COMPUTATION OF NONLINEAR STRUCTURES

COMPUTATION OF NONLINEAR STRUCTURES

EXTREMELY LARGE ELEMENTS FOR FRAMES, PLATES AND SHELLS

Debabrata Ray
PhD (Univ. of California, Berkeley)
ME, BE (Bengal Engineering College, Shibpur)
Principal, Institute for Dynamic Response

WILEY

This edition first published 2016

Registered office
John Wiley & Sons Ltd, The Atrium, Southern Gate, Chichester, West Sussex, PO19 8SQ, United Kingdom

For details of our global editorial offices, for customer services and for information about how to apply for permission to reuse the copyright material in this book please see our website at www.wiley.com.

Library of Congress Cataloging-in-Publication Data applied for.

A catalogue record for this book is available from the British Library.

ISBN: 9781118996959

Set in 10/12pt Times by Aptara Inc., New Delhi, India

Printed in Singapore by C.O.S. Printers Pte Ltd

1 2016

To the memory of my dad & mom, Provanshu and Sushila Roy, with whose altruistic love it was nurtured

&

the best thing ever happened to me, my wife, Anjana Ray, M.D., with whose infinite patience and eternal support it blossomed

&

the budding analysts and researchers like my sons, Dipanjan Ray, PhD and Shonket Ray, PhD, to whom it is offered

Contents

Acknowledgements

I would like to sincerely thank my friend, **Anil K. Chopra**, Johnson Professor of Structural Engineering, Department of Civil and Environmental Engineering, University of California, Berkeley, who introduced me to structural dynamics, for his encouragement and generous support in presenting my work to **Eric Willner**, Executive Commissioning Editor of John Wiley & Sons Ltd, who graciously agreed to publish the book, for which I am deeply appreciative. I must also thank **Anne Hunt**, Associate Commissioning Editor, Mechanical Engineering, for her unstinting support, and, **Clive Lawson**, Project Editor, Content Capture, Natural Sciences, Engineering & Stats, Professional Practice and Learning, Wiley. My special thanks for meticulous scrutiny of the manuscript to copy editor, Paul Beverley, LCGI, and Baljinder Kaur, Project Manager, Professional Publishing, Aptara.

During the writing of the manuscript for the book, I was paralysed as a result of a botched surgical procedure; however, from the initial period of rehabilitation to the date of this writing, I have been extremely fortunate to be surrounded by innumerable friends and well-wishers. I would like to thank them all; especially, my deep appreciation goes to my friend, Prof. Amitabha Basu of Dept. of South Asian Studies, University of California, Berkeley, and, to my care-givers, Levi Soler and John Viray.

Finally, however, the faults and mistakes, if any, are entirely mine.

DR
August 31, 2014

1

Introduction: Background and Motivation

1.1 What This Book Is All About

The book introduces linear and nonlinear structural analysis through a combination of of mesh generation, solid mechanics and a new finite element methodology called ***c-type finite element method*** (Ray, 1999, 2003, 2004, 2005, 2007, 2008). The ultimate objective is to present the largest possible (curved) beam, plate and shell elements undergoing extremely large displacement and rotation, and to apply these to solve standard industrial problems. Any finite element method is only as strong as its weakest link. In other words, the book is not just about unification of mesh generation and finite element methodology but it strives to serve as a reference for budding researchers, engineers, analysts, upper division and graduate students and teachers by demonstrating what various interdisciplinary machinery has to be accurately harnessed to devise a solid and conducive theoretical framework upon which to build a robust, reliable and efficient numerical methodology for linear and nonlinear static and dynamic analysis of beams, plates and shells. As indicated, the principal goal of the book is to produce the largest possible arbitrary shaped elements (a) defined and restricted solely by the requirements of geometry, material, loading and support conditions, (b) avoiding computational problems such as locking in the conventional finite element methods and (c) presenting new, accurate and explicit expressions for resolution of the symmetry issue of the tangent operator for beams, plates and shells in areas of extreme nonlinearity. The 'mega-sized' elements may result in substantial cost saving and reduced bookkeeping for the subsequent finite element analysis, and a reduced engineering manpower requirement for the final quality assurance. For example, the explicit algebraic and symmetric expressions of the tangent operator, as presented in the book, are an absolute necessity for computational cost efficiency, especially in repetitive calculations that are commonly associated with nonlinear problems. It must be recognized that the requirements for numerical convergence should be purely adaptive and subservient to the main delineating factors already mentioned. However, this strategy of computer generation of mega-elements of arbitrary shape, as it turns out, takes its toll on the analyst. Firstly, only accurate theoretical formulation can be used for the underlying continuum or solid mechanics principles without unnecessary 'short-circuiting' by proliferation of ad hoc numerical manipulations. Secondly, it demands that the applicable finite element method be devised to successfully accept computer generated elements with arbitrarily distorted shapes, with edges (or faces) consisting of up to truly 3D curved boundaries (or surfaces)

Computation of Nonlinear Structures: Extremely Large Elements for Frames, Plates and Shells, First Edition. Debabrata Ray.

with natural twist and bend (e.g. for shell elements). Thirdly, for these 'hyperelements' with conformity, the finite element method must be able to accommodate effortlessly and naturally C^0, C^1 or C^2 inter-element continuity on demand.

1.2 A Brief Historical Perspective

Every meaningful structural analysis is an exercise in abstraction about a structural system in the real world, so just as with any other natural or man-made phenomenon, the viability and safety of the structural system is intricately associated with the methodology underpinning such abstraction. More specifically, for a structural system, abstractions lie in the geometric modeling of its material body, its relevant support conditions and its imposed loadings, and finally its material properties; we associate the sum total of these abstractions with a structural theory. Moreover, each of these abstractions defines the extent to which a particular structural theory can efficiently and logically predict and control the response of a structural system. To paraphrase Einstein's incisive comment to Heisenberg that led to the latter's discovery of the uncertainty principle, every theory, like a mirror or a horse's blinder, filters and determines what we can see of the real world. Naturally, lest we miss out on important real-world phenomena, as structural analysts, we have to critically evaluate structural theories, propounded in both the distant and the recent past, so that we can be successful in the ultimate goal of our exercise in abstraction, namely, the prediction and control of structural response to external stimuli. Translated into actual methodology for solid bodies, these abstractions reduce to two fundamentally complementary disciplines: solid mechanics and numerical analysis – each determining and harnessing the strength of the other.

1.2.1 Operational Mechanics

For hundreds of years, even before the digital age, the basic theoretical premises of linear and nonlinear solid mechanics involving the study of the deformation of a body, transmission of force through it and the characterization of its material properties, have been well established, and they were perfected during the twentieth century (Eringen, 1962; Green and Zerna, 1968; Ogden, 1997; Malvern, 1969). However, without the computational power of modern computers, the forms of the various equations in solid mechanics, while accurately describing real world problems, were not of much use for finding numerical solutions for material bodies of very complicated geometry and intricate support conditions, subjected to complex loading systems. Thus, before the computer era, solid mechanics had no unifying numerical methodology, and so structural problems were solved on a case-by-case basis using a variety of different analytical methods.

1.2.2 Conventional Finite Element Methods

With the coming of the digital era, numerical methods became much more dominant. Over 60 years ago, an energy-based methodology called the finite element method (Turner, *et al.*, 1956) made its triumphant entry into the realm of numerical structural analysis, its theory having been established earlier (Courant, 1943). Its chief advance was to choose triangular elements with a complete set of basis functions following Pascal's triangle, but soon after, the application of similar basis functions to quadrilateral elements proved to be a poor choice because arbitrarily curtailed and incomplete polynomials resulted in interpolation problems requiring various ad hoc numerical artifacts such as under integration and reduced integration, to rectify numerous locking issues that resulted. Both the conventional finite element methods – h-type with Lagrangian and

Hermite basis functions (Zienkiewicz and Taylor, 2000) and p-type with Legendre basis functions (Szabo and Babuska, 1991), – suffered from these ill-conceived ideas.

While the general theoretical formulation of the displacement finite element method of analysis is based on rigorous variational method, the practice of finite element method is another story – a harrowing experience because, in an ad-hoc manner, it tried in vain to tear apart the fundamental and inalienable dictum of the Rayleigh–Ritz–Galerkin method that says: "... method is a 'package deal', and neither requires nor permits the user to make independent decisions about different parts of the problem" (Strang and Fix, 1973, p. 33). Based on the most common construction mechanism of global basis or Ritz functions from appropriate mapping of the chosen elemental or local basis functions, the ultimate success or failure of a practiced or computer-implemented finite element method essentially depends on the choice of the local basis functions (also known as the elemental shape functions), the adequacy of the mapping functions and the evaluations of the integrals. It is no wonder then that for more than 30 years, the primary focus of all notable fundamental researches in the field of finite element analysis has been on devising the "best" shape functions and associated mapping functions. It is instructive to briefly review the interpolation failure and concomitant patchy, ad hoc remedies with the ill side effects and to present some extremely important conceptual problems that remained unsolved by the conventional finite element methods, insofar as interpolation is concerned.

h-type Methodology: Hermitian shape functions, C^1 and C^2, are used for situations where flexural strain exists such as in the case of beam. The nodes in these cases are what can be considered as multiple nodes; that is, apart from the values, the slopes and/or curvatures are also taken as degrees of freedom at a physical location. The simplest element in the family has two end nodes with two degrees of freedom each for a total of four degrees of freedom. A cubic Hermite is used for this. The inter-element continuity that can be imposed is that the slope or the first derivatives of the functions are continuous, that is, C^1 continuity. To speed up convergence where the solution is smooth or has curved flexural elements, one can use a quintic Hermite with curvature or bending moment continuity, that is, C^2 continuity. Note that because of the inclusion of the derivative components as degrees of freedom, the interpolation loses its barycentric nature or convexity. The isoparametric elements are obtained following the assumption and procedure that the functional representation of the deformational behavior is employed in representation of the element geometry. In other words, the displacement vector components and the geometry use the same shape functions. This, in turn, forces the conventional element to have as many internal nodes as necessary to make geometry description isoparametric to the displacement function. For example, for a quadratic interpolation function for a truss element, it becomes a three-node element with one internal node. For a conventional isoparametric element, the accuracy increases with the corresponding element size measured by h, the diameter of the smallest circle encompassing the element, and, hence, also identified as h-type elements. The conventional h-type elements in two dimensions are interpolated with polynomials of degree one, two and sometimes three. The main elements, that have been known and used for quite some time, can be categorized into two major groups: Lagrangian and serendipity. These are obtained by taking the tensor product of the one-dimensional Lagrangian elements. The mapping functions are linear. For quadratic elements, say, the Lagrangian tensor product element consists naturally of nine nodes – four corner nodes, four edge nodes and one interior node. For the serendipity element, in this case, the interior node is dropped in favor of incomplete interpolation. The higher-order h-type elements are obtained analogously and may be found in any of the standard texts already mentioned.

p-type Methodology: From later developments come the so-called hierarchical elements of increasing polynomial degree $\boldsymbol{p}$, and hence, identified as p-type elements. The main idea behind these elements is that the next improved element of higher degree is achieved by retaining the earlier expressions of all degrees less than the improved one. In other words, to obtain an element of degree $p, p \geq 2$, only higher-order additional terms are introduced to the shape functions of the element of $p - 1$ degree. Moreover, these additional shape functions must vanish at old nodes, then since the pth derivative of an old $(p - 1)$ degree shape function always vanishes, this condition is chosen to insert a new node, and finally the new shape function is scaled so that it assumes the value of unity at this generalized displacement node. The combination is not barycentric or convex, and the degrees of freedom consist of both functions and their derivatives as in the mixed formulations. For p-type quadrilateral elements, there is more than one variant of the polynomial spaces. The shape functions are categorized in those spaces into three groups: nodal, side and interior modes. The basic idea behind the shape functions is the same as in one dimension, that is, choosing Legendre polynomials as the primary functions. However, in order to enforce the fundamental property of basis functions, namely that each assumes a value of unity at one node and vanishes at the rest of the nodes, blending functions similar to h-type elements are used.

Triangular Elements: The triangular elements, by contrast, have long been described in the barycentric coordinates. The root geometry has been described in three coordinates, suggesting a three-dimensional figure. However, a closer look will reveal that the root element is still planar, that is, a subset of two-dimensional space, $\mathbb{R}^2$, because only two of the coordinates are independent and all three are related; the barycentric coordinates are essentially the area coordinates in two dimensions. In other words, any point in two dimensions can be expressed as convex combinations of the vertex of a triangle containing the point. The coefficients of combinations are the barycentric coordinates of the point. As shown in this book, the three coordinates are proportional to the three smaller triangles generated by joining the point to each vertex of the triangle. All shape functions can be defined on the standard triangle in terms of these barycentric coordinates. For higher-degree h-type triangular elements, edge nodes (e.g. for degree 2) and inside nodes (e.g. for degree three) are included in the element. For the hierarchic p-type elements, for $p = 1, 2$, the shape functions for the triangular elements are exactly the same as those for the h-type elements. The mapping functions are chosen as linear for straight edge triangles. For large curved triangles, mapping functions are isoparametric for $p = 2$. For higher-order triangular elements, mapping functions are developed based on a blending method. For the purpose of critical evaluation to be introduced later, this concludes a brief presentation of the characteristics of the shape functions and the mapping functions for both families, namely the h-type and the p-type, of conventional finite elements.

1.2.2.1 Problems of the Conventional Finite Element Methods

We start by referring to the comment by MacNeal (1994) that the "days of pioneering" or the days of "heuristics, hunches and experimental data to guide . . . design choices" is over, and that we should take design of finite elements ". . . less as an art and more as a science". The noteworthy applied mathematicians such as Strang and Fix appeared to be too apologetic, maybe because the engineers beat the mathematicians and discovered the finite element method; they also appeared to have been temptingly permissive by announcing that the serious violations of the theory are mere "variational crimes": ". . . it [the rule] is broken every day, and for good reason" (MacNeal, 1994). No wonder then that MacNeal is forced to believe that "mathematical rigor falls well short of the goal of converting finite element design into science". In any case, because of these conflicting signals, scores of books keep appearing that simply repeat the same old ("variational crime"-ridden) information relating to the shape functions. The main goal, therefore, will be first to identify the various problems that crept into the mainstream conventional finite element method. However, in order for us to recognize the violations of the conventional finite element method, we need to reiterate the main dictates of the variational theorems: geometry must be accurately interpolated with extreme care for curved boundaries; the trial functions must have adequate derivatives across the element boundaries required by the inter-element continuity requirement; essential boundary conditions must be honored; and the integrals must be computed accurately.

Power Series as Basis Functions: One look at the history of the development of finite elements will reveal that primarily the power series made up of monomials of Cartesian position coordinates constituted the basis functions. Naturally, the Pascal triangle had been the guiding post for determining the completeness of the polynomials made up of such monomials. But, innocuous and familiar as these are, the polynomials constructed from these monomials, in their indiscriminate use as basis functions, produced some very distressing phenomena.

Induced Anisotropy: Once such commitment to power series has been made – the notion of completeness of the polynomials making up the basis functions became important for both convergence and suitability of the shape functions. Depending on the displacement solution function, convergence of incomplete polynomial basis functions to the solution is not guaranteed. Moreover, the basis functions based on incomplete polynomials can easily induce anisotropy for triangular or tetrahedral elements in the description of, say, the stiffness matrix, that is, the definition has to depend on the orientation of the coordinate axes. Similar loss of symmetry and hence induced anisotropy may result for arbitrary quadrilateral or hexahedral elements.

Incomplete Strain States: Incomplete polynomials as basis functions fail to invoke all the internal strain states that may result due to all the possible imposed conditions. For example, a four-node linear conventional quadrilateral element can only represent two internal linear strain states as opposed to required six. In fact, this condition afflicts all serendipity and Lagrange elements in two or three dimensions. Absence of complete strain states gives rise to what is known as ***locking*** which we shall deal with in the chapter on linear application. Based on the preceding discussion, polynomial basis functions made up of monomials do not seem to be a good choice. In fact, p-type hierarchical elements dropped monomials in favor of Legendre polynomials and thus avoided these problems. However, completeness is not the only dominating issue in the choice of the basis functions and mapping functions. We now embark on one such important issue.

Inter-element Continuity: As indicated in this book about the theory of Rayleigh–Ritz–Galerkin or finite element method, the trial and test spaces are finite dimensional subspaces of the solution space. The accuracy of the solution increases with the corresponding increase in the dimensionality of the subspaces. Furthermore, the energy in the error is equal to the error in the energy, and the approximated strain energy corresponding to any finite dimensional approximation is always less than that of the actual strain energy associated with exact solution. This, of course, is easy to understand when one considers that the description of an infinite dimensional displacement function by a finite set of basis functions only imposes additional constraints on the system. The additional stiffness, in turn, reduces the displacement and hence the strain energy of the system. The strain energy increases with increase in the number of basis functions. For this to continue to happen, however, the basis functions must satisfy some continuity requirements. For example, for plane stress problems in two dimensions, the trial function, for the Rayleigh–Ritz–Galerkin method, must be such that its first derivative has finite energy, that is, that it must be at least continuous, so that the virtual work expression is meaningful. In general, the trial functions must have finite energy (in the energy norm) in their m^{th} derivatives when the system-governing differential equation is of order $2m$. In finite element analysis, the geometry is reconstructed as an assemblage of finite elements. The above condition, then, becomes that the basis functions must be of class C^{m-1} across the element boundary. For plane stress analysis, $m = 1$, so the trial functions must be at least C^0 across the boundary, whereas for plate analysis, $m = 2$ and the corresponding continuity requirement is C^1, that is, the first derivative of the displacement function must be continuous. Any element that violates this requirement is known as the ***non-conforming*** element (Taylor *et al.*, 1976); otherwise, it is ***conforming***. In general, to establish, say, C^0 continuity at any edge between any two elements, the shape functions for a node connected to the edge must be identical for the two elements, and the shape functions for the rest of the nodes must vanish at the edge. For example, it can easily be shown that the constant strain, three-node triangle is conforming, and so are triangular or tetrahedral elements with edge nodes, as long as the edges are straight and all edges have same number of nodes. However, for the general four-node quadrilateral element, a simple coordinate transformation by rotation will show that the element is non-conforming. The only quadrilateral element that is conforming is the four-node rectangular element. All other general quadrilateral elements with edge nodes on the straight edges are non-conforming. Finally, if the edges of the elements, triangular or quadrilateral, are curved, the elements are non-conforming. All these elements made up of basis functions by incomplete power series of position coordinates lack the barycentric or convexity property of interpolation, and thus are devoid of invariance under affine – for example, rotational – transformations. In fact, as will be shown later, the problem associated with the general curved conventional elements of today's finite element method is one of interpolation that translates into lack of inter-element continuity in general. Historically speaking, however, these earlier problematic h-type elements found their partial savior in what is known as parametric mapping and rigid body rotation.

Parametric Mapping and Rigid Body Rotation: In the theoretical formulation of the isoparametric finite element method, it is assumed that the elemental basis functions are expressed in a root element, and the displacement functions are then mapped isoparametrically with the geometry. For the case where the geometry is straight and can be expressed by a linear root element, the displacement functions could be of higher order, needing conventionally edge or interior nodes; here, the mapping of the root element is known as ***subparametric***. In the reverse situation of curved geometry with simple strain states, the number of basis functions necessary to completely describe the geometry may be higher than that needed for displacement description; this mapping goes by the name of ***superparametric***. For superparametric elements, the representation of rigid body rotation and constant strain states is affected adversely. The history of construction of conventional finite elements with variable nodes appeared to have failed to realize one fundamental mathematical relationship: *the conversion in description of non-parametric functions to parametric description automatically and inescapably fixes all the internal nodes*. The displacement functions or the geometry coordinates, to start with, are non-parametric functions. To generate ***isoparametry*** between them, both needed to be first transformed into parametric entities. This then, by the principle just stated, fixes the intermediate nodes, be it just the edge nodes for one dimension, or the edge, face or internal nodes in their higher dimensional counterpart. Any discussion on distortion of the intermediate nodes is irrelevant and futile. In fact, if complete isoparametry is established, guaranteeing a barycentric or convex combination of the basis functions of the root element, the ensuing finite element, irrespective of any elemental distortion, will be conforming and complete. Short of guaranteed completeness of the polynomials, we would like to have a device by which various conventional elements can be measured for their acceptance. We now engage in discussing one such tool.

Patch Test: In earlier days, when terms were being added or dropped for designing new finite elements in ad hoc fashion, a standard numerical test was necessary for evaluating their worth. One such test is the patch test. The idea behind the patch test is simple. A chunk of elements representing a system is plucked out and subjected some known boundary conditions for which the internal solutions are known. It is expected, then, for acceptance of the elements, that the corresponding computed internal quantities should be the same as those known expectations. In other words, if a linear black box (the element) is correct, the output and input must correspond. To make calculations easy, the boundary is taken simple as well, such as square or cube for two and three dimensions, respectively. The constant strain patch test is considered standard for two- dimensional in-plane load conditions, because as the elements reduce sufficiently in size, the state of the strain (or derivative of the displacement) becomes constant in the limit. So, for convergence, the element should be able to represent such a state of strain. For plate bending, this strain becomes the bending curvature. In any case, linear displacement-equivalent forces are applied on the external nodes and solved for internal node displacements. The derivatives of this solution must match the known constant strains. There are several versions of patch tests. A patch test was initially conceived as a numerical experimentation for element validation. For smooth solution problems, the patch test was later shown mathematically to be a necessary and sufficient test for convergence of h-type elements. It appears quite clear to the author that the patch test is, in the main, an interpolation check. However, it serves to detect other failures such as integration or equilibrium failures. Because of the conformity (satisfying equilibrium requirements) and completeness (satisfying interpolation requirements) of the conventional linear isoparametric elements having only corner nodes, generation of constant strain or linear displacement state is extremely easy. Similarly, the conventional variable-node higher-order isoparametric elasticity elements with straight edges or faces with evenly spaced nodes can pass a higher-order patch test. As rectangular elements with evenly spaced edge nodes, the serendipity elements fail the patch test while its counterpart full-noded Lagrangian elements do not. Because of incomplete quadratic terms, isoparametric three- or four-noded bending elements cannot pass the constant curvature or quadratic displacement patch test. For curved elements, triangular (and tetrahedral) or quadrilateral (and hexahedral) with variable nodes arbitrarily placed cannot pass the quadratic displacement or constant patch test as necessary for plate and shell elements. As will be shown in the subsequent chapters, this is clearly an interpolation failure associated with elements with variable nodes under a curved situation. Finally, of course, the linear elements fail to pass the constant curvature patch test. In other words, these elements should not be used to model, say, plate or shell systems.

Locking Problems: The discussion on the patch test has been included merely to show the innate interpolation problems that have been afflicting conventional finite elements. More specifically, these translated into other serious numerical maladies known as ***locking***. We will therefore now look at the phenomenon of locking only in conventional rectangular elements. Other elements such as triangular elements experience the same locking phenomena, so we will dscribe the remedial methods that have been tried, but not altogether successfully. The futility crept in because none of these so-called tricks ever tried to correct the innate interpolation problems. In fact, these remedies, which include reduced and selective integration, incompatible modes and so on, do not appear to be numerically robust in the classical sense of consistency, convergence and stability. As a result, we are burdened with the additional problem of spurious modes introduced by reduced or selective integration. Simple solutions based on exact interpolation theory that allow the reader to appreciate the strength of the new c-type finite element method are presented in Chapter 9 about linear applications.

Shear and Membrane Locking and Shape Dependence: The major debacle of interpolation failure is locking, that is, elements showing unacceptably high stiffness, almost bordering on rigidity, in certain displacement states. In other words, the low order or variable node elements are failing to respond flexibly enough for or failing to recognize as it should, certain imposed conditions. Moreover, it was recognized that the severity of this phenomenon of locking depends on the shape of the element. Finally, the shape may be boiled down to a specific parameter that allegedly perpetrates such locking.

Now let us briefly examine the conventional state-of-the-art remedies.

Reduced/Selective Integration, Bubble Mode and Drilling Freedom: The linear four-node element failed to interpolate the quadratic term, and thus experienced dilatation locking.

Likewise, inability to handle quadratic term gave rise to spurious shear strain. This, in turn, contributed to the shear locking. As we already know, the numerical integration to evaluate, say, the stiffness matrix, is performed by selecting Gauss or other quadrature points. Thus, an immediate remedy to shear locking was offered by ***selective under-integration*** (Zienkiewicz *et al.*, 1971) by evaluating the integral involving shear at only one Gauss point, which effectively wiped out the shear contribution to the stiffness matrix. The problem with this remedy is that the reduced integration, is obviously less accurate. More importantly, it creates undesirable ***spurious modes***, that is, artificial singularities or eigenmodes induced by possible reduction in the rank of the stiffness matrix. Another ad hoc remedy was to include quadratic interpolation terms: so-called ***bubble modes*** (Wilson, 1973), without introducing any additional node, to handle both shear and dilatation locking. This, of course, renders the element non-conforming as two nodes cannot possibly represent a quadratic term. In line with the addition of modes of flexibility just described, modes known as ***drilling freedoms*** have been devised to resolve locking problems. Drilling freedoms (Allman, 1984) are nodal rotational degrees of freedom about an axis perpendicular to the plane of the element. These in-plane rotational degrees of freedom, in improving membrane performance, may create problems for bending performance of curved shell elements, where these degrees of freedom are already coupled with the bending modes. But, in general, most of these patchy remedies that worked on regular rectangular shaped element, fail to eradicate locking when applied to general h-type isoparametric curved elements of arbitrary shape. Moreover, these varieties of so-called remedial elements for specific displacement states increase the number of elements available commercially to such an extent that an ordinary user can be easily confused as to their use. It is extremely important that the element library be simple and user-friendly.

Plate and Shell Elements: For our purpose, we need only to note that for arbitrary shaped plate and curved shell elements, the interpolation problems and proposed solutions are similar to those already described. The remedies that have been offered for offsetting interpolation failure are not robust and are highly dependent on the element shapes. For example, trapezoidal plate or shell elements can still create problems of transverse shear locking.

1.2.3 Essential Mesh Generation: Curves and Surfaces

At around the same time as the introduction of the finite element method to structural analysis, a different numerical revolution, unnoticed by the finite element analysts, was underway primarily in the car industry (Coons and Herzog, 1967; Bezier, 1972; de Casteljau, 1986; Gordon and Riesenfeld 1974a,b), to improve upon the production of automotives with arbitrarily curved bodies: namely, **computer-aided geometric design** (CAGD). Of all the new methods of CAGD, **solid modeling** (Mantyla, 1988; Hoffmann, 1989) and **boundary representation** (B-Rep) methods (Barnhill, 1974; Faux and Pratt, 1979; Bohm, 1981, 1984; Foley and van Dam, 1982; Yamaguchi, 1988; Farin, 1992; Piegl and Tillman, 1995) took the leading role. Solid modeling is based on using geometrically simple elements as building blocks with set-theoretic operations of union, intersection, complementarity and so on, but without any information on inter-element continuity. On the other hand, B-Rep methods are all about building the geometry from simple shapes, from curves to surfaces to solids, with information and control over the inter-element boundaries. The foundations of these methods are based on complete Bernstein polynomials (Davis, 1975) as shape factors in Bezier control vector form and Schoenberg B-spline (Schoenberg, 1946) curves in de Boor–Cox (de Boor, 1972; Cox, 1972) control forms. The most important aspect of this revolution, in so far as it could relate to possible application to finite element analysis, is that B-splines are constructed geometrically as opposed to algebraically (Prenter, 1975) and thus provide the necessary local support for each finite element; in the ultimate analysis, these polynomials are complete for interpolation and approximation with local support.

1.3 Symbiotic Structural Analysis

To solve nonlinear structural problems such as beams, plates and shells under large displacement and rotation, it is not enough to engage the first part, namely, the mesh generation technique, and, the second part, namely, the c-type finite element method, free of 'variational crimes' that lead to locking, ad hoc numerical artefacts, and so on, without reformatting the fundamental axioms of solid mechanics to corresponding computational forms. In other words, the transformed governing equations of solid mechanics applied to beams, plates and shells form the third essential part of the system for possible production of largest finite elements. The reason for using such a transformation is primarily that finite rotations, unlike displacements of Euclidean space, traverse a configuration space that is a Riemannian curve space necessitating ***covariant derivatives*** as opposed to ***Gateaux derivatives***; this book tries to draw together these three components in developing the largest possible nonlinear beam, plate and shell static and dynamic finite elements.

1.4 Linear Curved Beams and Arches

The study of curved beams in finite element development is extremely important for anticipating problems for a given proposed method, when extended to the more general case of curved elements, namely, shells. In the linear regime, for many years, Timoshenko- or Mindlin-type curved beam and arch elements reportedly suffered from what is commonly known as shear and membrane locking in the thin regime (Ashwell & Sabir, 1971; Ashwell *et al.*, 1971). This, has triggered, for thirty years, a range of interpolationally inadequate approaches with limited success in terms of accuracy, rate of convergence and general applicability. Unfortunately, there are textbooks (e.g. Cook, 1994; Dow, 1999), that constantly refer to these methods as incurable observations, and this is misleading for the aspiring analysts. Nonetheless, we list a few such attempts with their associated pitfalls: the application of reduced integration of shear and membrane energy producing spurious modes of deformations in thin limits (Stolarski and Belytschko, 1982; Panadian, 1989); the truncation of so-called field-inconsistent strains resulting in low convergence rate (Babu and Prathap, 1986); the application of mixed polynomial-trigonometric displacement field restricting its general applicability to non-circular curved beams (Cantin & Clough, 1968; Ashwell, 1971); the utilization of higher-order quintic-quintic independent displacement fields (Dawe, 1974); the problem-specific penalty relaxation method (Tessler & Spiridigliozzi, 1986); the application of so-called quasi-conforming technique (Shi and Voyiadjis, 1991); and the circular curve-specific application of quadratic curvature interpolation (Lee and Sin, 1994). Finally, there is a method in which so-called material finite elements (MFE) are generated by analysis applicable only to linear circularly curved beams resulting in polynomial distributions for the field variables that are consistent with the equilibrium equations (Raveendranath *et al.*, 1999). The elements are lock-free with good predictability for field variables but unfortunately the shear strain distribution is restricted to being constant, resulting in only approximately constant shear force over each element. All these methods mentioned are either numerically ineffective or problem-specific. Also, there continues to be an unnecessary general consensus that to be locking free, a consistent formulation must have the order of interpolation for tangential displacement one order higher than that of normal displacement (Meck, 1980). For formulation in the linear regime of truly three-dimensional curved beams, we refer to the original works of Love (1944,

1972). For warping with seven degrees of freedom – three translations, three rotations and one twist, we refer to Washizu (1964).

1.5 Geometrically Nonlinear Curved Beams and Arches

One main focus of this book, as already indicated, is to demonstrate the versatility, simplicity and accuracy of the c-type method as applied to three-dimensional, beams and arches. The beams may be shear-deformable and extensible, straight or truly three-dimensional space curved with natural twist and bend, and finally geometrically nonlinear with extreme rotations and displacements without warping. The three-dimensional curved beam problem in the geometrically nonlinear regime – that is, for finite deformations – was first reported by Reissner (1973) and then in (1981) with an improved formulation. However, the analysis is restricted to second-order approximations on the rotation parameter and presents no numerically explicit expressions for stiffness, and so on. The work of Argyris (1982) is, in an unduly long and confusing sense (Spring, 1986), an introduction to theoretical kinematics for the mechanics community. For formulation in the extreme geometrically nonlinear regime, see the work of Simo (1985). In the analytical formulation, the possibility of a naturally twisted and bent situation is suggested. However, the numerical examples presented in the referenced work modeled curved beams as an assemblage of straight elements, resulting in a larger than necessary number of elements as will be demonstrated later. Simo and Vu-Quoc (1986) succeeded in enriching this method by considering finite rotation as an algebraic structure, namely, a ***Lie group***. Unfortunately, this paper missed the well-known role of covariant derivatives in parameters that travel on a differentiable manifold, that is, generalized multi-dimensional surfaces. However, this otherwise outstanding paper was taken as a standard by several other authors to justify their conclusions about the symmetry property of the tangent stiffness matrix. For example, Crisfield (1990), following a co-rotational formulation, wrongly asserted that even for a conservative system, the tangent stiffness matrix is unsymmetrical everywhere except at an equilibrium state. In view of this, we refer later to a simple lemma from Kreyszig (1959) to bring out the essence of the matter: in summary, the Gateaux derivatives are only appropriate on linear vector spaces, not on manifolds, for linearization and symmetry of the virtual functional, because, unlike 'flat-spaces' (e.g. Euclidean) where vectors are 'free' vectors, the Riemannian and differentiable manifolds accommodate only 'bound' vectors. An explicit proof for the multi-dimensional surface was eventually reported by Simo (1992). The tangent spaces are linear vector spaces, thus restriction of the rotation parameterization to elements of these spaces can avoid issues relating to covariant derivatives and allow the use of the familiar Gateaux derivatives needed for linearization. Following this idea, a series of papers appeared, starting with Cardona and Geradin (1988), where the symmetry of the geometric stiffness was correctly assumed, but without any proof. Similarly, in a related paper (Ibrahimbegovic, 1995), because of a less than instructive explicit expression of the same, the symmetry could only be verified but not proved. Another paper (Gruttmann *et al.*, 2000) presented a proof but because of the Cartesian nature of the analysis, the expression for the geometric stiffness remained rather convoluted. Of course, covariant derivatives are always needed for a curvilinear formulation, say, in the development of the equilibrium equations as we will show. We attend to this problem by choosing all the parameters as vector fields measured in curvilinear coordinates for various reasons. The first is rooted in our intention to unite the geometric mesh generation and the subsequent finite element analysis. Thus, the information – Bezier geometric controls, curvature and twist in Frenet frames – produced by a mesh generation provides the starting point for our analysis. Secondly, applications of the covariant derivatives for the stress resultant and stress couples in deriving the equilibrium equations and subsequent virtual work principle result in a direct proof and an elegant yet explicit expression for both the material and the geometric stiffness. Thirdly, the elements with only six degrees of freedom per node can be directly incorporated into the standard direct stiffness based finite element packages. Finally, in the numerical examples presented in the literature, almost invariably, curved beams are modeled as an assemblage of straight elements resulting in more than the necessary number of elements. We treat curved beams as mega-elements with truly three-dimensional attributes of natural bend and twist, and thus use a minimum number of elements for analysis, as will be demonstrated. The book presents the nonlinear beam theory for both quasi-static and dynamic loading conditions. For the quasi-static problems a Newton-type iteration scheme has been employed, similar to (Crisfield, 1981), to solve various benchmark problems; no attempt has been made in this book to apply the theory to dynamic problems.

1.6 Geometrically Nonlinear Plates and Shells

Another important focus of the book, lies in a demonstration of the versatility, simplicity and accuracy of the c-type method as applied to shells. The shells may be shear-deformable and extensible, and geometrically cylindrical, spherical or truly arbitrary surfaces, and finally geometrically nonlinear with extreme rotations and displacements and subjected to quasi-static or dynamic loading. Computationally speaking, the methodologies for modeling and solution of shell structures can be generally categorized into four groups: (1) the earliest finite element approach consisted of facet-type plate elements with loss of slope continuity for Love–Kirchhoff-type shells that need -continuity, and that are plagued by the shear locking problem when dealing with Mindlin–Reissner-type shells with shear strains; (2) three-dimensional solid (brick) elements are used as shell elements based on 3D elasticity equation with a large number of nodes and degrees of freedoms. For thinner shells, these elements broke down with erratic behavior and locking problem and ill-conditioning; (3) A finite element modeling known as degenerative shell elements from three-dimensional shell brick elements based on Mindlin–Reissner-type shell theory with reduced integration; and (4) shell elements modeled after various old classical shell theories (Wempner, 1989; Ahmad *et al.*, 1970; Zienkiewicz *et al.*, 1971; Pawsey and Clough, 1971).

For theoretical developments, we can divide the various efforts into two main groups: (1) in the ***direct approach*** the shell is taken at the very start as a 2D object and all the laws are developed for this, including the constitutive laws; Cosserat shells belong to this category (Naghdi, 1972); (2) the ***derived approach***, where various suitable constraints are applied to the three-dimensional shell-like body. If the main assumption is that the normals to the surface remain normal during deformation and inextensible, this results in what is known as the Love–Kirchhoff theory of shells with only three degrees of freedom at each point; otherwise, if normals just remain straight and inextensible, we have the Mindlin–Reissner model of shells with five degrees of freedom. Finally, inclusion of extensibility accommodates the full set of six degrees of freedom and goes by the name of Timoshenko-type shell theory (Pietraszkiewicz, 1979; Simo *et al.*, 1990). All these approaches suffer from adhocism in either their theoretical underpinning or their numerical evaluation. In this book, a modified method is developed following Reissner (1960) and Libai & Simmonds (1998). In this concept, we start from the exact three-dimensional balance laws and derive the variational functional exactly; any approximation is contained where it should naturally belong, that is the constitutive theory, which is at best empirical. The variational functional is linearized for numerical evaluation by a Newton-type iterative method by recognizing that the rotation tensor travels on a curve space and hence the absolute necessity of the covariant derivative of curve spaces over the Gateaux derivative of so-called 'flat' Euclidean space; for the corresponding constitutive theory, we follow Chroscielewski *et al.*, (1992).

1.7 Symmetry of the Tangent Operator: Nonlinear Beams and Shells

In nonlinear computational mechanics the prevalent Newton-type methods require linearization of the virtual functional for subsequent finite element formulation. The ad hoc practice of symmetrizing the tangent operator away from equilibrium, based on an application of Gateaux derivatives (a simple generalization of partial derivatives) in the linearization of parameters on a Riemannian surface, resulted in the claim that the geometric stiffness is non-symmetric away from equilibrium even for conservative systems. From a computational point of view, even for feasibility of producing a convergent solution without shear locking, and so on with *truly curved and large* beam or shell element, it is imperative that the concomitant theoretical formulation be derived in the most accurate and cost-effective sense. For beam and shell formulations in the geometrically nonlinear regimes, this brings us to the discussion of the following topics that show the essential motivation of our book.

Symmetry of the Geometric Stiffness – Conservative Systems: The three-dimensional curved beam formulation with a rotating or spinning coordinate system, extended in shell theory by Simmonds and Danielson (1970), in the geometric nonlinear regime (i.e. for finite deformations) was first reported by Reissner (1973) and subsequently again in (1981), with an improved formulation. However, the analysis is restricted to second-order approximations on the rotation parameter and presents no numerically explicit expressions for the residual and stiffness needed for a finite element development. Simo and Vu-Quoc (1986) enriched this by consideration of finite rotation as an algebraic structure, namely, a Lie group. Unfortunately, their paper overlooked the well-known role of the covariant (absolute) derivative in consistent linearization with respect to parameters that travel on a smooth manifold, that is, generalized multi-dimensional surfaces. To quote Simo and Vu-Quoc (1986 p. 80): "we show ... the global geometric stiffness arising from the (consistently) linearized weak form is non-symmetric, *even for conservative loading*, at a non equilibrium configuration ... Argyris and coworkers [5–8] point out that this lack of symmetry *inevitably* arises at the element (local) level" (emphasis added). The mistake is rooted in the application of the Gateaux derivative (a generalized partial derivative applicable only to Euclidean or "flat" surfaces) to define and compute the Hessian from the virtual work functional defined on Riemannian or "curved" surfaces. As a result, to overcome the inevitable dilemma, they introduced an ad hoc procedure that has since become known as the ***symmetrization of the geometric stiffness***. Simo recovered from this adhocism in a subsequent paper (1992) published four years later. However, Simo's original assertion was taken as a standard by several other authors to justify their mistaken conclusions about the symmetry property of the tangent stiffness operator. For example, Crisfield stated, (p. 148, Crisfield, 1990): "The tangent matrix is ... non-symmetric. ... *for the conservative problems* analyzed, symmetry is *recovered* as an equilibrium state is reached" (emphasis added). Based on this, the paper goes off at a tangent to recommend future research: "a study of the geometric stiffness matrix in order to discover which terms may be reasonably neglected without serious cost to the performance" (Crisfield, 1990 p. 148).

Symmetry of the Geometric Stiffness – Non-conservative Systems: For general three-dimensional situations, it will be recognized that, as indicated by Ziegler (1977), a moment loading (i.e. either distributed or end moments) makes the system non-conservative and so the resulting geometric stiffness is non-symmetric. But, as we will show, an identification of the terms responsible for the asymmetry suggests that the non-symmetrical structure may be computationally rather weak. Thus, in the absence of moment loading or where the end boundary conditions are purely kinematic with no distributed body moments, the system is conservative, and thus the geometric stiffness is definitely symmetric as discussed in the previous section. For two-dimensional or planar systems with planar deformations, the reduction of the 3D expression of the geometric stiffness to the 2D situation will reveal that all moment-related terms, including the reactive and externally applied distributed or end moments, vanish identically, leaving the geometric stiffness unconditionally symmetric.

Theoretical motivation: One important result of our theoretical motivation, *is that we show that the symmetry of the tangent operator is* ***always*** *guaranteed for a conservative system, in equilibrium or away from the equilibrium, if, instead of the Gateaux derivative, the concept of the covariant derivative is applied in the definition of consistent linearization.* Note that in any repetitive numerical solution (as in nonlinear problems): (a) the computational saving in working with a symmetric matrix, as opposed to a non-symmetric matrix, could be considerable and thus, always optimal, (b) an exact explicit definition, as opposed to an implicit one such as a derivative form or an approximate one, is always advantageous computationally. Thus, for a finite element formulation in the nonlinear regime with very large elements, it is not enough to prove that the tangent stiffness is symmetric for a conservative system. It is also imperative that the tangent stiffness present itself in its most simple and computationally effective form. In this sense, all papers published in the literature to date are not optimally suitable. Simo and Vu-Quoc (1986) and other similar papers presented the tangent stiffness in differential form. Cardona and Geradin (1988), where the symmetry of the geometric stiffness was correctly assumed, had to settle for an approximate expression of the tangent stiffness (on p. 2428, the comment after equation (146)). Similarly, in a related paper (Ibrahimbegovic, 1995), because of the disordered nature of the derived expression, the symmetry could not be proved. Gruttmann *et al.* (2000) presented a proof, but because of the Cartesian nature of the analysis, the expression for the geometric stiffness remained somewhat convoluted.

Computational motivation: As a principal result of our computational motivation, we present in this book a crucial ***central lemma*** (Section 10.1) *involving the virtual rotation and incremental rotation of a beam that avoids awkward derivation* [which forced Cardona and Geradin (1988) to settle for an approximation] *of second derivatives of tensors on a Riemannian surface. With the help of this lemma, along with other results, we present an accurate expression for the tangent stiffness in an optimal, cost-effective operational form.* In this book, for several reasons, we choose all the parameters to be vector or tensor fields measured in curvilinear coordinates. This is first rooted in our intention of uniting the geometric mesh generation and the subsequent finite element analysis. Thus, the information, namely, Bezier geometric controls, curvature and twist in Frenet frames, produced by a mesh generation provides the starting point of our analysis. Secondly, applications of the covariant derivatives for the stress resultant and stress couples in deriving the equilibrium equations and subsequent linearization of the virtual work functional result in a direct proof of symmetry and an elegant yet explicit expression for both the material and the geometric part of the stiffness. Thirdly, the elements with only six degrees of freedom per node can be directly incorporated into a standard displacement-based finite element system. In conclusion, with the goal of developing a truly three-dimensional, geometrically nonlinear curved finite element, a geometrically exact curvilinear, nonlinear formulation for the truly three-dimensional curved beams/arches and shells is presented in the book. The symmetry question of the geometric stiffness matrix is fully treated and shows that the tangent operator is symmetric for a conservative system. The moment loading is discussed thoroughly by identification of specific terms of non-symmetry. The two-dimensional specialization establishes the unconditional symmetry of the operator. Of utmost importance, considering the requirement of repetitive generation of the tangent operator, is the presentation of a closed form expression for the associated geometric stiffness in its most simple, direct and explicit form compared to those available in the published literature.

Finally, a few words about the chronological history is in order. The original work by the author of geometrically applying Bernstein–Bezier and B-spline bases to beams was initiated in 1995, and a paper (Ray, 1999) contained the c-type method, noting the importance of choosing appropriate basis functions for a finite element method. Note that about 13 years before any paper containing a similar finite element method (the iso-geometric method) was published, a single conference paper (Reus, 1992) was found to exist through an extensive literature search; the present author, having been completely unaware of the paper, developed the c-type method independently.

1.8 Road Map of the Book

Based on the discussions of this chapter on background and motivations, the book is designed to accomplish its goals through a logical sequence of chapters, the content of each of the chapters is as follows:

Chapter 1: this introductory chapter.

Chapter 2: presents some essential mathematical entities from real analysis such as sets, group, algebra and functions and continuity, normed spaces, the Sobolev space, the Vainberg principle, and Gram–Schmidt orthogonalization.

Chapter 3: introduces tensors in general, and then particularizes to second-order tensors; it identifies tensors both as linear functionals and dyadics or polyadics; it presents various properties and the differential and integral calculus of tensors so that all structural equations can be written in compact absolute notation devoid of cumbersome indicial description except where absolutely necessary.

Chapter 4: discusses in detail the representation and properties of rotational tensors necessary for analysis of structures with finite rotational responses; it covers both the theoretical and computational details pertaining to rotation tensor, and thus introduces the mathematical object called the quaternion.

Chapter 5: deals with the theory and computation of real curves in anticipation of its application to beams; in particular, it presents curves in the form of Bernstein polynomials and Bezier controls. It then includes B-spline representation geometrically by composite Bezier curves for local support.

Chapter 6: deals with the theory and computation of real surfaces in anticipation of its application to plates and shells; more specifically, it presents surfaces in the form of Bernstein polynomials and Bezier controls. It extends further to include B-spline representation geometrically by composite Bezier surfaces for local support.

Chapter 7: presents briefly the essential elements of nonlinear solid mechanics: deformation (strain), transmission of force (stress), balance laws and the constitutive theory of hyperelastic materials.

Chapter 8: introduces finite element theory as an energy method, and it gives a presentation of the new c-type finite element method.

Chapter 9: applies c-type finite element method to linear structural problems involving rods, straight and curved beams, plane stress and plane strain elements; it also provides solutions for various locking problems.

Chapter 10: develops the theory for nonlinear beams subjected to quasi-static and dynamic loading, resulting in extremely large displacement and rotation responses, and applies the c-type finite element method to solving numerical problems with extremely large beam elements.

Chapter 11: develops the theory for nonlinear plates and shells subjected to quasi-static and dynamic loading resulting in extremely large displacement and rotation responses and applies c-type finite element method to solve numerical problems with extremely large plate and shell elements.

References

Ahmad *et al.* (1970) Analysis of thick and thin shell structures by curved finite elements, *International Journal for Numerical Methods in Engineering* **2**(3), 419–451

Allman, D.J. (1984) A compatible triangular element including vertex rotations for plane elasticity analysis. *Computers & Structures*, **19**, 1–8.

Argyris, J.H. (1982) An excursion into large rotations. *Computer Methods in Applied Mechanics and Engineering*, **32**, 85–155.

Ashwell, D.G. and Sabir, A.B. (1971) Limitations of certain curved finite elements when applied to arches. *International Journal of Mechanical Sciences*, **13**, 133–139.

Ashwell, D.G. *et al.* (1971) Further study the application of curved finite element to circular arches. *International Journal of Mechanical Sciences*, **13**, 507–517.

Babu, C.R. and Prathap, G. (1986) A linear thick curved beam element. *International Journal for Numerical Methods in Engineering*, **23**, 1313–1328.

Barnhill, R.E. (1974) *Smooth Interpolation Over Triangles, in Computer Aided Geometric Design*, Academic Press.

Bathe, K.J. and Bolourchi, S. (1979) Large displacement analysis of three-dimensional beam structures. *International Journal for Numerical Methods in Engineering*, **14**, 961–986.

Bezier, P.E. (1972) Numerical control in automobile design and manufacture of curved surfaces, in *Curved Surfaces in Engineering*, IPC Science and Technology Press, London.

Bohm, W. (1981) Generating the Bezier points of B-spline curves and surfaces. *Computer Aided Design*, **13**(6), 365–366

Bohm, W. *et al.* (1984) A survey of curve and surface methods in CAGD. *Computer-Aided Geometric Design*, **I**, 1–60.

Cantin, G. and Clough, R.W. (1968) A curved cylindrical shell finite element. *AIAA Journal*, **6**, 1057–1062.

Cardona, A. and Geradin, M. (1988) A beam finite element non-linear theory with finite rotations. *International Journal for Numerical Methods in Engineering*, **26**, 2403–2438.

Chroscielewski, J. *et al.* (1992) Genuinely resultant shell finite elements accounting for geometric and material non-linearity. *International Journal for Numerical Methods in Engineering*, **35**, 63–94.

Cook, R.D. *et al.* (1994) *Concepts and Applications of Finite Element Analysis*, 3rd edn, John Wiley & Sons, Inc., NY, pp. 343–350.

Coons, S.A. and Herzog, B. (1967) *Surfaces for computer-aided aircraft design*. Paper 67-895, AIAA.

Courant, R. (1943) Variational methods for the solution of problems of equilibrium and vibration. *Bulletin of the American Mathematical Society*, **49**, 1–23.

Cox, M.G. (1972) The Numerical Evaluation of B-splines. *Journal of the Institute of Mathematics and its Application*, **10**, 134–149.

Crisfield, M.A. (1981) A fast incremental/iterative solution procedure that handles 'snap-through'. *Computers & Structures*, **13**, 55–62.

Crisfield, M.A. (1990) A consistent co-rotational formulation for non-linear, three-dimensional, beam-elements. *Computer Methods in Applied Mechanics and Engineering*, **81**, 131–150.

Davis, P.J. (1975) *Interpolation & Approximation*, Dover Publications, Inc., NY.

Dawe, D.J. (1974) Numerical studies using circular arch finite elements. *Computers & Structures*, **4**, 729–740.

De Boor, C. (1972) On calculating with B-splines. *Journal of Approximation Theory*, **6**, 50–62.

de Casteljau, P. (1986) *Shape Mathematics and CAD*, Kogan Page, London.

Dow, J.O. (1999) *A Unified Approach to the Finite Element Method and Error Analysis Procedures*, Academic Press.

Eringen, A.C. (1962) *Nonlinear Theory of Continuous Media*, McGraw-Hill, Inc.

Farin, G. (1992) *Curves and Surfaces for Computer Aided Geometric Design - A Practical Guide*, 3rd edn, Academic Press.

Faux, L. and Pratt, M. (1979) *Computational Geometry for Design & Manufacture*, Ellis Horwood.

Foley, J. and van Dam, A. (1982) *Fundamentals of Interactive Computer Graphics*, Addison-Wesley.

Gordon, W.J. and Riesenfeld, R.F. (1974a) Bernstein-Bezier methods for the computer-aided design of free-form curves and surfaces. *Journal of the Association for Computing Machinery*, **21**(2), 293–310.

Gordon, W.J. and Riesenfeld, R.F. (1974b) B-spline curves and surfaces, in *Computer Aided Geometric Design*, Academic Press, NY.

Green, A.E. and Zerna, W. (1968) *Theoretical Elasticity*, 2nd edn, Oxford University Press.

Gruttmann, F. *et al.* (2000) Theory and numerics of three-dimensional beams with elastoplastic material behavior. *International Journal for Numerical Methods in Engineering*, **48**, 1675–1702.

Hoffmann, C. (1989) *Geometric & Solid Modeling – An Introduction*, Morgan Kaufmann Publishers, Inc., CA.

Huddleston, J.V. (1968) Finite deflections and snap-through of high circular arches. *Journal of Applied Mechanics*, **35**(4), 763–769.

Ibrahimbegovic, A. *et al.* (1995) Computational aspects of vector-like parametrization of three-dimensional finite rotations. *International Journal for Numerical Methods in Engineering*, **38**, 3653–3673.

Kreyszig, E. (1959) *Differential Geometry*, University of Toronto Press.

Lang, S. (1968) *Linear Algebra*, 3rd Printing, Addison-Wesley Publishing Co., Inc.

Lee, P.G. and Sin, H.C. (1994) Locking-free curved beam element based on curvature. *International Journal for Numerical Methods in Engineering*, **37**, 989–1007.

Libai, A. and Simmonds, J.G. (1998) *Nonlinear Theory of Elastic Shells*, 2nd edn, Cambridge University Press.

Love, A.E.H. (1944) *The Mathematical Theory of Elasticity*, Dover, New York.

MacNeal, R.H. (1994) *Finite Elements: Their Design and Performance*, Mercel Dekker, Inc.

Malvern, L.E. (1969) *Introduction to the Mechanics of Continuous Medium*, Prentice Hall, Inc., NJ.

Mantyla, M. (1988) *An Introduction to Solid Modeling*, Computer Science Press, Helsinki.

Meck, H.R. (1980) An accurate polynomial displacement function for finite ring elements. *Computers & Structures*, **11**, 265–269.

Naghdi, P.M. (1972) The theory of plates and shells, in *Handbuch der Physik Band VIa/2*, Springer-Verlag, pp. 425–640.

Ogden, R.W. (1997) *Non-Linear Elastic Deformations*, Dover Publications, Inc., NY.

Palazotto, A.N. and Dennis, S.T. (1992) *Nonlinear analysis of shell structures*. AIAA Education series, Washington, DC, pp. 205–206.

Panadian, N. *et al.* (1989) Studies on performance of curved beam finite elements for analysis of thin arches. *Computers & Structures*, **31**, 997–1002.

Pawsey, S.E. and Clough, RW. (1971) Improved numerical integration of thick shell finite elements. *International Journal for Numerical Methods in Engineering*, **3**, 545–586.

Piegl, L. and Tiller, W. (1995) *The NURBS Book*, Springer-Verlag, Berlin.

Pietraszkiewicz, W. (1979) *Finite Rotations and Lagrangian Description in the Nonlinear Theory of Shells*, PWN, Warszawa=Poznan.

Prathap, G. (1985) The curved beam/deep arch finite ring element revisited. *International Journal for Numerical Methods in Engineering*, **21**, 389–407.

Prathap, G. and Shashirekha, B.R. (1993) Variationally correct assumed strain field for the simple Curved beam Element. *International Journal for Numerical Methods in Engineering*, **37**, 989–1007.

Prenter, P.M. (1975) *Splines and Variational Methods*, John Wiley & Sons.

Raveendranath, P. *et al.* (1999) A two-noded locking-free shear flexible curved beam element. *International Journal for Numerical Methods in Engineering*, **44**, 265–280.

Ray, D. (1999) Beam flexure and plane stress (unpublished)

Ray, D. (2003) C-type method of unified CAMG & FEA. Part I: Beam and arch mega-elements – 3D linear and 2D non-linear. *International Journal for Numerical Methods in Engineering*, **58**, 1297–1320.

Ray, D. (2004) c-Type Method of Unified CAMG & FEA: Part II: Beam & Arch Mega-Elements – True 3D Curved, and Geometrically Nonlinear: An Exact Total & Incremental Curvilinear Formulation. (unpublished)

Ray, D. (2004) c-Type Method of Unified CAMG & FEA: Geometrically Nonlinear Curved Beams, Plates & Shells, Indian Association for Computational Mechanics. International Congress on Computational Mechanics and Simulation. (ICCMS-04), December, IIT-Kanpur.

Ray, D. (2004) Part III: On the Symmetry of the Tangent Operator of Nonlinear Beams: An Exact Curvilinear Formulation, (unpublished)

Ray, D. (2007) Optimal Nonlinear Beam, Plate & Shell Elements & c-type FEM. 9th U.S. Congress on Computational Mechanics, San Francisco.

Ray, D. (2008) Largest Geometrically Exact Nonlinear Beam, Plate & Shell Elements & c-type FEM. Proceedings of 6th International Conference, IASS-IACM 2008 – 6th ICCSSS, Cornell Univ., NY.

Reddy, J.N. and Liu, C.F. (1985) A higher order shear deformation theory of laminated elastic shells. *International Journal of Engineering Science*, **23**(3), 319–330

Reissner, E. (1960) On some problems in shell theory. Proceedings of First Symposium Naval Structural Mech., pp. 74–114.

Reissner, E. (1972) On one-dimensional finite strain beam theory: The plane problem. *Journal of Applied Mathematics and Physics*, **23**, 795–804.

Reissner, E. (1973) On one-dimensional large displacement finite strain beam theory. *Studies in Applied Mathematics*, **52**, 87–95.

Reissner, E. (1981) On finite deformations of space-curved beams. *Journal of Applied Mathematics and Physics*, **32**, 734–744.

Reus, J. (1992) Mechanical Deformations of Hyperpatch Solids. Second. International Conference On Computer Graphics and Visual Technology, December, pp. 147–158.

Schoenberg, I.J. (1946) Contributions to the problem of approximation of equidistant data by analytic functions. *Quarterly Applied Mathematics*, **4**, 45–99.

Shi, G. and Voyiadjis, G.Z. (1991) Simple and efficient shear flexible Two-node arch/beam and four-node cylindrical shell/plate elements. *International Journal for Numerical Methods in Engineering*, **31**, 759–776.

Simmonds, J.G. and Danielson, D.A. (1970) Nonlinear shell theory with finite rotation vector. *Proceedings of Koninklijke Nederlandse Akademie, Wetenschappen, B*, **73**, 460–478.

Simo, JC. (1985) A finite strain beam formulation. The three-dimensional dynamic problem. *Part I. Computer Methods in Applied Mechanics and Engineering*, **49**, 55–70.

Simo, J.C. and Vu-Quoc, L. (1986) A three-dimensional finite strain rod model. Part II: Computational Aspects. *Computer Methods in Applied Mechanics and Engineering*, **58**(1), 79–116.

Simo, J.C. *et al.* (1990) On a stress resultant geometrically exact shell model. Part III: Computational aspects of the nonlinear theory. *Computer Methods in Applied Mechanics and Engineering*, **79**, 21–70.

Simo, J.C. (1992) The (symmetric) Hessian for geometrically nonlinear models in solid mechanics: Intrinsic definition and geometric interpretation. *Computer Methods in Applied Mechanics and Engineering*, **58**(1), 189–200.

Spring, KW. (1986) Euler parameters and the use of quaternion algebra in the manipulation of finite rotations: a review. *Mechanism and Machine Theory*, **21**(5), 365–373.

Stolarski, H. and Belytschko, T. (1982) Membrane locking and reduced integration for curved elements. *Journal of Applied Mechanics*, **49**, 172–178

Strang, G. and Fix, G. (1973) *An Analysis of the Finite Element Method*, Prentice Hall.

Szabo, B. and Babuska, I. (1991) *Finite Element Analysis*, John Wiley & Sons, Inc.

Taylor, RL, *et al.* (1976) A nonconforming element for stress analysis. *International Journal for Numerical Methods in Engineering*, **10**, 1211–1219.

Taylor, R.L. and Ray, D. (2005) Rods – Some Developments for Large Displacements. 5th International Conference on Computation of Shell and Spatial Structures, Salzburg, Austria.

Tessler, A. and Spiridigliozzi, L. (1986) Curved beam elements with penalty relaxation. *International Journal for Numerical Methods in Engineering*, **23**, 2245–2262.

Timoshenko, S and Goodier, J, (1951) *Theory of Elasticity*, 2nd Edition, McGraw-Hill, Inc.

Turner, M.J., Clough, R.W., Martin, H.C., and Topp, LJ. (1956) Stiffness & deflection analysis of complex structures. *Journal of Aerospace Science*, **23**, 805–823.

Vainberg, M.M. (1964) *Variational Methods for the Study of Nonlinear Operators*, Holden-Day. Inc.

Washizu, K. (1964) Some considerations on a naturally curved and twisted slender beam. *Journal of Mathematics and Physics*, **43**(2), 111–116.

Wempner, G. (1989) Mechanics and finite elements of shells. *Applied Mechanics Reviews*, **42**, 129–142.

Wilson, E. *et al.* (1973) Incompatible displacement models, in *Numerical and Computer Methods in Structural Mechanics* (eds S.T. Fenves *et al.*), Academic Press, pp. 43–57.

Yamaguchi, F. (1988), *Curves and Surfaces in Computer Aided Geometric Design*, Springer-Verlag, Berlin.

Ziegler, H. (1977) *Principles of Structural Stability*, 2nd edn, Berkhauser, Basel.

Zienkiewicz, O.C, Taylor, R.L., and Too, J.M. (1971) Reduced integration techniques in general analysis of plates and shells. *International Journal for Numerical Methods in Engineering*, **3**, 275–290.

Zienkiewicz, O.C. and Taylor, R.L. (2000) *The Finite Element Method - Basic Formulattion and Fundamentals*, Butterworth.

Part I

Essential Mathematics

2

Mathematical Preliminaries

2.1 Essential Preliminaries

In this chapter, we will briefly touch upon some of the mathematical ideas that are important for our subsequent investigation of problems in structural engineering, structural mechanics and computer-aided geometric design. Various mathematics and numerical analysis books can be also consulted.

2.1.1 Groups

First, let us introduce some operations and structures needed for the definition of groups:

- ***Binary Operation*** on set S: $f : S \times S \rightarrow S, S \neq \{\phi\}$; the + operation on Z is an example, while the − operation on Z^+ is not; z, z^+ are integer and positive integer sets, respectively.
- ***Commutative binary operation*** on set S: the binary operation is commutative if $\forall a, b \in S$, $a \bullet b = b \bullet a$; clearly, + operation is commutative but the − operation is not since $a - b \neq b - a$.
- ***Binary Structure***: $(S, \bullet)$, set S with binary operation; $(Z, +)$ is an example, while $(Z^+, -)$ is not.
- ***Semi-group***: $(S, \bullet)$ with associative property: $\forall a, b, c \in S, (a \bullet b) \bullet c = a \bullet (b \bullet c)$; $(Z, +), (Z^+, +)$ is an example, while $(Z^+, -)$ is not; $(a - b) - c \neq a - (b - c)$.
- ***Identity element***: $\forall a \in S$, and $(S, \bullet)$, a binary structure, e is an identity if there exists $e \in S$ such that $a \bullet e = e \bullet a = a$. For $(Z, +)$, $e = 0$ is the (additive) identity element because: $\forall a \in Z, \quad a + 0 = 0 + a = a$; for $(Z, \times)$, $e = 1$ is the (multiplicative) identity element because: $\forall a \in Z, \quad a \times 1 = 1 \times a = a$.
- ***Monoid***: a semi-group with an identity. $(Z, +)$ with $e = 0$, and $(Z, \times)$ with $e = 1$ are monoids, while $(Z^+, +)$ with $e = 0$ is not, because $0 \notin (Z^+, +)$.
- ***Inverse element***: $\forall a \in S$, and $(S, \bullet)$, a monoid with e as the identity element, $\tilde{a}$ is an inverse if there exists $\tilde{a} \in S$ such that $a \bullet \tilde{a} = \tilde{a} \bullet a = e$. For $(Z, +)$ with $e = 0$, -2 is the inverse of 2 since $2 + (-2) = -2 + 2 = 0$.

Now, we are in a position to define a group; a ***group*** is a monoid, $(S, \bullet)$ with identity element e such that $\forall a \in S$, there exists an inverse element, $\tilde{a}$. Thus, a group is a set of elements with a binary operation satisfying four properties: ***closure***, ***associativity***, existence of ***identity element*** and ***inverse elements***. Clearly, $(Z, +)$ with $e = 0$ form a group; but, $(Z, \times)$ with $e = 1$ does not form a

Computation of Nonlinear Structures: Extremely Large Elements for Frames, Plates and Shells, First Edition. Debabrata Ray.

group because $\tilde{a} \notin Z$ for $\forall a \in Z$. Similarly, $(Q^*, +)$ is not a group since $e = 0 \notin Q^*(Q^* = Q - \{0\}$ *with* Q, rational number set). In any event, in order to check for a group, we must see that the identity element and the inverse elements for all elements of the binary structure with associative property exist. A commutative group is known as an ***Abelian group***. Clearly, $(Z, +)$ and $(Q^*, \times)$ are Abelian groups. We will be interested in matrix groups, and particularly in rotational matrix groups under multiplication, denoted as special orthogonal group, $SO(3)$, (see Chapter 4 on rotation tensors); this belongs to what is known as the ***Lie group***, which has the geometry of a smooth manifold (a topological space resembling points and their neighborhoods as Euclidean spaces) with the algebraic structure of smooth continuous mapping; thus, the global group property can be constructed from the local group on the linear space known as ***Lie algebra***.

2.1.2 Linear Vector Space

A vector space is defined as follows: given any two (members of the defining set) vectors, $\mathbf{a}$, $\mathbf{b}$, of a vector space, and scalars, α, β, the linear (i.e. homogeneity: scalar multiplication and additivity) combination of vectors given by: $\alpha\mathbf{a} + \beta\mathbf{b}$ is also a vector (***closure*** property) belonging to the vector space; a vector space contains an origin, the vector $\mathbf{0}$, that follows by choosing: $\mathbf{a} = \mathbf{b}$; $\alpha = 1$; $\beta = -1$. A vector addition is both ***commutative*** $(\mathbf{a} + \mathbf{b} = \mathbf{b} + \mathbf{a})$ and ***associative*** $(\mathbf{a} + (\mathbf{b} + \mathbf{c}) = (\mathbf{a} + \mathbf{b}) + \mathbf{c})$. For example, the forces can be represented as vectors, but the confounding nature of finite rotations does not permit them to be vectors: although a single rotation can be described by a vector-like directed segment called an ***axial vector*** along the axis of rotation, it does not obey the vector addition property; in fact, it takes a tensor (see Chapter 4 on rotation) to represent rotation as will be seen in the discussion on rotations. A set of vectors are ***linearly dependent*** if they can be linearly combined to add up to the zero vector without all of the coefficients of combination being simultaneously zero: in an $n + 1$-dimensional vector space, if $c_0\mathbf{v}_0 + c_1\mathbf{v}_1 + \cdots + c_n\mathbf{v}_n = \mathbf{0}$ with not all c_i equal to zero, then the set of vectors $\{\mathbf{v}_i\}$ are linearly dependent; otherwise, the set of vectors is known as ***linearly independent***: in an $n + 1$-dimensional vector space. A set of $n + 1$ vectors that are linearly independent forms what is known as a ***basis vector set*** with each vector known as a ***base vector***. A basis vector set spans the space, that is, every vector of the space can be expressed as a linear combination of this basis vector set. Let us consider a vector, $\mathbf{v}$, of a vector space of $n + 1$ dimensions such that $\mathbf{v} = c_0\mathbf{v}_0 + c_1\mathbf{v}_1 + \cdots + c_n\mathbf{v}_n = c_i\mathbf{v}_i$ with not all coefficients,c_i, $i = 0, 1, 2, \ldots, n$, simultaneously zero, then the vector, $\mathbf{v}$, and the set of $n + 1$ vectors form a linearly dependent set of vectors; the set of vectors, $\mathbf{v}_i$, $\quad i = 0, 1, \ldots, n$ form a basis vector set.

2.1.3 Real Coordinate Space, $\mathbb{R}^n$

The real coordinate space, $\mathbb{R}^n$, consisting of $n-$ tuples of real numbers is an example of a n-dimensional linear vector space over a real number field.

Indicial notation: all scalar components of a vector, $\mathbf{a}$, are denoted in indicial notations for the purpose of both analysis and computation. For example, instead of designating components as $x, y, z, \ldots$, these will be subscripted such as $a_1, a_2, a_3, \ldots$, or in compact form as: a_i, $\ i = 1, n$; as will be seen later, for ***curvilinear coordinate systems***, these could be either subscripted or superscripted indices such as: a^j, $\ j = 1, n$. Thus, a column vector, $\mathbf{x}$, is written component-wise as: $\mathbf{x} = (x_1, x_2, \ldots, x_n)^T$; the superscript T stands for vector transposition. The origin is the zero vector, $\mathbf{0}$, with all its components zero: $\mathbf{O} = (0, 0, \ldots, 0)^T$. The linear combination of two vectors, $\mathbf{x}$ and $\mathbf{y}$ is given component-wise as: $\alpha\mathbf{x} + \beta\mathbf{y} = (\alpha x_1 + \beta y_1, \alpha x_2 + \beta y_1, \ldots, \alpha x_n + \beta y_n)^T$.

2.1.4 Convex Set and Convex Hull

A subset C of a linear space X is called ***convex*** if C contains all the elements on the line segment joining any two elements on the subset, that is, $\forall \mathbf{a}, \mathbf{b} \in C \Rightarrow \mathbf{c} \in C$ where $\mathbf{c} = t\mathbf{a} + (1-t)\mathbf{b}$, $t \in [0,1]$. For example, all circles are a convex set in $\mathbb{R}_2$. The ***convex hull*** of a set S is the intersection of all the convex sets containing S.

Einstein summation convention: for the example in the previous paragraph, in compact form, we could introduce and use the convention of summing over repeated indices. In Cartesian coordinate systems, the repeated indices are all subscripted; thus: $c_i\mathbf{v}_i = c_0\mathbf{v}_0 + c_1\mathbf{v}_1 + \cdots + c_n\mathbf{v}_n$, or $a_ja_j = a_1^2 + a_2^2 + a_3^2$ in three dimensions. However, for curvilinear coordinate systems, the repeated indices must be pair-wise diagonally placed, that is, if one is superscripted, the other must be subscripted; for example, $a^jb_j = a^1b_1 + a^2b_2 + a^3b_3$, and $a^{\alpha\beta}b_{\eta\beta} = \sum_{\beta=1}^{3} a^{\alpha\beta}b_{\eta\beta} = a^{\alpha 1}b_{\eta 1} + a^{\alpha 2}b_{\eta 2} + a^{\alpha 3}b_{\eta 3}$ in three dimensions. Finally, in compact form, a single equation representing the length, l, of a vector, $\mathbf{a} = (a_1, a_2, \dots, a_n)^T$ can be written as: $l^2 = a_ia_i$ whereas, in three dimensions, $a_i = mb_i$ hides three equations in it.

The real coordinate space, $\mathbb{R}^n$, has n standard bases given as: $\mathbf{i}_1 = (1,0,0,\dots,0)^T$, $\mathbf{i}_2 = (0,1,0,\dots,0)^T, \dots, \mathbf{i}_n = (0,0,0,\dots,1)^T$. Thus, a vector, $\mathbf{x}$, can be written, using the summation convention, as: $\mathbf{x} = \sum_{i=1}^{n} x_i\mathbf{i}_i = x_i\mathbf{i}_i$ where x_i's are the components of the vector $\mathbf{x}$.

2.1.5 Classes of Functions: Continuity and Differentiability

For both interpolations and approximations, we will deal almost exclusively with functions of real variables over finite intervals. The degree of smoothness of these functions determines in general the strength with which the analytical properties of interpolation and approximation can validate the process. In other words, the smoother the function, the better is the approximation. In the mathematical literature, functions of varying smoothness abound. Thus, we may organize functions depending on how smooth they are: a continuous function, $f(x)$, $x \in [a,b]$, belongs to the class of $C^0[a,b]$ functions. It is important to note that we have implied both the interior and the boundary of the domain in the definition; if the property of the function holds strictly in the interior of the domain, we will denote it by $C^0(a,b)$. For any $n \geq 0$ integer, if the same function and its derivatives up to and including order n are continuous, then it belongs to $C^n[a,b]$; of course, this classification holds for any vector-valued functions as long as all the component functions individually satisfy the property. An infinitely smooth function belongs to the class denoted by $C^\infty[a,b]$. Clearly, every polynomial function is a member of the class, $C^\infty[a,b]$. A step function belongs to $C^{-1}[a,b]$. In finite element analysis, as we shall see, the constructed global basis functions are infinitely smooth polynomials in the interior of a single element but may be zero on the boundary and over all other elements, that is, all derivatives on the boundary are zero; these functions are denoted as class of C_0^∞. Another class of important functions, for our purpose, belong to the set, $L_p[a,b]$, for which $\int_a^b |f(x)|^p \, dx < \infty$, where the integration is a Lebesgue integral (defined by $\int_a^b f(x)\, dx = \lim_{n\to\infty} \int_a^b f_n(x)\, dx$) as opposed to a Riemann integral. Of particular interest is the situation where $p = 2$, that is the $L_2[a,b]$ space; all continuous functions belong to this class, However, more importantly, continuous functions which have discontinuities at a finite number of points belong to this class. To sum up, there are several reasons to include this class: (a) classes of functions which are not Riemann integrable but are Lebesgue integrable, that is, functions that are continuous except at a finite number of points, can still be investigated, (b) while $C[a,b]$ is not complete (i.e. does not contain all the limit points of Cauchy sequences of functions on them), $L_2[a,b]$ is; in other words, the class of continuous functions are completized by inclusion into $L_2[a,b]$. This is important because it ensures that in any adaptive approximation procedure such as the finite element methods, any sequence of continuous functions approximating the solution will be guaranteed to converge eventually to the actual solution. One of the most important and 'decent' class of functions is the space of polynomial functions, $P_n[a,b]$, of degree n or less in the interval $[a,b]$; they are continuously infinitely differentiable and nicely structured so that term-by-term operations of calculus and algebra are easy. Now, for our purposes here, we will need linear spaces with more structure than just the linear vector spaces defined above; thus, we will briefly introduce several spaces that we will refer to in the finite element methodology described here.

2.1.6 Inner Product Space: Hilbert Space

A real linear space X will be called an inner product space if for any $\mathbf{x}, \mathbf{y} \in X$, there exists a real-valued function, designated by $(\mathbf{x}, \mathbf{y})$ with the following properties: (1) Positivity: $(\mathbf{x}, \mathbf{x}) \geq 0, \quad (\mathbf{x}, \mathbf{x}) = 0,$ *iff* $\mathbf{x} = \mathbf{0}$, (2) Symmetry: $(\mathbf{x}, \mathbf{y}) = (\mathbf{y}, \mathbf{x})$, (3) Linearity: $(\mathbf{x} + \mathbf{y}, \mathbf{z}) = (\mathbf{x}, \mathbf{z}) + (\mathbf{y}, \mathbf{z}), \quad \forall \mathbf{z} \in X$, (4) Homogeneity: $(\beta\mathbf{x} + \mathbf{y}) = \beta(\mathbf{x}, \mathbf{y}), \quad \forall \beta$ real; $(\mathbf{x}, \mathbf{y})$ is called the ***inner product*** of $\mathbf{x}, \mathbf{y} \in X$. A complete inner product space is called a ***Hilbert space***.

Examples: It can be easily seen that the n-dimensional Euclidean space, E^n, is an inner product space with inner product defined as: $(\mathbf{x}, \mathbf{y}) = \sum_{i=1}^{n} x_i\, y_i$, $\mathbf{x} = (x_1, \ldots, x_n) \in E^n$, and so on. So is the space, say, $C^2[a,b]$ with inner product defined as: $(f,g) = \int_a^b f(t)\, g(t)\, dt$, or $(f,g) = \int_a^b \left(f\, g + f_{,t}\; g_{,t}\right)\, dt, f, g \in C^2[a,b]$, and so on, Now, for an inner product space, each inner product defines a ***norm***: $\|\mathbf{x}\| = \sqrt{(\mathbf{x}, \mathbf{x})}$; that all norm properties are satisfied is easily shown by Schwarz (triangular) inequality. Finally, an inner product space encompasses all the geometric properties such as the parallelogram theorem, angular symmetry, the Pythagoras theorem, projection, and so on. For example, we can define ***orthogonality***, a geometric concept, of vectors: iff $(\mathbf{x}, \mathbf{y}) = 0$, we say $\mathbf{x}$ is orthogonal to $\mathbf{y}$; a set $S \subset X$, X inner product space, is called an orthonormal set if $\forall \mathbf{x}, \mathbf{y} \in S$, $(\mathbf{x}, \mathbf{y}) = 0$ if $\mathbf{x} \neq \mathbf{y}$, and equal to 1 if $\mathbf{x} = \mathbf{y}$. One important theorem is that any finite set of non-zero orthogonal elements are linearly independent and thus can form the basis set of the subspace known as an orthogonal subspace. Now, the converse is not true, but every finite set of linearly independent elements can be orthogonalized (Lang, 1968), which we apply in our study of mesh generation; we briefly describe in 3D, without any loss of generalization, the construction of the process algorithmically.

2.1.7 Gram–Schmidt Orthogonalization

Step 0: given linearly independent $\mathbf{x}_1, \mathbf{x}_2, \mathbf{x}_3 \in X$, an inner product space.

Step 1: set $\mathbf{y}_1 = \mathbf{x}_1$ and $\mathbf{z}_1 = \mathbf{y}_1 / \|\mathbf{y}_1\|$;

Step 2: set $\mathbf{y}_2 = \mathbf{x}_2 - \frac{(\mathbf{y}_1, \mathbf{x}_2)}{(\mathbf{y}_1, \mathbf{y}_1)}\mathbf{y}_1$, that is, $\mathbf{y}_2$ is $\mathbf{x}_2$ minus the projection of $\mathbf{x}_2$ on $\mathbf{y}_1$, and $\mathbf{z}_2 = \mathbf{y}_2 / \|\mathbf{y}_2\|$; thus, $\mathbf{y}_2 = \mathbf{x}_2 - \frac{\cancel{\|\mathbf{y}_1\|}(\mathbf{z}_1, \mathbf{x}_2)}{\cancel{\|\mathbf{y}_1\|^2}}\cancel{\|\mathbf{y}_1\|}\mathbf{z}_1 = \mathbf{x}_2 - (\mathbf{z}_1, \mathbf{x}_2)\, \mathbf{z}_1$; note that: $(\mathbf{y}_2, \mathbf{y}_1) = (\mathbf{x}_2, \mathbf{x}_1) - \frac{(\mathbf{x}_1, \mathbf{x}_2)}{\cancel{(\mathbf{x}_1, \mathbf{x}_1)}}\cancel{(\mathbf{x}_1, \mathbf{x}_1)} = 0$, that is, $\mathbf{y}_2$ is perpendicular to $\mathbf{y}_1$.

Step 3: set $\mathbf{y}_3 = \mathbf{x}_3 - \frac{(\mathbf{y}_1, \mathbf{x}_3)}{(\mathbf{y}_1, \mathbf{y}_1)}\mathbf{y}_1 - \frac{(\mathbf{y}_2, \mathbf{x}_3)}{(\mathbf{y}_2, \mathbf{y}_2)}\mathbf{y}_2 = \mathbf{x}_3 - (\mathbf{z}_1 - \mathbf{x}_3)\,\mathbf{z}_1 - (\mathbf{z}_2 - \mathbf{x}_3)\,\mathbf{z}_2$, that is, $\mathbf{y}_3$ is $\mathbf{x}_3$ minus $\mathbf{x}_3$'s projection on the plane containing $\mathbf{y}_1$ and $\mathbf{y}_2$; and $\mathbf{z}_3 = \mathbf{y}_3 / \|\mathbf{y}_3\|$. Then, $\mathbf{z}_1$, $\mathbf{z}_2$, $\mathbf{z}_3$ are completely orthonormalized; since orthogonalized vectors are linearly independent, they form a basis set for the space spanned by them; of course, orthonormal vectors depend on the definition of the inner product of the space.

Remark

- A complete inner product space is called a ***Hilbert space***.
- Every space, clearly, is not an inner product space. There are spaces in which the distance measure, that is, norm, cannot be induced by an inner product; these are known as normed linear spaces, thus, every inner product space is a normed linear space but the converse is not true. For example, a normed linear space with, say, max – norm, cannot define an inner product. In our finite element analysis, we will encounter normed linear spaces, thus we first introduce a normed linear space and then briefly describe a special and important normed linear space: Sobolev space.

2.1.8 Normed Linear Space: Banach Space

A linear space, X, is called a normed linear space if for each element $\mathbf{x} \in X$, there exists a real number, designated $\|\mathbf{x}\|$ and called the ***norm*** of $\mathbf{x}$ with properties: (1) positivity: $\|\mathbf{x}\| \geq 0$, (2) definiteness: $\|\mathbf{x}\| = 0$, iff $\mathbf{x} = \mathbf{0}$, (3) homogeneity: $\|\beta\mathbf{x}\| = |\beta|\,\|\mathbf{x}\|$, $\forall\beta$ scalar, (4) triangular inequality: $\|\mathbf{x} + \mathbf{y}\| \leq \|\mathbf{x}\| + \|\mathbf{y}\|$. It is easily seen that a normed linear space is a metric space with the metric, $d(\mathbf{x}, \mathbf{y}) = \|\mathbf{x} - \mathbf{y}\|$. Examples of normed linear spaces abound: $\mathbb{R}^n$ with $\|\mathbf{x}\|^2 = \sum_{i=1}^{n} x_i^2$, the square or Euclidean norm; another example is $C[a, b]$ with $\|f\| = \underbrace{\max}_{x\in[a,b]} f(x)$. A complete linear space is called a ***Banach space***.

2.1.9 Sobolev Space

In finite element analysis, we encounter varieties of functions, for example, displacement functions, strain functions as derivatives of the displacement functions, and so on. We need to understand these functions as to how they are different from and similar to each other. Mathematically, this tantamount to devising an appropriate norm, that is, a distance measure or a measuring scale. One natural way is to devise a norm, known as the L_2-norm and denoted by $\|\bullet\|_0$, that measures the distance between a function and the identically zero function in an average sense over the domain of the function:

$$\|f\|_0 = \left(\int f^2\,dx\right)^{\frac{1}{2}} \tag{2.1}$$

However, we may have a function, for example, a small saw-tooth function, which may have values very close to the zero function and hence appear similar, that is, a short distance apart under L_2-norm, but their derivatives may be wide apart pointwise, making them in reality very different, that is, far apart from each other. Thus, the L_2-norm which has no information on the derivatives may not be the most suitable norm (or sieve) for the situation. Looking from another angle, instead of solving a mechanics problem, for example, the Dirichlet problem, involving differential equations in the interior and derivatives conditions at the boundary of a material body, a finite element method, which is a variational method, as we shall see in later chapters, converts it into a problem involving an integral functional and then finding the solution as a minimization process; the main idea behind this is that instead of seeking the solution function satisfying pointwise the differential equation and the boundary conditions, from the stringent C^n class, we plan to make it easier by weakening the derivative (averaging or smoothing by an integral operation) requirements by choosing the function from a wider class. This motivates devising what is known as the ***Sobolev norm***, $\|\bullet\|_n$, of degree $n \geq 0$, in one dimension, and the space defined by this norm is known as the ***Sobolev space*** denoted by $\mathcal{H}^n$ whereby we define:

$$\begin{aligned}\|f\|_n &= \left(\int \left(f^2 + \left(\frac{\partial f}{\partial x}\right)^2 + \bullet\bullet\bullet + \left(\frac{\partial^n f}{\partial x^n}\right)^2\right) dx\right)^{\frac{1}{2}} \\ &= \left(\|f\|_0^2 + \left\|\left(\frac{\partial f}{\partial x}\right)\right\|_0^2 + \bullet\bullet\bullet + \left\|\left(\frac{\partial^n f}{\partial x^n}\right)\right\|_0^2\right)^{\frac{1}{2}} \end{aligned} \tag{2.2}$$

Clearly then $L_2 = \mathcal{H}^0$. The Sobolev space, $\mathcal{H}^n[a, b]$, can be thought of as the completion of the class $C^n[a, b]$ with the Sobolev norm, $\|\bullet\|_n$. In other words, every function belonging to a Sobolev space either belongs to $C^n[a, b]$ space or it can be generated as a sequence of functions in $C^n[a, b]$ space. Finally, a function that is a good approximation to the zero function $\mathcal{H}^0$-norm may not be a bad approximation in $\mathcal{H}^1$-norm.

2.1.10 Linear Transformations: Functionals and Operators

The motivation for the introduction of the linear functionals and linear operators arises from our need to deal with real-valued functions whose arguments could be functions or field variables themselves. For example, in the calculus of variations dealing with solid mechanics, the strain energy of a body may be expressed as a function of an associated displacement field or a function that defines the deformed body. Similarly, in mathematics, the definite integral of a function will serve as an example. In linear structural mechanics, the differential equations of equilibrium may be viewed as a linear operator equation involving a ***differential operator***. In other words, any function that operates on its arguments (domain) that are functions to produce (range) real numbers is called a functional. For computational purposes, we are primarily interested in linear functionals defined over normed linear spaces.

2.1.11 Linear Functionals

Let us, then, briefly summarize the following definitions and results about ***functionals***:

- $f(\mathbf{x})$ is a ***functional*** on a ***linear vector space***, $\mathbb{Z}$, if a scalar is assigned to every vector $\mathbf{x} \in \mathbb{Z}$. For example, the space may be a vector function space where every point in the space represents a vector function, say, $\mathbf{x} = \mathbf{x}(t), \quad t \in [a, b]$. In other words, a functional is a function of functions. Moreover, it is a ***linear functional*** if $f(\alpha\mathbf{x} + \beta\mathbf{y}) = \alpha f(\mathbf{x}) + \beta f(\mathbf{y})$ for any $\mathbf{x}, \mathbf{y} \in \mathbb{Z}$ and α, β arbitrary scalars,

- Given any linear vector space, $\mathbb{Z}$, and the definition of ***inner products*** on it, every inner product, $(\mathbf{z}, \mathbf{x}) \equiv \mathbf{z} \bullet \mathbf{x}$, where $\mathbf{z} \in \mathbb{Z}$ is a fixed vector and $\mathbf{x} \in Z$ any argument vector, is a continuous (hence bounded) linear functional, say, $f(\mathbf{x})$,
- The converse is not trivial but true: every continuous linear functional, $f(\mathbf{x})$, has an inner product representation, that is, there exists a vector $\mathbf{z} \in \mathbb{Z}$ such that $f(\mathbf{x}) = (\mathbf{z}, \mathbf{x})$.
- The set of linear functionals on a linear space, X, themselves form a linear space, X^*, called an ***algebraic conjugate space*** defined by: $L_1, L_2 \in X^* \Rightarrow (aL_1 + bL_2)\mathbf{x} = aL_1(\mathbf{x}) + bL_2(\mathbf{x})$ for any a, b scalar, and $\forall \mathbf{x} \in X$.
- ***Orthogonal transformation***: a linear transformation, $L : X \to X$, where X is an inner product space, is called an orthogonal transformation if $(L\mathbf{x}, L\mathbf{y}) = (\mathbf{x}, \mathbf{y})$, where $\mathbf{x}, \mathbf{y} \in X$; the rotational transformation serves as an example (see Chapter 4 on rotation tensors).
- **Energy inner product**: every positive definite (i.e. $(L\mathbf{x}, \mathbf{x}) > 0$ unless $\mathbf{x} = 0$), symmetric (i.e. $(L\mathbf{x}, \mathbf{y}) = (\mathbf{x}, L\mathbf{y})$) linear operator, $L : X \to Y$, X an inner product space and $Y \subseteq X$, provides a second inner product, designated by $(\bullet, \bullet)_L$, defined by $(\mathbf{x}, \mathbf{y})_L = (L\mathbf{x}, \mathbf{y}), \forall \mathbf{x}, \mathbf{y} \in X$.

Example LF1: Vectorial linear functional

Let T be a linear transformation: $T \in (\mathbb{R}^2, \mathbb{R}^3)$ such that $\mathbf{b} = T(\mathbf{a}) = (a_1 + 2a_2,\ a_1 - a_2,\ 2a_1 + 3a_2)$ with $\mathbf{a}(= a_i\mathbf{E}_i) \in^2$ and $\mathbf{b}(= (a_1 + 2a_2)\mathbf{E}_1 + (\ a_1 - a_2)\mathbf{E}_2 + (2a_1 + 3a_2)\mathbf{E}_3) \in \mathbb{R}^3$ under standard bases, $\mathbf{E}_i$, for both the domain ($i = 1, 2$) and range ($i = 1, 2, 3$), with: $\mathbf{E}_1 = (1, 0, 0)$, $\mathbf{E}_2 = (0, 1, 0)$, $\mathbf{E}_3 = (0, 0, 1)$.

Example LF2: Integral linear functional

Let the linear vector space $X = C[a, b]$, the class of continuous functions on $[a, b] \in \mathbb{R}^1$, the real line. Then, $L(f) = \int_a^b x^2 f(x)dx$ is a linear functional, because, for all $f, g \in C[a, b]$, and for all $[a, b] \in \mathbb{R}^1$, we have, by definition of functions and the linearity property of integration:

$$\begin{aligned} L(\alpha f + \beta g) &= \int_a^b x^2(\alpha f + \beta g)(x)dx = \int_a^b x^2(\alpha f(x) + \beta g(x))dx = \int_a^b x^2 f(x)dx + \int_a^b x^2 g(x)dx \\ &= \alpha L(f) + \beta L(g). \end{aligned}$$

Example LF3: Differential linear functional

Let the linear vector space $X = p^2[a, b]$, the class of polynomials of degree two or less on $[a, b] \in \mathbb{R}^1$, the real line. Then, $L(p) = (tD + 2)p = t\frac{\partial p(t)}{\partial t} + 2$, $p \in X, t \in [a, b]$ is a linear functional, because, for all $f, g \in p^2[a, b]$, and for all $\alpha, \beta \in \mathbb{R}^1$, we have, by definition of functions and linearity property of differentiation:

$$\begin{aligned} L(\alpha f + \beta g) &= \left(\frac{\partial}{\partial t} + 2\right)(\alpha f(t) + \beta g(t)) = \alpha\left\{\left(\frac{\partial}{\partial t} + 2\right)f(t)\right\} + \beta\left\{\left(\frac{\partial}{\partial t} + 2\right)g(t)\right\} \\ &= \alpha L(f) + \beta L(g) \end{aligned}$$

2.1.12 Linear Operators

We note that the functionals are merely scalar-valued functions of generally vector arguments which are themselves functions of some vector argument. In other words, functionals are scalar functions of functions. The ***operators***, on the other hand, are vector functions of functions. Let us, then, briefly summarize the following definitions and results about ***operators***:

- A function $L : X \to Y$, X, Y linear vector spaces, is called a ***linear operator***, if, for all $\mathbf{x}, \mathbf{y} \in X$, and for all α, β scalars, $L(\alpha\mathbf{x} + \beta\mathbf{y}) = aL(\mathbf{x}) + \beta L(\mathbf{y})$. In other words, the operator or mapping is linear if the image of a linear combination of any set of vectors is the linear combination of the image vectors.
- ***Positive definite Linear Operator***: a function $L : X \to X, X$ an inner product space, is called positive definite if $(L\mathbf{x}, \mathbf{x}) > 0$ unless $\mathbf{x} = \mathbf{0}, \mathbf{x} \in X$.
- ***Symmetric Linear Operator***: a function $L : X \to X$, X an inner product space, is called symmetric if $(L\mathbf{x}, \mathbf{y}) = (L\mathbf{y}, \mathbf{x}) > 0, \forall \mathbf{x}, \mathbf{y} \in X$.
- ***Energy inner product***: every positive definite, symmetric linear operator, $L : X \to X$, X linear space, and $Y \supset X$, having an inner product, provides a second inner product, called energy inner product and designated as: $(\bullet, \bullet)_L$, and defined by $(\mathbf{x}, \mathbf{y})_L = (L\mathbf{x}, \mathbf{y}), \forall \mathbf{x}, \mathbf{y} \in X$.

Example LO1: Matrix linear operator

Let the linear vector spaces: $X = E^n$, $Y = E^m$, be the Euclidean spaces of dimension n and m, respectively. Then, a matrix, $\mathbf{A} = (a_{ij})$, is a linear operator from $X \to Y$.

Example LO2: Integral linear operator

Let $X = Y = L_2[a, b]$, $K(s, t) \in C[a, b] \times C[a, b]$. If $u(s) \in L_2[a, b]$, then $I(\mathbf{u})$ as defined under is a linear operator: $I(u) = \int_a^b K(s, t)\, u(s)\, ds$

Example LO3: Energy inner product

Let $X = \{u \in C^2[0, 1] \mid u(0) = 0,\ u(1) = 0\}$ and $Y = C[0, 1]$ with inner product: $(f, g) = \int_0^1 f(t)\, g(t)\, dt,\ f, g \in Y$. Now, define the differential operator, $Du \overset{\circ}{=} -\frac{\partial^2 u}{\partial t^2} \equiv -u_{,tt}(t)$. Then, X is linear subspace of Y because $u_{tt}(t)$ is continuous and $u(t) = 0$ identically is contained in X. Now, $\forall u, v \in X$, $(Du, v) = -\int_0^1 u_{,tt}(t)\, v(t)\, dt = \int_0^1 u_{,t}(t)\, v_{,t}(t)\, dt$, by integration by parts and boundary conditions. So, $(Du, v) = (Dv, u)$, that is, is symmetric. Now, $(Du, u) = \int_0^1 (u_{,t}(t))^2\ dt > 0$ unless $u(t) = c$, a constant. But, then, by boundary conditions, $u(t)$ is identically 0. Thus, D is a symmetric, positive definite operator, and D defines an inner product on X, known as ***energy inner product***.

2.1.13 Gateaux or Directional Derivative and Rate of Change of Fields and Functionals

We recall from our previous discussion that functionals are merely scalar-valued functions of generally vector or tensor arguments which are themselves functions of some vector or scalar argument. In other words, functionals are scalar functions of functions. The operators are, likewise, vector or tensor functions of functions, that is, multi-dimensional functionals. Like the scalar functions, the functionals may also assume extrema (maxima or minima) for certain argument functions. In order for us to determine such extrema, we have to be able to track the rate of change of the functional with respect to its argument function. It is just like finding the extrema of a scalar function – the vanishing of the first derivative determines an extremum location. Similarly, the corresponding second derivative determines the nature of the extremum – stationary, maximum or minimum. The situation is very similar for functionals except that we have to have proper definitions for differentiations of functionals.

2.1.14 *Gateaux or Directional Derivative*

Directional derivative is a generalization of the concept of derivatives of ordinary scalar-valued functions with scalar arguments and differentials. Consider an ordinary scalar-valued vector function, $f(\mathbf{x})$, $\mathbf{x} \in \mathbb{R}^3$. Now, if we move away from vector $\mathbf{x} = (x_1, x_2, x_3)^T$ along a vector $\bar{\mathbf{x}} = (\bar{x}_1, \bar{x}_2, \bar{x}_3)^T$, the approximate change df in f is given by:

2.1.14.1 Method 1: "∇-method"

The partial differentiations as:

$$df = f_{,x_1}\bar{x}_1 + f_{,x_2}\bar{x}_2 + f_{,x_3}\bar{x}_3 = \nabla f(\mathbf{x}) \bullet \bar{\mathbf{x}}$$
$$\text{with} \quad f_{,x_j} \equiv \frac{\partial f}{\partial x_j} \tag{2.3}$$

where,

$$\boxed{\nabla f(\mathbf{x}) \equiv \begin{pmatrix} \dfrac{\partial f}{\partial x_1} & \dfrac{\partial f}{\partial x_2} & \dfrac{\partial f}{\partial x_3} \end{pmatrix}^T \equiv \frac{\partial f}{\partial \mathbf{x}}} \tag{2.4}$$

is the ***gradient*** of the function f. We also note that if $f(\mathbf{x})$ is linear in $\mathbf{x}$, that is, $f(\mathbf{x}) = a\mathbf{x} + b$, $a,\ b$ constants, then $df = \frac{\partial f}{\partial \mathbf{x}}\,\mathbf{dx} = a\,\mathbf{dx}$, that is, the differential df does not depend on $\mathbf{x}$, but only on $\mathbf{dx}$. Interpreted from another angle, if we move along a vector $\bar{\mathbf{x}} = (\bar{x}_1, \bar{x}_2, \bar{x}_3)^T$ such that we always stay on a surface given by: $f(\mathbf{x}) = 0$, that is, $\bar{\mathbf{x}}$ is a tangent to the surface, then we have $0 = df = \nabla f(\mathbf{x}) \bullet \bar{\mathbf{x}}$, which implies that $\nabla f(\mathbf{x})$ is normal to the surface; if $f(\mathbf{x}) = c$ depicts a contour of a mountainous surface, then the gradient points toward the direction of"steepest ascent".

2.1.14.2 Method 2: "ε- Method"

However, for the purpose of further generalization to the case of functionals and operators, we may consider another scheme to define differentials based on the Taylor series expansion of $f(\mathbf{x} + \varepsilon\bar{\mathbf{x}})$ in the neighborhood of $\mathbf{x}$ along the vector direction $\bar{\mathbf{x}}$, then, we will have:

$$f(\mathbf{x} + \varepsilon\bar{\mathbf{x}}) = f(\mathbf{x}) + \varepsilon(\nabla f(\mathbf{x}) \bullet \bar{\mathbf{x}}) + O(\varepsilon^2) \tag{2.5}$$

Now, if we maintain $\mathbf{x}$ and $\bar{\mathbf{x}}$ fixed and thus consider $f(\mathbf{x} + \varepsilon\bar{\mathbf{x}})$ as an one-parameter function of ε alone, the first derivative, as defined as follows, gives the same expression for the differential, df:

$$\begin{aligned} \left[\frac{d}{d\varepsilon} f(\mathbf{x} + \varepsilon\bar{\mathbf{x}})\right]_{\varepsilon=0} &\equiv \lim_{\varepsilon \to 0} \left\{ \frac{f(\mathbf{x} + \varepsilon\bar{\mathbf{x}}) - f(\mathbf{x})}{\varepsilon} \right\} \\ &= \lim_{\varepsilon \to 0} \frac{1}{\varepsilon} \{ f(\mathbf{x}) + \varepsilon(\nabla f(\mathbf{x}) \bullet \bar{\mathbf{x}}) + O(\varepsilon^2) - f(\mathbf{x}) \} \\ &= \nabla f(\mathbf{x}) \bullet \bar{\mathbf{x}} \\ &= df \end{aligned} \tag{2.6}$$

Evidently, the second definition of the derivative produces the same differential without requiring any predefinition such as gradient, for the derivatives of the function with respect to its vector argument, $\mathbf{x}$, as in the previous definition of df.

Now, let us replace f by its generalization: ***functional***, $G(\mathbf{u}(\mathbf{x}))$, $\mathbf{x} \in \mathbb{R}^m$, $\mathbf{u} : \mathbb{R}^m \to \mathbb{R}^n$. That is, G is a scalar function of its vector argument, $\mathbf{u}$, which itself is a n-dimensional vector function of its m-dimensional vector argument, $\mathbf{x}$. But, $dG = G_{,\mathbf{u}}\ \mathbf{du}$ has no meaning because the derivative of a functional with respect to its argument vector functions, $G_{,\mathbf{u}} \equiv \frac{\partial G}{\partial \mathbf{u}}$, has no meaning as yet. But we can still invoke the second scheme, that is, "ε -method", to define the rate of change of functionals. Thus:

- The ***Gateaux*** or ***directional derivatives*** of a scalar, vector or tensor (see Chapter 3 on tensors) field in Cartesian coordinate system are defined as:

$$\boxed{\begin{aligned} D\mathbf{T}(\mathbf{x}) \bullet \bar{\mathbf{x}} &\equiv \bar{\mathbf{T}}(\mathbf{x}, \bar{\mathbf{x}}) \\ &= \lim_{\varepsilon \to 0} \frac{d}{d\varepsilon}\{\mathbf{T}(\mathbf{x} + \varepsilon\bar{\mathbf{x}})\} = \lim_{\varepsilon \to 0} \frac{\{\mathbf{T}(\mathbf{x} + \varepsilon\bar{\mathbf{x}})_{\text{parallel transported}} - \mathbf{T}(\mathbf{x})\}}{\varepsilon} \end{aligned}} \tag{2.7}$$

where $D\mathbf{T}(\mathbf{x}) \bullet \bar{\mathbf{x}} \equiv \bar{\mathbf{T}}(\mathbf{x}, \bar{\mathbf{x}})$ is the ***directional*** or ***Gateaux derivative of a tensor field***, $\mathbf{T}(\mathbf{x})$, at $\mathbf{x}$ along the vector direction $\bar{\mathbf{x}}$.

- As to functionals, let $G(\mathbf{u}(\mathbf{x}))$ be a functional defined for an argument function $\mathbf{u}(\mathbf{x}) \in \mathbb{R}^n$ at any $\mathbf{x} \in \mathbb{R}^m$. Now, let $\bar{\mathbf{u}}(\mathbf{x}) \in \mathbb{R}^n$ be another function at $\mathbf{x}$. Then, the functional defined by the ***directional*** or ***Gateaux derivative of the functional***, denoted as $G_{,\mathbf{u}}(\mathbf{u}, \bar{\mathbf{u}})$, or $DG(\mathbf{u}) \bullet \bar{\mathbf{u}}$, of G at $\mathbf{u}$ along the direction $\bar{\mathbf{u}}$ is defined as:

$$\boxed{DG(\mathbf{u}) \bullet \bar{\mathbf{u}} \equiv G_{,u}(\mathbf{u}, \bar{\mathbf{u}}) \stackrel{\Delta}{=} \left[\frac{d}{d\varepsilon} G(\mathbf{u} + \varepsilon\mathbf{u})\right]_{\varepsilon=0} = \lim_{\varepsilon -> 0} \left\{ \frac{G(\mathbf{u} + \varepsilon\bar{\mathbf{u}}) - G(\mathbf{u})}{\varepsilon} \right\}} \tag{2.8}$$

where $DG(\mathbf{u})$ is an operator. As with the Taylor series expansion of functions, if the operator has an expansion such as:

$$G(\mathbf{u} + \varepsilon\bar{\mathbf{u}}) = G(\mathbf{u}) + \varepsilon\ dG + \varepsilon^n\ R(\mathbf{u}, \bar{\mathbf{u}}, \varepsilon) n \gtrsim 2$$

with the property that the residual: $\lim_{\varepsilon -> 0} \varepsilon^{n-1} R(\mathbf{u}, \bar{\mathbf{u}}, \varepsilon) = 0$, then,

$$\begin{aligned} DG(\mathbf{u}) \bullet \bar{\mathbf{u}} \equiv G_{,\mathbf{u}}(\mathbf{u}, \bar{\mathbf{u}}) &\equiv \left[\frac{d}{d\varepsilon} G(\mathbf{u} + \varepsilon\bar{\mathbf{u}})\right]_{\varepsilon=0} \\ &= \left[\frac{d}{d\varepsilon}\left\{G(\mathbf{u}) + \varepsilon\ dG + \varepsilon^n\ R(\mathbf{u}, \bar{\mathbf{u}}, \varepsilon) - G(\mathbf{u})\right\}\right]_{\varepsilon=0} = dG \end{aligned} \tag{2.9}$$

justifies the name differential.

Remarks

- $G_{,\mathbf{u}}(\mathbf{u}, \bar{\mathbf{u}})$ is also sometimes denoted in the literature as operator $\Delta G(\mathbf{u}, \bar{\mathbf{u}})$.
- In the context of ***virtual functionals*** used in finite element analysis, notations found include $\boxed{\delta G(\mathbf{u}, \delta\mathbf{u}) \equiv DG(\mathbf{u}) \bullet \delta\mathbf{u}}$ to indicate the first variation of $G(\mathbf{u}(\mathbf{x}))$ due to $\delta\mathbf{u}(\mathbf{x})$, the variation of $\mathbf{u}$ at point $\mathbf{x}$ of a domain.

- As with the linear functions discussed above, if a functional is linear, that is, $G(\mathbf{u}+\varepsilon\mathbf{v}) = G(\mathbf{u}) + \varepsilon\, G(\mathbf{v})$, then $G_{,\mathbf{u}} = G(\mathbf{v})$ and thus does not depend on $\mathbf{u}$.
- As in the case of the functions, the product and chain rules of differentiation hold for functionals, that is, if, in turn, $\mathbf{u} = \mathbf{u}(t)$ and $\mathbf{v} = \mathbf{v}(t)$ with $t \in \mathbb{R}$, then we have $\dot{G}(\mathbf{u},\mathbf{v}) \equiv \frac{\partial}{\partial t} G(\mathbf{u}(t),\mathbf{v}(t)) = G_{,\mathbf{u}}(\mathbf{u},\mathbf{v}) \bullet \frac{\partial \mathbf{u}}{\partial t} = G_{,\mathbf{u}}(\mathbf{u},\mathbf{v}) \bullet \dot{\mathbf{u}}$.

2.1.15 Vainberg Principle: Symmetric Operator and Potentialness

For problems in nonlinear structural mechanics and engineering, we encounter equations of equilibrium described by nonlinear differential operators. In variational methods such as finite element methods, as we shall see, a problem of such nature is solved in weak form by converting the equations into integral functional form – in an inner product space, say, Hilbert space, (see Chapter 8, on finite element method) – as a virtual work functional; then seeking the extremum or stationary point of this functional having, say, a Gateaux differential, and setting the differential to zero usually gives us the nonlinear equations (known as ***Euler equations***) in operator form. For structural and mechanical problems, the virtual functional always exist for any problems. It is always beneficial to know whether there exists a functional whose gradient is the virtual work functional – but it does not always exist. If it does, it reveals what is known as the ***energy principle***, and the functional is known as the ***energy functional***. The ***Vainberg principle*** (Vainberg, 1964) provides the necessary and sufficient condition under which we can guarantee the existence of the energy functional; moreover, it provides the rule from which to construct the energy functional from a corresponding virtual work functional. According to this principle, the operator in the operator equation must be symmetric for existence of the energy functional known as the potential functional, and the virtual functional is the gradient of the potential functional. Before we introduce the construction algorithm, let us try to understand the Vainberg principle by using an example.

Let us first consider a linear operator equation, in structural engineering and mechanics, given by: $\mathbf{A\,d} = \mathbf{f}$, where $\mathbf{A}$ may be a differential operator from a normed linear space, E, to its conjugate space,E^*; $\mathbf{d} \in E$, the displacement vector, and $\mathbf{f} \in E$ is the force vector. Then, with the virtual displacement designated by $\bar{\mathbf{d}}$, the virtual functional, $G(\mathbf{d},\bar{\mathbf{d}}) \in H$, a Hilbert space, can be represented by the inner product: $G(\mathbf{d},\bar{\mathbf{d}}) = (\mathbf{A\,d} - \mathbf{f},\, \bar{\mathbf{d}})$. $\forall\bar{\mathbf{d}}$. Now, the Gateaux derivative functional, δG, of the functional, G, along the direction, $\hat{\mathbf{d}}$, is given by: $\delta G(\mathbf{d},\bar{\mathbf{d}};\hat{\mathbf{d}}) = \lim_{\beta\to 0} \frac{d}{d\beta}\left\{(\mathbf{A}(\mathbf{d}+\beta\hat{\mathbf{d}}),\bar{\mathbf{d}}) - (\mathbf{f},\bar{\mathbf{d}})\right\} = (\mathbf{A}\,\hat{\mathbf{d}},\,\bar{\mathbf{d}}) = (\mathbf{A}^*\,\bar{\mathbf{d}},\,\hat{\mathbf{d}})$, where $\mathbf{A}^*$ is the adjoint operator of $\mathbf{A}$. Now, if the operator, $\mathbf{A}$, is ***symmetric***, that is, self-adjoint: $\mathbf{A} = \mathbf{A}^*$, then, from the previous expression and the definition of Gateaux derivative, we get: $\delta G(\mathbf{d},\bar{\mathbf{d}};\hat{\mathbf{d}}) = (\mathbf{A}^T\,\bar{\mathbf{d}},\,\hat{\mathbf{d}}) = (\mathbf{A}\,\bar{\mathbf{d}},\,\hat{\mathbf{d}}) = \delta G(\mathbf{d},\hat{\mathbf{d}};\bar{\mathbf{d}})$, $\forall\bar{\mathbf{d}},\,\hat{\mathbf{d}}$. In other words, the functional, δG, is symmetric with respect to $\bar{\mathbf{d}}$ and $\hat{\mathbf{d}}$. In this situation, there exists a functional, $I(\mathbf{d})$, called the ***potential*** (or, in structural engineering, the energy functional), whose gradient turns out to be $G(\mathbf{d},\bar{\mathbf{d}})$, $\forall\bar{\mathbf{d}}$. For this, let us define: $I(\mathbf{d}) = \int_0^1 G(t\mathbf{d},\mathbf{d})\,dt = \int_0^1 (t\mathbf{A}\mathbf{d},\mathbf{d})\,dt - \int_0^1 (\mathbf{f},\mathbf{d})\,dt = \frac{1}{2}(\mathbf{A}\mathbf{d},\mathbf{d}) - (\mathbf{f},\mathbf{d})$. Using the definition for the Gateaux derivative and the symmetry property of the operator, $\mathbf{A}$, we can prove this as: $\delta I(\mathbf{d}) = \lim_{\beta\to 0}\frac{d}{d\beta}\left\{\frac{1}{2}(\mathbf{A}(\mathbf{d}+\beta\,\bar{\mathbf{d}}),\mathbf{d}+\beta\,\bar{\mathbf{d}}) - (\mathbf{f},\mathbf{d}+\beta\,\bar{\mathbf{d}})\right\} = (\mathbf{A\,d}-\mathbf{f})\,\bar{\mathbf{d}} = G(\mathbf{d},\,\bar{\mathbf{d}})$. The Vainberg principle works even for a nonlinear operator (for which it was originally proved) equation of structural engineering and mechanics problems. For our purposes, it basically says that if the derivative of the virtual functional is symmetric, then a functional, called potential or energy functional, exists whose gradient turns out to be the virtual functional itself. In other words, there exists a potential functional, I, such that the virtual functional is given by $\delta I = G$, and its second derivative, $\delta^2 I = \delta G$ is symmetric. For curved spaces, the Gateaux derivative must be replaced by the appropriate ***covariant derivative***. Later, in our development of beam and shell theories, we will apply the Vainberg theorem to recover real strain tensors from its virtual counterparts. In this connection, and as a converse, a theorem from differential geometry is worth mentioning: the tensor obtained from the covariant derivative of a vector field that is the gradient of a scalar field is symmetric. In the language of structural mechanics, for conservative systems, an energy principle exists and the tangent operator (stiffness matrix), obtained by linearization (first-order derivatives) of the virtual functional which itself is a gradient of the energy functional, is symmetric. Algorithmically speaking, in order to apply the Vainberg principle to a virtual functional in virtual displacement method, the real field, $\mathbf{d}$, is replaced by $t\,\mathbf{d}$, $t \in [0, 1]$, and the virtual field, $\bar{\mathbf{d}}$, is replaced by the real field, $\mathbf{d}$, and then we integrate over $t \in [0, 1]$.

2.1.16 Successive Forward Difference Operator

There will be occasions where we need to apply a forward difference scheme and notations in the construction of curves and surfaces.

Definition: Given a sequence of values $f_0, f_1, \ldots$, the n-th forward difference is defined recursively as:

$$
\begin{aligned}
\Delta f_i &= f_{i+1} - f_i \\
\Delta^n f_i &= \Delta(\Delta^{n-1} f_i) = \Delta^{n-1} f_{i+1} - \Delta^{n-1} f_i \\
\text{with } \Delta^0 f_i &= f_i
\end{aligned}
$$

where Δ is called the forward difference operator. We can particularize this, to write down the following useful expressions:

$$
\begin{aligned}
\Delta^1 f_i &= \Delta^0 f_{i+1} - \Delta^0 f_i = f_{i+1} - f_i \\
\Delta^2 f_i &= \Delta^1 f_{i+1} - \Delta^1 f_i = f_{i+2} - 2f_{i+1} + f_i \\
\Delta^3 f_i &= \Delta^2 f_{i+1} - \Delta^2 f_i = f_{i+3} - 3f_{i+2} + 3f_{i+1} - f_i
\end{aligned}
$$

In general, it can be written as:

$$
\Delta^n f_i = \sum_{r=0}^{n} (-1)^{n-r} \binom{n}{r} f_{i+r} \quad \text{with} \quad \binom{n}{r} \stackrel{\Delta}{=} \frac{n!}{r!(n-r)!}
$$

Remarks

- ***Linearity***: the difference operator, Δ^n, is linear in the sense that $\{f_i\}$ and $\{g_i\}$ are two sequences, and a, b scalars, then, $\Delta^n(a\,f_i + b\,g_i) = a\,\Delta^n f_i + b\,\Delta^n g_i$, $\forall n \in \{0, 1, 2, \ldots\}$.
- ***Annihilation***: with evenly spaced knots, Δ^n annihilates all polynomials of degree $(n-1)$; that is, $\Delta^n p_{n-1}(x)$ with evenly spaced knots of any polynomial of degree $(n-1)$ is identically equal to zero.

2.2 Affine Space, Vectors and Barycentric Combination

If we bring in the ***real coordinate space*** to Euclidean geometry, we give algebraic structure to the geometry that creates what is known as the Euclidean space. Euclidean geometry embeds such geometric concepts as the distance (length) between two points of a space and the angle between two sides (lines) obtained from the points in the space. There are two allowable transformations that affect the points in Euclidean space: translation and orthogonal transformation. Under ***translation***, or shift of all points in the space, the distance between any two points remains invariant; under orthogonal transformation such as ***rotation*** about the origin, the angle between two lines remain invariant. Considering that a space is a set of points with a metric (distance function) defined on it, a Euclidean space is thus a real coordinate space equipped with operations such as the ***inner product*** (depicting the angle between two lines) that induces a ***metric*** and a ***norm***, the ***vector product*** (signifying a special kind of rotation of lines) and ***tensor product*** delineating general transformation of lines or vectors. Thus, we proceed to summarize these operations as follows:

- ***Scalar product*** or ***dot product*** or ***inner product***: a first kind of vector multiplication that yields a scalar: $\mathbf{a} \bullet \mathbf{b} = a_i b_i$; it is commutative ($\mathbf{a} \bullet \mathbf{b} = \mathbf{b} \bullet \mathbf{a}$) and distributive ($\mathbf{a} \bullet (\mathbf{b} + \mathbf{c}) = \mathbf{a} \bullet \mathbf{b} + \mathbf{a} \bullet \mathbf{c}$); this makes a Euclidean space an inner product space (i.e. a Hilbert space).
- ***Norm or length of a vector***: the scalar product allows us to define the Euclidean length or ***norm***, $\|\bullet\|$, of a vector as: $\|\mathbf{a}\| = \sqrt{\mathbf{a} \bullet \mathbf{a}} = \sqrt{a_i a_i}$; this makes a Euclidean space a normed space.
- ***Euclidean metric,*** $d(\mathbf{a}, \mathbf{b})$: the norm, in turn, defines the metric (distance) as: $d(\mathbf{a}, \mathbf{b}) \equiv \|\mathbf{a} - \mathbf{b}\| = \sqrt{\sum_{i=1}^{n} (a_i - b_i)^2}$; this makes a Euclidean space a metric space.
- ***Vector product*** or ***cross product***: the second kind of vector multiplication yields a vector: $\mathbf{c} = \mathbf{a} \times \mathbf{b} \neq \mathbf{b} \times \mathbf{a}$ where $\mathbf{c}$ is a vector normal to both $\mathbf{a}$ and $\mathbf{b}$; thus, it is not commutative but distributive ($\mathbf{a} \times (\mathbf{b} + \mathbf{c}) = \mathbf{a} \times \mathbf{b} + \mathbf{a} \times \mathbf{c}$). A vector product is not associative: $\mathbf{a} \times (\mathbf{b} \times \mathbf{c}) \neq (\mathbf{a} \times \mathbf{b}) \times \mathbf{c}$. The cross product can be seen as an operation that transforms one vector to another: in the above example, the $\mathbf{a}\times$ operation transforms vector $\mathbf{b}$ into vector $\mathbf{c}$; as it will turn out, seen from this viewpoint of transformation, the cross product fits the definition of a tensor, and as such has a tensor representation. Finally, since we will use these quite often, we introduce the following important identities:

$$\boxed{\begin{aligned} \mathbf{a} \times (\mathbf{b} \times \mathbf{c}) = (\mathbf{a} \bullet \mathbf{c})\mathbf{b} - (\mathbf{a} \bullet \mathbf{b})\mathbf{c} \\ (\mathbf{a} \times \mathbf{b}) \times \mathbf{c} = (\mathbf{a} \bullet \mathbf{c})\mathbf{b} - (\mathbf{b} \bullet \mathbf{c})\mathbf{a} \end{aligned}} \tag{2.10}$$

- ***Tensor product***: there is a third kind of vector multiplication that yields tensor; later, the definition and the properties of a tensor product will be introduced in great detail.
- ***Vector triple product***: yields a scalar: $\mathbf{a} \times \mathbf{b} \bullet \mathbf{c} = \mathbf{b} \times \mathbf{c} \bullet \mathbf{a} = \mathbf{c} \times \mathbf{a} \bullet \mathbf{b}$, which gives the volume of a parallelepiped formed by three vectors: $\mathbf{a}, \mathbf{b}$ and $\mathbf{c}$ as the three sides.
- ***Permutation symbol***: in order that the vector products in three dimensions can be represented in indicial notation, we need what is called the permutation symbol, e_{ijk}, with the following properties:
  - If any two indices repeat: $e_{ijk} = 0$; example: $e_{113} = 0$,
  - If indices are cyclic in $1, 2, 3$: $e_{ijk} = +1$; example: $e_{123} = e_{231} = e_{312} = +1$,
  - If indices are not cyclic in $1, 2, 3$: $e_{ijk} = -1$; example: $e_{213} = e_{213} = e_{321} = -1$,

  Now, the vector product: $\mathbf{c} = \mathbf{a} \times \mathbf{b}$ can be written, in a right-handed coordinate system, component-wise as: $c_i = e_{ijk} a_j b_k$.

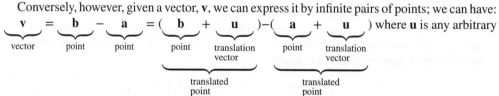

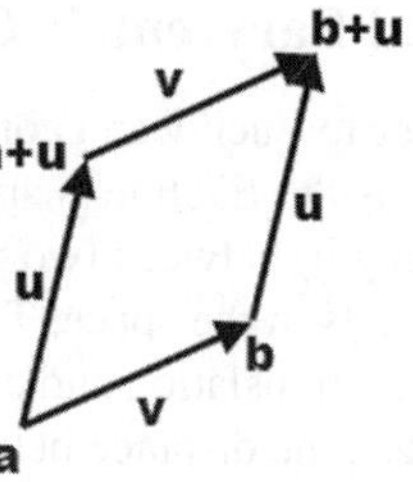

Figure 2.1 Free vector: translation.

2.2.1 Vectors in Affine Space

Now, in order that we can mathematically describe physical entities in a coordinate-independent way, we resort to what is known as ***affine space***, a Euclidean space that has forgotten, so to say, its origin by adding translation to linear maps. Unlike vector spaces, in an affine space:

- no special significance is attached to origin – it is like any other ***point*** that contains n real numbers as components
- only the subtraction operation on the elements of the space, that is the points, are allowed
- a unique ***vector*** in an affine space is then defined by the difference of two points; a ***directed*** line segment beginning with one point called the initial point and ending in the other point known as the final point; for example, if **a** and **b** are two points, then we can define vector, **v**, such that: $\mathbf{v} \equiv \mathbf{b} - \mathbf{a}$, with **v** pointing from **a** to **b**
- the addition of a point to a vector to obtain another point is a valid operation known as ***translation*** (Figure 2.1).

Conversely, however, given a vector, **v**, we can express it by infinite pairs of points; we can have:

$$\underbrace{\mathbf{v}}_{\text{vector}} = \underbrace{\mathbf{b}}_{\text{point}} - \underbrace{\mathbf{a}}_{\text{point}} = \underbrace{(\underbrace{\mathbf{b}}_{\text{point}} + \underbrace{\mathbf{u}}_{\substack{\text{translation}\\ \text{vector}}})}_{\substack{\text{translated}\\ \text{point}}} - \underbrace{(\underbrace{\mathbf{a}}_{\text{point}} + \underbrace{\mathbf{u}}_{\substack{\text{translation}\\ \text{vector}}})}_{\substack{\text{translated}\\ \text{point}}}$$

where **u** is any arbitrary translation vector. Thus, *the vectors are invariant under translation while the points are not*; in other words, the vector components are position independent under translation, that is, the initial point of the vector can be arbitrarily chosen. Under translation alone (***parallel transport*** in Euclidean space), vector components, and hence the vector length, stay invariant in a Euclidean (more accurately, affine) space, so the vectors in a Euclidean space are ***free vectors*** in the sense that they are position independent, that is, the initial point of a vector can be arbitrarily determined by translation. In other words, two vectors in a Euclidean space are identical if and only if they are parallel and have equal length and sense of direction. As we will see, the term "free" in free vectors of Euclidean space is introduced to distinguish these from bound vectors of curved spaces where vector initial point is position dependent and cannot be arbitrarily chosen or moved.

Then, the question is: given that addition of points is not a valid operation, can we express, as with the vectors in a vector space, a point, **a**, of an affine space by a linear combination of other points, known collectively as ***affinely dependent*** points, such as: $\mathbf{a} = c_0\mathbf{a}_0 + c_1\mathbf{a}_1 + \cdots + c_n\mathbf{a}_n = c_i\mathbf{a}_i$? The answer is that although the addition operation is not defined for points, for a special kind of linear combination – ***barycentric*** or ***affine combinations*** where the coefficients of combinations

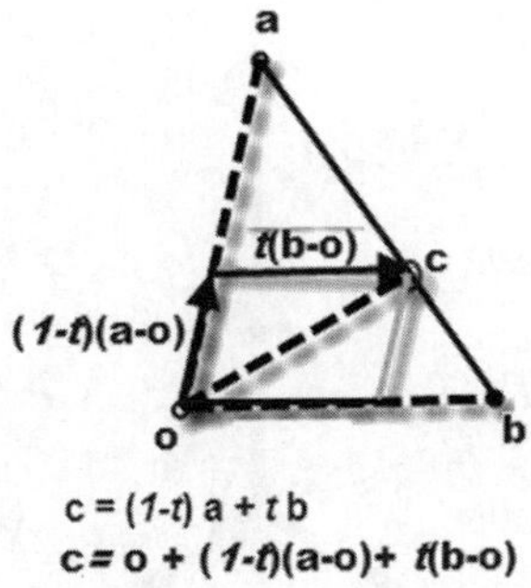

Figure 2.2 Barycentric combination as point transmission.

all add up to unity – the points can be made to look as if they admit addition. So let us consider the following expression, where a point is allowably added to vectors:

$$\mathbf{a} = \underbrace{\mathbf{a}_0}_{\text{point}} + \underbrace{\sum_{i=1}^{n} c_i(\mathbf{a}_i - \mathbf{a}_0)}_{\text{vector}}, \quad \sum_{i=0}^{n} c_i = 1 \tag{2.11}$$

As indicated below, we can express equation (2.11), equivalently, as a kind of summation of points:

$$\begin{aligned} \mathbf{a} &= \mathbf{a}_0 + \sum_{i=1}^{n} c_i(\mathbf{a}_i - \mathbf{a}_0) = \mathbf{a}_0 + \sum_{i=1}^{n} c_i\mathbf{a}_i - \mathbf{a}_0 \sum_{i=1}^{n} c_i \\ &= \mathbf{a}_0\left(1 - \sum_{i=1}^{n} c_i\right) + \sum_{i=1}^{n} c_i\mathbf{a}_i = \mathbf{a}_0 c_0 + \sum_{i=1}^{n} c_i\mathbf{a}_i = \sum_{i=0}^{n} c_i\mathbf{a}_i, \quad \text{with} \quad \sum_{i=0}^{n} c_i = 1 \end{aligned} \tag{2.12}$$

There are numerous examples of summation of points found in mathematics and science: an arbitrary point **c** on a line segment from point **a** to point **b** is given by the barycentric combination of the points as: $\mathbf{c}(t) = (1-t)\mathbf{a} + t\mathbf{b}$, with coefficients adding up to unity: $(1-t) + t = 1$; clearly, the expression is independent of any choice of an origin. To obtain an equivalent allowable operation of summation of a point and a vector, we can rewrite it as (Figure 2.2): $\mathbf{c}(t) = \underbrace{\mathbf{a}}_{\text{point}} \underbrace{+t(\mathbf{b}-\mathbf{a})}_{\text{vector}}$.

We may also note that if we choose any arbitrary point **o** as origin, then we can express: $\mathbf{c}(t) = (1-t)\mathbf{a} + t\mathbf{b} = \underbrace{\underbrace{\underbrace{\mathbf{o}}_{\text{point}} \underbrace{+(1-t)(\mathbf{a}-\mathbf{o})}_{\text{vector}}}_{\text{translated point}} \underbrace{+t(\mathbf{b}-\mathbf{o})}_{\text{vector}}}_{\text{translated point}}$, which is the scaled successive translations of the point **o**; thus, a barycentric or affine combination is independent of the choice of the origin.

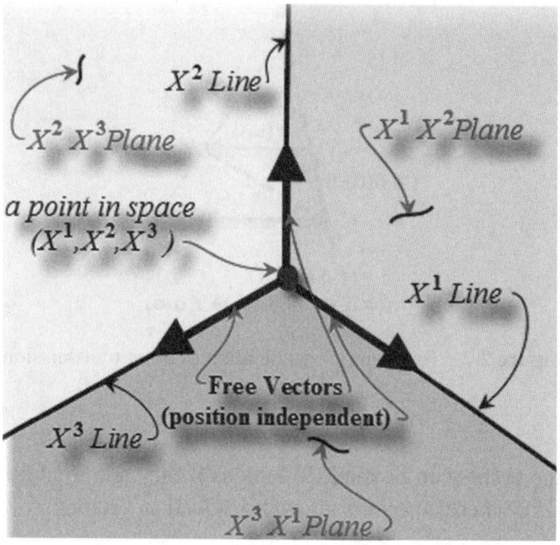

Figure 2.3 Euclidean geometry.

2.2.2 *Base or Coordinate Vectors for Cartesian Coordinate Systems*

The vectors and tensors are coordinate-independent objects, and as such, the form of equations written in these objects are invariant, but for computational purposes we will have to resort to specific component descriptions of these objects. This, then, brings us to a basis vector set of a Euclidean space that determines the components of a vector for a specific coordinate system.

Referring to the Figure 2.3, in the Cartesian coordinate system of a Euclidean space, a line segment can be considered as intersections of two planes, and points can be recognized as the intersection of three planes. For Cartesian coordinate systems, where the coordinate planes are mutually normal at any point, the normal to a coordinate plane is identical to the tangent along the coordinate line created by the intersections of the other two of these planes – giving rise to only one kind of base vector, $\{\mathbf{i}_i\}$, $i = 1, 2, 3$, for a three-dimensional space:

let $\mathbf{r} = \mathbf{r}(\mathbf{X}) = X_1\ \mathbf{i}_1 + X_2\ \mathbf{i}_2 + X_3\ \mathbf{i}_3$ is a radius vector at any point, $\mathbf{X}$ with $\mathbf{i}_i$, $\quad i = 1, 2, 3$ as the standard base vectors in three dimensions. Then, the derivatives along the coordinate directions, $\mathbf{r}_{,i} \equiv \frac{\partial \mathbf{r}}{\partial X^i} = \mathbf{i}_i$, $\quad i = 1, 2, 3$, are position independent. The base vectors of any other Cartesian coordinate system can be obtained by the rotation of the standard base vectors; referring to the Figure 2.4, then, we have: $\mathbf{e}_i = \mathbf{A}\ \mathbf{i}_i$, $\quad i = 1, 2, 3$. Note that we have conventionally chosen to describe these with subscripted indices.

Now that we have the definition of vectors in place, we will move directly to the curved spaces to introduce these two kinds of sets of base vectors. Finally, we avoid a very similar introduction of tensors in Cartesian coordination system in favor of a more general discussion in the curvilinear coordinate system.

2.3 Generalization: Euclidean to Riemannian Space

Let us now examine how the concepts in Euclidean geometry are directly generalized to Riemannian spaces, that is, curved spaces with adequate differentiability. We start with Euclidean curvilinear coordinate system to establish connection between the two spaces. However, expressions for the equilibrium equations and solution processes are rather cumbersome for curved structures such as curved beams and shell surfaces when expressed in Euclidean coordinate

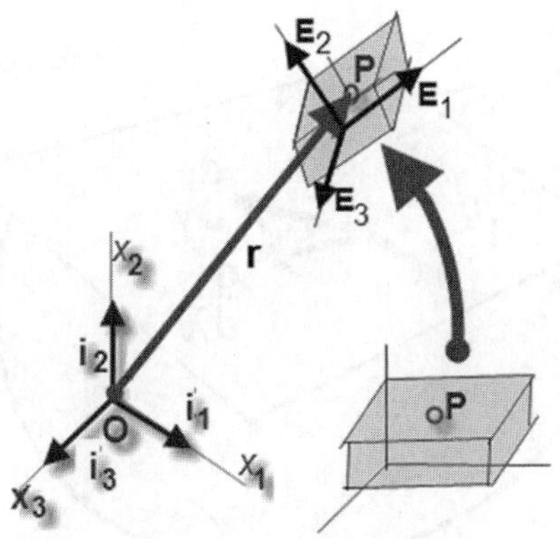

Figure 2.4 Cartesian basis transformation.

systems. So we quickly abandon such Euclidean connection in favor of a more general parametric representation for curved spaces. Thus, as shown in the Figure 2.5, for three-dimensional spaces, any point P of a curved space may be represented parametrically by a three-parameter set, $\{\xi^i\}$, $i = 1, 2, 3$. As indicated above, it can be functionally related to the position vector, $\mathbf{r}$, of a Euclidean Cartesian coordinate system as:

$$\mathbf{r} = \mathbf{r}(\xi^1, \xi^2, \xi^3) = \left(X^1(\xi^1, \xi^2, \xi^3), X^2(\xi^1, \xi^2, \xi^3), X^3(\xi^1, \xi^2, \xi^3)\right)^T \tag{2.13}$$

For the parametric representation expressed in equation (2.13) to be allowable, it must be arbitrarily smooth and must satisfy that the scaling matrix, or ***Jacobian matrix***, $\mathbf{J}$, as defined below, must be of rank 3; that is, there must be at least one non-vanishing determinant of order three to be invertible at any point in the space:

$$\mathbf{J} \equiv \begin{bmatrix} \dfrac{\partial X^1}{\partial \xi^1} & \dfrac{\partial X^1}{\partial \xi^2} & \dfrac{\partial X^1}{\partial \xi^3} \\ \dfrac{\partial X^2}{\partial \xi^1} & \dfrac{\partial X^2}{\partial \xi^2} & \dfrac{\partial X^2}{\partial \xi^3} \\ \dfrac{\partial X^3}{\partial \xi^1} & \dfrac{\partial X^3}{\partial \xi^2} & \dfrac{\partial X^3}{\partial \xi^3} \end{bmatrix} \tag{2.14}$$

In other words, the function given by equation (2.13) is invertible and one-to-one.

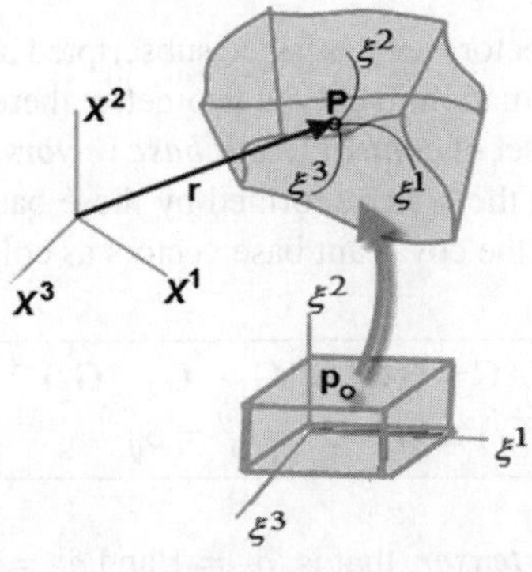

Figure 2.5 Allowable parametric representation.

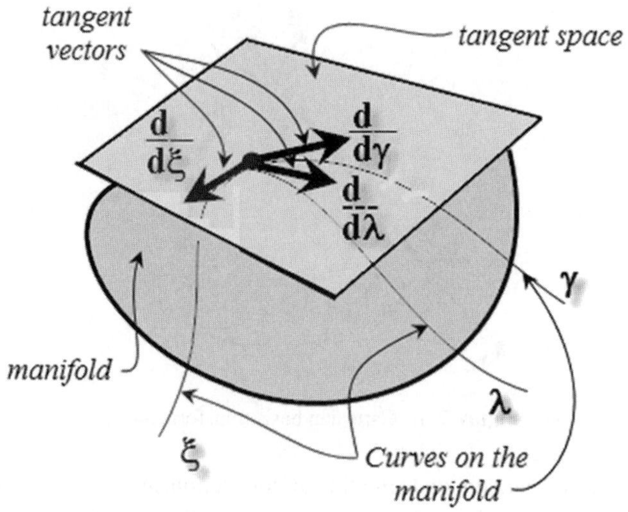

Figure 2.6 Curved space and tangent vectors.

Example: We examine the two-dimensional polar coordinate system, (r, θ), where $\xi^1 \equiv r$ and $\xi^2 \equiv \theta$, and equation (2.13) becomes: $\mathbf{r} = \left(X^1, X^2\right)^T = (r\cos\theta, r\sin\theta)^T$, then, from equation (2.14), we have: $\mathbf{J} = \begin{bmatrix} \cos\theta & -r\sin\theta \\ \sin\theta & r\cos\theta \end{bmatrix}$.

2.3.1 Dual Base vectors and Vector Components

As shown in the Figure 2.6, we can think of affine curves at a point in the space, and the tangents to these curves at a point form a local tangent space. As in the case of Euclidean space, taking the tangents along the affine coordinate curves, we can locally construct the set of base or coordinate vectors at a point that are generally not mutually orthogonal and known as the set of ***covariant base vectors***, $\{\mathbf{G}_i\}$, where:

$$\boxed{\mathbf{G}_i \equiv \frac{\partial \mathbf{r}}{\partial \xi^i}, i = 1,2,3} \tag{2.15}$$

Note that the covariant base vectors are described subscripted because, as opposed to Euclidean geometry, we have a different situation in curved geometry: there exists a dual set of bound base vectors at a point, known as the set of ***contravariant base vectors***, denoted superscripted as $\{\mathbf{G}^i\}$, $i = 1, 2, 3$ and defined such that the matrix, formed by these base vectors as the columns is the inverse of the matrix formed by the covariant base vectors as columns, In other words:

$$\boxed{\begin{aligned} &[\mathbf{G}^1 \quad \mathbf{G}^2 \quad \mathbf{G}^3] = [\mathbf{G}_1 \quad \mathbf{G}_2 \quad \mathbf{G}_3]^{-1}, \text{ or,} \\ &(\mathbf{G}_i, \mathbf{G}^j) \equiv \mathbf{G}_i \bullet \mathbf{G}^j = \delta_i^{\bullet j} = \delta_{ij} \end{aligned}} \tag{2.16}$$

where δ_{ij} is the ***Kronecker delta tensor***, that is, $\delta_{ii} = 1$ and $\delta_{ij} = 0,\ i \neq j$.

Remarks

- Because of the definition of the covariant base vectors, as given by equation (2.15), the Jacobian matrix, **J**, as given by equation (2.14), can be rewritten as the matrix, $\begin{bmatrix}\mathbf{G}_1 & \mathbf{G}_2 & \mathbf{G}_3\end{bmatrix}$, made up of the covariant base vectors as the columns; its determinant, known as the ***Jacobian determinant***, J, is given as:

$$\boxed{J \equiv \det[\mathbf{G}_1 \quad \mathbf{G}_2 \quad \mathbf{G}_3] = \mathbf{G}_1 \bullet (\mathbf{G}_2 \times \mathbf{G}_3) = \mathbf{G}_2 \bullet (\mathbf{G}_3 \times \mathbf{G}_1) = \mathbf{G}_3 \bullet (\mathbf{G}_1 \times \mathbf{G}_2)} \tag{2.17}$$

- From equation (2.17) and the definition of the contravariant base vectors, as in equation (2.16), we get:

$$\boxed{\mathbf{G}^1 = \frac{1}{J}\mathbf{G}_2 \times \mathbf{G}_3, \mathbf{G}^2 = \frac{1}{J}\mathbf{G}_3 \times \mathbf{G}_1, \mathbf{G}^3 = \frac{1}{J}\mathbf{G}_1 \times \mathbf{G}_2} \tag{2.18}$$

- The differential, **dr**, at a point with the Einstein summation convention, is given by:

$$\mathbf{dr} = \frac{\partial \mathbf{r}}{\partial \xi^i}\, d\xi^i = \mathbf{G}_i\, d\xi^i \tag{2.19}$$

- The invariant metric or the distance along an arc of a curve, ds, from equation (2.19), is then given by:

$$\boxed{ds^2 = \mathbf{dr} \bullet \mathbf{dr} = \mathbf{G}_i \bullet \mathbf{G}_j d\xi^i d\xi^j \equiv \mathbf{G}_{ij} d\xi^i d\xi^j} \tag{2.20}$$

where we have introduced the definition of a tensor, ($\mathbf{G}_{ij}$), as the ***covariant metric tensor***, given by:

$$\boxed{\mathbf{G}_{ij} \equiv \mathbf{G}_i \bullet \mathbf{G}_j} \tag{2.21}$$

- We will present details of this very important tensor shortly. In the mean time, for the Euclidean Cartesian coordinate system, the base vectors, $\mathbf{E}_i$'s, are given by: $\mathbf{E}_i = \mathbf{A}\, \mathbf{i}_i$, where **A** is a fixed rotation tensor and the $\mathbf{i}_i$'s are fixed standard mutually perpendicular base vectors, $\mathbf{E}_i \bullet \mathbf{E}_j = \mathbf{E}_i = \mathbf{A}\, \mathbf{i}_i \bullet \mathbf{A}\, \mathbf{i}_j = \mathbf{i}_i \bullet \mathbf{A}^T\, \mathbf{A}\, \mathbf{i}_j = \mathbf{I}_{ij}$. In other words, for a Cartesian coordinate system, the metric tensor is the second-order identity tensor, **I**.
- Unlike the Euclidean metric, the metric in the equation (2.20) is of quadratic differential form and known as the ***Riemannian metric*** or distance; the geometry with a Riemannian metric is known as the ***Riemannian geometry*** – after E. Riemann, its originator – and the metric space is known as the ***Riemannian space***.
- Geometrically speaking, as a generalization of the ***Euclidean spaces***, the planes are replaced by curved surfaces and lines, in general by curved lines in ***Riemannian spaces*** as shown in Figure 2.7.
- From Figure 2.7, the covariant base vectors can be identified by the tangents to the coordinate curves, and the contravariant base vectors with the normals to the coordinate surfaces. Moreover, note that, unlike in a Euclidean space, the normals to the coordinate surfaces are generally not coincident with the tangents to the coordinate curves, and thus two types of base vectors.

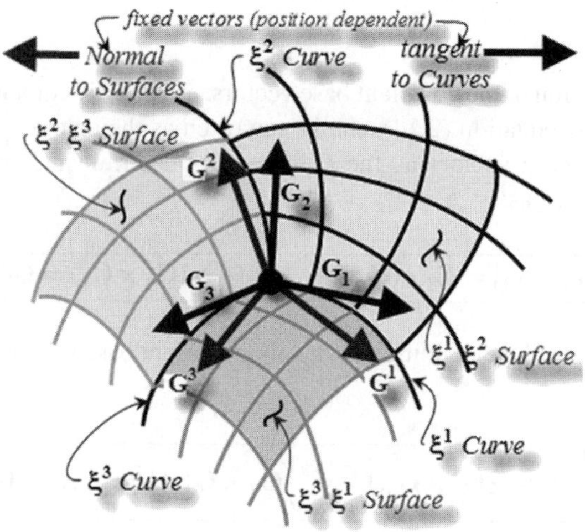

Figure 2.7 Riemannian geometry.

With all these materials available to us, we are ready to introduce the concepts and the properties related to tensors. In particular, we will first identify tensors as linear combinations followed by their coordinate-independent definition based on how the components respond to allowable coordinate transformations; next, we bring in various relevant properties along with important special tensors that are frequently used in all phases of our investigation. Finally, tensor calculus is introduced for our later use.

Example: Continuing on with the two-dimensional polar coordinate system, we have: $\mathbf{G}_1 \equiv \frac{\partial \mathbf{r}}{\partial \xi^1} = \frac{\partial \mathbf{r}}{\partial r} = (\cos\theta,\ \sin\theta)^T$ and $\mathbf{G}_2 \equiv \frac{\partial \mathbf{r}}{\partial \xi^2} = \frac{\partial \mathbf{r}}{\partial \theta} = r(-\sin\theta,\ \cos\theta)^T$ implying: $\|\mathbf{G}_1\| = 1$, $\widehat{\mathbf{G}}_1 = \frac{\mathbf{G}_1}{\|\mathbf{G}_1\|} = \mathbf{G}_1$ and $\|\mathbf{G}_2\| = r$, $\widehat{\mathbf{G}}_2 = \frac{\mathbf{G}_2}{\|\mathbf{G}_2\|} = \frac{1}{r}\mathbf{G}_2$. Now, using the relations: $\mathbf{G}^1 \bullet \mathbf{G}_1 = 1$, $\mathbf{G}^1 \bullet \mathbf{G}_2 = 0$, and $\mathbf{G}^2 \bullet \mathbf{G}_1 = 0$, $\mathbf{G}^2 \bullet \mathbf{G}_2 = 1$ solving for unknowns $\mathbf{G}^1$ and $\mathbf{G}^2$, we get: $\mathbf{G}^1 = (\cos\theta,\ \sin\theta)^T$ and $\mathbf{G}^2 = \frac{1}{r}(-\sin\theta,\ \cos\theta)^T$, implying: $\|\mathbf{G}^1\| = 1$, $\widehat{\mathbf{G}^1} = \frac{\mathbf{G}^1}{\|\mathbf{G}^1\|} = \mathbf{G}^1$ and $\|\mathbf{G}^2\| = \frac{1}{r}$, $\widehat{\mathbf{G}^2} = \frac{\mathbf{G}^2}{\|\mathbf{G}^2\|} = r\mathbf{G}^2$. Thus, $\mathbf{G}^1 = \mathbf{G}_1$ and$\mathbf{G}^2 = \frac{1}{r^2}\mathbf{G}_2$. The determinant, J, of the Jacobian matrix is given as: $J = r$; the differential: $\mathbf{dr} = (\cos\theta,\ \sin\theta)^T\, dr + r(-\sin\theta,\ \cos\theta)^T\, d\theta$, implying that the metric is given by: $ds^2 = dr^2 + r^2 d\theta^2$.

2.4 Where We Would Like to Go

In this chapter, we have briefly presented some of the mathematical preliminaries that are important for our purposes. However, there are two other topics that need to be immediately investigated with a more detailed exposition for which we should look up:

- Tensor analysis and calculus (Chapter 3)
- Rotational tensors (Chapter 4)

3

Tensors

3.1 Introduction

3.1.1 What We Should Already Know

Lincar vector space, base vectors, real coordinate space, linear functional or operator.

3.1.2 Why We Are Here

Nature is oblivious to any chosen coordinate system when it comes to its behavior; thus, the laws of nature – or more specifically, the equations encoding these laws – should not depend on any preferred coordinate system to describe them. In other words, the objects representing the physical entities and the mathematical ***form*** of the equations describing the natural phenomena must be independent of all possible coordinate systems. For any scalars involved, we know that we do not need to introduce any coordinate system, that is, the value of a scalar is the same in any coordinate system. In other words, a scalar is what is known as an ***invariant*** under transformation of coordinate systems. In the case of vectors (the geometric objects, not to be confused with the namesake members of a vector space), the question is: is it possible to talk about vectors without worrying about the coordinate system? The answer is yes. In fact, we can obtain coordinate-free mathematical objects such as vectors by defining in appropriate space: ***affine space*** (a Euclidean space with no preferred point for origin, as will be seen later) of Euclidean geometry; equations revealing natural laws that are written using vectors have invariant forms that do not depend on the coordinate system in this space. For practical computational purposes, however, we have to choose a specific coordinate system: the vectors are defined on the basis of how their components transform as we change from one coordinate system to another by allowable coordinate transformation: affine transformation. For computational purposes where we need these component description, we can put an equation in any coordinate system of our choice. Thus, from the practical point of view, the use of vectors in invariant or absolute form allows us to describe a physical phenomenon in a coordinate-independent invariant form (i.e. using mathematical objects in absolute notation), from which we can easily derive the component relations for any chosen coordinate system.

However, there are phenomena such as the description of surfaces, the constitutive relations of materials, and Einstein's relativity theory, for which vectors alone are not sufficient to describe them in a coordinate-independent invariant form; we have to deal with multidimensional arrays of scalars (equivalently, columns of vectors) as opposed to single rows or columns of scalars

Computation of Nonlinear Structures: Extremely Large Elements for Frames, Plates and Shells, First Edition. Debabrata Ray.

(equivalently, a single row or column vector). From this point of view, we also need to generalize the idea of vectors to higher-order mathematical entities: tensors (which, as we shall see shortly, are of great effectiveness); defining vectors based on the component transformation properties not only renders these objects coordinate-independent but also provides the added advantage of generalization of vectors to tensors; we can talk about tensors in a coordinate-independent way, and just as with the vectors, a tensor can be defined, among other possibilities, on the basis of how its components transform under allowable coordinate transformations.

Before we delve into tensors, we may recall that a vector function is an ordered one-dimensional array of n scalar functions of variables, with n determining the dimensionality of the vector function. Each scalar function constitutes what is known as a component of the vector function which, in turn, can be a scalar (resulting in a scalar-valued vector function) or a vector (resulting in a vector-valued vector function) or a combination thereof. As a generalization of the concept of vectors or vector functions, a tensor function will turn out to be an ordered multi-dimensional matrix array of scalar or scalar functions of variables that can be, in turn, a scalar (resulting in scalar-valued tensor function) or a vector (resulting in vector-valued tensor function) or a tensor (resulting in tensor-valued tensor function) or a combination thereof. The fact that we can talk about a tensor as a matrix array of functions is another way of saying that a tensor is a ***linear transformation (or operator)***. More specifically, a tensor function can be seen as an operator whose components are ***functionals***, that is, functions of a function. As we will see, an important property of linear functionals allows us to have an operationally powerful representation of tensors: the dyadic notation. Most importantly, however, any arbitrary one-dimensional collection of scalar functions cannot be a vector unless its components satisfy the condition of coordinate transformation. From this, we can relate to a scalar as simply an one-dimensional, (i.e. $n = 1$) vector. Similarly, any arbitrary multi-dimensional array of scalar functionals cannot be a tensor unless its components satisfy the condition of allowable coordinate transformation. Thus it will follow that a ***vector is a tensor of order*** 1 (i.e. one row or column of components), and that a ***scalar is a tensor of order*** 0 (i.e. just one component); thus, tensor analysis is a unifying concept.

Moreover, the components of vectors and tensors themselves undergo generalizations as we move from the ***Cartesian coordinate systems*** of the Euclidean 'flat' space to ***curvilinear coordinate systems*** and more general spaces such as Riemannian or differentiable 'curved' spaces. We often have to deal with such coordinate systems or spaces: for example, cylindrical or spherical geometry, more general curved surfaces, or, in structural mechanics, the displacements related to, say, the rigid body cross-sectional rotations of beams, and curved surface geometries of shell structures. By visualizing a curved space as the union of patches of deformed Euclidean spaces, the differential geometry of curved spaces (for example, surfaces) is locally related to tensors, the linear operators. However, unlike the Cartesian coordinate systems of an Euclidean space where we have only one set of components for the basis vectors and basis tensors, in curvilinear coordinate systems, there exist dual basis vectors and three types of basis tensors, so the components of the vectors and tensors as coefficients of linear combination of these basis vectors and tensors may be either covariant, contravariant or mixed (for tensors) as we will see shortly. It is important to recall here that the ***fifth or the parallel postulate of Euclid's geometry*** ceases to hold in curved geometry: the sum of the two internal angles subtended by a transversal does not stay at 180°, and that the parallel lines may meet at some point in the space, as shown in Figure 3.1, for example in the case of a spherical surface.

In a Euclidean Cartesian coordinate system, parallelism is central to translation of vectors in the space giving rise to what is known as parallel transport. More specifically, as we will see later, a vector remains as an invariant geometric object under translation because the components of a vector are preserved by parallel transport. Now, we can think of another geometric property that defines parallelism in Cartesian coordinate system; the invariance of the angle between any

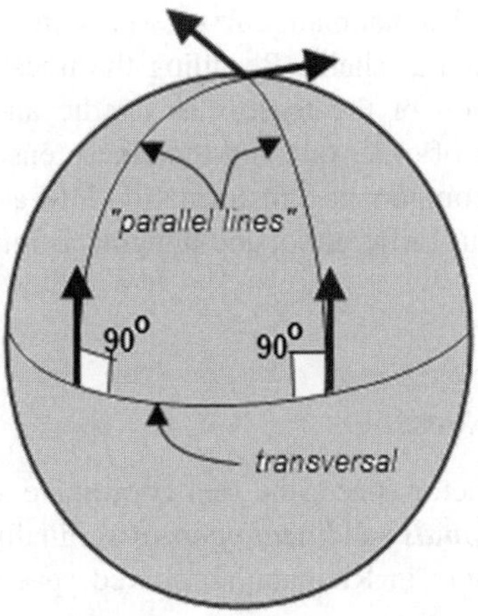

Figure 3.1 Spherical surface.

transversal and a vector can induce parallel translation. In a curved space, such as a spherical surface, translation is not well defined and components of a vector are not preserved because the coordinate system itself is changing from one point to a neighboring point; thus, the concept of parallel transport can only be achieved by maintaining the angle between an infinitesimal vector on the tangent space at a point of a curve on the surface and the coordinate at the point of the curve. Geometrically speaking, the curve acts as a transversal, and thus, parallel transport depends on the curve along which a vector is translated.

Finally, as applications of tensors in mechanics, we will find that the ***general deformations*** of, and the ***force transmission*** through, a material body need to be locally described effectively by tensors called the ***strain tensor*** and the ***stress tensor***, respectively. These tensors, that involve transformation of vectors to vectors, belong to one of the most important classes of tensors for our purposes: ***second-order tensors***. On the other hand, for the ***constitutive theory*** of materials in mechanics that involves transformation of the second-order tensors such as stress and strain tensors, we will have to consider higher-order tensors – more specifically – tensors of order 4.

Clearly, then, we need to do the following:

- define tensors as linear transformations and introduce the tensor product,
- define tensors based on how the components must transform under coordinate transformation, so that we can test the validity of an array of scalars as a tensor: the geometric interpretation of tensors,
- organize the domain of a tensor function in terms of one-, two- or multi-dimensional mathematical entities such as scalars, vectors or tensors, respectively,
- know about the structure of a tensor so that we can evaluate its (range) value for any point in its domain; this brings us to the definition of tensors in absolute, component-free notation through the introduction of an operation known as the tensor product of component vectors of a tensor,
- define various algebraic operations,
- introduce tensor calculus.

3.1.3 How We Get It

Recalling real coordinate space and Euclidean geometry, a geometric object such as a vector is defined in a coordinate-independent way by introducing an affine space; the Euclidean 'flat' space is further generalized into Riemannian 'curved' space in anticipation of tackling problems relating to curved structures such as shells. Recalling the linear functionals and operators, the coordinate-independent definition of the tensors as dyadic and polyadic are introduced; the vectors are identified as tensors of order one and scalars as tensors of order zero. Alternatively, tensors are described by their component transformation law, and various properties of tensors are noted. Finally, tensor calculus (differential and integral) is introduced for both Euclidean and Riemannian spaces.

3.1.4 What We Need to Recall

We need to recall the linear vector space, the real coordinate space. For tensor development, likewise, we recall ***linear functionals*** and ***linear operators***. Finally, for generalization to covariant derivatives of vectors and tensors in Riemannian curved spaces, we evoke the directional or Gateaux derivative of Euclidean space.

3.2 Tensors as Linear Transformation

A tensor may be defined as a continuous linear transformation (and hence bounded), transforming a set of ordered "vectors" of a vector space to another set of ordered "vectors"; we have put quotation mark around the word "vector" because it is intended to be interpreted in its most general sense, which includes tensors themselves, as will be seen later. Let us again note that the linearity of transformations guarantees the existence of a linear space formed by the domain vector spaces and the range vector spaces. In an abstract sense, by definition, if $\mathbf{T}$ is a tensor, i.e., a linear transformation, we may write its operational relation as: $\mathbf{T} : \mathbf{a} \to \mathbf{b}$, simply as: $\mathbf{b} = \mathbf{T}\,\mathbf{a}$, where $\mathbf{a}$ and $\mathbf{b}$ are vectors belonging to the domain and range, respectively. Noting that linear transformations themselves form vector spaces, the domain and the range could be vector spaces of linear transformations. In other words, $\mathbf{a}$ and $\mathbf{b}$ vectors belonging to the domain and range are used in the general sense of members belonging to vector spaces. Obviously, we need more structure to our definition of tensors for computational purposes. The abundance of linear transformation is experienced in all phases of linear or nonlinear (locally linear) analysis in engineering – strain to stress transformation (constitutive relations), displacement to strain transformation (compatibility relations), displacement to force transformation, or temperature to heat flow transformation to name but a few.

3.2.1 Second-Order Tensor

The simplest kind of tensor (generally adequate for our purposes, with some exceptions in the need for higher-order tensors for, say, constitutive theory) is envisioned with both the domain and the range of transformation consisting of vector spaces of one-dimensional column vectors, that is, $\mathbf{a}$ and $\mathbf{b}$ mentioned above are one-dimensional vectors; the domain and range can belong to the same vector space. This gives rise to what is known as ***second-order tensors***. We have already encountered a very special second-order tensor, the ***metric tensor***, and the ***rotation*** mentioned above, which transforms a vector to another vector, is also a second-order tensor. For another example, the familiar ***vector cross product*** is a linear transformation of the form, $\mathbf{T} \equiv (\mathbf{a}\times)$, and given a vector, $\mathbf{a}$, it transforms any other vector, say $\mathbf{b}$, into another vector, $\mathbf{c} = \mathbf{T}\,\mathbf{b} \equiv (\mathbf{a}\times)\,\mathbf{b}$, or the zero vector as it tries to transform itself, that is, $\mathbf{b} = \mathbf{a} \Rightarrow (\mathbf{a}\times)\,\mathbf{a} = \mathbf{0}$. Note that the vector cross product is clearly a linear transformation. Thus, the vector cross product is an example of a tensor.

But, it is clearly a very restrictive and special kind of transformation as it only produces vectors that are normal to the plane containing the given vector and any other vector being transformed. So, in what follows, we proceed to identify more general tensors.

3.2.2 Tensor Product (Dyadic) Notation

Now, let $\mathbf{T}$ be a linear operator over vector space, $\mathbb{Z}$. Then, for any $\mathbf{x} \in \mathbb{Z}$, $\mathbf{Tx} \in \mathbb{Z}$, and we can express the image vector as linear combinations of, say, the covariant base vectors, $\{\mathbf{G}_i\}$ as:

$$\mathbf{Tx} = x^1(\mathbf{x})\,\mathbf{G}_1 + x^2(\mathbf{x})\,\mathbf{G}_2 + \cdots + x^n(\mathbf{x})\,\mathbf{G}_n \tag{3.1}$$

where n is the dimensionality of the vector space. Note that we have indicated by $x^i(\mathbf{x})$ in equation (3.1) the fact that the scalar coefficients of combinations are all functions of $\mathbf{x}$. Thus, the coefficients $x^i(\mathbf{x})$ are functionals, that is, mapping vector functions to scalar functions. We will now show that the coefficients are, in fact, linear functionals:

Note that for $\mathbf{x}, \mathbf{y} \in \mathbb{Z}$ and arbitrary scalars, α, β, we have, from equation (3.1):

$$\mathbf{T}(\alpha\mathbf{x} + \beta\mathbf{y}) = x^1(\alpha\mathbf{x} + \beta\mathbf{y})\,\mathbf{G}_1 + x^2(\alpha\mathbf{x} + \beta\mathbf{y})\,\mathbf{G}_2 + \cdots + x^n(\alpha\mathbf{x} + \beta\mathbf{y})\,\mathbf{G}_n \tag{3.2}$$

and because of the linearity of $\mathbf{T}$:

$$\begin{aligned}\mathbf{T}(\alpha\mathbf{x} + \beta\mathbf{y}) &= \alpha\mathbf{T}(\mathbf{x}) + \beta\mathbf{T}(\mathbf{y}) \\ &= \alpha x^1(\mathbf{x})\,\mathbf{G}_1 + \alpha x^2(\mathbf{x})\,\mathbf{G}_2 + \cdots + \alpha x^n(\mathbf{x})\,\mathbf{G}_n \\ &\quad + \beta x^1(\mathbf{y})\,\mathbf{G}_1 + \beta x^2(\mathbf{y})\,\mathbf{G}_2 + \cdots + \beta x^n(\mathbf{y})\,\mathbf{G}_n\end{aligned} \tag{3.3}$$

Thus, it follows from equation (3.2) and equation (3.3):

$$x^i(\alpha\mathbf{x} + \beta\mathbf{y}) = \alpha x^i(\mathbf{x}) + \beta x^i(\mathbf{y}), \quad i = 1, 2, \dots, n \tag{3.4}$$

In other words, the coefficient functionals are linear functionals. Then, they have a scalar product representation, that is, there exist vectors, $z^i \in \mathbb{Z}$ for $i = 1, 2, \dots, n$ such that: $x^i(\mathbf{x}) = (\mathbf{z}^i, \mathbf{x})$. Thus, we can rewrite equation (3.1) as:

$$\mathbf{Tx} = \mathbf{G}_1(\mathbf{z}^1, \mathbf{x}) + \mathbf{G}_2(\mathbf{z}^2, \mathbf{x}) + \cdots + \mathbf{G}_n(\mathbf{z}^n, \mathbf{x}) \tag{3.5}$$

Now, if we represent $\mathbf{x}$ and base vectors $\mathbf{G}_i$ as column vectors and denote them as $\mathbf{x}$ and $\mathbf{G}_i$, and row vectors as $\mathbf{z}^i$, we can rewrite equation (3.5) again as:

$$\begin{aligned}\mathbf{Tx} &= \mathbf{G}_1(\mathbf{z}^1, \mathbf{x}) + \mathbf{G}_2(\mathbf{z}^2, \mathbf{x}) + \cdots + \mathbf{G}_n(\mathbf{z}^n, \mathbf{x}) \\ &= \mathbf{G}_1(\mathbf{z}^1 + \mathbf{G}_2)(\mathbf{z}^2 + \cdots + \mathbf{G}_n)(\mathbf{z}^n, \mathbf{x})\end{aligned} \tag{3.6}$$

Noting that $\mathbf{x}$) is any arbitrary vector, we finally have Dirac's dyad notation for tensors as:

$$\boxed{\mathbf{T} = \mathbf{G}_1(\mathbf{z}^1 + \mathbf{G}_2)(\mathbf{z}^2 + \cdots + \mathbf{G}_n)\mathbf{z}^n} \tag{3.7}$$

3.2.3 Tensor Product

Now, instead, we can introduce a rule of operation as:

$$\boxed{(\mathbf{a} \otimes \mathbf{b})\mathbf{c} = (\mathbf{b} \bullet \mathbf{c})\mathbf{a}} \tag{3.8}$$

where $\mathbf{a}, \mathbf{b}, \mathbf{c} \in \mathbb{Z}$ and operation $\otimes$ is known as the tensor product of any two vectors, $\mathbf{a}, \mathbf{b} \in \mathbb{Z}$ that transforms vector $\mathbf{c} \in \mathbb{Z}$ to a vector $\mathbf{a} \in \mathbb{Z}$ scaled by $(\mathbf{b} \bullet \mathbf{c})$. Then, clearly from equation (3.5), we have:

$$\begin{aligned} \mathbf{Tx} &= (\mathbf{G}_1 \otimes \mathbf{z}^1)\mathbf{x} + (\mathbf{G}_2 \otimes \mathbf{z}^2)\mathbf{x} + \cdots + (\mathbf{G}_n \otimes \mathbf{z}^n)\mathbf{x} \\ &= \{(\mathbf{G}_1 \otimes \mathbf{z}^1) + (\mathbf{G}_2 \otimes \mathbf{z}^2) + \cdots + (\mathbf{G}_n \otimes \mathbf{z}^n)\}\,\mathbf{x} \end{aligned} \tag{3.9}$$

Finally, noting that $\mathbf{x}$ is arbitrary, from equation (3.9), we get the tensor product definition of the tensor as:

$$\boxed{\mathbf{T} = (\mathbf{G}_1 \otimes \mathbf{z}^1) + (\mathbf{G}_2 \otimes \mathbf{z}^2) + \cdots + (\mathbf{G}_n \otimes \mathbf{z}^n)} \tag{3.10}$$

To sum up:

- a tensor product, denoted by $\otimes$, of two vectors, **a** and **b**, defines a simple ***second-order tensor*** $\mathbf{T} \equiv \mathbf{a} \otimes \mathbf{b}$ with the rule that for any vector **c** belonging to the same vector space as **b**, we have: $\mathbf{Tc} = (\mathbf{a} \otimes \mathbf{b})\mathbf{c} \equiv (\mathbf{b} \bullet \mathbf{c})\mathbf{a}$.

Clearly, by definition, $\mathbf{a} \otimes \mathbf{b} \neq \mathbf{b} \otimes \mathbf{a}$ that is it does not satisfy the commutative property. However, because of the linearity property of the dot product, the tensor product demonstrates both the distributive $(\mathbf{a} \otimes (\mathbf{b} + \mathbf{c}) = \mathbf{a} \otimes \mathbf{b} + \mathbf{a} \otimes \mathbf{c})$ and the associative $(\alpha\mathbf{a} \otimes \mathbf{b} = \mathbf{a} \otimes (\alpha\mathbf{b}) = \alpha(\mathbf{a} \otimes \mathbf{b})$ properties for any scalar α and vectors **a**, **b**, **c**. If we settle for column vectors only, as is customary in mechanics, then the above tensor product can also be expressed simply as $\mathbf{T} \equiv \mathbf{ab}^T$ since by this definition, we also get: $\mathbf{Tc} \equiv \mathbf{ab}^\mathbf{T}\mathbf{c} = (\mathbf{b}^\mathbf{T}\mathbf{c})\mathbf{a} = (\mathbf{b} \bullet \mathbf{c})\mathbf{a}$. Note that only **b** and **c** need to belong to the same vector space in the above example.

Finally, tensor product between two vectors (dyad) may be seen as the backbone of second-order tensors (various second-order tensors are presented later) in deriving various properties. Similarly generalizing the idea, simple arbitrary-order (polyad) tensors can be formed out of more than two vectors with appropriate rules of operation. For example, a third-order tensor, **S**, may be generated as: $\mathbf{S} \equiv \mathbf{a} \otimes \mathbf{b} \otimes \mathbf{c}$.

3.3 General Tensor Space

As indicated in the case of ***general tensors***, a formal definition comes from a generalization of the representation of vectors in a vector space. Just as a vector space is spanned by its ***base vectors***, a tensor space (i.e. ***a linear vector space*** whose elements are tensors) can be spanned by its base tensors. Also, just as the vector components are dependent on the choice of the base vector set, the tensor components depend on the choice of base tensor set; given a base tensor set, the components yield the matrix representation of a tensor. Of course, as indicated before, we have assumed that the ***linear transformations*** form a linear vector space as they do. The base tensors are formed by the tensor product of underlying ***covariant*** and/or ***contravariant*** base vectors; then, the base tensors are of the form: $\mathbf{g_i} \otimes \mathbf{g_j}, \mathbf{g_i} \otimes \mathbf{g^j}, \mathbf{g^i} \otimes \mathbf{g_j}, \mathbf{g^i} \otimes \mathbf{g^j}$. This can easily be seen, for example, if we substitute for any arbitrary vectors $\mathbf{a} = a^i\mathbf{g_i}$ and $\mathbf{b} = b^j\mathbf{g}_j$ in the tensor definition: $\mathbf{T} \equiv \mathbf{a} \otimes \mathbf{b}$, then by the associative law and definition $T^{ij} \equiv a^i b^j$, we have $\mathbf{T} \equiv T^{ij}\mathbf{g}_i \otimes \mathbf{g}_j$, where we have used the ***Einstein summation convention***. We may note that just like the vectors, **T** is an invariant object. Clearly, in three dimensions with nine base tensors, a second-order tensor has nine components. In general, then, depending on pure covariant or contravariant or mixed base tensors, component-wise we can describe by: $\mathbf{T} \equiv T^{ij}\mathbf{g}_i \otimes \mathbf{g}_j = T_{ij}\mathbf{g}^i \otimes \mathbf{g}^j = T^i_j\mathbf{g}_i \otimes \mathbf{g}^j = T_i^j\mathbf{g}^i \otimes \mathbf{g}_j$ where $T^{ij} \equiv a^i b^j, T_{ij} \equiv a_i b_j, T^i_j \equiv a^i b_j, T_i^j \equiv a_i b^j$. Thus, just as with the vectors, a general second-order tensor is represented by the base tensors of the underlying tensor space.

Definition: A tensor is an invariant object which may be described as a linear combination of the base tensors with the constants of combination recognized as the components of the tensor, yielding a matrix representation of the tensor corresponding to the chosen set of base tensors.

3.3.1 Component Description: Matrix Representation

The physical significance of any component, say T_{ij}, can be seen by first using the definition of the ***tensor product*** to operate the tensor on a covariant base vector $\mathbf{g}_j$, and then performing the vector scalar product with the resulting transformed vector and $\mathbf{g}_i$, that is $\mathbf{g}_i \bullet (\mathbf{T}\mathbf{g}_j) = \mathbf{g}_i \bullet T_{mn}(\mathbf{g}^n \bullet \mathbf{g}_j)\mathbf{g}^m = T_{mn}\delta^n_j(\mathbf{g}_i \bullet \mathbf{g}^m) = T_{mj}\delta^m_i = T_{ij}$, where $\delta_i^j = \delta_{ij} = \delta^{ij} = \delta^i_j$ are the components of a special tensor called an ***identity tensor***. In other words, we can see that T_{ij} is the ith component of the vector $\mathbf{T}\mathbf{g}_j$ and is known as the covariant component of a tensor. These equivalences are also written in more compact form without the scalar product symbol: $\mathbf{g}_i\mathbf{T}\mathbf{g}_j \equiv \mathbf{g}_i \bullet (\mathbf{T}\mathbf{g}_j)$. Similar operations can be performed with base tensors formed by pure covariant base vectors or a mix of covariant and contravariant base vectors to describe contravariant component (T^{ij}) or mixed components (T^i_j, T_i^j) of a tensor.

Now, we are ready to see, component-wise, how a tensor operates on any arbitrary vector, say, $\mathbf{a} = a^j\mathbf{g}_j$. Following the pattern above, we can easily verify that $\mathbf{T}\mathbf{a} = T_{mn}a^j(\mathbf{g}^n \bullet \mathbf{g}_j)\mathbf{g}^m = T_{mn}a^j\delta^n_j\mathbf{g}^m = (T_{mn}a^n)\mathbf{g}^m$, where we have used the summation convention. For example, for a transformation of an N-dimensional vector space to an M-dimensional vector space, m runs from 1 to M, and n runs from 1 to N. Clearly, the resulting entity is a M-dimensional vector. Because of our choice of the base tensors and hence the components of the tensor, the resulting vector turned out to be expressed as a linear combination of the contravariant base vectors. If we express for the resulting vector: $\mathbf{b} \equiv \mathbf{T}\mathbf{a} = \mathbf{b}_m\mathbf{g}^m$, then the components of the transformed vector are given by: $\mathbf{b}_m = T_{mn}a^n$ with summation implied for n over the range.

- Thus, every tensor, **T**, as a continuous linear operator, has a matrix representation, that is, there exist vectors, $\mathbf{G}_i, \mathbf{G}_j, \mathbf{G}^i, \mathbf{G}^j \in \mathbb{Z}$ such that $T_{ij} = \mathbf{G}_i \bullet \mathbf{T}\,\mathbf{G}_j$ or $T^{ij} = \mathbf{G}^i \bullet \mathbf{T}\,\mathbf{G}^j$ or $T^i_j = \mathbf{G}^i \bullet \mathbf{T}\,\mathbf{G}_j$ or $T_i^j = \mathbf{G}_i \bullet \mathbf{T}\,\mathbf{G}^j$, where T_{ij}, T^{ij} are the covariant and the contravariant components, respectively, and T_i^j, T^i_j are mixed components of the matrix.
- The matrix elements are nothing other than the components of a tensor when both the domain and the range base vectors or base tensors are already chosen; thus, the matrices are coordinate dependent while the corresponding tensors are not.
- Because the matrix representation of a tensor is dependent on both the domain and the range basis sets, any change in either of these will change the components of the matrix. Let **T** be a second-order tensor such that $\mathbf{b} = \mathbf{T}\,\mathbf{a}$ where **a** and **b** are vectors belonging to the domain and range, respectively; with domain and range bases chosen, let $\mathbf{b}^r = \mathbf{T}^m\,\mathbf{a}^d$, where $\mathbf{T}^m$ is the

matrix representation of the tensor, $\mathbf{a}^d$ is the domain coordinate vector and $\mathbf{b}^r$ is the range coordinate vector. Now suppose that the basis sets of both the domain and range are changed such that: $\mathbf{a}^d = \mathbf{Q}^m\, \overline{\mathbf{a}^d}$ and $\mathbf{b}^r = \mathbf{P}^m\, \overline{\mathbf{b}^r}$ where $\overline{\mathbf{a}^d}$, $\overline{\mathbf{b}^r}$ are the new domain and range coordinate vectors, respectively, and $\mathbf{Q}^m$, $\mathbf{P}^m$ are the corresponding matrices reflecting the basis changes. Then, using the new relationships, we have: $\mathbf{P}^m\, \overline{\mathbf{b}^r} = \mathbf{b}^r = \mathbf{T}^m\, \mathbf{Q}^m\, \overline{\mathbf{a}^d}$. Thus, the change of bases result in a new matrix representation: $\overline{\mathbf{b}^r} = \overline{\mathbf{T}^m}\, \overline{\mathbf{a}^d}$, given by:

$$\boxed{\overline{\mathbf{T}^m} = (\mathbf{P}^m)^{-1}\, \mathbf{T}^m\, \mathbf{Q}^m} \tag{3.11}$$

Example MR1: Matrix representation of a tensor

Continuing from ***Example LF1*** (Chapter 2), for $\mathbf{T}$, the linear transformation or tensor, we would like to get (a) its matrix representation, $\mathbf{T}^{\mathbf{E}}_{\mathbf{E}}$, under standard bases, $\mathbf{E}_i$, for both the domain $(i = 1, 2)$ and range $(i = 1, 2, 3)$, and (b) its matrix representation, $\mathbf{T}^{\mathbf{g}}_{\mathbf{G}}$, under domain bases: $\mathbf{G}_1 = (1, 1)$, $\mathbf{G}_2 = (0, -1)$, and range bases: $\mathbf{g}_1 = (1, 1, 1)$, $\mathbf{g}_2 = (1, 0, 1)$, $\mathbf{g}_3 = (0, 0, 1)$.

(a) For standard bases, it is easy to get the matrix representation by observation of the tensor definition:

$$\mathbf{T}^{\mathbf{E}}_{\mathbf{E}} = \begin{bmatrix} 1 & 2 \\ 1 & -1 \\ 2 & 3 \end{bmatrix} \quad \mathbf{b}(= (a_1 + 2a_2)\mathbf{E}_1 + (a_1 - a_2)\mathbf{E}_2 + (2a_1 + 3a_2)\mathbf{E}_3) \in \mathbb{R}^3$$

(b) Following equation (3.11), $\mathbf{T}^m = \mathbf{T}^{\mathbf{E}}_{\mathbf{E}} = \begin{bmatrix} 1 & 2 \\ 1 & -1 \\ 2 & 3 \end{bmatrix}$, $\mathbf{Q}^m = \begin{bmatrix} 1 & 0 \\ 1 & -1 \end{bmatrix}$, and $\mathbf{P}^m = \begin{bmatrix} 1 & 1 & 0 \\ 1 & 0 & 0 \\ 1 & 1 & 1 \end{bmatrix}$.

Now, noting that $(\mathbf{P}^m)^{-1} = \begin{bmatrix} 0 & 1 & 0 \\ 1 & -1 & 0 \\ -1 & 0 & 1 \end{bmatrix}$, it follows from equation (3.11):

$$\overline{\mathbf{T}^m} = (\mathbf{P}^m)^{-1}\, \mathbf{T}^m\, \mathbf{Q}^m = \begin{bmatrix} 0 & 1 & 0 \\ 1 & -1 & 0 \\ -1 & 0 & 1 \end{bmatrix} \begin{bmatrix} 1 & 2 \\ 1 & -1 \\ 2 & 3 \end{bmatrix} \begin{bmatrix} 1 & 0 \\ 1 & -1 \end{bmatrix} = \begin{bmatrix} 0 & 1 \\ 3 & -3 \\ 2 & -1 \end{bmatrix}.$$

Example MR2: Matrix representation of differential linear functional

Continuing from ***Example LF3*** (Chapter 2), for the linear vector space $X = p^2[a, b]$, the class of polynomials of degree two or less on $[a, b] \in \mathbb{R}^1$, the real line, let $\mathbf{b} = \mathbf{T}\,\mathbf{a}$ where $\mathbf{a}, \mathbf{b} \in X$, are polynomials of degree two or less, and $\mathbf{T}(\bullet) = (tD + 2)(\bullet) = t\frac{\partial(\bullet)(t)}{\partial t} + 2$ is the tensor function. Now the problem is (a) to find vector $\mathbf{b}$ if $\mathbf{a} = 1 + 2t^2$, and the matrix representation of $\mathbf{T}$ and (b) when both the domain and range basis sets are given by the monomials of up to degree two: $(1 \quad t \quad t^2)$, (c) when both the basis sets are changed to Bernstein bases (used extensively in computer-aided geometric design): $((1-t)^2 \quad 2t(1-t) \quad t^2)$.

(a) $\mathbf{b} = \mathbf{T}\,\mathbf{a} = \left(t\frac{\partial}{\partial t} + 2\right)(1 + 2t^2) = 2 + 8t^2$

(b) with monomial bases: $\mathbf{a} = (1 \quad t \quad t^2)\mathbf{a}^{mon}$ where the coordinate vector: $\mathbf{a}^{mon} = \begin{Bmatrix} 1 \\ 0 \\ 2 \end{Bmatrix}$;

similarly, $\mathbf{b} = (1 \quad t \quad t^2)\mathbf{b}^{mon}$ where the range coordinate vector, $\mathbf{b}^{mon} = \begin{Bmatrix} ? \\ ? \\ ? \end{Bmatrix}$ yet to be determined. Now,

$$\mathbf{b} = (1 \quad t \quad t^2)\mathbf{b}^{mon} = \mathbf{T}\,\mathbf{a} = \left(t\frac{\partial}{\partial t} + 2\right)(1 \quad t \quad t^2)\mathbf{a}^{mon}$$

$$= (2 \quad 3t \quad 4t^2)\mathbf{a}^{mon} = (1 \quad t \quad t^2)\begin{bmatrix} 2 & 0 & 0 \\ 0 & 3 & 0 \\ 0 & 0 & 4 \end{bmatrix}\mathbf{a}^{mon}$$

Comparing both sides of the above expression, we get: $\mathbf{b}^{mon} = \mathbf{T}^{mon}_{mon}\,\mathbf{a}^{mon}$ where $\mathbf{T}^{mon}_{mon}$ is the matrix representation, under monomial bases, of the tensor, $\mathbf{T}$, given by: $\mathbf{T}^{mon}_{mon} \equiv \begin{bmatrix} 2 & 0 & 0 \\ 0 & 3 & 0 \\ 0 & 0 & 4 \end{bmatrix}$. We can check the invariance of the range vector, $\mathbf{b}$, as: $\mathbf{b} = (1 \quad t \quad t^2)\mathbf{b}^{mon} =$

$$(1 \quad t \quad t^2)\begin{bmatrix} 2 & 0 & 0 \\ 0 & 3 & 0 \\ 0 & 0 & 4 \end{bmatrix}\mathbf{a}^{mon} = (1 \quad t \quad t^2)\begin{bmatrix} 2 & 0 & 0 \\ 0 & 3 & 0 \\ 0 & 0 & 4 \end{bmatrix}\begin{Bmatrix} 1 \\ 0 \\ 2 \end{Bmatrix} = 2 + 8t^2, \text{ as before.}$$

(c) Now we can relate Bernstein polynomial bases with the monomial bases as: $((1-t)^2 \quad 2t(1-t) \quad t^2) = (1 \quad t \quad t^2)\mathbf{P}^{Ber}_{mon}$ where the basis transformation matrix, $\mathbf{P}^{Ber}_{mon}$, is given as: $\mathbf{P}^{Ber}_{mon} = \begin{bmatrix} 1 & 0 & 0 \\ -2 & 2 & 0 \\ 1 & -2 & 1 \end{bmatrix}$ with $\left(\mathbf{P}^{Ber}_{mon}\right)^{-1} = \begin{bmatrix} 1 & 0 & 0 \\ 1 & \frac{1}{2} & 0 \\ 1 & 1 & 1 \end{bmatrix}$. Using the change of basis formula as in equation (3.11), we then have the new matrix representation, $\mathbf{b}^{Ber} = \mathbf{T}^{Ber}\,\mathbf{a}^{Ber}$ where:

$$\mathbf{T}^{Ber} = (\mathbf{P}^{Ber}_{mon})^{-1}\,\mathbf{T}^{mon}_{mon}\,\mathbf{P}^{Ber}_{mon} = \begin{bmatrix} 1 & 0 & 0 \\ 1 & \frac{1}{2} & 0 \\ 1 & 1 & 1 \end{bmatrix}\begin{bmatrix} 2 & 0 & 0 \\ 0 & 3 & 0 \\ 0 & 0 & 4 \end{bmatrix}\begin{bmatrix} 1 & 0 & 0 \\ -2 & 2 & 0 \\ 1 & -2 & 1 \end{bmatrix} = \begin{bmatrix} 2 & 0 & 0 \\ -1 & 3 & 0 \\ 0 & -2 & 4 \end{bmatrix}.$$

We can now express $\mathbf{a}^{Ber}$, the coordinate vector of $\mathbf{a}$, in Bernstein polynomial bases from the relation between the monomial bases and the Bernstein polynomial bases:

$$\mathbf{a} = (1 \quad t \quad t^2)\mathbf{a}^{mon} = (1 \quad t \quad t^2)\begin{Bmatrix} 1 \\ 0 \\ 2 \end{Bmatrix} = ((1-t)^2 \quad 2t(1-t) \quad t^2)\mathbf{a}^{Ber} = (1 \quad t \quad t^2)\mathbf{P}^{Ber}_{mon}\,\mathbf{a}^{Ber}$$

implying:

$$\mathbf{a}^{Ber} = \left(\mathbf{P}_{mon}^{Ber}\right)^{-1} \mathbf{a}^{mon} = \begin{bmatrix} 1 & 0 & 0 \\ 1 & \frac{1}{2} & 0 \\ 1 & 1 & 1 \end{bmatrix} \begin{Bmatrix} 1 \\ 0 \\ 2 \end{Bmatrix} = \begin{Bmatrix} 1 \\ 1 \\ 3 \end{Bmatrix}$$

So, we get:

$$\mathbf{b}^{Ber} = \mathbf{T}^{Ber}\, \mathbf{a}^{Ber} = \begin{bmatrix} 2 & 0 & 0 \\ -1 & 3 & 0 \\ 0 & -2 & 4 \end{bmatrix} \begin{Bmatrix} 1 \\ 1 \\ 3 \end{Bmatrix} = \begin{Bmatrix} 2 \\ 2 \\ 10 \end{Bmatrix}$$

We can again check for the invariance geometric object, **b**, from the Bernstein component vector, $\mathbf{b}^{Ber}$, as:

$$\mathbf{b} = (\,(1-t)^2 \quad 2t(1-t) \quad t^2\,)\mathbf{b}^{Ber} = (\,(1-t)^2 \quad 2t(1-t) \quad t^2\,) \begin{Bmatrix} 2 \\ 2 \\ 10 \end{Bmatrix} = 2 + 8t^2,$$

as expected.

3.3.2 *Pseudo-vector Representation of Tensors*

For second-order tensors, vector-type notation of components and base vectors – that is, as a linear combination of base vectors with components as vectors – greatly simplifies tensor analysis. For example, a second-order tensor, $\mathbf{T} \equiv T^{ij}\mathbf{g}_i \otimes \mathbf{g}_j$, can also be represented as: $\mathbf{T} \equiv \mathbf{T}^j \otimes \mathbf{g}_j$, where $\mathbf{T}^j$'s are the pseudo-vectors defined by $\mathbf{T}^j \equiv T^{ij}\mathbf{g}_i$.

3.4 Tensor by Component Transformation Property

As desired earlier, we can define tensors without reference to any chosen coordinate system. That is, we can define general tensors as invariant entities by identifying how their components respond to coordinate transformation. From this point of view, the scalars and the vectors can also be considered as tensors as follows:

- A scalar, with only one component that is invariant under coordinate transformation is identified as a tensor of order 0.
- A vector, with only one column of components that transforms under coordinate transformation under a certain law as described below, is identified as a tensor of order 1.
- A tensor, with multi-dimensional components that transforms under coordinate transformation under a certain law as described below, is identified as a second-order tensor if the dimension is 2, as a third-order tensor if the dimension is 3, and so on.

Thus, in order to understand the component transformation properties that determine tensors, we first need to discuss the relevant notions about allowable coordination transformations.

3.4.1 Allowable Coordinate Transformation

We can have parameterization by introduction of suitable, generally nonlinear, transformation such as:

$$\xi^i = \xi^i(\bar{\xi}^i, \bar{\xi}^2, \bar{\xi}^3), \quad i = 1, 2, 3 \tag{3.12}$$

with the assumption that the Jacobian (i.e. the determinant of the Jacobian matrix), as defined below, is non-vanishing at all points:

$$D = \frac{\partial(\xi^1, \xi^2, \xi^3)}{\partial(\bar{\xi}^1, \bar{\xi}^2, \bar{\xi}^3)} \equiv \det \begin{bmatrix} \frac{\partial \xi^1}{\partial \bar{\xi}^1} & \frac{\partial \xi^1}{\partial \bar{\xi}^2} & \frac{\partial \xi^1}{\partial \bar{\xi}^3} \\ \frac{\partial \xi^2}{\partial \bar{\xi}^1} & \frac{\partial \xi^2}{\partial \bar{\xi}^2} & \frac{\partial \xi^2}{\partial \bar{\xi}^3} \\ \frac{\partial \xi^3}{\partial \bar{\xi}^1} & \frac{\partial \xi^3}{\partial \bar{\xi}^2} & \frac{\partial \xi^3}{\partial \bar{\xi}^3} \end{bmatrix} \neq 0 \tag{3.13}$$

Remarks

- We have restricted our discussion to three dimensions; the arguments and derivations for any other dimensions are exactly the same, and the results are obtained in a similar way.
- Because of the invertibility of the transformation given by equation (3.12), there exists the following transformation:

$$\bar{\xi}^i = \bar{\xi}^i(\xi^i, \xi^2, \xi^3), \quad i = 1, 2, 3 \tag{3.14}$$

- Because of the generalization of the allowable coordinate transformations to nonlinear mapping, the base vectors (defined either by tangents or normals which are coordinate-dependent) in a curved space turn out to be ***bound vectors*** in the sense that they are position dependent; for example, in the tangent, vectors to the curves are pointwise fixed or bounded.
- Now, considering ξ^i and $\bar{\xi}^j$ independent, we note that by chain rule:

$$\begin{aligned} \frac{\partial \xi^i}{\partial \xi^j} &= \frac{\partial \xi^i}{\partial \bar{\xi}^k}\frac{\partial \bar{\xi}^k}{\partial \xi^j} = \delta^i_j, \\ \frac{\partial \bar{\xi}^i}{\partial \bar{\xi}^j} &= \frac{\partial \bar{\xi}^i}{\partial \xi^k}\frac{\partial \xi^k}{\partial \bar{\xi}^j} = \delta^i_j \end{aligned} \tag{3.15}$$

For linear transformations, we have to resort to transformations locally of differentials $\mathbf{d}\bar{\xi}$ to $\mathbf{d}\xi$ and vice versa; from equation (3.12), we get:

$$\boxed{d\xi^i = \frac{\partial \xi^i}{\partial \bar{\xi}^j} d\bar{\xi}^j, \quad \text{or,} \quad d\bar{\xi}^i = \frac{\partial \bar{\xi}^i}{\partial \xi^j} d\xi^j} \tag{3.16}$$

where we have used the Einstein summation convention for repeated diagonal indices. Thus the allowable transformation as given by equation (3.12) induces the linear transformation properties of the differentials as in equation (3.16). In other words, Riemann spaces are made of locally deformed patches of Euclidean spaces.

Remarks

- We note for the covariant base vectors, the invariant transformation relation as:

$$\boxed{\mathbf{G}_i \equiv \frac{\partial}{\partial \xi^i} = \frac{\partial}{\partial \bar{\xi}^j}\frac{\partial \bar{\xi}^j}{\partial \xi^i} = \frac{\partial \bar{\xi}^j}{\partial \xi^i}\overline{\mathbf{G}}_j, \quad \text{and,} \quad \overline{\mathbf{G}}_j \equiv \frac{\partial}{\partial \bar{\xi}^j} = \frac{\partial}{\partial \xi^k}\frac{\partial \xi^k}{\partial \bar{\xi}^j} = \frac{\partial \xi^k}{\partial \bar{\xi}^j}\mathbf{G}_k} \tag{3.17}$$

- Considering the infinitesimal arc length, ds, to be an invariant geometric object, we have from coordinate transformation given by equations (3.12) and (3.14), and inserting equation (3.16) into the expression for the infinitesimal arc length, ds:

$$\begin{aligned} ds^2 &= \mathbf{G}_{ij}\, d\xi^i\, d\xi^j = \mathbf{G}_{ij}\frac{\partial \xi^i}{\partial \bar{\xi}^m}\frac{\partial \xi^j}{\partial \bar{\xi}^n} d\bar{\xi}^m d\bar{\xi}^n = \bar{\mathbf{G}}_{mn}\, d\bar{\xi}^m\, d\bar{\xi}^n \\ &\Rightarrow \boxed{\bar{\mathbf{G}}_{mn} = \mathbf{G}_{ij}\frac{\partial \xi^i}{\partial \bar{\xi}^m}\frac{\partial \xi^j}{\partial \bar{\xi}^n}} \end{aligned} \tag{3.18}$$

- Similarly, we have:

$$\begin{aligned} ds^2 &= \bar{\mathbf{G}}_{ij}\, d\bar{\xi}^i\, d\bar{\xi}^j = \bar{\mathbf{G}}_{ij}\frac{\partial \bar{\xi}^i}{\partial \xi^m}\frac{\partial \bar{\xi}^j}{\partial \xi^n} d\xi^m d\xi^n = \mathbf{G}_{mn}\, d\xi^m\, d\xi^n \\ &\Rightarrow \boxed{\mathbf{G}_{mn} = \bar{\mathbf{G}}_{ij}\frac{\partial \bar{\xi}^i}{\partial \xi^m}\frac{\partial \bar{\xi}^j}{\partial \xi^n}} \end{aligned} \tag{3.19}$$

3.4.2 Contravariant Components of a Vector (First-order Tensor) Transformation Law

Let $\mathbf{a} = a^i\mathbf{G}_i = \bar{a}^j\overline{\mathbf{G}}_j$ be a vector with as yet undefined type components, $a^i, \bar{a}^j$, described in two different coordinate systems; then from equation (3.17), we have:

$$a^i\mathbf{G}_i = \bar{a}^j\overline{\mathbf{G}}_j = \bar{a}^j\frac{\partial \xi^i}{\partial \bar{\xi}^j}\mathbf{G}_i, \quad \text{and,} \quad \bar{a}^i\overline{\mathbf{G}}_i = a^j\mathbf{G}_j = a^j\frac{\partial \bar{\xi}^i}{\partial \xi^j}\overline{\mathbf{G}}_i \tag{3.20}$$

Thus, from equation (3.20), we finally get the transformation relations for the contravariant components of a vector, as:

$$\boxed{a^i = \frac{\partial \xi^i}{\partial \bar{\xi}^j}\bar{a}^j, \quad \text{and,} \quad \bar{a}^i = \frac{\partial \bar{\xi}^i}{\partial \xi^j}a^j} \tag{3.21}$$

Now, comparing the transformation relation given by equation (3.21) with the transformation relations for the differentials as in equation (3.16), we are ready to define these components as follows.

- For an object to be a first-order tensor, that is, a vector, the contravariant components of the vector must transform similarly to the differentials of the coordinates, that is, according to equation (3.16).

3.4.3 Covariant Base Vector Transformation Law

For covariant base vectors, the transformation relationships follow from those of the contravariant components of a vector, as follows.

Let $\mathbf{a} = a^i \mathbf{G}_i = \bar{a}^j \overline{\mathbf{G}}_j$ be a vector with contravariant components, $a^i, \bar{a}^j$, described in two different coordinate systems; then from equation (3.21), we have:

$$\mathbf{a} = \bar{a}^j \overline{\mathbf{G}}_j = a^i \frac{\partial \bar{\xi}^j}{\partial \xi^i} \overline{\mathbf{G}}_j = a^i \mathbf{G}_i, \quad \text{and,} \quad \mathbf{a} = a^j \mathbf{G}_j = \bar{a}^i \frac{\partial \xi^j}{\partial \bar{\xi}^i} \mathbf{G}_j = \bar{a}^i \overline{\mathbf{G}}_i,$$
$$\Rightarrow \boxed{\overline{\mathbf{G}}_i = \frac{\partial \xi^j}{\partial \bar{\xi}^i} \mathbf{G}_j, \quad \text{and,} \quad \mathbf{G}_i = \frac{\partial \bar{\xi}^j}{\partial \xi^i} \overline{\mathbf{G}}_j} \tag{3.22}$$

3.4.4 Length of a Vector

Considering an invariant geometric object such as a vector, $\mathbf{u}$, written in terms of the covariant base vectors, $\{\mathbf{G}_i\}$, and the contravariant base vectors, $\{\mathbf{G}^i\}$, we have: $\mathbf{u} = u_i \mathbf{G}^i = u^i \mathbf{G}_i$, where $u_i = u \bullet \mathbf{G}_i$'s and $u^i = \mathbf{u} \bullet \mathbf{G}^i$'s are the covariant and the contravariant components, respectively, of the vector. Finally, for the length of the vector, $\|\bullet\|$, we have:

$$\|\mathbf{u}\|^2 = \mathbf{u} \bullet \mathbf{u} = u_i \mathbf{G}^i \bullet u^j \mathbf{G}_j = u_i u^j \delta^i_j = u_i \, u^i \tag{3.23}$$

As a scalar object, the length of a vector is invariant under coordinate transformation, which helps us to define the covariant component transformation law for vectors as follows.

3.4.5 Covariant Components of a Vector (First-order Tensor) Transformation Law

Similar transformation properties for the ***covariant components*** of a vector may be obtained by considering scalar-valued vector functions; let us consider the invariant scalar function, $\phi(\xi^1, \xi^2, \xi^3)$, such that, under allowable coordinate transformation, we have:

$$\phi(\xi^1, \xi^2, \xi^3) = \phi(\bar{\xi}^1, \bar{\xi}^2, \bar{\xi}^3) \tag{3.24}$$

Then, the gradients of the function transform as:

$$\frac{\partial \phi}{\partial \xi^i} = \frac{\partial \bar{\xi}^j}{\partial \xi^i} \frac{\partial \phi}{\partial \bar{\xi}^j}, \quad \text{and,} \quad \frac{\partial \phi}{\partial \bar{\xi}^i} = \frac{\partial \xi^j}{\partial \bar{\xi}^i} \frac{\partial \phi}{\partial \xi^j} \tag{3.25}$$

Now, inserting the contravariant component transformation relation given by equation (3.21) in the expression for the invariant length of a vector, **a**, as in equation (3.23), we have:

$$a_i a^i = a_i \frac{\partial \xi^i}{\partial \bar{\xi}^j} \bar{a}^j = \bar{a}_j \bar{a}^j, \quad \text{and,} \quad \bar{a}_i \bar{a}^i = \bar{a}_i \frac{\partial \bar{\xi}^i}{\partial \xi^j} a^j = a_j a^j$$
$$\Rightarrow \boxed{a_j = \frac{\partial \bar{\xi}^i}{\partial \xi^j} \bar{a}_i, \quad \text{and,} \quad \bar{a}_j = \frac{\partial \xi^i}{\partial \bar{\xi}^j} a_i} \tag{3.26}$$

Now, comparing the transformation relation given by equation (3.26) with the transformation relations for the gradients as in equation (3.25), we are ready to define these components as follows.

- For an object to be a first-order tensor, that is, a vector, the covariant components of the vector must transform similarly to the gradient of a scalar function, that is, according to equation (3.25).

3.4.6 Contravariant Base Vector Transformation Law

We note that, for contravariant base vectors, the transformation relationships follow from those of the covariant components of a vector, as follows.

Let $\mathbf{a} = a_i \mathbf{G}^i = \bar{a}_j \overline{\mathbf{G}}^j$ be a vector with covariant components, $a_i, \bar{a}_j$, described in two different coordinate systems; then from equation (3.26), we have:

$$\mathbf{a} = \bar{a}_j \overline{\mathbf{G}}^j = a_i \frac{\partial \xi^i}{\partial \bar{\xi}^j} \overline{\mathbf{G}}^j = a_i \mathbf{G}^i, \quad \text{and,} \quad \mathbf{a} = a_j \mathbf{G}^j = \bar{a}_i \frac{\partial \bar{\xi}^i}{\partial \xi^j} \mathbf{G}^j = \bar{a}_i \overline{\mathbf{G}}^i,$$
$$\Rightarrow \boxed{\mathbf{G}^i = \frac{\partial \xi^i}{\partial \bar{\xi}^j} \overline{\mathbf{G}}^j, \quad \text{and,} \quad \overline{\mathbf{G}}^i = \frac{\partial \bar{\xi}^i}{\partial \xi^j} \mathbf{G}^j} \tag{3.27}$$

3.4.7 Second-orderOrder Tensor Component Transformation Laws

We can describe various transformation laws pertaining to T_{ij}, the covariant, T^{ij}, the contravariant, and $T^i_{\ j}, T_i^{\ j}$, the mixed components of a second-order tensor, $\mathbf{T} = \mathbf{a} \otimes \mathbf{b}$, given as: $\mathbf{T} \equiv T^{ij} \mathbf{g}_i \otimes \mathbf{g}_j = T_{ij} \mathbf{g}^i \otimes \mathbf{g}^j = T^i_{\ j} \mathbf{g}_i \otimes \mathbf{g}^j = T_i^{\ j} \mathbf{g}^i \otimes \mathbf{g}_j$ where $T^{ij} \equiv a^i b^j, \quad T_{ij} \equiv a_i b_j, \quad T^i_{\ j} \equiv a^i b_j, \quad T_i^{\ j} \equiv a_i b^j$.

3.4.7.1 Covariant Components of a Second-order Tensor Transformation Law

Noting that $T_{ij} \equiv a_i b_j$, and using the transformation law for covariant components of a vector as in equation (3.26), we get:

$$\boxed{\overline{T_{ij}} \equiv \overline{a_i}\, \overline{b_j} = \frac{\partial \xi^m}{\partial \bar{\xi}^i} \frac{\partial \xi^n}{\partial \bar{\xi}^j} a_i\, b_j = \frac{\partial \xi^m}{\partial \bar{\xi}^i} \frac{\partial \xi^n}{\partial \bar{\xi}^j} T_{ij}} \tag{3.28}$$

3.4.7.2 Contravariant Components of a Second-order Tensor Transformation Law

Noting that $T^{ij} \equiv a^i b^j$, and using transformation law for contravariant components of a vector as in equation (3.21), we get:

$$\overline{T^{ij}} \equiv \overline{a^i}\ \overline{b^j} = \frac{\partial \overline{\xi}^i}{\partial \xi^m} \frac{\partial \overline{\xi}^j}{\partial \xi^n} a^m\ b^n = \frac{\partial \overline{\xi}^i}{\partial \xi^m} \frac{\partial \overline{\xi}^j}{\partial \xi^n} T^{mn} \tag{3.29}$$

3.4.7.3 Mixed Components of a Second-order Tensor Transformation Law

Noting that $T^i_j \equiv a^i b_j$, $\quad T^j_i \equiv a_i b^j$, and using the transformation law for covariant and contravariant components of a vector as in equations (3.21) and (3.26), we get:

$$\overline{T}^i_{\ j} = \frac{\partial \overline{\xi}^i}{\partial \xi^m} \frac{\partial \xi^n}{\partial \overline{\xi}^j} T^m_{\ n}, \quad \text{and,} \quad \overline{T}^j_{\ i} \equiv \frac{\partial \xi^m}{\partial \overline{\xi}^i} \frac{\partial \overline{\xi}^j}{\partial \xi^n} T^n_{\ m} \tag{3.30}$$

3.4.8 Higher-order Tensor Component Transformation Laws

The second-order tensor component transformation laws can be readily generalized to higher-order tensor component transformation law as follows:

$$\overline{T}^{ij\bullet\bullet\bullet}_{ab\bullet\bullet} = \left(\frac{\partial \overline{\xi}^i}{\partial \xi^m} \frac{\partial \overline{\xi}^j}{\partial \xi^n} \bullet\bullet\bullet \frac{\partial \xi^p}{\partial \overline{\xi}^a} \frac{\partial \xi^q}{\partial \overline{\xi}^b} \bullet\bullet\bullet \right) T^{mn\bullet\bullet\bullet}_{pq\bullet\bullet\bullet} \tag{3.31}$$

Remark

- For Cartesian coordinate systems, the difference between the covariant and contravariant components disappears, and the mixed components do not exist.

3.4.9 Tensor Operations

Given tensors of various orders, new tensors of different orders can be generated by tensor operations such as simple contraction and double contraction, as follows.

3.4.9.1 Simple Contraction

Given a tensor $\mathbf{A} \equiv \mathbf{a} \otimes \mathbf{b}$ and another tensor $\mathbf{B} \equiv \mathbf{c} \otimes \mathbf{d}$, the composite second-order tensor $\mathbf{C}$ by ***simple contraction*** is given by: $\mathbf{C} = \mathbf{AB} = (\mathbf{a} \otimes \mathbf{b})(\mathbf{c} \otimes \mathbf{d}) = (\mathbf{b} \bullet \mathbf{c})\mathbf{a} \otimes \mathbf{d}$. The simple contraction is also denoted by: $\mathbf{C} = \mathbf{A} \bullet \mathbf{B}$. For column vectors, it can be easily seen that: $\mathbf{C} = \mathbf{AB} = (\mathbf{ab}^T)(\mathbf{cd}^T) = \mathbf{a}(\mathbf{b}^T\mathbf{c})\mathbf{d}^T = (\mathbf{b}^T\mathbf{c})\ \mathbf{ad}^T = (\mathbf{b} \bullet \mathbf{c})\mathbf{a} \otimes \mathbf{d}$.

- ***A simple contraction between any two tensors of arbitrary order reduces the order of the resulting tensor by two less than the order of the tensor product of the argument tensors.***

For example, a simple contraction between a tensor of order 3 and a tensor of order 4 will result in a tensor of order 5 (= 3 + 4 − 2). In terms of component-wise descriptions, if $\mathbf{A} = A_{ij}\mathbf{g}^i \otimes \mathbf{g}^j$ and $\mathbf{B} = B^{kl}\mathbf{g}_k \otimes \mathbf{g}_l$, then, we have: $\mathbf{C} = \mathbf{AB} = A_{ij}\mathbf{g}^i \otimes \mathbf{g}^j \bullet B^{kl}\mathbf{g}_k \otimes \mathbf{g}_l = A_{ij}B^{kl}\delta_{jk}\mathbf{g}^i \otimes \mathbf{g}_l = A_{ij}B^{jl}\mathbf{g}^i \otimes \mathbf{g}_l$; now, using the indices lowering or raising properties of the metric tensor, $\{g_{ij}\}$, we have: $\mathbf{C} = \mathbf{AB} = A_{im}B^{ml}\mathbf{g}^i \otimes g_{jl}\mathbf{g}^j = A_{im}B^m_{\ j}\ \mathbf{g}^i \otimes \mathbf{g}^j$. Thus, if we express $\mathbf{C} = C_{ij}\mathbf{g}^i \otimes \mathbf{g}^j$, we get the covariant components of $\mathbf{C}$ as: $C_{ij} = A_{im}B^m_{\ j}$.

It is possible to apply simple contraction to a higher-order tensor from both sides. For example, if $\mathbf{a}$ and $\mathbf{b}$ are vectors (i.e. tensors of order 1), then a simple contraction, say, to a second-order tensor, $\mathbf{C} = C_{ij}\mathbf{g}^i \otimes \mathbf{g}^j$, from both sides will result in a scalar: $\mathbf{aCb} = \mathbf{a} \bullet \mathbf{C} \bullet \mathbf{b} = a^i\mathbf{g}_i \bullet C_{jk}\mathbf{g}^j \otimes \mathbf{g}^k \bullet b^l\mathbf{g}_l = a^iC_{jk}b^l(\mathbf{g}_i \bullet \mathbf{g}^j)(\mathbf{g}^k \bullet \mathbf{g}_l) = a^iC_{jk}b^l\delta_{ij}\delta_{kl} = a^jC_{jk}b^k$. Conversely, we note that given a scalar, $a^jC_{jk}b^k$, an equivalent expression in absolute notation is: $a^jC_{jk}b^k = \mathbf{aCb} = \mathbf{a} \bullet \mathbf{C} \bullet \mathbf{b}$.

3.4.9.2 Double Contraction

Assuming column vectors, and noting that $trace(\mathbf{ab}^T) = (\mathbf{ab}) = \mathbf{a}^T\mathbf{b}$, the ***double contraction*** is given by: $C = \mathbf{A} : \mathbf{B} = \mathbf{a} \otimes \mathbf{b} : \mathbf{c} \otimes \mathbf{d} = (\mathbf{a} \bullet \mathbf{c})(\mathbf{b} \bullet \mathbf{d}) = trace(\mathbf{ac}^T)(\mathbf{b}^T\mathbf{d}) = trace(\mathbf{ab}^T\mathbf{dc}^T) = trace(\mathbf{AB}^T)$. We may note that C is a scalar as expected of a double contraction which is a generalization of the vector scalar product. A double contraction of second-order tensors results in a scalar.

- ***A double contraction between any two tensors of arbitrary order reduces the order of the resulting tensor by four less than the order of the tensor product of the argument tensors.***

For example, a double contraction between a tensor of order 3 and a tensor of order 4 will result in a tensor of order 3 (= 3 + 4 − 4). Thus:

$$
\begin{aligned}
\mathbf{P} : \mathbf{Q} &= P_{ijkl}\mathbf{g}^i \otimes \mathbf{g}^j \otimes \mathbf{g}^k \otimes \mathbf{g}^l : Q^{mns}\mathbf{g}_m \otimes \mathbf{g}_n \otimes \mathbf{g}_s \\
&= P_{ijkl}Q^{mns}(\mathbf{g}^k \bullet \mathbf{g}_m)(\mathbf{g}^l \bullet \mathbf{g}_n)\mathbf{g}^i \otimes \mathbf{g}^j \otimes \mathbf{g}_s \\
&= P_{ijkl}Q^{mns}\delta_{km}\delta_{ln}\mathbf{g}^i \otimes \mathbf{g}^j \otimes \mathbf{g}_s = P_{ijml}Q^{mls}\mathbf{g}^i \otimes \mathbf{g}^j \otimes \mathbf{g}_s
\end{aligned}
$$

A double contraction is usually not commutative but obeys both distributive and associative rules.

Finally, just as in simple contractions, a scalar-valued function can be written in a symbolic notation by double contraction: $u = C^{ijrs}e_{ij}e_{rs}$, $i, j, r, s = 1, 2, 3$, is equivalent to: $u = C^{ijrs}e_{ij}e_{rs} = \mathbf{e} : \mathbf{C} : \mathbf{e} = e_{ij}\mathbf{g}^i \otimes \mathbf{g}^j : C^{klmn}\mathbf{g}_k \otimes \mathbf{g}_l \otimes \mathbf{g}_m \otimes \mathbf{g}_n : e_{rs}\mathbf{g}^r \otimes \mathbf{g}^s$.

Fact 1: for an arbitrary second-order tensor, $\mathbf{T}$, and another second-order tensor, $\mathbf{c} \otimes \mathbf{d}$, we have: $\mathbf{T} : (\mathbf{c} \otimes \mathbf{d}) = \mathbf{cTd}$; that is, a double contraction can be replaced by simple contractions from both sides of a second-order tensor.

Proof. Let $\mathbf{T} = \mathbf{a} \otimes \mathbf{b}$; then,

$$
\mathbf{T} : (\mathbf{c} \otimes \mathbf{d}) = (\mathbf{a} \otimes \mathbf{b}) : (\mathbf{c} \otimes \mathbf{d}) = (\mathbf{a} \bullet \mathbf{c})(\mathbf{b} \bullet \mathbf{d}) = \mathbf{c} \bullet (\mathbf{a} \otimes \mathbf{b})\mathbf{d} = \mathbf{c} \bullet\ (\mathbf{Td}) = \mathbf{cTd} \qquad (3.32)
$$

3.4.10 Quotient Rule and Invariance

As indicated earlier, a simple collection or matrix array of functionals does not form a tensor unless the transformation property of the tensor components is manifested by such an array. However, we can invoke the invariance property to determine the tensor characteristic of an array and hence bypass any investigation into its transformation property; such an invariance property is contained in what is known as the quotient rule: if a simple contraction with a tensor and an as yet undetermined array results in a tensor, then the array must also be a tensor. In other words, if a simple contraction with an invariant object and an array produces an invariant object, then the array must also be an invariant object.

For example, the length, $\|\mathbf{u}\|^2 = u_i\, u^i$, an invariant property of a vector, $\mathbf{u}$, as given by equation (3.23), is obtained by simple contraction of u_i, the covariant components, and u^i, the contravariant components of the vector. More, importantly, if one object is covariant, the other must be contravariant. As a practical application, for work, $w = u_i\, f^i$, to be invariant, if u_i, the displacement is taken as covariant, then f^i, the force effecting such displacement must be contravariant.

As another example, let $\mathbf{q}$ be a tensor of order 1, that is, a vector with q^i, its contravariant component, and $\mathbf{P}(i,j)$, be a two-dimensional array such that a multiplication results in a tensor, $\mathbf{P}_j$, as given by Einstein's summation convention:

$$\mathbf{P}(i,j)\, q^i = P_j \tag{3.33}$$

then, we can show that $\mathbf{P}(i,j)$ is a tensor of type P_{ij}, that is, a second-order tensor.

Proof. Since $\mathbf{P}(i,j)\, q^i$ is a tensor of order 1 (i.e. a vector) with covariant component type, it must satisfy the transformation law given by equation (3.26):

$$\overline{\mathbf{P}}(i,j)\, \bar{q}^i = \bar{P}_j = \frac{\partial \xi^n}{\partial \bar{\xi}^j} P_n = \frac{\partial \xi^n}{\partial \bar{\xi}^j} \mathbf{P}(m,n)\, q^m \tag{3.34}$$

Now, noting that q^m, the contravariant component transforms according to equation (3.21) such that $q^m = \frac{\partial \xi^m}{\partial \bar{\xi}^i} \bar{q}^i$, from equation (3.34), we get:

$$\left(\overline{\mathbf{P}}(i,j) - \frac{\partial \xi^m}{\partial \bar{\xi}^i} \frac{\partial \xi^n}{\partial \bar{\xi}^j} \mathbf{P}(m,n) \right) \bar{q}^i = 0 \tag{3.35}$$

Now, with $\bar{q}^i \neq 0$ an arbitrary tensor, we have:

$$\overline{\mathbf{P}}(i,j) = \frac{\partial \xi^m}{\partial \bar{\xi}^i} \frac{\partial \xi^n}{\partial \bar{\xi}^j} \mathbf{P}(m,n) \tag{3.36}$$

which is a covariant component transformation law for a second-order tensor of P_{ij} type as given by equation (3.28). Thus, $\mathbf{P}(m,n) = P_{mn}$ are components of a tensor.

3.5 Special Tensors

We list below several special tensors that will be used quite frequently in our structural analysis.

3.5.1 Metric or Fundamental Tensor

Following up on the definition given by equation (3.15), $G_{ij} \equiv \mathbf{G}_i \bullet \mathbf{G}_j,\ i,j = 1,2,3$ are the covariant components of what is known as the second-order metric or fundamental tensor; it is so called as it is capable of providing all the geometric information such as the distances, the curvatures and the angular separations of vectors in curved spaces.

Now, considering a vector, $\mathbf{u}$, written in terms of the covariant base vectors, $\{\mathbf{G}_i\}$, and the contravariant base vectors, $\{\mathbf{G}^i\}$, we have: $\mathbf{u} = u_i\mathbf{G}^i = u^i\mathbf{G}_i$, where $u_i = u \bullet \mathbf{G}_i$'s and $u^i = \mathbf{u} \bullet \mathbf{G}^i$'s are the covariant and the contravariant components, respectively, of the vector; then, $u^i = \mathbf{u} \bullet \mathbf{G}^i = \mathbf{G}^i \bullet u_j\mathbf{G}^j = \mathbf{G}^i \bullet \mathbf{G}^j u_j \equiv G^{ij}u_j$ where we have defined the contravariant components of the metric tensor as:

$$\boxed{G^{ij} \equiv \mathbf{G}^i \bullet \mathbf{G}^j} \tag{3.37}$$

Fact 1: The metric tensors are symmetric: $G_{ij} = G_{ji}$ and $G^{ij} = G^{ji}$.

Proof. Since the inner product of two vectors is symmetric, we have: $G_{ij} \equiv \mathbf{G}_i \bullet \mathbf{G}_j = \mathbf{G}_j \bullet \mathbf{G}_i \equiv G_{ji}$. Similarly, $G^{ij} \equiv \mathbf{G}^i \bullet \mathbf{G}^j = \mathbf{G}^j \bullet \mathbf{G}^i = G^{ji}$.

Fact 2: The metric tensor has the property of raising and lowering of indices. For example, $v_i = G_{ij}v^j$ and $T_{ij} = G_{ik}T^k_j$.

Proof. For a vector, $\mathbf{v}$: $v_i = \mathbf{v} \bullet \mathbf{G}_i = v^j\mathbf{G}_j \bullet \mathbf{G}_i = G_{ij}v^j$; thus, the covariant component of a vector is obtained by lowering the index of the contravariant component by multiplying with the covariant components of the metric tensor. Similarly, for a second-order tensor, $\mathbf{T}$:

$$T_{ij} = \mathbf{G}_i \bullet \mathbf{T} \bullet \mathbf{G}_j = \mathbf{G}_i \bullet T^k_{.l}\mathbf{G}_k \otimes \mathbf{G}^l \bullet \mathbf{G}_j = T^k_{.l}(\mathbf{G}_i \bullet \mathbf{G}_k)(\mathbf{G}^l \bullet \mathbf{G}_j) = T^k_{.l}G_{ik}\delta_{lj} = G_{ik}T^k_{.j}$$

Fact 3: $G_{ij}G^{ik} = \delta^{.k}_i$ shows the connection between the covariant and the contravariant components of the metric tensor.

Proof. Using definition of G_{ij} as in equation (3.18) and raising of index property of the metric tensor as in

Fact 2: we have:

$$\boxed{G_{ij}G^{ik} = (\mathbf{G}_i \bullet \mathbf{G}_j)G^{ik} = \underbrace{\mathbf{G}_i G^{ik}}_{\text{raising of index}} \bullet \mathbf{G}_j = \mathbf{G}^k \bullet \mathbf{G}_j = \delta^{.k}_j} \tag{3.38}$$

Fact 4: If $G \equiv \det(G_{ij})$, then $\det(G^{ij}) = \frac{1}{G}$ where det stands for the determinant of the corresponding tensors.

Proof. Using the relation as in ***Fact 3***, we have:

$$\boxed{\begin{aligned}&\det(G_{ij}G^{ik}) = \det(G_{ij})\det(G^{ik}) = \det(\delta_i^{\cdot k}) = 1\\ &\Rightarrow \det(G^{ik}) = \frac{1}{\det(G_{ij})} = \frac{1}{G}\end{aligned}} \tag{3.39}$$

Fact 5: $G = \det(G_{ij}) = J^2$, where $J = \det[\,\mathbf{G}_1 \quad \mathbf{G}_2 \quad \mathbf{G}_3\,] = \mathbf{G}_1 \bullet (\mathbf{G}_2 \times \mathbf{G}_3)$ is the Jacobian determinant of the Jacobian matrix made up of the covariant base vectors as the columns of the matrix.

Proof. We can express the metric tensor $[G_{ij}]$ in matrix form by the columnar covariant base vectors as:

$$\begin{bmatrix} \mathbf{G}_1 \bullet \mathbf{G}_1 & \mathbf{G}_1 \bullet \mathbf{G}_2 & \mathbf{G}_1 \bullet \mathbf{G}_3 \\ \mathbf{G}_2 \bullet \mathbf{G}_1 & \mathbf{G}_2 \bullet \mathbf{G}_2 & \mathbf{G}_2 \bullet \mathbf{G}_3 \\ \mathbf{G}_3 \bullet \mathbf{G}_1 & \mathbf{G}_3 \bullet \mathbf{G}_2 & \mathbf{G}_3 \bullet \mathbf{G}_3 \end{bmatrix} = \begin{Bmatrix} \mathbf{G}_1^T \\ \mathbf{G}_2^T \\ \mathbf{G}_3^T \end{Bmatrix} [\,\mathbf{G}_1 \quad \mathbf{G}_2 \quad \mathbf{G}_3\,] \tag{3.40}$$

Then, using the result from equation, the product rule for determinants of matrices and the fact that, for any matrix, $\mathbf{A} : \det \mathbf{A} = \det \mathbf{A}^T$, we get:

$$\boxed{G = \det(G_{ij}) = \det \begin{Bmatrix} \mathbf{G}_1^T \\ \mathbf{G}_2^T \\ \mathbf{G}_3^T \end{Bmatrix} \det[\,\mathbf{G}_1 \quad \mathbf{G}_2 \quad \mathbf{G}_3\,] = J^2, \quad \text{or,} \quad J = \sqrt{G}} \tag{3.41}$$

3.5.2 *Metric Tensor Transformation Laws*

From equation (3.17), we get the transformation relation for the covariant components of the metric tensor as:

$$G_{ij} \equiv \mathbf{G}_i \bullet \mathbf{G}_j = \frac{\partial \bar{\xi}^k}{\partial \xi^i}\overline{\mathbf{G}}_k \bullet \frac{\partial \bar{\xi}^l}{\partial \xi^j}\overline{\mathbf{G}}_l \Rightarrow \boxed{G_{ij} = \frac{\partial \bar{\xi}^k}{\partial \xi^i}\frac{\partial \bar{\xi}^l}{\partial \xi^j}\overline{G}_{kl}} \tag{3.42}$$

In fact, equation (3.42) depicts the transformation law for the covariant components of any second-order tensor. In other words,

- ***for an object to be a second-order tensor, the covariant components of the object must transform according to equation*** (3.42); this, then, serves to define a second-order tensor.

Similarly, using equation (3.22), we have the transformation relation for the contravariant components of the metric tensor as:

$$G^{ij} \equiv \mathbf{G}^i \bullet \mathbf{G}^j = \frac{\partial \xi^i}{\partial \bar{\xi}^k} \bar{\mathbf{G}}^k \bullet \frac{\partial \xi^j}{\partial \bar{\xi}^l} \bar{\mathbf{G}}^l \Rightarrow \boxed{G^{ij} = \frac{\partial \xi^i}{\partial \bar{\xi}^k} \frac{\partial \xi^j}{\partial \bar{\xi}^l} \bar{G}^{kl}} \tag{3.43}$$

3.5.3 Second-order Identity Tensor, **I**

Consider a tensor **I** such that, for any vector **x**, we have: $\mathbf{I}\,\mathbf{x} = \mathbf{x}$. Now, from vector properties, the vector **x** can be expressed by its ***contravariant components*** as: $\mathbf{x} = x^i\,\mathbf{G}_i$ where $\mathbf{G}_i$'s are the ***covariant base vectors*** and $x^i = \mathbf{G}^i \bullet \mathbf{x}$ where $\mathbf{G}^i$'s are the reciprocal ***contravariant base vectors***. Then, we clearly have: $\mathbf{I}\,\mathbf{x} = \mathbf{x} = x^i\,\mathbf{G}_i = (\mathbf{G}^i \bullet \mathbf{x})\mathbf{G}_i = (\mathbf{G}_i \otimes \mathbf{G}^i)\,\mathbf{x}$. Thus, $\mathbf{I} = (\mathbf{G}_i \otimes \mathbf{G}^i)$. This is known as the second-order identity tensor. Note that we have used the Einstein diagonal summation convention. Alternatively, considering that the vector **x** can also be expressed by its ***covariant components*** as: $\mathbf{x} = x_i\,\mathbf{G}^i$ where $\mathbf{G}^i$'s are the reciprocal ***contravariant base vectors*** and $x_i = \mathbf{G}_i \bullet \mathbf{x}$ where $\mathbf{G}_i$'s are the ***covariant base vectors***, we have: $\mathbf{I}\,\mathbf{x} = \mathbf{x} = x_i\,\mathbf{G}^i = (\mathbf{G}_i \bullet \mathbf{x})\mathbf{G}^i = (\mathbf{G}^i \otimes \mathbf{G}_i)\,\mathbf{x}$ leading to $\mathbf{I} = (\mathbf{G}^i \otimes \mathbf{G}_i)$. Now, using the index lowering and raising property of the ***metric tensor***, $\mathbf{G} = \{G_{ij} = \mathbf{G}_i \bullet \mathbf{G}_j\}$, we get:

$$\boxed{\mathbf{I} = (\mathbf{G}_i \otimes \mathbf{G}^i) = (\mathbf{G}_i \otimes G^{ij}\mathbf{G}_j) = G^{ij}(\mathbf{G}_i \otimes \mathbf{G}_j) = (\mathbf{G}^i \otimes \mathbf{G}_i)} \tag{3.44}$$

Using the tensor product of the mixed base vectors, other equivalent expressions for the identity tensor can be generated.

Fact 1: For any second-order tensor, **A**, we get: $\mathbf{IA} = \mathbf{AI} = \mathbf{A}$

Proof. with $\mathbf{A} = A^{mn}\mathbf{G}_m \otimes \mathbf{G}_n$, we have:

$$\mathbf{IA} = (\mathbf{G}_i \otimes \mathbf{G}^i)A^{mn}\mathbf{G}_m \otimes \mathbf{G}_n = A^{mn}\delta^i_m \mathbf{G}_i \otimes \mathbf{G}_n = A^{in}\mathbf{G}_i \otimes \mathbf{G}_n = \mathbf{A}$$

Fact 2: For any two vectors, **a** and **b**, the scalar product can be expressed as double contraction through the second-order identity tensor: $\mathbf{a} \bullet \mathbf{b} = \mathbf{I} : (\mathbf{a} \otimes \mathbf{b}) = \mathbf{a}\,\mathbf{I}\,\mathbf{b}$, that is, a scalar product can be expressed as a double contraction through the identity tensor, **I**.

Proof. Using equation (3.32), $\mathbf{I} : (\mathbf{a} \otimes \mathbf{b}) = \mathbf{a}\,\mathbf{I}\,\mathbf{b} = \mathbf{a} \bullet (\mathbf{I}\,\mathbf{b}) = \mathbf{a} \bullet \mathbf{b} = a^i b_i$. Alternatively, from the definition of the identity tensor:

$$\mathbf{I} : (\mathbf{a} \otimes \mathbf{b}) = (\mathbf{g}^i \otimes \mathbf{g}_i) : (\mathbf{a} \otimes \mathbf{b}) = (\mathbf{g}^i \bullet \mathbf{a})\,(\mathbf{g}_i \bullet \mathbf{b}) = a^i b_i = (\mathbf{a} \bullet \mathbf{b})$$

Fact 3: For any second-order tensor, **A**, the trace is defined as: $tr\,\mathbf{A} \equiv \mathbf{A} : \mathbf{I}$. Then, we get: $tr\,\mathbf{A} = A_i^{\cdot i}$

Proof. with $\mathbf{A} = A^{mn}\mathbf{G}_m \otimes \mathbf{G}_n$, we have:

$$tr\,\mathbf{A} \equiv \mathbf{A} : \mathbf{I} = (\mathbf{G}_i \otimes \mathbf{G}^i) : A^{mn}\mathbf{G}_m \otimes \mathbf{G}_n = A^{mn}G_{im}\delta^i_n = A_i^{\cdot i} \tag{3.45}$$

3.5.4 Fourth-order Identity Tensor, $\overset{4}{\mathbf{I}}$

As a generalization of the second-order identity tensor, we have the fourth-order identity tensor, $\overset{4}{\mathbf{I}}$, defined as:

$$\boxed{\overset{4}{\mathbf{I}} \equiv \mathbf{I} \otimes \mathbf{I} = \mathbf{G}_i \otimes \mathbf{G}^i \otimes \mathbf{G}_j \otimes \mathbf{G}^j} \tag{3.46}$$

$\overset{4}{\mathbf{I}}$ is extensively utilized in material modeling as in the stress–strain transformation relation of continuum mechanics. Now, as indicated earlier, the tensor $\mathbf{A}$ can be expressed by its ***pseudo-contravariant components***, $\mathbf{A}^i$, as: $\mathbf{A} = A^{ij}\mathbf{G}_i \otimes \mathbf{G}_j = \mathbf{A}^j \otimes \mathbf{G}_j$ where $\mathbf{G}_i$'s are the ***covariant base vectors*** and $\mathbf{A}^j \equiv A^{ij}\mathbf{G}_i$. Then, we clearly have:

$$\begin{aligned}\overset{4}{\mathbf{I}} : A &= \mathbf{G}_i \otimes \mathbf{G}^i \otimes \mathbf{G}_j \otimes \mathbf{G}^j : A^{kl}\,\mathbf{G}_k \otimes \mathbf{G}_l \\ &= A^{kl}(\mathbf{G}_j \bullet \mathbf{G}_k)(\mathbf{G}^j \bullet \mathbf{G}_l)\mathbf{G}_i \otimes \mathbf{G}^i \\ &= G_{jk}A^{kl}\delta^j_l\,\mathbf{I} = A_j^{\cdot l}\delta^j_l\,\mathbf{I} = A_j^{\cdot j}\,\mathbf{I} = (tr\,\mathbf{A})\,\mathbf{I}\end{aligned} \tag{3.47}$$

where $tr\,\mathbf{A}$ is the abbreviation for the trace of a second-order tensor, $\mathbf{A}$, defined by the double contraction with $\mathbf{I}$, the second-order identity tensor, that is,

$$\boxed{tr\,\mathbf{A} \equiv \mathbf{A} : \mathbf{I} = A_j^{\cdot j}} \tag{3.48}$$

Fact 1: For any second-order tensor, $\mathbf{A}$, we get: $\mathbf{A} : \overset{4}{\mathbf{I}} : A = (tr\,\mathbf{A})^2$

Proof. Using equations (3.47) and (3.48), $\mathbf{A} : \overset{4}{\mathbf{I}} : A = \mathbf{A} : (tr\,\mathbf{A})\mathbf{I} = (tr\,\mathbf{A})^2$

3.5.5 Third Order Permutation Tensor, E

In vector calculus, the cross product of any two vectors, **a** and **b**, is given by:

$$\begin{gathered}\mathbf{a} \times \mathbf{b} = a^i b^j \mathbf{g}_i \times \mathbf{g}_j = \varepsilon_{ijk}a^i b^j \mathbf{g}^k = \varepsilon^{ijk}a_i b_j \mathbf{g}_k \\ \text{where } \mathbf{g}_i \times \mathbf{g}_j = \varepsilon_{ijk}\mathbf{g}^k, \quad \text{and,} \quad \mathbf{g}^i \times \mathbf{g}^j = \varepsilon^{ijk}\mathbf{g}_k\end{gathered} \tag{3.49}$$

and ε_{ijk} and ε^{ijk} are the covariant and contravariant components, respectively, of the third-order permutation tensor, $\mathbf{E}$, that is,

$$\boxed{\mathbf{E} \equiv \varepsilon_{ijk}\mathbf{g}^i \otimes \mathbf{g}^j \otimes \mathbf{g}^k = \varepsilon^{ijk}\mathbf{g}_i \otimes \mathbf{g}_j \otimes \mathbf{g}_k, \quad \text{with} \quad \varepsilon_{ijk} = -\varepsilon_{jik} = \varepsilon_{jki} = -\varepsilon_{kji}} \tag{3.50}$$

Fact 1:

$$\mathbf{g}_i \times \mathbf{g}_j = \varepsilon_{ijk}\mathbf{g}^k = \mathbf{E} : (\mathbf{g}_i \otimes \mathbf{g}_j) = -\mathbf{g}_i\ \mathbf{E}\ \mathbf{g}_j,$$

that is, the vector cross product has a tensorial representation.

Proof. Using equations (3.32), (3.49) and (3.50), we have:

$$\mathbf{E} : (\mathbf{g}_i \otimes \mathbf{g}_j) = \varepsilon_{ijk}\mathbf{g}^i \otimes \mathbf{g}^j \otimes \mathbf{g}^k : (\mathbf{g}_i \otimes \mathbf{g}_j) = \varepsilon_{ijk}\delta^k_j \delta^j_i \mathbf{g}^i = \varepsilon_{ijk}\mathbf{g}^k$$

and

$$\begin{aligned} -\mathbf{g}_i\ \mathbf{E}\ \mathbf{g}_j &= -\mathbf{g}_i \bullet \varepsilon_{ijk}\mathbf{g}^i \otimes \mathbf{g}^j \otimes \mathbf{g}^k \bullet \mathbf{g}_j = -\varepsilon_{ijk}\delta^k_j \mathbf{g}^j \\ &= -\varepsilon_{ikj}\mathbf{g}^k = \varepsilon_{ijk}\mathbf{g}^k = \mathbf{g}_i \times \mathbf{g}_j = \mathbf{E}:(\mathbf{g}_i \otimes \mathbf{g}_j) \end{aligned}$$

3.6 Second-order Tensors

As indicated in our discussion on tensors as linear transformations, a simple second-order tensor space can be defined in an invariant component-independent form from what is known as the ***tensor product*** (or dyadic) of two vectors:

Definition:

- A tensor product of any two vectors, **a** and **b**, may define a second-order tensor $\mathbf{T} \equiv \mathbf{a} \otimes \mathbf{b}$ such that, for any vector **c**, the transformed vector **d** is determined by: $\mathbf{d} = \mathbf{Tc} = (\mathbf{a} \otimes \mathbf{b})\mathbf{c} = (\mathbf{b} \bullet \mathbf{c})\mathbf{a}$.

In other words, according to this rule of transformation, the vector **c** is transformed by the tensor into a scaled vector along the direction of **a**, as shown in Figure 3.2: If we settle for column vectors only, as is customarily done in mechanics, then the above tensor product can be expressed as $\mathbf{T} \equiv \mathbf{ab}^T$ since by this representation we also get: $\mathbf{Tc} \equiv \mathbf{ab}^T c = (\mathbf{b}^T c)\mathbf{a} = (\mathbf{b} \bullet \mathbf{c})\mathbf{a}$ as above. Note that only **b** and **c** need to belong to the same ***vector space*** to have an operational definition of tensors by tensor product.

However, as indicated in the case of ***general tensors***, a formal definition comes from a generalization of the representation of vectors in a vector space. Let us recall that just as a vector space is spanned by its ***base vectors***, tensor spaces (i.e. ***linear vector spaces*** whose elements are tensors) are spanned by their base tensors. Of course, as indicated earlier, we have assumed that the ***linear transformations*** form a linear vector space as they do. The base tensors are formed by the tensor product of underlying the ***covariant*** and/or ***contravariant*** base vectors.

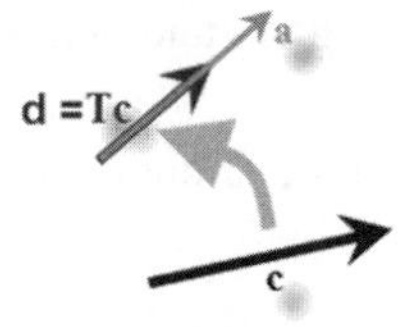

Figure 3.2 Second-order tensor.

That is the base tensors are of the form: $\mathbf{g}_i \otimes \mathbf{g}_j, \mathbf{g}_i \otimes \mathbf{g}^j, \mathbf{g}^i \otimes \mathbf{g}_j, \mathbf{g}^i \otimes \mathbf{g}^j$. This can be easily seen, for example, if we substitute for any arbitrary vectors $\mathbf{a} = a^i\mathbf{g}_i$ and $\mathbf{b} = b^j\mathbf{g}_j$ in the tensor definition, then by associative law, and defining $T^{ij} \equiv a^ib^j$, we have $\mathbf{T} \equiv T^{ij}\mathbf{g}_i \otimes \mathbf{g}_j$, where we have used the ***Einstein summation convention*** of summing over the repeated diagonal superscripts and subscripts. We may recall, from our discussion of tensors as linear transformations, that just like the vectors, **T** is an invariant object independent of the choice of coordinate systems. Clearly, in three dimensions with nine base tensors, a second-order tensor has nine components. In general, then, depending on pure covariant or contravariant or mixed base tensors, component-wise we can describe it by: $\mathbf{T} \equiv T^{ij}\mathbf{g}_i \otimes \mathbf{g}_j = T_{ij}\mathbf{g}^i \otimes \mathbf{g}^j = T^i_{\ j}\mathbf{g}_i \otimes \mathbf{g}^j = T_i^{\ j}\mathbf{g}^i \otimes \mathbf{g}_j$ where $T^{ij} \equiv a^ib^j, T_{ij} \equiv a_ib_j, T^i_{\ j} \equiv a^ib_j, T_i^{\ j} \equiv a_ib^j$. Thus, just like the vectors, a general second-order tensor is represented by the base tensors of the underlying tensor space.

So any second-order tensor is an invariant object described as a linear combination of the base tensors with the constants of combination recognized as the matrix components of the tensor corresponding to the chosen set of base tensors.

With the tensor components thus identified, we can easily see that for any $\mathbf{d} = \mathbf{Tc}$, we have for ***covariant components*** of the transformed vector: $d_i\mathbf{g}^i = (T_{ij}\mathbf{g}^i \otimes \mathbf{g}^j)(c^k\mathbf{g}_k) = T_{ij}c^k(\mathbf{g}^j \bullet \mathbf{g}_k)\mathbf{g}^i = T_{ij}c^k\delta^j_k\mathbf{g}^i = (T_{ij}c^j)\mathbf{g}^i$. Thus, $d_i = T_{ij}c^j$. Similarly, for mixed components of tensors: $d_i = T_i^{\ j}\, c_j$, $d^i = T^i_{\ j}\, c^j$, and for contravariant components of tensors: $d^i = T^{ij}c_j$.

3.6.1 *Inverse Tensors*

The second-order inverse tensor, denoted as $\mathbf{A}^{-1}$, of a tensor, **A**, is defined such that:

$$\mathbf{A}^{-1}\,\mathbf{A} = \mathbf{A}\,\mathbf{A}^{-1} = \mathbf{I} \tag{3.51}$$

where **I** is the second-order identity tensor.

- clearly: $\mathbf{I}^{-1}\,\mathbf{I} = \mathbf{I}\,\mathbf{I}^{-1} = \mathbf{I}$, that is, the identity tensor is the inverse of itself,
- $(\mathbf{A}^{-1})^{-1}\,\mathbf{A}^{-1} = \mathbf{I} \Rightarrow (\mathbf{A}^{-1})^{-1} = \mathbf{A}$,
- $(\mathbf{AB})^{-1}\,\mathbf{AB} = \mathbf{I} \Rightarrow (\mathbf{AB})^{-1}\,\underbrace{\mathbf{ABB}^{-1}\mathbf{A}^{-1}}_{=\mathbf{I}} = \mathbf{B}^{-1}\mathbf{A}^{-1} \Rightarrow (\mathbf{AB})^{-1} = \mathbf{B}^{-1}\mathbf{A}^{-1}$

3.6.2 *Transposed Tensors*

The second-order tensor transposition is an operation that is routinely utilized to determine symmetry, or lack of it, for tensors analysis.

The transposed tensor, denoted by $\mathbf{T}^T$, of a simple second-order tensor $\mathbf{T} \equiv \mathbf{a} \otimes \mathbf{b}$ is defined by $\mathbf{T}^T \equiv \mathbf{b} \otimes \mathbf{a}$. For column vectors: $\mathbf{T} \equiv \mathbf{ab}^T \Rightarrow \mathbf{T}^T = (\mathbf{ab}^T)^T = \mathbf{ba}^T$.

This definition can be easily extended to a general second-order tensor. Let $\mathbf{a} \equiv \mathbf{t}_i = T_{ij}\mathbf{g}^i$ and $\mathbf{b} \equiv \mathbf{g}^j$. Then, we have a general second-order tensor: $\mathbf{T} \equiv \mathbf{a} \otimes \mathbf{b} = \mathbf{t}_j \otimes \mathbf{g}^j = T_{ij}\mathbf{g}^i \otimes \mathbf{g}^j$ with covariant components. The transposed tensor is, then, defined by:

$$\mathbf{T}^T \equiv (\mathbf{T}^T)_{ij}\mathbf{g}^i \otimes \mathbf{g}^j = \mathbf{b} \otimes \mathbf{a} = T_{ij}\mathbf{g}^j \otimes \mathbf{g}^i = T_{ji}\mathbf{g}^i \otimes \mathbf{g}^j \tag{3.52}$$

Note that in the last equality, we have only switched the dummy indices. Thus, the transposed tensor components are given by: $(\mathbf{T}^T)_{ij} = T_{ji}$. Similarly, we can get the transposed contravariant and mixed components: for mixed: $(\mathbf{T}^T)_i^{\ j} = T^j_{\ i}$ and $(\mathbf{T}^T)^i_{\ j} = T_j^{\ i}$, and for contravariant: $(\mathbf{T}^T)^{ij} = T^{ji}$.

Fact 1: Given any two vectors, **c** and **d**, and a tensor $\mathbf{T} = \mathbf{t}_j \otimes \mathbf{g}^j$, we have:

$$\mathbf{cTd} = \mathbf{c} \bullet (\mathbf{Td}) = \mathbf{dT}^T\mathbf{c} = \mathbf{d} \bullet (\mathbf{T}^T\mathbf{c}) \tag{3.53}$$

Proof. $\mathbf{cTd} = \mathbf{c}(\mathbf{t}_j \otimes \mathbf{g}^j)\mathbf{d} = (\mathbf{c} \bullet \mathbf{t}_j)(\mathbf{g}^j \bullet \mathbf{d}) = (\mathbf{d} \bullet \mathbf{g}^j)(\mathbf{t}_j \bullet \mathbf{c}) = \mathbf{d}(\mathbf{g}^j \otimes \mathbf{t}_j)\mathbf{c} = \mathbf{dT}^T\mathbf{c}$. In the equalities, we used the properties of ***vector scalar product*** and definition of a transposed tensor.

Fact 2: Given any three tensors, **A**, **B** and **C**, we have:

$$\mathbf{A} : \mathbf{BC} = \mathbf{B}^T\mathbf{A} : \mathbf{C} = \mathbf{AC}^T : \mathbf{B} \tag{3.54}$$

Proof. Let $\mathbf{A} = \mathbf{a}^i \otimes \mathbf{g}_i$, $\mathbf{B} = \mathbf{b}^j \otimes \mathbf{g}_j$ and $\mathbf{C} = \mathbf{c}^k \otimes \mathbf{g}_k$. Then, $\mathbf{A} : \mathbf{BC} = \mathbf{A} : (\mathbf{b}^j \otimes \mathbf{g}_j)(\mathbf{c}^k \otimes \mathbf{g}_k) = (\mathbf{a}^i \otimes \mathbf{g}_i) : (\mathbf{b}^j \otimes \mathbf{g}_k)(\mathbf{g}_j \bullet \mathbf{c}^k) = (\mathbf{a}^i \bullet \mathbf{b}^j)(\mathbf{g}_i \bullet \mathbf{g}_k)(\mathbf{g}_j \bullet \mathbf{c}^k)$. Now, we can rewrite: $(\mathbf{a}^i \bullet \mathbf{b}^j)(\mathbf{g}_i \bullet \mathbf{g}_k)(\mathbf{g}_j \bullet \mathbf{c}^k) = (\mathbf{b}^j \bullet \mathbf{a}^i)(\mathbf{g}_j \otimes \mathbf{g}_i) : (\mathbf{c}^k \otimes \mathbf{g}_k) = (\mathbf{g}_j \otimes \mathbf{b}^j)(\mathbf{a}^i \otimes \mathbf{g}_i) : \mathbf{C} = \mathbf{B}^T\mathbf{A} : \mathbf{C}$. In the equalities, we used the definition of transposed tensor, ***single contraction*** and ***double contraction***.

- By similar arguments, we get: $\mathbf{A} : \mathbf{BC} = \mathbf{AC}^T : \boldsymbol{B}$
- By utilizing the linear properties of vectors, we get: $(\mathbf{A} + \mathbf{B})^T = \mathbf{A}^T + \mathbf{B}^T$ and $(\mathbf{AB})^T = \mathbf{B}^T\mathbf{A}^T$

Fact 3: For the ***identity tensor***, $\mathbf{I} = \mathbf{g}_i \otimes \mathbf{g}^i$, we have $\mathbf{I} = \mathbf{I}^T = \mathbf{g}^i \otimes \mathbf{g}_i$.

Proof. for any vector: $\mathbf{h} = h_j g^j$, we get $\mathbf{I}^T\mathbf{h} = (\mathbf{g}^i \otimes \mathbf{g}_i)h_j\mathbf{g}^j = h_j(\mathbf{g}_i \bullet \mathbf{g}^j)\mathbf{g}^i = h_j\delta_i^j\mathbf{g}^i = h_j\mathbf{g}^j = \mathbf{h} = \mathbf{Ih}$. Thus, $\mathbf{I} = \mathbf{I}^T$.

Fact 4: The ***inverse*** and transpose operations on a tensor are order independent, that is, for any tensor **A**, we have: $(\mathbf{A}^T)^{-1} = (\mathbf{A}^{-1})^T = \mathbf{A}^{-T}$.

Proof. with the definition of inverse, we have: $\mathbf{AA}^{-1} = \mathbf{I}$. Now, taking transpose of both sides and noting the transpose property of a composite tensor, we have: $(\mathbf{AA}^{-1})^T = (\mathbf{A}^{-1})^T\mathbf{A}^T = \mathbf{I}^T = \mathbf{I}$. Thus, we find that $(\mathbf{A}^{-1})^T$ is inverse of $\mathbf{A}^T$. In other words, $(\mathbf{A}^{-1})^T = (\mathbf{A}^T)^{-1}$. So, the parenthesis is irrelevant.

3.6.3 Orthogonal and Rotation Tensors

A tensor **Q** is called an Orthogonal tensor if:

$$\boxed{\mathbf{Q}^T = \mathbf{Q}^{-1} \Rightarrow \mathbf{QQ}^T = \mathbf{Q}^T\mathbf{Q} = \mathbf{Q}^{-1}\mathbf{Q} = \mathbf{QQ}^{-1} = \mathbf{I}} \tag{3.55}$$

The orthogonality imposes 6 conditions (3 equations due to orthogonality: $\mathbf{QQ}^T = \mathbf{I}$, and 3 due to symmetry: $\mathbf{QQ}^T = \mathbf{Q}^T\mathbf{Q}$, the tensor **Q** is left with only 3 independent variables.

An orthogonal transformation preserves both the lengths and in-between angle of any two vectors. To wit, for any two vectors, **a** and **b** with $\hat{\mathbf{a}} = \mathbf{Qa}$ and $\hat{\mathbf{b}} = \mathbf{Qb}$, we get: $\hat{\mathbf{a}} \bullet \hat{\mathbf{b}} = \mathbf{Qa} \bullet \mathbf{Qb} = \mathbf{a} \bullet \mathbf{Q}^T\mathbf{Qb} = \mathbf{a} \bullet \mathbf{b}$. This property, clearly, demonstrates that a rotational transformation, that preserves both the lengths and in-between angle of two vectors, is an orthogonal tensor.

For a ***detailed presentation on rotation tensor*** that is of utmost importance in our geometrically nonlinear analysis of beams, plates and shells, we would follow up in Chapter 4.

Another important orthogonal tensor is the reflection tensor, say, **P**, defined as:

$$\boxed{\mathbf{P} = \mathbf{I} - \frac{2}{(\mathbf{a} \bullet \mathbf{a})}(\mathbf{a} \otimes \mathbf{a})} \tag{3.56}$$

where **I** is the identity tensor, **a** is the normal to the plane of reflection of a vector. Let $\mathbf{b} = \mathbf{c} + \mathbf{d}$ be a vector such that it is represented by its components: $\mathbf{c} = \alpha\mathbf{a}$ parallel to the normal **a** and **d** perpendicular to it, that is, contained in the plane; then,

$$\begin{aligned}\mathbf{P}\mathbf{b} &= \left(\mathbf{I} - \frac{2}{(\mathbf{a} \bullet \mathbf{a})}(\mathbf{a} \otimes \mathbf{a})\right)(\alpha\mathbf{a} + \mathbf{d}) \\ &= \alpha\mathbf{a} + \mathbf{d} - \frac{2}{(\mathbf{a} \bullet \mathbf{a})}\{\alpha(\mathbf{a} \bullet \mathbf{a})\mathbf{a} + \cancel{(\mathbf{a} \bullet \mathbf{d})\mathbf{a}}\} = \alpha\mathbf{a} + \mathbf{d} - 2\alpha\mathbf{a} = -\alpha\mathbf{a} + \mathbf{d}\end{aligned}$$

Thus, the component perpendicular to the plane is reversed and the one along the plane is not disturbed, that is, the vector **b** is reflected by the plane with normal given by the vector, **a**.

3.6.4 Symmetric Tensors

A tensor is symmetric if its ***transposed*** is same as itself, that is, if $\mathbf{A} = \mathbf{A}^T$. Symmetric tensors abound in mechanics. For example, for non-polar mechanics, various stress ***tensors*** and ***strain tensors*** are symmetric; the ***identity tensor*** is symmetric and so on. Symmetric tensors can be formed by various ways from a general non-symmetric tensor. For example, a ***simple contraction*** of a tensor by its transposed generates a symmetric tensor. For example, for any **A**, we have $\mathbf{A}\mathbf{A}^T$ symmetric since by definition of transpose of composite tensors: $(\mathbf{A}\mathbf{A}^T)^T = (\mathbf{A}^T)^T\mathbf{A}^T = \mathbf{A}\mathbf{A}^T$.

Fact 1: As an example, in continuum mechanics, the symmetric ***Cauchy stress tensor***, $\boldsymbol{\sigma}$ is related to the ***second Piola–Kirchhoff stress tensor***, **S**, through the ***deformation tensor***, **F**, by the formula: $\mathbf{S} \equiv \mathbf{F}^{-1}\boldsymbol{\sigma}\mathbf{F}^{-T}$ to a constant. If $\boldsymbol{\sigma}$ is symmetric, so is **S**.

Proof. We have $\mathbf{S}^T = (\mathbf{F}^{-1}\boldsymbol{\sigma}\mathbf{F}^{-T})^T = (\mathbf{F}^{-T})^T\boldsymbol{\sigma}^T(\mathbf{F}^{-1})^T$ by transposition properties of composite tensors. Now, using symmetry of $\boldsymbol{\sigma}$ and order independence of inverse and transpose, we get: $\mathbf{S}^T = (\mathbf{F}^{-T})^T\boldsymbol{\sigma}^T(\mathbf{F}^{-1})^T = \mathbf{F}^{-1}\boldsymbol{\sigma}\mathbf{F}^{-T} = \mathbf{S}$.

Fact 2: For any symmetric tensor **A**, the covariant components have the property: $A_{ij} = A_{ji}$.

Proof. From the component relations with transposed tensor, we get for the covariant components of any symmetric tensor $\mathbf{A} = A_{ij}\mathbf{g}^i \otimes \mathbf{g}^j = \mathbf{A}^T = (A^T)_{ij}\mathbf{g}^i \otimes \mathbf{g}^j = A_{ji}\mathbf{g}^i \otimes \mathbf{g}^j$. With base tensors arbitrary, we get: $A_{ij} = A_{ji}$.

Similarly, it is easy to show that for mixed: $A_i^j = A_i^{\,j} = A^j_{\,i}$, and for contravariant: $A^{ij} = A^{ji}$ for **A** symmetric. Note also that the order of superscripted and subscripted indices is immaterial for mixed components of a symmetric tensor.

Finally, for any **A**, $\mathbf{B} \equiv (\mathbf{A} + \mathbf{A}^T)$ is symmetric.

3.6.4.1 Eigen Properties of Symmetric Tensors

By definition, for a symmetric tensor, **S**, with covariant components, the off-diagonal components, $S_{ij},\ i \neq j$, are identical for any interchange of the indices.

Then, for ***eigenanalysis*** of **S**, we have to solve for the ***eigenvalues***, λ, and subsequently determine the corresponding ***eigenvectors***, **x**, in the ***eigenequation***:

$$\mathbf{S}\,\mathbf{x} = \lambda\mathbf{x} \Leftrightarrow\ (\mathbf{S}-\lambda\mathbf{I})\mathbf{x} = \mathbf{0} \tag{3.57}$$

For a non-trivial solution of equation (3.65), we must have the determinant $|\mathbf{S} - \lambda\mathbf{I}|$ vanish identically. Following the Cayley–Hamilton theorem, the characteristic polynomial is given by:

$$\boxed{p(\lambda) \equiv \det(\mathbf{S} - \lambda\mathbf{I}) = -\lambda^3 + I_1\lambda^2 - I_2\lambda + I_3 = 0} \tag{3.58}$$

where, the three invariants, $I_1,\ I_2,\ I_3$ of a second-order symmetric tensor are given as:

$$\boxed{\begin{aligned} I_1 &= tr\ \mathbf{S} = S^i_i \\ I_2 &= \frac{1}{2}\{(tr\ \mathbf{S})^2 - tr\ \mathbf{S}^2\} = S^i_i S^j_j - S^i_j S^j_i \\ I_3 &= \det\ \mathbf{S} = \left|S^i_j\right| \end{aligned}} \tag{3.59}$$

From equation (3.57), we have, for a symmetric tensor:

$$\mathbf{x}_i\mathbf{S}\,\mathbf{x}_i = \mathbf{x}_i \bullet \lambda_i\mathbf{x}_i = \lambda_i\|\mathbf{x}_i\|^2 \geq \mathbf{0} \tag{3.60}$$

This implies that the eigenvalues, λ_i, of a symmetric tensor are real and positive.

Again from equation (3.57), we have, for a symmetric tensor:

$$(\lambda_i - \lambda_j)\mathbf{x}_i\ \mathbf{x}_j = \mathbf{x}_j\mathbf{S}\,\mathbf{x}_i - \mathbf{x}_i\mathbf{S}\,\mathbf{x}_j = 0 \tag{3.61}$$

This implies that because $\lambda_i \neq \lambda_j$, the eigenvectors of a symmetric tensor are orthogonal to each other.

Thus, the ***eigenvalues of a general symmetric tensor are real and positive, and the eigenvectors are mutually orthogonal***.

3.6.5 Skew-symmetric Tensors

A tensor is defined as skew-symmetric if its negative ***transpose*** is the same as itself. A tensor **A** is skew-symmetric if $\mathbf{A} = -\mathbf{A}^T$, that is, $\mathbf{A} + \mathbf{A}^T = \mathbf{0}$ where **0** is the null tensor.

Clearly, $A_{ij} + A_{ji} = 0$ or $A^{ij} + A^{ji} = 0$ implies that the diagonal elements of a skew-symmetric tensor under covariant or contravariant component representation, are identically zero. Thus, in these cases, the number of independent components is given by $\frac{1}{2}n(n-1)$ where n is the underlying dimension of the domain and range spaces. However, for the mixed components of a skew symmetric tensor, we have: $A^j_i = -A^j_i$. Noting that in general, $A^i_i \neq A^i_i$, the diagonal elements of a skew-symmetric tensor under mixed component representation is generally non-zero. Now, given an arbitrary tensor, $\mathbf{C} = \mathbf{a} \otimes \mathbf{b}$, we can extract the skew-symmetric part of the tensor as: $\boldsymbol{\Omega} \equiv \frac{1}{2}(\mathbf{C} - \mathbf{C}^T) = \frac{1}{2}(\mathbf{a} \otimes \mathbf{b} - \mathbf{b} \otimes \mathbf{a})$ since it can be easily shown that $\boldsymbol{\Omega} + \boldsymbol{\Omega}^T = \mathbf{0}$.

3.6.5.1 Axial Vectors

An axial vector is intimately related to the second-order skew-symmetric tensors in three dimensions. If we restrict the space of skew-symmetric tensors to the non-relativistic physical world, that is, an $n = 3$ dimensional space, the number of independent covariant or contravariant components of a skew-symmetric tensor becomes also 3 $(= \frac{1}{2}n(n-1))$. This makes it possible to treat equivalently operations by skew-symmetric tensors by means of vector algebra. That is, for any three-dimensional vector **c** and a given skew-symmetric tensor: $\mathbf{\Omega} = \frac{1}{2}(\mathbf{a} \otimes \mathbf{b} - \mathbf{b} \otimes \mathbf{a})$, we have $\mathbf{\Omega c} = \boldsymbol{\omega} \times \mathbf{c}$, where $\boldsymbol{\omega} \equiv \frac{1}{2}(\mathbf{b} \times \mathbf{a})$. Using the definition of a skew-symmetric tensor, we can easily show:

$$\boxed{\begin{aligned}\mathbf{\Omega c} &= \frac{1}{2}(\mathbf{a} \otimes \mathbf{b} - \mathbf{b} \otimes \mathbf{a})\mathbf{c} = \frac{1}{2}\{(\mathbf{c} \bullet \mathbf{b})\mathbf{a} - (\mathbf{c} \bullet \mathbf{a})\mathbf{b}\} \\ &= \frac{1}{2}\mathbf{c} \times (\mathbf{a} \times \mathbf{b}) = \frac{1}{2}(\mathbf{b} \times \mathbf{a}) \times \mathbf{c} = \boldsymbol{\omega} \times \mathbf{c}\end{aligned}} \tag{3.62}$$

where we have used the well-known vector identity: $\mathbf{a} \times (\mathbf{b} \times \mathbf{c}) = (\mathbf{a} \bullet \mathbf{c})\mathbf{b} - (\mathbf{a} \bullet \mathbf{b})\mathbf{c}$ for any three vectors. Clearly, then, a simple contraction $\mathbf{\Omega c}$ is equivalent to the vector operation $\boldsymbol{\omega} \times \mathbf{c}$. Thus, there is a one-to-one correspondence between the tensor operation by $\mathbf{\Omega}$ and vector cross product operation by $\boldsymbol{\omega}$. This equivalence is denoted by $\boxed{\mathbf{\Omega} = \mathbf{\Omega}\times = [\mathbf{\Omega}]}$, and tensor algebra anywhere for a skew-symmetric tensor can be replaced by a corresponding vector operation. Hence, for orthonormal bases with components of $\boldsymbol{\omega}$ given as: $\boldsymbol{\omega} = \{\,\omega_1 \quad \omega_2 \quad \omega_3\,\}^T$, the component matrix of the tensor $\mathbf{\Omega}$ has only three independent components and the following representation:

$$\mathbf{\Omega} = \begin{bmatrix} 0 & -\omega_3 & \omega_2 \\ \omega_3 & 0 & -\omega_1 \\ -\omega_2 & \omega_1 & 0 \end{bmatrix} \tag{3.63}$$

As an example, for a ***shell coordinate system***, if $\mathbf{\Omega}$ is applied to a fixed surface normal, $\mathbf{g}_3$, then $\boldsymbol{\omega}$ has only two components since $\mathbf{\Omega g}_3 = \boldsymbol{\omega} \times \mathbf{g}_3 = \omega^i \mathbf{g}_i \times \mathbf{g}_3 = \omega^\sigma \mathbf{g}_\sigma$, where the Latin indices run over 1 to 3, and the Greek indices over 1 and 2. Recall the ***summation*** and ***notation convention*** that Greek indices run through only two in three-dimensional tensors.

For $n > 3$, this correspondence between a skew-symmetric tensor and a vector does not exist and hence we do not have any other option than to carry on with tensor algebra. For $n = 2$, a skew-symmetric tensor is equivalent to a scalar.

Note also that by the definition of the ***permutation tensor*** E, we have a symbolic tensor representation:

$$\boxed{\boldsymbol{\omega} \equiv \frac{1}{2}(\mathbf{b} \times \mathbf{a}) = \frac{1}{2}\mathbf{E} : (\mathbf{b} \otimes \mathbf{a})} \tag{3.64}$$

3.6.5.2 Eigenproperties of Skew-symmetric Tensors

By definition, for a skew-symmetric tensor, **S**, with covariant components, all the diagonal components $\{S_{ii}\}$ are zero, and the off-diagonal elements have the property: $S_{ij} + S_{ji} = 0,\ i \neq j$. Now, let $\mathbf{s} = (\,s_1 \quad s_2 \quad s_3)^T$ be the corresponding axial vector such that $\mathbf{S} = \begin{bmatrix} 0 & -s_3 & s_2 \\ s_3 & 0 & -s_1 \\ -s_2 & s_1 & 0 \end{bmatrix}$.

Then, for ***eigenvalues*** of **S**, we have to solve for λ in the eigenequation:

$$\mathbf{S}\,\mathbf{x} = \lambda\mathbf{x} \Leftrightarrow (\mathbf{S}-\lambda\mathbf{I})\mathbf{x} = \mathbf{0} \tag{3.65}$$

For a non-trivial solution of equation (3.65), we must have the determinant $|\mathbf{S} - \lambda\mathbf{I}|$ vanish identically. In other words, the characteristic polynomial is given by: $p(\lambda) \equiv \det(\mathbf{S} - \lambda\mathbf{I}) = 0$. Upon simplification, we have:

$$\boxed{\begin{aligned}&\lambda^3 + \left(s_1^2 + s_2^2 + s_3^2\right)\lambda = 0, \quad \text{or,} \quad \lambda^3 + \|\mathbf{s}\|^2\lambda = 0\\ &\Rightarrow \lambda\left\{\lambda^2 + \left(s_1^2 + s_2^2 + s_3^2\right)\right\} = 0\end{aligned}} \tag{3.66}$$

Clearly, then, ***the eigenvalues of a general skew-symmetric tensor are either zero or purely imaginary numbers***.

3.6.5.3 Important Identities of Skew-symmetric Tensors

For later use, in finite element analyses, we list here some important identities for skew-symmetric tensors.

Since every tensor satisfies its own characteristic polynomial, we have, from equation (3.66), the important identity:

$$\boxed{\begin{aligned}&\mathbf{S}^3 + s^2\mathbf{S} = 0, \quad \text{where} \quad s \equiv \|\mathbf{s}\| = \sqrt{s_1^2 + s_2^2 + s_3^2}\\ &\Rightarrow \mathbf{S}^3 = -\|\mathbf{s}\|^2\mathbf{S}\end{aligned}} \tag{3.67}$$

Now, for all the following identities, we assume **A** and **B** as skew-symmetric tensors with **a**, **b** as the corresponding axial vectors, that is, $\mathbf{A} = [\mathbf{a}]$ and $\mathbf{B} = [\mathbf{b}]$. $h \in \mathbb{R}^3$ is any arbitrary vector. Note that $\mathbf{Aa} = \mathbf{a} \times \mathbf{a} = \mathbf{0}$ and $\mathbf{Bb} = \mathbf{b} \times \mathbf{b} = \mathbf{0}$. Additionally, taking all vectors as column vectors, we have: $(\mathbf{b} \otimes \mathbf{a}) = \mathbf{b}\,\mathbf{a}^T$, where $(\mathbf{b} \otimes \mathbf{a})$ is recalled as the ***tensor product*** of two vectors, **a** and **b**.

Fact 1: $\mathbf{Ab} + \mathbf{Ba} = \mathbf{0}$

Proof. $\mathbf{Ab} + \mathbf{Ba} = \mathbf{a} \times \mathbf{b} + \mathbf{b} \times \mathbf{a} = \mathbf{a} \times \mathbf{b} - \mathbf{a} \times \mathbf{b} = \mathbf{0}$

Fact 2: $\mathbf{AB} = (\mathbf{b} \otimes \mathbf{a}) - (\mathbf{a} \bullet \mathbf{b})\mathbf{I}$

Proof. $\mathbf{ABh} = \mathbf{a} \times (\mathbf{b} \times \mathbf{h}) = (\mathbf{a} \bullet \mathbf{h})\mathbf{b} - (\mathbf{a} \bullet \mathbf{b})\mathbf{h} = [(\mathbf{b} \otimes \mathbf{a}) - (\mathbf{a} \bullet \mathbf{b})\mathbf{I}]\mathbf{h}$

Fact 3: $\mathbf{A}^2 = (\mathbf{a} \otimes \mathbf{a}) - ||\mathbf{a}||^2\mathbf{I}$

Proof. substitution of **B** by **A** in **Fact 2** completes the proof.

Fact 4: $\mathbf{A}^2\,\mathbf{B} + \mathbf{B}\,\mathbf{A}^2 = -(\mathbf{a} \bullet \mathbf{b})\mathbf{A} - ||\mathbf{a}||^2\mathbf{B}$

Proof.

$$\begin{aligned}\mathbf{A}^2\,\mathbf{B} + \mathbf{B}\,\mathbf{A}^2 &= \mathbf{A}(\mathbf{AB}) + \mathbf{B}(\mathbf{a}\otimes\mathbf{a}) - ||\mathbf{a}||^2\mathbf{B} = \mathbf{A}(\mathbf{b}\otimes\mathbf{a}) - (\mathbf{a}\bullet\mathbf{b})\mathbf{A} + \mathbf{B}(\mathbf{a}\otimes\mathbf{a}) - ||\mathbf{a}||^2\mathbf{B}\\ &= \mathbf{A}(\mathbf{b}\otimes\mathbf{a}) + \mathbf{B}(\mathbf{a}\otimes\mathbf{a}) - (\mathbf{a}\bullet\mathbf{b})\mathbf{A} - ||\mathbf{a}||^2\mathbf{B}\\ &= (\{\mathbf{Ab}+\mathbf{Ba}\}\otimes\mathbf{a}) - (\mathbf{a}\bullet\mathbf{b})\mathbf{A} - ||\mathbf{a}||^2\mathbf{B} = -(\mathbf{a}\bullet\mathbf{b})\mathbf{A} - ||\mathbf{a}||^2\mathbf{B}\end{aligned}$$

Fact 5: $\mathbf{ABA} = -(\mathbf{a}\bullet\mathbf{b})\mathbf{A}$.

Proof. Using ***Fact 2***, we have: $\mathbf{ABA} = \mathbf{A}[(\mathbf{a}\otimes\mathbf{b}) - (\mathbf{a}\bullet\mathbf{b})\mathbf{I}] = (\mathbf{Aa}\otimes\mathbf{b}) - (\mathbf{a}\bullet\mathbf{b})\mathbf{A} = -(\mathbf{a}\bullet\mathbf{b})\mathbf{A}$

Fact 6: $\mathbf{AB}\,\mathbf{A}^2 = \mathbf{A}^2\,\mathbf{BA} = -(\mathbf{a}\bullet\mathbf{b})\mathbf{A}^2$.

Proof. Using ***Fact 5***, we have: $\mathbf{AB}\,\mathbf{A}^2 = (\mathbf{ABA})\mathbf{A} = -(\mathbf{a}\bullet\mathbf{b})\mathbf{A}$. Similarly, for the second equality.

Fact 7: $\mathbf{A}^2\,\mathbf{B}\,\mathbf{A}^2 = ||\mathbf{a}||^2(\mathbf{a}\bullet\mathbf{b})\mathbf{A}$.

Proof. Using **Fact 6**, and noting that $\mathbf{A}$ satisfies its characteristic equations, that is, $||\mathbf{a}||^2\mathbf{A} + \mathbf{A}^3 = \mathbf{0}$, we have: $\mathbf{A}^2\,\mathbf{B}\,\mathbf{A}^2 = \mathbf{A}(\mathbf{A}\,\mathbf{B}\,\mathbf{A}^2) = -(\mathbf{a}\bullet\mathbf{b})\mathbf{A}^3 = ||\mathbf{a}||^2(\mathbf{a}\bullet\mathbf{b})\mathbf{A}$.

Fact 8: $\mathbf{AB} - \mathbf{BA} = (\mathbf{b}\otimes\mathbf{a}) - (\mathbf{a}\otimes\mathbf{b}) = [\mathbf{a}\times\mathbf{b}]$.

Proof.

$$\begin{aligned}(\mathbf{AB}-\mathbf{BA})\mathbf{h} &= \mathbf{a}\times(\mathbf{b}\times\mathbf{h}) - \mathbf{b}\times(\mathbf{a}\times\mathbf{h}) = (\mathbf{a}\bullet\mathbf{h})\mathbf{b} - (\mathbf{a}\bullet\mathbf{b})\mathbf{h} - (\mathbf{b}\bullet\mathbf{h})\mathbf{a} + (\mathbf{a}\bullet\mathbf{b})\mathbf{h}\\ &= \{(\mathbf{b}\otimes\mathbf{a}) - (\mathbf{a}\otimes\mathbf{b})\}\mathbf{h}, \quad \forall\mathbf{h}\in\mathbb{R}^3\end{aligned}$$

Also, it follows that:

$$(\mathbf{AB}-\mathbf{BA})\mathbf{h} = (\mathbf{a}\bullet\mathbf{h})\mathbf{b} - (\mathbf{b}\bullet\mathbf{h})\mathbf{a} = (\mathbf{h}\bullet\mathbf{a})\mathbf{b} - (\mathbf{h}\bullet\mathbf{b})\mathbf{a} = \mathbf{h}\times(\mathbf{b}\times\mathbf{a}) = (\mathbf{a}\times\mathbf{b})\times\mathbf{h} = [\mathbf{a}\times\mathbf{b}]\mathbf{h}, \quad \forall\mathbf{h}\in\mathbb{R}^3$$

Fact 9: $\mathbf{hAh} = \mathbf{0}, \quad \forall\mathbf{h}\in\mathbb{R}^3$.

Proof. $\mathbf{hAh} = \mathbf{A}^T\mathbf{h}\,\mathbf{h} = -\mathbf{Ah}\,\mathbf{h} = -\mathbf{h}\,\mathbf{Ah}, \quad \forall\mathbf{h}\in\mathbb{R}^3$ completes the proof.

3.6.6 Contractions

Given a tensor $\mathbf{A}\equiv\mathbf{a}\otimes\mathbf{b}$ and another tensor $\mathbf{B}\equiv\mathbf{c}\otimes\mathbf{d}$, the composite second-order tensor $\mathbf{C}$ by ***simple contraction*** is given by: $\mathbf{C} = \mathbf{AB} = (\mathbf{a}\otimes\mathbf{b})(\mathbf{c}\otimes\mathbf{d}) = (\mathbf{b}\bullet\mathbf{c})\mathbf{a}\otimes\mathbf{d}$. The simple contraction is also denoted by: $\mathbf{C} = \mathbf{A}\bullet\mathbf{B}$. A simple contraction of second-order tensors results in a second-order tensor.

Moreover, with the column vector restriction and noting that $trace(\mathbf{a}\bullet\mathbf{b}^T) = (\mathbf{ab}) = \mathbf{a}^T\mathbf{b}$, the ***double contraction*** is given by: $C = \mathbf{A}:\mathbf{B} = \mathbf{a}\otimes\mathbf{b}:\mathbf{c}\otimes\mathbf{d} = (\mathbf{a}\bullet\mathbf{c})(\mathbf{b}\bullet\mathbf{d}) = trace(\mathbf{ac}^T)(\mathbf{b}^T\mathbf{d}) = trace(\mathbf{ab}^T\mathbf{dc}^T) = trace(\mathbf{AB}^T)$. We may note that C is a scalar, as expected of a double contraction, which is a generalization of vector scalar product. A double contraction of second-order tensors results in a scalar.

3.6.7 Trace, Norm and Determinant

As it turns out, the ***trace*** and ***determinant*** of a tensor appear as two of the three invariant scalars that define the associated characteristic polynomial resulting from the ***eigenanalysis*** of a tensor. The trace in turn helps to generalize the length concept of a vector through the definition of the ***norm***, an invariant measure of a tensor.

3.6.7.1 Trace

Denoting a tensor, $\mathbf{A}$, in the dyadic form: $\mathbf{A} = \mathbf{A}^j \otimes \mathbf{g}_j = \mathbf{A}_k \otimes \mathbf{g}^k$ where $\mathbf{A}^j \equiv \mathbf{A}^{ij}\mathbf{g}_i$ and $\mathbf{A}_k = A_{mk}\mathbf{g}^m$ are the contravariant and covariant vector representation, respectively, the ***trace*** of $\mathbf{A}$, denoted by $tr\mathbf{A}$, is defined as:

$$\boxed{tr\mathbf{A} \equiv \mathbf{I} : \mathbf{A}} \tag{3.68}$$

where $\mathbf{I}$ is the three-dimensional identity tensor. Thus,

$$\boxed{\begin{aligned} tr\mathbf{A} &\equiv \mathbf{I} : \mathbf{A} = \mathbf{g}_m \otimes \mathbf{g}^m : A^n \otimes \mathbf{g}_n = \mathbf{g}_m \bullet \mathbf{A}^n \delta^m_n = \mathbf{g}_m \bullet \mathbf{A}^m \\ &= \mathbf{g}_m \bullet \mathbf{A}^{nm}\mathbf{g}_n = \mathbf{g}_{mn}\mathbf{A}^{nm} = A^m_{\ m} = A^n_{\ n} \end{aligned}} \tag{3.69}$$

Thus, denoting a tensor in dyadic form, we have from the application point of view: $tr\mathbf{A} = \mathbf{g}_m \bullet \mathbf{A}^m = A^m_{\ m} = A^n_{\ n}$. Moreover, with the column vector restriction, we have $tr\mathbf{A} = tr(\mathbf{a} \otimes \mathbf{b}) = tr(\mathbf{a}\mathbf{b}^T) = \mathbf{a}^T\mathbf{b} = (\mathbf{a} \bullet \mathbf{b})$.

- $tr\mathbf{A}^T = tr(\mathbf{b} \otimes \mathbf{a}) = (\mathbf{b} \bullet \mathbf{a}) = (\mathbf{a} \bullet \mathbf{b}) = tr\mathbf{A}$,
- The trace is a linear function of tensors: $tr(\mathbf{A} + \mathbf{B}) = tr\mathbf{A} + tr\mathbf{B}$.

Fact 1: $tr(\mathbf{A}\mathbf{B}^T) = tr(\mathbf{B}\mathbf{A}^T) = \mathbf{A} : \mathbf{B} = \mathbf{B}^T : A^T$

Proof. Let $\mathbf{A} = \mathbf{a} \otimes \mathbf{b}$ and $\mathbf{B} = \mathbf{c} \otimes \mathbf{d}$, then, $\mathbf{A}\mathbf{B}^T = (\mathbf{a} \otimes \mathbf{b}) \bullet (\mathbf{d} \otimes \mathbf{c}) = (\mathbf{b} \bullet \mathbf{d})(\mathbf{a} \otimes \mathbf{c})$. Similarly, $\mathbf{B}\mathbf{A}^T = (\mathbf{b} \bullet \mathbf{d})(\mathbf{c} \otimes \mathbf{a})$. Thus, $tr(\mathbf{A}\mathbf{B}^T) = (\mathbf{b} \bullet \mathbf{d})\, tr(\mathbf{a} \otimes \mathbf{c}) = (\mathbf{b} \bullet \mathbf{d})\,(\mathbf{a} \bullet \mathbf{c}) = (\mathbf{b} \bullet \mathbf{d})\,(\mathbf{c} \bullet \mathbf{a}) = (\mathbf{d} \bullet \mathbf{b})\, tr(\mathbf{c} \otimes \mathbf{a}) = tr(\mathbf{B}\mathbf{A}^T)$. Note that we have used the definition of ***simple contraction*** for composite tensors. The trace can be directly verified by application of the concept of ***double contraction*** since: $\mathbf{A} : \mathbf{B} = (\mathbf{a} \otimes \mathbf{b}) : (\mathbf{c} \otimes \mathbf{d}) = (\mathbf{b} \bullet \mathbf{d})(\mathbf{a} : \mathbf{c})$. Note also that $tr(\mathbf{A}\mathbf{B}) \neq tr\mathbf{A}\; tr\mathbf{B}$.

Fact 2: for any ***symmetric*** $\mathbf{A}$, $(tr\mathbf{A})^2 - tr(\mathbf{A}^2) = \mathbf{A}^i_{\ i}\mathbf{A}^j_{\ j} - \mathbf{A}^j_{\ i}\mathbf{A}^i_{\ j}$

Proof. we have: $(tr\mathbf{A})^2 = tr\mathbf{A}\; tr\mathbf{A} = \mathbf{A}^i_{\ i}\mathbf{A}^j_{\ j}$.

Now, $tr(\mathbf{A}^2) = tr(\mathbf{A}\mathbf{A}) = tr(\mathbf{A}\mathbf{A}^T) = \mathbf{A} : \mathbf{A} = \mathbf{A}_j \otimes \mathbf{g}^j : A^k \otimes \mathbf{g}_k = \mathbf{A}_j \bullet \mathbf{A}^k \delta^j_k = \mathbf{A}_j \bullet \mathbf{A}^j = A^i_{\ j}\mathbf{g}_i \bullet A^j_{\ m}\, \mathbf{g}^m = A^i_{\ j}A^j_{\ i}$. Now, switching the dummy indices completes the proof.

3.6.7.2 Norm

Just as in vectors (first-order tensors), the norm of a second-order tensor is a measure of its 'generalized length' and, as in Hilbert space, may be induced from the definition of the underlying scalar product. The scalar product for tensors is generalized to ***double contraction***. Thus, the norm, denoted by $\| \cdot \|$, of a general second-order tensor **A** may be defined as:

$$\boxed{\|\mathbf{A}\| = \sqrt{\mathbf{A}:\mathbf{A}} = \sqrt{tr(\mathbf{A}\mathbf{A}^T)} = \sqrt{tr(\mathbf{A}^T\mathbf{A})}} \tag{3.70}$$

Fact 1: The Schwarz inequality: $\|\mathbf{AB}\| \le \|\mathbf{A}\|\,\|\mathbf{B}\| = \sqrt{tr(\mathbf{A}\mathbf{A}^T)}\sqrt{tr(\mathbf{B}\mathbf{B}^T)} = \sqrt{\mathbf{A}:\mathbf{A}}\sqrt{\mathbf{B}:\mathbf{B}}$

Fact 2: From ***invariants*** study, the ***trace*** of a composite tensors in simple contraction is given by: $tr(\mathbf{A}\mathbf{B}^T) = \mathbf{A}:\mathbf{B}$. Then, for a symmetric tensor **A**, we have:

$$tr(\mathbf{A}^2) = \mathbf{A}_i^j\mathbf{A}_j^i \tag{3.71}$$

Proof.

$$tr(\mathbf{A}^2) = tr(\mathbf{A}\mathbf{A}) = tr(\mathbf{A}\mathbf{A}^T) = \mathbf{A}:\mathbf{A}^T = A_i^j\mathbf{g}^i \otimes \mathbf{g}_j : A_l^k\mathbf{g}_k \otimes \mathbf{g}^l = A_i^j\delta_k^i\delta_j^l A_l^k = A_i^j A_j^i.$$

Fact 3: for any ***symmetric*** **A**, $(tr\mathbf{A})^2 - tr(\mathbf{A}^2) = \mathbf{A}_i^i\mathbf{A}_j^j - \mathbf{A}_i^j\mathbf{A}_j^i$

Proof. $(tr\mathbf{A})^2 = (tr\mathbf{A})(tr\mathbf{A}) = \mathbf{A}_i^i\mathbf{A}_j^j$, and using equation (3.71): $tr(\mathbf{A}^2) = tr(\mathbf{A}\mathbf{A}) = tr(\mathbf{A}\mathbf{A}^T) = \mathbf{A}:\mathbf{A}^T = \mathbf{A}_i^j\mathbf{A}_j^i$ completes the proof.

3.6.7.3 Determinant

The definition of the determinant, also known as the third invariant, of a second-order tensor depends on the choice of the tensor components. To facilitate connection with the vector cross product through the permutation symbol, we may choose to define i:t by the mixed components of a tensor.

Thus, for a second-order tensor $\mathbf{A} = A_j^i\,\mathbf{g}_i \otimes \mathbf{g}^j$, the determinant is defined as:

$$\boxed{\det\mathbf{A} \equiv \left|A_j^i\right| = e_{ijk}A_1^i\,A_2^j\,A_3^k} \tag{3.72}$$

where e_{ijk} is the third-order ***permutation tensor*** associated with the Cartesian coordinate systems. Given $|A_i^j|$ as the determinant of a coefficient matrix with A_i^j as the ith row and jth column element, this definition is motivated by the verifiable relations: $e^{ijk}A_i^r A_j^s A_k^t = |A_i^j|e^{rst}$ and $e_{ijk}A_r^i A_s^j A_t^k = |A_i^j|e_{rst}$.

3.6.8 *Decomposition of Tensors*

As it turns out, the ***double contraction*** of a ***symmetric*** and a ***skew-symmetric*** tensor results in the scalar zero, that is, it annihilates it. To wit: by definition, if **A** is symmetric, then $A^{ij} = A^{ji}$, and for a skew-symmetric tensor **B**, we have: $B_{kk} = 0$, and $B_{kl} = -B_{lk}$ for $k \neq l$. Thus,

$$\mathbf{A} : \mathbf{B} = A^{ij}\mathbf{g}_i \otimes \mathbf{g}_j : B_{kl}\mathbf{g}^k \otimes \mathbf{g}^l = A^{ij}B_{kl}(\mathbf{g}_j \bullet \mathbf{g}^l)(\mathbf{g}_i \bullet \mathbf{g}^k) = A^{ij}B_{kl}\delta^l_j\delta^k_i = A^{ij}B_{ij} = 0 \qquad (3.73)$$

This result is especially useful in tensor algebraic manipulations that arise frequently in mechanics. Thus, there is a need for an additive decomposition of a general second-order tensor for simplification as well as insight into the nature of the tensor. Also, the polar decomposition serves to identify the translational and rotational aspect of deformations, say, in strain tensors. Finally, for both analysis as in constitutive theory and numerical formulations such as mixed finite element analysis, stress and strain tensors can be profitably viewed from the volumetric (spherical) and the shearing (deviatoric) stresses and deformations.

3.6.8.1 Additive Decomposition: Symmetric and Skew-symmetric Tensors

A general second-order tensor **A** can always be represented as an additive sum of a symmetric tensor and a skew-symmetric tensor: $\mathbf{A} = A^j \otimes \mathbf{g}_j = \frac{1}{2}(A^j \otimes \mathbf{g}_j + \mathbf{g}_j \otimes A^j) + \frac{1}{2}(A^j \otimes \mathbf{g}_j - \mathbf{g}_j \otimes A^j) = \frac{1}{2}(\mathbf{A} + \mathbf{A}^T) + \frac{1}{2}(\mathbf{A} - \mathbf{A}^T)$. Recall that by our notation convention about contravariant vectors, $\mathbf{A}^j \equiv A^{ij}\mathbf{g}_i$. Now, by definition of a ***transposed tensor***, the first part of the decomposition is a symmetric tensor, and we use this to denote: $\boxed{sym\mathbf{A} \equiv \frac{1}{2}(\mathbf{A} + \mathbf{A}^T)}$. Likewise, by the same argument, the second part is a skew-symmetric tensor and we denote: $\boxed{skew\mathbf{A} \equiv \frac{1}{2}(\mathbf{A} - \mathbf{A}^T)}$. Thus, we have a representation: $\boxed{\mathbf{A} = sym\mathbf{A} + skew\mathbf{A}}$.

3.6.8.2 Polar Decomposition

A general second-order tensor **F** can always be represented as a multiplicative sum of another second-order tensor **U** or **V** and an orthogonal tensor **R** as: $\mathbf{F} = \mathbf{RU} = \mathbf{VR}$ Recalling that for any orthogonal tensor **R**, we have: $\mathbf{RR}^T = \mathbf{R}^T\mathbf{R} = \mathbf{I}$, it can easily be verified that: $\mathbf{F} = \mathbf{IF} = (\mathbf{RR}^T)\mathbf{F} = \mathbf{R}(\mathbf{R}^T\mathbf{F}) = \mathbf{RU}$. Similarly, $\mathbf{F} = \mathbf{FI} = \mathbf{F}(\mathbf{R}^T\mathbf{R}) = (\mathbf{FR}^T)\mathbf{R} = \mathbf{VR}$. We have clearly used the definitions: $\mathbf{U} = \mathbf{R}^T\mathbf{F}$ and $\mathbf{V} = \mathbf{FR}^T$.

However, for a total of 12 independent variables for the decomposed tensors, we only have 9 equations due to the original tensor **F**. Thus, the decomposition is under-determined and hence, non-unique. In order to make the decomposition unique, we may impose three more symmetry constraints on **U** or **V** in the form: $\mathbf{U} = \mathbf{U}^T$ or $\mathbf{V} = \mathbf{V}^T$ for the determination of **R**. Thus, a general second-order tensor **F** can always be represented as a multiplicative sum of a symmetric second-order tensor **U** or **V** and an orthogonal tensor **R** as: $\boxed{\mathbf{F} = \mathbf{RU} = \mathbf{VR}}$.

3.6.8.3 Spherical and Deviatoric Decomposition

A general second-order tensor **E** can always be represented as an additive sum of two tensors known as the spherical tensor *sph* **E** and the deviatoric tensor *dev* **E** as: $\boxed{\mathbf{E} = sph\mathbf{E} + dev\mathbf{E}}$. The equality can be easily verified by recalling the meaning of trace and double contraction along with the following definitions:

$$\boxed{sph\,\mathbf{E} \equiv \tfrac{1}{3}(tr\,\mathbf{E})\mathbf{I} = \tfrac{1}{3}(\mathbf{E}:\mathbf{I})\mathbf{I} = \tfrac{1}{3}E^k_k\, g_{ij}\,\mathbf{g}^i \otimes \mathbf{g}^j} \tag{3.74}$$

and

$$\boxed{dev\,\mathbf{E} \equiv \mathbf{E} - sph\,\mathbf{E} = (E_{ij} - E^k_k\, g_{ij})\mathbf{g}^i \otimes \mathbf{g}^j} \tag{3.75}$$

Clearly, $sph\,\mathbf{E}$ is a symmetric tensor.

Fact 1: For any $\mathbf{A}$, $sph\,(sph\,\mathbf{A}) = sph\,\mathbf{A}$.

Proof. Recalling that $tr\,\mathbf{I} = \mathbf{I}:\mathbf{I} = \delta^i_i = 3$, and applying the definitions, we have: $sph\,(sph\,\mathbf{A}) = \frac{1}{3}(tr\,\mathbf{A})sph(\mathbf{I}) = \frac{1}{3}(tr\,\mathbf{A})\frac{1}{3}(tr\,\mathbf{I})\mathbf{I} = \frac{1}{3}(tr\,\mathbf{A})\underbrace{\frac{1}{3}(tr\,\mathbf{I})}_{=1}\mathbf{I} = sph\,\mathbf{A}$

Fact 2: For any $\mathbf{A}$, $sph\,(dev\,\mathbf{A}) = \mathbf{0}$.

Proof. Using **Fact 1**: $sph\,(dev\,\mathbf{A}) = sph\,(\mathbf{A} - sph\,\mathbf{A}) = sph\,\mathbf{A} - sph\,(sph\,\mathbf{A}) = sph\,\mathbf{A} - sph\,\mathbf{A} = \mathbf{0}$

Fact 3: For any $\mathbf{A}$ and $\mathbf{B}$, $sph\,\mathbf{B} : dev\,\mathbf{A} = \mathbf{0}$.

Proof. Recalling that $tr\,\mathbf{I} = \mathbf{I}:\mathbf{I} = \delta^i_i = 3$, and applying the definitions, we have: $sph\,\mathbf{B} : dev\,\mathbf{A} = \frac{1}{3}(tr\,\mathbf{B})\mathbf{I} : (\mathbf{A} - \frac{1}{3}(tr\,\mathbf{A})\mathbf{I}) = \frac{1}{3}(tr\,\mathbf{B})(tr\,\mathbf{A}) - \frac{1}{9}(tr\,\mathbf{B})(tr\,\mathbf{A})(tr\,\mathbf{I}) = 0$

Fact 4: For any $\mathbf{A}$ and $\mathbf{B}$, $sph\,\mathbf{B} : \mathbf{A} = sph\,\mathbf{B} : sph\,\mathbf{A} = \mathbf{B} : sph\,\mathbf{A}$.

Proof. From **Fact 1** and **Fact 3**, we have: $sph\,\mathbf{B} : \mathbf{A} = sph\,\mathbf{B} : (sph\,\mathbf{A} + dev\,\mathbf{A}) = sph\,\mathbf{B} : sph\,\mathbf{A}$. Finally, $sph\,\mathbf{B} : \mathbf{A} = sph\,\mathbf{B} : sph\,\mathbf{A} = (sph\,\mathbf{B} + dev\,\mathbf{B}) : sph\,\mathbf{A} = \mathbf{B} : sph\,\mathbf{A}$

Fact 5: Show that for any $\mathbf{A}$ and $\mathbf{B}$, $dev\,\mathbf{B} : \mathbf{A} = \mathbf{B} : dev\,\mathbf{A} = dev\,\mathbf{B} : dev\,\mathbf{A}$ V. Using **Fact 3**, we have: $dev\,\mathbf{B} : \mathbf{A} = dev\,\mathbf{B} : (sph\,\mathbf{A} + dev\,\mathbf{A}) = dev\,\mathbf{B} : dev\,\mathbf{A}$.

Fact 6: Show that for any $\mathbf{A}$ and $\mathbf{B}$, $\mathbf{B} : \mathbf{A} = dev\,\mathbf{B} : dev\,\mathbf{A} + sph\,\mathbf{B} : sph\,\mathbf{A}$.

Proof. Using **Fact 3**, **Fact 4** and **Fact 5**, we have: $\mathbf{B} : \mathbf{A} = (sph\,\mathbf{B} + dev\,\mathbf{B}) : A = sph\,\mathbf{B} : sph\,\mathbf{A} + dev\,\mathbf{B} : dev\,\mathbf{A}$.

3.7 Calculus Tensor

Now that we have some idea about the geometry and algebraic operations of tensors along with vectors (tensors of order 1) and scalars (tensors of order 0), we conclude our treatment with the calculus of tensors.

As a start, we define a field as a function over some continuous region or domain. A scalar field is defined when a scalar is assigned at each point of the region; similarly, a vector field is defined when a vector is assigned at each point of the region, and finally, a tensor field is obtained when the range is a tensor space. On the other hand, the domain itself could be a scalar, vector or tensor space. A scalar field over a vector is exemplified by, say, the temperature distribution over a three-dimensional body; a force distribution likewise defines a vector field over a two- or three-dimensional body. A tensor field over a tensor domain appears in the constitutive theory of mechanics where the second-order stress tensor is a function of the second-order strain tensor. In any case, for any or all of the above fields, we will be interested in the rate at which these fields change, leading us to differential calculus, or in the determination of the average or summing operation resulting in integral calculus. Furthermore, we need to recognize that our discussion must anticipate curved surfaces such as shells where, as will be revealed shortly, we will encounter derivatives other than the familiar partial derivatives – namely, covariant or absolute derivatives.

3.8 Partial Derivatives of Tensors

Partial derivatives arise practically in all three aspects of mechanics: for example, the ***balance principles***, the ***equations of motion***, the ***compatibility conditions*** and ***constitutive theory***. In fact, partial derivatives with respect to scalars, vectors and tensors are needed in developing these theories.

3.8.1 Partial Derivatives of Base Vectors

Partial derivatives of the covariant base vectors, denoted as $\mathbf{G}_{i,j} \equiv \frac{\partial^2}{\partial \xi^i \partial \xi^j}$, with respect to the coordinate variables, are also vectors, and hence have a vector representation:

$$\boxed{\mathbf{G}_{i,j} \equiv \frac{\partial^2}{\partial \xi^i \partial \xi^j} = \Gamma^k_{ij}\mathbf{G}_k, \qquad i,j,k = 1,2,3} \tag{3.76}$$

where we have introduced: $\{\Gamma^k_{ij}\}$, $i,j,k = 1,2,3$, known as the Christoffel symbols of the second kind.

Remarks

- For Euclidean Cartesian coordinate system, the base vectors, $\mathbf{E}_i$'s are given by: $\mathbf{E}_i = \mathbf{A}\ \mathbf{i}_i$ where $\mathbf{A}$ is a fixed rotation tensor and the $\mathbf{i}_i$'s are fixed standard base vector; thus, $\mathbf{E}_{i,j} = \mathbf{0}$. In other words, the Christoffel symbols vanish identically: $\Gamma^k_{ij} = \mathbf{0}$ for all i,j,k.
- As an example, $\mathbf{G}_{1,2} \equiv \frac{\partial^2}{\partial \xi^1 \partial \xi^2} = \Gamma^1_{12}\mathbf{G}_1 + \Gamma^2_{12}\mathbf{G}_2 + \Gamma^3_{12}\mathbf{G}_3$.
- By definition, Christoffel symbols are symmetric in lower indices: $\mathbf{G}_{i,j} \equiv \mathbf{G}_{j,i} \Rightarrow \Gamma^k_{ij} = \Gamma^k_{ji}$,
- From equations (2.16) and (3.76), : $\boxed{\Gamma^k_{ij} = \mathbf{G}_{i,j} \bullet \mathbf{G}^k}$

- Again using equations (2.16) and (3.76), we also get the partial derivatives of the contravariant base vectors:

$$\boxed{\mathbf{G}_{i,k} \bullet \mathbf{G}^j + \mathbf{G}_i \bullet \mathbf{G}^j{}_{,k} = \mathbf{0} \quad \Rightarrow \mathbf{G}_i \bullet \mathbf{G}^j{}_{,k} = -\mathbf{G}_{i,k} \bullet \mathbf{G}^j = -\Gamma^m_{ik}\mathbf{G}_m \bullet \mathbf{G}^j = -\Gamma^j_{ik}} \quad (3.77)$$

implying:

$$\boxed{\mathbf{G}^j{}_{,k} = -\Gamma^j_{ik}\mathbf{G}^i} \quad (3.78)$$

3.8.2 Remark

Although the Christoffel symbols appear as components of a third-order tensor, the do not form a tensor as they do not satisfy the transformation properties of a tensor.

Example: Using two-dimensional polar coordinate system, we have: $\mathbf{G}^1 = (\cos\theta,\ \sin\theta)^T \Rightarrow \mathbf{G}^1{}_{,1} = (0,0)^T$, and $\mathbf{G}^2 = \frac{1}{r}(-\sin\theta,\ \cos\theta)^T \Rightarrow \mathbf{G}^2{}_{,1} = \frac{1}{r^2}(\sin\theta,\ -\cos\theta)^T$; similarly, $\mathbf{G}^1{}_{,2} = (-\sin\theta, \cos\theta)^T$ and $\mathbf{G}^2{}_{,2} = -\frac{1}{r}(\cos\theta,\ \sin\theta)^T$. Thus, the Christoffel symbols are given as: $\Gamma^1_{11} \equiv -\mathbf{G}^1{}_{,1} \bullet \mathbf{G}_1 = (0,0) \bullet (\cos\theta,\ \sin\theta) = 0$, $\Gamma^1_{12} = \Gamma^1_{21} \equiv -\mathbf{G}^1{}_{,1} \bullet \mathbf{G}_2 = (0,0) \bullet r(-\sin\theta,\ \cos\theta) = 0$, and $\Gamma^1_{22} \equiv -\mathbf{G}^1{}_{,2} \bullet \mathbf{G}_2 = (-\sin\theta,\ \cos\theta) \bullet r(-\sin\theta,\ \cos\theta) = r$; similarly, $\Gamma^2_{12} \equiv -\mathbf{G}^2{}_{,2} \bullet \mathbf{G}_1 = \frac{1}{r}$, $\Gamma^2_{22} \equiv -\mathbf{G}^2{}_{,2} \bullet \mathbf{G}_2 = 0$ and $\Gamma^2_{11} \equiv -\mathbf{G}^2{}_{,1} \bullet \mathbf{G}_1 = 0$.

3.9 Covariant or Absolute Derivative

Referring to the definition given by equation (2.18), to successfully complete the difference operation: $G(\mathbf{u} + \epsilon\bar{\mathbf{u}}) - G(\mathbf{u})$, we implied that $\mathbf{u}$ and $\mathbf{u} + \epsilon\bar{\mathbf{u}}$, belong to the same vector space – an assumption that is only valid for an Euclidean 'flat' space: we can ***translate*** or make a parallel transport (i.e. dragging a vector by keeping each component length of the vector intact) of $\bar{\mathbf{u}}$ to $\mathbf{u}$ and then perform the ϵ-scaled addition because (i) vectors are invariant under translation in an Euclidean space and (ii) both $\mathbf{u}$ and $\bar{\mathbf{u}}$ belong to the same vector space, as discussed before. Similar assumptions have also been made for the Gateaux derivative of tensor fields as given by equation (2.7). But, these assumptions are not generally true for Riemannian 'curved' spaces: the tangent space, which serves locally as a linear vector space at each point on a curved space, is different from point to point, and as such, makes both $\mathbf{u} + \epsilon\bar{\mathbf{u}}$ and $G(\mathbf{u} + \epsilon\bar{\mathbf{u}}) - G(\mathbf{u})$ operations generally invalid unless some connection is devised among the points in a neighborhood, that is, a parallel transport needs to be redefined.

One such redefinition of parallelism is offered by the Levi–Civita connection: instead of preserving the coordinate components of a vector intact for parallel transport as in Euclidean Cartesian coordinate systems, here we require that another invariant property of parallel lines be retained: that a transversal subtends equal angles with a line that is being moved in parallel fashion as in Figure 3.3. For the curved spaces, a curve generalizes the transversal with the angle defined between the tangents to the curve and the vector being moved. Thus, while the Gateaux derivative is intimately related to tensor calculus in Cartesian coordinate systems of a Euclidean space, for curvilinear coordinate systems or curved spaces, the concept of the Gateaux derivative fails to preserve the tensor characteristic of a geometric object; for example, while a differentiation

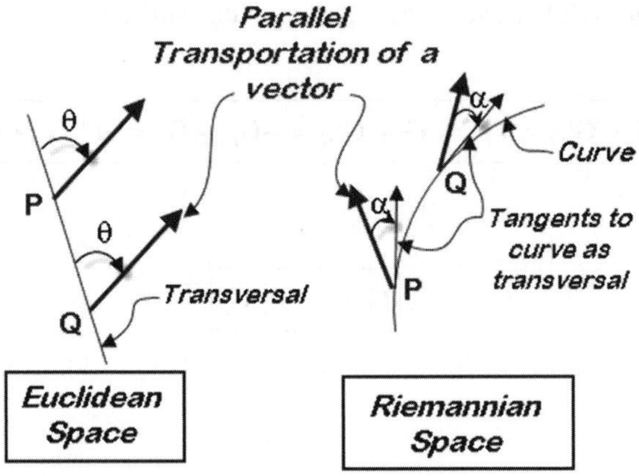

Figure 3.3 Parallel transport.

(gradient) of a scalar (tensor of order 0) field is a covariant vector field, the partial differentiation of a vector (tensor of order 1) field is not a tensor field. So, in the light of Levi–Civita connection, it is required to be further generalized to what is known as the covariant derivative – covariant in the sense that it satisfies that objects in geometry such as vectors and tensors should remain so after derivative operations, independent of their description in a particular coordinate system. In other words, the covariant derivative is the generalization of the Gateaux directional derivative of the Euclidean Cartesian coordinate systems to the curvilinear coordinate systems, or the curved spaces where $\mathbf{v}(\mathbf{x} + \rho\bar{\mathbf{x}})$ and $\mathbf{v}(\mathbf{x})$ of the definition do not belong to the same vector (tangent) space, and hence the difference operation does not have a meaning. In fact, the vector, itself, does not have a meaning other than as a tangent operation to an affine curve. Thus, the generalization is effected through the choice of affinely parameterized unique curves on the curved surface; this makes the definition of the covariant derivative path-dependent. Finally, in the analysis, by describing the vectors and tensors in coordinate-independent absolute form, we can effectively avoid use of the derivatives of the base vectors and stay with the familiar partial derivatives of a functional or vector or tensor field, since the covariant derivatives of the base vectors are identically zero. Here is the coordinate independent definition of the covariant derivative:

- The ***covariant derivatives***, denoted by $\nabla_{\mathbf{x}}\mathbf{T}(\mathbf{x}) \bullet \bar{\mathbf{x}}$, of a vector or tensor field on a ***curved space*** or ***manifold*** are given by:

$$\boxed{\begin{aligned} \nabla_{\mathbf{x}}\mathbf{T}(\mathbf{x}) \bullet \bar{\mathbf{x}} &\equiv \bar{\mathbf{T}}\left(\mathbf{x} \equiv \frac{\partial}{\partial \lambda}, \bar{\mathbf{x}} \equiv \frac{\partial}{\partial \xi}\right) \\ &= \lim_{\varepsilon \to 0} \frac{d}{d\varepsilon}\{\mathbf{T}(\mathbf{x} + \varepsilon\bar{\mathbf{x}})\} = \lim_{\varepsilon \to 0} \frac{\{\mathbf{T}(\lambda_0 + \varepsilon\xi)_{\substack{\text{transported along} \\ \text{an affine curve to } \lambda_0}} - \mathbf{T}(\lambda_0)\}}{\varepsilon} \end{aligned}} \tag{3.79}$$

where $\nabla_{\mathbf{x}}\mathbf{T}(\mathbf{x}) \bullet \bar{\mathbf{x}}$ is the ***covariant derivative*** of $\mathbf{T}(\mathbf{x})$ at $\mathbf{x}$ along the vector direction $\bar{\mathbf{x}}$; note that $\mathbf{T}(\mathbf{x})$ could be a vector or a tensor field.
- The ***covariant derivatives***, $\nabla_{\mathbf{u}}G(\mathbf{u}, \bar{\mathbf{u}})$, of a functional on a ***curved space*** or ***manifold*** are given by:

Let $G(\mathbf{u}(\mathbf{x}))$ be a functional or operator defined for an argument function $\mathbf{u}(\mathbf{x}) =$ at any $\mathbf{x} \in \mathbb{R}^m$. Now, let $\bar{\mathbf{u}}(\mathbf{x}) \in \mathbb{R}^n$ be another function at $\mathbf{x}$. Then, the covariant derivative, denoted as $G_{,\mathbf{u}}(\mathbf{u}, \bar{\mathbf{u}})$, or $DG(\mathbf{u}) \bullet \bar{\mathbf{u}}$, of G at $\mathbf{u}$ along the direction $\bar{\mathbf{u}}$ is defined as:

$$\boxed{\nabla_{\mathbf{u}}(\mathbf{u}) \bullet \bar{\mathbf{u}} \equiv \left[\frac{d}{d\epsilon} G(\mathbf{u} + \epsilon\bar{\mathbf{u}})\right]_{\epsilon=0}} \tag{3.80}$$

However, as a practical scheme for computing the covariant derivatives of the components of a vector field (or tensor field), all we need to do is to include a correction in the form of the partial derivatives of the base vectors (or base tensors) as they are the tangents (or the tensor product of these) to the coordinate curves on the manifold. Thus, in order to preserve the tensor characteristic of a tensor, a translation operation must respect the change of basis rule reflected in the derivative of the base vectors, as in equation (3.76) and (3.78), through introduction of the Christoffel symbols.

3.9.1 Covariant or Absolute Derivatives of the Components of Vectors

With the definition of the partial derivatives of the covariant base vectors, $\{\mathbf{G}_i\}$, as given by equation (3.76), given a vector function $\mathbf{a}(\xi^1, \xi^2, \cdots) = a^i(\xi^1, \xi^2, \cdots)\, \mathbf{G}_i(\xi^1, \xi^2, \cdots)$ on a curved space, the absolute or covariant derivative of the contravariant components of a vector, denoted by $a^i|_j$, is given as:

$$\boxed{a^i|_j \equiv \underbrace{\underbrace{a^i{}_{,j}}_{\text{not a tensor}} + \underbrace{a^m\, \Gamma^i_{mj}}_{\text{not a tensor}}}_{\text{whole expression is a tensor}}} \tag{3.81}$$

Thus:

$$\mathbf{a}|_j = a^i{}_{,j}\, \mathbf{G}_i + a^i\, \mathbf{G}_{i,j} = a^i{}_{,j}\, \mathbf{G}_i + a^i\, \Gamma^k_{ij}\mathbf{G}_k = \left(a^i{}_{,j} + a^m\, \Gamma^i_{mj}\right) \mathbf{G}_i \equiv a^i|_j\, \mathbf{G}_i \tag{3.82}$$

Similarly, expressing $\mathbf{a}(\xi^1, \xi^2, \cdots) = a_i(\xi^1, \xi^2, \cdots)\, \mathbf{G}^i(\xi^1, \xi^2, \cdots)$, and using equation (3.77), the absolute or covariant derivative of the covariant components of a vector is denoted by $a_i|_j$ and given as:

$$\boxed{a_i|_j \equiv a_{i,j} - a_k\, \Gamma^k_{ij}} \tag{3.83}$$

There are other alternative notations for the covariant derivative: $D_j a_i \equiv a_i;j \equiv a_i|_j$, and so on. Similarly, the Christoffel symbols of the second kind is alternatively written as: $\left\{ \begin{matrix} k \\ i \quad j \end{matrix} \right\} \equiv \Gamma^k_{ij}$.

The covariant derivatives of higher-order tensors are similarly given as:

$$\mathbf{T}^{jk\ldots}_{qr\ldots}\Big|_m \equiv \mathbf{T}^{jk\ldots}_{qr\ldots,m} + \Gamma^j_{im}\, \mathbf{T}^{ik\ldots}_{qr\ldots} + \Gamma^k_{im}\, \mathbf{T}^{ji\ldots}_{qr\ldots} + - \Gamma^i_{qm}\, \mathbf{T}^{jk\ldots}_{ir\ldots} - \Gamma^i_{rm}\, \mathbf{T}^{jk\ldots}_{qi\ldots} - \tag{3.84}$$

Note that in equation (3.84), for each contravariant index a Christoffel symbol term is added, while for each covariant index, a Christoffel symbol term is subtracted. Finally, note that for the differential geometry of thin surface-like structures such as shells, we can develop expressions without explicitly resorting to the Christoffel symbols.

3.9.2 Covariant Derivative of the Base Vectors and the Metric Tensor

Generalizing on the covariant derivatives of vector components as in equation (3.81), for, $\mathbf{T}^j$, the contravariant pseudo-vector components of a tensor, $\mathbf{T} \equiv \mathbf{T}^j \otimes \mathbf{g}_j$, we can express the covariant derivatives as:

$$\boxed{\mathbf{T}^j\big|_k \equiv \mathbf{T}^j{}_{,k} + \mathbf{T}^m\,\Gamma^j_{km}} \tag{3.85}$$

Similarly, for, $\mathbf{T}_j$, the covariant pseudo-vector components of a tensor, $\mathbf{T} \equiv \mathbf{T}_j \otimes \mathbf{g}^j$, and generalizing equation (3.83), we have the covariant derivatives as:

$$\boxed{\mathbf{T}_j\big|_k \equiv \mathbf{T}_{j,k} - \mathbf{T}_m\,\Gamma^m_{jk}} \tag{3.86}$$

Now, applying equation (3.86) to the base vectors themselves and noting the partial derivatives of the base vectors as in equation (3.76), we get the absolute or covariant derivative of the covariant base vectors, $\{\mathbf{G}_i\}$, we get $\mathbf{G}_i\big|_j = \mathbf{0}$. Thus:

$$\boxed{\mathbf{G}_i\big|_j = \mathbf{G}_{i,j} - \mathbf{G}_m\,\Gamma^m_{ij} = \Gamma^k_{ij}\mathbf{G}_k - \mathbf{G}_m\,\Gamma^m_{ij} = \mathbf{0}} \tag{3.87}$$

Now, noting the definition of the metric tensor, the covariant derivatives, $G_{ij}\,|_k$, of the components, G_{ij}, of the metric tensor are identically zero; so, using equation (3.87), we have:

$$\boxed{G_{ij}\,|_k = \mathbf{G}_i\,|_k \bullet \mathbf{G}_j + \mathbf{G}_i \bullet \mathbf{G}_j\,|_k = 0} \tag{3.88}$$

In summary:

- ***The covariant derivatives of both the base vectors and the metric tensor are identically zero; that is, under covariant derivative, they both act like constants.***

3.10 Riemann–Christoffel Tensor: Ordered Differentiation

Using the covariant derivatives of a vector, as in equations (3.83) and (3.84), and taking again the covariant derivative of the covariant derivative of the covariant component of a vector, we have:

$$\begin{aligned} D_i D_j\, a_k &= D_i\left(a_{k,j} - \Gamma^m_{kj}\,a_m\right) \\ &= \left(a_{k,j} - \Gamma^m_{kj}\,a_m\right)_{,i} - \Gamma^n_{ki}\left(a_{n,j} - \Gamma^m_{nj}\,a_m\right) - \Gamma^n_{ji}\left(a_{k,n} - \Gamma^m_{kn}\,a_m\right) \\ &= a_{k,ji} - \Gamma^m_{kj}\,a_{m,i} - \Gamma^m_{kj,i}\,a_m - \Gamma^n_{ki}\,a_{n,j} + \Gamma^n_{ki}\,\Gamma^m_{nj}\,a_m \\ &\quad - \Gamma^n_{ji}\,a_{k,n} + \Gamma^n_{ji}\,\Gamma^m_{kn}\,a_m \end{aligned} \tag{3.89}$$

where we have used D_i as the covariant derivative with respect to the i coordinate direction. Now, by permutation of i and j in equation (3.89), we get:

$$D_i D_j\, a_k - D_j D_i\, a_k = \underbrace{\left\{\Gamma^n_{ki}\,\Gamma^m_{nj} - \Gamma^n_{kj}\,\Gamma^m_{ni} + \Gamma^m_{ki},_j - \Gamma^m_{kj},_i\right\}}_{\overset{\Delta}{=} R^m_{ijk}} a_m \equiv R^m_{ijk}\, a_m \tag{3.90}$$

where we have defined, $\boxed{R^m_{ijk} \overset{\Delta}{=} \Gamma^n_{ki}\,\Gamma^m_{nj} - \Gamma^n_{kj}\,\Gamma^m_{ni} + \Gamma^m_{ki},_j - \Gamma^m_{kj},_i}$, the components of a fourth-order tensor known as the Riemann–Christoffel curvature tensor.

Remarks

- It can be shown that the R^m_{ijk}'s are the components of a tensor, that is, they obey the transformation law for a tensor.
- Since the Christoffel symbols depend solely on the metric tensor, the Riemann–Christoffel tensor belongs to the class of fundamental tensors.
- Since the Christoffel symbols, Γ^k_{ij}, identically vanish in Cartesian coordinate systems, the Riemann–Christoffel tensor is identically zero for all Cartesian coordinate systems in Euclidean space.
- Obviously, in Riemannian space, the Riemann–Christoffel tensor does not vanish, so ***the order of differentiation cannot be interchanged***.
- From equation (3.90), the difference between the covariant derivatives of the covariant components of a second-order tensor can be given as:

$$D_i D_j\, a_{kl} - D_j D_i\, a_{kl} = R^m_{kji}\, a_{ml} + R^m_{lji}\, a_{km} \tag{3.91}$$

3.11 Partial (PD) and Covariant (C.D.) Derivatives of Tensors

Using the derivatives of the base vectors as in equation (3.76), the partial derivatives of a tensor, say, a second-order tensor, $\mathbf{T} \equiv T^{ij}\mathbf{g}_i \otimes \mathbf{g}_j$, can be expressed as:

$$\boxed{\begin{aligned} \mathbf{T},_k &\equiv \frac{\partial \mathbf{T}}{\partial \xi^k} = T^{ij},_k\, \mathbf{g}_i \otimes \mathbf{g}_j + T^{ij}\mathbf{g}_i,_k \otimes \mathbf{g}_j + T^{ij}\mathbf{g}_i \otimes \mathbf{g}_j,_k \\ &= T^{ij},_k\, \mathbf{g}_i \otimes \mathbf{g}_j + T^{ij}\Gamma^m_{ik}\mathbf{g}_m \otimes \mathbf{g}_j + T^{ij}\mathbf{g}_i \otimes \Gamma^m_{jk}\mathbf{g}_m \\ &= \left(T^{ij},_k + T^{mj}\Gamma^i_{mk} + T^{im}\Gamma^j_{mk}\right)\mathbf{g}_i \otimes \mathbf{g}_j = T^{ij}\big|_k\, \mathbf{g}_i \otimes \mathbf{g}_j \\ &\text{where } T^{ij}\big|_k \equiv T^{ij},_k + T^{mj}\Gamma^i_{mk} + T^{im}\Gamma^j_{mk} \end{aligned}} \tag{3.92}$$

Now, because the covariant derivative of the base vectors are zero as given by equation (3.87), the covariant derivatives, $\mathbf{T}|_k$, of the tensor, $\mathbf{T}$, are identical to the partial derivatives of the tensor:

$$\boxed{\mathbf{T},_k = \mathbf{T}|_k = T^{ij}\big|_k\, \mathbf{g}_i \otimes \mathbf{g}_j = \left(T^{ij},_k + T^{mj}\Gamma^i_{mk} + T^{im}\Gamma^j_{mk}\right)\mathbf{g}_i \otimes \mathbf{g}_j} \tag{3.93}$$

Thus, let us note that:

The partial derivatives of a tensor are identical to the corresponding covariant derivatives of the tensor (equation 3.93) but the partial derivatives of the components of a tensor are different from the corresponding covariant derivatives of the components of the tensor (equation 3.92).

3.12 Partial Derivatives of Scalar Functions of Tensors

Now, we may consider a scalar (tensor of order zero) valued function $f(\mathbf{A})$ of a tensor. Then, the partial derivative of f with respect to second-order tensor, $\mathbf{A}$, is a second-order tensor defined as:

$$\boxed{f,_{\mathbf{A}} \equiv \frac{\partial f}{\partial \mathbf{A}} = \frac{\partial f}{\partial A_{ij}}\mathbf{g}_i \otimes \mathbf{g}_j = \frac{\partial f}{\partial A^{ij}}\mathbf{g}^i \otimes \mathbf{g}^j} \tag{3.94}$$

A similar expression can be obtained in terms of the mixed components of $\mathbf{A}$. In any event, the definition ensures that the partial derivative tensor is invariant, irrespective of the choice of components. The partial derivatives of various scalar functions – for example, the ***trace***, one of the ***invariants***, of a second-order tensor – are routinely needed in analysis, as follows.

Fact 1: For any $\mathbf{A} \equiv \mathbf{A}_j \otimes \mathbf{g}^j$, $\boxed{(tr\mathbf{A}),_{\mathbf{A}} = \mathbf{I}}$.

Proof. Let us recall that we have: $tr\mathbf{A} = \mathbf{A} : \mathbf{I} = \mathbf{A}_i \otimes \mathbf{g}^i : \mathbf{g}_j \otimes \mathbf{g}^j = g^{ij}(\mathbf{A}_i \bullet \mathbf{g}_j) = g^{ij}A_{ij} = A^j_{\ j}$. Then, $(tr\mathbf{A}),_{\mathbf{A}} = \frac{\partial A^j_{\ j}}{\partial A^p_{\ q}}\mathbf{g}^p \otimes \mathbf{g}_q = \delta^j_p \delta^q_j \mathbf{g}^p \otimes \mathbf{g}_q = \mathbf{g}^j \otimes \mathbf{g}_j = \mathbf{I}$.

Fact 2: For any $\mathbf{A} \equiv \mathbf{A}_j \otimes \mathbf{g}^j$, $\boxed{(tr\mathbf{A})^2,_{\mathbf{A}} = 2(tr\mathbf{A})\mathbf{I}}$.

Proof. Clearly, by **Fact 1** and the chain rule: $(tr\mathbf{A})^2,_{\mathbf{A}} = 2(tr\mathbf{A})\mathbf{I}$.

Fact 3: For any $\mathbf{A} \equiv \mathbf{A}_j \otimes \mathbf{g}^j$, $\boxed{tr(\mathbf{A}^2),_{\mathbf{A}} = 2\mathbf{A}^T}$.

Proof. Recalling that $tr(\mathbf{A}^2) = tr(A^i_{\ j}\mathbf{g}_i \otimes \mathbf{g}^j \bullet A^k_{\ l}\mathbf{g}_k \otimes \mathbf{g}^l) = tr(A^i_{\ j}A^k_{\ l}\delta^j_k\mathbf{g}_i \otimes \mathbf{g}^l) = A^i_{\ j}A^j_{\ l}\delta^l_i = A^i_{\ j}A^j_{\ i}$, we have: $tr(\mathbf{A}^2),_{\mathbf{A}} = \frac{\partial(A^i_{\ j}A^j_{\ i})}{\partial A^p_{\ q}}\mathbf{g}^p \otimes \mathbf{g}_q = \delta^i_p\delta^q_j A^j_{\ i}\mathbf{g}^p \otimes \mathbf{g}_q + A^i_{\ j}\delta^j_p\delta^q_i\mathbf{g}^p \otimes \mathbf{g}_q = 2A^q_{\ p}\mathbf{g}^p \otimes \mathbf{g}_q = 2\mathbf{A}^q \otimes \mathbf{g}_q$. Now, noting the definition: $\mathbf{A} \equiv \mathbf{A}_j \otimes \mathbf{g}^j$, we clearly have: $tr(\mathbf{A}^2),_{\mathbf{A}} = 2\mathbf{A}^T$.

Exercise: Show that for any $\mathbf{A}$, $\boxed{tr(\mathbf{A}^n),_{\mathbf{A}} = n\mathbf{A}^T}$

3.13 Partial Derivatives of Tensor Functions of Tensors

Now, the above definition can be extended to the second-order tensor with the coefficients being functions of another tensor. Let $\mathbf{B} = \mathbf{B}^{ij}(\mathbf{A})\,\mathbf{g}_i \otimes \mathbf{g}_j$ be the tensor-valued tensor function. Then, the partial derivative of **B** with respect to **A** is given by a fourth-order tensor:

$$\boxed{\mathbf{B}_{,\mathbf{A}} \equiv \frac{\partial \mathbf{B}}{\partial \mathbf{A}} = \frac{\partial B^{ij}}{\partial A_{kl}}\,\mathbf{g}_i \otimes \mathbf{g}_j \otimes \mathbf{g}_k \otimes \mathbf{g}_l} \tag{3.95}$$

With $\mathbf{B} \equiv \mathbf{A}$, we have: $\mathbf{A}_{,\mathbf{A}} \equiv \frac{\partial \mathbf{A}}{\partial \mathbf{A}} = \frac{\partial A^{ij}}{\partial A_{kl}}\,\mathbf{g}_i \otimes \mathbf{g}_j \otimes \mathbf{g}_k \otimes \mathbf{g}_l = \delta^{ik}\delta^{jl}\mathbf{g}_i \otimes \mathbf{g}_j \otimes \mathbf{g}_k \otimes \mathbf{g}_l = \mathbf{g}_i \otimes \mathbf{g}_j \otimes \mathbf{g}_i \otimes \mathbf{g}_j$. Now, recalling from equation (3.46) a fourth-order identity tensor from equation that is defined by only the base vectors: $\overset{4}{\tilde{\mathbf{I}}} \equiv \mathbf{g}_i \otimes \mathbf{g}_j \otimes \mathbf{g}_i \otimes \mathbf{g}_j$, we may have a compact symbolical definition: $\boxed{\mathbf{A}_{,\mathbf{A}} \equiv \frac{\partial \mathbf{A}}{\partial \mathbf{A}} = \overset{4}{\tilde{\mathbf{I}}}}$. A repeat partial differentiation of a scalar valued tensor function will result in a fourth-order tensor:

$$\begin{aligned} f_{,\mathbf{A}\otimes\mathbf{A}} &\equiv \frac{\partial^2 f}{\partial \mathbf{A}\,\partial \mathbf{A}} = \frac{\partial}{\partial \mathbf{A}}\left(\frac{\partial f}{\partial \mathbf{A}}\right) = \frac{\partial^2 f}{\partial A_{ij}\partial A_{pq}}\mathbf{g}_i \otimes \mathbf{g}_j \otimes \mathbf{g}_p \otimes \mathbf{g}_q \\ &= B^{ijpq}\mathbf{g}_i \otimes \mathbf{g}_j \otimes \mathbf{g}_p \otimes \mathbf{g}_q \end{aligned} \tag{3.96}$$

where $B^{ijpq} \equiv \frac{\partial^2 f}{\partial A_{ij}\partial A_{pq}}$. As before, similar expressions can be written in terms of the covariant and mixed components or their combinations.

3.14 Partial Derivatives of Parametric Functions of Tensors

For a scalar function of a tensor with the tensor having parametric dependence, the derivative of the function with respect to the parameter is obtained by a double contraction, equivalently describing the usual chain rule of differentiation. Thus, with $f = f(\mathbf{A}(t))$, we have:

$$\boxed{\dot{f} \equiv \frac{\partial f}{\partial t} = f_{,\mathbf{A}} : \dot{\mathbf{A}}} \tag{3.97}$$

Fact 1: From the definition of ***norm*** of a tensor, we have $||\mathbf{A}||^2 \equiv \mathbf{A} : \mathbf{A} = tr(\mathbf{A}\mathbf{A}^T)$. Then, for a symmetric tensor **A**, the partial derivative of the norm with respect to **A** is given by:

$$\boxed{||\mathbf{A}||_{,\mathbf{A}} \equiv \frac{\partial\, ||\mathbf{A}||}{\partial \mathbf{A}} = \frac{\mathbf{A}}{||\mathbf{A}||}} \tag{3.98}$$

Proof. $||\mathbf{A}||^2 = tr(\mathbf{A}\mathbf{A}^T) = tr(\mathbf{A}^2)$. Now, noting from ***invariants*** study that $(tr\mathbf{A}^2)_{,\mathbf{A}} \equiv 2\mathbf{A}$, and taking derivatives of both sides with respect to **A**, we get: $2||\mathbf{A}||\,||\mathbf{A}||_{,\mathbf{A}} = 2\mathbf{A}$. Now, diving both sides by $2||\mathbf{A}||$ completes the proof.

3.15 Differential Operators

The differential operators are a generalization of derivatives of scalar functions of scalars (zero-order tensors) to higher-order arguments. Interestingly, just like the vector multiplications that come under three different flavors – dot product (resulting in scalar), cross product (resulting in vector) and tensor product (resulting in tensor) – the vector differentiations manifest themselves as either divergence (resulting in a scalar field), curl (resulting in a vector field) or gradients (resulting in a tensor field). These, in turn, are conveniently employed through the introduction of the so-called Nabla operator.

3.15.1 Del or Nabla Operator: ∇

From the differential geometry of ***surfaces***, let $\phi(\xi^1, \xi^2, \xi^3) = const$ define a surface. For example, a spherical surface with ($\xi^1 \equiv r, \xi^2 \equiv \theta, \xi^3 \equiv \varphi$) such that $x_1 = \xi^1 \sin\xi^3 \cos\xi^2$, $x_2 = \xi^1 \sin\xi^3 \sin\xi^2$, $x_3 = \xi^1 \cos\xi^3$ may be defined by $\phi(\xi^1, \xi^2, \xi^3) = \xi^1 = r = const$. Then, if we move a bit $\mathbf{d}\xi$ on the surface, the change in $\mathbf{d}\phi = \frac{\partial\phi}{\partial\xi^i} d\xi^i = 0$. We define a pseudo-vector operator:

$$\boxed{\nabla(\bullet) \equiv \mathbf{G}^i \frac{\partial(\bullet)}{\partial\xi^i}} \tag{3.99}$$

With this definition, the above change in $\mathbf{d}\phi$ may be effectively represented by a scalar product of a vector, $\mathbf{d}\xi = d\xi^k \mathbf{G}_k$ and a pseudo-vector, $\nabla\phi = \frac{\partial\phi}{\partial\xi^i}\mathbf{G}^i$ resulting in $\boxed{\mathbf{d}\phi = \nabla\phi \bullet \mathbf{d}\xi = 0}$. Thus, if $\mathbf{d}\xi$ is a change on the surface, more precisely, on the tangent plane at the point on the surface, then $\nabla\phi$ must be a vector along the normal to the surface.

Thus, ∇, the pseudo-vector operator changes a scalar function on to a vector function along the normal to the surface. $\nabla\varphi$ is known as the gradient of the scalar function ϕ. For the spherical surface example above, $\nabla\phi = \mathbf{G}^1$ is the (radial) base vector normal to the so-called $\xi^1 = const$ surface.

Example: Continuing with the two-dimensional polar coordinate system with $\mathbf{G}^1 = \mathbf{G}_1$ and $\mathbf{G}^2 = \frac{1}{r^2}\mathbf{G}_2$, we have: $\nabla \equiv \mathbf{G}^i \frac{\partial}{\partial\xi^i} = \mathbf{G}^1 \frac{\partial}{\partial r} + \mathbf{G}^2 \frac{\partial}{\partial\theta} = \mathbf{G}_1 \frac{\partial}{\partial r} + \frac{1}{r^2}\mathbf{G}_2 \frac{\partial}{\partial\theta}$, implying that in polar coordinates:

$$\boxed{\nabla = \widehat{\mathbf{G}}_r \frac{\partial}{\partial r} + \frac{1}{r}\widehat{\mathbf{G}}_\theta \frac{\partial}{\partial\theta}} \tag{3.100}$$

where $\widehat{\mathbf{G}}_r, \widehat{\mathbf{G}}_\theta$ are the unit covariant base vectors.

3.16 Gradient Operator: *GRAD*$(\bullet)$ or $\nabla(\bullet)$

For scalar-valued vector functions, the ∇ operator is a logical generalization of the definition of differentiation of a scalar-valued scalar function. However, when we try to extend this to a vector- and tensor-valued function, or to other forms of differentiation (such as divergence, to be introduced shortly), then the ∇-operator, with column vectors in mind, is a bit confusing because it involves the transposed tensors. For this reason, we will introduce definitions that are more uniform for all orders of tensors.

3.16.1 Gradient of Scalar Functions: *GRAD* ϕ

The gradient of a scalar function $\phi(\xi^i)$, $i = 1, 2, \ldots$ is a vector function, denoted as $GRAD\,\phi$, defined by:

$$\boxed{grad\,\phi \equiv \frac{\partial \phi}{\partial \xi^i}\mathbf{G}^i = \nabla \phi} \tag{3.101}$$

Thus, the gradient of a zero-order tensor (scalar) function results in a first-order tensor (vector) function. Also, by definition, the *GRAD* and ∇ operators turn out to be the same.

Remark

- In nonlinear finite element analysis, which involves numerical minimization of an energy expression, we often need to determine the second derivative (i.e. the Hessian) of this scalar entity. For numerical efficiency, we further need to know whether the Hessian tensor is symmetric. For this information, we include the following important lemma.

3.16.2 Hessian Symmetry Lemma

The Hessian tensor obtained by taking the covariant derivative of the covariant components of the gradient vector of a scalar function is always symmetric.

Proof. Let $a_i = \frac{\partial \phi}{\partial \xi^i}$'s be the covariant components of the gradient vector of a scalar function, $\phi(\xi^i)$. Then, by equation (3.83), a_{ij}, the components of a Hessian obtained by the covariant derivative of a_i is given by: $a_{ij} = a_i\big|_j \equiv a_{i},_j - a_k\,\Gamma^k_{ij}$. Now, by symmetry of the ***Christoffel symbols*** in lowercase indices, that is, $\Gamma^k_{ij} = \Gamma^k_{ji}$, and noting the interchangeability of the partial derivatives, that is, $a_{i},_j = \frac{\partial^2 \phi}{\partial \xi^i \partial \xi^j} = \frac{\partial^2 \phi}{\partial \xi^j \partial \xi^i} = a_{j},_i$ we get: $a_{ij} - a_{ji} = a_{i},_j - a_k\,\Gamma^k_{ij} - a_{j},_i + a_k\,\Gamma^k_{ji} = 0$, that is, noting that any tensor can be decomposed as a sum of a symmetric and an anti-symmetric tensor, the anti-symmetric part of the Hessian is identically zero, which completes the proof.

3.16.3 Gradient of Vector Functions: *GRAD* $\boldsymbol{\psi}$

The gradient of a vector function $\psi(\xi^i)$, $i = 1, 2, \ldots$ is a second-order tensor function, denoted as $GRAD\,\boldsymbol{\psi}$, defined by:

$$\boxed{GRAD\,\boldsymbol{\psi} \equiv \frac{\partial \boldsymbol{\psi}}{\partial \xi^j} \otimes \mathbf{G}^j = \boldsymbol{\psi},_j \otimes \mathbf{G}^j} \tag{3.102}$$

Now, noting that the ***covariant derivatives*** of the base vectors are zero by equation (3.87), we have:

$$\boxed{\underbrace{GRAD\,\boldsymbol{\psi}}_{\text{invariant}} = \underbrace{\boldsymbol{\psi}|_j}_{\text{covariant}} \otimes \overbrace{\mathbf{G}^j}^{\text{contravariant}}} \tag{3.103}$$

Thus, by the ***quotient rule***, $GRAD\,\boldsymbol{\psi}$ is invariant and hence a tensor. We can write the component-wise description as: $GRAD\,\boldsymbol{\psi} = \psi_i|_j\,\mathbf{G}^i \otimes \mathbf{G}^j$. Finally, with the definition of the del operator as above, we have: $\nabla\boldsymbol{\psi} = \mathbf{G}^i \otimes \frac{\partial \boldsymbol{\psi}}{\partial \xi^i} = \mathbf{G}^i \otimes \boldsymbol{\psi}_{,i}$. Thus, comparing with $GRAD\,\boldsymbol{\psi}$, we find that: $\boxed{GRAD\,\boldsymbol{\psi} = \nabla\boldsymbol{\psi}^T}$.

3.16.4 Gradient of Tensor Functions: GRAD A

The gradient of a second-order tensor function, $\mathbf{A}(\xi^i),\ i = 1, 2, \ldots$, is a third-order tensor function, denoted as $GRAD\,\mathbf{A}$, defined by:

$$\boxed{GRAD\,\mathbf{A} \equiv \frac{\partial \mathbf{A}}{\partial \xi^j} \otimes \mathbf{G}^j = \mathbf{A}_{,j} \otimes \mathbf{G}^j} \tag{3.104}$$

Now, noting that the ***covariant derivatives*** of the base vectors are zero by equation (3.87), we have:

$$\boxed{\underbrace{GRAD\,\mathbf{A}}_{\text{invariant}} = \underbrace{\mathbf{A}|_j}_{\text{covariant}} \otimes \overbrace{\mathbf{G}^j}^{\text{contravariant}}} \tag{3.105}$$

Thus, by ***quotient rule***, $GRAD\,\mathbf{A}$ is invariant and hence a tensor. We can write the component-wise description as: $GRAD\,\mathbf{A} = A_{ij}|_k\,\mathbf{G}^i \otimes \mathbf{G}^j \otimes \mathbf{G}^k$. Finally, comparing with the definition of del operator, we find that: $GRAD\,\mathbf{A} = \nabla\mathbf{A}^T$.

Fact 2: For a scalar function φ and a vector function $\mathbf{v}$:

$$\boxed{\begin{aligned} GRAD(\varphi\mathbf{v}) &= \varphi\ GRAD\,\mathbf{v} + \mathbf{v} \otimes GRAD\varphi \\ \nabla(\varphi\mathbf{v}) &= \varphi\ \nabla\mathbf{v} + \mathbf{v} \otimes \nabla\varphi \end{aligned}} \tag{3.106}$$

Proof.

$$\begin{aligned} GRAD(\varphi\mathbf{v}) &= (\varphi\mathbf{v})_{,k} \otimes \mathbf{G}^k = (\varphi_{,k}\,\mathbf{v} + \varphi\ \mathbf{v}_{,k}) \otimes \mathbf{G}^k \\ &= \mathbf{v} \otimes \varphi_{,k}\ \mathbf{G}^k + \varphi\ \mathbf{v}_{,k} \otimes \mathbf{G}^k = \mathbf{v} \otimes GRAD\varphi + \varphi\ GRAD\,\mathbf{v} \end{aligned}$$

3.17 Divergence Operator: *DIV* or ∇•

Because of the argument presented above as to the confusion related to the ∇ operator, we follow the gradient operator in introducing the definitions below that are consistent for tensors of order one and above.

3.17.1 Divergence of Vector Functions: DIV ψ

The divergence of a vector function $\boldsymbol{\psi}(\xi^i),\ i = 1, 2, \ldots$ is a scalar or zero-order tensor function, denoted by $DIV\,\boldsymbol{\psi}$, defined as:

$$\boxed{DIV\,\boldsymbol{\psi} \equiv GRAD\,\boldsymbol{\psi} : \boldsymbol{I}} \tag{3.107}$$

Fact 1: For any vector function ψ:

$$DIV\,\psi = \psi_{,j}\,\mathbf{G}^j, \quad \text{or,} \quad \nabla \bullet \psi = \psi_{,j}\,\mathbf{G}^j \tag{3.108}$$

Proof. From the definition of gradient and divergence of vector functions above and ***double contraction***, we have:

$$DIV\,\psi = GRAD\,\psi : \mathbf{I} = (\psi_{,j} \otimes \mathbf{G}^j) : (\mathbf{G}_i \otimes \mathbf{G}^i) = (\mathbf{G}^j \bullet \mathbf{G}^i)(\psi_{,j} \bullet \mathbf{G}_i) = G^{ij}(\psi_{,j} \bullet \mathbf{G}_i) = \psi_{,j} \bullet \mathbf{G}^j$$

Fact 2: For any vector function ψ, and a scalar function, u, we have:

$$\begin{aligned} DIV\,(\psi u) &= u\,DIV\,\psi + \psi \bullet GRAD\,u \\ \nabla \bullet (\psi u) &= u\,(\nabla \bullet \psi) + \psi \bullet \nabla u \end{aligned} \tag{3.109}$$

Proof. From equations (3.101) and (3.108), we have:

$$DIV\,(u\psi) = (u\psi)_{,j} \bullet \mathbf{G}^j = u\,(\psi_{,j} \bullet \mathbf{G}^j) + \psi\,u_{,j} \bullet \mathbf{G}^j = u\,(\nabla \bullet \psi) + \psi \bullet u_{,j}\,\mathbf{G}^j = u\,(\nabla \bullet \psi) + \psi \bullet \nabla u$$

Fact 3: For any vector function ψ:

$$DIV\,\psi = \psi^j|_j \tag{3.110}$$

Proof. From the definition of the gradient above anda ***double contraction***, and raising of indices by the ***contravariant*** components of the ***metric tensor***, it follows that:

$$DIV\,\psi = \psi_{,j} \otimes \mathbf{G}^j \colon \mathbf{G}_k \otimes \mathbf{G}^k = (\mathbf{G}^i \bullet \mathbf{G}^k)(\psi_i|_j \mathbf{G}^i \bullet \mathbf{G}_k) = G^{jk}(\psi_i|_j \delta^i_k) = G^{jk}\psi_k|_j = (G^{jk}\psi_k)|_j = \psi^j|_j$$

Note that in the final inequality, we have used the fact that the covariant derivative of the components of the metric tensor is identically zero, that is, $G^{ij}\big|_k = 0$.

Let us also note here that the divergence operator reduces the order of the argument tensor by one. Finally, noting the definition of the ∇ operator, we have:

$$DIV\,\psi = \nabla \bullet \psi \tag{3.111}$$

Example: Continuing with the two-dimensional polar coordinate system, we first note that the nabla operator, ∇, written in terms of the unit base vectors, $\widehat{\mathbf{G}}_r, \widehat{\mathbf{G}}_\theta$ as in equation (3.100), is not suitable for application of the Christoffel symbols, Γ^k_{ij}'s. So, representing a vector, **v**, as: $\mathbf{v} = v_i \widehat{\mathbf{G}}^i = v_r \widehat{\mathbf{G}}^r + v_\theta \widehat{\mathbf{G}}^\theta$, and using the relations: $\widehat{\mathbf{G}^r} = \mathbf{G}^r$ and $\widehat{\mathbf{G}}^\theta = r\mathbf{G}^\theta$, we have, from equation (3.100): $\nabla \bullet \mathbf{v} = \widehat{\mathbf{G}}_r \bullet \frac{\partial \mathbf{v}}{\partial r} + \frac{1}{r}\widehat{\mathbf{G}}_\theta \bullet \frac{\partial \mathbf{v}}{\partial \theta} = \widehat{\mathbf{G}}_1 \bullet \frac{\partial \mathbf{v}}{\partial \xi^1} + \frac{1}{r}\widehat{\mathbf{G}}_2 \bullet \frac{\partial \mathbf{v}}{\partial \xi^2}$ with

$$\begin{aligned}
\widehat{\mathbf{G}}_1 \bullet \frac{\partial \mathbf{v}}{\partial \xi^1} &= \left(v_{i},_{1} \widehat{\mathbf{G}}^i + v_i \widehat{\mathbf{G}}^i,_{1} \right) \bullet \widehat{\mathbf{G}}_1 \\
&= v_{1},_{1} \mathbf{G}^1 \bullet \mathbf{G}_1 + v_1 \mathbf{G}^1,_{1} \bullet \mathbf{G}_1 + v_{2,1} \cancel{\widehat{\mathbf{G}}^2 \bullet \widehat{\mathbf{G}}_1} + v_2 (r\mathbf{G}^2),_{1} \bullet \mathbf{G}_1 \\
&= v_{1,1} + v_1(-\cancel{\Gamma^1_{11}}) + v_2 \left(\mathbf{G}^2 + r\mathbf{G}^2,_{1} \right) \bullet \mathbf{G}_1 = v_{1},_{1} + v_2(\cancel{\mathbf{G}^2 \bullet \mathbf{G}_1} + r\cancel{\Gamma^2_{11}}) \\
&= v_{r},_{r}
\end{aligned}$$

and

$$\begin{aligned}
\widehat{\mathbf{G}_2} \bullet \frac{1}{r}\frac{\partial \mathbf{v}}{\partial \xi^2} &= \frac{1}{r} \left(v_{i},_{2} \widehat{\mathbf{G}}^i + v_i \widehat{\mathbf{G}}^i,_{2} \right) \bullet \widehat{\mathbf{G}}_2 \\
&= \frac{1}{r} \left(\cancel{v_{1,2} \mathbf{G}^1 \bullet \frac{1}{r}} \mathbf{G}_2 + v_1 \mathbf{G}^1,_{2} \bullet \frac{1}{r}\mathbf{G}_2 + v_{2},_{2}\, r\mathbf{G}^2 \bullet \frac{1}{r}\mathbf{G}_2 + v_2 (r\mathbf{G}^2),_{2} \bullet \frac{1}{r}\mathbf{G}_2 \right) \\
&= v_1 \frac{1}{r} \left(-\frac{1}{r}\Gamma^1_{22} \right) + \frac{1}{r} v_{2},_{2} + \frac{1}{r} v_2 \cancel{\Gamma^2_{22}} = \frac{1}{r} v_{2},_{2} + \frac{1}{r} v_1 \left(-\frac{1}{r}(-r) \right) \\
&= \frac{1}{r} \left(v_{\theta},_{\theta} + v_r \right)
\end{aligned}$$

Thus, for the divergence of a vector in polar coordinates, we finally have:

$$\boxed{\nabla \bullet \mathbf{v} = v_{r},_{r} + \frac{1}{r}(v_{\theta,\theta}) = v_{r},_{r} + \frac{1}{r} v_r + \frac{1}{r} v_{\theta},_{\theta} = \frac{1}{r}\frac{\partial}{\partial r}(r v_r) + \frac{1}{r}\frac{\partial v_\theta}{\partial \theta}} \tag{3.112}$$

3.17.2 *Divergence of Tensor Functions: DIV* A

The divergence of a tensor function $\mathbf{A}(\xi^i)$, $i = 1, 2, \ldots$ is a vector or first-order tensor function, denoted as *DIV* **A**, defined by:

$$\boxed{DIV\ \mathbf{A} \equiv GRAD\ \mathbf{A} : \mathbf{I}} \tag{3.113}$$

Fact 3: For any tensor function **A**, $\boxed{DIV\ \mathbf{A} = \mathbf{A},_{j}\ \mathbf{g}^j}$.

Proof. From the definitions of gradient and divergence of tensor functions above and using double contraction, we have:

$$\boxed{\begin{aligned} DIV\ \mathbf{A} &= GRAD\ \mathbf{A} : \mathbf{I} = (\mathbf{A},_{j} \otimes \mathbf{G}^j) : (\mathbf{G}_i \otimes \mathbf{G}^i) \\ &= (\mathbf{G}^j \bullet \mathbf{G}^i)(\mathbf{A},_{j} \bullet \mathbf{G}_i) = G^{ij}(\mathbf{A},_{j} \bullet \mathbf{G}_i) = \mathbf{A},_{j} \bullet \mathbf{G}^j \end{aligned}} \tag{3.114}$$

Fact 4: For any tensor function **A**, $\boxed{div\ \mathbf{A} = \mathbf{A}^j|_j}$, where $\mathbf{A}^j$ are the contravariant pseudo-vector components of the tensor obtained by ***vector-like notation*** for the second-order tensor: $\mathbf{A} = \mathbf{A}^j \otimes \mathbf{G}_j$

Proof. From the definition of gradient above and using double contraction, and raising of indices by the contravariant components of the metric tensor, it follows that:

$$\boxed{\begin{aligned} DIV\ \mathbf{A} &= \mathbf{A}_{,j} \otimes \mathbf{G}^j : \mathbf{G}_k \otimes \mathbf{G}^k = (\mathbf{G}^i \bullet \mathbf{G}^k)(\mathbf{A}_i|_j\ \mathbf{G}^i \bullet \mathbf{G}_k) \\ &= G^{jk}(\mathbf{A}_i|_j \delta^i_k) = G^{jk}\mathbf{A}_k|_j = (G^{jk}\mathbf{A}_k)|_j = \mathbf{A}^j|_j \end{aligned}} \tag{3.115}$$

Note that in the final inequality, we have used the fact that the covariant derivative of the components of the metric tensor is identically zero, that is, $G^{ij}|_k = 0$.

Let us also note that the divergence operator reduces the order of the argument tensor by one. Finally, noting the definition of the ∇ operator, we have: $\boxed{DIV\ \mathbf{A} = \nabla\ \mathbf{A}^T}$.

Fact 5: For a vector function. **a**, and a second-order tensor function, **B**:

$$\boxed{DIV(\mathbf{aB}) = \mathbf{B} : GRAD\ \mathbf{a} + \mathbf{a}\ \bullet\ DIV\ \mathbf{B}} \tag{3.116}$$

Proof. Using **Fact 1**, we have: $DIV(\mathbf{aB}) = (\mathbf{a}\ \bullet\ \mathbf{B})_{,j} \bullet \mathbf{G}^j = (\mathbf{a}_{,j} \bullet \mathbf{B}) \bullet \mathbf{G}^j + (\mathbf{a}\ \bullet\ \mathbf{B}_{,j})\ \bullet\ \mathbf{G}^j$. Now, from **Fact 3**, we get $(\mathbf{a}\ \bullet\ \mathbf{B}_{,j})\ \bullet\ \mathbf{G}^j = \mathbf{a}\ \bullet\ (\mathbf{B}_{,j} \bullet \mathbf{G}^j) = \mathbf{a}\ DIV\ \mathbf{B}$. Similarly with the definition of double contraction, we have:

$$\begin{aligned} (\mathbf{a}_{,j} \bullet \mathbf{B})\ \bullet\ \mathbf{G}^j &= \mathbf{a}_{,j} \bullet \mathbf{B}\ \bullet\ \mathbf{G}^j = \mathbf{a}_{,j} \bullet \mathbf{B}^i \otimes \mathbf{G}_i \bullet \mathbf{G}^j = (\mathbf{a}_{,j} \bullet \mathbf{B}^i)(\mathbf{G}_i \bullet \mathbf{G}^j) \\ &= (\mathbf{B}^i \bullet\ \mathbf{a}_{,j})(\mathbf{G}_i \bullet\ \mathbf{G}^j) = (\mathbf{B}^i \otimes \mathbf{G}_i) : (\mathbf{a}_{,j} \otimes \mathbf{G}^j) = \mathbf{B} : GRAD\ \mathbf{a} \end{aligned}$$

3.18 Integral Transforms: Green–Gauss Theorems

For virtual integral transforms in the finite element analysis, we will occasionally encounter scalar-, vector- and tensor-valued functions or fields, or their derivatives such as gradients, divergence or curls, defined over a body, Ω, or its boundary, Γ. Thus, there will be need for relating their integrals over the body with those over the corresponding boundary. These relationships go by the general name of Green–Gauss theorems. Integral transforms range from where the body is one-dimensional with two end points as boundary, and the functions are scalar-valued to cases where they are, correspondingly volume, surface and tensor-valued; then the result for the simplest case is familiarly known as the integration by parts.

3.18.1 Integration by Parts: One-dimensional Scalar-valued Functions

Let $u(x),\ \bar{u}(x) \in C^1[a, b]$ be one-dimensional scalar-valued fields; then, by the chain rule for the differential: $d(u\ \bar{u}) = u\ d\bar{u} + \bar{u}\ du$; we also note that: $du = \frac{\partial u}{\partial x}\ dx = u_{,x}\ dx$, and $d\bar{u} = \frac{\partial \bar{u}}{\partial x}\ dx = \bar{u}_{,x}\ dx$. Now, integrating: $\int_a^b d(u\ \bar{u}) = \int_a^b u\ d\bar{u} + \int_a^b \bar{u}\ du$ implying that $u\ \bar{u}|_a^b = \int_a^b u\ d\bar{u} + \int_a^b \bar{u}\ du$, where $u\ \bar{u}|_a^b \equiv u(b)\ \bar{u}(b) - u(a)\ \bar{u}(a)$. Thus, finally:

$$\boxed{\begin{aligned} \int_a^b u\ d\bar{u} &= u\ \bar{u}|_a^b - \int_a^b \bar{u}\ du \\ \int_a^b u\ \bar{u}_{,x}\ dx &= u\ \bar{u}|_a^b - \int_a^b \bar{u}\ u_{,x}\ dx \end{aligned}} \tag{3.117}$$

3.18.2 Two- or Three-dimensional Scalar-valued Functions

Let $f = u\,\bar{u}$, where $u(x_i),\ \bar{u}(x_i) \in C^1[\overline{\Omega}],\ i = 1, 2, 3$ be scalar-valued fields on a volume, Ω, and boundary surface, Γ, with $\overline{\Omega}$ as the closure; x_i's are the covariant components of the coordinate vectors, $\mathbf{x} \in \overline{\Omega}$, and let n_i be the covariant components of the normal field, $\mathbf{n}$, on the surface such that: $\mathbf{n} = n_i\,\mathbf{G}^i$. Then, it can be easily shown by generalization of equation (3.117) that:

$$\boxed{\int_\Omega f_{,i}\ d\Omega = \int_\Gamma f\,n_i\ d\Gamma \Rightarrow \int_\Omega u\,\bar{u}_{,i}\ d\Omega = \int_\Gamma u\,\bar{u}\,n_i\ d\Gamma - \int_\Omega \bar{u}\,u_{,i}\ d\Omega} \tag{3.118}$$

3.18.3 Vector-valued Functions: Divergence or Gauss Theorem

Let $\mathbf{f} \in C^1(\overline{\Omega})$ be a vector-valued field on a volume, Ω, and boundary surface, Γ, with $\overline{\Omega}$ as the closure, as shown in Figure 3.4; x_i's are the covariant components of the coordinate vectors, and let n_i be the covariant components of the normal field, $\mathbf{n}$, on the surface such that: $\mathbf{n} = n_i\,\mathbf{G}^i$.

Then, by generalizing the first expression of equation (3.118), we get: $\int_\Omega (f_{i},_i\ d\Omega = \int_\Gamma f_i\,n_i\ d\Gamma$, implying:

$$\boxed{\int_\Omega DIV\,\mathbf{f}\ d\Omega = \int_\Gamma (\mathbf{f} \bullet \mathbf{n})\ d\Gamma, \quad \text{or,} \quad \int_\Omega (\nabla \bullet \mathbf{f})\ d\Omega = \int_\Gamma (\mathbf{f} \bullet \mathbf{n})\ d\Gamma} \tag{3.119}$$

Now, if $\mathbf{f} = \bar{u}\,\mathbf{u}$, where $\mathbf{u}$ is a vector field and $\bar{u}$ is a scalar field, from equations (3.109) and (3.119), we get: $\int_\Omega DIV\,(\bar{u}\,\mathbf{u})\ d\Omega = \int_\Omega (\bar{u}\ DIV\,\mathbf{u} + \mathbf{u} \bullet GRAD\ \bar{u})\ d\Omega = \int_\Gamma (\bar{u}\,\mathbf{u}) \bullet \mathbf{n}\ d\Gamma$, implying:

$$\boxed{\begin{aligned} \int_\Omega (DIV\ \mathbf{u})\ \bar{u}\ d\Omega &= \int_\Gamma \bar{u}\,(\mathbf{u} \bullet \mathbf{n})\ d\Gamma - \int_\Omega (\mathbf{u} \bullet GRAD\ \bar{u})\ d\Omega \\ \int_\Omega (\nabla \bullet \mathbf{u})\ \bar{u}\ d\Omega &= \int_\Gamma \bar{u}\,(\mathbf{u} \bullet \mathbf{n})\ d\Gamma - \int_\Omega (\mathbf{u} \bullet \nabla\bar{u})\ d\Omega \end{aligned}} \tag{3.120}$$

3.18.4 Tensor-valued Functions: Green–Gauss Theorem

Let $\mathrm{T} \in C^1(\overline{\Omega})$ be a second-order tensor-valued field on a volume, Ω, and boundary surface, Γ, with $\overline{\Omega}$ as the closure; x_i's are the covariant components of the coordinate vectors, and let n_i be

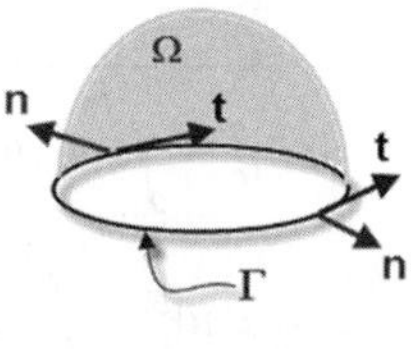

Figure 3.4 Green–Gauss theorem.

the covariant components of the normal field, **n**, on the surface such that: $\mathbf{n} = n_i\,\mathbf{G}^i$. Then, we have:

$$\boxed{\int_{\Omega} DIV\,\mathbf{T}\,d\Omega = \int_{\Gamma} (\mathbf{T}\,\mathbf{n})\,d\Gamma, \quad \text{or,} \quad \int_{\Omega} (\nabla \bullet \mathbf{T})\,d\Omega = \int_{\Gamma} (\mathbf{T}\,\mathbf{n})\,d\Gamma} \tag{3.121}$$

Proof. Let **h** be a constant vector; then, applying the divergence theorem for vectors as in equation (3.119), we have:

$$\int_{\Gamma} \mathbf{h} \bullet (\mathbf{Tn})\,d\Gamma = \int_{\Gamma} \mathbf{n} \bullet (\mathbf{T}^T\mathbf{h})\,d\Gamma = \int_{\Omega} (\nabla \bullet (\mathbf{T}^T\mathbf{h}))\,d\Omega = \int_{\Omega} \mathbf{h} \bullet (\nabla \bullet \mathbf{T})\,d\Omega;$$

noting that the statement is true, for any constant vector, completes the proof.

Now, noting from equation (3.115) that for any tensor function **T**, $\boxed{DIV\,\mathbf{T} = \mathbf{T}^j|_j}$, where $\mathbf{T}^j$ are the contravariant vector components of the tensor obtained by ***vector-like notation*** for the second-order tensor: $\mathbf{T} = \mathbf{T}^j \otimes \mathbf{G}_j$, we get from equation (3.121):

$$\boxed{\int_{\Omega} \mathbf{T}^j|_j\,d\Omega = \int_{\Gamma} (\mathbf{T}^j n_j)\,d\Gamma} \tag{3.122}$$

3.18.5 An Important Lemma

Let $\mathrm{T} \in C^1(\overline{\Omega})$ be a tensor field and $\bar{\mathbf{u}} \in C^1(\overline{\Omega})$ be a vector function, then we have:

$$\boxed{\begin{aligned} \int_{\Omega} DIV\,\mathbf{T} \bullet \bar{\mathbf{u}}\,d\Omega &= \int_{\Gamma} \mathbf{Tn} \bullet \bar{\mathbf{u}}\,d\Gamma - \int_{\Omega} (\mathbf{T}{:}GRAD\,\bar{\mathbf{u}})\,d\Omega \\ \int_{\Omega} (\nabla \bullet \mathbf{T}) \bullet \bar{\mathbf{u}}\,d\Omega &= \int_{\Gamma} \mathbf{Tn} \bullet \bar{\mathbf{u}}\,d\Gamma - \int_{\Omega} (\mathbf{T}{:}\nabla\bar{\mathbf{u}})\,d\Omega \end{aligned}} \tag{3.123}$$

Proof. Integrating the expression in equation (3.116) and applying divergence or Gauss relation of equation (3.119), and noting that, as a scalar by simple contraction of a second-order tensor by vectors from both sides: $\bar{\mathbf{u}}\mathbf{T} \bullet \mathbf{n} = \mathbf{Tn} \bullet \bar{\mathbf{u}}$ we get:

$$\int_{\Omega} DIV\,\mathbf{T} \bullet \bar{\mathbf{u}}\,d\Omega = \int_{\Omega} DIV(\bar{\mathbf{u}}\mathbf{T})\,d\Omega - \int_{\Omega} (\mathbf{T}{:}GRAD\,\bar{\mathbf{u}})\,d\Omega = \int_{\Gamma} \bar{\mathbf{u}}\mathbf{T} \bullet \mathbf{n}\,d\Gamma - \int_{\Omega} (\mathbf{T}{:}GRAD\,\bar{\mathbf{u}})\,d\Omega$$

$$= \int_{\Gamma} \mathbf{Tn} \bullet \bar{\mathbf{u}}\,d\Gamma - \int_{\Omega} (\mathbf{T}{:}GRAD\,\bar{\mathbf{u}})\,d\Omega$$

3.19 Where We Would Like to Go

The treatment of the geometry and mechanics of structures such as plates and shells become much more manageable by application of tensor analyses. There remain various other tensor topics such as Ricci or curvature tensor, Bianchi identity, and so on, that we have not included here, but they are of considerable importance for the geometric analysis of curved structures. Thus, for these and general geometric properties of curved structures such as shells, we must later look into ***the differential geometry of surfaces*** (Chapter 6).

Finally, the finite rotation tensor that inherently belongs to a curved space poses special problems in both the theoretical and computational treatment of geometrically (i.e. with large displacement and rotation) nonlinear structures; these, in turn, need more detailed inspection of the nature of rotational tensor for which we must turn to: the ***rotation tensor*** (Chapter 4).

4

Rotation Tensor

4.1 Introduction

4.1.1 What We Should Already Know

Configuration space, vector, matrices, determinants, traces, tensors, covariant base vectors, contravariant base vectors, linear transformation, orthogonal transformation, metric tensor, Cayley–Hamilton theorem, characteristic polynomial, eigenvalues, eigenvectors, skew-symmetric tensor, axial vectors, matrix Lie group, manifold, vector triple product, Gateaux derivative, covariant derivative.

4.1.2 Why We Are Here

The main goal here is to present a ***definition*** for rigid body rotations for use in structural and continuum mechanics such as ***finite rotations*** in ***geometrically nonlinear beam*** or ***shell analysis*** or ***polar decomposition of the deformation gradient***. As an example of rigid body rotation in structural mechanics, we may cite finite deformations of a structural element such as a thin beam, where the cross-sections perpendicular to the centroidal curve of the beam are often modeled as rigid. This simplifies the *configuration space* of a beam to be characterized by the displacements (a translational movement) of the reference centroidal curve coupled with rigid body rotations of the corresponding cross-sections. The displacement and the rotation together at a point define what is known as the generalized displacement in computational mechanics. We present various ways that a rotation can be mathematically realized and represented; these offer various comparative possibilities for finite rotation parameterizations suitable for subsequent numerical evaluations in structural mechanics. As we shall see, ***Rodrigues parametrization*** ingeniously represents a rotation matrix with only three parameters; however, in this process, we have a singularity as the *trace* of the rotation matrix reaches -1. If we need to represent the rotation matrix in an unconditionally non-singular fashion, we have ***Euler–Rodrigues parametrization***, but a vectorial parametrization of the rotation matrix is most important for our numerical purposes in the sense that the number of parameters (namely, 3) is the same as the usual rotational degrees of freedom of the structural elements such as beams, plates and shells, for computational schemes such as the *finite element method*. Although it is singular at some rotation, for most structural problems involving moderate finite rotations, it suffices. In the event that it is not, a numerical strategy based on a user-transparent combination of four-parameter parameterization – such

Computation of Nonlinear Structures: Extremely Large Elements for Frames, Plates and Shells, First Edition. Debabrata Ray.

as ***Euler–Rodrigues*** or ***Hamilton–Rodrigues quaternions***, in the background with the three-parameter Rodrigues vector – will be able to make rotation representation completely singularity-free.

4.1.3 How We Get It

Rotation is tractably associated with and recognized as an orthogonal transformation; in fact, as indicated in our tensor analysis, it is a linear transformation, that is, a *tensor*. As such, we get to discuss the tensorial description of a rotation tensor. This allows us to identify a rotation matrix with the components of a corresponding rotation tensor. We also determine that a pure rotation has the spherical displacement property. In other words, while a translational displacement belongs to a *Euclidean space*, the rotation travels on a Riemannian curved space: on the 3D surface of a 4D sphere. Realization of this property will eventually help us to devise a numerical scheme for finite rotational motions of structural elements. In order to help us with the parameterization of rotation for subsequent computational efficiency, we will look into the eigenproperties of a rotation matrix. Finally, we recognize rotation as forming a non-commutative group – that is, a matrix Lie group – and derive both algebraic and geometric insights into rotational transformation.

Remark

- ***Unless otherwise mentioned, we will restrict our attention to the more interesting physical three-dimensional (3D) space for our purposes.***

4.1.4 Orthogonal Transformation and the Rotation Tensor

If we consider the most general motion of a rigid body, it can be broken down into a ***translation*** and a ***rotation*** component. The two together, known as the ***generalized displacement***, as indicated above, are also referred to as the ***screw motion*** in theoretical kinematics. For the motion of a rigid body, generalized displacement is an invariant distance-preserving (i.e. distance measured between two points in the body) and hence an angle-(measured between two lines in the body)-preserving transformation. Now, in the absence of any translational displacement, a point in the configuration space (i.e. the space that contains the location of any point on the rigid body at all times of the motion) is completely characterized by only rotational transformation. In other words, as shown in Figure 4.1, if a point *P* in the initial state with coordinates given by the vector

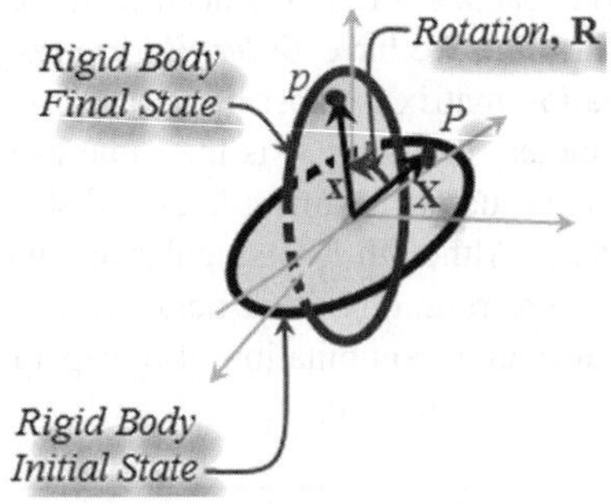

Figure 4.1 Rigid body rotation.

X, measured by a system fixed in the initial state of a rigid body, is rotated to a point p in the final state with coordinates given by the vector **x** measured by the same system in the initial state, we may represent the rotation of the vector by:

$$\mathbf{x} = \mathbf{R}\,\mathbf{X} \tag{4.1}$$

where **R** is the rotational transformation.

Fact 1: Rotational transformation, **R**, is an *orthogonal transformation*, that is, $\mathbf{R}^T\mathbf{R} = \mathbf{R}\,\mathbf{R}^T = \mathbf{I}$ or $\mathbf{R}^{-1} = \mathbf{R}^T$.

Proof. To guarantee that equation (4.1) represents a rigid transformation, the Euclidean distance between any two points on the body, with coordinates **X** and **Y** that are rotated to **x** and **y**, respectively, by **R**, must remain the same. In other words, $|\mathbf{x}-\mathbf{y}| = |\mathbf{R}(\mathbf{X}-\mathbf{Y})| = \{(\mathbf{X}-\mathbf{Y})^T(\mathbf{R}^T\mathbf{R})(\mathbf{X}-\mathbf{Y})\}^{\frac{1}{2}}$ must be same as $|\mathbf{X}-\mathbf{Y}| = \{(\mathbf{X}-\mathbf{Y})^T(\mathbf{X}-\mathbf{Y})\}^{\frac{1}{2}}$. The only way this is satisfied is when

$$\boxed{\mathbf{R}^T\mathbf{R} = \mathbf{I} \Leftrightarrow \mathbf{R}^T = \mathbf{R}^{-1}} \tag{4.2}$$

that is, the rotational motion of a rigid body is completely characterized mathematically by an orthogonal transformation.

Fact 2: The *determinant* of an orthogonal transformation, **R**, is ±1.

Proof. $1 = \det(\mathrm{I}) = \det(\mathbf{R}^T\mathbf{R}) = \det^2(\mathbf{R}) \Rightarrow \det(\mathbf{R}) = \pm 1$

We now introduce two definitions.

Definitions

- Orthogonal tensors with determinant +1 are called the ***direct or proper rotation tensors***,
- Orthogonal tensors with determinant −1 are called the indirect rotation tensors, or ***reflection tensors***.

Note that although reflections satisfy the rigid transformation constraint, only direct rotations are utilized to fix the position of a rigid body. Thus, from now on, any reference to a rotation tensor will mean a direct rotation tensor, that is, an orthogonal tensor with the determinant given by +1.

Fact 3: Rotational transformation, **R**, is linear, that is, **R** is a tensor.

Proof. A pure rotational operation, **R**, on a rigid body preserves its size, shape and orientation. Mathematically expressed, it preserves dot product and cross product of any two vectors on a body, as they represent length, angle and orientation, and hence the determinant of a matrix. To wit, let vectors $\mathbf{a}, \mathbf{b}, \bar{\mathbf{a}}, \bar{\mathbf{b}}$ be such that $\bar{\mathbf{a}} = \mathbf{R}\,\mathbf{a}$ and $\bar{\mathbf{b}} = \mathbf{R}\,\mathbf{b}$. Now, $\mathbf{Ra} \bullet \mathbf{Rb} = \mathbf{a} \bullet \mathbf{R}^T\mathbf{Rb} = \mathbf{a} \bullet \mathbf{b}$; so we need the following two corollaries to complete the proof:

Corollary 1: For any matrix, $\mathbf{A}$, and any three vectors, $\mathbf{b}_1$,$\mathbf{b}_2$,$\mathbf{b}_3$, we have:

$$\mathbf{Ab}_1 \times \mathbf{Ab}_2 \bullet \mathbf{Ab}_3 = (\det\ \mathbf{A})(\mathbf{b}_1 \times \mathbf{b}_2 \bullet \mathbf{b}_3) \tag{4.3}$$

Proof. Let the three columns of a 3×3 matrix, $\mathbf{B} \equiv (\,\mathbf{b}_1\ \mathbf{b}_2\ \mathbf{b}_3\,)$ whose determinant is clearly given by a triple product: $\det \mathbf{B} = \mathbf{b}_1 \times \mathbf{b}_2 \bullet \mathbf{b}_3$. Now, $\det(\mathbf{AB}) = \det \mathbf{A} \det \mathbf{B} = (\det \mathbf{A})(\mathbf{b}_1 \times \mathbf{b}_2 \bullet \mathbf{b}_3)$. Again, $\det(\mathbf{AB}) = \det(\mathbf{A}(\,\mathbf{b}_1\ \mathbf{b}_2\ \mathbf{b}_3\,)) = \det(\,\mathbf{Ab}_1\ \mathbf{Ab}_2\ \mathbf{Ab}_3\,) = \mathbf{Ab}_1 \times \mathbf{Ab}_2 \bullet \mathbf{Ab}_3$. The equality of the two previous expressions completes the proof.

Corollary 2: For any matrix, $\mathbf{A}$, and any two vectors, $\mathbf{b}_1$, $\mathbf{b}_2$, we have:

$$\boxed{\mathbf{A}^T(\mathbf{Ab}_1 \times \mathbf{Ab}_2) = (\det\ \mathbf{A})(\mathbf{b}_1 \times \mathbf{b}_2)} \tag{4.4}$$

Proof. From **Corollary 1**, we have: $\mathbf{A}^T(\mathbf{Ab}_1 \times \mathbf{Ab}_2) \bullet \mathbf{b}_3 = (\det\ \mathbf{A})(\mathbf{b}_1 \times \mathbf{b}_2) \bullet \mathbf{b}_3$ for any arbitrary vector $\mathbf{b}_3$. Hence, $\mathbf{A}^T(\mathbf{Ab}_1 \times \mathbf{Ab}_2) = (\det\ \mathbf{A})(\mathbf{b}_1 \times \mathbf{b}_2)$.

Now, continuing on with our proof of **Fact 3**, we note that for a proper rotation tensor,$\mathbf{R}$, $\det \mathbf{R} = +1$, we have, from **Corollary 2**, $\mathbf{R}^T(\mathbf{Ra} \times \mathbf{Rb}) = (\mathbf{a} \times \mathbf{b}) \Rightarrow (\mathbf{Ra} \times \mathbf{Rb}) = \mathbf{R}(\mathbf{a} \times \mathbf{b})$. So, the cross product is preserved under rotation. Finally, from **Corollary 1**, $\det(\mathbf{R}(\,\mathbf{a}\ \mathbf{b}\ \mathbf{c}\,)) = \det(\,\mathbf{Ra}\ \mathbf{Rb}\ \mathbf{Rc}\,) = \mathbf{Ra} \bullet \mathbf{Rb} \times \mathbf{Rc} = \det(\mathbf{R})(\mathbf{a} \bullet \mathbf{b} \times \mathbf{c}) = (\mathbf{a} \bullet \mathbf{b} \times \mathbf{c})$. Thus, the determinant is preserved under the rotation operation. Since these preserved operations are linear, the rotation transformation is also linear, and hence by definition, a second-order tensor.

4.1.4.1 Tensorial Description of Rotational Transformation

If we consider a set of covariant base vectors, $\{\mathbf{G}_i\}$, that undergoes a rigid body rotation to become the set of base vectors, $\{\mathbf{g}_i\}$, in the final state, then we have, by definition: $\mathbf{g}_i = \mathbf{R}\,\mathbf{G}_i$, for all $i \in \{1, 2, 3\}$ for a three-dimensional space. By the relation: $\mathbf{R}^{-1} = \mathbf{R}^T$, we also have: $\mathbf{G}_i = \mathbf{R}^{-1}\,\mathbf{g}_i = \mathbf{R}^T\,\mathbf{g}_i$, We can immediately recognize that the tensorial description of a rotation tensor through the tensor product is given by:

$$\boxed{\mathbf{R} = \mathbf{g}_i \otimes \mathbf{G}^i \quad \text{and} \quad \mathbf{R}^T = \mathbf{G}_i \otimes \mathbf{g}^i} \tag{4.5}$$

where $\mathbf{G}^i$ and $\mathbf{g}^i$ are the contravariant base vectors corresponding to the initial and final states, respectively. Similarly, for the rotation of the contravariant base vectors: $\mathbf{g}^i = \mathbf{R}\,\mathbf{G}^i$ and $\mathbf{G}^i = \mathbf{R}^T\,\mathbf{g}^i$. By the relation: $\mathbf{R}^{-1} = \mathbf{R}^T$, we then have: $\mathbf{G}^i == \mathbf{R}^{-1}\,\mathbf{g}^i = \mathbf{R}^T\,\mathbf{g}^i$, and:

$$\boxed{\mathbf{R} = \mathbf{g}^i \otimes \mathbf{G}_i \quad \text{and} \quad \mathbf{R}^T = \mathbf{G}^i \otimes \mathbf{g}_i} \tag{4.6}$$

4.1.4.2 Rotation Matrix: Components of a Rotation Tensor

Recalling that the matrix corresponding to a linear transformation is identified by the choice of the base vectors, we may have the rotation matrix as $\{R_{ij}\}$ or $\{r_{ij}\}$, based on the component definition:

$$\mathbf{R} = R_{ij}\,\mathbf{G}^i \otimes \mathbf{G}^j = r_{ij}\,\mathbf{g}^i \otimes \mathbf{g}^j \tag{4.7}$$

The components of a rotation tensor are invariant with respect to the base vectors, that is, we have the following.

Fact 4: $\{R_{ij}\}$ and $\{r_{ij}\}$ are identical, that is, independent of the original or the rotated set of base vectors:

$$R_{ij} = r_{ij} \tag{4.8}$$

Proof. Using expression (4.5), $\mathbf{R} = R_{ij}\,\mathbf{G}^i \otimes \mathbf{G}^j \Rightarrow R_{ij} = \mathbf{G}_i\,\mathbf{R}\,\mathbf{G}_j = \mathbf{G}_i\,(\mathbf{g}_k \otimes \mathbf{G}^k)\mathbf{G}_j = \mathbf{G}_i\,\mathbf{g}_j = \mathbf{R}^T\mathbf{g}_i\,\mathbf{g}_j = \mathbf{g}_i\,\mathbf{R}\,\mathbf{g}_j = r_{ij}$ completes the proof.

Thus, ***the matrix representation is one and the same, whether it is defined in the unrotated or rotated base vectors***.

4.1.5 The Eigenproperties of a Rotation Tensor

The set of points in a body that do not rotate under the action of rotation, $\mathbf{R}$, may be described by the equation:

$$\boxed{\mathbf{R}\mathbf{X} = \lambda\mathbf{X} \Leftrightarrow (\mathbf{R} - \lambda\mathbf{I})\mathbf{X} = \mathbf{0}} \tag{4.9}$$

that is, a point P, with coordinates given by vector $\mathbf{X}$, retains its direction under the action of the rotation. For a non-trivial solution of equation (4.9), we must have the determinant $|\mathbf{R} - \lambda\mathbf{I}|$ vanish identically. Now, for a 3×3 matrix or second-order tensor in 3D, we can use ***Cayley–Hamilton theorem*** to describe its ***characteristic polynomial*** as:

$$\boxed{\begin{aligned} &p(\lambda) \equiv \lambda^3 - I_R\,\lambda^2 + II_R\,\lambda - III_R = 0,\\ &I_R \equiv tr\,\mathbf{R} = \mathbf{R} : \mathbf{I} = R_{11} + R_{22} + R_{33} = \lambda_1 + \lambda_2 + \lambda_3\\ &II_R \equiv \tfrac{1}{2}\{(tr\,\mathbf{R})^2 - tr\,\mathbf{R}^2\} = \lambda_1\lambda_2 + \lambda_2\lambda_3 + \lambda_3\lambda_1\\ &III_R \equiv \det\mathbf{R} = +1 = \lambda_1\lambda_2\lambda_3 \end{aligned}} \tag{4.10}$$

We can immediately observe that $\lambda_1 = 1$ is the only real root or ***eigenvalue***. Accordingly we can factorize equation (4.10) and represent as:

$$p(\lambda) \equiv (\lambda - 1)\{\,\lambda^2 - \,\lambda(R_{11} + R_{22} + R_{33} \;-\; 1) + 1\} = 0 \tag{4.11}$$

The other two roots may, then, be given as complex conjugates:

$$\boxed{\begin{aligned} &\lambda_2 = \cos\theta + i\sin\theta = e^{i\theta}\\ &\lambda_3 \equiv \lambda_2^* \;= \cos\theta - i\sin\theta = e^{-i\theta} \end{aligned}} \tag{4.12}$$

with scalar θ determined by:

$$\boxed{\cos\theta = \tfrac{1}{2}(\lambda_2 + \lambda_3) = \tfrac{1}{2}\{(R_{11} + R_{22} + R_{33}) - 1\} = \tfrac{1}{2}(tr\,\mathbf{R} - 1)} \tag{4.13}$$

Now, let us introduce a new definition.

Definition

- **s** ***as the real eigenvector associated with the real root,*** $\lambda_1 = 1$

Clearly, then, by equation (4.9): $\mathbf{R}\,\mathbf{s} = \lambda \mathbf{s} = \mathbf{s}$, that is, all points along **s** vector are fixed in position under rotation. In other words, **s** is the instantaneous axis of rotation of the body. Now, let the complex eigenvectors **p** and $\mathbf{p}^*$ be associated with the complex eigenvalues λ_2 and λ_3. Then, by equation (4.9): $\mathbf{s} \bullet \lambda_2 \mathbf{p} = \mathbf{R}\,\mathbf{s} \bullet \mathbf{R}\,\mathbf{p} = \mathbf{s} \bullet (\mathbf{R}^T\mathbf{R})\,\mathbf{p} = (\mathbf{s} \bullet \mathbf{p}) \Rightarrow (\lambda_2 - 1)(\mathbf{s} \bullet \mathbf{p}) = \mathbf{0}$. Now, since $\lambda_2 \neq 1$, we must have: $\mathbf{s} \bullet \mathbf{p} = \mathbf{0}$. In other words, **p** is perpendicular to the axis of rotation, **s**. Similarly, $\mathbf{p}^*$ is also perpendicular to the axis, **s**. Thus, the **p** and $\mathbf{p}^*$ vectors define a plane perpendicular to the axis of rotation, **s**. We can easily construct a pair of real orthogonal vectors, $\mathbf{q}_1$ and $\mathbf{q}_2$, from the complex conjugate pairs, **p** and $\mathbf{p}^*$, by setting:

$$
\begin{aligned}
\mathbf{q}_1 &\equiv \frac{1}{2}(\mathbf{p} + \mathbf{p}^*) \\
\mathbf{q}_2 &\equiv \frac{1}{2}i(\mathbf{p} - \mathbf{p}^*)
\end{aligned}
\tag{4.14}
$$

The final rotated coordinates, $\mathbf{Q}_1$ and $\mathbf{Q}_2$, measured in initial unrotated frame, of $\mathbf{q}_1$ and $\mathbf{q}_2$ is given by:

$$
\begin{aligned}
\mathbf{Q}_1 &= \mathbf{R}\,\mathbf{q}_1 = \frac{1}{2}(\mathbf{R}\,\mathbf{p} + \mathbf{R}\,\mathbf{p}^*) = \frac{1}{2}(\lambda_2\,\mathbf{p} + \lambda_2^*\,\mathbf{p}^*) = \cos\theta\,\mathbf{q}_1 + \sin\theta\,\mathbf{q}_2 \\
\mathbf{Q}_2 &= \mathbf{R}\,\mathbf{q}_2 = \frac{1}{2}i(\mathbf{R}\,\mathbf{p} - \mathbf{R}\,\mathbf{p}^*) = \frac{1}{2}i(\lambda_2\,\mathbf{p} - \lambda_2^*\,\mathbf{p}^*) = -\sin\theta\,\mathbf{q}_1 + \cos\theta\,\mathbf{q}_2
\end{aligned}
\tag{4.15}
$$

or, in matrix form:

$$
\boxed{\begin{Bmatrix} \mathbf{Q}_1 \\ \mathbf{Q}_2 \end{Bmatrix} = \begin{bmatrix} \cos\theta & \sin\theta \\ -\sin\theta & \cos\theta \end{bmatrix} \begin{Bmatrix} \mathbf{q}_1 \\ \mathbf{q}_2 \end{Bmatrix}}
\tag{4.16}
$$

The above equation clearly shows a pure planar rotation of θ about **s**, of the real plane (see Figure 4.2) defined by the real eigenvectors, $\mathbf{q}_1$ and $\mathbf{q}_2$ and perpendicular to axis **s**.

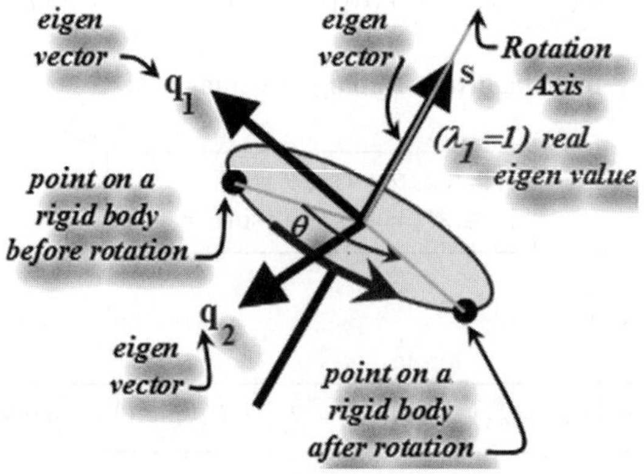

Figure 4.2 Eigenproperties of rotation.

4.1.5.1 Summary: The Eigenproperties

Let us then conclude that for all eigenvalues, λ_k, $k = 1, 2, 3$ of an orthogonal matrix:

- $|\lambda_k| = 1$,
- If λ_k is real, then $\lambda_k = +1$ or -1,
- If λ_k is imaginary, $\lambda_k = e^{i\theta}$ and $e^{-i\theta}$,
- An invariant line, representing the axis of rotation and corresponding to a real eigen value, is orthogonal to all invariant planes containing other eigenvectors.
- Finally, we see that our eigenanalysis tells us that: instead of the nine coefficients of matrix $\mathbf{R}$, we may equivalently keep track of four elements: the angle of rotation, θ, and the vector, $\mathbf{s}$, the axis of rotation.

Remark

- This brings us to the various ways of parameterizing rotation tensor for optimum numerical efficiency in our subsequent finite element analysis of various structural elements such as beams, plates and shells. If we want to work with the original nine parameters of a rotation matrix, we need to carry six side constraints on the nine parameters: three for the lengths of the columns and three for the orthogonality of the columns of the matrix, so that the rotation matrix always stays a rotation matrix upon multiplication. This is rather cumbersome and awkward numerically, hence the need for various other parameterizations.

4.1.5.2 Spherical Displacement

Now, as shown Figure 4.3, note that under pure rigid body rotation, because of length preservation of a vector in rotation, a point P of a body always remains on the ***surface of a sphere***. This is why a pure rotation is often termed a spherical displacement in kinematics. We will discuss it more later; in the meantime, let us sum up:

- Pure rotational rigid body transformation can be identified as spherical kinematics through an orthogonal transformation.

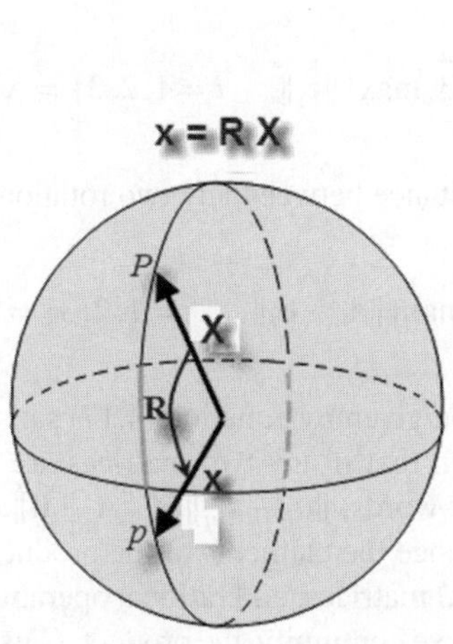

Figure 4.3 Spherical displacement.

4.1.6 Special Orthogonal Group, $SO(3)$, Lie Group and Rotation Matrix

Motivated by the need to identify entities such as the tangent operator of the rotation transformation, the angular velocities, and so on, and to analyze and numerically treat rotational motion, we introduce here minimally the concept of the ***special orthogonal group***, $SO(3)$, and of the ***Lie group***, and continue our discussion.

4.1.6.1 Special Orthogonal Group, $SO(3)$

Recall the algebraic structure of a ***group***: a set with four properties: existence of identity, existence of inverse, associative property of products and closure. The 3×3 rotation matrix as orthogonal transformations clearly satisfies all the above properties under multiplication and thus forms a group known as the ***orthogonal group***, $O(3)$. To wit, $\mathbf{I}$ such that $\mathbf{X} = \mathbf{I}\,\mathbf{X}$, that is, no rotation of any point $\mathbf{X}$ of a body, serves as the identity transformation; every $\mathbf{R}$ has an inverse given by $\mathbf{R}^{-1} = \mathbf{R}^T$. Given any three rotational transformations: $\mathbf{R}_1,\ \mathbf{R}_2,\ \mathbf{R}_3$, we have: $\mathbf{R}_1(\mathbf{R}_2 + \mathbf{R}_3) = \mathbf{R}_1\mathbf{R}_2 + \mathbf{R}_1\mathbf{R}_3$; finally, for closure: if $\mathbf{R}_1,\ \mathbf{R}_2 \in O(3)$, under multiplication, $\mathbf{R} \stackrel{\Delta}{=} \mathbf{R}_2\mathbf{R}_1$ belongs to $O(3)$ since $\mathbf{R}^T\,\mathbf{R} = (\mathbf{R}_2\mathbf{R}_1)^T\mathbf{R}_2\mathbf{R}_1 = \mathbf{R}_1^T\mathbf{R}_2^T\mathbf{R}_2\mathbf{R}_1 = \mathbf{I} \in O(3)$. Moreover, since the determinant is $+1$, the rotation matrices in three dimensions under multiplication belong to what is known as the ***special orthogonal group***, $SO(3)$. To wit, if the rotation matrices $\mathbf{R}_1,\ \mathbf{R}_2 \in SO(3)$ such that $\mathbf{R} \stackrel{\Delta}{=} \mathbf{R}_2\mathbf{R}_1$, then $\det(\mathbf{R}) = \det(\mathbf{R}_2\mathbf{R}_1) = \det(\mathbf{R}_2)\det(\mathbf{R}_1) = 1 \Rightarrow \mathbf{R} \in SO(3)$.

4.1.6.2 Matrix Lie Group, $SO(3)$

Recall that a continuous manifold which is also an algebraic group is called a ***Lie group***, bearing the name of its discoverer, Sophus Lie. A Lie group is endowed with both the algebraic structure of a group and the geometry of the Riemannian manifold; it is a non-commutative group. The rotation matrices form a Lie group under non-commutative multiplication, that is, $\mathbf{R}_2\mathbf{R}_1 \neq \mathbf{R}_1\mathbf{R}_2$. If we can show that the matrix product is continuous, we will have $SO(3)$ as a Lie group. For establishing continuity, we need to define a suitable norm, $||\mathbf{R}||$, for a rotation matrix, $\mathbf{R} = \{\mathbf{r}_1\ \ \mathbf{r}_2\ \ \mathbf{r}_3\}$, where $\mathbf{r}_1,\ \mathbf{r}_2,\ \mathbf{r}_3$ are the column vectors. We choose the norm by replacing each column by the largest column, that is:

$$||\mathbf{R}|| \equiv \sqrt{3}\ \max\{||\mathbf{r}_i||,\quad i = 1,2,3\} = \sqrt{3}\ ||\mathbf{r}_{\max}|| \tag{4.17}$$

Then, the metric – that is, the distance between any two rotation matrices, $\mathbf{A}$, $\mathbf{B}$ – is given by:

$$||\mathbf{A} - \mathbf{B}|| \equiv \sqrt{3}\ \max\{||\mathbf{a}_i - \mathbf{b}_i||,\quad i = 1,2,3\} = \sqrt{3}\ ||\mathbf{a}_i - \mathbf{b}_i||_{\max} \tag{4.18}$$

It can be shown that the definition given by equation (4.17) satisfies all the properties of a norm. Using this norm, it can be shown that if any two matrices are close together, then so are their corresponding elements. In other words, $||a_{ij} - b_{ij}|| \leq ||\mathbf{A} - \mathbf{B}||,\quad i,j = 1,2,3$, meaning that the elements are continuous. Now, since the elements of the product matrices, $\mathbf{C} = \mathbf{AB}$, are rational functions of each of the individual matrices, and rational operations such as addition, subtraction, multiplication and division preserve continuity, the product, $\mathbf{C}$, is a continuous function of the two matrices, $\mathbf{A}$, $\mathbf{B}$. Similarly, $\det\ \mathbf{C}$ and C^{-1} varies continuously as $\mathbf{C}$. Thus, $SO(3)$ is a continuous

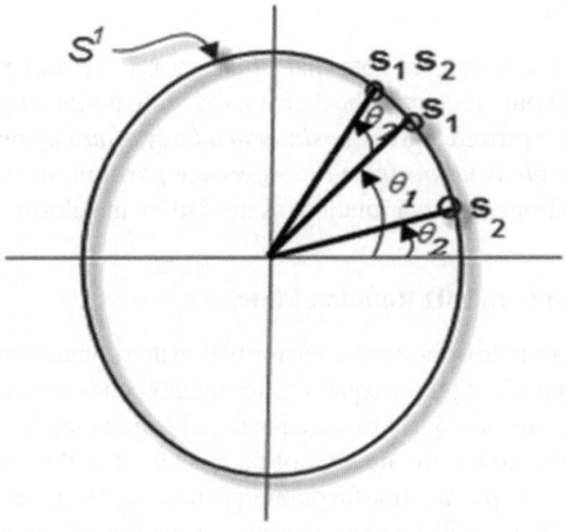

Figure 4.4 Simplest Lie group: circle.

manifold and forms a ***Lie subgroup***. Now, with both the algebraic structure and the geometry of $SO(3)$, we can begin to visualize, say, curves made up of rotation matrices, themselves as always belonging to or traveling on the 3D surface (denoted as 3-***sphere***) of a 4D spherical ball. In other words, a rotation matrix belongs to a curved surface as opposed to a flat Euclidean surface. In this context, let us recall that a 3D point under pure rotation travels on a 2D surface (2-***sphere***) of a 3D spherical ball. As will be shown, this way of looking into the rotation matrices helps us in devising various numerically efficient parameterizations of a rotation matrix and more accurate numerical integrations of the rotational motions.

Example: The simplest of the Lie groups is the unit circle, S^1 as shown in Figure 4.4. Clearly, S^1 depicts a one-dimensional manifold. There are several ways to depict points on the circle:

- Points can be parameterized in the complex plane as: $\mathbf{s} = \cos\theta + i\sin\theta,\ \ \theta \in [0, 2\pi]$; under complex multiplication, S^1 forms an algebraic group.
- Points can be expressed through a plane rotation matrix parametrized by θ as: $\mathbf{R}(\theta) = \begin{bmatrix} \cos\theta & -\sin\theta \\ \sin\theta & \cos\theta \end{bmatrix}$; the set of $\mathbf{R}$ forms a matrix group, $SO(2)$, under matrix multiplication.
- Points can also be expressed through a scalar exponential map: $\mathbf{s} = e^{i\theta}$; in matrix form, it is given as: $\mathbf{s} = \exp\begin{bmatrix} 0 & -\theta \\ \theta & 0 \end{bmatrix} = e^{\Theta},\ \ \Theta \equiv \begin{bmatrix} 0 & -\theta \\ \theta & 0 \end{bmatrix}$ where Θ, the ***skew-symmetric matrix*** belongs to what is known as ***Lie algebra***, that is, the tangent space, $so(2)$, with $e^{\Theta} = \mathbf{I} + \Theta + \frac{\Theta^2}{2} + ...$

Note that $\theta = 0 \Rightarrow e^{\Theta} = \mathbf{I}$, meaning that Lie algebra, $so(2)$, is the tangent space at identity, and S^1 has both the algebraic group structure and the geometry of a manifold, and thus forms a Lie group. In what follows, we use the above introductory notions, the algebraic structure and the geometry of $SO(3)$ to shed light on the meaning of various parameterizations of the rotational motions, for example, as in ***Cayley's*** parameterization.

4.1.6.3 Euler Theorem

In this connection, it is interesting to note that in the year 1752, Euler discovered the closure property of the group without, of course, the then non-existing notion of groups. By his ingenious geometric construction, he proved that ***the position of a body after any two consecutive rotations can be obtained by a single rotation from the reference position***, meaning that the product of any two consecutive rotations is also a rotation, as needed by the closure property of a group.

4.1.6.4 Representation of the 3D Rotation Matrix

We have identified rotation as an orthogonal transformation that contains nine components with the orthogonality condition providing three equations for interdependency of the parameters. But, for computational efficiency, we would like to seek any possible parameterization of a rotation matrix that results in further reduction in the number of parameters. For this, we have a representation theorem by ***Cayley*** that will put us in a favorable position to evaluate various parametrization possibilities. Concurrently, we will invoke Lie group properties of rotational transformations to eventually help us identify the best strategies for numerical treatment of nonlinear static and dynamic structural problems involving large finite rotations that travels on a curved Riemann surface: a 3-sphere, as we shall see. Clearly, then, for a fundamental representation of a rotation matrix, we must explore Cayley's representation.

4.2 Cayley's Representation

We introduce here a very important theorem by ***Cayley*** that allows us immediately to parameterize a rotation matrix by a vector: the ***Rodrigues vector***. This opportunity profoundly increases numerical efficiency by considering only a three-dimensional vector (i.e. three parameters) instead of the computationally cumbersome nine components of a rotation tensor. Finally, the geometry of the theorem by Cayley that provides valuable insights for an eventual numerical strategy is appreciated by invoking the $SO(3)$ and Lie group properties of the rotational transformations. As indicated above, we present an algebraic representation and a geometric interpretation for finite rotation suitable for subsequent numerical evaluations in structural mechanics. We may recall that rotation is tractably associated with and interpreted to be an orthogonal transformation belonging to $SO(3)$ group. This, then, leads us to representation of rotation by an associated ***skew-symmetric matrix*** given by an important theorem by Cayley. Finally, we introduce ***axial vectors***, corresponding to the skew-symmetric matrices in 3D, giving us the desired possibility of parametrization of rotation matrices by 3D vector quantities for optimum computational efficiency. The associated skew symmetric matrices can be identified as belonging to the ***tangent space at identity,*** $so(3)$, which is known as the ***Lie algebra***.

4.2.1 Cayley's Theorem for Representation of an Orthogonal Matrix

Every orthogonal matrix, **R**, can be represented by a skew-symmetric matrix, **S** such that:

$$\boxed{\mathbf{R} = (\mathbf{I} - \mathbf{S})^{-1}(\mathbf{I} + \mathbf{S})} \tag{4.19}$$

Proof. Let $\mathbf{p} = \mathbf{R}\,\mathbf{P}$ define a rigid body rotation. For rigid body rotation, length is preserved, that is, $\mathbf{p} \bullet \mathbf{p} - \mathbf{P} \bullet \mathbf{P} = 0 \Rightarrow (\mathbf{p} + \mathbf{P}) \bullet (\mathbf{p} - \mathbf{P}) = 0$, that is,as shown Figure 4.5, the diagonals $\mathbf{p} + \mathbf{P}$ and $\mathbf{p} - \mathbf{P}$ of a rhombus with edges **P** and **p** intersect at right angles. So, if we define: $\mathbf{m} \equiv \mathbf{p} + \mathbf{P}$ and $\mathbf{n} \equiv \mathbf{p} - \mathbf{P}$ we have: $\mathbf{m} \bullet\ \mathbf{n} = (\mathbf{p} + \mathbf{P}) \bullet (\mathbf{p} - \mathbf{P}) = 0$ for all **p**. Eliminating **p** by application of

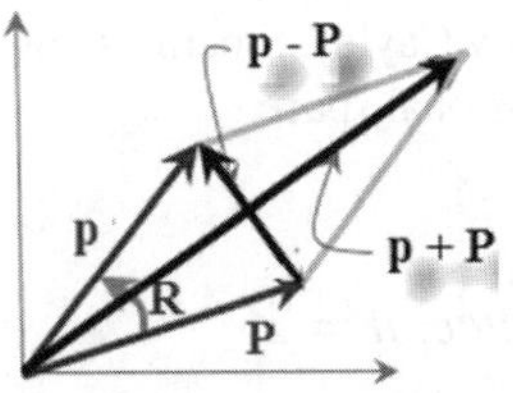

Figure 4.5 Cayley's theorem.

the rotation relation, we have: $\mathbf{m} = (\mathbf{R} + \mathbf{I})\mathbf{P}$ and $\mathbf{n} = (\mathbf{R} - \mathbf{I})\mathbf{P}$. Now substituting $\mathbf{P}$ from $\mathbf{m}$ into $\mathbf{n}$, we have: $\mathbf{n} = (\mathbf{R} - \mathbf{I})(\mathbf{R} + \mathbf{I})^{-1}\mathbf{m}$. Let us define:

$$\mathbf{S} \equiv [\mathbf{R} - \mathbf{I}][\mathbf{R} + \mathbf{I}]^{-1} \tag{4.20}$$

So, $\mathbf{n} = \mathbf{S}\,\mathbf{m}$. Noting from the above that: $\mathbf{m} \bullet \mathbf{n} = \mathbf{n} \bullet \mathbf{m} = 0$ and applying the relation: $\mathbf{n} = \mathbf{S}\,\mathbf{m}$, we have: $\mathbf{m} \bullet \mathbf{n} = \mathbf{m}^T\mathbf{S}\,\mathbf{m} = 0$ and $\mathbf{n} \bullet \mathbf{m} = \mathbf{m}^T\mathbf{S}^T\,\mathbf{m} = 0$. Now, adding these two relations, we have: $\mathbf{m}^T(\mathbf{S} + \mathbf{S}^T)\,\mathbf{m} = 0$ for all $\mathbf{m} \in \mathbb{R}^3$. Thus, finally, we get: $\mathbf{S} + \mathbf{S}^T = \mathbf{0}$ which means that $\mathbf{S}$ ***is a skew-symmetric matrix by definition***. Now, solving $\mathbf{S}$ for $\mathbf{R}$ completes the proof.

Remarks

- The inverse of $(\mathbf{I} - \mathbf{S})$ in equation (4.19) always exist. Otherwise, it will imply that $\det(\mathbf{I} - \mathbf{S}) = 0$ which means: for a $\mathbf{h} \in \mathbb{R}_3$, $(\mathbf{I} - \mathbf{S})\,\mathbf{h} = 0 \Rightarrow \mathbf{S}\,\mathbf{h} = \mathbf{h}$, implying that $\lambda = 1$ is an eigenvalue of the skew-symmetric matrix,$\mathbf{S}$. But, the only real eigenvalue of a skew-symmetric matrix is 0; so, this is a contradiction, and hence, matrix $(\mathbf{I} - \mathbf{S})$ is always invertible.
- The matrices: $(\mathbf{I} - \mathbf{S})$ and $(\mathbf{I} + \mathbf{S})$ commute. To wit, $(\mathbf{I} + \mathbf{S})(\mathbf{I} - \mathbf{S}) = (\mathbf{I} - \mathbf{S})(\mathbf{I} + \mathbf{S}) = (\mathbf{I} - \mathbf{S}^2)$. Now, pre-mumultiplying and post-multiplying these relations by $(\mathbf{I} - \mathbf{S})^{-1}$, we can see that $(\mathbf{I} - \mathbf{S})^{-1}(\mathbf{I} + \mathbf{S}) = (\mathbf{I} + \mathbf{S})(\mathbf{I} - \mathbf{S})^{-1}$, that is, $(\mathbf{I} - \mathbf{S})^{-1}$ and $(\mathbf{I} + \mathbf{S})$ also commute. Thus, we have an equivalent expression for Cayley's theorem:

$$\boxed{\mathbf{R} = (\mathbf{I} + \mathbf{S})(\mathbf{I} - \mathbf{S})^{-1}} \tag{4.21}$$

- Now, conversely, we would like to show that given a skew-symmetrix matrix, $\mathbf{S} \equiv (\mathbf{R} - \mathbf{I})(\mathbf{R} + \mathbf{I})^{-1}$, we must have $\mathbf{R}$ as an orthogonal matrix. We have: $\mathbf{R} = (\mathbf{I} - \mathbf{S})^{-1}(\mathbf{I} + \mathbf{S}) = (\mathbf{I} + \mathbf{S})(\mathbf{I} - \mathbf{S})^{-1}$ (because they commute as shown above). Now, from the definition of $\mathbf{S}$, we have $\mathbf{R} = (\mathbf{I} - \mathbf{S})^{-1}(\mathbf{I} + \mathbf{S}) \Rightarrow \mathbf{R}^T = (\mathbf{I} + \mathbf{S}^T)(\mathbf{I} - \mathbf{S}^T)^{-1} = (\mathbf{I} - \mathbf{S})(\mathbf{I} + \mathbf{S})^{-1}$ (in the last equality, we have used the property of skew-symmetric matrix: $\mathbf{S} = -\mathbf{S}^T$). Now, multiplying, we get: $\mathbf{R}\,\mathbf{R}^T = (\mathbf{I} + \mathbf{S})(\mathbf{I} - \mathbf{S})^{-1}(\mathbf{I} - \mathbf{S})(\mathbf{I} + \mathbf{S})^{-1} = \mathbf{I}$. Hence, $\mathbf{R}$ is an orthogonal matrix.

4.2.1.1 Summary: Cayley's Theorem

- Every rotation matrix, $\mathbf{R}$, is in a one-to-one correspondence with a skew-symmetric matrix, $\mathbf{S}$, and related by: $\mathbf{R} = (\mathbf{I} - \mathbf{S})^{-1}\,(\mathbf{I} + \mathbf{S})$.

Remarks

- For direct rotation, that is, $\det(\mathbf{R}) = +1$, the formula (4.19) excludes the situation where the real eigenvalue of $\mathbf{R}$ becomes -1. However, the representation can be obtained by a limiting procedure as developed by Cayley. **S** being skew-symmetric, we need only $\frac{1}{2}n(n-1)$ parameters for a general n-dimensional space.

4.2.2 *Three-dimensional Space, $n = 3$*

Of particular interest to us is the case of the 3D space, i.e. $n = 3$. In this situation, the skew-symmetric matrix is completely determined by three coefficients above the diagonal. Thus, the number of parameters needed to describe a rotation matrix also becomes 3 $(= \frac{1}{2}(3)(3-1))$. This gives us an opportunity to associate the skew matrices with the corresponding three-dimensional vectors. Recall that a vector, **a**, is called the *axial vector* of a skew-symmetric matrix, **A**, if for all three-dimensional vectors, $\mathbf{h} \in \mathbb{R}^3$, we have:

$$\boxed{\mathbf{A\,h} = \mathbf{a} \times \mathbf{h}, \quad \forall \mathbf{h} \in \mathbb{R}^3} \tag{4.22}$$

Note

- From now on, if **a** is an axial vector, the corresponding skew-symmetric matrix will be shown by [**a**]. Operationally, it is understood that $[\mathbf{a}] \equiv (\mathbf{a}\,\times)$.

4.2.3 *Closed-form Explicit Description of Rotation Matrix*

For future numerical efficiency, we can use equation (4.21) to describe the rotation matrix, **R**, in a closed and explicit form in terms of the components of the skew-symmetric matrix, **S**. To this end, we need the inverse of the matrix $(\mathbf{I} - \mathbf{S})$. From direct solution, we have:

$$(\mathbf{I}-\mathbf{S})^{-1} = \frac{1}{\Delta}(\Delta\,\mathbf{I}+\mathbf{S}+\mathbf{S}^2) = \frac{1}{\Delta}\begin{bmatrix} 1+s_1^2 & s_1\,s_2 - s_3 & s_3\,s_1 + s_2 \\ s_1\,s_2 - s_3 & 1+s_2^2 & s_2\,s_3 - s_1 \\ s_3\,s_1 - s_2 & s_2\,s_3 + s_1 & 1+s_3^2 \end{bmatrix} \tag{4.23}$$

where, $\Delta \equiv 1 + \|\mathbf{s}\|^2 = 1 + s_1^2 + s_2^2 + s_3^2$. Now, by direct multiplication of matrices in equation (4.21), we have:

$$\begin{gathered}\mathbf{R} \equiv (\mathbf{I}-\mathbf{S})^{-1}\,(\mathbf{I}+\mathbf{S}) = \frac{1}{\Delta}\begin{bmatrix} 1+s_1^2 & s_1\,s_2 - s_3 & s_3\,s_1 + s_2 \\ s_1\,s_2 + s_3 & 1+s_2^2 & s_2\,s_3 - s_1 \\ s_3\,s_1 - s_2 & s_2\,s_3 + s_1 & 1+s_3^2 \end{bmatrix}\begin{bmatrix} 1 & -s_3 & s_2 \\ s_3 & 1 & -s_1 \\ -s_2 & s_1 & 1 \end{bmatrix}, \\ \Delta \equiv 1 + \|\mathbf{s}\|^2 = 1 + s_1^2 + s_2^2 + s_3^2 \end{gathered} \tag{4.24}$$

leading to:

$$\boxed{\begin{gathered}\mathbf{R} = \frac{1}{\Delta}\begin{bmatrix} 1+s_1^2-s_2^2-s_3^2 & 2(s_1\,s_2 - s_3) & 2(s_3\,s_1 + s_2) \\ 2(s_1\,s_2 + s_3) & 1-s_1^2+s_2^2-s_3^2 & 2(s_2\,s_3 - s_1) \\ 2(s_3\,s_1 - s_2) & 2(s_2\,s_3 + s_1) & 1-s_1^2-s_2^2+s_3^2 \end{bmatrix}, \\ \Delta \equiv 1 + \|\mathbf{s}\|^2 = 1 + s_1^2 + s_2^2 + s_3^2 \end{gathered}} \tag{4.25}$$

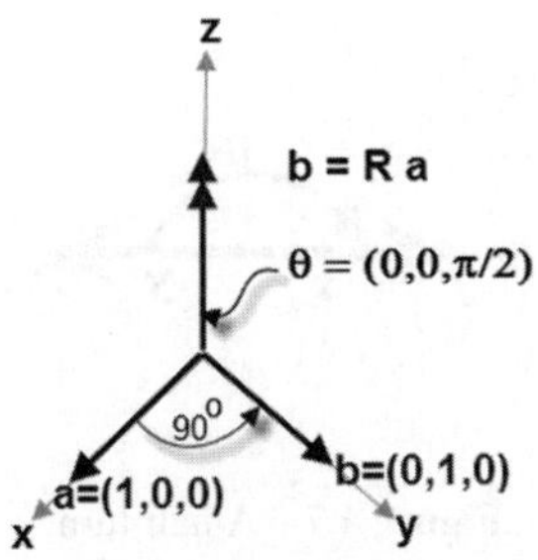

Figure 4.6 Example: Cayley's method.

Thus, we do not need to worry about the original and computationally cumbersome nine-component parameterization with six constraints of a 3D rotation matrix. The equivalent 3D skew-symmetric matrix, and hence 3D axial vector, representation helps us to consider only three parameters for the rotation matrix suitable for numerical efficiency.

Example: As shown in Figure 4.6, a vector, $\mathbf{a} = (1, 0, 0)$, along the x-axis is rotated 90° about z-axis to produce the vector, $\mathbf{b} = (0, 1, 0)$ along the y-axis such that $\mathbf{b} = \mathbf{R}\ \mathbf{a}$. We are interested in the expression for the rotation matrix, $\mathbf{R}$, according to Cayley's representation. Now, the instantaneous axial vector is given by $\mathbf{s} = (0, 0, 1)$. So, the corresponding skew-symmetric matrix $\mathbf{S} = \begin{bmatrix} 0 & -1 & 0 \\ 1 & 0 & 0 \\ 0 & 0 & 0 \end{bmatrix}$; $(\mathbf{I} - \mathbf{S}) = \begin{bmatrix} 1 & 1 & 0 \\ -1 & 1 & 0 \\ 0 & 0 & 1 \end{bmatrix}$ with $(\mathbf{I} - \mathbf{S})^{-1} = \frac{1}{2}\begin{bmatrix} 1 & -1 & 0 \\ 1 & 1 & 0 \\ 0 & 0 & 2 \end{bmatrix}$ and $(\mathbf{I} + \mathbf{S}) = \begin{bmatrix} 1 & -1 & 0 \\ 1 & 1 & 0 \\ 0 & 0 & 1 \end{bmatrix}$. Thus, finally, we have: $\mathbf{R} = (\mathbf{I} - \mathbf{S})^{-1}(\mathbf{I} + \mathbf{S}) = \frac{1}{2}\begin{bmatrix} 1 & -1 & 0 \\ 1 & 1 & 0 \\ 0 & 0 & 2 \end{bmatrix}\begin{bmatrix} 1 & -1 & 0 \\ 1 & 1 & 0 \\ 0 & 0 & 1 \end{bmatrix} = \begin{bmatrix} 0 & -1 & 0 \\ 1 & 0 & 0 \\ 0 & 0 & 1 \end{bmatrix}$ as expected. Given that the $\theta = \frac{\pi}{2}$, we can compare with the familiar expression for the rotation matrix. $\mathbf{R} = \begin{bmatrix} \cos\frac{\pi}{2} & -\sin\frac{\pi}{2} & 0 \\ \sin\frac{\pi}{2} & \cos\frac{\pi}{2} & 0 \\ 0 & 0 & 1 \end{bmatrix} = \begin{bmatrix} 0 & -1 & 0 \\ 1 & 0 & 0 \\ 0 & 0 & 1 \end{bmatrix}$.

4.2.4 Instantaneous Rotation

It can easily be seen by the definition of a skew-symmetric matrix that for two-dimensional (planar) space, that is, $n = 2$, we only need one parameter, a, to describe a rotation. Thus, for planar rotation, $\mathbf{S} = \begin{bmatrix} 0 & -a \\ a & 0 \end{bmatrix}$, $(\mathbf{I} - \mathbf{S})^{-1} = \frac{1}{1+a^2}\begin{bmatrix} 1 & -a \\ a & 1 \end{bmatrix}$ where $\det(\mathbf{I} - \mathbf{S}) \equiv \Delta = 1 + a^2$. Then the rotation matrix is given by: $\mathbf{R} = (\mathbf{I} - \mathbf{S})^{-1}\,(\mathbf{I} + \mathbf{S}) = \frac{1}{\Delta}\begin{bmatrix} 1 - a^2 & -2a \\ 2a & 1 - a^2 \end{bmatrix}$. Now, if we describe a planar rotation about the third axis by θ such that

$$a \equiv \tan\left(\frac{\theta}{2}\right) \tag{4.26}$$

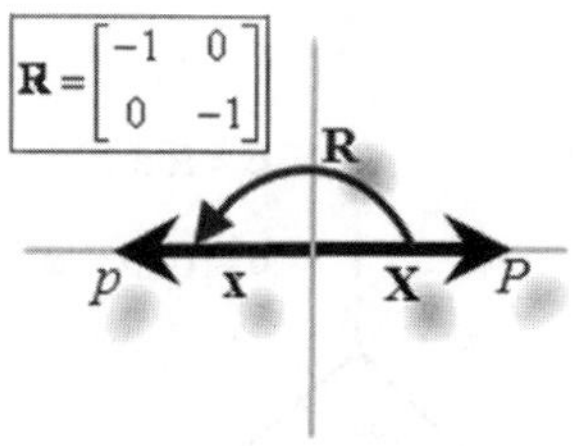

Figure 4.7 A half turn.

then: $\frac{1-a^2}{1+a^2} = 2\cos\frac{\theta}{2} - 1 = \cos\theta$ and $\frac{2a}{1+a^2} = \sin\theta$. With these relationships in the background, we get our familiar expression for planar rotation as:

$$\boxed{\mathbf{R} = \begin{bmatrix} \cos\theta & -\sin\theta \\ \sin\theta & \cos\theta \end{bmatrix}} \tag{4.27}$$

4.2.4.1 Limiting Case: $a \to \infty$

Following from this, as the parameter $a \to \infty$, we have: $\frac{1-a^2}{1+a^2} \to -1$ and $\frac{2a}{1+a^2} \to 0$ resulting in $\boxed{\mathbf{R} = \begin{bmatrix} -1 & 0 \\ 0 & -1 \end{bmatrix}}$. This rotation, $\mathbf{R}$, corresponds to a half turn with eigenvalue $\lambda = -1$, as shown in Figure 4.7.

4.2.5 *Lie Group, Tangent Operators, Exponential Map and Adjoint Map*

Recall that a Lie group is endowed with both the algebraic structure of a group and the geometry of a Riemannian surface, that is, a differentiable curved surface. Recall also that the rotational transformations belong to $SO(3)$, the special orthogonal group in three dimension, and form a matrix Lie group. In other words, $SO(3)$ is a differentiable manifold.

4.2.5.1 The Tangent Operators of $SO(3)$

Let us consider on $SO(3)$, a continuously differentiable affinely parametrized curve of rotational transformations, $\mathbf{R}(\lambda)$, $\lambda \in [0, 1]$ with $\mathbf{R}(0) = \mathbf{I}$ where $\mathbf{I}$ is the 3×3 identity matrix. Then, the rotational matrix function, $\mathbf{R}(\lambda) :\to SO(3)$, generates a continuous set of points, $\mathbf{X}(\lambda) \in \mathbb{R}_3$, of any point, $\mathbf{x} \in \mathbb{R}_3$ through the relation: $\mathbf{X}(\lambda) = \mathbf{R}(\lambda)\,\mathbf{x}$. The tangent direction to this curve is given by the derivative, denoted by $(\bullet)'$, with respect to λ, that is, $\mathbf{X}'(\lambda) = \mathbf{R}'(\lambda)\,\mathbf{x} = (\mathbf{R}'(\lambda)\,\mathbf{R}^{-1}(\lambda))\,\mathbf{X}(\lambda) = (\mathbf{R}'(\lambda)\,\mathbf{R}^T(\lambda))\,\mathbf{X}(\lambda)$ where we have used for $\mathbf{R}(\lambda)$ the orthogonality condition: $\mathbf{R}(\lambda)^T\,\mathbf{R}(\lambda) = \mathbf{I}$. Noting the identity: $\mathbf{R}'(\lambda) = (\mathbf{R}'(\lambda)\,\mathbf{R}^T(\lambda))\,\mathbf{R}(\lambda)$, we see that $(\mathbf{R}'(\lambda)\,\mathbf{R}^T(\lambda))$ also transforms the rotation matrix, $\mathbf{R}(\lambda)$, to give its derivative, $\mathbf{R}'(\lambda)$. Now, also note that: $(\mathbf{R}'(\lambda)\,\mathbf{R}^T(\lambda)) + (\mathbf{R}'(\lambda)\,\mathbf{R}^T(\lambda))^T = \mathbf{R}'(\lambda)\,\mathbf{R}^T(\lambda) + \mathbf{R}(\lambda)(\mathbf{R}'(\lambda)\,)^T = (\mathbf{R}(\lambda)\,\mathbf{R}^T(\lambda))' = \mathbf{0}$, because of the orthogonality condition: $\mathbf{R}(\lambda)\,\mathbf{R}^T(\lambda) = \mathbf{I}$. Thus the matrix, $(\mathbf{R}'(\lambda)\,\mathbf{R}^T(\lambda))$, is a skew-symmetric matrix and known as the ***tangent operator*** on $SO(3)$.

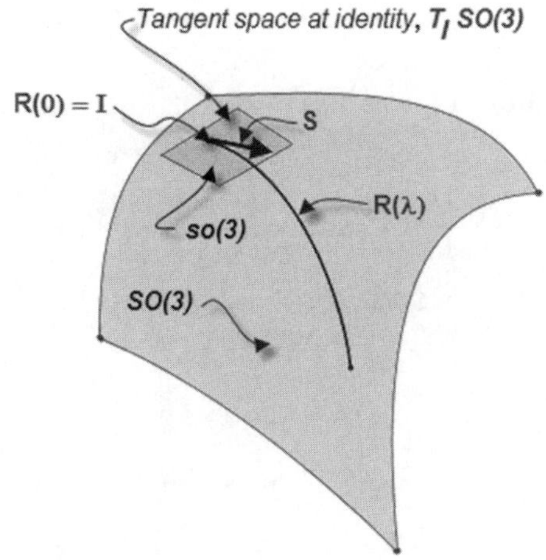

Figure 4.8 Cayley's theorem: $SO(3)$ and $so(3)$.

4.2.5.2 Exponential Map: the Case of Constant Tangent Operator

Let us consider the case where the tangent operator of $\mathbf{R}(\lambda)$ is constant and given by the skew-symmetric matrix, $\mathbf{S}$. Then, we have:

$$\mathbf{R}'(\lambda) = \mathbf{S}\,\mathbf{R}(\lambda) \tag{4.28}$$

Note that if the initial condition is given as $\mathbf{R}(0) = \mathbf{I}$, the identity matrix, then, from equation (4.28), we have: $\mathbf{S} = \mathbf{R}'(0)$. In other words, $\mathbf{S}$ is the tangent operator at identity. Thus, the set of tangent operators on $SO\,(3)$ is identical to the set of tangent directions at identity.

The tangent space at identity, denoted by $T_I SO(3)$, is a vector space of skew-symmetric matrices (see Figure 4.8). Recalling that for any two skew-symmetric matrices, $\mathbf{A}$, $\mathbf{B}$, with $\mathbf{a}$, $\mathbf{b}$ as the corresponding axial vectors, we have the identity: $\mathbf{AB} - \mathbf{BA} = (\mathbf{b} \otimes \mathbf{a}) - (\mathbf{a} \otimes \mathbf{b}) = [\mathbf{a} \times \mathbf{b}]$, we define a product, known as ***Lie product*** or ***Lie bracket***, as: $[\mathbf{A},\ \mathbf{B}] = \mathbf{AB} - \mathbf{BA}$. Recalling that a vector space with a product is known as algebra, the tangent space at identity is also recognized as a ***Lie algebra*** and is denoted by $so(3)$. Now, focusing on equation (4.28) with initial condition: $\mathbf{R}(0) = \mathbf{I}$, we have the solution:

$$\mathbf{R}(\lambda) = e^{\lambda \mathbf{S}} = \mathbf{I} + \lambda\mathbf{S} + \frac{1}{2}(\lambda\mathbf{S})^2 + \cdots \tag{4.29}$$

Thus, as shown Figure 4.9, ***a rotation matrix can be generated from the corresponding skew-symmetric matrix through an exponential map:***

$$\boxed{\mathbf{R} = e^{\mathbf{S}} = \mathbf{I} + \mathbf{S} + \frac{1}{2}(\mathbf{S})^2 + \cdots \quad \text{and} \quad \mathbf{R}^T = \mathbf{R}^{-1} = e^{-\mathbf{S}}} \tag{4.30}$$

If we normalize $\mathbf{S}$ by dividing it by the length $s = \|\mathbf{s}\|$, of its axial vector, $\mathbf{s}$, then we have: $\boxed{\mathbf{S} = s\mathbf{N}}$ such that $\|\mathbf{N}\| = 1$, that is, the axial vector, $\mathbf{n}$, corresponding to the skew-symmetric matrix, $\mathbf{N}$, is

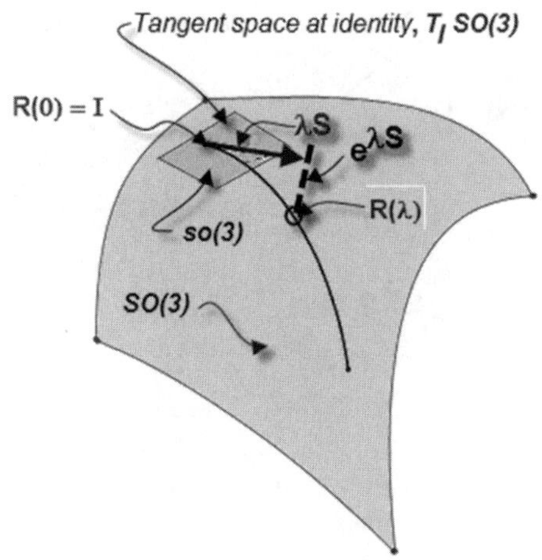

Figure 4.9 Cayley's theorem: exponential map.

an unit vector. Noting the *identity* that a matrix satisfies its own characteristic polynomial as in the case of the skew-symmetric matrix, **S**:

$$\boxed{\mathbf{S}^3 + s^2\mathbf{S} = 0 \quad \text{where} \quad s \equiv \|\mathbf{s}\| = \sqrt{s_1^2 + s_2^2 + s_3^2}} \tag{4.31}$$

we have:

$$\boxed{\mathbf{N}^3 + \mathbf{N} = 0} \tag{4.32}$$

Now, using identity (4.32) in equation (4.29), we get:

$$\begin{aligned}\mathbf{R}(\lambda) = e^{\lambda\mathbf{S}} &= \mathbf{I} + \lambda\mathbf{S} + \tfrac{1}{2}(\lambda\mathbf{S})^2 + \cdots = \mathbf{I} + \lambda s\mathbf{N} + \tfrac{1}{2}(\lambda s\mathbf{N})^2 + \cdots \\ &= \mathbf{I} + \left(\lambda s - \frac{(\lambda s)^3}{3!} + \cdots\right)\mathbf{N} + \left(1 - \left\{1 - \tfrac{1}{2}(\lambda s)^2 + \frac{(\lambda s)^4}{4!} - \cdots\right\}\right)\mathbf{N}^2 \\ \Rightarrow \mathbf{R}(\lambda) &= \mathbf{I} + (\sin\lambda s)\,\mathbf{N} + (1 - \cos\lambda s)\mathbf{N}^2 \Rightarrow \boxed{\mathbf{R} = \mathbf{I} + (\sin s)\,\mathbf{N} + (1 - \cos s)\mathbf{N}^2}\end{aligned} \tag{4.33}$$

Remark

- From our later discussion on ***Rodrigues parametrization*** of a rotation matrix by three parameters, we will be able to see the geometry of equation (4.33): the unit axial vector, **n**, of the unit skew-symmetric matrix, **N**, represents the axis of rotation with λs as the amount of rotation about it.

4.2.5.3 Adjoint Map

If $\mathbf{R} \in SO(3)$ and $\mathbf{r} \in so(3)$, then $\mathbf{R}\,[\mathbf{r}]\,\mathbf{R}^{-1} = \mathbf{R}\,[\mathbf{r}]\,\mathbf{R}^T \in so(3)$ where $\mathbf{R}\,[\mathbf{r}]\,\mathbf{R}^T = [\mathbf{R}\mathbf{r}]$ with the notation: $[\mathbf{a}] = \mathbf{a}\times$ for any vector $\mathbf{a} \in so(3)$.

Proof. Recall that the cross product is preserved under rotation, that is, $\mathbf{Ra} \times \mathbf{Rb} = \mathbf{R}(\mathbf{a} \times \mathbf{b})$ for any $\mathbf{a}, \mathbf{b} \in \mathbb{R}_3$. So, $(\mathbf{R}[\mathbf{r}]\mathbf{R}^T)\mathbf{h} = \mathbf{R}(\mathbf{r} \times \mathbf{R}^T\mathbf{h}) = \mathbf{Rr} \times \mathbf{RR}^T\mathbf{h} = \mathbf{Rr} \times \mathbf{h} = [\mathbf{Rr}]\mathbf{h}$ for any $\mathbf{h} \in \mathbb{R}_3$ completes the proof.

4.2.5.4 Discussion

As indicated before, ***Rodrigues parametrization*** ingeniously represents a rotation matrix with only three parameters; thus, for our subsequent computational descriptions of the rotation matrix, we now explore the Rodrigues vector and rotation parameterization.

4.3 Rodrigues Parameters

We take advantage of the earlier construction of the important theorem by Cayley that allows us to parameterize the rotation matrix by a vector, the ***Rodrigues vector***. We now introduce the unit ***axial vectors*** corresponding to the skew-symmetric matrices in 3D giving us the possibility of parametrization of rotation matrices by vector quantities for optimum computational efficiency. Alternatively, we introduce Rodrigues parametrization through geometrical construction. This assertion greatly increases numerical efficiency by considering only a three-dimensional vector (i.e. three parameters) instead of the computationally cumbersome nine parameters of a rotation matrix with six constraints.

Remark

- The Rodrigues vector was devised by Olinde Rodrigues in 1840, and Cayley's theorem appeared in 1845. Thus, there is no need to resort to Cayley's theorem to describe Rodrigues parameterization; this is clearly demonstrated towards the end, that is, where we see the geometry of the Rodrigues vector without resorting to Cayley's theorem. In what follows, however, we choose the same geometric construction as we did in developing Cayley's theorem as the starting point for introducing the Rodrigues vector. In so doing, we present our primary method for finite rotation parameterizations suitable for subsequent numerical evaluations in structural mechanics.

4.3.1 Rodrigues Vector, $\mathbf{n}_\theta$, in Three-dimensional Space

From Figure 4.10 and our discussion on Cayley's theorem, recall that for $\mathbf{x} = \mathbf{R}\mathbf{X}$, we find that $\mathbf{x} - \mathbf{X}$ is perpendicular to $\mathbf{x} + \mathbf{X}$, and it follows that:

$$\mathbf{x} - \mathbf{X} = \mathbf{S}(\mathbf{x} + \mathbf{X}) \quad \text{where} \quad \mathbf{S} \equiv [\mathbf{R} - \mathbf{I}][\mathbf{R} + \mathbf{I}]^{-1} \tag{4.34}$$

Now, from $\mathbf{s}$ as the axial vector of $\mathbf{S}$ it follows, by definition:

$$\mathbf{x} - \mathbf{X} = \mathbf{s} \times (\mathbf{x} + \mathbf{X}) \tag{4.35}$$

If we consider the projection of $\mathbf{X}$ and $\mathbf{x}$ on a plane perpendicular to $\mathbf{s}$, equation (4.35) still remains true if we replace $\mathbf{X}$ and $\mathbf{x}$ by their respective projections, that is,

$$proj(\mathbf{x} - \mathbf{X}) = \mathbf{s} \times \; proj(\mathbf{x} + \mathbf{X}) \tag{4.36}$$

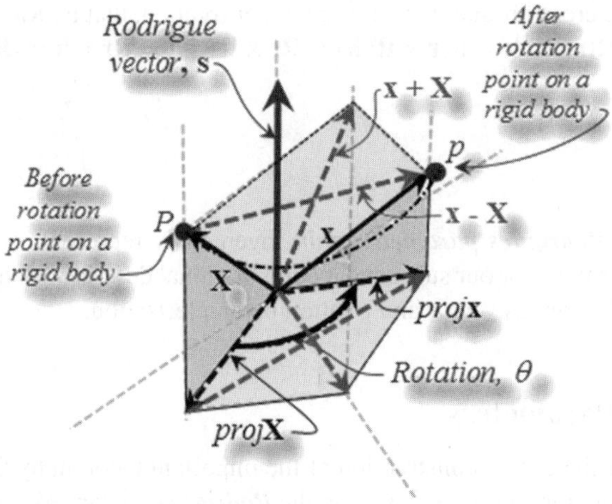

Figure 4.10 Rodrigues vector.

By definition of a cross product and noting that **s** is perpendicular to $proj(\mathbf{x}+\mathbf{X})$, we have the following magnitude relation:

$$\|proj(\mathbf{x}-\mathbf{X})\| = \|\mathbf{s}\| \, \|proj(\mathbf{x}+\mathbf{X})\| \tag{4.37}$$

Then, from Figure 4.10, we have:

$$\tan\left(\frac{\theta}{2}\right) = \frac{\frac{1}{2}\|proj(\mathbf{x}-\mathbf{X})\|}{\frac{1}{2}\|proj(\mathbf{x}+\mathbf{X})\|} = \frac{\|proj(\mathbf{x}-\mathbf{X})\|}{\|proj(\mathbf{x}+\mathbf{X})\|} \tag{4.38}$$

So, from equations (4.37) and (4.38), we get for the magnitude of **s**:

$$\boxed{s = \|\mathbf{s}\| = \tan\left(\frac{\theta}{2}\right)} \tag{4.39}$$

Thus, Cayley's axial vector, **s**, may be written as:

$$\boxed{\mathbf{s} = \tan\left(\frac{\theta}{2}\right)\mathbf{n}_\theta, \quad \|\mathbf{n}_\theta\| = 1} \tag{4.40}$$

Definition

- The unit axial vector, $\mathbf{n}_\theta$, that parameterizes a rotation matrix, is known as the ***Rodrigues vector***.

4.3.2 3D Rotation Vector, $\boldsymbol{\Theta}$, and the Rodrigues Vector, $\mathbf{n}_\theta$

For numerical computations in our subsequent finite element analysis and following the magnitude of **s** from equation (4.39), we introduce:

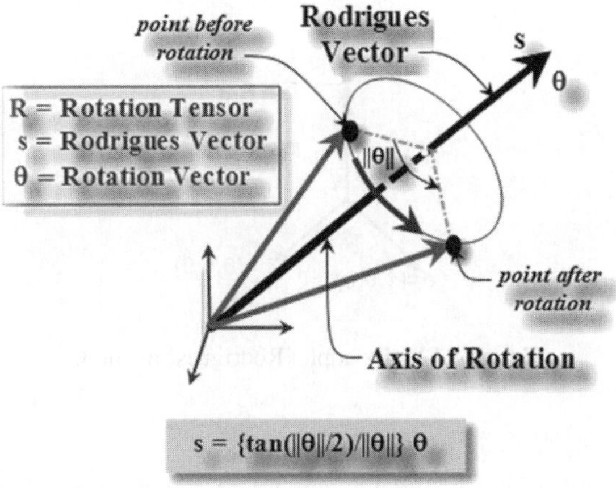

Figure 4.11 Rotation tensor and Rodrigues vector.

Definition A ***rotation vector***, $\boldsymbol{\theta}$, along the Rodrigues vector is defined such that:

$$\boxed{\mathbf{s} = \frac{1}{\theta}\tan\left(\frac{\theta}{2}\right)\boldsymbol{\theta}} \tag{4.41}$$

where $\theta \equiv \|\boldsymbol{\theta}\| = \sqrt{(\theta \bullet \theta)}$ is the magnitude of the rotation vector. We can easily verify that equation (4.41) implies equation (4.39). As shown in Figure 4.11, the rotation vector is the instantaneous axis of rotation of a rigid body. From equations (4.40) and (4.41), we have:

$$\boxed{\mathbf{S} = \frac{1}{\theta}\tan\left(\frac{\theta}{2}\right)\Theta = \tan\left(\frac{\theta}{2}\right)\mathbf{N}_\theta \quad \text{where} \quad \mathbf{N}_\theta \equiv \frac{\Theta}{\theta} = [\mathbf{n}_\theta], \quad \Theta = [\boldsymbol{\theta}], \quad \mathbf{n}_\theta = \frac{\boldsymbol{\theta}}{\theta},} \tag{4.42}$$

4.3.3 *Rotation Matrix,* R*, and Rodrigues Vector,* $\mathbf{n}_\theta \equiv \frac{\boldsymbol{\theta}}{\theta}$

Now, inserting expression (4.42) for $\mathbf{S}$ in terms of the components of the Rodrigues vector, $\mathbf{n}_\theta \equiv \frac{\boldsymbol{\theta}}{\theta}$, of equation (4.42), in equation (4.25), we get our final representation of a rotation matrix in terms of the components of the Rodrigues vector:

$$\boxed{\begin{gathered}\mathbf{R}(\mathbf{s}(\boldsymbol{\theta})) = \frac{1}{\rho}\begin{bmatrix} 1+s_1^2-s_2^2-s_3^2 & 2(s_1s_2-s_3) & 2(s_3s_1+s_2) \\ 2(s_1s_2+s_3) & 1-s_1^2+s_2^2-s_3^2 & 2(s_2s_3-s_1) \\ 2(s_3s_1-s_2) & 2(s_2s_3+s_1) & 1-s_1^2-s_2^2+s_3^2 \end{bmatrix} \\ \text{with } \rho \equiv 1+s_1^2+s_2^2+s_3^2 = 1+\|\mathbf{s}\|^2\text{, and,} \\ \mathbf{s} = \frac{1}{\theta}\tan\left(\frac{\theta}{2}\right)\boldsymbol{\theta}, \quad \theta \equiv \|\boldsymbol{\theta}\|, \quad \|\mathbf{s}\| = \tan\left(\frac{\theta}{2}\right)\end{gathered}} \tag{4.43}$$

Example: As shown in Figure 4.12, a vector, $\mathbf{a} = (1, 0, 0)$, along the x-axis is rotated 90° about z-axis to produce the vector, $\mathbf{b} = (0, 1, 0)$ along the y-axis such that $\mathbf{b} = \mathbf{R}\,\mathbf{a}$. We are interested in the expression for the rotation matrix, $\mathbf{R}$, according to the Rodrigues formula. Now, with $\theta = \frac{\pi}{2}$ and $\mathbf{n}_\theta = (0, 0, 1)$, the Rodrigues vector is given by: $\mathbf{s} = \tan\frac{\theta}{2}\,\mathbf{n}_\theta = (0, 0, 1)$. So, from

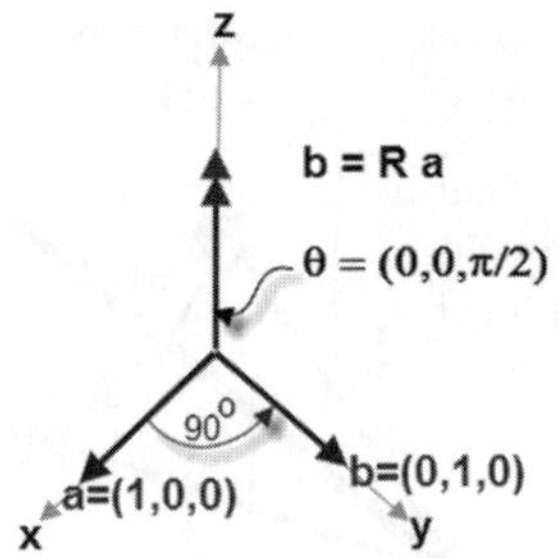

Figure 4.12 Example: Rodrigues' method.

equation (4.43), we have: $\|\mathbf{s}\| = \tan\left(\frac{\theta}{2}\right) = 1 \Rightarrow \rho = 1 + \|\mathbf{s}\|^2 = 2$ and $\mathbf{R} = \frac{1}{2}\begin{bmatrix} 0 & -2 & 0 \\ 2 & 0 & 0 \\ 0 & 0 & 2 \end{bmatrix} = \begin{bmatrix} 0 & -1 & 0 \\ 1 & 0 & 0 \\ 0 & 0 & 1 \end{bmatrix}$ as expected.

4.3.4 Geometry of the Rotation Matrix and the Rodrigues Vector

Let (X, Y, Z) be a global frame of reference. Now, referring to Figure 4.13, we choose – for simplicity and without any loss of generality – a local frame, (x, y, z), such that z-axis is aligned with the axis of rotation: $\boldsymbol{\theta} = (\theta_X\ \theta_Y\ \theta_Z)$, and xz-plane contains the initial unrotated vector, $\mathbf{P}$ that is rotated by rotation matrix, $\mathbf{R}$, to the final rotated vector, $\mathbf{p}$. Clearly, $\mathbf{P}_x$, the projection vector of the initial unrotated vector, $\mathbf{P}$, along x-axis is given by:

$$\mathbf{P}_x = \mathbf{P} - (\mathbf{P} \bullet \mathbf{n})\,\mathbf{n} = \mathbf{P} - (\mathbf{n} \otimes \mathbf{n})\,\mathbf{P} = (\mathbf{I} - (\mathbf{n} \otimes \mathbf{n}))\,\mathbf{P} \tag{4.44}$$

Noting that $\mathbf{n}$ is a unit vector, we have: $\|\mathbf{P}_x\| = \|\mathbf{p}_{xy}\| = \|\mathbf{n} \times \mathbf{P}\|$. So, using equation (4.44), the projection vector, $\mathbf{p}_{xy}$, of the final rotated vector, $\mathbf{p}$, is given by the components as:

$$\begin{aligned} \mathbf{p}_{xy} &= \mathbf{p}_x + \mathbf{p}_y = \mathbf{P}_x \cos\theta + (\mathbf{n} \times \mathbf{P}) \sin\theta = \mathbf{P}_x \cos\theta + (\mathbf{n} \times \mathbf{P}) \sin\theta \\ &= \{(\mathbf{I} - (\mathbf{n} \otimes \mathbf{n})) \cos\theta + [\mathbf{n}] \sin\theta\}\,\mathbf{P} \end{aligned} \tag{4.45}$$

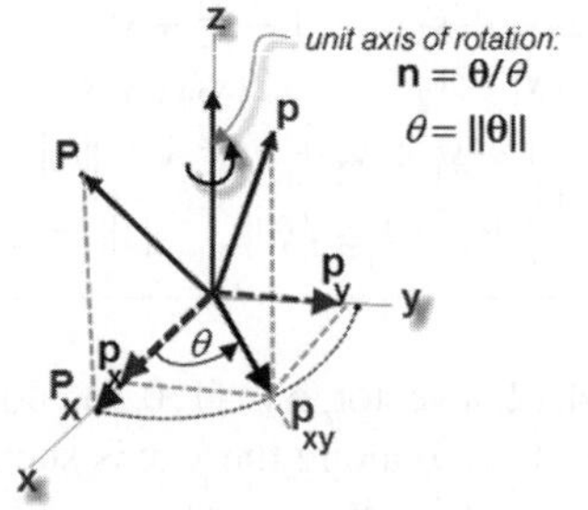

Figure 4.13 Geometry of Rodrigues vector.

Now, by using equation (4.45), the final rotated vector, $\mathbf{p}$, can be expressed in terms of the unrotated initial vector, $\mathbf{P}$, as:

$$\begin{aligned}\mathbf{p} &= \mathbf{p}_{xy} + (\mathbf{P} \bullet \mathbf{n})\,\mathbf{n} \\ &= \{(\mathbf{I} - (\mathbf{n} \otimes \mathbf{n}))\cos\theta + [\mathbf{n}]\sin\theta\}\,\mathbf{P} + (\mathbf{n} \otimes \mathbf{n})\,\mathbf{P} \\ &= \{\mathbf{I}\cos\theta + [\mathbf{n}]\sin\theta + (1 - \cos\theta)(\mathbf{n} \otimes \mathbf{n})\}\,\mathbf{P}\end{aligned} \tag{4.46}$$

If we define the rotation matrix, $\mathbf{R}$, as: $\mathbf{p} = \mathbf{R}\,\mathbf{P}$, then by equation (4.46), we have:

$$\mathbf{R} = \mathbf{I}\cos\theta + [\mathbf{n}]\sin\theta + (1 - \cos\theta)(\mathbf{n} \otimes \mathbf{n}) \tag{4.47}$$

Finally, recalling that for any $\mathbf{h} \in \mathbb{R}_3$, $\{\mathbf{I} + \mathbf{N}_\theta^2\}\,\mathbf{h} = \mathbf{h} + \mathbf{n} \times \mathbf{n} \times \mathbf{h} = \not{\mathbf{h}} + (\mathbf{n} \bullet \mathbf{h})\,\mathbf{n} - (\mathbf{n} \bullet \mathbf{n})\mathbf{h} = (\mathbf{n} \otimes \mathbf{n})\,\mathbf{h}$, that is, $\mathbf{I} + \mathbf{N}_\theta^2 = (\mathbf{n} \otimes \mathbf{n})$, where $\mathbf{N}_\theta$, as before, is the *skew-symmetric matrix* corresponding to the axial vector, $\mathbf{n}_\theta$, from equation (4.47), we have:

$$\boxed{\mathbf{R} = \mathbf{I} + \sin\theta\,\mathbf{N}_\theta + (1 - \cos\theta)\,\mathbf{N}_\theta^2} \tag{4.48}$$

If all the vectors are column vectors, we have equivalently, from equation (4.48):

$$\boxed{\mathbf{R} = \mathbf{I} + \sin\theta\,(\mathbf{n}_\theta \times) + (1 - \cos\theta)\left(\mathbf{n}_\theta\,\mathbf{n}_\theta^T - \mathbf{I}\right)} \tag{4.49}$$

Thus, the rotation is parameterized by three parameters: the components of the rotation axis or the Rodrigues vector. As we will see below, the three-parameter parametrization is singular at some rotations.

4.3.5 2D (Planar): Rodrigues Vector and Rotation Vector

It is instructive to bring out the familiar expression for rotation matrix for planar, that is, 2D rotation about the third axis with $\theta_1 = \theta_2 \equiv 0 \Rightarrow \|\boldsymbol{\theta}\| = \theta_3 \equiv \theta$. From expression (4.41), we immediately have: $s_1 = s_2 \equiv 0 \Rightarrow s_3 \equiv \left(\tan\frac{\theta}{2}\right)\frac{\theta}{\theta} = \tan\frac{\theta}{2}$. Now, let us note: $\frac{1}{\rho}(1 - s_3^2) = \dfrac{1 - \tan^2\frac{\theta}{2}}{1 + \tan^2\frac{\theta}{2}} = 2\cos^2\frac{\theta}{2} - 1 = \cos\theta$, and $\frac{1}{\rho}(2s_3) = \dfrac{2\tan\frac{\theta}{2}}{1 + \tan^2\frac{\theta}{2}} = 2\sin\frac{\theta}{2}\cos\frac{\theta}{2} = \sin\theta$. Thus, the expression (4.43) reduces to the familiar instantaneous rotation matrix for 2D:

$$\boxed{\begin{aligned}&\mathbf{R} = \frac{1}{\rho}\begin{bmatrix} 1 - s_3^2 & -2s_3 & 0 \\ 2s_3 & 1 - s_3^2 & 0 \\ 0 & 0 & 1 + s_3^2 \end{bmatrix} = \begin{bmatrix} \cos\theta & -\sin\theta & 0 \\ \sin\theta & \cos\theta & 0 \\ 0 & 0 & 1 \end{bmatrix} \\ &\text{with } \rho \equiv 1 + s_3^2\end{aligned}} \tag{4.50}$$

4.3.6 *Singularity of the Rodrigues Rotation Matrix at* $\|\theta\| = (2n+1)\pi, \quad n = 0, 1, ...$

For numerical computation, it is important to note that a rotation matrix as expressed by equation (4.43) is singular whenever $tr\,\mathbf{R} \to -1$, where tr stands for the ***trace of a matrix***, or equivalently, $\|\mathbf{s}\| \to \infty$, or $\|\theta\| = (2n+1)\pi, \quad n = 0, 1, ...$ To wit, let us first note that: $\|\mathbf{s}\| = \mathbf{s} \bullet \mathbf{s} = \left(\tan^2 \frac{\theta}{2}\right) \frac{(\boldsymbol{\theta} \bullet \boldsymbol{\theta})}{\|\boldsymbol{\theta}\|^2} = \tan^2 \frac{\theta}{2}$, where, recall that: $\theta \equiv \|\boldsymbol{\theta}\|$. Clearly, then, $\|\mathbf{s}\| \to \infty \Leftrightarrow \tan \frac{\theta}{2} \to \infty \Leftrightarrow \theta \to (2n+1)\pi$ for $n = 0, 1, ...$ Now, from equation (4.43): $\frac{1}{2}(tr\,\mathbf{R} - 1) = \frac{1}{2}\left\{\frac{1}{\rho}(3 - \|s\|^2) - 1\right\} = \frac{1}{2}\frac{(1 - \|s\|^2)}{(1 + \|s\|^2)} = \dfrac{1 - \tan^2 \frac{\theta}{2}}{1 + \tan^2 \frac{\theta}{2}} = \cos\theta$. Finally, knowing that $\cos\theta \to -1$ as $\theta \to (2n+1)\pi$ with $n = 0, 1, ...,$ completes our results.

4.3.6.1 Discussion

A vectorial parameterization of the rotation matrix is most important for our purposes because the number of parameters (namely, three) is the same as that of the usual rotational degrees of freedom of structural elements such as beams, plates and shells, for computational schemes such as the finite element method. Although it is singular at some rotation, as shown above, for most structural problems with moderate rotations, it suffices. In the event that it is not, various other four-parameter parameterizations, such as the ***Euler–Rodrigues parameterization***, can be combined transparently to make it singularity-free. Thus, for singularity-free parameterization, we consider next the Euler–Rodrigues parameters and rotation parameterization.

4.4 Euler – Rodrigues Parameters

The main goal here is to present a computationally singularity-free parameterization of rigid body rotations. More specifically, we concentrate on a very important parameterization of a rotation matrix by the four-dimensional ***Euler–Rodrigues parameter vector***. This parameterization eliminates the singularity problems of parameterization by the ***Rodrigues vectors***.

4.4.1 *4D Euler–Rodrigues Vector,* $\mathbf{X} \equiv \{X_1 \quad X_2 \quad X_3 \quad X_4\}^T$

Recalling our discussion on the Rodrigues rotation parameterization, the ***Rodrigues rotation matrix*** becomes ***singular*** at those $\boldsymbol{\theta}$'s for which $\|\theta\| = (2n+1)\pi, \quad n = 0, 1, ...$ Furthermore, recalling that, although there are only three independent parameters (nine minus six due to orthonormality constraints) in a rotation tensor, it is not possible for a three-parameter set such as the Rodrigues parameter vector, $\boxed{\mathbf{s} = \tan\left(\frac{\theta}{2}\right)\mathbf{n}_\theta, \quad \|\mathbf{n}_\theta\| = 1}$, $\mathbf{n}_\theta \equiv \frac{\boldsymbol{\theta}}{\theta}$, $\theta = \|\boldsymbol{\theta}\|$, to uniquely describe all possible rotations. A non-singular one-to-one mapping takes at least five parameters while a non-singular many-to-one mapping requires at least four parameters.

4.4.1.1 Cayley's Representation and Euler–Rodrigues Parameters

Using the Rodrigues vector from equation (4.43) and inserting in Cayley's representation as in equation (4.19), we have:

$$\mathbf{R} = (\mathbf{I}-\mathbf{S})^{-1}(\mathbf{I}+\mathbf{S}) = (\mathbf{I} - \left(\frac{\sin\frac{\theta}{2}}{\cos\frac{\theta}{2}}\right)\mathbf{N}_\theta)^{-1}(\mathbf{I} + \left(\frac{\sin\frac{\theta}{2}}{\cos\frac{\theta}{2}}\right)\mathbf{N}_\theta)$$

$$= \left(\cos\tfrac{\theta}{2}\mathbf{I} - \sin\tfrac{\theta}{2}\mathbf{N}_\theta\right)^{-1}\left(\cos\tfrac{\theta}{2}\mathbf{I} + \sin\tfrac{\theta}{2}\mathbf{N}_\theta\right) \tag{4.51}$$

where $\mathbf{N}_\theta = [\mathbf{n}_\theta] = \mathbf{n}_\theta \times$. Now, we define:

Definition

- The four-dimensional unit Euler–Rodrigues parameter vector, $\mathbf{X} = \{X_1 \quad X_2 \quad X_3 \quad X_4\}^T$, defined as:

$$\boxed{\begin{array}{l} X_1 \equiv \sin\frac{\theta}{2}\, n_1, X_2 \equiv \sin\frac{\theta}{2}\, n_2, X_3 \equiv \sin\frac{\theta}{2}\, n_3, X_4 \equiv \cos\frac{\theta}{2}, \\ \|\mathbf{X}\| \equiv 1, \text{with } \mathbf{n}_\theta \equiv (n_1 \quad n_2 \quad n_3) \quad \text{and} \quad [\mathbf{n}_\theta] = \mathbf{N}_\theta \end{array}} \tag{4.52}$$

Noting that $\left(\cos\frac{\theta}{2}\mathbf{I} - \sin\frac{\theta}{2}\mathbf{N}_\theta\right)^{-1} = \cos\frac{\theta}{2}\mathbf{I} + \sin\frac{\theta}{2}\mathbf{N}_\theta + \dfrac{\sin^2\frac{\theta}{2}}{\cos\frac{\theta}{2}}\left(\mathbf{I}+\mathbf{N}_\theta^2\right)$, we have, from equation (4.51):

$$\boxed{\mathbf{R} = \mathbf{I} + 2\sin\tfrac{\theta}{2}\cos\tfrac{\theta}{2}\mathbf{N}_\theta + 2\sin^2\tfrac{\theta}{2}\mathbf{N}_\theta^2 = \mathbf{I} + \sin\theta\,\mathbf{N}_\theta + (1-\cos\theta)\,\mathbf{N}_\theta^2} \tag{4.53}$$

Note that we have used the ***characteristic polynomial identity***: $\boxed{\mathbf{N}_\theta^3 + \mathbf{N}_\theta = 0}$ in the above derivation. The representations of a rotation matrix as given by equation (4.53) are identical to what was obtained through Rodrigues parameterization.

Exercise: prove that $\left(\cos\frac{\theta}{2}\mathbf{I} - \sin\frac{\theta}{2}\mathbf{N}_\theta\right)^{-1} = \cos\frac{\theta}{2}\mathbf{I} + \sin\frac{\theta}{2}\mathbf{N}_\theta + \dfrac{\sin^2\frac{\theta}{2}}{\cos\frac{\theta}{2}}\left(\mathbf{I}+\mathbf{N}_\theta^2\right)$

An orthogonal rotation matrix, **R**, with Euler–Rodrigues parameters, is characterized by a point on a three-dimensional unit spherical surface embedded in four dimensions, that is, a 3-sphere of radius 1, as shown in Figure 4.14, and is given by:

$$\boxed{\begin{array}{l} \mathbf{R} = \begin{bmatrix} X_1^2 - X_2^2 - X_3^2 + X_4^2 & 2(X_1X_2 - X_3X_4) & 2(X_1X_3 + X_2X_4) \\ 2(X_1X_2 + X_3X_4) & -X_1^2 + X_2^2 - X_3^2 + X_4^2 & 2(X_2X_3 - X_1X_4) \\ 2(X_1X_3 - X_2X_4) & 2(X_2X_3 + X_1X_4) & -X_1^2 - X_2^2 + X_3^2 + X_4^2 \end{bmatrix}, \\ \text{with} \quad \|\mathbf{X}\|^2 \equiv X_1^2 + X_2^2 + X_3^2 + X_4^2 = 1 \end{array}} \tag{4.54}$$

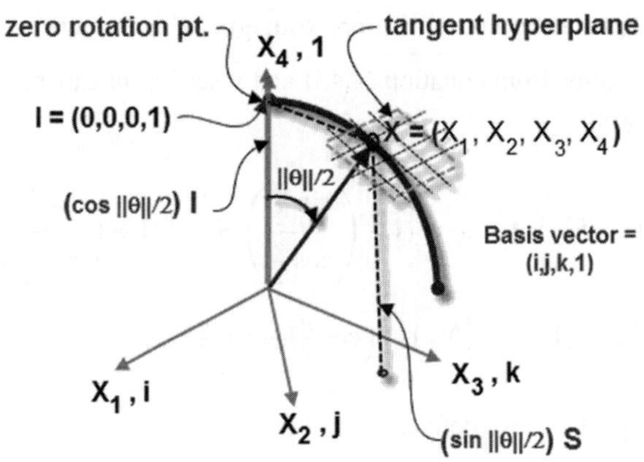

Figure 4.14 Euler–Rodrigues parameters.

with

$$\boxed{\begin{array}{l} X_1 \equiv \left(\sin\dfrac{\theta}{2}\right)\dfrac{\theta_1}{\theta}, X_2 \equiv \left(\sin\dfrac{\theta}{2}\right)\dfrac{\theta_2}{\theta}, X_3 \equiv \left(\sin\dfrac{\theta}{2}\right)\dfrac{\theta_3}{\theta}, X_4 \equiv \cos\dfrac{\theta}{2}, \\ \boldsymbol{\theta} = \{\theta_1 \quad \theta_2 \quad \theta_3\}^T, \quad \theta \equiv \|\boldsymbol{\theta}\| = \sqrt{\theta_1^2 + \theta_2^2 + \theta_3^2} \end{array}} \tag{4.55}$$

From equations (4.54) and (4.55), rotations are non-singular two-to-one maps where globally both $+\mathbf{X}$ and $-\mathbf{X}$ represent the same rotation. Thus, a four-parameter Euler–Rodrigues vector is adequate against singularity instead of a full five-parameter representation. Finally, geometrically speaking, the finite rotations, **R**, move around on a 3D unit sphere; the set of instantaneous rotations define a tangent hyperplane at a point on this unit sphere.

Example: As shown in Figure 4.15, a vector, $\mathbf{a} = (1, 0, 0)$, along the x-axis is rotated 90° about z-axis to produce the vector, $\mathbf{b} = (0, 1, 0)$ along the y-axis such that $\mathbf{b} = \mathbf{R}\,\mathbf{a}$. We are interested in the expression for the rotation matrix, **R**, according to the Euler–Rodrigues formula. Now, with $\theta = \frac{\pi}{2}$, we have, from equation (4.55), the Euler–Rodrigues parameters: $X_1 = 0, X_2 = 0, X_3 = \frac{1}{\sqrt{2}}, X_4 = \frac{1}{\sqrt{2}}$ and $\|\mathbf{X}\| = 1$. Thus, finally using equation (4.54), we get $\boxed{\mathbf{R} = \begin{bmatrix} 0 & -1 & 0 \\ 1 & 0 & 0 \\ 0 & 0 & 1 \end{bmatrix}}$ as expected.

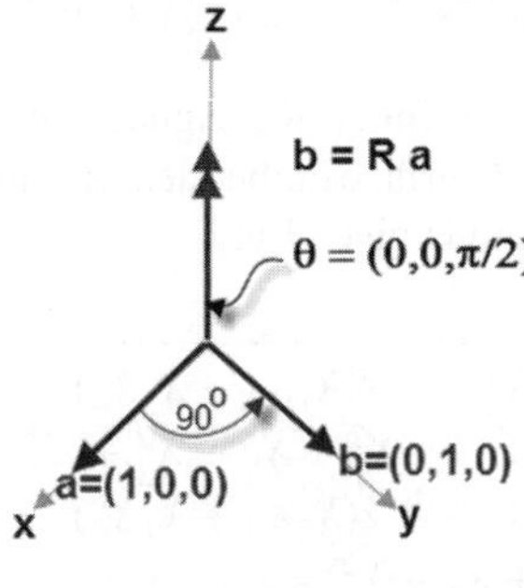

Figure 4.15 Example: Euler–Rodrigues method.

4.4.1.2 Discussion

As an alternative to Euler–Rodrigues parameterization, four-parameter ***Hamilton–Rodrigues quaternion*** non-singular parameterizations can be combined transparently to make numerical treatment singularity-free. But, before we can introduce Hamilton–Rodrigues quaternions and rotation parameterization, we will first familiarize ourselves with Hamilton's quaternion.

4.5 Hamilton's Quaternions

The main goal here is to introduce the concept of the quaternion as a precursor to its application as a locally singularity-free ***parameterization*** of rigid body rotations in structural and continuum mechanics. We introduce and discuss the important properties of the quaternions as devised by Hamilton. Particularly, the definitions for a ***conjugate quaternion***, the ***norm of a quaternion***, ***inverse quaternion***, ***pure or vector quaternions***, and for a ***unit quaternion*** are introduced. Finally, recognizing that quaternions form a non-commuting group, that is, a Lie group, a particular group operation reveals that quaternions are similar to rotation matrices, giving us the possibility of parameterization of rotation matrices by quaternions for optimum computational efficiency.

4.5.1 Four-dimensional Quaternions: $\mathbf{Q} \equiv \{q_1 \quad q_2 \quad q_3 \quad q_4\}^T$

William Rowan Hamilton was searching for the next higher dimensional extension to the two-dimensional complex number: $x = a + \mathbf{i}b$ with the rule $\mathbf{i}^2 = -1$; he was unsuccessful for 15 years as he was concentrating on three dimensions. On an October stroll in 1843, he realized that the next successor dimension is four and not three; he named his mathematical entity a ***quaternion*** and described a quaternion, **Q**, as:

$$\boxed{\mathbf{Q} \equiv a + b\mathbf{i} + c\mathbf{j} + d\mathbf{k} \quad \text{with} \quad \mathbf{i}^2 = \mathbf{j}^2 = \mathbf{k}^2 = \mathbf{ijk} = -1} \tag{4.56}$$

It follows then: $\mathbf{ij} = \mathbf{k}$, $\mathbf{ji} = -\mathbf{k}$ and so on. To wit, $\mathbf{ijk} = -1 \Rightarrow \mathbf{ijk}^2 = -\mathbf{k} \Rightarrow \mathbf{ij} = \mathbf{k}$ because $\mathbf{k}^2 = -1$, and $\mathbf{ji} = -\mathbf{ji}(\mathbf{ijk}) = -\mathbf{j}(\mathbf{i}^2)\mathbf{jk} = \mathbf{j}^2\mathbf{k} = -\mathbf{k}$. The multiplication of two quaternions with above rules as a four-dimensional entity is similar to that of complex numbers of two dimensions.

For our purpose of geometrical interpretation and parameterization of the rotation matrices, we may regroup the parameters in the above description of a quaternion to rewrite it as a "scalar plus a vector":

$$\boxed{\mathbf{Q} = (\underbrace{V}_{3D\ vector}, \underbrace{s}_{scalar}) \quad \text{with} \quad \mathbf{v} = (b\mathbf{i} + c\mathbf{j} + d\mathbf{k}) \quad \text{and} \quad s = a} \tag{4.57}$$

In order to explain the "scalar plus a vector" oddity, we may think of it as an addition of two underlying quaternions: $\mathbf{Q} = (\mathbf{v}, s) = (\mathbf{v}, 0) + (\mathbf{0}, s) = \mathbf{Q}_v + \mathbf{Q}_s$. Note that although, in modern mathematics, vectors are used far more than almost forgotten quaternions, historically speaking, quaternions preceded vectors. Only after the recognition of ***Euler–Rodrigues parameterization*** for rotational transformation has the quaternion received renewed interest among analysts.

4.5.1.1 Quaternion Properties

- It follows, from Hamilton's rules and equation (4.57), that notationally a quaternion is expressed component-wise as:

$$\boxed{\mathbf{Q} = (\mathbf{v} \quad s)^T = (q_1 \equiv v_1 \quad q_2 \equiv v_2 \quad q_3 \equiv v_3 \quad q_4 \equiv s)^T} \tag{4.58}$$

- Two quaternions are added (or subtracted) by simply adding (or subtracting) vector to vector and scalar to scalar.
- The ***multiplication rule*** is given by:

$$\boxed{\mathbf{Q}_1 = (\mathbf{v}_1, s_1) \textit{ and } \mathbf{Q}_2 = (\mathbf{v}_2, s_2) \Rightarrow \mathbf{Q} \equiv \mathbf{Q}_1\ \mathbf{Q}_2 = (s_1\mathbf{v}_2 + s_2\mathbf{v}_1 + \mathbf{v}_1 \times \mathbf{v}_2,\ s_1 s_2 - \mathbf{v}_1 \bullet \mathbf{v}_2)} \tag{4.59}$$

  From equation (4.59), we can see that, except for the presence of the cross product, a quaternion multiplication has similar characteristics to the familiar complex number multiplication: $(a_1 + \mathbf{i}b_1)(a_2 + \mathbf{i}b_2) = \mathbf{i}(a_1b_2 + a_2b_1) + (a_1a_2 - b_1b_2) = \left(a_1b_2 + a_2b_1,\ a_1a_2 - b_1b_2\right)$.
- A ***pure or vector quaternion*** is a quaternion with the scalar part identically zero:

$$\boxed{\mathbf{Q} \equiv (\mathbf{v}, 0)} \tag{4.60}$$

  As will be seen later, this allows smooth transition from vector to quaternion and back.
- The ***conjugate***, $\tilde{\mathbf{Q}}$, of any quaternion, $\mathbf{Q} \equiv (\mathbf{v}, s)$, is given by:

$$\boxed{\tilde{\mathbf{Q}} \equiv (-\mathbf{v}, s)} \tag{4.61}$$

- The ***norm***, $\|\mathbf{Q}\|$ of any quaternion, $\mathbf{Q} \equiv (\mathbf{v}, s)$, is given by:

$$\boxed{\|\mathbf{Q}\| = \sqrt{\mathbf{v} \bullet \mathbf{v} + s^2}} \tag{4.62}$$

  It follows that: $\|\mathbf{Q}\tilde{\mathbf{Q}}\| = \|(\cancel{-s\mathbf{v} + s\mathbf{v}} - \cancel{\mathbf{v} \times \mathbf{v}},\quad s^2 + \mathbf{v} \bullet \mathbf{v})\| = \|(\mathbf{0},\quad \mathbf{v} \bullet \mathbf{v} + s^2)\| = \|\mathbf{Q}\|$.
- The ***inverse quaternion***, $\mathbf{Q}^{-1}$ of any quaternion, $\mathbf{Q} \equiv (\mathbf{v}, s)$, is defined as:

$$\boxed{\mathbf{Q}^{-1} \equiv \frac{\tilde{\mathbf{Q}}}{\|\mathbf{Q}\|} = \frac{1}{\|\mathbf{Q}\|}(-\mathbf{v}, s)} \tag{4.63}$$

  $\|\mathbf{Q}\mathbf{Q}^{-1}\| = \left\|\mathbf{Q}\frac{\tilde{\mathbf{Q}}}{\|\mathbf{Q}\|}\right\| = \frac{\|\mathbf{Q}\|}{\|\mathbf{Q}\|} = 1$ justifies the multiplicative inverse definition.
- A ***unit quaternion***, $\mathbf{Q}$, such that $\|\mathbf{Q}\| = \sqrt{\mathbf{v} \bullet \mathbf{v} + s^2} = 1$, converts to a ***rotation matrix*** under quaternion group operation, $\boxed{\mathbf{Q}\ \mathbf{Q}_a\ \tilde{\mathbf{Q}}}$ where the ***vector quaternion*** $\mathbf{Q}_a$ embeds the vector $\mathbf{a}$ that is undergoing rotation such that $\mathbf{Q}_a \equiv (\mathbf{a}, 0)$.

Note that ***under this quaternion operation, the resulting quaternion is also a vector quaternion***, that is, $\mathbf{Q}\ \mathbf{Q}_a\ \tilde{\mathbf{Q}} = (\mathbf{b}, 0)$ for some vector $\mathbf{b}$. To wit, let us recall the identity: $\|\mathbf{v}\|^2\,\mathbf{I} + \mathbf{V}^2 = (\mathbf{v} \otimes \mathbf{v})$ where $\mathbf{v}$ is the axial vector corresponding to the skew-symmetric matrix, $\mathbf{V}$, that is, $\mathbf{V} = [\mathbf{v}]$ (because, for any $\mathbf{h} \in \mathbb{R}_3$ $\{\|\mathbf{v}\|^2\mathbf{I} + \mathbf{V}^2\}\mathbf{h} = \|\mathbf{v}\|^2\mathbf{h} + \mathbf{v} \times (\mathbf{v} \times \mathbf{h})$

$= \|\mathbf{v}\|^2\mathbf{h} + (\mathbf{v} \bullet \mathbf{h})\,\mathbf{v} - (\mathbf{v} \bullet \mathbf{v})\mathbf{h} = (\mathbf{v} \otimes \mathbf{v})\,\mathbf{h})$ with $(\mathbf{v} \otimes \mathbf{v})$ as the *tensor product* of two vectors (for column vectors, $(\mathbf{v} \otimes \mathbf{v}) = \mathbf{v}\,\mathbf{v}^T$), and $\mathbf{V}^2 = \mathbf{V}\,\mathbf{V} = \begin{bmatrix} 0 & -v_3 & v_2 \\ v_3 & 0 & -v_1 \\ -v_2 & v_1 & 0 \end{bmatrix}\begin{bmatrix} 0 & -v_3 & v_2 \\ v_3 & 0 & -v_1 \\ -v_2 & v_1 & 0 \end{bmatrix} =$ $\begin{bmatrix} -v_2^2 - v_3^2 & v_1 v_2 & v_3 v_1 \\ v_1 v_2 & -v_3^2 - v_1^2 & v_2 v_3 \\ v_3 v_1 & v_2 v_3 & -v_1^2 - v_2^2 \end{bmatrix}$.

So, now, noting that $\|\mathbf{v}\|^2 + s^2 = 1$ for the unit quaternion, $\mathbf{Q} = (\mathbf{v}, s)$, we get:

$$\begin{aligned}
\mathbf{Q}\;\mathbf{Q}_a\;\widetilde{\mathbf{Q}} &= (\mathbf{v}, s)(\mathbf{a}, 0)(-\mathbf{v}, s) = (s\mathbf{a} + \mathbf{v} \times \mathbf{a},\ -(\mathbf{v} \bullet \mathbf{a}))(-\mathbf{v}, s) \\
&= ((\mathbf{v} \bullet \mathbf{a})\mathbf{v} + s^2\mathbf{a} + 2s(\mathbf{v} \times \mathbf{a}) + \mathbf{v} \times (\mathbf{v} \times \mathbf{a}), \\
&\quad -s(\mathbf{v} \bullet \mathbf{a}) + s(\mathbf{v} \bullet \mathbf{a}) + (\mathbf{v} \times \mathbf{a}) \bullet \mathbf{v} \\
&= (((\mathbf{v} \otimes \mathbf{v}) + 2s\mathbf{V} + s^2\mathbf{I} + \mathbf{V}^2)\mathbf{a},\ 0) = ((\|\mathbf{v}\|^2\,\mathbf{I} + 2\mathbf{V}^2 + 2s\mathbf{V} + s^2\mathbf{I})\mathbf{a},\ 0) \\
&= (((\|\mathbf{v}\|^2 + s^2)\mathbf{I} + 2s\mathbf{V} + 2\mathbf{V}^2)\mathbf{a},\ 0) = ((\mathbf{I} + 2s\mathbf{V} + 2\mathbf{V}^2)\mathbf{a},\ 0) = (\mathbf{b},\ 0) \\
&= \mathbf{Q}_b
\end{aligned} \tag{4.64}$$

where we have defined:

$$\boxed{\mathbf{b} \equiv (\mathbf{I} + 2s\mathbf{V} + \mathbf{V}^2)\mathbf{a} = \mathbf{R}\;\mathbf{a}} \tag{4.65}$$

with

$$\boxed{\begin{aligned} \mathbf{R} &\equiv \mathbf{I} + 2s\mathbf{V} + 2\mathbf{V}^2 \\ &= \begin{bmatrix} 1 - 2v_2^2 - 2v_3^2 & 2(v_1 v_2 - s v_3) & 2(v_1 v_3 + s v_2) \\ 2(v_1 v_2 + s v_3) & 1 - 2v_1^2 - 2v_3^2 & 2(v_2 v_3 - s v_1) \\ 2(v_1 v_3 - s v_2) & 2(v_2 v_3 + s v_1) & 1 - 2v_1^2 - 2v_2^2 \end{bmatrix} \end{aligned}} \tag{4.66}$$

Notationally, this is equivalent to:

$$\boxed{\mathbf{R} = \begin{bmatrix} 1 - 2q_2^2 - 2q_3^2 & 2(q_1\,q_2 - q_3\,q_4) & 2(q_1\,q_3 + q_2\,q_4) \\ 2(q_1\,q_2 + q_3\,q_4) & 1 - 2q_1^2 - 2q_3^2 & 2(q_2\,q_3 - q_1\,q_4) \\ 2(q_1\,q_3 - q_2\,q_4) & 2(q_2\,q_3 - q_1\,q_4) & 1 - 2q_1^2 - 2q_2^2 \end{bmatrix}} \tag{4.67}$$

Or, noting that the quaternion, $\mathbf{Q}$, is a unit quaternion, that is, $\|\mathbf{Q}\|^2 = \|\mathbf{v}\|^2 + s^2 = q_1^2 + q_2^2 + q_3^2 + q_4^2 = 1$, we get:

$$\mathbf{R} = \begin{bmatrix} q_1^2 - q_2^2 - q_3^2 + q_4^2 & 2(q_1\,q_2 - q_3\,q_4) & 2(q_1\,q_3 + q_2\,q_4) \\ 2(q_1\,q_2 + q_3\,q_4) & -q_1^2 + q_2^2 - q_3^2 + q_4^2 & 2(q_2\,q_3 - q_1\,q_4) \\ 2(q_1\,q_3 - q_2\,q_4) & 2(q_2\,q_3 - q_1\,q_4) & -q_1^2 - q_2^2 + q_3^2 + q_4^2 \end{bmatrix} \tag{4.68}$$

Thus, from equations (4.64)–(4.68), we see that:

The quaternion group operation, $\boxed{\mathbf{Q}\;\mathbf{Q}_a\;\widetilde{\mathbf{Q}}}$, where $\mathbf{Q}$ is a unit *quaternion*, transforms the vector $\mathbf{a}$ to a vector $\mathbf{b}$ such that $\mathbf{b} = \mathbf{R}\;\mathbf{a}$.

Lemma: Prove that the matrix, **R**, is, in fact, a rotation matrix.

Proof. Let us first recall that the characteristic polynomial, $p(\lambda)$, of the skew symmetric matrix, $\mathbf{V} \equiv [\mathbf{v}] \equiv (\mathbf{v}\times)$ of an axial vector, $\mathbf{v}$, with length, $\|\mathbf{v}\|$, is given by:

$$p(\lambda) = \lambda^3 + \|\mathbf{v}\|^2 \lambda = 0 \tag{4.69}$$

and that the matrix, $\mathbf{V} \equiv [\mathbf{v}] \equiv (\mathbf{v}\times)$, satisfies its own characteristic polynomial, that is,

$$\mathbf{V}^3 = -\|\mathbf{v}\|^2 \mathbf{V} \tag{4.70}$$

From equation (4.70), it follows that,

$$\mathbf{V}^4 = -\|\mathbf{v}\|^2 \mathbf{V}^2 \tag{4.71}$$

Now, noting that the transpose of a skew-symmetric matrix, $\mathbf{V} \equiv [\mathbf{v}] \equiv (\mathbf{v}\times)$, is given by: $\mathbf{V}^T = -\mathbf{V}$, we have, from equation (4.66): $\mathbf{R}^T \equiv \mathbf{I} - 2s\mathbf{V} + 2\mathbf{V}^2$. Thus, finally, using the relation given by equation (4.71) and noting that for the unit quaternion: $\|\mathbf{Q}\|^2 = \|\mathbf{v}\|^2 + s^2 = 1$, we get:

$$\begin{aligned}
\mathbf{R}\,\mathbf{R}^T &= (\mathbf{I} + 2s\mathbf{V} + 2\mathbf{V}^2)(\mathbf{I} - 2s\mathbf{V} + 2\mathbf{V}^2) \\
&= \mathbf{I} - \cancel{2s\mathbf{V}} + 2\mathbf{V}^2 + \cancel{2s\mathbf{V}} - 4s^2\mathbf{V}^2 + \cancel{4s\mathbf{V}^3} + 2\mathbf{V}^2 \cancel{- 4s\mathbf{V}^3} + 4\mathbf{V}^4 \\
&= \mathbf{I} - \cancel{(1 - s^2 - \|\mathbf{v}\|^2)}4\mathbf{V}^2 = \mathbf{I}
\end{aligned}$$

Thus, **R** is an orthogonal transformation, which completes the proof.

Later, when we introduce the ***Hamilton–Rodrigues quaternion***, we will be able to recognize that **R** is indeed a rotation matrix with the quaternion vector component, **v**, and the quaternion scalar part, *s*, associated with the axis and the amount of rotation, respectively.

- Also, just like a rotation matrix, a unit quaternion preserves dot products and cross products embedded in the quaternion product. To wit, given any two *vector quaternions*, $\mathbf{Q}_a \equiv (\mathbf{a}, 0)$ and $\mathbf{Q}_b \equiv (\mathbf{b}, 0)$, we have: $\boxed{(\mathbf{Q}\ \mathbf{Q}_a\ \widetilde{\mathbf{Q}})(\mathbf{Q}\ \mathbf{Q}_b\ \widetilde{\mathbf{Q}}) = \mathbf{Q}\ (\mathbf{Q}_a\ \mathbf{Q}_b)\ \widetilde{\mathbf{Q}}}$. It is interesting to note that it is the cross product in the quaternion multiplication rule that allows the rotation to confound axes.
- The quaternions form a non-commuting group, that is, a Lie group, under quaternion multiplication. However, quaternions with same vector parts commute because the cross product vanishes. To wit, if $\mathbf{Q}_1 \equiv (\mathbf{v}, s_1)$ and $\mathbf{Q}_2 \equiv (\mathbf{v}, s_2)$, then $\mathbf{Q}_1\mathbf{Q}_2 = ((s_1 + s_2)\mathbf{v},\quad s_1 s_2 - \|\mathbf{v}\|^2) = \mathbf{Q}_2\mathbf{Q}_1$. As we will be see, this tells us that any two consecutive rotations about the same axis are commutative, as expected.

Example: Show that if **v** in a unit quaternion, and $\mathbf{Q} \equiv (\mathbf{v}, s)$, is an axis of rotation, then any rotation of a vector $\lambda\mathbf{v}$, a scaled vector of **v**, about **v** will keep $\lambda\mathbf{v}$ undisturbed. To wit, let $\mathbf{Q}_\lambda \equiv (\lambda\mathbf{v}, 0)$, then, noting that $\|\mathbf{Q}\|^2 = \|\mathbf{v}\|^2 + s^2 = 1$, we have:

$$\begin{aligned}
\mathbf{Q}\ \mathbf{Q}_\lambda\ \widetilde{\mathbf{Q}} &= (\mathbf{v}, s)(\lambda\mathbf{v}, 0)(-\mathbf{v}, s) = (\lambda s\mathbf{v},\ -\lambda\,\|\mathbf{v}\|^2)(-\mathbf{v}, s) \\
&= \left((\|\mathbf{v}\|^2 + s^2)\lambda\mathbf{v}, -\cancel{\lambda s\|\mathbf{v}\|^2} + \cancel{\lambda s\|\mathbf{v}\|^2}\right) = (\lambda\mathbf{v}, 0)
\end{aligned}$$

Thus, $\mathbf{Q}_\lambda \equiv (\lambda\mathbf{v}, 0)$ stays invariant under the quaternion operation, as expected.

Discussion: A vectorial parameterization of a rotation matrix is most important for our purposes in the sense that the number of parameters (three) is same as the usual rotational degrees of freedom of structural elements such as beams, plates and shells for computational schemes like the finite element method. Although it is singular at some rotation, for most structural problems, it suffices. In the event that it is not, ***Hamilton–Rodrigues quaternions*** and parameterizations, as already mentioned, can be combined transparently to make it singularity free.

4.6 Hamilton–Rodrigues Quaternion

The main goal here is to present a computationally singularity-free parameterization of rigid body rotations. More specifically, we concentrate on a very important parameterization of a rotation matrix by the four-dimensional Hamilton–Rodrigues quaternion; this parameterization eliminates the singularity problems of parameterization by the Rodrigues vectors. Also, this possibility profoundly increases numerical efficiency and robustness by incorporating a strategy of transparently combining only a three-dimensional vector, say, Rodrigues vector, with this four-parameter Hamilton–Rodrigues parameterization of a rotation matrix instead of the computationally cumbersome nine parameters of a rotation matrix with six constraints.

4.6.1 4D Hamilton–Rodrigues Quaternion, $\mathbf{Q} = \{q_1 \quad q_2 \quad q_3 \quad q_4\}^T$

Recalling the discussion on the Rodrigues rotation parameterization, the Rodrigues rotation matrix becomes singular at those $\boldsymbol{\theta}$'s for which $\|\theta\| = (2n+1)\pi, \quad n = 0, 1, \ldots$ Furthermore, recalling the discussion on rotation matrix, although there are only three independent parameters (nine minus six due to orthonormality constraints) in a rotation tensor, it is not possible for a three-parameter set such as the Rodrigues vector, $\boxed{\mathbf{s} = \tan(\frac{\theta}{2})\, \mathbf{n}_\theta, \quad \|\mathbf{n}_\theta\| = 1}$, $\mathbf{n}_\theta \equiv \frac{\boldsymbol{\theta}}{\theta}$, $\theta = \|\boldsymbol{\theta}\|$, to uniquely describe all possible rotations; a non-singular one-to-one mapping takes at least five parameters while a non-singular many-to-one mapping requires at least four parameters.

Definition

- The four-dimensional *unit Hamilton–Rodrigues quaternion* (Figure 4.16), $\mathbf{Q} = \{q_1 \quad q_2 \quad q_3 \quad q_4\}^T$, defined as:

$$\boxed{\begin{array}{l} q_1 \equiv \sin\frac{\theta}{2}\,\mathbf{n}_1, \quad q_2 \equiv \sin\frac{\theta}{2}\,\mathbf{n}_2, \quad q_3 \equiv \sin\frac{\theta}{2}\,\mathbf{n}_3, \quad q_4 \equiv \cos\frac{\theta}{2}, \\ \|\mathbf{Q}\| \equiv 1 \end{array}} \tag{4.72}$$

where the unit vector, $\mathbf{n}_\theta = (n_1 \quad n_2 \quad n_3)^T = \frac{1}{\theta}(\theta_1 \quad \theta_2 \quad \theta_3)^T$, is the ***Rodrigues axial vector*** along the rotation axis, $\boldsymbol{\theta}$, is a non-singular two-to-one map where globally both $+\mathbf{Q}$ and $-\mathbf{Q}$ represent the same rotation. Thus, a four-parameter Hamilton–Rodrigues quaternion will be adequate instead of a full five-parameter representation. The components of a point on the unit hypersphere (3-sphere) can be viewed as the components of a unit ***Hamilton–Rodrigues quaternion***,$\mathbf{Q}$. A

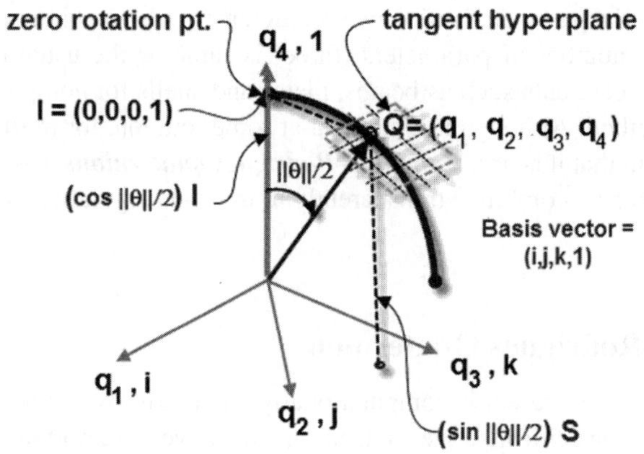

Figure 4.16 Hamilton–Rodrigues quaternions.

comparison between the definitions of Euler–Rodrigues vector, **X**, and of Hamilton–Rodrigues quaternion, **Q**, as given by equation (4.72), shows that $\mathbf{Q} \equiv (q_1 \equiv X_1, q_2 \equiv X_2, q_3 \equiv X_3, q_4 \equiv X_4)$, that is,

$$\boxed{\mathbf{Q} \equiv \{(\sin\frac{\theta}{2})\,\frac{\boldsymbol{\theta}}{\theta} + \cos\frac{\theta}{2}\}, \quad \|\mathbf{Q}\|^2 \equiv \mathbf{Q}\widetilde{\mathbf{Q}} = 1} \tag{4.73}$$

with $\widetilde{\mathbf{Q}} \equiv (\tilde{q}_1 \equiv -X_1, \tilde{q}_2 \equiv -X_2, \tilde{q}_3 \equiv -X_3, \tilde{q}_4 \equiv X_4)$ as the conjugate quaternion of **Q**.

Remark

- With the above assignment, we see, as indicated, that the vector part of a quaternion, $\mathbf{Q} \equiv (\mathbf{v}, s)$, represents some function of the axis of rotation, that is, $\boxed{\mathbf{v} \equiv (\sin\frac{\theta}{2})\,\frac{\boldsymbol{\theta}}{\theta}}$, and the scalar part represents some function of the amount of rotation, that is, $\boxed{s \equiv \cos\frac{\theta}{2}}$; and the matrix **R** given by the equation (4.66) turns out to be a true rotation matrix.

Then, any three-dimensional Euclidean point $\mathbf{x} \in \mathbb{R}_3$ with the corresponding four-dimensional vector or pure quaternion, $\boxed{\mathbf{Q}_x \equiv \left(\underbrace{\mathbf{x}}_{\in\mathbb{R}_3}, 0\right)}$, under quaternion operation given by $\boxed{\mathbf{Q}\,\mathbf{Q}_x\,\widetilde{\mathbf{Q}}}$, will have an image $\mathbf{y} \in \mathbb{R}_3$ lifted similarly to the four-dimensional vector quaternion, $\mathbf{Q}_y \equiv \left(\underbrace{\mathbf{y}}_{\in\mathbb{R}_3}, 0\right)$ that lie on a 3-sphere (recall that a 3-sphere means a three-dimensional spherical surface) embedded in a four-dimensional space such that:

$$\boxed{\mathbf{Q}_y = \mathbf{Q}\,\mathbf{Q}_x\,\widetilde{\mathbf{Q}}} \tag{4.74}$$

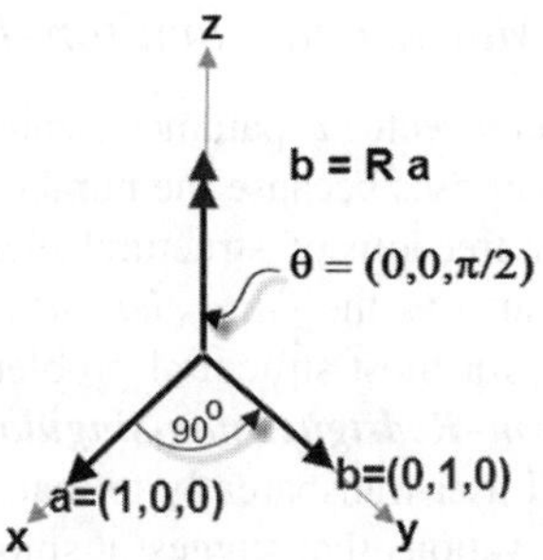

Figure 4.17 Example: Hamilton–Rodrigues quaternion method.

with

$$\begin{Bmatrix} y_1 \\ y_2 \\ y_3 \end{Bmatrix} = \begin{bmatrix} q_1^2 - q_2^2 - q_3^2 + q_4^2 & 2(q_1\,q_2 - q_3\,q_4) & 2(q_1\,q_3 + q_2\,q_4) \\ 2(q_1\,q_2 + q_3\,q_4) & -q_1^2 + q_2^2 - q_3^2 + q_4^2 & 2(q_2\,q_3 - q_1\,q_4) \\ 2(q_1\,q_3 - q_2\,q_4) & 2(q_2\,q_3 - q_1\,q_4) & -q_1^2 - q_2^2 + q_3^2 + q_4^2 \end{bmatrix} \begin{Bmatrix} x_1 \\ x_2 \\ x_3 \end{Bmatrix}, \qquad \mathbf{y} = \mathbf{R}\,\mathbf{x} \tag{4.75}$$

Note that the matrix given in equation (4.75) is exactly the same as that given by equation (4.68) derived in the Hamilton quaternion section and the rotation matrix defined by the Euler–Rodrigues parameters. Thus, ***a spherical map***, that is, a ***pure rotation***, can always be described by the quaternion map given in equation (4.74). Additionally, note that the finite rotations, $\mathbf{R}$, move around on a 3D unit spherical surface, i.e., a 3-sphere; the set of instantaneous rotations define a tangent hyperplane at a point on this unit sphere.

Now, noting that $\|\mathbf{Q}\|^2 \equiv q_1^2 + q_2^2 + q_3^2 + q_4^2 = 1$, we can express the rotation matrix, $\mathbf{R}$, of equation (4.75), alternatively by:

$$\mathbf{R} = \begin{bmatrix} 1 - 2q_2^2 - 2q_3^2 & 2(q_1\,q_2 - q_3\,q_4) & 2(q_1\,q_3 + q_2\,q_4) \\ 2(q_1\,q_2 + q_3\,q_4) & 1 - 2q_1^2 - 2q_3^2 & 2(q_2\,q_3 - q_1\,q_4) \\ 2(q_1\,q_3 - q_2\,q_4) & 2(q_2\,q_3 - q_1\,q_4) & 1 - 2q_1^2 - 2q_2^2 \end{bmatrix} \tag{4.76}$$

Example As shown in Figure 4.17, a vector, $\mathbf{a} = (1, 0, 0)$, along the x-axis is rotated 90° about the z-axis to produce the vector, $\mathbf{b} = (0, 1, 0)$ along the y-axis such that $\mathbf{b} = \mathbf{R}\,\mathbf{a}$. We are interested in the expression for the rotation matrix, $\mathbf{R}$, according to the Hamilton–Rodrigues formula. Now, with $\theta = \frac{\pi}{2}$, we have, from equation (4.72), the Hamilton–Rodrigues quaternions as: $\mathbf{Q}_x = \begin{pmatrix} 0 & 0 & \frac{1}{\sqrt{2}} & \frac{1}{\sqrt{2}} \end{pmatrix}$, $\|\mathbf{Q}\| = 1$, $\widetilde{\mathbf{Q}}_x = \begin{pmatrix} 0 & 0 & -\frac{1}{\sqrt{2}} & \frac{1}{\sqrt{2}} \end{pmatrix}$ and $\mathbf{Q}_a = (1 \quad 0 \quad 0 \quad 0)$. Thus, using the quaternion multiplication rule, as in equation (4.59), and $\mathbf{i}, \mathbf{j}, \mathbf{k}$ as the the unit base vectors, we get: $\mathbf{Q}_x\,\mathbf{Q}_a = \left(\frac{1}{\sqrt{2}}\mathbf{i} + \frac{1}{\sqrt{2}}\mathbf{j} + 0\mathbf{k},\ 0 \right)$, and finally, $\mathbf{Q}_b = \mathbf{Q}_x\,\mathbf{Q}_a\widetilde{\mathbf{Q}}_x = (0\mathbf{i} + \mathbf{j} + 0\mathbf{k},\ 0)$, a unit vector quaternion, implying that $\mathbf{b} = (0 \quad 1 \quad 0)^T$ as expected. Using the expression for $\mathbf{Q}_x$ in equation (4.76), we get the rotation matrix as: $\mathbf{R} = \begin{bmatrix} 0 & -1 & 0 \\ 1 & 0 & 0 \\ 0 & 0 & 1 \end{bmatrix}$ as expected.

4.6.2 Transition: Rotation Matrix and Hamilton–Rodrigues Quaternion

As indicated before, a three-parameter vectorial parameterization, such as Rodrigues, of a rotation matrix is most important for our purposes because the number of parameters (three) is the same as the usual rotational degrees of freedom of structural elements such as beams, plates and shells for computational scheme such as the *finite element method.* Although it is singular at some rotation, as indicated above, for most structural problems, it suffices. In the event that it does not, a four-parameter ***Hamilton–Rodrigues non-singular quaternion*** parameterization, as presented above, can be combined user-transparently to make numerical treatment singularity-free. These are precisely the observations that suggest a smooth transition from the Rodrigues three-parameter $\boldsymbol{\theta}$-space to the Hamilton–Rodrigues four-parameter $\mathbf{Q}$-quaternion space and back to obtain a singularity-free updating scheme for the rotation matrix for a typical nonlinear numerical analysis in structural engineering and mechanics.

4.6.2.1 Hamilton–Rodrigues Quaternion to Rotation Matrix

Given a unit Hamilton–Rodrigues quaternion, $\mathbf{Q}$, we can construct the corresponding rotation matrix, $\mathbf{R}$, by equation (4.76).

4.6.2.2 Rotation Matrix to Hamilton–Rodrigues Quaternion

Given the rotation matrix, $\mathbf{R}$, as in equation (4.76), recovering the corresponding quaternion, $\mathbf{Q}$, is a bit more involved but still a straightforward process. Algorithmically, we can describe it as follows:

Step 0: Compute the corresponding scalar part,$Q_4 \equiv s \equiv X_4$, of the quaternion as: $Q_4^2 \equiv s^2 \equiv X_4^2 = \frac{1}{4}(1 + trace(\mathbf{R}))$. If $Q_4^2 > 0$, then, go to ***Step 1***, else go to ***Step 2***.

Step 1: (It is immaterial whether we choose the positive or negative square root as both will represent the same rotation matrix; we will choose the positive square root), and

$$Q_4 \equiv s \equiv X_4 = \frac{1}{2}\sqrt{(1 + trace(\mathbf{R}))} \quad \text{and} \quad Q_1 \equiv v_1 \equiv X_1 = \frac{1}{4Q_4}(R_{32} - R_{23})$$
$$Q_2 \equiv v_2 \equiv X_2 = \frac{1}{4Q_4}(R_{13} - R_{31}) \quad \text{and} \quad Q_3 \equiv v_3 \equiv X_3 = \frac{1}{4Q_4}(R_{21} - R_{12})$$

STOP

Step 2: Compute $Q_1^2 \equiv v_1^2 \equiv X_1^2 = -\frac{1}{2}(R_{22} + R_{33})$. If $Q_1^2 > 0$, then, go to ***Step 3***. else go to ***Step 4***.

Step 3:

$$Q_4 \equiv s \equiv X_4 = 0 \quad \text{and} \quad Q_1 \equiv v_1 \equiv X_1 = \frac{1}{\sqrt{2}}\sqrt{(-R_{22} - R_{33})}$$
$$Q_2 \equiv v_2 \equiv X_2 = \frac{1}{2Q_1}R_{12} \quad \text{and} \quad Q_3 \equiv v_3 \equiv X_3 = \frac{1}{2Q_1}R_{13}$$

STOP

Step 4: Compute $Q_2^2 \equiv v_2^2 \equiv X_2^2 = \frac{1}{2}(1 - R_{33})$. If $Q_2^2 > 0$, then, go to ***Step 5***. else go to ***Step 6***.
Step 5.

$$
\begin{aligned}
&Q_4 \equiv s \equiv X_4 = 0 \quad \text{and} \quad Q_1 \equiv v_1 \equiv X_1 = 0 \\
&Q_2 \equiv v_2 \equiv X_2 = \frac{1}{\sqrt{2}}\sqrt{1 - R_{33}} \quad \text{and} \quad Q_3 \equiv v_3 \equiv X_3 = \frac{1}{2Q_2}R_{23}
\end{aligned}
$$

STOP
Step 6:

$$
\begin{aligned}
&Q_4 \equiv s \equiv X_4 = 0 \quad \text{and} \quad Q_1 \equiv v_1 \equiv X_1 = 0 \\
&Q_2 \equiv v_2 \equiv X_2 = 0 \quad \text{and} \quad Q_3 \equiv v_3 \equiv X_3 = 1
\end{aligned}
$$

STOP

4.6.3 Interpolation and Approximation on a Rotational 3-Sphere

In interpolating the continuous rotational history from a few key rotations, or approximating the rotational history in a time series analysis in structural dynamics, we need to deal with the interpolation or approximation of a rotation tensor on a curved surface: a 3-sphere. We will be interested only in the linear interpolation of rotations for a nonlinear static problem or, say, in the linear approximation of rotational accelerations for nonlinear dynamic time history analysis.

4.6.3.1 Linear Interpolation or Approximation on 3-Sphere

Let **A**, **B** represent the unit Hamilton–Rodrigues quaternions for rotations at parameters $t = 0$ and $t = 1$, respectively, as shown in Figure 4.18.

In The figure , the rotations are shown as points on a four-dimensional space; the rotation from the reference state, **A**, to a final state, **B**, is taken as a single rotation, which is always

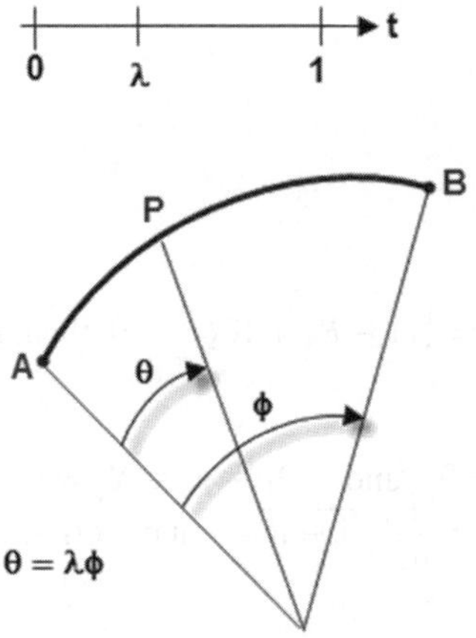

Figure 4.18 Spherically linear interpolation.

possible by ***Euler's theorem***. Now, let $\mathbf{P}(t)$, with $||\mathbf{P}|| = 1$, represents a rotation on this curve at an affine parameter $t = \lambda \in [0, 1]$. We are interested in expressing $\mathbf{P}$ in terns of $\mathbf{A}$ and $\mathbf{B}$. Clearly, with $||\mathbf{A}|| = ||\mathbf{B}|| = ||\mathbf{P}|| = 1$, we have: $\mathbf{A} \bullet \mathbf{P} = \cos\theta$ and $\mathbf{A} \bullet \mathbf{B} = \cos\phi$ with parametric relation: $\theta = \lambda\phi$.

Now, let us assume that $\mathbf{P}$ is given as a linear combination of $\mathbf{A}$ and $\mathbf{B}$ by: $\mathbf{P} = \alpha\mathbf{A} + \beta\mathbf{B}$. Then,

$$\mathbf{A} \bullet \mathbf{P} = \alpha(\mathbf{A} \bullet \mathbf{A}) + \beta(\mathbf{A} \bullet \mathbf{B}) \Rightarrow \cos\theta = \alpha + \beta\cos\phi \tag{4.77}$$

With $||\mathbf{P}|| = 1$, we have: $(\alpha\mathbf{A} + \beta\mathbf{B}) \bullet (\alpha\mathbf{A} + \beta\mathbf{B}) = 1 \Rightarrow \alpha^2 + 2\alpha\beta\cos\phi + \beta^2 = 1$. Replacing expression for α in the previous expression by equation (4.77), we get:

$$(\cos\theta - \beta\cos\phi)^2 + 2\beta(\cos\theta - \beta\cos\phi)\cos\phi + \beta^2 = 1 \Rightarrow \boxed{\beta = \frac{\sin\theta}{\sin\phi}} \tag{4.78}$$

Applying β of equation (4.78) in equation (4.77), we get: $\boxed{\alpha = \frac{\sin(\phi - \theta)}{\sin\phi}}$. Thus, finally:

$$\boxed{\mathbf{P} = \frac{\sin(1-\lambda)\phi}{\sin\phi}\mathbf{A} + \frac{\sin\lambda\phi}{\sin\phi}\mathbf{B}, \quad \lambda \in [0.1]} \tag{4.79}$$

Remark

- For nonlinear structural dynamic problems involving finite rotations, we may choose the rotational acceleration distribution according to equation (4.79) for a predictor–corrector type of numerical analysis; the rotational velocity and rotation distribution may then be obtained by corresponding integration of the acceleration.

4.6.3.2 Discussion

For numerical computations in the finite element phase, we will need the derivatives and the variational properties of a rotation matrix and the corresponding vector representations. Thus, for our subsequent computational needs, we must become familiar with the ***spatial and temporal derivatives and variations of rotation tensor field***.

4.7 Derivatives, Angular Velocity and Variations

Here, our main goal is to determine derivative and variational properties of a rotation tensor field. In the beam finite rotation nonlinear analysis, we will encounter a one-parameter-rotation tensor field: $\mathbf{R}(S)$ where $\mathbf{R}$ is the rotation tensor with S as the parameter for the beam centroidal axis curve. Similarly, In the plate or shell finite rotation nonlinear analysis, we will have to deal with a two-parameter-rotation tensor field: $\mathbf{R}(S^1, S^2)$ where $\mathbf{R}$ is the rotation tensor with S^α, $\alpha = 1,2$ as the arc length parameters along the lines of curvature of a plate or shell reference surface.

- These, in turn, necessitate evaluation of spatial derivatives: $\frac{\partial \mathbf{R}}{\partial S} \equiv \mathbf{R}_{,S} \equiv \mathbf{R}'$ or $\frac{\partial \mathbf{R}}{\partial S^\alpha} \equiv \mathbf{R}_{,S^\alpha}$, $\alpha = 1,2$, of the rotation tensor field on a curved surface: $SO(3)$, generating ***curvature tensor field***.

Additionally, in the case of dynamic analysis, with rotation field given as: $\mathbf{R}(S,t)$ or $\mathbf{R}(S^1, S^2, t)$, where $t \in \mathbb{R}_+$ as the time variable,

- we will need $\frac{\partial \mathbf{R}}{\partial t} \equiv \mathbf{R}_{,t} \equiv \dot{\mathbf{R}}$, of the rotation tensor field generating rotational or ***angular velocity tensor field***.

Finally, for a variational method of numerical analysis, such as the finite element method, we will be required to define the admissible virtual space, corresponding to the beam, plate or shell ***configuration space*** consisting of the displacement vector field and the rotational tensor field, that identifies the spatial virtual strain fields and the corresponding spatial real strain fields in the weak form of equations of a beam centroidal axis or a shell reference surface.

- This, in turn, will again require, in essence, the one-parameter derivative of the rotation tensor field defining the variation of the rotation tensor.

In each of these above situations, we will need the covariant derivative of the rotation tensor such that we can compose or combine rotations, so that the resulting object is still a rotation. Noting that the rotation field travels on a curved surface, the derivatives and the variations of the rotation field are carefully constructed with the help of the covariant derivative concept and the tangent space at identity, $so(3)$ (which is also known as Lie algebra), the necessary differential geometric and the group properties of the rotation tensor as an orthogonal transformation of the special orthogonal matrix group, $SO(3)$ which, because of its continuous topology and the algebraic structure, is also a Lie group.

4.7.1 Tangent Operator: Revisited

Let us recall that on a curved space, geometrical objects such as vectors and tensors must be defined locally instead of using a bi-local definition in the neighborhood, as in the flat Euclidean space. In other words, in an Euclidean ('flat') space, we can think of, say, a vector as joined by two neighboring points, but, in a curved space, a similar definition will yield a vector that may not even completely belong to the space. This necessitated the generalization from the definition of a vector as an arrow link between two neighboring points to a tangent operation to a curve (to be more specific, an affine curve for invariance or coordinate independence) at a point on the manifold. Thus, let the differentiable rotation tensor (or matrix) function $\mathbf{R}(\rho) :\rightarrow SO(3)$ generates a continuous set of points defining a curve on $SO(3)$, such that it passes through $\mathbf{R}(0) = \mathbf{R}$ at $\rho = 0$ where $\mathbf{R} \in SO(3)$. Then, $\mathbf{R},_\rho$, defined by the covariant derivative: $\mathbf{R},_\rho \equiv \left.\frac{d\mathbf{R}(\rho)}{d\rho}\right|_{\rho=0}$, is the tangent tensor to $SO(3)$ at $\mathbf{R}$. The set of all tangent tensors at $\mathbf{R}$, corresponding to various smooth curves passing though it, forms a vector space, denoted by $\mathbf{T_R}SO(3)$. Now, by taking the derivative of the orthogonality property, $\mathbf{R}^T\,\mathbf{R} = \mathbf{I}$, and rearranging, we have:

$$\boxed{\mathbf{R},_\rho = \mathrm{K}\,\mathbf{R} \Rightarrow \mathrm{K} \equiv \mathbf{R},_\rho\,\mathbf{R}^T = \mathbf{R},_\rho\,\mathbf{R}^{-1}} \tag{4.80}$$

where we have introduced the following definition.

Definition

- The ***curvature tensor***, K, the skew-symmetric matrix (because, with $\mathbf{R}\mathbf{R}^T = \mathbf{I}$, we have: $\mathrm{K} + \mathrm{K}^T = \mathbf{R},_\rho\,\mathbf{R}^T + \mathbf{R}\,\mathbf{R},_\rho^T = (\mathbf{R}\mathbf{R}^T),_\rho = \mathbf{0}$); K is thus the instantaneous and linearized skew-symmetric tensor at a rotation, $\mathbf{R} \in SO(3)$.

Because the tangent space is populated with the skew-symmetric matrices, the dimension of the tangent space is three (with only $n(n-1)/2$ elements in the matrix with $n = 3$ for the three-dimensional space). Note also that the tangent space is a linear vector space.

4.7.1.1 Tangent Space at Identity: Revisited

Recall that the tangent space at identity, that is, at $\mathbf{R} = \mathbf{I}$, is given the special name, ***Lie algebra*** (recall that a vector space with a vector multiplication, $[\bullet\,,\,\bullet]$, defined on it is an algebra) and denoted by $so(3)$. The reason for this special treatment of the neighborhood at identity, $\mathbf{I}$, is that the local structure of $SO(3)$, the continuous rotation group, can be constructed out of this infinitesimal tangent space, $so(3)$, by the exponential map given by equation (4.30); in other words, all the information of the group is contained in the algebra.

4.7.2 Adjoint Map and Left and Right Translations: Revisited

We have already seen that the elements of the Lie group, $SO(3)$, are related to the corresponding elements of its Lie algebra, $so(3)$, by the exponential map. Thus, we may define rotational variation through the exponential map in two different ways: $\bar{\mathbf{R}} = \lim_{\varepsilon\to 0}\frac{d}{d\varepsilon}(\mathbf{R}_\varepsilon) = \lim_{\varepsilon\to 0}\frac{d}{d\varepsilon}(\mathbf{R}\,e^{\varepsilon\,\bar{\Phi}}) = \mathbf{R}\,\bar{\Phi}$, or $\bar{\mathbf{R}} = \lim_{\varepsilon\to 0}\frac{d}{d\varepsilon}(\mathbf{R}_\varepsilon) = \lim_{\varepsilon\to 0}\frac{d}{d\varepsilon}(e^{\varepsilon\,\bar{\Phi}}\,\mathbf{R}) = \bar{\Phi}\,\mathbf{R}$ where $\bar{\Phi} \in so(3)$ is the instantaneous or infinitesimal variation and $\bar{\mathbf{R}} \in SO(3)$ is the final rotation due to this variation. Recall now that another map known as the ***adjoint map*** also relates the elements of $SO(3)$ with the corresponding elements of $so(3)$. If an arbitrary rotation matrix, $\mathbf{R} \in SO(3)$, and its corresponding skew-symmetric matrix, $\mathbf{S} \in so(3)$, are given, then we have the skew-symmetric matrix,$\mathbf{T}$, defined by the adjoint map: $\mathbf{T} \equiv \mathbf{R}\,\mathbf{S}\,\mathbf{R}^T$, also as an element of $so(3)$. We can easily see that because ($\mathbf{T} + \mathbf{T}^T = \mathbf{R}\,\mathbf{S}\,\mathbf{R}^T + \mathbf{R}\,\mathbf{S}^T\,\mathbf{R}^T = \mathbf{R}\,\mathbf{S}\,\mathbf{R}^T - \mathbf{R}\,\mathbf{S}\,\mathbf{R}^T = \mathbf{0}$) the symmetric part vanishes identically; it is a member of $so(3)$. Thus, we have: $\mathbf{T}\,\mathbf{R} = \mathbf{R}\,\mathbf{S}$. Now, similar to equation (4.80), we have two ways to ***superpose*** to represent the virtual rotation, $\bar{\mathbf{R}}$, by: $\bar{\mathbf{R}} = \bar{\Phi}\,\mathbf{R} = \mathbf{R}\,\bar{\Psi}$. In the first equality, known as the ***left translation***, the virtual rotation, $\bar{\Phi}$, is superposed on the current rotation; in the second equality, known as the ***right translation***, the current rotation is superposed on the virtual rotation, $\bar{\Psi}$.

4.7.3 Rotational Variation: $\bar{R}$ and $\bar{\Phi}$

We choose the following form of superposition in all our subsequent applications:

$$\boxed{\begin{array}{c} \bar{\mathbf{R}} = \bar{\Phi}\,\mathbf{R} \quad \Rightarrow \quad \bar{\Phi} = \bar{\mathbf{R}}\,\mathbf{R}^T \\ \text{where } \bar{\mathbf{R}}\,(\mathbf{R}, \delta\mathbf{R}) \equiv \left(\nabla_\rho\,\mathbf{R} \bullet \delta\mathbf{R}\right) \quad \text{and} \quad \bar{\Phi}\,(\mathbf{R}, \delta\Phi) \equiv (\nabla_\rho\,\mathbf{R} \bullet \delta\Phi) \end{array}} \tag{4.81}$$

4.7.4 Material Curvature Tensor: $\mathbf{R},_\rho$ and $\mathbf{K}$

Similarly, we can define the curvature tensor in two different ways as: $\mathbf{K} \equiv \mathbf{R},_\rho\,\mathbf{R}^T$ or $\mathbf{L} \equiv \mathbf{R}^T\,\mathbf{R},_\rho$, and thus the spatial derivatives of the variables can have two representations.

We choose the following form of superposition in all our subsequent applications:

$$\boxed{\mathbf{R},_\rho = \mathbf{K}\,\mathbf{R} \quad \Rightarrow \quad \mathbf{K} = \mathbf{R},_\rho\,\mathbf{R}^T} \tag{4.82}$$

4.7.5 Angular Velocity Tensor: $\dot{\mathbf{R}}$ and Ω

Likewise, we can define the rotational or angular velocity tensor in two different ways as: $\Omega_S \equiv \dot{\mathbf{R}}\,\mathbf{R}^T$ or $\Omega_M \equiv \mathbf{R}^T\,\dot{\mathbf{R}}$, and thus the temporal derivatives of the variables can have two representations. We choose the following form of superposition in some of our subsequent applications:

$$\dot{\mathbf{R}} = \Omega_S\,\mathbf{R} \Rightarrow \Omega_S \equiv \dot{\mathbf{R}}\,\mathbf{R}^T \tag{4.83}$$

4.7.6 Important Computational Identities

For future application in nonlinear analysis of beams, plates and shell structures, we introduce a collection of some useful results related to rotational transformation, that is parameterized by the Rodrigues vector, $\mathbf{s} = \frac{1}{\theta}\tan(\frac{\theta}{2})\,\Theta$, where Θ is the familiar rotation vector.

Fact D1: Given **k**, ϖ and $\bar{\boldsymbol{\phi}}$ as the axial vectors of the curvature tensor,**K**, the angular velocity tensor, Ω, and the variation tensor,$\bar{\Phi}$, respectively, we have:

$$\boxed{\begin{aligned} &\mathbf{K}(\rho) = \frac{2}{\Sigma}(\mathbf{S}\,\mathbf{S},_\rho - \mathbf{S},_\rho\,\mathbf{S} + \mathbf{S},_\rho) \quad \text{and} \quad \mathbf{k}(\rho) = \mathbf{L}_1(\mathbf{s})\,\mathbf{s},_\rho \\ &\bar{\Phi}(\mathbf{s},\bar{\mathbf{s}}) = \frac{2}{\Sigma}(\mathbf{S}\,\bar{\mathbf{S}} - \bar{\mathbf{S}}\,\mathbf{S} + \bar{\mathbf{S}}) \quad \text{and} \quad \bar{\boldsymbol{\phi}} = \mathbf{L}_1(\mathbf{s})\,\bar{\mathbf{s}} \\ &\Omega(t) = \frac{2}{\Sigma}(\mathbf{S}\,\dot{\mathbf{S}} - \dot{\mathbf{S}}\,\mathbf{S} + \dot{\mathbf{S}}), \quad \text{and} \quad \varpi(t) = \mathbf{L}_1(\mathbf{s})\,\dot{\mathbf{s}} \end{aligned}} \tag{4.84}$$

where

$$\boxed{\mathbf{L}_1(\mathbf{s}) \equiv \frac{2}{\Sigma}(\mathbf{I} + \mathbf{S})} \tag{4.85}$$

Proof D1. By taking direct one parameter, ρ, derivative of the rotation matrix in equation (4.43), applying the definition of **K** from equation (4.82), and through some algebra, we have the first of the equations. Then, in axial vector form, we have:

$$\mathbf{k} = \frac{2}{\Sigma}(\mathbf{s} \times \mathbf{s},_\rho + \mathbf{s},_\rho) = \frac{2}{\Sigma}(\mathbf{I} + \mathbf{S})\,\mathbf{s},_\rho = \mathbf{L}_1(\mathbf{s})\,\mathbf{s},_\rho$$

as required. Similarly, replacing $(),_\rho$ by $\dot{()}$ and $\overline{()}$ in the above expressions, respectively, we get: $\varpi(t) = \mathbf{L}_1(\mathbf{s})\,\dot{\mathbf{s}}$ and $\bar{\boldsymbol{\phi}} = \mathbf{L}_1(\mathbf{s})\,\bar{\mathbf{s}}$, which completes the proof.

Fact D2: $\mathbf{L}_1$ defined in equation (4.85) is given in terms of the rotation vector, $\boldsymbol{\theta}$, with the Rodrigues vector, $\mathbf{s} = \frac{1}{\theta}\tan(\frac{\theta}{2})\,\boldsymbol{\theta}$, as:

$$\boxed{\begin{aligned} &\mathbf{L}_1(\boldsymbol{\theta}) \equiv (1+\cos\theta)\left(\mathbf{I} + \frac{\sin\theta}{(1+\cos\theta)}\mathbf{N}\right) \\ &\text{where} \quad \mathbf{N} \equiv \frac{1}{\theta}\Theta, \quad \Theta = [\boldsymbol{\theta}], \quad \theta = \|\boldsymbol{\theta}\| \end{aligned}} \tag{4.86}$$

Proof D2. Following from equation (4.43): $\frac{2}{\Sigma} = \frac{2}{1+\mathbf{s}\cdot\mathbf{s}} = \frac{2}{1+\tan^2(\theta/2)} = 2\cos^2(\theta/2) = 1 + \cos\theta$, and noting that: $\tan(\theta/2) = \frac{2\sin(\theta/2)\cos(\theta/2)}{2\cos^2(\theta/2)} = \frac{\sin\theta}{1+\cos\theta}$, and getting: $\mathbf{S} = (\tan(\theta/2))\mathbf{N} = \frac{\sin\theta}{1+\cos\theta}\mathbf{N}$, completes the proof.

Fact D3:

$$\boxed{\begin{aligned} &\mathbf{s},_\rho = \mathbf{L}_2(\boldsymbol{\theta})\,\boldsymbol{\theta},_\rho, \quad \dot{\mathbf{s}} = \mathbf{L}_2(\boldsymbol{\theta})\,\dot{\boldsymbol{\theta}}, \quad \bar{\mathbf{s}} = \mathbf{L}_2(\boldsymbol{\theta})\,\bar{\boldsymbol{\theta}}, \\ &\text{where} \quad \mathbf{L}_2(\boldsymbol{\theta}) = \frac{1}{1+\cos\theta}\left(\mathbf{I} - \frac{\sin\theta - \theta}{\theta}\mathbf{N}^2\right) \end{aligned}} \tag{4.87}$$

Proof D3. Using the definition in equation (4.43) and taking direct one parameter, ρ, derivative of the Rodrigues vector, $\mathbf{s} = \frac{1}{\theta}\tan(\frac{\theta}{2})\,\boldsymbol{\theta}$, and noting that $\|\boldsymbol{\theta}\|,_\rho \equiv \|\boldsymbol{\theta}\|' = \frac{(\boldsymbol{\theta}\cdot\boldsymbol{\theta},_\rho)}{\|\boldsymbol{\theta}\|}$ and $\mathbf{n} = \frac{\boldsymbol{\theta}}{\|\boldsymbol{\theta}\|} = \frac{\boldsymbol{\theta}}{\theta}$, we have:

$$\mathbf{s}_{,\rho} = \frac{1}{2}\left\{\sec^2\left(\frac{\theta}{2}\right)\frac{(\boldsymbol{\theta}\bullet\boldsymbol{\theta}_{,\rho})}{\|\boldsymbol{\theta}\|^2}\right\}\boldsymbol{\theta} - \left\{\frac{(\boldsymbol{\theta}\bullet\boldsymbol{\theta}_{,\rho})}{\|\boldsymbol{\theta}\|^3}\tan\left(\frac{\theta}{2}\right)\right\}\boldsymbol{\theta} + \left\{\tan\left(\frac{\theta}{2}\right)\frac{\boldsymbol{\theta}_{,\rho}}{\|\boldsymbol{\theta}\|}\right\}$$
$$= \left(\frac{1}{\theta}\tan\left(\frac{\theta}{2}\right)\right)\left\{\boldsymbol{\theta}_{,\rho} - \left(1 - \frac{\theta}{\sin\theta}\right)(\mathbf{n}\bullet\boldsymbol{\theta}_{,\rho})\,\mathbf{n}\right\} = \left(\frac{1}{\theta}\tan\left(\frac{\theta}{2}\right)\right)$$
$$\left\{\mathbf{I} - \left(1 - \frac{\theta}{\sin\theta}\right)(\mathbf{n}\otimes\mathbf{n})\right\}\boldsymbol{\theta}_{,\rho}$$

where $\mathbf{n}$ is the unit axial vector of $\mathbf{N}$. Note that for column vectors: $(\mathbf{n}\otimes\mathbf{n}) = \mathbf{n}\,\mathbf{n}^T$. Now, using the identity for skew-symmetric matrices: $\mathbf{N}^2 = [\mathbf{n}\otimes\mathbf{n}] - \mathbf{I}$ and $\tan\frac{\theta}{2} = \frac{\sin\theta}{1+\cos\theta}$ completes the proof of the first part. Similarly, replacing $()_{,\rho}$ by $\dot{()}$ and $\overline{()}$ in the above expressions, respectively, we get: $\dot{\mathbf{s}} = \mathbf{L}_2(\boldsymbol{\theta})\,\dot{\boldsymbol{\theta}}$ and $\bar{\mathbf{s}} = \mathbf{L}_2(\boldsymbol{\theta})\,\bar{\boldsymbol{\theta}}$, which completes the proof.

Lemma L1: Given **Fact D2** and **Fact D3** as above, we have:

$$\boxed{\mathbf{k}(\rho) = \mathbf{W}(\boldsymbol{\theta})\,\boldsymbol{\theta}_{,\rho}, \quad \boldsymbol{\varpi}(t) = \mathbf{W}(\boldsymbol{\theta})\,\dot{\boldsymbol{\theta}}, \quad \bar{\boldsymbol{\phi}}(\boldsymbol{\theta},\bar{\boldsymbol{\theta}}) = \mathbf{W}(\boldsymbol{\theta})\,\bar{\boldsymbol{\theta}}} \tag{4.88}$$

where

$$\boxed{\begin{aligned} &\mathbf{W}(\Theta) \equiv \mathbf{L}_1\,\mathbf{L}_2 = \mathbf{I} + c_1\,\Theta + c_2\,\Theta^2 \\ &\mathbf{W}(\boldsymbol{\theta}) \equiv \mathbf{I} + c_1\,[\boldsymbol{\theta}] + c_2\left\{(\boldsymbol{\theta}\otimes\boldsymbol{\theta}) - \|\boldsymbol{\theta}\|^2\mathbf{I}\right\} \\ &\text{where} \quad c_1 \equiv \frac{1-\cos\theta}{\theta^2} \quad \text{and} \quad c_2 \equiv \frac{\theta - \sin\theta}{\theta^3} \end{aligned}} \tag{4.89}$$

Proof L1. We directly pre-multiply by $\mathbf{L}_1$, of equation (4.86), $\mathbf{L}_2$, of equation (4.87), and recall that $\mathbf{N}$, a skew-symmetric matrix with $\|\,\mathbf{n}\,\| = 1$, has the characteristic equation given by: $\lambda^3 + \lambda = 0$. Now, noting that every matrix satisfies its characteristic equations, substitution of the relation: $\mathbf{N}^3 + \mathbf{N} = \mathbf{0}$, and definition: $\mathbf{N} \equiv \frac{1}{\theta}\Theta$ from **Fact D2** completes the first expression. Now, with the use of the identity for the skew-symmetric tensor: $\Theta^2 = (\boldsymbol{\theta}\otimes\boldsymbol{\theta}) - \|\boldsymbol{\theta}\|^2\mathbf{I}$ where $\mathbf{I}$ is the three-dimensional identity tensor completes the proof.

Lemma LD1: Defining $\mathbf{k}$ and $\bar{\boldsymbol{\phi}}$ as the axial vectors of $\mathbf{K}$ and $\bar{\Phi}$, respectively, we have:

$$\boxed{\begin{aligned} \bar{\Phi}(s)_{,s} &= \bar{\mathbf{K}} + [\mathbf{K}\,\bar{\Phi} - \bar{\Phi}\,\mathbf{K}] \\ \bar{\boldsymbol{\phi}}(s)_{,s} &= \bar{\mathbf{k}} + \mathbf{k}\times\bar{\boldsymbol{\phi}} \end{aligned}} \tag{4.90}$$

Proof LD1: Using relations (4.81) and (4.82), we have:

$$\bar{\Phi}_{,s} = \bar{\mathbf{R}}_{,s}\;\mathbf{R}^T + (\bar{\Phi}\;\mathbf{R})\;\mathbf{R}^T{}_{,s}$$
$$= \bar{\mathbf{K}}\;\mathbf{R}\;\mathbf{R}^T + \mathbf{K}\;\bar{\Phi}\;\mathbf{R}\;\mathbf{R}^T + \bar{\Phi}\;\mathbf{R}\;\mathbf{R}^T\;\mathbf{K}^T$$

Now, noting the orthogonality property: $\mathbf{R}\,\mathbf{R}^T = \mathbf{I}$, and the skew-symmetry property: $\mathbf{K}^T = -\mathbf{K}$ completes the proof of the first part. Recognition of the Lie bracket property: for $\mathbf{A}, \mathbf{B}$ skew-symmetric tensor, and $\mathbf{a}, \mathbf{b}$, corresponding axial vectors, $[\mathbf{AB} - \mathbf{BA}] = [\mathbf{a} \times \mathbf{b}]$, for the second part, completes the proof.

Remark

- $\mathbf{W}(\theta)$ in equation (4.89), fundamentally describes the first derivative of a rotation matrix parameterized by Rodrigues vector; for our future application, we will also need the second derivative (or the Hessian) of a rotation tensor, which leads us to the following lemma.

Lemma LD2: Given $\mathbf{W}(\Theta)$ as in equation (4.89), and defining one parameter spatial derivative: $(\bullet)' \equiv \frac{\partial}{\partial \rho}(\bullet)$, variation: $\bar{\mathbf{a}}(\mathbf{x}, \bar{\mathbf{x}}) \equiv \left[\frac{d}{d\rho}\mathbf{a}(\mathbf{x} + \rho\bar{x})\right]_{\rho=0}$ and time derivative: $\dot{(\bullet)} \equiv \frac{\partial}{\partial t}(\bullet)$, we have:

$$
\boxed{
\begin{aligned}
&\mathbf{W}'(\theta) \equiv c_1\,\Theta' + c_2\,\{(\theta \otimes \theta') + (\theta' \otimes \theta)\} + \{(c_2 - c_1)\mathbf{I} + c_3\Theta + c_4(\theta \otimes \theta)\}(\theta \bullet \theta')\\
&\overline{\mathbf{W}}(\theta, \overline{\theta}) \equiv c_1\,\overline{\Theta} + c_2\,\{(\theta \otimes \overline{\theta}) + (\overline{\theta} \otimes \theta)\} + \{(c_2 - c_1)\mathbf{I} + c_3\overline{\Theta} + c_4(\theta \otimes \theta)\}(\theta \bullet \overline{\theta})\\
&\dot{\mathbf{W}}(\theta) \equiv c_1\,\dot{\Theta} + c_2\,\{(\theta \otimes \dot{\theta}) + (\dot{\theta} \otimes \theta)\} + \{(c_2 - c_1)\mathbf{I} + c_3\Theta + c_4(\theta \otimes \theta)\}(\theta \bullet \dot{\theta})\\
&c_1 \equiv \frac{1}{\theta^2}(1 - \cos\theta) \qquad c_2 \equiv \frac{1}{\theta^3}(\theta - \sin\theta)\\
&c_3 \equiv \frac{1}{\theta^2}(1 - 2c_1 - \theta^2 c_2) \qquad c_4 \equiv \frac{1}{\theta^2}(c_1 - 3c_2)
\end{aligned}
}
\tag{4.91}
$$

Proof LD2. First noting that $\overrightarrow{\theta^2} \equiv \overline{\|\theta\|^2} = \overline{(\theta \bullet \theta)^2} \Rightarrow \bar{\theta} = \frac{1}{\theta}(\theta \bullet \bar{\theta})$ and taking variation of c_1 and c_2, we have: $\bar{c}_1 = c_3(\theta \bullet \bar{\theta})$ and $\bar{c}_2 = c_4(\theta \bullet \bar{\theta})$ and taking variation of $\mathbf{W}$ in equation (4.89), we get: $\bar{\mathbf{W}}(\Theta,\ \bar{\theta}) = -c_1\bar{\Theta} + c_2(\bar{\Theta}\Theta + \Theta\bar{\Theta}) + (c_3\Theta + c_4\Theta^2)(\theta \bullet \bar{\theta})$. Now, recalling the identities for skew-symmetric matrices: $\bar{\Theta}\Theta + \Theta\bar{\Theta} = [\theta \otimes \bar{\theta}] + [\bar{\theta} \otimes \theta] - 2(\theta\bar{\theta})\mathbf{I}$, and replacing all superscripted barred variables, $\overline{(\bullet)}$, by superscripted dot, $\dot{(\bullet)}$, and primed, $(\bullet)'$, variables completes the proof.

Finally, we look at the central lemma for the nonlinear finite element analysis by Newton-type method.

Central Duality Lemma: Given $\mathbf{W}$, c_1 and c_2 as in equation (4.89), we have:

$$
\boxed{
\begin{aligned}
\bar{\mathbf{W}}(\theta,\ \bar{\theta})\ \theta_{,\rho} &= \mathbf{X}(\theta,\ \theta_{,\rho})\ \bar{\theta}\\
\Delta\mathbf{W}(\theta,\ \Delta\theta)\ \theta_{,\rho} &= \mathbf{X}(\theta,\ \theta_{,\rho})\ \Delta\theta
\end{aligned}
}
\tag{4.92}
$$

where

$$
\boxed{
\begin{aligned}
&\mathbf{X} \equiv -c_1\Theta,_\rho + c_2\{(\theta \otimes \theta,_\rho) - (\theta,_\rho \otimes \theta) + (\theta \bullet \theta,_\rho)\mathbf{I}\} + \mathbf{V}(\theta,_\rho \otimes \theta) \\
&\text{where} \quad \mathbf{V} \equiv -c_2\mathbf{I} + c_3\Theta + c_4\Theta^2 \\
&c_3 \equiv \frac{1}{\theta^2}\{1 - 2c_1 - \theta^2 c_2\} \\
&c_4 \equiv \frac{1}{\theta^2}(c_1 - 3c_2)
\end{aligned}
}
\tag{4.93}
$$

where $\bar{\theta}$ and $\Delta\theta$ are ***virtual (variational) rotation*** and ***incremental rotation*** vectors, respectively.

Proof of Central Lemma: Noting the expression for $\overline{\mathbf{W}}(\theta, \overline{\theta})$ from equation (4.91) and recalling the identities for skew-symmetric matrices: $\Theta\bar{\Theta} + \Theta\bar{\Theta} = [\theta \otimes \bar{\theta}] + [\bar{\theta} \otimes \theta] - 2(\theta \bullet \bar{\theta})\mathbf{I}$, and $(\theta \otimes \bar{\theta})\theta,_\rho = (\theta \otimes \theta,_\rho)\bar{\theta}$, we have:

$$
\begin{aligned}
\bar{\mathbf{W}}(\theta, \bar{\theta})\, \theta,_\rho &= (c_1\,\overline{\Theta} + c_2\,\{(\theta \otimes \overline{\theta}) + (\overline{\theta} \otimes \theta)\} + \{(c_2 - c_1)\mathbf{I} + c_3\overline{\Theta} + c_4(\theta \otimes \theta)\}(\theta \bullet \overline{\theta}))\, \theta,_\rho \\
&= (-c_1\,\Theta,_\rho + c_2\{(\theta \otimes \theta,_\rho) - (\theta,_\rho \otimes \theta) + (\theta \bullet \theta,_\rho)\mathbf{I}\} \\
&\quad + \{-c_2\mathbf{I} + c_3\Theta + c_4\Theta^2\}(\theta,_\rho \otimes \theta))\, \bar{\theta} \\
&= \mathbf{X}(\theta,\, \theta,_\rho)\, \bar{\theta}
\end{aligned}
$$

proves the first part of the equation (4.92). Now, replacing the $\overline{()}$ terms by the $\Delta()$ completes the proof.

4.7.7 Where We Would Like to Go

In the foregoing, we have introduced various parameterizations of rotation tensor and derived all the necessary relationships to be used in our beam, plate and shell analyses. To get familiarized with the overall computational strategy for these structures, we need to explore:

- the road map: nonlinear beams (Chapter 10)
- the road map: nonlinear plates and shells (Chapter 11)

Part II

Essential Mesh Generation

5

Curves: Theory and Computation

5.1 Introduction

5.1.1 What We Should Already Know

Vectors, derivatives, vector dot product, vector cross product, vector triple product, basis functions, interpolation, approximation, conics, continuous functions, polynomials, forward difference operators, linear vector space, power basis.

5.1.2 Why We Are Here

The main goal here is to present a topical discussion of the basic theory and computation of curves, and define relevant curve properties necessary for both ***mesh generation*** and ***finite element analysis*** phases of curved beams. However, as a start, let us remind ourselves that all of our theoretical discussions are primarily aimed at achieving robust computational methods and to guide us with useful practical tips. More specifically, we define and discuss the properties of parametric curves as opposed to their implicit description – to investigate computational Bezier–Bernstein–de Casteljau and Schoenberg–de Boor–Cox B-spline parametric curves. In the interactive ***mesh generation*** phase, we are interested in developing the final model of structures. Starting from a few key points of information, we would like to create the subassemblies of the model, and with the appropriate continuity conditions among the subassemblies eventually we will blend them into one integral mesh. In the ***finite element analysis*** phase, the mesh is used for structural integrity under various load conditions by determining key entities or field quantities such as displacements, stresses, and so on. Finally, for cost effectiveness in numerical analysis in the form of ***interpolations*** in the mesh generation phase and ***approximations*** in the finite element phases, we strive for elements that are as large as possible without compromising accuracy. Thus, in order that the interpolation and approximation algorithms can be easily understood, here we continue with the fundamental geometric entity: ***curves***. There are two basic ways to represent a curve – ***implicit form*** or ***parametric form***. For various reasons, the implicit description appears less promising than the parametric form for our purpose. Within the parametric domain, for several reasons, we prefer to represent curves in the Bezier–Bernstein–de Casteljau and Schoenberg-de Boor B-spline forms to *various other possible forms* such as piecewise Lagrangian and Hermite polynomials. The common shapes in ***conics*** such as circle, parabola and ellipse, need special attention under our choice of the parametric form. It turns out that in order to describe these shapes exactly, we must resort to ***rational Bezier*** and ***rational B-spline curves***. Implementationally,

Computation of Nonlinear Structures: Extremely Large Elements for Frames, Plates and Shells, First Edition. Debabrata Ray.

Bezier curves and B-splines by composite Bezier construction act as the cornerstone of our computer-aided geometric modeling algorithms. We make these also the building blocks of our proposed finite element analysis:

- We begin with the ***differential geometry of real curves***.
- For an introduction to the fundamental concepts and properties of Bezier–Bernstein–de Casteljau, we introduce ***non-rational Bezier–Bernstein–de Casteljau curves***.
- For parametric descriptions of conics and similar curves, we investigate rational Bezier and ***rational B-spine curves and NURBs***.
- For an introduction to the fundamental concepts and properties of de Boor–B-Spline forms of representation of curves, we conclude with ***deBoor-Cox B-spline curves***.
- For a unified implementation on the computer of Bezier curves and B-spine curves in Bezier form, we conclude with ***computer implementation of Bezier and B-spline curves***.

5.1.3 How We Get It

We introduce some fundamental concepts in differential geometry of parametric curves, such as the Frenet frame and Frenet–Serret derivatives with the definitions for curvature and twist; then we specialize in Bernstein–Bezier curves and de Boor–Cox B-spline curves, rational and non-rational, for computational purposes.

5.1.4 What We Need to Recall

In what follows, we touch upon some fundamental ideas and concepts that will be eventually utilized to introduce the concepts of computational curves. As indicated above, we need to recall the ***linear vector space,*** the ***real coordinate space*** and ***Euclidean 'flat' space*** for the introduction of an ***affine space***.

5.2 Affine Transformation and Ratios

5.2.1 Affine Transformation

Let us recall that the two-point (more specifically, the difference of two points) definition of vectors is easy to grasp geometrically, but it does not help generalization to higher-order invariant entities such as tensors. Thus, for an alternative definition, we resort to appropriate transformations and identify the transformation properties of vectors for their definition under such transformations. For this, we consider the ***affine transformation***s in a Euclidean geometry of the type:

$$\boxed{\mathbf{X}' = \mathbf{A}\,\mathbf{X} + \mathbf{b} \quad \text{and} \quad X_i' = \mathbf{A}_{ij}\,X_j + b_i} \tag{5.1}$$

where $\mathbf{A}$ is interpreted as a constant orthogonal transformation such that $\mathbf{A}\,\mathbf{A}^T = \mathbf{I}$ and $\det\ \mathbf{A} = 1$, and $\mathbf{b}$ is a translation; $\mathbf{X}$ and $\mathbf{X}'$ are the points before and after the transformation, respectively, but an alternative interpretation could be that $\mathbf{X}$ and $\mathbf{X}'$ are the coordinates of the same point in two different Cartesian coordinate systems, before and after the above transformation. Clearly, equation (5.1) is not a linear map because of the presence of the translation term, $\mathbf{b}$. However, since a vector remains invariant under translation, the linear transformation defining vectors may be written as:

$$\boxed{\mathbf{X}' = \mathbf{A}\,\mathbf{X}} \tag{5.2}$$

Component-wise, equation (5.2) can be expressed, with the Einstein summation convention, as:

$$\boxed{X_i' = \mathbf{A}_{ij} X_j} \tag{5.3}$$

Incidentally, for the component forms in equations (5.1) and (5.3), we have somewhat resorted to concepts that appeared earlier in the book: $\mathbf{A}_{ij}$'s are the matrix components of tensor $\mathbf{A}$ expressed in Cartesian coordinate base tensors in dyadic notation as: $\mathbf{A} = \mathbf{A}_{ij}\,\mathbf{i}_i \otimes \mathbf{i}_j$; in other words, the matrices to a tensor are similar to what the vector components are to a vector.

Examples of affine transformations abound in computer-aided geometric design (CAGD) (referring to equation (5.1)):

- ***Translation:*** $\mathbf{A} = \mathbf{0}$ and $\mathbf{b}$ non-zero.
- ***Identity***: $\mathbf{A} = \mathbf{I}$ and $\mathbf{b} = \mathbf{0}$.
- ***Rotation***: $\mathbf{A} = \mathbf{R}$ and $\mathbf{b} = \mathbf{0}$ with $\mathbf{R}$ as a rotation matrix: $\mathbf{R}\mathbf{R}^T = \mathbf{I}$ and $\det \mathbf{R} = 1$.
- ***Scaling***: $\mathbf{b} = \mathbf{0}$ and $\mathbf{A}$ is a diagonal matrix with entries defining scaling of a vector.
- ***Shear***: $\mathbf{b} = \mathbf{0}$ and $\mathbf{A}$ is a matrix with diagonal entries all unity and the off-diagonal terms defining shearing of a vector.

Now, we are in a position to redefine a vector based on transformation property:

- ***an n-dimensional vector*** is an ordered system of n numbers called its components that transfer according to equation (5.3) when subjected to an affine transformation given by equation (5.1).

For example, if we are given two points, $\mathbf{p}$ and $\mathbf{q}$, and transform them to points, $\mathbf{p}'$ and $\mathbf{q}'$, respectively, such that: $p_i' = \mathbf{A}_{ij}\,p_j + b_i$ and $q_i' = \mathbf{A}_{ij}\,q_j + b_i$, and take the difference, we get: $p_i' - q_i' = \mathbf{A}_{ij}\,(p_j - q_j)$, that is, since the component transformations satisfy equation (5.3), $\mathbf{p} - \mathbf{q}$ must be a vector by the given definition; in other words, the difference of two points that previously defined a vector is identical to the current definition of a vector based on the transformation property. As we have seen, the vector definition by transformation can be generalized to higher-order invariant geometric objects such as tensors.

It is important to note that a barycentric combination of points is preserved, that is, remains invariant, under the affine transformation given by equation (5.1). This is very important since geometric shapes, usually described by the barycentric combination of points in CAGD, remain invariant under affine coordinate transformation. To wit, let us consider a point, $\mathbf{a}$, given as a barycentric combination as $\mathbf{a} = c_0\mathbf{a}_0 + c_1\mathbf{a}_1 + \cdots + c_n\mathbf{a}_n = c_i\mathbf{a}_i$, $\sum_{i=0}^{n} c_i = 1$. Now, applying the affine transformation, we will get a point, $\bar{\mathbf{a}}$, such that: $\bar{\mathbf{a}} = c_0\overline{\mathbf{a}_0} + c_1\overline{\mathbf{a}_1} + \cdots + c_n\overline{\mathbf{a}_n} = c_i\overline{\mathbf{a}_i}$, $\sum_{i=0}^{n} c_i = 1$, because:

$$\begin{aligned}\bar{\mathbf{a}} &= \mathbf{A}\mathbf{a} + \mathbf{b} = \mathbf{A}\left(\sum c_i\mathbf{a}_i\right) + \mathbf{b} = \mathbf{A}\left(\sum c_i\mathbf{a}_i\right) + \sum c_i\mathbf{b} \quad \text{since} \quad \sum c_i = 1\\ &= \sum c_i\mathbf{A}\mathbf{a}_i + \sum c_i\mathbf{b} = \sum c_i(\mathbf{A}\mathbf{a}_i + \mathbf{b}) = \sum c_i\overline{\mathbf{a}_i}\end{aligned}$$

- ***Position*** or ***radius vector***: Finally, for analytical and computational purposes, if we fix a coordinate system by choosing a point with all components identically zero, then the vector formed by the difference of any point and this point is known as the ***position*** or ***radius vector***; the components of the position vector are identical to those of the other point. Thus, given a coordinate system, any point of an affine space is given by a position vector. We note too that in the absence of translation, the position vectors transform according to transformation of the type given by equation (5.3) and not by equation (5.1).

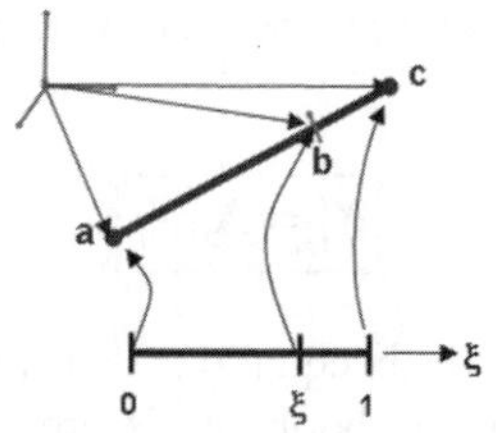

Figure 5.1 Linear interpolation and ratio.

5.2.2 *Linear Interpolation and Ratio Invariance*

One of the simplest barycentric combinations is exemplified by the linear interpolation, as shown in Figure 5.1; given any two points **a** and **b** as end points of a straight line segment, any point $\mathbf{c}(\xi)$ may be generated as a ***linear interpolation*** of the end points, and parametrically expressed as: $\mathbf{c}(\xi) = (1-\xi)\mathbf{a} + \xi\mathbf{b}$ with parameter $\xi \in [0, 1]$. We can also see this as a map of a point ξ on the real line segment belonging to one-dimensional real space with end points 0 and 1: $\xi = (1-\xi)0 + \xi 1$, to the three-dimensional affine space; clearly, the barycentric combination is preserved under this map. Thus, linear interpolation is an affine map; in other words, linear interpolation is affinely invariant.

Now, for any three collinear points, **a**, **b**, **c**, the barycentric coefficients, $\alpha_1 \stackrel{\Delta}{=} (1-\xi)$ and $\alpha_2 \stackrel{\Delta}{=} \xi$ such that $\mathbf{b} = \alpha_1\,\mathbf{a} + \alpha_2\,\mathbf{c}, \quad \alpha_1 + \alpha_2 = 1$, can also be given as:

$$\alpha_1 \equiv \frac{\text{area}\,(\mathbf{b},\mathbf{c})}{\text{area}\,(\mathbf{a},\mathbf{c})} = \frac{\|\mathbf{c}-\mathbf{b}\|}{\|\mathbf{c}-\mathbf{a}\|} \quad \text{and} \quad \alpha_2 \equiv \frac{\text{area}\,(\mathbf{a},\mathbf{b})}{\text{area}\,(\mathbf{a},\mathbf{c})} = \frac{\|\mathbf{b}-\mathbf{a}\|}{\|\mathbf{c}-\mathbf{a}\|} \tag{5.4}$$

where "area" is included to accommodate definitions for planar geometry and stands for absolute length in one-dimensional objects. Then, what is known as ***ratio*** is defined as:

$$\boxed{\text{ratio}(\mathbf{a},\mathbf{b},\mathbf{c}) \equiv \frac{\text{area}\,(\mathbf{a},\mathbf{b})}{\text{area}\,(\mathbf{b},\mathbf{c})} = \frac{\alpha_2}{\alpha_1} = \frac{\xi}{1-\xi}} \tag{5.5}$$

Because, as shown above, the barycentric coefficients are invariant under affine transformation, the equation (5.5) states that ratios are preserved under barycentric combination. As will become clear later, the concept of ratio forms the backbone of the de Casteljau algorithm of non-rational curve generation by linear interpolation.

5.2.3 *Four-dimensional Homogeneous Coordinate System*

As already indicated, an affine map is not linear because of the translational vector; in other words, given a function or map, $f(\mathbf{x})$, and scalars, a, b, c, it does not satisfy homogeneity, that is, $f(c\mathbf{x}) = c\,f(\mathbf{x})$, or additivity, that is, $f(a\mathbf{x} + b\mathbf{y}) = a\,f(\mathbf{x}) + b\,f(\mathbf{y})$. To wit, let the affine map be given by equation (5.1) with $\mathbf{X}' = \mathbf{A}\,\mathbf{X} + \mathbf{b}$ and $\mathbf{Y}' = \mathbf{A}\,\mathbf{Y} + \mathbf{b}$; clearly, additivity is violated: $\mathbf{A}\,(\mathbf{X}+\mathbf{Y}) + \mathbf{b} \neq \mathbf{X}' + \mathbf{Y}'\ (= \mathbf{A}\,\mathbf{X} + \mathbf{A}\,\mathbf{Y} + 2\mathbf{b})$, and so is homogeneity. In view of this drawback, computer graphics that depend on matrix manipulations (i.e. linear operations) cannot be efficiently performed. A clever and ingenious device to circumvent this problem is to use a dimensionally augmented coordinate system known as a ***4D homogeneous coordinate system***;

thus, the affine map given by equation (5.1) can be rewritten in homogeneous coordinate system, $(\mathbf{X}, w)$ as:

$$\begin{Bmatrix} X_1' \\ X_2' \\ X_3' \\ w \end{Bmatrix} = \begin{bmatrix} A_{11} & A_{12} & A_{13} & b_1 \\ A_{21} & A_{22} & A_{23} & b_2 \\ A_{31} & A_{32} & A_{33} & b_3 \\ 0 & 0 & 0 & 1 \end{bmatrix} \begin{Bmatrix} X_1 \\ X_2 \\ X_3 \\ w \end{Bmatrix} \tag{5.6}$$

As it turns out, for both non-rational and rational curves that we are about to delve into, we will utilize this 4D homogeneous coordinate system in computer coding as a most natural artifice or apparatus.

5.3 Real Parametric Curves: Differential Geometry

Let us consider three-dimensional Euclidean space (more precisely, ***affine space***), and real curves. If an origin is chosen, then every point in space can be identified with a position vector, $\mathbf{C} = \left(X \equiv X_1 \; Y \equiv X_2 \; Z \equiv X_3 \right)^T$.

5.3.1 Real Curve

A real curve is defined by a real vector function, $\mathbf{C}(\xi)$, where $\xi \in [a, b]$, that is, $a \leq \xi \leq b$ is a real parameter, as shown in Figure 5.2. In component curve form:

$$\mathbf{C}(\xi) = \begin{Bmatrix} X_1(\xi) \\ X_2(\xi) \\ X_3(\xi) \end{Bmatrix}, \quad \xi \in [a, b] \tag{5.7}$$

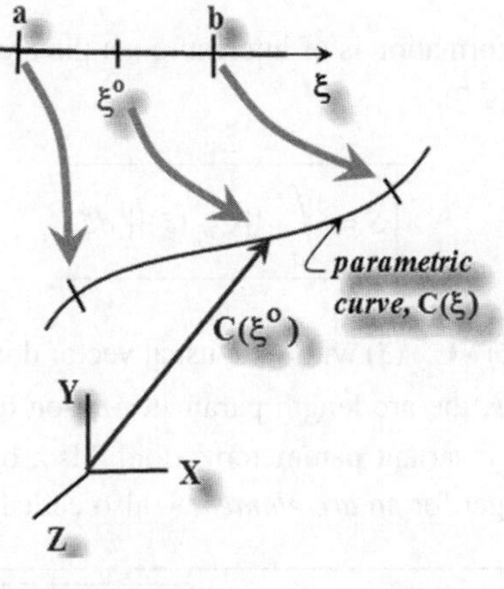

Figure 5.2 Real parametric curve.

5.3.2 *Derivative of a Curve*

We will define the derivative of the curve at any parameter, ξ, by:

$$\mathbf{C},_\xi(\xi) \equiv \frac{d\mathbf{C}(\xi)}{d\xi} = \left(\frac{dX_1(\xi)}{d\xi} \; \frac{dX_2(\xi)}{d\xi} \; \frac{dX_3(\xi)}{d\xi} \right)^T \tag{5.8}$$

5.3.3 *Regular Curve*

A regular curve will be defined as the curve, $\mathbf{C}(\xi)$, where the derivatives of the component curves do not all simultaneously vanish at any parameter, $\xi \in [a, b]$, that is, $\mathbf{C},_\xi(\xi) \neq \mathbf{0}$ vector. A point where a curve is regular is called an ordinary point, otherwise a singular point. As a counter example to regular curves, we may consider a curve: $\mathbf{C}(\xi) = (1 - f(\xi))\mathbf{C}(\xi_0) + f(\xi)\,\mathbf{C}(\xi_1), \quad \xi \in [\xi_0, \xi_1]$ such that the function $f(\xi)$ has stationary points, that is, $\frac{df}{d\xi} = 0$ at some $\xi \in [\xi_0, \xi_1]$; then, $\frac{d\mathbf{C}}{d\xi} = (\mathbf{C}(\xi_1) - \mathbf{C}(\xi_0))\frac{df}{d\xi} = 0$. As a general rule, if a curve has a cusp, then at that point, $\frac{d\mathbf{C}}{d\xi}$ is either discontinuous or zero.

5.3.4 *Allowable Parametric Representation*

A representation where a curve is regular and continuously differentiable $r(\geq 1)$ many times, depending on a problem, is called an allowable parametric representation. The curve is said to be of class C^r.

Thus, from now on, we will always assume our curves to be regular and of class C^r, $r(\geq 1)$. In other words, we will assume that our curves will always be as smooth as necessary.

5.3.5 *Allowable Parameter Transformations: Regular Reparameterization*

We define an allowable parameter transformation as of type: $\tau = \tau(\xi)$, where τ is a differentiable function of ξ, and $\frac{d\tau}{d\xi} \neq 0$ for the entire interval that will not change the shape of the curve. Clearly, then, $\tau(\xi)$ is an invertible function and we can find: $\xi = \xi(\tau)$.

5.3.6 *Invariant Arc Length Parameterization*

A particular parameter transformation is of interest to us: the arc length parameterization where S, the arc length is related to ξ by:

$$\boxed{S = \int_a^\xi ||\mathbf{C},_\xi(\xi)||\, d\xi} \tag{5.9}$$

where, $||\mathbf{C},_\xi(\xi)|| \equiv \sqrt{\mathbf{C},_\xi(\xi) \bullet \mathbf{C},_\xi(\xi)}$ with $\bullet$ the usual vector dot product. Because, $\mathbf{C},_\xi(\xi)\,d\xi \equiv \frac{d\mathbf{C}}{d\xi}d\xi = \frac{d\mathbf{C}}{d\tau}\frac{d\tau}{d\xi}d\xi = \mathbf{C},_\tau(\tau)\,d\tau$, the arc length parameterization does not depend on the original parameter, and hence, is an invariant parameterization. Also, by considering the above representation as: $S = \int_0^S dS$, we get for an ***arc element*** – also called a linear element – of a curve:

$$\boxed{dS = ||\mathbf{C},_\xi(\xi)||\, d\xi \equiv \sqrt{\mathbf{C},_\xi(\xi) \bullet \mathbf{C},_\xi(\xi)}\, d\xi} \tag{5.10}$$

Remarks

- Equation (5.10) gives us, component-wise, the usual relation:

$$dS = \sqrt{dX^2 + dY^2 + dZ^2}\, d\xi \tag{5.11}$$

- In order to maintain that $\mathbf{C}_{,\xi}(\xi) \neq 0$ at all ξ, the arc length parameter, S, must monotonously increase, that is, $\xi_1 > \xi_0 \Rightarrow S(\xi_1) > S(\xi_0)$.
- As a counter example to regular curves, we may consider a curve: $\mathbf{C}(\xi) = (1 - f(\xi))\mathbf{C}(\xi_0) + f(\xi)\,\mathbf{C}(\xi_1),\quad \xi \in [\xi_0, \xi_1]$ such that the function $f(\xi)$ has stationary points, that is, $\frac{df}{d\xi} = 0$ at some $\xi \in [\xi_0, \xi_1]$; then, $\frac{d\mathbf{C}}{d\xi} = (\mathbf{C}(\xi_1) - \mathbf{C}(\xi_0)) = 0$.

5.3.7 Curve Tangent Vector

As shown in Figure 5.3, let us consider two points, $\mathbf{C}(\xi)$ and $\mathbf{C}(\xi + \Delta\xi)$, on a curve at parameters, ξ and $\xi + \Delta\xi$. Now, if the curve is regular at ξ and as $\Delta\xi \to 0$ in the limit, we get, by definition: $\mathbf{C}_{,\xi}(\xi) = \lim_{\Delta\xi \to 0} \frac{\mathbf{C}(\xi+\Delta\xi) - \mathbf{C}(\xi)}{\Delta\xi} = \frac{d\mathbf{C}}{d\xi}$ is the tangent vector to the curve at ξ.

5.3.8 Unit Tangent Vector, **T**

Now, as shown in Figure 5.4, considering the arc length parameter, S, and relation (5.10), we get:

$$C'(S) \equiv \frac{d\mathbf{C}}{dS} = \frac{d\mathbf{C}}{d\xi}\frac{d\xi}{dS} = \frac{\frac{d\mathbf{C}}{d\xi}}{\frac{dS}{d\xi}} = \frac{\mathbf{C}_{,\xi}}{||\mathbf{C}_{,\xi}||} = \text{unit vector along } \mathbf{C}_{,\xi} \tag{5.12}$$

In other words, the unit tangent vector at a point on a curve is given by the derivative of the position vector at the point with respect to the arc length parameter. The unit tangent vector will be denoted by **T**.

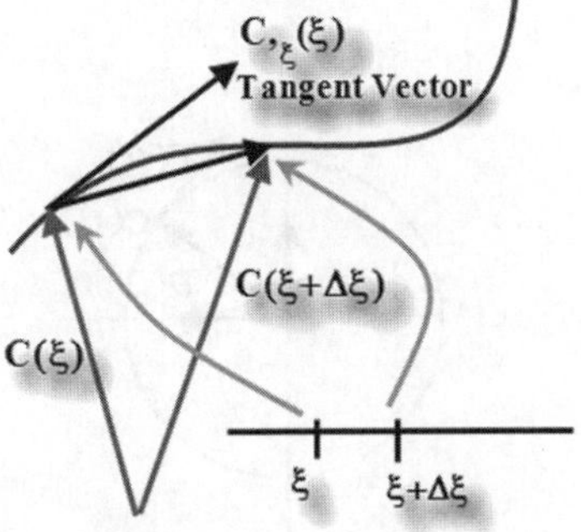

Figure 5.3 Tangent vector.

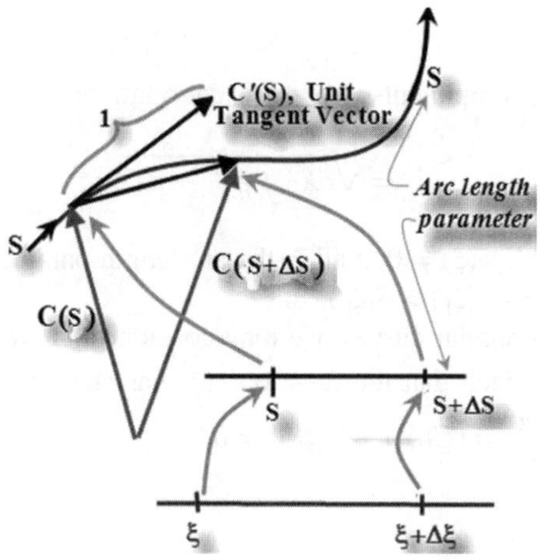

Figure 5.4 Unit tangent vector.

Remarks

- From now on, notationally, we will use: $(\bullet)' \equiv \frac{d(\bullet)}{dS}$, that is, the derivative at a point on a curve with respect to arc length parameter.

Example: Let us consider the circle, shown in Figure 5.5, with points on it that are given by: $\mathbf{C}(\theta) = a\ \cos\theta\ \mathbf{e}_1 + a\ \sin\theta\ \mathbf{e}_2$ where $\mathbf{e}_1$ and $\mathbf{e}_2$ are two-dimensional unit Cartesian base vectors; then, the unit tangent vector is given as:

$$\mathrm{T}(\theta) \stackrel{\Delta}{=} \frac{C_{,\theta}}{\left\| C_{,\theta} \right\|} = \frac{-a\sin\theta\ \mathbf{e}_1 + a\cos\theta\mathbf{e}_2}{\sqrt{a^2\sin^2\theta + a^2\cos^2\theta}} = -\sin\theta\mathbf{e}_1 + \cos\theta\mathbf{e}_2$$

5.3.9 Normal Plane

As shown in Figure 5.6, a plane that passes through a curve at a point, say, ξ^0, and whose normal is the tangent to the curve at the point, ξ^0, is called a ***normal plane***. Clearly, then, if $\mathbf{P}$ is a point on the plane, the normal plane is given by equation: $(\mathbf{P} - \mathbf{C}(\xi^0)) \bullet \mathbf{C}_{,\xi}\,(\xi^0) = 0$.

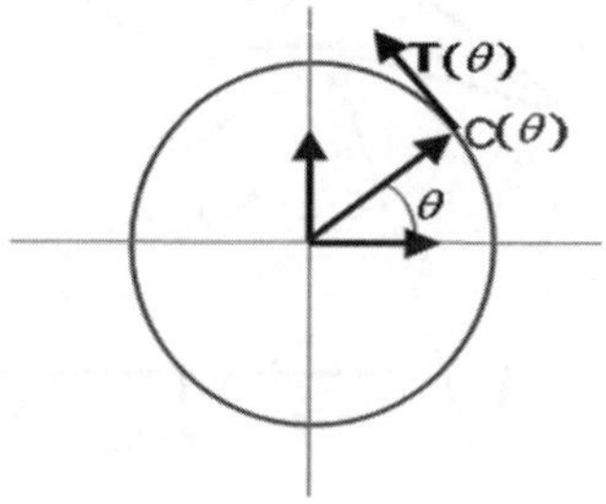

Figure 5.5 Example of a tangent vector of circle.

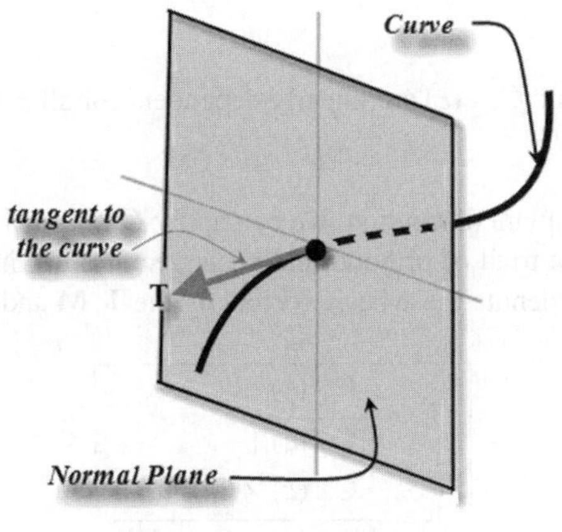

Figure 5.6 Normal plane.

5.3.10 *The Frenet Frame: Principal Normal Vector,* **M**, *and Bi-normal Vector,* **B**

The tangent at each point to a curve (we assume C^r, $r \geq 2$) continuously changes as we move from point to point along the curve. It will be interesting to see how the tangents change along the curve. In fact, as we will see shortly, this will enable us to define a very important and suitable local Cartesian frame at a point on the curve.

Based on Figure 5.7, by applying Taylor series approximation in the neighborhood of a point, $\mathbf{C}(\xi)$, on the curve, we have:

$$\mathbf{C}(\xi + \Delta\xi) = \mathbf{C}(\xi) + (\Delta\xi)\,\mathbf{C},_{\xi}(\xi) + \frac{1}{2}(\Delta\xi)^2\,\mathbf{C},_{\xi\xi}(\xi) + \frac{1}{6}(\Delta\xi)^3\,\mathbf{C},_{\xi\xi\xi}(\xi) + \textit{higher order terms} \tag{5.13}$$

If we assume that $\mathbf{C},_{\xi}(\xi)$, $\mathbf{C},_{\xi\xi}(\xi)$ and $\mathbf{C},_{\xi\xi\xi}(\xi)$ are linearly independent then they can form an affine coordinate system.

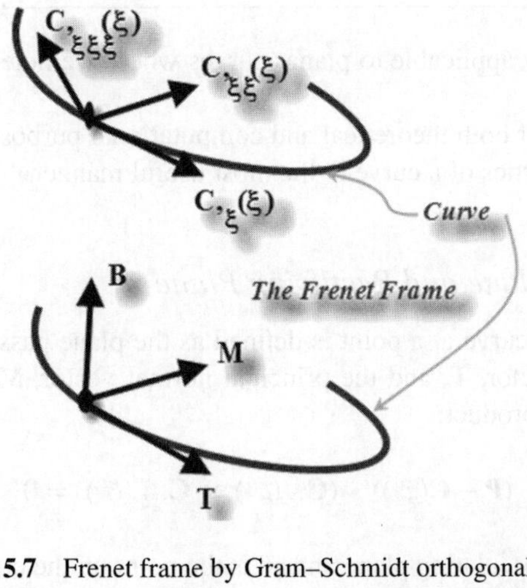

Figure 5.7 Frenet frame by Gram–Schmidt orthogonalization.

Remarks

- We note that if $\mathbf{C}_{,\xi}(\xi)$ and $\mathbf{C}_{,\xi\xi}(\xi)$ are linearly dependent for all ξ, then the curve is a straight line.

Now, going back to our main discussion, we can apply ***Gram–Schmidt orthogonalization*** to these basis vectors to get a triad of orthonormal basis vectors forming a local Cartesian frame with origin at $\mathbf{C}(\xi)$. Let us denote these basis vectors by the **T**, **M** and **B** defined by:

$$\boxed{\begin{aligned} \mathbf{T} &= \frac{\mathbf{C}_{,\xi}(\xi)}{||\mathbf{C}_{,\xi}(\xi)||} \\ \mathbf{B} &= \frac{\mathbf{C}_{,\xi}(\xi) \times \mathbf{C}_{,\xi\xi}(\xi)}{||\mathbf{C}_{,\xi}(\xi) \times \mathbf{C}_{,\xi\xi}(\xi)||} \\ \mathbf{M} &= \mathbf{B} \times \mathbf{T} \end{aligned}} \tag{5.14}$$

Definition

- The local Cartesian frame defined by **T**, **M** and **B** is known as the ***Frenet frame***.
- **T** is easily recognized as the unit ***tangent vector*** defined earlier.
- **M** is known as the principal ***normal vector***.
- **B** is known as the ***bi-normal vector***.

Remarks

- From the orthogonalization process, we can write the direct expression for normal vector in terms of the curve properties as follows:

$$\boxed{\mathbf{M} = \frac{(\mathbf{C}_{,\xi} \bullet \mathbf{C}_{,\xi})\mathbf{C}_{,\xi\xi} - (\mathbf{C}_{,\xi\xi} \bullet \mathbf{C}_{,\xi})\mathbf{C}_{,\xi}}{||(\mathbf{C}_{,\xi} \bullet \mathbf{C}_{,\xi})\mathbf{C}_{,\xi\xi} - (\mathbf{C}_{,\xi\xi} \bullet \mathbf{C}_{,\xi})\mathbf{C}_{,\xi}||}} \tag{5.15}$$

This equation is equally applicable to planar curves where we note that **B** coincides with the normal to the plane.

As we shall see later, for both theoretical and computational purposes, the Frenet frame helps to bring out various properties of a curve in the most useful manner.

5.3.11 Osculating Plane and Rectifying Plane

An ***osculating plane*** of a curve at a point is defined as the plane passing through the point and containing the tangent vector, **T**, and the principal normal vector, **M**. The osculating plane is given by the vector triple product:

$$(\mathbf{P} - \mathbf{C}(\xi^o)) \bullet (\mathbf{C}_{,\xi}(\xi^o) \times \mathbf{C}_{,\xi\xi}(\xi^o)) = 0 \tag{5.16}$$

where **P** is a point on the osculating plane and ξ^o is the point on the curve.

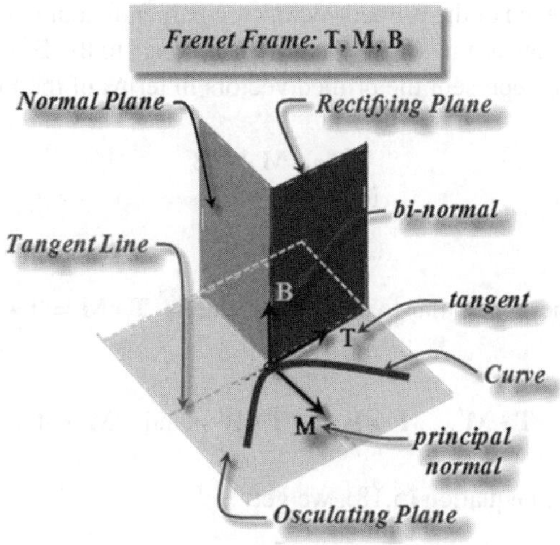

Figure 5.8 Local axes on a curve.

As shown in Figure 5.8, a plane passing through a point on the curve and spanned by the normal vector, **M** and the bi-normal vector, **B**, is known as the ***rectifying plane***.

5.3.12 Discussion

- In the foregoing, we have covered the definition and some of the basic properties of parametric curves and associated Frenet frames. However, for both finite element analysis and mesh generation, we would also be interested in how the Frenet frames are related as we move along the curve points, in other words, the derivatives of the Frenet frame. This gives us some very important properties, such as curvature and twist, of a general three-dimensional curve. Also, for computational purposes, both in mesh generation and finite element phases, a constructive description of curves is necessary that is easy to manipulate and calculate on computers.

5.4 Frenet–Serret Derivatives

Here, our main goal is to get acquainted with additional important properties – curvature and twist – of a curve, This, in turn, requires us to investigate the derivatives, known as the ***Frenet–Serret derivatives***, of the Frenet frames of a general three-dimensional curve. In what follows, we will use the notation: $(\bullet)' \stackrel{\Delta}{=} \frac{d(\bullet)}{dS}$.

5.4.1 Frenet–Serret Derivatives

Let us consider the Frenet frame, $\mathbf{T}(S), \mathbf{M}(S), \mathbf{B}(S)$, as given by equation (5.14), with S as the arc length parameter, and all three axes being unit in length, that is, $\|\mathbf{T}\| = \|\mathbf{M}\| = \|\mathbf{B}\| = 1$; clearly, then:

$$\|\mathrm{T}\|' = (\mathrm{T} \bullet \mathrm{T})' = \mathrm{T}' \bullet \mathrm{T} = 0 \Rightarrow \mathrm{M}' \bullet \mathrm{M} = 0 \text{ and } \mathrm{B}' \bullet \mathrm{B} = 0 \tag{5.17}$$

From equation (5.17), each of the primed vectors are perpendicular to the corresponding Frenet axis; specifically, $\mathbf{T}'$ belongs to the $\mathbf{M}-\mathbf{B}$ plane, $\mathbf{M}'$ belongs to the $\mathbf{B}-\mathbf{T}$ plane and $\mathbf{B}'$ belongs to $\mathbf{T}-\mathbf{M}$ plane. So, let us represent the primed vectors in terms of the Frenet frame as:

$$\begin{aligned}\mathbf{T}' &= \kappa\mathbf{M} + \lambda\mathbf{B}\\ \mathbf{M}' &= \tau\mathbf{B} + \xi\mathbf{T}\\ \mathbf{B}' &= \gamma\mathbf{T} + \eta\mathbf{M}\end{aligned} \tag{5.18}$$

With the Frenet axes perpendicular to each other, that is, $\mathbf{T}\bullet\mathbf{M} = \mathbf{T}\bullet\mathbf{B} = \mathbf{M}\bullet\mathbf{B} = 0$, we get:

$$\mathbf{T}'\bullet\mathbf{M} = -\mathbf{T}\bullet\mathbf{M}', \quad \mathbf{T}'\bullet\mathbf{B} = -\mathbf{T}\bullet\mathbf{B}' \quad \text{and} \quad \mathbf{M}'\bullet\mathbf{B} = -\mathbf{M}\bullet\mathbf{B}' \tag{5.19}$$

Using equation (5.19) in equation (5.18), we get:

$$\xi = -\kappa, \quad \eta = -\tau \quad \text{and} \quad \gamma = -\lambda \tag{5.20}$$

Now, applying equation (5.18) in equation (5.18) with $\lambda \equiv 0$, that is, choosing $\mathbf{T}'$ along the principal normal, $\mathbf{M}$, we get:

$$\boxed{\begin{aligned}\mathbf{T}' &= \kappa\mathbf{M}\\ \mathbf{M}' &= \tau\mathbf{B} - \kappa\mathbf{T}\\ \mathbf{B}' &= -\tau\mathbf{M}\end{aligned}} \tag{5.21}$$

In matrix form, we can rewrite equation (5.21) as:

$$\boxed{\frac{d}{dS}\begin{Bmatrix}\mathbf{T}\\ \mathbf{M}\\ \mathbf{B}\end{Bmatrix} = \begin{Bmatrix}\mathbf{T}\\ \mathbf{M}\\ \mathbf{B}\end{Bmatrix}' \equiv \begin{Bmatrix}\mathbf{T}'\\ \mathbf{M}'\\ \mathbf{B}'\end{Bmatrix} = \begin{bmatrix}0 & k & 0\\ -\kappa & 0 & \tau\\ 0 & -\tau & 0\end{bmatrix}\begin{Bmatrix}\mathbf{T}\\ \mathbf{M}\\ \mathbf{B}\end{Bmatrix}} \tag{5.22}$$

It is clear, from equation (5.22), that the matrix is skew-symmetric and hence has a corresponding axial vector given by:

$$\mathbf{d}(S) \equiv \begin{Bmatrix}\tau(S)\\ 0\\ \kappa(S)\end{Bmatrix} \tag{5.23}$$

where the vector $\mathbf{d}(S)$ is known as ***Darboux's vector***; with $\mathbf{a} = \mathbf{T}, \mathbf{M}, \mathbf{B}$, we can express the derivatives of the Frenet axes as: $\frac{d\mathbf{a}}{dS} = \mathbf{d} \times \mathbf{a} = [\mathbf{d}\times]\,\mathbf{a}$.

5.4.1.1 Curvature and Radius of Curvature

In equation (5.21), κ is known as the ***curvature*** at a point of a curve depicting ***the rate at which the curve departs from its tangent,*** as shown in Figure 5.9.

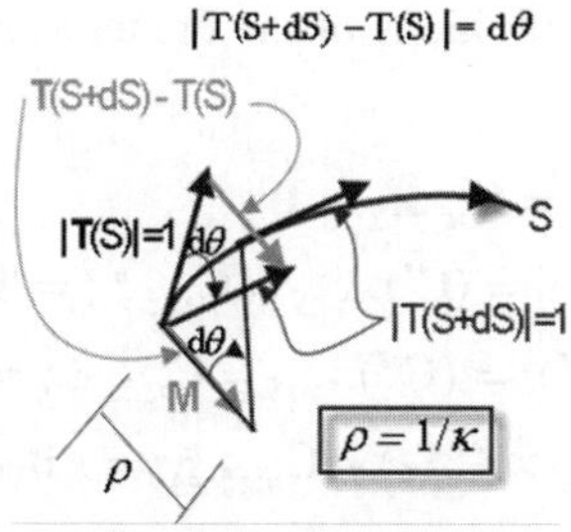

Figure 5.9 Curvature of a curve.

The derivative of the unit tangent vector, **T**, by definition, is given by: $\mathbf{T}'(S) = \mathbf{C}'' = \lim\limits_{dS \to 0} \frac{T(S+dS)-T(S)}{dS} = \lim\limits_{dS \to 0} \frac{d\theta}{dS} = \frac{1}{\rho}$. Now, from Figure 5.9 and the Frenet frame relation, as in equation (5.21), we get the relation between the ***curvature***, κ, and the ***radius of curvature***, ρ, at a point on the curve as:

$$\boxed{\kappa = \|\kappa\mathbf{M}\| = \|\mathbf{T}'(S)\| = \|\mathbf{C}''\| = \frac{d\theta}{dS} = \frac{1}{\rho}} \tag{5.24}$$

Given the relationship between the arc length parameter, S, and the coordinate parameter, ξ, as: $S = S(\xi)$, and performing simple differentiation, and applying the definitions of the Frenet axes as given by equation (5.21), we get:

$$\begin{aligned} &\mathbf{C}' = \mathbf{C}_{,S} = \mathbf{C}_{,\xi}\, \xi_{,S} = \mathbf{T} \quad \text{and} \quad \mathbf{C}'' = (\mathbf{C}')' = (\mathbf{C}')_{,\xi}\xi_{,S} = \mathbf{C}_{,\xi\xi}\, \xi_{,S}^{2} = \mathbf{T}' = \kappa\mathbf{M} \\ &\Rightarrow \kappa = \|\kappa\mathbf{B}\| = \|\mathbf{T} \times \kappa\mathbf{M}\| = \|\mathbf{C}' \times \mathbf{C}''\| = \left\|(\mathbf{C}_{,\xi}\, \xi_{,S}) \times (\mathbf{C}_{,\xi\xi}\, \xi_{,S}^{2})\right\| \\ &= \left\|(\mathbf{C}_{,\xi} \times \mathbf{C}_{,\xi\xi})\right\| \left\|\xi_{,S}^{3}\right\| \end{aligned} \tag{5.25}$$

Now, from equation (5.10), noting that: $\xi_{,S} \equiv \frac{d\xi}{dS} = \frac{1}{\|\mathbf{C}_{,\xi}(\xi)\|}$, we also get, from equation (5.25), an expression for the curvature, κ, in terms of the original parameter, ξ, as:

$$\boxed{\kappa = \frac{\|(\mathbf{C}_{,\xi} \times \mathbf{C}_{,\xi\xi})\|}{\|\mathbf{C}_{,\xi}\|^{3}}} \tag{5.26}$$

Remarks

- For a straight line, **T** is constant, implying that the curvature is zero, that is, $\kappa \equiv 0$, and $\rho = \infty$.

5.4.1.2 Torsion or Twist

In equation (5.21), τ is known as the ***torsion*** at a point of a curve depicting ***the rate at which a twisted curve departs from its osculating plane***. Now, given the relationship between the arc length parameter, S, and the coordinate parameter, ξ, as: $S = S(\xi)$, and performing simple differentiation, and applying the definitions of the Frenet axes as given by equation (5.21), we get:

$$\begin{aligned}
\mathbf{C}' &= \mathbf{C}_{,S} = \mathbf{C}_{,\xi}\,\xi_{,S} = \mathbf{T}\\
\mathbf{C}'' &= (\mathbf{C}')' = (\mathbf{C}')_{,\xi}\xi_{,S} = \mathbf{C}_{,\xi\xi}\,\xi_{,S}^{2} = \mathbf{T}' = \kappa\mathbf{M}\\
\mathbf{C}''' &= (\mathbf{C}'')' = (\mathbf{C}'')_{,\xi}\xi_{,S} = \mathbf{C}_{,\xi\xi\xi}\,\xi_{,S}^{3} = \kappa'\mathbf{M} + \kappa(\tau\mathbf{B} - \kappa\mathbf{T})\\
(\mathbf{C}'\times\mathbf{C}'')\bullet\mathbf{C}''' &= (\mathbf{C}_{,\xi}\times\mathbf{C}_{,\xi\xi})\bullet\mathbf{C}_{,\xi\xi\xi}\,\xi_{,S}^{6} = \kappa\mathbf{B}\bullet(\cancel{\kappa'\mathbf{M}} + \kappa(\tau\mathbf{B} - \cancel{\kappa\mathbf{T}}))\\
&= \kappa^{2}\tau
\end{aligned} \tag{5.27}$$

From equations (5.26) and (5.27), we finally get the expression for torsion, τ, in triple product form as:

$$\boxed{\tau = \frac{(\mathbf{C}'\times\mathbf{C}'')\bullet\mathbf{C}'''}{\kappa^{2}} = \frac{(\mathbf{C}_{,\xi}\times\mathbf{C}_{,\xi\xi})\bullet\mathbf{C}_{,\xi\xi\xi}}{||\mathbf{C}_{,\xi}(\xi)||^{6}\,\kappa^{2}}} \tag{5.28}$$

Remarks

- For a plane curve, twist is zero, that is, $\tau \equiv 0$, implying that the osculating plane does not change from point to point on the curve, From equation (5.28), the condition for a plane curve is given by the requirement that the triple product vanishes, that is, $(\mathbf{C}_{,\xi}\times\mathbf{C}_{,\xi\xi})\bullet\mathbf{C}_{,\xi\xi\xi} = 0$. In matrix form, equivalently, we get:

$$(\mathbf{C}_{,\xi}\times\mathbf{C}_{,\xi\xi})\bullet\mathbf{C}_{,\xi\xi\xi} = \det\begin{bmatrix} X_{1,\xi}(\xi) & X_{1,\xi\xi}(\xi) & X_{1,\xi\xi\xi}(\xi)\\ X_{2,\xi}(\xi) & X_{1,\xi\xi}(\xi) & X_{2,\xi\xi\xi}(\xi)\\ X_{3,\xi}(\xi) & X_{3,\xi\xi}(\xi) & X_{3,\xi\xi\xi}(\xi)\end{bmatrix} = 0 \tag{5.29}$$

where det stands for the determinant of the matrix.

5.4.1.3 Discussion

This investigation has established the theoretical basis for parametric curves. However, for computational accuracy, cost-effectiveness and smoothness, we are specifically interested in some special classes of approximating and interpolating curves: polynomial and piecewise polynomial curves. In what follows, we prefer and investigate ***Bernstein polynomials*** and ***spline functions*** with special attention to a class of piecewise polynomials known as ***Schoenberg B-splines***, over many other polynomials and piecewise polynomials such as Lagrangian and Hermite.

5.5 Bernstein Polynomials

Here, our main goal is to present a short introduction to the Bernstein polynomials insofar as it is useful for both geometric and displacement modeling in mesh generation and finite element analysis phases, respectively. Bernstein polynomials form the backbone of modern-day CAGD through the boundary representation method and c-type finite element method which we introduced. These form the polynomial basis functions for a Bezier curve. Bernstein polynomials show up in a constructive proof of the Weierstrass approximation theorem on continuous functions. The inverse problem gives rise to the Bernstein–Bezier curves:

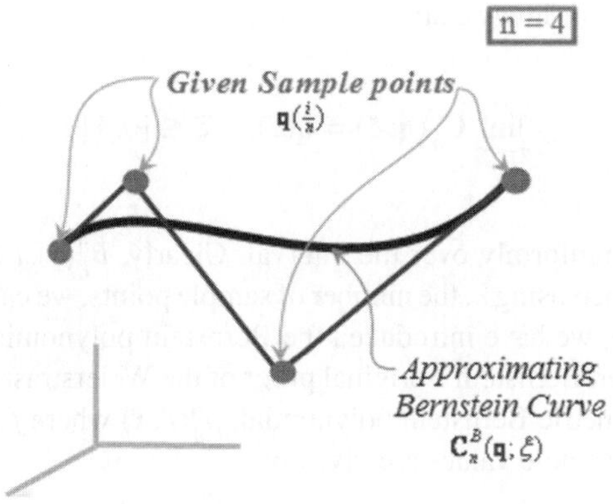

Figure 5.10 A Bernstein polynomial.

5.5.1 Weierstrass Approximation Theorem

One of the most fundamental theorems in approximation theory is the Weierstrass approximation theorem of 1885. In plane language, it states that every ***continuous function*** can be uniformly approximated by a sequence of ***polynomials*** over a closed interval. Of all the different proofs, Bernstein's constructive proof is most relevant for us. The net contention of the proof is established through the following definition of the Bernstein polynomials.

5.5.2 Bernstein Polynomials

Let $\mathbf{q}(\xi)$ be a continuous curve as shown in Figure 5.10. Then the approximating n^{th} Bernstein polynomial, $\mathbf{C}^B_n(\mathbf{q};\xi)$ for $\mathbf{q}(\xi)$ is defined by the linear combination of the Bernstein basis functions with $n+1$ sample points, $\mathbf{q}(\frac{i}{n})$, at $(\frac{i}{n})$, $i \in \{0, 1, \ldots, n\}$, that is, equidistant points in the interval $[0, 1]$, taken as the coefficients of combination. In other words:

$$\boxed{\mathbf{C}^B_n(\mathbf{q};\xi) = \sum_{i=0}^{n} \mathbf{q}\left(\frac{i}{n}\right) B^n_i(\xi)} \tag{5.30}$$

where, we have used the following definition.

Definition

- The ith ***Bernstein basis function***, $B^n_i(\xi)$, of the n-dimensional polynomial subspace of the class of continuous functions as:

$$\boxed{\begin{aligned} B^n_i(\xi) &= \binom{n}{i}\, \xi^i\,(1-\xi)^{n-i}, i \in \{0, 1, \ldots, n\}, \\ \text{where, } \binom{n}{i} &\equiv \begin{cases} \dfrac{n!}{i!(n-i)!} & 0 \le i \le n \\ 0, & \textit{otherwise} \end{cases} \end{aligned}} \tag{5.31}$$

Then, the Bernstein proof shows that:

$$\lim_{n\to\infty} \mathbf{C}_n^B(\mathbf{q};\xi) = \mathbf{q}(\xi), \quad \xi \in [0,1]$$

where the limit holds uniformly over the interval. Clearly, $B_i^n(\xi), i \in \{0, 1, \dots, n\}$ are all n^{th} degree polynomials. By increasing n, the number of sample points, we can get as close to the curve, $\mathbf{q}(\xi)$, as we choose. Here, we have introduced the Bernstein polynomial in terms of ***parametric curves***, $\mathbf{C}_n^B(\mathbf{q};\xi)$. However, Bernstein's original proof of the Weierstrass theorem was constructed in terms of the non-parametric Bernstein polynomial, $p_n^B(f;x)$ where $f(x)$ is a target function, to be approximated, whose sample values are given by:

$$\boxed{p_n^B(f;x) = \sum_{i=0}^{n} f\left(\frac{i}{n}\right) B_i^n(x)} \tag{5.32}$$

In the example shown by Figure 5.11, we are interested in approximating a continuous function $f(x)$ by a quadratic Bernstein polynomial, $p_2^B(f;x)$. Thus, from equation (5.32) with $n = 2$, we must have the sample points of f at locations $\frac{i}{2}$, $i = 0, 1, 2$. From Figure 5.11, it is easily seen that we can always establish the connection between a non-parametric function and its corresponding parametric form as a curve because of the ***linear precision*** property – $\sum_0^n (\frac{i}{n}) B_i^n(\xi) = \xi$ – of the Bernstein polynomials. This, as we shall see later, is the underlying principal motivation for introducing the Bernstein–Bezier or B-spline-de Boor parametric curve form for the ***non-parametric displacement functions*** in finite element analysis. The most relevant information to remember about the proof is that there is a theoretical guarantee of uniform convergence in the approximation. With these in the background, we can now elaborate on the various properties of Bernstein polynomial curves.

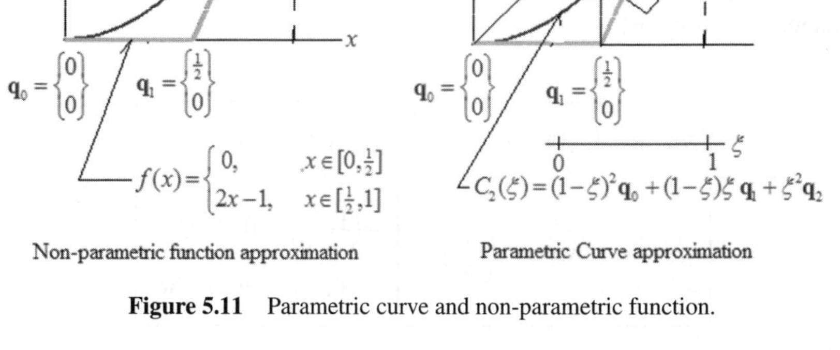

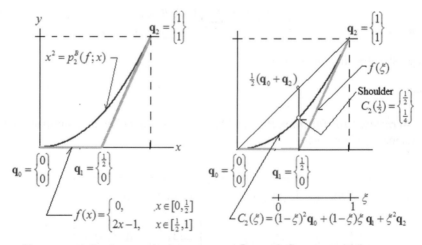

Figure 5.11 Parametric curve and non-parametric function.

5.5.3 *End Point Interpolation*

$$\mathbf{C}_n^B(\mathbf{q};0) = \mathbf{q}(0), \quad \text{and} \quad \mathbf{C}_n^B(\mathbf{q};1) = \mathbf{q}(1)$$

This follows from the end point property of the Bernstein basis functions. This property is fundamentally important and tremendously advantageous in both geometric modeling and finite element space definition.

5.5.4 *Forward Difference Form*

By applying the general definition of the ***forward difference operators***, $\Delta^i\mathbf{q}$, we can write the forward difference form for the approximating Bernstein polynomial, $\mathbf{C}_n^B(\mathbf{q};\xi)$, as:

$$\mathbf{C}_n^B(\mathbf{q};\xi) = \sum_{i=0}^{n} \Delta^i\mathbf{q}\left(\frac{i}{n}\right)\binom{n}{i}\xi^i$$

Exercise: Prove the above equality.

5.5.5 *Simultaneous Approximation*

The Bernstein polynomials simultaneously approximate both a function and its derivatives if it is differentiable. In other words, the Bernstein polynomials follow the behavior of a function with a remarkable degree of closeness.

5.5.6 *Partition of Unity*

If $\mathbf{q}(\xi) = 1$, using the property of successive forward difference ($\Delta^0\mathbf{q}(0) = 1; \Delta^1\mathbf{q}(0) = \Delta^2\mathbf{q}(0) = \cdots = 0$), and the property of Bernstein basis functions, we can show that

$$\mathbf{C}_n^B(1;\xi) = \sum_{i=0}^{n} B_i^n(\xi) \equiv 1$$

In other words, it is the binomial expansion for $1^n = \{\xi + (1-\xi)\}^n$. This property has profound implication in the finite element method of approximation.

Exercise: Prove the above equality.

5.5.7 *Linear Precision*

If $\mathbf{q}(\xi) = \xi$, the property of successive forward difference ($\Delta^0\mathbf{q}(0) = 0; \Delta^1\mathbf{q}(0) = \frac{1}{n}, \Delta^2\mathbf{q}(0) = \cdots = 0$) and the property of Bernstein basis functions show us:

$$\mathbf{C}_n^B(\xi;\xi) = \sum_{i=0}^{n}\left(\frac{i}{n}\right) B_i^n(\xi) \equiv \xi$$

In other words, the initial linear function is reproduced.

Exercise: Prove the above equality.

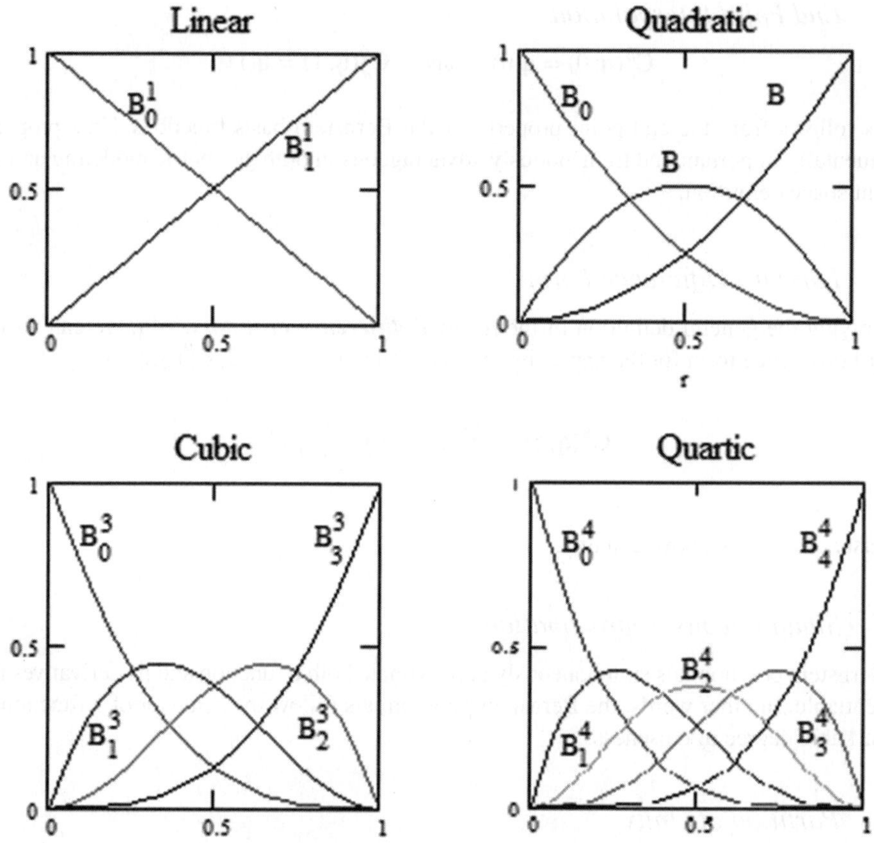

Figure 5.12 Bezier curves of various degrees.

5.5.8 *Bernstein Basis Functions: Properties*

Here, our main goal is to present the essential properties of the Bernstein basis functions. The modern-day CAGD relies heavily on these. Our c-type finite element analysis, likewise, utilizes these properties.

Example: Based on the definition given in equation (5.31), here are a few explicit expressions for the Bernstein basis functions for $n = 1, 2, 3, 4$. The corresponding graphs are also shown in Figure 5.12:

Linear ($n = 1$) basis functions: $B_0^1(\xi) = 1 - \xi$, $B_1^1(\xi) = \xi$

Quadratic ($n = 2$) basis functions: $B_0^2(\xi) = (1 - \xi)^2$, $B_1^2(\xi) = 2\,(1 - \xi)\,\xi$, $B_2^2(\xi) = \xi^2$

Cubic ($n = 3$) basis functions: $B_0^3(\xi) = (1 - \xi)^3$, $B_1^3(\xi) = 3\,(1 - \xi)^2\,\xi$, $B_2^3(\xi) = 3\,(1 - \xi)\,\xi^2$, $B_3^3(\xi) = \xi^3$

Quartic ($n = 4$) basis functions: $B_0^4(\xi) = (1 - \xi)^4$, $B_1^4(\xi) = 4\,(1 - \xi)^3\,\xi$, $B_2^4(\xi) = 6\,(1 - \xi)^2\,\xi^2$, $B_3^4(\xi) = 4\,(1 - \xi)\,\xi^3$, and $B_4^4(\xi) = \xi^4$,

5.5.8.1 Non-negativity and Unique Maximum

Bernstein basis functions, $B_i^n(\xi)$, are all non-negative and have one maximum, at $\xi = \frac{i}{n}$ on the interval, $\xi \in [0, 1]$.

5.5.8.2 Partition of Unity

For Bernstein basis functions, for any *n*,

$$\sum_{i=0}^{n} B_i^n(\xi) = 1, \quad \xi \in [0,1] \tag{5.33}$$

5.5.8.3 Recursive Property

The Bernstein basis function of differing degrees are recursively connected:

$$B_i^n(\xi) = \begin{cases} (1-\xi)\, B_i^{n-1}(\xi) \;+\; \xi\, B_{i-1}^{n-1}(\xi), & i \in \{0, 1, \dots, n\} \\ 0, & i \notin \{0, 1, \dots, n\} \end{cases}, \tag{5.34}$$
with $B_0^0(\xi) \equiv 1$

It can be seen that $B_i^n(\xi)$ is obtained as the ***convex combination*** $B_i^{n-1}(\xi)$ and $B_{i-1}^{n-1}(\xi)$ for all n and for all $i \in \{0, 1, \dots, n\}$.

5.5.8.4 Power Basis Connection

Power basis functions, $\{\xi^i\}$, and the Bernstein basis functions are connected to each other:

$$\xi^i = \sum_{j=i}^{n} \frac{\binom{j}{i}}{\binom{n}{i}} B_j^n(\xi) \quad \text{and} \quad B_i^n(\xi) = \sum_{j=i}^{n} (-1)^{j-i} \binom{n}{j} \binom{j}{i} \xi^j \tag{5.35}$$

5.5.8.5 Subdivision Property

$$B_i^n(c\xi) = \sum_{i=0}^{n} B_i^j(c)\, B_j^n(\xi) \tag{5.36}$$

5.5.8.6 Product Rule

This property is useful in the discretization of a specified traction to its nodal counterparts in the finite element analysis phase:

$$B_i^m(\xi)\, B_j^n(\xi) = \frac{\binom{m}{i}\binom{n}{j}}{\binom{m+n}{i+j}} B_{i+j}^{m+n}(\xi) \tag{5.37}$$

5.5.8.7 Degree Elevation Formulas

$$
\begin{aligned}
(1-\xi)\,B_i^n(\xi) &= \frac{n+1-i}{n+1}\,B_i^{n+1}(\xi)\\
\xi\,B_i^n(\xi) &= \frac{i+1}{n+1}\,B_{i+1}^{n+1}(\xi)\\
B_i^n(\xi) &= \frac{n+1-i}{n+1}\,B_i^{n+1}(\xi) + \frac{i+1}{n+1}\,B_{i+1}^{n+1}(\xi)
\end{aligned} \tag{5.38}
$$

5.5.8.8 The First Derivative

The first derivative is easily obtained as:

$$
B_{i,1}^n(\xi) = n\,\{B_{i-1}^{n-1}(\xi) - B_i^{n-1}(\xi)\} \tag{5.39}
$$

5.5.8.9 Symmetry Property

$$
B_i^n(\xi) = B_{n-i}^n(\xi) \tag{5.40}
$$

5.5.8.10 Integral Property

$$
\begin{aligned}
&\int_0^{\xi} B_i^n(\varsigma)\,d\varsigma = \frac{1}{n+1}\sum_{j=i+1}^{n+1} B_j^{n+1}(\xi)\\
&\text{and,}\\
&\int_0^{1} B_i^n(\varsigma)\,d\varsigma = \frac{1}{n+1}
\end{aligned} \tag{5.41}
$$

5.6 Non-rational Curves Bezier–Bernstein–de Casteljau

Remarks

- As to Bernstein polynomial, suppose that instead of the original curve, only a finite set of $n+1$ sample points, $\{\mathbf{q}_i\}$, are given such that, $\mathbf{q}_i \equiv \mathbf{q}(\frac{i}{n})$, $\quad i \in \{0, 1, \ldots, n\}$. Then, because of the properties of the Bernstein basis functions, the approximating curve defined as above will closely follow the polygon through the given sample points. Finally, by changing the location of this finite set of the sample points, we will be able to visually control the shape of the approximating curve on a viewing surface while it will still mimic the polygon all the time. This is precisely the motivation for the definition of a Bezier curve, and serves to establish the connection between the Bernstein polynomials and the Bezier curves.
- As already indicated, most curves are amenable to representation by the Bernstein polynomials with the coefficients of combination being ***non-rational*** except for the conics, for which we will need ***rational*** coefficients,
- Thus, for a detailed discussion on Bezier curves, the underpinning tool of modern-day CAGD, we investigate first the ***non-rational Bernstein–Bezier curves*** followed by the ***rational Bernstein–Bezier curves***,
- For more complicated curves that require composition of Bezier curves with certain smoothness, we present ***Bezier form of B-spline curves***.

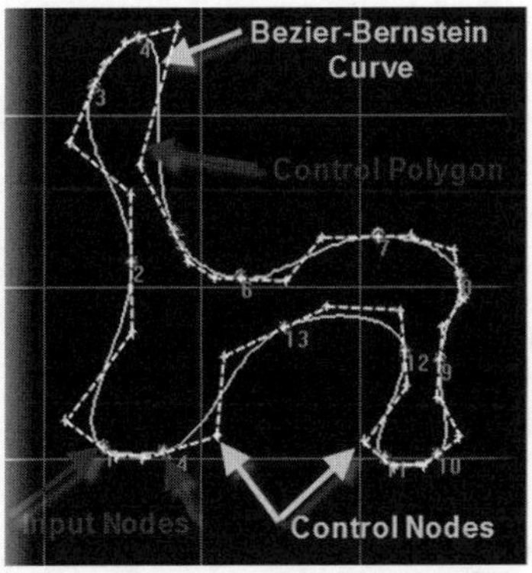

Figure 5.13 Definition of an inverse problem. (See colour plate section)

- It is easy to identify a curve represented by a given control polygon, but in most practical applications where we are given points on a curve and required to generate the corresponding control polygon, as shown in Figure 5.13, things can get really involved; these are known as the inverse problems, and we will deal with them later.
- Finally, noting that the affine transformation is not linear because of the translation part, in computer graphics all computations are performed where the three-dimensional space is embedded into a four-dimensional ***homogeneous coordinate space***; all our code snippets will be given in these coordinates.

5.6.1.1 Bezier–Bernstein Curves

Here, our main goal is to introduce Bezier curves in terms of the ***Bernstein polynomials*** – motivated by the need to discover several very important properties of a Bezier form of description of ***parametric curves***. These properties have a pivotal role in both the algebraic and geometric understanding of the representation. Furthermore, these are essential for an application of the Bezier representation to ***interpolation*** in the ***mesh generation*** phase and ***approximation*** in the ***finite element*** phase of analysis. Let us recall that the ***Bezier curves*** are parametric polynomial curves. Thus, definition utilizes an important property that says that any ***Euclidean points*** belonging to an ***affine space*** can only be added meaningfully in ***barycentric combinations***. We identify the ***Bernstein polynomials*** as the barycentric coefficients, and control points as the Euclidean points. In a non-parametric polynomial representation of continuous functions, as in the finite element analysis phase, the Bernstein polynomials can be seen as the basis functions of the underlying linear vector space. So, we will use these notions rather less rigorously and interchangeably, as long as we are aware of the context.

5.6.1.2 Bezier Curves with Bernstein Polynomials

We introduce here a Bezier curve as in the following definitions.

Definition

- An nth ***degree Bezier curve***, $\mathbf{C}_n(\xi)$, is defined as:

$$\boxed{\mathbf{C}_n(\xi) = \sum_{i=0}^{n} B_i^n(\xi)\,\mathbf{q}_i, \quad 0 \le \xi \le 1} \tag{5.42}$$

where the basis (blending) functions, $\{B_i^n(\xi)\}$, are the nth degree Bernstein polynomials:

$$\boxed{B_i^n(\xi) = \binom{n}{i} \xi^i\,(1-\xi)^{n-i}, i \in \{0,1,\ldots,n\} \quad \text{where} \quad \binom{n}{i} \equiv \frac{n!}{i!(n-i)!}} \tag{5.43}$$

- The coefficients of linear combination or Euclidean points, $\{\mathbf{q}_i\}$, are known as the ***control points*** of the curve.
- The polygon containing the control points, $\{\mathbf{q}_i\}$, is known as the ***control polygon*** of the curve.

Let us note that $\xi \in [0,1]$. Also, from the definition of a parametric curve, recall that in two dimensions: $\mathbf{C}(\xi) = \{X(\xi)\; Y(\xi)\}^T$, and in three dimensions: $\mathbf{C}(\xi) = \{X(\xi) \quad Y(\xi) \quad Z(\xi)\}^T$. Let us elucidate the above definitions by some simple examples. In mesh generation phase, usually curves with degree up to and including three are considered for numerical efficiency. For finite element analysis with mega-elements, it appears that up to quintic should be considered for, say, shell analysis.

Example: Bezier curves of degree one – linear

Let us consider the case of $n = 1$, that is, Bezier curves of degree one.

We have, from equation (5.31), $B_0^1(\xi) = 1 - \xi$ and $B_1^1(\xi) = \xi$. Thus, from equation (5.42), we get: $\mathbf{C}_1(\xi) = (1-\xi)\,\mathbf{q}_0 + \xi\,\mathbf{q}_1, \quad 0 \le \xi \le 1$. In other words, $\mathbf{C}_1(\xi)$ describes a straight line segment between points $\mathbf{q}_0$ and $\mathbf{q}_1$. Let us observe that the curve passes through the end control points.

Example: Bezier curves of degree two – quadratics

Let us consider the case of $n = 2$, that is, Bezier curves of degree two.

We have, from equation (5.31), $B_0^2(\xi) = (1-\xi)^2$, $B_1^2(\xi) = 2(1-\xi)\,\xi$ and $B_2^2(\xi) = \xi^2$. Thus, from equation (5.42), we get: $\mathbf{C}_2(\xi) = (1-\xi)^2\,\mathbf{q}_0 + 2\,(1-\xi)\,\xi\,\mathbf{q}_1 + \xi^2\,\mathbf{q}_2, \quad 0 \le \xi \le 1$. In other words, $\mathbf{C}_2(\xi)$ describes a parabolic arc between points $\mathbf{q}_0$ and $\mathbf{q}_2$. Let us observe again that the curve passes through the end control points.

Example: Bezier curves of degree three – cubics

Let us consider the case of $n = 3$, that is, Bezier curves of degree three.

We have, from equation (5.31), $B_0^3(\xi) = (1-\xi)^3, B_2^3(\xi) = 3(1-\xi)^2\,\xi, B_1^3(\xi) = 3(1-\xi)\,\xi^2$ and $B_2^2(\xi) = \xi^2$. Thus, from equation (5.42), we get: $\mathbf{C}_3(\xi) = (1-\xi)^2\,\mathbf{q}_0 + 3\xi\,(1-\xi)\,\mathbf{q}_1 + 3\xi\,(1-\xi)^2\,\mathbf{q}_2 + \xi^3\,\mathbf{q}_3, \quad 0 \le \xi \le 1$. In other words, $\mathbf{C}_3(\xi)$ describes a cubic arc between points $\mathbf{q}_0$ and $\mathbf{q}_3$, as shown in Figure 5.14. Let us observe again that the curve passes through the end control points.

Remarks

- Curves of degree up to two are planar curves, that is, a plane containing the curves fully in it can be identified.
- Curves of degree greater than two are generally true three-dimensional curves, that is, in general, a plane cannot be identified that fully contains a curve.
- The end point interpolation by the Bezier curves is a very suitable feature for its application to finite element analysis in any dimension.

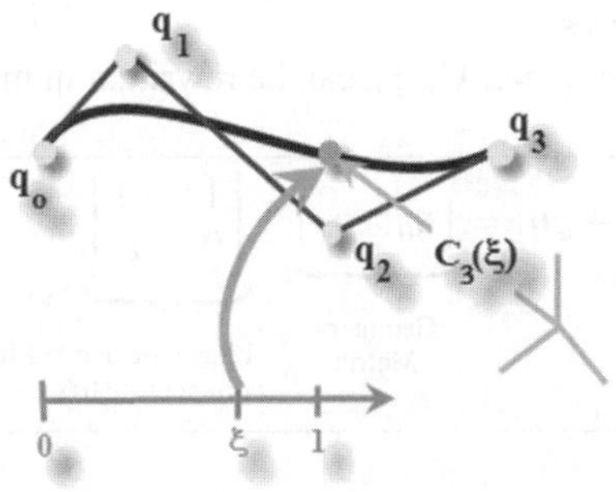

Figure 5.14 A cubic Bezier curve.

5.6.1.3 Horner's Algorithm

Let us consider a cubic polynomial curve, $\mathbf{C}_3(\xi)$, expressed in monomial basis functions, $\{1,\ \xi,\ \xi^2,\ \xi^3\}$, such that:

$$\mathbf{C}_3(\xi) = A + B\,\xi + C\,\xi^2 + D\,\xi^3 \tag{5.44}$$

The exponential powers associated with the basis functions make evaluation of the curve points numerically inaccurate as the degree of the curve increases. Computationally, a robust solution is offered by what is known as the ***Horner's algorithm*** of nested multiplications by rewriting the expression (5.44) as:

$$\boxed{\mathbf{C}_3(\xi) = A + \xi(B + \xi(C + D\,\xi))} \tag{5.45}$$

and then, calculating the expression by evaluating the multiplication and addition operations from the innermost bracket outward. The matrix form of the Bernstein–Bezier curves as introduced next can also be cast in monomial basis functions with appropriate ***Bezier coefficient matrices*** for computational stability, accuracy and efficiency; this approach will be algorithmically documented when we deal with computational aspects of curves.

5.6.1.4 Matrix Form of a Bezier Curve

By collecting the control vectors into a matrix and the Bernstein polynomials into a vector, we can express a curve in matrix form. Thus, in the matrix form, the same Bezier curve, $\mathbf{C}_n(\xi)$, may be described as:

$$\boxed{\underset{m\times 1}{\mathbf{C}_n(\xi)} = \underset{m\times n}{\mathbf{Q}}\ \underset{n\times 1}{\mathbf{B}(\xi)}, \quad 0 \le \xi \le 1} \tag{5.46}$$

For three-dimensional space, $m = 3$, equation (5.46) represents a Bezier curve, $\mathbf{C}_n(\xi)$, with Bezier control matrix: $\mathbf{Q} = \begin{bmatrix} \mathbf{q}_0 & \mathbf{q}_1 & \cdots & \mathbf{q}_n \end{bmatrix}$, and Bernstein basis matrix: $\mathbf{B} = \begin{Bmatrix} B_0^n(\xi) & B_1^n(\xi) & \cdots & B_n^n(\xi) \end{Bmatrix}^T$. As indicated before, we would rather like the expression (5.46) further reduced to monomial form for computational purposes with applications in manipulation of Bezier curves on a digital computer.

5.6.1.4.1 *Linear Bezier Curves*

For $n = 1$, that is, linear Bezier curve, $\mathbf{C}_1(\xi)$, can be rewritten in matrix form as:

$$\mathbf{C}_1(\xi) = (1-\xi)\mathbf{q}_0 + \xi\mathbf{q}_1 = \underbrace{\begin{bmatrix} \mathbf{q}_0 & \mathbf{q}_1 \end{bmatrix}}_{\substack{\text{Geometry}\\ \text{Matrix}}} \underbrace{\begin{bmatrix} 1 & -1 \\ 0 & 1 \end{bmatrix}}_{\substack{\text{Linear Bezier}\\ \text{coefficient matrix}}} \underbrace{\begin{Bmatrix} 1 \\ \xi \end{Bmatrix}}_{\substack{\text{Monomial basis functions}\\ \text{of degree 1}}} \tag{5.47}$$

where $\begin{bmatrix} \mathbf{q}_0 & \mathbf{q}_1 \end{bmatrix}$ is appropriately known as the ***geometry matrix*** as it determines the geometry of a linear curve, and $\begin{bmatrix} 1 & -1 \\ 0 & 1 \end{bmatrix}$, as the ***linear Bezier Coefficient matrix***.

5.6.1.4.2 *Quadratic Bezier Curves*

For $n = 2$, that is, quadratic Bezier curve, $\mathbf{C}_2(\xi)$, can be rewritten in matrix form as:

$$\begin{aligned} \mathbf{C}_2(\xi) &= (1-\xi)^2\mathbf{q}_0 + \xi(1-\xi)\mathbf{q}_1 + \xi^2\mathbf{q}_2 \\ &= \underbrace{\begin{bmatrix} \mathbf{q}_0 & \mathbf{q}_1 & \mathbf{q}_2 \end{bmatrix}}_{\substack{\text{Geometry}\\ \text{matrix}}} \underbrace{\begin{bmatrix} 1 & -2 & 1 \\ 0 & 2 & -2 \\ 0 & 0 & 1 \end{bmatrix}}_{\substack{\text{Quadratic Bezier}\\ \text{coefficient matrix}}} \underbrace{\begin{Bmatrix} 1 \\ \xi \\ \xi^2 \end{Bmatrix}}_{\substack{\text{Monomial basis functions}\\ \text{of degree 2}}} \end{aligned} \tag{5.48}$$

where $\begin{bmatrix} \mathbf{q}_0 & \mathbf{q}_1 & \mathbf{q}_2 \end{bmatrix}$ is appropriately known as the ***geometry matrix*** as it determines the geometry of a quadratic curve, and $\begin{bmatrix} 1 & -2 & 1 \\ 0 & 2 & -2 \\ 0 & 0 & 1 \end{bmatrix}$, as the ***quadratic Bezier coefficient matrix***.

5.6.1.4.3 *Cubic Bezier Curves*

For $n = 3$, that is, cubic Bezier curve, $\mathbf{C}_3(\xi)$, can be rewritten in matrix form as:

$$\begin{aligned} \mathbf{C}_3(\xi) &= (1-\xi)^3\mathbf{q}_0 + 3\xi(1-\xi)^2\mathbf{q}_1 + 3\xi^2(1-\xi)\mathbf{q}_2 + \xi^3\mathbf{q}_3 \\ &= \underbrace{\begin{bmatrix} \mathbf{q}_0 & \mathbf{q}_1 & \mathbf{q}_2 & \mathbf{q}_3 \end{bmatrix}}_{\substack{\text{Geometry}\\ \text{matrix}}} \underbrace{\begin{bmatrix} 1 & -3 & 3 & 1 \\ 0 & 3 & -6 & 3 \\ 0 & 0 & 3 & -3 \\ 0 & 0 & 0 & 1 \end{bmatrix}}_{\substack{\text{Cubic Bezier}\\ \text{coefficient matrix}}} \underbrace{\begin{Bmatrix} 1 \\ \xi \\ \xi^2 \\ \xi^3 \end{Bmatrix}}_{\substack{\text{Monomial basis functions}\\ \text{of degree 3}}} \end{aligned} \tag{5.49}$$

where $\begin{bmatrix} \mathbf{q}_0 & \mathbf{q}_1 & \mathbf{q}_2 & \mathbf{q}_3 \end{bmatrix}$ is appropriately known as the ***geometry matrix*** as it determines the geometry of a cubic curve, and $\begin{bmatrix} 1 & -3 & 3 & 1 \\ 0 & 3 & -6 & 3 \\ 0 & 0 & 3 & -3 \\ 0 & 0 & 0 & 1 \end{bmatrix}$, as the ***cubic Bezier coefficient matrix***.

Remarks

- In computer programming, it is also possible to consider the monomial basis function vectors in reverse order. For example, in the cubic case above, we can express the vector as: $\{\xi^3, \xi^2, \xi, 1\}^T$; of course, in those cases, the ***columns*** of the ***Bezier coefficient matrices*** also need to be reversed. In our example code snippet on curves, we will adopt this convention.

5.6.1.5 Sneak Preview of the de Casteljau Algorithm

Let us dig a bit further into the example above with $n = 2$, that is, the parabolic curve. We can rewrite the expression for the curve as:

$$\begin{aligned} \mathbf{C}_2(\xi) &= (1-\xi)^2\,\mathbf{q}_0 + 2\,(1-\xi)\,\xi\,\mathbf{q}_1 + \xi^2\,\mathbf{q}_2, \quad 0 \le \xi \le 1 \\ &= \underbrace{(1-\xi)\left(\overbrace{(1-\xi)\,\mathbf{q}_0 + \xi\,\mathbf{q}_1}^{\text{Linear combination}}\right) + \xi\left(\overbrace{(1-\xi)\,\mathbf{q}_1 + \xi\,\mathbf{q}_2}^{\text{Linear combination}}\right)}_{\text{Linear combination}} \end{aligned} \tag{5.50}$$

Now, we can clearly see that a parabolic curve, $\mathbf{C}_2(\xi)$, is obtained by the linear combinations of curves one degree less (namely of degree one) than the degree of the intended curve. In other words, the parabolic curves are constructed by repeated linear interpolations of linear curves on the control polygons. This is not an isolated observation! In fact, it forms the basis of a very elegant geometric algorithm for construction of any Bezier curve and its other related properties known as ***de Casteljau algorithm***.

Remarks

- Here we have introduced a definition of Bezier curves in terms of Bernstein polynomials. This definition is purely algebraic and suitable for both analytical and computational use. However, it is of little use for geometric visualization. As suggested, for a geometric understanding to aid visualization and, equally importantly, for computational efficiency in the mesh generation phase, we need to look at the ***Bezier–de Casteljau–Bernstein connection***.
- Note that here we have restricted ourselves merely to the definition of a Bezier curve. For computational efficiency in mesh generation and finite element analysis, we will need to utilize the various important properties of the Bezier curves. For a detailed discussion on the properties of a Bezier curve, we need to investigate the ***properties of Bezier curves***,
- Finally, for manipulations of curves in the power basis, we need to be familiar with the ***matrix form of a Bezier Curve in power basis***

5.6.1.6 Bezier–Bernstein–de Casteljau Connection

Here, our main goal is to present the de Casteljau algorithm for constructing Bezier or Bernstein polynomial curves. It is numerically more efficient and less error prone than computation in power basis form of the Bernstein basis functions obtained even through the ***Horner-type algorithm***. In the mesh generation phase, the de Casteljau algorithm forms the backbone of modern-day CAGD through a boundary representation method that utilizes Bezier curves and surfaces. Let us consider the following simple case of quadratic Bernstein polynomial curve with $n = 2$, that is, the parabolic curve. Based on the Bernstein basis functions, we can write the expression for the curve as:

$$\mathbf{C}_2(\xi) = (1-\xi)^2\,\mathbf{q}_0 + 2\,(1-\xi)\,\xi\,\mathbf{q}_1 + \xi^2\,\mathbf{q}_2, \quad 0 \le \xi \le 1 \tag{5.51}$$

However, by viewing it from a geometric point of view, we can rewrite the above equation as:

$$\mathbf{C}_2(\xi) = \underbrace{(1-\xi)\left(\overbrace{(1-\xi)\,\mathbf{q}_0 + \xi\,\mathbf{q}_1}^{\text{Convex combination}}\right) + \xi\left(\overbrace{(1-\xi)\,\mathbf{q}_1 + \xi\,\mathbf{q}_2}^{\text{Convex combination}}\right)}_{\text{Convex combination}} \tag{5.52}$$

Now, we can clearly see that the parabolic curve, $\mathbf{C}_2(\xi)$, is obtained by ***convex combination*** of curves of degree one less (i.e. of degree one) than the degree of the original curve. In other words, the parabolic curves are constructed by repeated linear interpolations of the segments of the ***control polygons*** which are linear curves. This is not an isolated observation! In fact, it forms the basis of the very elegant geometric algorithm for the construction of any Bezier curve and its other related properties such as tangents and curvatures, known as the ***de Casteljau algorithm***. The central scheme is based on the recognition of the ***recursive relation of repeated linear interpolations for the Bernstein polynomials***; for finding a point on a curve at any parameter point, we start with the given Bezier control polygon describing the curve; we locate a new control point in each segment of the polygon by linear interpolation (in the same ratio as the parameter point to the parameter interval) of the end control points of the corresponding segments. By continuing recursively a number of times equal to the degree of the curve, we recover the intended point on the curve. Before we present the de Casteljau algorithm, let us reintroduce:

Definition

- The set $\{\mathbf{q}_i\}$, $i \in \{0, 1, \ldots, n\}$ are called the ***Bezier points or control points***.
- The polygon formed by joining successively $\mathbf{q}_i$ and $\mathbf{q}_{i+1}$, $i \in \{0, 1, \ldots, n\}$ is called the ***Bezier polygon or control polygon*** of the curve.

5.6.1.7 de Casteljau Algorithm

Let $\{\mathbf{q}_i\}$, $i \in \{0, 1, \ldots, n\}$ be a finite set of control points in ^{3} and $\xi \in$. Then, de Casteljau algorithm is defined by the following recursive relation via repeated ***linear interpolations***:

$$\mathbf{q}_i^d(\xi) = \underbrace{(1-\xi)\,\mathbf{q}_i^{d-1}(\xi) + \xi\,\mathbf{q}_{i+1}^{d-1}(\xi)}_{\text{Convex combination}}, \quad \forall \xi \in, \quad d \in \{1, 2, \ldots, n\}$$
$$i \in \{0, 1, \ldots, n-d\}$$
$$\text{with } \mathbf{q}_i^0(\xi) = \mathbf{q}_i \tag{5.53}$$

Then, $\mathbf{q}_i^n(\xi)$, $\forall \xi \in \mathbb{R}$ defines a parametric Bezier curve, $\mathbf{C}_n(\xi)$ of degree n. What better way to understand the details than actually going through an example:

Example: Cubic ($n = 3$) Bezier curve

We have: $d \in \{1, 2, 3\}$ in equation (5.53). Now, let us explicitly record the steps:

Step 0:	$\mathbf{q}_0^0 = \mathbf{q}_0,\quad \mathbf{q}_1^0 = \mathbf{q}_1,\quad \mathbf{q}_2^0 = \mathbf{q}_2$ and $\mathbf{q}_3^0 = \mathbf{q}_3$, for a given $\xi \in [0, 1]$
Recursion, $d = 1$:	$\begin{cases} \mathbf{q}_0^1 = (1-\xi)\,\mathbf{q}_0 + \xi\,\mathbf{q}_1, \\ \mathbf{q}_1^1 = (1-\xi)\,\mathbf{q}_1 + \xi\,\mathbf{q}_2 \\ \mathbf{q}_2^1 = (1-\xi)\,\mathbf{q}_2 + \xi\,\mathbf{q}_3 \end{cases}$
Recursion, $d = 2$:	$\mathbf{q}_0^2 = (1-\xi)\,\mathbf{q}_0^1 + \xi\,\mathbf{q}_1^1,\quad \mathbf{q}_1^2 = (1-\xi)\,\mathbf{q}_1^1 + \xi\,\mathbf{q}_2^1,$
Recursion, $d = 3$:	$C_3(\xi) = \mathbf{q}_0^3 = (1-\xi)\,\mathbf{q}_0^2 + \xi\,\mathbf{q}_1^2$, as shown in Figure 5.15.

The whole process can be seen schematically as in Figure 5.15.

Note that the control point sets $\{\mathbf{q}_0, \mathbf{q}_0^1, \mathbf{q}_0^2, \mathbf{q}_0^3\}$ and $\{\mathbf{q}_0^3, \mathbf{q}_1^2, \mathbf{q}_2^1, \mathbf{q}_3\}$ subdivide the original cubic curve into two cubic curves at the parameter point: one on the left and one on the right. This aspect of the de Casteljau algorithm gives rise to the powerful notion of partition or ***subdivision*** of a Bezier curve, essential for various efficient applications such as computation of ***tangents***, ***curvatures*** and ***twist***. Note too that the entire algorithm is a collection of series of linear interpolations, that is, scaling and additions of control points. Thus, numerical errors are minimal and the process is fast. We can quickly reverse the process to see the Bernstein connection of a Bezier curve:

$$\begin{aligned} C_3(\xi) = \mathbf{q}_0^3 &= (1-\xi)\,\mathbf{q}_0^2 + \xi\,\mathbf{q}_1^2 \\ &= (1-\xi)\,\{(1-\xi)\,\mathbf{q}_0^1 + \xi\,\mathbf{q}_1^1\} + \xi\{(1-\xi)\,\mathbf{q}_1^1 + \xi\,\mathbf{q}_2^1\} \\ &= \ldots, = \sum_{i=0}^{3} B_i^3(\xi)\,\mathbf{q}_i \end{aligned}$$

This being a cubic expression in $-\infty \le \xi \le \infty$, $C_3(\xi)$ is a cubic curve.

Exercise: Fill in the intermediate steps for the reverse process of a cubic.

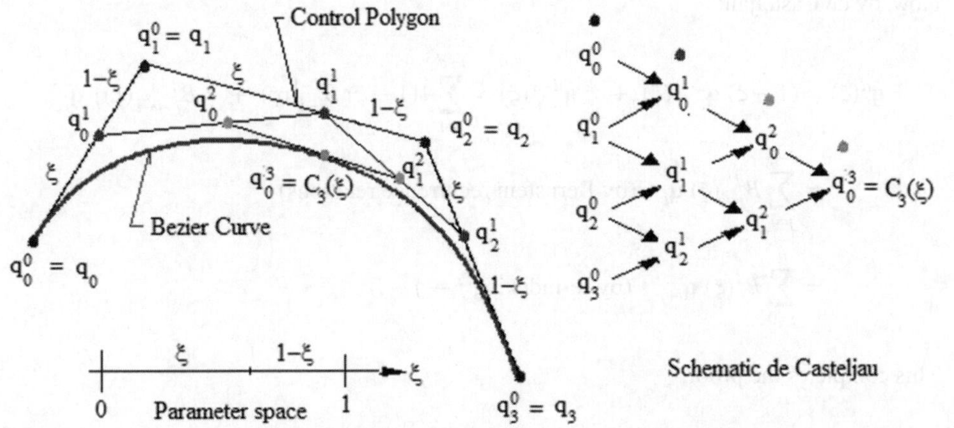

Figure 5.15 de Casteljau method for a cubic curve.

5.6.1.8 Intermediate de Casteljau Curves

The intermediate de Casteljau curves have a suitable alternative expression that implies that these are indeed Bezier curves. Now, we present the expressions for these curves in the form of a lemma with proof by induction.

Lemma

$$\mathbf{q}_i^d(\xi) = \sum_{j=0}^{d} B_j^d(\xi)\, \mathbf{q}_{i+j}, \quad d \in \{0, 1, \ldots, n\}, \quad i \in \{0, 1, \ldots, n-d\} \tag{5.54}$$

Proof. Let us recognize that by the de Casteljau algorithm for $d = 1$, we have: $\mathbf{q}_i^1(\xi) = \sum_{j=0}^{1} B_j^1(\xi)\, \mathbf{q}_{i+j}$, $i \in \{0, 1, \ldots, n-1\}$, that is, a set of pair-wise linear combinations of the original control points. Now, we will show that if it is true for $d - 1$ then it is also true for d. Then, by induction, the proof will be complete: Thus, assuming it is true for $d - 1$, we have:

$$\begin{aligned}
\mathbf{q}_i^{d-1}(\xi) &= \sum_{j=0}^{d-1} B_j^{d-1}(\xi)\, \mathbf{q}_{i+j} && \text{(by hypothesis)}\\
&= \sum_{j=i}^{i+d-1} B_{j-i}^{d-1}(\xi)\, \mathbf{q}_j && \text{(by setting } k = i + j \text{ and renaming } k \to j)\\
&= \sum_{j=i}^{i+d} B_{j-i}^{d-1}(\xi)\, \mathbf{q}_j && \text{(added } j = i \text{ term since } B_d^{d-1}(\xi) \equiv 0 \text{ for}\\
& && \text{all } d \notin \{0, 1, \ldots, d-1\} \text{ by definition)}
\end{aligned}$$

Similarly, we can show:

$$\mathbf{q}_{i+1}^{d-1}(\xi) = \sum_{j=i}^{i+d} B_{j-i-1}^{d-1}(\xi)\, \mathbf{q}_j$$

Now, by de Casteljau:

$$\begin{aligned}
\mathbf{q}_i^d(\xi) &= (1-\xi)\, \mathbf{q}_i^{d-1}(\xi) \;+\; \xi\, \mathbf{q}_{i+1}^{d-1}(\xi) = \sum_{j=i}^{i+d} \{(1-\xi)\, B_{j-i}^{d-1}(\xi) \;+\; \xi\, B_{j-i-1}^{d-1}(\xi)\}\, \mathbf{q}_j\\
&= \sum_{j=i}^{i+d} B_{j-i}^{d}(\xi)\, \mathbf{q}_j \quad \text{(by Bernstein recurrence relations)}\\
&= \sum_{j=0}^{d} B_j^d(\xi)\, \mathbf{q}_{i+j} \quad \text{(by re-indexing } j \to j - i)
\end{aligned}$$

This completes the proof.

Exercise: In the above proof, show that: $\mathbf{q}_{i+1}^{d-1}(\xi) = \sum_{j=i}^{i+d} B_{j-i-1}^{d-1}(\xi)\, \mathbf{q}_j$

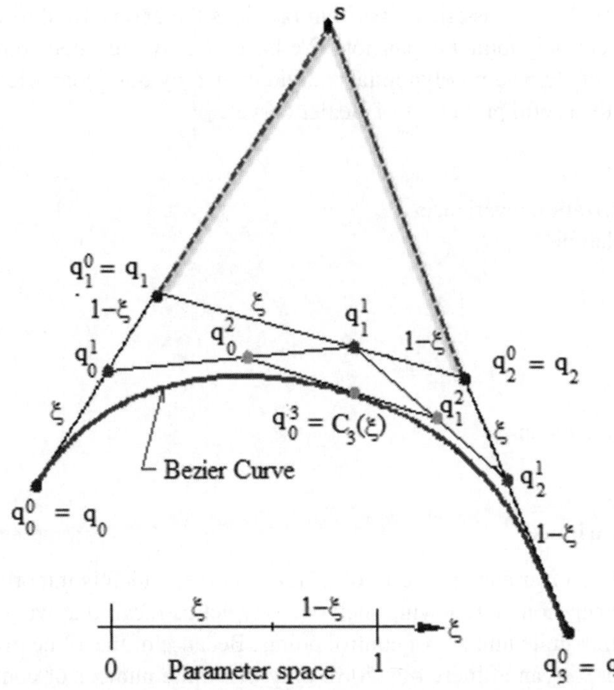

Figure 5.16 de Casteljau construction of a spline from cubic curve.

Remarks

- Let us note that each of these intermediate de Casteljau points defines a Bezier curve, that is, $\mathbf{q}_i, \mathbf{q}_{i+1}, \ldots, \mathbf{q}_{i+d}$ define $\mathbf{q}_i^d(\xi)$, a Bezier curve of degree d. For example, $\mathbf{q}_0, \mathbf{q}_1$ and $\mathbf{q}_2$ define a parabola, $\mathbf{q}_0^2(\xi)$, $\xi \in [0, 1]$. This observation is central to several crucial theorems that eventually lead up to generalization of the Bezier curves to a wider class of curves – B-spline curves that are of utmost importance both in computer-aided geometry modeling and finite element analysis.

5.6.1.9 Sneak Preview of B-spline Curve Construction

While we will have some detailed discussion on B-splines later, let us for now go back to the cubic example. We note that if we set: $3\mathbf{q}_1 - \mathbf{q}_0 = 3\mathbf{q}_2 - \mathbf{q}_3 = \mathbf{s}$, we may, as shown in Figure 5.16, describe the same curve as a curve of degree two, defined by the control points, $\{\mathbf{q}_0, \mathbf{s}, \mathbf{q}_3\}$, as: $\mathbf{C}_3(\xi) = (1-\xi)^2\mathbf{q}_0 + 2(1-\xi)\,\xi\,\mathbf{s} + \xi^2\,\mathbf{q}_3$. It turns out that $\mathbf{s}$ represents a control point of a more versatile class of curves – the ***B-spline curves*** in de Boor geometric form.

5.6.2 Properties: Non-rational Bezier Curves

Here, we provide various useful and relevant properties of a Bezier curve essential for both mesh generation and the finite element method. Our two different yet connected introductions to Bezier curves through the de Casteljau algorithm and the Bernstein polynomial basis functions nicely tie in with the two complementary ways of looking at any computational entity: by geometry and by algebra. Bezier by the de Casteljau algorithm provides the geometric flavor of the algebraic structures of Bernstein polynomial expansion. We have already indicated some of the relevant algebraic properties of Bernstein polynomials; strengthened by our geometric perspective, here we present most of the useful properties of Bezier curves:

- affine invariance
- affine reparameterization invariance
- end point interpolation
- convex hull
- symmetry
- derivatives
- integrals
- degree elevation
- partitioning or subdivision

5.6.2.1 Affine Invariance

To introduce the affine invariance property of a Bezier curve, which is important in developing an efficient mesh generation and viewing methodology, let us recall that we generate our finite element models from a finite number of control points. Because of the affine invariance property as will be explained, we can affinely transform only this finite number of control points rather than the entire infinite-dimensional curves or digitally finite but vast number of points on them to obtain the same transformed geometric object for, say, our viewing needs. We already know that barycentric combinations are invariant under affine transformations such as translation, rotations, scaling and reflection. Once we identify that Bernstein expansion is barycentric, our job is practically done – invariance proof utilizes the matrix form of a Bezier curve. Let us recall that an nth Bezier curve, $\mathbf{C}_n(\xi)$, is defined as:

$$\mathbf{C}_n(\xi) = \sum_{i=0}^{n} B_i^n(\xi)\, \mathbf{q}_i, \quad 0 \leq \xi \leq 1 \tag{5.55}$$

where the basis (blending) functions, $\{B_i^n(\xi)\}$, are the nth degree Bernstein polynomials:

$$B_i^n(\xi) = \binom{n}{i} \xi^i\,(1-\xi)^{n-i}, i \in \{0, 1, \dots, n\} \quad \text{where} \quad \binom{n}{i} \equiv \frac{n!}{i!(n-i)!} \tag{5.56}$$

with the coefficients of linear combination, $\{\mathbf{q}_i\}$, known as the control points of the curve.

Let us note that the expansion given in equation (5.42) is a barycentric combination, since from the properties of Bernstein polynomials we have $\sum_{i=0}^{n} B_i^n(\xi) = 1$. In the matrix form, the same Bezier curve, $\mathbf{C}_n(\xi)$, may be described as:

$$\underset{m\times 1}{\mathbf{C}_n(\xi)} = \underset{m\times n}{\mathbf{Q}}\ \underset{n\times 1}{\mathbf{B}(\xi)}, \quad 0 \leq \xi \leq 1 \tag{5.57}$$

For three-dimensional space, $m = 3$, equation (5.46) represent a Bezier curve, $\mathbf{C}^n(\xi)$, with Bezier control matrix: $\mathbf{Q} = \begin{bmatrix} \mathbf{q}_0 & \mathbf{q}_1 & \cdots & \mathbf{q}_n \end{bmatrix}$, and Bernstein basis matrix: $\mathbf{B} = \begin{Bmatrix} B_0^n(\xi) & B_1^n(\xi) & \cdots & B_n^n(\xi) \end{Bmatrix}^T$. Now, suppose that we define by an $m \times m$ matrix of affine, say, rotation, transformation, **T**, such that, after application of the rotation to the curve directly, we obtain $\mathbf{C}_n^*(\xi)$ as the transformed curve. We claim that:

the same curve can be obtained by simply transforming the Bezier control matrix, **Q** to $\mathbf{Q}^* = \mathbf{T}\,\mathbf{Q}$, and using this new control matrix, $\mathbf{Q}^*$, to describe the curve, $\mathbf{C}_n^*(\xi)$.

Because,

$$\underset{m\times 1}{\mathbf{C}_n^*(\xi)} = \underset{m\times n}{\mathbf{Q}^*}\ \underset{n\times 1}{\mathbf{B}(\xi)} = \underset{m\times n}{\mathbf{T}\,\mathbf{Q}}\ \underset{n\times 1}{\mathbf{B}(\xi)} = \mathbf{T}\,\mathbf{C}_n(\xi), \quad 0 \leq \xi \leq 1 \tag{5.58}$$

as proposed. This is what is known as the ***affine invariance*** property of a Bezier curve.

Example: A two-dimensional Bezier curves of degree 1: linear.

Let us consider the case of $m = 2$ and $n = 1$, that is, a two-dimensional line segment. Suppose, the segment is directed along the X-axis at the origin and of length unity. Then, clearly, $\mathbf{q}_0 = \begin{Bmatrix} 0 \\ 0 \end{Bmatrix}$ and $\mathbf{q}_1 = \begin{Bmatrix} 1 \\ 0 \end{Bmatrix}$, which makes the control matrix: $\mathbf{Q} = \begin{bmatrix} 0 & 1 \\ 0 & 0 \end{bmatrix}$. The Bernstein basis matrix is given as: $\mathbf{B} = \begin{Bmatrix} B_0^0(\xi) = 1 - \xi \\ B_0^1(\xi) = \xi \end{Bmatrix}$. Thus, $\underset{m\times 1}{\mathbf{C}_n(\xi)} = \begin{bmatrix} 0 & 1 \\ 0 & 0 \end{bmatrix} \begin{Bmatrix} 1-\xi \\ \xi \end{Bmatrix} = \begin{Bmatrix} \xi \\ 0 \end{Bmatrix}$, $0 \leq \xi \leq 1$. Now, suppose that we want the transformed description of the line segment when it is rotated by, say, 90° towards the positive Z-axis. The transformation matrix is given by: $\mathbf{T} = \begin{bmatrix} 0 & -1 \\ 1 & 0 \end{bmatrix}$. We could apply the transformation directly to $\mathbf{C}_n(\xi)$, that is, transform for all $0 \leq \xi \leq 1$, to draw on the screen the new transformed line: $\mathbf{C}_n^*(\xi) = \begin{bmatrix} 0 & -1 \\ 1 & 0 \end{bmatrix} \begin{Bmatrix} \xi \\ 0 \end{Bmatrix} = \begin{Bmatrix} 0 \\ \xi \end{Bmatrix}$, $0 \leq \xi \leq 1$. But why? Instead, for far greater efficiency, we only need to transform the two control vectors that define the line segment to get new control vectors: $\mathbf{Q}^* = \begin{bmatrix} 0 & -1 \\ 1 & 0 \end{bmatrix} \begin{bmatrix} 0 & 1 \\ 0 & 0 \end{bmatrix} = \begin{bmatrix} 0 & 0 \\ 0 & 1 \end{bmatrix}$, and we have: $\mathbf{C}_n^*(\xi) = \begin{bmatrix} 0 & 0 \\ 0 & 1 \end{bmatrix} \begin{Bmatrix} 1-\xi \\ \xi \end{Bmatrix} = \begin{Bmatrix} 0 \\ \xi \end{Bmatrix}$, $0 \leq \xi \leq 1$. Note that its new control points will be given by: $\mathbf{q}_0 = \begin{Bmatrix} 0 \\ 0 \end{Bmatrix}$ and $\mathbf{q}_1 = \begin{Bmatrix} 0 \\ 1 \end{Bmatrix}$ as expected. Obviously, the higher the degree of the curve, the more is the computational efficiency and hence more real-time response.

5.6.2.2 Affine Reparameterization Invariance

Here, our main goal is to introduce the affine reparameterization invariance property of a Bezier curve – motivated by the need to highlight the so-called "standard curves" instead of dealing with an arbitrarily given parameter range. Quite often, for convenience, a Bezier curve is given in what is known as the ***standard form***, where the parameter ξ travels from 0 to 1. The need for the convenient standard form arises for various reasons. Referring to Figure 5.17, if we are to describe an extracted part of a curve between $\xi = \xi^1$ to $\xi = \xi^2$, for, say, numerical processing, it is most convenient to reparameterize with $\sigma \in [0, 1]$, that is, $0 \leq \sigma \leq 1$, like the original curve, for the sake of uniformity of application of the standard formulas already developed. For the extracted curve, $\xi \in [\xi^1, \xi^2]$ is known as the ***global parameter*** and $\sigma \in [0, 1]$ is called the ***local parameter***. We can think of another numerical situation: the ***de Casteljau algorithm*** for construction of a

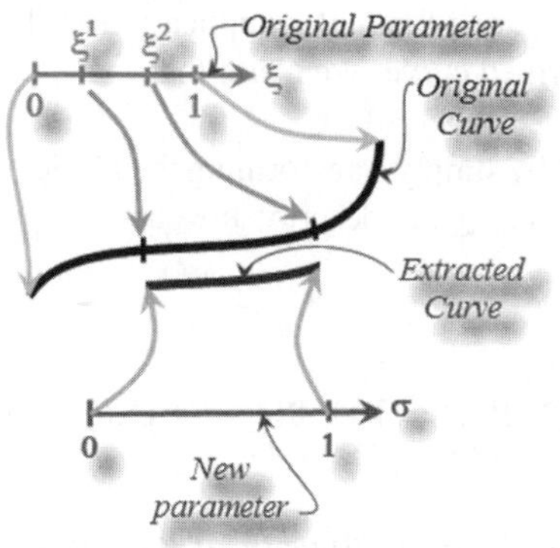

Figure 5.17 Reparameterization of curves.

Bezier curve is usually given assuming that the ***"standard" domain parameter*** runs from 0 to 1, although the actual domain parameter of the curve may be from a to b. Clearly, we would need a domain parameter transformation from an arbitrary range to the standard domain. So, the question is how do we reparameterize so that the curve degree and the shape remain invariant? In other words, how is an arbitrary domain parameter range related to the standard domain parameter range of [0, 1], so that the degree and the shape of the curve are not disturbed, that is, stays invariant. This is exactly what we set ourselves to achieve in the following. Let us recall that the ***barycentric or affine combinations*** are invariant under the ***affine transformations*** of translation, rotation, shearing, scaling and reflection. Once we identify that the Bernstein expansion of a Bezier curve is barycentric, and that we impose that the required transformation from the global to the local parameter (or, an arbitrary range to the standard range) is an affine map, our job is practically done, that is, the invariance of a Bezier curve under domain reparameterization is proved. Another way to see this is to recognize that a Bezier curve can be constructed by the de Casteljau algorithm, which is essentially a sequence of linear interpolations. But, a linear interpolation is an affine map, because the barycentric combination is invariant under linear interpolation, and hence a Bezier curve is invariant if the reparameterization itself is an affine map.

5.6.2.3 Affine Reparameterization Invariance: Standard Bezier Curve

Consider the Bezier curve as shown in Figure 5.18. The global parameter, σ, runs from a to b. We would be interested to find the local (or standard) parameter, $\xi \in [0, 1]$, and its relationship with

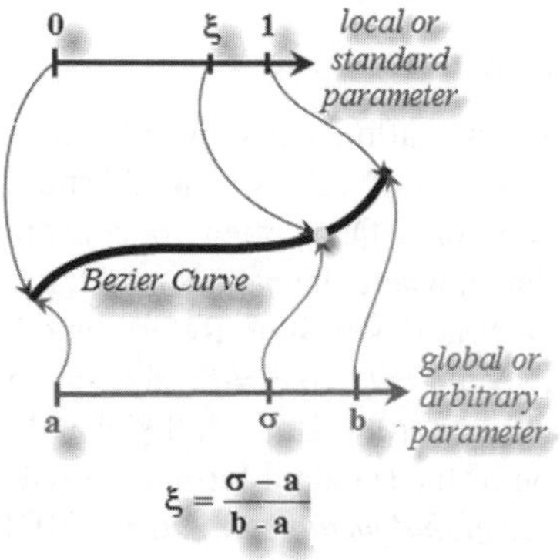

Figure 5.18 Global–local coordinate relationship.

the global (or arbitrary) parameter, σ. We define the standard parameter, ξ, such that $0 \leq \xi \leq 1$, to describe a curve in the standard form. For invariance under this reparameterization, we seek to describe the curve with the standard parameter by the same degree Bezier curve as is done by the arbitrary parameter, $\sigma \in [a, b]$. In order that the degree and the shape of the curve remain unchanged, we must have the linear relationship between any global parameter, σ, and the corresponding local parameter, ξ, as follows:

$$\sigma = (1 - \xi)a + \xi b = a + (b - a)\xi \tag{5.59}$$

Clearly then $\xi = 0$ and $\xi = 1$ correspond to a and b, respectively. Thus, from equation (5.59), it follows that we may define the local parameter, ξ, in terms of the global parameter, σ, as:

$$\boxed{\xi = \frac{(\sigma - a)}{(b - a)} = \frac{(1)}{(b - a)}\sigma + \frac{(-a)}{(b - a)}} \tag{5.60}$$

and ξ assumes values from 0 and 1, that is, $0 \leq \xi \leq 1$. From the second equality, we may now recognize that the parameter transformation given by expression (5.60) is an ***affine map***, that is, of the form: $\bar{x} = cx + d$ where x is transformed to $\bar{x}$ with c, d constants.

Note that expression (5.60) is not a linear map as it contains an element of translation indicated by the part $d = \frac{(-a)}{(b-a)}$ above.

Now, recall that an affine map is defined by the requirement that the ***barycentric combination*** remains invariant. Bezier curves are clearly given as a barycentric combination, as in equation (5.42), because of the ***partition property of the Bernstein basis functions***, that is, $\sum_{i=0}^{n} B_i^n(\xi) = 1, \quad \xi \in [0, 1]$. Thus, under affine transformation given by expression (5.60), the barycentric combination remains invariant. Thus, the degree and the shape of the curve remain invariant.

5.6.2.4 Effect of Reparameterization on the de Casteljau Algorithm

Based on our discussion above, for construction of a Bezier curve by de Casteljau algorithm, we simply need to plug in the local parameter, ξ, computed from a global (or arbitrary) to local (or standard) parameter transformation given by equation (5.60), and continue with the de Casteljau algorithm. Now, we will elaborate on this with a couple of examples.

5.6.2.4.1 Linear Interpolation: de Casteljau Scheme

Referring to Figure 5.19, suppose that the arbitrary parameter range is given by: $\sigma \in [a, b]$. The two Bezier control points, for a linear curve, are $\mathbf{q_0} \equiv \mathbf{q_0^0}$ and $\mathbf{q_1} \equiv \mathbf{q_1^0}$.

Now, following the de Casteljau algorithm in terms of domain parameter, $\sigma \in [a, b]$, we have:

$$\mathbf{C}(\sigma) \equiv \mathbf{q_0^1} = \left\{ \frac{b - \sigma}{b - a}\mathbf{q_0} + \frac{\sigma - a}{b - a}\mathbf{q_1} \right\} \tag{5.61}$$

Note that the curve is given by a barycentric combination, because: $\frac{b-\sigma}{b-a} + \frac{\sigma-a}{b-a} = 1$. Clearly, however, expression (5.61) is rather awkward and cumbersome. So, if we plug in the parameter

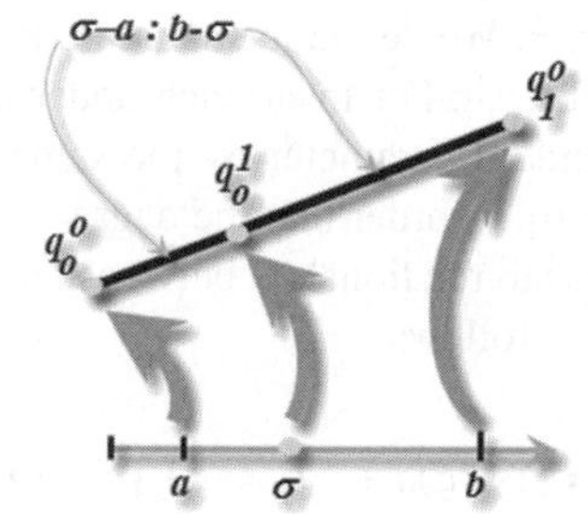

Figure 5.19 Linear interpolation.

transformation given by equation (5.59) in expression (5.61), we get de Casteljau in standard parameter, $\xi \in [0, 1]$:

$$\begin{aligned}\mathbf{C}(\xi) &= \frac{1}{b-a}\{(b-\sigma)\mathbf{q}_0 + (\sigma - a)\mathbf{q}_1\} \\ &= \frac{1}{b-a}[\{(1-\xi)b - (1-\xi)\}\mathbf{q}_0 + (b-a)\xi\mathbf{q}_1] = \underline{\underline{(1-\xi)\mathbf{q}_0 + \xi\mathbf{q}_1}}\end{aligned} \tag{5.62}$$

The description of the same curve by equation (5.62) in terms of the standard parameter is obviously a barycentric combination. But, clearly, this time the expression is much nicer and simpler.

5.6.2.4.2 Quadratic Interpolation: de Casteljau Scheme

Referring to Figure 5.20, suppose that the arbitrary parameter range is given by: $\sigma \in [a, b]$. The three Bezier control points, for a quadratic curve, are $\mathbf{q}_0 \equiv \mathbf{q}_0^0$, $\mathbf{q}_1 \equiv \mathbf{q}_1^0$ and $\mathbf{q}_2 \equiv \mathbf{q}_2^0$.

Now, following the de Casteljau algorithm in terms of domain parameter, $\sigma \in [a, b]$, we have, for first iteration:

$$\mathbf{q}_0^1(\sigma) = \left\{\frac{b-\sigma}{b-a}\mathbf{q}_0^0 + \frac{\sigma-a}{b-a}\mathbf{q}_1^0\right\}, \quad \mathbf{q}_1^1(\sigma) = \left\{\frac{b-\sigma}{b-a}\mathbf{q}_1^0 + \frac{\sigma-a}{b-a}\mathbf{q}_2^0\right\} \tag{5.63}$$

By applying the transformation given by equation (5.59) in equations (5.63), as we did in the previous linear interpolation, we get:

$$\mathbf{q}_0^1(\xi) = \left\{(1-\xi)\mathbf{q}_0^0 + \xi\mathbf{q}_1^0\right\}, \mathbf{q}_1^1(\xi) = \left\{(1-\xi)\mathbf{q}_1^0 + \xi\mathbf{q}_2^0\right\} \tag{5.64}$$

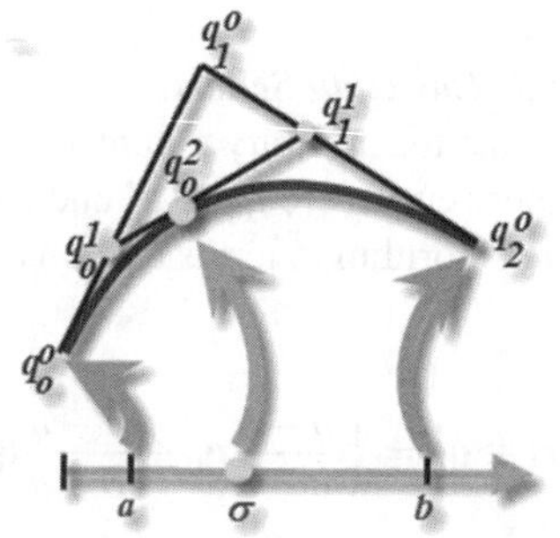

Figure 5.20 Quadratic interpolation.

Now, for the second iteration of the de Casteljau algorithm, we have:

$$\mathbf{C}(\sigma) \equiv \mathbf{q}_0^2 = \left\{ \frac{b-\sigma}{b-a}\mathbf{q}_0^1(\sigma) + \frac{\sigma-a}{b-a}\mathbf{q}_1^1(\sigma) \right\} \tag{5.65}$$

Note that the curve is given by a barycentric combination, because: $\frac{b-\sigma}{b-a} + \frac{\sigma-a}{b-a} = 1$. Clearly, however, expression (5.65) is rather awkward and cumbersome. So, if we plug in the parameter transformation given by equation (5.59) in expression (5.65), we get de Casteljau in standard parameter, $\xi \in [0, 1]$:

$$\begin{aligned}
\mathbf{C}(\xi) &= \frac{1}{b-a}\{(b-\sigma)\mathbf{q}_0^1(\sigma) + (\sigma-a)\mathbf{q}_1^1(\sigma)\} \\
&= \frac{1}{b-a}[\{(1-\xi)b - (1-\xi)\}\mathbf{q}_0^1(\xi) + (b-a)\xi\mathbf{q}_1^1(\xi)] \\
&= (1-\xi)\mathbf{q}_0^1(\xi) + \xi\mathbf{q}_1^1(\xi) = \underline{\underline{(1-\xi)^2\mathbf{q}_0 + 2(1-\xi)\xi\mathbf{q}_1 + \xi^2\mathbf{q}_2}}
\end{aligned} \tag{5.66}$$

The description of the same curve by equation (5.66) in terms of the standard parameter is obviously a barycentric combination. But, clearly, this time the expression is much nicer and simpler.

Hence, if we are given an arbitrary domain parameter range, we would simply apply the affine transformation to the standard parameter and then continue with the de Casteljau algorithm of construction of Bezier curves.

5.6.2.5 End Point Interpolation

The end point interpolation property tells us that a Bezier curve, given an ordered sequence of ***control points***, will always pass through the two end points, that is, the first and the last points. This property puts Bezier curves at a great advantage compared to ordinary splines which do not interpolate the end points without a multiplicity of ***knots*** or ***nodes***. Accordingly, in finite element method, the end Bezier control points clearly provide a natural definition of nodes of the displacement functions. Let us recall that an nth standard Bezier curve, $\mathbf{C}^n(\xi)$, is defined as:

$$\mathbf{C}_n(\xi) = \sum_{i=0}^{n} B_i^n(\xi)\,\mathbf{q}_i, \quad 0 \le \xi \le 1 \tag{5.67}$$

where the basis (blending) functions, $\{B_i^n(\xi)\}$, are the nth degree Bernstein polynomials. Since, $B_i^n(0) = B_i^n(1) = 1$ if and only if $i = n$, and because the Bernstein polynomials are non-negative and make a partition of unity: $\sum_{i=0}^{n} B_i^n(\xi) = 1$, we have $\mathbf{C}_n(0) = \mathbf{q}_0$ as one end point and $\mathbf{C}_n(1) = \mathbf{q}_n$ as another end point, for any degree, that is, n.

5.6.2.6 Convex Hull

The convex hull property has an enormously cost-effective application in an otherwise very costly ***interference checking*** between objects. One can easily form the so-called ***min–max boxes*** wrapping around the ***control points*** of each object and check if these boxes interfere. If they do not, the curves certainly do not. The saving could be tremendous if we have thousands of objects to check against. Of course, if they do, further checks need to be made. Because of repeated

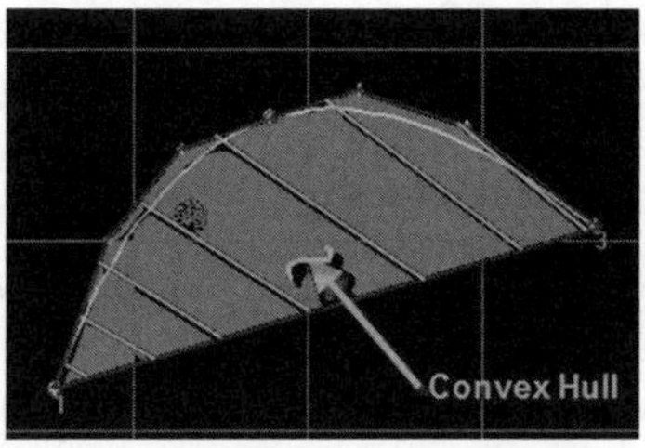

Figure 5.21 Convex hull. (See colour plate section)

convex combinations of control points, as shown in the construction of a Bezier curve by de Casteljau algorithm, the resulting curve is fully contained within the convex hull of its control points. Because of this property, if the control points are planar, so is the curve as shown in Figure 5.21.

5.6.2.7 Symmetry of the Bezier Curve Representation

The symmetry property enables the description of a Bezier curve in terms of its control points, independent of the way the control polygon is traversed. This has its application in joining two or more Bezier curves or in the subdivision of a Bezier curve. It follows from the symmetry property of the Bernstein polynomials. Suppose that we are given an $(n+1)$ ordered set of control points: $\{\mathbf{q}_0 \quad \mathbf{q}_1 \quad \cdots \quad \mathbf{q}_{n-1} \quad \mathbf{q}_n\}$ describing a Bezier curve. We can describe the same curve by reversing the order of the control point sequence to: $\{\mathbf{q}_n \quad \mathbf{q}_{n-1} \quad \cdots \quad \mathbf{q}_1 \quad \mathbf{q}_0\}$. However, noting the symmetry property of the Bernstein polynomials as: $B_i^n(\xi) = B_{n-i}^n(1-\xi)$, we must have the curve characterized as:

$$\mathbf{C}_n(\xi) = \sum_{i=0}^{n} B_i^n(\xi)\,\mathbf{q}_i = \sum_{i=0}^{n} B_i^n(1-\xi)\,\mathbf{q}_{n-i}, \quad 0 \le \xi \le 1 \tag{5.68}$$

Recall that the Bernstein polynomials are symmetric about $\xi = \frac{1}{2}$.

5.6.3 *Derivatives of Bezier Curves*

Our main goal here is to provide a gateway to various ways of calculating the derivatives (i.e. tangents, curvatures, etc.) of a ***Bezier curve*** essential for both mesh generation and finite element method. Our two different yet connected introductions to Bezier curves through de Casteljau algorithm and Bernstein polynomial basis functions nicely tie up the two complementary ways of looking at any computational entity – by geometry and by algebra. Bezier by the de Casteljau algorithm provides the geometric flavor over the algebraic structures of the Bernstein polynomial expansion. Corresponding to each of these two different ways of describing a Bezier curve, naturally there is a method to find the derivatives of a Bezier curve:

- Method 1: Algebraically, by the application of the ***forward difference operators***, resulting from the ***recurrence relations of Bernstein polynomials***, to the ***control points*** of a Bezier curve.
- Method 2: As a by-product of the geometric de Casteljau algorithm by application of the difference operator to the control points of the ***intermediate auxiliary Bezier curves*** and then taking their difference.

Furthermore, as we shall see, it is clearly more economical to ***subdivide*** (or partition or split) a curve at the point of interest and evaluate the end point derivatives by either of the two above methods of any one of the resulting subdivided curves. Finally, for most practical applications, we need the first derivatives (related to the tangents) and the second derivatives (related to the curvatures). However, for twist computation, we need the third derivative of a Bezier curve. Thus, we will now present two different ways of evaluating the derivatives of, and subdivision algorithm for, a curve.

5.6.3.1 Derivatives: Bernstein Form

Here, we present the derivative formulas for a Bezier curve in terms of the Bernstein basis functions. If we are interested only in the derivatives as in the finite element phase of element stiffness matrix generation with as yet unknown displacement control vectors, these forms seem appropriate for use. We apply directly the derivative formulas of the Bernstein polynomials to the definition of a Bezier curve of a given order and obtain the derivatives of a Bezier curve in a recurrence form involving lower-order Bezier curves.

5.6.3.1.1 First Derivative

Let us recall the first derivative property of a Bernstein polynomial:

$$B^n_{i,\xi}(\xi) = n\left(B^{n-1}_{i-1}(\xi) - B^{n-1}_i(\xi)\right), \quad 0 \le \xi \le 1 \tag{5.69}$$

Let us also recall, from the Bernstein–Bezier connection, that a Bezier curve, $\mathbf{C}_n(\xi)$, of degree n is given by:

$$\mathbf{C}_n(\xi) = \sum_{i=0}^{n} B^n_i(\xi)\,\mathbf{q}_i, \quad 0 \le \xi \le 1 \tag{5.70}$$

Now applying equation (5.69) directly in equation (5.42), we get the first-order derivative, $\mathbf{C}_{n,1}(\xi)$, of a Bezier curve as:

$$\begin{aligned}
\mathbf{C}_{n,1}(\xi) &= n\sum_{i=0}^{n}\left(B^{n-1}_{i-1}(\xi) - B^{n-1}_i(\xi)\right)\mathbf{q}_i \\
&= n\sum_{i=1}^{n}\left(B^{n-1}_{i-1}(\xi)\,\mathbf{q}_i - n\sum_{i=0}^{n-1} B^{n-1}_i(\xi)\right)\mathbf{q}_i \qquad \left(\because B^{n-1}_{i-1}(\xi), \quad i \notin \{1,2,\dots,n-1\}\right) \\
&= n\sum_{i=0}^{n-1}\left(B^{n-1}_i(\xi)\,\mathbf{q}_{i+1} - n\sum_{i=0}^{n-1} B^{n-1}_i(\xi)\right)\mathbf{q}_i \qquad (\text{Re-index: } i-1 \to i) \\
&= n\sum_{i=0}^{n-1}(\mathbf{q}_{i+1} - \mathbf{q}_i)\,B^{n-1}_i(\xi)
\end{aligned} \tag{5.71}$$

Let us note the following.

- The first derivative of a Bezier curve is also a Bezier curve. Its degree is one less than that of the original one. Its control points are obtained by simply taking the difference of the original control points in a forward difference manner.
- The derivative expression greatly improves computational efficiency: we never have to take the actual derivative of the curve; we simply use the forward difference of the control points, and lower the degree of the Bernstein polynomial by one.

In order to write equation (5.71) in a compact form, we can recall our definition of the first forward difference operator: $\Delta^1 \mathbf{q}_i \equiv \mathbf{q}_{i+1} - \mathbf{q}_i$. Thus, we can rewrite the first derivative of a Bezier curve as:

$$\boxed{\mathbf{C}_{n,1}(\xi) = n \sum_{i=0}^{n-1} (\Delta^1 \mathbf{q}_i)\, B_i^{n-1}(\xi)} \tag{5.72}$$

5.6.3.1.2 End Point First Derivatives

The end point derivatives are of great importance because derivatives of a Bezier curve at any $\xi \in [0, 1]$ can be reduced to finding the corresponding end point derivative by subdividing the curve at ξ and considering either of the two resulting curves. Noting the value of $B_i^{n-1}(0)$ and $B_i^{n-1}(1)$, we can easily write the expressions for the end point derivatives:

$$\boxed{\begin{aligned} \mathbf{C}_{n,1}(0) &= n\,(\Delta^1 \mathbf{q}_0) = n\,(\mathbf{q}_1 - \mathbf{q}_0) \\ \mathbf{C}_{n,1}(1) &= n\,(\Delta^1 \mathbf{q}_{n-1}) = n\,(\mathbf{q}_n - \mathbf{q}_{n-1}) \end{aligned}} \tag{5.73}$$

Let us observe the following.

- The first derivatives at end points involve one more control point than the end control points themselves.
- The tangents of the curve at end points will point along the direction of the end sides of the Bezier polygon.

Example: For $n = 3$, that is, a cubic Bezier curve

$$\mathbf{C}_{3,1}(0) = 3\,(\mathbf{q}_1 - \mathbf{q}_0) \qquad \mathbf{C}_{3,1}(1) = 3\,(\mathbf{q}_3 - \mathbf{q}_2)$$

5.6.3.1.3 General Derivatives

The higher-order derivatives can be similarly expressed through the use of higher-order forward difference operators and repeated application of the first-order derivatives. Thus, the dth derivative, $\mathbf{C}_{n,d}(\xi)$, of a Bezier curve is:

$$\boxed{\mathbf{C}_{n,d}(\xi) = \frac{n!}{(n-d)!} \sum_{i=0}^{n-d} (\Delta^d \mathbf{q}_i)\, B_i^{n-d}(\xi), \qquad d \in \{0, 1, \ldots, n\}} \tag{5.74}$$

Example: For a cubic Bezier curve, $\mathbf{C}_3(\xi)$, we would like to find the point, $\mathbf{C}_3(\frac{1}{3})$, and the first derivative, $\mathbf{C}_{3,1}(\xi)$, at $\xi = \frac{1}{3}$. With $n = 3$, $d = 1$ and $\xi = \frac{1}{3}$, for the point on the curve at $\xi = \frac{1}{3}$, we get:

$$\begin{aligned} \mathbf{C}_3\left(\frac{1}{3}\right) \equiv \mathbf{C}_3(\xi)\Big|_{\xi=\frac{1}{3}} &= \left\{(1-\xi)^3\, \mathbf{q}_0 + 3(1-\xi)^2\, \xi\, \mathbf{q}_1 + 3\,(1-\xi)\xi^2\, \mathbf{q}_2 + \xi^3\, \mathbf{q}_3\right\}_{\xi=\frac{1}{3}} \\ &= \underline{\underline{\left(\frac{8}{27}\right) \mathbf{q}_0 + \left(\frac{4}{9}\right) \mathbf{q}_1 + \left(\frac{2}{9}\right) \mathbf{q}_2 + \left(\frac{1}{27}\right) \mathbf{q}_3}} \end{aligned}$$

Then, using equation (5.74) with $n = 3$, $d = 1$ and $\xi = \frac{1}{3}$, we have:

$$\begin{aligned}\mathbf{C}_{3,1}(\xi)\Big|_{\xi=\frac{1}{3}} &= 3\sum_{i=0}^{2}(\Delta^1\mathbf{q}_i)\,B_i^2(\xi)\\ &= 3\{(1-\xi)^2(\mathbf{q}_1-\mathbf{q}_0)+2\xi(1-\xi)(\mathbf{q}_2-\mathbf{q}_1)+\xi^2(\mathbf{q}_3-\mathbf{q}_2)\}_{\xi=\frac{1}{3}}\\ &= 3\{\tfrac{4}{9}(\mathbf{q}_1-\mathbf{q}_0)+\tfrac{4}{9}(\mathbf{q}_2-\mathbf{q}_1)+\tfrac{1}{9}(\mathbf{q}_3-\mathbf{q}_2)\} = \underline{\underline{-\tfrac{4}{3}\mathbf{q}_0+\mathbf{q}_2+\tfrac{1}{3}\mathbf{q}_3}}\end{aligned}$$

5.6.3.1.4 End Point Higher Derivatives

As indicated before the end point derivatives are extremely important for smoothness definition between adjoining Bezier curves, or inversely for subdivision (i.e. splitting of Bezier curves) algorithms central to both interactive modeling and finite element mesh adaptation and refinement. The end point derivatives are given as:

$$\boxed{\mathbf{C}_{n,d}(0) = \frac{n!}{(n-d)!}\,(\Delta^d\mathbf{q}_0),\quad \mathbf{C}_{n,1}(1) = \frac{n!}{(n-d)!}\,(\Delta^d\mathbf{q}_{n-d})} \tag{5.75}$$

Let us note the following.

- The dth end point derivatives of a Bezier curve depend only on $(d+1)$ control points from the corresponding ends. This follows directly from the definition of the dth degree forward difference which we may recall as: $\Delta^d\mathbf{q}_i \equiv \sum_{j=0}^{d}(-1)^{d-j}\binom{d}{j}\mathbf{q}_{i+j}$

Example: For $n = 3$, that is, a cubic Bezier curve, the first ($d = 1$) and second derivatives ($d = 2$) at starting end point ($\xi = 0$) are given as:

$$\boxed{\mathbf{C}_{3,1}(0) = 3\,(\mathbf{q}_1-\mathbf{q}_0),\quad \mathbf{C}_{3,2}(0) = 6\,\Delta^2\mathbf{q}_0 = 6\,(\mathbf{q}_2-2\mathbf{q}_1+\mathbf{q}_0)} \tag{5.76}$$

Thus the tangent at $\xi = 0$ is defined by $\mathbf{q}_0$ and $\mathbf{q}_1$. The curvature, likewise, is proportionally defined by only $\mathbf{q}_0$, $\mathbf{q}_1$ and $\mathbf{q}_2$.

5.6.3.1.5 Discussion

- Here we have presented the derivative expressions for a Bezier curve with special reference to the first- and second-order derivatives, which are of more practical importance. However, in order to determine them simultaneously along with the Bezier curve points, a better alternative may be a utilization of the de Casteljau algorithm. This method of the derivative evaluation is introduced next.

5.6.3.2 Derivatives: de Casteljau Form

Next we will look at the simultaneous evaluation of a point and the derivatives of a Bezier curve through the de Casteljau algorithm. In the mesh generation phase, we often need information on both the point on a curve and its derivatives, as in the smooth joining of multiple Bezier curves. In such situations, the de Casteljau algorithm seems more suitable than the Bernstein form of the derivatives. We first observe that the forward difference operator is linear in control points. Then, starting from the derivative definition in terms of the Bernstein polynomials, we can eventually identify the derivatives with the difference of the intermediate de Casteljau control points.

5.6.3.2.1 General Derivatives

Let us recall that ***the Bernstein form*** of the dth derivative, $\mathbf{C}_{n,d}(\xi)$, for an n-degree Bezier curve is given as:

$$\boxed{\mathbf{C}_{n,d}(\xi) = \frac{n!}{(n-d)!}\sum_{i=0}^{n-d}(\Delta^d \mathbf{q}_i)\, B_i^{n-d}(\xi), \qquad d \in \{0, 1, \dots, n\}} \tag{5.77}$$

Now noting the linearity property of a forward difference operator, we can rewrite $\mathbf{C}_{n,d}(\xi)$ as:

$$\begin{aligned}
\mathbf{C}_{n,d}(\xi) &= \frac{n!}{(n-d)!}\sum_{i=0}^{n-d}(\Delta^d \mathbf{q}_i)\, B_i^{n-d}(\xi), && d \in \{0, 1, \dots, n\}\\
&= \frac{n!}{(n-d)!}\Delta^d \left\{\sum_{i=0}^{n-d} \mathbf{q}_i\, B_i^{n-d}(\xi)\right\} && \text{(linear operators } \Sigma \text{ and } \Delta\text{commute)}\\
&= \frac{n!}{(n-d)!}\Delta^d\, \mathbf{C}_{n-d}(\xi) && \text{(by definition)}\\
&= \frac{n!}{(n-d)!}\Delta^d\, \mathbf{q}_0^{n-d}(\xi) && \text{(by de Casteljau algorithm)}\\
&= \frac{n!}{(n-d)!}(\Delta^{d-1}\, \mathbf{q}_1^{n-d}(\xi) \;-\; \Delta^{d-1}\, \mathbf{q}_0^{n-d}(\xi)) && \text{(definition of } \Delta^d\text{)}
\end{aligned} \tag{5.78}$$

Thus, the de Casteljau form of the derivatives for a Bezier curve is given by:

$$\boxed{\mathbf{C}_{n,d}(\xi) = \frac{n!}{(n-d)!}(\Delta^{d-1}\, \mathbf{q}_1^{n-d}(\xi) \;-\; \Delta^{d-1}\, \mathbf{q}_0^{n-d}(\xi))} \tag{5.79}$$

where, Δ is the forward-difference operator, $\mathbf{q_i}$'s are the control points and B_i^n is the ith Bernstein basis functions for Bernstein polynomials of degree n.

Let us note that

- In the de Casteljau algorithm, the polygons, $\mathbf{q}_i^{n-d}(\xi), \quad i \in \{0, 1, .., n\}$ are generated as intermediate auxiliary polygons.

5.6.3.2.2 First Derivatives

The above observation provides an excellent opportunity to use the de Casteljau algorithm to compute curve points and the first derivatives at those points, if required simultaneously. Thus, for the computation of the first derivative, or equivalently, the ***tangent vectors*** at any $\xi \in [0, 1]$, we get with $d = 1$:

$$\boxed{\mathbf{C}_{n,1}(\xi) = n\left(\mathbf{q}_1^{n-1} - \mathbf{q}_0^{n-1}\right)} \tag{5.80}$$

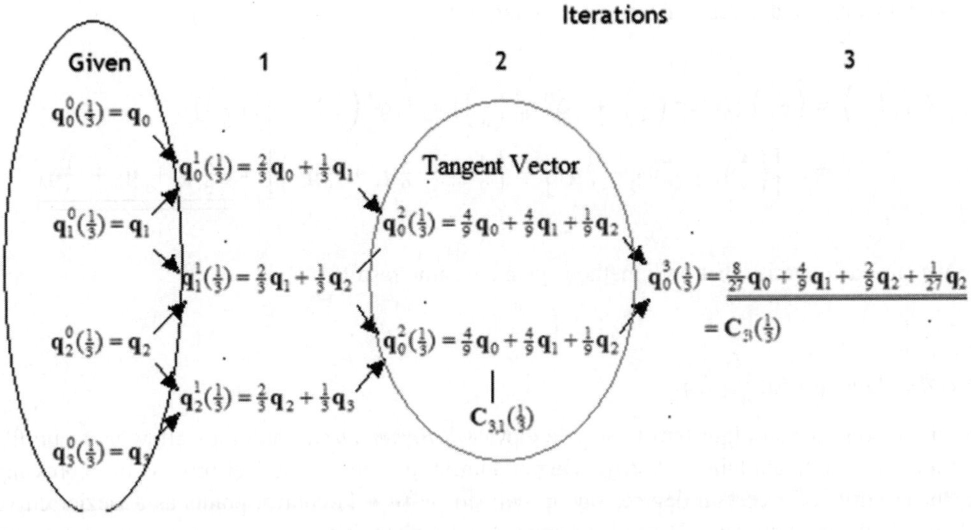

Figure 5.22 de Casteljau construction of a point on cubic curve.

- The tangent vectors are given by the difference of the intermediate de Casteljau points scaled by n. In other words, among the intermediate polygons that are created successively towards evaluation of a point on the curve, the final polygon, which is a line segment, coincides with the tangent at that point.

5.6.3.2.3 *Geometric Feel*

The above observation may be grasped better by presenting the de Casteljau algorithm in a geometric form. Let us now consider the de Casteljau algorithm as applied to the case of a cubic ($n = 3$) Bezier curve.

Example: Recalling the same example that was solved by the Bernstein form of derivative computation, for a cubic Bezier curve, $\mathbf{C}_3(\xi)$, we would like to find the point, $\mathbf{C}_3(\frac{1}{3})$, and the first derivative, $\mathbf{C}_{3,1}(\xi)$, at $\xi = \frac{1}{3}$ by the de Casteljau method. As shown in Figure 5.22, and using the de Casteljau algorithm with $n = 3$, $d = 1$ and $\xi = \frac{1}{3}$:

Initialize: $\mathbf{q}_0^0\left(\frac{1}{3}\right) = \mathbf{q}_0 \quad \mathbf{q}_1^0\left(\frac{1}{3}\right) = \mathbf{q}_1 \quad \mathbf{q}_2^0\left(\frac{1}{3}\right) = \mathbf{q}_2\mathbf{q}_3^0\left(\frac{1}{3}\right) = \mathbf{q}_3$

Iteration 1: $\mathbf{q}_0^1\left(\frac{1}{3}\right) = \frac{2}{3}\mathbf{q}_0 + \frac{1}{3}\mathbf{q}_1, \quad \mathbf{q}_1^1\left(\frac{1}{3}\right) = \frac{2}{3}\mathbf{q}_1 + \frac{1}{3}\mathbf{q}_2, \quad \mathbf{q}_2^1\left(\frac{1}{3}\right) = \frac{2}{3}\mathbf{q}_2 + \frac{1}{3}\mathbf{q}_3$

Iteration 2: $\mathbf{q}_0^2\left(\frac{1}{3}\right) = \frac{2}{3}\mathbf{q}_0^1\left(\frac{1}{3}\right) + \frac{1}{3}\mathbf{q}_1^1\left(\frac{1}{3}\right) = \frac{4}{9}\mathbf{q}_0 + \frac{4}{9}\mathbf{q}_1 + \frac{1}{9}\mathbf{q}_2, \quad \mathbf{q}_1^2\left(\frac{1}{3}\right) = \frac{2}{3}\mathbf{q}_1^1\left(\frac{1}{3}\right) + \frac{1}{3}\mathbf{q}_2^1\left(\frac{1}{3}\right) = \frac{4}{9}\mathbf{q}_1 + \frac{4}{9}\mathbf{q}_2 + \frac{1}{9}\mathbf{q}_3$

Iteration 3: $\mathbf{C}_3\left(\frac{1}{3}\right) \equiv \mathbf{q}_0^3\left(\frac{1}{3}\right) = \frac{2}{3}\mathbf{q}_0^2\left(\frac{1}{3}\right) + \frac{1}{3}\mathbf{q}_1^2\left(\frac{1}{3}\right) = \underline{\underline{\frac{8}{27}\mathbf{q}_0 + \frac{4}{9}\mathbf{q}_1 + \frac{2}{9}\mathbf{q}_2 + \frac{1}{27}\mathbf{q}_2}}$

And, finally, the derivative from terms in iteration 2:

$$\mathbf{C}_{3,1}\left(\frac{1}{3}\right) = \left(\frac{3!}{2!}\right)\{\Delta^0\mathbf{q}_1^2\left(\frac{1}{3}\right) + \Delta^0\mathbf{q}_0^2\left(\frac{1}{3}\right)\} = 3\{\mathbf{q}_1^2\left(\frac{1}{3}\right) - \mathbf{q}_0^2\left(\frac{1}{3}\right)\}$$
$$= 3\left[\left\{\frac{4}{9}\mathbf{q}_1 + \frac{4}{9}\mathbf{q}_2 + \frac{1}{9}\mathbf{q}_3\right\} - \left\{\frac{4}{9}\mathbf{q}_0 + \frac{4}{9}\mathbf{q}_1 + \frac{1}{9}\mathbf{q}_2\right\}\right] = \underline{\underline{-\frac{4}{3}\mathbf{q}_0 + \mathbf{q}_2 + \frac{1}{3}\mathbf{q}_3}}$$

We can easily verify that both methods give the same results.

5.6.4 Degree Elevation

Next we consider an algorithm to degree elevate a ***Bezier curve***, and thus allow us to profile a curve as will be explained shortlyl. Degree elevation is the numerical process of expressing a Bezier curve of a certain degree, say, n, defined by $(n + 1)$ control points as a Bezier curve of $(n + 1)$ degree with $(n + 2)$ control points without changing the shape of the original curve. Degree elevation is central to both interactive modeling and finite element analysis. In the modeling phase based on boundary representation, the surfaces are formed by boundary curves of the same degree. Thus, if any of these curves is previously generated with a lower degree than the highest degree boundary curve, the lower degree curve needs to be degree elevated, that is, redefined in terms of the highest degree without changing the shape of the curve in any way for the surface definition. In another situation, where, for example, a curve of degree 1 (i.e. a line segment) is joined with a curve of degree 2 (i.e. a parabola) to generate a boundary curve of a surface, the line segment needs to be expressed as a special degenerate quadratic curve so that the entire profile curve is quadratic; an example of curve blending by degree elevation is shown in Figure 5.23.

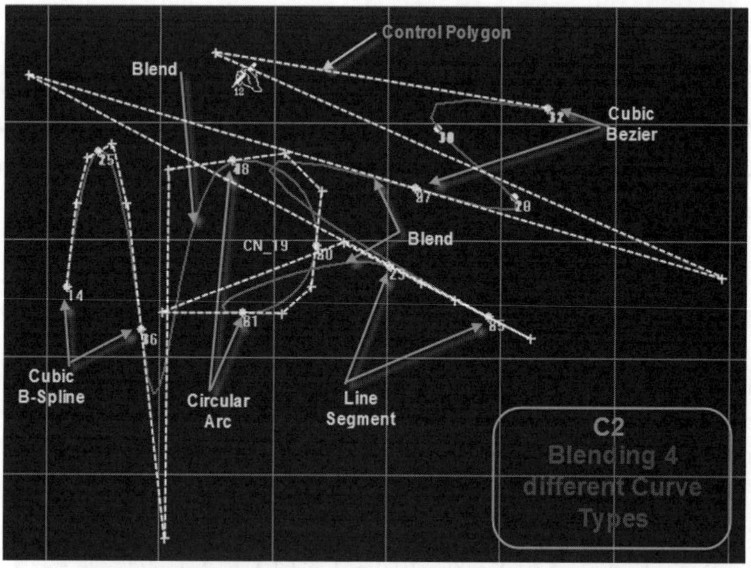

Figure 5.23 Curve blending by degree elevation. (See colour plate section)

5.6.4.1 Degree Elevation Proposition

Let $\{\mathbf{q}_i\}, \quad i \in \{0, 1, 2, \bullet\bullet\bullet, n\}$ be the control vertices of a Bezier polygon defining a Bezier curve, $\mathbf{C}_n(\xi)$, of degree n on $\xi \in [0, 1]$. Then, let us define: $\{\mathbf{q}_i^e\}, \quad i \in \{0, 1, 2, \bullet\bullet\bullet, n+1\}$, the new set of $(n+2)$ control vertices such that:

$$\boxed{\mathbf{q}_i^e = \left(\frac{i}{n+1}\right)\mathbf{q}_{i-1} + \left(1 - \frac{i}{n+1}\right)\mathbf{q}_i, \quad i \in \{0, 1, 2, \bullet\bullet\bullet, n+1\}} \tag{5.81}$$

that is, $\{\mathbf{q}_i^e\}, \quad i \in \{0, 1, 2, \bullet\bullet\bullet, n+1\}$ are obtained by piecewise linear interpolation of $\mathbf{q}_{i-1}$ and $\mathbf{q}_i$ at parameter value: $\xi = \frac{i}{n+1}, \forall i \in \{0, 1, \bullet\bullet\bullet, n+1\}$. Then, both the sets of control points define the same Bezier curve, $\mathbf{C}_n(\xi)$.

Proof Let $\mathbf{C}^e_{n+1}(\xi)$ be the curve generated by $\{\mathbf{q}_i^e\}, \quad i \in \{0, 1, 2, \bullet\bullet\bullet, n+1\}$; we will show that $\mathbf{C}^e_{n+1}(\xi) = \mathbf{C}_n(\xi)$. By definition and using index $j = i - 1$, we get:

$$\begin{aligned}
\mathbf{C}^e_{n+1}(\xi) &= \sum_{i=0}^{n+1} \mathbf{q}_i^e\, B_i^{n+1}(\xi) = \sum_{i=0}^{n+1} \left(\frac{i}{n+1}\right) \mathbf{q}_{i-1}\, B_i^{n+1}(\xi) + \sum_{i=0}^{n+1} \left(1 - \frac{i}{n+1}\right) \mathbf{q}_i\, B_i^{n+1}(\xi) \\
&= \sum_{i=0}^{n+1} \left(\frac{i}{n+1}\right) \frac{(n+1)!}{(n+1-i)!\, i!} \xi^i (1-\xi)^{n+1-i} \mathbf{q}_{i-1} + \sum_{i=0}^{n+1} \left(\frac{n+1-i}{n+1}\right) \frac{(n+1)!}{(n+1-i)!\, i!} \xi^i (1-\xi)^{n+1-i} \mathbf{q}_i \\
&= \xi \left\{ \sum_{j=-1}^{n+1} \frac{n!}{(n-j)!\, j!} \xi^j (1-\xi)^{n-j} \mathbf{q}_j \right\} + (1-\xi) \left\{ \sum_{i=0}^{n+1} \left(\frac{n+1-i}{n+1}\right) \frac{n!}{(n-i)!\, i!} \xi^i (1-\xi)^{n+1-i} \mathbf{q}_i \right\} \\
&= \xi \left\{ \sum_{i=-1}^{n} \mathbf{q}_i\, B_i^n(\xi) \right\} + (1-\xi) \left\{ \sum_{i=0}^{n+1} \mathbf{q}_i\, B_i^n(\xi) \right\}
\end{aligned}$$

Now, noting that $B_i^n(\xi) \equiv 0, \quad \forall i \notin \{0, 1, \ldots, n\}$, following through above expression, we have: $\mathbf{C}^e_{n+1}(\xi) = \xi \left\{ \sum_{i=0}^{n} \mathbf{q}_i\, B_i^n(\xi) \right\} + (1-\xi) \left\{ \sum_{i=0}^{n+1} \mathbf{q}_i\, B_i^n(\xi) \right\} = \sum_{i=0}^{n} \mathbf{q}_i\, B_i^n(\xi) = \underline{\underline{\mathbf{C}_n(\xi)}}$ completes the proof.

Example DE1: Degree elevate the line segment shown in Figure 5.24 to the nth degree.

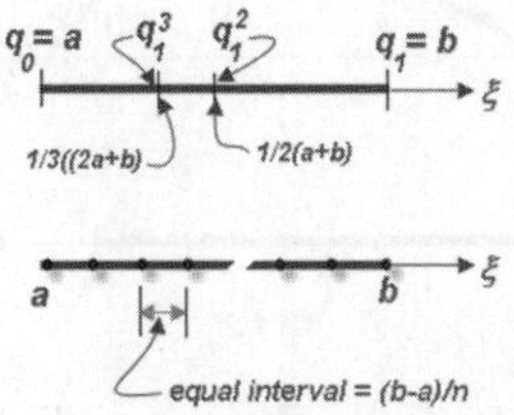

Figure 5.24 Example: degree elevation applied to a linear curve.

We have: $n = 1, \mathbf{q}_0^0 = \mathbf{a}, \mathbf{q}_1^0 = \mathbf{b}$, and the forward difference operator: $\Delta\mathbf{q}_0 = (\mathbf{q}_1 - \mathbf{q}_0) = \mathbf{b} - \mathbf{a}$. Thus, application of the degree elevation algorithm once results in a quadratic, that is, $n = 2$, curve as:

$$\mathbf{q}_0^2 = \mathbf{q}_0$$
$$\mathbf{q}_1^2 = \frac{1}{2}\mathbf{q}_0 + \frac{1}{2}\mathbf{q}_1 = \mathbf{q}_0 + \frac{1}{2}\Delta\mathbf{q}_0$$
$$\mathbf{q}_2^2 = \mathbf{q}_1$$

Now, application of another degree elevation will result in a cubic ($n = 3$) curve as:

$$\mathbf{q}_0^3 = \mathbf{q}_0$$
$$\mathbf{q}_1^3 = \frac{1}{3}\mathbf{q}_0^2 + \frac{2}{3}\mathbf{q}_1^2 = \mathbf{q}_0 + \frac{1}{3}\Delta\mathbf{q}_0$$
$$\mathbf{q}_2^3 = \frac{2}{3}\mathbf{q}_1^2 + \frac{1}{3}\mathbf{q}_2^2 = \mathbf{q}_0 + \frac{2}{3}\Delta\mathbf{q}_0$$
$$\mathbf{q}_3^3 = \mathbf{q}_1$$

By repeated applications of the degree elevation procedure up to degree n, it can be shown to be given by: $\mathbf{q}_i^n = \mathbf{q}_0 + \frac{i}{n}\Delta\mathbf{q}_0$. The resulting control points are shown in Figure 5.24.

Example DE2: The Bezier curve shown in Figure 5.25 has been expressed as a quadratic curve as well as a cubic curve; establish their equivalence.

Let the quadratic, as shown in Figure 5.25, be represented as: $\mathbf{C}_2(\xi) = \sum_{i=0}^{2} \mathbf{q}_i\, B_i^2(\xi) = (1-\xi)^2\mathbf{q}_0 + 2\xi(1-\xi)\mathbf{q}_1 + \xi^2\mathbf{q}_2$. We now let: $\mathbf{q}_0^e = \mathbf{q}_0$, $\mathbf{q}_3^e = \mathbf{q}_2$, and $\mathbf{q}_1^e = \frac{1}{3}(\mathbf{q}_0 + 2\mathbf{q}_1)$, $\mathbf{q}_3^e = \frac{1}{3}(2\mathbf{q}_1 + \mathbf{q}_2)$, then:

$$\begin{aligned}\mathbf{C}_3^e(\xi) &= \sum_{i=0}^{2} \mathbf{q}_i^e\, B_i^3(\xi)\\ &= (1-\xi)^3\mathbf{q}_0 + 3\xi(1-\xi)^2\frac{1}{3}(\mathbf{q}_0 + 2\mathbf{q}_1) + 3\xi^2(1-\xi)\frac{1}{3}(2\mathbf{q}_1 + \mathbf{q}_2) + \xi^3\mathbf{q}_2\\ &= (1-\xi)^2\mathbf{q}_0 + 2\xi(1-\xi)\mathbf{q}_1 + \xi^2\mathbf{q}_2 = \underline{\underline{\mathbf{C}_2(\xi)}}\end{aligned}$$

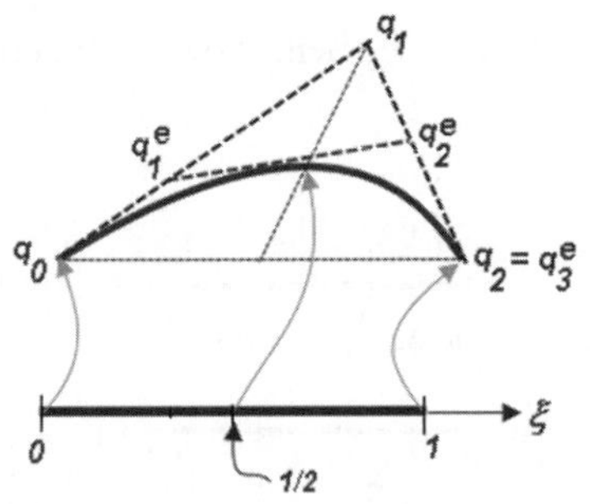

Figure 5.25 Example: degree elevation applied to a quadratic curve.

The process of degree elevation can be applied repeatedly to obtain higher-degree representation of the same Bezier curve. It can be proved by induction that if $\{\mathbf{q}_i^{e_m}\}$, $i \in \{0, 1, \ldots, n+m\}$ are the control vertices obtained after m repeated degree elevations, then:

$$\mathbf{q}_i^{e_m} = \sum_{j=0}^{n} \left[\binom{n}{j} \frac{\binom{m}{i-j}}{\binom{n+m}{i}} \right] \mathbf{q}_j, \quad \forall i \in \{0, 1, \ldots, n+m\} \tag{5.82}$$

5.6.5 Subdivision Algorithm

Here, we offer an algorithm to subdivide or partition a Bezier curve at any given parameter point, and thus produce new Bezier curves. As an example, a point P, corresponding to any $\xi \in [0, 1]$ on a cubic Bezier curve, $\mathbf{C}_3(\xi)$, divides the curve into two curves as shown in Figure 5.26.

Looking at Figure 5.26, the question is: given the Bezier representation of the entire curve, and given the control points, do we have a Bezier representation, that is, can we find the Bezier control points of each part of the curve, $\mathbf{L}_n(\xi)$ on the left of P, and $\mathbf{R}_n(\xi)$ on the right of P? The answer is yes and the process of finding such representations is given by what is known as the ***subdivision algorithm***. It has various applications. For example, in the model generation phase, the formula for evaluating the tangent vector at any arbitrary parameter point on the curve is rather involved and cumbersome. However, the formula for the tangents at the two ends of a curve is very simple and is given by the scaled difference of the two consecutive control points at the end. Thus, if we subdivide the curve at the parameter point, then we can evaluate the tangent vector by applying the end point formula to any of the partitioned curves. The subdivision algorithm is equally important in the ***adaptive refinement*** phase, say in contact–impact problems of ***finite element method*** where we may produce smaller elements from the original element on demand. For convenience and without any loss of generalization, we restrict ourselves to the de Casteljau algorithm for a cubic curve in the following discussion. Then, we explain the subdivision algorithm as a by-product of the de Casteljau algorithm. We provide a simple example of a cubic curve to demonstrate the various properties of the algorithm.

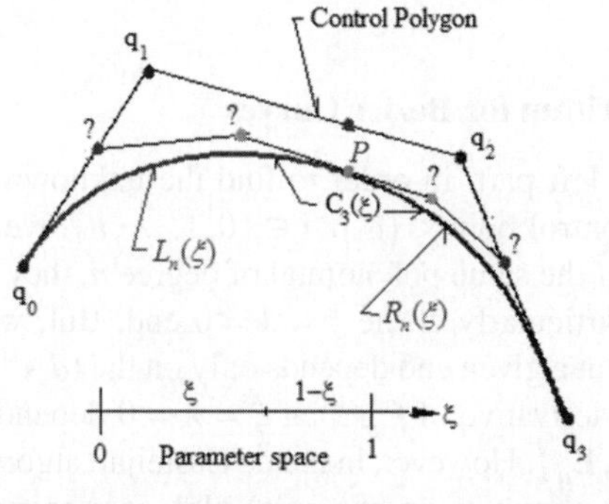

Figure 5.26 Subdivision: left and right half of a cubic curve.

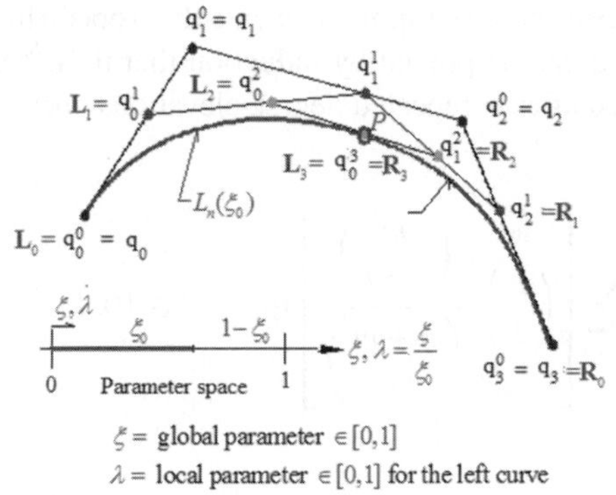

Figure 5.27 Subdivision: left and right control polygons of a cubic curve.

5.6.5.1 de Casteljau Algorithm for a Cubic Bezier

As shown in Figure 5.27, for a cubic ($n = 3$) Bezier Curve, we can explicitly go through the steps of the de Casteljau algorithm as follows:

Step 0: $\mathbf{q}_0^0 = \mathbf{q}_0, \quad \mathbf{q}_1^0 = \mathbf{q}_1, \quad \mathbf{q}_2^0 = \mathbf{q}_2$ and $\mathbf{q}_3^0 = \mathbf{q}_3$, for a given $\xi \in [0, 1]$

Recursion 1: $\mathbf{q}_0^1 = (1 - \xi)\,\mathbf{q}_0 + \xi\,\mathbf{q}_1, \mathbf{q}_1^1 = (1 - \xi)\,\mathbf{q}_1 + \xi\,\mathbf{q}_2, \quad \mathbf{q}_2^1 = (1 - \xi)\,\mathbf{q}_2 + \xi\,\mathbf{q}_3$

Recursion 2: $\mathbf{q}_0^2 = (1 - \xi)\,\mathbf{q}_0^1 + \xi\,\mathbf{q}_1^1, \quad \mathbf{q}_1^2 = (1 - \xi)\,\mathbf{q}_1^1 + \xi\,\mathbf{q}_2^1,$

Recursion 3: $\mathbf{C}_3(\xi) = \mathbf{q}_0^3 = (1 - \xi)\,\mathbf{q}_0^2 + \xi\,\mathbf{q}_1^2$, as shown in Figure 5.27.

Note that the control point sets $\{\mathbf{q}_0, \mathbf{q}_0^1, \mathbf{q}_0^2, \mathbf{q}_0^3\}$ and $\{\mathbf{q}_0^3, \mathbf{q}_1^2, \mathbf{q}_2^1, \mathbf{q}_3\}$ subdivide the original cubic curve into two cubic curves at the parameter point: $\mathbf{L}_n(\xi)$ curve on the left and $\mathbf{R}_n(\xi)$ curve on the right with $n = 3$. Thus, we see that, for a cubic, the de Casteljau algorithm provides us with the control points for the two subdivided curves. Thus, it can be seen as another important by-product of the de Casteljau algorithm that generalizes into what is known as a ***subdivision algorithm*** and is fundamental to both interactive mesh generation (e.g. tangent and curvature evaluation) and finite element analysis (e.g. adaptive refinement).

5.6.5.2 Subdivision Algorithm for Bezier Curves

We first concentrate on the left part: in order to find the unknown control points, $\{\mathbf{L}_i\}$, of the left curve from the given control points, $\{q_i\}$, $i \in \{0, 1, \dots, n\}$, we note that since both curves, $\mathbf{C}_n(\xi)$ and $\mathbf{L}_n(\xi)$, are parts of the same polynomial of degree n, they must be identical in all their derivatives everywhere – particularly at the $\xi = \lambda = 0$ end. But, we already know that the dth ***derivative of a Bezier curve*** at a given end depends only on the $(d + 1)$ consecutive control points from the end. Thus the dth derivative of $\mathbf{C}_n(\xi)$ at $\xi = \lambda = 0$ depends on $\{\mathbf{q}_i\}$, $i \in \{0, 1, \dots, d\}$, and of $\mathbf{L}_n(\xi)$ on $\{\mathbf{L}_0, \mathbf{L}_1, \dots, \mathbf{L}_d\}$. However, in the de Casteljau algorithm, the first $(d + 1)$ control points (see the list of steps above as an example) also correspondingly define the $\mathbf{q}_0^d(\xi)$ and

$\mathbf{L}_0^d(\lambda)$ curves; these two curves are identical since they have the same derivatives of all degrees at $\xi = \lambda = 0$. Now, evaluating at $\xi = \xi_0$ or $\lambda = 1$, we get: $\mathbf{L}_0^d(1) = \mathbf{q}_0^d(\xi_0)$. Finally, noting that $\mathbf{L}_0^d(1) = \mathbf{L}_d$, we have:

$$\mathbf{L}_d = \mathbf{q}_0^d(\xi_0), \quad d \in \{0, 1, \ldots, n\} \tag{5.83}$$

Thus, $\mathbf{L}_d$, the dth control point of the left part, is obtained by the de Casteljau intermediate control points, $\mathbf{q}_0^d(\xi_0)$, of the original curve evaluated at the partition point, $\xi = \xi_0$.

For the right curve, recalling the symmetry property of the Bezier curve, we can easily deduce that $\mathbf{R}_d$, the dth control point of the right part, counted from the right end point towards the partition point, is given by the de Casteljau intermediate control points, $\mathbf{q}_{n-d}^d$, evaluated at the partition point, $\xi = \xi_0$. Thus, we have:

$$\mathbf{R}_d = \mathbf{q}_{n-d}^d(\xi_0), \quad d \in \{0, 1, \ldots, n\} \tag{5.84}$$

In other words, $\mathbf{R}_d$, the dth control point of the right part is obtained by the de Casteljau intermediate control points, $\mathbf{q}_{n-d}^d(\xi_0)$ of the original curve evaluated at the partition point, $\xi = \xi_0$. The expressions given by equations (5.83) and (5.84) are known as the ***subdivision formulas*** for a Bezier curve.

This subdivision process is widely useful. For example:

- For rendering of curved surfaces generated by Bezier patches, they eventually have to be appropriately subdivided for application of planar algorithms of color, light reflection, and so on.
- For derivative evaluations at any point on a curve, the object may be momentarily subdivided to convert subject point to end point of a curve, for which the derivative expressions are very simple and thus cost-effective.
- For various intersection algorithms, the object may need to be subdivided repeatedly.
- For mesh refinement during adaptive finite element scheme such as in contact-impact problems, subdivision is very important as it allows change in element size on demand.

5.7 Composite Bezier–Bernstein Curves

A general mesh representation of a physical continuum of a structural system is expected to contain complex curves to start with. While Bezier representation is very good at mimicking the behavior of a given curve, the degree requirement on the interpolating Bezier curves for high complexity can be prohibitively large. For such cases, it is numerically wise to use the following.

Definition

- A ***composite Bezier curve*** is a set of piecewise Bezier parametric polynomial curves of a particular degree between the junction points with the desired continuity at junction points (known as ***knots***).
- Each component Bezier curve defines a ***Bezier element***, or simply ***element***. Thus far in our discussion we have almost exclusively defined a single Bezier curve as a transformation of a

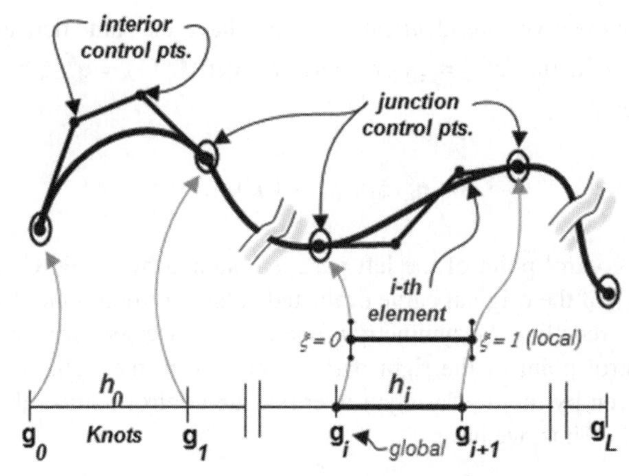

Figure 5.28 Composite Bezier curve.

unit real line segment, $\xi \in [0, 1]$, because of the invariance property of an affine parameter transformation. However, we have piecewise Bezier curves that we can still define the ***element*** or ***local parameter***, $\xi \in [0, 1]$, defines the domain of the individual Bezier curve component.

- However, the composite Bezier curve may be defined over some union of intervals, $\bigcup_{i=0}^{L} [g_i, g_{i+1}]$, in which we designate the ***global parameter*** such that $g \in \bigcup_{i=0}^{L} [g_i, g_{i+1}]$ defines the domain of the entire Bezier curve with the following definitions.
- ***Knots*** are defined as those parameter values, $g_i, \quad i \in \{0, 1, \dots, L\}$ that define the intervals.
- A ***knot sequence*** or ***knot vector*** is defined as the ordered set, $\pi = \{g_0, g_1, \dots, g_L\}$.

For interpolation problems (to be discussed later) in structural mesh generation, representations by composite quadratic C^1 curves for $2D$ geometry, and composite cubic C^1 or C^2 curves for $3D$ geometry seem reasonably adequate. Therefore, from now on, we will restrict our discussion primarily to focusing on cubic Bezier curves unless we need to touch upon general results. Figure 5.28 shows a composite cubic Bezier curve with local and global parameters, and knots.

Obviously, for any i^{th} Bezier element, $i \in \{0, 1, \dots, L\}$, the local parameter, ξ_i, and the global parameter, g, are related by:

$$\xi_i = \frac{g - g_i}{h_i}, \quad i \in \{0, 1, \dots, L\} \tag{5.85}$$

where $h_i = g_{i+1} - g_i, \quad i \in \{0, 1, \dots, L\}$, is the ith knot interval. This clearly makes $\xi_i \in [0, 1]$, $i \in \{0, 1, \dots, L\}$, as desired. For a composite Bezier curve written as $\mathbf{C}(g(\xi))$ or $\mathbf{C}(\xi(g))$, the derivative at a point on the curve, with respect to g or ξ, is related by chain rule as:

$$\boxed{\frac{d\mathbf{C}}{dg} = \left(\frac{1}{h_i}\right) \frac{d\mathbf{C}(\xi_i)}{d\xi_i} \quad \Rightarrow \quad \frac{d\mathbf{C}}{d\xi} = h_i \frac{d\mathbf{C}(g)}{dg}} \tag{5.86}$$

Example: To find the end point first derivatives of a cubic composite Bezier curve with respect to the global parameter, g, we recall from equation (5.76) that for the first segment:

$\frac{d\mathbf{C}_3(0)}{d\xi} = 3(\mathbf{q}_1 - \mathbf{q}_0)$, and for the last segment: $\frac{d\mathbf{C}_3(1)}{d\xi} = 3(\mathbf{q}_{3L-1} - \mathbf{q}_{3L})$, so using equation (5.86), we get:

$$\boxed{\frac{d\mathbf{C}_3(0)}{dg} = \frac{3}{h_0}(\mathbf{q}_1 - \mathbf{q}_0), \quad \frac{d\mathbf{C}_3(1)}{dg} = \frac{3}{h_{L-1}}(\mathbf{q}_{3L-1} - \mathbf{q}_{3L})} \tag{5.87}$$

5.7.1.1 Inter-element Continuity Class, C^d

In our discussion on the subdivision algorithm for a Bezier curve, we reinterpret each piece of the two subdivided Bezier curves as an element joined at the junction point of subdivision, and the entire Bezier curve as the composite curve of these two Bezier elements, and so we can safely deduce that these elements, being parts of the same polynomial curve of degree n which is at least C^n, themselves, in particular, are C^n, that is, n-times continuously differentiable, at all $\xi \in [0, 1]$. Now, let us consider two elements of a composite Bezier curve of degree n, as shown in Figure 5.28. Let us define a local parameter, ξ, such that $\xi = \frac{g-g_i}{h_i}$; then the knot, g_{i+2}, with respect to ξ, is given by extrapolation as: $\xi = \frac{h_i+h_{i+1}}{h_i}$. Let $\{\mathbf{q}_i\}, \quad i \in \{0, 1, \ldots, n\}$, and $\{\mathbf{q}_i\}, \quad i \in \{n+1, n+2, \ldots, 2n\}$ be the control nets of the ith and the $(i+1)^{\text{th}}$ elements, respectively. Then, for C^d continuity, that is, d times continuous differentiability, at $g = g_{i+1}$, we must have, by the chain rule of differentiation, the following conditions:

$$\frac{d^d\mathbf{C}^i}{dg^d} = \frac{d^d\mathbf{C}^{i+1}}{dg^d} \Rightarrow \left(\frac{1}{h_i}\right)^j (\Delta^j \mathbf{q}_{n-j}) = \left(\frac{1}{h_{i+1}}\right)^j (\Delta^j \mathbf{q}_n), \quad j \in \{0, 1, \ldots, d\} \tag{5.88}$$

where Δ^j is the jth finite difference operator. In actual applications, however, we will be interested only in C^0, C^1 and C^2 continuity:

Case C^0 continuity: For $d = 0, j = 0 \Rightarrow \Delta^j \mathbf{q}_l = \mathbf{q}_l$, so we will have:

$$\mathbf{q}_n = \mathbf{q}_n \tag{5.89}$$

as expected, that is, both the elements should have the same control points at $g = g_i$.

Case C^1 continuity: For $d = 1$, with $j = 1 \Rightarrow \Delta^j \mathbf{q}_l = \mathbf{q}_{l+1} - \mathbf{q}_l$, so we will have, apart from the C^0 continuity condition given by equation (5.89), the following additional condition:

$$\left(\frac{1}{h_i}\right)(\mathbf{q}_n - \mathbf{q}_{n-1}) = \left(\frac{1}{h_{i+1}}\right)(\mathbf{q}_{n+1} - \mathbf{q}_n) \tag{5.90}$$

We recognize, from equation (5.89), that both the elements should have the same control points at $g = g_i$, and from equation (5.90), that not only the directions of the tangent vectors – $(\mathbf{q}_n - \mathbf{q}_{n-1})$ of element i and $(\mathbf{q}_{n+1} - \mathbf{q}_n)$ of element $(i+1)$, – at both sides of the knot, g_i, must be parallel, but also their lengths must be proportional to the interval length, that is, to $h_i{:}h_{i+1}$; for equal intervals, that is, for $h_i = h_{i+1}$, the tangent vectors are equal. Now, for future application of representing ***cubic spline curves*** by piecewise Bezier curves, as will be done shortly, for C^1

continuity between two adjoining pieces of Bezier curves at ***junction control point***, $\mathbf{q}_n$, we must satisfy, from equation (5.90), the relation:

$$\boxed{\mathbf{q}_n = \frac{h_{i+1}}{h_i + h_{i+1}}\mathbf{q}_{n-1} + \frac{h_i}{h_i + h_{i+1}}\mathbf{q}_{n+1}} \tag{5.91}$$

Case C^2 continuity: For $d = 2$, with $j = 2 \Rightarrow \Delta^2\mathbf{q}_n = \mathbf{q}_{n+2} - 2\mathbf{q}_{n+1} + \mathbf{q}_n$, and $\Delta^2\mathbf{q}_{n-2} = \mathbf{q}_{n-2} - 2\mathbf{q}_{n-1} + \mathbf{q}_n$, in equation (5.88), so we will have, apart from the C^0 and C^1 continuity conditions given by equation (5.89) and equation (5.90), respectively, the following additional condition:

$$\left(\frac{1}{h_i}\right)^2 (\mathbf{q}_n - 2\mathbf{q}_{n-1} + \mathbf{q}_{n-2}) = \left(\frac{1}{h_{i+1}}\right)^2 (\mathbf{q}_{n+2} - 2\mathbf{q}_{n+1} + \mathbf{q}_n) \tag{5.92}$$

We recognize, from equation (5.89), that both the elements should have the same control points at $g = g_i$, and from equation (5.90), both the elements should have same control points at $g = g_i$, and that the tangent vectors must be collinear and with the same ratio, as before, and additionally from equation (5.92), we find, as expected, that C^2 continuity is determined by two consecutive ***inner control points*** on either side of the ***junction control point***. Although the above formula, as given by equation (5.92), for C^2 continuity condition derived by derivative method is important on its own right algebraically, for our purposes of an eventual generalized and geometrically constructive representation of the B-spline curves by composite Bezier curves it is rather less useful. For this, we turn instead to the concept behind the subdivision algorithm and the de Casteljau construction presented earlier. Let us recall that for a Bezier curve of degree n, $\{\mathbf{q}_{n-d}, \mathbf{q}_{n-d+1}, \cdots, \mathbf{q}_n\}$ on the left-hand side, and $\{\mathbf{q}_n, \mathbf{q}_{n+1}, \cdots, \mathbf{q}_{n+d}\}$ for the right-hand side represent two Bezier curves of degree d on $[g_{i-1}, g_i]$ and $[g_i, g_{i+1}]$, respectively. And, they must together define a global Bezier curve of degree d on $[g_{i-1}, g_{i+1}]$, and thus, guarantee the existence of $(d + 1)$ control points that include end control points, $\mathbf{q}_{n-d}$ and $\mathbf{q}_{n+d}$. Now, for C^2 continuity, with $d = 2$, there must exist a quadratic Bezier curve with three control points with two of them as, $\mathbf{q}_{n-2}$ and $\mathbf{q}_{n+2}$. Let us denote the third control point as $\mathbf{s}$, as shown in Figure 5.29.

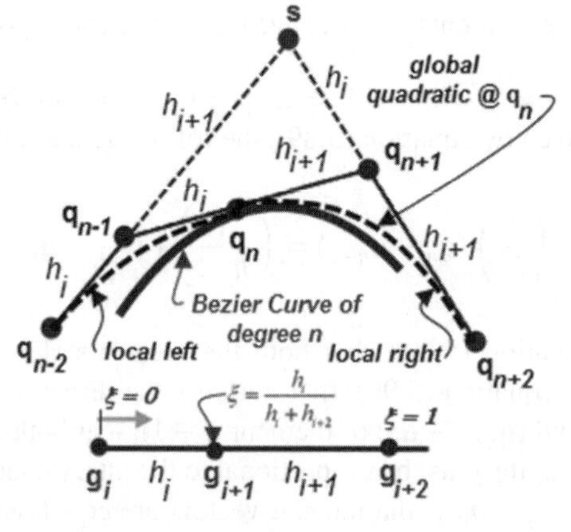

Figure 5.29 C^2 continuous composite Bezier curves.

Then, the two polygons, $\{\mathbf{q}_{n-2}, \mathbf{q}_{n-1}, \mathbf{q}_n\}$ and $\{\mathbf{q}_n, \mathbf{q}_{n+1}, \mathbf{q}_{n+2}\}$ define a quadratic Bezier polygon, $\{\mathbf{q}_{n-2}, \mathbf{s}, \mathbf{q}_{n+2}\}$ such that:

$$\boxed{\mathbf{q}_{n-1} = (1-\xi_i)\,\mathbf{q}_{n-2} + \xi_i \quad \mathbf{s}, \mathbf{q}_{n+1} = (1-\xi_i)\,\mathbf{s} + \xi_i\ \mathbf{q}_{n+2}} \tag{5.93}$$

or, in other words,

$$\boxed{\mathbf{q}_{n-1} = \frac{h_{i+1}}{h_i + h_{i+1}}\ \mathbf{q}_{n-2} + \frac{h_i}{h_i + h_{i+1}}\ \mathbf{s}, \quad \mathbf{q}_{n+1} = \frac{h_{i+1}}{h_i + h_{i+1}}\mathbf{s} + \frac{h_i}{h_i + h_{i+1}}\ \mathbf{q}_{n+2}} \tag{5.94}$$

5.7.2 Discussions

- For higher-degree inter-element continuity, such explicit expressions as in equations (5.93) and (5.94) are effectively impossible to generate. For this, we need to turn to a more general class of curves with inter-element continuity built into them – splines, which are a natural generalization of the composite Bezier curves
- So we study next the explicit expressions as in equations (5.93) and (5.94), but they provide the crucial connection for representing ***cubic B-splines*** by composite Bezier curves where $\mathbf{s}$ stands for what will be described as one of the ***de Boor control points*** defining the B-spline curves.

Thus, we now turn to curves or functions – splines and B-splines – that are a natural generalization of the composite Bezier curves or functions. In our implementation phase, however, we always come back to representing the ***cubic B-splines*** by the ***composite Beziers*** using the C^0, C^1 and C^2 continuities introduced earlier.

5.8 Splines: Schoenberg B-spline Curves

In the chapter introduction, we alluded to limitations of piecewise Lagrange and Hermite polynomials, and even of Bezier spaces. A generalization that includes and improves upon these spaces as special cases will be discussed here that tries to avoid the limitations by converting the mixed problem of interpolation and smoothness to one of just interpolations; as it turns out, this is achieved by defining piecewise polynomials with junction smoothness embedded in them. Thus, in what follows, we introduce and investigate these polynomials after we briefly summarize the historical background of their genesis.

5.8.1 Physical Splines

Historically speaking, a spline is a long thin flexible elastic strip of metal or wood used to design smooth curves passing through some given points describing parts in the ship or automobile industry, as shown in Figure 5.30.

In Figure 5.30, the arrows represent what are known as ***weights*** or ***ducks***. For an extremely thin strip, with small deformation, the shape of the deformed strip is such that the bending moment, $M(x)$, between any two consecutive supports varies linearly and obeys the Bernoulli–Euler law, that is, $M(x) = \frac{d^2y}{dx^2} = y''$. Thus, the deformed shape is cubic at each interval of the supports with not only the slopes but the curvatures being continuous at the weights; but, of course, the shear, $V(x)$, which is proportional to the third derivative, $y'''(x)$, is discontinuous at the weights. Thus, what we have is a deformed shape that could possibly be described by a function of class C^2 that

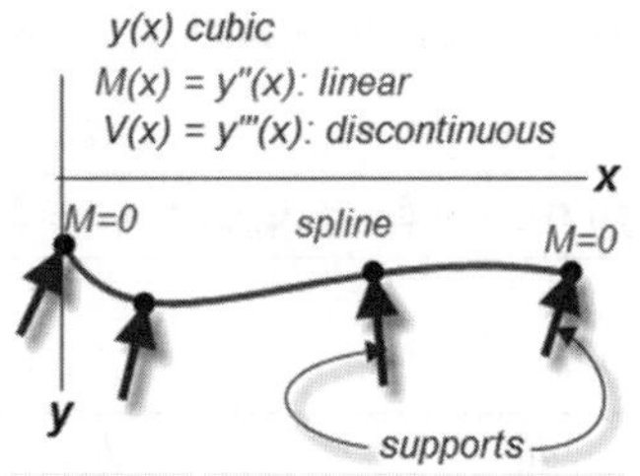

Figure 5.30 Physical spline with supports.

reduces to cubic polynomials (i.e. of degree 3) in each of the subintervals created by the weights. Note also that because these are simple supports with no clamping, the curvatures as well as the bending moments at both ends are zero, and hence the slopes at both ends are linear. This physical description of a spline can thus be translated and generalized into their mathematical counterparts known as the ***spline functions***.

5.8.1.1 Spline Space, $S(\pi, n, k)$

Apart from building on the physical splines, the need for a special class of functions arises in a natural way in trying to improve upon piecewise Lagrange, $L(\pi, n)$, or piecewise Hermite, $H(\pi, n)$, polynomials of degree n, with π known as the ***partition*** or ***knot sequence***. In our attempt to identify this class of functions, and thus the space thereof, the primary goal is to convert the mixed constraint problems of approximation – that is, containing both interpolations and smoothness constraints – to those having pure smoothness constraints. Thus, as we will be see, it is achieved by constructing the basic functions for the space with smoothness conditions already built into them. With this in mind, let us first introduce the spline functions and the space thereof, and then finally the basis functions of the space, with the clear understanding that the problem of approximation still remains as one of mixed constraints for which a solution is being sought.

Definition: A ***polynomial spline function*** of degree n (or order $n+1$) with a partition or knot sequence, $\pi = \{g_0, g_1, \dots, g_L\} \in [a, b]$, is any function $\in C^1\,[a, b]$ that reduces to a polynomial of degree n in each interval, $\pi = \{g_i, g_{i+1}\}$, $i \in \{1, \dots, g_{L-1}\}$. Thus, a spline function consists of a number of polynomials of degree n connected together at knots where they are k-times continuously differentiable. Clearly, these functions form a linear space which we will denote as $S(\pi, n, k)$. By definition, every polynomial of degree n defined on $[a, b]$ belongs to this space; it may readily be recognized that $S(\pi, n, 0) = L(\pi, n)$, the ***piecewise Lagrange polynomial space***, and that $S(\pi, n, 1) = H(\pi, n)$, the ***piecewise Hermite polynomial space*** of degree n. Finally, $S(\pi, n, -1)$ is the space of all step functions with discontinuity at knots.

Remarks

- However, given our main objective of constructing spline functions by generalization of the Bezier functions, to be introduced and detailed later, we will restrict our attention to $S(\pi, n, n-1)$, that is, the space of functions $\in C^{n-1}[a, b]$ that reduces to polynomials of degree n in each interval of π. In other words, we will consider only piecewise polynomials of degree n with smoothness condition on each knot defined by $(n-1)$ times continuous differentiability. We know that, from now on, the degree of smoothness is one less than that of the component polynomials, so we may rename our space as, $B(\pi, n)$:
- Definition: $\boxed{S(\pi, n, n-1) \equiv B(\pi, n)}$.

We would now like to know various relevant properties of $B(\pi, n)$ – its dimensions, variants, constructive characterization, the basis functions that span it, and so on.

5.8.1.2 Dimension of $B(\pi, n)$

Suppose the partition, $\pi = \{g_0, g_1, \dots, g_L\}$, consists of $(L + 1)$ distinct knots defining L subintervals of $[a, b]$. Now, it follows from the definition of $B(\pi, n)$ that there are L many n-degree polynomials connected at $(L - 1)$ internal knots where they are $(n - 1)$ times continuously differentiable. Then, L many n-degree polynomials, each requiring $(n + 1)$ parameters to define them, need a total of $L(n + 1)$ parameters. But $(n - 1)$ differentiability constraints at each $(L - 1)$ internal knot reduce the number of free (or independent) parameters by $n(L - 1)$. Thus, the net-free degrees of freedom defining the dimension of the space is given by: $L(n + 1) - n(L - 1) = L + n$. Clearly, then, the dimension is given as:

$$\dim B(\pi, n) = L + n \tag{5.95}$$

that is, it takes $(L + n)$ parameters to uniquely determine a member of $B(\pi, n)$. In the above definitions, we have assumed that the knots are distinct, that is, ***simple knots***. However, the definition of $B(\pi, n)$ does not restrict knots of multiplicity. In fact, as we shall see, multiplicity of knots crucially determines the end interpolation and intermittent smoothness: the more the multiplicity, the less is the smoothness in the interior knots. We will get back to this shortly, but first let us characterize its effect on the dimension of $B(\pi, n)$. Thus, suppose the polynomials are joined at knots in such a way that at knot, $g_i, \quad i \in \{0, 1, \dots, L\}$, the function is only $(n - m_i)$ times continuously differentiable, then, by a similar argument as above, we have:

$$\boxed{\dim B(\pi, n) = (n + 1) + \sum_{i=1}^{L-1} m_i} \tag{5.96}$$

We can easily verify that if $m_i = 1, \quad \forall i \in \{1, \dots, L - 1\}$ – that is, the functions are $(n - 1)$ continuously differentiable at all internal knots – the dimension of the space reduces to $(L + n)$ as before in equation (5.95). Now, several observations are in order.

Remarks

- The smoothness condition being already built into the spline function, there is no derivative data required for internal knots, and this eliminates one of the limitations of the members of $H(\pi, n)$, that is, the space of piecewise Hermites of degree n,
- For $n = 2j + 1$, that is, odd-degree polynomials with $(L + 1)$ distinct knots, $\dim B(\pi, 2j + 1) = L + 2j + 1$. Thus, of $(L + 2j + 1)$ free degrees of freedom, only $(L + 1)$ interpolatory constraints are supplied by the data at $(L + 1)$ knots. However, to uniquely determine a member function, we will still need $2j$ constraints; these are generally supplied as j-interpolatory derivative constraints at each of the end knots. In fact, depending on the problem at hand, variants of spline functions can be generated based on these end knot interpolatory conditions. For example, for $n = 3$, that is, a cubic spline function, $j = 1$; thus, we will still need either slope or curvature constraints at, say, each end to uniquely determine an interpolation spline that belongs to $B(\pi, 3)$.

We will very soon present some variants based on this remark. But, for now, let us try to characterize $B(\pi, n)$.

5.8.1.3 Spline Function Representation

Before we set out to accomplish our stated goal, let us first extend the partition, π, to $\bar{\pi}$ such that:

$$\bar{\pi} \stackrel{\Delta}{=} \{-\infty, g_{-1}\} \cup \pi \cup \{g_{L+1}, \infty\} \tag{5.97}$$

and introduce a special class of functions belonging to $B(\bar{\pi}, n)$, as follows.

Definition: A ***truncated power function***, $(g - g_i)_+^n$ is defined as:

$$(g - g_i)_+^n \stackrel{\Delta}{=} \begin{cases} (g - g_i)^n, & g \geq g_i \\ 0, & \text{otherwise} \end{cases} \tag{5.98}$$

It can be easily verified that a truncated power function for a given g_i is continuous at $g = g_i$, and is $(n-1)$ times continuously differentiable; thus $(g - g_i)_+^n \in B(\hat{\pi}, n)$, where $\hat{\pi} \stackrel{\Delta}{=} \{-\infty, g_i, \infty\} \subset [-\infty, \infty]$.

5.8.1.3.1 Schoenberg–Whitney Theorem

Every spline function, $f(g) \in B(\bar{\pi}, n)$, can be presented as:

$$f(g) = p(g) + \sum_{i=0}^{L+1} \alpha_i \, (g - g_i)_+^n \tag{5.99}$$

where the α_i's are constants, and $p(g)$ is a polynomial of degree n.

Proof. Let $p_i(g)$ be a polynomial of degree n for $g \in \{g_i, g_{i+1}\}$, with $p_i^d(g)$ as the dth derivative for $d \in \{0, 1, \ldots, n-1\}$, that is, $p_i(g) = p_i^0(g)$, and by definition, $p_{i-1}^d(g_i) = p_i^d(g_i)$ for all $d \in \{0, 1, \ldots, n-1\}$ and $i \in \{0, 1, \ldots, L-1\}$ with $p_{-1}(g) = p(g)$, an nth degree polynomial. Then, the difference function, $p_i(g) - p_{i-1}(g)$, is clearly of degree n, with n roots at g_i, $i \in \{0, 1, \ldots, L-1\}$. Thus, the difference function may be represented as: $p_i(g) - p_{i-1}(g) = \alpha_i(g - g_i)^n$, or equivalently $p_i(g) = p_{i-1}(g) + \alpha_i(g - g_i)^n$. Applying this recurrence relation for each $i \in \{0, 1, \ldots, L-1\}$, and noting that $p_{-1}(g) = p(g)$, we have: $p_i(g) = p(g) + \sum_{j=0}^{L-1} \alpha_i(g - g_j)^n$, $g \in [g_i, g_{i+1}]$. Now, utilizing the definition for the power function and recognizing that $p_i(g)$'s are $f(g)$ completes the proof.

5.8.2 Variants of Spline Functions

As remarked earlier, the specification of the end conditions determines varieties of spline functions. Of course, the specifications depend on the problem at hand, that is, on the boundary value problem in differential forms. We describe later just a few among others that are relevant to interactive modeling and finite element analysis. For this, we will restrict ourselves to odd degree

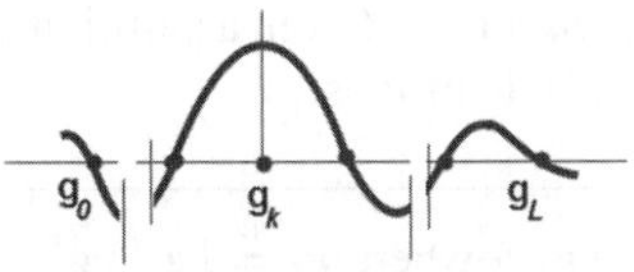

Figure 5.31 Cardinal spline.

splines, that is, $n = 2m - 1$ implying $\dim B(\pi, 2m-1) = L + 2m - 1 = (L+1) + (2m-2)$. This means that, apart from $(L+1)$ interpolatory constraints at $(L+1)$ knots, there need to be $(2m-2)$ additional constraints.

Natural Splines: This class of splines is the exact generalization of the physical splines introduced earlier. Thus, we define this class as: a spline which is of $(2j-1)$ degree over partition, π, and at most $(j-1)$ degree at two end intervals of $\bar{\pi}$, is called a natural spline. For example, for $j = 2$, we have a cubic natural spline over partition, π, and linear at two infinite end intervals outside π. We have already noted that the curvature at two ends is identically zero, thus the end slopes are linear and the spline stretches out linearly at each end. Hence, for cubic natural splines, an equivalent definition will be to impose free end conditions at each end, that is, the curvature requiring to be zero at each end. For the cubic example, the natural spline clearly describes the equilibrating shape of the physical splines. Thus, from the minimum potential energy principle, the cubic natural spline reflects the minimum potential energy configuration, that is, the natural shape of a physical spline; this justifies the name.

Periodic Splines: A spline of degree n, which has equal slope and curvature at both end knots of π, is called a ***periodic spline***. The usefulness of periodic splines for our purposes is in building closed curves where the end knots are equated to each other; this will be detailed later in our discussion on computational curves.

Cardinal Splines: This class of splines is designed to mimic the Lagrange polynomial basis functions. Thus, given a partition, $\pi = \{g_0, g_1, \ldots, g_L\}$, a cardinal spline, $c_k^n(\pi)$, is defined as a Kronecker delta function, δ_{ik}, over knots, that is,

$$c_k^n(\pi) = \begin{cases} 1, & \text{if } k = i, \quad k, i \in \pi \\ 0, & \text{, otherwise} \end{cases}$$

with suitable end conditions. One such cardinal spline is shown in Figure 5.31.

Although these oscillate less than the corresponding Lagrange polynomials of degree n, the cardinal splines are non-zero over every subinterval of the partition, and thus are unsuitable for any numerical application because the resulting matrices – in both interpolation problems of computer modeling and approximation problems of finite element analysis – will be full and costly. Thus, for numerical effectiveness, we need basis functions with minimum (compact) support. It turns out, fortunately, that there exist just such polynomial splines – known as ***B-splines*** (for basis splines) – that form suitable basis functions for the polynomial spline space.

5.8.2.1 B-splines (Basis Functions)

Before we move on to discuss the most important class of spline functions for interpolations in computer-aided mesh generation and approximations in finite element analysis, we will introduce two more definitions.

Extended Knot Sequence or Partition, π^e: Given a partition, $\pi = \{g_0, g_1, \dots, g_L\}$, and a spline of degree n, the ***extended partition*** is defined as:

$$\pi^e \overset{\Delta}{=} \pi_L \cup \pi \cup \pi_R \quad \text{where } \pi_L \overset{\Delta}{=} \{g_{-n}, g_{-(n-1)}, \dots, g_{-1}\} \\ \pi_R \overset{\Delta}{=} \{g_{L+1}, g_{L+2}, \dots, g_{L+n}\} \tag{5.100}$$

that is, extended partition, π^e, consists of n additional knots on either side of the original partition, π.

Knots of Multiplicity: If a knot g_i equals $(m-1)$ other consecutive knots in value, then the knot g_i is said to have a multiplicity of m; thus, a ***simple knot*** is said to have a multiplicity of 1. We reiterate that we are interested in the class of splines that help us to define a spline space with basis functions that span the space with minimum local support; B-splines turned out to be these basis functions. The B-splines can be defined in various different ways, based on the method of construction. One method based on forward difference and the truncated power function turned out to be numerically unsatisfactory. The method based on recurrence relation (due to deBoor, Cox and Bohm) is introduced here because it is most useful for generating piecewise Bezier curves and functions.

Definition: Given a knot sequence, $\{g_i\}$, the ith ***B-Spline***, N_i^n, of degree n, is defined by the following four properties:

- convex or barycentric combination (i.e. partition of unity): $\sum_i N_i^n(g) = 1$,
- positive: $N_i^n(g) \geq 0. \quad \forall i$,
- local support: $N_i^n(g) \equiv 0. \quad \text{if} \quad g \notin [g_i, g_{i+n+1}]$,
- continuity class: $N_i^n(g) \in C^{n-1}$, that is, $N_i^n(g)$ is $(n-1)$ times continuously differentiable.

5.8.2.1.1 Recursion Relation

The B-splines are related by the following recursion relations:

$$N_i^d(g) = \beta_i^d \, N_i^{d-1}(g) + (1 - \beta_{i+1}^d) \, N_{i+1}^{d-1}(g), \\ \text{where } \beta_i^d \equiv \frac{g - g_i}{g_{i+d} - g_i} \quad \text{with} \quad N_i^0(g) = \begin{cases} 1, & \text{if } g \in [g_i, g_{i+1}) \\ 0, & \text{otherwise} \end{cases} \tag{5.101}$$

We can visualize the above recursion relation for cubic B-splines as shown in Figure 5.32.

Example BS1: In Figure 5.33, we represent graphically the B-splines of degrees $0, 1, 2, 3$.

As is evident from the definition and the examples given, a B-spline, $N_i^n(g)$, is defined with $(g_{i+n+1} - g_i)$ as its support, consisting of $(n+1)$ consecutive subintervals starting at knot i.

5.8.2.2 B-splines and Knot Multiplicity

If m knots, $g_i = \dots = g_{i+m-1}$, coincide, that is, m denotes the ***knot multiplicity***, the B-spline curve may become only C^{n-m} continuous at g_i, as shown in Figure 5.34. Thus, in order that a B-spline curve should have non-vanishing support, we must have $m \leq n+1$ so that the recursion formula applies.

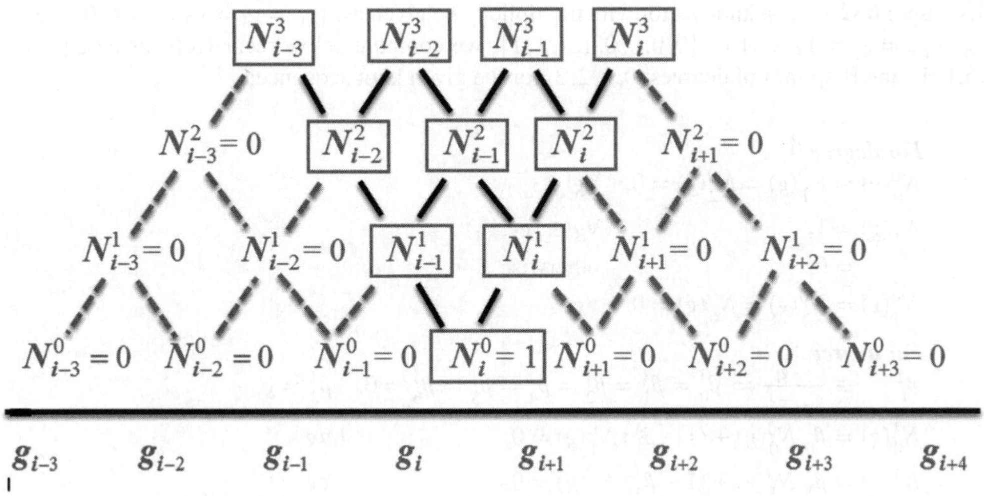

Figure 5.32 Cubic B-spline recursions.

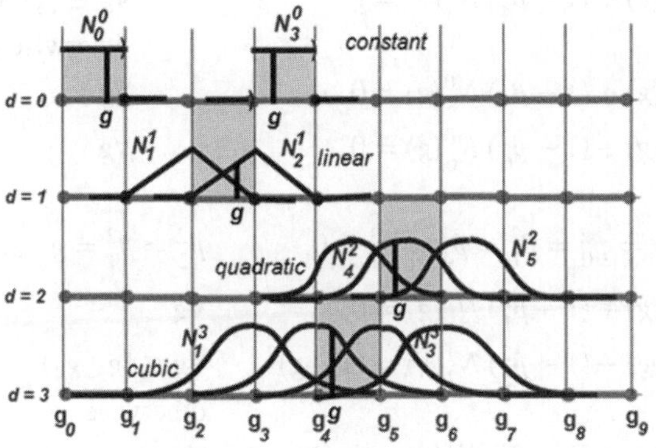

Figure 5.33 Example: B-splines.

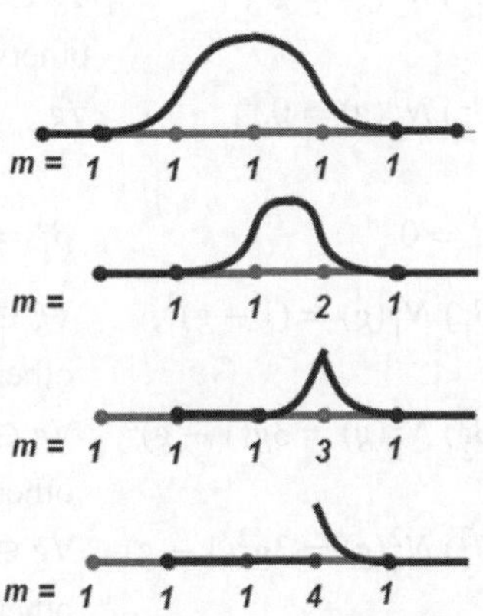

Figure 5.34 B-spline: effect of multiple knots.

Example BS2: Let a knot vector with multiplicity be given as: $\{g_0 = g_1 = g_2 = g_3 = 0, \quad g_4 = g_5 = g_6 = g_7 = 1\}$, that is, $\{0, 0, 0, 0, 1, 1, 1, 1\}$; we compute below recursively, using equation (5.101), the B-splines of degrees: $0, 1, 2, 3$, for the given knot sequence:

For degree 0:

$$
\begin{aligned}
&N_0^0(g) = N_1^0(g) = N_2^0(g) = 0, && \forall g\\
&N_3^0(g) = 1, && \forall g \in [g_3, g_4)\\
&\qquad\quad = 0, && \text{otherwise}\\
&N_4^0(g) = N_5^0(g) = N_6^0(g) = 0, && \forall g
\end{aligned}
$$

For degree 1:

$$
\begin{aligned}
&\beta_i^1 \equiv \tfrac{g-g_i}{g_{i+1}-g_i} \Rightarrow \beta_0^1 = \beta_1^1 = \beta_2^1 = \beta_4^1 = \beta_5^1 = \beta_6^1 = 0, \quad \beta_3^1 = g\\
&N_0^1(g) = \beta_0^1\, N_0^0(g) + (1 - \beta_1^1)\, N_1^0(g) = 0, && \forall g\\
&N_1^1(g) = \beta_1^1\, N_1^0(g) + (1 - \beta_2^1)\, N_2^0(g) = 0, && \forall g\\
&N_2^1(g) = \beta_2^1\, N_2^0(g) + (1 - \beta_3^1)\, N_3^0(g) = (1 - g), && \forall g \in [g_3, g_4)\\
&\qquad\quad = 0, && \text{otherwise}\\
&N_3^1(g) = \beta_3^1\, N_3^0(g) + (1 - \beta_4^1)\, N_4^0(g) = g, && \forall g \in [g_3, g_4)\\
&\qquad\quad = 0, && \text{otherwise}\\
&N_4^1(g) = \beta_4^1\, N_4^0(g) + (1 - \beta_5^1)\, N_5^0(g) = 0, && \forall g\\
&N_5^1(g) = \beta_5^1\, N_5^0(g) + (1 - \beta_6^1)\, N_6^0(g) = 0, && \forall g
\end{aligned}
$$

For degree 2:

$$
\begin{aligned}
&\beta_i^2 \equiv \tfrac{g-g_i}{g_{i+2}-g_i} \Rightarrow \beta_0^2 = \beta_1^2 = \beta_4^2 = \beta_5^2 = 0, && \beta_2^2 = \beta_3^2 = g\\
&N_0^2(g) = \beta_0^2\, N_0^1(g) + (1 - \beta_1^2)\, N_1^1(g) = 0, && \forall g\\
&N_1^2(g) = \beta_1^2\, N_1^1(g) + (1 - \beta_2^2)\, N_2^1(g) = (1 - g)^2, && \forall g \in [g_3, g_4)\\
&\qquad\quad = 0, && \text{otherwise}\\
&N_2^2(g) = \beta_2^2\, N_2^1(g) + (1 - \beta_3^2)\, N_3^1(g) = 2g(1 - g), && \forall g \in [g_3, g_4)\\
&\qquad\quad = 0, && \text{otherwise}\\
&N_3^2(g) = \beta_3^2\, N_3^1(g) + (1 - \beta_4^2)\, N_4^1(g) = g^2, && \forall g \in [g_3, g_4)\\
&\qquad\quad = 0, && \text{otherwise}\\
&N_4^2(g) = \beta_4^2\, N_4^1(g) + (1 - \beta_5^2)\, N_5^1(g) = 0, && \forall g
\end{aligned}
$$

For degree 3:

$$
\begin{aligned}
&\beta_i^3 \equiv \tfrac{g-g_i}{g_{i+3}-g_i} \Rightarrow \beta_0^3 = \beta_4^3 = 0, && \beta_1^3 = \beta_2^3 = \beta_3^3 = g\\
&N_0^3(g) = \beta_0^3\, N_0^2(g) + (1 - \beta_1^3)\, N_1^2(g) = (1 - g)^3, && \forall g \in [g_3, g_4)\\
&\qquad\quad = 0, && \text{otherwise}\\
&N_1^3(g) = \beta_1^3\, N_1^2(g) + (1 - \beta_2^3)\, N_2^2(g) = 3g(1 - g)^2, && \forall g \in [g_3, g_4)\\
&\qquad\quad = 0, && \text{otherwise}\\
&N_2^3(g) = \beta_2^3\, N_2^2(g) + (1 - \beta_3^3)\, N_3^2(g) = 3g^2(1 - g), && \forall g \in [g_2, g_3)\\
&\qquad\quad = 0, && \text{otherwise}\\
&N_3^3(g) = \beta_3^3\, N_3^2(g) + (1 - \beta_4^3)\, N_4^2(g) = g^3, && \forall g \in [g_3, g_4)\\
&\qquad\quad = 0, && \text{otherwise}
\end{aligned}
$$

Remark

- From the results of degree 3 above, we can easily recognize that the B-splines, $N_0^3(g), N_1^3(g), N_2^3(g), N_3^3(g)$, with multiplicity of 4 at both ends, are the same as the cubic Bezier basis functions, $B_i^3, \quad i \in \{0, 1, 2, 3\}$. This has a profound implication in the sense that the spline curves with appropriate multiplicity can be represented by piecewise Bezier curves; the influence of knot multiplicity is at the heart of the representation of splines by composite Bezier curves for interpolation as will be seen later.

Now, two important properties of the B-splines lend themselves to the usefulness of these curves, that is, allow B-splines to pass as basis functions of the spline space: minimal support and linear independence property (without proof), as we will introduce next.

5.8.2.3 Minimal Support Property

Proposition: Any piecewise polynomial, with the same smoothness properties over the same knot sequence and less support than $N_i^n(g)$, is identically zero, that is, no such non-trivial piecewise polynomial exists, making the B-spline with minimal support.

Proof. For any piecewise polynomial defined over $[g_i, g_{i+n+1}]$, the support for $N_i^n(g)$ are elements of the function space of dimension $(2n+1)$ – because $(n+1)$ polynomials of degree n freedoms reduced by $(n-1)$ differentiability and one value-matching condition at n internal nodes gives $(n+1)(n+1)-(n)(n)$. Now, with one shorter interval, that is, with n polynomials of degree n reduced by $n-1$ differentiability and one value-matching condition at $(n-1)$ internal nodes, we have dimension of $2n$ (i.e. $n(n+1)-n(n-1)$). So, for any polynomial with $(n-1)$ differentiability at two ends puts $2n$ restraints, leaving no room for any non-trivial value inside; thus, the polynomial must be identically zero completing our proof.

5.8.2.4 Linear Independence Property

Proposition: Given a partition, $\pi = \{g_0, g_1, \cdots, g_L\}$, and the extended partition defined as: $\pi^e \stackrel{\Delta}{=} \pi_L \cup \pi \cup \pi_R$ where $\pi_L \stackrel{\Delta}{=} \{g_{-n}, g_{-(n-1)}, \cdots, g_{-1}\}$, and $\pi_R \stackrel{\Delta}{=} \{g_{L+1}, g_{L+2}, \cdots, g_{L+n}\}$, the B-splines of degree n, $N_i^n(g), \quad g \in \pi,\ i \in \{0, \ldots, L+n-1\}$, are linearly independent.

Exercise: Prove the above proposition.

We are now ready to express any spline function of degree n, of linear space, $B(\pi, n)$, as a linear combination of its basis functions – the B-splines, as given by the following theorem due to Curry and Schoenberg.

Theorem: Any spline function, $p_n^S(\pi) \in B(\pi, n)$, with $\pi = \{g_0, g_1, \cdots, g_L\}$, can be expressed as a linear combination of the B-splines defined over the ***extended knot set***, $\pi^e \stackrel{\Delta}{=} \pi \cup \pi_R$,with $\pi_R \stackrel{\Delta}{=} \{g_{L+1}, g_{L+2}, \cdots, g_{L+n}\}$, as:

$$\boxed{p_n^S(g) = \sum_{i=l}^{l+n} N_i^n(g)\ \mathbf{s}_i, \quad g \in [g_l, g_{l+1}] \subset \pi,\ l \in [0, \ldots, L-1]} \tag{5.102}$$

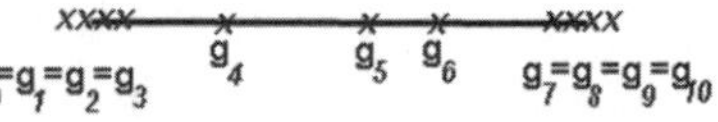

Figure 5.35 Cubic Bezier by repeated knots in B-splines.

Remarks

- To compute a point on a spline at parameter point, $g \in [g_l, g_{l+1}]$, with given knot vector and control points, we will have to first compute the relevant B-spline basis functions, over its support, $N_i^n(g)$, $i \in [l, \ldots, l+n]$, as in **Example BS2** (above), and then perform the linear combinations as in equation (5.102).
- Because of local support, a particular segment of a B-spline curve may be changed without disturbing the entire curve, as in Bezier curves.
- For $\pi = \{g_0, g_1, \cdots, g_L\}$, we have dim $B(\pi, n) = L + n$, with:

Number of B-spline control points $= L + n$,

Support for each Basis function $= [g_i, g_{i+n}]$

Support for $N_0^n(g)$, the first Basis function $= [g_0, g_n]$

Support for $N_{L+n}^n(g)$, the last Basis function $= [g_{L+n}, g_{L+2n}]$

Number of total knots $= L + 2n + 1$

Number of distinct curve segments $= L$

Number of distinct knots $= L + 1$

Number of repeated knots = total knots − distinct knots $= (L + 2n + 1) - (L + 1) = 2n$

Remark: Normally, n repeated knots are used at each end, so that the end control points interpolate the curve ends, that is, they reside on the curve.

Example: consider a cubic (i.e. $n = 3$) B-spline curve with $L = 4$ segments with $L + n$ (= 7) control points given by: $\{\mathbf{s}_0, \mathbf{s}_1, \mathbf{s}_2, \mathbf{s}_3, \mathbf{s}_4, \mathbf{s}_5, \mathbf{s}_6\}$. The number of distinct knots is $L + 1$ (= 5) and the total number of knots is $L + 2n + 1$ (= 11); the number of repeated knots is n (= 3) at each end, as shown in Figure 5.35, so that the control points $\mathbf{s}_0$ and $\mathbf{s}_6$ interpolate the curve ends.

- For our purposes, however, we will always represent a spline function by an appropriate composite (piecewise) Bezier function as will be seen later. As suggested by **Example BS2**, and as a special case for a single span cubic ($n = 3$) spline curve, with $4(= n + 1)$ multiple knots at each ends, we will obtain a single span cubic Bezier curve.

5.8.2.5 Splines, Differential Equations and Error Analysis

All the discussions presented so far about spline curves and functions, have their counterparts as the solution of ordinary differential equations when restricted to odd-degree polynomials.

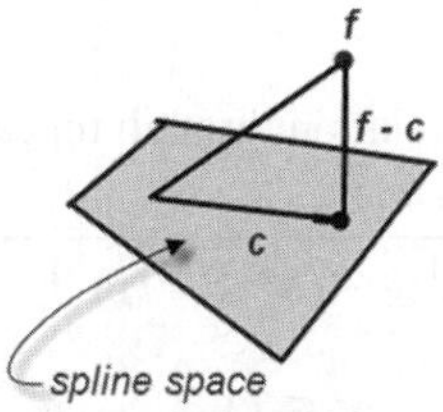

Figure 5.36 Best approximations by spline space.

The correspondence is easily established through interpolatory and smoothness conditions of splines as boundary conditions on the partition over which the differential equation is defined. For example, a cubic spline, $c_3(g)$, can be viewed as an interpolate to a given function, $f(g)$, related by the following 4th-order differential equation and the boundary conditions:

$$
\begin{aligned}
&D^4(c_3(g)) = 0, \text{on } \Omega = \bigcup_{i=0}^{L-1} [g_i, g_{i+1}) \subset [a, b],\\
&c_{3,g}(g_0) = f_{,g}(g_0), c_3(g_i) = f(g_i), \quad i \in \{0, 1, \ldots, L\}\\
&c_{3,g}(g_L) = f_{,g}(g_L)
\end{aligned}
$$

where partition: $\pi = \{g_0, g_1, \ldots, g_L\}$. The proof due independently to Ahlberg and Greville that $c_3(g)$ is, in fact, a cubic spline is omitted here. In fact, all the different splines introduced before – natural splines, periodic splines, and so on – have their corresponding differential equation definitions. It is interesting to note that the spline interpolates to functions are the best approximation in the sense that they satisfy the Pythagorean relationship under a particular definition of an inner product space with pseudo-norm or semi-norm, as follows.Let $H = K_2^m[a, b]$, be the space of real-valued functions, f, defined over $[a, b]$ such that f is absolutely $(m-1)$ times continuously differentiable with Lebesgue square integrability; H has an inner product defined on it: $(f, g) = \int_a^b f^{(m)}\, g^{(m)}\, dx$ where $f^{(m)}$, $g^{(m)}$ are the mth derivatives of $f, g \in H$. The norm, $|f|$, is defined as: $|f| = \sqrt{(f,f)}$ such that while $|f|$ satisfies all other properties of a norm, $|f| = 0$ does not imply $f \equiv 0$, that is, $|\bullet|$ is a pseudo- or semi-norm. Then, a spline interpolate, $c_{2m-1}(x) \in B(\pi, 2m-1)$ to $f \in H$, satisfies the Pythagorean relation: $|f|^2 = |f - c|^2 + |c|^2$ as shown in Figure 5.36.

5.8.2.6 Discussion

- This investigation establishes the theoretical basis for parametric curves. However, for practical curve constructions that are numerically robust and cost-effective on a computer, we need to discuss various numerical aspects and computational representation of the above parametric curves as applied to CAGD and the finite element method (FEM).

5.9 Recursive Algorithm: de Boor–Cox Spline

As introduced earlier in equation (5.100), let $\pi^e\ \Delta\pi_L\ \text{U}\pi\ \text{U}\pi_R$ be an ***extended partition*** or ***knot sequence***, $\{s_i\}$, $i \in \pi^e$ be a finite set of points in $\mathbb{R}_3$, and $g \in [g_l, g_{l+1}] \subset \pi = \{g_0, g_1, \ldots, g_L\}$, $l \in [0, L-1]$.

5.9.1 Proposition

Let us define the following recursive relation through repeated linear interpolation:

$$\boxed{\begin{array}{l} \mathbf{s}_i^d(g) = \left(1-\alpha_i^d\right)\,\mathbf{s}_{i-1}^{d-1}(g) + \alpha_i^d\,\mathbf{s}_i^{d-1}(g) \\ \text{with } \alpha_i^d \stackrel{\Delta}{=} \dfrac{g-g_i}{g_{i+n+1-d}-g_i}, \quad d \in \{1,2,\dots,n-m\}, i \in \{l-n+d,\dots,l\} \\ \text{and } \mathbf{s}_i^0(g) = \mathbf{s}_i \end{array}} \tag{5.103}$$

where m, the ***multiplicity*** of g, is given by:

$$\boxed{m = \begin{cases} 0, & \text{if g is not a knot} \\ m_i, & \text{if } \mathrm{g} = \mathrm{g}_i, m_i \text{ is the multiplicity of knot } \mathrm{g}_i \end{cases}} \tag{5.104}$$

Then, $\mathbf{s}_l^n(g), \quad \forall l \in [0, L-1]$, defines a parametric spline curve, $\mathbf{C}_n^S(g)$, of degree n.

Definitions

- The polygon, D, formed by joining $\mathbf{s}_i$ to $\mathbf{s}_{i+1}$, $i \in \{-n,\dots,L-1\}$ is called the ***de Boor polygon***, or ***spline control polygon***, of the curve, $\mathbf{C}_n^S(g)$, of degree n and is defined over π.
- $\{\mathbf{s}_i\}, \quad i \in \{-n,\dots,L-1\}$, are known as the ***de Boor points***, or ***spline control points***.

Remarks

- $\mathbf{C}_n^S(g)$ is, in fact, a spline curve, that is, it is a piecewise polynomial of degree n, belonging to $\mathbf{C}^{n-1}[\pi]$, which can be proved.
- The de Boor definition of the spline curve by recursion can easily be seen as repeated linear interpolations applied to successive segments resulting from the original de Boor polygon; linear interpolation is invariant under affine transformation and so is repeated linear interpolation. Accordingly, a spline curve is invariant under affine transformation of its domain, that is, the real line intervals.

Example: Let us look at the interval, $g \in [g_2, g_3]$, of a quadratic spline curve, $\mathbf{C}_n^S(g)$ (i.e. of degree 2), with no multiplicity of knots, that is, $n = 2, \quad l = 2, \quad m = 0$, as shown in Figure 5.37.

Then, $\mathbf{s}_0^0 = \mathbf{s}_0$, $\mathbf{s}_1^0 = \mathbf{s}_1$ and $\mathbf{s}_2^0 = \mathbf{s}_2$ with $d \in \{1,2\}$ and $i \in \{d,\dots,2\}$ resulting in:

$$\begin{aligned} \mathbf{s}_1^1(g) &= \left(1-\alpha_1^1\right)\mathbf{s}_0 + \alpha_1^1\mathbf{s}_1, \quad \alpha_1^1 = \frac{g-g_1}{g_2-g_1} \\ \mathbf{s}_2^1(g) &= \left(1-\alpha_2^1\right)\mathbf{s}_1 + \alpha_2^1\mathbf{s}_2, \quad \alpha_2^1 = \frac{g-g_2}{g_3-g_2} \\ \mathbf{C}_2(g) = \mathbf{s}_2^2(g) &= \left(1-\alpha_2^2\right)\mathbf{s}_1^1(g) + \alpha_2^2\mathbf{s}_1^2(g), \quad \alpha_2^2 = \frac{g-g_2}{g_3-g_2} \end{aligned} \tag{5.105}$$

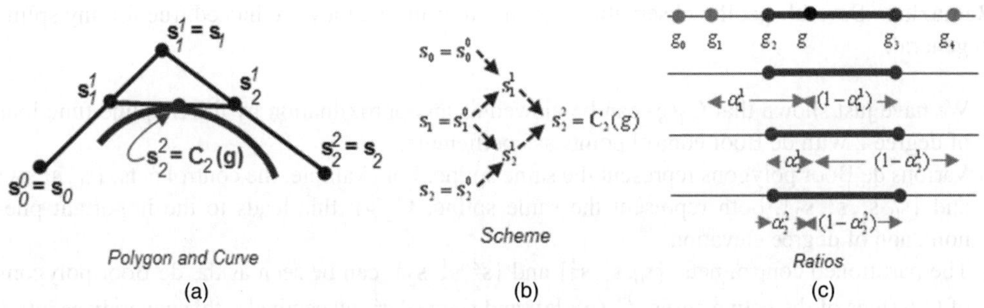

Figure 5.37 (a) Example: (B-spline as composite Beziers), (b) Scheme: (B-spline as composite Beziers), (c) Ratio: (B-spline as composite Beziers).

Let us look closer at $\mathbf{C}_2(g)$, as given by equation (5.105), by tracing back the recursive relations in reverse order, from which we get:

$$\mathbf{C}_2(g) = \left[\left(1-\alpha_2^2\right)\left(1-\alpha_1^1\right)\right]\mathbf{s}_0 + \left[\alpha_1^1\left(1-\alpha_2^2\right) + \alpha_2^2\left(1-\alpha_2^1\right)\right]\mathbf{s}_1 + \alpha_2^2\,\alpha_2^1\,\mathbf{s}_2 \tag{5.106}$$

Now, substituting for α_2^2, α_1^1 in equation (5.106) from their values in equation (5.105), we get:

$$\left(1-\alpha_2^2\right)\left(1-\alpha_1^1\right) = \frac{(g_3-g)^2}{(g_3-g_2)} \tag{5.107}$$

Looking at the recursion relations of B-splines, as given by equation (5.101), we get: $N_0^0(g) = N_1^0(g) = 0$ because $g \notin [g_0, g_2)$, and $N_2^0(g) = 1$ because $g \in [g_2, g_3)$, and

$$N_0^1(g) = \frac{g-g_0}{g_1-g_0}\,N_0^0(g) + \frac{g_2-g}{g_2-g_1}\,N_1^0(g) = 0,$$
$$N_1^1(g) = \frac{g-g_1}{g_2-g_1}\,N_1^0(g) + \frac{g_3-g}{g_3-g_2}\,N_2^0(g) = \frac{g_3-g}{g_3-g_2}$$

And, finally,

$$N_0^2(g) = \frac{g-g_0}{g_2-g_0}\,N_0^1(g) + \frac{g_3-g}{g_3-g_1}\,N_1^1(g) = \frac{(g_3-g)^2}{(g_3-g_1)(g_3-g_2)} \tag{5.108}$$

which, when compared to equation (5.107), shows that $N_0^2(g)$ matches the coefficient of $\mathbf{s}_0$ in equation (5.106); similarly, we can show that $N_1^2(g)$ and $N_2^2(g)$ match the coefficients of $\mathbf{s}_1$ and $\mathbf{s}_2$ in equation (5.106), respectively. Thus, we can express $\mathbf{C}_2(g)$ of equation (5.106) as:

$$\boxed{\mathbf{C}_2(g) = \sum_{i=0}^{2} N_i^2(g)\,\mathbf{s}_i} \tag{5.109}$$

that is, as the linear combinations of the B-spline or basis curves with the de Boor control points, $\{\mathbf{s}_i\}$, as the coefficients of combination.

Remarks: Several specific observations are in order in that they are indeed true for any spline in general:

- We have just shown that $\mathbf{C}_2(g)$ can be viewed as an approximation by the B-spline functions of degree 2 with de Boor control points as coefficients.
- Various de Boor polygons represent the same spline. For example, the control nets, $\{\mathbf{s}_0, \mathbf{s}_1, \mathbf{s}_2\}$ and $\{\mathbf{s}_0, \mathbf{s}_1^1, \mathbf{s}_2^1, \mathbf{s}_2\}$, both represent the same spline, $\mathbf{C}_2(g)$; this leads to the important phenomenon of degree elevation.
- The partitioned control nets, $\{\mathbf{s}_0, \mathbf{s}_1^1, \mathbf{s}_2^2\}$ and $\{\mathbf{s}_2^2, \mathbf{s}_2^1, \mathbf{s}_2\}$, can be seen as the de Boor polygons of two sides of the spline curve, $\mathbf{C}_2(g)$: left and right of $\mathbf{s}_2^2$, respectively; this generalizes into a subdivision algorithm similar to the Bernstein–Bezier curves.
- A general spline possesses most of the properties of Bezier curves discussed before – linear precision, convex hull property, and so on.
- For our purposes of representing spline curves by composite Bezier curves, as will be discussed later, the end knots of a spline curve of degree n, will always be assumed to have multiplicity of $n+1$.

5.10 Rational Bezier Curves: Conics and Splines

5.10.1 Projective Map and Cross Ratio Invariance

As indicated earlier, for non-rational curves, algorithms for curve generation based on repeated linear interpolation utilize the concepts of ratio and affine map for their invariance. However, for conics and other rational curves, we will need a more general map, known as a ***projective map***, such as perspective in geometry and a generalization of the concept of ratio known as the ***cross ratio***.

To start with, for a projection, we have a center of projection, O, as shown in Figure 5.38, which projects, for example, a line segment, $ABCD$, to another line segment, $abcd$, onto a plane known as the image plane. We can clearly see that under the projective map of one line segment to another, ratios are not preserved because individual lengths are distorted. However, the cross ratio, denoted as cr $(\bullet, \bullet, \bullet, \bullet)$, of four collinear points is defined as:

$$\boxed{\text{cr}\,(a,b,c,d) = \frac{\text{ratio}\,(a,b,d)}{\text{ratio}\,(a,c,d)}} \tag{5.110}$$

This can be shown to be preserved under projective map, that is, cr $(a,b,c,d) =$ cr (A,B,C,D). Now, referring to Figure 5.38, we are going to show that under the projective map, a pre-image

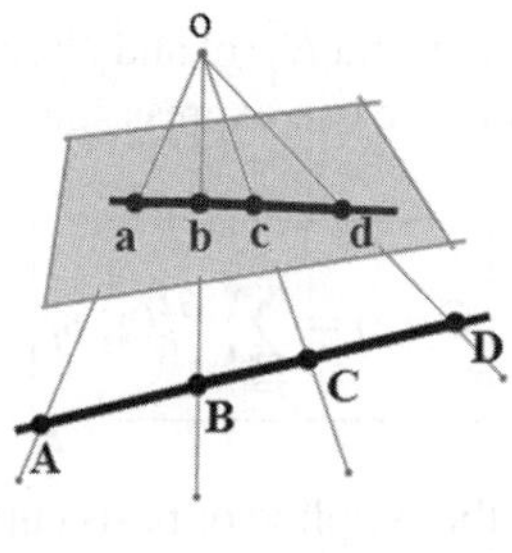

Figure 5.38 Projection map and cross ratio.

point, Θ, on line segment, *ABCD*, and the corresponding image point, θ, on the line segment, *abcd*, are connected by a ***rational linear transformation***. For this, we first note that unlike the affine map of line segments that is determined by two points – the end points – a projective map is determined by three pre-image and three image points. So, by invariance of cross ratios, we have: $\mathrm{cr}(a, b, \theta, c) = \mathrm{cr}(A, B, \Theta, C)$. For simplicity and without loss of generality, we may choose, $a = A = 0$ and $c = C = 1$; then, by applying the definition as in equation (5.110), we get: $\frac{\mathrm{ratio}(0,B,1)}{\mathrm{ratio}(0,\Theta,1)} = \frac{\mathrm{ratio}(0,b,1)}{\mathrm{ratio}(0,\theta,1)}$. Now, letting: $\mathrm{ratio}(0, B, 1) \stackrel{\Delta}{=} \Pi$ and $\mathrm{ratio}(0, b, 1) \stackrel{\Delta}{=} \pi$, we get from previous expression: $\frac{\pi\Theta}{(\Theta-1)} = \left\{ \frac{\Pi\theta}{(\theta-1)} \right\}$. Finally, solving for Θ, we get:

$$\boxed{\theta = \frac{\Pi\Theta}{\Pi\Theta + \pi(1 - \Theta)}} \tag{5.111}$$

The transformation of a pre-image point, Θ, to the image point, θ, as given by equation (5.111), is clearly a rational linear map; this is where the name ***rational Bezier curves*** came from, and they are based on the projective map which we will investigate next.

5.10.1.1 Rational Bezier Curves

It may be recalled from our derivation of the Bernstein basis representation of non-rational Bezier curves that the dimensionality of the underlying vector space containing a curve, $\mathbf{C}_n(\xi)$, was not crucial. In fact, generalization of the Bernstein–Bezier representation to rational Bezier curves follows from this observation; accordingly, we have the following definition.

Definition:

- A rational Bezier curve of degree n in three-dimensional real space, $\mathbb{R}_3$, is the central projection, as shown in Figure 5.39, of a non-rational Bezier curve of the same degree in a four-dimensional real space, $\mathbb{R}_4$.

Thus, we are now in a position to characterize a rational Bezier curve in 3D in terms of its 4D reimage as a non-rational Bezier curve, which we have already discussed. Incidentally, note that conveniently for uniformity in computer coding, both the rational and the non-rational are represented in 4D as in a ***homogenous coordinate system***.

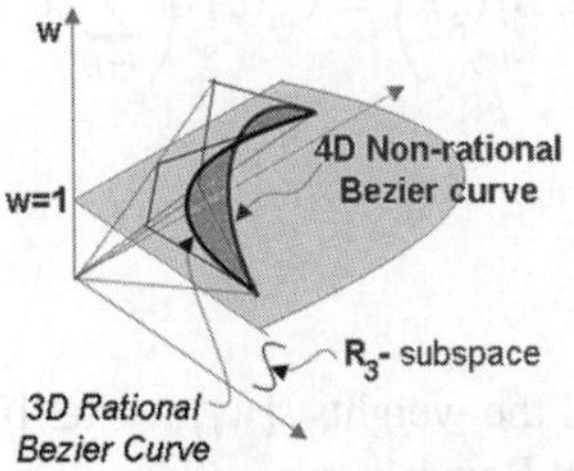

Figure 5.39 Rational Bezier by projection from 4D to 3D.

5.10.1.2 Representation Theorem

Let $\mathbf{C}_n(\xi) \in \mathbb{R}_3$ be a point on a rational Bezier curve of degree n; then, there exists $\{w_i \in \mathbb{R}\}$ and $\{\mathbf{q}_i \in \mathbb{R}_3\}$ for $\forall i \in \{0, 1, \dots, n\}$, known as the ***weights*** and the ***control points*** of the curve, respectively, such that:

$$\mathbf{C}_n(\xi) = \frac{\sum_{i=0}^{n} w_i \, \mathbf{q}_i \, B_i^n(\xi)}{\sum_{i=0}^{n} w_i \, B_i^n(\xi)} = \sum_{i=0}^{n} \underbrace{\left(\frac{w_i \, B_i^n(\xi)}{\sum_{i=0}^{n} w_i \, B_i^n(\xi)} \right)}_{\text{Bernstein basis functions for rational Bezier}} \mathbf{q}_i = \sum_{i=0}^{n} \underbrace{B_i^{wn}(\xi)}_{\text{Bernstein basis functions for rational Bezier}} \mathbf{q}_i, \quad \forall \xi \tag{5.112}$$

Proof. $\mathbf{C}_n(\xi) \in \mathbb{R}_3$ can be seen as a point $\begin{Bmatrix} \mathbf{C}_n(\xi) \\ 1 \end{Bmatrix} \in \mathbb{R}_4$. Now, because of above definition of a rational Bezier curve, $\begin{Bmatrix} \mathbf{C}_n(\xi) \\ 1 \end{Bmatrix}$ can also be seen as a point on the curve, $\mathbf{C}_n^{4D}(\xi) \equiv \begin{Bmatrix} \mathbf{C}_n(\xi)\, w(\xi) \\ w(\xi) \end{Bmatrix}$ with $w(\xi)$ so chosen that it is a non-rational Bezier curve of degree n. Then, $w(\xi)$ must have a representation as: $w(\xi) = \sum_{i=0}^{n} w_i \, B_i^n(\xi)$ where $\{w_i\}, \quad i \in \{0, 1, \dots, n\}$ are the Bezier coefficients. So,

$$\mathbf{C}_n^{4D}(\xi) \equiv \begin{Bmatrix} \mathbf{C}_n(\xi) \sum_{i=0}^{n} w_i \, B_i^n(\xi) \\ \sum_{i=0}^{n} w_i \, B_i^n(\xi) \end{Bmatrix} \tag{5.113}$$

But, $\mathbf{C}_n^{4D}(\xi)$ itself is a non-rational Bezier curve of degree n, and hence, has a representation:

$$\mathbf{C}_n^{4D}(\xi) \equiv \sum_{i=0}^{n} \begin{Bmatrix} \hat{\mathbf{q}}_i \\ w_i \end{Bmatrix} B_i^n(\xi) \text{ with } \hat{\mathbf{q}}_i \in \mathbb{R}_3 \tag{5.114}$$

Thus, comparing equations (5.113) and (5.114), we have:

$$\left(\sum_{i=0}^{n} \hat{\mathbf{q}}_i \, B_i^n(\xi) \right) = \mathbf{C}_n(\xi) \left(\sum_{i=0}^{n} w_i \, B_i^n(\xi) \right) \tag{5.115}$$

which completes the proof.

Remarks

- It can be easily shown that if all the weights, $\{w_i\}, \quad i \in \{0, 1, \dots, n\}$, are same, then $\mathbf{C}_n^{4D}(\xi)$ degenerates into the non-rational Bezier curve.
- To avoid singularity, we will require that the weights be all non-negative.

- All the properties of a non-rational Bezier curve introduced before apply to its rational counterparts.
- The rational Bezier curve, as described, can be thought of as a representation by Bernstein basis functions defined by: $\left(\frac{w_i\, B_i^n(\xi)}{\sum_{i=0}^{n} w_i\, B_i^n(\xi)} \right)$, $\forall i \in \{0, 1, \ldots, n\}$; it can be proved that they actually form a basis set.
- The basis functions sum to unity, that is, they form partitions of unity, indicating a barycentric combination, and thus, invariant under affine transformations.
- Because of the restrictions that all weights be non-negative reals, the convex hull property is preserved.
- Rational Bezier curves retain end point interpolations.
- Invariance under affine parameter transformation is, likewise, is maintained.

5.10.1.3 Reparameterization, Weights and Standard form

A rational linear parameter transformation (equivalently, the projective map) does not change the shape and the degree of a curve, only its control points and weights. So, we give the following definition.

Definition: A rational Bezier curve is in ***standard form*** if its end point weights are unity, that is, $\boxed{w_0 = w_n = 1}$

A rational Bezier curve can always be transformed to its standard form by using rational linear parameter transformation without changing the shape and the degree of the curve. Equivalently, if we reparameterize a rational Bezier curve by changing the weights, $\{w_i\}$, to new weights, $\{\hat{w}_i\}$, such that:

$$\boxed{\hat{w}_i = \left(\frac{w_0}{w_n} \right)^{i/n} w_i, \quad i \in \{0, 1, \ldots, n\}} \tag{5.116}$$

then, from equation (5.116), we obtain: $\hat{w}_n = \hat{w}_0 = w_0$; now, dividing all weights by w_0 gives us the curve in standard form.

5.10.2 Properties of Rational Bezier Curves

As expected, all properties of a rational Bezier curve are very much analogous to their non-rational counterpart when the rational Bezier curve is treated as its 4D non-rational pre-image. In any event, we list below the important properties to show the effect of the weights in their characterization, and thus how they can be implemented with slight variations to their non-rational counterparts.

5.10.2.1 Subdivision Property

The subdivision algorithm for a rational Bezier curve is analogous to its non-rational counterpart, that is, the de Casteljau algorithm is used in 4D pre-image space of a 3D rational Bezier curve, $\mathbf{C}_n(\xi)$. In other words, the algorithm is applied on both $\{w_i\, \mathbf{q}_i\}$ and $\{w_i\}$ curves for all $i \in \{0, 1, \ldots, n\}$. The intermediate control points are projected onto the hyperplane, $w = 1$, to provide us with the left and right subdivided Bezier polygons.

5.10.2.2 Derivative Property

Given: $\mathbf{C}_n(\xi) = \frac{N(\xi)}{D(\xi)}$ with $\underbrace{N(\xi) = \sum_{i=0}^{n} w_i\, q_i\, B_i^n(\xi)}_{\text{Numerator}}$ and $\underbrace{D(\xi) = \sum_{i=0}^{n} w_i\, B_i^n(\xi)}_{\text{Denominator}}$

The first derivative, $\mathbf{C}_{n,1}(\xi)$, is given by:

$$\mathbf{C}_{n,1}(\xi) = \frac{N_{,1}(\xi)\, D(\xi) - N(\xi)\, D_{,1}(\xi)}{D^2(\xi)} \tag{5.117}$$

Now, noting that: at $\xi = 0$, $N(0) = w_0\, \mathbf{q}_0$, $D(0) = w_0$, $N_{,1}(0) = w_0\, \mathbf{q}_0 - w_1\, \mathbf{q}_1$, and $D_{,1}(0) = w_1 - w_0$, we get the first derivative at $\xi = 0$ end:

$$\boxed{\mathbf{C}_{n,1}(0) = \frac{w_1}{w_0}(\mathbf{q}_1 - \mathbf{q}_0) = \frac{w_1}{w_0}\Delta \mathbf{q}_0} \tag{5.118}$$

with Δ as the ***forward difference operator***. The derivatives at all other parameter points can be obtained by first subdividing the curve at the parameter point and then evaluating the first derivative, using equation (5.118), at the left end point of the right subdivision, or the right end point of the left subdivision.

For higher derivatives such as curvature or torsion, it seems cost-effective to apply the subdivision algorithm and evaluate these by end point formulas which can be proved to be:

$$\boxed{\begin{aligned} \text{curvature: } \kappa &= \frac{n(n-1)}{n}\frac{w_0}{w_1\, w_2} = \frac{\text{area}[\mathbf{q}_0, \mathbf{q}_1, \mathbf{q}_2]}{\|\mathbf{q}_1 - \mathbf{q}_0\|^3} \\ \text{torsion: } \tau &= \frac{3(n-2)}{2n}\frac{w_0\, w_3}{w_1\, w_2} = \frac{\text{volume}[\mathbf{q}_0, \mathbf{q}_1, \mathbf{q}_2, \mathbf{q}_3]}{(\text{area}[\mathbf{q}_0, \mathbf{q}_1, \mathbf{q}_2])^2} \end{aligned}} \tag{5.119}$$

5.10.2.2.1 C++ Code: Tangent at Any Point on a Rational Bezier

We present below the C++ code excerpts for computing the tangent at any point $t \in [0, 1]$ of a rational Bezier curve:

```
/////////////////////////////////////////////////////////////
typedef int        *pINT;
typedef double     *pDOUBLE;
typedef     struct
      {
            double x;
            double y;
            double z;
      } WORLD, *pWORLD;
//////////////////////////////////////////////////////////////////////////
int CCurve::Rat_Tangent_t(pWORLD Cons, pDOUBLE Wts, double t, int nDegree,
WORLD& tangent)
{
      /////////////////////////////////////////////////////////////////////
      //   Returns tangent vector of rational Bezier curve at parameter t
      //   Always:
      //          Larger polygon selected for subdivision
```

```
    //          and its beginning tangent computed
    ////////////////////////////////////////////////////////////////////
    pWORLD   NewC = new WORLD[nDegree+1];
    pDOUBLE NewW = new double[nDegree+1];
    ///////////////////////////////////
    double sign = 1.0;
    if(t<0.5)      // subdivide Right
         Rat_SubDivide_t(Cons, Wts, t, NewC, NewW, FALSE/*bLeft*/, Degree);
    else           // subdivide Left
    {
         Rat_SubDivide_t(Cons, Wts, t, NewC, NewW, TRUE/*bLeft*/, nDegree);
         sign = -1.0;   //negative <- left polygon order reversed in Subdivision
    }
    ///////////////////////////////////////////////////////// Tangent
    Rat_Tangent_0(NewC, NewW, nDegree, tangent);//
    tangent.x = tangent.x * sign; //
    tangent.y = tangent.y * sign; //
    tangent.z = tangent.z * sign; //
    /////////////////
    delete [] NewC;
    delete [] NewW;
    /////////////////
    return 0;
}

int CCurve::Rat_SubDivide_t(pWORLD Cons, pDOUBLE Wts, double t,
          pWORLD NewC, pDOUBLE NewW, BOOL bLeft, int nDegree)
{
    ////////////////////////////////////////////////////
    //    Subdivide rational Bezier Curve at parameter t
    //          Left is reversed ordered
    ////////////////////////////////////////////////////
    int r,i;
    double tm1 = 1.0 - t,w1,w2;
    /////////////////////////////////////
    if(!bLeft)                      // Right Polygon by Casteljau
    {
          for(i=0;i<= nDegree;i++)
          {
                NewW[i]     = Wts[i];
                NewC[i].x   = Cons[i].x;
                NewC[i].y   = Cons[i].y;
                NewC[i].z   = Cons[i].z;
          }
          for(r=1;r<= nDegree;r++)
          {
                for(i=0;i<= nDegree-r;i++)
                {
                      w1                = NewW[i];
                      w2                = NewW[i+1];
                      ////////////////////////
                      NewW[i]    = tm1 * NewW[i] + t * NewW[i+1];
                      NewC[i].x    = (tm1 * w1 * NewC[i].x + t * w2 *
```

```
      NewC[i+1].x)/NewW[i];
                        NewC[i].y   = (tm1 * w1 * NewC[i].y + t * w2 *
NewC[i+1].y)/NewW[i];
                        NewC[i].z   = (tm1 * w1 * NewC[i].z + t * w2 *
      NewC[i+1].z)/NewW[i];
                  }
            }
      }
      else                              // Left polygon by Casteljau
      {
            tm1 = t;            // Reverse ordering to get left side
            t   = 1. - tm1;
            /////////////////////////
            for(i=0;i<= nDegree;i++)
            {
            NewW[nDegree-i]   = Wts[i];
            NewC[nDegree-i].x = Cons[i].x;
            NewC[nDegree-i].y = Cons[i].y;
            NewC[nDegree-i].z = Cons[i].z;
      }
      for(r=1;r<= nDegree;r++)
      {
            for(i=0;i<= nDegree-r;i++)
            {
                  w1                = NewW[i];
                  w2                = NewW[i+1];
                  ////////////////////////
                  NewW[i]    = tm1 * NewW[i] + t * NewW[i+1];
                  NewC[i].x   = (tm1 * w1 * NewC[i].x + t * w2 *
NewC[i+1].x)/NewW[i];
                  NewC[i].y   = (tm1 * w1 * NewC[i].y + t * w2 *
NewC[i+1].y)/NewW[i];
                  NewC[i].z   = (tm1 * w1 * NewC[i].z + t * w2 *
NewC[i+1].z)/NewW[i];
            }
      }
    }
    //////////
    return 0;

}

int CCurve::Rat_Tangent_0(WORLD q0, WORLD q1, double Wt0, double Wt1,
int Degree, WORLD& tangent)
{
//////////////////////////////////////////
      // Computes tangent of rational Bezier:
      //    at the beginning side ( t = 0)
//////////////////////////////////////////
//
//          tangent at 0 = nDegree * k1 * delq0
//    where
//          k1   = w1/w0
```

```
//          delq0 = q1-q0
/////////////////////////////////////////////
        C3DMath Math3D;
        ///////////////
double k1;
k1      = Wt1/Wt0;
        Math3D.Sub3DPts(q1, q0, tangent);       // subtract: q1 - q0
        /////////////////////////////////////////
        Math3D.Scale3DPts(nDegree * k1, tangent);   // scale by: nDegree*k1
        ///////////////////
        return 0;
}
```

5.10.2.3 Conics as Implicit Functions in $\mathbb{R}_2$

A conic is defined as the set of points, $\mathbf{q} \in \mathbb{R}_2$, satisfying an implicit equation in quadratic form:

$$c(\mathbf{q}) = \mathbf{q}^T \mathbf{A}\ \mathbf{q} = 0, \tag{5.120}$$

where $\mathbf{A}$ is a 3×3 symmetric matrix; in component form, for $(x, y) \in \mathbb{R}_2$, and $a, b, c, d, e, f \in \mathbb{R}$,

$$c(x, y) = a\,x^2 + 2b\,xy + 2c\,x + d\,y^2 + 2e\,y + f = 0 \tag{5.121}$$

The examples of conics include single lines (degenerate conics), double lines (degenerate conics), circles, ellipses, parabolas and hyperbolas. It may be noted that for a unique conic, only five out of the six scalars need to be determined; geometrically, for example, five distinct points in $\mathbb{R}_2$ (or a plane), uniquely determine a conic. However, for interactive geometry modeling, and more importantly, for the finite element displacement field definition, the implicit form of the conics is not suitable for various reasons apart from the obvious inefficiency and ill-conditioning involved in the determinant computation necessary for conic coefficient determinations.

5.10.2.4 Discussions

- We have introduced the rational Bezier curve in 3D as a central projective (i.e. a perspective) map or transformation of a corresponding 4D non-rational Bezier curve through the invariance requirement that both the image and the pre-image retain the same degree in the Bernstein–Bezier expansion. Of particular importance is the case when the degree is 2, that is, the quadratic rational Bezier curves, that most effectively represents the geometric objects known as ***conics***, as defined earlier; historically, introduction of the rational Bezier curves in the field of CAGD is due to S. A. Coons; since these curves encompass polynomial curves as well as the conic sections, they have been included in aircraft geometry design.
- Finally, the study of the quadratic rational Bezier curve separately is further predicated by the question of numerical accuracy of representation of conics; for the subsequent ***finite element analysis*** of, say, ***spherical shell structures***, with mega-size elements in mind, numerical accuracy in the geometric representation is an absolute necessity. Except for the degenerate conics and parabolas, non-rational Bezier curves fail to accurately capture a conic, as will be indicated in the following example.

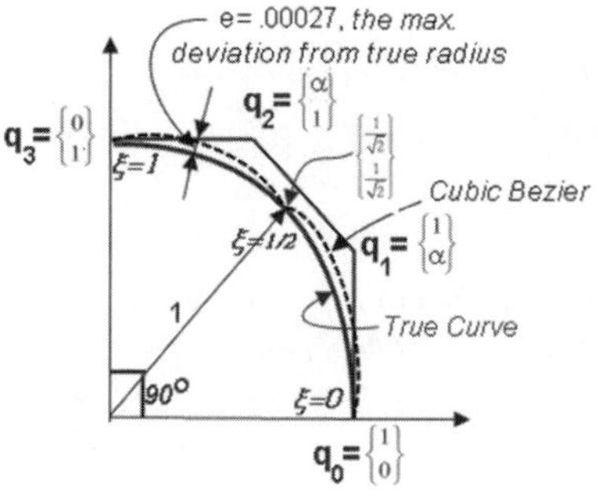

Figure 5.40 Quadrant of a circle approximation by non-rational cubic.

Example: Consider the problem of representation of one quadrant of a circle of unit radius, as shown in Figure 5.40, by a non-rational cubic Bezier curve.

Let $\mathbf{C}_3(\xi) = \sum_{i=0}^{3} B_i^3(\xi)\,\mathbf{q}_i, \quad 0 \le \xi \le 1$ be a cubic curve that tries to approximate the true circular quadrant with $\mathbf{q}_i$'s as shown in Figure 5.40; the undetermined scalar α has to be determined by the symmetry condition, that is, $\mathbf{C}_3(\frac{1}{2}) = (\frac{1}{\sqrt{2}}\ \frac{1}{\sqrt{2}})^T$. Upon solving, we get $\alpha = \frac{4}{3}(\sqrt{2}-1)$. The cubic Bezier, however, fails to trace out the circular arc identically. The radius of the approximate arc varies between 1 and $1 + e$, as shown in Figure 5.40, with e depending on the central angle of the circle.

5.10.3 Quadratic Rational Bezier Curves and Conics

Following the general definition of a rational Bezier curve of degree n, as in equation (5.112), we can now particularize and reiterate it with $n = 2$ for its special significance to conics, as:

$$\mathbf{C}_2(\xi) = \frac{a(\xi)}{b(\xi)} = \frac{\sum_{i=0}^{2} w_i\,\mathbf{q}_i\,B_i^2(\xi)}{\sum_{i=0}^{2} w_i\,B_i^2(\xi)} = \sum_{i=0}^{2} \underbrace{\left(\frac{w_i\,B_i^2(\xi)}{\sum_{i=0}^{2} w_i\,B_i^2(\xi)}\right)}_{\text{Bernstein basis functions for rational Bezier}} \mathbf{q}_i, \quad \forall\xi \tag{5.122}$$

where, as before, $\{\mathbf{q}_0, \mathbf{q}_1, \mathbf{q}_2\}$ are the control points and $\{w_0, w_1, w_2\}$ are known as the ***weights*** associated with the corresponding control points. It is easily seen that the coefficients add up to unity, that is, $\sum_{i=0}^{2}\left(\frac{w_i\,B_i^2(\xi)}{\sum_{i=0}^{2} w_i\,B_i^2(\xi)}\right) = 1$.

5.10.3.1 Barycentric Combination and Convex Hull

If $\{w_i\}$'s are all positive, then the Bezier coefficients are also positive, and since they add up to unity, the rational Bezier coefficients define a barycentric combination of the control points. Since $\mathbf{C}_2(\xi)$ lies in the convex hull of $\mathbf{q}_0, \mathbf{q}_1, \mathbf{q}_2$, which defines a plane, $\mathbf{C}_2(\xi)$ is planar curve.

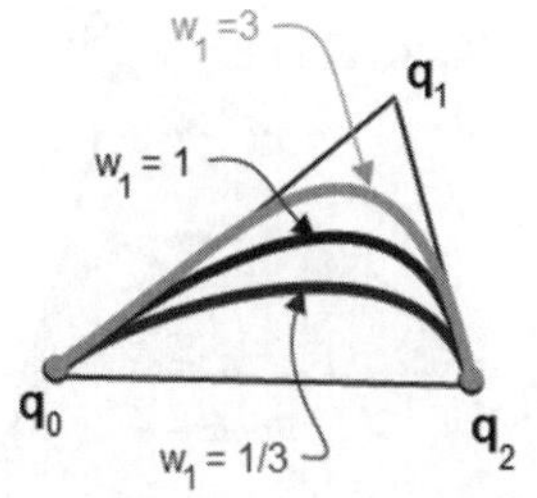

Figure 5.41 Dependence of rational Bezier on weight, w1.

5.10.3.2 Standard Form

Following the previous definition, the standard form of a quadratic Bezier is defined with the end points as unity, that is, $w_0 = w_2 = 1$. Thus, in standard form, all control points remaining the same, the shape of a quadratic Bezier curve is solely determined by w_1.

For a quadratic rational Bezier in standard form, the variations of the shapes are shown in Figure 5.41 for a few values of w_1; an increasing w_1 pulls the curve towards $\mathbf{q}_1$ and a decreasing one pushes it away.

5.10.3.3 Conics as Quadratic Rational Bezier

For the case where we set w_1 to unity, because $\sum_{i=0}^{2} B_i^2(\xi) = 1$, we can readily see that the quadratic rational Bezier curve degenerates to its non-rational counterpart, that is, quadratic non-rational Bezier curve, on a plane defined by the three control points, that, we know, describes a unique parabola – a conic section in 2D. But, the quadratic rational Bezier defined only by three control points, $(\mathbf{q}_i, 1), \quad i = 0, 1, 2$, also determines the same parabola by considering the plane in 3D. Thus, a parabola in $\mathbb{R}_2$ is the central projection of the parabola viewed in $\mathbb{R}_3$. As it turns out, this is true for general conics, that is,

- a conic in $\mathbb{R}_2$ is the projection of a parabola in $\mathbb{R}_3$ with an appropriate value of w_1.

The connection between a conic section and a quadratic rational Bezier can be established both algebraically and geometrically; the geometrical interpretation provides an insight through an introduction of the homogeneous coordinate system and the projective map, which also form the basis for computer implementation. In what follows, we first establish the algebraic connection and then provide the geometric interpretation. The expression for a quadratic Bezier, as given by equation (5.122), may be rewritten as:

$$\mathbf{C}_2(\xi) = \sum_{i=0}^{2} \left(\frac{w_i}{w(\xi)} \right) B_i^2(\xi)\, \mathbf{q}_i \quad \text{where} \quad w(\xi) = \sum_{i=0}^{2} w_i\, B_i^2(\xi) \quad \forall \xi \tag{5.123}$$

Now, if $\mathbf{C}_2(\xi)$ is a conic, that is, a planar curve defined in the plane defined by co-planar $\mathbf{q}_0$, $\mathbf{q}_1$, $\mathbf{q}_2$, then every point on it can be expressed as a barycentric combination with coordinates, say ξ_0, ξ_1, ξ_2 with respect to a triangle in the plane with $\mathbf{q}_0$, $\mathbf{q}_1$, $\mathbf{q}_2$ as its vertices, as shown in Figure 5.42.

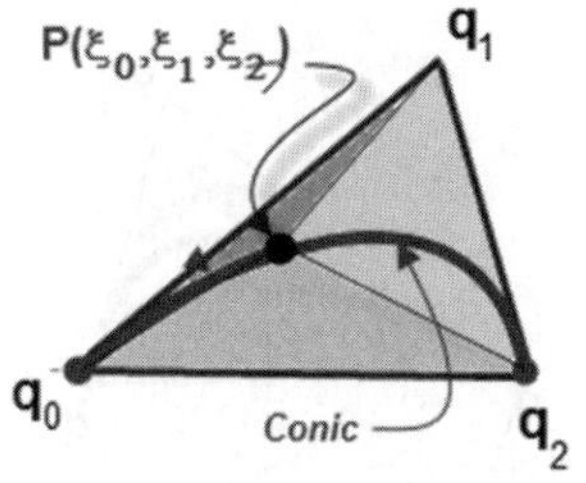

Figure 5.42 Barycentric coordinates.

We have:

$$\begin{aligned} \mathbf{C}_2(\xi) &= \xi_0\mathbf{q}_0 + \xi_1\mathbf{q}_1 + \xi_2\mathbf{q}_2, \quad \forall\xi \in [0,1] \\ &\text{with } \xi_0(\xi) + \xi_1(\xi) + \xi_2(\xi) = 1 \end{aligned} \tag{5.124}$$

where $\left(\xi_0(\xi),\ \xi_1(\xi),\ \xi_2(\xi)\right)$ are the barycentric coordinates of a point $\mathbf{P}(\xi)$ on a conic, $\mathbf{C}_2(\xi)$, given geometrically by:

$$\begin{aligned} \xi_0(\xi) &= \frac{area\Delta\mathbf{P}\mathbf{q}_1\mathbf{q}_2}{area\Delta\mathbf{q}_0\mathbf{q}_1\mathbf{q}_2}, \\ \xi_1(\xi) &= \frac{area\Delta\mathbf{P}\mathbf{q}_2\mathbf{q}_0}{area\Delta\mathbf{q}_0\mathbf{q}_1\mathbf{q}_2}, \\ \xi_2(\xi) &= \frac{area\Delta\mathbf{P}\mathbf{q}_0\mathbf{q}_1}{area\Delta\mathbf{q}_0\mathbf{q}_1\mathbf{q}_2} \end{aligned} \tag{5.125}$$

Comparing the barycentric coordinates of equation (5.125) with those of the quadratic rational Bezier as in equation (5.123), we have:

$$\begin{aligned} \xi_0(\xi) &= \frac{w_0(1-\xi)^2}{w}, \\ \xi_1(\xi) &= \frac{2w_1\xi(1-\xi)}{w}, \\ \xi_2(\xi) &= \frac{w_2\xi^2}{w} \end{aligned} \tag{5.126}$$

Now, solving for $(1-\xi)$ and ξ in the first and third of the equations, and plugging into the second, we finally get the connection between the barycentric coordinates of a conic and its quadratic rational Bezier representation as:

$$\boxed{\frac{\xi_1^2}{\xi_0\ \xi_2} = \frac{4w_1^2}{w_0\ w_2} = \lambda = \text{constant}} \tag{5.127}$$

Thus, by knowing the barycentric coordinates of a conic, the implicit equation (5.127) provides the weight w_1 that determine the shape of the curve represented in standard form, that is, with $w_0 = w_2 = 1$; the barycentric coordinates of any point other than the end points of a curve are easily obtained by solving a 3×3 matrix equation.

5.10.3.4 Shoulder Point and Parameterization

Now, suppose the control points are given and fixed so that control points, $\mathbf{q}_0$, $\mathbf{q}_2$, interpolate two end points of a conic, and $(\mathbf{q}_1 - \mathbf{q}_0)$ and $(\mathbf{q}_2 - \mathbf{q}_1)$ vectors represent to end tangent directions, respectively, with $\mathbf{q}_1$ as the tangent intersection point. We have so far fixed only four of the five necessary parameters, as given by equation (5.121) and a theorem by Pascal using implicit description for conics, to uniquely determine a conic as reflected by the general polynomial implicit equation of degree 2. Thus, choosing only $\{\mathbf{q}_i\}$, $i = 0, 1, 2$ represents a one-parameter family of conics, each of which is interpolated at the end points and in the end tangent directions; this parameter is given by λ in equation (5.127); that is, fixing λ determines a specific conic. This can be achieved by defining a fourth point $\bar{\mathbf{s}}$ on the conic known as the ***shoulder point***. However, although the shape of a conic is determined by the introduction of the shoulder point (which fixes λ), the weights – that is, the parameterization – of a conic is not fixed. In fact, various parameterizations with different choices of w_0, w_1, w_2 determine the same conic as long as $\frac{4w_1^2}{w_0\, w_2}$ equals λ of the conic. To fix the parameterization, we will also need a point, $\hat{\xi}$, on the conic such that $\bar{\mathbf{s}} = \mathbf{C}_2(\hat{\xi})$. In order to facilitate the computation, we describe a rational Bezier in standard form, that is, $w_0 = w_2 = 1$. Thus, for a quadratic rational Bezier in standard form, fixing w_1 alone completes parameterization of a conic. This is achieved by first determining the barycentric coordinates of the conic at $\bar{\mathbf{s}} = \mathbf{C}_2(\hat{\xi})$, then, we get:

$$\boxed{w_1 = \frac{\xi_1(\hat{\xi})}{2\sqrt{\xi_0(\hat{\xi}),\ \xi_2(\hat{\xi})}}} \tag{5.128}$$

where $\left(\xi_0(\hat{\xi}),\ \xi_1(\hat{\xi}),\ \xi_2(\hat{\xi})\right)$ are the barycentric coordinates of the shoulder point, $\bar{\mathbf{s}} = \mathbf{C}_2(\hat{\xi})$, on a conic, $\mathbf{C}_2(\xi)$, with respect to the control points, $\{\mathbf{q}_i\}$, $i = 0, 1, 2$, that is, $\bar{\mathbf{s}} = \mathbf{C}_2(\hat{\xi}) = \sum_{i=0}^{2} \xi_i(\hat{\xi})\,\mathbf{q}_i$.

Remarks

- We can think of w_1 as the conic shape factor.
- For $\xi_0 = 0$, that is, $\bar{\mathbf{s}}$ on $\mathbf{q}_1 - \mathbf{q}_2$ line, or for $\xi_2 = 0$, that is, $\bar{\mathbf{s}}$ on $\mathbf{q}_0 - \mathbf{q}_1$ line, there is no solution.
- For a conic in standard form, the shoulder point is defined by setting $\hat{\xi} = \frac{1}{2}$, that is, $\bar{\mathbf{s}} = \mathbf{C}_2(\frac{1}{2})$.
- The tangent to a conic at the shoulder point is parallel to the $\mathbf{q}_0 - \mathbf{q}_2$ line of the triangle determining the barycentric coordinates of the conic.

5.10.3.5 Determining w_1

Lemma: For conics in standard form, w_1 is given by:

$$\boxed{w_1 = \text{ratio}(\mathbf{m}, \mathbf{s}, \mathbf{q}_1) = \frac{\|\mathbf{s} - \mathbf{m}\|}{\|\mathbf{q}_1 - \mathbf{s}\|}} \tag{5.129}$$

where, as shown in Figure 5.43, $\mathbf{m}$ is the mid-point of the line $\mathbf{q}_0 - \mathbf{q}_2$, and $\mathbf{s} = \mathbf{C}_2(\frac{1}{2})$ is the ***shoulder point*** of the conic, $\mathbf{C}_2(\xi)$, at $\hat{\xi} = \frac{1}{2}$.

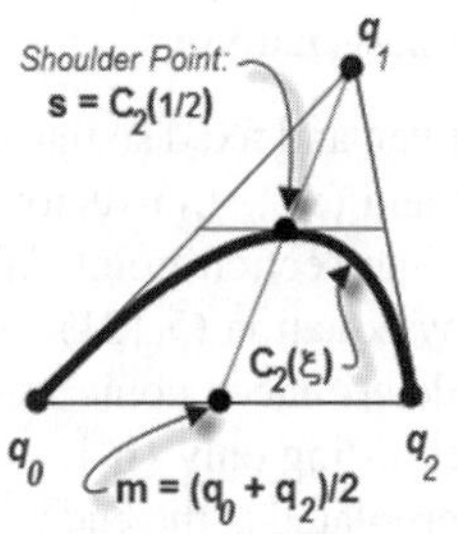

Figure 5.43 Determination of weight, w_1.

Proof. $\mathbf{s} = \mathbf{C}_2(\frac{1}{2}) = \frac{\frac{1}{4}\mathbf{q}_0 + \frac{1}{2}w_1\,\mathbf{q}_1 + \frac{1}{4}\mathbf{q}_2}{\frac{1}{4} + \frac{1}{2}w_1 + \frac{1}{4}} = \frac{\frac{1}{2}(\mathbf{q}_0 + \mathbf{q}_2) + w_1\,\mathbf{q}_1}{1 + w_1} = \frac{\mathbf{m} + w_1\,\mathbf{q}_1}{1 + w_1}$. So, $\mathbf{s} - \mathbf{m} = \frac{w_1(\,\mathbf{q}_1 - \mathbf{m})}{1 + w_1}$. Now, $\mathbf{q}_1 - \mathbf{s} = \frac{\mathbf{q}_1 - \mathbf{m}}{1 + w_1}$. Thus, ratio($\mathbf{m}$, $\mathbf{s}$, $\mathbf{q}_1$) $= \frac{\|\mathbf{s} - \mathbf{m}\|}{\|\mathbf{q}_1 - \mathbf{s}\|} = \frac{w_1\,\|\mathbf{q}_1 - \mathbf{m}\|}{1 + w_1}\,\frac{1 + w_1}{\|\mathbf{q}_1 - \mathbf{m}\|} = w_1$ completes the proof.

5.10.3.6 Classification of Conics and Complementary Segment

Finally, we would like to classify conics, as represented by the rational Bezier curves, into familiar types – ellipse, parabola and hyperbola. However, in order to accomplish this, we need to introduce the following concept.

Complementary Segment of a Conic: For a conic, $\mathbf{C}_2(\xi)$, in standard form, that is, with $w_0 = w_2 = 1$, the complementary segment, $\widetilde{\mathbf{C}}_2(\xi)$, of the conic is defined with $w_1 = -1$.

We note that:

- The complementary segment, $\widetilde{\mathbf{C}}_2(\xi)$, satisfies the implicit relation given by equation (5.127), since w_1 appears as squared in the equation; thus, the conic and the complementary segment both belong to the same conic.
- Moreover, $\widetilde{\mathbf{C}}_2(\xi) - \mathbf{q}_1 = \frac{w(\xi)}{\tilde{w}(\xi)}(\mathbf{C}_2(\xi) - \mathbf{q}_1)$, $\forall \xi \in [0, 1]$ where $\tilde{w}(\xi) \stackrel{\Delta}{=} w_0 B_0^2(\xi) - w_1 B_1^2(\xi) + w_2 B_2^2(\xi)$; clearly, then, $\mathbf{C}_2(\xi)$, $\mathbf{q}_1$ and $\widetilde{\mathbf{C}}_2(\xi)$ are collinear for all $\xi \in [0, 1]$, as shown in Figure 5.44.
- Finally, the end tangent directions of the complementary segment are opposite to their counterparts in the standard segment.

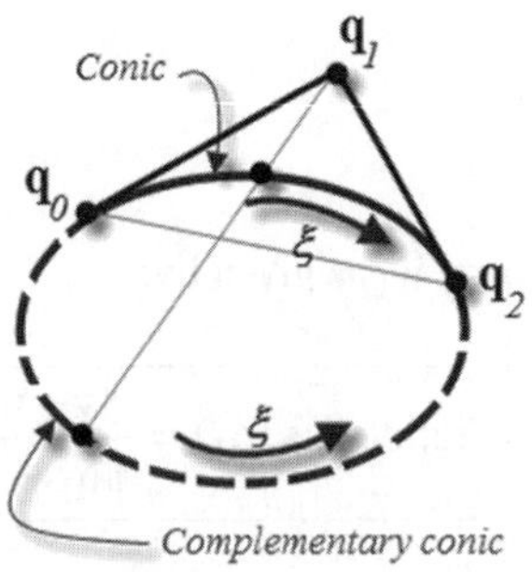

Figure 5.44 Complementary segment of a rational curve.

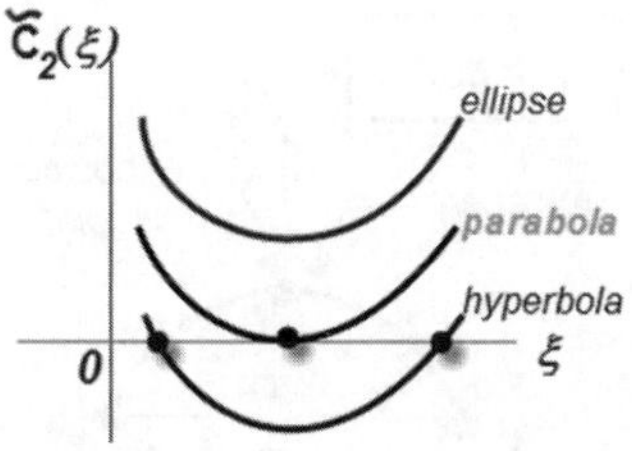

Figure 5.45 Roots of the complementary curve.

Thus, both the standard and the complementary segments of a conic are swept by the same rational quadratic Bezier curve with appropriate signs for w_1. However, as noted earlier, $\widetilde{\mathbf{C}}_2(\xi)$ is entirely depended on the denominator, $\tilde{w}(\xi)$. For example, if $\tilde{w}(\xi) \neq 0$, for all $\xi \in [0, 1]$, then, $\widetilde{\mathbf{C}}_2(\xi)$ is bounded. So, let us find out the conditions for real roots of $\tilde{w}(\xi)$: by expanding and rewriting, we have, for standard form, $\tilde{w}(\xi) = 2(1 + w_1)\xi^2 - 2(1 + w_1)\xi + 1$, and its roots, ρ_1, ρ_2 are given by:

$$\rho_{1,2} = \frac{(1 + w_1) \pm \sqrt{w_1 - 1}}{2(1 + w_1)} \tag{5.130}$$

Based on the type and number of roots, as shown in Figure 5.45, we have:

- ***Ellipse***: No singularity – $\boxed{w_1 < 1}$, no real roots, so $\tilde{w}(\xi) \neq 0$, for all $\xi \in [0, 1]$, then, $\widetilde{\mathbf{C}}_2(\xi)$ is bounded.
- ***Parabola***: One singularity – $\boxed{w_1 = 1}$, two equal real roots of the same sign, so $\widetilde{\mathbf{C}}_2(\xi)$ is unbounded but does not change sign.
- ***Hyperbola***: Two singularities – $\boxed{w_1 > 1}$, two unequal real roots of the same sign, so $\widetilde{\mathbf{C}}_2(\xi)$ is unbounded and changes sign for $\xi \in [\rho_1, \rho_2]$, and then regains back.

Finally, we noted before that as w_1 increases, the curve is pulled towards the control point, $\mathbf{q}_1$. Thus, for all control points remaining same, a hyperbola ($\boxed{w_1 > 1}$), a parabola ($\boxed{w_1 = 1}$) and an ellipse ($\boxed{w_1 < 1}$) will appear as shown in Figure 5.46.

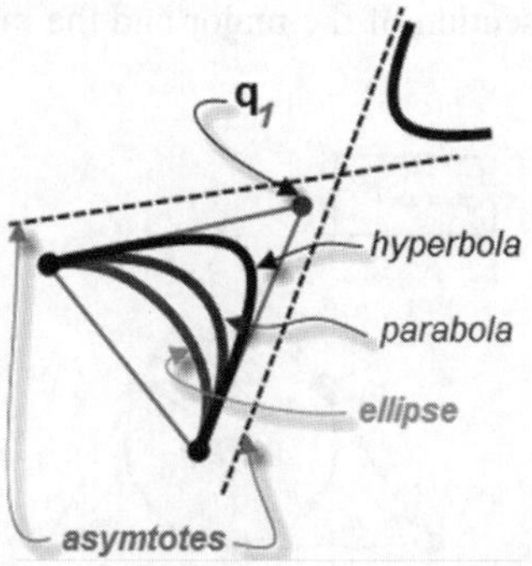

Figure 5.46 Classification of rational curves.

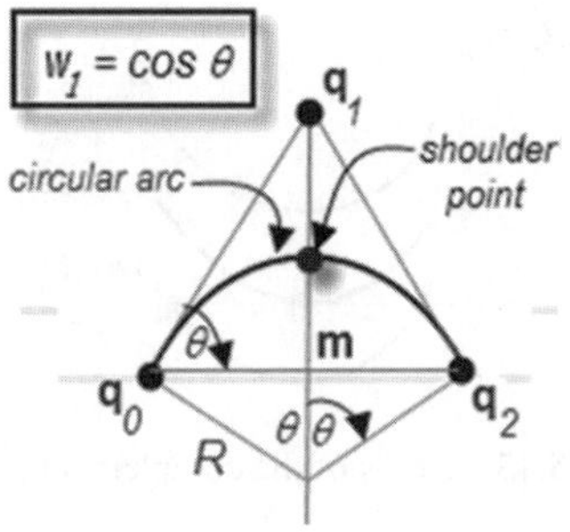

Figure 5.47 Weight (w_1) for a circular arc.

Example: *Circle*

Let us find out w^1 for a circular arc, in ***standard form***, that is, $w_0 = w_2 = 0$, with control points, $\mathbf{q}_0, \mathbf{q}_1, \mathbf{q}_2$, radius, R, and central angle, 2θ, are known as shown in Figure 5.47.

From Figure 5.47: $\|\mathbf{q}_1 - \mathbf{m}\| = \frac{R}{\cos\theta} - R\cos\theta$, and $\|\mathbf{s} - \mathbf{m}\| = R(1 - \cos\theta)$; so, $\|\mathbf{q}_1 - \mathbf{s}\| = \|\mathbf{q}_1 - \mathbf{m}\| - \|\mathbf{s} - \mathbf{m}\| = \frac{R}{\cos\theta} - \cancel{R\cos\theta} - R + \cancel{R\cos\theta} = \frac{R(1-\cos\theta)}{\cos\theta}$. Finally, using equation (5.129), we get:

$$\boxed{w_1 = \text{ratio}\ (\mathbf{m}, \mathbf{s}, \mathbf{q}_1) = \frac{\|\mathbf{s} - \mathbf{m}\|}{\|\mathbf{q}_1 - \mathbf{s}\|} = \cos\theta} \tag{5.131}$$

Remarks

- Clearly, if the central angle:$2\theta > 120°$, the triangle encompassing the circular arc cannot be formed; in such cases, the circular arc has to be generated in pieces, such that for each arc the central angle:$2\theta > 120°$,
- The minimum number of control points for a closed circle is obtained by considering the vertices of the equilateral triangle encompassing the circle, as can be seen from Figure 5.48, as the three control points, $\mathbf{q}_1$, for three circular arcs that can be represented by quadratic rational Beziers with $w_1 = \cos 60^0 = 0.5 < 1$:

Example: *Ellipse*

Let us find w_1 and $\mathbf{q}_1$ for an ellipse, in standard form, that is, $w_0 = w_2 = 0$, with control points, $\mathbf{q}_0, \mathbf{q}_2$, origin at the intersection of the major and the minor semi-axes, $a = 2$ and $b = 1$,

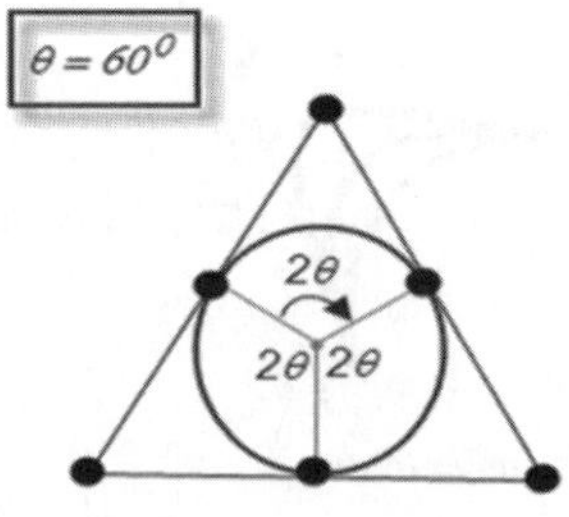

Figure 5.48 Bezier control polygon for a full circle.

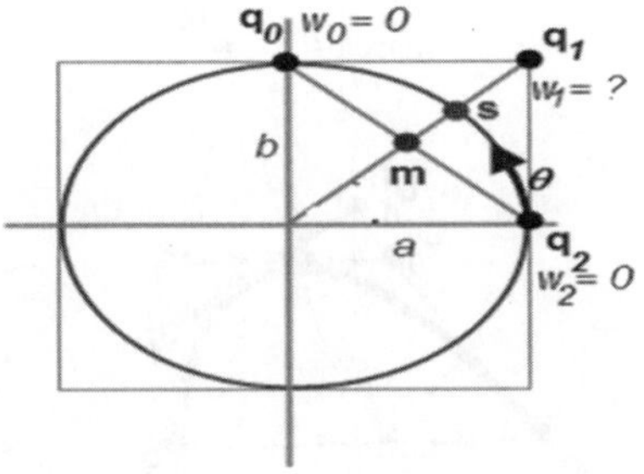

Figure 5.49 Weight (w_1) for an ellipse.

respectively, the shoulder point, $\mathbf{s} = \mathbf{C}_2(\frac{1}{2})$, and the encompassing rectangle, known as shown in Figure 5.49.

Note that $\mathbf{q}_1 = (a \quad b)^T, \mathbf{m} = \frac{1}{2}(\mathbf{q}_0 + \mathbf{q}_2) = 0.5(a \quad b)^T$ and $\mathbf{C}_2(\theta) = (a\cos\theta \quad b\sin\theta)^T$ with $\theta \in [0, \pi/2]$ for the first quadrant. Using reparameterization, $\xi = 2\theta/\pi$ such that $\xi \in [0, 1]$, in standard Bezier form for the same quadrant, we get: $\mathbf{s} = \mathbf{C}_2(\frac{1}{2}) = \left(a\cos\frac{\pi}{4} \quad b\sin\frac{\pi}{4}\right)^T = \frac{1}{\sqrt{2}}(a \quad b)^T = 0.7071(a \quad b)^T$. Then, for $a = 2$ and $b = 1$, we get: $\|\mathbf{q}_1 - \mathbf{s}\| = (1 - 0.7071)\|(2 \quad 1)^T\| = 0.2929\sqrt{5}$, and $\|\mathbf{s} - \mathbf{m}\| = (0.7071 - 0.5)\|(2 \quad 1)^T\| = 0.2021\sqrt{5}$. Thus, $\boxed{w_1 = ratio\,(\mathbf{m}, \mathbf{d}, \mathbf{q}_1) = \frac{\|\mathbf{s}-\mathbf{m}\|}{\|\mathbf{q}_1-\mathbf{s}\|} = 0.6899}$ which is less than 1, as it should be for an ellipse.

5.10.3.7 Subdivision of General Conic

We have already seen that a general conic can be expressed both as a rational quadratic Bezier curve given by equation (5.123), and as a barycentric combination shown in equation (5.124); in standard form, the preserved relationships given by equation (5.126) reduce to:

$$\begin{aligned} \xi_0(\xi) &= \frac{(1-\xi)^2}{w}, \\ \xi_1(\xi) &= \frac{2w_1\xi(1-\xi)}{w}, \\ \xi_2(\xi) &= \frac{\xi^2}{w} \end{aligned} \tag{5.132}$$

where w_1 for the original conic is given by equation (5.128). Then, from equation (5.132), we get:

$$\frac{(1-\xi)}{\xi} = \sqrt{\frac{\xi_0(\xi)}{\xi_2(\xi)}} \tag{5.133}$$

Now, suppose we need to subdivide a conic at any ξ as shown in Figure 5.50, with the de Casteljau algorithm applied to both the numerator and denominator of the quadratic rational Bezier as conic with intermediate control points, q_0^1 and q_1^1 to finally get the point $\mathbf{C}_2(\xi)$ on the curve.

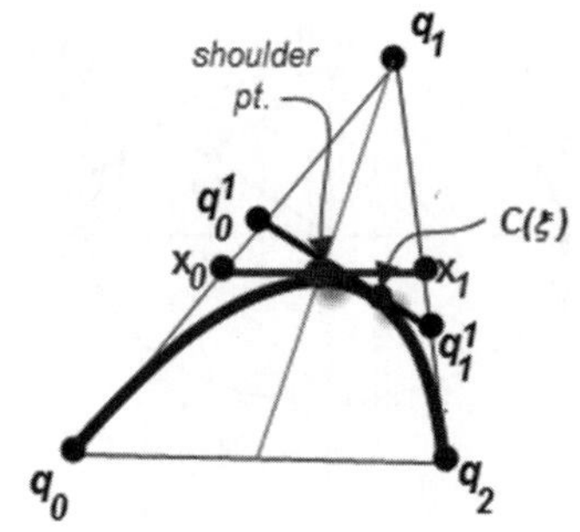

Figure 5.50 Subdivision of a conic.

We know that the tangent segment, $\mathbf{x}_0\ \mathbf{x}_1$, at the shoulder point, $\mathbf{s} = \mathbf{C}(\frac{1}{2})$, is parallel to the line segment, $\mathbf{q}_0\ \mathbf{q}_2$; thus, following equation (5.129) and similarity condition, we have:

$$
\begin{aligned}
ratio(\mathbf{q}_0, \mathbf{x}_0, \mathbf{q}_1) &\stackrel{\Delta}{=} \frac{\|\mathbf{x}_0 - \mathbf{q}_0\|}{\|\mathbf{q}_1 - \mathbf{x}_0\|} = w_1 \\
ratio(\mathbf{q}_1, \mathbf{x}_1, \mathbf{q}_2) &\stackrel{\Delta}{=} \frac{\|\mathbf{x}_1 - \mathbf{q}_1\|}{\|\mathbf{q}_2 - \mathbf{x}_1\|} = \frac{1}{w_1}
\end{aligned}
\tag{5.134}
$$

As we have already indicated, for rational Bezier polygon under projection map, the cross ratio is preserved and is equal to $\frac{1-\xi}{\xi}$, that is, using equations (5.110) and (5.133), we get:

$$
\begin{aligned}
cr(\mathbf{q}_0, \mathbf{x}_0, \mathbf{q}_0^1, \mathbf{q}_1) &= \frac{ratio(\mathbf{q}_0, \mathbf{x}_0, \mathbf{q}_1)}{ratio(\mathbf{q}_0, \mathbf{q}_0^1, \mathbf{q}_1)} = \frac{1-\xi}{\xi} = \sqrt{\frac{\xi_0(\xi)}{\xi_2(\xi)}} \\
cr(\mathbf{q}_1, \mathbf{x}_1, \mathbf{q}_1^1, \mathbf{q}_2) &= \frac{ratio(\mathbf{q}_1, \mathbf{x}_1, \mathbf{q}_2)}{ratio(\mathbf{q}_1, \mathbf{q}_1^1, \mathbf{q}_2)} = \frac{1-\xi}{\xi} = \sqrt{\frac{\xi_0(\xi)}{\xi_2(\xi)}}
\end{aligned}
\tag{5.135}
$$

Let us define the ratios involving as yet unknown control points, $\mathbf{q}_0^1$, $\mathbf{q}_1^1$, of the subdivided conics, as:

$$
\begin{aligned}
r_0 &\stackrel{\Delta}{=} ratio(\mathbf{q}_0, \mathbf{q}_0^1, \mathbf{q}_1) \\
r_1 &\stackrel{\Delta}{=} ratio(\mathbf{q}_1, \mathbf{q}_1^1, \mathbf{q}_2)
\end{aligned}
\tag{5.136}
$$

Using equations (5.134) and (5.136) in equation (5.135), we get unknown ratios in terms of the known weight, w_1, and the barycentric coordinates, $\xi_0(\xi)$, $\xi_2(\xi)$ of the subdivision point, $\mathbf{C}_2(\xi)$, on the conic, as:

$$
\begin{aligned}
r_0 &= w_1 \sqrt{\frac{\xi_2(\xi)}{\xi_0(\xi)}} \\
r_1 &= \frac{1}{w_1} \sqrt{\frac{\xi_2(\xi)}{\xi_0(\xi)}}
\end{aligned}
\tag{5.137}
$$

Finally, knowing the ratios, r_0, r_1, we can locate the unknown control points, $\mathbf{q}_0^1$, $\mathbf{q}_1^1$ and weights, w_0^1, w_1^1, of the two standard form subdivided conics: $(\mathbf{q}_0, \mathbf{q}_0^1, \mathbf{C}_2(\xi))$ and $(\mathbf{C}_2(\xi), \mathbf{q}_1^1, \mathbf{q}_2)$, as:

$$\boxed{\begin{aligned} \mathbf{q}_0^1 &= \frac{\mathbf{q}_0 + r_0\mathbf{q}_1}{1+r_0}, \quad w_0^1 = \frac{1 + r_0 w_1}{1+r_0} \\ \mathbf{q}_1^1 &= \frac{\mathbf{q}_1 + r_1\mathbf{q}_2}{1+r_1}, \quad w_1^1 = \frac{w_1 + r_1}{1+r_1} \end{aligned}} \tag{5.138}$$

5.10.4 *Rational B-spline Curves*

A 3D rational B-spline curve, like the 3D rational Bezier curves, is obtained by projection through origin of a 4D non-rational B-spline curve on the hyperplane $w = 1$. Thus, Let $\mathbf{C}_n(\xi) \in \mathbb{R}_3$ be a point on a 3D ***rational B-spline curve*** of degree n; then there exists $\{w_i \in\}$ and $\{\mathbf{q}_i \in \mathbb{R}_3\}$ for $\forall i \in \{0, 1, \dots, L+n-1\}$, known as the ***weights*** and the ***control points*** of the curve, respectively, of L segments, such that:

$$\boxed{\begin{aligned} \mathbf{C}_n(\xi) &= \frac{\sum_{i=0}^{L+n-1} w_i\, \mathbf{s}_i\, N_i^n(\xi)}{\sum_{i=0}^{L+n-1} w_i\, N_i^n(\xi)} = \sum_{i=0}^{L+n-1} \underbrace{\left(\frac{w_i\, N_i^n(\xi)}{\sum_{i=0}^{L+n-1} w_i\, N_i^n(\xi)} \right)}_{\text{de Boor–Cox basis functions for rational B–spline}} \\ \mathbf{s}_i &= \sum_{i=0}^{n} \quad \underbrace{N_i^{wn}(\xi)}_{\text{deBoor–Cox basis functions for rational Bezier}} \quad \mathbf{s}_i, \quad \forall \xi\pi \end{aligned}} \tag{5.139}$$

Remarks

- For computational purposes, a 3D rational B-spline curve may be generated by considering the 3D numerator and the 1D denominator and then carrying on the division operation as indicated in equation (5.139); we consider 4D vectors such as $(w_i\ \mathbf{s}_i,\ w_i)$ for the development of algorithms, as we will see shortly.

5.11 Composite Bezier Form: Quadratic and Cubic B-spline Curves

Although we have introduced before the general class of non-rational and rational B-spline curves, from now on, we will restrict ourselves to the case of C^1 ***quadratic*** and C^2 ***cubic B-splines*** for our purposes of geometric model development of a structural system and subsequent finite element analysis. Since, internally in computer implementation, we intend to store all curves, non-rational and/or rational, uniformly as ***composite Bezier curves***, here we set out to represent these curves as composite Bezier curves.

5.11.1 C^1 Quadratic B-spline Curves

Suppose, we are given the following.

- We have a quadratic B-spline curve (i.e. degree $n = 2$ or order $k = n + 1 = 3$) belonging to $B(\pi, n = 2)$ space with knot sequence or partition, $\pi = \{g_0, g_1, \dots, g_L\}$ such that $\dim B(\pi, 2) = L + k - 1 = L + n = L + 2$.
- The $(L + 2)$ de Boor control points are $\{\mathbf{s}_0, \mathbf{s}_1, \dots, \mathbf{s}_L, \mathbf{s}_{L+1}\}$; the polygon, D, formed by joining $\mathbf{s}_i$ to $\mathbf{s}_{i+1}$, $i \in \{0, \dots, L\}$ is the de Boor control polygon.
- The number of B-spline internal simple knots equals $L - 1$ with two end knots of multiplicity $m = 3$ at the ends, so that the curve ends are interpolated; that is, the end control points are located on the curve.

Problem: We want to represent the above C^1 quadratic B-spline curve by a composite Bezier curve.

Solution: with L segments, the number of Bezier control points needed is $2L + 1\ \{\underbrace{L}_{\text{interior}} + \underbrace{2}_{\text{ends}} + (\underbrace{L-1}_{\text{junction}})\}$. Let us designate these as: $\{\mathbf{q}_i\}, \quad i \in \{0, 1, \dots, 2L-1, 2L\}$.

Of these, L control points, known as ***Bezier interior control points***, are designated as $\mathbf{q}_1, \mathbf{q}_3, \mathbf{q}_5, \dots, \mathbf{q}_{2L-3}, \mathbf{q}_{2L-1}$; 2 control points, known as ***Bezier end control points***, are designated as $\mathbf{q}_0, \mathbf{q}_{2L}$; and finally, $(L - 1)$ control points, known as ***Bezier junction control points***, are designated as $\{\mathbf{q}_{2i}\}, \quad i \in \{1, \dots, L-1\}$. Let us now, describe the representation of the quadratic B-spline curve by a quadratic composite Bezier curve algorithmically as follows.

5.11.1.1 Algorithm: C1BS2BZ

Step 0: Identify two Bezier end control points with two known de Boor end control points such that: $\mathbf{q}_0, \mathbf{q}_{2L}$ correspond to $\mathbf{s}_0, \mathbf{s}_L$, respectively.

Step 1: Identify the remaining L Bezier interior control points with L known de Boor interior control points such that: $\mathbf{q}_1, \mathbf{q}_3, \mathbf{q}_5, \dots, \mathbf{q}_{2L-3}, \mathbf{q}_{2L-1}$ correspond to $\mathbf{s}_1, \dots, \mathbf{s}_L$, respectively; in other words, $\mathbf{q}_{2i-1} \equiv \mathbf{s}_i$ for all $i \in \{1, 2, \dots, L\}$.

Step 2: The $(L - 1)$ Bezier junction control points, $\{\mathbf{q}_{2i}\}, \quad i \in \{1, \dots, L-1\}$, are obtained by the C^1 continuity condition given by equation (5.91) with $n = 2i$ such that:

$$\boxed{\mathbf{q}_{2i} = \frac{h_{i+1}}{h_i + h_{i+1}}\mathbf{q}_{2i-1} + \frac{h_i}{h_i + h_{i+1}}\mathbf{q}_{2i+1}, \quad \forall i \in \{1, \dots, L-1\}} \tag{5.140}$$

5.11.2 C^2 Cubic B-spline Curves

Suppose, we are given the following.

- We have a cubic B-spline curve (i.e. degree $n = 3$ or order $k = n + 1 = 4$) belonging to $B(\pi, n = 3)$ space with knot sequence or partition, $\pi = \{g_0, g_1, \dots, g_L\}$ such that $\dim B(\pi, 3) = L + k - 1 = L + n = L + 3$, where L stands for the number of segments.
- The number of B-spline knots equals $L + n = L + 3$ with knot multiplicity $m = 4$ at the ends, so that the curve ends are interpolated, that is, end control points reside on the curve.
- The $(L + 3)$ de Boor control points: $\{\mathbf{s}_0, \mathbf{s}_1, \dots, \mathbf{s}_L, \mathbf{s}_{L+2}\}$; the polygon, D, formed by joining $\mathbf{s}_i$ to $\mathbf{s}_{i+1}$, $i \in \{0, \dots, L+2\}$ is the de Boor polygon.

Problem: We want to represent the above C^2 cubic B-spline curve by a cubic composite Bezier curve.

Solution: With L segments, the number of Bezier control points needed is $3L+1\{\underbrace{2L}_{\text{interior}}+\underbrace{2}_{\text{ends}}+(\underbrace{L-1}_{\text{junction}})\}$. Let us designate these as: $\{\mathbf{q}_i\},\quad i\in\{0,1,\dots,3L-1,3L\}$.

Of these, there are $2L$ control points – known as ***Bezier interior control points*** – that are designated as $\mathbf{q}_1,\mathbf{q}_2,\mathbf{q}_4,\dots,\mathbf{q}_{3L-4},\mathbf{q}_{3L-2},\mathbf{q}_{3L-1}$; two control points, known as ***Bezier end control points***, and designated as $\mathbf{q}_0,\mathbf{q}_{3L}$; and finally, $(L-1)$ control points, known as ***Bezier junction control points***, designated as $\{\mathbf{q}_{2i}\},\quad i\in\{1,\dots,L-1\}$. Now, let us describe the representation of the cubic B-spline curve by a cubic composite Bezier curve algorithmically as follows.

5.11.2.1 Algorithm: C2BS2BZO

Step 0: The $(L-1)$ junction control points, $\{\mathbf{q}_{2i}\},\quad i\in\{1,\dots,L-1\}$, are obtained by the C^1 continuity condition given by equation (5.91) with $n=3i$ such that:

$$\mathbf{q}_{3i}=\frac{h_i}{h_{i-1}+h_i}\mathbf{q}_{3i-1}+\frac{h_{i-1}}{h_{i-1}+h_i}\mathbf{q}_{3i+1},\quad \forall i\in\{1,\dots,L-1\} \tag{5.141}$$

Step 1: identify two end Bezier control points with two known de Boor control points such that:

$$\mathbf{q}_0=\mathbf{s}_0 \quad\text{and}\quad \mathbf{q}_{3L}=\mathbf{s}_{L+2} \tag{5.142}$$

Step 2: the $2L$ interior Bezier control points, $\mathbf{q}_1,\mathbf{q}_2,\dots,\mathbf{q}_{3L-2},\mathbf{q}_{3L-1}$, that is, $\{\mathbf{q}_{3i-2},\mathbf{q}_{3i-1}\},\quad \forall i\in\{1,\dots,L\}$, are obtained by the C^2 continuity condition given by equation (5.94) with $n=3i$, applied on two successive segments, as shown in Figure 5.51, such that:

$$\begin{aligned}\mathbf{q}_{3i-1}&=\frac{h_i}{h_{i-1}+h_i}\mathbf{q}_{3i-2}+\frac{h_{i-1}}{h_{i-1}+h_i}\mathbf{s}_i\\ \mathbf{q}_{3i-2}&=\frac{h_{i-1}}{h_{i-2}+h_{i-1}}\mathbf{s}_{i-1}+\frac{h_{i-2}}{h_{i-2}+h_{i-1}}\mathbf{q}_{3i-1}\end{aligned} \tag{5.143}$$

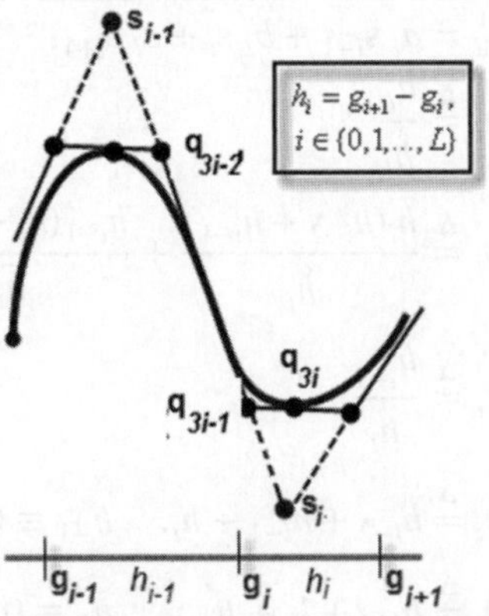

Figure 5.51 C^2 cubic B-spline to composite Bezier curve.

Now, inserting the expression for $\mathbf{q}_{3i-2}$ from the second of equation (5.143) into the first, we get an expression for $\mathbf{q}_{3i-1}$, and then inserting that in the second expression, we get an expression for $\mathbf{q}_{3i-2}$ in terms of the de Boor control points as follows:

$$\boxed{\mathbf{q}_{3i-1} = \frac{h_i}{\hat{h}_i}\mathbf{s}_{i-1} + \frac{h_{i-2}+h_{i-1}}{\hat{h}_i}\mathbf{s}_i \quad \text{and} \quad \mathbf{q}_{3i-2} = \frac{h_{i-1}+h_i}{\hat{h}_i}\mathbf{s}_{i-1} + \frac{h_{i-2}}{\hat{h}_i}\mathbf{s}_i,} \tag{5.144}$$

where

$$\boxed{\hat{h}_i \stackrel{\Delta}{=} h_{i-2} + h_{i-1} + h_i, \quad \hat{h}_{-1} = \hat{h}_L \equiv 0, \forall i \in \{1, \dots, L\}} \tag{5.145}$$

Exercise: Derive equation (5.144) following the hints given above in ***Step 2***.

Remark: For rational cubic B-spline curve conversion to rational cubic Bezier curves, algorithm C2RBS2RBZO, can be constructed by considering additionally the one-dimensional weights similar to control points as above in algorithm C2BS2BZO.

5.11.3 Open Curve Interpolation: Cubic B-spline and Cubic Composite Bezier Connection

In the curve fitting or interpolation by C^2 cubic B-spline curves (to be introduced shortly), we will also need an expression for $\mathbf{q}_{3i+1}$ in terms of the B-spline de Boor control points; once we observe that index $3(i+1) - 2 = 3i + 1$, we get from the second of equation (5.144):

$$\boxed{\mathbf{q}_{3i+1} = \frac{h_i + h_{i+1}}{\bar{h}_i}\mathbf{s}_i + \frac{h_{i-1}}{\bar{h}_i}\mathbf{s}_{i+1},} \tag{5.146}$$

where

$$\boxed{\bar{h}_i \stackrel{\Delta}{=} h_{i-1} + h_i + h_{i+1}, \forall i \in \{1, \dots, L-2\}} \tag{5.147}$$

Now, substituting for $\mathbf{q}_{3i-1}$ of equation (5.143), and $\mathbf{q}_{3i+1}$ of equation (5.146) into the C^1 continuity condition given by equation (5.141), we get:

$$\boxed{\begin{aligned}
&(h_{i-1} + h_i)\mathbf{q}_{3i} = a_i\,\mathbf{s}_{i-1} + b_i\,\mathbf{s}_i + c_i\,\mathbf{s}_{i+1}, \quad \forall i \in \{1, \dots, L-1\} \\
&\text{where} \quad a_i \stackrel{\Delta}{=} \frac{h_i^2}{\hat{h}_i} \\
&\qquad b_i \stackrel{\Delta}{=} \frac{h_i(h_{i-2} + h_{i-1})}{\hat{h}_i} + \frac{h_{i-1}(h_i + h_{i+1})}{\bar{h}_i} \\
&\qquad c_i \stackrel{\Delta}{=} \frac{h_{i-1}^2}{\bar{h}_i} \\
&\text{with} \quad \hat{h}_i \stackrel{\Delta}{=} h_{i-2} + h_{i-1} + h_i, \quad h_{-1} \equiv 0 \\
&\qquad \bar{h}_i \stackrel{\Delta}{=} h_{i-1} + h_i + h_{i+1}, \quad h_L \equiv 0
\end{aligned}} \tag{5.148}$$

Step 3: the two sets of end interior Bezier control points, $\mathbf{q}_1, \mathbf{q}_2$, and $\mathbf{q}_{3L-2}, \mathbf{q}_{3L-1}$ that is, $\{\mathbf{q}_{3i-2}, \mathbf{q}_{3i-1}\}, \quad \forall i \in \{1, L\}$, are obtained by the C^2 continuity condition given by equation (5.144), except that we set $\boxed{h_{-1} \equiv 0}$ to get for $i = 1$:

$$\boxed{\begin{aligned} &\mathbf{q}_1 = \frac{h_0 + h_1}{\hat{h}_1}\mathbf{s}_1 = \mathbf{s}_1, \quad \mathbf{q}_2 = \frac{h_1}{\hat{h}_1}\mathbf{s}_1 + \frac{h_0}{\hat{h}_1}\mathbf{s}_2, \\ &\text{where } \hat{h}_1 \stackrel{\Delta}{=} h_0 + h_1 \end{aligned}} \tag{5.149}$$

and that we set $\boxed{h_L \equiv 0}$ to get for $i = L$:

$$\boxed{\begin{aligned} &\mathbf{q}_{3L-2} = \frac{h_{L-1}}{\bar{h}_L}\mathbf{s}_L + \frac{h_{L-2}}{\bar{h}_L}\mathbf{s}_{L+1}, \quad \mathbf{q}_{3L-1} = \mathbf{s}_{L+1}, \\ &\text{where} \quad \bar{h}_L \stackrel{\Delta}{=} h_{L-2} + h_{L-1}, \end{aligned}} \tag{5.150}$$

Example: Consider the left half of a cubic B-spline basis function as shown in Figure 5.52, where the end knots have a multiplicity of 4, and thus interpolate the end points of the curve; in structural engineering terms, these are fixed end conditions, say, of a beam. Assuming that we already know the B-spline control points, the problem is to represent this curve in C^2 composite cubic Bezier form, that is, we need to determine Bezier control points.

Note that the degree of the B-spline, $n = 3$, and the number of segments, $L = 2$ with parameters $\{g_0 = 0, g_1 = 1, g_2 = 2\}$ and $h_i = g_{i+1} - g_i$ for $i = 0, 1$; we have: $h_0 = h_1 = 1$. The number of B-spline control points $= L + n = 5$, denoted by known $\{\mathbf{s}_0 = 0, \mathbf{s}_1 = 0, \mathbf{s}_2 = 0, \mathbf{s}_3 = 4, \mathbf{s}_5 = 4\}$; the number of composite Bezier control points $= 3L + 1 = 7$, identified but as yet unknown, $\{\mathbf{q}_i\}, \quad i = 0, \ldots, 6$.

Now, using ***Step 1*** of algorithm C2BS2BZO, the Bezier end control point, $\mathbf{q}_0 = \mathbf{s}_0 = 0, \mathbf{q}_6 = \mathbf{s}_4 = 4$. Noting that $h_2 \equiv 0$, from equation (5.145), we get: $h = h_0 + h_1 = 2$, and from equation (5.144), the Bezier interior control point, $\mathbf{q}_3 = \frac{1}{2}\mathbf{s}_1 + \frac{1}{2}\mathbf{s}_2 = 0$ and $\mathbf{q}_4 = \frac{1}{2}\mathbf{s}_2 + \frac{1}{2}\mathbf{s}_3 = 2$. From equation (5.149), $\mathbf{q}_1 = \mathbf{s}_1 = 0$ and $\mathbf{q}_2 = \frac{h_1}{h}\mathbf{s}_1 + \frac{h_0}{h}\mathbf{s}_2 = 0$; similarly, from equation (5.150),

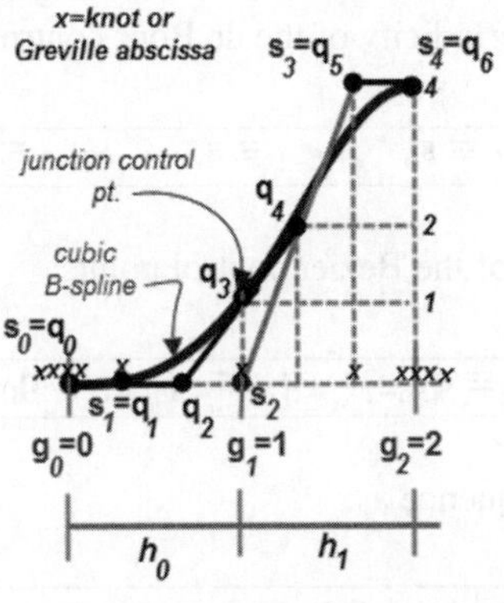

Figure 5.52 One half of cubic B-spline to Bezier control polygon.

$\mathbf{q}_5 = \mathbf{s}_3 = 4$, and $\mathbf{q}_4 = \frac{h_1}{h}\mathbf{s}_2 + \frac{h_0}{h}\mathbf{s}_3 = 2$, as before. Now, the Bezier junction control point, by equation (5.141), is given as: $\mathbf{q}_3 = \frac{h_1}{h_0+h_1}\mathbf{q}_2 + \frac{h_0}{h_0+h_1}\mathbf{q}_4 = 1$.

We can write the relation between the B-spline and Bezier control points in matrix form as:

$$\begin{Bmatrix} \mathbf{q}_0 \\ \mathbf{q}_1 \\ \mathbf{q}_2 \\ \mathbf{q}_3 \\ \mathbf{q}_4 \\ \mathbf{q}_5 \\ \mathbf{q}_6 \end{Bmatrix} = \begin{bmatrix} 1 & 0 & 0 & 0 & 0 \\ 0 & 1 & 0 & 0 & 0 \\ 0 & b_0 & a_0 & 0 & 0 \\ 0 & b_0^2 & 2a_0b_0 & a_0^2 & 0 \\ 0 & 0 & b_0 & a_0 & 0 \\ 0 & 0 & 0 & 1 & 0 \\ 0 & 0 & 0 & 0 & 1 \end{bmatrix} \begin{Bmatrix} \mathbf{s}_0 \\ \mathbf{s}_1 \\ \mathbf{s}_2 \\ \mathbf{s}_3 \\ \mathbf{s}_4 \end{Bmatrix} \tag{5.151}$$

where

$$a_0 \stackrel{\Delta}{=} \frac{h_0}{h_0 + h_1} \quad \text{and} \quad b_0 \stackrel{\Delta}{=} \frac{h_1}{h_0 + h_1} \tag{5.152}$$

- The above matrix form may be useful for converting an equilibrium equation written in Bezier form into B-spline form in finite element analysis for fixed end conditions.

5.11.4 *Closed Curve Interpolation: Cubic B-spline and Cubic Composite Bezier Connection*

Step 0: For a closed curve, we have the following additional C^0, C^1, C^2 continuity conditions to be satisfied at $i = 0$ and $i = L$:

$$\boxed{\mathbf{q}_0 = \mathbf{q}_{3L}, \quad \mathbf{q}_{0,g} = \mathbf{q}_{3L,g}, \quad \mathbf{q}_{0,gg} = \mathbf{q}_{3L,gg}} \tag{5.153}$$

Note that C^2 continuity conditions of equation (5.148) may be extended for $i = 0$ and $i = L$, if we recognize the following periodicity of the de Boor control points:

$$\boxed{\mathbf{s}_L \equiv \mathbf{s}_0, \quad \mathbf{s}_{L-1} \equiv \mathbf{s}_{-1}, \quad \mathbf{s}_{L-2} \equiv \mathbf{s}_{-2}} \tag{5.154}$$

and the following periodicity of the Bezier control points:

$$\boxed{\mathbf{q}_{-2} \equiv \mathbf{q}_{3L-2}, \quad \mathbf{q}_{-1} \equiv \mathbf{q}_{3L-1}, \quad \mathbf{q}_0 \equiv \mathbf{q}_{3L}} \tag{5.155}$$

and the periodicity of knot sequence as:

$$\boxed{h_0 \equiv h_L, \quad h_{L+1} \equiv h_1, \quad h_{-1} \equiv h_{L-1}, \quad h_{-2} \equiv h_{L-2}} \tag{5.156}$$

Step 1: We specialize equation (5.148) for $i = 0$ with periodicity of equations (5.154)–(5.156) to get:

$$\boxed{\begin{array}{ll} & (h_{L-1} + h_0)\mathbf{q}_0 = a_0\ \mathbf{s}_{L-1} + b_0\ \mathbf{s}_0 + c_0\ \mathbf{s}_1, \\ \text{where} & a_0 \stackrel{\Delta}{=} \dfrac{h_0^2}{\hat{h}_0}, b_0 \stackrel{\Delta}{=} \dfrac{h_0(h_{L-2} + h_{L-1})}{\hat{h}_0} + \dfrac{h_{L-1}(h_0 + h_1)}{\bar{h}_0}, c_0 \stackrel{\Delta}{=} \dfrac{h_{L-1}^2}{\bar{h}_0} \\ \text{with} & \hat{h}_0 \stackrel{\Delta}{=} h_{L-2} + h_{L-1} + h_0, \bar{h}_0 \stackrel{\Delta}{=} h_{L-1} + h_0 + h_1 \end{array}} \tag{5.157}$$

and for $i = L$ with periodicity as in equation (5.154)–(5.156) to get:

$$\boxed{\begin{array}{ll} & (h_{L-1} + h_0)\mathbf{q}_{3L} = a_L\ \mathbf{s}_{L-1} + b_L\ \mathbf{s}_0 + c_L\ \mathbf{s}_1 \\ \text{where} & a_L \stackrel{\Delta}{=} \dfrac{h_0^2}{\hat{h}_L}, b_L \stackrel{\Delta}{=} \dfrac{h_0(h_{L-2} + h_{L-1})}{\hat{h}_L} + \dfrac{h_{L-1}(h_0 + h_1)}{\bar{h}_L}, c_L \stackrel{\Delta}{=} \dfrac{h_{L-1}^2}{\bar{h}_L} \\ \text{with} & \hat{h}_L \stackrel{\Delta}{=} h_{L-2} + h_{L-1} + h_0, \bar{h}_L \stackrel{\Delta}{=} h_{L-1} + h_0 + h_1 \end{array}} \tag{5.158}$$

Remark:

- Equations (5.157) and (5.158) already satisfy the C^0, C^1 and C^2 conditions at $i = 0 = L$ of equation (5.153) by using equation (5.154) for periodicity of the de Boor control points.
- Considering that $\mathbf{q}_0 = \mathbf{q}_{3L}$ because of the periodicity of a closed curve, and with $\hat{h}_L = \hat{h}_0$, $\bar{h}_L = \bar{h}_0$, equations (5.157) and (5.158) are identical as they should be.

Step 3: finally, for $i = L - 1$ with periodicity as in equation (5.156) we get:

$$\boxed{\begin{array}{ll} & (h_{L-2} + h_{L-1})\mathbf{q}_{3(L-1)} = a_{L-1}\ \mathbf{s}_{L-2} + b_{L-1}\ \mathbf{s}_{L-1} + c_{L-1}\ \mathbf{s}_0 \\ \text{where} & a_{L-1} \stackrel{\Delta}{=} \dfrac{h_0^2}{\hat{h}_{L-1}}, b_{L-1} \stackrel{\Delta}{=} \dfrac{h_0(h_{L-3} + h_{L-2})}{\hat{h}_{L-1}} + \dfrac{h_{L-2}(h_{L-1} + h_0)}{\bar{h}_{L-1}}, c_{L-1} \stackrel{\Delta}{=} \dfrac{h_{L-2}^2}{\bar{h}_{L-1}} \\ \text{with} & \hat{h}_{L-1} \stackrel{\Delta}{=} h_{L-3} + h_{L-2} + h_{L-1}, \bar{h}_{L-1} \stackrel{\Delta}{=} h_{L-2} + h_{L-1} + h_0 \end{array}} \tag{5.159}$$

5.11.5 *C++ Code: B-spline to Bezier (Rational or Non-rational): Open and Close*

We present below the C++ code excerpt for conversion of B-spline control points and weights to Bezier control points and weights according to C2RBS2RBZO and C2RBS2RBZC:

```
////////////////////////////////////////////////////////////
typedef int         *pINT;
typedef double      *pDOUBLE;
typedef     struct
```

```
	{
		double x;
		double y;
		double z;
	} WORLD, *pWORLD;

typedef	double (*PDMA3)[4];
//////////////////////////////////////////////////////////////////////////////

Convert_BS2BZ(PDMA3 Sol, World Data, Double w_DT,pDOUBLE
Knot,pDOUBLE w_BZ,pWORLD Bez,int L,BOOL bClosed,int nDim)
{
	// Converts cubic B-spline to Bezier -> C2 class
	//////////////////////////////////////////////////////////////////////////////
///////////////////////////
	INPUT:
	// Sol		= 4D augmented solution for cubic B-spline Problem
	// Data		= input points
	// w_DT	= input weights
	// Knot		= input knots
	// w_BZ	= converted Bezier weights
	// Bez		= converted Bezier control points
	// L	= number of segments
	// bClosed	= open or close
	// nDim	= 3 for non-rational; 4 for rational
	////////////////////////////////////////////////////////////////
	// OPEN: L = nData - 1 = intervals
	//		Spline Dimension : L+3: (0, ... ,L+2)
	//		Bezier Dimension : 3L+1: (0, ... ,3L)
	//
	// CLOSE: L = nData = intervals: Periodic( Period = L)
	//		Spline Dimension : L: (0, ... ,L-1)
	//		Bezier Dimension : 3L: (0, ... ,3L - 1)
	////////////////////////////////////////////////////////////////
	//
	PDMA3 pBS,pBZ;
	pBS	= new double[L+2+1][4]; // reserve for rational
	pBZ = new double[3*L+1][4];	// reserve for rational
	////////////////////////////////////////////////////////////////
Spline: 4D Pre-image
	Set4DPre-image_BS(Sol,Data,w_DT,pBS,L,bClosed,nDim);
	////////////////////////////////////////////////////////////////
4D_BS -> 4D_BZ
	Convert_BS2BZ(pBS,Knot,pBZ,L,bClosed,nDim);
	////////////////////////////////////////////////////////////////
Bezier: 3D Image & Weights
	FillUpBezier(pBZ,Bez,w_BZ,L,bClosed,nDim);
	//////////////////////////////////////////
	delete [] pBS;
	delete [] pBZ;
	//////////////
	return 0;
```

```
}

Set4DPre-image_BS(PDMA3 Sol,pWORLD Data,pDOUBLE w_DT,PDMA3 BS,
                                int L,BOOL bClosed,int nDim)
{
      //////////////////////////////////////////////// Augment
      int k;
      ///////////
      if(!bClosed)
      // OPEN
      {
            ////////////////////////////////////////// First: 1 of them
            BS[0][0]    = Data[0].x;
            BS[0][1]    = Data[0].y;
            BS[0][2]    = Data[0].z;
            if(nDim>3)   //
            BS[0][3]    = w_DT[0];
            ////////////////////////////////////////// In-between: L of them
            for(k=0;k<=L;k++)
            {
                  BS[k+1][0]    = Sol[k][0];
                  BS[k+1][1]    = Sol[k][1];
                  BS[k+1][2]    = Sol[k][2];
                  if(nDim>3)
                  BS[k+1][3]    = Sol[k][3];
            }          ///
            ////////////////////////////////////////// Last: 1 of them {-}{-}{-}{-}{-}{-
}{-}{-}{-}{-}{-}/
            BS[L+2][0]    = Data[L].x;
            BS[L+2][1]    = Data[L].y;
            BS[L+2][2]    = Data[L].z;
            if(nDim>3) //   //
            BS[L+2][3]    = w_DT[L]; //      /////////////
      }
      else                    // CLOSED
      {
            ////////////////////////////////////////// First: 1 of them ——-|
            BS[0][0]    = Sol[L-1][0];   // d[-1] = d[L-1] // |
            BS[0][1]    = Sol[L-1][1]; // |
            BS[0][2]    = Sol[L-1][2];   // |
            if(nDim>3) //   //
            BS[0][3]    = Sol[L-1][3]; //
            ////////////////////////////////////////// In-between: L of them ——|
            for(k=0;k<L;k++)    // |
            {     // |-- (L+2) of them
                  BS[k+1][0]    = Sol[k][0];      // |-- L = nData
                  BS[k+1][1]    = Sol[k][1];      // |
                  BS[k+1][2]    = Sol[k][2];      //
                  if(nDim>3) //   //
                  BS[k+1][3]    = Sol[k][3];    //
            }          ///
            ////////////////////////////////////////// Last: 1 of them ——-/
            BS[L+1][0]    = Sol[0][0];   // d[L] = d[0]
```

```
            BS[L+1][1]    = Sol[0][1];
            BS[L+1][2]    = Sol[0][2];
/////////////////////////
            if(nDim>3) //   //
            BS[L+1][3]    = Sol[0][3]; //
}
       //////////
       return 0;
}

Convert_BS2BZ(PDMA3 BS,pDOUBLE Knot,PDMA3 BZ,int L,BOOL bClosed,int nDim)
{
       // NON-RATIONAL COUNTERPART
       ///////////////////////////////////////////////////////////////
       //            Converts cubic B-spline to Bezier -> still C2 class
       //            1. Spline can be rational(nDim = 4) or non-rational(nDim = 3)
       //            2. Spline can be open or closed.
       //                                      KNOTS ARE SIMPLE
       ///////////////////////////////////////////////////////////////
       // OPEN: L = nData - 1 = intervals
       //          Spline Dimension : L+3: (0, ...,L+2)
       //          Bezier Dimension : 3L+1: (0, ...,3L)
       //
       // CLOSE: L = nData = intervals: Periodic( Period = L)
       //           Spline Dimension : L: (0, ...,L-1)
       //           Bezier Dimension : 3L: (0, ...,3L - 1)
       ///////////////////////////////////////////////////////////////
       //
       //           h.i = Knot[i+1] - Knot[i] = Forward Difference
       //
       //           BZ[0] = BS[0]; BZ[1] = BS[1];
       //           BZ[2] = (h.1*BS[0]+h.0*BS[1])/h.0+h.1
       //
       //           BZ[3i-1] = (h.i*BS[i]+(h.i-2+h.i-1)*BS[i+1])/(h.i-2+h.i-1+h.i)
       //           BZ[3i+1] = (h.i*BS[i]+(h.i-1+h.i )*BS[i+1])/(h.i-2+h.i-1+h.i)
       //           BZ[3i ] = h.i*BZ[3i-1]/(h.i-1+h.i)+h.i-1*BZ[3i+1]/(h.i-1+h.i)
       //
       //           OPEN:
       //                       BZ[3L-2] = (h.L-1*BS[L]+h.L-2*BS[L+1])/(h.L-2+h.L-1)
       //                       BZ[3L-1] = BS[L+1]
       //                       BZ[3L ] = BS[L+2]
       //
       //           CLOSED: L MUST BE >= 3
// BZ[3L-2] = (h.L-1*BS[L%Period]+h.L-2*BS[(L+1)%Period])/(h.L-2+h.L-1)
       //                       BZ[3L-1] = BS[(L+1)%Period]
       //                       BZ[3L ] = modulo operation done in generation time
       //
       ///////////////////////////////////////////////////////////////////////////
////////////////
       int k,j,Lt3;
        ///////////////////////////////////////////////////////////////
First points
        if(!bClosed)
```

```
            // OPEN
        {
            for(j=0;j<nDim;j++)
            {
                BZ[0][j]    = BS[0][j];
                BZ[1][j]    = BS[1][j];
            }
            ////////////////
            double hi,hip1,sum,sum0,sum1;//,him1,sum2,sum3,sum4;
            if(L>1)
            {
                hip1 = Knot[2] - Knot[1];
                hi = Knot[1] - Knot[0];
                sum0 = Knot[2] - Knot[0];
                ///////////////////////////
               for(j=0;j<nDim;j++)
               {
                   BZ[2][j] = (hip1 * BS[1][j] + hi * BS[2][j] ) / sum0;
               }
            }
            ////////////////
            if(L>2)
            {
                sum1 = Knot[3] - Knot[1];
                sum = Knot[3] - Knot[0];
                ///////////////////////////
                for(j=0;j<nDim;j++)
                {
                    BZ[4][j] = (sum1 * BS[2][j] + hi * BS[3][j] ) / sum;
                    BZ[3][j] = (hip1 * BZ[2][j] + hi * BZ[4][j] ) / sum0;
                }
            }
        }
        else                                    // CLOSED (ALWAYS L>2)
        {
            ////////////////
            double hi,him1,him2,hip1,sum0,sum1,sum2,sum3,sum4;
            ////////////
            him2 = Knot[L-1]      - Knot[L-2];   // h[-2] = h[L-2]
            him1 = Knot[L]        - Knot[L-1];   // h[-1] = h[L-1]
            hip1 = Knot[2]        - Knot[1];
            hi = Knot[1]          - Knot[0];     // i = 0
            sum0 = him2 + him1;
            sum1 = sum0 + hi;
            sum2 = hi + hip1;
            sum3 = him1 + sum2;
            sum4 = him1 + hi;
            ///////////////////////////
            for(j=0;j<nDim;j++)
            {
                BZ[3*L-1][j]    = (hi * BS[0 ][j] + sum0 * BS[1][j])/sum1;
// BZ[-1] = BZ[3L-1]
                BZ[1] [j]    = (sum2 * BS[1 ][j] + him1 * BS[2][j])/sum3;
```

```
                BZ[0] [j]   = (hi * BZ[3*L-1][j] + him1 * BZ[1][j])/sum4;
            }

            him2 = Knot[L]    - Knot[L-1];      // h[-1] = h[L-1]
            him1 = Knot[1]    - Knot[0];
            hip1 = Knot[3]    - Knot[2];
            hi = Knot[2]      - Knot[1];      // i = 1
            sum0 = him2 + him1;
            sum1 = sum0 + hi;
            sum2 = hi + hip1;
            sum3 = him1 + sum2;
            sum4 = him1 + hi;
            ///////////////////////////
            for(j=0;j<nDim;j++)
            {
                BZ[2][j] = (hi * BS[1][j] + sum0 * BS[2][j] ) / sum1;
                BZ[4][j] = (sum2 * BS[2][j] + him1 * BS[3][j] ) / sum3;
                BZ[3][j] = (hi * BZ[2][j] + him1 * BZ[4][j] ) / sum4;
            }
        }
        ///////////////////////////////////////////////////////////////////
In-between points
        for (k=2;k<L-1;k++)
        {
            Lt3 = 3 * k;
            ////////////////
            double hi,him1,hip1,sum0,sum1,sum2,sum3,sum4;
            ///////////
            him1 = Knot[k]      - Knot[k-1];
            hip1 = Knot[2]        - Knot[1];
            hi = Knot[k+1]      - Knot[k];
            sum0 = Knot[k+2]      - Knot[k];
            sum1 = Knot[k]      - Knot[k-2];
            sum2 = Knot[k+1]    - Knot[k-2];
            sum3 = Knot[k+2]      - Knot[k-1];
            sum4 = Knot[k+1]    - Knot[k-1];
            ///////////////////////////
            for(j=0;j<nDim;j++)
            {
                BZ[Lt3-1][j] = ( hi * BS[k ][j]   + sum1 * BS[k+1 ][j] ) / sum2;
                BZ[Lt3+1][j] = ( sum0 * BS[k+1 ][j] + him1 * BS[k+2 ][j] ) / sum3;
                BZ[Lt3 ][j] = ( hi * BZ[Lt3-1][j]   + him1 * BZ[Lt3+1 ][j] ) / sum4;
            }
        }
        ///////////////////////////////////////////////////////////////////
Last points
        Lt3   = 3 * L;
        ///////////
        if(!bClosed)
            // OPEN
        {
            ////////////////
            double hi,him1,sum0,sum1,sum2;
```

```
            ////////////
            if(L>2)
            {
                  him1 = Knot[L]    - Knot[L-1];
                  sum0 = Knot[L-1]  - Knot[L-3];
                  sum1 = Knot[L]    - Knot[L-3];
                  ///////////////////////////
                  for(j=0;j<nDim;j++)
                        BZ[Lt3-4][j] = ( him1 * BS[L-1][j] + sum0 * BS[L][j] ) / sum1;
            }
            ////////////////
            if(L>1)
            {
                  him1   = Knot[L] - Knot[L-1];
                  hi     = Knot[L-1]- Knot[L-2];
                  sum1 = Knot[L]    - Knot[L-3];
                  sum2 = Knot[L]    - Knot[L-2];
                  ///////////////////////////
                  for(j=0;j<nDim;j++)
                  {
                        BZ[Lt3-2][j] = ( him1 * BS[L ][j] + hi * BS[L+1 ][j] ) / sum2;
                        BZ[Lt3-3][j] = ( him1 * BZ[Lt3-4][j] + hi * BZ[Lt3-
2][j]  ) / sum2;
                  }
            }
            ////////////////
            for(j=0;j<nDim;j++)
            {
                  BZ[Lt3-1][j] = BS[L+1][j];
                  /////////////////////////// Open
                  BZ[Lt3 ][j] = BS[L+2][j];
                  ////////////
            }
      }
      else                   // CLOSED (ALWAYS L>2)
      {

            ////////////////
            double hi,him1,him2,hip1,sum0,sum1,sum2,sum3,sum4;
            ////////////
            him2    = Knot[L-2] - Knot[L-3];
            him1    = Knot[L-1] - Knot[L-2];
            hi      = Knot[L]   - Knot[L-1];
            hip1    = Knot[1]   - Knot[0];    // h[L] = h[0];
            sum0    = him2 + him1;
            sum1    = sum0 + hi;
            sum2    = hi + hip1;
            sum3    = sum2 + him1;
            sum4    = hi   + him1;
            ////////////////////////////
            for(j=0;j<nDim;j++)
            {
                  BZ[Lt3-4][j] = ( hi * BS[L-1 ][j] + sum0 * BS[L ][j]) / sum1;
```

```
                        BZ[Lt3-2][j] = ( sum2 * BS[L ][j]    + him1 * BS[L+1 ][j]) / sum3;
                        BZ[Lt3-3][j] = ( hi * BZ[Lt3-4][j]    + him1 * BZ[Lt3-2][j]) / sum4;
                }
                        ////////////////
//                      BZ[Lt3-1][j] already done before
                        /////////////
        }
        /////////////
        return 0;

}

FillUpBezier(PDMA3 BZ4,pWORLD Con,pDOUBLE Wts,int L,BOOL bClosed,int nDim)
{
      ///////////////////////////////////////////////////// fill up
Controls and wts (Bezier)
      int k,Lt3;
      //////////
      Lt3 = 3 * L + 1;
      if(bClosed)
            Lt3-;
      //////////////////////////////////////////
      if(nDim>3)                // RATIONAL
      {
            for(k=0;k<Lt3;k++)
            {
                  double r = 1./BZ4[k][3];
                  Con[k].x    = BZ4[k][0] * r;
                  Con[k].y    = BZ4[k][1] * r;
                  Con[k].z    = BZ4[k][2] * r;
                  Wts[k]    = BZ4[k][3];
            }
      }
      else                      // Non-rational
      {
            for(k=0;k<Lt3;k++)
            {
                  Con[k].x    = BZ4[k][0];
                  Con[k].y    = BZ4[k][1];
                  Con[k].z    = BZ4[k][2];
            }
      }
      /////////////
      return 0;
}
```

5.11.6 Discussion

- Although the theory of general splines of degree *n* was introduced earlier, we are primarily interested in cubic B-splines. For cubic B-splines, we have just seen that they may be represented as composite Bezier curves; this connection between the Bernstein–Bezier curves/functions with the de Boor–B-spline curves/functions will be most useful for curve fitting and interpolation problems which we will explore next

5.12 Curve Fitting: Interpolations

So far, we have discussed various curves and their representations by control polygons; in other words, given:

- parameterization and knot sequence,
- the control polygon,
- the weights (for rational counterparts) of a curve and
- the boundary or end conditions,

we can quite effectively compute any point on the curve, be it non-rational or rational Bezier, or B-splines or non-uniform rational B-splines (NURB). (Incidentally, as mentioned before, we prefer to store internally in the computer all curves in Bezier form.) However, for mesh generations of structural continuums, we will only be given the input points and some end or boundary conditions, and be asked to interpolate these points to come up with appropriate curves, that is, fitting or passing a curve through all these input points satisfying the given set of boundary conditions. Thus, we will also have to make decisions on choosing appropriate parameterization, as we have seen that it factors in and influences the shape of a curve for the same set of control points and weights. Next, we intend to do just that by applying our foregoing discussions about curves to develop numerically efficient and robust algorithms for interpolation schemes. Before so doing, however, let us remind ourselves that for the sake of uniformity in computer coding under affine space and rational curve with weights considerations, all algorithms are set in 4D homogeneous coordinates. As will be seen later, for non-rational curves, the fourth coordinate, that stores weight, is trivially set to 1.

5.12.1 Parameterization and Knot Sequence

Since, for our purposes, we restrict ourselves to cubic non-rational and rational B-spline curves in corresponding Bezier forms, we only consider simple knots for internal junctions and multiple knots at the ends with multiplicity equal to the order of the curve; also, the knots are located at segments or junction points. There are several ways to parameterize for a curve, but the ones that factor in the given input points seem to be more reasonable. For the definitions below, we suppose that we are given input points: $\mathbf{x}_i,\ i \in \{0, \dots L\}$.

5.12.1.1 Uniform Parameterization

- In this simplest of schemes, we assign:

$$g_i = i, \quad \forall i \in \{0, \dots, L\} \tag{5.160}$$

This way of defining parameters does not incorporate the intended geometry, and thus, may produce unsatisfactory curves. It is not recommended.

5.12.1.2 Chord Length Parameterization

- In this important scheme, we assign:

$$\begin{aligned} g_0 &= 0 \\ g_i &= g_{i-1} + \left\| \mathbf{x}_i - \mathbf{x}_{i-1} \right\|, \quad \forall i \in \{1, \dots, L\} \end{aligned} \tag{5.161}$$

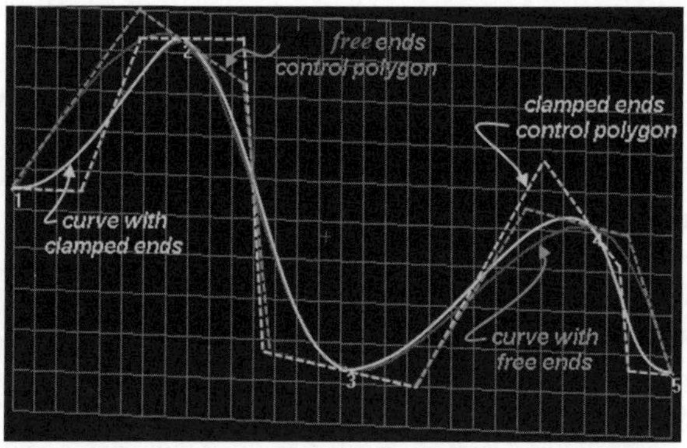

Figure 5.53 Effect of boundary conditions on a Bezier curve. (See colour plate section)

This is the most widely used scheme, but it involves length measures that are not invariant under affine transformation of the input points. In any case, this way of defining parameters incorporates the intended geometry behind the input points and produces satisfactory curves, and hence is recommended.

5.12.1.3 Centripetal Parameterization

- In this important scheme, we assign:

$$\begin{aligned} g_0 &= 0 \\ g_i &= g_{i-1} + \sqrt{\|\mathbf{x}_i - \mathbf{x}_{i-1}\|}, \quad \forall i \in \{1, \dots, L\} \end{aligned} \tag{5.162}$$

As in the previous case, it involves length measures that are not invariant under affine transformation of the input points. In any case, this way of defining parameters incorporates the intended geometry, and thus produces satisfactory curves, and hence is recommended.

5.12.2 Boundary or End Conditions

As indicated before, we restrict ourselves to cubic non-rational and rational B-spline curves. For curve interpolation, we will have fewer input data points than the total number of control points needed to be determined. Thus, we will require additional information in the form of boundary conditions to solve for all the control points and weights (see Figure 5.53).

5.12.2.1 Boundary Condition for C^1 Quadratic B-spline

For C^1 quadratic (degree, $n = 2$, and order, $k = n + 1 = 3$) B-spline curves with, say, input points, $\mathbf{x}_i$, $i = 0, \dots, L$, and input weights, $\bar{w}_i$, $i = 0, \dots, L$, over parameters, g_i, $i = 0, \dots, L$, we consider the following.

- ***Open curve***: We need to store $(L + k - 1) = L + 2$ B-spline control points, $\{\mathbf{s}_0, \mathbf{s}_1, \dots, \mathbf{s}_L, \mathbf{s}_{L+1}\}$, because there are L curve segments with support over $k = 3$ spans; for example, the first segment will have: $\mathbf{s}_0, \mathbf{s}_1, \mathbf{s}_2$ control points, and the Lth curve segment will have: $\mathbf{s}_{L-1}, \mathbf{s}_L, \mathbf{s}_{L+1}$ control points – similarly for weights. Since the number of control points to be determined $(L+2)$ is more than the input points, $(L+1)$ we will need one additional relationship in the form of a boundary condition, either left or right end.

- ***Close curve***: We need to store L control points and weights, because there are L curve segments with support over $k = 3$ spans; the additional control points are determined by the ***indices modulo period*** $(= L)$ operations. For example, the first segment will have: $\mathbf{s}_0, \mathbf{s}_1, \mathbf{s}_2$ control points, and the Lth curve segment will have: $\mathbf{s}_{L-1}, \mathbf{s}_0, \mathbf{s}_1$ control points – similarly for weights – so we will need no additional relationship in the form of boundary conditions.

5.12.2.2 Boundary Conditions for C^2 Cubic B-spline

For C^2 cubic (degree, $n = 3$, and order, $k = n + 1 = 4$) B-spline curves with, say, input points, $\mathbf{x}_i,\quad i = 0, \ldots, L$, and input weights, $\bar{w}_i,\quad i = 0, \ldots, L$, over parameters, $g_i,\quad i = 0, \ldots, L$, we consider the following.

- ***Open curve***: We need to store $(L + k - 1) = L + 3$ B-spline control points, $\{\mathbf{s}_0, \mathbf{s}_1, \ldots, \mathbf{s}_{L+1}, \mathbf{s}_{L+2}\}$, because there are L curve segments with support over $k = 4$ spans; for example, the first segment will have: $\mathbf{s}_0, \mathbf{s}_1, \mathbf{s}_2, \mathbf{s}_3$ control points, and the Lth curve segment will have: $\mathbf{s}_{L-1}, \mathbf{s}_L, \mathbf{s}_{L+1}, \mathbf{s}_{L+2}$ control points – similarly for weights. Since the number of control points to be determined $(L + 3)$ is more than the input points, $(L + 1)$ we will need two additional relationships in the form of boundary conditions at both left and right ends.
- ***Close curve***: We need to store L control points and weights, because there are L curve segments with support over $k = 4$ spans; the additional control points are determined by the ***indices modulo period*** $(= L)$ operation. For example, the first segment will have: $\mathbf{s}_0, \mathbf{s}_1, \mathbf{s}_2, \mathbf{s}_3$ control points, and the Lth curve segment will have: $\mathbf{s}_{L-1}, \mathbf{s}_0, \mathbf{s}_1, \mathbf{s}_2$ control points – similarly for weights; so we will need no additional relationship in the form of boundary conditions.

Remark: However, as we ultimately represent all B-spline curves of our interest by Bezier curves, we will need to determine them in standard form:

- From C^1 quadratic B-splines, $(2L + 1)$ Bezier control points, $\{\mathbf{q}_0, \mathbf{q}_1, \ldots, \mathbf{q}_{2L}\}$, and weights, $\{w_0, w_1, \ldots, w_{2L}\}$, assuming $w_0 = w_{2L} = 1$ at two end control points;
- From C^2 cubic B-splines, $(3L + 1)$ Bezier control points, $\{\mathbf{q}_0, \mathbf{q}_1, \ldots, \mathbf{q}_{3L}\}$, and weights, $\{w_0, w_1, \ldots, w_{3L}\}$, assuming $w_0 = w_{2L} = 1$ at two end control points;
- Thus, let us recall some Bezier curve information that will be needed for the prescription of the boundary conditions: at $i = 0$: the end tangent vectors, $\mathbf{t_0}$ and $\mathbf{t}_L$, are related by:

$$\boxed{\begin{aligned} @i = 0: &\quad \mathbf{q}_1 - \mathbf{q}_0 = \frac{h_0}{2}\mathbf{t_0} \\ @i = L: &\quad \mathbf{q}_{2L} - \mathbf{q}_{2L-1} = \frac{h_{L-1}}{2}\mathbf{t}_L \end{aligned}} \tag{5.163}$$

- With these additional relationships, we will develop and identify various boundary conditions that could be assigned to curve interpolation problems as will be seen shortly.

5.12.3 *Bezier Curve Interpolations*

We will consider below cases from degree 1 (equivalently, of order 2) to degree 3 (equivalently, of order 4):

5.12.3.1 Case: Linear Bezier Interpolation of Single Segment

Problem: Given two input points, $\mathbf{x}_i$, $\quad i = 0, 1$, determine a non-rational linear Bezier curve through these points.

5.12.3.1.1 Algorithm NR1S1

Step 0: Set control points: $\mathbf{q}_i = \mathbf{x}_i$, and $w_i = 1, \quad i = 0, 1$.

5.12.3.2 Case: Quadratic Non-Rational Bezier Interpolation of Single Segment: Parabola

Problem: Problem: Given three input points, $\mathbf{x}_i$, $\quad i = 0, 1, 2$, determine a non-rational quadratic Bezier curve through these points (see Figure 5.54), that is, determine three Bezier control points, $\mathbf{q}_i$, $\quad i = 0, 1, 2$.

5.12.3.2.1 Algorithm NR2S1

Step 0: Set end control points: $\mathbf{q}_i = \mathbf{x}_i$, $\quad i = 0, 2$ and weights: $w_i = 1, \quad i = 0, 1, 2$.
Step 1: Set shoulder point: $\mathbf{s} = \mathbf{x}_1$; set mid. pt. between $\mathbf{x}_0$ and $\mathbf{x}_2$ as: mid.pt. $= 0.5(\mathbf{x}_0 + \mathbf{x}_2)$.
Step 2: $\mathbf{q}_1 =$ mid.pt. $+ 2(\mathbf{s} -$ mid.pt.$) = 2\mathbf{x}_1 - 0.5(\mathbf{x}_0 + \mathbf{x}_2)$ and $w_i = 1$

5.12.3.3 Case: Cubic Non-Rational Bezier Interpolation of Single Segment

Problem: Given four input points, $\mathbf{x}_i$, $\quad i = 0, 1, 2, 3$, determine a non-rational cubic Bezier curve through these points (see Figure 5.55), that is, determine four Bezier control points, $\mathbf{q}_i$, $\quad i = 0, 1, 2, 3$.

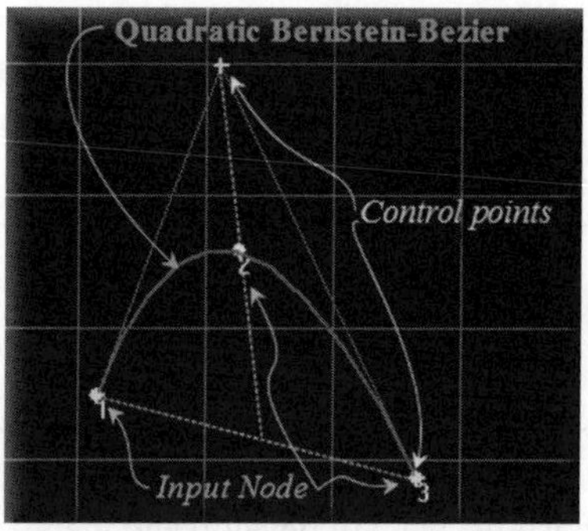

Figure 5.54 Quadratic curve fitting input points. (See colour plate section)

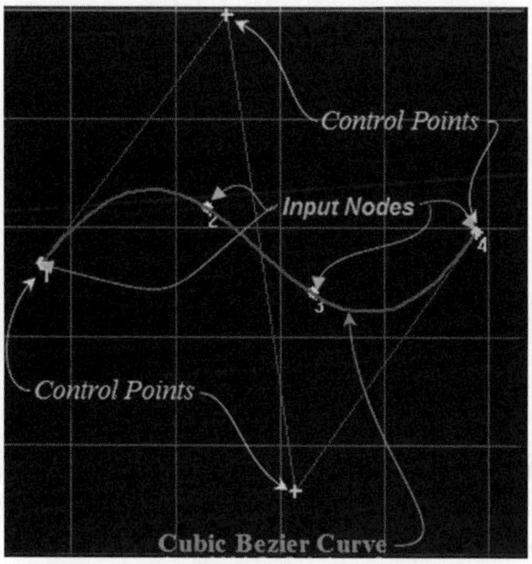

Figure 5.55 Cubic curve fitting input points. (See colour plate section)

5.12.3.3.1 Algorithm NR3S1

Step 0: Set end control points: $\mathbf{q}_i = \mathbf{x}_i, \quad i = 0, 3$ and weights: $w_i = 1, \quad i = 0, 1, 2, 3$.

Step 1: Set chord length parameters: $\overline{\xi_0} = 0, \overline{\xi_{i+1}} = \overline{\xi_i} + \|\mathbf{x}_{i+1} - \mathbf{x}_i\|, \quad i = 0, 1, 2$.

Step 2: Set chord length parameters in standard form: $\xi_0 = \overline{\xi_0} = 0, \xi_3 = 1. \ \xi_i = \frac{\overline{\xi_i}}{\overline{\xi_3}}, \quad i = 1, 2$.

Step 3: A point $\mathbf{x}$ on a cubic Bezier corresponding to a parameter, $\xi \in [0, 1]$, as: $\mathbf{x} = \sum_{i=0}^{3} B_i^3(\xi)\, \mathbf{q}_i$; so, set:

$$\begin{aligned} B_1^3(\xi_1)\mathbf{q}_1 + B_2^3(\xi_1)\mathbf{q}_2 &= \mathbf{r} \stackrel{\Delta}{=} \underbrace{\mathbf{x}_1 - B_0^3(\xi_1)\mathbf{x}_0 - B_3^3(\xi_1)\mathbf{x}_3}_{\text{Known}} \\ B_1^3(\xi_2)\mathbf{q}_1 + B_2^3(\xi_2)\mathbf{q}_2 &= \mathbf{s} \stackrel{\Delta}{=} \underbrace{\mathbf{x}_2 - B_0^3(\xi_2)\mathbf{x}_0 - B_3^3(\xi_2)\mathbf{x}_3}_{\text{Known}} \end{aligned} \tag{5.164}$$

Step 4: Solve equation (5.164) to set in-between control points:

$$\mathbf{q}_1 = \frac{B_2^3(\xi_2)\mathbf{r} - B_2^3(\xi_1)\mathbf{s}}{B_1^3(\xi_1)B_2^3(\xi_2) - B_1^3(\xi_2)B_2^3(\xi_1)}$$

$$\mathbf{q}_2 = \frac{B_1^3(\xi_1)\mathbf{r} - B_1^3(\xi_2)\mathbf{s}}{B_1^3(\xi_1)B_2^3(\xi_2) - B_1^3(\xi_2)B_2^3(\xi_1)}$$

5.12.3.4 Case: Quadratic Rational Bezier Interpolation: Circle

Problem: Given three input points, $\mathbf{x}_i, i = 0, 1, 2$, determine a circular arc, that is, a rational quadratic Bezier curve through these points in standard form, that is, determine three Bezier control points, $\mathbf{q}_i, i = 0, 1, 2$, and weight, w_1 at $\mathbf{q}_1$, assuming $w_0 = w_2 = 1$ at two end control points.

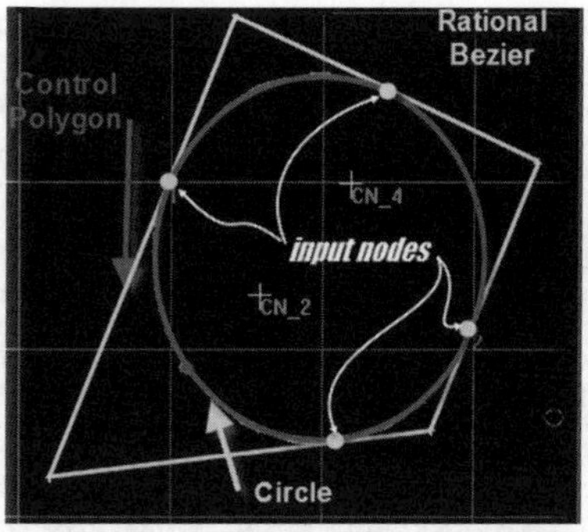

Figure 5.56 Exact full circle by rational Bezier. (See colour plate section)

Remarks

- Using the following algorithm, we can generate full circles from given input nodes as shown in Figure 5.56; however, an additional minimum number of nodes on the arcs are sometimes generated internally to keep the angle subtended at the center by any two consecutive nodes, less than 120° for tangent intersection to be possible.
- In the following algorithm, we need to identify or compute both the center and the radius of a circular arc; in actual practice, a circular arc is generated by various alternative sets of input information, as shown in the following.

5.12.3.4.1 Algorithm RCIRC1

Step 0: Set the end control points: $\mathbf{q}_i = \mathbf{x}_i$, and end weights: $w_i = 1, \quad i = 0, 2$.
Step 1: Set the shoulder point: $\mathbf{s} = \mathbf{x}_1$; set mid. pt. between $\mathbf{x}_0$ and $\mathbf{x}_2$ as: mid.pt. $= 0.5(\mathbf{x}_0 + \mathbf{x}_2)$.
Step 2: Set the mid point of the chord: $\mathbf{r}_m = 0.5(\mathbf{x}_0 + \mathbf{x}_2)$.
Step 3: Calculate the center, $\mathbf{c}$, and radius, R, and length: $\|\mathbf{r}_m - \mathbf{c}\|$; if 2θ is the angle subtended by input nodes, $\mathbf{x}_i, \quad i = 0, 2$, then, $\cos\theta = \frac{\|\mathbf{r}_m - \mathbf{c}\|}{R}$.
Step 4: Set the in-between weight: $w_1 = \cos\theta$ as given by equation (5.131).
Step 5: Length from the center, $\mathbf{c}$, to $\mathbf{q}_1$ is given as: set weight $\|\mathbf{q}_1 - \mathbf{c}\| = \frac{R}{\cos\theta}$; so, set the in-between control point, $\mathbf{q}_1 = \mathbf{c} + = \left(\frac{R}{\cos\theta}\right)\frac{(\mathbf{r}_m - \mathbf{c})}{\|\mathbf{r}_m - \mathbf{c}\|}$.

5.12.4 B-spline Curve Interpolations and Bezier Representations

We will now consider two very useful cases: rational quadratic (i.e. degree 2, or equivalently of order 3) and rational cubic (i.e. degree 3, or equivalently of order 4) of multiple segments:

Remark: As indicated before, all control points and weights are stored in the computer as Bezier control points and weights; thus, in the algorithm we use the following.

- First, the input points, weights, knot sequence and boundary conditions are utilized to solve for the interpolating 4D pre-image of the B-spline curve.,
- Then the 4D pre-image of the Bezier curve is constructed from the 4D pre-image of the B-spline curve.
- Finally, the corresponding 3D Bezier control points and weights are recovered.

Number of B-spline control points = $L+3=(L+1)+2=\#\text{of input points}+2$,

Suport for each Basis function = $[g_i, g_{i+3}]$

Suport for $N_0^3(g)$, the first Basis function = $[g_0, g_n]=[g_0, g_3]$

Suport for $N_{L+3}^3(g)$, the last Basis function = $[g_{L+n}, g_{L+2n}]=[g_{L+3}, g_{L+6}]$

Number of total knots = $L+(2)(3)+1=L+7$

Number of distinct curve segments = L

Number of distinct knots = $L+1$

Number of repeated knots

= total knots - distinct knots = $(L+2n+1)-(L+1)=2n=6$

Figure 5.57 Properties of C^2 B-spline curve.

5.12.4.1 Case: Cubic Rational B-spline (NURB) Interpolation of Multiple Segments

Problem: Given input points, $\mathbf{x}_i$, $i = 0, \ldots, L$, and input weights, $\bar{v}_i$, $i = 0, \ldots, L$, over the knot sequence, g_i, $i = 0, \ldots, L$, determine a rational cubic Bezier curve through these points in standard form, that is, for open curves: determine $(3L + 1)$ Bezier control points, $\{\mathbf{q}_0, \mathbf{q}_1, \ldots, \mathbf{q}_{3L}\}$, and weights, $\{w_0, w_1, \ldots, w_{3L}\}$, with $w_0 = w_{2L} = 1$ at two end control points; similarly, for closed curves: determine $3L$ Bezier control points, $\{\mathbf{q}_0, \mathbf{q}_1, \ldots, \mathbf{q}_{3L-1}\}$, and weights, $\{w_0, w_1, \ldots, w_{3L-1}\}$.

Remarks

- As indicated earlier, since we first generate the B-spline de Boor control points, let us recapture the main characteristics of a B-spline curve, particularized for cubics, that is, $n = 3$, from the remark after equation (5.102), as in Figure 5.57.
- All intermediate knots are simple knots, situated at segment junction points corresponding to the input points; the two end knots are multiple knots of multiplicity 4, that is, 3 additional knots each at two ends, equaling 6 as indicated above.

Remarks: ***Control Points and Weights for Bezier Curve***
As we ultimately represent all curves of our interest by composite Bezier curves, we will need to determine $(3L + 1)$ Bezier control points, $\{\mathbf{q}_0, \mathbf{q}_1, \ldots, \mathbf{q}_{3L}\}$, and weights, $\{w_0, w_1, \ldots, w_{3L}\}$, in standard form with $w_0 = w_{2L} = 1$ at two end control points; thus, let us recall some Bezier curve information that will be needed for the prescription of the boundary conditions.

- ***Open curve***: We need to store $(k-1)L + 1 = 3L + 1$ control points, because there are L curve segments; for example, the first segment will have: $\mathbf{q}_0, \mathbf{q}_1, \mathbf{q}_2, \mathbf{q}_3$ control points, and the Lth curve segment will have: $\mathbf{q}_{3L-3}, \mathbf{q}_{3L-2}, \mathbf{q}_{3L-1}, \mathbf{q}_{3L}$ control points – similarly for weights.
- ***Close curve***: We need to store $(k-1)L = 3L$ control points and weights, because there are L curve segments; the additional control points are determined by the indices modulo period $(= L)$ operation. For example, the first segment will have: $\mathbf{q}_0, \mathbf{q}_1, \mathbf{q}_2, \mathbf{q}_3$ control points, and the Lth curve segment will have: $\mathbf{q}_{3L-3}, \mathbf{q}_{3L-2}, \mathbf{q}_{3L-1}, \mathbf{q}_0$ control points – similarly for weights.

- For open curves, the end tangent vectors, $\mathbf{t_0}$ and $\mathbf{t}_L$, according to equation (5.87) are related by:

$$\boxed{\begin{aligned} @i=0: &\quad \mathbf{q}_1 - \mathbf{q}_0 = \frac{h_0}{3}\mathbf{t_0} \\ @i=L: &\quad \mathbf{q}_{3L} - \mathbf{q}_{3L-1} = \frac{h_{L-1}}{3}\mathbf{t}_L \end{aligned}} \tag{5.165}$$

Now, we are ready to describe the algorithm:

5.12.4.1.1 Algorithm C2RBS3O: Open Curves

Step 0: Set the 3D B-spline end control points: $\mathbf{s}_0 = \mathbf{q}_0 = \mathbf{x}_0$, that is, the left-most input point, and $\mathbf{s}_{L+2} = \mathbf{x}_L$, that is, the right-most input point, and the end weights as input weights: $v_0 = \bar{v}_0$ and $v_{L+2} = \bar{v}_L$.

Step 1: Set the 3D Bezier curve segment ends to input points: $\mathbf{q}_{3i} = \mathbf{x}_i, \quad \forall i \in \{0, \ldots, L\}$.

Step 2: Now, for computer implementation, define a 4D input vector, $\bar{\mathbf{x}}$, of dimension $(L+1)$ such that $\begin{Bmatrix} \bar{\mathbf{x}}_i(i,0) \\ \bar{\mathbf{x}}_i(i,1) \\ \bar{\mathbf{x}}_i(i,2) \\ \bar{\mathbf{x}}_i(i,3) \end{Bmatrix} = \begin{Bmatrix} \mathbf{x}_i(0) * \bar{v}_i \\ \mathbf{x}_i(1) * \bar{v}_i \\ \mathbf{x}_i(2) * \bar{v}_i \\ \bar{v}_i \end{Bmatrix}, \quad \forall i \in \{0, \ldots, L\}.$

Step 3: Using equation (5.148), we can set up the following $(L+1) \times (L+1)$ tri-diagonal matrix equations to be solved for the $(L+1)$ 4D B-spline augmented and as yet unknown control points, $\bar{\mathbf{e}}_i \stackrel{\Delta}{=} (\underbrace{\mathbf{e}_i}_{1\times 3} \; e_4), \forall i \in \{1, \ldots, L+1\}$:

$$\boxed{\begin{array}{l}
0|\,[-\,-]\ \{\bar{\mathbf{s}}_0\ \} = \{\bar{\mathbf{x}}_0\}(\text{control} = \text{left-most data}) \\
{*} \\
\begin{array}{c} \\ 1 \\ 2 \\ \bullet \\ \bullet \\ \bullet \\ L \\ L+1 \end{array}
\left\|\begin{bmatrix} b_0 & c_0 & 0 & 0 & 0 & 0 & 0 \\ a_1 & b_1 & c_1 & 0 & 0 & 0 & 0 \\ 0 & \bullet & \bullet & \bullet & 0 & 0 & 0 \\ 0 & 0 & \bullet & \bullet & \bullet & 0 & 0 \\ 0 & 0 & 0 & \bullet & \bullet & \bullet & 0 \\ 0 & 0 & 0 & 0 & a_{L-1} & b_{L-1} & c_{L-1} \\ 0 & 0 & 0 & 0 & 0 & a_L & b_L \end{bmatrix}\right.
\begin{Bmatrix} \bar{\mathbf{e}}_0 \\ \bar{\mathbf{e}}_1 \\ \bullet \\ \bullet \\ \bullet \\ \bar{\mathbf{e}}_{L-1} \\ \bar{\mathbf{e}}_L \end{Bmatrix}
=
\begin{Bmatrix} \mathbf{p}_0 \\ \mathbf{p}_1 \\ \bullet \\ \bullet \\ \bullet \\ \mathbf{p}_{L-1} \\ \mathbf{p}_L \end{Bmatrix}
\begin{array}{l} \leftarrow \text{determined by left BC} \\ --\backslash \\ --| \\ --| \leftarrow \text{as under} \\ --| \\ --/ \\ \leftarrow \text{determined by right BC} \end{array} \\
\qquad {*}{*}{*}{*}{*}{*}{*}{*}{*}{*}\ (L+1) \times (L+1)\ {*}{*}{*}{*}{*}{*} \\
L+2|\,[-\,-\,-\,-\,-\,-\,-\,-\,-\,-\,-\,-\,-\,-\,-\,-\,-\,-\,-]\ \{\bar{\mathbf{s}}_{L+2}\} = \{\bar{\mathbf{x}}_L\ \}(\text{control} = \text{right-most data})
\end{array}} \tag{5.166}$$

where again using equation (5.148) for all $i \in \{1, 2, \ldots, L-1\}$:

$$\boxed{\begin{array}{l}
\quad a_0 = c_L = 0 \\
\quad b_0, c_0 \leftarrow \text{depends on left boundary condition} \\
\quad a_L, b_L \leftarrow \text{depends on left boundary condition} \\
\text{and } \ \forall i \in \{1, 2, \ldots, L-1\} \\
\quad a_i \stackrel{\Delta}{=} \dfrac{h_i^2}{\widehat{h}_i}, \quad b_i \stackrel{\Delta}{=} \dfrac{h_i(h_{i-2} + h_{i-1})}{\widehat{h}_i} + \dfrac{h_{i-1}(h_i + h_{i+1})}{\bar{h}_i}, \quad c_i \stackrel{\Delta}{=} \dfrac{h_{i-1}^2}{\bar{h}_i} \\
\text{with } h_{-1} = h_L = 0, \quad h_i \stackrel{\Delta}{=} g_{i+1} - g_i, \quad \widehat{h}_i \stackrel{\Delta}{=} h_{i-2} + h_{i-1} + h_i, \quad \bar{h}_i \stackrel{\Delta}{=} h_{i-1} + h_i + h_{i+1}
\end{array}} \tag{5.167}$$

and the 4D auxiliary vectors, $\mathbf{p}_i$, $i \in \{0, \dots, L\}$ are defined as:

$$\boxed{\begin{aligned}\mathbf{p}_0 &= \text{determined by left BC}\\ \mathbf{p}_i &= (h_{i-1} + h_i)\,\bar{\mathbf{x}}_i, \quad i = 1, \dots, L-1\\ \mathbf{p}_L &= \text{determined by right BC}\end{aligned}} \tag{5.168}$$

Step 4: We cannot solve the tri-diagonal matrix equation (5.166) until we determine b_0, c_0, a_L, b_L in equation (5.167), and $\mathbf{p}_0$ and $\mathbf{p}_L$ in equation (5.168) through prescription of the boundary conditions.

Step 5: ***Boundary conditions***: with assignments of B-spline end controls to input points: $\mathbf{s}_0 = \mathbf{x}_0$ in ***Step 0*** and Bezier Control point, $\mathbf{q}_3 = \mathbf{x}_1$ in ***Step 1***, and expression for $\mathbf{q}_2$ in terms of $\mathbf{e}_0$, $\mathbf{e}_1$ as in equation (5.149), we get from the end point tangent relations of equation (5.165):

$$\boxed{\mathbf{e}_0 = \mathbf{x}_0 + \frac{h_0}{3}\mathbf{t}_0, \quad \mathbf{x}_1 = \frac{h_1\mathbf{e}_0 + h_0\mathbf{e}_1}{h_0 + h_1} + \frac{h_0}{3}\mathbf{t}_1} \tag{5.169}$$

With this in the background, we can now develop various boundary conditions:

Free end BC: Defined by natural ends with linear shape, that is, zero curvatures; thus, for the left end: $\frac{d^2\mathbf{x}(g_0)}{dg^2} = \frac{d^2\mathbf{x}(g_L)}{dg^2} = \mathbf{0}$. Now, with notation, $\mathbf{x}_{,gg}(g) = \frac{d^2\mathbf{x}(g)}{dg^2}$, and considering linear interpolation between g_0 and g_1, that is, $\mathbf{x}_{,gg}(g) = \frac{g-g_0}{h_0}\mathbf{x}_{,gg}(g_1) = \frac{g-g_0}{h_0}\mathbf{x}_{1,gg}$, and integrating twice with two constants evaluated by application of equation (5.169), we get:

$$\begin{aligned}\mathbf{x}(g) &= \frac{g-g_0}{3}(\mathbf{x}_{,g}(g) - \mathbf{t}_0) + (g - g_0)\mathbf{t}_0 + \mathbf{x}_0\\ \Rightarrow \Delta\mathbf{x}_0 &\stackrel{\Delta}{=} \mathbf{x}_1 - \mathbf{x}_0 = \frac{h_0}{3}(\mathbf{t}_1 + 2\mathbf{t}_0)\end{aligned} \tag{5.170}$$

Now, eliminating the tangents in equation (5.170) by their expressions from equation (5.169), we get:

$$\boxed{\left(2 - \frac{h_1}{h_0 + h_1}\right)\mathbf{e}_0 - \frac{h_0}{h_0 + h_1}\mathbf{e}_1 = \mathbf{x}_0} \tag{5.171}$$

Thus, with similar derivation for the right end, we set:

$$\boxed{\begin{aligned}b_0 &\stackrel{\Delta}{=} 2 - \frac{h_1}{h_0 + h_1}, \quad c_0 \stackrel{\Delta}{=} -\frac{h_0}{h_0 + h_1}, \quad \mathbf{p}_0 \stackrel{\Delta}{=} \mathbf{x}_0\\ b_L &\stackrel{\Delta}{=} 2 - \frac{h_{L-2}}{h_{L-2} + h_{L-1}}, \quad a_L \stackrel{\Delta}{=} -\frac{h_{L-1}}{h_{L-2} + h_{L-1}}, \quad \mathbf{p}_L \stackrel{\Delta}{=} \mathbf{x}_L\end{aligned}} \tag{5.172}$$

Fixed end or applied moment BC: Defined by both end tangents having been specified, that is, $\mathbf{t}_0 = \overline{\mathbf{t}_0}$ and $\mathbf{t}_L = \overline{\mathbf{t}_L}$; thus, from equation (5.169) and similar expression for the right end, we get:

$$\boxed{\mathbf{e}_0 = \mathbf{x}_0 + \frac{h_0}{3}\overline{\mathbf{t}_0}, \quad \mathbf{e}_L = \mathbf{x}_L + \frac{h_{L-1}}{3}\overline{\mathbf{t}_L}} \tag{5.173}$$

Thus, we set:

$$\boxed{\begin{aligned} b_0 &\stackrel{\Delta}{=} 1, \quad c_0 \stackrel{\Delta}{=} 0, \quad \mathbf{p}_0 \stackrel{\Delta}{=} \mathbf{x}_0 + \frac{h_0}{3}\overline{\mathbf{t}_0} \\ b_L &\stackrel{\Delta}{=} 1, \quad a_L \stackrel{\Delta}{=} 0, \quad \mathbf{p}_L \stackrel{\Delta}{=} \mathbf{x}_L + \frac{h_{L-1}}{3}\overline{\mathbf{t}_L} \end{aligned}} \tag{5.174}$$

Bessel end BC: Defined by passing a parabola through three end input points at each end, that is, passing a parabola through $(\mathbf{x}_0, \mathbf{x}_1, \mathbf{x}_2)$, such that $\mathbf{t}_0$ is taken as the tangent to the parabola, and similarly, passing a parabola through $(\mathbf{x}_{L-2}, \mathbf{x}_{L-1}, \mathbf{x}_L)$, such that $\mathbf{t}_L$ is tangent to the parabola. Thus, for the left end, let $\mathbf{x} = \mathbf{a} + (g - g_0)\,\mathbf{b} + (g - g_0)^2\,\mathbf{c}, \quad g \in [g_0, g_2]$, be the interpolating parabola; clearly, then, $\mathbf{x}_0 = \mathbf{a}$, $\mathbf{x}_1 = \mathbf{x}_0 + \mathbf{b}\,h_0 + \mathbf{c}\,h_0^2$ and $\mathbf{x}_2 = \mathbf{x}_0 + \mathbf{b}\,(h_0 + h_1) + \mathbf{c}\,(h_0 + h_1)^2$ leading to: $\mathbf{t}_0 \stackrel{\Delta}{=} \mathbf{x}_{0,g} = \mathbf{b} = -\frac{(2h_0+h_1)}{h_0\,(h_0+h_1)}\mathbf{x}_0 + \frac{(h_0+h_1)}{h_0\,h_1}\mathbf{x}_1 - \frac{h_0}{h_1(h_0+h_1)}\mathbf{x}_2$. So, eliminating $\mathbf{t}_0$ from the first of equation (5.169), we get:

$$\boxed{\begin{aligned} \mathbf{e}_0 &= \mathbf{x}_0 + \frac{h_0}{3}\mathbf{t}_0 \\ &= \frac{1}{3}\left\{\left(3 - \frac{(2h_0 + h_1)}{(h_0 + h_1)}\right)\mathbf{x}_0 + \frac{(h_0 + h_1)}{h_1}\mathbf{x}_1 - \frac{h_0^2}{h_1(h_0 + h_1)}\mathbf{x}_2\right\} \\ &= \frac{1}{3}\left\{(1 + \alpha)\mathbf{x}_0 + \frac{1}{\alpha}\mathbf{x}_1 - \frac{\beta^2}{\alpha}\mathbf{x}_2\right\} \\ \text{where} \quad \alpha &\stackrel{\Delta}{=} \frac{h_0}{(h_0 + h_1)}, \quad \beta \stackrel{\Delta}{=} (1 - \alpha) = \frac{h_1}{(h_0 + h_1)} \end{aligned}} \tag{5.175}$$

Thus, with a similar expression for the right end, we set:

$$\boxed{\begin{aligned} b_0 &\stackrel{\Delta}{=} 1, \quad c_0 \stackrel{\Delta}{=} 0, \quad \mathbf{p}_0 \stackrel{\Delta}{=} \frac{1}{3}\left\{(1 + \alpha)\mathbf{x}_0 + \frac{1}{\alpha}\mathbf{x}_1 - \frac{\beta^2}{\alpha}\mathbf{x}_2\right\} \\ b_L &\stackrel{\Delta}{=} 1, \quad a_L \stackrel{\Delta}{=} 0, \quad \mathbf{p}_L \stackrel{\Delta}{=} \frac{1}{3}\left\{(1 + \alpha)\mathbf{x}_L + \frac{1}{\alpha}\mathbf{x}_{L-1} - \frac{\beta^2}{\alpha}\mathbf{x}_{L-2}\right\} \end{aligned}} \tag{5.176}$$

Quadratic end BC: Defined by constant curvature at two end input points at each end, that is, $\mathbf{x}_{,gg}(g_0) = \mathbf{x}_{,gg}(g_1)$ for the left end, and $\mathbf{x}_{,gg}(g_L) = \mathbf{x}_{,gg}(g_{L-1})$ for the right end.
Thus, for the left end: $\mathbf{x}_{,g}(g), \quad g \in [g_0, g_1]$ is linear, and we can represent it as: $\mathbf{x}_{,g}(g) = \frac{g_1-g}{h_0}\,\mathbf{t}_0 + \frac{g-g_0}{h_0}\,\mathbf{t}_1, \quad g \in [g_0, g_1]$; now, integrating this expression once, evaluating at two input points, g_0 and g_1, and solving for the tangents, we get: $\mathbf{t}_0 + \mathbf{t}_1 = \frac{2}{h_0}(\Delta\mathbf{x}_0)$ where $\Delta\mathbf{x}_0 \stackrel{\Delta}{=} \mathbf{x}_1 - \mathbf{x}_0$.
Now, substituting for $\mathbf{t}_0, \mathbf{t}_1$ from equation (5.169), we get:

$$\boxed{\left(\frac{h_0}{h_0 + h_1}\right)(\mathbf{e}_0 + \mathbf{e}_1) = -\frac{1}{3}(\mathbf{x}_1 - \mathbf{x}_0)} \tag{5.177}$$

Thus, with a similar expression for the right end, we set:

$$\boxed{\begin{aligned} b_0 &\stackrel{\Delta}{=} \frac{h_{L-1}}{h_{L-2}+h_{L-1}}, \quad c_0 \stackrel{\Delta}{=} b_0, \quad \mathbf{p}_0 \stackrel{\Delta}{=} -\frac{1}{3}(\mathbf{x}_1 - \mathbf{x}_0) \\ b_L &\stackrel{\Delta}{=} \frac{h_0}{h_0+h_1}, \quad a_L \stackrel{\Delta}{=} b_L, \quad \mathbf{p}_L \stackrel{\Delta}{=} -\frac{1}{3}(\mathbf{x}_{L-1} - \mathbf{x}_L) \end{aligned}} \tag{5.178}$$

Step 6: From the 4D solution vectors, $\bar{\mathbf{s}}_i$, $\quad i \in \{1, \ldots, L+1\}$, we may get the 3D B-spline control vectors as: $\mathbf{s}_i(j) = \frac{\bar{\mathbf{s}}_i(j)}{\bar{\mathbf{s}}_i(3)}$, $\quad j \in \{0, 1, 2\}$, and the weights as: $v_i = \bar{\mathbf{s}}_i(3)$, $\quad \forall i \in \{1, \ldots, L+1\}$. However, since we always store internally as Bezier curves, using the algorithm C2RBS2RBZ, we first compute the 4D pre-image Bezier $(3L+1)$ vectors from the 4D B-spline $(L+3)$ augmented vectors for open curves; similarly, 4D pre-image Bezier $3L$ vectors from the 4D B-spline L augmented vectors for closed curves. Finally, we recover the Bezier control vectors, $\mathbf{q}_i$, and the weights, w_i, for $\forall i \in \{0, \ldots, 3L\}$; a computer code snippet is shown for the algorithm C2RBS3O following the example below:

Example: To determine the matrix equation (5.166) we solve for an open rational B-spline curve.

1. Uniform parameter intervals, that is, $h_i = g_{i+1} - g_i \equiv h$ for $\forall i \in \{0, \ldots, L-1\}$.
2. Fixed ends with zero tangent lengths, that is, $\mathbf{s}_0 = \mathbf{s}_1 = \bar{\mathbf{x}}_0$ and $\mathbf{s}_L = \mathbf{s}_{L-1} = \bar{\mathbf{x}}_L$, we get from equation (5.167):

$$\boxed{\begin{aligned} & a_0 = c_L = 0, \\ & h_{-1} = h_L = 0, \quad h_i \stackrel{\Delta}{=} g_{i+1} - g_i \equiv h \\ & \hat{h}_i \stackrel{\Delta}{=} h_{i-2} + h_{i-1} + h_i = \begin{cases} 2h, & \text{for } i = 1 \\ 3h, & \text{otherwise} \end{cases} \\ & \bar{h}_i \stackrel{\Delta}{=} h_{i-1} + h_i + h_{i+1} = \begin{cases} 2h, & \text{for } i = L-1 \\ 3h, & \text{otherwise} \end{cases} \end{aligned}} \tag{5.179}$$

implying:

$$\boxed{\begin{aligned} & a_1 = c_{L-1} = \frac{1}{2}h \\ & b_1 = b_{L-1} = \frac{h}{2} + \frac{2h}{3} = \frac{7}{6}h \\ & c_1 = a_{L-1} = \frac{1}{3}h \end{aligned}} \tag{5.180}$$

and:

$$\boxed{\begin{aligned} \forall i \in \{2,\ldots,L-2\},\quad a_i &= \frac{1}{3}h \\ b_i &= \frac{2}{3}h + \frac{2}{3}h = \frac{4}{3}h \\ c_i &= \frac{1}{3}h \end{aligned}} \tag{5.181}$$

with the right-hand side, from equation (5.168), as:

$$\boxed{\mathbf{p}_i = (h_{i-1} + h_i)\,\bar{\mathbf{x}}_i = 2h\bar{\mathbf{x}}_i, \quad i = 1,\ldots,L-1} \tag{5.182}$$

For fixed end boundary conditions with zero tangent lengths, we get, from equation (5.174):

$$\boxed{\begin{aligned} b_0 &\stackrel{\Delta}{=} 1, \quad c_0 \stackrel{\Delta}{=} 0, \quad \mathbf{p}_0 = \mathbf{x}_0 \\ b_L &\stackrel{\Delta}{=} 1, \quad a_L \stackrel{\Delta}{=} 0, \quad \mathbf{p}_L = \mathbf{x}_L \end{aligned}} \tag{5.183}$$

Thus, using equations (5.179)–(5.183), we can finally express the 4D matrix equation with weights included as:

$$\boxed{\begin{array}{c} [------------]\ \{\bar{\mathbf{s}}_0\ \} = \{\bar{\mathbf{x}}_0\} \\ ************************************ \\ \dfrac{1}{12}\begin{bmatrix} 12 & 0 & 0 & 0 & 0 & 0 & 0 \\ 3 & 7 & 2 & 0 & 0 & 0 & 0 \\ 0 & 2 & 8 & 2 & 0 & 0 & 0 \\ 0 & 0 & \bullet & \bullet & \bullet & 0 & 0 \\ 0 & 0 & 0 & 2 & 8 & 2 & 0 \\ 0 & 0 & 0 & 0 & 2 & 7 & 3 \\ 0 & 0 & 0 & 0 & 0 & 0 & 12 \end{bmatrix} \begin{Bmatrix} \bar{\mathbf{e}}_0 \\ \bar{\mathbf{e}}_1 \\ \bullet \\ \bullet \\ \bullet \\ \bar{\mathbf{e}}_{L-1} \\ \bar{\mathbf{e}}_L \end{Bmatrix} = \begin{Bmatrix} \bar{\mathbf{x}}_0 \\ \bar{\mathbf{x}}_1 \\ \bar{\mathbf{x}}_2 \\ \bullet \\ \bar{\mathbf{x}}_{L-2} \\ \bar{\mathbf{x}}_{L-1} \\ \bar{\mathbf{x}}_L \end{Bmatrix} \\ *****\ (L+1)\times(L+1)\ ***** \\ [------------]\ \{\bar{\mathbf{s}}_{L+2}\} = \{\bar{\mathbf{x}}_L\} \end{array}} \tag{5.184}$$

5.12.4.1.2 C++ Code: B-spline to Bezier (Rational or Non-rational): Open

We present below the C++ code excerpt for generation of B-spline control points and weights according to C2RBS3O:

```
SetupSystem_Open(pDOUBLE a,pDOUBLE b,pDOUBLE c,pDOUBLE Knot,int L)
{

     //////////////////////////////////////////////////////////////////
     //   KNOTS MUST BE SIMPLE i.e. NO MULTIPLICITY OTHER THAN 1
     // WITH END KNOTS OF MULTIPLICITY 4
     //////////////////////////////////////////////////////////////////
     //     L    = nData - 1;    // no. of intervals
     /////////////////////////////////////////////////////////////////////
```

```
// First row:   b[0],c[0] depends on BC -> will be done in BC routine
// Last row:    a[L],b[L] depends on BC -> will be done in BC routine
//////////////////////////////////////////////////////////////////////
double   hi, him1, him2, hip1, sumim1, sumi, sum;
///////////////////////////////////////////////////
a[0] = 0.0;
c[L] = 0.0;
/////////////////////////////////////////////////// Case: One interval
if( L == 1)          // b[0],c[0],a[L],b[L] will be done in
     return 0;   // BC routine
/////////////////////////////////////////////////// Case: Two intervals
int i;
//////////
if( L == 2)      // b[0],c[0],a[L],b[L] will be done in
{                            // BC routine

     i = 1;      // 2nd row of matrix
     //////
     him2    = 0.0;          // h[i-2] = 0.0 for i = 1
     him1    = Knot[i]   - Knot[i-1];      // h[i-1]
     hi      = Knot[i+1] - Knot[i];        // h[i]
     hip1    = 0.0;            // h[i+1] = 0.0 for i = 1

     sumim1    = him2 + him1 + hi;
     sumi      = him1 + hi + hip1;
     sum       = him1 + hi;
     ////////////////////////
     a[i] = hi*hi/sumim1;
     b[i] = hi*(him2+him1)/sumim1 + him1*(hi+hip1)/sumi;
     c[i] = him1*him1/sumi;
     //////////////// Corresponding RHS will be divided by sum in
BC routine, so
     a[i] /= sum;
     b[i] /= sum;
     c[i] /= sum;
     //////////////
     return 0;

}
/////////////////////////////////////////////////// Case: More than
two intervals
for (i=1;i<L;i++)    // 2rd to Last-but-1 Rows
{

     if(i == 1)
     {
          him2    = 0.0;       // h[i-2] = 0.0
          hip1    = Knot[i+2] - Knot[i+1];    // h[i+1]
     }
     else
     if(i == L-1)
     {
          him2    = Knot[i-1] - Knot[i-2];    // h[i-2]
          hip1    = 0.0;          // h[i+1] = 0.0
```

```
        }
        else
        {
             him2    = Knot[i-1] - Knot[i-2];    // h[i-2]
             hip1    = Knot[i+2] - Knot[i+1];    // h[i+1]
        }
        //////
        him1    = Knot[i]    - Knot[i-1];      // h[i-1]
        hi    = Knot[i+1] - Knot[i];        // h[i]

        sumim1    = him2 + him1 + hi;
        sumi    = him1 + hi + hip1;
        sum      = him1 + hi;
        ////////////////////////
        a[i] = hi*hi/sumim1;
        b[i] = hi*(him2+him1)/sumim1 + him1*(hi+hip1)/sumi;
        c[i] = him1*him1/sumi;
        ///////////// Corresponding RHS will be divided by sum in
BC routine, so
       a[i] /= sum;
       b[i] /= sum;
       c[i] /= sum;
       ///////////////

      }
      //////////
      return 0;

}
```

5.12.4.1.3 Algorithm C2RBS3C: Closed Curves

Step 0: Set the 3D B-spline end control point: $\mathbf{s}_0 = \mathbf{q}_0 = \mathbf{x}_0$, that is, the left-most input point, and the end weights as input weight: $v_0 = \bar{v}_0$.

Step 1: Set the 3D Bezier curve segment ends to input points: $\mathbf{q}_{3i} = \mathbf{x}_i, \quad \forall i \in \{1, \ldots, L-1\}$.

Step 2: Now, for computer implementation, define a 4D input vector, $\bar{\mathbf{x}}$, of dimension $(L+1)$ such that $\begin{Bmatrix} \bar{\mathbf{x}}_i(i,0) \\ \bar{\mathbf{x}}_i(i,1) \\ \bar{\mathbf{x}}_i(i,2) \\ \bar{\mathbf{x}}_i(i,3) \end{Bmatrix} = \begin{Bmatrix} \mathbf{x}_i(0) * \bar{v}_i \\ \mathbf{x}_i(1) * \bar{v}_i \\ \mathbf{x}_i(2) * \bar{v}_i \\ \bar{v}_i \end{Bmatrix}, \quad \forall i \in \{0, \ldots, L-1\}$.

Step 3: ***Closed (periodic) curves***: defined by: $\mathbf{x}_0 = \mathbf{x}_L$ with $\mathbf{s}_0 = \mathbf{s}_L$, $\mathbf{s}_1 = \mathbf{s}_{L-1}$ and so on. Thus, from equations (5.157) and (5.159), the matrix equation to solve is:

$$\begin{array}{c} \text{**} \\ \begin{array}{c} 1 \\ 2 \\ \bullet \\ \bullet \\ \bullet \\ L-1 \\ L \end{array} \left[\begin{array}{ccccccc} b_0 & c_0 & 0 & 0 & 0 & 0 & a_0 \\ a_1 & b_1 & c_1 & 0 & 0 & 0 & 0 \\ 0 & \bullet & \bullet & \bullet & 0 & 0 & 0 \\ 0 & 0 & \bullet & \bullet & \bullet & 0 & 0 \\ 0 & 0 & 0 & \bullet & \bullet & \bullet & 0 \\ 0 & 0 & 0 & 0 & a_{L-2} & b_{L-2} & c_{L-2} \\ c_{L-1} & 0 & 0 & 0 & 0 & a_{L-1} & b_{L-1} \end{array}\right] \begin{Bmatrix} \bar{\mathbf{s}}_0 \\ \bar{\mathbf{s}}_1 \\ \bullet \\ \bullet \\ \bullet \\ \bar{\mathbf{s}}_{L-2} \\ \bar{\mathbf{s}}_{L-1} \end{Bmatrix} = \begin{Bmatrix} \mathbf{p}_0 \\ \mathbf{p}_1 \\ \bullet \\ \bullet \\ \bullet \\ \mathbf{p}_{L-2} \\ \mathbf{p}_{L-1} \end{Bmatrix} \\ \text{*********}\ (L) \times (L)\ \text{******} \end{array} \tag{5.185}$$

with

$$
\boxed{\begin{aligned}
a_i &\stackrel{\Delta}{=} \frac{h_i^2}{\widehat{h}_i} \\
b_i &\stackrel{\Delta}{=} \frac{h_i(h_{i-2}+h_{i-1})}{\widehat{h}_i} + \frac{h_{i-1}(h_i+h_{i+1})}{\bar{h}_i} \\
c_i &\stackrel{\Delta}{=} \frac{h_{i-1}^2}{\bar{h}_i} \\
h_{-1} &= h_{L-1} \\
h_{-2} &= h_{L-2} \\
h_i &\stackrel{\Delta}{=} g_{i+1} - g_i \\
\widehat{h}_i &\stackrel{\Delta}{=} h_{i-2} + h_{i-1} + h_i \\
\bar{h}_i &\stackrel{\Delta}{=} h_{i-1} + h_i + h_{i+1}
\end{aligned}} \tag{5.186}
$$

and the 4D auxiliary vectors, $\mathbf{p}_i, \quad i \in \{0, \dots, L-1\}$ are defined as:

$$
\boxed{\mathbf{p}_i = (h_{i-1} + h_i)\,\bar{\mathbf{x}}_i, \quad i = 0, \dots, L-1} \tag{5.187}
$$

Step 4: From the 4D solution vectors, $\bar{\mathbf{s}}_i, \quad i \in \{1, \dots, L+1\}$, we may get the 3D B-spline control vectors as: $\mathbf{s}_i(j) = \frac{\bar{\mathbf{s}}_i(j)}{\bar{\mathbf{s}}_i(3)}, \quad j \in \{0, 1, 2\}$, and the weights as: $v_i = \bar{\mathbf{s}}_i(3), \quad \forall i \in \{0, \dots, L-1\}$.

However, since we always store internally as Bezier curves, using the algorithm C2RBS2RBZ, we first compute the 4D pre-image Bezier $3L$ vectors from the 4D B-spline L augmented vectors for closed curves. Finally, we recover the Bezier control vectors, $\mathbf{q}_i$, and the weights, w_i, for $\forall i \in \{0, \dots, 3L-1\}$,

5.12.4.1.4 C++ Code: B-spline to Bezier (Rational or Non-rational): CLOSE

We present below the C++ code excerpt for generation of B-spline control points and weights according to ***C2RBS3C***:

```
SetupSystem_Close(pDOUBLE a,pDOUBLE b,pDOUBLE c,pDOUBLE Knot,int L)
{

//////////////////////////////////////////////////////////////////////////////////
//    LAST DATA IS NOT FIRST DATA, i.e. FIRST DATA PT. IS NOT TAGGED ONTO BACK    //
      //   nData = No. of Data WITHOUT REPETITION       //
      //     Knot[i],Wts[i], i={0, ...nData-1}
      //   However, L = Intervals = #of Curve Segments = nData = Periodicity
      //   Knot[L] = Knot[0] etc.
//////////////////////////////////////////////////////////////////////////////////
      double   hi, him1, him2, hip1, sumim1, sumi, sum;
      ////////////////////////////////////////////////////// Case: One or two
interval: ERROR
```

```
      if( L < 3)
      {
//"Error:SetupSystem_Close\nToo Few Number of Intervals to be Closed!"
            return;
      }
      ////////////////////////////////////////////////// Case: More than two
intervals
      int i;
      for (i=0;i<L;i++)      // 2rd to Last-but-1 Rows
      {

            if(i == 0)
            {
                  him1  = Knot[L]   - Knot[L-1];      // h[-1] = h{L-1]
                  him2  = Knot[L-1] - Knot[L-2];      // h[-2] = h[L-2]
            }
            else
            if(i == 1)
            {
                  him1  = Knot[i]   - Knot[i-1];      / h[i-1]
                  him2  = Knot[L]   - Knot[L-1];      / h[-1] = h[L-1]
            }
            else
            {
                  him1  = Knot[i]   - Knot[i-1];      / h[i-1]
                  him2  = Knot[i-1] - Knot[i-2];      / h[i-2]
            }
            if(i == (L-1) )
                  hip1  = Knot[1]   - Knot[0];      / h[L] = h[0]
            else
                  hip1  = Knot[i+2] - Knot[i+1];      // h[i+1]
            //////
            hi   = Knot[i+1]      - Knot[i];      // h[i]

            sumim1   = him2 + him1 + hi;
            sumi   = him1 + hi + hip1;
            sum     = him1 + hi;
            ////////////////////////
            a[i] = hi*hi/sumim1;
            b[i] = hi*(him2+him1)/sumim1 + him1*(hi+hip1)/sumi;
            c[i] = him1*him1/sumi;
            ////////////////// Corresponding RHS will be divided by sum in
BC routine, so
            a[i] /= sum;
            b[i] /= sum;
            c[i] /= sum;
            ///////////////
      }
      /////////
      return 0;

}
```

5.12.4.2 Case: C^1 Quadratic Rational B-spline Interpolation of Multiple Segments

Exercise: Given input points, $\mathbf{x}_i$, $\quad i = 0, \ldots, L$, and input weights, $\bar{v}_i$, $\quad i = 0, \ldots, L$, over knot sequence, g_i, $\quad i = 0, \ldots, L$, determine a rational quadratic Bezier curve through these points in standard form, that is, determine $(2L + 1)$ Bezier control points, $\{\mathbf{q}_0, \mathbf{q}_1, \ldots, \mathbf{q}_{2L}\}$, and weights, $\{w_0, w_1, \ldots, w_{2L}\}$, with $w_0 = w_{2L} = 1$ at two end control points.

Hint: Follow the logic behind the cubic B-spline interpolation algorithm detailed above, noting that we need only one boundary condition for completing the task.

5.13 Where We Would Like to Go

We have presented here in some detail a topical discussion of the basic theory and computation of curves as a building block to study differential geometry and computation of surfaces and to apply this to structural analysis of beams by the finite element method. Thus, we may now proceed to:

- beam analysis (Chapter 10)
- the theory and computation of surfaces (Chapter 6).

6

Surfaces: Theory and Computation

6.1 Introduction

6.1.1 What We Should Already Know

Vectors, derivatives, vector dot product, vector cross product, vector triple product, basis functions, interpolation, approximation, conics, continuous functions, polynomials, forward-difference operators, linear vector space, power basis.

6.1.2 Why We Are Here

The main goal here is to present a topical discussion of the basic theory and computation of surfaces and to define relevant surface properties necessary for both ***mesh generation*** and ***finite element analysis*** phases of plates and shells. However, before we start, let us remind ourselves that all of our theoretical discussions are primarily motivated to achieve robust computational methods and to guide us with useful practical tips. More specifically, we define and discuss the properties of parametric surfaces as opposed to their implicit description – to investigate computational Bezier–Bernstein–de Casteljau and Schoenberg–de Boor–Cox B-spline parametric surfaces. In the interactive ***mesh generation*** phase, we are interested in developing the final model of structures. Starting from a few key points of information, we would like to create the subassemblies of the model, and with the appropriate continuity conditions among the subassemblies eventually we will blend them all together into one integral mesh. In the finite element analysis phase, on the other hand, the mesh is used for structural integrity under various load conditions by determining key entities or field quantities such as displacements and stresses.

Finally, for cost effectiveness in numerical analysis in the form of ***interpolations*** in the mesh generation phase and ***approximations*** in the finite element phases, we strive for elements that are as large as possible without compromising accuracy. Thus, in order that the interpolation and the approximation algorithms can be easily understood, here we continue with the fundamental geometric entity: ***surfaces***. There are two basic ways to represent a surface – ***implicit form*** or ***parametric form***. For various reasons, the implicit description appears less promising than the parametric form for our purposes. Within the parametric domain, for several reasons, we prefer to represent surfaces in the Bezier–Bernstein–de Casteljau and Schoenberg–deBoor B-spline forms to various other possible forms such as piecewise Lagrangian and Hermite polynomials. The common shapes in ***conics of revolution***, such as spheres, paraboloids and ellipsoids, need special attention under our choice of the parametric form. It turns out that in order to describe these shapes

Computation of Nonlinear Structures: Extremely Large Elements for Frames, Plates and Shells, First Edition. Debabrata Ray.

exactly, we must resort to ***rational Bezier*** and ***rational B-spline surfaces***. Implementationally, Bezier surfaces and B-spline surfaces by composite Bezier construction act as the cornerstone of our computer-aided geometric modeling algorithms. We make these also the building blocks of our proposed finite element analysis.

We begin with:

- Section 6.2: the differential geometry of real parametric surfaces
- Section 6.3: the Gauss–Weingarten formula: optimal coordinate system

For an introduction to the fundamental concepts and properties of Bezier–Bernstein–de Casteljau surfaces and de Boor B-spline forms of representation of surfaces, we introduce:

- Section 6.4: Cartesian product Bernstein–Bezier surfaces
- Section 6.5: Control node generation: Cartesian product surfaces
- Section 6.6: Quadratic and cubic B-splines as composite Bezier surfaces

Finally, for an introduction to the fundamental concepts and properties of triangular surfaces, we conclude with:

- Section 6.7: Triangular surfaces

6.1.3 What We Get Here

Following the stated aims of this book, we get to know, in concrete terms, the various useful definitions and assumptions specific to surfaces, as indicated in the adjoining document map or bookmarks.

6.1.4 How We Get It

We introduce some fundamental concepts in differential geometry of parametric surfaces: Gauss frame, Gauss–Weingarten derivatives with the definitions for curvature and twist, and so on; then we specialize on Bernstein–Bezier surfaces and de Boor–Cox B-spline surfaces, rational and non-rational, for computational purposes.

6.1.5 What We Need To Recall

In what follows, we touch upon some fundamental ideas and concepts that will be eventually taken advantage of, using the concepts of tensors introduced earlier. As indicated above, we need to recall the ***linear vector space***, ***real coordinate space*** and ***Euclidean 'flat' space*** and ***affine space***. For the use of tensors, likewise, we need to recall linear functionals and linear operators. Finally, for generalization to ***covariant derivatives*** of vectors and tensors in ***Riemannian curved spaces***, we invoke ***directional or Gateaux derivatives*** of Euclidean space.

6.2 Real Parametric Surface: Differential Geometry

Let us consider three-dimensional Euclidean space (more precisely, ***affine space***), and real surfaces embedded in it. If we assume a Cartesian coordinate system, then every point in space can be identified by a position vector, $\mathbf{r} = (X \equiv X_1 \quad Y \equiv X_2 \quad Z \equiv X_3)^T$.

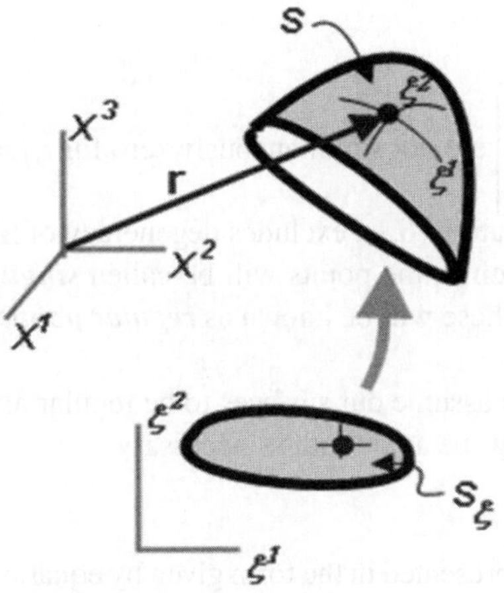

Figure 6.1 Real parametric surface.

6.2.1.1 Real Surface

A real surface is defined by some ***allowable transformation*** (to be identified soon) as a set of points, $\mathbf{S} = \{\mathbf{r} \in \mathbb{R}_3 | \, (\xi^1, \xi^2) \in \mathbf{S}_\xi\}$, where ξ^1, ξ^2 are real parameters, as shown in Figure 6.1. In component form:

$$\mathbf{r}(\xi^1,\ \xi^2) = \begin{Bmatrix} X_1(\xi^1,\ \xi^2) \\ X_2(\xi^1,\ \xi^2) \\ X_3(\xi^1,\ \xi^2) \end{Bmatrix} \tag{6.1}$$

As shown in Figure 6.1 for two-dimensional surfaces, any point on a surface may be represented parametrically by a two-parameter set, $\{\xi^\alpha\}$, $\alpha = 1, 2$. Assuming the surface to be embedded in three-dimensional Euclidean space, it can be functionally related to the position vector, **r**, of a Euclidean Cartesian coordinate system as given by equation (6.1).

6.2.1.2 Allowable Parametric Representation

A representation where a surface is regular and continuously differentiable $r(\geq 1)$ many times, depending on a problem, is called an ***allowable parametric representation***. The surface is said to be of class C^r. For the parametric representation expressed in equation (6.1) to be allowable, it must be arbitrarily smooth and must satisfy that the scaling matrix or ***Jacobian matrix***, **J**, as defined later, must be rank 2 in **S**, that is, there must be at least one non-vanishing determinant of order two to be invertible at any point in the space:

$$\mathbf{J} \equiv \begin{bmatrix} \dfrac{\partial X^1}{\partial \xi^1} & \dfrac{\partial X^1}{\partial \xi^2} \\ \dfrac{\partial X^2}{\partial \xi^1} & \dfrac{\partial X^2}{\partial \xi^2} \\ \dfrac{\partial X^3}{\partial \xi^1} & \dfrac{\partial X^3}{\partial \xi^2} \end{bmatrix} \tag{6.2}$$

In other words,

- $\mathbf{J}_{ij} \equiv \frac{\partial(X^i, X^j)}{\partial(\xi^1, \xi^2)} = \begin{bmatrix} \frac{\partial X^i}{\partial \xi^1} & \frac{\partial X^i}{\partial \xi^2} \\ \frac{\partial X^j}{\partial \xi^1} & \frac{\partial X^j}{\partial \xi^2} \end{bmatrix}$ are not simultaneously zero for $i, j = 1, 2, 3$.
- The function, given by equation (6.1), excludes degeneracy of **S** to curve.
- For rank < 2 at any point on **S**, the points will be called ***singular points*** with respect to the representation, otherwise, these will be known as ***regular points***.

From now on, we will always assume our surfaces to be regular and of class C^r, $r(\geq 1)$; in other words, the surfaces will always be as smooth as necessary.

1. The X^1, X^2 plane can be represented in the form given by equation (6.1) with $\xi^1 = X^1$, $\xi^2 = X^2$ as the ***Cartesian coordinates***; then: $\mathbf{J} \equiv \begin{bmatrix} 1 & 0 \\ 0 & 1 \\ 0 & 0 \end{bmatrix}$
2. If $\xi^1 = r$, $\xi^2 = \theta$ are the ***polar coordinates*** such that $\mathbf{r} = (\xi^1 \cos \xi^2 \quad \xi^1 \sin \xi^2 \quad 0)^T$; then: $\mathbf{J} \equiv \begin{bmatrix} \cos \xi^2 & -\xi^1 \sin \xi^2 \\ \sin \xi^2 & \xi^1 \cos \xi^2 \\ 0 & 0 \end{bmatrix}$
3. If $\xi^1 = \theta$, $\xi^2 = \phi$ are the ***spherical coordinates*** for a spherical surface (see Figure 6.2) of radius R with center at $\mathbf{r}(0, 0, 0)$ such that the position vector is given as: $\mathbf{r} = (R \cos \xi^1 \cos \xi^2 \quad R \sin \xi^1 \cos \xi^2 \quad R \sin \xi^2)^T$; then: $\mathbf{J} \equiv R \begin{bmatrix} -\sin \xi^1 \cos \xi^2 & -\cos \xi^1 \sin \xi^2 \\ \cos \xi^1 \cos \xi^2 & \sin \xi^1 \sin \xi^2 \\ 0 & \cos \xi^2 \end{bmatrix}$

6.2.2 Tangent Plane, Iso-curves, Vectors on a Surface

The isoparametric curves: $\xi^1 =$ constant, and $\xi^2 =$ constant curves can be seen in Figure 6.3. The portion of the tangent plane at $\mathbf{r}(\xi^1, \xi^2)$ is also shown in Figure 6.3. Let t be a parameter describing a regular curve, $(\hat{\xi}_1(t), \hat{\xi}_2(t))$, on the $\hat{\xi}_1$, $\hat{\xi}_2$ plane; we can think of affine curves at a

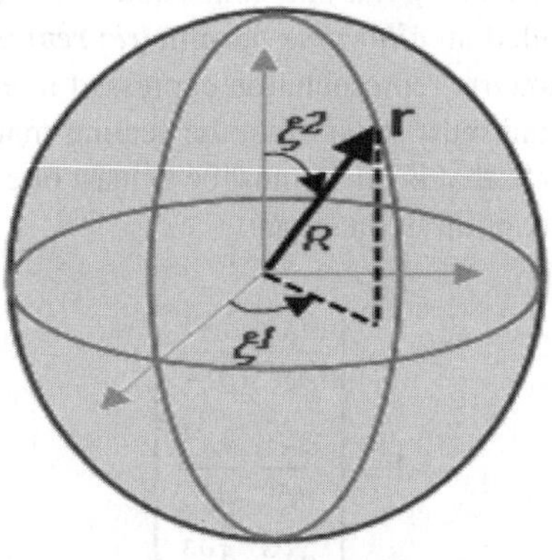

Figure 6.2 Spherical surface.

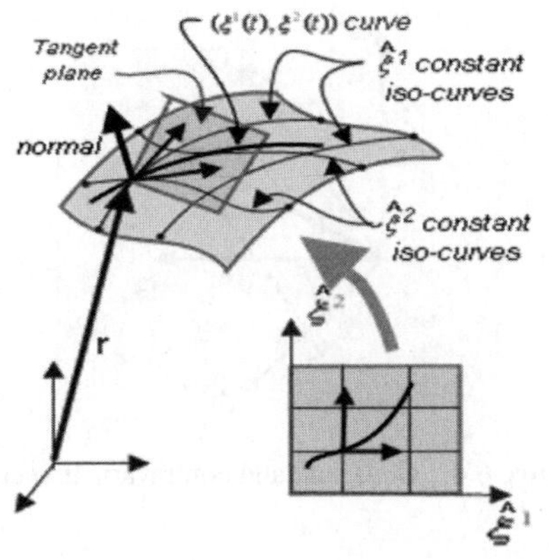

Figure 6.3 Iso-curves, tangent plane and tangent vector.

point on the surface such as the image curve given by $\mathbf{r}(\hat{\xi}_1(t),\ \hat{\xi}_2(t))$, and the tangents to these curves at a point form a local tangent space.

6.2.2.1 Covariant Base Vectors

As in the case of Euclidean space, taking the tangents along the affine coordinate curves: $\hat{\xi}^1 =$ constant and $\hat{\xi}^2 =$ constant curves, we can locally construct two of the bases or coordinate vectors at a point that are generally not mutually orthogonal and are known as the set of ***covariant base vectors***, $\{\hat{\mathbf{G}}_\alpha\}$:

$$\boxed{\hat{\mathbf{G}}_\alpha \equiv \frac{\partial \mathbf{r}}{\partial \hat{\xi}^\alpha},\ \alpha = 1,2 \quad \text{with} \quad \hat{\mathbf{G}}_1 \times \hat{\mathbf{G}}_2 \neq 0} \tag{6.3}$$

Then, for any arbitrary $\mathbf{r}(\hat{\xi}_1(t),\ \hat{\xi}_2(t))$ curve, the tangent defines a vector on the surface given by:

$$\frac{d\mathbf{r}}{dt} = \frac{\partial \mathbf{r}}{\partial \hat{\xi}^\alpha}\frac{d\hat{\xi}^\alpha}{dt} = \frac{d\hat{\xi}^\alpha}{dt}\hat{\mathbf{G}}_\alpha,\quad \alpha = 1,2 \tag{6.4}$$

So, the tangent plane at $\mathbf{r}$ is defined by all such tangents and is spanned by $\hat{\mathbf{G}}_\alpha$, $\alpha = 1,2$. If $\mathbf{y}$ is any point on the tangent plane, then $(\mathbf{y}-\mathbf{r})$, $\hat{\mathbf{G}}_\alpha$, $\alpha = 1,2$ must be linearly dependent, and thus the implicit equation of the tangent plane is given by:

$$\det[\,(\mathbf{y}-\mathbf{r}) \quad \hat{\mathbf{G}}_1 \quad \hat{\mathbf{G}}_2\,] = 0 \tag{6.5}$$

The parametric equation of the tangent plane is given by:

$$\mathbf{y} = \mathbf{r} + \Delta\xi^\alpha \hat{\mathbf{G}}_\alpha \tag{6.6}$$

Note that the covariant base vectors are described subscripted because, as opposed to a Euclidean geometry, we have a different situation in a curved geometry; there exists a dual set of bound base vectors at a point, known as the set of ***contravariant base vectors***, denoted superscripted as $\{\mathbf{G}^\alpha\}$, $\alpha = 1,2$ as follows.

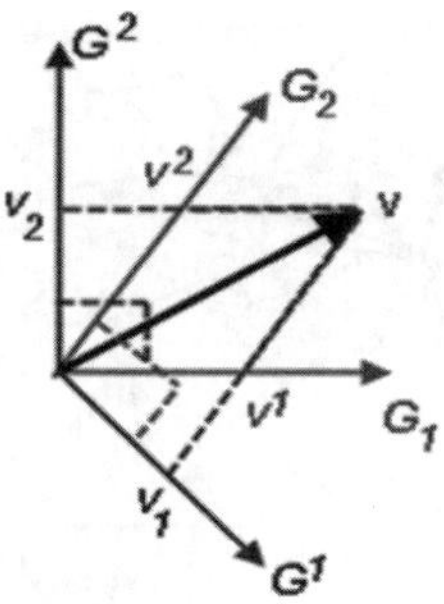

Figure 6.4 Covariant and contravariant vectors.

6.2.2.2 Contravariant Base Vectors

The set of contravariant base vectors is defined by:

$$(\mathbf{G}_\alpha, \mathbf{G}^\beta) \equiv \mathbf{G}_\alpha \bullet \mathbf{G}^\beta = \delta_\alpha^\beta = \delta_{\alpha\beta}, \quad \alpha, \beta = 1, 2 \tag{6.7}$$

where $\delta_{\alpha\beta}$ is the ***Kronecker delta tensor***, that is, $\delta_{\alpha\alpha} = 1$ and $\delta_{\alpha\beta} = 0,\ \alpha \neq \beta$.

6.2.2.3 Covariant and Contravariant Components of Vectors on a Surface

Thus, a vector **v** on a tangent plane has components, as shown in Figure 6.4, represented in both contravariant and covariant coordinate systems related by:

$$\mathbf{v} = v^\alpha \mathbf{G}_\alpha = v_\alpha \mathbf{G}^\alpha, \quad \alpha = 1, 2 \tag{6.8}$$

where $\mathbf{v}_\alpha$'s are ***covariant components*** measured in the parallel direction with the corresponding ***contravariant base vectors***, $\mathbf{G}^\alpha$, and $\mathbf{v}^\alpha$'s are ***contravariant components*** measured in the perpendicular direction to the corresponding ***covariant base vectors***, $\mathbf{G}_\alpha$.

6.2.3 First Fundamental form: Riemannian Metric

Unlike Euclidean geometry, the metric or distance to be measured is for a curved, invariant geometric object: an infinitesimal arc length, dS, as follows. From equation (6.4), we have: $d\mathbf{r} = \frac{\partial \mathbf{r}}{\partial \xi^\alpha} d\xi^\alpha = \hat{\mathbf{G}}_\alpha d\xi^\alpha$, where we have applied Einstein's summation convention for $\alpha = 1, 2$; so, for arc length, we get:

$$dS^2 \equiv d\mathbf{r} \bullet d\mathbf{r} = (\hat{\mathbf{G}}_\alpha \bullet \hat{\mathbf{G}}_\beta) d\xi^\alpha\, d\xi^\beta = \hat{G}_{\alpha\beta} d\xi^\alpha\, d\xi^\beta \quad \text{where} \quad \hat{G}_{\alpha\beta} \equiv \hat{\mathbf{G}}_\alpha \bullet \hat{\mathbf{G}}_\beta \tag{6.9}$$

- Equation (6.9) defines what is known as the ***first fundamental form*** of a surface.
- The tensor $\hat{G}_{\alpha\beta}$ is known as the ***metric tensor*** because it helps measure distance, angle between two vectors and areas on a surface, as we will see shortly.
- The metric defined by equation (6.9) is of quadratic differential form and as such is known as a ***Riemannian metric***.

- A geometry with the Riemannian metric is known as a ***Riemannian geometry***.
- The space containing the Riemannian metric is known as ***Riemannian space***.
- The first fundamental form is due to Carl Gauss, and in his notation, we get:

$$\boxed{\begin{aligned} &dS^2 \equiv E(d\xi^1)^2 + 2F\, d\xi^1\, d\xi^2 + G(d\xi^2)^2 \\ \text{where}\quad &E \equiv \hat{G}_{11},\quad F \equiv \hat{G}_{12} = \hat{G}_{21},\quad G \equiv \hat{G}_{22} \end{aligned}} \tag{6.10}$$

Example: Following on the previous sphere, if $\xi^1 = \theta,\ \xi^2 = \phi$ are the ***spherical coordinates*** for a spherical surface of radius R with center at $\mathbf{r}(0,0,0)$, such that the position vector is given as: $\mathbf{r} = (R\cos\xi^1\cos\xi^2 \quad R\xi^1\sin\xi^1\cos\xi^2 \quad R\sin\xi^2)^T$, then, an arc element, dS, on the surface is given by:

$$\begin{aligned} \hat{G}_1 &= (-R\sin\xi^1\cos\xi^2 \quad R\cos\xi^1\cos\xi^2 \quad 0)^T \\ \hat{G}_2 &= (-R\cos\xi^1\sin\xi^2 \quad -R\sin\xi^1\sin\xi^2 \quad R\cos\xi^2)^T \\ \Rightarrow \hat{G}_{11} &= R^2\cos^2\xi^2, \quad \hat{G}_{22} = R^2 \quad \hat{G}_{12} = \hat{G}_{21} = 0 \\ \Rightarrow &\boxed{dS^2 = R^2\cos^2\xi^2(d\xi^1)^2 + R^2(d\xi^2)^2} \end{aligned}$$

6.2.3.1 Properties of Metric or Fundamental Tensor

Based on equation (6.9), $\hat{G}_{\alpha\beta} \equiv \hat{\mathbf{G}}_\alpha \bullet \hat{\mathbf{G}}_\beta,\ \alpha,\beta = 1,2$ are the ***covariant components*** of the second-order ***metric or fundamental tensor***; as indicated earlier, it is so called because it is capable of providing all the geometric information such as distances, curvatures, angular separations of vectors and surface areas in curved surfaces. Now, considering a vector, $\mathbf{u}$, on the ***tangent space*** of a surface, written in terms of the covariant base vectors, $\{\hat{\mathbf{G}}_\alpha\}$, and the contravariant base vectors, $\{\hat{\mathbf{G}}^\alpha\}$, for $\alpha = 1,2$, we have: $\mathbf{u} = u_\alpha\hat{\mathbf{G}}^\alpha = u^\alpha\hat{\mathbf{G}}_\alpha$ where $u_\alpha = u \bullet \hat{\mathbf{G}}_\alpha$'s and $u^\alpha = \mathbf{u} \bullet \hat{\mathbf{G}}^\alpha$'s are the covariant and the contravariant components, respectively, of the vector; then, $u^\alpha = \mathbf{u} \bullet \hat{\mathbf{G}}^\alpha = \hat{\mathbf{G}}^\alpha \bullet u_\beta\hat{\mathbf{G}}^\beta = \hat{\mathbf{G}}^\alpha \bullet \hat{\mathbf{G}}^\beta u_\beta \equiv \hat{G}^{\alpha\beta}u_\beta$ where we have defined the ***contravariant components of the metric tensor*** as:

$$\boxed{\hat{G}^{\alpha\beta} \equiv \hat{\mathbf{G}}^\alpha \bullet \hat{\mathbf{G}}^\beta} \tag{6.11}$$

Fact 1: The metric tensors are symmetric: $\hat{G}_{12} = \hat{G}_{21}$ and $\hat{G}^{12} = \hat{G}^{21}$.

Proof. Since the inner product of two vectors is symmetric, we have: $\hat{G}_{12} \equiv \hat{\mathbf{G}}_1 \bullet \hat{\mathbf{G}}_2 = \hat{\mathbf{G}}_2 \bullet \hat{\mathbf{G}}_1 \equiv \hat{G}_{21}$. Similarly, $G^{12} \equiv \mathbf{G}^1 \bullet \mathbf{G}^2 = \mathbf{G}^2 \bullet \mathbf{G}^1 = G^{21}$.

Fact 2: The metric tensor has the property of raising and lowering of indices. For example, $v_\alpha = \hat{G}_{\alpha\beta}v^\beta$ and $T_{\alpha\beta} = \hat{G}_{\alpha\chi}T^\chi_\beta$.

Proof. For a vector, $\mathbf{v}$, $v_\alpha = \mathbf{v} \bullet \hat{\mathbf{G}}_\alpha = v^\beta\hat{\mathbf{G}}_\beta \bullet \hat{\mathbf{G}}_\alpha = G_{\alpha\beta}v^\beta$; thus, the covariant component of a vector is obtained by lowering the index of the contravariant component by multiplying with the covariant components of the metric tensor. Similarly, for a 2×2 second-order tensor, $\mathbf{T}$:

$$T_{\alpha\beta} = \hat{\mathbf{G}}_\alpha \bullet \mathbf{T} \bullet \hat{\mathbf{G}}_\beta = \hat{\mathbf{G}}_\alpha \bullet T^\gamma_\chi\hat{\mathbf{G}}_\gamma \otimes \hat{\mathbf{G}}^\chi \bullet \hat{\mathbf{G}}_\beta = T^\gamma_\chi(\hat{\mathbf{G}}_\alpha \bullet \hat{\mathbf{G}}_\gamma)(\hat{\mathbf{G}}^\chi \bullet \hat{\mathbf{G}}_\beta) = T^\gamma_\chi\hat{G}_{\alpha\gamma}\delta_{\chi\beta} = \hat{G}_{\alpha\gamma}T^\gamma_\beta$$

Fact 3: $G_{\alpha\beta}G^{\alpha\chi} = \delta_{\beta}^{\chi}$ shows the connection between the covariant and the contravariant components of the metric tensor.

Proof. Using definition of $\hat{G}_{\alpha\beta}$, as in equation (6.9), and the raising of index property of the metric tensor as in **Fact 2**, we have:

$$\boxed{\hat{G}_{\alpha\beta}\hat{G}^{\alpha\chi} = (\hat{\mathbf{G}}_{\alpha} \bullet \hat{\mathbf{G}}_{\beta})\hat{G}^{\alpha\chi} = \underbrace{\hat{\mathbf{G}}_{\alpha}\hat{G}^{\alpha\chi}}_{\text{raising of index}} \bullet \hat{\mathbf{G}}_{\beta} = \hat{\mathbf{G}}^{\chi} \bullet \hat{\mathbf{G}}_{\beta} = \delta_{\beta}^{\chi}} \tag{6.12}$$

Fact 4: If $\hat{G} \equiv \det(\hat{G}_{\alpha\beta})$, then $\det(\hat{G}^{\alpha\beta}) = \frac{1}{\hat{G}}$ where det stands for the determinant of the corresponding tensors.

Proof. Using the relation as in **Fact 3**, we have:

$$\boxed{1 = \det(\delta_{\alpha}^{\chi}) = \det(\hat{G}_{\alpha\beta}\hat{G}^{\alpha\chi}) = \det(\hat{G}_{\alpha\beta})\det(\hat{G}^{\alpha\chi}) \Rightarrow \det(\hat{G}^{\alpha\chi}) = \frac{1}{\det(\hat{G}_{\alpha\beta})} = \frac{1}{\hat{G}}} \tag{6.13}$$

Fact 5: $\sqrt{\hat{G}} = \sqrt{\det(\hat{G}_{\alpha\beta})} = \|\hat{\mathbf{G}}_1 \times \hat{\mathbf{G}}_2\|$.

Proof. Let us first note that:

$$\hat{G} = \det(\hat{G}_{\alpha\beta}) = \det \begin{bmatrix} \hat{G}_{11} & \hat{G}_{12} \\ \hat{G}_{21} & \hat{G}_{22} \end{bmatrix} = \hat{G}_{11}\hat{G}_{22} - \hat{G}_{12}^2 \tag{6.14}$$

Now, let θ be the angle between the base vectors, $\hat{\mathbf{G}}_1$ and $\hat{\mathbf{G}}_2$; then, from elementary calculus:

$$\|\hat{\mathbf{G}}_1 \times \hat{\mathbf{G}}_2\|^2 = \|\hat{\mathbf{G}}_1\|^2\|\hat{\mathbf{G}}_2\|^2 \sin^2\theta = \hat{G}_{11}\,\hat{G}_{22}(1 - \cos^2\theta) = \hat{G}_{11}\,\hat{G}_{22} - \hat{G}_{12}^2 \tag{6.15}$$

So, by comparing equations (6.14) and (6.15), we get:

$$\boxed{\|\hat{\mathbf{G}}_1 \times \hat{\mathbf{G}}_2\| = \sqrt{\det(\hat{G}_{\alpha\beta})} = \sqrt{\hat{G}}} \tag{6.16}$$

Fact 6: $\hat{G}^{11} = \frac{\hat{G}_{22}}{\hat{G}}$, $\hat{G}^{22} = \frac{\hat{G}_{11}}{\hat{G}}$ and $\hat{G}^{12} = -\frac{\hat{G}_{12}}{\hat{G}}$

Proof. Let us first note that:

$$\begin{bmatrix} \hat{G}_{11} & \hat{G}_{12} \\ \hat{G}_{21} & \hat{G}_{22} \end{bmatrix} \begin{bmatrix} \hat{G}^{11} & \hat{G}^{2} \\ \hat{G}_{21} & \hat{G}^{22} \end{bmatrix} = \begin{bmatrix} 1 & 0 \\ 0 & 1 \end{bmatrix} \tag{6.17}$$

Now, taking the inverse of the metric tensor with covariant components, we get:

$$\begin{bmatrix} \hat{G}^{11} & \hat{G}^{2} \\ \hat{G}_{21} & \hat{G}^{22} \end{bmatrix} = \frac{1}{\hat{G}} \begin{bmatrix} \hat{G}_{22} & -\hat{G}_{12} \\ -\hat{G}_{21} & \hat{G}_{11} \end{bmatrix} \tag{6.18}$$

which completes the proof.

Problem: show by applying the ***Lagrange identity*** that:

$$\|\mathbf{G}_1 \times \mathbf{G}_2\| = \sqrt{G} \tag{6.19}$$

Solution: Applying Lagrange identity: $(\mathbf{a} \times \mathbf{b}) \bullet (\mathbf{c} \times \mathbf{d}) = (\mathbf{a} \bullet \mathbf{c})(\mathbf{b} \bullet \mathbf{d}) - (\mathbf{a} \bullet \mathbf{d})(\mathbf{b} \bullet \mathbf{c})$, $\|\hat{\mathbf{G}}_1 \times \hat{\mathbf{G}}_2\|^2 = (\hat{\mathbf{G}}_1 \times \hat{\mathbf{G}}_2) \bullet (\hat{\mathbf{G}}_1 \times \hat{\mathbf{G}}_2) = (\hat{\mathbf{G}}_1 \bullet \hat{\mathbf{G}}_1)(\hat{\mathbf{G}}_2 \bullet \hat{\mathbf{G}}_2) - (\hat{\mathbf{G}}_1\, \hat{\mathbf{G}}_2)^2 = \hat{\mathbf{G}}_{11}\, \hat{\mathbf{G}}_{22} - (\hat{\mathbf{G}}_{12})^2 \equiv G$

6.2.3.2 Allowable Coordinate Transformation

The parameterization given by equation (6.1) is not the only possibility; in fact, we can have another representation by introduction of suitable, generally nonlinear, transformation such as:

$$\xi^\alpha = \xi^\alpha\,(\bar{\xi}^i, \bar{\xi}^2), \quad \alpha = 1, 2 \tag{6.20}$$

with the assumption that the ***surface Jacobian*** (i.e. the determinant of the Jacobian matrix), as defined later, is non-vanishing at all points:

$$D = \frac{\partial(\xi^1, \xi^2)}{\partial(\bar{\xi}^1, \bar{xi}^2)} \equiv \det \begin{bmatrix} \dfrac{\partial \xi^1}{\partial \bar{\xi}^1} & \dfrac{\partial \xi^1}{\partial \bar{\xi}^2} \\ \dfrac{\partial \xi^2}{\partial \bar{\xi}^1} & \dfrac{\partial \xi^2}{\partial \bar{\xi}^2} \end{bmatrix} \neq 0 \tag{6.21}$$

Remarks

- Because of the invertibility of the transformation given by equation (6.21), there exists the following transformation:

$$\bar{\xi}^\alpha = \bar{\xi}^\alpha\,(\xi^1, \xi^2), \quad \alpha = 1, 2 \tag{6.22}$$

- Because of the generalization of the allowable coordinate transformations to nonlinear mapping, the base vectors (defined by first derivatives, which are coordinate dependent) in a curved space turn out to be ***bound vectors*** in the sense that they are position dependent. For example, in Figure 6.5 the tangent vectors to the curves are pointwise fixed or bounded.

  Now, considering ξ^i and ξ^j independent, we note that, by the chain rule:

$$\begin{aligned} \frac{\partial \xi^\alpha}{\partial \xi^\beta} &= \frac{\partial \xi^\alpha}{\partial \bar{\xi}^\chi} \frac{\partial \bar{\xi}^\chi}{\partial \xi^\beta} = \delta^\alpha_\beta, \\ \frac{\partial \bar{\xi}^\alpha}{\partial \bar{\xi}^\beta} &= \frac{\partial \bar{\xi}^\alpha}{\partial \xi^\chi} \frac{\partial \xi^\chi}{\partial \bar{\xi}^\beta} = \delta^\alpha_\beta \end{aligned} \tag{6.23}$$

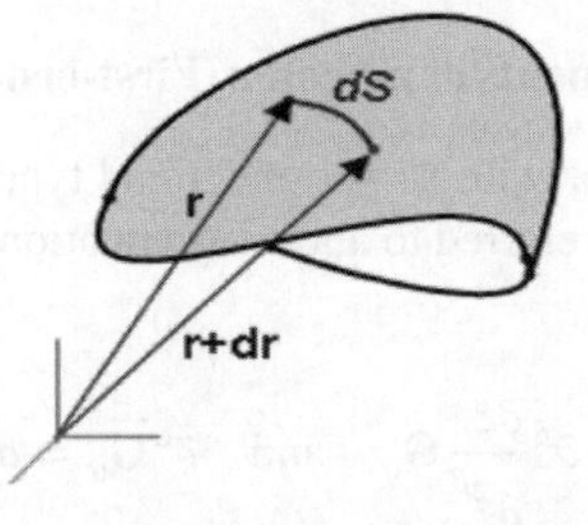

Figure 6.5 Arc (distance) element.

For linear transformations, we have to resort to transformations locally of differentials $\mathbf{d\xi}$ to $\mathbf{d\bar{\xi}}$ and vice versa. From equations (6.20) and (6.22), we get:

$$\boxed{d\xi^{\alpha} = \frac{\partial \xi^{\alpha}}{\partial \bar{\xi}^{\beta}} d\bar{\xi}^{\beta} \quad \text{and} \quad d\bar{\xi}^{\alpha} = \frac{\partial \bar{\xi}^{\alpha}}{\partial \xi^{\beta}} d\xi^{\beta}} \tag{6.24}$$

where we have used the Einstein summation convention for repeated diagonal indices. Thus, the allowable transformation, as given by equation (6.20), induces the linear transformation properties of the differentials as in equation (6.24). In other words, Riemann spaces are made of locally deformed patches of Euclidean spaces.

Remarks

- We note that for the covariant base vectors, the invariant transformation relation is:

$$\boxed{\begin{aligned} \mathbf{G}_{\alpha} &\equiv \frac{\partial}{\partial \xi^{\alpha}} = \frac{\partial}{\partial \bar{\xi}^{\beta}} \frac{\partial \bar{\xi}^{\beta}}{\partial \xi^{\alpha}} = \frac{\partial \bar{\xi}^{\beta}}{\partial \xi^{\alpha}} \overline{\mathbf{G}}_{\beta} \\ \overline{\mathbf{G}}_{\alpha} &\equiv \frac{\partial}{\partial \bar{\xi}^{\alpha}} = \frac{\partial}{\partial \xi^{\beta}} \frac{\partial \xi^{\beta}}{\partial \bar{\xi}^{\alpha}} = \frac{\partial \xi^{\beta}}{\partial \bar{\xi}^{\alpha}} \mathbf{G}_{\beta} \end{aligned}} \tag{6.25}$$

- Considering that the infinitesimal arc length, ds, to be an invariant geometric object, we have from coordinate transformation given by equations (6.20) and (6.22), and inserting equation (6.24) into the expression for the infinitesimal arc length as given by equation (6.9), we get:

$$\begin{aligned} ds^2 &= G_{\alpha\beta}\, d\xi^{\alpha}\, d\xi^{\beta} = G_{\alpha\beta} \frac{\partial \xi^{\alpha}}{\partial \bar{\xi}^{\lambda}} \frac{\partial \xi^{\beta}}{\partial \bar{\xi}^{\mu}} d\bar{\xi}^{\lambda} d\bar{\xi}^{\mu} = \bar{G}_{\lambda\mu}\, d\bar{\xi}^{\lambda}\, d\bar{\xi}^{\mu} \\ &\Rightarrow \boxed{\bar{G}_{\lambda\mu} = G_{\alpha\beta} \frac{\partial \xi^{\alpha}}{\partial \bar{\xi}^{\lambda}} \frac{\partial \xi^{\beta}}{\partial \bar{\xi}^{\mu}}} \end{aligned} \tag{6.26}$$

- Similarly, we have:

$$\begin{aligned} ds^2 &= \bar{G}_{\alpha\beta}\, d\bar{\xi}^{\alpha}\, d\bar{\xi}^{\beta} = \bar{G}_{\alpha\beta} \frac{\partial \bar{\xi}^{\alpha}}{\partial \xi^{\lambda}} \frac{\partial \bar{\xi}^{\beta}}{\partial \xi^{\mu}} d\xi^{\lambda} d\xi^{\mu} = G_{\lambda\mu}\, d\xi^{\lambda} d\xi^{\mu} \\ &\Rightarrow \boxed{G_{\lambda\mu} = \bar{G}_{\alpha\beta} \frac{\partial \bar{\xi}^{\alpha}}{\partial \xi^{\lambda}} \frac{\partial \bar{\xi}^{\beta}}{\partial \xi^{\mu}}} \end{aligned} \tag{6.27}$$

6.2.3.3 Contravariant Components of a Vector (First-order Tensor) Transformation Law

Let $\mathbf{a} = a^{\alpha}\mathbf{G}_{\alpha} = \bar{a}^{\beta}\overline{\mathbf{G}}_{\beta}$ be a vector with as yet undefined type components, $a^{\alpha}, \bar{a}^{\beta}$, described in two different coordinate systems referred to above by equation (6.20), then from equation (6.25), we have:

$$a^{\alpha}\mathbf{G}_{\alpha} = \bar{a}^{\beta}\overline{\mathbf{G}}_{\beta} = \bar{a}^{\beta} \frac{\partial \xi^{\alpha}}{\partial \bar{\xi}^{\beta}} \mathbf{G}_{\alpha} \quad \text{and} \quad \bar{a}^{\alpha}\overline{\mathbf{G}}_{\alpha} = a^{\beta}\mathbf{G}_{\beta} = a^{\beta} \frac{\partial \bar{\xi}^{\alpha}}{\partial \xi^{\beta}} \overline{\mathbf{G}}_{\alpha} \tag{6.28}$$

Thus, from equation (6.28), we finally get the transformation relations for the contravariant components of a vector, as:

$$\boxed{a^{\alpha} = \frac{\partial \xi^{\alpha}}{\partial \overline{\xi}^{\beta}} \overline{a}^{\beta} \quad \text{and} \quad \overline{a}^{\alpha} = \frac{\partial \overline{\xi}^{\alpha}}{\partial \xi^{\beta}} a^{\beta}} \tag{6.29}$$

Now, comparing the transformation relation given by equation (6.29) with the transformation relations for the differentials as in equation (6.24), we are ready to define these components:

- The contravariant components of the vector, that is, a first-order tensor, must transform similar to the differentials of the coordinates, that is, according to equation (6.24).

6.2.3.4 Covariant Base Vector Transformation Law

For covariant base vectors, the transformation relationships follow from those of the contravariant components of a vector, as follows: let $\mathbf{a} = a^{\alpha}\mathbf{G}_{\alpha} = \overline{a}^{\beta}\overline{\mathbf{G}}_{\beta}$ be a vector with contravariant components, $a^{\alpha}, \overline{a}^{\beta}$, described in two different coordinate systems; then from equation (6.29), we have:

$$\begin{aligned} &\mathbf{a} = \overline{a}^{\beta}\overline{\mathbf{G}}_{\beta} = a^{\alpha}\frac{\partial \overline{\xi}^{\beta}}{\partial \xi^{\alpha}}\overline{\mathbf{G}}_{\beta} = a^{\alpha}\mathbf{G}_{\alpha} \quad \text{and} \quad \mathbf{a} = a^{\beta}\mathbf{G}_{\beta} = \overline{a}^{\alpha}\frac{\partial \xi^{\beta}}{\partial \overline{\xi}^{\alpha}}\mathbf{G}_{\beta} = \overline{a}^{\alpha}\overline{\mathbf{G}}_{\alpha}, \\ &\Rightarrow \boxed{\mathbf{G}_{\alpha} = \frac{\partial \overline{\xi}^{\beta}}{\partial \xi^{\alpha}}\overline{\mathbf{G}}_{\beta} \quad \text{and} \quad \overline{\mathbf{G}}_{\alpha} = \frac{\partial \xi^{\beta}}{\partial \overline{\xi}^{\alpha}}\mathbf{G}_{\beta}} \end{aligned} \tag{6.30}$$

6.2.3.5 Length of a Vector on a Surface

Considering an invariant geometric object such as a vector, $\mathbf{u}$, written in terms of the covariant base vectors, $\{\mathbf{G}_{\alpha}\}$, and the contravariant base vectors, $\{\mathbf{G}^{\alpha}\}$, we have: $\mathbf{u} = u_{\alpha}\mathbf{G}^{\alpha} = u^{\alpha}\mathbf{G}_{\alpha}$, where $u_{\alpha} = u \bullet \mathbf{G}_{\alpha}$'s and $u^{\alpha} = \mathbf{u} \bullet \mathbf{G}^{\alpha}$'s are the covariant and the contravariant components, respectively, of the vector. Finally, for the length of the vector, $\|\mathbf{u}\|$, we have:

$$\|\mathbf{u}\|^2 = \mathbf{u} \bullet \mathbf{u} = u_{\alpha}\mathbf{G}^{\alpha} \bullet u^{\beta}\mathbf{G}_{\beta} = u_{\alpha}u^{\beta}\delta^{\alpha}_{\beta} = u_{\alpha}\, u^{\alpha} \tag{6.31}$$

As a scalar object, the length of a vector is invariant under coordinate transformation, which helps us to define the covariant component transformation law for vectors, as follows.

6.2.3.6 Covariant Components of a Vector (First-order Tensor) Transformation Law

Similar transformation properties for the covariant components of a vector may be obtained by considering scalar-valued vector functions. Let us consider the invariant scalar function, $\phi(\xi^1, \xi^2)$, such that under allowable coordinate transformation, we have:

$$\phi(\xi^1, \xi^2) = \phi(\overline{\xi}^1, \overline{\xi}^2) \tag{6.32}$$

Then, the gradients of the function transforms are given as:

$$\frac{\partial \phi}{\partial \xi^{\alpha}} = \frac{\partial \bar{\xi}^{\beta}}{\partial \xi^{\alpha}} \frac{\partial \phi}{\partial \bar{\xi}^{\beta}} \quad \text{and} \quad \frac{\partial \phi}{\partial \bar{\xi}^{\alpha}} = \frac{\partial \xi^{\beta}}{\partial \bar{\xi}^{\alpha}} \frac{\partial \phi}{\partial \xi^{\beta}} \tag{6.33}$$

Now, inserting the contravariant component transformation relation given by equation (6.29) in the expression for invariant length of a vector, **a**, as in equation (6.31), we have:

$$a_{\alpha} a^{\alpha} = a_{\alpha} \frac{\partial \xi^{\alpha}}{\partial \bar{\xi}^{\beta}} \bar{a}^{\beta} = \bar{a}_{\beta} \bar{a}^{\beta} \quad \text{and} \quad \bar{a}_{\alpha} \bar{a}^{\alpha} = \bar{a}_{\alpha} \frac{\partial \bar{\xi}^{\alpha}}{\partial \xi^{\beta}} a^{\beta} == a_{\beta} a^{\beta}$$

$$\Rightarrow \boxed{\bar{a}_{\beta} = \frac{\partial \xi^{\alpha}}{\partial \bar{\xi}^{\beta}} a_{\alpha} \quad \text{and} \quad a_{\beta} = \frac{\partial \bar{\xi}^{\alpha}}{\partial \xi^{\beta}} \bar{a}_{\alpha},} \tag{6.34}$$

Now, comparing the transformation relation given by equation (6.34) with the transformation relations for the gradients as in equation (6.33), we are ready to define these components:

- The covariant components of the vector, that is, a first-order tensor, must transform similar to the gradient of a scalar function, that is, according to equation (6.33).

6.2.3.7 Contravariant Base Vector Transformation Law

We note for contravariant base vectors, the transformation relationships follow from those of the covariant components of a vector, as follows: let $\mathbf{a} = a_{\alpha} \mathbf{G}^{\alpha} = \bar{a}_{\beta} \overline{\mathbf{G}}^{\beta}$ be a vector with covariant components, $a_{\alpha}, \bar{a}_{\beta}$, described in two different coordinate systems; then from equation (6.34), we have:

$$\mathbf{a} = \bar{a}_{\beta} \overline{\mathbf{G}}^{\beta} = a_{\alpha} \frac{\partial \xi^{\alpha}}{\partial \bar{\xi}^{\beta}} \overline{\mathbf{G}}^{\beta} = a_{\alpha} \mathbf{G}^{\alpha} \quad \text{and} \quad \mathbf{a} = a_{\beta} \mathbf{G}^{\beta} = \bar{a}_{\alpha} \frac{\partial \bar{\xi}^{\alpha}}{\partial \xi^{\beta}} \mathbf{G}^{\beta} = \bar{a}_{\alpha} \overline{\mathbf{G}}^{\alpha},$$

$$\Rightarrow \boxed{\mathbf{G}^{\alpha} = \frac{\partial \xi^{\alpha}}{\partial \bar{\xi}^{\beta}} \overline{\mathbf{G}}^{\beta} \quad \text{and} \quad \overline{\mathbf{G}}^{\alpha} = \frac{\partial \bar{\xi}^{\alpha}}{\partial \xi^{\beta}} \mathbf{G}^{\beta}} \tag{6.35}$$

6.2.3.8 Metric Tensor Transformation Laws

From equations (6.9) and (6.30), we get the transformation relation for the covariant components of the metric tensor as:

$$G_{\alpha\beta} \equiv \mathbf{G}_{\alpha} \bullet \mathbf{G}_{\beta} = \frac{\partial \bar{\xi}^{\mu}}{\partial \xi^{\alpha}} \overline{\mathbf{G}}_{\mu} \bullet \frac{\partial \bar{\xi}^{\nu}}{\partial \xi^{\beta}} \overline{\mathbf{G}}_{\nu} \Rightarrow \boxed{G_{\alpha\beta} = \frac{\partial \bar{\xi}^{\mu}}{\partial \xi^{\alpha}} \frac{\partial \bar{\xi}^{\nu}}{\partial \xi^{\beta}} \overline{\mathbf{G}}_{\mu\nu}} \tag{6.36}$$

Similarly, using equations (6.9) and (6.35), we have the transformation relation for the contravariant components of the metric tensor as:

$$G^{\alpha\beta} \equiv \mathbf{G}^{\alpha} \bullet \mathbf{G}^{\beta} = \frac{\partial \xi^{\alpha}}{\partial \bar{\xi}^{\mu}} \bar{\mathbf{G}}^{\mu} \bullet \frac{\partial \xi^{\beta}}{\partial \bar{\xi}^{\nu}} \overline{\mathbf{G}}^{\nu} \Rightarrow \boxed{G^{\alpha\beta} = \frac{\partial \xi^{\alpha}}{\partial \bar{\xi}^{\mu}} \frac{\partial \xi^{\beta}}{\partial \bar{\xi}^{\nu}} \overline{\mathbf{G}}^{\mu\nu}} \tag{6.37}$$

6.2.3.9 Metric Tensor: Curve Length, Angle between Tangents and Area on a Surface

As indicated earlier, the metric tensor helps measure all the geometric properties of a surface. Let $\mathbf{C}{:}\xi^1(t), \xi^2(t), \quad t \in [a, b]$ be a parametric curve on a surface, $\mathbf{S} = \{\mathbf{r} \in \mathbf{R}_3 | \ (\xi^1, \xi^2) \in \mathbf{S}_\xi\}$; then:

6.2.3.9.1 Length of a Curve:

$$\boxed{s = \int_a^b \sqrt{\mathbf{r}_{,t} \bullet \mathbf{r}_{,t}} dt = \int_a^b \sqrt{\mathbf{G}_\alpha \xi^\alpha{}_{,t} \bullet \mathbf{G}_\beta \xi^\beta{}_{,t}} dt = \int_a^b \sqrt{\mathbf{G}_{\alpha\beta}\, \xi^\alpha{}_{,t}\, \xi^\beta{}_{,t}} dt} \tag{6.38}$$

6.2.3.9.2 Angle between Vectors

Let **u** and **v** be two vectors; then, the angle between them:

$$\boxed{\mathbf{u} \bullet \ \mathbf{v} = u^\alpha \mathbf{G}_\alpha \bullet v_\beta \mathbf{G}^\beta = u^\alpha v_\alpha = G_{\alpha\beta} u^\alpha v^\beta = G^{\alpha\beta} u_\alpha v_\beta} \tag{6.39}$$

Now, considering $\mathbf{G}_1$ and $\mathbf{G}_2$, and the angle, θ, between them, we get: $\cos\theta = \frac{\mathbf{G}_{12}}{\|\mathbf{G}_1\| \|\mathbf{G}_2\|}$. So, if they are orthogonal, we must have: $\boxed{\mathbf{G}_{12} \equiv 0}$.

6.2.3.9.3 Surface Area

Let us recall that: $d\mathbf{r}_1 = \frac{\partial \mathbf{r}}{\partial \xi^1} d\xi^1 = \mathbf{G}_1 \ d\xi^1$ and $d\mathbf{r}_2 = \frac{\partial \mathbf{r}}{\partial \xi^2} d\xi^2 = \mathbf{G}_2 \ d\xi^2$; then, a differential surface area, dA (see Figure 6.6), is given by:

$$dA = \|d\mathbf{r}_1 \times d\mathbf{r}_2\| = \|\mathbf{G}_1 \times \ \mathbf{G}_2\| d\xi^1 \ d\xi^2 \tag{6.40}$$

Now, by using the Lagrange identity: $(\mathbf{a} \times \mathbf{b}) \bullet (\mathbf{c} \times \mathbf{d}) = (\mathbf{a} \bullet \mathbf{c})(\mathbf{b} \bullet \mathbf{d}) - (\mathbf{a} \bullet \mathbf{d})(\mathbf{b} \bullet \mathbf{c})$ for equation (6.40), we get:

$$\boxed{\begin{aligned} dA &= \|\mathbf{G}_1 \times \ \mathbf{G}_2\| \ d\xi^1 \ d\xi^2 = \sqrt{(\mathbf{G}_1 \times \ \mathbf{G}_2) \bullet (\mathbf{G}_1 \times \ \mathbf{G}_2)} d\xi^1 \ d\xi^2 \\ &= \sqrt{G_{11} G_{22} - G_{12}^2} d\xi^1 \ d\xi^2 = \sqrt{G} \ d\xi^1 \ d\xi^2 \end{aligned}} \tag{6.41}$$

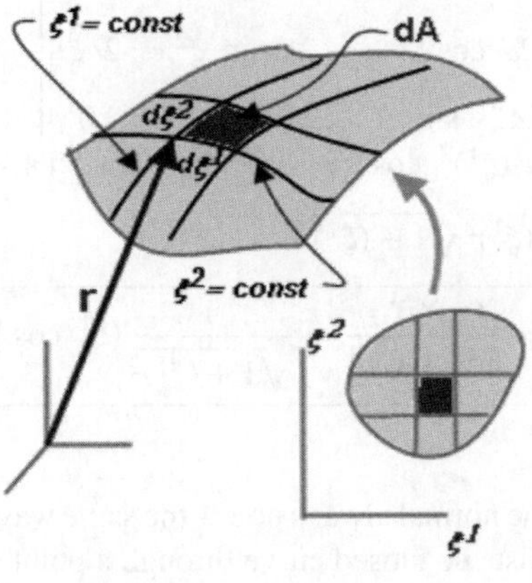

Figure 6.6 Surface area.

where G is the determinant of the surface Jacobian. Then, the total surface area is given by integration as:

$$\text{Surface area} = \iint_{\text{surface}} \sqrt{G}d\xi^1 d\xi^2 \tag{6.42}$$

Example: To find the surface area of a sphere of radius R, let us recall that: $G_{11} = R^2 \cos^2 \xi^2$, $G_{22} = R^2$ and $G_{12} = 0$; thus, $\sqrt{G} = R^2 \cos^2 \xi^2$, and by equation (6.41), $dA = R^2 \cos^2 \xi^2 \, d\xi^1 \, d\xi^2$. Finally, using equation (6.42), the surface area of the sphere is given by:

$$\text{Spherical surface area} = 2 \int_{\xi^1=0}^{2\pi} \int_{\xi^2=0}^{\pi/2} R^2 \cos^2 \xi^2 \, d\xi^1 \, d\xi^2 = 4\pi R^2 \tag{6.43}$$

6.2.3.10 Unit Surface Normal

The unit normal, $\hat{\mathbf{N}}(\hat{\xi}^1, \hat{\xi}^2)$, at $\mathbf{r}(\hat{\xi}_1(t), \ \hat{\xi}_2(t))$ to the tangent plane is also the normal to the surface at the point and is defined as:

$$\hat{\mathbf{N}}(\hat{\xi}^1, \hat{\xi}^2) = \frac{\hat{\mathbf{G}}_1(\hat{\xi}^1, \hat{\xi}^2) \times \hat{\mathbf{G}}_2(\hat{\xi}^1, \hat{\xi}^2)}{\|\hat{\mathbf{G}}_1(\hat{\xi}^1, \hat{\xi}^2) \times \hat{\mathbf{G}}_2(\hat{\xi}^1, \hat{\xi}^2)\|} \tag{6.44}$$

Now, using equation (6.16), we can also express the normal at a point on a surface as:

$$\hat{\mathbf{N}} = \frac{1}{\sqrt{\hat{G}}}(\hat{\mathbf{G}}_1 \times \hat{\mathbf{G}}_2) \tag{6.45}$$

Example: A paraboloid surface is given by: $r(\xi^1, \xi^2) = 2a\xi^1 \cos \xi^2 \, \mathbf{i} + 2a\xi^1 \sin \xi^2 \, \mathbf{j} + a(\xi^1)^2 \, \mathbf{k}$; we try to determine the normal to the surface:

$$\begin{aligned}
\mathbf{r}_{,1} \times \mathbf{r}_{,2} &= \begin{vmatrix} \mathbf{i} & \mathbf{j} & \mathbf{k} \\ 2a \cos \xi^2 & 2a \sin \xi^2 & 2a\xi^1 \\ -2a\xi^1 \sin \xi^2 & 2a\xi^1 \cos \xi^2 & 0 \end{vmatrix} \\
&= -4a^2(\xi^1)^2 \cos \xi^2 \mathbf{i} - 4a^2(\xi^1)^2 \sin \xi^2 \mathbf{j} + 4a^2 \xi^1 \mathbf{k} \\
\Rightarrow \|\mathbf{r}_{,1} \times \mathbf{r}_{,2}\| &= 4a^2(\xi^1)^2 \sqrt{1 + (\xi^1)^2} \\
&\Rightarrow \mathbf{n} = \frac{\mathbf{r}_{,1} \times \mathbf{r}_{,2}}{\|\mathbf{r}_{,1} \times \mathbf{r}_{,2}\|} = \frac{1}{\sqrt{1 + (\xi^1)^2}}(\xi^1 \cos \xi^2 \mathbf{i} - \xi^1 \sin \xi^2 \mathbf{j} - \mathbf{k})
\end{aligned}$$

- On an ***orientable surface***, the normal always points the same way on going around any curve on the surface, that is, there exists no closed curve through a point such that the normal direction switches by going once around. A counter example is the Mobius strip.

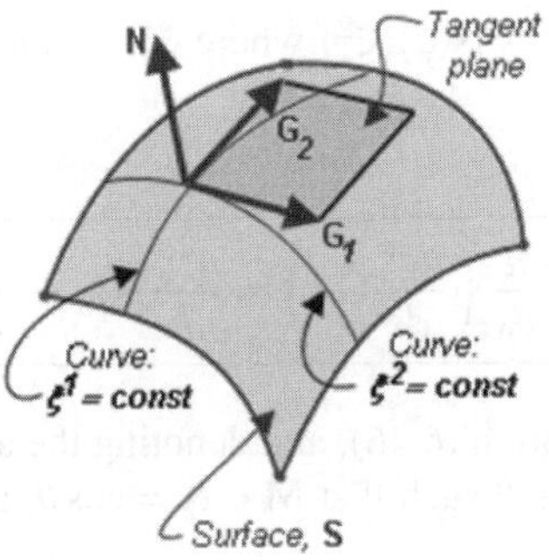

Figure 6.7 Local Gauss frame.

6.2.3.11 Local Gauss Frame

A Gauss frame may be defined by the set $\{\, \hat{\mathbf{G}}_1 \quad \hat{\mathbf{G}}_2 \quad \hat{\mathbf{N}} \,\}$, the set of covariant base vectors and the normal at a point on the surface.

Local Gauss frames (see Figure 6.7) at points on a surface serve the same important task as the ***Frenet frames*** of a curve; however, unlike Frenet frames, a local Gauss frame depends on parameterization.

6.2.4 Second Fundamental form: Surface Curvatures

Referring to Figure 6.8, let us recall that for a curve, $\mathbf{C}(s)$, where s is the arc length parameter, the unit ***tangent***, $\mathbf{T}(s)$, to the curve, and the unit ***principal normal***, $\mathbf{M}(s)$, are related to each other by the ***Frenet–Serret*** relation:

$$\boxed{\mathbf{T}' \overset{\Delta}{=} \frac{d\mathbf{T}}{ds} = \kappa \mathbf{M} = \frac{1}{R}\mathbf{M}} \tag{6.46}$$

where $(\bullet)' \equiv \frac{d}{ds}$, κ is the curvature of the curve, and R is the radius of curvature at the point on the curve. Now, to adapt this formula for curves on a surface, we have for the curve arc length parameter: $s = s(\xi^1, \xi^2)$. Then, for a curve on a surface: $\mathbf{S} = \{\mathbf{r} \in \mathbf{R}_3 |\ (\xi^1, \xi^2) \in \mathbf{S}_\xi\}$, with

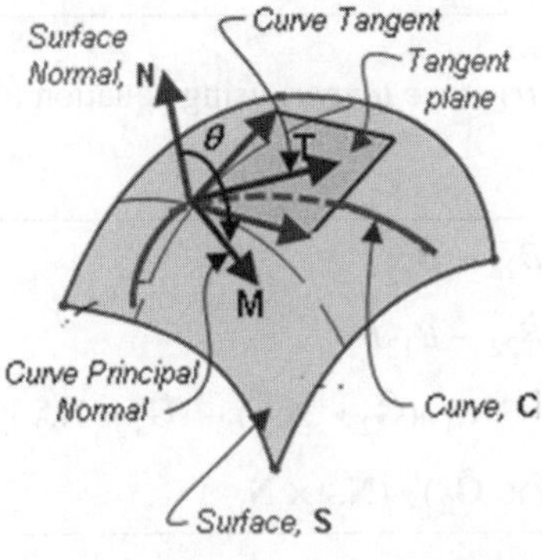

Figure 6.8 Curvature of a curve on a surface.

the surface unit normal denoted as $\mathbf{N}(\xi^1, \xi^2)$ where ξ^1, ξ^2 are real parameters, with $\mathbf{T} = \frac{d\mathbf{r}}{ds} = \frac{d\mathbf{r}}{d\xi^\alpha}\frac{\partial \xi^\alpha}{\partial s} = \frac{\partial \xi^\alpha}{\partial s}\hat{\mathbf{G}}_\alpha$ we have:

$$\boxed{\mathbf{T}' = \frac{d^2\mathbf{r}}{ds^2} = \frac{d}{ds}\left(\frac{d\xi^\alpha}{ds}\hat{\mathbf{G}}_\alpha\right) = \frac{d^2\xi^\alpha}{ds^2}\hat{\mathbf{G}}_\alpha + \frac{d\xi^\alpha}{ds}\frac{d\xi^\beta}{ds}\hat{\mathbf{G}}_{\alpha,\beta}} \tag{6.47}$$

So, using equation (6.47) in equation (6.46), and denoting the angle between the surface normal and the curve principal normal as θ such that $\mathbf{M} \bullet \mathbf{N} = \cos\theta$, and noting that: $\mathbf{N} \bullet \hat{\mathbf{G}}_\alpha = 0,\ \alpha = 1, 2$, we get:

$$\boxed{\kappa \cos\theta = \frac{1}{R}\cos\theta = \mathbf{T}' \bullet \mathbf{N} = \frac{d\xi^\alpha}{ds}\frac{d\xi^\beta}{ds}(\hat{\mathbf{G}}_{\alpha,\beta} \bullet \mathbf{N})} \tag{6.48}$$

Now, defining the covariant components of the curvature tensor, $\hat{B}_{\alpha\beta}$, as:

$$\boxed{\hat{B}_{\alpha\beta} \stackrel{\Delta}{=} \hat{\mathbf{G}}_{\alpha,\beta} \bullet \hat{\mathbf{N}} = \hat{\mathbf{G}}_{\alpha,\beta} \bullet \frac{\hat{\mathbf{G}}_1 \times \hat{\mathbf{G}}_2}{\sqrt{\hat{\mathbf{G}}_1 \times \hat{\mathbf{G}}_2}} = \frac{1}{\sqrt{G}}|\, \hat{\mathbf{G}}_{\alpha,\beta} \quad \hat{\mathbf{G}}_1 \quad \hat{\mathbf{G}}_2 \,|} \tag{6.49}$$

where $|\, \hat{\mathbf{G}}_{\alpha,\beta} \quad \hat{\mathbf{G}}_1 \quad \hat{\mathbf{G}}_2 \,|$ is the determinant; the quadratic form $\boxed{\hat{B}_{\alpha\beta} d\xi^\alpha d\xi^\beta}$ is known as the ***second fundamental form*** of a surface. Thus, we can rewrite equation (6.48) as:

$$\boxed{\kappa \cos\theta = \frac{1}{R}\cos\theta = \frac{\hat{B}_{\alpha\beta} d\xi^\alpha d\xi^\beta}{ds^2} = \frac{\hat{B}_{\alpha\beta} d\xi^\alpha d\xi^\beta}{\hat{G}_{\alpha\beta} d\xi^\alpha d\xi^\beta} = \frac{\text{Second fundamental form}}{\text{First fundamental form}}} \tag{6.50}$$

- Since $\hat{\mathbf{G}}_{\alpha,\beta} = \dfrac{\partial \mathbf{r}}{\partial \xi^\alpha \partial \xi^\beta} = \dfrac{\partial \mathbf{r}}{\partial \xi^\beta \partial \xi^\alpha} = \hat{\mathbf{G}}_{\beta,\alpha}$, $\hat{B}_{\alpha\beta}$ is symmetric.
- Since $\hat{\mathbf{G}}_\alpha \bullet \mathbf{N} = 0$, then, by differentiating, we have:

$$\boxed{\hat{B}_{\alpha\beta} \stackrel{\Delta}{=} \hat{\mathbf{G}}_{\alpha,\beta} \bullet \mathbf{N} = -\hat{\mathbf{G}}_\alpha \bullet \mathbf{N}_{,\beta}} \tag{6.51}$$

- The ***determinant***, $\hat{B}$, of the ***curvature tensor***, using equation (6.51) and the Lagrange identity, is given by:

$$\boxed{\begin{aligned} \hat{B} &= \hat{B}_{11}\hat{B}_{22} - \hat{B}_{12}^2 \\ &= \hat{B}_{11}\hat{B}_{22} - \hat{B}_{12}\hat{B}_{21} \\ &= (\hat{\mathbf{G}}_1 \bullet \mathbf{N}_{,1})(\hat{\mathbf{G}}_2 \bullet \mathbf{N}_{,2}) - (\hat{\mathbf{G}}_1 \bullet \mathbf{N}_{,2})(\hat{\mathbf{G}}_2 \bullet \mathbf{N}_{,1}) \\ &= (\hat{\mathbf{G}}_1 \times \hat{\mathbf{G}}_2) \bullet (\mathbf{N}_{,1} \times \mathbf{N}_{,2}) \end{aligned}} \tag{6.52}$$

- While the first fundamental form is positive definite, the second fundamental form may vanish.

- Given $\cos\theta$, that is, the angle between the curve principal normal and the surface normal, and ds^2 from the first fundamental form, we can use equation (6.50) to recover curvature κ from the second fundamental form.

6.2.4.1 Meusnier's Theorem: Normal Section of a Surface

Since the first fundamental form, $\hat{G}_{\alpha\beta}$, and the second fundamental form, $\hat{B}_{\alpha\beta}$, depend only on the surface geometry at a point, that is, are independent of the choice of a surface curve, the curvature, κ, of a surface curve, given by equation (6.50), depends only on the tangent $(d\xi^\alpha d\xi^\beta)$ and the principal normal $(\cos\theta)$. Now, knowing that the tangent to the curve and the principal normal of a surface curve belong to the osculating plane, we can state the following:

- The curvature of a surface curve at a fixed point on a surface depends solely on the osculating plane of the curve.

6.2.4.1.1 Meusnier's Theorem

- All surface curves at a point on a surface with the same osculating plane have the same curvature at the point.

This includes plane curves; thus, we may consider only the plane curves from now on. Following equation (6.50), it is clear that at a fixed point on a surface, the tangent direction, as determined by:

$$\lambda = \tan\alpha = \frac{d\xi^2}{d\xi^1} \tag{6.53}$$

fixes the right-hand side of the equation. Then, the curvature of the curve depends entirely on the angle between the normal to the surface and the principal normal of the curve. Let us, then, define the ***normal section*** of a surface (see Figure 6.9) for a given $(\mathbf{r}, \mathbf{T})$, that is, the position vector and the tangent vector, as follows:

- The ***plane curve*** of intersection between the surface and the plane contains the normal to surface and the tangent to the curve.

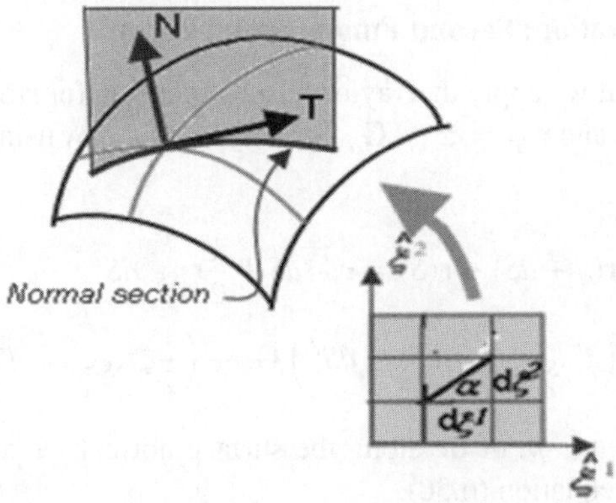

Figure 6.9 Normal section.

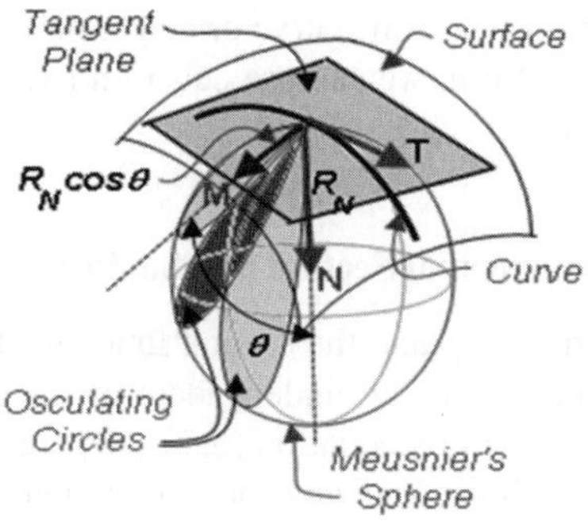

Figure 6.10 Meusnier's sphere.

For a normal section, $\cos\theta = \pm 1$, that is, the curve principal normal coincides ($\theta = 0° \Rightarrow \kappa = \kappa_N$) with the surface normal, or opposite ($\theta = \pi \Rightarrow \kappa = -\kappa_N$) to it, then, it follows, from equation (6.50), that:

$$\boxed{\kappa_N(\mathbf{r},\mathbf{T}) = \frac{1}{R_N(\mathbf{r},\mathbf{T})} = \frac{\hat{B}_{\alpha\beta} d\xi^\alpha d\xi^\beta}{\hat{G}_{\alpha\beta} d\xi^\alpha d\xi^\beta} = \frac{\text{Second fundamental form}}{\text{First fundamental form}}} \tag{6.54}$$

In other words, when the osculating plane of the curve containing the curve principal normal and the tangent to the curve coincides with plane containing the normal to the surface and the tangent to the curve, the curvature, known as the ***normal curvature***, is given by equation (6.54). Now comparing equation (6.50) with equation (6.54), we get:

$$\boxed{\begin{aligned} \kappa \cos\theta &= \kappa_N \\ R &= R_N \cos\theta \end{aligned}} \tag{6.55}$$

- From equation (6.55) which is another form of the ***Meusnier's Theorem***, the osculating circles of all the surface curves at a point on the surface with the same tangent direction form a sphere which has the same tangent plane at the point as the tangent plane to the surface at the point, as shown in Figure 6.10.

6.2.4.2 Alternate Derivation: Second Fundamental Form

As shown in Figure 6.11, if we apply the Taylor series expansion for $\mathbf{r}(S + dS)$ at $\mathbf{S}$, and note that: $\mathbf{r}_{,S} = \xi^\alpha{}_{,S}\ \mathbf{r}_{,\alpha} = \xi^\alpha{}_{,S}\ \mathbf{G}_\alpha$, and $\mathbf{r}_{,SS} = \xi^\alpha{}_{,SS}\,\mathbf{G}_\alpha + \xi^\alpha{}_{,S}\,\xi^\beta{}_{,S}\ \mathbf{G}_{\alpha,\beta}$, by using the Einstein summation convention, we get:

$$\begin{aligned} \mathbf{dr}(S) &= \mathbf{r}(S + dS) - \mathbf{r}(S) = \mathbf{r}_{,S}\ dS + \frac{1}{2}\mathbf{r}_{,SS}\ dS^2 \\ &= \left(\xi^\alpha{}_{,S}\, dS + \frac{1}{2}\xi^\alpha{}_{,SS}\ dS^2\right)\mathbf{G}_\alpha + \left(\frac{1}{2}\xi^\alpha{}_{,S}\,\xi^\beta{}_{,S}\ dS^2\right)\mathbf{G}_{\alpha,\beta} \end{aligned} \tag{6.56}$$

Now, taking the component, dp, of $\mathbf{dr}$ along the surface normal, $\mathbf{N}$, and noting that $\mathbf{N} \bullet \mathbf{G}_\alpha = 0,\ \alpha = 1, 2$, we get, from equation (6.56):

$$dp = \left(\frac{1}{2}\xi^\alpha{}_{,S}\,\xi^\beta{}_{,S}\ dS^2\right)(\mathbf{G}_{\alpha,\beta} \bullet \mathbf{N}) = \frac{1}{2}\hat{B}_{\alpha\beta}\ d\xi^\alpha\ d\xi^\beta \tag{6.57}$$

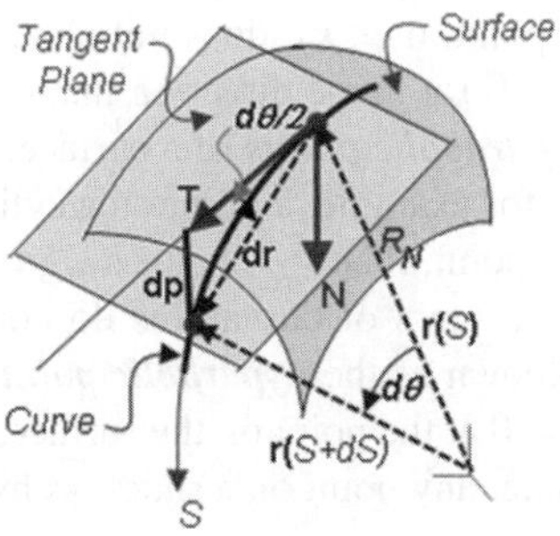

Figure 6.11 Second fundamental form.

where, as before, we have defined $\boxed{\hat{B}_{\alpha\beta} \overset{\Delta}{=} \mathbf{G}_{\alpha},_{\beta} \bullet \mathbf{N}}$ as the ***curvature tensor***. We can directly choose normal section for a given $(\mathbf{r}, \mathbf{T})$, that is, position vector and tangent vector, as a consequence of the Meusnier's Theorem; noting that $\frac{1}{R_N} = \frac{d\theta}{dS}$, and $\frac{1}{2}d\theta = \frac{dp}{ds}$, we get: $\kappa_N \overset{\Delta}{=} \frac{1}{R_N} = \lim_{dS \to 0} \frac{1}{dS}(2\frac{dp}{ds})$. Thus, using equation (6.57) for dp, we get:

$$\boxed{\kappa_N(\mathbf{r}, \mathbf{T}) = \frac{1}{R_N(\mathbf{r}, \mathbf{T})} = \frac{\hat{B}_{\alpha\beta}\, d\xi^\alpha\, d\xi^\beta}{\hat{G}_{\alpha\beta}\, d\xi^\alpha\, d\xi^\beta} = \frac{\text{Second fundamental form}}{\text{First fundamental form}}} \tag{6.58}$$

as before. Equation (6.54), or equation (6.58), can be rewritten as:

$$\boxed{(\hat{B}_{\alpha\beta} - \kappa_N \hat{G}_{\alpha\beta})\, d\xi^\alpha\, d\xi^\beta = 0 \quad \text{or} \quad (\hat{B}_{\alpha\beta} - \frac{1}{R_N}\hat{G}_{\alpha\beta})\, d\xi^\alpha\, d\xi^\beta = 0} \tag{6.59}$$

- As indicated, κ_N depends on the tangent direction of a normal section; the normal section for which $\kappa_N = 0$ is called an ***asymptotic curve***. For example, the generator of a cylindrical surface is an asymptotic curve.
- In fact, any straight line on a surface is an asymptotic curve.
- The coordinate curves of a coordinate system are asymptotic curves if and only if $\hat{B}_{11} = \hat{B}_{22} = 0$.

6.2.4.3 Elliptical, parabolic and Hyperbolic Points on a Surface

Since the first fundamental form is always positive definite, the sign of κ_N depends on the second fundamental form which, in turn, depends on the sign of the determinant, $\hat{B} = \hat{B}_{11}\hat{B}_{22} - \hat{B}_{12}$ of the curvature tensor:

- ***Elliptical Point***: If $\hat{B} > 0$ at a point, then $\kappa_N > 0$ for every normal section at the point on a surface, meaning that the centers of curvatures always lie on the same side of the surface. These points are known as the ***elliptic points*** of a surface, and there is no asymptotic line at the point on the surface; for example, a point on a sphere is elliptical.

- ***Parabolic Point***: If $\hat{B} = 0$ at a point, then κ_N does not change sign for any normal section at the point on a surface, but $\kappa_N = 0$ for some direction that corresponds to an asymptotic line. These points are known as the ***parabolic points*** of a surface, and there is only one asymptotic line at the point on the surface; for example, a point on a cylinder or a cone is parabolic.
- ***Hyperbolic Point***: If $\hat{B} < 0$ at a point, then κ_N changes sign over normal sections at the point on a surface, meaning that the centers of curvatures do not always lie on the same side of the surface. These points are known as the ***hyperbolic points*** of a surface, and there are two asymptotic lines for which $\kappa_N = 0$ at the point on the surface, and κ_N switches sign every time it crosses these lines; for example, any point on a saddle is hyperbolic.

6.2.5 The Lines of Curvature from the Iso-curves

The ***normal curvature*** of a curve on a surface, $\kappa_N(\mathbf{r}, \mathbf{T})$, as given by equation (6.58), depends on $\mathbf{r}, \mathbf{T}$, that is, the point on the surface and a curve with a fixed tangent direction. We can now investigate the extremum property of this curvature at a point by varying the tangent direction at the fixed point.

6.2.5.1 Extremum Directions of Curvature: Lines of Curvature

Referring to Figure 6.12, the tangent direction is determined by:

$$\lambda = \tan \alpha = \frac{d\hat{\xi}^2}{d\hat{\xi}^1} \tag{6.60}$$

The normal curvature can be rewritten using equation (6.60) in terms of λ, by dividing both the numerator and denominator of equation (6.58) by $(d\hat{\xi}^1)^2$, as:

$$\boxed{\kappa_N(\lambda) = \frac{\hat{B}_{11} + 2\hat{B}_{12}\ \lambda + \hat{B}_{22}\ \lambda^2}{\hat{G}_{11} + 2\hat{G}_{12}\ \lambda + \hat{G}_{22}\ \lambda^2}} \tag{6.61}$$

- If $\hat{B}_{11} : \hat{B}_{12} : \hat{B}_{22} = \hat{G}_{11} : \hat{G}_{12} : \hat{G}_{22}$ in equation (6.61) are proportional at a point, the normal curvature, κ_N, becomes independent of λ; then, the point is called an ***umbilical point***. For example, all points of a sphere are umbilical points.

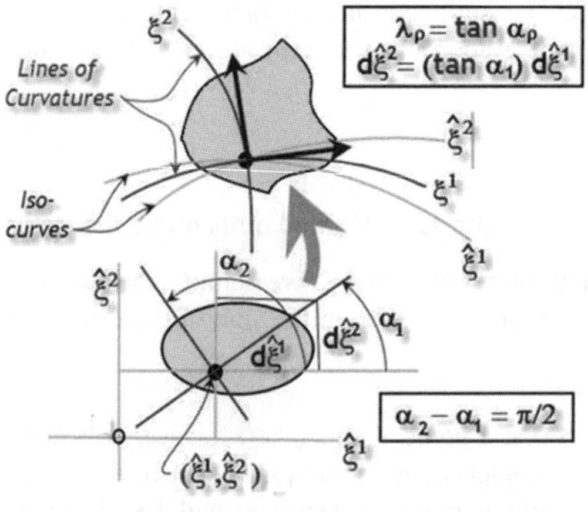

Figure 6.12 Lines of curvature.

Now, the condition for the extremum property of the normal curvature is given by: $\frac{d\kappa_N(\lambda)}{d\lambda} = 0$. Thus, by taking derivatives of $\kappa_N(\lambda)$ as given in equation (6.61), we get:

$$\begin{aligned}
&2(\hat{G}_{12} + \hat{G}_{22}\,\lambda)\kappa_N + (\hat{G}_{11} + 2\hat{G}_{12}\,\lambda + \hat{G}_{22}\,\lambda^2)\frac{d\kappa_N}{d\lambda} = 2(\hat{B}_{12} + \hat{B}_{22}\,\lambda)\\
&\Rightarrow \frac{\hat{B}_{11} + 2\hat{B}_{12}\,\lambda + \hat{B}_{22}\,\lambda^2}{\hat{G}_{11} + 2\hat{G}_{12}\,\lambda + \hat{G}_{22}\,\lambda^2} = \frac{\hat{B}_{12} + \hat{B}_{22}\,\lambda}{\hat{G}_{12} + \hat{G}_{22}\,\lambda}\\
&\Rightarrow \boxed{(\hat{G}_{12}\hat{B}_{22} - \hat{G}_{22}\hat{B}_{12})\lambda^2 + (\hat{G}_{11}\hat{B}_{22} - \hat{G}_{22}\hat{B}_{11})\lambda + (\hat{G}_{11}\hat{B}_{12} - \hat{G}_{12}\hat{B}_{11}) = 0}
\end{aligned} \tag{6.62}$$

Finally, equation (6.62) for the condition of extrema of curvature can be rewritten as:

$$\boxed{\det\begin{bmatrix} \lambda^2 & -\lambda & 1 \\ \hat{G}_{11} & \hat{G}_{12} & \hat{G}_{22} \\ \hat{B}_{11} & \hat{B}_{12} & \hat{B}_{22} \end{bmatrix} = 0} \tag{6.63}$$

where, $\hat{G}_{\alpha\beta}$ and $\hat{B}_{\alpha\beta}$, with $\alpha, \beta = 1, 2$, are the components of the ***surface metric tensor*** and the ***curvature tensor***, respectively, at any point on the surface, described by the $\{\hat{\xi}^\alpha\}$, $\alpha = 1, 2$ coordinate system. The two roots, λ_α, $\alpha = 1, 2$, of equation (6.63), determine the ***principal directions***, that is, maximum and minimum curvature directions, at a point;

- It can be shown that the roots are always real.
- A surface curve with its tangents pointing in a principal direction at all points on the curve is known as a ***line of curvature*** on a surface.

6.2.5.2 Extremum Values of Curvature: Gauss and Mean Curvatures

Now, the principal (i.e. along the lines of curvature) radii of curvature, R_α, themselves, at a point on the surface, are determined by the stationarity of $\frac{1}{R_N}$ or κ_N in equation (6.59), since the condition of extremum radius of curvature is $d(\frac{1}{R_\alpha}) = d\kappa_N = 0$; so, differentiating equation (6.59) with respect to $d\xi^\sigma$, and noting the symmetry of both $\hat{G}_{\alpha\beta}$, the metric tensor, and $\hat{B}_{\alpha\beta}$, the curvature tensor, we get:

$$\begin{aligned}
&(\hat{B}_{\alpha\beta} - \kappa_N\hat{G}_{\alpha\beta})\left(\frac{d\hat{\xi}^\alpha}{d\xi^\sigma}\,d\xi^\beta + d\xi^\alpha\,\frac{d\hat{\xi}^\beta}{d\xi^\sigma}\right) = 0\\
&\Rightarrow (\hat{B}_{\alpha\beta} - \kappa_N\hat{G}_{\alpha\beta})\,(\delta^\alpha_\sigma\,d\hat{\xi}^\beta + d\hat{\xi}^\alpha\,\delta^\beta_\sigma) = 0\\
&\Rightarrow (\hat{B}_{\sigma\beta} - \kappa_N\hat{G}_{\sigma\beta})\,d\hat{\xi}^\beta + (\hat{B}_{\alpha\sigma} - \kappa_N\hat{G}_{\alpha\sigma})\,d\hat{\xi}^\alpha = 0\\
&\Rightarrow (\hat{B}_{\sigma\alpha} - \kappa_N\hat{G}_{\sigma\alpha})\,d\hat{\xi}^\alpha + (\hat{B}_{\sigma\alpha} - \kappa_N\hat{G}_{\sigma\alpha})\,d\hat{\xi}^\alpha = 0\\
&\Rightarrow \boxed{\begin{array}{l}(\hat{B}_{\sigma\alpha} - \kappa_N\hat{G}_{\sigma\alpha})\,d\hat{\xi}^\alpha = 0, \quad \sigma = 1, 2\\ \left(\hat{B}_{\sigma\alpha} - \dfrac{1}{R_N}\hat{G}_{\sigma\alpha}\right)\,d\hat{\xi}^\alpha = 0, \quad \sigma = 1, 2\end{array}}
\end{aligned} \tag{6.64}$$

Now, for a non-trivial solution of the linear equations (6.64), we must have:

$$\det\begin{vmatrix} \hat{B}_{11}-\frac{1}{R_N}\hat{G}_{11} & \hat{B}_{12}-\frac{1}{R_N}\hat{G}_{12} \\ \hat{B}_{21}-\frac{1}{R_N}\hat{G}_{21} & \hat{B}_{22}-\frac{1}{R_N}\hat{G}_{22} \end{vmatrix}=0$$

$$\Rightarrow \left(\hat{B}_{11}-\frac{1}{R_N}\hat{G}_{11}\right)\left(\hat{B}_{22}-\frac{1}{R_N}\hat{G}_{22}\right)-\left(\hat{B}_{12}-\frac{1}{R_N}\hat{G}_{12}\right)^2=0 \tag{6.65}$$

$$\Rightarrow \hat{B}-\frac{1}{R_N}(\hat{G}_{11}\hat{B}_{22}+\hat{G}_{22}\hat{B}_{11}-2\hat{G}_{12}\hat{B}_{12})+\hat{G}\left(\frac{1}{R_N}\right)^2=0$$

where we have used: $\hat{B}\stackrel{\Delta}{=}\det|\hat{B}_{\alpha\beta}|=\hat{B}_{11}\hat{B}_{22}-\hat{B}_{12}^2$ and $\hat{G}\stackrel{\Delta}{=}\det|\hat{G}_{\alpha\beta}|=\hat{G}_{11}\hat{G}_{22}-\hat{G}_{12}^2$. Now, using the contravariant–covariant relationship, as illustrated in equation (6.18), of the surface metric tensor: $\hat{G}_{11}=\hat{G}\,\hat{G}^{22}$, $\hat{G}_{22}=\hat{G}\,\hat{G}^{11}$ and $\hat{G}_{12}=-\hat{G}\,\hat{G}^{12}$, we get from equation (6.65):

$$\boxed{\left(\frac{1}{R_\alpha}\right)^2-\hat{B}_{\alpha\beta}\,\hat{G}^{\alpha\beta}\left(\frac{1}{R_\alpha}\right)+\frac{\hat{B}}{\hat{G}}=0} \tag{6.66}$$

where $\hat{G}^{\alpha\beta}$ are the components of the tensor reciprocal to the surface metric tensor, defined by $\hat{\mathbf{G}}^\alpha\,\hat{\mathbf{G}}_\beta=\delta^\alpha_\beta$, with δ^α_β as the components of the identity tensor and $\hat{G}$ and $\hat{B}$ the determinants of the surface metric tensors and the curvature tensors at a point on the surface, respectively.

- Each of the roots, $\frac{1}{R_1}$ and $\frac{1}{R_2}$ corresponding to the directions $d\hat{\xi}^\alpha_{(1)}$, $\alpha=1,2$ and $d\hat{\xi}^\alpha_{(2)}$, $\alpha=1,2$, respectively, satisfy equation (6.64) such that:

$$\left(\hat{B}_{\alpha\beta}-\frac{1}{R_1}\hat{G}_{\alpha\beta}\right)d\hat{\xi}^\alpha_{(1)}=0 \quad\text{and}\quad \left(\hat{B}_{\alpha\beta}-\frac{1}{R_2}\hat{G}_{\alpha\beta}\right)d\hat{\xi}^\alpha_{(2)}=0 \tag{6.67}$$

- Now, multiplying the first of equation (6.67) by $d\hat{\xi}^\beta_{(2)}$, and the second by $d\hat{\xi}^\beta_{(1)}$, and subtracting, we get:

$$\left(\frac{1}{R_1}-\frac{1}{R_2}\right)\hat{G}_{\alpha\beta}\,d\hat{\xi}^\alpha_{(1)}\,d\hat{\xi}^\beta_{(2)}=0 \tag{6.68}$$

- Thus, if $(\frac{1}{R_1}-\frac{1}{R_2})\neq 0$, then, for solution of equation (6.68), the directions given by $d\hat{\xi}^\alpha_{(1)}$ and $d\hat{\xi}^\beta_{(2)}$ must be orthogonal (i.e. perpendicular to each other).
- Thus, it is always possible to construct a mutually orthogonal coordinate system following the directions of principal curvatures, that is, the directions of the ***lines of curvature***.
- In these principal coordinates, both the metric tensor, $G_{\alpha\beta}$, and the curvature tensor, $B_{\alpha\beta}$, become of diagonal form, that is, $G_{12}=B_{12}\equiv 0$. In other words, any allowable coordinate system on a surface coincides with the lines of curvature if and only if:

$$\boxed{G_{12}=0 \quad\text{and}\quad B_{12}=0} \tag{6.69}$$

Problem: show that:

$$\boxed{\frac{\hat{B}}{\hat{G}} = \hat{B}_{1}^{\cdot 1}\hat{B}_{2}^{\cdot 2} - \hat{B}_{2}^{\cdot 1}\hat{B}_{1}^{\cdot 2}} \tag{6.70}$$

Proof we have: $\hat{B}_{11} = \hat{B}_{1}^{\cdot\sigma}G_{\sigma 1}$, $\hat{B}_{22} = \hat{B}_{2}^{\cdot\sigma}G_{\sigma 2}$, $\hat{B}_{12} = \hat{B}_{1}^{\cdot\sigma}G_{\sigma 2}$ and $\hat{B}_{21} = \hat{B}_{2}^{\cdot\sigma}G_{\sigma 1}$; now, $\hat{B} \triangleq \hat{B}_{11}\hat{B}_{22} - \hat{B}_{12}\hat{B}_{21} = \hat{B}_{1}^{\cdot\sigma}\hat{B}_{2}^{\cdot\tau}(G_{\sigma 1}G_{\tau 2} - G_{\sigma 2}G_{\tau 1})$. For $\sigma = \tau$, $G_{\sigma 1}G_{\tau 2} - G_{\sigma 2}G_{\tau 1} = 0$, and for $\sigma \neq \tau$, $\hat{B} = \hat{B}_{1}^{\cdot 1}\hat{B}_{2}^{\cdot 2}(\hat{G}_{11}\hat{G}_{22} - \hat{G}_{12}^{2}) + \hat{B}_{1}^{\cdot 2}\hat{B}_{2}^{\cdot 1}(\hat{G}_{12}^{2} - \hat{G}_{11}\hat{G}_{22}) = \hat{G}(\hat{B}_{1}^{\cdot 1}\hat{B}_{2}^{\cdot 2} - \hat{B}_{1}^{\cdot 2}\hat{B}_{2}^{\cdot 1})$ completes the proof.

Problem: show that along the lines of curvature:

$$\boxed{B_{\alpha\beta}G^{\alpha\beta} = B_{1}^{\cdot 1} + B_{2}^{\cdot 2}} \tag{6.71}$$

Proof we have: $B_{\alpha\beta}G^{\alpha\beta} = B_{11}G^{11} + B_{22}G^{22} + \cancel{B_{12}G^{12}} + \cancel{B_{21}G^{21}} = B_{1}^{\cdot 1} + B_{2}^{\cdot 2}$

6.2.5.3 Gaussian Curvature and Mean Curvature

Let $\frac{1}{R_1}$ and $\frac{1}{R_2}$ be the roots of equation (6.66) corresponding to slope λ_1 and λ_2, respectively; then the ***mean curvature***, H, and the ***Gaussian curvature***, K, are, defined as:

$$\boxed{\begin{aligned} H &\equiv \frac{1}{2}\left(\frac{1}{R_1} + \frac{1}{R_2}\right) = \hat{B}_{\alpha\beta}\,\hat{G}^{\alpha\beta} = \frac{1}{2}\hat{B}_{\alpha}^{\cdot\alpha} \\ K &\equiv \frac{1}{R_1}\frac{1}{R_2} = \hat{B}_{1}^{\cdot 1}\hat{B}_{2}^{\cdot 2} - \hat{B}_{1}^{\cdot 2}\hat{B}_{2}^{\cdot 1} = \frac{\hat{B}}{\hat{G}} \end{aligned}} \tag{6.72}$$

where we have used the relationships given by equations (6.70) and (6.71).

- When the coordinate curves are the ***lines of curvature***, we can find the coordinate radii curvatures from equation (6.64) by using $d\xi^2 = 0$ for $\frac{1}{R_1}$, and $d\xi^1 = 0$ for $\frac{1}{R_2}$ such that:

$$\begin{aligned} &\left(B_{11} - \frac{1}{R_1}G_{11}\right)d\xi^1 + \left(B_{12} - \frac{1}{R_1}G_{12}\right)\cancel{d\xi^2} = 0 \\ &\left(B_{21} - \frac{1}{R_1}G_{21}\right)\cancel{d\xi^1} + \left(B_{22} - \frac{1}{R_2}G_{22}\right)d\xi^2 = 0 \\ &\Rightarrow \boxed{\frac{1}{R_1} = \frac{B_{11}}{G_{11}} = B_{1}^{\cdot 1} \text{ and } \frac{1}{R_2} = \frac{B_{22}}{G_{22}} = B_{2}^{\cdot 2}} \end{aligned} \tag{6.73}$$

- When the coordinate curves are the ***lines of curvature***, we have $B_{12} = B_{21} = G_{12} = G_{21} = 0$, and accordingly:

$$\boxed{\frac{\hat{B}}{\hat{G}} = \hat{B}_{1}^{\cdot 1}\hat{B}_{2}^{\cdot 2}} \tag{6.74}$$

- Thus, when the coordinate curves are the lines of curvature, the expressions for the mean curvature and the Gaussian curvature, using equations (6.73) and (6.74) in equation (6.72), reduce to:

$$\boxed{\begin{aligned} H &= \frac{1}{2}\left(\frac{1}{R_1}+\frac{1}{R_2}\right)=\frac{1}{2}\left(\frac{B_{11}}{G_{11}}+\frac{B_{22}}{G_{22}}\right) \\ K &= \frac{1}{R_1}\frac{1}{R_2}=\frac{B_{11}}{G_{11}}\frac{B_{22}}{G_{22}} \end{aligned}} \tag{6.75}$$

- Clearly, the ***Gaussian curvature***, K, is positive for an elliptic point, vanishes for a parabolic point and negative for a hyperbolic or saddle point.

6.2.5.4 Invariant Arc Length Parameterization

A particular parameter transformation is of profound importance for computational purposes (as will be seen shortly): the ***arc length parameterization*** where S^α, the arc length is related to ξ^α for $\alpha = 1, 2$ by:

$$\boxed{S^1(\xi^1,\xi^2)=\int_0^{\xi^1}\|\mathbf{r}(\xi,\xi^2),_\xi\|d\xi \quad \text{and} \quad S^2(\xi^1,\xi^2)=\int_0^{\xi^2}\|\mathbf{r}(\xi^1,\xi),_\xi\|d\xi} \tag{6.76}$$

where, $\|\mathbf{r}(\xi,\xi^2),_\xi\| \equiv \sqrt{\mathbf{r},_\xi(\xi,\xi^2)\bullet\mathbf{r},_\xi(\xi,\xi^2)}$, and $\|\mathbf{r}(\xi^1,\xi),_\xi\| \equiv \sqrt{\mathbf{r},_\xi(\xi^1,\xi)\bullet\mathbf{r},_\xi(\xi^1,\xi)}$ with $\bullet$ being the usual vector dot product. Because $\mathbf{r},_\xi(\xi,\xi^2)\,d\xi \equiv \frac{d\mathbf{r}}{d\xi}d\xi = \frac{d\mathbf{r}}{d\tau}\frac{d\tau}{d\xi}d\xi = \mathbf{r},_\tau(\tau)\,d\tau$, the arc length parameterization does not depend on the original parameter, and hence, is an invariant parameterization. Also, by considering the above representation as: $S^\alpha = \int_0^{S^\alpha} dS^\alpha, \quad \alpha = 1, 2$, we get for ***arc elements***, also called ***linear elements***, of a surface:

$$\boxed{\begin{aligned} dS^1 &= \|\mathbf{r}(\xi^1,\xi^2),_{\xi^1}\|d\xi^1 \equiv \sqrt{\mathbf{r},_{\xi^1}(\xi^1,\xi^2)\bullet\mathbf{r},_{\xi^1}(\xi^1,\xi^2)}\,d\xi^1 \\ dS^2 &= \|\mathbf{r}(\xi^1,\xi^2),_{\xi^2}\|d\xi^2 \equiv \sqrt{\mathbf{r},_{\xi^2}(\xi^1,\xi^2)\bullet\mathbf{r},_{\xi^2}(\xi^1,\xi^2)}\,d\xi^2 \end{aligned}} \tag{6.77}$$

Remark

- Equation (6.77) gives us, component-wise, the usual relation:

$$dS^\alpha = \sqrt{dX_\alpha^2 + dY_\alpha^2 + dZ_\alpha^2}\,d\xi, \quad \alpha = 1, 2 \tag{6.78}$$

- In order to maintain that $\mathbf{r}(\xi,\xi^2),_\xi \neq \mathbf{0}$ and $\mathbf{r},_\xi(\xi^1,\xi) \neq \mathbf{0}$ at all $\xi^\alpha, \quad \alpha = 1, 2$, the arc length parameters, $S^\alpha, \quad \alpha = 1, 2$, must monotonically increase along the corresponding $\xi^\alpha, \quad \alpha = 1, 2$.

Finally, with the arc length parameters along the lines of curvature, we have, using equation (6.77), the coordinate unit base vectors, $\mathbf{G}_1$ and $\mathbf{G}_2$ because:

$$\|\mathbf{G}_1\| \equiv \left\|\frac{\partial \mathbf{r}}{\partial S^1}\right\| = \left\|\frac{\partial \mathbf{r}}{\partial \xi^1}\frac{d\xi^1}{dS^1}\right\| = \frac{\|\mathbf{r}_{,\xi^1}\|}{\|\mathbf{r}_{,\xi^1}\|} = 1 \text{ and } \|\mathbf{G}_2\| \equiv \left\|\frac{\partial \mathbf{r}}{\partial S^2}\right\| = \left\|\frac{\partial \mathbf{r}}{\partial \xi^2}\frac{d\xi^2}{dS^2}\right\| = \frac{\|\mathbf{r}_{,\xi^2}\|}{\|\mathbf{r}_{,\xi^2}\|} = 1 \tag{6.79}$$

Thus, with $G_{11} = G_{22} = 1$, for arc length parameter along the lines of curvature, we have:

$$\boxed{\begin{aligned} H &= \frac{1}{2}(B_{11} + B_{22}) \\ K &= B_{11}\, B_{22} \end{aligned}} \tag{6.80}$$

6.2.5.5 The Transformations: The Iso-curves to the Lines of Curvature

In our subsequent numerical evaluations of various quantities of interest by the finite element method, we would like to resort to the descriptions of these entities in terms of the optimal normal coordinate system: $\{S^1, S^2\}$, because of the most desired form of simplicity in expressions, and hence, computational efficiency. For example, the elements ***stiffness matrices*** are numerically evaluated through integrations by ***quadrature rules*** at specific quadrature points. At these points on the shell reference surface, we can determine the lines of curvature as above for the simplest expressions for the involved derivatives for numerical efficiency (see shell analysis for relevant details). But our geometry of the shell surface is presented to us in terms of the non-orthonomal coordinate system: $(\hat{\xi}_1, \hat{\xi}_2)$. Thus, we need to know the necessary transformations for the derivatives in the above two coordinate systems: $\hat{\mathbf{G}}_\alpha \equiv \frac{\partial \mathbf{r}}{\partial \hat{\xi}^\alpha} = \frac{\partial \mathbf{r}}{\partial S^\beta}\frac{\partial S^\beta}{\partial \hat{\xi}^\alpha} = (\frac{\partial S^\beta}{\partial \hat{\xi}^\alpha})\, \mathbf{G}_\alpha$. Thus, because of the orthonormality of $\mathbf{G}_\alpha$, that is, with $\mathbf{G}^\alpha = \mathbf{G}_\alpha$:

$$\boxed{\frac{\partial S^\beta}{\partial \hat{\xi}^\alpha} = \hat{\mathbf{G}}_\alpha \bullet \mathbf{G}^\beta = \hat{\mathbf{G}}_\alpha \bullet \mathbf{G}_\beta} \tag{6.81}$$

Now, from equation (6.81), we have the desired relations:

$$\frac{\partial(\bullet)}{\partial \hat{\xi}^\alpha} = \frac{\partial S^\beta}{\partial \hat{\xi}^\alpha}\frac{\partial(\bullet)}{\partial S^\beta} = (\hat{\mathbf{G}}_\alpha \bullet \mathbf{G}_\beta)\frac{\partial(\bullet)}{\partial S^\beta} \Rightarrow \frac{\partial(\bullet)}{\partial S^\beta} = (\hat{\mathbf{G}}_\alpha \bullet \mathbf{G}_\beta)^{-1}\frac{\partial(\bullet)}{\partial \hat{\xi}^\alpha} \tag{6.82}$$

In matrix form, we can express the desired transformation relations as:

$$\boxed{\begin{Bmatrix} \dfrac{\partial(\bullet)}{\partial S^1} \\ \dfrac{\partial(\bullet)}{\partial S^2} \end{Bmatrix} = \begin{bmatrix} \hat{\mathbf{G}}_1 \bullet \mathbf{G}_1 & \hat{\mathbf{G}}_1 \bullet \mathbf{G}_2 \\ \hat{\mathbf{G}}_2 \bullet \mathbf{G}_1 & \hat{\mathbf{G}}_2 \bullet \mathbf{G}_2 \end{bmatrix}^{-1} \begin{Bmatrix} \dfrac{\partial(\bullet)}{\partial \hat{\xi}^1} \\ \dfrac{\partial(\bullet)}{\partial \hat{\xi}^2} \end{Bmatrix}} \tag{6.83}$$

6.3 Gauss–Weingarten Formulas: Optimal Coordinate System

Among the most important formulas in the description of surfaces are the ***Gauss–Weingarten formulas*** which relate the base vector derivatives to the base vectors themselves – these are the surface counterparts of the ***Frenet–Serret formulas*** related to the curves. However, for computational ease and efficiency for shell structures, we must express these in the optimal coordinate system, that is, in terms of the base vectors along the lines of curvature and parameterized in terms of the arc length parameters. Thus, our main goal is to get acquainted with additional important properties – curvature and twist – of a surface, This, in turn, requires us to investigate the derivatives, known as the ***Gauss–Weingarten derivatives***, of the Gauss frame of a general two-dimensional surface; particularly, we will be interested to find out whether we can express the derivatives of the base vectors, $(\mathbf{G}_1, \mathbf{G}_2, \mathbf{N})$ defining the local Gaussian frame along the lines of curvature with arc length parameterization, (S^1, S^2). In what follows, we will use the notation: $(\bullet)_\alpha \overset{\Delta}{=} \frac{d(\bullet)}{dS^\alpha}$.

6.3.1.1 Normal Coordinate System

First, let us note that for a unit normal, $\mathbf{N}(\xi^1, \xi^2)$, such that $\mathbf{N} \bullet \mathbf{N} = 1$, we have: $\mathbf{N}_{,\alpha} \bullet \mathbf{N} = 0$, $\alpha = 1, 2$, with $(\bullet)_\alpha \overset{\Delta}{=} \frac{d(\bullet)}{d\xi^\alpha}$, that is, $\mathbf{N}_{,\alpha} \perp \mathbf{N}$, $\alpha = 1, 2$, and that $\mathbf{N}_{,\alpha}$ $\alpha = 1, 2$ belong to the tangent space at a point on the surface. Thus, we can express it as:

$$\mathbf{N}_{,\alpha} = (\mathbf{N}_{,\alpha} \bullet \hat{\mathbf{G}}^\gamma)\hat{\mathbf{G}}_\gamma, \quad \alpha = 1, 2 \tag{6.84}$$

But, for $\alpha = 1, 2$, using definition of the curvature tensor as in equation (6.51), we have: $\mathbf{N}_{,\alpha} \bullet \hat{\mathbf{G}}^\gamma = (\mathbf{N}_{,\alpha} \bullet \hat{\mathbf{G}}_\eta)\hat{G}^{\gamma\eta} = -\hat{B}_{\alpha\eta}\hat{G}^{\gamma\eta} = -\hat{B}_\alpha^{\cdot\gamma}$; thus, we get:

6.3.1.1.1 Weingarten Formula

The derivative of the normal is related to the curvature tensor:

$$\boxed{\mathbf{N}_{,\alpha} = -\hat{B}_\alpha^{\cdot\beta}\hat{\mathbf{G}}_\beta, \quad \alpha = 1, 2 \quad \text{with} \quad \hat{B}_\alpha^{\cdot\beta} = \hat{G}^{\sigma\beta}\hat{B}_{\alpha\sigma}} \tag{6.85}$$

6.3.1.1.2 Gauss Formula

Now, with $\mathbf{N} \bullet \hat{\mathbf{G}}_{\alpha,\beta} = \hat{B}_{\alpha\beta}$ in equation (6.51), the derivatives of $\hat{\mathbf{G}}_\alpha$, $\alpha = 1, 2$, can be expressed, in terms of the three base vectors, as:

$$\boxed{\hat{\mathbf{G}}_{\alpha,\beta} = \Gamma^\gamma_{\alpha\beta}\hat{\mathbf{G}}_\gamma + \hat{B}_{\alpha\beta}\mathbf{N}, \quad \alpha = 1, 2 \quad \text{with} \quad \Gamma^\gamma_{\alpha\beta} = \hat{\mathbf{G}}_{\alpha,\beta} \bullet \hat{\mathbf{G}}^\gamma} \tag{6.86}$$

where we have introduced what is known as the ***Christoffel symbol of the second kind***:

$$\boxed{\Gamma^\gamma_{\alpha\beta} = \hat{\mathbf{G}}_{\alpha,\beta} \bullet \hat{\mathbf{G}}^\gamma, \quad \alpha, \beta, \gamma = 1, 2} \tag{6.87}$$

- Since the Christoffel symbol of the second kind vanishes in a rectangular coordinate system (i.e. $\hat{\mathbf{G}}^\gamma = \hat{\mathbf{G}}_\gamma$, $\gamma = 1, 2 \Rightarrow \hat{\mathbf{G}}_{\alpha,\beta} \bullet \hat{\mathbf{G}}_\gamma = 0$), it is coordinate dependent and hence is not a ***tensor***.
- Let us also introduce the ***Christoffel symbol of the first kind*** as:

$$\boxed{\Gamma_{\alpha\beta\gamma} = \hat{\mathbf{G}}_{\alpha,\beta} \bullet \hat{\mathbf{G}}_\gamma, \quad \alpha, \beta, \gamma = 1, 2 \quad \text{with} \quad \Gamma_{\alpha\beta\gamma} = \Gamma^\lambda_{\alpha\beta}\,\hat{G}_{\gamma\lambda} \quad \text{and} \quad \Gamma^\gamma_{\alpha\beta} = \Gamma_{\alpha\beta\lambda}\,\hat{G}^{\gamma\lambda}} \tag{6.88}$$

The Christoffel symbols are symmetric in two covariant indices, that is,

$$\boxed{\Gamma^{\gamma}_{\alpha\beta} = \hat{\mathbf{G}}_{\alpha},_{\beta} \bullet \hat{\mathbf{G}}^{\gamma} = \hat{\mathbf{G}}_{\beta},_{\alpha} \bullet \hat{\mathbf{G}}^{\gamma} = \Gamma^{\gamma}_{\beta\alpha}, \quad \alpha, \beta, \gamma = 1, 2 \quad \text{and} \quad \Gamma_{\alpha\beta\gamma} = \Gamma_{\beta\alpha\gamma}} \tag{6.89}$$

- The total numbers of symbols for a 2-surface is $2^3 = 8$; however, because of the symmetry in the first two indices of the first kind, we have only six distinct symbols: $\Gamma_{111}, \Gamma_{112}, \Gamma_{121}, \Gamma_{122}, \Gamma_{221}, \Gamma_{222}$.
- The derivative of the determinant, G, of the metric tensor is related to the Christoffel symbols:

$$\begin{aligned}
G,_{\gamma} &= \left(G_{11}G_{22} - G_{12}^2\right),_{\gamma} = G_{11},_{\gamma} G_{22} + G_{11}G_{22},_{\gamma} - 2G_{12}G_{12},_{\gamma} \\
&= G(G^{11}G_{11},_{\gamma} + G^{22}G_{22},_{\gamma} - 2G^{12}G_{12},_{\gamma}) = GG^{\alpha\beta}G_{\alpha\beta},_{\gamma} \\
&= GG^{\alpha\beta}(\Gamma_{\alpha\gamma\beta} + \Gamma_{\beta\gamma\alpha}) = GG^{\alpha\beta}\left(G_{\lambda\beta}\Gamma^{\lambda}_{\alpha\gamma} + G_{\lambda\alpha}\Gamma^{\lambda}_{\beta\gamma}\right) \\
&= G\left(\delta^{\alpha}_{\lambda}\Gamma^{\lambda}_{\alpha\gamma} + \delta^{\beta}_{\lambda}\Gamma^{\lambda}_{\beta\gamma}\right) = G\left(\Gamma^{\alpha}_{\alpha\gamma} + \Gamma^{\beta}_{\beta\gamma}\right) = 2G\Gamma^{\alpha}_{\alpha\gamma} \Rightarrow \boxed{G,_{\gamma} = 2G\Gamma^{\alpha}_{\alpha\gamma}}
\end{aligned} \tag{6.90}$$

- From equation (6.90), we get:

$$\boxed{\Gamma^{\alpha}_{\alpha\gamma} = \frac{1}{2G}G,_{\gamma} = \left(\log\left(\frac{1}{\sqrt{G}}\right)\right),_{\gamma}} \tag{6.91}$$

- We can express the Christoffel symbols in terms of only the metric tensor and its derivatives; let us first note that, by using equation (6.88), the derivative of a metric tensor is completely described by the Christoffel symbols of the first kind:

$$\boxed{G_{\alpha\gamma} = G_{\alpha} \bullet G_{\gamma} \Rightarrow G_{\alpha\gamma},_{\beta} = G_{\alpha},_{\beta} \bullet G_{\gamma} + G_{\gamma},_{\beta} \bullet G_{\alpha} = \Gamma_{\alpha\beta\gamma} + \Gamma_{\gamma\beta\alpha}} \tag{6.92}$$

Similarly,

$$G_{\gamma\beta},_{\alpha} = \Gamma_{\gamma\alpha\beta} + \Gamma_{\beta\alpha\gamma} \tag{6.93}$$

and

$$G_{\beta\alpha},_{\gamma} = \Gamma_{\beta\gamma\alpha} + \Gamma_{\alpha\gamma\beta} \tag{6.94}$$

Now, adding equations (6.92) and (6.93) together, and subtracting equation (6.94) from it, and noting the symmetry of the Christoffel symbols in the first two indices, we get:

$$\boxed{\Gamma_{\alpha\beta\gamma} = \frac{1}{2}(G_{\beta\gamma},_{\alpha} + G_{\alpha\gamma},_{\beta} - G_{\alpha\beta},_{\gamma})} \tag{6.95}$$

Thus, the Christoffel symbols of the first kind depend only on the ***first fundamental form***.

6.3.1.2 Gauss–Weingarten Formula

Like the Frenet–Serret formulas of the curves, the second partial derivatives of a vector function, $\mathbf{r}(\hat{\xi}^1, \hat{\xi}^2)$, by which a 2-surface can be represented, are linear combinations of the base vectors, $\{\hat{\mathbf{G}}_1 \quad \hat{\mathbf{G}}_2 \quad \hat{\mathbf{G}}_3 \equiv \mathbf{N}\}$:

$$\boxed{\begin{aligned} \hat{\mathbf{G}}_{\beta,\hat{\alpha}} &= \hat{\Gamma}^{\gamma}_{\beta\alpha}\,\hat{\mathbf{G}}_\gamma + B_{\beta\alpha}\,\mathbf{N} \quad \text{(Gauss)} \\ \mathbf{N}_{,\hat{\alpha}} &= -B^{\cdot\beta}_{\alpha}\,\hat{\mathbf{G}}_\beta \qquad\qquad \text{(Weingarten)} \end{aligned}} \tag{6.96}$$

In matrix form, we can express this as:

$$\boxed{\frac{\partial}{\partial\alpha}\begin{Bmatrix}\hat{\mathbf{G}}_1\\ \hat{\mathbf{G}}_2\\ \mathbf{N}\end{Bmatrix} \stackrel{\Delta}{=} \begin{Bmatrix}\hat{\mathbf{G}}_1\\ \hat{\mathbf{G}}_2\\ \mathbf{N}\end{Bmatrix}_{,\alpha} = \begin{bmatrix}\hat{\Gamma}^1_{1\alpha} & \hat{\Gamma}^2_{1\alpha} & \hat{B}_{1\alpha}\\ \hat{\Gamma}^1_{2\alpha} & \hat{\Gamma}^2_{2\alpha} & \hat{B}_{2\alpha}\\ -\hat{B}^{\cdot 1}_{\alpha} & -\hat{B}^{\cdot 2}_{\alpha} & 0\end{bmatrix}\begin{Bmatrix}\hat{\mathbf{G}}_1\\ \hat{\mathbf{G}}_2\\ \mathbf{N}\end{Bmatrix}} \tag{6.97}$$

where the Christoffel symbols of the second kind, $\hat{\Gamma}^{\gamma}_{\beta\alpha}$, are given as the functions of the components of the metric tensor as:

$$\boxed{\hat{\Gamma}^{\sigma}_{\alpha\beta} = \frac{1}{2}\hat{G}^{\sigma\gamma}(\hat{G}_{\alpha\sigma,\hat{\beta}} + \hat{G}_{\gamma\beta,\hat{\alpha}} - \hat{G}_{\alpha\beta,\hat{\gamma}}), \hat{\Gamma}^{3}_{\alpha\beta} \equiv \hat{B}_{\alpha\beta} \quad \text{and} \quad \hat{B}^{\cdot\beta}_{\alpha} = \hat{G}^{\sigma\beta}\hat{B}_{\alpha\sigma}} \tag{6.98}$$

- Noting that the Christoffel symbols of the second kind are related to those of the first kind by the surface metric tensor, as given by equation (6.88), which, in turn, is related to the first fundamental form, we see that the Gauss–Weingarten formula, as given by equation (6.96), is completely defined by the first fundamental form and the second fundamental form.

6.3.1.3 Normal Coordinate System Along the Lines of Curvature

If we now consider the normal coordinate system, with base vectors set: $\{\bar{\mathbf{G}}_1 \equiv \frac{\partial \mathbf{r}}{\partial \bar{\xi}^1} \quad \bar{\mathbf{G}}_2 \equiv \frac{\partial \mathbf{r}}{\partial \bar{\xi}^2} \quad \bar{\mathbf{G}}_3 \equiv \mathbf{N}\}$, and parameters $(\bar{\xi}^1, \bar{\xi}^2)$ along the lines of curvature (which are, of course, perpendicular to each other), we have: $\hat{G}^{\alpha\beta} \to \bar{G}^{\alpha\beta} = 0, \hat{B}^{\alpha\beta} \to \bar{B}^{\alpha\beta} = 0,\ \alpha \neq \beta$; also, with the determinant, $\bar{G}$, of the 2×2 surface metric tensor, we have:

$$\bar{G} = \bar{G}_{11}\,\bar{G}_{22} - \cancel{\bar{G}_{12}\bar{G}_{21}} \Rightarrow \boxed{\begin{aligned} \bar{G}^{(11)} &= \frac{\bar{G}_{(22)}}{\bar{G}} = \frac{1}{\bar{G}_{(11)}} \\ \bar{G}^{(22)} &= \frac{1}{\bar{G}_{(22)}} \end{aligned}} \tag{6.99}$$

with $(\alpha\alpha)$ means that no sum is intended. Now, noting from equation (6.73) that: $\frac{1}{R_1} = B^{\cdot 1}_{1} = \bar{G}^{\gamma 1}\bar{B}_{1\gamma} = \bar{G}^{(11)}\bar{B}_{(11)} = \frac{\bar{B}_{(11)}}{\bar{G}_{(11)}}$, and letting $\bar{G}_{\alpha\alpha} = \bar{\mathbf{G}}_\alpha \bullet \bar{\mathbf{G}}_\alpha = \gamma_\alpha \mathbf{G}_\alpha \bullet \gamma_\alpha \mathbf{G}_\alpha = \gamma^2_\alpha$ where we have introduced unit base vector set:

$$\boxed{\mathbf{G}_\alpha \equiv \frac{\partial \mathbf{r}}{\partial S^\alpha}, \quad \alpha = 1, 2} \tag{6.100}$$

with γ_α as the length of $\bar{\mathbf{G}}_\alpha$, from the third of equation (6.97), we get:

$$\frac{\partial \mathbf{N}}{\partial \xi^\alpha} = -B_{\alpha}^{\cdot\,\beta}\,\bar{\mathbf{G}}_\beta = -B_{\alpha}^{\cdot\,\alpha}\,\bar{\mathbf{G}}_\alpha - \underbrace{B_{\alpha}^{\cdot\,\beta}\bar{\mathbf{G}}_\beta}_{\alpha\neq\beta} = -\frac{1}{R_\alpha}\bar{\mathbf{G}}_\alpha = -\frac{\gamma_\alpha}{R_\alpha}\mathbf{G}_\alpha \tag{6.101}$$

and from $\bar{\mathbf{G}}_\alpha = \gamma_\alpha \mathbf{G}_\alpha$, we get:

$$\frac{\partial \mathbf{G}_\alpha}{\partial \xi^\alpha} = \frac{1}{\gamma_\alpha}\bar{\mathbf{G}}_{\alpha,\alpha} - \frac{1}{\gamma_\alpha^2}\gamma_{\alpha,\alpha}\,\bar{\mathbf{G}}_\alpha \tag{6.102}$$

Now, from first of the relations in equation (6.97) with $\bar{G}_{\alpha\beta} = \bar{B}_{\alpha\beta} = 0, \quad \alpha \neq \beta$, and $\bar{G}^{(\alpha\alpha)} = \frac{1}{\bar{G}_{(\alpha\alpha)}}$ and so on, and noting that ${}^{\xi}\bar{\Gamma}^{\sigma}_{\alpha\beta}$ described in $\{\xi^\alpha\}$-coordinate system, we get:

$${}^{\xi}\bar{\Gamma}^{\sigma}_{\alpha\beta} = \frac{1}{2}\bar{G}^{\sigma\gamma}\left(\frac{\partial \bar{G}_{\alpha\sigma}}{\partial \xi^\beta} + \frac{\partial \bar{G}_{\gamma\beta}}{\partial \xi^\alpha} - \frac{\partial \bar{G}_{\alpha\beta}}{\partial \xi^\gamma}\right) \tag{6.103}$$

Particularizing equation (6.103): $\bar{\Gamma}^1_{11} = \frac{1}{2}\{\bar{G}^{11}\bar{G}_{11,\bar{1}}\} = \frac{1}{2}\frac{1}{\bar{G}_{11}}\bar{G}_{11,\bar{1}} = \frac{1}{\gamma_1}\gamma_{1,\bar{1}}$; similarly, we get: $\bar{\Gamma}^2_{11} = -\frac{\gamma_1}{\gamma_2^2}\gamma_{1,\bar{2}}$. Now, using equations (6.101) and (6.102), we get from the first of equation (6.97):

$$\begin{aligned}
\mathbf{G}_\alpha &= \frac{1}{\gamma_\alpha}\bar{\mathbf{G}}_{\alpha,\bar{\alpha}} - \frac{1}{\gamma_\alpha^2}\gamma_{\alpha,\bar{\alpha}}\,\bar{\mathbf{G}}_\alpha = \frac{1}{\gamma_\alpha}\left(\bar{\Gamma}^{\alpha}_{\alpha\alpha}\,\bar{\mathbf{G}}_\alpha + B_{(\alpha\alpha)}\,\mathbf{N}\right) - \frac{1}{\gamma_\alpha^2}\gamma_{\alpha,\bar{\alpha}}\,\bar{\mathbf{G}}_\alpha \\
&= \frac{1}{\gamma_\alpha}\left(\bar{\Gamma}^{\alpha}_{\alpha\alpha}\,\bar{\mathbf{G}}_\alpha + \bar{\Gamma}^{\beta}_{\alpha\alpha}\,\bar{\mathbf{G}}_\beta + \frac{\gamma_\alpha^2}{R_\alpha}\,\mathbf{N}\right) - \frac{1}{\gamma_\alpha^2}\gamma_{\alpha,\bar{\alpha}}\,\bar{\mathbf{G}}_\alpha \\
&= \frac{1}{\gamma_\alpha^2}\gamma_{\alpha,\bar{\alpha}}\bar{\mathbf{G}}_\alpha - \frac{1}{\gamma_\beta^2}\,\gamma_{\alpha,\bar{\beta}}\,\bar{\mathbf{G}}_\beta + \frac{\gamma_\alpha}{R_\alpha}\,\mathbf{N} - \frac{1}{\gamma_\alpha^2}\gamma_{\alpha,\bar{\alpha}}\bar{\mathbf{G}}_\alpha \\
&= -\frac{1}{\gamma_\beta}\gamma_{\alpha,\bar{\beta}}\,\mathbf{G}_\alpha + \frac{\gamma_\alpha}{R_\alpha}\,\mathbf{N}, \quad \alpha \neq \beta
\end{aligned} \tag{6.104}$$

Now, using equations (6.101) and (6.104), we can rewrite the Gauss–Weingarten formula as:

$$\boxed{\frac{\partial}{\partial \xi^1}\begin{Bmatrix}\mathbf{G}_1\\ \mathbf{G}_2\\ \mathbf{N}\end{Bmatrix} = \begin{bmatrix} 0 & -\frac{1}{\gamma_2}\gamma_{1,\bar{2}} & \frac{\gamma_1}{R_1} \\ \frac{1}{\gamma_2}\gamma_{1,\bar{2}} & 0 & 0 \\ -\frac{\gamma_1}{R_1} & 0 & 0 \end{bmatrix}\begin{Bmatrix}\mathbf{G}_1\\ \mathbf{G}_2\\ \mathbf{N}\end{Bmatrix}} \tag{6.105}$$

and

$$\frac{\partial}{\partial \xi^2}\begin{Bmatrix}\mathbf{G}_1\\ \mathbf{G}_2\\ \mathbf{N}\end{Bmatrix}=\begin{bmatrix}0 & \frac{1}{\gamma_1}\gamma_{2,\bar{1}} & 0\\ \frac{1}{\gamma_1}\gamma_{2,\bar{1}} & 0 & \frac{\gamma_2}{R_2}\\ 0 & -\frac{\gamma_2}{R_2} & 0\end{bmatrix}\begin{Bmatrix}\mathbf{G}_1\\ \mathbf{G}_2\\ \mathbf{N}\end{Bmatrix} \tag{6.106}$$

6.3.1.4 Optimal Normal Coordinate System

Now, let us consider the optimal normal coordinate system, that is, coordinates along the lines of curvature with arc length parameterization; let (S^1, S^2) define a point on the surface.

Noting that $\frac{\partial\gamma_1}{\partial S^1}=\frac{\partial\gamma_1}{\partial\xi^1}\frac{\partial\xi_1}{\partial S^1}=\frac{1}{\gamma_1}\frac{\partial\gamma_1}{\partial\xi^1}$, we get: $\gamma_{1,\bar{2}}\equiv\frac{\partial\gamma_1}{\partial\xi^2}=\frac{\partial\gamma_1}{\partial S^\alpha}\frac{\partial S^\alpha}{\partial\xi^2}=\gamma_2\frac{\partial\gamma_1}{\partial S^2}\equiv\gamma_2\,\gamma_{1,2}$

$$\mathbf{G}_{1,\bar{1}}\equiv\frac{\partial\mathbf{G}_1}{\partial\xi^1}=\frac{\partial\mathbf{G}_1}{\partial S^\alpha}\frac{\partial S^\alpha}{\partial\xi^1}=\gamma_1\mathbf{G}_{1,1} \tag{6.107}$$

Now, using the first row of the relation in equation (6.105), we get:

$$\begin{aligned}\mathbf{G}_{1,1}\equiv\frac{\partial\mathbf{G}_1}{\partial S^1}&=\frac{1}{\gamma_1}\mathbf{G}_{1,\bar{1}}\equiv\frac{1}{\gamma_1}\frac{\partial\mathbf{G}_1}{\partial\xi^1}=\frac{1}{\gamma_1}\left(-\frac{1}{\cancel{\gamma_1}}\cancel{\gamma_2}\frac{1}{\ }\frac{\partial\gamma_1}{\partial S^2}\right)\mathbf{G}_2+\frac{1}{\cancel{\gamma_1}}\frac{\cancel{\gamma_1}}{R_1}\mathbf{N}\\ &=-\frac{1}{\gamma_1}\frac{\partial\gamma_1}{\partial S^2}\mathbf{G}_2+\frac{1}{R_1}\mathbf{N}\end{aligned} \tag{6.108}$$

Modifications similar to equation (6.108) can be developed for each term of the equations (6.105) and (6.106). Thus, from equations (6.101) and (6.104), we get the desired Gauss–Weingarten formulas for the derivatives of the triad of base vectors in matrix form with respect to the arc length parameters, S^α, $\alpha=1,2$, is given by:

$$\begin{Bmatrix}\mathbf{G}_{1,1}\\ \mathbf{G}_{2,1}\\ \mathbf{N}_{,1}\end{Bmatrix}\equiv\frac{\partial}{\partial S^1}\begin{Bmatrix}\mathbf{G}_1\\ \mathbf{G}_2\\ \mathbf{N}\end{Bmatrix}=\begin{bmatrix}0 & -\frac{1}{\gamma_1}\gamma_{1,2} & \frac{1}{R_1}\\ \frac{1}{\gamma_1}\gamma_{1,2} & 0 & 0\\ -\frac{1}{R_1} & 0 & 0\end{bmatrix}\begin{Bmatrix}\mathbf{G}_1\\ \mathbf{G}_2\\ \mathbf{N}\end{Bmatrix}\equiv\overset{0}{\widehat{\mathrm{K}}}_1\begin{Bmatrix}\mathbf{G}_1\\ \mathbf{G}_2\\ \mathbf{N}\end{Bmatrix} \tag{6.109}$$

and

$$\begin{Bmatrix}\mathbf{G}_{1,2}\\ \mathbf{G}_{2,2}\\ \mathbf{N}_{,2}\end{Bmatrix}\equiv\frac{\partial}{\partial S^2}\begin{Bmatrix}\mathbf{G}_1\\ \mathbf{G}_2\\ \mathbf{N}\end{Bmatrix}=\begin{bmatrix}0 & \frac{1}{\gamma_2}\gamma_{2,1} & 0\\ -\frac{1}{\gamma_2}\gamma_{2,1} & 0 & \frac{1}{R_2}\\ 0 & -\frac{1}{R_2} & 0\end{bmatrix}\begin{Bmatrix}\mathbf{G}_1\\ \mathbf{G}_2\\ \mathbf{N}\end{Bmatrix}\equiv\overset{0}{\widehat{\mathrm{K}}}_2\begin{Bmatrix}\mathbf{G}_1\\ \mathbf{G}_2\\ \mathbf{N}\end{Bmatrix} \tag{6.110}$$

where $\widehat{\mathbf{K}}^0_\alpha, \alpha = 1,2$ are the surface curvature matrices with $R_\alpha(S^1, S^2), \alpha = 1,2$, are the radii of curvatures at any point, $\{S^\alpha\}$, of the surface, and $\gamma_\alpha,\ \alpha = 1,2$, are the lengths of the base vectors: $\gamma_\alpha \equiv \frac{dS^\alpha}{d\xi^\alpha} = \|\bar{\mathbf{G}}_\alpha\| = \sqrt{\bar{\mathbf{G}}_{\alpha\alpha}}$ where $\bar{\mathbf{G}}_\alpha(\xi^1,\xi^2) \equiv \frac{\partial \mathbf{r}(\xi^1,\xi^2)}{\partial \xi^\alpha}$, with no sum on α.

- The curvature matrices are skew-symmetric.
- The individual coefficients of the matrices can be easily identified with the Christoffel symbols of the second kind, Γ^i_{jk}. However, in our subsequent shell analysis, we will completely avoid explicit use of these obscuring symbols by adhering to familiar geometric entities such as the radii of curvature, and so on, as we have done earlier, for ease of interpretation and clarity.

6.3.2 *Surface Curvature Tensors and Vectors*

On the other hand, the Gauss–Weingarten relation for the derivatives of the triad of base vectors on a surface in invariant tensor form with respect to the arc length parameters, S^α, $\alpha = 1,2$, may be obtained by first introducing:

Definition: Surface curvature tensors, $\mathbf{K}^0_\alpha$, $\alpha = 1,2$:

$$\boxed{\mathbf{K}^0_\alpha \equiv \left(\widehat{\mathbf{K}}^0_\alpha\right)^T_{ij} \mathbf{G}_i \otimes \mathbf{G}_j = -\mathbf{G}_i \otimes \mathbf{G}_{i,\alpha}, \quad \alpha = 1,2; \quad i,j = 1,2,3} \tag{6.111}$$

To wit, for the second equality in equation (6.111), we note that because of anti-symmetry:

$$\left(\widehat{\mathbf{K}}^0_\alpha\right)^T_{ij} = -\left(\widehat{\mathbf{K}}^0_\alpha\right)_{ij}.$$

Then, we have:

$$\begin{aligned}\mathbf{K}^0_\alpha &\equiv \left(\widehat{\mathbf{K}}^0_\alpha\right)^T_{ij} \mathbf{G}_i \otimes \mathbf{G}_j = -\left(\widehat{\mathbf{K}}^0_\alpha\right)_{ij} \mathbf{G}_i \otimes \mathbf{G}_j = -\mathbf{G}_i \otimes \left(\widehat{\mathbf{K}}^0_\alpha\right)_{ij} \mathbf{G}_j \\ &= -\mathbf{G}_i \otimes \mathbf{G}_{i,\alpha}, \quad \alpha = 1,2; \quad i,j = 1,2,3\end{aligned}$$

where we have used the result from equation (6.111).

Note that the curvature tensor is skew-symmetric and thus: $(\mathbf{K}^0_\alpha)^T = -\mathbf{K}^0_\alpha$. This can be easily verified:

$$\begin{aligned}(\mathbf{K}^0_\alpha)^T &\equiv \left(\widehat{\mathbf{K}}^0_\alpha\right)^T_{ij} \mathbf{G}_j \otimes \mathbf{G}_i \underbrace{=}_{Switch\ Indices} \left(\widehat{\mathbf{K}}^0_\alpha\right)^T_{ji} \mathbf{G}_i \otimes \mathbf{G}_j \\ &\underbrace{=}_{Skew-symmetric} -\left(\widehat{\mathbf{K}}^0_\alpha\right)^T_{ij} \mathbf{G}_i \otimes \mathbf{G}_j \underbrace{=}_{definition} -\mathbf{K}^0_\alpha, \quad \alpha = 1,2;\end{aligned}$$

Definition: Surface curvature vector, $\mathbf{k}^0_\alpha, \alpha = 1,2$:

The axial vectors corresponding to the initial curvature tensors, $\mathbf{K}^0_\alpha$, $\alpha = 1, 2$, are given by:

$$\boxed{\begin{aligned}\mathbf{k}^0_1 &\equiv \left\{ 0 \quad -\frac{1}{R_1} \quad -\frac{1}{\gamma_1}\gamma_{1},_{2} \right\}^T \\ \mathbf{k}^0_2 &\equiv \left\{ \frac{1}{R_2} \quad 0 \quad \frac{1}{\gamma_2}\gamma_{2},_{1} \right\}^T\end{aligned}} \tag{6.112}$$

Now, we can easily get from equation (6.111), the invariant tensor form as:

$$\boxed{\begin{aligned}\mathbf{G}_{i},_{\alpha} &= \mathbf{K}^0_\alpha \, \mathbf{G}_i, \qquad \alpha = 1, 2;\ i = 1, 2, 3 \\ &= \mathbf{k}^0_\alpha \times \ \mathbf{G}_i\end{aligned}} \tag{6.113}$$

Note that, in tensor form as in expression (6.111), unlike the matrix representations of (6.109) and (6.110), each unit base vector derivative is given as a transformation of itself. As we shall see, this allows us to write expressions in absolute compact form free of matrix indices.

Exercise: Verify the tensor form of the derivatives of base vectors given by equation (6.113).

Solution:

$$\boxed{\begin{aligned}\mathbf{K}^0_\alpha \, \mathbf{G}_i \underbrace{\quad = \quad}_{Skew-symmetry} -(\mathbf{K}^0_\alpha)^T \, \mathbf{G}_i \underbrace{\quad = \quad}_{Definition} -(-\mathbf{G}_{k},_{\alpha} \otimes \mathbf{G}_k)\mathbf{G}_i \\ = \mathbf{G}_{i},_{\alpha}, \qquad \alpha = 1, 2;\ i = 1, 2, 3\end{aligned}}$$

- We have so far described representation of surfaces by the $G_{\alpha\beta}$, the first fundamental form, and $B_{\alpha\beta}$, the second fundamental form, However, the inverse question of whether we can construct a surface from any two given tensors, $G_{\alpha\beta}$ and $B_{\alpha\beta}$, is equally intriguing and noteworthy; the answer is generally negative unless these tensors satisfy some integrability conditions: known as the ***Mainardi–Codazzi conditions*** along the surface normal direction, and the ***Riemann curvature tensor*** arising out of $\mathbf{G}_\alpha$, $\alpha = 1, 2$ directions; these constitute all the integrability conditions for the Gauss–Weingarten formulas given by equation (6.96), and are obtained by asserting that:

$$G_{\alpha\beta,\gamma} = G_{\alpha\gamma,\beta} \tag{6.114}$$

where, by taking derivative of the Gauss equation (6.96), we get:

$$\boxed{\begin{aligned}G_{\alpha\beta,\gamma} &= \left[\frac{\partial \Gamma^\rho_{\alpha\beta}}{\partial \xi^\gamma} + \Gamma^\eta_{\alpha\gamma}\Gamma^\rho_{\eta\beta} - B_{\alpha\beta}B^\rho_\gamma\right]\mathbf{G}_\rho + \left[\Gamma^\eta_{\alpha\beta}B_{\eta\gamma} + \frac{\partial B_{\alpha\beta}}{\partial \xi^\gamma}\right]\mathbf{N} \\ G_{\alpha\gamma,\beta} &= \left[\frac{\partial \Gamma^\rho_{\alpha\gamma}}{\partial \xi^\beta} + \Gamma^\eta_{\alpha\beta}\Gamma^\rho_{\eta\gamma} - B_{\alpha\gamma}B^\rho_\beta\right]\mathbf{G}_\rho + \left[\Gamma^\eta_{\alpha\gamma}B_{\eta\beta} + \frac{\partial B_{\alpha\gamma}}{\partial \xi^\beta}\right]\mathbf{N}\end{aligned}} \tag{6.115}$$

- Since $\mathbf{G}_\alpha$, $\alpha = 1, 2$ and $\mathbf{N}$ are linearly independent, the application of equations (6.115) into equation (6.114) will yield an equation where the coefficient of each of these base vectors must be individually zero; so, taking the coefficients for $\mathbf{G}_\alpha$, $\alpha = 1, 2$, we have:

$$R^{\rho}_{\alpha\gamma\beta} = B_{\alpha\beta} B^{\cdot\rho}_{\gamma} - B_{\alpha\gamma} B^{\cdot\rho}_{\beta} \tag{6.116}$$

where we have introduced the ***mixed Riemann curvature tensor*** of order four, as given by:

$$\boxed{R^{\rho}_{\alpha\gamma\beta} \overset{\Delta}{=} \frac{\partial \Gamma^{\rho}_{\alpha\beta}}{\partial \xi^{\gamma}} - \frac{\partial \Gamma^{\rho}_{\alpha\gamma}}{\partial \xi^{\beta}} + \Gamma^{\eta}_{\alpha\beta}\Gamma^{\rho}_{\eta\gamma} - \Gamma^{\eta}_{\alpha\gamma}\Gamma^{\rho}_{\eta\beta}} \tag{6.117}$$

- Clearly, the ***covariant Riemann curvature tensor***, $R_{\delta\alpha\beta\gamma}$, is given by:

$$\boxed{R_{\delta\alpha\beta\gamma} = G_{\delta\rho} R^{\rho}_{\alpha\gamma\beta}} \tag{6.118}$$

- The covariant Riemann curvature tensor, $R_{\delta\alpha\beta\gamma}$, can be shown to be given as:

$$\boxed{R_{\delta\alpha\beta\gamma} = B_{\alpha\beta} B_{\gamma\delta} - B_{\alpha\gamma} B_{\beta\delta}} \tag{6.119}$$

- It can be seen, from equation (6.119), that $R_{\delta\alpha\beta\gamma}$ is ***symmetric*** in the sense that the first pair can be interchanged with the second pair of indices without changing the expression, that is,

$$R_{\delta\alpha\beta\gamma} = R_{\beta\gamma\delta\alpha} \tag{6.120}$$

- $R_{\delta\alpha\beta\gamma}$ is skew-symmetric in both the first two and the second two indices, that is,

$$R_{\delta\alpha\beta\gamma} = -R_{\alpha\delta\beta\gamma} \quad \text{and} \quad R_{\delta\alpha\beta\gamma} = -R_{\delta\alpha\gamma\beta} \tag{6.121}$$

- For a 2-surface, that is, two-dimensional surface, the total number of terms in the fourth-order covariant Riemann tensor is supposed to be $2^4 = 16$; however, because of the above skew-symmetry conditions, the non-zero terms are obtained only when: $\gamma \neq \beta$ and $\delta \neq \alpha$. Thus, only four components are different from zero:

$$\boxed{\begin{aligned} R_{1212} &= R_{2121} = B_{11}B_{22} - B_{12}^2 = B \\ R_{2112} &= R_{1221} = B_{12}^2 - B_{11}B_{22} = -B \end{aligned}} \tag{6.122}$$

In view of equations (6.117), (6.118) and (6.122), the Riemann curvature tensor and hence, the determinant of the curvature tensor, B, is independent of the second fundamental form. Thus:

Gauss's theorema egregium: the Gaussian curvature, K, is dependent only on the first fundamental form and its first and second derivative, as given by:

$$\boxed{K = \frac{B}{G} = \frac{R_{1212}}{G}} \tag{6.123}$$

- For surfaces such as cones, cylinders and planes which have vanishing curvatures, all components of the Riemann curvature also vanish.

6.4 Cartesian Product Bernstein–Bezier surfaces

Here, our main goal is to introduce Cartesian product Bezier surfaces in terms of the Bernstein polynomials – we are motivated by the need to demonstrate several very important properties of a Bezier form of description of parametric surfaces. These properties occupy a pivotal role in both algebraic and geometric understanding of the representation. Furthermore, these are essential for an application of the Bezier representation of interpolation in the mesh generation phase and approximation in the finite element phase of analysis. Let us recall that the Bezier curves are parametric polynomial curves. Thus, definition has an important property that says that any points belonging to an affine space can only be added meaningfully in barycentric combinations. We identify the Bernstein polynomials as the barycentric coefficients, and control points as the Euclidean points. In non-parametric polynomial representations of continuous functions as in the finite element analysis phase, the Bernstein polynomials can be seen as the basis functions of the underlying linear vector space. So, we will use these notions rather less rigorously and interchangeably, as long as we are aware of the context.

- We will first introduce ***Cartesian product patches*** or surfaces, that is, surfaces generated by mapping two-dimensional rectangles to three-dimensional surfaces; later, we will discuss another important class of surfaces known as ***triangular patches or surfaces***.
- As to the Bernstein polynomial, suppose that instead of the original surface, only a finite set of $(m+1) \times (n+1)$ sample points, $\{\mathbf{q}_{ij}\}$, are given with $\mathbf{q}_{ij} \equiv \mathbf{q}(\frac{i}{m}, \frac{j}{n}), \quad i \in \{0, 1, \dots, m\} \quad j \in \{0, 1, \dots, n\}$. Then, because of the properties of the Bernstein basis functions, the approximating surface defined as above will closely follow the polygonal net through the given sample points. Finally, by changing the location of this finite set of the sample points, we will be able to visually control the shape of the approximating surface on a viewing surface while it will still mimic the polygonal net all the time. This is precisely the motivation for definition of a Bezier surface and serves to establish the connection between the Bernstein polynomials and the Bezier surfaces.
- As indicated earlier, most surfaces are amenable to representation by the ***Cartesian product*** of Bernstein polynomials with the coefficients of combination being non-rational except for those as are obtained by revolution of conics for which we will need rational coefficients,
- Thus, for a detailed discussion on Bezier surfaces, the underpinning tool of modern day computer-aided geometry development, we investigate first the ***non-rational Bernstein–Bezier surfaces*** followed by the ***rational Bernstein–Bezier surfaces***.
- For more complicated surfaces that require ***composition*** of Bezier surfaces with certain smoothness requirement, we present the ***Bezier form of B-spline surfaces***,
- It is easy to identify a surface represented by a given control polygonal net, but in most practical applications where we are given points on a surface and required to generate the corresponding control polygonal net, things can get really involved; these are known as the inverse problems, which we will deal with later.
- Finally, noting that the affine transformation is not linear because of the translation part, in computer graphics all computations are performed where the three-dimensional space is embedded into a four-dimensional ***homogeneous coordinate space***; all our code snippets will be given in these coordinates.

6.4.1 Non-rational Bezier Surfaces with Bernstein Polynomials

Based on our understanding of the ***non-rational Bezier curves***, we define here the non-rational Bezier Cartesian product surfaces as follows.

Definitions:

- An $m \times n$ non-rational Bezier surface, $\mathbf{S}(\xi^1, \xi^2)$, is defined as:

$$\boxed{\mathbf{S}(\xi^1, \xi^2) = \sum_{i=0}^{m} \sum_{j=0}^{n} B_i^m(\xi^1)\, B_j^n(\xi^2)\, \mathbf{q}_{ij}, \quad 0 \le \xi^1, \xi^2 \le 1} \tag{6.124}$$

where the basis (blending) function sets, $\{B_i^m(\xi^1)\}$, $\{B_j^n(\xi^2)\}$, are the nth and m^{th} degree Bernstein polynomials, respectively; for example:

$$\boxed{B_j^n(\xi^2) = \binom{n}{j} (\xi^2)^j\, (1-\xi^2)^{n-j}, j \in \{0, 1, \ldots, n\}, \quad \text{where} \quad \binom{n}{j} \equiv \frac{n!}{j!(n-j)!}} \tag{6.125}$$

- The coefficients of linear combination, or Euclidean points, $\{\mathbf{q}_{ij}\}$, are known as the ***control points*** of the surface.
- The polygonal net containing the control points, $\{\mathbf{q}_{ij}\}$, is known as the ***control polygonal net*** of the surface.

Let us elucidate the above definitions by examples. In the mesh generation phase, usually surfaces with degree up to and including three are considered for numerical efficiency. For finite element analysis with mega-elements, it appears that up to quintic should be considered for, say, shell analysis.

Example CP1: Dealing with a Bezier surface of degree one in both directions, that is, a bilinear surface, let us consider the case of $m = n = 1$, that is, Bezier surfaces of degree one.

We have, from equation (6.125), $B_0^1(\xi^X) = (1 - \xi^X)$ and $B_1^1(\xi^X) = \xi^X$ with $\xi^X \in [0, 1]$ for $X = 1, 2$. Thus, from equation (6.124), rewritten in matrix form, we get the representation of any point, $\mathbf{S}(\xi^1, \xi^2) \in \mathbf{R}_3$, on the surface as:

$$\boxed{\underbrace{\mathbf{S}(\xi^1, \xi^2)}_{3\times 1} = \underbrace{\mathbf{T}(\xi^1, \xi^2)}_{3\times 12}\; \underbrace{\mathbf{q}}_{12\times 1}} \tag{6.126}$$

where the control net in vector form is given by:

$$\boxed{\underbrace{\mathbf{q}}_{12\times 1} = \left\{ \underbrace{q_{00}^x\; q_{10}^x\; q_{01}^x\; q_{11}^x}_{1\times 4}\; \underbrace{q_{00}^y\; q_{10}^y\; q_{01}^y\; q_{11}^y}_{1\times 4}\; \underbrace{q_{00}^z\; q_{10}^z\; q_{01}^z\; q_{11}^z}_{1\times 4} \right\}^T} \tag{6.127}$$

and the Bezier transformation matrix is given by:

$$\mathbf{T}(\xi^1, \xi^2) = \underbrace{\begin{bmatrix} B_0^1B_0^1\; B_1^1B_0^1\; B_0^1B_1^1\; B_1^1B_1^1 & 0 & 0 \\ 0 & B_0^1B_0^1\; B_1^1B_0^1\; B_0^1B_1^1\; B_1^1B_1^1 & 0 \\ 0 & 0 & B_0^1B_0^1\; B_1^1B_0^1\; B_0^1B_1^1\; B_1^1B_1^1 \end{bmatrix}}_{3\times 12} \tag{6.128}$$

The matrix form given by equation (6.126) is convenient for approximations in finite element analysis, but for conceptual understanding and interpolations in mesh generation, the following matrix form is more intuitive for each coordinate direction, $D = x, y, z$:

$$\boxed{\underbrace{\mathbf{S}^D(\xi^1,\xi^2)}_{1\times1} = \underbrace{\left\langle B_0^1 = 1-\xi^1 \quad B_1^1 = \xi^1 \right\rangle}_{1\times2} \underbrace{\begin{bmatrix} \mathbf{q}_{00}^D & \mathbf{q}_{01}^D \\ \mathbf{q}_{10}^D & \mathbf{q}_{11}^D \end{bmatrix}}_{2\times2} \underbrace{\begin{Bmatrix} B_0^1 = 1-\xi^2 \\ B_1^1 = \xi^2 \end{Bmatrix}}_{2\times1}} \tag{6.129}$$

From equation (6.129), we can algorithmically think of a point on the surface as being generated by a two-stage already studied Bezier curve generation processes:

Step 1: Compute, by linear interpolation, a point on the curve along the ξ^2-direction at a fixed ξ^2 point as:

$$\boxed{\begin{aligned} \mathbf{q}_0^D(\xi^2) &= (B_0^1 = 1-\xi^2)\mathbf{q}_{00}^D + (B_1^1 = \xi^2)\mathbf{q}_{01}^D \\ \mathbf{q}_1^D(\xi^2) &= (B_0^1 = 1-\xi^2)\mathbf{q}_{10}^D + (B_1^1 = \xi^2)\mathbf{q}_{11}^D \end{aligned}} \tag{6.130}$$

Step 2: Compute, by linear interpolation, the point on the surface by evaluating the curve along the ξ^1-direction at a fixed ξ^1 point with the computed points $\mathbf{q}_0^D(\xi^2)$ and $\mathbf{q}_1^D(\xi^2)$ on the curves of ***Step 1*** as the control points for the current curve:

$$\boxed{\underbrace{\mathbf{S}^D(\xi^1,\xi^2)}_{1\times1} = (B_0^1 = 1-\xi^1)\mathbf{q}_0^D\mathbf{q}_1^D(\xi^2) + (B_1^1 = \xi^1)\mathbf{q}_1^D\mathbf{q}_1^D(\xi^2)} \tag{6.131}$$

- The surface may be thought of as the ***range*** of a ***linear map*** of a ***root element*** or ***domain*** as a unit square defined by $\xi^D \in [0, 1]$ for $D = 1, 2$.
- The surface is linear in both directions, that is, bilinear, and hence a ***ruled surface***.
- The curves on the surface generated by the map of lines along each coordinate direction are known as ***isoparametric curves***.
- The boundary curves on the surface, being linear, interpolate all four input control points.
- As shown in Figure 6.13, at some other rotated angle from the coordinate directions, the curves are parabolic and hyperbolic, which is why the surface is known as a ***hyperbolic paraboloid surface***.
- As will be seen shortly, the directional linear interpolations indicated above by equations (6.130) and (6.131) can be algorithmically extended for higher-degree surfaces by application of repeated linear interpolations already introduced in our study of Bernstein–Bezier–de Casteljau curves.

Example CP2: Looking at a Bezier surface of degree three in both directions (see Figure 6.14), that is, a cubic surface, let us consider the case of $m = n = 3$, that is, Bezier surfaces of degree three.

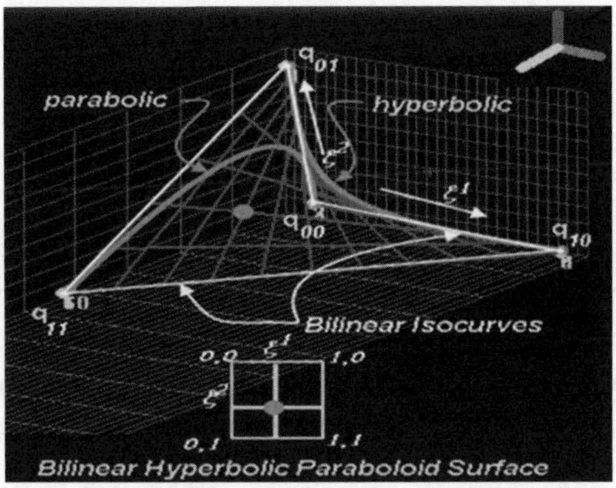

Figure 6.13 Bilinear hyperbolic paraboloid surface. (See colour plate section)

We have, from equation (6.125), $B_0^3(\xi^X) = (1 - \xi^X)^3$, $B_2^3(\xi^X) = 3(1 - \xi^X)^2\, \xi^X$, $B_1^3(\xi^X) = 3(1 - \xi^X)\,(\xi^X)^2$ and $B_3^3(\xi^X) = (\xi^X)^3$ for $X = 1, 2$. Thus, from equation (6.124), rewritten in matrix form, we get the representation of any point, $\mathbf{S}(\xi^1, \xi^2)$, on the surface as:

$$\boxed{\underbrace{\mathbf{S}(\xi^1, \xi^2)}_{3\times1} = \underbrace{\mathbf{T}(\xi^1,\, \xi^2)}_{3\times48}\ \underbrace{\mathbf{q}}_{48\times1}} \tag{6.132}$$

where the control net in vector form is given by:

$$\underbrace{\mathbf{q}}_{48\times1} = \Big\{ \underbrace{q_{00}^x \cdots q_{30}^x\ q_{01}^x \cdots q_{31}^x\ q_{03}^x \cdots q_{33}^x}_{1\times16}\ \underbrace{q_{00}^y \cdots q_{30}^y\ q_{01}^y \cdots q_{31}^y\ q_{03}^y \cdots q_{33}^y}_{1\times16} \\ \underbrace{q_{00}^z \cdots q_{30}^z\ q_{01}^z \cdots q_{31}^z\ q_{03}^z \cdots q_{33}^z}_{1\times16} \Big\}^T \tag{6.133}$$

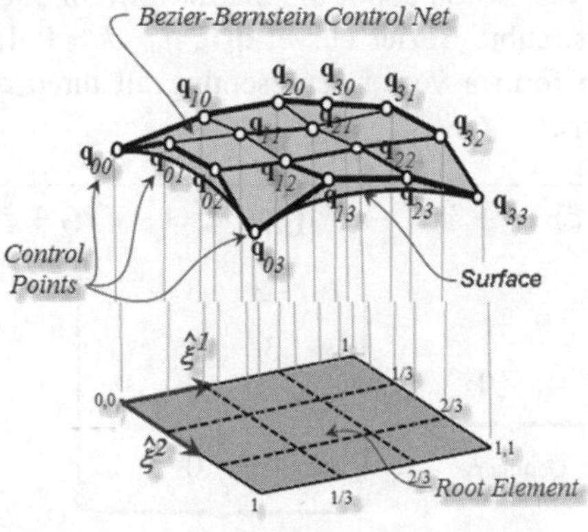

Figure 6.14 Cubic Bezier non-rational surface.

and the Bezier transformation matrix is given by:

$$\mathbf{T}(\xi^1,\xi^2) = \underbrace{\begin{bmatrix} B_0^3B_0^3 & B_1^3B_0^3\dots & B_3^3B_3^3 & 0 & 0 & \dots & \dots & \dots & 0 \\ 0 & 0 & 0 & B_0^3B_0^3 & B_1^3B_0^3\dots & B_3^3B_3^3 & 0 & \dots & 0 \\ 0 & 0 & \dots & \dots & 0 & 0 & B_0^3B_0^3 & B_1^3B_0^3\dots & B_3^3B_3^3 \end{bmatrix}}_{3\times 48} \tag{6.134}$$

The matrix form given by equation (6.132) is convenient for approximations in finite element analysis; however, for conceptual understanding and interpolations in mesh generation, the following matrix form is more intuitive for each coordinate direction, $D = x, y, z$:

$$\underbrace{\mathbf{S}^D(\xi^1,\xi^2)}_{1\times1} = \underbrace{\left\langle B_0^3 \quad B_1^3 \quad B_2^3 \quad B_3^3 \right\rangle}_{1\times4} \underbrace{\begin{bmatrix} \mathbf{q}_{00}^D & \mathbf{q}_{01}^D & \mathbf{q}_{02}^D & \mathbf{q}_{03}^D \\ \mathbf{q}_{10}^D & \mathbf{q}_{11}^D & \mathbf{q}_{12}^D & \mathbf{q}_{13}^D \\ \mathbf{q}_{20}^D & \mathbf{q}_{21}^D & \mathbf{q}_{22}^D & \mathbf{q}_{23}^D \\ \mathbf{q}_{30}^D & \mathbf{q}_{31}^D & \mathbf{q}_{32}^D & \mathbf{q}_{33}^D \end{bmatrix}}_{4\times4} \underbrace{\begin{Bmatrix} B_0^3 \\ B_1^3 \\ B_2^3 \\ B_3^3 \end{Bmatrix}}_{4\times1} \tag{6.135}$$

From equation (6.135), we can algorithmically think of a point on the surface as being generated by the already-studied two-stage Bezier curve generation processes:

Step 1: Compute, by cubic interpolation, a point for a fixed ξ^2 on each of the four curves along the ξ^2-direction as:

$$\begin{aligned} \mathbf{q}_0^D(\xi^2) &= B_0^3(\xi^2)\,\mathbf{q}_{00}^D + B_1^3(\xi^2)\,\mathbf{q}_{01}^D + B_2^3(\xi^2)\,\mathbf{q}_{02}^D + B_3^3(\xi^2)\,\mathbf{q}_{03}^D \\ \mathbf{q}_1^D(\xi^2) &= B_0^3(\xi^2)\,\mathbf{q}_{10}^D + B_1^3(\xi^2)\,\mathbf{q}_{11}^D + B_2^3(\xi^2)\,\mathbf{q}_{12}^D + B_3^3(\xi^2)\,\mathbf{q}_{13}^D \\ \mathbf{q}_2^D(\xi^2) &= B_0^3(\xi^2)\,\mathbf{q}_{20}^D + B_1^3(\xi^2)\,\mathbf{q}_{21}^D + B_2^3(\xi^2)\,\mathbf{q}_{22}^D + B_3^3(\xi^2)\,\mathbf{q}_{23}^D \\ \mathbf{q}_3^D(\xi^2) &= B_0^3(\xi^2)\,\mathbf{q}_{30}^D + B_1^3(\xi^2)\,\mathbf{q}_{31}^D + B_2^3(\xi^2)\,\mathbf{q}_{32}^D + B_3^3(\xi^2)\,\mathbf{q}_{33}^D \end{aligned} \tag{6.136}$$

Now, as indicated in our discussion about the matrix form of curve followed by appropriate remark, for $n = 3$, that is, cubic Bezier curve, $\mathbf{q}_k(\xi^2)$, $\quad k = 0, 1, 2, 3$, as above in equation (6.136) (but collected to form a vector representing all three physical directions) can be rewritten in matrix form as:

$$\begin{aligned} \mathbf{C}_3(\xi) &= (1-\xi)^3\mathbf{q}_0 + 3\xi(1-\xi)^2\mathbf{q}_1 + 3\xi^2(1-\xi)\mathbf{q}_2 + \xi^3\mathbf{q}_3 \\ &= \underbrace{[\mathbf{q}_0 \quad \mathbf{q}_1 \quad \mathbf{q}_2 \quad \mathbf{q}_3]}_{\substack{\text{Geometry} \\ \text{matrix}}} \underbrace{\begin{bmatrix} 1 & -3 & 3 & 1 \\ 0 & 3 & -6 & 3 \\ 0 & 0 & 3 & -3 \\ 0 & 0 & 0 & 1 \end{bmatrix}}_{\substack{\text{Cubic Bezier} \\ \text{coefficient matrix}}} \underbrace{\begin{Bmatrix} 1 \\ \xi \\ \xi^2 \\ \xi^3 \end{Bmatrix}}_{\substack{\text{Monomial basis functions} \\ \text{of degree 3}}} \end{aligned} \tag{6.137}$$

where $[\,\mathbf{q}_0 \quad \mathbf{q}_1 \quad \mathbf{q}_2 \quad \mathbf{q}_3\,]$ is appropriately known as the ***geometry matrix*** as it determines the geometry of a cubic curve, and $\begin{bmatrix} 1 & -3 & 3 & 1 \\ 0 & 3 & -6 & 3 \\ 0 & 0 & 3 & -3 \\ 0 & 0 & 0 & 1 \end{bmatrix}$, as the ***cubic Bezier coefficient matrix***.

Step 2: Compute, by cubic interpolation, the point on the surface by evaluating the curve along the ξ^1-direction with the computed points, $\mathbf{q}_0^D(\xi^2)$, $\mathbf{q}_1^D(\xi^2)$, $\mathbf{q}_2^D(\xi^2)$ and $\mathbf{q}_3^D(\xi^2)$, on the curve of ***Step 1*** as the control points:

$$\underbrace{\mathbf{S}^D(\xi^1,\xi^2)}_{1\times 1} = B_0^3(\xi^1)\,\mathbf{q}_0^D(\xi^2) + B_1^3(\xi^1)\,\mathbf{q}_1^D(\xi^2) + B_2^3(\xi^1)\,\mathbf{q}_2^D(\xi^2) + B_3^3(\xi^1)\,\mathbf{q}_3^D(\xi^2) \tag{6.138}$$

- The surface may be thought of as the range of a cubic map of a root element or domain as a unit square defined by $\xi^D \in [0,1]$ for $D = 1, 2$.
- The surface is cubic in both directions, that is, bi-cubic.
- As before, the curves on the surface generated by the map of lines along each coordinate direction are known as ***isoparametric curves***.
- The boundary curves on the surface, being cubic, interpolate only corner input control points.
- As will be seen shortly, the coordinate directional cubic interpolations indicated above by equations (6.136) and (6.138) can be algorithmically generalized for surfaces with different degrees in each coordinate direction.

Based on the above examples, an $m \times n$ Cartesian product Bezier surface, $\mathbf{S}(\xi^1,\xi^2)$, $\xi^1,\xi^2 \in [0,1]$, as defined by equation (6.124), can be rewritten for algorithmic development as follows:

Step 1: Compute a point at a fixed ξ^2 on each of the m Bezier curves of degree n along ξ^2 direction, according to:

$$\mathbf{q}_i(\xi^2) = \sum_{j=0}^{n} B_j^n(\xi^2)\,\mathbf{q}_{ij}, \quad i = 0, \ldots, m \tag{6.139}$$

Step 2: Use the curve points of ***Step 1*** as control points of a Bezier curve of degree m in the ξ^1 direction to obtain the points on the surface according to:

$$\mathbf{S}(\xi^1,\xi^2) = \sum_{i=0}^{m} B_i^m(\xi^1)\,\mathbf{q}_i(\xi^2) \tag{6.140}$$

- Because of the above algorithmic presentation of Bezier surface points generated by bidirectional Bezier curves, all the properties of ***Bezier curve***s such as affine invariance, convex hull, linear precision, and so on, apply to Bezier tensor product surfaces.

6.4.2 Derivatives of Non-rational Cartesian Product Bezier Surfaces

We need to consider various ways of calculating the derivatives (i.e. tangents, twist, curvatures, etc.) of a Bezier surface essential for both mesh generation and the finite element method.

As we have seen, Bezier Cartesian product surfaces can be algorithmically generated by Bezier curves in two coordinate directions. So discussions on derivatives of curves generally apply to those for surfaces, where we refer to the section on the theory and computation of curves. However, we did not deal with the question of cross derivatives, that is, ***twist***, which is only germane to surfaces. Furthermore, as we shall see, it is clearly more economical to ***subdivide*** (partition or split) a surface at the point of interest and evaluate the end point derivatives by either of the two methods for Bezier curves to the resulting coordinate curves at the subdivided surface. Finally, for most practical applications, we usually need the first derivatives (related to tangents), the second derivatives (related to curvatures) and the cross derivative (related to twist) of a Cartesian product surface. Thus, we will now look at two different ways of evaluating the derivatives of and subdivision algorithm for a surface.

6.4.2.1 Derivatives: Bernstein Form

So let's look at the derivative formulas for a Bezier surface in terms of the Bernstein basis functions. If we are interested only in the derivatives as in the finite element phase of element stiffness matrix generation with as yet unknown displacement control vectors, then these forms seem appropriate for use. We apply directly the derivative formulas of the ***Bernstein polynomials*** to the definition of a Bezier curve of a given order and obtain the derivatives of a Bezier curve in a recurrence form involving lower-order Bezier curves.

6.4.2.1.1 First and Second Derivatives

First, let us recall the following:

- The first derivative of a Bezier curve is also a Bezier curve. Its degree is one less than that of the original one. Its control points are obtained by simply taking the difference of the original control points in a forward-difference manner.
- The derivative expression has tremendous ramifications in computational efficiency: we never have to take an actual derivative of the curve; we simply use the forward-difference of the control points, and lower the degree of the Bernstein polynomial by one.

In order to write the derivative expression in a compact form, we can recall our definition of the first forward-difference operator and rewrite it to accommodate the two coordinate directions of a surface, $\mathbf{S}(\xi^1, \xi^2), \quad \xi^1, \xi^2 \in [0, 1]$.

For later use in coordinate directional partial derivatives, $\frac{\partial \mathbf{S}}{\partial \xi^i}, \frac{\partial^2 \mathbf{S}}{\partial \xi^i \partial \xi^i}, \quad i = 1, 2$, we define the following forward-difference operators:

$$\boxed{\begin{aligned} \Delta^{10}\mathbf{q}_{ij} &\equiv \mathbf{q}_{(i+1)j} - \mathbf{q}_{ij} \\ \Delta^{20}\mathbf{q}_{ij} &\equiv \Delta^{10}\mathbf{q}_{(i+1)j} - \Delta^{10}\mathbf{q}_{ij} \\ \Delta^{01}\mathbf{q}_{ij} &\equiv \mathbf{q}_{i(j+1)} - \mathbf{q}_{ij} \\ \Delta^{02}\mathbf{q}_{ij} &\equiv \Delta^{01}\mathbf{q}_{i(j+1)} - \Delta^{01}\mathbf{q}_{ij} \end{aligned}} \tag{6.141}$$

and for later use in mixed partial derivatives, $\frac{\partial^2 \mathbf{S}}{\partial \xi^1 \partial \xi^2}$, we define the following forward-difference operators:

$$\boxed{\begin{aligned}\Delta^{11}\mathbf{q}_{ij} &\equiv \Delta^{01}\mathbf{q}_{(i+1)j} - \Delta^{01}\mathbf{q}_{ij}\\ &= (\mathbf{q}_{(i+1)(j+1)} - \mathbf{q}_{(i+1)j}) - (\mathbf{q}_{i(j+1)} - \mathbf{q}_{ij})\end{aligned}} \tag{6.142}$$

or,

$$\begin{gathered}\boxed{\begin{aligned}\Delta^{11}\mathbf{q}_{ij} &\equiv \Delta^{10}\mathbf{q}_{i(j+1)} - \Delta^{10}\mathbf{q}_{ij}\\ &= (\mathbf{q}_{(i+1)(j+1)} - \mathbf{q}_{i(j+1)}) - (\mathbf{q}_{(i+1)j} - \mathbf{q}_{ij})\\ &= \text{constant}\end{aligned}}\\ \Rightarrow\\ \mathbf{q}_{(i+1)(j+1)} = (\mathbf{q}_{(i+1)j} - \mathbf{q}_{ij}) + \mathbf{q}_{i(j+1)}) + \Delta^{11}\mathbf{q}_{ij}\end{gathered} \tag{6.143}$$

Now, following the derivations for Bezier curves and the above defined forward-difference operators, we can write the various important derivatives of a Bezier surface as follows:

6.4.2.1.2 *Tangents or First Partial Derivative Along Coordinate Directions*

$$\boxed{\frac{\partial}{\partial \xi^1}\mathbf{S}(\xi^1,\xi^2) = \sum_{j=0}^{n}\underbrace{\left(\sum_{i=0}^{m}\left(\frac{\partial B_i^m(\xi^1)}{\partial \xi^1}\,\mathbf{q}_{ij}\right)\right.}_{\substack{\text{Tangents along } \xi^1 \text{ are the}\\ \text{control pts for } \xi^2\text{-curves}}} B_j^n(\xi^2) = m\sum_{j=0}^{n}\underbrace{\left(\sum_{i=0}^{m-1}\left(B_i^{m-1}(\xi^1)\,\Delta^{10}\mathbf{q}_{ij}\right)\right.}_{\substack{\text{Tangents along } \xi^1 \text{ are the}\\ \text{control pts for } \xi^2\text{-curves}}} B_j^n(\xi^2)\Bigg)} \tag{6.144}$$

and

$$\boxed{\frac{\partial}{\partial \xi^2}\mathbf{S}(\xi^1,\xi^2) = \sum_{i=0}^{m}\underbrace{\left(\sum_{j=0}^{n}\left(\frac{\partial B_j^n(\xi^2)}{\partial \xi^2}\,\mathbf{q}_{ij}\right)\right.}_{\substack{\text{Tangents along } \xi^2 \text{ are the}\\ \text{control pts for } \xi^1\text{-curves}}} B_i^m(\xi^1) = n\sum_{i=0}^{m}\underbrace{\left(\sum_{j=0}^{n-1}\left(B_j^{n-1}(\xi^2)\,\Delta^{01}\mathbf{q}_{ij}\right)\right.}_{\substack{\text{Tangents along } \xi^2 \text{ are the}\\ \text{control pts for } \xi^1\text{ -curves}}} B_i^m(\xi^1)\Bigg)} \tag{6.145}$$

6.4.2.1.3 *Curvatures or Second Partial Derivative Along Coordinate Directions*

$$\boxed{\begin{aligned}\frac{\partial^2}{\partial \xi^1 \partial \xi^1}\mathbf{S}(\xi^1,\xi^2) &= \sum_{j=0}^{n}\underbrace{\left(\sum_{i=0}^{m}\left(\frac{\partial^2 B_i^m(\xi^1)}{\partial \xi^1 \partial \xi^1}\,\mathbf{q}_{ij}\right)\right)}\, B_j^n(\xi^2)\\ &= m(m-1)\sum_{j=0}^{n}\underbrace{\left(\sum_{i=0}^{m-2}(B_i^{m-2}(\xi^1)\,\Delta^{20}\mathbf{q}_{ij})\right.}\, B_j^n(\xi^2)\Bigg)\end{aligned}} \tag{6.146}$$

and

$$\frac{\partial^2}{\partial\xi^2\partial\xi^2}\mathbf{S}(\xi^1,\xi^2) = \sum_{i=0}^{m}\underbrace{\left(\sum_{j=0}^{n}\left(\frac{\partial^2 B_j^n(\xi^2)}{\partial\xi^2\partial\xi^2}\mathbf{q}_{ij}\right)B_i^m(\xi^1)\right.}$$
$$= n(n-1)\sum_{i=0}^{m}\underbrace{\left(\sum_{j=0}^{n-2}(B_j^{n-2}(\xi^2)\,\Delta^{02}\mathbf{q}_{ij})\,B_i^m(\xi^1)\right.} \tag{6.147}$$

6.4.2.1.4 *Twist or Second Mixed Partial Derivative*

$$\frac{\partial^2}{\partial\xi^1\partial\xi^2}\mathbf{S}(\xi^1,\xi^2) = \sum_{j=0}^{n}\underbrace{\sum_{i=0}^{m}\frac{\partial B_i^m(\xi^1)}{\partial\xi^1}\frac{\partial B_j^n(\xi^2)}{\partial\xi^1}\mathbf{q}_{ij}}$$
$$= mn\sum_{j=0}^{n-1}\underbrace{\left(\sum_{i=0}^{m-1}(B_i^{m-1}(\xi^1)\,\Delta^{11}\mathbf{q}_{ij})\,B_j^{n-1}(\xi^2)\right.} \tag{6.148}$$

6.4.2.1.5 *End Point First Derivatives (Tangents)*

In practice, the end point derivatives are of great importance because derivatives of a Bezier surface at any $\xi^1, \xi^2 \in [0,1]$ can be reduced to finding the corresponding end point derivative by subdividing the isoparametric curves at (ξ^1, ξ^2) and retaining the left or right parts and considering the subdivided patch with a new control net.

Algorithm: Given two surface points: (ξ_1^1, ξ_1^2) and (ξ_2^1, ξ_2^2), compute the derivatives along the isoparametric curves (see Figure 6.15):

- Interpret all rows of the control net as control points of Bezier curves;
- Subdivide each of these curves at $\xi^1 = \xi_1^1$, and retain the right part of the subdivided curve with the new set of control points;

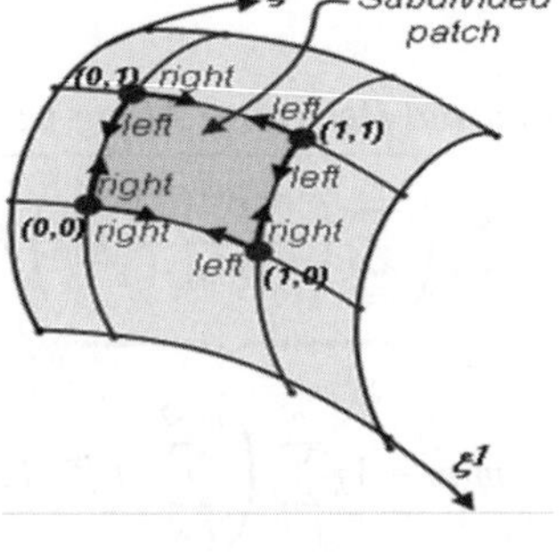

Figure 6.15 End derivatives by subdivision.

- Subdivide each of these curves at $\xi^1 = \xi^1_2$, and retain the left part of the subdivided curve with the new set of control points;
- Interpret all the columns of the control net as control points of Bezier curves;
- Subdivide each of these curves at $\xi^2 = \xi^2_1$, and retain the right part of the subdivided curve with the new set of control points;
- Subdivide each of these curves at $\xi^2 = \xi^2_2$, and retain the left part of the subdivided curve with the new set of control points;

Noting the value of $B^n_0(0) = B^n_n(1) = 1, \quad \forall n$, and $B^n_i(0) = B^n_i(1) = 0, \quad \forall i \neq 0, n$, we can easily write the expressions for the end point derivatives with the subdivided patch, $\overline{\mathbf{S}}$ with new control net, $\{\bar{\mathbf{q}}_{ij}\}$. Denoting the end point tangents as $\mathbf{Tan}^{00}, \mathbf{Tan}^{10}, \mathbf{Tan}^{01}, \mathbf{Tan}^{11}$, from equations (6.144) and (6.145), we get:

$$\boxed{\begin{aligned}
\mathbf{Tan}^{\mathbf{00}}_1 &= \frac{\partial}{\partial \xi^1}\overline{\mathbf{S}}(0,0) = m\Delta^{10}\bar{\mathbf{q}}_{00} = m(\bar{\mathbf{q}}_{10} - \bar{\mathbf{q}}_{00})\\
\mathbf{Tan}^{\mathbf{00}}_2 &= \frac{\partial}{\partial \xi^2}\overline{\mathbf{S}}(0,0) = n\Delta^{01}\bar{\mathbf{q}}_{00} = n(\bar{\mathbf{q}}_{01} - \bar{\mathbf{q}}_{00})\\
\mathbf{Tan}^{\mathbf{10}}_1 &= \frac{\partial}{\partial \xi^1}\overline{\mathbf{S}}(1,0) = m\Delta^{10}\bar{\mathbf{q}}_{(m-1)0} = m(\bar{\mathbf{q}}_{m0} - \bar{\mathbf{q}}_{(m-1)0})\\
\mathbf{Tan}^{\mathbf{10}}_2 &= \frac{\partial}{\partial \xi^2}\overline{\mathbf{S}}(1,0) = n\Delta^{01}\bar{\mathbf{q}}_{m0} = n(\bar{\mathbf{q}}_{m1} - \bar{\mathbf{q}}_{m0})\\
\mathbf{Tan}^{01}_1 &= \frac{\partial}{\partial \xi^1}\overline{\mathbf{S}}(0,1) = m\Delta^{10}\bar{\mathbf{q}}_{0n} = m(\bar{\mathbf{q}}_{1n} - \bar{\mathbf{q}}_{0n})\\
\mathbf{Tan}^{01}_2 &= \frac{\partial}{\partial \xi^2}\overline{\mathbf{S}}(0,1) = n\Delta^{01}\bar{\mathbf{q}}_{0(n-1)} = n(\bar{\mathbf{q}}_{0n} - \bar{\mathbf{q}}_{0(n-1)})\\
\mathbf{Tan}^{11}_1 &= \frac{\partial}{\partial \xi^1}\overline{\mathbf{S}}(1,1) = m\Delta^{10}\bar{\mathbf{q}}_{(m-1)n} = m(\bar{\mathbf{q}}_{mn} - \bar{\mathbf{q}}_{(m-1)n})\\
\mathbf{Tan}^{11}_2 &= \frac{\partial}{\partial \xi^2}\overline{\mathbf{S}}(1,1) = n\Delta^{01}\bar{\mathbf{q}}_{m(n-1)} = n(\bar{\mathbf{q}}_{mn} - \bar{\mathbf{q}}_{m(n-1)})
\end{aligned}} \tag{6.149}$$

Similarly, denoting the end point twists as $\mathbf{Twist}^{00}, \mathbf{Twist}^{10}, \mathbf{Twist}^{01}, \mathbf{Twist}^{11}$, from equation (6.148), we get:

$$\boxed{\begin{aligned}
\mathbf{Twist}^{\mathbf{00}} &= \frac{\partial^2}{\partial \xi^1 \partial \xi^2}\overline{\mathbf{S}}(0,0) = mn\Delta^{11}\bar{\mathbf{q}}_{00} = mn[(\bar{\mathbf{q}}_{11} - \bar{\mathbf{q}}_{10}) - (\bar{\mathbf{q}}_{01} - \bar{\mathbf{q}}_{00})]\\
\mathbf{Twist}^{\mathbf{10}} &= \frac{\partial^2}{\partial \xi^1 \partial \xi^2}\overline{\mathbf{S}}(1,0) = mn\Delta^{11}\bar{\mathbf{q}}_{(m-1)0} = mn[(\bar{\mathbf{q}}_{m1} - \bar{\mathbf{q}}_{m0}) - (\bar{\mathbf{q}}_{(m-1)1} - \bar{\mathbf{q}}_{(m-1)0})]\\
\mathbf{Twist}^{01} &= \frac{\partial^2}{\partial \xi^1 \partial \xi^2}\overline{\mathbf{S}}(0,1) = mn\Delta^{11}\bar{\mathbf{q}}_{0(n-1)} = mn[(\bar{\mathbf{q}}_{1n} - \bar{\mathbf{q}}_{1(n-1)}) - (\bar{\mathbf{q}}_{0n} - \bar{\mathbf{q}}_{0(n-1)})]\\
\mathbf{Twist}^{11} &= \frac{\partial^2}{\partial \xi^1 \partial \xi^2}\overline{\mathbf{S}}(1,1) = mn\Delta^{11}\bar{\mathbf{q}}_{(m-1)(n-1)} = mn[(\bar{\mathbf{q}}_{mn} - \bar{\mathbf{q}}_{m(n-1)}) - (\bar{\mathbf{q}}_{(m-1)n} - \bar{\mathbf{q}}_{(m-1)(n-1)})]
\end{aligned}} \tag{6.150}$$

Note:

- The first derivatives at end points involve one more control point than the end control points themselves of the subdivided patch.
- The tangents of the surface at end points point along the direction of the end sides of the subdivided Bezier polygonal net.

6.4.2.2 Surface Normal Vectors

In the process of illuminating patches in the rendering phase of mesh generation, and computational phase of the finite element analysis of, say, shells, we will need the normals at any point on the surface. Let us recall from the theory of surfaces, the normal vector, $\mathbf{N}(\xi^1, \xi^2)$ at any point, (ξ^1, ξ^2), on a surface is given by:

$$\boxed{\mathbf{N}(\xi^1, \xi^2) = \frac{\dfrac{\partial \mathbf{S}(\xi^1, \xi^2)}{\partial \xi^1} \times \dfrac{\partial \mathbf{S}(\xi^1, \xi^2)}{\partial \xi^2}}{\left\| \dfrac{\partial \mathbf{S}(\xi^1, \xi^2)}{\partial \xi^1} \times \dfrac{\partial \mathbf{S}(\xi^1, \xi^2)}{\partial \xi^2} \right\|}} \tag{6.151}$$

where the surface coordinate tangents, $\frac{\partial \mathbf{S}(\xi^1,\xi^2)}{\partial \xi^1}$ and $\frac{\partial \mathbf{S}(\xi^1,\xi^2)}{\partial \xi^2}$ are obtained from equation (6.149) of a subdivided patch at point, (ξ^1, ξ^2).

6.4.2.2.1 Normal Vector for a Degenerate Patch

If a triangular patch (which we will describe shortly in detail) is described by a ***degenerate Cartesian product patch*** where two end control points coalesce, the simple cross product of the surface tangents may be undefined. In such cases, a limiting process to determine the cross product is necessary which may involve the following mixed partial.

Let us consider a triangular surface modeled as a degenerate Cartesian product patch with $m = 3$ and $n = 2$, as shown in Figure 6.16. For the patch, $\mathbf{q}_{01} - \mathbf{q}_{00} = 0$ and $\mathbf{q}_{02} - \mathbf{q}_{01} = 0$ at degeneracy point, $\mathbf{P}$. Thus, following equation (6.149), the surface tangents at $\mathbf{P}$ along ξ^2-isoparametric curve are zero: $\frac{\partial \mathbf{P}(0,0)}{\partial \xi^2} = 2(\mathbf{q}_{01} - \mathbf{q}_{00}) = 0$. $\frac{\partial \mathbf{P}(0,1)}{\partial \xi^2} = 2(\mathbf{q}_{02} - \mathbf{q}_{01}) = 0$. So, we evaluate the tangents at an infinitesimal distance $\Delta\xi^1$ and let $\Delta\xi^1 \to 0$ determine the normal vector in the limiting process as follows.

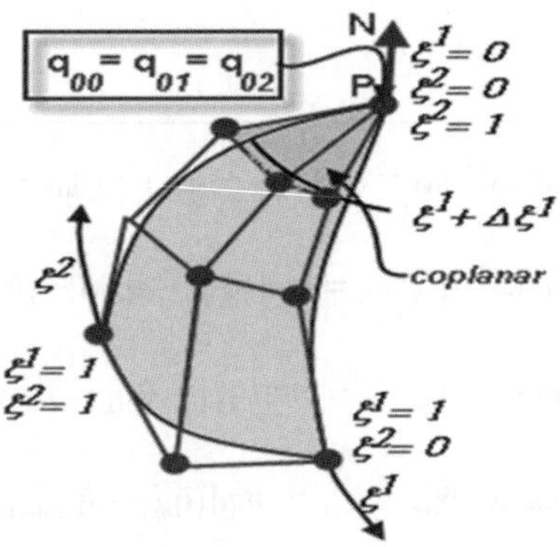

Figure 6.16 Degenerate Cartesian product surface.

Noting the first-order approximation:$\mathbf{P}(\Delta\xi^1, \xi^2) \simeq \mathbf{P}(0, \xi^2) + \Delta\xi^1 \frac{\partial \mathbf{P}(0,\xi^2)}{\partial \xi^1}$, we get:

$$\begin{aligned}
\mathbf{T}_1 &\triangleq \frac{\partial \mathbf{S}(\Delta\xi^1, \xi^2)}{\partial \xi^1} \simeq \frac{\partial \mathbf{S}(0, \xi^2)}{\partial \xi^1} + \Delta\xi^1 \frac{\partial^2 \mathbf{S}(0, \xi^2)}{\partial \xi^1 \partial \xi^1} \\
\mathbf{T}_2 &\triangleq \frac{\partial \mathbf{S}(\Delta\xi^1, \xi^2)}{\partial \xi^2} \simeq \cancel{\frac{\delta \mathbf{S}(0, \xi^2)}{\delta \xi^2}} + \Delta\xi^1 \frac{\partial^2 \mathbf{S}(0, \xi^2)}{\partial \xi^1 \partial \xi^2} = \Delta\xi^1 \frac{\partial^2 \mathbf{S}(0, \xi^2)}{\partial \xi^1 \partial \xi^2}
\end{aligned} \tag{6.152}$$

Thus, as $\Delta\xi^1 \to 0$, for the normal vector, $\mathbf{N}(0, \xi^2)$ given by equation (6.151), we have:

$$\begin{aligned}
\mathbf{N}(0, \xi^2) &= \lim_{\Delta\xi^1 \to 0} \frac{\mathbf{T}_1 \times \mathbf{T}_2}{\|\mathbf{T}_1 \times \mathbf{T}_2\|} \\
&= \lim_{\Delta\xi^1 \to 0} \frac{\cancel{\Delta\xi^1} \left(\frac{\partial \mathbf{S}(0, \xi^2)}{\partial \xi^1} \times \frac{\partial^2 \mathbf{S}(0, \xi^2)}{\partial \xi^1 \partial \xi^2} + \Delta\xi^1 \frac{\partial^2 \mathbf{S}(0, \xi^2)}{\partial \xi^1 \partial \xi^1} \right)}{\cancel{\Delta\xi^1} \left\| \frac{\partial \mathbf{S}(0, \xi^2)}{\partial \xi^1} \times \frac{\partial^2 \mathbf{S}(0, \xi^2)}{\partial \xi^1 \partial \xi^2} + \cancel{\Delta\xi^1} \frac{\partial^2 \mathbf{S}(0, \xi^2)}{\partial \xi^1 \partial \xi^1} \right\|} \\
&= \frac{\frac{\partial \mathbf{S}(0, \xi^2)}{\partial \xi^1} \times \frac{\partial^2 \mathbf{S}(0, \xi^2)}{\partial \xi^1 \partial \xi^2}}{\left\| \frac{\partial \mathbf{S}(0, \xi^2)}{\partial \xi^1} \times \frac{\partial^2 \mathbf{S}(0, \xi^2)}{\partial \xi^1 \partial \xi^2} \right\|}
\end{aligned} \tag{6.153}$$

6.4.2.3 Degree Elevation

Degree elevation is the numerical process of expressing a Bezier surface of a certain degrees, say, $m \times n$, defined by $(m + 1) \times (n + 1)$ control points as a Bezier surface of $(m + 1) \times n$ degrees with $(m + 2) \times (n + 1)$ control points without changing the shape of the original surface. Degree elevation is central to both interactive modeling and finite element analysis. In the modeling phase based on boundary representation, the surfaces are formed by boundary curves of the same degree. Thus, if any of these curves is previously generated with lower degree than the highest degree boundary curve, the lower degree curve needs to be degree elevated, that is, redefined in terms of the highest degree without changing the shape of the curve in any way for the surface definition. In another situation, where, for example, a curve of degree one (i.e. a line segment) is joined with a curve of degree two (i.e. a parabola) to generate a boundary curve of a surface, the line segment needs to be expressed as a special degenerate quadratic curve so that the entire profile curve is quadratic. Again, in the process of generation of patches, say, a ruled surface, the two end curves could be of different degrees, in which case the curve with the lower degree must undergo ***degree elevation***. As in the case of Bezier curves, the degree elevation of Bezier surfaces can be achieved by sequentially degree elevating isoparametric curves along each directions.

Suppose, we are given a Bezier surface of degree $m \times n$, as indicated by equation (6.124), and we want a degree-elevated Bezier surface of degree $(m + 1) \times n$ such that:

$$\boxed{\mathbf{S}(\xi^1, \xi^2) = \sum_{j=0}^{n} B_j^n(\xi^2) \left(\sum_{i=0}^{m+1} B_i^{m+1}(\xi^1)\, \mathbf{q}_{ij}^{10} \right), \quad 0 \leq \xi^1, \xi^2 \leq 1} \tag{6.154}$$

6.4.2.3.1 *Algorithm*

Step 0: Define $\mathbf{q}_{ij}^{10}$ as the control points of the degree-elevated Bezier surface after elevating degree m by 1.

Step 1: Compute $\mathbf{q}_{ij}^{10}$ as:

$$\mathbf{q}_{ij}^{10} = \left(1 - \frac{i}{m+1}\right)\mathbf{q}_{ij} + \frac{i}{m+1}\mathbf{q}_{(i-1)j}, \quad i \in \{0, 1, \dots, m+1\}; \quad j \in \{0, 1, \dots, n\} \tag{6.155}$$

- From equations (6.154) and (6.155), we can easily see that degree elevation is obtained by considering the surface as $(n + 1)$ curves of degree m in the ξ^2-direction and then applying the formula for the Bezier curves.
- Degree elevation in both direction can be effected by first elevating in the ξ^1-direction and then using the new control net for the ξ^2-direction.

6.4.3 *Rational Bezier Surfaces with Bernstein Polynomials*

Based on our discussion of the rational Bezier curves, we define here the rational Bezier Cartesian product surfaces.

Definitions:

- An $m \times n$ rational Cartesian product Bezier surface, $\mathbf{S}(\xi^1, \xi^2)$, is defined as:

$$\mathbf{S}(\xi^1, \xi^2) = \sum_{i=0}^{m}\sum_{j=0}^{n}\left(\frac{w_{ij}B_i^m(\xi^1)\,B_j^n(\xi^2)}{\sum_{i=0}^{m}\sum_{j=0}^{n} w_{ij}B_i^m(\xi^1)\,B_j^n(\xi^2)}\right)\mathbf{q}_{ij}, \quad 0 \le \xi^1, \xi^2 \le 1 \tag{6.156}$$

  where the basis (blending) function sets, $\{B_i^m(\xi^1)\}, \{B_j^n(\xi^2)\}$, are the nth and m^{th} degree Bernstein polynomials, respectively;
- w_{ij} are known as the ***weights*** at the control points, $\mathbf{q}_{ij}$, $i \in [0, m]$, $j \in [0, n]$.
- The coefficients of linear combination or Euclidean points, $\{\mathbf{q}_{ij}\}$, are known as the ***control points*** of the surface.
- The polygonal net containing the control points, $\{\mathbf{q}_{ij}\}$, is known as the ***control polygonal net*** of the surface.
- Rational Bezier surfaces are obtained as projections of ***4D Cartesian product non-rational patches***; rational patches themselves are not generally Cartesian product patches as these cannot be, in general, expressed in $\sum\sum c_{ij}A_i(\xi^1)B_j(\xi^2)$ form.

6.4.4 *Derivatives of Rational Cartesian Product Bezier Surfaces*

There are various ways of calculating the derivatives (i.e. tangents, twist, curvatures, etc.) of a Rational Bezier surface essential for both mesh generation and finite element method. As we have already seen, rational Bezier Cartesian product surfaces can be algorithmically generated by Bezier curves in two coordinate directions. So discussions on derivatives of curves generally apply to those for surfaces. Furthermore, as we shall see, it is more economical to subdivide (partition or split) a surface at the point of interest and evaluate the end point derivatives to the resulting coordinate curves at the subdivided surface. Finally, for most practical applications, we usually need the first derivatives (related to tangents), the second derivatives (related to curvatures) and the cross derivative (related to twist) of a rational Cartesian product surface. So we now present two different ways of evaluating the derivatives of and subdivision algorithm for a surface.

6.4.4.1 Derivatives: Bernstein Form

We turn now to the derivative formulas for a Bezier surface in terms of the Bernstein basis functions. If we are interested only in the derivatives as in the finite element phase of element stiffness matrix generation with as yet unknown displacement control vectors, these forms seem appropriate for use. We apply directly the derivative formulas of the Bernstein polynomials to the definition of a Bezier curve of a given order and obtain the derivatives of a Bezier curve in a recurrence form involving lower-order Bezier curves.

6.4.4.1.1 Tangents or First Partial Derivative Along Coordinate Directions
First, let us introduce the compact notations in equation (6.156):

$$\begin{aligned} w(\xi^1,\xi^2) &\triangleq \sum_{i=0}^{m}\sum_{j=0}^{n} w_{ij} B_i^m(\xi^1)\, B_j^n(\xi^2) \\ \mathbf{b}(\xi^1,\xi^2) &\triangleq \sum_{i=0}^{m}\sum_{j=0}^{n} \mathbf{q}_{ij} B_i^m(\xi^1)\, B_j^n(\xi^2) \\ &\Rightarrow \\ \mathbf{b}(\xi^1,\xi^2) &= \mathbf{S}(\xi^1,\xi^2)\, w(\xi^1,\xi^2), \quad 0 \le \xi^1,\xi^2 \le 1 \text{ (4D preimage)} \end{aligned} \tag{6.157}$$

Now, taking the derivatives of both sides of the third expression of (6.157), and readjusting the terms, we get the expressions for surface coordinate tangents:

$$\begin{aligned} \frac{\partial \mathbf{S}(\xi^1,\xi^2)}{\partial \xi^1} &= \frac{1}{w(\xi^1,\xi^2)} \left\{ \frac{\partial \mathbf{b}(\xi^1,\xi^2)}{\partial \xi^1} - \mathbf{S}(\xi^1,\xi^2)\, \frac{\partial w(\xi^1,\xi^2)}{\partial \xi^1} \right\} \\ \frac{\partial \mathbf{S}(\xi^1,\xi^2)}{\partial \xi^2} &= \frac{1}{w(\xi^1,\xi^2)} \left\{ \frac{\partial \mathbf{b}(\xi^1,\xi^2)}{\partial \xi^2} - \mathbf{S}(\xi^1,\xi^2)\, \frac{\partial w(\xi^1,\xi^2)}{\partial \xi^2} \right\} \end{aligned} \tag{6.158}$$

6.4.4.1.2 Twist or Second Mixed Partial Derivative
Similarly, taking the mixed partials of the both sides of the third expression of equation (6.157), and readjusting the terms, we get the expression for twist:

$$\frac{\partial^2 \mathbf{S}}{\partial \xi^1 \partial \xi^2} = \frac{1}{w} \left\{ \frac{\partial^2 \mathbf{b}}{\partial \xi^1 \partial \xi^2} - \frac{\partial \mathbf{S}}{\partial \xi^1}\, \frac{\partial w}{\partial \xi^2} - \frac{\partial \mathbf{S}}{\partial \xi^2}\, \frac{\partial w}{\partial \xi^1} - \mathbf{S}\, \frac{\partial^2 w}{\partial \xi^1 \partial \xi^2} \right\} \tag{6.159}$$

6.4.4.1.3 End Point First Derivatives (Tangents)

In practice, the end point derivatives are of great importance because derivatives of a Bezier surface at any $\xi^1, \xi^2 \in [0,1]$ can be reduced to finding the corresponding end point derivative by subdividing the isoparametric curves at (ξ^1, ξ^2) and retaining left or right parts and considering the subdivided patch, as in the non-rational algorithm presented before, with new control net and weights.

As before, noting the value of $B_0^n(0) = B_n^n(1) = 1, \quad \forall n$, and $B_i^n(0) = B_i^n(1) = 0, \quad \forall i \neq 0, n$, we can easily write the expressions for the end point derivatives with the subdivided patch, $\overline{\mathbf{S}}$ with new control net, $\{\bar{\mathbf{q}}_{ij}\}$, new weights, $\{\bar{w}_{ij}\}$. Denoting the end point tangents as $\mathbf{Tan}^{00}, \mathbf{Tan}^{10}, \mathbf{Tan}^{01}, \mathbf{Tan}^{11}$, from equation (6.158), we get:

$$\boxed{\begin{aligned}
\mathbf{Tan}_1^{00} &= \frac{\partial}{\partial \xi^1}\overline{\mathbf{S}}(0,0) = m\left(\frac{\bar{w}_{10}}{\bar{w}_{00}}\right)\Delta^{10}\bar{\mathbf{q}}_{00} = m\left(\frac{\bar{w}_{10}}{\bar{w}_{00}}\right)(\bar{\mathbf{q}}_{10} - \bar{\mathbf{q}}_{00})\\
\mathbf{Tan}_2^{00} &= \frac{\partial}{\partial \xi^2}\overline{\mathbf{S}}(0,0) = n\left(\frac{\bar{w}_{01}}{\bar{w}_{00}}\right)\Delta^{01}\bar{\mathbf{q}}_{00} = n\left(\frac{\bar{w}_{01}}{\bar{w}_{00}}\right)(\bar{\mathbf{q}}_{01} - \bar{\mathbf{q}}_{00})\\
\mathbf{Tan}_1^{10} &= \frac{\partial}{\partial \xi^1}\overline{\mathbf{S}}(1,0) = m\left(\frac{\bar{w}_{(m-1)0}}{\bar{w}_{m0}}\right)\Delta^{10}\bar{\mathbf{q}}_{(m-1)0} = m\left(\frac{\bar{w}_{(m-1)0}}{\bar{w}_{m0}}\right)(\bar{\mathbf{q}}_{m0} - \bar{\mathbf{q}}_{(m-1)0})\\
\mathbf{Tan}_2^{10} &= \frac{\partial}{\partial \xi^2}\overline{\mathbf{S}}(1,0) = n\left(\frac{\bar{w}_{m1}}{\bar{w}_{m0}}\right)\Delta^{01}\bar{\mathbf{q}}_{m0} = n\left(\frac{\bar{w}_{m1}}{\bar{w}_{m0}}\right)(\bar{\mathbf{q}}_{m1} - \bar{\mathbf{q}}_{m0})\\
\mathbf{Tan}_1^{01} &= \frac{\partial}{\partial \xi^1}\overline{\mathbf{S}}(0,1) = m\left(\frac{\bar{w}_{1n}}{\bar{w}_{0n}}\right)\Delta^{10}\bar{\mathbf{q}}_{0n} = m\left(\frac{\bar{w}_{1n}}{\bar{w}_{0n}}\right)(\bar{\mathbf{q}}_{1n} - \bar{\mathbf{q}}_{0n})\\
\mathbf{Tan}_2^{01} &= \frac{\partial}{\partial \xi^2}\overline{\mathbf{S}}(0,1) = n\left(\frac{\bar{w}_{0(n-1)}}{\bar{w}_{0n}}\right)\Delta^{01}\bar{\mathbf{q}}_{0(n-1)} = n\left(\frac{\bar{w}_{0(n-1)}}{\bar{w}_{0n}}\right)(\bar{\mathbf{q}}_{0n} - \bar{\mathbf{q}}_{0(n-1)})\\
\mathbf{Tan}_1^{11} &= \frac{\partial}{\partial \xi^1}\overline{\mathbf{S}}(1,1) = m\left(\frac{\bar{w}_{(m-1)n}}{\bar{w}_{mn}}\right)\Delta^{10}\bar{\mathbf{q}}_{(m-1)n} = m\left(\frac{\bar{w}_{(m-1)n}}{\bar{w}_{mn}}\right)(\bar{\mathbf{q}}_{mn} - \bar{\mathbf{q}}_{(m-1)n})\\
\mathbf{Tan}_2^{11} &= \frac{\partial}{\partial \xi^2}\overline{\mathbf{S}}(1,1) = n\left(\frac{\bar{w}_{m(n-1)}}{\bar{w}_{mn}}\right)\Delta^{01}\bar{\mathbf{q}}_{m(n-1)} = n\left(\frac{\bar{w}_{m(n-1)}}{\bar{w}_{mn}}\right)(\bar{\mathbf{q}}_{mn} - \bar{\mathbf{q}}_{m(n-1)})
\end{aligned}} \quad (6.160)$$

6.4.4.1.4 End Point Mixed Derivatives (Twists)

To calculate twist at the four corners of the subdivided patch with new control net and weights, we first compute $\frac{\partial^2 \mathbf{b}}{\partial \xi^1 \partial \xi^2}$ at the four corners and then apply the expression in equation (6.159) with the aid of expressions for tangents from equation (6.160), and the derivatives of the new weights of the subdivided patch as given below:

$$\boxed{\begin{aligned}
\frac{\partial \bar{w}(0,0)}{\partial \xi^1} &= m\Delta^{10}\bar{w}_{00} = m(\bar{w}_{10} - \bar{w}_{00})\\
\frac{\partial \bar{w}(0,0)}{\partial \xi^2} &= n\Delta^{01}\bar{w}_{00} = n(\bar{w}_{01} - \bar{w}_{00})\\
\frac{\partial^2 \bar{w}(0,0)}{\partial \xi^1 \partial \xi^2} &= mn[(\bar{w}_{11} - \bar{w}_{10}) - (\bar{w}_{01} - \bar{w}_{00})]
\end{aligned}} \quad (6.161)$$

We, then, get the mixed partial derivatives of the 4D pre-image at the four corners of the subdivided patch as:

$$\begin{aligned}
\frac{\partial^2 \mathbf{b}(0,0)}{\partial\xi^1\partial\xi^2} &= mn\Delta^{11}\bar{\mathbf{b}}_{00} = mn[(\bar{w}_{11}\bar{\mathbf{q}}_{11} - \bar{w}_{10}\bar{\mathbf{q}}_{10}) - (\bar{w}_{01}\bar{\mathbf{q}}_{01} - \bar{\mathbf{q}}_{00})] \\
\frac{\partial^2 \mathbf{b}(1,0)}{\partial\xi^1\partial\xi^2} &= mn\Delta^{11}\bar{\mathbf{b}}_{(m-1)0} = mn[(\bar{w}_{m1}\bar{\mathbf{q}}_{m1} - \bar{w}_{m0}\bar{\mathbf{q}}_{m0}) - (\bar{w}_{(m-1)1}\bar{\mathbf{q}}_{(m-1)1} - \bar{w}_{(m-1)0}\bar{\mathbf{q}}_{(m-1)0})] \\
\frac{\partial^2 \mathbf{b}(0,1)}{\partial\xi^1\partial\xi^2} &= mn\Delta^{11}\bar{\mathbf{b}}_{0(n-1)} = mn[(\bar{w}_{1n}\bar{\mathbf{q}}_{1n} - \bar{w}_{1(n-1)}\bar{\mathbf{q}}_{1(n-1)}) - (\bar{w}_{0n}\bar{\mathbf{q}}_{0n} - \bar{w}_{0(n-1)}\bar{\mathbf{q}}_{0(n-1)})] \\
\frac{\partial^2 \mathbf{b}(1,1)}{\partial\xi^1\partial\xi^2} &= mn\Delta^{10}\bar{\mathbf{b}}_{(m-1)(n-1)} = mn\Delta^{11}\bar{w}_{(m-1)(n-1)}\bar{\mathbf{q}}_{(m-1)(n-1)} \\
&= mn[(\bar{w}_{mn}\bar{\mathbf{q}}_{mn} - \bar{w}_{m(n-1)}\bar{\mathbf{q}}_{m(n-1)}) - (\bar{w}_{(m-1)n}\bar{\mathbf{q}}_{(m-1)n} - \bar{w}_{(m-1)(n-1)}\bar{\mathbf{q}}_{(m-1)(n-1)})]
\end{aligned} \tag{6.162}$$

Now, denoting the end point twists as $\mathbf{Twist}^{00}, \mathbf{Twist}^{10}, \mathbf{Twist}^{01}, \mathbf{Twist}^{11}$, we get these from equations (6.159), (6.161) and (6.162), and note the following:

- The first derivatives at end points involve one more control point and weight than the end control points and weights themselves of the subdivided patch.
- The tangents of the surface at end points point along the direction of the end sides of the subdivided rational Bezier polygonal net.

Example: We will be interested in describing spheres as rational Bezier surfaces (see Figure 6.17).

A sphere may be thought of as a surface of revolution: a semicircle, a planar rational Bezier curve, known as the ***generatrix***, is rotated around an axis of revolution belonging to the same plane as the curve, through 360°, generating a rational Bezier surface; details on how to generate surfaces of revolution will be presented shortly.

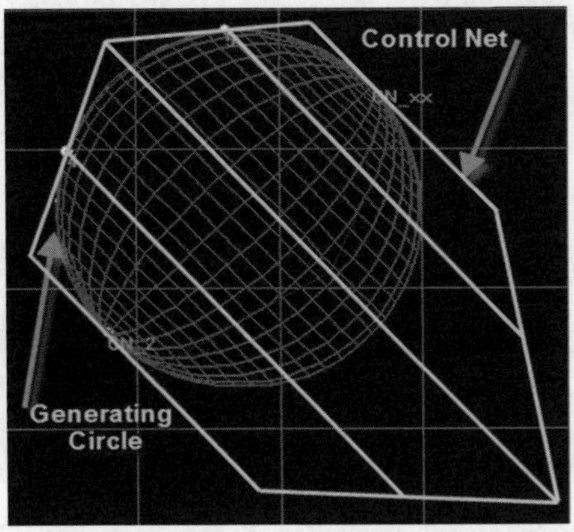

Figure 6.17 Exact sphere as rational Bezier surface. (See colour plate section)

6.5 Control Net Generation: Cartesian Product Surfaces

So far, we have introduced expressions for the general class of Cartesian product non-rational and rational Bezier and B-spline surfaces as in equations (6.124) and (6.156); we can compute the coordinates of any point on these surface only if the corresponding control net is known. Here, we present various ways to generate the control net of a surface; all of these methods generate surfaces starting from one or more boundary curves.

6.5.1 *Patch by Extrusion of One Boundary Curve*

This is the simplest of the surfaces (see Figure 6.18) generated from one curve: also known as ***ruled surfaces***.

Problem: Given a curve with control points, $\{\mathbf{q}_{i0}\}, \quad i = 0, 1, \ldots, m$, and a unit extrusion vector, $\mathbf{v}$, and an extrusion length, λ, generate the patch control net.

6.5.1.1.1 Algorithm EPC1

Step 0: By curve construction methodology, identify the control points, $\{\mathbf{q}_{i0}\}, \quad i = 0, 1, \ldots, m$.

Step 1: Extrude the curve control points to generate the necessary control points such that $\mathbf{q}_{i1} = \mathbf{q}_{i0} + \lambda \mathbf{v}, \quad i = 0, 1, \ldots, m$.

6.5.2 *Patch by Lofting of Two Boundary Curves*

Lofted surfaces are generated from two curves (see Figure 6.19) acting as two opposite boundary curves of the surface: also known as ***ruled surfaces***.

Problem: Given two curves with control points, $\{\mathbf{q}_{i0}\}, \{\mathbf{q}_{i1}\}, \quad i = 0, 1, \ldots, m$, generate the patch control net.

6.5.2.1.1 Algorithm LPC2

Step 0: By the curve construction methodology, identify the control points for each of the curves: $\{\mathbf{q}_{i0}\}, \quad i = 0, 1, \ldots, m$, and $\{\mathbf{q}_{i1}\}, \quad i = 0, 1, \ldots, m$

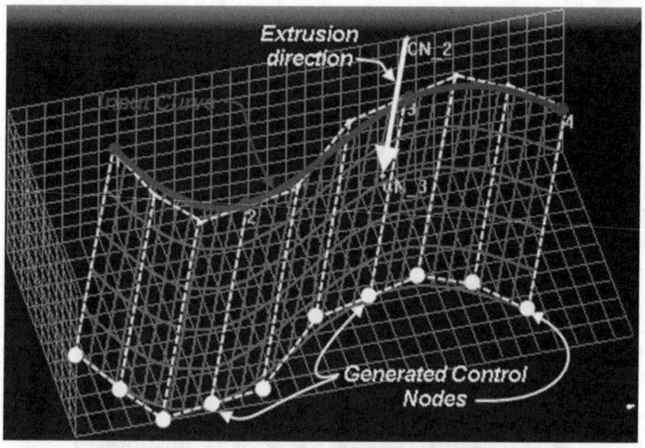

Figure 6.18 Patch by extrusion with Bezier control net. (See colour plate section)

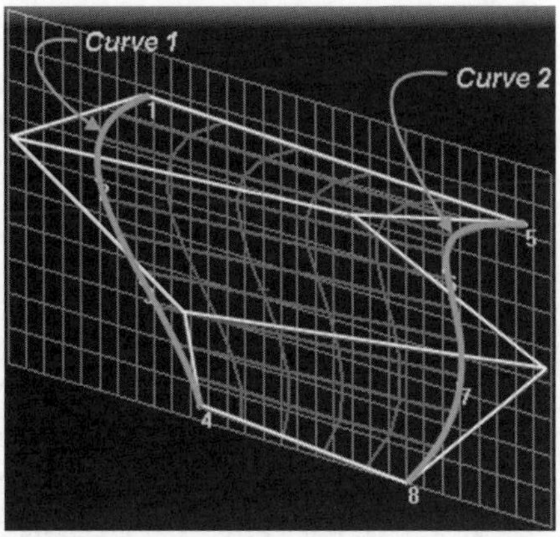

Figure 6.19 Patch by lofting of two curves with Bezier control net. (See colour plate section)

Step 1: Joining by a straight line the corresponding control points, generate the necessary control net of the resulting patch.

- The concept of lofting can be extended to more than two curves, as shown in Figure 6.20:

6.5.3 Coons' Patch by Bilinear Blending of Four Boundary Curves

A ***Coons' surface*** (due to Steven Coons) is generated from four given curves acting as the four boundary curves of the surface.

Problem: Given four curves with control points, the two sets of opposite boundary curves are given as: $\mathbf{C}_0 : \{\mathbf{q}_{0j}\}, \mathbf{C}_1 : \{\mathbf{q}_{1j}\}, \quad j = 0, 1, \ldots, n$, and $\mathbf{C}_2 : \{\mathbf{q}_{i0}\}, \mathbf{C}_3 : \{\mathbf{q}_{i1}\}, \quad i = 0, 1, \ldots, m$, as shown in Figure 6.21; generate the patch control net by bilinear blending.

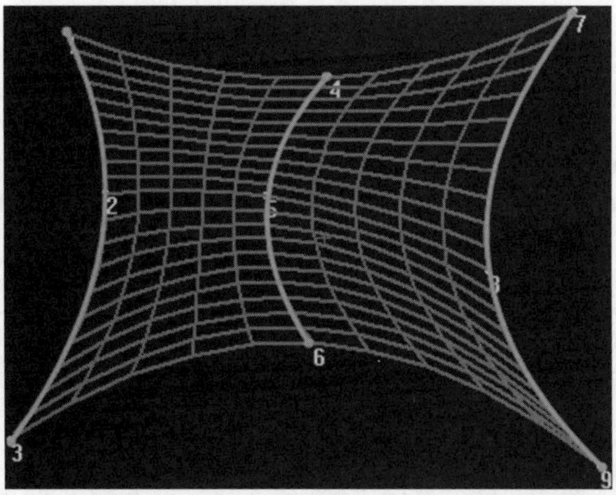

Figure 6.20 Patch by lofting of three curves with Bezier control net. (See colour plate section)

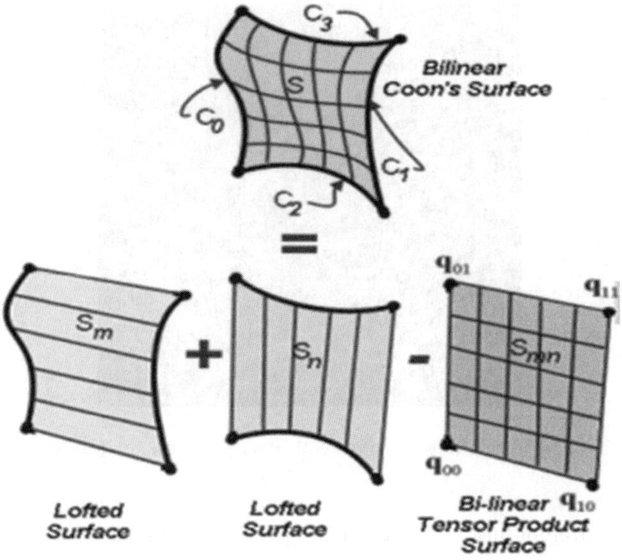

Figure 6.21 Coons' patch construction.

6.5.3.1.1 Algorithm

Step 0: Generate the ruled surface, $\mathbf{S}_m$ by lofting the control points of each of the opposite boundary curves, $\mathbf{C}_0 : \{\mathbf{q}_{0j}\}$, and $\mathbf{C}_1 : \{\mathbf{q}_{1j}\}$, $\quad j = 0, 1, \ldots, n$ such that any point, (ξ^1, ξ^2), on the surface, $\mathbf{S}_m$, is given by:

$$\boxed{\mathbf{S}_m(\xi^1, \xi^2) = (1 - \xi^1)\mathbf{C}_0(\xi^2) + \xi^1\mathbf{C}_1(\xi^1)} \tag{6.163}$$

Step 1: Generate the ruled surface, $\mathbf{S}_n$ by lofting the control points of each of the opposite boundary curves, $\mathbf{C}_2 : \{\mathbf{q}_{i0}\}$, and $\mathbf{C}_3 : \{\mathbf{q}_{i1}\}$, $\quad i = 0, 1, \ldots, m$ such that any point, (ξ^1, ξ^2), on the surface, $\mathbf{S}_n$, is given by:

$$\boxed{\mathbf{S}_n(\xi^1, \xi^2) = (1 - \xi^2)\mathbf{C}_2(\xi^1) + \xi^2\mathbf{C}_3(\xi^1)} \tag{6.164}$$

Step 2: Generate the bilinear Cartesian product surface, $\mathbf{S}_{mn}$, as rewritten from equation (6.129), through the four corner control points, such that any point, (ξ^1, ξ^2), on the surface, $\mathbf{S}_n$, is given by:

$$\boxed{\mathbf{S}_{mn}(\xi^1, \xi^2) = (1 - \xi^1)(1 - \xi^2)\mathbf{q}_{00} + \xi^1(1 - \xi^2)\mathbf{q}_{10} + (1 - \xi^1)\xi^2\mathbf{q}_{01} + \xi^1\xi^2\mathbf{q}_{11}} \tag{6.165}$$

Step 3: the bilinearly blended Coons' surface, $\mathbf{S}_{mn}$, is then given by:

$$\boxed{\mathbf{S}(\xi^1, \xi^2) = \mathbf{S}_m(\xi^1, \xi^2) + \mathbf{S}_n(\xi^1, \xi^2) - \mathbf{S}_{mn}(\xi^1, \xi^2)} \tag{6.166}$$

- We can easily show that the bilinearly blended Coons' surface does satisfy all four boundary curves; let us prove it for the curve, $\mathbf{C}_0(\xi^2) = \mathbf{S}(0, \xi^2)$:

$$\boxed{\begin{aligned}\mathbf{S}(0,\xi^2) &= \mathbf{S}_m(0,\xi^2) + \mathbf{S}_n(0,\xi^2) - \mathbf{S}_{mn}(0,\xi^2)\\ &= \mathbf{C}_0(\xi^2) + (1-\xi^2)\mathbf{C}_2(0) + \xi^2\mathbf{C}_3(0) - (1-\xi^2)\mathbf{q}_{00} - \xi^2\mathbf{q}_{01}\\ &= \mathbf{C}_0(\xi^2) + \cancel{(1-\xi^2)\mathbf{q}_{00}} + \cancel{\xi^2\mathbf{q}_{01}} - \cancel{(1-\xi^2)\mathbf{q}_{00}} - \cancel{\xi^2\mathbf{q}_{01}}\\ &= \mathbf{C}_0(\xi^2)\end{aligned}} \tag{6.167}$$

- In spite of the name, a bilinearly blended Coons' patch is clearly not a bilinear Cartesian product surface.
- The general Coons' surfaces, say, bi-cubically blended surfaces can be generated in a similar fashion; the bilinearly or bi-cubically blended Coons' patch approximations are used to estimate twists of more complicated surfaces such as bi-cubic B-spline surfaces or composite Bezier surfaces as will be seen shortly.

6.5.4 *Patch by Revolution of a Curve*

Surfaces with axial symmetry can be generated exactly, by revolving a planar curve around an axis belonging to the same plane, as shown in Figure 6.22.

Problem: Given a planar (non-rational or rational) curve, $\mathbf{C}(\xi^2)$, $\xi^2 \in [0, 1]$, and an axis of revolution as vector, $\mathbf{v}(\xi^2)$, as shown in Figure 6.22, generate the patch control net and weight net by revolution of the curve around the axis. Note that the radius of revolution, $r(\xi^2)$, at any $\xi^2 \in [0, 1]$ is obtained by taking the difference between the point on the axis at ξ^2 and the point on the curve by intersection of the curve with the plane perpendicular to the axis of revolution at ξ^2. The curve is generally known as a generatrix; the surface may be described as:

$$\boxed{\mathbf{S}(\xi^1, \xi^2) = \begin{pmatrix} r(\xi^2)\cos\xi^1 & r(\xi^2)\sin\xi^1 & z(\xi^2)\end{pmatrix}^T} \tag{6.168}$$

with the generatrix expressed as:

$$\boxed{\mathbf{C}(\xi^2) = \mathbf{S}(0, \xi^2) = \begin{pmatrix} r(\xi^2) & 0 & z(\xi^2)\end{pmatrix}^T} \tag{6.169}$$

It is assumed that the Bezier control points and the corresponding Bezier standard weights of the curve are known. If it is a rational B-spline curve, it is assumed that the curve is already converted into a composite Bezier curve with appropriate Bezier control points and weights. Finally, it is intended that each parallel circle at any ξ^2 of the resulting surface be composed of three equal circular arcs; in other words, three control points for the circles form the vertices of an equilateral triangle, and the three control points midway between the control points at the vertices belong to the circles.

6.5.4.1.1 Algorithm

Step 0: For each control point of the generatrix, redefine the control point as a vertex of the equilateral triangle.

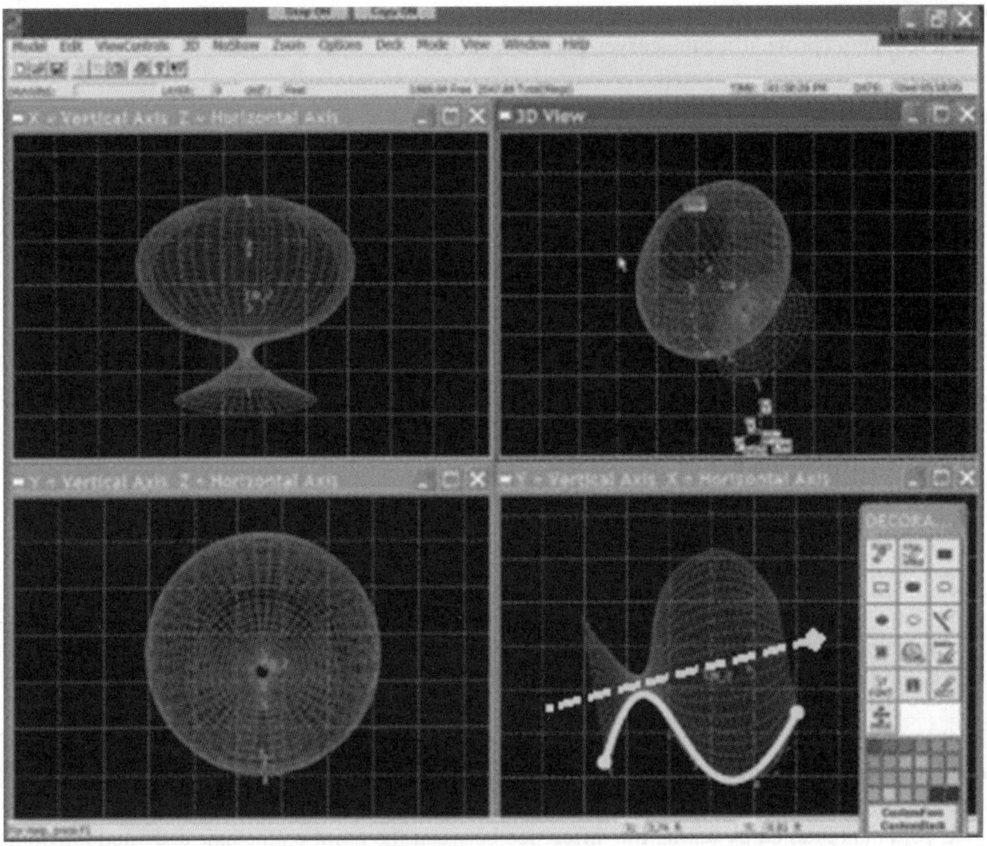

Figure 6.22 Patch by revolution of a generating curve. (See colour plate section)

Step 1: Construct the equilateral triangle for the circle using the curve generation technology; this generates the control net of the surface of revolution.

Step 2: The remaining two vertices of the equilateral triangle are assigned the same weights as the weight corresponding to the control point of the generatrix.

Step 3: The three mid-points between each pair of vertices of the equilateral triangle are assigned half the weights ($\cos\ 60 = \frac{1}{2}$ with $60°$ being the angle subtended by two consecutive mid-points with the in-between vertex of the triangle) as the weight corresponding to the control point of the generatrix.

6.6 Composite Bezier Form: Quadratic and Cubic B-splines

Although we have introduced before the general class of non-rational and rational B-spline surfaces, we will now restrict ourselves to the case of quadratic and cubic B-spline surfaces for our purposes of geometric model development of a structural system. Since, internally in computer implementation, we intend to store all B-spline surfaces, non-rational and/or rational, uniformly as composite Bezier surfaces, here we set out to represent these surfaces as composite Bezier surfaces.

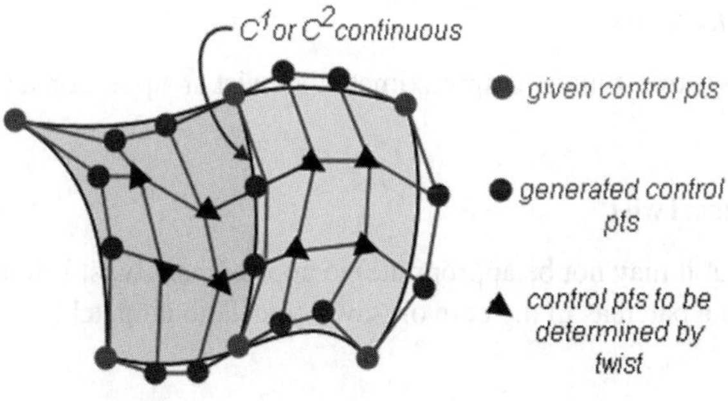

Figure 6.23 Twist problem for composite Bezier surface.

As shown in Figure 6.23, we are given (which is what is expected normally) only the corner control points (and needed boundary conditions) for individual patches of non-rational B-spline surfaces or rational B-spline surfaces. These surfaces, as indicated above, are stored internally in the computer by transforming these into composite Bezier surfaces with Bezier control net and weights (for the rational surfaces). This transformation is achieved by treating each row of the spline control net as control points of a corresponding spline curve and then applying the algorithm discussed relating to spline curve to Bezier composite curve transformation. The resulting boundary composite Bezier curves in both directions through these Bezier control points can be constructed following the method discussed for composite Bezier curves; this, in turn, then generates all the edge control points of any individual patch which amount to a total of 12 out of 16 needed for defining a ***bi-cubic Bezier patch***. The remaining four internal control points for each Bezier patch remain undefined, unless we apply additional conditions: mixed partial derivatives or twist. Thus, next we discuss how to estimate these twists.

In order to understand how twist condition helps us to determine the unknown control points, let us consider four control points as shown in Figure 6.24. Based on equation (6.143), we can easily see that if three control points are known, the fourth one can be determined if the twist at the corner control point is known as the twist $\Delta^{11}\mathbf{q}_{ij}$ at the corner control point, $\mathbf{q}_{ij}$, is merely the deviation of the corner control point, $\mathbf{q}_{(i+1)(j+1)}$, from the tangent plane defined by the known three corner control points, namely. $\mathbf{q}_{ij}, \mathbf{q}_{i(j+1)}, \mathbf{q}_{(i+1)j}$ of the patch.

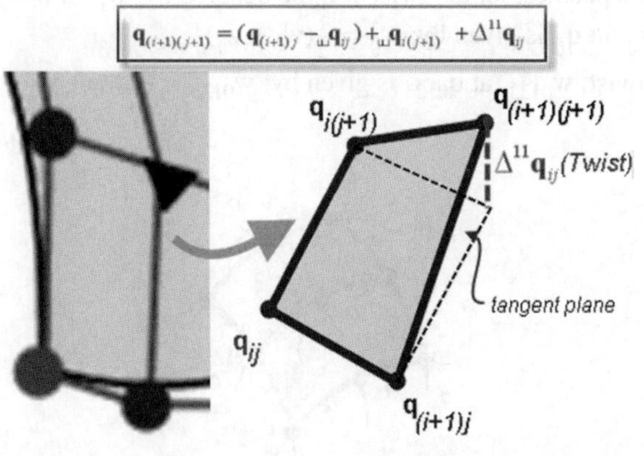

Figure 6.24 Twist and determination of internal control points.

6.6.1 Twist Conditions

There are several ways to estimate or approximate the twist at a point on a surface.

6.6.1.1 Case 1: Zero Twist

The simplest way (but it may not be appropriate) to approximate twist is to assume that the twist is identically zero on a patch as in the case of, say, a translational patch.

6.6.1.2 Case 2: Method Using Bilinear Surface Approximation

Recalling equation (6.143), for a bilinear surface given in matrix form as:

$$\boxed{S(\xi^1,\xi^2) =< (1-\xi^1) \quad \xi^1 > \begin{bmatrix} \mathbf{q}_{00} & \mathbf{q}_{01} \\ \mathbf{q}_{10} & \mathbf{q}_{11} \end{bmatrix} \begin{Bmatrix} (1-\xi^2) \\ \xi^2 \end{Bmatrix}} \tag{6.170}$$

the twist is constant over the surface, and is given by:

$$\boxed{\begin{aligned} S_{,12}(\xi^1,\xi^2) \overset{\Delta}{=} \frac{\partial^2 S}{\partial \xi^1\, \partial \xi^2} &= < -1 \quad 1 > \begin{bmatrix} \mathbf{q}_{00} & \mathbf{q}_{01} \\ \mathbf{q}_{10} & \mathbf{q}_{11} \end{bmatrix} \begin{Bmatrix} -1) \\ 1 \end{Bmatrix} \\ &= \mathbf{q}_{00} - \mathbf{q}_{10} - \mathbf{q}_{01} + \mathbf{q}_{11} = \text{constant} \end{aligned}} \tag{6.171}$$

As shown in Figure 6.25, given a surface made up of a collection of patches, we consider four patches that intersect at a corner where we intend to compute the twist.

Problem: We are interested in estimating twist at $\mathbf{q}_{0123}$.

6.6.1.2.1 Algorithm

Step 0: According to equation (6.171), twist at each of the four bilinearly approximated patches is constant over the patches; let the twist at $\mathbf{q}_i$ be denoted by: $\mathbf{w}_i$ for patch i, $\quad i = 1,2,3,4$.

Step 1: The twist, $\mathbf{w}_{ij}$, at $\mathbf{q}_{ij}$ is given by: $\mathbf{w}_{ij} = \frac{1}{2}(\mathbf{w}_i + \mathbf{w}_j)$.

Step 2: Finally, the twist, $\mathbf{w}_{0123}$, at $\mathbf{q}_{0123}$ is given by: $\mathbf{w}_{0123} = \frac{1}{4}(\mathbf{w}_0 + \mathbf{w}_1 + \mathbf{w}_2 + \mathbf{w}_3)$

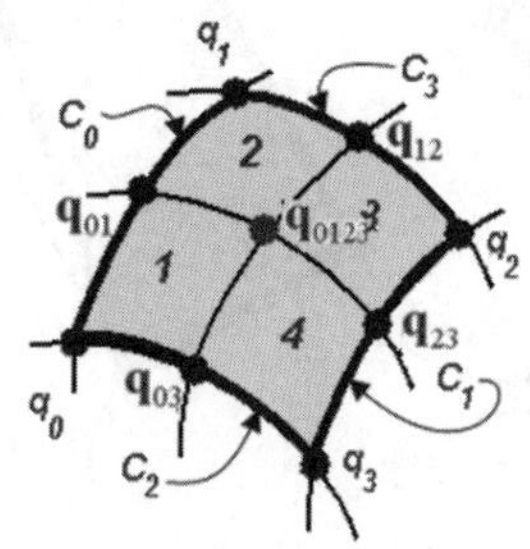

Figure 6.25 Twist by Coons' patch fitting.

6.6.1.3 Case 3: Method Using Bilinear Coons' Surface Approximation

As shown in Figure 6.25, given a surface made up of a collection of patches, we consider four patches that intersect at a corner where we intend to compute the twist, but in this method we pass a bilinearly blended Coons' patch through the surface bounded by the four outer boundary curves of the four patches.

We rewrite equations (6.163)–(6.166) for easy reference; then, the bilinearly blended Coons' surface, $\mathbf{S}$, over four patches with outer boundary curves denoted by $\mathbf{C}_0, \mathbf{C}_1, \mathbf{C}_2, \mathbf{C}_3$, and corner control points indicated by $\mathbf{q}_0, \mathbf{q}_1, \mathbf{q}_2, \mathbf{q}_3$, is given by:

$$\boxed{\mathbf{S}(\xi^1, \xi^2) = \mathbf{S}_1(\xi^1, \xi^2) + \mathbf{S}_2(\xi^1, \xi^2) - \mathbf{S}_{12}(\xi^1, \xi^2)} \tag{6.172}$$

where

$$\boxed{\begin{aligned} \mathbf{S}_1(\xi^1, \xi^2) &= (1 - \xi^1)\mathbf{C}_0(\xi^1) + \xi^1\mathbf{C}_1(\xi^1) \\ \mathbf{S}_2(\xi^1, \xi^2) &= (1 - \xi^2)\mathbf{C}_2(\xi^2) + \xi^2\mathbf{C}_3(\xi^2) \\ \mathbf{S}_{12}(\xi^1, \xi^2) &= (1 - \xi^1)(1 - \xi^2)\mathbf{q}_0 + \xi^1(1 - \xi^2)\mathbf{q}_3 + (1 - \xi^1)\xi^2\mathbf{q}_1 + \xi^1\xi^2\mathbf{q}_2 \end{aligned}} \tag{6.173}$$

Problem: We are interested in estimating twist at $\mathbf{q}_{0123}$.

6.6.1.3.1 Algorithm

Step 0: According to equations (6.172) and (6.173), twist, $\mathbf{S}_{,12}(\xi^1, \xi^2) \triangleq \frac{\partial^2 \mathbf{S}(\xi^1,\xi^2)}{\partial \xi^1 \, \partial \xi^2}$ at any point, (ξ^1, ξ^2), is given by:

$$\boxed{\begin{aligned} \mathbf{S}_{,12}(\xi^1, \xi^2) &= \mathbf{S}_{1,12}(\xi^1, \xi^2) + \mathbf{S}_{2,12}(\xi^1, \xi^2) - \mathbf{S}_{12,12}(\xi^1, \xi^2) \\ &= \mathbf{C}_1(\xi^2)_{,2} - \mathbf{C}_0(\xi^2)_{,2} + \mathbf{C}_2(\xi^1)_{,1} - \mathbf{C}_3(\xi^1)_{,1} - c \\ \text{where} \quad & \\ c &= \mathbf{q}_0 - \mathbf{q}_3 - \mathbf{q}_1 + \mathbf{q}_2 = \text{constant} \end{aligned}} \tag{6.174}$$

Step 1: From equation (6.174), the corner twists, $\mathbf{w}_1 \triangleq \mathbf{S}_{,12}(0, 1)$, $\mathbf{w}_2 \triangleq \mathbf{S}_{,12}(1, 1)$, $\mathbf{w}_3 \triangleq \mathbf{S}_{,12}(1, 0)$ and $\mathbf{w}_4 \triangleq \mathbf{S}_{,12}(0, 0)$, for each of the patch outer corners, are given by:

$$\boxed{\begin{aligned} \mathbf{w}_1 &\triangleq \mathbf{S}_{,12}(0, 1) = \mathbf{C}_1(1)_{,2} - \mathbf{C}_0(1)_{,2} + \mathbf{C}_2(0)_{,1} - \mathbf{C}_3(0)_{,1} - c \\ \mathbf{w}_2 &\triangleq \mathbf{S}_{,12}(1, 1) = \mathbf{C}_1(1)_{,2} - \mathbf{C}_0(1)_{,2} + \mathbf{C}_2(1)_{,1} - \mathbf{C}_3(1)_{,1} - c \\ \mathbf{w}_3 &\triangleq \mathbf{S}_{,12}(1, 0) = \mathbf{C}_1(0)_{,2} - \mathbf{C}_0(0)_{,2} + \mathbf{C}_2(1)_{,1} - \mathbf{C}_3(1)_{,1} - c \\ \mathbf{w}_4 &\triangleq \mathbf{S}_{,12}(0, 0) = \mathbf{C}_1(0)_{,2} - \mathbf{C}_0(0)_{,2} + \mathbf{C}_2(0)_{,1} - \mathbf{C}_3(0)_{,1} - c \end{aligned}} \tag{6.175}$$

Step 2: The twist, $\mathbf{w}_{ij}$, at junction points between the corresponding two patches is given by: $\mathbf{w}_{ij} = \frac{1}{2}(\mathbf{w}_i + \mathbf{w}_j)$.

Step 3: Finally, the twist, $\mathbf{w}_{0123}$, at $\mathbf{q}_{0123}$ is given by: $\mathbf{w}_{0123} = \frac{1}{4}(\mathbf{w}_0 + \mathbf{w}_1 + \mathbf{w}_2 + \mathbf{w}_3)$

6.6.1.4 Case 4: Method Using Bilinear Coons' Surface Approximation of Control Net

In this method, the entire control net (describing a surface) itself is treated as a piecewise bilinear surface and thus approximated by a bilinear Coons' surface to compute the missing four internal control points of each of the patches; this method appears to be the simplest of the all. Assuming that we already have the control points of the Cartesian product cubic B-spline curves converted into piecewise cubic Bezier form for both the control points and the weights (for rational B-spline surface), we provide below the C++ code for this method of completing the entire control net by generating along the way the four missing internal control points of each patch.

6.6.1.4.1 C++ Code: Generating Missing Control Points

```
/////////////////////////////////////////////////////////////
typedef int *pINT;
typedef double    *pDOUBLE;
typedef      struct
      {
            double x;
            double y;
            double z;
      } WORLD, *pWORLD;
////////////////////////////////////////////////////////////////////////////////
int CoonNetGenerate
(
/////////////////////////
// Input data points should be ALREADY TRANSFORMED into cubic Bezier
// Interior four control points are created by Bilinear Coons' patch fit of the
// control net
/////////////////////////
      short        nPts_S,      //    number of input control points in S-dir
      short        nPts_T,      //    number of input control points in T-dir
      pWORLD       pInC1,       //    input data point for Curve 1
      pDOUBLE      pWts1,       //    input data weight for Curve 1
      pWORLD       pInC2,       //    input data point for Curve 2
      pDOUBLE      pWts2,       //    input data weight for Curve 2
      pWORLD       pInC3,       //    input data point for Curve 3
      pDOUBLE      pWts3,       //    input data weight for Curve 3
      pWORLD       pInC4,       //    input data point for Curve 4
      pDOUBLE      pWts4,       //    input data weight for Curve 4
      pWORLD       pOut, //     Output pts array
      pDOUBLE      pOutWts //   Output wts array
)
{
      int        nV,i,j,k;
      ////////////////
      if( pOut )
      {
```

```
        nV          = nPts_S*nPts_T;
        ////////////////////////////
        double rS = 1./(nPts_S-1);
        double rT = 1./(nPts_T-1);
        double u,v,u1,v1;
        //////////////////////
        for( j = 0; j<nPts_T; j++ )
        {
            for( i = 0; i<nPts_S; i++ )
            {
                u    = rS * i;
                v    = rT * j;
                u1   = 1. - u;
                v1   = 1. - v;
                ////////////////
                k = j*nPts_S + i;
                ////////////////
                pOut[k].x =
pInC4[nPts_T-j-1].x* u1 + pInC2[j].x * u      +
                      pInC1[i].x * v1         +
pInC3[nPts_S-i-1].x * v    -
    pInC1[0].x * u1 * v1      - pInC4[0].x * u1 * v    -
                      pInC2[0].x * u * v1     -
pInC3[0].x * u * v      ;
                /////////
                pOut[k].y =
pInC4[nPts_T-j-1].y * u1    + pInC2[j].y * u      +
                    pInC1[i].y * v1         +
pInC3[nPts_S-i-1].y * v    -
                    pInC1[0].y * u1 * v1      -
pInC4[0].y * u1 * v       -
                    pInC2[0].y * u * v1       -
pInC3[0].y * u * v        ;
                /////////
                pOut[k].z =
pInC4[nPts_T-j-1].z * u1    + pInC2[j].z * u      +
                  pInC1[i].z * v1         +
pInC3[nPts_S-i-1].z * v    -
                pInC1[0].z * u1 * v1      - pInC4[0].z * u1 * v      -
                pInC2[0].z * u * v1       - pInC3[0].z * u * v       ;
                ////////////////////////////////////////
                pOutWts[k]=
pWts4[nPts_T-j-1] * u1      + pWts2[j] * u         +
                pWts1[i] * v1        +
pWts3[nPts_S-i-1] * v - -
                pWts1[0] * u1 * v1       - pWts4[0] * u1 * v      -
                pWts2[0] * u * v1      - pWts3[0] * u * v       ;
                ////////////////////////////////////////
    }
        }
    }
    return nV;
}
```

6.6.2 Cartesian Product Composite Cubic Bezier Surface

Now that we have the means to determine twists of a surface, we finally provide the algorithm to generate a Cartesian product cubic B-spline surface converted as a composite cubic Bezier surface using the curve technology in each direction:

6.6.2.1.1 Algorithm

Step 0: Assuming that the iso-curves have been generated, using curve technology along ξ^1 and ξ^2 directions, produce the control points and weights (in case of rational curves).

Step 1: These curves are all appropriately degree elevated for compatibility in terms of number of patches along each coordinate direction and rational/non-rational behavior. For example, if there are two curves – one rational cubic and one linear along ξ^1 direction, then the linear curve is degree elevated and treated as rational. Finally, if any curve is rational in any direction, all curves are treated as rational and treated in the 4D homogeneous pre-image space.

Step 2: The curves in each direction are converted into composite Bezier curves producing Bezier control points and weights.

Step 3: Use Case 4, that is, the method using bilinear Coons' surface approximation of control net above for twist estimation to determine the unknown internal control points of each patch of the surface.

6.7 Triangular Bezier–Bernstein Surfaces

6.7.1 Triangles and Barycentric Coordinates

Three vertices of a triangle are always co-planar; every point in it can be expressed as a ***barycentric combination*** with coordinates, say $(\xi_0,\ \xi_1,\ \xi_2)$, with respect to a triangle in the plane with $\mathbf{q}_{001},\ \mathbf{q}_{010},\ \mathbf{q}_{100}$, as its vertices, as shown in Figure 6.26.

We have:

$$\boxed{\mathbf{P}(\xi_0, \xi_1, \xi_2) = \xi_0\mathbf{q}_{100} + \xi_1\mathbf{q}_{010} + \xi_2\mathbf{q}_{001} \quad \text{with} \quad \xi_0 + \xi_1 + \xi_2 = 1} \tag{6.176}$$

where $(\xi_0,\ \xi_1,\ \xi_2)$ are the ***barycentric coordinates*** of a point, $\mathbf{P}$, with respect to vertices, $\mathbf{q}_{100},\ \mathbf{q}_{010}$ and $\mathbf{q}_{001}$, of the triangle, given geometrically by:

$$\boxed{\begin{aligned} \xi_0 &= \frac{area\Delta\mathbf{P}\mathbf{q}_{010}\mathbf{q}_{001}}{area\Delta\mathbf{q}_{100}\mathbf{q}_{010}\mathbf{q}_{001}}, \\ \xi_1 &= \frac{area\Delta\mathbf{P}\mathbf{q}_{100}\mathbf{q}_{001}}{area\Delta\mathbf{q}_{100}\mathbf{q}_{010}\mathbf{q}_{001}}, \\ \xi_2 &= \frac{area\Delta\mathbf{P}\mathbf{q}_{100}\mathbf{q}_{010}}{area\Delta\mathbf{q}_{100}\mathbf{q}_{010}\mathbf{q}_{001}} \end{aligned}} \tag{6.177}$$

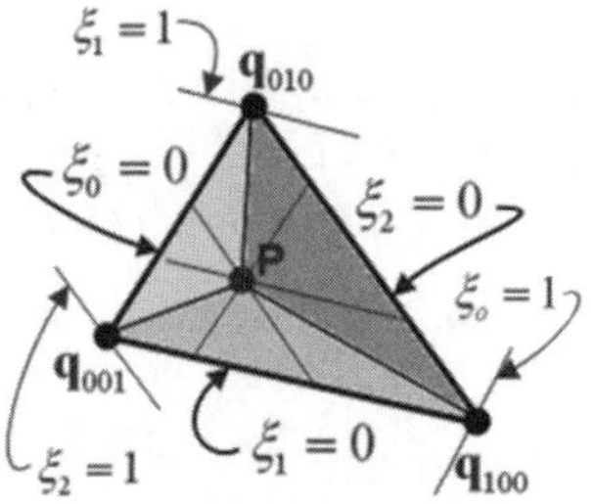

Figure 6.26 Barycentric coordinates.

For example, it is easy to see from equations (6.176) and (6.177) that when $area\Delta \mathbf{P}\mathbf{q}_{010}\mathbf{q}_{001} = area\Delta \mathbf{q}_{100}\mathbf{q}_{010}\mathbf{q}_{001}$, that is, $\xi_0 = 1$, $\xi_1 = \xi_2 = 0$, the point, **P**, becomes the vertex $\mathbf{q}_{100}$.

- Although from equation (6.176), the representation of a point seems as if it were tri-variate, it is clearly bi-variate because of the constraint condition: $\xi_0 + \xi_1 + \xi_2 = 1$.

6.7.1.1 Linear Interpolation

As indicated in our discussion on curves, barycentric combinations are affinely invariant, meaning that under affine transformation, the coordinates are maintained; for example, under an affine map, A, of all the four points, **P**, $\mathbf{q}_{100}$, $\mathbf{q}_{010}$ and $\mathbf{q}_{001}$ of equation (6.176) into A**P**, A$\mathbf{q}_{100}$, A$\mathbf{q}_{010}$ and A$\mathbf{q}_{001}$, the barycentric coordinates of the point, A**P**, will still be given by: (ξ_0, ξ_1, ξ_2) with respect to vertices A$\mathbf{q}_{100}$, A$\mathbf{q}_{010}$ and A$\mathbf{q}_{001}$ of the transformed triangle. As suggested by equation (6.176), the point **P** is clearly obtained by linear interpolation of a point in some specific triangle in 2D, known as the ***domain***, corresponding to the triangle, known as the ***range***, given by the vertices, $\mathbf{q}_{100}$, $\mathbf{q}_{010}$ and $\mathbf{q}_{001}$, in 3D, and, as such, linear interpolation is an affine transformation. In other words, given $\mathbf{q}_{100}$, $\mathbf{q}_{010}$ and $\mathbf{q}_{001}$ in 3D as the affine map of vertices of some triangle in 2D, and any point **p** in 2D with barycentric coordinates, (ξ_0, ξ_1, ξ_2), with respect to the same triangle in 2D, its image point **P** in 3D can be obtained by linear combination given by equation (6.176).

6.7.1.2 Triangle Number

The triangle number, T, which determines the dimension of the vector space, and hence the number of basis functions, is defined as:

$$\boxed{\begin{aligned} T &= \frac{(n+1)(n+2)}{2} \\ &= \frac{(\#\text{ of rows})(\#\text{ of rows} + 1)}{2} \end{aligned}} \tag{6.178}$$

where n is the degree of the triangle; for linear triangle, $n = 1, \#\text{ of rows} = 2$, implying $T = 3$, for quadratic triangle, $n = 2, \#\text{ of rows} = 3$, implying $T = 6$, and for cubic triangle, $n = 3, \#\text{ of rows} = 4$, implying $T = 10$, and so on.

6.7.1.3 Higher Degree Interpolation

Just as in the case of curves, and as shown in Figure 6.27, we can think of higher-degree interpolation in two different ways:

- as repeated linear interpolations as given by the ***de Casteljau algorithm*** specialized for triangular barycentric coordinates,
- as a bi-variate ***Bernstein polynomial*** expansion.

6.7.2 Triangular Bernstein Polynomials

Just as in the case of Bezier curves of degree n, the Bernstein polynomials form the basis functions resulting from the univariate binomial expansion or partition of unity: $1^n = (\xi_0 + (1 - \xi_0))^n$, we

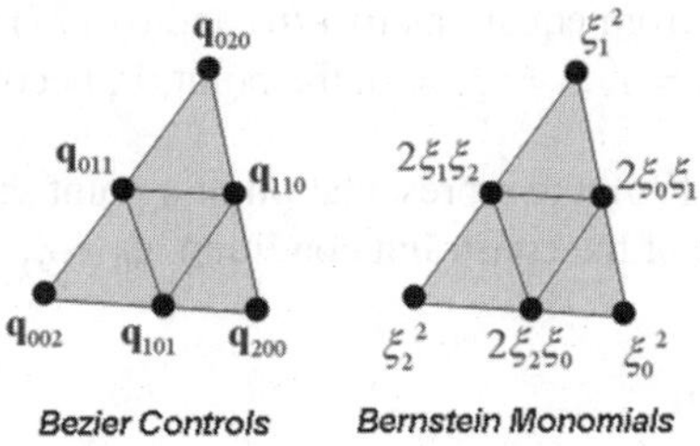

Figure 6.27 Bezier controls and Bernstein monomials for a triangle.

can initiate a bi-variate binomial expansion for a triangular surface, noting that $\xi_0 + \xi_1 + \xi_2 = 1$, with the restriction: $i + j + k = n$, as:

$$\begin{aligned}
1^n &= (\xi_0 + (1 - \xi_0))^n = (\xi_0 + (\xi_1 + \xi_2))^n = \sum_{i=0}^{n} \binom{n}{i} \xi_0^i (\xi_1 + \xi_2)^{n-i} \\
&= \sum_{i=0}^{n} \binom{n}{i} \xi_0^i \sum_{j=0}^{n-i} \binom{n-i}{j} \xi_1^j \, \xi_2^{n-i-j} = \sum_{i=0}^{n} \sum_{j=0}^{n-i} \binom{n}{i} \binom{n-i}{j} \xi_0^i \, \xi_1^j \, \xi_2^{n-i-j} \\
&= \sum_{i=0}^{n} \sum_{j=0}^{n-i} \frac{n!}{i!\cancel{(n-i)!}} \frac{\cancel{(n-i)!}}{j!(n-i-j)!} \xi_0^i \, \xi_1^j \, \xi_2^{n-i-j} = \sum_{i=0}^{n} \sum_{j=0}^{n-i} \left(\frac{n!}{i!j!k!} \xi_0^i \, \xi_1^j \, \xi_2^k \right)
\end{aligned} \tag{6.179}$$

Let us introduce some shorthand vector notations for simplicity in expression: 3D point ξ and a pseudo-vector $\mathbf{i}$ such that:

$$\boxed{\begin{aligned}
\xi &\stackrel{\Delta}{=} (\xi_0 \quad \xi_1 \quad \xi_2)^T \\
\mathbf{i} &\stackrel{\Delta}{=} ijk = (i, j, k) \\
|\mathbf{i}| &\stackrel{\Delta}{=} i + j + k
\end{aligned}} \tag{6.180}$$

With the above binomial expansion as in equation (6.179), and notations in equation (6.180), we can now define the Bernstein basis functions for barycentric coordinates as follows:

Definition

- The $\mathbf{i}^{\text{th}}$ triangular Bernstein basis functions, $B_{\mathbf{i}}^n(\xi)$, of the n-dimensional polynomial subspace of the class of continuous functions are defined as:

$$\boxed{\begin{aligned}
&B_{\mathbf{i}}^n(\xi) = \binom{n}{\mathbf{i}} \xi_0^i \, \xi_1^j \, \xi_2^k, \quad \mathbf{i} \stackrel{\Delta}{=} ijk, \; |\mathbf{i}| \stackrel{\Delta}{=} i + j + k = n, \\
&\textit{where} \; \binom{n}{\mathbf{i}} \equiv \begin{cases} 0, & \text{if any of } i, j, k < 0 \\ \dfrac{n!}{i!j!k!} & \text{otherwise} \end{cases}
\end{aligned}} \tag{6.181}$$

Example: Based on the definition given in equation (6.181), we give below, in triangular form, a few explicit expressions for the Bernstein basis functions for $n = 1, 2, 3$.

Linear ($n = 1$): 3 (triangle number, $T = \frac{(1+1)(1+2)}{2}$) basis functions.

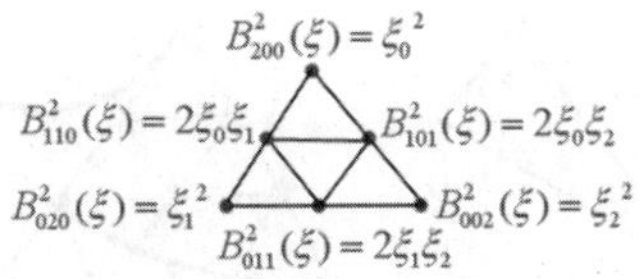

Figure 6.28 Bezier basis functions for quadratic triangles.

Remark

- For later use in the definition of directional derivatives over a triangular surface with respect to the barycentric coordinates, let us note the following.

$$\boxed{\begin{aligned} B^1_{100} &= \xi_0 \\ B^1_{010} &= \xi_1 \\ B^1_{001} &= \xi_2 \end{aligned}} \tag{6.182}$$

Quadratic ($n = 2$): 6 (triangle number, $T = \frac{(2+1)(2+2)}{2}$) basis functions in Figure 6.28.
Cubic ($n = 3$): 10 (triangle number, $T = \frac{(3+1)(3+2)}{2}$) basis functions in Figure 6.29.

- The distributions of these basis functions over the corresponding triangle can be easily visualized as we concentrate on their expressions and the barycentric coordinates at the points of interest; for example, Figure 6.30 shows graphically the distributions for B^1_{100} and B^3_{300}.

6.7.2.1 Non-negativity and Unique Maximum

Bernstein basis functions, $B^n_{\mathbf{i}}(\xi)$, are all non-negative and attain a maximum for $\xi_0 \geq 0$, $\xi_1 \geq 0$ and $\xi_2 \geq 0$.

6.7.2.2 Partition of Unity

For Bernstein basis functions, for any n,

$$\sum_{i=0}^{n} B^n_{\mathbf{i}}(\xi) = 1, \quad \xi \stackrel{\Delta}{=} (\xi_0, \xi_1, \xi_2) \in [0,1] \times [0,1] \times [0,1] \tag{6.183}$$

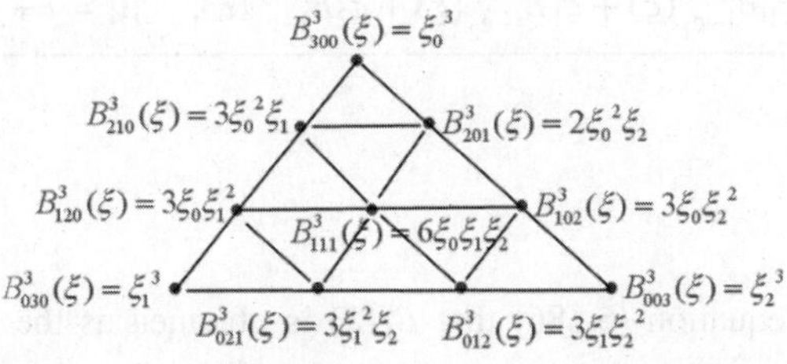

Figure 6.29 Bezier basis functions for cubic triangles.

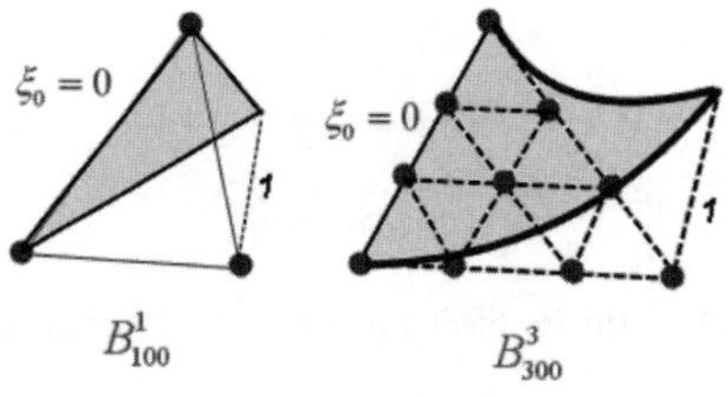

Figure 6.30 B_{100} and B_{300} of a linear and cubic triangles, respectively.

6.7.2.3 Recursive Property

The Bernstein basis functions of differing degrees are recursively connected; from the definition of the Bernstein polynomial given by equation (6.181), we get:

$$\begin{aligned}
B^n_{\mathbf{i}}(\xi) &= \frac{n!}{i!j!k!}\xi_0^i\,\xi_1^j\,\xi_2^k, \quad |\mathbf{i}| \triangleq i+j+k=n \\
&= \frac{n(n-1)!}{i!j!k!}\xi_0^i\,\xi_1^j\,\xi_2^k = \left(\frac{i(n-1)!}{i!j!k!}+\frac{j(n-1)!}{i!j!k!}+\frac{k(n-1)!}{i!j!k!}\right)\xi_0^i\,\xi_1^j\,\xi_2^k \\
&= \left(\frac{(n-1)!}{(i-1)!j!k!}+\frac{(n-1)!}{i!(j-1)!k!}+\frac{(n-1)!}{i!j!(k-1)!}\right)\xi_0^i\,\xi_1^j\,\xi_2^k \\
&= \xi_0\left(\frac{(n-1)!}{(i-1)!j!k!}\xi_0^{(i-1)}\,\xi_1^j\,\xi_2^k\right)+\xi_1\left(\frac{(n-1)!}{i!(j-1)!k!}\xi_0^i\,\xi_1^{(j-1)}\,\xi_2^k\right)+\xi_2\left(\frac{(n-1)!}{i!j!(k-1)!}\xi_0^i\,\xi_1^j\,\xi_2^{(k-1)}\right)
\end{aligned} \tag{6.184}$$

Now, introducing the notation for the set of natural basis:

$$\boxed{\begin{aligned} \mathbf{e}_1 &\triangleq (1,0,0) \\ \mathbf{e}_2 &\triangleq (0,1,0) \\ \mathbf{e}_3 &\triangleq (0,0,1) \end{aligned}} \tag{6.185}$$

we get from equations (6.181), (6.184) and (6.185), the recursion relation for the triangular Bezier polynomials as:

$$\boxed{B^n_{\mathbf{i}}(\xi) = \xi_0 B^{n-1}_{\mathbf{i}-\mathbf{e}_1}(\xi) + \xi_1 B^{n-1}_{\mathbf{i}-\mathbf{e}_2}(\xi) + \xi_2 B^{n-1}_{\mathbf{i}-\mathbf{e}_3}(\xi), \quad |\mathbf{i}| \triangleq i+j+k=n} \tag{6.186}$$

Remark

- It can be seen from equation (6.186) that $B^n_{\mathbf{i}}(\xi)$ is obtained as the convex combination of $B^{n-1}_{\mathbf{i}-\mathbf{e}_1}(\xi)$, $B^{n-1}_{\mathbf{i}-\mathbf{e}_2}(\xi)$ and $B^{n-1}_{\mathbf{i}-\mathbf{e}_3}(\xi)$ for all n and for all $|\mathbf{i}| \triangleq i+j+k=n$.

6.7.2.4 Derivative Property

First, let us introduce the multivariate definition and notation for the derivative of a function, $f(\xi_0, \xi_1, \xi_2)$, with respect to coordinates, ξ_0, ξ_1, ξ_2, as:

$$\boxed{\partial^{\mathbf{i}} f(\xi_0, \xi_1, \xi_2) \stackrel{\Delta}{=} \frac{\partial^{|\mathbf{i}|} f}{\partial \xi_0^i \, \partial \xi_1^j \, \partial \xi_2^k}, \quad \mathbf{i} \stackrel{\Delta}{=} ijk, \ |\mathbf{i}| = i + j + k} \tag{6.187}$$

We can show that the $\mathbf{j}^{th}$ derivative, of the $\mathbf{i}^{\text{th}}$ triangular Bernstein basis functions, $B_{\mathbf{i}}^n(\xi)$, of Bernstein basis functions, as given by equation (6.181), with respect to the barycentric coordinates, is given by:

$$\boxed{\begin{aligned} \partial^{\mathbf{j}} B_{\mathbf{i}}^n(\xi) &= \frac{n!}{(n-s)!} B_{\mathbf{i}-\mathbf{j}}^{n-s}(\xi), \quad |\mathbf{i}| = i + j + k = n, \ |\mathbf{j}| = p + q + r = s \\ &= \frac{n!}{(n-s)!} B_{(i-p)(j-q)(k-r)}^{n-s}(\xi) \end{aligned}} \tag{6.188}$$

Proof. Following the definition given by equation (6.181), we have:

$$\partial^{\mathbf{i}} f(\xi_0, \xi_1, \xi_2) \stackrel{\Delta}{=} \frac{\partial^{|\mathbf{i}|} f}{\partial \xi_0^i \, \partial \xi_1^j \, \partial \xi_2^k} B_{(i-p)(j-q)(k-r)}^{n-s} = \frac{(n-s)!}{(i-p)!(j-q)!(k-r)!} \xi_0^{(i-p)} \ \xi_1^{(j-q)} \ \xi_2^{(k-r)} \tag{6.189}$$

Now, using the derivative definition given by equation (6.187), and the definition given by equation (6.189), we get:

$$\begin{aligned} \partial^{\mathbf{j}} B_{\mathbf{i}}^n(\xi) &= \frac{\partial^{|\mathbf{j}|}}{\partial \xi_0^p \, \partial \xi_1^q \, \partial \xi_2^r} \left\{ \frac{n!}{i!j!k!} \xi_0^i \ \xi_1^j \ \xi_2^k \right\}, \quad |\mathbf{j}| = p + q + r = s \\ &= \frac{n!}{(i-p)!(j-q)!(k-r)!} \xi_0^{(i-p)} \ \xi_1^{(j-q)} \ \xi_2^{(k-r)} = \frac{n!}{(n-s)!} B_{(i-p)(j-q)(k-r)}^{n-s} \\ &= \frac{n!}{(n-s)!} B_{\mathbf{i}-\mathbf{j}}^{n-s} \end{aligned}$$

This completes the proof.

Example: Let us consider a quadratic triangular Bernstein polynomial, that is, $n = 2$; now, let $i = 2$, then with $i + j + k = n = 2$, we get: $j = k = 0$. So, the Bernstein polynomial is $B_{200}^2 = \xi_0^2$, and its first derivative with $|\mathbf{j}| = p + q + r = s = 1 \Rightarrow p = 1, q = r = 0$, using equation (6.188), is given by: $\partial^1 B_{200}^2(\xi_0) = \frac{2!}{(2-1)!} B_{(2-1)(0-0)(0-0)}^{2-1} = 2B_{100}^1(\xi_0)$; it is easily verified to be true because: $B_{100}^1 = \xi_0$, and $\frac{\partial}{\partial \xi_0} B_{200}^2 = 2\xi_0$.

6.7.3 *Triangular Bezier Surfaces with Bernstein Polynomials*

A point on a triangular Bezier surface can be obtained by the triangular Bernstein polynomials as the basis functions and the Bezier control points as follows.

Definitions

- An nth degree triangular Bezier surface, $\mathbf{S}(\xi)$, is defined as:

$$\boxed{\begin{aligned}\mathbf{S}^n(\xi) &= \sum_{|\mathbf{j}|=n} B^n_{\mathbf{j}}(\xi)\,\mathbf{q_j}, \quad \mathbf{j} \stackrel{\Delta}{=} ijk, \quad |\mathbf{j}| = i+j+k, \quad \xi = (\xi_0, \xi_1, \xi_2)\\ &= \sum_{|\mathbf{j}|=n} \left(\frac{n!}{i!j!k!}\xi_0^i\, \xi_1^j\, \xi_2^k\right)\mathbf{q_j}\end{aligned}} \tag{6.190}$$

- Based on the relation given by equations (6.179) and (6.181), we can express an nth degree triangular Bezier surface, $\mathbf{S}^n(\xi_0, \xi_1, \xi_2)$, in bi-variate summation, as:

$$\boxed{\begin{aligned}\mathbf{S}^n(\xi_0, \xi_1, \xi_2) &= \sum_{i=0}^{n}\sum_{j=0}^{n-i} B^n_{ijk}(\xi_0, \xi_1, \xi_2)\,\mathbf{q}_{ijk}, \quad i+j+k=n\\ &= \sum_{i=0}^{n}\sum_{j=0}^{n-i} \left(\frac{n!}{i!j!k!}\xi_0^i\, \xi_1^j\, \xi_2^k\right)\mathbf{q}_{ijk}\end{aligned}} \tag{6.191}$$

- The basis (blending) function sets, $B^n_{\mathbf{j}}(\xi)$, are the nth degree triangular Bernstein polynomial given by equation (6.181).
- The coefficients of linear combination, $\{\mathbf{q_j}\}$, are known as the control points of the surface.
- The polygonal net containing the control points, $\{\mathbf{q_j}\}$, is known as the control polygonal net of the surface.

Remark

- For any of the barycentric coordinates identically zero, the triangular surface boundary becomes a Bezier curve; for example, for $\xi_0 = 0$, implying $\xi_2 = 1 - \xi_1$, from equation (6.190), we get the typical expression for a Bezier curve, $\mathbf{C}^n(0, \xi_1)$, as:

$$\boxed{\mathbf{C}^n(0, \xi_1) = \sum_{j=0}^{n}\left(\frac{n!}{j!(n-j)!}\xi_1^j\, \xi_1^{(n-j)}\right)\mathbf{q}_{0j(n-j)},} \tag{6.192}$$

Now, let us elucidate the above definitions by examples. In the mesh generation phase, usually surfaces with degree up to and including three are considered for numerical efficiency. For finite element analysis with mega-elements, it appears that up to quintic should be considered for, say, shell analysis. The examples below also provide the formulas to determine points on surfaces of degree $n = 1, 2, 3$.

Example TR1: ***Linear Triangle***

From equation (6.190) or (6.191), for $n = 1$, we get the linear surface:

$$\boxed{\mathbf{S}^1(\xi_0, \xi_1, \xi_2) = \xi_0\,\mathbf{q}_{100} + \xi_1\,\mathbf{q}_{010} + \xi_2\,\mathbf{q}_{001}} \tag{6.193}$$

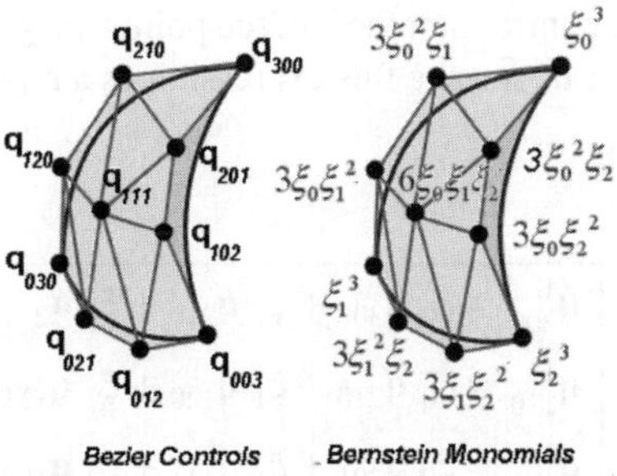

Figure 6.31 Cubic Bezier triangle with control net.

Example TR2: ***Quadratic Triangle***

From equation (6.190) or (6.191), for $n = 2$, we get the quadratic surface:

$$\boxed{\begin{aligned}\mathbf{S}^2(\xi_0, \xi_1, \xi_2) = {} & \xi_0^2\ \mathbf{q}_{200} + \xi_1^2\ \mathbf{q}_{020} + \xi_2^2\ \mathbf{q}_{002} \\ & + 2\xi_0\xi_1\ \mathbf{q}_{110} + 2\xi_2\xi_0\ \mathbf{q}_{101} + 2\xi_1\xi_2\ \mathbf{q}_{011}\end{aligned}} \tag{6.194}$$

Example TR3: ***Cubic Triangle***

From equation (6.190) or (6.191), for $n = 3$, we get the cubic surface (Figure 6.31):

$$\boxed{\begin{aligned}\mathbf{S}^3(\xi_0, \xi_1, \xi_2) = {} & \xi_0^3\ \mathbf{q}_{300} + \xi_1^3\ \mathbf{q}_{030} + \xi_2^3\ \mathbf{q}_{003} \\ & + 3\xi_0^2\ \xi_1\ \mathbf{q}_{210} + 3\xi_0\ \xi_2^2\ \mathbf{q}_{102} + 3\xi_1^2\ \xi_2\ \mathbf{q}_{021} \\ & + 3\xi_0^2\ \xi_2\ \mathbf{q}_{201} + 3\xi_1\ \xi_2^2\ \mathbf{q}_{012} + 3\xi_0\ \xi_1^2\ \mathbf{q}_{120} \\ & + 6\xi_0\ \xi_1\ \xi_2\ \mathbf{q}_{111}\end{aligned}} \tag{6.195}$$

6.7.4 Triangular Bezier Surfaces by the de Casteljau Algorithm

As in the univariate case of curves, given the Bezier control points, a point on a triangular Bezier surface can also be obtained by a de Casteljau algorithm.

Let us consider a triangular cubic control net as shown in Figure 6.32. We describe triangular linear interpolation as a map of a triangle from 2D to 3D with a barycentric combination of the

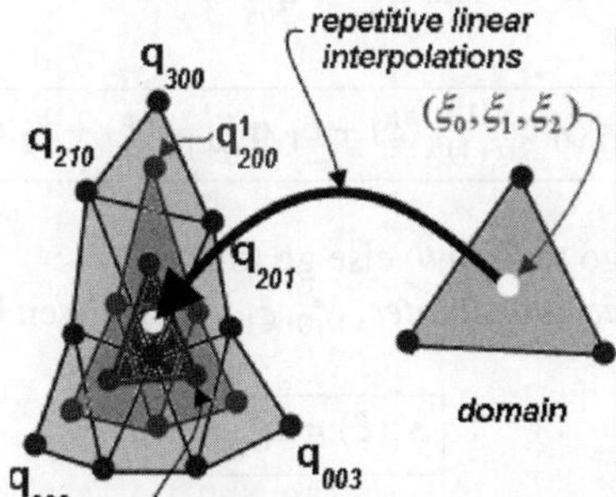

Figure 6.32 de Casteljau algorithm for triangle.

triangle vertices. For the cubic control net, the surface points are given by Bernstein polynomials as given in equation (6.195); let us rewrite this expression as a repetitive linear interpolation as:

Step 0: Linear interpolations:

$$\boxed{\begin{aligned}\mathbf{q}^1_{200} &= \xi_0\ \mathbf{q}_{300} + \xi_1\ \mathbf{q}_{210} + \xi_2\ \mathbf{q}_{201}\\ \mathbf{q}^1_{110} &= \xi_0\ \mathbf{q}_{210} + \xi_1\ \mathbf{q}_{120} + \xi_2\ \mathbf{q}_{111}\\ \mathbf{q}^1_{101} &= \xi_0\ \mathbf{q}_{201} + \xi_1\ \mathbf{q}_{111} + \xi_2\ \mathbf{q}_{102}\\ \mathbf{q}^1_{020} &= \xi_0\ \mathbf{q}_{120} + \xi_1\ \mathbf{q}_{030} + \xi_2\ \mathbf{q}_{021}\\ \mathbf{q}^1_{011} &= \xi_0\ \mathbf{q}_{111} + \xi_1\ \mathbf{q}_{021} + \xi_2\ \mathbf{q}_{012}\\ \mathbf{q}^1_{002} &= \xi_0\ \mathbf{q}_{102} + \xi_1\ \mathbf{q}_{012} + \xi_2\ \mathbf{q}_{003}\end{aligned}} \tag{6.196}$$

Step 1: Linear interpolations:

$$\boxed{\begin{aligned}\mathbf{q}^2_{100} &= \xi_0\ \mathbf{q}^1_{200} + \xi_1\ \mathbf{q}^1_{110} + \xi_2\ \mathbf{q}^1_{101}\\ \mathbf{q}^2_{010} &= \xi_0\ \mathbf{q}^1_{110} + \xi_1\ \mathbf{q}^1_{020} + \xi_2\ \mathbf{q}^1_{011}\\ \mathbf{q}^2_{001} &= \xi_0\ \mathbf{q}^1_{101} + \xi_1\ \mathbf{q}^1_{011} + \xi_2\ \mathbf{q}^1_{002}\end{aligned}} \tag{6.197}$$

Step 2: Linear interpolations:

$$\boxed{\mathbf{S}^3(\xi_0, \xi_1, \xi_2) = \mathbf{q}^3_{000} = \xi_0\ \mathbf{q}^2_{100} + \xi_1\ \mathbf{q}^2_{010} + \xi_2\ \mathbf{q}^2_{001}} \tag{6.198}$$

It can be easily verified that repeated triangular linear interpolations as in equations (6.196)–(6.198) provide the same result for a point on the cubic triangular surface by Bernstein polynomials as given by equation (6.195). Now, we can generalize the above observations to describe ***de Casteljau algorithm*** of repeated triangular interpolations for a triangular surface of degree n; given $\mathbf{q}^n_{ijk}$, $i+j+k=n$, $i,j,k \geq 0$.

6.7.5 de Casteljau Algorithm

Step 0: set $s = 1$ and $i+j+k = n-s$ and $\mathbf{q}^0_{ijk} = \mathbf{q}_{ijk}$

$$\boxed{\mathbf{q}^s_{ijk}(\boldsymbol{\xi}) = \xi_0\ \mathbf{q}^{s-1}_{(i+1)jk}(\boldsymbol{\xi}) + \xi_1\ \mathbf{q}^{s-1}_{i(j+1)k}(\boldsymbol{\xi}) + \xi_2\ \mathbf{q}^{s-1}_{ij(k+1)}(\boldsymbol{\xi})} \tag{6.199}$$

Step 1: set $s = s+1$; if $s < n$, go to ***Step 0***, else go to ***Step 2***.
Step 2: a point, with ***barycentric coordinates***, (ξ_0, ξ_1, ξ_2), is given by:

$$\boxed{\mathbf{S}^n(\boldsymbol{\xi}) = \mathbf{q}^n_{000}(\boldsymbol{\xi})} \tag{6.200}$$

Let us look at some intermediate points in de Casteljau algorithm for a quadratic triangle, that is, $n = 2$; for $s = 1$, we get, from ***de Casteljau algorithm*** as given by equations (6.199) and

(6.200), $i + j + k = n - s = 2 - 1 = 1$: let $i = 1$, $j = k = 0$, then, noting equation (6.182), and introducing: $\mathbf{j} \equiv (l, m, n)$ such that $|\mathbf{j}| \equiv l + m + n$, we can express the intermediate point, $\mathbf{q}^1_{100}$, in terms of summation over Bernstein polynomials as under:

$$\begin{aligned}
\mathbf{q}^s_{\mathbf{i}} = \mathbf{q}^1_{100} &= \xi_0\, \mathbf{q}^0_{200} + \xi_1\, \mathbf{q}^0_{110} + \xi_2\, \mathbf{q}^0_{101} \\
&= \xi_0\, \mathbf{q}_{200} + \xi_1\, \mathbf{q}_{110} + \xi_2\, \mathbf{q}_{101} \\
&= B^1_{100}\, \mathbf{q}_{200} + B^1_{010}\, \mathbf{q}_{110} + B^1_{001}\, \mathbf{q}_{101} \\
&= B^s_{l00}\, \mathbf{q}_{i00+l00} + B^s_{0m0}\, \mathbf{q}_{i00+0m0} + B^s_{00n}\, \mathbf{q}_{i00+00n} \\
&= \sum_{|\mathbf{j}|=s=1} B^s_{lmn}\, \mathbf{q}_{(ijk)+(lmn)} = \sum_{|\mathbf{j}|=s=1} B^s_{\mathbf{j}}\, \mathbf{q}_{\mathbf{i}+\mathbf{j}}
\end{aligned} \tag{6.201}$$

In other words, the intermediate points at each iteration form a triangular surface.

6.7.6 Bernstein Polynomials and de Casteljau Algorithm

Since triangular surfaces can be expressed by both Bernstein polynomials and de Casteljau algorithm as presented above, we expect a connection between the intermediate points of the de Casteljau algorithm and the Bernstein polynomials; this relationship can be shown, by inductive reasoning and an application of the recursion relation of the Bernstein polynomials as given by equation (6.186), to be:

$$\begin{aligned}
\mathbf{q}^s_{\mathbf{i}}(\xi) &= \sum_{|\mathbf{j}|=s} B^s_{\mathbf{j}}(\xi)\, \mathbf{q}_{\mathbf{i}+\mathbf{j}}, \quad |\mathbf{i}| = n - s \\
\mathbf{q}^s_{ijk}(\xi) &= \sum_{p+q+r=s} B^s_{pqr}(\xi)\, \mathbf{q}_{(i+p)(j+q)(k+r)}, \quad i + j + k = n - s
\end{aligned} \tag{6.202}$$

Now, taking, $s = n$ in equation (6.202), we can express any point on a triangular Bezier patch as:

$$\begin{aligned}
\mathbf{S}^n(\xi) \stackrel{\Delta}{=} \mathbf{q}^n_{\mathbf{0}}(\xi) &= \sum_{|\mathbf{j}|=n} B^n_{\mathbf{j}}(\xi)\, \mathbf{q}_{\mathbf{j}}, \\
\mathbf{q}^n_{ijk}(\xi) &= \sum_{p+q+r=n} B^n_{pqr}(\xi)\, \mathbf{q}_{pqr}
\end{aligned} \tag{6.203}$$

6.7.7 Derivatives and Normal to Triangular Surface

The derivatives at a point on a triangular surface is required for various reasons; particularly, the first derivatives and the cross derivatives are necessary for evaluating the ***normal*** to the surface. Assuming that a triangular surface is embedded in a three-dimensional Euclidean space, we will be interested in ***directional derivatives***. Suppose, any two points on the surface are given by the barycentric coordinates, $\boldsymbol{\xi} = (\xi_0, \xi_1, \xi_2)^T$ and $\boldsymbol{\eta} = (\eta_0, \eta_1, \eta_2)^T$, respectively; and we are interested in finding the derivative at $\boldsymbol{\xi}$ along $\boldsymbol{\gamma} = (\gamma_0, \gamma_1, \gamma_2)^T = (\boldsymbol{\eta} - \boldsymbol{\xi})$ direction. Clearly, $\gamma_0 + \gamma_1 + \gamma_2 = 0$ since for ***barycentric coordinates***, $\xi_0 + \xi_1 + \xi_2 = 1$ and $\eta_0 + \eta_1 + \eta_2 = 1$.

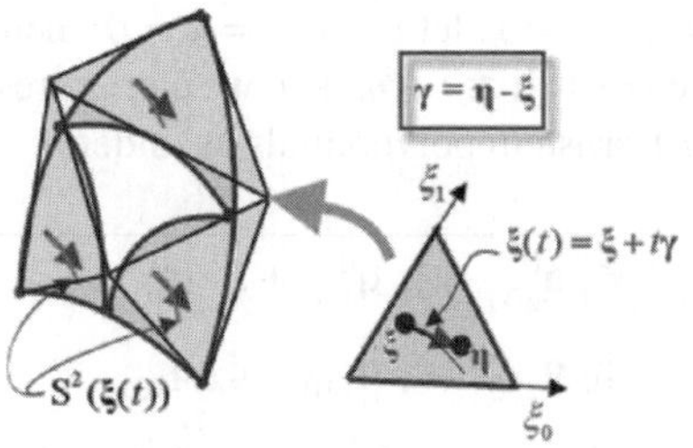

Figure 6.33 First derivative of a quadrilateral triangle.

6.7.8 The First Derivatives: Tangents

Following the ε- method for the ***Gateaux or directional derivative***, $\nabla_{\gamma}\mathbf{S}^n(\boldsymbol{\xi})$, of a triangular surface, $\mathbf{S}^n(\boldsymbol{\xi}) = (S_0^n(\boldsymbol{\xi}), S_1^n(\boldsymbol{\xi}), S_2^n(\boldsymbol{\xi}))^T$, at $\boldsymbol{\xi}$ along $\boldsymbol{\gamma}$ direction is given by the ***Jacobian*** or scaling matrix as:

$$\nabla_{\gamma}\mathbf{S}^n(\boldsymbol{\xi}) = \lim_{\varepsilon\to 0}\left\{\frac{\mathbf{S}^n(\boldsymbol{\xi}+\varepsilon\boldsymbol{\gamma}) - \mathbf{S}^n(\boldsymbol{\xi})}{\varepsilon}\right\} = \frac{\partial \mathbf{S}^n(\boldsymbol{\xi})}{\partial \boldsymbol{\xi}} \bullet \boldsymbol{\gamma} = \begin{bmatrix} \dfrac{\partial S_0^n}{\partial \xi_0} & \dfrac{\partial S_0^n}{\partial \xi_1} & \dfrac{\partial S_0^n}{\partial \xi_2} \\ \dfrac{\partial S_1^n}{\partial \xi_0} & \dfrac{\partial S_1^n}{\partial \xi_1} & \dfrac{\partial S_1^n}{\partial \xi_2} \\ \dfrac{\partial S_2^n}{\partial \xi_0} & \dfrac{\partial S_2^n}{\partial \xi_1} & \dfrac{\partial S_2^n}{\partial \xi_2} \end{bmatrix} \begin{Bmatrix} \gamma_0 \\ \gamma_1 \\ \gamma_1 \end{Bmatrix} \tag{6.204}$$

Geometrically, as shown in Figure 6.33, the directional derivative defines the image, $\mathbf{S}(\xi(t))$, on a three-dimensional triangular surface, of a vector function, $\boldsymbol{\xi}(t) = \boldsymbol{\xi} + t\boldsymbol{\gamma}$, of t with $\boldsymbol{\gamma} \overset{\Delta}{=} \boldsymbol{\eta} - \xi$ as a vector in the two-dimensional triangle with beginning and end points having barycentric coordinates, $\boldsymbol{\xi} = (\xi_0, \xi_1, \xi_2)^T$ and $\boldsymbol{\eta} = (\eta_0, \eta_1, \eta_2)^T$, respectively; the sum of the components of $\boldsymbol{\gamma}$ is given by:

$$\begin{aligned} \gamma_0 + \gamma_1 + \gamma_2 &= (\eta_0 - \xi_0) + (\eta_1 - \xi_1) + (\eta_2 - \xi_2) \\ &= (\eta_0 + \eta_1 + \eta_2) - (\xi_0 + \xi_1 + \xi_2) \\ &= 0 \end{aligned} \tag{6.205}$$

Noting, from equations (6.181) and (6.205), that:

$$\begin{aligned} B_{100}^1(\boldsymbol{\gamma}) &= \gamma_0 = \xi_0 - \eta_0 \\ B_{010}^1(\boldsymbol{\gamma}) &= \gamma_1 = \xi_1 - \eta_1 \\ B_{001}^1(\boldsymbol{\gamma}) &= \gamma_2 = \xi_2 - \eta_2 \end{aligned} \tag{6.206}$$

we can rewrite the first derivative expression given by equation (6.204) as:

$$\boxed{\nabla_{\gamma}\mathbf{S}^n(\xi) = \sum_{|\mathbf{j}|=1} \partial^{\mathbf{j}}\mathbf{S}^n(\xi)\, B^1_{\mathbf{j}}(\gamma)} \tag{6.207}$$

where $\partial^{\mathbf{j}}(\bullet)$ is defined as in equation (6.187).

6.7.9 *The Higher Order Derivatives*

We can find the expressions for the higher-order derivatives by repeated differentiation of equation (6.207); for example, for second derivative:

$$\boxed{\begin{aligned}
\nabla^2_{\gamma}\mathbf{S}^n(\xi) &= \nabla_{\gamma}(\nabla_{\gamma}\mathbf{S}^n(\xi)) \\
&= \left\{\begin{matrix}
\gamma_0\left\{\gamma_0\dfrac{\partial^2}{\partial\xi_0^2} + \gamma_1\dfrac{\partial^2}{\partial\xi_0\partial\xi_1} + \gamma_2\dfrac{\partial^2}{\partial\xi_0\partial\xi_2}\right\} + \\
\gamma_1\left\{\gamma_0\dfrac{\partial^2}{\partial\xi_0\partial\xi_1} + \gamma_1\dfrac{\partial^2}{\partial\xi_1^2} + \gamma_2\dfrac{\partial^2}{\partial\xi_1\partial\xi_2}\right\} + \\
\gamma_2\left\{\gamma_0\dfrac{\partial^2}{\partial\xi_0\partial\xi_2} + \gamma_1\dfrac{\partial^2}{\partial\xi_1\partial\xi_2} + \gamma_2\dfrac{\partial^2}{\partial\xi_2^2}\right\}
\end{matrix}\right\}\mathbf{S}^n(\xi) \\
&= \sum_{|\mathbf{j}|=2} \partial^{\mathbf{j}}\mathbf{S}^n(\xi)\, B^2_{\mathbf{j}}(\gamma)
\end{aligned}} \tag{6.208}$$

where we have used the definition of the Bernstein polynomials, $B^2_{\mathbf{j}}(\gamma)$, with argument set at $\boldsymbol{\gamma}$ as given by equation (6.181); the equation (6.208), clearly can be written in matrix form, with the matrix known as the ***Hessian matrix***, as:

$$\boxed{\nabla^2_{\gamma}\mathbf{S}^n(\xi) = \begin{Bmatrix}\gamma_0\\ \gamma_1\\ \gamma_2\end{Bmatrix}^T \underbrace{\begin{bmatrix}
\dfrac{\partial^2}{\partial\xi_0^2} & \dfrac{\partial^2}{\partial\xi_0\partial\xi_1} & \dfrac{\partial^2}{\partial\xi_0\partial\xi_2} \\
\dfrac{\partial^2}{\partial\xi_1\partial\xi_0} & \dfrac{\partial^2}{\partial\xi_1^2} & \dfrac{\partial^2}{\partial\xi_1\partial\xi_2} \\
\dfrac{\partial^2}{\partial\xi_2\partial\xi_0} & \dfrac{\partial^2}{\partial\xi_2\partial\xi_1} & \dfrac{\partial^2}{\partial\xi_2^2}
\end{bmatrix}}_{\text{Hessian Matrix}} \begin{Bmatrix}\gamma_0\\ \gamma_1\\ \gamma_2\end{Bmatrix}\mathbf{S}^n(\xi)} \tag{6.209}$$

We can generalize equation (6.209) to express any s^{th} ***directional derivatives*** as:

$$\boxed{\nabla^s_{\gamma}\mathbf{S}^n(\xi) = \sum_{|\mathbf{j}|=s} \partial^{\mathbf{j}}\mathbf{S}^n(\xi)\, B^s_{\mathbf{j}}(\gamma)} \tag{6.210}$$

Additionally, If we apply equation (6.210) to Bernstein polynomials along with equation (6.188), we will get:

$$\nabla_\gamma^s B_\mathbf{i}^n(\xi) = \sum_{|\mathbf{j}|=s} \partial^\mathbf{j} B_\mathbf{i}^n(\xi)\, B_\mathbf{j}^s(\gamma) = \frac{n!}{(n-s)!} \sum_{|\mathbf{j}|=s} B_{\mathbf{i}-\mathbf{j}}^{n-s}(\xi)\, B_\mathbf{j}^s(\gamma) \tag{6.211}$$

Using definition for the points, $\mathbf{S}^n(\xi)$, from equation (6.203) and applying equation (6.211) in equation (6.210), we get:

$$\begin{aligned}\nabla_\gamma^s \mathbf{S}^n(\xi) &= \sum_{|\mathbf{j}|=s} \partial^\mathbf{j} \mathbf{S}^n(\xi)\, B_\mathbf{j}^s(\gamma) = \sum_{|\mathbf{j}|=s} \partial^\mathbf{j} \Big(\sum_{|\mathbf{k}|=n} B_\mathbf{k}^n(\xi)\, \mathbf{q}_\mathbf{k} \Big) B_\mathbf{j}^s(\gamma) \\ &= \sum_{|\mathbf{j}|=s} \sum_{|\mathbf{k}|=n} \partial^\mathbf{j} B_\mathbf{k}^n(\xi)\, B_\mathbf{j}^s(\gamma)\, \mathbf{q}_\mathbf{k} = \frac{n!}{(n-s)!} \sum_{|\mathbf{k}|=n} \sum_{|\mathbf{j}|=s} B_{\mathbf{k}-\mathbf{j}}^{n-s}(\xi)\, B_\mathbf{j}^s(\gamma)\, \mathbf{q}_\mathbf{k}\end{aligned} \tag{6.212}$$

In equation (6.212), letting $\mathbf{i} \stackrel{\Delta}{=} \mathbf{k} - \mathbf{j}$, with $|\mathbf{i}| = n - s$, and applying equation (6.202), we finally get for the ***directional derivative of a Bezier triangular surface***:

$$\nabla_\gamma^s \mathbf{S}^n(\xi) = \frac{n!}{(n-s)!} \sum_{|\mathbf{j}|=s} B_\mathbf{j}^s(\gamma) \left(\sum_{|\mathbf{i}|=n-s} B_\mathbf{i}^{n-s}(\xi)\, \mathbf{q}_{\mathbf{i}+\mathbf{j}} \right) = \frac{n!}{(n-s)!} \sum_{|\mathbf{j}|=s} B_\mathbf{j}^s(\gamma)\, \mathbf{q}_\mathbf{j}^{n-s}(\xi) \tag{6.213}$$

Now, equation (6.202) holds equally true if we replace ξ by γ; in this case, we get the dual expression for the ***directional derivative of a Bezier triangular surface*** from equation (6.212) as:

$$\begin{aligned}\nabla_\gamma^s \mathbf{S}^n(\xi) &= \frac{n!}{(n-s)!} \sum_{|\mathbf{i}|=n-s} B_\mathbf{i}^{n-s}(\xi) \left(\sum_{|\mathbf{j}|=s} B_\mathbf{j}^s(\gamma)\, \mathbf{q}_{\mathbf{i}+\mathbf{j}} \right) \\ &= \frac{n!}{(n-s)!} \sum_{|\mathbf{i}|=n-s} B_\mathbf{i}^{n-s}(\xi)\, \mathbf{q}_\mathbf{i}^s(\gamma)\end{aligned} \tag{6.214}$$

Remarks

- For $s = 1$ in equation (6.214), we get the first directional derivative as:

$$\nabla_\gamma^1 \mathbf{S}^n(\xi) = n \sum_{|\mathbf{i}|=n-1} B_\mathbf{i}^{n-1}(\xi)\, \mathbf{q}_\mathbf{i}^1(\gamma) \tag{6.215}$$

where, by de Casteljau algorithm given by equation (6.199) with ξ replaced by γ, we have:

$$\mathbf{q}_\mathbf{i}^1(\gamma) \stackrel{\Delta}{=} \mathbf{q}_{ijk}^1(\gamma) = \gamma_0\, \mathbf{q}_{(i+1)jk} + \gamma_1\, \mathbf{q}_{i(j+1)k} + \gamma_2\, \mathbf{q}_{ij(k+1)} \tag{6.216}$$

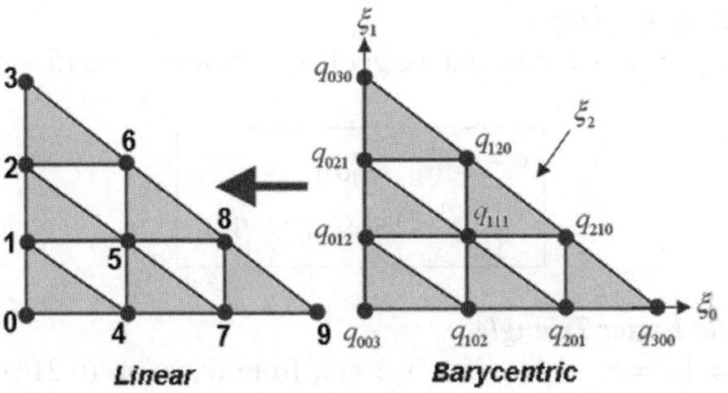

Figure 6.34 Barycentric to linear correspondence for a cubic triangle.

- From equation (6.216), $\mathbf{q}_\mathbf{i}^1(\boldsymbol{\gamma})$ are affine map of the two-dimensional vector, $\boldsymbol{\gamma}$, to the three-dimensional triangle formed by $\mathbf{q}_{(i+1)jk}, \gamma_1\ \mathbf{q}_{i(j+1)k}$ and $+\gamma_2\ \mathbf{q}_{ij(k+1)}$.
- From equation (6.215), the first directional derivative of a triangular Bezier surface is a triangular Bezier patch whose coefficients are the images of $\boldsymbol{\gamma}$ on each subtriangle in the control net,
- Computationally speaking, we first compute $n\ \mathbf{q}_\mathbf{i}^1(\boldsymbol{\gamma})$ and then use these as the input for a direct Bernstein or iterative de Casteljau algorithm of Bezier patch evaluation.

6.7.9.1 Algorithm TBT: Tangents for a Bezier Triangle

The tangents for various ***barycentric coordinate***s are stored in a linear format as in a vector for simple and efficient computer memory management; we indicate this below as ***Bary2Lin*** storage. For example, Figure 6.34 depicts the connection between the linear positions and the corresponding barycentric coordinates of control net of a cubic patch:

Now, we can set up the algorithm as follows:

Step 0: for each triangular patch, collect the control net
Step 1: form the tangent control net, using equation (6.216)
Step 2: store in ***Bary2Lin*** array.

6.7.10 *Normals on a Bezier Triangle*

As indicated before, the normals on a triangular surface taken as a degenerate Cartesian product patch is given by equation **(6.153)**; for computation of normals on its own right on the basis of barycentric coordinates of a Bezier triangular surface, we will need first derivatives, $\nabla_{\gamma_{\xi_1}} \mathbf{S}^n(\xi)$ and $\nabla_{\gamma_{\xi_2}} \mathbf{S}^n(\xi)$ along ξ_1 and ξ_2 directions, respectively, where:

$$\boxed{\begin{aligned} \boldsymbol{\gamma}_{\xi_0} &= (1, 0, -1)^T \text{ along } \xi_0 \\ \boldsymbol{\gamma}_{\xi_1} &= (0, 1, -1)^T \text{ along } \xi_1 \end{aligned}} \tag{6.217}$$

6.7.10.1 Axis-Directional Tangents

Using equation (6.216), we can get the tangents along $\boldsymbol{\gamma}_{\xi_1}$ and $\boldsymbol{\gamma}_{\xi_2}$ defined by equation (6.217); let us indicate these for linear, quadratic and cubic triangles:

6.7.10.1.1 Linear Bezier Triangle

In this case, $n = 1 \Rightarrow |\mathbf{i}| = n - 1 = 0$, and we get from equations (6.215)–(6.217):

$$\boxed{\begin{aligned} c &= \mathbf{q}_{100} - \mathbf{q}_{001} \\ \nabla^1_{\gamma_{\xi_1}} \mathbf{S}^1(\xi) &= \mathbf{q}_{010} - \mathbf{q}_{001} \end{aligned}} \tag{6.218}$$

6.7.10.1.2 Quadratic Bezier Triangle

In this case, $n = 2 \Rightarrow |\mathbf{i}| = n - 1 = 1$, and we get, from equation (6.216), the tangent net for direction γ_{ξ_0} as:

$$\boxed{\begin{aligned} 2\mathbf{q}^1_{100}(\gamma_{\xi_0}) &= 2(\mathbf{q}_{200} - \mathbf{q}_{101}) \\ 2\mathbf{q}^1_{010}(\gamma_{\xi_0}) &= 2(\mathbf{q}_{110} - \mathbf{q}_{011}) \\ 2\mathbf{q}^1_{001}(\gamma_{\xi_0}) &= 2(\mathbf{q}_{101} - \mathbf{q}_{002}) \end{aligned}} \tag{6.219}$$

and for direction γ_{ξ_1} as:

$$\boxed{\begin{aligned} 2\mathbf{q}^1_{100}(\gamma_{\xi_1}) &= 2(\mathbf{q}_{110} - \mathbf{q}_{101}) \\ 2\mathbf{q}^1_{010}(\gamma_{\xi_1}) &= 2(\mathbf{q}_{020} - \mathbf{q}_{011}) \\ 2\mathbf{q}^1_{001}(\gamma_{\xi_1}) &= 2(\mathbf{q}_{011} - \mathbf{q}_{002}) \end{aligned}} \tag{6.220}$$

Finally, we get the first directional derivatives, $\nabla^1_{\gamma_{\xi_0}} \mathbf{S}^2(\xi)$ and $\nabla^1_{\gamma_{\xi_1}} \mathbf{S}^2(\xi)$, of the Bezier triangle by using the tangent nets given by equation (6.219) and (6.220), respectively, in equation (6.215) for linear patch.

6.7.10.1.3 Cubic Bezier Triangle

In this case, $n = 3 \Rightarrow |\mathbf{i}| = n - 1 = 2$, and we get, from equation (6.216), the tangent net (see Figure 6.35) for direction γ_{ξ_0} as:

$$\boxed{\begin{aligned} 3\mathbf{q}^1_{200}(\gamma_{\xi_0}) &= 3(\mathbf{q}_{300} - \mathbf{q}_{201}) \\ 3\mathbf{q}^1_{110}(\gamma_{\xi_0}) &= 3(\mathbf{q}_{210} - \mathbf{q}_{111}) \\ 3\mathbf{q}^1_{101}(\gamma_{\xi_0}) &= 3(\mathbf{q}_{201} - \mathbf{q}_{102}) \\ 3\mathbf{q}^1_{011}(\gamma_{\xi_0}) &= 3(\mathbf{q}_{111} - \mathbf{q}_{012}) \\ 3\mathbf{q}^1_{020}(\gamma_{\xi_0}) &= 3(\mathbf{q}_{1201} - \mathbf{q}_{021}) \\ 3\mathbf{q}^1_{002}(\gamma_{\xi_1}) &= 3(\mathbf{q}_{102} - \mathbf{q}_{021}) \end{aligned}} \tag{6.221}$$

and similar expressions can be obtained for direction γ_{ξ_1}. Finally, we get the first directional derivatives, $\nabla^1_{\gamma_{\xi_0}} \mathbf{S}^3(\xi)$ and $\frac{\partial}{\partial \gamma_{\xi_1}} \mathbf{S}^3_{ijk}$, of the Bezier triangle by using the tangent nets given by equation (6.219) and (6.220), respectively, in equation (6.215) for quadratic patch.

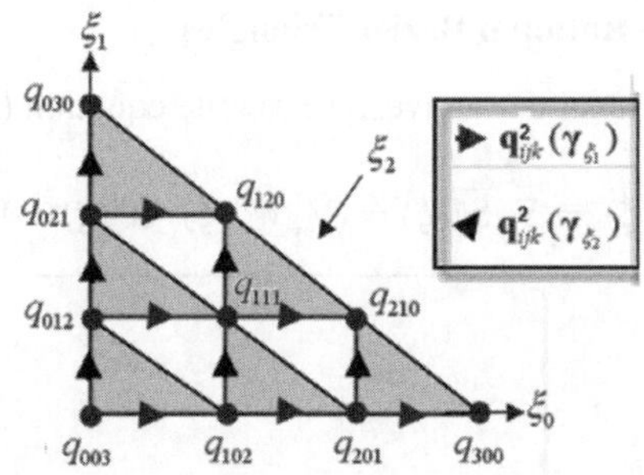

Figure 6.35 Tangent directions for a cubic triangle.

6.7.10.2 Normals to the Bezier Triangle

Knowing the first axis-directional derivatives as given above, we can determine the expression for the normal, $\mathbf{N}(\xi)$, at a point, ξ, on a Bezier triangle as:

$$\boxed{\mathbf{N}(\xi) = \frac{\nabla^1_{\gamma_{\xi_0}} \mathbf{S}^n(\xi) \times \nabla^1_{\gamma_{\xi_1}} \mathbf{S}^n(\xi)}{\left\| \nabla^1_{\gamma_{\xi_0}} \mathbf{S}^n(\xi) \times \nabla^1_{\gamma_{\xi_1}} \mathbf{S}^n(\xi) \right\|}} \tag{6.222}$$

6.7.11 *Rational Bezier Triangle and Derivatives*

Following along the line of rational Bezier curves and rational Cartesian product Bezier surfaces, we can think of rational Bezier triangular surface as projection of 4D non-rational Bezier triangular surface to 3D as:

$$\boxed{\begin{aligned} \mathbf{R}^n(\xi) = \frac{\overrightarrow{\mathbf{S}^n}(\xi)}{W(\xi)} &= \frac{\sum\limits_{|\mathbf{j}|=n} w_{\mathbf{j}} B^n_{\mathbf{j}}(\xi)\, \mathbf{q}_{\mathbf{j}}}{\sum\limits_{|\mathbf{j}|=n} w_{\mathbf{j}} B^n_{\mathbf{j}}(\xi)}, \quad \mathbf{j} \stackrel{\Delta}{=} ijk, \quad |\mathbf{j}| = i+j+k, \quad \xi = (\xi_0, \xi_1, \xi_2) \\ &= \underbrace{\left(\frac{\sum\limits_{|\mathbf{j}|=n} w_{\mathbf{j}} B^n_{\mathbf{j}}(\xi)}{\sum\limits_{|\mathbf{j}|=n} w_{\mathbf{j}} B^n_{\mathbf{j}}(\xi)} \right)}_{\substack{\text{Basis functions for} \\ \text{rational Bezier triangle}}} \mathbf{q}_{\mathbf{j}} \end{aligned}} \tag{6.223}$$

where $w_{\mathbf{j}}$'s are, as before, the weights corresponding to the control points, $\mathbf{q}_{\mathbf{j}}$.

6.7.11.1 Derivatives of the Rational Bezier Triangles

For computation of the directional derivatives, we rewrite equation (6.223) such that:

$$\overrightarrow{\mathbf{S}^n}(\xi) = W(\xi)\,\mathbf{R}^n(\xi) \Rightarrow \nabla_\gamma \overrightarrow{\mathbf{S}^n}(\xi) = (\nabla_\gamma W^n(\xi))\,\mathbf{R}^n(\xi) + W^n(\xi)\,(\nabla_\gamma \mathbf{R}^n(\xi))$$

$$\Rightarrow \boxed{\nabla_\gamma \mathbf{R}^n(\xi) = \frac{1}{W^n(\xi)}\left\{ \underbrace{\nabla_\gamma \overrightarrow{\mathbf{S}^n}(\xi)}_{\substack{\text{Computed as in}\\ \text{non-rational case}\\ \text{with } w_j B_j^n(\xi) \text{ as}\\ \text{coefficients}}} - \underbrace{(\nabla_\gamma W^n(\xi))}_{\substack{\text{Computed as}\\ \text{one-dimensional}\\ \text{non-rational case}}}\ \mathbf{R}^n(\xi)\right\}} \tag{6.224}$$

- Instead of computing the directional derivative at any particular barycentric coordinates, say, (ξ_0, ξ_1, ξ_2), we may apply subdivision using de Casteljau algorithm at the point to break it into three subpatches and evaluate the derivative from one of the subdivided triangles with barycentric coordinates at the corner point given by $(0, 0, n)$.

6.7.12 Where We Would Like To Go

We have presented here in some detail a topical discussion of the basic theory and computation of surfaces as a building block to apply to structural analysis of plates and shells by finite element method. Thus, we may now proceed to:

- Essential Mechanics (Chapter 7)
- plates and shell analysis (Chapter 11)

Part III

Essential Mechanics

7

Nonlinear Mechanics: A Lagrangian Approach

7.1 Introduction

7.1.1 What We Should Already Know

Introduction and motivation, essential mathematics, tensors, rotation tensor

7.1.2 Why We Are Here

The main goal here is to review the essential aspects of nonlinear solid mechanics in so far as these help us to understand our subsequent ***c-type finite element method*** and its application to nonlinear beam, plate and shell analyses. The nonlinear solid mechanics may be visualized essentially to be built on three "pillars" (Figure 7.1): (1) the ***geometry of deformation*** or ***strains***, (2) the ***balance principles*** for ***transmission of force*** or ***stress***, and (3) the ***constitutive theory*** relating stress and strain of a material body. The geometry of deformation and the balance principles produce exact relationships that are applicable to any material body while the constitutive theory, by its very nature of fitting the experimental data, is approximate at best, but it distinguishes or establishes similarities in different materials. The nonlinearity in solid mechanics may be identified under two different categories: (1) the ***geometric nonlinearity*** due to finite displacement and rotation and (2) the ***material nonlinearity*** due to nonlinear stress–strain law. For applications covered here, we restrict our discussions to only geometric nonlinearity; the material behavior is assumed to be what may be termed a ***linear***, ***homogeneous***, ***isotropic***, ***hyperelastic*** response under external stimuli. In any case, as opposed to linear mechanics, nonlinear mechanics must recognize the difference between the undeformed and deformed geometry of a material body. This, in turn, results in the description of all entities measured either in terms of the undeformed, reference configuration known as the Lagrangian frame of reference, or deformed, spatial configuration known as the Eulerian frame of analysis. Throughout the entire book, we will follow exclusively the Lagrangian description of a material body with minimal connection to the Eulerian description such as the Cauchy stress tensor for later utilization in the constitutive theory.

First, we introduce an analysis of the geometry of deformation that gives rise to the local concept of the ***deformation gradient tensor***, which will help us to define the Lagrangian strain tensor: the symmetric ***Green–Lagrange strain tensor***. On the other hand, the balance principles, that include, under non-relativistic setting, the ***principle of mass conservation***, the ***principle of***

Computation of Nonlinear Structures: Extremely Large Elements for Frames, Plates and Shells, First Edition. Debabrata Ray.

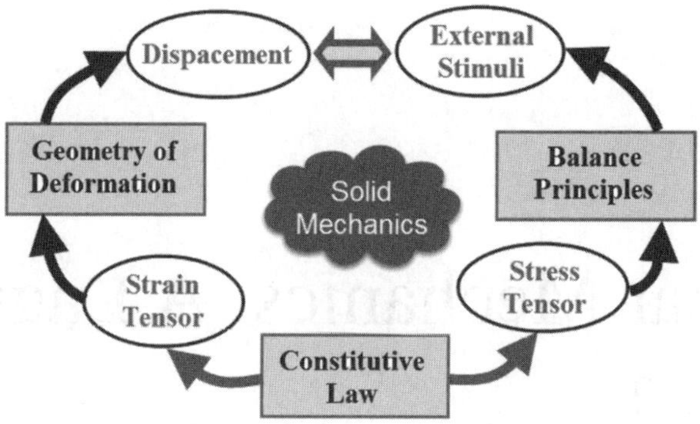

Figure 7.1 Three pillars of continuum mechanics.

momentum balance, the ***principle of moment of momentum balance*** and the ***principle of energy balance***, are discussed to culminate in defining the symmetric Lagrangian stress tensors: the first ***Piola–Kirchhoff (PK I)*** and the ***second Piola–Kirchhoff (PK II) stress tensors***, and the ***energy conjugate*** stress–strain pairs. Finally, the constitutive theory of nonlinear mechanics is introduced with the ultimate goal of developing the classical elasticity relationship between the Lagrangian strains and stresses of a material body; this is achieved through presentation of the ***principle of determinism***, the ***principle of local action***, and two major restrictions on the constitutive theory: the ***condition of frame indifference*** and the ***condition of symmetry*** through representation of what is known as the ***isotropic tensor functions***, to generate material stress–strain relationships that try to mimic the properties of real materials.

7.1.3 Where We Would Like to Go

Following the above introduction, we would like to explore the essential aspects of nonlinear solid mechanics as follows:

- Section 7.2: Lagrangian geometry of deformation: strain tensors
- Section 7.3: Balance principles and transmission of forces: stress tensors
- Section 7.4: Constitutive theory: stress–strain relationship

7.2 Deformation Geometry: Strain Tensors

7.2.1 Why We Are Here

Here we describe the first metaphorical pillar: the deformation patterns, for example, extension, contraction, shear and twist, of solid deformable bodies, subjected to external stimulations such as forces, moments and heat, that was alluded to in the introduction for nonlinear solid mechanics, to help us understand our subsequent c-type finite element method and its application to nonlinear beam, plate and shell analyses. One of the important aspects to recognize about the scope of this study is that the deformation of a body is all about its geometry and its physical compatibility without any reference to how the causal forces transmit through the body. Thus, in short, it is the differential geometry (as introduced in Chapter 3 on tensors) of deformations that we focus on.

Briefly, the deformation of a material body may be defined as the history of separation over time of any two points in the body, the points being the map of the material particles of the body, defined by its volume and surface over time, that occupies a continuous region on, say, Euclidean or Riemannian point space. Accordingly, the aim is to establish suitable space and coordinate systems in the space and characterize the geometry (i.e. volume and surface area) of the body and its possible deformation pattern over time in terms of such coordinate systems. Additionally, any deformation, apart from being relative motion between the material particles of a body, may include rigid body motion meaning the distance between any two points of a body remains the same over time. For infinitesimal, that is, linear motions of a body, we are not interested in the rigid body motion part of the deformation history; and as it will turn out, the strain measure with what is known as the ***strain tensor*** suffices to avoid inclusion of any rigid body motion. However, for finite, that is, nonlinear motions of a body deforming with large displacements and large rotations, it is important to devise a measure that can capture and reflect both these deformations. For this, we introduce what are known as the ***deformation map*** and the ***deformation gradient***. For infinitesimal, that is, linear, motion of a body all field quantities of solid mechanics are referred to the undeformed reference coordinate system, but for finite displacements and rotations, that is, nonlinear motions, of a body, we can either name a material particle and track its motion with reference material coordinate system known as the ***Lagrangian*** or ***material coordinate system***, or concentrate on a point in space, and associate that point with a particle in the deformed body that happens to occupy the space, with reference to the current coordinate system known as the ***Eulerian*** or ***spatial coordinate system***. The latter seems more useful for fluid bodies; in what follows, for deformable solid bodies, we choose to select and restrict ourselves strictly to the Lagrangian material frame of reference. It is important to note that in fact a body may not necessarily ever assume the reference state, but it helps analytically and computationally to identify the reference state with the undeformed state of the body. Henceforth, all field quantities will be expressed with reference to the undeformed, material, Lagrangian frame of reference. Finally, under the Lagrangian frame of reference, we can think of two coordinate systems: the Cartesian coordinate system and the curvilinear (convected) coordinate system, as referred in the Chapter 3 on tensors, Chapter 5 on curves and Chapter 6 on surfaces; the name ***convected***, or sometime ***intrinsic***, refers to the requirement that the numerical values of the coordinates of a material point remain the same before and after the deformation over all times of interest. For curved beam and shell geometries, it seems more suitable to describe the field quantities in curvilinear coordinates. However, for our purposes of representing the large displacement and rotational motion of beams, plates and shells with shear deformations, it turned out that the most desirable and efficient strategy is to leave behind both the Cartesian coordinate system and the convected curvilinear coordinate system and concentrate on what may be called a ***rotating coordinate system*** in the case of static loading, or a ***spinning coordinate system*** in the case of dynamic loading. We will have ample opportunities to have detailed discussion of these coordinate systems when we introduce the chapters on applications to nonlinear beams, plates and shells. So, in what follows, we engage in a discussion of the geometry of deformation, primarily in the curvilinear coordinate system in Lagrangian material framework that sets the stage for future discussion of the deformation map and deformation gradient in rotating or spinning coordinate systems.

7.2.2 Deformation Map

Both the Cartesian coordinate system and the curvilinear coordinate system with corresponding definition of base vectors, and so on, were detailed in Chapter 3 on tensors. For the curvilinear

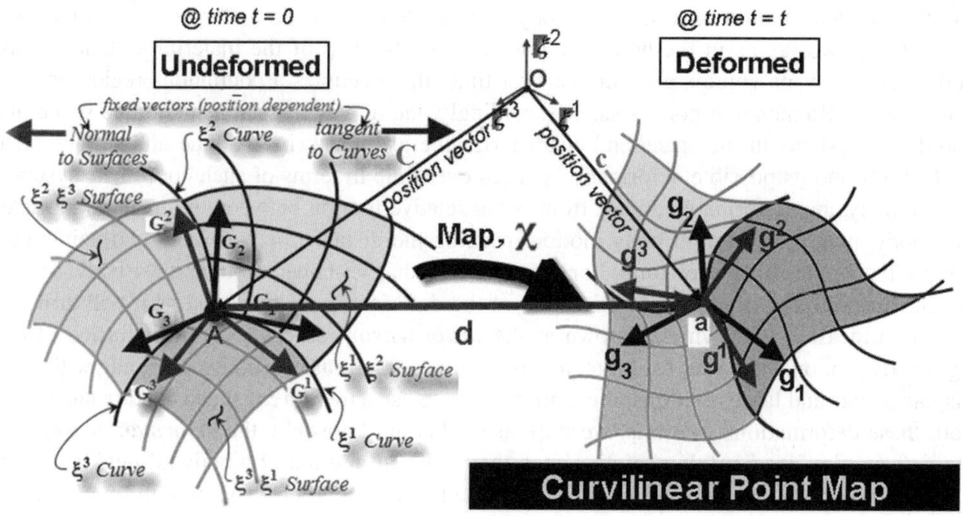

Figure 7.2 Curvilinear point map.

coordinate system, as shown in Figure 7.2, let us consider a material particle, $A \in \Omega_0$, with its position vector, $\mathbf{C}$, defined by the triplet of real numbers as its coordinates, $\xi^i, i = 1, 2, 3$, that is, $\mathbf{C} = \xi^i\, \mathbf{G}_i$ with $\{\mathbf{G}_i = \frac{\partial \mathbf{C}}{\partial \xi^i}\}$, $i = 1, 2, 3$, as a set of covariant base vectors in the curvilinear coordinate system in the undeformed body, Ω_0, at material particle, A.

After deformation, the same point assumes the location marked as a, with the position vector given as $\mathbf{c}$, defined by the same (convective) triplet of real numbers as its coordinates, $\xi^i, i = 1, 2, 3$, that is, $\mathbf{c} = \xi^i\, \mathbf{g}_i$ with $\{\mathbf{g}_i = \frac{\partial \mathbf{c}}{\partial \xi^i}\}$, $i = 1, 2, 3$, as a set of covariant base vectors in the curvilinear coordinate system in the deformed body, Ω, at material particle, a. Although, in its most general form, the origins can be chosen as separate points for the undeformed and deformed position vector, for our purposes and without loss of much generality, both the origins are chosen to coincide; we recall that the position vectors depend on the choice of origin. The point, a, after deformation is separated from its undeformed location, A, by a vector, $\mathbf{d}$, known as the ***displacement vector***. We again recall that all vectors are affine in the sense that they do not depend on the choice of origin. Clearly, the displacement vector ties up both the undeformed and the deformed space, and hence is known as a two-point vector, and as such can be expressed in both the base vector sets, belonging to undeformed and deformed spaces. In all our subsequent application, we choose to express all physical fields, including the displacement field, in terms of the undeformed reference material coordinate system, referred to before as the Lagrangian frame of reference. We can express this deformation pattern by an one-to-one mapping, χ, known as the deformation map:

$$\begin{gathered} \mathbf{c} = \chi(\mathbf{C}) \\ \mathbf{c}(\xi^1,\ \xi^2,\ \xi^3) = \chi(\mathbf{C}(\xi^1,\ \xi^2,\ \xi^3)) \end{gathered} \tag{7.1}$$

A similar definition for the deformation map for the Cartesian coordinate system can be easily developed.

As shown in Figure 7.3, we only have to recall that the covariant derivatives coincide with the Gateaux or ordinary partial derivatives as the Christoffel symbols vanish identically, and the covariant and the contravariant base vectors coincide for the Cartesian coordinate system.

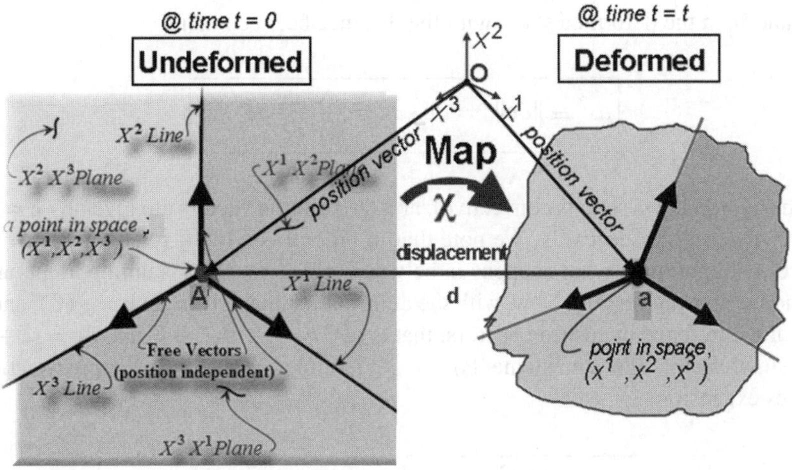

Figure 7.3 Cartesian point map.

7.2.3 *Deformation Gradient*

In order to study the deformation pattern local to, that is, in the neighborhood of, any material point, we need to introduce what is known as the ***deformation gradient*** at a material point.

7.2.3.1 Curvilinear (Convected) Coordinate System

Let us consider, as in Figure 7.4, this material particle, $A \in \Omega_0$, with position vector, $\mathbf{C}$.

Let $B \in \Omega_0$ be another material particle in the neighborhood of A separated by an infinitesimal distance, dS, where:

$$\boxed{dS^2 \equiv \|\mathbf{dC}\|^2 = \mathbf{dC} \bullet \mathbf{dC} = \frac{\partial \mathbf{C}}{\partial \xi^i}\frac{\partial \mathbf{C}}{\partial \xi^j} d\xi^i\, d\xi^j} \tag{7.2}$$

where $\mathbf{dC} = \frac{\partial \mathbf{C}}{\partial \xi^i} d\xi^i$ is a vector from A to B, in other words, the position vector and the coordinates of B are given by, $\mathbf{C} + d\mathbf{C}$, and $\xi^i + d\xi^i$, $i = 1, 2, 3$, respectively, that is, $\mathbf{C} + \mathbf{dC} = (\xi^i + d\xi^i)\, \mathbf{G}_i$.

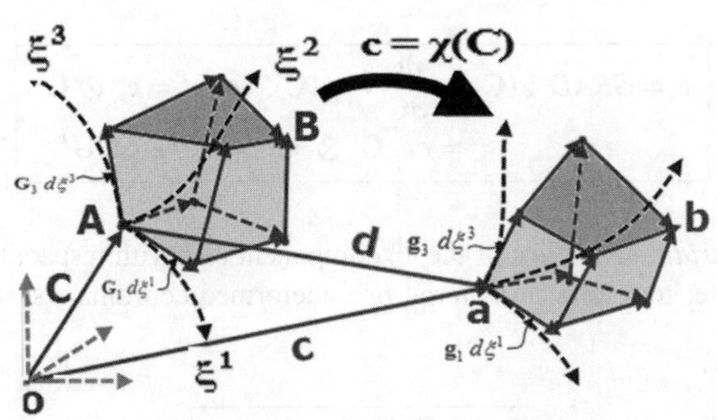

Figure 7.4 Curvilinear deformation map.

In a deformed state, the body Ω_0 becomes Ω; the particles A and B of the undeformed state become **a** and **b**, in the deformed state, with the distance between them:

$$\boxed{ds^2 \equiv \|d\mathbf{c}\|^2 = d\mathbf{c} \bullet d\mathbf{c} = \frac{\partial \mathbf{c}}{\partial \xi^i}\frac{\partial \mathbf{c}}{\partial \xi^j} d\xi^i \, d\xi^j} \tag{7.3}$$

(because the coordinate system is convective), and the position vectors are given by $\mathbf{c} = \xi^i\ \mathbf{g}_i$ and $\mathbf{c} + \mathbf{dc} = (\xi^i + d\xi^i)\ \mathbf{g}_i$, respectively. We note that in order to study the relationship between any two states of configuration, what happens in between these two states is unimportant, and so the time dependence is unnecessary. Now, with the deformation map given as: $\mathbf{c} = \chi(\mathbf{C})$, and vectors defined by the difference in position vectors, that is, $\mathbf{dC} = (\mathbf{C} + \mathbf{dC}) - \mathbf{C}$ and $\mathbf{dc} = (\mathbf{c} + \mathbf{dc}) - \mathbf{c}$, for infinitesimal vectors (i.e. line elements), we get as a first-order (linear) approximation in the Taylor series expansion:

$$\boxed{\begin{aligned} &\mathbf{d}c = \frac{\partial \chi(\mathbf{C})}{\partial C}\mathbf{d}C = \frac{\partial \mathbf{c}}{\partial \mathbf{C}}\mathbf{dC} \Rightarrow \mathbf{g}_\mathrm{i}\ d\xi^i = \frac{\partial \mathbf{c}}{\partial \mathbf{C}}\ \mathbf{G}_\mathrm{i}\ d\xi^i \\ \Rightarrow\quad &\mathbf{g}_\mathrm{i} = \frac{\partial \mathbf{c}}{\partial \mathbf{C}}\ \mathbf{G}_\mathrm{i} = \mathbf{F}\ \mathbf{G}_\mathrm{i} \\ \text{where}\quad &\mathbf{F} \triangleq \frac{\partial \mathbf{c}}{\partial \mathbf{C}} = \mathbf{g}_\mathrm{i} \otimes \mathbf{G}^\mathrm{i}\ \text{(in dyadic form)} \end{aligned}} \tag{7.4}$$

where we have introduced the following.

Definition: F, a second-order two-point (i.e. generally the base tensors made up of base vectors from two different spaces) tensor, known as the ***deformation gradient tensor***, defined by:

$$\boxed{\mathbf{F} \equiv \frac{\partial \chi(\mathbf{C})}{\partial \mathbf{C}} = \frac{\mathbf{dc}}{\mathbf{dC}} = \mathbf{g}_i \otimes \mathbf{G}^i \Rightarrow \mathbf{g}_i = \mathbf{F}\ \mathbf{G}_i} \tag{7.5}$$

If we apply the definition of *GRAD* (•), from Chapter 3, as the material gradient (or, Lagrangian gradient with reference to the undeformed coordinate system) functional: $\boxed{GRAD\ \chi(\mathbf{C}) \equiv \frac{\partial \chi(\mathbf{C})}{\partial \xi^j} \otimes \mathbf{G}^j = \chi(\mathbf{C})_{,j} \otimes \mathbf{G}^j}$, we get the component description for the deformation gradient as:

$$\boxed{\begin{aligned} \mathbf{F} = GRAD\ \chi(\mathbf{C}) = \frac{\partial \mathbf{c}}{\partial \mathbf{C}} &= \chi(\mathbf{C})_{,j} \otimes \mathbf{G}^j = \mathbf{c}_{,j} \otimes \mathbf{G}^j \\ &= c_{i|j}\ \mathbf{G}^i \otimes\ \mathbf{G}^j = F_{ij}\ \mathbf{G}^i \otimes\ \mathbf{G}^j \end{aligned}} \tag{7.6}$$

where $c_{i|j}$ is the ***covariant derivative*** of the ith component of **c** with respect to the jth component with both indices refer to reference material or undeformed coordinate system. From equation (7.6), we get:

$$\boxed{F_{ij} = \mathbf{G}_i\ \mathbf{F}\ \mathbf{G}_j = \mathbf{G}_i \bullet\ \mathbf{g}_j} \tag{7.7}$$

In terms of the displacement vector, $\mathbf{d}$, with the covariant derivative component, in the undeformed coordinate frame, given as: $\mathbf{d},_j = d_{k|j}\ \mathbf{G}^k$ and $\mathbf{g}_j = \mathbf{G}_j + \mathbf{d},_j$, we get from equations (7.5) and (7.7):

$$\boxed{\begin{array}{l} \mathbf{F} = \mathbf{g}_i \otimes \mathbf{G}^i = (\mathbf{G}_i + \mathbf{d},_i) \otimes \mathbf{G}^i = \mathbf{I} + GRAD\ \mathbf{d} \\ F_{ij} = \mathbf{G}_i \bullet (\mathbf{G}_j + \mathbf{d},_j) = G_{ij} + d_{k|j}\ \mathbf{G}_i \bullet \mathbf{G}^k = G_{ij} + d_{i|j} \end{array}} \tag{7.8}$$

where $\mathbf{I}$ is the 3×3 identity tensor in the curvilinear coordinate system.

From equation (7.5), the deformation gradient can be seen as a local map from the base vectors at a point in the undeformed state to the base vectors at the same point in the deformed state. As indicated earlier, the deformation map, $\chi : \mathbf{C} \to \mathbf{c}$, is assumed to be one-to-one to exclude unrealistic penetration of one part of the material by any other, and as such, the Jacobian, $J = \left|\frac{\partial \mathbf{c}}{\partial \mathbf{C}}\right| > 0$, and $\frac{\partial \mathbf{c}}{\partial \mathbf{C}}$ is invertible; clearly, the transpose and the inverse of the deformation gradient tensor can be obtained as:

$$\boxed{\begin{array}{l} \mathbf{g}_i = \mathbf{F}\ \mathbf{G}_i \Rightarrow \mathbf{F}^{-1}\ \mathbf{g}_i = \mathbf{G}_i \Rightarrow \mathbf{F}^{-1} = \mathbf{G}_i \otimes \mathbf{g}^i \Rightarrow \mathbf{F}^{-T} = \mathbf{g}^i \otimes \mathbf{G}_i \\ \mathbf{F} = \mathbf{g}_i \otimes \mathbf{G}^i \Rightarrow \mathbf{F}^T = \mathbf{G}^i \otimes \mathbf{g}_i \Rightarrow \mathbf{F}^T\ \mathbf{g}^i = \mathbf{G}^i \Rightarrow \mathbf{g}^i = \mathbf{F}^{-T}\ \mathbf{G}^i \Rightarrow \mathbf{F}^{-T} = \mathbf{g}^i \otimes \mathbf{G}_i \end{array}} \tag{7.9}$$

7.2.3.2 Deformation Gradient as the Tangent Map

There is an alternative way to define the deformation map and deformation gradient: we may define them by using the deformation gradient as the tangent map on the configuration manifold. A motion may be viewed as a curve on the configuration space passing through the undeformed state as well as other and current deformation states. We can then view the deformation gradient which carries local information as the tangent map that relates the tangent to the curve at the undeformed state to that at the current deformed state.

For example, let us briefly consider the deformation of a shell body; elaboration of its importance will be detailed in the Chapter 11 on nonlinear shells. If we denote $\mathbf{C}(S^1, S^2, \eta) = \boldsymbol{\vartheta}_0(S^1, S^2, \eta)$ and $\mathbf{c}(S^1, S^2, \eta) = \boldsymbol{\vartheta}(S^1, S^2, \eta)$ as the values of the functions $\boldsymbol{\vartheta}_0$ and $\boldsymbol{\vartheta}$ as shown in Figure 7.5, we get χ the deformation map from the undeformed shell-like body to the current

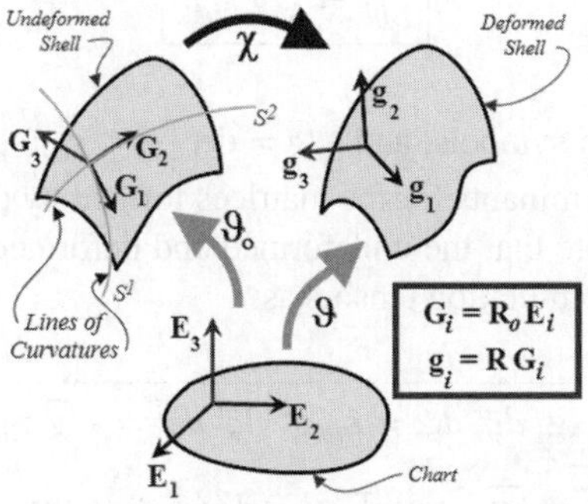

Figure 7.5 Alternate deformation map for shell surface.

deformed body as:

$$\chi \equiv \boldsymbol{\vartheta} \circ \boldsymbol{\vartheta}_0^{-1} \tag{7.10}$$

Noting that, $\nabla\boldsymbol{\vartheta} = \mathbf{g}_i \otimes \mathbf{E}^i$, we have: $\mathbf{g}_i = (\nabla\boldsymbol{\vartheta})\ \mathbf{E}_i = (\nabla\boldsymbol{\vartheta})\ \left(\nabla\boldsymbol{\vartheta}_0\right)^{-1}\ \mathbf{G}_i = \mathbf{F}\ \mathbf{G}_i$ where we have used the definition of the deformation gradient as: $\mathbf{F} \equiv \mathbf{g}_i \otimes \mathbf{G}^i$. Then, using the definition of the deformation map for the shell body, we have: $\nabla\boldsymbol{\vartheta} = \boldsymbol{\vartheta}_{,\alpha} \otimes \mathbf{E}^\alpha + \boldsymbol{\vartheta}_{,\eta} \otimes \mathbf{E}^3 = \left\{\mathbf{c}_{0,\alpha} + \eta\, \mathbf{g}_{3,\alpha}\right\} \otimes \mathbf{E}^\alpha + \mathbf{g}_3 \otimes \mathbf{E}^3$. The tangent map of the deformation map is then given as:

$$\mathbf{F} \equiv (\nabla\boldsymbol{\vartheta})\left(\nabla\boldsymbol{\vartheta}_0\right)^{-1} \tag{7.11}$$

The deformation gradient is also an extremely useful computational tool in relating objects in the undeformed and deformed configurations; here we give two important examples with proof.

7.2.3.3 Differential Area Element – Nanson's Formula (Piola Transformation)

Let $\mathbf{da} = \mathbf{n}\ da$ be the axial vector corresponding to the area, da, of a parallelogram with the sides given by $\mathbf{g}_i\ d\xi^i$ and $\mathbf{g}_j\ d\xi^j$, and $\mathbf{n}$ as the unit normal vector perpendicular to the differential area, da, in the deformed state; similarly, let $\mathbf{dA} = \mathbf{N}\ dA$ be the axial vector corresponding to the area, dA, of the same parallelogram with the side given by $\mathbf{G}_i\ d\xi^i$ and $\mathbf{G}_j\ d\xi^j$, and $\mathbf{N}$ as the unit normal vector perpendicular to the differential area, dA, in the undeformed state. We can prove the following transformation, known as ***Piola transformation*** or ***Nanson's formula***, that relates the undeformed area to the deformed area under deformation gradient, $\mathbf{F}$:

$$\boxed{\begin{aligned} &\mathbf{n}\ da = (\det\mathbf{F})\ \mathbf{N}\ \mathbf{F}^{-1}\ dA = (\det\mathbf{F})\ \mathbf{F}^{-T}\ \mathbf{N}\ dA \\ \text{or,}\quad &\mathbf{da} = (\det\mathbf{F})\ \mathbf{F}^{-T}\ \mathbf{dA} = J\ \mathbf{F}^{-T}\ \mathbf{dA} \end{aligned}} \tag{7.12}$$

Solution: Let us first recall the third-order permutation tensors, $\mathbf{E} = \left\{\varepsilon_{ijk}\right\}$, and $\hat{\mathbf{E}} = \left\{\hat{\varepsilon}_{ijk}\right\}$ (see Chapter 3), for the undeformed and the deformed coordinate system, respectively, with components given as:

$$\boxed{\begin{aligned} \varepsilon_{ijk} &= \sqrt{G}\ e_{ijk} \\ \hat{\varepsilon}_{ijk} &= \sqrt{g}\ e_{ijk} \end{aligned}} \tag{7.13}$$

where e_{ijk} are the permutation symbols, and $\sqrt{g} = \det\left[\mathbf{g}_1 \quad \mathbf{g}_2 \quad \mathbf{g}_3\right] = \mathbf{g}_1 \times \mathbf{g}_2 \bullet \mathbf{g}_3$ and $\sqrt{G} = \det\left[\mathbf{G}_1 \quad \mathbf{G}_2 \quad \mathbf{G}_3\right]$, the determinants for the matrices formed by the deformed and undeformed base vectors, respectively. Note that the undeformed and deformed differential area vectors can be expressed through these permutation tensors as:

$$\boxed{\begin{aligned} \mathbf{n}\ da &= \mathbf{g}_i \times\ \mathbf{g}_j\ d\xi^i\ d\xi^j = \hat{\varepsilon}_{ijk}\mathbf{g}^k\ d\xi^i\ d\xi^j = \sqrt{g}\ e_{ijk}\ \mathbf{g}^k\ d\xi^i\ d\xi^j \\ &= \sqrt{\frac{g}{G}}\ \sqrt{G}\ e_{ijk}\ \mathbf{F}^{-T}\ \mathbf{G}^k\ d\xi^i\ d\xi^j = (\det\mathbf{F})\ \ \mathbf{F}^{-T}\ \mathbf{N}\ dA \end{aligned}} \tag{7.14}$$

where we have used equations (7.9) and (7.13). Now, using equation (7.9), we see that:

$$\begin{aligned} &\mathbf{F}^{-T}\,\mathbf{N} = \mathbf{g}^i \otimes \mathbf{G}_i \bullet N_j\,G^j = N_i\,\mathbf{g}^i \\ \text{and,}\quad &\mathbf{N}\,\mathbf{F}^{-1} = N_j\,G^j \bullet \mathbf{G}_i \otimes \mathbf{g}^i = N_i\,\mathbf{g}^i \\ \Rightarrow\quad &\mathbf{F}^{-T}\,\mathbf{N} = \mathbf{N}\,\mathbf{F}^{-1} \end{aligned} \tag{7.15}$$

Finally, using equation (7.15) in equation (7.14), we get equation (7.12). Thus, we see that the deformation gradient helps express the differential area in the deformed state in terms of those in the undeformed or reference state.

7.2.3.4 Differential Volume Element

As before let $\sqrt{g} = \det\left[\mathbf{g}_1 \quad \mathbf{g}_2 \quad \mathbf{g}_3\right]$ and $\sqrt{G} = \det\left[\mathbf{G}_1 \quad \mathbf{G}_2 \quad \mathbf{G}_3\right]$ be determinants for the matrices formed by the deformed and undeformed base vectors, respectively, connected by the deformation gradient, $\mathbf{F}$; note that $\sqrt{g} = \mathbf{g}_1 \times \mathbf{g}_2 \bullet \mathbf{g}_3$ is the volume of a differential parallelepiped element with sides $\mathbf{g}_1$, $\mathbf{g}_2$ and $\mathbf{g}_3$ in the deformed state, and similarly, $\sqrt{G} = \mathbf{G}_1 \times \mathbf{G}_2 \bullet \mathbf{G}_3$ is the volume of the same differential parallelepiped element with sides $\mathbf{G}_1$, $\mathbf{G}_2$ and $\mathbf{G}_3$ in the undeformed state. We can show that:

$$\sqrt{g} = J\sqrt{G}, \quad J \equiv \det\mathbf{F} \quad \mathbf{d}v = J\,\mathbf{d}V \tag{7.16}$$

Solution: Referring to Chapter 3, for any invertible matrix, $\mathbf{A}$, and any two vectors, $\mathbf{u}$ and $\mathbf{v}$, we have: $\mathbf{Au} \times \mathbf{Av} = (\det\mathbf{A})\mathbf{A}^{-T}(\mathbf{u} \times \mathbf{v})$; so we get:

$$\begin{aligned} \sqrt{g} &= \mathbf{g}_1 \times \mathbf{g}_2 \bullet \mathbf{g}_3 = \mathbf{FG}_1 \times \mathbf{FG}_2 \bullet \mathbf{FG}_3 = J\mathbf{F}^{-T}(\mathbf{G}_1 \times \mathbf{G}_2) \bullet \mathbf{FG}_3 \\ &= J\mathbf{F}^T\mathbf{F}^{-T}(\mathbf{G}_1 \times \mathbf{G}_2 \bullet \mathbf{G}_3) = J\sqrt{G} \end{aligned} \tag{7.17}$$

Now, we know: $dV = \sqrt{G}\,d\xi^1\,d\xi^2\,d\xi^3$ and $dv = \sqrt{g}\,d\xi^1\,d\xi^2\,d\xi^3$ are the volumes of a parallelepiped element in undeformed and deformed state, respectively; then, from equation (7.17), we get:

$$\frac{dv}{dV} = \sqrt{\frac{g}{G}} = J = \det\mathbf{F} = \left|\frac{\partial \mathbf{c}}{\partial \mathbf{C}}\right| \tag{7.18}$$

Thus, we see again that the deformation gradient helps us to express the differential volume in the deformed state in terms of that in the undeformed or reference state.

Remarks

- Clearly, if $J = 1$ in equation (7.18), there is no change in volume element before and after deformation, a condition that is known as ***isochoric deformation***.

- The local relationship, given by equation (7.18), is equally valid globally for a body in isochoric deformation, and can be expressed by taking integrals over the undeformed and deformed body:

$$\boxed{\int_{\Omega} \mathbf{d}v = \int_{\Omega_0} J\,\mathbf{d}V} \tag{7.19}$$

7.2.3.5 Polar Decomposition of the Deformation Gradient Tensor

As indicated, the deformation gradient is most suitable for tracking large rotations in a deformable body; in what follows, this becomes obvious when we express deformation gradient under polar decomposition (recall that every second-order tensor has a polar decomposition). Additionally, in order for us to introduce suitable definitions for objective strain tensors that remain unaffected by rigid body motions, that is, rigid body translations and rotations by definition, we need to consider polar decomposition of the deformation gradient tensor. An ***objective deformation variable*** is defined as one that vanishes identically for rigid body translation and rotation; the deformation gradient is not objective. To wit, if $\mathbf{g_i} = \mathbf{R}\,\mathbf{G_i}$ where $\mathbf{R}$ is the rigid body rotation, that is, any two material points maintain the distance between them, then, by equation (7.5), we get: $\boxed{\mathbf{F} = \mathbf{g}_i \otimes \mathbf{G}^i = \mathbf{R}\,\mathbf{G}_i \otimes \mathbf{G}^i = \mathbf{R} \neq \mathbf{0}}$; similarly, for rigid body displacement, $\mathbf{d}$, (i.e. independent of position) such that $\mathbf{g_i} = \mathbf{G_i} + \not{\mathbf{d}}_{,i} = \mathbf{G_i}$, we get: $\boxed{\mathbf{F} = \mathbf{g}_i \otimes \mathbf{G}^i = \mathbf{G}_i \otimes \mathbf{G}^i = \mathbf{I} \neq \mathbf{0}}$. We may define the polar decomposition of the deformation gradient, $\mathbf{F}$, as:

$$\boxed{\mathbf{F} = \mathbf{R}\,\mathbf{U} = \mathbf{v}\,\mathbf{R}} \tag{7.20}$$

where $\mathbf{U}$ and $\mathbf{v}$ are known as the ***right stretch tensor*** and ***left stretch tensor***, respectively; these are both symmetric and positive definite tensors.

Remark

- As indicated, for our subsequent application to analyses of beams, plates and shells, we will be interested in Lagrangian or material undeformed coordinate systems for which we only need to look at the first relation in equation (7.20), that is, $\boxed{\mathbf{F} = \mathbf{R}\,\mathbf{U}}$.

  Now, because of the orthonormality of the rotation tensor, we note that:

$$\boxed{\mathbf{F}^T\,\mathbf{F} = \mathbf{U}^T(\mathbf{R}^T\,\mathbf{R})\,\mathbf{U} = \mathbf{U}^2 \triangleq \boldsymbol{C}} \tag{7.21}$$

where the symmetric tensor, $\boldsymbol{C}$, is known as the ***right Cauchy–Green tensor***; it is important to note that the deformation tensor, $\boldsymbol{C}$, is independent of any pure rotational motion, as this observation serves as a key to defining objective strain tensor as follows.

7.2.3.6 Green–Lagrange Strain Tensor

From equations (7.3) and (7.21), we get:

$$\boxed{ds^2 \equiv \|d\mathbf{c}\|^2 = d\mathbf{c} \bullet d\mathbf{c} = \mathbf{F}\,d\mathbf{C} \bullet \mathbf{F}\,d\mathbf{C} = d\mathbf{C} \bullet (\mathbf{F}^T\,\mathbf{F})\,d\mathbf{C} = d\mathbf{C} \bullet \boldsymbol{C}\,d\mathbf{C}} \tag{7.22}$$

Now, using equations (7.2) and (7.22), we can define the ***Green–Lagrange Strain tensor***, **E**, from:

$$ds^2 - dS^2 = d\mathbf{C} \bullet (\mathcal{C} - \mathbf{I})\, d\mathbf{C} = d\mathbf{C} \bullet (2\mathbf{E})\, d\mathbf{C} \tag{7.23}$$

as:

$$\boxed{\mathbf{E} = \frac{1}{2}(\mathcal{C} - \mathbf{I}) = \frac{1}{2}(\mathbf{F}^T\,\mathbf{F} - \mathbf{I}) = \frac{1}{2}(\mathbf{U}^2 - \mathbf{I})} \tag{7.24}$$

From equation (7.24), we see that the strain tensor, **E**, is symmetric and allows us to measure the change in the square of the undeformed length, dS^2, in the material coordinate system, and hence it is known as the ***material strain tensor*** with components, E_{ij}, given by:

$$\boxed{\begin{aligned} &\mathbf{E} = E_{ij}\mathbf{G}^i \otimes \mathbf{G}^j \\ \text{and,}\quad &E_{ij} = \mathbf{G}_i\,\mathbf{E}\,\mathbf{G}_j \end{aligned}} \tag{7.25}$$

We can express the components, E_{ij}, of the material strain tensor, **E**, explicitly; for this, we use equation (7.8) in equation (7.24), and note that $\mathbf{I} = \mathbf{G}_j \otimes \mathbf{G}^j = G_{ij}\mathbf{G}^i \otimes \mathbf{G}^j = (\mathbf{G}_i \bullet \mathbf{G}_j)\,\mathbf{G}^i \otimes \mathbf{G}^j$ to get:

$$\boxed{\begin{aligned} \mathbf{E} &= \frac{1}{2}(\mathbf{F}^T\,\mathbf{F} - \mathbf{I}) = \frac{1}{2}[(\mathbf{G}^i \otimes \mathbf{g}_i)(\mathbf{g}_j \otimes \mathbf{G}^j) - \mathbf{I}] \\ &= \frac{1}{2}(\mathbf{g}_i \bullet \mathbf{g}_j - \mathbf{G}_i \bullet \mathbf{G}_j)\,\mathbf{G}^i \otimes \mathbf{G}^j \end{aligned}} \tag{7.26}$$

With equation (7.25), we get the component descriptions as:

$$\boxed{E_{ij} = \frac{1}{2}(\mathbf{g}_i \bullet \mathbf{g}_j - \mathbf{G}_i \bullet \mathbf{G}_j)} \tag{7.27}$$

We can express the material strain tensor in terms of the displacement vector: knowing that: $\mathbf{c} = \mathbf{C} + \mathbf{d}$, we get: $\mathbf{F} = \frac{\partial \mathbf{c}}{\partial \mathbf{C}} = \mathbf{I} + \frac{\partial \mathbf{d}}{\partial \mathbf{C}} = \mathbf{I} + GRAD\ \mathbf{d}$; defining $\mathbf{H} \equiv GRAD\ \mathbf{d}$, we get, in terms of the displacement vector, the expression for **E**, the material strain tensor, from equation (7.24), as:

$$\boxed{\begin{aligned} \mathbf{E} &= \frac{1}{2}\{(\mathbf{I} + \mathbf{H}(\mathbf{d}))^T\,(\mathbf{I} + \mathbf{H}(\mathbf{d})) - \mathbf{I}\} \\ &= \underbrace{\frac{1}{2}\{\mathbf{H}(\mathbf{d}) + \mathbf{H}(\mathbf{d})^T\}}_{\text{Linear and symmetric}} + \underbrace{\frac{1}{2}\,\mathbf{H}(\mathbf{d})^T\,\mathbf{H}(\mathbf{d})}_{\text{Nonlinear}} \end{aligned}} \tag{7.28}$$

For sufficiently small strain, we get from equation (7.28) the expression for the symmetric linear strain as:

$$\boxed{\mathbf{E} = \frac{1}{2}\{\mathbf{H}(\mathbf{d}) + \mathbf{H}(\mathbf{d})^T\}} \tag{7.29}$$

Finally, we can get the components, E_{ij}, in terms of the displacement vector, $\mathbf{d}$, and its components. To wit, let us note that: $\mathbf{g}_i = \mathbf{G}_i + \mathbf{d}_{,i}$, $\mathbf{d}_{,j} = d_{i|j}\ \mathbf{G}^i \Rightarrow \mathbf{G}_i \bullet\ \mathbf{d}_{,j} = d_{i|j}$ and $\mathbf{d}_{,i} \bullet \mathbf{d}_{,j} = d_{k|i}\ \mathbf{G}^k \bullet\ d_{m|j}\ \mathbf{G}^m = G^{km}\ d_{k|i}\ d_{m|j}\ = d^m_{|i}\ d_{m|j}$; thus, from equation (7.27), we get:

$$\boxed{\begin{aligned} E_{ij} &= \frac{1}{2}\{(\mathbf{G}_i + \mathbf{d}_{,i}) \bullet\ (\mathbf{G}_j + \mathbf{d}_{,j}) - \mathbf{G}_i \bullet\ \mathbf{G}_j\} \\ &= \frac{1}{2}\{\mathbf{G}_i \bullet\ \mathbf{d}_{,j} + \mathbf{G}_j \bullet\ \mathbf{d}_{,i} + \mathbf{d}_{,i} \bullet \mathbf{d}_{,j}\} \\ &= \frac{1}{2}\left\{d_{i|j} +\ d_{j|i} + d^m_{|i} \bullet\ d_{m|j}\right\} \end{aligned}} \tag{7.30}$$

Remarks

- If $\boldsymbol{C} = \mathbf{F}^T\ \mathbf{F} = \mathbf{I}$ in equation (7.24), the body is clearly unstrained, that is, $\mathbf{E} = \mathbf{0}$. For example, for rigid body rotation, that is, $\mathbf{F} = \mathbf{R}$, we get: $\mathbf{E} = \mathbf{F}^T\ \mathbf{F} - \mathbf{I} = \mathbf{R}^T\ \mathbf{R} - \mathbf{I} = \mathbf{I} - \mathbf{I} = \mathbf{0}$, that is, it is unstrained as expected.
- There exist several other strain measures: Eulerian Almansi strain tensor, Hencky strain tensor, Seth–Hill strain tensors, and so on; but, for our applications here, we are only interested in the Green–Lagrange material strain tensor.
- All of the above expressions have been deduced using curvilinear coordinate system.

These can be easily reduced for Cartesian coordinate system (see Figure 7.6) if we recognize that the Christoffel symbols are identically zero for Cartesian frame of reference, and thus, all covariant differentiations reduce to ordinary partial differentiations, and, for Cartesian coordinate systems, the covariant and the contravariant components of a vector coincide. For example, the components of the Lagrangian strain tensor, from equation (7.30), become:

$$\boxed{E_{ij} = \frac{1}{2}\{d_{i,j} + d_{j,i} + d_{m,i} \bullet d_{m,j}\}} \tag{7.31}$$

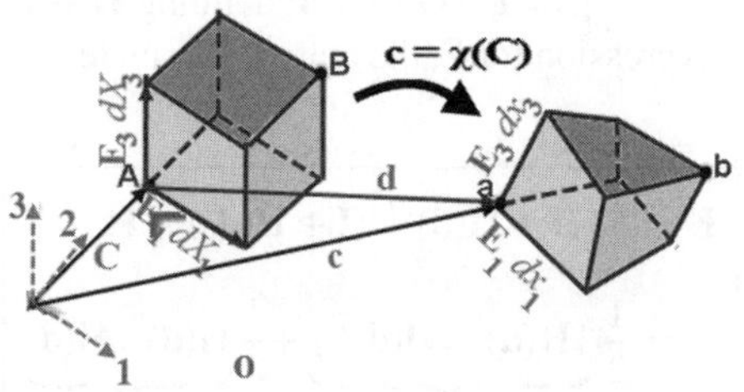

Figure 7.6 Cartesian deformation map.

where $d_{i,j} = \frac{\partial d_i}{\partial X_j}$, and so on. In the linear regime, where the strains are sufficiently small, we get the familiar strain tensor, $\mathbf{E} = E_{ij}\ \mathbf{E}_i \otimes \mathbf{E}_j$ with the $[E_{jj}]$ coefficient matrix given as:

$$[E_{ij}] = \begin{bmatrix} E_{11} \triangleq \dfrac{\partial u}{\partial X} = \varepsilon_{xx} & E_{12} \triangleq \dfrac{1}{2}\left(\dfrac{\partial u}{\partial X} + \dfrac{\partial u}{\partial Y}\right) = \dfrac{1}{2}\gamma_{xy} & E_{13} \triangleq \dfrac{1}{2}\left(\dfrac{\partial u}{\partial Z} + \dfrac{\partial w}{\partial X}\right) = \dfrac{1}{2}\gamma_{xz} \\ & E_{22} \triangleq \dfrac{\partial v}{\partial Y} = \varepsilon_{yy} & E_{23} \triangleq \dfrac{1}{2}\left(\dfrac{\partial v}{\partial Z} + \dfrac{\partial w}{\partial Y}\right) = \dfrac{1}{2}\gamma_{yz} \\ \text{sym} & & E_{33} \triangleq \dfrac{\partial w}{\partial Z} = \varepsilon_{zz} \end{bmatrix} \tag{7.32}$$

where we have replaced the coordinates X_1, X_2, X_3 by x, y, z, respectively, and displacement components d_1, d_2, d_3 by u, v, w, respectively, with ε_{xx}, ε_{yy}, ε_{zz} as the familiar axial strains and γ_{xy}, γ_{yz} γ_{zx} as the shear strains of engineering terminology.

- For curvilinear coordinates, we have deduced all the above relationships based on the natural coordinates, $\{\xi^i\}$, $i = 1, 2, 3$; however, as indicated in Chapter 5 on curves (used for beam analysis) and Chapter 6 on surfaces (used for plates and shell analyses), the most efficient coordinate system is given instead by the arc length parameters, $\{S^i\}$, $i = 1, 2, 3$, and also, these along the lines of curvature of shell structures; we will introduce the modified relationships in the Chapter 10 on beams and the Chapter 11 on shells.
- For rotating or spinning coordinate systems, the deformation gradient is best left in the fundamental form:

$$\mathbf{F} \equiv \frac{\partial \mathbf{c}}{\partial \mathbf{C}} \tag{7.33}$$

  as $\mathbf{c}_{,i} \neq \mathbf{g}_i$; more on this in the application chapters on nonlinear beams, plates and shells.
- Finally, for the rotation tensor and its properties, we refer to Chapter 4 on the rotation tensor.

7.2.4 *Where We Would Like to Go*

With this first pillar, the deformation patterns, having been essentially described above, the next stage is to identify the second pillar, in nonlinear mechanics of deformable solid bodies:

- stress tensors and the transmission of force

7.3 Balance Principles: Stress Tensors

Her we look at the balance principles, the conservation laws and transmission of forces for the motion of deformable solid bodies subjected to external stimulations such as forces, moments, and heat, to help us understand our subsequent c-type finite element method and its application to nonlinear beam, plate and shell analyses. One of the important aspects to recognize about

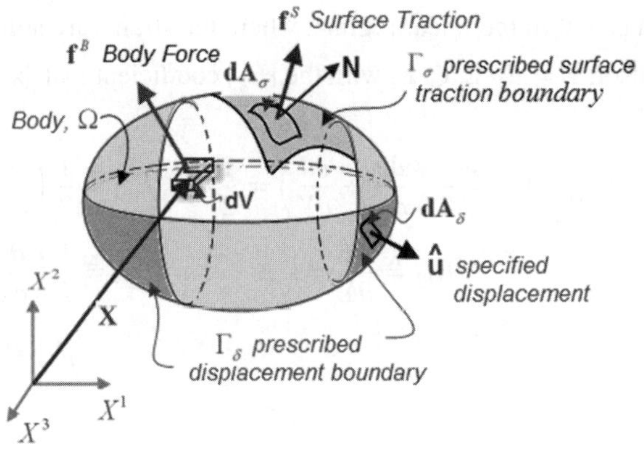

Figure 7.7 Body deformation and load definition.

the scope of this study is that an axiomatic approach helps us to build a unified theory of the global behavior of deformable solid bodies irrespective of the type of medium. As indicated earlier, we cover the relevant axioms that describe the conservation laws of solid mechanics: the conservation of mass, the balance of momentum, the balance of moment of momentum and the conservation of energy. The balance principles give rise to what is known as the ***stress tensor*** which manifests itself as a unique measure at a point for the transmission of force through a deformable body. The energy conservation principle leads to the definition of what is known as the ***internal energy*** of a deformable body in terms of the stress tensor and the corresponding strain measure. This correspondence, in turn, is expressed through the concept of ***energy conjugate*** stress and deformation measure pair. Finally, for later applications in the book, these principles, among other possibilities such as the Eulerian framework, are strictly expressed in the Lagrangian or material frame of reference, i.e. all physical entities are as defined shown in Figure 7.7. In what follows, therefore, we engage in the discussion of the relevant axioms primarily in the curvilinear coordinate system in Lagrangian material framework which helps define the stress tensors at a point, its relationship with the ***surface traction vector*** and its importance in the description of the internal energy in a deformable body.

7.3.1 Axiom #1: Conservation of Mass – Global and Local

Principle: Globally, the total mass of a body remains invariant under any motion; local mass of an infinitesimal portion of a body is conserved under any motion.

In order to express this principle in its mathematical form, we must define ***mass***: the mass of a body, in classical mechanics, is a non-negative, additive measure associated with the body. If the mass of the body is absolutely continuous, there exists what is known as its ***mass density per unit volume***, designated $\rho_0 = \rho_0(\mathbf{C}, t_0)$, in the undeformed body with material coordinate system at a point with position vector, $\mathbf{C}$, and at time, t_0, and $\rho = \rho(\mathbf{c}, t)$ in the current or deformed body with same origin (chosen without any loss of generality) and material coordinate system of the same point in deformed configuration with position vector, $\mathbf{c} = \chi(\mathbf{C}, t)$, and at time, t, with χ as the ***deformation map*** such that for a mass. dM, of an infinitesimal volume, dV, of a body with total volume, B, in the undeformed state that becomes the mass. dm, the volume, dv, and the body with total volume, b, in the deformed state, we have: $dM = \rho_0\, dV$ and $dm = \rho\, dv$; then, according to ***the principle of global mass conservation***, we get;

Material equation of continuity:

$$\begin{aligned} &\int dM = \int_B \rho_0\, dV = \int_b \rho\, dv = \int dm \\ \Rightarrow \\ &\int_B (\rho_0 - \rho\, J) dV = 0 \qquad \text{(Global mass conservation)} \\ \Rightarrow \\ &\rho_0 = \rho\, J \qquad \text{(Local mass conservation)} \end{aligned} \tag{7.34}$$

where we recall that $J = \left| \frac{\partial c^i}{\partial C^j} \right| = \det \mathbf{F}$ is the Jacobian, a strain invariant, relating the infinitesimal undeformed and deformed volume of body as: $dv = J\, dV$ with $\mathbf{F}$ as the deformation gradient at the point with position vector, $\mathbf{C}$. Clearly, ρ_0 being independent of time, $\rho \ \det \mathbf{F}$ does not change under the deformation process, that is, $\overset{\bullet}{\overbrace{\rho \ \det \mathbf{F}}} = 0$. To be sure, this constancy of the mass of a body is only applicable in non-relativistic mechanics; when the effect of relativity is taken into consideration (i.e. at or near the speed of light), imparting of any energy into the body by external stimuli is equivalent to increasing in the mass of the body, and vice versa. We restrict ourselves to only non-relativistic mechanics and thus the mass is conserved under deformation.

7.3.2 Axiom #2: Balance of Momentum: Global

Principle: The time rate of change of momentum of a body is equal to the resultant force acting on the body for any motion.

First of all, it is important to recognize that, in nonlinear mechanics, all forces and moments acting on a body must be considered on the current deformed state of a material body, although, as will soon be seen, these variables can be measured using the undeformed material coordinate system known as the Lagrangian framework. Let us, then, consider a body in its undeformed configuration, as shown in Figure 7.8, that is subsequently deformed by external stimuli such as forces, moments, and so on.

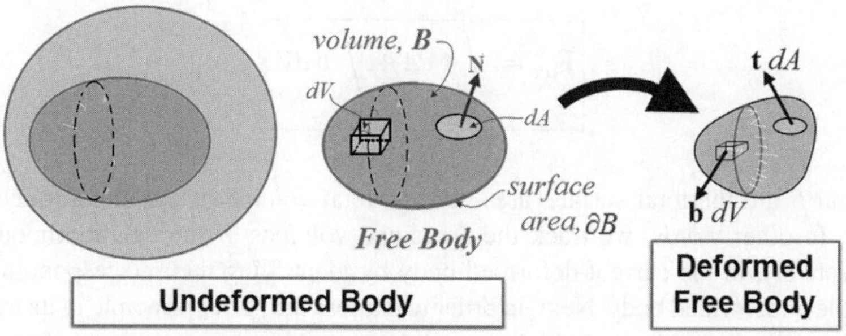

Figure 7.8 Traction definition.

If we extract a bounded ***free-body***, with undeformed volume, B, bounded by undeformed surface area, ∂B, as shown in Figure 7.8, the surface of the free-body must represent the effects of the remaining portion on it; these interaction effects will manifest themselves in the form of forces and moments on every particle on the bounding surface of the free-body.

Assumption: Here, we make the assumption of excluding representation of any distributed moment on the surface particles; in other words, as a principle of solid mechanics, the couple stresses such as dipoles in, say, dislocation theory, are neglected, and it is assumed that only the force vectors are sufficient to represent the interaction between the free-body and its complement.

With this assumption in place, we are left with two kinds of basic forces to deal with in solid mechanics: the ***body forces per unit of volume of the body*** and the ***surface (contact) forces per unit of surface area of the body***, also known as the ***surface tractions***; the volume and the surface area, for a Lagrangian framework, refer to the undeformed body. Clearly, the surface of the free-body common to the surface of its complement of an entire body must represent the traction forces that represent the interaction between the two. The ***surface force***, $\mathbf{t}\, dA$, where the vector, $\mathbf{t}$, is known as the ***surface traction***, acting on the surface of the current deformed body but measured per unit of the undeformed surface area, dA, and the ***volume force***, $\mathbf{b}\, dV$, where the vector, $\mathbf{b}$, is known as the ***body force***, acting in the current deformed body but measured per unit of the undeformed volume, dV. In other words, all the geometric attributes of a body are measured in the undeformed configuration of the body, so the outward unit normal, $\mathbf{N}$, at the point of pre-image of the deformed surface, is also measured in the Lagrangian coordinate system, and is not necessarily coincident with the direction of the traction, $\mathbf{t}$, that acts at the image point. A theorem due to Cauchy that helps to establish the orientation of the surface traction with respect to the normal at a point on the surface, is as follows.

Cauchy Reciprocal Theorem: It can be shown, similar to ***Newton's third law*** that every action has an equal and opposite reaction, that the traction, $\mathbf{t}$, at a point on the surface with the pre-image normal, $\mathbf{N}$, is equal and opposite to the traction with normal, $-\mathbf{N}$. Thus, if we designate by $\mathbf{t}_N$ the traction vector at a point on the surface with the pre-image normal $\mathbf{N}$, then, by Cauchy reciprocal theorem, we get:

$$\boxed{\mathbf{t}_N = -\mathbf{t}_{-N}} \tag{7.35}$$

Then, the ***total force***, $\mathbf{F}_{tot}$, acting on the deformed free-body can be expressed as:

$$\boxed{\mathbf{F}_{tot} = \int_{\partial B} \mathbf{t}\, dA + \int_{B} \mathbf{b}\, dV} \tag{7.36}$$

where ∂B and B are the total surface area and the total volume of the undeformed free-body, respectively. In other words, we track the areas and volumes of the neighborhoods around, $\mathbf{t}$ and $\mathbf{b}$, respectively, in the current deformed body by identifying their corresponding areas and volumes in the undeformed body. Next, in order to express the above principle in its mathematical form, we must define the velocity field, the acceleration field and the momentum at a point on the deformed body. So, suppose the position vector, $\mathbf{C}(\xi^1,\ \xi^2,\ \xi^3, t_0)$, in the undeformed body and the position vector, $\mathbf{c}(\xi^1, \xi^2, \xi^3, t)$, in the deformed body for the same material point are related through the displacement vector, $\mathbf{d}(\xi^1, \xi^2, \xi^3, t)$, as:

$$\boxed{\mathbf{c}(\xi^1, \xi^2, \xi^3, t) = \mathbf{C}(\xi^1, \xi^2, \xi^3, t_0) + \mathbf{d}(\xi^1, \xi^2, \xi^3, t)} \tag{7.37}$$

Then, the ***velocity vector field***, $\mathbf{v}(\xi^1, \xi^2, \xi^3, t)$, at the current location, $\mathbf{c}(\xi^1, \xi^2, \xi^3, t)$, can be expressed in the material frame of reference as:

$$\boxed{\mathbf{v}(\xi^1, \xi^2, \xi^3, t) = \frac{d}{dt}\mathbf{c}(\xi^1, \xi^2, \xi^3, t) = \frac{d}{dt}\mathbf{d}(\xi^1, \xi^2, \xi^3, t) \equiv \dot{\mathbf{d}}(\xi^1, \xi^2, \xi^3, t)} \tag{7.38}$$

Similarly, the ***acceleration vector field***, $\mathbf{a}(\xi^1, \xi^2, \xi^3, t)$, at the same current location, can be expressed in the material frame of reference as:

$$\boxed{\mathbf{a}(\xi^1, \xi^2, \xi^3, t) = \frac{d^2}{dt^2}\mathbf{d}(\xi^1, \xi^2, \xi^3, t) \equiv \ddot{\mathbf{d}}(\xi^1, \xi^2, \xi^3, t)} \tag{7.39}$$

Then, the ***total momentum***, $\mathcal{P}(t)$, of the deformed body at any time, t, may be defined as:

$$\boxed{\mathcal{P}(t) = \int \mathbf{v}\, dM = \int_V \rho_0\, \mathbf{v}\, dV} \tag{7.40}$$

and the ***rate of total momentum***, $\frac{d}{dt}\mathcal{P}(t)$, of the deformed body at any time, t, may be defined as:

$$\boxed{\begin{aligned}\frac{d}{dt}\mathcal{P}(t) &= \frac{d}{dt}\int_V \mathbf{v}\, dM = \int_V \rho_0\, \mathbf{a}\, dV \\ &= \int_V \rho_0\, \ddot{\mathbf{d}}\, dV\end{aligned}} \tag{7.41}$$

where we have used the principle of conservation of mass: $\frac{d}{dt}dM = \frac{d}{dt}(\rho_0\, dV) = 0$. Now, we are in a position to express the principle of momentum balance mathematically as: the three-dimensional equations of motion (or the ***balance of momentum***) in Lagrangian (material) formulation for a 3D body, with undeformed volume, B, and the undeformed surface area, ∂B, for dynamic loading conditions, can be written as:

$$\boxed{\int_{\partial B} \mathbf{t}\, dA + \int_B \mathbf{b}\, dV = \frac{d}{dt}\int_B \mathbf{v}\, \rho_0\, dV = \int_B \rho_0\, \ddot{\mathbf{d}}\, dV \quad \text{(Momentum balance)}} \tag{7.42}$$

where $\mathbf{t}$ and $\mathbf{b}$ are the surface traction vector measured per unit of the surface area and the body force vector measured per unit volume, of the undeformed body, respectively, but acting on the current deformed body. In other words, we track the areas and the volumes of the neighborhoods around, $\mathbf{t}$ and $\mathbf{b}$, respectively, in the current deformed body by identifying their corresponding areas and volumes in the undeformed body.

7.3.3 Axiom #3: Balance of Moment of Momentum: Global

Principle: The time rate of change of moment of momentum of a body about a fixed point is equal to the resultant moment about the same point acting on the body for any motion.

The total moment, $\mathbf{M}_{tot}$, acting on the deformed free-body can be expressed as:

$$\boxed{\mathbf{M}_{tot} = \int_{\partial B} \mathbf{c} \times \mathbf{t}\, dA + \int_{B} \mathbf{c} \times \mathbf{b}\, dV} \tag{7.43}$$

where, again, ∂B and B are the total surface area and the total volume of the undeformed free-body, respectively. Next, in order for us to express the above principle in its mathematical form, we must define moment of momentum at a point on the deformed body. The ***total moment of momentum***, $\mathcal{M}(t)$, of the deformed body at any time, t, about the origin, may be defined as:

$$\boxed{\mathcal{M}(t) = \int \mathbf{c} \times \mathbf{v}\, dM = \int_{B} \rho_0 \;\; \mathbf{c} \times \mathbf{v}\, dV} \tag{7.44}$$

and the ***rate of total moment of momentum***, $\frac{d}{dt}\mathcal{M}(t)$, of the deformed body at any time, t, may be defined as:

$$\boxed{\begin{aligned} \frac{d}{dt}\mathcal{M}(t) &= \frac{d}{dt}\int \mathbf{c} \times \mathbf{v}\, dM = \frac{d}{dt}\int_{B} \mathbf{c} \times \dot{\mathbf{d}}\; \rho_0\, dV \\ &= \cancel{\int_{B} \dot{\mathbf{d}} \times \dot{\mathbf{d}}\, \rho_0\, dV} + \int_{B} \mathbf{c} \times \mathbf{a}\, \rho_0\, dV \\ &= \int_{B} \mathbf{c} \times \ddot{\mathbf{d}}\; \rho_0\, dV \end{aligned}} \tag{7.45}$$

where we have used equations (7.38) and (7.39), and the principle of conservation of mass: $\frac{d}{dt}dM = \frac{d}{dt}(\rho_0\; dV) = 0$. Now, we are in a position to express the principle of moment of momentum balance mathematically as: the three-dimensional equations of motion (or, the ***balance of moment of momentum***) in Lagrangian (material) formulation for a $3D$ body, with undeformed volume, B, and the undeformed surface area, ∂B, for dynamic loading conditions, can be written as:

$$\boxed{\int_{\partial B} \mathbf{c} \times \mathbf{t}\, dA + \int_{B} \mathbf{c} \times \mathbf{b}\, dV = \int_{B} \mathbf{c} \times \ddot{\mathbf{d}}\; \rho_0\, dV \quad \textit{(Moment of momentum balance)}} \tag{7.46}$$

where $\mathbf{t}$ and $\mathbf{b}$ are the surface traction vector measured per unit surface area and the body force vector measured per unit volume of the undeformed body, respectively, but acting on the current deformed body. The momentum balance and the moment of momentum balance given by equations (7.42) and (7.46), respectively, are expressed in global forms; for developing the corresponding local forms, we first need to define appropriate stress tensor.

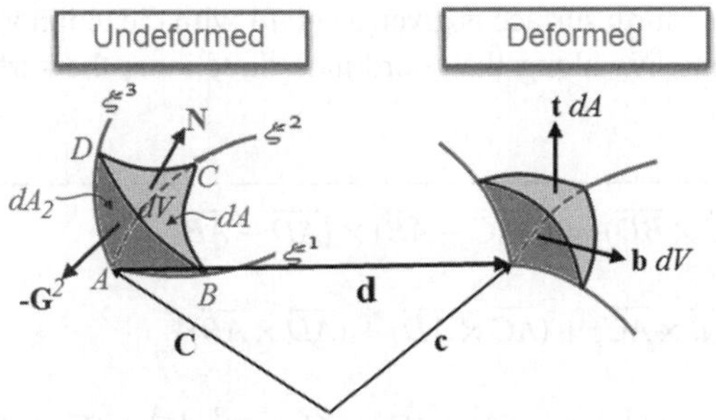

Figure 7.9 Piola tetrahedron.

7.3.4 Piola–Kirchhoff Stress Tensors and Local Balance Principles

Before we can focus on the local balance of momentum, and of moment of momentum principles and stress tensors in the Lagrangian frame of reference, we need to establish some preliminary geometrical relationships that are necessary to arrive at such a construction. As already indicated, all physical and mathematical entities are measured in terms of the undeformed, material frame of reference in the Lagrangian description of the laws of motion; on the other hand, we must remember that the forces and moments, in fact, always act on the current, deformed body.

7.3.4.1 Geometry of Piola–Kirchhoff Tetrahedron

Let us consider an arbitrary infinitesimal tetrahedron, bounded by iso-curves (see Chapter 6 on surface) of lengths, $\mathbf{G}_i\, d\xi^i$, with $\mathbf{G}_i$, $i = 1, 2, 3$, as the covariant base vectors, in the undeformed reference frame at a point, A, with position vector, $\mathbf{C}(\xi^1, \xi^2, \xi^3, t_0)$, that is made up of three coordinate surfaces denoted: ABD, ACD and ABC in Figure 7.9; the fourth surface is given by BCD. The same tetrahedron deforms into the current shape with the material point A having the new position vector shown as $\mathbf{c}(\xi^1, \xi^2, \xi^3, i)$ in Figure 7.9. The outward unit normal to the undeformed surface, BCD, is given by the vector, $\mathbf{N}$, with the surface area denoted by dA; the area vector, we may recall, is given by $\mathbf{N}\, dA$. The corresponding outward normals, not necessarily of unit length, for the coordinate surfaces, ABD, ACD and ABC, are given by the contravariant base vectors, $-\mathbf{G}^2$ (shown in Figure 7.9), $-\mathbf{G}^1$ and $-\mathbf{G}^3$, respectively; let us denote the corresponding surface areas as dA_2, dA_1 and dA_3, respectively. Now, considering the coordinate surface area vectors, $\mathbf{dA}_i$, for $i = 1, 2, 3$, we have

$$\boxed{\begin{aligned}\mathbf{dA}_i &= \frac{1}{2}\mathbf{G}_j \times \mathbf{G}_k\, d\xi^j\, d\xi^k \quad (\text{no sum on } j, k \text{ and } j \neq k)\\ &= \frac{1}{2}\varepsilon_{jki}\, \mathbf{G}^i\, d\xi^j\, d\xi^k \quad (i \neq j \neq k) = \frac{1}{2}\sqrt{G}\, \mathbf{G}^i\, d\xi^j\, d\xi^k\end{aligned}} \tag{7.47}$$

where recall, from Chapter 3 on tensors, that $\sqrt{G}$ is the square root of the determinant of the matrix formed by the covariant base vectors as columns, that is, $G = \det\begin{bmatrix}\mathbf{G}_1 & \mathbf{G}_2 & \mathbf{G}_3\end{bmatrix} = \mathbf{G}_1 \times \mathbf{G}_2 \bullet \mathbf{G}_3$; the corresponding areas for $i = 1, 2, 3$, are given as:

$$\boxed{dA_i = \frac{1}{2}\sqrt{G}\,\|\mathbf{G}^i\|\; d\xi^j\, d\xi^k = \frac{1}{2}\sqrt{G\, G^{ii}}\, d\xi^j\, d\xi^k} \tag{7.48}$$

The area vector for the fourth surface is given as $\mathbf{N}\,dA$ with the unit normal, $\mathbf{N} = N_i\,\mathbf{G}^i$, having the ***covariant*** components, N_i, along the coordinate directions; the surface area vector can be expressed as:

$$\begin{aligned}
\mathbf{N}\,dA &= \frac{1}{2}(\overrightarrow{BC}\times\overrightarrow{BD}) = \frac{1}{2}(\overrightarrow{AC}-\overrightarrow{AB})\times(\overrightarrow{AD}-\overrightarrow{AB})\\
&= \frac{1}{2}\{(\overrightarrow{AB}\times\overrightarrow{AC}) + (\overrightarrow{AC}\times\overrightarrow{AD}) + (\overrightarrow{AD}\times\overrightarrow{AB})\}\\
&= \frac{1}{2}\{(\mathbf{G}_1\times\mathbf{G}_2)\,d\xi^1\,d\xi^2 + (\mathbf{G}_2\times\mathbf{G}_3)\,d\xi^2\,d\xi^3 + (\mathbf{G}_3\times\mathbf{G}_1)\,d\xi^3\,d\xi^1\}\\
&= \frac{1}{2}\{\varepsilon_{123}\mathbf{G}^3\,d\xi^1\,d\xi^2 + \varepsilon_{231}\mathbf{G}^1\,d\xi^2\,d\xi^3 + \varepsilon_{312}\mathbf{G}^2\,d\xi^3\,d\xi^1\}\\
&= \sum_{\substack{i=1\\ i\neq j\neq k}}^{3}\left(\frac{1}{2}\sqrt{GG^{ii}}\,d\xi^j\,d\xi^k\right)\frac{\mathbf{G}^i}{\sqrt{G^{ii}}}
\end{aligned} \tag{7.49}$$

Now, using equation (7.48) with the normal, $\mathbf{N} = N_i\,\mathbf{G}^i$, with N_i as the covariant components, in equation (7.49), we finally get the desired geometric relations between the coordinate surface areas and the surface area with unit normal, $\mathbf{N}$, for future use, as:

$$N_i\,\sqrt{G^{ii}}\,dA = dA_i \tag{7.50}$$

Let us assume that the height of the tetrahedron is Δh; then, the volume of the undeformed tetrahedron is given by, $dV = \frac{1}{3}\Delta h\,dA$. Next, in order for us to express the principles of momentum balance and moment of momentum balance in its mathematical form, the total force, $\mathbf{F}_{tot}$, acting on the deformed tetrahedron can be expressed as:

$$\mathbf{F}_{tot} = \int_A \mathbf{t}\,dA + \int_V \mathbf{b}\,dV \tag{7.51}$$

where $A = dA + dA_1 + dA_2 + dA_3$ and $V = dV$ are the total surface area and the total volume of the undeformed tetrahedron, respectively. The total momentum, $\boldsymbol{\mathcal{P}}$, of the deformed tetrahedron at any time, t, may be defined as:

$$\boldsymbol{\mathcal{P}} = \int_V \mathbf{v}\,dM = \int_V \rho_0\,\dot{\mathbf{d}}\,dV \tag{7.52}$$

Now, we are in a position to express the ***local principle of momentum balance*** mathematically for the deformed tetrahedron as:

$$\begin{aligned}
&\mathbf{t}\,dA - \hat{\mathbf{t}}^1\,dA_1 - \hat{\mathbf{t}}^2\,dA_2 - \hat{\mathbf{t}}^3\,dA_3 + \frac{1}{3}\mathbf{b}\,\Delta h\,dA = \frac{1}{3}\rho_0\,\ddot{\mathbf{d}}\,\Delta h\,dA\\
\Rightarrow\quad &\{\mathbf{t} - N_1\,\hat{\mathbf{t}}^1 - N_2\,\hat{\mathbf{t}}^2 - N_3\,\hat{\mathbf{t}}^3\} = \frac{\Delta h}{3}(\rho_0\,\ddot{\mathbf{d}} - \mathbf{b})
\end{aligned} \tag{7.53}$$

where we have considered the fact that dA is arbitrary, and the outward normals on the coordinate surfaces of the tetrahedron have negative signs, and application of the Cauchy reciprocal theorem results in the negative signs for the coordinate surface tractions, $\hat{\mathbf{t}}^i$, $i = 1, 2, 3$, in the equation. Now, taking $\Delta h \to 0$ in the limit, using equation (7.53), Einstein summation convention and dyadic form of a second-order tensor, the traction at A can be expressed as:

$$\boxed{\begin{aligned}\mathbf{t} &= (\hat{\mathbf{t}}^i \sqrt{G^{ii}})\, N_i = \mathbf{t}^i\, N_i \\ &= \mathbf{t}^i\, (\mathbf{N} \bullet \mathbf{G}_i) = (\mathbf{t}^i \otimes \mathbf{G}_i)\, \mathbf{N}\end{aligned}} \tag{7.54}$$

In equation (7.54), we have defined: $\mathbf{t}^i \equiv \hat{\mathbf{t}}^i \sqrt{G^{ii}}$, $i = 1, 2, 3$, because $\mathbf{t}^i$ is a contravariant vector by the quotient rule: since N_i's are covariant components and $\mathbf{t}$ is an invariant vector, we must have $\mathbf{t}^i$ as contravariant vectors. Now, knowing that the traction vector, $\mathbf{t}$, works on the deformed configuration of the body, it is natural to express its components in terms of the covariant base vector set, $\{\mathbf{g}_i\}$, $i = 1, 2, 3$, in the deformed body at the mapping of the point, A, of the undeformed body, and inserting in equation (7.54), we get:

$$\boxed{\mathbf{t} = (P^{ij}\, \mathbf{g}_i \otimes \mathbf{G}_j)\, \mathbf{N} = \mathbf{P}\, \mathbf{N}} \tag{7.55}$$

where we have introduced what is known as the ***first Piola–Kirchhoff stress tensor (PK I)***, $\mathbf{P}$, in mixed bases, as:

$$\boxed{\mathbf{P} = P^{ij}\, \mathbf{g}_i \otimes \mathbf{G}_j} \tag{7.56}$$

where $\mathbf{P}^j = P^{ij}\, \mathbf{g}_i$ is known as the ***contravariant pseudo-vector components*** for the first Piola–Kirchhoff tensor, $\mathbf{P}$.

Remarks

- It may be noted that the stress tensor at a point is uniquely determined by the normal to the surface at the point; thus, given the traction vector at a point, for all different surfaces with the same normal (or, equivalently, the same tangent plane) will exhibit the same stress tensor at the point.
- The first Piola–Kirchhoff stress tensor (PK I) can also be defined fully in terms of the undeformed base vector dyadics, $\mathbf{G}_i \otimes \mathbf{G}_j$; but, such a description usually results in asymmetric tensor components.
- By describing in the mixed basis dyadic, shown in equation (7.56), we will soon see, as a consequence of the local moment of momentum balance principle, that the components of the first Piola–Kirchhoff stress tensor are symmetric. This helps considerably in computational efficiency when applied to beam, plate and shell analyses as will be seen later.
- Of course, we can devise other stress tensors, both in the Lagrangian framework such as the ***second Piola–Kirchhoff stress tensor (PK II)***, and in the Eulerian framework such as the symmetric ***Cauchy stress tensor***, denoted by, $\boldsymbol{\sigma}$. Given a traction vector, $\mathbf{t}$, at a point on the deformed body with the surface normal given by $\mathbf{n}$ at the point, we get:

$$\boxed{\mathbf{t} = \boldsymbol{\sigma}\, \mathbf{n}} \tag{7.57}$$

For later use in the constitutive theory, and with the help of Piola transformation relating the areas on the undeformed and the deformed surfaces, the first Piola–Kirchhoff and the Cauchy stress tensors can be shown to be related by:

$$\boxed{\mathbf{P} = (\det\mathbf{F})\, \sigma\, \mathbf{F}^{-T}} \tag{7.58}$$

- For our purposes here, we are primarily interested in the Lagrangian stress tensors; thus, the PK II stress tensor, **S**, whose components are defined with respect to the undeformed base vector dyadic, is given as:

$$\boxed{\begin{array}{l} \mathbf{S} = S^{ij}\, \mathbf{G}_i \otimes \mathbf{G}_j \text{ or, } \mathbf{S} = \mathbf{S}^j \otimes \mathbf{G}_j \\ \text{and,} \quad \mathbf{t} = \mathbf{S}\,\mathbf{N} \end{array}} \tag{7.59}$$

7.3.4.2 Local Momentum Balance Principle: Lagrange Equations of Motion

By incorporating equation (7.55) in the global momentum balance equation (7.42), and using the Green–Gauss theorem (Chapter 3 on tensors), we get, knowing that dV is arbitrary, the local momentum balance equation as:

$$\begin{array}{l} \displaystyle\int_{\partial B} \mathbf{P}\,\mathbf{N}\, dA + \int_B (\mathbf{b} - \rho_0\, \ddot{\mathbf{d}}) dV = \mathbf{0} \Rightarrow \int_B DIV\, \mathbf{P}\, dV + \int_B (\mathbf{b} - \rho_0\, \ddot{\mathbf{d}}) dV = \mathbf{0} \\ \Rightarrow \\ \boxed{DIV\, \mathbf{P} + \mathbf{b} - \rho_0\, \ddot{\mathbf{d}} = \mathbf{0}} \end{array} \tag{7.60}$$

Recalling that $DIV\, \mathbf{P} = \mathbf{P}^i\,|_i$ where $\mathbf{P}^i\,|_i$ is the covariant derivative of the contrvariant pseudo-vector components, $\mathbf{P}^i$, of the first Piola–Kirchhoff stress tensor, **P**, as in equation (7.56), we can also rewrite the local momentum balance equation as:

$$\boxed{\mathbf{P}^i\,|_i + \mathbf{b} - \rho_0\, \ddot{\mathbf{d}} = \mathbf{0}} \tag{7.61}$$

Equations (7.60) and (7.61) describe what are called the ***Lagrange equations of motion***.

7.3.4.3 Local Moment of Momentum Balance Principle: Symmetry of PK I and PK II

By incorporating equation (7.55) in the global moment of momentum balance equation (7.46), we get:

$$\boxed{\int_{\partial B} \mathbf{c} \times \mathbf{P}\,\mathbf{N}\, dA + \int_B \mathbf{c} \times \mathbf{b}\, dV = \int_B \mathbf{c} \times \ddot{\mathbf{d}}\, \rho_0\, dV} \tag{7.62}$$

The surface integral on left side of equation (7.62) must be converted to a volume integral in order to arrive at the local balance principle; for this, we again apply the Green–Gauss theorem as follows:

$$\begin{aligned}\int_{\partial B} \mathbf{c}\times\mathbf{P}\,\mathbf{N}\,dA &= \int_{\partial B} \mathbf{c}\times(\mathbf{P}^i\otimes\mathbf{G}_i)\bullet(N_j\otimes\mathbf{G}^j)\,dA = \int_{\partial B}(\mathbf{c}\times\mathbf{P}^i)N_i\,dA\\ &= \int_B (\mathbf{c}\times\mathbf{P}^i)\,|_i\,dV = \int_B (\mathbf{c},_i\times\mathbf{P}^i)\,dV + \int_B \mathbf{c}\times\mathbf{P}^i\,|_i\,dV\\ &= \int_B (\mathbf{c},_i\times\mathbf{P}^i)\,dV + \int_B \mathbf{c}\times(DIV\,\mathbf{P})\,dV\end{aligned} \tag{7.63}$$

where we have used the definition of the contravariant pseudo-vector, $\mathbf{P}^i$, of the second-order stress tensor, $\mathbf{P}$, and the relationship: $DIV\,\mathbf{P} = \mathbf{P}^i\,|_i$; also, we have used the identity: $\mathbf{c}\,|_i = \mathbf{c},_i$, that is, the covariant derivative of a vector is identical with its partial derivative because the covariant derivative of the base vectors vanishes (see Chapter 3 on tensors). Now, incorporating equation (7.63) in (7.62), we get:

$$\begin{aligned}&\int_B (\mathbf{c},_i\times\mathbf{P}^i)\,dV + \int_B \mathbf{c}\times\underbrace{(DIV\mathbf{P}+\mathbf{b}-\rho_0\,\ddot{\mathbf{d}})}_{\text{Momentum balance}}\,dV = \mathbf{0}\\ \Rightarrow\quad &\int_B (\mathbf{c},_i\times\mathbf{P}^i)\,dV = \mathbf{0}\end{aligned} \tag{7.64}$$

Noting that the volume, B, is arbitrary and that $\mathbf{c},_i = \mathbf{g}_i$, where we recall that $\mathbf{g}_i$, $i = 1,2,3$ are the base vector set on the deformed body, we finally get the ***principle of local balance of moment of momentum***, with Einstein summation convention, as:

$$\boxed{g_i\times\mathbf{P}^i = \mathbf{0}} \tag{7.65}$$

More importantly, however, using the fact that $\mathbf{P}^i = P^{ij}\,\mathbf{g}_j$, equation (7.65) can be reduced to:

$$\boxed{\sqrt{g}\{(P^{23}-P^{32})g_1 + (P^{31}-P^{13})g_2 + (P^{12}-P^{21})g_3\} = \mathbf{0}} \tag{7.66}$$

where $g = \det\begin{bmatrix}\mathbf{g}_1 & \mathbf{g}_2 & \mathbf{g}_3\end{bmatrix}$. Finally, noting that the base vectors, $\{\mathbf{g}_i\}$, by definition, are independent, we get the symmetry of the first Piola–Kirchhoff stress tensor in mixed bases. That is,

$$\boxed{\mathbf{P} = \mathbf{P}^T \quad\Rightarrow\quad P^{ij} = P^{ji}} \tag{7.67}$$

Now, knowing that $\mathbf{g}_i = \mathbf{F}\,\mathbf{G}_i$, where $\mathbf{F}$ is the deformation gradient, and using equations (7.56) and (7.59), we get:

$$\boxed{\mathbf{P} = P^{ij}\,\mathbf{g}_i\otimes\mathbf{G}_j = \mathbf{F}\,P^{ij}\,\mathbf{G}_i\otimes\mathbf{G}_j} \tag{7.68}$$

Thus, if we choose the components of PK I in mixed bases identical to those of PK II, that is, in equation (7.59), $S^{ij} \equiv P^{ij}$, then by equations (7.68) and (7.59), we get:

$$\boxed{\mathbf{P} = \mathbf{F}\,\mathbf{S}} \tag{7.69}$$

and PK II is clearly symmetric. Now, using equations (7.58) and (7.69), the second Piola–Kirchhoff stress tensor and the Cauchy stress tensor are related by

$$\boxed{\mathbf{S} = (\mathbf{detF})\,\mathbf{F}^{-1}\,\sigma\,\mathbf{F}^{-T}} \tag{7.70}$$

7.3.5 Axiom #4: Conservation of Energy: Energy Conjugated Stress–strain

Principle: The time rate of change of the kinetic and internal energy is equal to the sum of the rate of work (i.e. power) of the external forces and all other energies that enter or leave the body per unit of time.

In order for us to express the above principle in its mathematical form, we must define the terms kinetic energy, internal energy and power of the external forces, and identify some of the energies that enter or leave a body. The various forms of energy may include electrical, magnetic, heat and sound; these are considered to be equivalent to each other and thus may be appropriately added together in an equation. The ***kinetic energy***, $\mathcal{K}(t)$ of a body in motion at any time, t, is expressed as:

$$\boxed{\mathcal{K}(t) = \frac{1}{2}\int_{\partial B} v^2\,dM = \frac{1}{2}\int_{\partial B} v^2\,\rho_0\,dV = \frac{1}{2}\int_{\partial B} (\mathbf{v}\bullet\mathbf{v})\,\rho_0\,dV} \tag{7.71}$$

where all other terms have been defined before in the text. The ***internal energy***, $\mathcal{E}(t)$, of a deformable body at any time, t, is defined as:

$$\boxed{\mathcal{E}(t) = \int_{\partial B} e\,\rho_0\,dV} \tag{7.72}$$

where e is known as the ***internal energy density*** or ***specific internal energy per unit of mass*** of a body. We will soon see, in isothermal processes, that $\rho_0\,\dot{e}$, that is, the weighted rate of the specific internal energy is equal to what is known as the ***stress power***. The internal energy is special in the sense that it is a state function, that is it is independent of the process by which a body goes from one state to another. In fact, it is a function of the material property of the body, that is, the constitutive variables. The ***power of the external forces*** is defined as the time rate of the external forces; it is, then, expressed, using equation (7.36), as:

$$\boxed{\mathcal{W}(t) = \int_{\partial B} \mathbf{t}\bullet\dot{\mathbf{d}}\,dA + \int_{B} \mathbf{b}\bullet\dot{\mathbf{d}}\,dV} \tag{7.73}$$

7.3.5.1 Kinetic Energy Balance

The kinetic energy term defined by equation (7.71) can be derived directly from the Lagrangian equation of motion given by equation (7.60) in local form; thus, taking the scalar product with velocity, $\mathbf{v} = \dot{\mathbf{d}}$, of the equation (7.60), we get:

$$\rho_0\ \dot{\mathbf{d}} \bullet \ddot{\mathbf{d}} = \dot{\mathbf{d}} \bullet DIV\,\mathbf{P} + \dot{\mathbf{d}} \bullet \mathbf{b} \tag{7.74}$$

Now, applying the relation: $DIV(\dot{\mathbf{d}} \bullet \mathbf{P}) = \mathbf{P} : GRAD\ \dot{\mathbf{d}} + \dot{\mathbf{d}} \bullet DIV\,\mathbf{P}$ (see Chapter 3 on tensors), and noting that: $\boxed{\mathbf{P} : GRAD\ \dot{\mathbf{d}} = \mathbf{P} : (\dot{\mathbf{d}},_i \otimes \mathbf{G}^i) = \mathbf{P} : (\dot{\mathbf{c}},_i \otimes \mathbf{G}^i) = \mathbf{P} : (\dot{\underset{i}{\mathbf{g}}} \otimes \mathbf{G}^i) = \mathbf{P} : \dot{\mathbf{F}}}$

where $\mathbf{c}(t) = \mathbf{C}(t_0) + \mathbf{d}(t)$, and that: $\rho_0\ (\mathbf{v} \bullet \dot{\mathbf{v}}) = \frac{1}{2}\rho_0\ \overbrace{(\mathbf{v} \bullet \mathbf{v})}^{\bullet} = \frac{d}{dt}\left\{\frac{1}{2}\rho_0\ (\mathbf{v} \bullet \mathbf{v})\right\}$, from equation (7.74), we get the local form of the kinetic energy as:

$$\boxed{\underbrace{\frac{d}{dt}\left\{\frac{1}{2}\rho_0\ (\mathbf{v} \bullet \mathbf{v})\right\}}_{\text{Kinetic energy}} = DIV(\dot{\mathbf{d}} \bullet \mathbf{P}) - \mathbf{P} : \dot{\mathbf{F}} + (\dot{\mathbf{d}} \bullet \mathbf{b})} \tag{7.75}$$

Now, integrating over the volume, B, noting that $\mathbf{t} = \mathbf{P}\,\mathbf{N}$ and applying the Green–Gauss theorem, we get the ***global form of the kinetic energy*** as:

$$\boxed{\begin{aligned} \underbrace{\frac{d}{dt}\left\{\int_B \frac{1}{2}\rho_0\ (\mathbf{v} \bullet \mathbf{v})\, dV\right\}}_{\text{Kinetic energy}} &= \int_B DIV(\dot{\mathbf{d}} \bullet \mathbf{P})\, dA - \int_B \mathbf{P} : \dot{\mathbf{F}}\, dV + \int_B (\dot{\mathbf{d}} \bullet \mathbf{b})\, dV \\ &= \underbrace{\int_{\partial B} (\dot{\mathbf{d}} \bullet \mathbf{t})\, dA + \int_B (\dot{\mathbf{d}} \bullet \mathbf{b})\, dV}_{\text{Power of the external forces}} - \underbrace{\int_B \underbrace{\mathbf{P} : \dot{\mathbf{F}}}_{\text{Stress power}}\, dV}_{\text{Internal energy}} \end{aligned}} \tag{7.76}$$

where the integrand in the final term on the right side of equation (7.75), that is, $\mathbf{P} : \dot{\mathbf{F}}$, is known as the ***stress power***. Thus, the difference in the power of the external forces and the stress power defines the rate of change of the kinetic energy of a body. Using the statement of the principle above for an isothermal process, we can identify the ***rate of specific energy*** with the stress power as:

$$\boxed{\rho_0\ \dot{e} = \mathbf{P} : \dot{\mathbf{F}}} \tag{7.77}$$

that is, the double contraction of the first Piola–Kirchhoff (PK I) stress tensor and the rate of deformation gradient gives us the specific energy rate, $\rho_0\ \dot{e}$; this condition of double contraction of a stress tensor and a deformation or strain tensor rate resulting in the specific energy rate is attributed to what is known as ***energy conjugation*** of the stress tensor and the rate of the deformation or strain tensor. Thus, PK I stress tensor, $\mathbf{P}$, is energy conjugate with the rate of the deformation gradient, $\overset{\bullet}{\mathbf{F}}$. We can use this definition to determine the deformation measure whose rate is the energy conjugate with the second Piola–Kirchhoff (PK II) stress tensor. From equations (7.69) and (7.77), we get:

$$
\begin{aligned}
\rho_0\, \overset{\bullet}{e} &= \mathbf{P} : \overset{\bullet}{\mathbf{F}} = \mathbf{F}\,\mathbf{S} : \overset{\bullet}{\mathbf{F}} = \mathbf{S} : \mathbf{F}^T\, \overset{\bullet}{\mathbf{F}} = \mathbf{S} : \mathbf{F}^T\, \overset{\bullet}{\mathbf{F}} = \frac{1}{2}(\mathbf{S} + \mathbf{S}^T) : \mathbf{F}^T\, \overset{\bullet}{\mathbf{F}} \\
&= \frac{1}{2}\mathbf{S} : \mathbf{F}^T\, \overset{\bullet}{\mathbf{F}} + \frac{1}{2}\mathbf{S} : \overset{\bullet}{\mathbf{F}}{}^T\, \mathbf{F} = \mathbf{S} : \frac{1}{2}(\mathbf{F}^T\, \overset{\bullet}{\mathbf{F}} + \overset{\bullet}{\mathbf{F}}{}^T\, \mathbf{F}) \\
&= \mathbf{S} : \overset{\bullet}{\mathbf{E}}
\end{aligned}
\tag{7.78}
$$

where we have used the symmetry of PK II such that $\mathbf{S} = \mathbf{S}^T \Rightarrow \mathbf{S} = \frac{1}{2}(\mathbf{S} + \mathbf{S}^T)$, Thus, we see, from equation (7.78), that the second Piola–Kirchhoff (PK II) stress tensor, $\mathbf{S}$, is energy conjugate with the rate of Green–Lagrange strain tensor, $\mathbf{E} = \frac{1}{2}(\mathbf{F}^T\,\mathbf{F} - \mathbf{I}) \Rightarrow \overset{\bullet}{\mathbf{E}} = \frac{1}{2}(\mathbf{F}^T\, \overset{\bullet}{\mathbf{F}} + \overset{\bullet}{\mathbf{F}}{}^T\, \mathbf{F})$. So, for an isothermal deformation, using equations (7.76) and (7.77), and definitions of the kinetic energy, $\mathcal{K}(t)$, as in equation (7.71), of internal energy, $\mathcal{E}(t)$, as in equation (7.72), and of the power of the external forces, $\mathcal{W}(t)$, as in equation (7.73), we can express the conservation of energy in mathematical form as:

$$
\begin{aligned}
&\frac{d}{dt}(\mathcal{K}(t) + \mathcal{E}(t)) = \mathcal{W}(t) \\
\text{or} \quad &\frac{d}{dt}\left(\int_B \frac{1}{2}\rho_0\,(\mathbf{v} \bullet \mathbf{v})\, dV\right) + \int_B \mathbf{P} : \overset{\bullet}{\mathbf{F}}\, dV = \int_{\partial B} (\overset{\bullet}{\mathbf{d}} \bullet \mathbf{t})\, dA + \int_B (\overset{\bullet}{\mathbf{d}} \bullet \mathbf{b})\, dV
\end{aligned}
\tag{7.79}
$$

For the non-isothermal case, we will need to incorporate terms due to heat energy into the principle. Accordingly, let us suppose that a body, also has a ***distributed heat source***, $\mathbf{h}$, per unit of the undeformed volume, and heat transfer across the bounding surface represented by the ***heat flux***, $\mathbf{q}$, per unit of the undeformed surface area. These are similar to the volume load and the surface traction, respectively. Then, the ***total heat energy input***, $\mathcal{Q}(t)$, can be given by:

$$
\mathcal{Q}(t) = -\int_{\partial B} \mathbf{q} \bullet \mathbf{N}\, dA + \int_B \mathbf{h}\, dV \tag{7.80}
$$

where $\mathbf{N}$ is the normal at a point on the undeformed bounding surface. We can come up with the final expression for the Axiom #4 in global form or what is known as the ***first law of thermodynamics*** as:

$$
\begin{aligned}
&\frac{d}{dt}(\mathcal{K}(t) + \mathcal{E}(t)) = \mathcal{W}(t) + \mathcal{Q}(t) \\
&\frac{d}{dt}\left\{\int_B \frac{1}{2}\rho_0\,(\mathbf{v} \bullet \mathbf{v})\, dV\right\} + \frac{d}{dt}\left\{\int_B \rho_0\, e\, dV\right\} = \int_{\partial B} (\overset{\bullet}{\mathbf{d}} \bullet \mathbf{t})\, dA + \int_B (\overset{\bullet}{\mathbf{d}} \bullet \mathbf{b})\, dV \\
&\qquad - \int_{\partial B} \mathbf{q} \bullet \mathbf{N}\, dA + \int_B \rho_0\, \mathbf{h}\, dV
\end{aligned}
\tag{7.81}
$$

Now, using the relation for the kinetic energy balance as in equation (7.76) in equation (7.81), we get:

$$\begin{aligned}\frac{d}{dt}\left\{\int_B \rho_0\, e\, dV\right\} &= \int_B \mathbf{P} : \dot{\mathbf{F}}\, dV - \int_{\partial B} \mathbf{q} \bullet \mathbf{N}\, dA + \int_B \mathbf{h}\, dV \\ &= \int_B \{\mathbf{P} : \dot{\mathbf{F}} - DIV\, \mathbf{q} + \mathbf{h}\}\, dV\end{aligned} \tag{7.82}$$

where we have used the Green–Gauss theorem on the second term on the right side of the equation. Finally, noting that the volume, dV, is arbitrary, and that because of conservation of mass principle: $\frac{d}{dt}(\rho_0\, dV) = 0$, we get the local form of the energy conservation equation as:

$$\boxed{\rho_0\, \dot{e} = \mathbf{P} : \dot{\mathbf{F}} - DIV\, \mathbf{q} + \mathbf{h}} \tag{7.83}$$

7.3.6 *Where We Would Like to Go*

Now with the second pillar in place – the balance principles, having been essentially identified above for our subsequent applications – the next logical place to explore will be the one that reveals the third and final pillar in nonlinear mechanics of deformable solid bodies:

- the constitutive theory

7.4 Constitutive Theory: Hyperelastic Stress–Strain Relation

Here, we introduce the elastic material properties, that is, the stress–strain relationship, or the constitutive laws, appropriate for a linear, hyperelastic, homogeneous, isotropic solid material body. Clearly, this will be an approximation: ***the entire subject of the constitutive theory by its very nature of fitting the real world experimental data is an empirical science at best in any dimension***. However, to arrive at some realistic constitutive theory of solid material bodies, we must have some rules and restrictions upon it. Accordingly, a constitutive modeling must satisfy four general principles of a consistent three-dimensional theory: (1) the principle of stress determinism, (2) the principle of local action, and two in the form of restrictions: (3) the principle of material frame indifference and (4) the principle of material symmetry. For later applications, we are primarily interested in the stress–strain relationship in the Lagrangian reference frame of analysis, that is, the relationship between the first or second Piola–Kirchhoff stress tensor and the deformation gradient or Green–Lagrange strain tensor.

7.4.1 *Frame Indifference and Objective Tensors*

A constitutive law must be independent of observers or frames of reference. In other words, the stress–strain relationship must be form-independent as a matter of principle to make natural properties observer independent.

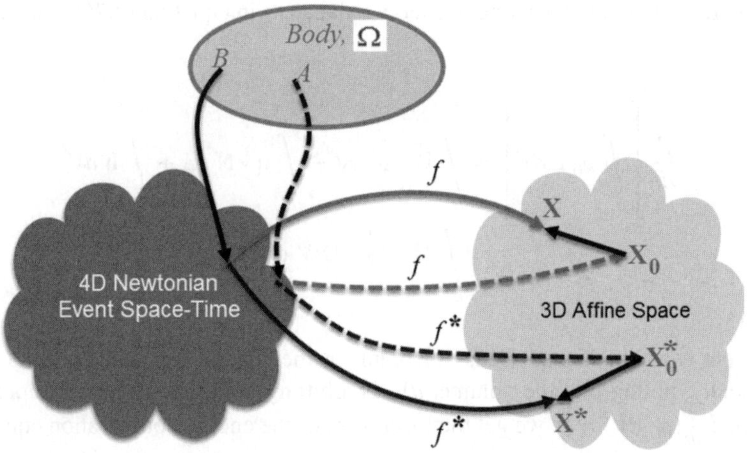

Figure 7.10 Frame of reference.

7.4.1.1 Frames of Reference

In order to express the principle of material objectivity, we must define what we mean by a frame of reference: generally, a change of frame of reference is confused as mere change in a coordinate system. We can talk about a frame of reference without any explicit consideration of a coordinate system. However, for an unambiguous description of frame indifference, we will have to restrict our discussion to affine space (see Chapter 3 on tensors) or Euclidean space; no counterpart appears to exist for curvilinear coordinates or curved spaces such as Riemannian space.

Let us consider a body, Ω_t, in motion; a particle, A, as shown in Figure 7.10, at any time, $t \in \mathbb{R}$, can be seen to be mapped as an ***event*** defined by an ordered pair, $(\mathbf{X}_0, t) \in \mathcal{E} \times \mathbb{R}$, of the 3D location, $\mathbf{X}_0$, and the instant of time, t, in a 4D ***Newtonian space-time event space***, $\mathcal{V}$, of classical mechanics, which may be seen as a disjoint union of sets, $\mathcal{V}_t$, of locations of each ***configuration***, Ω_t, of the body at different instant of time such that: $\mathcal{V} = \bigcup_{t \in \mathbb{R}_+} \mathcal{V}_t$; the configuration is defined as the placement of a body at a certain instant of time. The event space, in turn, can be mapped one-to-one as function, $f : \mathcal{V} \to \mathcal{E} \times \mathbb{R}$, to a 3D affine space (see Chapter 3 on tensors), $\mathcal{E}$, at location, $\mathbf{X}_0$, of the particle at each instant of time, t, such that: $f_t : \mathcal{V}_t \to \mathcal{E}$. We will call the function, f, the ***frame of reference***. Clearly, then, if we think of each frame of reference as associated with what may be called an ***observer***, then any two independently and relatively moving observers, having their clock synchronized such that simultaneous events can be recognized, will end up identifying two different locations of an event at the same instant of time through two independent frames of reference. In other words, it is all relative and there is no absolute space of mapping by which a motion of a body or the configuration of a body can be determined, because it depends on the observers of the events. Let us consider two such frames of reference with two observers, one belonging to each, given by, say, f and f^*, and accordingly, allowing us to measure the distance between any two simultaneous events. We call the composite mapping $f^* \circ f^{-1} : \mathcal{E} \times \mathbb{R} \to \mathcal{E} \times \mathbb{R}$ as the ***change of frame of reference*** such that $(\mathbf{X}_0{}^*, t^*) = (f^* \circ f^{-1})(\mathbf{X}_0, t)$. In other words, both $(\mathbf{X}_0, t)$ and $(\mathbf{X}_0^*, t^*)$ of the event space depict the same event as seen by two different observers, and similarly for particle B of the body, $(\mathbf{X}, t)$ and $(\mathbf{X}^*, t^*)$ of the event space. For a body in motion, with $(\mathbf{x}, t)$ and $(\mathbf{x}^*, t^*)$ being the descriptions of the same event observed simultaneously by the frames of reference, f and f^*, respectively, the agreement between the observers under the most general affine transformation has to be:

$$\boxed{\mathbf{x}^* = \mathbf{Q}(t)\,\mathbf{x}(t) + \mathbf{b}(t), \quad t^* = t + a} \tag{7.84}$$

where $\mathbf{Q} \in SO(3)$ is an orthogonal transformation (i.e. $\mathbf{Q}^T\,\mathbf{Q} = \mathbf{I}$), rotation tensor, $\mathbf{b}$ is a translational vector and a, is a constant.

7.4.1.2 Reference Configuration: Lagrangian Description

A body is an assemblage of material particles; the description of the history of these particles in motion under external stimuli by itself is not very useful for analysis purposes in continuum mechanics. In space-time, a one-to-one mapping, $\chi_t : \Omega \to \mathcal{V}_t$, defines an event; a continuous sequence of events defines the motion of a body. For an observer, f, with $f_t : \mathcal{V}_t \to \mathcal{E}$, the composite mapping: $\chi_{f_t} \equiv (f_t \circ \chi_t) : \Omega \to \mathcal{E}$ such that $\mathbf{x} = \chi_{f_t}(B) = f_t(\chi_t(B)), \quad B \in \Omega$ defines the motion of the body. Thus, for the entire body: $\Omega_{f_t} = \chi_{f_t}(\Omega) \subset \mathcal{E}$ with the mapping, χ_{f_t}, defining a position for the body at time, t, in the affine space, is known as the configuration of the body at time t under observer, f (i.e. it locates in affine space, $\mathcal{E}$, and at each instant of time, $t \in \mathbb{R}$, the particles of a body, Ω, under a frame of reference); the sequence of configurations can then be interpreted as the motion of a body. However, the motion can be reinterpreted as a sequence of deformations by choosing a particular configuration, c_f, as the ***reference configuration*** such that: $c_f : \Omega \to \mathcal{E}, \quad \mathbf{X} = c_f(B), \quad B \in \Omega, \quad \Omega_c = c_f(\Omega)$. The motion with respect to the reference configuration can then be redefined as: $\chi_{c_f}(\bullet, t) : \Omega_c \to \mathcal{E}, \quad \mathbf{x} = \chi_{c_f}(\mathbf{X}, t) = \chi_{c_f}(c_f^{-1}(\mathbf{X}))$. The mapping, χ_{c_f}, is known as the ***deformation*** of the body with respect to the reference configuration c_f under the frame of reference (or the observer), f. It is important to note that it is not necessary for the body to actually occupy the reference configuration, but for ease of analysis and computation, the unstressed or undeformed configuration of the body is usually chosen as the reference configuration in practical applications. Finally, the description of motion with respect to a reference configuration is known as the ***Lagrangian description*** in continuum mechanics. In the absence of any ambiguity related to any other frame of reference configuration or frame of reference, the Lagrangian description may be simply given by: $\mathbf{x} = \chi(\mathbf{X}, t)$.

7.4.1.3 Objective Tensors of Order 0 (Scalars), 1 (Vectors) and 2 (Second-order Tensors)

Under the most general affine transformation, the positions, $\mathbf{X}_0$ and $\mathbf{X}$, assume new positions, $\mathbf{x}_0(t)$ and $\mathbf{x}(t)$, respectively, related at any instant of time, t, by:

$$\boxed{\begin{aligned} \mathbf{x}_0(t) &= \mathbf{Q}(t)\,\mathbf{X}_0 + \mathbf{b}(t) \\ \mathbf{x}(t) &= \mathbf{Q}(t)\,\mathbf{X} + \mathbf{b}(t) \end{aligned}} \tag{7.85}$$

Clearly, with $\mathbf{Q}^T = \mathbf{Q}^{-1}$ by definition, for static cases, that is, equivalently for an instant of time, we have the defining transformation property of a vector (see Chapter 3 on tensors):

$$\boxed{\mathbf{c} \equiv \mathbf{x} - \mathbf{x}_0 = \mathbf{Q}\,(\mathbf{X} - \mathbf{X}_0) \equiv \mathbf{Q}\,\mathbf{C}} \tag{7.86}$$

whereby the distances, that is, scalars, are preserved:

$$\begin{aligned}\|\mathbf{c}\|^2 &\equiv \|\mathbf{x}-\mathbf{x}_0\|^2 = (\mathbf{x}-\mathbf{x}_0)\bullet(\mathbf{x}-\mathbf{x}_0)\\ &= \mathbf{Q}\,(\mathbf{X}-\mathbf{X}_0)\bullet\mathbf{Q}\,(\mathbf{X}-\mathbf{X}_0) = (\mathbf{X}-\mathbf{X}_0)\bullet\mathbf{Q}^T\,\mathbf{Q}\,(\mathbf{X}-\mathbf{X}_0)\\ &= \|\mathbf{X}-\mathbf{X}_0\|^2 = \|\mathbf{C}\|^2\end{aligned} \tag{7.87}$$

The transformation property given by equation (7.86) defines an entity as an ***invariant vector*** (i.e. a tensor of order 1). Now, let us suppose that a difference vector, say, $\mathbf{c}(f) = \mathbf{x} - \mathbf{x}_0$, in a frame of reference, f, becomes $\mathbf{c}^*(f^*) = \mathbf{x}^* - \mathbf{x}_0^*$ for a change of frame of reference to f^*, then, from equation (7.85) transformation and following the same argument as in equation (7.86), we get:

$$\mathbf{c}^*(f^*) = Q(t)\,\mathbf{c}(f) \tag{7.88}$$

Because equation (7.88) reflects the same intrinsic transformation property of an invariant vector given in equation (7.86), we say that any vector that transforms according to equation (7.88) for a change of frame in affine space, is an ***objective vector*** with respect to affine or Euclidean transformation. A velocity vector, $\dot{\mathbf{x}} \equiv \frac{\partial \mathbf{x}}{\partial t}$, or an acceleration vector, $\ddot{\mathbf{x}} \equiv \frac{\partial^2 \mathbf{x}}{\partial t^2}$, according to this definition, is not an objective vector because taking, for a body in motion, the time derivative of $\mathbf{x}^*(t) = \mathbf{Q}(t)\,\mathbf{x} + \mathbf{b}(t)$, we get:

$$\begin{aligned}\dot{\mathbf{x}}^* &= \frac{\partial \mathbf{x}^*}{\partial t} = \mathbf{Q}(t)\,\dot{\mathbf{x}}(t) + \dot{\mathbf{Q}}(t)\,\mathbf{x}(t) + \dot{\mathbf{b}}(t)\\ \ddot{\mathbf{x}}^* &= \frac{\partial^2 \mathbf{x}^*}{\partial t^2} = \mathbf{Q}(t)\,\ddot{\mathbf{x}}(t) + 2\,\dot{\mathbf{Q}}(t)\,\dot{\mathbf{x}}(t) + \ddot{\mathbf{Q}}(t)\,\mathbf{x}(t) + \ddot{\mathbf{b}}(t)\end{aligned} \tag{7.89}$$

whereas the base vectors, $\mathbf{g}_i = \mathbf{x},_i$, $i = 1, 2, 3$, are objective vectors because:

$$\mathbf{g}_i^*(f^*) = g_i^* = \frac{\partial \mathbf{x}^*}{\partial \xi^i} = \mathbf{Q}\,\mathbf{x},_i = \mathbf{Q}\,\mathbf{g}_i = \mathbf{Q}\,\mathbf{g}_i(f) \tag{7.90}$$

Similarly, recalling (see Chapter 3 on tensors) the intrinsic transformation property of a second-order tensor, we define, $\mathbf{T}$, to be an ***objective second-order tensor***, under a change of frame, if it transforms according to:

$$\mathbf{T}^*(f^*) = \mathbf{Q}(t)\,\mathbf{T}(f)\,\mathbf{Q}(t)^T \tag{7.91}$$

The deformation gradient tensor, $\mathbf{F} = \mathbf{g}_i \otimes \mathbf{G}^i$, is not an objective tensor because:

$$\mathbf{F}^*(f^*) = \mathbf{g}_i^* \otimes \mathbf{G}^i = \mathbf{Q}\,\mathbf{g}_i \otimes \mathbf{G}^i = \mathbf{Q}\,\mathbf{F}(f) \tag{7.92}$$

where we have used equation (7.90) for objectivity of the base vectors. However, the right Cauchy–Green tensor, $\mathbf{C} = \mathbf{F}^T\,\mathbf{F}$, respects frame indifference since:

$$\mathbf{C}^*(f^*) = \overset{*}{\mathbf{F}}{}^T\,\overset{*}{\mathbf{F}} = \mathbf{F}^T\,\mathbf{Q}^T\,\mathbf{Q}\,\mathbf{F} = \mathbf{F}^T\,\mathbf{F} = \mathbf{C}(f) \tag{7.93}$$

where we have used equation (7.92); similar argument also shows that the Green–Lagrange strain tensor, $\mathbf{E}(\mathbf{C}) = \frac{1}{2}(\mathbf{C} - \mathbf{G})$, and the right stretch tensor, $\mathbf{U} = \mathbf{R}^T\ \mathbf{F} = \mathbf{C}^{\frac{1}{2}}$, are also frame indifferent since:

$$\boxed{\begin{aligned} &\mathbf{E}^*(f^*) = \frac{1}{2}(\mathbf{C}^* - \mathbf{G}) = \frac{1}{2}(\mathbf{C} - \mathbf{G}) = \mathbf{E}(f) \\ &\mathbf{U}^*(f^*) = \mathbf{C}^{*\frac{1}{2}} = \mathbf{C}^{\frac{1}{2}} = \mathbf{U}(f) \end{aligned}} \tag{7.94}$$

7.4.1.4 Frame-indifferent or Objective Functions

Scalar-valued, $\phi : \mathbb{R} \to \mathbb{R}$, vector-valued, $\mathbf{r} : \mathcal{V} \to \mathbb{R}$, and tensor-valued, $\mathbf{T} : \mathcal{L}(\mathcal{V}) \to \mathbb{R}$, functions of tensor argument (where $\mathcal{L}(\mathcal{V})$ are the space of linear transformations (second-order tensors) on the vector space, $\mathcal{V}$, as the translation space of an affine space), are objective (Euclidean or affine objective) or frame indifferent scalar, vector and tensor, respectively, as long as the following conditions are met:

$$\boxed{\begin{aligned} \phi(f^*) &= \phi(f) \\ \mathbf{r}(f^*) &= \mathbf{Q}\ \mathbf{r}(f) \\ \mathbf{T}(f^*) &= \mathbf{Q}\ \mathbf{T}(f)\ \mathbf{Q}^T \end{aligned}} \tag{7.95}$$

As an example, the scalar function, $J = \det \mathbf{F}$, the Jacobian, is frame indifferent because:

$$\boxed{J^*(f^*) = \det \mathbf{F}^* = \det\ (\mathbf{Q}\ \mathbf{F}) = \det\ (\mathbf{Q})\ \det\ (\mathbf{F}) = \det\ (\mathbf{F}) = J(f)} \tag{7.96}$$

where we have used the property: $\det\ (\mathbf{Q}) = 1$. The objectivity condition of a tensor-valued function follows from that of the scalar-valued and the vector-valued functions. To wit, let $\phi = \mathbf{a} \bullet \mathbf{T} \bullet \mathbf{b}$ be an objective scalar such that:

$$\phi(f^*) = \mathbf{a}(f^*) \bullet \mathbf{T}(f^*) \bullet \mathbf{b}(f^*) = \phi(f) = \mathbf{a}(f) \bullet \mathbf{T}(f) \bullet \mathbf{b}(f) \tag{7.97}$$

under two observers, f and f^*. Now, if $\mathbf{a}$ and $\mathbf{b}$ are objective vector functions, then

$$\begin{aligned} \phi^*(f^*) &= \mathbf{a}^*(f^*) \bullet \mathbf{T}^*(f^*) \bullet \mathbf{b}^*(f^*) \\ &= \mathbf{Q}\ \mathbf{a}(f) \bullet \mathbf{T}^*(f^*) \bullet \mathbf{Q}\ \mathbf{b}(f) \\ &= \mathbf{a}(f) \bullet (\mathbf{Q}^T\ \mathbf{T}^*(f^*)\ \mathbf{Q}) \bullet \mathbf{b}(f) \end{aligned} \tag{7.98}$$

Now, comparing equations (7.97) and (7.98), we get the objectivity condition:

$$\boxed{\mathbf{T}^*(f^*) = \mathbf{Q}\ \mathbf{T}(f)\ \mathbf{Q}^T} \tag{7.99}$$

The affine or Euclidean objective quantities are also known as ***frame indifferent*** with respect to the Euclidean transformation. Based on the above discussions, we finally summarize the transformation properties that must be respected so that we can call these specific functions of interest frame indifferent or objective:

$$
\begin{array}{|ll|}
\hline
\mathbf{v}^* = \mathbf{Q(t)}\ \mathbf{v} & \text{vectors} \\
\mathbf{T}^* = \mathbf{Q(t)}^T\ \mathbf{T}\ \mathbf{Q(t)} & \text{tensors} \\
\mathbf{F}^* = \mathbf{Q(t)}\ \mathbf{F} & \text{deformation gradient} \\
\mathbf{C}^* = \mathbf{C} & \text{right Cauchy–Green tensor} \\
\mathbf{U}^* = \mathbf{U} & \text{right stretch tensor} \\
\mathbf{E}^* = \mathbf{E} & \text{Green–Lagrange strain tensor} \\
\hline
\end{array}
\tag{7.100}
$$

7.4.2 Isotropic Tensor Functions: Representation and Invariance

In Lagrangian (material or referential) analysis, the constitutive properties of a material may be different for different ***reference configurations***. In other words, these types of materials respond to the same external stimuli in different ways, depending upon its orientation relative to the applied disturbances. In particular, if the reference configuration is rotated by a proper orthogonal that is, rotation tensor, the material behavior may change. For example, wood constitutive behavior along its fiber direction is different from what it is in the directions transverse to the fibers. However, there are materials (for example, construction steel) that manifest the same constitutive properties, that is, they respond equally to the same external stimuli, in all orientations with respect to the reference configuration of choice. In order to analyze and categorize these behaviors, we need to study what is known as the ***isotropic functions***. Let us define a domain space: $\mathcal{L}(\mathcal{V})$ where $\mathcal{V}$ and $\mathcal{L}(\mathcal{V})$ are the translation vector space in an affine space and the linear transformation (second-order tensor) space on the vector space, respectively. Then, a tensor function, $\mathbf{T} : \mathcal{L}(\mathcal{V}) \to \mathcal{L}(\mathcal{V})$, is called an ***isotropic tensor function*** if the following holds for any tensor, $\mathrm{H} \in \mathcal{L}(\mathcal{V})$, and for any proper orthogonal (rotation) tensor, $\mathbf{Q} \in SO3$:

$$\boxed{\mathbf{T}(\mathbf{Q}\ \mathbf{H}\ \mathbf{Q}^T) = \mathbf{Q}\ \mathbf{T}(\mathbf{H})\ \mathbf{Q}^T} \tag{7.101}$$

For example, let us suppose that the reference configuration is rotated by $\mathbf{Q}$, a rotation tensor. If we choose the undeformed base vectors, $\{\mathbf{G}_i\}$, $i = 1, 2, 3$, in the original reference configuration, we get the base vectors, $\{\hat{\mathbf{G}}_i\}$, for the rotated reference configuration, as $\hat{\mathbf{G}}_i = \mathbf{Q}\ \mathbf{G}_i$; then, the deformation gradient, $\hat{\mathbf{F}}$, and the right Cauchy–Green tensor, $\hat{\mathbf{C}}$, in the rotated configuration, are given by:

$$
\boxed{
\begin{aligned}
&\hat{\mathbf{F}} = \mathbf{g}_i \otimes \hat{\mathbf{G}}_i = \mathbf{g}_i \otimes \mathbf{Q}\ \mathbf{G}_i = (\mathbf{g}_i \otimes \mathbf{G}_i)\ \mathbf{Q}^T = \mathbf{F}\ \mathbf{Q}^T \\
&\hat{\mathbf{C}} = \hat{\mathbf{F}}^T\ \hat{\mathbf{F}} = \mathbf{Q}\ \mathbf{F}^T\ \mathbf{F}\ \mathbf{Q}^T = \mathbf{Q}\ \mathbf{C}\ \mathbf{Q}^T
\end{aligned}
}
\tag{7.102}
$$

where we have used the tensor property: $(\mathbf{a} \otimes \mathbf{Q}\ \mathbf{b}) = (\mathbf{a} \otimes\ \mathbf{b})\ \mathbf{Q}^T$ since $\forall \mathbf{h} \in \mathbb{R}$, we have: $(\mathbf{a} \otimes \mathbf{Q}\ \mathbf{b})\ \mathbf{h} = (\mathbf{Q}\ \mathbf{b} \bullet\ \mathbf{h})\mathbf{a} = (\mathbf{b} \bullet \mathbf{Q}^T\ \mathbf{h})\mathbf{a} = (\mathbf{a} \otimes\ \mathbf{b})\ \mathbf{Q}^T\ \mathbf{h}$. We find that the right Cauchy–Green tensor, $\hat{\mathbf{C}}$, is an isotropic tensor function while the deformation gradient, $\hat{\mathbf{F}}$, is not. Furthermore, the components of both $\mathbf{C}$ and $\hat{\mathbf{C}}$ are the same. To wit, using the transformation property of the undeformed base vectors for the reference configuration rotation and equation (7.102):

$$\boxed{\hat{C}_{ij} = \hat{\mathbf{G}}_i\ \hat{\mathbf{C}}\ \hat{\mathbf{G}}_j = \mathbf{G}_i\ \mathbf{Q}^T\ (\mathbf{Q}\ \mathbf{C}\ \mathbf{Q}^T)\ \mathbf{Q}\ \mathbf{G}_j = \mathbf{G}_i\ \mathbf{C}\ \mathbf{G}_j = C_{ij}} \tag{7.103}$$

Next, we introduce a theorem from linear algebra for later use in the representation theorem for isotropic tensor functions.

7.4.2.1 Cayley–Hamilton Theorem

Let the characteristic polynomial, $p(\lambda) = \det(\mathbf{T} - \lambda\mathbf{I})$, of a second-order symmetric tensor, $\mathbf{T} \in \mathcal{L}_{SYM}(\mathcal{V})$, where $\mathcal{L}_{SYM}(\mathcal{V})$ is the space of all symmetric second-order tensors or linear transformations, be given by:

$$\boxed{\begin{aligned} p(\lambda) &= \det(\mathbf{T} - \lambda\mathbf{I}) \\ &= \lambda^3 - I_T\,\lambda^2 + II_T\,\lambda - III_T \end{aligned}} \tag{7.104}$$

where, $\boxed{I_{\mathbf{T}}\,, II_{\mathbf{T}},\ III_{\mathbf{T}}}$ are the scalar invariants of the symmetric tensor, $\mathbf{T}$, given by:

$$\boxed{\begin{aligned} I_{\mathbf{T}} &= tr(\mathbf{T}) = \lambda_1 + \lambda_2 + \lambda_3 \\ II_{\mathbf{T}} &= \frac{1}{2}\left\{tr(\mathbf{T})^2 - tr(\mathbf{T}^2)\right\} = \lambda_1\,\lambda_2 + \lambda_2\,\lambda_3 + \lambda_3\,\lambda_1 \\ III_{\mathbf{T}} &= \det(\mathbf{T}) = \lambda_1\,\lambda_2\,\lambda_3 \end{aligned}} \tag{7.105}$$

where λ_1, λ_2, λ_3 are the distinct and real eigenvalues of the symmetric tensor, $\mathbf{T}$, that is, are the solution of the root equation: $p(\lambda) = \det(\mathbf{T} - \lambda\mathbf{I}) = 0$. Then, the tensor, $\mathbf{T}$, itself satisfies its characteristic polynomial equation; in other words, we have:

$$\boxed{\mathbf{T}^3 - I_T\,\mathbf{T}^2 + II_T\,\mathbf{T} - III_T = 0} \tag{7.106}$$

7.4.2.2 Representation Theorem: Symmetric Isotropic Tensor Functions of Symmetric Tensors

If $\mathbf{H} : \mathcal{L}_{SYM}(\mathcal{V}) \to \mathcal{L}_{SYM}(\mathcal{V})$, is an isotropic tensor function, then for any symmetric tensor, $\mathbf{T}$, we must be able to represent the function by a quadratic polynomial given by:

$$\boxed{\mathbf{H}(\mathbf{T}) = a_0\mathbf{I} + a_1\mathbf{T} + a_2\mathbf{T}^2} \tag{7.107}$$

where a_0, a_1, a_2 are functions of $\boxed{I_{\mathbf{T}}\,, II_{\mathbf{T}},\ III_{\mathbf{T}}}$, the scalar invariants of the symmetric tensor, $\mathbf{T}$, as given by equation (7.105). Moreover, if the isotropic tensor function, $\mathbf{H}(\mathbf{T})$, is linear, then it is given as:

$$\boxed{\mathbf{H}(\mathbf{T}) = \lambda(tr\mathbf{T})\mathbf{I} + 2\mu\mathbf{T}} \tag{7.108}$$

where λ and μ are some constants independent of the tensor, $\mathbf{T}$.

7.4.3 Elastic Materials

Despite all the mathematical support introduced so far, constitutive theory will still be, at best, an approximate model of a real material; it is based on experimental measurements and hence is approximate and empirical. So, to come up with a realistic model with properties as close as possible to those of its purported real material, several restrictions or rules must be imposed on the constitutive theory. In other words, it must guarantee some general principles to be physically plausible. In what follows, we discuss these general principles for purely mechanical phenomena, that is, with the stress tensor as the only dependent variable in the constitutive relationship, of an isotropic, homogeneous, linear elastic material.

7.4.3.1 Principle of Determinism

Principle: The stresses in a body are determined by the past history of the motion of the body.

According to this principle, the history of deformation up to the present time determines the properties of a material body. For elastic material, behaviors depend solely on the present state of deformation. The principle, thus, allows inclusion of elastic materials whose only dependence on history is characterized by the natural state to which a body will return upon unloading; in other words, for elastic materials, a body recovers its original state, independent of any history of its deformation. We will restrict our discussion from now on exclusively to elastic materials.

7.4.3.2 Principle of Local Action

Principle: The stress at a point in a material body is determined by the motion in the immediate small neighborhood of the body surrounding the point.

Under the previous principle of determinism alone, the mechanical history of any part of a material body is allowed to affect the response at a point the body. According to this principle, such a global influence is irrelevant for all practical purposes. Thus, the stress at a point in a body is determined by the local motion restricted to a small neighborhood of the point. Let A be such a point, with position vector, $\mathbf{X} \in \Omega$, and $N = \{\mathbf{Y} \in \Omega \,|\, \|\mathbf{Y} - \mathbf{X}\| \leq \varepsilon\}$ is a small neighborhood of an undeformed material body, Ω; then, the local motion, by first-order Taylor series expansion around $\mathbf{X}$, can be approximated by: $\mathbf{dx} = \mathbf{F}\,\mathbf{dX}$ where $\mathbf{dx}$ is the deformed image of the undeformed line element, $\mathbf{dX}$, completely contained in the neighborhood, N, with $\mathbf{F}(\mathbf{X}, t) = \frac{\partial \mathbf{x}}{\partial \mathbf{X}}$ as the deformation gradient at the point, $\mathbf{X}$, at any time, $t \in \mathbb{R}$. Then, according to this principle, the stress at a point is determined by a local deformation which, in turn, is completely characterized in the limit by the deformation gradient at the point. Thus, the Cauchy stress, $\boldsymbol{\sigma}$, for an elastic material can be described as a tensor-valued function:

$$\boxed{\boldsymbol{\sigma} = \boldsymbol{\sigma}(\mathbf{F}, \mathbf{X}, t)} \tag{7.109}$$

For $\boldsymbol{\sigma}(\mathbf{F})$ to be frame indifferent for arbitrary orthogonal transformation, $\mathbf{Q}$, we must have, according to equation (7.100):

$$\boxed{\mathbf{Q}\,\boldsymbol{\sigma}(\mathbf{F}, \mathbf{X}, t)\,\mathbf{Q}^T = \boldsymbol{\sigma}(\mathbf{Q}\,\mathbf{F}, \mathbf{X}, t)} \tag{7.110}$$

We may particularly choose: $\mathbf{Q} = \mathbf{R}^T$, where $\mathbf{R}$ is the rotation tensor that makes up for the polar decomposition of the deformation gradient as: $\mathbf{F} = \mathbf{R}\,\mathbf{U}$, with $\mathbf{U}$ as the right stretch tensor. Applying this to equation (7.110), we get:

$$\boxed{\boldsymbol{\sigma}(\mathbf{F}, \mathbf{X}, t) = \mathbf{R}\,\boldsymbol{\sigma}(\mathbf{U}, \mathbf{X}, t)\,\mathbf{R}^T} \tag{7.111}$$

In other words, for frame indifference, the Cauchy stress tensor cannot depend on the deformation gradient in any arbitrary manner; it must follow the pattern of the constituting polar components given by equation (7.111). Now, considering $\mathbf{P}(\mathbf{X}, t)$, the first Piola–Kirchhoff stress tensor (PK I), one of the Lagrangian stress tensors, and the Piola transformation (see Section 7.2.3.3 on deformation) relating it to the Cauchy stress tensor, we get, from equation (7.111):

$$\boxed{\begin{aligned}\mathbf{P} &= J\,\boldsymbol{\sigma}(\mathbf{F}, \mathbf{X}, \mathbf{t})\,\mathbf{F}^{-T} = (\det \mathbf{F})\,\mathbf{R}\,\boldsymbol{\sigma}(\mathbf{U}, \mathbf{X}, t)\,\mathbf{R}^T\,\mathbf{R}\,\mathbf{U}^{-T}\\ &= \mathbf{R}\,(\det \mathbf{U})\,\boldsymbol{\sigma}(\mathbf{U}, \mathbf{X}, t)\,\mathbf{U}^{-T} \equiv \mathbf{R}\,\mathbf{P}(\mathbf{U}, \mathbf{X}, t)\end{aligned}} \tag{7.112}$$

where we have used: $\det \mathbf{R} = 1$, and defined:

$$\boxed{\mathbf{P}(\mathbf{U}, \mathbf{X}, t) \equiv (\det \mathbf{U})\,\boldsymbol{\sigma}(\mathbf{U}, \mathbf{X}, t)\,\mathbf{U}^{-T}} \tag{7.113}$$

Finally, noting that: $\mathbf{P} = \mathbf{F}\,\mathbf{S}$ where $\mathbf{S}$ is the second Piola–Kirchhoff stress tensor (PK II), we get from equation (7.112):

$$\boxed{\begin{aligned}\mathbf{S} &= \mathbf{F}^{-1}\,\mathbf{R}\,\mathbf{P}(\mathbf{U}, \mathbf{X}, t) = \mathbf{U}\,\mathbf{P}(\mathbf{U}, \mathbf{X}, t)\\ &= \hat{\mathbf{S}}(\mathbf{U}, \mathbf{X}, t) = \hat{\mathbf{S}}(\mathbf{C}^{\frac{1}{2}}, \mathbf{X}, t)\\ &= \tilde{\mathbf{S}}(\mathbf{C}, \mathbf{X}, t) = \tilde{\mathbf{S}}(2\mathbf{E} + \mathbf{I}, \mathbf{X}, t) = S(\mathbf{E}, \mathbf{X}, t)\end{aligned}} \tag{7.114}$$

Thus, the second Piola–Kirchhoff stress tensor, $\mathbf{S}$, is a function of its energy conjugate strain measure: the Green–Lagrange strain tensor, $\mathbf{E}$, representing the constitutive relationship for an elastic material.

7.4.3.3 Homogeneous Elastic Material

The tensor function, $\mathbf{S}(\mathbf{E}, \mathbf{X}, \mathbf{t})$, in equation (7.114), is normally position dependent, that is, it is different for different points in a material body; however, from now on, we will restrict discussion to a homogeneous material body such that the constitutive relation becomes:

$$\boxed{\mathbf{S} = \mathbf{S}(\mathbf{E})} \tag{7.115}$$

that is, it is independent of the position of a point in the body; we have also taken out the explicit dependence on time for simplicity.

7.4.3.4 Isotropic, Homogeneous, Elastic Material: Symmetry

In addition to the condition of frame indifference, another fundamental restriction is imposed on the constitutive relationship: the condition of material symmetry. In general, the constitutive functions are dependent on the reference configuration. In other words, the responses of a material body under some stimuli could be different under different configurations. If we consider two reference configurations, c and $\bar{c}$, then a motion with respect to these reference configurations is related by: $\mathbf{x} = \chi_c(\mathbf{X}, t) = \chi_{\bar{c}}(\bar{\mathbf{X}}, t)$ where, say, $\bar{\mathbf{X}} = \beta(\mathbf{X})$ such that $\bar{Z} = GRAD\ \bar{\mathbf{X}} = \frac{\partial \beta(\mathbf{X})}{\partial \mathbf{X}} \in \mathcal{L}(\mathcal{V})$. Thus, by the chain rule: $\frac{\partial \chi_c(\mathbf{X},t)}{\partial \mathbf{X}} = \frac{\partial \chi_{\bar{c}}(\bar{\mathbf{X}},t)}{\partial \bar{\mathbf{X}}} \frac{\partial \beta(\mathbf{X})}{\partial \mathbf{X}}$ implying $\boxed{\mathbf{F} = \bar{\mathbf{F}}\, \bar{\mathbf{Z}}}$, relating the deformation gradients with reference to the two different configurations. Thus, a constitutive function, $\mathcal{T}_c(\mathbf{F})$ in reference configuration, c, is related to $\bar{\mathcal{T}}_{\bar{c}}(\bar{\mathbf{F}})$, the constitutive function in reference configuration, $\bar{c}$, by: $\bar{\mathcal{T}}_{\bar{c}}(\bar{\mathbf{F}}) \triangleq \mathcal{T}_c(\bar{\mathbf{F}}\, \bar{\mathbf{Z}}) = \mathcal{T}_c(\mathbf{F})$. Normally, $\bar{\mathcal{T}}_{\bar{c}}$ and $\mathcal{T}_c$ will be different; but for some materials, there are reference configurations, defining what is known as ***material symmetry***, that are indistinguishable, that is, there is a form invariance, implying that the constitutive functions are the same at these configurations. Thus, for the material symmetry condition, we must have:

$$\boxed{\mathcal{T}_c(\mathbf{F}) = \mathcal{T}_c(\mathbf{F}\,\mathbf{Z}), \quad \forall \mathbf{F}} \tag{7.116}$$

where $\mathbf{Z}$ is a linear transformation known as the ***material symmetry transformation***. In order to avoid arbitrarily large volume change due to successive changes in the reference configurations, we must have: $\det(\mathbf{Z}) = 1$. The set of all material symmetry transformations belong to a group, $\mathcal{Z}_c$, known as the ***material symmetry group*** such that $\mathbf{Z} \in \mathcal{Z}_c$. Now, following equation (7.116), for $\mathbf{S}$, the second Piola–Kirchhoff stress tensor, $\mathbf{S} = (\det \mathbf{F})\ \mathbf{F}^{-1}\ \boldsymbol{\sigma}(\mathbf{F})\ \mathbf{F}^{-T}$, with $\boldsymbol{\sigma}(\mathbf{F})$, the Cauchy constitutive function, we have the following material symmetry condition:

$$\begin{aligned} \mathbf{S}(\mathbf{C}) &= \mathbf{S}(\mathbf{F}^T\ \mathbf{F}) = \mathbf{S}((\mathbf{F}\mathbf{Z})^T\ (\mathbf{F}\mathbf{Z})) = \mathbf{S}(\mathbf{Z}^T \mathbf{C}\ \mathbf{Z}) \\ &= (\det \mathbf{F})\ \mathbf{F}^{-1}\ \boldsymbol{\sigma}(\mathbf{F})\ \mathbf{F}^{-T} = (\det \mathbf{F}\mathbf{Z})\ (\mathbf{F}\mathbf{Z})^{-1}\ \boldsymbol{\sigma}(\mathbf{F}\mathbf{Z})\ (\mathbf{F}\mathbf{Z})^{-T} \\ &= \mathbf{Z}^{-1}\{(\det \mathbf{F})\ \mathbf{F}^{-1}\ \boldsymbol{\sigma}(\mathbf{F})\ \mathbf{F}^{-T}\}\mathbf{Z}^{-T} \\ &= \mathbf{Z}^{-1}\ \mathbf{S}(\mathbf{C})\ \mathbf{Z}^{-T}, \qquad \forall \mathbf{Z} \in \mathcal{Z}_c \end{aligned} \tag{7.117}$$

With the relationship between the right Cauchy–Green tensor and the Green–Lagrange tensor, we get from equation (7.117):

$$\mathbf{S}(\mathbf{Z}^T \mathbf{E}\ \mathbf{Z}) = \mathbf{Z}^{-1}\ \mathbf{S}(\mathbf{E})\ \mathbf{Z}^{-T}, \quad \forall \mathbf{Z} \in \mathcal{Z} \tag{7.118}$$

Now, if $\mathcal{Z}_c \subseteq SO3$, that is the material symmetry transformations are proper rotations, then we have: $\mathbf{Z}^{-1} = \mathbf{Z}^T$, and if we particularly choose, $\mathbf{R}^T = \mathbf{Z}$, where $\mathbf{R}$ is the rotation tensor in the polar decomposition of the deformation gradient, $\mathbf{F}$, then, from equation (7.118), we finally get the material symmetry condition as:

$$\boxed{\mathbf{S}(\mathbf{R}\mathbf{E}\ \mathbf{R}^T) = \mathbf{R}\ \mathbf{S}(\mathbf{E})\ \mathbf{R}^T} \tag{7.119}$$

If for a material body, $\mathcal{Z}_c = SO3$, that is, the entire rotation group, then, the material is called an ***isotropic material***; that is, the material response is the same, irrespective of the direction of the reference configuration; otherwise, it is ***anisotropic***. From now on we will restrict ourselves to isotropic materials. Let us note that the isotropic symmetry condition given by equation (7.119) implies that $\mathbf{S}(\mathbf{E})$ is an isotropic tensor function according to equation (7.101). Then, using equation (7.107), the most general constitutive function of an isotropic material can be represented by:

$$\boxed{\mathbf{S}(\mathbf{C}) = a_0\mathbf{I} + a_1\mathbf{C} + a_2\mathbf{C}^2} \tag{7.120}$$

where $a_0,\ a_1,\ a_2$ are functions of $\boxed{I_{\mathbf{T}},\ II_{\mathbf{T}},\ III_{\mathbf{T}}}$, the scalar invariants of the symmetric tensor, **C**.

7.4.3.5 Hyperelastic, Isotropic, Homogeneous *Material*

The class of hyperelastic materials is characterized by the existence of a strain energy functional (i.e. elastic potential), $W = \rho\,\Phi$, per unit of undeformed volume, that depends on the deformation gradient or the strain tensor; ρ is the current density and Φ is the ***specific strain energy*** (i.e. strain energy per unit mass), also known as the ***Helmholtz free energy***; for Lagrangian analysis, the elastic potential depends on **E**, the Green–Lagrange strain tensor, with its material time derivative equal to the stress power, that is,

$$\boxed{\dot{W} = \rho_0\,\dot{\varepsilon} = \mathbf{S} : \dot{\mathbf{E}}} \tag{7.121}$$

But, $\dot{W} = \frac{\partial W(\mathbf{E})}{\partial \mathbf{E}} : \dot{\mathbf{E}}$, so we get, from equation (7.121), the second Piola–Kirchhoff stress tensor as:

$$\boxed{\mathbf{S} = \frac{\partial W(\mathbf{E})}{\partial \mathbf{E}}} \tag{7.122}$$

By considering the total potential, it can be shown that the primary characteristic of a hyperelastic material is that the internal work performed by the stress is independent of the path of the contributing conjugate strain and that it only depends on the beginning and end strains.

7.4.3.6 Material Tensor

By taking the time derivative of equation (7.122), we get:

$$\boxed{\dot{\mathbf{S}} = \frac{\partial^2 W(\mathbf{E})}{\partial \mathbf{E}\,\partial \mathbf{E}} : \dot{\mathbf{E}} = W_{,\mathbf{E}\otimes\mathbf{E}} : \dot{\mathbf{E}} \triangleq \mathbb{C} : \dot{\mathbf{E}}} \tag{7.123}$$

where $W_{,\mathbf{E}\otimes\mathbf{E}}$ is a fourth-order tensor, $\mathbb{C}$, and known as the ***material tensor***; in component form, in material coordinate frame, we can express it as:

$$\boxed{\mathbb{C} = \mathbb{C}^{ijkl}\ \mathbf{G}_i \otimes \mathbf{G}_j \otimes \mathbf{G}_k \otimes \mathbf{G}_l} \tag{7.124}$$

The isotropic hyperelastic material is characterized by the isotropic function which, in turn, has a representation as a quadratic polynomial as noted in equation (7.107).

7.4.3.7 Linear, Hyperelastic, Isotropic, Homogeneous Material

In our applications later, we are interested in linear, isotropic, homogeneous, hyperelastic materials; thus, here we discuss linearization of the constitutive function, $\mathbf{S}(\mathbf{E})$, relating $\mathbf{S}$, the second Piola–Kirchhoff (PK II) stress tensor to $\mathbf{E}$, the Green–Lagrange strain tensor through the displacement function, $\mathbf{d}(\mathbf{X}, t) = \mathbf{x} - \mathbf{X}$ at $\mathbf{X}$. Let us recall that with $\mathbf{H}(\mathbf{d}) \triangleq GRAD\ \mathbf{d} = \frac{\partial \mathbf{d}}{\partial \mathbf{X}}$ such that the deformation gradient, $\mathbf{F}(\mathbf{d}) = \mathbf{I} + \mathbf{H}(\mathbf{d})$, we get for the Green–Lagrange strain tensor:

$$\boxed{\mathbf{E}(\mathbf{d}) = \frac{1}{2}\underbrace{\{\mathbf{H}(\mathbf{d}) + \mathbf{H}(\mathbf{d})^T\}}_{\text{Linear and symmetric}} + \frac{1}{2}\underbrace{\mathbf{H}(\mathbf{d})^T\ \mathbf{H}(\mathbf{d})}_{\text{Nonlinear}}} \tag{7.125}$$

Recalling that the directional derivative, $\bar{\mathbf{E}}(\mathbf{d}, \bar{\mathbf{d}})$, of a tensor-valued function, $\mathbf{E}(\mathbf{d})$, along the direction, $\bar{\mathbf{d}}$, is defined (see Chapter 3 on tensors) by: $\bar{\mathbf{E}}(\mathbf{d}, \bar{\mathbf{d}}) = \lim_{\varepsilon \to 0} \frac{d}{d\varepsilon}\mathbf{E}(\mathbf{d} + \varepsilon\bar{\mathbf{d}})$, we get, by using the representation in equation (7.125), the expression for $L\mathbf{E}(\mathbf{d}, \bar{\mathbf{d}})$, the linearized tensor-valued function from $\mathbf{E}(\mathbf{d})$ at $\mathbf{d}$ along the direction $\bar{\mathbf{d}}$ as:

$$\boxed{L\mathbf{E}(\mathbf{d}, \bar{\mathbf{d}}) = \mathbf{E}(\mathbf{d}) + \bar{\mathbf{E}}(\mathbf{d}, \bar{\mathbf{d}})} \tag{7.126}$$

Now, evaluating at $\mathbf{d} = \mathbf{0}$, that is, at the reference configuration, and assuming that $\mathbf{E}(\mathbf{0}) = \mathbf{0}$, that is, the material body is unstrained at the reference configuration, and noting that the linear constitutive function, $\mathbf{S} = \mathbf{S}(\mathbf{E})$, is an isotropic symmetric tensor-valued function of symmetric tensor argument, we get, from the representation theorem given by equation (7.108), its representation as:

$$\boxed{\mathbf{S}(\mathbf{E}) = \lambda(tr\mathbf{E})\mathbf{I} + 2\mu\mathbf{E}} \tag{7.127}$$

where λ and μ are known as the ***Lame constants***. Recalling that $\mathbf{I} = \mathbf{G}_i \otimes \mathbf{G}^i$ and $\overset{4}{\mathbf{I}} = \mathbf{G}_i \otimes \mathbf{G}_j \otimes \mathbf{G}^i \otimes \mathbf{G}^j = G^{ik}G^{jl}\ \mathbf{G}_i \otimes \mathbf{G}_j \otimes \mathbf{G}_k \otimes \mathbf{G}_l$ as the second-order and the fourth-order identity tensors, respectively, we have:

$$\boxed{\begin{aligned}\mathbf{I} \otimes \mathbf{I} : \mathbf{E} &= \mathbf{G}_i \otimes \mathbf{G}^i \otimes \mathbf{G}_j \otimes \mathbf{G}^j : E_{mn}\ \mathbf{G}^m \otimes \mathbf{G}^n = \delta_j^m G^{jn} E_{mn}\ \mathbf{G}_i \otimes \mathbf{G}^i = E_j^{\cdot j}\mathbf{I}\\ &= (tr\ \mathbf{E})\mathbf{I}\end{aligned}} \tag{7.128}$$

and

$$\boxed{\begin{aligned}\overset{4}{\mathbf{I}} : \mathbf{E} &= \mathbf{G}_i \otimes \mathbf{G}_j \otimes \mathbf{G}^i \otimes \mathbf{G}^j : E_{mn}\ \mathbf{G}^m \otimes \mathbf{G}^n = G^{im}G^{jn}E_{mn}\ \mathbf{G}_i \otimes \mathbf{G}_j = E_{mn}\ \mathbf{G}^m \otimes \mathbf{G}^n\\ &= \mathbf{E}\end{aligned}} \tag{7.129}$$

So, applying equations (7.128) and (7.129) in equation (7.127), we get:

$$\boxed{\mathbf{S}(\mathbf{E}) = \left\{\lambda(\mathbf{I} \otimes \mathbf{I}) + 2\mu\overset{4}{\mathbf{I}}\right\} : \mathbf{E}} \tag{7.130}$$

Finally, we define the fourth-order constitutive or material tensor, $\mathbb{C}$, as:

$$\boxed{\mathbb{C} \triangleq \lambda(\mathbf{I} \otimes \mathbf{I}) + 2\mu \overset{4}{\mathbf{I}} = C^{ijkl}\, \mathbf{G}_i \otimes \mathbf{G}_j \otimes \mathbf{G}_k \otimes \mathbf{G}_l} \tag{7.131}$$

to get the desired constitutive relation between second Piola–Kirchhoff stress tensor and the Green–Lagrange strain tensor, using equation (7.131) in equation (7.130), as:

$$\boxed{\begin{aligned} \mathbf{S} &= \mathbb{C} : \mathbf{E} \\ S^{ij} &= C^{ijkl}\, E_{kl} \end{aligned}} \tag{7.132}$$

Clearly, since both the stress and strain tensors are symmetric, the material matrix is symmetric in the first two and the last two indices; also, as can be seen from equation (7.131), the material tensor is completely defined by just two independent parameters – the Lame constants. We record below some of the more useful definitions, to be used in the applications later in the book, that follow from these constants:

$$\boxed{\begin{array}{ll} \mu = \dfrac{E}{2(1+\nu)} = G & \text{:Lame constant or shear modulus} \\ \lambda = \dfrac{E\nu}{2(1+\nu)(1-2\nu)} & \text{:Lame constant} \\ E = 2\mu(1+\nu) & \text{:Elasticity or Youngs modulus} \\ \nu = \dfrac{\lambda}{2(\lambda+\mu)} & \text{:Poissons ratio} \\ \kappa = \dfrac{E}{3(1-2\nu)} = \dfrac{1}{3}(3\lambda + 2\mu) & \text{:Bulk or volume modulus} \end{array}} \tag{7.133}$$

Now, using this symmetry of the material tensor, noting the definitions of the second-order and fourth-order identity tensors, and of the Lame constants in equation (7.133), we get from equation (7.131):

$$\boxed{C^{ijkl} = \mu \left\{ G^{ik}\, G^{jl} + G^{il}\, G^{jk} + \frac{2\nu}{1-2\nu} G^{ij}\, G^{kl} \right\}} \tag{7.134}$$

7.4.4 *Where We Would Like to Go*

With the final pillar – the constitutive theory – having been essentially described for our purposes, the next logical stage will be the application and furtherance by transformation of the concepts of solid mechanics to computational mechanics, that is

- finite element methodology: a new finite element space (Chapter 8)

Part IV

A New Finite Element Method

8

C-type Finite Element Method

8.1 Introduction

8.1.1 What We Should Already Know

Introduction and motivation, essential mechanics, essential mathematics, strain–displacement, stress–strain, equilibrium equations, curves, Bernstein–Bezier functions, B-spline functions, vector, matrices, determinants, traces, tensors, stress tensor, strain tensor, constitutive law, Sobolev space, inner product spaces, vector spaces, best approximation, linear variety, integration by parts, Green's identity, divergence theorem, linear space, subspace, distribution, bilinear form, linear form, linear functional, basis functions, positive definite matrix, self-adjoint matrix, potential energy, strain energy, Euler equation, state vector, displacement, temperature, the principle of minimum potential energy, square integrable, plain strain, plane stress, axisymmetric, Euler–Bernoulli beam, plate, shell, undeformed (Lagrangian) state, deformed (Eulerian) state, surfaces, reduced integration, selective integration, bubble mode, constructive solid modeling, boundary representation method, Cartesian product form, Bezier control vector, barycentric property, convex combination, Jacobian, linear precision, degree elevation, conductivity, heat flow, de Boor vertices, C^n-continuity

8.1.2 Why We Are Here

This chapter introduces our ***c-type finite element method***; however, before we can shed any light on this method, we must ask ourselves: what is a finite element method? In the S-form, that is, the strong differential equation form, of boundary value problem (BVP) in which most mechanics problems are described, it is immensely difficult, if not impossible, to satisfy the governing partial differential equations at every point in the body at all time even for a mildly complicated problem. The aim is then to transform these equations so that effective numerical methods can be applied to solve problems approximately but with the requirement to converge as close as possible to the exact solutions. Integration (anti-derivative) being the opposite of differentiation and S-form being of differential character, the most intuitive expectation is that the new form should have the appearance of integral equations. This idea is a winner, and the new forms (in fact, two forms as will be seen later) are all about integration. This, in turn, makes the new form more amenable to numerical determination. We can suggest one possible argument in support of this from our elementary idea of calculus: as far as smoothness (a helping attribute of the ***trial solutions***, to be detailed soon, for effective numerical treatment) is concerned, operation of differentiation in

Computation of Nonlinear Structures: Extremely Large Elements for Frames, Plates and Shells, First Edition. Debabrata Ray.

S-form only weakens it, while integration in the new forms strengthens it. But, then, again intuitively, smoothing by integration is also a process of averaging and thus is prone to losing important pointwise information that is otherwise retained by the differential form. So there a price is attached to this transformation. Given the computer age, perhaps the most important reason in support of ***variational formulation*** – a type of formulation that relies on integral transform – lies in its ability to accommodate approximations by various powerful numerical concepts that are necessary for solving practical problems of any complexity. The finite element method has turned out to be the most popular of such approximation methods, particularly in structural and continuum mechanics, and all branches of engineering in general. Thus, a finite element method, in theory, is a variational method. However, given this problem, in solid mechanics, which generally involves a material body with complicated geometry supported in a complicated manner and subjected to complicated loading system, the triumph of a finite element method lies in simplifying the problem by breaking the geometry into a ***finite*** set of mathematically easily tractable pieces (***local basis functions***), building the unknown solution ***locally*** on these pieces and finally ***globally*** constructing the required state (***global basis function***). In other words, a finite element method numerically constructs the global solution by piecing together the local description on a finite set of simple shapes, usually determined by the geometry of the material body, the support and the loading conditions. In the past, the finite element methods arbitrarily violated, as indicated in Chapter 1 on introduction and motivation, the underlying variational principles and thus, erroneously fell prey to ad hoc numerical artifacts with nagging undesirable consequences. As will be seen later, the c-type finite element method can strictly adhere to the theoretical underpinning by choosing the appropriate local basis functions and thus, numerically becomes the most efficient, robust and accurate method with "mega-sized" finite elements compared to all other hitherto published methods. In what follows, then, we present:

Section 8.2: We start by looking at the general theory underlying the variational formulations of boundary value problems, followed by the concepts of energy space, trial space and test space. In so doing, we identify various formulations as simply variants of the general theory arising out of the choice of the test space. We then establish the connection between the different forms: strong, weak or Galerkin, variational and residual forms.

Section 8.3: Here we touch on the different variational theorems: along with the potential energy functional, the minimum potential energy principle is detailed followed by the virtual work principle for both linear and nonlinear problems in structural engineering and structural mechanics.

Section 8.4: This section starts with a description of the classical Rayleigh–Ritz–Galerkin technique serving as a basis for the finite element method, then a new finite element space – c-type – which allows the generation of large or mega-sized curved elements is introduced. These follow strictly the displacement-based iso- or subparametric Rayleigh–Ritz–Galerkin method; the method is then particularized through application of the virtual work principle for linear elasticity and heat conduction problems; the section ends with a discussion of inter-element continuity conditions-on-demand and the c-type method.

Section 8.5: This details the Newton-type iteration method with arc length constraint, as needed for nonlinear force–deformation path computation.

Section 8.6: Finally, we discuss numerical integration by Gauss quadrature and other formulas.

8.2 Variational Formulations

8.2.1 Strong or Classical Form: (S-Form)

In its most general form in linear regime, we may introduce the inhomogeneous boundary value problem, (IBVP) as follows.

Given:

$$\boxed{\begin{array}{lll} \mathbf{A}\,\mathbf{d}=\mathbf{f} & \text{in }\Omega, & \mathbf{f}\in\mathbb{S}^{\mathbf{f}} \\ \mathbf{B}\,\mathbf{d}=\mathbf{g} & \text{on }\Gamma, & \mathbf{g}\in\mathbb{S}^{\mathbf{g}} \end{array}} \tag{8.1 and 8.2}$$

Find:

$$\mathbf{d}\in\mathbb{S}^{\mathbf{d}}_{\mathbf{B}} \tag{8.3}$$

where,

$\mathbb{S}^{\mathbf{d}}_{\mathbf{B}}$	=	a normed linear space of state vectors, **d**, that satisfies the boundary conditions
$\mathbb{S}^{\mathbf{f}}$	=	a normed linear space for the given data, **f**
$\mathbb{S}^{\mathbf{g}}$	=	a normed linear space for the given data, **g**
$\mathbf{A}:\mathbb{S}^{\mathbf{d}}_{\mathbf{B}}\to\mathbb{S}^{\mathbf{f}}$	=	a linear differential operator of order $2m$, that is, the highest degree of derivatives are of order $2m$
$\mathbf{B}:\mathbb{S}^{\mathbf{d}}_{\mathbf{B}}\to\mathbb{S}^{\mathbf{g}}$	=	a linear (possibly, differential) operator of order, $2m-1$, that is, the highest degree of derivatives are of order $2m-1$,

Remarks

- For linear structural engineering and structural mechanics problems, equations (8.1) and (8.2) represent the state equilibrium equations and boundary conditions, respectively. For **d** to be the displacement vector serving as the only state vector, the other independent physical entities, σ, the stress tensor, and ε, the strain tensor, have been replaced by application of appropriate constitutive law and ***strain–displacement compatibility conditions***. In all our future references to both linear and nonlinear problems, we will stay with the displacements as the only constituting entities of the state vector; this will eventually lead us to the displacement-based finite element method which is our main interest in this book.
- For nonlinear problems, we replace $\mathbf{A}\,\mathbf{d}=\mathbf{f}$ by, say, $\mathbf{A}\,(\mathbf{d})=\mathbf{0}$, where **A** is a nonlinear operator of the displacement vector, **d**.

Example 1: At one extreme, in its most general and complex linear form, we may present, for elliptic equations in an n-dimensional region, Ω, the differential operator, **A**, as: $\mathbf{A}(\bullet)=\sum_{|\mathbf{p}|,|\mathbf{q}|\le k}(-1)^{|\mathbf{q}|}\,\mathbf{D}^{\mathbf{q}}(a_{\mathbf{pq}}(\mathbf{x})\mathbf{D}^{\mathbf{p}}(\bullet)$ where, $\mathbf{x}=(x_1,\dots,x_n)\in\Omega$, $\mathbf{p}=(p_1,\dots,p_n)$ and $\mathbf{q}=(q_1,\dots,q_n)$ are positive multi-integer vectors with $|\mathbf{p}|=p_1+\dots+p_n$, $|\mathbf{q}|=q_1+\dots+q_n$, and $\mathbf{D}^{\mathbf{p}}(\bullet)=\frac{\partial^{|\mathbf{p}|}(\bullet)}{\partial x_1^{p_1}\dots\partial x_n^{p_n}}$.

The operator, A, is said to be ***elliptic***, if:

$$\sum_{|\mathbf{p}|,|\mathbf{q}|\le k} a_{\mathbf{pq}}(\mathbf{x})\mathbf{x}^{\mathbf{p}}\,\mathbf{x}^{\mathbf{q}} \;\ge C\sum_{|\mathbf{p}|\le k}|\mathbf{x}^{\mathbf{p}}|^2,\forall\mathbf{x}\in\Omega \tag{8.4}$$

where $\mathbf{x}^{\mathbf{p}}=\{x_1^{p_1},\dots,x_n^{p_n}\}$ and C, a constant.

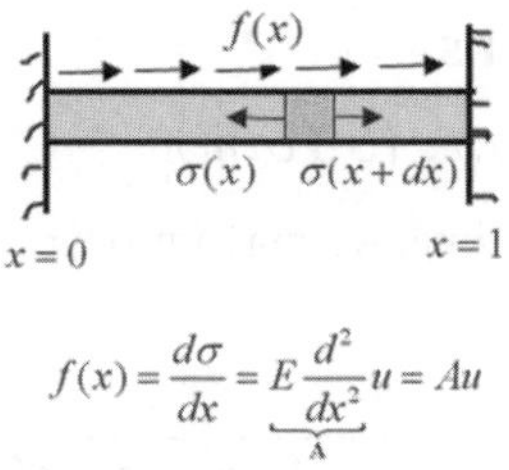

Figure 8.1 Example 2.

Remark

- This example is presented only to show the ease with which a problem of considerable complexity with multivariate calculus can be put into vector space format. Accordingly, the theory presented next will equally be applicable to the problem with all its generality. Of course, to work out the actual details will be a mathematically formidable task, but fortunately, for our purposes we have no need to delve into it any further. Instead, we will now present an example of extreme simplicity that is still of interest for structural analysis.

Example 2: Consider the case of a simple one-dimensional bar. For this, referring to Figure 8.1, we have: $\Omega = \{x \in (0, L) \subset \mathbf{R}^1\}$, that is, the bar excluding the end points, $\Gamma = \{0\} \cup \{L\}$, that is, the two end points of the bar, $\mathbb{S}^{\mathbf{f}} = \{f | f \in C^0(\Omega)\}$ and $\mathbb{S}^{\mathbf{d}}_{\mathbf{B}} = \{u | u \in C^2(\Omega),\ u(0) = 0,\ A(L)E(L)\ u_{,x}(L) = P_L\}$ with $\mathbf{A} : \mathbb{S}^{\mathbf{d}}_{\mathbf{B}} \mathbf{x} \mathbb{S}^{\mathbf{f}} = \mathbf{D}^{\mathbf{1}}[a_{11}(x)\mathbf{D}^{\mathbf{1}}(\bullet)] = \frac{\partial}{\partial x}[A(x)E(x)\frac{\partial(\bullet)}{\partial x}]$, that is, $m = 1$ where, $A(x)$ and $E(x)$ are cross-sectional area and elasticity modulus, respectively, at any $\mathbf{x} \in \Omega$, and finally, **B** is represented by: $u(0) = 0$ and A(L)E(L) $u_{,x}(L) = P_L$.

Remarks

- With the strong form, the full set of boundary conditions is imposed on the solution, In other words, if **f** is just continuous or has finite energy, that is, $\mathbf{f} \in \mathcal{H}^0$,(recall that the definition of general Sobolev spaces, $\mathcal{H}^s$, of degree s is the complete normed linear ***spaces*** that contain functions whose s th derivative has finite energy), then **d** must be sought from a space such that $\mathbf{d} \in \mathcal{H}^{2m}_{\mathbf{B}}$, where the subscript **B** indicates that boundary conditions are satisfied. For the bar in the example above, we will have $\mathbf{d} \in \mathcal{H}^2_{\mathbf{B}}$, that is, **d** has to have its second derivative with finite energy.
- In general, an exact closed form solution is not possible for problems stated in strong form that have any degree of complexity in their description of the differential operator, the region and/or the boundary conditions. We simply need to settle for the ***best approximation*** in the sense introduced earlier for the solutions to these problems.

Fortunately, there are very many numerical methods available to us, all of which can be effectively categorized into two major classes, depending on what we choose as a starting point of the numerical formulation:

- Stay with the original strong form of the statement of the problem; these methods include the finite difference method and the residual error method.

- Devise some new form by judiciously yet equivalently transforming the original strong (differential) form of the statement of the problem into a weak (integral) form; these methods are collectively known as ***variational methods***.

8.2.2 Finite Difference Methods

This immensely important class of numerical methods arises out of the decision to deal directly with the strong form of the statement of the problem. Generally speaking, the infinite-dimensional problem is transformed into a finite set of simultaneous equations first by discretizing the domain through identification of some finite set of locations, and then, replacing the differential operators by their corresponding finite difference operators at those discrete set of point of the domain; hence, the name ***finite difference methods***. The boundary conditions are handled similarly. Of course, the accuracy and convergence characteristics of this class of methods depend on both the domain discretization and the finite difference operators. Finally, the simultaneous set of algebraic equations is solved by computers by application of any of the familiar equation solvers available. We shall not pursue this class of methods any further in this book; interested readers may consult a textbook on numerical analysis. Now, to facilitate our future discussion on variational methods we introduced a variant of the strong form

8.2.3 Residual Error Form: (R-Form)

In its general form, we have the IBVP as:
Given:

$$\mathbf{R}(\mathbf{d}) \equiv \mathbf{A}\,\mathbf{d} - \mathbf{f} = \mathbf{0} \tag{8.5}$$

Find:

$$\mathbf{d} \in \mathbb{S}_{\mathbf{B}}^{\mathbf{d}} \tag{8.6}$$

Remarks

- Clearly, for R-form, the solution belongs to the null space of $\mathbf{R}$, the residual error transformation. This differential form is similar to the strong form except that forcing function has been moved to the left side of the equation. For solutions based directly on differential equation, this form does not change the situation from that of its strong form counterpart. However, this form essentially sets the stage for variational formulations by possibly changing the problem to one of minimization of the residual error, $\mathbf{R}(\mathbf{d})$, over some appropriate solution space. The weighted residual methods derive their name from this form of the statement of the problem.

8.2.4 Variational Methods

These belong to an alternative and effective class of numerical methods. They are aimed at transforming the strong form description of the problem. Here lies the root of several variants of what is known as the class of ***variational methods*** of which, given the modern computer resources, the finite element method is the most popular one. Generally speaking, in these methods, we seek to approximate the solution by minimization of the residual error, $\mathbf{R(d)}$, over a solution set consisting of a linear variety of a subspace, known as the ***trial space***, of an appropriate normed linear space. The dimensionality of the trial space is dependent on another auxiliary normed linear space, known as a ***test space***. As we will soon seen, it is the choice of the test space that determines the different variational methods that exist in the literature today. Additionally, a particular choice of test space which makes it one and the same as the trial space, necessitates the introduction of ***energy space*** and the splitting of the boundary conditions into two types: the ***essential*** and ***natural*** boundary conditions, for the problem at hand. Finally, all these methods apply a basic tool of the differential operator theory: the familiar integration by parts for one dimension or its higher-dimensional counterparts – ***Green's identity*** or the ***divergence theorem***. In so doing, it gives rise to the desired transformed form that is most amenable for generating numerical algorithm suitable for computer solution of problems of an enormous variety of complexity.

Remark

- Because the strong form is expressed in differential form, the underlying spaces of the transformed form are, in a sense, an inversion of the strong form. Accordingly, the spaces are expected to be normed linear spaces or inner product spaces with bilinear forms or inner product, respectively, expressed naturally in integral – that is, anti-derivative – form. The transformed form referred to here is known as the ***weak or Galerkin form***.

8.2.5 Weak or Galerkin Form: (W- Form)

In view of the cursory description of the weak or Galerkin form presented above, a formal statement of the form needs to wait until we have finished our digression to define various spaces and boundary conditions introduced therein.

Remarks

- Before we delve into any discussion, note that we will have ample opportunities to revisit these concepts, revealing their innermost characteristics through concrete examples involving practical problems of interest in structural and solid mechanics. However, for now, we refer to our simple one-dimensional model from time to time, to elaborate on these concepts.
- Finally, for the following discussions, we will, without any loss of generality, assume that the $\mathbf{f}$, the forcing function in equation (8.5), is only continuous, that is, $\mathbf{f}$ has finite energy or $\mathbf{f} \in \mathcal{H}^0(\Omega)$, that is, $\|f\|_0 \equiv \int_\Omega f^2 \, d\Omega < \infty$.

8.2.5.1 Test Space: $\mathbf{V}_0(\Omega)$

The test space, $\mathbf{V}_0(\Omega)$, is a subspace of the Sobolev space, $\mathcal{H}^s(\Omega)$. The subscript 0 in the notation indicates that the elements belonging to this space satisfy the homogeneous counterparts of all prescribed boundary conditions (to be defined shortly) that generally involve derivatives of less than or equal to s. The superscript s for the space may vary from -1 to $2m$, depending on the type of the variational method. As we will see, the most intriguing test space that we are interested in is when $s = m$, in which case, the test space could be chosen to be one and the same as the trial space, a subspace of an energy space, to be defined shortly.

8.2.5.2 Solution Set: $\mathbf{U}^{Sol}(\Omega)$

The solution set, $\mathbf{U}^{Sol}(\Omega)$, is a subset of the Sobolev space, $\mathcal{H}^{2m-s}(\Omega)$, where s in the superscript is same as that defines a test space. The *Sol* in the superscript indicates that elements of the set satisfies all boundary conditions that involve derivatives of degree less than $(2m - s)$. For the case of our ultimate interest, that is, when $s = m$, any element of the set satisfies all essential boundary conditions whose derivatives are of degree less than or equal to $(m - 1)$. It may be generally noted that by definition, for inhomogeneous boundary conditions, the solution set does not constitute a linear space or subspace. This brings us to the following definition.

8.2.5.3 Trial Space: $\mathbf{U}_0(\Omega)$

The trial space, $\mathbf{U}_0(\Omega)$, is a subspace of the Sobolev space, $\mathcal{H}^{2m-s}(\Omega)$, such that elements of the subspace satisfy all the homogeneous counterparts of the boundary conditions that involve derivatives of degree less than $(2m - s)$. In other words, the solution set, $\mathbf{U}^{Sol}(\Omega)$, is a linear variety of the trial space, $\mathbf{U}_0(\Omega)$. Applying the translation theorem on linear variety, we can express any element $\mathbf{d}_{Sol} \in \mathbf{U}^{Sol}(\Omega)$ as a summation of $\mathbf{d_0} \in \mathbf{U_0}(\Omega)$ and an arbitrary fixed element $\hat{\mathbf{d}} \in \mathbf{U}^{Sol}(\Omega)$, that is,

$$\boxed{\mathbf{d}_{Sol} = \hat{\mathbf{d}} + \mathbf{d_0}} \tag{8.7}$$

8.2.5.4 Admissible Boundary Conditions

As to the role of boundary conditions for various formulations, any boundary condition that makes sense in a variational formulation must involve derivatives of degree less than $2m - s$, where the trial space is a subspace of $\mathcal{H}^{2m-s}(\Omega)$.

Example 3: For the one-dimensional problem introduced earlier, the end condition at the $x = 0$ side, that is, $u(0) = u_0$ constitutes an essential boundary condition, since the degree of the derivative is $0 < m$ where for both the trial space and the test space, $s = m = 1$.

Remarks

- The solution to an IBVP, can be obtained by first dealing with the corresponding BVP with homogeneous boundary conditions (HBVP), and then adding a fixed vector determined by the inhomogeneous boundary conditions.
- As indicated above, if $s = -1$, in the test space, that is, the test space is the space of δ-functions, then the trial space becomes $\mathcal{H}^{2m}(\Omega)$, that is, the solution is sought in the strongest sense. In other words, all the essential boundary conditions of degree less than or equal to $(2m - 1)$ are imposed on the solution set.
- At the other extreme, if the test space contains infinitely smooth functions, the solution can be sought in the weakest form, that is, in the distribution sense.

We are finally ready to define the weak or Galerkin form (W-form) of the statement of the problem corresponding to the strong form expressed by the equations (8.1)–(8.3), as follows:

Given: $\Omega, \mathbf{A}, \mathbf{f}, \mathbf{B}$ and $\mathbf{g}$ as in equations (8.1) and (8.2)

Find: $\mathbf{d} \in \mathbf{U_0}(\Omega)$ such that

$$(\mathbf{Ad}, \mathbf{v}) = (\mathbf{f}, \mathbf{v}), \quad \forall \mathbf{v} \in \mathbf{V_0}(\Omega) \tag{8.8}$$

or

$$(\mathbf{R}(\mathbf{d}), \mathbf{v}) = \mathbf{0}, \quad \forall \mathbf{v} \in \mathbf{V_0}(\Omega) \tag{8.9}$$

where $\mathbf{U_0}(\Omega)$ is the trial space, that is, the space from which all solutions are being sought, $\mathbf{V_0}(\Omega)$ is the test space (generally known as the virtual displacement space), $(\mathbf{Ad}, \mathbf{v})$ is a bilinear form of $\mathbf{U_0}(\Omega) \times \mathbf{V_0}(\Omega)$ and $(\mathbf{f}, \mathbf{v})$ is a linear form or linear functional of $\mathbf{V_0}(\Omega)$.

Remark

- As indicated above, because the W-form has to hold for all $\mathbf{v} \in \mathbf{V_0}(\Omega)$, the choice of the test space, $\mathbf{V_0}(\Omega)$, clearly determines everything about the weak form definition of HBVP including the computational algorithm that follows for the desired solution, $\mathbf{d} \in \mathrm{U}_0(\Omega)$. In other words, the numerical algorithm – and hence the type of variational method – is pretty much determined by the test space.

Equation (8.9) provides us with a geometrical insight that resembles a similar perspective introduced in the discussion on the best approximation. This, in turn, is very useful in appreciating the underlying scheme of things for the weak form: the residual error is perpendicular to every element of the test space, as shown in Figure 8.2.

This perpendicularity, viewed as the shortest distance between the solution point and the test space, suggests some sort of minimization algorithm involving the distance between the image of the elements of the trial space under residual error transformation, $\mathbf{R}$, and the elements of the test space. Before we concentrate on the type of variational method of our interest that serves as a precursor to the finite element method, let us briefly touch upon a few other variants that arise out of specific choices of the test space.

8.2.5.5 Collocation Method ($s = -1$)

This method is based on the extreme choice of the test space, $\mathbf{V}_0 \equiv \mathcal{H}^{-1}(\Omega)$, that is, only infinite integrals of $\mathbf{v} \in \mathbf{V}_0$ having finite energy. In other words, the test space consists of only δ-functions, with their basis functions given as:

$$\Phi_j(\mathbf{c}) = \delta(\mathbf{c} - \mathbf{c_j}), \quad \forall \mathbf{c}, \mathbf{c_j} \in \Omega \tag{8.10}$$

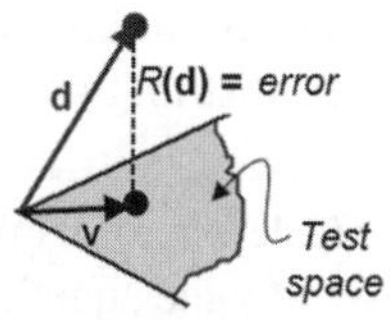

Figure 8.2 Best approximation.

The corresponding numerical method is an obvious discretization based on the choice of the test space as a finite N-dimensional subspace, $\mathbf{V}_0^{\mathbf{N}} \subset \mathbf{V}_0$. This then provides the numerical algorithm in the following W-form:

$$\mathbf{A}\ \mathbf{d}(\mathbf{c_j}) = \mathbf{f}(\mathbf{c_j}), \quad \mathbf{j} = \mathbf{j}_1, \mathbf{j}_2, \dots, \mathbf{j}_N \tag{8.11}$$

In other words, $\mathbf{d} \in \mathbf{U}_0(\Omega) \equiv \mathcal{H}_0^{2m}$ must satisfy all the boundary conditions and must satisfy them at all discretization points, that is, **d** satisfies the problem in the strong sense and the W-form satisfies the problem pointwise.

Remark

- For one-dimensional Ω, all quantities in equation (8.11) are scalar. However, for $\Omega \subset \mathbf{R}^m$, $m \geq 2$, the corresponding items are vector quantities belonging to $\mathbf{R}^m$. For example, $\mathbf{j} = (s, s, \dots, m\text{–many})$ where s is a scalar.

8.2.5.6 Green's Function Method ($s = \infty$)

This method is based on the other extreme choice of the test space, $\mathbf{V}_0 \equiv \mathcal{H}^\infty(\Omega)$, that is, it contains infinitely smooth functions which vanish on the boundary, Γ. A formal integration by parts, or an application of its two- or three-dimensional counterparts, of the Green's identity or the divergence theorem to equation (8.8) or (8.9) transfers all the derivatives from $\mathbf{d} \in \mathbf{U}_0$, the trial space on to $\mathbf{v} \in \mathbf{V}_0$, the test space. This results in the following new W-form:

$$(\mathbf{d}, \mathbf{A}^*\ \mathbf{v}) = (\mathbf{f}, \mathbf{v}), \quad \forall \mathbf{v} \in \mathbf{V}_0 \tag{8.12}$$

where $\mathbf{A}^*$ is the formal adjoint of **A**. If **A** is self-adjoint, that is, $\mathbf{A} = \mathbf{A}^*$, then we have:

$$(\mathbf{d}, \mathbf{A}\mathbf{v}) = (\mathbf{f}, \mathbf{v}), \quad \forall \mathbf{v} \in \mathbf{V}_0 \tag{8.13}$$

Clearly, then, **d**, the solution, in this weakest form has to satisfy the equations only in the distribution sense with the trial space, $\mathbf{U}_0 \equiv \mathcal{H}^{-1}(\Omega)$.

8.2.5.7 Galerkin Methods ($-1 < s < m$)

For any choice of the test space, say $\mathbf{V}_0 \equiv \mathcal{H}^s(\Omega)$, in between those above extreme choices, the boundary conditions that are meaningful are obviously of derivatives of degree $< 2m - s$, while the solution space, from which **d** is sought, is of degree $2m - s$, that is, $\mathbf{U}_0 \equiv \mathcal{H}^{2m-s}(\Omega)$. This is accomplished by shifting s many derivatives from **d** on to $\mathbf{v} \in \mathbf{V}_0$ by application of integration by parts in one dimension or ***Green's identity*** in higher dimension. Numerically, it is difficult to handle these infinite-dimensional subspaces. The ***Galerkin method*** is thus an obvious discretization of this situation where the solution space is a N-dimensional subspace, $\mathbf{U}_0^{\mathbf{N}} \subset \mathbf{U}_0$, of the solution space, and the test space is a M-dimensional subspace, $\mathbf{V}_0^{\mathbf{M}} \subset \mathbf{V}_0$ such that the weak form becomes as follows:

Find: $\mathbf{d}^{\mathbf{N}} \in U_0^N(\Omega)$ such that

$$(\mathbf{A}\mathbf{d}^{\mathbf{N}}, \mathbf{v}^{\mathbf{M}}) = (\mathbf{f}, \mathbf{v}^{\mathbf{M}}), \quad \forall \mathbf{v}^{\mathbf{M}} \in \mathbf{V}_0^{\mathbf{M}}(\Omega) \tag{8.14}$$

The left side of the equation (8.14) needs s integration by parts. If the solution subspace and the test subspace are of equal dimensions, say, $N = M$, then assuming $\Phi = (\phi_1, \phi_2, \ldots, \phi_{\mathbf{N}})$ as the set of basis functions for the solution subspace and $\Psi = (\psi_1, \psi_2, \ldots, \psi_{\mathbf{N}})$ as the set of basis functions for the test subspace, we have from equation (8.14), with $N = M$, that the solution, in matrix form, defined by:

$$\mathbf{d} = \Phi\mathbf{q} \tag{8.15}$$

must satisfy the matrix equation:

$$\boxed{\begin{array}{ll} & \mathbf{G}\,\mathbf{q} = \mathbf{F} \\ \text{with} & \mathbf{G} \equiv \Psi^{\mathbf{T}}\mathbf{A}\Phi, \\ & \mathbf{F} = \Psi^{\mathbf{T}}\mathbf{f} \end{array}} \tag{8.16}$$

In component form:

$$\boxed{\begin{array}{ll} & \mathbf{d} = \sum_{i=1}^{N} q_i \Phi_i \\ \text{with} & \mathbf{q} = \{q_1, q_2, \ldots, q_N\}^T \end{array}} \tag{8.17}$$

and

$$\boxed{\mathbf{G}_{ij} = (\psi_i, \mathbf{A}\phi_j) = \psi_i^{\mathbf{T}}\mathbf{A}\phi_j} \tag{8.18}$$

and

$$\boxed{\mathbf{F}_i = (\mathbf{f}, \psi_i) = \psi_i^{\mathbf{T}}\mathbf{f}} \tag{8.19}$$

8.2.5.8 Rayleigh–Ritz–Galerkin Method ($s = m$)

The most important choice of the test space, for our purposes, is when $\boxed{s = m}$. For this case, the trial space and the test space become one and the same and are a subspace of an underlying space known as the energy space when the differential operator, **A**, is positive definite and self-adjoint (symmetric). The boundary conditions are then split into what are known as essential and natural boundary conditions. Accordingly, let us first characterize these new entities before we define the weak form of the Rayleigh–Ritz–Galerkin method.

8.2.6 Energy Space: E(Ω)

The energy space, $\mathbf{E}(\Omega)$, is a Sobolev space, $\mathcal{H}^m(\Omega)$, for a problem with the differential operator, **A**, as symmetric and positive definite of order $2m$. The functions belonging to this space, and their derivatives up to degree m have finite energy under the norm arising out of the ***energy inner product*** induced by the operator, **A**; the energy inner product is so termed because it is used to minimize the potential energy of the physical systems. In fact, they are the underlying inner product for all finite element methods. The energy inner product, $\mathcal{Q}(\mathbf{d}, \mathbf{v})$, is defined by:

$$\boxed{\mathcal{Q}(\mathbf{d}, \mathbf{v}) = (\mathbf{d}, \mathbf{v})_{\mathbf{A}} = (\mathbf{Ad}, \mathbf{v}), \quad \forall \mathbf{d}, \mathbf{v} \in \mathbf{E_0}} \tag{8.20}$$

where $\mathbf{E_0} \subset \mathbf{E}(\Omega)$ is the subspace of the energy space acting as both the trial space and the test space, and that $(\mathbf{Ad}, \mathbf{v})$ needs m integration by parts for one-dimensional problems, or application of ***Green's identity*** or ***Green–Gauss divergence theorem*** for higher dimensions; $(\bullet, \bullet)$ is the ***natural inner product*** of the space.

Example 4: Consider our bar problem in **Example 2** with the following changes: $A(x) = E(x) \equiv 1, \quad \forall x \in [0, 1]$, with $L = 1$. Let us assume, u(0) = u(L) = 0, and that a continuous axial force is applied along the length. Then, $\bar{\Omega} = \{x \in [0, 1] \subset \mathbb{R}^1\}, \mathbb{S}^{\mathbf{f}} = \{f | f \in C^0(\bar{\Omega})\}, \mathbb{S}^{\mathbf{d}}_{\mathbf{B}} = \{u | u \in C^2(\bar{\Omega}),\ u(0) = 0,\ u(1) = 0\}$ and $\mathbf{A} : \mathbb{S}^{\mathbf{d}}_{\mathbf{B}} \times \mathbb{S}^{\mathbf{f}} = -\frac{\partial^2(\bullet)}{\partial x^2} = -(\bullet)_{,xx}$, that is, $m = 1$. Now suppose $\mathbb{S}^{\mathbf{f}}$ has the inner product defined by $(f, g) = \int_0^1 f(x)g(x)dx$, then, let us, (a) show that $\mathbf{A}$ is symmetric and positive definite, and (b) find the energy inner product induced by it.

Solution: Note first that $\mathbb{S}^{\mathbf{d}}_{\mathbf{B}}$ is a linear subspace of $\mathbb{S}^{\mathbf{f}}$ and $\mathbf{A}u \in \mathbb{S}^{\mathbf{f}}, \ \forall u \in \mathbb{S}^{\mathbf{d}}_{\mathbf{B}}$, that is $u_{,xx}$ is continuous. Also, for $\forall u, v \in \mathbb{S}^{\mathbf{d}}_{\mathbf{B}}, \quad (Au, v) = \int_0^1 u_{,xx}(x)\, v(x)\, dx$. Integrating this by parts $m = 1$ times, and noting the homogeneous boundary conditions for elements of $\mathbb{S}^{\mathbf{d}}_{\mathbf{B}}$, we have:

$$(\mathbf{A}u, v) = \int_0^1 u_{,x}(x)\, v_{,x}(x)\, dx = (\mathbf{A}v, u) \tag{8.21}$$

that is, $\mathbf{A}$ is symmetric. Now, to see that $\mathbf{A}$ is positive definite, note that

$$(\mathbf{A}u, u) = \int_0^1 (u_{,x}(x))^2\, dx \tag{8.22}$$

Thus, for $(\mathbf{A}u, u) = 0$ we must have $u_{,x}(x) \equiv 0$, since $u_{,x}(x)$ is continuous. This, in turn, means that $u(x)$ is constant. But, because of the homogeneous boundary condition, $u(x) \equiv 0$. This proves that $\mathbf{A}$ is positive definite. We define the induced energy inner product, $\mathcal{Q}(u,v)$ as:

$$\boxed{\mathcal{Q}(u, v) \equiv (u, v)_{\mathbf{A}} = (\mathbf{A}u, v) = \int_0^1 u_{,x}(x)\, v_{,x}(x)\, dx} \tag{8.23}$$

For the bar problem, it can be easily seen from the above definition that $\frac{1}{2}\mathcal{Q}(u, u)$ is the strain energy of the system which is given by: $\frac{1}{2}\int_0^1 \varepsilon(x)\, \sigma(x)\, dx = \frac{1}{2}\int_0^1 u^2(x)\, dx = \frac{1}{2}\mathcal{Q}(u, u)$.

Remark

- When the boundary conditions are prescribed in a way that contains linear combination terms with derivatives of degree below m and those with derivatives of degree greater than or equal to m, the energy inner product must reflect these peculiarities as shown in the following example.

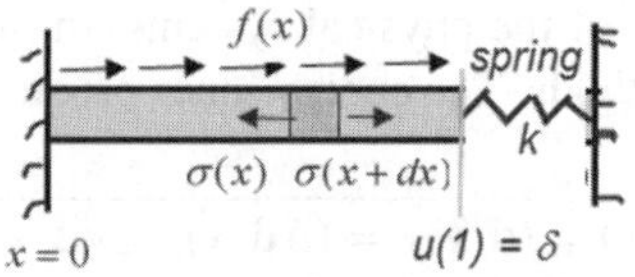

Figure 8.3 Example 5.

Example 5: Dependence of energy inner product on boundary condition.

Considering the problem in **Example 4**, everything else remaining same except that the boundary condition at $x = 1$ is given by:

$$k(\delta - u(1)) = AE\ u_{,x}(1) \Rightarrow k\,u(1) + u_{,x}(1) = k\delta$$

that is, a spring with stiffness k is attached to the $x = 1$ end with an imposed displacement δ, as shown in Figure 8.3. Then, we need to find the induced energy inner product, $Q(u, v)$.

Solution: Integrating by parts $m = 1$ times, we have:

$$(\mathbf{A}u, v) = \int_0^1 u_{,x}(x)\, v_{,x}(x)\, dx - [u_{,x}(x)\, v(x)]_0^1 \tag{8.24}$$

Now, applying the boundary conditions to the integrated term in equation (8.24), and substituting for $u_{,x}(1)$ from the expression for boundary condition at $x = 1$, we get,

$$(\mathbf{A}u, v) = [\int_0^1 u_{,x}(x)\, v_{,x}(x)\, dx + k\,u(1)\,v(1)] - (k\delta)\,v(1) \tag{8.25}$$

Recognizing that the boundary condition has introduced both a bilinear term and a linear term, we define the induced energy inner product, $Q(u,v)$, in this case, as:

$$\boxed{Q(u, v) = (u, v)_{\mathbf{A}} = \int_0^1 u_{,x}(x)\, v_{,x}(x)\, dx + k\,u(1)\,v(1)} \tag{8.26}$$

Thus, the spring-type boundary conditions affect the definition of the energy inner product. As will be shown shortly, the remaining linear term enters the definition of the functional in the variational form of formulation.

8.2.6.1 Solution Set: $\mathbf{E}^{Sol}(\Omega)$

The solution set, $\mathbf{E}^{Sol}(\Omega)$, is a subset of the energy space, $\mathbf{E}(\Omega)$. The *Sol* in the superscript indicates that elements of the set satisfy all the essential boundary conditions that involve derivatives of degree $< m$. It may be noted, by definition, that in general for inhomogeneous essential boundary conditions, the solution set does not constitute a linear space or subspace. Thus, we define $\mathbf{E}^{Sol}(\Omega)$ as:

$$\boxed{\mathbf{E}^{Sol}(\Omega) \stackrel{\Delta}{=} \{\mathbf{d} \in \mathbf{E}(\Omega) | \mathbf{d} \text{ satisfies all essential BC}\}} \tag{8.27}$$

8.2.6.2 Essential Boundary Conditions: Dirichlet Type, Kinematic, Admissible

The usual test for essential boundary conditions is that they involve only derivatives of degree $< m$; although, this condition is neither necessary nor sufficient. In terms of structural mechanics or elasticity problems, the essential boundary conditions are primarily the kinematic boundary conditions, that is, the constraints on the geometry, such as generated by support conditions. Thus, the essential boundary conditions are also known as kinematic or admissible boundary conditions.

Example 6: For the one-dimensional problem, **Example 2**, the end condition at the $x = 0$ side, that is, $u(0) = u_0$, constitutes an essential boundary condition, since the degree of the derivative is $0 < m$, where for both the trial space and the test space, $s = m = 1$. The applied force condition on the $x = L$ side, that is, $A(L)\, E(L)\, \frac{\partial u(L)}{\partial x} = P_L$, does not qualify as an essential boundary condition since the degree of the derivative is $m = 1$. In fact, this latter type of boundary condition belongs to what is known as the natural boundary conditions.

8.2.6.3 Natural Boundary Conditions: Neumann Type

All meaningful boundary conditions other than the essential boundary conditions are known as the natural boundary conditions. As we can recognize, the natural boundary conditions are not included in any of the variational spaces introduced so far. In fact, as we shall see shortly, these are not considered in any weak or Galerkin type formulation. However, if the theories are mathematically consistent, then it is expected that these natural boundary conditions will be concurrently satisfied when the variational problems are solved. Now, we are in a position to define our Rayleigh–Ritz–Galerkin weak form, as follows.

Given: $\Omega, \mathbf{A}, \mathbf{f}, \mathbf{B}$ and $\mathbf{g}$ as in equations (8.1) and (8.2)

Find: $\mathbf{d_0} \in \mathbf{U_0}(\Omega)$ such that

$$\boxed{\mathcal{Q}(\mathbf{d_0}, \mathbf{v}) = (\mathbf{f}, \mathbf{v}) - \mathcal{Q}(\hat{\mathbf{d}}, v), \quad \forall \mathbf{v} \in \mathbf{V_0}(\Omega)} \tag{8.28}$$

where

$$\boxed{\mathbf{U_0}(\Omega) = \mathbf{V_0}(\Omega) \equiv \mathbf{E_0}(\Omega) \subset \mathbf{E}^{Sol}(\Omega)} \tag{8.29}$$

then $\mathbf{d}_{Sol} \in \mathbf{E}^{Sol}(\Omega)$ is given as the summation of $\mathbf{d_0} \in \mathbf{U_0}(\Omega)$ and an arbitrary fixed element $\hat{\mathbf{d}} \in \mathbf{E}^{Sol}(\Omega)$, that is,

$$\boxed{\mathbf{d}_{Sol} = \mathbf{d_0} + \hat{\mathbf{d}}} \tag{8.30}$$

The Rayleigh–Ritz–Galerkin method is then an obvious discretization of equation (8.28) with trial space and test space taken as an N-dimensional finite subspace, $\mathbf{E}_\mathbf{0}^\mathbf{N}(\Omega)$ of $\mathbf{E_0}(\Omega)$ such that the weak form becomes:

Find: $\mathbf{d}^\mathbf{N} \in \mathbf{E}_\mathbf{0}^\mathbf{N}(\Omega)$ such that

$$\boxed{\mathcal{Q}(\mathbf{d}^\mathbf{N}, \mathbf{v}^\mathbf{N}) = (\mathbf{f}, \mathbf{v}^\mathbf{N}), \quad \forall \mathbf{v}^\mathbf{N} \in \mathbf{E}_\mathbf{0}^\mathbf{N}(\Omega)} \tag{8.31}$$

Now, assuming $\Phi = (\phi_1, \phi_2, \dots, \phi_N)$ as the set of basis functions for the trial space and the test space, the solution, in matrix form, defined by:

$$\mathbf{d} = \Phi \mathbf{q} \tag{8.32}$$

must satisfy the matrix equation:

$$\boxed{\begin{aligned} &\mathbf{K}\mathbf{q} = \mathbf{F} \\ \text{with} \quad &\mathbf{K} \equiv \Phi^{\mathbf{T}} \mathbf{A} \Phi, \\ &\mathbf{F} = \Phi^{\mathbf{T}} \mathbf{f} \end{aligned}} \tag{8.33}$$

In component form:

$$\boxed{\begin{aligned} &\mathbf{d} = \sum_{i=1}^{N} q_i \Phi_i \\ \text{with} \quad &\mathbf{q} = \{q_1, q_2, \dots, q_N\}^T \end{aligned}} \tag{8.34}$$

and

$$\boxed{\mathbf{K}_{ij} = (\phi_i, \mathbf{A}\phi_j) = \phi_i^{\mathbf{T}} \mathbf{A} \phi_j} \tag{8.35}$$

and

$$\boxed{\mathbf{F}_i = (\mathbf{f}, \phi_i) = \phi_i^{\mathbf{T}} \mathbf{f}} \tag{8.36}$$

In equation (8.33), for elasticity and solid mechanics problems, it may be recognized that **K** is the stiffness matrix and **F** is the generalized load vector.

8.2.7 Equivalence of S-form and W-form

The equivalence of strong (S-form) and weak form (W-form) is trivially established as follows:

- Take account of the basic tool of differential operator: the integration by parts in one dimension and its higher-dimensional counterparts such as the Green's identity or the Divergence theorem.
- Construct the W-form from the S-form and vice versa.

Thus, we will briefly summarize the construction of each form from the other.

8.2.8 Construction of W-form from S-form

Based on the discussions so far, we simply take the following steps:

Step 0: Based on the problem in S-form, choose the test space with degree, say, s, the differential operator being of order $2m$.

Step 1: Multiply both sides of the differential equation in S-form by any arbitrary element of the test space and integrate over the domain of the problem.

Step 2: Integrate by parts s times to transfer derivatives off the trial space element and onto the test space element, so that the trial space element's highest degree of derivative is $2m - s$. Integration by parts must recognize and apply appropriate boundary conditions.

8.2.9 Construction of S-form from W-form

Using similar arguments, we simply take the following steps:

Step 0: Based on the problem in W-form we identifiy the test space with appropriate degree, say, s, with the trial space degree identified as $2m - s$. Further notice that the solution set satisfies all the boundary conditions. This will be obvious from the integrand of the bilinear form associated with the W-form.

Step 1: Integrate by parts s times to transfer derivatives off the test space element onto the trial space element, so that the trial space element highest degree of derivative is $2m$. In so doing, all the boundary integrals will drop off, because the test space only satisfies the homogeneous counterparts of the relevant boundary conditions.

Step 2: All integrands associated with the test space element are collected together. Noting that the test space element is arbitrary, the kernel of the integral must vanish identically, giving rise to the differential equations of the associated S-form of the problem.

Remarks

- It may be pointed out that for physical systems – either the S-form or the W-form or both – may be known. In situations where the differential operator representing a physical phenomenon is known to be not self-adjoint or symmetric, we can only describe the phenomenon either in strong form or weak form. Most transient phenomena in elasto-dynamics fall under this category, although these problems can be manipulated to produce a self-adjoint (symmetric) operator.
- However, for most steady-state problems (e.g. in solid mechanics and elasticity), the differential operator is self-adjoint. In such situations, another very important form equivalent to W-form secures itself as the starting point of the numerical analysis based on the variational method instead of W-form. We introduce this form next.

8.2.10 Variational form (V-form)

As already noted, when the differential operator, **A**, is self-adjoint, it is always possible to define a quadratic functional on the domain, the minimization (if **A** is also positive definite) or by finding the stationary point (if **A** is only positive semi-definite) which gives us back the Galerkin or weak form of the problem statement. Thus, we may convert the problem of inverting the differential operator in strong form into one of minimization of a functional, giving rise to what is known as the variational form. There is no unique description of the functional. However, a particular choice appears to be both conceptually and computationally more suitable than others (e.g. least square form which we will also introduce later) for solid and structural mechanics problems.

The variational form for the boundary value problem defined in equations (8.1)–(8.3) is given as follows:

Given: $\Omega, \mathbf{A}, \mathbf{f}, \mathbf{B}$ and $\mathbf{g}$ as in equations (8.1) and (8.2), and $\mathbf{U_0}(\Omega) = \mathbf{V_0}(\Omega) \equiv \mathbf{E_0}(\Omega) \subset \mathbf{E}^{Sol}(\Omega)$, and the quadratic functional, $I(\mathbf{d})$ as:

$$\boxed{I(\mathbf{d}) \stackrel{\Delta}{=} \frac{1}{2}(\mathbf{Ad}, \mathbf{d}) - (\mathbf{f}, \mathbf{d}), \quad \forall \mathbf{d} \in \mathbf{E}^{Sol}(\Omega)} \tag{8.37}$$

Find: $\mathbf{d}_{Sol} \in \mathbf{E}^{Sol}(\Omega)$ such that

$$\boxed{I(\mathbf{d}_{Sol}) = \min_{\mathbf{d} \in \mathbf{E}^{Sol}} I(\mathbf{d})} \tag{8.38}$$

Remark

- Equation (8.37) can be easily recognized as: $I(\mathbf{d}) \stackrel{\Delta}{=} \frac{1}{2}(\mathcal{Q}(\mathbf{d}, \mathbf{d}) - \mathcal{L}(\mathbf{d})$, where $\mathcal{Q}(\mathbf{d}, \mathbf{d})$ is the bilinear energy inner product induced by $\mathbf{A}$, and $\mathcal{L}(\mathbf{d})$ is the linear functional induced by $\mathbf{f}$ and which depends on the boundary conditions.

We can readily show that the variational form is equivalent to the strong form through weak form, that is, minimization of the functional and inversion of the differential operator leads to the same solution of a boundary value problem.

8.2.11 Equivalence of S-form, W-form and V-form

The equivalence of the strong (S-form) and the weak form (W-form) has already been established. Let us now show the implication of the W-form by the V-form.

Obviously, if $I(\mathbf{d})$ has a minimum or stationary value at $\mathbf{d}_{Sol} \in \mathbf{E}^{Sol}(\Omega)$, then for any $\mathbf{v_0} \in \mathbf{V_0}(\Omega)$ and ε, real and small, we must have:

$$I(\mathbf{d}_{Sol} + \varepsilon \mathbf{v_0}) \geq I(\mathbf{d}_{Sol})$$

Now, using equation (8.37),

$$\begin{aligned} I(\mathbf{d}_{Sol} + \varepsilon \mathbf{v_0}) &= \frac{1}{2}(\mathbf{A}(\mathbf{d}_{Sol} + \varepsilon \mathbf{v_0}), \mathbf{d}_{Sol} + \varepsilon \mathbf{v_0}) - (\mathbf{f}, \mathbf{d}_{Sol} + \varepsilon \mathbf{v_0}) \\ &= \frac{1}{2}(\mathbf{Ad}_{Sol}, \mathbf{d}_{Sol}) - (\mathbf{f}, \mathbf{d}_{Sol}) + \varepsilon \left[\frac{1}{2}(\mathbf{Ad}_{Sol}, \mathbf{v_0}) + \frac{1}{2}(\mathbf{Av_0}, \mathbf{d}_{Sol}) - (\mathbf{f}, \mathbf{v_0})\right] \\ &+ \frac{1}{2}\varepsilon^2(\mathbf{Av_0}, \mathbf{v_0}) \end{aligned}$$

For $\mathbf{A}$ self-adjoint, that is, $\mathbf{A} = \mathbf{A}^*$, and by symmetry of the inner product:

$$(\mathbf{Av_0}, \mathbf{d}_{Sol}) = (\mathbf{v_0}, \mathbf{A}^*\mathbf{d}_{Sol}) = (\mathbf{v_0}, \mathbf{Ad}_{Sol}) = (\mathbf{Ad}_{Sol}, \mathbf{v_0})$$

and noting the definition of $I(\mathbf{d}_{Sol})$, we get,

$$I(\mathbf{d}_{Sol} + \varepsilon \mathbf{v}) = I(\mathbf{d}_{Sol}) + \varepsilon[(\mathbf{Ad}_{Sol}, \mathbf{v_0}) - (\mathbf{f}, \mathbf{v_0})] + \frac{1}{2}\varepsilon^2(\mathbf{Av_0}, \mathbf{v_0}) \tag{8.39}$$

Now, because of the positive definiteness of $\mathbf{A}$, and square of ε makes the last term on the right side of the last expression always positive. Finally, ε being arbitrary in sign, for $I(\mathbf{d}_{Sol})$ to be minimum, the coefficient of ε must vanish. In other words, the first variation must be zero for any $\mathbf{v}_0 \in \mathbf{V}_0$. Thus, from equation (8.39), we get:

$$(\mathbf{A}\mathbf{d}_{Sol}, \mathbf{v}_0) - (\mathbf{f}, \mathbf{v}_0) = \mathbf{0}, \quad \forall \mathbf{v}_0 \in \mathbf{V}_0 \tag{8.40}$$

Now, $\mathbf{d}_{Sol} \in \mathbf{E}^{Sol}(\Omega)$ can be obtained as the summation of $\mathbf{d}_0 \in \mathbf{U}_0(\Omega)$ and an arbitrary fixed element $\hat{\mathbf{d}} \in \mathbf{E}^{Sol}(\Omega)$, that is,

$$\mathbf{d}_{Sol} = \hat{\mathbf{d}} + \mathbf{d}_0 \tag{8.41}$$

Applying equation (8.41) in equation (8.40), we get:

$$(\mathbf{A}\mathbf{d}_0, \mathbf{v}_0) = (\mathbf{f}, \mathbf{v}_0) - (\mathbf{A}\hat{\mathbf{d}}, \mathbf{v}_0), \quad \forall \mathbf{v}_0 \in \mathbf{V}_0 \tag{8.42}$$

Noting that for the Rayleigh–Ritz–Galerkin case, that is, for $s = m$, the self-adjoint and positive-definite $\mathbf{A}$ induces the energy inner product $\mathcal{Q}(\mathbf{d}, \mathbf{v})$ defined by equation (8.20), we finally get:

$$\boxed{\mathcal{Q}(\mathbf{d}_0, \mathbf{v}_0) = (\mathbf{f}, \mathbf{v}_0) - \mathcal{Q}(\hat{\mathbf{d}}, \mathbf{v}_0), \quad \forall \mathbf{v}_0 \in \mathbf{V}_0} \tag{8.43}$$

which is nothing else than the weak form that we presented earlier, except that it is formatted to show the modification necessitated by the inhomogeneous boundary conditions. This is the form which is actually used for numerical solutions, where both $\mathbf{d}_0$ and $\mathbf{v}_0$ are chosen from the same subspace $\mathbf{E}_0(\Omega) \subset \mathbf{E}^{Sol}(\Omega)$. We will elaborate more on this in the next section. Now, rewriting the equation (8.40) as:

$$(\mathbf{A}\mathbf{d}_{Sol} - \mathbf{f}, \mathbf{v}_0) = \mathbf{0}, \quad \forall \mathbf{v}_0 \in \mathbf{V}_0 \tag{8.44}$$

and noting that equation (8.44) holds for $\forall \mathbf{v}_0 \in \mathbf{V}_0$, we get the Euler equation or the strong form equation:

$$\mathbf{A}\mathbf{d}_{Sol} = \mathbf{f}$$

and the boundary conditions follow from the definition and subsequent integration by parts of the functional. This is demonstrated by the example problem that follows. Thus, the various forms are equivalent:

$$\text{S} - \text{form} \Leftrightarrow \text{W} - \text{form} \Leftrightarrow \text{V} - \text{form}$$

The numerical method arising out of the V-form is an obvious discretization of the functional by choosing, as before, the trial space and the test space as the finite-dimensional subspaces of the trial and test spaces. Finally, integration by parts m times gives us the simultaneous set of algebraic equations ready to be solved by computer.

Remarks

- The variational form consisting of a quadratic bilinear form and a linear form, as presented above, or the equivalent weak or Galerkin form are resorted to for the generation of a numerical algorithm for the finite element method. The primary reason for this is that for solid and structural mechanics problems, the variational form is virtually equivalent to the potential energy minimization principle, and the corresponding weak or Galerkin form represents the virtual work principle. In other words, these forms provide us with the physical meaning by links to the familiar physical entities such as stresses, strains, strain energy, and so on.

- The definition of the quadratic bilinear form and the linear form in the V-form depends not only on the differential equation of the S-form but is affected by the inhomogeneous boundary conditions. In order to gain some insight, consider the following example.

Example 7: Continuing with **Example 5**, we would like to construct the functional for the variational form and recover from it the strong form and the boundary conditions. Noting that, from this example, the energy inner product is given by:

$$Q(u, v) = (\mathbf{A}u, v) = \int_0^1 u_{,x}(x)\, v_{,x}(x)\, dx + k u(1)\, v(1)$$

and the linear term due to the boundary as: $-k\delta\, v(1)$, we propose our functional as:

$$I(u) = \frac{1}{2}\left\{\int_0^1 (u_{,x}(x))^2\, dx + k(u(1))^2\right\} - \delta\, u(1)$$

Now, for $I(u)$ to be minimum, for any real and small ε, and for any element of the test space, $v_0 \in \mathbf{V_0}$, we must have: $I(u + \varepsilon v_0) \geq I(u)$. But,

$$I(u + \varepsilon v_0) = \tfrac{1}{2}\{\textstyle\int_0^1 (u_{,x} + \varepsilon v_{0,x})^2 dx + k((u(1) + \varepsilon v_0(1))^2 - k\delta\,(u(1) + \varepsilon v_0(1))$$

After algebraic manipulation and integration by parts, and noting $v_0(0) = 0$, we get the coefficient of ε, that is, the first variation vanishing condition as:

$$[-\int_0^1 u_{,xx} v_0(x) dx] + [u_{,x}(1) + ku(1) - k\delta] v_0(1) = 0$$

Now, noting that v_0 is arbitrary on the entire domain, we must have that each term vanish independently without v_0. This gives us the following:

$$\boxed{\begin{aligned} \text{differential equation} &: \Rightarrow u_{,xx} = 0, \\ \text{boundary condition} &: \Rightarrow u_{,x}(1) + ku(1) = k\delta \end{aligned}}$$

Note that u as an element of the trial space already satisfies the essential boundary condition: $u(0) = 0$. Thus, we achieve what we wanted to accomplish in this example. As pointed out before, the functional of our choice, as presented here, is not the only one. Other functionals exist for statement of the problem in variational form. We now present one such variant.

8.2.12 *Least Square Form* (L–form)

Although it is obviously a variational statement, to distinguish from the previous V-form, we identify it as L-form and state as it as follows:

Given: $\Omega, \mathbf{A}, \mathbf{f}, \mathbf{B}$ and $\mathbf{g}$ as in equations (8.1) and (8.2), and $\mathbf{U_0}(\Omega) = \mathbf{V_0}(\Omega) \equiv \mathbf{E_0}(\Omega) \subset \mathbf{E}(\Omega)$, and the functional $I(\mathbf{d})$ as residual error norm:

$$\boxed{I(\mathbf{d}) = \|\mathbf{Ad} - \mathbf{f}\| = (\mathbf{Ad} - \mathbf{f}, \mathbf{Ad} - \mathbf{f})} \tag{8.45}$$

Find: $\mathbf{d_0} \in \mathbf{U_0}(\Omega)$ such that

$$I(\mathbf{d_0}) = \min_{\mathbf{d} \in \mathbf{U_0}} I(\mathbf{d}) \tag{8.46}$$

then, $\mathbf{d}_{Sol} \in \mathbf{E}^{Sol}(\Omega)$ is given as the summation of $\mathbf{d_0} \in \mathbf{U_0}(\Omega)$ and an arbitrary fixed element $\hat{\mathbf{d}} \in \mathbf{E}^{Sol}(\Omega)$, that is,

$$\mathbf{d}_{Sol} = \hat{\mathbf{d}} + \mathbf{d_0} \tag{8.47}$$

Following similar arguments as in V-form, we can easily show that the vanishing of the first variation requirement is the same, as follows:

$$(\mathbf{A^T A\, d_0}, \mathbf{v}) = (\mathbf{A^T f}, \mathbf{v}), \quad \forall \mathbf{v} \in \mathbf{V_0} \tag{8.48}$$

The least square method is an obvious discretization of equation (8.48) by choosing both the trial and test spaces as finite-dimensional subspaces of the trial and test spaces. Finally, integration by parts m times derives the simultaneous set of algebraic equations ready to be solved by computer.

Remark

- For structural and solid mechanics problems, this form appears to be computationally (e.g. for stiffness matrix generation) awkward as well as lacking a direct interpretation with the physical entities mentioned earlier. Accordingly as far as finite element is concerned, this form will not be discussed any more here.

8.2.13 *Discussion*

In this section, we have presented, in a unifying manner, most of the important variational forms for linear problems in structural engineering and structural mechanics. Throughout the rest of the book, however, we will restrict ourselves only to the Rayleigh–Ritz–Galerkin type variational form and its corresponding weak or Galerkin form, which serves as a precursor to most finite element methods. In other words, from now on, we will mostly deal with positive definite, self-adjoint operators, and with situations where the trial space and the test space are one and the same subspace of some energy space with appropriate energy inner product and norm induced in it by the differential operator. With these in the background, we now try to specialize the variational formulation as it applies to the field of linear and nonlinear structural and solid mechanics, eventually bringing out the basic theory behind the linear and nonlinear finite element method, particularly restricting to what is known as the ***virtual displacement method***.

8.3 Energy Precursor to Finite Element Method

In the previous section, of necessity, the presentation was a little bit abstract mathematically; nonetheless, we have introduced the general theories behind variational methods. In so doing, the state vector, **d**, was never specifically identified, although, the examples presented therein suggested the unknown displacement vector to be the ***state vector***. As will be obvious through the subsequent sections, the main strength of finite element methods lies in their ability to accommodate arbitrarily complicated:

- geometry
- loading conditions
- boundary conditions.

This section acts as a precursor which elaborates, in more definitive yet robust terms, all the necessary concepts, introduced in the previous section, that will eventually, allow us to:

- unravel the full strength and promise of the finite element method,
- critically evaluate the old finite element method, insofar as it departs from its theoretical counterpart,
- offer our new robust finite element method that strictly adheres to the theory, and thus eradicates all the problems created by the old finite element method.

For the scope of our book, that is, dealing with applications primarily in solid and structural mechanics, and simple heat conduction, the fundamental state variables will be the displacement quantities (and temperature for heat conduction) in the sense that all other variables can be derived from it, at least in the computational sense. However, historically, two other fundamental entities – stress and strain tensors or vectors – are treated as independent state variables. Of course, this has its justification. For example, we have presented equilibrium equations without ever recognizing the displacement vector. Similarly, the strain energy can be defined only by the strain and the stress vectors or tensors. However, we have already seen that stresses and strains lose their independent status as soon as we adopt some constitutive stress–strain relationship and impose the strain compatibility conditions. Nevertheless, this loss of independence has not hindered the formulation of variational methods, commonly known as the ***mixed formulations***, based on variational theorems that include stresses and strains tensors or vectors along with displacement vectors as the underlying state vectors. In what follows, therefore, for the sake of completeness, we will briefly introduce two well-known energy or variational theorems due to Hu–Washizu and Hellinger–Reissner that include stresses and/or strains as state vectors. Then, we will quickly concentrate on the Rayleigh–Ritz–Galerkin energy functional in variational form, establishing its connection with the principle of minimum potential energy of continuum and structural mechanics, which includes displacement as the only state vector of the system. This, in turn will enable us to tie the principle of virtual work to the corresponding weak or Galerkin form that flows as the underlying basis for all displacement-based finite element methods. Finally, for ease of understanding of the basic principles, we will assume that a solid body is undergoing linear deformations; the effect of nonlinearity, material or geometric, will be dealt with later, and handled later in full detail in specific cases: the geometric nonlinearity of beams, plates and shells.

Before we engage in stating the energy theorems, let us recall the definition of the following entities for linear elasticity: Ω, the finite continuum or body, $\mathbf{c} \in \bar{\Omega}$; Γ, the boundary of Ω, with $\Gamma = \Gamma_d \cup \Gamma_\sigma \cup \Gamma_C$, where Γ_d, the boundary where displacement or kinematic condition is prescribed; Γ_σ, the boundary where stress related condition is prescribed, Γ_C, the boundary where convective conditions are prescribed; $\bar{\Omega}$, the closure of Ω that is, $\bar{\Omega} = \Omega \cup \Gamma$; $\mathbf{c} = (x, y, z)^T$ the column vector of the material points or Cartesian coordinates of Ω; $\mathbf{d}$, the column vector of the nodal generalized temperatures defining the displacement state of $\bar{\Omega} = \Omega \cup \Gamma$, the closure of Ω, under boundary conditions on Γ; $\sigma(c, d)$, stress vector function, $\sigma(\mathbf{c}, \mathbf{d})$, $\mathbf{c} \in \bar{\Omega}$, $\sigma : \bar{\Omega} \times E_{Ess}^{Sol} \rightarrow \mathbf{R}^{n(n+1)/2}$; $\varepsilon(\mathbf{c}, \mathbf{d})$, the strain vector function, $\varepsilon(\mathbf{c}, \mathbf{d})$, $\mathbf{c} \in \bar{\Omega}$, $\varepsilon : \bar{\Omega} \times E_{sol} \rightarrow \mathbf{R}^{n(n+1)/2}$. The strain–displace linear differential operator or matrix represents the fundamental kinematic or strain compatibility relationship in elasticity:

$$\varepsilon(\mathbf{c}, \mathbf{d}) = \mathcal{S}\ \mathbf{d}(\mathbf{c}) \tag{8.49}$$

The elements of the matrix obviously depend on the dimensionality of the problem. For one dimension: $\mathcal{S} = \frac{\partial(\bullet)}{\partial x} = (\bullet)_{,x}$; For two dimensions:

$$\mathcal{S} = \begin{bmatrix} \frac{\partial}{\partial x} & 0 \\ 0 & \frac{\partial}{\partial y} \\ \frac{\partial}{\partial y} & \frac{\partial}{\partial x} \end{bmatrix} \quad \text{or,} \quad \begin{bmatrix} \frac{\partial}{\partial x_1} & 0 \\ 0 & \frac{\partial}{\partial x_2} \\ \frac{\partial}{\partial x_2} & \frac{\partial}{\partial x_1} \end{bmatrix} \quad \text{or,} \quad \begin{bmatrix} (\cdot)_{,x} & 0 \\ (\cdot)_{,y} & 0 \\ (\cdot)_{,y} & (\cdot)_{,x} \end{bmatrix} \quad \text{or,} \quad \begin{bmatrix} (\cdot)_{,1} & 0 \\ (\cdot)_{,2} & 0 \\ (\cdot)_{,2} & (\cdot)_{,1} \end{bmatrix};$$

For three dimensions: $\mathcal{S} = \begin{bmatrix} \frac{\partial}{\partial x} & 0 & 0 \\ 0 & \frac{\partial}{\partial y} & 0 \\ 0 & 0 & \frac{\partial}{\partial z} \\ \frac{\partial}{\partial y} & \frac{\partial}{\partial x} & 0 \\ 0 & \frac{\partial}{\partial z} & \frac{\partial}{\partial y} \\ \frac{\partial}{\partial z} & 0 & \frac{\partial}{\partial x} \end{bmatrix}$ or, $\begin{bmatrix} \frac{\partial}{\partial x_1} & 0 & 0 \\ 0 & \frac{\partial}{\partial x_2} & 0 \\ 0 & 0 & \frac{\partial}{\partial x_3} \\ \frac{\partial}{\partial x_2} & \frac{\partial}{\partial x_1} & 0 \\ 0 & \frac{\partial}{\partial x_3} & \frac{\partial}{\partial x_2} \\ \frac{\partial}{\partial x_3} & 0 & \frac{\partial}{\partial x_1} \end{bmatrix}$, or similar to other two-dimensional forms.

Remarks

- It can be easily seen that for two dimensions the displacement distribution of: $\mathbf{d} = a_1\{0 \quad 1\}^T$, or $\mathbf{d} = a_2\{0 \quad 1\}^T$ or $\mathbf{d} = a_3\{y \quad -x\}^T$ induces zero strain condition in the system. In other words, rigid body movement is predicated by these distributions.
- Similarly, for three dimensions, the displacement distribution of:

$$\boxed{\begin{array}{l} \mathbf{d} = a_1\{1 \quad 0 \quad 0\}^T, \text{ or, } \mathbf{d} = a_2\{0 \quad 1 \quad 0\}^T, \text{ or, } \mathbf{d} = a_3\{0 \quad 0 \quad 1\}^T \\ \mathbf{d} = a_4\{y \quad -x \quad 0\}^T, \text{or, } \mathbf{d} = a_5\{0 \quad y \quad -z\}^T, \text{ or, } \mathbf{d} = a_6\{x \quad 0 \quad -z\}^T \end{array}}$$

  induces the zero strain condition in the system. In other words, a body experiences only rigid body movements when subjected to these displacement distributions, where a_i, $i = 1, \ldots, 6$ are arbitrary constants.

The initial strains, ϵ^I, could be due to thermal expansion, such as defined by $\epsilon^I = \alpha(\theta - \theta^I)$, where α is the vector representing the coefficients of thermal expansion, θ^I is the ambient stress-free temperature and θ is the operating temperature. Similarly, initial stress, σ^I, could be the stresses induced by forced fitting different components of the body having lack of fit. The stress vector σ is then the actual stress state of a body. The ϵ is the total strain state while $\epsilon - \epsilon^I$ denotes the mechanical strain induced by the deformation under some imposed loading. The stress–strain elasticity matrix, $\mathcal{D}$, is given by:

$$\boxed{\sigma = \mathcal{D}(\epsilon - \epsilon^I) + \sigma^I} \tag{8.50}$$

The stress–strain matrix reflects the constitutive law governing the material behavior in linear elasticity. Of course, the elasticity matrix depends on the dimensionality and the state of stress of the problem.

For one dimension: $\mathcal{D} = \text{E}$, the one-dimensional elasticity or Young's modulus. For two and three dimensions, by applying equation (8.49) in equation (8.50), we get the alternate expression relating displacement vector as:

$$\boxed{\sigma = \mathcal{D}\mathcal{S}\,\mathbf{d} - \mathcal{D}\epsilon^I + \sigma^I} \tag{8.51}$$

Remarks

- Whenever the context is clear, the arguments of all the scalar, vector, tensor or matrix quantities may be arbitrarily removed, in the interest of clutter-free readable mathematical statements.
- $\mathcal{D}$ is a symmetric and positive definite matrix, that is, if there exists a displacement, $\mathbf{d}$, not identically zero, then, $\mathbf{d}^T\mathcal{D}\,\mathbf{d} > 0$.
- In the virtual work principle presented later, it will be seen that the virtual displacements are the variations in ***total*** displacements; hence, the need for ***total*** virtual strains and the ***actual*** stress, given by equations (8.49) and (8.51) respectively, for the expression for the virtual work.
- For material nonlinearity, of course, the stress–strain relationship will in general be functionally described based on material types.

As indicated earlier, one of the main strengths of the finite element method lies in its ability to accommodate complicated loading conditions. There are several kinds of load commonly applied to structural finite element models. In general, the loads fall in three categories: (1) body loads, represented by the $\mathbf{f}^B$ vector (force/volume); (2) surface loads or tractions, represented by the $\mathbf{f}^S$ vector (force/area); (3) concentrated loads, represented by the $\mathbf{f}^C$ vector (force units). The body loads are generally induced by gravity or other inertial effect. The surface tractions are assumed to be applied to the boundary of the body in one or more patches. The collective area of the boundary for all surface tractions defines Γ_σ. The surface tractions can be in coordinate directions or in normal and tangential directions to the surface. In any case, they can be expressed by simple trigonometric matrix relationships. The surface tractions are required to be balanced by the body's internally generated stresses at those surfaces where they are prescribed. The concentrated loads are similar to surface tractions, except that they are supposed to be applied at an infinitesimal surface. Thus, they can assume any direction.

8.3.1 Equilibrium Equations in Linear Elasticity

Following the presentation of the previous section, in its most compacted form, the equilibrium equations can be expressed in operational form as:

$$\boxed{\begin{array}{ll} \mathcal{A}\,\mathbf{d} = f, & \forall \mathbf{c} \in \Omega \\ \mathcal{B}\,\mathbf{d} = \mathbf{g}, & \forall \mathbf{c} \in \Gamma_\sigma \cup \Gamma_{\sigma d} \end{array}} \tag{8.52}$$

where

$$
\begin{aligned}
\mathcal{A} &\equiv \mathcal{S}^T \mathcal{D}\, \mathcal{S}, \\
\mathbf{f} &\equiv -\mathbf{f}^B + \mathcal{S}^T \mathcal{D} \epsilon^I - \mathcal{S}^T \sigma^I, \\
\mathcal{B}\mathbf{d} &\equiv \mathcal{D}\, \mathcal{S}\mathbf{d} - \mathcal{D}\epsilon^I + \sigma^I, \\
g &\equiv \mathbf{f}^S
\end{aligned}
\tag{8.53}
$$

However, in terms of the familiar stress vector, noting that σ is given by equation (8.51), we get from equations (8.52) and (8.53), in a component independent form:

$$
\begin{aligned}
\mathcal{S}^T \sigma + \mathbf{f}^B = \mathbf{0}, &\quad \forall \mathbf{c} \in \Omega \\
\sigma = \mathbf{f}^S, &\quad \forall \mathbf{c} \in \Gamma_\sigma
\end{aligned}
\tag{8.54}
$$

Finally, applying the definitions above, we can easily rewrite equation (8.54) component-wise. For example, for $n = 2$, that is, two dimensions, with uniform thickness, the equation of equilibrium becomes:

$$
\begin{aligned}
\frac{\partial \sigma_x(\mathbf{d})}{\partial x} + \frac{\partial \tau_{xy}(\mathbf{d})}{\partial y} + f_x^B = 0 \\
\frac{\partial \tau_{xy}(\mathbf{d})}{\partial x} + \frac{\partial \sigma_y(\mathbf{d})}{\partial y} + f_y^B = 0
\end{aligned}
\tag{8.55}
$$

Remark

- For non-uniform, variable thickness, two-dimensional problems, the equilibrium equations are written in terms of forces, namely in-plane membrane forces, by including the thickness parameter as a multiplier for the stresses, instead of stresses alone as shown in equation (8.55). Note that although the thicknesses may be discontinuous, the membrane forces are continuous in the body.

Example EP1: Assuming a linear, isotropic elastic material, for plane stress conditions in two dimensions, let us find the expanded expression for the equilibrium equation from equation (8.52). For plane stress with Poisson's ratio, ν, and elasticity modulus, E, we have the elasticity matrix, $\mathcal{D}$, given by:

$$
\mathcal{D} = \frac{E}{1-\nu^2} \begin{bmatrix} 1 & \nu & 0 \\ \nu & 1 & 0 \\ 0 & 0 & \frac{1-\nu}{2} \end{bmatrix}
$$

Now, applying the definition of $\mathcal{S}$ and $\mathcal{D}$ in $\mathcal{A}$, we get after some necessary matrix multiplications:

$$
\frac{E}{1-\nu^2} \begin{bmatrix} \frac{\partial^2}{\partial x^2} + \left(\frac{1-\nu}{2}\right) \frac{\partial^2}{\partial y^2} & \left(\frac{1+\nu}{2}\right) \frac{\partial^2}{\partial x \partial y} \\ \left(\frac{1+\nu}{2}\right) \frac{\partial^2}{\partial x \partial y} & \left(\frac{1-\nu}{2}\right) \frac{\partial^2}{\partial x^2} + \frac{\partial^2}{\partial y^2} \end{bmatrix} \begin{Bmatrix} u \\ v \end{Bmatrix}
$$

$$
= -\begin{Bmatrix} f_x^B \\ f_y^B \end{Bmatrix} + \begin{Bmatrix} \frac{\partial}{\partial x}\left(\epsilon_x^I + \nu \epsilon_y^I\right) + \left(\frac{1-\nu}{2}\right) \frac{\partial \gamma_{xy}^I}{\partial y} \\ \frac{\partial}{\partial y}\left(\nu \epsilon_x^I + \epsilon_y^I\right) + \left(\frac{1-\nu}{2}\right) \frac{\partial \gamma_{xy}^I}{\partial x} \end{Bmatrix} + \begin{Bmatrix} \frac{\partial \sigma_x^I}{\partial x} + \frac{\partial \tau_{xy}^I}{\partial y} \\ \frac{\partial \tau_{xy}^I}{\partial x} + \frac{\partial \sigma_y^I}{\partial y} \end{Bmatrix}
$$

This is the same equation of equilibrium as equation (8.55), except expressed in terms of the displacement vectors and initial stress and strain components. The matrix on the left-hand side of the expression may be identified as $\mathcal{A}$.

8.3.2 The Boundary Conditions

As discussed and exemplified in the previous section, the inhomogeneous boundary conditions play an important role in the definition of both the energy inner product and the linear functional of the underlying spaces of the variational representations of the problems. Thus, we need to identify various possibilities in the prescription of the boundary conditions of the problems before the energy theorem can be stated. In general, the boundary conditions may be prescribed in general Cartesian coordinate directions or normal and tangential directions at a point on the boundary, Γ. The boundary conditions are generally prescribed as normal and tangential components of: traction vectors or displacement vectors or a linear combination of the traction and displacement vectors. In the previous section, we provided an example that shows how a boundary condition of category 3 (concentrated loads) representing springs affects both the energy inner product – that is, the quadratic form – and the linear functional. Similar effects reveal itself for higher-dimensional problems as well for boundary conditions of category 3, which represents distributed springs.

The boundary Γ of the body may be considered to be union of: Γ_δ, the collective boundary where displacements, that is, category 1 above, are prescribed, Γ_σ, the collective boundary where the tractions, that is, category 2 above, are specified, and Γ_{sp}, the collective boundary where the distributed springs with imposed displacements, that is, category 3 above, are specified. These boundary conditions are treated in detail, when we discuss problems individually in one, two and three dimensions in subsequent chapters. For now, let us just assume that the tractions, $\mathbf{f}^s \in \Gamma_\sigma$, and displacements, $\bar{\mathbf{d}} \in \Gamma_\delta$, are prescribed with no distributed springs assigned.

8.3.3 Hu–Washizu Variational Theorem

Given these notational and elasticity relationships, we are now in a position to state what is known as the Hu–Washizu variational theorem:

The Hu–Washizu functional is given by:

$$\boxed{\begin{aligned} I(\mathbf{d}, \sigma, \varepsilon) = {} & \frac{1}{2}\int_\Omega \varepsilon^T D\, \varepsilon\, d\Omega - \int_\Omega \varepsilon^T D\, \varepsilon^I\, d\Omega + \int_\Omega \varepsilon^T \sigma^I\, d\Omega \\ & \int_\Omega \sigma^T(S\mathbf{d} - \varepsilon)\, d\Omega - \int_\Omega \mathbf{d}^T \bar{\mathbf{f}}^B d\Omega \ - \int_{\Gamma_\sigma} \mathbf{d}^T \bar{\mathbf{f}}^s d\Gamma \ - \int_{\Gamma_\delta} \mathbf{f}^{sT}(\mathbf{d} - \bar{\mathbf{d}}) d\Gamma \end{aligned}} \tag{8.56}$$

where the barred quantities are prescribed or given. The solution, $\mathbf{d}_{Sol}, \sigma_{Sol}, \varepsilon_{Sol}$, is the state at which the functional becomes stationary.

Remark

- The stationary condition is given by the first variation of the functional vanishing with respect to its arguments $\mathbf{d}$, σ and ε. Variation of the functional over the appropriate vector space easily generates the weak or Galerkin form, and eventually application of Green's identity or divergence theorem; that is, the higher-dimensional counterparts of integration by parts of one dimension provide the governing strong-form differential equations and the boundary conditions. The Galerkin-type variational methods from this functional are based on an obvious discretization by a finite-dimensional subspace containing $\mathbf{d}$, σ and ε.

8.3.4 Hellinger–Reissner Variational Theorem

In this theorem, the strain vector, ε, ceases to be a state vector, having been replaced by the stress vector, $\boldsymbol{\sigma}$, through the application of the constitutive law of linear elasticity:

$$\varepsilon = \varepsilon^I + \mathcal{D}^{-1}(\sigma - \sigma^I) \tag{8.57}$$

Now, introducing equation (8.57) into equation (8.56), we define the ***Hellinger–Reissner functional*** as follows:

$$\begin{aligned} I(\mathbf{d}, \sigma) = &-\frac{1}{2}\int_\Omega (\bar{\varepsilon}^I)^T \mathcal{D}\, \bar{\varepsilon}^I \, d\Omega - \frac{1}{2}\int_\Omega (\bar{\sigma}^I)^T \mathcal{D}^{-1}\, \bar{\sigma}^I \, d\Omega - \frac{1}{2}\int_\Omega \sigma^T \mathcal{D}^{-1} \sigma \, d\Omega \\ &- \int_\Omega \sigma^T \bar{\varepsilon}^I \, d\Omega + \int_\Omega (\bar{\varepsilon}^I)^T \bar{\sigma}^I \, d\Omega + \int_\Omega \sigma^T \mathcal{D}^{-1}\, \bar{\sigma}^I \, d\Omega + \int_\Omega \mathcal{S}\mathbf{d} \, d\Omega \\ &- \int_\Omega \mathbf{d}^T \mathbf{f}^B d\Omega \; - \int_{\Gamma_\sigma} \mathbf{d}^T \bar{\mathbf{f}}^s d\Gamma \; - \int_{\Gamma_\delta} \mathbf{f}^{s^T} (\mathbf{d} - \bar{\mathbf{d}}) d\Gamma \end{aligned} \tag{8.58}$$

Remarks

- In this equation, the superscripted barred quantities indicate that they are specified or given as an input and, as such, are known.
- This theorem is primarily used for what is commonly known as ***mixed-formulation***-based finite element methods. As discussed in the next section, the use of low-order elements fails to solve problems of a limiting nature, such as one of incompressibility, by the conventional displacement-based finite element method. A diversion to mixed formulation based on this theorem has been resorted to in an attempt to avoid such failures.
- As we shall see later, for our new displacement-based finite element method we will have no need for any mixed formulation. So we will not discuss any further either the Hu–Washizu or the Hellinger–Reissner theorems.
- Our interest is strictly centered on the variational theorem that has the displacement as the only state vector. We now proceed to present such a variational theorem.

8.3.5 Minimum Potential Energy Theorem and Virtual Work Principle

The minimum potential energy theorem seeks to eliminate both the stress tensor, σ, and the strain tensor, ε, from being members of the state vector. Thus the state vector becomes synonymous with the displacement vector, **d**, that solely, completely and uniquely defines any state of a deformable elastic body.

This is achieved as follows.

- The strain is completely defined by the strain–displacement relationship:

$$\varepsilon(\mathbf{c}) = \mathcal{S}\, \mathbf{d}(\mathbf{c}), \quad \forall \mathbf{c} \in \Omega \tag{8.59}$$

  The actual stress, σ, is completely defined by equation (8.51).
- The displacement boundary conditions are considered as constraints on the principle, that is, the elements in the solution space must a priori satisfy all the displacement boundary conditions known as the essential boundary conditions defined in the previous section.

Remarks

- With the above assumptions, first we will state the potential energy functional and, in so doing, justify the name by identifying it with familiar concepts such as potential energy and strain energy of the system. This will also enable us to associate operator-induced energy inner product of the previous section with the strain energy of the system.
- Second, the minimization principle known as the ***principle of minimum potential energy*** that leads to weak or Galerkin form is stated. This, in turn, allows us to associate the soon-to-be-introduced principle of virtual work to this form. Finally, we recover the equilibrium equations and the boundary conditions, that is, the Euler equations of the system.

8.3.6 Potential Energy Functional, $I(\mathbf{d}) \equiv \Pi(\mathbf{d})$

The potential energy functional may be expressed in any one of the following ways:

- In its most compact form:

$$\boxed{I(\mathbf{d}) = \frac{1}{2}\mathcal{Q}(\mathbf{d},\mathbf{d}) - \mathcal{L}(\mathbf{d})} \tag{8.60}$$

- In the expanded matrix form:

$$\boxed{\mathcal{Q}(\mathbf{d},\mathbf{v}) = \int_\Omega (S\mathbf{v})^T D\,(S\mathbf{d})\,d\Omega, \quad \forall \mathbf{d},\mathbf{v} \in \mathbf{E}^{Sol}_{Ess}(\Omega) \subseteq \mathfrak{S}^m(\Omega)} \tag{8.61}$$

  with the subscript *Ess* in the solution space, $\mathbf{E}^{Sol}_{Ess}(\Omega)$, indicates that the displacement solution will be sought from among the elements of the space that satisfies all the essential or admissible boundary conditions.
- In compact inner-product form:

$$\boxed{\mathcal{Q}(\mathbf{d},\mathbf{v}) = (\mathcal{S}\mathbf{v}, \mathcal{D}\mathcal{D}\mathcal{S}\mathbf{d}), \quad \forall \mathbf{d},\mathbf{v} \in \mathbf{E}^{Sol}_{Ess}(\Omega)} \tag{8.62}$$

  where $(\bullet,\bullet)$ is the natural inner product, and $\mathcal{Q}(\mathbf{d},\mathbf{v})$ is the energy product induced in $\mathfrak{S}^m(\Omega)$ by the operator, A, defined in equation (8.53), and assuming that $\mathbf{f} \in^0 (\Omega)$, that is, square integrable: for plane strain, plane stress and axisymmetric problems: $m = 1$ that is, $\mathbf{d}$ and $\mathbf{v}$ must have square-integrable first derivatives, and for plate bending and shell problems: $m = 2$, that is, $\mathbf{d}$ and $\mathbf{v}$ must have square-integrable second derivatives (more on this later).
- Using the definition of total strain vector in equations (8.61) and (8.62), we get the alternative expressions:

$$\boxed{\mathcal{Q}(\mathbf{d},\mathbf{v}) = \int_\Omega \varepsilon(\mathbf{v})^T D\varepsilon(\mathbf{d})\,d\Omega, \quad \forall \mathbf{d},\mathbf{v} \in \mathbf{E}^{Sol}_{Ess}(\Omega) \subseteq \mathfrak{S}^m(\Omega)} \tag{8.63}$$

  and

$$\boxed{\mathcal{Q}(\mathbf{d},\mathbf{v}) = (\varepsilon(\mathbf{v}), D\varepsilon(\mathbf{d})), \quad \forall \mathbf{d},\mathbf{v} \in \mathbf{E}^{Sol}_{Ess}(\Omega)} \tag{8.64}$$

  Now, as to the last term of the equation (8.60):

- In expanded matrix form:

$$\boxed{\mathcal{L}(\mathbf{d}) = \int_\Omega \mathbf{d}^T\mathbf{f}^B\, d\Omega + \int_\Omega (S\mathbf{d})^T D\epsilon^I\, d\Omega - \int_\Omega (S\mathbf{d})^T \sigma^I\, d\Omega + \int_{\Gamma_\sigma} \mathbf{d}^T\mathbf{f}^S\, d\Gamma} \tag{8.65}$$

- In compact inner product form:

$$\boxed{\mathcal{L}(\mathbf{d}) = (\mathbf{d},\ \mathbf{f}^B) + (S\mathbf{d}, D\varepsilon^I) - (S\mathbf{d}, \sigma^I) + (\mathbf{d}, \mathbf{f}^S)_{\Gamma_\sigma}} \tag{8.66}$$

Remarks

- Recall that the strain vector is $\epsilon(\mathbf{d}) = S\,\mathbf{d}$, and the stress vector is given by $\sigma(\mathbf{d}) = \mathcal{D}S\,\mathbf{d} - \mathcal{D}\epsilon^I + \sigma^I$, as in equations (8.49) and (8.51), respectively.
- So it is immediately seen by applying the definition of $\mathcal{Q}(\mathbf{d},\mathbf{d})$ from equation (8.63), that the first term in equation (8.60) is exactly equal to the strain energy, $\mathcal{U}(\mathbf{d})$, of the system under deformation, $\mathbf{d}$, defined by the product of the strain times the corresponding stress due to straining alone:

$$\begin{aligned} U(\mathbf{d}) &= \frac{1}{2}\int_\Omega e\epsilon(\mathbf{d})^T\ [(\sigma - \sigma^I) + D\epsilon^I]\, d\Omega \\ &= \frac{1}{2}\int_\Omega e(S\mathbf{d})^T\ D\,(S\,\mathbf{d})\, d\Omega \\ &\equiv \frac{1}{2}\mathcal{Q}(\mathbf{d},\mathbf{d}) \end{aligned} \tag{8.67}$$

Note that in equation (8.67), we have used the definition given in equation (8.51).
- From the definition of the linear term $\mathcal{L}(\mathbf{d})$ from equation (8.65), we see that the first term is the work done by the body forces. The second and third terms represent the work done by the initial strains and stresses, respectively, and the final term denotes the work done by surface tractions. Accordingly, the second term $\mathcal{L}(\mathbf{d})$ in equation (8.60) represents the total work performed by the specified input data, denoted $\mathcal{W}(\mathbf{d})$, and defined by:

$$\begin{aligned} \mathcal{W}(\mathbf{d}) &= \int_\Omega \epsilon(\mathbf{d})^T D\epsilon^I\, d\Omega - \int_\Omega \epsilon(\mathbf{d})^T \sigma^I\, d\Omega + \int_\Omega \mathbf{d}^T\mathbf{f}^B\, d\Omega \\ &\quad + \int_{\Gamma_\sigma} \mathbf{d}^T\mathbf{f}^S\, d\Gamma \\ &= \int_\Omega (S\mathbf{d})^T D\epsilon^I\, d\Omega - \int_\Omega (S\mathbf{d})^T \sigma^I\, d\Omega + \int_\Omega \mathbf{d}^T\mathbf{f}^B\, d\Omega \\ &\quad + \int_{\Gamma_\sigma} \mathbf{d}^T\mathbf{f}^S\, d\Gamma \\ &\equiv \mathcal{L}(\mathbf{d}) \end{aligned} \tag{8.68}$$

- Finally, the functional is easily seen to be the potential energy, $\Pi(\mathbf{d})$, of the system under deformation $\mathbf{d}$, defined by:

$$\begin{aligned} \Pi(\mathbf{d}) &= \mathcal{U}(\mathbf{d}) - \mathcal{W}(\mathbf{d}) \\ &\equiv I(\mathbf{d}) \end{aligned} \tag{8.69}$$

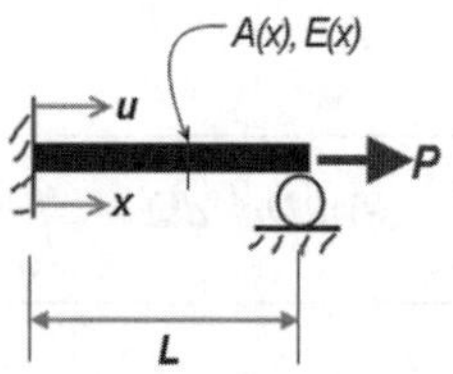

Figure 8.4 Example EP2.

- Thus, from now on, for the sake of notational convention in engineering terms, we will use $\mathbf{\Pi}(d)$ to denote the potential energy functional.

Before presenting the principle of minimum potential energy functional, we propose to elaborate on the potential energy functional through some examples. For simplicity, these are restricted to one-dimensional bar and the Euler–Bernoulli beam. We will have ample demonstrative examples of higher dimensions and complexities in the later sections.

Example EP2: We are interested in finding the potential energy functional, $\Pi(u)$, of a one-dimensional bar of cross-sectional area, $A(x)$, with material elasticity modulus, $E(x)$, and pulled axially by a load, P, with boundary conditions as shown in Figure 8.4.

Note that, for this example, $\mathbf{d}(\mathbf{c})$ is simply $u(x)$ and $\Omega = \{x|x \in (0, L)\}$ and $\bar{\Omega} = \{x|x \in [0, L]\}$. The axial strain, $\varepsilon(x) = \frac{\partial u}{\partial x} = u,_x$. Thus, the strain energy becomes:

$$\mathcal{U}(u) = \frac{1}{2} \int_0^L A(x)\, E(x)\, (u,_x)^2\, dx$$

The external work, $W(u)$, is due only to P and given by:

$$\mathcal{W}(u) = P\, u(L)$$

Thus, the potential energy functional, $\mathbf{\Pi}(u)$, of the bar is defined as:

$$\boxed{\Pi(u) = \frac{1}{2} \int_0^L A(x)\, E(x)\, (u,_x)^2\, dx - P\, u(L)} \tag{8.70}$$

Example EP3: We now intend to find the potential energy functional of a Bernoulli–Euler (i.e. thin, long and flexural only) beam subjected to arbitrary lateral load as shown in Figure 8.5.

For this case, $\mathbf{d}(\mathbf{c})$ is simply $v(y)$ along the neutral axis, which is assumed inextensible and $\Omega = \{x|x \in (0, L),\ d\Omega = dA(x)\, dx\}$. The moment of inertia about z axis is given by: $I_{zz} = \int_A y^2 dA$.

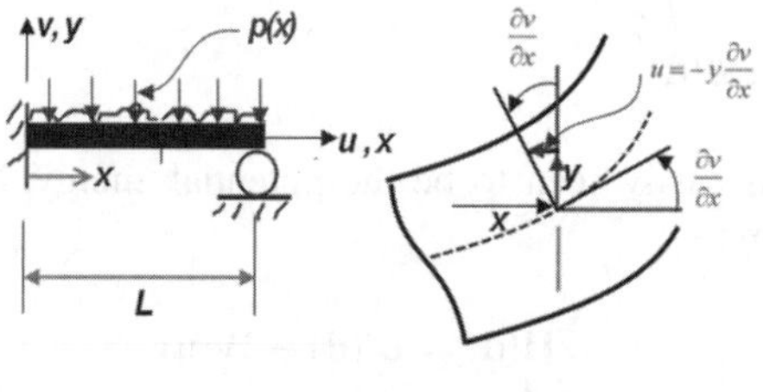

Figure 8.5 Example EP3.

For the Euler–Bernoulli beam, the plane section normal to the neutral axis before bending remains normal after bending, that is, there is no shear strain in the cross-sections along the length. Only strain considered is the axial strain, $\varepsilon(x)$, due to bending. The kinematics of the beam relate the displacement, $u(x, y)$, along the x axis with the displacement, $v(x)$, normal to x axis by:

$$u(x, y) = -y\ v_{,x}$$

The axial strain, $\varepsilon_x(x, y) \equiv u_{,x}(x, y)$, at any height y from the neutral axis is given by:

$$\varepsilon_x(x, y) \equiv u_{,x}(x, y) = -y\ v_{,xx}$$

Thus, the strain energy, $\mathcal{U}(v)$, is given by:

$$\boxed{\mathcal{U}(v) = \frac{1}{2}\int_{\Omega} \varepsilon_x(\mathbf{c})^T \sigma(\mathbf{c})\, d\Omega = \frac{1}{2}\int_0^L [\int_A y^2 dA(x)] E(x)\,(v_{,xx})^2\, dx = \frac{1}{2}\int_0^L E(x) I(x)\,(v_{,xx})^2\, dx}$$

The external work, $\mathcal{W}(v)$ is given by:

$$\mathcal{W}(v) = \int_0^L p(x)\, v(x)\, dx$$

Finally, the potential energy functional of the beam can be written as:

$$\boxed{\mathcal{U}(v) = \frac{1}{2}\int_0^L E(x)\, I(x)\,(v_{,xx})^2\, dx - \int_0^L p(x)\, v(x)\, dx} \tag{8.71}$$

8.3.7 *Minimum Potential Energy Principle*

Now that we have some idea about both the theoretical and practical nature of the potential energy functional, we can state the principle in words as:

Of all the displacement states that a system can experience (i.e. all admissible displacement functions), the one, and the only one, that minimizes the potential energy of the system determines the equilibrium state of the system under the imposed loading conditions.

Mathematically, this is the variational form of the Rayleigh–Ritz–Galerkin method of the previous section, applied to the problems of structural mechanics. Thus, we can write the principle as follows:

Given: the quadratic functional, $\mathbf{\Pi}(d)$ as:

$$\boxed{\mathbf{\Pi}(\mathbf{d}) = \frac{1}{2}\mathcal{Q}(\mathbf{d}, \mathbf{d}) - \mathcal{L}(\mathbf{d})} \tag{8.72}$$

where in expanded matrix form:

$$\boxed{\mathcal{Q}(\mathbf{d}, \mathbf{v}) = \int_{\Omega} (\mathcal{S}\mathbf{v})^T D\,(\mathcal{S}\mathbf{d})\, d\Omega, \quad \forall \mathbf{d}, \mathbf{v} \in \mathbf{E}^{Sol}_{Ess}(\Omega) \subseteq \mathfrak{S}^m(\Omega)} \tag{8.73}$$

with the subscript *Ess* in the solution space, $\mathbf{E}^{Sol}_{Ess}(\Omega)$, indicates that the body in the space satisfies all the essential or admissible boundary conditions. In compact inner product form:

$$\boxed{\mathcal{Q}(\mathbf{d},\mathbf{v}) = (\mathcal{S}\mathbf{v}, \mathcal{D}\mathcal{S}\mathbf{d}), \quad \forall \mathbf{d},\mathbf{v} \in \mathbf{E}^{Sol}_{Ess}(\Omega)} \tag{8.74}$$

where $(\bullet,\bullet)$ is the natural inner product, and $\mathcal{Q}(\mathbf{d},\mathbf{v})$ is the quadratic form or the energy product induced in $\mathcal{H}^m(\Omega)$ by the operator $\mathcal{A}$ defined in equation (8.53), and is related to strain energy, $\mathcal{U}(\mathbf{d})$ by:

$$\mathcal{U}(\mathbf{d}) = \frac{1}{2}\mathcal{Q}(\mathbf{d},\mathbf{d}) \tag{8.75}$$

and $\mathcal{L}(\mathbf{d})$, the linear functional, is related to the work done by the imposed loading and initial conditions by:

$$\mathcal{L}(\mathbf{d}) = \mathcal{W}(\mathbf{d}) \tag{8.76}$$

with, for plane strain, plain stress and axisymmetric problems: $m = 1$, that is, $\mathbf{d}$ and $\mathbf{v}$ must have square-integrable first derivatives, and for plate bending and shell problems: $m = 2$, that is, $\mathbf{d}$ and $\mathbf{v}$ must have square-integrable second derivatives.

Find: $\mathbf{d}_{Sol} \in \mathbf{E}^{Sol}_{Ess}(\Omega)$ such that

$$\boxed{\Pi(\mathbf{d}_{Sol}) = \min_{\mathbf{d}\in\mathbf{E}^{Sol}_{Ess}} \Pi(\mathbf{d})} \tag{8.77}$$

We can quickly establish the minimization condition by using our standard variational technique, as follows. In so doing, let us first recall some spaces that we will be working with for consistent development: $\mathbf{E}^{Sol}(\Omega)$, the energy space with $\|\mathbf{d}\|_{\mathcal{A}} = \sqrt{\mathcal{U}(\mathbf{d})} = \sqrt{\frac{1}{2}\mathcal{Q}(\mathbf{d},\mathbf{d})}$ and $\mathbf{E}^{Sol}_{Ess}(\Omega) \subset \mathbf{E}^{Sol}(\Omega)$; let $\mathbf{v} \in \mathbf{V}_0$ where $\mathbf{V}_0(\Omega) \subset \mathbf{E}^{Sol}_0(\Omega)$ be such that the elements of $\mathbf{V}_0(\Omega)$ satisfy all the essential boundary conditions in their homogeneous state, that is, the elements of $\mathbf{V}_0(\Omega)$ vanish at all locations where the essential boundary conditions are prescribed. Also, let ε be a small real number. Then, for $\mathbf{\Pi}(\mathbf{d}_{Sol})$ to be minimum, we must have:

$$\mathbf{\Pi}(\mathbf{d}_{Sol} + \varepsilon\mathbf{v}_0) \geq \mathbf{\Pi}(\mathbf{d}_{Sol})$$

Now, using equations (8.72) and (8.74),

$$\begin{aligned}\mathbf{\Pi}(\mathbf{d}_{Sol} + \varepsilon\mathbf{v}_0) &= \frac{1}{2}(\mathcal{S}(\mathbf{d}_{Sol} + \varepsilon\mathbf{v}_0), \mathcal{D}\mathcal{S}(\mathbf{d}_{Sol} + \varepsilon\mathbf{v}_0)) - \mathcal{L}(\mathbf{d}_{Sol} + \varepsilon\mathbf{v}_0)\\ &= \frac{1}{2}(\mathcal{S}\mathbf{d}_{Sol}, \mathcal{D}\mathcal{S}\mathbf{d}_{Sol}) - \mathcal{L}(\mathbf{d}_{Sol})\\ &\quad + \varepsilon\left[\frac{1}{2}(\mathcal{S}\mathbf{d}_{Sol}, \mathcal{D}\mathcal{S}\mathbf{v}_0) + \frac{1}{2}(\mathcal{S}\mathbf{v}_0, \mathcal{D}\mathcal{S}\mathbf{d}_{Sol}) - \mathcal{L}(\mathbf{v}_0)\right] +\\ &\quad \frac{1}{2}\varepsilon^2(\mathcal{S}\mathbf{v}_0, \mathcal{D}\mathcal{S}\mathbf{v}_0)\end{aligned} \tag{8.78}$$

For D symmetric, that is, $D = D^T$, and symmetry of the inner product:

$$(\mathcal{S}\mathbf{d}_{Sol}, \mathcal{D}\mathcal{S}\mathbf{v}_0) = (D^T\mathcal{S}\mathbf{d}_{Sol}, \mathcal{S}\mathbf{v}_0) = (\mathcal{D}\mathcal{S}\mathbf{d}_{Sol}, \mathcal{S}\mathbf{v}_0) = (\mathcal{S}\mathbf{v}_0, \mathcal{D}\mathcal{S}\mathbf{d}_{Sol}) \tag{8.79}$$

and noting the definition of $\mathbf{\Pi}(d_{Sol})$, we get,

$$\mathbf{\Pi}(\mathbf{d}_{Sol} + \varepsilon\mathbf{v_0}) \geq \mathbf{\Pi}(\mathbf{d}_{Sol}) + \varepsilon[(\mathcal{D}\mathcal{S}\mathbf{d}_{Sol}, \mathcal{S}\mathbf{v_0}) - \mathcal{L}(\mathbf{v_0})] + \frac{1}{2}\varepsilon^2(\mathcal{D}\mathcal{S}\mathbf{v_0}, \mathcal{S}\mathbf{v_0}) \tag{8.80}$$

Now, positive definiteness of $\mathcal{S}^T\mathcal{D}\mathcal{S}$ with $\mathbf{v}_0 = 0$ on Γ_δ, and square of ε make the last term on the right side of the last expression always positive. Finally, ε being arbitrary in sign, for $\Pi(\mathbf{d}_{Sol})$ to be minimum, the coefficient of ε must vanish. In other words, the first variation must be zero for any $\mathbf{v}_0 \in \mathbf{V}_0$. Thus, we get:

$$(\mathcal{D}\mathcal{S}\mathbf{d}_{Sol}, \mathcal{S}\mathbf{v_0}) = \mathcal{L}(\mathbf{v_0}), \quad \forall\mathbf{v_0} \in \mathbf{V_0} \tag{8.81}$$

Using the definition of $\mathcal{Q}(\mathbf{d}, \mathbf{v})$ in equation (8.74), we finally get the weak or Galerkin form:

$$\boxed{\mathcal{Q}(\mathbf{d}_{Sol}, \mathbf{v_0}) = \mathcal{L}(\mathbf{v_0}), \quad \forall\mathbf{v_0} \in \mathbf{V_0}} \tag{8.82}$$

8.3.8 *Virtual Work Principle for Linear Systems*

Using the definitions $\mathcal{Q}(\mathbf{d}_{Sol}, \mathbf{v_0})$ and $\mathcal{L}(\mathbf{v_0})$ from equations (8.63) and (8.65), respectively, we can express Galerkin or weak form given by equation (8.82) in expanded matrix integral form as:

$$\begin{aligned}\int_\Omega (\mathcal{S}\mathbf{v}_0)^T(\mathcal{D}\mathcal{S}\mathbf{d}_{Sol})\, d\Omega &= \int_\Omega \mathbf{v}_0^T\mathbf{f}^B\, d\Omega + \int_{\Gamma_\sigma} \mathbf{v}_0^T\mathbf{f}^S\, d\Gamma \\ &+ \int_\Omega (\mathcal{S}\mathbf{v}_0)^T\mathcal{D}\epsilon^I\, d\Omega - \int_\Omega (\mathcal{S}\mathbf{v}_0)^T\boldsymbol{\sigma}^I\, d\Omega, \\ &\forall\mathbf{v}_0 \in V_0(\Omega)\end{aligned} \tag{8.83}$$

Now, transferring the third and the final term of the right-hand side to the left and invoking the definition of $\boldsymbol{\sigma}$ from equation (8.51), we get:

$$\boxed{\int_\Omega \epsilon(\mathbf{v}_0)^T\boldsymbol{\sigma}(\mathbf{d}_{Sol})\, d\Omega = \int_\Omega \mathbf{v}_0^T\mathbf{f}^B\, d\Omega + \int_{\Gamma_\sigma} \mathbf{v}_0^T\mathbf{f}^S\, d\Gamma, \quad \forall\mathbf{v}_0 \in V_0(\Omega)} \tag{8.84}$$

If $\mathbf{v}_0 \in V_0(\Omega)$ is considered as the virtual or kinematically admissible displacement, then the left-hand side of the equation (8.84) clearly represents the internal virtual work of the system and the right-hand side reflect the external virtual work done by the imposed loading. Thus, the equations (8.82), (8.83) and (8.84) are the mathematical equivalents of the ***principle of virtual work*** which states:

> ***When a body in equilibrium is perturbed by any kinematically admissible virtual displacements, the internal virtual work performed by the body is equal to the external virtual work by the imposed loading and initial conditions.***

Accordingly, the Galerkin or weak form is nothing other than the principle of virtual work of structural and solid mechanics. We are almost ready to embark on the finite element method, except for one final modification. From the translation theorem on linear variety and subspace (see Chapter 2 on essential mathematics), $\mathbf{d}_{Sol} \in \mathbf{E}^{Sol}_{Ess}(\Omega)$ can be obtained as the summation of $\mathbf{d_0} \in \mathbf{U_0}(\Omega)$ and an arbitrary fixed element $\hat{\mathbf{d}} \in \mathbf{E}^{Sol}(\Omega)$, that is,

$$\mathbf{d}_{Sol} = \hat{\mathbf{d}} + \mathbf{d_0} \tag{8.85}$$

where $\mathbf{U_0}(\Omega) = \mathbf{V_0}(\Omega) \equiv \mathbf{E_0}(\Omega) \subset \mathbf{E}^{Sol}_{Ess}(\Omega)$, that is, $\mathbf{U_0}(\Omega)$ is the space of displacement functions that vanish at all boundaries where the inhomogeneous essential boundary conditions are specified.

Applying equation (8.85) to equations (8.82) and (8.83), we get:

$$\mathcal{Q}(\mathbf{d}_0, \mathbf{v_0}) = \mathcal{L}(\mathbf{v_0}) - \mathcal{Q}(\hat{\mathbf{d}}, \mathbf{v_0}), \quad \forall \mathbf{v_0} \in \mathbf{V_0} \tag{8.86}$$

and

$$\begin{aligned}\int_\Omega (\mathcal{S}\mathbf{v}_0)^T(\mathcal{D}\mathcal{S}\mathbf{d}_0)\, d\Omega = &\int_\Omega \mathbf{v}_0^T\mathbf{f}^B\, d\Omega + \int_{\Gamma_\sigma} \mathbf{v}_0^T\mathbf{f}^S\, d\Gamma \\ &+ \int_\Omega (\mathcal{S}\mathbf{v}_0)^T\mathcal{D}\epsilon^I\, d\Omega - \int_\Omega (\mathcal{S}\mathbf{v}_0)^T\boldsymbol{\sigma}^I\, d\Omega, \\ &- \int_\Omega (\mathcal{S}\mathbf{v}_0)^T(\mathcal{D}\mathcal{S}\hat{\mathbf{d}})\, d\Omega, \quad \forall \mathbf{v}_0 \in V_0(\Omega)\end{aligned} \tag{8.87}$$

The form given by equation (8.87) is actually used for numerical solutions where both $\mathbf{d_0}$ and $\mathbf{v_0}$ are chosen from the same subspace $\mathbf{E_0}(\Omega) \subset \mathbf{E}^{Sol}(\Omega)$. We will deal with this in full detail later. In the meantime, however, we still need to show that the weak form given by equation (8.82) is equivalent to strong form given by equation (8.52). Thus, first noting the definition of $\boldsymbol{\sigma}$ given by equation (8.51), and applying Green's identity:

$$\int_\Omega (\mathcal{S}\mathbf{v})^T\boldsymbol{\sigma}\, d\Omega = -\int_\Omega \mathbf{v}^T(\mathcal{S}^T\boldsymbol{\sigma})\, d\Omega + \int_{\Gamma_\sigma} \mathbf{v}^T\boldsymbol{\sigma}\, d\Gamma \tag{8.88}$$

we can rewrite equation (8.84) and get:

$$\int_\Omega \mathbf{v}_0^T(\mathcal{S}^T\boldsymbol{\sigma} + \mathbf{f}^B)\, d\Omega - \int_{\Gamma_\sigma} \mathbf{v}_0^T(\boldsymbol{\sigma} - \mathbf{f}^S)\, d\Gamma = \mathbf{0}, \quad \forall \mathbf{v}_0 \in V_0(\Omega) \tag{8.89}$$

Now, noting that equation (8.89) holds for $\forall \mathbf{v_0} \in \mathbf{V_0}$, the coefficients of $\mathbf{v_0}$ in each integrand must be identically zero, that is,

$$\boxed{\begin{array}{rll}\mathcal{S}^T\boldsymbol{\sigma} + \mathbf{f}^B = \mathbf{0}, & \forall c \in \Omega & \text{(Euler equation)} \\ \boldsymbol{\sigma} = \mathbf{f}^S & \forall c \in \Gamma_\sigma & \text{(Boundary condition)}\end{array}} \tag{8.90}$$

Equation (8.90) can be compared to equation (8.54) and found to be the same as the strong form of the problem.

Remarks

- First of all, for $\mathbf{d}_{Sol}$ to have finite energy to make any sense, care must be taken to verify that the various input data – the material properties, boundary conditions and loading functions – has finite strain energy and thus can be considered admissible.
- However, concentrated loading in all other situations is very important to consider in practice. If we express the concentrated force, $\mathbf{f}^c$ at $\bar{\mathbf{c}} \in \Gamma_\sigma$, as distributions such as $\delta(\mathbf{c} - \bar{\mathbf{c}})\, \mathbf{f}^c$, then we

can split the surface traction terms in equation (8.90) with the one containing the concentrated force for which, after performing the integral on $\delta(\mathbf{c}-\bar{\mathbf{c}})\,\mathbf{f}^c$, we get $\mathbf{v}_0(\bar{c})^T\,\mathbf{f}^c$. If there is more than one specified, we add the term denoted by $\sum_{n=1}^{No.\,of\,loads} \mathbf{v}_0(\bar{c}_n)^T\,\mathbf{f}^{c_n}$ to the equation; more on this in the subsequent chapters.

- The energy norm induced by the operator, A, is readily seen to be given by:

$$||\mathbf{d}|| \equiv \sqrt{\mathcal{U}(\mathbf{d})} \tag{8.91}$$

This norm is known as the energy norm along with the natural norm of the energy space.

- The elements, $\mathbf{d} \in \mathbf{E}^{Sol}_{Ess}$, do not necessarily have to be continuous for problems in more than one dimension. In other words, the displacement functions are not necessarily bounded by the energy norm, that is, infinite displacement distribution under certain conditions of the order of the space and differentiability can still produce finite strain energy.
- The virtual work principle (the weak or Galerkin form) or the minimum potential energy principle (Rayleigh–Ritz–Galerkin form) forms the theoretical basis of all finite element methods. However, the practical implementations, that is conventional finite element methods, often violate the basic premises of the underlying energy theorems and of course, not without paying a heavy price for it.
- Finally, let us note that so far in our development we have assumed linear mechanics. However, for various structural problems, we may not have an equivalent energy theorem, but the virtual work principle always exists for all structural problems, linear or nonlinear, and this principle can be directly constructed from the equilibrium equations of a system, so we introduce in the following the direct construction of the virtual work principle for nonlinear elasto-dynamic problems.

8.3.9 Virtual (Displacement) Work Principle for Nonlinear Systems

As indicated before in Chapter 7 on essential mathematics, for nonlinear mechanics we have to decide on an appropriate or suitable choice of analysis coordinate system, between the undeformed (Lagrangian) state or on the deformed (Eulerian) state of a body. For all of our future analyses and presentations, we prefer to measure all physical entities and the balance principles of mechanics in coordinate space in reference to the undeformed or Lagrangian state of a body. In the Lagrangian or reference coordinate system, $\mathbf{X} = (X^1, X^2, X^3)^T$, let Ω be a body in its undeformed state with body force, $\mathbf{f}^B(\mathbf{X})$ per unit volume of the body with volume density given by, ρ_0; $\mathbf{E}$, the Green strain tensor, and $\mathbf{S}$, the second Piola–Kirchhoff stress tensor are the energy conjugate stress–strain pair appropriate for the Lagrangian coordinate system; $\mathbf{u}(\mathbf{X})$ is the displacement vector at any point, $\mathbf{X} \in \Omega$, induced by the body force, and the prescribed surface traction, $\mathbf{f}^S$ per unit area on the traction boundary, Γ_σ. Finally, there are specified displacements, $\hat{\mathbf{u}}$, on the displacement boundary, Γ_δ. Given this data, the corresponding nonlinear elasto-dynamic problem may be completely (with the body's elastic state given by the triplet: $(\mathbf{E}, \mathbf{S}, \mathbf{u})$) described by the following set of equations in strong form:

$$\boxed{\begin{aligned} \mathbf{E}(\mathbf{u}) - \mathbf{E} = \mathbf{0}, &\quad \text{(strain-displacement)}\\ \frac{\partial}{\partial \mathbf{E}} U - \mathbf{S} = \mathbf{0}, &\quad \text{(stress-strain)}\\ \nabla \bullet (\mathbf{I} + \nabla \mathbf{u})\mathbf{S} + \rho_0 \mathbf{f}^B - \rho_0\,\ddot{\mathbf{x}} = \mathbf{0}, &\quad \text{(momentum balance)} \end{aligned}} \tag{8.92}$$

where, for clarity in material or Lagrangian description, let us recall the component-wise description in the undeformed coordinate system with $\mathbf{G}_i$ and $\mathbf{G}^i$ for $i = 1, 2, 3$, being the covariant and the contravariant base vectors, respectively, and $(\bullet)|_j$ being the covariant derivative with respect to the base vectors, $\mathbf{G}_i$:

$$
\boxed{
\begin{array}{ll}
\mathbf{u} = U_i\, \mathbf{G}^i & \ddot{\mathbf{x}} = \ddot{U}_i(t)\, \mathbf{G}^i \\
\mathbf{f}^B = f_B^i\, \mathbf{G}_i & \mathbf{E} = E_{ij}\, \mathbf{G}^i \otimes \mathbf{G}^j \\
\mathbf{S} = S^{ij}\, \mathbf{G}_i \otimes \mathbf{G}_j & \mathbf{E}(\mathbf{u}) = \dfrac{1}{2}\{\nabla \mathbf{u} + \nabla \mathbf{u}^T + \nabla \mathbf{u}^T\, \nabla \mathbf{u}\} \\
U(\mathbf{E}) = \dfrac{1}{2}\mathbf{C}(\mathbf{E}) \bullet \mathbf{E} & \mathbf{S} = \mathbf{C}(\mathbf{E}) \\
\mathbf{E} = \mathbf{C}^{-1}(\mathbf{S}) & \nabla(\bullet) = GRAD(\bullet) = \dfrac{\partial(\bullet)}{\partial X^k}\mathbf{G}^k \\
\nabla \bullet (\bullet) = DIV(\bullet) = \dfrac{\partial}{\partial X^k}\mathbf{G}^k \bullet (\bullet) & \nabla \bullet \mathbf{S} = DIV\ \mathbf{S} = S^{ij}|_j\ \mathbf{G}_i
\end{array}
}
\tag{8.93}
$$

For a complete description of a well-posed boundary value problem, the state equation (8.92) is supplemented by the following boundary conditions:

$$
\boxed{
\begin{array}{lll}
\mathbf{u} = \hat{\mathbf{u}}, & \text{on } \Gamma_\delta & \text{(essential boundary condition)} \\
\mathbf{f}^S = (\mathbf{I} + \nabla \mathbf{u})\mathbf{S} \bullet \mathbf{N}, & \text{on } \Gamma_\sigma & \text{(natural boundary condition)}
\end{array}
}
\tag{8.94}
$$

Now, for the derivation of the virtual displacement principle, we will assume that both the strain–displacement and the stress–strain relationships have been satisfied, and thus the state is reduced to only the displacement vector, $\mathbf{u}$, which completely describes the system.

Furthermore, let us recall that the Gateaux directional derivative helps define the virtual function:

$$
\boxed{\bar{\mathbf{f}}(\mathbf{u}, \bar{\mathbf{u}}) \overset{\Delta}{=} \delta \mathbf{f}(\mathbf{u}, \delta \mathbf{u}) = \lim_{\varepsilon \to 0} \frac{d}{d\varepsilon}\mathbf{f}(\mathbf{u} + \varepsilon\ \delta \mathbf{u})}
\tag{8.95}
$$

where we have indicated that the variation or virtual quantities may be denoted by a superscripted bar, $\overline{(\bullet)}$, or as $\delta(\bullet)$. With these in the background, we can devise algorithmically a standard procedure for deriving the virtual displacement equations in linear or nonlinear mechanics.

8.3.9.1 Algorithm VWP

Step 0: Start with the equilibrium equation expressed in the Lagrangian coordinate system:

$$
\boxed{\nabla \bullet (\mathbf{I} + \nabla \mathbf{u})\mathbf{S} + \rho_0 \mathbf{f}^B - \rho_0\, \ddot{\mathbf{x}} = \mathbf{0}, \quad \text{(momentum balance)}}
\tag{8.96}
$$

Step 1: Take the inner product with the variation, $\bar{\mathbf{u}}$, and volume integrate over the body, Ω:

$$
\begin{aligned}
\iiint (\nabla \bullet (\mathbf{I} + \nabla \mathbf{u})\mathbf{S} + \rho_0 \mathbf{f}^B - \rho_0\, \ddot{\mathbf{x}}) \bullet \bar{\mathbf{u}} d\Omega = 0 \\
\Rightarrow I_1 + I_2 + I_3 = 0
\end{aligned}
\tag{8.97}
$$

where,

$$\begin{aligned} I_1 &\stackrel{\Delta}{=} \iiint_{\Omega} (\nabla \bullet (\mathbf{I} + \nabla \mathbf{u})\mathrm{S}) \bullet \bar{\mathbf{u}} d\Omega \\ I_2 &\stackrel{\Delta}{=} \iiint_{\Omega} \rho_0 \mathbf{f}^B \bullet \bar{\mathbf{u}} d\Omega \\ I_3 &\stackrel{\Delta}{=} \iiint_{\Omega} \rho_0 \ddot{\mathbf{x}} \bullet \bar{\mathbf{u}} d\Omega \end{aligned} \tag{8.98}$$

Step 2: For the integral, I_1, let us first recall the divergence of vector times tensor formula:

$$\begin{aligned} &\nabla \bullet (\delta \mathbf{u}\ \mathbf{S}) = \delta \mathbf{u} \bullet (\nabla \bullet \mathbf{S}) + \mathbf{S} : \quad \nabla \delta \mathbf{u} \\ &\Rightarrow \\ &(\nabla \bullet \mathbf{S}) \bullet \delta \mathbf{u} = \nabla \bullet (\delta \mathbf{u}\ \mathbf{S}) - \mathbf{S} : \quad \nabla \delta \mathbf{u} \end{aligned} \tag{8.99}$$

where**:** is the notation for double contraction of a second-order tensor; additionally, we recall the Green–Gauss integral formula:

$$\iiint_{\Omega} (\nabla \bullet \{\bar{\mathbf{u}} \bullet (\mathbf{I} + \nabla \mathbf{u})\mathrm{S}\}) d\Omega = \iint_{\Gamma} (\{\bar{\mathbf{u}} \bullet (\mathbf{I} + \nabla \mathbf{u})\mathrm{S}\} \bullet \mathrm{N}) d\Gamma \tag{8.100}$$

where **N** is the normal at a point on the surface, Γ. So, using equations (8.99) and (8.100), we get:

$$\begin{aligned} I_1 &\stackrel{\Delta}{=} (\nabla \bullet \iiint_{\Omega} \bullet (\mathbf{I} + \nabla \mathbf{u})\mathrm{S}) \bullet \bar{\mathbf{u}} d\Omega \\ &= -\iiint_{\Omega} \nabla \bullet (\bar{\mathbf{u}}\{(\mathbf{I} + \nabla \mathbf{u})\mathrm{S}\}) d\Omega - \iiint_{\Omega} \{(\mathbf{I} + \nabla \mathbf{u})\mathrm{S}{:}\nabla \bar{bfu}\} d\Omega \\ &= \iint_{\Gamma} (\{\bar{\mathbf{u}} \bullet (\mathbf{I} + \nabla \mathbf{u}) \bullet \mathbf{N})\} d\Gamma - \iiint_{\Omega} \{(\mathbf{I} + \nabla \mathbf{u})\mathrm{S} : \nabla \bar{\mathbf{u}}\} d\Omega \end{aligned} \tag{8.101}$$

The kernel of the second term in equation (8.101) can be rewritten as:

$$\begin{aligned} &(\mathbf{I} + \nabla \mathbf{u})\mathbf{S}{:}\nabla \bar{\mathbf{u}} = (\mathbf{I} + \nabla \mathbf{u}){:}\nabla \bar{\mathbf{u}}\ \mathbf{S}^T = \nabla \bar{\mathbf{u}}^T (\mathbf{I} + \nabla \mathbf{u}){:}\mathbf{S} = (\nabla \bar{\mathbf{u}}^T + \nabla \bar{\mathbf{u}}^T \nabla \mathbf{u}){:}\mathbf{S} \\ &= \left\{ \begin{array}{l} \underbrace{\tfrac{1}{2}\left(\nabla \bar{\mathbf{u}}^T + \nabla \bar{\mathbf{u}}\right)}_{\text{Symmetric}} + \underbrace{\tfrac{1}{2}\left(\nabla \bar{\mathbf{u}}^T - \nabla \bar{\mathbf{u}}\right)}_{\text{Antisymmetric}} + \\ \underbrace{\tfrac{1}{2}\left(\nabla \bar{\mathbf{u}}^T \nabla \mathbf{u} + \nabla \mathbf{u}^T \nabla \bar{\mathbf{u}}\right)}_{\text{Symmetric}} + \underbrace{\tfrac{1}{2}\left(\nabla \bar{\mathbf{u}}^T \nabla \mathbf{u} - \nabla \mathbf{u}^T \nabla \bar{\mathbf{u}}\right)}_{\text{Antisymmetric}} \end{array} \right\} : \underbrace{\mathbf{S}}_{\text{Symmetric}} \\ &= \tfrac{1}{2} \left\{ \underbrace{\nabla \bar{\mathbf{u}}^T + \nabla \bar{\mathbf{u}} + \nabla \bar{\mathbf{u}}^T \nabla \mathbf{u} + \nabla \mathbf{u}^T \nabla \bar{\mathbf{u}}}_{\text{Symmetric}} \right\} : \underbrace{\mathbf{S}}_{\text{Symmetric}} \end{aligned} \tag{8.102}$$

where we have used the fact that the second Piola–Kirchhoff stress tensor, **S**, is symmetric, and that the double contraction between a second-order symmetric tensor and an anti-symmetric tensor is zero.

Step 3: We take the variation of the actual strain tensor, $\mathbf{E}(\mathbf{u})$, of equation (8.93), to get:

$$\begin{aligned}\delta\mathbf{E}(\mathbf{u}) &= \bar{\mathbf{E}}(\mathbf{u},\bar{\mathbf{u}}) = \lim_{\varepsilon\to 0}\frac{d}{d\varepsilon}\mathbf{E}(\mathbf{u}+\varepsilon\,\bar{\mathbf{u}})\\ &= \frac{1}{2}\lim_{\varepsilon\to 0}\frac{d}{d\varepsilon}\{\nabla(\mathbf{u}+\varepsilon\,\bar{\mathbf{u}})+\nabla(\mathbf{u}+\varepsilon\,\bar{\mathbf{u}})^T+\nabla(\mathbf{u}+\varepsilon\,\bar{\mathbf{u}})^T\,\nabla(\mathbf{u}+\varepsilon\,\bar{\mathbf{u}})\}\\ &= \frac{1}{2}\{\nabla\bar{\mathbf{u}}+\nabla\bar{\mathbf{u}}^T+\nabla\bar{\mathbf{u}}^T\,\nabla\mathbf{u}+\nabla\mathbf{u}^T\,\nabla\bar{\mathbf{u}}\}\end{aligned} \tag{8.103}$$

Step 4: Now, comparing the expression for $\delta\mathbf{E}(\mathbf{u})$ of equation (8.103) with that of equation (8.102), and plugging it into equation (8.101), and inserting the expression for the natural boundary condition as given by equation (8.94) into the kernel of the first integral of I_1, we get:

$$I_1 \overset{\Delta}{=} \iiint_\Omega \mathbf{S}:\delta\mathbf{E}\,d\Omega - \iint_\Gamma \mathbf{f}_s\bullet\bar{\mathbf{u}}\,d\Gamma \tag{8.104}$$

Step 5: Finally, inserting the expression of equation (8.104) into equation (8.97), the ***virtual (displacement) work principle*** is given by:

$$\underbrace{\iiint_\Omega \mathbf{S}:\delta\mathbf{E}\,d\Omega}_{\text{Internal virtual work}} + \underbrace{\iiint_\Omega \rho_0\,\ddot{\mathbf{x}}\bullet\bar{\mathbf{u}}\,d\Omega}_{\text{Inertial virtual work}} - \underbrace{\iiint_\Omega \rho_0\mathbf{f}^B\bullet\bar{\mathbf{u}}d\Omega + \iint_\Gamma \mathbf{f}_s\bullet\bar{\mathbf{u}}\,d\Gamma}_{\text{External virtual work}} = 0 \tag{8.105}$$

Remarks

- All nonlinear displacement-based finite element methods use the virtual work principle expressed in equation (8.105) as the starting point for analysis.
- The virtual work principle always exists, even though an energy theorem may not, for some problems.
- The strain–displacement compatibility and the constitutive relationships do not enter directly but act as constraints on the principle.
- The nonlinear equations of the virtual work principle cannot generally be solved in closed form; a numerical iteration scheme which we present at the end of this chapter is always needed to solve nonlinear quasi-static or dynamic problems.
- Before we get involved with our ultimate goal of discussing nonlinear problems, we will next explore the linear displacement-based finite element method.

8.4 c-type FEM: Linear Elasticity and Heat Conduction

Here we extend the Rayleigh–Ritz–Galerkin weak form or the principle of virtual work of the previous section to introduce the general formulation of what is known as the c-type finite element methodology in linear systems. In so doing, we examine the source of its strength and the requirements for its convergence to an actual solution. This, then, allows us, in the following chapter, to take a critical look at how the actual practice of the method has deviated, more often than not opportunistically, far beyond the bounds dictated by the theory. Thus, in what follows, we present the theoretical basis of the finite element method as a Rayleigh–Ritz–Galerkin method. The applications of the c-type finite element method to nonlinear problems will be presented later for beams, plates and shell structures.

8.4.1 The Classical Rayleigh–Ritz–Galerkin Method

As introduced in previous sections, the energy space and its linear variety and subspaces are infinite-dimensional spaces, that is, they contain infinitely many linearly independent functions, known as ***basis functions***, that span the spaces. In other words, any displacement solution sought from these spaces, for example, will be characterized as an infinite sum of these basis displacement functions. For computer application with decently manageable data handling capabilities, this is simply prohibitively costly and hence, unusable. We would ideally like some finite sequence of approximate yet ***good*** solutions, preferably of increasing monotonic accuracy, that can be pushed as close to the exact solution as desired. We may present this underlying theme of all Rayleigh–Ritz–Galerkin-type methods as deceptively innocuous steps as follows:

Step 0: Choose finite subspaces of all the relevant spaces by selecting an appropriate finite set of basis functions spanning and characterizing these finite subspaces.

Step 1: Describe the unknown approximate displacement solution and the virtual displacement functions, each as a linear combination of this finite number of basis functions with the finite number of as yet unknown coefficients of linear combinations.

Step 2: Apply the approximate displacement functions and the virtual displacement functions in the equations representing the weak or Galerkin form, more commonly known as the principle of virtual work, as introduced in the previous section.

Step 3: Solve for the unknown coefficients and construct from them the required solutions of interest, that is, displacement, strain and stress, etc.

Step 4: Construct the sequence of improved solutions by increasing either the dimension of the subspace or the quality of the subspace.

Remarks

- The classical Rayleigh–Ritz–Galerkin methods have been in use for a long time. This is because they provided a systematic method for otherwise intractable problems or where the governing differential equations of the systems were not known a priori. However, in the past, an attempt to choose appropriate global basis functions – also known as the ***Ritz functions*** – restricted applications severely, and the geometry of the body, as a result, needed to be very simple and regular, such as squares, cubes, quadrics, etc.
- To satisfy the dictates of variational theory, the essential boundary conditions could only be very simple.
- For similar practical reasons, that is, the evaluation of the necessary surface and volume integrals, the load conditions could not include more complicated ones of practical interest.
- It must be recognized here, that these restrictions are not just logistical. Because, even with the advent of computers, the inability or limitation of this classical approach of selecting Ritz functions lay in its global character, which invariably turned out to be a conceptual shortcoming limiting practical applications. This is clear when it is compared computationally with what came out triumphantly as a remedy by constructing a global Ritz function from simple, locally

supported, piecewise polynomials. Of course, we mean the emergence of the finite element method.

As we shall see later, the practice of the finite element method is another story, since it tried in vain to tear apart the fundamental dictum of the Rayleigh–Ritz–Galerkin method that still remains as invincible as ever:

The Ritz method is a "package deal" and neither requires and permits the user to make independent decisions about different parts of the problem.

In what follows, therefore, we first rewrite and annotate the above steps of the classical Rayleigh–Ritz–Galerkin method in light of the displacement finite element method.

Step 0: Choose finite subspaces of all the relevant spaces by constructing an appropriate finite set of basis functions spanning and characterizing these finite subspaces in following way.
Step 1: Break up the geometry into appropriate finite-sized, geometrically simple pieces known as ***finite elements*** so that the assemblage, known as the ***mesh***, of all these finite elements covers the original geometry exactly.
Step 2: Establish appropriate regular standard elements, known as the ***root elements***.
Step 3: Choose simple polynomials on these root elements as local basis functions.
Step 4: Establish mapping functions or transformations that relate both the geometry and the basis function of the root elements to the corresponding entities on the actual finite elements creating generally non-polynomial basis functions on the actual finite elements.
Step 5: Finally, construct the required global Ritz functions by appropriately combining these mapped, local elemental basis functions; of course, restrictions apply, as we shall see shortly, to inter-element behavior to satisfy the conditions of the underlying energy theorem.

All other remaining steps are the same as described for the classical method.

Before we elaborate on these steps, let us recap on some of the major definitions and equations that will facilitate our subsequent discussions.

- The ***solution space***, $\mathbb{S}^{Sol} \subset \mathfrak{S}^m$, is the space from whose appropriate subspace all displacement solutions will be sought; it is defined as:

$$\boxed{\mathbb{S}^{Sol} = \{\mathbf{d}|\text{ strain energy, } \mathcal{U}(\mathbf{d}) < \infty\}} \tag{8.106}$$

with the energy norm, induced by a symmetric, positive-definite operator, $\mathcal{A} \equiv \mathcal{S}^T \mathcal{D} \mathcal{S}$, defined by:

$$\boxed{\|\mathbf{d}\|_A = \sqrt{\mathcal{U}(\mathbf{d})}} \tag{8.107}$$

$\mathcal{S}$ is the strain–displacement differential operator, and $\mathcal{D}$ is the stress–strain elasticity matrix representing the constitutive law governing the material. In other words, the displacement functions must have finite energy when measured in the energy norm; that is, the space can be equivalently redefined using equation (8.107) as:

$$\mathbb{S}^{Sol} = \{\mathbf{d} \in \mathfrak{S}^m | \|\mathbf{d}\|_{\mathcal{A}} < \infty\} \tag{8.108}$$

The ***solution linear variety***, $S^{Sol}_{Ess} \subset S^{Sol}$, is a subset of the energy space defined as:

$$\boxed{\mathbb{S}^{Sol}_{Ess} = \{\hat{\mathbf{d}} \in \mathbb{S}^{Sol} | \hat{\mathbf{d}} \text{ satisfies all essential boundary conditions}\}} \tag{8.109}$$

- The ***trial subspace***, $\mathbf{U}_0 \subset \mathbb{S}^{Sol}$, of the displacement functions is defined by:

$$\mathbf{U}_0 = \{\mathbf{d}_0 \in S^{Sol}\} \tag{8.110}$$

that is, $\mathbf{U_0}$ is the displacement solution space for a corresponding boundary value problem with homogeneous essential boundary conditions.
- The ***virtual or test subspace***, $\mathbf{V_0} \equiv \mathbf{U_0} \subset \mathbb{S}^{Sol}$, of the virtual displacement functions, that is, the virtual displacements and solution displacements are sought from the same subspace.

The exact solution, $\mathbf{d}_{Sol} \in \mathbb{S}^{Sol}_{Ess}(\Omega)$ can be obtained as the direct sum of the homogeneous solution, $\mathbf{d_0} \in \mathbf{U_0}(\Omega)$, and an arbitrary fixed element $\hat{\mathbf{d}} \in \mathbb{S}^{Sol}(\Omega)$, that is,

$$\boxed{\mathbf{d}_{Sol} = \hat{\mathbf{d}} + \mathbf{d_0}} \tag{8.111}$$

- The ***principle of virtual work*** is defined by:

$$\boxed{\mathcal{Q}(\mathbf{d_0}, \mathbf{v_0}) = \mathcal{L}(\mathbf{v_0}) - \mathcal{Q}(\hat{\mathbf{d}}, \mathbf{v_0}), \quad \forall \mathbf{v_0} \in \mathbf{V_0}} \tag{8.112}$$

or,

$$\boxed{\begin{aligned} \int_{\Omega} (S\mathbf{v}_0)^T D(S\mathbf{d}_0)\, d\Omega = & \int_{\Omega} \mathbf{v}_0^T \mathbf{f}^B \, d\Omega + \int_{\Gamma_\sigma} \mathbf{v}_0^T \mathbf{f}^S \, d\Gamma + \mathbf{v}_0^T \mathbf{F}^C \\ & + \int_{\Omega} (S\mathbf{v}_0)^T D\epsilon^I \, d\Omega - \int_{\Omega} (S\mathbf{v}_0)^T \sigma^I \, d\Omega, \\ & - \int_{\Omega} (S\mathbf{v}_0)^T D(S\hat{\mathbf{d}})\, d\Omega, \quad \forall \mathbf{v}_0 \in V_0(\Omega) \end{aligned}} \tag{8.113}$$

where, $\mathcal{Q}(\mathbf{d}, \mathbf{d})$, the inner product induced by the symmetric, positive definite operator, *calA*, related to strain energy, $\mathcal{U}(\mathbf{d})$, by:

$$\boxed{\mathcal{U}(\mathbf{d}) = \tfrac{1}{2}\mathcal{Q}(\mathbf{d}, \mathbf{d})} \tag{8.114}$$

that is referring to the definition of the norm of the solution space, S^{Sol}; we see that the solution space is a complete inner product space. With the above steps and definitions in place, let us now continue to explain the finite element methodology as it relates to the new expansion of our element library.

8.4.2 New Finite Element Method: c-type

Here we revisit the general theory of the finite element methodology as a means of defining and constructing our new finite element space, exemplify anding presenting the attendant finite element library:

- As seen from our presentation earlier of curves (Chapter 4) and surfaces (Chapter 5), the use of ***Bernstein–Bezier polynomials*** in computer-aided geometry design (CAGD) is very useful and routinely used.
- Similarly, the isoparametric direct stiffness formulation of the finite element method has been used since the early days of the method. However, as indicated earlier, it is the underlying ***finite element space*** and the associated ***element library*** used for the formulation that identify a specific variant of the general method. We have alluded to two such examples: h-type and p-type methods in Chapter 1 on introduction and motivation.
- Now, considering the entire published history of the finite element methodology, we believe that the use of the Bezier control vectors as the generalized displacements, and the Bernstein polynomials as the elemental basis functions, towards the definition of the finite element space is new and unique. Likewise, the linked idea of ***coerced isoparametry*** through the degree elevation algorithm applied to the mesh describing the geometry and inter-element continuity on demand is novel, and it seamlessly unifies modeling and the finite element analysis.
- For practical solution methodology of most structural elements such as beams, plates and shells, ***subparametry*** – that is, the highest degree of polynomial to describe the geometry is less than that for the displacement distribution – seems more efficient numerically without any loss of accuracy, as will be seen later in our examples dealing with linear and nonlinear finite element solutions.
- The new finite element space with the associated new elements jointly gives rise to what is termed the ***c-type finite element method***.
- Finally, from what follows, it will be clear that the c-type method is a displacement-based finite element method; as such, the state of deformation is determined by the displacement distribution, a non-parametric function, while the c-type finite element method is motivated by the Bernstein–Bezier parametric curves. We may now recall from Chapter 4 on curves that any non-parametric function can be represented by parametric Bernstein–Bezier curves because of the linear precision property: $\sum_{0}^{n} (\frac{i}{n}) B_i^n(\xi) = \xi$. Thus, a displacement function, $d(\xi)$, can be transformed into parametric form as:

$$\boxed{\begin{aligned} \mathbf{c}(\xi) &= \begin{Bmatrix} x(\xi) \\ y(\xi) \end{Bmatrix} = \begin{Bmatrix} \xi \\ d(\xi) \end{Bmatrix} = \sum_{0}^{n} B_i^n(\xi)\, \mathbf{q}_i \\ \mathbf{q}_i &\equiv \begin{Bmatrix} \left(\dfrac{i}{n}\right) \\ q_i \end{Bmatrix} \text{ where, } \quad d(\xi) = \sum_{0}^{n} B_i^n(\xi)\, q_i \end{aligned}} \tag{8.115}$$

  that is, the control vertices are located at equal intervals.

In what follows, we could easily recognize that, in our new c-type method, we have retained in a modified way the strong points: isoparametry (or subparametry) and direct stiffness formulation of the old FEM. At the same time, we have offered a new finite element space that does not have the weaknesses of the conventional or old finite element formulation. Some examples carrying the essence of our new space have been previewed for solving some of the nagging problems of the earlier finite element methods. Thus, we detail what is new, and we touch upon those parts of the formulation that are necessary but not new, for the presentation of our theory. Wherever necessary, we elaborate the structure, sequence and organization of our new concept and algorithm with examples. Finally, we conclude by presenting the new c-type element library.

In the finite element approximation for the displacement distribution, a solution is sought in a finite subspace of an ***admissible space*** whose ***linear variety*** is the solution set. The ***finite subspace*** is defined by a finite set of suitable basis functions that spans the subspace. These basis functions, in turn, are determined by:

- the finite element mesh:

$$\Delta = \{\Omega^e, \mathrm{e} = 1, 2, \ldots, \mathrm{N}(\Delta)\}, \ni \Omega = \bigcup_{e \in N(\Delta)} \Omega^e \tag{8.116}$$

  of the domain, Ω, where the Ω^e 's are known as the finite elements,
- the root or standard element, Ω^R, and
- the set of transformations:

$$\mathbf{T} = \{\mathbf{T}^e : \Omega^{\mathrm{R}} \rightarrow \Omega^{\mathrm{e}}, \mathrm{e} = 1, 2, \ldots, \mathrm{N}(\Delta)\} \tag{8.117}$$

Now we will define each of these entities in relation to our general theory. In that following chapters, we will deal with these again in more specific quantized form by specializing them and solving real problems for individual cases in one-, two- and three-dimensional situations.

8.4.3 Finite Element Mesh, Δ: Element Set

Finite element mesh generation is performed in the interactive modeling phase. As indicated before, the success of a finite element method based on the Rayleigh–Ritz–Galerkin technique comes as a "package deal", that is, the choice of basis functions defining the elements in a method primarily determines the success and failure of the method. This pre-empts any ad hoc manipulations such as ***reduced*** or ***selective integration***, or so-called ***bubble modes***, based on violation of the underlying variational principles. Thus, for our theory, the choice of basis functions further permeates the mesh generation phase, that is, the mesh generation and finite element analysis procedure is fully unified. In other words, mesh generation methodology predetermines the outcome of our finite element analysis methodology. In the conventional mesh generation technique, geared for subsequent finite element analysis, once the coordinates of the individual elements are generated, its task is assumed complete. In other words, otherwise profitable global information of the model geometry in terms of continuity among the elements is lost forever by the time the data presents itself to the finite element analysis procedure. Of course, if the mesh generation is performed by ***constructive solid modeling,*** as is done in several of the commercially available CAD programs that are used as third-party mesh generators for providers of general-purpose finite element programs, the useful global information is lacking to start with. Accordingly, based on the ***boundary representation method***, the basis functions of our mesh generation scheme chosen to describe the geometry also serve as the de facto basis functions for our finite element displacement space. This firmly establishes the iso- or subparametric relationship between the geometry and the solution space. Finally, note that for the sake of ultimate integrity of the continuum in finite element analysis, we cannot allow the generation of

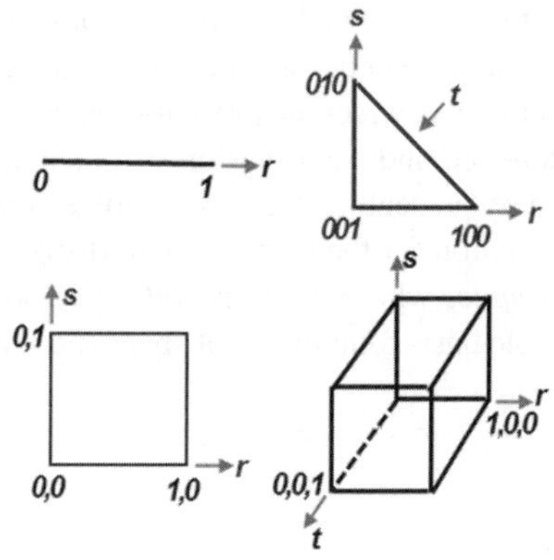

Figure 8.6 Examples of root elements.

mesh with mid-point dangling nodes in any edge or face of our finite elements. In other words, in our mesh generation phase expositions, we have the following restrictions: the physical nodes are located only at the ends of an edge or corners of the faces with no mid-point or intermediate nodes. This guarantees generation of only a regular mesh, preventing possibility of a so-called irregular mesh. Let us now detail how we obtain what constitutes the necessary outcome in the interactive modeling phase. The mesh is generated by first securing a root or standard element.

8.4.4 Root Element, Ω^R: Parameter Domain

The root element (see Figure 8.6) depends on the real geometry of the continuum, and as such, on the dimensionality of the problem. In any case, the root element consists of the simplest of the geometry possible for the problem. The ***local*** Cartesian coordinates of the root elements for general three-dimensional continuum is denoted by:

> $\xi = (r, s, t)^T \in \Omega^R$ = the parameter column vector defining a point on the root element, also known as the ***naturalized coordinates***, where $r \in [0, 1]$, $s \in [0, 1]$ and $t \in [0, 1]$.

For elements of one dimension, such as bars and thin beams, the root element consists of a straight line element of unit length; for two-dimensional elements, the root elements could be an equilateral triangle or an isosceles triangle of unit side or a square of unit side. Finally, for elements in three dimensions, the root element may be a unit regular tetrahedron, pentahedron or hexahedron; our root elements ***locally*** are similar to those of the conventional finite element, except for the definition of the coordinate boundaries – that is, our coordinates in each direction belongs to $[0, 1]$ instead of the conventional $[-1, 1]$. However, our root elements are also described by a ***global*** coordinate system. This is determined at the model mesh generation time by the input geometry data. This aspect of the root elements is important for the imposition of inter-element continuity on demand by the user in the finite element analysis, and accordingly discussed later in the section.

Remarks

- It is assumed at this point that a geometric assembly has been successfully defined by interactive modeling. In other words, all the geometric properties of the mesh along with the Bezier control vectors, $\mathbf{q}^c$, defining the geometry of each of the elements are already available to us.

- However, we must emphasize here, that the discretization of the geometry in the modeling phase cannot and should not be done without regard to the loading and support conditions that define a particular problem. The loading conditions – as will be seen shortly – are also isoparametrically developed along with the mesh for input standardization.

Having secured the root domain, the primary task of mesh generation is to establish the one-to-one transformations between the root domain and the various finite elements based on interpolation. This is, in essence, the starting point of our departure from the old finite element method.

8.4.5 Domain Transformation Set: $\{T^e : \Omega^R \rightarrow \Omega^e,\ e = 1, 2, \ldots, N(\Delta)\}$

In the modeling phase, we essentially develop and determine the domain transformations. For general three-dimensional hexahedral elements, the transformations, relating the root elements to the physical geometry, can be expressed in triple Cartesian product form:

$$
\begin{aligned}
x(r,s,t) &= \sum_{i=0}^{L}\sum_{j=0}^{M}\sum_{k=0}^{N} \widehat{B}_i^L(r)\widehat{B}_j^M(s)\widehat{B}_k^N(t)\ \widehat{\mathbf{q}}_{ijk}^{x_e}\\
y(r,s,t) &= \sum_{i=0}^{L}\sum_{j=0}^{M}\sum_{k=0}^{N} \widehat{B}_i^L(r)\widehat{B}_j^M(s)\widehat{B}_k^N(t)\ \widehat{\mathbf{q}}_{ijk}^{y_e}\\
z(r,s,t) &= \sum_{i=0}^{L}\sum_{j=0}^{M}\sum_{k=0}^{N} \widehat{B}_i^L(r)\widehat{B}_j^M(s)\widehat{B}_k^N(t)\ \widehat{\mathbf{q}}_{ijk}^{z_e}
\end{aligned}
\tag{8.118}
$$

where,

$\widehat{\mathbf{q}}_{ijk}^{\mathbf{c}}$ is the ijk^{th} component of the Bezier control vector defining the geometry of the element Ω^e, as indicated; the vector is known after the interactive model generation phase.

Now, let us express the indices of the Bezier controls in the column vector form:

$$
\begin{aligned}
\widehat{\mathbf{q}}^{\mathbf{c}} = \{&(\widehat{\mathrm{q}}_{000}^{\mathrm{x}}, \widehat{\mathrm{q}}_{100}^{\mathrm{x}}, \ldots, \widehat{\mathrm{q}}_{\mathrm{L00}}^{\mathrm{x}}, \ldots, \widehat{\mathrm{q}}_{\mathrm{LM0}}^{\mathrm{x}}, \ldots \widehat{\mathrm{q}}_{\mathrm{LMN}}^{\mathrm{x}}),\\
&(\widehat{\mathrm{q}}_{000}^{\mathrm{x}}, \widehat{\mathrm{q}}_{100}^{\mathrm{x}}, \ldots, \widehat{\mathrm{q}}_{\mathrm{L00}}^{\mathrm{x}}, \ldots, \widehat{\mathrm{q}}_{\mathrm{LM0}}^{\mathrm{x}}, \ldots \widehat{\mathrm{q}}_{\mathrm{LMN}}^{\mathrm{x}}),\\
&(\widehat{\mathrm{q}}_{000}^{\mathrm{x}}, \widehat{\mathrm{q}}_{100}^{\mathrm{x}}, \ldots, \widehat{\mathrm{q}}_{\mathrm{L00}}^{\mathrm{x}}, \ldots, \widehat{\mathrm{q}}_{\mathrm{LM0}}^{\mathrm{x}}, \ldots \widehat{\mathrm{q}}_{\mathrm{LMN}}^{\mathrm{x}})\}^{\mathrm{T}}
\end{aligned}
\tag{8.119}
$$

where, the $\{(L+1)(M+1)(N+1)\} \times 1$ column vector of known Bezier controls defines the general three-dimensional geometry, Ω^e. Similarly, let us introduce the $1 \times \{(L+1)(M+1)(N+1)\}$ row transformation matrix,

$$
\begin{aligned}
\widehat{\bar{\mathbf{T}}}^{\mathbf{c}} &= \{(\mathrm{T}_{000}, \mathrm{T}_{100}, \ldots, \mathrm{T}_{\mathrm{L00}}, \mathrm{T}_{\mathrm{L10}}, \ldots, \mathrm{T}_{\mathrm{LM0}}, \mathrm{T}_{\mathrm{LM1}}, \ldots \mathrm{T}_{\mathrm{LMN}})\},\\
\text{with}\quad & \forall i \in \{\mathbf{1}, \mathbf{2}, \ldots, \boldsymbol{L}\},\ \forall j \in \{\mathbf{1}, \mathbf{2}, \ldots, \boldsymbol{M}\},\ \forall k \in \{\mathbf{1}, \mathbf{2}, \ldots, \boldsymbol{N}\}\\
\mathrm{T}_{\mathrm{ijk}} &= \widehat{B}_i^L(r)\widehat{B}_j^M(s)\widehat{B}_k^N(t)
\end{aligned}
\tag{8.120}
$$

where

$\widehat{B}_i^L(r)$ = the ith Bernstein polynomial of degree L defined along r direction of the root element,

$\widehat{B}_j^M(s)$ = the jth Bernstein polynomial of degree M defined along s direction of the root element,

$\widehat{B}_k^N(t)$ = the kth Bernstein polynomial of degree N defined along t direction of the root element.

Recall that $\widehat{B}_i^N(r)$, the ith Bernstein polynomial of degree n is defined as:

$$\boxed{\begin{aligned} B_i^n(r) &= \binom{n}{n-i} r^i\,(1-r)^{n-i}, i \in \{0,1,\ldots,n\},\\ &= 0, i \notin \{0,1,\ldots,n\},\\ \text{where,}\quad \binom{n}{n-i} &= \frac{n!}{i!(n-i)!}\end{aligned}} \tag{8.121}$$

Now, for three-dimensional hexahedral element, the $3 \times \{(L+1)(M+1)(N+1)\}$ transformation matrix, $\widehat{\bar{T}}^e$, will then be defined as:

$$\widehat{\bar{T}}^e = \begin{bmatrix} \widehat{\bar{T}}^e & \mathbf{0} & \mathbf{0} \\ \mathbf{0} & \widehat{\bar{T}}^e & \mathbf{0} \\ \mathbf{0} & \mathbf{0} & \widehat{\bar{T}}^e \end{bmatrix} \tag{8.122}$$

where $\widehat{\bar{T}}^e$ is as defined in equation (8.120), and $\mathbf{0} = \{0\ 0\ \ldots\ \text{M-}times\}$. Finally, we can represent equation (8.118) in matrix form as:

$$\boxed{\mathbf{c}^e(\xi) = \widehat{\bar{T}}^e(\xi)\ \widehat{\mathbf{q}}^c} \tag{8.123}$$

where, $\widehat{\mathbf{q}}^c$ is defined by equation (8.119), and $\widehat{\bar{\mathbf{T}}}^e(\xi)$ is given by equation (8.122).

Remarks

- It may be observed that the Bezier control vectors are defined in such a way that all the x-coordinates are packed first, then the y-coordinates and finally the z-coordinates in the vectors. Of course, it could have been done in alternate way such as by collecting x-, y- and z- directions of each control location, and then repeating for each such location.
- For hexahedral element with plane faces, for example, the degree of the corresponding Bernstein polynomials will be 1; for cubic curved faces, it will be 3 and for faces with one direction linear and another quadratic, the corresponding degrees will be 1 and 2, respectively, and so on.
- Because of the end point interpolation property of piecewise Bernstein–Bezier polynomials, our elements consist of only corner physical nodes of conventional or old finite elements and

other generalized 'non-physical' (in the conventional node sense) nodes represented by Bezier control vectors, but never of any intermediate or mid-point nodes. This, of course, has profound implications for inter-element continuity, which will be discussed shortly.

From the second remark above, we see that, unlike the old finite elements where shape functions directly relate the finite element nodal coordinates to corresponding properties inside the element, our elements are indirectly defined by Bezier control vectors with corner node interpolation providing the nodal attributes and Bernstein polynomials serving as the shape functions. Because of the barycentric properties of Bernstein polynomials, that is,

$$\boxed{\sum_{i=0}^{p} B_i^p(r) = 1, \qquad p = L, M, N} \tag{8.124}$$

the element geometry is characterized by convex combination of Bezier controls guaranteeing a non-singular Jacobian which is crucially important for iso- or subparametric formulations. Because of the linear precision and the degree elevation properties of Bernstein–Bezier curves, the degrees of shape functions can be freely manipulated that enable us to formulate what we term as coerced isoparametry, which we discuss next.

8.4.6 Coerced Isoparametry and Transformations

Consider a hexahedral geometry with plane faces. In this situation, referring to equation (8.118), it is easy to see that the choice of Bernstein polynomials of degree 1, that is, $L = M = N = 1$, in all naturalized coordinate directions, r, s and t, will exactly replicate the geometry. However, for general loading conditions imposed on the element, the displacement distributions over the element can hardly be expected to be linear in any parameter direction. Suppose we need the element to represent the cubic displacement distribution across r, quadratic in s and cubic in t parametric directions. Although the geometry has been represented by linear faces, we can easily achieve the displacement distribution as above by first coercing up the degrees in each parametric direction by applying the ***degree elevation algorithm*** repeatedly an appropriate number of times on the geometry representation. Then we represent the displacement distribution isoparametric to this new geometry representation. Because of the linear precision and degree elevation properties of Bernstein–Bezier representation, geometry shape is left intact by this coercion, while the displacement is represented isoparametrically with the geometry. This is what we call ***coerced isoparametry***.

Remark

- Note that coercion is not necessarily needed for most practical problems; in such situations, subparametry – that is, the geometry highest degree being less than the displacement highest degree for the chosen polynomial – is adequate for recovering efficiently geometric properties such as Jacobians etc., in the numerical treatment in this new finite element procedure. We will continue, however, with the concept of coerced isoparametry for completeness.

Example CIP 1: Let us consider a bar, straight in geometry with only one element. In this situation, as indicated in the first remark earlier, in modeling phase, a linear domain mapping will be the exact representation of the bar:

$$x(r) = (1 - r)\tilde{q}_0^{x^1} + r\tilde{q}_1^{x^1} \tag{8.125}$$

with, : $\tilde{q}_0^{x^1} \equiv x(0) = x_1$, and $\tilde{q}_{\tilde{p}^1}^{x^1} \equiv x(1) = x_2$, and in matrix form:

$$x(r) = \mathbf{T}^1(r)\,\tilde{\mathbf{q}}^{x^1} \tag{8.126}$$

where, $\mathbf{T} \equiv \mathbf{T}^1(r) = [(1 - r) \quad r] = (1 \times 2)$ transformation matrix, $\tilde{\mathbf{q}}^x \equiv \tilde{\mathbf{q}}^{x^1} = \{x_1 \quad x_2\}^T = (2 \times 1)$ coordinate control column vector. However, under general loading imposed on the bar, an inspection of the governing differential equation will quickly reveal that the displacement distribution along the bar will be far from linear, depending on the forcing function. In this typical situation, for conventional finite element methodologies, it is common practice to include as many linear elements to ensure convergence, or to introduce more nodes in the element, or to increase the degree of the element. For the new method, suppose the displacement distribution is of degree d, and the geometry degree is g. Then, in order to establish isoparametry, we use the ***linear precision*** property of the Bezier curves, that is, we degree elevate the geometry representation by $(d - g)$ repeated applications. This in turn will increase the number of Bezier coordinate control vertices by $(d - g)$ and the parameters of the new vertices will be given by $(\frac{i}{d})$, $i = 1, 2, \ldots, (d - g)$, for $d = 3$ and $g = 1$, that is, the displacement distribution is cubic and the geometry is linear. Given d and g, the coordinate locations are fixed. Of course, for this straight bar situation, this is achieved trivially by inspection. For $d = 3$ and $g = 1$, $\widehat{\mathbf{q}}^x$, the (2×1) column vector will be replaced by $\mathbf{q}^x$, a (4×1) column vector as follows:

$$\mathbf{q}^x = \{q_0^x \equiv \widehat{q}_0^x \quad q_1^x \quad q_2^x \quad q_3^x \equiv \widehat{q}_1^x\}^T \text{ with,}$$
$$q_1^x = \widehat{q}_0^x + \tfrac{1}{3}(\widehat{q}_1^x - \widehat{q}_0^x), \text{ and,}$$
$$q_2^x = \widehat{q}_0^x + \tfrac{2}{3}(\widehat{q}_1^x - \widehat{q}_0^x),$$

In other words, $\mathbf{q}^x$ is fully defined. For the general case of $g \neq 1$, the relationship is given by degree elevation algorithm presented in the Chapter 4 on curves.

Remarks

- The number of physical nodes still remains as two, that is, the two end nodes.
- As remarked before, we emphasize that, for most problems, subparametry will suffice; that is, there is no need for isoparametry because the geometry is primarily used for computing the Jacobian, as will be seen by various examples later. The coerced isoparametry is introduced only to understand the role of degree elevation in the new methodology.
- For complicated geometry, it is always advisable to break it down into such detail that no more than cubic piecewise Bezier, or equivalently, cubic B-spline curves or surfaces are required to describe it geometrically while the displacement distribution can be of higher degree.

In any event, because of this coercion technique or not, let us for now assume that the transformations for both geometry and displacements are the same and, as such, are given by

$\mathbf{T}^e(\boldsymbol{\xi})$; that is, $\widehat{T}^e(\boldsymbol{\xi})$ of equation (8.122) is modified by appropriate Bernstein polynomials. Now, we are ready to represent the element displacement distribution isoparametrically as:

$$\mathbf{d}(x(r,s,t),y(r,s,t),z(r,s,t)) = \mathbf{T}^e(r,s,t)\mathbf{q}^{d_e} \tag{8.127}$$

or in compacted coordinate-independent form:

$$\mathbf{d}(\mathbf{c}(\boldsymbol{\xi})) = \mathbf{T}^e(\boldsymbol{\xi})\mathbf{q}^{d_e} \tag{8.128}$$

where, $\mathbf{q}^{\mathbf{d}}_{ijk}$ = the ijk^{th} Bezier control vector defining the displacements at generalized coordinates; they are the desired unknown quantities, or

$$\mathbf{q}^{\mathbf{d}} = \{(\mathrm{q}^{\mathrm{u}}_{000}, \mathrm{q}^{\mathrm{u}}_{100}, \cdots, \mathrm{q}^{\mathrm{u}}_{\mathrm{LMN}}), (\mathrm{q}^{\mathrm{v}}_{000}, \mathrm{q}^{\mathrm{v}}_{100}, \cdots, \mathrm{q}^{\mathrm{v}}_{\mathrm{LMN}}), (\mathrm{q}^{\mathrm{w}}_{000}, \mathrm{q}^{\mathrm{w}}_{100}, \cdots, \mathrm{q}^{\mathrm{w}}_{\mathrm{LMN}})\}^{\mathrm{T}} \tag{8.129}$$

the $(L+1)(M+1)(N+1)\times 1$ column vector of unknown Bezier controls for displacements. Everything else is as defined earlier. Similarly, we can readily write all the required transformations for imposed loading functions having same degree of representation as:

$$\mathbf{f}(x(r,s,t),y(r,s,t),z(r,s,t)) = \mathbf{T}^e(r,s,t)\mathbf{q}^{\mathbf{f}^e} \tag{8.130}$$

or, in compacted, coordinate-independent form:

$$\mathbf{f}(\mathbf{c}) = \mathbf{T}^e(x)\mathbf{q}^{\mathbf{f}^e} \tag{8.131}$$

where,

$$\mathbf{f}^{\mathbf{c}} = \{(\mathrm{f}^{\mathrm{x}}_{000}, \mathrm{f}^{\mathrm{x}}_{100}, \cdots, \mathrm{f}^{\mathrm{x}}_{\mathrm{LMN}}), (\mathrm{f}^{\mathrm{y}}_{000}, \mathrm{f}^{\mathrm{y}}_{100}, \cdots, \mathrm{f}^{\mathrm{y}}_{\mathrm{LMN}}), (\mathrm{f}^{\mathrm{z}}_{000}, \mathrm{f}^{\mathrm{z}}_{100}, \cdots, \mathrm{f}^{\mathrm{z}}_{\mathrm{LMN}})\}^{\mathrm{T}} \tag{8.132}$$

the $(L+1)(M+1)(N+1)\times 1$ column vector of known Bezier controls for forces. Here, it is intended to represent generally all the different force types, $\mathbf{q}^{\mathbf{f}}_{ijk} = ijk^{th}$ Bezier control vector defining the loadings at generalized coordinates; they are the known quantities generated during interactive modeling phase.

Remarks

- As indicated before, for cases where both the coordinates and displacement spaces of a particular element are of the same degree in any particular coordinate direction i, $i \in \{1,2,3\}$,

$$\bar{\mathbf{T}}^e_i = \mathbf{T}^e_i \tag{8.133}$$

  where we have introduced the notation: $\mathbf{T}^e = \{\mathbf{T}^e_x\ \mathbf{T}^e_y\ \mathbf{T}^e_z\}^T$ and $\bar{\mathbf{T}}^e = \{\bar{\mathbf{T}}^e_x\ \bar{\mathbf{T}}^e_y\ \bar{\mathbf{T}}^e_z\}^T$, that is, it degenerates into isoparametry of the old finite element procedures except with the crucial difference that for our method, it is still the unknown Bezier displacement control vectors that are our generalized coordinates instead of conventional displacements at the nodes.
- Although there is no theoretical restrictions, a indicated before, for practical general-purpose finite element programs, the geometry representation is recommended to be limited to piecewise

cubic Beziers of C^0, C^1 and C^2 (B-spline) representation. Thus, it is assumed that during interactive modeling phase:

- Concentrated forces or moments should have the input node positioned at the point of their applications.
- For complicated distributed loading, the geometry should be partitioned into a mesh, such that cubic or lesser-degree Bezier or B-spline functions should be able to model the loading.
- A node is positioned wherever there is a support.
- Finally, highly curved geometry should also be partitioned such that piecewise Beziers or B-splines of degree three or less should be able to model the geometry with any desired accuracy.

- In the above description of the transformations, it is assumed that the loading functions are also interpolated isoparametrically to the displacement distribution. This, of course, allows standardization of loading input for any general-purpose finite element computer program. Arbitrary cross-sectional and material properties, likewise, can be interpolated. However, for general-purpose programming, numerical integration is performed by appropriate quadrature rules that treat these properties in a discretized way at the quadrature points.
- It follows that at each geometry control point that is needed to describe an element, there is one corresponding displacement and/or loading control point.

8.4.7 Element (Local) Level Formulation

As presented in Sections 8.3.8 and 8.3.9 on the theory of finite element method, we know that the principle of virtual work provides us with the equilibrium conditions for the continuum. Then, satisfaction of the strain–displacement and constitutive relationships generates the necessary finite element algebraic simultaneous set of equations in unknown Bezier displacement control vectors relating the stiffness matrix and the discretized imposed load vectors. Just as in the old finite element methodology, to express the virtual work in terms of the unknown displacement control vectors, we need to develop the strain–displacement matrix relating element strains and unknown Bezier displacement control vectors, as follows.

8.4.7.1 Strain–Displacement Control Matrix: Linear Elasticity

The strain tensors, $\epsilon(\mathbf{d}(c^e))$, at any point $c \in \Omega^e$, as a function of displacement vector, $\mathbf{d}$, are given by:

$$\epsilon(\mathbf{d}(\mathbf{c}^e)) = \mathcal{S}\,\mathbf{d}(\mathbf{c}^e) \tag{8.134}$$

where $\mathcal{S}$ is a differential operator defined as:

$$\mathcal{S} = \begin{bmatrix} \frac{\partial}{\partial x} & 0 & 0 \\ 0 & \frac{\partial}{\partial y} & 0 \\ 0 & 0 & \frac{\partial}{\partial z} \\ \frac{\partial}{\partial y} & \frac{\partial}{\partial x} & 0 \\ 0 & \frac{\partial}{\partial z} & \frac{\partial}{\partial y} \\ \frac{\partial}{\partial z} & 0 & \frac{\partial}{\partial x} \end{bmatrix} \tag{8.135}$$

Let us define the Jacobian transformation matrix, $\mathbf{J}(x) \equiv \frac{\partial \mathbf{c}}{\partial x}$, that is, the scaling matrix relating the derivatives of the generalized coordinates, $\mathbf{c} \in \Omega^e$, with respect to the naturalized coordinates, $x \in \Omega^R$, such that:

$$\frac{\partial(\bullet)}{\partial x} = \mathbf{J}(x)\frac{\partial(\bullet)}{\partial \mathbf{c}} \tag{8.136}$$

or, in expanded operator form for general three dimensions:

$$\begin{Bmatrix} \frac{\partial(\bullet)}{\partial r} \\ \frac{\partial(\bullet)}{\partial s} \\ \frac{\partial(\bullet)}{\partial t} \end{Bmatrix} = \begin{bmatrix} \frac{\partial x}{\partial r} & \frac{\partial y}{\partial r} & \frac{\partial z}{\partial r} \\ \frac{\partial x}{\partial s} & \frac{\partial y}{\partial s} & \frac{\partial z}{\partial s} \\ \frac{\partial x}{\partial t} & \frac{\partial y}{\partial t} & \frac{\partial z}{\partial t} \end{bmatrix} \begin{Bmatrix} \frac{\partial(\bullet)}{\partial x} \\ \frac{\partial(\bullet)}{\partial y} \\ \frac{\partial(\bullet)}{\partial z} \end{Bmatrix} \tag{8.137}$$

Then, assuming the Jacobian to be non-singular, we have:

$$\frac{\partial(\bullet)}{\partial \mathbf{c}} = (\mathbf{J}(x))^{-1}\frac{\partial(\bullet)}{\partial x} \tag{8.138}$$

or, in expanded operator form for general three dimensions:

$$\begin{Bmatrix} \frac{\partial(\bullet)}{\partial x} \\ \frac{\partial(\bullet)}{\partial y} \\ \frac{\partial(\bullet)}{\partial z} \end{Bmatrix} = \begin{bmatrix} \frac{\partial r}{\partial x} & \frac{\partial s}{\partial x} & \frac{\partial t}{\partial x} \\ \frac{\partial r}{\partial y} & \frac{\partial s}{\partial y} & \frac{\partial t}{\partial y} \\ \frac{\partial r}{\partial z} & \frac{\partial s}{\partial z} & \frac{\partial t}{\partial z} \end{bmatrix} \begin{Bmatrix} \frac{\partial(\bullet)}{\partial x} \\ \frac{\partial(\bullet)}{\partial y} \\ \frac{\partial(\bullet)}{\partial z} \end{Bmatrix} \tag{8.139}$$

Thus, from equation (8.128) and equations (8.134) and (8.138), we get the derivatives of the displacement vector, $\mathbf{d}$, with respect to the generalized coordinate vector, $\mathbf{c}$, in terms of the naturalized coordinate vector, ξ, and the unknown displacement control vector, $\mathbf{q}$, as:

$$\frac{\partial \mathbf{d}(\mathbf{c}(\xi))}{\partial \mathbf{c}} = (\mathbf{J}(\xi))^{-1}\frac{\partial \mathbf{T}^e(\xi)}{\partial \xi}\mathbf{q}^{d_e} \tag{8.140}$$

Now, knowing the differential operator, $\mathcal{S}$, as in equation (8.135) for general three dimensions, and the displacement derivatives as described in equation (8.140), we can finally construct, by appropriate collection of the derivative components, the desired strain–displacement control matrix, $\mathcal{B}(\xi)$, such that:

$$\boxed{\epsilon(\mathbf{c}_e(\xi)) = \mathcal{B}(\xi)\,\mathbf{q}^{d_e}, \qquad \forall e \in \{1, 2, \ldots, N(\Delta)\}} \tag{8.141}$$

It may be noted that $\mathcal{B}(\xi)$ relates the strain distribution over the element to a finite number of nodal displacements of the element. Following the same methodology, we get for the virtual strain:

$$\mathcal{S}\,\mathbf{v} = \mathcal{B}(\xi)\,\mathbf{q}^{\mathrm{v}_e}, \qquad \forall e \in \{1, 2, \ldots, N(\Delta)\} \tag{8.142}$$

where $\mathbf{q}^{\mathrm{v}_e}$ is the corresponding virtual displacement control vector. It may be noted that, unlike the old methodology, $\mathcal{B}(\xi)$ relates the strain distribution over the element to a finite number of displacement controls defined as components of control vector, $\mathbf{q}$.

8.4.7.2 Virtual Work Principle: Linear Elasticity

We are now in a position to formulate the equilibrium equations arising out of the virtual work principle. Writing only the virtual work principle for a single element following the old methodology, we get for $e \in \{1, 2, \ldots, N(\Delta)\}$:

$$\begin{aligned}\mathbf{q}_0^{\mathrm{v}_{e}T}\left(\int_{\Omega^e} \mathcal{B}^T \mathcal{D}\mathcal{B}\, d\Omega^e\right)\mathbf{q}_0^{d_e} = \mathbf{q}_0^{\mathrm{v}_e^T}\{&\int_{\Omega^e} \mathbf{T}^{e^T}\mathbf{f}^B\, d\Omega^e + \int_{\Gamma_\sigma} \mathbf{T}^{e^T}\mathbf{f}^S\, d\Gamma + \mathbf{F}^C \\ &+ \int_{\Omega^e} \mathcal{B}^T \mathcal{D}\epsilon^I\, d\Omega^e - \int_{\Omega^e} \mathcal{B}^T \sigma^I\, d\Omega^e, \\ &- \int_{\Omega^e} \mathcal{B}^T \mathcal{D}\mathcal{S}\hat{\mathbf{d}}\, d\Omega^e\},\end{aligned} \tag{8.143}$$

8.4.8 Assembly (Global) Level Formulation

For equation (8.143) to be true for arbitrary $\mathbf{v}_0^{e_{Nod}}$, $\forall e \in \{1, 2, \ldots, N(\Delta)\}$, we must have by performing the appropriate global assembly-level summation over the elements:

$$\boxed{\mathbf{K}\,\mathbf{q}_0^d = \mathbf{F} - \hat{\mathbf{F}}} \tag{8.144}$$

where, $\mathbf{K}$, the stiffness control matrix for the entire body, Ω, is given by:

$$\boxed{\mathbf{K} \equiv \sum_e \mathbf{K}^e = \sum_e \int_{\Omega^e} (\mathcal{B}^e)^T \mathcal{D}^e\, \mathcal{B}^e\, d\Omega^e} \tag{8.145}$$

and $\mathbf{F}$, the total generalized nodal load control vector defined by:

$$\boxed{\mathbf{F} \equiv \mathbf{F}^B + \mathbf{F}^S + \mathbf{F}^C + \mathbf{F}^{\epsilon^I} + \mathbf{F}^{\sigma^I}} \tag{8.146}$$

where,

$\mathbf{F}^B$, the generalized nodal body load control vector is given by:

$$\mathbf{F}^B \equiv \left[\sum_e \int_{\Omega^e} (\mathbf{T}^e)^T \mathbf{T}^e\, d\Omega^e\right] \mathbf{q}_e^{fB} \tag{8.147}$$

$\mathbf{F}^S$, the generalized nodal surface traction load control vector is given by:

$$\mathbf{F}^S \equiv \left[\sum_e \int_{\Gamma^e} (\mathbf{T}^e)^T \mathbf{T}^e\, d\Gamma^e\right] \mathbf{q}_e^{fS} \tag{8.148}$$

$\mathbf{F}^C$, the generalized nodal concentrated load control vector at nodes,

$\mathbf{F}^{\sigma^I}$, the generalized nodal initial stress load control vector is given by:

$$\mathbf{F}^{\sigma^I} \equiv \sum_e \int_{\Omega^e} (\mathcal{B}^e)^T \sigma^{I^e}\, d\Omega^e \tag{8.149}$$

$\mathbf{F}^{\epsilon^I}$, the generalized nodal initial strain load control vector is given by:

$$\mathbf{F}^{\epsilon^I} \equiv \sum_e \int_{\Omega^e} (B^e)^T D \epsilon^{I^e} \, d\Omega^e \tag{8.150}$$

and $\hat{\mathbf{F}}$, the generalized nodal inhomogeneous load control vector is given by:

$$\hat{\mathbf{F}} \equiv \sum_e \int_{\Omega^e} (B^e)^T D (S \hat{\mathbf{d}}^e) \, d\Omega^e \tag{8.151}$$

and the finite element displacement solution, $\mathbf{d}^{FEM}$, is given by:

$$\mathbf{d}^{FEM} = \mathbf{d}_0^{FEM} + \hat{\mathbf{d}} \tag{8.152}$$

where $\mathbf{d}_0^{FEM}$ at any point in the body is obtained from the solved displacement control vectors, $\mathbf{q}_0^d$, by application of the transformation or mapping functions. The stiffness control matrix, $\mathbf{K}$, and the load control vectors can be further characterized for numerical evaluation by noting that the differential element, $d\Omega^e (= dx\, dy\, dz)$, on the element, Ω^e, is related to the differential element, $d\Omega^R (= dr\, ds\, dt)$, of the root element, Ω^R, as:

$$d\Omega^e = [\det \mathbf{J}^e(\xi)] \, d\Omega^R \tag{8.153}$$

where, $[\det \mathbf{J}^e(\xi)]$ is the determinant of the Jacobian. Then, we will have, for example:

$$\boxed{\mathbf{K} \equiv \sum_e \int_{\Omega^e} (B^e)^T D^e \; B^e \; [\det \mathbf{J}^e(\xi)] \, d\Omega^R} \tag{8.154}$$

Remarks

- In the derivation of the stiffness matrix and load vectors of the entire body, Ω, the summation of elements, of course, implied an appropriate rule of combination for the corresponding elemental quantities.
- We have assumed above that the body force vector, $\mathbf{f}^B$, and the surface force vector, $\mathbf{f}^S$, are also interpolated in terms of the control vectors.
- The determinant, $[\det \mathbf{J}^e(\xi)]$, of the Jacobian matrix, a geometric property, is most efficiently computed at the mesh generation phase without any regard to isoparametry.
- Note that the stiffness matrix and the load vector are given in terms of integrals that do not generally have closed form solutions; thus, these must be evaluated by numerical integration such as Gauss quadrature formulas, which we present at the end of the chapter.

Let us now get familiar with this new concept by working out a real yet simple example problem.

Example NFEM 1: Let us consider the extension of a bar as shown in Figure 8.7.

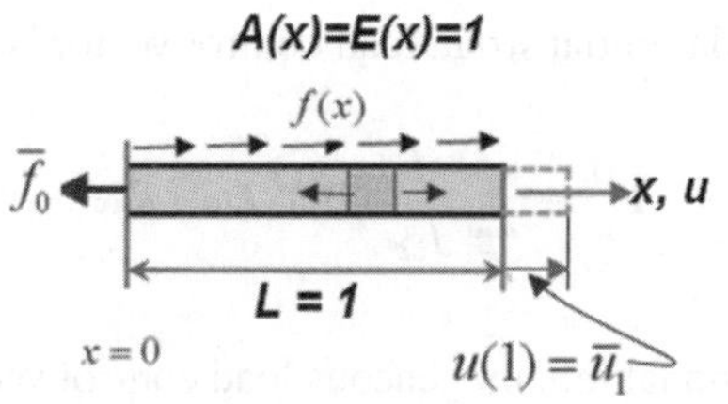

Figure 8.7 Example NFEM1.

8.4.8.1 Geometry and Material Property: Ω

For the geometry of the bar which is considered one-dimensional, the coordinate vector, c, contains only one component, x, that is, $\mathbf{c} = \{x\}$. Note that $\bar{\Omega} = \{x \in [0,1] \subset \mathbb{R}^1\}$ that is, $L = 1$. Let $A(x) = E(x) \equiv 1, \quad \forall x \in \Omega$.

8.4.8.2 Loading

Let us also suppose that the bar is subjected to an axial force along the bar described by a smooth function, $f^S(x)$, that is, $f \in C^\infty(\Omega)$.

8.4.8.3 Momentum Balance Equation and Boundary Conditions

Then, if the axial displacement is denoted by $u(x)$, we have the familiar governing differential equation of equilibrium as:

$$\begin{aligned} &((\underbrace{A(x)E(x)}_{=1}\ u,_x(x)),_x + f^S(x) = 0 \quad \text{in} \quad \Omega = (0, \mathrm{L} = 1) \\ &\text{or,} \qquad \frac{\partial^2 u(x)}{\partial x^2} + f^S(x) = 0, \quad \forall x \in \Omega \end{aligned} \tag{8.155}$$

Now, for the uniqueness of the solution, let us assume the following boundary conditions:

$$\begin{aligned} &\text{Essential boundary condition: } \boldsymbol{u}(L) = \bar{u}_1 \quad \text{on} \quad \Gamma_d = (\boldsymbol{x} = \boldsymbol{L} = \mathbf{1}) \\ &\text{Natural boundary condition: } (\underbrace{A(L)E(L)}_{=1}\,\boldsymbol{u},_x(0) = -\bar{f}_0 \quad \text{on} \quad \Gamma_\sigma = (x = 0) \end{aligned} \tag{8.156}$$

8.4.8.4 Virtual Work Principle and Functionals

As can be seen in equation (8.156), the highest degree of differentiation of the displacement-based equilibrium equation is $2(= 2m)$, that is, $m = 1$. Now, multiplying the displacement-based equilibrium equation given in the assumption by a virtual function $\hat{u}(x)$, with the property $\hat{u}(1) = 0$, that is, with homogeneous essential boundary condition satisfied, and integrating over the domain, we get:

$$\int_0^1 \frac{d^2u(x)}{dx^2}\, \hat{u}(x)\, dx + \int_0^1 f^B(x)\, \hat{u}(x)\, dx = 0 \tag{8.157}$$

Now, integrating by parts, and applying the boundary conditions and the property of the virtual function, and rearranging, we get:

$$\boxed{\int_0^1 \left(\frac{du(x)}{dx}\right)\left(\frac{d\hat{u}(x)}{dx}\right) dx - \left(\int_0^1 f^B(x)\,\hat{u}(x)\,dx - \overline{f}_0\,\hat{u}(0)\right) = 0} \tag{8.158}$$

Thus, if we define:

$$\boxed{\begin{aligned}
\mathcal{G}(u,\hat{u}) &= \mathcal{Q}(u,\hat{u}) - \mathcal{L}(\hat{u}) && : \text{Virtual work functional}\\
\mathcal{Q}(u,\hat{u}) &= \int_0^1 \left(\frac{du(x)}{dx}\right)\left(\frac{d\hat{u}(x)}{dx}\right) dx && : \text{Internal virtual work functional}\\
\mathcal{L}(\hat{u}) &= \int_0^1 f^B(x)\,\hat{u}(x)\,dx - \overline{f}_0\,\hat{u}(0) && : \text{External virtual work functional}
\end{aligned}} \tag{8.159}$$

then, we have the statement of the virtual work principle for the problem:

$$\boxed{\mathcal{G}(u,\hat{u}) = 0, \qquad \forall \hat{u} \in \left\{\hat{u} \in \mathcal{H}^1(\Omega) \mid \hat{u}(1) = 0\right\}} \tag{8.160}$$

Remark

- $\mathcal{Q}(u,\hat{u})$ is symmetric in its first and second arguments.

8.4.8.5 Minimum Potential Energy Principle: $\mathcal{U}(u)$ and $\mathcal{W}(u)$

Let us compute the first variation of the virtual work functional through the definition of the Gateaux differential:

$$\boxed{\begin{aligned}
\delta\mathcal{G}(u,\tilde{u};\hat{u}) &= \lim_{\varepsilon\to 0} \frac{d}{d\varepsilon}\mathcal{G}(u+\varepsilon\tilde{u},\hat{u})\\
&= \int_0^1 \left(\frac{d\tilde{u}(x)}{dx}\right)\left(\frac{d\hat{u}(x)}{dx}\right) dx\\
&= \mathcal{Q}(\tilde{u},\hat{u})\\
&= \text{symmetric in } \tilde{u} \text{ and } \hat{u}
\end{aligned}} \tag{8.161}$$

Thus, an energy principle exists. Therefore, by the Vainberg Principle, the energy functional, $\Pi^{Tot}(u)$, known as the total potential energy functional is given by:

$$\boxed{\begin{aligned}
\Pi^{Tot}(u) &= \int_0^1 \mathcal{G}(\rho u, u)\,d\rho\\
&= \frac{1}{2}\int_0^1 \left(\frac{du(x)}{dx}\right)^2 dx - \left(\int_0^1 f^B(x)\,u(x)\,dx - \overline{f}_0\,u(0)\right)\\
&= \frac{1}{2}\mathcal{Q}(u,u) - \mathcal{L}(u)
\end{aligned}} \tag{8.162}$$

Accordingly, for the problem at hand, we can identify the strain energy functional, $\mathcal{U}(u)$, and the external work functional, $\mathcal{W}(u)$, as the first and second term, respectively, of the total potential energy functional, in equation (8.162), as:

$$\boxed{\begin{aligned}\mathcal{U}(u) &= \frac{1}{2}\mathcal{Q}(u,u) = \frac{1}{2}\int_0^1 \left(\frac{du(x)}{dx}\right)^2 dx\\ \mathcal{W}(u) &= \mathcal{L}(u) = \textstyle\int_0^1 f^B(x)\, u(x)\, dx - \overline{f}_0\, u(0)\end{aligned}} \tag{8.163}$$

Remark

- The first variation of the total potential energy functional is equal to the virtual work functional, that is,

$$\boxed{\delta\Pi^{Tot}(u,\hat{u}) = \mathcal{G}(u,\hat{u}) = 0} \tag{8.164}$$

- The second variation of the total potential energy functional is:

$$\boxed{\begin{aligned}\delta^2\Pi^{Tot}\left(u,\hat{u}\right) &= \lim_{\varepsilon\to 0}\frac{d^2}{d\varepsilon^2}\Pi^{Tot}\left(u+\varepsilon\hat{u}\right) = \int_0^1\left(\frac{d\hat{u}(x)}{dx}\right)^2 dx\\ &= \mathcal{Q}(\hat{u},\hat{u}) > 0\end{aligned}} \tag{8.165}$$

- So, the principle of minimum total potential functional energy exists for the problem.

8.4.8.6 The Essential and Natural Boundary Conditions

The essential boundary condition corresponding to the derivative $\leq\ 0\,(= m-1)$ is given by $u(1) = \overline{u}_1$. The solution set does not need to satisfy the other boundary condition, $-\frac{du(0)}{dx} = \overline{f}(0)$, as the displacement function contains a derivative of degree $1 >\ 0\,(= m-1)$. Thus, this boundary condition acts as a natural boundary condition.

8.4.8.7 The Energy Space and Norm: $E(\Omega)$ and $||u||_{energy}$

Since the highest derivative in the integral expression for virtual work principle is 1, that is, $m = 1$, we may define the ***energy space***, a normed linear space for the problem from where all the displacement functions, real or virtual, may be sought, as:

$$\boxed{E(\Omega) = \left\{u \in \mathcal{H}^1(\Omega) \mid ||u||_{energy} < \infty\right\}} \tag{8.166}$$

with energy norm, $||u||_{energy}$, defined by the strain energy functional, $\mathcal{U}(u)$, as:

$$\boxed{||u||_{energy} \overset{def}{=} \sqrt{\mathcal{U}(u)} = \frac{1}{2}\mathcal{Q}(u,u) = \frac{1}{2}\int_0^1\left(\frac{du(x)}{dx}\right)^2 dx} \tag{8.167}$$

For this problem, with symmetric Q, the energy space is easily seen as an inner product space with the norm induced by the following, energy inner product $(\cdot\ ,\ \cdot)_Q$:

$$\boxed{(u\ ,\ v)_Q \stackrel{def}{=} \int_0^1 \left(\frac{du}{dx}\right)\left(\frac{dv}{dx}\right)\, dx, \forall u, v \in E(\Omega)} \tag{8.168}$$

8.4.8.8 The Solution Set and Trial and Virtual Subspace: $E_{Ess}^{Sol}(\Omega)$ and $E_0^{Sol}(\Omega)$

Thus, the solution set, may be defined as:

$$\boxed{E_{Ess}^{Sol}(\Omega) = \left\{u \in E(\Omega) \mid u(1) = \overline{u}_1\right\}} \tag{8.169}$$

Clearly, $E_{Ess}^{Sol}(\Omega)$ is not a subspace – in fact, not even a vector space. It is a linear manifold. We apply the translation theorem of linear manifolds to represent u as the direct sum of u^0 and any arbitrary $u^\star \in E_{Ess}^{Sol}(\Omega)$, that is,

$$\boxed{\begin{array}{ll} & u(x) = u^0(x) \oplus u^\star(x) \\ \text{where} & u^0 \in E_0^{Sol}(\Omega) \quad \text{with} \quad E_0^{Sol}(\Omega) = \left\{u^0 \in E_{Ess}^{Sol}(\Omega) \mid u^0(1) = 0\right\} \end{array}} \tag{8.170}$$

Remark

- $E_0^{Sol}(\Omega)$ contains the displacement functions that satisfy essential boundary condition in its homogeneous state. However, this happens to be the precise definition of the virtual space according to the virtual work principle. Thus, the direct sum representation allows us to make the trial space and the virtual space one and the same. In the Rayleigh–Ritz–Galerkin finite element formulation, this in turn allows us to choose the trial and virtual finite-dimensional subspace to be the same. Of course, for the final solution, we still have to determine the arbitrary $u^\star$ as a function to satisfy the inhomogeneous boundary condition; this is normally done in a way that is trivially simple and numerically efficient, as we shall see shortly.

For convenience, let us now summarize the necessary entities for the given infinite-dimensional problem: the energy inner product, the strain energy, the energy space, the solution set which is the linear variety of the trial space and the virtual space, as:

$$\boxed{\begin{array}{ll} \text{Energy inner product:} & Q(u, \mathrm{v}) \equiv \int_0^1 (\frac{\partial u}{\partial x})(\frac{\partial \hat{u}}{\partial x})\, dx, \quad \forall u, \hat{u} \in \mathcal{H}^1(\Omega) \\ \text{Strain energy:} & \mathcal{U}(u) \equiv \frac{1}{2} Q(u, u), \\ \text{Energy space:} & ||u||_e \equiv \sqrt{\mathcal{U}(u)}, \\ & E(\Omega) \equiv \{u \in \mathcal{H}^1(\Omega) |\ ||u||_e < \infty\}, \\ \text{Solution set or linear variety:} & E_{Ess}^{Sol}(\Omega) \equiv \{u \in E(\Omega) |\ u(1) = \bar{u}_1,\ ||u||_e < \infty\}, \\ \text{Trial subspace:} & U_0 \equiv \{u | u(1) = 0\} \subset E_{Ess}^{Sol}, \\ \text{Virtual subspace:} & V_0 \equiv U_0 \end{array}} \tag{8.171}$$

8.4.8.9 Finite Element Mesh: $\Delta = \bigcup_{e=1,2,\ldots,N} \Omega^e,\ \Omega^R, \mathcal{T}$

For this problem, we choose $N = 1$, that is, one element only. Thus, the geometric element, Ω^e, coincides with the entire geometry, Ω. Since the bar is straight, and the length is unity, the root element, Ω^R, and the physical element, Ω^e, coincide. We denote the root element natural coordinates as ξ. Clearly, $\xi = x$, that is, $\xi \in [0, 1]$. Also, the bar being a straight element requires only linear Bernstein polynomials for geometry representation. In other words, the transformation, $\mathcal{T}$, relating the root element to the physical element is given by:

$$\boxed{\begin{aligned} x(\xi) &= \overline{\mathbf{T}}(\xi)\mathbf{q}^x && : \text{in matrix form} \\ \overline{\mathbf{T}}(\xi) &= \left[B_0^1(\xi)\ \ B_1^1(\xi)\right] \\ \mathbf{q}^x &= \left\{q_0^x\ \ q_1^x\right\}^T \\ x(\xi) &= (1-\xi)\, q_0^x + \xi\, q_1^x : && \text{in component form} \end{aligned}} \tag{8.172}$$

For the problem at hand, clearly, $q_0^x = 0$ and $q_1^x = 1$. Notice, that for a trivial geometry such as this, we have, by inspection, $x = \xi$. However, we choose to express this relation in the above elaborate form for the sake of uniformity that aids our subsequent exposition.

The transformation set trivially contains one element, that is, $\mathcal{T} = \{\overline{\mathbf{T}}\}$.

8.4.8.10 Finite Element Trial and Virtual Subspace: $S^P\left(\Omega,\Delta,\mathcal{T},C\right) \subset E_0^{Sol}(\Omega)$

The forcing function (we can think of it as a distributed surface traction), $f^S(\xi)$, may be, for our example problem, an identically zero function or at a best linear continuous function. We will seek our trial and virtual displacements from among the set $\mathbb{S}^2$ of piecewise quadratic functions ($p = 2$) which are continuous at all nodes and satisfy the essential boundary condition at its homogeneous state. Now, each quadratic function describes a parabola for each element, requiring three free parameters for its definition. However, the continuity constraint at each physical end node reduces the number of free control vertices of each parabola ending in the node, to two. Thus, for the problem, the dimension of the finite element trial and virtual subspace is $2N$, where N is the number of elements. For the above choice of one finite element, the number of free parameters should be $2N_e = 2$.

8.4.8.11 Finite Element Basis Functions and Displacement Control Vector: $B_i^2(\xi)$ and q_i^u

For a c-type finite element method, the local or elemental basis functions are Bernstein polynomials with Bezier control polygon. For quadratic functions, these become the quadratic Bernstein polynomials, $\left\{B_i^2(\xi), i = 0, 1, 2\right\}$. Now, the parameters serving as the coefficients of linear combination of these basis functions are the as-yet-undetermined Bezier controls, $\left\{q_i^u, i = 0, 1, 2\right\}$. In other words, any displacement function, $u(\xi)$, is given by a linear combination of this finite set of basis functions as:

$$\boxed{u(\xi) = \sum_{i=0}^{2} B_i^2(\xi)\, q_i^u} \tag{8.173}$$

Remark

- The actual number of physical nodes is two, with q_0^u and q_2^u serving as the two end nodes. Knowing that $B_0^2(0) = 1, B_1^2(0) = B_2^2(0) = 0$, we have $u(0) = q_0^u$. Accordingly, for satisfaction of the homogeneous counterpart of the essential boundary conditions, we only need to set $q_0^u = 0$. As already explained, we allow the displacement trial function to satisfy the essential boundary conditions only in their homogeneous state, so that both the displacement trial subspace and the virtual subspace (functions of which, by choice, must vanish wherever essential boundary conditions are prescribed) can be one and the same with same basis functions spanning the subspace. This, in turn, allows generation of a numerically efficient symmetric stiffness matrix, etc. Thus, the as-yet-unknown parameters are q_1^u and q_2^u.

8.4.8.12 Coerced Isoparametry: $u^0(\xi)$, $\hat{u}^0(\xi)$, and q^x, q^u and $\hat{\mathbf{q}}^u$

Recall that the bar required only linear Bernstein polynomials, that is, of degree one, for geometry representation, that is, mesh generation. Recall also that the displacement solution function is already indicated to be quadratic, that is, of degree two. Thus, we may, although not necessary for computation, conceptually coerce the geometry transformation up by one application of the degree elevation technique. This then makes the geometry representation – that is, the transformation from root to physical element – to be:

$$\boxed{\begin{aligned} &x(\xi) = \mathbf{T}(\xi)\mathbf{q}^x \\ &\mathbf{T}(\xi) = \left[B_0^2(\xi)\ B_1^2(\xi)\ B_2^2(\xi)\right] \\ &\mathbf{q}^x = \left\{0 \tfrac{1}{2} 1\right\}^T \end{aligned}} \tag{8.174}$$

Most importantly, note that coercing did not introduce any new physical node.

The element trial displacement function (see Figure 8.8 for control vertices) can be given isoparametrically as:

$$\boxed{\begin{aligned} &u^0(\xi) = \mathbf{T}(\xi)\mathbf{q}^u \\ &\mathbf{q}^u = \left\{q_0^u\ q_1^u \quad q_2^u \equiv 0\right\}^T \end{aligned}} \tag{8.175}$$

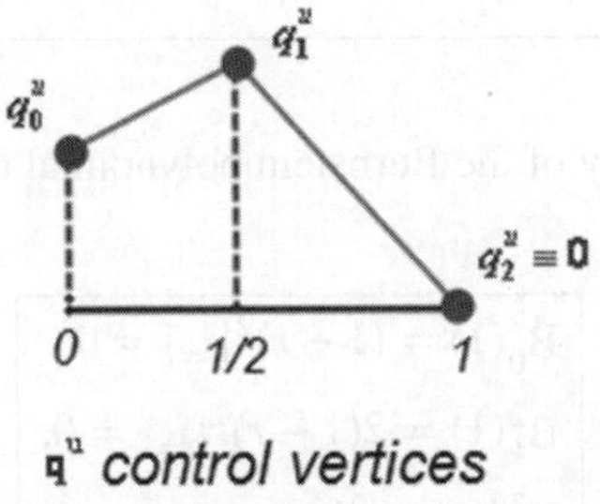

Figure 8.8 Displacement control vertices.

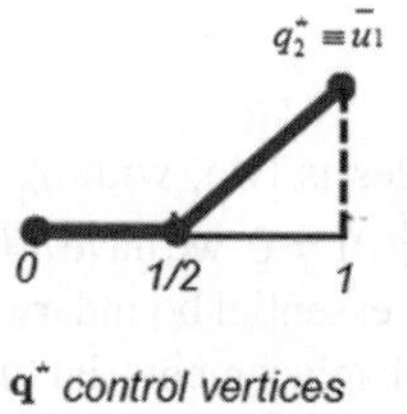

Figure 8.9 Trivial solution for control vertices.

The virtual displacement function, likewise, can be given by:

$$\boxed{\begin{aligned} \widehat{u}^0(\xi) &= \mathbf{T}(\xi)\widehat{\mathbf{q}}^u \\ \widehat{\mathbf{q}}^u &= \left\{\widehat{q}_0^u \; \widehat{q}_1^u \quad \widehat{q}_2^u \equiv 0\right\}^T \end{aligned}} \tag{8.176}$$

8.4.8.13 Inhomogeneous Essential Condition: $u^0 \oplus u^\star$

We may construct the trivial solution, $u^\star(\xi)$, due to the inhomogeneous boundary condition (see Figure 8.9 for trivial control vertices) as:

$$\boxed{\begin{aligned} u^\star(\xi) &= \mathbf{T}(\xi)\, \mathbf{q}^\star \\ \mathbf{q}^\star &= \left\{0 \;\; 0 \quad q_2^\star \equiv \bar{u}_1\right\}^T \end{aligned}} \tag{8.177}$$

Considering description of $u^0(\xi)$ in equation (8.175) and $\widehat{u}^0(\xi)$ in equation (8.177), we see, from equation (8.170), that $u^{FEM}(\xi)$, indeed, belongs to $E_{Ess}^{Sol}(\Omega)$, as follows:

$$\boxed{\begin{aligned} u^{FEM}(1) &= u^0(1) + u^\star(1) \\ &= \{0 \;\; 0 \;\; 1\}\begin{Bmatrix} q_0^u \\ q_1^u \\ 0 \end{Bmatrix} + \{0 \;\; 0 \;\; 1\}\begin{Bmatrix} 0 \\ 0 \\ \bar{u}_1 \end{Bmatrix} \\ &= \bar{u}_1 \end{aligned}} \tag{8.178}$$

where, we have used the property of the Bernstein polynomial of degree two:

$$\boxed{\begin{aligned} &\mathrm{B}_0^2(1) = (1-r)^2|_{r=1} = 0, \\ &\mathrm{B}_1^2(1) = 2(1-r)r|_{r=1} = 0, \\ &\mathrm{B}_2^2(1) = r^2|_{r=1} = 1, \end{aligned}} \tag{8.179}$$

8.4.8.14 Strain–displacement Control Matrix: $\mathcal{B}(\xi)$

The strain–displacement control matrix is defined by the finite element strain and displacement as:

$$\begin{aligned} \int^{FEM}(\xi) &= \frac{du^{FEM}(\xi)}{d\xi} = \mathcal{B}(\xi)\, \mathbf{q}^u \\ \mathcal{B}(\xi) &= \frac{d\mathbf{T}(\xi)}{d\xi} \end{aligned} \tag{8.180}$$

8.4.8.15 c-type Virtual Work Principle and Equilibrium Equation

Now, using definitions of u^0, $\hat{u}^0$ and $\mathcal{Q}(u^0, \hat{u}^0)$ in the equation of the virtual work principle, we get, in compact form:

$$\left(\hat{\mathbf{q}}^u\right)^T \mathbf{K}\, \mathbf{q}^u = \left(\hat{\mathbf{q}}^u\right)^T \left[\mathbf{F}^S \quad -\mathbf{F}^0 \quad -\mathbf{F}^a\right], \quad \forall \hat{\mathbf{q}}^u \tag{8.181}$$

where, $\mathbf{K}$ is the (3 × 3) ***unconstrained*** or ***full singular stiffness control matrix*** given by:

$$\begin{aligned} \mathbf{K} &= \int_0^1 \mathcal{B}(\xi)^T\, \mathcal{B}(\xi)\, d\xi = \int_0^1 \left(\frac{d\mathbf{T}(\xi)}{d\xi}\right)^T \left(\frac{d\mathbf{T}(\xi)}{d\xi}\right) d\xi \\ &= \int_0^1 \begin{Bmatrix} B^2_{0,\xi} \\ B^2_{1,\xi} \\ B^2_{2,\xi} \end{Bmatrix} \left\{B^2_{0,\xi} \quad B^2_{1,\xi} \quad B^2_{2,\xi}\right\} d\xi = \frac{2}{3}\begin{bmatrix} 2 & -1 & -1 \\ -1 & 2 & -1 \\ -1 & -1 & 2 \end{bmatrix} \end{aligned} \tag{8.182}$$

and the load control vectors are:

$$\begin{aligned} \mathbf{F}^S &= \int_0^1 T(\xi)^T f^S(\xi)\, d\xi \\ \mathbf{F}^0 &= T(0)^T \bar{f}^0 \\ \mathbf{F}^\star &= \left[\int_0^1 \mathcal{B}(\xi)^T\, \mathcal{B}(\xi)\, d\xi\right] \mathbf{q}^\star = \mathbf{K}\, \mathbf{q}^\star = \frac{2}{3}\begin{Bmatrix} -\bar{u}_1 \\ -\bar{u}_1 \\ 2\bar{u}_1 \end{Bmatrix} \end{aligned} \tag{8.183}$$

Finally, noting that the third element for both the trial and virtual displacement vectors to be identically zero, and that the virtual work principle is true for any ***kinematically admissible virtual displacement***, we can write the ***constrained finite element equation*** of equilibrium from equation (8.181) in expanded matrix form as:

$$\begin{aligned} &\frac{2}{3}\begin{bmatrix} 2 & -1 \\ -1 & 2 \end{bmatrix} \begin{Bmatrix} q^u_0 \\ q^u_1 \end{Bmatrix} = \begin{Bmatrix} F_1 \\ F_2 \end{Bmatrix} \\ &\text{where, } \begin{Bmatrix} F_1 \\ F_2 \end{Bmatrix} = \begin{Bmatrix} \int_0^1 f^S(\xi)\,(1-\xi)^2\, d\xi + \bar{f}^0 + \frac{2}{3}\bar{u}_1 \\ 2\int_0^1 f^S(\xi)\,\xi(1-\xi)\, d\xi + \frac{2}{3}\bar{u}_1 \end{Bmatrix} \end{aligned} \tag{8.184}$$

The equation (8.184) can be solved for the desired Bezier control vertices only if the body force, $f^B(\xi)$, is given explicitly. Let us, then, consider two cases as under:

Case I: $f^S(\xi) \equiv 0$, that is, homogeneous differential equation:
In this case, the load vector, the right-hand side of equation (8.184), becomes:

$$\boxed{\begin{Bmatrix} F_1 \\ F_2 \end{Bmatrix} = \begin{Bmatrix} \bar{f}^0 + \frac{2}{3}\bar{u}_1 \\ \frac{2}{3}\bar{u}_1 \end{Bmatrix}} \tag{8.185}$$

Now, solving equation (8.184) with the load vector given by equation (8.185), we get:

$$\boxed{\begin{Bmatrix} q_0^u \\ q_1^u \end{Bmatrix} = \frac{1}{2}\begin{Bmatrix} 2\bar{f}^0 + 2\bar{u}_1 \\ \bar{f}^0 + 2\bar{u}_1 \end{Bmatrix}} \tag{8.186}$$

Then, using equation (8.186) for the known Bezier controls for $u^0(\xi)$ as in equation (8.175) with $u^\star(\xi)$ as in equation (8.177) and $\xi = x$ in the expression for the finite element displacement, we finally get the desired displacement solution as:

$$\boxed{\begin{aligned} u^{FEM}(x) &= (1-x)^2\, q_0^u + 2x(1-x)\, q_1^u + x^2\bar{u}_1 \\ &= \tfrac{1}{2}(2\bar{f}_0 + 2\bar{u}_1)(1-x)^2 + (\bar{f}_0 + 2\bar{u}_1)x(1-x) + x^2\bar{u}_1 \\ &= (1-x)\bar{f}_0 + \bar{u}_1 \\ &= \text{exact solution!} \end{aligned}} \tag{8.187}$$

The equation (8.187) represents the exact homogeneous part of the solution that may be simply obtained by a mere sequence of integrations of the equilibrium equation.

Case II: $f^S(\xi) \equiv \bar{f}^S$ = constant, that is, inhomogeneous differential equation:
In this case, we show only the additional part of the solution, that is, the particular solution of the differential equation due to this inhomogeneous term. The homogeneous part of the solution remains the same as expressed by equation (8.187). The generalized load control vector due to this is obtained from the integral expressions in equation (8.184) as:

$$\boxed{\begin{Bmatrix} F_1^{part} \\ F_2^{part} \end{Bmatrix} = \frac{1}{3}\bar{f}^S\begin{Bmatrix} 1 \\ 1 \end{Bmatrix}} \tag{8.188}$$

Now, solving equation (8.184) for this particular load vector given by equation (8.188), we get:

$$\boxed{\begin{Bmatrix} {}^{part}q_0^u \\ {}^{part}q_1^u \end{Bmatrix} = \frac{1}{2}\bar{f}^S\begin{Bmatrix} 1 \\ 1 \end{Bmatrix}} \tag{8.189}$$

Then, we finally get the desired particular part of the displacement solution as:

$$\boxed{\begin{aligned} u_{\text{part}}^{FEM}(x) &= (1-x)^2\, q_0^{u^{\text{part}}} + 2x(1-x)\, q_1^{u^{\text{part}}} \\ &= \tfrac{1}{2}\bar{f}^S\{(1-x)^2 + 2x(1-x)\} = \frac{1}{2}(1-x^2)\bar{f}^S \\ &= \text{exact solution!} \end{aligned}} \tag{8.190}$$

The equation (8.190) represents the exact particular solution of the differential equation; this completes the presentation of the current example.

8.4.9 Linear Elasto-dynamic Problems

For linear elasto-dynamic problems, we can continue as in the old finite element method presented in the appendix. The only thing that must be kept in mind, however, is that, unlike the old method, the generalized displacement consists of Bezier displacement control vectors.

Let, ρ^e be the effective mass density, and ς^e the effective damping coefficients of the element, Ω^e. Then, the generalized nodal mass, $\mathbf{M}$, and damping, $\mathbf{C}$, matrices can be similarly obtained if the acceleration and velocity distributions are assumed to be of the same isoparametric form as the coerced geometry and the displacement, respectively. They are formally defined as follows:

$$\mathbf{M} \equiv \left[\sum_e \int_{\Omega^e} \rho^e (\mathbf{T}^e)^T \mathbf{T}^e \, d\Omega^e \right] \mathbf{q}^{de} \tag{8.191}$$

and

$$\mathbf{C} \equiv \left[\sum_e \int_{\Omega^e} \varsigma^e (\mathbf{T}^e)^T \mathbf{T}^e \, d\Omega^e \right] \mathbf{q}^{de} \tag{8.192}$$

In this case, the equilibrium equations become:

$$\boxed{\mathbf{M}\,\ddot{\mathbf{q}}^d + \mathbf{C}\,\dot{\mathbf{q}}^d + \mathbf{K}\,\mathbf{q}^d = \mathbf{F}} \tag{8.193}$$

where, $\dot{\mathbf{q}}^d$ is the generalized nodal velocity control vector and $\ddot{\mathbf{q}}^d$ is the generalized nodal acceleration control vector, and all other terms are as described earlier, except for $\mathbf{F}^B$, the generalized nodal body load vector as follows:

Remark

- It may be seen that for our unknown generalized nodal vectors, we have $\mathbf{q^d}$, the Bezier displacement controls which involve only the corner nodes of the old finite element methodologies. Furthermore, we have no need to include the ad hoc rotational degrees of freedom usually associated with the plate and shell analysis, irrespective of the kinematic relationship that makes these degrees of freedom necessary in the old finite element methodologies.

In the derivation of the stiffness control matrix and load control vectors of the entire body, Ω, the summation of elements, of course, implied an appropriate rule of combination for the corresponding elemental quantities. In this section, we explain these rules and thus introduce the new methodology for inter-element smoothness or continuity control. This phase of finite element formulation is widely known as the ***global assembly*** process. For practical applications to structural or continuum mechanics problems, we generally need to consider only C^0, C^1 and C^2 continuity across the element boundaries, which we discuss here.

8.4.10 Heat Conduction and Potential Flow Problems

The Rayleigh–Ritz–Galerkin method, and hence finite element methods, apply equally successfully to heat conduction and other potential flow problems. Obviously, our theory can be applied equally, with the exception that the state variables are represented as linear combinations of Bernstein polynomials with generalized unknown nodal coefficients as Bezier control vertices. For example, let us consider the heat conduction analysis where the state variable is the one-dimensional, steady-state temperature distribution, θ. We can apply everything discussed so far for linear elasticity to the heat conduction analysis, when we redefine various terms as follows: Ω the finite continuum or body, Γ, = the boundary of Ω, with $\Gamma = \Gamma_\theta \cup \Gamma_H \cup \Gamma_C$, where Γ_θ, the boundary where temperature is prescribed, Γ_H, the boundary where heat flux is prescribed, Γ_C, the boundary where convective conditions are prescribed, $\bar{\Omega}$ the closure of Ω that is, $\bar{\Omega} = \Omega \cup \Gamma$; $\mathbf{d} \stackrel{\Delta}{=} \boldsymbol{\theta}$, the column vector of the nodal generalized temperatures defining the temperature state of $\bar{\Omega}$, under imposed boundary conditions: $\mathbf{f} = (f_x, f_y, f_z)^T$, the imposed heat flow inputs along coordinate directions at any point, $\mathbf{c} \in \Gamma$; $\mathbf{f}$ may consist of any combination of: $\mathbf{f}^B = (f_x^B, f_y^B, f_z^B)^T$, the body heat transfer per unit volume generated (or absorbed) by the distributed heat source (or sink) at any, $\mathbf{c} \in \Omega$, the surface heat transfer per unit area at any $\mathbf{c} \in \Gamma$, $\mathbf{f}^C = (f_x^C, f_y^C, f_z^C)^T$, the concentrated heat flow input at any $\mathbf{c} \in \Gamma$; $q^H = \{q_x^H,\ q_y^H,\ q_z^H\}^T$ = the heat flux per unit volume column vector; $\mathcal{S}$, the heat–temperature differential matrix (operator), $\mathcal{D}$, the heat conductivity matrix and $\mathcal{B}$, the heat–temperature control matrix.

Now, the temperature distribution is expressed as:

$$\boxed{\theta(r,s,t) = \sum_{i=0}^{L} \sum_{j=0}^{M} \sum_{k=0}^{N} \widehat{B}_i^L(r)\widehat{B}_j^M(s)\widehat{B}_k^N(t)\ \widehat{\mathbf{q}}_{ijk}^{\theta_e}} \tag{8.194}$$

where $\widehat{\mathbf{q}}_{ijk}^{\theta_e}$ is the Bezier generalized nodal temperature control vertex at location *ijk*, and all other terms are as defined in linear elasticity. Then, applying the appropriate virtual work principle described in the previous sections, and the methodology used as in the linear elasticity, we get the global heat balance equation in the matrix form as:

$$\mathbf{K}\ \mathbf{q}_0^\theta = \mathbf{F} - \hat{\mathbf{F}} \tag{8.195}$$

where, $\mathbf{K}$, the conductivity control matrix for the entire body, Ω, is given by:

$$\mathbf{K} \equiv \sum_e \mathbf{K}^e = \sum_e \int_{\Omega^e} (\mathcal{B}^e)^T \mathcal{D}^e\ \mathcal{B}^e\ d\Omega^e \tag{8.196}$$

and $\mathbf{F}$, the total generalized nodal heat flow control vector is defined by:

$$\mathbf{F} \equiv \mathbf{F}^B + \mathbf{F}^S + \mathbf{F}^C \tag{8.197}$$

where, $\mathbf{F}^B$, the generalized nodal body heat flow control vector is given by:

$$\mathbf{F}^B \equiv \left[\sum_e \int_{\Omega^e} (\mathbf{T}^e)^T \mathbf{T}^e \, d\Omega^e \right] \mathbf{q}_e^{fB} \tag{8.198}$$

$\mathbf{F}^S$, the generalized nodal surface heat flow control vector is given by:

$$\mathbf{F}^S \equiv \left[\sum_e \int_{\Gamma^e} (\mathbf{T}^e)^T \mathbf{T}^e \, d\Gamma^e \right] \mathbf{q}_e^{fS} \tag{8.199}$$

$\mathbf{F}^C$, the generalized nodal concentrated heat flow control vector at nodes, and finally, $\hat{\mathbf{F}}$, the generalized nodal inhomogeneous heat flow control vector is given by:

$$\hat{\mathbf{F}} \equiv \sum_e \int_{\Omega^e} (B^e)^T D(calS\hat{\theta}^e) \, d\Omega^e \tag{8.200}$$

Finally, the finite element temperature solution, θ^{FEM}, is given by:

$$\theta^{FEM} = \theta_0^{FEM} + \hat{\theta} \tag{8.201}$$

where, θ_0^{FEM} at any point in the body is obtained from the solved temperature control vectors, $\mathbf{q}_0^{\theta}$, by application of the transformation or mapping functions.

Remark

- The convective heat boundary condition, like the springs in the elasticity situation, will contribute to both the global conductivity matrix and the heat flux parts of the equation (8.195).
- We have only considered the steady state solution of the heat transfer problem; the transient solution follows the theory presented earlier.
- Similar finite element solutions using the Bernstein elemental basis function and the Bezier control vertices representation of the state variables can be obtained for other field problems such as fluid flow, seepage and general flow problems.

8.4.11 Inter-element Continuity (IEC): Bezier to B-spline

A general mesh representation of a real physical continuum or a structural system can be expected to involve complex shapes. While Bezier representation amazingly mimics the behavior of a given shape, from the standpoint of the finite element philosophy which demands assemblage of simple elements, and from the mesh generation viewpoint, interpolation with Bezier elements with a very high degree is prohibitive. In such cases, the most natural choice is to introduce the concept of what may be called a composite ***Bezier super-element*** which is a composition of piecewise Bezier parametric elements with desired continuity at the junction with the neighboring elements. In other words, the finite element mesh defined by equation (8.116) can be imagined as one composite Bezier super-element. For finite element analysis, we will generally be interested in at most C^1 continuity for, say, plate bending and shell problems involving fourth-order differential equations. However, higher continuity such as C^2 will accelerate convergence wherever applicable. For elaboration and details on ideas relating to inter-element continuity, we refer to Chapter 5 on curves. For example, with the one-dimensional situation, to achieve these goals constructively in finite element analysis, we will need to recast the inter-element continuity equations, as follows.

8.4.11.1 Linear Elements and C^0 Continuity

For connection between linear elements, only C^0 continuity is meaningful, which is trivially achieved by renaming the junction displacement control vectors of each element so that they are one and the same.

8.4.11.2 Quadratic Elements and C^1 Continuity

For connection between quadratic elements, additionally C^1 continuity is meaningful. For example, let there be two elements, element 1 on the left and element 2 on the right, that are defined by the two displacement control vector sets, $\{q_0^d, q_1^d, q_2^d\}$ and $\{q_2^d, q_3^d, q_4^d\}$, respectively. Notice that the C^0 continuity between the two is already satisfied. Let h_1 and h_2 be the corresponding lengths measured in global parameter of the root elements; notice that if the elements are straight, these quantities represent their respective lengths. Now the C^1 continuity is given by:

$$\boxed{\begin{aligned} & q_2^d = \beta\, q_1^d + \alpha\, q_3^d, \\ \text{where,} \\ & \alpha \equiv \frac{h_1}{h_1 + h_2}, \text{ and, } \beta \equiv \frac{h_2}{h_1 + h_2} \end{aligned}} \tag{8.202}$$

For $\mathbf{C}^1$ continuity, we introduce the de Boor polygon vertices as:

$$\boxed{\mathbf{s}^{d_1} \equiv \left\{\mathbf{s}_{-1}^{d_1} \equiv q_0^d,\ \mathbf{s}_0^{d_1} \equiv q_1^d,\ \mathbf{s}_1^{d_1} \equiv q_3^d,\ \mathbf{s}_2^{d_1} \equiv q_4^d\right\}^T} \tag{8.203}$$

It can be shown that with de Boor vertices constructed as above, and the equation (8.202) for C^1 continuity at q_2^d particularized as:

$$q_2^d = \beta\, \mathbf{s}_0^{d_1} + \alpha\, \mathbf{s}_1^{d_1} \tag{8.204}$$

guarantees that two quadratic Bezier displacement elements are converted to one quadratic B-spline displacement element. In vector-matrix form, the constraint defined by equation (8.204) along with the de Boor vector definition (8.203) can be expressed as:

$$\underbrace{\begin{Bmatrix} q_0^d \\ q_1^d \\ q_2^d \\ q_3^d \\ q_4^d \end{Bmatrix}}_{\substack{\text{Bezier–Bernstein} \\ \text{control vectors}}} = \begin{bmatrix} 1 & 0 & 0 & 0 \\ 0 & 1 & 0 & 0 \\ 0 & \beta & \alpha & 0 \\ 0 & 0 & 1 & 0 \\ 0 & 0 & 0 & 1 \end{bmatrix} \underbrace{\begin{Bmatrix} \mathbf{s}_{-1}^{d_1} \\ \mathbf{s}_0^{d_1} \\ \mathbf{s}_1^{d_1} \\ \mathbf{s}_2^{d_1} \end{Bmatrix}}_{\substack{\text{B-spline–de Boor} \\ \text{control vectors}}} \quad \text{or, } \mathbf{q}^{d_{Bezier}} = \mathbf{T}^{BS_1}\, \mathbf{s}^{d_1} \tag{8.205}$$

In other words, inclusion of the C^1 constraint converts the two quadratic Bezier elements into one quadratic B-spline element. The transformation matrix relating the unknown displacements in terms of the B-spline generalized values to those in terms of the Bezier control vectors is given by $\mathbf{T}^{BS}$. We can apply this matrix to both the global stiffness matrix and the load vector to obtain the equilibrium equation in terms of the B-spline control vectors as:

$$\boxed{\begin{aligned} \mathbf{K}^{BS_1}\,\mathbf{b}^{d_1} &= \mathbf{F}^{BS_1} \\ \text{where} \qquad \mathbf{K}^{BS_1} &\equiv (\mathbf{T}^{BS_1})^T\,\mathbf{K}\,\mathbf{T}^{BS_1}, \\ \mathbf{F}^{BS_1} &= (\mathbf{T}^{BS_1})^T\,\mathbf{F} \end{aligned}} \tag{8.206}$$

where $\mathbf{K}$ and $\mathbf{F}$ are our standard stiffness control matrix and load control vector, respectively, and defined in equation (8.144).

Remark

- The transformation matrix, $\mathbf{T}^{BS_1}$, is extremely sparse, and in actual implementation, the constraints could be incorporated by means of sparse matrix technology, avoiding multiplication by zeros. Every nodal continuity constraint reduces the total assembled degrees of freedom by one in the system. Recall that our Bezier end nodes only interpolate the physical nodes.

8.4.11.3 Cubic Elements and C^2 Continuity

For connection between cubic elements, additionally C^2 continuity is meaningful, and when applied, defines a B-spline element in terms of the de Boor polygon vertices. For example, let there be two elements, element 1 on the left and element 2 on the right that are defined by the two displacement control vector sets, $\{q_0^d, q_1^d, q_2^d, q_3^d\}$ and $\{q_3^d, q_4^d, q_5^d, q_6^d\}$, respectively. Notice that the C^0 continuity between the two is already satisfied. Let h_1 and h_2, as before, be the corresponding lengths measured in global parameter of the root elements; notice that if the elements are straight, these quantities represent their respective lengths. Now the C^1 continuity, following equation (8.202), is given by:

$$q_3^d = \beta\, q_2^d + \alpha\, q_4^d \quad \text{where} \quad \alpha \equiv \frac{h_1}{h_1 + h_2} \text{ and } \beta \equiv \frac{h_2}{h_1 + h_2} \tag{8.207}$$

For C^2 continuity, we introduce the de Boor polygon vertices as:

$$\mathbf{s}^{d_2} \equiv \left\{\mathbf{s}_{-1}^{d_2} \equiv q_0^d,\ \mathbf{s}_0^{d_2} \equiv q_1^d,\ \mathbf{s}_1^{d_2},\ \mathbf{s}_2^{d_2} \equiv q_5^d,\ \mathbf{s}_3^{d_2} \equiv q_6^d\right\}^T \tag{8.208}$$

It can be shown that, with the de Boor vertices constructed as above, the C^2 continuity at q_3^d, particularized as:

$$q_2^d = \beta\, \mathbf{s}_0^{d_2} + \alpha\, \mathbf{s}_1^{d_2} \quad \text{and} \quad q_4^d = \beta\, \mathbf{s}_1^{d_2} + \alpha\, \mathbf{s}_2^{d_2} \tag{8.209}$$

guarantees that two cubic Bezier displacement elements are converted to one cubic B-spline displacement element. Now, substituting expressions for q_2^d and q_4^d from equation (8.209) above in equation (8.207) for C^1 continuity, we get for the junction node:

$$q_3^d = \beta^2\, \mathbf{s}_0^{d_2} + 2\alpha\beta\, \mathbf{s}_1^{d_2} + \alpha^2\, \mathbf{s}_2^{d_2} \tag{8.210}$$

Finally, we can write in matrix form, expanded or compact, the constraint conditions imposed by the C^2 condition as:

$$\underbrace{\begin{Bmatrix} q_0^d \\ q_1^d \\ q_2^d \\ q_3^d \\ q_4^d \\ q_5^d \\ q_6^d \end{Bmatrix}}_{\substack{\text{Bezier–Bernstein} \\ \text{control vectors}}} = \begin{bmatrix} 1 & 0 & 0 & 0 & 0 \\ 0 & 1 & 0 & 0 & 0 \\ 0 & \beta & \alpha & 0 & 0 \\ 0 & \beta^2 & 2\alpha\beta & \alpha^2 & 0 \\ 0 & 0 & \beta & \alpha & 0 \\ 0 & 0 & 0 & 1 & 0 \\ 0 & 0 & 0 & 0 & 1 \end{bmatrix} \underbrace{\begin{Bmatrix} \mathbf{s}_{-1}^{d_2} \\ \mathbf{s}_0^{d_2} \\ \mathbf{s}_1^{d_2} \\ \mathbf{s}_2^{d_2} \\ \mathbf{s}_3^{d_2} \end{Bmatrix}}_{\substack{\text{B-spline–de Boor} \\ \text{control vectors}}} \tag{8.211}$$

or,

$$\mathbf{q}^{d_{Bezier}} = \mathbf{T}^{BS_2}\, \mathbf{s}^{d_2}$$

In other words, inclusion of the C^2 constraint converts the two quadratic Bezier elements into one quadratic B-spline element. The transformation matrix relating the unknown displacements in terms of the B-spline generalized values to those in terms of the Bezier control vectors is given by $\mathbf{T}^{BS_2}$. We can apply this matrix to both the global stiffness matrix and the load vector to obtain the equilibrium equation in terms of the B-spline control vectors as:

$$\begin{aligned} &\mathbf{K}^{BS_2}\, \mathbf{b}^{d_2} = \mathbf{F}^{BS_2} \\ &\text{where}\quad \mathbf{K}^{BS_2} \equiv (\mathbf{T}^{BS_2})^T\, \mathbf{K}\, \mathbf{T}^{BS_2} \quad \text{and} \quad \mathbf{F}^{BS_2} = (\mathbf{T}^{BS_2})^T\, \mathbf{F} \end{aligned} \tag{8.212}$$

where **K** and **F** are our standard stiffness control matrix and load control vector, respectively, and defined in equation (8.144).

Remark

- As before, the transformation matrix, $\mathbf{T}^{BS_2}$, is extremely sparse, and in actual implementation, the constraints could be incorporated by means of sparse matrix technology avoiding multiplication by zeros. Every nodal continuity constraint reduces the total assembled degrees of freedom by two in the system. Recall, as before, that the Bezier end nodes only interpolate the physical nodes.

Let us now summarize the mechanisms by which inter-element continuity conditions are implemented for general situations: let Ω^i and $\Omega^j, i, j \in N(\Delta)$ be two adjacent finite elements, sharing a boundary with each other. As to boundary, for the one-dimensional case, it is the common node between them, for two- and three-dimensional cases; they are the shared edge or face, respectively. As discussed earlier, depending on the degree of the Bernstein polynomials across the shared boundary, the element displacements are described by rows of hitherto unknown

Bezier displacement control vectors parallel to the shared boundary. Now, we are ready to state the inter-element continuity conditions in term of these displacement controls. In actual application, these are as follows:

- For C^0 continuity of displacements between elements, we simply rename the displacement control nodes that determine the displacements at the adjoining boundary of Ω^i and Ω^j.
- For C^1 continuity of displacements between elements, in addition to imposition of C^0 continuity, we introduce the constraint conditions in the form of linear combinations of those displacement control vectors that are on the first row parallel and nearest to the adjoining boundary control vectors of both Ω^i and Ω^j, such that the tangent or tangent plane, depending on the dimensionality of the problem, across the boundary is continuous.

For C^2 continuity of displacements between elements, in addition to the imposition of C^0 and C^1 continuity as above, we introduce additional constraint conditions. These take the form of linear combinations of those displacement control vectors that are on the first two rows parallel and nearest to the adjoining boundary control vectors of both Ω^i and Ω^j. Thus, the curvature of the surface, depending on the dimensionality of the problem, across the boundary is secured continuous.

Remark

- All of these conditions are, in practice, tantamount to a pre- and post-multiplication of the assembled stiffness control matrix, **K**, and pre-multiplication of the assembled load control vector, **F**, by a very sparse matrix. These can be performed very effectively by sparse matrix technology at the assembled (global) level. However, element (local) level implementation is also possible, since we will have a priori knowledge of the continuity conditions and the topology of the entire structure or continuum. In other words, the method of incorporation of these conditions is exactly the same as that of any constraint conditions.

Now, we elaborate on the concept of inter-element continuity on demand by an example. For the old finite element method, the rate of convergence to the exact solution of a problem may be improved, normally, by:

- increasing the number of elements (h-type),
- increasing the degree of the element (p-type).

For a smooth solution, or where applicable, our new method introduces another control for improvement of the rate of convergence:

- inter-element continuity condition (c-type).

We can now, finally, summarize to define our basis functions and the new finite element space.

8.4.12 *The Basis Functions:* $\Phi^{P,Q,R}(\Omega, \Delta, T, C)$

The basis functions in c-type finite element space are conceptually generated by first defining Bernstein basis functions on the root element of some degree P, Q and R for three global directions, and then mapping them to the ith physical elements, Ω^e_i, through transformation, $\mathbf{T}_i$, for $i = 1, 2, ...$ Finally, the basis functions for Ω can be thought to be as these elemental basis functions joined appropriately, so that certain continuity, C^x, ($x = 0, 1, 2$ normally) is maintained at the joins. Thus, the basis functions depend on the mesh, Δ, the transformations, $\mathbf{T}_i$, the Bernstein degree p, and finally, the continuity, C^x. We denote a typical member of the basis functions set as $\Phi^{P,Q,R}(\Omega, \Delta, T, C)$. Then, of course, the displacement vectors, as usual, are given as linear combinations of this set of basis functions, except that, in our case, the coefficients of combination are Bezier displacement control vectors, $\mathbf{q}^d$.

8.4.13 The c-type Finite Element Space: $S^{P,Q,R}(\Omega, \Delta, T, C)$

Finally, we are in a position to formally define the c-type finite element space:

A finite element space consists of:

- a partitioning or mesh, Δ, of the body, Ω, into a finite set of N bodies, Ω^e, such that $\bigcup_{e\in\{1,2,...N\}} \Omega^e \equiv \Omega$,
- a set of N transformations, $\mathbf{T}^e : \Omega^R \rightarrow \Omega^e, \quad e \in \{1, 2, \ldots, N(\Delta)\}$, where the $\mathbf{T}^e$'s represent the complete set of Bernstein polynomials of degree, P, Q and R in three global directions,
- minimally and exactly required continuity conditions, C, between the adjacent finite elements.

Remarks

- For the first time, we have, through our new method, brought together the geometry (mesh generation) and algebra (analysis), and local properties (displacement control vectors) of the finite difference and global properties (variational theory) of the finite element to solve general problems.
- The only way in which this finite element space with selective continuity fails to improve over all other the old methods is in the approximation of solutions which do not contain continuous derivatives, or solutions which have singularities. In the case of internal singularities, of course, it is essential not to impose excess smoothness of the trial space. This aspect will be further elucidated through examples, such as a typical beam problem loaded with nodal moment.
- The cubic finite elements, with C^2 continuity across the element boundary, is the equivalent to B-spline space as described above.
- Most importantly, the assembly level continuity control is the ultimate in combining the global nature of the finite element with the local nature of the finite difference scheme. The use of the Bezier form with piecewise Bernstein polynomials as the finite element space uniquely provides the ground on which the finite element and finite difference scheme implicitly combine to achieve a solution of high accuracy with derivatives of high-order concomitantly approximated.
- A finite element space, $S^{P,Q,R}(\Omega, \Delta, T, C)$ is exactly and minimally conforming; that is, it constitutes the largest admissible subspace of the associated energy space. Later, we will show that it is necessary for it to be minimally conforming to arrive at the best approximation, as defined by the underlying variational principle.
- To be a finite element space, it must consist of finite-dimensional subspaces formed by the piecewise Bezier polynomials of certain degree, say, p, formed in Bernstein basis functions pieced together with C^k, $k = 0, 1, 2...$, continuity based on the solution space.
- c-type finite element space is totally different from all spaces published to date, yet it subsumes the old methodologies both conceptually and computationally. As a result, conventional, general purpose computer programs currently in use can be readily adapted to produce a problem-free convergent, controlled and cost-efficient procedure that unifies modeling and analysis.

- Because of its novelty and the profound difference from the old methods, we demonstrate, now, in great detail, the promises that it presents, by examining the procedure in one, two and three dimensions – and in both linear and nonlinear regimes. We suggest that the usefulness of the new methodology will be greatest in problems with curved boundaries. Accordingly, we will look at solving one-dimensional problems that involve bodies with curved boundaries, that is, curved bars and beams along with their usual straight counterparts. However, we are more interested in seeing its usefulness in two- and three-dimensional problems – e.g. plate and shell problems – which, of course, happen to be the primary targets of the bulk of the finite element applications.

8.4.14 The c-type Finite Element Library

Finally, we are in a position to formally reveal our new finite element library. Here, it will be quickly seen that, unlike the old type finite elements with confusing arrays of elements in commercially available codes, our list of elements is very simple and short. The following general observations may be made.

- The elements depend primarily on the dimensionality and the degree of the basis functions.
- All elements, curved or otherwise, are described in Cartesian coordinates.
- The only physical nodes present are the end nodes in one dimension and the corner nodes in higher dimensions. In other words, there are no physical edge nodes nor interior nodes.

So, let us go ahead and formally describe them in increasing dimensionality.

8.4.14.1 One-dimensional Elements

An element is characterized by:

- the root element, which is a unit straight line in geometry,
- the transformation, which is subparametric or coerced isoparametric and defined by:

$$\boxed{\begin{aligned} \mathbf{c}(r) &= \sum_{i=0}^{M} \mathrm{B}_i^M(r)\, q_i^c, \\ \text{and, } \mathbf{d}(r) &= \sum_{i=0}^{N} \mathrm{B}_i^N(r)\, q_i^d, \end{aligned}} \tag{8.213}$$

All terms are as defined earlier in the text.

8.4.14.2 Two-dimensional Elements

An element is characterized by:

- the root element which is either:
  - an unit square, or
  - an isosceles triangle

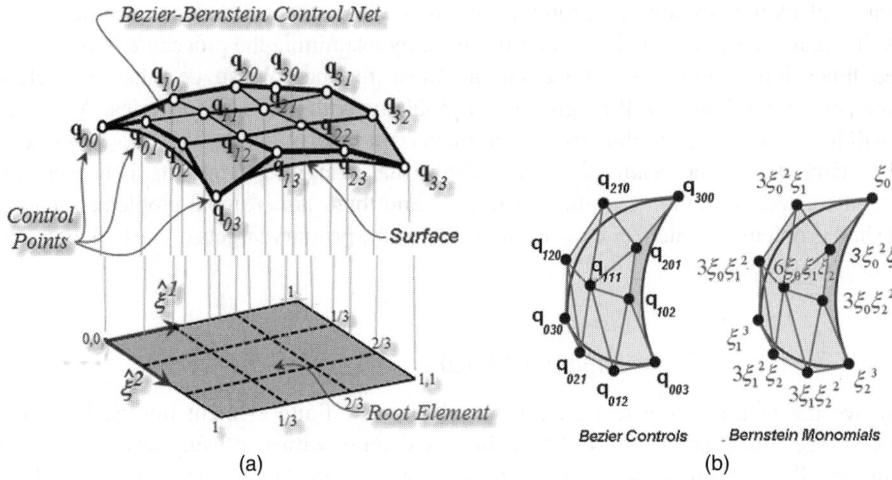

Figure 8.10 (a) Cubic Bezier rectangular element; (b) Cubic Bezier triangular element.

- the transformation, which is subparametric or coerced isoparametric and defined by:
  - rectangular elements:

$$\begin{aligned} \mathbf{c}(r,s) &= \sum_{i=0}^{K}\sum_{j=0}^{L} \mathrm{B}_i^K(r)\mathrm{B}_j^L(s)\, q_{ij}^c, \\ \text{and,}\quad \mathbf{d}(r,s) &= \sum_{i=0}^{M}\sum_{j=0}^{N} \mathrm{B}_i^M(r)\mathrm{B}_j^N(s)\, q_{ij}^d, \end{aligned} \tag{8.214}$$

  all terms are as defined earlier in the text; an example is shown in Figure 8.10(a).
  - triangular elements:

$$\begin{aligned} \mathbf{c}(r,s,t) &= \sum_{i+j+k=M} \mathrm{B}_{ijk}^M(r,s,t)\, q_{ijk}^c, \\ \text{and,}\quad \mathbf{d}(r,s,t) &= \sum_{i+j+k=N} \mathrm{B}_{ijk}^N(r,s,t)\, q_{ijk}^d \end{aligned} \tag{8.215}$$

  where, $B_{ijk}^X(r,s,t)$, $X = M, N$ are the bi-variate triangular Bernstein polynomials in barycentric coordinates, given by:

$$\begin{aligned} B_{ijk}^N(r,s,t) &= \binom{N}{ijk} r^i s^j t^k \quad \text{where} \quad i+j+k=N, r+s+t=1, \\ \text{and,}\quad \binom{N}{ijk} &= \frac{N!}{i!j!k!} \end{aligned} \tag{8.216}$$

  An example is shown in Figure 8.10(b).

8.4.14.3 Three-dimensional Elements

An element is characterized by:

- the root element which is either:
  - a right unit hexahedron, or
  - a right unit tetrahedron, or
  - a right unit pentahedron
- the transformation, which is subparametric or coerced isoparametric and defined by:
  - hexahedral elements:

$$\begin{aligned} \mathbf{c}(r,s,t) &= \sum_{i=0}^{P}\sum_{j=0}^{Q}\sum_{k=0}^{R} \mathrm{B}_i^P(r)\mathrm{B}_j^Q(s)\mathrm{B}_k^R(t)\, q_{ijk}^c, \\ \text{and,}\quad \mathbf{d}(r,s,t) &= \sum_{i=0}^{L}\sum_{j=0}^{M}\sum_{k=0}^{N} \mathrm{B}_i^L(r)\mathrm{B}_j^M(s)\mathrm{B}_k^N(t)\, q_{ijk}^d, \end{aligned} \tag{8.217}$$

  all terms are as defined earlier in the text.
  - tetrahedral elements:

$$\begin{aligned} \mathbf{c}(r,s,t,p) &= \sum_{i+j+k+l=M} \mathrm{B}_{ijkl}^M(r,s,t,p)\, q_{ijkl}^c, \\ \text{and,}\quad \mathbf{d}(r,s,t,p) &= \sum_{i+j+k+l=N} \mathrm{B}_{ijkl}^N(r,s,t,p)\, q_{ijkl}^d \end{aligned} \tag{8.218}$$

  where, $\mathrm{B}_{ijkl}^X(r,s,t,p)$, $X = M, N$ are the trivariate Bernstein polynomials, given by:

$$\begin{aligned} &\mathrm{B}_{ijkl}^N(r,s,t,p) = \begin{pmatrix} N \\ ijkl \end{pmatrix} r^i s^j t^k p^l, \\ &\text{where,}\quad i+j+k+l = N, r+s+t+p = 1, \\ &\text{and,}\quad \begin{pmatrix} N \\ ijkl \end{pmatrix} = \frac{N!}{i!j!k!l!} \end{aligned} \tag{8.219}$$

  - pentahedral elements:

$$\begin{aligned} \mathbf{c}(r,s,t,p) &= \sum_{i+j+k=K}\sum_{l=0}^{L} \mathrm{B}_{ijk}^K(r,s,t)\mathrm{B}_l^L(p)\, q_{ijkl}^c, \\ \text{and,}\quad \mathbf{d}(r,s,t,p) &= \sum_{i+j+k=N}\sum_{l=0}^{M} \mathrm{B}_{ijk}^N(r,s,t)\mathrm{B}_l^M(p)\, q_{ijkl}^c \end{aligned} \tag{8.220}$$

  where, all terms are defined as before.

Discussion:

- In the above description of the transformations, it is assumed that the loading functions are also interpolated isoparametrically to the geometry. This, of course, allows standardization of the loading input for any general-purpose finite element computer program. Arbitrary cross-sectional and material properties can likewise be interpolated. However, for general-purpose programming, numerical integration is performed by appropriate quadrature rules that treat these properties as discretized at the quadrature points. It follows that to each geometry control point necessary to describe an element, there corresponds one displacement and loading control points.

- The above solutions with c-type bar or truss elements, straight and curved beam elements, and plane stress and plate elements indicate that use of the c-type elements will exactly represent all the element properties: displacements, strains, stresses and strain energies. Moreover, because of the existence of the physical nodes of our elements as only corner nodes, the conformity can be achieved on demand. Although the examples above demonstrate many specificities, these should not be construed as limiting the scope of the c-type finite element method but as mere illustrations of the strength and superiority of the c-type method over the old h-type and p-type finite element methods.
- In this section, for simplicity in application of the c-type method, the example problems are restricted to the linear regime, that is, small displacement, rotations and strains with homogeneous, isotropic material behavior. In later chapters, we remove some of these assumptions. For kinematic behavior, we eventually assume that a plane section remains plane but not necessarily perpendicular to the centroidal axis or plane, to include shear deformations. More importantly, we apply our c-type finite element method to complicated problems of beams, plates and shells in extreme geometrical nonlinearity, that is, very large displacements and rotations. The applications of the c-type method to beam, plates and shell problems in both linear and nonlinear regime are equally robust, stable and accurate; later, we will demonstrate these applications in detail.

8.5 Newton Iteration and Arc Length Constraint

Our main aim here is to devise a numerical scheme to trace nonlinear force–deformation path of a structural system subjected to quasi-static loads. For nonlinear structural problems, complicated or not, closed form solutions, more often than not, are non-existent. To trace the equilibrium (i.e. the force–deformation) path of a structural system subjected to quasi-static loadings, we need to resort to some kind of numerical method that is robust, stable and computationally cost-effective, and mimics the closed form solution. Most algorithms build on Newton's method of solving the linearized (tangent operator) equations of the original set of nonlinear equations, and then drive the solution to its true value by some sort of iteration scheme. Here, we discuss and present one such method that we use extensively in subsequent application problems to our new c-type finite element method.

As to the peculiarity of nonlinear structural equilibrium path tracing, generally there are two fundamentally different scenarios for loss of stability of a structure, as shown in Figure 8.11. These are related to the limit point associated with the snap-through phenomena, or the bifurcation point associated with the buckling situation. In the case of limit points, a basic Newton-type method will either overshoot near a limit point that is about to signal an unloading or undershoot near a limit point heralding a switch to a loading situation with no chance of correction or recovery. Similarly, out of two possible controls of tracing a load–displacement curve, we prefer load control to displacement control. That is, during numerical iteration, the load is incremented and the corresponding displacement increment is determined from the solution of the equilibrium equation; our discussion will be restricted to load control methods. Finally, for a practical problem,

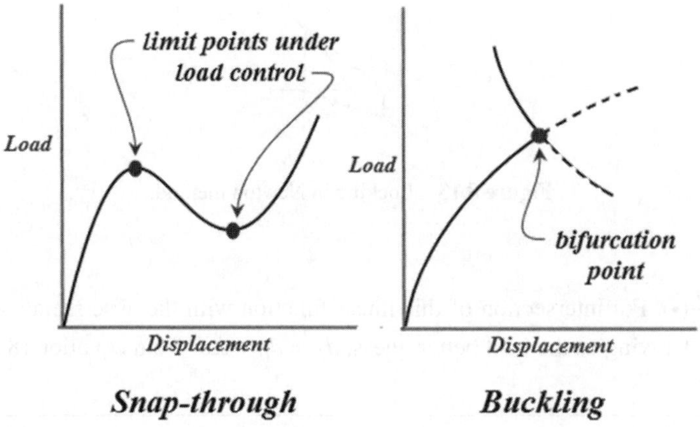

Figure 8.11 Snap-through and bifurcation buckling.

all that may be necessary is the first limit point, that is, the collapse load; even, in this situation, the simple Newton-type iteration method must be equipped with a mechanism to converge or predict the collapse load. Later, we introduce what is known as the ***arc length constraint mechanism*** for all numerical example of quasi-static proportional loading presented for nonlinear beams and shells. With these in the background, let us start by briefly reviewing the Newton's method and the arc length constraint algorithm.

8.5.1 Newton's Method

Let us consider a scalar-valued scalar function, $G(d)$; we are interested, say, in the zeros of this function as shown in Figure 8.12.

Geometrically speaking, we make an initial guess, say, d_0, and determine the tangent to the curve at this point; the intersection of this tangent with the abscissa gives us the next better guess for the zero of the function; the equivalent algebraic operations will be to determine the first-order approximation, $\bar{G}(d)$, to the function at the initial guess as:

$$\boxed{\bar{G}(d) = G(d_0) + \Delta d \; G'(d_0), \quad \Delta d = d - d_0} \tag{8.221}$$

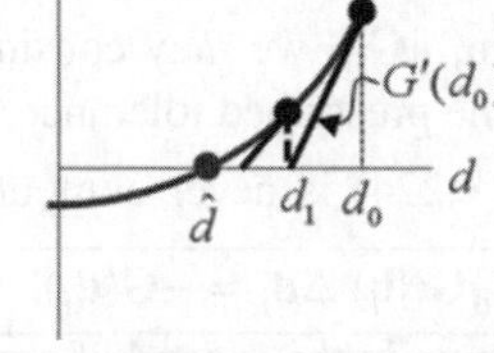

Figure 8.12 Newton's method for scalar function.

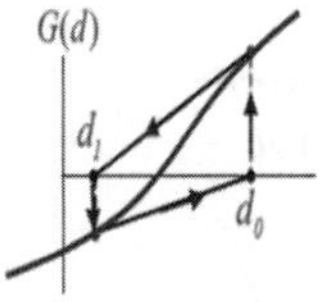

Figure 8.13 Locking in Newton method.

where $(\bullet)' \stackrel{\Delta}{=} \frac{\partial}{\partial d}(\bullet)$. For intersection of this linear function with the abscissa at, say, $d = d_1$, we take : $\bar{G}(d_1) = 0$, giving us the next better guess, $d_1 = d_0 + \Delta d$, from equation (8.221):

$$\boxed{\bar{G}(d_1) = G(d_0) + \Delta d\ G'(d_0) = 0 \quad \Rightarrow d_1 \stackrel{\Delta}{=} d_0 + \Delta d = d_0 - \frac{G(d_0)}{G'(d_0)}} \tag{8.222}$$

Of course, we can keep on doing it iteratively and hope that it will get closer and closer to the zero of the function with a tolerance, $\varepsilon > 0$, for error prescribed to us a priori, such that we can stop the iteration process whenever $|G(d_n)| < \varepsilon$ for some finite n, but note that for this to happen, we need to beware of the following.

- The function should not have a point, d_i, at which the derivative vanishes.
- We must have the starting guess close to the appropriate zero of the function, otherwise, we may settle into a situation that locks the iteration into an infinite loop, as shown in Figure 8.13.
- The Newton iteration method is easily generalized for a system of nonlinear equations; let us suppose we have N nonlinear equations in N variables, given below in vector form as a vector-valued vector function:

$$\boxed{\mathbf{G}(\mathbf{d}) = \mathbf{0}, \quad \mathbf{G} \in \mathbb{R}_N, \quad \mathbf{d} \in \mathbb{R}_N} \tag{8.223}$$

then, the Newton method at the ith iteration is given as:

$$\boxed{\begin{array}{l} \bar{\mathbf{G}}(\mathbf{d}_{i+1}) = \mathbf{G}(\mathbf{d}_i) + \Delta\mathbf{d}_i\ \nabla_{\mathbf{d}}\mathbf{G}(\mathbf{d}_i) = \mathbf{0}, \quad i = 0, 1, 2, \ldots \\ \Rightarrow \mathbf{d}_{i+1} \stackrel{\Delta}{=} \mathbf{d}_i + \Delta\mathbf{d}_i = \mathbf{d}_i - \left(\nabla_{\mathbf{d}}\mathbf{G}(\mathbf{d}_i)\right)^{-1} \mathbf{G}(\mathbf{d}_i) \end{array}} \tag{8.224}$$

where $\nabla_{\mathbf{d}}\mathbf{G}(\mathbf{d}_i) = \frac{\partial G_m(\mathbf{d}_i)}{\partial d_n} E_m \otimes E_n$ is the $N \times N$ gradient tensor of the function, $\mathbf{G}(\mathbf{d}_i)$, at point $\mathbf{d}_i$. Now, assuming a suitable norm, $\|\mathbf{G}\|$, we may consider that the iteration has converged whenever $\|\mathbf{G}\| < \varepsilon$ with $\varepsilon > 0$ as the prescribed tolerance. Note that the inverse of the matrix, $\left[\frac{\partial G_m(\mathbf{d}_i)}{\partial d_n}\right]$ suggested by equation (8.224) is never computed in practical systems with large N; instead, the linear equation: $\boxed{\nabla_{\mathbf{d}}\mathbf{G}(\mathbf{d}_i)\ \Delta\mathbf{d}_i = -\mathbf{G}(\mathbf{d}_i), \quad i = 0, 1, 2, \ldots}$ is solved for $\Delta\mathbf{d}_i$ to update the iteration variable, as shown in the second of equation (8.224).
- As will be seen by the problems solved in nonlinear beams and nonlinear shells, the Newton method of iteration, if successful, converges quadratically.

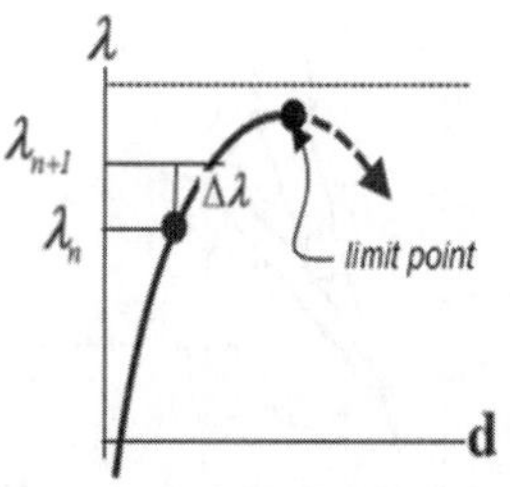

Figure 8.14 Load control in Newton method.

8.5.1.1 Algorithm For Newton Method of Iteration

We can sum up the previous presentation in the form of an algorithm:

Step 0: Given $\mathbf{G}(\mathbf{d})$, set $i = 0$, and choose: starting point, $\mathbf{d}_i$, and tolerance, $\varepsilon > 0$
Step 1: If the residual: $\|\mathbf{G}(\mathbf{d}_i)\| < \varepsilon$, go to Step 3 ; else go to Step 2.
Step 2: Solve incremental tangent equation: $\boxed{\nabla_{\mathbf{d}}\mathbf{G}(\mathbf{d}_i)\,\Delta\mathbf{d}_i = -\mathbf{G}(\mathbf{d}_i), \quad i = 0, 1, 2, \ldots}$; update the estimate: $\mathbf{d}_{i+1} \stackrel{\Delta}{=} \mathbf{d}_i + \Delta\mathbf{d}_i$ and set the iteration counter: $i = i + 1$; go to Step 1.
Step 3: $\mathbf{d}_i$ is the desired solution; stop.

8.5.2 *Non-symmetric Arc Length Constraint Method*

The Newton method as presented above works fine as long as we are away from limit points. Particularly, for load–deformation equilibrium path tracing with load control, using the ordinary Newton scheme as above, we may overshoot the load increment around the limit points with no possible intersection of the tangent with the load–deformation curve, and thus we may not recover thereafter as suggested by Figure 8.14.

Let us recall the nonlinear equilibrium equation obtained for a structural system under quasi-static proportional loading:

$$\mathbf{G}(\mathbf{d}, \lambda) = \mathbf{0} \tag{8.225}$$

where $\mathbf{G}$ represents a system of N equations defined by N-dimensional generalized displacement vector and scalar λ, the load proportionality factor. Suppose now that at each iteration we choose $\Delta\lambda$ as a known increment from the previously equilibrium condition; then, we are left with N equations in N variables, given by equation (8.225), to solve for $\mathbf{d}$. Of course, using the Newton algorithm, we can try to make ad hoc decisions to arbitrarily adjust the incremental load control to get to the limit point of which, however, we have no prior knowledge. Moreover, as we get closer and closer to a limit point, the gradient tensor become more and more ill-conditioned and eventually becomes singular (because it ends up having zero eigenvalue) at the limit point signaling the failure of the Newton method. On the contrary, we may treat load control, λ, as an independent parameter and take note of the fact that there are N equations to solve for $(N + 1)$ independent variables. Thus, for a well-determined equation solution, we may introduce one more equation, $c(\mathbf{d}, \lambda) = 0$, known as the constraint equation, such that we have the following $(N + 1)$ system of equations in $(N + 1)$ variables:

$$\boxed{\begin{aligned} \mathbf{G}(\mathbf{d}, \lambda) &= \mathbf{0} \quad \text{Original equations} \\ c(\mathbf{d}, \lambda) &= 0 \quad \text{Constraint equation} \end{aligned}} \tag{8.226}$$

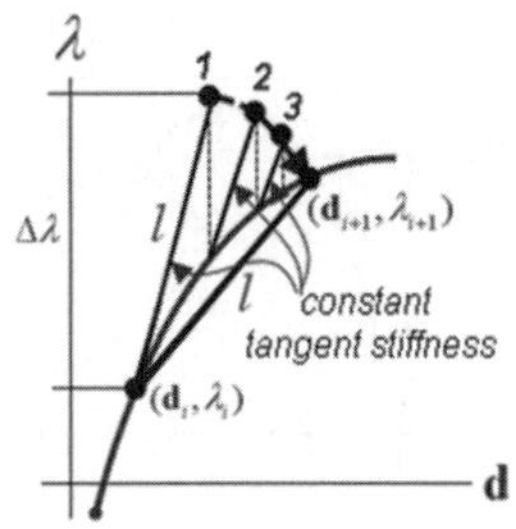

Figure 8.15 Arc length constraint.

8.5.2.1 Choosing the Constraint Equation

The constraint equation may be chosen in several ways: generally speaking, all constraint algorithms are designed to constrain iteration steps by what is known as the ***arc length constraint***; given the ith converged equilibrium state, $(\mathbf{d}_i, \lambda_i)$, the next guess (see Figure 8.15) in the current iteration step, $(\mathbf{d}, \lambda)$, is always constrained by a fixed length, l, expressed by a parameter, β, as:

$$\boxed{\begin{aligned} c(\mathbf{d}, \lambda) &= \beta\,\{\Delta\lambda\}^2 + \|\Delta\mathbf{d}\|^2 - l^2, \quad l > 0,\ \beta = 0, 1 \\ \Delta\lambda &= \lambda - \lambda_i \quad \text{and} \quad \Delta\mathbf{d} = \mathbf{d} - \mathbf{d}_i \end{aligned}} \tag{8.227}$$

where assuming a Euclidean norm, we have: $\|\mathbf{d}\|^2 = \mathbf{d} \bullet \mathbf{d} = \mathbf{d}^T\,\mathbf{d}$. With $\beta = 1$, we get what is known as ***spherical constraint***, because it can be seen as a spherical ball of constraint determined by the load and displacement increment at each iteration step with l as the spherical radius given by:

$$c(\mathbf{d}, \lambda) = \{\Delta\lambda\}^2 + \|\Delta\mathbf{d}\|^2 - l^2, \quad l > 0 \tag{8.228}$$

With $\beta = 0$, we get what may be characterized as the ***cylindrical constraint*** which involves only the displacement increment and may be expressed accordingly as:

$$c(\mathbf{d}) = \|\Delta\mathbf{d}\|^2 - l^2, \quad l > 0 \tag{8.229}$$

Now, Newton's method is applied, assuming that we have an $(N + 1)$ system of equations in $(N + 1)$ independent variables: N dimensional displacement vector, $\mathbf{d}$, and a scalar load parameter, λ, as given by equation (8.226). If the constraint radius, l, is small enough, it is expected to intersect the equilibrium path at two points, one going backward and one forward of the last converged state; clearly, we are seeking the forward point on the path. To start with, at each iteration step, as before, we need to linearize (i.e. find the tangent operator) of both equations in equation (8.226) as:

$$\begin{aligned} \bar{\mathbf{G}}(\mathbf{d}_{i+1}, \lambda_{i+1}) &= \mathbf{G}(\mathbf{d}_i^\alpha, \lambda_i^\alpha) + \Delta\mathbf{d}_i^\alpha\ \nabla_{\mathbf{d}}\mathbf{G}(\mathbf{d}_i^\alpha, \lambda_i^\alpha) + \Delta\lambda_i^\alpha\ \nabla_\lambda\mathbf{G}(\mathbf{d}_i^\alpha, \lambda_i^\alpha) = \mathbf{0} \\ \bar{c}(\mathbf{d}_{i+1}, \lambda_{i+1}) &= c(\mathbf{d}_i^\alpha, \lambda_i^\alpha) + \Delta\mathbf{d}_i^\alpha\ \nabla_{\mathbf{d}}c(\mathbf{d}_i^\alpha, \lambda_i^\alpha) + \Delta\lambda_i^\alpha\ \nabla_\lambda c(\mathbf{d}_i^\alpha, \lambda_i^\alpha) = \mathbf{0} \end{aligned} \tag{8.230}$$

where $\nabla_{\mathbf{d}}\mathbf{G}(\mathbf{d}_i^\alpha, \lambda_i^\alpha) = \frac{\partial G_m(\mathbf{d}_i^\alpha, \lambda_i^\alpha)}{\partial d_n} E_m \otimes E_n$ is the $N \times N$ gradient tensor of the function, $\mathbf{G}(\mathbf{d}_i, \lambda_i)$. with respect to displacement vector, $\mathbf{d}$, and $\nabla_\lambda\mathbf{G}(\mathbf{d}_i^\alpha, \lambda_i^\alpha) = \frac{\partial G_m(\mathbf{d}_i^\alpha, \lambda_i^\alpha)}{\partial\lambda} E_m$ is the $N \times 1$ gradient vector of the function, $\mathbf{G}(\mathbf{d}_i, \lambda_i)$. with respect to load parameter, λ, at an intermediate point $\mathbf{d}_i^\alpha$, λ_i^α

at an intermediate point $\mathbf{d}_i^\alpha, \lambda_i^\alpha$, $\alpha = 0, 1, 2, ...$, in the main converged step i and the next desired converged step $(i + 1)$; similar definitions apply to $\nabla_{\mathbf{d}} c(\mathbf{d}_i^\alpha, \lambda_i^\alpha)$ and $\nabla_\lambda\, c(\mathbf{d}_i^\alpha, \lambda_i^\alpha)$. If we collect appropriately, the matrix form of equation (8.230) to be solved at each intermediate step may be given as:

$$\underbrace{\begin{bmatrix} \nabla_{\mathbf{d}}\mathbf{G}(\mathbf{d}_i^\alpha, \lambda_i^\alpha) & \nabla_\lambda \mathbf{G}(\mathbf{d}_i^\alpha, \lambda_i^\alpha) \\ \nabla_{\mathbf{d}} c(\mathbf{d}_i^\alpha, \lambda_i^\alpha) & \nabla_\lambda c(\mathbf{d}_i^\alpha, \lambda_i^\alpha) \end{bmatrix} \begin{Bmatrix} \Delta\mathbf{d}_i^\alpha \\ \Delta\lambda_i^\alpha \end{Bmatrix}}_{(N+1)\times(N+1)} = -\underbrace{\begin{Bmatrix} \mathbf{G}(\mathbf{d}_i^\alpha, \lambda_i^\alpha) \\ c(\mathbf{d}_i^\alpha, \lambda_i^\alpha) \end{Bmatrix}}_{(N+1)\times 1} \tag{8.231}$$

Remark

- If the original tangent operator is symmetric, its symmetry is clearly destroyed by forming the augmented matrix given by (8.231) making it inefficient for computation when we are dealing with numbers of displacement degrees of freedom as high as the hundreds or thousands. It is possible to devise a separate numerical scheme using a similar idea, that is, arc length constraint, such that the symmetry of the tangent stiffness is preserved; this is what we use in our later example problems relating to nonlinear beams, plates and shells, since the tangent stiffness (i.e. material stiffness and geometric stiffness) is generally symmetric, and this we introduce next.

8.5.3 Symmetry Preserving Arc Length Constraint Iteration Method

In this scheme, the solution involving the tangent stiffness and that for the arc length constraint equation are kept separate, that is, at an intermediate point $\mathbf{d}_i^\alpha, \lambda_i^\alpha$, $\alpha = 0, 1, 2, ...$, in the main converged step i and the next desired converged step $(i + 1)$. Let us recall that the governing covariantly linearized equation for a structural system undergoing quasi-static loading may be given by:

$$\begin{aligned} \mathbf{G}(\mathbf{d}_i + \Delta\mathbf{d}_i^\alpha, \lambda_i + \Delta\lambda_i^\alpha) &= \mathbf{G}(\mathbf{d}_i, \lambda_i) + \nabla_{\mathbf{d}}\mathbf{G}(\mathbf{d}_i, \lambda_i)\, \Delta\mathbf{d}_i^\alpha + \Delta\lambda_i^\alpha\, \nabla_\lambda \mathbf{G}(\mathbf{d}_i, \lambda_i) \\ &= \mathbf{G}(\mathbf{d}_i, \lambda_i) + \mathbf{K}_T\, \Delta\mathbf{d}_i^\alpha - \Delta\lambda_i^\alpha\, \mathbf{P} = \mathbf{0} \\ \Rightarrow \qquad \mathbf{K}_T\, \Delta\mathbf{d}_i^\alpha &= \Delta\lambda_i^\alpha\, \mathbf{P} - \mathbf{G}(\mathbf{d}_i, \lambda_i), \quad \alpha = 1, 2, 3, ... \end{aligned} \tag{8.232}$$

where $\mathbf{K}_T$, $\mathbf{P}$ and $\mathbf{G}(\mathbf{d}_i, \lambda_i)$ are the tangent stiffness, the external load vector and the unbalanced force vector, respectively. In this scheme, the solution, $\Delta\mathbf{d}_i^\alpha$, for the iteration step is divided into two parts: $\Delta\mathbf{d}_i$, involving the unbalanced force and $\Delta\mathbf{d}_P$ due to the unscaled external load vector such that:

$$\begin{array}{lll} \Delta\mathbf{d}_i = -\mathbf{K}_T^{-1}\, \mathbf{G}(\mathbf{d}_i, \lambda_i) & & \text{Out-of-balance displacement} \\ \Delta\mathbf{d}_P = \mathbf{K}_T^{-1}\, \mathbf{P} & & \text{Tangential displacement} \\ \Delta\mathbf{d}_i^\alpha = \Delta\lambda_i^\alpha\, \Delta\mathbf{d}_P + \Delta\mathbf{d}_i, & \alpha = 0, 1, 2, ... & \text{Total incremental displacement} \end{array} \tag{8.233}$$

Remarks

- It is important to note that in this scheme the original tangent stiffness matrix at the beginning of each load step is used for all intermediate iteration steps, and $\Delta\mathbf{d}_P$ is solved only once in the beginning of a step; this scheme results in more iteration steps and fewer tangent stiffness computations.

- Depending on the severity of nonlinearity, we may choose to update the tangent stiffness at each iteration step; this normally results in fewer iteration steps.

Now, the displacement and load parameter increments are updated as:

$$\boxed{\begin{aligned} \Delta\mathbf{d}_{i+1} &= \Delta\mathbf{d}_i + \Delta\mathbf{d}_i^{\alpha}, \quad \alpha = 1,2,3,... \\ \Delta\lambda_{i+1} &= \Delta\lambda_i + \Delta\lambda_i^{\alpha} \end{aligned}} \tag{8.234}$$

Remarks

- The arc length constraint given by equation (8.228) may be replaced by the simplified version for practical problems at each intermediate iteration step and thus equation (8.229) may be used instead.
- Because the constraint equation is quadratic, there is a distinct possibility that some iteration steps may back up, doubling the previous solution; a strategy must be adopted to avoid such situation.

Using equation (8.233) in equation (8.229), we get:

$$\boxed{\begin{aligned} & \Delta\mathbf{d}_{i+1}^T \Delta\mathbf{d}_{i+1} = l^2 \Rightarrow a\,(\Delta\lambda_i^{\alpha})^2 + 2b\;\Delta\lambda_i^{\alpha} + c = 0 \\ \text{where} \quad & a \stackrel{\Delta}{=} \Delta\mathbf{d}_P^T\,\Delta\mathbf{d}_P \\ & b \stackrel{\Delta}{=} (\Delta\mathbf{d}_i^{\alpha} + \Delta\mathbf{d}_i)^T\,\Delta\mathbf{d}_P \\ & c \stackrel{\Delta}{=} (\Delta\mathbf{d}_i^{\alpha} + \Delta\mathbf{d}_i)^T\,(\Delta\mathbf{d}_i^{\alpha} + \Delta\mathbf{d}_i) - l^2 \end{aligned}} \tag{8.235}$$

Remarks

- The solution of the quadratic equation (8.235) will have two roots, say, $\Delta\lambda_{i1}^{\alpha}$ and $\Delta\lambda_{i2}^{\alpha}$:
  - If any of the roots is imaginary, we would cut the load increment as it is too large to intersect with the equilibrium path, so cut the arc length, l, in half and start the iteration again.
  - In order to avoid doubling back on some previous point on the equilibrium path, we would expect that the angle between the incremental displacement after iteration, $\Delta\mathbf{d}_{i+1}$, and the displacement at the start of the iteration, $\Delta\mathbf{d}_i$, will be acute. So, we compute two estimates, $\Delta\mathbf{d}_{i+1}^1$ and $\Delta\mathbf{d}_{i+1}^2$, corresponding to the roots, $\Delta\lambda_{i1}^{\alpha}$ and $\Delta\lambda_{i2}^{\alpha}$, respectively; we compute: $\cos\theta_k = (\Delta\mathbf{d}_{i+1}^k \bullet \Delta\mathbf{d}_i)$, $k = 1,2$, and we choose $\Delta\lambda_i^{\alpha} = \Delta\lambda_{ik}^{\alpha}$ based on the positive θ_k that is closest to the solution of the linear equation corresponding to equation (8.235), that is,

$$\boxed{2b\;\Delta\lambda_i^{\alpha} + c = 0 \Rightarrow \Delta\lambda_i^{\alpha} = -\frac{c}{2b}} \tag{8.236}$$

- Now, the iteration can be initiated from the beginning by assuming an initial load increment, $\Delta\lambda_0$; then, the arc length, l, can be chosen as:

$$\boxed{l_0 = \Delta\lambda_0\sqrt{\Delta\mathbf{d}_P \bullet \Delta\mathbf{d}_P}} \tag{8.237}$$

- For any subsequent iteration, we would like to choose the arc length in such a way that the number of iterations remain more or less the same at each load step; we would normally expect that the less the severity of nonlinearity in equilibrium path, the less is the number of iterations in each load step, and vice versa. So, if n_{i-1} and N_S are the number of iterations at load steps, $i-1$ and the assigned maximum number of iterations, respectively, and l_{i-1} is the arc length at the $(i-1)^{th}$ load step, then the arc length, l_i, at load step i may be set in inverse proportion as:

$$\boxed{l_i = \left(\frac{N_{\max}}{n_{i-1}}\right) l_{i-1}} \tag{8.238}$$

- For any load step other than the first one, we will choose the increment of load step as:

$$\boxed{\Delta\lambda_i = \pm\frac{l_i}{\sqrt{\Delta\mathbf{d}_P \bullet \Delta\mathbf{d}_P}}} \tag{8.239}$$

- The sign of the load increment in equation (8.239) depends on whether we are loading up on the equilibrium path or unloading after, say, a limit point; the indication comes from the determinant of the tangent stiffness matrix: during the loading cycle, the determinant should be positive as the tangent stiffness is positive definite, and it is negative during the unloading cycle for negative definitive tangent stiffness, The positive sign should be chosen for the loading cycle and negative for unloading; exactly at the limit point, the stiffness is singular but it is a very low probability of occurrence numerically.
- A tolerance for error of convergence criterion, $\varepsilon > 0$, and an arc length constraint parameter are input by the user of a computer code.
- Finally, for any load step i, the total displacement, $\mathbf{d}_{i+1}$, and the total load increment, λ_{i+1}, after convergence, are given as:

$$\boxed{\mathbf{d}_{i+1} = \mathbf{d}_i + \Delta\mathbf{d}_i \quad \text{and} \quad \lambda_{i+1} = \lambda_i + \Delta\lambda_i} \tag{8.240}$$

Based on these discussions, we construct and choose the following algorithm for numerical iteration in all succeeding nonlinear problems.

8.5.3.1 Algorithm: Numerical Iteration

Step 0: Choose, an initial load increment, $\Delta\lambda_0$, maximum iteration count, $N_{\max}$, maximum load step count, $L_{\max}$, a maximum load increment, $\lambda_{\max}$, and a tolerance for convergence, ε; set load step counter, $i = 0$, iteration counter, $\alpha = 0$.

Step 1: If $i \neq 0$, that is, not the first load step, go to ***Step 2***; if the iteration counter, $\alpha \neq 0$, that is, not first iteration, go to ***Step 3*** else: compute $\Delta\mathbf{d}_P$ from equation (8.233); compute l_i from equation (8.237); compute $\Delta\mathbf{d}_i^\alpha$ from equation (8.233); go to ***Step 3***.

Step 2: If the iteration counter, $\alpha \neq 0$, that is, not first iteration, go ***Step 4*** else: compute l_i from equation (8.238),

Step 3: Compute $\Delta\mathbf{d}_P$ from equation (8.233); set $\Delta\lambda_i$ using equation (8.239) with sign chosen based on the determinant of the current tangent stiffness, $\mathbf{K}_T$; go to ***Step 4***.

Step 4: If tangent stiffness, $\mathbf{K}_T$, is always kept constant during iteration, go to ***Step 5*** else recompute $\Delta\mathbf{d}_P$ from equation (8.233), and go to ***Step 5***.

Step 5: compute the out-of-balance displacement, $\Delta\mathbf{d}_i$, from equation (8.233); solve the quadratic equation (8.235) for two roots, say, $\Delta\lambda^{\alpha}_{i1}$ and $\Delta\lambda^{\alpha}_{i2}$: if any of the roots is imaginary, set $l_i = \frac{l_i}{2}$ and $\alpha = 0$, and go to ***Step 3***; else, compute: $\cos\theta_k = (\Delta\mathbf{d}^k_{i+1} \bullet \Delta\mathbf{d}_i)$, $k = 1, 2$; if $\theta_1 > 0$, $\Delta\lambda^{\alpha}_i = \Delta\lambda^{\alpha}_{i1}$, and $\Delta\mathbf{d}_{i+1} = \Delta\mathbf{d}^1_{i+1}$, where $\Delta\mathbf{d}^1_{i+1}$ corresponds to the root, $\Delta\lambda^{\alpha}_{i1}$; if $\theta_2 > 0$, $\Delta\lambda^{\alpha}_i = \Delta\lambda^{\alpha}_{i2}$, and $\Delta\mathbf{d}_{i+1} = \Delta\mathbf{d}^2_{i+1}$, where $\Delta\mathbf{d}^1_{i+1}$ corresponds to the root, $\Delta\lambda^{\alpha}_{i2}$; if both θ_k, $k = 1, 2$, are positive, then choose $\Delta\lambda^{\alpha}_i = \Delta\lambda^{\alpha}_{ik}$ based on positive θ_k that is closest to the solution of the linear equation (8.236); update total displacement, $\mathbf{d}^{\alpha}_i = \mathbf{d}_i + (\Delta\mathbf{d}^{\alpha}_i - \Delta\mathbf{d}^{\alpha-1}_i)$. If $\left\|\mathbf{d}^{\alpha}_i - \mathbf{d}^{\alpha-1}_i\right\| \le \varepsilon \left\|\mathbf{d}^{\alpha}_i\right\|$, set $\mathbf{d}_{i+1} = \mathbf{d}^{\alpha}_i$, $n_i = n_\alpha$ and update $\Delta\lambda_{i+1}$ using equation (8.234), and go to ***Step 6***; else set: $\alpha = \alpha + 1$, and go to ***Step 4***.

Step 6: If load step count, $i \ge L_{\max}$, or load increment, $\lambda_i \ge \lambda_{\max}$, stop, else set: $\alpha = 0$, and go to ***Step 2***.

8.6 Gauss–Legendre Quadrature Formulas

Here, our main goal is to introduce numerical schemes to evaluate integrals of functions; as indicated in previous sections on the finite element method, in computing the stiffness matrix, mass matrix and load vectors, integrals of the following general form need to be integrated:

$$I = \int_a^b w(x)\, f(x)\, dx \tag{8.241}$$

where $w(x)$ are known as the weights; with user-supplied variable geometric properties such as thickness and area, and the inverse of the Jacobian function in the form of even rational polynomial as part of the integrand, $f(r)$, the integration is invariably evaluated numerically.

8.6.1 Interpolatory Quadratures for Numerical Integration

We have already seen that a numerical integration scheme relies on approximating integration by finite summation; the interval of the integration is partitioned into a finite number of points, and the function values at these point serve as summands for the approximation by summation. In the most general scheme, the partition locations are variable, but in relatively simpler schemes such as the Newton–Cotes formulas, the partition points are fixed in position. So, considering one-dimensional integration, suppose $(n + 1)$ distinct partition points are ordered as:

$$x_0 < x_1, \bullet\bullet\bullet, < x_{n-1} < x_n \tag{8.242}$$

With these interpolatory points, we can construct the interpolatory polynomial, $P_n(x)$, of degree n, such that:

$$f(x_j) = P_n(x_j), \quad j = 0, 1, \dots, n \tag{8.243}$$

Then, as an approximation of the integration given by equation (8.241), we set:

$$I_{\text{approx}} = \int_a^b w(x)\, P_n(x)\, dx + error(f) \tag{8.244}$$

Now, if we choose the Lagrange form for the polynomial, that is,

$$P_n(x) = \sum_{j=0}^{n} \frac{l_n(x)}{(x - x_j)\, l_{n,x}(x_j)} f(x_j) \quad \text{with} \quad l_n(x) \equiv \prod_{i=0}^{n} (x - x_i) \tag{8.245}$$

where Π is the multiplicative symbol. Upon integration, we get from equation (8.244):

$$I_{\text{approx}} = \sum_{j=0}^{n} A_j f(x_j) \quad \text{where} \quad A_j = \int_a^b w(x) \frac{l_n(x)}{(x - x_j)\, l_{n,x}(x_j)} dx \tag{8.246}$$

where A_j's are known as the weights of the quadrature.

Remark

- From the definition of $w_n(x)$ given in equation (8.245) and the Lagrangian error estimate, it can be shown that the error in integration vanishes if $f(x)$ is a polynomial of degree $\leq n$; thus, a $(n+1)$ point interpolatory quadrature has degree of precision of at least n.

8.6.2 Newton–Cotes Formulas

These include end points as interpolatory points, and as such known as "closed" type formulas; in what follows, we assume: $w(x) = 1$, as indicated before, if $f(x) = P_m(x), \quad m \leq n$, the rules are exact.

8.6.2.1 Trapezoidal Rule: Two End Points

$$\int_a^b f(x)\, dx = \frac{h}{2} f(a) + \frac{h}{2} f(b) - \frac{h^3}{12} f^{(2)}(\xi), \quad h \equiv b - a \tag{8.247}$$

8.6.2.2 Simpson's Rule: Two End Points and the Mid-point

$$\int_a^b f(x)\, dx = \left(\frac{h}{3}\right) \left\{ f(a) + 4f\left(\frac{a+b}{2}\right) + f(b) \right\} - \frac{h^5}{90} f^{(4)}(\xi), \quad h \equiv \frac{b-a}{2} \tag{8.248}$$

8.6.2.3 Simpson's $\frac{3}{8}$th Rule: Four Equidistant Points Including Two End Points

$$\int_a^b f(x)\, dx = \left(\frac{3h}{8}\right) \{f(a) + 3f(a+h) + 3f(a+2h) + f(b)\} - \frac{3h^5}{80} f^{(4)}(\xi), \quad h \equiv \frac{b-a}{3} \tag{8.249}$$

Remarks

- It is interesting to note that there exist $f \in C^0[a, b]$ for which as $n \to \infty$, the Newton–Cotes formulas do not converge; in other word, the above rules not necessarily converge as $n \to \infty$.

8.6.3 Linear Transformation of Interval

Suppose two sets of intervals: $x \in [a, b]$ and $u \in [c, d]$ are given, and we know the rules for the interval, $[a, b]$ such that:

$$\boxed{\int_a^b w(x) f(x)\, dx = \sum_{k=1}^{n} A_k f(x_k)} \tag{8.250}$$

Now, suppose $x = \gamma u + \beta \Rightarrow u = \frac{1}{\gamma}(x - \beta)$ where $\gamma = \frac{b-a}{d-c}$ and $\beta = \frac{ad-bc}{d-c}$; we would like to know the rule such that:

$$\boxed{\int_a^b w(x) f(x)\, dx = \int_c^d \hat{w}(u)\, g(u)\, du} \tag{8.251}$$

where

$$\boxed{g(u) = f(\gamma u + \beta), \quad \hat{w}(\gamma u + \beta)} \tag{8.252}$$

The rule can be easily found to be:

$$\boxed{\begin{aligned} \int_c^d \hat{w}(u)\, g(u)\, du &= \sum_{k=1}^{n} \hat{A}_k\, g(u_k) \\ \text{where} \qquad \hat{A}_k &= \frac{1}{\gamma} A_k \\ u_k &= \frac{1}{\gamma}(x_k - \beta) \end{aligned}} \tag{8.253}$$

8.6.4 Composite Formulas

We have already noted that the previous formulas do not necessarily converge, but if the interval is subdivided and the simple formulas are applied repeatedly, the convergence is achieved as $n \to \infty$.

8.6.4.1 Repeated Trapezoidal Rule

With $(n - 1)$ subintervals, we get:

$$\boxed{\int_a^b f(x)\, dx = \frac{h}{2} f(a) + h \sum_{k=2}^{n-1} f(a + (k-1)h) + \frac{h}{2} f(b) + error(f), \quad h \equiv \frac{b-a}{n-1}} \tag{8.254}$$

8.6.4.2 Repeated Mid-point Rule: "Open", i.e. End Points Are Not Nodes

Based on the single interval mid-point. rule:

$$\int_a^b f(x)\,dx = (b-a)\,f\left(\frac{a+b}{2}\right) \tag{8.255}$$

With n subintervals, for repeated mid-point. rule, we get:

$$\boxed{\int_a^b f(x)\,dx = h\sum_{k=1}^{n} f\left(a+\left(k-\frac{1}{2}\right)h\right) + error(f), \quad h \equiv \frac{b-a}{n}} \tag{8.256}$$

8.6.4.3 Repeated Simpson's Rule

With $(n-1)$ subintervals, we get:

$$\begin{aligned}\int_a^b f(x)\,dx = {} & \left(\frac{h}{3}\right)f(a) + \left(\frac{4h}{3}\right)f(a+h) + \left(\frac{2h}{3}\right)f(a+2h)\\ & + \left(\frac{4h}{3}\right)f(a+3h) + \left(\frac{2h}{3}\right)f(a+4h) + \bullet\bullet\bullet + \left(\frac{2h}{3}\right)f(b-2h)\\ & + \left(\frac{4h}{3}\right)f(b-h) + \left(\frac{h}{3}\right)f(b) + error(f), \quad h \equiv \frac{b-a}{n-1}\end{aligned} \tag{8.257}$$

Remark

- The above repeated rules constitute what is known as the ***Riemann sum***; thus, $\sum \to \int$ as $n \to \leq \infty$

8.6.5 Gauss – Legendre Quadratures

Let us recall that the Newton–Cotes formulas presented earlier with, say, n nodes, are exact for polynomial functions of only degree $\leq n-1$. The question is whether there exist quadratures that are exact for polynomials of higher degrees, and, if so, what is the highest degree of polynomial for which it remains exact. The answer is yes: if we make both the node locations of the nodes and the weights, that is, a total of $2n$ quantities, variable, then the quadrature constructed optimally can be exact for polynomials of degree $\leq 2n-1$: in other words, for a special choice of nodes, $x_i, \quad i = 1, 2, \ldots, n$, we can find $A_k, \quad k = 1, 2, \ldots, n$ so that

$$\boxed{\int_a^b w(x)\,f(x)\,dx = \sum_{k=1}^{n} A_k\,f(x_k) + \text{error}(f)} \tag{8.258}$$

is exact for polynomials of degree $\leq 2n-1$.

Remarks

- It is important to note that $(2n-1)$ is the highest degree of a polynomial that can be exactly represented by any quadrature formula with n nodes.
- These families of quadratures are known as ***Gaussian quadratures***.
- The ***degree of precision*** is defined as the highest degree of polynomials for which a particular quadrature is exact; the degree of precision for an n-point Gaussian quadrature is $(2n-1)$, and thus the Gaussian quadrature formulas have the highest degree of precision.

The basic idea behind a Gaussian quadrature is to locate the nodes of quadrature, x_i, $i = 1, 2, \ldots, n$, at the zeros of an appropriate polynomial, $P_n(x)$, of degree n, that is orthogonal to the interval of the integral with respect to the weighting function, $w(x)$, to all polynomials, $Q_{n-1}(x)$, of degree $\leq n-1$, that is,

$$\boxed{\int_a^b w(x)\, P_n(x)\, Q_{n-1}(x)\, dx = 0} \tag{8.259}$$

Remarks

- It can be shown that the zeros of $P_n(x)$ are real and distinct and reside in the open interval (a, b).
- For $w(x) = 1$, and $[a, b] = [-1, 1]$, the orthogonal polynomials are known as ***Legendre polynomials***.
- All weights, A_k, $k = 1, 2, \ldots, n$, of the Gaussian quadrature are positive.
- Gaussian quadratures converge to an exact integral as $n \to \infty$.
- Geometrically speaking, the Legendre polynomials, say, being orthogonal to $Q_{n-1}(x)$, are at the shortest distance from the space of all polynomials of degree $\leq n-1$, and thus minimize error and provide the best approximation.
- Similarly, if the integrand is a polynomial of degree d, then $(2n-1)$ should be at least equal to d, that is, the quadrature must have: $n \geq \frac{1}{2}(d+1)$; for example, if $d = 5$, say, a quintic polynomial, the integration must be evaluated with at least a three-point Gaussian quadrature formula.
- The tables listing the Gauss–Legendre quadrature nodes and weights are widely available in any standard literature on numerical analysis, but the ones that we use here for c-type finite element analysis of beams, shells, etc., are listed later.
- The Gaussian quadratures are interpolatory; thus, the weights, A_k, $k = 1, 2, \ldots, n$, may be expressed, among other ways, as:

$$\boxed{A_k = \frac{1}{P_{n,x}(x_k)} \int_a^b \frac{P_n(x)}{(x - x_k)} dx, \quad k = 1, 2, \ldots, n} \tag{8.260}$$

8.6.5.1 Linear Transformation of Interval

For the c-type finite element method, $x \in [0, 1]$, instead of $x \in [-1, 1]$ as in the old finite element methods, and this linear transformation is easily handled by Gaussian quadratures as before.

With $x \in [a,b] = [-1,1]$ and $u \in [c,d] = [0,1]$, we get: $\gamma = \frac{b-a}{d-c} = 2$ and $\beta = \frac{ad-bc}{d-c} = -1$; so it follows from equation (8.253):

$$\begin{aligned} &\int_{c=0}^{d=1} \hat{w}(u)\, g(u)\, du = \sum_{k=1}^{n} \hat{A}_k\, g(u_k) \\ &\text{where} \quad \hat{A}_k = \frac{1}{\gamma} A_k = \frac{1}{2} A_k \\ &u_k = \frac{1}{\gamma}(x_k - \beta) = \frac{1}{2}(x_k + 1) \end{aligned} \tag{8.261}$$

An example will clarify these concepts.

Example: Consider a function, $f(x)$, that is a quintic polynomial, that is, of degree five, defined as:

$$f(x) = \alpha_0 + \alpha_1 x + \alpha_2 x^2 + \alpha_3 x^3 + \alpha_4 x^4 + \alpha_5 x^5 \tag{8.262}$$

The exact integration, I, is given by:

$$I = \int_{-1}^{1} f(x)\, dx = 2\alpha_0 + \frac{2}{3}\alpha_2 + \frac{2}{5}\alpha_4 \tag{8.263}$$

The three parameters, $\alpha_0, \alpha_2, \alpha_4$ completely define the integral; we can construct the underlying Legendre polynomial, $P_3(x)$, of degree three whose roots define the quadrature as follows: we know that $P_3(x)$ is orthogonal to all polynomials of degree ≤ 2; thus, letting:

$$P_3(x) = x^3 + \beta_2 x^2 + \beta_1 x + \beta_0 \tag{8.264}$$

and noting that the linear combination of the monomials, $1 = x^0, x, x^2$, spans the subspace of polynomials of degree ≤ 2, and we have for orthogonality with x^k, $k = 0, 1, 2$

$$\begin{aligned} &\int_{-1}^{1} P_2(x)\, dx = 0 \quad \Rightarrow 2\beta_0 + \frac{2}{3}\beta_2 = 0 \\ &\int_{-1}^{1} x\, P_2(x)\, dx = 0 \quad \Rightarrow \frac{2}{5} + \frac{2}{3}\beta_1 = 0 \\ &\int_{-1}^{1} x^2\, P_2(x)\, dx = 0 \quad \Rightarrow \frac{2}{3}\beta_0 + \frac{2}{5}\beta_2 = 0 \end{aligned} \tag{8.265}$$

Now, solving equation (8.265), we get: $\beta_0 = \beta_2 = 0$ and $\beta_1 = -\frac{3}{5}$; thus, the interpolating Legendre polynomial is given by:

$$\boxed{P_3(x) = x^3 - \frac{3}{5}x} \tag{8.266}$$

and its roots are given by:

$$\boxed{x = 0,\ +\sqrt{\frac{3}{5}},\ -\sqrt{\frac{3}{5}}} \tag{8.267}$$

that form the interpolating points, $x_k,\ k = 0, 1, 2$, respectively. Using equation (8.260), for the weights, $A_k,\ k = 0, 1, 2$, we get:

$$\boxed{\begin{aligned} A_0 &= \frac{1}{\left(-\frac{3}{5}\right)} \int_{-1}^{1} \frac{\left(x^3 - \frac{3}{5}x\right)}{x} dx = \frac{8}{9} \\ A_1 = A_2 &= \frac{1}{\left(\frac{6}{5}\right)} \int_{-1}^{1} \frac{\left(x^3 - \frac{3}{5}x\right)}{\left(x \pm \sqrt{\frac{3}{5}}\right)} dx = \left(\frac{5}{6}\right) \int_{-1}^{1} \left(x^2 \pm \sqrt{\frac{3}{5}}x\right) dx = \frac{5}{9} \end{aligned}} \tag{8.268}$$

Finally, we can express the Gauss–Legendre quadrature as:

$$\boxed{\int_{-1}^{1} f(x)\, dx = \left(\frac{8}{9}\right) f(0) + \left(\frac{5}{9}\right) f\left(\sqrt{\frac{3}{5}}\right) + \left(\frac{5}{9}\right) f\left(-\sqrt{\frac{3}{5}}\right)} \tag{8.269}$$

Remarks

- This example has been worked out using the general principle of evaluating the nodes and the weights of Gauss–Legendre quadratures.
- The weights can be seen to be real, positive and distinct, and they reside inside the interval of integration.
- A polynomial of degree five has been exactly integrated by evaluating the function values at three points.

8.6.6 Higher Dimensional Gauss–Legendre Nodes and Weights

The numerical integration for higher dimensions – for example, two or three – can be achieved by straightforward extensions to the one-dimensional procedure described above. To wit, let us write the two-dimensional quadrature rule as:

$$\boxed{\int_a^b \int_c^d f(x, y)\, dx\, dy = \sum_{k=1}^{n} \int_c^d A_k f(x_k, y)\, dy = \sum_{l=1}^{m} \sum_{k=1}^{n} A_k A_l f(x_k, y_l)} \tag{8.270}$$

8.6.7 Tabular Examples of Gauss–Legendre Nodes and Weights

For future reference, we present below some values of the nodes and the weights for quadratures discussed above for various degrees that we actually use later in our applications:

```
Real(kind=8) Function GaussPtVal(index,Nint,iPtVal)
!
!
!     Notes:  Nint should be at least 2
!             index       = Index of Columns of Pts: 1 -> Nint
!             Nint        = Number of Gauss Points
!             iPtVal      = 1: Gauss Pt/ 2:Gauss Value
!     Outputs
!             GaussPtVal  = Result based on iPtVal
!
      Implicit Real(kind=8) (a-h,o-z)
!     Input Variables & Arrays
!-----------------------------------------------------------------------
!     Rk(8,8)             = Gauss-Legendre Nodes       Upto 8 Points
!     Ak(8,8)             = Gauss-Legendre Weights     Upto 8 Points
!     Recall:
!           Our integration, r=[0,1]
!           Conventional  x=[-1,1]
!           i.e. x = (2r-1)
!     So, following values modified as:
!           Rk  = (1/2)(xk+1)
!           Ak  = (1/2)Axk
!-----------------------------------------------------------------------
      Real(kind=8) GLO16,GLO5,GLO7
      Dimension GLO16(16,2),GLO5(5,2),GLO7(7,2)
      Real(kind=8) Rk,Ak,D,B,Add,DB
      Dimension Rk(10,9),Ak(10,9),D(3,3),B(3,32),Add(32),DB(4)
      DATA Rk/  0.5D0,                              0.0D0,        ! 1-point
& 0.0D0,                      0.0D0,
& 0.0D0,                      0.0D0,
& 0.0D0,                      0.0D0,
& 0.0D0,                      0.0D0,
&         0.211324865405187D0, 0.788675134594813D0,         ! 2-point
&         0.0D0,                      0.0D0,
&         0.0D0,                      0.0D0,
&         0.0D0,                      0.0D0,
&         0.0D0,                      0.0D0,
&         0.112701665379258D0, 0.5D0,                       ! 3-point
&         0.887298334620742D0, 0.0D0,
&         0.0D0,                      0.0D0,
&         0.0D0,                      0.0D0,
&         0.0D0,                      0.0D0,
&         0.069431844202974D0, 0.330009478207572D0,         ! 4-point
&         0.669990521792428D0, 0.930568155797026D0,
&         0.D0,                   0.0D0,
&         0.D0,                   0.0D0,
&         0.D0,                   0.0D0,
&         0.0D0, 0.230765344947158D0,                       ! 5-point
&         0.769234655052841D0, 0.046910077030668D0,
```

```
&          0.953089922969332D0, 0.0D0,
&          0.D0,                  0.0D0,
&          0.D0,                  0.0D0,
&          0.966234757101576D0, 0.033765242898424D0,      ! 6-point
&          0.830604693233132D0, 0.169395306766868D0,
&          0.619309593041598D0, 0.380690406958402D0,
&          0.D0,                  0.0D0,
&          0.D0,                  0.0D0,
&          0.5D0, 0.297077424311301D0,                    ! 7-point
& 0.702922575688699D0, 0.129234407200303D0,
& 0.870765592799697D0, 0.025446043828620D0,
& 0.974553956171379D0, 0.D0,
&          0.D0,             0.0D0,
&          0.019855071751232D0, 0.980144928248768D0,      ! 8-point
& 0.101666761293187D0, 0.898333238706813D0,
& 0.237233795041836D0, 0.762766204958164D0,
& 0.408282678752175D0, 0.591717321247825D0,
&          0.D0,             0.0D0,
&          0.5D0, 0.337873288298095D0,         ! 9-point
& 0.662126711701904D0, 0.193314283649705D0,
& 0.806685716350295D0, 0.081984446336682D0,
& 0.918015553663318D0, 0.015919880246187D0,
&          0.984080119753813D0, 0.0D0
& /
!
      DATA Ak/    2.0D0,              0.0D0,             ! 1-point
& 0.0D0,                   0.0D0,
& 0.0D0,                   0.0D0,
& 0.0D0,                   0.0D0,
& 0.0D0,                   0.0D0,
&          0.5D0,          0.5D0,                         ! 2-point
&          0.0D0,          0.0D0,
& 0.0D0,                   0.0D0,
& 0.0D0,                   0.0D0,
& 0.0D0,                   0.0D0,
&          0.277777777777778D0, 0.444444444444444D0,      ! 3-point
&          0.277777777777778D0, 0.0D0,
& 0.0D0,                   0.0D0,
& 0.0D0,                   0.0D0,
& 0.0D0,                   0.0D0,
&          0.173927422568727D0, 0.326072577431273D0,      ! 4-point
&          0.326072577431273D0, 0.173927422568727D0,
&          0.0D0,                     0.0D0,
&          0.0D0,                     0.0D0,
& 0.0D0,                   0.0D0,
& 0.284444444444444D0, 0.239314335249683D0,              ! 5-point
&          0.239314335249683D0, 0.118463442528095D0,
& 0.118463442528095D0, 0.0D0,
&          0.0D0,          0.0D0,
& 0.0D0,                   0.0D0,
&          0.085662246189585D0, 0.085662246189585D0,      ! 6-point
&          0.180380786524069D0, 0.180380786524069D0,
& 0.233956967286346D0, 0.233956967286346D0,
```

```
&             0.0D0,                0.0D0,
& 0.0D0,                            0.0D0,
&             0.208979591836734D0, 0.190915025252560D0,     ! 7-point
&             0.190915025252560D0, 0.139852695744638D0,
& 0.139852695744638D0, 0.064742483084435D0,
&             0.064742483084435D0, 0.0D0,
& 0.0D0,                            0.0D0,
&             0.050614268145188D0, 0.050614268145188D0,     ! 8-point
&             0.111190517226687D0, 0.111190517226687D0,
& 0.156853322938943D0, 0.156853322938943D0,
&             0.181341891689181D0, 0.181341891689181D0,
& 0.0D0,                            0.0D0,
&             0.165119677500630D0, 0.156173538520001D0,     ! 9-point
&             0.156173538520001D0, 0.130305348201468D0,
& 0.130305348201468D0, 0.090324080347429D0,
&             0.090324080347429D0, 0.040637194180787D0,
& 0.040637194180787D0, 0.0D0
& /
!-----------------------------------------------------------------------
DATA ((GLO5(i,j),j=1,2),i=1,5)/
& 0.9530899229693319, 0.1184634425280945,
& 0.0469100770306680, 0.1184634425280945,
& 0.7692346550528415, 0.2393143352496832,
& 0.2307653449471584, 0.2393143352496832,
& 0.5000000000000000, 0.2844444444444444
& /
DATA ((GLO7(i,j),j=1,2),i=1,7)/
& 0.9745539561713792, 0.0647424830844349,
& 0.0254460438286208, 0.0647424830844349,
& 0.8707655927996972, 0.1398526957446383,
& 0.1292344072003028, 0.1398526957446383,
& 0.7029225756886985, 0.1909150252525595,
& 0.2970774243113014, 0.1909150252525595,
& 0.5000000000000000, 0.2089795918367347
& /
DATA ((GLO16(i,j),j=1,2),i=1,16)/
& 0.9947004674958250D0, 0.0135762297058770D0,
& 0.0052995325041750D0, 0.0135762297058770D0,
& 0.9722875115366163D0, 0.0311267619693239D0,
& 0.0277124884633837D0, 0.0311267619693239D0,
& 0.9328156011939159D0, 0.0475792558412464D0,
& 0.0671843988060841D0, 0.0475792558412464D0,
& 0.8777022041775016D0, 0.0623144856277669D0,
& 0.1222977958224985D0, 0.0623144856277669D0,
& 0.8089381222013219D0, 0.0747979944082884D0,
& 0.1910618777986781D0, 0.0747979944082884D0,
& 0.7290083888286136D0, 0.0845782596975013D0,
& 0.2709916111713863D0, 0.0845782596975013D0,
& 0.6408017753896295D0, 0.0913017075224618D0,
& 0.3591982246103705D0, 0.0913017075224618D0,
& 0.5475062549188188D0, 0.0947253052275343D0,
& 0.4524937450811813D0, 0.0947253052275343D0
& /
!-----------------------------------------------------------------------
```

```
      if (iPtVal == 1) Then
            if (Nint == 5) Then
                  GaussPtVal = GLO5(index,1)
            else if (Nint == 7) Then
                  GaussPtVal = GLO7(index,1)
            else if (Nint == 16) Then
                  GaussPtVal = GLO16(index,1)
            else
                  GaussPtVal = Rk(index,Nint)
            end if
      else if (iPtVal == 2) Then
            if (Nint == 5) Then
                  GaussPtVal = GLO5(index,2)
            else if (Nint == 7) Then
                  GaussPtVal = GLO7(index,2)
            else if (Nint == 16) Then
                  GaussPtVal = GLO16(index,2)
            else
                  GaussPtVal = Ak(index,Nint)
            endif
      end if
!
      return
      end
```

8.6.8 Where We Would Like to Go

We have introduced Gauss–Legendre quadrature for integrations that are routinely needed in structural system analysis as one of the final pieces of the puzzle for applying the c-type finite element method. So, to check the efficacy and robustness of the method, we should check out:

- applications of c-type finite element method (Chapter 9, Chapter 10 and Chapter 11)

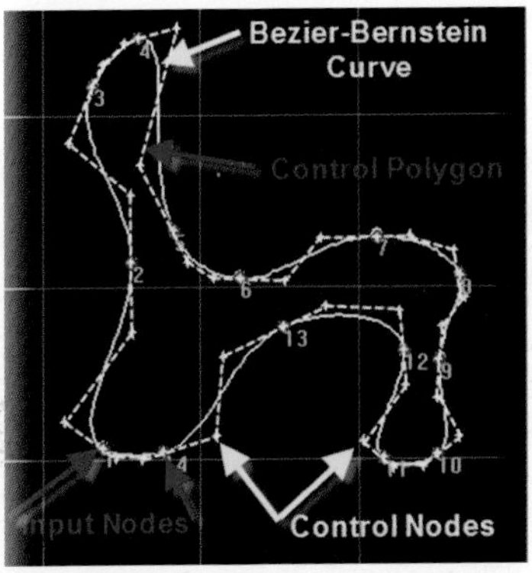

Figure 5.13 Definition of an inverse problem.

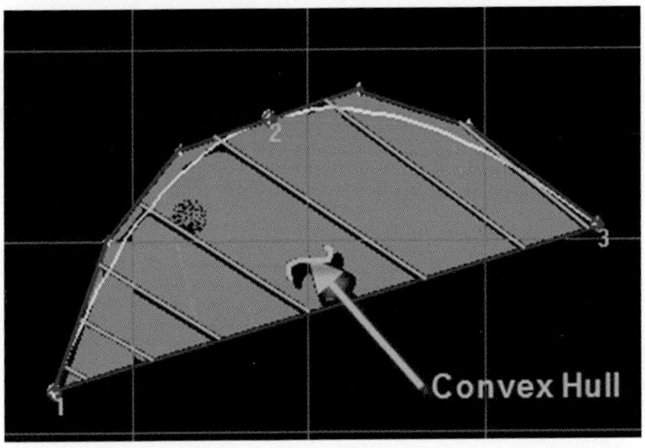

Figure 5.21 Convex hull.

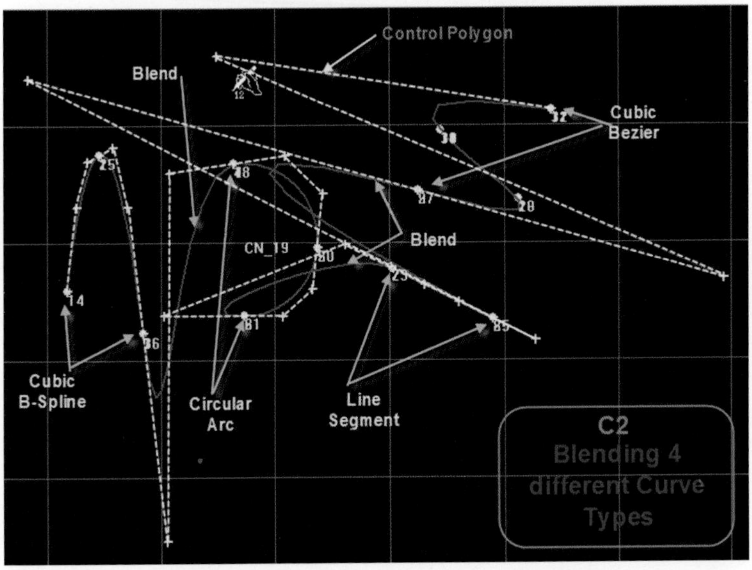

Figure 5.23 Curve blending by degree elevation.

Computation of Nonlinear Structures: Extremely Large Elements for Frames, Plates and Shells, First Edition. Debabrata Ray.

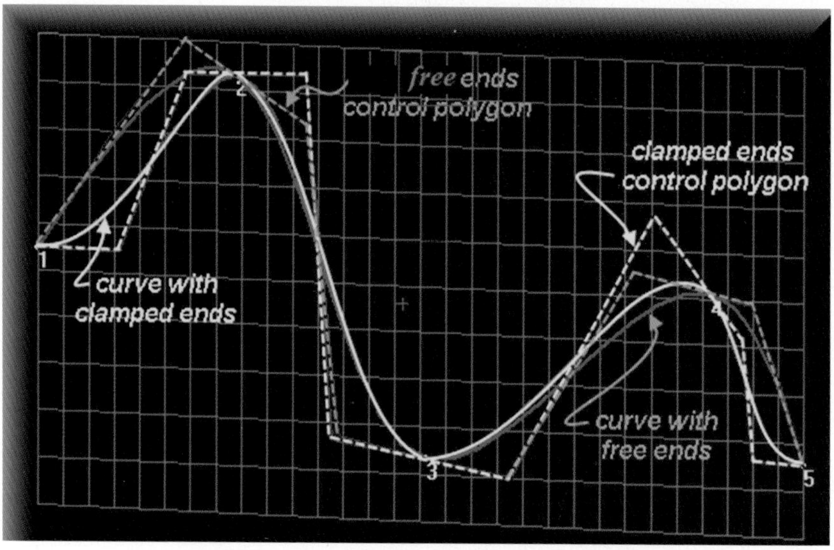

Figure 5.53 Effect of boundary conditions on a Bezier curve.

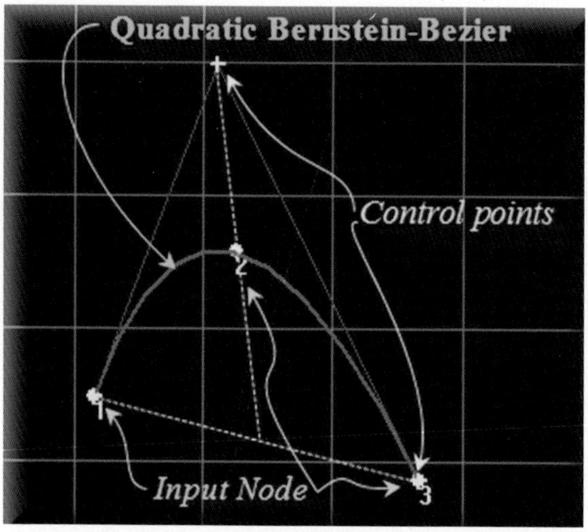

Figure 5.54 Quadratic curve fitting input points.

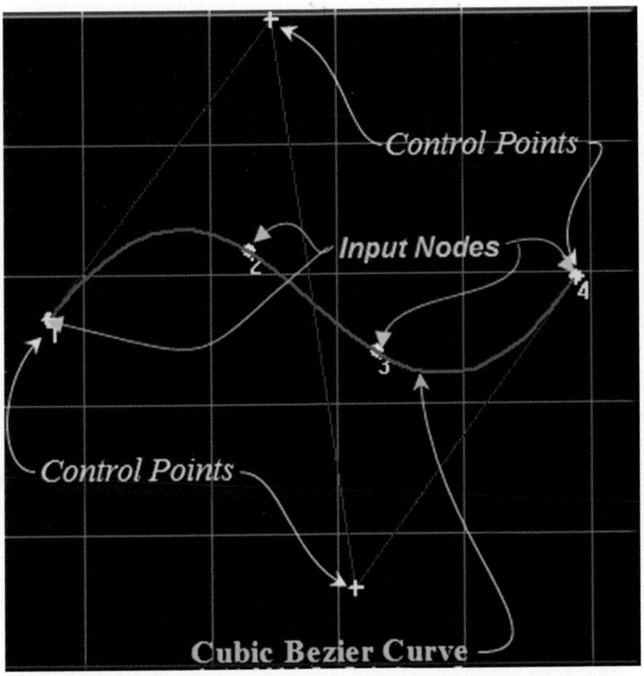

Figure 5.55 Cubic curve fitting input points.

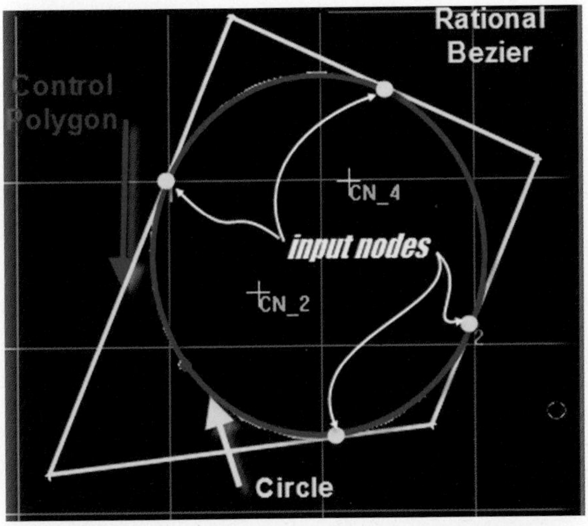

Figure 5.56 Exact full circle by rational Bezier.

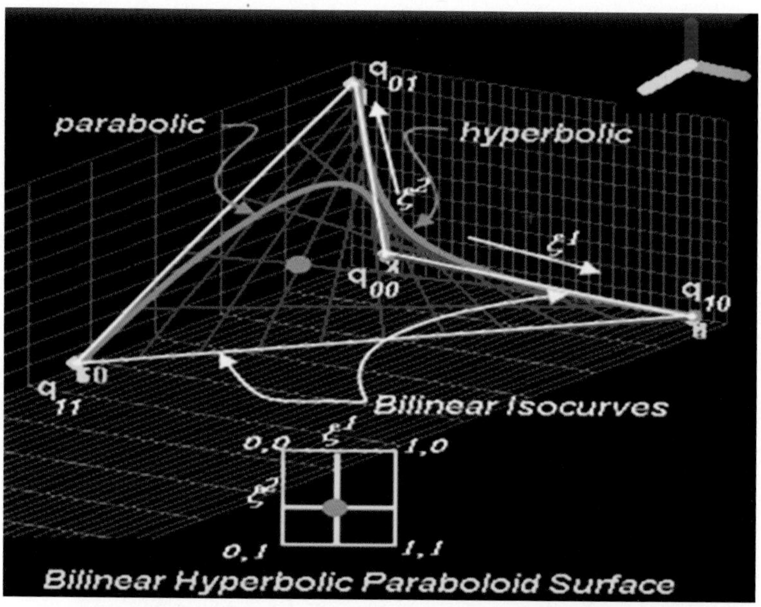

Figure 6.13 Bilinear hyperbolic paraboloid surface.

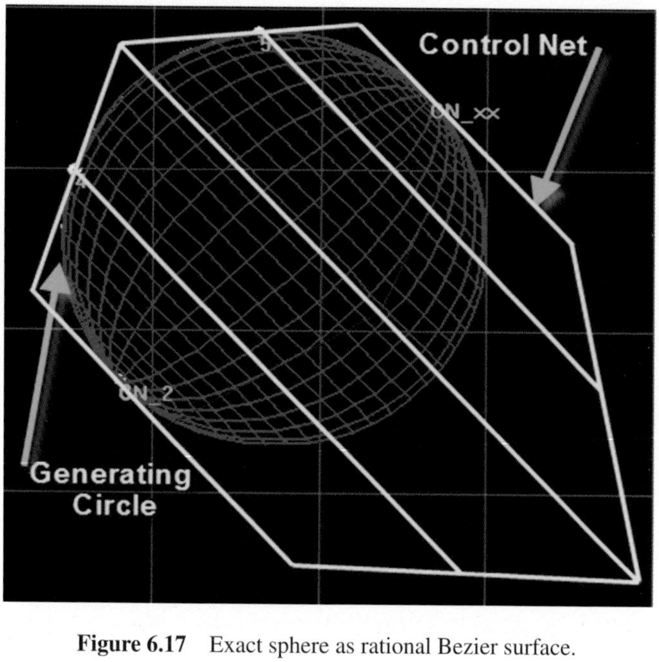

Figure 6.17 Exact sphere as rational Bezier surface.

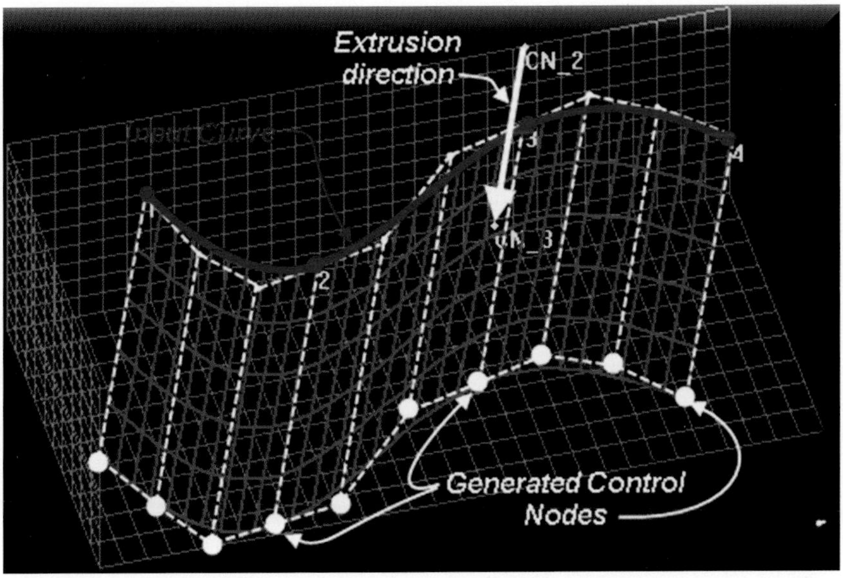

Figure 6.18 Patch by extrusion with Bezier control net.

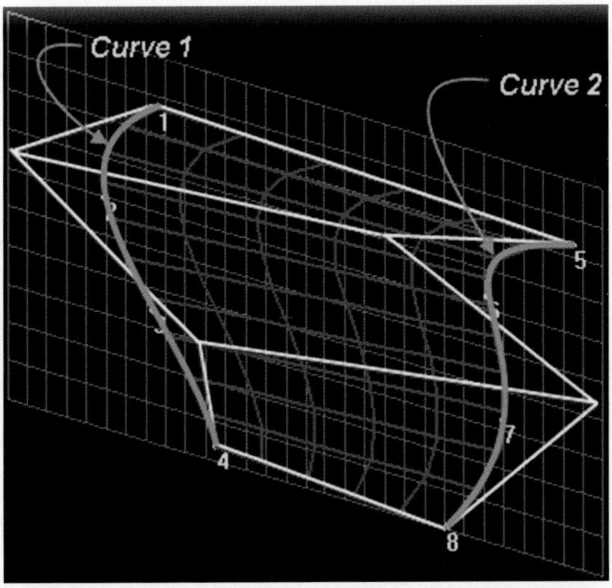

Figure 6.19 Patch by lofting of two curves with Bezier control net.

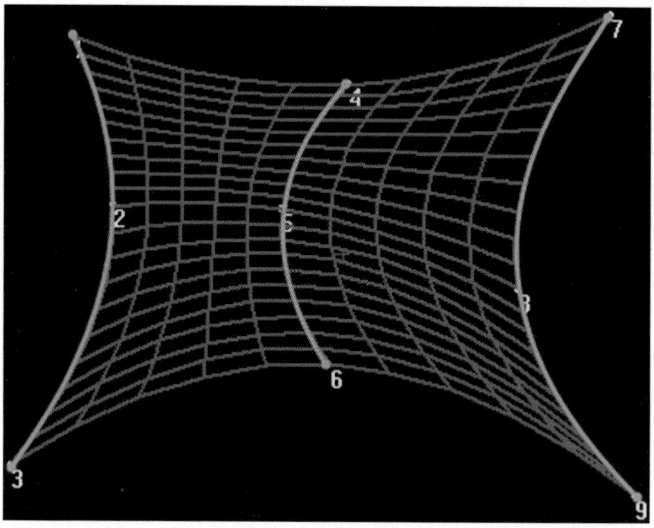

Figure 6.20 Patch by lofting of three curves with Bezier control net.

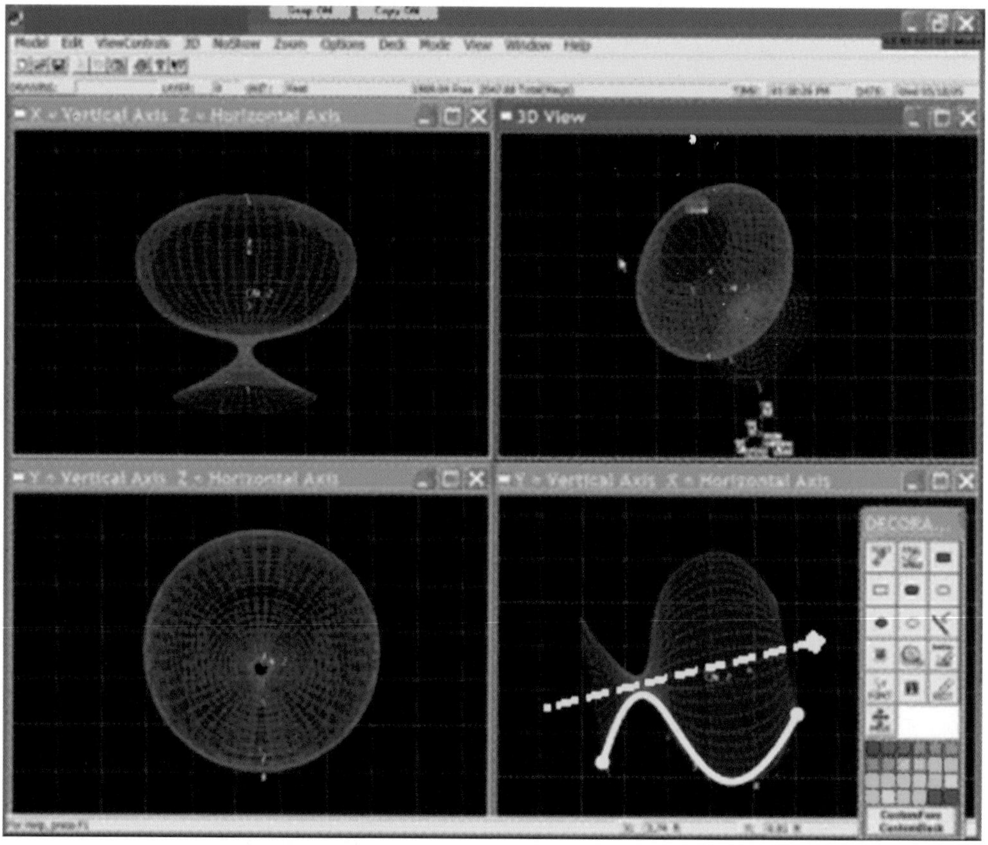

Figure 6.22 Patch by revolution of a generating curve.

Part V

Applications: Linear and Nonlinear

9

Application to Linear Problems and Locking Solutions

9.1 Introduction

9.1.1 Why We Are Here

Thus far we have presented in great detail our new c-type finite element method – a displacement-based finite element method, but nothing can make it more amenable for comprehension or more reliable for industrial usage than testing it with example problems of interest in structural engineering and structural mechanics. In each of these applications, the state is determined by the displacement distribution, a non-parametric function, while c-type finite element method is motivated by the Bernstein–Bezier parametric curves. We may now recall from our previous presentation that any non-parametric function can be represented by parametric Bernstein–Bezier curves because of the linear precision property: $\sum_0^n (\frac{i}{n}) B_i^n(\xi) = \xi$. Thus, we recall that a displacement function, $d(\xi)$, can be transformed into parametric form as:

$$\mathbf{c}(\xi) = \begin{Bmatrix} x(\xi) \\ y(\xi) \end{Bmatrix} = \begin{Bmatrix} \xi \\ d(\xi) \end{Bmatrix} = \sum_0^n B_i^n(\xi)\, \mathbf{q}_i$$

$$\mathbf{q}_i \equiv \begin{Bmatrix} \left(\dfrac{i}{n}\right) \\ q_i \end{Bmatrix} \tag{9.1}$$

$$\text{where, } d(\xi) = \sum_0^n B_i^n(\xi)\, q_i$$

that is, the control vertices are located at equal intervals. In order to evaluate the proposed c-type element library, several test problems, most of which are generally available in the literature, are solved and the results are presented and compared to demonstrate its numerical accuracy, robustness and stability. In all the numerical examples presented below, normal exact full Gauss quadratures were used for stiffness matrix computations avoiding ad hoc reduced or selective integration, and so on, of old methodologies. It may also be noticed that all problems are solved with the minimum number (compared to the old finite element methods) of mega-sized arbitrary

Computation of Nonlinear Structures: Extremely Large Elements for Frames, Plates and Shells, First Edition. Debabrata Ray.

shaped (wherever applicable) elements that are central to the main motivation of the c-type method. In this chapter, we restrict applications to only linear problems; in the subsequent chapters, extreme nonlinearity is considered.

Thus, in this chapter we will present the following.

> ***Section 2***: The proposed c-type method with special reference to inter-element continuity on demand is exemplified by its application to one-dimensional straight truss or straight rod problems.
>
> ***Section 3***: This discusses applications of c-type application to straight Euler–Bernoulli beams.
>
> ***Section 4***: Here we investigate application of c-type Bezier and B-spline elements to curved beam problems, including pinched rings. The unification of mesh generation and finite element analysis in c-type is exemplified by a linear, truly three-dimensional curved beam element with natural twist and bend (the treatment of beams, plates and shells in extreme geometric nonlinearity is elaborated in the following two chapters).
>
> ***Section 5***: The versatility, straightforward nature and ease of use of the method are demonstrated by looking at the problem of arbitrarily distorted and near-incompressible plane strain and plane stress deep beam elements that have, otherwise, remained in the literature and recent textbooks as an annoying nuisance for all other isoparametric finite element methods.
>
> ***Section 6***: The solutions to usual locking problems: shear locking, dilatational locking, transverse shear locking, and so on, are generated with the c-type methodology.

9.2 c-type Truss and Bar Element

One-dimensional finite element analysis, although less consequential for practical applications, brings out most of the salient features of our c-type finite element formulation presented in the previous chapter; thus, we specialize our method by applying it to one-dimensional bars or rods (truss element) that involve second-order differential equations as the momentum balance equation. It is assumed at this point that a one-dimensional geometric assemblage has been generated at the interactive modeling phase; we must emphasize right at the outset that the discretization of a geometry in the modeling phase should not be performed without due regard to the loading and support conditions that define a particular problem for which the geometry is being discretized. For a one-dimensional truss element, the geometric property, the cross-sectional area, $A(x)$, the material property, the elasticity modulus, $E(x)$, the loading, the body forces, $f^B(x)$, per unit volume will generally depend only on the axial coordinate, x. For linear truss elements, a hierarchical set of elements may be conceived through the polynomial degree allowed for displacement up to cubic for numerical efficiency. Given this limit, however, we will build two main categories of elements: Bernstein–Bezier elements characterized by Bezier control polygons, B-spline–de Boor elements formed by cubic composite Bernstein–Bezier polynomials connected by C^2 continuity. With these in the background, let us consider the following one-dimensional bar example problem.

Example CFEM 1: One-dimensional bar with variable elasticity modulus

Consider the bar or truss element of length, L, with constant cross-sectional area, A, as shown in Figure 9.1, but whose elasticity constant varies linearly along its length as $E(x) = E_0(1 + \frac{x}{L})$ with $\boxed{A = E_0 = L = P = 1}$; the bar is subjected to an axial load, P, at $x = L$ end. We will try to identify for the problem: (a) stiffness control matrix, (b) load control vector, (c) displacement

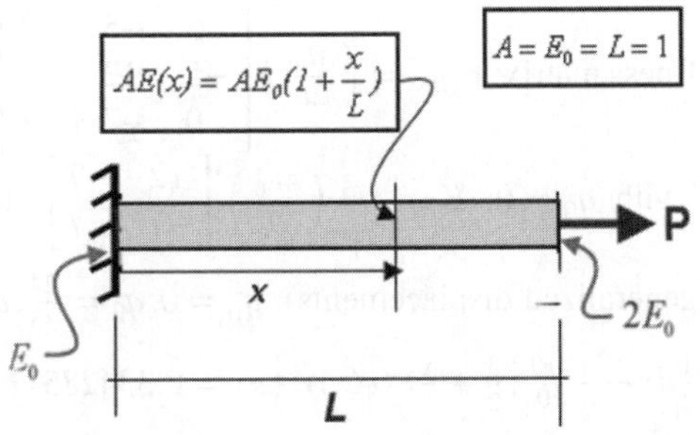

Figure 9.1 Example CFEM1: 1D bar with variable elasticity.

distribution, (d) strain distribution and (e) strain energy of the bar, obtained under a hierarchy of c-type bar elements, along with varying inter-element continuity. For this problem, let us recognize that: $\Omega = (0, L)$; $\Gamma_d = \{0\}$; $\Gamma_\sigma = \{L\}$; $\Gamma = \Gamma_d \cup \Gamma_\sigma$; $\bar{\Omega} = \Omega \cup \Gamma = [0, L]$, and the associated boundary value problem can be posed as:

Differential equation:	$(AE(x)u_{,x}(x))_{,x} = 0 \text{ in } x \in \Omega$
Essential boundary condition:	$u(0) = 0 \text{ on } \Gamma_d$
Natural boundary condition:	$AE(L)\, u_{,x}(L) = P \text{ on } \Gamma_\sigma$

(9.2)

The exact solution can be easily determined to be:

$$\text{Displacements: } u_{Exact}(x) = \left(\frac{PL}{AE_0}\right) \ln\left(1 + \frac{x}{L}\right) \text{ with } u_{Exact}(L) = \ln 2 \left(\frac{PL}{AE_0}\right) < \left(\frac{PL}{AE_0}\right)$$

$$\text{Strains: } \varepsilon_{Exact}(x) = \left(\frac{P}{AE_0}\right) \frac{1}{\left(1 + \frac{x}{L}\right)} \text{ with}$$

$$\varepsilon_{Exact}(0) = \left(\frac{P}{AE_0}\right) \text{ and } \varepsilon_{Exact}(L) = \frac{1}{2}\left(\frac{P}{AE_0}\right)$$

$$\text{Strain energy: } U_{Exact} = \frac{1}{2} \ln 2 \left(\frac{P^2 L}{AE_0}\right) = 0.3465736 \left(\frac{P^2 L}{AE_0}\right)$$

(9.3)

9.2.1.1 Case 1: One Linear Element – Constant Strain (CSL)

Full singular stiffness matrix: $\mathbf{K} = 1.5\left(\frac{AE_0}{L}\right)\begin{bmatrix} 1 & -1 \\ -1 & 1 \end{bmatrix}$, Bezier controls (generalized Displacements): $q_0 = 0\ q_1 = \frac{2}{3}$; strain energy: $U = \frac{1}{2}\int_0^1 (1+x)\,(\varepsilon(x))^2 dx = 0.333333$.

9.2.1.2 Case 2: Two Linear Elements – C^0 Continuity

Singular stiffness matrices: Elem #1: $\mathbf{K}_1 = \frac{5}{2}\left(\frac{AE_0}{L}\right)\begin{bmatrix} 1 & -1 \\ -1 & 1 \end{bmatrix}$, Elem #2: $\mathbf{K}_2 = \frac{7}{2}\left(\frac{AE_0}{L}\right)\begin{bmatrix} 1 & -1 \\ -1 & 1 \end{bmatrix}$.

Assembled singular stiffness matrix: $\mathbf{K}_{SIN} = \left(\frac{AE_0}{2L}\right)\begin{bmatrix} 5 & -5 & 0 \\ -5 & 12 & -7 \\ 0 & -7 & 7 \end{bmatrix}$; assembled non-singular stiffness control matrix: with $q_0 = 0$: $\mathbf{K}_{NON} = \left(\frac{AE_0}{2L}\right)\begin{bmatrix} 12 & -7 \\ -7 & 7 \end{bmatrix}$; load control vector: $P = \begin{Bmatrix} 0 \\ 1 \end{Bmatrix}$; Bezier controls (generalized displacements): $q_0 = 0, q_2 = \frac{24}{35}, q_1 = \frac{7}{12}q_2$; strain energy: $U = \frac{1}{2}\int_0^1 (1+\frac{x}{2})\,(\varepsilon(x))^2\frac{1}{2}dx + \frac{1}{2}\int_0^1 (\frac{3}{2}+\frac{x}{2})\,(\varepsilon(x))^2\frac{1}{2}dx = 0.33428571429.$

9.2.1.3 Case 3: One Quadratic Element

Full singular stiffness matrix: $\mathbf{K}_{SIN} = \left(\frac{AE_0}{3L}\right)\begin{bmatrix} 5 & -2 & -3 \\ -2 & 6 & -4 \\ -3 & -4 & 7 \end{bmatrix}$; non-singular stiffness control matrix: with $q_0 = 0$: $\mathbf{K}_{NON} = \left(\frac{AE_0}{3L}\right)\begin{bmatrix} 6 & -4 \\ -4 & 7 \end{bmatrix}$; s control vector: $P = \begin{Bmatrix} 0 \\ 1 \end{Bmatrix}$; Bezier controls (generalized displacements): $q_0 = 0 \quad q_2 = \frac{9}{13} \quad q_1 = \frac{2}{3} \quad q_2$; strain energy: $U = \int_0^1 2(1+x)\ \left[(1-2x)q_1 + xq_2\right]^2 dx = 0.3461538462.$

9.2.1.4 Case 4: Two Quadratic Elements – C^0 Continuity

Assembled non-singular stiffness control matrix: with $q_0 = 0$: $\mathbf{K}_{NON} = \left(\frac{AE_0}{3L}\right)\begin{bmatrix} 10 & -6 & 0 & 0 \\ -6 & 24 & -6 & -7 \\ 0 & -6 & 14 & -8 \\ 0 & -7 & -8 & 15 \end{bmatrix}$ load control vector: $P = \begin{Bmatrix} 0 & 0 & 0 & 1 \end{Bmatrix}^T$; Bezier controls: $q_0 = 0, q_4 = \frac{1872}{2701}, q_2 = \left(\frac{365}{624}\right)q_4, q_3 = \left(\frac{513}{624}\right)q_4, q_1 = \left(\frac{3}{5}\right)q_2$; strain energy: $U = \int_0^1 (2+x)\ \left[(1-2x)q_1 + xq_2\right]^2 dx + \int_0^1 (3+x)\ \left[-(1-x)q_2 + (1-2x)q_3 + xq_4\right]^2 dx = 0.3465383191.$

9.2.1.5 Case 5: Two Quadratic Elements – C^1 Continuity

Assembled singular stiffness control matrix and C^1 continuity transformation matrix:

$$\mathbf{T}_1 = \begin{bmatrix} 1 & 0 & 0 & 0 \\ 0 & 1 & 0 & 0 \\ 0 & .5 & .5 & 0 \\ 0 & 0 & 1 & 0 \\ 0 & 0 & 0 & 1 \end{bmatrix} \quad \hat{\mathbf{K}}_{SIN} = \begin{bmatrix} 9 & -4 & -5 & 0 & 0 \\ -4 & 10 & -6 & 0 & 0 \\ -5 & -6 & 24 & -6 & -7 \\ 0 & 0 & -6 & 14 & -8 \\ 0 & 0 & -7 & -8 & 15 \end{bmatrix}$$

$$\mathbf{K}_{SIN} = \mathbf{T}_1^T \hat{\mathbf{K}}_{SIN} \mathbf{T}_1 = \begin{bmatrix} 9 & -6.5 & -2.5 & 0 \\ -6.5 & 10 & 0 & -3.5 \\ -2.5 & 0 & 14 & -11.5 \\ 0 & -3.5 & -11.5 & 15 \end{bmatrix}$$

load control vector: $P = \left\{ 0 \quad 0 \quad 0 \quad 1 \right\}^T$; assembled non-singular stiffness control matrix: with $q_0 = 0$: $\mathbf{K}_{NON} = \left(\frac{AE_0}{6L}\right)\begin{bmatrix} 20 & 0 & -7 \\ 0 & 28 & -23 \\ -7 & -23 & 30 \end{bmatrix}$; Bezier controls: $q_0 = 0 q_4 = \frac{70}{101} q_2 = \left(\frac{41}{70}\right) q_4 q_3 = \left(\frac{23}{28}\right) q_4 q_1 = \left(\frac{7}{20}\right) q_2$; strain energy: $U = \int_0^1 (2+x)\,\left[(1-2x)q_1 + xq_2\right]^2 dx + \int_0^1 (3+x)\,\left[-(1-x)q_2 + (1-2x)q_3 + xq_4\right]^2 dx = 0.3465346535$.

9.2.1.6 Case 6: One Cubic Element

Full singular stiffness control matrix: $\mathbf{K}_{SIN} = \left(\frac{9AE_0}{60L}\right)\begin{bmatrix} 14 & -6 & -5 & -3 \\ -6 & 10 & 3 & -7 \\ -5 & 3 & 14 & -12 \\ -3 & -7 & -12 & 22 \end{bmatrix}$. Non-singular stiffness control matrix: with $q_0 = 0$: $\mathbf{K}_{NON} = \left(\frac{9AE_0}{60L}\right)\begin{bmatrix} 10 & 3 & -7 \\ 3 & 14 & -12 \\ -7 & -12 & 22 \end{bmatrix}$.

Bezier controls: $q_0 = 0 q_3 = \frac{131}{189} q_2 = \frac{99}{131} q_3 q_1 = \frac{62}{131} q_3$; strain energy: $U = \frac{1}{2}\int_0^1 (1+x)\,\left[3[-(1-x)^2 q_0 + (1-4x+3x^2)q_1 + (2x-3x^2)q_2 + x^2 q_3]\right]^2 dx \Rightarrow$
$U = 0.3465608466$.

9.2.1.7 Case 7: Two Cubic Elements_ C^0- Continuity

Assembled non-singular stiffness control matrix:

$$\mathbf{K}_{NON} = \left(\frac{3AE_0}{20L}\right)\begin{bmatrix} 18 & 5 & -11 & 0 & 0 & 0 \\ 5 & 22 & -18 & 0 & 0 & 0 \\ -11 & -18 & 72 & -18 & -13 & -7 \\ 0 & 0 & -18 & 26 & 7 & -15 \\ 0 & 0 & -13 & 7 & 30 & -24 \\ 0 & 0 & -7 & -15 & -24 & 46 \end{bmatrix}$$

$$\mathbf{P} = \begin{Bmatrix} 0 \\ 0 \\ 0 \\ 0 \\ 0 \\ 1 \end{Bmatrix} \Rightarrow \text{Bezier controls: } \mathbf{q} = \begin{Bmatrix} 0 \\ 0.1661 \\ 0.2940 \\ 0.4055 \\ 0.5164 \\ 0.6097 \\ 0.6931 \end{Bmatrix}$$

strain energy: $U_{Elem\_1} = 0.2027322404 \quad U_{Elem\_2} = 0.1438410075$ $\boxed{U = 0.3465732479}$.

9.2.1.8 Case 8: Two Cubic Elements – C^1 Continuity

Assembled non-singular stiffness control matrix:

$$\mathbf{K}_{NON} = \left(\frac{3AE_0}{20L}\right)\begin{bmatrix} 18 & 5 & -11 & 0 & 0 & 0 \\ 5 & 22 & -18 & 0 & 0 & 0 \\ -11 & -18 & 72 & -18 & -13 & -7 \\ 0 & 0 & -18 & 26 & 7 & -15 \\ 0 & 0 & -13 & 7 & 30 & -24 \\ 0 & 0 & -7 & -15 & -24 & 46 \end{bmatrix} \quad \mathbf{P}_{NON} = \begin{Bmatrix} 0 \\ 0 \\ 0 \\ 0 \\ 0 \\ 1 \end{Bmatrix}$$

$$\mathbf{T} = \begin{bmatrix} 1 & 0 & 0 & 0 & 0 \\ 0 & 1 & 0 & 0 & 0 \\ 0 & .5 & .5 & 0 & 0 \\ 0 & 0 & 1 & 0 & 0 \\ 0 & 0 & 0 & 1 & 0 \\ 0 & 0 & 0 & 0 & 1 \end{bmatrix}$$

$$\mathbf{K} = \mathbf{T}^T\mathbf{K}_{NON}\mathbf{T} = \begin{bmatrix} 2.7 & -0.075 & -0.825 & 0 & 0 \\ -0.075 & 3.3 & 0 & -0.975 & -0.525 \\ -0.825 & 0 & 3.9 & 0.075 & -2.775 \\ 0 & -0.975 & 0.075 & 4.5 & -3.6 \\ 0 & -0.525 & -2.775 & -3.6 & 6.9 \end{bmatrix}$$

$$\mathbf{P} = \mathbf{T}^T\mathbf{P}_{NON} = \begin{Bmatrix} 0 \\ 0 \\ 0 \\ 0 \\ 1 \end{Bmatrix}; \mathbf{q} = \begin{Bmatrix} 0 \\ 0.1660 \\ 0.2942 \\ 0.4054 \\ 0.5166 \\ 0.6096 \\ 0.6931 \end{Bmatrix}$$

strain energy: $U_{Elem\_1} = 0.2026499611 \quad U_{Elem\_2} = 0.1439231038 \quad \boxed{U = 0.3465730648}$.

9.2.1.9 Case 9: Two Cubic Elements – C^2 Continuity: One Cubic B-spline Element

$$\mathbf{K}_{NON} = \left(\frac{3AE_0}{20L}\right)\begin{bmatrix} 18 & 5 & -11 & 0 & 0 & 0 \\ 5 & 22 & -18 & 0 & 0 & 0 \\ -11 & -18 & 72 & -18 & -13 & -7 \\ 0 & 0 & -18 & 26 & 7 & -15 \\ 0 & 0 & -13 & 7 & 30 & -24 \\ 0 & 0 & -7 & -15 & -24 & 46 \end{bmatrix}$$

$$\mathbf{P}_{NON} = \begin{Bmatrix} 0 \\ 0 \\ 0 \\ 0 \\ 0 \\ 1 \end{Bmatrix} \quad \mathbf{T}_{BS} = \begin{bmatrix} 1 & 0 & 0 & 0 \\ .5 & .5 & 0 & 0 \\ .25 & .5 & .25 & 0 \\ 0 & .5 & .5 & 0 \\ 0 & 0 & 1 & 0 \\ 0 & 0 & 0 & 1 \end{bmatrix}$$

$$\mathbf{K}_{BS} = \mathbf{T}^T\mathbf{K}_{NON}\mathbf{T} = \begin{bmatrix} 3.45 & 0.375 & -0.9 & -0.2625 \\ 0.375 & 1.8 & 0.525 & -1.65 \\ -0.9 & 0.525 & 5.55 & -4.9875 \\ -0.2625 & -1.65 & -4.987 & 6.9 \end{bmatrix} \quad \mathbf{P}_{BS} = \mathbf{T}_{BS}^T\mathbf{P}_{NON} => \begin{Bmatrix} 0 \\ 0 \\ 0 \\ 1 \end{Bmatrix}$$

$$\Rightarrow \text{de Boor controls}: \mathbf{s} = \mathbf{K}_{BS}^{-1}\,\mathbf{P}_{BS} = \begin{Bmatrix} 0 \\ 0.1658 \\ 0.4230 \\ 0.6098 \\ 0.6931 \end{Bmatrix} \text{ Bezier controls:}\mathbf{q} = \mathbf{T}_{BS}\,\mathbf{s} = \begin{Bmatrix} 0 \\ 0.1658 \\ 0.2944 \\ 0.4054 \\ 0.5164 \\ 0.6098 \\ 0.6931 \end{Bmatrix}$$

strain energy: $U_{Elem\_1} = 0.2026622896 \quad U_{Elem\_2} = 0.1439105656 \quad \boxed{U = 0.3465728552}$.

We show in Figures 9.2(a) and (b) the displacement and the strain distribution for various cases; the strains are not as easily satisfied as the displacements.

9.3 c-type Straight Beam Element

Here we present the element level stiffness control matrix and load control vector for an Euler–Bernoulli beam:

The Euler–Bernoulli beam equation for the beam, shown in Figure 9.3, is given by:

$$\boxed{\begin{aligned} \text{Momentum balance equation: } EI\, w,_{ssss}(s) &= 0, \quad s \in (0, L) \\ \text{Boundary conditions: } EI\, w,_{sss}(0) &= P_1, \\ EI\, w,_{sss}(L) &= -P_2, \\ EI\, w,_{ss}(0) &= -M_1, \\ EI\, w,_{ss}(L) &= M_2 \end{aligned}} \tag{9.4}$$

Now, using the virtual work principle with $\bar{w}(s) \equiv \delta w(s)$ as the virtual displacement, and applying integration by parts twice, we get:

$$\begin{aligned} \int_0^L \bar{w},_{ss}\, EI\, w,_{ss}\, ds &= -\bar{w}\, EI\, w,_{sss}\big|_0^L + \bar{w}\, EI\, w,_{ss}\big|_0^L \\ &= \bar{w}(0)\, P_1 + \bar{w}(L)\, P_2 + \bar{w},_s(0)\, M_1 + \bar{w},_s(L)\, M_2 \end{aligned} \tag{9.5}$$

Let us reparameterize the domain as: $\xi = \frac{s}{L}$ such that $\xi \in [0, 1]$ with $ds = L\, d\xi$ and $\frac{\partial}{\partial s} = \frac{d\xi}{ds}\frac{\partial}{\partial \xi} = \frac{1}{L}\frac{\partial}{\partial \xi}$ implying $w,_s = \frac{1}{L}\, w,_\xi$ and $w,_{ss} = \frac{1}{L^2}\, w,_{\xi\xi}$. So, equation (9.5) becomes:

$$\boxed{\frac{EI}{L^3}\int_0^L \bar{w},_{\xi\xi}\; w,_{\xi\xi}\; d\xi = \bar{w}(0)\, P_1 + \bar{w}(1)\, P_2 + \bar{w},_\xi(0)\,\frac{M_1}{L} + \bar{w},_\xi(1)\,\frac{M_2}{L} = \bar{\mathbf{w}}^T\,\mathbf{P}} \tag{9.6}$$

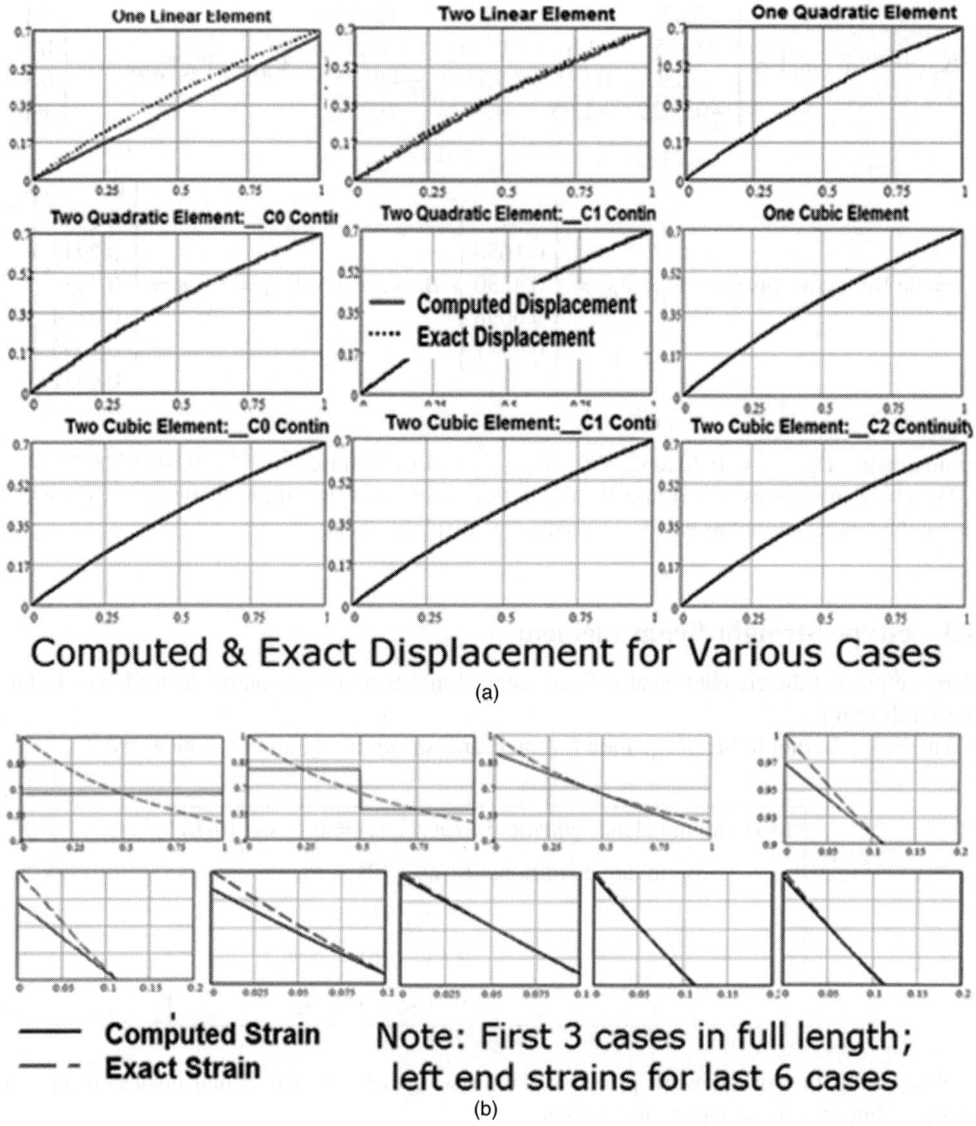

Figure 9.2 (a) Example CFEM1: computed and exact displacements for various cases. (b) Example CFEM1: computed and exact strains for various cases.

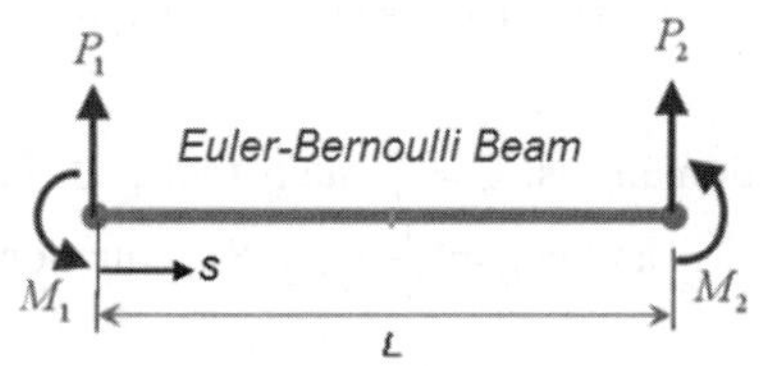

Figure 9.3 Bernstein–Bezier beam element.

where

$$\bar{\mathbf{w}} \equiv \begin{Bmatrix} \bar{w}(0) \\ \bar{w},_{\xi}(0) \\ \bar{w}(1) \\ \bar{w},_{\xi}(1) \end{Bmatrix}, \quad \text{and} \quad \mathbf{P} \equiv \frac{1}{L} \begin{Bmatrix} P_1 L \\ M_1 \\ P_2 L \\ M_2 \end{Bmatrix} \tag{9.7}$$

We now represent both the displacement and the virtual displacement by cubic Bernstein–Bezier functions as:

$$w(\xi) = \sum_0^3 B_i^3(\xi)\, q_i \quad \text{and} \quad \bar{w}(\xi) \equiv \delta w(\xi) = \sum_0^3 B_i^3(\xi)\, \delta q_i \tag{9.8}$$

In matrix form, let $\mathbf{q} = (q_0\ q_1\ q_2\ q_3)^T$, $\delta\mathbf{q} = (\delta q_0\ \delta q_1\ \delta q_2\ \delta q_3)^T$, and from equation (9.8), we get:

$$\begin{aligned} \mathbf{B}(\xi) &\overset{\Delta}{=} \frac{d}{d\xi}\{B_0^3\ \ B_1^3\ \ B_2^3\ \ B_3^3\}^T = 3\underbrace{\{-(1-\xi)^2\quad (1-4\xi+3\xi^2)\quad (2\xi-3\xi^2)\quad \xi^2\}^T}_{1\times 4} \\ \mathbf{C}(\xi) &\overset{\Delta}{=} \frac{d^2}{d\xi^2}\{B_0^3\ \ B_1^3\ \ B_2^3\ \ B_3^3\}^T = 6\underbrace{\{(1-\xi)\quad -(2-3\xi)\quad (1-3\xi)\quad \xi\}^T}_{1\times 4} \end{aligned} \tag{9.9}$$

Applying definitions given in equations (9.8) and (9.9) in equations (9.6) and (9.7), and by performing the integration on the left-hand side of the equation (9.6), we get in Bezier controls:

$$\begin{aligned} &\delta\mathbf{q}^T(\mathbf{K}_B\, \mathbf{q}) = \delta\mathbf{q}^T\, \mathbf{P}_B \\ &\text{where } \mathbf{K}_B = \frac{6EI}{L^3}\begin{bmatrix} 2 & -3 & 0 & 1 \\ -3 & 6 & -3 & 0 \\ 0 & -3 & 6 & -3 \\ 1 & 0 & -3 & 2 \end{bmatrix} \quad \text{and} \quad \mathbf{P}_B = \frac{1}{L}\begin{Bmatrix} P_1 L - 3M_1 \\ 3M_1 \\ -3M_2 \\ P_2 L + 3M_2 \end{Bmatrix} \end{aligned} \tag{9.10}$$

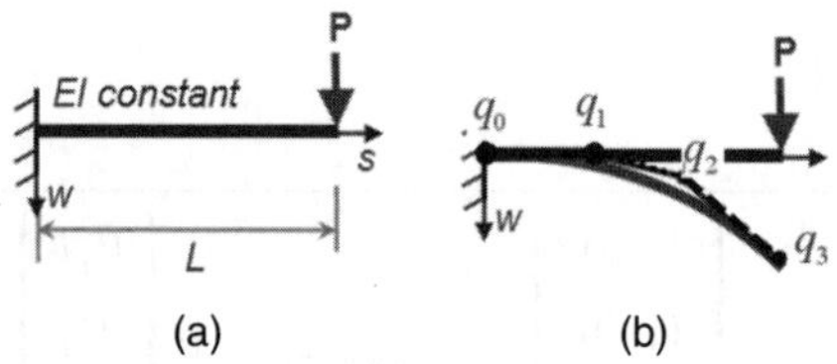

Figure 9.4 (a) Example CFEM2: Cantilevered Euler–Bernoulli beam. (b) Example CFEM2: Control polygon.

Definition: $\mathbf{K}_B$ is the unconstrained or singular Bernstein–Bezier beam element stiffness control matrix for an Euler–Bernoulli beam, $\mathbf{P}_B$ is the Bernstein–Bezier beam element load control vector. Now, with $\delta\mathbf{q}$ arbitrary, we must have the following c-type finite element equation:

$$\mathbf{K}_B\,\mathbf{q} = \mathbf{P}_B$$
$$\text{where } \mathbf{K}_B = \frac{6EI}{L^3}\begin{bmatrix} 2 & -3 & 0 & 1 \\ -3 & 6 & -3 & 0 \\ 0 & -3 & 6 & -3 \\ 1 & 0 & -3 & 2 \end{bmatrix}, \quad \text{and} \quad \mathbf{P}_B = \frac{1}{L}\begin{Bmatrix} P_1L - 3M_1 \\ 3M_1 \\ -3M_2 \\ P_2L + 3M_2 \end{Bmatrix} \tag{9.11}$$

Remarks: Later, we will use equation (9.11) that describes a Bernstein–Bezier beam element to develop a B-spline beam element. But next let us look at a problem that we can solve in two ways: by directly applying the above element derived by the virtual work principle and by using the minimum potential energy principle.

Example CFEM 2: One cubic Bernstein–Bezier element

Let us consider the cantilevered Euler–Bernoulli beam, that is, bending in flexure with no shear strain, subjected to a tip load, P, as shown in Figure 9.4(a).

Let $w(s)$ be the deflection at any point $s \in [0, L]$; however, for the Bernstein–Bezier approximation, we reparameterize by $\xi = \frac{s}{L}$ so that $\xi \in [0, 1]$, $d\xi = \frac{1}{L}ds$, and $\frac{\partial}{\partial s} = \frac{\partial}{\partial \xi}\frac{d\xi}{ds} = \frac{1}{L}\frac{\partial}{\partial \xi}$. Thus, $\frac{d^2w}{ds^2} \equiv w,_{ss} = \frac{1}{L^2}w,_{\xi\xi}$. Now, the potential energy, Π, in flexure exists and is given by:

$$\Pi = \frac{1}{2}\int_0^L EI(w,_{ss})^2\,ds - P\,w(L) = \frac{EI}{2L^3}\int_0^1 (w,_{\xi\xi})^2\,d\xi - P\,w(1) \tag{9.12}$$

Applying the minimum potential energy principle, that is, the first variation of the potential energy vanishing to zero, to equation (9.12) with $\bar{w} \equiv \delta w$ as the virtual displacement, and by Gateaux derivative, $\bar{w},^2_{\xi\xi} \equiv \delta(w,^2_{\xi\xi}) = \lim_{\varepsilon\to 0}\frac{d}{d\varepsilon}\{(w + \varepsilon\delta w),_{\xi\xi}\}^2 = 2w,_{\xi\xi}\,\delta w,_{\xi\xi} = 2w,_{\xi\xi}\,\bar{w},_{\xi\xi}$, we get:

$$\delta\Pi = \frac{EI}{2L^3}\int_0^L \delta(w,_{\xi\xi})^2\,d\xi - P\,\delta w(1) = \frac{EI}{L^3}\int_0^L w,_{\xi\xi}\,\delta w,_{\xi\xi}\,d\xi - P\,\delta w(1) = 0 \tag{9.13}$$

We choose to approximate both the displacement function, $w(\xi)$, and the virtual displacement function, $\delta w(\xi)$, by the elements of the same polynomial space: the space of cubic Bernstein polynomials in Bezier representation. Thus,

$$w(\xi) = \sum_{0}^{3} B_i^3(\xi)\, q_i \quad \text{and} \quad \delta w(\xi) = \sum_{0}^{3} B_i^3(\xi)\, \delta q_i \tag{9.14}$$

In matrix form, let $\mathbf{q} = (q_0\; q_1\; q_2\; q_3)^T$, $\delta\mathbf{q} = (\delta q_0\; \delta q_1\; \delta q_2\; \delta q_3)^T$, and $\mathbf{B}(\xi) \stackrel{\Delta}{=} \frac{d^2}{d\xi^2}\{B_0^3(\xi)\; B_1^3(\xi)\; B_2^3(\xi)\; B_3^3(\xi)\}^T = 6\underbrace{\{(1-\xi) \quad -(2-3\xi) \quad (1-3\xi) \quad \xi\}^T}_{1\times 4}$. Then, applying equation (9.14) to equation (9.13) and upon integrating, we get:

$$\delta\mathbf{q}^T(\mathbf{K}_B\,\mathbf{q}) = \delta\mathbf{q}^T\,(\mathbf{P}_B) \quad \text{where} \quad \mathbf{K}_B = \frac{EI}{L^3}\int_0^L (\mathbf{B}(\xi)^T\,\mathbf{B}(\xi))\, d\xi$$
$$= \frac{6EI}{L^3}\begin{bmatrix} 2 & -3 & 0 & 1 \\ -3 & 6 & -3 & 0 \\ 0 & -3 & 6 & -3 \\ 1 & 0 & -3 & 2 \end{bmatrix} \quad \text{and} \quad \mathbf{P}_B = \begin{Bmatrix} 0 & 0 & 0 & P \end{Bmatrix}^T \tag{9.15}$$

Definition: $\mathbf{K}_B$ is the unconstrained or singular Bernstein–Bezier beam element stiffness control matrix for an Euler–Bernoulli beam, $\mathbf{P}_B$ is the Bernstein–Bezier beam element load control vector. Now, noting that $\delta\mathbf{q}$ in equation (9.15) is arbitrary, we must have the finite element equation to solve for $\mathbf{q}$ as:

$$\frac{6EI}{L^3}\begin{bmatrix} 2 & -3 & 0 & 1 \\ -3 & 6 & -3 & 0 \\ 0 & -3 & 6 & -3 \\ 1 & 0 & -3 & 2 \end{bmatrix}\mathbf{q} = \begin{Bmatrix} 0 \\ 0 \\ 0 \\ 1 \end{Bmatrix} P \tag{9.16}$$

With the fixed end essential boundary conditions: $w(0) \Rightarrow q_0 = 0$, and $w_{,s}(s) = 3(q_1 - q_0) = 0 \Rightarrow q_1 = 0$, we get, from equation (9.16):

$$\frac{6EI}{L^3}\begin{bmatrix} 6 & -3 \\ -3 & 2 \end{bmatrix}\begin{Bmatrix} q_2 \\ q_3 \end{Bmatrix} = \begin{Bmatrix} 0 \\ P \end{Bmatrix} \tag{9.17}$$

resulting in (see Figure 9.4(b) for the control polygon):

$$\begin{Bmatrix} q_2 \\ q_3 \end{Bmatrix} = \frac{PL^3}{EI}\begin{Bmatrix} \frac{1}{6} \\ \frac{1}{3} \end{Bmatrix} \tag{9.18}$$

Finally, applying equation (9.18), we get:

$$\begin{aligned} \text{Tip displacement} &= w(1) = q_3 = \frac{PL^3}{3EI}(\text{Exact!}) \\ \text{Tip slope} &= w,_s(1) = \frac{1}{L}w,_\xi(1) = \frac{3}{L}(q_3 - q_2) = \frac{PL^2}{2EI}(\text{Exact!}) \end{aligned} \tag{9.19}$$

Discussion: For this problem, we have arrived at the solution by application of the minimum potential energy principle because the potential energy functional exists for the problem which involves a conservative system. However, not all systems possess a potential energy functional, but all systems do possess a virtual work principle. Thus, we could have used directly the result developed earlier to arrive at the same solution. Thus, in equation (9.10) with: $P_2 = P$ and $P_1 = M_1 = M_2 = 0$, we get:

$$\delta\mathbf{q}^T(\mathbf{K}_B\,\mathbf{q}) = \delta\mathbf{q}^T\,\mathbf{P}_B \quad \text{where} \quad \mathbf{K}_B = \frac{6EI}{L^3}\begin{bmatrix} 2 & -3 & 0 & 1 \\ -3 & 6 & -3 & 0 \\ 0 & -3 & 6 & -3 \\ 1 & 0 & -3 & 2 \end{bmatrix} \quad \text{and} \quad \mathbf{P}_B = \begin{Bmatrix} 0 \\ 0 \\ 0 \\ P \end{Bmatrix} \tag{9.20}$$

which is exactly same as given by equation (9.15).

9.3.1 Two-node C^1 Bernstein–Bezier Element

Here we present the solution for an Euler–Bernoulli beam, as in Figure 9.5, by selecting two two-node Bernstein–Bezier elements, as presented before, with C^1 continuity between them.

Example CFEM 3: Two cubic Bernstein–Bezier elements with C^1 continuity

The inter-element boundary condition with imposed moment M makes the beam curvature discontinuous at $x = x_1$, which in turn allows us only to consider up to C^1 inter-element continuity; in fact the non-trivial natural boundary condition is given by:

$$EI\,w,_{xx}\left(x_1^+\right) - EI\,w,_{xx}\left(x_1^-\right) = -M \tag{9.21}$$

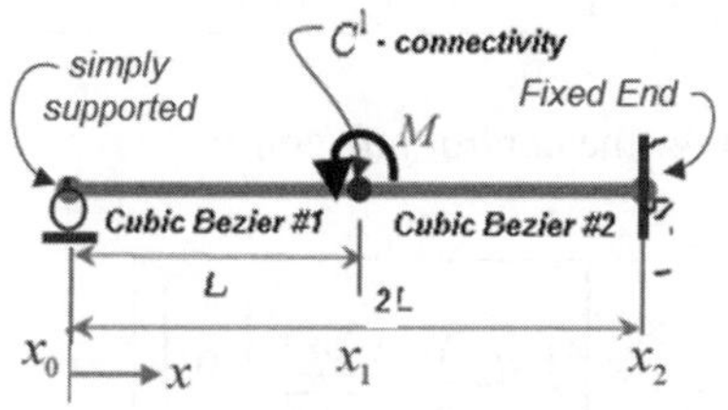

Figure 9.5 Example CFEM3: Two beams with junction moment loading.

The essential boundary conditions are:

$$w(x_0) = w(x_2) = w_{,x}(x_2) = 0 \tag{9.22}$$

The virtual work equation for the entire beam, as before, is given by:

$$\begin{aligned}\sum_{i=0}^{1}\int_0^L \bar{w}_{,xx}\, EI\, w_{,xx}\, dx &= \cancel{-\sum_{i=0}^{1} \bar{w} EI w_{,xxx}}\Bigg| + \sum_{i=0}^{1} \bar{w}_{,x}\; EI\; w_{,xx}\Bigg|_0^L \\ &= \bar{w}_{,x}(x_1^-)\; EI\; w_{,xx} - \bar{w}_{,x}(x_1^+)\; EI\; w_{,xx}\end{aligned} \tag{9.23}$$

where we have already accommodated the trivial applied shear condition. Based on the Bernstein–Bezier element stiffness matrix given by equation (9.11), the (7×7) C^0-assembled unconstrained Bezier stiffness matrix, $\mathbf{K}_B^0$, for the entire beam, following the left-hand side of equation (9.23), is given as:

$$\mathbf{K}_B^0 = \frac{6EI}{L^3} \underbrace{\begin{bmatrix} 2 & -3 & 0 & 1 & 0 & 0 & 0 \\ -3 & 6 & -3 & 0 & 0 & 0 & 0 \\ 0 & -3 & 6 & -3 & 0 & 0 & 0 \\ 1 & 0 & -3 & 4 & -3 & 0 & 1 \\ 0 & 0 & 0 & -3 & 6 & -3 & 0 \\ 0 & 0 & 0 & 0 & -3 & 6 & -3 \\ 0 & 0 & 0 & 1 & 0 & -3 & 2 \end{bmatrix}}_{7\times 7} \tag{9.24}$$

Before we can proceed similarly with the right-hand side of equation (9.23) for the assembled load vector, we need to establish the connection between the applied moments and the control vertices; for the first derivative of the Bernstein–Bezier function we have:

$$w_{,x}(x_1^-) = \frac{3}{L}(q_3 - q_2) \quad \text{and} \quad w_{,x}(x_1^+) = \frac{3}{L}(q_4 - q_3) \tag{9.25}$$

If we apply the slope continuity condition to equation (9.25), that is,

$$w_{,x}(x_1) = w_{,x}(x_1^-) = w_{,x}(x_1^+) \Rightarrow q_3 = \frac{1}{2}(q_4 - q_2) \tag{9.26}$$

then, for the right-hand side of equation (9.23), we get the assembled Bezier load vector, $\mathbf{P}_B$, for the entire beam as:

$$\underbrace{\mathbf{P}_B^0}_{7\times 1} = \frac{1}{L}\begin{bmatrix} 0 & 0 & -\frac{3}{2}M & 0 & \frac{3}{2}M & 0 & 0 \end{bmatrix}^T \tag{9.27}$$

and the assembled C^0 continuous c-type finite element control matrix equation is given as:

$$\mathbf{K}_B^0\, \mathbf{q}^0 = \mathbf{P}_B^0 \tag{9.28}$$

Now, for C^1 continuity at the junction for the control vertices, we have the transformation matrix, $\mathbf{T}_{BB}^{01}$ relating $\mathbf{q}^0$, the (7×1) C^0 control vertices with $\mathbf{q}^1$, the (6×1) C^1 control vertices as:

$$\mathbf{q}^0 = \mathbf{T}_{BB}^{01}\,\mathbf{q}^1, \text{ in expanded form: } \underbrace{\begin{Bmatrix} q_0^0 \\ q_1^0 \\ q_2^0 \\ q_3^0 \\ q_4^0 \\ q_5^0 \\ q_6^0 \end{Bmatrix}}_{7\times1} = \underbrace{\begin{bmatrix} 1 & 0 & 0 & 0 & 0 & 0 \\ 0 & 1 & 0 & 0 & 0 & 0 \\ 0 & 0 & 1 & 0 & 0 & 0 \\ 0 & 0 & .5 & .5 & 0 & 0 \\ 0 & 0 & 0 & 1 & 0 & 0 \\ 0 & 0 & 0 & 0 & 1 & 0 \\ 0 & 0 & 0 & 0 & 0 & 1 \end{bmatrix}}_{7\times6} \underbrace{\begin{Bmatrix} q_0^1 \\ q_1^1 \\ q_2^1 \\ q_3^1 \\ q_4^1 \\ q_5^1 \end{Bmatrix}}_{6\times1} \tag{9.29}$$

Applying equation (9.29) to equation (9.28), we get $\mathbf{K}_B^1$, the (6×6) C^1 continuous unconstrained Bernstein–Bezier stiffness control matrix as:

$$\underbrace{\mathbf{K}_B^1}_{6\times6} = \underbrace{\mathbf{T}_{BB}^{01T}}_{6\times7}\,\underbrace{\mathbf{K}_B^0}_{7\times7}\,\underbrace{\mathbf{T}_{BB}^{01}}_{7\times6} = \frac{6EI}{L^3}\begin{bmatrix} 2 & -3 & .5 & .5 & 0 & 0 \\ -3 & 6 & -3 & 0 & 0 & 0 \\ .5 & -3 & 4 & -2 & 0 & .5 \\ .5 & 0 & -2 & 4 & -3 & .5 \\ 0 & 0 & 0 & -3 & 6 & -3 \\ 0 & 0 & .5 & .5 & -3 & 2 \end{bmatrix} \tag{9.30}$$

and $\mathbf{P}_B^1$, the (6×1) Bernstein–Bezier control vector as:

$$\underbrace{\mathbf{P}_B^1}_{6\times1} = \underbrace{\mathbf{T}_{BB}^{01T}}_{6\times7}\,\underbrace{\mathbf{P}_B^0}_{7\times1} = \frac{1}{L}\begin{bmatrix} 0 & 0 & -1.5M & 1.5M & 0 & 0 \end{bmatrix}^T \tag{9.31}$$

with the C^1- continuous unconstrained Bernstein–Bezier equation as:

$$\mathbf{K}_B^1\,\mathbf{q}^1 = \mathbf{P}_B^1 \tag{9.32}$$

Now, we can impose the essential boundary conditions as given in equation (9.22) to control the vertices: $q_0^1 = q_4^1 = q_5^1 = 0$, so the reduced equation, from equation (9.32), is given as:

$$\frac{6EI}{L^2}\begin{bmatrix} 6 & -3 & 0 \\ -3 & 4 & -2 \\ 0 & -2 & 4 \end{bmatrix}\begin{Bmatrix} q_1^1 \\ q_2^1 \\ q_3^1 \end{Bmatrix} = \frac{3M}{2}\begin{Bmatrix} 0 \\ -1 \\ +1 \end{Bmatrix} \tag{9.33}$$

which upon solution for the C^1 control vertices, and application of the transformation relating the C^1 to the C^0 control vertices, as given by equation (9.29), gives:

$$\begin{Bmatrix} q_0^0 \\ q_1^0 \\ q_2^0 \\ q_3^0 \\ q_4^0 \\ q_5^0 \\ q_6^0 \end{Bmatrix} = \frac{ML^2}{96EI} \begin{Bmatrix} 0 \\ -4 \\ -8 \\ -3 \\ 2 \\ 0 \\ 0 \end{Bmatrix} \tag{9.34}$$

With the help of the Bernstein–Bezier functions and the computed control vertices as shown in equation (9.34), we can compute the displacements and slopes along the beam:

$$\begin{aligned}
&w(x_0) = q_0 = 0 && w,_x(x_0) = \frac{1}{L} w,_{\xi_1}(0) = \frac{3}{L}(q_1 - \cancel{q_0}) = -\frac{12ML}{96EI} \\
&w(x_1^-) = q_3 = -\frac{3ML^2}{96EI} && w,_x(x_1^-) = \frac{1}{L} w,_{\xi_1}(1) = \frac{3}{L}(q_3 - \cancel{q_2}) = \frac{15ML}{96EI} \\
&w(x_1^+) = q_3 = -\frac{3ML^2}{96EI} && w,_x(x_1^+) = \frac{1}{L} w,_{\xi_2}(0) = \frac{3}{L}(q_4 - q_3) = \frac{15ML}{96EI} \\
&w(x_2) = q_6 = 0 && w,_x(x_2) = \frac{1}{L} w,_{\xi_2}(0) = \frac{3}{L}(q_6 - q_5) = 0
\end{aligned} \tag{9.35}$$

We present the exact deformation as indicated by equation (9.35) graphically as shown in Figure 9.6.

Now, for the force and moment distribution, let us first note the second (curvature) and third derivatives of the Bernstein–Bezier polynomials:

$$\begin{aligned}
w,_{\xi_1\xi_1} &= 6\{(1-\xi_1)(q_2 - 2q_1 + \cancel{q_0}) + \xi_1(q_3 - 2q_2 + q_1)\}, \quad \xi_1 L \in [x_0, x_1] \\
w,_{\xi_1\xi_1\xi_1} &= 6\{3q_1 - 3q_2 + q_3\} \\
w,_{\xi_2\xi_2} &= 6\{(1-\xi_2)(\cancel{q_5} - 2q_4 + q_3) + \xi_2(\cancel{q_6} - \cancel{2q_5} + q_4)\}, \quad \xi_2 L \in [x_1, x_2] \\
w,_{\xi_2\xi_2\xi_2} &= 6\{-q_3 + 3q_4\}
\end{aligned} \tag{9.36}$$

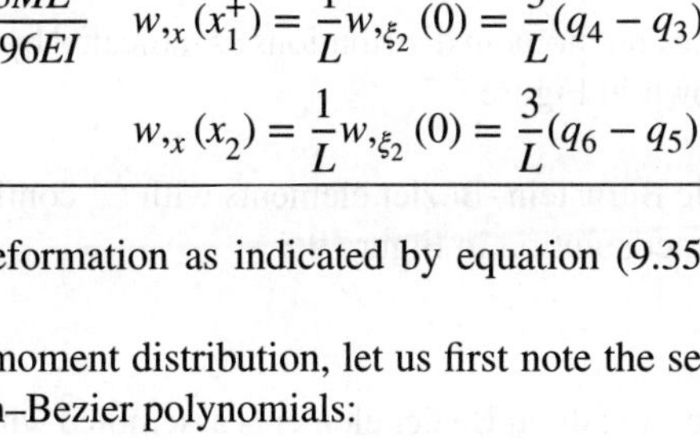

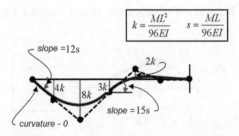

Figure 9.6 Example CFEM3: Exact deformed shape.

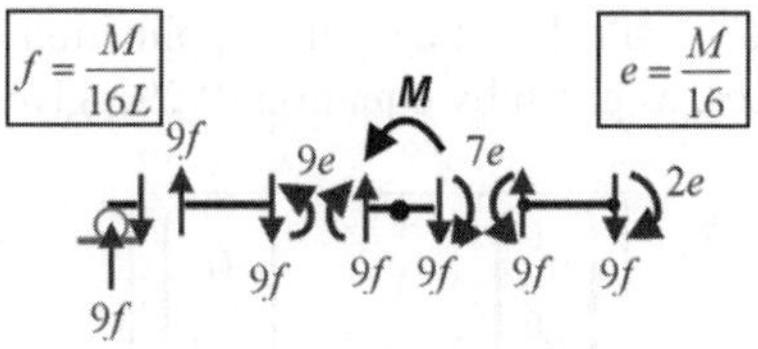

Figure 9.7 Example CFEM3: Exact force and moment distribution.

Using the expressions in equation (9.36), we can determine the force and moment distributions as:

$$\begin{aligned}
&M(x_0 = 0) = EIw,_{xx}(x_0) = \frac{EI}{L^2}w,_{\xi\xi}(0) = 0\\
&M(x_1^- = L^-) = EIw,_{xx}(x_1^-) = \frac{EI}{L^2}w,_{\xi\xi}(\xi_1 = 1) = \frac{9M}{16}\\
&M(x_1^+ = L^+) = EIw,_{xx}(x_1^+) = \frac{EI}{L^2}w,_{\xi\xi}(\xi_2 = 0) = -\frac{7M}{16}\\
&M(x_2 = 2L) = EIw,_{xx}(x_2) = \frac{EI}{L^2}w,_{\xi\xi}(\xi_2 = 1) = \frac{2M}{16}\\
&F(x) = EIw,_{xxx}(x) = \frac{EI}{L^3}w,_{\xi\xi\xi}(\xi) = \frac{9M}{16L}, \quad x \in [0, 2L]
\end{aligned} \tag{9.37}$$

We present the exact force and moment distributions as indicated by equation (9.37) graphically as a free-body diagram shown in Figure 9.7.

Exercise: Apply two cubic Bernstein–Bezier elements with C^1 continuity as before to solve the problem shown with the exact solution in Figure 9.8.

Remark

- Let us note that two cubic Bernstein–Bezier elements assembled with C^1 continuity reproduced the exact deformations, forces and moments; imposition of any additional continuity such as C^2 continuity will spell disaster because the curvature is obviously discontinuous at the junction where a moment has been applied. However, we can refine our elements by considering B-spline elements, which we consider next.

9.3.2 Two-node B-spline Beam: Composite Bernstein–Bezier Form

We may recall from our discussion in Chapter 5 on curves that a cubic B-spline can be constructed by two piecewise cubic Bernstein–Bezier curves joined together by C^2 connectivity. Let us

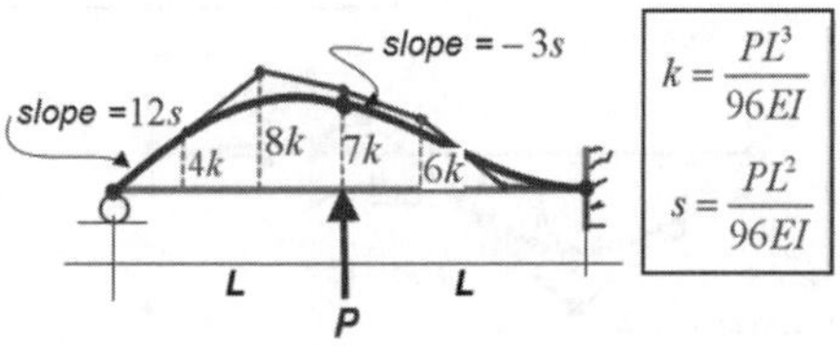

Figure 9.8 Exercise problem.

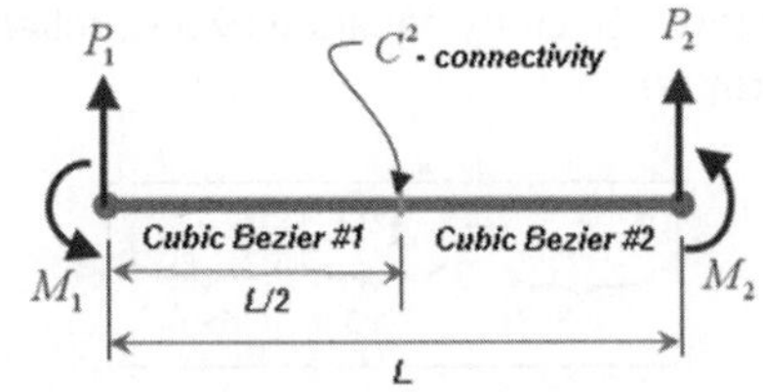

Figure 9.9 B-spline element in Bernstein–Bezier composite form.

consider the free-body of an Euler–Bernoulli beam, that is, bending in flexure with no shear strain, as shown in Figure 9.9.

Let the control points of the Bezier element # 1 be given by: $\{\mathbf{q}_0, \mathbf{q}_1, \mathbf{q}_2, \mathbf{q}_3\}$, and for the Bezier element # 2: $\{\mathbf{q}_4, \mathbf{q}_5, \mathbf{q}_6, \mathbf{q}_7\}$, and for the resulting B-spline controls: $\{\mathbf{s}_0, \mathbf{s}_1, \mathbf{s}_2, \mathbf{s}_3, \mathbf{s}_4\}$. Recalling that $k = 4$, with $k = n + 1$ where n, k are the degree and the order of a cubic curve, respectively, the original two cubic Bezier curves have $lk = 2 \times 4 = 8$ controls where $l = 2$ is the number of Bezier segments; the B-spline, correspondingly, has: $l + k - 1 = l + n = 2 + 3 = 5$ control points which may be explained as: C^0 continuity at the junction amounting to one constraint, and thus reduces the Bezier control number to $l(k - 1) + 1 = 2 \times 3 + 1 = 7$; C^1 continuity adds one extra constraint, reducing the number to $7 - 1 = 6$. Finally, C^2 continuity gives the number for the B-spline controls as $6 - 1 = 5$ (which could have been directly obtained for B-spline controls as indicated earlier and as shown in Figure 9.10:

Let us now recapture the effect of the constraints due and up to C^2- continuity, described before in Chapter 5 on curves, as follows:

$$\boxed{\begin{aligned}
&q_0 = s_0 \quad q_1 = s_1 \\
&q_2 = \frac{h_1}{h_0 + h_1} s_1 + \frac{h_0}{h_0 + h_1} s_2 \quad (C^2 \text{ continuity}) \\
&q_3 = \frac{h_1\, q_2 + h_0\, q_4}{h_0 + h_1} \qquad\qquad (\text{junction pt. } C^1 \text{ continuity}) \\
&q_4 = \frac{h_1}{h_0 + h_1} s_2 + \frac{h_0}{h_0 + h_1} s_3 \quad (C^2 \text{ continuity}) \\
&q_5 = s_3 \quad q_1 = s_4
\end{aligned}} \tag{9.38}$$

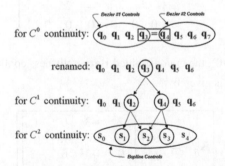

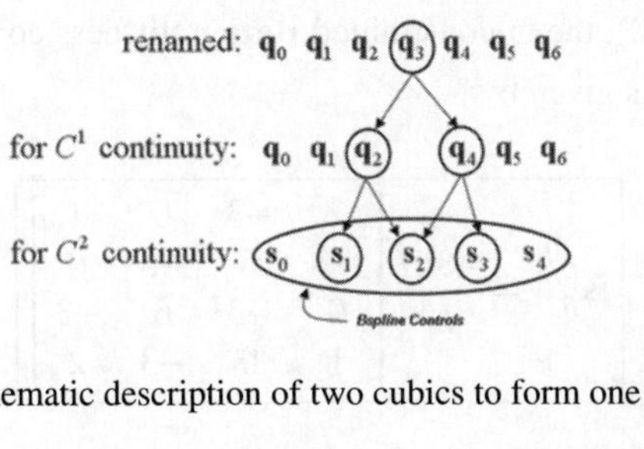

Figure 9.10 Schematic description of two cubics to form one B-spline element.

We can write the transformation, given by equation (9.38), relating the Bezier controls to the B-spline controls in matrix form as:

$$\boxed{\underbrace{\mathbf{q}}_{7\times1} = \underbrace{\mathbf{T}^{BS}}_{7\times5}\underbrace{\mathbf{s}}_{5\times1}} \tag{9.39}$$

where in expanded form:

$$\underbrace{\begin{Bmatrix} q_0 \\ q_1 \\ q_2 \\ q_3 \\ q_4 \\ q_5 \\ q_6 \end{Bmatrix}}_{7\times1} = \underbrace{\begin{bmatrix} 1 & 0 & 0 & 0 & 0 \\ 0 & 1 & 0 & 0 & 0 \\ 0 & \beta_1 & \beta_0 & 0 & 0 \\ 0 & \beta_1^2 & 2\beta_1\beta_0 & \beta_0^2 & 0 \\ 0 & 0 & \beta_1 & \beta_0 & 0 \\ 0 & 0 & 0 & 1 & 0 \\ 0 & 0 & 0 & 0 & 1 \end{bmatrix}}_{7\times5} \underbrace{\begin{Bmatrix} s_0 \\ s_1 \\ s_2 \\ s_3 \\ s_4 \end{Bmatrix}}_{5\times1} \tag{9.40}$$

With

$$\boxed{\beta_0 \stackrel{\Delta}{=} \frac{h_0}{h_0 + h_1} \quad \text{and} \quad \beta_1 \stackrel{\Delta}{=} \frac{h_1}{h_0 + h_1}} \tag{9.41}$$

where for the interval length of each of the Bezier beam segments, we have: $h_0 = h_1 = \frac{L}{2}$. Then we get, the transformation matrix, $\mathbf{T}^{BS}$, from B-spline controls, $\mathbf{s}$, to Bezier controls, $\mathbf{q}$, as:

$$\underbrace{\mathbf{T}^{BS}}_{7\times5} = \begin{bmatrix} 1 & 0 & 0 & 0 & 0 & 0 & 0 \\ 0 & 1 & \frac{1}{2} & \frac{1}{4} & 0 & 0 & 0 \\ 0 & 0 & \frac{1}{2} & \frac{1}{2} & \frac{1}{2} & 0 & 0 \\ 0 & 0 & 0 & \frac{1}{4} & \frac{1}{2} & 1 & 0 \\ 0 & 0 & 0 & 0 & 0 & 0 & 1 \end{bmatrix}^T \tag{9.42}$$

Now, from equation (9.10), the unconstrained Bezier stiffness control matrix, $\mathbf{K}_B^{L/2}$, modified for $\frac{1}{2}L$ long flexural beam, is given by:

$$\boxed{\mathbf{K}_B^{L/2} = \frac{48EI}{L^3}\begin{bmatrix} 2 & -3 & 0 & 1 \\ -3 & 6 & -3 & 0 \\ 0 & -3 & 6 & -3 \\ 1 & 0 & -3 & 2 \end{bmatrix}} \tag{9.43}$$

Thus, the (7×7) C^0-assembled unconstrained Bezier stiffness matrix, $\mathbf{K}_B$, for the entire beam is given as:

$$\mathbf{K}_B = \frac{48EI}{L^3}\underbrace{\begin{bmatrix} 2 & -3 & 0 & 1 & 0 & 0 & 0 \\ -3 & 6 & -3 & 0 & 0 & 0 & 0 \\ 0 & -3 & 6 & -3 & 0 & 0 & 0 \\ 1 & 0 & -3 & 4 & -3 & 0 & 1 \\ 0 & 0 & 0 & -3 & 6 & -3 & 0 \\ 0 & 0 & 0 & 0 & -3 & 6 & -3 \\ 0 & 0 & 0 & 1 & 0 & -3 & 2 \end{bmatrix}}_{7\times 7} \tag{9.44}$$

Similarly, from equation (9.10), the Bezier load control vector, $\mathbf{P}_B^{L/2}$, modified for $\frac{1}{2}L$ long flexural beam, is given by:

$$\mathbf{P}_B^{L/2} = \frac{1}{L}\left[P_1L - 6M_1 \quad 6M_1 \quad -6M_2 \quad P_2L + 6M_2\right]^T \tag{9.45}$$

Thus, the (7×7) C^0-assembled Bezier load vector, $\mathbf{P}_B$, for the entire beam, with junction load and moments identically zero, is given as:

$$\underbrace{\mathbf{P}_B}_{7\times 1} = \frac{1}{L}\left[P_1L - 6M_1 \quad 6M_1 \quad 0 \quad 0 \quad 0 \quad -6M_2 \quad P_2L + 6M\right]^T \tag{9.46}$$

Thus, for the beam, we have the C^0-assembled Bezier element equation as:

$$\mathbf{K}_B\,\mathbf{q} = \mathbf{P}_B$$

$$\text{where } \mathbf{K}_B = \frac{48EI}{L^3}\underbrace{\begin{bmatrix} 2 & -3 & 0 & 1 & 0 & 0 & 0 \\ -3 & 6 & -3 & 0 & 0 & 0 & 0 \\ 0 & -3 & 6 & -3 & 0 & 0 & 0 \\ 1 & 0 & -3 & 4 & -3 & 0 & 1 \\ 0 & 0 & 0 & -3 & 6 & -3 & 0 \\ 0 & 0 & 0 & 0 & -3 & 6 & -3 \\ 0 & 0 & 0 & 1 & 0 & -3 & 2 \end{bmatrix} \text{ and } \underbrace{\mathbf{P}_B}_{7\times 1} = \frac{1}{L}\begin{Bmatrix} P_1L - 6M_1 \\ 6M_1 \\ 0 \\ 0 \\ 0 \\ -6M_2 \\ P_2L + 6M_2 \end{Bmatrix}}_{7\times 7} \tag{9.47}$$

Now, by pre-multiplying equation (9.47) by $\mathbf{T}_{BS}^T$ and applying the transformation equation, we finally get the c-type equation for the two-node B-spline flexural element:

$$\boxed{\underbrace{\mathbf{K}_S}_{5\times5}\ \underbrace{\mathbf{s}}_{5\times1} = \underbrace{\mathbf{P}_S}_{5\times1}} \tag{9.48}$$

where $\underbrace{\mathbf{K}_S}_{5\times5}$ is the unconstrained or singular B-spline element stiffness control matrix given as:

$$\boxed{\underbrace{\mathbf{K}_S}_{5\times5} = \underbrace{\mathbf{T}_{BS}^T}_{5\times7}\underbrace{\mathbf{K}_B}_{7\times7}\underbrace{\mathbf{T}_{BS}}_{7\times5} = \frac{12EI}{L^3}\begin{bmatrix} 8 & -11 & 2 & 1 & 0 \\ -11 & 16 & -4 & -2 & 1 \\ 2 & -4 & 4 & -4 & 2 \\ 1 & -2 & -4 & 16 & -11 \\ 0 & 1 & 2 & -11 & 8 \end{bmatrix}} \tag{9.49}$$

and $\underbrace{\mathbf{P}_S}_{5\times1}$ is the B-spline element load control vector given as:

$$\boxed{\underbrace{\mathbf{P}_S}_{5\times1} = \underbrace{\mathbf{T}_{BS}^T}_{5\times7}\underbrace{\mathbf{P}_B}_{7\times1} = \frac{1}{L}\left[P_1L - 6M_1 \quad 6M_1 \quad 0 \quad -6M_2 \quad P_2L + 6M_2\right]^T} \tag{9.50}$$

Example CFEM 4: Two cubic B-spline elements with C^1 connectivity

Let us consider the beam in flexure with no shear strain as shown in Figure 9.11.

Let the de Boor control polygon: $\{s_0^0, s_1^0, s_2^0, s_3^0, s_4^0\}$ describes the B-spline element # 1, and the B-spline element # 2 is described by: $\{s_0^1, s_1^1, s_2^1, s_3^1, s_4^1\}$. Using the B-spline stiffness matrix equation (9.48) and applying direct stiffness assembly to two B-spline elements with $P_1 = 0$, we get, for the entire system, the C^0 control polygon, $\{s_0, s_1, s_2, s_3, s_4, s_5, s_6, s_7, s_8\}$, stiffness

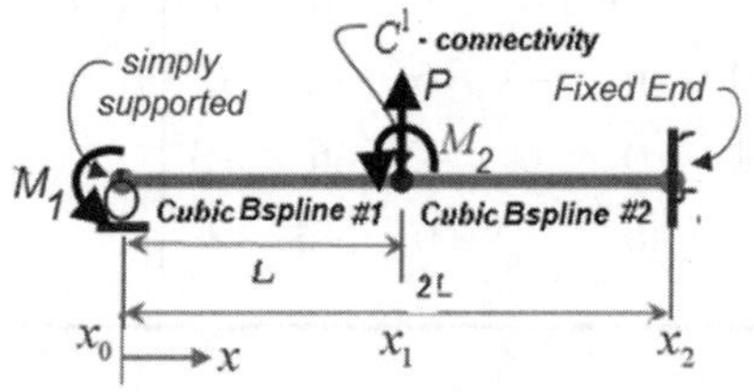

Figure 9.11 Example CFEM4: Two B-splines with C1-junction for shear and moment.

matrix, $\underbrace{\mathbf{K}_S^0}_{9\times 9}$, and load vector, $\underbrace{\mathbf{P}_S^0}_{9\times 1}$, related by:

$$\boxed{\underbrace{\mathbf{K}_S^0}_{9\times 9}\ \underbrace{\mathbf{s}^0}_{9\times 1} = \underbrace{\mathbf{P}_S^0}_{9\times 1}}$$

$$\begin{bmatrix} 8 & -11 & 2 & 1 & 0 & 0 & 0 & 0 & 0 \\ -11 & 16 & -4 & -2 & 1 & 0 & 0 & 0 & 0 \\ 2 & -4 & 4 & -4 & 2 & 0 & 0 & 0 & 0 \\ 1 & -2 & -4 & 16 & -11 & 0 & 0 & 0 & 0 \\ 0 & 1 & 2 & -11 & 16 & -11 & 2 & 1 & 0 \\ 0 & 0 & 0 & 0 & -11 & 16 & -4 & -2 & 1 \\ 0 & 0 & 0 & 0 & 2 & -4 & 4 & -4 & 2 \\ 0 & 0 & 0 & 0 & 1 & -2 & -4 & 16 & -11 \\ 0 & 0 & 0 & 0 & 0 & 1 & 2 & -11 & 16 \end{bmatrix} \begin{Bmatrix} s_0 = s_0^0 \\ s_1 = s_1^0 \\ s_2 = s_2^0 \\ s_3 = s_3^0 \\ s_4 = s_4^0 = s_0^1 \\ s_5 = s_1^1 \\ s_6 = s_2^1 \\ s_7 = s_3^1 \\ s_8 = s_4^1 \end{Bmatrix} = \begin{Bmatrix} -6M_1 \\ 6M_1 \\ 0 \\ -3M_2 \\ P_2L \\ 3M_2 \\ 0 \\ 0 \\ 0 \end{Bmatrix} \tag{9.51}$$

Now, imposing the C^1 continuity matrix, $\mathbf{T}_{SS}^{01}$ – with $s_4 = \frac{1}{2}(s_3 + s_5)$ and thus eliminating control vertex, s_4 – to equation (9.51), similar to the previous example, we get the corresponding C^1 control polygon, $\mathbf{s}^1 = \{s_0, s_1, s_2, s_3, s_5, s_6, s_7, s_8\}^T$, stiffness matrix, $\underbrace{\mathbf{K}_S^1}_{8\times 8} = \underbrace{\mathbf{T}_{SS}^{01T}}_{8\times 9}\ \underbrace{\mathbf{K}_S^0}_{9\times 9}\ \underbrace{\mathbf{T}_{SS}^{01}}_{9\times 8}$, and load vector, $\underbrace{\mathbf{P}_S^1}_{8\times 1} = \underbrace{\mathbf{T}_{SS}^{01T}}_{8\times 9}\ \underbrace{\mathbf{P}_S^0}_{9\times 1}$, related by:

$$\boxed{\underbrace{\mathbf{K}_S^1}_{9\times 9}\ \underbrace{\mathbf{s}^1}_{9\times 1} = \underbrace{\mathbf{P}_S^1}_{9\times 1}}$$

$$\frac{12EI}{L^2}\begin{bmatrix} 8 & -11 & 2 & 1 & 0 & 0 & 0 & 0 \\ -11 & 16 & -4 & -1.5 & .5 & 0 & 0 & 0 \\ 2 & -4 & 4 & -3 & 1 & 0 & 0 & 0 \\ 1 & -1.5 & -3 & 9 & -7 & 1 & .5 & 0 \\ 0 & .5 & 1 & -7 & 9 & -3 & -1.5 & 1 \\ 0 & 0 & 0 & 1 & -3 & 4 & -4 & 2 \\ 0 & 0 & 0 & .5 & -1.5 & -4 & 16 & -11 \\ 0 & 0 & 0 & 0 & 1 & 2 & -11 & 8 \end{bmatrix} \begin{Bmatrix} s_0 \\ s_1 \\ s_2 \\ s_3 \\ s_5 \\ s_6 \\ s_7 \\ s_8 \end{Bmatrix} = \begin{Bmatrix} -6M_1 \\ 6M_1 \\ 0 \\ .5P_2L - 3M_2 \\ .5P_2L + 3M_2 \\ 0 \\ 0 \\ 0 \end{Bmatrix} \tag{9.52}$$

Applying the essential boundary conditions: $s_0 = 0$ for the simply supported end, and $s_7 = s_8 = 0$ for the fixed end to equation (9.52), we get:

$$\frac{12EI}{L^2}\begin{bmatrix} 16 & -4 & -1.5 & .5 & 0 \\ -4 & 4 & -3 & 1 & 0 \\ -1.5 & -3 & 9 & -7 & 1 \\ .5 & 1 & -7 & 9 & -3 \\ 0 & 0 & 1 & -3 & 4 \end{bmatrix}\begin{Bmatrix} s_1 \\ s_2 \\ s_3 \\ s_5 \\ s_6 \end{Bmatrix} = \frac{1}{2}P_2L\begin{Bmatrix} 0 \\ 0 \\ 1 \\ 1 \\ 0 \end{Bmatrix} - 6M_1\begin{Bmatrix} 1 \\ 0 \\ 0 \\ 0 \\ 0 \end{Bmatrix} + 3M_2\begin{Bmatrix} 0 \\ 0 \\ -1 \\ 1 \\ 0 \end{Bmatrix} \tag{9.53}$$

We can solve for the control vertices in equation (9.53) to get:

$$\begin{Bmatrix} s_1 \\ s_2 \\ s_3 \\ s_5 \\ s_6 \end{Bmatrix} = \frac{L^2}{96EI}\left[P_2L\begin{Bmatrix} 2 \\ 6 \\ 7.5 \\ 6.5 \\ 3 \end{Bmatrix} - M_1\begin{Bmatrix} 8 \\ 16 \\ 14 \\ 10 \\ 4 \end{Bmatrix} + 3M_2\begin{Bmatrix} -2 \\ -6 \\ -1.5 \\ -.5 \\ 1 \end{Bmatrix}\right] \tag{9.54}$$

and additionally, from the C^1 continuity condition, we get:

$$s_4 = \frac{1}{2}(s_3 + s_5) = \frac{L^2}{96EI}\{7P_2L - 12M_1 - 3M_2\} \tag{9.55}$$

We can now use Bernstein–Bezier and B-spline connection as given in equations (9.39) and (9.42), into equations (9.54) and (9.55) to describe each element in local Bernstein–Bezier control vertices:

Element #1: Bernstein–Bezier control polygon for displacements and rotations

$$\begin{Bmatrix} q_0^1 \\ q_1^1 \\ q_2^1 \\ q_3^1 \\ q_4^1 \\ q_5^1 \\ q_6^1 \end{Bmatrix} = \frac{L^2}{96EI}\left[P_2L\begin{Bmatrix} 0 \\ 2 \\ 4 \\ 5\frac{3}{8} \\ 6\frac{1}{4} \\ 7\frac{1}{2} \\ 7 \end{Bmatrix} - M_1\begin{Bmatrix} 0 \\ 8 \\ 12 \\ 13\frac{1}{2} \\ 15 \\ 14 \\ 12 \end{Bmatrix} + M_2\begin{Bmatrix} 0 \\ -2 \\ -4 \\ -4\frac{7}{8} \\ -5\frac{3}{4} \\ -5\frac{1}{2} \\ -3 \end{Bmatrix}\right] \tag{9.56}$$

Element #2: Bernstein–Bezier control polygon for displacements and rotations

$$\begin{Bmatrix} q_0^1 \\ q_1^1 \\ q_2^1 \\ q_3^1 \\ q_4^1 \\ q_5^1 \\ q_6^1 \end{Bmatrix} = \frac{L^2}{96EI}\left[P_2 L \begin{Bmatrix} 0 \\ 2 \\ 4 \\ 5\frac{3}{8} \\ 6\frac{1}{4} \\ 7\frac{1}{2} \\ 7 \end{Bmatrix} - M_1 \begin{Bmatrix} 0 \\ 8 \\ 12 \\ 13\frac{1}{2} \\ 15 \\ 14 \\ 12 \end{Bmatrix} + M_2 \begin{Bmatrix} 0 \\ -2 \\ -4 \\ -4\frac{7}{8} \\ -5\frac{3}{4} \\ -5\frac{1}{2} \\ -3 \end{Bmatrix} \right] \tag{9.57}$$

9.3.2.1 Moment Distributions

Let us first recall that each of the B-spline elements is composed of two Bernstein–Bezier elements; thus, for the second half, that is, $x \in (\frac{L}{2}, L]$, of the B-spline element #1, the curvature is given by: $w,_{xx} = \left(\frac{1}{L/2}\right)^2 w,_{\xi_1\xi_1}$, and we get:

Case M_1: Moment distribution

$$\begin{aligned}
&M(0) = EIw,_{xx}(x_1 = 0) = \cancel{EI}\left(\frac{M_1\cancel{L^2}}{96\cancel{EI}}\right)\left(\frac{4}{\cancel{L^2}}\right)\{6\,(q_2^1 - 2q_1^1 + q_0^1) = -M_1\\
&M\left(\frac{L}{2}\right) = EIw,_{xx}\left(x_1 = \frac{L}{2}\right) = \cancel{EI}\left(\frac{M_1\cancel{L^2}}{96\cancel{EI}}\right)\left(\frac{4}{\cancel{L^2}}\right)\{6\,(q_3^1 - 2q_2^1 + q_1^1) = -\left(\frac{5}{8}\right)M_1\\
&M(L^-) = EIw,_{xx}(x_1 = L^-) = \cancel{EI}\left(\frac{M_1\cancel{L^2}}{96\cancel{EI}}\right)\left(\frac{4}{\cancel{L^2}}\right)\left\{6\left(q_4^1 - 2q_5^1 + q_6^1\right) = -\left(\frac{1}{4}\right)M_1\right.\\
&\qquad\qquad\qquad\qquad\qquad\qquad\qquad\qquad\qquad\text{Curvature matched } @x = L\\
&M(L^+) = EIw,_{xx}(x_2 = 0^+) = \cancel{EI}\left(\frac{M_1\cancel{L^2}}{96\cancel{EI}}\right)\left(\frac{4}{\cancel{L^2}}\right)\{6\,(q_0^2 - 2q_1^2 + q_2^2) = -\left(\frac{1}{4}\right)M_1\\
&M(2L) = EIw,_{xx}(x_2 = L) = \cancel{EI}\left(\frac{M_1\cancel{L^2}}{96\cancel{EI}}\right)\left(\frac{4}{\cancel{L^2}}\right)\left\{6\left(q_4^2 - 2q_5^2 + q_6^2\right) = \left(\frac{1}{8}\right)M_1\right.
\end{aligned} \tag{9.58}$$

Case M_2: Moment distribution
Similarly, we get:

$$M(0) = EIw_{,xx}(x_1 = 0) = \cancel{EI}\left(\frac{M_2\cancel{L^2}}{96\cancel{EI}}\right)\left(\frac{4}{\cancel{L^2}}\right)\{6\,(q_2^1 - 2q_1^1 + q_0^1) = 0$$

$$M\left(\frac{L}{2}\right) = EIw_{,xx}\left(x_1 = \frac{L}{2}\right) = \cancel{EI}\left(\frac{M_2\cancel{L^2}}{96\cancel{EI}}\right)\left(\frac{4}{\cancel{L^2}}\right)\{6\,(q_3^1 - 2q_2^1 + q_1^1) = \left(\frac{9}{32}\right)M_2$$

$$M(L^-) = EIw_{,xx}(x_1 = L^-) = \cancel{EI}\left(\frac{M_2\cancel{L^2}}{96\cancel{EI}}\right)\left(\frac{4}{\cancel{L^2}}\right)\left\{6\left(q_4^1 - 2q_5^1 + q_6^1\right) = \left(\frac{9}{16}\right)M_2\right.$$

Curvature discontinuous @$x = L$

$$M(L^+) = EIw_{,xx}(x_2 = 0^+) = \cancel{EI}\left(\frac{M_2\cancel{L^2}}{96\cancel{EI}}\right)\left(\frac{4}{\cancel{L^2}}\right)\{6\,(q_0^2 - 2q_1^2 + q_2^2) = \left(\frac{7}{16}\right)M_2$$

$$M(2L) = EIw_{,xx}(x_2 = L) = \cancel{EI}\left(\frac{M_2\cancel{L^2}}{96\cancel{EI}}\right)\left(\frac{4}{\cancel{L^2}}\right)\left\{6\left(q_4^2 - 2q_5^2 + q_6^2\right) = \left(\frac{1}{8}\right)M_2\right.$$

(9.59)

Case P_2: Moment distribution
Similarly, we get:

$$M(0) = EIw_{,xx}(x_1 = 0) = \cancel{EI}\left(\frac{P_2L^3}{96\cancel{EI}}\right)\left(\frac{4}{\cancel{L^2}}\right)\{6\,(q_3^1 - 2q_2^1 + q_1^1) = 0$$

$$M\left(\frac{L}{2}\right) = EIw_{,xx}\left(x_1 = \frac{L}{2}\right) = \cancel{EI}\left(\frac{M_2\cancel{L^2}}{96\cancel{EI}}\right)\left(\frac{4}{\cancel{L^2}}\right)\{6\,(q_3^1 - 2q_2^1 + q_1^1) = -\left(\frac{5}{32}\right)P_2L$$

$$M(L^-) = EIw_{,xx}(x_1 = L^-) = \cancel{EI}\left(\frac{P_2L^3}{96\cancel{EI}}\right)\left(\frac{4}{\cancel{L^2}}\right)\left\{6\left(q_4^1 - 2q_5^1 + q_6^1\right) = -\left(\frac{5}{16}\right)P_2L\right.$$

Curvature matched @$x = L$

$$M(L^+) = EIw_{,xx}(x_2 = 0^*) = \cancel{EI}\left(\frac{P_2L^3}{96\cancel{EI}}\right)\left(\frac{4}{\cancel{L^2}}\right)\{6\,(q_0^2 - 2q_1^2 + q_2^2) = -\left(\frac{5}{16}\right)P_2L$$

$$M(2L) = EIw_{,xx}(x_2 = L) = \cancel{EI}\left(\frac{P_2L^3}{96\cancel{EI}}\right)\left(\frac{4}{\cancel{L^2}}\right)\left\{6\left(q_4^2 - 2q_5^2 + q_6^2\right) = \left(\frac{3}{8}\right)P_2L\right.$$

(9.60)

9.3.2.2 Shear Distributions

Let us recall again that each of the B-spline elements are composed of two Bernstein–Bezier elements; thus, for each Bezier element, the shear is constant is given by: $w_{,xxx} = \left(\frac{1}{L/2}\right)^3 w_{,\xi_1\xi_1\xi_1} = \frac{8}{L^3} w_{,\xi_1\xi_1\xi_1}$; let the shear be denoted as $V(x)$, then, for moment loading:

$$\begin{aligned}
&\text{Applied moment:}\\
&V(x) = EIw_{,xxx} = \frac{8}{L^3} w_{,\xi_1\xi_1\xi_1} = \left(\frac{8}{L^3}\right)\left(\frac{ML^2}{96EI}\right) 6\{q_3 - 3q_2 + 3q_1 - q_0\}\\
&\quad = \frac{M}{2L}\{q_3 - 3q_2 + 3q_1 - q_0\}\\
&\text{Applied force:}\\
&V(x) = EIw_{,xxx} = \frac{8}{L^3} w_{,\xi_1\xi_1\xi_1} = \left(\frac{8}{L^3}\right)\left(\frac{PL^3}{96EI}\right) 6\{q_3 - 3q_2 + 3q_1 - q_0\}\\
&\quad = \frac{P}{2}\{q_3 - 3q_2 + 3q_1 - q_0\}
\end{aligned} \tag{9.61}$$

Case M_1: Shear distribution

$$\begin{aligned}
V(0) &= EIw_{,xxx}(x_1 = 0) = \frac{M_1}{2L}\left\{\frac{27}{2} - 36 + 24 - 0\right\} = \left(\frac{3}{4}\right)\frac{M_1}{L}\\
V(L^-) &= EIw_{,xxx}(x_1 = L^-) = \frac{M_1}{2L}\left\{-12 + 42 - 45 + \frac{27}{2}\right\} = \left(\frac{3}{4}\right)\frac{M_1}{L}\\
V(L^+) &= EIw_{,xxx}(x_2 = 0^+) = \frac{M_1}{2L}\left\{\frac{9}{2} - 21 + 30 - 12\right\} = \left(\frac{3}{4}\right)\frac{M_1}{L}\\
V(2L) &= EIw_{,xxx}(x_2 = L) = \frac{M_1}{2L}\left\{-\frac{9}{2} + 21 - 30 + 12\right\} = -\left(\frac{3}{4}\right)\frac{M_1}{L}
\end{aligned} \tag{9.62}$$

Case M_2: Shear distribution

Similarly, we get:

$$\begin{aligned}
V(0) &= EIw_{,xxx}(x_1 = 0) = \frac{M_2}{2L}\left\{\frac{39}{8} + 12 - 6 - 0\right\} = \left(\frac{9}{16}\right)\frac{M_2}{L}\\
V(L^-) &= EIw_{,xxx}(x_1 = L^-) = \frac{M_2}{2L}\left\{-3 + \frac{33}{2} - \frac{69}{4} + \frac{39}{8}\right\} = \left(\frac{9}{16}\right)\frac{M_2}{L}\\
V(L^+) &= EIw_{,xxx}(x_2 = 0^+) = \frac{M_2}{2L}\left\{\frac{3}{8} - \frac{3}{4} - \frac{3}{2} + 3\right\} = \left(\frac{9}{16}\right)\frac{M_2}{L}\\
V(2L) &= EIw_{,xxx}(x_2 = L) = \frac{M_2}{2L}\left\{-\frac{3}{8} + \frac{3}{2} - 0 + 0\right\} = \left(\frac{9}{16}\right)\frac{M_2}{L}
\end{aligned} \tag{9.63}$$

Case P_2: Shear distribution

Similarly, we get:

$$\begin{aligned}
V(0) &= EIw_{,xxx}(x_1=0) = \frac{P_2}{2}\left\{\frac{43}{8} - 12 + 6 - 0\right\} = -\left(\frac{5}{16}\right)P_2\\
V(L^-) &= EIw_{,xxx}(x_1=L^-) = \frac{P_2}{2}\left\{7 - \frac{45}{2} + \frac{8}{4} - \frac{43}{8}\right\} = -\left(\frac{5}{16}\right)P_2\\
V(L^+) &= EIw_{,xxx}(x_2=0^+) = \frac{P_2}{2}\left\{-\frac{25}{8} + \frac{57}{4} - \frac{39}{2} + 7\right\} = -\left(\frac{11}{16}\right)P_2\\
V(2L) &= EIw_{,xxx}(x_2=L) = \frac{P_2}{2}\left\{0 - 0 + \frac{9}{2} - \frac{25}{8}\right\} = -\left(\frac{11}{16}\right)P_2
\end{aligned} \tag{9.64}$$

Remarks

- It can be easily verified that all distributions – displacement, rotation, shear and moment – are exact.
- From the free-body diagram, it can be seen that the equilibrium is preserved everywhere.

9.4 c-type Curved Beam Element

In the following example, we demonstrate imposition of the inter-element continuity conditions. In so doing, we show improvement in accuracy by introduction of a hierarchy of our c-type elements. It will also be shown constructively that lack of adequate inter-element continuity, otherwise required by the degree of the associated differential equation, spells disaster.

9.4.1 3D Linear Curved Beams with Natural Twist and Bend

For linear problems, as shown in Figure 9.12, the undeformed basis set, **t**, **m**, **b**, and the deformed basis set, **n**, $\mathbf{m}_n$, $\mathbf{b}_n$, are identical. It is assumed here that the curved beams are embedded in the general three-dimensional space with principal axes in the cross-sections aligned along the **m** and **b** directions. Let the generalized displacement and rotation components at the centroidal axis

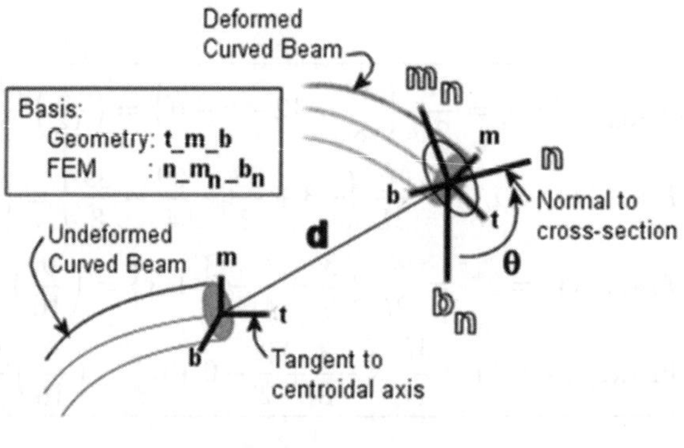

Figure 9.12 Curved beam: Frenet frame for geometry and shear deformed frame for c-type FEM.

along **t**, **m**, **b** axes be given by $u(\xi)$, $v(\xi)$, $w(\xi)$, $\theta_t(\xi)$ $\theta_m(\xi)$ $\theta_b(\xi)$, as functions only of $\xi \in [0, 1]$. Under these assumptions and neglecting the warping effect, the Green's strain components at any point $(m,\ b)$ of the cross-section at ξ are given by:

$$\boxed{\begin{aligned} &e_{mm} = e_{bb} = e_{mb} = 0 \\ &\sqrt{g}e_{tt} = \varepsilon_t + b\varpi_m - m\varpi_b \\ &2\sqrt{g}e_{tm} = \varepsilon_m - \theta_b - b\varpi_s \\ &2\sqrt{g}e_{tb} = \varepsilon_b + \theta_m + m\varpi_s \end{aligned}} \tag{9.65}$$

where, $\sqrt{g} = 1 - m\kappa$ with g as the determinant of the metric tensor of the geometry, and

$$\boxed{\begin{aligned} &\varepsilon_m = u,_s - \tau v + \kappa w \quad && \chi_m = \theta_m,_s - \tau\theta_b + \kappa\theta_t \\ &\varepsilon_b = v,_s + \tau u && \chi_b = \theta_b,_s + \tau\theta_m \\ &\varepsilon_t = w,_s - \kappa u && \chi_t = \theta_t,_s - \kappa\theta_m \end{aligned}} \tag{9.66}$$

where, $(\bullet),_s = (\bullet),_\xi / J$. For comparison with published results in the literature, some of the following examples are planar curves and circular in shape with the plane of the circle coinciding with the osculating plane. With the additional assumption of cross-section symmetry, planar applied load and $m\kappa \ll 1$, we can reduce the above expressions by setting:

$$\boxed{\begin{aligned} &\tau = 0, \quad \theta_m = \theta_t = 0, \quad \theta_b \equiv \theta \\ &v = 0, \ \psi = 0, \quad e_{tb} = 0, \quad \sqrt{g} \simeq 1 \end{aligned}} \tag{9.67}$$

If we denote, for simplicity, the uniform axial strain as: $\varepsilon \equiv e_{tt}$, bending strain as: $\chi = \theta,_s$ and the shear strain as: $\gamma = 2e_{tm}$, we get:

$$\boxed{\varepsilon = w,_s - \frac{u}{R}, \quad \chi = \theta,_s, \quad \gamma = u,_s + \frac{w}{R} - \theta} \tag{9.68}$$

For inextensional circular beam with no shear strain deformation, we have further simplification for the bending strain which now depends only on the radial displacement and is given by:

$$\boxed{\chi = w,_{ss} + \frac{w}{R}} \tag{9.69}$$

Likewise, let N_t, Q_m, Q_b, M_t, M_m, M_b be the corresponding generalized axial force along the **t** axis, transverse shear forces in the **m** and **b** directions, torsional moment about the **t** axis, and bending moments about the **m** and **b** axes, respectively. The constitutive relations for a beam element are then given as:

$$\begin{Bmatrix} N_t \\ Q_b \\ Q_m \\ M_t \\ M_b \\ M_m \end{Bmatrix} = \begin{bmatrix} EA & 0 & 0 & 0 & 0 & 0 \\ 0 & kG & 0 & 0 & 0 & 0 \\ 0 & 0 & kG & 0 & 0 & 0 \\ 0 & 0 & 0 & GJ & 0 & 0 \\ 0 & 0 & 0 & 0 & EI_b & 0 \\ 0 & 0 & 0 & 0 & 0 & EI_m \end{bmatrix} \begin{Bmatrix} e_{tt} \\ \gamma_{tb} = 2e_{tb} \\ \gamma_{tm} = 2e_{tm} \\ \chi_t \\ \chi_b \\ \chi_m \end{Bmatrix} \tag{9.70}$$

where, E, G, A, I_b, I_m and J are the usual elasticity and shear moduli, cross-sectional area, moments of inertia about the **b** and **m** axes and polar moment of inertia, respectively; k is the shear factor. In the absence of body forces and surface traction, the equilibrium equations involving the above stress resultants and stress couples, obtained by the application of the virtual work principle, are given as:

$$\boxed{\begin{aligned} N_{t},_{s} - \kappa Q_m &= 0 \\ Q_{m},_{s} + \kappa N_t - \tau Q_b &= 0 \\ Q_{b},_{s} + \tau Q_m &= 0 \\ M_{t},_{s} - \kappa M_m &= 0 \\ M_{m},_{s} + \kappa M_t - \tau M_b - Q_b &= 0 \\ M_{b},_{s} - \tau M_m + Q_m &= 0 \end{aligned}} \tag{9.71}$$

The corresponding simplified expression for a circular beam is given by:

$$\boxed{\begin{aligned} M,_{s} - Q &= 0 \\ Q,_{s} - \frac{N}{R} &= 0 \\ \frac{M,_{s}}{R} + N,_{s} &= 0 \end{aligned}} \tag{9.72}$$

9.4.1.1 Finite Element Interpolations

In general, the cubic Bernstein–Bezier polynomials in root variable, ξ, define the finite element interpolation fields for generalized displacements and rotations for straight beams. In matrix form, the finite element displacement distributions may be denoted, in matrix form and component-wise, respectively, as:

$$\boxed{\mathbf{d}(\xi) = \mathbf{T}(\xi)\,\mathbf{q}^{\mathbf{d}} \quad \text{and} \quad \mathbf{d}_j(\xi) = \sum_{i=0}^{N} \mathrm{B}_i^N(\xi)\, q_i^d, \quad j = 1, \ldots, 6} \tag{9.73}$$

where $\mathbf{T}(\xi)$ is the interpolation transformation matrix containing appropriate Bernstein polynomials, $\mathbf{q}^{\mathbf{d}} = \{\, \mathbf{q}^u \quad \mathbf{q}^v \quad \mathbf{q}^w \quad \mathbf{q}^{\theta_t} \quad \mathbf{q}^{\theta_b} \quad \mathbf{q}^{\theta_m} \,\}^T$ is the displacement control vector, and N denotes the appropriate degree of the polynomial. It may be reemphasized that for computational efficiency in the mesh generation phase, the geometry may be modeled subparametrically depending on the curve. For example, for the general three-dimensional space-curved beam, geometry may be represented at most by cubic Bernstein–Bezier with as many elements as needed. Similarly, for circular or conic shape other than parabolic beam, only the quadratic rational Bezier or cubic Bernstein–Bezier is needed. Finally, a quadratic polynomial exactly represents a parabolic beam. For linear curved beams, c-type quartic Bernstein–Bezier elements suffice. However, for extreme nonlinear problems considered next, quintic Bernstein–Bezier elements are necessary for so-called mega-elements.

9.4.1.2 Element Post-processing and Stress Recovery

A few words relating to stress recovery are in order. The displacements and rotations converge in the norm sense. Accordingly, the uncoupled rotations can recover the bending moment

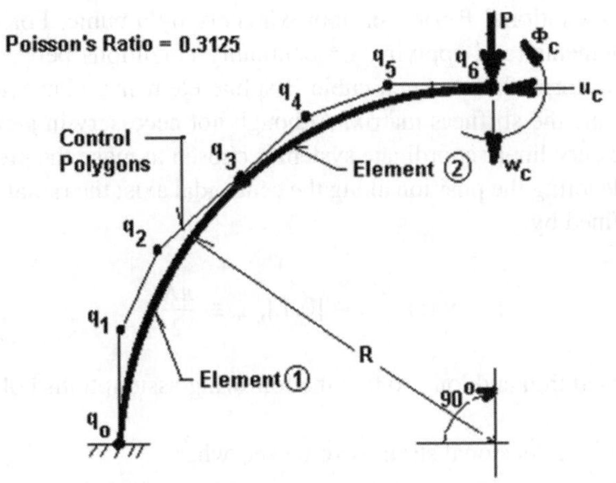

Figure 9.13 Example CFEM5: Curved beam with inter-element continuity: control nodes for two cubics are shown.

distributions by constitutive relationship exactly. Because of the constraints imposed by the kinematic assumptions, the shear force and axial force cannot be recovered from the first three constitutive relations given by equations (9.70). For problems considered here, a simpler method based on the integration of the bending moments satisfying the equilibrium conditions given above by equations (9.71) produces accurate results for both elemental axial force and shear force distributions. For example, for each three-dimensional space-curved beam element in the following, after recovering the moment components by the usual application of the constitutive relations and the nodal reactions, the axial force computation preceded shear force distribution through integration of the equilibrium equations as given by:

$$\boxed{\begin{aligned} N_t(\xi) &= N_t(0) - \int_0^{\xi} \kappa\, M_{b},_{\xi}\, d\xi - \int_0^{\xi} \kappa\tau\, M_m\, J\, d\xi \\ Q_b(\xi) &= Q_b(0) + \int_0^{\xi} \tau\, M_{b},_{\xi}\, d\xi + \int_0^{\xi} \tau^2\, M_m\, J\, d\xi \\ Q_m(\xi) &= Q_m(0) - \int_0^{\xi} \tau\, Q_b\, J\, d\xi - \int_0^{\xi} \kappa\, N_t\, J\, d\xi \end{aligned}} \tag{9.74}$$

With these in the background, let us consider the application of inter-element continuity (IEC) on the following example of interest in structural engineering and solid mechanics.

Example CFEM 5: Quadrant of a circular beam element

For demonstration of inter-element continuity condition imposition, we choose the curved (circular) cantilever beam shown in Figure 9.13. The beam is assumed fixed at one end, and loaded vertically down at the other end. The beam is assumed very thin and long so that the extensional strain and shear strains are zero. One cubic c-type Bernstein–Bezier element suffices, but for demonstration of inter-element continuity, we continue to improve by increasing the number of cubic elements. The results of the hierarchic solutions for: (a) one Bernstein–Bezier cubic element, (b) two Bernstein–Bezier cubic elements with C^1 continuity, and finally, (c) two cubic c-type B-spline elements with C^1 continuity. The original circular arc can be geometrically exactly generated by a rational Bezier, or approximately by a cubic. For the last case, taking two cubic Bezier elements, and applying C^2 continuity conditions between them, a cubic B-spline element is first formed. Two such cubic B-spline elements, connected by C^1 continuity between them, generate the stiffness matrix. Although not necessary in general applications, as indicated before, the curvilinear coordinate system is chosen to make the presentation simple for our purpose with s denoting the position along the centroidal axis; the radial displacement, w, for circular beam, is defined by:

$$\mathrm{w}(s), \quad s \in [0, L], \; L \equiv \frac{\pi R}{2} \tag{9.75}$$

The beam is assumed thin and long so that the following assumptions hold:

- inextensional, that is, extensional strain is zero everywhere:

$$\epsilon_\theta \equiv \mathrm{u},_s + \frac{\mathrm{w}}{R} = 0 \quad \Rightarrow \quad \mathrm{u},_s = -\frac{\mathrm{w}}{R} \tag{9.76}$$

where, $u(s)$ represents the hoop displacement.

- no shear strain, that is, $\gamma \equiv 0$

Now, let the bending strain, or the curvature, $\kappa = \psi,_s$, where ψ denotes the rotation of the normal to the neutral axis at any point s. Then, we have, applying equation (9.69), for bending strain:

$$\kappa(s) = \mathrm{w},_{ss} - \frac{\mathrm{u},_s}{R} = \mathrm{w},_{ss} + \frac{\mathrm{w}}{R^2} \tag{9.77}$$

The exact solution, for radial load P applied at $s = L$, is given by:

$$\boxed{\mathrm{w}_{exact}(L) = \frac{\pi R^3 P}{4EI}, \quad \mathrm{u}_{exact}(L) = \frac{PR^3}{2EI}, \quad \psi_{exact}(L) = \frac{PR^2}{EI}} \tag{9.78}$$

The root element is defined by: $r = \frac{s}{L} \Rightarrow r \in [0, 1]$. The Jacobian, $\mathrm{J}(r) = L$, is constant, and its determinant is, obviously, the same. Then, the potential energy functional, $\Pi(w)$, exists for the system for a displacement state, w, and is given by:

$$\Pi(w) = \frac{EI}{2} \int_0^L \left[w,_{ss} + \frac{w}{R^2} \right]^2 ds \; - Pw(L) = \frac{EI}{2L^3} \int_0^1 \left[w,_{rr} + \frac{\pi^2}{4} \frac{w}{R^2} \right]^2 dr \; - Pw(1) \tag{9.79}$$

Now, let the displacement function be as described in matrix form:

$$\begin{aligned} w(r) &= \mathbf{T}\,\mathbf{q}, \\ w,_r(r) &= \mathbf{N}\,\mathbf{q}, \quad \mathbf{N} \equiv \mathbf{T},_r \\ w,_{rr}(r) &= \mathbf{H}\,\mathbf{q}, \quad \mathbf{H} \equiv \mathbf{T},_{rr} \end{aligned} \tag{9.80}$$

where, **T** and **q** are the transformation matrix and Bezier displacement control vector, respectively, dependent on the choice of our c-type element. Let us consider several cases based on the number and type of element as under:

9.4.1.3 Case 1: One c-type Cubic Bernstein–Bezier Element

For this case, first note that the geometry in the curvilinear coordinate system is linear, while the displacement function is assumed cubic. Thus, although not necessary,we may coerce up the geometry representation to cubic also by requiring the location of the geometry Bezier control vector, $\mathbf{q}^c$, as following:

$$\mathbf{q}^c = \frac{\pi R}{2}\{0 \quad \tfrac{1}{3} \quad \tfrac{2}{3} \quad 1\}^T \tag{9.81}$$

So with the transformation, **T**, defined as follows, we have a coerced isoparametric representation of the geometry and the displacement function:

$$\begin{aligned}\mathbf{q}^d &= \{q_0^d, q_1^d, q_2^d, q_3^d\}^T\\ \mathbf{T} &\equiv [(1-r)^3 \quad 3(1-r)^2 r \quad 3(1-r)r^2 \quad r^3]\\ \mathbf{N} &\equiv 3[-(1-r)^2 \quad (1-4r+3r^2) \quad (2r-3r^2) \quad r^2]\\ \mathbf{H} &\equiv 6[(1-r) \quad -(2-3r) \quad (1-3r) \quad r],\\ w(1) &= T(1)q^d = \{0 \quad 0 \quad 0 \quad 1\}q^d \equiv T_1 q^d\end{aligned} \tag{9.82}$$

Now, taking the variation of the potential energy functional given by equation (9.79), we have the virtual work principle:

$$\mathbf{K}\,\mathbf{q}^d = \mathbf{T}_1^T P, \quad \text{where} \quad \mathbf{K} \equiv \frac{EI}{L^3}\int_0^1 \left[\mathbf{H} + \frac{\pi^2}{4}\mathbf{T}\right]^T \left[\mathbf{H} + \frac{\pi^2}{4}\mathbf{T}\right] dr \tag{9.83}$$

where, **K**, is the system-full, unconstrained stiffness matrix, and all other terms are defined as before; given the matrices, it can be easily computed to be:

$$\mathbf{K} = \begin{bmatrix} 18.791 & -20.526 & 3.135 & 7.524\\ -20.526 & 30.6 & -19.089 & 3.135\\ 3.135 & -19.089 & 30.6 & -20.526\\ 7.524 & 3.135 & -20.526 & 18.791 \end{bmatrix} \tag{9.84}$$

Now, for solving for the Bezier displacement vector, $\mathbf{q}^d$ in equation (9.83), we impose the essential boundary conditions (i.e. the fixed end boundary conditions) as follows:

$$\begin{aligned} \mathrm{w}(0) &= q_0^d = 0,\\ \mathrm{w},_r(0) &= \frac{3}{L}\left(q_1^d - q_0^d\right) = 0 \Rightarrow q_1^d = 0 \end{aligned} \tag{9.85}$$

We can solve for the remaining components of the unknown displacement control vectors, by considering only the lower (2×2) submatrix of the stiffness matrix, and get:

$$\begin{Bmatrix} q_2^d\\ q_3^d \end{Bmatrix} = \begin{Bmatrix} 0.134\\ 0.199 \end{Bmatrix} \frac{PL^3}{EI} \tag{9.86}$$

We can then extract the desired radial displacement, w(L), at load point as:

$$\boxed{\mathrm{w}(L)=\left[\frac{0.199\pi^2}{2}\right]\left(\frac{\pi R^3P}{4EI}\right)=0.982\left(\frac{\pi R^3P}{4EI}\right)} \tag{9.87}$$

Now, for axial displacement, u(L), under the load point, after noting the boundary condition that u(0) = 0, we have, from equation (9.76), upon integration:

$$\boxed{\begin{aligned}\mathrm{u}(L)&=-\frac{L}{R}\int_0^1\mathrm{w}(r)\,dr\\&=-\left[\frac{\pi^4}{8}\{q_0^d+q_1^d+q_2^d+q_3^d\}\right]\left(\frac{PR^3}{2EI}\right)=-1.014\left(\frac{PR^3}{2EI}\right)\end{aligned}} \tag{9.88}$$

In equation (9.88) above, we have used the integration property of Bernstein polynomials:

$$\boxed{\int_0^1\sum_{i=0}^{n}\mathbf{T}(r)\,\mathbf{q}^d\,dr=\frac{1}{n+1}\sum_{i=0}^{n}\mathbf{q}_i^d,\quad n=3} \tag{9.89}$$

Finally, for rotation of the normal to the centroidal axis, $\psi(L)$, under the load point, we have:

$$\boxed{\psi(L)=\mathrm{w}_{,s}(L)-\frac{\mathrm{u}(L)}{R}=0.99\left(\frac{PR^2}{2EI}\right)} \tag{9.90}$$

Remark

- We see that with only one element, we have achieved a pretty good estimate of the solution by our c-type method. However, we can continue to improve it by increasing the number of cubic elements. Let us try the following case.

9.4.1.4 Case 2: Two Cubic Bernstein–Bezier Elements with C^1 continuity

Following the method described above, we can show that the assembled finite element equilibrium equations with:

- C^0 continuity imposed between the elements
- dropping off the first two rows and columns to reflect the fixed end boundary conditions, is given by:

$$\boxed{\begin{aligned}&\mathbf{K}^0\,\tilde{\mathbf{q}}^d=\mathbf{P}^0\\&\text{where }\mathbf{K}^0=\left(\frac{8EI}{L^3}\right)\begin{bmatrix}34.552&-18.713&0&0&0\\-18.713&2(13.535)&-18.713&.751&6.373\\0&-18.713&34.552&-18.346&.751\\0&.751&-18.346&34.552&-18.713\\0&6.373&0.751&-18.713&13.535\end{bmatrix}\\&\tilde{\mathbf{q}}^d=\{q_2^d\quad q_3^d\quad q_4^d\quad q_5^d\quad q_6^d\}^T,\mathbf{P}^0=\{0\quad 0\quad 0\quad 0\quad 1\}^TP\end{aligned}} \tag{9.91}$$

Remark

- Any attempt to solve equation (9.91) will spell disaster because, for a variational solution, a system governed by fourth-order differential equation, as in our case, needs at least C^1 continuity between the elements.

So, let's go ahead and impose the C^1 continuity condition, between the elements. We apply to the C^0-assembled stiffness control matrix, K^0, and the load control vector, P^0, and get the corresponding C^1-assembled stiffness control matrix and the load ontrol vector, denoted by K^1 and P^1, respectively, as given by:

$$\mathbf{K}^1 = \mathbf{C}^T\,\mathbf{K}^0\,\mathbf{C} = \begin{bmatrix} 22.606 & -11.946 & 0.375 & 3.186 \\ -11.946 & 22.606 & -17.970 & 3.937 \\ 0.375 & -17.970 & 34.552 & -18.713 \\ 3.186 & 3.937 & -18.713 & 13.535 \end{bmatrix}, \qquad (9.92)$$
$$\mathbf{P}^1 = \mathbf{C}^T\,\mathbf{P}^0 = \{0 \quad 0 \quad 0 \quad 1\},$$

where, the C^1 continuity condition matrix, $\mathbf{C}$, is given by:

$$\tilde{\mathbf{q}}^d = \mathbf{C}\tilde{q}^d,$$
$$\text{or, in expanded form: } \begin{Bmatrix} q_2^d \\ q_3^d \\ q_4^d \\ q_5^d \\ q_6^d \end{Bmatrix} = \begin{bmatrix} 1 & 0 & 0 & 0 \\ .5 & .5 & 0 & 0 \\ 0 & 1 & 0 & 0 \\ 0 & 0 & 1 & 0 \\ 0 & 0 & 0 & 1 \end{bmatrix} \begin{Bmatrix} q_2^d \\ q_4^d \\ q_5^d \\ q_6^d \end{Bmatrix} \qquad (9.93)$$

Remark

- The total number of degrees of freedom is reduced by one, that is, the junction degree of freedom, q_3^d, between the two elements is eliminated from the equations by the condition of C^1 continuity.

Now, solving equation (9.91), and continuing with the procedure as before, element by element, we get the desired solution as:

$$\mathrm{w}(L) = (0.9984)\frac{\pi R^3 P}{4EI},$$
$$\mathrm{u}(L) = -(0.99995)\frac{PR^3}{2EI}, \qquad (9.94)$$
$$\psi(L) = (0.99932)\frac{PR^2}{EI}$$

Remarks

- The above results almost match the exact solutions! It can be shown that improving the continuity condition does not improve the results anymore.

- However, as noted earlier, we can further increase the accuracy by changing the elements from simple cubic Beziers to next hierarchical elements: cubic B-spline elements connected by the C^1 continuity condition. That is exactly what we propose to do next.

9.4.1.5 Case 3: Two c-type Cubic B-spline Elements with C^1 Continuity

For this case, we do not intend to present our analysis in detail. We only highlight the major ideas as follows:

- As suggested by the definition, a cubic B-spline element is formed by taking two cubic Bezier elements, and applying C^2 continuity conditions between them.
- Each cubic B-spline element will, then, have five degrees of freedom, that is, three less than the eight degrees of freedom of two cubic elements.
- The two cubic B-spline element when connected by C^1 continuity between them, will have a total of eight degrees of freedom, that is, two less than ten for the two elements.
- For the problem at hand with fixed end boundary condition, the system total significant degrees of freedom are six.

Finally, by carrying out the solution as before, we get:

$$\boxed{\begin{aligned} \mathrm{w}(L) &= (1.000176)\frac{\pi R^3 P}{4EI}, \\ \mathrm{u}(L) &= -(1.00029)\frac{PR^3}{2EI}, \\ \psi(L) &= (1.000219)\frac{PR^2}{EI} \end{aligned}} \tag{9.95}$$

Remark

- The results, of course, speak for themselves – we have achieved almost the exact solution. We summarize below, in tabular form (see Figure 9.14), the tip response comparison between the analytical results and the solutions obtained by various hierarchic c-type finite elements.

Case	Element Type	$w_c/(\frac{\pi R^3 P}{4EI})$	$u_c/(\frac{PR^3}{2EI})$	$\Phi_c/(\frac{PR^2}{EI})$
Exact	Analytic	1.0	-1.0	1.0
I	1 Cubic	0.982	-1.014	0.99
II	2Cubic (C^1 contnuity)	0.9984	-0.99995	0.99932
III	2Cubic (C^2 contnuity)	1.000716	-1.00029	1.00022

Figure 9.14 Curved beam: c-type responses for various cases.

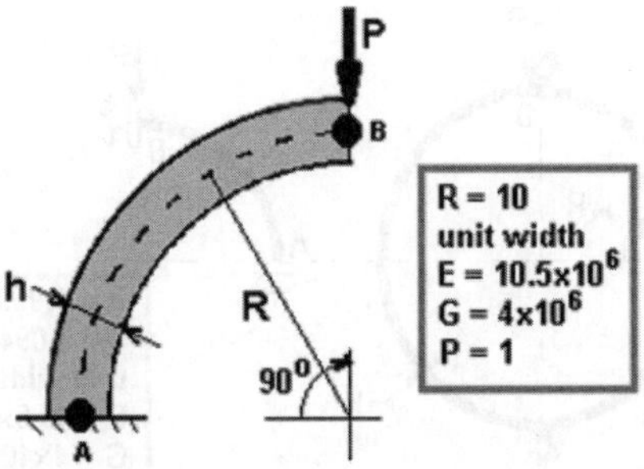

Figure 9.15 Example CFEM 6: The 90° circular cantilevered beam with tip radial load.

Example CFEM 6: A 90° cantilevered circular ring with tip radial load

As shown in Figure 9.15, and as a prelude to application of the c-type method to general shell analysis, we consider here the curved ring problem with both axial and shear deformation possibilities. The geometric control polygons are generated for circular curve by both cubic Bernstein–Bezier polynomials with $k_i = \frac{4}{3}R_i(\sqrt{2}-1), \quad i = 1, 2$, and rational quadratic Bezier polynomials. The results obtained by c-type formulation are shown for various aspect ratio, $\frac{R}{h}$, in a tabular form as shown in Figure 9.16:

Remark

- Considering the remarkable accuracy of the responses, we may conclude that the interpolation problems known as ***shear, membrane and dilatation locking*** that severely and continuously plague the old finite element methodologies are absent in the results presented above. The results are compared with the corresponding tip responses obtained by application of ***Castigliano's energy theorem***. In comparing, however, it must be recognized that for a very thick beam, analytical solutions obtained by Castigliano's theorem are only approximate when the model is treated as a simple beam.

Example CFEM 7: Thin circular pinched ring

As shown in Figure 9.17, a thin ring pinched by two diametrically opposite and identical forces is analyzed to assess the c-type curved beam element performance for modeling a deep arch

Slenderness Ratio	$\frac{w_B}{w_{Exact}}$	$\frac{u_B}{u_{Exact}}$	$\frac{\Phi_B}{\Phi_{Exact}}$
4	1.01660	1.01693	1.00022
10	1.00290	1.00307	1.00022
20	1.00093	1.00110	1.00022
50	1.00038	1.00054	1.00022
100	1.00030	1.00047	1.00022
200	1.00028	1.00045	1.00022
500	1.00028	1.00044	1.00022
1000	1.00028	1.00044	1.00022

Figure 9.16 Example CFEM 6: c-type responses for several slenderness ratios.

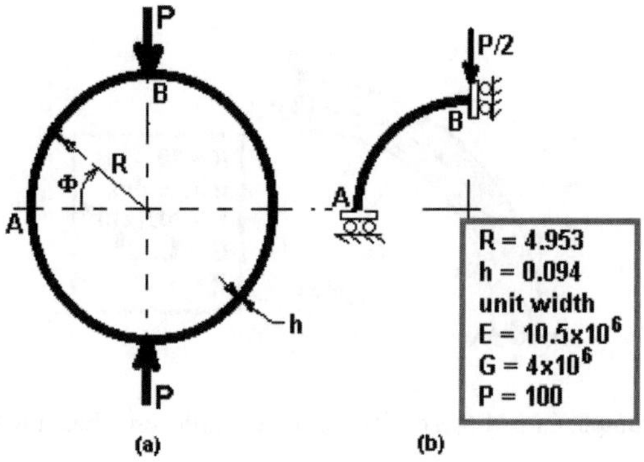

Figure 9.17 Example CFEM 7: Pinched ring problem (a) geometry (b) finite elemnt model.

configuration. Because of symmetry, only one quadrant of the ring is modeled with appropriate boundary conditions and then analyzed twice: once with only one element and then with two-element discretization. The same problem analyzed by many researchers has been reported in the literature where ad hoc treatments were needed to deal with the locking problems in curved beams.

In Figure 9.18, the tip radial normalized displacement generated by the c-type method is compared with that given by others reported in the literature. For most cases, significantly more elements are needed to model the problem to obtain the same level of accuracy by the c-type curved beam element. The distributions of the bending moment, axial force and shear force are shown in Figures 9.19(a)–(c):

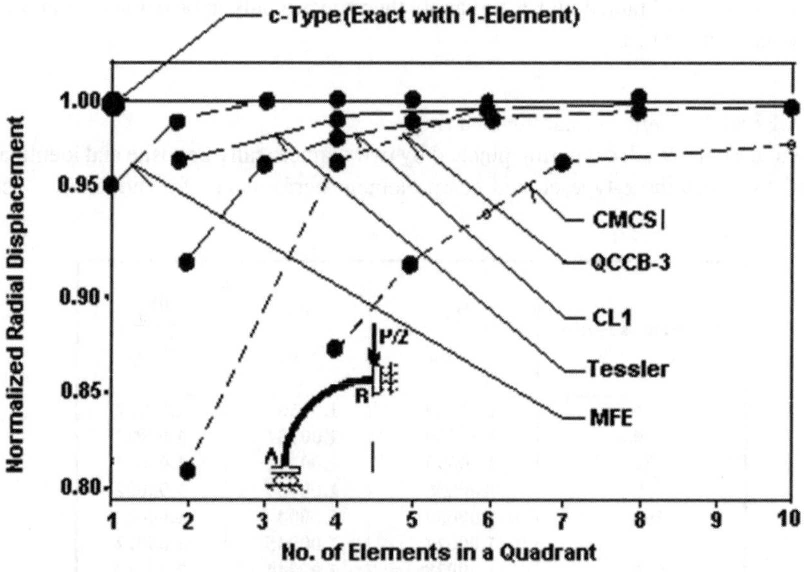

References are cited in the introduction; curved beams

Figure 9.18 Example CFEM 7: Convergence comparison with various published results.

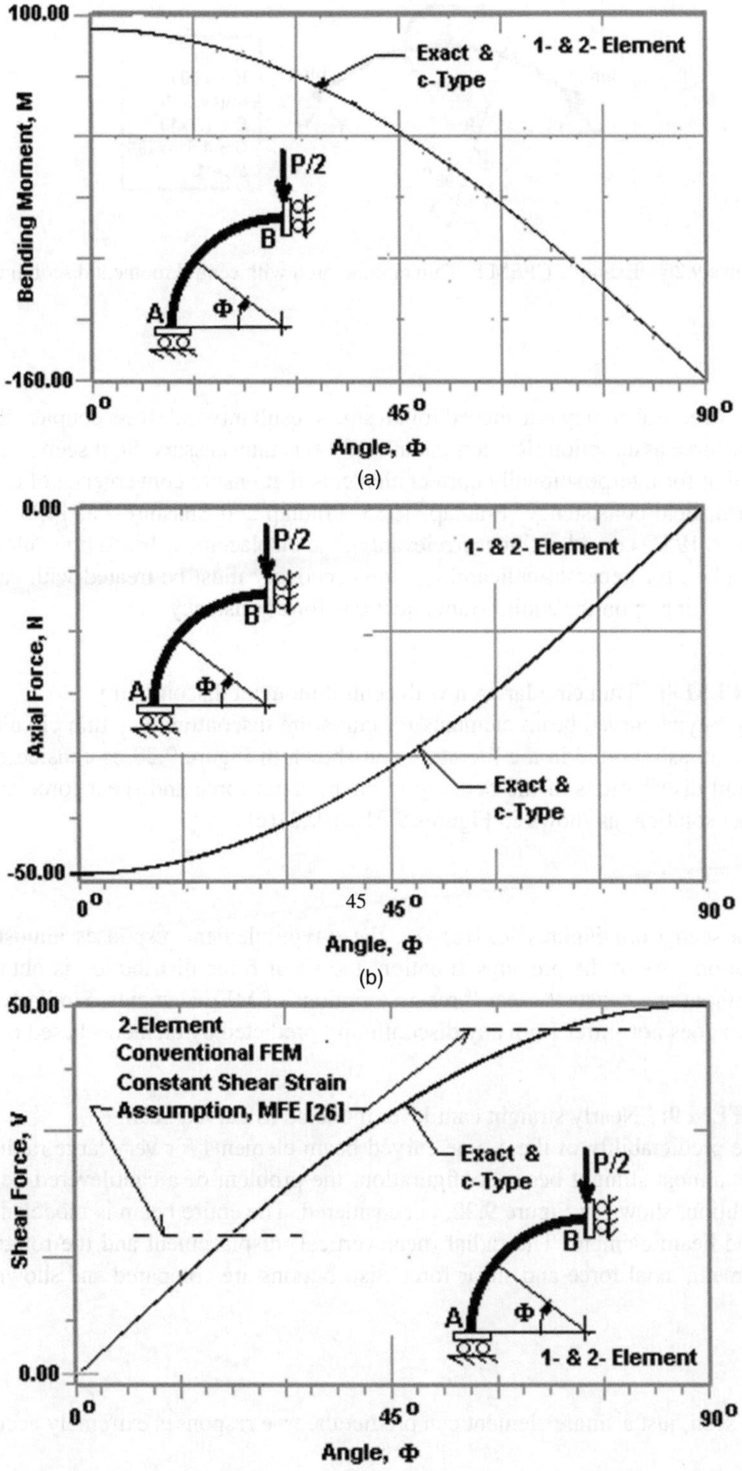

Figure 9.19 (a) Example CFEM 7: Bending moment distribution. (b) Example CFEM 7: Axial force distribution. (c) Example CFEM 7: Shear force distribution.

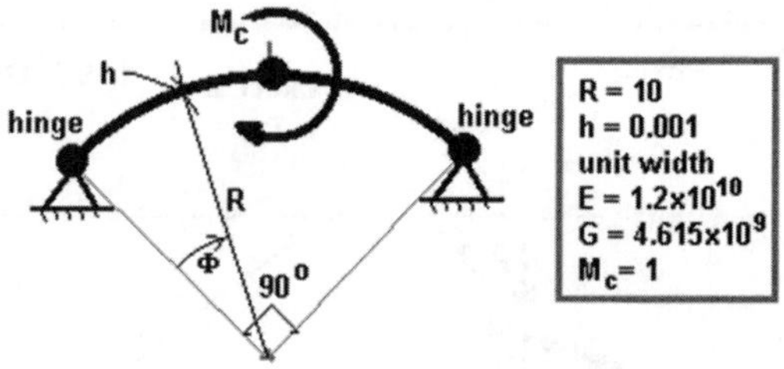

Figure 9.20 Example CFEM 8: Thin circular arch with central moment discontinuity.

Remark

- Note that extreme accuracy is achieved for all stress resultants and stress couples; thus, the constant shear force assumption (Raveendranath, 1999) is unnecessary. So it seems appropriate to point out that for interpolationally correct elements that ensure convergence of displacements in the norm, field consistency (Prathap, 1985; Prathap and Shashirekha, 1993; Timoshenko and Goodier, 1951) consideration is irrelevant in the displacement-based finite element formulations. Rather, for general applicability, stress recovery must be treated with care, as noted earlier, by calling upon the equilibrium equations for consistency.

Example CFEM 8: Thin circular arch with central moment discontinuity

To qualify c-type curved beam elements for curvature discontinuity, a thin circular arch with central moment, as reported in the literature and shown in Figure 9.20, is considered here. The c-type element distributions of the bending moment, axial force and shear force are compared with the exact solution, as shown in Figures 9.21(a)–9.21(c):

Remark

- As can be seen from Figures 9.21(a)–(c), the c-type element responses almost mimic the exact solutions. As in the previous situation, the shear force distribution is obtained almost exactly without any constant shear force assumptions of MFE elements. Similarly, axial force distribution does not suffer from any discontinuity predicted by elements based on curvature,

Example CFEM 9: Nearly straight cantilever modeled as curved beam

To test the predictability of the c-type curved beam elements for very large radius and short span, that is, almost straight beam configuration, the problem of a cantilevered beam with tip loading condition, shown in Figure 9.22, is considered. The entire beam is modeled by a single c-type curved beam element. The radial (near vertical) displacement and the rotation, and the bending moment, axial force and shear force distributions are compared and shown in Figures 9.23(a)–(c):

Remark

- As can be seen, just a single element can predict the true responses extremely accurately.

Example CFEM 10: Truly three-dimensional cantilevered curved beam element with natural twist and bend

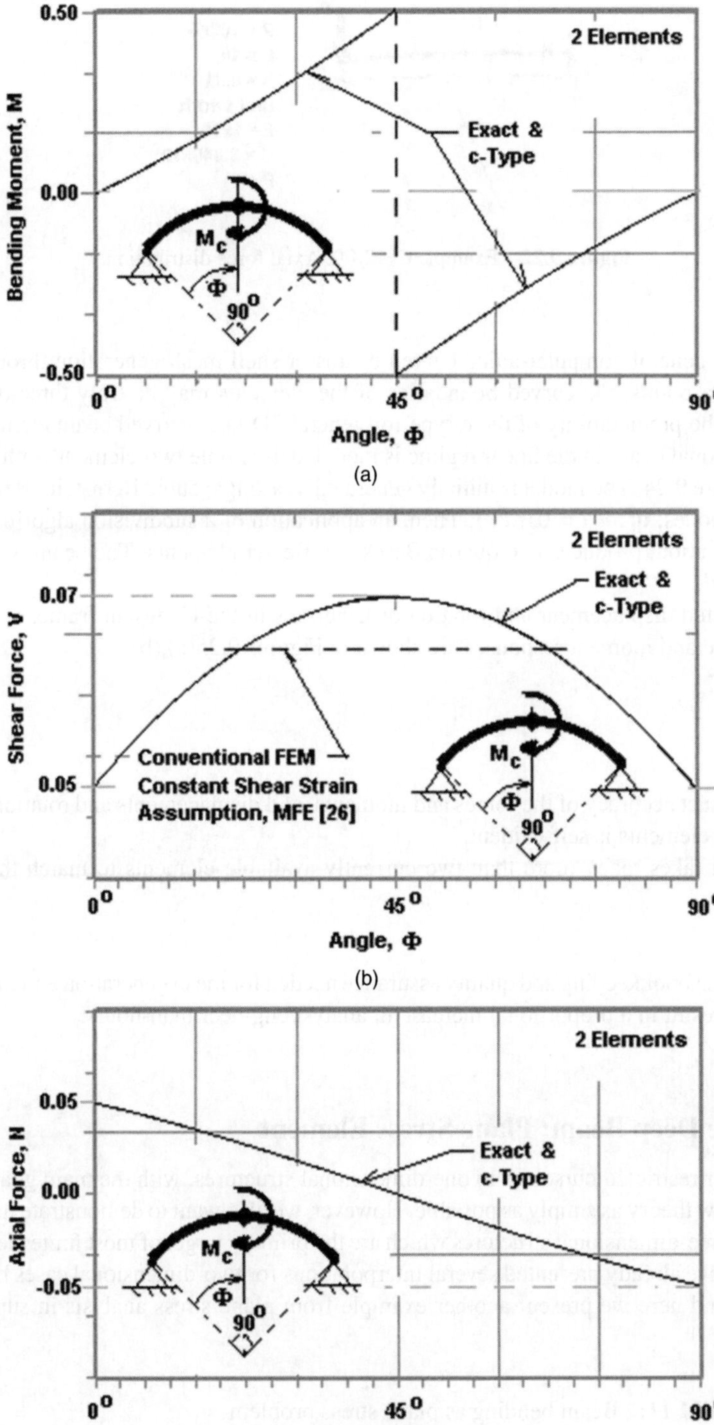

Figure 9.21 (a) Example CFEM 8: Bending moment distribution. (b) Example CFEM 8: Shear force distribution. (c) Example CFEM 8: Axial force distribution.

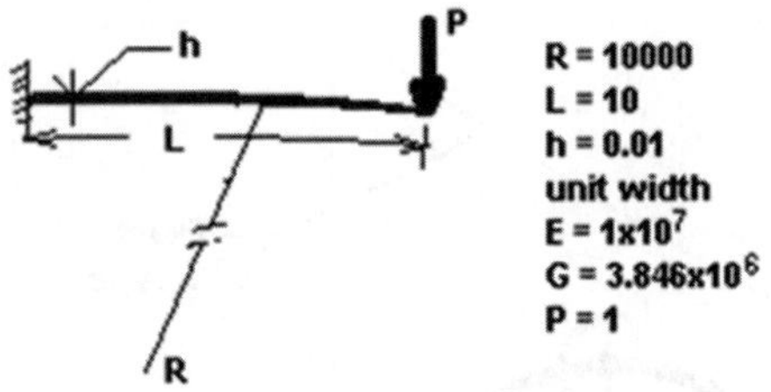

Figure 9.22 Example CFEM 9: Axial force distribution.

Finally, for general computer-aided curved beams or shell mesh generation through input of a few key data points, the curved boundaries of the elements may be truly three-dimensional. Thus, to test the predictability of the c-type for general 3D large curved beam elements, a truly three-dimensional beam in the linear regime is modeled as simple two-element configurations as shown in Figure 9.24. The model is initially generated as a single cubic Bernstein–Bezier element with control nodes, $\mathbf{q}_i^1$, $i = 0, \ldots, 3$. Then, an application of a subdivision algorithm followed by degree elevation, produces two quartic Bernstein–Bezier elements. The beam is subjected to simple tip load.

The computed displacement and rotation distributions in the Cartesian frame, and the corresponding force and moment responses are shown in Figures 9.25(a),(b).

Remarks

- The near-exact accuracy of the forces and moments, and displacements and rotations with only two quartic elements is self-evident.
- Note that it takes many more than two currently available elements to match the presented accuracy.

Likewise, the bookkeeping and quality assurance needed for mesh generation and finite element analysis will result in a proportional increase in analyst/engineer man-hours.

9.5 c-type Deep Beam: Plane Stress Element

We have so far restricted ourselves to one-dimensional structures, with the main goal of demonstrating the new theory as simply as possible. However, we also want to demonstrate its usefulness to two- and three-dimensional structures which are the primary target of most finite element applications. We have already presented several interpolations for two-dimensional cases based on our new theory, and here we present another example from plane stress analysis in support of our theory.

Example CFEM 11: Beam bending as plane stress problem

For the beam shown in Figure 9.26, we consider one of our cubic–cubic elements, but note that the edges of the beam geometry are straight, requiring only linear–linear element for the representation of its geometry. So, although not necessary, we can apply degree-elevation and coerce up the geometry representation by cubic–cubic Bernstein polynomials so that it becomes

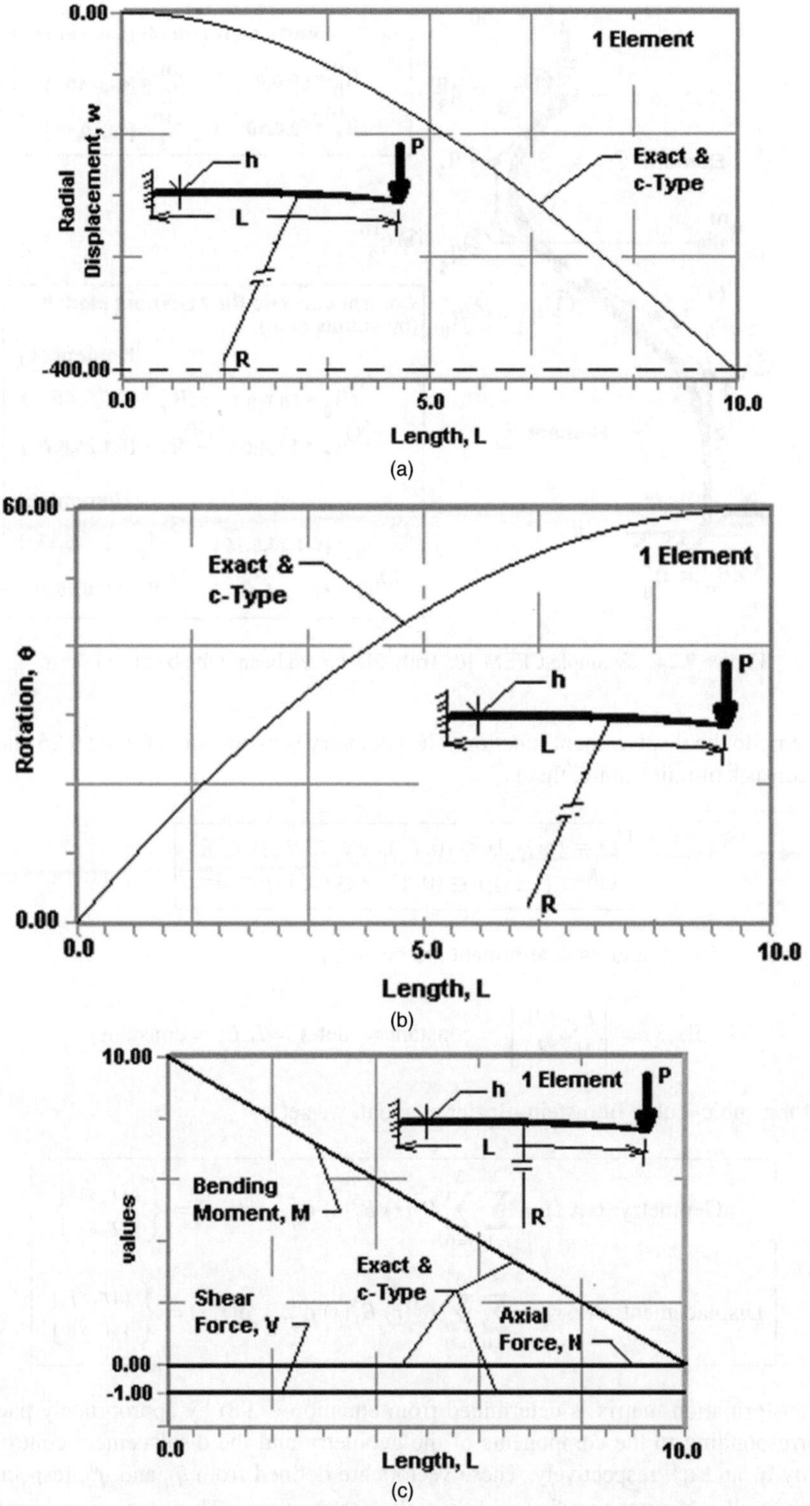

Figure 9.23 (a) Example CFEM 9: Vertical displacement distribution. (b) Example CFEM 9: Rotation distribution. (c) Example CFEM 9: Bending moment, shear and axial force distribution.

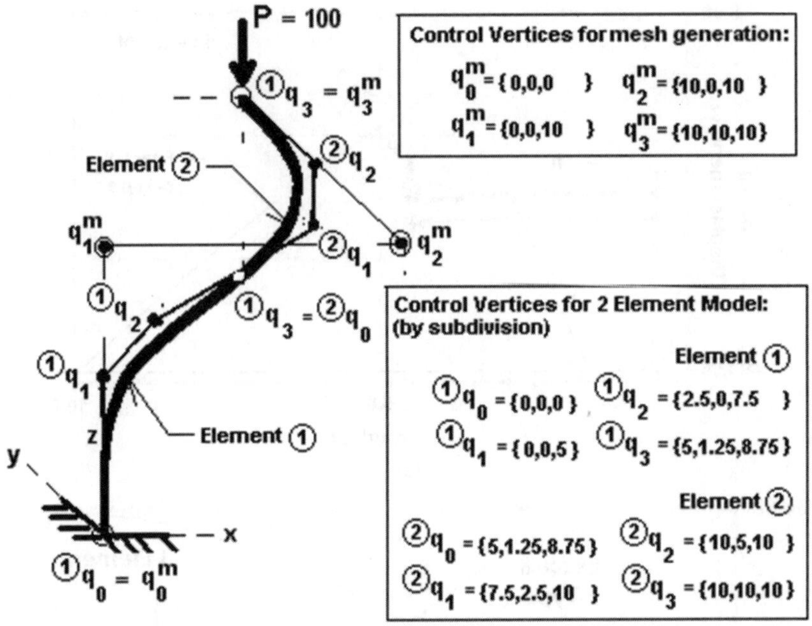

Figure 9.24 Example CFEM 10: Truly 3D curved beam with bend and twist.

isoparametric to the displacement function. The geometry is as shown in Figure 9.26 and the root element consists of unit square, that is,

$$\begin{array}{|l|}\hline \Omega \equiv \{(x, y) | x \in (0, L_x),\ y \in (0, L_y)\} \subset \mathbb{R}^2 \\ \Omega^R \equiv \{(r, s) | r \in (0, 1),\ s \in (0, 1)\} \subset \mathbb{R}^2 \\ \hline \end{array} \tag{9.96}$$

The Jacobian, $\mathbf{J}(r, s)$, and its determinant are given by:

$$\mathbf{J}(r, s) = \begin{bmatrix} L_x & 0 \\ 0 & L_y \end{bmatrix} = \text{constant} \Rightarrow \det \mathbf{J} = L_x L_y = \text{constant} \tag{9.97}$$

Now, for a cubic–cubic Bernstein–Bezier element, we set:

$$\begin{array}{|l|}\hline \text{Geometry: } \mathbf{c}(r, s) = \sum_{i=0}^{3} \sum_{j=0}^{3} B_i^3(r)\, B_j^3(s)\, \tilde{\mathbf{q}}_{ij}^c, \quad \mathbf{c}(r, s) = \begin{Bmatrix} x(r, s) \\ y(r, s) \end{Bmatrix} \\ \text{Displacement: } \mathbf{d}(r, s) = \sum_{i=0}^{3} \sum_{j=0}^{3} B_i^3(r)\, B_j^3(s)\, \tilde{\mathbf{q}}_{ij}^d, \quad \mathbf{d}(r, s) = \begin{Bmatrix} u(r, s) \\ v(r, s) \end{Bmatrix} \\ \hline \end{array} \tag{9.98}$$

The transformation matrix is determined from equation (9.98) by appropriately packing the terms corresponding to the components of the geometry and the displacement control vectors denoted by $\mathbf{q}^c$ and $\mathbf{q}^d$, respectively. These vectors are defined from $\tilde{q}_{ij}^c$ and $\tilde{q}_{ij}^d$, respectively, by packing their all x-components first and then all y-components. Then, the strain–displacement control expressions can be easily found. Use of internal virtual work gives, as usual, the stiffness

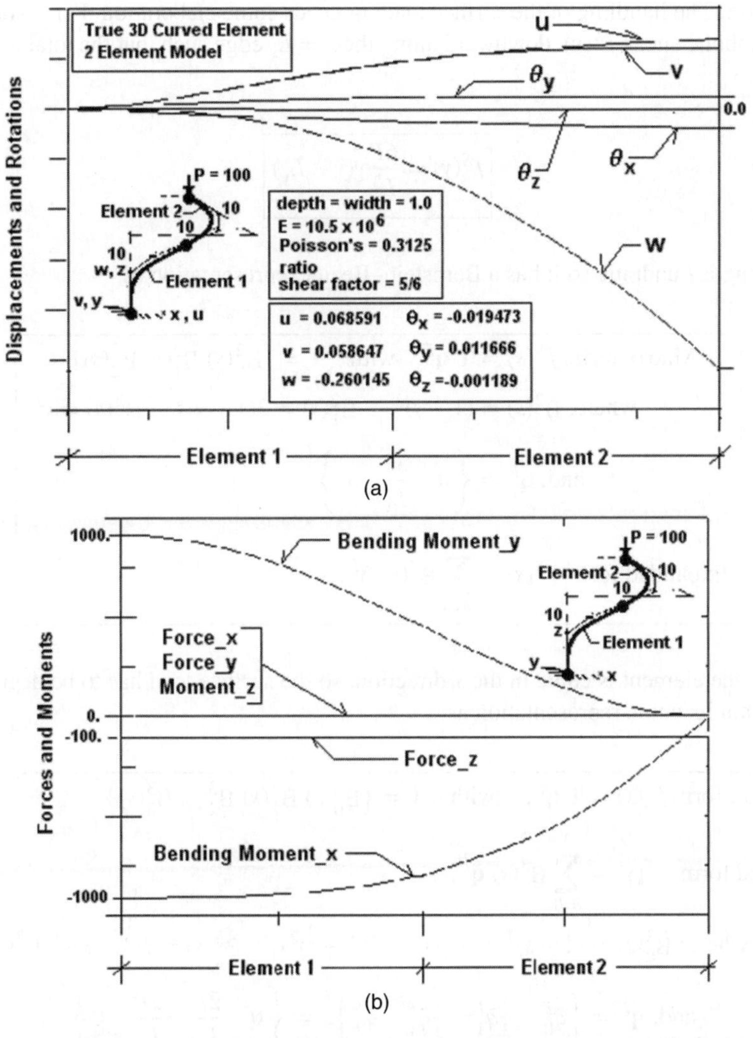

Figure 9.25 (a) Example CFEM 10: Displacement and rotation distribution. (b) Example CFEM 10: Force and moment distribution.

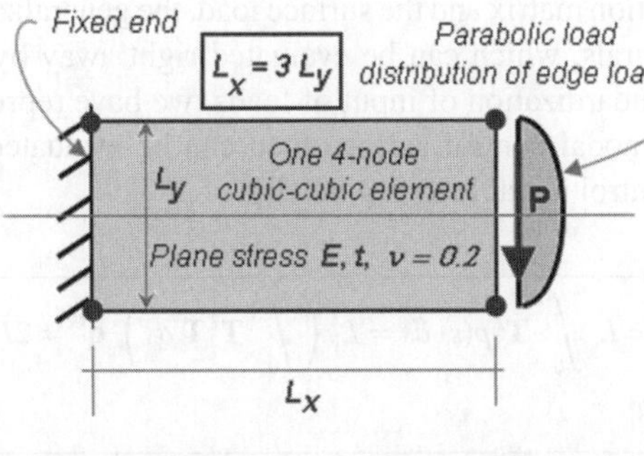

Figure 9.26 Example CFEM 11: Deep beam bending as plane stress problem.

control matrix. The handling of the surface loading needs some elaboration. For surface loading, we have parabolic shear, $f^S(y)$, downward along the $x = L_x$ edge such that the total load amounts to P, that is,

$$\boxed{f^S(y) = \frac{6P}{L_y^3} y(y - L_y)} \tag{9.99}$$

The loading is quadratic so it has a Bernstein–Bezier representation as:

$$\begin{aligned}
&\text{Matrix form: } f^S(s) = \mathbf{T}\,\mathbf{q}^F \quad \text{with} \quad \mathbf{T} = \{\mathrm{B}_0^2(s)\ \mathrm{B}_1^2(s)\ \mathrm{B}_2^2(s)\}\\
&\text{where, } \mathrm{B}_0^2(s) = (1-s)^2, \quad \mathrm{B}_1^2(s) = 2(1-s)s, \quad \mathrm{B}_2^2(s) = s^2,\\
&\text{and, } \bar{\mathbf{q}}^F = \left\{0 \quad \frac{3P}{L_y} \quad 0\right\}^T\\
&\text{Expanded form: } f^S(s) = \sum_{i=0}^{2} \mathrm{B}_i^2(s)\,\bar{\mathbf{q}}_i^F,
\end{aligned} \tag{9.100}$$

However, the element is cubic in the s-direction, so the surface load has to be degree elevated once, to obtain its cubic representation as:

$$\begin{aligned}
&\text{Matrix form } f^S(s) = \mathbf{T}\,\mathbf{q}^F, \quad \text{with} \quad \mathbf{T} = \left\{\mathrm{B}_0^3(s)\ \mathrm{B}_1^3(s)\ \mathrm{B}_2^3(s)\ \mathrm{B}_3^3(s)\right\}\\
&\text{Expanded form: } f^S(s) = \sum_{i=0}^{3} \mathrm{B}_i^3(s)\,\mathbf{q}_i^F,\\
&\text{where: } \mathrm{B}_0^3(s) = (1-s)^3, \quad \mathrm{B}_1^3(s) = 3(1-s)^2 s, \quad \mathrm{B}_2^3(s) = 3(1-s)s^2, \mathrm{B}_3^3(s) = s^3\\
&\text{and: } \mathbf{q}^F = \left\{\bar{q}_0^F \quad \tfrac{2}{3}\bar{q}_1^F \quad \tfrac{2}{3}\bar{q}_1^F \quad \bar{q}_2^F\right\}^T = \left\{0 \quad \frac{2P}{L_y} \quad \frac{2P}{L_y} \quad 0\right\}^T
\end{aligned} \tag{9.101}$$

Now, given the transformation matrix and the surface load, the generalized nodal surface control loads are given by line integrals, which can be evaluated right away by numerical integration. However, for the sake of standardization of input of loads, we have represented the surface load isoparametrically. Then the nodal control surface load can be evaluated in closed form as we explain next. The surface control nodal load is given by:

$$\begin{aligned}
&\mathbf{F}^S = \int_{\Gamma_\sigma} \mathbf{T}^T f^s\, d\Gamma_\sigma = L_y \int_0^1 \mathbf{T}^T p(s)\, ds = L_y \left(\int_0^1 \mathbf{T}^T\mathbf{T}\, ds\right) \mathbf{q}^F = 2P \left(\int_0^1 \mathbf{T}^T\ \mathbf{T}\, ds\right) \hat{\mathbf{q}}^F,\\
&\text{where, } \hat{\mathbf{q}}^F = \{0 \quad 1 \quad 1 \quad 0\}^T
\end{aligned} \tag{9.102}$$

This integration can be evaluated numerically by Gaussian quadrature formulas. But, we apply the useful properties of Bernstein polynomials for a closed form expression, as follows:

$$\boxed{\begin{aligned} &\mathrm{B}_i^m(s)\,\mathrm{B}_j^n(s) = \frac{\binom{m}{i}\binom{n}{j}}{\binom{m+n}{i+j}}\mathrm{B}_{i+j}^{m+n}(s) \quad \text{where} \quad \binom{m}{i} = \frac{m!}{i!(m-i)!} \\ &\text{and,} \int_0^1 \mathrm{B}_i^n(s)\,ds = \frac{1}{n+1} \end{aligned}} \tag{9.103}$$

Finally, using formulas (9.103) in equation (9.102), we get the generalized surface control load in two-dimensional notation as:

$$\boxed{F_{ij}^S = \int_0^1 \mathbf{T}_i^T\,\mathbf{T}_j\,ds = \int_0^1 B_i^3(s)\,B_j^3(s)\,ds = \frac{\binom{3}{i}\binom{3}{j}}{\binom{6}{i+j}}\int_0^1 B_i^6(s)\,ds = \left(\frac{1}{7}\right)\frac{\binom{3}{i}\binom{3}{j}}{\binom{6}{i+j}}} \tag{9.104}$$

The boundary condition is taken to be completely fixed at the $x = 0$ edge, that is, no warping is allowed. Finally, the problem is treated as a plane stress problem with the Poisson's ratio, $\nu = 0.20$. Now, solving for the particular case of:

$$\frac{L_x}{L_y} = 3 \tag{9.105}$$

we get the tip displacement downward in the y-direction at mid-point of the loaded surface as:

$$\boxed{\begin{aligned} \mathrm{v}\left(L_x, \tfrac{1}{2}L_y\right) &= 115.54\left(\frac{P}{Et}\right)\text{(computed)} \\ &= 116.00\left(\frac{P}{Et}\right)\text{(exact)} \end{aligned}} \tag{9.106}$$

where, E and t are the material elasticity modulus and the thickness, respectively, of the beam. For one element and large aspect ratio, the solution seems quite remarkable. Of course, the accuracy can be significantly improved with two elements.

Example CFEM 12: Beam bending as plane stress problem (distortion and incompressibility)

The beam shown in Figure 9.27 is supported at one end with warp possibility and is shear loaded vertically up on the other end. Let us consider a c-type cubic–cubic element in u and v displacements. Use of the virtual work principle, gives the stiffness control matrix and load control vector, as usual. However, the handling of the imposed surface load (likewise, the balancing shear load on the supported end) needs elaboration. The parabolic shear, $f^S(y)$, along the $x = L_x$ edge, such that the total load amounts to P, that is, $f^S(y) = \frac{6P}{L_y^3}y(y - L_y)$, is treated in the same way as in the previous example to find the nodal load controls corresponding to cubic distribution. To

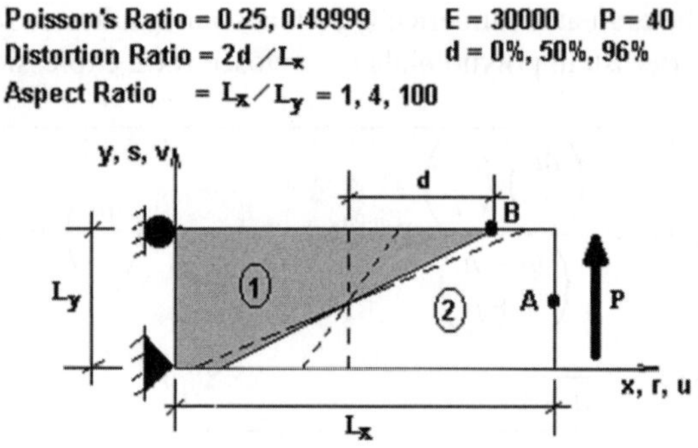

Figure 9.27 Example CFEM 12: Deep beam bending as plane stress problem – distortion and incompressibility.

further demonstrate the effectiveness of the c-type method, we continue with the two-element configuration of the same problem as introduced before. The tip displacements and extensional stresses of the extreme fiber at the common boundary of the two elements for various aspect ratios, distortion ratios and Poisson's ratio are recorded in tabular form, as shown in Figure 9.28. The straight ***Timoshenko beam*** with the rotation of the normal to the neutral axis and the transverse displacement taken for cubic–cubic representation has also been considered. The results, although are not reproduced here, are equally excellent.

Aspect Ratio $a_r = \frac{L_x}{L_y}$	Poisson's Ratio μ	Distortion Ratio $d_r = \frac{2d}{L_x}$	Displacement v^A		Axial Stress σ^B	
			Computed	Exact	Computed	Exact
1**	0.25	0 %	0.0091	0.0088		
4	0.25	0 %	0.3524		1.000	1.000
		33.33 %	0.3528	0.3553*	1.327	1.333
		50.00 %	0.3531		1.501	1.500
		91.67 %	0.3538		1.904	1.917
	0.49999	0 %	0.3510		1.000	1.000
		33.33 %	0.3522	0.3586*	1.323	1.333
		50.00 %	0.3529		1.500	1.500
		91.67 %	0.3550		1.879	1.917
100	0.25	0 %	3990935 $		297.766	300.00
		66.67 %	3994340	4000256	448.954	450.00
		96.00 %	3995810		585.461	576.00
	0.49999	0 %	3954158		290.500	300.00
		66.67 %	3967103	4000325	444.257	450.00
		96.00 %	3974440		575.314	576.00
$ For Fixed End Condition: Typical Computed v^A = 4000238; ** 1-element configuration						

Figure 9.28 Example CFEM 12: Tip displacement and extensional stresses for various aspect ratio.

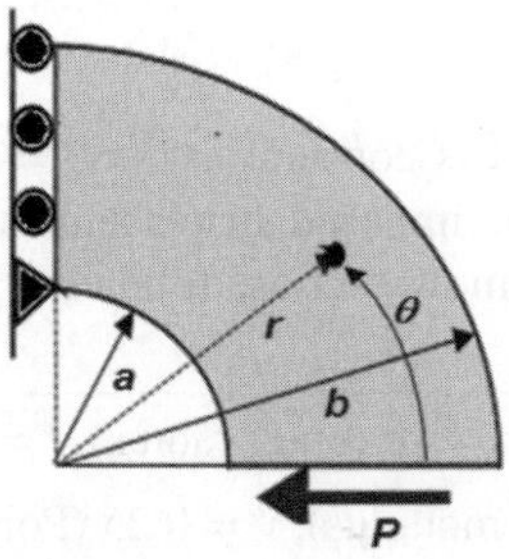

Figure 9.29 Example CFEM 13: Curved deep beam as a plane stress problem.

Example CFEM 13: Curved deep beam bending as a plane stress problem.

Problem: Let us consider the deep beam in the form of one quadrant of a circle, shown in Figure 9.29, in plane stress. The problem is solved in Timoshenko and Goodier (1951) by way of using stress functions. The stresses are found to be:

$$\boxed{\begin{aligned} \sigma_{rr} &= \frac{P}{N}\left[r + \frac{a^2b^2}{r^3} - \frac{a^2+b^2}{r}\right]\sin\theta \\ \sigma_{\theta\theta} &= \frac{P}{N}\left[3r - \frac{a^2b^2}{r^3} - \frac{a^2+b^2}{r}\right]\sin\theta \\ \tau_{r\theta} &= \frac{P}{N}\left[r - \frac{a^2b^2}{r^3} - \frac{a^2+b^2}{r}\right]\cos\theta \\ \text{where}\quad N &\equiv a^2 - b^2 + (a^2+b^2)\log\frac{b}{a} \end{aligned}} \tag{9.107}$$

For the boundary conditions shown in Figure 9.29, the displacement solutions are found to be:

$$\boxed{\begin{aligned} u_r &= \frac{P}{NE}\left\{\begin{array}{l}\left[\begin{array}{l}\frac{1}{2}(1-3\nu)r^2 - \frac{a^2b^2(1+\nu)}{2r^2} \\ -(a^2+b^2)(1-\nu)\log r\end{array}\right]\sin\theta \\ +(a^2+b^2)(2\theta-\pi)\cos\theta\end{array}\right\} - K\sin\theta \\ u_\theta &= -\frac{P}{NE}\left\{\begin{array}{l}\left[\begin{array}{l}\frac{1}{2}(5+\nu)r^2 - \frac{a^2b^2(1+\nu)}{2r^2} \\ +(a^2+b^2)[(1-\nu)\log r + (1+\nu)]\end{array}\right]\cos\theta \\ -(a^2+b^2)(2\theta-\pi)\sin\theta\end{array}\right\} - K\cos\theta \\ &\text{where } u_r(a,\pi/2) = 0 \Rightarrow K \equiv \frac{P}{NE}\left[\frac{1}{2}(1-3\nu)a^2 - \frac{b^2(1+\nu)}{2} - (a^2+b^2)(1-\nu)\log a\right] \\ &\text{with } u_r(r,0) = -\frac{\pi P}{NE}(a^2+b^2) = \text{constant} \end{aligned}} \tag{9.108}$$

Remark

- Since the radial displacement, u_r, is constant at the $\theta = 0$ shearing load boundary, we may choose to solve this problem for imposed displacement only rather than one of imposed tractions. For our example problem, we choose following properties:

$$\boxed{\begin{array}{l} a = 5 \text{ (inner radius)}, \ b = 10 \text{ (outer radius)}, \ t = 1 \text{ (thickness)} \\ E = 10000 \text{ (elasticity modulus)}, \ \nu = 0.25 \text{ (Poissons ratio)} \\ \text{Boundary conditions: } v = -0.01 \text{ on all points at the shear load edge} \\ u(5, \pi/2) = 0, \ v(r, \pi/2) = 0 \\ \text{Exact total energy: } E_{\text{exact}} = \dfrac{1}{\pi}(\log 2 - 0.6) = 0.02964966844238 \end{array}} \tag{9.109}$$

In the mesh generation phase, the geometry of the structure can be represented either approximately by a non-rational cubic Bernstein–Bezier curve, or exactly by a rational quadratic Bernstein–Bezier curve. The strain–displacement relationship, likewise, can be developed in both Cartesian and polar coordinates, but for this problem we choose Cartesian coordinates as our basis of development. Finally, the problem is solved subparametrically, as opposed to isoparametry, as will be seen shortly; we will elaborate on the solution in some details because the problem may be taken as a prelude to more complicated membrane problems associated with structural surfaces such as shells, and such like.

Solution We determine the accuracy of the c-type finite element method by choosing to solve this problem with increasing complexity in terms of geometry representation and element displacement distribution selection.

Mesh generation phase: For non-rational geometry representation, let us recall, from Chapter 5 on curves, that a quadrant of a circle of unit radius can be represented approximately by a cubic Bernstein–Bezier curve.

Thus, the geometry of the surface, $\mathbf{S}(\xi^1, \xi^2) = \begin{pmatrix} X(\xi^1, \xi^2) & Y(\xi^1, \xi^2) \end{pmatrix}^T$, is represented always by the linear–cubic Bernstein–Bezier root element, linear in the radial direction and cubic in the circumferential direction, shown in Figure 9.30, as:

- a 1×3 non-rational Bernstein–Bezier surface, $\mathbf{S}(\xi^1, \xi^2)$, defined by:

$$\boxed{\mathbf{S}(\xi^1, \xi^2) = \begin{Bmatrix} X(\xi^1, \xi^2) \\ Y(\xi^1, \xi^2) \end{Bmatrix} == \sum_{i=0}^{1} \sum_{j=0}^{3} B_i^1(\xi^1) B_j^3(\xi^2) (\mathbf{q}_{ij}^c = \begin{Bmatrix} q_{ij}^X \\ q_{ij}^Y \end{Bmatrix}), \quad 0 \leq \xi^1, \xi^2 \leq 1} \tag{9.110}$$

where the basis (blending) function sets, $\{B_i^1(\xi^1)\}, \{B_j^3(\xi^2)\}$, are the Bernstein polynomials of degree one and three, respectively; for example:

$$\boxed{B_j^n(\xi^2) = \begin{pmatrix} n \\ j \end{pmatrix} (\xi^2)^j \, (1 - \xi^2)^{n-j}, j \in \{0, 1, \ldots, n\} \text{ where } \begin{pmatrix} n \\ j \end{pmatrix} \equiv \frac{n!}{j!(n-j)!}} \tag{9.111}$$

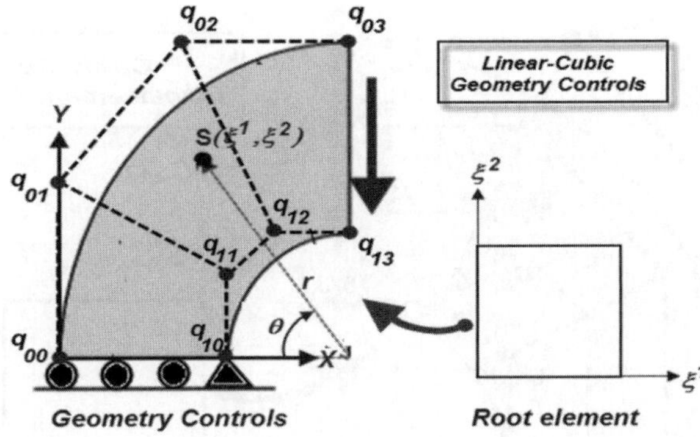

Figure 9.30 Example CFEM 13: Geometry control polygon and root element.

with $\alpha = \frac{4}{3}(\sqrt{2} - 1)$, the control vertices are:

$$\mathbf{q}^c_{00} = \begin{Bmatrix} 0 \\ 0 \end{Bmatrix} \quad \mathbf{q}^c_{01} = \begin{Bmatrix} 0 \\ \alpha b \end{Bmatrix} \quad \mathbf{q}^c_{02} = \begin{Bmatrix} b(1-\alpha) \\ b \end{Bmatrix} \quad \mathbf{q}^c_{03} = \begin{Bmatrix} b \\ b \end{Bmatrix}$$
$$\mathbf{q}^c_{10} = \begin{Bmatrix} b-a \\ 0 \end{Bmatrix} \quad \mathbf{q}^c_{11} = \begin{Bmatrix} b-a \\ \alpha a \end{Bmatrix} \quad \mathbf{q}^c_{12} = \begin{Bmatrix} b-\alpha a) \\ a \end{Bmatrix} \quad \mathbf{q}^c_{13} = \begin{Bmatrix} b \\ a \end{Bmatrix} \tag{9.112}$$

The Jacobian, $\mathbf{J} = \frac{\partial(X,Y)}{\partial(\xi^1,\xi^2)}$, and the determinant of the Jacobian, $\det \mathbf{J}$, from equation (9.110), are computed in the mesh generation phase as follows:

Step 0:

$$\begin{Bmatrix} \frac{\partial}{\partial \xi^1} \\ \frac{\partial}{\partial \xi^2} \end{Bmatrix} = \begin{bmatrix} \frac{dX}{d\xi^1} & \frac{dY}{d\xi^1} \\ \frac{dX}{d\xi^2} & \frac{dY}{d\xi^2} \end{bmatrix} \begin{Bmatrix} \frac{\partial}{\partial X} \\ \frac{\partial}{\partial Y} \end{Bmatrix} \Leftrightarrow \begin{Bmatrix} \frac{\partial}{\partial \xi^1} \\ \frac{\partial}{\partial \xi^2} \end{Bmatrix} = \mathbf{J} \begin{Bmatrix} \frac{\partial}{\partial X} \\ \frac{\partial}{\partial Y} \end{Bmatrix} \tag{9.113}$$

Note that, in actual implementation, the derivatives of the Bernstein–Bezier functions are computed by computationally economic difference operator.

Step 1:

$$\det \mathbf{J} = \frac{dX}{d\xi^1}\frac{dY}{d\xi^2} - \frac{dX}{d\xi^2}\frac{dY}{d\xi^1} \tag{9.114}$$

Step 2: The inverse of the Jacobian, needed for the strain–displacement relationship, is also stored during the mesh generation phase as:

$$\begin{Bmatrix} \frac{\partial}{\partial X} \\ \frac{\partial}{\partial Y} \end{Bmatrix} = \frac{1}{\det \mathbf{J}} \begin{bmatrix} \frac{dY}{d\xi^2} & -\frac{dY}{d\xi^1} \\ -\frac{dX}{d\xi^2} & \frac{dX}{d\xi^1} \end{bmatrix} \begin{Bmatrix} \frac{\partial}{\partial \xi^1} \\ \frac{\partial}{\partial \xi^2} \end{Bmatrix} \tag{9.115}$$

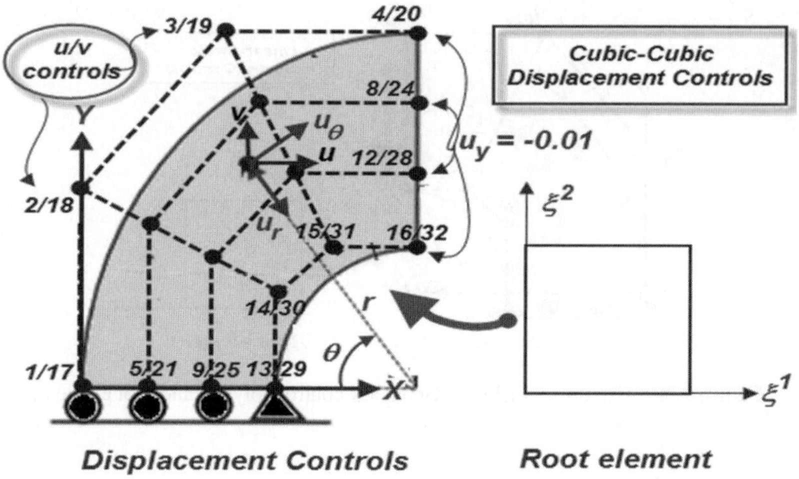

Figure 9.31 Example CFEM 13: Displacement control polygon and root element.

c-type finite element phase: Likewise, the displacement distribution, as shown in Figure 9.31, is represented by cubic–cubic non-rational Bernstein–Bezier functions, with as yet unknown control nct, as:

$$\boxed{\mathbf{d}(\xi^1,\xi^2) = \left\{ \begin{array}{c} u(\xi^1,\xi^2) \\ v(\xi^1,\xi^2) \end{array} \right\} == \sum_{i=0}^{3}\sum_{j=0}^{3} B_i^3(\xi^1)\, B_j^3(\xi^2) \left(\mathbf{q}_{ij}^d = \left\{ \begin{array}{c} q_{ij}^u \\ q_{ij}^v \end{array} \right\} \right), \quad 0 \leq \xi^1,\xi^2 \leq 1} \tag{9.116}$$

The (3×32) Cartesian strain–displacement matrix is obtained by computing $u_{,X}, u_{,Y}, v_{,X}, v_{,Y}$ from equation (9.116) with the help of equation (9.115). With the constitutive matrix for plane stress and the strain–displacement matrix, the element singular stiffness control matrix may be computed. However, as indicated in the theory of finite element formulation, the integral is seldom calculated in closed form, but is numerically integrated by Gaussian quadrature formula. For the cubic–cubic displacement distribution, the full (4×4) Gaussian points are selected, and integration is performed by summation over these quadrature points. Before solving it, we need to impose the displacement constraints at appropriate degrees of freedom to generate the non-singular equation as follows:

$$\boxed{\begin{array}{l} u_{13} = u_{20} = u_{24} = u_{28} = u_{32} = 0, \\ v_{17} = v_{21} = v_{25} = v_{29} = 0 \end{array}} \tag{9.117}$$

The stiffness equation is then solved for the unknown displacement controls; the displacements, u and v are computed for any (ξ^1,ξ^2) using equation (9.116). Finally, the radial displacement, u_r, and the circumferential displacement, u_θ, at any (r,θ), are obtained by simple rotation of the coordinate relationship. The problem is solved for two choices of geometry representation and various choices of displacement distributions:

> **Case 1:** One element: linear–cubic non-rational Bernstein–Bezier geometry with cubic–cubic non-rational Bernstein–Bezier displacement distribution, as described above,
>
> **Case 2:** Two equal elements: linear–cubic non-rational Bernstein–Bezier geometry elements with cubic–cubic non-rational Bernstein–Bezier displacement distribution connected circumferentially by C^2- condition, that is, a single B-spline element; the two cubic elements are obtained by subdivision algorithm.
>
> **Case 3:** Two equal elements: linear–cubic non-rational Bernstein–Bezier geometry with cubic–cubic non-rational Bernstein–Bezier displacement distribution,
>
> **Case 4:** A linear-quadratic Rational Bernstein–Bezier geometry with quintic-quintic non-rational Bernstein–Bezier displacement distribution.
>
> **Case 5:** A linear–cubic non-rational Bernstein–Bezier geometry with quartic-quartic non-rational Bernstein–Bezier displacement distribution; solved by 5×5 Gauss quadrature points.
>
> **Case 6:** A linear–cubic non-rational Bernstein–Bezier geometry with quintic-quintic non-rational Bernstein–Bezier displacement distribution; solved by 6×6 Gauss quadrature points.

The radial and circumferential displacement distributions for the quintic–quintic case are shown in Figure 9.32(a), and, the total energy for the various cases mentioned above are shown and compared with the exact value in tabular form as in Figure 9.32(b).

9.6 c-type Solutions: Locking Problems

Recall that interpolation problems have been the nagging nuisance to all old (namely, h-type and p-type) finite element methods. Thus, as a further demonstration of superiority of the c-type finite element method, we now offer some solutions to outstanding and nagging locking problems of the these methods as presented in Chapter 1's introduction and motivation, which may render old safety calculations dubious and unacceptably risky; we will use some practical problems that routinely appear in industry, and that are posed and considered solved by ad hoc means such as under-integration, reduced integration and arbitrary drilling degree of freedom in the old finite element methods. So, before we present our solutions for these locking problems, let us briefly recall the conventional state-of-the-art remedies.

9.6.1.1 Reduced/Selective Integration, Bubble Mode and Drilling Freedom

As we already know, the numerical integration used to evaluate, say, the stiffness matrix, is performed by selecting Gauss or other quadrature points. Thus, an immediate remedy to shear locking was offered by ***selective under-integration*** by arbitrarily evaluating the integral involving shear at only $\mathbf{x} = \mathbf{0}$, that is, one point Gauss, which effectively wipes out the shear contribution to the stiffness matrix. The problem with this remedy is that the reduced integration is obviously less

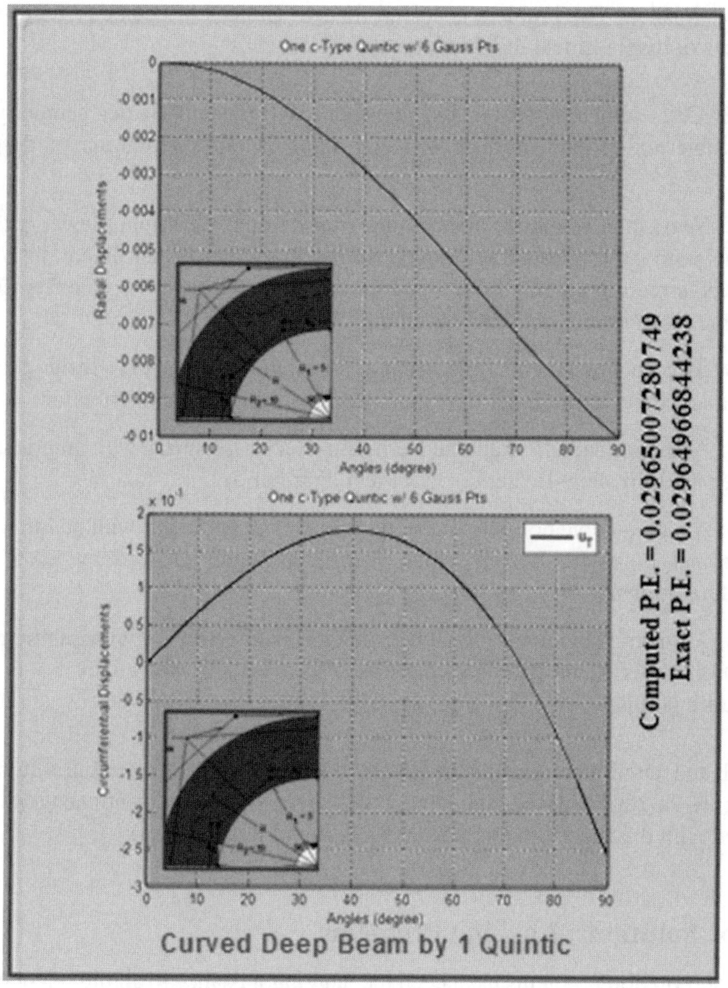

(a)

Case	Computed Total Energy	Exact Total Energy
1	0.03077147592990	0.02964966844238
2	0.02978271529071	
3	0.02974391245645	
4	0.02968587309359	
5	0.02967249805952	
6	0.02965007280749	

(b)

Figure 9.32 (a) Example CFEM 13: Radial and circumferential displacement for a quintic-quintic case. (b) Example CFEM 13: Total energy for various cases compared to exact value.

accurate. More importantly, it creates undesirable ***spurious modes***, that is, artificial singularities or eigenmodes, induced by possible reduction in the rank of the stiffness matrix. Another ad hoc remedy was to include quadratic interpolation terms, the so-called ***bubble modes***, without introducing any additional node, to handle both shear and dilatation locking. This, of course, renders the element non-conforming as two nodes cannot possibly represent a quadratic term. In line with the addition of modes of flexibility just described, modes known as the ***drilling freedoms*** have been devised to resolve locking problems. The drilling freedoms are nodal rotational degrees of freedom about an axis perpendicular the plane of the element. These in-plane rotational degrees of freedom, in improving membrane performance, may create a problem for bending performance of the curved shell elements where these degrees of freedoms are already coupled with the bending modes. In general, most of these patchy remedies that were used on regular rectangular elements, fail to eradicate locking when applied to general h-type isoparametric curved element of arbitrary shape. Moreover, these varieties of so-called remedial elements for specific displacement states increase the number of elements available commercially to such an extent that an ordinary user can be easily confused as to their usefulness – it is important that the element library be simple and user-friendly.

Thus, in what follows, it will be recognized that these locking problems are interpolational ones that crop up due to an incomplete set of basis functions, and hence are easily remedied by a complete set of ingenious polynomial functions with geometric representation such as the Bernstein–Bezier functions, which contain all the above modes (bubble, etc.) of deformation naturally.

Example RSL: Remedy for shear locking: c-type four-node quadratic element

For shear locking (as discussed in Chapter 1), consider the following state of stress imposed on a four-node old (i.e. h-type or p-type) rectangular membrane element: in-plane bending load due to a couple of moment applied at opposite ends of the element in plane stress situation, as shown in Figure 9.33.

The state of the actual displacement is given by:

$$\boxed{u = cxy \quad \text{and} \quad v = -\frac{1}{2}cx^2} \tag{9.118}$$

and shown in Figure 9.34.

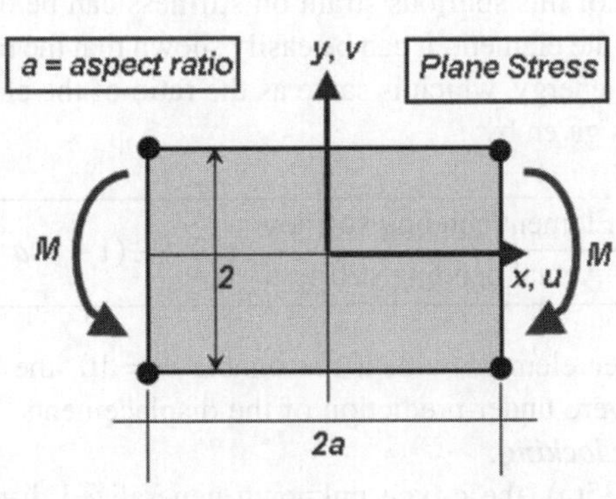

Figure 9.33 Example RSL: Remedy for shear locking problem.

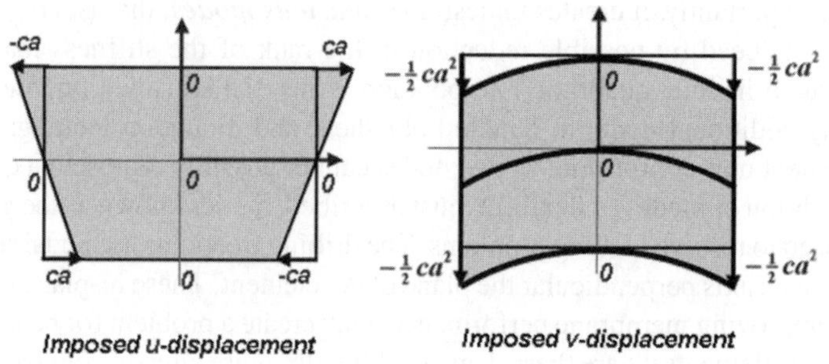

Figure 9.34 Example RSL: Imposed **u** and **v** displacements.

The corresponding states of actual strain and stress for the plane stress situation with E, ν as the elasticity modulus and Poisson's ratio, respectively, are given by:

$$\begin{array}{ll} \epsilon_x \equiv \dfrac{\partial u}{\partial x} = cy & \sigma_x = \dfrac{cEy}{1-\nu^2} \\ \epsilon_y \equiv \dfrac{\partial v}{\partial y} = 0 & \sigma_y = \dfrac{\nu cEy}{1-\nu^2} \\ \gamma_{xy} \equiv \dfrac{\partial u}{\partial y} + \dfrac{\partial v}{\partial x} = 0 & \tau_{xy} = 0 \end{array} \tag{9.119}$$

An application of the displacement state, given by the equation (9.118), to the joint nodes of the element and then subsequent calculation (by inspection) yields the element internal state of stress as:

$$\begin{array}{ll} \epsilon_x = cy & \sigma_x = \dfrac{cEy}{1-\nu^2} \\ \epsilon_y = 0 & \sigma_y = \dfrac{\nu cEy}{1-\nu^2} \\ \gamma_{xy} = cx & \tau_{xy} = \dfrac{cEx}{2(1+\nu)} \end{array} \tag{9.120}$$

From equation (9.120), we can easily see that the element has generated undesirable, spurious shear strain. The ill effect of this spurious strain on stiffness can be evaluated by observing the computed strain energy of the element. It can be easily shown that the ratio of the computed strain energy to the actual strain energy, which is same as the ratio of the element bending stiffness to actual bending stiffness, is given by:

$$\frac{\text{Element bending stiffness}}{\text{Actual bending stiffness}} = 1 + \frac{1}{2}(1-\nu)a^2 \tag{9.121}$$

Obviously, for a slender element with a reasonable $a = 10$, the element stiffness will be excessive and result in severe under-prediction of the displacements. This spurious shear strain brought in the name ***shear locking***.

As shown in Figure 9.35(a), the c-type unknown generalized displacement vectors are the Bezier control vectors, $\mathbf{q}^u$ and $\mathbf{q}^v$ in x and y directions, respectively. The mapping function is

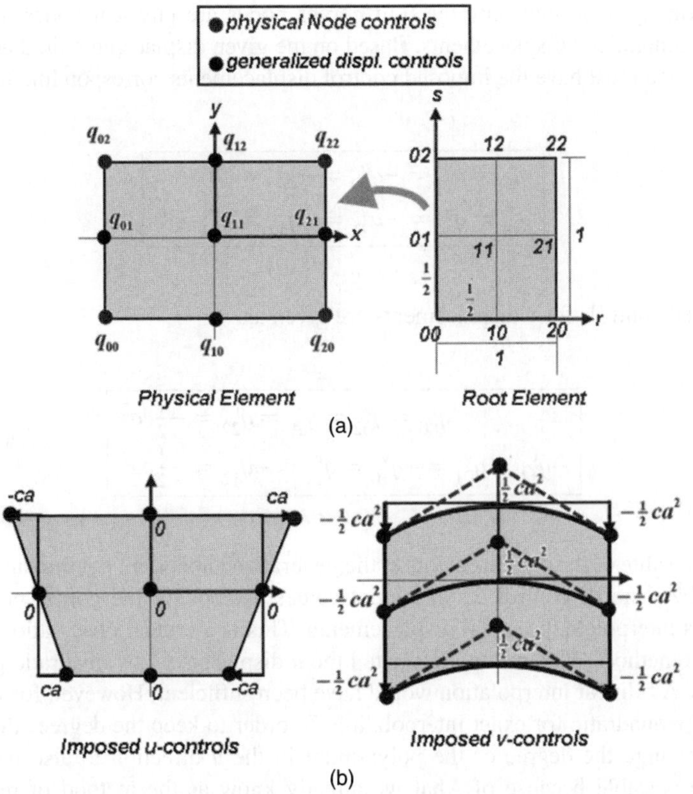

Figure 9.35 (a) Example RSL: c-type quadratic membrane physical and root element. (b) Example RSL: Imposed displacement **u**-controls and **v**-controls.

composed of Bernstein polynomials; the root element is a square with r and s as the coordinate directions. The root to physical element inverse mapping is given by:

$$\boxed{r = \frac{x+a}{2a} \quad \text{and} \quad s = \frac{y+1}{2}} \tag{9.122}$$

In component form, we express the displacement vectors as a linear combination of the basis functions as:

$$\boxed{\mathrm{u}(r,s) = \sum_{i=0}^{2}\sum_{j=0}^{2} B_i^2(r)B_j^2(s)\, q_{ij}^u \quad \text{and} \quad \mathrm{v}(r,s) = \sum_{i=0}^{2}\sum_{j=0}^{2} B_i^2(r)B_j^2(s)\, q_{ij}^v} \tag{9.123}$$

where, the second degree Bernstein polynomials for, say, r, are given by:

$$\boxed{B_0^2(r) = (1-r)^2, B_1^2(r) = 2(1-r)r, B_2^2(r) = r^2} \tag{9.124}$$

Most importantly, note that only the four corner nodes are physical nodes; the rest can be considered as generalized displacements. Based on the given displacement field and as shown in Figure 9.35(b), we must have the imposed control displacements corresponding to u, as:

$$\boxed{\begin{aligned} q^u_{00} &= q^u_{22} = -q^u_{20} = -q^u_{02} = ca \\ q^u_{01} &= q^u_{10} = -q^u_{21} = -q^u_{12} = q^u_{11} = 0 \end{aligned}} \tag{9.125}$$

and the imposed controls for v displacements are given as:

$$\boxed{\begin{aligned} q^v_{00} = q^v_{02} = q^v_{20} = q^v_{22} &= -\tfrac{1}{2}ca^2 \\ -q^v_{11} = q^v_{01} = -q^v_{10} = q^v_{21} = -q^v_{12} &= -\tfrac{1}{2}ca^2 \end{aligned}} \tag{9.126}$$

Note that the value of the displacements at the generalized nodes are not the same as the values of the generalized nodal control displacement values, except for the corner nodes where the control vectors interpolate the nodal displacements. This is a crucial observation to understand the new c-type method. We have approximated the u displacement by quadratic polynomials in two directions. A bilinear interpolation would have been sufficient. However, for v displacement field, we need a quadratic for exact interpolation. In order to keep the degrees the same for the example, we change the degree of the polynomial in the u direction to also being quadratic. This has been possible because of what we already know as the method of repeated ***degree elevation*** guaranteed by the linear precision property of the Bernstein polynomials. This is one of the cornerstones of the theory behind the c-type method. Most importantly, note that the linear combination given by equation (9.123) is barycentric because of the barycentric property of the Bernstein polynomials, that is,

$$\boxed{\sum_{i=0}^{n} B_i^n = 1} \tag{9.127}$$

Now, for u displacement, substituting the values of the imposed generalized control displacements, given by equation (9.125), and the definition of Bernstein polynomials, as given by equation (9.124), in equation (9.123), we get:

$$\boxed{\begin{aligned} u(r,s) &= \begin{bmatrix} \{(1-r)^2(1-2s+s^2-s^2)\} \\ -\{r^2(1-2s+s^2-s^2)\} \end{bmatrix} q^u_{00} \\ &= (1-2r)(1-2s)q^u_{00} = cxy = \text{Exact solution!} \end{aligned}} \tag{9.128}$$

Thus, u ***displacement is interpolated exactly, and so will be the strains, stresses and element strain energy***.

Similarly, for v displacement, substituting the values of generalized displacements, given by equation (9.126), and definition of Bernstein polynomials, as given by equation (9.124), in equation (9.123), we get:

$$\begin{aligned} v(r,s) &= (1-r)^2\left\{1-2s+s^2+2s-2s^2+s^2\right\}q^v_{00} \\ &\quad +2r(1-r)\left\{1-2s+s^2+2s-2s^2+s^2\right\}q^v_{10} \\ &\quad +r^2\left\{1-2s+s^2+2s-2s^2+s^2\right\}q^v_{00} \\ &= (1-2r+2r^2-2r+2r^2)q^v_{00} \\ &= (1-2r)^2 q^v_{00} = \frac{x^2}{a^2}q^v_{00} = -\tfrac{1}{2}cx^2 = \text{Exact solution!} \end{aligned} \tag{9.129}$$

Thus, v ***displacement is interpolated exactly, and so will be the strains, stresses and element strain energy, irrespective of the aspect ratio representing shape sensitivity***.

Example RDL: ***Remedy for Dilatation Locking***: c-type four-node cubic element incompressible state ($\nu \to \frac{1}{2}$)

Incompressibility is associated, for example, with a practical bridge support made up of elastomeric rubber materials, or fluid flow problems as in modeling, say, reservoir water in dam safety computations. The rubber material and the fluid are nearly incompressible. For the dilatation locking problem arising out of it, consider the following state of stress imposed on a old four-node rectangular membrane element: in-plane pure bending state due to a couple of moment applied at the opposite end of the element in plane strain situation. The state of actual displacement is given by:

$$\begin{aligned} &u = cxy \quad \text{and} \quad v = c_x x^2 + c_y y^2 \\ &\text{where, } c_x \equiv -\tfrac{1}{2}c \quad \text{and} \quad c_y \equiv -\frac{\nu}{2(1-\nu)}c \end{aligned} \tag{9.130}$$

The corresponding states of actual strain and stress for the plane strain situation (i.e. $\sigma_z \neq 0$) with E, ν as the elasticity modulus and Poisson's ratio, respectively, are given by:

$$\begin{aligned} &\epsilon_x = cy \qquad &&\sigma_x = \frac{Ecy}{1-\nu^2} \\ &\epsilon_y = -\frac{\nu}{1-\nu}cy \qquad &&\sigma_y = 0 \\ &\gamma_{xy} = 0 \qquad &&\tau_{xy} = 0 \end{aligned} \tag{9.131}$$

For future reference, let us recall, from Chapter 1, the following definitions related to volumetric strains and stresses:

Dilatation, e:

$$\begin{aligned} &e \equiv \epsilon_x + \epsilon_y + \epsilon_z = \left(\frac{1-2\nu}{1-\nu}\right)(cy), \\ &\nu \to \tfrac{1}{2} \Rightarrow e \to 0 \end{aligned} \tag{9.132}$$

Bulk modulus, K:

$$\boxed{K = \frac{E}{3(1-2\nu)}} \tag{9.133}$$

Mean pressure, p:

$$\boxed{\begin{aligned} p &= -(\sigma_x + \sigma_y + \sigma_z) = -K\,e \\ &= \frac{-Ecy}{3(1-\nu)}, \quad \nu \to \tfrac{1}{2} \Rightarrow p = \frac{-2Ecy}{3} < \infty \end{aligned}} \tag{9.134}$$

Actual strain energy density of the element, $\mathcal{U}$:

$$\boxed{\mathcal{U} \equiv \tfrac{1}{2}\sigma^T \epsilon = \frac{E}{2(1-\nu^2)}(cy)^2} \tag{9.135}$$

An application of the displacement state, given by the equation (9.130), to the joint nodes of the element and then subsequent calculation yields the element internal state of stress as:

$$\boxed{\begin{aligned} \epsilon_x &= y \quad \sigma_x = \frac{E(1-\nu)y}{(1+\nu)(1-2\nu)} \\ \epsilon_y &= 0 \quad \sigma_y = \frac{E\nu y}{(1+\nu)(1-2\nu)} \\ \gamma_{xy} &= x \quad \tau_{xy} = \frac{Ex}{2(1+\nu)} \end{aligned}} \tag{9.136}$$

It can be easily seen that as ν approaches $\frac{1}{2}$, the dilatation, e, does not go to zero as it should for incompressibility. The computed element strain energy density reflecting its stiffness is given by:

$$\boxed{\mathcal{U}^{Computed} = \frac{E}{2(1+\nu)}\left[\left(\frac{1-\nu}{1-2\nu}\right)y^2 + \frac{x^2}{2}\right]} \tag{9.137}$$

A comparison with equation (9.135), will immediately reveal that the computed strain energy density given by equation (9.137) is inaccurate. More importantly, the first term in the square bracket, in equation (9.137), shows that as ν approaches $\frac{1}{2}$, the computed strain energy density, and hence, the bending stiffness of the element will be extremely high, severely underestimating the displacement state.; this is what is known as ***dilatation locking***. The second term, as before, shows the effect of spurious shear strain. Both of these effects jeopardize the validity of the old FEM calculations for design safety; additionally, the remedy was ad hoc in the terms of including non-conforming terms.

The c-type unknown generalized displacement vectors are the Bezier control vectors, $\mathbf{q}^u$ and $\mathbf{q}^v$ in x and y directions, respectively (see Figure 9.36). The mapping function is composed of

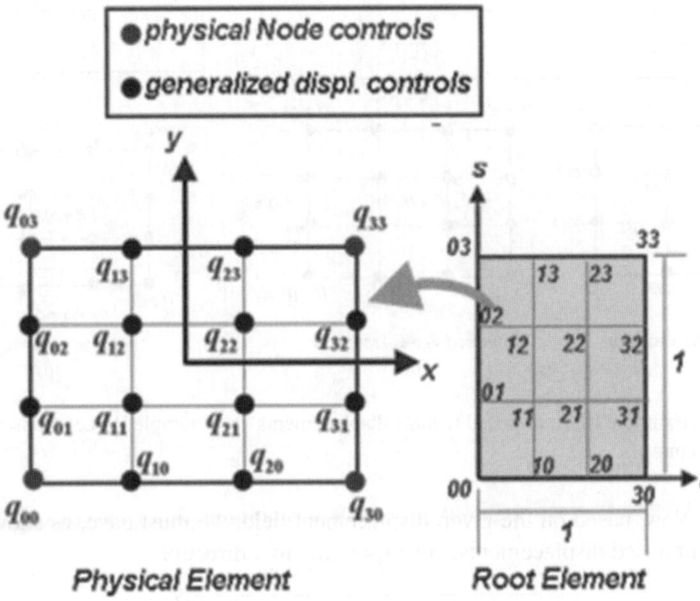

Figure 9.36 Example RDL: Remedy for dilatational locking problem with four-node cubic membrane physical and root element.

Bernstein polynomials and the root element is a square with **r** and **s** as the coordinate directions. The root to physical element inverse mapping, as before, is given by:

$$r = \frac{x+a}{2a} \quad \text{and} \quad s = \frac{y+1}{2} \tag{9.138}$$

In component form, applying c-type cubic–cubic element, we express the displacement vectors as linear combination of the basis functions as:

$$u(r,s) = \sum_{i=0}^{3}\sum_{j=0}^{3} B_i^3(r)B_j^3(s)\, q_{ij}^u \quad \text{and} \quad v(r,s) = \sum_{i=0}^{3}\sum_{j=0}^{3} B_i^3(r)B_j^3(s)\, q_{ij}^v \tag{9.139}$$

where the Bernstein polynomials are given by:

$$B_0^3(r) = (1-r)^3, \quad B_1^3(r) = 3(1-r)^2 r, B_2^3(r) = 3(1-r)r^2, \quad B_3^3(r) = r^3 \tag{9.140}$$

Note that as in our solution of Example RSL, we actually needed bilinear interpolation for the u displacement. This is equally applicable to the present solution, as the u displacement field is the same. However, for the v displacement field, we need a cubic for exact interpolation. In order to keep the degrees the same, we change the degree of the polynomial in the u direction to also be cubic. This has been possible by the method of repeated degree elevation guaranteed by the linear precision property of the Bernstein polynomials. Most importantly, note that only the four corner nodes are physical nodes and the rest of them can be considered as generalized control

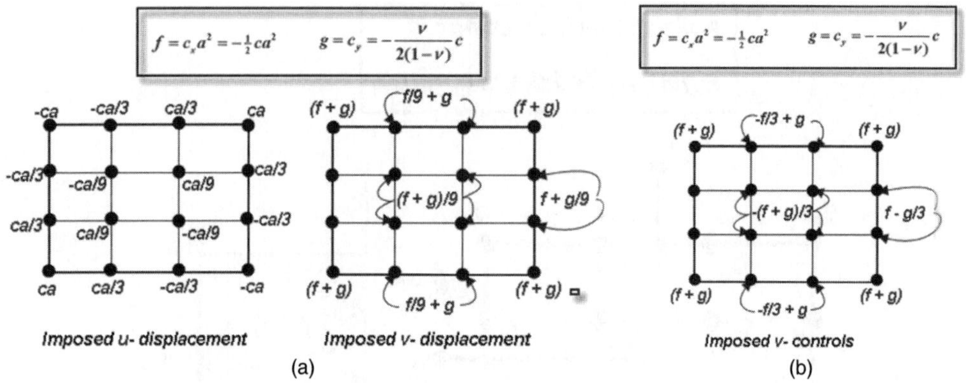

Figure 9.37 (a) Example RDL: imposed **u** and **v** displacements. (b) Example RDL: Imposed displacement **u** controls and **v** controls.

displacements. Now, based on the given displacement field, we must have, as shown in Figures 9.37(a),(b),the imposed displacements, corresponding to u direction:

$$\boxed{\begin{aligned} u_{00} &= -u_{03} = -u_{30} = u_{33} = ca \\ u_{10} &= -u_{20} = -u_{13} = u_{23} = \tfrac{1}{3}ca \\ u_{01} &= -u_{02} = -u_{31} = u_{32} = \tfrac{1}{3}ca \\ u_{11} &= -u_{12} = -u_{21} = u_{22} = \tfrac{1}{9}ca \end{aligned}} \tag{9.141}$$

Similarly, based on the given displacement field, we must have, as shown in Figure 9.37(a), the imposed v displacements as:

$$\boxed{\begin{aligned} v_{00} &= v_{03} = v_{30} = v_{33} = c_x a^2 + c_y \\ v_{10} &= v_{20} = v_{13} = v_{23} = \tfrac{1}{9}c_x a^2 + c_y \\ v_{01} &= v_{02} = v_{31} = v_{32} = c_x a^2 + \tfrac{1}{9}c_y \\ v_{11} &= v_{12} = v_{21} = v_{22} = \tfrac{1}{9}(c_x a^2 + c_y) \end{aligned}} \tag{9.142}$$

Figure 9.37(b) shows the imposed Bezier control vectors for v displacement, derived from equation (9.142), that is given by:

$$\boxed{\begin{aligned} q^v_{00} &= q^v_{03} = q^v_{30} = q^v_{33} = c_x a^2 + c_y \\ q^v_{10} &= q^v_{20} = q^v_{13} = q^v_{23} = -\tfrac{1}{3}c_x a^2 + c_y \\ q^v_{01} &= q^v_{02} = q^v_{31} = q^v_{32} = c_x a^2 - \tfrac{1}{3}c_y \\ q^v_{11} &= q^v_{12} = q^v_{21} = q^v_{22} = -\tfrac{1}{3}(c_x a^2 + c_y) \end{aligned}} \tag{9.143}$$

To show how we obtain expression (9.143) from expression (9.142), let us look into the following auxiliary degree elevation problem:

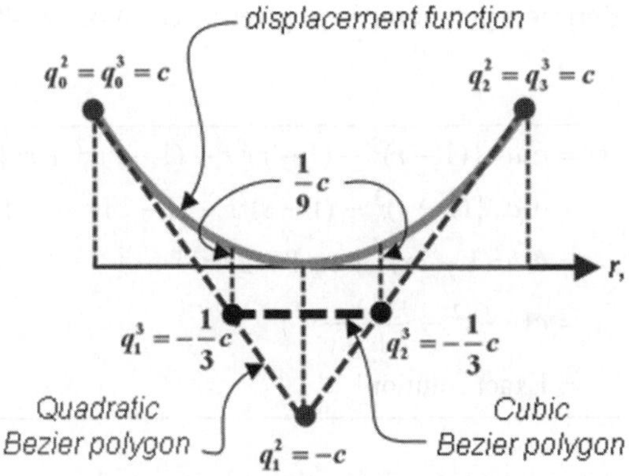

Figure 9.38 Auxiliary degree elevation problem.

9.6.1.2 Auxiliary Degree Elevation Problem

Problem: Given a one-dimensional quadratic displacement distribution, in parametric domain, for arbitrary constant, c, as:

$$\boxed{\{u(r), \quad r \in [0,1] | u(0) = c, u\left(\frac{1}{2}\right) = 0, u(1) = c\}} \tag{9.144}$$

As shown in Figure 9.38, show the quadratic Bezier control displacements degree elevate and show the corresponding cubic Bezier controls. Let q_0^2, q_1^2 and q_2^2 are the required quadratic controls. From the property of the parabolic distribution and the Bernstein–Bezier interpolation, we get:

$$\boxed{q_0^2 = c, \quad q_1^2 = -c, \quad q_2^2 = c} \tag{9.145}$$

Now, let q_0^3, q_1^3, q_2^3 and q_3^3 be the corresponding cubic controls that describe the same displacement distribution. Using the formula for degree elevation and applying expressions in equation (9.145), we get:

$$\boxed{\begin{array}{ll} q_0^3 = q_0^2 = c, & q_1^3 = \frac{1}{3}q_0^2 + \frac{2}{3}q_1^2 = -\frac{1}{3}c \\ q_2^3 = \frac{2}{3}q_1^2 + \frac{1}{3}q_2^2 = -\frac{1}{3}c, & q_0^3 = q_3^3 = c \end{array}} \tag{9.146}$$

completing our demonstration.

Equipped with our description of the imposed control displacements supported by the above auxiliary problem, let us now go back to the original problem. For u displacement, substituting the values of imposed generalized control displacements, given by equation (9.141), and definition of Bernstein polynomials, as given by equation (9.140), in equation (9.139), we can show that ***the computed displacement field matches exactly the given imposed one***. Similarly, for **v** displacement, substituting the values of generalized control displacements, given by equation

(9.142), and definition of Bernstein polynomials, as given by equation (9.140), in equation (9.139), we get:

$$\begin{aligned} v(r,s) &= c_x a^2 \left\{(1-r)^3 - (1-r)^2 r - (1-r)r^2 + r^3\right\} \\ &\quad + c_y \left\{(1-s)^3 - (1-s)^2 r - (1-s)s^2 + s^3\right\} \\ &= c_x a^2 (1-2r)^2 + c_y (1-2s)^2 \\ &= c\{-\tfrac{1}{2}x^2 - \tfrac{1}{2}\frac{\nu}{1-\nu}y^2\} \\ &= \text{Exact solution!} \end{aligned} \tag{9.147}$$

Thus, the v displacement is interpolated exactly as before and irrespective of the Poisson's ratio. In other words, incompressible materials can be treated just as easily.

Discussion: The plate or shell elements are used to deal with problems related to, say, design of floors and dome roofs of buildings, ship hull design, and so on. Particularly, trapezoidal plate or shell elements can create problems of transverse shear locking. To demonstrate this problem and its solution by the c-type method, let us look at another example.

Example RTSL: Remedy for transverse shear locking: c-type four-node plate element

Problem: Given the element in the xy plane and the normal quadratic displacement in z direction as:

$$w = x^2 \tag{9.148}$$

Find the transverse shear strains, γ_{xz} and γ_{yz}, for the isosceles trapezoid shown in Figure 9.39(a).

The c-type unknown generalized displacement vector is the Bezier control vectors, $\mathbf{q}^w$ in the z direction. The mapping function is composed of Bernstein polynomials, and the root element is a square with $\boldsymbol{r}$ and $\boldsymbol{s}$ as the coordinate directions. The root to physical element mapping is given by:

$$x = \lambda c(2r-1)\{1 - a(2s-1)\} \quad \text{and} \quad y = c(2s-1) \tag{9.149}$$

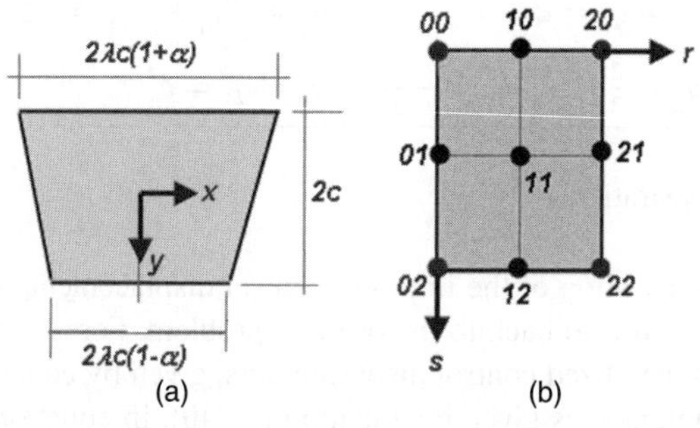

Figure 9.39 (a) Example RTSL: Remedy for transverse shear locking problem with four-node plate element. (b) Example RTSL: c-type control net locations.

In component form, applying our quadratic element, we express the normal displacement vector as a linear combination of the basis functions as:

$$w(r,s) = \sum_{i=0}^{2}\sum_{j=0}^{2} B_i^2(r)B_j^2(s)\, q_{ij}^w \tag{9.150}$$

where the Bernstein polynomials, as before, for, say, r, are given by:

$$B_0^2(r) = (1-r)^2, \quad B_1^2(r) = 2(1-r)r, \quad B_2^2(r) = r^2 \tag{9.151}$$

The locations of the Bezier control locations for components, q_{ij}^w are shown in Figure 9.39(b). Most importantly, note that only the four corner nodes are physical nodes and the rest of them can be considered as generalized control displacements. Now, based on the displacement field given by the equation (9.149), we get, as shown in Figure 9.40, the imposed displacements corresponding to w as:

$$\begin{aligned} w_{00}^w &= w_{20}^u = \lambda^2c^2(1+a)^2 \\ w_{02}^w &= w_{22}^u = \lambda^2c^2(1-a)^2 \\ w_{01}^w &= w_{21}^w = \lambda^2c^2 \\ w_{10}^u &= w_{11}^u = w_{12}^w = 0 \end{aligned} \tag{9.152}$$

Similarly, based on the displacement field given by equation (9.149), we get after trivial computation, as shown in Figure 9.40, the imposed control displacements corresponding to w as:

$$\begin{aligned} q_{00}^w &= -q_{10}^u = q_{20}^u = \lambda^2c^2(1+a)^2 \\ q_{02}^w &= -q_{12}^u = q_{22}^u = \lambda^2c^2(1-a)^2 \\ q_{01}^w &= -q_{11}^w = q_{21}^w = \lambda^2c^2(1-a^2) \end{aligned} \tag{9.153}$$

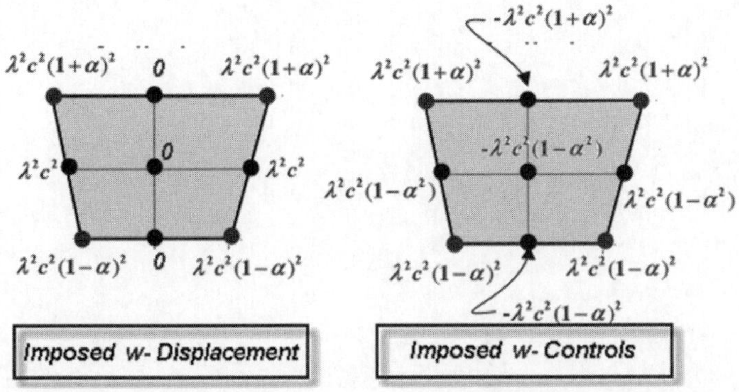

Figure 9.40 Example RTSL: Imposed **w** displacement and **w** control nets.

For w displacement, substituting the values of imposed generalized control displacements, given by equation (9.153), and definition of Bernstein polynomials, as given by equation (9.151), in equation (9.150), we can show that the computed displacement field matches exactly the imposed one given by equation (9.152). Thus, w displacement is interpolated exactly as before and irrespective of the shape of the element. In other words, shape sensitivity of the old finite elements is basically interpolation problem that can be eradicated just as easily by the c-type elements.

Discussions: The above solutions with c-type bar or truss elements, straight and curved beam elements, and plane stress and plate elements suggest that use of the c-type elements will exactly represent all the element properties: displacements, strains, stresses and strain energies. Moreover, because of the existence of the physical nodes of our elements as only corner nodes, the conformity can be achieved on demand. Although the examples we have given demonstrate many specificities, these should not be construed as limiting the scope of the c-type finite element method but as mere illustrations of the strength and superiority of the c-type method over the old h-type and p-type finite element methods. In this section, for simplicity in application of the c-type method, the example problems are restricted to the linear regime, that is, small displacement, rotations and strains with homogeneous, isotropic material behavior. In the next two chapters, we remove some of these assumptions. For kinematic behavior, we eventually assume that a plane section remains plane but not necessarily perpendicular to the centroidal axis or plane to include shear deformations; More importantly, we apply our c-type finite element method to complicated problems of beams, plates and shells in extreme geometrical nonlinearity, that is, very large displacements and rotations. The applications of the c-type method to beam, plates and shell problems in both linear and nonlinear regime are equally robust, stable and accurate. We will now demonstrate these applications in detail.

9.6.2 Where We Would Like to Go

In this chapter, we have introduced our c-type finite element method and solved various linear problems with it. As indicated, we intend to apply the method to beams, plates and shells undergoing extreme geometric nonlinearity. To become familiar with the computational strategy for these structures, we need to explore:

- Linear and nonlinear beams (Chapter 10)
- Linear and nonlinear plates and shells (Chapter 11)

10

Nonlinear Beams

10.1 Introduction

10.1.1 Why We Are Here

In this chapter, we will develop and examine the nonlinear theory of beams for both nonlinear quasi-static and dynamic loadings; we specialize the theory for thin hyperelastic beams subjected to extremely large displacements and rotations by deriving the most efficient and robust computational form to accommodate and devise extremely large c-type finite elements:

> ***Section 10.2:*** We define various geometric properties of an undeformed structural beam-like body for our subsequent analysis and indicate the assumptions that are implicit in the definition.
>
> ***Section 10.3:*** We develop a simple and direct way (as opposed to reduction of the equations of motion of continuum mechanics from 3D to 1D) of applying the momentum and moment of momentum (i.e. angular momentum) balance principles to establish the beam static and dynamic equations of motion. We remind ourselves that because we want to include the case of large (i.e. finite) displacements and rotations resulting in the nonlinear responses, all equilibrium equations must be considered in the deformed geometry of a beam. Hence, we will develop the balance equations on the current geometry for the nonlinear responses, and then particularize by linearization for the case of infinitesimal deformations, that is, in the linear regime. Furthermore, in order to arrive at our stated goal, at the very outset we need to view the beam as a representative but fictitious one-dimensional physical entity, as opposed to the three-dimensional continuum it really is, identified geometrically as being composed of a reference curve and a rigid cross-section with equivalent and physical properties assigned to simulate real deformations that is expected of a 3D continuum. We will then see that this way of viewing a beam as a one-dimensional continuum produces the exact equations of motion obtained otherwise through an elaborate and rigorous reduction from the 3D continuum viewpoint to an equivalent 1D curve except that the 3D connection provides insight into the real character of the forces

Computation of Nonlinear Structures: Extremely Large Elements for Frames, Plates and Shells, First Edition. Debabrata Ray.

and moments, and the linear and angular momenta, introduced here. Thus, while the direct approach provides an easy and simple overview of the balance laws, we must remember that a consistent interpretation of the terms developed here must await the derivations that involve reduction from the 3D continuum balance principles. We recognize that a beam, by definition, is reduced to a geometric curve and rigid cross-sections with representative and appropriate deformational properties and applied quantities, assigned to every point on the curve. Clearly, then, by treating it as a curve, we will be able to characterize the geometry (undeformed and deformed): the centroidal axis of a beam by a single parameter along the curve representing the beam. More specifically, the arc length parameter along the curve will be our natural choice in a curvilinear framework for description of a beam viewed as a curve with rigid cross-section.

Section 10.4: We reduce the dynamic and static balance equations of 3D continuum mechanics to 1D beam equations of motion. In so doing, we derive the beam dynamic and quasi-static vector momentum and moment of momentum balance equations. The main reason behind the starting point as a reduction from the 3D equations is to lend support to the accuracy, consistency and validity of the beam equations of motion that may also be derived directly and simply by following an engineering approach by stating a beam-like body geometrically as a simple 1D curve. Furthermore, it connects the beam kinematic and force boundary conditions to the actually prescribed 3D counterparts that is essential to establish the equivalence between the real problems of a 3D nature and the supposedly corresponding problems to be solved by a beam theory – a theory that models a real body effectively as a nonlinear, thin, one-dimensional object.

Section 10.5: We identify the most suitable beam kinematic variables and the configuration space, that is, the set of kinematic parameters containing all the information about the motion history (i.e. the configurations) of the body, from the beam weak form of the balance equations. Our goal is motivated by the requirement that for the determination of a beam configuration path or the history of motion for a beam numerically by an energy-type procedure such as the c-type finite element method, we must first identify the equivalent one-dimensional configuration parameters of a 3D beam-like body that is energetically consistent with the exact balance laws. Towards this, we start from the derived 1D exact equations of the momentum and the angular momentum balance laws of motion of a beam and by the method of selecting test functions (see Chapter 8 on the finite element), develop the weak form of equations which in turn allows us to identify the kinematic parameters appropriate for the conjugate virtual strains. However, alternatively, it is possible to start directly with the integral equations of the stress or deformation power and an assumed kinematics and the deformation map for the beam-like body, and define equivalently both the forces and moments and identify the conjugate virtual strains. For the beam-like body, we will perform these operations as an example. Philosophically, however, it seems to make more sense to adhere to the first method because the various beam theories with different kinematics crop up naturally in an exact fashion.

Section 10.6: We define the curvature vector and tensor, the angular velocity vector and tensor and the admissible virtual space, corresponding to the beam configuration space consisting of the displacement vector field and the rotational tensor field, necessary to identify the spatial strain rate fields and the corresponding spatial real strain fields in the weak form of equations (and later for the computational virtual work equation) of a beam reference curve. In so doing, we describe in terms of the rotation field, the curvature vector and tensor, and the angular velocity vector and tensor field, and the variations of the kinematic variables, that is, the displacement vector field and the rotation tensor field, constituting the configuration space with respect to the arc length parameter along the centroidal axis of a beam reference curve.

Section 10.7: We identify the force and moment virtual strain rate fields from the internal virtual work functional, and the corresponding real strain rate fields whose variations validate the presumptions that the strain fields, derived from the weak form, in the internal virtual work expressions are actually virtual in nature, that is, there exist functions (real fields) whose variations are these fields. As will be apparent later in our discussion on the beam constitutive theory, we are also motivated by the need to develop the appropriate constitutive laws for the one-dimensional beam by establishing the existence of an equivalent one-dimensional strain energy functional.

Section 10.8: We define the component or operational forms of various vectors associated with the beam vector balance equations. These definitions allow us to convert the invariant vector form of the beam equations of motion to the component form suitable for subsequent computational development. Let us recall that our interest in any theory is solely governed by its relevance to our computational endeavors. In other words, here we will develop a computationally efficient component, or operational vector forms, by first collecting the curvilinear components of displacements and rotations, of forces and moments and of linear and angular velocities and momenta, and then describe these as components of corresponding vectors that help to define both the undeformed and deformed states of an arbitrary beam. To be sure, we are interested ultimately in the computation of the components of displacements and rotations, and of the reactive forces and moments at any point on a beam that has been subjected to some loading conditions. This, in turn, motivates us to directly describe the relevant parameters and equations of a beam in component vectors. Accordingly, we then develop the notion of component vectors. More specifically, we define the displacement and rotation, and the force and moment component vectors. Finally, we establish the relations that exist between the force and moment component vectors as defined in the undeformed frame and a rotating or spinning deformed frame.

Section 10.9: Here, we define the covariant derivatives of the component or operational vectors necessary to convert the beam equations of motion in invariant vector form to a suitable component vector form for ease in subsequent computational development.

Section 10.10: We convert the invariant vector form of the balance equations of motion of a beam (as obtained by the reduction of the equations of motion of the continuum mechanics from 3D to 1D or by direct 1D beam balance equations) to a modified and computationally efficient component vector or operational form suitable for subsequent computational tasks such as the virtual work equation development which is the cornerstone of a finite element procedure. For finite rotation – a situation for which we will develop our theory, the beam real configuration space being a Riemannian or curved surface – we will need to work with the covariant derivatives of the component vectors. Accordingly, we then develop the beam equations of motion in component vector or operator form in curvilinear coordinates suitable for subsequent generation of the virtual work equations necessary for solution by a finite element method such as our c-type finite element procedure.

Section 10.11: We define the computational curvature tensor and vector, the computational angular velocity tensor and vector and their linearization (as a precursor to covariant linearization of the dynamic virtual functional necessary for a Newton-type method of iterative solution technique), and the computational virtual space, corresponding to the beam configuration space consisting of the displacement vector field and the rotational tensor field necessary to derive the computational virtual work functional for our subsequent numerical treatment by the c-type finite element method. In so doing, we describe the curvature and angular velocity vector and tensor fields in terms of the rotation field, and the variations of the kinematic variables, that is, the displacement vector field and the rotation tensor field, constituting the configuration space with respect to the arc length parameter along the beam reference curve. This is also a prerequisite for the identification of the virtual strain fields and the development of the linearized virtual work functionals necessary for a Newton-type iterative procedure for the numerical solution of beam problems by a variational scheme such as the c-type finite element method.

Section 10.12: We develop the thin beam computational virtual work equations from the weak form of the balance equations which we have specialized for thin beams and described as the computational beam balance equation. We do this to determine a beam configuration path or the history of motion for a beam numerically by an energy-type procedure such as the c-type finite element method, and we will have to derive, from the differential strong form of the beam equations, the integral virtual weak form of equations. Finally, we look at a semi-discrete finite element formulation in which the beam geometry and the spatial distribution of the deformations are discretized by the c-type finite element method and thus, reduce the virtual work equation eventually into a matrix second-order ordinary nonlinear differential equation. Under this type of formulation, the virtual space is taken as time-independent as the time is viewed only as an indexing parameter of the configuration space variables: the displacement vector and the rotation tensor, and thus, the time marching essentially traces a curve on the configuration space.

Section 10.13: We express the virtual strains in a computationally efficient matrix form of description, leading to a helpful description of the virtual work functional, and we also define various generalized entities of relevance for notational compactness. In so doing, we rewrite the virtual work functional of a beam in the generalized form ready for our subsequent computational endeavor for the case of beams subjected to proportional loadings in the case of the quasi-static or static analyses, and time history of loadings for dynamic analyses.

Section 10.14: We develop compatible real strain fields from the virtual strain fields that we have already derived as energetically conjugate to the real stress resultant and stress couple fields of a beam. The real strain field is necessary as one of the essential ingredients in the computational procedure; the real strains with the appropriate constitutive relationship allow us to compute the real internal reactive force (the stress resultants and the stress couples) fields of a beam. The contradiction, if any, between the internal and the applied force system drives an iterative method such as a Newton-type algorithm, in computational structural engineering and solid mechanics.

Section 10.15: We introduce the elastic material properties, that is, the stress–strain relationship, or the constitutive laws, appropriate for an elastic beam. We need to reduce the three-dimensional stress–strain laws to a one-dimensional form that could be suitably incorporated in the internal virtual work functional of a beam. Clearly, this will be an approximation, but then the entire subject of the constitutive theory by its very nature of fitting the real world experimental data is an empirical science, at best. Although the virtual work principle is exact, irrespective of the material properties, here we restrict ourselves to the case of elastic beams.

Section 10.16: Here we deal with the covariant linearization of the nonlinear virtual work functional that is necessary for the computation of the configuration history of a beam, digitally, by a Newton-type method of iterative solution for the nonlinear quasi-static (proportional loading), and by a numerical integration method of solution for the nonlinear dynamic (time history of loading) cases under semi-discrete formulation. In other words, we would like to obtain the linearized equations for both the quasi-static case under the proportional loading system (for incremental load step) identified by the coefficient of load proportionality, and the dynamic case under any time history of applied loading system (for incremental time step). However, since the virtual work equation is non-linear in the displacement, velocity and acceleration states of a beam, a modified Newton-type iterative scheme is used for each time step; as indicated earlier, this sometimes involves linearization or finding of the tangent operator of the virtual work equation at each step. Also, both the time integration method and the iterative scheme must be cognizant of the fact that the unit rotational vector is constrained to travel on a curved space at each time step of the evolution, and thus, in a time discrete formulation, belongs to a tangent space in the beginning of a time step that is different at the end of the time step. Thus, linearization requires special treatment: parallel transport or covariant differentiation, as we have already accomplished. In this context, the special treatment relates to updating the rotation vector in the time-marching algorithm, which will be tackled later. Now, for the internal virtual work functional, the linearization (or, equivalently, obtaining the tangent operator) consisted of two parts, the linearization of the generalized reactive forces resulting in the material tangent stiffness, and the linearization of the virtual strain functional (i.e. due to change in the geometry of a beam) culminating in the geometric tangent stiffness. Thus the linearization of the first part followed from an application of the definition of the virtual or the first variation of the real strain along the virtual generalized displacement vector. The incremental strain mimics the virtual strain whereby the virtual variables are correspondingly replaced by the incremental variables. This allows us to extract the definition of the symmetric material stiffness matrix from the functional. As to the second part, it holds one of the most important and essential roles in the finite element formulation for nonlinear analysis involving finite deformation of structural systems. Physically, it is an additional stiffness or flexibility that arises purely because of the finite change in the geometry of a beam undergoing finite deformations. However, even for a simple materially hyperelastic beam behavior to which we have restricted our discussion here, a closed form description of the geometric stiffness for computational ease is painstakingly involved. As a result, all work reported in the literature is either in differential form left to be evaluated numerically or compromised some form of approximation (see Chapter 1). While the proof of symmetry theoretically is rather simple (by having the instantaneous configuration space variables belonging to a vector space, namely the tangent spaces of the configuration space and the symmetry of the dot product), the explicit expression of a symmetric geometric stiffness matrix is quite a challenging and involved task. However, the successful derivation we presented earlier will reward us with the cleanest, most simple yet exact description of the geometric stiffness matrix suitable for repetitive numerical evaluation that is natural and accurate in the tracing of the equilibrium configurations, especially with a minimum number of "mega" elements, as we try to accomplish by the c-type finite element method.

Section 10.17: We present a simple yet exact definitive description of a symmetric material stiffness matrix of a beam. Starting from the linearized virtual work functional, here we intend to transform a part of it, specifically the material part of the expression, to its matrix form. The transformed matrix part will subsequently give rise to the material stiffness matrix of a beam; this matrix form, however, is not yet suitable for subsequent finite element formulation since the sought-after independent incremental variables appear only in an implicit way hidden behind the incremental strains and the virtual strain–displacement functions; we would like to have a form of the linearized virtual equation that is explicit in the incremental generalized displacement vector so that the virtual work equation can be solved digitally by a Newton-type iterative method.

Section 10.18: We present an exact definitive description of a symmetric geometric stiffness matrix of a beam. Starting from the linearized virtual work functional, here we transform a part of it, specifically the second-order geometric part of the expression, to its matrix form. The transformed matrix part will subsequently give rise to what is known as the geometric stiffness matrix of a beam. However, this matrix form is not yet suitable for subsequent finite element formulation since the sought-after independent incremental variables appear only in an implicit way, hidden behind the incremental strains and the virtual strain–displacement functions. We would like to have a form of the linearized virtual equation that is explicit in the incremental generalized displacement vector so that the virtual work equation can be solved digitally by a Newton-type iterative method. The geometric stiffness holds one of the most important roles in the finite element formulation of a finite deformation, nonlinear analysis of structural systems. It arises as a part of the tangent operator resulting from the linearization of the virtual functional necessary for digital computation based on a Newton-type method. However, even for a materially hyperelastic beam behavior to which we have restricted our discussion here, a closed form description for the kernel of the geometric stiffness for computational ease is painstakingly involved. As a result, all work reported in the literature is either in differential form, left to be evaluated numerically, or compromises some form of approximation (see Chapter 1). While the proof of symmetry is theoretically rather simple, the explicit expression, of a symmetric geometric stiffness matrix is quite formidable. The successful derivation, however, will reward us with the cleanest, most simple yet exact description of the geometric stiffness tensor suitable for repetitive numerical evaluation that is natural in the tracing of the equilibrium configurations.

Section 10.19: We present the semi-discrete c-type finite element formulation for beams with dynamic time history loading systems; in so doing, we also present the most simple, explicit yet exact definitive description of the symmetric material and geometric stiffness matrices for a beam. We extend this section with a description of the residual forces with the introduction of the c-type finite elements and the notion of control vectors, and complete it with a computationally efficient exposition of the incremental virtual work equation for a time step of the motion of a beam finite element. Starting from the already derived fully linearized virtual work functional, first we intend to transform a part, specifically the internal linearized virtual work functional, to its suitable matrix representation; we focus on the part known as the inertial linearized virtual work functional and transform it to the most suitable matrix representation; the transformed matrix representations will subsequently give rise to what is known as the mass matrix, gyroscopic (damping) matrix and the centrifugal (stiffness) matrix of a beam undergoing dynamic motion. Using these developments, we present the c-type finite element formulation for the dynamic loading systems and generate the final form of the incremental dynamic virtual work equation; the previous form recalled later is not suitable for a subsequent finite element implementation since the sought-after independent incremental variables appear only in an implicit way, hidden behind the incremental strains and the virtual strain–displacement functions for the incremental internal virtual work functional. Also, the mass matrix, the gyroscopic matrix and the centrifugal matrix were not fully identified in that form for the incremental inertial virtual work functional. For the semi-discrete finite element formulation, we would like to have a form of the linearized virtual equation that is explicit in the incremental generalized displacement vector, the incremental generalized velocity vector, and the incremental generalized acceleration vector so that the virtual work equation can be used digitally by a numerical time integration method in conjunction with a Newton-type iterative algorithm for tracing the dynamic motion history of a beam structure.

Section 10.20: We present the c-type finite element formulation for beams with quasi-static proportional loading systems and its numerical implementation, followed by several numerical examples; we extend this section with a description of the residual and applied proportional forces with the introduction of the c-type finite elements and the notion of control vectors, and complete it with a computationally efficient exposition of the incremental equation for each finite element. Using these developments, we present the c-type finite element formulation for the quasi-static loading systems and generate the final form of the incremental virtual work equation. We get a form of the linearized virtual equation that is explicit in the incremental generalized displacement vector so that the virtual work equation can be solved digitally by a Newton-type iterative method; then, we discuss the c-type beam finite element computer programming details on the finite element formulation of the material control stiffness matrix, the geometric control stiffness matrix, the applied loads and the residual loads, and so on, of a beam for subsequent direct stiffness method of assembly and solution by Newton-type algorithm to ensure solutions without any locking phenomena – shear, membrane, and so on. Finally, we present solutions for several nonlinear beam problems of practical importance by c-type finite element method.

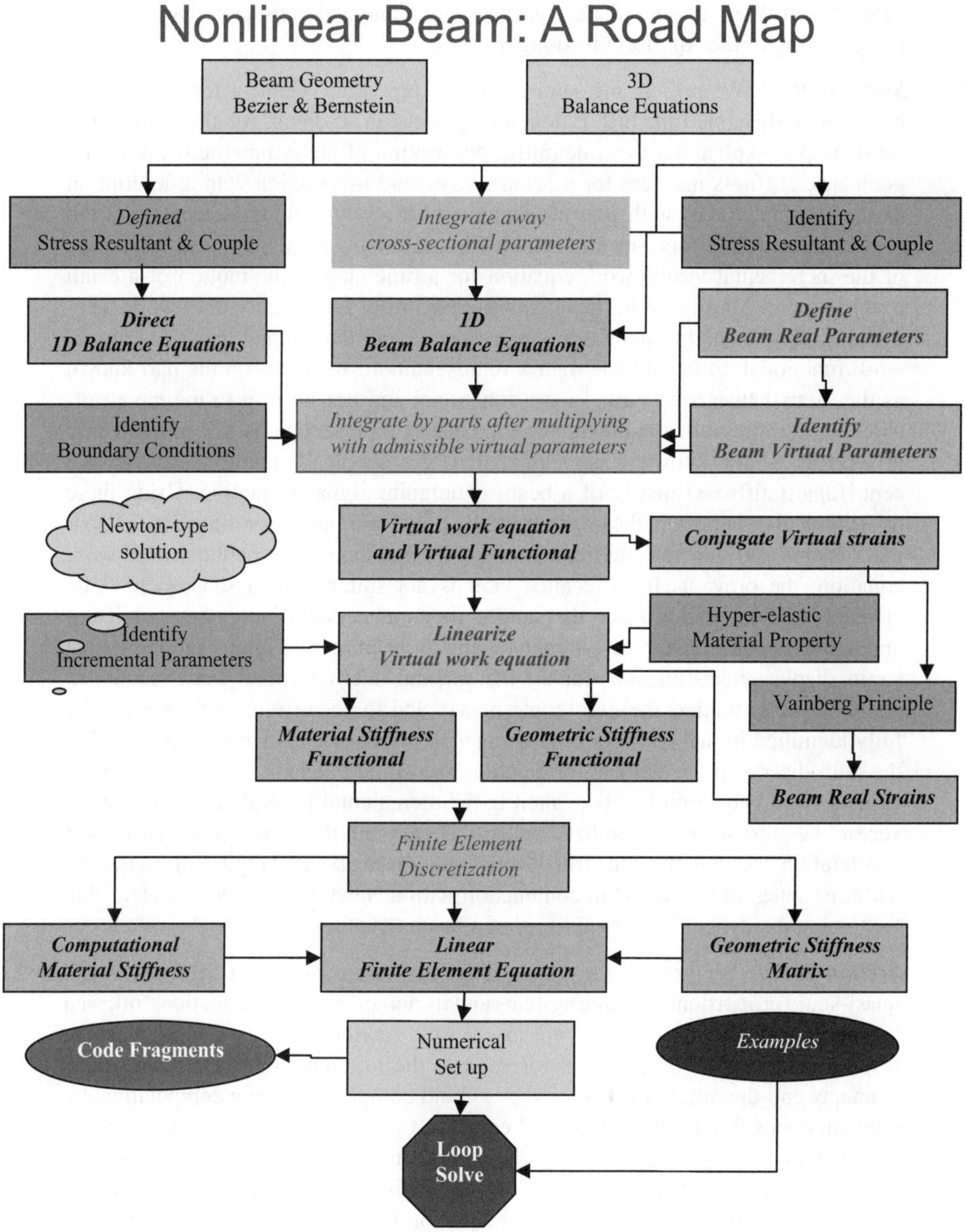

10.2 Beam Geometry: Definition and Assumptions

10.2.1 What We Need to Recall

As will be seen here, because thin beam geometry is essentially captured by a tortuous curve, we must recall from our study on curve theory various geometric properties of a curve:

- Curve local axes are given by the ***Frenet frame***.

10.2.2 Assumptions

Beam representation depends on the assumptions implicit in the method of analysis. Note that for a computationally optimal nonlinear and generally 3D tortuous curved beam treatment with our ***c-type beam "mega-element"***, we have to be especially mindful of the accuracy of the theoretical formulation. Thus, for applicability of our method, we would like to identify various assumptions that encompass our analysis. In classical thin beam theory, the beam kinematic parameters are restricted to only six components – three for displacement and three for rotation – at any cross-section of the beam. Clearly, this was possible with various assumptions on the definition and behavior of a beam. For example, to include more accurately the warping phenomenon in the beam cross-sections at various locations along the length, it is simply not possible with only six degrees of freedom. The out-of-plane deformation of the cross-sections entails at least one other degree of freedom with an assumed warp function. In our nonlinear theory, we intend to hold on to these same six components, and also for engineering applications. Naturally, we have to resort to reasonable assumptions to our definition of a beam that is thin and long. As an example, for kinematic behavior it has been assumed that a plane section remains plane (i.e. no warping) but not necessarily perpendicular to the centroidal axis. Moreover, the plane cross-sections remain rigid in their own plane. This is commonly known as the ***plane-sections hypothesis***. However, we must guarantee that this hypothesis is natural. In other words, adoption of this assumption is contingent upon the fact that the true three-dimensional effects may be acceptably captured in an average sense over the cross-sections. The long and thin beams with length dimension conspicuously dominant tend to behave in a way that makes this assumption acceptably reasonable. That is, with this assumption, we can still predict quite comfortably all the general behaviors of a beam: the axial, shear and bending characteristics. Thus, in our subsequent analysis, we assume the following.

- The beam is thin and long.
- The beam is capable of undergoing flexure, shear and axial deformations.
- The beam cross-section is rigid in its own plane and remains plane during the entire history of deformation; thus, beam cross-sectional warping is neglected as it is considered to be of secondary importance for a long thin beam.
- The beam cross-section is initially perpendicular to the tangent to the centroidal axis at the point of attachment but does not necessarily remain so during the course of the beam deformations. In other words, a beam cross-section is capable of rotation (which will be identified later as shear) independent of the corresponding rotation (flexural) of the tangent to the centroidal axis.
- The beam can undergo any finite (nonlinear) displacements and rotations as opposed to just infinitesimal (linear) displacements and rotations.
- For simplicity, and without much loss of generalization, it is further assumed that the curved beam geometry is embedded in the general three-dimensional space with principal axes in the cross-sections aligned along $\mathbf{M}$, the principal normal, and $\mathbf{B}$, the bi-normal directions.

With these assumptions, the beam kinematics for geometrical nonlinearity reduce to the study of large displacement and rotational motion of a series of rigid bodies (beam cross-sections) constrained by certain material connections between them.

10.2.3 Beam: Undeformed Geometry

An undeformed structural beam is a 3D continuum with two of its dimensions relatively small compared to its third dimension. In other words, for our computation, a structural beam is a long thin 3D continuum.

The first dimension (ξ^1) defines its length while the last two dimensions (ξ^2, ξ^3) are identified with the beam's cross-section. The centroidal axis is then given by the parametric map, $\mathbf{C}(\xi) \equiv \mathbf{C}(\xi^1 \equiv \xi, 0, 0)$, $\xi \in [0, 1]$. The usual invariant arc length parameter, S, is related to the parameter ξ by the relation: $S(\xi) = \int_0^{\xi} ||\mathbf{C}_{,\eta}||d\eta$, where $\mathbf{C}(\xi)$ describes the geometric curve representing the beam in the undeformed configuration in terms of the root parameter.

As we shall see in the derivation of the equilibrium equations, we can develop beam equations of equilibrium by directly representing a beam as a 1D curved element as in engineering analysis. Alternatively, a beam's 1D equations can be deduced by assuming that a beam is a thin long continuum with side surfaces and cross-sectional areas and then, by a reduction process from 3D equations to 1D equations. Of course, this method helps define the applied forces in a consistent manner. Thus, representation of a beam is dependent on the nature of analysis to be used.

10.2.4 Beam 1D Direct Engineering Representation

For direct 1D equilibrium equation derivation, a beam is defined geometrically as a tortuous 3D space curve with a rigid cross-section attached to each of its points along its length. The curve itself is identified as the centroidal axis of the beam and the rigid cross-sections are connected to each other by applicable material properties as shown in Figure 10.1. As we can see from our list of assumptions above, to handle shear deformations, rigid cross-sections are assumed to be capable of rotations, independent of that of the tangents to the curve. Thus, for subsequent finite element analysis, we will need two sets of coordinate frames with one attached to the curve tracking its movement and the other to the cross-sections monitoring its rotations as shown in Figure 10.2.

10.2.5 Beam 3D Continuum Representation

A beam is defined geometrically as a thin, long continuum with the interior of the body and enclosing surfaces defined as follows.

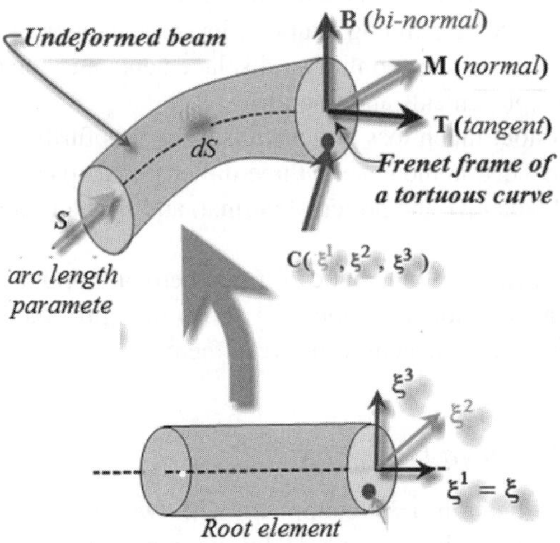

Figure 10.1 Geometric Curvilinear map.

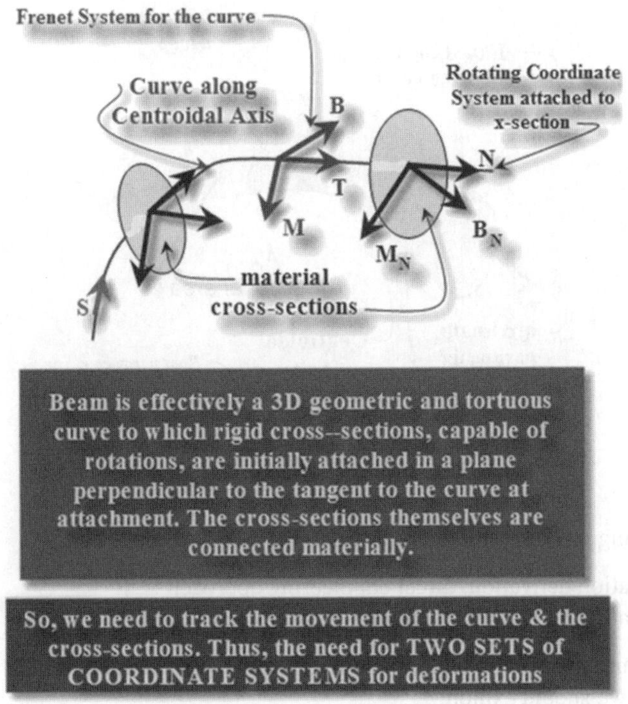

Figure 10.2 Beam definition.

10.2.5.1 Beam Body: Interior and Boundary Surfaces

Recall that we have separate equations for the interior and boundary of a 3D continuum: a set of balance equations for the interior and a set for applied kinematic and/or traction boundary conditions. Thus, in order to reduce the 3D integral balance equations to a set of beam balance equations, we must characterize the beam as a continuum and as such identify its interior and boundaries. Furthermore, the boundaries, in turn, need to be further categorized to be able to subsequently specify various applicable kinematic and traction boundary conditions. As shown in Figure 10.3, a beam is bounded by the side surface and the end cross-sectional area. The corresponding differential areas are given by $dA_S = d\Gamma\, dS$ and dA_c, and the differential volume of the interior of the beam is given by $dV = dA_c\, dS$ where dS is the differential arc length of a beam.

10.2.6 Where We Would Like to Go

So far, we have defined beams suitable for computation based on the given assumptions. In so doing, we have made extensive use of the general theory of curves. Clearly, then, we should check out the theory of curves:

- Theory of curves

For numerical computations, we need a clear measure for deformations – translation of the centroidal axis and rigid body rotation of the cross – sections. Thus, for details on kinematic parameters, and associated rotation tensor and its parameterization, we would like to visit:

- Kinematic parameters

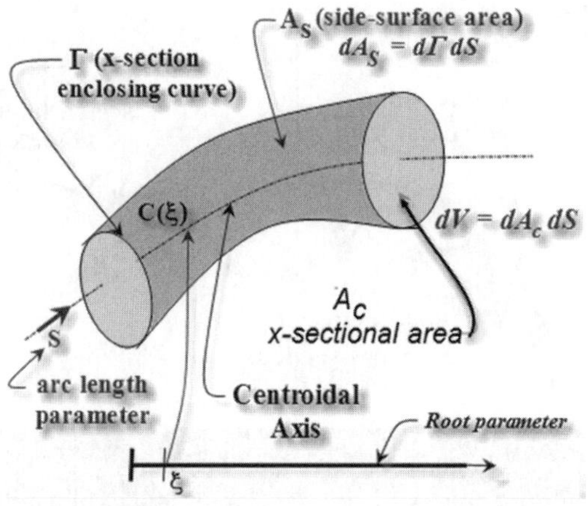

Figure 10.3 Beam 3D continuum representation.

Now, for proceeding onward with the beam analysis, we must move on to:

- Thin beam equation derivation: direct engineering approach
- Thin beam equation derivation: 3D to 1D

Finally, to get familiarized with and be conscious about the overall computational strategy for nonlinear beams, we should explore:

- the road map: nonlinear beams

10.3 Static and Dynamic Equations: Engineering Approach

10.3.1 What We Need To Recall

Let us recall the following.

- The undeformed coordinate system, that is, the ***Frenet frame***, hereinafter known as the ***tangent coordinate system***, corresponding to the centroidal axis with arc length parameterization, $S \in [0, L]$, the normal and the bi-normal to the beam reference curve, with the base vectors: $\{\mathbf{T}(S) \equiv \mathbf{G}_1(S), \mathbf{M}(S) \equiv \mathbf{G}_2(S), \mathbf{B}(S) \equiv \mathbf{G}_3(S)\}$, is an orthonormal system of coordinates; arc length parameter, S, is essentially a smooth reparameterization of the root parameter, ξ, through the relationship: $S(\xi) = \int_0^{\xi} \| C_{0},_{\eta}(\eta) \| \, d\eta$ with the notation: $(\bullet),_{\eta} \equiv \frac{d(\bullet)}{d\eta}$.
- For the ensuing vector form of the balance equations, no kinematic assumptions are necessary.

10.3.2 Deformed Geometry and Stress Resultants and Stress Couples

As indicated, because we are interested in including in our analysis the case of finite displacements and rotations, all equilibrium equations must be considered for the current deformed geometry of a beam. Thus, as a starting point, we will attend to the following.

10.3.2.1 Deformed Geometry

Let us consider a typical differential element of the beam reference curve; we will parameterize the beam reference curve by the arc length parameterization S along this centroidal axis. Referring to Figure 10.4, the position vector, $\mathbf{c}_0(S, t)$ locates a point a on a deformed curve (representing

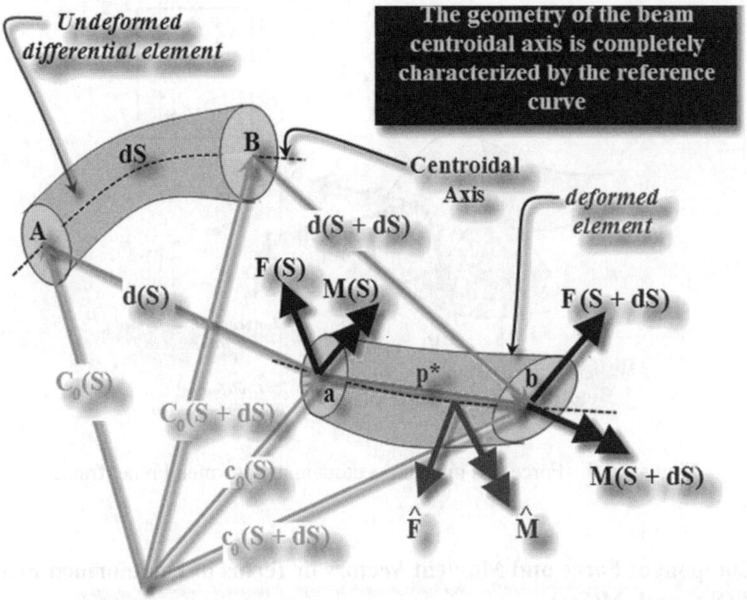

Figure 10.4 Direct engineering equilibrium derivation.

the beam in a deformed state at time $t \in \mathbb{R}_+$) whose corresponding location on the curve $\mathbf{C}_0(S)$ (representing the undeformed beam) at S is denoted by A. Similarly, considering a differential element dS at A on the curve, $\mathbf{c}_0(S + dS, t)$ locates a point b on the same deformed curve whose corresponding location on the undeformed curve $\mathbf{C}_0(S + dS)$ at $S + dS$ is denoted by B. In other words, a differential element AB on the undeformed beam becomes ab on a deformed state of the beam.

10.3.2.2 Force (Stress Resultant) and Moment (Stress Couple) Vectors: $\mathbf{F}(S, t)$ and $\mathbf{M}(S, t)$

Again, referring to Figure 10.4, we introduce the force and moment vectors acting on the deformed beam curve as follows.

- $\mathbf{F}(S, t)$ = the force vector per unit length of the undeformed curve acting outward on a cross-section of the deformed beam curve (which was the $S = const.$ cross-section at the S-point of the undeformed beam reference curve) at any time $t \in \mathbb{R}_+$.
- $\mathbf{M}(S, t)$ = the moment axial vector per unit length of the undeformed curve acting outward on that cross-section of the deformed beam curve (which was the $S = const.$ cross-section at the S-point of the undeformed beam reference curve) at any time $t \in \mathbb{R}_+$.

- Note that in the absence of any kinematic assumptions, we do not have any computationally useful stress-level interpretations and hence expressions for either the invariant force (i.e. stress resultant) vector, $\mathbf{F}(S, t)$, or the moment (i.e. stress couple) vector, $\mathbf{M}(S, t)$. However, the names suggested in parenthesis allude to an assumed beam through-the-thickness deformation distribution, which we will refrain from discussing further until our derivation of the balance laws from the fundamental principles of continuum mechanics.

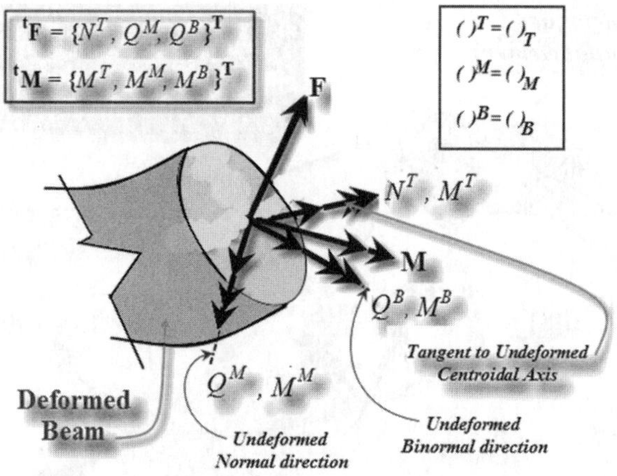

Figure 10.5 Force and moment vectors in undeformed Frenet frame.

10.3.2.3 Component Force and Moment Vectors in Terms of Undeformed Frame: ${}^t\mathbf{F}(S,t)$ and ${}^t\mathbf{M}(S,t)$

The invariant force vector, $\mathbf{F}(S,t)$, and the moment vector, $\mathbf{M}(S,t)$, for any $t \in R_+$, can have component representation in any triad of base vectors, Cartesian or curvilinear. At this point in the analysis, we choose to describe the components in the undeformed curvilinear coordinate system (henceforth, referred to as the "tangent coordinate system") with the triad of base vectors, $\{\mathbf{T} \equiv \mathbf{G_1}, \mathbf{M} \equiv \mathbf{G_2}, \mathbf{B} \equiv \mathbf{G}_3\}$ on the beam reference curve; in differential geometry parlance for the theory of curves, the system is known as the Frenet frame. Thus, referring to Figure 10.5, we can write:

$$\boxed{\begin{aligned} \mathbf{F}(S,t) &= {}^tF^1(\equiv N_T)\,\mathbf{G_1} + {}^tF^2(\equiv Q_M)\,\mathbf{G}_2 + {}^tF^3(\equiv Q_B)\,\mathbf{G}_3 \\ &\equiv N_T\ \mathbf{T} + Q_M\ \mathbf{M} + Q_B\ \mathbf{B} = {}^tF^i(S,t)\ \mathbf{G}_i(S), \quad i = 1,2,3 \end{aligned}} \tag{10.1}$$

where, N_T is known as the axial force, Q_M, Q_B as the shear forces in the engineering community; and

$$\boxed{\begin{aligned} \mathbf{M}(S,t) &= {}^tM^1(\equiv M_T)\,\mathbf{T} + {}^tM^2(\equiv M_M)\,\mathbf{M} + {}^tM^3(\equiv M_B)\,\mathbf{B} \\ &= {}^tM^i(S,t)\ \mathbf{G}_i(S), \qquad i = 1,2,3 \end{aligned}} \tag{10.2}$$

where, M_T is known as the ***torque***, and M_M, M_B are known as the bending moments on a beam cross-section.

10.3.2.4 (Effective) Linear Momentum and (Effective) Angular Momentum Vectors: $\mathbf{L}(S,t)$ and $\mathbf{A}_S(S,t)$

We introduce the linear momentum and the angular moment vectors acting on the deformed beam curve as follows.

- $\mathbf{L}(S,t)$ = the linear momentum vector per unit length of the undeformed curve acting at a cross-section of the deformed beam curve (which was the $S = const.$ cross-section at the S-point of the undeformed beam reference curve) at any time $t \in \mathbb{R}_+$.
- $\mathbf{A}_S(S,t)$ = the angular momentum or the moment of momentum per unit length of the undeformed curve acting at a cross-section of the deformed beam curve (which was the $S = const.$ cross-section at the S-point of the undeformed beam reference curve) at any time $t \in \mathbb{R}_+$.

- Note that in the absence of any kinematic assumptions, we do not have any computationally useful deformation-level interpretations, and hence, expressions for either the invariant linear momentum (i.e. effective) vector, $\mathbf{L}(S,t)$, or the angular momentum (i.e. effective) vector, $\mathbf{A}_S(S,t)$. However, the qualifiers suggested in the parenthesis allude to an assumed beam through-the-thickness rate of deformation distribution, which we will refrain from discussing until after the determination of the balance laws.

10.3.3 Dynamic Balance Equations

Let us again emphasize that the forces and moments, and the linear and the angular momenta, are acting on the current deformed beam reference curve but defined per unit length in the undeformed beam reference curve along the centroidal axis at the point they are acting with the components given in the local basis directions of the undeformed curvilinear coordinate system as shown in Figures 10.6(a),(b). With the above preliminaries as the background and referring to Figure 10.4, we can directly apply the principles of momentum balance and the moment of momentum balance (moment being taken about the point b of the deformed differential element ds), that is, the angular momentum balance, and we get the following.

10.3.3.1 Momentum Balance Equations (Force Equilibrium)

$$\sum \mathbf{F} = \mathbf{F}(S+dS) - \mathbf{F} + \hat{\mathbf{F}}\, dS = \cancel{\mathbf{F}} + \mathbf{F}_{,S}\, dS - \cancel{\mathbf{F}} + \hat{\mathbf{F}}\, dS = \frac{d}{dt}\mathbf{L}(S,t)\, dS$$

or

$$\boxed{\mathbf{F}_{,S} + \hat{\mathbf{F}} = \dot{\mathbf{L}}} \tag{10.3}$$

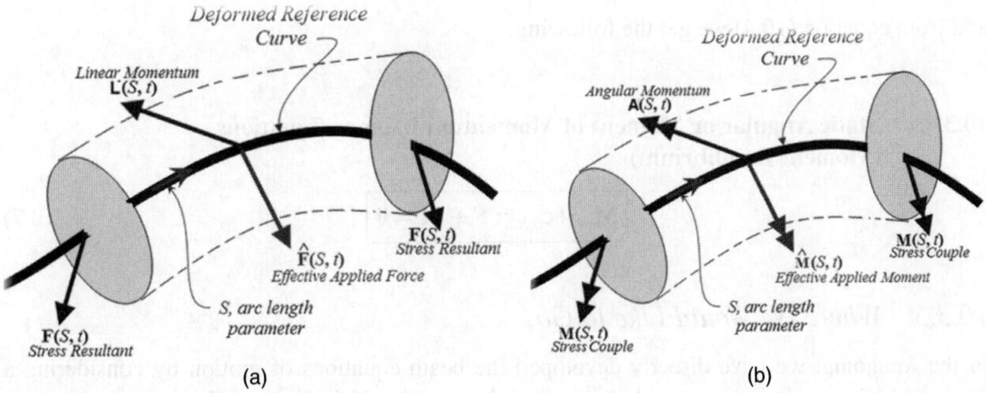

Figure 10.6 (a) 1D beam forces and linear momentum. (b) 1D beam moments and angular momentum.

10.3.3.2 Angular or Moment of Momentum Balance Equations (Moment Equilibrium)

$$
\begin{aligned}
\sum \mathbf{M} &= \mathbf{M}(S + dS, t) - \mathbf{M}(S, t) + \mathbf{p}^*(S, t) \times \mathbf{F}(S, t) + \hat{\mathbf{M}}(S, t)\, dS \\
&= \cancel{\mathbf{M}(S, t)} + \mathbf{M}(S, t)_{,S}\, dS - \cancel{\mathbf{M}(S, t)} + \mathbf{p}^*(S, t) \times \mathbf{F}(S, t) + \hat{\mathbf{M}}(S, t)\, dS \\
&= \mathbf{M}(S, t)_{,S}\, dS + \mathbf{p}^*(S, t) \times \mathbf{F}(S, t) + \hat{\mathbf{M}}(S, t)\, dS = \frac{d}{dt}\mathbf{A}_S(S, t)\, dS
\end{aligned}
$$

Now, from Figure 10.4, we have the following vector identity:

$$
\cancel{\mathbf{c}_0(S, t)} + \mathbf{p}^*(S, t) = \mathbf{c}_0(S + dS, t) = \cancel{\mathbf{c}_0(S, t)} + \mathbf{c}_{0,S}(S, t)\, dS
$$

or

$$
\mathbf{p}^*(S, t) = \mathbf{c}_{0,S}(S, t)\, dS
$$

Thus, we finally get:

$$
\boxed{\mathbf{M}_{,S} + \mathbf{c}_{0,S} \times \mathbf{F} + \hat{\mathbf{M}} = \dot{\mathbf{A}}} \tag{10.4}
$$

10.3.4 *Static Balance Equations*

The quasi-static balance laws can be easily obtained from their dynamic counterparts by assuming that the linear and the angular momentum vectors identically vanish, that is,

$$
\boxed{\begin{aligned} \mathbf{L}(S, t) &\equiv \mathbf{0} \\ \mathbf{A}_S(S, t) &\equiv \mathbf{0} \end{aligned}} \tag{10.5}
$$

Thus, we get the following from equation (10.3).

10.3.4.1 Static Momentum Balance Equations (Force Equilibrium)

$$
\boxed{\mathbf{F}_{,S} + \hat{\mathbf{F}} = \mathbf{0}} \tag{10.6}
$$

and from equation (10.4) we get the following.

10.3.4.2 Static Angular or Moment of Momentum Balance Equations (Moment Equilibrium)

$$
\boxed{\mathbf{M}_{,S} + \mathbf{c}_{0,S} \times \mathbf{F} + \hat{\mathbf{M}} = \mathbf{0}} \tag{10.7}
$$

10.3.5 *Where We Would Like to Go*

In the foregoing, we have directly developed the beam equations of motion by considering a beam as a 1D geometric object with the force and moment, and the linear and angular momentum fields. However, to validate the consistency of these equations of motion, and to gain insight into the true nature of these quantities, by establishing the connection between the beam equations and the more fundamental balance equations of 3D continuum mechanics, these equations will be derived by starting from and reducing the 3D equations of motion of solid mechanics. For this development, we must check out the following.

- beam balance equations: 3D to 1D

While the beam equations presented in equations (10.3), (10.4), (10.6) and (10.7) are very compact and thus useful for analysis, we will, as suggested before, use the modified component or operational vector forms as those are most suitable for computational purposes where the components of displacements, rotations, forces, moments and so on, are needed to be determined as the primary variables of interest. As prerequisites to such a task in the case of reduction of beam balance equations from 3D to 1D, we should check out:

- configuration space and kinematics
- component or operational vectors
- derivatives of component vectors

Then, with these background materials, we will be able to have the fully interpreted, final and computationally most effective balance equations:

- computational balance equations

10.4 Static and Dynamic Equations: Continuum Approach – 3D to 1D

10.4.1 What We Need to Recall

- A structural beam, rod and so on, by definition, is a thin, long continuum. In other words, a beam, viewed as a 3D continuum, is biased to have one dimension – the length – much more dominant compared to the other two: – those that make up the transverse cross-section of a beam. That is, a continuum locally in the form of a 1D curve called the ***beam reference curve***; the reference curve is the locus of the mass centroids along the beam.
- The root coordinates are given by: $\xi^1 \in [0,1]$, ξ^2 and ξ^3.
- $S \in \{0, L\}$ is the arc length parameter associated with ξ^1 by the relation: $S(\xi^1) = \int_0^{\xi^1} ||\mathbf{C}(\eta, 0, 0)_{,\eta}||\, d\eta$, where $\mathbf{C}(\xi^1, 0, 0)$ is the mass centroidal axis of the beam in terms of the root parameters.
- We characterize the beam undeformed reference curve, M, by the arc length parameter, S, so that the position vectors on the beam reference curve may be given as $\mathbf{C}_0(S) \equiv \mathbf{C}(S, 0, 0)$. Thus, we will eventually be able to characterize the configuration (undeformed and deformed) of a beam by a single parameter, namely the arc length parameter, $S \in [0, L]$, along the reference curve of the beam for quasi-static loading, and by two parameters, $(S, t) \in [0, L] \times \mathbb{R}_+$, for dynamic loading.
- A thin beam is modeled as a mass centroidal axis represented by a geometric curve, $\mathbf{C}(S, t)$, to which a rigid plane cross-section is attached that is initially normal to the tangent of the curve but at any subsequent time does not necessarily remain normal to the tangent to the deformed curve; that is, shear and extensional strains are permitted in the beam.
- The centroidal curve is attached to the curvilinear Frenet frame at every point.

- On the undeformed centroidal axis, the Frenet frame may be denoted by a triad of orthonormal vectors **T**, **M** and **B** as the tangent, principal normal and bi-normal, respectively (see Chapter 5 on curves), given by:

$$\boxed{\begin{aligned} \mathbf{G}_1 &\equiv \mathbf{T} = \frac{\mathbf{C}_{,\xi}(\xi)}{\|\mathbf{C}_{,\xi}(\xi)\|} \\ \mathbf{G}_2 &\equiv \mathbf{M} = \mathbf{B} \times \mathbf{T} \\ \mathbf{G}_3 &\equiv \mathbf{B} = \frac{\mathbf{C}_{,\xi}(\xi) \times \mathbf{C}_{,\xi\xi}(\xi)}{\|\mathbf{C}_{,\xi}(\xi) \times \mathbf{C}_{,\xi\xi}(\xi)\|} \end{aligned}} \tag{10.8}$$

where, ξ, in our current notation is related to arc length parameter, S. As shown in equation (10.8), for later use in compact form: $\mathbf{G}_k$, $k = 1, 2, 3$; let the base vectors for the curvilinear coordinate system be denoted by: $\mathbf{G}_1 \equiv \mathbf{T}$, $\mathbf{G}_2 \equiv \mathbf{M}$ and $G_3 \equiv \mathbf{B}$. We also note that the unit normal, **N**, to the undeformed cross-sectional area is identical to the unit tangent, **T**, to the centroidal axis, and that the principal normal **M** is along the curvature vector of the beam.
- On the deformed centroidal axis, the Frenet triad of orthonormal base vectors, $\mathbf{g}_i$, $i = 1, 2, 3$ is given by:

$$\boxed{\mathbf{g}_i = \mathbf{R}\,\mathbf{G}_i, \quad i = 1, 2, 3} \tag{10.9}$$

- The undeformed 3D beam interior, B, is completely enclosed by the boundary ∂B which, in turn is made up of three disjoint surfaces: the end cross-sectional planes with the normals that are tangents to the reference curve with the beginning plane cross-sectional area denoted by M^{C-}, the ending cross-sectional area by M^{C+} and a side surface denoted by M^S. Any undeformed plane cross-section normal to the tangent, **T**, of the reference curve at a point, S, is fully parameterized by two parameters (ξ^2, ξ^3) along two mutually perpendicular directions on the cross-section at a point, S, on the reference curve, defined by the principal normal, **M**, and the bi-normal, **B**, respectively; thus, any point on the beam body is identified by the coordinates (S, ξ^2, ξ^3) at any time $t \in \mathbb{R}_+$.
- The 3D equations of motion in Lagrangian (material) formulation for a 3D body, B, with the surface undeformed surface area, ∂B, for dynamic loading conditions, can be written (see Chapter 7 on nonlinear mechanics) as:

$$\boxed{\begin{aligned} &\int_{\partial B} \mathbf{t}\, dA + \int_B \mathbf{b}\, dV = \frac{d}{dt}\int_B \mathbf{v}\,\rho_0\, dV && \text{(Momentum balance)} \\ &\int_{\partial B} \mathbf{c} \times \mathbf{t}\, dA + \int_B \mathbf{c} \times \mathbf{b}\, dV = \frac{d}{dt}\int_B \mathbf{c} \times \mathbf{v}\,\rho_0\, dV && \text{(Moment of momentum balance)} \end{aligned}} \tag{10.10}$$

where **t** and **b** are the surface traction vector, measured per unit surface area, and the body force vector, measured per unit volume of the undeformed body, respectively, but acting on the current deformed body. In other words, we track the areas and the volumes of the neighborhoods around, **t** and **b**, respectively, in the current deformed body by identifying their corresponding areas and volumes in the undeformed body.

- $\mathbf{v}(S, \xi^2, \xi^3, t)$ is the current velocity at any point (S, ξ^2, ξ^3) and at the current time $t \in \mathbb{R}_+$; ρ_0 is the mass density per unit volume of the undeformed body; $\mathbf{c}$ is the position vector of the point (S, ξ^2, ξ^3) on the body in the deformed configuration.
- On the undeformed beam, the differential area on any cross-sectional planes is given by: $dA^c = d\xi^2\, d\xi^3$
- In the undeformed beam interior around any point (S, ξ^2, ξ^3), the differential volume, dV, is given by: $dV = \mathbf{dA}^c \bullet \mathbf{dS} = \mathbf{T}\, dA^c \bullet \mathbf{T}\, dS = d\xi^2\, d\xi^3\, dS$
- The beam side surface differential area with the differential perimeter around it denoted by $d\Gamma$ (Γ being the arc length around the perimeter of a curve enclosing the side surface) and of any length, dS, along the beam reference curve, is given by: $dA^s = d\Gamma\, dS$
- For any tensor, $\mathbf{P}$, with representation as: $\mathbf{P} = \mathbf{F}^i \otimes \mathbf{G}_i$, we have:

$$DIV\,\mathbf{P} = \mathbf{F}^i\,|_i \tag{10.11}$$

- For any vector $\mathbf{a}$ and tensor $\mathbf{B}$, we have:

$$DIV(\mathbf{a} \times \mathbf{B}) = \mathbf{a} \times DIV\,\mathbf{B} + \mathbf{a},_i \times \mathbf{B} \bullet \mathbf{G}^i \tag{10.12}$$

- As a notational reminder, in what follows, all lowercase Greek indices assume values 2 or 3, and the lowercase Latin indices runs through $1, 2, 3$; also, for repeated diagonal indices, Einstein's summation convention of summing over their ranges holds.

10.4.2 Thin and Very Flexible Beam: Deformation Map

The theory we are trying to detail is expected to handle thin and very flexible beams. So, let us assume that $\chi : \mathrm{B} \to \mathrm{E}^3$ defines a sufficiently smooth deformation map for the beam body. In other words, the deformed position vector, $\mathbf{c}(S, \xi^2, \xi^3, t)$, of a point that had a position vector, $\mathbf{C}(S, \xi^2, \xi^3, 0)$, in the undeformed reference state is related by: $\mathbf{c} = \chi(\mathbf{C})$. On the beam deformed reference curve, $\chi(M)$, the position vectors are denoted by: $\mathbf{c}_0(S, t) \equiv \mathbf{c}(S, 0, 0, t)$.

Now, we will assume that the deformation of a thin 3D beam-like body may be given by:

$$\boxed{\begin{aligned} &\mathbf{c}(S, \xi^2, \xi^3, t) = \mathbf{c}_0(S, t) + \mathbf{p}(S, \xi^2, \xi^3, t), \\ &\mathbf{p}(S, 0, 0, t) \equiv 0 \end{aligned}} \tag{10.13}$$

where

$$\boxed{\mathbf{p}(S, \xi^2, \xi^3, t) = \mathbf{R}(S, t)\, \mathbf{P}(\xi^2, \xi^3, 0)} \tag{10.14}$$

with $\mathbf{R}(S, t)$, a rotation tensor, defining the rotation of the cross-section at the current time, $t \in \mathbb{R}_+$ and $\mathbf{P}(\xi^2, \xi^3, 0)$ is a point on the original cross-section at any point, S, on the undeformed reference curve. Clearly, the assumption implies that a plane cross-section remains rigid, but not necessarily perpendicular to the deformed reference curve during deformation (because it rotates due to shear deformations. This is a commonly accepted underpinning of theories for very thin and flexible beams with shear deformations (reflected in the rotation of the cross-sections) and goes by the name: ***Reissner–Mindlin***-type kinematic assumptions.

- Alternatively, we could choose: $\mathbf{p}(S, \xi^2, \xi^3, t)$ always remaining perpendicular to the tangent to the deformed reference curve at point $S \in [0, L]$; clearly, this assumption excludes any shear deformation and includes only flexural deformation of a beam and goes by the name: ***Love–Kirchhoff***-type kinematic assumption.
- Yet another choice would be to include a multiplicative, through-the-thickness variable to introduce a warping effect.
- Thus, the choice of $\mathbf{p}(S, \xi^2, \xi^3, t)$ essentially helps to categorize the various beam theories with different kinematic assumptions.

Thus, the cross-section of the beam model being rigid, the position vector $\mathbf{c}(S, \xi^2, \xi^3, t)$ locates a point p (with the location from the deformed centroidal axis denoted by the local vector $\mathbf{p}(\xi^2, \xi^3)$) on the deformed cross-section of the body whose corresponding undeformed location is denoted by P (with the location from the undeformed centroidal axis denoted by the local vector $\mathbf{P}(\xi^2, \xi^3)$) with the same (ξ^2, ξ^3) as the curvilinear coordinates. As shown in Figure 10.7, the configuration space of a thin beam is completely characterized by a displacement vector, $\mathbf{d}(S, t)$, of the centroidal curve, and a rotation tensor, $\mathbf{R}(S, t)$, of the rigid cross-section; thus, for $\mathbf{c}(S, \xi^2, \xi^3, t)$, the position vector of the point, p, on the body in the deformed configuration at time t, we have:

$$\boxed{\begin{aligned} \mathbf{c}_0(S,t) &= \mathbf{C}_0(S) + \mathbf{d}(S,t) \\ \mathbf{p}(S,\xi^2,\xi^3,t) &= \mathbf{P}(\xi^2,\xi^3)\,\mathbf{R}(S,t) \\ \mathbf{c}(S,\xi^2,\xi^3,t) &= \mathbf{c}_0(S,t) + \mathbf{p}(S,\xi^2,\xi^3,t) = \mathbf{C}_0(S) + \mathbf{d}(S,t) + \mathbf{P}(\xi^2,\xi^3)\,\mathbf{R}(S,t) \end{aligned}} \tag{10.15}$$

Now, the rotational tensor, $\mathbf{R}$, is orthonormal, that is, $\mathbf{R}^T\,\mathbf{R} = \mathbf{I}$. Thus, the material angular velocity tensor, $\mathbf{\Omega}_M$, is skew-symmetric, and is given by:

$$\boxed{\mathbf{\Omega}_M(S,t) = \mathbf{R}(S,t)^T\,\dot{\mathbf{R}}(S,t) \quad \Rightarrow \quad \dot{\mathbf{R}}(S,t) = \mathbf{R}(S,t)\,\mathbf{\Omega}_M(S,t)} \tag{10.16}$$

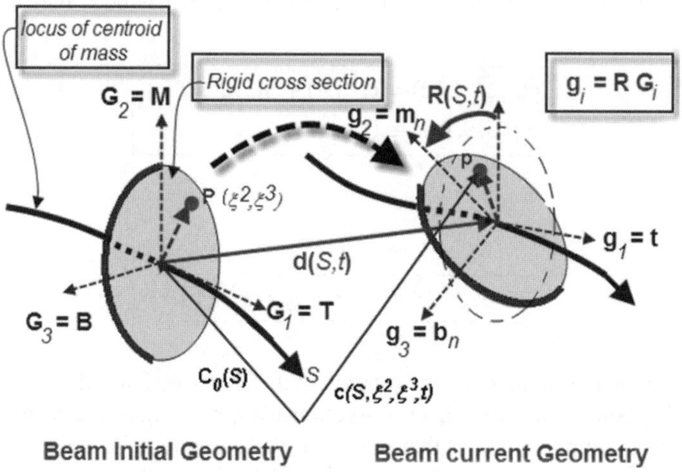

Figure 10.7 Beam initial and current geometry.

Then, $\mathbf{v}(S,\xi^2,\xi^3,t) \equiv \dot{\mathbf{c}}(S,\xi^2,\xi^3,t)$ is the current velocity at any point, $\mathbf{p}(\xi^2,\xi^3)$, on the cross-section, at the current time $t \in \mathbb{R}_+$, and is given, using equations (10.15) and (10.16) by:

$$\boxed{\begin{aligned} \mathbf{v}(S,\xi^2,\xi^3,t) \equiv \dot{\mathbf{c}}(S,\xi^2,\xi^3,t) &= \overset{\bullet}{\mathbf{d}}(S,t) + \mathbf{P}(\xi^2,\xi^3)\ \overset{\bullet}{\mathbf{R}}(S,t) \\ &= \overset{\bullet}{\mathbf{d}}(S,t) + \mathbf{P}(\xi^2,\xi^3)\ \mathbf{R}(S,t)\ \mathbf{\Omega}_M(S,t) \end{aligned}} \tag{10.17}$$

Let ρ_0 be the mass density per unit volume of the undeformed body; it follows from the definition of the mass centroidal axis as the locus of the origins of the cross-sectional coordinate system:

$$\boxed{\int_{A_C} \rho_0\ \mathbf{P}(\xi^2,\xi^3)\ dA_C = 0} \tag{10.18}$$

Finally, in what follows, the surface area integral, of the 3D equations of motion in Lagrangian (material) formulation as given by equation (10.10) for the beam free-body, is separated for the undeformed end cross-sectional area, A_c, for contact traction, and the side surface area, A_s, for surface traction.

10.4.3 Dynamic Balance Equations

As indicated, we would like to establish the connection between the one-dimensional beam balance equations and the balance equations of a beam viewed as a three-dimensional continuum or body. For this, if we extract, from the undeformed reference body, a completely arbitrary and closed free-body with the interior given by $\Omega \subset \mathrm{B}$, and the enclosing boundary denoted by $\partial\Omega$, we must have the Cauchy momentum balance and the Cauchy angular momentum balance laws satisfied for each equilibrium configuration of the free-body. We are free to choose the shape of the free-body with arbitrary boundary description, but the choice is usually made such that the resulting derivation is as simple as possible, and thus, naturally follows the undeformed geometry definition in the Lagrangian or material form of analysis.

Accordingly, in the undeformed reference body (see Figure 10.8), the boundary surface, $\partial\Omega$, is made up of three disjointed surfaces: two plane cross-sectional areas, $\partial\Omega^{C-}$, for the beginning plane cross-section and $\partial\Omega^{C+}$, for the ending plane cross-section, both perpendicular to the tangent base vector, $\mathbf{T}$, to the reference curve, and the length-wise side surface, $\partial\Omega^S \subset M^S$; all these reference surfaces are defined as such at the reference time $t = 0$.

Now, let us recognize the surface traction, $\mathbf{t}$, at any point, (S,ξ^2,ξ^3) at the current time $t \in \mathbb{R}_+$, on the deformed boundary of the free-body as: $\mathbf{t} = \mathbf{P} \bullet \mathbf{A}^j$ with $\mathbf{P}(S,\xi^2,\xi^3,t)$ as the two-point first Piola–Kirchhoff stress tensor (PK1) defined by: $\mathbf{P} = \mathbf{P}^j \otimes \mathbf{A}_j$ with $\{\mathbf{A}_i(S,\xi^2,\xi^3,t=0)\}$ as the set of covariant base vectors at a point, in the undeformed body, which when deformed to the current state becomes the traction point in consideration on the deformed body. Note that $\mathbf{t} \equiv \mathbf{P}^j = \mathbf{P} \bullet \mathbf{A}^j$ is the traction at a point on the currently deformed surface that had $\mathbf{A}^j$, the contravariant base vector at the point (S,ξ^2,ξ^3), as the normal in the undeformed state. Note that PK1 is expressed in a mixed basis form with the domain basis set as $\{\mathbf{A}_j\}$, and the range basis set in the deformed body, yet to be explicitly specified; the intended and computationally useful set is identified only after we recognize the nature of the underlying configuration space as shown later. Considering that the beam bounding surface is made up of three surfaces, the tractions on each of these deformed surfaces can be identified as shown in Figure 10.9. With these definitions, we can specialize in

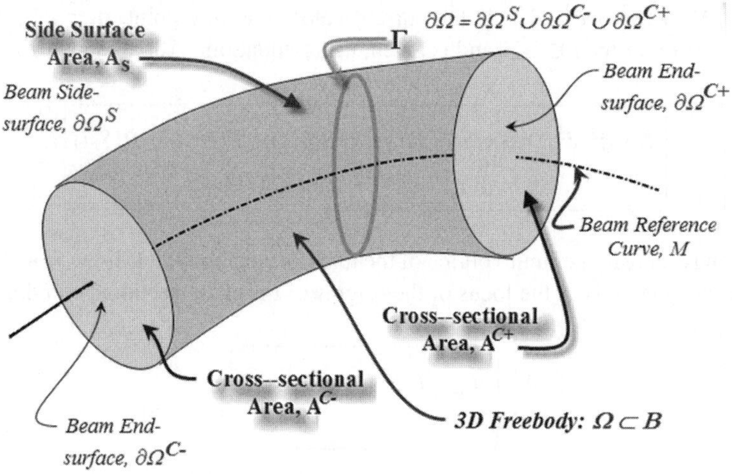

Figure 10.8 Beam undeformed free-body.

the beam space the balance laws of equation (10.10) as:

$$
\begin{aligned}
&\int_{\partial\Omega^S} (\mathbf{t}^S \equiv \mathbf{P}\,\nu^S)\,dA^S + \int_{\partial\Omega^{C+}} (\mathbf{t}^{C+} \equiv \mathbf{P}\,\mathbf{T}^{C+})\,dA^{C+} - \int_{\partial\Omega^{C-}} (\mathbf{t}^{C-} \equiv \mathbf{P}\,\mathbf{T}^{C-})\,dA^{C-} \\
&\qquad + \int_{\Omega} \mathbf{b}\,dV = \frac{d}{dt}\int_{\Omega} \mathbf{v}\,\rho_0\,dV \\
&\int_{\partial\Omega^S} \mathbf{c}\times\mathbf{P}\,\nu^S\,dA^s + \int_{\partial\Omega^{C+}} \mathbf{c}\times\mathbf{P}\,\mathbf{T}^{C+}\,dA^{C+} - \int_{\partial\Omega^{C-}} \mathbf{c}\times\mathbf{P}\,\mathbf{T}^{C-}\,dA^{C-} \\
&\qquad + \int_{\Omega} \mathbf{c}\times\mathbf{b}\,dV = \frac{d}{dt}\int_{\Omega} \mathbf{c}\times\mathbf{v}\,\rho_0\,dV
\end{aligned}
\tag{10.19}
$$

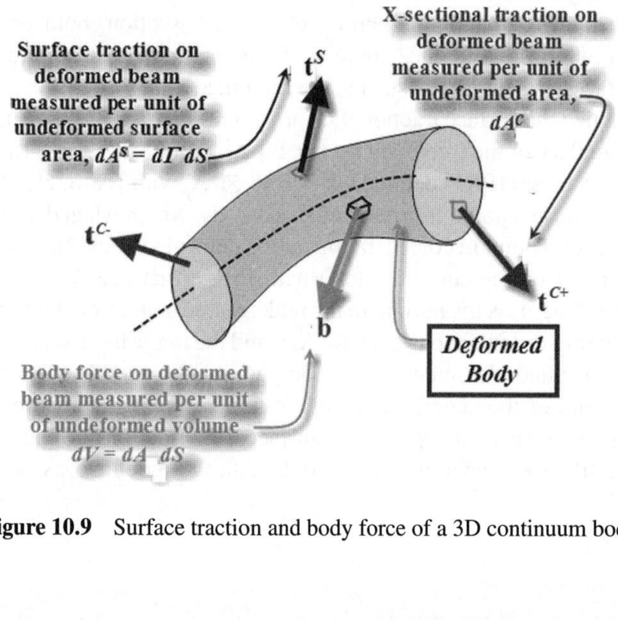

Figure 10.9 Surface traction and body force of a 3D continuum body.

where ν^S is the normal at a point on the side surface, $\partial\Omega^S$, of the undeformed body; **T** is the tangent vector (which is also normal to the undeformed cross-sections) at the beam reference curve. Then dA^{C+}, dA^{C-} and dA^S are the differential areas in the neighborhood of the point on the beginning cross-section, $\partial\Omega^{C-}$, the end cross-section, $\partial\Omega^{C+}$, and the side surface, $\partial\Omega^S$, respectively, of the undeformed body.

10.4.4 Dynamic Momentum Balance Equations: Integral and Differential Forms

Using the beam geometry and the deformation map as discussed above and rearranging the terms, we can rewrite the first of equation (10.19) to get:

$$\int_{\partial\Omega^S} \mathbf{P}\,\nu^S\,dA^S = \int_M \left\{ \int_{\Gamma} \mathbf{P}\,\nu^S\,d\Gamma \right\} dS \tag{10.20}$$

For the second and third integrals of the first of equations (10.19), we first complete the surface integrals by fictitiously adding a side surface with no tractions on it (the applied external tractions have already been accounted for in equation (10.20)); then we apply divergence theorem to transform these into a volume integral. Now, noting the property of a tensor representation as given by equation (10.11), we have:

$$\begin{aligned}\int_{\partial\Omega^{C+}} \mathbf{P}\,\mathbf{T}\,dA^{C+} - \int_{\partial\Omega^{C-}} \mathbf{P}\,\mathbf{T}\,dA^{C-} &= \int_{\partial\Omega} \mathbf{P}\,\nu\,dA = \int_{\Omega} DIV\,\mathbf{P}\,dV \\ &= \int_M \left\{ \int_{A^C} {}^S\mathbf{F}^1|_S\,dA^C \right\} dS = \int_M \left\{ \int_{A^C} {}^S\mathbf{F}^1\,dA^C \right\} |_S dS \\ &= \int_M \left\{ \int_{A^C} {}^S\mathbf{F}^1\,dA^C \right\}_{,S} dS = \int_M \left\{ \int_{A^C} {}^S\mathbf{F}^1\,dA^C \right\}' dS\end{aligned} \tag{10.21}$$

where ${}^S\mathbf{F}^1(S, \xi^1, \xi^2, t)$ is the surface traction per unit of the undeformed cross-sectional area acting on the deformed cross-section at any time $t \in \mathbb{R}_+$ at a material point with undeformed coordinates (S, ξ^2, ξ^3); note that for an invariant vector, the partial derivative is identical to the covariant derivative, and that we have used the notation: $(\bullet)' \equiv \frac{d}{dS}(\bullet)$. In order to get the final beam integral or differential equation for the momentum balance principle, we introduce the following definitions for the beam stress resultant, linear momentum and the external applied forces.

Internal resistive stress resultant, $\mathbf{F}(S, t)$, on an end cross-section:

$$\boxed{\mathbf{F}(S,t) \equiv \int_{A_c} \mathbf{T}_c\,dA_c = \int_{A^C} {}^S\mathbf{F}^1(S, \xi^2, \xi^3, t)\;d\xi^2\,d\xi^3} \tag{10.22}$$

10.4.4.1 (Effective) Linear Momentum, $\mathbf{L}(S,t)$, and Total Linear Momentum, $\mathbf{L}_{TOT}(t)$

From the integral on the right-hand side of the first of equation (10.19), we have:

$$\begin{aligned} \dot{\mathbf{L}}_{TOT}(t) &\equiv \frac{d}{dt}\int_{\Omega} \mathbf{v}\,\rho_0\,dV = \frac{d}{dt}\int_{M}\left\{\int_{A^C} \mathbf{v}(S,\xi^2,\xi^3,t)\,\rho_0(S,\xi^2,\xi^3)\;d\xi^2\,d\xi^3\right\} dS \\ &= \frac{d}{dt}\int_{M} \mathbf{L}(S,t)\,dS = \int_{M} \dot{\mathbf{L}}(S,t)\,dS \end{aligned} \tag{10.23}$$

where we have used the notation: $\dot{(\bullet)} \equiv \frac{d}{dt}(\bullet)$, and introduced:

Beam total linear momentum, $\mathbf{L}_{TOT}(t)$:

$$\boxed{\mathbf{L}_{TOT}(t) \equiv \int_{M} \mathbf{L}(S,t)\,dS} \tag{10.24}$$

with

Beam (effective) linear momentum, $\mathbf{L}(S,t)$, per unit length of the beam reference curve:

$$\boxed{\begin{aligned} \mathbf{L}(S,t) &\equiv \int_{A^C} \mathbf{v}(S,\xi^2,\xi^3,t)\,\rho_0(S,\xi^2,\xi^3)\;d\xi^2\,d\xi^3 \\ &= \int_{A^C} \dot{\mathbf{c}}(S,\xi^2,\xi^3,t)\,\rho_0(S,\xi^2,\xi^3)\;d\xi^2\,d\xi^3 \end{aligned}} \tag{10.25}$$

where $\mathbf{v}(S,\xi^2,\xi^3,t) \equiv \dot{\mathbf{c}}(S,\xi^2,\xi^3,t)$ is, as indicated earlier, the velocity at current time $t \in \mathbb{R}_+$ at a point with undeformed coordinates (S,ξ^2,ξ^3) in the interior of the beam-like body. Now, let us define:

- the cross-sectional mass, $M_0(S)$, per unit of undeformed length, as:

$$\boxed{M_0(S) \stackrel{\Delta}{=} \int_{\partial\Omega} \rho_0(S,\xi^2,\xi^3)\;dA_c} \tag{10.26}$$

Then, using equations (10.17), (10.18) and (10.26) in equation (10.25), we get the final expression for the effective linear momentum, $\mathbf{L}(S,t)$:

$$\boxed{\begin{aligned} \mathbf{L}(S,t) &= \int_{A^C} \dot{\mathbf{c}}(S,\xi^2,\xi^3,t)\,\rho_0(S,\xi^2,\xi^3)\;d\xi^2\,d\xi^3 \\ &= \dot{\mathbf{d}}(S,t)\int_{A^C} \rho_0(S,\xi^2,\xi^3)\;d\xi^2\,d\xi^3 \\ &\quad + \dot{\mathbf{c}}_0(S,t)\left(\int_{A^C} \rho_0(S,\xi^2,\xi^3)\mathbf{P}(\xi^2,\xi^3)\;d\xi^2\,d\xi^3\right)\dot{\mathbf{R}}(S,t) \\ &= M_0(S)\;\dot{\mathbf{d}}(S,t) \end{aligned}} \tag{10.27}$$

10.4.4.2 Effective External Applied Force

We have from the fourth integral (involving body forces) of the first of equation (10.19) and the surface traction equation (10.20):

$$\boxed{\hat{\mathbf{F}}(S,t) \equiv \int_{\Gamma} \mathbf{t}\, d\Gamma + \int_{A^C} \mathbf{b}\; d\xi^2\, d\xi^3} \tag{10.28}$$

Note that $\hat{\mathbf{F}}(S,t)$ is the statically equivalent beam surface forces to the prescribed 3D surface tractions, $\mathbf{t}(S,\Gamma,t)$, on the side surface, and body forces, $\mathbf{b}(S,\xi^2,\xi^3,t)$. Thus, we can always apply the prescribed forces of a 3D beam-like body as equivalent forces on its 1D beam model.

10.4.4.3 Dynamic Momentum Balance Equation

Now, incorporating definitions (10.22), (10.24), (10.25) and (10.28) in equation (10.19), we get the integral form of the momentum balance equation form the first of equation (10.19) as:

$$\boxed{\int_M \{\mathbf{F}|_S + \hat{\mathbf{F}}\}\, dS = \dot{\mathbf{L}}_{TOT}(t) = \int_M \dot{\mathbf{L}}\, dS} \tag{10.29}$$

Noting that beam free-body curve length dS can be arbitrarily chosen, we have the invariant form of the differential equation as:

$$\boxed{\mathbf{F}|_S + \hat{\mathbf{F}} = \dot{\mathbf{L}} = M_0\, \ddot{\mathbf{d}}} \tag{10.30}$$

- Recall our notations for derivatives: $(\bullet)_{,S}$ represents partial derivatives with respect to S, and $(\bullet)|_S$ represents covariant derivatives with respect to S.

***Exercise*:** Show that if $\mathbf{P}(\xi^i) = \mathbf{A}^k(\xi^i) \otimes \mathbf{G}_k(\xi^i)$, then $DIV\,\mathbf{P} = \mathbf{A}^k|_k$

> We can now recognize equation (10.30) as identical to what is obtained by the direct engineering approach except that, as intended, we now have consistent interpretations for the stress resultant, $\mathbf{F}(S,t)$, the effective linear momentum, $\mathbf{L}(S,t)$, and the effective external applied load, $\hat{\mathbf{F}}(S,t)$, by equations (10.22), (10.25) and (10.28), respectively; this is not possible in the direct engineering approach whereby we could, at best, resort to ad hoc assumptions for interpretations of the above beam curve entities.

10.4.5 Angular Momentum Equations: Integral and Differential Forms

Using the beam geometry and the deformation map as discussed, and rearranging the terms, and noting that: $\mathbf{c} = \mathbf{c}_0 + \mathbf{p}$, as in equation (10.15), we can rewrite the first and fourth integrals in the second of equations (10.19):

$$
\begin{aligned}
\int_{\partial\Omega^S} \mathbf{c}\times\mathbf{P}\,\nu^S\,dA^s + \int_{\Omega} \mathbf{c}\times\mathbf{b}\,dV &= \int_M \left\{ \int_\Gamma \mathbf{c}\times\mathbf{P}\,\nu^S\,d\Gamma + \int_{A_C} \mathbf{c}\times\mathbf{b}\,dA_C \right\} dS \\
&= \int_M \left\{ \mathbf{c}_0 \times \left(\int_\Gamma \mathbf{P}\,\nu^S\,d\Gamma + \int_{A_C} \mathbf{b}\,dA_C \right) \right\} dS + \int_M \left\{ \int_\Gamma \mathbf{p}\times\mathbf{P}\,\nu^S\,d\Gamma + \int_{A_C} \mathbf{p}\times\mathbf{b}\,dA_C \right\} dS \\
&= \int_M \left\{ \mathbf{c}_0 \times \left(\int_\Gamma \mathbf{t}\,d\Gamma + \int_{A_C} \mathbf{b}\,dA_C \right) \right\} dS + \int_M \left\{ \int_\Gamma \mathbf{p}\times\mathbf{t}\,d\Gamma + \int_{A_C} \mathbf{p}\times\mathbf{b}\,dA_C \right\} dS \\
&= \int_M \{\mathbf{c}_0 \times \hat{\mathbf{F}}\} dS + \int_M \left\{ \int_\Gamma \mathbf{p}\times\mathbf{t}\,d\Gamma + \int_{A_C} \mathbf{p}\times\mathbf{b}\,dA_C \right\} dS
\end{aligned}
\tag{10.31}
$$

For the second and third integrals of the second of equations (10.19), as in equation (10.21) for the momentum balance principle, we first complete the surface integrals by fictitiously adding a side surface with no tractions on it (the applied external tractions have already been accounted for in equation (10.31)); then we apply the divergence theorem to transform these into a volume integral. Now, noting the property of a tensor representation as given by equation (10.11), we have:

$$
\begin{aligned}
\int_{\partial\Omega^{C+}} \mathbf{c}\times\mathbf{P}\,\mathbf{T}\,dA^{C+} - \int_{\partial\Omega^{C-}} \mathbf{c}\times\mathbf{P}\,\mathbf{T}\,dA^{C-} &= \int_{\partial\Omega} \mathbf{c}\times\mathbf{P}\,\nu\,dA = \int_{\partial\Omega} (\mathbf{c}\times {}^S\mathbf{F}^k)\nu_k\,dA \\
&= \int_{\Omega} DIV(\mathbf{c}\times {}^S\mathbf{F}^k)_{,k}\,dV = \int_M \left\{ \int_{A^C} (\mathbf{c}\times {}^S\mathbf{F}^1)|_S\,dA^C \right\} dS \\
&= \int_M \left\{ \mathbf{c}_0|_S \times \int_{A^C} {}^S\mathbf{F}^1|_S\,dA^C + \mathbf{c}_0 \times \int_{A^C} {}^S\mathbf{F}^1|_S\,dA^C + \int_{A^C} (\mathbf{p}\times {}^S\mathbf{F}^1)\,|_S\,dA^C \right\} dS \\
&= \int_M \left\{ (\mathbf{c}_{0},_S \times \mathbf{F} + \mathbf{c}_0 \times \mathbf{F},_S) + \int_{A^C} (\mathbf{p}\times {}^S\mathbf{F}^1)\,|_S\,dA^C \right\} dS
\end{aligned}
\tag{10.32}
$$

where ${}^S\mathbf{F}^1(S, \xi^1, \xi^2, t)$ is the surface traction per unit of the undeformed cross-sectional area acting on the deformed cross-section at any time $t \in \mathbb{R}_+$ at a material point with undeformed coordinates (S, ξ^2, ξ^3). In order to get the final beam integral or differential equation for the angular momentum balance principle, we introduce the following definitions for the beam stress couple, the effective angular momentum and the effective external applied moments:

$$
\boxed{\mathbf{M}(S,t) \equiv \int_{A^C} (\mathbf{p}(S,\xi^2,\xi^3,t) \times {}^S\mathbf{F}^1(S,\xi^2,\xi^3,t))|_S\,d\xi^2\,d\xi^3}
\tag{10.33}
$$

10.4.5.1 (Effective) Angular Momentum, $\mathbf{A}(S,t)$ and Total Angular Momentum, $\mathbf{A}_{TOT}(t)$

From the integral on the right-hand side of the first of equation (10.19), we have:

$$\begin{aligned}
\dot{\mathbf{A}}(S,t) &\equiv \frac{d}{dt}\int_{\Omega} \mathbf{c}\times\mathbf{v}\,\rho_0\,dV \\
&= \frac{d}{dt}\int_{M}\left\{\int_{A^C} \mathbf{c}(S,\xi^2,\xi^3,t)\times\dot{\mathbf{c}}(S,\xi^2,\xi^3,t)\;\rho_0(S,\xi^2,\xi^3)\,dA^C\right\}dS \\
&= \frac{d}{dt}\int_{M}\left\{\begin{array}{l}\mathbf{c}_0(S,t)\times\left(\int_{A^C}\dot{\mathbf{c}}(S,\xi^2,\xi^3,t)\;\rho_0(S,\xi^2,\xi^3)\,d\xi^2\,d\xi^3\right)\\ +\left(\int_{A^C}\mathbf{p}(S,\xi^2,\xi^3,t)\times\dot{\mathbf{c}}(S,\xi^2,\xi^3,t)\;\rho_0(S,\xi^2,\xi^3)\,d\xi^2\,d\xi^3\right)\end{array}\right\}dS \\
&= \frac{d}{dt}\int_{M}\left\{\mathbf{c}_0(S,t)\times\mathbf{L}(S,t)+\mathbf{A}(S,t)\right\}dS
\end{aligned} \tag{10.34}$$

where we have used the definition for the effective linear momentum, $\mathbf{L}(S,t)$, as in equation (10.25), and introduced:

Total angular momentum, $\mathbf{A}_{TOT}(t)$:

$$\boxed{\mathbf{A}_{TOT}(t) \equiv \int_{M}\left\{\mathbf{c}_0(S,t)\times\mathbf{L}(S,t)+\mathbf{A}(S,t)\right\}dS} \tag{10.35}$$

with

Effective angular momentum, $\mathbf{A}(S,t)$, per unit length of the beam undeformed reference curve:

$$\boxed{\begin{aligned}\mathbf{A}(S,t) &\equiv \int_{A^C}\mathbf{p}(S,\xi^2,\xi^3,t)\times\mathbf{v}(S,\xi^2,\xi^3,t)\;\rho_0(S,\xi^2,\xi^3)\,dA^C \\ &= \int_{A^C}\mathbf{p}(S,\xi^2,\xi^3,t)\times\dot{\mathbf{c}}(S,\xi^2,\xi^3,t)\;\rho_0(S,\xi^2,\xi^3)\,dA^C\end{aligned}} \tag{10.36}$$

where $\mathbf{v}(S,\xi^2,\xi^3,t)\equiv\dot{\mathbf{c}}(S,\xi^2,\xi^3,t)$ is, as indicated earlier, the velocity at current time $t\in\mathbb{R}_+$ at a point with undeformed coordinates (S,ξ^2,ξ^3) in the interior of the beam-like body.

10.4.5.2 Rigid Body Dynamics

In order to reduce the right-hand side of the second of equation (10.19), that represents the angular momentum of the cross-section of a beam, we must have some preliminary results relating to rigid body dynamics.

- Note that:

$$\begin{aligned}\mathbf{R\,P}\times\dot{\mathbf{R}}\;\mathbf{P} &= \mathbf{R\,P}\times\,\boldsymbol{\Omega}_S\,\mathbf{R\,P} = \mathbf{R\,P}\times\boldsymbol{\omega}_S\times\,\mathbf{R\,P}\\ &= (\mathbf{R\,P}\bullet\mathbf{R\,P})\,\boldsymbol{\omega}_S - (\mathbf{R\,P}\bullet\boldsymbol{\omega}_S)\,\mathbf{R\,P}\}\\ &= \{\|\mathbf{P}\|^2\,\mathbf{I}-(\mathbf{R\,P}\otimes\mathbf{R\,P})\}\,\boldsymbol{\omega}_S\\ &= \{\|\mathbf{P}\|^2\,\mathbf{R\,R}^T - \mathbf{R}\,(\mathbf{P}\otimes\mathbf{P})\mathbf{R}^T\}\,\boldsymbol{\omega}_S\\ &= \mathbf{R}\,\{\|\mathbf{P}\|^2\mathbf{I}-(\mathbf{P}\otimes\,\mathbf{P})\}\mathbf{R}^T\,\boldsymbol{\omega}_S\end{aligned}\tag{10.37}$$

  where we have used the facts:
  - $\boldsymbol{\Omega}_S$ is the skew-symmetric current (spatial) angular velocity tensor measured in Lagrangian coordinates defined by: $\boxed{\boldsymbol{\Omega}_S \stackrel{\Delta}{=} \dot{\mathbf{R}}\;\mathbf{R}^T \Leftrightarrow \dot{\mathbf{R}} \stackrel{\Delta}{=} \boldsymbol{\Omega}_S\,\mathbf{R}}$, with its corresponding axial vector, denoted by $\boldsymbol{\omega}_S$, such that: $\boldsymbol{\Omega}_S\,\mathbf{h} = \boldsymbol{\omega}_S\times\mathbf{h}$ for $\forall\mathbf{h}\in\mathbb{R}_3$; and
  - the Lagrangian identity for the vector triple product: $\mathbf{a}\times\mathbf{b}\times\mathbf{c} = (\mathbf{a}\bullet\mathbf{c})\mathbf{b} - (\mathbf{a}\bullet\mathbf{b})\mathbf{c}$.
- Instead of defining as above, we could choose the undeformed (material) counterpart: $\boldsymbol{\Omega}_M$ as the skew-symmetric angular velocity tensor measured in Lagrangian coordinates defined by: $\boxed{\boldsymbol{\Omega}_M \stackrel{\Delta}{=} \mathbf{R}^T\;\dot{\mathbf{R}} \Leftrightarrow \dot{\mathbf{R}} \stackrel{\Delta}{=} \mathbf{R}\,\boldsymbol{\Omega}_M}$, with its corresponding axial vector, denoted by $\boldsymbol{\omega}_M$, such that: $\boldsymbol{\Omega}_M\,\mathbf{h} = \boldsymbol{\omega}_M\times\mathbf{h}$ for $\forall\mathbf{h}\in\mathbb{R}_3$.
- Recall the adjoint map: $\boxed{\dot{\mathbf{R}} \stackrel{\Delta}{=} \mathbf{R}\boldsymbol{\Omega}_M = \boldsymbol{\Omega}_S\mathbf{R} \Rightarrow [\boldsymbol{\omega}_S] = \boldsymbol{\Omega}_S = \mathbf{R}\boldsymbol{\Omega}_M\mathbf{R}^T = [\mathbf{R}\boldsymbol{\omega}_M] \Rightarrow \boldsymbol{\omega}_S = \mathbf{R}\boldsymbol{\omega}_M}$; then:

$$\begin{aligned}\mathbf{R\,P}\times\dot{\mathbf{R}}\;\mathbf{P} &= \mathbf{R\,P}\times\mathbf{R}\,\boldsymbol{\Omega}_M\;\mathbf{P} = \mathbf{R}\,(\mathbf{P}\times\boldsymbol{\omega}_M\times\mathbf{P})\\ &= \mathbf{R}\,\{(\mathbf{P}\bullet\mathbf{P})\,\boldsymbol{\omega}_M - (\mathbf{P}\bullet\boldsymbol{\omega}_M)\,\mathbf{P}\}\\ &= \mathbf{R}\,\{\|\mathbf{P}\|^2\mathbf{I}-(\mathbf{P}\otimes\mathbf{P})\}\,\boldsymbol{\omega}_M\\ &= \mathbf{R}\,\{\|\mathbf{P}\|^2\mathbf{I}-(\mathbf{P}\otimes\mathbf{P})\}\,\mathbf{R}^T\,\boldsymbol{\omega}_S\end{aligned}\tag{10.38}$$

  which relationship is same as in equation (10.37). Now, the right-hand side of the second of equation (10.19) can be expressed, using equation (10.37), as:

$$\begin{aligned}&\int_{A_c}\mathbf{c}\times\mathbf{v}\,\rho_0\,dA = \int_{A_c}\mathbf{c}\times\dot{\mathbf{c}}\;\rho_0\,dA = \int_{A_c}(\mathbf{C}_0+\mathbf{d}+\mathbf{R\,P})\times(\dot{\mathbf{d}}+\dot{\mathbf{R}}\;\mathbf{P})\,\rho_0\,dA\\ &= (\mathbf{C}_0+\mathbf{d})\times\left\{\left(\int_{A_c}\rho_0\,dA\right)\dot{\mathbf{d}}+\dot{\mathbf{R}}\left(\cancel{\int_{A_c}\rho_0\mathbf{P}\,dA}\right)\right\}\\ &+\mathbf{R}\times\left(\cancel{\int_{A_c}\rho_0\mathbf{P}\,dA}\right)\dot{\mathbf{d}}+\left(\int_{A_c}\rho_0\,\mathbf{R\,P}\times\dot{\mathbf{R}}\;\mathbf{P}\,dA\right)\\ &= \mathbf{c}_0\times M_0\;\dot{\mathbf{d}}+\mathbf{R}\left(\int_{A_c}\rho_0\,\{\|\mathbf{P}\|^2-(\mathbf{P}\otimes\mathbf{P})\}\,dA\right)\mathbf{R}^T\,\boldsymbol{\omega}_S\\ &= \mathbf{c}_0\times\mathbf{L}+(\mathbf{R}\,\mathbf{I}_M\,\mathbf{R}^T)\boldsymbol{\omega}_S = \mathbf{c}_0\times\mathbf{L}+\mathbf{I}_S\,\boldsymbol{\omega}_S = \underbrace{\mathbf{c}_0(S,t)\times\mathbf{L}(S,t)}_{\text{Moment of linear momentum}} + \underbrace{\mathbf{A}_S(S,t)}_{\text{Angular momentum}}\end{aligned}\tag{10.39}$$

where we have used the definitions:

- the ***Lagrangian inertia tensor***, $\mathbf{I}_M(S)$, per unit of undeformed length, as:

$$\boxed{\mathbf{I}_M(S) \stackrel{\Delta}{=} \int_{A_c} \rho_0(S,\xi^2,\xi^3)\ \{\|\mathbf{P}(\xi^2,\xi^3)\|^2\mathbf{I} - (\mathbf{P}(\xi^2,\xi^3)\otimes\ \mathbf{P}(\xi^2,\xi^3))\}\ dA} \tag{10.40}$$

In solid mechanics, it is also referred to as the material inertia tensor in the sense that the coordinates refer to the initial undeformed configuration.

Now, $\|\mathbf{P}(\xi^2,\xi^3)\|^2 = (\mathbf{P}\bullet\mathbf{P}) = \xi^\alpha\ \xi^\beta(\mathbf{G}_\alpha\bullet\mathbf{G}_\beta) = \xi^\alpha\ \xi^\beta\delta_{\alpha\beta}$ and $(\mathbf{P}\otimes\mathbf{P}) = \xi^\alpha\ \xi^\beta(\mathbf{G}_\alpha\otimes\mathbf{G}_\beta)$ for $\alpha,\beta = 2,3$ with the Einstein summation convention implied. Thus, we have:

$$\boxed{\begin{aligned}\mathbf{I}_M(S) &\stackrel{\Delta}{=} \left(\int_{A_c} \rho_0(S,\xi^2,\xi^3)\ \xi^\alpha\ \xi^\beta\ dA_C\right)\{\delta_{\alpha\beta}\mathbf{I} - (\mathbf{G}_\alpha\otimes\mathbf{G}_\beta)\}\\ &= I_{\alpha\beta}\{\delta_{\alpha\beta}\mathbf{I} - (\mathbf{G}_\alpha\otimes\mathbf{G}_\beta)\},\quad \alpha,\beta = 2,3\end{aligned}} \tag{10.41}$$

with $\mathbf{I} = \mathbf{G}_\alpha\otimes\mathbf{G}_\alpha$, $\alpha = 2,3$. Now, note that if the material coordinates, (S,ξ^2,ξ^3), are aligned along the principal axes of inertia, then the inertia tensor will be diagonal, that is, $\mathbf{I}_M = diag(I_2,I_3)$ with the polar moment of inertia: $J = I_2 + I_3$ where $I_2 = \int_{A_C}(\xi^3)^2\ \rho_0(S,\xi^2,\xi^3)\ dA_C$ and $I_3 = \int_{A_C}(\xi^2)^2\ \rho_0(S,\xi^2,\xi^3)\ dA_C$; then, the inertia tensor gets the familiar look:

$$\boxed{\mathbf{I}_M(S) = J(\mathbf{G}_1\otimes\mathbf{G}_1) + I_2(\mathbf{G}_2\otimes\mathbf{G}_2) + I_3(\mathbf{G}_3\otimes\mathbf{G}_3)} \tag{10.42}$$

- Spatial (in the sense that the components refer to current body) inertia tensor, $\mathbf{I}_S$,:

$$\boxed{\mathbf{I}_S \stackrel{\Delta}{=} \mathbf{R}\ \mathbf{I}_M\ \mathbf{R}^T} \tag{10.43}$$

Incorporating equation (10.40) in equation (10.43), we get:

$$\boxed{\begin{aligned}\mathbf{I}_S \stackrel{\Delta}{=} \mathbf{R}\ \mathbf{I}_M\ \mathbf{R}^T &= \left(\int_{A_c} \rho_0(S,\xi^2,\xi^3)\ \xi^\alpha\ \xi^\beta\ dA\right)\{\delta_{\alpha\beta}\mathbf{I} - \mathbf{R}(\mathbf{G}_\alpha\otimes\mathbf{G}_\beta)\mathbf{R}^T\}\\ &= \left(\int_{A_c} \rho_0(S,\xi^2,\xi^3)\ \xi^\alpha\ \xi^\beta\ dA\right)\{\delta_{\alpha\beta}\mathbf{I} - (\mathbf{R}\mathbf{G}_\alpha\otimes\mathbf{R}\mathbf{G}_\beta)\}\\ &= \left(\int_{A_c} \rho_0(S,\xi^2,\xi^3)\ \xi^\alpha\ \xi^\beta\ dA\right)\{\delta_{\alpha\beta}\mathbf{I} - (\mathbf{g}_\alpha\otimes\mathbf{g}_\beta)\}\end{aligned}} \tag{10.44}$$

We note that the coefficients of inertia tensor with respect to the spinning frame remain the same and independent of time. As in the material coordinate system, we get for alignment with the principal coordinates:

$$\boxed{\mathbf{I}_S(S) = J(\mathbf{g}_1 \otimes \mathbf{g}_1) + I_2(\mathbf{g}_2 \otimes \mathbf{g}_2) + I_3(\mathbf{g}_3 \otimes \mathbf{g}_3)} \tag{10.45}$$

- the effective spatial angular momentum, $\mathbf{A}_S(S,t)$, as:

$$\boxed{\mathbf{A}_S(S,t) \stackrel{\Delta}{=} \left(\mathbf{R}(S,t)\,\mathbf{I}_M(S)\,\mathbf{R}^T(S,t)\right)\boldsymbol{\omega}_S(S,t) = \mathbf{I}_S(S,t)\,\boldsymbol{\omega}_S(S,t)} \tag{10.46}$$

- the effective material angular momentum, $\mathbf{A}_M(S,t)$, as:

$$\boxed{\mathbf{A}_M(S,t) \stackrel{\Delta}{=} \mathbf{I}_M(S)\,\boldsymbol{\omega}_M(S,t)} \tag{10.47}$$

From equations (10.46) and (10.47), we get:

$$\boxed{\mathbf{A}_S(S,t) = \mathbf{R}(S,t)\,\mathbf{A}_M(S,t)} \tag{10.48}$$

Now, taking the time derivative of the angular momentum, $\dot{\mathbf{A}}_S(S,t)$, we get:

$$\boxed{\begin{aligned}\dot{\mathbf{A}}_S &= \dot{\mathbf{R}}\,\mathbf{I}_M\,\mathbf{R}^T\,\boldsymbol{\omega}_S + \mathbf{R}\,\mathbf{I}_M\,\dot{\mathbf{R}}^T\,\boldsymbol{\omega}_S + \mathbf{R}\,\mathbf{I}_M\,\mathbf{R}^T\,\dot{\boldsymbol{\omega}}_S \\ &= \boldsymbol{\Omega}_S\,\mathbf{R}\,\mathbf{I}_M\,\mathbf{R}^T\,\boldsymbol{\omega}_S - \mathbf{R}\,\mathbf{I}_M\,\mathbf{R}^T\,\cancel{\boldsymbol{\Omega}_S\,\boldsymbol{\omega}_S} + \mathbf{R}\,\mathbf{I}_M\,\mathbf{R}^T\,\dot{\boldsymbol{\omega}}_S \\ &= \mathbf{I}_S\,\dot{\boldsymbol{\omega}}_S + \boldsymbol{\omega}_S \times \mathbf{A}_S\end{aligned}} \tag{10.49}$$

and taking the time derivative of the angular momentum, $\dot{\mathbf{A}}_M(S,t)$, we get:

$$\boxed{\dot{\mathbf{A}}_M = \mathbf{I}_M\,\dot{\boldsymbol{\omega}}_M} \tag{10.50}$$

Clearly, then from equations (10.47) and (10.48), we get:

$$\boxed{\begin{aligned}\dot{\mathbf{A}}_S &= \dot{\mathbf{R}}\,\mathbf{A}_M + \mathbf{R}\,\dot{\mathbf{A}}_M = \mathbf{R}\,\boldsymbol{\omega}_M \times \mathbf{A}_M + \mathbf{R}\,\dot{\mathbf{A}}_M \\ &= \mathbf{R}\left(\boldsymbol{\omega}_M \times \mathbf{A}_M + \dot{\mathbf{A}}_M\right) = \mathbf{R}\left(\mathbf{I}_M\,\dot{\boldsymbol{\omega}}_M + \boldsymbol{\omega}_M \times \mathbf{A}_M\right)\end{aligned}} \tag{10.51}$$

Now, going back to the right-hand side of the second of equation (10.19), we get:

$$\boxed{\frac{d}{dt}\int_{A_c} \mathbf{c} \times \mathbf{v}\,\rho_0\,dA = \dot{\mathbf{c}}_0(S,t) \times \mathbf{L}(S,t) + \mathbf{c}_0(S,t) \times \dot{\mathbf{L}}(S,t) + \dot{\mathbf{A}}_S(S,t)} \tag{10.52}$$

10.4.5.3 Effective External Applied Moment

We have from equation (10.31):

Effective external applied moment, $\hat{\mathbf{M}}(S,t)$, per unit of undeformed reference curve length:

$$\boxed{\hat{\mathbf{M}}(S,t) \equiv \int_{\Gamma} \mathbf{p}(S,\Gamma,t) \times \mathbf{t}(S,\Gamma,t)\ d\Gamma + \int_{A^C} \mathbf{p}(S,\xi^2,\xi^3,t) \times \mathbf{b}(S,\xi^2,\xi^3,t)\ \ d\xi^2\ d\xi^3} \tag{10.53}$$

Note that $\hat{\mathbf{M}}(S,t)$ is the statically equivalent beam curve moment to the prescribed 3D surface tractions, $\mathbf{t}(S,\Gamma,t)$, on the side surface, and body forces, $\mathbf{b}(S,\xi^2,\xi^3,t)$. Thus, we can always apply the prescribed moments of a 3D beam-like body as equivalent moments on its 1D beam model with the cross-sectional kinematics known. Now, incorporating definitions (10.53) in equation (10.31):

$$\int_{\partial\Omega^S} \mathbf{c} \times \mathbf{P}\, \nu^S\, dA^s + \int_{\Omega} \mathbf{c} \times \mathbf{b}\, dV = \int_M \{\mathbf{c}_0 \times \hat{\mathbf{F}}\} dS + \int_M \hat{\mathbf{M}}\, dS \tag{10.54}$$

and incorporating definitions (10.33) in equation (10.32)

$$\int_{\partial\Omega^{C+}} \mathbf{c} \times \mathbf{P}\,\mathbf{T}\, dA^{C+} - \int_{\partial\Omega^{C-}} \mathbf{c} \times \mathbf{P}\,\mathbf{T}\, dA^{C-} = \int_M \left\{\mathbf{c}_{0,S} \times \mathbf{F} + \mathbf{c}_0 \times \mathbf{F}_{,S} + \mathbf{M}_{,S}\right\} dS \tag{10.55}$$

10.4.5.4 Beam Angular Momentum Balance Equations

Now, using equations (10.54), (10.55) and (10.52), the second of the equation (10.19) becomes:

$$\begin{aligned}
&\int_S \{\mathbf{M}|_S + \mathbf{c}_0|_S \times\ \mathbf{F} + \hat{\mathbf{M}}\} dS + \int_S \mathbf{c}_0 \times (\mathbf{F}|_S + \hat{\mathbf{F}})\ dS \\
&= \int_S \{\cancel{\dot{\mathbf{c}}_0(S,t) \times \mathbf{L}(S,t)} + \mathbf{c}_0(S) \times \dot{\mathbf{L}}(S,t) + \dot{\mathbf{A}}_S(S,t)\}\ dS \\
&\Rightarrow \int_S \{\mathbf{M}|_S + \ \mathbf{c}_0|_S \times\ \mathbf{F} + \hat{\mathbf{M}}\} dS + \int_S \mathbf{c}_0 \times \underbrace{\cancel{(\mathbf{F}|_S + \hat{\mathbf{F}} - \dot{\mathbf{L}})}}_{\text{Momentum balance}} \}dS \\
&= \int_S \{\mathbf{M}\ \cancel{\dot{\mathbf{d}}(S,t) \times \dot{\mathbf{d}}(S,t)} + \dot{\mathbf{A}}_S(S,t)\} dS \\
&\Rightarrow \boxed{\int_S \{\mathbf{M}|_S + \ \mathbf{c}_0|_S \times\ \mathbf{F} + \hat{\mathbf{M}}\} dS = \int_S \dot{\mathbf{A}}_S(S,t) dS}
\end{aligned} \tag{10.56}$$

where we have used the fact that the linear momentum balance, as given by equation (10.30), is simultaneously satisfied. Finally, noting that equation (10.56) is true for all arc lengths, S, we get from the kernel of the integral form, the local form of the vector equations for a beam under dynamic forces and moments as:

$$\boxed{\mathbf{M}|_S + \mathbf{c}_0|_S \times \mathbf{F} + \hat{\mathbf{M}} = \dot{\mathbf{A}}_S} \tag{10.57}$$

with

$$\begin{aligned}
\dot{\mathbf{A}}_S(S,t) &= \mathbf{I}_S\ \dot{\boldsymbol{\omega}}_S(S,t) + \boldsymbol{\omega}_S(S,t) \times \mathbf{A}_S(S,t) \\
&= \mathbf{R}\big(\mathbf{I}_M\ \dot{\boldsymbol{\omega}}_M(S,t) + \boldsymbol{\omega}_M(S,t) \times \mathbf{A}_M(S,t)\big) = \mathbf{R}\big(\boldsymbol{\omega}_M \times \mathbf{A}_M + \dot{\mathbf{A}}_M\big) \\
\text{where}\quad & \mathbf{A}_S = \mathbf{I}_S\ \boldsymbol{\omega}_S = \mathbf{R}\ \mathbf{A}_M, \quad \mathbf{A}_M = \mathbf{I}_M\ \boldsymbol{\omega}_M \\
\mathbf{I}_S(S,t) &\stackrel{\Delta}{=} \mathbf{R}(S,t)\ \mathbf{I}_M(S)\ \mathbf{R}^T(S,t) \\
\mathbf{I}_M(S) &\stackrel{\Delta}{=} \left(\int_{A_c} \rho_0(S,\xi^2,\xi^3)\ \xi^\alpha\ \xi^\beta\ dA_C\right)\{\delta_{\alpha\beta}\mathbf{I} - (\mathbf{G}_\alpha \otimes \mathbf{G}_\beta)\}, \quad \alpha,\beta = 2,3 \\
\mathbf{I}_S(S,t) &= \left(\int_{A_c} \rho_0(S,\xi^2,\xi^3)\ \xi^\alpha\ \xi^\beta\ dA\right)\{\delta_{\alpha\beta}\mathbf{I} - (\mathbf{g}_\alpha(t) \otimes \mathbf{g}_\beta(t))\} \\
\boldsymbol{\Omega}_M &= [\boldsymbol{\omega}_M] = \mathbf{R}^T\ \dot{\mathbf{R}}, \boldsymbol{\Omega}_S = [\boldsymbol{\omega}_S] = \dot{\mathbf{R}}\ \mathbf{R}^T = \mathbf{R}\ \boldsymbol{\Omega}_M\ \mathbf{R}^T, \boldsymbol{\omega}_S = \mathbf{R}\ \boldsymbol{\omega}_M
\end{aligned} \tag{10.58}$$

We can now recognize equation (10.57) as identical to what is obtained by the direct engineering approach except that, as intended, now we have consistent interpretations for the stress couple, $\mathbf{M}(S,t)$, the effective angular momentum, $\mathbf{A}(S,t)$, and the effective external applied moment, $\hat{\mathbf{M}}(S,t)$, by equations (10.33), (10.36) and (10.53), respectively; this is not possible in the direct engineering approach where we could, at best, resort to ad hoc assumptions for interpretations of the beam curve entities.

10.4.6 Beam Balance Equations of Motion: Dynamic

For easy reference, here we copy and collect the derived balance equations:

$$\begin{array}{lr}
\mathbf{F}|_S + \hat{\mathbf{F}} = \dot{\mathbf{L}} & \text{Momentum balance} \\
\mathbf{M}|_S + \mathbf{c}_0|_S \times \mathbf{F} + \hat{\mathbf{M}} = \dot{\mathbf{A}}_S & \text{Moment of momentum balance}
\end{array} \tag{10.59}$$

where

$$
\begin{aligned}
\overset{\bullet}{\mathbf{A}}_S(S,t) &= \mathbf{I}_S\ \overset{\bullet}{\boldsymbol{\omega}}_S(S,t) + \boldsymbol{\omega}_S(S,t) \times \mathbf{A}_S(S,t) \\
&= \mathbf{R}\big(\mathbf{I}_M\ \overset{\bullet}{\boldsymbol{\omega}}_M(S,t) + \boldsymbol{\omega}_M(S,t) \times \mathbf{A}_M(S,t)\big) = \mathbf{R}\big(\boldsymbol{\omega}_M \times \mathbf{A}_M + \overset{\bullet}{\mathbf{A}}_M\big) \\
\text{where}\quad & \mathbf{A}_S = \mathbf{I}_S\ \boldsymbol{\omega}_S = \mathbf{R}\ \mathbf{A}_M, \quad \mathbf{A}_M = \mathbf{I}_M\ \boldsymbol{\omega}_M \\
\mathbf{I}_S(S,t) &\stackrel{\Delta}{=} \mathbf{R}(S,t)\ \mathbf{I}_M(S)\ \mathbf{R}^T(S,t) \\
\mathbf{I}_M(S) &\stackrel{\Delta}{=} \left(\int_{A_c} \rho_0(S,\xi^2,\xi^3)\ \xi^\alpha\ \xi^\beta\ dA_C \right) \{\delta_{\alpha\beta}\mathbf{I}\ - (\mathbf{G}_\alpha \otimes \mathbf{G}_\beta)\}, \quad \alpha,\beta = 2,3 \\
\mathbf{I}_S(S,t) &= \left(\int_{A_c} \rho_0(S,\xi^2,\xi^3)\ \xi^\alpha\ \xi^\beta\ dA \right) \{\delta_{\alpha\beta}\mathbf{I}\ - (\mathbf{g}_\alpha(t) \otimes \mathbf{g}_\beta(t))\} \\
\boldsymbol{\Omega}_M &= [\boldsymbol{\omega}_M] = \mathbf{R}^T\ \overset{\bullet}{\mathbf{R}}, \quad \boldsymbol{\Omega}_S = [\boldsymbol{\omega}_S] = \overset{\bullet}{\mathbf{R}}\ \mathbf{R}^T = \mathbf{R}\,\boldsymbol{\Omega}_M\,\mathbf{R}^T, \quad \boldsymbol{\omega}_S = \mathbf{R}\ \boldsymbol{\omega}_M
\end{aligned}
\tag{10.60}
$$

- Note that, as shown in equation (10.59), the vector beam balance equations of motion do not depend on the rotation parameters, whether shear deformation is considered or not, although the displacement parameter is implicit in the second term of the moment of momentum balance equation.

10.4.7 Where We Would Like to Go

So far, we have derived the beam balance equations by reducing the 3D equations of motion of elasticity. However, the beam equations of motion can be directly developed by considering a beam as a 1D object along the axis of the beam. For this simple method, we would like to check out the following.

- ***Direct 1D beam balance equations***
  While the beam equations presented is very compact and thus useful for analysis, we will modify its current vector form to component or operational vector form suitable for computational purposes where the components of the forces and moments need to be determined. As prerequisites to such a task, we should check out:
- ***Component or operational vectors***
- ***Derivatives of component vectors***
  With these background materials, we will be able to modify the equations in terms of what may be called the component vector or operator form as described in:
- ***Computational balance equations***

10.5 Weak Form: Kinematic and Configuration Space

10.5.1 What We Need to Recall

Let us recall the following.

- Thin beam vector balance equations:

$$\begin{array}{|lr|}\hline \mathbf{F}_{,S} + \hat{\mathbf{F}} = \dot{\mathbf{L}} & \text{Momentum balance} \\ \mathbf{M}_{,S} + \mathbf{c}_{0,S} \times \mathbf{F} + \hat{\mathbf{M}} = \dot{\mathbf{A}} & \text{Moment of momentum balance} \\ \hline \end{array} \tag{10.61}$$

- The internal resistive stress resultant, $\mathbf{F}$, is defined by:

$$\boxed{\mathbf{F}(S,t) \equiv \int_{A^C} {}^{S}\mathbf{F}^1(S,\xi^2,\xi^3,t)\; d\xi^2\, d\xi^3} \tag{10.62}$$

- the effective linear momentum, $\mathbf{L}(S,t)$, is defined by:

$$\boxed{\begin{aligned} \mathbf{L}(S,t) &\equiv \int_{A^C} \mathbf{v}(S,\xi^2,\xi^3,t)\, \rho_0(S,\xi^2,\xi^3)\; d\xi^2\, d\xi^3 \\ &= \int_{A^C} \dot{\mathbf{c}}(S,\xi^2,\xi^3,t)\, \rho_0(S,\xi^2,\xi^3)\; d\xi^2\, d\xi^3 \end{aligned}} \tag{10.63}$$

where $\mathbf{v}(S,\xi^2,\xi^3,t) \equiv \dot{\mathbf{c}}(S,\xi^2,\xi^3,t)$ is the velocity at current time $t \in \mathbb{R}_+$ at a point with undeformed coordinates (S,ξ^1,ξ^2) in the interior of the beam-like body.

- The external applied force, $\hat{\mathbf{F}}(S,t)$, is defined by:

$$\boxed{\hat{\mathbf{F}}(S,t) \equiv \int_{\Gamma} \mathbf{t}\, d\Gamma + \int_{A^C} \mathbf{b}\; d\xi^2\, d\xi^3} \tag{10.64}$$

- The internal resistive stress couple, $\mathbf{M}$, is defined by:

$$\boxed{\mathbf{M}(S,t) \equiv \int_{A^C} \left(b(S,\xi^2,\xi^3,t) \times {}^{S}\mathbf{F}^1(S,\xi^2,\xi^3,t)\right)|_S\; d\xi^2\, d\xi^3} \tag{10.65}$$

- The angular momentum, $\mathbf{A}(S,t)$, is defined by:

$$\boxed{\begin{aligned} \mathbf{A}(S,t) &\equiv \int_{A^C} b(S,\xi^2,\xi^3,t) \times \mathbf{v}(S,\xi^2,\xi^3,t)\; \rho_0(S,\xi^2,\xi^3)\; dA^C \\ &= \int_{A^C} b(S,\xi^2,\xi^3,t) \times \dot{\mathbf{c}}(S,\xi^2,\xi^3,t)\; \rho_0(S,\xi^2,\xi^3)\; dA^C \end{aligned}} \tag{10.66}$$

- The vector triple product property: $\mathbf{a} \times \mathbf{b} \bullet \mathbf{c} = \mathbf{b} \times \mathbf{c} \bullet \mathbf{a} = \mathbf{c} \times \mathbf{a} \bullet \mathbf{b}$, $\forall \mathbf{a}, \mathbf{b}, \mathbf{c}$ vectors.
- In what follows, a barred quantity signifies that it is the variation of a corresponding real entity.
- The definition of a rotation tensor, its matrix representation and the algebra, geometry and topology of rotation.

10.5.2 Weak Form of the Vector Balance Equations

Let us consider the beam reference curve, $C \equiv M(S)$, with two ends given by $S = 0$ and $S = L$. Let $(\bar{\mathbf{v}} \equiv \dot{\bar{\mathbf{d}}}, \bar{\boldsymbol{\omega}} \equiv \dot{\bar{\boldsymbol{\varphi}}})$ be any sufficiently smooth vector fields defined on $C \times \mathbb{R}_+$. Now, following the standard model of the weak form generation, we multiply equation (10.61) appropriately by $\bar{\mathbf{v}} = \dot{\bar{\mathbf{d}}}$ and $\bar{\boldsymbol{\omega}} = \dot{\bar{\boldsymbol{\varphi}}}$, and integrate over the beam reference curve, C, to get the beam weak form functional, $\hat{G}_{Dyn}$, given as:

$$\hat{G}_{Dyn} \equiv -\int_C \left\{ \begin{array}{l} (\mathbf{F},_S + \hat{\mathbf{F}} - \dot{\mathbf{L}}) \bullet \bar{\mathbf{v}} + \\ (\mathbf{M},_S + \mathbf{c}_0,_S \times \mathbf{F} + \hat{\mathbf{M}} - \dot{\mathbf{A}}) \bullet \bar{\boldsymbol{\omega}} \end{array} \right\} dS = \mathbf{0} \tag{10.67}$$

Now, by an application of integration by parts and recalling the commutative property of vector triple product, we get the virtual functional, $\hat{G}_{Dyn}$, as:

$$\begin{aligned} \hat{G}_{Dyn} \equiv & \overbrace{\int_C \left\{ \left(\mathbf{F} \bullet (\dot{\bar{\mathbf{d}}},_S + \mathbf{c}_0,_S \times \dot{\bar{\boldsymbol{\varphi}}}\right) + \mathbf{M} \bullet \dot{\bar{\boldsymbol{\varphi}}},_S \right\} dS}^{\text{Internal stored power}} + \overbrace{\int_C \left\{ \dot{\mathbf{L}} \bullet \dot{\bar{\mathbf{d}}} + \dot{\mathbf{A}} \bullet \dot{\bar{\boldsymbol{\varphi}}} \right\} dS}^{\text{Intertial power}} \\ & - \overbrace{\left\{ \left(\hat{\mathbf{F}}_\nu \bullet \dot{\bar{\mathbf{d}}} + \hat{\mathbf{M}}_\nu \bullet \dot{\bar{\boldsymbol{\varphi}}}\right)_{S=0}^{S=L} + \int_C \left(\hat{\mathbf{F}} \bullet \dot{\bar{\mathbf{d}}} + \hat{\mathbf{M}} \bullet \dot{\bar{\boldsymbol{\varphi}}}\right) dS \right\}}^{\text{External applied power}} \\ & = \bar{W}_{Internal} + \bar{W}_{Inertia} - \bar{W}_{External} = \mathbf{0} \end{aligned} \tag{10.68}$$

where we have identified the rates of stored internal virtual work, the inertial work and the external virtual work in the context of the energy conjugate integral given by equation (10.68), as follows.

Internal virtual work rate, $\bar{W}_{Internal}$, as:

$$\bar{W}_{Internal} \equiv \int_C \left\{ \left(\mathbf{F} \bullet \left(\dot{\bar{\mathbf{d}}},_S + \mathbf{c}_0,_S \times \dot{\bar{\boldsymbol{\varphi}}} \right) + \mathbf{M} \bullet \dot{\bar{\boldsymbol{\varphi}}},_S \right) \right\} dS \tag{10.69}$$

Inertial virtual work rate, $\bar{W}_{Inertia}$, as:

$$\bar{W}_{Inertia} \equiv \int_C \left\{ \dot{\mathbf{L}} \bullet \dot{\bar{\mathbf{d}}} + \dot{\mathbf{A}} \bullet \dot{\bar{\boldsymbol{\varphi}}} \right\} dS \tag{10.70}$$

External virtual work rate, $\bar{W}_{External}$, as:

$$\bar{W}_{External} \equiv \left(\hat{\mathbf{F}}_\nu \bullet \dot{\bar{\mathbf{d}}} + \hat{\mathbf{M}}_\nu \bullet \dot{\bar{\boldsymbol{\varphi}}}\right)_{S=0}^{S=L} + \int_C \left(\hat{\mathbf{F}} \bullet \dot{\bar{\mathbf{d}}} + \hat{\mathbf{M}} \bullet \dot{\bar{\boldsymbol{\varphi}}}\right) dS \tag{10.71}$$

10.5.3 Kinematic Parameters

The weak form of the equation for the one-dimensional beam, that is, the beam undeformed reference curve, given by equation (10.68), holds whenever the strong form, given by equation (10.61), is true. In fact, the weak form is just as exact as the strong form. Moreover, to be energy conjugate in the internal virtual work, we may call $\bar{\mathbf{d}}_{,S} + \mathbf{c}_{0,S} \times \bar{\varphi}$, and $\bar{\varphi}_{,S}$, in equation (10.68), as the force strain vector and the moment strain vector, respectively, as the 1D virtual strain fields over the beam reference curve as functions of the virtual parameters, $(\bar{\mathbf{d}}(S,t), \bar{\varphi}(S,t))$, presuming that these functions exist necessarily as the variations of some as yet undetermined 1D real strain fields over the reference curve as functions of some as yet undetermined real parameters. Naturally, then, this way of going about the problem will make sense only if we are able to determine those real parameters and corresponding real strain fields. In other words, we have set our problem as one of determining the kinematics of a 1D beam reference curve. As it will turn out, in the material coordinate system, that is, in the event we describe the invariant force, $\mathbf{F}$, and moment, $\mathbf{M}$, vectors in terms of the undeformed basis vectors, $\{\mathbf{G}_i\}$. The kinematic translational expression, $\bar{\mathbf{d}}_{,S} + \mathbf{c}_{0,S} \times \bar{\varphi}$, and rotational expression, $\bar{\varphi}_{,S}$, that comprise the internal virtual work, cannot be the expressions for the virtual strains since there exist no functions whose variations could produce these expressions. It will be found later that the expressions $\bar{\mathbf{d}}_{,S} + \mathbf{c}_{0,S} \times \bar{\varphi}$, and $\bar{\varphi}_{,S}$ in equation (10.68), ***are truly virtual in nature only when these are interpreted in, as yet unknown, some spinning orthogonal basis set*** that will be called, in the case of static loading, the ***co-rotated bases***, or in the case of dynamic loading, the ***co-spinning bases*** of a ***spatial coordinate system***. To be sure, we must understand that this coordinate system is not what is commonly referred to as the convected coordinate system. Now, this in turn will enable us to identify the additional kinematic variables in the form of a rotation tensor, $\mathbf{R}(S,t)$. Clearly, for $\bar{\mathbf{d}}$, we can choose the kinematic variables, $\mathbf{d}(S,t)$, as the displacement vector of the translational movement.

Thus, our configuration space consists of: the history of an ordered pair $(\mathbf{d}(S,t), \mathbf{R}(S,t))$ with $\mathbf{d}(S,t)$, as the displacement vector, and $\mathbf{R}(S,t)$, as the rotation tensor describing the co-spinning bases, for all points of the 1D beam reference curve over all times of interest.

Of course, this does not mean that we cannot choose to express the components of the force and the moment vectors, and the linear and angular momentum vectors in material coordinates; in fact, we would do just that in our component or operational form of the virtual work equations later. However, in that case, we have to rotate the internal work equations to express these again in the co-rotated (or co-spun) spatial coordinate system. Now, in connection with the choice of the kinematic variables, note the following.

- Transparently, one more decision has been made that has overwhelming importance for computational ease and accuracy. It is about the choice of independent kinematic variables. We could settle for the as yet unknown real translational and bending strain tensors as the independent kinematic variables for the numerical formulation of a subsequent finite element procedure. The major pitfall of this choice is that the real strain tensors must guarantee that a real deformed curve can be constructed with these tensors; in other words, these will then need to additionally satisfy the strain compatibility equations for the reference curve; this can get too complicated for numerical comfort. Thus, from now on, our sought-after kinematic variables will be the real counterparts, $(\mathbf{d}(S,t), \mathbf{R}(S,t))$, of the virtual entities, $\bar{\mathbf{d}}$, $\bar{\varphi}$. The real compatible strains, as will be seen later, can then be easily derived and computed from these parameters satisfying the strain compatibility a priori.

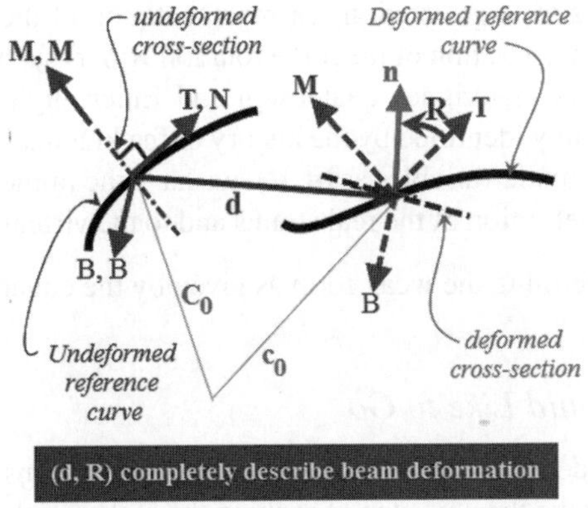

Figure 10.10 Kinematic and configuration space.

10.5.4 Configuration Space, C

So, given a beam problem with the beam having been defined by the beam reference curve and rigid cross-section, the beam kinematics is completely characterized (see Figure 10.10) by a displacement vector field, $\mathbf{d}(S,t)$, tracking each point on the reference curve, and an orthonormal rotational tensor field, $\mathbf{R}(S,t)$, that monitors rotations of the points on the beam cross-section where S is the arc length parameter along the undeformed beam reference curve for any time $t \in \mathbb{R}_+$. Keeping the assumptions listed earlier in mind, we can ascertain the location of any point on the beam at any time during its deformation history with these two fields. Thus, we may define the beam ***configuration space*** as the space defined by $(\mathbf{d}(S,t),\ \mathbf{R}(S,t))$, $S \in C$, $t \in \mathbb{R}_+$. In other words, a configuration space tells us all about the possible deformations that a beam can undergo, and thus consists of all ordered pairs of $(\mathbf{d}(S,t),\ \mathbf{R}(S,t))$ for all points on the beam reference curve for the entire deformation history; a particular motion of a beam reference curve traverses a unique curve on the configuration space. Additionally, assuming adequate smoothness properties on the fields defining the configuration space, we will also be able to establish later the corresponding virtual space that dictates what admissible variations, $(\bar{\mathbf{v}} = \dot{\bar{\mathbf{d}}},\ \bar{\boldsymbol{\omega}} = \dot{\bar{\boldsymbol{\varphi}}})$, of the configuration space we are allowed to consider in our subsequent computational virtual work principle application for our c-type finite element method. We just note that because of the rotational tensor field, $\mathbf{R}(S,t)$ (algebraically, a member of $SO(3)$, the 3D special orthogonal group), the configuration space is not a linear space, but rather a geometrically curved space and what mathematicians call a ***Riemannian*** (we assume $\mathbf{R}(S,t)$ as smooth as necessary) ***differentiable manifold***. As we shall see later, this has major implications in our treatment of the derivatives and the variations of the rotation tensor, and for the linearization of the virtual work equations for our intended computational scheme. The dimension of the configuration space comprises three displacement components and three independent rotational components (because there are six constraints of orthonormality of the rotational tensor) for every point on the beam for all times of interest – a huge dimension.

10.5.5 Kinematic Parameterization: Displacements and Rotations

We note that because of the choice of orthonormality (i.e. $\mathbf{R}^T\mathbf{R} = \mathbf{I}$ where $\mathbf{I}$ is the identity tensor) for $\mathbf{R}(S,t)$, only three degrees of freedom for rotations can be chosen for parameterization. As discussed in the kinematics of large rotation, for representation of the rotational motions, many possibilities exist for parameterization of the finite rotation $\mathbf{R}(S,t)$ such as Rodrigues parameters, Euler–Rodrigues parameters, Rodrigues quaternion and Euler angles; thus, our configuration space for a beam is completely identified by the history of the kinematic parameters, $(\mathbf{d},\mathbf{R})$. Now, by accepting the beam kinematic variables as $(\mathbf{d},\mathbf{R})$, we have the immediate task of relating these real variables through the definition of the real strains and to the virtual strains through the virtual entities $(\bar{\mathbf{v}} = \dot{\bar{\mathbf{d}}},\ \bar{\omega} = \dot{\bar{\varphi}})$ defining the weak form as given by the equation (10.68).

10.5.6 Where We Would Like to Go

As discussed earlier, in order to describe the presumed virtual strains as the variations of some real strain fields, we must take the next step of seeking the real spatial strain fields, the variations of which will guarantee the virtual nature of the energy conjugate strain expressions in the internal virtual work given by equation (10.69). However, before we can do so, we need to equip ourselves with adequate mathematical tools relating to the definitions of curvatures, variations and time derivatives. Thus, we must visit next:

- curvatures, variations and time derivatives
- spatial virtual and real strains.

10.6 Admissible Virtual Space: Curvature, Velocity and Variation

10.6.1 What We Need to Recall

- The vector triple product property: $\mathbf{a}\times\mathbf{b}\bullet\mathbf{c} = \mathbf{b}\times\mathbf{c}\bullet\mathbf{a} = \mathbf{c}\times\mathbf{a}\bullet\mathbf{b}$, $\forall\mathbf{a},\mathbf{b},\mathbf{c}$ vectors
- The vector triple cross product property: $\mathbf{a}\times\mathbf{b}\times\mathbf{c} = (\mathbf{a}\bullet\mathbf{c})\mathbf{b} - (\mathbf{a}\bullet\mathbf{b})\mathbf{c}$, $\forall\mathbf{a},\mathbf{b},\mathbf{c}$ vectors.
- The definition of a rotation tensor, its matrix representation, and algebra, geometry and topology of rotation.
- A rotation matrix, $\mathbf{R}$, corresponding to the three-dimensional special orthogonal matrix group, $SO(3)$, may be parameterized by the Rodrigues vector, $\mathbf{s}$, the axis of the instantaneous rotation.
- The Gateaux derivatives on a vector space are:

$$\boxed{\bar{\mathbf{v}}(\mathbf{x},\bar{\mathbf{x}}) = \lim_{\rho\to 0}\frac{d}{d\rho}\{\mathbf{v}(\mathbf{x}+\rho\bar{\mathbf{x}})\} = \lim_{\rho\to 0}\frac{\{\mathbf{v}(\mathbf{x}+\rho\bar{\mathbf{x}}) - \mathbf{v}(\mathbf{x})\}}{\rho}} \tag{10.72}$$

  where $\bar{\mathbf{v}}(\mathbf{x},\bar{\mathbf{x}})$ is the one-parameter variation of a vector or tensor field, say $\mathbf{v}(\mathbf{x})$, where $\mathbf{x}$ and $\bar{\mathbf{x}}$ belong to a linear vector space; in other words, $\bar{\mathbf{v}}(\mathbf{x},\bar{\mathbf{x}})$ is the directional derivative of $\mathbf{v}(\mathbf{x})$ at $\mathbf{x}$ along the vector direction $\bar{\mathbf{x}}$.
- The ***covariant derivatives*** on a ***curved space*** or ***manifold*** are:

$$\boxed{\bar{\mathbf{v}}\left(\mathbf{x}\equiv\frac{d}{d\xi},\bar{\mathbf{x}}\equiv\frac{d}{d\lambda}\right) = \lim_{\rho\to 0}\frac{\{\mathbf{v}(\lambda_0+\rho)_{parallel\ transported\ to\ \lambda_0} - \mathbf{v}(\lambda_0)\}}{\rho}} \tag{10.73}$$

  where $\bar{\mathbf{v}}(\mathbf{x},\bar{\mathbf{x}})$ is the covariant derivative of $\mathbf{v}(\mathbf{x})$ at $\mathbf{x}$ along the vector direction $\bar{\mathbf{x}}$. The covariant derivative is the generalization of the Gateaux directional derivative of the vector spaces to the curved spaces where $\mathbf{v}(\mathbf{x}+\rho\bar{\mathbf{x}})$ and $\mathbf{v}(\mathbf{x})$ of the definition (10.72) do not belong to the same vector (tangent) space, and hence the difference operation does not have a meaning. In fact, the vector, itself, does not have a meaning other than as a tangent operation to a curve. Thus, and for invariance, the generalization is effected through the choice of affinely parameterized unique curves; this makes the definition of the covariant derivative path-dependent. However, as a practical scheme for computing the covariant derivatives of vector fields (or tensor fields), all we need to do is to include the derivatives of the base vectors (or base tensors) as they are the tangents (or the tensor product of these) to the coordinate curves on the manifold.

- For notational uniformity, we denote all variations of entities by a bar superscripted on it. Furthermore all entities measured in $\mathbf{G}_1 = \mathbf{T}, \mathbf{G}_2 = \mathbf{M}, \mathbf{G}_3 = \mathbf{B}$ frame are superscripted as ${}^t(\bullet)$, and those in the rotating $\mathbf{g}_1 = \mathbf{t}, \mathbf{g}_2 = \mathbf{m}_n, \mathbf{g}_3 = \mathbf{b}_n$ coordinate system as ${}^n(\bullet)$. However, for the sake of uncluttered readability, sometimes the superscripts may be dropped, but the context should help determine the appropriate underlying coordinate system.

10.6.2 Curvature and Angular Velocity: The Admissible Virtual Space

In the development of the beam virtual work principle necessary for numerical analysis by a variational method such as the c-type finite element method, we will need the derivatives and the variations of the parameters of the configuration space. The configuration space parameters are the continuous displacement vector field, $\mathbf{d}(S, t)$, and the continuous rotation tensor field, $\mathbf{R}(S, t)$, of the centroidal axis representing the beam reference curve where S is the arc length parameter along the centroidal axis as the beam reference curve, and $t \in \mathbb{R}_+$ is the time parameter. For the dynamic motion, we will need both the temporal and the spatial derivatives, and the variations of these fields. For quasi-static motion, the fields are independent of time, that is, $\mathbf{d} = \mathbf{d}(S)$ and $\mathbf{R} = \mathbf{R}(S)$, in which case we will need the variations and the spatial derivatives of the fields for the subsequent real and the virtual strain measures, and so on.

10.6.2.1 Real Displacement Vector Field and Admissible Virtual Space

The real displacement vector field, $\mathbf{d}(S, t)$, describing the translational movement of the points on an arbitrary configuration of the beam curve, for all times of interest, belongs to the three-dimensional Euclidean space. Thus, we can construct the displacement portion of the virtual space as the vector space containing the perturbed or the infinitesimal displacement vector field that is 'superimposed' on the real displacement vector field to obtain the perturbed or virtual configuration; it will be identified as $\bar{\mathbf{d}}(S, t)$. The method of superposition is clearly defined by the vector addition operation.

10.6.2.2 Real Rotation Tensor, Manifold Lie Group, *SO*(3) and Admissible Virtual Space

The real rotation field, describing the independent rotation of the triad of base vectors associated with the beam reference curve belongs algebraically to the special orthogonal group, $SO(3)$, and geometrically to a curved space or manifold. Thus, the admissible virtual space for rotation is considerably complicated compared to its displacement counterpart described in the previous section. We will need to define the infinitesimal (later, they will also turn out to be the instantaneous) rotations that can serve as the required virtual rotations which can then be 'superposed' on the real rotations to described the perturbed or virtual configurations. This clearly needs identification of the space to which these infinitesimal rotations belong, and additionally, the method of superposition. In other words, in an Euclidean ('flat') space, we have identified, for the displacement vector field, superposition simply as the vector addition operation, the question for us to resolve is if we can do the same for the rotation tensor field. It turns out that with appropriate parameterization for the rotation field, the answer is yes, although certain singularity issues must be tackled. Moreover, we will need for future use the curvature and angular velocity fields, that is, the spatial and temporal derivatives of the rotation field. In what follows, we try to precisely meet these goals. Now, recall that the $SO(3)$ group is the set of all 3×3 real proper orthogonal matrices, that is, $\mathbf{R} \in SO(3) \Rightarrow \mathbf{R}^T \mathbf{R} = \mathbf{I}$, with $\mathbf{I}$ as the identity matrix, and the rotation matrix has positive unit determinant, that is, $\det \mathbf{R} = +1$. The four properties of a group: the existence of the identity, the existence of the inverse element, the associative property of the product and closure, are all satisfied by the rotation matrices under the product rule given by the usual matrix multiplication. The identity element: $\mathbf{I}$ with $\det \mathbf{I} = 1$, the inverse element: $\mathbf{R}^{-1} = \mathbf{R}^T$ for each rotation $\mathbf{R}$ belonging to the group, and the associative property of the matrices, satisfy the first three conditions. Now for closure: with $\mathbf{R}, \mathbf{S} \in SO(3)$, two proper orthogonal matrices, we have $\mathbf{R}\,\mathbf{S}$ also orthonormal because $(\mathbf{R}\,\mathbf{S})^T\,(\mathbf{R}\,\mathbf{S}) = \mathbf{S}^T(\mathbf{R}^T\mathbf{R})\,\mathbf{S} = \mathbf{I}$, and proper because: $\det(\mathbf{R}\,\mathbf{S}) = \det\,\mathbf{R}\,\det\,\mathbf{S} = 1$; in other words, $\mathbf{R} \in SO(3),\ \mathbf{S} \in SO(3) \Rightarrow \mathbf{R}\,\mathbf{S} \in SO(3)$. Additionally, a set of invertible tensor field is a continuous manifold. Since a group which is also a continuous manifold is called a Lie group, $SO(3)$ is a Lie group.

10.6.3 *Tangent Operator and Exponential Map*

Let us recall that on a curved space, geometrical objects such as vectors and tensors, must be defined locally instead of bi-local definition in the neighborhood as in the flat Euclidean space. In other words, in an Euclidean ('flat') space, we can think of, say, a vector as joined by two neighboring points, but in a curved space a similar definition will yield a vector that may not even completely belong to the space. This necessitated the generalization from the definition of a vector as an arrow link between two neighboring points to a tangent operation to a curve (to be more specific, an affine curve for invariance or coordinate independence) at a point on the manifold.

10.6.3.1 Spatial tangent operator

Let the one spatial parameter family of the rotation tensor (or matrix) function $\mathbf{R}(\rho) : \mathbb{R} \rightarrow SO(3)$ generate a continuous set of points defining a curve on $SO(3)$, such that it passes through $\mathbf{R}(0) = \mathbf{R}$ at $\rho = 0$ where $\mathbf{R} \in SO(3)$. Then, $\mathbf{R},_\rho$, defined by the covariant derivative: $\mathbf{R},_\rho \equiv \frac{d}{d\rho}\mathbf{R}(\rho)|_{\rho=0}$, is a tangent vector to $SO(3)$ at $\mathbf{R}$. The set of all tangent vectors at $\mathbf{R}$, corresponding to various smooth curves passing though it forms a vector space, denoted by $\mathbf{T_R}SO(3)$. Now, by taking the derivative of the orthogonality property, $\mathbf{R}^T\,\mathbf{R} = \mathbf{I}$, and rearranging, we have:

$$\boxed{\mathbf{R},_\rho = \mathrm{K}\,\mathbf{R} \quad \Rightarrow \mathrm{K} \equiv \mathbf{R},_\rho\,\mathbf{R}^T = \mathbf{R},_\rho\,\mathbf{R}^{-1}} \tag{10.74}$$

where we have introduced

Definition

- ***the tangent (linearized) operator***, K, the skew-symmetric matrix (because, with $\mathbf{R}\mathbf{R}^T = \mathbf{I}$, we have: $\mathrm{K} + \mathrm{K}^T = \mathbf{R},_\rho\,\mathbf{R}^T + \mathbf{R}\,\mathbf{R},_\rho^T = (\mathbf{R}\mathbf{R}^T),_\rho = \mathbf{0}$); K is the instantaneous and linearized skew-symmetric tensor at a rotation, $\mathbf{R} \in SO(3)$.

Now, if we consider a constant tangent operator, K, that is, independent of the parameter ρ, we have from equation (10.74), $\mathbf{R}(\rho)_{,\rho} = \mathrm{K}\,\mathbf{R}(\rho)$, which has a solution given by:

$$\boxed{\mathbf{R}(\rho) = e^{\rho \mathrm{K}} \equiv \left(\mathbf{I} + \rho \mathrm{K} + \frac{(\rho \mathrm{K})^2}{2!} + \dots\right)} \tag{10.75}$$

Thus, each element of the tangent space is related to a corresponding element of the curve space by an exponential map given by equation (10.75). In other words, there always exists a skew-symmetric matrix that can represent a rotation matrix by this exponential map. More specifically, for a rotation matrix, $\mathbf{R} \in SO(3)$, there exists a linearized skew-symmetric matrix, $\mathbf{K} \in so(3)$ related by:

$$\boxed{\mathbf{R} = e^{\mathbf{K}}} \tag{10.76}$$

10.6.3.2 Temporal tangent operator

Similarly, let the one temporal (time) parameter family of the rotation tensor (or matrix) function $\mathbf{R}(t) : \mathbb{R} \rightarrow SO(3)$ generate a continuous set of points defining a curve on $SO(3)$, such that it passes through $\mathbf{R}(0) = \mathbf{R}$ at $t = 0$ where $\mathbf{R} \in SO(3)$. Then, $\dot{\mathbf{R}} \equiv \frac{d}{dt}\mathbf{R}$, defined by the covariant derivative: $\dot{\mathbf{R}} \equiv \frac{d\mathbf{R}(t)}{dt}|_{t=0}$, is the tangent vector to $SO(3)$ at $\mathbf{R}$. The set of all tangent vectors at $\mathbf{R}$, corresponding to various smooth curves passing through it forms a vector space, denoted by $\mathbf{T_R}SO(3)$. Now, by taking the temporal derivative of the orthogonality property, $\mathbf{R}^T\,\mathbf{R} = \mathbf{I}$, and rearranging, we have:

$$\boxed{\dot{\mathbf{R}} = \boldsymbol{\Omega}\,\mathbf{R} \quad \Rightarrow \boldsymbol{\Omega} \equiv \dot{\mathbf{R}}\,\mathbf{R}^T = \dot{\mathbf{R}}\,\mathbf{R}^{-1}} \tag{10.77}$$

where we have introduced

Definition

- the ***tangent (linearized) operator***, $\boldsymbol{\Omega}$, the skew-symmetric matrix (because, with $\mathbf{R}\mathbf{R}^T = \mathbf{I}$, we have: $\boldsymbol{\Omega} + \boldsymbol{\Omega}^T = \dot{\mathbf{R}}\,\mathbf{R}^T + \mathbf{R}\,\dot{\mathbf{R}}^T = \frac{d}{dt}(\mathbf{R}\mathbf{R}^T) = \mathbf{0}$); $\boldsymbol{\Omega}$ is the instantaneous and linearized skew-symmetric tensor at a rotation, $\mathbf{R} \in SO(3)$.

Because the tangent space is populated with the skew-symmetric matrices, the dimension of the tangent space is three (with only $n(n-1)/2$ elements in the matrix with $n = 3$ for the 3D space). Note also that the tangent space is a linear vector space. Now, if we consider a constant tangent operator, $\boldsymbol{\Omega}$, that is, independent of the parameter t, we have from equation (10.77), $\dot{\mathbf{R}}(t) = \boldsymbol{\Omega}\,\mathbf{R}(t)$, which has a solution given by:

$$\boxed{\mathbf{R}(t) = e^{t\boldsymbol{\Omega}} \equiv \left(\mathbf{I} + t\boldsymbol{\Omega} + \frac{(t\boldsymbol{\Omega})^2}{2!} + \dots\right)} \tag{10.78}$$

Thus, each element of the tangent space is related to a corresponding element of the curve space by an exponential map given by equation (10.78). In other words, there always exists a skew-symmetric matrix that can represent a rotation matrix by this exponential map. More specifically, for a rotation matrix, $\mathbf{R} \in SO(3)$, there exists a linearized skew-symmetric matrix, $\boldsymbol{\Omega} \in so(3)$ related by:

$$\boxed{\mathbf{R} = e^{\boldsymbol{\Omega}}} \tag{10.79}$$

Remark

- Although equations (10.76) and (10.79) relate the rotation matrix with the corresponding skew-symmetric matrix of the tangent space through the exponential map, there is, as derived earlier, a better way of relating the two by expressing the exponential map by the Rodrigues vector or Rodrigues quaternion parameters of the rotation matrix through Cayley's theorem.

10.6.3.3 Tangent Space at Identity

The tangent space at identity, that is, at $\mathbf{R} = \mathbf{I}$, is given the special name, ***Lie algebra*** (recall that a vector space with a vector multiplication, $[\bullet, \bullet]$, defined on it is an algebra) and denoted by $so(3)$. The reason for this special treatment of the neighborhood at identity, $\mathbf{I}$, is that the local structure of $SO(3)$, the continuous rotation group, can be constructed out of this infinitesimal tangent space, $so(3)$, by the exponential map given by equation (10.78); in other words, all the information of the group is contained in the algebra. Thus, $so(3)$ consists of 3×3 skew-symmetric matrices; as any 3×3 skew-symmetric matrix has a corresponding 3×1 axial vector, $so(3)$ is isomorphic to the 3D Euclidean vector space. More generally, a Lie algebra is a vector space, $\mathcal{V}$, with a bilinear map, $[\bullet, \bullet] : \mathcal{V} \times \mathcal{V} \to \mathcal{V}$, known as the ***Lie bracket***, so that for every $a, b, c \in \mathcal{V} \Rightarrow$, 1) $[a\ ,\ a] = 0$, 2) $[a, [b, c]]\ +\ [b, [c, a]]\ +\ [c, [a, b]] = 0$ (the Jacobi identity). From condition (1) and the bilinearity of the map, $[a + b, a + b] = [a, b] + [b, a] = 0$ implying $[a, b] = -[b, a]$. For square matrix Lie group and Lie algebra, the Lie bracket reduces to the matrix commutator: $[\mathbf{A}, \mathbf{B}] = \mathbf{AB} - \mathbf{BA}, \quad \forall \mathbf{A}, \mathbf{B} \in so(3)$. With the notation, $[\mathbf{a}] \in so(3)$ as the skew-symmetric matrix corresponding to the axial vector, $\mathbf{a}$, we can easily show that $\mathbf{AB} - \mathbf{BA} = [\mathbf{a} \times \mathbf{b}]$.

10.6.4 The Admissible Virtual Space

Based on our discussion about the tangent space at identity, $so(3)$, the admissible virtual space is defined by this linear space of the configuration space made up of the union set of all the possible configurations of the beam reference curve with rigid cross-section defined by the ordered pair $(\mathbf{d}, \mathbf{R})$, that is, the displacement vector field and the rotation tensor field, respectively. In other words, every ordered pair, $(\bar{\mathbf{d}}, \bar{\mathbf{R}})$ of the virtual displacement vector, $\bar{\mathbf{d}}$, and the infinitesimal, skew-symmetric rotation tensor, $\bar{\mathbf{R}}$, belongs to this linear space. This, then, takes care of the part of our original questions about how to define admissible virtual space. But, we still have not determined how to "superimpose" the virtual rotations onto the current real rotation to obtain the perturbed or virtual configuration space necessary for the subsequent virtual work principle. As it turns out, there are two possible way to accomplish this and we have to conveniently choose one. This aspect of the study we will discuss next.

10.6.5 Adjoint Map and Material and Spatial Variations

We have already seen through equation (10.78) that the elements of the Lie group, $SO(3)$, are related to the corresponding elements of its Lie algebra, $so(3)$, by the exponential map. Another map known as the ***adjoint map*** also relates the elements of $SO(3)$ with the corresponding elements of $so(3)$: If an arbitrary rotation matrix, $\mathbf{R} \in SO(3)$, and its corresponding skew-symmetric matrix,

$\bar{\boldsymbol{\Psi}} \in so(3)$, are given then we have $\bar{\boldsymbol{\Phi}}$ defined by the adjoint map: $\bar{\boldsymbol{\Phi}} \equiv \mathbf{R}\,\bar{\boldsymbol{\Psi}}\,\mathbf{R}^T$, also an element of $so(3)$. We can easily see that because $(\bar{\boldsymbol{\Phi}} + \bar{\boldsymbol{\Phi}}^T = \mathbf{R}\,\bar{\boldsymbol{\Psi}}\,\mathbf{R}^T + \mathbf{R}\,\bar{\boldsymbol{\Psi}}^T\,\mathbf{R}^T == \mathbf{R}\,\bar{\boldsymbol{\Psi}}\,\mathbf{R}^T - \mathbf{R}\,\bar{\boldsymbol{\Psi}}\,\mathbf{R}^T = \mathbf{0})$, the symmetric part vanishes identically; it is a member of $so(3)$. Thus, we have: $\bar{\boldsymbol{\Phi}}\,\mathbf{R} = \mathbf{R}\,\bar{\boldsymbol{\Psi}}$, and using similar to equation (10.74), we have two ways to "superpose" to represent the virtual rotation, $\bar{\mathbf{R}}$, by:

$$\bar{\mathbf{R}} = \bar{\boldsymbol{\Phi}}\,\mathbf{R} = \mathbf{R}\,\bar{\boldsymbol{\Psi}} \tag{10.80}$$

In the first equality of equation (10.80), known as the ***left translation***, the virtual rotation, $\bar{\boldsymbol{\Phi}}$, is being superposed on the current rotation to obtain the updated rotation; we use this relationship in our subsequent analysis, but the components are measured in the ***undeformed and fixed material frame***, $\mathbf{G_1} = \mathbf{T}, \mathbf{G_2} = \mathbf{M}, \mathbf{G_3} \equiv \mathbf{B}$. In the second equality of equation (10.80), known as the ***right translation***, the current rotation is being superposed on the virtual rotation, $\bar{\boldsymbol{\Psi}}$, to obtain the updated rotation. We do not use this relationship further in our subsequent analysis and hence will not discuss it any further. Finally, then, the finite rotation, $\mathbf{R}$, belongs to the spherical curved space, $SO(3)$, and the instantaneous and infinitesimal virtual rotation, $\bar{\boldsymbol{\Phi}}$, represented by a skew symmetric matrix, belongs to the tangent space at identity, the Lie algebra.

10.6.6 Rotational Variation: $\bar{\boldsymbol{\Phi}}$ and $\bar{\boldsymbol{\varphi}}$

As indicated above, we choose the following form for rotational variation in all subsequent applications:

$$\boxed{\bar{\mathbf{R}} = \bar{\boldsymbol{\Phi}}\,\mathbf{R} \quad \Leftrightarrow \quad \bar{\boldsymbol{\Phi}} \equiv \bar{\mathbf{R}}\,\mathbf{R}^T} \tag{10.81}$$

For the 3D rotation tensors, the corresponding skew-symmetric tensors only have three significant variables (i.e. the number of variables equal the dimension of the space), and thus can have axial vector representation. Let $\bar{\boldsymbol{\varphi}} \equiv \{\bar{\varphi}_1 \quad \bar{\varphi}_2 \quad \bar{\varphi}_3\}^T$ be such an axial vector corresponding to the skew-symmetric tensor, $\bar{\boldsymbol{\Phi}}$, with $\bar{\boldsymbol{\Phi}}\,\bar{\boldsymbol{\varphi}} = \mathbf{0}$. Knowing that $\bar{\boldsymbol{\Phi}} = \begin{bmatrix} 0 & -\bar{\varphi}_3 & \bar{\varphi}_2 \\ \bar{\varphi}_3 & 0 & -\bar{\varphi}_1 \\ -\bar{\varphi}_2 & \bar{\varphi}_1 & 0 \end{bmatrix}$, the equality can be readily verified.

10.6.7 Rotational Spatial Derivatives, that is, Curvatures: $\mathbf{K}$ and $\mathbf{k}$

Next, we can define the curvature tensor in two different ways as: $\mathbf{K} \equiv \mathbf{R}_{,S}\ \mathbf{R}^T$ or $\mathbf{L} \equiv \mathbf{R}^T\,\mathbf{R}_{,S}$, and thus, the spatial derivatives of variables of a beam, in the form of a reference curve and rigid cross-sections, can have two representations. We choose the following spatial form in all subsequent applications:

$$\boxed{\mathbf{R}_{,S} = \mathbf{K}\,\mathbf{R} \quad \Leftrightarrow \quad \mathbf{K} \equiv \mathbf{R}_{,S}\ \mathbf{R}^T} \tag{10.82}$$

As before, let $\mathbf{k}$ be the axial vector corresponding to the skew-symmetric tensor, $\mathbf{K}$, with $\mathbf{K}\,\mathbf{k} = \mathbf{0}$.

10.6.8 Rotational Time Derivative, Angular Velocity: $\boldsymbol{\Omega}$ and $\boldsymbol{\omega}$

Similarly, we can define the angular velocity tensor in two different ways as: $\boldsymbol{\Omega} \equiv \dot{\mathbf{R}}\,\mathbf{R}^T$ or $\Xi \equiv \mathbf{R}^T\,\dot{\mathbf{R}}$, where $\dot{(\bullet)} \equiv \frac{d}{dt}(\bullet)$, and thus, the time derivatives of variables of a beam, in the form of a reference curve and rigid cross-sections, can have two representations. We choose the following form in all subsequent applications:

$$\boxed{\dot{\mathbf{R}} = \mathbf{\Omega}\,\mathbf{R} \quad \Leftrightarrow \quad \mathbf{\Omega} = \dot{\mathbf{R}}\,\mathbf{R}^T} \tag{10.83}$$

As before, let $\boldsymbol{\omega}$ be the axial vector corresponding to the skew-symmetric tensor, $\mathbf{\Omega}$, with $\mathbf{\Omega}\,\boldsymbol{\omega} = \mathbf{0}$.

10.6.9 *Coordinate Frame Rotation and Objective Variations and Derivatives*

For future application, let us suppose that an initial orthonormal coordinate frame, $\{\mathbf{G}_i(S)\}$, $i = 1, 2, 3$, attached at a material point, S, to a material curve in space; let us also suppose that at time t, a current coordinate system, $\{\mathbf{g}_i(S, t)\}$, $i = 1, 2, 3$, is obtained from the initial coordinate frame by pure rotation such that $\mathbf{g}_i = \mathbf{R}\,\mathbf{G}_i$, $i = 1, 2, 3$ where $\mathbf{R}(S, t)$ is the rotation tensor. Now, let $\mathbf{x}(S, t)$ be any vector field attached to the body at point S. If an observer is sitting on the body such that it also co-spins as above, we would be interested in how it sees the variations and time rate of change in $\mathbf{x}(S, t)$; we will call this the ***objective variations***, and the ***objective time rate***, respectively, of vector fields. As it turns out, it is also intimately connected with what we have described above as the ***Lie derivative***. However, before we delve into the derivations of objective rate, we note the following.

10.6.9.1 Variation of Current Base Vectors: $\{\mathbf{g}_i(S,t)\}, i = 1,2,3$

From equation (10.81), the variation of $\mathbf{g}_i$ is given as:

$$\overline{\mathbf{g}_i} = \bar{\mathbf{R}}\,\mathbf{G}_i = \bar{\mathbf{\Phi}}\,\mathbf{R}\,\mathbf{G}_i = \bar{\mathbf{\Phi}}\,\mathbf{g}_i = \bar{\boldsymbol{\varphi}} \times \mathbf{g}_i \tag{10.84}$$

10.6.9.2 Spatial Derivative of Current Base Vectors: $\{\mathbf{g}_i(S,t)\}, i = 1,2,3$

From equation (10.82), the spatial derivative of $\mathbf{g}_i$ is given as:

$$\mathbf{g}_i,_S = \mathbf{R},_S\,\mathbf{G}_i = \mathbf{K}\,\mathbf{R}\,\mathbf{G}_i = \mathbf{K}\,\mathbf{g}_i = \mathbf{k} \times \mathbf{g}_i \tag{10.85}$$

10.6.9.3 Time Derivative of Current Base Vectors: $\{\mathbf{g}_i(S,t)\}, i = 1,2,3$

From equation (10.83), the time derivatives of $\mathbf{g}_i$ is given as:

$$\dot{\mathbf{g}}_i = \dot{\mathbf{R}}\,\mathbf{G}_i = \mathbf{\Omega}\,\mathbf{R}\,\mathbf{G}_i = \mathbf{\Omega}\,\mathbf{g}_i = \boldsymbol{\omega} \times \mathbf{g}_i \tag{10.86}$$

10.6.9.4 Objective Variation, $\overset{\mathrm{v}}{(\bullet)}$

Now, let us also observe that for any vector field $\mathbf{x}$ described in the current rotated coordinate system, that is, $\mathbf{x} = x^i\,\mathbf{g}_i$, we have, using equation (10.84), for variation in a spatial coordinate system:

$$\underbrace{\bar{\mathbf{x}}}_{\text{Total variation}} = \bar{x}^i\,\mathbf{g}_i + x^i\,\bar{\mathbf{g}}_i = \bar{x}^i\,\mathbf{g}_i + \bar{\boldsymbol{\varphi}} \times x^i\,\mathbf{g}_i = \bar{x}^i\,\mathbf{g}_i + \bar{\boldsymbol{\varphi}} \times \mathbf{x}$$

$$\Rightarrow \quad \overset{\mathbf{v}}{\mathbf{x}} \equiv \underbrace{\bar{x}^i\,\mathbf{g}_i}_{\substack{\text{Variation in rotated}\\ \text{coordinate system, i.e.}\\ \text{objective variation}}} = \underbrace{\bar{\mathbf{x}}}_{\text{Total variation}} - \underbrace{\bar{\boldsymbol{\varphi}} \times \mathbf{x}}_{\text{Spinning variation}}$$

$$= \mathbf{R}\left\{\mathbf{R}^T\bar{\mathbf{x}} + \left(-\mathbf{R}^T\bar{\boldsymbol{\Phi}}\right)\mathbf{x}\right\} = \mathbf{R}\left\{\mathbf{R}^T\bar{\mathbf{x}} + \bar{\mathbf{R}}^T\,\mathbf{x}\right\} \tag{10.87}$$

$$= \underbrace{\mathbf{R}}_{\text{Push forward}} \underbrace{\left(\overline{\underbrace{\mathbf{R}^T}_{\text{Pull back}}\;\mathbf{x}}\right)}_{\text{Variation}}$$

From equation (10.87), we introduce below what is known as the ***objective variation***, or the ***co-rotated variation***.

Definition: ***Objective variation***, $\overset{\mathbf{v}}{(\bullet)}$:

$$\boxed{\overset{\mathbf{v}}{(\bullet)} \equiv \overline{(\bullet)} - \bar{\boldsymbol{\varphi}} \times (\bullet)} \tag{10.88}$$

10.6.9.5 Objective Time Rate, $\overset{\mathbf{t}}{(\bullet)}$

Similarly, using equation (10.84), for the time derivative in the current rotated coordinate system, we have:

$$\underbrace{\dot{\mathbf{x}}}_{\text{Total derivative}} = \dot{x}^i\,\mathbf{g}_i + x^i\,\dot{\mathbf{g}}_i = \dot{x}^i\,\mathbf{g}_i + \boldsymbol{\omega} \times x^i\,\mathbf{g}_i = \dot{x}^i\,\mathbf{g}_i + \boldsymbol{\omega} \times \mathbf{x}$$

$$\Rightarrow \quad \overset{\mathbf{t}}{\mathbf{x}} \equiv \underbrace{\dot{x}^i\,\mathbf{g}_i}_{\substack{\text{Derivative in rotated}\\ \text{coordinate system, i.e.}\\ \text{objective derivative}}} = \underbrace{\dot{\mathbf{x}}}_{\text{Total derivative}} - \underbrace{\boldsymbol{\omega} \times \mathbf{x}}_{\substack{\text{Spinning observer}\\ \text{correction}}}$$

$$= \mathbf{R}\left\{\mathbf{R}^T\dot{\mathbf{x}} + \left(-\mathbf{R}^T\boldsymbol{\Omega}\right)\mathbf{x}\right\} \tag{10.89}$$

$$= \mathbf{R}\left\{\mathbf{R}^T\dot{\mathbf{x}} + \dot{\mathbf{R}}^T\,\mathbf{x}\right\} = \underbrace{\mathbf{R}}_{\text{Push forward}} \underbrace{\left(\overset{\bullet}{\overbrace{(\underbrace{\mathbf{R}^T}_{\text{Pull back}}\;\mathbf{x})}}\right)}_{\text{Derivative}}$$

From equation (10.89), we introduce below what is known as the ***objective time rate***, or the ***co-rotated time rate***.

Definition: ***Objective time rate***, $\overset{\mathbf{t}}{(\bullet)}$:

$$\boxed{\overset{\mathbf{t}}{(\bullet)} \equiv \frac{\partial}{\partial t}(\bullet) = (\bullet) - \omega \times (\bullet)} \tag{10.90}$$

Remark

- The idea behind the objective variation (or objective time derivative) in the current rotated body attached coordinate system is that it is the net variation (or time derivative) in the co-rotated coordinate system, that is, the total (material) variation minus the variation of the spinning coordinate system itself. In other words, it is the variation (or time derivative) observed by an observer attached to the spinning coordinate system given by the base vectors, $\{\mathbf{g}_i\}$.
- The objective variation (or time derivative) may also be seen as the ordered sequence of operations as the "pulling back" to the material coordinate system, taking the variation, and finally "pushing forward" to the current coordinate system, as in the definition of the Lie derivative.
- Note, however, that, as we will see later, the components of all involved vector and tensor quantities are measured in terms of the undeformed coordinate system.

10.6.9.6 Relations between $\bar{\boldsymbol{\Phi}}$ and $\bar{\varphi}$ and K and k

From equation (10.81) and equation (10.82), and noting that $\mathbf{R}\,\mathbf{R}^T = \mathbf{I}$, $\mathbf{I}$ being the 3D identity tensor, we have for the variation of the curvature:

$$\begin{aligned}
\bar{\boldsymbol{\Phi}}_{,S} &= (\bar{\mathbf{R}}\,\mathbf{R}^T)_{,S} = \bar{\mathbf{R}}_{,S}\,\mathbf{R}^T + \bar{\mathbf{R}}\,\mathbf{R}^T{}_{,S} = \overline{\mathbf{K}\,\mathbf{R}}\,\mathbf{R}^T + \bar{\boldsymbol{\Phi}}\,\mathbf{R}\,(\mathbf{R}^T\,\mathbf{K}^T) \\
&= \bar{\mathbf{K}}\,\mathbf{R}\,\mathbf{R}^T + \mathbf{K}\,\bar{\boldsymbol{\Phi}}\,\mathbf{R}\,\mathbf{R}^T - \bar{\boldsymbol{\Phi}}\,\mathbf{R}\,\mathbf{R}^T\,\mathbf{K} = \bar{\mathbf{K}} + \underbrace{\mathbf{K}\,\bar{\boldsymbol{\Phi}} - \bar{\boldsymbol{\Phi}}\,\mathbf{K}}_{\textit{Lie Bracket}}
\end{aligned} \tag{10.91}$$

implying, with the application of the definition of objective variation from equation (10.88) to equation (10.91), and the definition of the axial vectors and the Lie brackets, that:

$$\boxed{\begin{aligned}
\bar{\varphi}_{,S} &= \bar{\mathbf{k}} - \bar{\varphi} \times \mathbf{k} = \mathbf{R}\{\mathbf{R}^T\bar{\mathbf{k}} + (-\mathbf{R}^T\bar{\boldsymbol{\Phi}})\mathbf{k}\} = \mathbf{R}\{\mathbf{R}^T\,\bar{\mathbf{k}} + \bar{\mathbf{R}}^T\,\mathbf{k}\} \\
&= \underbrace{\mathbf{R}}_{\text{Push forward}}\ \underbrace{\left(\overline{\underbrace{\mathbf{R}^T}_{\text{Pull back}}\ \mathbf{k}}\right)}_{\text{Variation}} \equiv \overset{\mathbf{v}}{\bar{\mathbf{k}}}
\end{aligned}} \tag{10.92}$$

10.6.9.7 Relations between Ω and ω and K and k

From equation (10.82) and equation (10.83), and noting that $\mathbf{R}\,\mathbf{R}^T = \mathbf{I}$, $\mathbf{I}$ being the 3D identity tensor, we have for the spatial derivative of the angular velocity:

$$\begin{aligned}
\boldsymbol{\Omega}_{,S} &= (\dot{\mathbf{R}}\,\mathbf{R}^T)_{,S} = \dot{\mathbf{R}}_{,S}\,\mathbf{R}^T + \dot{\mathbf{R}}\,\mathbf{R}^T{}_{,S} = \overset{\bullet}{\overbrace{\mathbf{K}\,\mathbf{R}}}\,\mathbf{R}^T + \boldsymbol{\Omega}\,\mathbf{R}\,(\mathbf{R}^T\,\mathbf{K}^T) \\
&= \dot{\mathbf{K}}\,\mathbf{R}\,\mathbf{R}^T + \mathbf{K}\,\boldsymbol{\Omega}\,\mathbf{R}\,\mathbf{R}^T - \boldsymbol{\Omega}\,\mathbf{R}\,\mathbf{R}^T\,\mathbf{K} = \dot{\mathbf{K}} + \underbrace{\mathbf{K}\,\boldsymbol{\Omega} - \boldsymbol{\Omega}\,\mathbf{K}}_{\text{Lie bracket}}
\end{aligned} \tag{10.93}$$

implying, with the application of the definition of objective time derivative from equation (10.90) to equation (10.93), and the definition of the axial vectors and the Lie brackets, that:

$$\boxed{\begin{aligned} \boldsymbol{\omega}_{,S} &= \overset{\bullet}{\mathbf{k}} - \boldsymbol{\omega} \times \mathbf{k} = \mathbf{R}\{\mathbf{R}^T \overset{\bullet}{\mathbf{k}} + (-\mathbf{R}^T\boldsymbol{\Omega})\mathbf{k}\} = \mathbf{R}\{\mathbf{R}^T \overset{\bullet}{\mathbf{k}} + \overset{\bullet}{\mathbf{R}^T} \mathbf{k}\} \\ &= \underbrace{\mathbf{R}}_{\text{Push forward}} \underbrace{\left(\underbrace{\mathbf{R}^T}_{\text{Pull back}} \mathbf{k}\right)}_{\text{Variation}} \equiv \overset{t}{\mathbf{k}} \end{aligned}} \tag{10.94}$$

Taking variation of $\boldsymbol{\omega}_{,S}$ in equation (10.94), and observing that both variation and derivative are linear operations (and hence the order of operation can be switched), we have:

$$\boxed{\bar{\boldsymbol{\omega}}_{,S} = \overline{\overset{t}{\mathbf{k}}}} \tag{10.95}$$

10.6.9.8 Relations between $\boldsymbol{\Omega}$ and $\boldsymbol{\omega}$ and $\bar{\boldsymbol{\Phi}}$ and $\bar{\boldsymbol{\varphi}}$

From equation (10.81) and equation (10.83), and noting that $\mathbf{R}\,\mathbf{R}^T = \mathbf{I}$, $\mathbf{I}$ being the 3D identity tensor, we have for the variation of the angular velocity:

$$\begin{aligned} \bar{\boldsymbol{\Omega}} &= \overline{(\overset{\bullet}{\mathbf{R}}\,\mathbf{R}^T)} = \overset{\bullet}{\bar{\mathbf{R}}}\,\mathbf{R}^T + \overset{\bullet}{\mathbf{R}}\,\bar{\mathbf{R}}^T = \overset{\bullet}{\overline{\bar{\boldsymbol{\Phi}}\,\mathbf{R}}}\,\mathbf{R}^T + \boldsymbol{\Omega}\,\mathbf{R}\,(\mathbf{R}^T\,\bar{\boldsymbol{\Phi}}^T) \\ &= \overset{\bullet}{\bar{\boldsymbol{\Phi}}}\,\mathbf{R}\,\mathbf{R}^T + \bar{\boldsymbol{\Phi}}\,\overset{\bullet}{\mathbf{R}}\,\mathbf{R}^T - \boldsymbol{\Omega}\,\mathbf{R}\,\mathbf{R}^T\,\bar{\boldsymbol{\Phi}} = \overset{\bullet}{\bar{\boldsymbol{\Phi}}} + \underbrace{\bar{\boldsymbol{\Phi}}\,\boldsymbol{\Omega} - \boldsymbol{\Omega}\,\bar{\boldsymbol{\Phi}}}_{\text{Lie bracket}} \end{aligned} \tag{10.96}$$

implying, with the application of the definition of objective time derivative from equation (10.90) to equation (10.96), and the definition of the axial vectors and the Lie brackets, that:

$$\boxed{\begin{aligned} \bar{\boldsymbol{\omega}} &= \overset{\bullet}{\bar{\boldsymbol{\varphi}}} - \boldsymbol{\omega} \times \bar{\boldsymbol{\varphi}} = \mathbf{R}\{\mathbf{R}^T\,\overset{\bullet}{\bar{\boldsymbol{\varphi}}} + (-\mathbf{R}^T\boldsymbol{\Omega})\bar{\boldsymbol{\varphi}}\} \\ &= \mathbf{R}\{\mathbf{R}^T\,\overset{\bullet}{\bar{\boldsymbol{\varphi}}} + \overset{\bullet}{\mathbf{R}^T}\,\bar{\boldsymbol{\varphi}}\} = \underbrace{\mathbf{R}}_{\text{Push forward}} \underbrace{\overset{\bullet}{\overbrace{(\underbrace{\mathbf{R}^T}_{\text{Pull back}}\,\bar{\boldsymbol{\varphi}})}}}_{\text{Derivative}} \equiv \overset{t}{\bar{\boldsymbol{\varphi}}} \end{aligned}} \tag{10.97}$$

Taking spatial derivative of $\bar{\boldsymbol{\omega}}$ in equation (10.97), and observing that both variation and derivative are linear operations (and hence the order of operation can be switched), we have:

$$\boxed{\bar{\boldsymbol{\omega}}_{,S} = \overset{t}{\bar{\boldsymbol{\varphi}}_{,S}}} \tag{10.98}$$

Thus, finally, from equation (10.95) and equation (10.98), we have:

$$\boxed{\bar{\omega}_{,S} = \overset{\mathbf{t}}{\bar{\varphi}}_{,S} = \overset{\overline{\mathbf{t}}}{\mathbf{k}}} \tag{10.99}$$

10.6.10 Where We Would Like to Go

So far, we have introduced various parameters and derived all the necessary relationships anticipating the next step of identifying the spatial real strain and the real strain rate fields for the beam reference curve. Clearly, then, we should proceed onward by embarking on the following.

- Spatial real strains and rates
  Finally, to get familiarized with and be conscious about the overall computational strategy for nonlinear beams, we should explore:
- the road map: nonlinear beams

10.7 Real Strain and Strain Rates from Weak Form

10.7.1 What We Need to Recall

Let us recall the following.

- The beam internal stored power, $P_{Internal}$, is:

$$P_{Internal} \equiv \int_M \{(\mathbf{F} \bullet (\mathbf{v}_{,S} + \mathbf{c}_{0,S} \times \boldsymbol{\omega}) + \mathbf{M} \bullet \boldsymbol{\omega}_{,S}\} \; dS \tag{10.100}$$

- The resistive stress resultant, $\mathbf{F}$, is defined by:

$$\boxed{\mathbf{F}(S,t) \equiv \int_{A_C} \mathbf{P}_1 \; dA_C} \tag{10.101}$$

- The resistive stress couple, $\mathbf{M}^{\alpha}$, is defined by:

$$\boxed{\mathbf{M}(S,t) \equiv \int_{A_C} (\mathbf{c}(S,\xi^2,\xi^3,t) - \mathbf{c}_0(S,t)) \times \mathbf{P}^1 \; dA_C} \tag{10.102}$$

- $(\mathbf{v} \equiv \dot{\mathbf{d}}, \boldsymbol{\omega})$ are the weak parameters.
- The gradient of a vector function $\boldsymbol{\psi}(\xi^i)$, $i = 1, 2, \ldots$ is a second-order tensor function, denoted as $GRAD\,\boldsymbol{\psi}$, defined by:

$$\boxed{GRAD\,\boldsymbol{\psi} \equiv \frac{\partial \boldsymbol{\psi}}{\partial \xi^j} \otimes \mathbf{G}^j = \boldsymbol{\psi}_{,j} \otimes \mathbf{G}^j} \tag{10.103}$$

- The beam configuration space is completely identified by the kinematic parameters, $(\mathbf{d}, \mathbf{R})$.

- The derivatives of the initial base vector set, $\{\mathbf{G}_i(S)\}$, $i = 1, 2, 3$, are given by:

$$\frac{d}{dS}\mathbf{G}_i(S) \equiv \mathbf{G}_i(S),_S = \mathbf{K}^0\ \mathbf{G}_i \tag{10.104}$$

- The temporal derivatives of the current base vector set, $\{\mathbf{g}_i(S,t)\}$, $i = 1, 2, 3$, are given by:

$$\dot{\mathbf{g}}_i = \boldsymbol{\omega} \times \mathbf{g}_i \tag{10.105}$$

- For any vector field $\mathbf{x}$ described in the rotated coordinate system, that is, $\mathbf{x} = x^i\ \mathbf{g}_i$, we have for the variation in the coordinate system:

$$\overset{\mathbf{v}}{\mathbf{x}} \equiv \underbrace{\bar{x}^i\ \mathbf{g}_i}_{\substack{\text{Variation in rotated}\\ \text{coordinate system, i.e.}\\ \text{objective variation}}} = \underbrace{\bar{\mathbf{x}}}_{\text{Total variation}} - \underbrace{\bar{\boldsymbol{\varphi}} \times \mathbf{x}}_{\text{Spinning variation}} \tag{10.106}$$

$\overset{\mathbf{v}}{\mathbf{x}}$ is the objective variation of $\mathbf{x}$, defined by: $\boxed{\overset{\mathbf{v}}{(\bullet)} \equiv \overline{(\bullet)} - \bar{\boldsymbol{\varphi}} \times (\bullet)}$.

- For any vector field $\mathbf{x}$ described in the rotated coordinate system, that is, $\mathbf{x} = x^i\ \mathbf{g}_i$, we have for the rate in the coordinate system:

$$\overset{\mathbf{t}}{\mathbf{x}} \equiv \underbrace{\dot{x}^i\ \mathbf{g}_i}_{\substack{\text{Derivative in rotated}\\ \text{coordinate system, i.e.}\\ \text{objective derivative}}} = \underbrace{\dot{\mathbf{x}}}_{\text{Total derivative}} - \underbrace{\boldsymbol{\omega} \times \mathbf{x}}_{\substack{\text{Spinning observer}\\ \text{correction}}} \tag{10.107}$$

- $\overset{\mathbf{t}}{\mathbf{x}}$ is the objective time rate of $\mathbf{x}$, defined by: $\boxed{\overset{\mathbf{t}}{(\bullet)} \equiv \frac{\partial}{\partial t}(\bullet) = \dot{(\bullet)} - \boldsymbol{\omega} \times (\bullet)}$.
- The relationship between the derivative of the angular velocity tensor, $\boldsymbol{\Omega}$, and the curvature tensor, $\mathbf{K}$:

$$\boldsymbol{\Omega},_S = \dot{\mathbf{K}} + \underbrace{\mathbf{K}\,\boldsymbol{\Omega} - \boldsymbol{\Omega}\,\mathbf{K}}_{\text{Lie bracket}} \Leftrightarrow \boldsymbol{\omega},_S = \dot{\mathbf{k}} - \boldsymbol{\omega} \times \mathbf{k} \tag{10.108}$$

- The vector triple product property: $\mathbf{a} \times \mathbf{b} \bullet \mathbf{c} = \mathbf{b} \times \mathbf{c} \bullet \mathbf{a} = \mathbf{c} \times \mathbf{a} \bullet \mathbf{b}$, $\forall \mathbf{a}, \mathbf{b}, \mathbf{c}$ vectors.

10.7.2 Real Strains and Strain Rates in Spinning Coordinate System

Note that with $\mathbf{g}_i(S,t) = \mathbf{R}(S,t)\ \mathbf{G}_i(S)$, where $\{\mathbf{g}_i\}$, and $\{\mathbf{G}_i\}$ are the base vector sets for the spinning coordinate system of the current deformed beam curve, and the undeformed coordinate beam reference curve, respectively. Now, with the relation given in equation (10.104), we can obtain for the derivatives of $\mathbf{g}_i$ as:

$$\begin{aligned}\mathbf{g}_i,_S &= \mathbf{R},_S\ \mathbf{G}_i + \mathbf{R}\ \mathbf{G}_i,_S = \mathbf{K}\,\mathbf{R}\ \mathbf{G}_i + \mathbf{R}\ \mathbf{K}^0\ \mathbf{G}_i\\ &= \mathbf{K}\ \mathbf{g}_i + \mathbf{R}\ \mathbf{K}^0\ \mathbf{R}^T\ \mathbf{R}\,\mathbf{G}_i \equiv (\mathbf{K} + \mathbf{R}\ \mathbf{K}^0\ \mathbf{R}^T)\,\mathbf{g}_i \equiv \mathbf{K}^{tot}\ \mathbf{g}_i = \mathbf{k}^{tot} \times \mathbf{g}_i\end{aligned} \tag{10.109}$$

where we introduce the following:

The total curvature tensor, $\mathbf{K}^{tot}$, and the corresponding axial vector, $\mathbf{k}^{tot}$, as:

$$\boxed{\mathbf{K}^{tot} \equiv \mathbf{K} + \mathbf{R}\,\mathbf{K}^0\,\mathbf{R}^T \Leftrightarrow \mathbf{k}^{tot} \equiv \mathbf{k} + \mathbf{R}\,\mathbf{k}^0} \tag{10.110}$$

The result is that for beams with initial curvature, the curvature must always be understood as the sum of one due to the current rotation, that is, $\mathbf{K} \equiv \mathbf{R},_S\ \mathbf{R}^T$, and another due to the initial curvature, that is, $\mathbf{k}^0$, of the undeformed beam reference curve where $\mathbf{K}^{tot}$ and $\mathbf{k}^{tot}$ are the total curvature tensor and the total curvature vector, respectively.

10.7.2.1 Force Real Strain Field

Note that with $\mathbf{c}_0(S,t),_S = \mathbf{G}_1(S) + \mathbf{d}(S,t),_S$, we have as time rate: $\overset{\bullet}{\overbrace{\mathbf{c}_0,_S}} = \dot{\mathbf{d}},_S = \mathbf{v},_S$, and applying equation (10.105): $\dot{\mathbf{d}},_S - \boldsymbol{\omega} \times \mathbf{g}_1 = \overset{\bullet}{\overbrace{\mathbf{c}_0,_S}} - \dot{\mathbf{g}}_1 = \overset{\bullet}{\overbrace{\mathbf{c}_0,_S - \mathbf{g}_1}}$. Now, with the results in equation (10.107) applied to the presumed force strain rate expression in the internal stored power, as given in equation (10.100), we get:

$$\boxed{\begin{aligned}
&\mathbf{v},_S + \mathbf{c}_0,_S \times \boldsymbol{\omega}\\
&\quad = \dot{\mathbf{d}},_S - \boldsymbol{\omega}\times\mathbf{g}_1 - \boldsymbol{\omega}\times\mathbf{c}_0,_S + \boldsymbol{\omega}\times\mathbf{g}_1 = \overset{\bullet}{\overbrace{(\mathbf{c}_0,_S - \mathbf{g}_1)}} - \boldsymbol{\omega}\times(\mathbf{c}_0,_S - \mathbf{g}_1)\\
&\quad \equiv \overset{t}{\overbrace{(\mathbf{c}_0,_S - \mathbf{g}_1)}} = \mathbf{R}\left\{\mathbf{R}^T\overset{\bullet}{\overbrace{(\mathbf{c}_0,_S - \mathbf{g}_1)}} + (-\mathbf{R}^T\boldsymbol{\Omega})(\mathbf{c}_0,_S - \mathbf{g}_1)\right\}\\
&\quad = \mathbf{R}\left\{\mathbf{R}^T\overset{\bullet}{\overbrace{(\mathbf{c}_0,_S - \mathbf{g}_1)}} + \dot{\mathbf{R}}^T(\mathbf{c}_0,_S - \mathbf{g}_1)\right\} = \underbrace{\mathbf{R}}_{\text{Push forward}}\underbrace{\left(\overset{\bullet}{\overbrace{\underbrace{\mathbf{R}^T}_{\text{Pull back}}(\mathbf{c}_0,_S - \mathbf{g}_1)}}\right)}_{\text{Time rate}} = \overset{t}{\overbrace{{}^R\boldsymbol{\beta}}}
\end{aligned}} \tag{10.111}$$

where we have introduced:

- The ***rotated force real strain***, ${}^R\boldsymbol{\beta}$:

$$\boxed{{}^R\boldsymbol{\beta} \equiv \mathbf{c}_0,_S - \mathbf{g}_1} \tag{10.112}$$

> Thus, we conclude that the expression $\mathbf{v},_S + \mathbf{c}_0,_S \times \boldsymbol{\omega}$, in equation (10.100), truly represents the rotated force strain rate, as it can be obtained from the time derivative of a function, the rotated force real strain, ${}^R\boldsymbol{\beta}$ defined by equation (10.112); it embodies translational strains, the real extensional strain and the real shear strain, along the rotated coordinate system with the base vector set given by $\{\mathbf{g}_i\}$.

Note that in the undeformed reference surface: as $\mathbf{d} \rightarrow \mathbf{0}$ and $\mathbf{R} \rightarrow \mathbf{I}$, the force real strain, ${}^R\beta \rightarrow \mathbf{0}$, as it should.

10.7.2.2 Moment Real Strain Field

By definition: the rotational velocity tensor, $\boldsymbol{\Omega} \equiv \dot{\mathbf{R}}\, \mathbf{R}^T$, the curvature, $\mathbf{K} \equiv \mathbf{R},_S\, \mathbf{R}^T$, the initial curvature of the undeformed beam reference curve, $\mathbf{K}^0$ with $\mathbf{G}_{i},_S = \mathbf{K}^0\, \mathbf{G}_i$, and finally, recall the property: for any skew-symmetric tensor $\hat{\mathbf{K}}^0$ with the corresponding axial vector, $\hat{\mathbf{k}}^0$, defined by: $\hat{\mathbf{K}}^0 = \mathbf{R}\, \mathbf{K}^0\, \mathbf{R}^T \Rightarrow \hat{\mathbf{k}}^0 = \mathbf{R}\, \mathbf{k}^0$, where $\mathbf{k}^0$ is the axial vector corresponding to the skew-symmetric tensor $\mathbf{K}^0$. Now, with $\mathbf{K}^{tot} \equiv \mathbf{K} + \mathbf{R}\, \mathbf{K}^0\, \mathbf{R}^T$ as the total curvature tensor with $\mathbf{k}^{tot} = \mathbf{k} + \mathbf{R}\, \mathbf{k}^0$ as the corresponding total axial curvature vector, we have:

$$\begin{aligned} \dot{\mathbf{K}} &= \dot{\mathbf{K}}^{tot} - \dot{\mathbf{R}}\, \mathbf{K}^0\, \mathbf{R}^T - \mathbf{R}\, \mathbf{K}^0\, \dot{\mathbf{R}}^T = \dot{\mathbf{K}}^{tot} - \boldsymbol{\Omega}\, \mathbf{R}\, \mathbf{K}^0\, \mathbf{R}^T + \mathbf{R}\, \mathbf{K}^0\, \mathbf{R}^T\, \boldsymbol{\Omega} \\ &= \dot{\mathbf{K}}^{tot} + \underbrace{(\mathbf{R}\, \mathbf{K}^0\, \mathbf{R}^T)\, \boldsymbol{\Omega} - \boldsymbol{\Omega}\, (\mathbf{R}\, \mathbf{K}^0\, \mathbf{R}^T)}_{\text{Lie bracket}} \end{aligned} \tag{10.113}$$

which implies in vector form:

$$\dot{\mathbf{k}} = \dot{\mathbf{k}}^{tot} + (\mathbf{R}\, \mathbf{k}^0) \times \boldsymbol{\omega} = \dot{\mathbf{k}}^{tot} - \boldsymbol{\omega} \times (\mathbf{R}\, \mathbf{k}^0) \tag{10.114}$$

Now, from the second of equation (10.108) and relation in equation (10.114), we have:

$$\boxed{\begin{aligned} \boldsymbol{\omega},_S &= \dot{\mathbf{k}} - \boldsymbol{\omega} \times \mathbf{k} \equiv \overbrace{\ \mathbf{k}\ }^{\mathbf{t}} = \mathbf{R} \left\{ \mathbf{R}^T\, \dot{\mathbf{k}} + (-\mathbf{R}^T \boldsymbol{\Omega})\mathbf{k} \right\} \\ &= \mathbf{R} \left\{ \mathbf{R}^T\, \dot{\mathbf{k}} + \dot{\mathbf{R}}^T\, \mathbf{k} \right\} = \underbrace{\mathbf{R}}_{\text{Push forward}} \underbrace{\left(\overbrace{\underbrace{\mathbf{R}^T}_{\text{Pull back}}\, \mathbf{k}}^{\bullet} \right)}_{\text{Time derivative}} = \overbrace{{}^R\chi}^{\mathbf{t}} \end{aligned}} \tag{10.115}$$

where we have introduced:

- The ***rotated moment real strain***:

$$\boxed{{}^R\chi \equiv \mathbf{k} = \mathbf{k}^{tot} - \mathbf{R}\, \mathbf{k}^0, \quad \mathbf{K} \equiv \mathbf{R},_S\, \mathbf{R}^T} \tag{10.116}$$

> Thus, we conclude that the expression $\boldsymbol{\omega},_S$ truly represents the rotated moment strain rate as it is obtained from the time derivative of a function, the rotated moment real strain, ${}^R\chi$; it embodies rotational strain, the real bending strain, along the rotated coordinate system with the base vector set given by $\{\mathbf{g}_i\}$.

Note that in the undeformed reference curve: $\mathbf{R} \rightarrow \mathbf{I}$, the moment real strain, ${}^R\chi \rightarrow \mathbf{0}$, as it should.

Thus, by inserting results from equation (10.111) and equation (10.115) into the internal stored power expression of equation (10.100), it can be rewritten as:

$$P_{Internal} \equiv \int_M \left\{ (\mathbf{F} \cdot \overbrace{{}^R\boldsymbol{\beta}}^{t} + \mathbf{M} \cdot \overbrace{{}^R\boldsymbol{\chi}}^{t} \right\} dS \tag{10.117}$$

- For our subsequent computational virtual work principle necessary for a c-type finite element formulation, we will develop all the rotating coordinate system by first expressing the invariant force and moment vectors in the material coordinate system and then applying the rotation tensor to the component representation for the expressions. This allows us to develop and track the most accurate numerical scheme with a cost-effective explicit identification of the relevant entities. Thus, for computational ease and accuracy, all the expressions pertaining to the weak form, the virtual strains, and so on, must be modified to most suitable ***operational*** or ***component vector form***, as we intend to do next.

However, before we get into the computational description of the beam behavior, we will take a short detour to present an alternative way to develop the weak form, the stress resultants and the stress couple and the associated conjugate strains for historical connection. We have developed the exact weak form by starting from the exact balance laws of motion. In this alternative way of thinking, we start by making some kinematic assumptions; then, the stress or deformation power (in case of static analysis, the internal strain energy variations) provides the conjugate real strains and real strain rates upon suitable definition of the stress resultants and the stress couples. So, let us look into it with a kinematic assumption that is appropriate and commonly used for thin beams as follows.

10.7.3 Alternative Derivation: Kinematics and Stress or Deformation Power

A simple 1D beam can never adequately approximate the behavior of a general 3D beam-like body; an additional mechanism must be embedded in the 1D beam to capture in some representative fashion the through-the-thickness response of the 3D beam-like body. It turns out that the 1D kinematics that are energetically consistent with the 3D beam-like body is best captured by the ***Reissner–Mindlin***-type model represented by the displacement vector and the rotation tensor with a coordinate system termed the ***co-rotating or spinning system***, the meaning of which will be clear shortly. We choose these coordinate system at the reference configuration to be coincident with the base vectors, $\{\mathbf{G}_i(S)\}$, $i = 1, 2, 3$, on the beam reference curve as before with S as the arc length parameter along the reference curve as introduced earlier. Then, the deformation map of a beam, with equation (10.109), is given by:

$$\begin{aligned} \mathbf{c}(S,t) &= \mathbf{c}_0(S,t) + \xi^\alpha \mathbf{g}_\alpha, \quad \alpha = 2,3 \\ \mathbf{g}_i(S,t) &= \mathbf{R}(S,t)\, \mathbf{G}_i(S), \quad i = 1,2,3 \\ \mathbf{g}_{i,S}(S,t) &= \mathbf{k} \times \mathbf{g}_i \end{aligned} \tag{10.118}$$

where $\mathbf{R}(S,t)$ is the rotational tensor field designating the rotation of the rigid cross-section independent of the deformation of the reference curve, and $\mathbf{k}^{tot}$ is the total curvature as given by equation (10.110).

Remark

- Note that the base vector set, $\{\mathbf{g}_i(S,t)\}, i = 1,2,3$ are not the convected coordinate bases; recall that the convected coordinate bases are given by $\{\mathbf{c}_0,_1, \mathbf{c}_0,_2, \mathbf{c}_0,_3\}$, if shear deformation is neglected.

10.7.3.1 Weak Form: Kinematic Assumption and Reduction from the 3D Beam-like Body

Contrary to what we have done before, we use the deformation map as our starting point. As an example, in what follows, we restrict our exposition to the case of beams with the basic kinematic assumption of rigid cross-section. Now, let us suppose that a particle with the position vector, $\mathbf{C}(S,\xi^2,\xi^3,0)$, in the 3D beam-like undeformed body assumes under deformation a location with the position vector given by $\mathbf{c}(S,\xi^2,\xi^3,t)$; let the corresponding position vectors on the beam reference curve be $\mathbf{C}_0(S) \equiv \mathbf{C}(S,0,0)$ and $\mathbf{c}_0(S,t) \equiv \mathbf{c}(S,0,0,t)$, respectively. Then, from equation (10.118), the deformation gradient, $\mathbf{F}_D$, is given by:

$$\begin{aligned}\mathbf{F}_D \equiv GRAD\,\mathbf{c} = \frac{\partial \mathbf{c}}{\partial \mathbf{C}} &= \mathbf{c},_i \otimes \mathbf{G}_i = \mathbf{c},_S \otimes \mathbf{G}_1 + \mathbf{c},_\alpha \otimes \mathbf{G}_\alpha, \quad \alpha = 2,3 \\ &= \left(\mathbf{c}_0,_S + \mathbf{k} \times (\mathbf{c} - \mathbf{c}_0)\right) \otimes \mathbf{G}_1 + \mathbf{g}_\alpha \otimes \mathbf{G}_\alpha, \quad \alpha = 2,3\end{aligned} \tag{10.119}$$

Now noting the time rate, from equation (10.105), we have, from equation (10.119), the time rate, $\overset{\bullet}{\mathbf{F}} \equiv \frac{d}{dt}\mathbf{F}$, of the deformation map as:

$$\begin{aligned}\overset{\bullet}{\mathbf{F}}_D &= \left\{ \overset{\bullet}{\mathbf{c}}_0,_S + \overset{\bullet}{\mathbf{k}} \times (\mathbf{c} - \mathbf{c}_0) + \mathbf{k} \times \overset{\bullet}{\overbrace{(\mathbf{c} - \mathbf{c}_0)}} \right\} \otimes \mathbf{G}_1 + \overset{\bullet}{\mathbf{g}}_\alpha \otimes \mathbf{G}_\alpha, \quad \alpha = 2,3 \\ &= \left\{ \overset{\bullet}{\mathbf{c}}_0,_S + \overset{\bullet}{\mathbf{k}} \times (\mathbf{c} - \mathbf{c}_0) + \mathbf{k} \times (\boldsymbol{\omega} \times (\mathbf{c} - \mathbf{c}_0)) \right\} \otimes \mathbf{G}_1 + (\boldsymbol{\omega} \times \mathbf{g}_\alpha) \otimes \mathbf{G}_\alpha, \quad \alpha = 2,3\end{aligned} \tag{10.120}$$

***Fact* 1:** As a supporting formula, note that for any $\mathbf{B} = \mathbf{a} \otimes \mathbf{b}$, a symmetric tensor, that is, $\mathbf{B} - \mathbf{B}^T = \mathbf{0}$, we have: $\mathbf{a} \times \mathbf{b} = \mathbf{0}$ because: $\forall \mathbf{h} \in \mathbb{R}_3$, $\mathbf{0} = (\mathbf{B} - \mathbf{B}^T)\mathbf{h} = (\mathbf{a} \otimes \mathbf{b})\mathbf{h} - (\mathbf{b} \otimes \mathbf{a})\mathbf{h} = (\mathbf{h} \bullet \mathbf{b})\mathbf{a} - (\mathbf{h} \bullet \mathbf{a})\mathbf{b} = (\mathbf{a} \times \mathbf{b}) \times \mathbf{h}$, where we have used the vector identity: $\mathbf{a} \times (\mathbf{b} \times \mathbf{c}) = (\mathbf{a} \bullet \mathbf{c})\mathbf{b} - (\mathbf{a} \bullet \mathbf{b})\mathbf{c}$.

Now, if $\mathbf{P} \equiv \mathbf{P}^j \otimes \mathbf{G}_j$ is the first Piola–Kirchhoff stress tensor, we get from equation (10.120):

$$\mathbf{P} : \overset{\bullet}{\mathbf{F}}_D = \mathbf{P}^1 \bullet \overset{\bullet}{\mathbf{c}}_0,_S + (\mathbf{c} - \mathbf{c}_0) \times \mathbf{P}^1 \bullet \overset{\bullet}{\mathbf{k}} + \mathbf{P}^1 \bullet \{\mathbf{k} \times (\boldsymbol{\omega} \times (\mathbf{c} - \mathbf{c}_0))\} + \boldsymbol{\omega}\,(\mathbf{g}_\alpha \times \mathbf{P}^\alpha), \quad \alpha = 2,3 \tag{10.121}$$

where we have used the definition that for any two tensors, $\mathbf{A} = \mathbf{a} \otimes \mathbf{b}$, and $\mathbf{B} = \mathbf{c} \otimes \mathbf{d}$, we have: $\mathbf{A} : \mathbf{B} = (\mathbf{a} \otimes \mathbf{b}) : (\mathbf{c} \otimes \mathbf{d}) = (\mathbf{a} \bullet \mathbf{c})(\mathbf{b} \bullet \mathbf{d})$. Now, let us recall that the angular momentum balance law gives the following symmetry relation:

$$\mathbf{F}_D\,\mathbf{P}^T = \mathbf{P}\,\mathbf{F}_D^T \tag{10.122}$$

whereby we get $\mathbf{F}_D\,\mathbf{P}^T = \mathbf{P}\,\mathbf{F}_D^T = \mathbf{c},_i \otimes \mathbf{G}^i \bullet \mathbf{G}_j \otimes \mathbf{P}^j = \mathbf{c},_i \otimes \mathbf{P}^i$ as symmetric, and by the above supporting ***Fact 1***, we conclude:

$$\mathbf{c},_i \times \mathbf{P}^i = \mathbf{0} \quad \Rightarrow \mathbf{c},_S \times \mathbf{P}^1 = -\mathbf{c},_\alpha \times \mathbf{P}^\alpha = -\mathbf{g}_\alpha \times \mathbf{P}^\alpha, \quad \alpha = 2,3 \tag{10.123}$$

So, from equation (10.123), we get the last term of equation (10.121) as:

$$\begin{aligned}\boldsymbol{\omega} \bullet (\mathbf{g}_\alpha \times \mathbf{P}^\alpha) &= -\boldsymbol{\omega} \bullet (\mathbf{c},_S \times \mathbf{P}^1) \\ &= -\mathbf{P}^1 \bullet \{\boldsymbol{\omega} \times \mathbf{c}_0,_S + \boldsymbol{\omega} \times (\mathbf{k} \times (\mathbf{c} - \mathbf{c}_0))\}\end{aligned} \tag{10.124}$$

Inserting equation (10.124) in equation (10.121), we get:

$$\begin{aligned}\mathbf{P} : \dot{\mathbf{F}}_D = \mathbf{P}^1 \bullet \left\{\dot{\mathbf{c}}_0,_S - \boldsymbol{\omega} \times \mathbf{c}_0,_S\right\} + (\mathbf{c} - \mathbf{c}_0) \times \mathbf{P}^1 \bullet \dot{\mathbf{k}} \\ + \mathbf{P}^1 \bullet \{\mathbf{k} \times (\boldsymbol{\omega} \times (\mathbf{c} - \mathbf{c}_0)) - \boldsymbol{\omega} \times (\mathbf{k} \times (\mathbf{c} - \mathbf{c}_0))\}\end{aligned} \tag{10.125}$$

Now, let us note the following supporting vector identity:

***Fact* 2:** for any three vectors, $\mathbf{a} \times (\mathbf{b} \times \mathbf{c}) - \mathbf{b} \times (\mathbf{a} \times \mathbf{c}) = (\mathbf{a} \times \mathbf{b}) \times \mathbf{c}$.

Proof. Let $\mathbf{Z} \equiv (\mathbf{b} \otimes \mathbf{a})$ be a tensor, then from the previous supporting fact, we have for the skew-symmetric part: $\mathbf{Z} - \mathbf{Z}^T \equiv (\mathbf{b} \otimes \mathbf{a}) - (\mathbf{a} \otimes \mathbf{b}) = [\mathbf{a} \times \mathbf{b}] = (\mathbf{a} \times \mathbf{b})\times$. Now, by the vector identity: $\mathbf{a} \times (\mathbf{b} \times \mathbf{c}) = (\mathbf{a} \bullet \mathbf{c})\mathbf{b} - (\mathbf{a} \bullet \mathbf{b})\mathbf{c}$, we complete the proof with: $\mathbf{a} \times (\mathbf{b} \times \mathbf{c}) - \mathbf{b} \times (\mathbf{a} \times \mathbf{c}) = (\mathbf{a} \bullet \mathbf{c})\mathbf{b} - (\mathbf{b} \bullet \mathbf{c})\mathbf{a} = \{(\mathbf{b} \otimes \mathbf{a}) - (\mathbf{a} \otimes \mathbf{b})\}\mathbf{c} = \{\mathbf{Z} - \mathbf{Z}^T\}\mathbf{c} = (\mathbf{a} \times \mathbf{b}) \times \mathbf{c}$.

So, applying this supporting fact, we get for the last term of equation (10.125) as:

$$\begin{aligned}&\mathbf{P}^1 \bullet \{\mathbf{k} \times (\boldsymbol{\omega} \times (\mathbf{c} - \mathbf{c}_0)) - \boldsymbol{\omega} \times (\mathbf{k} \times (\mathbf{c} - \mathbf{c}_0))\} \\ &= \mathbf{P}^1 \bullet \{(\mathbf{k} \times \boldsymbol{\omega}) \times (\mathbf{c} - \mathbf{c}_0)\} = -(\mathbf{c} - \mathbf{c}_0) \times \mathbf{P}^1(\boldsymbol{\omega} \times \mathbf{k})\end{aligned} \tag{10.126}$$

Now, inserting equation (10.126) in equation (10.125), we finally get, with help of equation (10.111) and equation (10.115). the expression for the stress or deformation power density as:

$$\boxed{\begin{aligned}\mathbf{P} : \dot{\mathbf{F}}_D &= \mathbf{P}^1 \bullet \left\{\dot{\mathbf{c}}_0,_S - \boldsymbol{\omega} \times \mathbf{c}_0,_S\right\} + (\mathbf{c} - \mathbf{c}_0) \times \mathbf{P}^1 \bullet \left\{\dot{\mathbf{k}} - (\boldsymbol{\omega} \times \mathbf{k})\right\} \\ &= \mathbf{P}^1 \bullet \overbrace{\mathbf{c}_0,_S}^{\mathbf{t}} + (\mathbf{c} - \mathbf{c}_0) \times \mathbf{P}^1 \bullet \overbrace{\mathbf{k}}^{\mathbf{t}} = \mathbf{P}^1 \bullet \overbrace{{}^R\boldsymbol{\beta}}^{\mathbf{t}} + (\mathbf{c} - \mathbf{c}_0) \times \mathbf{P}^1 \bullet \overbrace{{}^R\boldsymbol{\chi}}^{\mathbf{t}}\end{aligned}} \tag{10.127}$$

Now, incorporating the definition of $\mathbf{F}$ and $\mathbf{M}$ as in equations (10.101) and (10.102), respectively, we get, using equation (10.127), the expression for the internal stored power as:

$$\boxed{P_{Internal} \equiv \int_M \left\{(\mathbf{F} \bullet \overbrace{{}^R\boldsymbol{\beta}}^{\mathbf{t}} + \mathbf{M} \bullet \overbrace{{}^R\boldsymbol{\chi}}^{\mathbf{t}}\right\} dS} \tag{10.128}$$

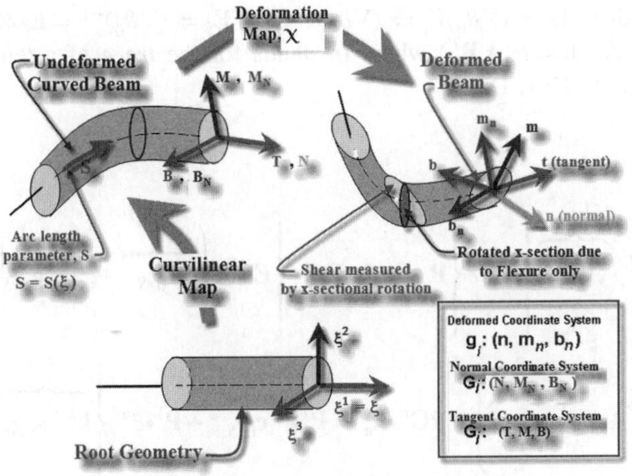

Figure 10.11 Beam deformation map.

which is exactly same as that given by equation (10.117). Another compact way of finding the relation given by equation (10.127) is to consider the beam motion in terms of the configuration space; this is what we will do next.

10.7.3.2 Deformation Gradient as the Tangent Map

We may choose to derive equation (10.127) by using the deformation gradient as the tangent map on the configuration manifold. A motion may be viewed as a curve on the configuration space passing through the undeformed state as well as other and current deformation states. We can then view the deformation gradient which carries local information as the tangent map that relates the tangent to the curve at the undeformed state to that at the current deformed state. In other words, If we denote $\mathbf{C}(S, \xi^2, \xi^3) = \vartheta_0(S, \xi^2, \xi^3)$ and $\mathbf{c}(S, \xi^2, \xi^3) = \vartheta(S, \xi^2, \xi^3)$ as the values of the functions ϑ_0 and ϑ as shown in Figure 10.11, we get χ the deformation map from the undeformed shell-like body to the current deformed body as:

$$\chi \equiv \vartheta \circ \vartheta_0^{-1} \tag{10.129}$$

Noting that, $\nabla\vartheta = \mathbf{g}_i \otimes \mathbf{E}^i$, we have: $\mathbf{g}_i = (\nabla\vartheta)\,\mathbf{E}_i = (\nabla\vartheta)\,(\nabla\vartheta_0)^{-1}\,\mathbf{G}_i = \mathbf{F}_D\,\mathbf{G}_i$ where we have used the definition of the deformation gradient as: $\mathbf{F}_D \equiv \mathbf{g}_i \otimes \mathbf{G}^i$. With the deformation map given by equation (10.118), we have: $\nabla\vartheta = \vartheta_{,S} \otimes \mathbf{E}^1 + \vartheta_{,\alpha} \otimes \mathbf{E}^\alpha = \{\mathbf{c}_{0,S} + \xi^\alpha\,\mathbf{g}_{\alpha,S}\} \otimes \mathbf{E}^1 + \mathbf{g}_\alpha \otimes \mathbf{E}^\alpha$, $\alpha = 2, 3$. The tangent map of the deformation map given by equation (10.129) is:

$$\mathbf{F}_D \equiv (\nabla\vartheta)(\nabla\vartheta_0)^{-1} \tag{10.130}$$

Then, from equation (10.130), by taking the time derivative, we have:

$$\begin{aligned}\dot{\mathbf{F}}_D &= \overset{\bullet}{\overbrace{\nabla\vartheta}}\;(\nabla\vartheta_0)^{-1}\\ &= \left\{\left(\overset{\bullet}{\overbrace{\mathbf{c}_{0,S}}} + \xi^\alpha\,\overset{\bullet}{\overbrace{\mathbf{g}_{\alpha,S}}}\right) \otimes \mathbf{E}^1 + \dot{\mathbf{g}}_\alpha \otimes \mathbf{E}^\alpha\right\}(\nabla\vartheta_0)^{-1}, \quad \alpha = 2, 3\end{aligned} \tag{10.131}$$

Now, noting that $\mathbf{G}_i = (\nabla\vartheta_0)\mathbf{E}_i \Rightarrow (\nabla\vartheta_0)^{-1}\mathbf{G}_i = \mathbf{E}_i \Rightarrow (\nabla\vartheta_0)^{-1} = \mathbf{E}_i \otimes \mathbf{G}^i \Rightarrow (\nabla\vartheta_0)^{-T} = \mathbf{G}^i \otimes \mathbf{E}_i$, and that $\mathbf{A} : \mathbf{B} = tr(\mathbf{A}\,\mathbf{B}^T)$ where tr stands for the ***trace*** of a tensor, by definition, we have with equation (10.131):

$$\mathbf{P} : \dot{\mathbf{F}}_D$$

$$= tr\left\{\mathbf{P}\,(\nabla\vartheta_0)^{-T}\,\overset{\bullet}{\overbrace{(\nabla\vartheta)^T}}\right\} = tr\left\{\mathbf{P}\,(\mathbf{G}^i \otimes \mathbf{E}_i)\bullet\left\{\mathbf{E}^1 \otimes \left(\overset{\bullet}{\overbrace{\mathbf{c}_{0,S}}} + \xi^\alpha\,\overset{\bullet}{\overbrace{\mathbf{g}_{\alpha,S}}}\right) + \mathbf{E}^\alpha \otimes \dot{\mathbf{g}}_\alpha\right\}\right\}$$

$$= \mathbf{P}\mathbf{G}^1\bullet\left(\overset{\bullet}{\overbrace{\mathbf{c}_{0,S}}} + \xi^\alpha\,\overset{\bullet}{\overbrace{\mathbf{g}_{\alpha,S}}}\right) + \mathbf{P}\mathbf{G}^\alpha\bullet\dot{\mathbf{g}}_\alpha = \mathbf{P}^1\bullet\overset{\bullet}{\overbrace{\mathbf{c}_{0,S}}} + \mathbf{P}^1\bullet\xi^\alpha\left\{\overset{\bullet}{\overbrace{\mathbf{k}^{tot}\times\mathbf{g}_\alpha}}\right\} + \mathbf{P}^\alpha\bullet(\boldsymbol{\omega}\times\mathbf{g}_\alpha)$$

(10.132)

where we have used the tensor trace property: $tr(\mathbf{a}\otimes\mathbf{b}) = \mathbf{a}\bullet\mathbf{b}$, and equations (10.105). Now, noting that: $\xi^\alpha\left\{\overset{\bullet}{\overbrace{\mathbf{k}^{tot}\times\mathbf{g}_\alpha}}\right\} = \overset{\bullet}{\overbrace{\mathbf{k}^{tot}}}\times\xi^\alpha\,\mathbf{g}_\alpha + \mathbf{k}^{tot}\times\xi^\alpha\,\overset{\bullet}{\overbrace{\mathbf{g}_\alpha}} = \overset{\bullet}{\overbrace{\mathbf{k}^{tot}}}\times(\mathbf{c}-\mathbf{c}_0) + \mathbf{k}^{tot}\times(\boldsymbol{\omega}\times(\mathbf{c}-\mathbf{c}_0))$, and using equation (10.124) for the third term, we get from equation (10.132):

$$\mathbf{P} : \dot{\mathbf{F}}_D = \mathbf{P}^1\bullet\overset{\bullet}{\overbrace{\mathbf{c}_{0,S}}} + \mathbf{P}^1\bullet\xi^\alpha\left\{\overset{\bullet}{\overbrace{\mathbf{k}^{tot}\times\mathbf{g}_\alpha}}\right\} + \mathbf{P}^\alpha\bullet(\boldsymbol{\omega}\times\mathbf{g}_\alpha)$$

$$= \mathbf{P}^1\bullet\overset{\bullet}{\overbrace{\mathbf{c}_{0,S}}} + \underbrace{(\mathbf{c}-\mathbf{c}_0)\times\mathbf{P}^1\bullet\overset{\bullet}{\overbrace{\mathbf{k}^{tot}}} + \mathbf{P}^1\bullet\mathbf{k}^{tot}\times(\boldsymbol{\omega}\times(\mathbf{c}-\mathbf{c}_0))}_{\text{Second term}} \tag{10.133}$$

$$+ \underbrace{\mathbf{P}^1\bullet\{-\boldsymbol{\omega}\times\mathbf{c}_{0,S} - \boldsymbol{\omega}\times(\mathbf{k}^{tot}\times(\mathbf{c}-\mathbf{c}_0))\}}_{\text{Third term}}$$

Finally, using ***Fact 1*** for the third term in equation (10.133), we get:

$$\boxed{\begin{aligned}\mathbf{P} : \dot{\mathbf{F}}_D &= \mathbf{P}^1\bullet\left\{\dot{\mathbf{c}}_{0,S} - \boldsymbol{\omega}\times\mathbf{c}_{0,S}\right\} + (\mathbf{c}-\mathbf{c}_0)\times\mathbf{P}^1\bullet\left\{\dot{\mathbf{k}}^{tot} - (\boldsymbol{\omega}\times\mathbf{k}^{tot})\right\}\\ &= \mathbf{P}^1\bullet\overset{\mathbf{t}}{\overbrace{\mathbf{c}_{0,S}}} + (\mathbf{c}-\mathbf{c}_0)\times\mathbf{P}^1\bullet\overset{\mathbf{t}}{\overbrace{\mathbf{k}^{tot}}} = \mathbf{P}^1\bullet\overset{\mathbf{t}}{\overbrace{{}^R\boldsymbol{\beta}}} + (\mathbf{c}-\mathbf{c}_0)\times\mathbf{P}^1\bullet\overset{\mathbf{t}}{\overbrace{{}^R\boldsymbol{\chi}}}\end{aligned}}$$

(10.134)

Clearly, the expressions in equation (10.127) and (10.134) are identical.

10.7.3.3 Reduced Internal Energy (Deformation or Stress Power): Conjugate Stresses and Strains

Using equation (10.127) or equation (10.134), we can identify the deformation or stress power, $\dot{U}$, of a 1D beam reference curve as the reduced 3D stress power, $P_{internal}$, of the corresponding beam-like body as:

$$\boxed{\begin{aligned} P_{\text{internal}} &= \int_M \left(\int_{A_C} \mathbf{P} : \dot{\mathbf{F}}_D \, dA_C \right) dS = \int_M \left(\int_{A_C} \mathbf{P}^1 \, dA_C \right) \left\{ \overbrace{\mathbf{c}_{0,S}}^{\bullet} - \boldsymbol{\omega} \times \mathbf{c}_{0,S} \right\} dS \\ &+ \int_M \left(\int_{A_C} (\mathbf{c} - \mathbf{c}_0) \times \mathbf{P}^1 \, dA_C \right) \bullet \left\{ \overbrace{\mathbf{k}^{tot}}^{\bullet} - (\boldsymbol{\omega} \times \mathbf{k}^{tot}) \right\} dS \\ &= \int_M \left\{ \mathbf{F} \bullet \overbrace{{}^R\boldsymbol{\beta}}^{\mathbf{t}} + \mathbf{M} \bullet \overbrace{{}^R\boldsymbol{\chi}}^{\mathbf{t}} \right\} dS = \dot{U}\left({}^R\boldsymbol{\beta}, {}^R\boldsymbol{\chi}\right) \end{aligned}} \tag{10.135}$$

where we have inserted the definitions as given in equations (10.101) and (10.102) for the stress resultant and the stress couple, respectively, and in equations (10.112) and (10.116) for the rotated translational and bending strains, respectively.

- Equation (10.135) establishes the existence of an equivalent 1D strain energy functional, $U({}^R\boldsymbol{\beta}, {}^R\boldsymbol{\chi})$, under conservative loading, or at least the existence of the strain energy functional as functions of the 1D beam reference curve strains only. This knowledge helps to free us from looking back to the originating 3D beam-like body ever again when we develop the appropriate constitutive theory for the beam reference curve as will soon become apparent.

10.7.4 Where We Would Like to Go

In conformity with our principal interest in computational structural engineering and structural mechanics, we need to bring every beam equation down to a form where we can easily solve the beam problems numerically by the use of "mega-beam elements" of our finite element scheme: c-type finite element method. This in turn, requires exact derivation of the compatible real strain fields in component or operational form along the configuration path from our knowledge of the above virtual strain fields. This endeavor will again validate our hunch that the presumed virtual strains are truly virtual, that is, these can be shown to exist as variations of some real strain fields. Thus, before anything else, the expressions for the virtual strain fields in the internal virtual work expression of equation (10.100) must all be modified to the component vector form. For this, we must now look at component or operational vector form.

10.8 Component or Operational Vector Form

10.8.1 What We Need to Recall

Our stated goal is achieved by use of the curvilinear Frenet frame description of a geometric curve in general 3D space. As a start, let us recall the following.

- For the undeformed tangent coordinate system (Frenet frame base vectors), $\mathbf{T}, \mathbf{M}, \mathbf{B}$, being an orthonormal system of coordinates, both the contravariant base vectors, $\{\mathbf{G}^i(S)\}$, and covariant base vectors, $\{\mathbf{G}_i(S)\}$, are one and the same.
- It is assumed that the current deformation of a beam is completely characterized by the displacement vector field: $\mathbf{d}(S,t)$, and a rotation tensor field: $\mathbf{R}(S,t)$, both measured in the undeformed coordinate system, $\{\mathbf{G_1} \equiv \mathbf{T}, \mathbf{G_2} \equiv \mathbf{M}, \mathbf{G}_3 \equiv \mathbf{B}\}$ on the beam reference curve at time $t \in \mathbb{R}_+$. However, we have another curvilinear co-spinning coordinate system (rotating synchronously with the rotation tensor, $\mathbf{R}(S,t)$) and denoted as the normal coordinate system, with base vectors, $\{\mathbf{G_1} \equiv \mathbf{N} = \mathbf{T}, \mathbf{G_2} \equiv \mathbf{M}, \mathbf{G}_3 \equiv \mathbf{B}\}$ in the undeformed configuration that becomes $\{\mathbf{g_1} \equiv \mathbf{n},\ \mathbf{g_2} \equiv \mathbf{m_n},\ \mathbf{g}_3 \equiv \mathbf{b_n}\}$ in the current deformed state at any point $S \in [0, L]$, the reference curve of a beam at time $t \in \mathbb{R}_+$. The undeformed and deformed base vectors have the relation: $\mathbf{n} = \mathbf{RN}$, $\mathbf{g}_\alpha = \mathbf{R}\,\mathbf{G}_\alpha$, $\alpha = 2, 3$. In compact tensor notation, we can rewrite these relations as: $\mathbf{g}_i = \mathbf{R}\,\mathbf{G}_i$, $i = 1, 2, 3$. Moreover, as can be seen from above definition, in the undeformed configuration of the beam, the base vector set of the tangent coordinate system coincides identically with that of the normal coordinate system.
- The rotation tensor, $\mathbf{R}$, and the Rodrigues vector, $\mathbf{s}$, are related by:

$$\boxed{\begin{aligned}&\mathbf{R} = \frac{1}{\rho}\begin{bmatrix} 1+s_1^2-s_2^2-s_3^2 & 2(s_1 s_2 - s_3) & 2(s_3 s_1 + s_2) \\ 2(s_1 s_2 + s_3) & 1-s_1^2+s_2^2-s_3^2 & 2(s_2 s_3 - s_1) \\ 2(s_3 s_1 - s_2) & 2(s_2 s_3 + s_1) & 1-s_1^2-s_2^2+s_3^2 \end{bmatrix}\\ &\text{with} \quad \rho \equiv 1+s_1^2+s_2^2+s_3^2\end{aligned}} \tag{10.136}$$

- The rotation vector, $\boldsymbol{\theta}$, and the Rodrigues vector, $\mathbf{s}$, are related such that:

$$\boxed{\mathbf{s} = \frac{1}{\theta}\tan\left(\frac{\theta}{2}\right)\boldsymbol{\theta}} \tag{10.137}$$

  where $\theta \equiv \ ||\boldsymbol{\theta}|| \ = \sqrt{(\theta \bullet \theta)}$ is the magnitude of the rotation vector.
- The angular velocity vector, $\boldsymbol{\omega}$, measured in the reference coordinate system is given by:

$$\boxed{\boldsymbol{\omega} \equiv \mathbf{W}(\boldsymbol{\theta})\ \dot{\boldsymbol{\theta}}} \tag{10.138}$$

  where the $\mathbf{W}(\boldsymbol{\theta})$ tensor is given by

$$\boxed{\begin{aligned}&\mathbf{W}(\boldsymbol{\theta}) \equiv \mathbf{I} + c_1\ \boldsymbol{\Theta} + c_2\ \boldsymbol{\Theta}^2,\\ &c_1(\boldsymbol{\theta}) = \frac{1-\cos\theta}{\theta^2}, \quad c_2(\boldsymbol{\theta}) = \frac{\theta - \sin\theta}{\theta^3}\end{aligned}} \tag{10.139}$$

10.8.2 Relevant Operational Matrix Representations

For our immediate use, we will represent the base vector sets in operational matrix form as:

$$\mathbf{G} \equiv [\mathbf{G}_1 \equiv \mathbf{T} \quad \mathbf{G}_2 \equiv \mathbf{M} \quad \mathbf{G}_3 \equiv \mathbf{B}], \quad \mathbf{g} \equiv [\mathbf{g}_1 \equiv \mathbf{n} \quad \mathbf{g}_2 \equiv \mathbf{m} \quad \mathbf{g}_2 \equiv \mathbf{b}] \tag{10.140}$$

where $\mathbf{G}^T\,\mathbf{G} = \mathbf{g}^T\,\mathbf{g} = \mathbf{I}$ with $\mathbf{I}$ as the 3×3 identity matrix. Now, the component representation of the rotation tensor, $\mathbf{R} = {}^t R_{ij}\,\mathbf{G}^i \otimes \mathbf{G}^j$, can be given as the matrix, ${}^t\mathbf{R}$, as:

$${}^t\mathbf{R} = \mathbf{G}^T\,\mathbf{R}\,\mathbf{G}, \qquad \mathbf{R} = \mathbf{G}\,{}^t\mathbf{R}\,\mathbf{G}^T \tag{10.141}$$

where the ij^{th}th component of ${}^t\mathbf{R}$ is given by ${}^tR_{ij}$. Finally, given the relations: $\mathbf{g}_i = \mathbf{R}\,\mathbf{G}_i$, $i = 1, 2, 3$, we can use equation (10.140) and equation (10.141) to express these relations in operational matrix form as:

$$\mathbf{g} \equiv [\mathbf{g}_1 \quad \mathbf{g}_2 \quad \mathbf{g}_3] = [\mathbf{R}\,\mathbf{G}_1 \quad \mathbf{R}\,\mathbf{G}_2 \quad \mathbf{R}\,\mathbf{G}_3] = \mathbf{R}\,\mathbf{G} = \mathbf{G}\,{}^t\mathbf{R}\,\mathbf{G}^T\,\mathbf{G} = \mathbf{G}\,{}^t\mathbf{R} \tag{10.142}$$

10.8.3 Component or Operational Vectors

As indicated in our stated goal, we define here the component vectors by first identifying the coefficients (i.e. the components) of linear combinations and the beam base vectors that make up the corresponding invariant vectors, and then collecting the components in an operational vector form.

10.8.3.1 Component Displacement and Rotation Vectors: ${}^t\mathbf{d}$ and ${}^t\boldsymbol{\theta}$

In terms of the linear combinations of parameters measured (see Figure 10.12) in the undeformed tangent coordinate system, $\mathbf{T}, \mathbf{M}, \mathbf{B}$, the displacement vector, $\mathbf{d}$, an invariant vector measuring the

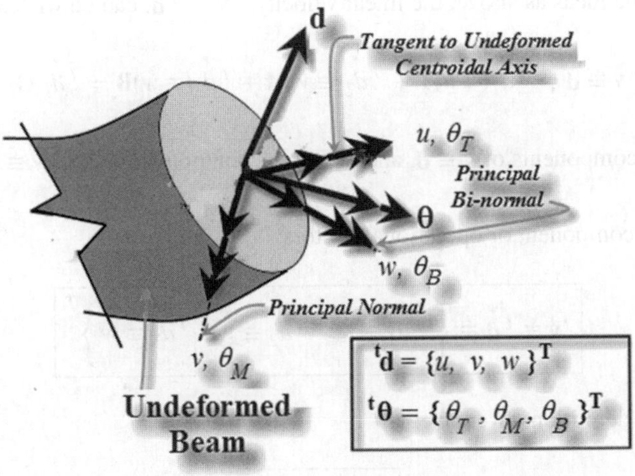

Figure 10.12 Displacement and rotation component vectors in undeformed Frenet frame.

movement of a point on the undeformed curve to its current position on the deformed curve, can be written as:

$$\mathbf{d} = {}^t d_1(\equiv u \equiv d_T)\mathbf{T} + {}^t d_2(\equiv v \equiv d_M)\mathbf{M} + {}^t d_3(\equiv w \equiv d_B)\mathbf{B} = {}^t d_i\, \mathbf{G}_i = \mathbf{G}\, {}^t\mathbf{d} \tag{10.143}$$

and similarly, the rotation vector, $\boldsymbol{\theta}$, as:

$$\boldsymbol{\theta} = {}^t\theta_1(\equiv \theta_T)\mathbf{T} + {}^t\theta_2(\equiv \theta_M)\mathbf{M} + {}^t\theta_3(\equiv \theta_B)\mathbf{B} = {}^t\theta_i\, \mathbf{G}_i = \mathbf{G}\, {}^t\boldsymbol{\theta} \tag{10.144}$$

where $\mathbf{G} \equiv [\mathbf{G}_1 \quad \mathbf{G}_2 \quad \mathbf{G}_3]$ is the orthonormal matrix with base vectors as its columns. Note that although it is immaterial for orthogonal frames, for the sake of consistency with the description in the curvilinear coordinates, the displacement and rotation components are shown as covariant components. We collect the components of $\mathbf{d}$ and $\boldsymbol{\theta}$, and define component vectors ${}^t\mathbf{d}$ and ${}^t\boldsymbol{\theta}$, respectively, as:

Definition: The component or operational vectors: ${}^t\mathbf{d}$ and ${}^t\boldsymbol{\theta}$

$$\boxed{{}^t\mathbf{d} \equiv \{d_1 \equiv d_T \equiv u \quad d_2 \equiv d_M \equiv v \quad d_3 \equiv d_B \equiv w\}^T} \tag{10.145}$$

$$\boxed{{}^t\boldsymbol{\theta} \equiv \{\theta_1 \equiv \theta_T \quad \theta_2 \equiv \theta_M \quad \theta_3 \equiv \theta_B\}^T} \tag{10.146}$$

Note that the superscript t reminds us that the components are measured in the undeformed tangent coordinate system. With the above definitions, we have:

$$\boxed{\mathbf{d} = \mathbf{G}\, {}^t\mathbf{d}, \qquad \boldsymbol{\theta} = \mathbf{G}\, {}^t\boldsymbol{\theta}} \tag{10.147}$$

where $\mathbf{G} \equiv [\mathbf{G}_1 \quad \mathbf{G}_2 \quad \mathbf{G}_3]$ is the orthonormal matrix with base vectors as its columns.

10.8.3.2 Component Linear Velocity Vectors: ${}^t\mathbf{v} \equiv {}^t\dot{\mathbf{d}}$

Following the same ideas as above, the linear velocity, ${}^t\mathbf{v} \equiv {}^t\dot{\mathbf{d}}$, can be written as:

$$\mathbf{v} \equiv \overset{*}{\mathbf{d}} = {}^t\dot{d}_1(\equiv \dot{u})\mathbf{T} + {}^t\dot{d}_2(\equiv \dot{v})\mathbf{M} + {}^t\dot{d}_3(\equiv \dot{w})\mathbf{B} = {}^t\dot{d}_i\, \mathbf{G}_i \tag{10.148}$$

We collect the components of $\mathbf{v} \equiv \overset{*}{\mathbf{d}}$, and define the component vector ${}^t\mathbf{v} \equiv {}^t\overset{*}{\mathbf{d}}$, as:

Definition: The component or operational vectors: ${}^t\dot{\mathbf{d}}$

$$\boxed{{}^t\mathbf{v} \equiv {}^t\overset{*}{\mathbf{d}} \equiv \left\{ {}^t\dot{d}_1 \equiv \dot{u} \quad {}^t\dot{d}_2 \equiv \dot{v} \quad {}^t\dot{d}_3 \equiv \dot{w} \right\}^T} \tag{10.149}$$

With the above definitions, we have:

$$\boxed{\mathbf{v} \equiv \overset{*}{\mathbf{d}} = \mathbf{G}\, {}^t\dot{\mathbf{d}}} \tag{10.150}$$

10.8.3.3 Component Angular Velocity Vectors: ${}^t\boldsymbol{\omega}$

The angular velocity, ${}^t\boldsymbol{\omega}$, can be written as:

$$\boldsymbol{\omega} = {}^t\omega_1\,\mathbf{T} + {}^t\omega_2\,\mathbf{M} + {}^t\omega_3\,\mathbf{B} = {}^t\omega_i\mathbf{G}_i = \mathbf{G}\,{}^t\boldsymbol{\omega} \tag{10.151}$$

We collect the components of ${}^t\boldsymbol{\omega}$, and define the component vector ${}^t\boldsymbol{\omega}$, as:

Definition: The ***component or operational angular velocity vectors***: ${}^t\boldsymbol{\omega}$

$$\boxed{{}^t\boldsymbol{\omega} \equiv \{\,{}^t\omega_1 \quad {}^t\omega_2 \quad {}^t\omega_3\}^T} \tag{10.152}$$

With the above definitions and equation (10.138), we have:

$$\boxed{{}^t\boldsymbol{\omega} \equiv \mathbf{G}\,{}^t\mathbf{W}\,\mathbf{G}^T\,\mathbf{G}\,{}^t\dot{\boldsymbol{\theta}} = \mathbf{G}\,{}^t\mathbf{W}\,{}^t\dot{\boldsymbol{\theta}}} \tag{10.153}$$

10.8.3.4 Component Force and Moment Vectors in Terms of Undeformed Frame: ${}^t\mathbf{F}$ and ${}^t\mathbf{M}$

In terms of the linear combinations of parameters measured in the undeformed tangent coordinate system, $\mathbf{T}, \mathbf{M}, \mathbf{B}$, the force vector, $\mathbf{F}$, an invariant vector measuring the current stress resultant on a deformed beam cross-section, and the moment vector, $\mathbf{M}$, an invariant vector measuring the current stress couple resultant on a deformed beam cross-section, can be written, respectively, as:

$$\mathbf{F} = {}^tF^1(\equiv N^T = N_T)\,\mathbf{T} + {}^tF^2(\equiv Q^M = Q_M)\,\mathbf{M} + {}^tF^3(\equiv Q^B = Q_B)\,\mathbf{B} = {}^tF^i\,\mathbf{G}_i \tag{10.154}$$

where, N_T is the axial force, and Q_M, Q_B are the shear forces; conventionally and notation-wise, these are described as such in engineering terms as shown in Figure 10.13, and

$$\mathbf{M} = {}^tM^1(\equiv M^T = M_T)\,\mathbf{T} + {}^tM^2(\equiv M^M = M_M)\,\mathbf{M} + {}^tM^3(\equiv M^B = M_B)\,\mathbf{B} = {}^tM^i\,\mathbf{G}_i \tag{10.155}$$

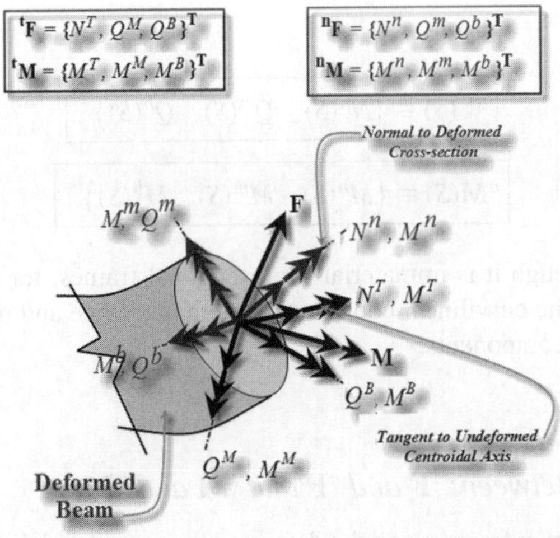

Figure 10.13 Force and moment component vectors in undeformed and deformed coordinate systems.

where, M_T is the twisting moment, M_M, M_B are the bending moments; conventionally and notation-wise, these are described as such in engineering terms, as shown in Figure 10.13. We collect the components of **F** and M, and define component vectors, respectively, as:

Definition

$$\boxed{{}^t\mathbf{F} \equiv \left\{F^1 \equiv N^T = N_T \quad F^2 \equiv Q^M = Q_M \quad F^3 \equiv Q^B = Q_B\right\}^T} \tag{10.156}$$

$$\boxed{{}^t\mathbf{M} \equiv \left\{M^1 \equiv M^T = M_T \quad M^2 \equiv M^M = M_M \quad M^3 \equiv M^B = M_B\right\}^T} \tag{10.157}$$

With the above definitions, we have:

$$\boxed{\mathbf{F} = \mathbf{G}\,{}^t\mathbf{F}, \qquad \mathbf{M} = \mathbf{G}\,{}^t\mathbf{M}} \tag{10.158}$$

where $\mathbf{G} \equiv [\mathbf{G}_1 \quad \mathbf{G}_2 \quad \mathbf{G}_3]$ is the orthonormal matrix with base vectors as its columns. Note that, although it is immaterial for orthogonal frames, for the sake of consistency with the description in the curvilinear coordinate system, the force and moment components are shown as both covariant and contravariant components.

10.8.3.5 Component Force and Moment Vectors in Terms of Deformed Frame: ${}^n\mathbf{F}$ and ${}^n\mathbf{M}$

In terms of the parameters measured in the current normal coordinate system, $\mathbf{n}, \mathbf{m}, \mathbf{b}$, let the current stress resultants (or forces): $\mathbf{F} = {}^nF^1(\equiv N^n)\,\mathbf{n} + {}^nF^2(\equiv Q^m)\,\mathbf{m} + {}^nF^3(\equiv Q^b)\,\mathbf{b} = {}^nF^i\,\hat{\mathbf{g}}_i$ and the current resultant stress couple (or moments):

$$\mathbf{M} = {}^nM^1(\equiv M^n)\,\mathbf{n} + {}^nM^2(\equiv M^m)\,\mathbf{m} + {}^nM^3(\equiv M^b)\,\mathbf{b} = {}^nM^i\,\mathbf{g}_i \tag{10.159}$$

We collect the components of **F** and **M**, and define component vectors, respectively, as:

Definition

$$\boxed{{}^n\mathbf{F}(S) \equiv \left\{N^n(S) \quad Q^m(S) \quad Q^b(S)\right\}^{\mathbf{T}}} \tag{10.160}$$

$$\boxed{{}^n\mathbf{M}(S) \equiv \left\{M^n(S) \quad M^m(S) \quad M^b(S)\right\}^{\mathbf{T}}} \tag{10.161}$$

Note again that, although it is immaterial for orthogonal frames, for the sake of consistency with the description in the curvilinear coordinate system, the force and moment components are shown as contravariant components.

10.8.4 Relations Between: ${}^t\mathbf{F}$ *and* ${}^n\mathbf{F}$ *and* ${}^t\mathbf{M}$ *and* ${}^n\mathbf{M}$

Considering that the $\mathbf{n}, \mathbf{m}, \mathbf{b}$ frame along the deformed beam centroidal axis, $\mathbf{c}_0(S(\xi))$, $\xi \in [0, 1]$, is defined by the orthogonal rotation tensor, $\mathbf{R}(S)$, measured in the $\mathbf{N} = \mathbf{T}, \mathbf{M}, \mathbf{B}$ frame of the

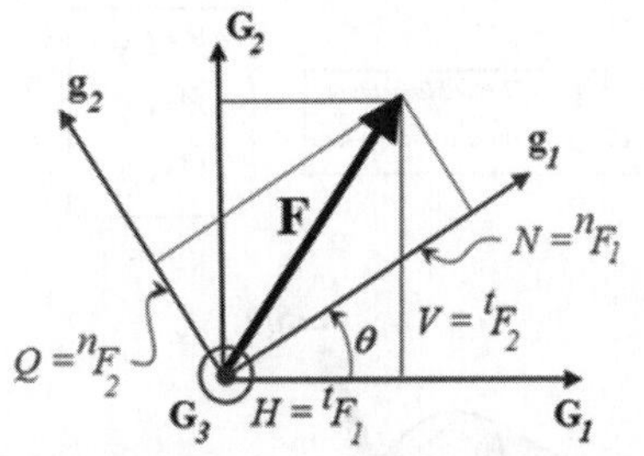

Figure 10.14 Component transformation due to rotation pf coordinate system.

undeformed beam, the two sets of the stress resultants and stress couples component vectors are related by the following relations:

$$\boxed{\mathbf{F} = \mathbf{G}\,{}^t\mathbf{F} = \mathbf{g}\,{}^n\mathbf{F} = \mathbf{G}\,{}^t\mathbf{R}\,{}^n\mathbf{F} \quad \text{and} \quad \mathbf{M} = \mathbf{G}\,{}^t\mathbf{M} = \mathbf{g}\,{}^n\mathbf{M} = \mathbf{G}\,{}^t\mathbf{R}\,{}^n\mathbf{M},} \tag{10.162}$$

where we have used equation (10.142). From equation (10.162), we can finally write the following relationships relating the components of forces and moments in undeformed and current spinning frames as:

$$\boxed{{}^t\mathbf{F} = \mathbf{R}\,{}^n\mathbf{F} \quad \text{and} \quad {}^t\mathbf{M} = \mathbf{R}\,{}^n\mathbf{M}} \tag{10.163}$$

To exemplify this, let us consider a planar rotation, θ, about, say, the basis vector $\mathbf{G}_3$, so that the rotation tensor has the matrix representation as:

$${}^t\mathbf{R} = \begin{bmatrix} \cos\theta & -\sin\theta & 0 \\ \sin\theta & \cos\theta & 0 \\ 0 & 0 & 1 \end{bmatrix} \tag{10.164}$$

Now, if an invariant force vector, $\mathbf{F}$, has component representations in column vectors, ${}^t\mathbf{F} \equiv \{H \equiv {}^tF_1 \quad V \equiv {}^tF_2 \quad h\}^T$ and ${}^n\mathbf{F} \equiv \{N \equiv {}^nF_1 \quad Q \equiv {}^nF_2 \quad h\}^T$ in basis sets $\{\mathbf{g}_i\}$ and $\{\mathbf{G}_i\}$, respectively, that is, $\mathbf{F} = {}^tF^i\,\mathbf{G}_i = {}^nF^j\,\mathbf{g}_j$, then, as shown in Figure 10.14, we have by simple application of the trigonometry:

$$\begin{Bmatrix} H \\ V \\ h \end{Bmatrix} = \begin{bmatrix} \cos\theta & -\sin\theta & 0 \\ \sin\theta & \cos\theta & 0 \\ 0 & 0 & 1 \end{bmatrix} \begin{Bmatrix} N \\ Q \\ h \end{Bmatrix}.$$
$$ {}^t\mathbf{F} = {}^t\mathbf{R}\,{}^n\mathbf{F} \quad \Rightarrow \quad {}^n\mathbf{F} = {}^t\mathbf{R}^T\,{}^t\mathbf{F} \tag{10.165}$$

A similar example can be devised for the components of a moment vector.

10.8.4.1 Components of Linear Momentum and Angular Momentum Vectors in Terms of the Undeformed Frame: ${}^t\mathbf{L}(S,t)$ and ${}^t\mathbf{A}(S,t)$

The invariant linear momentum vector, $\mathbf{L}(S,t)$, and the angular momentum vector, $\mathbf{A}(S,t)$, for any $t \in \mathbb{R}_+$, can have component representation (see Figure 10.15) in any triad of base vectors,

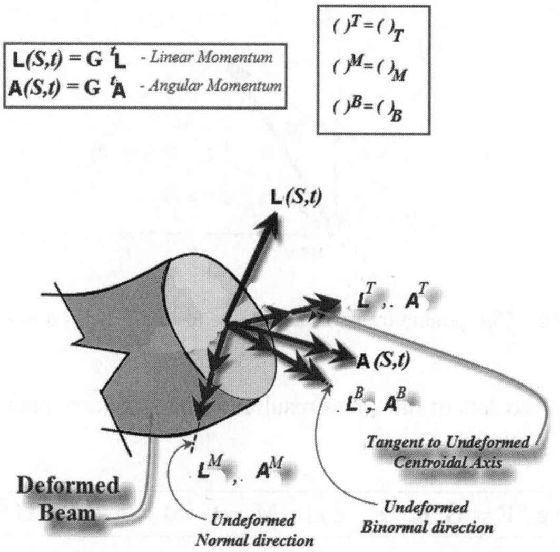

Figure 10.15 Linear and angular momentum component vectors in undeformed Frenet frame.

Cartesian or curvilinear. Thus, we can write:

$$\boxed{\begin{aligned} \mathbf{L}(S,t) &= {}^tL^T(\equiv L_T)\ \mathbf{G_1} + {}^tL^M(\equiv L_M)\ \mathbf{G_2} + {}^tL^B(\equiv L_B)\ \mathbf{G}_3 \\ &= {}^tL^i(S,t)\ \mathbf{G}_i(S), \quad i = 1,2,3 \end{aligned}} \tag{10.166}$$

and

$$\boxed{\begin{aligned} \mathbf{A}(S,t) = &= {}^tA^T(\equiv A_T)\ \mathbf{G_1} + {}^tA^M(\equiv A_M)\ \mathbf{G_2} + {}^tA^B(\equiv A_B)\ \mathbf{G}_3 \\ &= {}^tA^i(S,t)\ \mathbf{G}_i(S), \quad i = 1,2,3 \end{aligned}} \tag{10.167}$$

We collect the components of **L** and **A**, and define the component vectors, respectively, as follows.

Definitions

$$\boxed{{}^t\mathbf{L}(S,t) \equiv \left\{ {}^tL^1 \quad {}^tL^2 \quad {}^tL^3 \right\}^T \quad \text{and} \quad {}^t\mathbf{A}(S,t) \equiv \left\{ {}^tA^1 \quad {}^tA^2 \quad {}^tA^3 \right\}^T} \tag{10.168}$$

With the above definitions, we can express these in the column vector form as:

$$\boxed{\mathbf{L}(S,t) = \mathbf{G}(S)\ {}^t\mathbf{L}(S,t) \quad \text{and} \quad \mathbf{A}(S,t) = \mathbf{G}(S)\ {}^t\mathbf{A}(S,t)} \tag{10.169}$$

Likewise, we have the time derivatives of the linear momentum and the angular momentum in terms of their material component vectors:

$$\boxed{\dot{\mathbf{L}}(S,t) = \mathbf{G}(S)\ {}^t\dot{\mathbf{L}}(S,t) \quad \text{and} \quad \dot{\mathbf{A}}(S,t) = \mathbf{G}(S)\ {}^t\dot{\mathbf{A}}(S,t)} \tag{10.170}$$

10.8.5 Where We Would Like to Go

We have now defined the component or operational vectors corresponding to the displacement and rotation vectors, the force and moment vectors, and the linear and angular momentums. These definitions play a vital role in converting the vector form of beam balance equations (3D to 1D) or beam balance equations (direct 1D) to computationally more suitable component vector form of balance equations. However, to understand this conversion, we would need:

- derivatives of component vectors
- computational curvatures, angular velocity and variations.

  Thus, with these background materials, we will be able to modify the equilibrium equations in terms of what may be called the component vector or operator form as described in:
- computational balance equations

10.9 Covariant Derivatives of Component Vectors

10.9.1 What We Need to Recall

- the Frenet–Serret relation for the derivatives of the triad of base vectors in matrix form with respect to the arc length parameter, S, is given by:

$$\boxed{\begin{Bmatrix}\mathbf{T}'\\ \mathbf{M}'\\ \mathbf{B}'\end{Bmatrix} \equiv \frac{\partial}{\partial S}\begin{Bmatrix}\mathbf{T}\\ \mathbf{M}\\ \mathbf{B}\end{Bmatrix} = \begin{bmatrix}0 & \kappa & 0\\ -\kappa & 0 & \tau\\ 0 & -\tau & 0\end{bmatrix}\begin{Bmatrix}\mathbf{T}\\ \mathbf{M}\\ \mathbf{B}\end{Bmatrix} \equiv \overset{\frown}{\mathbf{K}}{}^{0}\begin{Bmatrix}\mathbf{T}\\ \mathbf{M}\\ \mathbf{B}\end{Bmatrix}} \tag{10.171}$$

  where $\overset{\frown}{\mathbf{K}}{}^{0}$ is the beam undeformed curvature matrix. Note that the curvature matrix is skew-symmetric. With the notation in curvilinear base vectors: $\mathbf{G}_1 \equiv \mathbf{T}$, $\mathbf{G}_2 \equiv \mathbf{M}$, $\mathbf{G}_3 \equiv \mathbf{B}$, we can rewrite the matrix form in compact notation as:

$$\boxed{\mathbf{G}_i' = \overset{\frown}{K}{}^{0}_{ij}\,\mathbf{G}_j,\ i = 1,2,3;\ ,\ j = 1,2,3} \tag{10.172}$$

  Clearly, this matrix form gives the derivative of a base vector as a linear combination of the base vectors.
- On the other hand, the Frenet–Serret relation for the derivatives of the triad of base vectors along a curve in invariant tensor form with respect to the arc length parameter, S, may be obtained by first introducing the following.

Definition: ***Beam initial curvature tensor***, $\mathbf{K}^0$:

$$\boxed{\mathbf{K}^0 \equiv (\overset{\frown}{K}{}^{0}_{ij})^T\,\mathbf{G}_i \otimes \mathbf{G}_j = -\mathbf{G}_i \otimes \mathbf{G}_i'} \tag{10.173}$$

where we have used the result from equation (10.172). Note that the curvature tensor is skew-symmetric and thus: $(\mathbf{K}^0)^T = -\mathbf{K}^0$.

Definition: ***Beam initial curvature vector***, $\mathbf{k}^0$:

The axial vector of the initial curvature tensor, $\mathbf{K}^0$, is given by $\mathbf{k}^0 \equiv \{\tau \quad 0 \quad \kappa\}$. Note that $\mathbf{k}^0$ is defined as the negative of the ***Darboux's vector*** of Frenet–Serret relations.

Now, we can easily get from equation (10.173), the invariant tensor form as:

$$\boxed{\mathbf{G}_i' = \mathbf{K}^0\,\mathbf{G}_i = \mathbf{k}^0 \times \mathbf{G}_i, \quad i = 1,2,3} \tag{10.174}$$

> Note that, in tensor form, unlike the matrix expression (10.172), each base vector derivative is given as a transformation of itself. As we shall see, this allows us to write expressions in absolute compact form, free of matrix indices.

***Exercise*:** Verify the tensor form of the derivatives of base vectors given by equation (10.174).

***Solution*:** $\boxed{\mathbf{K}^0\,\mathbf{G}_i = -(\mathbf{K}^0)^T\,\mathbf{G}_i = -(-\mathbf{G}_i' \otimes \mathbf{G}_i)\mathbf{G}_i = \mathbf{G}_i', \quad i = 1,2,3}$

- Finally, the covariant derivative of a base vector is the zero vector, that is, $\mathbf{G}_i|_S = \mathbf{0}$, $i = 1,2,3$. Thus, the covariant derivative of a tensor function of any order is identical with its partial derivative. If we take a vector (a tensor of order one) function, $\mathbf{a}(S)$, we, then, have: $\mathbf{a}|_S = \mathbf{a},_S = {}^t a_i|_S \mathbf{G}_i = \mathbf{G}\,{}^t\mathbf{a}|_S$ where $\mathbf{G} \equiv [\mathbf{G}_1 \quad \mathbf{G}_2 \quad \mathbf{G}_3]$ is the orthonormal matrix with base vectors as its columns, and ${}^t\mathbf{a}(S)$ is the component vector corresponding to the vector function, $\mathbf{a}(S)$.

***Exercise*:** Show that the partial derivative of a tensor function of any order is identical with its covariant derivative with respect to its arguments. Hint: for vector function: $\mathbf{d}\,|_S = {}^t d_i\,|_S\,\mathbf{G}_i + {}^t d_i\,\cancel{\mathbf{G}_i|_S} = {}^t d_i\,|_S\,\mathbf{G}_i = \mathbf{d},_S$, why do you think $\mathbf{G}_i|_S = \mathbf{0}$?

10.9.2 Covariant Derivatives of Component or Operational Vectors

Now, having found the above results we are in a position to express the derivatives of various component vectors, with the help of the following lemma for a beam.

Lemma: Given any vector function, $\mathbf{a}(S)$, on a beam, and its corresponding component vector, ${}^t\mathbf{a}(S)$, the covariant derivative of the component vector is given as:

$$\boxed{{}^t\mathbf{a}|_S = {}^t\mathbf{a},_S + \mathbf{k}^0 \times {}^t\mathbf{a} \quad \Rightarrow \quad {}^t\mathbf{a}|_S = {}^t\mathbf{a},_S + \mathbf{K}^0\,{}^t\mathbf{a}} \tag{10.175}$$

where $\times$ denotes the vector cross product operation.

Proof. Using results from above, we have:

$$\boxed{\mathbf{a}\,|_S = \mathbf{a},_S = {}^t a_i,_S\ \mathbf{G}_i + {}^t a_i\ \mathbf{G}_i,_S = \mathbf{G}(\,{}^t\mathbf{a},_S + \mathbf{K}^0\,{}^t\mathbf{a}) = \mathbf{G}({}^t\mathbf{a},_S + \mathbf{k}^0 \times {}^t\mathbf{a})} \tag{10.176}$$

Now, noting that $\mathbf{a}\,|_S = \mathbf{a},_S = \mathbf{G}\,{}^t\mathbf{a}\,|_S$ completes the proof.

10.9.2.1 Covariant Derivatives of Component Displacement and Rotation Vectors: ${}^t\mathbf{d}$ and ${}^t\boldsymbol{\theta}$

Based on equation (10.175) of the lemma, we have:

$$\boxed{{}^t\mathbf{d}|_S = {}^t\mathbf{d},_S + \mathbf{k}^0 \times {}^t\mathbf{d}, \quad \text{and} \quad {}^t\boldsymbol{\theta}|_S = {}^t\boldsymbol{\theta},_S + \mathbf{k}^0 \times {}^t\boldsymbol{\theta}} \tag{10.177}$$

Now, with the definition: $\boxed{{}^t\mathbf{d} \equiv \{d_1 \equiv d_T \equiv u \quad d_2 \equiv d_M \equiv v \quad d_3 \equiv d_B \equiv w\}^T}$, we can express the covariant derivative of the displacement vector in the component form as:

$$\boxed{{}^t\mathbf{d}|_S = \begin{Bmatrix} u|_S \\ v|_S \\ w|_S \end{Bmatrix} = \begin{Bmatrix} u,_S - \kappa v \\ v,_S + \kappa u - \tau w \\ w,_S + \tau v \end{Bmatrix}} \tag{10.178}$$

Similarly, results can be obtained for the rotation vector, ${}^t\boldsymbol{\theta}$.

***Exercise*:** Verify the component form given by equation (10.178) for ${}^t\boldsymbol{\theta}$.

10.9.2.2 Covariant Derivatives of Component Force and Moment Vectors: ${}^t\mathbf{F}$ and ${}^t\mathbf{M}$

Based on equation (10.175) of the lemma, we have:

$$\boxed{{}^t\mathbf{F}|_S = {}^t\mathbf{F},_S + \mathbf{k}^0 \times {}^t\mathbf{F} \quad \text{and} \quad {}^t\mathbf{M}|_S = {}^t\mathbf{M},_S + \mathbf{k}^0 \times {}^t\mathbf{M}} \tag{10.179}$$

Now, with the definition: $\boxed{{}^t\mathbf{F}(S) \equiv \{N^T(S) \quad Q^M(S) \quad Q^B(S)\}^{\mathrm{T}}}$, we can express the covariant derivative of the force vector in the component form as:

$$\boxed{{}^t\mathbf{F}|_S = \begin{Bmatrix} N^T|_S \\ Q^M|_S \\ Q^B|_S \end{Bmatrix} = \begin{Bmatrix} N^T,_S - \kappa\, Q^M \\ Q^M,_S + \kappa\, N^T - \tau\, Q^B \\ Q^B,_S + \tau\, Q^M \end{Bmatrix}} \tag{10.180}$$

Similarly, results can be obtained for the moment vector, ${}^t\mathbf{M}$.

***Exercise*:** Verify the component form given by equation (10.180) for ${}^t\mathbf{M}$.

10.9.3 Where We Would Like to Go

We have now defined the covariant derivatives of the component or operational vectors corresponding to the displacement and rotation vectors, and the force and moment vectors. These definitions play a vital role in converting the vector form of:

- beam balance equations, 3D to 1D
- beam 1D direct balance equations into a computationally more suitable component vector form of balance equations. Thus, with these background materials, we will be able to modify the

equations in terms of what may be called the component vector or operator form as described in the:

- computational balance equations.

10.10 Computational Equations of Motion: Component Vector Form

10.10.1 What We Need to Recall

- The invariant vector beam dynamic balance equations are given by:

$$\boxed{\begin{array}{rl} \mathbf{F}|_S + \hat{\mathbf{F}} = \dot{\mathbf{L}} & \text{Momentum balance} \\ \mathbf{M}|_S + \mathbf{c}_0|_S \times \mathbf{F} + \hat{\mathbf{M}} = \dot{\mathbf{A}} & \text{Moment of momentum balance} \end{array}} \tag{10.181}$$

- We have for the component or operational vectors: ${}^t\mathbf{d}$ and ${}^t\boldsymbol{\theta}$

$$\boxed{{}^t\mathbf{d} \equiv \{d_1 \equiv u \quad d_2 \equiv v \quad d_3 \equiv w\}^T} \tag{10.182}$$

$$\boxed{{}^t\boldsymbol{\theta} \equiv \{\theta_1 \quad \theta_2 \quad \theta_3\}^T} \tag{10.183}$$

with:

$$\boxed{\mathbf{d} = \mathbf{G}\,{}^t\mathbf{d} \qquad \text{and} \qquad \boldsymbol{\theta} = \mathbf{G}\,{}^t\boldsymbol{\theta}} \tag{10.184}$$

- Similarly, for the forces and moments, we have:

$$\boxed{\begin{aligned} {}^t\mathbf{F} &= {}^tF^1(\equiv N_T)\,\mathbf{G}_1(=\mathbf{T}) + {}^tF^2(\equiv Q_M)\,\mathbf{G}_2(=\mathbf{M}) + {}^tF^3(\equiv Q_B)\,\mathbf{G}_3(=\mathbf{B}) \\ &= {}^tF^i\,\mathbf{G}_i, \quad i = 1,2,3 \end{aligned}} \tag{10.185}$$

$$\boxed{\begin{aligned} {}^t\mathbf{M} &= {}^tM^1(\equiv M_T)\,\mathbf{G}_1(=\mathbf{T}) + {}^tM^2(\equiv M_M)\,\mathbf{G}_2(=\mathbf{M}) + {}^tM^3(\equiv M_B)\,\mathbf{G}_3(=\mathbf{B}) \\ &= {}^tM^i\,\mathbf{G}_i, \quad i = 1,2,3 \end{aligned}} \tag{10.186}$$

with:

$$\boxed{\mathbf{F} = \mathbf{G}\,{}^t\mathbf{F} \qquad \text{and} \qquad \mathbf{M} = \mathbf{G}\,{}^t\mathbf{M}} \tag{10.187}$$

- For the covariant derivative of displacement component vectors, we have:

$$\mathbf{d}|_S = \mathbf{d}_{,S} = \mathbf{G}\ {}^t\mathbf{d}|_S \tag{10.188}$$

with:

$$\boxed{{}^t\mathbf{d}|_S = {}^t\mathbf{d}_{,S} + \mathbf{k}^0 \times {}^t\mathbf{d} \qquad \Leftrightarrow \qquad {}^t\mathbf{d}|_S = {}^t\mathbf{d}_{,S} + \mathbf{K}^0\ {}^t\mathbf{d}} \tag{10.189}$$

Similar results follow for the rotation component vector.

- Likewise, we have the covariant derivatives of forces and moments:

$$\mathbf{F}_{,S} = \mathbf{G}\ {}^t\mathbf{F}|_S \tag{10.190}$$

with:

$$\boxed{{}^t\mathbf{F}|_S = {}^t\mathbf{F}_{,S} + \mathbf{k}^0 \times {}^t\mathbf{F} \quad \Leftrightarrow \quad {}^t\mathbf{F}|_S = {}^t\mathbf{F}_{,S} + \mathbf{K}^0\ {}^t\mathbf{F}} \tag{10.191}$$

and

$$\mathbf{M}_{,S} = \mathbf{G}\ {}^t\mathbf{M}|_S \tag{10.192}$$

with:

$$\boxed{{}^t\mathbf{M}|_S = {}^t\mathbf{M}_{,S} + \mathbf{k}^0 \times {}^t\mathbf{M} \quad \Leftrightarrow \quad {}^t\mathbf{M}|_S = {}^t\mathbf{M}_{,S} + \mathbf{K}^0\ {}^t\mathbf{M}} \tag{10.193}$$

In the above expressions, $\mathbf{k}^0$ is the initial curvature vector, that is, the axial vector corresponding to the initial curvature tensor, $\mathbf{K}^0$, and is given by $\mathbf{k}^0 \equiv \{\tau \quad 0 \quad \kappa\}$. Finally, $\mathbf{G} = [\mathbf{G}_1 \equiv \mathbf{T} \quad \mathbf{G}_2 \equiv \mathbf{M} \quad \mathbf{G}_3 \equiv \mathbf{B}]$ is the 3×3 orthonormal matrix with the base vectors stored as its columns.
- Likewise, we have the time derivatives of the linear momentum and the angular momentum in terms of the components measured in the undeformed coordinate system:

$$\boxed{\mathbf{L}(S,t) = \mathbf{G}\ {}^t\mathbf{L}(S,t) \quad \text{and} \quad \mathbf{A}(S,t) = \mathbf{G}\ {}^t\mathbf{A}(S,t)} \tag{10.194}$$

10.10.1.1 Covariant Derivative of the Deformed Position Vector

The derivative of the deformed position vector, $\mathbf{c}_0$, in the second of the balance equation (10.181), can be rewritten as:

$$\boxed{\begin{aligned}\mathbf{c}_0|_S = \mathbf{c}_{0,S} &= \mathbf{C}_{0,S} + \mathbf{d}_{,S} = \mathbf{G}_1 + \mathbf{G}\ {}^t\mathbf{d}|_S \\ &= \mathbf{G}\,(\hat{\mathbf{1}} + {}^t\mathbf{d}_{,S} + \mathbf{k}^0 \times {}^t\mathbf{d}) = \mathbf{G}\,(\mathbf{e} + \mathbf{k}^0 \times {}^t\mathbf{d}) = \mathbf{G}\,\mathbf{a}\end{aligned}} \tag{10.195}$$

where we have introduced:

$$\boxed{\begin{aligned}\mathbf{a} &\equiv (\hat{\mathbf{1}} + {}^t\mathbf{d}_{,S}) + \mathbf{k}^0 \times {}^t\mathbf{d} = \mathbf{e} + \mathbf{k}^0 \times {}^t\mathbf{d} \\ &= \{(1 + u' - \kappa v) \quad (v' + \kappa u - \tau w) \quad (w' + \tau v)\}^T\end{aligned}} \tag{10.196}$$

Definitions

$$\boxed{\hat{\mathbf{1}} \equiv \{1 \quad 0 \quad 0\}^T} \tag{10.197}$$

is the unit pseudo-vectors along the base vectors, $\{\mathbf{G}_i\}$, $i = 1, 2, 3$, and

$$\boxed{\mathbf{e} \equiv \hat{\mathbf{1}} + {}^t\mathbf{d}' = \{(1 + u') \quad v' \quad w'\}^T} \tag{10.198}$$

Then, from the second term in the angular momentum balance equation (10.181), by using the first relation in equation (10.187), we have:

$$\mathbf{c}_{0},_{S} \times \mathbf{F} = \mathbf{G}\,\mathbf{a} \times \mathbf{G}\,{}^{t}\mathbf{F} = (\det\,\mathbf{G})\,\mathbf{G}(\mathbf{a} \times\,{}^{t}\mathbf{F}) = \mathbf{G}(\mathbf{a} \times\,{}^{t}\mathbf{F}) \tag{10.199}$$

Exercise: Show that for any two non-collinear vector $\mathbf{b}_1, \mathbf{b}_2$, and an orthonormal tensor $\mathbf{A}$: $\mathbf{A}\mathbf{b}_1 \times \mathbf{A}\mathbf{b}_2 = (\det \mathbf{A})\,\mathbf{A}(\mathbf{b}_1 \times \mathbf{b}_2)$.

Solution: Let $\mathbf{b}_3$ be any arbitrary vector non-collinear with $\mathbf{b}_1, \mathbf{b}_2$. Let us define a tensor $\mathbf{B} \equiv \mathbf{b}_i \otimes \mathbf{G}^i$ where $\mathbf{G}^i$ are the contravariant base vectors. Note that $\mathbf{b}_i \equiv \mathbf{B}\,\mathbf{G}_i$; that is, the matrix representation of $\mathbf{B}$ has three columns as $\mathbf{b}_1, \mathbf{b}_2$ and $\mathbf{b}_3$. Thus, $\det \mathbf{B} = \mathbf{b}_1 \times \mathbf{b}_2\ \mathbf{b}_3$. Note that $\mathbf{AB} = (\mathbf{A}\mathbf{b}_j \otimes \mathbf{G}^j)$ because for $\forall \mathbf{h} \in \mathbb{R}_3$, $\mathbf{ABh} = \mathbf{A}(\mathbf{b}_j \otimes \mathbf{G}^j)\mathbf{h} = \mathbf{A}(\mathbf{G}^j \bullet \mathbf{h})\mathbf{b}_j = (\mathbf{A}\mathbf{b}_j)(\mathbf{G}^j \bullet \mathbf{h}) = (\mathbf{A}\mathbf{b}_j \otimes \mathbf{G}^j)\mathbf{h}$. So, $\det(\mathbf{AB}) = \mathbf{A}\mathbf{b}_1 \times \mathbf{A}\mathbf{b}_2 \bullet \mathbf{A}\mathbf{b}_3$. But, by the product rule of determinants, we have: $\det(\mathbf{AB}) = \det(\mathbf{A})\det(\mathbf{B}) = \det(\mathbf{A})(\mathbf{b}_1 \times \mathbf{b}_2 \bullet \mathbf{b}_3)$. So, we get: $\mathbf{A}^T(\mathbf{A}\mathbf{b}_1 \times \mathbf{A}\mathbf{b}_2) \bullet \mathbf{b}_3 = \det(\mathbf{A})(\mathbf{b}_1 \times \mathbf{b}_2) \bullet \mathbf{b}_3$. Now, noting that $\mathbf{b}_3$ is arbitrary and that $\mathbf{A}\mathbf{A}^T = \mathbf{I}$ because $\mathbf{A}$ is orthonormal, completes the proof.

Now, evaluating the expressions in equation (10.196), we have:

$$\mathbf{a} = \begin{Bmatrix} 1 + u,_s - \kappa v \\ v,_s + \kappa u - \tau w \\ w,_s + \tau v \end{Bmatrix} \tag{10.200}$$

10.10.1.2 Component Vector Form of Dynamic Balance Equations

Now, by applying equations (10.170), (10.190) and (10.192) in the equation (10.181), we get with equation (10.199):

$$\boxed{\begin{aligned} \mathbf{G}({}^{t}\mathbf{F}|_S + {}^{t}\hat{\mathbf{F}} - {}^{t}\dot{\mathbf{L}}) = 0 \\ \mathbf{G}({}^{t}\mathbf{M}|_S + \mathbf{a} \times {}^{t}\mathbf{F} + {}^{t}\hat{\mathbf{M}} - {}^{t}\dot{\mathbf{A}}) = 0 \end{aligned}} \tag{10.201}$$

10.10.1.3 Expanded Component-wise Force and Moment Covariant Derivatives

As a preliminary result, we may expand expressions in equation (10.191) component-wise:

$$\boxed{\begin{aligned} {}^{t}\mathbf{F}|_S &= {}^{t}\mathbf{F},_S + \mathbf{k}^0 \times {}^{t}\mathbf{F} \\ &= \begin{Bmatrix} N_T' - \kappa Q_M \\ Q_M' + \kappa N_T - \tau Q_B \\ Q_B' + \tau Q_M \end{Bmatrix} = \begin{Bmatrix} N_T' \\ Q_M' \\ Q_B' \end{Bmatrix} + \begin{bmatrix} 0 & -\kappa & 0 \\ \kappa & 0 & -\tau \\ 0 & \tau & 0 \end{bmatrix} \begin{Bmatrix} N_T \\ Q_M \\ Q_B \end{Bmatrix} \\ &= {}^{t}\mathbf{F}' + \mathbf{K}^0\,{}^{t}\mathbf{F} \end{aligned}} \tag{10.202}$$

Similarly, from equation (10.193) with the definition of $\mathbf{a}_\alpha$ as above, we get:

$$
\begin{aligned}
{}^t\mathbf{M}|_s + \mathbf{a} \times {}^t\mathbf{F} &= {}^t\mathbf{M}' + \kappa^0 \times {}^t\mathbf{M} + \{\kappa^0 \times \mathbf{d} + \mathbf{e}\} \times {}^t\mathbf{F} \\
&= \begin{Bmatrix} M_T' - \kappa M_M + (w' - \tau v + \kappa u)Q_B - (v' + \tau w)Q_M \\ M_M' + \kappa M_T - \tau M_B + (v' + \tau w) - (1 + u' - \kappa w)Q_B \\ M_B' + \tau M_M + (1 + u' - \kappa w)Q_M - (w' - \tau v + \kappa u)N_T \end{Bmatrix} \\
&= {}^t\mathbf{M}' + \mathrm{K}^0\,{}^t\mathbf{M} + \{(\mathrm{K}^0\,\mathbf{D} - \mathbf{D}\,\mathrm{K}^0) + \mathbf{E}\}\,{}^t\mathbf{F}
\end{aligned}
\tag{10.203}
$$

In the above expression, we have used the following:

- $\mathbf{D} \equiv [\mathbf{d}]$ is the skew-symmetric tensor with the displacement vector, $\mathbf{d} = \{u \equiv d_1 \quad v \equiv d_2 \quad w \equiv d_3\}^T$ taken as its axial vector.
- $\mathbf{E} \equiv [\mathbf{e}]$ is the skew-symmetric tensor corresponding to the special displacement derivative vector $\mathbf{e} \equiv \{(1 + u') \quad v' \quad w'\}^T = \hat{\mathbf{1}} + \mathbf{d}'$ taken as its axial vector.

Finally, we have used the Lie bracket definition, a (commutator) property relating a skew-symmetric tensor to its axial vector.

***Exercise*:** Given any two skew-symmetric tensors, $\mathbf{A}$ and $\mathbf{B}$ with the corresponding axial vectors, $\mathbf{a}$ and $\mathbf{b}$, respectively, we have the commutator (or Lie bracket) relationship: $\mathbf{AB} - \mathbf{BA} = [\mathbf{a} \times \mathbf{b}] = (\mathbf{a} \times \mathbf{b})\times$.

***Solution*:** for any $\mathbf{h} \in \mathbb{R}_3$, $(\mathbf{AB} - \mathbf{BA})\mathbf{h} = \mathbf{a} \times (\mathbf{b} \times \mathbf{h}) - \mathbf{b} \times (\mathbf{a} \times \mathbf{h}) = (\mathbf{a} \bullet \mathbf{h})\mathbf{b} - \cancel{(\mathbf{a} \bullet \mathbf{b})\mathbf{h}} - (\mathbf{b} \bullet \mathbf{h})\mathbf{a} + \cancel{(\mathbf{a} \bullet \mathbf{b})\mathbf{h}} = (\mathbf{a} \bullet \mathbf{h})\mathbf{b} - (\mathbf{b} \bullet \mathbf{h})\mathbf{a} = \{(\mathbf{b} \otimes \mathbf{a}) - (\mathbf{a} \otimes \mathbf{b})\}\mathbf{h}$ completes the first equality; also, $(\mathbf{a} \bullet \mathbf{h})\mathbf{b} - (\mathbf{b} \bullet \mathbf{h})\mathbf{a} = \mathbf{h} \times (\mathbf{b} \times \mathbf{a}) = (\mathbf{a} \times \mathbf{b}) \times \mathbf{h} = [\mathbf{a} \times \mathbf{b}]\mathbf{h}$ establishes the rest of the equalities. In all of the derivations, we have used the vector identity: $\mathbf{a} \times (\mathbf{b} \times \mathbf{c}) = (\mathbf{a} \bullet \mathbf{c})\mathbf{b} - (\mathbf{a} \bullet \mathbf{b})\mathbf{c}$.

***Exercise*:** Verify expressions given by equation (10.202) and equation (10.203).

10.10.1.4 Component-wise Dynamic Balance Equations

Now, using equations (10.202) and (10.203) in equation (10.201), we can express the combined balance equations component-wise as:

$$
\begin{aligned}
\mathbf{G}({}^t\mathbf{F},_S + \kappa^0 \times {}^t\mathbf{F} + {}^t\hat{\mathbf{F}} - {}^t\dot{\mathbf{L}}) = \mathbf{0} \\
\mathbf{G}({}^t\mathbf{M},_S + \kappa^0 \times {}^t\mathbf{M} + \{\kappa^0 \times \mathbf{d} + \mathbf{e}\} \times {}^t\mathbf{F} + {}^t\hat{\mathbf{M}} - {}^t\dot{\mathbf{A}}) = \mathbf{0}
\end{aligned}
\tag{10.204}
$$

10.10.2 Static Balance Equations

The quasi-static balance laws can be easily obtained from their dynamic counterparts by assuming that the linear and the angular momentum vectors identically vanish, that is,

$$
\begin{aligned}
\mathbf{L}(S, t) \equiv \mathbf{0} \\
\mathbf{A}(S, t) \equiv \mathbf{0}
\end{aligned}
\tag{10.205}
$$

We get the following from equation (10.201).

10.10.2.1 Component Vector Form of Static Balance Equations

$$\boxed{\begin{aligned} \mathbf{G}({}^t\mathbf{F}|_S + {}^t\hat{\mathbf{F}}) = 0 \\ \mathbf{G}({}^t\mathbf{M}|_S + \mathbf{a} \times {}^t\mathbf{F} + {}^t\hat{\mathbf{M}}) = 0 \end{aligned}} \tag{10.206}$$

as the static component vector form of the linear momentum and angular momentum balance equations.

10.10.2.2 Component-wise Beam Static Balance Equations

Now, using equations (10.202) and (10.203) in equation (10.206), we can express the combined balance equations component-wise as:

$$\boxed{\begin{aligned} N_T' - \kappa Q_M + \hat{N}_T = 0 \\ Q_M' + \kappa N_T - \tau Q_B + \hat{Q}_M = 0 \\ Q_B' + \tau Q_M + \hat{Q}_B = 0 \\ M_T' - \kappa M_M + (w' - \tau v + \kappa u)Q_B - (v' + \tau w)Q_M + \hat{M}_T = 0 \\ M_M' + \kappa M_T - \tau M_B + (v' + \tau w)N_T - (1 + u' - \kappa w)Q_B + \hat{M}_M = 0 \\ M_B' + \tau M_M + (1 + u' - \kappa w)Q_M - (w' - \tau v + \kappa u)N_T + \hat{M}_B = 0 \end{aligned}} \tag{10.207}$$

10.10.2.3 Matrix Form of Beam Static Balance Equations

Equation (10.207) may be expressed in matrix form as follows.

10.10.2.4 Momentum Balance Equations

$$\begin{Bmatrix} N_T' \\ Q_M' \\ Q_B' \end{Bmatrix} + \begin{bmatrix} 0 & -\kappa & 0 \\ \kappa & 0 & -\tau \\ 0 & \tau & 0 \end{bmatrix} \begin{Bmatrix} N_T \\ Q_M \\ Q_B \end{Bmatrix} + \begin{Bmatrix} \hat{N}_T \\ \hat{Q}_M \\ \hat{Q}_B \end{Bmatrix} = \begin{Bmatrix} 0 \\ 0 \\ 0 \end{Bmatrix} \tag{10.208}$$

and

10.10.2.5 Angular (or Moment of) Momentum Balance Equations

$$\boxed{\begin{Bmatrix} M_T' \\ M_M' \\ M_B' \end{Bmatrix} + \begin{bmatrix} 0 & -\kappa & 0 \\ \kappa & 0 & -\tau \\ 0 & \tau & 0 \end{bmatrix} \begin{Bmatrix} M_T \\ M_M \\ M_B \end{Bmatrix} + \left(\begin{aligned} \begin{bmatrix} 0 & -\kappa & 0 \\ \kappa & 0 & -\tau \\ 0 & \tau & 0 \end{bmatrix} \begin{bmatrix} 0 & -v & w \\ v & 0 & -u \\ -w & u & 0 \end{bmatrix} \\ - \begin{bmatrix} 0 & -v & w \\ v & 0 & -u \\ -w & u & 0 \end{bmatrix} + \begin{bmatrix} 0 & -\kappa & 0 \\ \kappa & 0 & -\tau \\ 0 & \tau & 0 \end{bmatrix} \end{aligned} \right) \begin{Bmatrix} N_T \\ Q_M \\ Q_B \end{Bmatrix} + \begin{Bmatrix} \hat{M}_T \\ \hat{M}_M \\ \hat{M}_B \end{Bmatrix} = \begin{Bmatrix} 0 \\ 0 \\ 0 \end{Bmatrix}} \tag{10.209}$$

Exercise: Starting from equation (10.203) verify expressions given by equation (10.209).

10.10.3 3D Linear Balance Equations

We can linearize equations (10.210) to get balance equations for 3D linear beams:

$$\boxed{\begin{aligned} N_T' - \kappa Q_M + \hat{N}_T &= 0 \\ Q_M' + \kappa N_T - \tau Q_B + \hat{Q}_M &= 0 \\ Q_B' + \tau Q_M + \hat{Q}_B &= 0 \\ M_T' - \kappa M_M + \hat{M}_T &= 0 \\ M_M' + \kappa M_T - \tau M_B - Q_B + \hat{M}_M &= 0 \\ M_B' + \tau M_M + Q_M + \hat{M}_B &= 0 \end{aligned}} \tag{10.210}$$

10.10.4 2D Balance Equations for Planar Beam under In-plane Loading

We can specialize equations (10.207) for 2D planar beam in the **T-M** plane by noting that the twist, $\tau \equiv 0$, shear force, $Q_B \equiv 0$, and out of plane moment, $M_M \equiv 0$, and torsion, $M_T \equiv 0$; thus, redefining: $Q \equiv Q_M$, $N \equiv N_T$ and $M \equiv M_B$, we get:

10.10.4.1 For Nonlinear Beam

$$\boxed{\begin{aligned} N' - \kappa Q + \hat{N}_T &= 0 \\ Q' + \kappa N + \hat{Q}_M &= 0 \\ M' + (1 + u')Q - (\kappa u)N + \hat{M}_B &= 0 \end{aligned}} \tag{10.211}$$

10.10.4.2 For Linear Beam

We can linearize equation (10.211) to get balance equations for 2D linear beams:

$$\boxed{\begin{aligned} N' - \kappa Q + \hat{N}_T &= 0 \\ Q' + \kappa N + \hat{Q}_M &= 0 \\ M' + Q + \hat{M}_B &= 0 \end{aligned}} \tag{10.212}$$

10.10.4.3 For Linear Circular Beam of Radius, *R*

In equation (10.211) we insert relationship: $\kappa = \frac{1}{R}$ to get balance equations for 2D linear circular beams:

$$\boxed{\begin{aligned} N' - \frac{1}{R}Q + \hat{N}_T &= 0 \\ Q' + \frac{1}{R}N + \hat{Q}_M &= 0 \\ M' + Q + \hat{M}_B &= 0 \end{aligned}} \tag{10.213}$$

10.10.5 Where We Would Like to Go

We have now derived the shell computational balance equations. These serve as the starting point in developing the:

- weak form or virtual work functional necessary for the numerical formulation based on a finite element method of solution.

10.11 Computational Derivatives and Variations

10.11.1 What We Need to Recall

- The vector triple product property: $\mathbf{a} \times \mathbf{b} \bullet \mathbf{c} = \mathbf{b} \times \mathbf{c} \bullet \mathbf{a} = \mathbf{c} \times \mathbf{a} \bullet \mathbf{b}$, $\forall \mathbf{a}, \mathbf{b}, \mathbf{c}$ vectors.
- The vector triple cross product property: $\mathbf{a} \times \mathbf{b} \times \mathbf{c} = (\mathbf{a} \bullet \mathbf{c})\mathbf{b} - (\mathbf{a} \bullet \mathbf{b})\mathbf{c}$, $\forall \mathbf{a}, \mathbf{b}, \mathbf{c}$ vectors.
- The definition of a rotation tensor, its matrix representation, and algebra, geometry and topology of rotation.
- A rotation matrix, $\mathbf{R}$, corresponding to the three-dimensional special orthogonal matrix group, $SO(3)$, may be parameterized by the Rodrigues vector, $\mathbf{s}$, the axis of the instantaneous rotation.
- The Rodrigues vector, $\mathbf{s} = \{s_1 \quad s_2 \quad s_3\}^T$, is orientated along the instantaneous axis of rotation at a point of the beam reference curve. The skew-symmetric matrix, $\mathbf{S}$, corresponding to the Rodrigues vector, $\mathbf{s}$, that is, $\mathbf{S}\,\mathbf{s} = \mathbf{0}$, is given by:

$$\mathbf{S} = \begin{bmatrix} 0 & -s_3 & s_2 \\ s_3 & 0 & -s_1 \\ -s_2 & s_1 & 0 \end{bmatrix} \tag{10.214}$$

- The rotation matrix, $\mathbf{R}(S,t)$, is related, through Cayley's theorem, to the Rodrigues vector, $\mathbf{s}(S,t) = \{s_1(S,t) \quad s_2(S,t) \quad s_3(S,t)\}^T$, with the components in the $\mathbf{T}, \mathbf{M}, \mathbf{B}$ frame at any point, S, parameterized by the arc length along the centroidal axis as a beam reference curve, at time $t \in \mathbb{R}_+$ by:

$$\begin{aligned} {}^t\mathbf{R} &= (\mathbf{I} - \mathbf{S})^{-1}\,(\mathbf{I} + \mathbf{S}) \\ &= \frac{1}{\Sigma}\begin{bmatrix} 1 + s_1^2 - s_2^2 - s_3^2 & 2(s_1 s_2 - s_3) & 2(s_1 s_3 + s_2) \\ 2(s_1 s_2 + s_3) & 1 - s_1^2 + s_2^2 - s_3^2 & 2(s_2 s_3 - s_1) \\ 2(s_1 s_3 - s_2) & 2(s_2 s_3 + s_1) & 1 - s_1^2 - s_2^2 + s_3^2 \end{bmatrix} \end{aligned} \tag{10.215}$$

where $\sum \equiv 1 + \| \mathbf{s} \|^2$ with $(\mathbf{I} - \mathbf{S})$ matrix always invertible for any rotation matrix, defined by:

$$(\mathbf{I} - \mathbf{S})^{-1} = \frac{1}{\Sigma}\begin{bmatrix} \left(1 + s_1^2\right) & (s_1\, s_2 - s_3) & (s_3\, s_1 + s_2) \\ (s_1\, s_2 + s_3) & \left(1 + s_2^2\right) & (s_2\, s_3 - s_1) \\ (s_3\, s_1 - s_2) & (s_2\, s_3 + s_1) & \left(1 + s_3^2\right) \end{bmatrix} \tag{10.216}$$

- Finally, the rotation vector, $\boldsymbol{\theta}$, is the scaled Rodrigues vector related by:

$$\boxed{\mathbf{s} = \frac{1}{\theta}\tan\left(\frac{\theta}{2}\right)\boldsymbol{\theta}} \tag{10.217}$$

where $\theta \equiv ||\boldsymbol{\theta}|| = (\boldsymbol{\theta} \bullet \boldsymbol{\theta})^{\frac{1}{2}}$ is the length measure of the rotation vector. Alternatively, defining the unit vector, $\mathbf{n} \equiv \frac{\boldsymbol{\theta}}{\theta}$, and the corresponding skew-symmetric matrix, $\mathbf{N}$, such that $\mathbf{N}\,\mathbf{h} = \mathbf{n} \times \mathbf{h}$, $\forall \mathbf{h} \in \mathbb{R}_3$, we have:

$$\boxed{\begin{aligned} &\mathbf{R} = (\mathbf{I} - \mathbf{S})^{-1}\,(\mathbf{I} + \mathbf{S}) \\ &\mathbf{s} = \tan\left(\frac{\theta}{2}\right)\mathbf{n}, \\ &\mathbf{S} = \tan\left(\frac{\theta}{2}\right)\mathbf{N} \end{aligned}} \tag{10.218}$$

- The Rodrigues quaternion and Euler–Rodrigues parameters.
- For notational uniformity, we denote all variations of entities by a bar superscripted on it. Furthermore all entities measured in $\mathbf{T}, \mathbf{M}, \mathbf{B}$ frame is superscripted as ${}^t(\bullet)$, and those in the rotating $t, \mathbf{m}_n, \mathbf{b}_n$ coordinate system as ${}^n(\bullet)$. However, for the sake of uncluttered readability, sometimes the superscripts may be dropped, but the context should help to determine the appropriate underlying coordinate system.

10.11.2 *Curvature Matrix, ${}^t\mathbf{K}$, Curvature Vector, ${}^t\mathbf{k}$*

Let $\mathbf{R}(S,t)$ be the rotation matrix at a point on the beam reference curve described in the arc length parameter, S, along the centroidal axis direction, at any fixed time $t \in \mathbb{R}_+$. Then, considering it as a one-parameter family along the direction, and taking the derivative, in the sense described above, of the orthonormality relationship: $\mathbf{R}\,\mathbf{R}^T = \mathbf{I}$, with respect to this parameter, we have:

$$(\mathbf{R}_{,S}\ \mathbf{R}^T) + (\mathbf{R}_{,S}\ \mathbf{R}^T)^T = \mathbf{0} \tag{10.219}$$

Now, we introduce the following definition.

Definition

- The curvature matrices, ${}^t\mathbf{K}$, as:

$$\boxed{{}^t\mathbf{K} \equiv {}^t\mathbf{R}_{,S}\ {}^t\mathbf{R}^{\mathbf{T}}} \tag{10.220}$$

By introducing the above definitions into equation (10.219), we have: ${}^t\mathbf{K} + ({}^t\mathbf{K})^T = \mathbf{0}$, signifying that ${}^t\mathbf{K}$ is 3D skew-symmetric matrix; we can then, define the corresponding axial vector, ${}^t\mathbf{k}$, such that ${}^t\mathbf{K}\,\mathbf{h} = {}^t\mathbf{k} \times \mathbf{h}$ for $\forall \mathbf{h} \in \mathbb{R}_3$. Moreover, from the matrix representation, as given by equation (10.215), of the rotation tensor, we have, by taking straightforward term-by-term derivatives, the corresponding the matrix representation of the curvature tensor,

${}^t\mathbf{K}$, with $\Sigma \equiv 1 + ||\mathbf{s}||^2$, as:

$$
\boxed{
\begin{aligned}
{}^t\mathbf{K} &= \frac{2}{\Sigma}\left\{ \underbrace{(\mathbf{S}\,\mathbf{S}_{,S} - \mathbf{S}_{,S}\,\mathbf{S})}_{\text{Commutator or Lie bracket}} + \mathbf{S}_{,S} \right\} \\
&= \frac{2}{\Sigma}\begin{bmatrix} 0 & (-s_1 s_{2,S} + s_{1,S}\, s_2 - s_{3,S}) & (-s_1 s_{3,S} + s_{1,S}\, s_3 + s_{2,S}) \\ (s_1 s_{2,S} - s_{1,S}\, s_2 + s_{3,S}) & 0 & (-s_2 s_{3,S} + s_{2,S}\, s_3 - s_{1,S}) \\ (s_1 s_{3,S} - s_{1,S}\, s_3 - s_{2,S}) & (s_2 s_{3,S} - s_{2,S}\, s_3 + s_{1,S}) & 0 \end{bmatrix}
\end{aligned}
}
\tag{10.221}
$$

where the third equality follows from the first equality by the fact: $(\mathbf{A}\,\mathbf{B} - \mathbf{B}\,\mathbf{A})\mathbf{h} = [\mathbf{a} \times \mathbf{b}]\mathbf{h}$, $\forall \mathbf{h} \in \mathbb{R}_3$ where $[(\bullet)]$ denotes a skew-symmetric matrix corresponding to an axial vector $(\bullet)$, and $\mathbf{A},\ \mathbf{B}$ are skew-symmetric matrices, and $\mathbf{a},\ \mathbf{b}$ are the corresponding axial vectors.

***Exercise*:** Show that matrix Lie bracket, $\mathbf{A}\,\mathbf{B} - \mathbf{B}\,\mathbf{A} = [\mathbf{a} \times \mathbf{b}]$

Solution:

$$
\begin{aligned}
(\mathbf{A}\,\mathbf{B} - \mathbf{B}\,\mathbf{A})\mathbf{h} &= \mathbf{a} \times \mathbf{b} \times \mathbf{h} - \mathbf{b} \times \mathbf{a} \times \mathbf{h} = (\mathbf{a} \bullet \mathbf{h})\mathbf{b} - \cancel{(\mathbf{a} \bullet \mathbf{b})\mathbf{h}} - (\mathbf{b} \bullet \mathbf{h})\mathbf{a} + \cancel{(\mathbf{b} \bullet \mathbf{a})\mathbf{h}} \\
&= \mathbf{h} \times (\mathbf{b} \times \mathbf{a}) = \mathbf{a} \times \mathbf{b} \times \mathbf{h} = [\mathbf{a} \times \mathbf{b}]\mathbf{h}, \quad \forall \mathbf{h} \in \mathbb{R}_3
\end{aligned}
$$

completes the proof.

***Exercise*:** Derive equation (10.221).

***Solution*:** By taking the derivative of equation (10.215) with respect to S, and using the definition of the curvature matrix as in equation (10.220), we have:

$$
\begin{aligned}
(\mathbf{I} - \mathbf{S})\,{}^t\mathbf{R}_{,S} - \mathbf{S}_{,S}\,{}^t\mathbf{R} = \mathbf{S}_{,S} &\Rightarrow (\mathbf{I} - \mathbf{S})\,{}^t\mathbf{R}_{,S}\,{}^t\mathbf{R}^T - \mathbf{S}_{,S}\,{}^t\mathbf{R}\,{}^t\mathbf{R}^T = \mathbf{S}_{,S}\,{}^t\mathbf{R}^T \\
&\Rightarrow {}^t\mathbf{K} = (\mathbf{I} - \mathbf{S})^{-1}\left\{\mathbf{S}_{,S} + \mathbf{S}_{,S}\,{}^t\mathbf{R}^T\right\}
\end{aligned}
$$

Now, using the inverse relation given by equation (10.216) and direct multiplication completes the proof.

10.11.2.1 Curvature Vector, ${}^t\mathbf{k}$

Now, the expressions for the axial curvature vector, ${}^t\mathbf{k}$ corresponding to the curvature matrix, ${}^t\mathbf{K}$, such that ${}^t\mathbf{K}\,{}^t\mathbf{k} = \mathbf{0}$, follow from equation (10.221), as:

$$
\boxed{{}^t\mathbf{k} = \frac{2}{\Sigma}\left(\mathbf{s} \times \mathbf{s}_{,S} + \mathbf{s}_{,S}\right) = \frac{2}{\Sigma}\left(\mathbf{S}\,\mathbf{s}_{,S} + \mathbf{s}_{,S}\right) = \left\{\frac{2}{\Sigma}(\mathbf{I} + \mathbf{S})\right\}\mathbf{s}_{,S} = \mathbf{W}_1(\mathbf{s})\,\mathbf{s}_{,S}}
\tag{10.222}
$$

where we have introduced, for future use, the definition of the matrix:

$$\mathbf{W}_1(\mathbf{s}) \equiv \frac{2}{\Sigma}(\mathbf{I}+\mathbf{S}) \tag{10.223}$$

with $\mathbf{I}$ as the 3×3 identity matrix. Now, noting that $\Sigma \equiv 1 + ||\mathbf{s}||^2 = 1 + \tan^2\frac{\theta}{2} = \frac{2}{1+\cos\theta}$ and $\tan\frac{\theta}{2} = \frac{\sin\theta}{1+\cos\theta}$, we have, using the second relation of equation (10.218) in equation (10.223):

$$\boxed{\mathbf{W}_1(\mathbf{s}) = (1+\cos\theta)\left\{\mathbf{I} + \frac{\sin\theta}{1+\cos\theta}\,\mathbf{N}\right\}} \tag{10.224}$$

Note that equation (10.224) gives the expression in terms of the rotation vector, $\boldsymbol{\theta}$.

10.11.3 Curvature Vector, ${}^t\mathbf{k}$ in Terms of the Rotation Vector ${}^t\boldsymbol{\theta}$

Note that from the fourth equality in equation (10.222), we only get ${}^t\mathbf{k}$ as the function of the derivatives, $\mathbf{s}_{,S}$, of the Rodrigues vector. But, we would like to get its dependence on the rotation vector, ${}^t\boldsymbol{\theta}$. For this, we take derivative of the relation given by equation (10.217):

$$\boxed{\begin{aligned}\mathbf{s}_{,S} &= \frac{1}{2}\left\{\left(\sec^2\frac{\theta}{2}\right)\frac{(\boldsymbol{\theta}\bullet\boldsymbol{\theta}_{,S})}{\theta^2}\right\}\boldsymbol{\theta} - \left\{\frac{(\boldsymbol{\theta}\bullet\boldsymbol{\theta}_{,S})}{\theta^3}\tan\frac{\theta}{2}\right\}\boldsymbol{\theta} + \left\{\frac{\boldsymbol{\theta}_{,S}}{\theta}\tan\frac{\theta}{2}\right\}\\ &= \left(\frac{1}{\theta}\tan\frac{\theta}{2}\right)\left\{\boldsymbol{\theta}_{,S} - \left(1-\frac{\theta}{\sin\theta}\right)(\mathbf{n}\bullet\boldsymbol{\theta}_{,S})\mathbf{n}\right\} = \left(\frac{1}{\theta}\tan\frac{\theta}{2}\right)\left\{\mathbf{I} - \left(1-\frac{\theta}{\sin\theta}\right)(\mathbf{n}\otimes\mathbf{n})\right\}\boldsymbol{\theta}_{,S}\\ &= \mathbf{W}_2\,\boldsymbol{\theta}_{,S}\end{aligned}} \tag{10.225}$$

where we have introduced the definition for the matrix:

$$\mathbf{W}_2 \equiv \left(\frac{1}{\theta}\tan\frac{\theta}{2}\right)\left\{\mathbf{I} - \left(1-\frac{\theta}{\sin\theta}\right)(\mathbf{n}\otimes\mathbf{n})\right\} \tag{10.226}$$

Alternatively, as previously done for equation (10.223), we can express the relationship in equation (10.226) in terms of the rotation vector, $\boldsymbol{\theta}$, as:

$$\boxed{\mathbf{W}_2 = \frac{1}{1+\cos\theta}\left\{\mathbf{I} - \left(\frac{\sin\theta-\theta}{\theta}\right)\mathbf{N}^2\right\}} \tag{10.227}$$

where we have additionally used the identity: $\mathbf{N}^2 = \{(\mathbf{n}\otimes\mathbf{n}) - \mathbf{I}\}$.

Exercise: show: $\mathbf{N}^2 = \{(\mathbf{n}\otimes\mathbf{n}) - \mathbf{I}\}$

Solution: $\mathbf{N}^2\,\mathbf{h} \equiv \mathbf{N}[\mathbf{n}\times\mathbf{h}] = \mathbf{n}\times\mathbf{n}\times\mathbf{h} = (\mathbf{n}\bullet\mathbf{h})\mathbf{n} - (\mathbf{n}\bullet\mathbf{n})\mathbf{h} = \{(\mathbf{n}\otimes\mathbf{n}) - \mathbf{I}\}\mathbf{h},\ \forall\mathbf{h}\in\mathbb{R}_3.$

Now, using equation (10.225) with definition (10.227) in equation (10.222) with definition (10.224), we have:

$$\boxed{{}^t\mathbf{k} = \mathbf{W}_1(\mathbf{s})\, \mathbf{s}_{,S} = \mathbf{W}_1\, \mathbf{W}_2\, \boldsymbol{\theta}_{,S} \equiv \mathbf{W}(\boldsymbol{\theta})\, {}^t\boldsymbol{\theta}_{,S}} \tag{10.228}$$

where we have introduced the following definitions.

Definitions

$$\boxed{\begin{aligned} \mathbf{W}(\boldsymbol{\theta}) &\equiv \mathbf{I} + c_1\, \boldsymbol{\Theta} + c_2\, \boldsymbol{\Theta}^2, \\ c_1(\boldsymbol{\theta}) &= \frac{1-\cos\theta}{\theta^2}, \\ c_2(\boldsymbol{\theta}) &= \frac{\theta - \sin\theta}{\theta^3} \end{aligned}} \tag{10.229}$$

Note that equation (10.228) with the definitions (10.139) describes the curvature vector ${}^t\mathbf{k}$ in terms of the derivative, $\boldsymbol{\theta}_{,S}$, of the rotation vector, where $\boldsymbol{\Theta}$ is the skew-symmetric matrix corresponding to rotation vector, $\boldsymbol{\theta}$, as the axial vector such that $\boldsymbol{\Theta}\,\boldsymbol{\theta} = \mathbf{0}$, and $\boldsymbol{\Theta}^2 \equiv \boldsymbol{\Theta}\boldsymbol{\Theta} = [\boldsymbol{\theta} \times \boldsymbol{\theta} \times]$.

***Exercise*:** Prove the relationships in equation (10.229).

***Solution*:** Let us note that $\mathbf{N}$, being a skew-symmetric matrix with the form given similar to equation (10.214), and that every matrix satisfies its characteristic equation; from its eigen or characteristic equation, $p(\lambda) \equiv \lambda^3 + \lambda = 0$, we have: $\mathbf{N} + \mathbf{N}^3 = 0 \Rightarrow \mathbf{N}^3 = -\mathbf{N}$. Now, using relations (10.224) and (10.227) in equation (10.228), we have:

$$\begin{aligned} \mathbf{W} \equiv \mathbf{W}_1\, \mathbf{W}_2 &= \left\{\mathbf{I} + \frac{\sin\theta}{1+\cos\theta}\mathbf{N}\right\}\left\{\mathbf{I} - \left(\frac{\sin\theta - \theta}{\theta}\right)\mathbf{N}^2\right\} \\ &= \mathbf{I} - \left(\frac{\sin\theta - \theta}{\theta}\right)\mathbf{N}^2 + \frac{\sin\theta}{1+\cos\theta}\mathbf{N} - \frac{\sin^2\theta - \theta\sin\theta}{\theta(1+\cos\theta)}\mathbf{N}^3 \\ &= \mathbf{I} + \left(\frac{\theta - \sin\theta}{\theta}\right)\mathbf{N}^2 + \left\{\frac{\sin\theta}{1+\cos\theta} + \frac{\sin^2\theta}{\theta(1+\cos\theta)} - \frac{\sin\theta}{1+\cos\theta}\right\}\mathbf{N} \\ &= \mathbf{I} + \underbrace{\left(\frac{1-\cos\theta}{\theta^2}\right)}_{\equiv c_1}\Theta + \underbrace{\left(\frac{\theta - \sin\theta}{\theta^3}\right)}_{\equiv c_2}\Theta^2 \end{aligned}$$

Noting that we have used the unit vector: $\mathbf{N} \equiv \frac{\Theta}{\theta}$, with Θ as the skew-symmetric matrix corresponding to $\boldsymbol{\theta}$ taken as the axial vector, completes the proof.

10.11.4 Angular Velocity Matrix, ${}^t\Omega_S$, Angular Velocity Vector, ${}^t\boldsymbol{\omega}_S$

This time we take the rotation tensor, $\mathbf{R}(S,t)$, to be fixed at a point on the beam reference curve described in the arc length parameter, S, along the centroidal axis direction for any time $t \in \mathbb{R}_+$. Then, considering it as a one-parameter family along the time coordinate, and following the steps as in the case of curvature derivation above but replacing the spatial derivative operations, $(\bullet)_{,S}$, everywhere by the time derivative operation, $\dot{(\bullet)} \equiv \frac{d}{dt}(\bullet)$, and recognizing that:

$$\boxed{{}^t\boldsymbol{\Omega}_S \equiv \dot{\mathbf{R}}\ \mathbf{R}^T} \tag{10.230}$$

we have from equation (10.230):

$$\boxed{\begin{aligned} {}^t\boldsymbol{\Omega}_S &= \frac{2}{\Sigma}\left\{\underbrace{\left(\mathbf{s}\ \dot{\mathbf{s}} - \dot{\mathbf{s}}\mathbf{s}\right)}_{\text{Commutator or Lie bracket}} + \dot{\mathbf{S}}\right\} \\ &= \frac{2}{\Sigma}\begin{bmatrix} 0 & (-s_1\dot{s}_2 + \dot{s}_1 s_2 - \dot{s}_3) & (-s_1\dot{s}_3 + \dot{s}_1 s_3 + \dot{s}_2) \\ (s_1\dot{s}_2 - \dot{s}_1 s_2 + \dot{s}_3) & 0 & (-s_2\dot{s}_3 + \dot{s}_2 s_3 - \dot{s}_1) \\ (s_1\dot{s}_3 - \dot{s}_1 s_3 - \dot{s}_2) & (s_2\dot{s}_3 - \dot{s}_2 s_3 + \dot{s}_1) & 0 \end{bmatrix} \end{aligned}} \tag{10.231}$$

10.11.4.1 Angular Velocity Vector, ${}^t\omega_S$

Now, from equation (10.231), we have:

$$\boxed{{}^t\boldsymbol{\omega}_S = \mathbf{W}_1(\mathbf{s})\ \dot{\mathbf{s}}} \tag{10.232}$$

where $\mathbf{W}_1(\mathbf{s})$ is given by equation (10.224).

10.11.4.2 Angular Velocity Vector, ${}^t\omega_S$, in Terms of the Rotation vector, ${}^t\theta$

Using equation (10.227) and replacing spatial derivatives by time derivatives in equation (10.228), we have:

$$\boxed{{}^t\boldsymbol{\omega}_S = \mathbf{W}_1(\mathbf{s})\ \dot{\mathbf{s}} = \mathbf{W}_1\ \mathbf{W}_2\ \dot{\boldsymbol{\theta}} \equiv \mathbf{W}(\boldsymbol{\theta})\ {}^t\dot{\boldsymbol{\theta}}} \tag{10.233}$$

where we have used the definition of $\mathbf{W}(\boldsymbol{\theta})$ as given by equation (10.139).

10.11.5 Variational Matrix, ${}^t\bar{\Phi} \equiv \delta\,{}^t\Phi$, Variational Vector, ${}^t\bar{\varphi} \equiv \delta\,{}^t\varphi$

This time we take the variation of the rotation tensor, $\mathbf{R}(S,t)$, at a point on the beam reference curve described in the arc length parameter, S, along the centroidal axis direction for any time $t \in \mathbb{R}_+$. Then, following the steps as in the case of angular velocity derivation above but replacing the time derivative operations, $\dot{(\bullet)}$, everywhere by the variation operation, $\overline{(\bullet)} \equiv \delta(\bullet)$, and recognizing that $\bar{\mathbf{\Phi}} \equiv \delta\,\mathbf{\Phi} \equiv \bar{\mathbf{R}}\,\mathbf{R}^{\mathbf{T}} \equiv \delta\mathbf{R}\,\mathbf{R}^{\mathbf{T}}$, we have from equation (10.231):

$$
\boxed{
\begin{aligned}
{}^t\bar{\mathbf{\Phi}} &= \frac{2}{\Sigma}\left\{\underbrace{(\mathbf{S}\,\bar{\mathbf{S}} - \bar{\mathbf{S}}\,\mathbf{S})}_{\text{Commutator or Lie bracket}} + \bar{\mathbf{S}}\right\} \\
&= \frac{2}{\Sigma}\begin{bmatrix} 0 & (-s_1\bar{s}_2 + \bar{s}_1 s_2 - \bar{s}_3) & (-s_1\bar{s}_3 + \bar{s}_1 s_3 + \bar{s}_2) \\ (s_1\bar{s}_2 - \bar{s}_1 s_2 + \bar{s}_3) & 0 & (-s_2\bar{s}_3 + \bar{s}_2 s_3 - \bar{s}_1) \\ (s_1\bar{s}_3 - \bar{s}_1 s_3 - \bar{s}_2) & (s_2\bar{s}_3 - \bar{s}_2 s_3 + \bar{s}_1) & 0 \end{bmatrix}
\end{aligned}
}
\tag{10.234}
$$

where a superscripted bar over the variable signifies variation of the variable by the definition: the variation of a along $\bar{a}$ is given by: $\bar{a} = \lim_{\varepsilon \to 0} \frac{d}{d\varepsilon}(a + \varepsilon\bar{a})$.

10.11.5.1 Rotational Variational Vector, ${}^t\bar{\boldsymbol{\varphi}} \equiv \delta\,{}^t\boldsymbol{\varphi}$

Now, the expressions for the axial rotational variational vector, $\bar{\boldsymbol{\varphi}}$, corresponding to the variation ${}^t\bar{\mathbf{\Phi}} \equiv \delta\,{}^t\mathbf{\Phi}$, follow from equation (10.234), as:

$$\boxed{{}^t\bar{\boldsymbol{\varphi}} = \mathbf{W}_1(\mathbf{s})\,\bar{\mathbf{s}}} \tag{10.235}$$

where we have $\mathbf{W}_1(\mathbf{s})$ as defined before in equation (10.223).

10.11.6 $\bar{\boldsymbol{\varphi}}$ in Terms of the Rotation Vector Variation, $\bar{\boldsymbol{\theta}}$

By substituting $\bar{\boldsymbol{\varphi}}$ for $\mathbf{k}_\alpha$, $\alpha = 1, 2$, and $\bar{\boldsymbol{\theta}}$ for $\boldsymbol{\theta}_{,\alpha}$ in equations (10.228) and (10.139), we get:

$$\boxed{{}^t\bar{\boldsymbol{\varphi}} = \mathbf{W}_1(\mathbf{s})\,\bar{\mathbf{s}} = \mathbf{W}_1\,\mathbf{W}_2\,\bar{\boldsymbol{\theta}} \equiv \mathbf{W}(\boldsymbol{\theta})\,{}^t\bar{\boldsymbol{\theta}}} \tag{10.236}$$

where $\mathbf{W}(\boldsymbol{\theta})$, $c_1(\boldsymbol{\theta})$, $c_2(\boldsymbol{\theta})$ have been defined earlier in equation (10.139).

10.11.7 Derivatives: $\mathbf{W}_{,S}(\boldsymbol{\theta})$ and $\dot{\mathbf{W}}(\boldsymbol{\theta})$, $\ddot{\mathbf{W}}(\boldsymbol{\theta})$ and Variation: $\bar{\mathbf{W}}(\boldsymbol{\theta}) \equiv \delta\mathbf{W}(\boldsymbol{\theta})$ of $\mathbf{W}(\boldsymbol{\theta})$

For future use, we will need the derivative: $\mathbf{W}_{,S}(\boldsymbol{\theta})$, the rates: $\dot{\mathbf{W}}(\boldsymbol{\theta})$, $\ddot{\mathbf{W}}(\boldsymbol{\theta})$, and the variation: $\bar{\mathbf{W}}(\boldsymbol{\theta}) \equiv \delta\mathbf{W}(\boldsymbol{\theta})$ of the matrix. $\mathbf{W}(\boldsymbol{\theta})$.

Upon differentiating by the expression for $\mathbf{W}(\theta)$ as given by equation (10.139), we have:

$$\begin{aligned}
\mathbf{W}_{,S} &= c_1\boldsymbol{\Theta}_{,S} + c_2(\boldsymbol{\Theta}_{,S}\,\boldsymbol{\Theta} + \boldsymbol{\Theta}\,\boldsymbol{\Theta}_{,S}) + c_{1,S}\,\boldsymbol{\Theta} + c_{2,S}\,\boldsymbol{\Theta}^2 \\
&= (c_2 - c_1)(\boldsymbol{\theta} \bullet \boldsymbol{\theta}_{,S})\mathbf{I} + c_1\boldsymbol{\Theta}_{,S} + c_2\{(\boldsymbol{\theta} \otimes \boldsymbol{\theta}_{,S}) + (\boldsymbol{\theta}_{,S} \otimes \boldsymbol{\theta})\} + c_{1,S}\,\boldsymbol{\Theta} + c_{2,S}\,(\boldsymbol{\theta} \otimes \boldsymbol{\theta}) \\
c_{1,S} &\equiv \frac{1}{\theta^2}(1 - 2c_1 - \theta^2 c_2)(\boldsymbol{\theta} \bullet \boldsymbol{\theta}_{,S}) = c_3(\boldsymbol{\theta} \bullet \boldsymbol{\theta}_{,S}), \\
c_{2,S} &\equiv \frac{1}{\theta^2}(c_1 - 3c_2)(\boldsymbol{\theta} \bullet \boldsymbol{\theta}_{,S}) = c_4(\boldsymbol{\theta} \bullet \boldsymbol{\theta}_{,S})
\end{aligned} \tag{10.237}$$

where we have introduced the following definitions.

Definitions

$$\begin{aligned}
c_3 &\equiv \frac{1}{\theta^2}(1 - 2c_1 - \theta^2 c_2), \\
c_4 &\equiv \frac{1}{\theta^2}(c_1 - 3c_2)
\end{aligned} \tag{10.238}$$

and applied the following identities:

$$\begin{aligned}
\boldsymbol{\Theta}^2 &= \{(\boldsymbol{\theta} \otimes \boldsymbol{\theta}) - \theta^2\mathbf{I}\} \\
(\boldsymbol{\Theta}_{,S}\ \boldsymbol{\Theta} + \boldsymbol{\Theta}\ \boldsymbol{\Theta}_{,S}) &= \left\{(\boldsymbol{\theta} \otimes \boldsymbol{\theta}_{,S}) + (\boldsymbol{\theta}_{,S} \otimes \boldsymbol{\theta}) - 2(\boldsymbol{\theta}\boldsymbol{\theta}_{,S})\mathbf{I}\right\} \\
\theta_{,S} &\equiv ||\boldsymbol{\theta}||_{,S} = \frac{1}{\theta}(\boldsymbol{\theta}\boldsymbol{\theta}_{,S})
\end{aligned} \tag{10.239}$$

***Exercise*:** Prove the identities (10.239).

***Solution*:** $\forall \mathbf{h} \in \mathbb{R}^3$,

$$\boldsymbol{\Theta}^2\ \mathbf{h} = \boldsymbol{\theta} \times \boldsymbol{\theta} \times \mathbf{h} = \{(\boldsymbol{\theta} \otimes \boldsymbol{\theta}) - \theta^2\mathbf{I}\}\mathbf{h}$$

and

$$\begin{aligned}
(\boldsymbol{\Theta}_{,\alpha}\,\boldsymbol{\Theta} + \boldsymbol{\Theta}\,\boldsymbol{\Theta}_{,\alpha})\mathbf{h} &= \boldsymbol{\theta}_{,\alpha} \times \boldsymbol{\theta} \times \mathbf{h} + \boldsymbol{\theta} \times \boldsymbol{\theta}_{,\alpha} \times \mathbf{h} \\
&= (\boldsymbol{\theta}_{,\alpha} \bullet \mathbf{h})\boldsymbol{\theta} - (\boldsymbol{\theta}_{,\alpha} \bullet \boldsymbol{\theta})\mathbf{h} + (\boldsymbol{\theta} \bullet \mathbf{h})\boldsymbol{\theta}_{,\alpha} - (\boldsymbol{\theta} \bullet \boldsymbol{\theta}_{,\alpha})\mathbf{h} \\
&= \left\{(\boldsymbol{\theta} \otimes \boldsymbol{\theta}_{,\alpha}) + (\boldsymbol{\theta}_{,\alpha} \otimes \boldsymbol{\theta}) - 2(\boldsymbol{\theta} \bullet \boldsymbol{\theta}_{,\alpha})\mathbf{I}\right\}\mathbf{h}
\end{aligned}$$

complete the proof.

***Exercise*:** Derive relations given in equation (10.237).

Solution: By taking the derivative with respect to S of $\mathbf{W}(\theta)$ in equation (10.139) and using relations in equation (10.239):

$$\begin{aligned}
\mathbf{W}_{,S} &= c_1\mathbf{\Theta}_{,S} + c_2(\mathbf{\Theta}_{,S}\mathbf{\Theta} + \mathbf{\Theta}\mathbf{\Theta}_{,S}) + c_{1,S}\mathbf{\Theta} + c_{2,S}\mathbf{\Theta}^2 \\
&= c_1\mathbf{\Theta}_{,S} + c_2\left\{(\theta\otimes\theta_{,S}) + (\theta_{,S}\otimes\theta) - 2\,(\theta\bullet\theta_{,S})\,\mathbf{I}\right\} \\
&\quad + c_{1,S}\mathbf{\Theta} + c_{2,S}\left\{(\theta\otimes\theta) - \theta^2\mathbf{I}\right\} \\
&= c_1\mathbf{\Theta}_{,S} + c_2\left\{(\theta\otimes\theta_{,S}) + (\theta_{,S}\otimes\theta)\right\} + c_{1,S}\mathbf{\Theta} + c_{2,S}(\theta\otimes\theta) \\
&\quad + \left\{-c_1 + 3c_2 - 2c_2\right\}(\theta\bullet\theta_{,S})\,\mathbf{I} \\
&= (c_2 - c_1)(\theta\bullet\theta_{,S})\mathbf{I} + c_1\mathbf{\Theta}_{,S} + c_2\{(\theta\otimes\theta_{,S}) + (\theta_{,S}\otimes\theta)\} + c_{1,S}\mathbf{\Theta} + c_{2,S}(\theta\otimes\theta)
\end{aligned}$$

completes the derivation.

Now, replacing $\theta_{,S}$ by $\dot{\theta}$ and $\mathbf{\Theta}_{,S}$ by $\dot{\mathbf{\Theta}}$ everywhere in equations (10.237)–(10.239), we have the expression for $\dot{\mathbf{W}}$, the time derivative of $\mathbf{W}$ matrix:

$$\boxed{\begin{aligned}
\dot{\mathbf{W}} &= c_1\dot{\mathbf{\Theta}} + c_2(\dot{\mathbf{\Theta}}\mathbf{\Theta} + \mathbf{\Theta}\dot{\mathbf{\Theta}}) + \dot{c}_1\mathbf{\Theta} + \dot{c}_2\mathbf{\Theta}^2 \\
&= (c_2 - c_1)\left(\theta\bullet\dot{\theta}\right)\mathbf{I} + c_1\dot{\mathbf{\Theta}} + c_2\left\{\left(\theta\otimes\dot{\theta}\right) + \left(\dot{\theta}\otimes\theta\right)\right\} + \dot{c}_1\mathbf{\Theta} + \dot{c}_2(\theta\otimes\theta) \\
\dot{c}_1 &\equiv \frac{1}{\theta^2}(1 - 2c_1 - \theta^2 c_2)(\theta\bullet\dot{\theta}) = c_3(\theta\bullet\dot{\theta}), \\
\dot{c}_2 &\equiv \frac{1}{\theta^2}(c_1 - 3c_2)(\theta\bullet\dot{\theta}) = c_4(\theta\bullet\dot{\theta})
\end{aligned}}$$

(10.240)

Taking the time derivative of $\dot{\mathbf{W}}$ in equation (10.240), we have the expression for $\ddot{\mathbf{W}}$, the second time derivative of $\mathbf{W}$ matrix:

$$\boxed{\begin{aligned}
\ddot{\mathbf{W}} &= c_1\ddot{\mathbf{\Theta}} + 2\left\{\dot{c}_1\dot{\mathbf{\Theta}} + \dot{c}_2(\dot{\mathbf{\Theta}}\mathbf{\Theta} + \mathbf{\Theta}\dot{\mathbf{\Theta}})\right\} + c_2(\ddot{\mathbf{\Theta}}\mathbf{\Theta} + 2\dot{\mathbf{\Theta}}^2 + \mathbf{\Theta}\ddot{\mathbf{\Theta}}) + \ddot{c}_1\mathbf{\Theta} + \ddot{c}_2\mathbf{\Theta}^2 \\
\ddot{c}_1 &\equiv c_5(\theta\bullet\dot{\theta})^2 + c_3\left\{\theta^2 + (\theta\bullet\dot{\theta})\right\} \\
\ddot{c}_2 &\equiv c_6(\theta\bullet\dot{\theta})^2 + c_4\left\{\theta^2 + (\theta\bullet\dot{\theta})\right\} \\
c_5 &\equiv -\frac{1}{\theta^2}(2c_2 + 4c_3 + \theta^2 c_4), \\
c_6 &\equiv -\frac{1}{\theta^2}(c_3 - 5c_4),
\end{aligned}}$$

(10.241)

where we have used the properties of the constants as given by equation (10.139) and equation (10.238).

Replacing $\theta,_S$ by $\bar{\theta}$ and $\Theta,_\alpha$ by $\bar{\Theta}$ everywhere in equations (10.237)–(10.239), we have the expression for $\bar{\mathbf{W}}$, the variation of $\mathbf{W}$ matrix:

$$\begin{aligned}
\bar{\mathbf{W}} &= c_1\bar{\Theta} + c_2(\bar{\Theta}\Theta + \Theta\bar{\Theta}) + \bar{c}_1\,\Theta + \bar{c}_2\,\Theta^2 \\
&= (c_2 - c_1)(\theta \bullet \bar{\theta})\mathbf{I} + c_1\bar{\Theta} + \ c_2\{(\theta \otimes \bar{\theta}) + (\bar{\theta} \otimes\ \theta)\} + \bar{c}_1\,\Theta + \bar{c}_2\,(\theta \otimes \theta) \\
\bar{c}_1 &\equiv \frac{1}{\theta^2}(1 - 2c_1 - \theta^2 c_2)(\theta \bullet \bar{\theta}) = c_3(\theta \bullet \bar{\theta}), \\
\bar{c}_2 &\equiv \frac{1}{\theta^2}(c_1 - 3c_2)(\theta \bullet \bar{\theta}) = c_4(\theta \bullet \bar{\theta})
\end{aligned} \tag{10.242}$$

10.11.8 *Angular Acceleration Vector, ${}^t\dot{\boldsymbol{\omega}}_S$, in Terms of the Rotation Vector, ${}^t\theta$*

Taking the time derivative of the angular acceleration given in equation (10.233), we have:

$$ {}^t\dot{\omega}_S = \dot{\mathbf{W}}(\theta)\,{}^t\dot{\theta}\ + \mathbf{W}(\theta)\,{}^t\ddot{\theta} \tag{10.243}$$

where $\mathbf{W}(\theta)$ is given by equation (10.139), and $\dot{\mathbf{W}}(\theta)$ by equation (10.240).

10.11.8.1 An Important Lemma

We will also need the following important result, relating the derivatives of the variational vector, ${}^t\bar{\varphi},_S$, and the curvature vectors, ${}^t\mathbf{k}$:

Lemma LD1: Defining ${}^t\mathbf{k}$ and ${}^t\bar{\varphi}$ as the axial vectors of ${}^t\mathbf{K}$ and ${}^t\bar{\mathbf{\Phi}}$, respectively, we have:

$$\begin{aligned}
{}^t\bar{\mathbf{\Phi}},_S &= {}^t\bar{\mathbf{K}} + [{}^t\mathbf{K}\ {}^t\bar{\mathbf{\Phi}} - {}^t\bar{\mathbf{\Phi}}\ {}^t\mathbf{K}] \\
\underbrace{{}^t\bar{\varphi},_S}_{\text{Effective material curvature vector variation}} &= \underbrace{{}^t\bar{\mathbf{k}}}_{\text{Material curvature vector variation}} + \underbrace{{}^t\mathbf{k}}_{\text{Material curvature vector}} \times \underbrace{{}^t\bar{\varphi}}_{\text{Material rotation vector variation}}
\end{aligned} \tag{10.244}$$

Proof LD1:. Using relations (10.220) and the definition of ${}^t\bar{\mathbf{\Phi}} \equiv \delta\,{}^t\mathbf{\Phi} \equiv \bar{\mathbf{R}}\,\mathbf{R}^\mathbf{T} \equiv \delta\mathbf{R}\,\mathbf{R}^\mathbf{T}$, we have:

$$\bar{\mathbf{\Phi}}^t,_S = \bar{\mathbf{R}},_S\ \mathbf{R}^T + ({}^t\bar{\mathbf{\Phi}}\,\mathbf{R})\,\mathbf{R}^T,_S = {}^t\bar{\mathbf{K}}\,\mathbf{R}\,\mathbf{R}^T + {}^t\mathbf{K}\ {}^t\bar{\mathbf{\Phi}}\,\mathbf{R}\,\mathbf{R}^T + {}^t\bar{\mathbf{\Phi}}\,\mathbf{R}\,\mathbf{R}^T\,({}^t\mathbf{K})^T$$

Now, noting the orthogonality property of a rotation tensor: $\mathbf{R}\,\mathbf{R}^T = \mathbf{I}$, and skew-symmetry property: $({}^t\mathbf{K})^T = -\,{}^t\mathbf{K}$ completes the proof of the first part. Recognition of the Lie bracket property: for $\mathbf{A}, \mathbf{B}$ skew, and $\mathbf{a}, \mathbf{b}$, the corresponding axial vectors, $[\mathbf{AB} - \mathbf{BA}] = [\mathbf{a} \times \mathbf{b}]$, for the second part, completes the proof.

***Exercise*:** Detail the steps in the above proof.

***Solution*:**

$$
\begin{aligned}
{}^t\bar{\mathbf{\Phi}}_{,S} &= {}^t\bar{\mathbf{K}}\,\mathbf{R}\,\mathbf{R}^T + {}^t\mathbf{K}\,{}^t\bar{\mathbf{\Phi}}\,\mathbf{R}\,\mathbf{R}^T + {}^t\bar{\mathbf{\Phi}}\,\mathbf{R}\,\mathbf{R}^T\,({}^t\mathbf{K})^T \\
&= {}^t\bar{\mathbf{K}} + \underbrace{{}^t\mathbf{K}\,{}^t\bar{\mathbf{\Phi}} - {}^t\bar{\mathbf{\Phi}}\,{}^t\mathbf{K}}_{\text{Lie bracket}} \quad \Rightarrow \quad \boxed{{}^t\bar{\boldsymbol{\varphi}}_{,S} = {}^t\bar{\mathbf{k}} - {}^t\bar{\boldsymbol{\varphi}} \times {}^t\mathbf{k}}
\end{aligned}
$$

10.11.8.2 The Central Lemma

We will need the following most important lemma, relating the derivative, $\boldsymbol{\theta}_{,S}$, and the variation, $\bar{\boldsymbol{\theta}}$, or the increment, $\Delta\boldsymbol{\theta}$, (which appears in the linearized virtual functional later, in the treatment of a Newton-type iterative algorithm) of the rotation vector, $\boldsymbol{\theta}$, for deriving the geometric stiffness operator and proving its symmetry under the conservative system of loading, presented later.

Lemma LD2: Given $\mathbf{W}$, c_1 and c_2 as in equation (10.139), we have:

$$
\boxed{\begin{aligned}
\bar{\mathbf{W}}\,\boldsymbol{\theta}_{,S} &= \mathbf{X}^S\,\bar{\boldsymbol{\theta}} \\
\Delta\mathbf{W}\,\boldsymbol{\theta}_{,S} &= \mathbf{X}^S\,\Delta\boldsymbol{\theta} \\
\Delta\mathbf{W}\,\dot{\boldsymbol{\theta}} &= \mathbf{X}^t\,\Delta\boldsymbol{\theta}
\end{aligned}} \tag{10.245}
$$

where

$$
\boxed{\begin{aligned}
\mathbf{X}^S &\equiv -c_1\mathbf{\Theta}_{,S} + c_2\{[\boldsymbol{\theta}\otimes\boldsymbol{\theta}_{,S}] - [\boldsymbol{\theta}_{,S}\otimes\boldsymbol{\theta}] + (\boldsymbol{\theta}\;\;\boldsymbol{\theta}_{,S})\mathbf{I}\} + \mathbf{V}\,[\boldsymbol{\theta}_{,S}\otimes\boldsymbol{\theta}] \\
\mathbf{X}^t &\equiv -c_1\,\dot{\mathbf{\Theta}} + c_2\{[\boldsymbol{\theta}\otimes\dot{\boldsymbol{\theta}}] - [\dot{\boldsymbol{\theta}}\otimes\boldsymbol{\theta}] + (\boldsymbol{\theta}\;\;\dot{\boldsymbol{\theta}})\mathbf{I}\} + \mathbf{V}\,[\dot{\boldsymbol{\theta}}\otimes\boldsymbol{\theta}] \\
\mathbf{V} &\equiv -c_2\mathbf{I} + c_3\mathbf{\Theta} + c_4\mathbf{\Theta}^2 \\
c_3 &\equiv \frac{1}{\theta^2}\{(1-\theta^2 c_2) - 2c_1 \\
c_4 &\equiv \frac{1}{\theta^2}(c_1 - 3c_2)
\end{aligned}} \tag{10.246}
$$

where $\bar{\boldsymbol{\theta}}$ and $\Delta\boldsymbol{\theta}$ are virtual (variational) and incremental rotation vectors, respectively.

Proof LD2:. First noting that $\bar{\theta} = \frac{1}{\theta}(\boldsymbol{\theta}\bullet\bar{\boldsymbol{\theta}})$ and taking one parameter variation of c_1 and c_2, we have, from equations (10.237), (10.240) and (10.246): $c_{1,S} = c_3(\boldsymbol{\theta}\bullet\boldsymbol{\theta}_{,S})$, $\bar{c}_1 = c_3(\boldsymbol{\theta}\bullet\bar{\boldsymbol{\theta}})$ and $c_{2,S} = c_4(\boldsymbol{\theta}\bullet\boldsymbol{\theta}_{,S})$, $\bar{c}_2 = c_4(\boldsymbol{\theta}\bullet\bar{\boldsymbol{\theta}})$. Now, rewriting the variation of $\mathbf{W}$ as defined in the first equality of equation (10.240), and multiplying both sides by $\boldsymbol{\theta}_{,S}$, we get:

$$
\bar{\mathbf{W}}\,\boldsymbol{\theta}_{,S} = c_1\bar{\mathbf{\Theta}}\,\boldsymbol{\theta}_{,S} + c_2(\bar{\mathbf{\Theta}}\mathbf{\Theta} + \mathbf{\Theta}\bar{\mathbf{\Theta}})\,\boldsymbol{\theta}_{,S} + (c_3\mathbf{\Theta} + c_4\mathbf{\Theta}^2)(\boldsymbol{\theta}\bullet\bar{\boldsymbol{\theta}})\,\boldsymbol{\theta}_{,S} \tag{10.247}
$$

Noting the identities: $\bar{\boldsymbol{\Theta}}\boldsymbol{\Theta} + \boldsymbol{\Theta}\bar{\boldsymbol{\Theta}} = [\theta \otimes \bar{\theta}] + [\bar{\theta} \otimes \theta] - 2(\theta \bullet \bar{\theta})\mathbf{I}$ and $[\theta \otimes \bar{\theta}]\theta_{,S} = [\theta \otimes \theta_{,S}]\bar{\theta}$, we have:

$$\begin{aligned} \bar{\boldsymbol{\Theta}}\,\theta_{,S} &= \bar{\theta} \times \theta_{,S} = -\theta_{,S} \times \bar{\theta} = -\boldsymbol{\Theta}_{,S}\,\bar{\theta}; \\ (\bar{\boldsymbol{\Theta}}\boldsymbol{\Theta} + \boldsymbol{\Theta}\bar{\boldsymbol{\Theta}})\,\theta_{,S} &= \bar{\theta} \times \theta \times \theta_{,S} + \theta \times \bar{\theta} \times \theta_{,S} \\ &= (\bar{\theta} \bullet \theta_{,S})\theta - (\bar{\theta} \bullet \theta)\theta_{,S} + (\theta \bullet \theta_{,S})\bar{\theta} - (\theta \bullet \bar{\theta})\theta_{,S} \\ &= \left\{(\theta \otimes \theta_{,S}) - (\theta_{,S} \otimes \theta) + (\theta \bullet \theta_{,S})\mathbf{I}\right\}\bar{\theta} - (\theta_{,S} \otimes \theta)\bar{\theta} \end{aligned} \tag{10.248}$$

Now, inserting relations (10.248) in equation (10.247), we get:

$$\bar{\mathbf{W}}\,\theta_{,S} = \left\{ \begin{array}{l} -c_1\boldsymbol{\Theta}_{,S} + c_2\left\{(\theta \otimes \theta_{,S}) - (\theta_{,S} \otimes \theta) + (\theta \bullet \theta_{,S})\mathbf{I}\right\} \\ + \left(-c_2\mathbf{I} + c_3\boldsymbol{\Theta} + c_4\boldsymbol{\Theta}^2\right)(\theta_{,S} \otimes \theta) \end{array} \right\}\bar{\theta} \tag{10.249}$$

Next, insertion of the definitions for $\mathbf{X}$ and $\mathbf{V}$ of equation (10.246) in equation (10.249) completes the proof for the first part of equation (10.245), and replacement of $\overline{(\bullet)}$ terms by the $\Delta(\bullet)$, completes the proof for the increment and spatial derivative. Finally, replacing $\mathbf{X}^S$ by $\mathbf{X}^t$, completes the proof for the increment and time rate.

10.11.9 Where We Would Like to Go

We have introduced various parameters and derived all the necessary relationships anticipating the next step of developing the virtual work equation for the beam reference curve. Clearly, then, we should proceed onward by embarking on:

- computational virtual work principle
  Finally, to get familiarized with, and be conscious about, the overall computational strategy for nonlinear beams, we should explore:
- the road map: nonlinear beams

10.12 Computational Virtual Work Equations

10.12.1 What We Need to Recall

- The real configuration space of a beam is characterized by $({}^t\mathbf{d}(S,t),\ {}^t\mathbf{R}(S,t))$.
- The virtual space for a beam may be described by the virtual displacement vector and the virtual rotational tensor fields, $(\bar{\mathbf{d}}(S) = \mathbf{G}\,{}^t\bar{\mathbf{d}},\ \bar{\mathbf{R}}(S) = \mathbf{G}\,{}^t\bar{\mathbf{R}}\,\mathbf{G}^T)$, with the component virtual entities defined as $({}^t\bar{\mathbf{d}} \equiv \delta\,{}^t\mathbf{d},\ {}^t\bar{\boldsymbol{\Phi}} \equiv \delta\,{}^t\boldsymbol{\Phi})$, where ${}^t\bar{\boldsymbol{\Phi}}(S)$, the instantaneous (variational) rotational skew-symmetric matrix describing the one-parameter variation ${}^t\bar{\mathbf{R}}(S) \equiv \delta\,{}^t\mathbf{R}(S)$ of the rotation matrix ${}^t\mathbf{R}$ defined by: ${}^t\bar{\boldsymbol{\Phi}} \equiv {}^t\bar{\mathbf{R}}\,{}^t\mathbf{R}^T$ with ${}^t\bar{\varphi}$ as its axial vector, that is, ${}^t\bar{\boldsymbol{\Phi}}\mathbf{h} = {}^t\bar{\varphi} \times \mathbf{h}, \forall\mathbf{h} \in \mathbb{R}^3$, and ${}^t\bar{\boldsymbol{\Phi}}\,{}^t\bar{\varphi} = \mathbf{0}$.
- The computational dynamic balance equations in component vector form:

$$\boxed{\begin{aligned} \mathbf{G}({}^t\mathbf{F}|_S + {}^t\hat{\mathbf{F}} - {}^t\dot{\mathbf{L}}) &= 0 \\ \mathbf{G}({}^t\mathbf{M}|_S + \mathbf{a} \times {}^t\mathbf{F} + {}^t\hat{\mathbf{M}} - {}^t\dot{\mathbf{A}}) &= 0 \end{aligned}} \tag{10.250}$$

or

$$\boxed{\begin{aligned} \mathbf{G}({}^t\mathbf{F}_{,S} + \boldsymbol{\kappa}^0 \times {}^t\mathbf{F} + {}^t\hat{\mathbf{F}} - {}^t\dot{\mathbf{L}}) = \mathbf{0} \\ \mathbf{G}({}^t\mathbf{M}_{,S} + \boldsymbol{\kappa}^0 \times {}^t\mathbf{M} + \{\boldsymbol{\kappa}^0 \times \mathbf{d} + \mathbf{e}\} \times {}^t\mathbf{F} + {}^t\hat{\mathbf{M}} - {}^t\dot{\mathbf{A}}) = \mathbf{0} \end{aligned}} \tag{10.251}$$

where $\mathbf{G} = [\mathbf{G}_1 \equiv \mathbf{T} \quad \mathbf{G}_2 \equiv \mathbf{M} \quad \mathbf{G}_3 \equiv \mathbf{B}]$ is the orthonormal matrix with the base vectors as the columns of the matrix such that $\mathbf{G}^T\,\mathbf{G} = \mathbf{I}$.

- The force and moment component vectors in undeformed and current deformed coordinate system are related by the force and moment rotation transformation:

$$\boxed{{}^t\mathbf{F} = {}^t\mathbf{R}\,{}^n\mathbf{F} \qquad \text{and} \qquad {}^t\mathbf{M} = {}^t\mathbf{R}\,{}^n\mathbf{M}} \tag{10.252}$$

where the superscript t over an object, which reminds us that the object is expressed in the undeformed $\mathbf{G}_1 \equiv \mathbf{T}$, $\mathbf{G}_2 \equiv \mathbf{M}$, $\mathbf{G}_3 \equiv \mathbf{B}$ orthonormal frame, may be dropped occasionally for clarity; a superscript n identifies variables in the rotated frame, $\mathbf{g}_1 = {}^t\mathbf{R}\,\mathbf{G}_1$, $\mathbf{g}_2 = {}^t\mathbf{R}\,\mathbf{G}_2$, $\mathbf{g}_3 = {}^t\mathbf{R}\,\mathbf{G}_3$, where ${}^t\mathbf{R}$ is the current rotation tensor measured in the undeformed coordinate system; the context should help us recall the appropriate frame of reference.

- The covariant derivatives of the displacement and rotation vectors:

$$\boxed{{}^t\mathbf{d}|_S = {}^t\mathbf{d}_{,S} + \mathbf{k}^0 \times {}^t\mathbf{d} \qquad \text{and} \qquad {}^t\boldsymbol{\theta}|_S = {}^t\boldsymbol{\theta}_{,S} + \mathbf{k}^0 \times {}^t\boldsymbol{\theta}} \tag{10.253}$$

Likewise, we have the time derivatives of the effective linear momentum, $\mathbf{L}(S,t)$, and the effective angular momentum, $\mathbf{A}(S,t)$, in terms of the components measured in the undeformed coordinate system:

$$\boxed{\begin{aligned} \dot{\mathbf{A}}_S(S,t) &= \mathbf{I}_S\;\dot{\boldsymbol{\omega}}_S(S,t) + \boldsymbol{\omega}_S(S,t) \times \mathbf{A}_S(S,t) \\ &= \mathbf{R}\big(\mathbf{I}_M\;\dot{\boldsymbol{\omega}}_M(S,t) + \boldsymbol{\omega}_M(S,t) \times \mathbf{A}_M(S,t)\big) = \mathbf{R}\big(\boldsymbol{\omega}_M \times \mathbf{A}_M + \dot{\mathbf{A}}_M\big) \\ &\text{where} \\ \mathbf{A}_S &= \mathbf{I}_S\;\boldsymbol{\omega}_S = \mathbf{R}\,\mathbf{A}_M, \qquad \mathbf{A}_M = \mathbf{I}_M\;\boldsymbol{\omega}_M \\ \mathbf{I}_S(S,t) &\stackrel{\Delta}{=} \mathbf{R}(S,t)\,\mathbf{I}_M(S)\,\mathbf{R}^T(S,t) \\ \mathbf{I}_M(S) &\stackrel{\Delta}{=} I_{\alpha\beta}\{\delta_{\alpha\beta}\mathbf{I} - (\mathbf{G}_\alpha \otimes \mathbf{G}_\beta)\}, \quad \alpha,\beta = 2,3 \\ \mathbf{I}_S(S,t) &= I_{\alpha\beta}\{\delta_{\alpha\beta}\mathbf{I} - (\mathbf{g}_\alpha(t) \otimes \mathbf{g}_\beta(t))\} \\ I_{\alpha\beta} &\stackrel{\Delta}{=} \int_{A_c} \rho_0(S,\xi^2,\xi^3)\;\xi^\alpha\;\xi^\beta\;dA \\ \boldsymbol{\Omega}_M &= [\boldsymbol{\omega}_M] = \mathbf{R}^T\,\dot{\mathbf{R}}, \quad \boldsymbol{\Omega}_S = [\boldsymbol{\omega}_S] = \dot{\mathbf{R}}\,\mathbf{R}^T = \mathbf{R}\,\boldsymbol{\Omega}_M\,\mathbf{R}^T, \quad \boldsymbol{\omega}_S = \mathbf{R}\,\boldsymbol{\omega}_M \end{aligned}} \tag{10.254}$$

- The definition of a rotation tensor, its matrix representation, and algebra, geometry and topology of rotation.
- A rotation matrix, ${}^t\mathbf{R}$, corresponding to the three-dimensional special orthogonal matrix group, $SO(3)$, may be parameterized by the Rodrigues vector, $\mathbf{s}$, the axis of the instantaneous rotation.

- the rotation matrix, ${}^t\mathbf{R}(S,t)$, is related, through Cayley's theorem, to the Rodrigues vector, $\mathbf{s}(S,t) = \{s_1(S,t) \quad s_2(S,t) \quad s_3(S,t)\}^T$, with the components in the $\mathbf{G}_1(\equiv \mathbf{T}) - \mathbf{G}_2(\equiv \mathbf{M}) - \mathbf{G}_3(\equiv \mathbf{B})$ frame at any point, S, and at a time, $t \in \mathbb{R}_+$, parameterized by the arc lengths along a beam reference curve by :

$$\mathbf{R} = \frac{1}{\Sigma}\begin{bmatrix} 1+s_1^2-s_2^2-s_3^2 & 2(s_1s_2-s_3) & 2(s_1s_3+s_2) \\ 2(s_1s_2+s_3) & 1-s_1^2+s_2^2-s_3^2 & 2(s_2s_3-s_1) \\ 2(s_1s_3-s_2) & 2(s_2s_3+s_1) & 1-s_1^2-s_2^2+s_3^2 \end{bmatrix} \tag{10.255}$$

where $\sum \equiv 1 + \| \mathbf{s} \|^2$.
- That the Rodrigues vector, $\mathbf{s}$, is orientated along the instantaneous axis of rotation at a point of the shell reference surface. The skew-symmetric matrix, $\mathbf{S}$, corresponding to the Rodrigues vector, $\mathbf{s}$, is given by:

$$\mathbf{S} = \begin{bmatrix} 0 & -s_3 & s_2 \\ s_3 & 0 & -s_1 \\ -s_2 & s_1 & 0 \end{bmatrix} \tag{10.256}$$

- Finally, the Rodrigues rotation vector, $\boldsymbol{\theta}$, is the scaled Rodrigues vector related by:

$$\boxed{\mathbf{s} = \frac{1}{\theta}\tan\left(\frac{\theta}{2}\right)\boldsymbol{\theta}} \tag{10.257}$$

where $\theta \equiv ||\boldsymbol{\theta}|| = (\boldsymbol{\theta} \bullet \boldsymbol{\theta})^{\frac{1}{2}}$ is the length measure of the rotation vector. Alternatively, defining the unit vector, $\mathbf{n} \equiv \frac{\boldsymbol{\theta}}{\theta}$, and the corresponding skew-symmetric matrix, $\mathbf{N}$, such that $\mathbf{N}\,\mathbf{h} = \mathbf{n} \times \mathbf{h},\ \forall \mathbf{h} \in \mathbb{R}_3$, we have:

$$\boxed{\begin{aligned} &\mathbf{R} = (\mathbf{I}-\mathbf{S})^{-1}\,(\mathbf{I}+\mathbf{S}) \\ &\mathbf{s} = \tan\left(\frac{\theta}{2}\right)\mathbf{n}, \qquad \mathbf{S} = \tan\left(\frac{\theta}{2}\right)\mathbf{N} \end{aligned}} \tag{10.258}$$

- The Rodrigues quaternion and Euler–Rodrigues parameters.
- The finite rotation, $\mathbf{R}$, belongs to the spherical curved space, $SO(3)$, and the instantaneous and infinitesimal virtual rotation, $\bar{\boldsymbol{\Phi}}$, represented by a skew-symmetric matrix, belongs to the tangent space at identity, Lie algebra, and they are related by an exponential map; but, computationally, it is more convenient to apply Cayley's formula:

$$\boxed{\mathbf{R} = (\mathbf{I}-\bar{\boldsymbol{\Phi}})^{-1}\,(\mathbf{I}+\bar{\boldsymbol{\Phi}})} \tag{10.259}$$

- For notational uniformity, we denote all variations of entities by a bar superscripted on it. Furthermore all entities measured in the $\mathbf{T}, \mathbf{M}, \mathbf{B}$ frame is superscripted as ${}^t(\bullet)$, and those in the rotating $\mathbf{g_1}(\equiv \mathbf{t})$, $\mathbf{g_2}(\equiv \mathbf{m}_n)$, $\mathbf{g}_3(\equiv \mathbf{b}_n)$ coordinate system as ${}^n(\bullet)$. However, for the sake of uncluttered readability, sometimes the superscripts may be dropped, but the context should help determine the appropriate underlying coordinate system.

- The material "curvature" tensors, ${}^t\mathbf{K}$, and "curvature" vector, ${}^t\mathbf{k}$, as:

$$\boxed{{}^t\mathbf{K} \equiv \mathbf{R},_S\, \mathbf{R}^{\mathrm{T}}} \tag{10.260}$$

$$\boxed{{}^t\mathbf{k} = \mathbf{W}(\boldsymbol{\theta})\ {}^t\boldsymbol{\theta},_S} \tag{10.261}$$

where we have:

$$\boxed{\begin{aligned} &\mathbf{W}({}^t\boldsymbol{\theta}) \equiv \mathbf{I} + c_1\,{}^t\boldsymbol{\Theta} + c_2\,{}^t\boldsymbol{\Theta}^2, \\ &c_1({}^t\boldsymbol{\theta}) = \frac{1-\cos\theta}{\theta^2}, \qquad c_2({}^t\boldsymbol{\theta}) = \frac{\theta - \sin\theta}{\theta^3} \end{aligned}} \tag{10.262}$$

with ${}^t\boldsymbol{\Theta}$ as the skew-symmetric matrix corresponding to the axial vector, ${}^t\boldsymbol{\theta}$, such that ${}^t\boldsymbol{\Theta}\,{}^t\boldsymbol{\theta} = \mathbf{0}$.

- The material first variation tensor, ${}^t\bar{\boldsymbol{\Phi}} \equiv \delta\,{}^t\boldsymbol{\Phi}$, and first variation vector, ${}^t\bar{\boldsymbol{\varphi}} \equiv \delta\,{}^t\boldsymbol{\phi}$, as:

$$\boxed{{}^t\bar{\boldsymbol{\Phi}} \equiv \delta\,{}^t\boldsymbol{\Phi} \equiv \bar{\mathbf{R}}\,\mathbf{R}^{\mathrm{T}} \equiv \delta\mathbf{R}\,\mathbf{R}^{\mathrm{T}}} \tag{10.263}$$

$$\boxed{{}^t\bar{\boldsymbol{\varphi}} = \mathbf{W}(\boldsymbol{\theta})\,{}^t\bar{\boldsymbol{\theta}}} \tag{10.264}$$

- The angular velocity component vector, ${}^t\boldsymbol{\omega}$, is related to the Rodrigues rotation component vector, ${}^t\boldsymbol{\theta}$, by:

$$\boxed{{}^t\boldsymbol{\omega} = \mathbf{W}(\boldsymbol{\theta})\,{}^t\dot{\boldsymbol{\theta}}} \tag{10.265}$$

- The angular acceleration component vector, ${}^t\dot{\boldsymbol{\omega}}$, is related to the Rodrigues rotation component vector, ${}^t\boldsymbol{\theta}$, by:

$$\boxed{{}^t\dot{\boldsymbol{\omega}} = \dot{\mathbf{W}}(\boldsymbol{\theta})\,{}^t\dot{\boldsymbol{\theta}} + \mathbf{W}(\boldsymbol{\theta})\,{}^t\ddot{\boldsymbol{\theta}}} \tag{10.266}$$

Lemma LD1: Defining ${}^t\mathbf{k}$ and ${}^t\bar{\boldsymbol{\varphi}}$ as the axial vectors of ${}^t\mathbf{K}$ and ${}^t\bar{\boldsymbol{\Phi}}$, respectively, we have:

$$\boxed{\begin{aligned} &{}^t\bar{\boldsymbol{\Phi}},_S = {}^t\bar{\mathbf{K}} + [\,{}^t\mathbf{K}\,{}^t\bar{\boldsymbol{\Phi}} - {}^t\bar{\boldsymbol{\Phi}}\,{}^t\mathbf{K}] \\ &\underbrace{{}^t\bar{\boldsymbol{\varphi}},_S}_{\substack{\text{Effective material}\\ \text{curvature vector}\\ \text{variation}}} = \underbrace{{}^t\bar{\mathbf{k}}}_{\substack{\text{Material}\\ \text{curvature vector}\\ \text{variation}}} + \underbrace{{}^t\mathbf{k}}_{\text{Material curvature vector}} \times \underbrace{{}^t\bar{\boldsymbol{\varphi}}}_{\substack{\text{Material}\\ \text{rotation vector}\\ \text{variation}}} \end{aligned}} \tag{10.267}$$

10.12.1.1 The Central Lemma

We will need the following most important lemma, relating the derivative, $\boldsymbol{\theta},_S$, and the variation, $\bar{\boldsymbol{\theta}}$, or the increment, $\Delta\boldsymbol{\theta}$ (which appears in the linearized virtual functional later in the treatment of a Newton-type iterative algorithm), of the rotation vector, $\boldsymbol{\theta}$, for deriving the geometric stiffness operator and proving its symmetry under the conservative system of loading, presented later:

Lemma LD2: Given $\mathbf{W}$, c_1 and c_2 as in equation (10.139), we have:

$$\boxed{\bar{\mathbf{W}}\,\boldsymbol{\theta}_{,S} = \mathbf{X}\,\bar{\boldsymbol{\theta}}, \qquad \Delta\mathbf{W}\,\boldsymbol{\theta}_{,S} = \mathbf{X}\,\Delta\boldsymbol{\theta}} \tag{10.268}$$

where

$$\boxed{\begin{aligned}
&\mathbf{X} \equiv -c_1\boldsymbol{\Theta}_{,S} + c_2\{[\boldsymbol{\theta}\otimes\boldsymbol{\theta}_{,S}] - [\boldsymbol{\theta}_{,S}\otimes\boldsymbol{\theta}] + (\boldsymbol{\theta}\bullet\boldsymbol{\theta}_{,S})\mathbf{I}\} + \mathbf{V}[\boldsymbol{\theta}_{,S}\otimes\boldsymbol{\theta}]\\
&\mathbf{V} \equiv -c_2\mathbf{I} + c_3\boldsymbol{\Theta} + c_4\boldsymbol{\Theta}^2\\
&c_3 \equiv \frac{1}{\theta^2}\{(1-\theta^2 c_2) - 2c_1 \qquad c_4 \equiv \frac{1}{\theta^2}(c_1 - 3c_2)
\end{aligned}} \tag{10.269}$$

where $\bar{\boldsymbol{\theta}}$ and $\Delta\boldsymbol{\theta}$ are virtual (variational) and incremental rotation vectors, respectively.

- Notationally, the bracket $[\bullet]$ denotes the skew-symmetric matrix corresponding to the argument, that is, $[\mathbf{a}] \equiv \mathbf{a}\times$ for an axial vector, $\mathbf{a}$, such that $[\mathbf{a}]\,\mathbf{a} = \mathbf{0}$.
- The vector triple product property: $\mathbf{a}\times\mathbf{b}\bullet\mathbf{c} = \mathbf{b}\times\mathbf{c}\bullet\mathbf{a} = \mathbf{c}\times\mathbf{a}\bullet\mathbf{b}$, $\forall\mathbf{a},\mathbf{b},\mathbf{c}$ vectors.

In what follows, notationally, all virtual entities are identified with a superscripted bar over them.

10.12.2 Boundary Conditions

Just as in the reduction of the 3D balance laws of a beam-like body by integration over the cross-section produced the exact counterparts for a 1D beam curve that are valid for the ***interior*** of $M \equiv [0, S]$, we need to determine the exact statically equivalent boundary conditions for the beam ends from the prescribed boundary conditions on the end cross-sections at $M^C = [S = 0] \cup [S = L]$ of a 3D beam-like body. The boundary conditions on a 3D beam-like body can basically be of the following types.

- ***Traction boundary condition***: Tractions are specified on all or part, M^C, of the end sections of a beam-like body. If the specification involves the entire end area, some sort of constraint must be imposed against the rigid body movement. For tractions specified partially on one or the other of the end sections, we designate the aggregate area as the set $M_f^C \subset M^C$. Let the specified 3D end traction, $\hat{\mathbf{t}}(S, \xi^2, \xi^3)$, be given which, of course, satisfies the relationship: at $(S, \xi^2, \xi^3) \in M \cap M_f^C \subset M \cap M^C$, $\hat{\mathbf{t}}(S, \xi^2, \xi^3) = \mathbf{P}(S, \xi^2, \xi^3)\,\mathbf{T}(S)$, where $\mathbf{T}(S)$ is the outward undeformed normal at any point, (ξ^2, ξ^3), on the rigid end cross-section with $\mathbf{G_1} \equiv \mathbf{T}$ on it. In this case, we may obtain the 1D counterparts by the over-the-cross-section integration as:

$$\boxed{\begin{aligned}
&\hat{\mathbf{F}}^S \equiv \int_{\partial A_c} \hat{\mathbf{t}}(S, \xi^2, \xi^3)\, dA_c, \quad S \in \partial M_f\\
&\hat{\mathbf{M}}^S \equiv \int_{\partial A_c} p(S, \xi^2, \xi^3) \times \hat{\mathbf{t}}(S, \xi^2, \xi^3)\, dA_c, \quad S \in \partial M_f
\end{aligned}} \tag{10.270}$$

where $\partial M_f = M \cap M_f^C$ is that part of the beam ends where the beam's statically equivalent tractions will be considered; for a thin beam with $\mathbf{p}(S, \xi^2, \xi^3) = \mathbf{R}(S)\, \mathbf{P}(\xi^2, \xi^3)$, from equation (10.270), we will have:

$$\begin{aligned} \hat{\mathbf{F}}^S &\equiv \int_{\partial A_c} \hat{\mathbf{t}}(S, \xi^2, \xi^3)\, dA_c, \quad S \in \partial M_f \\ \hat{\mathbf{M}}^S &\equiv \mathbf{R}(S) \int_{\partial A_c} \mathbf{P}(\xi^2, \xi^3) \times \hat{\mathbf{t}}(S, \xi^2, \xi^3)\, dA_c, \quad S \in \partial M_f \end{aligned} \tag{10.271}$$

- ***Kinematic boundary condition***: Displacements are specified on one or both of the end cross-sections. For displacements specified partially on one of the cross-sections, we set $M_d^C \subset M^C$. We will admit only those displacements that have finite strain energy. Now, if the displacements are homogeneous, we simply set: ${}^t\mathbf{d} = \mathbf{0}$, ${}^t\mathbf{R} = \mathbf{0}$, $({}^t\mathbf{d},\, {}^t\mathbf{R}) \in \partial M_d$ where $\partial M_d \subset \partial M$ is that portion of the end cross-sections, ∂M, that is also common with M_d^S. For inhomogeneous displacement specifications, we will set: ${}^t\mathbf{d} = \hat{\mathbf{d}}$, ${}^t\mathbf{R} = \hat{\mathbf{R}}$, $({}^t\mathbf{d},\, {}^t\mathbf{R}) \in \partial M_d$; the hat superscript means the quantities are prescribed. In this case, the kinematically admissible virtual displacements, $({}^t\bar{\mathbf{d}} \equiv \delta\, {}^t\mathbf{d},\ \bar{\mathbf{R}} \equiv \delta\, {}^t\mathbf{R})$, will be identically zero on ∂M_d. Thus, in the subsequent virtual work equation, we must have the boundary integral on ∂M_d identically vanished. The actual finite element solution for a beam with an inhomogeneous displacement condition will be given as an additive sum of two displacement functions with one part as the solution of the relevant virtual work equation and the other part as an arbitrary fixed admissible displacement function.
- ***Mixed traction and displacement boundary conditions***: In the event that the end sections of the three-dimensional beam-like body is in contact with other elastic bodies such as, say, elements modeled as distributed springs, the boundary conditions will be expressed as a linear combination of both the traction and displacement vectors with the spring stiffness as given. This condition is essentially a combination of the previous two boundary conditions.

10.12.3 Initial Conditions

For a beam semi-discrete finite element formulation (which we alluded to in Chapter 1 on motivation), the virtual work equation eventually reduces to a matrix second-order ordinary differential equations. Clearly, the typical initial conditions will be the prescribed initial displacement and velocity at the finite element nodes for calculating the time-evolution of the beam deformations and internal reactive stresses. Then, the initial conditions may be written as:

$$\begin{aligned} &\mathbf{d}(S,0) = \widehat{\mathbf{d}_0} && \theta(S,0) = \widehat{\theta_0}. \\ &\mathbf{v}(S,0) \equiv \dot{\mathbf{d}}(S,0) = \widehat{\mathbf{v}_0}, && \omega(S,0) \equiv \dot{\theta}(S,0) = \widehat{\omega_0} \end{aligned} \tag{10.272}$$

10.12.4 Virtual Work or Weak Form of Balance Equations

Let $(\bar{\mathbf{d}}, \bar{\varphi})$ be a virtual vector set belonging to the ***admissible test***, or ***virtual space***. If $({}^t\bar{\mathbf{d}},\, {}^t\bar{\varphi})$ is the corresponding component vectors, then we have $\bar{\mathbf{d}} = \mathbf{G}\, {}^t\bar{\mathbf{d}}$ and $\bar{\varphi} = \mathbf{G}\, {}^t\bar{\varphi}$; note that $({}^t\bar{\mathbf{d}},\, {}^t\bar{\varphi})$ is time-independent. Now, with the above background in mind, we multiply equation (10.204) by any admissible virtual state vectors ${}^t\bar{\mathbf{d}}$ and ${}^t\bar{\boldsymbol{\varphi}}$, and integrate over the range, $S \in [0, L]$, of the beam arc length parameter, S, to get the beam virtual work functional, $G(\mathbf{d}, ; \bar{\mathbf{d}}, \bar{\boldsymbol{\varphi}})$, given as:

$$G(\mathbf{d}, \boldsymbol{\varphi}; \bar{\mathbf{d}}, \bar{\boldsymbol{\varphi}}) \equiv \int_0^L \left\{ \begin{array}{l} ({}^t\mathbf{F},_s + \mathbf{k}^0 \times {}^t\mathbf{F} + {}^t\hat{\mathbf{F}} - {}^t\dot{\mathbf{L}}) \bullet {}^t\bar{\mathbf{d}} + \\ ({}^t\mathbf{M},_s + \mathbf{k}^0 \times {}^t\mathbf{M} + (\mathbf{k}^0 \times \mathbf{d} + \mathbf{e}) \times {}^t\mathbf{F} + {}^t\hat{\mathbf{M}} - {}^t\dot{\mathbf{A}}_S) \bullet {}^t\bar{\boldsymbol{\varphi}} \end{array} \right\} dS$$
$$= \mathbf{0}, \quad \forall \text{ admissible } {}^t\bar{\mathbf{d}}, {}^t\bar{\boldsymbol{\varphi}} \tag{10.273}$$

where we have applied the fact that $\mathbf{G}^T\,\mathbf{G} = \mathbf{I}$. Now, recalling the vector triple product property, and integrating by parts, we get from equation (10.273):

$$\boxed{\begin{aligned} G_{Dyn}({}^t\mathbf{d}, {}^t\mathbf{R}; {}^t\bar{\mathbf{d}}, {}^t\bar{\boldsymbol{\varphi}}) &\equiv G_{Inertial} + G_{Internal} - G_{External} \\ &= \overbrace{\int_0^L \{(M_0\,{}^t\ddot{\mathbf{d}} \bullet {}^t\bar{\mathbf{d}} + {}^t\dot{\mathbf{A}}_S \bullet {}^t\bar{\boldsymbol{\varphi}}\}\, dS}^{\text{Inertial}} \\ &\quad \overbrace{+ \int_0^L \left\{ \begin{array}{l} [{}^t\mathbf{F} \bullet [(\bar{\mathbf{d}},_s + \mathbf{k}^0 \times \bar{\mathbf{d}}) - \bar{\boldsymbol{\varphi}} \times (\mathbf{e} + \mathbf{k}^0 \times \mathbf{d})]] + \\ [{}^t\mathbf{M} \bullet [\bar{\boldsymbol{\varphi}},_s - \bar{\boldsymbol{\varphi}} \times \mathbf{k}^0]] \end{array} \right\} dS}^{\text{Internal}} \\ &\quad \overbrace{- \int_0^L \{{}^t\hat{\mathbf{F}}^B \bullet \bar{\mathbf{d}} + {}^t\hat{\mathbf{M}}^B \bullet \bar{\boldsymbol{\varphi}}\}\, dS + \{[{}^t\mathbf{F}^S \bullet \bar{\mathbf{d}}]_0^L + [{}^t\mathbf{M}^S \bullet \bar{\boldsymbol{\varphi}}]_0^L\}}^{\text{External}} \\ &= \mathbf{0}, \quad \forall \text{ admissible } {}^t\bar{\mathbf{d}}, {}^t\bar{\boldsymbol{\varphi}} \end{aligned}}$$
$$\tag{10.274}$$

where we have the angular velocity component vector, ${}^t\boldsymbol{\omega}$, related to the Rodrigues' rotation component vector, ${}^t\boldsymbol{\theta}$, and identified the inertial virtual work, the internal virtual work and the external virtual work of the virtual work principle as:

Inertial virtual work, $G_{Inertial}$, as:

$$\boxed{G_{Inertial} \equiv \int_0^L \left\{ (M_0\,{}^t\ddot{\mathbf{d}} \bullet {}^t\bar{\mathbf{d}} + (\mathbf{I}_S\,\dot{\boldsymbol{\omega}}_S + \boldsymbol{\omega}_S \times \mathbf{A}_S) \bullet {}^t\bar{\boldsymbol{\varphi}} \right\} dS} \tag{10.275}$$

Internal virtual work, $G_{Internal}$, as:

$$\boxed{G_{Internal} \equiv \int_0^L \{[{}^t\mathbf{F} \bullet [(\bar{\mathbf{d}},_s + \mathbf{k}^0 \times \bar{\mathbf{d}}) - \bar{\boldsymbol{\varphi}} \times (\mathbf{e} + \mathbf{k}^0 \times \mathbf{d})]] + [{}^t\mathbf{M} \bullet [\bar{\boldsymbol{\varphi}},_s - \bar{\boldsymbol{\varphi}} \times \mathbf{k}^0]]\}\, dS}$$
$$\tag{10.276}$$

External virtual work, $G_{External}$, as:

$$G_{external} \equiv \int_S \left[{}^t\hat{\mathbf{F}}^B \bullet {}^t\bar{\mathbf{d}} + {}^t\hat{\mathbf{M}}^B \bullet {}^t\bar{\boldsymbol{\varphi}}\right] dS + \left[{}^t\hat{\mathbf{F}}^S \bullet {}^t\bar{\mathbf{d}}\right]_{S=0}^{S=L} + \left[{}^t\hat{\mathbf{M}}^S \bullet {}^t\bar{\boldsymbol{\varphi}}\right]_{S=0}^{S=L} \tag{10.277}$$

where, ${}^t\hat{\mathbf{F}}^B \equiv {}^t\hat{\mathbf{F}}$ and ${}^t\hat{\mathbf{M}}^B \equiv {}^t\hat{\mathbf{M}}$ are the specified distributed forces and moments along the length of the beam (analogous to the body forces), and ${}^t\hat{\mathbf{F}}^S \equiv {}^t\mathbf{F}$ and ${}^t\hat{\mathbf{M}}^S \equiv {}^t\mathbf{M}$ are the specified end forces and moments at $S = 0$ and $S = L$ end cross-sections (analogous to the surface traction) of the beam. Note that the expression for the external virtual work depends on the nature of the kinematically admissible virtual functions which, in turn, depends on the boundary conditions.

***Exercise*:** Fill in the details and verify the weak form in equation (10.274).

10.12.5 Where We Would Like to Go

Let us note that the virtual functional of equations (10.274) is expressed in terms of rotational variation vector, ${}^t\bar{\boldsymbol{\varphi}}$, the variational axial vector corresponding to the instantaneous variational matrix, ${}^t\bar{\boldsymbol{\Phi}}$, of the rotation matrix, ${}^t\mathbf{R}$. Since we parameterize the rotation matrix by the Rodrigues rotation vector, ${}^t\boldsymbol{\theta}$, we must express, for computational purposes, the virtual strains, angular accelerations and the virtual functional as functions of this rotation vector and its virtual counterpart, ${}^t\bar{\boldsymbol{\theta}}$. Thus, we must check out:

- computational virtual work: revisited.

10.13 Computational Virtual Work Equations and Virtual Strains: Revisited

10.13.1 What we need to recall

- The virtual work equation is given as:

$$\begin{aligned}
G_{Dyn}({}^t\mathbf{d}, {}^t\mathbf{R}; {}^t\bar{\mathbf{d}}, {}^t\bar{\boldsymbol{\varphi}}) &\equiv G_{Inertial} + G_{Internal} - G_{External} \\
&= \overbrace{\int_0^L \left\{ (M_0\, {}^t\ddot{\mathbf{d}} \bullet {}^t\bar{\mathbf{d}} + {}^t\dot{\mathbf{A}}_S \bullet {}^t\bar{\boldsymbol{\varphi}} \right\} dS}^{\text{Inertial}} \\
&\quad + \overbrace{\int_0^L \left\{ \begin{matrix} [{}^t\mathbf{F}\bullet[(\bar{\mathbf{d}}_{,s} + \mathbf{k}^0 \times \bar{\mathbf{d}}) - \bar{\boldsymbol{\varphi}} \times (\mathbf{e} + \mathbf{k}^0 \times \mathbf{d})]] + \\ [{}^t\mathbf{M}\bullet[\bar{\boldsymbol{\varphi}}_{,s} - \bar{\boldsymbol{\varphi}} \times \mathbf{k}^0]] \end{matrix} \right\} dS}^{\text{Internal}} \\
&\quad - \overbrace{\int_0^L \left\{ {}^t\hat{\mathbf{F}}^B\bullet\bar{\mathbf{d}} + {}^t\hat{\mathbf{M}}^B\bullet\bar{\boldsymbol{\varphi}} \right\} dS + \left\{ [{}^t\mathbf{F}^S\bullet\bar{\mathbf{d}}]_0^L + [{}^t\mathbf{M}^S\bullet\bar{\boldsymbol{\varphi}}]_0^L \right\}}^{\text{External}} \\
&= \mathbf{0}, \quad \forall \text{ admissible } {}^t\bar{\mathbf{d}}, {}^t\bar{\boldsymbol{\varphi}}
\end{aligned} \tag{10.278}$$

- Inertial virtual work, $G_{Inertial}$, as:

$$G_{Inertial} \equiv \int_0^L \{(M_0\, {}^t\ddot{\mathbf{d}} \bullet {}^t\bar{\mathbf{d}} + (\mathbf{I}_S\, \dot{\boldsymbol{\omega}}_S + \boldsymbol{\omega}_S \times \mathbf{A}_S) \bullet {}^t\bar{\boldsymbol{\varphi}}\}\, dS \tag{10.279}$$

- Internal virtual work, $G_{Internal}$, as:

$$G_{Internal} \equiv \int_0^L \{[{}^t\mathbf{F} \bullet [(\bar{\mathbf{d}}_{,s} + \mathbf{k}^0 \times \bar{\mathbf{d}}) - \bar{\boldsymbol{\varphi}} \times (\mathbf{e} + \mathbf{k}^0 \times \mathbf{d})]] + [{}^t\mathbf{M} \bullet [\bar{\boldsymbol{\varphi}}_{,s} - \bar{\boldsymbol{\varphi}} \times \mathbf{k}^0]]\}\, dS \tag{10.280}$$

- External virtual work, $G_{External}$, as:

$$G_{External} \equiv \int_0^L \left[{}^t\hat{\mathbf{F}}^B \bullet {}^t\bar{\mathbf{d}} + {}^t\hat{\mathbf{M}}^B \bullet {}^t\bar{\boldsymbol{\varphi}}\right] dS + \left[{}^t\hat{\mathbf{F}}^S \bullet {}^t\bar{\mathbf{d}}\right]_{S=0}^{S=L} + \left[{}^t\hat{\mathbf{M}}^S \bullet {}^t\bar{\boldsymbol{\varphi}}\right]_{S=0}^{S=L} \tag{10.281}$$

- The translational virtual strain field, ${}^n\bar{\boldsymbol{\beta}}$, is given as:

$$\begin{aligned} {}^n\bar{\boldsymbol{\beta}} &= {}^t\mathbf{R}^T\left({}^t\bar{\mathbf{d}}_{,S} + {}^t\mathbf{k}^0 \times {}^t\bar{\mathbf{d}}\right) - {}^t\mathbf{R}^T\, {}^t\bar{\boldsymbol{\Phi}}\left({}^t\mathbf{e} + {}^t\mathbf{k}^0 \times {}^t\mathbf{d}\right) \\ &= \overline{{}^t\mathbf{R}^T(\hat{1} + {}^t\mathbf{d}|_S)} \end{aligned} \tag{10.282}$$

- The bending virtual strain field, ${}^n\bar{\boldsymbol{\chi}}$, is given as:

$$\begin{aligned} {}^n\bar{\boldsymbol{\chi}} &\equiv {}^t\mathbf{R}^T\, {}^t\bar{\boldsymbol{\varphi}}|_S = {}^t\mathbf{R}^T\{{}^t\bar{\boldsymbol{\varphi}}_{,S} + \mathbf{k}^0 \times {}^t\bar{\boldsymbol{\varphi}}\} \\ &= \overline{{}^t\mathbf{R}^T\,(\mathbf{k}^0 + \mathbf{k})} \end{aligned} \tag{10.283}$$

- The force and moment component vectors in undeformed and current deformed coordinate system are related by the force and moment rotation transformation:

$$ {}^t\mathbf{F} = {}^t\mathbf{R}\, {}^n\mathbf{F} \quad \text{and} \quad {}^t\mathbf{M} = {}^t\mathbf{R}\, {}^n\mathbf{M} \tag{10.284}$$

- The covariant derivatives of the displacement and rotation vectors:

$$ {}^t\mathbf{d}|_S = {}^t\mathbf{d}_{,S} + \mathbf{k}^0 \times {}^t\mathbf{d} \quad \text{and} \quad {}^t\boldsymbol{\theta}|_S = {}^t\boldsymbol{\theta}_{,S} + \mathbf{k}^0 \times {}^t\boldsymbol{\theta} \tag{10.285}$$

- The definition of a rotation tensor, its matrix representation, and algebra, geometry and topology of rotation.
- A rotation matrix, ${}^t\mathbf{R}$, corresponding to the three-dimensional special orthogonal matrix group, $SO(3)$, may be parameterized by the Rodrigues vector, $\mathbf{s}$, the axis of the instantaneous rotation.

- The rotation matrix, ${}^t\mathbf{R}(S,t)$, is related, through Cayley's theorem, to the Rodrigues vector, $\mathbf{s}(S,t) = \{s_1(S,t) \quad s_2(S,t) \quad s_3(S,t)\}^T$, with the components in the $\mathbf{T}, \mathbf{M}, \mathbf{B}$ frame at any point, S, and at a time, $t \in \mathbb{R}_+$, parameterized by the arc lengths on a beam reference curve by:

$$\mathbf{R} = \frac{1}{\Sigma}\begin{bmatrix} 1+s_1^2-s_2^2-s_3^2 & 2(s_1s_2-s_3) & 2(s_1s_3+s_2) \\ 2(s_1s_2+s_3) & 1-s_1^2+s_2^2-s_3^2 & 2(s_2s_3-s_1) \\ 2(s_1s_3-s_2) & 2(s_2s_3+s_1) & 1-s_1^2-s_2^2+s_3^2 \end{bmatrix} \tag{10.286}$$

where $\sum \equiv 1+ \| \mathbf{s} \|^2$

- The Rodrigues vector, $\mathbf{s}$, is orientated along the instantaneous axis of rotation at a point of the beam reference curve. The skew-symmetric matrix, $\mathbf{S}$, corresponding to the Rodrigues vector, $\mathbf{s}$, is given by:

$$\mathbf{S} = \begin{bmatrix} 0 & -s_3 & s_2 \\ s_3 & 0 & -s_1 \\ -s_2 & s_1 & 0 \end{bmatrix} \tag{10.287}$$

- Finally, Rodrigues rotation vector, $\boldsymbol{\theta}$, is the scaled Rodrigues vector related by:

$$\boxed{\mathbf{s} = \frac{1}{\theta}\tan\left(\frac{\theta}{2}\right)\boldsymbol{\theta}} \tag{10.288}$$

where $\theta \equiv ||\boldsymbol{\theta}|| = (\boldsymbol{\theta} \bullet \boldsymbol{\theta})^{\frac{1}{2}}$ is the length measure of rotation vector. Alternatively, defining the unit vector, $\mathbf{n} \equiv \frac{\boldsymbol{\theta}}{\theta}$, and the corresponding skew-symmetric matrix, $\mathbf{N}$, such that $\mathbf{N}\,\mathbf{h} = \mathbf{n} \times \mathbf{h}$, $\forall \mathbf{h} \in \mathbb{R}_3$, we have:

$$\boxed{\begin{aligned} &\mathbf{R} = (\mathbf{I}-\mathbf{S})^{-1}\,(\mathbf{I}+\mathbf{S}) \\ &\mathbf{s} = \tan\left(\frac{\theta}{2}\right)\mathbf{n}, \\ &\mathbf{S} = \tan\left(\frac{\theta}{2}\right)\mathbf{N} \end{aligned}} \tag{10.289}$$

- The Rodrigues quaternion and Euler–Rodrigues parameters.
- The finite rotation, $\mathbf{R}$, belongs to the spherical curved space, $SO(3)$, and the instantaneous and infinitesimal virtual rotation, $\bar{\boldsymbol{\Phi}}$, represented by a skew-symmetric matrix, belongs to the tangent space at identity, Lie algebra, and they are related by an exponential map. However, computationally, it is more convenient to apply Cayley's formula:

$$\boxed{\mathbf{R} = (\mathbf{I}-\bar{\boldsymbol{\Phi}})^{-1}\,(\mathbf{I}+\bar{\boldsymbol{\Phi}})} \tag{10.290}$$

- For notational uniformity, we denote all variations of entities by a bar superscripted on it. Furthermore all entities measured in $\mathbf{G}_1, \mathbf{G}_2, \mathbf{G}_3$ frame are superscripted as ${}^t(\bullet)$, and those in the rotating $\mathbf{g}_1, \mathbf{g}_2, \mathbf{g}_3$ coordinate system as ${}^n(\bullet)$. However, for the sake of uncluttered readability, sometimes the superscripts may be dropped, but the context should help determine the appropriate underlying coordinate system.

- The material "curvature" tensor, ${}^t\mathbf{K}$, and "curvature" vector, ${}^t\mathbf{k}$, as:

$$\boxed{{}^t\mathbf{K} \equiv \mathbf{R}_{,S}\,\mathbf{R}^{\mathrm{T}}} \tag{10.291}$$

$$\boxed{{}^t\mathbf{k} = \mathbf{W}(\boldsymbol{\theta})\ {}^t\boldsymbol{\theta}_{,S}} \tag{10.292}$$

where we have:

Definitions

$$\boxed{\begin{aligned} &\mathbf{W}({}^t)\boldsymbol{\theta} \equiv \mathbf{I} + c_1\,{}^t\boldsymbol{\Theta} + c_2\,{}^t\boldsymbol{\Theta}^2, \\ &c_1({}^t)\boldsymbol{\theta} = \frac{1-\cos\theta}{\theta^2}, \\ &c_2({}^t)\boldsymbol{\theta} = \frac{\theta-\sin\theta}{\theta^3} \end{aligned}} \tag{10.293}$$

with ${}^t\boldsymbol{\Theta}$ is the skew-symmetric matrix corresponding to the axial vector, ${}^t\boldsymbol{\theta}$, such that ${}^t\boldsymbol{\Theta}\,{}^t\boldsymbol{\theta} = \mathbf{0}$.

- The material first variation tensor, ${}^t\bar{\boldsymbol{\Phi}} \equiv \delta\,{}^t\boldsymbol{\Phi}$, and first variation vector, ${}^t\bar{\boldsymbol{\varphi}} \equiv \delta\,{}^t\boldsymbol{\varphi}$, as:

$$\boxed{{}^t\bar{\boldsymbol{\Phi}} \equiv \delta\,{}^t\boldsymbol{\Phi} \equiv \bar{\mathbf{R}}\,\mathbf{R}^{\mathrm{T}} \equiv \delta\mathbf{R}\,\mathbf{R}^{\mathrm{T}}} \tag{10.294}$$

$$\boxed{{}^t\bar{\boldsymbol{\varphi}} = \mathbf{W}(\boldsymbol{\theta})\ {}^t\bar{\boldsymbol{\theta}}} \tag{10.295}$$

- The angular velocity component vector, ${}^t\boldsymbol{\omega}$, is related to the Rodrigues rotation component vector, ${}^t\boldsymbol{\theta}$, by:

$$\boxed{{}^t\boldsymbol{\omega}_S = \mathbf{W}(\boldsymbol{\theta})\,{}^t\dot{\boldsymbol{\theta}}} \tag{10.296}$$

- The angular acceleration component vector, ${}^t\dot{\boldsymbol{\omega}}_S$, is related to the Rodrigues rotation component vector, ${}^t\boldsymbol{\theta}$, by:

$$\boxed{{}^t\dot{\boldsymbol{\omega}}_S = \dot{\mathbf{W}}(\boldsymbol{\theta})\,{}^t\dot{\boldsymbol{\theta}} + \mathbf{W}(\boldsymbol{\theta})\,{}^t\ddot{\boldsymbol{\theta}}} \tag{10.297}$$

Lemma LD1:

- Defining ${}^t\mathbf{k}$ and ${}^t\bar{\boldsymbol{\varphi}}$ as the axial vectors of ${}^t\mathbf{K}$ and ${}^t\bar{\boldsymbol{\Phi}}$, respectively, we have:

$$\boxed{\begin{aligned} {}^t\bar{\boldsymbol{\Phi}}_{,S} &= {}^t\bar{\mathbf{K}} + [{}^t\mathbf{K}\,{}^t\bar{\boldsymbol{\Phi}} - {}^t\bar{\boldsymbol{\Phi}}\,{}^t\mathbf{K}] \\ \underbrace{{}^t\bar{\boldsymbol{\varphi}}_{,S}}_{\substack{\text{Effective material}\\ \text{curvature vector}\\ \text{variation}}} &= \underbrace{{}^t\bar{\mathbf{k}}}_{\substack{\text{Material}\\ \text{curvature vector}\\ \text{variation}}} + \underbrace{{}^t\mathbf{k}}_{\substack{\text{Material}\\ \text{curvature vector}}} \times \underbrace{{}^t\bar{\boldsymbol{\varphi}}}_{\substack{\text{Material}\\ \text{rotation vector}\\ \text{variation}}} \end{aligned}} \tag{10.298}$$

10.13.1.1.1 The Central Lemma

We will need the following most important lemma, relating the derivatives, $\boldsymbol{\theta}_{,\alpha}$, and the variation, $\bar{\boldsymbol{\theta}}$, or the increment, $\Delta\boldsymbol{\theta}$, (which appears in the linearized virtual functional later in the treatment of a Newton-type iterative algorithm), of the rotation vector, $\boldsymbol{\theta}$, for deriving the geometric stiffness operator and proving its symmetry under the conservative system of loading, presented later:

Lemma LD2: Given $\mathbf{W}$, c_1 and c_2 as in equation (10.139), we have:

$$\boxed{\begin{aligned} \bar{\mathbf{W}}\,\boldsymbol{\theta}_{,S} &= \mathbf{X}^S\,\bar{\boldsymbol{\theta}} \\ \Delta\mathbf{W}\,\boldsymbol{\theta}_{,S} &= \mathbf{X}^S\,\Delta\boldsymbol{\theta} \\ \Delta\mathbf{W}\,\dot{\boldsymbol{\theta}} &= \mathbf{X}^t\,\Delta\boldsymbol{\theta} \end{aligned}} \tag{10.299}$$

where

$$\boxed{\begin{aligned} \mathbf{X}^S &\equiv -c_1\boldsymbol{\Theta}_{,S} + c_2\{[\boldsymbol{\theta}\otimes\boldsymbol{\theta}_{,S}] - [\boldsymbol{\theta}_{,S}\otimes\boldsymbol{\theta}] + (\boldsymbol{\theta}\bullet\boldsymbol{\theta}_{,S})\mathbf{I}\} + \mathbf{V}\,[\boldsymbol{\theta}_{,S}\otimes\boldsymbol{\theta}] \\ \mathbf{X}^t &\equiv -c_1\,\dot{\boldsymbol{\Theta}} + c_2\{[\boldsymbol{\theta}\otimes\dot{\boldsymbol{\theta}}] - [\dot{\boldsymbol{\theta}}\otimes\boldsymbol{\theta}] + (\boldsymbol{\theta}\bullet\dot{\boldsymbol{\theta}})\mathbf{I}\} + \mathbf{V}\,[\dot{\boldsymbol{\theta}}\otimes\boldsymbol{\theta}] \\ \mathbf{V} &\equiv -c_2\mathbf{I} + c_3\boldsymbol{\Theta} + c_4\boldsymbol{\Theta}^2 \\ c_3 &\equiv \frac{1}{\theta^2}\{(1-\theta^2 c_2) - 2c_1 \\ c_4 &\equiv \frac{1}{\theta^2}(c_1 - 3c_2) \end{aligned}} \tag{10.300}$$

where $\boldsymbol{\theta}_{,S}$, $\bar{\boldsymbol{\theta}}$, $\dot{\boldsymbol{\theta}}$ and $\Delta\boldsymbol{\theta}$ are the spatial derivative, the time rate, the virtual (variational) and the incremental rotation vectors, respectively.

- Notationally, the bracket $[\bullet]$ denotes the skew-symmetric matrix corresponding to the argument, that is, $[\mathbf{a}] \equiv \mathbf{a}\times$ for an axial vector, $\mathbf{a}$, such that $[\mathbf{a}]\,\mathbf{a} = \mathbf{0}$.

10.13.2 Virtual Functional and Virtual Strains Revisited

Here we aim to eliminate ${}^t\bar{\boldsymbol{\varphi}}$ from the virtual functional and all the virtual strains, and replace them with ${}^t\bar{\boldsymbol{\theta}}$, the Rodrigues rotation vector.

10.13.3 Inertial Virtual Functional: Revisited

Using the relationships in equation (10.295) and equation (10.297), we get:

$$\boxed{G_{Inertial} \equiv \int_0^L \{(M_0\,{}^t\ddot{\mathbf{d}}\bullet{}^t\bar{\mathbf{d}} + \mathbf{W}^T\,(\mathbf{I}_S\,{}^t\dot{\boldsymbol{\omega}}_S + {}^t\boldsymbol{\omega}_S\times\mathbf{A}_S)\bullet{}^t\bar{\boldsymbol{\theta}}\}\,dS} \tag{10.301}$$

10.13.4 Internal Virtual Functional and Virtual Strains Revisited

We would also like to express the translational and the bending virtual strains in computationally suitable matrix form. From an apparent energy conjugation in the internal virtual work equation (10.278), the expressions, $(\bar{\mathbf{d}},_s + \mathbf{k}^0 \times \bar{\mathbf{d}}) - \bar{\varphi} \times (\mathbf{e} + \mathbf{k}^0 \times \mathbf{d})$ and $\bar{\varphi},_S - \bar{\varphi} \times \mathbf{k}^0$ present themselves as the force and moment virtual strains, but it turns out that there exists no function whose variations can possibly be identified with these functions; in other words, these are not truly the virtual strains derivable from the real strains. Fortunately, however, if we rotate the force and moment component vectors by pre-multiplying with ${}^t\mathbf{R}^T$ to obtain forces and moments as the components in the rotating deformed frame, we achieve our goal of establishing the true virtual nature of the newly generated and energy conjugated strain functions; this is precisely what we intend to do next. However, it is important to note that all relevant quantities are measured in Lagrangian or material coordinate system.

10.13.4.1.1 Virtual Translational and Rotational Strain Vectors

Now, we introduce the rotation matrix, ${}^t\mathbf{R}$, to express the internal virtual work in the rotated frame, and by applying equation (10.285) in equation (10.280) and noting that ${}^t\mathbf{R}$ is orthonormal, that is, ${}^t\mathbf{R}^T\,{}^t\mathbf{R} = \mathbf{I}$, we get:

$$
\begin{aligned}
G_{Internal} &= \int_0^L \{{}^t\mathbf{R}^T\,{}^t\mathbf{F} \bullet {}^t\mathbf{R}^T({}^t\bar{\mathbf{d}}|_S - {}^t\bar{\varphi} \times (\hat{1} + {}^t\mathbf{d}|_S)) + {}^t\mathbf{R}^T\,{}^t\mathbf{M} \bullet {}^t\mathbf{R}\,{}^t\bar{\varphi}|_S\}\, dS \\
&= \int_0^L \left\{ \underbrace{{}^n\mathbf{F}}_{\substack{\text{Real component}\\ \text{force vector along}\\ \text{rotated frame}}} \quad \underbrace{\{{}^t\mathbf{R}^T\,{}^t\bar{\mathbf{d}}|_S - ({}^t\mathbf{R}^T\,{}^t\bar{\Phi})\,(\hat{1} + {}^t\mathbf{d}|_S)\}}_{\text{Virtual translational strain vector}} + \underbrace{{}^n\mathbf{M}}_{\substack{\text{Real component}\\ \text{moment vector}\\ \text{along rotated frame}}} \quad \underbrace{{}^t\mathbf{R}^T\,{}^t\bar{\varphi}|_S}_{\substack{\text{Virtual rotational}\\ \text{strain vector}}} \right\} dS
\end{aligned}
\tag{10.302}
$$

Next, applying the definition of the covariant derivatives as given by equation (10.285) and interpreting ${}^t\bar{\Phi}$ (being skew-symmetric: ${}^t\bar{\Phi}^T = -{}^t\bar{\Phi}$) as the variation of the rotation matrix, ${}^t\mathbf{R}$, related by: ${}^t\bar{\mathbf{\Phi}} \equiv {}^t\bar{\mathbf{R}}\,{}^t\mathbf{R}^T \Rightarrow ({}^t\bar{\mathbf{R}})^T = ({}^t\bar{\mathbf{\Phi}}\,{}^t\mathbf{R})^T = {}^t\mathbf{R}^T\,{}^t\bar{\mathbf{\Phi}}^T = -\,{}^t\mathbf{R}^T\,{}^t\bar{\mathbf{\Phi}}$, we can rewrite and define the virtual strain vectors as follows.

The virtual translational (axial & shear) component strain vector, ${}^n\bar{\beta}$:

$$
\begin{aligned}
{}^n\bar{\beta} &\equiv {}^t\mathbf{R}^T\,{}^t\bar{\mathbf{d}}|_S - ({}^t\mathbf{R}^T\,{}^t\bar{\Phi})\,(\hat{1} + {}^t\mathbf{d}|_S) = {}^t\mathbf{R}^T\,\overline{(\hat{1} + {}^t\mathbf{d}|_S)} + \overline{{}^t\mathbf{R}^T}\,(\hat{1} + {}^t\mathbf{d}|_S) \\
&= \overline{{}^t\mathbf{R}^T(\hat{1} + {}^t\mathbf{d}|_S)}
\end{aligned}
\tag{10.303}
$$

The virtual rotational (bending) component strain vector:

$$
\begin{aligned}
{}^n\bar{\chi} &\equiv {}^t\mathbf{R}^T\,{}^t\bar{\varphi}|_S = {}^t\mathbf{R}^T\{{}^t\bar{\varphi},_S + \mathbf{k}^0 \times {}^t\bar{\varphi}\} \\
&= {}^t\mathbf{R}^T\{\bar{\mathbf{k}} - {}^t\bar{\Phi}\,(\mathbf{k}^0 + \mathbf{k})\} = {}^t\mathbf{R}^T\,\bar{\mathbf{k}} + (-\,{}^t\mathbf{R}^T\,{}^t\bar{\Phi})\,(\mathbf{k}^0 + \mathbf{k}) \\
&= {}^t\mathbf{R}^T\,\overline{(\mathbf{k}^0 + \mathbf{k})} + \overline{{}^t\mathbf{R}^T}\,(\mathbf{k}^0 + \mathbf{k}) \\
&= \overline{{}^t\mathbf{R}^T\,(\mathbf{k}^0 + \mathbf{k})}
\end{aligned}
\tag{10.304}
$$

In equation (10.304), we have used the virtual derivative relation: ${}^t\bar{\varphi}_{,S} = \bar{\mathbf{k}} - {}^t\bar{\boldsymbol{\Phi}}\,\mathbf{k}$ with $\mathbf{k}^0$ as the initial curvature vector, and $\mathbf{k}$ as the current curvature vector along S direction; also, recall that all entities with barred superscript are virtual in nature. Finally, introducing equations (10.302)–(10.304) in equation (10.280), we get the desired beam internal virtual functional as:

$$G_{Internal}({}^t\mathbf{d}, {}^t\mathbf{R}; {}^t\bar{\mathbf{d}}, {}^t\bar{\varphi}) = \int_0^L \{({}^n\mathbf{F} \bullet {}^n\bar{\beta} + {}^n\mathbf{M} \bullet {}^n\bar{\chi}\} \, dS$$

with

$$\begin{aligned} {}^n\bar{\beta} &\equiv \overline{{}^t\mathbf{R}^T(\hat{1} + {}^t\mathbf{d}|_S)} \\ {}^n\bar{\chi} &\equiv \overline{{}^t\mathbf{R}^T(\mathbf{k}^0 + \mathbf{k})} \end{aligned} \tag{10.305}$$

Note that in the internal virtual work, the forces and the moments, and the corresponding conjugate translational and bending virtual strain vectors, although measured in undeformed coordinates, are all expressed as the components in the current deformed (rotated) local coordinate system, that is, the co-spinning coordinate system. One of the main consequences is that the constitutive relationship must also be considered in this deformed coordinate system which is the co-spinning coordinate system: $\mathbf{t}, \mathbf{m}_n, \mathbf{b}_n$, in the current deformed configuration.

10.13.4.1.2 Computational Translational Virtual Strain in Matrix Form

Now, we would also like to express the translational and bending virtual strains in computationally suitable matrix form. Using the relationship, ${}^t\bar{\varphi} = \mathbf{W}\,{}^t\bar{\theta}$ from equation (10.295), and using definitions $\mathbf{a} \equiv \mathbf{e} + \mathbf{k}^0 \times \mathbf{d}$ where $\mathbf{e} \equiv \mathbf{d}_{,S} + \hat{\mathbf{1}}$ with $\hat{\mathbf{1}} \equiv \{1 \quad 0 \quad 0\}^T$, we have: $-\mathbf{R}^T\,{}^t\bar{\boldsymbol{\Phi}}({}^t\mathbf{e} + {}^t\mathbf{k}^0 \times {}^t\mathbf{d}) = \mathbf{R}^T[\mathbf{a}]\,{}^t\bar{\varphi} = \mathbf{R}^T[\mathbf{a}]\mathbf{W}\,{}^t\bar{\theta}$; then, we can rewrite the expression for the translational virtual strain field, ${}^n\bar{\beta}$, of equation (10.303) as:

$${}^n\bar{\beta} = {}^t\mathbf{R}^T[[\mathbf{k}^0] \quad [\mathbf{a}]\mathbf{W} \quad \mathbf{I} \quad \mathbf{0}]\begin{Bmatrix} {}^t\bar{\mathbf{d}} \\ {}^t\bar{\theta} \\ {}^t\bar{\mathbf{d}}_{,S} \\ {}^t\bar{\theta}_{,S} \end{Bmatrix} \tag{10.306}$$

10.13.4.1.3 Computational Rotational Virtual Strain in Matrix Form

Using the relationship, ${}^t\bar{\varphi}_{,S} = {}^t\bar{\mathbf{k}} + {}^t\mathbf{k} \times {}^t\bar{\varphi}$ from the second of equation (10.298), ${}^t\mathbf{k}_\alpha = \mathbf{W}\,\theta_{,\alpha}$, and the central lemma: $\bar{\mathbf{W}}\,\theta_{,S} = \mathbf{X}\,\bar{\theta}$ of equation (10.299), we have:

$$\begin{aligned} {}^n\bar{\chi} &= {}^t\mathbf{R}^T\,{}^t\bar{\varphi}_{,S} - {}^t\mathbf{R}^T\,{}^t\bar{\boldsymbol{\Phi}}\,{}^t\mathbf{k}^0 = {}^t\mathbf{R}^T\{{}^t\bar{\mathbf{k}} + ({}^t\mathbf{k} + {}^t\mathbf{k}^0) \times {}^t\bar{\varphi}\} \\ &= {}^t\mathbf{R}^T\left\{\bar{\mathbf{W}}\,\theta_{,S} + \mathbf{W}\,\bar{\theta}_{,S} + [{}^t\mathbf{k} + {}^t\mathbf{k}^0]\mathbf{W}\,\bar{\theta}\right\} \\ &= {}^t\mathbf{R}^T\left\{\mathbf{X}\,\bar{\theta} + \mathbf{W}\,\bar{\theta}_{,S} + [{}^t\mathbf{k} + {}^t\mathbf{k}^0]\mathbf{W}\,\bar{\theta}\right\} \\ &= {}^t\mathbf{R}^T\left\{(\mathbf{X} + [{}^t\mathbf{k} + {}^t\mathbf{k}^0]\mathbf{W})\bar{\theta} + \mathbf{W}\,\bar{\theta}_{,S}\right\} \end{aligned} \tag{10.307}$$

Now, we introduce the following definition.

Definition: The ***total current "curvature" vectors***, ${}^t\mathbf{k}^c$,

$$\boxed{{}^t\mathbf{k}^c \equiv {}^t\mathbf{k} + {}^t\mathbf{k}^0} \tag{10.308}$$

so we have, in matrix form:

$$\boxed{{}^n\bar{\chi} = {}^t\mathbf{R}^T[\mathbf{0} \quad \mathbf{X}^S + [{}^t\mathbf{k}^c]\,\mathbf{W} \quad \mathbf{0} \quad \mathbf{W}]\begin{Bmatrix} {}^t\bar{\mathbf{d}} \\ {}^t\bar{\theta} \\ {}^t\bar{\mathbf{d}}_{,S} \\ {}^t\bar{\theta}_{,S} \end{Bmatrix}} \tag{10.309}$$

Exercise: Deduce the virtual strains for planar beams from equations (10.306) and (10.309) with $\mathbf{d} = (u \quad w \quad 0)^T$, $\theta = (0 \quad 0 \quad \theta)^T$ and $\kappa = (0 \quad 0 \quad \kappa)^T$, that is, $\tau \equiv 0$, as:

$$\begin{Bmatrix} {}^n\bar{\varepsilon} \\ {}^n\bar{\gamma} \\ {}^n\bar{\chi} \end{Bmatrix} = \begin{bmatrix} \kappa\sin\theta & -\kappa\cos\theta & {}^n\gamma & \cos\theta & \sin\theta & 0 \\ \kappa\cos\theta & \kappa\sin\theta & {}^n\varepsilon & -\sin\theta & \cos\theta & 0 \\ 0 & 0 & 0 & 0 & 0 & 1 \end{bmatrix}\begin{Bmatrix} \bar{u} \\ \bar{w} \\ \bar{\theta} \\ \bar{u}' \\ \bar{w}' \\ \bar{\theta}' \end{Bmatrix}$$

where (10.310)

$$\begin{Bmatrix} {}^n\varepsilon \\ {}^n\gamma \end{Bmatrix} = \begin{bmatrix} \kappa\sin\theta & -\kappa\cos\theta & 0 & \cos\theta & \sin\theta & 0 \\ \kappa\cos\theta & \kappa\sin\theta & 0 & -\sin\theta & \cos\theta & 0 \end{bmatrix}\begin{Bmatrix} u \\ w \\ \theta \\ u' \\ w' \\ \theta' \end{Bmatrix} - \begin{Bmatrix} 1 \\ 0 \end{Bmatrix}$$

Hint: It follows that: $\mathbf{W} = \dfrac{1}{\theta}\begin{bmatrix} \sin\theta & -(1-\cos\theta) & 0 \\ (1-\cos\theta) & \sin\theta & 0 \\ 0 & 0 & \theta \end{bmatrix}$ and $\mathbf{k} = \mathbf{W}\theta' = (0 \quad 0 \quad \theta')$

10.13.4.1.4 Virtual Strain–Displacement

We introduce the following definitions.

Definitions

- The (12×1)virtual generalized deformation function is $\mathbf{H}(\hat{\bar{\theta}}) \equiv \{\bar{\mathbf{d}} \quad \bar{\theta} \quad \bar{\mathbf{d}}_{,S} \quad \bar{\theta}_{,S}\}^T$.
- The total current "curvature" skew-symmetric matrices are $[\mathbf{k}^c] \equiv [\mathbf{k}^0] + [{}^t\mathbf{k}]$.
- The virtual generalized strain vector is $\breve{\bar{\varepsilon}} \equiv \{{}^n\bar{\beta} \quad {}^n\bar{\chi}\}^T$.

- The generalized reactive force field is ${}^n\breve{\mathbf{F}}$, as: $\underbrace{{}^n\breve{\mathbf{F}}}_{6\times1} \equiv \{{}^n\mathbf{F} \quad {}^n\mathbf{M}\}^T$.

So now we can express the virtual strain–displacement relations in compact matrix form as:

$$\underbrace{\breve{\varepsilon}(\mathbf{d},\theta;\bar{\mathbf{d}},\bar{\theta}) \equiv \{{}^n\bar{\beta} \quad {}^n\bar{\chi}\}^T}_{6\times1} = \underbrace{\mathbf{E}(\mathbf{d},\theta)}_{6\times12}\ \underbrace{\mathbf{H}(\bar{\mathbf{d}},\bar{\theta})}_{12\times1} \tag{10.311}$$

where, we have introduced the following definitions.

Definition: The virtual strain–displacement matrix, $\mathbf{E}(\mathbf{d},\theta)$:

$$\underbrace{\mathbf{E}(\mathbf{d},\theta)}_{6\times12} = \underbrace{\begin{bmatrix} \mathbf{R}^{\mathrm{T}} & \mathbf{0} \\ \mathbf{0} & \mathbf{R}^{\mathrm{T}} \end{bmatrix}}_{6\times6} \underbrace{\begin{bmatrix} [\mathbf{k}^0] & [\mathbf{a}]\mathbf{W} & \mathbf{I} & \mathbf{0} \\ \mathbf{0} & \mathbf{X}^S + [\mathbf{k}^c]\mathbf{W} & \mathbf{0} & \mathbf{W} \end{bmatrix}}_{6\times12} \tag{10.312}$$

where, $\mathbf{R}(\theta)$, $\mathbf{W}(\theta)$, $\mathbf{X}(\theta)$ matrices are as defined in equations (10.286) with (10.288), (10.293), and (10.300), respectively, and $\mathbf{I}$, is the 3x3 identity matrix. With these in mind, we can express the internal virtual work of equation (10.305) in generalized matrix form as:

$$G_{Internal}({}^t\mathbf{d},{}^t\mathbf{R};{}^t\bar{\mathbf{d}},{}^t\bar{\varphi}) = \int_M \underbrace{\mathbf{H}(\bar{\mathbf{d}},\bar{\theta})^T}_{1\times12}\ \underbrace{\mathbf{E}(\mathbf{d},\theta)^T}_{12\times6}\ \underbrace{{}^n\breve{\mathbf{F}}}_{6\times1}\ dS \tag{10.313}$$

10.13.5 External Virtual Functional Revisited

A definitive form of the external virtual work depends on the basic nature of the externally applied loading system. In what follows, we will consider two types of loading system.

10.13.5.1.1 Quasi-static Proportional Loading

This type of loading system essentially results in the static analysis of a beam structure in the nonlinear regime; in this case, the linear and the angular momentum of a beam is assumed identically zero, that is, the inertial virtual work vanishes identically: $G_{Inertia} \equiv 0$. Thus, in order that we can prepare for a concrete solution strategy for the quasi-static analysis of a beam, we need to characterize the external virtual work more precisely. As indicated above, the exact expression for the external virtual work depends on the forces applied to a beam. We may take the distributed forces and the moments along the beam body, and the forces and moments along the end boundary cross-sections (end forces and moments at $S = 0$ and at $S = L$) to be purely proportional with the former being only functions of the beam arc length parameter, $S \in [0, L]$. In other words, we assume that the prescribed forces and moments are functions of a single real parameter, $\lambda \in \mathbb{R}$. Clearly, within this assumption of one parameter family of forces and moments, various possibilities exist such as the combination of a constant, $\mathbf{F}_1^B(S)$, and a proportional distributed load, $\mathbf{F}_2^B(\lambda, S)$, that is, $\mathbf{F}^B(\lambda, S) \equiv \mathbf{F}_1^B(S) + \mathbf{F}_2^B(\lambda, S)$ where λ is the constant of proportionality for the distributed loading. Similarly, displacement state dependent loadings, that is, ***non-conservative*** loadings such as ***follower forces*** and such like, can also be considered through an additional term reflecting the appropriate differentiations. We will only consider a conservative loading system for static analysis here. The utility of the choice of a one-parameter family of forces and moments is that by specializing the external virtual work equations for this family, we can nonlinearly trace the configuration path of a beam. With this in mind, we may include the load proportionality parameter, λ, as a formal parameter, and rewrite the external virtual work functional, $G_{External}(\lambda; {}^t\bar{\mathbf{d}}, {}^t\bar{\theta})$, as:

$$\begin{aligned} G_{External}(\lambda; \bar{\mathbf{d}}, \bar{\varphi}) \equiv & \int_0^L \left[{}^t\hat{\mathbf{F}}^B(\lambda, S) \bullet {}^t\bar{\mathbf{d}} + \mathbf{W}^T\, {}^t\hat{\mathbf{M}}^B(\lambda, S) \bullet {}^t\bar{\theta} \right] dS \\ & + \lambda \left[{}^t\hat{\mathbf{F}}^S \bullet {}^t\bar{\mathbf{d}} \right]_{S=0}^{S=L} + \lambda \left[\mathbf{W}^T\, {}^t\hat{\mathbf{M}}^S \bullet {}^t\bar{\theta} \right]_{S=0}^{S=L} \end{aligned} \tag{10.314}$$

where the superscript t and n indicate that all quantities are expressed in the undeformed frame and the deformed rotated frame, respectively.

10.13.5.1.2 Time-dependent Dynamic Loading

This type of loading system essentially results in the full dynamic analysis of a beam structure in the linear and nonlinear regime. For this case also, we will only consider conservative loading system here. For these situations and using the relationships in equation (10.295), we rewrite the external virtual work functional as:

$$G_{External}(\bar{\mathbf{d}}, \bar{\varphi}) \equiv \int_0^L \left[{}^t\hat{\mathbf{F}}^B(S) \bullet {}^t\bar{\mathbf{d}} + \mathbf{W}^T\, {}^t\hat{\mathbf{M}}^B(S) \bullet {}^t\bar{\theta} \right] dS + \left[{}^t\hat{\mathbf{F}}^S \bullet {}^t\bar{\mathbf{d}} \right]_{S=0}^{S=L} + \left[\mathbf{W}^T\, {}^t\hat{\mathbf{M}}^S \bullet {}^t\bar{\theta} \right]_{S=0}^{S=L} \tag{10.315}$$

10.13.5.2 Dynamic Computational Virtual Work Functional

Now, we can express the virtual functional of equation (10.305) in terms of the real displacement, ${}^t\mathbf{d}$, and rotation, ${}^t\theta$, and their virtual counterparts for dynamic analysis as:

$$\begin{aligned} G({}^t\mathbf{d}, {}^t\theta; {}^t\bar{\mathbf{d}}, {}^t\bar{\theta}) \equiv\ & G_{Inertial} + G_{Internal} - G_{External} \\ =\ & \overbrace{\int_0^L \left\{ (M_0\, {}^t\ddot{\mathbf{d}} \bullet {}^t\bar{\mathbf{d}} + \mathbf{W}^T (\mathbf{I}_S\, {}^t\dot{\omega}_S + {}^t\omega_S \times \mathbf{A}_S) \bullet {}^t\bar{\theta} \right\} dS}^{\text{Inertial}} \\ & + \overbrace{\int_0^L \{ ({}^n\mathbf{F} \bullet {}^n\bar{\beta} + {}^n\mathbf{M} \bullet {}^n\bar{\chi} \} \, dS}^{\text{Internal}} \\ & - \overbrace{\int_0^L \left[{}^t\hat{\mathbf{F}}^B(S) \bullet {}^t\bar{\mathbf{d}} + \mathbf{W}^T\, {}^t\hat{\mathbf{M}}^B(S) \bullet {}^t\bar{\theta} \right] dS + \left[{}^t\hat{\mathbf{F}}^S \bullet {}^t\bar{\mathbf{d}} \right]_{S=0}^{S=L} + \left[\mathbf{W}^T\, {}^t\hat{\mathbf{M}}^S \bullet {}^t\bar{\theta} \right]_{S=0}^{S=L}}^{\text{External}} \\ =\ & \mathbf{0}, \quad \forall \text{ admissible } {}^t\bar{\mathbf{d}}, {}^t\bar{\theta} \end{aligned} \tag{10.316}$$

with the initial conditions given as:

$$\boxed{\begin{aligned} \mathbf{d}(0) &= \mathbf{d}_0. \\ \boldsymbol{\theta}(0) &= \boldsymbol{\theta}_0. \\ \mathbf{v}(0) \equiv \dot{\mathbf{d}}(0) &= \mathbf{v}_0, \\ \boldsymbol{\omega}(0) \equiv \dot{\boldsymbol{\theta}}(0) &= \boldsymbol{\omega}_0 \end{aligned}} \tag{10.317}$$

10.13.5.3 Static and Quasi-static Computational Virtual Work Functional

Similarly, we can express the virtual functional of equation (10.305) in terms of the real displacement, ${}^t\mathbf{d}$, and rotation, ${}^t\boldsymbol{\theta}$, and their virtual counterparts for static and quasi-static analysis as:

$$\boxed{\begin{aligned} G(\lambda, {}^t\mathbf{d}, {}^t\boldsymbol{\theta}; {}^t\bar{\mathbf{d}}, {}^t\bar{\boldsymbol{\theta}}) &\equiv G_{Internal} - G_{External} \\ &= \overbrace{\int_0^L \{({}^n\mathbf{F} \bullet {}^n\bar{\boldsymbol{\beta}} + {}^n\mathbf{M} \bullet {}^n\bar{\boldsymbol{\chi}}\}\, dS}^{\text{Internal}} \\ &\quad \overbrace{- \int_0^L \left[{}^t\hat{\mathbf{F}}^B(\lambda, S) \bullet {}^t\bar{\mathbf{d}} + \mathbf{W}^T\, {}^t\hat{\mathbf{M}}^B(\lambda, S) \bullet {}^t\bar{\boldsymbol{\theta}}\right] dS}^{\text{External}} \\ &\quad + \lambda \left[{}^t\hat{\mathbf{F}}^S\, {}^t\bar{\mathbf{d}}\right]_{S=0}^{S=L} + \lambda \left[\mathbf{W}^T\, {}^t\hat{\mathbf{M}}^S \bullet {}^t\bar{\boldsymbol{\theta}}\right]_{S=0}^{S=L} \\ &= \mathbf{0}, \quad \forall \text{ admissible } {}^t\bar{\mathbf{d}}, {}^t\bar{\boldsymbol{\theta}} \end{aligned}} \tag{10.318}$$

with appropriate prescribed boundary conditions.

10.13.5.4 Generalized Definitions

For notational compactness and as a follow-up on our discussion of the beam configuration space, we may combine both the real vector displacement and rotation fields, ${}^t\mathbf{d}$ and ${}^t\boldsymbol{\theta}$, and the virtual vector displacement and rotation fields, ${}^t\bar{\mathbf{d}}$ and ${}^t\bar{\boldsymbol{\theta}}$, respectively, for the following definitions.

Definitions

- (6×1) the generalized real state field, ${}^t\breve{\mathbf{d}}$ as: ${}^t\breve{\mathbf{d}} \equiv \{{}^t\mathbf{d} \quad {}^t\boldsymbol{\theta}\}^T$
- (6×1) the generalized virtual state field, ${}^t\breve{\bar{\mathbf{d}}}$ as: ${}^t\breve{\bar{\mathbf{d}}} \equiv \{{}^t\bar{\mathbf{d}} \quad {}^t\bar{\boldsymbol{\theta}}\}^T$
- (6×1) the generalized real velocity field, ${}^t\dot{\breve{\mathbf{d}}}$ as: ${}^t\dot{\breve{\mathbf{d}}} \equiv \{\mathbf{v} \quad \boldsymbol{\omega}\}^T$
- (6×1) the generalized real acceleration field, ${}^t\ddot{\breve{\mathbf{d}}}$ as: ${}^t\ddot{\breve{\mathbf{d}}} \equiv \{\dot{\mathbf{v}} \quad \dot{\boldsymbol{\omega}}\}^T$

- (6×1) the generalized real inertial force field, ${}^t\breve{\mathbf{F}}_{Iner}$ as:

$$ {}^t\breve{\mathbf{F}}_{Iner} \equiv \left\{ M_0\, {}^t\ddot{\breve{\mathbf{d}}} \quad \mathbf{W}^T\,(\mathbf{I}_S\, {}^t\dot{\boldsymbol{\omega}}_S + {}^t\boldsymbol{\omega}_S \times \mathbf{A}_S) \right\}^T $$

Similarly, we have:

Definitions

- (6×1) the generalized real strain, ${}^n\breve{\boldsymbol{\varepsilon}}$, as ${}^n\breve{\boldsymbol{\varepsilon}} \equiv \{{}^n\boldsymbol{\beta} \quad {}^n\boldsymbol{\chi}\}^T$
- (6×1) the generalized virtual strain, ${}^n\breve{\bar{\boldsymbol{\varepsilon}}}$, as: ${}^n\breve{\bar{\boldsymbol{\varepsilon}}}(\mathbf{d}, \boldsymbol{\phi}; \bar{\mathbf{d}}, \bar{\boldsymbol{\varphi}}) \equiv \{{}^n\bar{\boldsymbol{\beta}} \quad {}^n\bar{\boldsymbol{\chi}}\}^T$
- (6×1) the generalized reactive force field, ${}^n\breve{\mathbf{F}}$, as: $\underbrace{{}^n\breve{\mathbf{F}}}_{6\times 1} \equiv \{{}^n\mathbf{F} \quad {}^n\mathbf{M}\}^T$
- (6×1) the generalized external force field, ${}^t\breve{\hat{\mathbf{F}}}$, as: $\underbrace{{}^t\breve{\hat{\mathbf{F}}}}_{6\times 1} \equiv \{{}^n\hat{\mathbf{F}} \quad {}^n\hat{\mathbf{M}}\}^T$

Similar generalizations apply for other force and moment terms.

10.13.6 Dynamic Virtual Work Functional in Generalized Form

With the above notational definitions and the load proportionality parameter, λ, we may rewrite the virtual work functional, $G({}^t\mathbf{d}, {}^t\boldsymbol{\theta}; {}^t\bar{\mathbf{d}}, {}^t\overleftarrow{\boldsymbol{\theta}})$, of equation (10.316) in compact generalized notation as:

$$ \begin{aligned} &G({}^t\breve{\mathbf{d}}; {}^t\breve{\bar{\mathbf{d}}}) \equiv G_{Internal} + G_{Inertial} - G_{External} = \mathbf{0}, \quad \forall \text{ admissible } {}^t\breve{\bar{\mathbf{d}}} \\ &G_{Internal} = \int_0^L ({}^n\breve{\bar{\boldsymbol{\varepsilon}}} \bullet {}^n\breve{\mathbf{F}})\, dS \\ &G_{Inertial} = \int_0^L ({}^t\breve{\bar{\mathbf{d}}}^T\, {}^t\breve{\mathbf{F}}_{Iner})\, dS \\ &G_{External} = \left\{ \int_0^L \hat{\bar{\mathbf{d}}}^{\mathbf{T}}\, \hat{\mathbf{F}}^B\, dS + \left[\hat{\bar{\mathbf{d}}}^{\mathbf{T}}\, \hat{\mathbf{F}}^S \right]_{S=0}^{S=L} \right\} \end{aligned} \tag{10.319} $$

with the applicable initial conditions on the dynamic state variables as:

$$ \begin{aligned} &\breve{\mathbf{d}}(0) = \breve{\mathbf{d}}_0, \\ &{}^t\dot{\breve{\mathbf{d}}}(0) = \breve{\mathbf{v}}_0, \end{aligned} \tag{10.320} $$

10.13.7 Static Virtual Work Functional in Generalized Form

With the above notational definitions and the load proportionality parameter, λ, we may rewrite the virtual work functional, $G(\lambda,\ {}^t\mathbf{d},\ {}^t\boldsymbol{\theta};\ {}^t\bar{\mathbf{d}},\ {}^t\overleftarrow{\boldsymbol{\theta}})$, of equation (10.318) in compact generalized notation as:

$$
\boxed{
\begin{aligned}
&G(\lambda,\ {}^t\breve{\mathbf{d}};\ {}^t\breve{\bar{\mathbf{d}}}) \equiv G_{Internal}({}^t\breve{\mathbf{d}};\ {}^t\breve{\bar{\mathbf{d}}}) - G_{external}(\lambda;\ {}^t\breve{\bar{\mathbf{d}}}) = 0, \quad \forall \text{ admissible } {}^t\breve{\bar{\mathbf{d}}} \\
&G_{Internal} = \int_0^L ({}^n\breve{\bar{\varepsilon}}\ {}^n\breve{\mathbf{F}})\, dS \\
&G_{External} = \left\{ \int_0^L \hat{\bar{\mathbf{d}}}^T\ \hat{\mathbf{F}}^B(\lambda, S)\, dS + \lambda \left[\hat{\bar{\mathbf{d}}}^T\ \hat{\mathbf{F}}^S\right]_{S=0}^{S=L} \right\}
\end{aligned}
}
\qquad (10.321)
$$

where the superscript t and n indicate that all quantities are expressed in the undeformed frame and the deformed frame, respectively.

10.13.8 Where We Would Like to Go

In conformity with our main interest in computational solid mechanics and structural engineering, we need to bring all the beam equations down to a form where we can easily solve beam problems numerically by our finite element scheme, the c-type finite element method. This, in turn, requires the derivation of the compatible real strain fields along the configuration path from our knowledge of the above virtual strain fields. This endeavor validates our assumption that the derived virtual strains are truly variational, that is, these can be shown to exist as variations of some real strain fields. Our expressions for the virtual strain fields given in equation (10.282) and equation (10.283) already suggest what the real strain fields will be. However, we would like to record formally the expressions for real strain fields through an application of what is known as ***the Vainberg principle***. Accordingly, the most logical route will be the following.

- ***The computational real strains in component form***
  As we can see from equations (10.319) and (10.321), the virtual functional, being generally highly nonlinear in displacements and rotations, may not be solved in a closed form in an analytical way. The solution, that is, the tracing of the beam motion or configuration path, will require some sort of iterative numerical method. Of all the possible schemes, the one that we intend to pursue is a Newton-type method. This, in turn, requires linearization of the equations at every iteration step. Clearly, then, we will have to further modify the virtual work equation for the ensuing computational strategy. Let us recall, however, that while the translational displacement of the beam reference curve travels on an Euclidean (flat) configuration space, the beam rotation tensor belongs to a Riemannian (curved) configuration space. More specifically, as we will see, we will need to recognize this by resorting to covariant derivatives as opposed to standard Gateaux derivatives of the Euclidean space in our linearization of the virtual work functional. Thus, the next logical place to go from here will be:
- covariant linearization of virtual work functionals.

10.14 Computational Real Strains

10.14.1 What We Need to Recall

Let us recall that we have defined the virtual strains in terms of the displacement and rotation fields in invariant (defined from the weak form of balance equations) and component vector forms (defined from the virtual work equation) by the following.

10.14.1.1.1 Internal Virtual Work Functional

$$\boxed{G_{Internal} \equiv \int_0^L \{ {}^n\mathbf{F} \bullet {}^n\bar{\boldsymbol{\beta}} + {}^n\mathbf{M} \bullet {}^n\bar{\boldsymbol{\chi}} \}\, dS} \tag{10.322}$$

10.14.1.1.2 Component Vector Form

- The translational virtual strain fields, ${}^n\bar{\boldsymbol{\beta}}$, are given as:

$$\boxed{{}^n\bar{\boldsymbol{\beta}} = {}^t\mathbf{R}^{\mathbf{T}} \left({}^t\bar{\mathbf{d}},_S + \mathbf{k}^0 \times {}^t\bar{\mathbf{d}} \right) - {}^t\mathbf{R}^{\mathbf{T}}\, {}^t\bar{\boldsymbol{\Phi}} \left(\mathbf{e} + \mathbf{k}^0 \times {}^t\mathbf{d} \right)} \tag{10.323}$$

Component-wise in spatial representation, that is, in rotated bases, we have:

$$\boxed{{}^n\bar{\boldsymbol{\beta}} = \left\{ {}^n\bar{\beta}_1 \quad {}^n\bar{\beta}_2 \quad {}^n\bar{\beta}_3 \right\}^T} \tag{10.324}$$

For equation (10.323), we have the following definitions:

$$\begin{aligned} \mathbf{e} &\equiv \hat{\mathbf{1}} + {}^t\mathbf{d},_S \\ {}^t\mathbf{d}|_S &= {}^t\mathbf{d},_S + \mathbf{k}^0 \times {}^t\mathbf{d} \end{aligned} \tag{10.325}$$

- The bending virtual strain fields, ${}^n\bar{\boldsymbol{\chi}}$, are given as:

$$\boxed{{}^n\bar{\boldsymbol{\chi}} = {}^t\mathbf{R}^{\mathbf{T}}\, {}^t\bar{\boldsymbol{\varphi}},_S - {}^t\mathbf{R}^{\mathbf{T}}\, {}^t\bar{\boldsymbol{\Phi}}\, \mathbf{k}^0} \tag{10.326}$$

Component-wise in spatial representation, that is, in rotated bases, we have:

$$\boxed{{}^n\bar{\boldsymbol{\chi}} = \left\{ {}^n\bar{\chi}_1 \quad {}^n\bar{\chi}_2 \quad {}^n\bar{\chi}_3 \right\}^T} \tag{10.327}$$

Furthermore, we need the following results:

$$\boxed{{}^t\mathbf{k} = \mathbf{W}(\boldsymbol{\theta})\, \boldsymbol{\theta},_S} \tag{10.328}$$

where:

$$\boxed{\begin{aligned} &\mathbf{W}(\boldsymbol{\theta}) \equiv \mathbf{I} + c_1\, \boldsymbol{\Theta} + c_2\, \boldsymbol{\Theta}^2, \\ &c_1(\theta) = \frac{1 - \cos\theta}{\theta^2}, \\ &c_2(\theta) = \frac{\theta - \sin\theta}{\theta^3} \end{aligned}} \tag{10.329}$$

Lemma LD1: Defining ${}^t\mathbf{k}$ and ${}^t\bar{\varphi}$ as the axial vectors of ${}^t\mathbf{K}_\alpha$, $\alpha = 1,2$ and ${}^t\bar{\mathbf{\Phi}}$, respectively, we have:

$$\begin{gathered}
{}^t\bar{\mathbf{\Phi}}_{,S} = {}^t\bar{\mathbf{K}} + [{}^t\mathbf{K}\,{}^t\bar{\mathbf{\Phi}} - {}^t\bar{\mathbf{\Phi}}\,{}^t\mathbf{K}] \\
\underbrace{{}^t\bar{\varphi}_{,S}}_{\substack{\text{Effective material}\\ \text{curvature vector}\\ \text{variation}}} = \underbrace{{}^t\bar{\mathbf{k}}}_{\substack{\text{Material}\\ \text{curvature vector}\\ \text{variation}}} + \underbrace{{}^t\mathbf{k}}_{\substack{\text{Material}\\ \text{curvature vector}}} \times \underbrace{{}^t\bar{\varphi}}_{\substack{\text{Material}\\ \text{rotation vector}\\ \text{variation}}}
\end{gathered} \tag{10.330}$$

10.14.2 Component Vector Form: Real Strains and the Vainberg Principle

10.14.2.1.1 Virtual Strain Symmetry

As indicated, in order for us to apply the Vainberg principle of determining the real strains from the virtual ones, we need to guarantee the symmetry of the virtual strains on a linear vector space, which is provided by the tangent space at identity; the components of the virtual strains are expressed in the rotated bases. Noting that $\breve{\bar{\varepsilon}} \equiv ({}^n\bar{\beta},\ {}^n\breve{\chi})$ belongs to a linear vector space (namely, the tangent space at identity), the following symmetry relations hold:

$$\breve{\bar{\varepsilon}}({}^t\mathbf{d},\,{}^t\theta;\,{}^t\tilde{\mathbf{d}},\,{}^t\tilde{\theta};\,{}^t\bar{\mathbf{d}},\,{}^t\bar{\theta}) = \breve{\bar{\varepsilon}}({}^t\mathbf{d},\,{}^t\theta;\,{}^t\bar{\mathbf{d}},\,{}^t\bar{\theta};\,{}^t\tilde{\mathbf{d}},\,{}^t\tilde{\theta}) \tag{10.331}$$

where the parameters ${}^t\theta$, ${}^t\tilde{\theta}$, ${}^t\bar{\theta}$ belong to the tangent space at identity with ${}^t\mathbf{d}$, ${}^t\theta$ as the real parameters, and ${}^t\tilde{\mathbf{d}}$, ${}^t\tilde{\theta}$; ${}^t\bar{\mathbf{d}}$, ${}^t\bar{\theta}$ are the two sets of the virtual parameters. In other words, the virtual strains are symmetric for the conservative loading, and thus, by the Vainberg principle, the potential functional exists. In other words, there exist functions (we will call them the real strains) the variation of which will yield the virtual strains.

10.14.2.1.2 Component Form: Real Translational (Extensional and Shear) Strain Field: ${}^n\beta$

The real translational strain field, ${}^n\boldsymbol{\beta}$, can be derived by an application of the integral expression in the definition of the Vainberg principle as follows (mnemonically, multiply the real variables by $\rho \in [0.1]$ such as $\mathbf{d} \to \rho\mathbf{d}$, and so on, and replace the virtual variables by the corresponding real one such as $\bar{\mathbf{d}} \to \mathbf{d}$, and so on, and integrate over $\rho \in [0.1]$) :

$$\begin{aligned}
{}^n\beta &= \int_0^1 {}^n\bar{\beta}(\rho\,{}^t\mathbf{d},\,\rho\,{}^t\varphi;\,{}^td,\,{}^t\varphi)\,d\rho \\
&= \int_0^1 \left\{{}^t\mathbf{R}_\rho^T\,{}^t\mathbf{d}|_S + (-{}^t\mathbf{R}_\rho^T\,{}^t\bar{\mathbf{\Phi}})(\hat{1} + \rho\,{}^t\mathbf{d}|_S)\right\}\,d\rho \\
&= \int_0^1 \frac{d}{d\rho}\left\{{}^t\mathbf{R}_\rho^T\,(\hat{1} + \rho\,{}^t\mathbf{d}|_S)\right\}\,d\rho = \left\{{}^t\mathbf{R}_\rho^T\,(\hat{1} + \rho\,{}^t\mathbf{d}|_S)\right\}_{\rho=0}^{\rho=1} \\
&= {}^t\mathbf{R}^T\,(\hat{1} + {}^t\mathbf{d}|_S) - \hat{1}
\end{aligned} \tag{10.332}$$

In this derivation, we have used relations in equation (10.325), and the following:

$${}^t\mathbf{R}_\rho \equiv {}^t\mathbf{R}(\rho\varphi)$$

Definitions

Facts:

- ${}^{t}\mathbf{R}_0 \equiv {}^{t}\mathbf{R}(0) = \mathbf{I} = {}^{t}\mathbf{R}^T(0) = {}^{t}\mathbf{R}_0^T$
- $\overline{{}^{t}\mathbf{R}^T} \equiv \delta({}^{t}\mathbf{R}^T) = -\,{}^{t}\mathbf{R}^T\,{}^{t}\mathbf{\Phi}$ (see discussions on the derivatives and variations of rotation).

10.14.2.1.3 Component Form: Real Bending Strain Field: ${}^{n}\chi$

Likewise, the real bending strains, ${}^{n}\chi$, can be derived by an application of the integral expression in the definition of the Vainberg principle, as in the exercise following: ${}^{n}\chi = {}^{t}\mathbf{R}^{\mathbf{T}}\,\mathbf{k}^c - \mathbf{k}^0$ where we have used the following definitions.

Definitions

- The current total curvature vector: $\mathbf{k}^c \equiv \mathbf{k}^0 + \mathbf{k}$, with $\mathbf{k} \equiv \mathbf{W}\ \boldsymbol{\theta}_{,S}$ where $\mathbf{W}(\boldsymbol{\theta}) \equiv I + c_1\,\mathbf{\Theta} + c_2\,\mathbf{\Theta}^2$, $c_1 \equiv \frac{1-\cos\theta}{\theta^2}$, $c_2 \equiv \frac{\theta-\sin\theta}{\theta^3}$ with $\mathbf{W}$ defined in equation (10.329) and introduced in the section on the derivatives and variations of rotation tensor, ${}^{t}\mathbf{R}$, in which $\mathbf{\Theta}$ is the skew-symmetric tensor corresponding to rotation vector, $\boldsymbol{\theta}$, and θ stands for the length of $\boldsymbol{\theta}$; $\mathbf{k}^0$ is the initial curvature vector of the beam reference curve.

Exercise: Following a similar procedure to that used for the real translational strains, derive the real bending strain as: ${}^{n}\chi = \mathbf{R}^{\mathbf{T}}\,\mathbf{k}^c - \mathbf{k}^0$.

Solution:

$$\begin{aligned}
{}^{n}\chi &= \int_0^1 {}^{n}\bar{\chi}(\rho\,{}^{t}\mathbf{d},\ \rho\,{}^{t}\boldsymbol{\phi};\ {}^{t}d,\ {}^{t}\boldsymbol{\phi})\,d\rho \\
&= \int_0^1 \frac{d}{d\rho}\left\{{}^{t}\mathbf{R}_\rho^T\,(\mathbf{k}^0 + \rho\,\mathbf{k})\right\}\,d\rho = \left\{{}^{t}\mathbf{R}_\rho^T\,(\mathbf{k}^0 + \rho\,\mathbf{k})\right\}_{\rho=0}^{\rho=1} \\
&= {}^{t}\mathbf{R}^T\,(\mathbf{k}^0 + \mathbf{k}) - \mathbf{k}^0 = {}^{t}\mathbf{R}^T\,\mathbf{k}^c - \mathbf{k}^0
\end{aligned} \tag{10.333}$$

where we have recalled the definition for the total curvature vector as: $\mathbf{k}^c \equiv \mathbf{k}^0 + \mathbf{k}$.

10.14.3 The Real Translational and Bending Strains

Thus, we have the real translational (axial and shear) and the bending strains given as:

$$\boxed{\begin{aligned}{}^{n}\boldsymbol{\beta} &= \mathbf{R}^{\mathbf{T}}(\mathbf{e} + \mathbf{k}^0 \times \mathbf{d}) - \hat{\mathbf{1}} \\ {}^{n}\chi &= \mathbf{R}^{\mathbf{T}}\mathbf{k}^c - \mathbf{k}^0\end{aligned}} \tag{10.334}$$

***Exercise*:** Deduce the real strains for planar beams from equation (10.334) with $\mathbf{d} = (u \quad w \quad 0)^T$, $\boldsymbol{\theta} = (0 \quad 0 \quad \theta)^T$ and $\mathbf{K} = (0 \quad 0 \quad k)^T$, that is, $\tau \equiv 0$, as:

$$\begin{Bmatrix} {}^n\varepsilon \\ {}^n\gamma \\ {}^n\chi \end{Bmatrix} = \begin{bmatrix} \kappa \sin\theta & -\kappa\cos\theta & 0 & \cos\theta & \sin\theta & 0 \\ \kappa\cos\theta & \kappa\sin\theta & 0 & -\sin\theta & \cos\theta & 0 \\ 0 & 0 & 0 & 0 & 0 & 1 \end{bmatrix} \begin{Bmatrix} u \\ w \\ \theta \\ u' \\ w' \\ \theta' \end{Bmatrix} - \begin{Bmatrix} 1 \\ 0 \\ 0 \end{Bmatrix} \tag{10.335}$$

Hint: It follows from: $\mathbf{R}^T = \begin{bmatrix} \cos\theta & \sin\theta & 0 \\ -\sin\theta & \cos\theta & 0 \\ 0 & 0 & 1 \end{bmatrix}$ and $\mathbf{k} = \mathbf{W}\theta' = (0 \quad 0 \quad \theta')$

10.14.4 Where We Would Like to Go

We have now derived the beam real strain fields by applying the Vainberg principle to the virtual strain fields. Now that both the real and the virtual strain fields are known, we must particularize the virtual functional further by determination of the reactive forces and moments through the application of a relevant constitutive law. Thus, we must check out the special material properties and the loading system for which the analysis will be specialized and developed later for numerical implementation. Let us also recall that the internal virtual work functional of equation (10.322) has been expressed in terms of the real and virtual displacements and rotations, and the reactive forces and moments. For notational compactness and to enable an uncluttered presentation, it is useful to generalize the expressions by lumping various translational entities with their corresponding rotational counterparts – such as the displacements with the rotations (real and virtual), and so on. Thus, a logical place to go from here must be:

- beam material property (constitutive laws)
- covariant linearization of virtual work equations

10.15 Hyperelastic Material Property

10.15.1 What We Need to Recall

- The invariant form of the real strain vectors in rotated (spatial) frame, and the components in rotating bases, are given as:

$$\boxed{{}^n\boldsymbol{\beta} = \mathbf{c}_{0},_{S} - \mathbf{g}_1 = {}^n\beta_i\,\mathbf{g}^i,\ i = 1,2,3} \tag{10.336}$$

$$\boxed{{}^n\boldsymbol{\chi} = \mathbf{k} = \mathbf{k}^c - \mathbf{R}\,\mathbf{k}^0 = {}^n\chi_i\,\mathbf{g}^i,\ i = 1,2,3} \tag{10.337}$$

- The resistive stress resultant, ${}^n\mathbf{F}$, is defined by:

$$\boxed{{}^n\mathbf{F} \equiv \int_{A_C} \mathbf{P}^1 \; dA_C} \tag{10.338}$$

- The resistive stress couple, $\mathbf{M}^{\alpha}$, is defined by:

$$\boxed{\mathbf{M} \equiv \int_{A_C} \mathbf{p}(\xi^1, \xi^2) \times \mathbf{P}^1 \; dA_C} \tag{10.339}$$

where $\mathbf{p}(\xi^1, \xi^2)$ defines the deformed location of a point on the beam-like body, and component-wise is denoted by:

$$\boxed{{}^n\mathbf{F} = {}^nF^i \; \mathbf{g}_i, \; i = 1, 2, 3 \qquad \text{and} \qquad {}^n\mathbf{M} = {}^nM^i \; \mathbf{g}_i} \tag{10.340}$$

- The component form of the real strain vectors, and the components in rotating bases, are given as:

$$\boxed{{}^n\boldsymbol{\beta} = \mathbf{R}^{\mathbf{T}}(\mathbf{e} + \mathbf{k}^0 \times \mathbf{d}) - \hat{\mathbf{1}} \qquad \text{and} \qquad {}^n\boldsymbol{\chi} = \mathbf{R}^{\mathbf{T}}\mathbf{k}^c - \mathbf{k}^0} \tag{10.341}$$

with components denoted by:

$$\boxed{{}^n\boldsymbol{\beta} = \left\{{}^n\beta_1 \quad {}^n\beta_2 \quad {}^n\beta_3\right\}^T \qquad \text{and} \qquad {}^n\boldsymbol{\chi} = \left\{{}^n\chi_1 \quad {}^n\chi_2 \quad {}^n\chi_3\right\}^T} \tag{10.342}$$

where a superscript n signifies that the object has a spatial representation in the rotating bases, $\{\mathbf{g}_i\}, \; i = 1, 2, 3$.

- The internal virtual work functional, $G_{Internal}(\mathbf{d}, \mathbf{R}; \bar{\mathbf{d}}, \bar{\boldsymbol{\varphi}})$, is given as:

$$\boxed{G_{Internal}(\mathbf{d}, \mathbf{R}; \bar{\mathbf{d}}, \bar{\boldsymbol{\varphi}}) \equiv \int_0^L \{({}^n\mathbf{F} \bullet {}^n\bar{\boldsymbol{\beta}} \; + \; {}^n\mathbf{M} \bullet {}^n\bar{\boldsymbol{\chi}}\} dS} \tag{10.343}$$

The beam reactive component or operational force vector, ${}^n\mathbf{F}$, and the beam reactive component or operational moment vector, ${}^n\mathbf{M}$, are component-wise denoted by:

$$\boxed{\begin{aligned} {}^n\mathbf{F} &= \{{}^nF^1 \quad {}^nF^2 \quad {}^nF^3\}^T = {}^t\mathbf{R}^T \, {}^t\mathbf{F} \\ {}^n\mathbf{M} &= \{{}^nM^1 \quad {}^nM^2 \quad {}^nM^3\}^T = {}^t\mathbf{R}^T \, {}^t\mathbf{M} \end{aligned}} \tag{10.344}$$

where the superscript t and n indicate that all quantities are measured and expressed in the undeformed $\mathbf{G}_1 \equiv \mathbf{T}, \mathbf{G}_2 \equiv \mathbf{M}, \mathbf{G}_3 \equiv \mathbf{B}$ frame, and the deformed and rotated $\mathbf{g}_1, \mathbf{g}_2, \mathbf{g}_3$ frame, respectively.

- The reduced internal energy variation is given as:

$$\boxed{\begin{aligned} \bar{W} \equiv \delta W &= \int_0^L \left(\int_{A_C} \mathbf{P} : \bar{\mathbf{F}} \; dA_C \right) dS = \int_0^L \{{}^n\mathbf{F} \bullet \overline{{}^n\boldsymbol{\beta}} + \mathbf{M} \bullet \overline{{}^n\boldsymbol{\chi}}\} dS \\ &= \bar{U} \equiv \delta U \end{aligned}} \tag{10.345}$$

where $\bar{U} \equiv \delta U$ is the strain energy variation of a one-dimensional beam reference curve identified as the reduced three-dimensional strain energy variation, $\bar{W} \equiv \delta W$, of the corresponding beam-like body.

10.15.2 Elastic Beam Material Property

As mentioned earlier, and in our discussion on the constitutive laws (see Chapter 7 on nonlinear mechanics), every constitutive theory is approximate by its very task of modeling and fitting experimental data. The beam constitutive properties will be no exception and, in fact, will induce additional approximations through the reduction of the 3D theory to an equivalent 1D beam material properties. In any case, as usual, the beam constitutive modeling must also satisfy the four general principles with two as common restrictions of a consistent 3D theory which we may recall are: (1) the principle of stress determinism, (2) the principle of local action, (3) the principle of material frame indifference and (4) the principle of material symmetry. As a reminder, note that the derived real translational and bending strains as given by equations (10.336) and (10.337), respectively, and the virtual work principle given by equation (10.343) are exact and do not depend on the material properties of a beam.

10.15.2.1 Principle of Equivalence

All the same, we need an equivalent 1D constitutive theory from a corresponding 3D one, that is, on the one hand, suitable for a beam, and replicates the response of its corresponding 3D counterparts, that is, a beam-like body, as close as possible, on the other. One way of measuring this equivalency is to require that the internal virtual work of a 3D beam-like body integrated over the cross-section (just like the way we have derived the 1D virtual work principle and the boundary conditions) be as close as possible to the 1D internal virtual work expressed in equation (10.343). This has been achieved elsewhere earlier in the text as shown in equation (10.345). Recall that for an elastic material, by definition, the first Piola–Kirchhoff stress tensor, $\mathbf{P} = \mathbf{P}^i \otimes \mathbf{G}_i$, can be expressed as a function of the deformation gradient tensor, $\mathbf{F}$, by:

$$\mathbf{P} = \bar{\mathcal{P}}(\mathbf{F}, \mathbf{X}) \tag{10.346}$$

where $\mathbf{X}$ denotes a particle on the 3D beam-like body in the undeformed reference state; recall that the imposition of the material frame indifference guarantees that equation (10.346) can be expressed in terms of energy conjugate stress and strain variables and that with $\mathbf{C} \equiv \mathbf{F}^T \mathbf{F}$ as the right Cauchy–Green stretch tensor, we can redefine equation (10.346) as:

$$\mathbf{P} = \mathcal{P}(\mathbf{C}, \mathbf{X}) \tag{10.347}$$

An application of the equivalence principle will generally result in the reduction of the 3D strain energy variation, $\bar{W} \equiv \delta W$, and as such establish the existence of the equivalent 1D strain energy variation, $\bar{U} \equiv \delta U$, as:

$$\boxed{\begin{aligned} \bar{W} \equiv \delta W &= \int_0^L \left(\int_{A_C} \mathbf{P} : \bar{\mathbf{F}}\, dA_C \right) dS \\ &= \int_0^L \{{}^n\mathbf{F} \bullet \overline{{}^n\boldsymbol{\beta}} + \mathbf{M} \bullet \overline{{}^n\boldsymbol{\chi}}\} dS + \text{possible error} \\ &= \bar{U} \equiv \delta U \end{aligned}} \tag{10.348}$$

Recall also that under the kinematic assumption of rigid cross-section, we have already established this reduction exactly, as shown by equation (10.345). However, if we generate the stress resultants and the stress couple, and the conjugate strains exactly, as we do in our theory, without any kinematic assumptions, the reduction in the form of equation (10.348) will generally entail possible error which conforms to our design of containing and restricting all the approximations in the constitutive theory of beams.

10.15.2.2 Isotropic, Hyperelastic Material

Now, if we restrict, as we do, our ensuing analysis to only isotropic hyperelastic or Green-elastic beams, then, recall that, by definition, there exists a strain energy functional, $W = \rho\,\Phi$, per unit of undeformed volume, that depends on the deformation gradient of the strain tensor; ρ is the current density and Φ is the specific strain energy (i.e. strain energy per unit mass), also known as the Helmholtz free energy, such that we have:

$$\mathcal{P}(\mathbf{C},\ \mathbf{X}) = \frac{\partial W}{\partial \mathbf{C}} = 2\mathbf{F}\frac{\partial W}{\partial \mathbf{F}} \tag{10.349}$$

Now, considering equation (10.348) with the beam real strain measures given by ${}^{n}\boldsymbol{\beta}$, the real translational (axial and shear) spatial strain, and ${}^{n}\boldsymbol{\chi}$, the real rotational (bending) strain, the beam strain energy, U, for the 1D beam curve, must be given by:

$$\boxed{\begin{aligned} U &\equiv U({}^{n}\boldsymbol{\beta}, {}^{n}\boldsymbol{\chi}, S, \xi^2, \xi^3) \\ &= \int_{A_C} W(\mathbf{C}, \mathbf{X})\, dA_C + \textit{possible error} \end{aligned}} \tag{10.350}$$

where we have included the explicit dependence on the position of a material particle by the arc length parameter, S along the curve, and ξ^2, ξ^3, the beam coordinates on the rigid cross-section. Then, by the postulate of hyperelasticity, there exist two vector-valued functions, $\mathcal{F}(\boldsymbol{\beta}, \boldsymbol{\chi}, S, \xi^2, \xi^3)$ and $\mathcal{M}(\boldsymbol{\beta}, \boldsymbol{\chi}, S, \xi^2, \xi^3)$ such that ${}^{n}\mathbf{F}$, the spatial stress resultant, and ${}^{n}\mathbf{M}$, the spatial stress couple, are given by:

$$\begin{aligned} {}^{n}\mathbf{F} &= \mathcal{F}({}^{n}\boldsymbol{\beta}, {}^{n}\boldsymbol{\chi}, S, \xi^2, \xi^3) \equiv \frac{\partial U}{\partial\, {}^{n}\boldsymbol{\beta}} \\ {}^{n}\mathbf{M} &= \mathcal{M}({}^{n}\boldsymbol{\beta}, {}^{n}\boldsymbol{\chi}, S, \xi^2, \xi^3) \equiv \frac{\partial U}{\partial\, {}^{n}\boldsymbol{\chi}} \end{aligned} \tag{10.351}$$

10.15.2.3 Homogeneous, Isotropic, Hyperelastic Material

If the material properties are independent of the position of a particle homogeneity, then, we can rewrite equation (10.351) as:

$$\begin{aligned} {}^{n}\mathbf{F} &= \mathcal{F}({}^{n}\boldsymbol{\beta}, {}^{n}\boldsymbol{\chi}) \equiv \frac{\partial U}{\partial\, {}^{n}\boldsymbol{\beta}} \\ {}^{n}\mathbf{M} &= \mathcal{M}({}^{n}\boldsymbol{\beta}, {}^{n}\boldsymbol{\chi}) \equiv \frac{\partial U}{\partial\, {}^{n}\boldsymbol{\chi}} \end{aligned} \tag{10.352}$$

10.15.2.4 Linear, Homogeneous, Isotropic, Hyperelastic Material

Here, we restrict the exposition only to linear, homogeneous, isotropic, hyperelastic materials. Let us recall that for such material properties, the strain energy is a quadratic function of the strains and thus, the constitutive law for a 3D body can be represented in tensor form, or component-wise by:

$$\begin{aligned} W &= \frac{1}{2}\mathbf{E}:\mathbf{C}:\mathbf{E} \\ &= \frac{1}{2}\left\{E_{ij}\,\mathbf{A}^i\otimes\mathbf{A}^j{:}\mathbb{C}^{mnpq}\,\mathbf{A}_m\otimes\mathbf{A}_n\otimes\mathbf{A}_p\otimes\mathbf{A}_q{:}E_{kl}\,\mathbf{A}^k\otimes\mathbf{A}^l\right\} \\ &= \frac{1}{2}\left\{\delta^i_m\,\delta^j_n\,\delta^k_p\,\delta^l_q\,\mathbb{C}^{mnpq}\,E_{ij}\,E_{kl}\right\} = \frac{1}{2}C^{ijkl}\,E_{ij}\,E_{kl} \\ \Rightarrow \mathbf{S} &= \frac{\partial W}{\partial \mathbf{E}} = \mathbf{C}:\mathbf{E} = \left\{C^{ijkl}\,E_{kl}\right\}\,\mathbf{A}_i\otimes\mathbf{A}_j, \\ S^{ij} &= C^{ijkl}\,E_{kl} \end{aligned} \tag{10.353}$$

where $\mathbf{S} = S^{ij}(S,\xi^2,\xi^3)\,\mathbf{A}_i(S,\xi^2,\xi^3)\otimes\,\mathbf{A}_j(S,\xi^2,\xi^3)$ is the second Piola–Kirchhoff stress tensor, and $\mathbf{E} = E_{kl}\,(S,\xi^2,\xi^3)\,\mathbf{A}^k(S,\xi^2,\xi^3)\otimes\,\mathbf{A}^l(S,\xi^2,\xi^3)$ is the Green–Lagrange strain tensor, $\{\mathbf{A}_i(S,\xi^2,\xi^3)\}$, $i = 1,2,3$ is the base vector set at any particle point (S,ξ^2,ξ^3) in the undeformed reference system. Now, recall that for linear isotropic material with the identity tensors: $\overset{4}{\mathbf{I}} \equiv \mathbf{A}_i\otimes\mathbf{A}_j\otimes\mathbf{A}^i\otimes\mathbf{A}^j$ and $\mathbf{I}\equiv\mathbf{A}_i\otimes\mathbf{A}^i$, the constitutive tensor, $\mathbf{C}$, is given by:

$$\mathbf{C} \equiv 2\mu\left(\overset{4}{\mathbf{I}}+\frac{\nu}{1-2\nu}\mathbf{I}\otimes\mathbf{I}\right) \tag{10.354}$$

resulting in the coefficients, C^{ijkl}, $i,j,k,l = 1,2,3$ of the fourth-order constitutive tensor, $\mathbf{C} = {}^{ijkl}\,\mathbf{A}_i\otimes\mathbf{A}_j\otimes\mathbf{A}_k\otimes\mathbf{A}_l$, in the curvilinear coordinates as given by:

$$C^{ijkl} \equiv \mu\left(A^{ik}A^{jl}+A^{il}A^{jk}+\frac{2\nu}{1-2\nu}A^{ij}A^{kl}\right) \tag{10.355}$$

where $\mu \equiv G = \frac{E}{2(1+\nu)}$ is the shear modulus with E as the elasticity modulus and ν as the Poisson's ratio.

10.15.2.5 First Piola–Kirchhoff Stress Tensor in Mixed Bases

The stress–strain relations expressed in equations (10.353)–(10.361) are all in terms of the second Piola–Kirchhoff stress tensor, $\mathbf{S}$, and the Green–Lagrange strain tensor, $\mathbf{E}$; however, we are interested in relations in terms of the first Piola–Kirchhoff stress tensor, $\mathbf{P} = P^{ij}\,\mathbf{a}_i\otimes\mathbf{A}_j$, and the deformation tensor, $\mathbf{F}$, defined by $\mathbf{F}\equiv\mathbf{a}_i\otimes\mathbf{A}^i$ where $\{\mathbf{a}_i\}$ and $\{\mathbf{A}_i\}$ are the deformed and the undeformed base vector sets, respectively. Recall that $\mathbf{P} = \mathbf{F}\,\mathbf{S}$. Note here that we have represented **P** in a mixed basis form, that is, the domain basis in the undeformed reference state and the range basis in the deformed state. Under the mixed basis representation, we have:

$$\begin{aligned}\mathbf{P} &= P^{ij}\,\mathbf{c}_{,i} \otimes \mathbf{A}_j \\ &= \mathbf{F}\,\mathbf{S} \\ &= \underbrace{\mathbf{c}_{,i} \otimes \mathbf{A}^i}_{\text{F, deformation gradient tensor}} \bullet \underbrace{S^{ij}\mathbf{A}_i \otimes \mathbf{A}_j}_{\text{S, second Piola–Kirchhoff tensor}}\end{aligned} \tag{10.356}$$

Thus, the components, P^{ij}, of **P** represented in the mixed basis are identical with the corresponding components, S^{ij}, of **S**, and hence, symmetric. Now, noting, that $\mathbf{P} = \mathbf{F}\,\mathbf{S}$, we can rewrite the constitutive relation of equation (10.353) as:

$$\begin{aligned}P^{ij}\,\mathbf{a}_i \otimes \mathbf{A}_j &= \mathbf{P} = \mathbf{F}\,\mathbf{S} = \mathbf{F}\,\mathbf{C} : \mathbf{E} \\ &= \mathbf{a}_i \otimes \mathbf{A}^i\, C^{kjmn}\, \mathbf{A}_k \otimes \mathbf{A}_j \otimes \mathbf{A}_m \otimes \mathbf{A}_n : E_{pq}\, \mathbf{A}^p \otimes \mathbf{A}^q \\ &= C^{ijmn}\,(E_{mn}\,\mathbf{a}_i \otimes \mathbf{A}_j) = (C^{ijmn}\,E_{mn})\,\mathbf{a}_i \otimes \mathbf{A}_j \\ &\Rightarrow \boxed{P^{ij} = C^{ijmn}\,E_{mn}}\end{aligned} \tag{10.357}$$

From equation (10.357), we see that if the first Piola–Kirchhoff stress tensor is represented in the mixed basis, then for the stress–strain relations, we can continue to use the same constitutive relations, as between the second Piola–Kirchhoff stress tensor and the Green–Lagrange strain tensor, with the effective strain tensor represented in the mixed form with the coefficients of the Green–Lagrange strain tensor.

10.15.3 Thin, Linear, Homogeneous, Isotropic, Hyperelastic Beam

Here, we restrict the exposition only to linear, homogeneous, isotropic, hyperelastic and very thin beams to derive the stress–strain relationship that will be used in our subsequent numerical simulations. In what follows for thin beams, we will assume that $\mathbf{p}(\xi^2, \xi^3) = \xi^\alpha \mathbf{g}_\alpha$, $\alpha = 2, 3$ in equation (10.339), keeping the cross-section rigid through the deformation history. Now, let us first recall that for our optimal normal beam coordinate system with the arc length as the coordinate parameter, for any particle with coordinates (S, ξ^2, ξ^3) in the undeformed beam-like body, we have:

$$\begin{aligned}\mathbf{C}(S, \xi^2, \xi^3) &= \mathbf{C}_0(S) + \xi^\alpha\,\mathbf{G}_\alpha(S) \\ &\Rightarrow \\ \mathbf{A}_\alpha &= \mathbf{G}_\alpha, \quad \alpha = 2, 3\end{aligned} \tag{10.358}$$

where $\mathbf{G}_\alpha$, $\alpha = 2, 3$ are the normal and the bi-normal base vectors of the Frenet frame of the undeformed beam curve.

10.15.3.1 Thin Beam Material Properties

The assumption of a thin beam with rigid cross-section may be mathematically expressed by the condition: $\mathbf{A}_\alpha \cong \mathbf{G}_\alpha$ for all practical purposes; additionally, we have for Poisson's ratio: $\nu \equiv 0$. Thus, we may approximate the coefficients of the constitutive tensor from equation (10.355) to:

$$C^{ijkl} \equiv \mu(G^{ik}G^{jl} + G^{il}G^{jk}) \tag{10.359}$$

and in the optimal normal beam coordinates, we can identify and particularize the coefficients of equation (10.359) to get:

$$\begin{aligned}
C^{11\gamma\delta} &= \mu\left(\cancel{G^{1\gamma}G^{1\delta}} + \cancel{G^{1\delta}G^{1\gamma}}\right) = 0. \quad \gamma, \delta = 2,3\\
C^{\alpha\beta 11} &= \mu\left(\cancel{G^{\alpha 1}G^{\beta 1}} + \cancel{G^{\alpha 1}G^{\beta 1}}\right) = 0, \quad \alpha, \beta = 2,3\\
C^{1\alpha 1\beta} &= \mu\left(\underbrace{G^{11}}_{=1} G^{\alpha\beta} + \cancel{G^{\alpha 1}G^{1\beta}}\right)\\
&= G\,\delta_{\alpha\beta}\\
\Rightarrow C^{1212} &= C^{1313} = G\\
C^{1111} &= \mu\,(2) = E
\end{aligned} \tag{10.360}$$

10.15.3.2 Explicit Stress Components

Using the relations given in equation (10.360), we can list the non-trivial stress components from equation (10.353) as:

$$\boxed{\begin{aligned} S^{11} &= E\,E_{11}\\ S^{12} &= G\,E_{12}\\ S^{13} &= G\,E_{13}\end{aligned}} \tag{10.361}$$

10.15.3.3 One-dimensional Strain Energy, $U(\beta_\alpha, \chi_\alpha)$

Before we can formally write down the expression for the 1D strain energy functional, $U(\beta_\alpha, \chi_\alpha)$ per unit of the beam length, we need to express the coefficients, E_{ij}, of the Green–Lagrange strain tensors in terms of the coefficients as given by equations (10.336) and (10.337) of ${}^n\beta$ and ${}^n\chi$, the spatial translational and bending strain vectors of the beam curve; similarly, the components of the stress resultant, ${}^n\mathbf{F}$, and those of the stress couple, ${}^n\mathbf{M}$, as given by equations (10.340) must be expressed in terms of the coefficients, S^{ij}, of the second Piola–Kirchhoff stress tensor. Let us recall some preliminaries: that $G = G_i \otimes G^i$ is the identity tensor in the undeformed optimal normal coordinate system; that for an arbitrary particle, the current particle position vector is given by $\mathbf{c} = \mathbf{c}_0 + \xi^\alpha\, \mathbf{g}_\alpha$, $\alpha = 2,3$; that, at an arbitrary particle, the deformed base vector set is given as:

$$\mathbf{g}_1 = \mathbf{c},_S = \mathbf{c}_0,_S + \xi^\alpha\, \mathbf{g}_\alpha,_S = \mathbf{c}_0,_S + \mathbf{k} \times\, \xi^\alpha\, \mathbf{g}_\alpha = \mathbf{c}_0,_S + \chi \times\, \xi^\alpha\, \mathbf{g}_\alpha \tag{10.362}$$

where we have used: $\mathbf{g}_{\alpha},_S = \mathbf{R},_S \mathbf{G}_\alpha = \mathbf{K}\,\mathbf{R}\,\mathbf{G}_\alpha = \mathbf{k}\times\ \mathbf{g}_\alpha$ with $\mathbf{k}$ as the current curvature vector; that the deformation gradient, $\mathbf{F}$, is given by:

$$\begin{aligned}\mathbf{F} &\equiv \mathbf{g}_i \otimes\ \mathbf{G}^i = \mathbf{c},_S \otimes \mathbf{G} + \mathbf{g}_\alpha \otimes\ \mathbf{G}^\alpha \\ &= \left(\mathbf{c}_0,_S + \mathbf{k}\times\ \xi^\alpha\, \mathbf{g}_\alpha\right) \otimes\ \mathbf{G}^1 + \mathbf{g}_\alpha \otimes\ \mathbf{G}^\alpha \\ &= \left(\boldsymbol{\beta} + \boldsymbol{\chi}\times\ \xi^\alpha\, \mathbf{g}_\alpha + \mathbf{g}_1\right) \otimes\ \mathbf{G}^1 + \mathbf{g}_\alpha \otimes\ \mathbf{G}^\alpha \\ \Rightarrow \mathbf{F}^T &= \mathbf{G}^1 \otimes\ \left(\boldsymbol{\beta} + \boldsymbol{\chi}\times\ \xi^\alpha\, \mathbf{g}_\alpha + \mathbf{g}_1\right) + \mathbf{G}^\alpha \otimes\ \mathbf{g}_\alpha\end{aligned} \tag{10.363}$$

where the deformed beam reference surface spin coordinate base vectors are given as: $\mathbf{g}_i = \mathbf{R}\,\mathbf{G}_i$ with $\{\mathbf{G}_i\}$ as the undeformed beam reference curve base vector set; that noting the strain expressions as given by equations (10.336) and (10.337), we have: $\mathbf{c}_0,_S = \boldsymbol{\beta} +\ \mathbf{g}_1$; from equation (10.363), noting that: $\mathbf{g}_i \bullet \mathbf{g}_j = \mathbf{R}\,\mathbf{G}_i \bullet \mathbf{R}\,\mathbf{G}_j = \mathbf{G}_i \bullet\ \mathbf{G}_j = G_{ij}$, we get:

$$\begin{aligned}\mathbf{F}^T\,\mathbf{F} = &\left\{\left(\boldsymbol{\beta} + \boldsymbol{\chi}\times\ \xi^\alpha\, \mathbf{g}_\alpha + \mathbf{g}_1\right) \bullet \left(\boldsymbol{\beta} + \boldsymbol{\chi}\times\ \xi^\alpha\, \mathbf{g}_\alpha + \mathbf{g}_1\right)\right\} \mathbf{G}^1 \otimes\ \mathbf{G}^1 \\ &+ \left\{\left(\boldsymbol{\beta} + \boldsymbol{\chi}\times\ \xi^\alpha\, \mathbf{g}_\alpha + \mathbf{g}_1\right) \bullet\ \mathbf{g}_\gamma\right\} \mathbf{G}^1 \otimes\ \mathbf{G}^\gamma \\ &+ \left\{\mathbf{g}_\delta \bullet \left(\boldsymbol{\beta} + \boldsymbol{\chi}\times\ \xi^\alpha\, \mathbf{g}_\alpha + \mathbf{g}_1\right)\right\} \mathbf{G}^\delta \otimes \mathbf{G}^1 \\ &+ (\mathbf{g}_\gamma \bullet\ \mathbf{g}_\delta) \mathbf{G}^\gamma \otimes\ \mathbf{G}^\delta\end{aligned} \tag{10.364}$$

10.15.3.4 Explicit Strain Components

Now, noting that the Greene strain tensor: $\mathbf{E} = \frac{1}{2}(\mathbf{C} - \mathbf{G}) = \frac{1}{2}(\mathbf{F}^T\ \mathbf{F} - \mathbf{G})$ and using equation (10.364), and retaining only the linear terms, we get component-wise:

$$\begin{aligned}E_{11} &= \frac{1}{2}\left\{\left(\boldsymbol{\beta} + \boldsymbol{\chi}\times\ \xi^\alpha\, \mathbf{g}_\alpha + \mathbf{g}_1\right) \bullet \left(\boldsymbol{\beta} + \boldsymbol{\chi}\times\ \xi^\beta\, \mathbf{g}_\beta + \mathbf{g}_1\right) - 1\right\} \\ &= \frac{1}{2}\left\{\begin{array}{l}\underbrace{\boldsymbol{\beta} \bullet \boldsymbol{\beta} + \boldsymbol{\chi}\times\ \xi^\alpha\, \mathbf{g}_\alpha \bullet\ \boldsymbol{\chi}\times\ \xi^\beta\, \mathbf{g}_\beta}_{\text{Nonlinear}} \\ \underbrace{+\,\boldsymbol{\beta} \bullet \boldsymbol{\chi}\times\ \xi^\alpha\, \mathbf{g}_\alpha + \boldsymbol{\beta} \bullet\ \boldsymbol{\chi}\times\ \xi^\beta\, \mathbf{g}_\beta}_{\text{Nonlinear}} \\ +2\beta_1 + 2\,\mathbf{g}_1 \bullet \boldsymbol{\chi}\times\ \xi^\alpha\, \mathbf{g}_\alpha + 1 - 1\end{array}\right\} \\ &= \beta_1 + e_{\beta\alpha}\ \xi^\alpha\ \chi_\beta, \quad \alpha, \beta = 2, 3\end{aligned} \tag{10.365}$$

where we have used the permutation matrix: $\begin{bmatrix} e_{22} & e_{23} \\ e_{32} & e_{33}\end{bmatrix} = \begin{bmatrix} 0 & 1 \\ -1 & 0\end{bmatrix}$, and

$$\begin{aligned}E_{12} &= \frac{1}{2}\left\{\left(\boldsymbol{\beta} + \boldsymbol{\chi}\times\ \xi^\alpha\, \mathbf{g}_\alpha + \mathbf{g}_1\right) \bullet\ \mathbf{g}_2\right\} \\ &= \frac{1}{2}\left\{\beta_2 + \boldsymbol{\chi}\times\ \xi^\alpha\, \mathbf{g}_\alpha \bullet\ \mathbf{g}_2\right\} = \frac{1}{2}\left\{\beta_2 + \xi^\alpha\, \mathbf{g}_\alpha \times\ \mathbf{g}_2 \bullet\ \boldsymbol{\chi}\right\} \\ &= \frac{1}{2}\left\{\beta_2 - \xi^3\, \mathbf{g}_1 \bullet\ \boldsymbol{\chi}\right\} = \frac{1}{2}\left\{\beta_2 - \xi^3 \chi_1\right\}\end{aligned} \tag{10.366}$$

Similarly,

$$E_{13} = \frac{1}{2}\left\{\beta_3 + \xi^2 \chi_1\right\} \tag{10.367}$$

Thus, from equations (10.365)–(10.367), we have for non-trivial strain components:

$$\boxed{\begin{aligned} E_{11} &= \beta_1 - \xi^2\ \chi_3 + \xi^3\ \chi_2 \\ E_{12} &= \frac{1}{2}\left\{\beta_2 - \xi^3 \chi_1\right\} \\ E_{13} &= \frac{1}{2}\left\{\beta_3 + \xi^2 \chi_1\right\} \end{aligned}} \tag{10.368}$$

10.15.3.5 Thin Beam 1D Strain Energy

For a thin beam, we will take the centers of the rigid cross-sectional area along the length as the locus of the beam reference curve implying: the area $A \equiv \int_{A_C} dA_C$, $\int_{A_C} \mathbf{p}(\xi^2, \xi^3)\ dA_C = 0$, the moments of inertia: $I_2 \equiv \int_{A_C} (\xi^3)^2\ dA_C$, $I_3 \equiv \int_{A_C} (\xi^2)^2\ dA_C$ and the polar moment of inertia: $J = I_2 + I_3$. Now, using equations (10.361) and (10.368) in equation (10.350), the one-dimensional strain energy may be expressed as:

$$\boxed{\begin{aligned} U &= \int_{A_C} W\ dA_C = \frac{1}{2}\int_{A_C} \left\{S^{11}E_{11} + 2S^{12}E_{12} + 2S^{13}E_{13}\right\}\ dA_C \\ &= \frac{1}{2}E\int_{A_C} \left\{(\beta_1 - \xi^2\ \chi_3 + \xi^3\ \chi_2)^2\right\}\ dA_C + \frac{1}{2}G\int_{A_C} \left\{(\beta_2 - \xi^3\ \chi_1)^2\right\}\ dA_C \\ &\quad + \frac{1}{2}G\int_{A_C} \left\{(\beta_3 + \xi^2\ \chi_1)^2\right\}\ dA_C \end{aligned}} \tag{10.369}$$

Using the relations given by equations (10.352) and (10.369), the stress resultants and the stress couples are given by:

$$\begin{aligned} {}^nN_T &= \frac{\partial U}{\partial^n \beta^1} = EA\ \beta_1 \\ {}^nQ_M &= \frac{\partial U}{\partial^n \beta^2} = GA\ \beta_2 \\ {}^nQ_B &= \frac{\partial U}{\partial^n \beta^3} = GA\ \beta_3, \\ {}^nM_T &= \frac{\partial U}{\partial^n \chi^1} = G(I_2 + I_3)\ \chi_1 = GJ\ \chi_1 \\ {}^nM_M &= \frac{\partial U}{\partial^n \chi^2} = GI_2\ \chi_2 \\ {}^nM_B &= \frac{\partial U}{\partial^n \chi^3} = GI_3\ \chi_3 \end{aligned} \tag{10.370}$$

Finally, writing equation (10.370) in matrix form, we can express the stress–strain matrix, ${}^n\mathbf{D}$, of a thin, linear, hyperelastic, homogeneous beam by:

$$\begin{Bmatrix} {}^nN_T \\ {}^nQ_M \\ {}^nQ_B \\ {}^nM_T \\ {}^nM_M \\ {}^nM_B \end{Bmatrix} = {}^n\breve{\mathbf{D}} \begin{Bmatrix} {}^n\beta_1 \\ {}^n\beta_2 \\ {}^n\beta_3 \\ {}^n\chi_1 \\ {}^n\chi_1 \\ {}^n\chi_1 \end{Bmatrix} \Rightarrow \qquad {}^n\breve{\mathbf{F}} = {}^n\breve{\mathbf{D}}\,{}^n\breve{\boldsymbol{\beta}} \tag{10.371}$$

where the (6 × 6) generalized constitutive matrix, ${}^n\breve{\mathbf{D}}$, is defined as:

$${}^n\breve{\mathbf{D}} = \begin{bmatrix} EA & 0 & 0 & 0 & 0 & 0 \\ 0 & GA & 0 & 0 & 0 & 0 \\ 0 & 0 & GA & 0 & 0 & 0 \\ 0 & 0 & 0 & GJ & 0 & 0 \\ 0 & 0 & 0 & 0 & GI_2 & 0 \\ 0 & 0 & 0 & 0 & 0 & GI_3 \end{bmatrix} \tag{10.372}$$

Note that the components of the stress resultants, the stress couples and the strains are all indicated in the rotated spin bases. Finally, note that without the approximations adopted above, the general constitutive relations for beams of simple linear isotropic hyperelastic materials is clearly rather involved.

10.15.4 Where We Would Like to Go

Now for the next step of the computational analysis of beams, as we can see from equation (10.343), the virtual functional in the integral or weak form of equation, being generally highly nonlinear in displacements and rotations, may not be solved in a closed form analytical way. The solution, that is, the tracing of the beam motion or the configuration path will require some sort of iterative numerical method. Of all the possible schemes, the one that we intend to pursue is a Newton-type method. This, in turn, requires linearization ("tangent") of the equations at every iteration step. Clearly, then, we will have to further modify the virtual work equation for computational strategy. Thus, a logical place to go from here will be:

- covariant linearization of virtual functional.

10.16 Covariant Linearization of Virtual Work

10.16.1 What We Need to Recall

- The generalized virtual work equation for quasi-static proportional loading, for all admissible generalized virtual displacement, ${}^t\breve{\bar{\mathbf{d}}} \equiv \{{}^t\bar{\mathbf{d}} \quad {}^t\bar{\boldsymbol{\theta}}\}^T$, is given as:

$$
\begin{aligned}
&G(\lambda, {}^t\breve{\mathbf{d}}; {}^t\breve{\bar{\mathbf{d}}}) \equiv G_{Internal}({}^t\breve{\mathbf{d}}; {}^t\breve{\bar{\mathbf{d}}}) - G_{external}(\lambda; {}^t\breve{\bar{\mathbf{d}}}) = 0, \quad \forall \text{ admissible } {}^t\breve{\bar{\mathbf{d}}}\\
&G_{Internal} = \int_0^L ({}^n\breve{\bar{\varepsilon}} \bullet {}^n\breve{\mathbf{F}})\, dS\\
&G_{External} = \left\{ \int_0^L \hat{\bar{\mathbf{d}}}^{\mathbf{T}}\, \hat{\mathbf{F}}^B(\lambda, S)\, dS + \lambda \left[\hat{\bar{\mathbf{d}}}^{\mathbf{T}}\, \hat{\mathbf{F}}^S\right]_{S=0}^{S=L} \right\}
\end{aligned}
\tag{10.373}
$$

- The generalized virtual work equation for dynamic loading, for all admissible generalized virtual displacement, ${}^t\breve{\bar{\mathbf{d}}} \equiv \{{}^t\bar{\mathbf{d}} \quad {}^t\bar{\theta}\}^T$, is given as:

$$
\begin{aligned}
&G({}^t\breve{\mathbf{d}}; {}^t\breve{\bar{\mathbf{d}}}) \equiv G_{Internal} + G_{Inertial} - G_{External} = \mathbf{0}, \quad \forall \text{admissible } {}^t\breve{\bar{\mathbf{d}}}\\
&G_{Internal} = \int_0^L ({}^n\breve{\bar{\varepsilon}}\ {}^n\breve{\mathbf{F}})\, dS\\
&G_{Inertial} = \int_0^L ({}^t\breve{\bar{\mathbf{d}}}^T\ {}^t\breve{\mathbf{F}}_{Iner})\, dS\\
&\qquad = \overbrace{\int_0^L \left\{(M_0\, {}^t\ddot{\mathbf{d}}\ {}^t\bar{\mathbf{d}} + \mathbf{W}^T\, (\mathbf{I}_S\, {}^t\dot{\omega}_S + {}^t\omega_S \times \mathbf{A}_S) \bullet {}^t\bar{\theta}\right\} dS}^{\text{Inertial}}\\
&G_{External} = \left\{ \int_0^L \hat{\bar{\mathbf{d}}}^{\mathbf{T}}\, \hat{\mathbf{F}}^B\, dS + \left[\hat{\bar{\mathbf{d}}}^{\mathbf{T}}\, \hat{\mathbf{F}}^S\right]_{S=0}^{S=L} \right\}
\end{aligned}
\tag{10.374}
$$

- The time derivatives of the effective linear momentum, $\mathbf{L}(S, t)$, and the effective angular momentum, $\mathbf{A}(S, t)$, in terms of the components measured in the undeformed coordinate system are:

$$
\begin{aligned}
\dot{\mathbf{A}}_S(S, t) &= \mathbf{I}_S\ \dot{\omega}_S(S, t) + \omega_S(S, t) \times \mathbf{A}_S(S, t)\\
&= \mathbf{R}\,(\mathbf{I}_M\ \dot{\omega}_M(S, t) + \omega_M(S, t) \times \mathbf{A}_M(S, t))\\
&= \mathbf{R}\,(\omega_M \times \mathbf{A}_M + \dot{\mathbf{A}}_M)\\
\text{where}\quad \mathbf{A}_S &= \mathbf{I}_S\ \omega_S = \mathbf{R}\,\mathbf{A}_M, \quad \mathbf{A}_M = \mathbf{I}_M\ \omega_M\\
\mathbf{I}_S(S, t) &\stackrel{\Delta}{=} \mathbf{R}(S, t)\ \mathbf{I}_M(S)\ \mathbf{R}^T(S, t)\\
\mathbf{I}_M(S) &\stackrel{\Delta}{=} I_{\alpha\beta}\{\delta_{\alpha\beta}\mathbf{I} - (\mathbf{G}_\alpha \otimes \mathbf{G}_\beta)\}, \quad \alpha, \beta = 2, 3\\
\mathbf{I}_S(S, t) &= I_{\alpha\beta}\{\delta_{\alpha\beta}\mathbf{I} - (\mathbf{g}_\alpha(t) \otimes \mathbf{g}_\beta(t))\}\\
I_{\alpha\beta} &\stackrel{\Delta}{=} \int_{A_c} \rho_0(S, \xi^2, \xi^3)\ \xi^\alpha\ \xi^\beta\ dA\\
\mathbf{\Omega}_M &= [\omega_M] = \mathbf{R}^T\ \dot{\mathbf{R}}, \quad \mathbf{\Omega}_S = [\omega_S] = \dot{\mathbf{R}}\ \mathbf{R}^T = \mathbf{R}\ \mathbf{\Omega}_M\ \mathbf{R}^T\\
\omega_S &= \mathbf{R}\ \omega_M
\end{aligned}
\tag{10.375}
$$

- ${}^{n}\breve{\boldsymbol{\varepsilon}}$, the generalized virtual strain field is given as:

$$\boxed{\begin{aligned}
&{}^{n}\breve{\boldsymbol{\varepsilon}} \equiv \{{}^{n}\bar{\boldsymbol{\beta}} \quad {}^{n}\bar{\boldsymbol{\chi}}\}^{T},\\
&{}^{n}\bar{\boldsymbol{\beta}} = {}^{t}\mathbf{R}^{T}\left({}^{t}\bar{\mathbf{d}},_{S} + {}^{t}\mathbf{k}^{0} \times {}^{t}\bar{\mathbf{d}}\right) - {}^{t}\mathbf{R}^{T}\,{}^{t}\bar{\boldsymbol{\Phi}}\left({}^{t}\mathbf{e} + {}^{t}\mathbf{k}^{0} \times {}^{t}\mathbf{d}\right) = \overline{{}^{t}\mathbf{R}^{T}(\hat{\mathrm{I}} + {}^{t}\mathbf{d}|_{S})},\\
&{}^{n}\bar{\boldsymbol{\chi}} \equiv {}^{t}\mathbf{R}^{T}\,{}^{t}\bar{\boldsymbol{\varphi}}|_{S} = {}^{t}\mathbf{R}^{T}\{{}^{t}\bar{\boldsymbol{\varphi}},_{S} + \mathbf{k}^{0} \times {}^{t}\bar{\boldsymbol{\varphi}}\} = \overline{{}^{t}\mathbf{R}^{T}(\mathbf{k}^{0} + \mathbf{k})},\\
&{}^{t}\bar{\boldsymbol{\varphi}} = \mathbf{W}\,{}^{t}\bar{\boldsymbol{\theta}}
\end{aligned}} \tag{10.376}$$

where $\mathbf{W}$ is given as:

$$\boxed{\begin{aligned}
&\mathbf{W}({}^{t}\boldsymbol{\theta}) \equiv \mathbf{I} + c_1\,{}^{t}\boldsymbol{\Theta} + c_2\,{}^{t}\boldsymbol{\Theta}^2,\\
&c_1({}^{t}\boldsymbol{\theta}) = \frac{1-\cos\theta}{\theta^2},\\
&c_2({}^{t}\boldsymbol{\theta}) = \frac{\theta - \sin\theta}{\theta^3}
\end{aligned}} \tag{10.377}$$

with ${}^{t}\boldsymbol{\Theta}$ being the skew-symmetric matrix corresponding to the axial vector, ${}^{t}\boldsymbol{\theta}$, such that ${}^{t}\boldsymbol{\Theta}\,{}^{t}\boldsymbol{\theta} = \mathbf{0}$.

- $\dot{\mathbf{W}}$ is given as:

$$\boxed{\begin{aligned}
\dot{\mathbf{W}} &= c_1\,\dot{\boldsymbol{\Theta}} + c_2(\dot{\boldsymbol{\Theta}}\,\boldsymbol{\Theta} + \boldsymbol{\Theta}\,\dot{\boldsymbol{\Theta}}) + \dot{c}_1\,\boldsymbol{\Theta} + \dot{c}_2\,\boldsymbol{\Theta}^2\\
&= (c_2 - c_1)(\boldsymbol{\theta} \bullet \dot{\boldsymbol{\theta}})\mathbf{I} + c_1\,\dot{\boldsymbol{\Theta}} + c_2\{(\dot{\boldsymbol{\theta}} \otimes \boldsymbol{\theta}) + (\boldsymbol{\theta} \otimes \dot{\boldsymbol{\theta}})\} + \dot{c}_1\,\boldsymbol{\Theta} + \dot{c}_2(\boldsymbol{\theta} \otimes \boldsymbol{\theta})\\
\dot{c}_1 &\equiv \frac{1}{\theta^2}(1 - 2c_1 - \theta^2 c_2)(\boldsymbol{\theta} \bullet \dot{\boldsymbol{\theta}}) = c_3(\boldsymbol{\theta} \bullet \dot{\boldsymbol{\theta}}),\\
\dot{c}_2 &\equiv \frac{1}{\theta^2}(c_1 - 3c_2)(\boldsymbol{\theta} \bullet \dot{\boldsymbol{\theta}}) = c_4(\boldsymbol{\theta} \bullet \dot{\boldsymbol{\theta}}),
\end{aligned}} \tag{10.378}$$

- $\ddot{\mathbf{W}}$ is given as:

$$\boxed{\begin{aligned}
&\ddot{\mathbf{W}} = c_1\,\ddot{\boldsymbol{\Theta}} + 2\{\dot{c}_1\,\dot{\boldsymbol{\Theta}} + \dot{c}_2(\dot{\boldsymbol{\Theta}}\,\boldsymbol{\Theta} + \boldsymbol{\Theta}\,\dot{\boldsymbol{\Theta}})\} + c_2(\ddot{\boldsymbol{\Theta}}\,\boldsymbol{\Theta} + 2\dot{\boldsymbol{\Theta}}^2 + \boldsymbol{\Theta}\,\ddot{\boldsymbol{\Theta}}) + \ddot{c}_1\,\boldsymbol{\Theta} + \ddot{c}_2\,\boldsymbol{\Theta}^2\\
&\ddot{c}_1 \equiv c_5(\dot{\boldsymbol{\theta}} \bullet \boldsymbol{\theta})^2 + c_3\{\theta^2 + (\boldsymbol{\theta} \bullet \dot{\boldsymbol{\theta}})\}, \qquad \ddot{c}_2 \equiv c_6(\boldsymbol{\theta} \bullet \dot{\boldsymbol{\theta}})^2 + c_4\{\theta^2 + (\boldsymbol{\theta} \bullet \dot{\boldsymbol{\theta}})\}\\
&c_5 \equiv -\frac{1}{\theta^2}(2c_2 + 4c_3 + \theta^2 c_4), \qquad c_6 \equiv -\frac{1}{\theta^2}(c_3 - 5c_4),
\end{aligned}} \tag{10.379}$$

- ${}^{n}\breve{\mathbf{F}}$, the generalized reactive force field is given as:

$$ {}^{n}\breve{\mathbf{F}} \equiv \{{}^{n}\mathbf{F} \quad {}^{n}\mathbf{M}\}^{T} \tag{10.380}$$

- ${}^t\breve{\mathbf{F}}_{Iner}$, the generalized inertial force field is given as:

$$\boxed{\begin{aligned} &{}^t\breve{\mathbf{F}}_{Iner} \equiv \left\{M_0\,{}^t\ddot{\mathbf{d}} \quad \mathbf{W}^T\,(\mathbf{I}_S\,{}^t\dot{\boldsymbol{\omega}}_S + {}^t\boldsymbol{\omega}_S \times \mathbf{A}_S)\right\}^T,\\ &{}^t\boldsymbol{\omega}_S \equiv \mathbf{W}\,{}^t\dot{\boldsymbol{\theta}},\quad {}^t\dot{\boldsymbol{\omega}}_S = (\dot{\mathbf{W}}\,{}^t\dot{\boldsymbol{\theta}} + \mathbf{W}\,{}^t\ddot{\boldsymbol{\theta}}),\quad \mathbf{I}_S = \mathbf{R}\,\mathbf{I}_M\,\mathbf{R}^T \end{aligned}} \tag{10.381}$$

where the spatial angular velocity vector, ${}^t\boldsymbol{\omega}_S$, is the axial vector corresponding to the skew-symmetric spatial angular velocity matrix, $\boldsymbol{\Omega}_S$, given by:

$$\boxed{\boldsymbol{\Omega}_S \equiv \dot{\mathbf{R}}\,\mathbf{R}^T} \tag{10.382}$$

- ${}^t\breve{\hat{\mathbf{F}}}$, the generalized external applied force field is given as:

$${}^t\breve{\hat{\mathbf{F}}} \equiv \{{}^t\hat{\mathbf{F}} \quad {}^t\hat{\mathbf{M}}\}^T \tag{10.383}$$

- Equations (10.373) and (10.374) are generally nonlinear in ${}^t\breve{\mathbf{d}} \equiv \{{}^t\mathbf{d} \quad {}^t\boldsymbol{\theta}\}^T$, the generalized displacement state of the beam; similarly, all entities in equations (10.376)–(10.383) are functions of generalized state variables, ${}^t\breve{\mathbf{d}} \equiv \{{}^t\mathbf{d} \quad {}^t\boldsymbol{\theta}\}^T$.
- For the quasi-static case, we have restricted our presentation to a weak form, involving only a proportional loading system with the assumption that the forces are truly external and hence independent of the displacement state, ${}^t\breve{\mathbf{d}} \equiv \{{}^t\mathbf{d} \quad {}^t\boldsymbol{\theta}\}^T$, and as such can be expressed with a multiplicative variable of proportionality, λ, as we have indicated in the above equation.
- Briefly, the basic characteristics of a Newton-type method of tracing a configuration path entail: starting from a known state or configuration (which is usually the unloaded and undeformed structure at the first step), we find the tangent to the equilibrium or configuration path (usually for every load step for a quasi-static system, or for every time step for a time history dynamic analysis) which helps us set up the instantaneous linear equations to be solved for the incremental and independent variables iteratively for each step; the state variables are updated by the incremental state variables and the process is repeated.
- Finally, the superscript n in the above expressions that helps to identify the relevant entities in the current, rotated or deformed configuration will occasionally be dropped for notational clarity, although the context will lend support to consistency.

10.16.2 Incremental State Variables

Thus, for numerical evaluation of the motion of a beam on a digital computer by a typical iterative Newton-type method, we will need to derive, for every step of the iteration, the instantaneous linearized form of the virtual equations as a function of the incremental generalized displacement state; from now on, we denote it as ${}^t\Delta\breve{\mathbf{d}} \equiv \{{}^t\Delta\mathbf{d} \quad {}^t\Delta\boldsymbol{\theta}\}^T$.

10.16.2.1 Quasi-static Proportional Loading

For quasi-static proportional load cases, in actual numerical evaluation with the incremental load step, denoted by $\Delta\lambda$, as a pre-assigned entity in a typical load-control incrementation method, a Newton-type method without any special treatment such as arc length constraint, is inherently disastrous at the limit points or at the bifurcation points of a configuration path. Thus, for computational purposes (i.e. to maintain symmetry of the stiffness matrices – more on this later), we consider the Newton-type method with an arc length constraint algorithm in which $\Delta\lambda$ is also treated as an additional independent variable as the displacement state variable, ${}^t\breve{\mathbf{d}} \equiv \{{}^t\mathbf{d} \quad {}^t\boldsymbol{\theta}\}^T$. Thus, the virtual work equation (10.373) must be linearized with respect to both the independent variable displacement sets: ${}^t\breve{\mathbf{d}} \equiv \{{}^t\mathbf{d} \quad {}^t\boldsymbol{\theta}\}^T$, and the coefficient of proportionality, λ.

10.16.2.2 Dynamic Time History Loading

For dynamic time history load cases, the virtual work equation (10.374) must be linearized with respect only to the independent variable displacement sets: ${}^t\breve{\mathbf{d}} \equiv \{{}^t\mathbf{d} \quad {}^t\boldsymbol{\theta}\}^T$, but we need to additionally linearize the inertial virtual work functional with particular attention to the angular acceleration, ${}^t\dot{\boldsymbol{\omega}}$.

Finally, in attempting to linearize the virtual equations, we must be aware that the rotation tensor belongs to a curved space; hence, the linearization must be performed with covariant differentiation in mind; the practical way to accommodate covariant differentiation is to simply include the differentiation of the associated base vectors (for the vector fields), or the base tensors (for the tensor field).

10.16.3 Linearization of the Virtual Work Equation

As already indicated, the virtual work – equation (10.373) for the quasi-static case or equation (10.374) for the dynamic case – must be linearized, that is, expressed in incremental form for incremental quasi-static load step or a dynamic time step, to apply a Newton-type iteration scheme for tracing the motion of a beam. Let the known – but not necessarily in equilibrium – configuration be given by ${}^t\breve{\mathbf{d}}^0$ at a load level λ^0, that is, $G(\lambda^0, {}^t\breve{\mathbf{d}}^0; {}^t\bar{\breve{\mathbf{d}}})$ may not satisfy equation (10.373) for the quasi-static case (similarly, $G({}^t\breve{\mathbf{d}}^0; {}^t\bar{\breve{\mathbf{d}}})$ for the dynamic analysis). We will try to seek equilibrium in the neighborhood of $(\lambda^0, {}^t\breve{\mathbf{d}}^0)$ for the quasi-static case (similarly, in the neighborhood of ${}^t\breve{\mathbf{d}}^0$ for the dynamic analysis). More specifically, at $(\lambda, {}^t\breve{\mathbf{d}})$ for the quasi-static case, that is, we want $G(\lambda, {}^t\breve{\mathbf{d}}; {}^t\bar{\breve{\mathbf{d}}}) = \mathbf{0}, \forall\, {}^t\bar{\breve{\mathbf{d}}}$ admissible (similarly, at ${}^t\breve{\mathbf{d}}$ for the dynamic analysis, that is, we want $G({}^t\breve{\mathbf{d}}; {}^t\bar{\breve{\mathbf{d}}}) = \mathbf{0}, \forall\, {}^t\bar{\breve{\mathbf{d}}}$ admissible). Expanding by Taylor series around $(\lambda^0, \breve{\mathbf{d}}^0)$ for the quasi-static case (similarly, around $\mathbf{d}^0$ for the dynamic case) and retaining only the first-order (linear) terms, that is, terms involving $\Delta^t\breve{\mathbf{d}}$ $(\equiv {}^t\breve{\mathbf{d}} - {}^t\breve{\mathbf{d}}^0)$ and $\Delta\lambda$ $(\equiv \lambda - \lambda^0)$, we have:

For quasi-static proportional loading:

$$
\begin{aligned}
&\tilde{G}(\lambda, {}^t\breve{\mathbf{d}}; {}^t\bar{\breve{\mathbf{d}}}; \Delta^t\breve{\mathbf{d}}, \Delta\lambda) \equiv G(\lambda^0, {}^t\breve{\mathbf{d}}^0; {}^t\bar{\breve{\mathbf{d}}}) + \mathcal{L}^{Quasi}_{\Delta^t\breve{\mathbf{d}},\Delta\lambda}[G(\lambda^0, {}^t\breve{\mathbf{d}}^0; {}^t\bar{\breve{\mathbf{d}}})] \\
&\qquad = \mathbf{0} \\
&\Rightarrow \mathcal{L}^{Quasi}_{\Delta^t\breve{\mathbf{d}},\Delta\lambda}[G(\lambda^0, {}^t\breve{\mathbf{d}}^0; {}^t\bar{\breve{\mathbf{d}}})] = -G(\lambda^0, {}^t\breve{\mathbf{d}}^0; {}^t\bar{\breve{\mathbf{d}}}; \Delta^t\breve{\mathbf{d}}) \\
&\mathcal{L}^{Quasi}_{\Delta^t\breve{\mathbf{d}},\Delta\lambda}[G(\lambda^0, {}^t\breve{\mathbf{d}}^0; {}^t\bar{\breve{\mathbf{d}}})] \equiv \nabla_{\lambda^0} G(\lambda^0, {}^t\breve{\mathbf{d}}^0; {}^t\bar{\breve{\mathbf{d}}}) \bullet \Delta\lambda + \nabla_{{}^t\breve{\mathbf{d}}^0} G(\lambda^0, {}^t\breve{\mathbf{d}}^0; {}^t\bar{\breve{\mathbf{d}}}) \bullet \Delta^t\breve{\mathbf{d}}
\end{aligned}
\tag{10.384}
$$

where $\nabla_{{}^t\breve{\mathbf{d}}^0} G$ and $\nabla_{\lambda^0} G$ are the covariant derivatives of G with respect to ${}^t\breve{\mathbf{d}}$ and λ evaluated at ${}^t\breve{\mathbf{d}}^0$ and λ^0, respectively, and the linear, that is, the first-order terms, part of the virtual functional is denoted by $\mathcal{L}^{Quasi}_{\Delta^t\breve{\mathbf{d}},\Delta\lambda}[G(\lambda^0, {}^t\breve{\mathbf{d}}^0; {}^t\bar{\breve{\mathbf{d}}})]$. In a non-equilibrium state, $G(\lambda^0, {}^t\breve{\mathbf{d}}^0; {}^t\bar{\breve{\mathbf{d}}})$, known as the total residual force functional, is non-zero. Noting that for conservative loading, the external loading does not depend on the current configuration, so we can express the residual as:

$$G_{Static}(\lambda^0, {}^t\breve{\mathbf{d}}^0; {}^t\bar{\breve{\mathbf{d}}}) \equiv G_{Internal}({}^t\breve{\mathbf{d}}^0; {}^t\bar{\breve{\mathbf{d}}}) - G_{External}(\lambda^0; {}^t\bar{\breve{\mathbf{d}}}) \tag{10.385}$$

Similarly, because of the coefficient of proportionality, λ, the linear operator, $\mathcal{L}^{Quasi}_{\Delta^t\breve{\mathbf{d}},\Delta\lambda}[G(\lambda^0, {}^t\breve{\mathbf{d}}^0; {}^t\bar{\breve{\mathbf{d}}})]$, also consists of two parts: the internal and the external linear virtual functional:

$$\begin{aligned}
&\mathcal{L}^{Quasi}_{\Delta^t\breve{\mathbf{d}},\Delta\lambda}[G(\lambda^0, {}^t\breve{\mathbf{d}}^0; {}^t\bar{\breve{\mathbf{d}}})] \equiv \Delta G_{Internal}({}^t\breve{\mathbf{d}}^0; {}^t\bar{\breve{\mathbf{d}}}; \Delta{}^t\breve{\mathbf{d}}) + \Delta G_{External}(\lambda^0; {}^t\bar{\breve{\mathbf{d}}}; \Delta\lambda)\\
&\Delta G_{Internal} \equiv \nabla_{{}^t\breve{\mathbf{d}}^0} G_{Internal}({}^t\breve{\mathbf{d}}^0; {}^t\bar{\breve{\mathbf{d}}}) \bullet \Delta{}^t\breve{\mathbf{d}}\\
&\Delta G_{External} \equiv \nabla_{\lambda^0} G_{External}(\lambda^0; {}^t\bar{\breve{\mathbf{d}}}) \bullet \Delta\lambda
\end{aligned} \tag{10.386}$$

For dynamic time history loading:

$$\begin{aligned}
&\tilde{G}({}^t\breve{\mathbf{d}}; {}^t\bar{\breve{\mathbf{d}}}) \equiv G({}^t\breve{\mathbf{d}}^0; {}^t\bar{\breve{\mathbf{d}}}) + \mathcal{L}^{Dyn}_{\Delta^t\breve{\mathbf{d}}}[G({}^t\breve{\mathbf{d}}^0; {}^t\bar{\breve{\mathbf{d}}})] = \mathbf{0}\\
&\Rightarrow \mathcal{L}^{Dyn}_{\Delta^t\breve{\mathbf{d}}}[G({}^t\breve{\mathbf{d}}^0; {}^t\bar{\breve{\mathbf{d}}})] = -G({}^t\breve{\mathbf{d}}^0; {}^t\bar{\breve{\mathbf{d}}})\\
&\mathcal{L}^{Dyn}_{\Delta^t\breve{\mathbf{d}}}[G({}^t\breve{\mathbf{d}}^0; {}^t\bar{\breve{\mathbf{d}}})] \equiv \nabla_{{}^t\breve{\mathbf{d}}^0} G({}^t\breve{\mathbf{d}}^0; {}^t\bar{\breve{\mathbf{d}}}) \bullet \Delta{}^t\breve{\mathbf{d}}
\end{aligned} \tag{10.387}$$

where $\nabla_{{}^t\breve{\mathbf{d}}^0} G$ is the covariant derivative of G with respect to ${}^t\breve{\mathbf{d}}$ evaluated at ${}^t\breve{\mathbf{d}}^0$, and the linear part, that is, the first-order terms, of the virtual functional is denoted by $\mathcal{L}^{Dyn}_{\Delta^t\breve{\mathbf{d}}}[G({}^t\breve{\mathbf{d}}^0; {}^t\bar{\breve{\mathbf{d}}})]$. In a non-equilibrium state, $G({}^t\breve{\mathbf{d}}^0; {}^t\bar{\breve{\mathbf{d}}})$, known as the ***residual force functional***, is non-zero. Noting that for conservative loading, the external loading does not depend on the current configuration, so we can express the residual as:

$$G_{Dyn}({}^t\breve{\mathbf{d}}^0; {}^t\bar{\breve{\mathbf{d}}}) \equiv G_{Internal}({}^t\breve{\mathbf{d}}^0; {}^t\bar{\breve{\mathbf{d}}}) + G_{Inertial}({}^t\breve{\mathbf{d}}^0; {}^t\bar{\breve{\mathbf{d}}}) - G_{External}({}^t\bar{\breve{\mathbf{d}}}) \tag{10.388}$$

The linear operator, $\mathcal{L}^{Dyn}_{\Delta^t\breve{\mathbf{d}}}[G(\lambda^0, {}^t\breve{\mathbf{d}}^0; {}^t\breve{\bar{\mathbf{d}}})]$, consists only of two parts: the internal and the inertial linear virtual functionals:

$$\boxed{\begin{aligned}&\mathcal{L}^{Dyn}_{\Delta^t\breve{\mathbf{d}}}[G({}^t\breve{\mathbf{d}}^0; {}^t\breve{\bar{\mathbf{d}}})] \equiv \Delta G_{Internal}({}^t\breve{\mathbf{d}}^0; {}^t\breve{\bar{\mathbf{d}}}; \Delta^t\breve{\mathbf{d}}) + \Delta G_{Inertial}({}^t\breve{\mathbf{d}}^0; {}^t\breve{\bar{\mathbf{d}}}; \Delta^t\breve{\mathbf{d}})\\ &\Delta G_{Internal} \equiv \nabla_{{}^t\breve{\mathbf{d}}^0} G_{Internal}({}^t\breve{\mathbf{d}}^0; {}^t\breve{\bar{\mathbf{d}}}) \bullet \Delta^t\breve{\mathbf{d}}\\ &\Delta G_{Inertial} \equiv \nabla_{{}^t\breve{\mathbf{d}}^0} G_{Inertial}({}^t\breve{\mathbf{d}}^0; {}^t\breve{\bar{\mathbf{d}}}) \bullet \Delta^t\breve{\mathbf{d}}\end{aligned}} \tag{10.389}$$

Now we will provide a complete description of the various parts mentioned.

10.16.4 Internal Virtual Functional: Covariant Linearization

Referring to equations (10.373) or (10.374), the internal virtual work functional is given as:

$$G_{Internal}({}^t\breve{\mathbf{d}}; {}^t\breve{\bar{\mathbf{d}}}) = \int_0^L {}^n\breve{\bar{\varepsilon}}({}^t\breve{\mathbf{d}}; {}^t\breve{\bar{\mathbf{d}}}) \bullet {}^n\breve{\mathbf{F}}({}^t\breve{\mathbf{d}})\, dS \tag{10.390}$$

Taking the covariant derivative of the expression in equation (10.390), we have:

$$\boxed{\begin{aligned}\Delta G({}^t\breve{\mathbf{d}}^0; {}^t\breve{\bar{\mathbf{d}}}; \Delta^t\breve{\mathbf{d}}) &\equiv \nabla_{{}^t\breve{\mathbf{d}}} G({}^t\breve{\mathbf{d}}^0; {}^t\breve{\bar{\mathbf{d}}})\, \Delta^t\hat{\mathbf{d}}\\ &= \int_0^L {}^n\breve{\bar{\varepsilon}}^T \Delta^n\breve{\mathbf{F}}\, dS + \int_0^L \Delta\, {}^n\breve{\bar{\varepsilon}}^T\, {}^n\breve{\mathbf{F}}\, dS\\ \Delta^n\breve{\mathbf{F}} &\equiv \nabla_{{}^t\breve{\mathbf{d}}}\, {}^n\breve{\mathbf{F}}^T\, \Delta^t\breve{\mathbf{d}},\\ \Delta^n\breve{\bar{\varepsilon}} &\equiv \nabla_{{}^t\breve{\mathbf{d}}}\, {}^n\breve{\bar{\varepsilon}}^T\, \Delta^t\breve{\mathbf{d}}\end{aligned}} \tag{10.391}$$

where $\nabla_{{}^t\breve{\mathbf{d}}} G$ are the covariant derivatives of G with respect to ${}^t\breve{\mathbf{d}}$, and we have introduced the following definitions.

Definitions: The incremental generalized force is $\Delta^n\breve{\mathbf{F}} \equiv (\nabla_{{}^t\breve{\mathbf{d}}}\, {}^n\breve{\mathbf{F}})^T\, \Delta^t\breve{\mathbf{d}}$, and the incremental generalized virtual strains: $\Delta\breve{\bar{\varepsilon}} \equiv [(\nabla_{{}^t\breve{\mathbf{d}}}\, {}^n\breve{\bar{\varepsilon}})^T\, \Delta^t\breve{\mathbf{d}}]$.

Now, noting that $\Delta^n\breve{\mathbf{F}} = {}^n\breve{\mathbf{d}}\ \Delta^n\breve{\varepsilon}$, for the generalized reactive forces in the current configuration of a linear, isotropic, homogeneous and hyperelastic material with ${}^n\mathbf{D}$ as the constitutive matrix, we can rewrite the above equation as:

$$\boxed{\nabla_{\breve{\mathbf{d}}^0} G(\lambda^0, {}^t\breve{\mathbf{d}}^0; {}^t\breve{\bar{\mathbf{d}}}) \bullet \Delta^t\breve{\mathbf{d}} = \underbrace{\int_0^L {}^n\breve{\bar{\varepsilon}}^T\, {}^n\breve{\mathbf{D}}\, \Delta^n\breve{\varepsilon}\, dS}_{\text{Material stiffness-specific}} + \underbrace{\int_0^L \Delta^n\breve{\bar{\varepsilon}}^T\, {}^n\breve{\mathbf{F}}\, dS}_{\text{Geometric stiffness-specific}}} \tag{10.392}$$

Remarks

- Note that the internal virtual functional linearization consists of two parts. The first part, $\int_0^L {}^n\breve{\bar{\varepsilon}}^T\, {}^n\breve{\mathbf{d}}\; \Delta\, {}^n\breve{\varepsilon}\; dS$, on the right-hand side of equation (10.392) involves the virtual strains and the incremental real strains derived from the linearization of the generalized reactive forces. More specifically, it is the general nature of the nonlinear formulation that the incremental and the first variations of the real strains eventually give rise to what is known as the ***material stiffness functional***, $\mathcal{K}_M$:

Definition: ***Material stiffness functional***, $\mathcal{K}_M$:

$$\boxed{\mathcal{K}_M\left({}^t\breve{\mathbf{d}},\, {}^t\breve{\bar{\mathbf{d}}};\, \Delta{}^t\breve{\mathbf{d}}\right) \equiv \int_0^L {}^n\breve{\bar{\varepsilon}}^T\; {}^n\breve{\mathbf{D}}\; \Delta\, {}^n\breve{\varepsilon}\; dS} \tag{10.393}$$

- Likewise, the second part, $\int_0^L \Delta\, {}^n\breve{\bar{\varepsilon}}^T\; {}^n\breve{\mathbf{F}}\; dS$, as a general characteristic of a geometrically nonlinear system, involves the second variation of the real strains or the first variation or increment of the virtual strains, derived from the linearization of the generalized strains and resulting in what is known as the ***geometric stiffness functional***, $\mathbf{K}_G$:

Definition: ***Geometric stiffness functional***, $\mathcal{K}_G$:

$$\boxed{\mathcal{K}_G\left({}^t\breve{\mathbf{d}},\, {}^t\breve{\bar{\mathbf{d}}};\, \Delta{}^t\breve{\mathbf{d}}\right) \equiv \int_0^L \Delta\, {}^n\breve{\bar{\varepsilon}}^T({}^t\breve{\mathbf{d}},\, {}^t\breve{\bar{\mathbf{d}}};\, \Delta{}^t\breve{\mathbf{d}})\; {}^n\breve{\mathbf{F}}\; dS} \tag{10.394}$$

- Clearly, if the real strain is linear, then the second variation (or the ***Hessian***), and thus, the second part vanishes identically; that is, $\mathcal{K}_G \equiv \mathbf{0}$. In other words, for linear beam responses, the geometric stiffness part drops out with the beam tangent stiffness, being purely due to the material stiffness contribution.

Definition: ***Tangent stiffness functional***, $\mathcal{K}_T$:

$$\boxed{\mathcal{K}_T \equiv \mathcal{K}_M + \mathcal{K}_G} \tag{10.395}$$

As indicated, if the geometric stiffness functional is absent, as in the case of a linear beam system, the tangent stiffness functional is populated solely by the material stiffness functional.

10.16.5 Inertial Virtual Functional: Covariant Linearization

Referring to equation (10.374), the inertial virtual work functional is given as:

$$G_{Inertial}({}^t\breve{\mathbf{d}};\, {}^t\breve{\bar{\mathbf{d}}}) = \int_0^L \overbrace{\{(M_0\, {}^t\ddot{\mathbf{d}} \bullet {}^t\bar{\mathbf{d}} + \mathbf{W}^T\, (\mathbf{I}_S\, {}^t\dot{\omega}_S + {}^t\omega_S \times \mathbf{A}_S) \bullet {}^t\bar{\theta}\}\, dS}^{\text{Inertial}} \tag{10.396}$$

10.16.5.1 Linearized Translational Acceleration Vector, $\Delta\, {}^t\ddot{\mathbf{d}}$

The translational acceleration vector, ${}^t\ddot{\mathbf{d}}$, is trivially linearized as $\Delta\, {}^t\ddot{\mathbf{d}}$, since it is a member of a Euclidean vector space. Now, linearizing the expression in equation (10.396), we have:

$$
\begin{aligned}
\Delta G_{Inertial}({}^t\breve{\mathbf{d}};\, {}^t\bar{\breve{\mathbf{d}}};\, \Delta\, {}^t\breve{\mathbf{d}}) &\equiv \{\nabla_{{}^t\breve{\mathbf{d}}}\, G_{Inertial}({}^t\breve{\mathbf{d}};\, {}^t\bar{\breve{\mathbf{d}}})\}^T\, \Delta\, {}^t\breve{\mathbf{d}} \\
&= \int_0^L {}^t\bar{\breve{\mathbf{d}}}^T\, \Delta\, {}^t\breve{\mathbf{F}}_{Iner}({}^t\breve{\mathbf{d}};\, \Delta\, {}^t\breve{\mathbf{d}})\, dS \\
&= \int_0^L [{}^t\bar{\mathbf{d}}^T\, (M_0\, \Delta\, {}^t\ddot{\mathbf{d}}) + {}^t\bar{\theta}^T\, (\Delta \mathbf{W}^T\, \dot{\mathbf{A}}_S + \mathbf{W}^T\, \Delta\, \dot{\mathbf{A}}_S)]\, dS
\end{aligned}
\qquad (10.397)
$$

Before we can generate the computationally efficient matrix form for the covariant linearization of the inertial virtual functional from equation (10.397), we need the following results.

10.16.5.2 Linearized Spatial Angular Momentum, $\Delta\, {}^t\mathbf{A}_S$

The skew-symmetric angular velocity matrix, $\mathbf{\Omega}_S$, is related to the rotation matrix, $\mathbf{R}$, that belongs to a curved space; hence, as before, linearization involves additional consideration as follows.

Let $\Delta\mathbf{\Psi}_S$ define the incremental spatial skew-symmetric matrix related to the linearized or incremental rotation matrix, $\Delta\mathbf{R}$, such that:

$$
\boxed{\Delta\mathbf{\Psi}_S \equiv \Delta\mathbf{R}\, \mathbf{R}^T} \qquad (10.398)
$$

whereby the axial vector, $\Delta\psi_S$, to the matrix, $\Delta\mathbf{\Psi}_S$, is related to the incremental rotation vector, $\Delta\theta$, by:

$$
\boxed{
\begin{aligned}
&\Delta\psi_S = \mathbf{W}({}^t\theta)\, \Delta\, {}^t\theta \\
&\Delta\, \dot{\psi}_S = \dot{\mathbf{W}}\, \Delta\theta + \mathbf{W}\, \Delta\, \dot{\theta} \\
&\Delta\, \ddot{\psi}_S = \ddot{\mathbf{W}}\, \Delta\theta + 2\, \dot{\mathbf{W}}\, \Delta\, \dot{\theta} + \mathbf{W}\, \Delta\, \ddot{\theta}
\end{aligned}
} \qquad (10.399)
$$

Now, starting from the relationship given by equation (10.382), and interchanging the order of the linear operations, the incremental or linearized angular velocity matrix, $\Delta\, {}^t\mathbf{\Omega}$, is given by:

$$
\boxed{
\begin{aligned}
\Delta\, {}^t\mathbf{\Omega}_S &= \Delta\, \dot{\mathbf{R}}\, \mathbf{R}^T + \dot{\mathbf{R}}\, \Delta\mathbf{R}^T = \overbrace{\Delta\mathbf{\Psi}\, \mathbf{R}}^{\bullet}\, \mathbf{R}^T + {}^t\mathbf{\Omega}_S\, \mathbf{R}\, \mathbf{R}^T\, \Delta\mathbf{\Psi}^T \\
&= \Delta\dot{\mathbf{\Psi}}\, \mathbf{R}\, \mathbf{R}^T + \Delta\mathbf{\Psi}\, \dot{\mathbf{R}}\, \mathbf{R}^T - {}^t\mathbf{\Omega}_S\, \Delta\mathbf{\Psi} \\
&= \Delta\dot{\mathbf{\Psi}} + \Delta\mathbf{\Psi}\, {}^t\mathbf{\Omega}_S\, \mathbf{R}\, \mathbf{R}^T - {}^t\mathbf{\Omega}_S\, \Delta\mathbf{\Psi} \\
&= \Delta\dot{\mathbf{\Psi}} + (\Delta\mathbf{\Psi}\, {}^t\mathbf{\Omega}_S - {}^t\mathbf{\Omega}_S\, \Delta\mathbf{\Psi})
\end{aligned}
} \qquad (10.400)
$$

where we have used the fact that $\mathbf{R}\,\mathbf{R}^T = \mathbf{I}$, with $\mathbf{I}$ as the 3×3 identity matrix. Noting the commutator or Lie-bracket property (i.e. for any two skew-symmetric matrices, $\mathbf{A}$, $\mathbf{B}$ with $\mathbf{a}$, $\mathbf{b}$ as the corresponding axial vectors in three dimensions, we have: $\mathbf{AB} - \mathbf{BA} = [\mathbf{a} \times \mathbf{b}]$ where the notation $[\mathbf{c}] \equiv \mathbf{c}\times$ for any vector $\mathbf{c}$ is implied) in equation (10.400), the incremental or the linearized angular velocity vector, $\Delta\,{}^t\boldsymbol{\omega}$, the axial vector of $\Delta\,{}^t\boldsymbol{\Omega}_S$, is given by:

$$\boxed{\Delta\,{}^t\boldsymbol{\omega}_S = \Delta\dot{\boldsymbol{\psi}}_S - {}^t\boldsymbol{\omega}_S \times \Delta\boldsymbol{\psi}_S = \Delta\dot{\boldsymbol{\psi}}_S - {}^t\boldsymbol{\Omega}_S\,\Delta\boldsymbol{\psi}_S} \tag{10.401}$$

From the definition of the inertia tensor in equation (10.381), and using equation (10.401), we get:

$$\begin{aligned}
\Delta\mathbf{A}_S &= \Delta\mathbf{I}_S\,\boldsymbol{\omega}_S + \mathbf{I}_S\,\Delta\boldsymbol{\omega}_S \\
&= (\Delta\mathbf{R}\,\mathbf{I}_M\,\mathbf{R}^T + \mathbf{R}\,\mathbf{I}_M\,\Delta\mathbf{R}^T)\,\boldsymbol{\omega}_S + \mathbf{I}_S\,(\Delta\dot{\boldsymbol{\psi}}_S - {}^t\boldsymbol{\Omega}_S\,\Delta\boldsymbol{\psi}_S) \\
&= (\Delta\boldsymbol{\Psi}_S\,\mathbf{R}\,\mathbf{I}_M\,\mathbf{R}^T + \mathbf{R}\,\mathbf{I}_M\,\mathbf{R}^T\,\Delta\boldsymbol{\Psi}_S^T)\,\boldsymbol{\omega}_S + \mathbf{I}_S\,(\Delta\dot{\boldsymbol{\psi}}_S - {}^t\boldsymbol{\Omega}_S\,\Delta\boldsymbol{\psi}_S) \\
&= (\Delta\boldsymbol{\Psi}_S\,\mathbf{I}_S - \mathbf{I}_S\,\Delta\boldsymbol{\Psi}_S)\,\boldsymbol{\omega}_S + \mathbf{I}_S\,(\Delta\dot{\boldsymbol{\psi}}_S - {}^t\boldsymbol{\Omega}_S\,\Delta\boldsymbol{\psi}_S) \\
&= \left(\cancel{\mathbf{I}_S\,\boldsymbol{\Omega}_S} - [\mathbf{A}_S] - \cancel{\mathbf{I}_S\,\boldsymbol{\Omega}_S}\right)\Delta\boldsymbol{\psi}_S + \mathbf{I}_S\,\Delta\dot{\boldsymbol{\psi}}_S = \mathbf{I}_S\,\Delta\dot{\boldsymbol{\psi}}_S - [\mathbf{A}_S]\,\Delta\boldsymbol{\psi}_S
\end{aligned} \tag{10.402}$$

where we have used the notation: $[\mathbf{a}] \equiv \mathbf{a}\times$.

10.16.5.3 Linearized Spatial Rate of Angular Momentum, $\Delta\,{}^t\dot{\mathbf{A}}_S$

Taking the time derivative of the expression in equation (10.402), and noting that: $\boxed{\dot{\mathbf{I}}_S = \boldsymbol{\Omega}_S\,\mathbf{I}_S - \mathbf{I}_S\,\boldsymbol{\Omega}_S}$, we get:

$$\begin{aligned}
\Delta\dot{\mathbf{A}}_S &= \mathbf{I}_S\,\Delta\dot{\boldsymbol{\psi}}_S - [\mathbf{A}_S]\,\Delta\boldsymbol{\psi}_S = \mathbf{I}_S\,\Delta\ddot{\boldsymbol{\psi}}_S + (\dot{\mathbf{I}}_S - [\mathbf{A}_S])\,\Delta\dot{\boldsymbol{\psi}}_S - [\dot{\mathbf{A}}_S]\,\Delta\boldsymbol{\psi}_S \\
&= \mathbf{I}_S\,\Delta\ddot{\boldsymbol{\psi}}_S + (\boldsymbol{\Omega}_S\,\mathbf{I}_S - \mathbf{I}_S\,\boldsymbol{\Omega}_S - [\mathbf{A}_S])\,\Delta\dot{\boldsymbol{\psi}}_S - [\dot{\mathbf{A}}_S]\,\Delta\boldsymbol{\psi}_S
\end{aligned} \tag{10.403}$$

Using the results in equation (10.403) in equation (10.397), we have the full linearization of the inertial virtual function in matrix form for $(\bar{\mathbf{d}}, \bar{\boldsymbol{\varphi}}; \Delta\mathbf{d}, \Delta\boldsymbol{\psi}_S)$ as:

$$\begin{aligned}
\Delta G_{Inertial}(\bar{\mathbf{d}},\ \bar{\boldsymbol{\varphi}};\Delta\mathbf{d},\ \Delta\boldsymbol{\psi}_S) &= \underbrace{\Delta G_M(\bar{\mathbf{d}},\ \bar{\boldsymbol{\varphi}};\Delta\mathbf{d},\ \Delta\boldsymbol{\psi}_S)}_{\text{Mass functional}} + \underbrace{\Delta G_G(\bar{\mathbf{d}},\ \bar{\boldsymbol{\varphi}};\Delta\mathbf{d},\ \Delta\boldsymbol{\psi}_S)}_{\text{Gyroscopic functional}} \\
&\quad + \underbrace{\Delta G_C(\bar{\mathbf{d}},\ \bar{\boldsymbol{\varphi}};\Delta\mathbf{d},\ \Delta\boldsymbol{\psi}_S)}_{\text{Centrifugal functional}}
\end{aligned} \tag{10.404}$$

where

$$\underbrace{\Delta G_M({}^t\breve{\mathbf{d}};\, {}^t\bar{\breve{\mathbf{d}}};\, \Delta\, {}^t\breve{\mathbf{d}})}_{\text{Mass functional}} = \int_0^L \begin{Bmatrix} \bar{\mathbf{d}} \\ \bar{\varphi} \end{Bmatrix} \bullet \begin{bmatrix} M_0\,\mathbf{I} & \mathbf{0} \\ \mathbf{0} & \mathbf{I}_S \end{bmatrix} \begin{Bmatrix} \Delta\ddot{\mathbf{d}} \\ \Delta\ddot{\psi}_S \end{Bmatrix} dS$$

$$\underbrace{\Delta G_G({}^t\breve{\mathbf{d}};\, {}^t\bar{\breve{\mathbf{d}}};\, \Delta\, {}^t\breve{\mathbf{d}})}_{\text{Gyroscopic functional}} = \int_0^L \begin{Bmatrix} \bar{\mathbf{d}} \\ \bar{\varphi} \end{Bmatrix} \bullet \begin{bmatrix} \mathbf{0} & \mathbf{0} \\ \mathbf{0} & \{\boldsymbol{\Omega}_S\,\mathbf{I}_S - \mathbf{I}_S\,\boldsymbol{\Omega}_S - [\mathbf{A}_S]\} \end{bmatrix} \begin{Bmatrix} \Delta\dot{\mathbf{d}} \\ \Delta\dot{\psi}_S \end{Bmatrix} dS \qquad (10.405)$$

$$\underbrace{\Delta G_C({}^t\breve{\mathbf{d}};\, {}^t\bar{\breve{\mathbf{d}}};\, \Delta\, {}^t\breve{\mathbf{d}})}_{\text{Centrifugal functional}} = \int_0^L \begin{Bmatrix} \bar{\mathbf{d}} \\ \bar{\varphi} \end{Bmatrix} \bullet \begin{bmatrix} \mathbf{0} & \mathbf{0} \\ \mathbf{0} & \{-[\dot{\mathbf{A}}_S]\} \end{bmatrix} \begin{Bmatrix} \Delta\mathbf{d} \\ \Delta\psi_S \end{Bmatrix} dS$$

Remark

- For direct computational purposes, we will be interested in modifying the expression in equations (10.404) and (10.405) in terms of $(\bar{\mathbf{d}}, \bar{\theta}; \Delta\mathbf{d}, \Delta\theta)$.

For this, we note that: $\bar{\varphi} = \mathbf{W}\,\bar{\theta}$ and $\Delta\varphi_S = \Delta\mathbf{W}\,\bar{\theta}$. Thus, we get:

$$\begin{aligned}
\Delta\dot{\mathbf{A}}_S \bullet \bar{\varphi} &= \mathbf{W}^T\ \Delta\dot{\mathbf{A}}_S \bullet \bar{\theta} \\
&= \begin{Bmatrix} \mathbf{W}^T\,\mathbf{I}_S\,(\mathbf{W}\,\Delta\,\ddot{\theta} + 2\,\dot{\mathbf{W}}\ \Delta\,\dot{\theta} + \ddot{\mathbf{W}}\ \Delta\theta) \\ +\,\mathbf{W}^T\,(\boldsymbol{\Omega}_S\,\mathbf{I}_S - \mathbf{I}_S\,\boldsymbol{\Omega}_S - [\mathbf{A}_S])(\dot{\mathbf{W}}\ \Delta\theta + \mathbf{W}\,\Delta\,\dot{\theta}) \\ -\,\mathbf{W}^T\,[\dot{\mathbf{A}}_S]\ \mathbf{W}\,\Delta\theta \end{Bmatrix} \bullet \bar{\theta} \\
&= \begin{Bmatrix} (\mathbf{W}^T\,\mathbf{I}_S\,\mathbf{W})\,\Delta\,\ddot{\theta} \\ +\,(2\mathbf{W}^T\,\mathbf{I}_S\ \dot{\mathbf{W}} + \mathbf{W}^T\,(\boldsymbol{\Omega}_S\,\mathbf{I}_S - \mathbf{I}_S\,\boldsymbol{\Omega}_S - [\mathbf{A}_S])\mathbf{W})\Delta\,\dot{\theta} \\ +\,(\mathbf{W}^T\,\mathbf{I}_S\ \ddot{\mathbf{W}} + \mathbf{W}^T\,(\boldsymbol{\Omega}_S\,\mathbf{I}_S - \mathbf{I}_S\,\boldsymbol{\Omega}_S - [\mathbf{A}_S])\ \dot{\mathbf{W}} - \mathbf{W}^T\,[\dot{\mathbf{A}}_S]\ \mathbf{W})\,\Delta\theta \end{Bmatrix} \bullet \bar{\theta}
\end{aligned} \qquad (10.406)$$

and

$$\begin{aligned}
\dot{\mathbf{A}}_S \bullet \Delta\bar{\varphi} &= \dot{\mathbf{A}}_S \bullet \Delta\mathbf{W}\ \bar{\theta} \\
&= \Delta\mathbf{W}^T\ \dot{\mathbf{A}}_S \bullet \bar{\theta} \\
&= \left\{-c_1\ \Delta\boldsymbol{\Theta} + c_2\,(\Delta\boldsymbol{\Theta}\ \boldsymbol{\Theta} + \boldsymbol{\Theta}\ \Delta\boldsymbol{\Theta}) + \left(c_3\boldsymbol{\Theta} + c_4\boldsymbol{\Theta}^2\right)(\theta \bullet \Delta\theta)\right\}\ \dot{\mathbf{A}}_S \bullet \bar{\theta} \\
&= \begin{Bmatrix} c_1\ [\dot{\mathbf{A}}_S] + (-c_2\ \mathbf{I} - c_3\ \boldsymbol{\Theta} + c_4\boldsymbol{\Theta}^2)(\dot{\mathbf{A}}_S \bullet \theta) \\ +c_2((\theta \otimes \dot{\mathbf{A}}_S) - (\dot{\mathbf{A}}_S \otimes \theta) + (\theta \bullet \dot{\mathbf{A}}_S))\mathbf{I} \end{Bmatrix} \Delta\theta \bullet\ \bar{\theta}
\end{aligned} \qquad (10.407)$$

***Exercise*:** Derive the expression in equation (10.407).

Finally, using the results in equation (10.406) and equation (10.407) in equation (10.397), we have the full and exact linearization of the inertial virtual function in matrix form in terms of $(\bar{\mathbf{d}},\ \bar{\boldsymbol{\theta}};\Delta\mathbf{d},\ \Delta\boldsymbol{\theta})$:

$$\Delta G_{Inertial}({}^t\breve{\mathbf{d}};\,{}^t\breve{\bar{\mathbf{d}}};\Delta\,{}^t\breve{\mathbf{d}}) = \underbrace{\Delta G_M({}^t\breve{\mathbf{d}};\,{}^t\breve{\bar{\mathbf{d}}};\Delta\,{}^t\breve{\mathbf{d}})}_{\text{Mass functional}} + \underbrace{\Delta G_G({}^t\breve{\mathbf{d}};\,{}^t\breve{\bar{\mathbf{d}}};\Delta\,{}^t\breve{\mathbf{d}})}_{\text{Gyroscopic functional}} + \underbrace{\Delta G_C({}^t\breve{\mathbf{d}};\,{}^t\breve{\bar{\mathbf{d}}};\Delta\,{}^t\breve{\mathbf{d}})}_{\text{Centrifugal functional}}$$

(10.408)

where

$$\underbrace{\Delta G_M({}^t\breve{\mathbf{d}};\,{}^t\breve{\bar{\mathbf{d}}};\Delta\,{}^t\breve{\mathbf{d}})}_{\text{Mass functional}} = \int_0^L \begin{Bmatrix}\bar{\mathbf{d}}\\ \bar{\boldsymbol{\theta}}\end{Bmatrix} \bullet \begin{bmatrix} M_0\,\mathbf{I} & \mathbf{0}\\ \mathbf{0} & (\mathbf{W}^T\,\mathbf{I}_S\,\mathbf{W})\end{bmatrix} \begin{Bmatrix}\Delta\ddot{\mathbf{d}}\\ \Delta\ddot{\boldsymbol{\theta}}\end{Bmatrix} dS$$

$$\underbrace{\Delta G_G({}^t\breve{\mathbf{d}};\,{}^t\breve{\bar{\mathbf{d}}};\Delta\,{}^t\breve{\mathbf{d}})}_{\text{Gyroscopic functional}} = \int_0^L \begin{Bmatrix}\bar{\mathbf{d}}\\ \bar{\boldsymbol{\theta}}\end{Bmatrix} \bullet \begin{bmatrix} \mathbf{0} & \mathbf{0}\\ \mathbf{0} & \{2\mathbf{W}^T\,\mathbf{I}_S\,\dot{\mathbf{W}} + \mathbf{W}^T\,(\boldsymbol{\Omega}_S\,\mathbf{I}_S - \mathbf{I}_S\,\boldsymbol{\Omega}_S - [\mathbf{A}_S])\mathbf{W}\}\end{bmatrix} \begin{Bmatrix}\Delta\dot{\mathbf{d}}\\ \Delta\dot{\boldsymbol{\theta}}\end{Bmatrix} dS$$

$$\underbrace{\Delta G_C({}^t\breve{\mathbf{d}};\,{}^t\breve{\bar{\mathbf{d}}};\Delta\,{}^t\breve{\mathbf{d}})}_{\text{Centrifugal functional}} = \int_0^L \begin{Bmatrix}\bar{\mathbf{d}}\\ \bar{\boldsymbol{\theta}}\end{Bmatrix} \bullet \begin{bmatrix} \mathbf{0} & \mathbf{0}\\ \mathbf{0} & \left\{\begin{matrix}\mathbf{U}(\dot{\mathbf{A}}_S) + \mathbf{W}^T\,\mathbf{I}_S\,\ddot{\mathbf{W}} + \mathbf{W}^T\,(\boldsymbol{\Omega}_S\,\mathbf{I}_S - \mathbf{I}_S\,\boldsymbol{\Omega}_S\\ -[\mathbf{A}_S])\,\dot{\mathbf{W}} - \mathbf{W}^T\,[\dot{\mathbf{A}}_S]\,\mathbf{W}\end{matrix}\right\}\end{bmatrix} \begin{Bmatrix}\Delta\mathbf{d}\\ \Delta\boldsymbol{\theta}\end{Bmatrix} dS$$

(10.409)

and from equation (10.407), we have defined operator, $\mathbf{U}$, as:

$$\begin{aligned}
&\mathbf{U}(\dot{\mathbf{A}}_S) \equiv \{c_1\,[\dot{\mathbf{A}}_S] + (-c_2\,\mathbf{I} - c_3\,\boldsymbol{\Theta} + c_4\boldsymbol{\Theta}^2)(\dot{\mathbf{A}}_S \bullet \boldsymbol{\theta}) + c_2((\boldsymbol{\theta}\otimes\dot{\mathbf{A}}_S) - (\dot{\mathbf{A}}_S\otimes\boldsymbol{\theta}) + (\boldsymbol{\theta}\bullet\dot{\mathbf{A}}_S)\mathbf{I}\},\\
&\dot{\mathbf{A}}_S \equiv \mathbf{I}_S\,\dot{\boldsymbol{\omega}}_S + \boldsymbol{\omega}_S \times \mathbf{A}_S\\
&\mathbf{A}_S \equiv \mathbf{I}_S\,\boldsymbol{\omega}_S\\
&\boldsymbol{\omega}_S \equiv \mathbf{W}\,\dot{\boldsymbol{\theta}}
\end{aligned}$$

(10.410)

where $\mathbf{W}$, $\dot{\mathbf{W}}$ and $\ddot{\mathbf{W}}$ are given by equations (10.377), (10.378) and (10.379), respectively.

10.16.6 Quasi-static External Virtual Functional: Covariant Linearization

Referring to equation (10.373), the quasi-static external virtual work functional for proportional loading is given as:

$$G_{External}(\lambda;\, {}^t\breve{\bar{\mathbf{d}}}) = \int_0^L \hat{\bar{\mathbf{d}}}^{\mathbf{T}}\, \hat{\mathbf{F}}^B(\lambda, S)\, dS + \lambda \left[\hat{\bar{\mathbf{d}}}^{\mathbf{T}}\, \hat{\mathbf{F}}^S\right]_{S=0}^{S=L} \tag{10.411}$$

Then, we have the following symbolic representation of the first-order terms:

$$\begin{aligned} \Delta G_{External}(\lambda,\, {}^t\breve{\mathbf{d}};\, {}^t\breve{\bar{\mathbf{d}}};\, \Delta\lambda) &\equiv \nabla_\lambda G_{External}(\lambda,\, {}^t\breve{\mathbf{d}};\, {}^t\breve{\bar{\mathbf{d}}})\ \Delta\lambda \\ &= \left\{\int_0^L \hat{\bar{\mathbf{d}}}^{\mathbf{T}}\, \hat{\mathbf{F}}^B{}_{,\lambda}(\lambda, S)\, dS + \left[\hat{\bar{\mathbf{d}}}^{\mathbf{T}}\, \hat{\mathbf{F}}^S\right]_{S=0}^{S=L}\right\}\ \Delta\lambda \end{aligned} \tag{10.412}$$

10.16.7 Quasi-static Linearized Virtual Work Equation

Thus, finally, using equation (10.392) and equation (10.412) in equation (10.384), we can introduce the following definition.

Definition: The linearized quasi-static virtual work equations:

$$\begin{aligned} &\overbrace{\int_0^L {}^n\breve{\bar{\boldsymbol{\varepsilon}}}^T\, {}^n\breve{\mathbf{D}}\ \Delta^n\breve{\boldsymbol{\varepsilon}}\, dS}^{\text{Material stiffness functional}} + \overbrace{\int_0^L \Delta^n\breve{\bar{\boldsymbol{\varepsilon}}}^{\,T}\, {}^n\breve{\mathbf{F}}\, dS}^{\text{Geometric stiffness functional}} \\ &= \underbrace{-\overbrace{G(\lambda^0,\, {}^t\breve{\mathbf{d}}^0;\, {}^t\breve{\bar{\mathbf{d}}})}^{\text{Unbalanced force functional}} + \overbrace{\left\{\int_0^L \hat{\bar{\mathbf{d}}}^{\mathbf{T}}\, \hat{\mathbf{F}}^B{}_{,\lambda}(\lambda, S)\, dS + \left[\hat{\bar{\mathbf{d}}}^{\mathbf{T}}\, \hat{\mathbf{F}}^S\right]_{S=0}^{S=L}\right\}\bullet\Delta\lambda}^{\text{Applied force functional}}}_{\text{Residual force functional}} \end{aligned} \tag{10.413}$$

Or,

$$\begin{aligned} &\overbrace{\overbrace{\mathcal{K}_M}^{\text{Material stiffness functional}} + \overbrace{\mathcal{K}_G}^{\text{Geometric stiffness functional}}}^{\text{Tangent functional}} \\ &= \underbrace{-\overbrace{G(\lambda^0,\, {}^t\breve{\mathbf{d}}^0;\, {}^t\breve{\bar{\mathbf{d}}})}^{\text{Unbalanced force functional}} + \overbrace{\left\{\int_0^L \hat{\bar{\mathbf{d}}}^{\mathbf{T}}\, \hat{\mathbf{F}}^B{}_{,\lambda}(\lambda, S)\, dS + \left[\hat{\bar{\mathbf{d}}}^{\mathbf{T}}\, \hat{\mathbf{F}}^S\right]_{S=0}^{S=L}\right\}\ \Delta\lambda}^{\text{Applied force functional}}}_{\text{Residual force functional}} \end{aligned} \tag{10.414}$$

where, we have, for convenience, introduced and collected for convenience the following definitions.

Definitions

- The ***unbalanced force functional***, $G(\lambda^0,\, {}^t\breve{\mathbf{d}}^0;\, {}^t\breve{\bar{\mathbf{d}}})$ is:

$$\boxed{G(\lambda^0, \breve{\mathbf{d}}^0; \breve{\bar{\mathbf{d}}}) \equiv \int_0^L ({}^n\breve{\bar{\varepsilon}}^T\; {}^n\breve{\mathbf{D}}\; {}^n\breve{\varepsilon})\, dS} \tag{10.415}$$

- The ***residual force functional*** is:

$$\boxed{\begin{aligned}\text{Residual force functional} \equiv -\; & \overbrace{G(\lambda^0,\, {}^t\breve{\mathbf{d}}^0;\, {}^t\breve{\bar{\mathbf{d}}})}^{\text{Unbalanced force functional}} \\ & + \overbrace{\left\{\int_0^L \hat{\bar{\mathbf{d}}}^{\mathbf{T}}\, \hat{\mathbf{F}}^B{}_{,\lambda}(\lambda^0, S)\, dS + \left[\hat{\bar{\mathbf{d}}}^{\mathbf{T}}\, \hat{\mathbf{F}}^S\right]_{S=0}^{S=L}\right\} \bullet \Delta\lambda}^{\text{Applied force functional}}\end{aligned}} \tag{10.416}$$

The material stiffness functional, $\mathcal{K}_M$ is:

$$\boxed{\mathcal{K}_{\mathbf{M}}({}^t\breve{\mathbf{d}},\, {}^t\breve{\bar{\mathbf{d}}};\, \Delta^t\breve{\mathbf{d}}) \equiv \int_0^L {}^n\breve{\bar{\varepsilon}}^T\, {}^n\breve{\mathbf{D}}\;\, \Delta^n\breve{\varepsilon}\, dS} \tag{10.417}$$

The geometric stiffness functional, $\mathcal{K}_G$ is:

$$\boxed{\mathcal{K}_G({}^t\breve{\mathbf{d}},\, {}^t\breve{\bar{\mathbf{d}}};\, \Delta\, {}^t\breve{\mathbf{d}}) \equiv \int_0^L \Delta^n\breve{\bar{\varepsilon}}\,{}^T({}^t\breve{\mathbf{d}},\, {}^t\breve{\bar{\mathbf{d}}};\, \Delta\, {}^t\breve{\mathbf{d}})\; {}^n\breve{\mathbf{F}}\, dS} \tag{10.418}$$

10.16.8 Dynamic Linearized Virtual Work Equation

Thus, finally, using equation (10.392) and equations (10.408)–(10.410) in equation (10.387), we can introduce and collect the following for convenience.

Definition: The linearized dynamic virtual work equations:

$$\boxed{\begin{aligned}&\overbrace{\int_0^L {}^n\breve{\bar{\varepsilon}}^T\, {}^n\breve{\mathbf{D}}\;\, \Delta^n\breve{\varepsilon}\, dS}^{\text{Material stiffness functional}} + \overbrace{\int_0^L \Delta^n\breve{\bar{\varepsilon}}\,{}^T\; {}^n\breve{\mathbf{F}}\, dS}^{\text{Geometric stiffness functional}} \\ &\overbrace{+ \underbrace{\Delta G_M({}^t\breve{\mathbf{d}};\, {}^t\breve{\bar{\mathbf{d}}};\, \Delta\, {}^t\breve{\mathbf{d}})}_{\text{Mass-specific}} + \underbrace{\Delta G_G({}^t\breve{\mathbf{d}};\, {}^t\breve{\bar{\mathbf{d}}};\, \Delta\, {}^t\breve{\mathbf{d}})}_{\text{Gyroscopic}} + \underbrace{\Delta G_C({}^t\breve{\mathbf{d}};\, {}^t\breve{\bar{\mathbf{d}}};\, \Delta\, {}^t\breve{\mathbf{d}})}_{\text{Centrifugal}}}^{\text{Inertial functional}} = \underbrace{-\; \overbrace{G({}^t\breve{\mathbf{d}}^0;\, {}^t\breve{\bar{\mathbf{d}}})}^{\text{Unbalanced force functional}}}_{\text{Residual force functional}}\end{aligned}}$$

(10.419)

or

$$\overbrace{\underbrace{\overbrace{\mathcal{K}_M}^{\text{Material stiffness functional}} + \overbrace{\mathcal{K}_G}^{\text{Geometric stiffness functional}}}_{}}^{\text{Tangent functional}} \overbrace{+ \underbrace{\Delta G_M({}^t\breve{\mathbf{d}};\, {}^t\bar{\breve{\mathbf{d}}}; \Delta\, {}^t\breve{\mathbf{d}})}_{\text{Mass-specific}} + \underbrace{\Delta G_G({}^t\breve{\mathbf{d}};\, {}^t\bar{\breve{\mathbf{d}}}; \Delta\, {}^t\breve{\mathbf{d}})}_{\text{Gyroscopic}} + \underbrace{\Delta G_C({}^t\breve{\mathbf{d}};\, {}^t\bar{\breve{\mathbf{d}}}; \Delta\, {}^t\breve{\mathbf{d}})}_{\text{Centrifugal}}}^{\text{Inertial functional}} = - \underbrace{\overbrace{G({}^t\breve{\mathbf{d}};\, {}^t\bar{\breve{\mathbf{d}}})}^{\text{Unbalanced force functional}}}_{\text{Residual force functional}} \tag{10.420}$$

where, for convenience, we collect the linearized internal and inertial terms for the following definitions.

Definitions

- The residual force functional, $G({}^t\breve{\mathbf{d}};\, {}^t\bar{\breve{\mathbf{d}}})$:

$$G(\breve{\mathbf{d}}; \bar{\breve{\mathbf{d}}}) \equiv \int_0^L ({}^n\bar{\breve{\varepsilon}}^T\, {}^n\breve{\mathbf{D}}\, {}^n\breve{\varepsilon}\,)\, dS - \left\{ \begin{array}{l} \displaystyle\int_0^L ({}^t\hat{\mathbf{F}}^B \bullet {}^t\bar{\mathbf{d}} + {}^t\hat{\mathbf{M}}^B \bullet \mathbf{W}\, {}^t\bar{\theta})\, dS \\ +[{}^t\hat{\mathbf{F}}^S \bullet {}^t\bar{\mathbf{d}} + {}^t\hat{\mathbf{M}}^S \bullet \mathbf{W}\, {}^t\bar{\theta}]_{S=0}^{S=L} \end{array} \right\} \tag{10.421}$$

- The *linearized symmetric mass functional*, $\Delta G_M({}^t\breve{\mathbf{d}};\, {}^t\bar{\breve{\mathbf{d}}}; \Delta\, {}^t\breve{\mathbf{d}})$:

$$\underbrace{\Delta G_M({}^t\breve{\mathbf{d}};\, {}^t\bar{\breve{\mathbf{d}}}; \Delta\, {}^t\breve{\mathbf{d}})}_{\text{Mass functional}} = \int_0^L \begin{Bmatrix} \bar{\mathbf{d}} \\ \bar{\theta} \end{Bmatrix} \bullet \begin{bmatrix} M_0\,\mathbf{I} & \mathbf{0} \\ \mathbf{0} & (\mathbf{W}^T\, \mathbf{I}_S\, \mathbf{W}) \end{bmatrix} \begin{Bmatrix} \Delta\ddot{\mathbf{d}} \\ \Delta\ddot{\theta} \end{Bmatrix} dS \tag{10.422}$$

- The *linearized gyroscopic (damping) functional*, $\Delta G_G({}^t\breve{\mathbf{d}};\, {}^t\bar{\breve{\mathbf{d}}}; \Delta\, {}^t\breve{\mathbf{d}})$:

$$\underbrace{\Delta G_G({}^t\breve{\mathbf{d}};\, {}^t\bar{\breve{\mathbf{d}}}; \Delta\, {}^t\breve{\mathbf{d}})}_{\text{Gyroscopic functional}} = \int_0^L \begin{Bmatrix} \bar{\mathbf{d}} \\ \bar{\theta} \end{Bmatrix} \bullet \begin{bmatrix} \mathbf{0} & \mathbf{0} \\ \mathbf{0} & \{2\mathbf{W}^T\, \mathbf{I}_S\, \dot{\mathbf{W}} + \mathbf{W}^T\, (\Omega_S\, \mathbf{I}_S - \mathbf{I}_S\, \Omega_S - [\mathbf{A}_S])\mathbf{W}\} \end{bmatrix} \begin{Bmatrix} \Delta\dot{\mathbf{d}} \\ \Delta\dot{\theta} \end{Bmatrix} dS \tag{10.423}$$

- The ***linearized centrifugal (stiffness) functional***, $\Delta G_C({}^t\breve{\mathbf{d}};\ {}^t\breve{\bar{\mathbf{d}}};\Delta\,{}^t\breve{\mathbf{d}})$:

$$\underbrace{\Delta G_C({}^t\breve{\mathbf{d}};\ {}^t\breve{\bar{\mathbf{d}}};\Delta\,{}^t\breve{\mathbf{d}})}_{\text{Centrifugal functional}} = \int_0^L \begin{Bmatrix} \bar{\mathbf{d}} \\ \bar{\theta} \end{Bmatrix} \bullet \begin{bmatrix} \mathbf{0} & \mathbf{0} \\ \mathbf{0} & \begin{Bmatrix} \mathbf{U}(\dot{\mathbf{A}}_S) + \mathbf{W}^T\,\mathbf{I}_S\,\ddot{\mathbf{W}} \\ +\mathbf{W}^T\,(\mathbf{\Omega}_S\,\mathbf{I}_S - \mathbf{I}_S\,\mathbf{\Omega}_S \\ -[\mathbf{A}_S])\,\dot{\mathbf{W}} - \mathbf{W}^T\,[\dot{\mathbf{A}}_S]\,\mathbf{W} \end{Bmatrix} \end{bmatrix} \begin{Bmatrix} \Delta\mathbf{d} \\ \Delta\theta \end{Bmatrix} dS$$

$$\mathbf{U}(\dot{\mathbf{A}}_S) \equiv \begin{Bmatrix} c_1\,[\dot{\mathbf{A}}_S] + (-c_2\,\mathbf{I} - c_3\,\mathbf{\Theta} + c_4\mathbf{\Theta}^2)(\dot{\mathbf{A}}_S\bullet\theta) \\ +c_2((\theta\otimes\dot{\mathbf{A}}_S) - (\dot{\mathbf{A}}_S\otimes\theta) + (\theta\bullet\dot{\mathbf{A}}_S))\mathbf{I} \end{Bmatrix},$$

$$\dot{\mathbf{A}}_S \equiv \mathbf{I}_S\,\dot{\omega}_S + \omega_S\times\mathbf{A}_S$$

$$\mathbf{A}_S \equiv \mathbf{I}_S\,\omega_S$$

$$\omega_S \equiv \mathbf{W}\,\dot{\theta} \tag{10.424}$$

with $[\mathbf{h}] \equiv \mathbf{h}\times$, $\forall\mathbf{h}\in\mathbb{R}_3$, and

$$\begin{aligned} \mathbf{W}({}^t\theta) &\equiv \mathbf{I} + c_1\,{}^t\mathbf{\Theta} + c_2\,{}^t\mathbf{\Theta}^2, \\ c_1({}^t\theta) &= \frac{1-\cos\theta}{\theta^2}, \\ c_2({}^t\theta) &= \frac{\theta-\sin\theta}{\theta^3} \end{aligned} \tag{10.425}$$

$$\begin{aligned} \dot{\mathbf{W}} &= c_1\,\dot{\mathbf{\Theta}} + c_2(\dot{\mathbf{\Theta}}\,\mathbf{\Theta} + \mathbf{\Theta}\,\dot{\mathbf{\Theta}}) + \dot{c}_1\,\mathbf{\Theta} + \dot{c}_2\,\mathbf{\Theta}^2 \\ &= (c_2 - c_1)(\theta\bullet\dot{\theta})\mathbf{I} + c_1\,\dot{\mathbf{\Theta}} + c_2\{(\theta\otimes\dot{\theta}) + (\dot{\theta}\otimes\theta)\} + \dot{c}_1\,\mathbf{\Theta} + \dot{c}_2(\theta\otimes\theta) \\ \dot{c}_1 &\equiv \frac{1}{\theta^2}(1 - 2c_1 - \theta^2 c_2)(\theta\bullet\dot{\theta}) = c_3(\theta\bullet\dot{\theta}),\ \dot{c}_2 \equiv \frac{1}{\theta^2}(c_1 - 3c_2)(\theta\bullet\dot{\theta}) = c_4(\theta\bullet\dot{\theta}) \end{aligned} \tag{10.426}$$

and

$$\begin{aligned} \ddot{\mathbf{W}} &= c_1\,\ddot{\mathbf{\Theta}} + 2\{\dot{c}_1\,\dot{\mathbf{\Theta}} + \dot{c}_2(\dot{\mathbf{\Theta}}\,\mathbf{\Theta} + \mathbf{\Theta}\,\dot{\mathbf{\Theta}})\} + c_2(\ddot{\mathbf{\Theta}}\,\mathbf{\Theta} + 2\,\dot{\mathbf{\Theta}}^2 + \mathbf{\Theta}\,\ddot{\mathbf{\Theta}}) + \ddot{c}_1\,\mathbf{\Theta} + \ddot{c}_2\,\mathbf{\Theta}^2 \\ \ddot{c}_1 &\equiv c_5(\theta\bullet\dot{\theta})^2 + c_3\{\dot{\theta}^2 + (\theta\bullet\ddot{\theta})\}, \qquad \ddot{c}_2 \equiv c_6(\theta\bullet\dot{\theta})^2 + c_4\{\dot{\theta}^2 + (\theta\bullet\ddot{\theta})\} \\ c_5 &\equiv -\frac{1}{\theta^2}(2c_2 + 4c_3 + \theta^2 c_4), \qquad c_6 \equiv -\frac{1}{\theta^2}(c_3 - 5c_4), \end{aligned} \tag{10.427}$$

10.16.9 Where We Would Like to Go

Clearly, the generalized form of the linearized equation (10.413) or equation (10.419) is presented in functional form and thus, still not suitable for computational purposes; that is, the expressions for the strains hide the sought-after incremental variable, $\Delta\,{}^t\breve{\mathbf{d}}$. We would like to have a form of the virtual linearized equation that is totally explicit in the incremental variables $\Delta\,{}^t\breve{\mathbf{d}} = (\Delta\,{}^t\mathbf{d} \quad \Delta\,{}^t\boldsymbol{\theta})^T$ and $\Delta\lambda$ for quasi-static cases, and $\Delta\,{}^t\breve{\mathbf{d}}$, $\Delta\,{}^t\dot{\breve{\mathbf{d}}}$ and $\Delta\,{}^t\ddot{\breve{\mathbf{d}}}$ for dynamic analyses, so that it can be easily solved digitally by a Newton-type iterative method. In other words, we would like the matrix forms of the equations that are easily solved on digital computers. For this, we need to make a few more logical stops as follows:

For matrix modification of the first part on the left side of equation (10.413) quasi-static/static loading:

- material stiffness matrix
  For matrix modification of the second part on the left side of equation (10.413)
- geometric stiffness matrix
  For matrix form of equation (10.413)
- linearized virtual incremental matrix equations
  For the matrix form of the incremental equation (10.419) for dynamic time history loading:
- finite element formulation: dynamic loading.

10.17 Material Stiffness Matrix and Symmetry

10.17.1 What We Need to Recall

- The virtual strain–displacement relations in compact matrix form are:

$$\boxed{\hat{\boldsymbol{\varepsilon}}(\hat{\mathbf{d}};\hat{\bar{\mathbf{d}}}) = \mathbf{E}(\hat{\mathbf{d}})\,\mathbf{H}(\hat{\bar{\mathbf{d}}})} \tag{10.428}$$

where

$$\boxed{\mathbf{E}(\hat{\mathbf{d}}) = \begin{bmatrix} [\mathbf{R^T}] & \mathbf{0} \\ \mathbf{0} & [\mathbf{R^T}] \end{bmatrix} \begin{bmatrix} [\mathbf{k}^0] & [\mathbf{a}]\mathbf{W} & \mathbf{I} & \mathbf{0} \\ \mathbf{0} & \mathbf{X} + [\mathbf{k}^c]\mathbf{W} & \mathbf{0} & \mathbf{W} \end{bmatrix}} \tag{10.429}$$

and $\mathbf{I}$, is the 3×3 identity matrix, the definition of $\mathbf{H}(\hat{\bar{\mathbf{d}}}) \equiv \{\bar{\mathbf{d}} \quad \bar{\boldsymbol{\theta}} \quad \bar{\mathbf{d}}_s \quad \bar{\boldsymbol{\theta}}_{,S}\}^T$, and from our knowledge of the relevant derivatives and variations, we have:

$$\boxed{\begin{aligned} c_1 &\equiv (1-\cos\theta)/\theta^2, \\ c_2 &\equiv (\theta-\sin\theta)/\theta^3, \\ \bar{c}_1 \equiv \delta c_1 \equiv c_3 &\equiv \{(1-\theta^2 c_2) - 2c_1\}/\theta^2, \\ \bar{c}_2 \equiv \delta c_2 \equiv c_4 &\equiv (c_1 - 3c_2)/\theta^2 \end{aligned}} \tag{10.430}$$

$$\boxed{\begin{aligned}&\mathbf{W} \equiv \mathbf{I} + c_1\boldsymbol{\Theta} + c_2\boldsymbol{\Theta}^2\\&\mathbf{X} \equiv -c_1\boldsymbol{\Theta}_{,s} + c_2\{[\boldsymbol{\theta} \otimes \boldsymbol{\theta}_{,s}] - [\boldsymbol{\theta}_{,s} \otimes \boldsymbol{\theta}] + (\boldsymbol{\theta} \bullet \boldsymbol{\theta}_{,s})\mathbf{I}\}\\&\quad + \mathbf{V}[\boldsymbol{\theta}_{,s} \otimes \boldsymbol{\theta}])\\&\mathbf{V} \equiv -c_2\mathbf{I} + c_3\boldsymbol{\Theta} + c_4\boldsymbol{\Theta}^2\end{aligned}} \tag{10.431}$$

- The (6×1) incremental generalized displacement state vector functions, $\Delta\hat{\mathbf{d}} \equiv (\Delta\mathbf{d} \quad \Delta\boldsymbol{\theta})^{\mathbf{T}}$.
- The symmetric constitutive matrix, ${}^n\hat{\mathbf{D}}$, is given by: ${}^n\hat{\mathbf{D}} = diag\{\mathbf{D}^F \quad \mathbf{D}^M\}$ with $\mathbf{D}^F \equiv diag\{EA \quad GA_m \quad GA_b\}$, and $\mathbf{D}^M \equiv diag\{GJ \quad GI_m \quad GI_b\}$ with E, G as the material elasticity and shear modulus, respectively. Obviously, EA is the axial stiffness, GA_m, GA_b are the cross-sectional shear stiffnesses along the principal normal and bi-normal directions, respectively, GJ is the torsional stiffness, and GI_m, GI_b are the bending stiffnesses about the normal and the bi-normal directions respectively.

10.17.2 Material Stiffness Functional Revisited: $\mathcal{K}_M$

As in the virtual expression above, we group together the incremental state variables and their derivatives with respect to the arc length parameter, S, into an augmented (12×1) incremental vector function; we can replace $\hat{\bar{\mathbf{d}}}$ by $\Delta\hat{\mathbf{d}}$ in the definition of $\mathbf{H}(\hat{\bar{\mathbf{d}}}) \equiv \{\bar{\mathbf{d}} \quad \bar{\boldsymbol{\theta}} \quad \bar{\mathbf{d}}_{,s} \quad \bar{\boldsymbol{\theta}}_{,S}\}^T$ to introduce the following definition.

Definition

- The incremental generalized displacement function is: $\mathbf{H}(\Delta\hat{\mathbf{d}}) \equiv \{\Delta\hat{\mathbf{d}} \quad \Delta\hat{\mathbf{d}}_{,S}\}^T = \{\Delta\mathbf{d} \quad \Delta\boldsymbol{\theta} \quad \Delta\mathbf{d}_{,S} \quad \Delta\boldsymbol{\theta}_{,S}\}^T$,

Now, replacing the variational variables in equation (10.428) with incremental variables with the above definition in mind, we get a similar expression.

Definition

- The incremental strain–displacement relation:

$$\boxed{\Delta^n\hat{\boldsymbol{\varepsilon}}(\hat{\mathbf{d}}; \Delta\hat{\mathbf{d}}) = \mathbf{E}(\hat{\mathbf{d}})\,\mathbf{H}(\Delta\hat{\mathbf{d}})} \tag{10.432}$$

Note that the incremental strain, $\Delta^n\hat{\boldsymbol{\varepsilon}}(\hat{\mathbf{d}}; \Delta\hat{\mathbf{d}})$, is clearly linear in the incremental generalized displacement, $\Delta\hat{\mathbf{d}}$.

By applying the virtual strain–displacement relation as in equation (10.428) and incremental strain–displacement relation of equation (10.432), we are now ready to transform equation (10.393) as:

$$\boxed{\mathcal{K}_M = \int_0^L {}^n\hat{\bar{\boldsymbol{\varepsilon}}}^T\, {}^n\hat{\mathbf{D}}\, \Delta^n\hat{\boldsymbol{\varepsilon}}\, dS = \int_0^L \mathbf{H}(\hat{\bar{\mathbf{d}}})^T\, \mathbf{E}(\hat{\mathbf{d}})^T\, {}^n\hat{\mathbf{D}}\, \mathbf{E}(\hat{\mathbf{d}})\, \mathbf{H}(\Delta\hat{\mathbf{d}})\, dS} \tag{10.433}$$

10.17.3 Material Stiffness Matrix

Based on equation (10.433), we can now redefine and present the material stiffness functional.

Definition: Material stiffness functional, $\mathcal{K}_M$:

$$\boxed{\mathcal{K}_M(\hat{\mathbf{d}}, \hat{\bar{\mathbf{d}}}; \Delta\hat{\mathbf{d}}) \equiv \int_0^L \{\mathbf{H}(\hat{\bar{\mathbf{d}}})\ \mathbf{K}(\hat{\mathbf{d}})\ \mathbf{H}(\Delta\hat{\mathbf{d}})\}\, dS} \tag{10.434}$$

where, we have introduced:

Definition: Material stiffness matrix, **K**, as:

$$\boxed{\mathbf{K} \equiv \mathbf{E}(\hat{\mathbf{d}})^T\ \hat{\mathbf{D}}^n\ \mathbf{E}(\hat{\mathbf{d}})} \tag{10.435}$$

***Exercise*:** Particularize the expression for the material stiffness, $\mathbf{K}_M$, for planar beam and motion, that is, for $\hat{\mathbf{d}} = \{d_1 \equiv u \quad d_2 \equiv w \quad d_3 \equiv 0 \quad \theta_1 \equiv 0 \quad \theta_2 \equiv 0 \quad \theta_3 \equiv \theta\}$, as:

$$\mathbf{E}^T\mathbf{D}\mathbf{E} \begin{bmatrix} (D_1s^2+D_2c^2)\kappa^2 & (D_2-D_1)\kappa^2 sc & \kappa(D_1\gamma s - D_2\epsilon c) & (D_1-D_2)\kappa sc & (D_1s^2+D_2c^2)\kappa & 0 \\ sym & (D_1c^2+D_2s^2)\kappa^2 & -\kappa(D_1\gamma c + D_2\epsilon s) & -(D_1c^2+D_2s^2)\kappa & -(D_1-D_2)\kappa sc & 0 \\ sym & sym & (D_1\gamma^2+D_2\epsilon^2) & (D_1\gamma c + D_2\epsilon s) & (D_1\gamma s - D_2\epsilon c) & 0 \\ sym & sym & sym & (D_1c^2+D_2s^2) & (D_1-D_2)sc & 0 \\ sym & sym & sym & sym & (D_1s^2+D_2c^2) & 0 \\ sym & sym & sym & sym & sym & D_3 \end{bmatrix}$$

where we have used the shorthand notations: $s \equiv \sin\theta$, $c \equiv \cos\theta$, $D_1 \equiv EA$, $D_2 \equiv GA$, $D_3 \equiv EI$, and $\gamma \equiv {}^n\gamma$, $e \equiv -(1 + {}^n\varepsilon)$.

Hint: It follows from: $\mathbf{d} = (u, w, 0)^T$, $\boldsymbol{\theta} = (0, 0, \theta)^T$ and $\mathbf{E} = \begin{bmatrix} \kappa s & -\kappa c & \gamma & c & s & 0 \\ \kappa c & \kappa s & -e & -s & c & 0 \\ 0 & 0 & 0 & 0 & 0 & 1 \end{bmatrix}$

with $\begin{Bmatrix} {}^n\varepsilon \\ {}^n\gamma \end{Bmatrix} = \begin{Bmatrix} [(1+u')c + w's - 1] + k[us - wc] \\ [-(1+u')s + w'c] + k[uc + ws] \end{Bmatrix}$ and $\mathbf{W} = \begin{bmatrix} s & -(1-c) & 0 \\ (1-c) & s & 0 \\ 0 & 0 & \theta \end{bmatrix}$ and $\boldsymbol{\kappa} = (0 \quad 0 \quad \kappa)^T$.

- Clearly, from equation (10.435), the material stiffness matrix, **K**, is symmetric in state or configuration space vector, $\hat{\mathbf{d}}$.

10.17.4 Where We Would Like to Go

For the modification of the second part of the linearized virtual work given by the following equation:

$$\boxed{\mathcal{K}_G(\hat{\mathbf{d}}, \hat{\hat{\mathbf{d}}}; \Delta\hat{\mathbf{d}}) \equiv \int_0^L \Delta\hat{\hat{\varepsilon}}^{n\ T}(\hat{\mathbf{d}}, \hat{\hat{\mathbf{d}}}; \Delta\hat{\mathbf{d}})\ \hat{\mathbf{F}}^n\, dS} \tag{10.436}$$

known also as the geometric stiffness part, we need to turn to the geometric stiffness functional part.

The material stiffness functional part expressed in equation (10.433) needs to be further modified in the subsequent finite element formulation for computational purpose. In so doing, we will be able to introduce the computational definition for beam material stiffness matrix. For this, we need to take a few more logical steps.

For finite element formulation:

- finite element formulation for quasi-static loading
- finite element formulation for dynamic loading.

  For final modification of equation (10.433) and stiffness matrix definition:
- material stiffness matrix.

  Finally, for the final computationally useful form of the linearized virtual work:
- final linearized virtual work equation.

10.18 Geometric Stiffness Matrix and Symmetry

10.18.1 What We Need to Recall

- In an expanded form, equation (10.394) may be expressed as follows:

$$\boxed{\mathcal{K}_G \equiv \overbrace{\int_0^L \Delta^n\bar{\beta} \bullet {}^n\mathbf{F}\, dS}^{\text{Translational part}} + \overbrace{\int_0^L \Delta^n\bar{\chi} \bullet {}^n\mathbf{M}\, dS}^{\text{Rotational part}}} \tag{10.437}$$

- where Δ, the incremental operator, involves the covariant derivatives of the virtual strain functions, and the superscript n reminds us that the relevant items are in rotated coordinate system, $\mathbf{n} - \mathbf{m}_n - \mathbf{b}_n$.
- $\mathbf{H}(\hat{\hat{\mathbf{d}}}) \equiv \{\bar{\mathbf{d}} \quad \bar{\theta} \quad \bar{\mathbf{d}}_{,s} \quad \bar{\theta}_{,S}\}^T$, and $\mathbf{H}(\Delta\hat{\mathbf{d}}) \equiv \{\mathbf{\Delta d} \quad \Delta\theta \quad \mathbf{\Delta d}_{,\mathbf{s}} \quad \Delta\theta_{,\mathbf{S}}\}^T$

Remark

- The expressions given in equation (10.394) or equation (10.437) involve the first variation (incremental) of the virtual strains or, equivalently, the second variation (Hessian) of the real strains. It is this second variation that renders the derivation of the geometric stiffness matrix extremely involved, and thus, tedious. However, the subsequent reward in terms of considerable computational ease and efficiency in situations of repetitive computations of the geometric stiffness, as we have in a nonlinear numerical analysis, makes it sufficiently worthwhile.

As indicated above, in view of the complexity of a lengthy derivation, and lest the readers get unduly distracted, we defer presentation of various detailed derivation; first, we write down the final result.

- Translational virtual work: linearization and symmetry
  We get as the final result:

$$\begin{aligned}\Delta\,{}^{n}\bar{\boldsymbol{\beta}}\bullet{}^{n}\mathbf{F} &= \bar{\mathbf{d}},_{s}\bullet([{}^{t}\mathbf{F}]^{\mathbf{T}}\,\mathbf{W})\Delta\theta+\Delta\mathbf{d},_{s}\bullet([{}^{t}\mathbf{F}]^{\mathbf{T}}\,\mathbf{W})\bar{\theta}\\ &\quad+\bar{\mathbf{d}}\bullet([\mathbf{k}^{0}]\,[{}^{t}\mathbf{F}]\,\mathbf{W})\Delta\theta+\Delta\mathbf{d}\bullet([\mathbf{k}^{0}]\,[{}^{t}\mathbf{F}]\,\mathbf{W})\bar{\theta}\\ &\quad+\bar{\theta}\bullet\mathbf{Z}_{\beta}^{Sym}\,\Delta\theta\\ \mathbf{Z}_{\beta}^{Sym} &= \frac{1}{2}\{\mathbf{W}^{\mathbf{T}}([{}^{t}\mathbf{F}\otimes\mathbf{a}]+[\mathbf{a}\otimes{}^{t}\mathbf{F}])\,\mathbf{W}+([\mathbf{h}^{f}\otimes\theta]+[\theta\otimes\mathbf{h}^{f}])\}\\ &\quad-({}^{t}\mathbf{F}\bullet\mathbf{a})\mathbf{W}^{\mathbf{T}}\,\mathbf{W}+c_{2}(\theta\bullet\mathbf{b}^{f})\mathbf{I}\end{aligned}\tag{10.438}$$

- Rotational virtual work: linearization and symmetry
  We get as the final result:

$$\begin{aligned}\Delta\,{}^{n}\bar{\chi}\bullet{}^{n}\mathbf{M} &= \Delta\theta\bullet\{\mathbf{W}^{\mathbf{T}}\,[{}^{t}\mathbf{M}]\,\mathbf{W}+\mathbf{Y}({}^{t}\mathbf{M})\}\bar{\theta},_{S}\\ &\quad+\bar{\theta}\bullet\{\mathbf{W}^{\mathbf{T}}\,[{}^{t}\mathbf{M}]\,\mathbf{W}+\mathbf{Y}({}^{t}\mathbf{M})\}\Delta\theta,_{S}\\ &\quad+\Delta\theta\bullet\mathbf{Z}_{\chi}^{Sym}\,\bar{\theta}\\ \mathbf{Z}_{\chi}^{Sym} &\equiv \frac{1}{2}\{\mathbf{W}^{\mathbf{T}}([{}^{t}\mathbf{M}\otimes\mathbf{k}^{c}]+[\mathbf{k}^{c}\otimes{}^{t}\mathbf{M}])\,\mathbf{W}+([\mathbf{h}^{m}\otimes\theta]+[\theta\otimes\mathbf{h}^{m}])\}\\ &\quad-({}^{t}\mathbf{M}\bullet\mathbf{k}^{c})\mathbf{W}^{\mathbf{T}}\,\mathbf{W}+c_{2}(\theta\bullet\mathbf{b}^{m})\mathbf{I}+\mathbf{W}^{\mathbf{T}}[{}^{t}\mathbf{M}]\mathbf{X}+\mathbf{X}^{\mathbf{T}}[{}^{t}\mathbf{M}]^{\mathbf{T}}\mathbf{W}\end{aligned}\tag{10.439}$$

10.18.2 *Geometric Stiffness Functional, $\mathcal{K}_G$ & Stiffness Matrix,* $\mathbf{G}(\hat{\mathbf{d}},\hat{\mathbf{F}})$

Based on equations (10.438) and (10.439), we can now redefine and present the geometric stiffness functional in its most uncluttered and explicit form compared to those published in the literature.

Definition: Geometric stiffness functional, $\mathcal{K}_G$, is:

$$\mathcal{K}_G(\hat{\mathbf{d}},\hat{\bar{\mathbf{d}}};\Delta\hat{\mathbf{d}})\equiv\int_C\{\mathbf{H}(\hat{\mathbf{d}})\;\mathbf{G}\;\mathbf{H}(\Delta\hat{\mathbf{d}})\}\,ds\tag{10.440}$$

where

Definition: Geometric stiffness matrix, $\mathbf{G}(\hat{\mathbf{d}},\hat{\mathbf{F}})$, is given as:

$$\mathbf{G}(\hat{\mathbf{d}},\hat{\mathbf{F}})\equiv\begin{bmatrix}\mathbf{0}&\mathbf{G}_{12}&\mathbf{0}&\mathbf{0}\\ \mathbf{G}_{12}^{\mathbf{T}}&\mathbf{G}_{22}&\mathbf{G}_{23}&\mathbf{G}_{24}\\ \mathbf{0}&\mathbf{G}_{23}^{\mathbf{T}}&\mathbf{0}&\mathbf{0}\\ \mathbf{0}&\mathbf{G}_{24}^{\mathbf{T}}&\mathbf{0}&\mathbf{0}\end{bmatrix}\tag{10.441}$$

where

$$\begin{aligned}
\mathbf{G}_{12} &\equiv [\mathbf{k}^0][{}^t\mathbf{F}]\,\mathbf{W}\\
\mathbf{G}_{22} &\equiv \tfrac{1}{2}\mathbf{W}^{\mathbf{T}}\{[{}^t\mathbf{F}\otimes\mathbf{a}]+[\mathbf{a}\otimes{}^t\mathbf{F}]-2({}^t\mathbf{F}\bullet\mathbf{a})\}\mathbf{W}+c_2(\boldsymbol{\theta}\bullet\mathbf{b}^f)\mathbf{I}\\
&\quad+\tfrac{1}{2}([\mathbf{h}^b\otimes\boldsymbol{\theta}]+[\boldsymbol{\theta}\otimes\mathbf{h}^b])\\
&\quad+\tfrac{1}{2}\mathbf{W}^{\mathbf{T}}\{[{}^t\mathbf{M}\otimes\mathbf{k}^c]+[\mathbf{k}^c\otimes{}^t\mathbf{M}]-2({}^t\mathbf{M}\bullet\mathbf{k}^c)\}\mathbf{W}+c_2(\boldsymbol{\theta}\bullet\mathbf{b}^m)\mathbf{I}\\
&\quad+\tfrac{1}{2}([\mathbf{h}^m\otimes\boldsymbol{\theta}]+[\boldsymbol{\theta}\otimes\mathbf{h}^m])+\mathbf{W}^{\mathbf{T}}[{}^t\mathbf{M}]\,\mathbf{X}+\mathbf{X}\,[{}^t\mathbf{M}]^{\mathbf{T}}\,\mathbf{W}\\
\mathbf{G}_{23} &\equiv \mathbf{W}^{\mathbf{T}}\,[{}^t\mathbf{F}]\\
\mathbf{G}_{24} &\equiv \mathbf{W}^{\mathbf{T}}\,[{}^t\mathbf{M}]\,\mathbf{W}+\mathbf{Y}({}^t\mathbf{M})
\end{aligned} \tag{10.442}$$

with

$$\begin{aligned}
c_1 &\equiv (1-\cos\theta)/\theta^2,\\
c_2 &\equiv (\theta-\sin\theta)/\theta^3,\\
\overline{c_1} \equiv \delta c_1 \equiv c_3 &\equiv \{(1-\theta^2 c_2)-2c_1\}/\theta^2,\\
\overline{c_2} \equiv \delta c_2 \equiv c_4 &\equiv (c_1-3c_2)/\theta^2
\end{aligned} \tag{10.443}$$

$$\begin{aligned}
\mathbf{a} &\equiv \mathbf{e}+[\mathbf{k}^0]\mathbf{d} = \{(\hat{\mathbf{1}}+\mathbf{d}_{,s})+[\mathbf{k}^0]\mathbf{d}\}\\
\mathbf{k}^t &\equiv \mathbf{W}\,\boldsymbol{\theta}_{,s}\\
\mathbf{k}^c &\equiv \mathbf{k}^0+\mathbf{k}^t\\
\mathbf{b}^f &\equiv \mathbf{F}^t\times\mathbf{a}\\
\mathbf{b}^m &\equiv \mathbf{M}^t\times\mathbf{k}^c
\end{aligned} \tag{10.444}$$

$$\begin{aligned}
\mathbf{W} &\equiv \mathbf{I}+c_1\boldsymbol{\Theta}+c_2\boldsymbol{\Theta}^2\\
\mathbf{X} &\equiv -c_1\boldsymbol{\Theta}_{,s}+c_2\{[\boldsymbol{\theta}\otimes\boldsymbol{\theta}_{,s}]-[\boldsymbol{\theta}_{,s}\otimes\boldsymbol{\theta}]+(\boldsymbol{\theta}\bullet\boldsymbol{\theta}_{,s})\mathbf{I}\}\\
&\quad+\mathbf{V}\,[\boldsymbol{\theta}_{,s}\otimes\boldsymbol{\theta}])\\
\mathbf{V} &\equiv -c_2\mathbf{I}+c_3\boldsymbol{\Theta}+c_4\boldsymbol{\Theta}^2\\
\mathbf{Y}(\mathbf{x}) &\equiv -c_1[\mathbf{x}]+c_2([\mathbf{x}\otimes\boldsymbol{\theta}]-[\boldsymbol{\theta}\otimes\mathbf{x}])+c_2(\boldsymbol{\theta}\bullet\mathbf{x})\mathbf{I}\\
&\quad+[\boldsymbol{\theta}\otimes\mathbf{V}^{\mathbf{T}}\mathbf{x}],\quad \forall\mathbf{x}\in R^3\\
\mathbf{h}^f &\equiv \mathbf{V}^{\mathbf{T}}\,\mathbf{b}^f\\
\mathbf{h}^m &\equiv \mathbf{V}^{\mathbf{T}}\,\mathbf{b}^m
\end{aligned} \tag{10.445}$$

***Exercise*:** Drawing upon the analyses referred to above for the translational and rotational geometric stiffness functional, derive the expression for **G** above. (A solution is given later.)

Remark

- Clearly, from equation (10.441), the geometric stiffness matrix, **G**, is symmetric in state or configuration space variables.

Exercise: Specialize the geometric stiffness matrix given by equation (10.442) for planar beams as:

$$\mathbf{G} = \begin{bmatrix} 0 & 0 & G_{13} & 0 & 0 & 0 \\ 0 & 0 & G_{23} & 0 & 0 & 0 \\ G_{31} & G_{32} & G_{33} & G_{34} & G_{35} & 0 \\ 0 & 0 & G_{43} & 0 & 0 & 0 \\ 0 & 0 & G_{53} & 0 & 0 & 0 \\ 0 & 0 & 0 & 0 & 0 & 0 \end{bmatrix} \text{ with } \boxed{\begin{aligned} G_{13} &\equiv G_{31} = \kappa\,{}^tN \\ G_{23} &\equiv G_{32} = \kappa\,{}^tQ \\ G_{33} &\equiv \kappa[w\,{}^tN - u\,{}^tQ] - [(1+u')\,{}^tN + w'\,{}^tQ] \\ G_{43} &\equiv G_{34} = -\,{}^tQ \\ G_{53} &\equiv G_{35} = {}^tN \end{aligned}} \tag{10.446}$$

Hint: It follows from: $\mathbf{d} = (u \quad w \quad 0)^T$, and $\boldsymbol{\theta} = (0 \quad 0 \quad \theta)^T$ implies that only the third row and column is non-zero; with $s \equiv \sin\theta,\ c \equiv \cos\theta$, we get: $c_1 = \frac{(1-c)}{\theta^2},\ c_2 = \frac{(\theta-s)}{\theta^3},\ \begin{Bmatrix} {}^n\varepsilon \\ {}^n\gamma \end{Bmatrix} = \begin{Bmatrix} [(1+u')c + w's - 1] + k[us - wc] \\ [-(1+u')s + w'c] + k[uc + ws] \end{Bmatrix}$ and $\mathbf{W} = \begin{bmatrix} s & -(1-c) & 0 \\ (1-c) & s & 0 \\ 0 & 0 & \theta \end{bmatrix}$ and $\boldsymbol{\kappa} = (0 \quad 0 \quad \kappa)^T$.

10.18.3 Symmetry: the Geometric Stiffness Matrix – Translational Part

This necessitates a collection of various vector and tensor algebraic and calculus results. Accordingly, we first break up this task into a number of pieces and finally make a summary presentation. Readers interested only in the end results can simply skip these details and inspect the geometric stiffness matrix itself explicitly given in equation (10.441) in its most simple and uncluttered form compared to those published in the literature.

10.18.3.1 Translational Virtual Work: Linearization & Symmetry

We would need several lemmas supporting the proof of the final theorem proving the symmetry of the translational part of the linearized virtual work functional, and hence the symmetry of the geometric stiffness; for these lemmas, however, we introduce few supporting mathematical relationships as facts:

Fact T1: Given the definition of c_1, c_2 and **W** in equations (10.443) and (10.445):

$$\mathbf{W}^T\,\mathbf{W} = (1 - \theta^2 c_2)^2\,\mathbf{I} + c_2\,(2 - \theta^2 c_2)\,[\boldsymbol{\theta} \otimes \boldsymbol{\theta}]$$

Proof T1. We use the fact that every matrix satisfies its characteristic polynomial. For any skew-symmetric matrix, $\boldsymbol{\Theta} \equiv [\boldsymbol{\theta}]$, the characteristic equation is given by: $\lambda^3 + (\theta_1 + \theta_2 + \theta_3)\,\lambda = 0$. Thus, $\boldsymbol{\Theta}^3 + \theta^2\,\boldsymbol{\Theta} = \mathbf{0}$. Additionally, making use of the relation: $\boldsymbol{\Theta}^2 = [\boldsymbol{\theta} \otimes \boldsymbol{\theta}] - \theta^2\,\mathbf{I}$, we have:

$$\begin{aligned}\mathbf{W}^T\,\mathbf{W} &= (\mathbf{I} - c_1\,\boldsymbol{\Theta} + c_2\,\boldsymbol{\Theta})(\mathbf{I} + c_1\,\boldsymbol{\Theta} + c_2\,\boldsymbol{\Theta})\\ &= \mathbf{I} + c_2\,(2 - \theta^2 c_2)\,\boldsymbol{\Theta}^2\\ &= (1 - \theta^2 c_2)^2\,\mathbf{I} + c_2\,(2 - \theta^2 c_2)\,[\boldsymbol{\theta} \otimes \boldsymbol{\theta}]\end{aligned}$$

which completes the proof.

Fact T2: Given: $\mathbf{U}^f \equiv [\mathbf{F} \otimes \mathbf{a}] - [\mathbf{a} \otimes \mathbf{F}]$ and $\mathbf{b}^f \equiv \mathbf{F} \times \mathbf{a} = -\mathbf{u}^f$ such that $\mathbf{U}^f\,\mathbf{x} = \mathbf{u}^f \times \mathbf{x}$, $\forall \mathbf{x} \in \mathbb{R}^3$,

$$\begin{aligned}\mathbf{W}^T\,\mathbf{U}^f\,\mathbf{W} &= (1 - \theta^2 c_2)([\mathbf{F} \otimes \mathbf{a}] - [\mathbf{a} \otimes \mathbf{F}])\\ &\quad - c_1([\boldsymbol{\theta} \otimes \mathbf{b}^f] - [\mathbf{b}^f \otimes \boldsymbol{\theta}])\\ &\quad - \{c_1 - c_2(1 - \theta^2 c_2)\}(\boldsymbol{\theta} \bullet \mathbf{b}^f)\boldsymbol{\Theta}\end{aligned}$$

Proof T2. We use properties: given two skew-symmetric matrices: $\mathbf{U}^f\boldsymbol{\Theta} - \boldsymbol{\Theta}\mathbf{U}^f = [\boldsymbol{\theta} \times \mathbf{u}^f] - [\mathbf{u}^f \times \boldsymbol{\theta}]$, $\boldsymbol{\Theta}\mathbf{U}^f\boldsymbol{\Theta} = -(\boldsymbol{\theta} \bullet \mathbf{u}^f)\boldsymbol{\Theta}$, $\boldsymbol{\Theta}^2\mathbf{U}^f\boldsymbol{\Theta}^2 = \theta^2(\boldsymbol{\theta} \bullet \mathbf{u}^f)\boldsymbol{\Theta}$, $\boldsymbol{\Theta}^2\mathbf{U}^f\boldsymbol{\Theta} - \boldsymbol{\Theta}\mathbf{U}^f\boldsymbol{\Theta}^2 = \mathbf{0}$, and $\boldsymbol{\Theta}^2\mathbf{U}^f + \mathbf{U}^f\boldsymbol{\Theta}^2 = -(\boldsymbol{\theta} \bullet \mathbf{u}^f)\boldsymbol{\Theta} - \theta^2\mathbf{U}^f$. Thus:

$$\begin{aligned}\mathbf{W}^T\,\mathbf{U}^F\,\mathbf{W} &= \mathbf{U}^f + c_1(\mathbf{U}^f\boldsymbol{\Theta} - \boldsymbol{\Theta}\mathbf{U}^f) - c_1^2\boldsymbol{\Theta}\mathbf{U}^f\boldsymbol{\Theta} + c_2^2\boldsymbol{\Theta}^2\mathbf{U}^f\boldsymbol{\Theta}^2\\ &\quad + c_1c_2(\boldsymbol{\Theta}^2\mathbf{U}^f\boldsymbol{\Theta} - \boldsymbol{\Theta}\mathbf{U}^f\boldsymbol{\Theta}^2) + c_2(\boldsymbol{\Theta}^2\mathbf{U}^f + \mathbf{U}^f\boldsymbol{\Theta}^2)\\ &= \mathbf{U}^f + c_1([\boldsymbol{\theta} \times \mathbf{u}^f] - [\mathbf{u}^f \times \boldsymbol{\theta}]) + c_1^2(\boldsymbol{\theta} \bullet \mathbf{u}^f)\boldsymbol{\Theta}\\ &\quad - c_2\{(\boldsymbol{\theta}\mathbf{u}^f)\boldsymbol{\Theta} + \theta^2\mathbf{U}^f\} + c_2^2\{\theta^2(\boldsymbol{\theta} \bullet \mathbf{u}^f)\boldsymbol{\Theta}\}\end{aligned}$$

Finally, collection of like terms completes the proof.

Fact T3: Given the definition of c_1, c_2 and $\mathbf{W}$ in equations (10.443) and (10.445):

$$\Delta\mathbf{W} = c_1\Delta\boldsymbol{\Theta} + c_2(\Delta\boldsymbol{\Theta}\,\boldsymbol{\Theta} + \boldsymbol{\Theta}\,\Delta\boldsymbol{\Theta}) + (\bar{c}_1\boldsymbol{\Theta} + \bar{c}_2\,\boldsymbol{\Theta}^2)(\boldsymbol{\theta} \bullet \Delta\boldsymbol{\theta})$$

where, $\bar{c}_1 \equiv \frac{1}{\theta^2}\{(1 - \theta^2 c_2) - 2c_1\}$ and $\bar{c}_2 \equiv \frac{1}{\theta^2}(c_1 - 3c_2)$

Proof T3. Taking incremental variations of c_1, c_2 and $\mathbf{W}$ completes the proof.

Fact T4: Given: $\mathbf{b}^f$ as before, $\boldsymbol{\Theta}[\mathbf{b}^f \otimes \boldsymbol{\theta}] = ([\mathbf{F} \otimes \mathbf{a}] - [\mathbf{a} \otimes \mathbf{F}])[\boldsymbol{\theta} \otimes \boldsymbol{\theta}]$

Proof T4.

$$\begin{aligned}\boldsymbol{\Theta}[\mathbf{b}^f \otimes \boldsymbol{\theta}]\mathbf{x} &= \{\boldsymbol{\theta} \times (\mathbf{F} \times \mathbf{a})\}(\boldsymbol{\theta} \bullet \mathbf{x}), \quad \forall \mathbf{x} \in R^3\\ &= ([\mathbf{F} \otimes \mathbf{a}] - [\mathbf{a} \otimes \mathbf{F}])[\boldsymbol{\theta} \otimes \boldsymbol{\theta}]\mathbf{x}\end{aligned}$$

This completes the proof.

Fact T5: Given: $\mathbf{b}^f$ as before, $[\boldsymbol{\Theta}\mathbf{b}^f \otimes \boldsymbol{\theta}] + [\boldsymbol{\theta} \otimes \boldsymbol{\Theta}\mathbf{b}^f] = (\boldsymbol{\theta} \bullet \mathbf{b}^f)\boldsymbol{\Theta} + \theta^2([\mathbf{F} \otimes \mathbf{a}] - [\mathbf{a} \otimes \mathbf{F}])$

Proof T5. We use identities: $\boldsymbol{\Theta}^2 = [\boldsymbol{\theta} \otimes \boldsymbol{\theta}] - \theta^2 \mathbf{I}$ and $\boldsymbol{\Theta}^2 \mathbf{U}^f + \mathbf{U}^f \boldsymbol{\Theta}^2 = -(\boldsymbol{\theta} \bullet \mathbf{u}^f)\boldsymbol{\Theta} - \theta^2 \mathbf{U}^f$ and that $\mathbf{b}^f \equiv \mathbf{F} \times \mathbf{a} = -\mathbf{u}^f$ from ***Fact T2***. Thus, first:

$$
\begin{aligned}
[\boldsymbol{\Theta}\mathbf{b}^f \otimes \boldsymbol{\theta}]\mathbf{x} &= (\boldsymbol{\theta}\mathbf{x})\{\boldsymbol{\theta} \times \mathbf{b}^f\} \qquad , \forall \mathbf{x} \in R^3 \\
&= ([\mathbf{F} \otimes \mathbf{a}] - [\mathbf{a} \otimes \mathbf{F}])[\boldsymbol{\theta} \otimes \boldsymbol{\theta}]\mathbf{x} \\
&= \mathbf{U}^f(\boldsymbol{\Theta}^2 + \theta^2 \mathbf{I})\mathbf{x}
\end{aligned}
$$

Then,

$$
\begin{aligned}
[\boldsymbol{\Theta}\mathbf{b}^f \otimes \boldsymbol{\theta}] + [\boldsymbol{\theta} \otimes \boldsymbol{\Theta}\mathbf{b}^f] &= (\mathbf{U}^f \boldsymbol{\Theta}^2 + \boldsymbol{\Theta}^2 \mathbf{U}^f + 2\theta^2 \mathbf{U}^f) \\
&= -(\boldsymbol{\theta} \bullet \mathbf{u}^f)\boldsymbol{\Theta} + \theta^2([\mathbf{F} \otimes \mathbf{a}] - [\mathbf{a} \otimes \mathbf{F}])
\end{aligned}
$$

Now, the relation: $\mathbf{b}^f = -\mathbf{u}^f$ completes the proof.

Fact T6: Given: $c_1, c_2, \bar{c}_1, \bar{c}_2, \mathbf{b}^f$ as defined before and letting: $\mathbf{h}^f \equiv [-c_2 \mathbf{I} - \bar{c}_1 \boldsymbol{\Theta} + \bar{c}_2 \boldsymbol{\Theta}^2]\mathbf{b}^f$, we have:

$$
\begin{aligned}
[\mathbf{h}^f \otimes \boldsymbol{\theta}] - [\boldsymbol{\theta} \otimes \mathbf{h}^f] &= (c_1 - 2c_2)\{[\boldsymbol{\theta} \otimes \mathbf{b}^f] - [\mathbf{b}^f \otimes \boldsymbol{\theta}]\} \\
&\quad - \bar{c}_1\{(\boldsymbol{\theta} \bullet \mathbf{b}^f)\boldsymbol{\Theta} + \theta^2([\mathbf{F} \otimes \mathbf{a}] - [\mathbf{a} \otimes \mathbf{F}])\}
\end{aligned}
$$

Proof T6. we use identities: $\boldsymbol{\Theta}^2 = [\boldsymbol{\theta} \otimes \boldsymbol{\theta}] - \theta^2 \mathbf{I}$ and ***Fact T5***. First, redefining $\mathbf{h}^f = \{-c_2 \mathbf{I} - \bar{c}_1 \boldsymbol{\Theta} + \bar{c}_2([\boldsymbol{\theta} \times \boldsymbol{\theta}] - \theta^2 \mathbf{I})\}\mathbf{b}^f = \{(2c_2 - c_1)\mathbf{b}^f - \bar{c}_1 \boldsymbol{\Theta}\mathbf{b}^f + \bar{c}_2(\boldsymbol{\theta} \bullet \mathbf{b}^f)\boldsymbol{\theta}$, we have:

$$
[\mathbf{h}^f \otimes \boldsymbol{\theta}] = \{(2c_2 - c_1)[\mathbf{b}^f \otimes \boldsymbol{\theta}] - \bar{c}_1[\boldsymbol{\Theta}\mathbf{b}^f \otimes \boldsymbol{\theta}] + \bar{c}_2(\boldsymbol{\theta} \bullet \mathbf{b}^f)[\boldsymbol{\theta} \otimes \boldsymbol{\theta}]
$$

Now, using ***Fact T5***, and noting symmetry of $[\boldsymbol{\theta} \otimes \boldsymbol{\theta}]$ completes the proof.

Fact T7: Given: $\mathbf{b}^f$ as before, $(\Delta\boldsymbol{\Theta}\,\boldsymbol{\Theta} + \boldsymbol{\Theta}\,\Delta\boldsymbol{\Theta})\mathbf{b}^f = \{(\boldsymbol{\theta} \bullet \mathbf{b}^f)\mathbf{I} + [\boldsymbol{\theta} \otimes \mathbf{b}^f] - 2[\mathbf{b}^f \otimes \boldsymbol{\theta}]\}\Delta\boldsymbol{\theta}$

Proof T7. We use the symmetry of the scalar product and the triple vector product identity indicated before. Thus,

$$
\begin{aligned}
(\Delta\boldsymbol{\Theta}\,\boldsymbol{\Theta} + \boldsymbol{\Theta}\,\Delta\boldsymbol{\Theta})\mathbf{b}^f &= \Delta\boldsymbol{\theta} \times (\boldsymbol{\theta} \times \mathbf{b}^f) + \boldsymbol{\theta} \times (\Delta\boldsymbol{\theta} \times \mathbf{b}^f) \\
&= (\mathbf{b}^f \bullet \Delta\boldsymbol{\theta})\boldsymbol{\theta} - (\boldsymbol{\theta} \bullet \Delta\boldsymbol{\theta})\mathbf{b}^f + (\boldsymbol{\theta} \bullet \mathbf{b}^f)\Delta\boldsymbol{\theta} - (\boldsymbol{\theta} \bullet \Delta\boldsymbol{\theta})\mathbf{b}^f
\end{aligned}
$$

Now, applying definition of the tensor product completes the proof.

Lemma LT1: Given ***Fact T2***, show that:

$$
\boxed{\{(\mathbf{a} \times \bar{\boldsymbol{\varphi}}) \times \Delta\boldsymbol{\phi}\} \bullet {}^t\mathbf{F} = \bar{\boldsymbol{\theta}} \bullet \mathbf{Z}_1\,\Delta\boldsymbol{\theta}} \tag{10.447}
$$

where,

$$
\boxed{\mathbf{Z}_1 \equiv \tfrac{1}{2}\{\mathbf{W}^{\mathbf{T}}([{}^t\mathbf{F} \otimes \mathbf{a}] + [\mathbf{a} \otimes {}^t\mathbf{F}])\,\mathbf{W}\} - ({}^t\mathbf{F} \bullet \mathbf{a})\mathbf{W}^{\mathbf{T}}\,\mathbf{W} + \tfrac{1}{2}\mathbf{W}^{\mathbf{T}}\mathbf{U}^f\,\mathbf{W}} \tag{10.448}
$$

Proof LT1. We use the cyclic permutation of the triple scalar product indicated earlier and the triple vector product identity: $\mathbf{a} \times (\mathbf{b} \times \mathbf{c}) = (\mathbf{a} \bullet \mathbf{c})\,\mathbf{b} - (\mathbf{a} \bullet \mathbf{b})\,\mathbf{c}$, tensor product definition: $[\mathbf{a} \otimes \mathbf{b}]\,\mathbf{x} = (\mathbf{b} \bullet \mathbf{x})\,\mathbf{a},\ \forall \mathbf{a}, \mathbf{b}, \mathbf{x} \in R^3$, and $\bar{\boldsymbol{\varphi}} = \mathbf{W}\,\bar{\boldsymbol{\theta}}$ and $\Delta\boldsymbol{\phi} = \mathbf{W}\,\Delta\boldsymbol{\theta}$. Thus,

$$\begin{aligned}\{(\mathbf{a} \times \bar{\boldsymbol{\varphi}}) \times \Delta\boldsymbol{\varphi}\} \bullet {}^t\mathbf{F}^t &= \Delta\boldsymbol{\varphi} \bullet \{({}^t\mathbf{F} \bullet \boldsymbol{\varphi})\,\mathbf{a} - ({}^t\mathbf{F} \bullet \mathbf{a})\,\bar{\boldsymbol{\varphi}})\} \\ &= \bar{\boldsymbol{\theta}} \bullet \mathbf{W}^{\mathbf{T}}\{[{}^t\mathbf{F} \otimes \mathbf{a}] - ({}^t\mathbf{F} \bullet \mathbf{a})\,\mathbf{I}\}\mathbf{W}\Delta\boldsymbol{\theta}\end{aligned}$$

Now, representing $\mathbf{W}^{\mathbf{T}}\{[{}^t\mathbf{F} \otimes \mathbf{a}] - ({}^t\mathbf{F} \bullet \mathbf{a})\,\mathbf{I}\}\mathbf{W}$ as sum of its symmetric and anti-symmetric parts and application of ***Fact T2*** completes the proof.

Exercise*:** Following the hints given, fill in the details for the proof of ***Lemma LT1.

Lemma LT2: Given $c_1, c_2, c_3, c_4, \mathbf{a}, \mathbf{b}^f$ as indicated before, ***Fact T3***, ***Fact T6*** and ***Fact T7***, show that:

$$\boxed{(\mathbf{a} \times \Delta\bar{\boldsymbol{\varphi}}) \bullet \mathbf{F}^t = \bar{\boldsymbol{\theta}} \bullet \mathbf{Z}_2\,\Delta\boldsymbol{\theta}} \tag{10.449}$$

where,

$$\boxed{\begin{aligned}\mathbf{Z}_2 \equiv\ & c_2(\boldsymbol{\theta} \bullet \mathbf{b}^f)\mathbf{I} + \tfrac{1}{2}([\mathbf{h}^f \otimes \boldsymbol{\theta}] + [\boldsymbol{\theta} \otimes \mathbf{h}^f]) - \left(c_1 + \tfrac{1}{2}\theta^2 c_3\right)([\mathbf{F} \otimes \mathbf{a}] - [\mathbf{a} \otimes \mathbf{F}]) \\ & + \tfrac{1}{2}c_1([\boldsymbol{\theta} \otimes \mathbf{b}^f] - [\mathbf{b}^f \otimes \boldsymbol{\theta}]) + \tfrac{1}{2}([\mathbf{h}^f \otimes \boldsymbol{\theta}] - [\boldsymbol{\theta} \otimes \mathbf{h}^f]) - \tfrac{1}{2}c_3(\boldsymbol{\theta} \bullet \mathbf{b}^f)\boldsymbol{\Theta}\end{aligned}} \tag{10.450}$$

Proof LT2. We use the cyclic permutation of the triple scalar product indicated earlier and the triple vector product identity: $\mathbf{a} \times (\mathbf{b} \times \mathbf{c}) = (\mathbf{a} \bullet \mathbf{c})\,\mathbf{b} - (\mathbf{a} \bullet \mathbf{b})\,\mathbf{c}$, tensor product definition: $[\mathbf{a} \otimes \mathbf{b}]\,\mathbf{x} = (\mathbf{b} \bullet \mathbf{x})\,\mathbf{a},\ \forall\,\mathbf{a}, \mathbf{b}, \mathbf{x} \in R^3$, and $\bar{\boldsymbol{\varphi}} = \mathbf{W}\,\bar{\boldsymbol{\theta}}$ and $\Delta\boldsymbol{\phi} = \mathbf{W}\,\Delta\boldsymbol{\theta}$:

$$\begin{aligned}(\mathbf{a} \times \Delta\bar{\boldsymbol{\varphi}}) \bullet \mathbf{F}^t &= \Delta\mathbf{W}\,\bar{\boldsymbol{\theta}} \bullet (\mathbf{F}^t \times \mathbf{a}) \\ &= \bar{\boldsymbol{\theta}} \bullet \{c_1(\mathbf{b}^f \times \Delta\boldsymbol{\theta}) + c_2(\Delta\boldsymbol{\Theta}\,\boldsymbol{\Theta} + \boldsymbol{\Theta}\,\Delta\boldsymbol{\Theta})\mathbf{b}^f + (-c_3\boldsymbol{\Theta} + c_4\boldsymbol{\Theta}^2)[\mathbf{b}^f \otimes \boldsymbol{\theta}]\Delta\boldsymbol{\theta} \\ &= \bar{\boldsymbol{\theta}} \bullet \{-c_1([\mathbf{F} \otimes \mathbf{a}] - [\mathbf{a} \otimes \mathbf{F}]) + c_2(\boldsymbol{\theta} \bullet \mathbf{b}^f)\mathbf{I} + c_2([\boldsymbol{\theta} \otimes \mathbf{b}^f] - [\mathbf{b}^f \otimes \boldsymbol{\theta}]) \\ &\quad +([-c_2\mathbf{I} - c_3\boldsymbol{\Theta} + c_4\boldsymbol{\Theta}^2]\mathbf{b}^f) \otimes \boldsymbol{\theta}\}\Delta\boldsymbol{\theta}\end{aligned}$$

Now recognizing the fourth term as $\mathbf{h}^f$ defined in ***Fact T6*** completes the proof.

Exercise*:** Following the hints given above, fill in the details for the proof of ***Lemma LT2.

Lemma LT3: Given ***Lemma LT1*** and ***Lemma LT2***, show that:

$$\boxed{\begin{aligned}\mathbf{Z}_{\beta}^{Sym} &= (\mathbf{Z}_1 + \mathbf{Z}_2)^{Sym} \equiv \tfrac{1}{2}\{(\mathbf{Z}_1 + \mathbf{Z}_2) + (\mathbf{Z}_1 + \mathbf{Z}_2)^T\} \\ &= \tfrac{1}{2}\{\mathbf{W}^{\mathbf{T}}([\mathbf{F} \otimes \mathbf{a}] + [\mathbf{a} \otimes \mathbf{F}])\,\mathbf{W} + ([\mathbf{h}^f \otimes \boldsymbol{\theta}] + [\boldsymbol{\theta} \otimes \mathbf{h}^f])\} \\ &\quad - (\mathbf{F} \bullet \mathbf{a})\mathbf{W}^{\mathbf{T}}\,\mathbf{W} + c_2(\boldsymbol{\theta} \bullet \mathbf{b}^f)\mathbf{I}\end{aligned}} \tag{10.451}$$

with $\mathbf{Z}^{Ant} = (\mathbf{Z}_1 + \mathbf{Z}_2)^{Ant} \equiv \frac{1}{2}\{(\mathbf{Z}_1 + \mathbf{Z}_2) - (\mathbf{Z}_1 + \mathbf{Z}_2)^T\}$ defined as under vanishes identically, that is:

$$\boxed{\begin{aligned}\mathbf{Z}^{Ant} &\equiv \frac{1}{2}\mathbf{W}^{\mathbf{T}}([\mathbf{F}\otimes\mathbf{a}] - [\mathbf{a}\otimes\mathbf{F}])\,\mathbf{W} - c_1([\mathbf{F}\otimes\mathbf{a}] - [\mathbf{a}\otimes\mathbf{F}])\\ &\quad + c_2([\boldsymbol{\theta}\otimes\mathbf{b}^f] - [\mathbf{b}^f\otimes\boldsymbol{\theta}]) + \frac{1}{2}([\mathbf{h}^f\otimes\boldsymbol{\theta}] - [\boldsymbol{\theta}\otimes\mathbf{h}^f])\\ &= \mathbf{0}\end{aligned}} \tag{10.452}$$

Proof LT3. For $\mathbf{Z}^{Ant}$, using ***Fact T2*** in the second term of $\mathbf{Z}_1$, we have:

$$\begin{aligned}\mathbf{Z}^{Ant} &= \frac{1}{2}\{(1-\theta^2 c_2 - 2c_1) - \theta^2\bar{c}_1\}([\mathbf{F}\otimes\mathbf{a}] - [\mathbf{a}\otimes\mathbf{F}])\\ &\quad + \frac{1}{2}(c_1 - c_1)([\boldsymbol{\theta}\otimes\mathbf{b}^f] - [\mathbf{b}^f\otimes\boldsymbol{\theta}]) - \frac{1}{2}\{c_1^2 - c_2(1-\theta^2 c_2) + \bar{c}_1\}(\boldsymbol{\theta}\bullet\mathbf{b}^f)\boldsymbol{\Theta}\end{aligned}$$

Applying the definitions of $c_1, c_2, \bar{c}_1 \equiv \delta c_1, \bar{c}_2 \equiv \delta c_2$, it is seen that all the coefficients in the above expression vanish identically. This completes the proof.

Exercise: Following the hints given above, fill in the details for the proof of ***Lemma LT3***.

Theorem TT1: Given, in the deformed coordinates, the translational virtual strain, ${}^n\bar{\boldsymbol{\beta}}$, and the conjugate force, ${}^n\mathbf{F}$, we have:

$$\boxed{\begin{aligned}\Delta\,{}^n\bar{\boldsymbol{\beta}}\bullet{}^n\mathbf{F} &= \bar{\mathbf{d}}_{,S}\bullet([{}^t\mathbf{F}]^{\mathbf{T}}\,\mathbf{W})\Delta\boldsymbol{\theta} + \Delta\mathbf{d}_{,S}\bullet([{}^t\mathbf{F}]^{\mathbf{T}}\,\mathbf{W})\bar{\boldsymbol{\theta}}\\ &\quad + \bar{\mathbf{d}}\bullet\;([\mathbf{k}^0]\,[{}^t\mathbf{F}]\,\mathbf{W})\Delta\boldsymbol{\theta} + \Delta\mathbf{d}\bullet\;([\mathbf{k}^0]\,[{}^t\mathbf{F}]\,\mathbf{W})\bar{\boldsymbol{\theta}}\\ &\quad + \bar{\boldsymbol{\theta}}\bullet\;\mathbf{Z}_{\boldsymbol{\beta}}^{Sym}\,\Delta\boldsymbol{\theta}\end{aligned}} \tag{10.453}$$

where $\mathbf{Z}_{\boldsymbol{\beta}}^{Sym}$, the symmetric part of $\mathbf{Z}_{\boldsymbol{\beta}}$ defined in equation (10.453) is given by:

$$\boxed{\begin{aligned}\mathbf{Z}_{\boldsymbol{\beta}}^{Sym} &\equiv \frac{1}{2}\mathbf{W}^T\{[{}^t\mathbf{F}\otimes\mathbf{a}] - [\mathbf{a}\otimes{}^t\mathbf{F}] - 2({}^t\mathbf{F}\bullet\mathbf{a})\}\,\mathbf{W} + \frac{1}{2}([\mathbf{h}^f\otimes\boldsymbol{\theta}] + [\boldsymbol{\theta}\otimes\mathbf{h}^f])\\ &\quad + c_2(\boldsymbol{\theta}\bullet\mathbf{b}^f)\mathbf{I}\end{aligned}} \tag{10.454}$$

and for completeness:

$$\boxed{\begin{aligned}&\mathbf{a} \equiv \mathbf{e} + [\mathbf{k}^0]\mathbf{d} = \{(\hat{\mathbf{1}} + \mathbf{d}_{,S}) + [\mathbf{k}^0]\mathbf{d}\},\\ &\mathbf{b}^f \equiv {}^n\mathbf{F}\times\mathbf{a},\\ &c_1 \equiv (1-\cos\theta)/\theta^2, \qquad c_2 \equiv (\theta - \sin\theta)/\theta^3,\\ &\mathbf{W} \equiv \mathbf{I} + c_1\boldsymbol{\Theta} + c_2\boldsymbol{\Theta}^2\\ &\bar{c}_1 \equiv \delta c_1 = \{(1-\theta^2 c_2) - 2c_1\}/\theta^2, \quad \bar{c}_2 \equiv \delta c_2 = (c_1 - 3c_2)/\theta^2,\\ &\mathbf{V} \equiv -c_2\mathbf{I} + \bar{c}_1\boldsymbol{\Theta} + \bar{c}_2\boldsymbol{\Theta}^2, \qquad \mathbf{h}^f \equiv \mathbf{V}^{\mathbf{T}}\,\mathbf{b}^f\end{aligned}} \tag{10.455}$$

Proof TT1. Using the definition of the virtual translational strain, ${}^{n}\bar{\boldsymbol{\beta}}$, the equation relating reactive forces in the undeformed and the deformed coordinates, the cyclic permutation of the triple scalar product, and the relations: $\bar{\boldsymbol{\varphi}} = \mathbf{W}\,\bar{\boldsymbol{\theta}}$ and $\Delta\boldsymbol{\phi} = \mathbf{W}\,\Delta\boldsymbol{\theta}$, we have:

$$\begin{aligned}
\Delta\,{}^{n}\bar{\boldsymbol{\beta}} \bullet {}^{n}\mathbf{F} &= \bar{\mathbf{d}}_{,S} \bullet ([{}^{t}\mathbf{F}]^{\mathrm{T}}\,\mathbf{W})\Delta\boldsymbol{\theta} + \Delta\mathbf{d}_{,S} \bullet ([{}^{t}\mathbf{F}]^{\mathrm{T}}\,\mathbf{W})\bar{\boldsymbol{\theta}} \\
&\quad + \bar{\mathbf{d}} \bullet ([\mathbf{k}^{0}]\,[{}^{t}\mathbf{F}]\,\mathbf{W})\Delta\boldsymbol{\theta} + \Delta\mathbf{d} \bullet ([\mathbf{k}^{0}]\,[{}^{t}\mathbf{F}]\,\mathbf{W})\bar{\boldsymbol{\theta}} \\
&\quad + {}^{t}\mathbf{F} \bullet \{(\mathbf{a}\times\bar{\boldsymbol{\varphi}})\times\Delta\boldsymbol{\varphi}\} + {}^{t}\mathbf{F} \bullet (\mathbf{a}\times\Delta\bar{\boldsymbol{\varphi}})
\end{aligned}$$

Now application of equations (10.447) and (10.449) with $\mathbf{Z}_{\beta}$ given by equation (10.451) for the sum of the last two parts of the above expression completes the proof.

Exercise*:** Following the hints given above, fill in the details for the proof of ***Theorem TT1.

10.18.3.2 Translational Contribution to the Geometric Stiffness Matrix

Note that the entire expression in the equation (10.453) is a function of the state displacement vectors, and is linear in its first variation and incremental counterparts, and thus amenable to matrix representation. Thus, we can write:

$$\boxed{\Delta\,{}^{n}\bar{\boldsymbol{\beta}} \bullet {}^{n}\mathbf{F} = \{\bar{\mathbf{d}} \quad \bar{\boldsymbol{\theta}} \quad \bar{\mathbf{d}}_{,s} \quad \bar{\boldsymbol{\theta}}_{,S}\}\begin{bmatrix} \mathbf{0} & [\mathbf{k}^{0}]\,[{}^{t}\mathbf{F}]\,\mathbf{W} & \mathbf{0} & \mathbf{0} \\ Sym & \mathbf{Z}_{\beta}^{Sym} & \mathbf{W}^{\mathrm{T}}[{}^{t}\mathbf{F}] & \mathbf{0} \\ \mathbf{0} & \mathbf{0} & \mathbf{0} & \mathbf{0} \\ \mathbf{0} & Sym & \mathbf{0} & \mathbf{0} \end{bmatrix}\begin{Bmatrix} \Delta\mathbf{d} \\ \Delta\boldsymbol{\theta} \\ \Delta\mathbf{d}_{,s} \\ \Delta\boldsymbol{\theta}_{,S} \end{Bmatrix}} \tag{10.456}$$

Or,

Introducing our definition of the generalized state vectors: $\mathbf{H}(\hat{\bar{\mathbf{d}}}) \equiv \{\bar{\mathbf{d}} \quad \bar{\boldsymbol{\theta}} \quad \bar{\mathbf{d}}_{,S} \quad \bar{\boldsymbol{\theta}}_{,S}\}^{T}$, and $\mathbf{H}(\Delta\hat{\mathbf{d}}) \equiv \{\Delta\mathbf{d} \quad \Delta\boldsymbol{\theta} \quad \Delta\mathbf{d}_{,S} \quad \Delta\boldsymbol{\theta}_{,S}\}^{T}$, and defining the matrix above as $\mathbf{G}^{Tran}$, the translational matrix contribution to the geometric stiffness is as follows.

Definition:

$$\boxed{\mathbf{G}^{Tran} \equiv \begin{bmatrix} \mathbf{0} & [\mathbf{k}^{0}]\,[{}^{t}\mathbf{F}]\,\mathbf{W} & \mathbf{0} & \mathbf{0} \\ Sym & \mathbf{Z}_{\beta}^{Sym} & \mathbf{W}^{\mathrm{T}}[{}^{t}\mathbf{F}] & \mathbf{0} \\ \mathbf{0} & \mathbf{0} & \mathbf{0} & \mathbf{0} \\ \mathbf{0} & Sym & \mathbf{0} & \mathbf{0} \end{bmatrix}} \tag{10.457}$$

We can have the final compact matrix representation as:

$$\boxed{\Delta\,{}^{n}\bar{\boldsymbol{\beta}}\ {}^{n}\mathbf{F} = \mathbf{H}(\hat{\bar{\mathbf{d}}})^{T}\ \mathbf{G}^{Tran}\ \mathbf{H}(\Delta\hat{\mathbf{d}})} \tag{10.458}$$

Clearly, the translational geometric stiffness matrix, $\mathbf{G}^{Tran}$, must be added to its counterpart, $\mathbf{G}^{Rot}$, the rotational geometric stiffness to obtain, finally, the complete geometric stiffness of the beam. Next, we construct the most computationally efficient symmetric form of the rotational geometric stiffness.

10.18.4 Symmetry: the Geometric Stiffness Matrix – Rotational Part

This turns out to be the most involved analysis needing a collection of various vector and tensor algebraic and calculus results. Accordingly, we first break this up into a number of pieces and finally make a summary presentation. Readers interested only in the end results can simply skip these details and inspect the geometric stiffness matrix itself explicitly given in equation (10.441) in its most simple and uncluttered form compared to those published in the literature.

10.18.4.1 Rotational Virtual Work: Linearization and Symmetry

We would need several lemmas supporting the proof of the final theorem proving the symmetry of the rotational part of the linearized virtual work functional, and hence the symmetry of the geometric stiffness. For these lemmas, however, we introduce a few supporting mathematical relationships as facts:

Fact R1: Given the definition of $\mathbf{W}$ in equation (10.445):

$$\Delta\mathbf{W}\,\theta_{,S} = \mathbf{X}\,\Delta\theta \tag{10.459}$$

Proof R1. Replacing $\overline{()}$ quantities by Δ in the central lemma completes the proof.

Fact R2: Given $c_2, \bar{c}_1, \bar{c}_2$ as before and $\mathbf{V} \equiv -c_2\mathbf{I} + \bar{c}_1\boldsymbol{\Theta} + \bar{c}_2\boldsymbol{\Theta}^2$, we have:

$$[\theta \otimes \Delta\theta_{,S}]\mathbf{V}^{\mathrm{T}}\,{}^t\mathbf{M} = [\theta \otimes {}^t\mathbf{M}]\mathbf{V}\,\Delta\theta_{,S} \tag{10.460}$$

Proof R2. For $\forall \mathbf{x} \in \mathbb{R}^3$,

$$\mathbf{x} \bullet [\theta \otimes \Delta\theta_{,S}]\mathbf{V}^{\mathrm{T}}\,{}^t\mathbf{M} = (\theta \bullet \mathbf{x})\,{}^t\mathbf{M} \bullet \mathbf{V}\,\Delta\theta_{,S}$$

An application of the tensor product definition completes the proof.

Fact R3: Given c_1, c_2 and $\bar{c}_1, \bar{c}_2$ in equation (10.443),

$$\mathbf{M} \bullet \bar{\mathbf{W}}\,\Delta\theta_{,S} = \bar{\theta} \bullet \mathbf{Y}(\mathbf{M})\,\Delta\theta_{,S} \tag{10.461}$$

where, for $\forall \mathbf{x} \in \mathbb{R}^3$,

$$\mathbf{Y}(\mathbf{x}) \equiv -c_1[\mathbf{x}] + c_2([\mathbf{x} \otimes \theta] - [\theta \otimes \mathbf{x}]) + c_2(\theta \bullet \mathbf{x})\mathbf{I} + [\theta \otimes \mathbf{V}^{\mathrm{T}}\mathbf{x}] \tag{10.462}$$

Proof R3.

$$\begin{aligned} {}^t\mathbf{M} \bullet \bar{\mathbf{W}}\,\Delta\theta_{,S} = \bar{\theta} \bullet \{&c_1\,\Delta\boldsymbol{\Theta}_{,S} + c_2([\Delta\theta_{,S} \otimes \theta] - [\theta \otimes \Delta\theta_{,S}]) \\ &+ c_2(\Delta\theta_{,S} \bullet \theta)\mathbf{I} + [\theta \otimes \Delta\theta_{,S}](-c_2\mathbf{I} - \bar{c}_1\boldsymbol{\Theta} + \bar{c}_2\boldsymbol{\Theta}^2)\}\,{}^t\mathbf{M} \end{aligned}$$

Replacing the last term by the result of ***Fact R2*** and noting that for $[\theta \otimes \mathbf{x}]\,\mathbf{V} = [\theta \otimes \mathbf{V}^{\mathrm{T}}\mathbf{x}]$ completes the proof.

Fact R4: Given: $c_1, c_2, \bar{c}_1, \bar{c}_2$ as defined before, and $\mathbf{k}^c \equiv \mathbf{k}^0 + \mathbf{k}$, $\mathbf{b}^m \equiv {}^t\mathbf{M} \times \mathbf{k}^c$, and $\mathbf{h}^m \equiv \mathbf{V}^{\mathrm{T}}\,\mathbf{b}^m$, we have:

$$[\mathbf{h}^m \otimes \theta] - [\theta \otimes \mathbf{h}^m] = (c_1 - 2c_2)\{[\theta \otimes \mathbf{b}^m] - [\mathbf{b}^m \otimes \theta]\} \\ - \bar{c}_1\{(\theta \bullet \mathbf{b}^m)\Theta + \theta^2([{}^t\mathbf{M} \otimes \mathbf{k}^c] - [\mathbf{k}^c \otimes {}^t\mathbf{M}])\}$$

Proof R4. A similar set of arguments as in ***Fact T6*** with $\mathbf{h}^b, \mathbf{b}, {}^t\mathbf{F}, \mathbf{a}$ replaced by $\mathbf{h}^m, \mathbf{b}^m, {}^t\mathbf{M}$ and $\mathbf{k}^c$, respectively, completes the proof.

Fact R5: Given $\mathbf{c}^t \equiv {}^t\mathbf{M} \times \mathbf{k}$ and $\mathbf{h}^m$ as in ***Fact R4***, we have:

$$\mathbf{M} \bullet (\mathbf{W}\,\theta,_S \times \Delta\mathbf{W}\,\bar{\theta}) = \Delta\theta \bullet \mathbf{Y}(\mathbf{c}^t)\,\bar{\theta} \tag{10.463}$$

Proof R5. An application of the cyclic permutation property of the scalar triple product, and noting that $\mathbf{k} = \mathbf{W}\,\theta,_S$ and $\Delta\mathbf{W}$ as given by equation (10.459), we have:

$$\begin{aligned}\mathbf{M} \bullet\ (\mathbf{W}\,\theta,_S \times \Delta\mathbf{W}\,\bar{\theta}) &= \mathbf{c}^t \bullet\ \Delta\mathbf{W}\,\bar{\theta} \\ &= \bar{\theta} \bullet\ \{-c_1\Delta\Theta + c_2([\Delta\theta \otimes \theta] - [\theta \otimes \Delta\theta]) \\ &\qquad + c_2\{(\Delta\theta \bullet\ \theta)\mathbf{I} + [\theta \otimes \Delta\theta]\mathbf{V}^{\mathrm{T}})\mathbf{c}^t \\ &= \bar{\theta} \bullet\ \{-c_1[\mathbf{c}^t] + c_2([\mathbf{c}^t \otimes \theta] - [\theta \otimes \mathbf{c}^t]) \\ &\qquad + c_2((\mathbf{c}^t \bullet \theta)\mathbf{I} + [\theta \otimes \mathbf{V}^{\mathrm{T}}\mathbf{c}^t])\}\Delta\theta\end{aligned}$$

Now, recalling the definition of $\mathbf{Y}(\mathbf{x})$ from equation (10.462) for $\mathbf{x} = \mathbf{c}^t$ completes the proof.

Fact R6: Given: $\mathbf{U}^m \equiv [{}^t\mathbf{M} \otimes \mathbf{k}^c] - [\mathbf{k}^c \otimes {}^t\mathbf{M}]$ and $\mathbf{b}^m$ as in ***Fact R4***, we have:

$$\mathbf{W}^T\,\mathbf{U}^m\,\mathbf{W} = (1 - \theta^2 c_2)([{}^t\mathbf{M} \otimes \mathbf{k}^c] - [\mathbf{k}^c \otimes {}^t\mathbf{M}]) \\ - c_1([\theta \otimes \mathbf{b}^m] - [\mathbf{b}^m \otimes \theta]) - \bar{c}_1(\theta \bullet\ \mathbf{b}^m)\Theta \tag{10.464}$$

Proof R6. A similar set of arguments as in ***Fact T2*** with $\mathbf{U}^f$, ${}^t\mathbf{F}$, $\mathbf{a}$, $\mathbf{b}^f$ replaced by $\mathbf{U}^m$, ${}^t\mathbf{M}$, $\mathbf{k}^c$ and $\mathbf{b}^m$ completes the proof.

Lemma LR1: Given ***Fact R5***, we have:

$$\boxed{\{(\bar{\varphi},_S + \mathbf{k}^0 \times \bar{\varphi}) \times \Delta\varphi\} \bullet {}^t\mathbf{M} = \Delta\theta \bullet \mathbf{Z}_3\bar{\theta} + \Delta\theta \bullet (\mathbf{W}^{\mathrm{T}}[\mathbf{M}^t]\mathbf{W})\bar{\theta},_S} \tag{10.465}$$

where,

$$\boxed{\begin{aligned}\mathbf{Z}_3 \equiv \tfrac{1}{2}\{\mathbf{W}^{\mathrm{T}}([{}^t\mathbf{M} \otimes \mathbf{k}^c] + \mathbf{k}^c \otimes {}^t\mathbf{M})\mathbf{W}\} - ({}^t\mathbf{M} \bullet \mathbf{k}^c)\mathbf{W}^{\mathrm{T}}\mathbf{W} \\ + \tfrac{1}{2}(\mathbf{W}^{\mathrm{T}}\mathbf{U}^m\mathbf{W})^{\mathrm{T}} + \mathbf{W}^{\mathrm{T}}[{}^t\mathbf{M}]\mathbf{X}\end{aligned}} \tag{10.466}$$

Proof LR1. We use the cyclic permutation of the triple scalar product, the triple vector product identity and the tensor product definition as indicated before, and $\bar{\varphi},_S =$

$\bar{\mathbf{W}}\,\theta,_s + \mathbf{W}\,\bar{\theta},_s + \mathbf{W}\,\theta,_s \times \mathbf{W}\,\bar{\theta}$ and $\Delta\phi = \mathbf{W}\,\Delta\theta$ from our discussion on derivatives and variations. Thus,

$$\begin{aligned}\{(\bar{\varphi},_s + \mathbf{k}^0 \times \bar{\varphi}) \times \Delta\phi\} \bullet {}^t\mathbf{M} &= \Delta\varphi \bullet \{[{}^t\mathbf{M}]\,(\bar{\varphi},_s + [\mathbf{k}^0]\bar{\varphi})\} \\ &= \Delta\theta \bullet \{(\mathbf{W}^T\,[{}^t\mathbf{M}]\,\bar{\mathbf{W}}\bar{\theta},_s) + (\mathbf{W}^T\,[{}^t\mathbf{M}]\,\mathbf{W})\bar{\theta},_s \\ &\quad + \mathbf{W}^T({}^t\mathbf{M} \times (\mathbf{W}\theta,_s \times \mathbf{W}\bar{\theta})) + \mathbf{W}^T\,[{}^t\mathbf{M}]\,[\mathbf{k}^0]\mathbf{W}\}\end{aligned}$$

We use: $\mathbf{W}^T\,\{{}^t\mathbf{M} \times (\mathbf{W}\theta,_s \times \mathbf{W}\bar{\theta})\} = \mathbf{W}^T\,\{[{}^t\mathbf{M} \otimes \mathbf{k}]^\mathbf{T} - ({}^t\mathbf{M} \bullet\ \mathbf{k})\mathbf{I}\}\mathbf{W}\bar{\theta}$ and represent the same as the sum of its symmetric and anti-symmetric parts, and use: $\bar{\mathbf{W}}\,\theta,_s = \mathbf{X}\,\bar{\theta},_s$ from the equation in the ***Central Lemma LD3***, and note the following identities for any two skew-symmetric matrices:

$$[\mathbf{k}^0]\,[{}^t\mathbf{M}] + [{}^t\mathbf{M}]\,[\mathbf{k}^0] = [{}^t\mathbf{M} \otimes \mathbf{k}^0] + [\mathbf{k}^0 \otimes {}^t\mathbf{M}] - 2({}^t\mathbf{M} \bullet\ \mathbf{k}^0)\mathbf{I}$$

and

$$[\mathbf{k}^0]\,[{}^t\mathbf{M}] - [{}^t\mathbf{M}]\,[\mathbf{k}^0] = [{}^t\mathbf{M} \otimes \mathbf{k}^0] - [\mathbf{k}^0 \otimes {}^t\mathbf{M}]$$

Now, with $\mathbf{k}^c$ defined as before in ***Fact R4***, and identification of $\mathbf{U}^m$ from ***Fact R6*** completes the proof.

Exercise: Following the hints given, fill in the details for the proof of ***Lemma LR1***.

Lemma LR2: Given $c_1, c_2, c_3, c_4, \mathbf{k}^c, \mathbf{b}^m$ as indicated before, and ***Fact T3***, ***Fact R3*** and ***Fact R4***,

$$\boxed{\begin{aligned}\{\Delta\bar{\varphi},_s + \mathbf{k}^0 \times \Delta\bar{\varphi}\} \bullet {}^t\mathbf{M} &= \Delta\theta \bullet \mathbf{Z}_4\,\bar{\theta} \\ &\quad + \bar{\theta} \bullet \{\mathbf{W}^T[{}^t\mathbf{M}]\mathbf{W} + [\mathbf{Y}({}^t\mathbf{M})]\}\Delta\theta,_s \\ &\quad + \bar{\theta},_s \bullet \{[\mathbf{Y}({}^t\mathbf{M})]^T\}\Delta\theta\end{aligned}} \tag{10.467}$$

where,

$$\boxed{\begin{aligned}\mathbf{Z}_4 \equiv\ & c_2(\theta \bullet \mathbf{b}^m)\mathbf{I} + \tfrac{1}{2}([\mathbf{h}^m \otimes \theta] + [\theta \otimes \mathbf{h}^m]) \\ & + \left(c_1 + \tfrac{1}{2}\theta^2 c_3\right)([{}^t\mathbf{M} \otimes \mathbf{k}^c] - [\mathbf{k}^c \otimes {}^t\mathbf{M}]) \\ & - \tfrac{1}{2}c_1([\theta \otimes \mathbf{b}^m] - [\mathbf{b}^m \otimes \theta]) + \tfrac{1}{2}c_3(\theta \bullet \mathbf{b}^m)\mathbf{\Theta}\end{aligned}} \tag{10.468}$$

Proof LR2. We rewrite $\mathbf{Y}(\mathbf{b}^m)$ in ***Fact R3*** with $\mathbf{h}^m$ as defined in ***Fact R4***, and note that:

$$\begin{aligned}\mathbf{Y}(\mathbf{b}^m) &\equiv \mathbf{Y}(\mathbf{b}^m)^{sym} + \mathbf{Y}(\mathbf{b}^m)^{ant} \\ \mathbf{Y}(\mathbf{b}^m)^{sym} &= \tfrac{1}{2}([\mathbf{h}^m \otimes \theta] + [\theta \otimes \mathbf{h}^m]) + c_2(\theta \bullet\ \mathbf{b}^m)\mathbf{I} \\ \mathbf{Y}(\mathbf{b}^m)^{ant} &= \left(c_1 + \tfrac{1}{2}\theta^2 c_3\right)([{}^t\mathbf{M} \otimes \mathbf{k}^c] - [\mathbf{k}^c \otimes {}^t\mathbf{M}]) \\ &\quad + \left(\tfrac{1}{2}c_1\right)([\mathbf{b}^m \otimes \theta] - [\theta \otimes \mathbf{b}^m]) \\ &\quad + \left(\tfrac{1}{2}c_3\right)(\theta \bullet\ \mathbf{b}^m)\mathbf{\Theta}\end{aligned} \tag{10.469}$$

Using the cyclic permutation of the triple scalar product and the triple vector product identity, tensor product definition as indicated before, and the relations: $\bar{\varphi} = \mathbf{W}\,\bar{\theta}$, $\Delta\varphi = \mathbf{W}\,\Delta\theta$, and $\mathbf{k} = \mathbf{W}\,\theta_{,S}$, $\bar{\varphi}_{,S} = \bar{\mathbf{k}} + \mathbf{k}\times\bar{\varphi}$, $\Delta\bar{\varphi}_{,S} = \Delta\bar{\mathbf{k}} + \Delta\mathbf{k}\times\bar{\varphi} + \mathbf{k}\times\Delta\bar{\varphi}$, with $\bar{\mathbf{k}} = \bar{\mathbf{W}}\,\theta_{,S} + \mathbf{W}\,\bar{\theta}_{,S}$ and $\Delta\bar{\mathbf{k}} = \bar{\mathbf{W}}\,\Delta\theta_{,S} + \Delta\mathbf{W}\,\bar{\theta}_{,S}$ we have:

$$\begin{aligned}\{\Delta\bar{\varphi}_{,S} + \mathbf{k}^0\times\Delta\bar{\varphi}\}\bullet {}^t\mathbf{M}^t &= \{(\Delta\mathbf{W}\,\theta_{,S} + \mathbf{W}\,\Delta\theta_{,S})\times\mathbf{W}\,\bar{\theta}\\ &\quad + \bar{\mathbf{W}}\,\Delta\theta_{,S} + (\mathbf{W}\,\theta_{,S})\times\Delta\mathbf{W}\,\bar{\theta} + \Delta\mathbf{W}\,\bar{\theta}_{,S}\}\\ &= \bar{\theta}\bullet\{\mathbf{W}^{\mathbf{T}}[{}^t\mathbf{M}]\mathbf{W} + \mathbf{Y}({}^t\mathbf{M})\}\Delta\theta_{,S} + \bar{\theta}_{,S}\bullet\{\mathbf{Y}({}^t\mathbf{M})^{\mathbf{T}}\}\Delta\theta\\ &\quad + \Delta\theta\bullet\{\mathbf{X}^{\mathbf{T}}[{}^t\mathbf{M}]^{\mathbf{T}}\mathbf{W} + \mathbf{Y}(\mathbf{b}^m)\}\bar{\theta}\end{aligned}$$

Now, representing the right-hand side as sum of its symmetric and anti-symmetric parts completes the proof.

Exercise: Following the hints given, fill in the details for the proof of ***Lemma LR2***.

Lemma LR3: Given ***Lemma LR1*** and ***Lemma LR2***: we have:

$$\boxed{\begin{aligned}\mathbf{Z}_{\mathrm{X}}^{Sym} &\equiv (\mathbf{Z}_3 + \mathbf{Z}_4)^{Sym}\\ &= \tfrac{1}{2}\{\mathbf{W}^T([{}^t\mathbf{M}\otimes\mathbf{k}^c] + [\mathbf{k}^c\otimes{}^t\mathbf{M}])\,\mathbf{W} + ([\mathbf{h}^m\otimes\theta] + [\theta\otimes\mathbf{h}^m])\}\\ &\quad - ({}^t\mathbf{M}\bullet\,\mathbf{k}^c)\mathbf{W}^T\,\mathbf{W} + c_2(\theta\bullet\,\mathbf{b}^m)\mathbf{I} + \mathbf{W}^T[{}^t\mathbf{M}]\mathbf{X} + \mathbf{X}^T[{}^t\mathbf{M}]^T\mathbf{W}\end{aligned}} \tag{10.470}$$

with $\mathbf{Z}^{ant}$ from $\mathbf{Z}_3$ and $\mathbf{Z}_4$ defined as under vanishes identically, that is:

$$\boxed{\mathbf{Z}^{Ant} \equiv \tfrac{1}{2}\mathbf{W}^T(\mathbf{U}^m)^T\,\mathbf{W} + \mathbf{Y}(\mathbf{b}^m)^{Ant} = \mathbf{0}} \tag{10.471}$$

where, $\mathbf{U}^m$ and $\mathbf{Y}(\mathbf{b}^m)^{ant}$ are as defined in Fact R6 and equation (10.469), respectively.

Proof LR3. For $\mathbf{Z}^{ant}$, using Fact R6 for the first term, we have:

$$\begin{aligned}\mathbf{Z}^{Ant} &= \tfrac{1}{2}\{2c_1 - (1-\theta^2 c_2) + \theta^2 c_3\}([{}^t\mathbf{M}\otimes\mathbf{k}^c] - [\mathbf{k}^c\otimes{}^t\mathbf{M}])\\ &\quad + \tfrac{1}{2}(c_1 - c_1)([\theta\otimes\mathbf{b}^m] - [\mathbf{b}^m\otimes\,\theta]) - \tfrac{1}{2}(c_3 - c_3)(\theta\bullet\,\mathbf{b}^m)\boldsymbol{\Theta}\end{aligned}$$

Applying the definitions of $c_1, c_2, \bar{c}_1 \equiv \delta c_1, \bar{c}_2 \equiv \delta c_2$, it is easily seen that all the coefficients in the above expression vanish identically. Then, annulment of $\mathbf{Z}^{ant}$ terms from $\mathbf{Z}_3$ and $\mathbf{Z}_4$ completes the proof.

Exercise: Following the hints given, fill in the details for the proof of ***Lemma LR3***.

Theorem TR1: Given the rotational virtual strain, ${}^n\bar{\chi}$, and the conjugate moment, ${}^n\mathbf{M}$, we have:

$$\boxed{\begin{aligned}{}^n\Delta\bar{\chi}\bullet{}^n\mathbf{M} &= \Delta\theta\bullet\{\mathbf{W}^T\,[{}^t\mathbf{M}]\,\mathbf{W} + \mathbf{Y}({}^t\mathbf{M})\}\bar{\theta}_{,S}\\ &\quad + \bar{\theta}\bullet\{\mathbf{W}^{\mathbf{T}}\,[{}^t\mathbf{M}]\,\mathbf{W} + \mathbf{Y}({}^t\mathbf{M})\}\Delta\theta_{,S}\\ &\quad + \Delta\theta\bullet\,\mathbf{Z}_{\mathrm{X}}^{Sym}\,\bar{\theta}\end{aligned}} \tag{10.472}$$

where $\mathbf{Z}_{\mathrm{X}}^{Sym}$ is defined as in equation (10.470).

Proof TR1. Using equation (10.283) defining ${}^n\chi$ from virtual bending strain definition, the equation relating moments in tangent and normal coordinates, the cyclic permutation of the triple scalar product, and the relations: $\bar{\boldsymbol{\varphi}} = \mathbf{W}\,\bar{\boldsymbol{\theta}}$, $\Delta\boldsymbol{\phi} = \mathbf{W}\,\Delta\boldsymbol{\theta}$, and $\mathbf{k} = \mathbf{W}\,\boldsymbol{\theta},_S$, $\bar{\boldsymbol{\varphi}},_s = \bar{\mathbf{k}} + \mathbf{k}\times\bar{\boldsymbol{\varphi}}$, $\Delta\bar{\boldsymbol{\varphi}},_S = \Delta\bar{\mathbf{k}} + \Delta\mathbf{k}\times\bar{\boldsymbol{\varphi}} + \mathbf{k}\times\Delta\bar{\boldsymbol{\varphi}}$, $\bar{\mathbf{k}} = \bar{\mathbf{W}}\,\boldsymbol{\theta},_S + \mathbf{W}\,\bar{\boldsymbol{\theta}},_S$ and $\Delta\bar{\mathbf{k}} = \bar{\mathbf{W}}\,\Delta\boldsymbol{\theta},_S + \Delta\mathbf{W}\,\bar{\boldsymbol{\theta}},_S$ we have: $\Delta\,{}^n\bar{\chi}\bullet{}^n\mathbf{M} = \{(\bar{\boldsymbol{\varphi}},_s + \mathbf{k}^0\times\bar{\boldsymbol{\varphi}})\times\Delta\boldsymbol{\phi}\}\bullet{}^t\mathbf{M} + \{\Delta\bar{\boldsymbol{\varphi}},_S + \mathbf{k}^0\times\Delta\bar{\boldsymbol{\varphi}}\}\bullet{}^t\mathbf{M}$. Now application of equations (10.465) and (10.467) with $\mathbf{Z}_\chi$ given by the equation (10.470) completes the proof.

Exercise*:** Following the hints given, fill in the details for the proof of ***Theorem TR1.

10.18.4.2 Rotational Geometric Stiffness Matrix

It may be remarked now that the entire expression in equation (10.472) is a function of the state vectors, and linear in its first variation and incremental counterparts, and thus amenable to matrix representation. Thus, we can write:

$$\boxed{\Delta\,{}^n\bar{\chi}\bullet{}^n\mathbf{M} = \{\bar{\mathbf{d}}\quad\bar{\boldsymbol{\theta}}\quad\bar{\mathbf{d}},_s\quad\bar{\boldsymbol{\theta}},_S\}\begin{bmatrix}\mathbf{0} & \mathbf{0} & \mathbf{0} & \mathbf{0}\\ \mathbf{0} & \mathbf{Z}_\chi^{Sym} & \mathbf{0} & \mathbf{W}^T[{}^t\mathbf{M}]\,\mathbf{W} + \mathbf{Y}({}^t\mathbf{M})\\ \mathbf{0} & \mathbf{0} & \mathbf{0} & \mathbf{0}\\ \mathbf{0} & Sym & \mathbf{0} & \mathbf{0}\end{bmatrix}\begin{Bmatrix}\Delta\mathbf{d}\\ \Delta\boldsymbol{\theta}\\ \Delta\mathbf{d},_s\\ \Delta\boldsymbol{\theta},_S\end{Bmatrix}} \tag{10.473}$$

Introducing our definition of the generalized state vectors: $\mathbf{H}(\hat{\bar{\mathbf{d}}}) \equiv \{\bar{\mathbf{d}}\quad\bar{\boldsymbol{\theta}}\quad\bar{\mathbf{d}},_s\quad\bar{\boldsymbol{\theta}},_S\}^T$, and $\mathbf{H}(\Delta\hat{\mathbf{d}}) \equiv \{\Delta\mathbf{d}\quad\Delta\boldsymbol{\theta}\quad\Delta\mathbf{d},_s\quad\Delta\boldsymbol{\theta},_S\}^T$, and defining the matrix above as $\mathbf{G}^{Rot}$, the translational matrix contribution to the geometric stiffness as follows.

Definition

$$\boxed{\mathbf{G}^{Rot} \equiv \begin{bmatrix}\mathbf{0} & \mathbf{0} & \mathbf{0} & \mathbf{0}\\ \mathbf{0} & \mathbf{Z}_\chi^{Sym} & \mathbf{0} & \mathbf{W}^T[{}^t\mathbf{M}]\,\mathbf{W} + \mathbf{Y}({}^t\mathbf{M})\\ \mathbf{0} & \mathbf{0} & \mathbf{0} & \mathbf{0}\\ \mathbf{0} & Sym & \mathbf{0} & \mathbf{0}\end{bmatrix}} \tag{10.474}$$

We can have the final compact matrix representation as:

$$\boxed{\Delta\,{}^n\bar{\chi}\bullet{}^n\mathbf{M} = \mathbf{H}(\hat{\bar{\mathbf{d}}})^T\,\mathbf{G}^{Rot}\,\mathbf{H}(\Delta\hat{\mathbf{d}})} \tag{10.475}$$

Clearly, the rotational geometric stiffness matrix, $\mathbf{G}^{Rot}$, must be added to its counterpart, $\mathbf{G}^{Tran}$, the translational geometric stiffness matrix to obtain, finally, the geometric stiffness matrix of the beam.

10.18.5 Geometric Stiffness Matrix with Distributed Moment Loading

For 3D beams, distributed external moment loading affects the symmetry of the geometric stiffness matrix. Noting that $\hat{\mathbf{M}} \bullet \bar{\varphi} = \hat{\mathbf{M}} \bullet \mathbf{W}\bar{\theta}$, we get as before in equation (10.445): $\Delta({}^t\hat{\mathbf{M}} \bullet \bar{\varphi}) = \Delta\mathbf{W}^T\, {}^t\hat{\mathbf{M}} \bullet \bar{\theta} = \bar{\theta} \bullet \mathbf{Y}^T({}^t\hat{\mathbf{M}})\, \Delta\theta$ with

$$\mathbf{Y}({}^t\hat{\mathbf{M}}) \equiv -c_1[{}^t\hat{\mathbf{M}}] + c_2([{}^t\hat{\mathbf{M}} \otimes \theta] - [\theta \otimes {}^t\hat{\mathbf{M}}]) + c_2(\theta \bullet {}^t\hat{\mathbf{M}})\mathbf{I} + [\theta \otimes \mathbf{V}^{\mathrm{T}}\, {}^t\hat{\mathbf{M}}] \tag{10.476}$$

where

$$\mathbf{V} \equiv -c_2\mathbf{I} + c_3\mathbf{\Theta} + c_4\mathbf{\Theta}^2$$

Clearly, $\mathbf{Y}^T(\hat{\mathbf{M}})$ is weakly unsymmetric. Thus, for distributed moment loading, the geometric stiffness matrix for a 3D beam is non-symmetric.

***Exercise*:** Investigate the effect of distributed external moment loading on the symmetry of the geometric stiffness matrix, $\mathbf{G}$, for planar motion, that is, for $\hat{\mathbf{d}} = \{d_1 \equiv u \quad d_2 \equiv w \quad d_3 \equiv 0 \quad \theta_1 \equiv 0 \quad \theta_2 \equiv 0 \quad \theta_3 \equiv \theta\}$.

***Solution*:** With $\theta = (0 \quad 0 \quad \theta)^T$ and ${}^t\hat{\mathbf{M}} = (0 \quad 0 \quad {}^tM)^T$, we get: $\mathbf{Y}({}^t\hat{\mathbf{M}}) \equiv -c_1[{}^t\hat{\mathbf{M}}] + c_2([{}^t\hat{\mathbf{M}} \otimes \theta] - [\theta \otimes {}^t\hat{\mathbf{M}}]) + c_2(\theta \bullet {}^t\hat{\mathbf{M}})\mathbf{I} + [\theta \otimes \mathbf{V}^{\mathrm{T}}\, {}^t\hat{\mathbf{M}}] = \begin{bmatrix} c_2\, {}^t\hat{\mathbf{M}}\theta & -c_1\, {}^t\hat{\mathbf{M}} & 0 \\ c_1\, {}^t\hat{\mathbf{M}} & c_2\, {}^t\hat{\mathbf{M}}\theta & 0 \\ 0 & 0 & 0 \end{bmatrix}$ implying $\bar{\theta} \bullet \mathbf{Y}({}^t\hat{\mathbf{M}})\, \Delta\theta = 0$. Thus, external moments in 2D have no effect on the symmetry of the geometric stiffness unlike in 3D beams.

10.18.6 Where We Would Like to Go

For the modification of the second part of linearized virtual work given by the following equation:

$$\mathcal{K}_M(\hat{\mathbf{d}}, \hat{\hat{\mathbf{d}}}; \Delta\hat{\mathbf{d}}) \equiv \int_0^L {}^n\hat{\bar{\varepsilon}}^T\, {}^n\hat{\mathbf{D}}\, \Delta\hat{\varepsilon}^n\, dS \tag{10.477}$$

known also as the material stiffness part, we need to turn to:

- material stiffness functional part
  The geometric stiffness functional part, expressed in equation (10.441), needs to be further modified in the subsequent finite element formulation for computational purposes. In so doing, we will be able to introduce the computational definition for beam geometric stiffness matrix. For this, we need to take a few more logical steps as under:
  For finite element formulations, we should look up:
- quasi-static finite element formulation
- dynamic finite element formulation

10.19 c-type FE Formulation: Dynamic Loading

10.19.1 What We Need to Recall

- The virtual work functional, $G_{Dyn}({}^t\mathbf{d}, {}^t\theta; {}^t\bar{\mathbf{d}}, {}^t\overleftarrow{\theta})$, in compact generalized notation:

$$\boxed{\begin{aligned}
&G_{Dyn}({}^t\breve{\mathbf{d}};\ {}^t\breve{\bar{\mathbf{d}}}) \equiv G_{Internal} + G_{Inertial} - G_{External} = \mathbf{0}, \quad \forall \text{ admissible } {}^t\breve{\bar{\mathbf{d}}}\\
&G_{Internal} = \int_0^L ({}^n\breve{\bar{\varepsilon}}\ {}^n\breve{\mathbf{F}})\, dS\\
&G_{Inertial} = \int_0^L ({}^t\breve{\bar{\mathbf{d}}}^T\ {}^t\breve{\mathbf{F}}_{Iner})\, dS\\
&\qquad = \overbrace{\int_0^L \left\{(M_0\ {}^t\ddot{\mathbf{d}} \bullet {}^t\bar{\mathbf{d}} + \mathbf{W}^T\ (\mathbf{I}_S\ {}^t\dot{\omega}_S + {}^t\omega_S \times \mathbf{A}_S)\ {}^t\bar{\theta}\right\} dS}^{\text{Inertial}}\\
&G_{External} = \left\{\int_0^L \breve{\bar{\mathbf{d}}}^{\mathbf{T}}\ \breve{\mathbf{F}}^B\, dS + \left[\breve{\bar{\mathbf{d}}}^{\mathbf{T}}\ \breve{\mathbf{F}}^S\right]_{S=0}^{S=L}\right\}
\end{aligned}} \tag{10.478}$$

with the applicable initial conditions on the dynamic state variables as:

$$\boxed{\begin{aligned}
&\breve{\mathbf{d}}(0) = \breve{\mathbf{d}}_0.\\
&{}^t\dot{\breve{\mathbf{d}}}(0) = \breve{\mathbf{v}}_0,
\end{aligned}} \tag{10.479}$$

the linearized dynamic virtual work equations: at a time $t \in \mathbb{R}_+$

$$\boxed{\begin{aligned}
&\overbrace{\int_0^L {}^n\breve{\bar{\varepsilon}}^T\ {}^n\breve{\mathbf{D}}\ \Delta^n\breve{\varepsilon}\, dS}^{\text{Material stiffness functional}} + \overbrace{\int_0^L \Delta^n\breve{\bar{\varepsilon}}^{\ T}\ {}^n\breve{\mathbf{F}}\, dS}^{\text{Geometric stiffness functional}}\\
&+ \overbrace{\underbrace{\Delta G_M({}^t\breve{\mathbf{d}};\ {}^t\breve{\bar{\mathbf{d}}}; \Delta\, {}^t\breve{\mathbf{d}})}_{\text{Mass-specific}} + \underbrace{\Delta G_G({}^t\breve{\mathbf{d}};\ {}^t\breve{\bar{\mathbf{d}}}; \Delta\, {}^t\breve{\mathbf{d}})}_{\text{Gyroscopic}} + \underbrace{\Delta G_C({}^t\breve{\mathbf{d}};\ {}^t\breve{\bar{\mathbf{d}}}; \Delta\, {}^t\breve{\mathbf{d}})}_{\text{Centrifugal}}}^{\text{Inertial functional}} = \underbrace{-\overbrace{G({}^t\breve{\mathbf{d}};\ {}^t\breve{\bar{\mathbf{d}}})}^{\text{Unbalanced force functional}}}_{\text{Residual force functional}}
\end{aligned}}$$

(10.480)

Or,

$$\overbrace{\overbrace{\mathcal{K}_M}^{\text{Material stiffness functional}} + \overbrace{\mathcal{K}_G}^{\text{Geometric stiffness functional}}}^{\text{Tangent functional}}$$
$$\overbrace{+\underbrace{\Delta G_M({}^t\breve{\mathbf{d}};\, {}^t\bar{\breve{\mathbf{d}}}; \Delta\, {}^t\breve{\mathbf{d}})}_{\text{Mass-specific}} + \underbrace{\Delta G_G({}^t\breve{\mathbf{d}};\, {}^t\bar{\breve{\mathbf{d}}}; \Delta\, {}^t\breve{\mathbf{d}})}_{\text{Gyroscopic}} + \underbrace{\Delta G_C({}^t\breve{\mathbf{d}};\, {}^t\bar{\breve{\mathbf{d}}}; \Delta\, {}^t\breve{\mathbf{d}})}_{\text{Centrifugal}}}^{\text{Inertial functional}} = -\underbrace{\overbrace{G({}^t\breve{\mathbf{d}};\, {}^t\bar{\breve{\mathbf{d}}})}^{\text{Unbalanced force functional}}}_{\text{Residual force functional}} \tag{10.481}$$

where the initial conditions on the generalized displacement and velocity are specified as:

$$\breve{\mathbf{d}}(0) \equiv \begin{Bmatrix} {}^t\mathbf{d}(0) \\ {}^t\theta(0) \end{Bmatrix} = \hat{\breve{\mathbf{d}}}_0 \equiv \begin{Bmatrix} {}^t\mathbf{d}_0 \\ {}^t\theta_0 \end{Bmatrix} \tag{10.482}$$

where, for convenience, we collect the linearized inertial terms for the following.

Definitions

- Definition: Material stiffness functional, $\mathcal{K}_M$:

$$\boxed{\mathcal{K}_M(\breve{\mathbf{d}},\ \bar{\breve{\mathbf{d}}};\ \Delta\breve{\mathbf{d}}) \equiv \int_C \{\mathbf{H}(\breve{\mathbf{d}})\ \mathbf{K}(\breve{\mathbf{d}})\ \mathbf{H}(\Delta\breve{\mathbf{d}})\}\ dS} \tag{10.483}$$

with the material stiffness matrix, $\mathbf{K}$, is given by:

$$\boxed{\mathbf{K} = \mathbf{E}(\breve{\mathbf{d}})^T\ {}^n\breve{\mathbf{D}}\ \mathbf{E}(\breve{\mathbf{d}})} \tag{10.484}$$

where,

$$\boxed{\mathbf{E}(\breve{\mathbf{d}}) = \begin{bmatrix} [\mathbf{R}^\mathbf{T}] & \mathbf{0} \\ \mathbf{0} & [\mathbf{R}^\mathbf{T}] \end{bmatrix} \begin{bmatrix} [\mathbf{k}^0] & [\mathbf{a}]\mathbf{W} & \mathbf{I} & \mathbf{0} \\ \mathbf{0} & \mathbf{X} + [\mathbf{k}^c]\mathbf{W} & \mathbf{0} & \mathbf{W} \end{bmatrix}} \tag{10.485}$$

and

Definition

- Geometric stiffness functional, $\mathcal{K}_G$:

$$\boxed{\mathcal{K}_G(\breve{\mathbf{d}},\ \bar{\breve{\mathbf{d}}};\ \Delta\breve{\mathbf{d}}) \equiv \int_C \{\mathbf{H}(\breve{\mathbf{d}})\ \mathbf{G}(\breve{\mathbf{d}}, \breve{\mathbf{F}})\ \mathbf{H}(\Delta\breve{\mathbf{d}})\}\ dS} \tag{10.486}$$

- The linearized inertial functional, $\Delta G_{Inertial}({}^t\breve{\mathbf{d}};\, {}^t\breve{\bar{\mathbf{d}}}; \Delta\, {}^t\breve{\mathbf{d}})$:

$$\boxed{\Delta G_{Inertial}({}^t\breve{\mathbf{d}};\, {}^t\breve{\bar{\mathbf{d}}}; \Delta\, {}^t\breve{\mathbf{d}}) = \underbrace{\Delta G_M({}^t\breve{\mathbf{d}};\, {}^t\breve{\bar{\mathbf{d}}}; \Delta\, {}^t\breve{\mathbf{d}})}_{\text{Mass functional}} + \underbrace{\Delta G_G({}^t\breve{\mathbf{d}};\, {}^t\breve{\bar{\mathbf{d}}}; \Delta\, {}^t\breve{\mathbf{d}})}_{\text{Gyroscopic functional}} + \underbrace{\Delta G_C({}^t\breve{\mathbf{d}};\, {}^t\breve{\bar{\mathbf{d}}}; \Delta\, {}^t\breve{\mathbf{d}})}_{\text{Centrifugal functional}}} \tag{10.487}$$

- The linearized mass functional, $\Delta G_M({}^t\breve{\mathbf{d}};\, {}^t\breve{\bar{\mathbf{d}}}; \Delta\, {}^t\breve{\mathbf{d}})$:

$$\boxed{\underbrace{\Delta G_M({}^t\breve{\mathbf{d}};\, {}^t\breve{\bar{\mathbf{d}}}; \Delta\, {}^t\breve{\mathbf{d}})}_{\text{Mass functional}} = \int_0^L \begin{Bmatrix} \bar{\mathbf{d}} \\ \bar{\theta} \end{Bmatrix} \bullet \begin{bmatrix} M_0\, \mathbf{I} & \mathbf{0} \\ \mathbf{0} & (\mathbf{W}^T\, \mathbf{I}_S\, \mathbf{W}) \end{bmatrix} \begin{Bmatrix} \Delta\ddot{\mathbf{d}} \\ \Delta\ddot{\theta} \end{Bmatrix} dS} \tag{10.488}$$

- The linearized gyroscopic functional, $\Delta G_G({}^t\breve{\mathbf{d}};\, {}^t\breve{\bar{\mathbf{d}}}; \Delta\, {}^t\breve{\mathbf{d}})$:

$$\boxed{\underbrace{\Delta G_G({}^t\breve{\mathbf{d}};\, {}^t\breve{\bar{\mathbf{d}}}; \Delta\, {}^t\breve{\mathbf{d}})}_{\text{Gyroscopic functional}} = \int_0^L \begin{Bmatrix} \bar{\mathbf{d}} \\ \bar{\theta} \end{Bmatrix} \bullet \begin{bmatrix} \mathbf{0} & \mathbf{0} \\ \mathbf{0} & \{2\mathbf{W}^T\, \mathbf{I}_S\, \dot{\mathbf{W}} + \mathbf{W}^T\, (\mathbf{\Omega}_S\, \mathbf{I}_S - \mathbf{I}_S\, \mathbf{\Omega}_S - [\mathbf{A}_S])\mathbf{W}\} \end{bmatrix} \begin{Bmatrix} \Delta\dot{\mathbf{d}} \\ \Delta\dot{\theta} \end{Bmatrix} dS} \tag{10.489}$$

- The linearized centrifugal functional, $\Delta G_C({}^t\breve{\mathbf{d}};\, {}^t\breve{\bar{\mathbf{d}}}; \Delta\, {}^t\breve{\mathbf{d}})$:

$$\boxed{\begin{aligned}
&\underbrace{\Delta G_C({}^t\breve{\mathbf{d}};\, {}^t\breve{\bar{\mathbf{d}}}; \Delta\, {}^t\breve{\mathbf{d}})}_{\text{Centrifugal functional}} = \int_0^L \begin{Bmatrix} \bar{\mathbf{d}} \\ \bar{\theta} \end{Bmatrix} \bullet \begin{bmatrix} \mathbf{0} & \mathbf{0} \\ \mathbf{0} & \begin{Bmatrix} \mathbf{U}(\dot{\mathbf{A}}_S) + \mathbf{W}^T\, \mathbf{I}_S\, \ddot{\mathbf{W}} + \mathbf{W}^T\, (\mathbf{\Omega}_S\, \mathbf{I}_S - \mathbf{I}_S\, \mathbf{\Omega}_S \\ -[\mathbf{A}_S])\, \dot{\mathbf{W}} - \mathbf{W}^T\, [\dot{\mathbf{A}}_S]\, \mathbf{W} \end{Bmatrix} \end{bmatrix} \begin{Bmatrix} \Delta\mathbf{d} \\ \Delta\theta \end{Bmatrix} dS \\
&\mathbf{U}(\dot{\mathbf{A}}_S) \equiv \begin{Bmatrix} c_1\, [\dot{\mathbf{A}}_S] + (-c_2\, \mathbf{I} - c_3\, \mathbf{\Theta} + c_4 \mathbf{\Theta}^2)(\dot{\mathbf{A}}_S \bullet \theta) \\ +c_2((\theta \otimes \dot{\mathbf{A}}_S) - (\dot{\mathbf{A}}_S \otimes \theta) + (\theta \bullet \dot{\mathbf{A}}_S))\mathbf{I} \end{Bmatrix}, \\
&\dot{\mathbf{A}}_S \equiv \mathbf{I}_S\, \dot{\omega}_S + \omega_S \times \mathbf{A}_S, \qquad \mathbf{A}_S \equiv \mathbf{I}_S\, \omega_S, \qquad \omega_S \equiv \mathbf{W}\, \dot{\theta}
\end{aligned}} \tag{10.490}$$

with $[\mathbf{h}] \equiv \mathbf{h}\times, \forall \mathbf{h} \in \mathbb{R}_3$, and

- The residual force functional, $R({}^t\breve{\mathbf{d}}(t);\, {}^t\breve{\bar{\mathbf{d}}}) = -G({}^t\breve{\mathbf{d}}(t);\, {}^t\breve{\bar{\mathbf{d}}})$ is given as:

$$
\begin{aligned}
R({}^t\breve{\mathbf{d}}(t);\, {}^t\breve{\bar{\mathbf{d}}}) &= -G({}^t\breve{\mathbf{d}}(t);\, {}^t\breve{\bar{\mathbf{d}}})\\
&= -G({}^t\mathbf{d}(t),\, {}^t\theta(t);\, {}^t\bar{\mathbf{d}},\, {}^t\bar{\theta})\\
&\equiv -\{G_{Internal} + G_{Inertial} - G_{External}\}\\
G_{Internal} &= \int_0^L ({}^n\breve{\varepsilon} \bullet {}^n\breve{\mathbf{F}})\, dS\\
G_{Inertial} &= \int_0^L ({}^t\breve{\bar{\mathbf{d}}}^T\, {}^t\breve{\mathbf{F}}_{Iner})\, dS\\
&= \int_0^L \overbrace{\left\{(M_0\, {}^t\ddot{\mathbf{d}} \bullet {}^t\bar{\mathbf{d}} + \mathbf{W}^T\, (\mathbf{I}_S\, {}^t\dot{\omega}_S + {}^t\omega_S \times \mathbf{A}_S) \bullet {}^t\bar{\theta}\right\}\, dS}^{\text{Inertial}}\\
G_{External} &= \left\{\int_0^L \breve{\bar{\mathbf{d}}}^{\mathrm{T}}\, \breve{\mathbf{F}}^B\, dS + \left[\breve{\bar{\mathbf{d}}}^{\mathrm{T}}\, \breve{\mathbf{F}}^S\right]_{S=0}^{S=L}\right\}
\end{aligned}
\tag{10.491}
$$

where $\mathbf{W}$ is given as:

$$
\begin{aligned}
\mathbf{W}({}^t)\theta &\equiv \mathbf{I} + c_1\, {}^t\mathbf{\Theta} + c_2\, {}^t\mathbf{\Theta}^2,\\
c_1({}^t)\theta &= \frac{1-\cos\theta}{\theta^2},\\
c_2({}^t)\theta &= \frac{\theta - \sin\theta}{\theta^3}
\end{aligned}
\tag{10.492}
$$

with ${}^t\mathbf{\Theta}$ as the skew-symmetric matrix corresponding to the axial vector, ${}^t\theta$, such that ${}^t\mathbf{\Theta}\,{}^t\theta = \mathbf{0}$.

- $\dot{\mathbf{W}}$ is given as:

$$
\begin{aligned}
\dot{\mathbf{W}} &= c_1\, \dot{\mathbf{\Theta}} + c_2(\dot{\mathbf{\Theta}}\, \mathbf{\Theta} + \mathbf{\Theta}\, \dot{\mathbf{\Theta}}) + \dot{c}_1\, \mathbf{\Theta} + \dot{c}_2\, \mathbf{\Theta}^2\\
&= (c_2 - c_1)(\theta \bullet \dot{\theta})\mathbf{I} + c_1\, \dot{\mathbf{\Theta}} + c_2\{(\theta \otimes \dot{\theta}) + (\dot{\theta} \otimes \theta)\} + \dot{c}_1\, \mathbf{\Theta} + \dot{c}_2(\theta \otimes \theta)\\
\dot{c}_1 &\equiv \frac{1}{\theta^2}(1 - 2c_1 - \theta^2 c_2)(\theta \bullet \dot{\theta}) = c_3(\theta \bullet \dot{\theta}),\ \dot{c}_2 \equiv \frac{1}{\theta^2}(c_1 - 3c_2)(\theta \bullet \dot{\theta}) = c_4(\theta \bullet \dot{\theta}),
\end{aligned}
\tag{10.493}
$$

- $\ddot{\mathbf{W}}$ is given as:

$$\boxed{\begin{aligned}&\ddot{\mathbf{W}} = c_1\,\ddot{\boldsymbol{\Theta}} + 2\{\dot{c}_1\,\dot{\boldsymbol{\Theta}} + \dot{c}_2(\dot{\boldsymbol{\Theta}}\,\boldsymbol{\Theta} + \boldsymbol{\Theta}\,\dot{\boldsymbol{\Theta}})\} + c_2(\ddot{\boldsymbol{\Theta}}\,\boldsymbol{\Theta} + 2\,\dot{\boldsymbol{\Theta}}^2 + \boldsymbol{\Theta}\,\ddot{\boldsymbol{\Theta}}) + \ddot{c}_1\,\boldsymbol{\Theta} + \ddot{c}_2\,\boldsymbol{\Theta}^2\\ &\ddot{c}_1 \equiv c_5(\boldsymbol{\theta}\bullet\dot{\boldsymbol{\theta}})^2 + c_3\{\theta^2 + (\boldsymbol{\theta}\bullet\ddot{\boldsymbol{\theta}})\}, \qquad \ddot{c}_2 \equiv c_6(\boldsymbol{\theta}\bullet\dot{\boldsymbol{\theta}})^2 + c_4\{\theta^2 + (\boldsymbol{\theta}\bullet\ddot{\boldsymbol{\theta}})\}\\ &c_5 \equiv -\frac{1}{\theta^2}(2c_2 + 4c_3 + \theta^2 c_4), \qquad c_6 \equiv -\frac{1}{\theta^2}(c_3 - 5c_4),\end{aligned}} \tag{10.494}$$

- ${}^t\breve{\mathbf{F}}_{Iner}$, the generalized inertial force field is given as:

$$\boxed{\begin{aligned}&{}^t\breve{\mathbf{F}}_{Iner} \equiv \{M_0\,{}^t\ddot{\mathbf{d}} \quad \mathbf{W}^T\,(\mathbf{I}_S\,{}^t\dot{\boldsymbol{\omega}}_S + {}^t\boldsymbol{\omega}_S \times \mathbf{A}_S)\}^T,\\ &{}^t\boldsymbol{\omega}_S \equiv \mathbf{W}\,{}^t\dot{\boldsymbol{\theta}}, \qquad \mathbf{A}_S \equiv \mathbf{I}_S\,{}^t\boldsymbol{\omega}_S\end{aligned}} \tag{10.495}$$

where the angular velocity vector, ${}^t\boldsymbol{\omega}_S$, is the axial vector corresponding to the skew-symmetric angular velocity matrix, $\boldsymbol{\Omega}_S$, given by:

$$\boxed{\boldsymbol{\Omega}_S \equiv \dot{\mathbf{R}}\,\mathbf{R}^T} \tag{10.496}$$

- The Jacobian determinant, $J(\xi)$, relating the arc length parameter, S, and the root coordinate, ξ, are defined as: $J(\xi) = ||\mathbf{C}_{,\xi}||$ where $C(\xi)$ is the curve representing the centroidal axis of the beam. Clearly, then:

$$\boxed{\begin{aligned}&dS = ||\mathbf{C}_{,\xi}||\;d\xi = J(\xi)\;d\xi\\ &\frac{\partial}{\partial S} = \frac{d\xi}{dS}\frac{\partial}{\partial \xi} = \frac{1}{J(\xi)}\frac{\partial}{\partial \xi}\end{aligned}} \tag{10.497}$$

- Superscript t reminds us that the relevant items are measured in the undeformed Frenet coordinate system, $\mathbf{G}_1 \equiv \mathbf{T}$, $\mathbf{G}_2 \equiv \mathbf{M}$, $\mathbf{G}_3 \equiv \mathbf{B}$; superscript n reminds us that the relevant items are in the rotated coordinate system, $\mathbf{g}_1(= \mathbf{n} = \mathbf{R}\,\mathbf{G}_1)$, $\mathbf{g}_2(= \mathbf{m}_n = \mathbf{R}\,\mathbf{G}_2)$, $\mathbf{g}_3(= \mathbf{b}_n = \mathbf{R}\,\mathbf{G}_3)$.

10.19.2 c-type Finite Element Formulation

Under the semi-discrete formulation and following the standard definition of a finite element discretization, a long beam is broken into finite number of intervals: $\bigcup_{e=1}^{e=no.\ of\ elements}$ *Finite Element*^{e}. The root element corresponding to a finite element is defined over the geometric root parameter, $\xi \in [0, 1]$. The root parameter, ξ, is related to the arc length parameter, S, of a beam by: $S(\xi) = \int_0^{\xi} ||\mathbf{C}^e{}_{,\eta}||d\eta$, where $\mathbf{C}^e(\xi)$, $\xi \in [0, 1]$ defines the centroidal axis part of the eth finite element, that is, *Finite Element*^{e}. Based on the assumption that the assembly of the finite elements will be performed by the direct stiffness assembly procedure, we can restrict our discussion under to a typical finite beam element of length, L, as before. For finite element discretization, we must express the generalized displacements, $\breve{\mathbf{d}}$, the variations in generalized displacements, $\breve{\bar{\mathbf{d}}}$, and the incremental generalized displacement, $\Delta\breve{\mathbf{d}}$, in terms of the Bernstein–Bezier interpolation polynomials, $B_i^n(\xi)$, $i = 0, \ldots, n$, where n is the degree of the polynomials. Let us define N as the order of the polynomial given by: $N \equiv n + 1$. We may recall that these polynomials serve as the local basis functions for the c-type beam element.

10.19.3 The c-type Beam Elements and State Control Vectors

Let us, then, introduce an n-degree c-type beam element as:

$$\begin{aligned}
\underbrace{\breve{\mathbf{d}}(\xi,t)}_{6\times1} &= \underbrace{{}^d\breve{\mathbf{T}}(\xi)}_{6\times6N}\ \underbrace{{}^d\breve{\mathbf{q}}(t)}_{6N\times1},\\
\underbrace{\dot{\breve{\mathbf{d}}}(\xi,t)}_{6\times1} &= \underbrace{{}^d\breve{\mathbf{T}}(\xi)}_{6\times6N}\ \underbrace{{}^d\dot{\breve{\mathbf{q}}}(t)}_{6N\times1},\\
\underbrace{\ddot{\breve{\mathbf{d}}}(\xi,t)}_{6\times1} &= \underbrace{{}^d\breve{\mathbf{T}}(\xi)}_{6\times6N}\ \underbrace{{}^d\ddot{\breve{\mathbf{q}}}(t)}_{6N\times1},\\
\breve{d}_j(\xi) &= \sum_{i=0}^{n} B_i^n(\xi)\, \breve{q}_i^{d_j}, \quad j = 1, \ldots, 6
\end{aligned} \tag{10.498}$$

where we have introduced the following:

Definitions

- $\breve{\mathbf{q}}^d = \{q_0^u \quad q_1^u \quad \cdots \quad q_n^u \quad \cdots \quad q_0^{\theta_b} \quad \cdots \quad q_n^{\theta_b}\}^T$ be the $(6N \times 1)$ displacement control vector.
- $\breve{\bar{\mathbf{q}}}^d$ be the $(6N \times 1)$ virtual displacement control vector, similar to $\breve{\mathbf{q}}^d$,
- $\Delta\breve{\mathbf{q}}^d$, be the $(6N \times 1)$ incremental control vector, similar to $\mathbf{q}^d$,
- $\breve{\mathbf{T}}^d(\xi)$ be the $(6 \times 6N)$ interpolation transformation matrix containing the Bernstein–Bezier polynomials, that is,

$$\breve{\mathbf{T}}^d(\xi) = \underbrace{\begin{bmatrix} B_0^n & \cdots & B_n^n & 0 & 0 & \cdots & \cdots & \cdots & 0\\ 0 & 0 & 0 & B_0^n & \cdots & B_n^n & 0 & \cdots & 0\\ & & \cdots & \cdots & & & & & \\ & & & \cdots & \cdots & & & & \\ 0 & 0 & \cdots & \cdots & & 0 & B_0^n & \cdots & B_n^n \end{bmatrix}}_{6\times6N} \tag{10.499}$$

In other words, each row of $\breve{\mathbf{T}}(\xi)$ is identical, except that the non-zero contents are shifted by $(1 \times 6N^2)$ zero entries with increasing row numbers. Now, let us further introduce the following.

Definition

- $\breve{\mathbf{B}}(\xi)$ be the $(12 \times 6N)$ augmented interpolation transformation matrix as:

$$\breve{\mathbf{B}}(\xi) \equiv \begin{bmatrix} \underbrace{\breve{\mathbf{T}}(\xi)}_{6\times6N} \\ ----- \\ \underbrace{\frac{1}{J}\breve{\mathbf{T}},_{\xi}(\xi)}_{6\times6N} \end{bmatrix} \tag{10.500}$$

Based on the above definitions, we also have:

$$\boxed{\begin{aligned} \breve{\bar{\mathbf{d}}} &= \breve{\mathbf{T}}^d(\xi)\, \breve{\bar{\mathbf{q}}}^d \\ \Delta\breve{\mathbf{d}} &= \breve{\mathbf{T}}^d(\xi)\, \Delta\breve{\mathbf{q}}^d \end{aligned}} \tag{10.501}$$

and

$$\boxed{\begin{aligned} \underbrace{\mathbf{H}(\breve{\bar{\mathbf{d}}})}_{12\times1} &= \underbrace{\breve{\mathbf{B}}(\xi)}_{12\times6N} \underbrace{\breve{\bar{\mathbf{q}}}^d}_{6N\times1} \\ \underbrace{\mathbf{H}(\Delta\breve{\mathbf{d}})}_{12\times1} &= \underbrace{\breve{\mathbf{B}}(\xi)}_{12\times6N} \underbrace{\Delta\breve{\mathbf{q}}^d}_{6N\times1} \end{aligned}} \tag{10.502}$$

Similarly, for incremental quantities, we have:

$$\boxed{\begin{aligned} \Delta\breve{\mathbf{d}}(\xi, t) &= {}^d\breve{\mathbf{T}}(\xi)\, \Delta\, {}^d\breve{\mathbf{q}}(t), \\ \Delta\, \dot{\breve{\mathbf{d}}}(\xi, t) &= {}^d\breve{\mathbf{T}}(\xi)\, \Delta\, {}^d\dot{\breve{\mathbf{q}}}(t). \\ \Delta\, \ddot{\breve{\mathbf{d}}}(\xi, t) &= {}^d\breve{\mathbf{T}}(\xi)\, \Delta\, {}^d\ddot{\breve{\mathbf{q}}}(t), \end{aligned}} \tag{10.503}$$

where we have similarly introduced the following.

Definitions

- $\Delta^d\breve{\mathbf{q}}(t)$, $\Delta^d\dot{\breve{\mathbf{q}}}(t)$ and $\Delta^d\ddot{\breve{\mathbf{q}}}(t)$, be the $(6N \times 1)$ time-dependent incremental generalized displacement, velocity and acceleration control vectors, with the components ordered in the same sequence as the components of ${}^d\breve{\mathbf{q}}(t)$, ${}^d\dot{\breve{\mathbf{q}}}(t)$ and ${}^d\ddot{\breve{\mathbf{q}}}(t)$, respectively.

Note that equation (10.498) implies that we have chosen to interpolate both the displacement components and the rotation components with the same degree control net.

10.19.4 Computational Element Symmetric Material Stiffness Matrix, ${}^{e}\mathbf{K}_M$

By substituting equation (10.502) in equation (10.483), we get for each finite element:

$$
\begin{aligned}
{}^{e}\mathcal{K}_M(\breve{\mathbf{d}}, \bar{\breve{\mathbf{d}}}; \Delta\breve{\mathbf{d}}) &\equiv \underbrace{{}^{d}\bar{\breve{\mathbf{q}}}^T}_{1\times 6N} \underbrace{\left[\int_0^1 \left\{ \underbrace{\breve{\mathbf{B}}(\xi)^T}_{6N\times 12} \underbrace{\overbrace{\mathbf{E}(\breve{\mathbf{d}})^T}^{12\times 6} \overbrace{{}^{n}\breve{\mathbf{d}}}^{6\times 6} \overbrace{\mathbf{E}(\breve{\mathbf{d}})}^{6\times 12}}_{12\times 12} \underbrace{\breve{\mathbf{B}}(\xi)}_{12\times 6N} \right\} J(\xi)\, d\xi \right]}_{6N\times 6N} \underbrace{{}^{d}\Delta\breve{\mathbf{q}}(t)}_{6N\times 1} \\
&= \underbrace{{}^{d}\bar{\breve{\mathbf{q}}}^T}_{1\times 6N} \underbrace{{}^{e}\mathbf{K}_M}_{6N\times 6N} \underbrace{{}^{d}\Delta\breve{\mathbf{q}}(t)}_{6N\times 1}
\end{aligned}
\tag{10.504}
$$

where, we have introduced the following.

Definition

- The ***computational element symmetric material stiffness matrix***, ${}^{e}\mathbf{K}_M$, as:

$$
\underbrace{{}^{e}\mathbf{K}_M}_{6N\times 6N} \equiv \underbrace{\int_0^1 \left\{ \underbrace{\breve{\mathbf{B}}(\xi)^T}_{6N\times 12} \underbrace{\overbrace{\mathbf{E}(\breve{\mathbf{d}})^T}^{12\times 6} \overbrace{{}^{n}\breve{\mathbf{D}}}^{6\times 6} \overbrace{\mathbf{E}(\breve{\mathbf{d}})}^{6\times 12}}_{12\times 12} \underbrace{\breve{\mathbf{B}}(\xi)}_{12\times 6N} \right\} J(\xi)\, d\xi}_{6N\times 6N}
\tag{10.505}
$$

10.19.5 Computational Element Symmetric Geometric Stiffness Matrix, ${}^{e}\mathbf{K}_G$

By substituting equation (10.502) into equation (10.486), we get for each finite element:

$$
\begin{aligned}
{}^{e}\mathcal{K}_G(\breve{\mathbf{d}}, \bar{\breve{\mathbf{d}}}; \Delta\breve{\mathbf{d}}) &\equiv \underbrace{{}^{d}\bar{\breve{\mathbf{q}}}^T}_{1\times 6N} \underbrace{\left[\int_0^1 \left\{ \underbrace{\breve{\mathbf{B}}(\xi)^T}_{6N\times 12} \underbrace{\mathbf{G}(\breve{\mathbf{d}}, {}^{t}\breve{\mathbf{F}})}_{12\times 12} \underbrace{\breve{\mathbf{B}}(\xi)}_{12\times 6N} \right\} J(\xi)\, d\xi \right]}_{6N\times 6N} \underbrace{{}^{d}\Delta\breve{\mathbf{q}}(t)}_{6N\times 1} \\
&= \underbrace{{}^{d}\bar{\breve{\mathbf{q}}}^T}_{1\times 6N} \underbrace{{}^{e}\mathbf{K}_G}_{6N\times 6N} \underbrace{{}^{d}\Delta\breve{\mathbf{q}}(t)}_{6N\times 1}
\end{aligned}
\tag{10.506}
$$

where, we have introduced the following.

Definition

- The ***computational element symmetric geometric stiffness matrix***, ${}^{e}K_G$, as:

$$\underbrace{{}^{e}\mathbf{K}_G}_{6N\times 6N} \equiv \underbrace{\int_0^1 \left\{ \underbrace{\breve{\mathbf{B}}(\xi)^T}_{6N\times 12} \; \underbrace{\mathbf{G}(\breve{\mathbf{d}}, {}^{t}\breve{\mathbf{F}})}_{12\times 12} \; \underbrace{\breve{\mathbf{B}}(\xi)}_{12\times 6N} \right\} J(\xi)\, d\xi}_{6N\times 6N} \tag{10.507}$$

10.19.6 Computational Element Symmetric Mass Matrix, ${}^{e}\mathbf{M}$

By substituting the expression for the generalized incremental acceleration vector, $\Delta\,\ddot{\breve{\mathbf{d}}}(t)$, and the generalized virtual displacement vector, $\bar{\breve{\mathbf{d}}}$, from equation (10.503) into equation (10.488), we get for each finite element:

$$\Delta G_M({}^{t}\breve{\mathbf{d}};\, {}^{t}\bar{\breve{\mathbf{d}}};\, \Delta\, {}^{t}\breve{\mathbf{d}}) \equiv {}^{t}\bar{\breve{\mathbf{q}}}^T\, {}^{e}\mathbf{M}\, \Delta\, {}^{t}\ddot{\breve{\mathbf{q}}} \tag{10.508}$$

where we have introduced the following.

Definition

- The ***computational element symmetric mass matrix***, ${}^{e}\mathbf{M}$, as:

$$\underbrace{{}^{e}\mathbf{M}}_{6N\times 6N} \equiv \int_0^1 \underbrace{{}^{d}\breve{\mathbf{T}}(\xi)^T}_{6N\times 6} \underbrace{\begin{bmatrix} M_0\, \mathbf{I} & \mathbf{0} \\ \mathbf{0} & (\mathbf{W}^T\, \mathbf{I}_S\, \mathbf{W}) \end{bmatrix}}_{6\times 6} \underbrace{{}^{d}\breve{\mathbf{T}}(\xi)}_{6\times 6N}\; J(\xi)\, d\xi \tag{10.509}$$

10.19.7 Computational Element Gyroscopic (Damping) Matrix, ${}^{e}\mathbf{C}_G$

By substituting the expression for the generalized incremental acceleration vector, $\Delta\,\ddot{\breve{\mathbf{d}}}(t)$, and the generalized virtual displacement vector, $\bar{\breve{\mathbf{d}}}$, from equation (10.503) into equation (10.489), we get for each finite element:

$$\Delta G_G({}^{t}\breve{\mathbf{d}};\, {}^{t}\bar{\breve{\mathbf{d}}};\, \Delta\, {}^{t}\breve{\mathbf{d}}) \equiv {}^{t}\bar{\breve{\mathbf{q}}}^T\, {}^{e}\mathbf{C}_G\, \Delta\, {}^{t}\dot{\breve{\mathbf{q}}} \tag{10.510}$$

where we have introduced the following.

Definition

- The ***computational element gyroscopic matrix***, ${}^e\mathbf{C}_G$, as:

$$\underbrace{{}^e\mathbf{C}_G}_{6N\times 6N} \equiv \int_0^1 \underbrace{{}^d\breve{\mathbf{T}}(\xi)^T}_{6N\times 6} \underbrace{\begin{bmatrix} \mathbf{0} & \mathbf{0} \\ \mathbf{0} & \left\{ \begin{matrix} 2\mathbf{W}^T\, \mathbf{I}_S\, \dot{\mathbf{W}} \\ +\mathbf{W}^T\, (\boldsymbol{\Omega}_S\, \mathbf{I}_S - \mathbf{I}_S\, \boldsymbol{\Omega}_S - [\mathbf{A}_S])\mathbf{W} \end{matrix} \right\} \end{bmatrix}}_{6\times 6} \underbrace{{}^d\breve{\mathbf{T}}(\xi)}_{6\times 6N}\, J(\xi)\, d\xi \tag{10.511}$$

10.19.8 Computational Element Centrifugal (Stiffness) Matrix, ${}^e\mathbf{K}_C$

By substituting the expression for the generalized incremental acceleration vector, $\Delta\, \ddot{\hat{\breve{\mathbf{d}}}}(t)$, and the generalized virtual displacement vector, $\bar{\breve{\mathbf{d}}}$, from equation (10.503) into equation (10.490), we get for each element:

$$\boxed{\Delta G_C({}^t\breve{\mathbf{d}};\, {}^t\bar{\breve{\mathbf{d}}};\, \Delta\, {}^t\breve{\mathbf{d}}) \equiv {}^t\bar{\breve{\mathbf{q}}}^T\, {}^e\mathbf{K}_C\, \Delta\, {}^t\breve{\mathbf{q}}} \tag{10.512}$$

where we have introduced the following.

Definition

- The ***computational element centrifugal stiffness matrix***, ${}^e\mathbf{K}_C$, as:

$$\underbrace{{}^e\mathbf{K}_C}_{6N\times 6N} \equiv \int_0^1 \underbrace{{}^d\breve{\mathbf{T}}(\xi)^T}_{6N\times 6} \underbrace{\begin{bmatrix} \mathbf{0} & \mathbf{0} \\ \mathbf{0} & \left\{ \begin{matrix} \mathbf{U}(\dot{\mathbf{A}}_S) + \mathbf{W}^T\, \mathbf{I}_S\, \ddot{\mathbf{W}} + \mathbf{W}^T\, (\boldsymbol{\Omega}_S\, \mathbf{I}_S - \mathbf{I}_S\, \boldsymbol{\Omega}_S \\ -[\mathbf{A}_S])\, \dot{\mathbf{W}} - \mathbf{W}^T\, [\dot{\mathbf{A}}_S]\, \mathbf{W} \end{matrix} \right\} \end{bmatrix}}_{6\times 6} \underbrace{{}^d\breve{\mathbf{T}}(\xi)}_{6\times 6N}\, J(\xi)\, d\xi \tag{10.513}$$

with

$$\mathbf{U}(\dot{\mathbf{A}}_S) \equiv \left\{ \begin{matrix} c_1\, [\dot{\mathbf{A}}_S] + (-c_2\, \mathbf{I} - c_3\, \boldsymbol{\Theta} + c_4 \boldsymbol{\Theta}^2)(\dot{\mathbf{A}}_S \bullet \theta) \\ + c_2((\theta \otimes \dot{\mathbf{A}}_S) - (\dot{\mathbf{A}}_S \otimes \theta) + (\theta \bullet \dot{\mathbf{A}}_S))\mathbf{I} \end{matrix} \right\},$$
$$\dot{\mathbf{A}}_S \equiv \mathbf{I}_S\, \dot{\omega}_S + \omega_S \times \mathbf{A}_S, \quad \mathbf{A}_S \equiv \mathbf{I}_S\, \omega_S, \quad \omega_S \equiv \mathbf{W}\, \dot{\theta} \tag{10.514}$$

where $[\mathbf{h}] \equiv \mathbf{h}\times$, $\forall \mathbf{h} \in \mathbb{R}_3$.

10.19.9 Computational Element Unbalanced Residual Force, ${}^{c}\mathbf{R}(\breve{\mathbf{d}})$

By substituting equation (10.502) into equation (10.491), we get for each finite element:

$$\begin{aligned} \overbrace{R(\breve{\mathbf{d}};\breve{\bar{\mathbf{d}}})}^{\text{Residual force}} &= -\overbrace{{}^{e}G(\breve{\mathbf{d}};\breve{\bar{\mathbf{d}}})}^{\text{Unbalanced force}} \\ &= -{}^{d}\breve{\bar{\mathbf{q}}}^{T}\,{}^{e}\breve{\mathbf{R}}(\breve{\mathbf{d}}) \end{aligned} \tag{10.515}$$

where, we have defined the following.

Definition

$$\underbrace{{}^{e}\breve{\mathbf{R}}(\breve{\mathbf{d}})}_{6N\times 1} \equiv \left\{ \int_0^L \left[\underbrace{\breve{\mathbf{B}}^{T}}_{6N\times 6}\ \underbrace{\mathbf{E}^{T}}_{6\times 6}\ \underbrace{{}^{n}\breve{\mathbf{F}}}_{6\times 1} + \underbrace{\breve{\mathbf{T}}^{T}}_{6N\times 6}\ (\underbrace{{}^{t}\breve{\mathbf{F}}_{Iner}}_{6\times 1} - \underbrace{\breve{\mathbf{F}}^{B}}_{6\times 1}) \right] dS + \left[\underbrace{\breve{\mathbf{T}}^{T}}_{6N\times 6}\ \underbrace{\breve{\mathbf{F}}^{S}}_{6\times 1} \right]_{S=0}^{S=L} \right\} \tag{10.516}$$

10.19.10 Geometrically Exact Element Incremental Dynamic Equation

Now substituting the expressions given in equations (10.504), (10.506), (10.509), (10.511), (10.513) and (10.515) in equation (10.480), we have:

$$ {}^{d}\breve{\bar{\mathbf{q}}}^{T}\{{}^{e}\mathbf{M}\,\Delta\,{}^{d}\ddot{\breve{\mathbf{q}}} + {}^{e}\mathbf{C}_G\,\Delta\,{}^{d}\dot{\breve{\mathbf{q}}} + ({}^{e}\mathbf{K}_M + {}^{e}\mathbf{K}_G + {}^{e}\mathbf{K}_C)\,\Delta{}^{d}\breve{\mathbf{q}}\} = -{}^{d}\breve{\bar{\mathbf{q}}}^{T}\,{}^{e}\breve{\mathbf{R}}(\breve{\mathbf{d}}) \tag{10.517}$$

Noting that this equation is true for any arbitrary admissible generalized virtual control displacement, ${}^{d}\breve{\bar{\mathbf{q}}}$, we can finally write the $6N \times 1$ geometrically exact fully linearized finite element equation for each element in unknown incremental generalized displacement control vector, $\Delta\,{}^{d}\breve{\hat{\mathbf{q}}}$, generalized velocity control vector, $\Delta\,{}^{d}\dot{\breve{\hat{\mathbf{q}}}}$, and generalized acceleration control vector, $\Delta\,{}^{d}\ddot{\breve{\hat{\mathbf{q}}}}$, for any time $t \in \mathbb{R}_{+}$, as:

$$\underbrace{{}^{e}\mathbf{M}}_{6N^2\times 6N^2}\ \underbrace{\Delta\,{}^{d}\ddot{\breve{\mathbf{q}}}}_{6N^2\times 1} + \underbrace{{}^{e}\mathbf{C}_G}_{6N^2\times 6N^2}\ \underbrace{\Delta\,{}^{d}\dot{\breve{\mathbf{q}}}}_{6N^2\times 1} + \underbrace{({}^{e}\mathbf{K}_C + {}^{e}\mathbf{K}_T)}_{6N^2\times 6N^2}\ \underbrace{\Delta\,{}^{d}\breve{\mathbf{q}}}_{6N^2\times 1} = -\underbrace{{}^{e}\breve{\mathbf{R}}(\breve{\mathbf{d}})}_{6N^2\times 1} \tag{10.518}$$

where, we have introduced the following.

Definition

- The computational tangent stiffness matrix, ${}^{e}\mathbf{K}_T^{c}$:

$$ {}^{e}\mathbf{K}_T \equiv {}^{e}\mathbf{K}_M + {}^{e}\mathbf{K}_G \tag{10.519}$$

of a beam, a concept familiar under quasi-static loading.

10.19.11 Direct Global Assembly

We have now detailed the development of the incremental second-order ordinary differential equation for a beam element at any time $t \in \mathbb{R}_+$ as given by equation (10.518). The corresponding incremental equation for the entire domain of a beam can be obtained by the standard direct global assembly procedure which is given as:

$$\boxed{\mathbf{M}\ \Delta^d \ddot{\breve{\mathbf{q}}}(t) + \mathbf{C}_G\ \Delta^d \dot{\breve{\mathbf{q}}}(t) + \mathbf{K}\ \Delta^d \breve{\mathbf{q}}(t) = -\breve{\mathbf{R}}(\breve{\mathbf{d}}(t))} \tag{10.520}$$

where the global matrices and vectors are obtained as:

$$\boxed{\begin{aligned} \mathbf{M} &\equiv \sum_{element} \underbrace{{}^e\mathbf{M}}_{6N^2\times 6N^2}, \\ \mathbf{C}_G &\equiv \sum_{element} \underbrace{{}^e\mathbf{C}_G}_{6N^2\times 6N^2}, \\ \mathbf{K} &\equiv \sum_{element} \underbrace{\left({}^e\mathbf{K}_M + {}^e\mathbf{K}_G + {}^e\mathbf{K}_C\right)}_{6N^2\times 6N^2}, \\ \breve{\mathbf{R}} &\equiv \sum_{element} \underbrace{{}^e\breve{\mathbf{R}}(\breve{\mathbf{d}})}_{6N^2\times 1} \end{aligned}} \tag{10.521}$$

An actual global assembly scheme, referred to in finite element jargon as the ***direct global assembly***, will be normally applied as we we will explain in the numerical implementation phase next.

10.19.12 Time Integration of the Semi-discrete Equation

So far, we have detailed the development of the global incremental matrix second-order ordinary differential equation as given by equation (10.520) under the semi-discrete c-type finite element formulation. The linearization has been introduced and devised with the understanding that a time integration algorithm for the ensuing matrix second-order ordinary differential equation will be developed in conjunction with a Newton-type iterative scheme that will require linearization of the nonlinear equations at each iterative step. For the time integration, we must realize that the rotation vector marching and updating must be given special attention as it is constrained to travel over a curved space: the 3-sphere.

- In this book, we have not implemented the dynamic treatment presented above; so, for the following sections, we restrict ourselves to implementation of beam analysis to static and quasi-static loading conditions.
- However, for nonlinear structural dynamic problems involving finite rotations, we conjecture that the rotational acceleration with a spherically linear distribution (see Chapter 4 on the rotation tensor), as we will see, for a predictor–corrector type of numerical analysis may be the way to go; the rotational velocity and rotation distribution may then be obtained by corresponding integration of the acceleration.

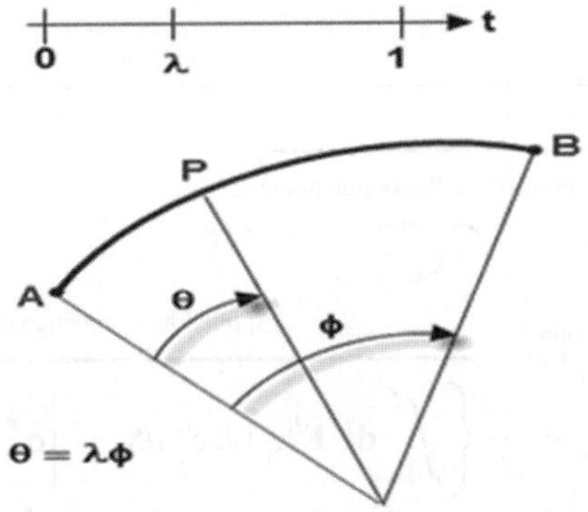

Figure 10.16 Spherical interpolation and approximation.

10.19.12.1 Linear Interpolation or Approximation on the 3-sphere

Let **A**, **B** represent the unit Hamilton–Rodrigues quaternion for rotations at parameters $t = 0$ and $t = 1$, respectively, as shown in Figure 10.16, where the rotations are shown as points on a four-dimensional space; the rotation from the reference state, **A**, to a final state, **B**, is taken as a single rotation, which is always possible by Euler's theorem. Now, let $\mathbf{P}(t)$, with $\|\mathbf{P}\| = 1$, represents a rotation on this curve at an affine parameter $t = \lambda \in [0, 1]$. We are interested in expressing **P** in terns of **A** and **B**:

$$\boxed{\mathbf{P} = \frac{\sin(1-\lambda)\phi}{\sin\phi}\mathbf{A} + \frac{\sin\lambda\phi}{\sin\phi}\mathbf{B}, \quad \lambda \in [0.1]} \tag{10.522}$$

10.19.13 Where We Would Like to Go

The main goal here was to present the finite element formulation of the dynamic fully linearized variational or virtual work equations. As indicated earlier, beam analysis subjected to dynamic loadings has not been implemented. We will continue with the implementational issues related to static and quasi-static analyses:

- finite element implementation for quasi-static analyses

10.20 c-type FE Implementation and Examples: Quasi-static Loading

10.20.1 What We Need to Recall

The linearized quasi-static virtual work equations:

$$\boxed{\begin{aligned} &\overbrace{\int_0^L {}^n\breve{\bar{\varepsilon}}^T\, {}^n\breve{\mathbf{D}}\, \Delta {}^n\breve{\varepsilon}\, dS}^{\text{Material stiffness functional}} + \overbrace{\int_0^L \Delta {}^n\breve{\bar{\varepsilon}}^T\, {}^n\breve{\mathbf{F}}\, dS}^{\text{Geometric stiffness functional}} \\ &\underbrace{= -\overbrace{G(\lambda^0, {}^t\breve{\mathbf{d}}^0; {}^t\breve{\bar{\mathbf{d}}})}^{\text{Unbalanced force functional}} - \overbrace{\left\{\int_0^L \hat{\bar{\mathbf{d}}}^{\mathrm{T}}\, \hat{\mathbf{F}}^B{}_{,\lambda}(\lambda, S)\, dS + \left[\hat{\bar{\mathbf{d}}}^{\mathrm{T}}\, \hat{\mathbf{F}}^S\right]_{S=0}^{S=L}\right\}}^{\text{Applied force functional}} \Delta\lambda}_{\text{Residual force functional}} \end{aligned}} \tag{10.523}$$

Or,

$$\overbrace{\overbrace{\mathcal{K}_M}^{\text{Material stiffness functional}} + \overbrace{\mathcal{K}_G}^{\text{Geometric stiffness functional}}}^{\text{Tangent functional}} = \underbrace{-\overbrace{G(\lambda^0, \breve{\mathbf{d}}^0; \breve{\bar{\mathbf{d}}})}^{\text{Unbalanced Force functional}} - \overbrace{\left\{ \int_0^L \breve{\bar{\mathbf{d}}}^T \, \hat{\mathbf{F}}^B{}_{,\lambda}(\lambda, S)\, dS + \left[\breve{\bar{\mathbf{d}}}^T \, \hat{\mathbf{F}}^S \right]_{S=0}^{S=L} \right\} \Delta\lambda}^{\text{Applied force functional}}}_{\text{Residual force functional}} \tag{10.524}$$

The linearized internal functional, $\Delta G_{Internal}$:

$$\begin{aligned} \Delta G_{Internal} &\equiv \overbrace{\overbrace{\int_M {}^n\breve{\bar{\varepsilon}}^T \, {}^n\breve{\mathbf{D}} \, \Delta^n\breve{\varepsilon} \, dA}^{\text{Material stiffness functional}} + \overbrace{\int_M \Delta^n\breve{\bar{\varepsilon}}^T \, {}^n\breve{\mathbf{F}} \, dA}^{\text{Geometric stiffness functional}}}^{\text{Linearized internal functional}} \\ &= \overbrace{\overbrace{\mathcal{K}_M}^{\text{Material stiffness functional}} + \overbrace{\mathcal{K}_G}^{\text{Geometric stiffness functional}}}^{\text{Tangent functional}} \end{aligned} \tag{10.525}$$

Definition: Material stiffness functional, $\mathcal{K}_M$:

$$\mathcal{K}_M(\breve{\mathbf{d}}, \breve{\bar{\mathbf{d}}}; \Delta\breve{\mathbf{d}}) \equiv \int_C \{\mathbf{H}(\breve{\bar{\mathbf{d}}}) \; \mathbf{K}_M(\breve{\mathbf{d}}) \; \mathbf{H}(\Delta\breve{\mathbf{d}})\} \, dS \tag{10.526}$$

with the material stiffness matrix, $\mathbf{K}_M$, given by:

$$\mathbf{K}_M = \mathbf{E}(\breve{\mathbf{d}})^T \, {}^n\breve{\mathbf{D}} \, \mathbf{E}(\breve{\mathbf{d}}) \tag{10.527}$$

where,

$$\mathbf{E}(\breve{\mathbf{d}}) = \begin{bmatrix} [\mathbf{R}^{\mathbf{T}}] & \mathbf{0} \\ \mathbf{0} & [\mathbf{R}^{\mathbf{T}}] \end{bmatrix} \begin{bmatrix} [\mathbf{k}^0] & [\mathbf{a}]\mathbf{W} & \mathbf{I} & \mathbf{0} \\ \mathbf{0} & \mathbf{X} + [\mathbf{k}^c]\mathbf{W} & \mathbf{0} & \mathbf{W} \end{bmatrix} \tag{10.528}$$

and

Definition: Geometric stiffness functional, $\mathcal{K}_G$:

$$\mathcal{K}_G(\breve{\mathbf{d}}, \breve{\bar{\mathbf{d}}}; \Delta\breve{\mathbf{d}}) \equiv \int_C \{\mathbf{H}(\breve{\bar{\mathbf{d}}}) \; \mathbf{G}(\breve{\mathbf{d}}, \breve{\mathbf{F}}) \; \mathbf{H}(\Delta\breve{\mathbf{d}})\} \, dS \tag{10.529}$$

- Additionally, for completeness, we have the unbalanced force at λ_0, $\breve{\mathbf{d}}_0$ as:

$$G(\lambda_0, \breve{\mathbf{d}}_0; \bar{\breve{\mathbf{d}}}) \equiv \int_0^L {}^n\bar{\hat{\boldsymbol{\varepsilon}}}^T \; {}^n\hat{\mathbf{F}}(\breve{\mathbf{d}}_0)\; dS - \left\{ \int_0^L \bar{\breve{\mathbf{d}}}^T\, \breve{\mathbf{F}}^B(\lambda_0)\, dS + \lambda_0 \left[\bar{\breve{\mathbf{d}}}^{\mathrm{T}}\, \breve{\mathbf{F}}^S \right]_{S=0}^{S=L} \right\} \qquad (10.530)$$

where we have used the hyperelastic constitutive relation: ${}^n\hat{\mathbf{F}}(\hat{\mathbf{d}}_0) = {}^n\hat{\mathbf{D}}\; {}^n\hat{\boldsymbol{\varepsilon}}(\hat{\mathbf{d}}_0)$.
- The definitions: $\mathbf{H}(\bar{\breve{\mathbf{d}}}) \equiv \{\bar{\mathbf{d}} \quad \bar{\boldsymbol{\theta}} \quad \bar{\mathbf{d}},_S \quad \bar{\boldsymbol{\theta}},_S\}^T$, and $\mathbf{H}(\Delta\breve{\mathbf{d}}) \equiv \{\Delta\mathbf{d} \quad \Delta\boldsymbol{\theta} \quad \Delta\mathbf{d},_S \quad \Delta\boldsymbol{\theta},_S\}^T$ and that $\breve{\mathbf{d}} = \{\mathbf{d} \quad \boldsymbol{\theta}\}^T$, with $\mathbf{d} = \{d_1 \equiv u \quad d_2 \equiv v \quad d_3 \equiv w\}^T$ and $\boldsymbol{\theta} = \{\theta_1 \equiv \theta_T \quad \theta_2 \equiv \theta_M \quad \theta_3 \equiv \theta_B\}^T$, is the generalized displacement vector field describing the deformation of a three-dimensional curved beam.
- The load factor and the incremental load factor are denoted by λ, and $\Delta\lambda$, respectively, for proportional loading.
- The Jacobian determinant, $J(\xi)$, relating the arc length parameter, S, and the root coordinate, ξ, is defined as: $J(\xi) = \;||\mathbf{C},_\xi||$ where $C(\xi)$ is the curve representing the centroidal axis of the beam. Clearly, then, $dS = ||\mathbf{C},_\xi||\; d\xi = J(\xi)\; d\xi$.
- Superscript t reminds us that the relevant items are measured in the undeformed Frenet coordinate system, $\mathbf{G}_1 \equiv \mathbf{T}$, $\mathbf{G}_2 \equiv \mathbf{M}$, $\mathbf{G}_3 \equiv \mathbf{B}$; superscript n reminds us that the relevant items are in the rotated coordinate system, $\mathbf{g}_1 (= \mathbf{n} = \mathbf{R}\,\mathbf{G}_1)$, $\mathbf{g}_2 (= \mathbf{m}_n = \mathbf{R}\,\mathbf{G}_2)$, $\mathbf{g}_3 (= \mathbf{b}_n = \mathbf{R}\,\mathbf{G}_3)$.

10.20.2 c-type Finite Element Discretization

Following the standard definition of a finite element discretization, the beam arc length $[0, L]$ is broken into a finite number of intervals: $[0, L] = \bigcup_{e=1}^{e=no.\ of\ elements}$ *Finite Element*^{e}. The root element corresponding to a finite element is defined over the geometric root parameter, $\xi \in [0, 1]$. The root parameter, ξ, is related to the arc length parameter, S, of a beam by: $S(\xi) = \int_0^\xi ||\mathbf{C}^e,_\eta||d\eta$, where $\mathbf{C}^e(\xi)$, $\xi \in [0, 1]$ defines the centroidal axis part of the eth finite element, that is, *Finite Element*^{e}. Based on the assumption that the assembly of the finite elements will be performed by the direct stiffness assembly procedure, we can restrict our discussion to a typical finite element for the beam. For finite element discretization, we must express the generalized displacements, $\breve{\mathbf{d}}$, the variations in generalized displacements, $\bar{\breve{\mathbf{d}}}$, and the incremental generalized displacement, $\Delta\breve{\mathbf{d}}$, in terms of the Bernstein–Bezier interpolation polynomials, $B_i^n(\xi)$, $i = 0, \ldots, n$ where n is the degree of the polynomials. Let us define N as the order of the polynomial given by: $N \equiv n + 1$. We may recall that these polynomials serve as the local basis functions for the c-type beam element.

10.20.3 c-type Beam Element Library

The library consists of curve elements determined simply by the degree of the underlying polynomial of interpolation or the so-called shape functions; both the non-rational Bernstein–Bezier and the rational Bernstein–Bezier interpolation types may be constructed. For general curves, non-rational types are applicable; for conics such as circles, ellipses, and so on, excepting the parabolas, we must use the rational counterparts for accuracy of representation. In fact, for large c-type mega elements, it is absolutely crucial. For example, it is well known that an acute-angled portion of a circle can be quite "closely" approximated geometrically by a cubic curve, but then the subsequent finite element analysis becomes erroneous for such elements of mega size.

10.20.3.1 Non-rational Elements

The root element of an n-degree typical non-rational c-type finite element interpolating the beam centroidal axis curve is a unit straight line, and the geometric interpolation of the element itself can be written as:

$$\boxed{\begin{aligned} \mathbf{C}_0(\xi) &= \sum_{j=0}^{L} \mathrm{B}_i^L(\xi)\ {}^c q_i \\ {}^t\breve{\mathbf{d}}(\xi) &= \sum_{j=0}^{M} \mathrm{B}_i^M(\xi)\ {}^d q_i \end{aligned}} \tag{10.531}$$

where $\mathbf{C}_0(\xi) \equiv \{X(\xi) \quad Y(\xi) \quad Z(\xi)\}^T$ defines the position vector of a point on the beam element, $\xi \in [0, 1]$; $\mathrm{B}_i^L(\xi)$ is the ith Bernstein–Bezier interpolation polynomials of degree L, with the control points ${}^c q_i$ for $i \in \{0, 1, \ldots, L\}$; $\mathrm{B}_i^L(\xi)$ is the ith non-rational basis function, and the control points ${}^c q_i$ for $i \in \{0, 1, \ldots, L\}$ form the Bernstein–Bezier net of control points. Similarly, for ${}^t\breve{\mathbf{d}}(\xi)$, the generalized displacement and rotation interpolation, $\mathrm{B}_i^M(\xi)$, is the ith Bernstein–Bezier interpolation polynomials of degree M, with the generalized displacement control points ${}^d q_i$ for $i \in \{0, 1, \ldots, M\}$; $\mathrm{B}_i^M(\xi)$ is the ith non-rational basis functions, and the control points ${}^d q_i$ for $i \in \{0, 1, \ldots, M\}$ form the Bernstein–Bezier set of generalized displacement control points.

Remark

- We must have $L \leq M$ with the equality for isoparametry, and the inequality for subparametry; superparamtery, that is, $L > M$ is computationally not recommended.

10.20.3.2 Rational Elements

For curve generation, the root element of an n-degree typical rational c-type geometric element interpolating the beam centroidal axis is a unit straight line and the geometric interpolation of the element itself can be written as:

$$\boxed{\begin{aligned} \mathbf{C}_0(\xi) &= \sum_{j=0}^{L} F_i(\xi)\ {}^c q_i \\ F_i(\xi) &\equiv \frac{w_i \mathrm{B}_i^L(\xi)}{\sum_{j=0}^{L} w_i \mathrm{B}_i^L(\xi)} \end{aligned}} \tag{10.532}$$

where $\mathbf{C}_0(\xi) \equiv \{X(\xi) \quad Y(\xi) \quad Z(\xi)\}^T$, as before, defines the position vector of a point on the beam element, $\xi \in [\mathbf{0}, \mathbf{1}]$; $F_i(\xi)$ for $i \in \{0, 1, \ldots, L\}$ is the ith rational basis functions; w_i for $i \in \{0, 1, \ldots, L\}$ are the weights at the control points.

- For generalized displacement interpolation, we still choose the non-rational curves as given in the equation (10.531) because, for unknown displacement distribution, we would like to avoid choosing ad hoc estimates for the corresponding required weights.
- In what follows, we will restrict our discussion to non-rational elements.

10.20.4 Bezier–Bernstein Geometric Curve Element

For a typical non-rational beam finite element, then, a n_c-degree (i.e. $N_c = n_c + 1$ as the order of the polynomial) typical c-type beam geometry can be written as:

$$\boxed{\underbrace{\mathbf{C}_0(\xi)}_{3\times 1} = \underbrace{\sum_{i=0}^{n_c} B_i^{n_c}(\xi)}_{3\times 3N_c} \underbrace{{}^c\mathbf{q}_i}_{3N_c\times 1},} \tag{10.533}$$

where $\mathbf{C}_0(\xi) \equiv \{X(\xi) \quad Y(\xi) \quad Z(\xi)\}^T$, $B_i^{n^c}(\xi)$, ${}^c\mathbf{q}_i$ and n^c are the geometric position vector, the Bernstein–Bezier basis functions, the geometry control vectors and the degree of the geometry element, respectively, at any point $\xi \in [0, 1]$. The equation (10.533) can be rewritten in transformation form as:

$$\boxed{\underbrace{\mathbf{C}_0(\xi)}_{3\times 1} = \underbrace{{}^c\mathbf{T}(\xi)}_{3\times 3N_c} \underbrace{{}^c\mathbf{q}}_{3N_c\times 1},} \tag{10.534}$$

where we have:

$$\boxed{\underbrace{{}^c\mathbf{q}}_{3N_c\times 1} = \{\underbrace{q_0^X \cdots q_{n_c}^X}_{1\times N_c} \underbrace{q_0^Y \cdots q_{n_c}^Y}_{1\times N_c} \underbrace{q_0^Z \cdots q_{n_c}^Z}_{1\times N_c}\}^T} \tag{10.535}$$

the $(3N_c \times 1)$ geometry control vector, and ${}^c\hat{\mathbf{T}}(\hat{\xi})$ as the $(3 \times 3N_c)$ geometry interpolation transformation matrix containing the Bernstein–Bezier polynomials, that is,

$${}^c\mathbf{T}(\xi) = \underbrace{\begin{bmatrix} B_0^{n_c} & B_1^{n_c} & B_{n_c}^{n_c} B_{n_c}^{n_c} & 0\ldots & & \cdots & & & \ldots 0 \\ 0\ldots & & & \ldots 0 & B_0^{n_c} & B_1^{n_c} & B_{n_c}^{n_c} & & \\ 0\ldots & & & & & & \ldots 0 & B_0^{n_c} & B_1^{n_c} & B_{n_c}^{n_c} \end{bmatrix}}_{3\times 3N_c} \tag{10.536}$$

Note that $N_c \equiv n_c + 1$ is the order of the geometric element with degree n_c.

Remarks

- It is recommended whenever possible to have geometry representation at most cubic, that is, $n^c \leq 3$ for ease and efficiency in the computation of tangents, curvatures and twists that are subsequently necessary. So, as an example, for a cubic Bernstein–Bezier geometry polygon, that is, $n_c = 3$, this results in the order of the geometry as: $N = 4$. Thus, we will have 12 (i.e. 3 sets of 4 geometry control nets) as the dimension of the geometry control vector, ${}^c\mathbf{q}$, and the interpolation transformation matrix is of dimension 3×4.

- Additionally, whenever conic curves (i.e. circles, ellipses, etc.) are involved, the geometric representation must be of rational Bernstein–Bezier form for accuracy.

Example: Let us consider a linear element, that is, $n = 1, N = 2$.

We have:

$$\underbrace{\mathbf{C}_0(\xi)}_{3\times1} = \underbrace{\sum_{i=0}^{1} B_i^1(\xi)}_{3\times2} \underbrace{{}^c\mathbf{q}_i}_{2\times1}, \tag{10.537}$$

with:

$$B_0^1(\xi) \equiv 1 - \xi, \qquad B_1^1(\xi) \equiv \xi \tag{10.538}$$

Implying

$$\underbrace{{}^c\mathbf{q}}_{6\times1} = \{\underbrace{q_0^X\ q_1^X}_{1\times2}\ \underbrace{q_0^Y\ q_1^Y}_{1\times2}\ \underbrace{q_0^Z\ q_1^Z}_{1\times2}\}^T \tag{10.539}$$

and

$${}^c\mathbf{T}(\xi) = \underbrace{\begin{bmatrix} \underbrace{\mathbf{T}}_{1\times2} & \underbrace{\mathbf{0}}_{1\times2} & \underbrace{\mathbf{0}}_{1\times2} \\ \underbrace{\mathbf{0}}_{1\times2} & \underbrace{\mathbf{T}}_{1\times2} & \underbrace{\mathbf{0}}_{1\times2} \\ \underbrace{\mathbf{0}}_{1\times2} & \underbrace{\mathbf{0}}_{1\times2} & \underbrace{\mathbf{T}}_{1\times2} \end{bmatrix}}_{3\times6} \text{ where } \mathbf{T} \equiv \{(1-\xi) \quad \xi\} \tag{10.540}$$

10.20.5 The c-type Beam Elements and State Control Vectors

Let us, then, introduce a n-degree c-type beam element as:

$$\begin{aligned} \breve{\mathbf{d}}(\xi) &= \breve{\mathbf{T}}^d(\xi)\, \breve{\mathbf{q}}^d, \\ \breve{d}_j(\xi) &= \sum_{i=0}^{n} B_i^n(\xi)\, \breve{q}_i^{d_j}, \quad j = 1, \ldots, 6 \end{aligned} \tag{10.541}$$

where we have introduced the following:

Definitions

- $\breve{\mathbf{q}}^d = \{q_0^u \quad q_1^u \quad \cdots \quad q_n^u \quad \cdots \quad q_0^{\theta_b} \quad \cdots \quad q_n^{\theta_b}\}^T$ be the $(6N \times 1)$ displacement control vector.
- $\bar{\breve{\mathbf{q}}}^d$ be the $(6N \times 1)$ virtual displacement control vector, similar to $\breve{\mathbf{q}}^d$,
- $\Delta\breve{\mathbf{q}}^d$, be the $(6N \times 1)$ incremental control vector, similar to $\breve{\mathbf{q}}^d$,
- $\breve{\mathbf{T}}^d(\xi)$ be the $(6 \times 6N)$ interpolation transformation matrix containing the Bernstein–Bezier polynomials, that is,

$$\breve{\mathbf{T}}^d(\xi) = \underbrace{\begin{bmatrix} B_0^n & \cdots & B_n^n & 0 & 0 & \cdots & \cdots & \cdots & 0 \\ 0 & 0 & 0 & B_0^n & \cdots & B_n^n & 0 & \cdots & 0 \\ & & \cdots & \cdots & & & & & \\ & & \cdots & \cdots & & & & & \\ 0 & 0 & \cdots & \cdots & 0 & B_0^n & \cdots & B_n^n \end{bmatrix}}_{6\times6N} \tag{10.542}$$

Now, let us further introduce the following.

Definition

- $\breve{\mathbf{B}}(\xi)$ be the $(12 \times 6N)$ augmented interpolation transformation matrix as:

$$\breve{\mathbf{B}}(\xi) \equiv \begin{bmatrix} \underbrace{\breve{\mathbf{T}}^d(\xi)}_{6\times6N} \\ ----- \\ \underbrace{\frac{1}{J}\breve{\mathbf{T}}^d{}_{,\xi}(\xi)}_{6\times6N} \end{bmatrix} \tag{10.543}$$

where $\mathbf{T}^d{}_{,\xi}$ is computed by the first derivative recursive formula of the Bernstein basis functions at Gauss points as:

$$\frac{d}{d\xi}B(\xi) = n\left(B_{i-1}^{n-1}(\xi) - B_i^{n-1}(\xi)\right) \tag{10.544}$$

Finally, we must express the infinitesimal length, $dS = J(\hat{\xi})\, d\hat{\xi}$, of the area integrals where $J(\hat{\xi})$ is the determinant of the Jacobian matrix expressed in terms of the geometry coordinate parameters, $\hat{\xi}$. Based on the above definitions, we have:

$$\boxed{\begin{aligned} \breve{\mathbf{d}} &= \breve{\mathbf{T}}^d(\xi)\, \breve{\mathbf{q}}^d \\ \bar{\breve{\mathbf{d}}} &= \breve{\mathbf{T}}^d(\xi)\, \bar{\breve{\mathbf{q}}}^d \\ \Delta\breve{\mathbf{d}} &= \breve{\mathbf{T}}^d(\xi)\, \Delta\breve{\mathbf{q}}^d \end{aligned}} \tag{10.545}$$

and

$$\boxed{\begin{aligned}\underbrace{\mathbf{H}(\breve{\bar{\mathbf{d}}})}_{12\times 1} &= \underbrace{\breve{\mathbf{B}}(\xi)}_{12\times 6N}\ \underbrace{\breve{\bar{\mathbf{q}}}^{d}}_{6N\times 1}\\ \underbrace{\mathbf{H}(\Delta\breve{\mathbf{d}})}_{12\times 1} &= \underbrace{\breve{\mathbf{B}}(\xi)}_{12\times 6N}\ \underbrace{\Delta\breve{\mathbf{q}}^{d}}_{6N\times 1}\end{aligned}} \qquad (10.546)$$

- Note that both the $\hat{\mathbf{T}}^{d}(\hat{\xi})$ and $\hat{\mathbf{B}}(\hat{\xi})$ matrices for all Gauss points can be computed and saved once, and stored away for repetitive use in the Newton-type iterative algorithm; of course, in adaptive refinements the locations of the Gauss points change, requiring recomputation.
- For convergence, we must have the elements at most isoparametric, that is, $n^c = n$; we must always avoid superparametry, that is, n^c is never allowed to exceed n.
- For computational efficiency such as tangent and curvature computations, and so on, a sub-parametric representation, that is, n^c strictly less than n, with $n^c = 3$, that is, up to cubic spline representation for the non-rational geometry, seems an optimal choice. Similarly, for rational geometry such as circles, ellipses, and so on, the quadratic rational spline is the desired geometric representation.
- It turns out that for general 3D beam finite element analysis, the optimum choice for the generalized displacement distribution must be with a quintic Bernstein–Bezier net, that is, $n = 5$, resulting in the order of the element as: $N = 6$. Thus, we will have 36 (i.e. 6×6 control points) as the dimension of the generalized displacement control vector, ${}^{d}\mathbf{q}$, and the interpolation transformation matrix is of dimension 6×36; however, if desired, all the control points other than the end control points can easily be statically condensed to produce compact matrices.

10.20.6 Computational Symmetric Material Stiffness Matrix, $\mathbf{K}_M^C$

By substituting equation (10.502) into equation (10.483), we get:

$$\boxed{\begin{aligned}\mathcal{K}_M(\breve{\mathbf{d}},\,\breve{\bar{\mathbf{d}}};\,\Delta\breve{\mathbf{d}}) \equiv \underbrace{\breve{\bar{\mathbf{q}}}^{dT}}_{1\times 6N} &\underbrace{\left[\int_0^1 \left\{\underbrace{\breve{\mathbf{B}}(\xi)^T}_{6N\times 12}\ \underbrace{\overbrace{\mathbf{E}(\breve{\mathbf{d}})^T}^{12\times 6}\ \overbrace{{}^{n}\mathbf{D}}^{6\times 6}\ \overbrace{\mathbf{E}(\breve{\mathbf{d}})}^{6\times 12}}_{12\times 12}\ \underbrace{\breve{\mathbf{B}}(\xi)}_{12\times 6N}\right\} J(\xi)\, d\xi\right]}_{6N\times 6N} \underbrace{\Delta\breve{\mathbf{q}}^{d}}_{6N\times 1}\\ = \underbrace{\breve{\bar{\mathbf{q}}}^{dT}}_{1\times 6N}\ &\underbrace{\mathbf{K}_M^C}_{6N\times 6N}\ \underbrace{\Delta\breve{\mathbf{q}}^{d}}_{6N\times 1}\end{aligned}}$$

(10.547)

where, we have introduced:

Definition: The computational material stiffness matrix, $\mathbf{K}_M^C$, as:

$$\underbrace{\mathbf{K}_M^C}_{6N\times 6N} \equiv \underbrace{\int_0^1 \left\{ \underbrace{\breve{\mathbf{B}}(\xi)^T}_{6N\times 12} \underbrace{\overbrace{\mathbf{E}(\breve{\mathbf{d}})^T}^{12\times 6} \overbrace{{}^n\breve{\mathbf{D}}}^{6\times 6} \overbrace{\mathbf{E}(\breve{\mathbf{d}})}^{6\times 12}}_{12\times 12} \underbrace{\breve{\mathbf{B}}(\xi)}_{12\times 6N} \right\} J(\xi)\, d\xi}_{6N\times 6N} \tag{10.548}$$

***Exercise*:** Derive expression given by equation (10.548)

10.20.6.1 Code Fragment for E Matrix

```
   Subroutine FormEMatrix(AMat,C0,nC,E,nE,W,R,AA,CK,nG,Iout)
! -------------------------------------------------------------------
--------------
!  Inputs
!        AMat(nC,nC)        = Rotation Matrix
!        nG                 = 12
!        C0(nC)             = init.Curvature vec = {torsion,0,curvature} = k0
!        nC                 = 3
!        nE                 = 6
!        W(nC,nC)           = W Matrix
!        R(nC,nC)           = R Matrix
!        AA(nC)             = e + [k0]d
!        CK(nC)             = k0 + k = k0 + W(thetaPrimed)
!
!  Outputs
!        E(nE,nG)           = E Matrix
! ----------------------------------------------------------------
   Implicit Real(kind=8) (a-h,o-z)
!
   Real(kind=8)            C0 ,E ,AMat
   Dimension               C0(nC),E(nE,nG),AMat(nC,nC)
   Real(kind=8)            E11 ,E12
   Dimension               E11(nC,nC),E12(nC,nC)
   Real(kind=8)            E13 ,E22 ,E23 ,E24
   Dimension               E13(nC,nC),E22(nC,nC),E23(nC,nC),E24(nC,nC)
   Real(kind=8)            CK ,AA
   Dimension               CK(nC),AA(nC)
   Real(kind=8)            W ,R
   Dimension               W(nC,nC),R(nC,nC)
!
   DATA zero/0.D0/,one/1.0D0/,two/2.0D0/,three/3.0D0/
!-------------------------------------------------------------------
------- Get EBlocks
   call FormEBlocks(C0,nC,W,R,AA,CK,AMat,E11,E12,E13,E22,E24,Iout)
```

```
!----------------------------------------------------------------------
-------- Form EMatrix
   E = zero
!  --------
   do 40 i = 1,3
   do 40 j = 1,3
   E(i,j)                   = E11(i,j)
   E(i,j+3)     = E12(i,j)
   E(i,j+6)     = E13(i,j)
   E(i+3,j+3)   = E22(i,j)
   E(i+3,j+9)   = E24(i,j)
40 continue
!-----------------------------------------------------------------
   return
   End

   Subroutine FormEBlocks(C0,nC,W,R,AA,CK,AMat,E11,E12,E13,E22,E24,
&                                                 Iout)
!  --------------------------------------------------------------------
--------------
!  Inputs
!       C0(nC)              = init. Curvature vector = {torsion,0,curvature}
!       nC                  = 3
!       AMat                = Rotation Matrix
!       W                   = matrix
!       R                   = matrix
!       AA                  = a Vector
!       CK                  = Kc Vector
!
!  Outputs
!       E11(nC,nC)          = E11 block
!       E12(nC,nC)          = E12 block
!       E13(nC,nC)          = E13 block
!       E22(nC,nC)          = E22 block
!       E24(nC,nC)          = E24 block
!  ------------------------------------------------------------------
   Implicit Real(kind=8) (a-h,o-z)
!
   Real(kind=8)       C0
   Dimension          C0(nC)
   Real(kind=8)       E11 ,E12 ,E13 ,E22
   Dimension          E11(nC,nC),E12(nC,nC),E13(nC,nC),E22(nC,nC)
   Real(kind=8)       CK ,AA ,AMat ,TAMat
   Dimension          CK(nC),AA(nC),AMat(nC,nC),TAMat(nC,nC)
   Real(kind=8)       W ,R ,E24 ,sAA ,sKc
   Dimension          W(nC,nC),R(nC,nC),E24(nC,nC),sAA(nC,nC),sKc(nC,nC)
!
   DATA zero/0.D0/,one/1.0D0/,two/2.0D0/,three/3.0D0/
!---------------------------------------------------------------------
-------------------------AAMatTran
   TAMat = TRANSPOSE(AMat)
!---------------------------------------------------------------------
-------------------------E11Block
```

```
   call FormSkew(C0,nC,E11,Iout)                    !skew[k0]
   E11 = MATMUL(TAMat, E11)                         !AMatTransposeE11
!-------------------------------------------------------------------------
--------------------------E12Block
   call FormSkew(AA,nC,sAA,Iout)                    !skew[a]
   E12 = MATMUL(sAA, W)                             ![a]W
   E12 = MATMUL(TAMat, E12)                         !AMatTransposeE12
!-------------------------------------------------------------------------
--------------------------E13Block
   E13 = TAMat                                      !AMatTransposeE13
!-------------------------------------------------------------------------
--------------------------E22Block
   call FormSkew(CK,nC,sKc,Iout)                    !skew[Kc]
   E22 = R + MATMUL(sKc,W)                    !R + [Kc]W
   E22 = MATMUL(TAMat, E22)                         !AMatTransposeE22
!-------------------------------------------------------------------------
--------------------------E24Block
   E24 = W
   E24 = MATMUL(TAMat, E24)                         !AMatTransposeE24
!-------------------------------------------------------------------
   return
   end
```

10.20.7 Computational Symmetric Geometric Stiffness Matrix, K_G^C

By substituting equation (10.546) into equation (10.529), we get:

$$\begin{aligned}\mathcal{K}_G(\breve{\mathbf{d}},\ \bar{\mathbf{d}};\ \Delta\breve{\mathbf{d}}) &\equiv \underbrace{\bar{\mathbf{q}}^{d\,T}}_{1\times 6N} \underbrace{\left[\int_0^1 \left\{ \underbrace{\breve{\mathbf{B}}(\xi)^T}_{6N\times 12}\ \underbrace{\mathbf{G}(\breve{\mathbf{d}},\breve{\mathbf{F}})}_{12\times 12}\ \underbrace{\breve{\mathbf{B}}(\xi)}_{12\times 6N} \right\} J(\xi)\, d\xi \right]}_{6N\times 6N} \underbrace{\Delta\breve{\mathbf{q}}^d}_{6N\times 1} \\ &= \underbrace{\bar{\mathbf{q}}^{d\,T}}_{1\times 6N}\ \underbrace{K_G^C}_{6N\times 6N}\ \underbrace{\Delta\breve{\mathbf{q}}^d}_{6N\times 1}\end{aligned} \tag{10.549}$$

where, we have introduced the following.

Definition: The computational geometric stiffness matrix, K_G^C, as:

$$\underbrace{K_G^C}_{6N\times 6N} \equiv \underbrace{\int_0^1 \left\{ \underbrace{\breve{\mathbf{B}}(\xi)^T}_{6N\times 12}\ \underbrace{\mathbf{G}(\breve{\mathbf{d}},\breve{\mathbf{F}})}_{12\times 12}\ \underbrace{\breve{\mathbf{B}}(\xi)}_{12\times 6N} \right\} J(\xi)\, d\xi}_{6N\times 6N} \tag{10.550}$$

where the symmetric geometric stiffness matrix, $\mathbf{G}$, is given by:

$$\mathbf{G}(\hat{\mathbf{d}}, \hat{\mathbf{F}}) \equiv \begin{bmatrix} \mathbf{0} & \mathbf{G}_{12} & \mathbf{0} & \mathbf{0} \\ \mathbf{G}_{12}^{\mathbf{T}} & \mathbf{G}_{22} & \mathbf{G}_{23} & \mathbf{G}_{24} \\ \mathbf{0} & \mathbf{G}_{23}^{\mathbf{T}} & \mathbf{0} & \mathbf{0} \\ \mathbf{0} & \mathbf{G}_{24}^{\mathbf{T}} & \mathbf{0} & \mathbf{0} \end{bmatrix} \tag{10.551}$$

with

$$\begin{aligned}
\mathbf{G}_{12} &\equiv [\mathbf{k}^0][{}^t\mathbf{F}]\,\mathbf{W} \\
\mathbf{G}_{22} &\equiv \tfrac{1}{2}\mathbf{W}^{\mathbf{T}}\{[{}^t\mathbf{F} \otimes \mathbf{a}] + [\mathbf{a} \otimes {}^t\mathbf{F}] - 2({}^t\mathbf{F}\ \ \mathbf{a})\}\mathbf{W} + c_2(\theta\ \ \mathbf{b}^f)\mathbf{I} \\
&\quad + \tfrac{1}{2}([\mathbf{h}^b \otimes \theta] + [\theta \otimes \mathbf{h}^b]) \\
&\quad + \tfrac{1}{2}\mathbf{W}^{\mathbf{T}}\{[{}^t\mathbf{M} \otimes \mathbf{k}^c] + [\mathbf{k}^c \otimes {}^t\mathbf{M}] - 2({}^t\mathbf{M}\ \ \mathbf{k}^c)\}\mathbf{W} + c_2(\theta\ \ \mathbf{b}^m)\mathbf{I} \\
&\quad + \tfrac{1}{2}([\mathbf{h}^m \otimes \theta] + [\theta \otimes \mathbf{h}^m]) + \mathbf{W}^{\mathbf{T}}[{}^t\mathbf{M}]\,\mathbf{X} + \mathbf{X}\,[{}^t\mathbf{M}]^{\mathbf{T}}\,\mathbf{W} \\
\mathbf{G}_{23} &\equiv \mathbf{W}^{\mathbf{T}}\,[{}^t\mathbf{F}] \\
\mathbf{G}_{24} &\equiv \mathbf{W}^{\mathbf{T}}\,[{}^t\mathbf{M}]\,\mathbf{W} + \mathbf{Y}({}^t\mathbf{M})
\end{aligned} \tag{10.552}$$

and

$$\begin{aligned}
c_1 &\equiv (1 - \cos\theta)/\theta^2, \\
c_2 &\equiv (\theta - \sin\theta)/\theta^3, \\
\bar{c}_1 \equiv \delta c_1 \equiv c_3 &\equiv \{(1 - \theta^2 c_2) - 2c_1\}/\theta^2, \\
\bar{c}_2 \equiv \delta c_2 \equiv c_4 &\equiv (c_1 - 3c_2)/\theta^2
\end{aligned} \tag{10.553}$$

$$\begin{aligned}
&\mathbf{a} \equiv \mathbf{e} + [\mathbf{k}^0]\mathbf{d} = \{(\hat{\mathbf{1}} + \mathbf{d},_s) + [\mathbf{k}^0]\mathbf{d}\} \\
&\mathbf{k}^t \equiv \mathbf{W}\,\theta,_s \qquad \mathbf{k}^c \equiv \mathbf{k}^0 + \mathbf{k}^t \\
&\mathbf{b}^f \equiv \mathbf{F}^t \times \mathbf{a} \qquad \mathbf{b}^m \equiv \mathbf{M}^t \times \mathbf{k}^c
\end{aligned} \tag{10.554}$$

$$\begin{aligned}
\mathbf{W} &\equiv \mathbf{I} + c_1\mathbf{\Theta} + c_2\mathbf{\Theta}^2 \\
\mathbf{X}_\alpha &\equiv -c_1\mathbf{\Theta},_\alpha + c_2\{[\theta \otimes \theta,_\alpha] - [\theta,_\alpha \otimes \theta] + (\theta \bullet \theta,_\alpha)\mathbf{I}\} \\
&\quad + \mathbf{V}\,[\theta,_\alpha \otimes \theta]) \\
\mathbf{V} &\equiv -c_2\mathbf{I} + c_3\mathbf{\Theta} + c_4\mathbf{\Theta}^2 \\
\mathbf{Y}(\mathbf{x}) &\equiv -c_1[\mathbf{x}] + c_2([\mathbf{x} \otimes \theta] - [\theta \otimes \mathbf{x}]) + c_2(\theta \bullet \mathbf{x})\mathbf{I} \\
&\quad + [\theta \otimes \mathbf{V}^{\mathbf{T}}\mathbf{x}], \quad \forall \mathbf{x} \in R^3 \\
\mathbf{h}_\alpha^f &\equiv \mathbf{V}^{\mathbf{T}}\,\mathbf{b}_\alpha^f \qquad \mathbf{h}_\alpha^m \equiv \mathbf{V}^{\mathbf{T}}\,\mathbf{b}_\alpha^m
\end{aligned} \tag{10.555}$$

Exercise: Derive expression given by equation (10.550).

10.20.7.1 Code Fragment for G Matrix

```
   Subroutine FormGMatrix(HD,nG,FM,nF,C0,nC,G,W,R,AA,CK,Iout)
!  -------------------------------------------------------------------
-------------
!  Inputs
!        HD(nG)             = {dis,rot,disPrimed,rotPrimed}
!        nG                 = 12
!        FM(nC)             = {force,moment}
!        nF                 = 6
!        C0(nC)             = init. Curvature vector = {torsion,0,curvature}
!        nC                 = 3
!
!  Outputs
!        G(nG,nG)           = GMatrix
!  ------------------------------------------------------------------
   Implicit Real(kind=8) (a-h,o-z)
!
   Real(kind=8)       HD ,FM ,C0 ,G
   Dimension          HD(nG),FM(nF),C0(nC),G(nG,nG)
   Real(kind=8)       G12 ,G22 ,G23 ,G24
   Dimension          G12(nC,nC),G22(nC,nC),G23(nC,nC),G24(nC,nC)
!
   DATA zero/0.D0/,one/1.0D0/,two/2.0D0/,three/3.0D0/
!---------------------------------------------------------------------
------- Get GBlocks
   call FormGBlocks(HD,nG,FM,nF,C0,nC,W,R,AA,CK,G12,G22,G23,G24,Iout)
!---------------------------------------------------------------------
------- Form GMatrix
   G = zero
!  -------------------------------------- Upper Triangle with Diagonals
   do 10 i = 1,3
   do 10 j = 1,3
   G(i,j+3)      = G12(i,j)
   G(i+3,j+3)    = G22(i,j)
   G(i+3,j+6)    = G23(i,j)
   G(i+3,j+9)    = G24(i,j)
10 continue
!------------------------------------------------ Symmetry:Lower Triangle
   do 50 i = 1,11
   do 50 j = i+1,12
   G(j,i) = G(i,j)
50 continue
!----------------------------------------------------------------
   return
   End

   Subroutine FormGBlocks(HD,nG,FM,nF,C0,nC,W,R,AA,CK,
&                         G12,G22,G23,G24,Iout)
```

```
! ------------------------------------------------------------------
--------------
!  Inputs
!       HD(nG)        = {dis,rot,disPrimed,rotPrimed}
!       nG            = 12
!       FM(nC)        = {force,moment}
!       nF            = 6
!       C0(nC)        = initCurvature vector = {torsion,0,curvature}
!       nC            = 3
!
!  Outputs
!       Gij(nC,nC)        = GBlocks
! ---------------------------------------------------------------
   Implicit Real(kind=8) (a-h,o-z)
!
   Real(kind=8)       HD ,FM ,C0
   Dimension          HD(nG),FM(nF),C0(nC)
   Real(kind=8)       G12 ,G22 ,G23 ,G24
   Dimension          G12(nC,nC),G22(nC,nC),G23(nC,nC),G24(nC,nC)
   Real(kind=8)       Di ,Dp ,Ro ,Rp ,Fo ,XM
   Dimension          Di(nC),Dp(nC),Ro(nC),Rp(nC),Fo(nC),XM(nC)
   Real(kind=8)       CK ,AA ,BB ,CC ,Hat1
   Dimension          CK(nC),AA(nC),BB(nC),CC(nC),Hat1(3)
   Real(kind=8)       W ,R ,SM ,SF ,SC
   Dimension          W(nC,nC),R(nC,nC),SM(nC,nC),SF(nC,nC),SC(nC,nC)
   Real(kind=8)       FA ,XMK ,TA ,ATA
   Dimension          FA(nC,nC),XMK(nC,nC),TA(nC,nC),ATA(nC,nC)
   Real(kind=8)       WMR ,TMT ,SMt
   Dimension          WMR(nC,nC),TMT(nC,nC),SMt(nC,nC)
!
   common/ThetaConstants/c1,c2,c1b,c2b
!
   DATA zero/0.D0/,one/1.0D0/,two/2.0D0/,three/3.0D0/
   DATA Hat1/1.0D0,0.D0,0.D0/
!  ------------------------------------------------------------------
----- H(dHat)
!   call Extract_Vector(HD,nG,Di,nC,1,Iout)  !displacement
   call Extract_Vector(HD,nG,Ro,nC,4,Iout)   !rotation
!   call Extract_Vector(HD,nG,Dp,nC,7,Iout)  !displacementPrimed
   call Extract_Vector(HD,nG,Rp,nC,10,Iout)!rotationPrimed
!
   call Extract_Vector(FM,nF,Fo,nC,1,Iout)          !Force
   call Extract_Vector(FM,nF,XM,nC,4,Iout)    !Moment
!------------------------------------------------------------------
--G22Block:Translational
!  ------------------------------------------------------------ FA =
WTran(Fxa + axF -2(F.a)I)W
   call FormF_a_PartMatrix(Fo,AA,FA,3,1,Iout)       !Fxa + axF -2(F.a)I
!   call Form_WMatrix(Ro,W,nC,c1,c2,c1b,c2b,Iout)
   FA = MATMUL(FA, W)                         !(Fxa + axF -2(F.a)I)W
   FA = MATMUL(TRANSPOSE(W), FA) !WTran(Fxa + axF -2(F.a)I)W
!
   call CrossProduct(Fo,AA,BB,Iout)                    ! b = F x a
```

```
   call DotProduct(Ro,BB,dot,nV,Iout)                    ! theta dot b
   C2TdotB = c2*dot                                      ! c2*(theta dot b)
!  ----------------------------------- .5*FA + c2(theta dot b)I
   FA = 0.5D0*FA
   FA(1,1) = FA(1,1) + C2TdotB
   FA(2,2) = FA(2,2) + C2TdotB
   FA(3,3) = FA(3,3) + C2TdotB
!  ----------------------------------------------------------
---------------- Rotational
!  ----------------------------------------XMK = WTran(Mxkc + kcxM -
2(M.kc)I)W
   call FormF_a_PartMatrix(XM,CK,XMK,3,2,Iout)!Mxkc + kcxM -2(M.kc)I
!
   XMK = MATMUL(XMK, W)            !(Mxkc + kcxM -2(M.kc)I)W
   XMK = MATMUL(TRANSPOSE(W), XMK)!WTran(Mxkc + kcxM -2(M.kc)I)W
!
   call CrossProduct(XM,CK,CC,Iout)                         ! c = M x kc
   call DotProduct(Ro,CC,dot,nV,Iout)                       ! theta dot c
   C2TdotC = c2*dot                                         ! c2*(theta dot c)
!  ----------------------------------- .5*XMK + c2(theta dot c)I
   XMK = 0.5D0*XMK
   XMK(1,1) = XMK(1,1) + C2TdotB
   XMK(2,2) = XMK(2,2) + C2TdotB
   XMK(3,3) = XMK(3,3) + C2TdotB
!  ----------------------------------------------------------
-AlfaCxRot + RotxAlfaC
   call FormThetaAlfaTrans_Matrix(Ro,CC,TA,nC,c1,c2,c1b,c2b,Iout)   !RotxAlfa
   ATA = 0.5D0*(TA + TRANSPOSE(TA))
!  ----------------------------------------------------------
-AlfaBxRot + RotxAlfaB
   call FormThetaAlfaTrans_Matrix(Ro,BB,TA,nC,c1,c2,c1b,c2b,Iout)   !RotxAlfa
   ATA = ATA + 0.5D0*(TA + TRANSPOSE(TA))
!  -------------------------------------------------------
WTranMR+RTranMTranW
   call FormSkew(XM,nC,SM,Iout)          !skew[M]
   call FormR_Matrix(Ro,Rp,R,nC,c1,c2,c1b,c2b,Iout) !R matrix
!
   WMR = MATMUL(TRANSPOSE(W),SM) !WTran[M]
   WMR = MATMUL(WMR,R)                   !WTran[M]R
!
   WMR = WMR + TRANSPOSE(WMR)
!  ----------------------------------------------------------
---------- G22Block
   G22 = FA + XMK + ATA + WMR
!-----------------------------------------------------------
-------------------------G12Block
!-----------------------------------------------------------
---------------------------------
   call FormSkew(Fo,nC,SF,Iout)           !skew[F]
   call FormSkew(C0,nC,SC,Iout)           !skew[k0]
!
   G12 = MATMUL(SC,SF)                    ![k0][Ft]
   G12 = MATMUL(G12,W)                    ![k0][Ft]W
```

```
!------------------------------------------------------------------
--------------------------G23Block
   G23 = MATMUL(TRANSPOSE(W),SF)        !WTran[Ft]
!------------------------------------------------------------------
--------------------------G24Block
   TMT = MATMUL(TRANSPOSE(W),SM) !WTran[Mt]
   TMT = MATMUL(TMT,W)                              !WTran[Mt]W
!
   call FormS_Matrix(Ro,XM,SMt,nC,c1,c2,c1b,c2b,Iout) !S(Mt) Matrix
!
   G24 = TMT + SMt
!------------------------------------------------------------
   return
   end
```

10.20.8 Unbalanced Force

By substituting equation (10.498) into equation (10.530), we get:

$$\overbrace{G(\lambda_0, \breve{\mathbf{d}}_0; \bar{\mathbf{d}})}^{\text{Unbalanced force}} = \bar{\mathbf{q}}^{dT}\, \breve{\mathbf{R}}(\lambda_0, \breve{\mathbf{d}}_0) \tag{10.556}$$

where, we have defined the following.

Definition

$$\boxed{\breve{\mathbf{R}}(\lambda_0, \breve{\mathbf{d}}_0) \equiv \left\{ \int_0^1 [\breve{\mathbf{B}}^T\, \breve{\mathbf{E}}^{T\,n}\breve{\mathbf{F}} - \breve{\mathbf{T}}^T\, \breve{\mathbf{F}}^B(\lambda_0, \xi)\,] J(\xi)\, d\xi - \lambda_0(\breve{\mathbf{T}}^T(1)\ \breve{\mathbf{F}}^S(1) - \ \breve{\mathbf{T}}^T(0)\ \breve{\mathbf{F}}^S(0)) \right\}} \tag{10.557}$$

10.20.9 Incremental Applied Force

By substituting equation (10.498) into the applied force part of equation (10.524), the applied external force for a step can be expressed as:

$$\overbrace{\left\{ \left(\int_0^1 \bar{\breve{\mathbf{d}}}^T\, \breve{\mathbf{F}}^B{}_{,\lambda}(\lambda)\, J(\xi)\, d\xi \right) + \left[\bar{\breve{\mathbf{d}}}^T\, \breve{\mathbf{F}}^S \right]_{\xi=0}^{\xi=1} \right\}}^{\text{Incremental applied force}} \Delta\lambda = \bar{\mathbf{q}}^{dT}\, \breve{\mathbf{p}}(\lambda)\, \Delta\lambda \tag{10.558}$$

where, we have defined the following.

Definition

$$\boxed{\breve{\mathbf{p}}(\lambda) \equiv \int_0^1 \breve{\mathbf{T}}^T(\xi)\, \breve{\mathbf{F}}^B(\lambda, \xi)_{,\lambda}\ J(\xi)\, d\xi\ +\ \breve{\mathbf{T}}^T(1)\ \breve{\mathbf{F}}^S(1) - \ \breve{\mathbf{T}}^T(0)\ \breve{\mathbf{F}}^S(0)} \tag{10.559}$$

10.20.10 Geometrically Exact Element Linearized Incremental Equation

Now substituting the expressions given in equations (10.548), (10.550), (10.557), and (10.559) in equation (10.524), we have:

$$\boxed{\breve{\bar{\mathbf{q}}}^{d\,T} \left\{ \mathbf{K}_M^C(\breve{\mathbf{q}}^d) + K_G^C(\breve{\mathbf{q}}^d) \right\} \Delta\breve{\mathbf{q}}^d = \breve{\bar{\mathbf{q}}}^{d\,T} \{ \breve{\mathbf{p}}(\lambda)\, \Delta\lambda - \breve{\mathbf{R}}(\lambda_0, \breve{\mathbf{d}}_0) \}} \tag{10.560}$$

Noting that equation (10.560) above is true for any arbitrary admissible generalized virtual control displacement, $\breve{\bar{\mathbf{q}}}^d$, we can finally write the geometrically exact linearized finite element equation in unknown incremental generalized displacement control, $\Delta\breve{\mathbf{q}}^d$, and the incremental load factor, $\Delta\lambda$, as:

$$\boxed{\mathbf{K}_T^c\, \Delta\breve{\mathbf{q}}^d + \breve{\mathbf{p}}(\lambda)\, \Delta\lambda = -\breve{\mathbf{R}}(\lambda_0, \breve{\mathbf{d}}_0)} \tag{10.561}$$

where, we have introduced the following.

Definition: The computational tangent stiffness matrix, $\mathbf{K}_T^c$:

$$\boxed{\mathbf{K}_T^c \equiv \mathbf{K}_M^C + K_G^C} \tag{10.562}$$

10.20.11 Implementation: Accuracy and Symmetry Preservation

Our main goal here is to discuss various important issues and solutions relating to economical, robust and stable numerical implementation, and to provide several illustrative and representative examples for beams under quasi-static loading resulting in nonlinear (finite displacement and rotation) responses. As to numerical implementation, the following topics are in order:

- Numerical integration of the material stiffness and the geometric stiffness: the solution requires application of Gaussian quadrature of appropriate degree to the integrals.
- Maintaining symmetry of the tangent operator during iteration for proportional loading: this is achieved by treating separately the linearized equations related to incremental displacement and rotations, $(\Delta\mathbf{d}, \Delta\boldsymbol{\theta})$, from that involving the incremental proportionality factor, $\Delta\lambda$.
- Newton-type iterative solution of nonlinear matrix equations.

10.20.12 Gauss Quadrature Points

For a quintic Bernstein–Bezier polygon, for example, over an element, that is, with $n = 5$, we have for the order of the element as: $N = 6$. Thus, for full integration, we will estimate a set of six Gauss quadrature points. With the locations of the Gauss quadrature points fixed, we need to construct the local Frenet frame at these points with the triad of base vectors, $\mathbf{T}, \mathbf{M}, \mathbf{B}$; by taking derivatives of equation (10.533) with respect to the curve parameter, we get:

$$\boxed{\begin{aligned}&\mathbf{T} = \frac{\mathbf{C}_{,\xi}(\xi)}{||\mathbf{C}_{,\xi}(\xi)||}\\ &\mathbf{B} = \frac{\mathbf{C}_{,\xi}(\xi) \times \mathbf{C}_{,\xi\xi}(\xi)}{||\mathbf{C}_{,\xi}(\xi) \times \mathbf{C}_{,\xi\xi}(\xi)||}\\ &\mathbf{M} = \mathbf{B} \times \mathbf{T}\end{aligned}} \tag{10.563}$$

where we use the following first-order forward difference operator, Δ^1 as:

$$\boxed{\Delta^1\,{}^c\mathbf{q}_i \equiv {}^c\mathbf{q}_{(i+1)} - {}^c\mathbf{q}_i} \tag{10.564}$$

Note that the great advantage of using difference operators is that we only need to know and use the differences of the original Bezier control points of the geometry. With this background, we can express all the components of the desired curvature vector: $\mathbf{k}^0 \equiv \{\tau \quad 0 \quad \kappa\}^T$ in terms of the coordinate parameters, ξ, with the local radius of curvature, $\rho = \frac{1}{\kappa}$, is given by $\boxed{\rho = \frac{||\mathbf{C}_{,\xi}||^3}{||\mathbf{C}_{,\xi} \times \mathbf{C}_{,\xi\xi}||}}$, κ, the curvature, is defined as the reciprocal of ρ. The twist, τ, of the curve is given by $\boxed{\tau = \frac{\mathbf{C}_{,\xi} \times \mathbf{C}_{,\xi\xi} \bullet \mathbf{C}_{,\xi\xi\xi}}{||\mathbf{C}_{,\xi} \times \mathbf{C}_{,\xi\xi}||^2}}$. Finally, $J(\xi)$, the determinant of the Jacobian matrix is evaluated at all the Gauss quadrature points of integration.

Remark

- Thus, note that the reference curve properties: the initial curvature vectors, $\mathbf{k}^0$ along with the determinant $J(\hat{\xi})$ of the Jacobian matrix, can be computed at all Gauss points and saved once, and stored away for repetitive use in the Newton-type iterative algorithm; of course, in adaptive refinement, the locations of the Gauss points change, requiring recomputation.

10.20.12.1 Code Fragment for Reference Geometry Info

```
      Subroutine Get_ReferenceInfo(Qc,nDim,nQc,RoCOld,CuCOld,nCInt,
&  Ri,jInt,R0,C0,CK,nC,T,B,
&  nDof,nDof2,nStif,Det,Iout,Nel,Nint)
!     -------------------------------------------------------------
-----------------
!     Get Reference Geometric Info.
!     -------------------------------------------------------------
-----------------
!     Inputs
!            Qc(nDim,nQc)       = Geometry Controls
!            nDim               = 3
!            nQc                = order of Bezier
!            RoCOld(nCInt)      = rot Controls:Old for all Gauss point
!            CuCOld(nCInt)      = Curvature:Old
!            nCInt              = no. of Gauss Integration points * nDim (=3)
!                               = nDim*nInt
```

```
!         Ri                    = Current Gauss location Pt.
!         jInt                  = Current Gauss Index
!         nDof                  = 6
!         nDof2                 = 12
!         nStif                 = 6*nQc
!
!     Outputs
!         R0(nDim)              = rot Controls:Old for the current Gauss Pt.
!      C0(nDim)                 = Ref Curvature:for the current Gauss Pt.
!      T(nDof,nStif)            = Dof Transformation Control matrix
!      B(nDof,nStif)            = Dof2 Transformation Control matrix
!      Det                      = Determinant (Jacobian)
!     ------------------------------------------------------------------
      Implicit Real(kind=8) (a-h,o-z)
!     Input Variables & Arrays
      Real(kind=8)        Qc
      Dimension           Qc(nDim,nQc)
      Real(kind=8)        RoCOld ,CuCOld
      Dimension           RoCOld(nCInt),CuCOld(nCInt)
!     Working Variables & Arrays
      Real(kind=8)       Br ,H1D ,Axes
      Dimension          Br(nQc),H1D(nQc),Axes(3,3)
!     Output Variables & Arrays
      Real(kind=8)       B ,T
      Dimension          B(12,nStif),T(6,nStif)
      Real(kind=8)       R0 ,C0 ,CK
      Dimension          R0(nDim),C0(nDim),CK(nDim)
      character*3        readstat
!==================================================================
      LOGICAL bReset
      common/LoadStepIter/iStep,iNut,bReset
      common/Scratch/iRotCur1,iRotCur2
!==================================================================
      DATA zero/0.D0/,one/1.0D0/,two/2.0D0/,three/3.0D0/
!     ------------------------------------------------------------------
      call TMB_CuLn(Qc,nQc,nOrderR,Ri,Br,H1D,Det,Cur,Tor,
&                           Axes,dxbardr,Nel,Iout,iPrt)
      C0(1) = Tor
      C0(2) = zero
      C0(3) = Cur
      AJI      = one/det
      kr = (jInt-1)*nDim+1
      call Extract_Vector(RoCOld,nCInt,R0,nC,kr,Iout) !rotationControlOld
      call Extract_Vector(CuCOld,nCInt,CK,nDIm,kr,Iout) CurvatureControlOld
!     ------------------- initialize with geometric curvature
      if(iStep == 1) Then
         CK = C0
      endif
!--------------------------------------------------------------------
Get !T(nDof,nDof*nQc),B(nDof2,nDof*nQc)
call Form_B_and_Tmatrices(Br,H1D,T,nDof,B,nDof2,nStif,AJI, nQc,Iout)
!--------------------------------------------------------------
return
```

```
end
      Subroutine TMB_CuLn(Qc,nQc,R,Det,Cur,Tor,Axes,dxbardr)
!     Inputs:
!           Qc(1) = Control X-Coordinates =    (qc0x qc1x...qc3x)
!           Qc(2) = Control Y-Coordinates =    (qc0y qc1y...qc3y)
!           Qc(3) = Control Z-Coordinates =    (qc0z qc1z...qc3z)
!           R     = root domain value
!     Outputs:
!           Axes      = T,M,B
!           Cur      = curvature
!           Tor      = torsion
!           Det      = determinant of The Jacobian
!
      Implicit Real(kind=8) (a-h,o-z)
      Real(kind=8)    Qc(3,nQc),Axes(3,3),Det,R,Rad
      Real(kind=8)    dum,delCx,delCy,delCz,xMult,pi
      Real(kind=8)    de2Cx,de2Cy,de2Cz,de3Cx,de3Cy,de3Cz
      Real(kind=8)    Br(nQc), B1Dr(nQc),dxbardr,H1D(nQc),,H2D(nQc),H3D(nQc)
      Real(kind=8)    xDot(3) ,xDo2(3) ,xDo3(3),AT(3),AM(3),AB(3),AJ, AJI
!------------------------------------------------
Displacement/Bernstein
      call Bernstein(Br,nQc,r,iOut)
!------------------------------------------------ Ist Derivative
      call Bernstein_IstDerivative(B1Dr,nQc,r,iOut)
!--------------------------------------- Derivative of H wrt r and s
      do 10 i = 1,nQc
      H1D(i) = B1Dr(i)
10 continue
!------------- Jacobian
      delCx = 0.0D0
      do 20 k = 1,nQc
      delCx = delCx + H1D(k)*Qc(1,k)
20 continue
      delCy = 0.0D0
      do 30 k = 1,nQc
      delCy = delCy + H1D(k)*Qc(2,k)
30 continue
      delCz = 0.0D0
      do 40 k = 1,nQc
      delCz = delCz + H1D(k)*Qc(3,k)
40 continue
      xDot(1) = delCx
      xDot(2) = delCy
      xDot(3) = delCz
!     ------------------------------------------ Determinant
of Jacobian
      det = dsqrt((delCx*delCx)+(delCy*delCy)+(delCz*delCz))
      if(det.gt.0.0000001) go to 45
      stop 'det 0'
45 continue
!------------------------------------------------ 2nd Derivative
      xn = Factorial(nQc-1)/Factorial(nQc-3)
      call Bernstein(B1Dr,nQc-2,r,iOut)
```

```
!       ---------------
      H2D    = 0.D0    ! ALL
      do 50 i = 1,nQc-2
      H2D(i) = B1Dr(i)
50 continue
      de2Cx = 0.0D0
      do 60 k = 1,nQc-2
      d2Qc = Qc(1,k+2) - 2.D0*Qc(1,k+1) + Qc(1,k)
      de2Cx = de2Cx + xn*H2D(k)*d2Qc
60 continue
      de2Cy = 0.0D0
      do 70 k = 1,nQc-2
      d2Qc = Qc(2,k+2) - 2.D0*Qc(2,k+1) + Qc(2,k)
      de2Cy = de2Cy + xn*H2D(k)*d2Qc
70 continue
      de2Cz = 0.0D0
      do 80 k = 1,nQc-2
      d2Qc = Qc(3,k+2) - 2.D0*Qc(3,k+1) + Qc(3,k)
      de2Cz = de2Cz + xn*H2D(k)*d2Qc
80 continue
      xDo2(1) = de2Cx
      xDo2(2) = de2Cy
      xDo2(3) = de2Cz
!--------------------------------------------------------- 3rd Derivative
(for torsion)
      xn = Factorial(nQc-1)/Factorial(nQc-4)
      call Bernstein(B1Dr,nQc-3,r,iOut)
!       --------------
      H3D    = 0.D0    ! ALL
      do 51 i = 1,nQc-3
      H3D(i) = B1Dr(i)
51 continue
      de3Cx = 0.0D0
      do 61 k = 1,nQc-3
      d2Qc = Qc(1,k+3) - 3.D0*Qc(1,k+2) + 3.D0*Qc(1,k+1) - Qc(1,k)
      de3Cx = de3Cx + xn*H3D(k)*d2Qc
61 continue
      de3Cy = 0.0D0
      do 71 k = 1,nQc-3
      d2Qc = Qc(2,k+3) - 3.D0*Qc(2,k+2) + 3.D0*Qc(2,k+1) - Qc(2,k)
      de3Cy = de3Cy + xn*H3D(k)*d2Qc
71 continue
      de3Cz = 0.0D0
      do 81 k = 1,nQc-3
      d2Qc = Qc(3,k+3) - 3.D0*Qc(3,k+2) + 3.D0*Qc(3,k+1) - Qc(3,k)
      de3Cz = de3Cz + xn*H3D(k)*d2Qc
81 continue
      xDo3(1) = de3Cx
      xDo3(2) = de3Cy
      xDo3(3) = de3Cz
!       --------------------------------------------------------------
---------------------- T
      AJ    = det
```

```
      AJI = 1.D0/AJ
!
      Axes(1,1) = xDot(1) * AJI
      Axes(1,2) = xDot(2) * AJI
      Axes(1,3) = xDot(3) * AJI
!     -----------------------------------------------------------------
----------------------- B
      call CrossProduct(xDot,xDo2,AB,Iout)
      AJ    = dsqrt(AB(1)*AB(1)+AB(2)*AB(2)+AB(3)*AB(3))
!-------------------------------------------------------
xDot & xDo2 not lin.indep
      if(AJ ==0.D0) Then
            Axes (3,1) = 0.D0
            Axes (3,2) = 0.D0
            Axes (3,3) = 0.D0
            return
      endif
      AJI = 1.D0/AJ
!
      AB(1) = AB(1) * AJI
      AB(2) = AB(2) * AJI
      AB(3) = AB(3) * AJI
!
      Axes (3,1) = AB(1)
      Axes (3,2) = AB(2)
      Axes (3,3) = AB(3)
!     -----------------------------------------------------------------
----------------------- M
      call CrossProduct(AB,AT,AM,Iout)
      AJ    = dsqrt(AM(1)*AM(1)+AM(2)*AM(2)+AM(3)*AM(3))
      AJI = 1.D0/AJ
!
      Axes (2,1) = AM(1) * AJI
      Axes (2,2) = AM(2) * AJI
      Axes (2,3) = AM(3) * AJI
!     ------ find dxbar/dr !   = [t|m|b]transpose.{dx/dr dy/dr dz/dr}transpose
      dxbardr = Axes(1,1)*delCx + Axes(1,2)*delCy + Axes(1,3)*delCz
!     -----------------------------------------------------------------
- Curvature & Torsion
      call CrossProduct(xDot,xDo2,AB,Iout)
      xNum = dsqrt(AB(1)*AB(1)+AB(2)*AB(2)+AB(3)*AB(3))
      xDen = dsqrt(xDot(1)*xDot(1)+xDot(2)*xDot(2)+xDot(3)*xDot(3))
      xDen = xDen*xDen*xDen
!
      Cur = xNum/xDen
!     --------------------------------- Torsion
      xDen = xNum*xNum
      xNum = AB(1)*xDo3(1)+AB(2)*xDo3(2)+AB(3)*xDo3(3)
!
      Tor = xNum/xDen
!     ---------------
      return
      end
```

10.20.13 Stiffness Matrices and Residual Forces

Both the instantaneous material stiffness matrix and the geometric stiffness matrix are formed with appropriate substitution of various fundamental vectors: $\mathbf{k}^0$, $\mathbf{a}$, $\mathbf{X}$ and so on. Moreover, generally for a c-type finite element, the internal Bezier control points can be easily eliminated by static condensation through partial Gauss elimination algorithm.

10.20.14 Incremental Equation Solution and Updates

Now let us recall the geometrically exact linearized finite element equation in unknown incremental generalized displacement, $\Delta\hat{\mathbf{q}}^d$, and the incremental load factor, $\Delta\lambda$, as:

$$\boxed{\mathbf{K}_T^c\ \Delta\hat{\mathbf{q}}^d + \hat{\mathbf{p}}(\lambda)\ \Delta\lambda = -\hat{\mathbf{R}}(\lambda_0, \hat{\mathbf{d}}_0)} \tag{10.565}$$

It may be recalled that the linearized equation (10.565) has been formulated to be solved by a Newton-type iterative solution procedure for the configuration path evaluation. The configuration path is sought for a set of discrete load level parameters: $\lambda = \lambda_0, \lambda_1, \lambda_2, \ldots$. Generally, a nonlinear configuration path may consist of both snap-through and snap-back buckling phenomena. A most suitable procedure for capturing these unstable situations is provided by embedding the arc length constraint algorithm in a Newton-type incremental/iterative method. At each step of an iterative solution procedure, both $\mathbf{d}$, the displacement vector, and $\mathbf{R}$, the rotation tensor, need to be updated for use in the next step of iteration, and for the determination of the beam configuration path, $C(\mathbf{d}, \mathbf{R})$, of interest. For the displacement vector, being a member of a vector space, the updating is additively straightforward as given by:

$$\mathbf{d}^{(i+1)} = \mathbf{d}^{(i)} + \Delta\mathbf{d}^{(i)} \tag{10.566}$$

where $\Delta\mathbf{d}^{(i)}$, $\mathbf{d}^{(i)}$, $\mathbf{d}^{(i+1)}$ are the incremental displacement vector obtained as a solution of the linearized incremental equation (10.565), the known starting displacement vector at iteration, i, and the updated displacement vector, respectively.

An additive update for the rotation tensor, being a member of a manifold, that is, a curved space (namely, $SO(3)$, the special orthogonal group), is not allowed. However, we can locally parameterize the rotation tensor by $\boldsymbol{\theta}$, the Rodrigues rotation vector that belong to the tangent spaces. A tangent space being a vector space, we can have an additive updating scheme for the rotation vector, without leaving the space, similar to that of the displacement update of the reference surface as:

$$\boldsymbol{\theta}^{(i+1)} = \boldsymbol{\theta}^{(i)} + \Delta\boldsymbol{\theta}^{(i)} \tag{10.567}$$

where $\Delta\boldsymbol{\theta}^{(i)}$, $\boldsymbol{\theta}^{(i)}$, $\boldsymbol{\theta}^{(i+1)}$ are the incremental rotation vector obtained as a solution of the linearized incremental equation (10.565), the known starting rotation vector at iteration, i, and the updated rotation vector, respectively. For most structural problems of interest, the additive updating procedure suffices, resulting in extremely accurate approximations, close to the exact solution for moderately large rotations.

Remark

- Recalling our discussion on the rotation tensor, the updating scheme given by equation (10.567) will not work for $||\theta|| = (2n+1)\pi,\ n = 0, 1, \ldots,$ as the Rodrigues rotation tensor becomes singular at those $\boldsymbol{\theta}$'s.

10.20.15 Singularity-free Update for the Configuration

The update for the translational configuration, that is, the displacement of the beam reference surface, is done as shown earlier by equation (10.566). Here, we are mainly concerned with a singularity-free rotational configuration update (see Chapter 4 on rotation tensors); although there are only three independent parameters in a rotation tensor, it is not possible for a three-parameter set such as the Rodrigues parameter vector, $\boldsymbol{\theta}$, to uniquely describe all possible rotations; a non-singular one-to-one mapping takes at least five parameters, while a non-singular many-to-one mapping requires at least four parameters. Recall further that the 4D unit Euler–Rodrigues parameter vector, $\mathbf{X} = \{X_1\quad X_2\quad X_3\quad X_4\}^T$, or the unit Rodrigues quaternion, $\mathbf{Q} = \{\mathbf{v}, s\} \equiv \{v_1\quad v_2\quad v_3\quad s\}^T$ with s as the scalar part, and $\mathbf{v}$ as the vector part, are non-singular two-to-one maps where globally both $+\mathbf{Q}$ and $-\mathbf{Q}$ represent the same rotation. Thus, a four-parameter Euler–Rodrigues vector or a quaternion will be adequate instead of a full five-parameter representation. Recall that an orthogonal rotation matrix, $\mathbf{R}$, with Euler–Rodrigues parameters, is characterized by a point on a unit spherical surface in 4D, that is, a 3-sphere of radius 1, and is given by:

$$\mathbf{R} = \begin{bmatrix} X_1^2 - X_2^2 - X_3^2 + X_4^2 & 2(X_1X_2 - X_3X_4) & 2(X_1X_3 + X_2X_4) \\ 2(X_1X_2 + X_3X_4) & -X_1^2 + X_2^2 - X_3^2 + X_4^2 & 2(X_2X_3 - X_1X_4) \\ 2(X_1X_3 - X_2X_4) & 2(X_2X_3 - X_1X_4) & -X_1^2 - X_2^2 + X_3^2 + X_4^2 \end{bmatrix},$$
$$||\mathbf{X}||^2 \equiv X_1^2 + X_2^2 + X_3^2 + X_4^2 = 1 \tag{10.568}$$

with

$$X_1 \equiv \left(\sin\frac{\theta}{2}\right)\frac{\theta_1}{\theta}, X_2 \equiv \left(\sin\frac{\theta}{2}\right)\frac{\theta_2}{\theta}, X_3 \equiv \left(\sin\frac{\theta}{2}\right)\frac{\theta_3}{\theta}, X_4 \equiv \cos\frac{\theta}{2},$$
$$\boldsymbol{\theta} = \{\theta_1\quad \theta_2\quad \theta_3\}^T,\quad \theta \equiv ||\boldsymbol{\theta}|| = \sqrt{\theta_1^2 + \theta_2^2 + \theta_3^2} \tag{10.569}$$

Note that the finite rotations, $\mathbf{R}$, move around on a unit sphere; the set of infinitesimal or virtual rotations define a tangent hyperplane at a point on this unit sphere. The components of a point on the unit hypersphere can also be viewed as the components of a unit Rodrigues quaternion, $\mathbf{Q}$. If we define the components of a unit quaternion by the Euler–Rodrigues parameter, $\mathbf{X}$, such that $\mathbf{Q} \equiv (Q_1 \equiv X_1, Q_2 \equiv X_2, Q_3 \equiv X_3, Q_4 \equiv X_4)$, that is,

$$\mathbf{Q} \equiv \left\{\left(\sin\frac{\theta}{2}\right)\frac{\boldsymbol{\theta}}{\theta}\quad \cos\frac{\theta}{2}\right\}^T,\quad ||\mathbf{Q}||^2 \equiv \mathbf{Q}\tilde{\mathbf{Q}} = 1 \tag{10.570}$$

with $\widetilde{\mathbf{Q}} \equiv (\widetilde{Q}_1 \equiv -X_1, \widetilde{Q}_2 \equiv -X_2, \widetilde{Q}_3 \equiv -X_3, \widetilde{Q}_4 \equiv X_4) = (-\mathbf{v}, s)$ as the conjugate quaternion of $\mathbf{Q}$, then any 3D Euclidean point $\mathbf{x} \in \mathbb{R}_3$ with the corresponding 4D vector quaternion, $\mathbf{Q}_x \equiv (\underbrace{\mathbf{x}}_{\in \mathbb{R}_3}, 0)$, under quaternion operation given by $\boxed{\mathbf{Q}\,\mathbf{Q}_x\,\widetilde{\mathbf{Q}}}$, will have an image $\mathbf{y} \in \mathbb{R}_3$ lifted similarly to the 4D vector quaternion, $\mathbf{Q}_y \equiv (\underbrace{\mathbf{y}}_{\in \mathbb{R}_3}, 0)$ that lie on a 4D hypersphere such that:

$$\boxed{\mathbf{Q}_y = \mathbf{Q}\,\mathbf{Q}_x\,\widetilde{\mathbf{Q}}} \tag{10.571}$$

with

$$\boxed{\begin{gathered}\begin{Bmatrix} y_1 \\ y_2 \\ y_3 \end{Bmatrix} = \begin{bmatrix} Q_1^2 - Q_2^2 - Q_3^2 + Q_4^2 & 2(Q_1\,Q_2 - Q_3\,Q_4) & 2(Q_1\,Q_3 + Q_2\,Q_4) \\ 2(Q_1\,Q_2 + Q_3\,Q_4) & -Q_1^2 + Q_2^2 - Q_3^2 + Q_4^2 & 2(Q_2\,Q_3 - Q_1\,Q_4) \\ 2(Q_1\,Q_3 - Q_2\,Q_4) & 2(Q_2\,Q_3 - Q_1\,Q_4) & -Q_1^2 - Q_2^2 + Q_3^2 + Q_4^2 \end{bmatrix} \begin{Bmatrix} x_1 \\ x_2 \\ x_3 \end{Bmatrix}, \\ \mathbf{y} = \mathbf{R}\,\mathbf{x} \end{gathered}} \tag{10.572}$$

Note that the matrix given in equation (10.572) is exactly the same as the rotation matrix defined by the Euler–Rodrigues parameters of equation (10.568). Thus, ***a spherical map***, that is, a ***pure rotation***, can always be described by the quaternion map given in equation (10.571). Now, noting that $||\mathbf{X}||^2 \equiv X_1^2 + X_2^2 + X_3^2 + X_4^2 = 1$, we can express the rotation matrix, $\mathbf{R}$, of equation (10.572), alternatively by:

$$\boxed{\mathbf{R} = \begin{bmatrix} 1 - 2Q_2^2 - 2Q_3^2 & 2(Q_1\,Q_2 - Q_3\,Q_4) & 2(Q_1\,Q_3 + Q_2\,Q_4) \\ 2(Q_1\,Q_2 + Q_3\,Q_4) & 1 - 2Q_1^2 - 2Q_3^2 & 2(Q_2\,Q_3 - Q_1\,Q_4) \\ 2(Q_1\,Q_3 - Q_2\,Q_4) & 2(Q_2\,Q_3 + Q_1\,Q_4) & 1 - 2Q_1^2 - 2Q_2^2 \end{bmatrix}} \tag{10.573}$$

Note that $trace(\mathbf{R}) = -1 + 4(Q_4)^2 \Rightarrow Q_4 = \frac{1}{2}\sqrt{(1 + trace(\mathbf{R}))}$. These are precisely the observations that suggest a smooth transition from the Rodrigues three-parameter $\boldsymbol{\theta}$-space to the Rodrigues four-parameter $\mathbf{Q}$-quaternion space and back to obtain a singularity-free updating scheme for the rotation matrix.

10.20.15.1 Rodrigues Quaternion to Rotation Matrix

Given a unit Rodrigues quaternion, $\mathbf{Q}$, we can construct the corresponding rotation matrix, $\mathbf{R}$, by equation (10.573).

10.20.15.2 Rotation Matrix to Rodrigues Quaternion

Given the rotation matrix, **R**, as in equation (10.573), recovering the corresponding quaternion, **Q**, is a bit more involved but is a straightforward process. Algorithmically, we can describe as follows.

Step 0: Compute the corresponding scalar part, $Q_4 \equiv s \equiv X_4$, of the quaternion as: $Q_4^2 \equiv s^2 \equiv X_4^2 = \frac{1}{4}(1 + trace(\mathbf{R}))$. If $Q_4^2 > 0$, then, go to ***Step 1***. else go to ***Step 2***.

Step 1: (It is immaterial whether we choose positive or negative square root as both will represent the same rotation matrix; we will choose positive square root), and

$$Q_4 \equiv s \equiv X_4 = \frac{1}{2}\sqrt{(1 + trace(\mathbf{R}))}$$

$$Q_1 \equiv v_1 \equiv X_1 = \frac{1}{4Q_4}(R_{32} - R_{23})$$

$$Q_2 \equiv v_2 \equiv X_2 = \frac{1}{4Q_4}(R_{13} - R_{31})$$

$$Q_3 \equiv v_3 \equiv X_3 = \frac{1}{4Q_4}(R_{21} - R_{12})$$

STOP

Step 2: Compute $Q_1^2 \equiv v_1^2 \equiv X_1^2 = -\frac{1}{2}(R_{22} + R_{33})$. If $Q_1^2 > 0$, then, go to Step, else go to ***Step 4***.

Step 3:

$$Q_4 \equiv s \equiv X_4 = 0$$

$$Q_1 \equiv v_1 \equiv X_1 = \frac{1}{\sqrt{2}}\sqrt{(-R_{22} - R_{33})}$$

$$Q_2 \equiv v_2 \equiv X_2 = \frac{1}{2Q_1}R_{12}$$

$$Q_3 \equiv v_3 \equiv X_3 = \frac{1}{2Q_1}R_{13}$$

STOP

Step 4: Compute $Q_2^2 \equiv v_2^2 \equiv X_2^2 = \frac{1}{2}(1 - R_{33})$. If $Q_2^2 > 0$, then, go to ***Step***, else go to ***Step 6***.

Step 5:

$$Q_4 \equiv s \equiv X_4 = 0$$

$$Q_1 \equiv v_1 \equiv X_1 = 0$$

$$Q_2 \equiv v_2 \equiv X_2 = \frac{1}{\sqrt{2}}\sqrt{1 - R_{33}}$$

$$Q_3 \equiv v_3 \equiv X_3 = \frac{1}{2Q_2}R_{23}$$

STOP

Step 6:

$$Q_4 \equiv s \equiv X_4 = 0$$
$$Q_1 \equiv v_1 \equiv X_1 = 0$$
$$Q_2 \equiv v_2 \equiv X_2 = 0$$
$$Q_3 \equiv v_3 \equiv X_3 = 1$$

STOP

10.20.15.3 Update for the Rotation Configuration

With the above conversions from the rotation matrix to the quaternion and back in place, let $\Delta\theta^{(i)}$ be the incremental Rodrigues rotation vector. We will have in storage the current quaternion, $\mathbf{Q}^{(i)}$, for the iteration step i; we can retrieve and convert it to the rotation matrix, $\mathbf{R}^{(i)}$, by using equations (10.572) or (10.573). Then, we can generate the quaternion, $\Delta\mathbf{Q}^{(i)}$, corresponding to $\Delta\theta^{(i)}$ by using equation (10.570) by replacing θ by $\Delta\theta^{(i)}$, and similarly for the norm; using equation (10.573), we can construct the incremental rotation matrix, $\Delta\mathbf{R}^{(i)}$ from the quaternion $\Delta\mathbf{Q}^{(i)}$. Now, we get the updated rotation matrix, $\mathbf{R}^{(i+1)}$, as $\mathbf{R}^{(i+1)} = \Delta\mathbf{R}^{(i)}\,\mathbf{R}^{(i)}$. Finally, by application of the algorithm for the conversion of a rotation matrix to the corresponding quaternion, we convert $\mathbf{R}^{(i+1)}$ to $\mathbf{Q}^{(i+1)}$ and store it in the database.

10.20.15.4 Code Fragment for Tangent Stiffness Matrix

```
   Subroutine Form_STIF(RoCOld,CuCOld,RoSav,CKSav,nCInt,Qd,Qc,nQc,
&                Nel,NelType,Nint,DMat,
&                Stif,BFors,StifK,nStif,Iout,Trace)
!-----------------------------------------------------------------------
---------------------------
!     Form Element Stiffness Matrix
!-----------------------------------------------------------------------
---------------------------
!     Notes:      Nint should be atleast 2
!
!     Inputs
!           Qd(nStif)     = Latest displacement controls: state vector
!           Qc(1,16)      = Bezier X-Coordinates =    (qc00x qc10x...qc33x)
!           Qc(2,16)      = Bezier Y-Coordinates =    (qc00y qc10y...qc33y)
!           Qc(3,16)      = Bezier Z-Coordinates =    (qc00z qc10z...qc33z)
!           Nel                 = Element Number
!           NelType             = Element Type: Beam
!           Nint                = Number of Gauss Points
!           E                   = Elasticity Modulus
!           PR                  = Poisson's ratio
!           Thk                 = Thickness
!           nStif               = Row or Col size of Stiffness matrix = nDof*nQc
!           Iout                = Error Output file#
!
!     Outputs
!           Stif(nStif,nStif) = Stiffness Matrix
```

```
!          bFors(nStif)             = Body force Vector Integrals
!          StifK(nStif)             = kLambda Stiffness last column Vector Inte-
grals
!          Trace               = trace of the matrix
!
      Implicit Real(kind=8) (a-h,o-z)
!
!     Input Variables & Arrays
!
      Real(kind=8)      Qc ,Qd
      Dimension         Qc(3,nQc),Qd(nStif)
      Real(kind=8)      RoSav ,CKSav
      Dimension         RoSav(nCInt),CKSav(nCInt)
      Real(kind=8)      RoCOld ,CuCOld
      Dimension         RoCOld(nCInt),CuCOld(nCInt)
      Real(kind=8)      DMat
      Dimension         DMat(6,6)
      integer                 Nel,NelType,Iout,nStif
!
!   Output Variables & Arrays
!
      Real(kind=8)      BFors ,StifK ,Stif            ,Trace
      Dimension         BFors(nStif),StifK(nStif),Stif(nStif,nStif)
      Real(kind=8)      EDG ,B ,DB ,T
      Dimension         EDG(12,12),B(12,nStif),DB(12),T(6,nStif)
      Real(kind=8)      Add ,HD ,RoOld ,Rp
      Dimension         Add(nStif),HD(12),HDOld(12),Rp(3)
      Real(kind=8)      BVEC ,SKLM ,R0 ,RK ,C0 ,CK
      Dimension         BVEC(nStif),SKLM(nStif),R0(3),RK(3),C0(3),CK(3)
      Real(kind=8) AM,BM,CM,TH,Ri,Si,Wt,sum
!
      LOGICAL bReset
!==================================================
      common/LoadStepIter/iStep,iNut,bReset
      common/Scratch/iRotCur1,iRotCur2
!==================================================
      DATA zero/0.D0/,one/1.0D0/
      DATA nC/3/,nDof/6/,nDof2/12/
!-------------------------------------------------------------
Stiffness Matrix Calculation
      BFors = 0.D0    !all
      StifK = 0.D0    !all
      Stif   = 0.D0    ! all elements = 0.
!============================================================ Restore
      inquire(iRotCur1,NEXTREC = nRec)
      if(nRec.GT.1) then
            nRec = Nel
            call PutGetRotCur(iRotCur1,RoCOld,CuCold,nCInt,nRec,2)    !Read
      endif
!========================================== Loop over Gauss Locations
      do 60 jInt = 1,Nint
      Ri = GaussPtVal(jInt,Nint,1)
!---------------------------------------------------------- Get
```

```
Geometric Curvature Vector etc.
      call Get_ReferenceInfo(Qc,nC,nQc,RoCOld,CuCOld,nCInt,
& Ri,jInt,R0,C0,CK,nC,T,B,
& nDof,nDof2,nStif,Det,Iout,Nel,Nint)
!--------------------------------------------------------------------
----- DisRots & Derivatives
      HD = MATMUL(B,Qd)
!-------------------------------------------------------------------
Get EDG(nDof2,nDof2)
      call Form_EtranDEplusGMatrix(R0,HD,nStif,DMat,C0,RK,CK,
&           T,B,BVEC,SKLM,EDG,3,nDof,nDof2,Iout)
!-------------------------------------------------------------------
Save Curvature
      do 10 ii = 1,nC
      ij = ii + (jInt -1)*nC
      RoCOld(ij) = RK(ii)
      CuCold(ij) = CK(ii)
10    continue
!----------------------------------------- Form: Btran_EDG_B & Add
to stiffness matrix
      Wt = GaussPtVal(jInt,Nint,2)*Det
!
      do 50 j = 1,nStif
      do 20 k = 1,nDof2       !
      DB(k) = 0.0
      do 20 l = 1,nDof2       !
      DB(k) = DB(k) + EDG(k,l)*B(l,j)
20 continue
      do 40 i = j,nStif
      sum = 0.0
      do 30 l = 1,nDof2       !
      sum = sum + B(l,i)*DB(l)
30    continue
      Stif(i,j) = Stif(i,j) + sum*Wt
40    continue
50 continue
!-------------------------------------------------------------- Load
Vector: Residual & Body force part
      do 53 i = 1,nStif
      BFors(i) = BFors(i) + Wt*BVec(i)
53 continue
!--------------------------------- Form: Last Column of stiffness:
KLambda Body Force part
      do 55 i = 1,nStif
      StifK(i) = StifK(i) + Wt*SKLM(i)
55 continue
60 continue
!
!     symmetry: Upper Triangle
!
      do 70 j = 1,nStif
      do 70 i = j,nStif
      Stif(j,i) = Stif(i,j)
```

```
70 continue
!============================================================= Save
!      nRec = (Nel-1)*Nint + jInt
       nRec = Nel
       call PutGetRotCur(iRotCur2,RoCOld,CuCold,nCInt,nRec,1)    !Write
!      Get Trace
       Trace = 0.D0
       do 80 i = 1,nStif
       Trace = Trace + Stif(i,i)
80 continue
!      Check each col. vector for Null
       Add = 0.D0
       do 100 j = 1,nStif
       sum = 0.0
       do 90 i = 1,nStif
       sum = sum + Stif(i,j)
90 continue
       Add(j) = sum
100 continue
!
       return
       end
```

10.20.16 Newton-type Iteration and Arc Length Constraint

Finally, a Newton-type method can bc devised to solve problems and perform stability or buckling analysis with quadratic convergence. For proportional loading, the coefficient of proportionality for the loading, λ, is gradually incremented through a Newton-type iterative method to trace the configuration path of the system. However, a straightforward application of the method fails near, at or beyond the stability limit point. Recall that there are generally two fundamentally different scenarios for loss of stability of a beam structure; these are related to the limit point associated with the snap-through phenomena, or the bifurcation point associated with the buckling situation. In the case of limit points, a basic Newton-type method will either overshoot near a limit point that is about to signal an unloading or undershoot near a limit point, heralding a switch to a loading situation with no chance of correction or recovery. Similarly, recalling two possible controls of tracing a load–displacement curve, we prefer load control to displacement control. Finally, for a practical problem, all that may be necessary is the first limit point, that is, the collapse load; even, in this situation, the simple Newton-type iteration method must be equipped with a mechanism to converge or predict the collapse load. Recalling our discussion about nonlinear iterative methods, we choose the arc length constraint mechanism for all the following numerical examples; usually, a tolerance for error of the convergence criterion and an arc length constraint parameter are input by the user of the computer code.

10.20.17 Numerical Examples

Now that we have the implementation details, we would like to apply these to various problems of interest, and examine and interpret the accuracy, robustness and numerical efficiency of the obtained results and compare with the published solutions by other methods, whenever available. It is important to note that in all of the following example full Gauss quadrature points were

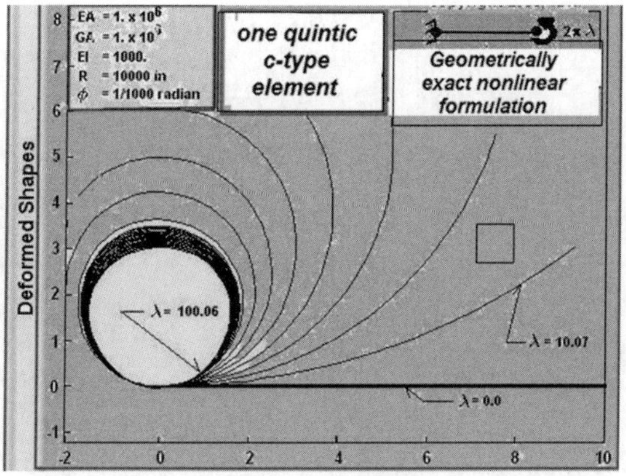

Figure 10.17 Example 1: Straight beam curling problem.

selected with absolute minimum number of mega c-type curved beam elements to show that ***locking*** is a non-issue and there is no need for any ad hoc so-called ***"under integration"*** or ***"reduced integration"***; this was only possible by c-type finite elements that are backed by a balanced exact theoretical formulation and a concomitant numerical implementation with Bernstein–Bezier polynomials as the basis functions.

Example 1: A 2D straight beam coiling to circle

A straight cantilevered beam, shown in Figure 10.17, of unit cross-sectional width and depth, subjected to a moment loading at the free end has been reported in the published literature. The elasticity modulus, E, and the shear modulus, G, are 10^6 and 0.5×10^6, respectively. The beam (displacement and rotation) is modeled as a single degenerate 3D quintic in X-Y plane with Z directional constraints. The deformed shapes for various tip moment loading are shown in Figure 10.17. Note that compared to the reported analyses that have used eight elements, a single mega quintic element suffices, and thus points towards great reduction in subsequent quality assurance efforts in practical endeavors.

Example 2: Geometrically exact nonlinear curved arches

To exemplify unification of geometry modeling and finite element analysis in both formulation and subsequent numerical solution, a geometrically exact nonlinear curved arch problem with shear deformation and extensibility is considered here. The load–deformation paths are shown in Figure 10.18 for two Huddleston compressibility parameters (Huddleston, 1968), $c = \frac{I}{A(2bd)}$ where, I is the moment of inertia, A represents cross-sectional area, and for b, d. One model analyzed with $c = 2.55 \times 10^{-6}$, that is, flexibility is so low that it renders the arch relatively incompressible. The other has $c = 0.01$, that is, has very high I, and thus, making the arch substantially compressible. The exact deformed shapes for various load levels for $c = 2.55 \times 10^{-6}$ are shown in Figure 10.18(c). It may be noted that it is solved with only one c-type quintic element to accurately model the problem, while it takes 40 Q-type shell elements of Palazotto and Dennis (1992), to seriously underestimate the limit point and track the wrong equilibrium path. The Q-type degenerate shell element, with Poisson's ratio vanishing, of Palazotto and Dennis is characterized by unnecessary Reddy-type quadratic transverse shear (Reddy and Liu, 1985) and quadratic in-plane displacement approximations resulting in 36 degree of freedoms. For flexible bodies, that is, beams, plates and shells, undergoing extreme geometric nonlinearity through large

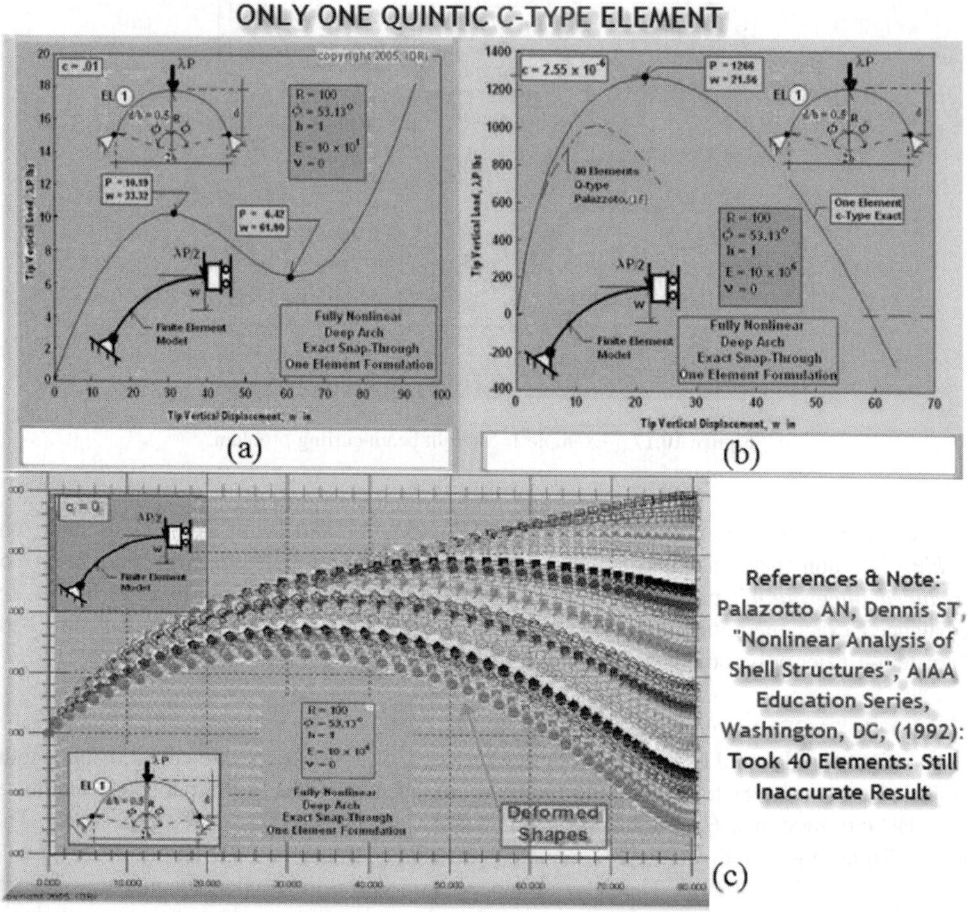

Figure 10.18 (a) Example 2: nonlinear arch with Huddleston parameter. (b) Example 2: nonlinear arch with Huddleston parameter. (c) Example 2: nonlinear arch deformations at various load levels for parameter.

rotations and displacement, it seems unwise to compromise analytical formulation prematurely for numerical solution early in the procedure.

Example 3: A 2D-curved beam element with 3D loading

A cantilevered 45° curved beam in X-Y plane loaded with a force in the Z direction at the free end invoking bending and torsion, is considered here. It was originally proposed by Bathe and Bolourchi (1979), and since analyzed by various other researchers. The model and the deformations for various load levels are shown in Figure 10.19(a) and (b), respectively. It took only five load steps to arrive at a load level of 600, compared to 60 equal load steps reported in Bathe and Bolourchi (1979). Figure 10.19(b) shows the load–displacement curves for various components of load point displacement. A comparison of deformed tip positions for reported load levels are shown, in tabular form, in Figure 10.19(c).

Note that while it took eight elements for most other researchers, a single mega c-type element suffices. Finally, there is no need for arbitrary under/reduced Gauss quadrature points for mythical shear locking.

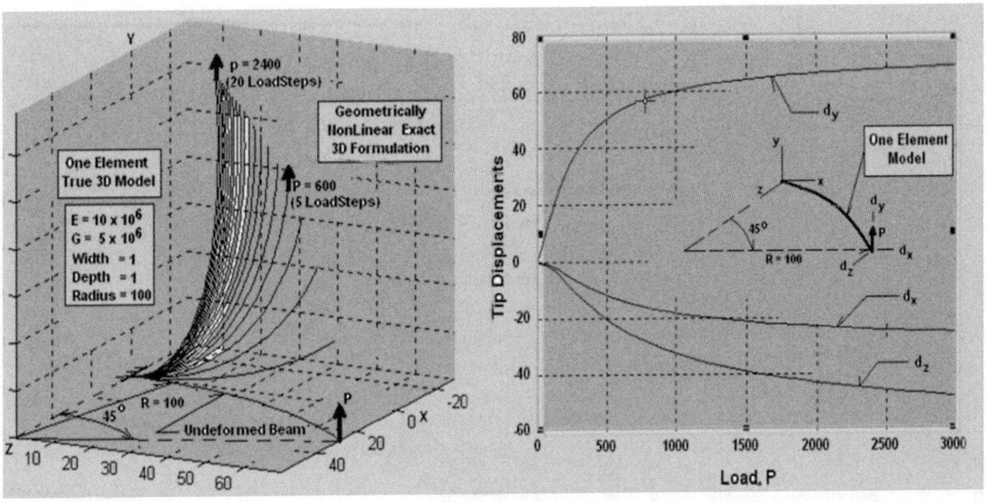

(a) (b)

+ see references in introduction Beam Model	Load Level						Number Of Elements Needed
	300			600			
	x	y	z	x	y	z	
Present c-Type	22.24	58.78	40.20	15.69	47.15	53.48	1
Simo & Vu-Quoc,+	22.33	58.84	40.08	15.79	47.23	53.37	8
Cardona & Geradin,+	22.14	58.64	40.35	15.55	47.04	53.50	8
Ibrahimbegovic et al, +	--	--	--	15.62	47.01	53.50	8
Bathe &Bolourchi,+	22.5	59.2	39.5	15.9	47.2	53.4	8

(c)

Figure 10.19 (a) Example 3: 2D rod problem with 3D loading. (b) Example 3: load–deformation curves. (c) Example 3: table of comparisons for number of elements needed.

Example 4: A 3D cantilevered curved beam element with natural twist and bend

Finally, we note that for general computer-aided curved beam mesh generation through the input of few key data points, the curved boundaries of the elements may be truly 3D. Thus, to test the robustness of the c-type finite element method for general large 3D curved beams, a 3D beam (see Chapter 9 for linear solution) is modeled by two truly three-dimensional elements, as shown in Figure 10.20(a). The geometric shape is initially generated as a single cubic with control nodes, $\mathbf{q}_{\mathbf{i}}^{\mathbf{1}}$, $i = 0, \ldots, 3$; then, an application of the subdivision algorithm produces two cubic Bernstein–Bezier geometric elements. The displacement finite elements, then, are made up of two quintic Bernstein–Bezier elements. The beam is subjected to simple tip load. The corresponding exact geometrically nonlinear deformations following the present formulation for various load levels are shown in Figure 10.20(b); the load-deformation curves are shown Figure 10.20(c). Note that it takes 100 currently available elements (FEAP computer program) compared to two c-type elements used here to match the presented accuracy as shown in Figure 10.20(d), which compares the present result with that by Simo and Vu-Quoc (1986); the quadratic convergence is shown in Figure 10.20(e).

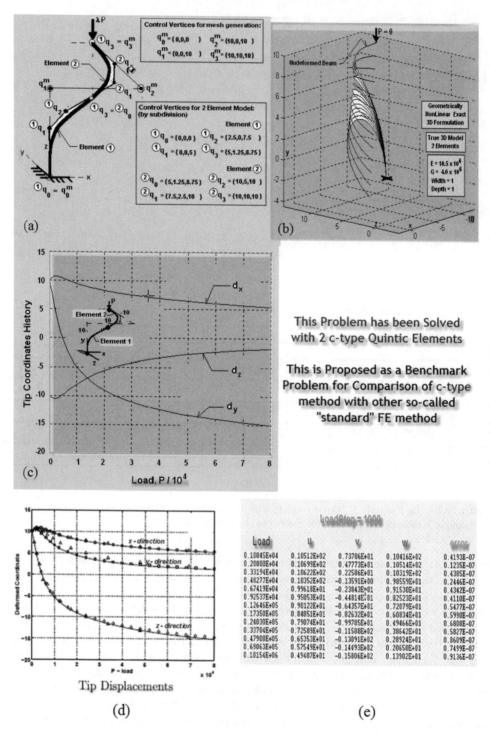

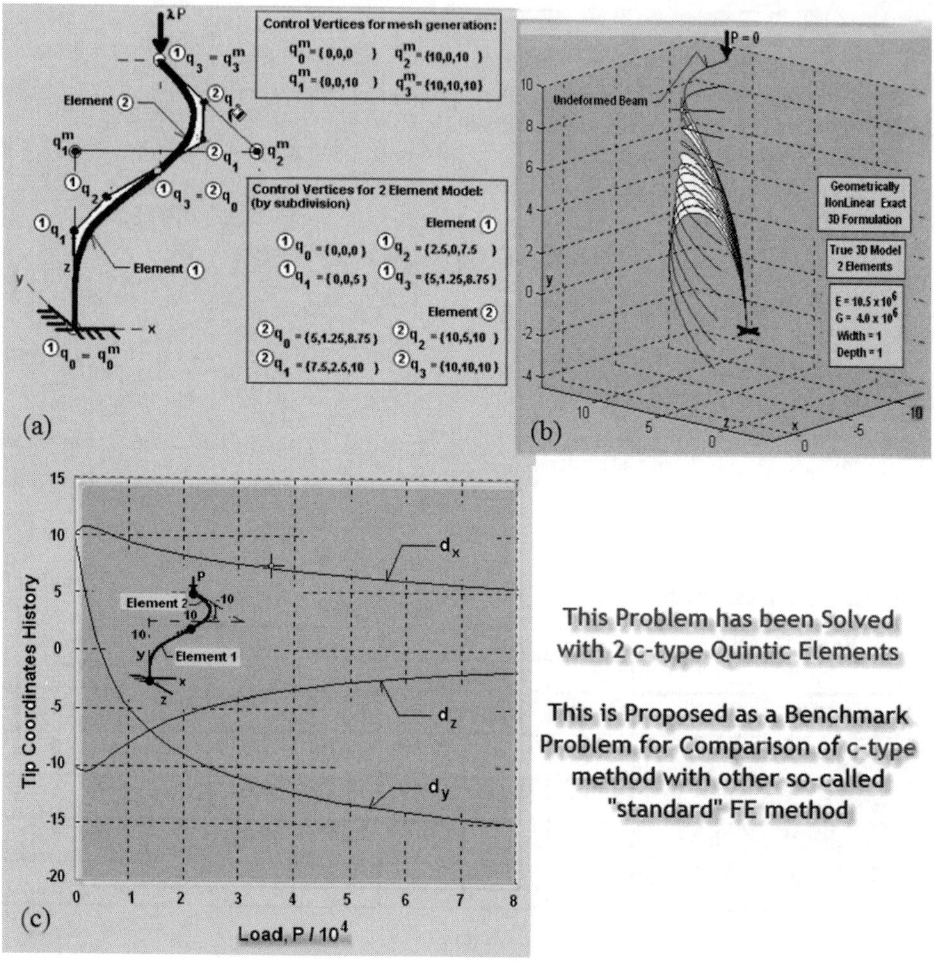

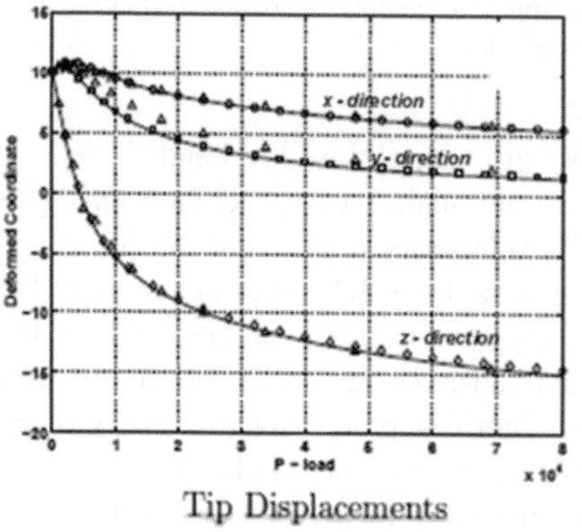

Tip Displacements

(d)

LoadStep = 1000

Load	u	v	w	error
0.10045E+04	0.10512E+02	0.73786E+01	0.10416E+02	0.4193E-07
0.20800E+04	0.10699E+02	0.47773E+01	0.10514E+02	0.1235E-07
0.33194E+04	0.10622E+02	0.22586E+01	0.10319E+02	0.4385E-07
0.48277E+04	0.10352E+02	-0.13591E+00	0.98559E+01	0.2446E-07
0.67419E+04	0.99618E+01	-0.23843E+01	0.91530E+01	0.4342E-07
0.92537E+04	0.95053E+01	-0.44814E+01	0.82523E+01	0.4110E-07
0.12646E+05	0.90122E+01	-0.64357E+01	0.72079E+01	0.5477E-07
0.17350E+05	0.84851E+01	-0.82632E+01	0.60834E+01	0.5990E-07
0.24030E+05	0.79074E+01	-0.99785E+01	0.49466E+01	0.6808E-07
0.33704E+05	0.72589E+01	-0.11588E+02	0.38642E+01	0.5827E-07
0.47908E+05	0.65353E+01	-0.13091E+02	0.28924E+01	0.8609E-07
0.69063E+05	0.57549E+01	-0.14493E+02	0.20650E+01	0.7499E-07
0.10154E+06	0.49487E+01	-0.15806E+02	0.13902E+01	0.9136E-07

(e)

Figure 10.20 (a) Example 4: 3D curved beam problem with natural twist and bend. (b) Example 4: nonlinear deformations of two quintic elements at various load levels. (c) Example 4: load–deformation curves. (d) Example 4: comparison of load–deformation histories with FEAP elements. (e) Example 4: quadratic convergence of displacement components.

In summary, the concomitant bookkeeping and quality assurance in mesh generation and finite element analysis will result in a proportional decrease in analyst/engineer man-hours if the presented formulation is used.

10.20.18 Where We Would Like to Go

This section concludes our nonlinear beam exposition; the next logical step will be to study nonlinear plates and shells for which we must check out:

- nonlinear plates and shells (Chapter 11)

11

Nonlinear Shell

11.1 Introduction

11.1.1 Why We Are Here

In this chapter we develop and examine the nonlinear theory of general shells for both nonlinear quasi-static and dynamic loadings; we, then, specialize the theory for thin hyperelastic shells subjected to extremely large displacements and rotations by deriving the most efficient and robust computational form to accommodate and devise extremely large c-type finite elements. We briefly detail below the contents of the various sections to achieve our goal:

> ***Section 11.2***: We define various geometric properties of an undeformed structural shell-like body for our subsequent analysis, and indicate the assumptions that are implicit in the definition.
>
> ***Section 11.3***: We reduce the dynamic and static balance of equations of 3D continuum mechanics to the shell equations of motion of two dimensions. In so doing, we derive the shell dynamic and quasi-static vector momentum balance and moment of momentum balance equations. The main reason behind this starting point as a reduction from the 3D equations is to lend support to the accuracy, consistency and validity of the shell equations of motion that may also be derived directly and simply by following an engineering approach of assuming a shell-like body as a mere 2D surface (see the direct engineering approach referred to in the Section 11.5). Furthermore, it connects the shell kinematic and force boundary conditions to the actually prescribed 3D counterparts that is pre-eminently essential to establish the equivalence between the real problems of three-dimensional nature and the supposedly corresponding problems to be solved by a shell theory – a theory that models real body effectively as a linear or nonlinear, thick or thin, two-dimensional surface.
>
> ***Section 11.4***: We specialize the dynamic and static balance equations of general shells to those of very thin and flexible shells. The case of thin and very flexible shells constitutes a class that is of great practical importance, and deserves special attention. The thinness attribute simplifies the shell equations for subsequent computational formulation.
>
> ***Section 11.5***: We develop a simple and direct engineering way (as opposed to the reduction of the equations of motion of continuum mechanics from 3D to 2D) of

applying the momentum balance and moment of momentum balance (i.e. angular momentum) principles to establish the shell static and dynamic equations of motion. Let us remind ourselves that because we are interested to include the case of extremely large (i.e. finite) displacements and rotations resulting in the nonlinear responses, all equilibrium equations must be considered in the deformed geometry of a shell. Hence, we will develop the balance equations on the current geometry for the nonlinear responses, and later, particularize by linearization for the case of infinitesimal deformations for the linear regime. Furthermore, in order to arrive at our stated goal, at the very outset we need to view a shell as a representative but fictitious thin two-dimensional curvilinear physical sheet, as opposed to the 3D continuum it really is, identified geometrically as the reference surface with equivalent cross-sectional and physical properties assigned to simulate real deformations that are expected of a shell-like 3D continuum. As it turns out that this way of viewing a shell as a two-dimensional surface produces the exact equations of motion obtained otherwise through an elaborate and rigorous reduction, from the 3D continuum viewpoint to an equivalent 2D surface except that the 3D connection provides insight into the real characters of the forces and moments, and the linear and angular momenta, introduced here. Thus, while the direct approach provides an easy and simple way to the balance laws, we must remember that consistent interpretation of the entities developed here must await the derivations that involve reduction from the 3D continuum balance principles.

Section 11.6: We identify the most suitable shell kinematic variables and the configuration space, that is, the set of kinematic parameters containing all the information about the motion history (i.e. the configurations) of the body, from the shell weak form of the balance equations. Our goal is motivated by the requirement that for the determination of a shell configuration path, or the history of motion for a shell numerically by an energy-type procedure such as the c-type finite element method, we must first identify the equivalent 2D configuration parameters of a 3D shell-like body that is energetically consistent with the exact balance laws. Towards this, we start from the derived 2D exact equations of the momentum and the angular momentum balance laws of motion of a shell and by the standard method of selecting test functions, develop the weak form of the equations which in turn allows us to identify the kinematic parameters appropriate for the conjugate virtual strains. However, alternatively, it is possible to start directly with the integral equations of the stress or deformation power and an assumed kinematics and the deformation map for the shell-like body, and define equivalently both the forces and moments and identify the conjugate virtual strains. For the shell-like body, we will have occasion to perform these operations as an example. Philosophically, however, it seems to make more sense to adhere to the first method as the various shell theories with different kinematics crop up naturally in an exact fashion.

Section 11.7: We define the curvature vectors and tensors, the angular velocity vectors and tensors and the admissible virtual space, corresponding to the shell configuration space consisting of the displacement vector field and the rotational tensor field, necessary to identify the spatial strain rate fields and the corresponding spatial real strain fields in the weak form of equations (and later for the computational virtual work equation) of a shell reference surface. In so doing, we describe in terms of the rotation field, the curvature vector and tensor, and the angular velocity vector and tensor field, and the variations of the kinematic variables, that is, the displacement vector field and the rotation tensor field, constituting the configuration space with respect to the arc length parameters along the lines of curvature of a shell reference surface.

Section 11.8: We identify the force and moment virtual strain rate fields from the internal virtual work functional, and the corresponding real strain rate fields whose variations validate the presumptions that the strain fields, derived from the weak form, in the internal virtual work expressions are actually virtual in nature, that is, there exist functions (real fields) whose variations are these fields. As will be apparent later in our discussion on the shell constitutive theory, our goal is also motivated by the need to develop the appropriate constitutive laws for the 2D shell surface by establishing the existence of an equivalent 2D surface strain energy functional.

Section 11.9: We define the component or operational forms of various vectors associated with the shell vector balance equations. These definitions allow us to convert the invariant vector form of the shell equations of motion to the component form suitable for subsequent computational development. Recall that our interest in any theory is solely governed by its relevance to our computational endeavors. In other words, here we will develop a computationally efficient component or operational vector forms by first collecting the curvilinear components of displacements and rotations, of forces and moments and of linear and angular velocities and momenta, and then describing these as components of corresponding vectors that help to define both the undeformed and the deformed states of an arbitrary shell. To be sure, we are interested ultimately in the computation of the components of displacements and rotations, and of the reactive forces and moments at any point on a shell that has been subjected to some loading conditions. This, in turn, motivates us to directly describe the relevant parameters and equations of a shell in component vectors. Accordingly, in what follows here, we develop the notion of component vectors. More specifically, we define the displacement and rotation, and the force and moment component vectors. Finally, we establish the relations that exist between the force and moment component vectors as defined in the undeformed frame and a rotating or spinning deformed frame.

Section 11.10: We define the covariant derivatives of the component or operational vectors necessary to convert the shell equations of motion in invariant vector form to a suitable component vector form for ease in subsequent computational development.

Section 11.11: We convert the invariant vector form of the balance equations of motion of a thin shell to a modified and computationally efficient component vector or operational form suitable for subsequent computational tasks such as the virtual work equation development which is the cornerstone of a finite element procedure such as the c-type finite element method. For finite rotation – a situation for which we will develop our theory, the shell real configuration space being a Riemannian or curved surface, we will need to work with the covariant derivatives of the component vectors.

Section 11.12: We define the computational curvature tensor and vector, the computational angular velocity tensor and vector and their linearization (as a precursor to covariant linearization of the dynamic virtual functional necessary for a Newton-type method of iterative solution technique) and the computational virtual space, corresponding to the shell configuration space consisting of the displacement vector field and the rotational tensor field necessary to derive the computational virtual work functional for our subsequent numerical treatment by the c-type finite element method. This is also aimed at the identification of the virtual strain fields and the development of the linearized virtual work functionals necessary for a Newton-type iterative procedure for the numerical solution of shell problems by a variational scheme such as the c-type finite element method.

Section 11.13: We develop the thin shell computational virtual work equations from the balance equations which we have specialized for thin shells and described as the computational shell balance equation. This is developed because, for the determination of a shell configuration path or the history of motion for a shell numerically by an energy-type procedure such as the c-type finite element method, we will have to derive, from the differential strong form of the shell equations, the integral virtual weak form of equations. Finally, we intend then to follow a semi-discrete finite element formulation in which the shell geometry and the spatial distribution of the deformations are discretized by the c-type finite element method and thus reduce the virtual work equation eventually into a matrix second-order ordinary differential equations. Under this type of formulation, the virtual space is taken as time-independent because the time is viewed only as an indexing parameter of the configuration space variables: the displacement vector and the rotation tensor, and thus, the time marching essentially traces a curve on the configuration space.

Section 11.14: We express the virtual strains in a computationally efficient matrix form of description leading to a desirable articulation of the virtual work functional, and also to a definition of various generalized entities of relevance for notational compactness. Furthermore, in so doing, we rewrite the virtual work functional of a shell in the generalized form to prepare for our subsequent computational work for shells subjected to proportional loadings in the case of the quasi-static or static analyses, and the time history of loadings for dynamic analyses.

Section 11.15: We develop the compatible real strain fields from the virtual strain fields that we have already derived as energetically conjugate to the real stress resultant and stress couple fields of a shell. The real strain field is necessary as one of the essential ingredients in the computational procedure; the real strains with the appropriate constitutive relationship allow us to compute the real internal reactive force (the stress resultants and the stress couples) fields of a shell. The contradiction, if any, between the internal and the applied force system drives an iterative method such as the Newton-type algorithm, in computational structural engineering and solid mechanics.

Section 11.16: We introduce the elastic material properties, that is, the stress–strain relationship, or the constitutive laws, appropriate for an elastic shell. We do this to reduce the 3D stress–strain laws to a 2D form that could be suitably incorporated in the internal virtual work functional of a shell. Clearly, this will be an approximation, but then the entire subject of the constitutive theory by its very nature of fitting real-world experimental data is an empirical science. Although the virtual work principle is exact, irrespective of the material properties, here we restrict ourselves to the case of the elastic shells.

Section 11.17: We accomplish the covariant linearization of the nonlinear virtual work functional that is necessary for the computation of the configuration history of a shell digitally by a Newton-type method of iterative solution for the nonlinear quasi-static (proportional loading), and by a numerical integration method of solution for the nonlinear dynamic (time history of loading) cases under semi-discrete formulation. In other words, we would like to obtain the linearized equations for both the quasi-static case under the proportional loading system (for incremental load step) identified by λ, the coefficient of load proportionality, and also the dynamic case under any time history of applied loading system (for incremental time step).

Section 11.18: We present the semi-discrete c-type finite element formulation for shells with dynamic time history loading systems. We extend this section with the description of the residual forces with the introduction of the c-type finite elements and the notion of control vectors, and complete it with a computationally efficient exposition of the incremental virtual work equation for a time step of the motion of a shell finite element. Starting from the already-derived fully linearized virtual work functional, first we aim to transform a part – specifically the internal linearized virtual work functional – to its aforesaid matrix representation. The transformed matrix representations will subsequently give rise to what is known as the material stiffness matrix and the geometric stiffness matrix of a shell. Next we focus on the part known as the inertial linearized virtual work functional, and transform it to the most suitable matrix representation, and the transformed matrix representations will subsequently give rise to what is known as the mass matrix, gyroscopic (damping) matrix and the centrifugal (stiffness) matrix of a shell undergoing dynamic motion. Using these developments, we present the c-type finite element formulation for the dynamic loading systems and generate the final form of the incremental dynamic virtual work equation; the previous form recalled later is not suitable for a subsequent finite element implementation since the sought-after independent incremental variables appear only in an implicit way, hidden behind the incremental strains and the virtual strain–displacement functions for the incremental internal virtual work functional. Also, the relevant matrices were not fully identified in that form for the incremental inertial virtual work functional. For the semi-discrete finite element formulation, we would like to have a form of the linearized virtual equation that is explicit in the incremental generalized displacement vector, the incremental generalized velocity vector, and the incremental generalized acceleration vector, so that the virtual work equation can be used digitally by a numerical time integration method in conjunction with a Newton-type iterative algorithm for tracing the dynamic motion history of a shell structure.

Section 11.19: We present the c-type finite element formulation for shells with quasi-static proportional loading systems; we then describe the residual and applied proportional forces with the introduction of the c-type finite elements and the notion of control vectors, and complete it with a computationally efficient exposition of the incremental equation for each finite element. Starting from the linearized virtual work functional, here we first intend to transform a part – specifically the internal linearized virtual work functional – to its suitable matrix representation. The transformed matrix representations will subsequently give rise to what is known as the material stiffness matrix and the geometric stiffness matrix of a shell. Using these developments, we present the c-type finite element formulation for the quasi-static loading systems and generate the final form of the incremental virtual work equation. The previous form recalled later is not yet suitable for a subsequent finite element implementation since the sought-after independent incremental variables appear only in an implicit way hidden behind the incremental strains and the virtual strain–displacement functions. We would like to have a form of the linearized virtual equation that is explicit in the incremental generalized displacement vector, so that the virtual work equation can be solved digitally by a Newton-type iterative method.

Section 11.20: We present the c-type shell finite element computer programming details on the finite element formulation of the material control stiffness matrix, the geometric control stiffness matrix, the applied loads and the residual loads, and so on, of a shell for subsequent direct stiffness method of assembly and solution by Newton-type algorithm to ensure solutions without any locking phenomena – shear, membrane, and so on. Finally, we present solutions for several nonlinear shell problems of practical importance by c-type finite element method.

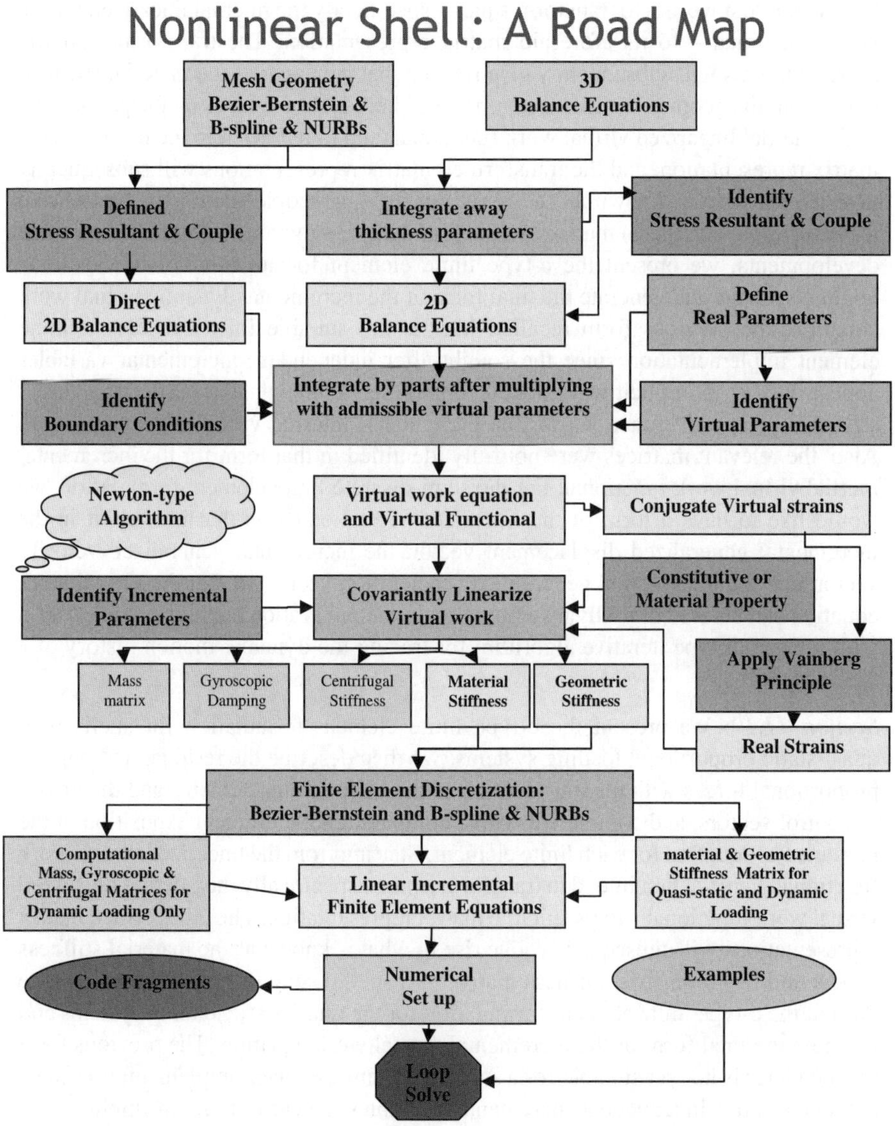

11.2 Shell Geometry: Definition and Assumptions

11.2.1 What We Need to Recall

As will be seen later, because thin shell geometry is essentially captured by a general 2D surface, we must recall, from Chapter 6 on surfaces, the various geometric properties of a surface.

11.2.2 3D Shell-like Body

For analysis and computational purposes, a 3D body may be called a **shell-like** body if it has any two of its dimensions conspicuously dominant compared to the remaining one. The two dominant dimensions will be identified with the **shell surfaces** and the remaining direction is associated with the **thickness** of a shell. Thus, as opposed to viewing the body just as a general 3D continuum identified in a Euclidean space, the body will then be equivalently perceived as the union or family of shell surfaces along the thickness direction that occupies the same 3D space, and henceforth, will be called a **shell space**. Our Euclidean space will be endowed with an inertial frame. Then, any point on the Euclidean space can be identified with a unique position vector with respect to this frame.

11.2.3 Shell Undeformed Geometry

In what follows, we are interested in the response of a shell to static and dynamic loads. So, let us suppose that, as shown in Figure 11.1, a three-dimensional shell material body in an undeformed reference state occupies a regular region, $\mathbf{B}$, in a Euclidean space with boundary denoted by $\partial\mathbf{B}$ at any time t_0. In other words, the shell body is considered embedded in a 3D Euclidean (flat) point space, although the shell surface itself constitutes a curved non-Euclidean space. We will assume that this shell space is differentiable to an order ≥ 2. We may introduce a curvilinear coordinate system, $\{\hat{\xi}^i\}$, $i = 1, 2, 3$ to describe the shell geometry and with which we may identify a position vector, $\hat{\mathbf{C}}(\hat{\boldsymbol{\xi}}) \equiv \hat{\mathbf{C}}(\hat{\xi}^1, \hat{\xi}^2, \hat{\xi}^3)$, at any point, $\hat{\boldsymbol{\xi}} \equiv \{\hat{\xi}^1, \hat{\xi}^2, \hat{\xi}^3\}^T$, of the body. Now, with the knowledge that geometrically the shell thickness direction is of different scale from those of the other two dimensions, and with an assignment of the first two curvilinear coordinates

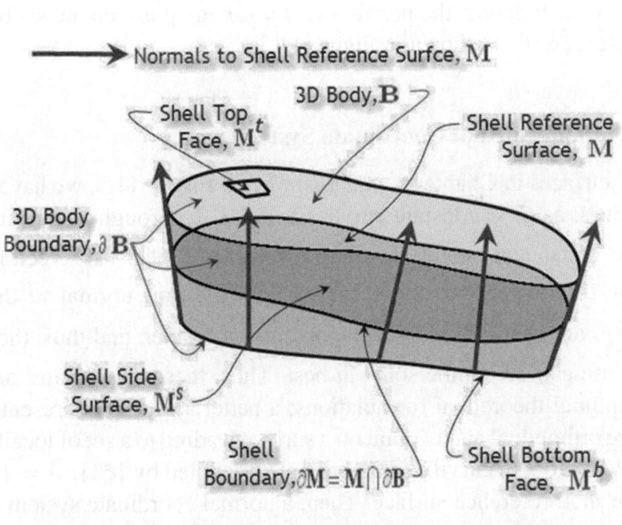

Figure 11.1 3D body and Shell geometric definition.

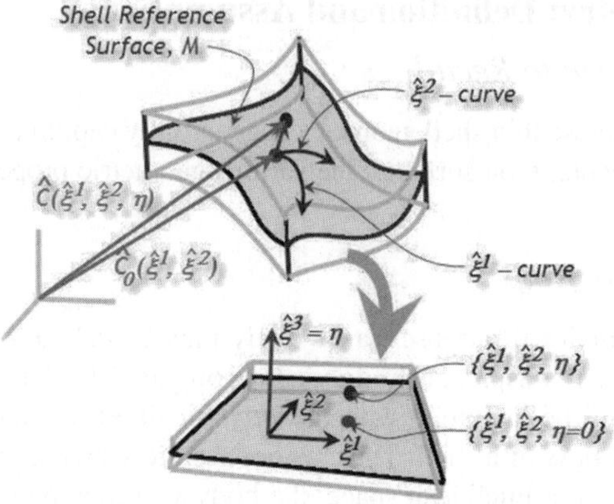

Figure 11.2 Shell space and map.

to the surface dimensions, we will redefine our curvilinear coordinates as: $\{\hat{\xi}^\alpha\}$, $\alpha = 1, 2$ to define surfaces, and $\eta \equiv \hat{\xi}^3$ to define the thickness direction. Thus, as shown in Figure 11.2, any material point on the shell body is given by $\{\hat{\xi}^1, \hat{\xi}^2, \eta\}$. Then, a coordinate system known as, a ***normal coordinate system***, can be introduced by defining a ***shell reference surface*** and taking the thickness direction, η, to be coincident along the normal at a point on this surface.

11.2.3.1 Shell Reference Surface

For reasons indicated above and for efficient and computationally optimal theoretical formulation, the shell body may be conceived as the union of surfaces given by $\{\hat{\xi}^1, \hat{\xi}^2, \eta = \text{const}\}$ along the thickness direction, η. While any one of these surfaces may serve as a shell reference surface, we choose our shell reference surface, denoted by $\mathbf{M} \subset \mathbf{B}$, as the one with $\eta = 0$. Thus, any point on the shell reference surface is given by $\{\hat{\xi}^1, \hat{\xi}^2, 0\}$.

Now, with the given arbitrary but fixed inertial frame and the position vectors, $\hat{\mathbf{C}}(\hat{\xi}^1, \hat{\xi}^2, \eta)$, at a point $\{\hat{\xi}^1, \hat{\xi}^2, \eta\}$, we will denote the position vector for any point on the shell reference surface as: $\hat{\mathbf{C}}_0(\hat{\xi}^1, \hat{\xi}^2) \equiv \hat{\mathbf{C}}(\hat{\xi}^1, \hat{\xi}^2, 0)$, as shown in Figure 11.3.

11.2.3.2 Shell Optimal Normal Coordinate System

From our study of surfaces in Chapter 6, and as shown in Figure 11.3, we have what is known as two sets of ***iso-curves***: as $\hat{\xi}^\alpha$ = constant curves for $\alpha = 1, 2$, through any point on these parallel surfaces. The triad of tangent vectors forming a set of the local base vectors: $\hat{\mathbf{G}}_\beta \equiv \frac{\partial \hat{\mathbf{C}}(\hat{\xi}^\alpha)}{\partial \hat{\xi}^\beta}$ at a point defined by the tangents to these curves (and the unit normal to these base vectors, $\hat{\mathbf{N}} \equiv \hat{\mathbf{G}}_3 = \frac{\hat{\mathbf{G}}_1 \times \hat{\mathbf{G}}_2}{\|\hat{\mathbf{G}}_1 \times \hat{\mathbf{G}}_2\|}$) are generally non-orthogonal to each other, and thus, their expressions are coupled and their use can be cumbersome at best. Thus, these coordinates are not suitable for computationally optimal theoretical formulations; a better scenario is presented by the lines of curvature which are orthogonal at any point on a surface leading to a set of locally orthogonal base vectors. So we may opt for the curvilinear coordinates, denoted by $\{\xi^\alpha\}$, $\alpha = 1, 2$, along the lines of curvature on the shell reference surface. Then, a normal coordinate system can be introduced locally by a triad of base vectors at a point: $\bar{\mathbf{G}}_\beta(\xi^1, \xi^2) \equiv \frac{\partial \mathbf{C}(\xi^1, \xi^2)}{\partial \xi^\beta}$ defined by the tangents to the

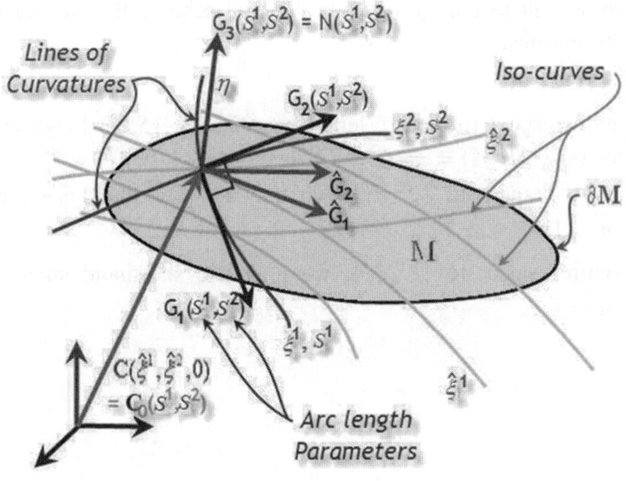

Figure 11.3 Shell reference surface.

lines of curvature, and the unit normal to these base vectors, $\bar{\mathbf{N}}(\xi^1, \xi^2) \equiv \bar{\mathbf{G}}_3(\xi^1, \xi^2) = \frac{\bar{\mathbf{G}}_1 \times \bar{\mathbf{G}}_2}{\|\bar{\mathbf{G}}_1 \times \bar{\mathbf{G}}_2\|}$. Then, given the fixed inertial frame and the position vectors, $\bar{\mathbf{C}}(\xi^1, \xi^2, \eta)$, defined by it, at any point $\{\xi^1, \xi^2, \eta\}$, we may denote the position vector for any point on the shell reference surface as: $\bar{\mathbf{C}}_0(\xi^1, \xi^2) \equiv \bar{\mathbf{C}}(\xi^1, \xi^2, 0)$.

However, parameterization of surfaces by $\{\xi^a\}, a = 1, 2$ along the lines of curvature is still not the best choice for either theoretical formulation or numerical evaluation. For example, the base vectors are not normalized, that is, not of unit lengths. Thus, the best choice is to replace $\{\xi^\alpha\}$, $\alpha = 1, 2$ of the shell reference surface by the corresponding invariant arc length parameterization, denoted by $\{S^\alpha\}$, $\alpha = 1, 2$, respectively, along the lines of curvature. Let us recall that the arc length parameter, S^α, is related to the parameter ξ^α by the relations: $S^1(\xi^1, \xi^2) = \int_0^{\xi^1} \|\bar{\mathbf{C}}(\rho, \xi^2, 0),_\rho\| d\rho$ and $S^2(\xi^1, \xi^2) = \int_0^{\xi^2} \|\bar{\mathbf{C}}(\xi^1, \rho, 0),_\rho\| d\rho$ where $\bar{\mathbf{C}}(\xi^1, \xi^2, 0)$ is the position vector of a point on the shell reference surface, Finally, any point on the shell reference surface may then be given by $\{S^1, S^2, 0\}$. We will denote the position vector for any point on the shell reference surface as: $\mathbf{C}_0(S^1, S^2) \equiv \mathbf{C}(S^1, S^2, 0)$. Thus, any material point on the shell body is given by $\{S^1, S^2, \eta\}$.

Now, we can define locally what we will call the ***optimal normal coordinate system*** by a triad of orthonormal base vectors at a point on the shell reference surface, $\mathbf{M}$, as:

- the two coordinate unit tangent vectors, $\mathbf{G}_\alpha(S^1, S^2) = C_0,_\alpha(S^1, S^2) \equiv \frac{\partial}{\partial S^\alpha} C_0(S^1, S^2)$, $\alpha = 1, 2$, along the lines of curvature on the shell reference surface, and the unit vector $\mathbf{N}(S^1, S^2) \equiv \mathbf{G}_3(S^1, S^2) = \frac{\mathbf{G}_1 \times \mathbf{G}_2}{\|\mathbf{G}_1 \times \mathbf{G}_2\|} = \frac{1}{\sqrt{G}} \mathbf{G}_1 \times \mathbf{G}_2$, normal to the shell reference surface, where G is the determinant of the 2×2 shell reference surface metric tensor given by: $\begin{bmatrix} G_{11} & G_{12} \\ G_{21} = G_{12} & G_{22} \end{bmatrix}$ with $\mathbf{G}_{\alpha\beta} = \mathbf{G}_\alpha \bullet \mathbf{G}_\beta$, $\alpha, \beta = 1, 2$. We have used the Lagrange identity: $(\mathbf{a} \times \mathbf{b}) \bullet (\mathbf{c} \times \mathbf{d}) = (\mathbf{a} \bullet \mathbf{c}) \bullet (\mathbf{b} \bullet \mathbf{d}) - (\mathbf{a} \bullet \mathbf{b})(\mathbf{c} \bullet \mathbf{d})$ to imply $\|\mathbf{G}_1 \times \mathbf{G}_2\| = \sqrt{G}$ in the above definition. Note that because of orthonormality: $\mathbf{G}_\alpha(S^1, S^2) = \mathbf{G}^\alpha(S^1, S^2)$, $\alpha = 1, 2$, and $G_{11} = G_{22} = 1$, $\mathbf{G}_{\alpha\beta} = 0$, $\alpha \neq \beta$ reducing the determinant to unity, that is, $\sqrt{G} = 1$.

Remarks: It is important to recognize the notational scheme that we have adopted for the derivatives of various entities.

- For derivatives with respect to the arc length parameters, $\{S^1, S^2\}$, we will sometimes use shorthand such as: $(\bullet),_{\alpha}(S^1, S^2) \equiv \frac{\partial}{\partial S^\alpha}(\bullet)(S^1, S^2), \quad \alpha = 1, 2.$
- For derivatives with respect to $\{\xi^1, \xi^2\}$, we may use shorthand such as: $(\bullet),_{\bar{\alpha}}(\xi^1, \xi^2) \equiv \frac{\partial}{\partial \xi^\alpha}(\bullet)(\xi^1, \xi^2), \quad \alpha = 1, 2.$
- For derivatives with respect to $\{\hat{\xi}^1, \hat{\xi}^2\}$, we may use shorthand such as: $(\bullet),_{\hat{\alpha}}(\hat{\xi}^1, \hat{\xi}^2) \equiv \frac{\partial}{\partial \hat{\xi}^\alpha}(\bullet)(\hat{\xi}^1, \hat{\xi}^2), \quad \alpha = 1, 2.$

Exercise: show that:

$$\boxed{\|\mathbf{G}_1 \times \mathbf{G}_2\| = \sqrt{G}} \tag{11.1}$$

Solution: Applying the Lagrange identity: $(\mathbf{a} \times \mathbf{b}) \bullet (\mathbf{c} \times \mathbf{d}) = (\mathbf{a} \bullet \mathbf{c})(\mathbf{b} \bullet \mathbf{d}) - (\mathbf{a} \bullet \mathbf{b})(\mathbf{c} \bullet \mathbf{d})$, $\|\mathbf{G}_1 \times \mathbf{G}_2\|^{\,2} = (\mathbf{G}_1 \times \mathbf{G}_2) \bullet (\mathbf{G}_1 \times \mathbf{G}_2) = (\mathbf{G}_1 \bullet \mathbf{G}_1)(\mathbf{G}_2 \bullet \mathbf{G}_2) - (\mathbf{G}_1 \mathbf{G}_2)^2 = \mathbf{G}_{11}\, \mathbf{G}_{22} - (\mathbf{G}_{12})^2$ $=$ det (surface metric tensor) $\equiv G$ completes the proof.

11.2.3.3 The Lines of Curvature from the Iso-curves on the Shell Reference Surface

Let us recall that given the iso-curves, the lines of curvature at a point for the shell reference surface can always be obtained through an application of the extremum principle leading to the following equations in λ_α, $\alpha = 1, 2$, for determining the directions of the lines of curvature:

$$\det \begin{bmatrix} \lambda^2 & -\lambda & 1 \\ \hat{G}_{11} & \hat{G}_{12} & \hat{G}_{22} \\ \hat{B}_{11} & \hat{B}_{12} & \hat{B}_{22} \end{bmatrix} = 0 \tag{11.2}$$

where, $\hat{G}_{\alpha\beta}$ and $\hat{B}_{\alpha\beta}$, $\quad \alpha, \beta = 1, 2$ are the components of the surface metric tensor and the curvature tensor, respectively, at any point on the shell reference surface described by the $\{\hat{\xi}^\alpha\}$, $\alpha = 1, 2$ coordinate system. Recall that the principal (i.e. along the lines of curvature) radii of curvatures, R_α, at a point on the shell reference surface, are given by:

$$\boxed{\left(\frac{1}{R_\alpha}\right)^2 - \hat{B}_{\alpha\beta}\,\hat{G}^{\alpha\beta}\left(\frac{1}{R_\alpha}\right) + \frac{\hat{B}}{\hat{G}} = 0} \tag{11.3}$$

where $\hat{G}^{\alpha\beta}$ are the components of the tensor reciprocal to the surface metric tensor defined by $\hat{\mathbf{G}}^\alpha \bullet \hat{\mathbf{G}}_\beta = \delta^\alpha_\beta$ with δ^α_β as the components of the identity tensor, and $\hat{G}$ and $\hat{B}$ as the determinants of the surface metric tensor and the curvature tensor on the shell reference surface, respectively. The mean curvature, H, and the Gaussian curvature, K, are then defined as:

$$\begin{aligned} H &\equiv \frac{1}{2}\left(\frac{1}{R_1}+\frac{1}{R_2}\right)=\frac{1}{2}\hat{B}_{\alpha}^{\cdot\alpha} \\ K &\equiv \frac{1}{R_1}\frac{1}{R_2}=\hat{B}_1^{\cdot 1}\hat{B}_2^{\cdot 2}-\hat{B}_1^{\cdot 2}\hat{B}_2^{\cdot 1}=\frac{\hat{B}}{\hat{G}} \end{aligned} \tag{11.4}$$

When the coordinate curves are the lines of curvature, we have $B_{12}=B_{21}=G_{12}=G_{21}=0$, and $G^{11}=\frac{1}{G_{22}}$ and $G^{22}=\frac{1}{G_{11}}$; thus, the expressions for the mean and the Gaussian curvatures reduce to:

$$\begin{aligned} H &= \frac{1}{2}\left(\frac{B_{11}}{G_{11}}+\frac{B_{22}}{G_{22}}\right)=\frac{1}{2}\left(\frac{1}{R_1}+\frac{1}{R_2}\right) \\ K &= \frac{B_{11}}{G_{11}}\frac{B_{22}}{G_{22}}=\frac{1}{R_1}\frac{1}{R_2} \end{aligned} \tag{11.5}$$

Finally, with the arc length parameters along the lines of curvature with $G_{11}=G_{22}=1$, we have: $H=\frac{1}{2}(B_{11}+B_{22})$ and $K=B_{11}\,B_{22}$.

11.2.3.4 Transformations: the Iso-curves to the Lines of Curvature

In our subsequent numerical evaluations of various quantities of interest by c-type finite element method, we would like to resort to the descriptions of these entities in terms of the optimal normal coordinate system: $\{S^1,S^2,\eta\}$, because of its most desired form of simplicity of expressions, and hence, computational efficiency. For example, the element stiffness matrices are numerically evaluated through integrations by quadrature rules at specific quadrature points; at these points on the shell reference surface, we can determine the lines of curvature as above for the simplest expressions for the involved derivatives with concomitant numerical efficiency, as will be detailed later. But, our geometry of the shell body is normally given to us in terms of the non-orthonormal coordinate system: $\{\hat{\xi}^1,\hat{\xi}^2,\eta\}$. Thus, we need to know the necessary transformations for the derivatives in the above two coordinate system. We have on the shell reference surface: $\hat{\mathbf{G}}_\alpha\equiv\frac{\partial\hat{\mathbf{C}}_0}{\partial\hat{\xi}^\alpha}=\frac{\partial\mathbf{C}_0}{\partial S^\beta}\frac{\partial S^\beta}{\partial\hat{\xi}^\alpha}=(\frac{\partial S^\beta}{\partial\hat{\xi}^\alpha})\,\mathbf{G}_\alpha$. Thus, because of orthonormality of $\mathbf{G}_\alpha$, that is, with $\mathbf{G}^\alpha=\mathbf{G}_\alpha$:

$$\frac{\partial S^\beta}{\partial\hat{\xi}^\alpha}=\hat{\mathbf{G}}_\alpha\bullet\mathbf{G}^\beta=\hat{\mathbf{G}}_\alpha\bullet\mathbf{G}_\beta \tag{11.6}$$

Now, from equation (11.6), we have the desired relations:

$$\frac{\partial(\bullet)}{\partial\hat{\xi}^\alpha}=\frac{\partial S^\beta}{\partial\hat{\xi}^\alpha}\frac{\partial(\bullet)}{\partial S^\beta}=(\hat{\mathbf{G}}_\alpha\bullet\mathbf{G}_\beta)\frac{\partial(\bullet)}{\partial S^\beta}\Rightarrow\frac{\partial(\bullet)}{\partial S^\beta}=(\hat{\mathbf{G}}_\alpha\bullet\mathbf{G}_\beta)^{-1}\frac{\partial(\bullet)}{\partial\hat{\xi}^\alpha} \tag{11.7}$$

In matrix form, we can express the desired transformation relations as:

$$\begin{Bmatrix} \dfrac{\partial(\bullet)}{\partial S^1} \\ \dfrac{\partial(\bullet)}{\partial S^2} \end{Bmatrix} = \begin{bmatrix} \hat{\mathbf{G}}_1 \bullet \mathbf{G}_1 & \hat{\mathbf{G}}_1 \bullet \mathbf{G}_2 \\ \hat{\mathbf{G}}_2 \bullet \mathbf{G}_1 & \hat{\mathbf{G}}_2 \bullet \mathbf{G}_2 \end{bmatrix}^{-1} \begin{Bmatrix} \dfrac{\partial(\bullet)}{\partial \hat{\xi}^1} \\ \dfrac{\partial(\bullet)}{\partial \hat{\xi}^2} \end{Bmatrix} \tag{11.8}$$

11.2.3.5 Shell Shifter Tensor

For a consistent reduction to, and development of, a 2D shell formulation in curved $\{\xi^1, \xi^2, \eta = 0\}$ shell reference surface parameters from the 3D continuum mechanics principles, it is necessary that all the variables of interest in the shell space, that is, $\{\xi^1, \xi^2, \eta \neq 0\}$ in particular, be described in terms of the variables on the shell reference surface. This is achieved by the introduction of a linear transformation, ${}^{\eta}\boldsymbol{\mu} \equiv \boldsymbol{\mu}(\eta) \equiv \frac{\partial \mathbf{C}}{\partial \mathbf{C}_0}$, known as the ***shifter tensor***. First, with $\boldsymbol{\xi} \equiv \{\, \xi^1 \; \xi^2 \; \eta \equiv \xi^3 \,\}^T$, let the undeformed shell space Jacobian, $\mathbf{J}(\eta)$, at any $\eta \neq 0 \in [\eta^t, \eta^b]$ be given by: $\mathbf{J}_\eta \equiv \frac{\partial \mathbf{C}(\xi^1, \xi^2, \eta)}{\partial \boldsymbol{\xi}} = \mathbf{A}_1 \otimes \mathbf{E}_i$ where $\mathbf{A}_\alpha = \frac{\partial \mathbf{C}(\xi^1, \xi^2, \eta)}{\partial \xi^\alpha}$ $\alpha = 1, 2$, and the normal, $\mathbf{N}(\xi^1, \xi^2) = \frac{\bar{\mathbf{G}}_1 \times \bar{\mathbf{G}}_2}{\|\bar{\mathbf{G}}_1 \times \bar{\mathbf{G}}_2\|}$, are the triad of base vectors at a point on the η-surface. In other words, we have $\mathbf{A}_i = \mathbf{J}_\eta\, \mathbf{E}_i$ by definition, and $\mathbf{J}_\eta$ may be represented in column form, that is, $\mathbf{J}_\eta = [\, \mathbf{A}_1 \; \mathbf{A}_2 \; \mathbf{A}_3 \equiv \mathbf{N} \,]$. Note that $\mathbf{N}$ is generally not normal to $\mathbf{A}_\alpha$, $\alpha = 1, 2$. Now, the reference surface Jacobian, $\mathbf{J}_0 = \mathbf{J}(\eta = 0)$, is given by: $\mathbf{J}_0 \equiv \frac{\partial \mathbf{C}_0}{\partial \boldsymbol{\xi}} = \bar{\mathbf{G}}_i \otimes \mathbf{E}_i$ where $\bar{\mathbf{G}}_\alpha = \frac{\partial \mathbf{C}_0(\xi^1, \xi^2)}{\partial \xi^\alpha}$ $\alpha = 1, 2$, and the normal, $\mathbf{N}$, forms a triad of orthogonal base vectors at a point on the shell reference surface. As before, we have $\bar{\mathbf{G}}_i = \mathbf{J}_0\, \mathbf{E}_i$ by definition, and $\mathbf{J}_0$ may be represented in column form, that is, $\mathbf{J}_0 = [\, \bar{\mathbf{G}}_1 \; \bar{\mathbf{G}}_2 \; \mathbf{N} \equiv \bar{\mathbf{G}}_3 \,]$; recall that $\{\mathbf{E}_i\}$ is the set of orthonormal bases associated with the fixed inertial reference frame.

Now, we have a representation for the shifter, ${}^{\eta}\boldsymbol{\mu} \equiv \boldsymbol{\mu}(\eta)$, by the chain rules of derivatives as:

$$\begin{aligned} &{}^{\eta}\boldsymbol{\mu} \equiv \frac{\partial \mathbf{C}}{\partial \mathbf{C}_0} = \frac{\partial \mathbf{C}}{\partial \boldsymbol{\xi}} \left(\frac{\partial \mathbf{C}_0}{\partial \boldsymbol{\xi}} \right)^{-1} = \mathbf{J}_\eta\, \mathbf{J}_0^{-1}. \\ &\mathbf{J}_\eta = {}^{\eta}\boldsymbol{\mu}\, \mathbf{J}_0 \end{aligned} \tag{11.9}$$

Alternatively, we may also obtain it as:

$$\mathbf{A}_i = \mathbf{J}_\eta\, \mathbf{E}_i = (\mathbf{J}_\eta\, \mathbf{J}_0^{-1})\, \bar{\mathbf{G}}_i = {}^{\eta}\boldsymbol{\mu}\, \bar{\mathbf{G}}_i \Rightarrow {}^{\eta}\boldsymbol{\mu} = \mathbf{A}_i \otimes \bar{\mathbf{G}}^i \tag{11.10}$$

Referring to the column definitions above of $\mathbf{J}_0$ and $\mathbf{J}_\eta$, from equation (11.9), we have a column representation of $\boldsymbol{\mu}$ as: $\boldsymbol{\mu} = [\, \boldsymbol{\mu}_1 \; \boldsymbol{\mu}_2 \; \boldsymbol{\mu}_3 \equiv \mathbf{N} \,]$ in the shell reference surface optimal normal bases with $\boldsymbol{\mu}_1$ and $\boldsymbol{\mu}_2$ yet to be explicitly determined. In other words, the covariant base vectors, $\mathbf{A}_\alpha$, $\alpha = 1, 2$ at any η-surface are related to those of the reference surface by the shifter, $\boldsymbol{\mu}$, and this has a tensor representation as: $\boldsymbol{\mu} = \mathbf{A}_i \otimes \bar{\mathbf{G}}^i$ through the contravariant base vectors, $\bar{\mathbf{G}}^i$, of the reference surface. Also, from equation (11.10), we have: $\bar{\mathbf{G}}_i = \boldsymbol{\mu}^{-1}\, \mathbf{A}_i \Rightarrow \boldsymbol{\mu}^{-1} = \bar{\mathbf{G}}_i \otimes \mathbf{A}^i \Rightarrow$

$(\boldsymbol{\mu}^{-1})^T = \mathbf{A}^i \otimes \bar{\mathbf{G}}_i$. Thus, the contravariant base vectors at any η-surface are related to those of the reference surface by the shifter through:

$$\boxed{\mathbf{A}^i = (\boldsymbol{\mu}^{-1})^T\,\bar{\mathbf{G}}^i} \tag{11.11}$$

Now, with the contravariant base vector $\bar{\mathbf{G}}^3 = \xi^3{,_{\mathbf{C}_0}} \equiv \frac{\partial \xi^3}{\partial \mathbf{C}_0} = \frac{\partial \eta}{\partial \mathbf{C}_0} \equiv \eta{,_{\mathbf{C}_0}}$, and the position vector of any point on a η-surface given by: $\mathbf{C}(\xi^1, \xi^2, \eta) = \mathbf{C}_0(\xi^1, \xi^2) + \eta\,\bar{\mathbf{G}}_3(\xi^1, \xi^2)$, we get from the definition of the shifter operator, $\boldsymbol{\mu}$:

$$\boxed{\begin{aligned} \boldsymbol{\mu}(\eta) \equiv \frac{\partial \mathbf{C}}{\partial \mathbf{C}_0} &= (\mathbf{C}_0 + \eta\,\bar{\mathbf{G}}_3){,_{\mathbf{C}_0}} = \bar{\mathbf{G}}_\alpha \otimes \bar{\mathbf{G}}^\alpha + \bar{\mathbf{G}}_3 \otimes \eta{,_{\mathbf{C}_0}} + \eta\,\bar{\mathbf{G}}_3{,_{\mathbf{C}_0}} \\ &= \bar{\mathbf{G}}_i \otimes \bar{\mathbf{G}}^i + \eta\,\bar{\mathbf{G}}_3{,_{\mathbf{C}_0}} = \mathbf{I} - \eta\,\mathbf{B} \end{aligned}} \tag{11.12}$$

where, $\mathbf{I} \equiv \bar{\mathbf{G}}_i \otimes \bar{\mathbf{G}}^i$, is the identity tensor, and $\mathbf{B}$ is the symmetric ***surface curvature tensor***, at a point on the reference surface, derived as follows: noting that the position vectors $\mathbf{C}_0(\xi^1, \xi^2) \equiv \mathbf{C}(\xi^1, \xi^2, 0)$ on the reference surface, and applying the surface curvature relationship – that is, the Weingarten formula – we have:

$$\boxed{\bar{\mathbf{G}}_{3,\,\alpha} \equiv \mathbf{N}_{,\,\alpha} = -b_\alpha^{\cdot\sigma}\,\bar{\mathbf{G}}_\sigma = -b_{\alpha\beta}\bar{G}^{\sigma\beta}\,\bar{\mathbf{G}}_\sigma = -b_{\alpha\beta}\,\bar{\mathbf{G}}^\beta} \tag{11.13}$$

By using equation (11.13) and switching indices, we define the desired reference surface curvature tensor, $\mathbf{B}$, as:

$$\boxed{\mathbf{B} \equiv -\bar{\mathbf{G}}_3{,_{\mathbf{C}_0}} = b_{\alpha\beta}\,\bar{\mathbf{G}}^\alpha \otimes \bar{\mathbf{G}}^\beta} \tag{11.14}$$

Now, by choosing our optimal coordinate system, that is, with arc length parameters, $\{S^\alpha\}$, along the lines of curvature, we have for the shifter, $\boldsymbol{\mu}$, the following matrix representation in curvilinear bases:

$$\boxed{\boldsymbol{\mu}(\eta) = \begin{bmatrix} 1 - \dfrac{\eta}{R_1} & 0 & 0 \\ 0 & 1 - \dfrac{\eta}{R_2} & 0 \\ 0 & 0 & 1 \end{bmatrix}} \tag{11.15}$$

In other words, we have:

$$\boxed{\begin{aligned} \mathbf{A}_1 &= \left(1 - \frac{\eta}{R_1}\right)\mathbf{G}_1, \\ \mathbf{A}_2 &= \left(1 - \frac{\eta}{R_2}\right)\mathbf{G}_2, \\ \mathbf{A}_3 &= \mathbf{N} \end{aligned}} \tag{11.16}$$

Remark

- We have taken time to get to equation (11.16) to establish it from the general curvilinear coordinate system. For coordinate systems along the lines of curvature with the arc length as the coordinate parameters, we could have directly used our specialized Weingarten formula to obtain the expression as:

$$\begin{aligned}
&\mathbf{C}(S^1, S^2, \eta) = \mathbf{C}_0(S^1, S^2) + \eta\, \mathbf{N}(S^1, S^2) \\
&\quad \Rightarrow \quad \mathbf{A}_\alpha \equiv \mathbf{C},_\alpha = \mathbf{G}_\alpha + \eta\, \mathbf{N},_\alpha = \mathbf{G}_\alpha - \eta\, B_{\alpha\beta} G^{\beta\sigma} \mathbf{G}_\sigma \\
&\qquad\qquad = \mathbf{G}_\alpha - \eta \frac{B_{(\alpha\alpha)}}{G_{(\alpha\alpha)}} \mathbf{G}_\sigma = \left(1 - \frac{\eta}{R_\alpha}\right) \mathbf{G}_\alpha \quad \text{(no sum)}
\end{aligned} \tag{11.17}$$

and

$$\mathbf{A}_3 \equiv \mathbf{C},_\eta = \mathbf{N} \tag{11.18}$$

where we have used: $G^{11} = \frac{1}{G_{22}}$; $G^{22} = \frac{1}{G_{11}}$; $G^{\alpha\beta} = 0,\ \alpha \neq \beta$, and $\frac{1}{R_\alpha} \equiv \frac{B_{(\alpha\alpha)}}{G_{(\alpha\alpha)}}$. Clearly the expressions in equations (11.17) and (11.18) are same as those in equation (11.16).

11.2.3.6 Shell Shifter (Translator) Determinant

In our subsequent effort to describe various infinitesimal area and volume elements in the shell space, as well as on the shell reference surface, we will need the shifter determinant, denoted by $\boldsymbol{\mu}$, of the shifter tensor, $\boldsymbol{\mu}$. From the definition of the shell space Jacobian, $\mathbf{J}$, and the shell reference surface Jacobian, $\mathbf{J}_0$, we have their determinants as:

$$\boxed{\begin{aligned}
&\det\, \mathbf{J}(\eta) = {}^\eta\mathbf{A}_1 \times {}^\eta\mathbf{A}_2 \bullet ({}^\eta\mathbf{A}_3 \equiv \mathbf{N}) = \varepsilon_{12} \mathbf{N} \bullet \mathbf{N} = \sqrt{{}^\eta A} \\
&\det\, \mathbf{J}_0 = \mathbf{G}_1 \times \mathbf{G}_2 \bullet (\mathbf{G}_3 \equiv \mathbf{N}) = \varepsilon_{12} \mathbf{N} \bullet \mathbf{N} = \sqrt{G}
\end{aligned}} \tag{11.19}$$

where, we have used the facts: $\mathbf{N}$ is orthonormal so that $\mathbf{A}_3 = \mathbf{A}^3 = \mathbf{G}_3 = \mathbf{G}^3 = \mathbf{N}$, and that $\varepsilon_{12}(\eta) = \sqrt{A(\eta)}\, e_{12}$, and $\varepsilon_{12}(\eta = 0) = \sqrt{A(\eta = 0)} = \sqrt{G}\, e_{12}$ are the components of the 2×2 surface covariant permutation tensors on the η-surface and the shell reference surface, respectively; e_{12} is the component of the 2×2 permutation matrix given by $\begin{bmatrix} 0 & 1 \\ 1 & 0 \end{bmatrix}$. Now, from the first of the equation (11.9), and equation (11.19), we get:

$$\boxed{{}^\eta\mu \equiv \det {}^\eta\boldsymbol{\mu} = \det(\mathbf{J}_\eta\, \mathbf{J}_0^{-1}) = \frac{\det\, \mathbf{J}_\eta}{\det\, \mathbf{J}_0} = \sqrt{\frac{{}^\eta A}{G}}} \tag{11.20}$$

The determinant, μ, of the shifter can also be given in terms of the mean curvature, H, and Gaussian curvature, K. To wit, noting that $\hat{\mathbf{C}}(\xi^1, \xi^2, \eta) = \mathbf{C}_0(\xi^1, \xi^2) + \eta\, \mathbf{N}(\xi^1, \xi^2)$, we have:

$\mathbf{A}_\alpha = \mathbf{G}_\alpha + \eta\,\mathbf{N},_\alpha = \mathbf{G}_\alpha - \eta\, B_{\alpha\beta}\,\mathbf{G}^\beta$. Thus, the components, $A_{\alpha\beta}$, of the metric tensor at the η-surface is given by:

$$\begin{aligned} A_{\alpha\beta} &= \mathbf{A}_\alpha \bullet \mathbf{A}_\beta = (\mathbf{G}_\alpha - \eta\, B_{\alpha\sigma}\,\mathbf{G}^\sigma) \bullet (\mathbf{G}_\beta - \eta\, B_{\beta\rho}\,\mathbf{G}^\rho) \\ &= G_{\alpha\beta} - 2\eta\, B_{\alpha\beta} + \eta^2\, B_\alpha^\rho\, B_{\beta\rho} = G^{\sigma\rho}(G_{\alpha\sigma} - \eta\, B_{\alpha\sigma})(G_{\beta\rho} - \eta\; B_{\beta\rho}) \end{aligned} \tag{11.21}$$

Now, let us note that for normal coordinates along the lines of curvature, the determinant of the contravariant metric tensor, $G^{\alpha\beta}$ is $\frac{1}{G}$ where $G = G_{11}\, G_{22}$ as $G_{12} = 0$. Similarly, the reference surface curvature tensor determinant is given by $B = B_{11}\, B_{22}$ with $B_{12} = 0$. Now, with:

$$\begin{aligned} \det(G_{\alpha\beta} - \eta\, B_{\alpha\beta}) &= (G_{11} - \eta\, B_{11})(G_{22} - \eta\, B_{22}) - \cancel{(G_{12} - \eta B_{12})^2} \\ &= G\left\{1 - 2\eta\left(\frac{B_{11}}{G_{11}} + \frac{B_{22}}{G_{22}}\right) + \eta^2 \tfrac{B}{G}\right\} = G(1 - 2\eta H + \eta^2 K) \end{aligned} \tag{11.22}$$

where, we have used equation (11.5) for the definition of H and K. Thus, the determinant, A, from equation (11.21), is given by:

$$\begin{aligned} {}^\eta A &= \det(G^{\alpha\beta})\,\{\det(G_{\alpha\beta} - \eta\, B_{\alpha\beta})\}^2 \\ &= \frac{1}{G}\, G^2(1 - 2\eta H + \eta^2 K)^2 = G\,(1 - 2\eta H + \eta^2 K)^2 \end{aligned} \tag{11.23}$$

Finally, the determinant, μ, using the above relation and equation (11.20), is given by:

$$\boxed{{}^\eta\mu \equiv \det {}^\eta\boldsymbol{\mu} = \det(\mathbf{J}_\eta\; \mathbf{J}_0^{-1}) = \frac{\det \mathbf{J}_\eta}{\det \mathbf{J}_0} = \sqrt{\frac{{}^\eta A}{G}} = 1 - 2\eta H + \eta^2 K} \tag{11.24}$$

Moreover, for arc length parameterization along the lines of curvature, the determinant, $\boldsymbol{\mu}$, in equation (11.24) can be used by direct application of H and K using the radii of curvatures as indicated in equation (11.15).

11.2.3.7 Shell Boundary

For a theoretical formulation of the momentum balance and the moment of momentum balance laws useful for subsequent numerical evaluation, it will be necessary to produce a closed free-body; thus, we need to identify and particularize various components of boundaries that make up the total boundary, $\partial\mathbf{B}$, of the 3D body in relation to the optimal normal coordinate systems on the shell reference surface; thus, we can think of it as being the union of three surfaces: shell top face, $\mathbf{M}^t$, shell bottom face, $\mathbf{M}^b$, and the shell side surface, $\mathbf{M}^S$.

11.2.3.8 Shell Top and Bottom Faces (Surfaces): $\mathbf{M}^t$ and $\mathbf{M}^b$

The shell top and bottom faces are defined to be the extreme surfaces along the thickness direction defined by $\eta = \eta^t(S^\alpha)$ and $\eta = \eta^b(S^\alpha)$, respectively. The position vectors to a point on the top and bottom faces, as shown in Figure 11.4, are given by $\mathbf{C}(S^1, S^2, \eta^t)$ and $\mathbf{C}(S^1, S^2, \eta^b)$, respectively,

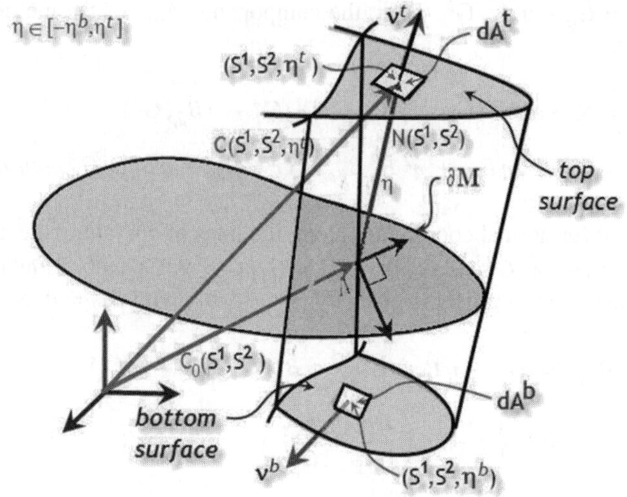

Figure 11.4 Shell top and bottom surface.

defined by:

$$\mathbf{C}(S^1, S^2, \eta^t) = \mathbf{C_0}(S^1, S^2) + \eta^t(S^1, S^2)\,\mathbf{N}, \quad (S^1, S^2) \in \mathbf{M} \tag{11.25}$$

and

$$\mathbf{C}(S^1, S^2, \eta^b) = \mathbf{C_0}(S^1, S^2) + \eta^b(S^1, S^2)\,\mathbf{N}, \quad (S^1, S^2) \in \mathbf{M} \tag{11.26}$$

11.2.3.9 Shell Side Surface: $\mathbf{M}^S$

In order to define shell side surface, we first define a curve, $\partial\mathbf{M}$, that completely encompasses and lies on the shell reference surface, $\mathbf{M}$. The shell side surface is, then, defined by the union of normals, $\mathbf{N}(S^1, S^2)$, $(S^1, S^2) \in \partial\mathbf{M}$, at and along the shell boundary, $\partial\mathbf{M}$, and bounded by the shell top and bottom faces, as shown in Figure 11.5. For subsequent analysis, we may parameterize the boundary curve, $\partial\mathbf{M}$, on the shell reference surface by an arc length parameter that will be denoted by S.

11.2.3.10 Shell Coordinates

With the above definitions, we can now sum up representation of the coordinates of all points in the 3D body as:

- a point $\{S^1, S^2, \eta\} \in \mathbf{B}$, $\eta \neq 0$:

$$\mathbf{C}(S^1, S^2, \eta) = \mathbf{C_0}(S^1, S^2) + \eta\,\mathbf{N}, \quad \eta \in [-\eta^b(S^1, S^2), \eta^t(S^1, S^2)], \ (S^1, S^2) \in \mathbf{M} \tag{11.27}$$

- a point on the shell reference surface:

$$\mathbf{C}_0(S^1, S^2) \equiv \mathbf{C}(S^1, S^2, 0)\,, \quad (S^1, S^2) \in \mathbf{M} \tag{11.28}$$

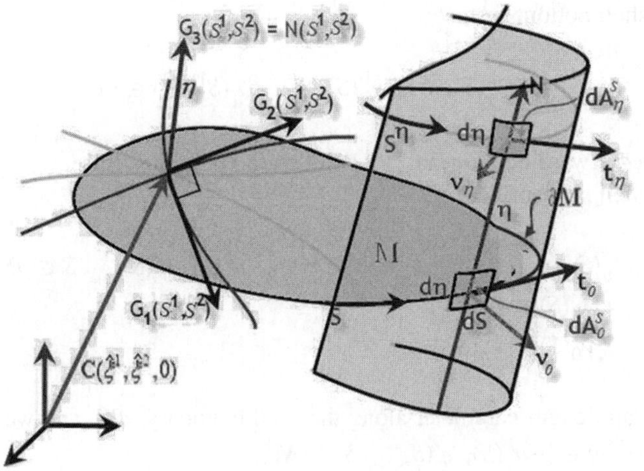

Figure 11.5 Shell side surface.

- a point on the top surface:

$$\mathbf{C}(S^1, S^2, \eta^t) = \mathbf{C_0}(S^1, S^2) \; + \; \eta^t(S^1, S^2)\,\mathbf{N}, \quad (S^1, S^2) \in \mathbf{M} \tag{11.29}$$

- a point on the bottom surface:

$$\mathbf{C}(S^1, S^2, \eta^b) = \mathbf{C_0}(S^1, S^2) \; + \; \eta^b(S^1, S^2)\,\mathbf{N}, \quad (S^1, S^2) \in \mathbf{M} \tag{11.30}$$

- a point on the shell side surface:

$$\mathbf{C}(S, \eta) = \mathbf{C_0}(S) \; + \; \eta(S)\,\mathbf{N}, \quad \eta \in [-\eta^b(S), \eta^t(S)], \qquad S \in \partial\mathbf{M} \tag{11.31}$$

where, S is the arc length parameter along the shell boundary, $\partial\mathbf{M}$.

11.2.3.11 Shell Triads of Covariant Base Vectors

With the definitions: $\mathbf{A}_\alpha(S^1, S^2, \eta) \equiv \frac{\partial \mathbf{C}(S^1,S^2,\eta)}{\partial S^\alpha} = \mathbf{G}_\alpha(S^1, S^2) + \eta\,\mathbf{N},_\alpha$ with $\mathbf{G}_\alpha(S^1, S^2) \equiv \frac{\partial \mathbf{C}_0(S^1,S^2)}{\partial S^\alpha}$, $\alpha = 1, 2$, $(S^1, S^2) \in \mathbf{M}$, and $\mathbf{N},_\alpha(S^1, S^2) \equiv \frac{\partial \mathbf{N}(S^1,S^2)}{\partial S^\alpha}$, we have for:

- a point $\{S^1, S^2, \eta\}$ on the parallel surfaces with $\eta \neq 0$:

$$\mathbf{A}^\eta_\alpha(S^1, S^2, \eta), \;\mathbf{N}(S^1, S^2), \quad \eta \in [-\eta^b(S^1, S^2), \eta^t(S^1, S^2)], \;(S^1, S^2) \in \mathbf{M} \tag{11.32}$$

- a point on the shell reference surface:

$$\mathbf{G}_\alpha(S^1, S^2), \;\mathbf{N}(S^1, S^2), \qquad (S^1, S^2) \in \mathbf{M} \tag{11.33}$$

- a point on the shell top surface:

$$\mathbf{A}^t_\alpha(S^1, S^2), \;\boldsymbol{\nu}^{\mathbf{t}}(S^1, S^2), \qquad (S^1, S^2) \in \mathbf{M} \tag{11.34}$$

where, $\boldsymbol{\nu}^{\mathbf{t}}$ is the outward unit normal at a point on the top surface,

- a point on the shell bottom surface:

$$\mathbf{A}^b_\alpha(S^1,S^2),\ \boldsymbol{\nu}^{\mathbf{b}}(S^1,S^2),\qquad (S^1,S^2)\in\mathbf{M} \tag{11.35}$$

where, $\boldsymbol{\nu}^{\mathbf{b}}$ is the outward unit normal at a point on the bottom surface,

- a point on the shell side surface:

$$\begin{aligned}&\mathbf{t}_\eta(S,\eta),\quad \boldsymbol{\nu}_\eta(S,\eta),\ \mathbf{N}(S),\quad \eta\in[-\eta^b(S),\ \eta^t(S)],\quad S\in\partial\mathbf{M}\\ &\mathbf{t}_0(S)\equiv\mathbf{t}_\eta(S,0),\\ &\boldsymbol{\nu}_0(S)\equiv\boldsymbol{\nu}_\eta(S,0),\end{aligned} \tag{11.36}$$

where, S is the arc length parameter along the shell boundary, $\partial\mathbf{M}$, and we have: $\boldsymbol{\nu}_\eta(S,\eta)\equiv \mathbf{t}_\eta(S,\eta)\times\ \mathbf{N}(S),\quad \eta\in[-\eta^b(S),\ \eta^t(S)],\quad S\in\partial\mathbf{M}$

11.2.3.12 Shell Reference Surface Differential Area and Interior Volume Elements

For our theoretical formulation in the Lagrangian form, we will subsequently need the description of a typical undeformed scalar differential area, dA, in the shell reference surface. Additionally, the area elements of the top, dA^t, bottom, dA^b, and side, dA^s, surfaces will have to be described in terms of the corresponding reference surface and boundary curve elements. Finally, we will also need the description of a typical differential volume, dV, element belonging the shell space interior.

11.2.3.12.1 Reference Surface Oriented Differential Area: $\mathbf{dA} = \mathbf{N}\,dA$

The line element, $\mathbf{dS}^{\bar\alpha}$, along a coordinate curve is given by $\mathbf{dS}^{\bar\alpha}=\frac{\partial\bar{\mathbf{C}}}{\partial\bar\xi^\alpha}\,d\bar\xi^\alpha=\bar{G}_{\bar\alpha}\,d\bar\xi^\alpha$ with $\bar\alpha$ not summed. The magnitude, clearly, is given by: $dS^{\bar\alpha}=\sqrt{G_{\bar\alpha\bar\alpha}}\,d\bar\xi^\alpha$. The line element, $\mathbf{dS}^\alpha$, in terms of the arc length parameter, S^α, is given by: $\mathbf{dS}^\alpha=\frac{\partial\mathbf{C}}{\partial S^\alpha}\,dS^\alpha=G_\alpha\,dS^\alpha$ with α not summed, and the magnitude by: dS^α with $G_{\alpha\alpha}$ being unity.

A typical differential area axial vector, using definition of the normal, $\mathbf{N}$, to the shell reference surface, and the fact that: $\|\mathbf{G}_1\times\mathbf{G}_2\| = \sqrt{G}$, is given by:

$$\begin{aligned}&\mathbf{dA}=\mathbf{N}\,dA\equiv\mathbf{dS}^1\times\ \mathbf{dS}^2=\mathbf{G}_1\,dS^1\times\ \mathbf{G}_2\,dS^2=\sqrt{G}\mathbf{N}\,dS^1dS^2\ =(\sqrt{G}\,dS^1dS^2)\,\mathbf{N}\\ &\Rightarrow \boxed{dA=\sqrt{G}\ dS^1dS^2}\end{aligned} \tag{11.37}$$

With $\mathbf{G}_\alpha$, $\alpha=1,2$ being unit vectors along the lines of curvature, we have: $\sqrt{G}=\sqrt{G_{11}G_{22}}=1$ with $G_{12}=G_{21}=0$.

We note, using equations (11.24) and (11.37), that the area element, dA^η, for any η in the shell interior, is given by:

$$\mathbf{dA}^\eta=\mathbf{N}\,dA^\eta=(\sqrt{{}^\eta A}\ dS^1dS^2)\,\mathbf{N}=\sqrt{\frac{{}^\eta A}{G}}(\sqrt{G}\ dS^1dS^2)\,\mathbf{N}=({}^\eta\mu\ dA)\,\mathbf{N} \tag{11.38}$$

where ${}^\eta\mu$ is defined as μ at any $\eta\in(-\eta^b,\ \eta^t)$.

11.2.3.12.2 Shell Interior Differential Volume: dV^η

Using equation (11.38) for $\mathbf{dA}^\eta$, we have:

$$dV^\eta = \mathbf{dA}^\eta \bullet \mathbf{d}\eta = \mathbf{N}\, dA^\eta \bullet \mathbf{N}\, d\eta = {}^\eta\mu\, dA\, d\eta \tag{11.39}$$

11.2.3.12.3 Top Surface Oriented Differential Area: $\mathbf{dA^t} = \boldsymbol{\nu}^\mathbf{t}\, dA^\mathbf{t}$

From equation (11.38), a typical differential area axial vector is given by: $\mathbf{dA}^t = \boldsymbol{\nu}^\mathbf{t}\, dA^t = (\mu^t\, dA)\, \boldsymbol{\nu}^\mathbf{t}$ where μ^t is defined as μ at $\eta = \eta^t$.

11.2.3.12.4 Bottom Surface Oriented Differential Area: $\mathbf{dA^b} = \boldsymbol{\nu}^\mathbf{b}\, \boldsymbol{dA}^\mathbf{b}$

From equation (11.38), a typical differential area axial vector is given by: $\mathbf{dA^b} = \boldsymbol{\nu}^\mathbf{b}\, dA^t = (\mu^b\, dA)\, \boldsymbol{\nu}^\mathbf{b}$ where μ^b is defined as μ at $\eta = \eta^b$.

11.2.3.12.5 Side Surface Differential Area: $\mathbf{dA^S} = \boldsymbol{\nu}_o\, dA^S$ and $\mathbf{dA}^{\mathbf{S}^\eta} = \boldsymbol{\nu}\, \mu^{S^\eta} dA^S$

There are several ways to relate the side surface area, $\mathbf{dA^S} = \boldsymbol{\nu}_o\, dA^S$, at $\eta = 0$ with the surface area, $\mathbf{dA}^{\mathbf{S}^\eta} = \boldsymbol{\nu}\, \mu^{S^\eta} dA^S$, at any $\eta \in [-\eta^b, \eta^t]$. Before we present two such methods, note that, by definition, a typical differential area axial vector $\mathbf{dA^S} = \boldsymbol{\nu}\, dA^S$ at $\eta = 0$ on the side surface, M^S, is given by: $\mathbf{dA}^{\mathbf{S}^\eta} = d\mathbf{C}^S \times\, d\mathbf{C}^\eta = \mathbf{C},_S \times \mathbf{C},_\eta\, dS\, d\eta = \mathbf{C},_S \times \mathbf{C},_\eta\, dA^S$ where $dA^S \equiv dS\, d\eta$. Moreover, $\mathbf{C}(S, \eta) = C(\mathbf{C}_0(S), \eta)$, $\mathbf{C}_0,_S = \mathbf{t}_0$, and as a triad of orthonormal bases, we have $\boldsymbol{\nu}_0 = \mathbf{t}_0 \times \mathbf{N}$. Finally, with the component representation of $\mathbf{t}_0$ by $\mathbf{t}_0 = t_0^\alpha\, \mathbf{G}_\alpha$, we have with $\mathbf{t}_0 = \mathbf{C}_0(S(S^\alpha)),_S = (\frac{dS}{dS^\alpha})\, \mathbf{C}_{0\cdot\alpha} = t_0^\alpha\, \mathbf{G}_\alpha$, the components as $t_0^\alpha \equiv \frac{dS}{dS^\alpha}$. Thus, $\mathbf{N},_S = \mathbf{N},_\alpha \frac{dS}{dS^\alpha} = t_0^\alpha\, \mathbf{N},_\alpha$; $\boldsymbol{\nu}_0 = \nu_{0\beta}\, \mathbf{G}^\beta$. Now, $\boldsymbol{\nu}_0 = \mathbf{t}_0 \times \mathbf{N} = S,_\alpha \mathbf{G}_\alpha \times \mathbf{N} = S,_\alpha \overset{0}{\varepsilon}_{\alpha\beta}\, \mathbf{G}^\beta = \nu_{0\beta}\, \mathbf{G}^\beta$ implying

$$S,_\alpha \overset{0}{\varepsilon}_{\alpha\beta} = \nu_{0\beta} \Rightarrow S,_\alpha \overset{0}{\varepsilon}_{\alpha\beta} \overset{0}{\varepsilon}{}^{\beta\alpha} = \overset{0}{\varepsilon}{}^{\beta\alpha}\, \nu_{0\beta} \Rightarrow \frac{\partial S}{\partial S^\alpha} \delta_\alpha^{\ \beta} = \overset{0}{\varepsilon}{}^{\beta\alpha}\, \nu_{0\beta} \tag{11.40}$$

11.2.3.12.6 Method 1 for Side Surface Area Transformation

Now, to have $\mathbf{dA}^{\mathbf{S}^\eta}$ in the form of $\boldsymbol{\nu}_\eta\, \mu^{S^\eta} dA^S$, we must have:

$$\boxed{{}^\eta\mu\, \boldsymbol{\nu} = {}^\eta\mathbf{T}\, \boldsymbol{\nu}_0} \tag{11.41}$$

where we have defined a transformation tensor, $\mathbf{T}$, as:

$$\boxed{{}^\eta\mathbf{T} \equiv \mathbf{I} + \eta\, [(\mathbf{N},_S \bullet \mathbf{t}_0)\mathbf{I} - (\mathbf{t}_0 \otimes \mathbf{N},_S)] = \mathbf{I} + \eta\, t_0^\alpha\, [(\mathbf{N},_\alpha \bullet \mathbf{t}_0)\mathbf{I} - (\mathbf{t}_0 \otimes \mathbf{N},_\alpha)]} \tag{11.42}$$

Exercise: Show that: ${}^\eta\mu\, \boldsymbol{\nu} = {}^\eta\mathbf{T}\, \boldsymbol{\nu}_0$

Solution

$$\begin{aligned} {}^\eta\mu\, \boldsymbol{\nu} &= \mathbf{C},_S \times \mathbf{C},_\eta = (\mathbf{t}_0 + \eta\, \mathbf{N},_S) \times\, \mathbf{N} = \boldsymbol{\nu}_0 + \eta\, (\mathbf{N},_S \times \boldsymbol{\nu}_0 \times\, \mathbf{t}_0) \\ &= \boldsymbol{\nu}_0 + \eta\{(\mathbf{N},_S \bullet \mathbf{t}_0)\boldsymbol{\nu}_0 - (\mathbf{N},_S \bullet \boldsymbol{\nu}_0)\mathbf{t}_0\} = \{\mathbf{I} + \eta\, [(\mathbf{N},_S \bullet \mathbf{t}_0)\mathbf{I} - (\mathbf{t}_0 \otimes \mathbf{N},_S)]\}\, \boldsymbol{\nu}_0 = {}^\eta\mathbf{T}\, \boldsymbol{\nu}_0 \end{aligned}$$

As a special case, if we have the side surfaces parallel to the lines of curvature, and use arc lengths, $\{S^\alpha\}$, as the reference surface parameter along the lines of curvature, then we have, say, for the side surface along the direction 1: $\mathbf{t}_0 \equiv \mathbf{G}_1$; $\boldsymbol{\nu}_0 \equiv \mathbf{G}_2$; implying $t_0^1 = 1$; $\mathbf{N},_1 = -\frac{1}{R_1}\mathbf{G}_1 = -\frac{1}{R_1}\mathbf{t}_0$;

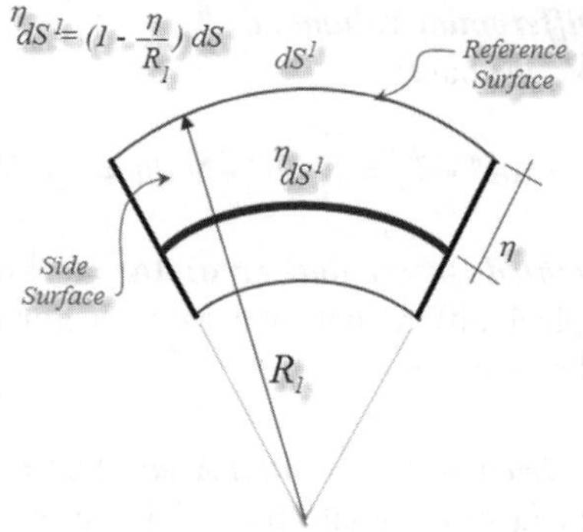

Figure 11.6 Shell surface along a line of curvature.

these imply that $t_0^1(\mathbf{N},_1 \bullet \mathbf{t}_0) = -\frac{1}{R_1}$, and $t_0^1(\mathbf{t}_0 \otimes \mathbf{N},_1)\, v_0 = -\frac{1}{R_1}(\mathbf{t}_0 \bullet v_0)\, \mathbf{t}_0 = \mathbf{0}$. In this special case, from equation (11.42), we have: ${}^{\eta}\mathbf{T} = (1 - \frac{\eta}{R_1})\mathbf{I}$ as shown in Figure 11.6. Similarly, for the side surface parallel to the other line of curvature, direction 2, we get: ${}^{\eta}\mathbf{T} = (1 - \frac{\eta}{R_2})\mathbf{I}$.

11.2.3.12.7 Method 2: For Side Surface Area Transformation

Now, a typical differential area axial vector $\mathbf{dA}^{S^{\eta}}$ at any $\eta \neq 0$ on the side surface, M^S, is obtained by focusing on the expression for $\mathbf{C}(S, \eta^t) = \mathbf{C_0}(S) + \eta(S)\,\mathbf{N}, \quad \eta \in [-\eta^b(S), \eta^t(S)], \quad S \in \partial\mathbf{M}$ as in equation (11.31). Note that ${}^{\eta}\mathbf{A}_{\alpha} \times \mathbf{N} = {}^{\eta}\varepsilon_{\alpha\gamma}{}^{\eta}\mathbf{A}^{\gamma}$ at any η, and ${}^{0}\mathbf{A}_{\alpha} \equiv \mathbf{G}_{\alpha}$ at $\eta = 0$ with ${}^{\eta}\varepsilon_{\alpha\gamma} = \sqrt{{}^{\eta}A}\, e_{\alpha\gamma} = \sqrt{\frac{{}^{\eta}A}{G}}\sqrt{G}\, e_{\alpha\gamma} = {}^{\eta}\mu\, {}^{0}\varepsilon_{\alpha\gamma}$. Furthermore, with the relation: ${}^{0}\varepsilon^{\beta\alpha}\, {}^{0}\varepsilon_{\alpha\gamma} = -\delta^{\beta}_{\gamma}$, we can show that, on the side surface at any η:

$$\boxed{v_{\eta}\, \mathbf{dA}^S_{\eta} = {}^{\eta}\mu\ v_{0\beta}\, {}^{\eta}\mathbf{A}^{\beta}\ dA^S_0} \tag{11.43}$$

Exercise: Prove the relation given by equation (11.43).

Solution: Using the relation given by equation (11.40), we get:

$$\begin{aligned} v_{\eta}\, \mathbf{dA}^S_{\eta} &= \mathbf{dC}(\eta, S) \times \mathbf{d}\eta = \mathbf{C},_S\, dS \times \mathbf{N}\, d\eta = S^{\alpha},_S\, \mathbf{C},_{\alpha} \times \mathbf{N}\ dA^S_0 \\ &= ({}^{0}\varepsilon^{\beta\alpha}\ v_{0\beta})(\mathbf{A}_{\alpha} \times \mathbf{N})\ dA^S_0 = ({}^{0}\varepsilon^{\beta\alpha}\ v_{0\beta})(-{}^{\eta}\varepsilon_{\alpha\gamma}{}^{\eta}\mathbf{A}^{\gamma})\ dA^S_0 \\ &= -v_{0\beta}({}^{0}\varepsilon^{\beta\alpha}\ {}^{\eta}\mu\ {}^{0}\varepsilon_{\alpha\gamma})\, {}^{\eta}\mathbf{A}^{\gamma}\ dA^S_0 = {}^{\eta}\mu\ v_{0\beta}\ \delta^{\beta}_{\gamma}\, {}^{\eta}\mathbf{A}^{\gamma}\ dA^S_0 = {}^{\eta}\mu\ v_{0\beta}\ {}^{\eta}\mathbf{A}^{\beta}\ dA^S_0 \end{aligned}$$

In what follows, we use both method 1 and method 2 relations to show alternative definitions for a shell side-surface tractions.

11.2.4 Gauss–Weingarten Relations: Along the Lines of Curvature with Arc Length Parameterization

Among the most important formulas in the description of surfaces are the ***Gauss–Weingarten formulas*** that relate the base vector derivatives to the base vectors themselves – these are the surface counterparts of the Frenet–Serret formulas for curves. However, for computational ease and efficiency, we must express these in the optimal coordinate system, that is, in terms of the base vectors along the lines of curvature, and parameterized in terms of the arc length parameters.

11.2.4.1 Normal Coordinate System

First, let us recall (see Chapter 6 on surfaces) the following Gauss–Weingarten formula in general shell coordinates system, with base vectors $\{\hat{\mathbf{G}}_1\ \hat{\mathbf{G}}_2\ \hat{\mathbf{G}}_3 \equiv \mathbf{N}\}$, and parameters $(\hat{\xi}_1, \hat{\xi}_2)$, of a 2-surface:

$$\boxed{\begin{aligned} \hat{\mathbf{G}}_{\beta,\hat{\alpha}} &= \hat{\Gamma}^{\gamma}_{\beta\alpha}\ \hat{\mathbf{G}}_\gamma + B_{\beta\alpha}\ \mathbf{N}\ \text{(Gauss)} \\ \mathbf{N}_{,\hat{\alpha}} &= -B_{\alpha}^{\cdot\,\beta}\ \hat{\mathbf{G}}_\beta \qquad \text{(Weingarten)} \end{aligned}} \tag{11.44}$$

In matrix form, we can express this as:

$$\boxed{\begin{Bmatrix} \hat{\mathbf{G}}_1 \\ \hat{\mathbf{G}}_2 \\ \mathbf{N} \end{Bmatrix}_{,\alpha} = \begin{bmatrix} \hat{\Gamma}^1_{1\alpha} & \hat{\Gamma}^2_{1\alpha} & \hat{B}_{1\alpha} \\ \hat{\Gamma}^1_{2\alpha} & \hat{\Gamma}^2_{2\alpha} & \hat{B}_{2\alpha} \\ -\hat{B}_{\alpha}^{\cdot\,1} & -\hat{B}_{\alpha}^{\cdot\,2} & 0 \end{bmatrix} \begin{Bmatrix} \hat{\mathbf{G}}_1 \\ \hat{\mathbf{G}}_2 \\ \mathbf{N} \end{Bmatrix}} \tag{11.45}$$

where recall that the Christoffel symbols of the second kind, $\hat{\Gamma}^{\gamma}_{\beta\alpha}$, are given as the functions of the components of the metric tensor as:

$$\boxed{\begin{aligned} \hat{\Gamma}^{\sigma}_{\alpha\beta} &= \tfrac{1}{2}\hat{G}^{\sigma\gamma}(\hat{G}_{\alpha\sigma,\hat{\beta}} + \hat{G}_{\gamma\beta,\hat{\alpha}} - \hat{G}_{\alpha\beta,\hat{\gamma}}) \\ \hat{\Gamma}^{3}_{\alpha\beta} &\equiv \hat{B}_{\alpha\beta} \\ \hat{B}_{\alpha}^{\cdot\,\beta} &= \hat{G}^{\sigma\beta}\hat{B}_{\alpha\sigma} \end{aligned}} \tag{11.46}$$

11.2.4.2 Normal Coordinate System Along the Lines of Curvature

If we now consider the normal coordinate system, with base vectors set: $\{\bar{\mathbf{G}}_1 \equiv \frac{\partial \mathbf{C}_0}{\partial \xi^1}\ \bar{\mathbf{G}}_2 \equiv \frac{\partial \mathbf{C}_0}{\partial \xi^2}\ \bar{\mathbf{G}}_3 \equiv \mathbf{N}\}$, and parameters $(\bar{\xi}^1, \bar{\xi}^2)$ along the lines of curvature (which are, of course, perpendicular to each other), we have: $\hat{G}^{\alpha\beta} \to \bar{G}^{\alpha\beta} = 0, \hat{B}^{\alpha\beta} \to \bar{B}^{\alpha\beta} = 0,\ \alpha \neq \beta$; also, with the determinant, $\bar{G}$, of the 2×2 surface metric tensor, we have: $\bar{G} = \bar{G}_{11}\ \bar{G}_{22} - \cancel{\bar{G}_{12}\ \bar{G}_{21}} \Rightarrow \bar{G}^{(11)} = \frac{\bar{G}_{(22)}}{\bar{G}} = \frac{1}{\bar{G}_{(11)}}$ and $\bar{G}^{(22)} = \frac{1}{\bar{G}_{(22)}}$ with $(\alpha\alpha)$ means no sum intended. Now, recalling that $\frac{1}{R_1} = B_{1}^{\cdot\,1} = \bar{G}^{\gamma 1}\bar{B}_{1\gamma} = \bar{G}^{(11)}\bar{B}_{(11)} = \frac{\bar{B}_{(11)}}{\bar{G}_{(11)}}$, and letting $\bar{G}_{\alpha\alpha} = \bar{\mathbf{G}}_\alpha \bullet \bar{\mathbf{G}}_\alpha = \gamma_\alpha \mathbf{G}_\alpha \bullet \gamma_\alpha \mathbf{G}_\alpha = \gamma_\alpha^2$ where we have introduced unit base vector set $\{\mathbf{G}_\alpha\} \equiv \{\frac{\partial \mathbf{C}}{\partial S^\alpha}\}$ with γ_α as the length of $\bar{\mathbf{G}}_\alpha$, from the third of equation (11.45), we get:

$$\frac{\partial \mathbf{N}}{\partial \xi^\alpha} = -B_{\alpha}^{\cdot\,\beta}\ \bar{\mathbf{G}}_\beta = -B_{\alpha}^{\cdot\,\alpha}\ \bar{\mathbf{G}}_\alpha - \cancel{\underbrace{B_{\alpha}^{\cdot\,\beta}\ \bar{\mathbf{G}}_\beta}_{\alpha\neq\beta}} = -\frac{1}{R_\alpha}\bar{\mathbf{G}}_\alpha = -\frac{\gamma_\alpha}{R_\alpha}\mathbf{G}_\alpha \tag{11.47}$$

and from $\bar{\mathbf{G}}_\alpha = \gamma_\alpha \mathbf{G}_\alpha$, we get:

$$\frac{\partial \mathbf{G}_\alpha}{\partial \xi^\alpha} = \frac{1}{\gamma_\alpha}\bar{\mathbf{G}}_{\alpha},_{\alpha} - \frac{1}{\gamma_\alpha^2}\gamma_{\alpha},_{\alpha}\bar{\mathbf{G}}_\alpha \tag{11.48}$$

Now, from the first of the relations in equation (11.45) with $\bar{G}_{\alpha\beta} = \bar{B}_{\alpha\beta} = 0, \quad \alpha \neq \beta$, and $\bar{G}^{(\alpha\alpha)} = \frac{1}{\bar{G}_{(\alpha\alpha)}}$ and so on, and noting that ${}^{\xi}\bar{\Gamma}^{\sigma}_{\alpha\beta}$ described in the $\{\xi^\alpha\}$-coordinate system, we get:

$$\boxed{{}^{\xi}\bar{\Gamma}^{\sigma}_{\alpha\beta} = \frac{1}{2}\bar{G}^{\sigma\gamma}\left(\frac{\partial \bar{G}_{\alpha\sigma}}{\partial \xi^\beta} + \frac{\partial \bar{G}_{\gamma\beta}}{\partial \xi^\alpha} - \frac{\partial \bar{G}_{\alpha\beta}}{\partial \xi^\gamma}\right)} \tag{11.49}$$

Particularizing equation (11.49): $\bar{\Gamma}^1_{11} = \frac{1}{2}\{\bar{G}^{11}\bar{G}_{11},_{\bar{1}}\} = \frac{1}{2}\frac{1}{\bar{G}_{11}}\bar{G}_{11},_{\bar{1}} = \frac{1}{\gamma_1}\gamma_1,_{\bar{1}}$; similarly, we get: $\bar{\Gamma}^2_{11} = -\frac{\gamma_1}{\gamma_2^2}\gamma_1,_{\bar{2}}$. Now, using equations (11.47) and (11.48), we get from the first of equation (11.45):

$$\begin{aligned}
\mathbf{G}_\alpha &= \frac{1}{\gamma_\alpha}\bar{\mathbf{G}}_{\alpha},_{\bar{\alpha}} - \frac{1}{\gamma_\alpha^2}\gamma_{\alpha},_{\bar{\alpha}}\bar{\mathbf{G}}_\alpha = \frac{1}{\gamma_\alpha}\left(\bar{\Gamma}^{\alpha}_{\alpha\alpha}\bar{\mathbf{G}}_\alpha + B_{(\alpha\alpha)}\mathbf{N}\right) - \frac{1}{\gamma_\alpha^2}\gamma_{\alpha},_{\bar{\alpha}}\bar{\mathbf{G}}_\alpha \\
&= \frac{1}{\gamma_\alpha}\left(\bar{\Gamma}^{\alpha}_{\alpha\alpha}\bar{\mathbf{G}}_\alpha + \bar{\Gamma}^{\beta}_{\alpha\alpha}\bar{\mathbf{G}}_\beta + \frac{\gamma_\alpha^2}{R_\alpha}\mathbf{N}\right) - \frac{1}{\gamma_\alpha^2}\gamma_{\alpha},_{\bar{\alpha}}\bar{\mathbf{G}}_\alpha \\
&= \frac{1}{\gamma_\alpha^2}\gamma_{\alpha},_{\bar{\alpha}}\bar{\mathbf{G}}_\alpha - \frac{1}{\gamma_\beta^2}\gamma_{\alpha},_{\bar{\beta}}\bar{\mathbf{G}}_\beta + \frac{\gamma_\alpha}{R_\alpha}\mathbf{N} - \frac{1}{\gamma_\alpha^2}\gamma_{\alpha},_{\bar{\alpha}}\bar{\mathbf{G}}_\alpha \\
&= -\frac{1}{\gamma_\beta}\gamma_{\alpha},_{\bar{\beta}}\mathbf{G}_\alpha + \frac{\gamma_\alpha}{R_\alpha}\mathbf{N}, \quad \alpha \neq \beta
\end{aligned} \tag{11.50}$$

Using equations (11.47) and (11.50), we can rewrite the Gauss–Weingarten formula as:

$$\boxed{\frac{\partial}{\partial \xi^1}\begin{Bmatrix}\mathbf{G}_1\\ \mathbf{G}_2\\ \mathbf{N}\end{Bmatrix} = \begin{bmatrix} 0 & -\frac{1}{\gamma_2}\gamma_1,_{\bar{2}} & \frac{\gamma_1}{R_1} \\ \frac{1}{\gamma_2}\gamma_1,_{\bar{2}} & 0 & 0 \\ -\frac{\gamma_1}{R_1} & 0 & 0 \end{bmatrix}\begin{Bmatrix}\mathbf{G}_1\\ \mathbf{G}_2\\ \mathbf{N}\end{Bmatrix}} \tag{11.51}$$

and

$$\boxed{\frac{\partial}{\partial \xi^2}\begin{Bmatrix}\mathbf{G}_1\\ \mathbf{G}_2\\ \mathbf{N}\end{Bmatrix} = \begin{bmatrix} 0 & \frac{1}{\gamma_1}\gamma_2,_{\bar{1}} & 0 \\ \frac{1}{\gamma_1}\gamma_2,_{\bar{1}} & 0 & \frac{\gamma_2}{R_2} \\ 0 & -\frac{\gamma_2}{R_2} & 0 \end{bmatrix}\begin{Bmatrix}\mathbf{G}_1\\ \mathbf{G}_2\\ \mathbf{N}\end{Bmatrix}} \tag{11.52}$$

11.2.4.3 Optimal Normal Coordinate System

Now, let us consider the optimal normal coordinate system, that is, coordinates along the lines of curvature with arc length parameterization; let (S^1, S^2) define a point on the shell undeformed reference surface. Noting that $\frac{\partial \gamma_1}{\partial S^1} = \frac{\partial \gamma_1}{\partial \xi^1}\frac{\partial \xi_1}{\partial S^1} = \frac{1}{\gamma_1}\frac{\partial \gamma_1}{\partial \xi^1}$, we get: $\gamma_{1,\bar{2}} \equiv \frac{\partial \gamma_1}{\partial \xi^2} = \frac{\partial \gamma_1}{\partial S^\alpha}\frac{\partial S^\alpha}{\partial \xi^2} = \gamma_2 \frac{\partial \gamma_1}{\partial S^2} \equiv \gamma_2\, \gamma_{1,2}$

$$\mathbf{G}_{1,\bar{1}} \equiv \frac{\partial \mathbf{G}_1}{\partial \xi^1} = \frac{\partial \mathbf{G}_1}{\partial S^\alpha}\frac{\partial S^\alpha}{\partial \xi^1} = \gamma_1 \mathbf{G}_{1,1} \tag{11.53}$$

Using the first row of the relation in equation (11.51), we get:

$$\begin{aligned} \mathbf{G}_{1,1} &\equiv \frac{\partial \mathbf{G}_1}{\partial S^1} = \frac{1}{\gamma_1}\mathbf{G}_{1,\bar{1}} \equiv \frac{1}{\gamma_1}\frac{\partial \mathbf{G}_1}{\partial \xi^1} \\ &= \frac{1}{\gamma_1}\left(-\frac{1}{\not{\gamma_2}}\not{\gamma_2}\frac{\partial \gamma_1}{\partial S^2}\right)\mathbf{G}_2 + \frac{1}{\not{\gamma_1}}\frac{\not{\gamma_1}}{R_1}\mathbf{N} \\ &= \frac{1}{\gamma_1}\frac{\partial \gamma_1}{\partial S^2}\mathbf{G}_2 + \frac{1}{R_1}\mathbf{N} \end{aligned} \tag{11.54}$$

Modifications similar to equation (11.54) can be developed for each term of the equations (11.51) and (11.52). Thus, from equations (11.47) and (11.50), we get the desired Gauss–Weingarten formulas for the derivatives of the triad of base vectors in matrix form with respect to the arc length parameters, S^α, $\alpha = 1, 2$, given by:

$$\begin{Bmatrix} \mathbf{G}_{1,1} \\ \mathbf{G}_{2,1} \\ \mathbf{N}_{,1} \end{Bmatrix} \equiv \frac{\partial}{\partial S^1}\begin{Bmatrix} \mathbf{G}_1 \\ \mathbf{G}_2 \\ \mathbf{N} \end{Bmatrix} = \begin{bmatrix} 0 & -\frac{1}{\gamma_1}\gamma_{1,2} & \frac{1}{R_1} \\ \frac{1}{\gamma_1}\gamma_{1,2} & 0 & 0 \\ -\frac{1}{R_1} & 0 & 0 \end{bmatrix}\begin{Bmatrix} \mathbf{G}_1 \\ \mathbf{G}_2 \\ \mathbf{N} \end{Bmatrix} \equiv \widehat{\mathbf{K}}_1^0 \begin{Bmatrix} \mathbf{G}_1 \\ \mathbf{G}_2 \\ \mathbf{N} \end{Bmatrix} \tag{11.55}$$

and

$$\begin{Bmatrix} \mathbf{G}_{1,2} \\ \mathbf{G}_{2,2} \\ \mathbf{N}_{,2} \end{Bmatrix} \equiv \frac{\partial}{\partial S^2}\begin{Bmatrix} \mathbf{G}_1 \\ \mathbf{G}_2 \\ \mathbf{N} \end{Bmatrix} = \begin{bmatrix} 0 & \frac{1}{\gamma_2}\gamma_{2,1} & 0 \\ -\frac{1}{\gamma_2}\gamma_{2,1} & 0 & \frac{1}{R_2} \\ 0 & -\frac{1}{R_2} & 0 \end{bmatrix}\begin{Bmatrix} \mathbf{G}_1 \\ \mathbf{G}_2 \\ \mathbf{N} \end{Bmatrix} \equiv \widehat{\mathbf{K}}_2^0 \begin{Bmatrix} \mathbf{G}_1 \\ \mathbf{G}_2 \\ \mathbf{N} \end{Bmatrix} \tag{11.56}$$

where $\widehat{\mathbf{K}}_\alpha^0$, $\alpha = 1, 2$ are the shell undeformed curvature matrices with $R_\alpha(S^1, S^2)$, $\alpha = 1, 2$, are the radii of curvatures at any point, $\{S^\alpha\}$, of the shell reference surface, and γ_α, $\alpha = 1, 2$, are the lengths of the base vectors: $\gamma_\alpha \equiv \frac{dS^\alpha}{d\xi^\alpha} = \|\bar{\mathbf{G}}_\alpha\| = \sqrt{\bar{\mathbf{G}}_{\alpha\alpha}}$ where $\bar{\mathbf{G}}_\alpha(\xi^1, \xi^2) \equiv \frac{\partial \mathbf{C}(\xi^1, \xi^2, 0)}{\partial \xi^\alpha}$, with no sum on α.

Remarks

- The curvature matrices are skew-symmetric.
- The individual coefficients of the matrices can be easily identified with the Christoffel symbols of second kind, Γ^i_{jk}.
- In our entire shell analysis, we will completely avoid explicit use of these obscuring symbols by adhering to familiar geometric entities such as the radii of curvature, and so on, as we have done before, for ease of interpretation and clarity.

11.2.5 Initial Curvature Tensors and Vectors

On the other hand, the Gauss–Weingarten relation for the derivatives of the triad of base vectors on a surface in invariant tensor form with respect to the arc length parameters, $S^\alpha, \quad \alpha = 1, 2$, may be obtained by first introducing the following.

Definition: ***Shell initial curvature tensors***, $\mathbf{K}^0_\alpha$, $\alpha = 1, 2$:

$$\boxed{\mathbf{K}^0_\alpha \equiv (\widehat{\mathbf{K}}^0_\alpha)^T_{ij}\, \mathbf{G}_i \otimes \mathbf{G}_j = -\mathbf{G}_i \otimes \mathbf{G}_{i,\alpha}, \quad \alpha = 1, 2; \quad i, j = 1, 2, 3} \tag{11.57}$$

To wit, for the second equality in equation (11.57), we note that because of anti-symmetry: $(\widehat{\mathbf{K}}^0_\alpha)^T_{ij} = -(\widehat{\mathbf{K}}^0_\alpha)_{ij}$. Then, we have:

$$\begin{aligned}\mathbf{K}^0_\alpha \equiv (\widehat{\mathbf{K}}^0_\alpha)^T_{ij}\, \mathbf{G}_i \otimes \mathbf{G}_j &= -(\widehat{\mathbf{K}}^0_\alpha)_{ij}\, \mathbf{G}_i \otimes \mathbf{G}_j = -\,\mathbf{G}_i \otimes (\widehat{\mathbf{K}}^0_\alpha)_{ij}\mathbf{G}_j \\ &= -\mathbf{G}_i \otimes \mathbf{G}_{i,\alpha}, \quad \alpha = 1, 2; \quad i, j = 1, 2, 3\end{aligned}$$

where we have used the result from equation (11.57). Note that the curvature tensor is skew-symmetric and thus: $(\mathbf{K}^0_\alpha)^T = -\mathbf{K}^0_\alpha$. This can be easily verified:

$$\begin{aligned}(\mathbf{K}^0_\alpha)^T \equiv (\widehat{\mathbf{K}}^0_\alpha)^T_{ij}\, \mathbf{G}_j \otimes \mathbf{G}_i &\underbrace{=}_{\text{Switch indices}} (\widehat{\mathbf{K}}^0_\alpha)^T_{ji}\, \mathbf{G}_i \otimes \mathbf{G}_j \\ = \underbrace{=}_{\text{Skew-symmetric}} -(\widehat{\mathbf{K}}^0_\alpha)^T_{ij}\mathbf{G}_i \otimes \mathbf{G}_j &\underbrace{=}_{\text{Definition}} -\mathbf{K}^0_\alpha, \quad \alpha = 1, 2;\end{aligned}$$

Definition: ***Shell initial curvature vector***, $\mathbf{k}^0_\alpha$, $\quad \alpha = 1, 2$:

The axial vectors corresponding to the initial curvature tensors, $\mathbf{K}^0_\alpha$, $\alpha = 1, 2$, are given by:

$$\boxed{\begin{aligned}\mathbf{k}^0_1 &\equiv \left\{0 \quad -\frac{1}{R_1} \quad -\frac{1}{\gamma_1}\gamma_{1,2}\right\}^T \\ \mathbf{k}^0_2 &\equiv \left\{\frac{1}{R_2} \quad 0 \quad \frac{1}{\gamma_2}\gamma_{2,1}\right\}^T\end{aligned}} \tag{11.58}$$

Now, we can easily get from equation (11.57), the invariant tensor form as:

$$\boxed{\begin{aligned}\mathbf{G}_{i},_{\alpha} &= \mathbf{K}^{0}_{\alpha}\,\mathbf{G}_{i}, \qquad \alpha = 1,2;\ i = 1,2,3\\ &= \mathbf{k}^{0}_{\alpha} \times\ \mathbf{G}_{i}\end{aligned}} \tag{11.59}$$

Note that, in tensor form as in expression (11.57), unlike the matrix representations (11.55) and (11.56), each unit base vector derivative is given as a transformation of itself. As we shall see, this allows us to write expressions in absolute compact form free of matrix indices.

Exercise: Verify the tensor form of the derivatives of base vectors given by equation (11.59).

Solution:

$$\boxed{\begin{aligned}\mathbf{K}^{0}_{\alpha}\,\mathbf{G}_{i} \underbrace{=}_{\text{Skew-symmetry}} -(\mathbf{K}^{0}_{\alpha})^{T}\,\mathbf{G}_{i} \underbrace{=}_{\text{Definition}} -(-\mathbf{G}_{k},_{\alpha} \otimes \mathbf{G}_{k})\mathbf{G}_{i}\\ = \mathbf{G}_{i},_{\alpha},\ \alpha = 1,2;\ i = 1,2,3\end{aligned}}$$

Exercise: Identify the components of the matrices in equations (11.55) and (11.56) as components of Christoffel symbol matrix, $\{\Gamma^{j}_{i\,\alpha}\},\ \ i = 1,2,3;\ j = 1,2,3;\ \alpha = 1,2.$

Discussions

- Shell geometry representation goes hand in hand with the assumptions that we build into our method of analysis. This, in turn, entails consideration of the range of both geometric (such as thin or thick shell) and material properties (such as small or finite strain level) that the shell may predictably undergo.
- In classical shell theory, the shell kinematic parameters are restricted to only five components – three for displacements and two for rotations about the surface coordinate vectors at any point on the shell. We will be interested in parameterization by all six degrees of freedom, that is, three degrees of freedom each for displacement and rotation. In the ensuing nonlinear theory, we intend to hold onto these same six components as well for engineering applications. As we shall see, we have to resort to reasonable kinematic and material assumptions in our definition of a shell as a thin structural element undergoing small material property changes (such as hyperelastic strain), or alternately, allow it to be thick and subjected to finite strain, and all appropriate approximations are lumped in the declaration of the constitutive theory, which is at best empirical anyway.
- As an example, for kinematic behavior, it has been assumed that a normal section remains plane (i.e. no warping) but not necessarily perpendicular to the shell reference surface. Moreover, the plane cross-sections remain rigid in their own plane. This is commonly known as the ***Reissner–Mindlin*** hypothesis in which linear through-the-thickness deformation distribution is permitted. However, we must guarantee that this hypothesis is natural. In other words, adoption of this assumption is contingent upon the fact that the true 3D effects may be acceptably captured on an average sense over the cross-sections. The thin shells with surface dimensions conspicuously

dominant tend to behave in a way that makes this assumption seem reasonable. That is, with this assumption, we can still predict quite comfortably all the general behaviors of a shell: the axial, shear and bending characteristics.

11.2.6 Where We Would Like to Go

In the foregoing, we have defined shells suitable for computation based on the assumptions delineated above. In so doing, we have made extensive use of the general theory of surfaces. Clearly, then, we should review the chapter on:

- the theory and computation of surfaces.
- Now, for proceeding onward with the shell analysis, we must move on to:
- shell equation derivation: 3D to 2D.
- Finally, to get familiarized with and be conscious about the overall computational strategy for nonlinear shells, we should explore:
- the road map: nonlinear shells.

11.3 Static and Dynamic Equations: Continuum Approach – 3D to 2D

11.3.1 What We Need to Recall

- A structural shell, by definition, is a 3D continuum with one dimension, namely the thickness, considerably smaller than the other two, that is, a continuum in the form of a 2D surface called the shell reference surface. Thus, as a surface, we will eventually be able to characterize the configurations (undeformed and deformed) of a shell as functions of two parameters of our choice: the arc lengths along the lines of curvature, on the reference surface of the shell.
- We characterize the shell undeformed reference surface by the arc length parameter S^1, S^2 so that the position vectors on the shell reference surface may be given as $\mathbf{C}_0(S^1, S^2) \equiv \mathbf{C}(S^1, S^2, 0)$.
- S^1, S^2, as smooth reparameterization, are associated with ξ^1, ξ^2, the natural coordinate parameters along the lines of curvature, respectively, by the relations: $S^1(\xi^1, \xi^2) = \int_0^{\xi^1} \|\mathbf{C}(\rho, \xi^2, 0),_\rho\| d\rho$ and $S^2(\xi^1, \xi^2) = \int_0^{\xi^2} \|\mathbf{C}(\xi^1, \rho, 0),_\rho\| d\rho$ where $\mathbf{C}(\xi^1, \xi^2, 0)$ is the position vector of a point on the shell reference surface.
- The undeformed 3D shell interior, B, is completely enclosed by the boundary ∂B which, in turn is made up of three disjoint surfaces: the top and bottom faces, denoted by M^t and M^b, respectively, and a side surface denoted by M^S. Any undeformed thickness direction point away from the undeformed reference surface is fully parameterized by a parameter, η, along the normal, $\mathbf{N}(S^1, S^2)$, to the reference surface at a point, (S^1, S^2), on it; thus, any point on the shell body is identified by the coordinates (S^1, S^2, η) at any time $t \in \mathbb{R}_+$.
- The 3D equations of motion in Lagrangian (material) formulation for a 3D body, B, with the surface undeformed surface area, ∂B, for dynamic loading conditions, can be written (see Chapter 7 on nonlinear mechanics) as:

$$\boxed{\begin{aligned} &\int_{\partial B} \mathbf{t}\, dA + \int_B \mathbf{b}\, dV = \frac{d}{dt}\int_B \mathbf{v}\,\rho_0\, dV && \text{(Momentum balance)} \\ &\int_{\partial B} \mathbf{c}\times\mathbf{t}\, dA + \int_B \mathbf{c}\times\mathbf{b}\, dV = \frac{d}{dt}\int_B \mathbf{c}\times\mathbf{v}\,\rho_0\, dV && \text{(Moment of momentum balance)} \end{aligned}} \tag{11.60}$$

where **t** and **b** are the surface traction vector measured per unit surface area and the body force vector measured per unit volume of the undeformed body, respectively, but acting on the current deformed body. In other words, we track the areas and the volumes of the neighborhoods around, **t** and **b**, respectively, in the current deformed body by identifying their corresponding areas and volumes in the undeformed body. $\mathbf{v}(S^1, S^2, \eta, t)$ is the current velocity at any point (S^1, S^2, η) and at the current time $t \in \mathbb{R}_+$; ρ_0 is the mass density per unit volume of the undeformed body; **c** is the position vector of the point (S^1, S^2, η) on the body in the deformed configuration.

- According to the theory of surfaces, on the undeformed reference surface, the optimal normal curvilinear frame along the lines of curvature and the normal to the surface may be denoted by a triad of orthonormal base vectors $\mathbf{G}_1, \mathbf{G}_2, \mathbf{G}_3 \equiv \mathbf{N}$. Additionally, for later use, we can equivalently write them in compact form as: $\mathbf{G}_k$, $k = 1, 2, 3$.
- On the shell reference surface, the differential area is given by: $dA = dS^1\, dS^2$
- In the undeformed shell interior around any point (S^1, S^2, η), the differential volume, dV^η, is given by: $dV^\eta = \mathbf{dA}^\eta \bullet \mathbf{d}\eta = \mathbf{N}\, dA^\eta \bullet \mathbf{N}\, d\eta = {}^\eta\mu\, dA\, d\eta$
- The shell top face differential area is given by: $dA^t = \mu^t\, dA$ where $\mu^t = \sqrt{\frac{^{\eta^t}A}{G}} = 1 - 2\eta^t H + (\eta^t)^2 K$ with μ^t defined as μ, the determinant of the shift tensor, at $\eta = \eta^t$, and H and K as the mean and Gaussian curvatures; A^{η^t} and G are the surface metric determinant at $\eta = \eta^t$ and $\eta = 0$, respectively.
- The shell bottom surface differential area is given by: $dA^b = \mu^b\, dA$ where $\mu^b = \sqrt{\frac{^{\eta^b}A}{G}} = 1 - 2\eta^b H + (\eta^b)^2 K$ with μ^t defined as μ at $\eta = \eta^b$.
- The shell side surface differential area at $\eta = 0$ is given by: $dA_0^s = dS\, d\eta$, and at any $\eta \neq 0$, we have: $\mathbf{v}_\eta\, dA_\eta^s = {}^\eta\mu^S\, \mathbf{v}_\eta\, dA_0^S$ where by
method 1:

$$\boxed{\begin{aligned} {}^\eta\mu^S\, \mathbf{v} &= {}^\eta\mathbf{T}\, \mathbf{v}_0 \\ {}^\eta\mathbf{T} &= \mathbf{I} + \eta\, t_0^\alpha\, [(\mathbf{N},_\alpha \bullet\, \mathbf{t}_0)\mathbf{I} - (\mathbf{t}_0 \otimes \mathbf{N},_\alpha)] \\ t_0^\alpha &= \mathbf{t}_0 \bullet \mathbf{G}^\alpha \end{aligned}} \tag{11.61}$$

and by
method 2:

$$\boxed{\begin{aligned} \mathbf{v}\, dA_\eta^s &= {}^\eta\mu^S\, v_{0\beta}\, {}^\eta\mathbf{A}^\beta\, dA_0^S \\ v_{0\beta} &= v_0 \bullet \mathbf{G}_\beta \\ {}^\eta\mathbf{A}^\beta &\equiv \mathbf{A}^\beta(S^1, S^2, \eta) \end{aligned}} \tag{11.62}$$

- With the representation of second-order tensor **A** as: $\mathbf{A} = \mathbf{A}^\alpha \otimes \mathbf{G}_\alpha$, we have:

$$DIV\, \mathbf{A} = \mathbf{A}^\alpha|_\alpha \tag{11.63}$$

where *DIV* is the surface divergence operator.

- For any vector **a** and second-order tensor **B**, with the contravariant base vector set $\{\mathbf{G}^\alpha\}$, we have:

$$DIV(\mathbf{a} \times \mathbf{B}) = \mathbf{a} \times DIV\,\mathbf{B} + \mathbf{a},_\alpha \times \mathbf{B} \bullet \mathbf{G}^\alpha \tag{11.64}$$

- Finally, as a notational reminder, in what follows, all lowercase Greek indices assume values 1 and 2, and the lowercase Latin indices runs through 1, 2, 3; also, for repeated diagonal indices, Einstein's summation convention of summing over their ranges holds.

11.3.2 Dynamic Balance Equations: General Theory

As indicated above, we would like to establish the connection between the 2D shell balance equations and the balance equations of a shell viewed as a 3D continuum or body. For this, if we extract from the undeformed reference body a completely arbitrary and closed free-body with the interior given by $\Omega \subset \mathrm{B}$, and the enclosing boundary denoted by $\partial\Omega$, we must have the Lagrange momentum balance and the Lagrange angular momentum balance laws satisfied for each equilibrium configuration of the free-body. We are free to choose the shape of the free-body with arbitrary boundary description; however, the choice is usually made such that the resulting derivation can be as simple as possible, and thus, naturally follows the undeformed geometry definition in the Lagrangian or material form of analysis.

Accordingly, in the undeformed reference body, as shown in Figure 11.7, the boundary surface, $\partial\Omega$, is made up of three disjointed surfaces: a top, $\partial\Omega^T \subset M^t$, and a bottom face, $\partial\Omega^B \subset M^b$, and a side surface, $\partial\Omega^S$. The side surface, denoted by $\partial\Omega^S$, is generated by the continuous set of normals, $\mathbf{N}(S, 0)$, at the reference time $t = 0$ drawn at any arc length parameter, S, along an arbitrary and closed curve, $\partial\Omega^M \subset M$, on the reference surface, perpendicular to the shell reference surface, $\Omega^M \subset M$, and intersecting the top and bottom faces. Now, let us recognize the surface traction, $\mathbf{t}$, at any point, (S^1, S^2, η) at the current time $t \in \mathbf{R}_+$, on the deformed boundary, as shown in Figure 11.8, of the free-body as: $\mathbf{t} = \mathbf{P} \bullet \mathbf{A}^j$ with $\mathbf{P}(S^1, S^2, \eta, t)$ as the two-point first Piola–Kirchhoff stress tensor (PK1) defined by: $\mathbf{P} = \mathbf{P}^j \otimes \mathbf{A}_j$with $\{\mathbf{A}_i(S^1, S^2, \eta, t = 0)\}$ as the set of covariant base vectors at a point, in the undeformed body, which when deformed to the current state becomes the traction point in consideration on the deformed surface. Note that: $\mathbf{t} \equiv \mathbf{P}^j = \mathbf{P} \bullet \mathbf{A}^j$ is the traction at a point on the currently deformed surface that had $\mathbf{A}^j$, the contravariant base vector at the point (S^1, S^2, η), as the normal in the undeformed state. Note that PK1 is expressed in a mixed basis form with the domain basis set as $\{\mathbf{A}_j\}$, and the range basis

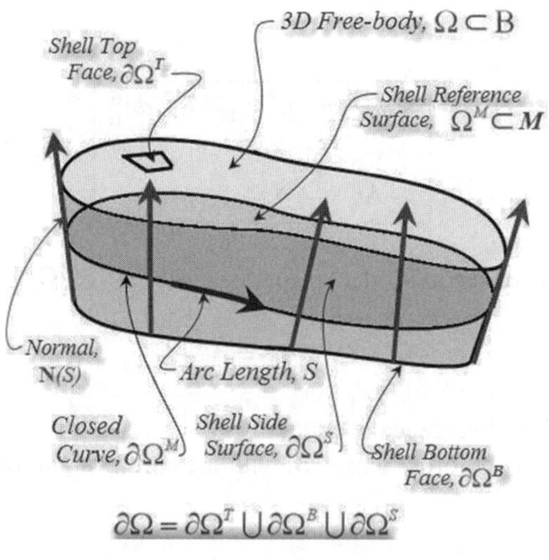

Figure 11.7 Shell undeformed free-body.

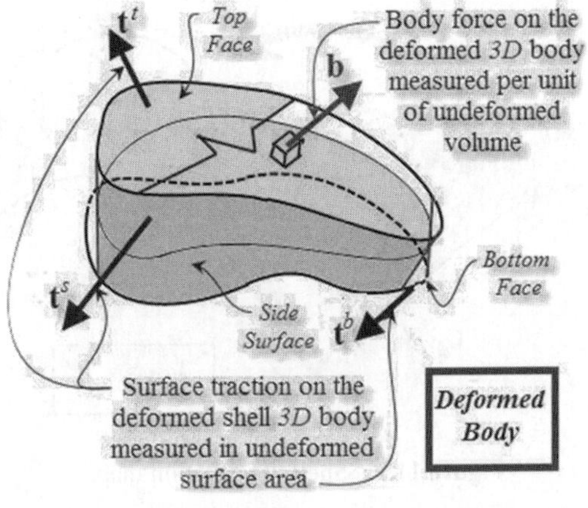

Figure 11.8 3D body: surface traction and body force.

set in the deformed body yet to be explicitly specified. The intended and computationally useful set is identified only after we recognize the nature of the underlying configuration space later. Considering that the shell bounding surface is made up of three surfaces, the tractions on each of these deformed surfaces can be identified as shown in Figure 11.7. With these definitions, we can specialize into the shell space the balance laws of equation (11.60) as:

$$\int_{\partial\Omega^S} (\mathbf{t}^S \equiv \mathbf{P}\,\nu^s)\, dA^s + \int_{\partial\Omega^T} (\mathbf{t}^t \equiv \mathbf{P}\,\nu^t)\, dA^t + \int_{\partial\Omega^B} (\mathbf{t}^b \equiv \mathbf{P}\,\nu^b)\, dA^b + \int_{\Omega} \mathbf{b}\, dV^\eta = \frac{d}{dt}\int_{\Omega} \mathbf{v}\,\rho_0\, dV^\eta$$

$$\int_{\partial\Omega^S} \mathbf{c}\times\mathbf{P}\,\nu^s\, dA^s + \int_{\partial\Omega^T} \mathbf{c}\times\mathbf{P}\,\nu^t\, dA^t + \int_{\partial\Omega^B} \mathbf{c}\times\mathbf{P}\,\nu^b\, dA^b + \int_{\Omega} \mathbf{c}\times\mathbf{b}\, dV^\eta = \frac{d}{dt}\int_{\Omega} \mathbf{c}\times\mathbf{v}\,\rho_0\, dV^\eta \tag{11.65}$$

where ν^t, ν^b and ν^s are the normals at a point on the top face, $\partial\Omega^T$, the bottom face, $\partial\Omega^B$, and the side surface, $\partial\Omega^S$, respectively, of the undeformed body; dA^t, dA^b and dA^s are the differential areas in the neighborhood of the point on the top face, the bottom face and the side surface, respectively, of the undeformed body; $\rho_0(S^1, S^2, \eta, 0)$ is the mass density per unit volume of the undeformed body, that is, at the time $t = 0$; $\mathbf{c}(S^1, S^2, \eta, t)$ is the position vector of the point (with undeformed coordinates (S^1, S^2, η)) on the body in the deformed configuration at the current time $t \in \mathbb{R}_+$; $\mathbf{v}(S^1, S^2, \eta, t) \equiv \dot{\mathbf{c}}(S^1, S^2, \eta, t)$ is the current velocity at any material point at the current time $t \in \mathbb{R}_+$that used to have the coordinates (S^1, S^2, η) in undeformed reference state; and $dV^\eta = {}^\eta\mu\; dA\, d\eta$ is the differential volume of the interior, Ω, of the undeformed body as the volumetric neighborhood of the body force vector and the momenta at the point with undeformed coordinates (S^1, S^2, η) on the body.

11.3.2.1 Deformation Map

The theory we are trying to detail is expected to handle a general deformation map. So, let us assume that $\chi : \mathrm{B} \rightarrow \mathrm{E}^3$ defines a sufficiently smooth deformation map for the shell body as

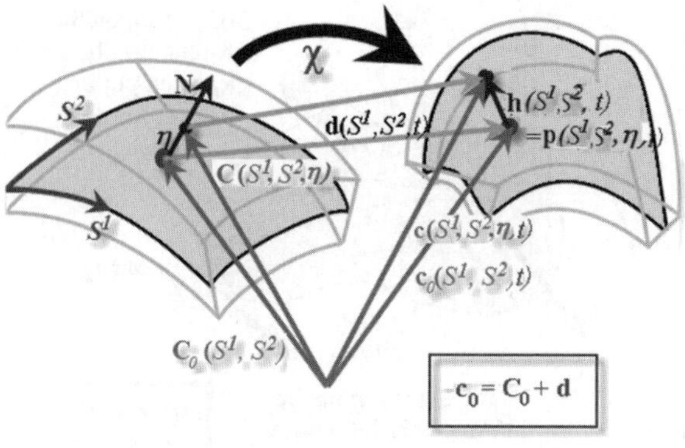

Figure 11.9 Shell deformation map.

shown in Figure 11.9. In other words, the deformed position vector, $\mathbf{c}(S^1, S^2, \eta, t)$, of a point that had a position vector, $\mathbf{C}(S^1, S^2, \eta, 0)$, in the undeformed reference state is related by: $\mathbf{c} = \chi(\mathbf{C})$. On the shell deformed reference surface, $\chi(M)$, the position vectors are denoted by: $\mathbf{c}_0(S^1, S^2, t) \equiv \mathbf{c}(S^1, S^2, 0, t)$. Now, we will assume that the deformation of the 3D shell-like body may be given by:

$$\boxed{\begin{aligned} &\mathbf{c}(S^1, S^2, \eta, t) = \mathbf{c}_0(S^1, S^2, t) + \mathbf{h}(S^1, S^2, \eta, t), \\ &\mathbf{h}(S^1, S^2, 0, t) \equiv 0 \\ &\mathbf{c}_0(S^1, S^2, t) = \mathbf{C}_0(S^1, S^2) + \mathbf{d}(S^1, S^2, t) \end{aligned}} \tag{11.66}$$

where, $\mathbf{h}(S^1, S^2, \eta, t)$ is an as yet unknown and arbitrary vector-valued through-the-thickness deformation function, with the usual requirement of invertibility, that locates any arbitrary point on the deformed body, given the shell reference surface map $\chi(M)$, and $\mathbf{d}(S^1, S^2, t)$ is the displacement of a point (S^1, S^2) on the reference surface at any time $t \in \mathbb{R}^+$. For example, we may take $\mathbf{h}(S^1, S^2, \eta, t) = \eta\ \mathbf{R}(S^1, S^2, t)\ \mathbf{N}(S^1, S^2)$ where $\mathbf{R}(S^1, S^2, t)$ is a rotation tensor at the current time, $t \in \mathbb{R}_+$ and $\mathbf{N}(S^1, S^2)$ is the normal at a point, (S^1, S^2), on the undeformed shell reference surface. Clearly, this assumption implies that the unit normal to the undeformed reference surface remains inextensible, but not necessarily perpendicular to the deformed reference surface during deformation (because it rotates due to shear deformations), which is a commonly accepted underpinning of theories for very thin and flexible shells with shear deformations (reflected in the rotation of the reference surface) and goes by the name: Reissner–Mindlin-type kinematic assumptions. Thus, the choice of $\mathbf{h}(S^1, S^2, \eta, t)$ essentially helps categorizing the various shell theories with differing kinematic assumptions; more on this later.

11.3.2.2 Dynamic Balance Equations: 3D Continuum to Shell Space

In what follows, we will reduce the 3D continuum mechanics balance laws: the momentum balance and the angular momentum balance laws, given by equation (11.65), to shell space balance laws. We will be able to recognize the reduced equations as identical to those obtained by the direct engineering approach and thus establish the latter's legitimacy as consistent with the laws of continuum mechanics.

11.3.3 Dynamic Momentum Balance Equations

Using the shell geometry and the deformation map as discussed above and rearranging the terms, we can rewrite the first of equation (11.65) as:

$$\int_{\partial\Omega^M}\left(\int_{\eta^b}^{\eta^t}\mathbf{P}\;\boldsymbol{\nu}\,{}^{\eta}\mu^S\,d\eta\right)dS+\int_{\Omega^M}\left(\int_{\eta^b}^{\eta^t}\mathbf{b}\,{}^{\eta}\mu\,d\eta+\mathbf{P}\,\boldsymbol{\nu}\,\mu\Big|_{\eta^b}^{\eta^t}\right)dA=\frac{d}{dt}\int_{\Omega^M}\left(\int_{\eta^b}^{\eta^t}\mathbf{v}\,\rho_0\,{}^{\eta}\mu\,d\eta\right)dA \tag{11.67}$$

In order to get the final shell integral or differential equation, we introduce the following definitions for the surface stress resultant, surface linear momentum and the surface external applied forces:

11.3.3.1 Stress Resultants, $\mathbf{F}^\alpha$, $\alpha = 1, 2$: by Method 1 Given by Equation (11.61)

From the first integral of equation (11.67), we have

$$\begin{aligned}\mathbf{F}_\nu(S,t) &\equiv \int_{\eta^b}^{\eta^t}\mathbf{P}(S,\eta,t)\,\boldsymbol{\nu}(S,t)\,{}^{\eta}\mu^S\,d\eta\\ &= \left(\int_{\eta^b}^{\eta^t}\mathbf{P}(S,\eta,t)\,\mathbf{T}(S,\eta,t)\,d\eta\right)\boldsymbol{\nu}_0(S)\\ &= \underbrace{{}^{f}\mathbf{P}(S,t)}_{\text{Shell resultant stress tensor}}\;\bullet\;\underbrace{\boldsymbol{\nu}_0(S)}_{\text{Shell boundary normal}}\end{aligned} \tag{11.68}$$

where, in the last equality above, we have used the following definition:
Stress resultant tensor:

$${}^{f}\mathbf{P}(S,t) \equiv \int_{\eta^b}^{\eta^t}\mathbf{P}(S,\eta,t)\;\mathbf{T}(S,\eta,t)\;d\eta \tag{11.69}$$

Remark

- Note the resemblance in the definition of shell edge traction between the last equality of equation (11.68) and the usual definition of a traction on a 3D body through the ***PK1***-like relationship: a traction at a point on the deformed surface edge is given by a two-point surface stress tensor, ${}^{f}\mathbf{P}$, that has domain in the undeforgmed surface and the range in the current deformed surface, that is, a linear transformation that maps a normal, $\boldsymbol{\nu}_0$, at a point on the undeformed surface edge that became the traction point in the current and deformed state to the traction, $\mathbf{F}_\nu(S,t)$.
- The unit normal, $\boldsymbol{\nu}_0$, is always defined in the tangent plane to any closed curve in the shell reference surface, that is, $\boldsymbol{\nu}_0 = \nu_0^\alpha\,\mathbf{G}_\alpha$, and has no component along the normal, $\mathbf{N}$.

So, the shell resultant stress tensor, ${}^f\mathbf{P}(S,t)$, may have a representation:

$$\boxed{{}^f\mathbf{P}(S,t) = \mathbf{F}_\nu(S,t) \otimes \boldsymbol{\nu}_0 = \nu_0^\alpha\, \mathbf{F}_\nu(S,t) \otimes \mathbf{G}_\alpha = \mathbf{F}^\alpha(S,t) \otimes \mathbf{G}_\alpha} \tag{11.70}$$

where, we have introduced the following.

Definition: ***Internal resistive stress resultant***, $\mathbf{F}^\alpha(S,t)$, $\alpha = 1,2$, per unit of undeformed length:

$$\boxed{\mathbf{F}^\alpha(S,t) \equiv \nu_0^\alpha(S)\, \mathbf{F}_\nu(S,t) = {}^f\mathbf{P}(S,t) \bullet \mathbf{G}^\alpha(S)} \tag{11.71}$$

acting on the deformed shell surface as the projection of $\mathbf{F}_\nu$ onto a tangent plane in the shell undeformed boundary containing the unit normal, $\boldsymbol{\nu}_0$. Noting that $1 = \boldsymbol{\nu}_0 \bullet \boldsymbol{\nu}_0 = \nu_{0\alpha}\nu_0^\alpha$ where $\nu_{0\alpha}$, ν_0^α are the covariant and the contravariant components of the unit normal, $\boldsymbol{\nu}_0$, from the first equality of equation (11.71), we have:

$$\boxed{\mathbf{F}^\alpha\, \nu_{0\alpha} \equiv \nu_{0\alpha}\nu_0^\alpha\, \mathbf{F}_\nu = \mathbf{F}_\nu} \tag{11.72}$$

11.3.3.2 Stress Resultant, $\mathbf{F}^\alpha(S,t)$, $\alpha = 1,2$: by Method 2 Given by Equation (11.62)

$$\boxed{\begin{aligned}\mathbf{F}_\nu(S,t) &= \int_{\eta^b}^{\eta^t} \mathbf{P}(S,\eta,t)\,(\boldsymbol{\nu}(S,t)\,{}^\eta\mu^S)\, d\eta = \left(\int_{\eta^b}^{\eta^t} \{\mathbf{P}(S,\eta,t) \bullet {}^\eta\mathbf{A}^\alpha\}\,{}^\eta\mu^S\, d\eta\right) \nu_{0\alpha} \\ &= \left(\int_{\eta^b}^{\eta^t} {}^\eta\mathbf{P}^\alpha(S,\eta,t)\,{}^\eta\mu^S\, d\eta\right) \nu_{0\alpha} = \mathbf{F}^\alpha(S,t)\, \nu_{0\alpha}\end{aligned}} \tag{11.73}$$

where, in the last equality above, we have used the following.

Definition: ***Internal resistive stress resultant***, $\mathbf{F}^\alpha(S,t)$, $\alpha = 1,2$, per unit of undeformed boundary length:

$$\boxed{\mathbf{F}^\alpha \equiv \int_{\eta^b}^{\eta^t} {}^\eta\mathbf{P}^\alpha\,{}^\eta\mu^S\, d\eta} \tag{11.74}$$

where, of course, ${}^\eta\mathbf{P}^\alpha = \mathbf{P} \bullet {}^\eta\mathbf{A}^\alpha$. Now, continuing further on equation (11.73), we have:

$$\boxed{\begin{aligned}\mathbf{F}_\nu(S,t) &= \mathbf{F}^\alpha(S,t)\, \nu_{0\alpha} = \mathbf{F}^\alpha \otimes \mathbf{G}_\alpha \bullet \nu_{0\alpha}\, \mathbf{G}^\alpha \\ &= \underbrace{{}^f\mathbf{P}(S,t)}_{\text{Shell resultant stress tensor}} \bullet \underbrace{\boldsymbol{\nu}_0(S)}_{\text{Shell boundary normal}}\end{aligned}} \tag{11.75}$$

and where, in the last equality above, we have used again the following definition:

Stress resultant tensor:

$$\boxed{{}^{f}\mathbf{P}(S,t) = \mathbf{F}^{\alpha}(S,t) \otimes \mathbf{G}_{\alpha}(S)} \tag{11.76}$$

whereby, we have:

$$\boxed{\mathbf{F}^{\alpha} = {}^{f}\mathbf{P} \bullet \mathbf{G}^{\alpha}} \tag{11.77}$$

Remarks

- Again, note the resemblance in the definition of shell edge traction between the last equality of equation (11.75) snd the usual definition of a traction on a 3D body through the Lagrange postulate-like relationship with a surface stress tensor, ${}^{f}\mathbf{P}$, and the normal, $\boldsymbol{\nu}_0$, to the boundary.
- Additionally, note that by equation (11.72) and the last equality of equation (11.73), the shell edge traction, $\mathbf{F}_{\nu}(S,t)$, is equivalently described in both the methods.

11.3.3.3 Surface Linear Momentum, $\mathbf{L}(S^1, S^2, t)$ and Total Linear Momentum, $\mathbf{L}_{TOT}(t)$

From the integral on the right-hand side of equation (11.67), we have:

$$\begin{aligned} \dot{\mathbf{L}}_{TOT}(t) &\equiv \frac{d}{dt}\int_{\Omega^M}\left(\int_{\eta^b}^{\eta^t} \mathbf{v}(S^1,S^2,\eta,t)\,\rho_0\,{}^{\eta}\mu\,d\eta\right) dA = \frac{d}{dt}\int_{\Omega^M} \mathbf{L}(S^1,S^2,t)\,dA \\ &= \int_{\Omega^M} \dot{\mathbf{L}}(S^1,S^2,t)\,dA \end{aligned} \tag{11.78}$$

where we have used the notation: $\dot{(\bullet)} \equiv \frac{d}{dt}(\bullet)$, and introduced:

Definitions

- ***Shell total linear momentum***, $\mathbf{L}_{TOT}(t)$:

$$\boxed{\mathbf{L}_{TOT}(t) \equiv \int_{\Omega^M} \mathbf{L}(S^1,S^2,t)\,dA} \tag{11.79}$$

with

- ***Surface linear momentum***, $\mathbf{L}(S^1,S^2,t)$, per unit of the undeformed reference surface area:

$$\boxed{\begin{aligned} \mathbf{L}(S^1,S^2,t) &\equiv \int_{\eta^b}^{\eta^t} \mathbf{v}(S^1,S^2,\eta,t)\,\rho_0(S^1,S^2,\eta)\,{}^{\eta}\mu(S^1,S^2,\eta)\,d\eta \\ &= \int_{\eta^b}^{\eta^t} \dot{\mathbf{c}}(S^1,S^2,\eta,t)\,\rho_0(S^1,S^2,\eta)\,{}^{\eta}\mu(S^1,S^2,\eta)\,d\eta \end{aligned}} \tag{11.80}$$

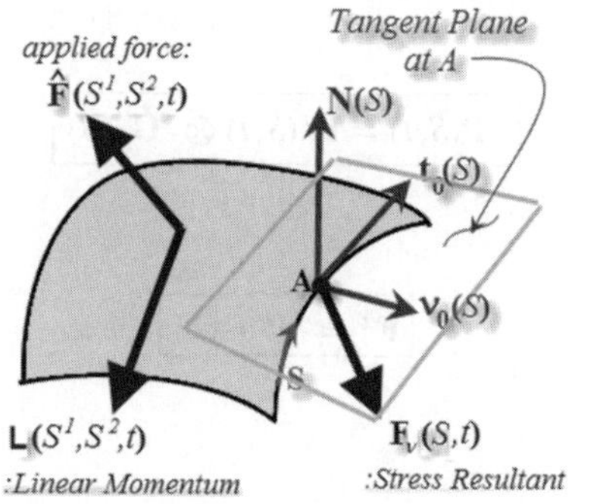

Figure 11.10 Shell surface forces and linear momentum.

where $\mathbf{v}(S^1, S^2, \eta, t) \equiv \dot{\mathbf{c}}(S^1, S^2, \eta, t) = \dot{\mathbf{d}}(S^1, S^2, t) + \dot{\mathbf{h}}(S^1, S^2, \eta, t)$ is, as indicated earlier, the velocity at current time $t \in \mathbb{R}_+$ at a point with undeformed coordinates (S^1, S^2, η) in the interior of the shell-like body.

Surface external applied force:
We have from the second integral of equation (11.67) the following.

Definition

- ***Surface external applied force***, $\hat{\mathbf{F}}(S^1, S^2, t)$, per unit of undeformed reference surface area:

$$\boxed{\hat{\mathbf{F}}(S^1, S^2, t) \equiv \int_{\eta^b}^{\eta^t} \mathbf{b}(S^1, S^2, \eta, t)\, \mu^\eta \, d\eta \; + \mathbf{t}(S^1, S^2, \eta^x, t)\, \mu|_{\eta^x=\eta^b}^{\eta^x=\eta^t}} \tag{11.81}$$

Note that $\hat{\mathbf{F}}(S^1, S^2, t)$ are the statically equivalent surface forces to the prescribed 3D surface tractions, $\mathbf{t}(S^1, S^2, \eta^x, t)$, $x = b, t$, on the top and bottom faces, and body forces, $\mathbf{b}(S^1, S^2, \eta, t)$. Thus, we can always apply the prescribed forces of a 3D shell-like body as equivalent forces on its 2D shell model, as shown in Figure 11.10.

11.3.3.4 Dynamic Momentum Balance Equations

Now, incorporating definitions (11.68) or (11.75), and (11.81) in equation (11.67), we get the integral form of the momentum balance equation:

$$\boxed{\begin{aligned} &\int_{\partial\Omega^M} \mathbf{F}_v(S, t)\, dS + \int_{\Omega^M} \hat{\mathbf{F}}(S^1, S^2, t)\, dA = \dot{\mathbf{L}}_{TOT}(t) \equiv \int_{\Omega^M} \dot{\mathbf{L}}(S^1, S^2, t)\, dA \\ \text{or, } &\int_{\partial\Omega^M} {}^f\mathbf{P}(S, t)\; \mathbf{v}_0(S)\, dS + \int_{\Omega^M} \hat{\mathbf{F}}(S^1, S^2, t)\, dA = \dot{\mathbf{L}}_{TOT}(t) \equiv \int_{\Omega^M} \dot{\mathbf{L}}(S^1, S^2, t)\, dA \end{aligned}} \tag{11.82}$$

Now, by applying the surface Green–Gauss or divergence theorem to the first part of the second of equation (11.82), we can rewrite it to get the invariant form of the integral equation as:

$$\boxed{\int_{\Omega^M} \{DIV^f\mathbf{P}(S^1,S^2,t) + \hat{\mathbf{F}}(S^1,S^2,t)\}\, dA = \int_{\Omega^M} \dot{\mathbf{L}}(S^1,S^2,t)\, dA} \tag{11.83}$$

where $DIV(\bullet) \equiv (\bullet),_{\alpha} \otimes \mathbf{G}^{\alpha}$ is the material surface divergence operator. Noting that the shell surface area dA can be arbitrarily chosen, we have the invariant form of the differential equation as:

$$\boxed{DIV\,^f\mathbf{P}(S^1,S^2,t) \;+\; \hat{\mathbf{F}}(S^1,S^2,t) = \dot{\mathbf{L}}(S^1,S^2,t)} \tag{11.84}$$

Remark

- Noting that with the representation of the stress resultant tensor, ${}^f\mathbf{P}(S,t) = \mathbf{F}^{\alpha} \otimes \mathbf{G}_{\alpha}$, as given by the last equality of equation (11.70), or equation (11.76), we have by equation (11.63):

$$\boxed{DIV\,^f\mathbf{P} \;= \mathbf{F}^{\alpha}\,|_{\alpha}} \tag{11.85}$$

Finally, using the relationship given by equation (11.85) in equation (11.84), we get the vector differential form of the momentum balance equations as:

$$\boxed{\mathbf{F}^{\alpha}|_{\alpha} + \hat{\mathbf{F}} = \dot{\mathbf{L}}} \tag{11.86}$$

We can now recognize equation (11.86) as identical to what is obtained by the direct engineering approach except that, as intended, we now have consistent interpretations for the stress resultant, $\mathbf{F}^{\alpha}(S,t)$, $\alpha = 1,2$, the surface linear momentum, $\mathbf{L}(S^1,S^2,t)$, and the surface external applied load, $\hat{\mathbf{F}}(S^1,S^2,t)$, by equations (11.74), (11.80) and (11.81), respectively; this was not possible in the direct engineering approach whereby we could, at best, resort to ad hoc assumptions for the interpretations of the above shell surface entities.

11.3.4 Dynamic Angular Momentum Balance Equations: Integral and Differential Forms

Using the shell geometry and the deformation map as discussed above and rearranging the terms, we can rewrite the second of equation (11.65) as:

$$\boxed{\begin{aligned}&\int_{\partial\Omega^M}\left(\int_{\eta^b}^{\eta^t} \mathbf{c}\times\mathbf{P}\ \mathbf{v}\ \mu^S_{\eta}\, d\eta\right) dS + \int_{\Omega^M}\left(\int_{\eta^b}^{\eta^t} \mathbf{c}\times\mathbf{b}^{\eta}\mu\, d\eta \;+\; \mathbf{c}\times\mathbf{P}\,\mathbf{v}\,\mu\Big|_{\eta^b}^{\eta^t}\right) dA\\ &= \frac{d}{dt}\int_{\Omega^M} (\mathbf{c}\times\mathbf{v}\,\rho_0{}^{\eta}\mu\, d\eta) dA\end{aligned}} \tag{11.87}$$

In order to get the final shell integral or differential equation, we introduce the following definitions for the shell surface stress couple, the surface angular momentum and the surface external applied moments.

11.3.4.1 Stress Couples: $\mathbf{M}^\alpha(S^1, S^2, t)$, $\alpha = 1, 2$ by Method 1 Given by Equation (11.61)

From the first integral of equation (11.87), we have, with $\mathbf{c} = \mathbf{c}_0 + \mathbf{h}$:

$$\begin{aligned}
&\mathbf{c}_0(S,t) \times \int_{\eta^b}^{\eta^t} \mathbf{P}(S,\eta,t)\,\mathbf{v}(S,t)\,\mu^\eta\,d\eta + \int_{\eta^b}^{\eta^t} \mathbf{h}(S,\eta,t) \times \mathbf{P}(S,\eta,t)\,\mathbf{v}(S,t)\,\mu^\eta\,d\eta \\
&= \mathbf{c}_0(S,t) \times \mathbf{F}_\nu(S,t) + \left(\int_{\eta^b}^{\eta^t} \mathbf{h}(S,\eta,t) \times \mathbf{P}(S,\eta,t)\;\mathbf{T}(S,\eta)\,d\eta \right) \boldsymbol{\nu}_0(S) \\
&= \mathbf{c}_0(S,t) \times \mathbf{F}_\nu(S,t) + \underbrace{{}^m\mathbf{P}(S,t)}_{\text{Shell boundary stress couple tensor}} \bullet \underbrace{\boldsymbol{\nu}_0(S)}_{\text{Shell boundary normal}} \\
&= \mathbf{c}_0(S,t) \times \mathbf{F}_\nu(S,t) + \mathbf{M}_\nu(S,t)
\end{aligned} \tag{11.88}$$

where, we have used equation (11.68) for the definition of $\mathbf{F}_\nu(S,t)$, and in the last equality above, we have used the following definition.

$$\begin{aligned}
\mathbf{M}_\nu(S,t) &\equiv \left(\int_{\eta^b}^{\eta^t} \mathbf{h}(S,\eta,t) \times \mathbf{P}(S,\eta,t)\;\mathbf{T}(S,\eta,t)\,d\eta \right) \boldsymbol{\nu}_0(S) \\
&= \underbrace{{}^m\mathbf{P}(S,t)}_{\text{Shell boundary stress couple tensor}} \bullet \underbrace{\boldsymbol{\nu}_0(S)}_{\text{Shell boundary normal}}
\end{aligned} \tag{11.89}$$

with the introduction of the following.

Definition: Stress couple tensor:

$$ {}^m\mathbf{P}(S,t) \equiv \int_{\eta^b}^{\eta^t} \mathbf{h}(S,\eta,t) \times \mathbf{P}(S,\eta,t)\;\mathbf{T}(S,\eta,t)\,d\eta \tag{11.90}$$

Remarks

- Note the resemblance in the definition of shell edge couple between the last equality of equation (11.89) and the usual definition of a couple on a 3D body through the Lagrange postulate-like relationship with a surface stress couple tensor, ${}^m\mathbf{P}$, and the normal, $\boldsymbol{\nu}_0$, to the boundary.
- Note also that the unit normal, $\boldsymbol{\nu}_0$, is always defined in the tangent plane to any closed curve in the shell reference surface, that is, $\boldsymbol{\nu}_0 = \nu_0^\alpha\,\mathbf{G}_\alpha$ and has no component along the normal, $\mathbf{N}$.

Now, with $\mathbf{M}(S,t) = {}^{m}\mathbf{P}(S,t) \bullet \boldsymbol{\nu}_0$ from equation (11.89), the shell boundary stress couple tensor, ${}^{m}\mathbf{P}(S,t)$, can have the representation:

$$\begin{aligned} {}^{m}\mathbf{P}(S,t) &= \mathbf{M}_{\nu}(S,t) \otimes \boldsymbol{\nu}_0(S) \\ &= \nu_0^{\alpha}\, \mathbf{M}_{\nu}(S,t) \otimes \mathbf{G}_{\alpha}(S) \\ &= \mathbf{M}^{\alpha}(S,t) \otimes \mathbf{G}_{\alpha}(S) \end{aligned} \tag{11.91}$$

where, we have the following.

Definition: ***Internal resistive stress couple***, $\mathbf{M}^{\alpha}(S,t),\ \alpha = 1,2$, per unit of undeformed length:

$$\begin{aligned} \mathbf{M}^{\alpha}(S,t) &\equiv \nu_0^{\alpha}(S)\, \mathbf{M}_{\nu}(S,t) \\ &= {}^{m}\mathbf{P}(S,t) \bullet \mathbf{G}^{\alpha}(S) \end{aligned} \tag{11.92}$$

as the projection of $\mathbf{M}_{\nu}$ on a tangent plane of the shell boundary containing the unit normal, $\boldsymbol{\nu}_0$. Noting that $1 = \boldsymbol{\nu}_0 \bullet \boldsymbol{\nu}_0 = \nu_{0\alpha}\, \nu_0^{\alpha}$ where $\nu_{0\alpha}$, ν_0^{α} are the covariant and the contravariant components of the unit normal, $\boldsymbol{\nu}_0$, from the first equality of equation (11.92), we have:

$$\mathbf{M}^{\alpha}\ \nu_{0\alpha} \equiv \nu_{0\alpha}\nu_0^{\alpha}\, \mathbf{M}_{\nu} = \mathbf{M}_{\nu} \tag{11.93}$$

11.3.4.2 Stress Couples, $\mathbf{M}^{\alpha}(S^1,S^2,t)$, $\alpha = 1,2$: by Method 2 Given by Equation (11.62)

$$\begin{aligned} \mathbf{M}_{\nu}(S,t) &= \int_{\eta^b}^{\eta^t} \mathbf{h}(S,\eta,t) \times \mathbf{P}(S,\eta,t)\ (\boldsymbol{\nu}(S,t)\, {}^{\eta}\mu^S)\ d\eta \\ &= \left(\int_{\eta^b}^{\eta^t} \mathbf{h}(S,\eta,t) \times \{\mathbf{P}(S,\eta,t)\ {}^{\eta}\mathbf{A}^{\alpha}\}\ {}^{\eta}\mu^S\ d\eta \right)\ \nu_{0\alpha} \\ &= \left(\int_{\eta^b}^{\eta^t} \mathbf{h}(S,\eta,t)\ \times\ {}^{\eta}\mathbf{P}^{\alpha}(S,\eta,t)\ {}^{\eta}\mu^S\ d\eta \right)\ \nu_{0\alpha} \\ &= \mathbf{M}^{\alpha}(S,t)\ \nu_{0\alpha}(S) \end{aligned} \tag{11.94}$$

where, in the last equality above, we have used the following.

Definition: ***Internal resistive stress couple***, $\mathbf{M}^{\alpha}(S,t),\ \alpha = 1,2$, per unit of undeformed boundary length:

$$\mathbf{M}^{\alpha}(S,t) \equiv \int_{\eta^b}^{\eta^t} \mathbf{h}(S,\eta,t) \times {}^{\eta}\mathbf{P}^{\alpha}(S,\eta,t)^{\eta}\mu^S(S,\eta,t)\ d\eta \tag{11.95}$$

where, of course, ${}^{\eta}\mathbf{P}^{\alpha} = \mathbf{P} \bullet {}^{\eta}\mathbf{A}^{\alpha}$.

Remarks

- Note that a specific choice as to the nature of the function $\mathbf{h}(S,\eta)$ identifies a specific shell theory. For example, in its simplest form: $\mathbf{h}(S,\eta) = \eta\ \mathbf{n}$ in equation (11.66), keeping the normal to the surface inextensible through the deformation history, Reissner–Mindlin shell theory may be developed. In this case, however, while we will have six degrees of freedoms, we end up with no so-called "drilling" couples in the deformed coordinate system: $M^{\alpha 3} = \mathbf{n} \bullet \mathbf{M}^{\alpha} = \mathbf{n} \bullet \int_{\eta^b}^{\eta^t} \eta\ \mathbf{n} \times {}^{\eta}\mathbf{P}^{\alpha}\ {}^{\eta}\mu^S\ d\eta = 0$ for $\alpha = 1, 2$. With the addition of one more through-the-thickness degree of freedom, $\lambda(\eta)$, we may relax into a higher-order shell theory by allowing the deformed normal extensible by choosing $\mathbf{h}(S,\eta) = \lambda(\eta)\ \mathbf{n}$. Clearly, other possibilities exist to model various materials such as rubber, and so on, in which cases appropriate constitutive theories demand representation of the more relaxed deformations.

Now, continuing further on equation (11.94), we have:

$$\boxed{\begin{aligned}\mathbf{M}_{\nu}(S,t) &= \mathbf{M}^{\alpha}(S,t)\ \nu_{0\alpha}(S) = \mathbf{M}^{\alpha} \otimes \mathbf{G}_{\alpha} \bullet \nu_{0\alpha}\ \mathbf{G}^{\alpha} \\ &= \underbrace{{}^{m}\mathbf{P}\,(S,t)}_{\text{Shell stress couple tensor}} \bullet \underbrace{\boldsymbol{\nu}_0(S)}_{\text{Shell boundary normal}}\end{aligned}} \tag{11.96}$$

and where, in the last equality above, we have again used the following.

Definition

- ***Stress couple tensor***:

$$\boxed{{}^{m}\mathbf{P}\,(S,t) = \mathbf{M}^{\alpha}(S,t) \otimes \mathbf{G}_{\alpha}(S)} \tag{11.97}$$

whereby, we have:

$$\boxed{\mathbf{M}^{\alpha}(S,t) = {}^{m}\mathbf{P}(S,t) \bullet \mathbf{G}^{\alpha}(S)} \tag{11.98}$$

Remark

- Again, we may note the resemblance in the definition of shell edge traction, between the last equality of equation (11.96), and the usual definition of a traction on a 3D body through the PK1-like relationship: a traction at a point on the deformed surface edge is given by a two-point surface stress couple tensor, ${}^{m}\mathbf{P}$, that has domain in the undeformed surface and the range in the current deformed surface, that is, a linear transformation that maps a normal, $\boldsymbol{\nu}_0$, at a point on the undeformed surface edge that became the traction point in the current and deformed state, to the traction, $\mathbf{M}_{\nu}(S,t)$. Also note that by equation (11.93) and the last equality of equation (11.94), the shell edge couple, $\mathbf{M}_{\nu}(S,t)$, is equivalently described in both the methods.

11.3.4.3 Surface Angular Momentum, $\mathbf{A}(S^1, S^2, t)$ and Total Angular Momentum, $\mathbf{A}_{TOT}(t)$

From the integral on the right-hand side of equation (11.87), we have, with $\mathbf{c} = \mathbf{c}_0 + \mathbf{h}$:

$$
\begin{aligned}
\dot{\mathbf{A}}_{TOT}(t) &\equiv \frac{d}{dt}\int_{\Omega^M}\left(\int_{\eta^b}^{\eta^t} \mathbf{c}\times\mathbf{v}\,\rho_0\,{}^{\eta}\mu\,d\eta\right) dA \\
&= \frac{d}{dt}\int_{\Omega^M} \mathbf{c}(S^1,S^2,t)\times\left(\int_{\eta^b}^{\eta^t} \dot{\mathbf{c}}(S^1,S^2,\eta,t)\,\rho_0(S^1,S^2,\eta)\,{}^{\eta}\mu(S^1,S^2,\eta)\,d\eta\right) dA \\
&= \frac{d}{dt}\int_{\Omega^M} \mathbf{c}(S^1,S^2,\eta,t)\times\left(\int_{\eta^b}^{\eta^t} \dot{\mathbf{c}}(S^1,S^2,\eta,t)\,\rho_0(S^1,S^2,\eta)\,{}^{\eta}\mu(S^1,S^2,\eta)\,d\eta\right) dA \\
&= \frac{d}{dt}\int_{\Omega^M} \left\{ \begin{aligned} &\mathbf{c}_0(S^1,S^2,t)\times\left(\int_{\eta^b}^{\eta^t} \dot{\mathbf{c}}(S^1,S^2,\eta,t)\,\rho_0(S^1,S^2,\eta)\,{}^{\eta}\mu(S^1,S^2,\eta)\,d\eta\right) \\ &+\int_{\eta^b}^{\eta^t} \mathbf{h}(S^1,S^2,\eta,t)\times\dot{\mathbf{c}}(S^1,S^2,\eta,t)\,\rho_0(S^1,S^2,\eta)\,{}^{\eta}\mu(S^1,S^2,\eta)\,d\eta \end{aligned}\right\} dA \\
&= \frac{d}{dt}\int_{\Omega^M} \left(\mathbf{c}_0(S^1,S^2,t)\,\times\,\mathbf{L}(S^1,S^2,t) + \mathbf{A}(S^1,S^2,t)\right) dA
\end{aligned}
\tag{11.99}
$$

where we have used the definition for the linear momentum, $\mathbf{L}(S^1, S^2, t)$, as given by equation (11.80), and introduced the total angular momentum, $\mathbf{A}_{TOT}(t)$, and the surface angular momentum, $\mathbf{A}(S^1, S^2, t)$, as follows:

Definitions

- ***Shell total angular momentum***, $\mathbf{A}_{TOT}(t)$:

$$
\boxed{\mathbf{A}_{TOT}(t) \equiv \int_{\Omega^M} \left(\mathbf{c}_0(S^1,S^2,t)\times\mathbf{L}(S^1,S^2,t) + \mathbf{A}(S^1,S^2,t)\right) dA}
\tag{11.100}
$$

with

- ***Surface angular momentum***, $\mathbf{A}(S^1, S^2, t)$, per unit of the undeformed reference surface area:

$$
\boxed{
\begin{aligned}
\mathbf{A}(S^1,S^2,t) &\equiv \int_{\eta^b}^{\eta^t} \mathbf{h}(S,\eta,t)\times\mathbf{v}(S^1,S^2,\eta,t)\,\rho_0(S^1,S^2,\eta)\,{}^{\eta}\mu(S^1,S^2,\eta)\;d\eta \\
&= \int_{\eta^b}^{\eta^t} \mathbf{h}(S,\eta,t)\times\dot{\mathbf{c}}(S^1,S^2,\eta,t)\,\rho_0(S^1,S^2,\eta)\,{}^{\eta}\mu(S^1,S^2,\eta)\;d\eta
\end{aligned}}
\tag{11.101}
$$

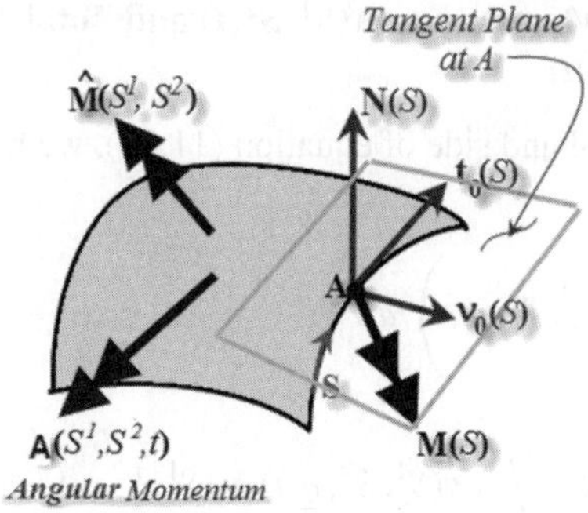

Figure 11.11 Shell surface moment and angular momentum.

where $\mathbf{v}(S^1, S^2, \eta, t) \equiv \dot{\mathbf{c}}(S^1, S^2, \eta, t) = \dot{\mathbf{d}}(S^1, S^2, t) + \dot{\mathbf{h}}(S^1, S^2, \eta, t)$ is the velocity at current time $t \in \mathbb{R}_+$ at a point with undeformed coordinates (S^1, S^2, η) in the interior of the shell-like body, as shown in Figure 11.11.

11.3.4.4 Surface External Applied Moment

From the second integral of equation (11.87), we have, with $\mathbf{c} = \mathbf{c}_0 + \mathbf{h}$:

$$\begin{aligned}
&\mathbf{c}_0(S^1, S^2, t) \times \int_{\eta^b}^{\eta^t} \mathbf{b}(S^1, S^2, \eta, t)\ \mu(S^1, S^2, \eta, t)\ d\eta + \int_{\eta^b}^{\eta^t} \mathbf{h}(S^1, S^2, \eta, t) \\
&\quad \times \mathbf{b}(S^1, S^2, \eta, t)\ \mu(S^1, S^2, \eta, t)\ d\eta + \mathbf{c}_0(S^1, S^2, t) \\
&\quad \times (\mathbf{t}(S^1, S^2, \eta, t)\ {}^{\eta}\mu)|_{\eta^b}^{\eta^t} + \ (\mathbf{h}(S^1, S^2, \eta, t) \times \mathbf{t}(S^1, S^2, \eta, t)\ {}^{\eta}\mu)|_{\eta^b}^{\eta^t} \\
&\quad = \mathbf{c}_0(S^1, S^2, t) \times \left(\int_{\eta^b}^{\eta^t} \mathbf{b}(S^1, S^2, \eta, t)\ \ {}^{\eta}\mu\ d\eta + \ (\mathbf{t}(S^1, S^2, \eta, t)\ {}^{\eta}\mu)|_{\eta^b}^{\eta^t} \right) \\
&\quad + \hat{\mathbf{M}}(S^1, S^2, t) = \mathbf{c}_0(S^1, S^2, t) \times \hat{\mathbf{F}}(S^1, S^2, t)\ +\ \hat{\mathbf{M}}(S^1, S^2, t)
\end{aligned} \tag{11.102}$$

where we have used equation (11.81) for the definition of $\hat{\mathbf{F}}(S^1, S^2)$, and in the last equality above, we have introduced the following.

Definition

- ***External applied moment***, $\hat{\mathbf{M}}(S^1, S^2, t)$, per unit of undeformed reference surface area:

$$\hat{\mathbf{M}}(S^1, S^2, t) \equiv \int_{\eta^b}^{\eta^t} \mathbf{h}(S^1, S^2, \eta, t) \times \mathbf{b}(S^1, S^2, \eta, t)\mu^{\eta} d\eta + \mathbf{h}(S^1, S^2, \eta, t) \times \mathbf{t}\mu|_{\eta^b}^{\eta^t} \tag{11.103}$$

Remark

- Note that $\hat{\mathbf{M}}(S^1, S^2)$ is the statically equivalent shell surface moments to the prescribed 3D surface tractions, **t**, on the top and bottom faces, and body forces, **b**. Thus, we can always apply the prescribed moments of a 3D shell-like body as equivalent moments on its 2D shell model.

Now, incorporating definitions (11.88) and (11.103) in equation (11.87), we get the integral form of the angular momentum balance equation:

$$\boxed{\begin{aligned}&\int_{\partial\Omega^M} (\mathbf{M}_\nu + \mathbf{c}_0 \times \mathbf{F}_\nu)\, dS + \int_{\Omega^M} (\mathbf{c}_0 \times \hat{\mathbf{F}} + \hat{\mathbf{M}})\, dA = \dot{\mathbf{A}}_{TOT}(t) \equiv \frac{d}{dt}\int_{\Omega^M} (\mathbf{c}_0 \times \mathbf{L} + \mathbf{A})\, dA \\ \text{or, } &\int_{\partial\Omega^M} ({}^m\mathbf{P} + \mathbf{c}_0 \times {}^f\mathbf{P}) \bullet \nu_0\, dS + \int_{\Omega^M} (\mathbf{c}_0 \times \hat{\mathbf{F}} + \hat{\mathbf{M}}) dA = \dot{\mathbf{A}}_{TOT}(t) \equiv \frac{d}{dt}\int_{\Omega^M} (\mathbf{c}_0 \times \mathbf{L} + \mathbf{A})\, dA\end{aligned}} \tag{11.104}$$

Noting that with the representation of ${}^m\mathbf{P}(S,t)$ as in equation (11.91), we have by equation (11.63):

$$\boxed{DIV\, {}^m\mathbf{P} = \mathbf{M}^\alpha |_\alpha} \tag{11.105}$$

and by equation (11.64), we have:

$$\boxed{DIV\,(\mathbf{c}_0 \times {}^f\mathbf{P}) = \mathbf{c}_0 \times DIV\, {}^f\mathbf{P} + \mathbf{c}_{0,\alpha} \times ({}^f\mathbf{P} \bullet \mathbf{G}^\alpha)} \tag{11.106}$$

Now, by applying surface Green–Gauss or divergence theorem to the second of equation (11.104) and applying relationships given by equation (11.105) and equation (11.106), and taking derivative of the right-hand side of second of the equation (11.104) and rearranging terms, we can rewrite it to get:

$$\begin{aligned}&\int_{\Omega^M} \left\{ DIV\, {}^m\mathbf{P} + \mathbf{c}_{0,\alpha} \times ({}^f\mathbf{P} \bullet \mathbf{G}^\alpha) + \mathbf{c}_0 \times \cancel{(DIV\, {}^f\mathbf{P} + \bar{\mathbf{F}} - \dot{\mathbf{L}})} + \hat{\mathbf{M}} \right\} dA \\ &= \int_{\Omega^M} \left(\dot{\mathbf{c}}_0 \times \mathbf{L} + \dot{\mathbf{A}} \right) dA\end{aligned} \tag{11.107}$$

If the momentum balance equation (11.84) is satisfied identically, the third term of equation (11.107) drops out as shown. Now, applying the relationship given by equation (11.77) in the second term, we get what we will call the ***invariant form of the integral equation***:

$$\boxed{\int_{\Omega^M} \{DIV\, {}^m\mathbf{P} + \mathbf{c}_{0,\alpha} \times \mathbf{F}^\alpha + \hat{\mathbf{M}}\}\, dA = \int_{\Omega^M} \left(\dot{\mathbf{c}}_0 \times \mathbf{L} + \dot{\mathbf{A}} \right) dA} \tag{11.108}$$

Noting that the shell free-body surface area Ω^M can be arbitrarily chosen, we have the invariant form of the differential equation as:

$$\boxed{DIV\,{}^m\mathbf{P} + \mathbf{c}_{0},_{\alpha} \times ({}^f\mathbf{P} \bullet \mathbf{G}^{\alpha}) + \hat{\mathbf{M}} = \dot{\mathbf{c}}_0 \times \mathbf{L} + \dot{\mathbf{A}}} \tag{11.109}$$

Finally, using relationship given by equation (11.105), we get the vector form of angular momentum balance equations as:

$$\boxed{\mathbf{M}^{\alpha}|_{\alpha} + \mathbf{c}_{0},_{\alpha} \times \mathbf{F}^{\alpha} + \hat{\mathbf{M}} = \dot{\mathbf{c}}_0 \times \mathbf{L} + \dot{\mathbf{A}}} \tag{11.110}$$

We can now recognize equation (11.110) as identical to what is obtained by the direct engineering approach except that, as intended, we now have consistent interpretations for the stress couple, $\mathbf{M}^{\alpha}(S,t)$, $\alpha = 1,2$, the surface angular momentum, $\mathbf{A}(S^1,S^2,t)$, and the surface external applied moment, $\hat{\mathbf{M}}(S^1,S^2,t)$, by equationS (11.95), (11.101) and (11.103), respectively. This was not possible in the direct engineering approach, where we could, at best, resort to ad hoc assumptions for interpretations of the shell surface entities.

Exercise: Show that if $\mathbf{P} = \mathbf{P}^{\beta} \otimes \mathbf{G}_{\beta}$, then $DIV\,\mathbf{P} = \mathbf{P}^{\beta}|_{\beta}$

Solution: Note that $\mathbf{G}_{\alpha}|_{\alpha} = \mathbf{0}$, that is, the covariant derivative of a base vector is $\mathbf{0}$ (why?). So, by definition: $DIV\,\mathbf{P} \equiv GRAD\;\mathbf{P}{:}\mathbf{I} = \mathbf{P}_{\alpha\beta|\gamma}\mathbf{G}^{\alpha} \otimes \mathbf{G}^{\beta} \otimes \mathbf{G}^{\gamma} : \mathbf{G}_{\sigma} \otimes \mathbf{G}^{\sigma} = \mathbf{P}_{\alpha\beta|\gamma}\mathbf{G}^{\gamma\beta}\,\mathbf{G}^{\alpha} = \mathbf{P}_{\alpha}{}^{\gamma}|_{\gamma}\,\mathbf{G}^{\alpha} = \mathbf{P}_{\alpha}{}^{\gamma}|_{\gamma}\,\mathbf{G}^{\rho\alpha}\,\mathbf{G}_{\rho} = \mathbf{P}^{\alpha\beta}|_{\beta}\mathbf{G}_{\alpha} = (\mathbf{P}^{\alpha\beta}\,\mathbf{G}_{\alpha})|_{\beta} = \mathbf{P}^{\beta}|_{\beta}$.

11.3.5 Static Balance Equations: General Theory

The quasi-static balance laws can be easily obtained from their dynamic counterparts by assuming that the linear and the angular momentum vectors identically vanish, that is,

$$\boxed{\begin{aligned} \mathbf{L}(S^1,S^2,t) &\equiv \mathbf{0} \\ \mathbf{A}(S^1,S^2,t) &\equiv \mathbf{0} \end{aligned}} \tag{11.111}$$

Thus, from equation (11.86), we get:

11.3.5.1 Static Momentum Balance Equations (Force Equilibrium)

$$\boxed{\mathbf{F}^{\alpha}|_{\alpha} + \hat{\mathbf{F}} = \mathbf{0}, \quad \alpha = 1,2} \tag{11.112}$$

and from equation (11.110):

11.3.5.2 Static Angular or Moment of Momentum Balance Equations (Moment Equilibrium)

$$\boxed{\mathbf{M}^{\alpha}|_{\alpha} + \mathbf{c}_{0},_{\alpha} \times \mathbf{F}^{\alpha} + \hat{\mathbf{M}} = \mathbf{0}, \quad \alpha = 1,2} \tag{11.113}$$

11.3.6 General Shell Vector Dynamic Balance Equations of Motion

For easy reference, here we copy and collect the derived dynamic balance equations as follows:

$$
\boxed{\begin{array}{lr}
\mathbf{F}^{\alpha},_{\alpha} + \hat{\mathbf{F}} = \dot{\mathbf{L}}, \quad \alpha = 1,2 & \text{Momentum balance} \\
\mathbf{M}^{\alpha},_{\alpha} + \mathbf{c}_0,_{\alpha} \times \mathbf{F}^{\alpha} + \hat{\mathbf{M}} = \dot{\mathbf{c}}_0 \times \mathbf{L} + \dot{\mathbf{A}}, \quad \alpha = 1,2 & \text{Moment of momentum balance}
\end{array}} \tag{11.114}
$$

11.3.7 General Shell Vector Static Balance Equations of Motion

Here we copy and collect the derived static balance equations:

$$
\boxed{\begin{array}{lr}
\mathbf{F}^{\alpha}|_{\alpha} + \hat{\mathbf{F}} = \mathbf{0}, \quad \alpha = 1,2 & \text{Momentum balance} \\
\mathbf{M}^{\alpha}|_{\alpha} + \mathbf{c}_0,_{\alpha} \times \mathbf{F}^{\alpha} + \hat{\mathbf{M}} = \mathbf{0}, \quad \alpha = 1,2 & \text{Moment of momentum balance}
\end{array}} \tag{11.115}
$$

Remark

- We have deduced the shell balance equations consistently by starting from the 3D continuum mechanics balance laws. In this respect, the equations are exact. They are the true equations for any shell-like body, linear or nonlinear, thick or thin.

11.3.8 Where We Would Like to Go

So far, we have derived the balance equations by reducing the 3D equations of motion of continuum mechanics for general shells. However, for our purposes in the rest of the book, we will restrict ourselves to an important subclass of shells: thin and very flexible shells. Thus, we must specialize the above balance equations to thin shells; for this, we should check out:

- thin shell dynamic and static balance equations

11.4 Static and Dynamic Equations: Continuum Approach – Revisited

11.4.1 What We Need to Recall

Let us recall the following for general shells.

- We characterize the shell undeformed reference surface by the arc length parameters S^1, S^2 so that the position vectors on the undeformed shell reference surface may be given as $\mathbf{C}_0(S^1, S^2) \equiv \mathbf{C}(S^1, S^2, 0, 0)$.
- S^1, S^2, as smooth reparameterizations, are associated with ξ^1, ξ^2, the natural coordinate parameters along the lines of curvature, respectively, by the relations: $S^1(\xi^1, \xi^2) = \int_0^{\xi^1} \|\mathbf{C}(\rho, \xi^2, 0),_{\rho}\| d\rho$ and $S^2(\xi^1, \xi^2) = \int_0^{\xi^2} \|\mathbf{C}(\xi^1, \rho, 0),_{\rho}\| d\rho$ where $\mathbf{C}(\xi^1, \xi^2, 0)$ is the position vector of a point on the shell reference surface.
- $\mathbf{c}(S^1, S^2, \eta, t)$ is the position vector of the point (with undeformed coordinates (S^1, S^2, η)) on the body in the deformed configuration at the current time $t \in \mathbb{R}_+$; $\mathbf{v}(S^1, S^2, \eta, t) \equiv \dot{\mathbf{c}}(S^1, S^2, \eta, t)$

is the current velocity at any material point at the current time $t \in \mathbb{R}_+$ that used to have the coordinates (S^1, S^2, η) in the undeformed reference state; on the shell deformed reference surface, $\chi(M)$, the position vectors are denoted by: $\mathbf{c}_0(S^1, S^2, t) \equiv \mathbf{c}(S^1, S^2, 0, t)$.

- The deformed position vector has a representation as:

$$\mathbf{c}_0(S^1, S^2, t) = \mathbf{C}_0(S^1, S^2) + \mathbf{d}(S^1, S^2, t) \tag{11.116}$$

where $\mathbf{d}(S^1, S^2, t)$ is the displacement vector of a point (S^1, S^2) on the mid-surface at current time $t \in \mathbb{R}_+$ and $\mathbf{C}_0(S^1, S^2)$ is the position vector of the same point on the undeformed mid-surface.

- $\mathbf{P}(S^1, S^2, \eta, t)$ as the two-point first Piola–Kirchhoff stress tensor (PK1) defined by: $\mathbf{P} = \mathbf{P}^j \otimes \mathbf{A}_j$with $\{\mathbf{A}_i(S^1, S^2, \eta, t = 0)\}$ as the set of covariant base vectors at a point in the undeformed body which, when deformed to the current state, become the traction point in consideration on the deformed surface.
- the deformation map of the 3D shell-like body may be given by:

$$\boxed{\begin{aligned}&\mathbf{c}(S^1, S^2, \eta, t) = \mathbf{c}_0(S^1, S^2, t) + \mathbf{h}(S^1, S^2, \eta, t),\\ &\mathbf{h}(S^1, S^2, 0, t) \equiv 0\end{aligned}} \tag{11.117}$$

where, $\mathbf{h}(S^1, S^2, \eta, t)$ is an as yet unknown and arbitrary vector-valued through-the-thickness deformation function, with the usual requirement of invertibility, that locates any arbitrary point on the deformed body, given the shell reference surface map $\chi(M)$.

- The stress resultant, $\mathbf{F}^\alpha(S, t),\ \alpha = 1, 2$, per unit of undeformed boundary length, is defined as:

$$\boxed{\mathbf{F}^\alpha \equiv \int_{\eta^b}^{\eta^t} {}^\eta\mathbf{P}^\alpha\ {}^\eta\mu^S\, d\eta} \tag{11.118}$$

where, of course, ${}^\eta\mathbf{P}^\alpha = \mathbf{P}\ {}^\eta\mathbf{A}^\alpha$.

- The effective surface linear momentum, $\mathbf{L}(S^1, S^2, t)$, per unit of undeformed surface area, is defined as:

$$\boxed{\begin{aligned}\mathbf{L}(S^1, S^2, t) &\equiv \int_{\eta^b}^{\eta^t} \mathbf{v}(S^1, S^2, \eta, t)\, \rho_0(S^1, S^2, \eta)\, {}^\eta\mu(S^1, S^2, \eta)\, d\eta\\ &= \int_{\eta^b}^{\eta^t} \dot{\mathbf{c}}(S^1, S^2, \eta, t)\ \rho_0(S^1, S^2, \eta)\, {}^\eta\mu(S^1, S^2, \eta)\, d\eta\end{aligned}} \tag{11.119}$$

where $\mathbf{v}(S^1, S^2, \eta, t) \equiv \dot{\mathbf{c}}(S^1, S^2, \eta, t)$ is, as indicated earlier, the velocity at current time $t \in \mathbb{R}_+$ at a point with undeformed coordinates (S^1, S^2, η) in the interior of the shell-like body.

- The surface external applied force, $\hat{\mathbf{F}}(S^1, S^2, t)$, per unit of undeformed reference surface area, is defined as:

$$\boxed{\hat{\mathbf{F}}(S^1, S^2, t) \equiv \int_{\eta^b}^{\eta^t} \mathbf{b}(S^1, S^2, \eta, t)\, \mu^\eta\, d\eta + \mathbf{t}(S^1, S^2, \eta^x, t)\, \mu\big|_{\eta^x = \eta^b}^{\eta^x = \eta^t}} \tag{11.120}$$

- The stress couple, $\mathbf{M}^\alpha(S^1, S^2, t)$, $\alpha = 1, 2$, per unit of undeformed boundary length, is defined as:

$$\boxed{\mathbf{M}^\alpha(S,t) \equiv \int_{\eta^b}^{\eta^t} \mathbf{h}(S,\eta,t) \times {}^\eta\mathbf{P}^\alpha(S,\eta,t)\, {}^\eta\mu^S(S,\eta,t)\, d\eta} \tag{11.121}$$

- The effective surface angular momentum, $\mathbf{A}(S^1, S^2, t)$, per unit of undeformed surface area, is defined as:

$$\boxed{\begin{aligned} \mathbf{A}(S^1,S^2,t) &\equiv \int_{\eta^b}^{\eta^t} \mathbf{h}(S,\eta,t) \times \mathbf{v}(S^1,S^2,\eta,t)\, \rho_0(S^1,S^2,\eta)\, {}^\eta\mu(S^1,S^2,\eta)\, d\eta \\ &= \int_{\eta^b}^{\eta^t} \mathbf{h}(S,\eta,t) \times \dot{\mathbf{c}}(S^1,S^2,\eta,t)\, \rho_0(S^1,S^2,\eta)\, {}^\eta\mu(S^1,S^2,\eta)\, d\eta \end{aligned}} \tag{11.122}$$

where $\mathbf{v}(S^1, S^2, \eta, t) \equiv \dot{\mathbf{c}}(S^1, S^2, \eta, t)$ is, as indicated earlier, the velocity at current time $t \in \mathbb{R}_+$ at a point with undeformed coordinates (S^1, S^2, η) in the interior of the shell-like body.

- The surface external applied moment, $\hat{\mathbf{M}}(S^1, S^2, t)$, per unit of undeformed reference surface area, is defined as:

$$\boxed{\hat{\mathbf{M}}(S^1,S^2,t) \equiv \int_{\eta^b}^{\eta^t} \mathbf{h}(S^1,S^2,\eta,t) \times \mathbf{b}(S^1,S^2,\eta,t)\, \mu^\eta\, d\eta \;+\; \mathbf{h}(S^1,S^2,\eta,t) \times \mathbf{t}\;\mu\big|_{\eta^b}^{\eta^t}} \tag{11.123}$$

- The vector dynamic momentum balance equations:

$$\boxed{\begin{array}{lr} \mathbf{F}^\alpha{}_{,\alpha} + \hat{\mathbf{F}} = \dot{\mathbf{L}}, \quad \alpha = 1,2 & \text{Momentum balance} \\ \mathbf{M}^\alpha{}_{,\alpha} + \mathbf{c}_0{}_{,\alpha} \times \mathbf{F}^\alpha + \hat{\mathbf{M}} = \dot{\mathbf{c}}_0 \times \mathbf{L} + \dot{\mathbf{A}}, \quad \alpha = 1,2 & \text{Moment of momentum balance} \end{array}} \tag{11.124}$$

- The vector static momentum balance equations:

$$\boxed{\begin{array}{lr} \mathbf{F}^\alpha|_\alpha + \hat{\mathbf{F}} = \mathbf{0}, \quad \alpha = 1,2 & \text{Momentum balance} \\ \mathbf{M}^\alpha|_\alpha + \mathbf{c}_0{}_{,\alpha} \times \mathbf{F}^\alpha + \hat{\mathbf{M}} = \mathbf{0}, \quad \alpha = 1,2 & \text{Moment of momentum balance} \end{array}} \tag{11.125}$$

- The spatial angular velocity, $\boldsymbol{\omega} \equiv \boldsymbol{\omega}_S$, is the axial vector corresponding to the skew-symmetric tensor, $\boldsymbol{\Omega} \equiv \boldsymbol{\Omega}_S$, defined by:

$$\boxed{\begin{gathered} \dot{\mathbf{R}} = \boldsymbol{\Omega}_S\, \mathbf{R} \Leftrightarrow \boldsymbol{\Omega}_S = \dot{\mathbf{R}}\, \mathbf{R}^T \\ \boldsymbol{\Omega}_S\, \boldsymbol{\omega}_S = \mathbf{0} \end{gathered}} \tag{11.126}$$

where $\mathbf{R}(S^1, S^2, t)$ is the rotation field at a point (S^1, S^2) on the shell reference surface at time $t \in \mathbb{R}_+$.

- Finally, as a notational reminder, in what follows, all lowercase Greek indices assume values 1 and 2, and the lowercase Latin indices runs through 1, 2, 3; also, for repeated diagonal indices, Einstein's summation convention of summing over their ranges holds.

11.4.2 Thin and Very Flexible Shell: Specialization

First of all, for very thin and flexible shells:

- The mid-surface is chosen as the shell undeformed reference surface for computational purposes; then, the locations through the thickness along the normal to the mid-surface are given by $\eta \in [-\frac{H}{2} \equiv H^-, +\frac{H}{2} \equiv H^+]$ where H is the constant thickness of the shall, that is, $\eta^t = +\frac{H}{2} \equiv H^+$ and $\eta^b = -\frac{H}{2} \equiv H^-$.
- It is assumed that the undeformed mid-surface is also the center of mass surface at time $t = 0$. So, from $\eta = 0$ on the mid-surface, and the definition of center of mass density, $\bar{\eta}$, we have:

$$\boxed{\begin{aligned} \bar{\eta} &= \frac{\displaystyle\int_{H^-}^{H^+} \eta\ \rho_0(S^1, S^2, \eta)\, {}^{\eta}\mu(S^1, S^2, \eta)\ d\eta}{\displaystyle\int_{H^-}^{H^+} \rho_0(S^1, S^2, \eta)\, {}^{\eta}\mu(S^1, S^2, \eta)\ d\eta} = 0 \\ &\Rightarrow \int_{H^-}^{H^+} \eta\ \rho_0(S^1, S^2, \eta)\, {}^{\eta}\mu(S^1, S^2, \eta)\ d\eta = 0 \end{aligned}} \tag{11.127}$$

11.4.2.1 Deformation Map: Thin Shell

For the most general case, the deformation map of a 3D shell-like body characterized by $\mathbf{h}(S^1, S^2, \eta, t)$ was defined in equation (11.117). The choice of $\mathbf{h}(S^1, S^2, \eta, t)$ essentially helps in categorizing the various shell theories with differing kinematic assumptions. For thin and very flexible shells, as shown in Figure 11.12, $\mathbf{h}(S^1, S^2, \eta, t)$ is taken as:

$$\boxed{\mathbf{h}(S^1, S^2, \eta, t) \equiv \mathbf{p}(S^1, S^2, \eta, t) = \ \eta\ \mathbf{n}(S^1, S^2, t) = \ \eta\ \mathbf{R}(S^1, S^2, t)\ \mathbf{N}(S^1, S^2)} \tag{11.128}$$

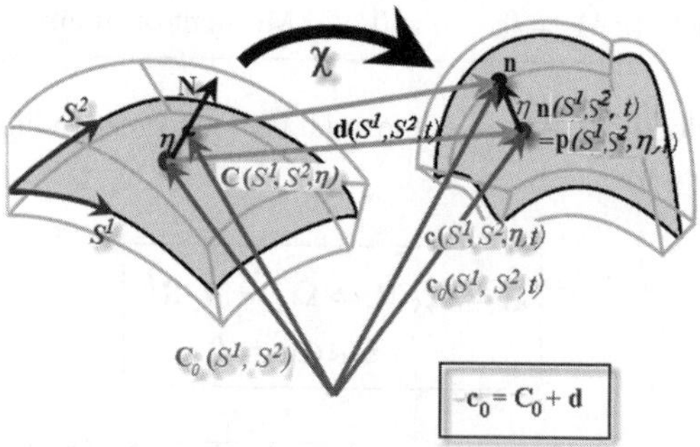

Figure 11.12 Thin shell deformation map.

where $\mathbf{n}(S^1, S^2, t)$ is the current unit normal obtained by rotation of $\mathbf{N}(S^1, S^2)$, the unit undeformed normal to the undeformed mid-surface, such that $\mathbf{n} = \mathbf{R}\,\mathbf{N}$ with $\mathbf{R}(S^1, S^2, t)$ the rotation at a point on the mid-surface at the current time, $t \in \mathbb{R}_+$. Noting that any point in the shell-like body at current time is obtained by pure rotation, the assumption implies that the unit normal to the undeformed mid-surface remains ***inextensible***, but not necessarily perpendicular to the deformed mid-surface during deformation (because it may rotate due to shear deformations), which is a commonly accepted underpinning of theories for very thin and flexible shells with shear deformations (reflected in the rotation of the mid-surface) and goes by the name of Reissner–Mindlin-type kinematic assumptions. For such cases, instead of equation (11.117), we will have deformation map as:

$$\boxed{\mathbf{c}(S^1, S^2, \eta, t) = \mathbf{c}_0(S^1, S^2, t) + \eta\, \mathbf{R}(S^1, S^2, t)\, \mathbf{N}(S^1, S^2),} \tag{11.129}$$

If we exclude considerations for all shear deformations and restrict ourselves only to flexural behaviors by requiring that the unit normal to the mid-surface always remains normal and inextensible to the deformed mid-surface, the involved theory is associated with the name: Love–Kirchhoff-type kinematic assumption. In what follows, we will be interested in specializing the balance equations to the Reissner–Mindlin-type shell configurations for computational purposes.

11.4.2.2 Effective Surface Linear Momentum

The effective surface linear momentum, $\mathbf{L}(S^1, S^2, t)$, defined in equation (11.119), can be rewritten as:

$$\begin{aligned}
\mathbf{L}(S^1, S^2, t) &= \int_{H^-}^{H^+} \dot{\mathbf{c}}(S^1, S^2, \eta, t)\; \rho_0(S^1, S^2, \eta)\, {}^{\eta}\mu(S^1, S^2, \eta)\, d\eta \\
&= \left(\int_{H^-}^{H^+} \eta\; \rho_0(S^1, S^2, \eta)\, {}^{\eta}\mu(S^1, S^2, \eta)\; d\eta \right) \dot{\mathbf{R}}(S^1, S^2, t)\, \mathbf{N}(S^1, S^2) \\
&\quad + \left(\int_{H^-}^{H^+} \rho_0(S^1, S^2, \eta)\, {}^{\eta}\mu(S^1, S^2, \eta)\; d\eta \right) \dot{\mathbf{c}}_0(S^1, S^2, t) \\
&= M_0(S^1, S^2)\; \mathbf{v}_0(S^1, S^2, t)
\end{aligned} \tag{11.130}$$

where we have dropped a term based on equation (11.127) and introduced:

Definition:

- Surface linear velocity, $\mathbf{v}_0(S^1, S^2, t)$:

$$\boxed{\mathbf{v}_0(S^1, S^2, t) = \dot{\mathbf{c}}_0(S^1, S^2, t) \equiv \dot{\mathbf{d}}(S^1, S^2, t)} \tag{11.131}$$

- Surface mass, $M_0(S^1, S^2)$, per unit surface area of the undeformed mid-surface:

$$\boxed{M_0(S^1,S^2) = \int_{H^-}^{H^+} \rho_0(S^1,S^2,\eta)\,{}^{\eta}\mu(S^1,S^2,\eta)\; d\eta} \tag{11.132}$$

A typical assumption for thin shell is that the mass density, $\rho_0(S^1, S^2, \eta)$, is constant through the thickness, that is, $\rho_0 \equiv \rho_0(S^1, S^2)$, and ${}^{\eta}\mu(S^1, S^2, \eta) \approx 1$. Thus, with the application of these assumptions to equation (11.132), we get:

$$\boxed{M_0(S^1,S^2) = \rho_0\; H} \tag{11.133}$$

- Effective surface linear momentum, $\mathbf{L}(S^1, S^2, t)$:

$$\boxed{\mathbf{L}(S^1,S^2,t) = M_0(S^1,S^2)\; \mathbf{v}_0(S^1,S^2,t) \approx \rho_0 H\; \dot{\mathbf{d}}} \tag{11.134}$$

11.4.2.3 Effective Surface Angular Momentum

Note that from the definition of the angular velocity, $\boldsymbol{\omega}$, from equation (11.126), we have:

$$\begin{array}{lll}
\mathbf{n} = \mathbf{R}\,\mathbf{N} \Rightarrow \mathbf{R} = \mathbf{n} \otimes \mathbf{N}, & & \\
\Rightarrow \dot{\mathbf{n}} = \dot{\mathbf{R}}\;\mathbf{N} = \dot{\mathbf{R}}\;\mathbf{R}^T\,\mathbf{R}\,\mathbf{N} = \boldsymbol{\Omega}\,\mathbf{n} & \Rightarrow & \boxed{\dot{\mathbf{n}} = \boldsymbol{\Omega}\,\mathbf{n} = \boldsymbol{\omega} \times \mathbf{n}}, \\
\boldsymbol{\Omega}\;\boldsymbol{\omega} = \mathbf{0} \Rightarrow (\dot{\mathbf{n}} \otimes \mathbf{n})\;\boldsymbol{\omega} = 0 \Rightarrow (\boldsymbol{\omega} \bullet \mathbf{n})\;\dot{\mathbf{n}} = 0 & \Rightarrow & \boxed{(\boldsymbol{\omega} \bullet \mathbf{n}) = 0}, \\
\mathbf{n} \times \dot{\mathbf{n}} = \mathbf{n} \times (\boldsymbol{\omega} \times \mathbf{n}) = (\mathbf{n} \bullet \mathbf{n})\;\boldsymbol{\omega} - \cancel{(\mathbf{n} \bullet \boldsymbol{\omega})}\,\mathbf{n} & \Rightarrow & \boxed{\mathbf{n} \times \dot{\mathbf{n}} = \boldsymbol{\omega}}
\end{array} \tag{11.135}$$

Then, with the last relation in equation (11.135), the effective (spatial) surface angular momentum, $\mathbf{A}(S^1, S^2, t)$, from equation (11.122), may be redefined as:

$$\boxed{\begin{aligned}
\mathbf{A}(S^1,S^2,t) &= \int_{\eta^b}^{\eta^t} \mathbf{h}(S,\eta,t) \times \dot{\mathbf{c}}(S^1,S^2,\eta,t)\,\rho_0(S^1,S^2,\eta)\,{}^{\eta}\mu(S^1,S^2,\eta)\; d\eta \\
&= \left(\int_{H^-}^{H^+} \eta\,\mathbf{n} \times \overbrace{(\mathbf{c}_0 + \eta\,\mathbf{n})}^{\bullet}\; \rho_0{}^{\eta}\mu\; d\eta\right) \\
&= \cancel{\left(\int_{H^-}^{H^+} \eta\rho_0(S^1,S^2,\eta)\,{}^{\eta}\mu(S^1,S^2,\eta)\; d\eta\right)}\; \mathbf{n} \times \dot{\mathbf{c}}_0 \\
&\quad + \left(\int_{H^-}^{H^+} \eta^2\rho_0(S^1,S^2,\eta)\,{}^{\eta}\mu(S^1,S^2,\eta)\; d\eta\right)\; \mathbf{n} \times \dot{\mathbf{n}} \\
&= I_0\;(\mathbf{n} \times \dot{\mathbf{n}}) = I_0\;\;\boldsymbol{\omega}
\end{aligned}} \tag{11.136}$$

where we have dropped a term based on equation (11.127) and introduced:

Definition

- Surface angular velocity, $\boldsymbol{\omega}(S^1, S^2, t)$:

$$\boxed{\begin{aligned} &\boldsymbol{\omega}(S^1, S^2, t) \equiv \mathbf{n}(S^1, S^2, t) \times \dot{\mathbf{n}}(S^1, S^2, t), \\ &\qquad\qquad \textit{with} \\ &\mathbf{n}(S^1, S^2, t) = \mathbf{R}(S^1, S^2, t)\, \mathbf{N}(S^1, S^2), \\ &\dot{\mathbf{n}}(S^1, S^2, t) = \dot{\mathbf{R}}(S^1, S^2, t)\, \mathbf{N}(S^1, S^2) = \dot{\mathbf{R}}\; \mathbf{R}^T\, \mathbf{R}\, \mathbf{N} \\ &\qquad\qquad = \boldsymbol{\Omega}(S^1, S^2, t)\, \mathbf{n}(S^1, S^2, t) = \boldsymbol{\omega} \times \mathbf{n}, \\ &\boldsymbol{\omega} \bullet \mathbf{n} = 0 \end{aligned}} \tag{11.137}$$

- Surface rotary inertia, $I_0(S^1, S^2)$, per unit surface area of the undeformed mid-surface:

$$\boxed{I_0(S^1, S^2) = \int_{H^-}^{H^+} \eta^2\, \rho_0(S^1, S^2, \eta)\, {}^{\eta}\mu(S^1, S^2, \eta)\, d\eta} \tag{11.138}$$

A typical assumption for thin shell is that the mass density, $\rho_0(S^1, S^2, \eta)$, is constant through the thickness, that is, $\rho_0 \equiv \rho_0(S^1, S^2)$, and ${}^{\eta}\mu(S^1, S^2, \eta) \approx 1$. Thus, with the application of these assumptions to equation (11.138), we get:

$$\boxed{I_0(S^1, S^2) = \rho_0\, \frac{H^3}{12}} \tag{11.139}$$

Effective surface angular momentum, $\mathbf{A}(S^1, S^2, t)$:

$$\boxed{\mathbf{A}(S^1, S^2, t) = I_0(S^1, S^2)\, \boldsymbol{\omega}(S^1, S^2, t) \approx \rho_0\, \frac{H^3}{12}\, \boldsymbol{\omega}} \tag{11.140}$$

11.4.2.4 Thin Shell Dynamic Momentum Balance Equations

Now, from the first of equation (11.124), with definitions in equation (11.134), we can rewrite the momentum balance equation as:

$$\boxed{\begin{aligned} &\mathbf{F}^{\alpha}{}_{,\alpha} + \hat{\mathbf{F}} = \dot{\mathbf{L}}, \quad \alpha = 1, 2, \\ &\dot{\mathbf{L}} = M_0\; \dot{\mathbf{v}}_0 = M_0\; \ddot{\mathbf{d}} \approx (\rho_0 H)\; \ddot{\mathbf{d}} \end{aligned}} \tag{11.141}$$

11.4.2.5 Thin Shell Dynamic Angular Momentum Balance Equations

Noting that: $\boxed{\dot{\mathbf{c}}_0 \times \mathbf{L} = \mathbf{v}_0 \times M_0\, \mathbf{v}_0 = \mathbf{0}}$, from the second of equation (11.124), and with definitions in equation (11.140), we can rewrite the angular momentum balance equation as:

$$\boxed{\begin{aligned} &\mathbf{M}^{\alpha}|_{\alpha} \;+\; \mathbf{c}_{0},_{\alpha} \times \mathbf{F}^{\alpha} \;+\; \hat{\mathbf{M}} = \dot{\mathbf{A}}, \quad \alpha = 1,2, \\ &\qquad \dot{\mathbf{A}}(S^1,S^2,t) = I_0(S^1,S^2)\; \dot{\boldsymbol{\omega}}(S^1,S^2,t) \approx \left(\rho_0\, \frac{H^3}{12}\right)\; \dot{\boldsymbol{\omega}}(S^1,S^2,t) \end{aligned}} \tag{11.142}$$

Finally, note that in equation (11.142), we have redefined the stress couple, $\mathbf{M}^{\alpha}(S,t),\ \alpha = 1,2$, of equation (11.121) and the external applied moment, $\hat{\mathbf{M}}(S^1,S^2,t)$, of equation (11.123), as follows.

Definitions

- Internal resistive stress couple, $\mathbf{M}^{\alpha}(S,t),\ \alpha = 1,2$, per unit of undeformed boundary length:

$$\boxed{\mathbf{M}^{\alpha}(S,t) \equiv \mathbf{n}(S,t) \;\times \int_{\eta^b}^{\eta^t} \eta\;\; {}^{\eta}\mathbf{P}^{\alpha}(S,\eta,t)\;\; {}^{\eta}\mu^{S}(S,\eta,t)\; d\eta} \tag{11.143}$$

- External applied moment, $\hat{\mathbf{M}}(S^1,S^2,t)$, per unit of undeformed reference surface area:

$$\boxed{\hat{\mathbf{M}}(S^1,S^2,t) \equiv \mathbf{n}(S,t) \;\times \int_{\eta^b}^{\eta^t} \eta\;\, \mathbf{b}(S^1,S^2,\eta,t)\, \mu^{\eta}\, d\eta \;+\; \eta\; \mathbf{n}(S,\eta,t) \;\times\; \mathbf{t}\; \mu|_{\eta^b}^{\eta^t}} \tag{11.144}$$

11.4.3 *Thin Shell Dynamic Balance Equations of Motion*

For easy reference, here we copy and collect the derived dynamic balance equations:

$$\boxed{\begin{aligned} &\mathbf{F}^{\alpha},_{\alpha} + \;\hat{\mathbf{F}} = \dot{\mathbf{L}}, \quad \alpha = 1,2 \qquad\qquad \text{Momentum balance} \\ &\dot{\mathbf{L}} = M_0\;\, \dot{\mathbf{v}}_0 = M_0\;\, \ddot{\mathbf{d}} \approx \rho_0 H\;\, \ddot{\mathbf{d}} \\ &\mathbf{M}^{\alpha},_{\alpha} \;\;+\; \mathbf{c}_{0},_{\alpha} \times \mathbf{F}^{\alpha} + \hat{\mathbf{M}} = \dot{\mathbf{A}}, \quad \alpha = 1,2 \quad \text{Moment of momentum balance} \\ &\dot{\mathbf{A}}(S^1,S^2,t) = I_0(S^1,S^2)\; \dot{\boldsymbol{\omega}}(S^1,S^2,t) \approx \rho_0\, \frac{H^3}{12}\;\, \dot{\boldsymbol{\omega}}(S^1,S^2,t) \end{aligned}} \tag{11.145}$$

11.4.4 *Where We Would Like to Go*

We have now derived the shell balance equations by reducing the 3D equations of motion of continuum mechanics. However, the shell equations of motion can be directly developed by considering a shell as a reference surface to start with. For this simple engineering method, we would like to check out the following.

The direct engineering approach to shell balance equations: while the shell equations presented in equation (11.141) and equation (11.143) are very compact and thus useful for analysis, we will modify the current vector form to a component or operational vector form suitable for computational purposes where the appropriate components of the forces and moments are of primary importance. As prerequisites to such a task, we should check out:

- component or operational vectors
- derivatives of component vectors.
  With these background materials, we will be able to modify the equations in terms of what may be called the component vector or operator form as described in:
- computational balance equations.

11.5 Static and Dynamic Equations: Engineering Approach

11.5.1 What We Need to Recall

- The undeformed coordinate system, hereinafter known as the ***tangent coordinate system***, corresponding to the lines of curvature with arc length parameterization, $\{S^1, S^2\}$ and to the unit normal to the shell reference surface, with base vectors: $\{\mathbf{G_1}(S^1, S^2), \mathbf{G_2}(S^1, S^2), \mathbf{G}_3(S^1, S^2) \equiv \mathbf{N}(S^1, S^2)\}$, is an orthonormal system of coordinates where both the sets of the contravariant base vectors, $\{\mathbf{G}^i\}$, and the covariant base vectors, $\{\mathbf{G}_i\}$, are one and the same; arc length parameters, $\{S^1, S^2\}$, are essentially a smooth reparameterization of the root parameters, $\{\xi^1, \xi^2\}$, through the relationship: $S^1(\xi^1, \xi^2) = \int_0^{\xi^1} \|\mathbf{C}_{0,\eta}(\eta, \xi^2)\|\, d\eta$ and $S^2(\xi^1, \xi^2) = \int_0^{\xi^2} \|\mathbf{C}_{0,\eta}(\xi^1, \eta)\|\, d\eta$.
- For the ensuing vector form of the balance equations, no kinematic assumptions are necessary; however, for the component form, we must bring in the displacement parameters for computationally meaningful expressions.
- The Gauss–Weingarten relation for the derivatives of the triad of base vectors in matrix form with respect to the arc length parameters, $S^\alpha, \quad \alpha = 1, 2$, is given by:

$$\begin{Bmatrix} \mathbf{G_{1,1}} \\ \mathbf{G}_{2,1} \\ \mathbf{N}_{,1} \end{Bmatrix} \equiv \frac{\partial}{\partial S^1} \begin{Bmatrix} \mathbf{G_1} \\ \mathbf{G}_2 \\ \mathbf{N} \end{Bmatrix} = \begin{bmatrix} 0 & -\frac{1}{\gamma_1}\gamma_{1,2} & \frac{1}{R_1} \\ \frac{1}{\gamma_1}\gamma_{1,2} & 0 & 0 \\ -\frac{1}{R_1} & 0 & 0 \end{bmatrix} \begin{Bmatrix} \mathbf{G_1} \\ \mathbf{G}_2 \\ \mathbf{N} \end{Bmatrix} \equiv \widehat{\mathbf{K}}_1^0 \begin{Bmatrix} \mathbf{G_1} \\ \mathbf{G}_2 \\ \mathbf{N} \end{Bmatrix} \tag{11.146}$$

and

$$\begin{Bmatrix} \mathbf{G_{1,2}} \\ \mathbf{G}_{2,2} \\ \mathbf{N}_{,2} \end{Bmatrix} \equiv \frac{\partial}{\partial S^2} \begin{Bmatrix} \mathbf{G_1} \\ \mathbf{G}_2 \\ \mathbf{N} \end{Bmatrix} = \begin{bmatrix} 0 & \frac{1}{\gamma_2}\gamma_{2,1} & 0 \\ -\frac{1}{\gamma_2}\gamma_{2,1} & 0 & \frac{1}{R_2} \\ 0 & -\frac{1}{R_2} & 0 \end{bmatrix} \begin{Bmatrix} \mathbf{G_1} \\ \mathbf{G}_2 \\ \mathbf{N} \end{Bmatrix} \equiv \widehat{\mathbf{K}}_2^0 \begin{Bmatrix} \mathbf{G_1} \\ \mathbf{G}_2 \\ \mathbf{N} \end{Bmatrix} \tag{11.147}$$

where $\widehat{\mathbf{K}}_\alpha^0, \quad \alpha = 1, 2$ are the shell undeformed curvature matrices with $R_\alpha(S^1, S^2), \quad \alpha = 1, 2$, are the radii of curvatures at any point, $\{S^\alpha\}$, of the shell reference surface, and $\gamma_\alpha, \quad \alpha = 1, 2$, are the lengths of the root base vectors, $\bar{\mathbf{G}}_\alpha$: $\gamma_\alpha \equiv \frac{dS^\alpha}{d\xi^\alpha} = \|\bar{\mathbf{G}}_\alpha\| = \sqrt{\bar{\mathbf{G}}_{\alpha\alpha}}$ where $\bar{\mathbf{G}}_\alpha(\xi^1, \xi^2) \equiv \frac{\partial \mathbf{C}(\xi^1,\xi^2,0)}{\partial \xi^\alpha}$, $\alpha = 1, 2$ with $\{\xi^1, \xi^2\}$ as the root parameters along the lines of curvature of the shell surface.

- On the other hand, the Gauss–Weingarten relation for the derivatives of the triad of base vectors on a surface in invariant tensor form with respect to the arc length parameters, S^α, $\alpha = 1, 2$, may be obtained by first introducing the following.

Definition: Shell initial curvature tensors, $\mathbf{K}^0_\alpha$, $\alpha = 1, 2$:

$$\boxed{\mathbf{K}^0_\alpha \equiv (\widehat{\mathbf{K}}^0_\alpha)^T_{ij}\,\mathbf{G}_i \otimes \mathbf{G}_j = -\mathbf{G}_i \otimes \mathbf{G}_{i},_\alpha\,, \qquad \alpha = 1, 2; \quad i, j = 1, 2, 3} \tag{11.148}$$

Definition: shell initial curvature vector, $\mathbf{k}^0_\alpha$, $\alpha = 1, 2$:
The axial vectors corresponding to the initial curvature tensors, $\mathbf{K}^0_\alpha$, $\alpha = 1, 2$, are given by:

$$\boxed{\begin{aligned}\mathbf{k}^0_1 &\equiv \left\{0 \quad -\frac{1}{R_1} \quad -\frac{1}{\gamma_1}\gamma_{1},_2\right\}^T \\ \mathbf{k}^0_2 &\equiv \left\{\frac{1}{R_2} \quad 0 \quad \frac{1}{\gamma_2}\gamma_{2},_1\right\}^T\end{aligned}} \tag{11.149}$$

Now, we can easily get from equation (11.148), the invariant tensor form as:

$$\boxed{\mathbf{G}_{i},_\alpha = \mathbf{K}^0_\alpha\,\mathbf{G}_i = \mathbf{k}^0_\alpha \times \,\mathbf{G}_i, \qquad \alpha = 1, 2;\ i = 1, 2, 3} \tag{11.150}$$

- Finally, the covariant derivative of a base vector is the zero vector, that is, $\mathbf{G}_i|_\alpha = \mathbf{0}$, $i = 1, 2, 3$. Thus, the covariant derivative of a tensor function of any order is identical to its partial derivative. If we take a vector (a tensor of order 1) function, $\mathbf{a}(S^\alpha)$, we then have: $\mathbf{a}\,|_{S^\alpha} \equiv \mathbf{a}\,|_\alpha = \mathbf{a},_\alpha = {}^t a_i\,|_\alpha \mathbf{G}_i = \mathbf{G}\,{}^t\mathbf{a}\,|_\alpha$ where $\mathbf{G} \equiv [\,\mathbf{G}_1 \quad \mathbf{G}_2 \quad \mathbf{G}_3\,]$ is the orthonormal matrix with base vectors as its columns, and ${}^t\mathbf{a}(S^\alpha)$ is the component vector corresponding to the vector function, $\mathbf{a}(S^\alpha)$.

Exercise: Show that the partial derivative of a tensor function of any order is identical with its covariant derivative with respect to its arguments. Hint: For vector function $\mathbf{d}\,|_\alpha = {}^t d_i\,|_\alpha\,\mathbf{G}_i + {}^t d_i\ \cancel{\mathbf{G}_i|_\alpha} = {}^t d_i\,|_\alpha\,\mathbf{G}_i = \mathbf{d},_\alpha$. Why do you think $\mathbf{G}_i|_\alpha = \mathbf{0}$? Because, by definition, $\mathbf{G}_i|_j = \mathbf{G}_{i},_j - \Gamma^k_{ij}\mathbf{G}_k = \Gamma^k_{ij}\mathbf{G}_k - \Gamma^k_{ij}\mathbf{G}_k = 0$.

11.5.2 Deformed Geometry, Stress Resultants and Stress Couples

As already indicated, because we are interested to include in our analysis the case of finite displacements and rotations, all equilibrium equations must be considered for the current deformed geometry of a shell. Thus, as a starting point, we will attend to deformed geometry.

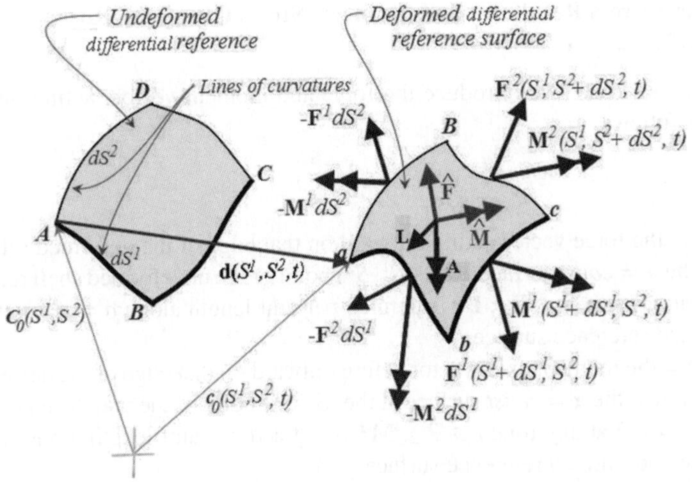

Figure 11.13 Direct engineering equilibrium derivation.

11.5.2.1 Deformed Geometry

Let us consider a typical differential element of the shell reference surface as shown in Figure 11.13, and also suppose that we have determined the lines of curvature directions from the given coordinate iso-curves of the undeformed shell surface, and that we will parameterize the shell reference surface by the arc length parameterization (S^1, S^2) along these lines of curvature (which are, of course, orthogonal to each other). Note first that we have intentionally chosen the shell element along the lines of curvature; this is not at all necessary for analysis and so we can choose any arbitrarily oriented element. However, it turns out that the equations are extremely efficient and simple for the most accurate computational development as we apply these expressions to our c-type finite element method. To jump ahead, as it will turn out computationally, all relevant information is needed only at the numerical integration points (i.e. the Gauss quadrature points) along the lines of curvature of the undeformed shell surface. Thus, without loss of any generality, we choose our elements along the lines of curvature to develop expressions for future computational use.

Referring to Figure 11.13, the position vector, $\mathbf{c}_0(S^1, S^2, t)$ locates a point a on a deformed reference surface (representing the shell in a deformed state) at any time $t \in \mathbb{R}_+$ whose corresponding location, on the reference surface $\mathbf{C}_0(S^1, S^2) \equiv \mathbf{c}_0(S^1, S^2, 0)$ (representing the undeformed shell) at initial reference time $t = 0$, is denoted by A. Similarly, considering a differential element $dS^1 \times dS^2$ along the lines of curvature at A on the surface, $\mathbf{c}_0(S^1 + dS^1, S^2, t)$ locates a point b on the same deformed reference surface whose corresponding location on the undeformed reference surface $\mathbf{C}_0(S^1 + dS^1, S^2)$ at $(S^1 + dS^1, S^2)$ is denoted by B. In other words, a differential element $ABCD$ on the undeformed shell reference surface becomes $abcd$ on a deformed state of the shell. For small enough dS^1 and dS^2, we may approximate to the first-order partial derivatives of the vector functions $\mathbf{C}_0(S^1, S^2)$ with respect to the arc length parameters, S^1 and S^2, respectively, as: $\mathbf{C}_0(S^1 + dS^1, S^2) \simeq \mathbf{C}_0(S^1, S^2) + \mathbf{C}_0(S^1, S^2)_{,S^1}\, dS^1$ and $\mathbf{C}_0(S^1, S^2 + dS^2) \simeq \mathbf{C}_0(S^1, S^2) + \mathbf{C}_0(S^1, S^2)_{,S^2}\, dS^2$. Finally, we indicate the displacement of the undeformed point A to its current point a by $\mathbf{d}(S^1, S^2, t)$ such that

$$\boxed{\mathbf{c}_0(S^1, S^2, t) = \mathbf{C}_0(S^1, S^2) + \mathbf{d}(S^1, S^2, t)} \tag{11.151}$$

11.5.2.2 Force (Stress Resultant) and Moment (Stress Couple) Vectors: $\mathbf{F}^\alpha \mathbf{M}^\alpha, \quad \alpha = 1, 2$

Referring to Figure 11.13, we introduce the force and moment vectors acting on the deformed shell surface as follows.

Definition:

- $\mathbf{F}^\alpha(S^1, S^2, t)$ = the force vector acting outward on that edge of the deformed reference surface (which was the $\alpha = const$ surface at the (S^1, S^2) point of the undeformed shell reference surface for $\alpha = 1, 2$ at any time $t \in \mathbb{R}_+$; $\mathbf{F}^\alpha$ is defined per unit length along $\alpha = const$cross-section of the undeformed reference surface.
- $\mathbf{M}^\alpha(S^1, S^2, t)$ = the moment axial vector acting outward on that edge of the deformed reference surface which was the $\alpha = const$ surface at the (S^1, S^2) point of the undeformed shell reference surface for $\alpha = 1, 2$ at any time $t \in \mathbb{R}_+$; $\mathbf{M}^\alpha$ is defined per unit length along $\alpha = const$cross-section of the undeformed reference surface.

Remark

- Let us note that at this time, in the absence of any kinematic assumptions, we do not have any computationally useful stress-level interpretations, and hence expressions, for either the invariant force (i.e. stress resultant) vector, $\mathbf{F}^\alpha(S^1, S^2, t)$, or the moment (i.e. stress couple) vector, $\mathbf{M}^\alpha(S^1, S^2, t)$. Clearly, the names suggested in parentheses allude to an assumed shell through-the-thickness deformation distribution which we refrain from discussing any further, and leave it for our derivation of the balance laws from the fundamental principles of continuum mechanics, as we have done in the section on 3D to 2D reduction.

11.5.2.3 Component Force and Moment Vectors in terms of Undeformed Frame: ${}^t\mathbf{F}^\alpha(S^1, S^2, t)$ and ${}^t\mathbf{M}^\alpha(S^1, S^2, t)$

The invariant force vectors, $\mathbf{F}^\alpha$, and the moment vectors, $\mathbf{M}^\alpha$, can have component representation in any triad of base vectors, Cartesian or curvilinear. At this point of the analysis, we choose to describe the components in the undeformed curvilinear coordinate system (henceforth, known as the tangent coordinate system) with the triad of base vectors, $\{\mathbf{G}_1, \mathbf{G}_2, \mathbf{G}_3 \equiv \mathbf{N}\}$ on the shell reference surface. Thus, referring to Figure 11.14, we can write:

$$\boxed{\begin{aligned} \mathbf{F}^\alpha(S^1, S^2, t) &= {}^tF^{\alpha 1}(\equiv N_{\alpha 1})\,\mathbf{G}_1 + {}^tF^{\alpha 2}(\equiv N_{\alpha 2})\,\mathbf{G}_2 + {}^tF^{\alpha 3}(\equiv Q_\alpha)\,\mathbf{N} \\ &= {}^tF^{\alpha i}(S^1, S^2, t)\,\mathbf{G}_i(S^1, S^2), \quad \alpha = 1, 2; \quad i = 1, 2, 3 \end{aligned}} \tag{11.152}$$

where, N_{11}, N_{22} are known as the in-plane extensional forces, N_{12}, N_{21} as the in-plane shear forces, and Q_1, Q_2 as the transverse shear forces, as shown in Figure 11.15; and

$$\boxed{\begin{aligned} \mathbf{M}^1(S^1, S^2, t) &= {}^tM^{11}(\equiv -M_{12})\,\mathbf{G}_1 + {}^tM^{12}(\equiv M_1)\,\mathbf{G}_2 + {}^tM^{13}(\equiv M_{13})\,\mathbf{N} \\ &= {}^tM^{1i}\,\mathbf{G}_i, \qquad i = 1, 2, 3 \\ \mathbf{M}^2(S^1, S^2, t) &= {}^tM^{21}(\equiv -M_2)\,\mathbf{G}_1 + {}^tM^{22}(\equiv M_{21})\,\mathbf{G}_2 + {}^tM^{23}(\equiv M_{23})\,\mathbf{N} \\ &= {}^tM^{2i}\,\mathbf{G}_i, \qquad i = 1, 2, 3 \\ \mathbf{M}^\alpha(S^1, S^2, t) &= {}^tM^{\alpha i}(S^1, S^2, t)\,\mathbf{G}_i(S^1, S^2), \quad \alpha = 1, 2, \quad i = 1, 2, 3 \end{aligned}} \tag{11.153}$$

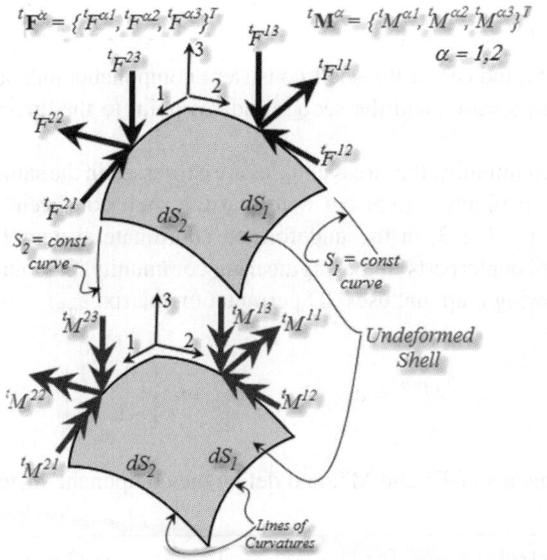

Figure 11.14 Shell force and moment components.

where, M_1, M_2 are known as the bending moments, M_{12}, M_{21} are known as the in-plane twisting moments, and $M_{\alpha 3}$ as the transverse twisting moments on a shell cross-section with normal to it pointing to α direction, $\alpha = 1, 2$ (for the classical shell theory, these components are identically zero. Conventionally and notationwise, these are described as such in engineering terminology as shown in Figure 11.14. In the final equalities of equations (11.152) and (11.153), we have introduced the following.

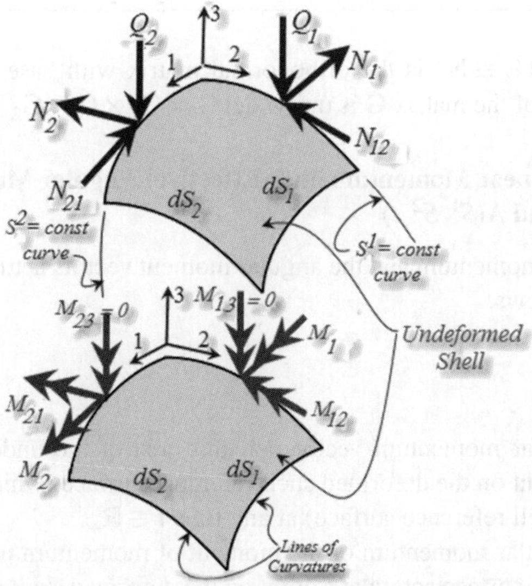

Figure 11.15 Engineering force and moment components.

Remarks

- Notationally, the first indices of the shell consistent components indicate the direction of the normals to the cross-section, and the second indices point to the direction of the forces and moments.
- In the engineering community, the stress couples are expressed in the same way as the associated stresses, as shown in Figure 11.15, but note that the shell consistent moment components, ${}^tM^{\alpha i},\quad \alpha = 1,2,\quad i = 1,2,3$, in the undeformed coordinate system $\{\mathbf{G}_1, \mathbf{G}_2, \mathbf{G}_3 \equiv \mathbf{N}\}$, are different from their counterparts in the engineering community. Of course, these can be easily related by the following map that uses 2D permutation matrix, $e_{\alpha\beta}$:

$$
{}^tM^{\alpha\beta} = e_{\eta\beta}\, M_{\alpha\eta}, \quad (e_{\eta\beta}) = \begin{bmatrix} 0 & 1 \\ -1 & 0 \end{bmatrix} \tag{11.154}
$$

We collect the components of $\mathbf{F}^\alpha$ and M^α, and define the component vectors, respectively, as:

$$
\boxed{\begin{aligned}
{}^t\mathbf{F}^1(S^1,S^2,t) &\equiv \left\{ {}^tF^{11} \equiv N_1 \quad {}^tF^{12} \equiv N_{12} \quad {}^tF^{13} \equiv Q_1 \right\}^T \\
{}^t\mathbf{F}^2(S^1,S^2,t) &\equiv \left\{ {}^tF^{21} \equiv N_{21} \quad {}^tF^{22} \equiv N_2 \quad {}^tF^{23} \equiv Q_2 \right\}^T
\end{aligned}} \tag{11.155}
$$

$$
\boxed{\begin{aligned}
{}^t\mathbf{M}^1(S^1,S^2,t) &\equiv \left\{ {}^tM^{11} \equiv -M_{12} \quad {}^tM^{12} \equiv M_1 \quad {}^tM^{13} \equiv M_{13} \right\}^T \\
{}^t\mathbf{M}^2(S^1,S^2,t) &\equiv \left\{ {}^tM^{21} \equiv -M_2 \quad {}^tM^{22} \equiv M_{21} \quad {}^tM^{23} \equiv M_{23} \right\}^T
\end{aligned}} \tag{11.156}
$$

With the above definitions, we can express these in the column vector form as:

$$
\boxed{\begin{aligned}
\mathbf{F}^\alpha(S^1,S^2,t) &= \mathbf{G}(S^1,S^2)\, {}^t\mathbf{F}^\alpha(S^1,S^2,t) \\
\mathrm{M}^\alpha(S^1,S^2,t) &= \mathbf{G}(S^1,S^2)\, {}^t\mathbf{M}^\alpha(S^1,S^2,t), \quad \alpha = 1,2
\end{aligned}} \tag{11.157}
$$

where $\mathbf{G} \equiv [\,\mathbf{G}_1 \quad \mathbf{G}_2 \quad \mathbf{G}_3 \equiv \mathbf{N}\,]$ is the orthonormal matrix with base vectors as its columns; clearly, the determinant of the matrix $\mathbf{G}$ is unity: $\det \mathbf{G} = \mathbf{G}_1 \times \mathbf{G}_2 \bullet \mathbf{G}_3 = 1$.

11.5.2.4 (Effective) Linear Momentum and (Effective) Angular Momentum Vectors: $\mathbf{L}(S^1,S^2,t)$ and $\mathbf{A}(S^1,S^2,t)$

We introduce the linear momentum and the angular moment vectors acting on the deformed shell reference surface as follows.

Definitions

- $\mathbf{L}(S^1,S^2,t)$ = the linear momentum vector per unit area of the undeformed shell reference surface acting at a point on the deformed shell reference surface (which was the (S^1,S^2) point of the undeformed shell reference surface) at any time $t \in \mathbb{R}_+$.
- $\mathbf{A}(S^1,S^2,t)$ = the angular momentum or the moment of momentum per unit area of the undeformed shell reference surface acting at a point on the deformed shell reference surface (which was the (S^1,S^2) point of the undeformed shell reference surface) at any time $t \in \mathbb{R}_+$.

Remark

- Let us note that, at this time, in the absence of any kinematic assumptions, we do not have any computationally useful deformation-level interpretations, and hence, expressions for either the invariant linear momentum (i.e. effective) vector, $\mathbf{L}(S^1, S^2, t)$, or the angular momentum (i.e. effective) vector, $\mathbf{A}(S^1, S^2, t)$. However, the qualifiers suggested in the parenthesis allude to an assumed shell through-the-thickness rate of deformation.

11.5.2.5 Material Components of Linear Momentum and Angular Momentum Vectors in Terms of the Undeformed Frame: ${}^t\mathbf{L}(S^1, S^2, t)$ and ${}^t\mathbf{A}(S^1, S^2, t)$

The invariant linear momentum vector, $\mathbf{L}(S^1, S^2, t)$, and the angular momentum vector, $\mathbf{A}(S^1, S^2, t)$, for any $t \in \mathbb{R}_+$, can have component representation in any triad of base vectors, Cartesian or curvilinear. At this point of the analysis, we choose to describe the components in the undeformed curvilinear coordinate system (the tangent coordinate system) with the triad of base vectors, $\{\mathbf{G}_1, \mathbf{G}_2, \mathbf{N} \equiv \mathbf{G}_3\}$ on the shell reference surface.

Thus, we can write:

$$\boxed{\begin{aligned}\mathbf{L}(S^1, S^2, t) &= {}^tL^1(\equiv P_1)\ \mathbf{G}_1 + {}^tL^2(\equiv P_2)\ \mathbf{G}_2 + {}^tL^3(\equiv P_N)\ \mathbf{G}_3\\ &= {}^tL^i(S^1, S^2, t)\ \mathbf{G}_i(S^1, S^2), \quad i = 1, 2, 3\end{aligned}} \tag{11.158}$$

and

$$\boxed{\begin{aligned}\mathbf{A}(S^1, S^2, t) = &= {}^tA^1(\equiv A_1)\ \mathbf{G}_1 + {}^tA^2(\equiv A_2)\ \mathbf{G}_2 + {}^tA^3(\equiv A_N)\ \mathbf{G}_3\\ &= {}^tA^i(S^1, S^2, t)\ \mathbf{G}_i(S^1, S^2), \quad i = 1, 2, 3\end{aligned}} \tag{11.159}$$

We collect the components of $\mathbf{L}$ and $\mathbf{A}$, and define the component vectors, respectively, as follows.

Definitions

$$\boxed{\begin{aligned}{}^t\mathbf{L}(S^1, S^2, t) &\equiv \{\,{}^tL^1 \quad {}^tL^2 \quad {}^tL^3 \equiv L_N\,\}^T\\ {}^t\mathbf{A}(S^1, S^2, t) &\equiv \{\,{}^tA^1 \quad {}^tA^2 \quad {}^tA^3 \equiv A_N\,\}^T\end{aligned}} \tag{11.160}$$

With the above definitions, we can express these in the column vector form as:

$$\boxed{\begin{aligned}\mathbf{L}(S^1, S^2, t) &= \mathbf{G}(S^1, S^2)\ {}^t\mathbf{L}(S^1, S^2, t)\\ \mathbf{A}(S^1, S^2, t) &= \mathbf{G}(S^1, S^2)\ {}^t\mathbf{A}(S^1, S^2, t)\end{aligned}} \tag{11.161}$$

Likewise, we have the time derivatives, $\dot{(\bullet)} \stackrel{\Delta}{=} \frac{\partial}{\partial t}$, of the of the linear momentum and the angular momentum in terms of their material component vectors:

$$\boxed{\begin{aligned}\dot{\mathbf{L}}(S^1, S^2, t) &= \mathbf{G}(S^1, S^2)\ {}^t\dot{\mathbf{L}}(S^1, S^2, t)\\ \dot{\mathbf{A}}(S^1, S^2, t) &= \mathbf{G}(S^1, S^2)\ {}^t\dot{\mathbf{A}}(S^1, S^2, t)\end{aligned}} \tag{11.162}$$

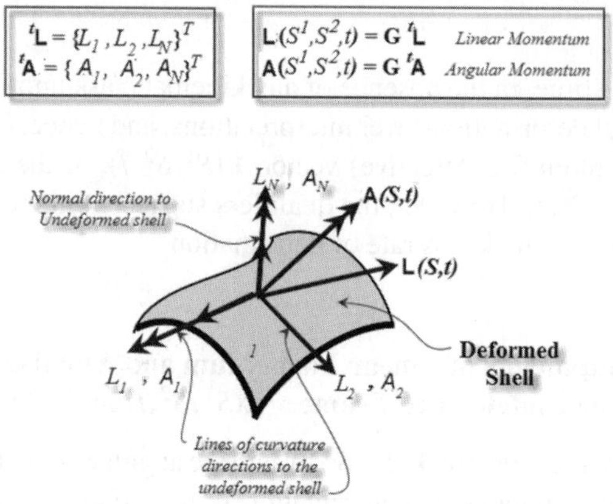

Figure 11.16 Linear and angular momentum in undeformed frame.

11.5.3 Dynamic Balance Equations

Let us again emphasize that the forces and moments, and the linear and the angular momenta, are acting on the current deformed shell reference surface but defined per unit length in the undeformed shell reference surface along the lines of curvature at the point where they are acting, with the components given in the local basis directions of the undeformed curvilinear coordinate system. With the above preliminaries as the background and referring to Figures 11.14 and 11.16, we can directly apply the momentum balance and the moment of momentum balance (moment being taken about the point b of the deformed differential element $ds^1 \times ds^2$), that is, the angular momentum balance, principles and we get the following.

11.5.3.1 Momentum Balance Equations (Force Equilibrium)

$$\begin{aligned}\sum \mathbf{F} &= \mathbf{F}^1(S^1+dS^1,S^2,t)dS^2 - \mathbf{F}^1(S^1,S^2,t)dS^2 + \mathbf{F}^2(S^1,S^2+dS^2,t)dS^1\\ &\quad -\mathbf{F}^2(S^1,S^2)dS^1 + \hat{\mathbf{F}}\,dS^1dS^2\\ &= \cancel{\mathbf{F}^1dS^2} + \mathbf{F}^1{}_{,S^1}\,dS^1dS^2 - \cancel{\mathbf{F}^1dS^2} + \cancel{\mathbf{F}^2dS^1} + \mathbf{F}^2{}_{,S^2}\,dS^1dS^2 - \cancel{\mathbf{F}^2dS^1} + \hat{\mathbf{F}}\,dS^1dS^2\\ &= \left(\mathbf{F}^1{}_{,S^1} + \mathbf{F}^2{}_{,S^2} + \hat{\mathbf{F}}\right)dS^1dS^2 = (\mathbf{F}^\alpha{}_{,\alpha} + \hat{\mathbf{F}})\,dS^1dS^2 = \frac{d}{dt}\mathbf{L}(S^1,S^2,t)\,dS^1dS^2\end{aligned}$$

Or, dividing the above equations all over by dS^1dS^2, we get

$$\boxed{\mathbf{F}^\alpha{}_{,\alpha} + \hat{\mathbf{F}} = \dot{\mathbf{L}}, \quad \alpha = 1,2} \tag{11.163}$$

where we have the following notation and relation for any vector function, $(\bullet)(S^1,S^2,t)$:

$$\begin{aligned}(\bullet)_{,\alpha} \equiv (\bullet)_{,S^\alpha} &= {}^t(\bullet)^i{}_{,\alpha}\mathbf{G}_i + {}^t(\bullet)^i\mathbf{G}_{i,\alpha} = {}^t(\bullet)^i{}_{,\alpha}\mathbf{G}_i + \boldsymbol{\kappa}^0_\alpha \times {}^t(\bullet)^i\,\mathbf{G}_i\\ &\equiv {}^t(\bullet)|_\alpha\end{aligned} \tag{11.164}$$

for the covariant derivative of the component vector, ${}^t(\bullet)$, corresponding to the invariant vector function, $(\bullet)(S^1,S^2,t)$, and $\mathbf{G}_{i,\alpha}$ depend on the shell surface metric tensor and the curvature tensor, as given by equations (11.146) and (11.147); we have also used Einstein's summation convention for the contravariant and the covariant repeated indices with $\alpha = 1, 2$. In fact, as indicated before, all Greek indices will be assumed to run over 1 and 2, and all Latin indices on 1,2 and 3. Note that the momentum balance equation of motion does not involve any kinematic parameter.

11.5.3.2 Angular or Moment of Momentum Balance Equations (Moment Equilibrium)

We intend to take moments about the origin of the fixed inertial frame, so we have:

$$\begin{aligned}
\sum \mathbf{M} &= \mathbf{M}^1(S^1 + dS^1, S^2, t)dS^2 - \mathbf{M}^1(S^1, S^2, t)dS^2 + \mathbf{M}^2(S^1, S^2 + dS^2)dS^1 \\
&\quad - \mathbf{M}^2(S^1, S^2, t)dS^2 + \mathbf{c}_0(S^1 + dS^1, S^2, t) \times \mathbf{F}^1(S^1 + dS^1, S^2, t)dS^2 \\
&\quad - \mathbf{c}_0(S^1, S^2, t) \times \mathbf{F}^1(S^1, S^2, t)dS^2 \\
&\quad + \mathbf{c}_0(S^1, S^2 + dS^2, t) \times \mathbf{F}^2(S^1, S^2 + dS^2, t)dS^1 \\
&\quad - \mathbf{c}_0(S^1, S^2, t) \times \mathbf{F}^2(S^1, S^2, t)dS^1 + \hat{\mathbf{M}}(S^1, S^2, t)\, dS^1 dS^2 \\
&\quad + \mathbf{c}_0(S^1 + \tfrac{1}{2}dS^1, S^2 + \tfrac{1}{2}dS^2, t) \times \hat{\mathbf{F}}(S^1 + \tfrac{1}{2}dS^1, S^2 + \tfrac{1}{2}dS^2, t)dS^1\, dS^2 \\
&= \left(\mathbf{M}^\alpha{}_{,\alpha} + \hat{\mathbf{M}}\right)\, dS^1 dS^2 \\
&= \frac{d}{dt}\left\{ \begin{array}{l} \mathbf{c}_0(S^1 + \tfrac{1}{2}dS^1, S^2 + \tfrac{1}{2}dS^2, t) \times \mathbf{L}(S^1 + \tfrac{1}{2}dS^1, S^2 \\ + \tfrac{1}{2}dS^2, t) + \mathbf{A}(S^1, S^2, t) \end{array} \right\} dS^1\, dS^2
\end{aligned} \tag{11.165}$$

Now, from equation (11.165), we intend to retain terms up to and including second differential order, such as $dS^1\, dS^2$, and neglect products of higher differential order terms. In order that we are consistent in our approximations, we must look into the following terms more closely:

Thus,

$$\begin{aligned}
&\mathbf{M}^1(S^1 + dS^1, S^2, t)dS^2 - \mathbf{M}^1(S^1, S^2, t)dS^2 + \mathbf{M}^2(S^1, S^2 + dS^2)dS^1 - \mathbf{M}^2(S^1, S^2, t)dS^2 \\
&\quad = \cancel{\mathbf{M}^1(S^1, S^2, t)dS^2} + \mathbf{M}^1(S^1, S^2, t)_{,1}\, dS^1 dS^2 - \cancel{\mathbf{M}^1(S^1, S^2, t)dS^2} + \quad \text{higher order} \\
&\quad\quad + \cancel{\mathbf{M}^2(S^1, S^2, t)dS^2} + \mathbf{M}^2(S^1, S^2)_{,2}\, dS^2 dS^1 - \cancel{\mathbf{M}^2(S^1, S^2, t)dS^2} \\
&\quad = \mathbf{M}^1(S^1, S^2, t)_{,1}\, dS^1 dS^2 + \mathbf{M}^2(S^1, S^2)_{,2}\, dS^2 dS^1 = \mathbf{M}^\alpha(S^1, S^2, t)_{,\alpha} dS^1 dS^2, \quad \alpha = 1, 2
\end{aligned} \tag{11.166}$$

and

$$\begin{aligned}
&\mathbf{c}_0(S^1 + dS^1, S^2, t) \times \mathbf{F}^1(S^1 + dS^1, S^2, t)dS^2 - \mathbf{c}_0(S^1, S^2, t) \times \mathbf{F}^1(S^1, S^2, t)dS^2 \\
&\quad = \left\{\mathbf{c}_0(S^1, S^2, t) + \mathbf{c}_{0,1}(S^1, S^2, t)dS^1 + \text{higher order}\right\} \\
&\quad\quad \times \left\{\mathbf{F}^1(S^1, S^2, t)dS^2 + \mathbf{F}^1{}_{,1}(S^1, S^2, t)dS^2\, dS^1\right\} \\
&\quad\quad - \mathbf{c}_0(S^1, S^2, t) \times \mathbf{F}^1(S^1, S^2, t)dS^2 + \text{higher order} \\
&\quad = \cancel{\mathbf{c}_0(S^1, S^2, t) \times \mathbf{F}^1(S^1, S^2, t)dS^2} + \mathbf{c}_0(S^1, S^2, t) \times \mathbf{F}^1{}_{,1}(S^1, S^2, t)dS^2\, dS^1 \\
&\quad\quad + \mathbf{c}_{0,1}(S^1, S^2, t)dS^1 \times \mathbf{F}^1(S^1, S^2, t)dS^2 \\
&\quad\quad + \underbrace{\mathbf{c}_{0,1}(S^1, S^2, t)dS^1 \times \mathbf{F}^1{}_{,1}(S^1, S^2, t)dS^2\, dS^1}_{\text{higher order}} - \cancel{\mathbf{c}_0(S^1, S^2, t) \times \mathbf{F}^1(S^1, S^2, t)dS^2} \\
&\quad = \mathbf{c}_0(S^1, S^2, t) \times \mathbf{F}^1{}_{,1}(S^1, S^2, t)dS^2\, dS^1 + \mathbf{c}_{0,1}(S^1, S^2, t)dS^1 \times \mathbf{F}^1(S^1, S^2, t)dS^2
\end{aligned} \tag{11.167}$$

Similarly, following the same arguments as in equation (11.167), we get:

$$\begin{aligned}&\mathbf{c}_0(S^1,S^2+dS^2,\ t)\times\ \mathbf{F}^2(S^1,S^2+dS^2,\ t)dS^1\ -\mathbf{c}_0(S^1,S^2,\ t)\times\ \mathbf{F}^2(S^1,S^2,\ t)dS^1\\&=\mathbf{c}_0(S^1,S^2,\ t)\times\ \mathbf{F}^2{}_{,2}(S^1,S^2,\ t)dS^1\,dS^2+\ \mathbf{c}_{0,2}(S^1,S^2,\ t)dS^1\times\ \mathbf{F}^2(S^1,S^2,\ t)dS^2\end{aligned} \tag{11.168}$$

By combining equations (11.167) and (11.168), we get for $\alpha = 1, 2$:

$$\begin{aligned}&\mathbf{c}_0(S^1+dS^1,S^2,\ t)\times\ \mathbf{F}^1(S^1+dS^1,S^2,\ t)dS^2\ -\mathbf{c}_0(S^1,S^2,\ t)\\&\quad\times\mathbf{F}^1(S^1,S^2,\ t)dS^2+\ \mathbf{c}_0(S^1,S^2+dS^2,\ t)\times\ \mathbf{F}^2(S^1,S^2+dS^2,\ t)dS^1\\&\quad-\mathbf{c}_0(S^1,S^2,\ t)\times\ \mathbf{F}^2(S^1,S^2,\ t)dS^1\\&=\left\{\mathbf{c}_0(S^1,S^2,\ t)\times\ \mathbf{F}^\alpha{}_{,\alpha}(S^1,S^2,\ t)+\ \mathbf{c}_{0,\alpha}(S^1,S^2,\ t)\times\ \mathbf{F}^\alpha(S^1,S^2,\ t)\right\}\,dS^1\,dS^2\end{aligned} \tag{11.169}$$

Next, we have:

$$\begin{aligned}&\mathbf{c}_0(S^1+\tfrac{1}{2}dS^1,S^2+\tfrac{1}{2}dS^2,\ t)\times\ \hat{\mathbf{F}}(S^1+\tfrac{1}{2}dS^1,S^2+\tfrac{1}{2}dS^2,\ t)dS^1\,dS^2\\&=\{\mathbf{c}_0(S^1,S^2,\ t)+\text{higher order}\}\times\ \{\hat{\mathbf{F}}(S^1,S^2,\ t)+\text{higher order}\}dS^1\,dS^2\\&=\mathbf{c}_0(S^1,S^2,\ t)\times\ \hat{\mathbf{F}}(S^1,S^2,\ t)\end{aligned} \tag{11.170}$$

Then, following the same arguments as in equation (11.170), we get:

$$\begin{aligned}&\mathbf{c}_0(S^1+\tfrac{1}{2}dS^1,S^2+\tfrac{1}{2}dS^2,\ t)\times\ \mathbf{l}(S^1+\tfrac{1}{2}dS^1,S^2+\tfrac{1}{2}dS^2,\ t)\ dS^1\,dS^2\\&=\{\mathbf{c}_0(S^1,S^2,\ t)+\text{higher order}\}\times\ \{\hat{\mathbf{F}}(S^1,S^2,\ t)+\text{higher order}\}\ dS^1\,dS^2\\&=\mathbf{c}_0(S^1,S^2,\ t)\times\ \hat{\mathbf{F}}(S^1,S^2,\ t)\ dS^1\,dS^2\end{aligned} \tag{11.171}$$

Now, using the relations in equations (11.166)–(11.171), we get from equation (11.165):

$$\sum\mathbf{M}=\mathbf{M}^\alpha{}_{,\alpha}\ +\ \mathbf{c}_{0,\alpha}\times\mathbf{F}^\alpha+\hat{\mathbf{M}}+\mathbf{c}_0\times\left\{\cancel{\mathbf{F}^\alpha{}_{,\alpha}+\hat{\mathbf{F}}-\dot{\mathbf{L}}}\right\}=\dot{\mathbf{c}}_0\times\mathbf{L}+\dot{\mathbf{A}} \tag{11.172}$$

where we have used the momentum balance equation (11.163) to strike out the terms in the second bracket and reused the notation for the time derivative: $\frac{\partial}{\partial t}(\bullet)\equiv\dot{(\bullet)}$. Thus, using equation (11.151), we finally get the angular momentum balance equations:

$$\boxed{\mathbf{M}^\alpha{}_{,\alpha}\ +\ \mathbf{c}_{0,\alpha}\ \times\ \mathbf{F}^\alpha\ +\ \hat{\mathbf{M}}\ =\dot{\mathbf{d}}\ \times\mathbf{L}+\dot{\mathbf{A}},\quad \alpha=1,2} \tag{11.173}$$

Remark

- If, for a thin shell, we define intuitively (will be derived in dynamic equation by reduction from 3D) the linear momentum, **L**, as:

$$\boxed{\mathbf{L}(S^1,S^2,t)=M_0(S^1,S^2)\ \dot{\mathbf{d}}(S^1,S^2,t)} \tag{11.174}$$

where $M_0(S^1, S^2)$ is the effective mass at the point, (S^1, S^2), measured per unit of the undeformed surface, then $\dot{\mathbf{d}} \times \mathbf{L} = \mathbf{0}$; thus the angular momentum balance equation (11.173) becomes:

$$\boxed{\mathbf{M}^\alpha,_\alpha + \mathbf{c}_0,_\alpha \times \mathbf{F}^\alpha + \hat{\mathbf{M}} = \dot{\mathbf{A}}, \quad \alpha = 1, 2} \tag{11.175}$$

11.5.4 Static Balance Equations

The quasi-static balance laws can be easily obtained from their dynamic counterparts by assuming that the linear and the angular momentum vectors identically vanish, that is,

$$\boxed{\begin{aligned} \mathbf{L}(S^1, S^2, t) &\equiv \mathbf{0} \\ \mathbf{A}(S^1, S^2, t) &\equiv \mathbf{0} \end{aligned}} \tag{11.176}$$

Thus, we get the following from equation (11.163).

11.5.4.1 Static Momentum Balance Equations (Force Equilibrium)

$$\boxed{\mathbf{F}^\alpha,_\alpha + \hat{\mathbf{F}} = \mathbf{0}, \quad \alpha = 1, 2} \tag{11.177}$$

and from equation (11.173):

11.5.4.2 Static Angular or Moment of Momentum Balance Equations (Moment Equilibrium)

$$\boxed{\mathbf{M}^\alpha,_\alpha + \mathbf{c}_0,_\alpha \times \mathbf{F}^\alpha + \hat{\mathbf{M}} = \mathbf{0}, \quad \alpha = 1, 2} \tag{11.178}$$

11.5.5 Vector Dynamic Balance Equations of Motion

For easy reference, here we copy and collect the derived dynamic balance equations:

$$\boxed{\begin{array}{lr} \mathbf{F}^\alpha,_\alpha + \hat{\mathbf{F}} = \dot{\mathbf{L}}, \quad \alpha = 1, 2 & \text{Momentum balance} \\ \mathbf{M}^\alpha,_\alpha + \mathbf{c}_0,_\alpha \times \mathbf{F}^\alpha + \hat{\mathbf{M}} = \dot{\mathbf{A}}, \quad \alpha = 1, 2 & \text{Moment of momentum balance} \end{array}} \tag{11.179}$$

Now, recalling that the partial derivatives and the covariant derivatives of a tensor of any order are the same, the above vector (tensor of order 1) equations can also be written as:

$$\boxed{\begin{array}{lr} \mathbf{F}^\alpha|_\alpha + \hat{\mathbf{F}} = \dot{\mathbf{L}}, \quad \alpha = 1, 2 & \text{Momentum balance} \\ \mathbf{M}^\alpha|_\alpha + \mathbf{c}_0,_\alpha \times \mathbf{F}^\alpha + \hat{\mathbf{M}} = \dot{\mathbf{A}}, \quad \alpha = 1, 2 & \text{Moment of momentum balance} \end{array}} \tag{11.180}$$

11.5.6 Vector Static Balance Equations of Motion

Likewise, here we copy and collect the derived static balance equations:

$$\boxed{\begin{array}{rll} \mathbf{F}^{\alpha},_{\alpha} + \hat{\mathbf{F}} = \mathbf{0}, & \alpha = 1,2 & \text{Momentum balance} \\ \mathbf{M}^{\alpha},_{\alpha} + \mathbf{c}_0,_{\alpha} \times \mathbf{F}^{\alpha} + \hat{\mathbf{M}} = \mathbf{0}, & \alpha = 1,2 & \text{Moment of momentum balance} \end{array}} \tag{11.181}$$

Now, recalling that the partial derivatives and the covariant derivatives of a tensor of any order are the same, the above vector (tensor of order 1) equations can also be written as:

$$\boxed{\begin{array}{rll} \mathbf{F}^{\alpha}|_{\alpha} + \hat{\mathbf{F}} = \mathbf{0}, & \alpha = 1,2 & \text{momentum balance} \\ \mathbf{M}^{\alpha}|_{\alpha} + \mathbf{c}_0,_{\alpha} \times \mathbf{F}^{\alpha} + \hat{\mathbf{M}} = \mathbf{0}, & \alpha = 1,2 & \text{moment of momentum balance} \end{array}} \tag{11.182}$$

Remarks

- Note that, as shown in equations (11.145) and (11.181), the vector shell balance equations of motions do not depend on the rotation parameters (which need to be introduced later) whether shear deformation is considered or not, although the displacement function is implicit in the second term of the moment of momentum balance equation.
- Finally, although we have already introduced in our analysis the force and moment components, and the momentum and the angular momentum, on a shell reference surface, we have deliberately avoided revealing their true character by connecting these to the kinematics and the stresses of a 3D shell-like body it really is supposed to represent. For a consistent treatment of this aspect of the analysis, we must deduce the balance equations from the basic balance laws of continuum mechanics which is, as we will show, detailed in the derivation of the balance equations by reduction from the 3D continuum mechanics counterparts to 2D shell balance equations.

11.5.7 Where We Would Like to Go

- We have directly developed the shell equations of motions by considering a shell as a 2D geometric object (i.e. a surface) with the forces and moments, and the linear and angular momentum functions. However, to validate the consistency of these equations of motion, and to gain insight into the true natures of these quantities, by establishing the connection between the shell equations and the more fundamental balance equations of 3D continuum mechanics, these equations will be derived by starting from and reducing the 3D equations of motion of solid mechanics. For this development, we must investigate:
  - shell balance equations: 3D to 2D.
- While the shell equations presented in equations (11.145) and (11.181) are very compact and thus useful for analysis, we will use the modified component or operational vector form that are most suitable for computational purposes where the components of displacements, rotations, forces and moments, and so on, need to be determined as the primary variables of interest. As prerequisites to such a task in the case of reduction of shell balance equations from 3D to 2D we should check out:
  - configuration space and kinematics

- component or operational vectors
- derivatives of component vectors.

  Then, with these background materials, we will be able to have the fully interpreted, final and computationally most effective balance equations:
- computational balance equations.

11.6 Weak Form: Kinematic and Configuration Space

11.6.1 What We Need to Recall

- The shell vector balance equations:

$$\boxed{\begin{array}{ll} \mathbf{F}^{\alpha},_{\alpha} + \hat{\mathbf{F}} = \dot{\mathbf{L}}, \quad \alpha = 1,2 & \text{momentum balance} \\ \mathbf{M}^{\alpha},_{\alpha} + \mathbf{c}_{0},_{\alpha} \times \mathbf{F}^{\alpha} + \hat{\mathbf{M}} = \dot{\mathbf{A}}, \quad \alpha = 1,2 & \text{moment of momentum balance} \end{array}} \tag{11.183}$$

The internal resistive stress resultant, $\mathbf{F}^{\alpha}$, is defined by:

$$\boxed{\mathbf{F}^{\alpha}(S,t) \equiv \int_{\eta^b}^{\eta^t} {}^{\eta}\mathbf{P}^{\alpha}(S,\eta,t)\ {}^{\eta}\mu^{S}(S,\eta,t)\ d\eta} \tag{11.184}$$

The internal resistive stress couple, $\mathbf{M}^{\alpha}$, is defined by:

$$\boxed{\mathbf{M}^{\alpha}(S,t) \equiv \int_{\eta^b}^{\eta^t} (S,\eta,t)\ \times\ {}^{\eta}\mathbf{P}^{\alpha}(S,\eta,t)\ {}^{\eta}\mu^{S}(S,\eta,t)\ d\eta} \tag{11.185}$$

The linear momentum, $\mathbf{L}(S^1,S^2,t)$, is defined by:

$$\boxed{\begin{aligned} \mathbf{L}(S^1,S^2,t) &\equiv \int_{\eta^b}^{\eta^t} \mathbf{v}(S^1,S^2,\eta,t)\,\rho_0(S^1,S^2,\eta)\,{}^{\eta}\mu(S^1,S^2,\eta)\,d\eta \\ &= \int_{\eta^b}^{\eta^t} \dot{\mathbf{c}}(S^1,S^2,\eta,t)\ \rho_0(S^1,S^2,\eta)\,{}^{\eta}\mu(S^1,S^2,\eta)\,d\eta \end{aligned}} \tag{11.186}$$

where $\mathbf{v}(S^1,S^2,\eta,t) \equiv \dot{\mathbf{c}}(S^1,S^2,\eta,t)$ is the velocity at current time $t \in \mathbb{R}_+$ at a point with undeformed coordinates (S^1,S^2,η) in the interior of the shell-like body.

The angular momentum, $\mathbf{A}(S^1,S^2,t)$, is defined by:

$$\boxed{\begin{aligned} \mathbf{A}(S^1,S^2,t) &\equiv \int_{\eta^b}^{\eta^t} \mathbf{c}(S^1,S^2,\eta,t) \times \mathbf{v}(S^1,S^2,\eta,t)\,\rho_0(S^1,S^2,\eta)\,{}^{\eta}\mu(S^1,S^2,\eta)\ d\eta \\ &= \int_{\eta^b}^{\eta^t} \mathbf{c}(S^1,S^2,\eta,t) \times \dot{\mathbf{c}}(S^1,S^2,\eta,t)\,\rho_0(S^1,S^2,\eta)\,{}^{\eta}\mu(S^1,S^2,\eta)\ d\eta \end{aligned}} \tag{11.187}$$

Definition: $\mathbf{F}_\nu$ as:

$$\boxed{\mathbf{F}_\nu \equiv {}^f\mathbf{P} \bullet \nu_0 \Leftarrow \mathbf{F}^\alpha \otimes \mathbf{G}_\alpha \bullet \nu_{0\beta}\, \mathbf{G}^\beta \Leftarrow \mathbf{F}^\alpha\, \nu_{0\alpha}} \tag{11.188}$$

Definition: $\mathbf{M}_\nu$ as:

$$\boxed{\mathbf{M}_\nu \equiv {}^m\mathbf{P} \bullet \nu_0 \Leftarrow \mathbf{M}^\alpha \otimes \mathbf{G}_\alpha \bullet \nu_{0\beta}\, \mathbf{G}^\beta \Leftarrow \mathbf{M}^\alpha\, \nu_{0\alpha}} \tag{11.189}$$

using the vector triple product property: $\mathbf{a} \times \mathbf{b} \bullet \mathbf{c} = \mathbf{b} \times \mathbf{c} \bullet \mathbf{a} = \mathbf{c} \times \mathbf{a} \bullet \mathbf{b},\ \forall \mathbf{a}, \mathbf{b}, \mathbf{c}$ vectors.

- In what follows, a barred quantity signifies that it is the variation of a corresponding real entity.
- The definition of a rotation tensor, its matrix representation, and the algebra, geometry and topology of rotation.

11.6.2 Weak Form of the Vector Balance Equations

Let us consider the shell reference surface, $\Omega^M \equiv M$, bounded by a smooth closed curve boundary, $\partial\Omega^M \equiv \partial M$. Let $(\mathbf{v}, \boldsymbol{\omega})$ be any sufficiently smooth vector fields defined on $M \times \mathbb{R}_+$. Now, following the standard model of the weak form generation, we multiply equation (11.183) appropriately by two vector fields: $\mathbf{v}$ and $\boldsymbol{\omega}$, and integrate over the arbitrary shell reference surface area, Ω, to get the shell weak form functional, $\hat{G}_{Dyn}$, given as:

$$\boxed{\begin{aligned}\hat{G}_{Dyn} &\equiv -\int_{\Omega^M} \left\{ \begin{aligned} &\left(\mathbf{F}^\alpha|_\alpha + \hat{\mathbf{F}} - \dot{\mathbf{L}}\right) \bullet \mathbf{v} + \\ &\left(\mathbf{M}^\alpha|_\alpha + \mathbf{c}_{0},_\alpha \times \mathbf{F}^\alpha + \hat{\mathbf{M}} - \left\{\dot{\mathbf{c}}_0 \times \mathbf{L} + \dot{\mathbf{A}}\right\}\right) \bullet \boldsymbol{\omega} \end{aligned} \right\} dA \\ &= \mathbf{0},\end{aligned}} \tag{11.190}$$

By an application of the Green–Gauss (or divergence) theorem for a surface: $\int_{\Omega^M} \mathbf{F}^\alpha|_\alpha\, dA = \int_{\Omega^M} DIV\,{}^f\mathbf{P}\, dA = \oint_{\partial\Omega^M} \mathbf{F}^\alpha \otimes \mathbf{G}_\alpha\, \nu_{0\beta}\, \mathbf{G}^\beta\, dS = \oint_{\partial\Omega^M} \mathbf{F}^\alpha\, \nu_{0\alpha}\, dS \equiv \oint_{\partial\Omega^M} \mathbf{F}_\nu\, dS$ where $\nu_0 = \nu_{0\alpha}\, \mathbf{G}^\alpha$ is the outward unit normal on the tangent plane on the boundary curve, Γ, to the undeformed shell reference surface area, Ω^M, and S is the arc length parameter along the boundary curve, Γ; $\mathbf{F}_\nu$ and $\mathbf{M}_\nu$ are given by equations (11.188) and (11.189), respectively.

Then, $\int_{\Omega^M} (\mathbf{F}^\alpha \bullet \mathbf{v})|_\alpha\, dA = \oint_{\partial\Omega^M} (\mathbf{F}^\alpha \bullet \mathbf{v})\, \nu_{0\alpha}\, dS = \oint_{\partial\Omega^M} (\mathbf{F}^\alpha\, \nu_{0\alpha} \bullet \mathbf{v})\, dS = \oint_{\partial\Omega^M} (\mathbf{F}_\nu \bullet \mathbf{v})\, dS$. Now, by applying the chain rule of covariant differentiation on the left-most term above and rearranging terms, we get:

$$-\int_{\Omega^M} (\mathbf{F}^\alpha|_\alpha \bullet \mathbf{v})\, dA = \int_{\Omega^M} (\mathbf{F}^\alpha \bullet \mathbf{v}|_\alpha)\, dA - \oint_{\partial\Omega^M} (\mathbf{F}_\nu \bullet \mathbf{v})\, dS \tag{11.191}$$

Similarly, for the moments, we have:

$$-\int_{\Omega^M} (\mathbf{M}^\alpha|_\alpha \bullet \boldsymbol{\omega})\, dA = \int_{\Omega^M} (\mathbf{M}^\alpha \bullet \boldsymbol{\omega}|_\alpha)\, dA - \oint_{\partial\Omega^M} (\mathbf{M}_\nu \bullet \boldsymbol{\omega})\, dS \tag{11.192}$$

Now, recalling the commutative property of a vector triple product, and applying equations (11.191) and (11.192) in equation (11.190), we get the weak form, $\hat{G}_{Dyn}$, as:

$$\hat{G}_{Dyn} \equiv \overbrace{\int_{\Omega^M} \{(\mathbf{F}^\alpha \bullet (\mathbf{v},_\alpha + \mathbf{c}_0,_\alpha \times \boldsymbol{\omega}) + \mathbf{M}^\alpha \bullet \boldsymbol{\omega},_\alpha\} \, dA}^{\text{Internal stored power}} + \overbrace{\int_{\Omega^M} \{\dot{\mathbf{L}} \bullet \mathbf{v} + (\dot{\mathbf{A}}) \bullet \boldsymbol{\omega}\} dA}^{\text{Internal power}} - \overbrace{\left\{ \oint_{\partial\Omega^M} (\hat{\mathbf{F}}_\nu \bullet \mathbf{v} + \hat{\mathbf{M}}_\nu \bullet \boldsymbol{\omega}) \, dS + \int_{\Omega^M} (\hat{\mathbf{F}} \bullet \mathbf{v} + \hat{\mathbf{M}} \bullet \boldsymbol{\omega}) \, dA \right\}}^{\text{External applied power}} = P_{Internal} + P_{Inertia} - P_{External} = \mathbf{0} \tag{11.193}$$

In the event that $\mathbf{v}$ and $\boldsymbol{\omega}$ are the actual linear velocity and the angular velocity of the shell reference surface, we can identify, in the integral given by equation (11.193), in the context of a mechanical energy conjugate relationship, the internal stored power (deformation or stress power), the inertial power and the external applied power as follows.

Definition: ***Internal deformation power***, $P_{Internal}$, as:

$$P_{Internal} \equiv \int_{\Omega^M} \{(\mathbf{F}^\alpha \bullet (\mathbf{v},_\alpha + \mathbf{c}_0,_\alpha \times \boldsymbol{\omega}) + \mathbf{M}^\alpha \bullet \boldsymbol{\omega},_\alpha\} \, dA \tag{11.194}$$

Inertial power, $P_{Inertia}$, as:

$$P_{Inertia} \equiv \int_{\Omega^M} \left\{ \dot{\mathbf{L}} \bullet \mathbf{v} + (\dot{\mathbf{A}}) \bullet \boldsymbol{\omega} \right\} dA \tag{11.195}$$

External applied power, $P_{External}$, as:

$$P_{External} \equiv \oint_{\partial\Omega^M} (\hat{\mathbf{F}}_\nu \bullet \mathbf{v} + \hat{\mathbf{M}}_\nu \bullet \boldsymbol{\omega}) \, dS + \int_{\Omega^M} (\hat{\mathbf{F}} \bullet \mathbf{v} + \hat{\mathbf{M}} \bullet \boldsymbol{\omega}) \, dA \tag{11.196}$$

11.6.3 *Kinematic Parameters*

The weak form of equation for the equivalent 2D shell – that is, the shell undeformed reference surface, given by equation (11.193) – holds whenever the strong form, given by equation (11.183), is true; in fact, the weak form is just as exact as the strong form.

11.6.3.1 New Moving or Spinning Coordinate System

Moreover, to be energy conjugate in the internal stored power, we may call $\mathbf{v},_{\alpha} + \mathbf{c}_0,_{\alpha} \times \boldsymbol{\omega}$, and $\boldsymbol{\omega},_{\alpha}$ for $\alpha = 1, 2$, in equation (11.193), (as the force strain rate vector and the moment strain rate vector, respectively) as the 2D strain rate fields over the shell reference surface as functions of the parameters, $(\mathbf{v}(S^1, S^2, t),\ \boldsymbol{\omega}(S^1, S^2, t))$, presuming that these functions exist necessarily as rates of some as yet undetermined 2D real strain fields over the reference surface as functions of some as yet undetermined real parameters. Naturally, then, this way of going about the problem will make sense only if we are able to determine those real parameters and 2D real strain fields. In other words, we have set our problem as one of determining the kinematics of a 2D shell reference surface. It will turn out, in the material coordinate system (i.e. in the event we describe the invariant force, $\mathbf{F}^{\alpha}$, and moment, $\mathbf{M}^{\alpha}$, vectors in terms of the undeformed basis vectors), $\{\mathbf{G}_i\}$, that the kinematic translational expression, $\mathbf{v},_{\alpha} + \mathbf{c}_0,_{\alpha} \times \boldsymbol{\omega}$, and rotational expression, $\boldsymbol{\omega},_{\alpha}$, for $\alpha = 1, 2$ (that comprise the internal stored power) cannot be the expressions for the real strain rates since there exist no functions whose rates could produce these expressions. However, It will be found later that the expressions $\mathbf{v},_{\alpha} + \mathbf{c}_0,_{\alpha} \times \boldsymbol{\omega}$, and $\boldsymbol{\omega},_{\alpha}$ for $\alpha = 1, 2$, in equation (11.193), are true rates only when these are interpreted in some, as yet unknown, spinning orthogonal basis set that will be called, in the case of static loading, the co-rotated bases, or in the case of dynamic loading, the co-spinning bases, defining, as shown in Figure 11.17, a new moving coordinate system; to be sure, we must understand that this new coordinate system is not what is commonly referred to as the convected coordinate system; more on this later. Now, the main point is that this, in turn, enables us to identify an additional kinematic variable in the form of a rotation tensor, $\mathbf{R}(S^1, S^2, t)$. Clearly, for the velocity vector, $\mathbf{v}$, we can choose the kinematic variables, $\mathbf{d}(S^1, S^2, t)$, as the displacement vector of the translational movement that belongs to the flat Euclidean space, that is, $\mathbf{v} \equiv \dot{\mathbf{d}}$. However, for the angular velocity vector, $\boldsymbol{\omega}$, it is not so obvious because of the very nature of the rotation tensor, $\mathbf{R}(S^1, S^2, t)$, that travels on a curved space, and thus we will have to wait until we consider the topic of the calculus of curved spaces.

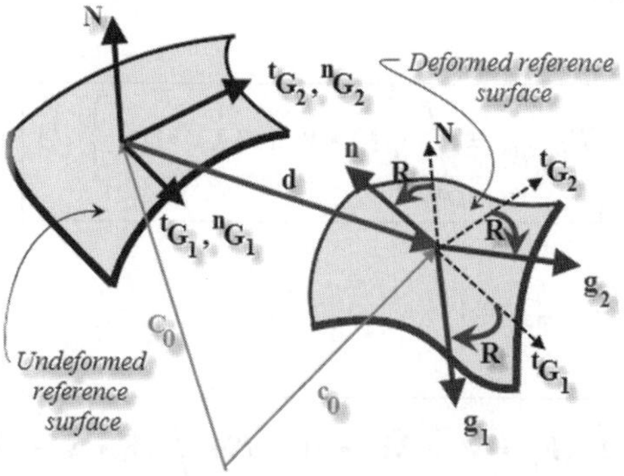

Figure 11.17 Shell Kinematics.

Thus, our configuration space consists of the history of an ordered pair $(\mathbf{d}(S^1, S^2, t), \mathbf{R}(S^1, S^2, t))$ with $\mathbf{d}(S^1, S^2, t)$, as the displacement vector, and $\mathbf{R}(S^1, S^2, t)$, as the rotation tensor describing the co-spinning bases, for all points of the 2D shell reference surface over all times of interest.

Of course, it does not mean that we cannot choose to express the components of the force and the moment vectors, and the linear and angular momentum vectors in material coordinate bases; in fact, we would do just that in our component or operational form of the virtual work equations. However, in that case, we have to rotate the internal work equations to express these again in the co-rotated (or co-spun) coordinate bases.

Remark

- Clearly, one more decision has been made that has overwhelming importance for the computational ease and accuracy. It is about the choice of independent kinematic variables. We can opt to settle, as has been done in the past, for the as yet unknown real translational and bending strain tensors as the independent kinematic variables for the numerical formulation of a subsequent finite element procedure. The major pitfall of this choice is that these real strain tensors must guarantee that a real deformed surface can be constructed with these tensors; in other words, these will then need to additionally satisfy the strain compatibility equations for the deformed reference surface, which is much like the integrability or the inverse problem for the metric and curvature tensors of constructing a 2D surface leading to the Gauss-Mainardi-Codazzi conditions. This can get too complicated, if not be impossible, for numerical comfort. Thus, from now on, our sought-after kinematic variables will be the real counterparts, $(\mathbf{d}(S^1, S^2, t), \mathbf{R}(S^1, S^2, t))$, of the real rate entities, $(\mathbf{v} \equiv \dot{\mathbf{d}},\ \boldsymbol{\omega})$. The real compatible strains and their rates, as we will see, can then be derived and computed from these parameters satisfying the strain compatibility a priori.

11.6.4 Configuration Space, C

Thus, given a shell problem with the shell having been defined by the equivalent and oriented shell reference surface, the shell kinematics is completely characterized by a displacement vector field, $\mathbf{d}(S^1, S^2, t)$, tracking each point on the reference surface, and an orthonormal rotational tensor field, $\mathbf{R}(S^1, S^2, t)$, that monitors rotations of the triad of base vectors (also referred to as the orientation defining directors) on the shell reference surface where $S^\alpha, \alpha = 1, 2$ are the arc length parameters along the lines of curvature of the undeformed shell reference surface for any time $t \in \mathbb{R}_+$. Keeping these assumptions in mind, we can ascertain the location of any point on the shell at any time during its deformation history with these two fields. Thus, we may define:

- the shell configuration space as the space defined by $(\mathbf{d}(S^1, S^2, t),\ \mathbf{R}(S^1, S^2, t))$, $(S^1, S^2) \in \Omega^M$, $t \in \mathbb{R}_+$.

In other words, a configuration space tells us all about the possible deformations that a shell can undergo, and thus consists of all ordered pairs of $(\mathbf{d}(S^1, S^2, t),\ \mathbf{R}(S^1, S^2, t))$ for all points on the shell reference surface for the entire deformation history; a particular history (induced by a specific set of loading system) of motion of a shell reference surface traces out a unique curve on the configuration space. Also, assuming adequate smoothness properties on the fields defining the configuration space, we will also be able to establish later the corresponding virtual space that defines the admissible variations, $(\bar{\mathbf{d}}, \bar{\mathbf{R}})$, of the configuration space that we are allowed to consider in our subsequent computational virtual work principle application for our c-type finite element method. We just note that because of the inclusion of the rotational tensor field, $\mathbf{R}(S^1, S^2, t)$ (algebraically, a member of $SO(3)$, the 3D special orthogonal group), the configuration space is not a linear space, but rather a geometrically curved space known to mathematicians as the Riemannian (we assume $\mathbf{R}(S^1, S^2, t)$ as smooth as necessary) differentiable manifold. As we shall see later, this has major implications in our treatment of the calculus (the derivatives and the variations) of the rotation tensor, and for the linearization of the virtual work equations for our intended computational scheme. The dimension of the configuration space comprises three displacement components, three independent rotational components (because there are six constraints of orthonormality of the rotational tensor) for every point on the shell for all times of interest – a huge dimension.

11.6.5 Kinematic Parameterization: Displacements and Rotations

Note that because of the choice of orthonormality (i.e. $\mathbf{R}^T \mathbf{R} = \mathbf{I}$ where $\mathbf{I}$ is the identity tensor) for $\mathbf{R}(S^1, S^2, t)$, only three degrees of freedom for rotations can be chosen for parameterization. As discussed in the kinematics of large rotation in Chapter 10, for representation of the rotational motions, many possibilities exists for parameterization of finite rotation $\mathbf{R}(S^1, S^2, t)$ such as Rodrigues parameters, Euler Rodrigues parameters, Hamilton–Rodrigues quaternions and Euler angles; more on this later. Now, by accepting the shell configuration space made up of the kinematic variables $(\mathbf{d}, \mathbf{R})$, we have the immediate task of defining the real strains in term of these real variables, and then relating these to the real strain rates through the rate entities $(\mathbf{v} = \dot{\mathbf{d}},\ \boldsymbol{\omega})$ defining the weak form as given by the equation (11.193).

11.6.6 Where We Would Like to Go

As discussed, in order that we can describe the presumed virtual strains as the variations of some real strain fields, we must take up the next step of seeking the real strain fields, the variations of which will guarantee the virtual nature of the energy conjugate strain expressions in the internal virtual work similar to that given by equation (11.194). However, before we can get on with it, we need to equip ourselves with adequate mathematical preliminaries relating to the definitions for the spatial derivatives (i.e. curvatures), temporal derivatives (i.e. rates) and variations. Thus, we must next visit:

- curvatures, angular velocity and variations
- real strains and strain rates

11.7 Admissible Virtual Space: Curvature, Velocity and Variation

11.7.1 What We Need to Recall

- The vector triple product property: $\mathbf{a} \times \mathbf{b} \bullet \mathbf{c} = \mathbf{b} \times \mathbf{c} \bullet \mathbf{a} = \mathbf{c} \times \mathbf{a} \bullet \mathbf{b},\ \forall \mathbf{a}, \mathbf{b}, \mathbf{c}$ vectors.
- The vector triple cross product property: $\mathbf{a} \times \mathbf{b} \times \mathbf{c} = (\mathbf{a} \bullet \mathbf{c})\mathbf{b} - (\mathbf{a} \bullet \mathbf{b})\mathbf{c},\ \forall \mathbf{a}, \mathbf{b}, \mathbf{c}$ vectors.
- The definition of a rotation tensor, its matrix representation, and algebra, geometry and topology of rotation.
- A rotation matrix, $\mathbf{R}$, corresponding to the three-dimensional special orthogonal matrix group, $SO(3)$, may be parameterized by the Rodrigues vector, $\mathbf{s}$, the axis of the instantaneous rotation.

- The Gateaux derivatives on a vector space are:

$$\bar{\mathbf{v}}(\mathbf{x},\bar{\mathbf{x}}) = \lim_{\rho\to 0}\frac{d}{d\rho}\{\mathbf{v}(\mathbf{x}+\rho\bar{\mathbf{x}})\} = \lim_{\rho\to 0}\frac{\{\mathbf{v}(\mathbf{x}+\rho\bar{\mathbf{x}})-\mathbf{v}(\mathbf{x})\}}{\rho} \tag{11.197}$$

  where $\bar{\mathbf{v}}(\mathbf{x},\bar{\mathbf{x}})$ is the one-parameter variation of a vector or tensor field, say $\mathbf{v}(\mathbf{x})$, where $\mathbf{x}$ and $\bar{\mathbf{x}}$ belong to a linear vector space; in other words, $\bar{\mathbf{v}}(\mathbf{x},\bar{\mathbf{x}})$ is the directional derivative of $\mathbf{v}(\mathbf{x})$ at $\mathbf{x}$ along the vector direction $\bar{\mathbf{x}}$.
- The covariant derivatives on a curved space are:

$$\bar{\mathbf{v}}\left(\mathbf{x}\equiv\frac{d}{d\xi},\bar{\mathbf{x}}\equiv\frac{d}{d\lambda}\right) = \frac{d\mathbf{x}}{d\lambda} = \lim_{\rho\to 0}\frac{\{\mathbf{v}(\lambda_0+\rho)_{\text{parallel transported to }\lambda_0}-\mathbf{v}(\lambda_0)\}}{\rho} \tag{11.198}$$

  where $\bar{\mathbf{v}}(\mathbf{x},\bar{\mathbf{x}})$ is the covariant derivative of $\mathbf{v}(\mathbf{x})$ at $\mathbf{x}$ along the vector direction $\bar{\mathbf{x}}$. The covariant derivative is the generalization of the Gateaux directional derivative of the vector spaces to the curved spaces where $\mathbf{v}(\mathbf{x}+\rho\bar{\mathbf{x}})$ and $\mathbf{v}(\mathbf{x})$ of the definition (11.197) do not belong to the same vector (tangent) space, and hence the difference operation do not have a meaning. In fact, the vector itself does not have a meaning other than as a tangent operation to a curve. Thus, and for invariance, the generalization is effected through the choice of affinely parameterized unique curves, but this makes the definition of the covariant derivative path-dependent. However, as a practical scheme for computing the covariant derivatives of vector fields (or tensor fields), all we need to do is to include the derivatives of the base vectors (or base tensors) as they are the tangents (or tensor product of these) to the coordinate curves on the manifold.
- For notational uniformity, we denote all variations of entities by using a superscripted bar. Furthermore all entities measured in the $\mathbf{G_1}, \mathbf{G_2}, \mathbf{N}$ frame are superscripted as ${}^{t}(\bullet)$, and those in the rotating $\mathbf{g_1}, \mathbf{g_2}, \mathbf{g_3}$ coordinate system as ${}^{n}(\bullet)$. However, for the sake of uncluttered readability, sometimes the superscripts may be dropped, but the context should help determine the appropriate underlying coordinate system.

11.7.2 Curvature and Angular Velocity Fields: the Admissible Virtual Space

In the development of the shell virtual work principle necessary for numerical analysis by a variational method such as the c-type finite element method, we will need the derivatives and the variations of the parameters of the configuration space. The configuration space parameters are the continuously differentiable displacement vector field, $\mathbf{d}(S^1, S^2, t)$, and the rotation tensor field, $\mathbf{R}(S^1, S^2, t)$, of the shell reference surface, where S^1, S^2 are the arc length parameters along the lines of curvature of the shell reference surface, and $t \in \mathbb{R}_+$ is the time parameter. For dynamic motion, we will need both the temporal and the spatial derivatives, and the variations of these fields; for quasi-static motion, the fields are independent of time, that is, $\mathbf{d} = \mathbf{d}(S^1, S^2)$ and $\mathbf{R} = \mathbf{R}(S^1, S^2)$, in which case we will need the variations and spatial derivatives of the fields for the subsequent real and the virtual strain measures, and so on.

11.7.2.1 Real Displacement Vector Field and Admissible Virtual Space

The real displacement vector field, $\mathbf{d}(S^1, S^2, t)$, describing the translational movement of the points on a arbitrary configuration of the surface, for all times of interest, belongs to the 3D Euclidean space, E_3. Thus, we can construct the displacement portion of the virtual space as the vector space containing the perturbed or the infinitesimal displacement vector field that is "superposed" on the real displacement vector field to obtain the perturbed or virtual configuration; it will be identified as $\bar{\mathbf{d}}(S^1, S^2, t) \equiv \delta\mathbf{d}(S^1, S^2, t)$; the method of superposition is clearly defined by the vector addition operation.

11.7.2.2 Real Rotation Tensor Field, Manifold Lie Group, *SO*(3) and Admissible Virtual Space

The real rotation field, describing the independent rotation of the triad of base vectors of a surface belongs algebraically to the special orthogonal group, $SO(3)$, and geometrically to a curved space. Thus, the admissible virtual space for rotation is considerably complicated compared to its displacement counterpart described in the previous section. We will need to define the infinitesimal (later, it will also be the instantaneous) rotations that can serve as the required virtual rotations which can then be superposed on the real rotations to describe the perturbed or virtual configurations. This, clearly, needs identification of the space to which these infinitesimal rotations belong to, and additionally, the method of superposition. In other words, in a Euclidean ('flat') space, we have identified, before for the displacement vector field, superposition simply as the vector addition operation, the question for us to resolve is if we can do the same for the rotation tensor field. It turns out that with appropriate parameterization for the rotation field, the answer is yes, although certain singularity issues must first be tackled. Moreover, we will need, for future use, the curvature and angular velocity fields, that is, the spatial and temporal derivatives of the rotation field. Next, we try to precisely meet these goals. Now, recall that the $SO(3)$ group is the set of all 3×3 real proper orthogonal matrices, that is, $\mathbf{R} \in SO(3) \Rightarrow \mathbf{R}^T\mathbf{R} = \mathbf{I}$, with $\mathbf{I}$ as the identity matrix and rotation matrix has positive unit determinant, that is, $\det \mathbf{R} = +1$. The four properties of a group: the existence of the identity, the existence of the inverse element, the associative property of the product and closure, are all satisfied by the rotation matrices under the product rule given by the usual matrix multiplication. The identity element: $\mathbf{I}$ with $\det \mathbf{I} = 1$, the inverse element: $\mathbf{R}^{-1} = \mathbf{R}^T$ for each rotation $\mathbf{R}$ belonging to the group, and the associative property of the matrices, satisfy the first three conditions. Now for closure: with $\mathbf{R}, \mathbf{S} \in SO(3)$, two proper orthogonal matrices, we have $\mathbf{R}\,\mathbf{S}$ also orthonormal because $(\mathbf{R}\,\mathbf{S})^T\,(\mathbf{R}\,\mathbf{S}) = \mathbf{S}^T(\mathbf{R}^T\mathbf{R})\,\mathbf{S} = \mathbf{I}$, and because: $\det(\mathbf{R}\,\mathbf{S}) = \det\,\mathbf{R}\,\det\,\mathbf{S} = 1$; in other words, $\mathbf{R} \in SO(3),\ \mathbf{S} \in SO(3) \Rightarrow \mathbf{R}\,\mathbf{S} \in SO(3)$. Also, a set of invertible tensor field is a continuous manifold. Since a group which is also a continuous manifold is called a Lie group, $SO(3)$ is a Lie group.

11.7.3 Tangent Operator and Exponential Map

Let us recall that on a curved space, geometrical objects such as vectors, tensors, and so on, must be defined locally instead of bi-local definition in the neighborhood as in the flat Euclidean space; in other words, in a Euclidean ('flat') space, we can think of, say, a vector as joined by two neighboring points, but in a curved space a similar definition will yield a vector that may not even completely belong to the space. This necessitated the generalization from the definition of a vector as an arrow link between two neighboring points to a tangent operation to a curve (to be more specific, an affine curve for invariance or coordinate independence) at a point on the manifold as shown next.

11.7.3.1 Spatial Tangent Operator

Let the one spatial parameter family of the rotation tensor (or matrix) function $\mathbf{R}(\rho) :\rightarrow SO(3)$ generate a continuous set of points defining a curve on $SO(3)$, such that it passes through $\mathbf{R}(0) = \mathbf{R}$

at $\rho = 0$ where $\mathbf{R} \in SO(3)$. Then, $\mathbf{R},_\rho$, defined by the covariant derivative: $\mathbf{R},_\rho \equiv \frac{d}{d\rho}\mathbf{R}(\rho)|_{\rho=0}$, is a tangent vector to $SO(3)$ at $\mathbf{R}$. The set of all tangent vectors at $\mathbf{R}$, corresponding to various smooth curves passing though it forms a vector space, denoted by $\mathbf{T_R}SO(3)$. Now, by taking the derivative of the orthogonality property, $\mathbf{R}^T\,\mathbf{R} = \mathbf{I}$, and rearranging, we have:

$$\boxed{\begin{aligned}&\mathbf{R},_\rho = \mathbf{K}\,\mathbf{R},\\ &\mathbf{K} \equiv \mathbf{R},_\rho\mathbf{R}^T = \mathbf{R},_\rho\mathbf{R}^{-1}\end{aligned}} \tag{11.199}$$

where we have introduced the following.

Definition

- The tangent (linearized) operator, $\mathbf{K}$, the skew-symmetric matrix (because, with $\mathbf{R}\mathbf{R}^T = \mathbf{I}$, we have: $\mathbf{K} + \mathbf{K}^T = \mathbf{R},_\rho\mathbf{R}^T + \mathbf{R}\,\mathbf{R},_\rho^T = (\mathbf{R}\mathbf{R}^T),_\rho = \mathbf{0}$); $\mathbf{K}$ is the instantaneous and linearized skew-symmetric tensor at a rotation, $\mathbf{R} \in SO(3)$.

Now, if we consider a constant tangent operator, $\mathbf{K}$, that is, independent of the parameter ρ, we have from equation (11.199), $\mathbf{R}(\rho),_\rho = \mathbf{K}\,\mathbf{R}(\rho)$, which has a solution given by:

$$\boxed{\mathbf{R}(\rho) = e^{\rho\mathbf{K}} \equiv \left(\mathbf{I} + \rho\mathbf{K} + \frac{(\rho\mathbf{K})^2}{2!} + \ldots\right)} \tag{11.200}$$

Thus each element of the tangent space is related to a corresponding element of the curve space by an exponential map given by equation (11.200). In other words, there always exists a skew-symmetric matrix that can represent a rotation matrix by this exponential map. More specifically, for a rotation matrix, $\mathbf{R} \in SO(3)$, there exists a linearized skew-symmetric matrix, $\mathbf{K} \in so(3)$ related by:

$$\mathbf{R} = e^{\mathbf{K}} \tag{11.201}$$

11.7.3.2 Temporal Tangent Operator

Similarly, let the one temporal (time) parameter family of the rotation tensor (or matrix) function $\mathbf{R}(t) :\rightarrow SO(3)$ generate a continuous set of points defining a curve on $SO(3)$, such that it passes through $\mathbf{R}(0) = \mathbf{R}$ at $t = 0$ where $\mathbf{R} \in SO(3)$. Then, $\dot{\mathbf{R}} \equiv \frac{d}{dt}\mathbf{R}$, defined by the covariant derivative: $\dot{\mathbf{R}} \equiv \frac{d\mathbf{R}(t)}{dt}|_{t=0}$, is the tangent vector to $SO(3)$ at $\mathbf{R}$. The set of all tangent vectors at $\mathbf{R}$, corresponding to various smooth curves passing though it forms a vector space, denoted by $\mathbf{T_R}SO(3)$. Now, by taking the temporal derivative of the orthogonality property, $\mathbf{R}^T\,\mathbf{R} = \mathbf{I}$, and rearranging, we have:

$$\boxed{\begin{aligned}&\dot{\mathbf{R}} = \omega\,\mathbf{R},\\ &\omega \equiv \dot{\mathbf{R}}\,\mathbf{R}^T = \dot{\mathbf{R}}\,\mathbf{R}^{-1}\end{aligned}} \tag{11.202}$$

where we have introduced the following.

Definition

- The tangent (linearized) operator, $\mathbf{\Omega}$, the skew-symmetric matrix (because, with $\mathbf{R}\mathbf{R}^T = \mathbf{I}$, we have: $\mathbf{\Omega} + \mathbf{\Omega}^T = \dot{\mathbf{R}}\mathbf{R}^T + \mathbf{R}\,\dot{\mathbf{R}}^T = \frac{d}{dt}(\mathbf{R}\mathbf{R}^T) = \mathbf{0}$); $\mathbf{\Omega}$ is the instantaneous and linearized skew-symmetric tensor at a rotation, $\mathbf{R} \in SO(3)$.

Because the tangent space is populated with the skew-symmetric matrices, the dimension of the tangent space is 3 (with only $n(n-1)/2$ elements in the matrix with $n = 3$ for the 3D space). Note also that the tangent space is a linear vector space.

Now, if we consider a constant tangent operator, $\mathbf{\Omega}$, that is, independent of the parameter t, we have from equation (11.202), $\dot{\mathbf{R}}(t) = \mathbf{\Omega}\,\mathbf{R}(t)$, which has a solution given by:

$$\boxed{\mathbf{R}(t) = e^{t\mathbf{\Omega}} \equiv \left(\mathbf{I} + t\mathbf{\Omega} + \frac{(t\mathbf{\Omega})^2}{2!} + \ldots\right)} \tag{11.203}$$

Thus, each element of the tangent space is related to a corresponding element of the curved space by an exponential map given by equation (11.203). In other words, there always exists a skew-symmetric matrix that can represent a rotation matrix by this exponential map. More specifically, for a rotation matrix, $\mathbf{R} \in SO(3)$, there exists a linearized skew-symmetric matrix, $\mathbf{\Omega} \in so(3)$ related by:

$$\mathbf{R} = e^{\mathbf{\Omega}} \tag{11.204}$$

Remark

- Although equations (11.201) and (11.204) relate the rotation matrix with the corresponding skew-symmetric matrix of the tangent space through the exponential map, there is, as derived before, a better way of relating the two by expressing the exponential map by Rodrigues vector or Rodrigues quaternion parameters of the rotation matrix through Cayley's theorem.

11.7.3.3 Tangent Space at Identity

The tangent space at identity, that is, at $\mathbf{R} = \mathbf{I}$, is given the special name, Lie algebra (recall that a vector space with a vector multiplication, $[\bullet, \bullet]$, defined on it is an algebra) and denoted by $so(3)$. The reason for this special treatment of the neighborhood at identity, $\mathbf{I}$, is that the local structure of $SO(3)$, the continuous rotation group, can be constructed out of this infinitesimal tangent space, $so(3)$, by the exponential map given by equation (11.203). In other words, all the information of the group is contained in the algebra. Thus, $so(3)$ consists of 3×3 skew-symmetric matrices; as any 3×3 skew-symmetric matrix has a corresponding 3×1 axial vector, $so(3)$ is isomorphic to the 3D Euclidean vector space. More generally, a Lie algebra is a vector space, $\mathcal{V}$, with a bilinear map, $[\bullet, \bullet] : \mathcal{V} \times \mathcal{V} \rightarrow \mathcal{V}$, known as the Lie bracket, so that for every $a, b, c \in \mathcal{V} \Rightarrow$, 1) $[a\ ,\ a] = 0$, 2) $[a, [b, c]]\ +\ [b, [c, a]]\ +\ [c, [a, b]] = 0$ (the Jacobi identity). From condition (1) and the bilinearity of the map, $[a + b, a + b] = [a, b] + [b, a] = 0$ implying $[a, b] = -[b, a]$. For square matrix Lie group and Lie algebra, the Lie bracket reduces to the matrix commutator: $[\mathbf{A}, \mathbf{B}] = \mathbf{AB} - \mathbf{BA}, \quad \forall \mathbf{A}, \mathbf{B} \in so(3)$. With the notation, $[\mathbf{a}] \in so(3)$ as the skew-symmetric matrix corresponding to the axial vector, $\mathbf{a}$, we can easily show that $\mathbf{AB} - \mathbf{BA} = [\mathbf{a} \times \mathbf{b}]$.

11.7.4 Admissible Virtual Space

Based on our discussion about the tangent space at identity, *so*(3), the admissible virtual space is defined by this linear space of the configuration space made up of the union set of all the possible configurations of the surface defined by the ordered pair $(\mathbf{d}, \mathbf{R})$, that is, the displacement vector field and the rotation tensor field, respectively. In other words, every ordered pair, $(\bar{\mathbf{d}} \equiv \delta\mathbf{d}, \bar{\mathbf{R}} \equiv \delta\mathbf{R})$ of the virtual of displacement vector, $\bar{\mathbf{d}}$, and the infinitesimal, skew-symmetric rotation tensor, $\bar{\mathbf{R}}$, belongs to this linear space. This, then, takes care of the part of our original questions about how to define admissible virtual space. But, we still have not determined how to superpose the virtual rotations to the current real rotation to obtain the perturbed or virtual configuration space necessary for the subsequent virtual work principle. As it turns out, there are two possible ways to accomplish this and we have to conveniently choose one; this aspect of the study we will discuss next.

11.7.5 Adjoint Map and Material and Spatial Variations

We have already seen through equation (11.203) that the elements of the Lie group, *SO*(3), is related to the corresponding elements of its Lie algebra, *so*(3), by the exponential map. Another map known as the adjoint map also relates the elements of *SO*(3) with the corresponding elements of *so*(3). If an arbitrary rotation matrix, $\mathbf{R} \in SO(3)$, and its corresponding skew-symmetric matrix, $\bar{\mathbf{\Psi}} \in so(3)$, are given, then, we have $\bar{\mathbf{\Phi}}$ defined by the adjoint map: $\bar{\mathbf{\Phi}} \equiv \mathbf{R}\,\bar{\mathbf{\Psi}}\,\mathbf{R}^T$, also an element of *so*(3). We can easily see that because $(\bar{\mathbf{\Phi}} + \bar{\mathbf{\Phi}}^T = \mathbf{R}\,\bar{\mathbf{\Psi}}\,\mathbf{R}^T + \mathbf{R}\,\bar{\mathbf{\Psi}}^T\,\mathbf{R}^T == \mathbf{R}\,\bar{\mathbf{\Psi}}\,\mathbf{R}^T - \mathbf{R}\,\bar{\mathbf{\Psi}}\,\mathbf{R}^T = \mathbf{0})$ the symmetric part vanishes identically, it is a member of *so*(3). Thus, we have: $\bar{\mathbf{\Phi}}\,\mathbf{R} = \mathbf{R}\,\bar{\mathbf{\Psi}}$, and using similar to equation (11.199), we have two ways to superpose to represent the virtual rotation, $\bar{\mathbf{R}}$, by:

$$\bar{\mathbf{R}} = \bar{\mathbf{\Phi}}\,\mathbf{R} = \mathbf{R}\,\bar{\mathbf{\Psi}} \tag{11.205}$$

In the first equality of equation (11.205), known as the left translation, the virtual rotation, $\bar{\mathbf{\Phi}}$, is being superposed on the current rotation to obtain the updated rotation; we use this relationship in our subsequent analysis, but the components are measured in the undeformed and fixed material frame, $\mathbf{G}_1, \mathbf{G}_2, \mathbf{G}_3 \equiv \mathbf{N}$. In the second equality of equation (11.205), known as the right translation, the current rotation is being superposed on the virtual rotation, $\bar{\mathbf{\Psi}}$, to obtain the updated rotation. We do not use this relationship further in our subsequent analysis and hence will not discuss it any further.

Finally, then, the finite rotation, $\mathbf{R}$, belongs to the spherical curved space, *SO*(3), and the instantaneous and infinitesimal virtual rotation, $\bar{\mathbf{\Phi}}$, represented by a skew-symmetric matrix, belongs to the tangent space at identity, the Lie algebra.

11.7.6 Rotational Variation: $\bar{\mathbf{\Phi}}$ and $\bar{\boldsymbol{\varphi}}$

As indicated above, we choose the following form for rotational variation in all subsequent applications:

$$\boxed{\bar{\mathbf{R}} = \bar{\mathbf{\Phi}}\,\mathbf{R} \Leftrightarrow \bar{\mathbf{\Phi}} \equiv \bar{\mathbf{R}}\,\mathbf{R}^T} \tag{11.206}$$

For the 3D rotation tensors, the corresponding skew-symmetric tensors only have three significant variables (i.e. the number of variables equal the dimension of the space), and thus can have axial vector representation. Let $\bar{\boldsymbol{\varphi}} \equiv \{ \bar{\varphi}_1 \bar{\varphi}_2 \bar{\varphi}_3 \ \}^T$ be such axial vector corresponding to the skew-symmetric tensor, $\bar{\boldsymbol{\Phi}}$, with $\bar{\boldsymbol{\Phi}}\ \bar{\boldsymbol{\varphi}} = \mathbf{0}$. Knowing that $\bar{\boldsymbol{\Phi}} = \begin{bmatrix} 0 & -\bar{\varphi}_3 & \bar{\varphi}_2 \\ \bar{\varphi}_3 & 0 & -\bar{\varphi}_1 \\ -\bar{\varphi}_2 & \bar{\varphi}_1 & 0 \end{bmatrix}$, the equality can be readily verified.

11.7.7 Rotational Spatial Derivatives – Curvatures: K *and* k

Next, we can define the curvature tensor in two different ways as: $\mathbf{K}_\alpha \equiv \mathbf{R},_\alpha\ \mathbf{R}^T$ or $\mathbf{L}_\alpha \equiv\ \mathbf{R}^T\ \mathbf{R},_\alpha$, and thus, the spatial derivatives of variables of the reference surface can have two representations; we choose the following spatial form in all subsequent applications:

$$\boxed{\mathbf{R},_\alpha = \mathbf{K}_\alpha\ \mathbf{R} \Leftrightarrow \mathbf{K}_\alpha \equiv \mathbf{R},_\alpha\ \mathbf{R}^T} \tag{11.207}$$

As before, let $\mathbf{k}$ be the axial vector corresponding to the skew-symmetric tensor, $\mathbf{K}$, with $\mathbf{K}\,\mathbf{k} = \mathbf{0}$.

11.7.8 Rotational Time Derivative – Angular Velocity: Ω *and* ω

Similarly, we can define the angular velocity tensor in two different ways as: $\boldsymbol{\Omega} \equiv \dot{\mathbf{R}}\ \mathbf{R}^T$ or $\Xi \equiv\ \mathbf{R}^T\ \dot{\mathbf{R}}$, where $\dot{(\bullet)} \equiv \frac{d}{dt}(\bullet)$, and thus, the time derivatives of variables of a Cosserat surface can have two representations. We choose the following form in all subsequent applications:

$$\dot{\mathbf{R}} = \boldsymbol{\Omega}\ \mathbf{R} \Leftrightarrow \boldsymbol{\Omega} = \dot{\mathbf{R}}\ \mathbf{R}^T \tag{11.208}$$

As before, let $\boldsymbol{\Omega}$ be the axial vector corresponding to the skew-symmetric tensor, $\boldsymbol{\Omega}$, with $\boldsymbol{\Omega}\ \boldsymbol{\omega} = \mathbf{0}$.

11.7.9 Coordinate Frame Rotation, Objective Variations and Derivatives

For future application, let us suppose an initial orthonormal coordinate frame, $\{\mathbf{G}_i(S^\alpha)\}$, $i = 1, 2, 3$, attached at a material point S^α, $\alpha = 1, 2$ to a material surface in space, and let us also suppose that in time t, a current coordinate system, $\{\mathbf{g}_i(S^\alpha, t)\}$, $i = 1, 2, 3$, is obtained from the initial coordinate frame by pure rotation such that $\mathbf{g}_i = \mathbf{R}\ \mathbf{G}_i$, $i = 1, 2, 3$ where $\mathbf{R}(S^\alpha, t)$ is the rotation tensor. Now, let $\mathbf{x}(S^\alpha, t)$ be any vector field attached to the body at point S^α, $\alpha = 1, 2$. If an observer is sitting on the body such that it also co-spins as above, we would be interested in how it sees the variations and time rate of change in $\mathbf{x}(S^\alpha, t)$; we will call this the ***objective variations***, and ***objective time rate***, respectively, of vector fields. However, before we delve into the derivations of the objective rate and note the following.

11.7.9.1 Variation of Current Base Vectors: $\{\mathbf{g}_i(S^\alpha, t)\}$, $i = 1, 2, 3$

From equation (11.206), the variation of $\mathbf{g}_i$ is given as:

$$\overline{\mathbf{g}_i} = \bar{\mathbf{R}}\,\mathbf{G}_i = \bar{\boldsymbol{\Phi}}\,\mathbf{R}\,\mathbf{G}_i = \bar{\boldsymbol{\Phi}}\;\mathbf{g}_i = \bar{\boldsymbol{\varphi}} \times \mathbf{g}_i \tag{11.209}$$

11.7.9.2 Spatial Derivative of Current Base Vectors: $\left\{\mathbf{g}_i(S^\alpha, t)\right\}$, $i = 1, 2, 3$

From equation (11.207), the spatial derivative of $\mathbf{g}_i$ is given as:

$$\mathbf{g}_{i},_{\alpha} = \mathbf{R},_{\alpha}\,\mathbf{G}_i = \mathbf{K}_\alpha\,\mathbf{R}\,\mathbf{G}_i = \mathbf{K}_\alpha\;\mathbf{g}_i = \mathbf{k}_\alpha \times \mathbf{g}_i \tag{11.210}$$

11.7.9.3 Time Derivative of Current Base Vectors: $\{\mathbf{g}_i(S^\alpha, t)\}$, $i = 1, 2, 3$

From equation (11.126), the time derivatives of $\mathbf{g}_i$ is given as:

$$\dot{\mathbf{g}}_i = \dot{\mathbf{R}}\;\mathbf{G}_i = \boldsymbol{\Omega}\,\mathbf{R}\,\mathbf{G}_i = \boldsymbol{\Omega}\;\mathbf{g}_i = \boldsymbol{\omega} \times \mathbf{g}_i \tag{11.211}$$

11.7.9.4 Objective Variation, $\check{(\bullet)}$

Now, let us also observe that for any vector field $\mathbf{x}$ described in the current rotated coordinate system, that is, $\mathbf{x} = x^i\,\mathbf{g}_i$, we have, using equation (11.209), for variation in spatial coordinate system:

$$\underbrace{\bar{\mathbf{x}}}_{\text{Total variation}} = \bar{x}^i\,\mathbf{g}_i + x^i\,\bar{\mathbf{g}}_i = \bar{x}^i\,\mathbf{g}_i + \bar{\boldsymbol{\varphi}} \times x^i\,\mathbf{g}_i = \bar{x}^i\,\mathbf{g}_i + \bar{\boldsymbol{\varphi}} \times \mathbf{x}$$

$$\Rightarrow \check{\mathbf{x}} \equiv \underbrace{\bar{x}^i\,\mathbf{g}_i}_{\substack{\text{Variation in rotated}\\ \text{coordinate system, i.e.}\\ \text{objective variation}}} = \underbrace{\bar{\mathbf{x}}}_{\text{Total variation}} - \underbrace{\bar{\boldsymbol{\varphi}} \times \mathbf{x}}_{\text{Spinning variation}} = \mathbf{R}\{\mathbf{R}^T\bar{\mathbf{x}} + (-\mathbf{R}^T\bar{\boldsymbol{\Phi}})\mathbf{x}\}$$

$$= \mathbf{R}\{\mathbf{R}^T\bar{\mathbf{x}} + \bar{\mathbf{R}}^T\,\mathbf{x}\} = \underbrace{\mathbf{R}}_{\text{Push forward}}\left(\underbrace{\overline{\underbrace{\mathbf{R}^T}_{\text{Pull back}}\;\mathbf{x}}}_{\text{Variation}}\right) \tag{11.212}$$

From equation (11.212), we now introduce what is known as the objective variation or the co-rotated variation.

Definition: ***Objective variation***, $\check{(\bullet)}$:

$$\boxed{\check{(\bullet)} \equiv \overline{(\bullet)} - \bar{\boldsymbol{\varphi}} \times (\bullet)} \tag{11.213}$$

11.7.9.5 Objective Time Rate, $\overset{t}{(\bullet)}$

Similarly, using equation (11.209), for the time derivative in the current rotated coordinate system, we have:

$$\underbrace{\dot{\mathbf{x}}}_{\text{Total derivative}} = \dot{x}^i\,\mathbf{g}_i + x^i\,\dot{\mathbf{g}}_i = \dot{x}^i\,\mathbf{g}_i + \boldsymbol{\omega} \times x^i\,\mathbf{g}_i = \dot{x}^i\,\mathbf{g}_i + \boldsymbol{\omega} \times \mathbf{x}$$

$$\Rightarrow \overset{t}{\mathbf{x}} \equiv \underbrace{\dot{x}^i\,\mathbf{g}_i}_{\substack{\text{Derivative in rotated}\\ \text{coordinate system,i.e.}\\ \text{objective derivative}}} = \underbrace{\dot{\mathbf{x}}}_{\text{Total derivative}} - \underbrace{\boldsymbol{\omega} \times \mathbf{x}}_{\substack{\text{Spinning observer}\\ \text{correction}}} = \mathbf{R}\{\mathbf{R}^T\dot{\mathbf{x}} + (-\mathbf{R}^T\boldsymbol{\omega})\mathbf{x}\}$$

$$= \mathbf{R}\{\mathbf{R}^T\dot{\mathbf{x}} + \dot{\mathbf{R}}^T\,\mathbf{x}\} = \underbrace{\mathbf{R}}_{\text{Push forward}}\left(\underbrace{\overbrace{(\underbrace{\mathbf{R}^T}_{\text{Pull back}}\ \mathbf{x})}^{\bullet}}_{\text{Derivative}}\right) \tag{11.214}$$

From equation (11.214), we now introduce what is known as the objective time rate, or the co-rotated time rate.

Definition: ***Objective time rate***, $\overset{t}{(\bullet)}$:

$$\boxed{\overset{t}{(\bullet)} \equiv \frac{\partial}{\partial t}(\bullet) = (\bullet) - \boldsymbol{\omega} \times (\bullet)} \tag{11.215}$$

Remarks

- The idea behind the objective variation (or objective time derivative) in the current rotated body attached coordinate system is that it is the net variation (or time derivative) in the co-rotated coordinate system, that is, the total (material) variation minus the variation of the spinning coordinate system itself; in other words, it is the variation (or time derivative) observed by an observer attached to the spinning coordinate system given by the base vectors, $\{\mathbf{g}_i\}$.
- The objective variation (or time derivative) may also be seen as the ordered sequence of operations as the "pulling back" to the material coordinate system, taking the variation, and finally "pushing forward" to the current coordinate system, as in the definition of the Lie derivative.
- Note, however, that, as will be seen in our subsequent computational efforts, the components of all involved vector and tensor quantities are measured in terms of the undeformed coordinate system.

11.7.9.6 Relations between $\bar{\mathbf{\Phi}}$ and $\bar{\varphi}$ and K and k

From equations (11.206) and (11.207), and noting that $\mathbf{R}\,\mathbf{R}^T = \mathbf{I}$, $\mathbf{I}$ being the 3D identity tensor, we have for the variation of the curvature:

$$\begin{aligned}\bar{\mathbf{\Phi}}_{,\alpha} &= (\bar{\mathbf{R}}\,\mathbf{R}^T)_{,\alpha} = \bar{\mathbf{R}}_{,\alpha}\,\mathbf{R}^T + \bar{\mathbf{R}}\,\mathbf{R}^T{}_{,\alpha} = \overline{\mathbf{K}_\alpha}\,\mathbf{R}\,\mathbf{R}^T + \bar{\mathbf{\Phi}} R\,(\mathbf{R}^T\,\mathbf{K}_\alpha^T)\\ &= \overline{\mathbf{K}_\alpha}\,\mathbf{R}\,\mathbf{R}^T + \mathbf{K}_\alpha\,\bar{\mathbf{\Phi}}\,\mathbf{R}\,\mathbf{R}^T - \bar{\mathbf{\Phi}}\,\mathbf{R}\,\mathbf{R}^T\,\mathbf{K}_\alpha = \overline{\mathbf{K}_\alpha} + \underbrace{\mathbf{K}_\alpha\,\bar{\mathbf{\Phi}} - \bar{\mathbf{\Phi}}\,\mathbf{K}_\alpha}_{\text{Lie bracket}}\end{aligned} \tag{11.216}$$

implying, with the application of the definition of objective variation from equation (11.213) to equation (11.216), and the definition of the axial vectors and the Lie brackets, that:

$$\begin{aligned}\bar{\varphi}_{,\alpha} &= \overline{\mathbf{k}_\alpha} - \bar{\varphi} \times \mathbf{k}_\alpha = \mathbf{R}\{\mathbf{R}^T\overline{\mathbf{k}_\alpha} + (-\mathbf{R}^T\bar{\mathbf{\Phi}})\mathbf{k}_\alpha\}\\ &= \mathbf{R}\{\mathbf{R}^T\overline{\mathbf{k}_\alpha} + \bar{\mathbf{R}}^T\,\mathbf{k}_\alpha\} = \underbrace{\underbrace{\mathbf{R}}_{\text{Push forward}}\,\overline{\underbrace{\mathbf{R}^T}_{\text{Pull back}}\,\mathbf{k}_\alpha}}_{\text{Variation}} \equiv \overset{\mathrm{v}}{\overline{\mathbf{k}_\alpha}}\end{aligned} \tag{11.217}$$

11.7.9.7 Relations between Ω and ω and K and k

From equations (11.207) and (11.126), and noting that $\mathbf{R}\,\mathbf{R}^T = \mathbf{I}$, $\mathbf{I}$ being the 3D identity tensor, we have for the spatial derivative of the angular velocity:

$$\begin{aligned}\mathbf{\Omega}_{,\alpha} &= (\dot{\mathbf{R}}\,\mathbf{R}^T)_{,\alpha} = \dot{\mathbf{R}}_{,\alpha}\,\mathbf{R}^T + \dot{\mathbf{R}}\,\mathbf{R}^T{}_{,\alpha} = \widehat{\dot{\mathbf{K}}_\alpha\,\mathbf{R}}\;\mathbf{R}^T + \mathbf{\Omega}\,\mathbf{R}\,(\mathbf{R}^T\,\mathbf{K}_\alpha^T)\\ &= \dot{\mathbf{K}}_\alpha\,\mathbf{R}\,\mathbf{R}^T + \mathbf{K}_\alpha\,\mathbf{\Omega}\,\mathbf{R}\,\mathbf{R}^T - \mathbf{\Omega}\,\mathbf{R}\,\mathbf{R}^T\,\mathbf{K}_\alpha\\ &= \dot{\mathbf{K}}_\alpha + \underbrace{\mathbf{K}_\alpha\,\mathbf{\Omega} - \mathbf{\Omega}\,\mathbf{K}_\alpha}_{\textit{Lie Bracket}}\end{aligned} \tag{11.218}$$

implying, with the application of the definition of objective time derivative from equation (11.215) to equation (11.218), and the definition of the axial vectors and the Lie brackets, that:

$$\begin{aligned}\omega_{,\alpha} &= \dot{\mathbf{k}}_\alpha - \omega \times \mathbf{k}_\alpha = \mathbf{R}\{\mathbf{R}^T\,\dot{\mathbf{k}}_\alpha + (-\dot{\mathbf{R}}^T\omega)\,\mathbf{k}_\alpha\}\\ &= \mathbf{R}\{\mathbf{R}^T\,\dot{\mathbf{k}}_\alpha + \dot{\mathbf{R}}^T\,\mathbf{k}_\alpha\} = \underbrace{\underbrace{\mathbf{R}}_{\text{Push forward}}\,\overset{\bullet}{\overbrace{\underbrace{\mathbf{R}^T}_{\text{Pull back}}\,\mathbf{k}_\alpha}}}_{\text{Variation}} \equiv \overset{\mathrm{t}}{\mathbf{k}_\alpha}\end{aligned} \tag{11.219}$$

Taking variation of $\omega,_\alpha$ in equation (11.219), and observing that both variation and derivative are linear operations (and hence the order of operation can be switched), we have:

$$\boxed{\bar{\omega},_\alpha = \overline{\overset{t}{\mathbf{k}}}_\alpha} \tag{11.220}$$

11.7.9.8 Relations Between $\boldsymbol{\Omega}$ and $\boldsymbol{\omega}$ *and* $\bar{\boldsymbol{\Phi}}$ and $\bar{\boldsymbol{\varphi}}$

From equations (11.206) and (11.126), and noting that $\mathbf{R}\,\mathbf{R}^T = \mathbf{I}$, $\mathbf{I}$ being the 3D identity tensor, we have for the variation of the angular velocity:

$$\begin{aligned}
\bar{\Omega} &= \overline{\left(\dot{\mathbf{R}}\;\mathbf{R}^T\right)} = \dot{\bar{\mathbf{R}}}\;\mathbf{R}^T + \dot{\mathbf{R}}\;\bar{\mathbf{R}}^T = \overset{\bullet}{\overline{\bar{\Phi}\,\mathbf{R}}}\;\mathbf{R}^T + \Omega\,\mathbf{R}\,(\mathbf{R}^T\,\bar{\Phi}^T) \\
&= \dot{\bar{\Phi}}\;\mathbf{R}\,\mathbf{R}^T + \bar{\Phi}\;\dot{\mathbf{R}}\;\mathbf{R}^T - \Omega\,\mathbf{R}\,\mathbf{R}^T\,\bar{\Phi} = \dot{\bar{\Phi}} + \underbrace{\bar{\Phi}\,\Omega - \Omega\,\bar{\Phi}}_{\text{Lie bracket}}
\end{aligned} \tag{11.221}$$

implying, with the application of the definition of objective time derivative from equation (11.215) to equation (11.221), and the definition of the axial vectors and the Lie brackets, that:

$$\boxed{\begin{aligned}
\bar{\omega} &= \dot{\bar{\varphi}} - \omega \times \bar{\varphi} = \mathbf{R}\left\{\mathbf{R}^T\,\dot{\bar{\varphi}} + \left(-\mathbf{R}^T\Omega\right)\bar{\Phi}\right\} \\
&= \mathbf{R}\left\{\mathbf{R}^T\;\dot{\bar{\varphi}} + \dot{\mathbf{R}}^T\;\bar{\varphi}\right\} = \underbrace{\mathbf{R}}_{\text{Push forward}}\underbrace{\left(\overset{\bullet}{\overbrace{(\underbrace{\mathbf{R}^T}_{\text{Pull back}}\;\bar{\varphi})}}\right)}_{\text{Derivative}} \equiv \overset{t}{\bar{\varphi}}
\end{aligned}} \tag{11.222}$$

Taking spatial derivative of $\bar{\omega}$ in equation (11.222), and observing that both variation and derivative are linear operations (and hence the order of operation can be switched), we have:

$$\boxed{\bar{\omega},_\alpha = \overset{t}{\bar{\varphi}},_\alpha} \tag{11.223}$$

Thus, finally, from equations (11.220) and (11.223), we have:

$$\boxed{\bar{\omega},_\alpha = \overset{t}{\bar{\varphi}},_\alpha = \overline{\overset{t}{\mathbf{k}}}_\alpha} \tag{11.224}$$

11.7.10 Where We Would Like to Go

Here, we have introduced various parameters and derived all the necessary relationships anticipating the next step of identifying the spatial real strain and the real strain rate fields for the shell reference surface. Clearly, then, we should proceed by embarking on:

- spatial real strains and rates.
  Finally, to get familiarized with and be conscious about the overall computational strategy for nonlinear shells, we should explore:
- the road map: nonlinear shells

11.8 Real Strain and Strain Rates from Weak Form

11.8.1 What We Need to Recall

- The shell internal stored power, $P_{Internal}$, as:

$$P_{Internal} \equiv \int_{\Omega} \{ (\mathbf{F}^{\alpha} \bullet (\mathbf{v},_{\alpha} + \mathbf{c}_0,_{\alpha} \times \boldsymbol{\omega}) + \mathbf{M}^{\alpha} \bullet \boldsymbol{\omega},_{\alpha} \} \, dA \tag{11.225}$$

The resistive stress resultant, $\mathbf{F}^{\alpha}$, is defined by:

$$\boxed{\mathbf{F}^{\alpha}(S,t) \equiv \int_{\eta^b}^{\eta^t} {}^{\eta}\mathbf{P}^{\alpha} \; {}^{\eta}\mu^S \, d\eta} \tag{11.226}$$

The resistive stress couple, $\mathbf{M}^{\alpha}$, is defined by:

$$\boxed{\mathbf{M}^{\alpha}(S,t) \equiv \int_{\eta^b}^{\eta^t} (\mathbf{c}(S,\eta,t) - \mathbf{c}_0(S,t)) \times \; {}^{\eta}\mathbf{P}^{\alpha} \; {}^{\eta}\mu^S \, d\eta} \tag{11.227}$$

where (S, η) defines the through-the-thickness deformation of the shell-like body.

$(\mathbf{v} \equiv \dot{\mathbf{d}}, \boldsymbol{\omega})$ are the weak parameters,

- The shell configuration space is completely identified by the history of the kinematic parameters, $(\mathbf{d}, \mathbf{R})$
- The derivatives of the initial base vector set, $\{\mathbf{G}_i(S^1, S^2)\}$, $\quad i = 1, 2, 3$, are given by:

$$\frac{d}{dS^{\alpha}} \mathbf{G}_i(S^1, S^2) \equiv \mathbf{G}_i(S^1, S^2),_{\alpha} = \mathbf{K}^0_{\alpha} \; \mathbf{G}_i, \quad \alpha = 1, 2 \tag{11.228}$$

The temporal derivatives of the current base vector set, $\{\mathbf{g}_i(S^1, S^2, t)\}$, $\quad i = 1, 2, 3$, are given by:

$$\dot{\mathbf{g}}_i = \boldsymbol{\omega} \times \mathbf{g}_i \tag{11.229}$$

- For any vector field $\mathbf{x}$ described in the rotated coordinate system, that is, $\mathbf{x} = x^i \, \mathbf{g}_i$, we have for the variation in the coordinate system:

$$\overset{\mathrm{v}}{\mathbf{x}} \equiv \underbrace{\bar{x}^i \, \mathbf{g}_i}_{\substack{\text{Variation in rotated} \\ \text{coordinate system, i.e.} \\ \text{objective variation}}} = \underbrace{\bar{\mathbf{x}}}_{\text{Total variation}} - \underbrace{\bar{\boldsymbol{\varphi}} \times \mathbf{x}}_{\text{Spinning variation}} \tag{11.230}$$

$\overset{\mathrm{v}}{\mathbf{x}}$ is the objective variation of $\mathbf{x}$, defined by: $\boxed{\overset{\mathrm{v}}{(\bullet)} \equiv \overline{(\bullet)} - \bar{\boldsymbol{\varphi}} \times (\bullet)}$.

- For any vector field $\mathbf{x}$ described in the rotated (coordinate system, i.e. $\mathbf{x} = x^i\,\mathbf{g}_i$, we have for the rate in this coordinate system:

$$\overset{\mathbf{t}}{\mathbf{x}} \equiv \underbrace{\dot{x}^i\;\mathbf{g}_i}_{\substack{\text{Derivative in rotated}\\ \text{coordinate system, i.e.}\\ \text{objective derivative}}} = \underbrace{\dot{\mathbf{x}}}_{\text{Total derivative}} - \underbrace{\boldsymbol{\omega}\times\mathbf{x}}_{\substack{\text{Spinning observer}\\ \text{correction}}} \tag{11.231}$$

- $\overset{\mathbf{t}}{\mathbf{x}}$ is the objective time rate of $\mathbf{x}$, defined by: $\boxed{\overset{\mathbf{t}}{(\bullet)} \equiv \frac{\partial}{\partial t}(\bullet) = \dot{(\bullet)} - \boldsymbol{\omega}\times(\bullet)}$.
- The relationship between the derivative of the angular velocity tensor, $\boldsymbol{\omega}$, and the curvature tensor, $\mathbf{K}_\alpha$:

$$\boldsymbol{\Omega},_\alpha = \dot{\mathbf{K}}_\alpha + \underbrace{\mathbf{K}_\alpha\,\boldsymbol{\Omega} - \boldsymbol{\Omega}\,\mathbf{K}_\alpha}_{\text{Lie bracket}} \Leftrightarrow \boldsymbol{\omega},_\alpha = \dot{\mathbf{k}}_\alpha - \boldsymbol{\omega}\times\mathbf{k}_\alpha \tag{11.232}$$

- The vector triple product property: $\mathbf{a}\times\mathbf{b}\bullet\mathbf{c} = \mathbf{b}\times\mathbf{c}\bullet\mathbf{a} = \mathbf{c}\times\mathbf{a}\bullet\mathbf{b}$, $\forall\mathbf{a},\mathbf{b},\mathbf{c}$ vectors.

11.8.2 Real Strains and Strain Rates in the Spinning Coordinate System

Let us note that $\mathbf{g}_i(S^1,S^2,t) = \mathbf{R}(S^1,S^2,t)\;\mathbf{G}_i(S^1,S^2)$, where $\{\mathbf{g}_i\}$, and $\{\mathbf{G}_i\}$ are the base vector sets for the spinning coordinate system of the current deformed shell surface, and the undeformed coordinate shell reference surface, respectively. Now, with the relation given in equation (11.228), we can obtain for the derivatives of $\mathbf{g}_i$ as:

$$\begin{aligned}\mathbf{g}_i,_\alpha &= \mathbf{R},_\alpha\,\mathbf{G}_i + \mathbf{R}\,\mathbf{G}_i,_\alpha = \mathbf{K}_\alpha\,\mathbf{R}\,\mathbf{G}_i + \mathbf{R}\,\mathbf{K}^0_\alpha\,\mathbf{G}_i\\ &= \mathbf{K}_\alpha\,\mathbf{g}_i + \mathbf{R}\,\mathbf{K}^0_\alpha\,\mathbf{R}^T\;\mathbf{R}\mathbf{G}_i \equiv (\mathbf{K}_\alpha + \mathbf{R}\,\mathbf{K}^0_\alpha\,\mathbf{R}^T)\,\mathbf{g}_i\\ &\equiv \mathbf{K}^{tot}_\alpha\,\mathbf{g}_i = \mathbf{k}^{tot}_\alpha\times\mathbf{g}_i\end{aligned} \tag{11.233}$$

where we introduce the following.

Definitions: The ***total curvature tensor***, $\mathbf{K}^{tot}_\alpha$, and the corresponding axial vector, $\mathbf{k}^{tot}_\alpha$, as:

$$\boxed{\mathbf{K}^{tot}_\alpha \equiv \mathbf{K}_\alpha + \mathbf{R}\,\mathbf{K}^0_\alpha\,\mathbf{R}^T \Rightarrow \mathbf{k}^{tot}_\alpha \equiv \mathbf{k}_\alpha + \mathbf{R}\,\mathbf{k}^0_\alpha} \tag{11.234}$$

The result is that, for shells with initial curvature, the curvature must always be understood as the sum of one due to the current rotation, that is, $\mathbf{K}_\alpha \equiv \mathbf{R},_\alpha\,\mathbf{R}^T$, and another due to the initial curvature, that is, $\mathbf{k}^0_\alpha$, of the undeformed shell reference surface with $\mathbf{K}^{tot}_\alpha$ and $\mathbf{k}^{tot}_\alpha$ being the total curvature tensor and the total curvature vector, respectively.

11.8.2.1 Force Real Strain Field

Note that with $\mathbf{c}_0(S^1,S^2,t),_\alpha = \mathbf{G}_\alpha(S^1,S^2) + \mathbf{d}(S^1,S^2,t),_\alpha$, we have as time rate: $\overset{\bullet}{\widehat{\mathbf{c}_0,_\alpha}} = \dot{\mathbf{d}},_\alpha = \mathbf{v},_\alpha$, and applying equation (11.229): $\dot{\mathbf{d}},_\alpha - \boldsymbol{\omega}\times\mathbf{g}_\alpha = \overset{\bullet}{\widehat{\mathbf{c}_0,_\alpha}} - \dot{\mathbf{g}}_\alpha = \overset{\bullet}{\widehat{\mathbf{c}_0,_\alpha - \mathbf{g}_\alpha}}$. Now, with the results in equation (11.231) applied to the presumed force strain rate expression in the internal stored power as given in equation (11.225), we get:

$$\begin{aligned}
&\mathbf{v},_{\alpha} + \mathbf{c}_{0},_{\alpha} \times \boldsymbol{\omega} \\
&= \dot{\mathbf{d}},_{\alpha} - \boldsymbol{\omega} \times \mathbf{g}_{\alpha} - \boldsymbol{\omega} \times \mathbf{c}_{0},_{\alpha} + \boldsymbol{\omega} \times \mathbf{g}_{\alpha} = \overbrace{(\mathbf{c}_{0},_{\alpha} - \mathbf{g}_{\alpha})}^{\bullet} - \boldsymbol{\omega} \times (\mathbf{c}_{0},_{\alpha} - \mathbf{g}_{\alpha}) \equiv \overbrace{(\mathbf{c}_{0},_{\alpha} - \mathbf{g}_{\alpha})}^{t} \\
&= \mathbf{R}\left\{\mathbf{R}^{T}\overbrace{(\mathbf{c}_{0},_{\alpha} - \mathbf{g}_{\alpha})}^{\bullet} + \left(-\mathbf{R}^{T}\boldsymbol{\Omega}\right)(\mathbf{c}_{0},_{\alpha} - \mathbf{g}_{\alpha})\right\} = \mathbf{R}\left\{\mathbf{R}^{T}\overbrace{(\mathbf{c}_{0},_{\alpha} - \mathbf{g}_{\alpha})}^{\bullet} + \dot{\mathbf{R}}^{T}(\mathbf{c}_{0},_{\alpha} - \mathbf{g}_{\alpha})\right\} \\
&= \underbrace{\mathbf{R}}_{\text{Push forward}} \left(\underbrace{\overbrace{\underbrace{\mathbf{R}^{T}}_{\text{Pull back}}(\mathbf{c}_{0},_{\alpha} - \mathbf{g}_{\alpha})}^{\bullet}}_{\text{Time rate}}\right) = \overbrace{{}^{R}\boldsymbol{\beta}_{\alpha}}^{t}
\end{aligned}$$

(11.235)

where we have introduced the following.

Definition

- The ***rotated force real strain***, ${}^{R}\boldsymbol{\beta}_{\alpha}$:

$$\boxed{{}^{R}\boldsymbol{\beta}_{\alpha} \equiv \mathbf{c}_{0},_{\alpha} - \mathbf{g}_{\alpha}} \tag{11.236}$$

Thus, we conclude that the expression $\mathbf{v},_{\alpha} + \mathbf{c}_{0},_{\alpha} \times \boldsymbol{\omega}$, in equation (11.225), truly represents the rotated force strain rate as it can be obtained from the time derivative of a function: the rotated force real strain, ${}^{R}\boldsymbol{\beta}_{\alpha}$ defined by equation (11.236); it embodies translational strains: the real extensional strain and the real shear strain, along the rotated coordinate system with the base vector set given by $\{\mathbf{g}_i\}$.

Note that in the undeformed reference surface: as $\mathbf{d} \to \mathbf{0}$ and $\mathbf{R} \to \mathbf{I}$, the force real strain, ${}^{R}\boldsymbol{\beta}_{\alpha} \to \mathbf{0}$ as it should be.

11.8.2.2 Moment Real Strain Field

By definition, the rotational velocity tensor, $\boldsymbol{\Omega} \equiv \dot{\mathbf{R}}\,\mathbf{R}^{T}$, the curvature, $\mathbf{K}_{\alpha} \equiv \mathbf{R},_{\alpha}\,\mathbf{R}^{T}$, the initial curvature of the undeformed shell reference surface, $\mathbf{K}_{\alpha}^{0}$ with $\mathbf{G}_{i},_{\alpha} = \mathbf{K}_{\alpha}^{0}\,\mathbf{G}_{i}$, and finally, recall the property: for any skew-symmetric tensor $\hat{\mathbf{K}}_{\alpha}^{0}$ with the corresponding axial vector, $\hat{\mathbf{k}}_{\alpha}^{0}$, defined by: $\hat{\mathbf{K}}_{\alpha}^{0} = \mathbf{R}\,\mathbf{K}_{\alpha}^{0}\,\mathbf{R}^{T} \Rightarrow \hat{\mathbf{k}}_{\alpha}^{0} = \mathbf{R}\,\mathbf{k}_{\alpha}^{0}$, where $\mathbf{k}_{\alpha}^{0}$ is the axial vector corresponding to the skew-symmetric tensor $\mathbf{K}_\alpha^0$. Now, with $\mathbf{K}_\alpha^{tot} \equiv \mathbf{K}_\alpha + \mathbf{R}\,\mathbf{K}_\alpha^0\,\mathbf{R}^T$ as the total curvature tensor with $\mathbf{k}_\alpha^{tot} = \mathbf{k}_\alpha + \mathbf{R}\,\mathbf{k}_\alpha^0$ as the corresponding total axial curvature vector, we have:

$$\begin{aligned}\dot{\mathbf{K}}_\alpha &= \dot{\mathbf{K}}_\alpha^{tot} - \dot{\mathbf{R}}\,\mathbf{K}_\alpha^0\,\mathbf{R}^T - \mathbf{R}\,\mathbf{K}_\alpha^0\,\dot{\mathbf{R}}^T = \dot{\mathbf{K}}_\alpha^{tot} - \boldsymbol{\Omega}\,\mathbf{R}\,\mathbf{K}_\alpha^0\,\mathbf{R}^T + \mathbf{R}\,\mathbf{K}_\alpha^0\,\mathbf{R}^T\,\boldsymbol{\Omega}\\ &= \dot{\mathbf{K}}_\alpha^{tot} + \underbrace{\left(\mathbf{R}\,\mathbf{K}_\alpha^0\,\mathbf{R}^T\right)\boldsymbol{\Omega} - \boldsymbol{\Omega}\left(\mathbf{R}\,\mathbf{K}_\alpha^0\,\mathbf{R}^T\right)}_{\text{Lie bracket}}\end{aligned} \tag{11.237}$$

implying in vector form:

$$\dot{\mathbf{k}}_\alpha = \dot{\mathbf{k}}_\alpha^{tot} + (\mathbf{R}\,\mathbf{k}_\alpha^0) \times \boldsymbol{\omega} = \dot{\mathbf{k}}_\alpha^{tot} - \boldsymbol{\omega} \times (\mathbf{R}\,\mathbf{k}_\alpha^0) \tag{11.238}$$

Now, from the second of equation (11.232) and relation in equation (11.238), we have:

$$\begin{aligned}\boldsymbol{\omega},_\alpha &= \dot{\mathbf{k}}_\alpha - \boldsymbol{\omega} \times \mathbf{k}_\alpha \equiv \overbrace{\mathbf{k}_\alpha}^{\mathbf{t}}\\ &= \mathbf{R}\left\{\mathbf{R}^T\,\dot{\mathbf{k}}_\alpha + \left(-\mathbf{R}^T\boldsymbol{\omega}\right)\mathbf{k}_\alpha\right\} = \mathbf{R}\left\{\mathbf{R}^T\,\dot{\mathbf{k}}_\alpha + \dot{\mathbf{R}}^T\,\mathbf{k}_\alpha\right\}\\ &= \underbrace{\mathbf{R}}_{\text{Push forward}}\left(\underbrace{\overbrace{\underbrace{\mathbf{R}^T}_{\text{Pull back}}\ \mathbf{k}_\alpha}^{\cdot}}_{\text{Time derivative}}\right) = \overbrace{{}^R\chi_\alpha}^{\mathbf{t}}\end{aligned} \tag{11.239}$$

where we have introduced:

Definition

- The ***rotated moment real strain***:

$$\boxed{{}^R\chi_\alpha \equiv \mathbf{k}_\alpha = \mathbf{k}_\alpha^{tot} - \mathbf{R}\,\mathbf{k}_\alpha^0, \quad \mathbf{K}_\alpha \equiv \mathbf{R},_\alpha\,\mathbf{R}^T} \tag{11.240}$$

Thus, we conclude that the expression $\boldsymbol{\omega},_\alpha$ truly represents the rotated moment strain rate as obtained from the time derivative of a function: the rotated moment real strain, ${}^R\chi_\alpha$; it embodies rotational strain: the real bending strain, along the rotated coordinate system with the base vector set given by $\{\mathbf{g}_i\}$.

Note that in the undeformed reference surface: $\mathbf{R} \to \mathbf{I}$, the moment real strain, ${}^{R}\chi_{\alpha} \to \mathbf{0}$ as it should. Thus, by inserting results from equations (11.235) and (11.239) into the internal stored power expression of equation (11.225), it can be rewritten as:

$$P_{Internal} \equiv \int_{\Omega} \left\{ (\mathbf{F}^{\alpha} \bullet \overbrace{{}^{R}\beta_{\alpha}}^{\mathrm{t}} + \mathbf{M}^{\alpha} \bullet \overbrace{{}^{R}\chi_{\alpha}}^{\mathrm{t}} \right\} dA \tag{11.241}$$

Remark

- For our subsequent computational virtual work principle necessary for a c-type finite element formulation, we will develop the rotating coordinate system by first expressing the invariant force and moment vectors in the material coordinate system and then applying the rotation tensor to the components. This allows us to develop and track the most accurate numerical scheme with a cost-effective explicit identification of the relevant entities. Thus, for computational ease and accuracy, all the expressions pertaining to the weak form, the virtual strains, and so on, must be modified to the most suitable operational or component vector form, which we intend to do next.

However, before we get into the computational description of the shell behavior, we will make a short detour to present an alternate way to develop the weak form, the stress resultants and the stress couple and the associated conjugate strains for historical connection. We have developed the exact weak form by starting from the exact balance laws of motion. In this alternative way of thinking, we instead starts by making some kinematic assumptions; then, the stress or deformation power (in the case of static analysis, the internal strain energy variations) provides the conjugate real strains and real strain rates upon suitable definition of the stress resultants and the stress couples. So, let us look into it with a kinematic assumption that is appropriate and commonly used for thin shells, as follows.

11.8.3 Alternative Derivation: Kinematics and Stress or Deformation Power

A simple 2D surface can never represent completely the behavior of a 3D shell-like body; an additional mechanism must be embedded in the 2D surface to capture, in some representative fashion, the through-the-thickness response of the 3D shell-like body. It turns out that the 2D kinematics that are energetically consistent with the 3D shell-like body is best captured by what is known as the Cosserat surface, which is an oriented surface, that is, a 2D surface with a triad of orthonormal vector fields, $\{\mathbf{g}_i(S^1, S^2)\}$, $i = 1, 2, 3$, known as the directors, acting as the base vectors of a coordinate system and identifying the deformed shell reference surface. This coordinate system will be termed as the co-rotating or spinning system, the meaning of which will be clear shortly. We choose these directors at the reference configuration to be coincident with the base vectors, $\{\mathbf{G}_i(S^1, S^2)\}$, $i = 1, 2, 3$, of the shell reference surface where $\mathbf{G}_1$, $\mathbf{G}_2$ are the unit tangent base vectors with S^{α}, $\alpha = 1, 2$ as the arc length parameters along the lines of curvature, and $\mathbf{G}_3 \equiv \mathbf{N}$ is the unit normal to the shell reference surface as introduced earlier.

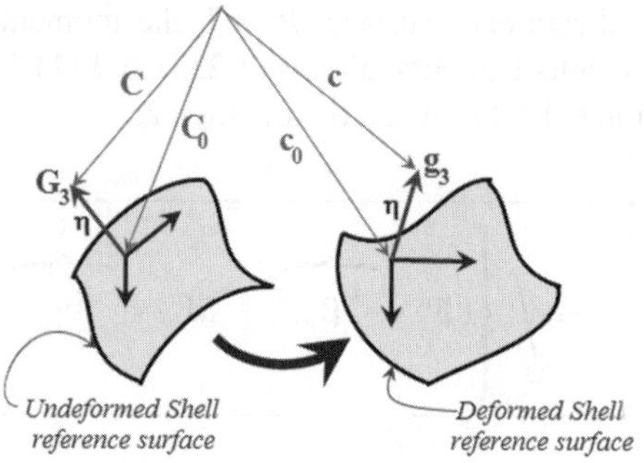

Figure 11.18 Kinematics and deformation map.

Then, as shown in Figure 11.18, the most general deformation map of a Cosserat shell surface is given by:

$$\boxed{\begin{aligned} \mathbf{c}_0(S^1,S^2) &= \mathbf{C}_0(S^1,S^2) + \mathbf{d}(S^1,S^2) \\ \mathbf{g}_i(S^1,S^2) &= \mathbf{R}(S^1,S^2)\ \mathbf{G}_i(S^1,S^2),\ i = 1,2,3 \end{aligned}} \tag{11.242}$$

where $\mathbf{d}(S^1,S^2)$ is the displacement vector field identifying the translational movement of the reference surface, $\mathbf{R}(S^1,S^2)$ is the rotational tensor field designating the rotation of the directors independent of the deformation of the reference surface; as introduced earlier, $\mathbf{C}_0(S^1,S^2)$ and $\mathbf{c}_0(S^1,S^2)$ are the corresponding position vectors in the reference surface and the current deformed surface, respectively.

Remark

- It is important to note that the base vector set, $\{\mathbf{g}_i(S^1,S^2)\}$, $i = 1,2,3$ is not the the set of convected coordinate bases; recall that the convected coordinate bases are given by $\{\mathbf{c}_{0,1}, \mathbf{c}_{0,2}, \mathbf{c}_{0,1} \times \mathbf{c}_{0,2}\}$, if shear deformation is neglected.

11.8.3.1 Weak Form: Kinematic Assumption and Reduction from the 3D Shell-like Body

Contrary to what we have done before, we use the deformation map as our starting point. As an example, in what follows, we restrict our exposition to the case of shells with a linear deformation distribution through the thickness of the shell body, that is, we take $\mathbf{h}(S^1,S^2,\eta) = \eta\ \mathbf{g}_3(S^1,S^2)$ in equation (11.227), except that we do not allow any thickness change. Thus, the basic kinematic assumption is the inextensible one-director Cosserat surface. Now, let us suppose that a particle with the position vector, $\mathbf{C}(S^1,S^2,\eta)$, in the 3D shell-like undeformed body assumes under deformation a location with the position vector given by $\mathbf{c}(S^1,S^2,\eta)$; and suppose the corresponding position vectors on the shell reference surface to be $\mathbf{C}_0(S^1,S^2) \equiv \mathbf{C}(S^1,S^2,0)$ and $\mathbf{c}_0(S^1,S^2) \equiv \mathbf{c}(S^1,S^2,0)$, respectively. Thus, any configuration of the shell-like body under the kinematic assumption is given by:

$$\mathbf{c}(S^1,S^2,\eta) = \mathbf{c}_0(S^1,S^2) + \eta\ \mathbf{g}_3 \tag{11.243}$$

The rotated base vector set, $\{\mathbf{g}_i\}$, on the deformed reference surface is related to the base vector set, $\{\mathbf{G}_i\}$, on the undeformed reference surface, as before, by: $\mathbf{g}_i = \mathbf{R}\,\mathbf{G}_i$. The reference configuration is exactly given by:

$$\mathbf{C}(S^1, S^2, \eta) = \mathbf{C}_0(S^1, S^2) + \eta\,\mathbf{G}_3 \tag{11.244}$$

Then, using equations (11.233) and (11.243), the deformation gradient, $\mathbf{F}_D$, is given by:

$$\begin{aligned}\mathbf{F}_D \equiv GRAD\,\mathbf{c} = \frac{\partial \mathbf{c}}{\partial \mathbf{C}} &= \mathbf{c},_i \otimes \mathbf{A}^i = \mathbf{c},_\alpha \otimes \mathbf{A}^\alpha + \mathbf{c},_\eta \otimes \mathbf{G}_3 \\ &= \left\{\mathbf{c}_0,_\alpha + \mathbf{k}_\alpha^{tot} \times (\mathbf{c} - \mathbf{c}_0)\right\} \otimes \mathbf{A}^\alpha + \mathbf{g}_3 \otimes \mathbf{G}_3\end{aligned} \tag{11.245}$$

where $\mathbf{A}^\alpha$, $\alpha = 1, 2$ are the contravariant base vectors at the particle of interest in the undeformed state.

Now noting the time rate, from equation (11.229), we have, from equation (11.245), the time rate, $\dot{\mathbf{F}} \equiv \frac{d}{dt}\mathbf{F}$, of the deformation map as:

$$\begin{aligned}\dot{\mathbf{F}}_D &= \left\{\dot{\mathbf{c}}_0,_\alpha + \dot{\mathbf{k}}_\alpha^{tot} \times (\mathbf{c} - \mathbf{c}_0) + \mathbf{k}_\alpha^{tot} \times \overbrace{(\mathbf{c} - \mathbf{c}_0)}^{\bullet}\right\} \otimes \mathbf{A}^\alpha + \dot{\mathbf{g}}_3 \otimes \mathbf{G}_3 \\ &= \left\{\dot{\mathbf{c}}_0,_\alpha + \dot{\mathbf{k}}_\alpha^{tot} \times (\mathbf{c} - \mathbf{c}_0) + \mathbf{k}_\alpha^{tot} \times (\boldsymbol{\omega} \times (\mathbf{c} - \mathbf{c}_0))\right\} \otimes \mathbf{A}^\alpha + (\boldsymbol{\omega} \times \mathbf{g}_3) \otimes \mathbf{G}_3\end{aligned} \tag{11.246}$$

Fact 1: As a supporting formula, let us note that for any $\mathbf{B} = \mathbf{a} \otimes \mathbf{b}$, a symmetric tensor, that is, $\mathbf{B} - \mathbf{B}^T = \mathbf{0}$, we have: $\mathbf{a} \times \mathbf{b} = \mathbf{0}$ because: $\forall \mathbf{h} \in \mathbb{R}_3$, $\mathbf{0} = (\mathbf{B} - \mathbf{B}^T)\mathbf{h} = (\mathbf{a} \otimes \mathbf{b})\mathbf{h} - (\mathbf{b} \otimes \mathbf{a})\mathbf{h} = (\mathbf{h} \bullet \mathbf{b})\mathbf{a} - (\mathbf{h} \bullet \mathbf{a})\mathbf{b} = (\mathbf{a} \times \mathbf{b}) \times \mathbf{h}$ where we have used the vector identity: $\mathbf{a} \times (\mathbf{b} \times \mathbf{c}) = (\mathbf{a} \bullet \mathbf{c})\mathbf{b} - (\mathbf{a} \bullet \mathbf{b})\mathbf{c}$.

Now, if $\mathbf{P} \equiv \mathbf{P}^j \otimes \mathbf{A}_j$ is the first Piola–Kirchhoff stress tensor, we get from equation (11.246):

$$\begin{aligned}\mathbf{P} : \dot{\mathbf{F}}_D = \mathbf{P}^\alpha \bullet \dot{\mathbf{c}}_0,_\alpha + (\mathbf{c} - \mathbf{c}_0) \times \mathbf{P}^\alpha \bullet \dot{\mathbf{k}}_\alpha^{tot} + \mathbf{P}^\alpha \left\{\mathbf{k}_\alpha^{tot} \times (\boldsymbol{\omega} \times (\mathbf{c} - \mathbf{c}_0))\right\} \\ + \boldsymbol{\omega} \bullet (\mathbf{g}_3 \times \mathbf{P}^3)\end{aligned} \tag{11.247}$$

where we have used the definition that for any two tensors, $\mathbf{A} = \mathbf{a} \otimes \mathbf{b}$, and $\mathbf{B} = \mathbf{c} \otimes \mathbf{d}$, we have: $\mathbf{A}{:}\mathbf{B} = (\mathbf{a} \otimes \mathbf{b}){:}(\mathbf{c} \otimes \mathbf{d}) = (\mathbf{a} \bullet \mathbf{c})(\mathbf{b} \bullet \mathbf{d})$. Now, let us recall that the angular momentum balance law gives the following symmetry relation:

$$\mathbf{F}_D\,\mathbf{P}^T = \mathbf{P}\,\mathbf{F}_D^T \tag{11.248}$$

whereby we get $\mathbf{F}_D\,\mathbf{P}^T = \mathbf{P}\,\mathbf{F}_D^T = \mathbf{c},_i \otimes \mathbf{A}^i \bullet \mathbf{A}_j \otimes \mathbf{P}^j = \mathbf{c},_i \otimes \mathbf{P}^i$ as symmetric, and by the above supporting fact, we conclude:

$$\mathbf{c},_i \times \mathbf{P}^i = \mathbf{0} \Rightarrow \mathbf{c},_\alpha \times \mathbf{P}^\alpha = -\mathbf{c},_\eta \times \mathbf{P}^3 = -\mathbf{g}_3 \times \mathbf{P}^3 \tag{11.249}$$

So, from equation (11.249), we get for the last term of equation (11.247):

$$\begin{aligned}\boldsymbol{\omega} \bullet (\mathbf{g}_3 \times \mathbf{P}^3) &= -\boldsymbol{\omega} \bullet (\mathbf{c},_\alpha \times \mathbf{P}^\alpha) \\ &= -\mathbf{P}^\alpha \bullet \left\{\boldsymbol{\omega} \times \mathbf{c}_{0},_\alpha + \boldsymbol{\omega} \times \left(\mathbf{k}_\alpha^{tot} \times (\mathbf{c} - \mathbf{c}_0)\right)\right\}\end{aligned} \tag{11.250}$$

Inserting equation (11.250) in equation (11.247), we get:

$$\begin{aligned}\mathbf{P} : \dot{\mathbf{F}}_D = \mathbf{P}^\alpha \bullet \left\{\dot{\mathbf{c}}_{0},_\alpha - \boldsymbol{\omega} \times \mathbf{c}_{0},_\alpha\right\} + (\mathbf{c} - \mathbf{c}_0) \times \mathbf{P}^\alpha \bullet \dot{\mathbf{k}}_\alpha^{tot} \\ + \mathbf{P}^\alpha \left\{\mathbf{k}_\alpha^{tot} \times (\boldsymbol{\omega} \times (\mathbf{c} - \mathbf{c}_0)) - \boldsymbol{\omega} \times \left(\mathbf{k}_\alpha^{tot} \times (\mathbf{c} - \mathbf{c}_0)\right)\right\}\end{aligned} \tag{11.251}$$

Now, let us note the following supporting vector identity.

Fact 2: For any three vectors, $\mathbf{a} \times (\mathbf{b} \times \mathbf{c}) - \mathbf{b} \times (\mathbf{a} \times \mathbf{c}) = (\mathbf{a} \times \mathbf{b}) \times \mathbf{c}$.

Proof. Let $\mathbf{Z} \equiv (\mathbf{b} \otimes \mathbf{a})$ be a tensor, then from the previous supporting fact, we have for the skew-symmetric part: $\mathbf{Z} - \mathbf{Z}^T \equiv (\mathbf{b} \otimes \mathbf{a}) - (\mathbf{a} \otimes \mathbf{b}) = [\mathbf{a} \times \mathbf{b}] = (\mathbf{a} \times \mathbf{b})\times$. Now, by the vector identity: $\mathbf{a} \times (\mathbf{b} \times \mathbf{c}) = (\mathbf{a} \bullet \mathbf{c})\mathbf{b} - (\mathbf{a} \bullet \mathbf{b})\mathbf{c}$, we complete the proof with: $\mathbf{a} \times (\mathbf{b} \times \mathbf{c}) - \mathbf{b} \times (\mathbf{a} \times \mathbf{c}) = (\mathbf{a} \bullet \mathbf{c})\mathbf{b} - (\mathbf{b} \bullet \mathbf{c})\mathbf{a} = \{(\mathbf{b} \otimes \mathbf{a}) - (\mathbf{a} \otimes \mathbf{b})\}\mathbf{c} = \{\mathbf{Z} - \mathbf{Z}^T\}\mathbf{c} = (\mathbf{a} \times \mathbf{b}) \times \mathbf{c}$.

So, applying this supporting fact, we get for the last term of equation (11.251) as:

$$\begin{aligned}&\mathbf{P}^\alpha \bullet \left\{\mathbf{k}_\alpha^{tot} \times (\boldsymbol{\omega} \times (\mathbf{c} - \mathbf{c}_0)) - \boldsymbol{\omega} \times \left(\mathbf{k}_\alpha^{tot} \times (\mathbf{c} - \mathbf{c}_0)\right)\right\} \\ &= \mathbf{P}^\alpha \bullet \left\{\left(\mathbf{k}_\alpha^{tot} \times \boldsymbol{\omega}\right) \times (\mathbf{c} - \mathbf{c}_0)\right\} = -(\mathbf{c} - \mathbf{c}_0) \times \mathbf{P}^\alpha \bullet \left(\boldsymbol{\omega} \times \mathbf{k}_\alpha^{tot}\right)\end{aligned} \tag{11.252}$$

Now, inserting equation (11.252) in equation (11.251), we finally get, with help of equations (11.235) and (11.239), the expression for the stress or deformation power density as:

$$\boxed{\begin{aligned}\mathbf{P} : \dot{\mathbf{F}}_D &= \mathbf{P}^\alpha \bullet \left\{\dot{\mathbf{c}}_{0},_\alpha - \boldsymbol{\omega} \times \mathbf{c}_{0},_\alpha\right\} + (\mathbf{c} - \mathbf{c}_0) \times \mathbf{P}^\alpha \left\{\dot{\mathbf{k}}_\alpha^{tot} - \left(\boldsymbol{\omega} \times \mathbf{k}_\alpha^{tot}\right)\right\} \\ &= \mathbf{P}^\alpha \bullet \overset{\mathbf{t}}{\overbrace{\mathbf{c}_{0},_\alpha}} + (\mathbf{c} - \mathbf{c}_0) \times \mathbf{P}^\alpha \bullet \overset{\mathbf{t}}{\overbrace{\mathbf{k}_\alpha^{tot}}} = \mathbf{P}^\alpha \bullet \overset{\mathbf{t}}{\overbrace{{}^R\boldsymbol{\beta}_\alpha}} + (\mathbf{c} - \mathbf{c}_0) \times \mathbf{P}^\alpha \bullet \overset{\mathbf{t}}{\overbrace{{}^R\boldsymbol{\chi}_\alpha}}\end{aligned}} \tag{11.253}$$

Now, incorporating the definition of $\mathbf{F}^\alpha$ and $\mathbf{M}^\alpha$ as in equations (11.226) and (11.227), respectively, we get, using equation (11.253), the expression for the internal stored power as:

$$\boxed{P_{Internal} \equiv \int_\Omega \left\{(\mathbf{F}^\alpha \bullet \overset{\mathbf{t}}{\overbrace{{}^R\boldsymbol{\beta}_\alpha}} + \mathbf{M}^\alpha \bullet \overset{\mathbf{t}}{\overbrace{{}^R\boldsymbol{\chi}_\alpha}}\right\} dA} \tag{11.254}$$

Which is exactly the same as that given by equation (11.241). Another compact way of finding the relation given by equation (11.253) is to consider the shell motion in terms of the configuration space; this is what we will do next.

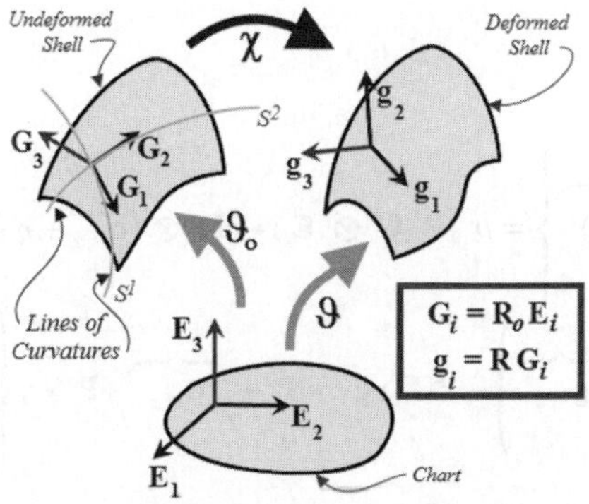

Figure 11.19 Shell deformation map.

11.8.3.2 Deformation Gradient as the Tangent Map

We may choose to derive equation (11.253) by using the deformation gradient as the tangent map on the configuration manifold. A motion may be viewed as a curve on the configuration space passing through the undeformed state as well as other and current deformation states. We can then view the deformation gradient which carries local information as the tangent map that relates the tangent to the curve at the undeformed state to that at the current deformed state. In other words, If we denote $\mathbf{C}(S^1, S^2, \eta) = \vartheta_0(S^1, S^2, \eta)$ and $\mathbf{c}(S^1, S^2, \eta) = \vartheta(S^1, S^2, \eta)$ as the values of the functions ϑ_0 and ϑ as shown in Figure 11.19, we get χ the deformation map from the undeformed shell-like body to the current deformed body as:

$$\chi \equiv \vartheta \circ \vartheta_0^{-1} \tag{11.255}$$

Noting that, $\nabla\vartheta = \mathbf{g}_i \otimes \mathbf{E}^i$, we have: $\mathbf{g}_i = (\nabla\vartheta)\ \mathbf{E}_i = (\nabla\vartheta)\ \left(\nabla\vartheta_0\right)^{-1}\ \mathbf{G}_i = \mathbf{F}\ \mathbf{G}_i$ where we have used the definition of the deformation gradient as: $\mathbf{F} \equiv \mathbf{g}_i \otimes \mathbf{G}^i$. With the deformation map given by equation (11.243), we have: $\nabla\vartheta = \vartheta,_\alpha \otimes \mathbf{E}^\alpha + \vartheta,_\eta \otimes \mathbf{E}^3 = \{\mathbf{c}_0,_\alpha + \eta\,\mathbf{g}_3,_\alpha\} \otimes \mathbf{E}^\alpha + \mathbf{g}_3 \otimes \mathbf{E}^3$. The tangent map of the deformation map given by equation (11.255) is:

$$\mathbf{F} \equiv (\nabla\vartheta)(\nabla\vartheta_0)^{-1} \tag{11.256}$$

Then, from equation (11.256), by taking the time derivative, we have:

$$\dot{\mathbf{F}} = \overbrace{\nabla\vartheta}^{\bullet}\ (\nabla\vartheta_0) = \left\{\left(\overbrace{\mathbf{c}_0,_\alpha}^{\bullet} + \eta\ \overbrace{\mathbf{g}_3,_\alpha}^{\bullet}\right) \otimes \mathbf{E}^\alpha + \dot{\mathbf{g}}_3 \otimes \mathbf{E}^3\right\}(\nabla\vartheta_0)^{-1} \tag{11.257}$$

Now, noting that $\mathbf{G}_i = (\nabla\vartheta_0)\mathbf{E}_i \Rightarrow (\nabla\vartheta_0)^{-1}\mathbf{G}_i = \mathbf{E}_i \Rightarrow (\nabla\vartheta_0)^{-1} = \mathbf{E}_i \otimes \mathbf{G}^i \Rightarrow (\nabla\vartheta_0)^{-T} = \mathbf{G}^i \otimes \mathbf{E}_i$, and that $\mathbf{A}{:}\mathbf{B} = tr(\mathbf{A}\,\mathbf{B}^T)$where *tr* stands for the trace of a tensor, by definition, we have with equation (11.257):

$$\mathbf{P}:\dot{\mathbf{F}} = tr\left\{\mathbf{P}\,(\nabla\vartheta_0)^{-T}\,\overbrace{(\nabla\vartheta)^T}^{\bullet}\right\} = tr\left\{\mathbf{P}\,(\mathbf{G}^i \otimes \mathbf{E}_i)\bullet\left\{\mathbf{E}^\alpha \otimes \left(\overline{\mathbf{c}_{0},_{\alpha}} + \eta\,\overline{\mathbf{g}_{3},_{\alpha}}\right) + \mathbf{E}^3 \otimes \overline{\mathbf{g}_3}\right\}\right\}$$

$$= \mathbf{PG}^\alpha \bullet \left(\overbrace{\mathbf{c}_{0},_{\alpha}}^{\bullet} + \eta\,\overbrace{\mathbf{g}_{3},_{\alpha}}^{\bullet}\right) + \mathbf{PG}^3 \bullet \dot{\mathbf{g}}_3 = \mathbf{P}^\alpha \bullet \overbrace{\mathbf{c}_{0},_{\alpha}}^{\bullet} + \mathbf{P}^\alpha \bullet \eta\left\{\overbrace{\mathbf{k}_\alpha^{tot} \times \mathbf{g}_3}^{\bullet}\right\} + \mathbf{P}^3 \bullet (\boldsymbol{\omega} \times \mathbf{g}_3) \tag{11.258}$$

where we have used the tensor trace property: $tr(\mathbf{a} \otimes \mathbf{b}) = \mathbf{a} \bullet \mathbf{b}$, and equations (11.229). Now, noting that: $\eta\left\{\overbrace{\mathbf{k}_\alpha^{tot} \times \mathbf{g}_3}^{\bullet}\right\} = \overbrace{\mathbf{k}_\alpha^{tot}}^{\bullet} \times \eta\,\mathbf{g}_3 + \mathbf{k}_\alpha^{tot} \times \eta\,\overbrace{\mathbf{g}_3}^{\bullet} = \overbrace{\mathbf{k}_\alpha^{tot}}^{\bullet} \times (\mathbf{c} - \mathbf{c}_0) + \mathbf{k}_\alpha^{tot} \times (\boldsymbol{\omega} \times (\mathbf{c} - \mathbf{c}_0))$, and using equation (11.250) for the third term, we get from equation (11.258):

$$\begin{aligned}\mathbf{P} : \dot{\mathbf{F}} &= \mathbf{P}^\alpha \bullet \overbrace{\mathbf{c}_{0},_{\alpha}}^{\bullet} + \mathbf{P}^\alpha \bullet \eta\left\{\overbrace{\mathbf{k}_\alpha^{tot} \times \mathbf{g}_3}^{\bullet}\right\} + \mathbf{P}^3 \bullet (\boldsymbol{\omega} \times \mathbf{g}_3) \\ &= \mathbf{P}^\alpha \bullet \overbrace{\mathbf{c}_{0},_{\alpha}}^{\bullet} + \underbrace{(\mathbf{c} - \mathbf{c}_0) \times \mathbf{P}^\alpha \bullet \overbrace{\mathbf{k}_\alpha^{tot}}^{\bullet} + \mathbf{P}^\alpha \bullet \mathbf{k}_\alpha^{tot} \times (\boldsymbol{\omega} \times (\mathbf{c} - \mathbf{c}_0))}_{2nd\ term} \\ &\quad + \underbrace{\mathbf{P}^\alpha \bullet \left\{-\boldsymbol{\omega} \times \mathbf{c}_{0},_{\alpha} - \boldsymbol{\omega} \times (\mathbf{k}_\alpha^{tot} \times (\mathbf{c} - \mathbf{c}_0))\right\}}_{3rd\ term}\end{aligned} \tag{11.259}$$

Finally, using Fact 1 for the third term in equation (11.259), we get:

$$\boxed{\begin{aligned}\mathbf{P} : \dot{\mathbf{F}} &= \mathbf{P}^\alpha \left\{\dot{\mathbf{c}}_{0},_{\alpha} - \boldsymbol{\omega} \times \mathbf{c}_{0},_{\alpha}\right\} + (\mathbf{c} - \mathbf{c}_0) \times \mathbf{P}^\alpha \bullet \left\{\dot{\mathbf{k}}_\alpha^{tot} - (\boldsymbol{\omega} \times \mathbf{k}_\alpha^{tot})\right\} \\ &= \mathbf{P}^\alpha \bullet \overbrace{\mathbf{c}_{0},_{\alpha}}^{t} + (\mathbf{c} - \mathbf{c}_0) \times \mathbf{P}^\alpha \bullet \overbrace{\mathbf{k}_\alpha^{tot}}^{t} = \mathbf{P}^\alpha \bullet \overbrace{{}^R\boldsymbol{\beta}_\alpha}^{t} + (\mathbf{c} - \mathbf{c}_0) \times \mathbf{P}^\alpha \bullet \overbrace{{}^R\boldsymbol{\chi}_\alpha}^{t}\end{aligned}} \tag{11.260}$$

Clearly, the expressions in equation (11.253) and equation (11.260) are identical.

11.8.3.3 Reduced Internal Energy (Deformation or Stress Power): Conjugate Stresses and Strains

Using equation (11.253) or equation (11.260), we can identify the deformation or stress power, $\overset{\bullet}{U}$, of a 2D shell reference surface as the reduced 3D stress power, $P_{internal}$, of the corresponding shell-like body as:

$$
\begin{aligned}
P_{internal} &= \int_M \left(\int_{\eta^b}^{\eta^t} \mathbf{P} : \dot{\mathbf{F}}\ {}^{\eta}\mu^S\, d\eta \right) dA \\
&= \int_M \left(\int_{\eta^b}^{\eta^t} \mathbf{P}^\alpha\ {}^{\eta}\mu^S\, d\eta \right) \bullet \left\{ \overbrace{\mathbf{c}_{0,\alpha}}^{\bullet} - \boldsymbol{\omega} \times \mathbf{c}_{0,\alpha} \right\} dA \\
&+ \int_M \left(\int_{\eta^b}^{\eta^t} (\mathbf{c} - \mathbf{c}_0) \times \mathbf{P}^\alpha\ {}^{\eta}\mu^S\, d\eta \right) \bullet \left\{ \overbrace{\mathbf{k}_\alpha^{tot}}^{\bullet} - \left(\boldsymbol{\omega} \times \mathbf{k}_\alpha^{tot} \right) \right\} dA \\
&= \int_M \left\{ \mathbf{F}^\alpha \bullet \overbrace{{}^R\boldsymbol{\beta}_\alpha}^{\mathbf{t}} + \mathbf{M}^\alpha \bullet \overbrace{{}^R\boldsymbol{\chi}_\alpha}^{\mathbf{t}} \right\} dA = \overset{\bullet}{U}\left({}^R\boldsymbol{\beta}_\alpha, {}^R\boldsymbol{\chi}_\alpha \right)
\end{aligned}
\qquad (11.261)
$$

where we have inserted the definitions as given in equations (11.226) and (11.227) for the stress resultant and the stress couple, respectively, and in equations (11.236) and (11.240) for the rotated translational and bending strains, respectively.

Remark

- Equation (11.261) establishes the existence of an equivalent 2D strain energy functional, $U\left({}^R\boldsymbol{\beta}_\alpha, {}^R\boldsymbol{\chi}_\alpha\right)$, under conservative loading, or at least the existence of the strain energy functional as functions of the 2D shell reference surface strains only. With this knowledge we never need to look back to the originating 3D shell-like body again, when we develop the appropriate constitutive theory for the shell reference surface as will become apparent soon.

11.8.4 Where We Would Like to Go

In conformity with our principal interest in computational structural engineering and structural mechanics, we need to bring every shell equation down to a form where we can easily solve the shell problems numerically by the use of mega-shell elements of our finite element scheme, the c-type finite element method. This in turn, requires exact derivation of the compatible real strain fields in component or operational form along the configuration path from our knowledge of the above virtual strain fields; this endeavor will again validate our belief that the presumed virtual strains are truly virtual, that is, these can be shown to exist as variations of some real strain fields. Thus, before anything else, the expressions for the virtual strain fields in the internal virtual work expression of equation (11.225) must all be modified to the ***component vector form***. For this, we must next look at:

- component or operational vector form.

11.9 Component or Operational Vector Form

11.9.1 What We Need to Recall

Our stated goal is achieved by use of the curvilinear description of a geometric surface embedded in a general 3D space, but first there are various facts to recall.

- For the undeformed coordinate system corresponding to the lines of curvature along and the normal to the shell reference surface with base vectors, $\{\mathbf{G_1}, \mathbf{G_2}, \mathbf{G}_3 \equiv \mathbf{N}\}$, being an orthonormal system of coordinates, both the sets of the contravariant base vectors, $\{\mathbf{G}^i(S^1, S^2)\}$, and the covariant base vectors, $\{\mathbf{G}_i(S^1, S^2)\}$, are one and the same.
- It is assumed that the current deformation of a shell is completely characterized by the displacement vector field: $\mathbf{d}(S^1, S^2, t)$, and a rotation tensor field: $\mathbf{R}(S^1, S^2, t)$, both measured in the undeformed coordinate system, $\{\mathbf{G_1}, \mathbf{G_2}, \mathbf{G}_3 \equiv \mathbf{N}\}$ on the shell reference surface at time $t \in \mathbb{R}_+$. However, we have another curvilinear co-spinning coordinate system (rotating synchronously with the rotation tensor, $\mathbf{R}(S^1, S^2, t)$) and denoted as the normal coordinate system, with base vectors, $\{{}^n\mathbf{G}_1, {}^n\mathbf{G}_2, \mathbf{N}\}$ in the undeformed configuration that becomes $\{{}^n\mathbf{g}_1, {}^n\mathbf{g}_2, \mathbf{n}\}$ in the current deformed state at any point $\{S^1, S^2\} \in \mathcal{M}$, the reference surface of a shell at time $t \in \mathbb{R}_+$. The undeformed and deformed base vectors have the relation: $\mathbf{n} = \mathbf{R}\mathbf{N}$, ${}^n\mathbf{g}_\alpha = \mathbf{R}\,{}^n\mathbf{G}_\alpha$, $\alpha = 1, 2$. In compact tensor notation, we can rewrite these relations as: ${}^n\mathbf{g}_i = \mathbf{R}\,\mathbf{G}_i$, $i = 1, 2, 3$. Moreover, in the undeformed configuration of the shell, $\{{}^n\mathbf{G}_1, {}^n\mathbf{G}_2, \mathbf{N}\}$ coincide identically with $\{\mathbf{G_1}, \mathbf{G_2}, \mathbf{G}_3 \equiv \mathbf{N}\}$, respectively,
- The rotation matrix, ${}^t\mathbf{R}$, measured in undeformed coordinate system, and the Rodrigues vector, $\mathbf{s} \equiv \{\, s_1 \quad s_2 \quad s_3 \,\}^T$, are related by:

$$
{}^t\mathbf{R} = \frac{1}{\rho}\begin{bmatrix} 1 + s_1^2 - s_2^2 - s_3^2 & 2(s_1 s_2 - s_3) & 2(s_3 s_1 + s_2) \\ 2(s_1 s_2 + s_3) & 1 - s_1^2 + s_2^2 - s_3^2 & 2(s_2 s_3 - s_1) \\ 2(s_3 s_1 - s_2) & 2(s_2 s_3 + s_1) & 1 - s_1^2 - s_2^2 + s_3^2 \end{bmatrix}
\quad \text{with } \rho \equiv 1 + s_1^2 + s_2^2 + s_3^2 \tag{11.262}
$$

- The rotation vector, $\boldsymbol{\theta}$, measured in the reference coordinate system, and the Rodrigues vector, $\mathbf{s}$, are related such that:

$$
\mathbf{s} = \frac{1}{\theta}\tan\left(\frac{\theta}{2}\right)\boldsymbol{\theta} \tag{11.263}
$$

  where $\theta \equiv \|\boldsymbol{\theta}\| = \sqrt{(\boldsymbol{\theta} \bullet \boldsymbol{\theta})}$ is the magnitude of the rotation vector.
- The angular velocity vector, $\boldsymbol{\omega}$, measured in the reference coordinate system is given by:

$$
\boldsymbol{\omega} \equiv \mathbf{W}(\boldsymbol{\theta})\,\dot{\boldsymbol{\theta}} \tag{11.264}
$$

where $\mathbf{W}(\boldsymbol{\theta})$ matrix is given by

$$\boxed{\begin{aligned} \mathbf{W}(\boldsymbol{\theta}) &\equiv \mathbf{I} + c_1\, \boldsymbol{\Theta} + c_2\, \boldsymbol{\Theta}^2, \\ c_1(\boldsymbol{\theta}) &= \frac{1-\cos\theta}{\theta^2}, \\ c_2(\boldsymbol{\theta}) &= \frac{\theta - \sin\theta}{\theta^3} \end{aligned}} \tag{11.265}$$

- In the tensor notation, for the rotating normal coordinate system, $\{{}^n\mathbf{G}_1, {}^n\mathbf{G}_2, \mathbf{N}\}$, in the undeformed configuration that becomes $\{{}^n\mathbf{g}_1, {}^n\mathbf{g}_2, \mathbf{n}\}$in the current deformed state at any point $\{S^1, S^2\} \in \mathcal{M}$, the reference surface of a shell, being orthonormal systems of coordinates, both the set of the contravariant base vectors, say, $\{\mathbf{G}^i\}$ for the undeformed and $\{\mathbf{g}^i\}$ for the deformed states, and the covariant base vectors, say, $\{\mathbf{G}_i\}$ for the undeformed and $\{\mathbf{g}_i\}$ for the deformed states, correspondingly, are one and the same.

11.9.2 Relevant Operational Matrix Representations

For our immediate use, we will represent the base vector sets in operational matrix form as:

$$\mathbf{G} \equiv [\,\mathbf{G}_1 \quad \mathbf{G}_2 \quad \mathbf{G}_3 \equiv \mathbf{N}\,], \qquad \mathbf{g} \equiv [\,{}^n\mathbf{g}_1 \quad {}^n\mathbf{g}_2 \quad \mathbf{n}\,] \tag{11.266}$$

where $\mathbf{G}^T\,\mathbf{G} = \mathbf{g}^T\,\mathbf{g} = \mathbf{I}$ with $\mathbf{I}$ as the 3×3 identity matrix. Now, the component representation of the rotation tensor, $\mathbf{R} = {}^tR_{ij}\; \mathbf{G}^i \otimes \mathbf{G}^j$, can be given as the matrix, ${}^t\mathbf{R}$, as:

$$\boxed{{}^t\mathbf{R} = \mathbf{G}^T\,\mathbf{R}\,\mathbf{G}, \Leftrightarrow \mathbf{R} = \mathbf{G}\,{}^t\mathbf{R}\,\mathbf{G}^T} \tag{11.267}$$

where ij^{th}-th component of ${}^t\mathbf{R}$ is given by ${}^tR_{ij}$. Finally, given the relations: ${}^n\mathbf{g}_i = \mathbf{R}\,\mathbf{G}_i$, $i = 1, 2, 3$, we can use equations (11.266) and (11.267) to express these relation in operational matrix form as:

$$\begin{aligned} &\mathbf{G} \equiv [\,\mathbf{G}_1 \quad \mathbf{G}_2 \quad \mathbf{G}_3 \equiv \mathbf{N}\,], \\ &\mathbf{g} \equiv [\,{}^n\mathbf{g}_1 \quad {}^n\mathbf{g}_2 \quad \mathbf{n}\,] = [\,\mathbf{R}\mathbf{G}_1 \quad \mathbf{R}\mathbf{G}_2 \quad \mathbf{R}\mathbf{G}_3 \equiv \mathbf{R}\mathbf{N}\,] \\ &\quad = \mathbf{R}\mathbf{G} = \mathbf{G}\,{}^t\mathbf{R}\,\mathbf{G}^T\,\mathbf{G} = \mathbf{G}\,{}^t\mathbf{R} \end{aligned} \tag{11.268}$$

11.9.3 Component or Operational Vectors

As indicated in our stated goal, and as shown in Figure 11.20, we define here the component vectors by first identifying the coefficients (i.e. the components) of linear combinations and the shell base vectors that make up the corresponding invariant vectors, and then collect the components in an operational vector form.

11.9.3.1 Component Displacement and Rotation Vectors: ${}^t\mathbf{d}$ and ${}^t\boldsymbol{\theta}$

In terms of the linear combinations of the parameters measured in the undeformed tangent coordinate system, $\{\mathbf{G}_1, \mathbf{G}_2, \mathbf{G}_3 \equiv \mathbf{N}\}$, the displacement vector, $\mathbf{d}$, an invariant vector measuring

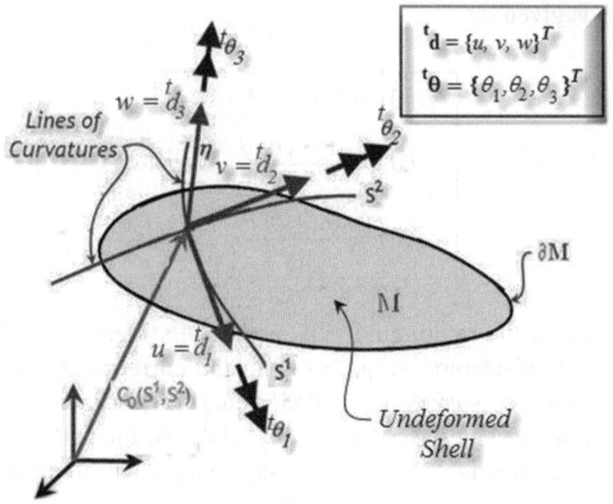

Figure 11.20 Displacement and rotation components.

the movement of a point on the undeformed reference surface to its current position on the deformed reference surface, can be written as:

$$\boxed{\mathbf{d} = {}^t d_1(\equiv u)\mathbf{G_1} + {}^t d_2(\equiv v)\mathbf{G_2} + {}^t d_3(\equiv w)\mathbf{N} = {}^t d_i\,\mathbf{G}_i} \tag{11.269}$$

and similarly, the rotation vector, $\boldsymbol{\theta}$, as:

$$\boxed{\boldsymbol{\theta} = ({}^t\theta_1 \equiv \theta_1)\mathbf{G_1} + ({}^t\theta_2 \equiv \theta_2)\mathbf{G}_2 + ({}^t\theta_3 \equiv \theta_3)\mathbf{N} = {}^t\theta_i\,\mathbf{G}_i} \tag{11.270}$$

where $\mathbf{G} \equiv [\,\mathbf{G}_1 \quad \mathbf{G}_2 \quad \mathbf{G}_3\,]$ is the orthonormal matrix with the base vectors as its columns. Note that, although it is immaterial for the orthogonal frames, for the sake of consistency with the description in the curvilinear coordinates, the displacement and rotation components are shown as covariant components. We collect the components of $\mathbf{d}$ and $\boldsymbol{\theta}$, and define the component vectors ${}^t\mathbf{d}$ and ${}^t\boldsymbol{\theta}$, respectively, as follows,

Definition: The component or operational vectors: ${}^t\mathbf{d}$ and ${}^t\boldsymbol{\theta}$

$$\boxed{{}^t\mathbf{d} \equiv \left\{\, {}^t d_1 \equiv u \quad {}^t d_2 \equiv v \quad {}^t d_3 \equiv w \,\right\}^T} \tag{11.271}$$

$$\boxed{{}^t\boldsymbol{\theta} \equiv \left\{\, {}^t\theta_1 \equiv \theta_1 \quad {}^t\theta_2 \equiv \theta_2 \quad {}^t\theta_3 \equiv \theta_3 \,\right\}^T} \tag{11.272}$$

Note that the superscript t reminds us that the components are measured in the undeformed tangent coordinate system. With the above definitions, we have:

$$\boxed{\begin{aligned} \mathbf{d} &= \mathbf{G}\,{}^t\mathbf{d} \\ \boldsymbol{\theta} &= \mathbf{G}\,{}^t\boldsymbol{\theta} \end{aligned}} \tag{11.273}$$

11.9.3.2 Component Linear Velocity Vectors: ${}^t\mathbf{v} \equiv {}^t\overset{\bullet}{\mathbf{d}}$

Following the same ideas as above, the linear velocity, ${}^t\mathbf{v} \equiv {}^t\overset{\bullet}{\mathbf{d}}$, can be written as:

$$\mathbf{v} \equiv \overset{*}{\mathbf{d}} = {}^t\overset{\bullet}{d_1}(\equiv \dot{u})\mathbf{G}_1 + {}^t\overset{\bullet}{d_2}(\equiv \dot{v})\mathbf{G}_2 + {}^t\overset{\bullet}{d_3}(\equiv \dot{w})\mathbf{N} = {}^t\overset{\bullet}{d_i}\,\mathbf{G}_i \tag{11.274}$$

We collect the components of $\mathbf{v} \equiv \overset{*}{\mathbf{d}}$, and define the component vector ${}^t\mathbf{v} \equiv {}^t\overset{*}{\mathbf{d}}$, as:

Definition: The component or operational vectors: ${}^t\overset{\bullet}{\mathbf{d}}$

$$\boxed{{}^t\mathbf{v} \equiv {}^t\overset{*}{\mathbf{d}} \equiv \left\{\; {}^t\overset{\bullet}{d_1} \equiv \dot{u} \quad {}^t\overset{\bullet}{d_2} \equiv \dot{v} \quad {}^t\overset{\bullet}{d_3} \equiv \dot{w} \;\right\}^T} \tag{11.275}$$

With the above definitions, we have:

$$\boxed{\mathbf{v} \equiv \overset{*}{\mathbf{d}} = \mathbf{G}\,{}^t\overset{\bullet}{\mathbf{d}}} \tag{11.276}$$

11.9.3.3 Component Angular Velocity Vectors: ${}^t\boldsymbol{\omega}$

The angular velocity, ${}^t\boldsymbol{\omega}$, can be written as:

$$\boxed{\boldsymbol{\omega} = {}^t\omega_1\,\mathbf{G}_1 + {}^t\omega_2\,\mathbf{G}_2 + {}^t\omega_3\,\mathbf{N} = {}^t\omega_i\mathbf{G}_i = \mathbf{G}\,{}^t\boldsymbol{\omega}} \tag{11.277}$$

We collect the components of ${}^t\boldsymbol{\omega}$, and define the component vector ${}^t\boldsymbol{\omega}$, as:

Definition: The component or operational vectors: ${}^t\boldsymbol{\omega}$

$$\boxed{{}^t\boldsymbol{\omega} \equiv \left\{\; {}^t\omega_1 \quad {}^t\omega_2 \quad {}^t\omega_3 \;\right\}^T} \tag{11.278}$$

With the above definitions and equation (11.264), we have:

$$\boxed{{}^t\boldsymbol{\omega} \equiv \mathbf{G}\,{}^t\mathbf{W}\,\mathbf{G}^T\,\mathbf{G}\,{}^t\overset{\bullet}{\boldsymbol{\theta}} = \mathbf{G}\,{}^t\mathbf{W}\,{}^t\overset{\bullet}{\boldsymbol{\theta}}} \tag{11.279}$$

11.9.3.4 Component Force and Moment Vectors in Undeformed Frame: ${}^t\mathbf{F}^\alpha$ and ${}^t\mathbf{M}^\alpha$

In terms of the linear combinations of parameters measured in the undeformed tangent coordinate system, $\{\mathbf{G}_1, \mathbf{G}_2, \mathbf{G}_3 \equiv \mathbf{N}\}$, the force vector, $\mathbf{F}^\alpha$, an invariant vector measuring the current stress resultant on an $\alpha = const.$ cross-section of the deformed shell, and the moment vector, $\mathbf{M}^\alpha$, an invariant vector measuring the current stress couple resultant on an $\alpha = const.$ cross-section of a deformed shell, can be written, respectively, as:

$$\boxed{\begin{aligned} \mathbf{F}^\alpha &= {}^tF^{\alpha 1}(\equiv N_{\alpha 1})\,\mathbf{G}_1 + {}^tF^{\alpha 2}(\equiv N_{\alpha 2})\,\mathbf{G}_2 + {}^tF^{\alpha 3}(\equiv Q_\alpha)\,\mathbf{N} \\ &= {}^tF^{\alpha i}\,\mathbf{G}_i, \quad \alpha = 1, 2; \quad i = 1, 2, 3 \end{aligned}} \tag{11.280}$$

where, N_{11}, N_{22} are known as the in-plane extensional forces, N_{12}, N_{21} as the in-plane shear forces, and Q_{13}, Q_{23} are the transverse shear forces; conventionally and notation-wise, these are described as such in engineering terminology, and

$$\boxed{\begin{aligned} \mathrm{M}^1 &= {}^tM^{11}(\equiv -M_{12})\,\mathbf{G_1} + {}^tM^{12}(\equiv M_1)\,\mathbf{G_2} + {}^tM^{13}(\equiv M_{13})\,\mathbf{N} \\ &= {}^tM^{1i}\,\mathbf{G}_i, \qquad i = 1,2,3 \\ \mathrm{M}^2 &= {}^tM^{21}(\equiv -M_2)\,\mathbf{G_1} + {}^tM^{22}(\equiv M_{21})\,\mathbf{G_2} + {}^tM^{23}(\equiv M_{23})\,\mathbf{N} \\ &= {}^tM^{2i}\,\mathbf{G}_i, \qquad i = 1,2,3 \\ \mathrm{M}^\alpha &= {}^tM^{\alpha i}\,\mathbf{G}_i, \quad \alpha = 1,2, \quad i = 1,2,3 \end{aligned}} \tag{11.281}$$

where, M_1, M_2 are known as the bending moments, M_{12}, M_{21} as the in-plane twisting moments, and $M_{\alpha 3}$ are the transverse twisting moments on a shell cross-section with the normal to it pointing in the α direction, $\alpha = 1, 2$ (for the classical shell theory, these components are identically zero); conventionally and notation-wise, these are described as such in engineering terminology.

Remarks

- Notationally, the first indices of the shell consistent components indicate the directions of the normals to the cross-section, and the second indices point to the directions of the forces and moments.
- In engineering, the stress couples are expressed in the same way as the associated stresses. However, we must reiterate that the shell consistent moment components, ${}^tM^{\alpha i}$, $\alpha = 1, 2$, $i = 1, 2, 3$, in the undeformed tangent coordinate system, $\{\mathbf{G_1}, \mathbf{G_2}, \mathbf{G_3} \equiv \mathbf{N}\}$, are different from their counterparts in engineering. Of course, these can be easily related by the following map that uses the 2D permutation matrix, $e_{\alpha\beta}$:

$$\boxed{{}^tM^{\alpha\beta} = e_{\eta\beta}\, M^{\alpha\eta}, \quad (e_{\eta\beta}) = \begin{bmatrix} 0 & 1 \\ -1 & 0 \end{bmatrix}} \tag{11.282}$$

We collect the components of $\mathbf{F}^\alpha$ and M^α, and define component vectors, respectively, as follows.

Definition:

$$\boxed{\begin{aligned} {}^t\mathbf{F}^1 &\equiv \left\{ {}^tF^{11} \equiv N_1 \quad {}^tF^{12} \equiv N_{12} \quad {}^tF^{13} \equiv Q_1 \right\}^T \\ {}^t\mathbf{F}^2 &\equiv \left\{ {}^tF^{21} \equiv N_{21} \quad {}^tF^{22} \equiv N_2 \quad {}^tF^{23} \equiv Q_2 \right\}^T \end{aligned}} \tag{11.283}$$

$$\boxed{\begin{aligned} {}^t\mathbf{M}^1 &\equiv \left\{ {}^tM^{11} \equiv -M_{12} \quad {}^tM^{12} \equiv M_1 \quad {}^tM^{13} \equiv M_{13} \right\}^T \\ {}^t\mathbf{M}^2 &\equiv \left\{ {}^tM^{21} \equiv -M_2 \quad {}^tM^{22} \equiv M_{21} \quad {}^tM^{23} \equiv M_{23} \right\}^T \end{aligned}} \tag{11.284}$$

With the above definitions, we have:

$$\boxed{\begin{aligned} \mathbf{F}^\alpha &= \mathbf{G}\, {}^t\mathbf{F}^\alpha \\ \mathrm{M}^\alpha &= \mathbf{G}\, {}^t\mathbf{M}^\alpha, \quad \alpha = 1,2 \end{aligned}} \tag{11.285}$$

where $\mathbf{G} \equiv [\,\mathbf{G}_1 \quad \mathbf{G}_2 \quad \mathbf{G}_3\,]$ is the orthonormal matrix with base vectors as its columns. Note that although it is immaterial for orthogonal frames, for the sake of consistency with the description in the curvilinear coordinate system, the force and moment components are shown as the contravariant components.

11.9.3.5 Component Force and Moment Vectors in Terms of Deformed Frame: ${}^n\mathbf{F}^\alpha$ and ${}^n\mathbf{M}^\alpha$

In terms of the parameters measured in the current normal coordinate system, $\{{}^n\mathbf{g}_1, {}^n\mathbf{g}_2, \mathbf{n}\}$, let the current stress resultants (or forces): ${}^n\mathbf{F}^\alpha = {}^nF^{\alpha 1}\,{}^n\mathbf{g}_1 + {}^nF^{\alpha 2}\,{}^n\mathbf{g}_2 + {}^nF^{\alpha 3}\,{}^n\mathbf{g}_3 = {}^nF^{\alpha i}\,{}^n\mathbf{g}_i$ and the current resultant stress couple (or moments): ${}^n\mathbf{M}^\alpha = {}^nM^{\alpha 1}\,{}^n\mathbf{g}_1 + {}^nM^{\alpha 2}\,{}^n\mathbf{g}_2 + {}^nM^{\alpha 3}\,{}^n\mathbf{g}_3 = {}^nM^{\alpha i}\,{}^n\mathbf{g}_i$.

We collect the components of $\mathbf{F}$ and $\mathbf{M}$, and define component vectors, respectively, as follows.

Definition:

$$\boxed{{}^n\mathbf{F}^\alpha(S^1,S^2) \equiv \left\{\, {}^nF^{\alpha 1}(S^1,S^2) \quad {}^nF^{\alpha 2}(S^1,S^2) \quad {}^nF^{\alpha 3}(S^1,S^2) \,\right\}^{\mathrm{T}}} \tag{11.286}$$

$$\boxed{{}^n\mathbf{M}^\alpha(S^1,S^2) \equiv \left\{\, {}^nM^{\alpha 1}(S^1,S^2) \quad {}^nM^{\alpha 2}(S^1,S^2) \quad {}^nM^{\alpha 3}(S^1,S^2) \,\right\}^{\mathrm{T}}} \tag{11.287}$$

11.9.4 Relations Between: ${}^t\mathbf{F}^\alpha$ and ${}^n\mathbf{F}^\alpha$, and ${}^t\mathbf{M}^\alpha$ and ${}^n\mathbf{M}^\alpha$, $\alpha = 1, 2$

Considering that the $\{{}^n\mathbf{g}_1, {}^n\mathbf{g}_2, \mathbf{n}\}$ frame in the deformed shell reference surface, is defined by the orthogonal rotation tensor, $\mathbf{R}(S^1, S^2)$, measured in the $\{\mathbf{G}_1, \mathbf{G}_2, \mathbf{G}_3 \equiv \mathbf{N}\}$ frame of the undeformed shell, the two sets of the stress resultants and stress couples component vectors are related by the following relations, for $\alpha = 1, 2$:

$$\boxed{\begin{aligned} {}^t\mathbf{F}^\alpha &= \mathbf{G}\,{}^t\mathbf{F}^\alpha = \mathbf{g}\,{}^n\mathbf{F}^\alpha = \mathbf{G}\,{}^t\mathbf{R}\,{}^n\mathbf{F}^\alpha \\ {}^t\mathbf{M}^\alpha &= \mathbf{G}\,{}^t\mathbf{M}^\alpha = \mathbf{g}\,{}^n\mathbf{M}^\alpha = \mathbf{G}\,{}^t\mathbf{R}\,{}^n\mathbf{M}^\alpha \end{aligned}} \tag{11.288}$$

where we have used equation (11.268). From equation (11.288), we can finally write the following relationships relating the components of forces and moments in undeformed and current spinning frames as:

$$\boxed{\begin{aligned} {}^t\mathbf{F}^\alpha &= {}^t\mathbf{R}\,{}^n\mathbf{F}^\alpha \\ {}^t\mathbf{M}^\alpha &= {}^t\mathbf{R}\,{}^n\mathbf{M}^\alpha, \quad \alpha = 1, 2 \end{aligned}} \tag{11.289}$$

11.9.4.1 Components of Linear Momentum and Angular Momentum Vectors in Terms of the Undeformed Frame: ${}^t\mathbf{L}(S^1, S^2, t)$ and ${}^t\mathbf{A}(S^1, S^2, t)$

The invariant linear momentum vector, $\mathbf{L}(S^1, S^2, t)$, and the angular momentum vector, $\mathbf{A}(S^1, S^2, t)$, for any $t \in \mathbb{R}_+$, can have component representation in any triad of base vectors, Cartesian or curvilinear.

Thus, we can write:

$$\boxed{\begin{aligned}\mathbf{L}(S^1,S^2,t) &= {}^tL^1(\equiv P_1)\ \mathbf{G_1} + {}^tL^2(\equiv P_2)\ \mathbf{G_2} + {}^tL^3(\equiv P_N)\ \mathbf{G}_3 \\ &= {}^tL^i(S^1,S^2,t)\ \mathbf{G}_i(S^1,S^2), \quad i = 1,2,3\end{aligned}} \tag{11.290}$$

and

$$\boxed{\begin{aligned}\mathbf{A}(S^1,S^2,t) = &= {}^tA^1(\equiv A_1)\ \mathbf{G_1} + {}^tA^2(\equiv A_2)\ \mathbf{G_2} + {}^tA^3(\equiv A_N)\ \mathbf{G}_3 \\ &= {}^tA^i(S^1,S^2,t)\ \mathbf{G}_i(S^1,S^2), \quad i = 1,2,3\end{aligned}} \tag{11.291}$$

We collect the components of **L** and **A**, and define the component vectors, respectively, as follows.

Definitions

$$\boxed{\begin{aligned}{}^t\mathbf{L}(S^1,S^2,t) &\equiv \left\{ {}^tL^1 \quad {}^tL^2 \quad {}^tL^3 \equiv L_N \right\}^T \\ {}^t\mathbf{A}(S^1,S^2,t) &\equiv \left\{ {}^tA^1 \quad {}^tA^2 \quad {}^tA^3 \equiv A_N \right\}^T\end{aligned}} \tag{11.292}$$

With the above definitions, we can express these in the column vector form as:

$$\boxed{\begin{aligned}\mathbf{L}(S^1,S^2,t) &= \mathbf{G}(S^1,S^2)\ {}^t\mathbf{L}(S^1,S^2,t) \\ \mathbf{A}(S^1,S^2,t) &= \mathbf{G}(S^1,S^2)\ {}^t\mathbf{A}(S^1,S^2,t)\end{aligned}} \tag{11.293}$$

Likewise, we have the time derivatives of the linear momentum and the angular momentum in terms of their material component vectors:

$$\boxed{\begin{aligned}\dot{L}(S^1,S^2,t) &= \mathbf{G}(S^1,S^2)\ {}^t\dot{\mathbf{L}}(S^1,S^2,t) \\ \dot{A}(S^1,S^2,t) &= \mathbf{G}(S^1,S^2)\ {}^t\dot{A}(S^1,S^2,t)\end{aligned}} \tag{11.294}$$

11.9.5 Where We Would Like to Go

- We have defined the component or operational vectors corresponding to the displacement, rotation and velocity vectors, and the force and moment vectors, and so on. These definitions play a vital role in converting the invariant vector form of the shell balance equations: 3D to 2D, or the shell balance equations: direct 2D to computationally more suitable component vector or operational form of balance equations. However, to understand this conversion fully, we would need and investigate next:
- derivatives of component vectors
- computational curvatures, angular velocity and variations

11.10 Covariant Derivatives of Component Vectors

11.10.1 What We Need to Recall

- The Gauss–Weingarten relation for the derivatives of the triad of base vectors in matrix form with respect to the arc length parameters, $S^\alpha, \quad \alpha = 1, 2$, is given by:

$$\begin{Bmatrix} \mathbf{G}_{1,1} \\ \mathbf{G}_{2,1} \\ \mathbf{N}_{,1} \end{Bmatrix} \equiv \frac{\partial}{\partial S^1}\begin{Bmatrix} \mathbf{G}_1 \\ \mathbf{G}_2 \\ \mathbf{N} \end{Bmatrix} = \begin{bmatrix} 0 & \frac{1}{\gamma_1}\gamma_{1,2} & -\frac{1}{R_1} \\ -\frac{1}{\gamma_1}\gamma_{1,2} & 0 & 0 \\ \frac{1}{R_1} & 0 & 0 \end{bmatrix}\begin{Bmatrix} \mathbf{G}_1 \\ \mathbf{G}_2 \\ \mathbf{N} \end{Bmatrix} \equiv \widehat{\mathbf{K}}_1^0 \begin{Bmatrix} \mathbf{G}_1 \\ \mathbf{G}_2 \\ \mathbf{N} \end{Bmatrix} \tag{11.295}$$

and

$$\begin{Bmatrix} \mathbf{G}_{1,2} \\ \mathbf{G}_{2,2} \\ \mathbf{N}_{,2} \end{Bmatrix} \equiv \frac{\partial}{\partial S^2}\begin{Bmatrix} \mathbf{G}_1 \\ \mathbf{G}_2 \\ \mathbf{N} \end{Bmatrix} = \begin{bmatrix} 0 & -\frac{1}{\gamma_2}\gamma_{2,1} & 0 \\ \frac{1}{\gamma_2}\gamma_{2,1} & 0 & -\frac{1}{R_2} \\ 0 & \frac{1}{R_2} & 0 \end{bmatrix}\begin{Bmatrix} \mathbf{G}_1 \\ \mathbf{G}_2 \\ \mathbf{N} \end{Bmatrix} \equiv \widehat{\mathbf{K}}_2^0 \begin{Bmatrix} \mathbf{G}_1 \\ \mathbf{G}_2 \\ \mathbf{N} \end{Bmatrix} \tag{11.296}$$

where $\widehat{\mathbf{K}}_\alpha^0, \quad \alpha = 1, 2$ are the shell undeformed curvature matrices with $R_\alpha(S^1, S^2), \quad \alpha = 1, 2$, are the radii of curvatures at any point, $\{S^\alpha\}$, of the shell reference surface, and $\gamma_\alpha, \quad \alpha = 1, 2$, are the lengths of the base vectors: $\gamma_\alpha \equiv \dfrac{dS^\alpha}{d\xi^\alpha} = \|\bar{\mathbf{G}}_\alpha\| = \sqrt{\bar{\mathbf{G}}_{\alpha\alpha}}$ where $\bar{\mathbf{G}}_\alpha(\xi^1, \xi^2) \equiv \dfrac{\partial \mathbf{C}(\xi^1, \xi^2, 0)}{\partial \xi^\alpha}$, with no sum on α. Note that the curvature matrices are skew-symmetric. With the notation for the unit base vectors: $\mathbf{G}_1, \ \mathbf{G}_2, \ \mathbf{G}_3 \equiv \mathbf{N}$, we can rewrite the matrix relations in compact notation as:

$$\mathbf{G}_{i},_{\alpha} = (\widehat{K}_\alpha^0)_{ij}\, \mathbf{G}_j, \quad \alpha = 1, 2; \ \ i = 1, 2, 3; \ \ j = 1, 2, 3 \tag{11.297}$$

Clearly, these matrix forms give the derivatives of a base vector as a linear combination of all the base vectors. The individual coefficients of the matrices can be easily identified with the Christoffel symbols of the second kind, Γ^i_{jk}. However, in our entire shell analysis, we will completely avoid explicit use of these obscuring symbols by adhering to familiar geometric entities such as the radii of curvature, and so on, as we have done above, for ease of interpretation and clarity.

- On the other hand, the Gauss–Weingarten relation for the derivatives of the triad of base vectors on a surface in invariant tensor form with respect to the arc length parameters, $S^\alpha, \quad \alpha = 1, 2$, may be obtained by first introducing the following.

Definition: ***Shell initial curvature tensors***, $\mathbf{K}_\alpha^0$, $\alpha = 1, 2$:

$$\mathbf{K}_\alpha^0 \equiv (\widehat{\mathbf{K}}_\alpha^0)^T_{ij}\, \mathbf{G}_i \otimes \mathbf{G}_j = -\mathbf{G}_i \otimes \mathbf{G}_{i},_{\alpha}, \quad \alpha = 1, 2; \quad i, j = 1, 2, 3 \tag{11.298}$$

To wit, for the second equality in equation (11.148), we note that because of anti-symmetry: $(\widehat{\mathbf{K}}^0_\alpha)^T_{ij} = -(\widehat{\mathbf{K}}^0_\alpha)_{ij}$. Then, we have:

$$\begin{aligned}\mathbf{K}^0_\alpha &\equiv (\widehat{\mathbf{K}}^0_\alpha)^T_{ij}\,\mathbf{G}_i \otimes \mathbf{G}_j = -(\widehat{\mathbf{K}}^0_\alpha)_{ij}\,\mathbf{G}_i \otimes \mathbf{G}_j \\ &= -\,\mathbf{G}_i \otimes (\widehat{\mathbf{K}}^0_\alpha)_{ij}\mathbf{G}_j = -\mathbf{G}_i \otimes \mathbf{G}_{i},_{\alpha}, \quad \alpha = 1,2; \quad i,j = 1,2,3\end{aligned}$$

where we have used the result from equation (11.297). Note that the curvature tensor is skew-symmetric and thus: $(\mathbf{K}^0_\alpha)^T = -\mathbf{K}^0_\alpha$. This can be easily verified:

$$(\mathbf{K}^0_\alpha)^T \equiv (\widehat{\mathbf{K}}^0_\alpha)^T_{ij}\,\mathbf{G}_j \otimes \mathbf{G}_i = (\widehat{\mathbf{K}}^0_\alpha)^T_{ji}\,\mathbf{G}_i \otimes \mathbf{G}_j = -\,(\widehat{\mathbf{K}}^0_\alpha)^T_{ij}\mathbf{G}_i \otimes \mathbf{G}_j = -\mathbf{K}^0_\alpha\,, \quad \alpha = 1,2;$$

Definition: ***Shell initial curvature vector***, $\mathbf{k}^0_\alpha$, $\alpha = 1,2$:
The axial vectors corresponding to the initial curvature tensors, $\mathbf{K}^0_\alpha$, $\alpha = 1,2$, are given by:

$$\boxed{\begin{aligned}\mathbf{k}^0_1 &\equiv \left\{0 \quad -\frac{1}{R_1} \quad -\frac{1}{\gamma_1}\gamma_{1},_{2}\right\}^T \\ \mathbf{k}^0_2 &\equiv \left\{\frac{1}{R_2} \quad 0 \quad \frac{1}{\gamma_2}\gamma_{2},_{1}\right\}^T\end{aligned}} \qquad (11.299)$$

Now, we can easily get from equation (11.148), the invariant tensor form as:

$$\boxed{\mathbf{G}_{i},_{\alpha} = \mathbf{K}^0_\alpha\,\mathbf{G}_i = \mathbf{k}^0_\alpha \times\ \mathbf{G}_i, \quad \alpha = 1,2;\ i = 1,2,3} \qquad (11.300)$$

Note that, in tensor form, unlike the matrix expression (11.297), each unit base vector derivative is given as a transformation of itself. As we shall see, this allows us to write expressions in absolute compact form free of matrix indices.

Exercise: Verify the tensor form of the derivatives of base vectors given by equation (11.150).

Solution: $\boxed{\mathbf{K}^0_\alpha\,\mathbf{G}_i = -(\mathbf{K}^0_\alpha)^T\,\mathbf{G}_i = -(-\mathbf{G}_{i},_{\alpha} \otimes \mathbf{G}_i)\mathbf{G}_i = \mathbf{G}_{i},_{\alpha}, \quad \alpha = 1,2;\ i = 1,2,3}$

- Finally, the covariant derivative of a base vector is the zero vector, that is, $\mathbf{G}_i|_\alpha = \mathbf{0}$, $i = 1,2,3$. Thus, the covariant derivative of a tensor function of any order is identical with its partial derivative. If we take a vector (a tensor of order 1) function, $\mathbf{a}(S^\alpha)$, we then have: $\mathbf{a}\,|_{S^\alpha} \equiv \mathbf{a}\,|_\alpha = \mathbf{a},_\alpha = {}^t a_i\,|_\alpha \mathbf{G}_i = \mathbf{G}\,{}^t\mathbf{a}\,|_\alpha$ where $\mathbf{G} \equiv [\,\mathbf{G}_1 \quad \mathbf{G}_2 \quad \mathbf{G}_3\,]$ is the orthonormal matrix with base vectors as its columns, and ${}^t\mathbf{a}(S^\alpha)$ is the component vector corresponding to the vector function, $\mathbf{a}(S^\alpha)$.

Exercise: Show that the partial derivative of a tensor function of any order is identical with its covariant derivative with respect to its arguments. Hint: For vector function $\mathbf{d}\,|_\alpha = {}^t d_i\,|_\alpha\,\mathbf{G}_i + {}^t d_i\,\cancel{\mathbf{G}_i|_\alpha} = {}^t d_i\,|_\alpha\,\mathbf{G}_i = \mathbf{d},_\alpha$. Why do you think $\mathbf{G}_i|_\alpha = \mathbf{0}$? Because, by definition, $\mathbf{G}_i|_j = \mathbf{G}_{i},_{j} - \Gamma^k_{ij}\mathbf{G}_k = \Gamma^k_{ij}\mathbf{G}_k - \ \Gamma^k_{ij}\mathbf{G}_k = 0$.

11.10.2 Covariant Derivatives of the Component or Operational Vectors

Now, with the above results in the background, we are in a position to express the derivatives of various component vectors with the help of the following lemma for a shell.

Lemma: Given any vector function, $\mathbf{a}(S^\alpha)$, on a shell reference surface, and its corresponding component vector, ${}^t\mathbf{a}(S^\alpha)$, the covariant derivative of the component vector is given as:

$$\boxed{\begin{aligned} {}^t\mathbf{a}|_\alpha &= {}^t\mathbf{a},_\alpha + \mathbf{k}^0_\alpha \times {}^t\mathbf{a}, \quad \alpha = 1,2 \\ {}^t\mathbf{a}|_\alpha &= {}^t\mathbf{a},_\alpha + \mathbf{K}^0_\alpha \, {}^t\mathbf{a} \end{aligned}} \tag{11.301}$$

where × denotes vector cross product operation.

Proof. Using results from above, we have:

$$\boxed{\begin{aligned} \mathbf{a}|_\alpha = \mathbf{a},_\alpha &= {}^t a_i,_\alpha \mathbf{G}_i + {}^t a_i \mathbf{G}_i,_\alpha = \mathbf{G}\left({}^t\mathbf{a},_\alpha + \mathbf{K}^0_\alpha \, {}^t\mathbf{a}\right) \\ &= \mathbf{G}\left({}^t\mathbf{a},_\alpha + \mathbf{k}^0_\alpha \times {}^t\mathbf{a}\right) \end{aligned}} \tag{11.302}$$

Now, noting that $\mathbf{a}|_\alpha = \mathbf{a},_\alpha = \mathbf{G}\,{}^t\mathbf{a}|_\alpha$ completes the proof.

11.10.2.1 Covariant Derivatives of Component Displacement and Rotation Vectors: ${}^t\mathbf{d}$ and ${}^t\theta$

Based on equation (11.301) of the lemma presented, we have:

$$\boxed{\begin{aligned} {}^t\mathbf{d}|_\alpha &= {}^t\mathbf{d},_\alpha + \mathbf{k}^0_\alpha \times {}^t\mathbf{d}, \quad \alpha = 1,2 \\ {}^t\theta|_\alpha &= {}^t\theta,_\alpha + \mathbf{k}^0_\alpha \times {}^t\theta \end{aligned}} \tag{11.303}$$

11.10.2.2 Covariant Derivatives of the Component Force and Moment Vectors: ${}^t\mathbf{F}^\alpha$ and ${}^t\mathbf{M}^\alpha$

Based on equation (11.301) of the lemma presented, we have:

$$\boxed{\begin{aligned} {}^t\mathbf{F}^\alpha|_\beta &= {}^t\mathbf{F}^\alpha,_\beta + \mathbf{k}^0_\beta \times {}^t\mathbf{F}^\alpha, \quad \alpha, \beta = 1,2 \\ {}^t\mathbf{M}^\alpha|_\beta &= {}^t\mathbf{M}^\alpha,_\beta + \mathbf{k}^0_\beta \times {}^t\mathbf{M}^\alpha \end{aligned}} \tag{11.304}$$

11.10.3 Where We Would Like to Go

We have defined the covariant derivatives of the component or operational vectors corresponding to the displacement and rotation vectors, and the force and moment vectors. These definitions play a vital role in converting the vector form of:

- the shell balance equations: 3D to 2D
- the shell 2D direct balance equations

to computationally more suitable component vector form of balance equations. Thus, with these background materials, we will be able to modify the equations in terms of what may be called the component vector or operator form as described in:

- the computational balance equations.

11.11 Computational Equations of Motion: Component Vector Form

11.11.1 What We Need to Recall

Let us recall the following for a thin shell.

- The invariant vector shell dynamic balance equations are given by:

$$\boxed{\begin{array}{ll} \mathbf{F}^{\alpha},_{\alpha} + \hat{\mathbf{F}} = \dot{\mathbf{L}}, \quad \alpha = 1,2 & \text{Momentum balance} \\ \dot{\mathbf{L}} = M_0\ \dot{\mathbf{v}}_0 = M_0\ \ddot{\mathbf{d}} \approx (\rho_0 H)\ \ddot{\mathbf{d}} & \\ \mathbf{M}^{\alpha},_{\alpha} + \mathbf{c}_0,_{\alpha} \times \mathbf{F}^{\alpha} + \hat{\mathbf{M}} = \dot{\mathbf{A}}, \quad \alpha = 1,2 & \text{Moment of momentum balance} \\ \dot{\mathbf{A}}(S^1,S^2,t) = I_0(S^1,S^2)\ \dot{\boldsymbol{\omega}}(S^1,S^2,t) \approx \left(\rho_0 \dfrac{H^3}{12}\right)\ \dot{\boldsymbol{\omega}}(S^1,S^2,t) & \end{array}} \tag{11.305}$$

- We have for the component or operational vectors: ${}^t\mathbf{d}$ and ${}^t\boldsymbol{\theta}$

$$\boxed{{}^t\mathbf{d} \equiv \{\, d_1 \equiv u \quad d_2 \equiv v \quad d_3 \equiv w \,\}^T} \tag{11.306}$$

$$\boxed{{}^t\boldsymbol{\theta} \equiv \{\, \theta_1 \quad \theta_2 \quad \theta_3 \,\}^T} \tag{11.307}$$

with:

$$\boxed{\begin{array}{l} \mathbf{d} = \mathbf{G}\ {}^t\mathbf{d} \\ \boldsymbol{\theta} = \mathbf{G}\ {}^t\boldsymbol{\theta} \end{array}} \tag{11.308}$$

- Similarly, for the forces and moments, we have:

$$\boxed{\begin{aligned} \mathbf{F}^{\alpha} &= {}^tF^{\alpha 1}(\equiv N_{\alpha 1})\ \mathbf{G_1} + {}^tF^{\alpha 2}(\equiv N_{\alpha 2})\ \mathbf{G_2} + {}^tF^{\alpha 3}(\equiv Q_{\alpha})\ \mathbf{N} \\ &= {}^tF^{\alpha i}\ \mathbf{G}_i, \quad \alpha = 1,2; \quad i = 1,2,3 \end{aligned}} \tag{11.309}$$

$$\boxed{\begin{aligned} \mathbf{M}^{\alpha} &= {}^tM^{\alpha 1}(\equiv -M_{\alpha 2})\ \mathbf{G_1} + {}^tM^{\alpha 2}(\equiv M_{\alpha 1})\ \mathbf{G_2} + {}^tM^{\alpha 3}(\equiv M_{\alpha 3})\ \mathbf{N} \\ &= {}^tM^{\alpha i}\ \mathbf{G}_i, \quad \alpha = 1,2, \quad i = 1,2,3 \end{aligned}} \tag{11.310}$$

with:

$$\boxed{\begin{array}{l} \mathbf{F}^{\alpha} = \mathbf{G}\ {}^t\mathbf{F}^{\alpha}, \quad \alpha = 1,2 \\ \mathbf{M}^{\alpha} = \mathbf{G}\ {}^t\mathbf{M}^{\alpha} \end{array}} \tag{11.311}$$

- $\mathbf{k}^0_\alpha$, $\alpha = 1, 2$, are the initial curvature vectors, that is, the axial vectors corresponding to the initial curvature tensors, $\mathbf{K}^0_\alpha$, $\alpha = 1, 2$, and are given by:

$$\boxed{\begin{aligned}\mathbf{k}^0_1 &\equiv \left\{0 \quad -\frac{1}{R_1} \quad -\frac{1}{\gamma_1}\gamma_{1,2}\right\}^T \equiv \left\{\tau^1 \quad \kappa^1_2 \quad \kappa^1_3\right\}^T \\ \mathbf{k}^0_2 &\equiv \left\{\frac{1}{R_2} \quad 0 \quad \frac{1}{\gamma_2}\gamma_{2,1}\right\}^T \equiv \left\{\kappa^2_3 \quad \tau^2 \quad \kappa^2_2\right\}^T\end{aligned}} \tag{11.312}$$

- For the covariant derivative of displacement component vectors, we have: $\mathbf{d}|_\alpha = \mathbf{d},_\alpha = \mathbf{G}\ {}^t\mathbf{d}|_\alpha$, $\alpha = 1, 2$ with:

$$\boxed{\begin{aligned}{}^t\mathbf{d}|_\alpha &= {}^t\mathbf{d},_\alpha + \mathbf{k}^0 \times {}^t\mathbf{d} \\ {}^t\mathbf{d}|_\alpha &= {}^t\mathbf{d},_\alpha + \mathbf{K}^0\ {}^t\mathbf{d}\end{aligned}} \tag{11.313}$$

  Similar results follow for the rotation component vector.
- Likewise, we have the covariant derivatives of forces and moments:

$$\boxed{\mathbf{F}^\alpha,_\alpha = \mathbf{G}\ {}^t\mathbf{F}^\alpha|_\alpha} \tag{11.314}$$

  with:

$$\boxed{\begin{aligned}{}^t\mathbf{F}^\alpha|_\alpha &= {}^t\mathbf{F}^\alpha,_\alpha + \mathbf{k}^0_\alpha \times {}^t\mathbf{F}^\alpha, \\ {}^t\mathbf{F}^\alpha|_\alpha &= {}^t\mathbf{F}^\alpha,_\alpha + \mathbf{K}^0_\alpha\ {}^t\mathbf{F}^\alpha\end{aligned}} \tag{11.315}$$

  and

$$\mathbf{M}^\alpha,_\alpha = \mathbf{G}\ {}^t\mathbf{M}^\alpha|_\alpha \tag{11.316}$$

  with:

$$\boxed{\begin{aligned}{}^t\mathbf{M}^\alpha|_\alpha &= {}^t\mathbf{M}^\alpha,_\alpha + \mathbf{k}^0_\alpha \times {}^t\mathbf{M}^\alpha \\ {}^t\mathbf{M}^\alpha|_\alpha &= {}^t\mathbf{M}^\alpha,_\alpha + \mathbf{K}^0_\alpha\ {}^t\mathbf{M}^\alpha\end{aligned}} \tag{11.317}$$

- Likewise, we have the time derivatives of the linear momentum and the angular momentum in terms of the components measured in the undeformed coordinate system:

$$\boxed{\begin{aligned}\mathbf{L}(S^1, S^2, t) &= \mathbf{G}(S^1, S^2)\ {}^t\mathbf{L}(S^1, S^2, t), \\ {}^t\dot{\mathbf{L}} &= M_0\ {}^t\ddot{\mathbf{d}} \approx (\rho_0 H)\ {}^t\ddot{\mathbf{d}} \\ \mathbf{A}(S^1, S^2, t) &= \mathbf{G}(S^1, S^2)\ {}^t\mathbf{A}(S^1, S^2, t), \\ {}^t\dot{\mathbf{A}} &= I_0\ {}^t\dot{\boldsymbol{\omega}} \approx \left(\rho_0 \frac{H^3}{12}\right)\ {}^t\dot{\boldsymbol{\omega}}\end{aligned}} \tag{11.318}$$

  Finally, $\mathbf{G} = [\,\mathbf{G}_1 \quad \mathbf{G}_2 \quad \mathbf{G}_3\,]$ is the 3×3 tensor that has a representation as an orthonormal matrix with the base vectors stored as its columns.

11.11.1.1 Covariant Derivative of the Deformed Position Vector

The derivative of the deformed position vector, $\mathbf{c}_0$, in the second of the balance equation (11.305), can be rewritten as:

$$\begin{aligned} \mathbf{c}_0|_\alpha = \mathbf{c}_0,_\alpha = & \ \mathbf{C}_0,_\alpha + \mathbf{d},_\alpha = \mathbf{G}_\alpha + \mathbf{G}\,{}^t\mathbf{d}|_\alpha \\ = & \ \mathbf{G}\,(\hat{\mathbf{1}}_\alpha + {}^t\mathbf{d},_\alpha + \mathbf{k}^0_\alpha \times {}^t\mathbf{d}) \\ = & \ \mathbf{G}\,(\mathbf{e}_\alpha + \mathbf{k}^0_\alpha \times {}^t\mathbf{d}) = \mathbf{G}\,\mathbf{a}_\alpha, \qquad \alpha = 1,2 \end{aligned} \tag{11.319}$$

where we have introduced the following.

Definitions

$$\begin{aligned} \hat{\mathbf{1}}_1 &\equiv \{1 \quad 0 \quad 0\}^T, \\ \hat{\mathbf{1}}_2 &\equiv \{0 \quad 1 \quad 0\}^T \end{aligned} \tag{11.320}$$

are the unit pseudo-vectors along the base vectors, $\{\mathbf{G}_i\}$, $i = 1, 2, 3$, and

$$\begin{aligned} \mathbf{e}_\alpha &\equiv \hat{\mathbf{1}}_\alpha + {}^t\mathbf{d},_\alpha, \quad \alpha = 1,2 \\ \mathbf{e}_1 &\equiv \hat{\mathbf{1}}_1 + {}^t\mathbf{d},_1 = \{(1 + u,_1) \quad v,_1 \quad w,_1\}^T, \\ \mathbf{e}_2 &\equiv \hat{\mathbf{1}}_2 + {}^t\mathbf{d},_2 = \{u,_2 \quad (1 + v,_2) \quad w,_2\}^T \end{aligned} \tag{11.321}$$

and

$$\begin{aligned} \mathbf{a}_\alpha &\equiv (\hat{\mathbf{1}}_\alpha + {}^t\mathbf{d}|_\alpha), \quad \alpha = 1,2 \\ &= (\hat{\mathbf{1}}_\alpha + {}^t\mathbf{d},_\alpha) + \mathbf{k}^0_\alpha \times {}^t\mathbf{d} \\ \mathbf{a}_1 &\equiv (\hat{\mathbf{1}}_1 + {}^t\mathbf{d},_1) + \mathbf{k}^0_1 \times {}^t\mathbf{d} = \mathbf{e}_1 + \mathbf{k}^0_1 \times {}^t\mathbf{d} \\ &= \left\{(1 + u,_1 - \kappa^1_3\, v + \kappa^1_2\, w) \quad (v,_1 + \kappa^1_3\, u) \quad (w,_1 - \kappa^1_2\, u)\right\}^T, \\ \mathbf{a}_2 &\equiv (\hat{\mathbf{1}}_2 + {}^t\mathbf{d},_2) + \mathbf{k}^0_2 \times {}^t\mathbf{d} = \mathbf{e}_2 + \mathbf{k}^0_2 \times {}^t\mathbf{d} \\ &= \left\{(u,_1 - \kappa^2_2\, v) \quad (1 + v,_1 + \kappa^2_2\, u - \kappa^2_3\, w) \quad (w,_1 + \kappa^2_3\, v)\right\}^T \end{aligned} \tag{11.322}$$

Then, from the second term in the angular momentum balance equation (11.305), by using the first relation in equation (11.311), we have:

$$\mathbf{c}_0,_\alpha \times \mathbf{F}^\alpha = \mathbf{G}\,\mathbf{a}_\alpha \times \mathbf{G}\,{}^t\mathbf{F}^\alpha = (\det\ \mathbf{G})\,\mathbf{G}(\mathbf{a}_\alpha \times {}^t\mathbf{F}^\alpha) = \mathbf{G}(\mathbf{a}_\alpha \times {}^t\mathbf{F}^\alpha), \quad \alpha = 1,2 \tag{11.323}$$

Exercise: Show that for any two non-collinear vectors $\mathbf{b}_1, \mathbf{b}_2$, and an orthonormal tensor $\mathbf{A}$: $\mathbf{A}\mathbf{b}_1 \times \mathbf{A}\mathbf{b}_2 = (\det \mathbf{A})\ \mathbf{A}(\mathbf{b}_1 \times \mathbf{b}_2)$.

Solution: Let $\mathbf{b}_3$ be any arbitrary vector non-collinear with $\mathbf{b}_1, \mathbf{b}_2$. Let us define a tensor $\mathbf{B} \equiv \mathbf{b}_i \otimes \mathbf{G}^i$ where $\mathbf{G}^i$ are the contravariant base vectors. Note that $\mathbf{b}_i \equiv \mathbf{B}\,\mathbf{G}_i$ that is, the matrix representation of $\mathbf{B}$ has three columns as $\mathbf{b}_1, \mathbf{b}_2$ and $\mathbf{b}_3$. Thus, $\det \mathbf{B} = \mathbf{b}_1 \times \mathbf{b}_2 \bullet \mathbf{b}_3$. Note that $\mathbf{AB} = (\mathbf{A}\mathbf{b}_j \otimes \mathbf{G}^j)$ because for $\forall \mathbf{h} \in \mathbb{R}_3$, $\mathbf{ABh} = \mathbf{A}(\mathbf{b}_j \otimes \mathbf{G}^j)\mathbf{h} = \mathbf{A}(\mathbf{G}^j \bullet \mathbf{h})\mathbf{b}_j = (\mathbf{A}\mathbf{b}_j)(\mathbf{G}^j \bullet \mathbf{h}) = (\mathbf{A}\mathbf{b}_j \otimes \mathbf{G}^j)\mathbf{h}$. So, $\det(\mathbf{AB}) = \mathbf{A}\mathbf{b}_1 \times \mathbf{A}\mathbf{b}_2 \bullet \mathbf{A}\mathbf{b}_3$. But, by the product rule of determinants, we have: $\det(\mathbf{AB}) = \det(\mathbf{A})\det(\mathbf{B}) = \det(\mathbf{A})(\mathbf{b}_1 \times \mathbf{b}_2 \bullet \mathbf{b}_3)$. So, we get: $\mathbf{A}^T(\mathbf{Ab}_1 \times \mathbf{Ab}_2) \bullet \mathbf{b}_3 = \det(\mathbf{A})(\mathbf{b}_1 \times \mathbf{b}_2) \bullet \mathbf{b}_3$. Now, noting that $\mathbf{b}_3$ is arbitrary and that $\mathbf{AA}^T = \mathbf{I}$ because $\mathbf{A}$ is orthonormal, completes the proof.

Now, evaluating the expressions in equation (11.322), we have:

$$\mathbf{a}_1 = \begin{Bmatrix} (1+u,_1) + \left(\frac{1}{\gamma_1}\gamma_{1},_2\right) v - \frac{1}{R_1} w \\ v,_1 - \left(\frac{1}{\gamma_1}\gamma_{1},_2\right) u \\ w,_1 + \frac{1}{R_1} u \end{Bmatrix}; \quad \mathbf{a}_2 = \begin{Bmatrix} u,_2 - \left(\frac{1}{\gamma_2}\gamma_{2},_1\right) v \\ (1+v,_2) + \left(\frac{1}{\gamma_2}\gamma_{2},_1\right) u - \frac{1}{R_2} w \\ w,_2 + \frac{1}{R_2} v \end{Bmatrix} \quad (11.324)$$

Then, noting in equation (11.323), $\mathbf{c}_0,_\alpha \times \mathbf{F}^\alpha$ in column vector form is given by $\mathbf{a}_\alpha \times {}^t\mathbf{F}^\alpha$ as:

$$\mathbf{a}_1 \times {}^t\mathbf{F}^1 = \begin{Bmatrix} \left(v,_1 - \left(\frac{1}{\gamma_1}\gamma_{1},_2\right) u\right) Q_1 - \left(w,_1 + \frac{1}{R_1} u\right) N_{12} \\ \left(w,_1 + \frac{1}{R_1} u\right) N_1 - \left((1+u,_1) + \left(\frac{1}{\gamma_1}\gamma_{1},_2\right) v - \frac{1}{R_1} w\right) Q_1 \\ \left((1+u,_1) + \left(\frac{1}{\gamma_1}\gamma_{1},_2\right) v - \frac{1}{R_1} w\right) N_{12} - \left(v,_1 - \left(\frac{1}{\gamma_1}\gamma_{1},_2\right) u\right) N_1 \end{Bmatrix} \quad (11.325)$$

and similarly,

$$\mathbf{a}_2 \times {}^t\mathbf{F}^2 = \begin{Bmatrix} \left((1+v,_2) + \left(\frac{1}{\gamma_2}\gamma_{2},_1\right) u - \frac{1}{R_2} w\right) Q_2 - \left(w,_2 + \frac{1}{R_2} v\right) N_2 \\ \left(w,_2 + \frac{1}{R_2} v\right) N_{21} - \left(u,_2 - \left(\frac{1}{\gamma_2}\gamma_{2},_1\right) v\right) Q_2 \\ \left(u,_2 - \left(\frac{1}{\gamma_2}\gamma_{2},_1\right) v\right) N_2 - \left((1+v,_2) + \left(\frac{1}{\gamma_2}\gamma_{2},_1\right) u - \frac{1}{R_2} w\right) N_{21} \end{Bmatrix} \quad (11.326)$$

11.11.1.2 Component Vector Form of Dynamic Balance Equations

Now, by applying equations (11.314), (11.316) and (11.162) in the equation (11.305), we get with equation (11.323):

$$\boxed{\begin{aligned} \mathbf{G}({}^t\mathbf{F}^\alpha|_\alpha + {}^t\hat{\mathbf{F}}^\alpha - {}^t\dot{\mathbf{L}}) &= 0 \\ \mathbf{G}({}^t\mathbf{M}^\alpha|_\alpha + \mathbf{a}_\alpha \times {}^t\mathbf{F}^\alpha + {}^t\hat{\mathbf{M}}^\alpha - {}^t\dot{\mathbf{A}}) &= 0 \end{aligned}} \quad (11.327)$$

11.11.1.3 Expanded Component-wise Force and Moment Covariant Derivatives

As a preliminary result, we may expand expressions in equation (11.315) component-wise:

$$
\begin{aligned}
{}^t\mathbf{F}^{\alpha}|_{\alpha} &= {}^t\mathbf{F}^{\alpha},_{\alpha} + \boldsymbol{\kappa}^0_{\alpha} \times {}^t\mathbf{F}^{\alpha} \\
&= \left\{ N_1,_1 + \frac{1}{\gamma_1}\gamma_1,_2 N_{12} - \frac{1}{\gamma_2}\gamma_2,_1 N_2 + N_{21},_2 - \frac{1}{R_1} Q_1 + \hat{F}^1 \right\} \mathbf{G}_1 \\
&\quad + \left\{ N_2,_2 + \frac{1}{\gamma_2}\gamma_2,_1 N_{21} - \frac{1}{\gamma_1}\gamma_1,_2 N_1 + N_{12},_1 - \frac{1}{R_2} Q_2 + \hat{F}^2 \right\} \mathbf{G}_2 \\
&\quad + \left\{ \frac{1}{R_1} N_1 + \frac{1}{R_2} N_2 + Q_1,_1 + Q_2,_2 + \hat{F}^3 \right\} \mathbf{N} \\
&= {}^t\mathbf{F}^{\alpha},_{\alpha} + \mathbf{K}^0_{\alpha}\, {}^t\mathbf{F}^{\alpha}
\end{aligned}
\qquad (11.328)
$$

Similarly, from equation (11.317) with the definition of $\mathbf{a}_{\alpha}$ as above, we get:

$$
\begin{aligned}
{}^t\mathbf{M}^{\alpha}|_{\alpha} + \mathbf{a}_{\alpha} \times {}^t\mathbf{F}^{\alpha} &= {}^t\mathbf{M}^{\alpha},_{\alpha} + \boldsymbol{\kappa}^0_{\alpha} \times {}^t\mathbf{M}^{\alpha} + \{\boldsymbol{\kappa}^0_{\alpha} \times \mathbf{d} + \mathbf{e}_{\alpha}\} \times {}^t\mathbf{F}^{\alpha} \\
&= \left\{ \begin{array}{c}
-M_2,_2 - M_{12},_1 - \frac{1}{R_1} M_{13} + \frac{1}{\gamma_1}\gamma_1,_2 M_1 + \frac{1}{\gamma_2}\gamma_2,_1 M_{21} + X_1 \\
M_1,_1 + M_{21},_2 - \frac{1}{R_2} M_{13} + \frac{1}{\gamma_1}\gamma_1,_2 M_{12} - \frac{1}{\gamma_2}\gamma_2,_1 M_2 + X_2 \\
M_{13},_1 + M_{23},_2 - \frac{1}{R_1} M_{12} + \frac{1}{R_2} M_{21} + X_3
\end{array} \right\} \\
&= {}^t\mathbf{M}^{\alpha},_{\alpha} + \mathbf{K}^0_{\alpha}\, {}^t\mathbf{M}^{\alpha} + \{(\mathbf{K}^0_{\alpha}\mathbf{D} - \mathbf{D}\,\mathbf{K}^0_{\alpha}) + \mathbf{E}_{\alpha}\}\, {}^t\mathbf{F}^{\alpha}
\end{aligned}
\qquad (11.329)
$$

where we have used the following new definitions:

$$
\begin{aligned}
X_1 &\equiv \left(v,_1 - \frac{1}{\gamma_1}\gamma_1,_2 u \right) Q_1 - \left(w,_1 + \frac{1}{R_1} u \right) N_{12} + \left(1 + v,_1 + \frac{1}{\gamma_2}\gamma_2,_1 u \right) Q_2 - \left(w,_1 + \frac{1}{R_2} v \right) N_2 \\
X_2 &\equiv \left(w,_1 + \frac{1}{R_1} u \right) N_1 - \left(1 + u,_1 + \frac{1}{\gamma_1}\gamma_1,_2 v - \frac{1}{R_1} w \right) Q_1 + \left(w,_1 + \frac{1}{R_2} v \right) N_{21} \\
&\quad - \left(u,_1 - \frac{1}{\gamma_2}\gamma_2,_1 v \right) Q_2 \\
X_3 &\equiv \left(1 + u,_1 + \frac{1}{\gamma_1}\gamma_1,_2 v - \frac{1}{R_1} w \right) N_{12} - \left(v,_1 - \frac{1}{\gamma_1}\gamma_1,_2 u \right) N_1 + \left(u,_1 - \frac{1}{\gamma_2}\gamma_2,_1 v \right) N_2 \\
&\quad - \left(1 + v,_1 + \frac{1}{\gamma_2}\gamma_2,_1 u \right) N_{21}
\end{aligned}
\qquad (11.330)
$$

In the above expression, we have used the following.

Definitions

- $\mathbf{D} \equiv [\mathbf{d}]$ is the skew-symmetric tensor with the displacement vector, $\mathbf{d} = \{\, u \equiv d_1 \quad v \equiv d_2 \quad w \equiv d_3 \,\}^T$ taken as its axial vector,
- $\mathbf{E}_1 \equiv [\mathbf{e}_1]$ is the skew-symmetric tensor corresponding to the special displacement derivative vector $\mathbf{e}_1 \equiv \{\,(1 + u,_1) \quad v,_1 \quad w,_1\,\}^T = \hat{\mathbf{1}}_1 + \mathbf{d},_1$ taken as its axial vector.
- $\mathbf{E}_2 \equiv [\mathbf{e}_2]$ is the skew-symmetric tensor corresponding to the special displacement derivative vector $\mathbf{e}_2 \equiv \{\, u,_2 \quad (1 + v,_2) \quad w,_2\,\}^T = \hat{\mathbf{1}}_2 + \mathbf{d},_2$ taken as its axial vector.

Finally, we have used the Lie bracket definition, a (commutator) property relating a skew-symmetric tensor to its axial vector.

Exercise: Given any two skew-symmetric tensors, **A** and **B** with the corresponding axial vectors, **a** and **b**, respectively, we have the commutator (or Lie bracket) relationship: $\mathbf{AB} - \mathbf{BA} = [\mathbf{a} \times \mathbf{b}] = (\mathbf{a} \times \mathbf{b})\times$.

Solution: For any $\mathbf{h} \in \mathbb{R}_3$, $(\mathbf{AB} - \mathbf{BA})\mathbf{h} = \mathbf{a} \times (\mathbf{b} \times \mathbf{h}) - \mathbf{b} \times (\mathbf{a} \times \mathbf{h}) = (\mathbf{a} \bullet \mathbf{h})\mathbf{b} - \cancel{(\mathbf{a} \bullet \mathbf{h})\mathbf{b}} - (\mathbf{b} \bullet \mathbf{h})\mathbf{a} + \cancel{(\mathbf{a} \bullet \mathbf{h})\mathbf{b}} = (\mathbf{a} \bullet \mathbf{h})\mathbf{b} - (\mathbf{b} \bullet \mathbf{h})\mathbf{a} = \{(\mathbf{b} \otimes \mathbf{a}) - (\mathbf{a} \otimes \mathbf{b})\}\,\mathbf{h}$ completes the first equality; also, $(\mathbf{a} \bullet \mathbf{h})\mathbf{b} - (\mathbf{b} \bullet \mathbf{h})\mathbf{a} = \mathbf{h} \times (\mathbf{b} \times \mathbf{a}) = (\mathbf{a} \times \mathbf{b}) \times \mathbf{h} = [\mathbf{a} \times \mathbf{b}]\mathbf{h}$ establishes the rest of the equalities. In all of the derivations, we have used the vector identity: $\mathbf{a} \times (\mathbf{b} \times \mathbf{c}) = (\mathbf{a} \bullet \mathbf{c})\mathbf{b} - (\mathbf{a} \bullet \mathbf{b})\mathbf{c}$.

Exercise: Verify expressions given by equations (11.328) and (11.329).

11.11.1.4 Component-wise Dynamic Balance Equations

Now, using equations (11.328) and (11.329) in equation (11.327), we can express the combined balance equations component-wise as follows:

$$\boxed{\begin{aligned} \mathbf{G}(\,{}^t\mathbf{F}^\alpha,_\alpha + \boldsymbol{\kappa}^0_\alpha \times {}^t\mathbf{F}^\alpha + {}^t\hat{\mathbf{F}}^\alpha - {}^t\dot{\mathbf{L}}) = \mathbf{0} \\ \mathbf{G}(\,{}^t\mathbf{M}^\alpha,_\alpha + \boldsymbol{\kappa}^0_\alpha \times {}^t\mathbf{M}^\alpha + \{\boldsymbol{\kappa}^0_\alpha \times \mathbf{d} + \mathbf{e}_\alpha\} \times {}^t\mathbf{F}^\alpha + {}^t\hat{\mathbf{M}}^\alpha - {}^t\dot{\mathbf{A}}) = \mathbf{0} \end{aligned}} \tag{11.331}$$

or, using the definition of the initial curvature vectors in equation (11.312) and of the force components in equation (11.309), we have for the first (momentum balance) of equation (11.331):

$$\begin{aligned} &\left\{ N_1,_1 + \frac{1}{\gamma_1}\gamma_1,_2 N_{12} - \frac{1}{\gamma_2}\gamma_2,_1 N_2 + N_{21},_2 - \frac{1}{R_1}Q_1 + \hat{F}^1 \right\} \mathbf{G}_1 \\ &+ \left\{ N_2,_2 + \frac{1}{\gamma_2}\gamma_2,_1 N_{21} - \frac{1}{\gamma_1}\gamma_1,_2 N_1 + N_{12},_1 - \frac{1}{R_2}Q_2 + \hat{F}^2 \right\} \mathbf{G}_2 \\ &+ \left\{ \frac{1}{R_1}N_1 + \frac{1}{R_2}N_2 + Q_1,_1 + Q_2,_2 + \hat{F}^3 \right\} \mathbf{N} = {}^t\dot{L}_i \mathbf{G}_i \end{aligned} \tag{11.332}$$

Now, in equation (11.332), noting that the base vectors are linearly independent, the coefficient of each has to independently vanish resulting in the following three momentum or force balance equations:

$$\begin{aligned}
N_{1},_{1}+\frac{1}{\gamma_1}\gamma_{1},_{2}N_{12}-\frac{1}{\gamma_2}\gamma_{2},_{1}N_2+N_{21},_{2}-\frac{1}{R_1}Q_1+\hat{F}^1&=\dot{L}_1\\
N_{2},_{2}+\frac{1}{\gamma_2}\gamma_{2},_{1}N_{21}-\frac{1}{\gamma_1}\gamma_{1},_{2}N_1+N_{12},_{1}-\frac{1}{R_2}Q_2+\hat{F}^2&=\dot{L}_2\\
\frac{1}{R_1}N_1+\frac{1}{R_2}N_2+Q_{1},_{1}+Q_{2},_{2}+\hat{F}^3&=\dot{L}_N
\end{aligned}\tag{11.333}$$

Next, using the similar arguments, we can find the expanded component-wise angular momentum balance equations. Thus, by combining the moment balance and the angular momentum balance equations, we have:

$$\boxed{\begin{aligned}
N_{1},_{1}+N_{21},_{2}+\frac{1}{\gamma_1}\gamma_{1},_{2}N_{12}-\frac{1}{\gamma_2}\gamma_{2},_{1}N_2-\frac{1}{R_1}Q_1+\hat{F}^1&=\dot{L}_1\\
N_{2},_{2}+N_{12},_{1}+\frac{1}{\gamma_2}\gamma_{2},_{1}N_{21}-\frac{1}{\gamma_1}\gamma_{1},_{2}N_1-\frac{1}{R_2}Q_2+\hat{F}^2&=\dot{L}_2\\
Q_{1},_{1}+Q_{2},_{2}+\frac{1}{R_1}N_1+\frac{1}{R_2}N_2+\hat{F}^3&=\dot{L}_N\\
-M_{2},_{2}-M_{12},_{1}-\frac{1}{R_1}M_{13}+\frac{1}{\gamma_1}\gamma_{1},_{2}M_1+\frac{1}{\gamma_2}\gamma_{2},_{1}M_{21}+X_1+\hat{M}^1&=\dot{A}_1\\
M_{1},_{1}+M_{21},_{2}-\frac{1}{R_2}M_{13}+\frac{1}{\gamma_1}\gamma_{1},_{2}M_{12}-\frac{1}{\gamma_2}\gamma_{2},_{1}M_2+X_2+\hat{M}^2&=\dot{A}_2\\
M_{13},_{1}+M_{23},_{2}-\frac{1}{R_1}M_{12}+\frac{1}{R_2}M_{21}+X_3+\hat{M}^3&=\dot{A}_N
\end{aligned}}\tag{11.334}$$

11.11.1.5 Matrix Form of Dynamic Balance Equations

Equation (11.334) may be expressed in matrix form as:

11.11.1.5.1 Momentum Balance Equations

$$\boxed{\begin{aligned}
&\begin{Bmatrix} N_{1},_{1}\\ N_{12},_{1}\\ Q_{1},_{1}\end{Bmatrix}+\begin{bmatrix} 0 & \frac{1}{\gamma_2}\gamma_{2},_{1} & 0\\ -\frac{1}{\gamma_2}\gamma_{2},_{1} & 0 & \frac{1}{R_2}\\ 0 & -\frac{1}{R_2} & 0\end{bmatrix}\begin{Bmatrix} N_1\\ N_{12}\\ Q_1\end{Bmatrix}+\\
&\begin{Bmatrix} N_{21},_{2}\\ N_{2},_{2}\\ Q_{2},_{2}\end{Bmatrix}+\begin{bmatrix} 0 & -\frac{1}{\gamma_1}\gamma_{1},_{2} & \frac{1}{R_1}\\ \frac{1}{\gamma_1}\gamma_{1},_{2} & 0 & 0\\ -\frac{1}{R_1} & 0 & 0\end{bmatrix}\begin{Bmatrix} N_{21}\\ N_2\\ Q_2\end{Bmatrix}=\begin{Bmatrix}\dot{L}_1\\ \dot{L}_2\\ \dot{L}_N\end{Bmatrix}
\end{aligned}}\tag{11.335}$$

and

11.11.1.5.2 Angular (or Moment of) Momentum Balance Equations

$$\left\{\begin{matrix} -M_{12},_1 \\ M_1,_1 \\ M_{13},_1 \end{matrix}\right\} + \begin{bmatrix} 0 & \frac{1}{\gamma_2}\gamma_2,_1 & 0 \\ -\frac{1}{\gamma_2}\gamma_2,_1 & 0 & \frac{1}{R_2} \\ 0 & -\frac{1}{R_2} & 0 \end{bmatrix} \left\{\begin{matrix} -M_{12},_1 \\ M_1,_1 \\ M_{13},_1 \end{matrix}\right\} + \mathbf{K}^0_{1D} \left\{\begin{matrix} N_1 \\ N_{12} \\ Q_1 \end{matrix}\right\}$$
$$+ \left\{\begin{matrix} -M_2,_2 \\ M_{21},_2 \\ M_{23},_2 \end{matrix}\right\} + \begin{bmatrix} 0 & -\frac{1}{\gamma_1}\gamma_1,_2 & \frac{1}{R_1} \\ \frac{1}{\gamma_1}\gamma_1,_2 & 0 & 0 \\ -\frac{1}{R_1} & 0 & 0 \end{bmatrix} \left\{\begin{matrix} -M_2,_2 \\ M_{21},_2 \\ M_{23},_2 \end{matrix}\right\} + \mathbf{K}^0_{2D} \left\{\begin{matrix} N_{21} \\ N_2 \\ Q_2 \end{matrix}\right\} + \left\{\begin{matrix} \hat{M}_1 \\ \hat{M}_2 \\ \hat{M}_N \end{matrix}\right\} = \left\{\begin{matrix} \dot{A}_1 \\ \dot{\mathbf{A}}_2 \\ \dot{\mathbf{A}}_N \end{matrix}\right\} \tag{11.336}$$

where $\mathbf{K}^0_D$, the matrix commutator, is given by:

$$\mathbf{K}^0_{1D} \equiv \mathbf{K}^0_1\,\mathbf{D} - \mathbf{D}\,\mathbf{K}^0_1 \text{ where}$$
$$\mathbf{K}^0_1 = \begin{bmatrix} 0 & \frac{1}{\gamma_2}\gamma_2,_1 & 0 \\ -\frac{1}{\gamma_2}\gamma_2,_1 & 0 & \frac{1}{R_2} \\ 0 & -\frac{1}{R_2} & 0 \end{bmatrix} \quad \text{and} \quad \mathbf{D} = \begin{bmatrix} 0 & -w & v \\ w & 0 & -u \\ -v & u & 0 \end{bmatrix} \tag{11.337}$$

and

$$\mathbf{K}^0_{2D} \equiv \mathbf{K}^0_2\,\mathbf{D} - \mathbf{D}\,\mathbf{K}^0_2 \text{ where}$$
$$\mathbf{K}^0_2 = \begin{bmatrix} 0 & -\frac{1}{\gamma_1}\gamma_1,_2 & \frac{1}{R_1} \\ \frac{1}{\gamma_1}\gamma_1,_2 & 0 & 0 \\ -\frac{1}{R_1} & 0 & 0 \end{bmatrix} \quad \text{and} \quad \mathbf{D} = \begin{bmatrix} 0 & -w & v \\ w & 0 & -u \\ -v & u & 0 \end{bmatrix} \tag{11.338}$$

Exercise: Starting from equation (11.329), verify expressions given by equation (11.336).

Remark

- It is interesting to note that the rotation tensor due to shell shear deformation does not appear in any of the balance equations. It is only when we develop, later, the virtual work functional in a co-rotated frame under the Reissner–Mindlin assumption that we begin to see the effect of rotation due to shear deformation. Under the Kirchhoff assumption of the normal at a point on the shell reference surface remaining straight and perpendicular to the shell deformed reference surface (i.e. without any shear deformation), the functional will be completely characterized by the displacements alone.

11.11.2 Static Balance Equations

The quasi-static balance laws can be easily obtained from their dynamic counterparts by assuming that the linear and the angular momentum vectors identically vanish, that is,

$$\boxed{\begin{aligned} &\mathbf{L}(S^1, S^2, t) \equiv \mathbf{0} \\ &\mathbf{A}(S^1, S^2, t \equiv \mathbf{0} \end{aligned}} \tag{11.339}$$

Thus, we get:

11.11.2.1 Static Momentum Balance Equations (Force Equilibrium)

$$\boxed{\mathbf{F}^{\alpha},_{\alpha} + \ \hat{\mathbf{F}} = \mathbf{0}, \quad \alpha = 1, 2} \tag{11.340}$$

and

11.11.2.2 Static Angular or Moment of Momentum Balance Equations (Moment Equilibrium)

$$\boxed{\mathbf{M}^{\alpha},_{\alpha} \ + \mathbf{c}_0,_{\alpha} \times \mathbf{F}^{\alpha} \ + \hat{\mathbf{M}} \ = \mathbf{0}, \quad \alpha = 1, 2} \tag{11.341}$$

11.11.2.3 Component Vector form of Static Balance Equations

We get from equation (11.327):

$$\boxed{\begin{aligned} \mathbf{G}(\,{}^t\mathbf{F}^{\alpha}|_{\alpha} + {}^t\hat{\mathbf{F}}^{\alpha}) = 0 \\ \mathbf{G}(\,{}^t\mathbf{M}^{\alpha}|_{\alpha} + \mathbf{a}_{\alpha} \times {}^t\mathbf{F}^{\alpha} + {}^t\hat{\mathbf{M}}^{\alpha}) = 0 \end{aligned}} \tag{11.342}$$

as the static component vector form of the linear momentum and angular momentum balance equations.

11.11.2.4 Component-wise Static Balance Equations

Now, from equation (11.331), we can express the static combined balance equations component-wise as:

$$\boxed{\begin{aligned} \mathbf{G}(\,{}^t\mathbf{F}^{\alpha},_{\alpha} + \ \boldsymbol{\kappa}^0_{\alpha} \times {}^t\mathbf{F}^{\alpha} + {}^t\hat{\mathbf{F}}^{\alpha}) = \mathbf{0} \\ \mathbf{G}(\,{}^t\mathbf{M}^{\alpha},_{\alpha} + \ \boldsymbol{\kappa}^0_{\alpha} \times \ {}^t\mathbf{M}^{\alpha} + \{\boldsymbol{\kappa}^0_{\alpha} \times \mathbf{d} \ + \mathbf{e}_{\alpha}\} \times {}^t\mathbf{F}^{\alpha} + {}^t\hat{\mathbf{M}}^{\alpha}) = \mathbf{0} \end{aligned}} \tag{11.343}$$

or

$$
\begin{aligned}
N_{1,1} + N_{21,2} + \frac{1}{\gamma_1}\gamma_{1,2}N_{12} - \frac{1}{\gamma_2}\gamma_{2,1}N_2 - \frac{1}{R_1}Q_1 + \hat{F}^1 &= 0 \\
N_{2,2} + N_{12,1} + \frac{1}{\gamma_2}\gamma_{2,1}N_{21} - \frac{1}{\gamma_1}\gamma_{1,2}N_1 - \frac{1}{R_2}Q_2 + \hat{F}^2 &= 0 \\
Q_{1,1} + Q_{2,2} + \frac{1}{R_1}N_1 + \frac{1}{R_2}N_2 + \hat{F}^3 &= 0 \\
-M_{2,2} - M_{12,1} - \frac{1}{R_1}M_{13} + \frac{1}{\gamma_1}\gamma_{1,2}M_1 + \frac{1}{\gamma_2}\gamma_{2,1}M_{21} + X_1 + \hat{M}^1 &= 0 \\
M_{1,1} + M_{21,2} - \frac{1}{R_2}M_{13} + \frac{1}{\gamma_1}\gamma_{1,2}M_{12} - \frac{1}{\gamma_2}\gamma_{2,1}M_2 + X_2 + \hat{M}^2 &= 0 \\
M_{13,1} + M_{23,2} - \frac{1}{R_1}M_{12} + \frac{1}{R_2}M_{21} + X_3 + \hat{M}^3 &= 0
\end{aligned}
\tag{11.344}
$$

11.11.2.5 Matrix Form of Shell Static Balance Equations

From equations (11.335) and (11.336), we get in matrix form as:

11.11.2.6 Momentum Balance Equations

$$
\begin{aligned}
&\begin{Bmatrix} N_{1,1} \\ N_{12,1} \\ Q_{1,1} \end{Bmatrix} + \begin{bmatrix} 0 & \frac{1}{\gamma_2}\gamma_{2,1} & 0 \\ -\frac{1}{\gamma_2}\gamma_{2,1} & 0 & \frac{1}{R_2} \\ 0 & -\frac{1}{R_2} & 0 \end{bmatrix} \begin{Bmatrix} N_1 \\ N_{12} \\ Q_1 \end{Bmatrix} \\
&+ \begin{Bmatrix} N_{21,2} \\ N_{2,2} \\ Q_{2,2} \end{Bmatrix} + \begin{bmatrix} 0 & -\frac{1}{\gamma_1}\gamma_{1,2} & \frac{1}{R_1} \\ \frac{1}{\gamma_1}\gamma_{1,2} & 0 & 0 \\ -\frac{1}{R_1} & 0 & 0 \end{bmatrix} \begin{Bmatrix} N_{21} \\ N_2 \\ Q_2 \end{Bmatrix} = \begin{Bmatrix} 0 \\ 0 \\ 0 \end{Bmatrix}
\end{aligned}
\tag{11.345}
$$

and

11.11.2.7 Angular (or Moment of) Momentum Balance Equations:

$$\begin{Bmatrix} -M_{12},_1 \\ M_1,_1 \\ M_{13},_1 \end{Bmatrix} + \begin{bmatrix} 0 & \frac{1}{\gamma_2}\gamma_2,_1 & 0 \\ -\frac{1}{\gamma_2}\gamma_2,_1 & 0 & \frac{1}{R_2} \\ 0 & -\frac{1}{R_2} & 0 \end{bmatrix} \begin{Bmatrix} -M_{12},_1 \\ M_1,_1 \\ M_{13},_1 \end{Bmatrix} + \mathbf{K}^0_{1D} \begin{Bmatrix} N_1 \\ N_{12} \\ Q_1 \end{Bmatrix}$$

$$+ \begin{Bmatrix} -M_2,_2 \\ M_{21},_2 \\ M_{23},_2 \end{Bmatrix} + \begin{bmatrix} 0 & -\frac{1}{\gamma_1}\gamma_1,_2 & \frac{1}{R_1} \\ \frac{1}{\gamma_1}\gamma_1,_2 & 0 & 0 \\ -\frac{1}{R_1} & 0 & 0 \end{bmatrix} \begin{Bmatrix} -M_2,_2 \\ M_{21},_2 \\ M_{23},_2 \end{Bmatrix} + \mathbf{K}^0_{2D} \begin{Bmatrix} N_{21} \\ N_2 \\ Q_2 \end{Bmatrix} + \begin{Bmatrix} \hat{M}_1 \\ \hat{M}_2 \\ \hat{M}_N \end{Bmatrix} = \begin{Bmatrix} 0 \\ 0 \\ 0 \end{Bmatrix} \quad (11.346)$$

11.11.3 Where We Would Like to Go

We have now derived the thin shell computational balance equations. These serve as the starting point in developing

- virtual work functional
  necessary for the numerical formulation based on a finite element method of solution.

11.12 Computational Derivatives and Variations

11.12.1 What We Need to Recall

- The vector triple product property: $\mathbf{a} \times \mathbf{b} \bullet \mathbf{c} = \mathbf{b} \times \mathbf{c} \bullet \mathbf{a} = \mathbf{c} \times \mathbf{a} \bullet \mathbf{b}$, $\forall \mathbf{a}, \mathbf{b}, \mathbf{c}$ vectors.
- The vector triple cross product property: $\mathbf{a} \times \mathbf{b} \times \mathbf{c} = (\mathbf{a} \bullet \mathbf{c})\mathbf{b} - (\mathbf{a} \bullet \mathbf{b})\mathbf{c}$, $\forall \mathbf{a}, \mathbf{b}, \mathbf{c}$vectors.
- The definition of a rotation tensor, its matrix representation and algebra, geometry and topology of rotation.
- A rotation matrix, $\mathbf{R}$, corresponding to the 3D special orthogonal matrix group, $SO(3)$, may be parameterized by the Rodrigues vector, $\mathbf{s}$, the axis of the instantaneous rotation.
- The Rodrigues vector, $\mathbf{s} = \{ s_1 \quad s_2 \quad s_3 \}^T$, is orientated along the instantaneous axis of rotation at a point of the shell reference surface. The skew-symmetric matrix, $\mathbf{S}$, corresponding to the Rodrigues vector, $\mathbf{s}$, that is, $\mathbf{S}\,\mathbf{s} = \mathbf{0}$, is given by:

$$\mathbf{S} = \begin{bmatrix} 0 & -s_3 & s_2 \\ s_3 & 0 & -s_1 \\ -s_2 & s_1 & 0 \end{bmatrix} \quad (11.347)$$

- The rotation matrix, $\mathbf{R}(S^1, S^2, t)$, is related, through Cayley's theorem, to the Rodrigues vector, $\mathbf{s}(S^1, S^2, t) = \{ s_1(S^1, S^2, t) \quad s_2(S^1, S^2, t) \quad s_3(S^1, S^2, t) \}^T$, with the components in the

$\mathbf{G}_1, \mathbf{G}_2, \mathbf{N}$ frame at any point, (S^1, S^2), parameterized by the arc lengths along the lines of curvature on a shell reference surface, at time $t \in \mathbb{R}_+$, by:

$$
\begin{aligned}
{}^t\mathbf{R} &= (\mathbf{I}-\mathbf{S})^{-1}\,(\mathbf{I}+\mathbf{S}) \\
&= \frac{1}{\Sigma}\begin{bmatrix} 1+s_1^2-s_2^2-s_3^2 & 2(s_1s_2-s_3) & 2(s_1s_3+s_2) \\ 2(s_1s_2+s_3) & 1-s_1^2+s_2^2-s_3^2 & 2(s_2s_3-s_1) \\ 2(s_1s_3-s_2) & 2(s_2s_3+s_1) & 1-s_1^2-s_2^2+s_3^2 \end{bmatrix}
\end{aligned} \tag{11.348}
$$

where $\sum \equiv 1 + \|\, \mathbf{s} \,\|^2$ with the $(\mathbf{I}-\mathbf{S})$ matrix always invertible for any rotation matrix, defined by:

$$
(\mathbf{I}-\mathbf{S})^{-1} = \frac{1}{\Sigma}\begin{bmatrix} (1+s_1^2) & (s_1\,s_2-s_3) & (s_3\,s_1+s_2) \\ (s_1\,s_2+s_3) & (1+s_2^2) & (s_2\,s_3-s_1) \\ (s_3\,s_1-s_2) & (s_2\,s_3+s_1) & (1+s_3^2) \end{bmatrix} \tag{11.349}
$$

- Finally, the rotation vector, $\boldsymbol{\theta}$, is the scaled Rodrigues vector related by:

$$
\boxed{\mathbf{s} = \frac{1}{\theta}\tan\left(\frac{\theta}{2}\right)\boldsymbol{\theta}} \tag{11.350}
$$

where $\theta \equiv \|\boldsymbol{\theta}\| = (\boldsymbol{\theta} \bullet \boldsymbol{\theta})^{\frac{1}{2}}$ is the length measure of rotation vector. Alternatively, defining the unit vector, $\mathbf{n} \equiv \dfrac{\boldsymbol{\theta}}{\theta}$, and the corresponding skew-symmetric matrix, $\mathbf{N}$, such that $\mathbf{N}\,\mathbf{h} = \mathbf{n} \times \mathbf{h}$, $\forall \mathbf{h} \in \mathbb{R}_3$, we have:

$$
\boxed{\begin{aligned} &\mathbf{R} = (\mathbf{I}-\mathbf{S})^{-1}\,(\mathbf{I}+\mathbf{S}) \\ &\mathbf{s} = \tan\left(\frac{\theta}{2}\right)\mathbf{n}, \\ &\mathbf{S} = \tan\left(\frac{\theta}{2}\right)\mathbf{N} \end{aligned}} \tag{11.351}
$$

- The Rodrigues quaternion and Euler–Rodrigues parameters.
- For notational uniformity, we denote all variations of entities by a superscripted bar. Furthermore all entities measured in $\mathbf{G_1}, \mathbf{G_2}, \mathbf{N}$ frame are superscripted as ${}^t(\bullet)$, and those in the rotating $\mathbf{g}_1, \mathbf{g}_2, \mathbf{g}_3$ coordinate system as ${}^n(\bullet)$. However, for the sake of uncluttered readability, sometimes the superscripts may be dropped, but the context should help determine the appropriate underlying coordinate system.

11.12.2 Curvature Matrices, ${}^t\mathbf{K}_\alpha$, $\alpha = 1, 2$, *Curvature Vectors,* ${}^t\mathbf{k}_\alpha$, $\alpha = 1, 2$

Let $\mathbf{R}(S^1,\ S^2, t)$ be the rotation matrix at a point on the shell reference surface described in the arc length parameters, $(S^1,\ S^2)$, along the lines of the curvature direction, at any fixed time $t \in \mathbb{R}_+$. Then, considering it as a one-parameter family along each direction, and taking the derivative, in the sense described above, of the orthonormality relationship: $\mathbf{R}\,\mathbf{R}^T = \mathbf{I}$, with respect to this parameter, we have:

$$\boxed{(\mathbf{R},_\alpha \mathbf{R}^T) + (\mathbf{R},_\alpha \mathbf{R}^T)^T = \mathbf{0}, \quad \alpha = 1,2} \tag{11.352}$$

Now, we introduce the following.

Definitions

- The curvature matrices, ${}^t\mathbf{K}_\alpha$, $\alpha = 1, 2$, as:

$$\boxed{{}^t\mathbf{K}_\alpha \equiv {}^t\mathbf{R},_\alpha {}^t\mathbf{R}^\mathbf{T}, \quad \alpha = 1,2} \tag{11.353}$$

By introducing above definitions into equation (11.352), we have: ${}^t\mathbf{K}_\alpha + ({}^t\mathbf{K}_\alpha)^T = \mathbf{0}$, $\alpha = 1, 2$, signifying that ${}^t\mathbf{K}_\alpha$, $\alpha = 1, 2$ are 3D skew-symmetric matrices; we can then define the corresponding axial vectors, ${}^t\mathbf{k}_\alpha$, $\alpha = 1, 2$, such that ${}^t\mathbf{K}_\alpha\,\mathbf{h} = {}^t\mathbf{k}_\alpha \times \mathbf{h}$, $\alpha = 1, 2$ for $\forall \mathbf{h} \in \mathbb{R}_3$.

Moreover, from the matrix representation, as given by equation (11.348), of the rotation tensor, we have, by taking straightforward term-by-term derivatives, the corresponding matrix representation of the curvature tensors, ${}^t\mathbf{K}_\alpha$, $\alpha = 1, 2$, as:

$$\boxed{\begin{aligned} {}^t\mathbf{K}_\alpha &= \frac{2}{\Sigma}\left\{ \underbrace{(\mathbf{S}\,\mathbf{S},_\alpha - \mathbf{S},_\alpha\mathbf{S})}_{\text{Commutator or Lie bracket}} + \mathbf{S},_\alpha \right\} \\ &= \frac{2}{\Sigma}\begin{bmatrix} 0 & (-s_1s_2,_\alpha + s_1,_\alpha s_2 - s_3,_\alpha) & (-s_1s_3,_\alpha + s_1,_\alpha s_3 + s_2,_\alpha) \\ (s_1s_2,_\alpha - s_1,_\alpha s_2 + s_3,_\alpha) & 0 & (-s_2s_3,_\alpha + s_2,_\alpha s_3 - s_1,_\alpha) \\ (s_1s_3,_\alpha - s_1,_\alpha s_3 - s_2,_\alpha) & (s_2s_3,_\alpha - s_2,_\alpha s_3 + s_1,_\alpha) & 0 \end{bmatrix} \end{aligned}} \tag{11.354}$$

where the third equality follows from the first equality by the following fact.

Fact: $(\mathbf{A}\,\mathbf{B} - \mathbf{B}\,\mathbf{A})\,\mathbf{h} = [\mathbf{a} \times \mathbf{b}]\,\mathbf{h}$, $\forall \mathbf{h} \in \mathbb{R}_3$ where $[(\bullet)]$ denotes a skew-symmetric matrix corresponding to an axial vector $(\bullet)$, and $\mathbf{A}$, $\mathbf{B}$ are skew-symmetric matrices, and $\mathbf{a}$, $\mathbf{b}$ are the corresponding axial vectors, respectively.

Exercise: Show that matrix Lie bracket, $\mathbf{A}\,\mathbf{B} - \mathbf{B}\,\mathbf{A} = [\mathbf{a} \times \mathbf{b}]$

Solution: $(\mathbf{A}\,\mathbf{B} - \mathbf{B}\,\mathbf{A})\mathbf{h} = \mathbf{a} \times \mathbf{b} \times \mathbf{h} - \mathbf{b} \times \mathbf{a} \times \mathbf{h}$, $\forall \mathbf{h} \in \mathbb{R}_3 = (\mathbf{a} \bullet \mathbf{h})\mathbf{b} - (\mathbf{a} \bullet \mathbf{b})\mathbf{h} - (\mathbf{b} \bullet \mathbf{h})\mathbf{a} - (\mathbf{b} \bullet \mathbf{a})\mathbf{h} = \mathbf{h} \times (\mathbf{b} \times \mathbf{a}) = \mathbf{a} \times \mathbf{b} \times \mathbf{h} = [\mathbf{a} \times \mathbf{b}]\,\mathbf{h}$ completes the proof.

Exercise: Derive equation (11.354).

Solution: By taking the derivative of equation (11.348) with respect to $\alpha = 1, 2$, and using the definition of the curvature matrix as in equation (11.353), we have:

$$\begin{aligned}(\mathbf{I}-\mathbf{S})\,{}^t\mathbf{R},_{\alpha} - \mathbf{S},_{\alpha}\,{}^t\mathbf{R} = \mathbf{S},_{\alpha} &\Rightarrow (\mathbf{I}-\mathbf{S})\,{}^t\mathbf{R},_{\alpha}\,{}^t\mathbf{R}^T - \mathbf{S},_{\alpha}\,{}^t\mathbf{R}\,{}^t\mathbf{R}^T = \mathbf{S},_{\alpha}\,{}^t\mathbf{R}^T \\ &\Rightarrow {}^t\mathbf{K}_{\alpha} = (\mathbf{I}-\mathbf{S})^{-1}\{\mathbf{S},_{\alpha} + \mathbf{S},_{\alpha}\,{}^t\mathbf{R}^T\}\end{aligned}$$

Now, using the inverse relation given by equation (11.349) and direct multiplication completes the proof.

11.12.2.1 Curvature Vector, ${}^t\mathbf{k}_{\alpha}$, $\alpha = 1, 2$

Now, the expressions for the axial curvature vectors, ${}^t\mathbf{k}_{\alpha}$, $\alpha = 1, 2$ corresponding to the curvature matrices, ${}^t\mathbf{K}_{\alpha}$, $\alpha = 1, 2$, such that ${}^t\mathbf{K}_{\alpha}\,{}^t\mathbf{k}_{\alpha} = \mathbf{0}$, $\alpha = 1, 2$, follow from equation (11.354), as:

$$\boxed{\begin{aligned}{}^t\mathbf{k}_{\alpha} &= \frac{2}{\Sigma}(\mathbf{s}\times\mathbf{s},_{\alpha} + \mathbf{s},_{\alpha}), \quad \alpha = 1, 2 \\ &= \frac{2}{\Sigma}(\mathbf{S}\,\mathbf{s},_{\alpha} + \mathbf{s},_{\alpha}) = \left\{\frac{2}{\Sigma}(\mathbf{I}+\mathbf{S})\right\}\mathbf{s},_{\alpha} = \mathbf{W}_1(\mathbf{s})\,\mathbf{s},_{\alpha}\end{aligned}} \tag{11.355}$$

where we have introduced, for future use, the definition of the matrix:

$$\boxed{\mathbf{W}_1(\mathbf{s}) \equiv \frac{2}{\Sigma}(\mathbf{I}+\mathbf{S})} \tag{11.356}$$

with $\mathbf{I}$ as the 3×3 identity matrix. Now, noting that $\Sigma \equiv 1 + \|\mathbf{s}\|^2 = 1 + \tan^2\frac{\theta}{2} = \frac{2}{1+\cos\theta}$ and $\tan\frac{\theta}{2} = \frac{\sin\theta}{1+\cos\theta}$, we have, using the second relation of equation (11.351) in equation (11.356):

$$\boxed{\mathbf{W}_1(\mathbf{s}) = (1+\cos\theta)\left\{\mathbf{I} + \frac{\sin\theta}{1+\cos\theta}\mathbf{N}\right\}} \tag{11.357}$$

Note that this equation gives the expression in terms of the rotation vector, $\boldsymbol{\theta}$.

11.12.2.2 Curvature Vectors, ${}^t\mathbf{k}_{\alpha}$, $\alpha = 1, 2$ in Terms of the Rotation vector ${}^t\boldsymbol{\theta}$

Note that from the fourth equality in equation (11.355) that we only get ${}^t\mathbf{k}_{\alpha}$ as the function of the derivatives, $\mathbf{s},_{\alpha}$, of the Rodrigues vector. But, we would like to get its dependence on the rotation vector, ${}^t\boldsymbol{\theta}$. For this, we take the derivative of the relation given by equation (11.350):

$$\boxed{\begin{aligned}\mathbf{s},_{\alpha} &= \frac{1}{2}\left\{\left(\sec^2\frac{\theta}{2}\right)\frac{(\boldsymbol{\theta}\bullet\boldsymbol{\theta},_{\alpha})}{\theta^2}\right\}\boldsymbol{\theta} - \left\{\frac{(\boldsymbol{\theta}\bullet\boldsymbol{\theta},_{\alpha})}{\theta^3}\tan\frac{\theta}{2}\right\}\boldsymbol{\theta} + \left\{\frac{\boldsymbol{\theta},_{\alpha}}{\theta}\tan\frac{\theta}{2}\right\} \\ &= \left(\frac{1}{\theta}\tan\frac{\theta}{2}\right)\left\{\boldsymbol{\theta},_{\alpha} - \left(1 - \frac{\theta}{\sin\theta}\right)(\mathbf{n}\bullet\boldsymbol{\theta},_{\alpha})\,\mathbf{n}\right\} \\ &= \left(\frac{1}{\theta}\tan\frac{\theta}{2}\right)\left\{\mathbf{I} - \left(1 - \frac{\theta}{\sin\theta}\right)(\mathbf{n}\otimes\mathbf{n})\right\}\boldsymbol{\theta},_{\alpha} = \mathbf{W}_2\,\boldsymbol{\theta},_{\alpha}, \quad \alpha = 1, 2\end{aligned}} \tag{11.358}$$

where we have introduced the definition for the matrix:

$$\mathbf{W}_2 \equiv \left(\frac{1}{\theta}\tan\frac{\theta}{2}\right)\left\{\mathbf{I} - \left(1 - \frac{\theta}{\sin\theta}\right)(\mathbf{n} \otimes \mathbf{n})\right\} \tag{11.359}$$

Alternatively, as previously done for equation (11.356), we can express the relationship in equation (11.359) in terms of the rotation vector, $\boldsymbol{\theta}$, as:

$$\boxed{\mathbf{W}_2 = \frac{1}{1+\cos\theta}\left\{\mathbf{I} - \left(\frac{\sin\theta - \theta}{\theta}\right)\mathbf{N}^2\right\}} \tag{11.360}$$

where we have additionally used the identity: $\mathbf{N}^2 = \{(\mathbf{n} \otimes \mathbf{n}) - \mathbf{I}\}$.

Exercise: Show: $\mathbf{N}^2 = \{(\mathbf{n} \otimes \mathbf{n}) - \mathbf{I}\}$

Solution: $\mathbf{N}^2\,\mathbf{h} \equiv \mathbf{N}[\mathbf{n} \times \mathbf{h}] = \mathbf{n} \times \mathbf{n} \times \mathbf{h} = (\mathbf{n} \bullet \mathbf{h})\mathbf{n} - (\mathbf{n} \bullet \mathbf{n})\mathbf{h} = \{(\mathbf{n} \otimes \mathbf{n}) - \mathbf{I}\}\,\mathbf{h}, \quad \forall \mathbf{h} \in \mathbb{R}_3$

Now, using equation (11.358) with definition (11.360) in equation (11.355) with definition (11.357), we have:

$$\boxed{{}^t\mathbf{k}_\alpha = \mathbf{W}_1(\mathbf{s})\,\mathbf{s},_\alpha = \mathbf{W}_1\,\mathbf{W}_2\,\boldsymbol{\theta},_\alpha \equiv \mathbf{W}(\boldsymbol{\theta})\,{}^t\boldsymbol{\theta},_\alpha, \quad \alpha = 1,2} \tag{11.361}$$

where we have introduced the following definitions:

Definitions

$$\boxed{\begin{aligned} &\mathbf{W}(\boldsymbol{\theta}) \equiv \mathbf{I} + c_1\,\boldsymbol{\Theta} + c_2\,\boldsymbol{\Theta}^2, \\ &c_1(\boldsymbol{\theta}) = \frac{1-\cos\theta}{\theta^2}, \\ &c_2(\boldsymbol{\theta}) = \frac{\theta - \sin\theta}{\theta^3} \end{aligned}} \tag{11.362}$$

Note that equation (11.361) with the definitions (11.265) describes the curvature vector ${}^t\mathbf{k}_\alpha$ in terms of the derivatives, $\boldsymbol{\theta},_\alpha$, of the rotation vector for $\alpha = 1,2$ where $\boldsymbol{\Theta}$ is the skew-symmetric matrix corresponding to rotation vector, $\boldsymbol{\theta}$, as the axial vector such that $\boldsymbol{\Theta}\,\boldsymbol{\theta} = \mathbf{0}$, and $\boldsymbol{\Theta}^2 \equiv \boldsymbol{\Theta}\boldsymbol{\Theta} = [\boldsymbol{\theta} \times \boldsymbol{\theta} \times]$.

Exercise: Prove the relationships in equation (11.265).

Solution: Let us note that $\mathbf{N}$, being a skew-symmetric matrix with the form given similar to equation (11.347), and that every matrix satisfies its characteristic equation; from its eigen or characteristic equation, $p(\lambda) \equiv \lambda^3 + \lambda = 0$, we have: $\mathbf{N} + \mathbf{N}^3 = 0 \Rightarrow \mathbf{N}^3 = -\mathbf{N}$. Now, using relations (11.357) and (11.360) in equation (11.361), we have:

$$\mathbf{W} \equiv \mathbf{W}_1\ \mathbf{W}_2 = \left\{\mathbf{I} + \frac{\sin\theta}{1+\cos\theta}\ \mathbf{N}\right\}\left\{\mathbf{I} - \left(\frac{\sin\theta - \theta}{\theta}\right)\mathbf{N}^2\right\}$$

$$= \mathbf{I} - \left(\frac{\sin\theta - \theta}{\theta}\right)\mathbf{N}^2 + \frac{\sin\theta}{1+\cos\theta}\mathbf{N} - \frac{\sin^2\theta - \theta\sin\theta}{\theta(1+\cos\theta)}\mathbf{N}^3$$

$$= \mathbf{I} + \left(\frac{\theta - \sin\theta}{\theta}\right)\mathbf{N}^2 + \left\{\cancel{\frac{\sin\theta}{1+\cos\theta}} + \frac{\sin^2\theta}{\theta(1+\cos\theta)} - \cancel{\frac{\sin\theta}{1+\cos\theta}}\right\}\mathbf{N}$$

$$= \mathbf{I} + \underbrace{\left(\frac{1-\cos\theta}{\theta^2}\right)}_{\equiv c_1}\Theta + \underbrace{\left(\frac{\theta - \sin\theta}{\theta^3}\right)}_{\equiv c_2}\Theta^2$$

Noting that we have used the unit vector: $\mathbf{N} \equiv \dfrac{\Theta}{\theta}$, with Θ as the skew-symmetric matrix corresponding to $\boldsymbol{\theta}$ taken as the axial vector, completes the proof.

11.12.3 Angular Velocity Matrix, ${}^t\Omega$, Angular Velocity Vector, ${}^t\omega$

This time we take the rotation tensor, $\mathbf{R}(S^1,\ S^2, t)$, to be fixed at a point on the shell reference surface described in the arc length parameters, $(S^1,\ S^2)$, along the lines of curvature directions for any time $t \in \mathbb{R}_+$. Then, considering it as a one-parameter family along the time coordinate, and following the steps used in the curvature derivation, but replacing the spatial derivative operations, $(\bullet)_{,\alpha}$, $\alpha = 1, 2$, everywhere by the time derivative operation, $\dot{(\bullet)} \equiv \dfrac{d}{dt}(\bullet)$, and recognizing that:

$$\boxed{{}^t\boldsymbol{\Omega} \equiv \dot{\mathbf{R}}\ \mathbf{R}^T} \qquad (11.363)$$

we have from equation (11.363):

$$\boxed{\begin{aligned} {}^t\boldsymbol{\Omega} &= \frac{2}{\Sigma}\left\{\underbrace{\left(\mathbf{S}\ \dot{\mathbf{S}} - \dot{\mathbf{S}}\ \mathbf{S}\right)}_{\text{Commutator or Lie bracket}} + \dot{\mathbf{S}}\right\} \\ &= \frac{2}{\Sigma}\begin{bmatrix} 0 & (-s_1\,\dot{s}_2 + \dot{s}_1\,s_2 - \dot{s}_3) & (-s_1\,\dot{s}_3 + \dot{s}_1\,s_3 + \dot{s}_2) \\ (s_1\,\dot{s}_2 - \dot{s}_1\,s_2 + \dot{s}_3) & 0 & (-s_2\,\dot{s}_3 + \dot{s}_2\,s_3 - \dot{s}_1) \\ (s_1\,\dot{s}_3 - \dot{s}_1\,s_3 - \dot{s}_2) & (s_2\,\dot{s}_3 - \dot{s}_2\,s_3 + \dot{s}_1) & 0 \end{bmatrix} \end{aligned}} \qquad (11.364)$$

11.12.3.1 Angular Velocity Vector, ${}^t\omega$

Now, from equation (11.364), we have:

$$\boxed{{}^t\omega = \mathbf{W}_1(\mathbf{s})\ \overset{\bullet}{\mathbf{s}}} \tag{11.365}$$

where $\mathbf{W}_1(\mathbf{s})$ is given by equation (11.357).

11.12.3.2 Angular Velocity Vector, ${}^t\omega$, in Terms of the Rotation Vector, ${}^t\theta$

Using equation (11.360) and replacing spatial derivatives by time derivatives in equation (11.361), we have:

$$\boxed{{}^t\omega = \mathbf{W}_1(\mathbf{s})\ \overset{\bullet}{\mathbf{s}} = \mathbf{W}_1\ \mathbf{W}_2\ \overset{\bullet}{\theta} \equiv \mathbf{W}(\theta)\ {}^t\overset{\bullet}{\theta}} \tag{11.366}$$

where we have used the definition of $\mathbf{W}(\theta)$ as given by equation (11.265).

11.12.4 Variational Matrix, ${}^t\bar{\Phi} \equiv \delta\, {}^t\Phi$, Variational Vector, ${}^t\bar{\varphi} \equiv \delta\, {}^t\varphi$

This time we take the variation of the rotation tensor, $\mathbf{R}(S^1,\ S^2, t)$, at a point on the shell reference surface described in the arc length parameters, $(S^1,\ S^2)$, along the lines of curvature directions for a time $t \in \mathbb{R}_+$. Then, following the steps used in the angular velocity derivation but replacing the time derivative operations, $\overset{\bullet}{(\bullet)}$, everywhere by the variation operation, $\overline{(\bullet)} \equiv \delta(\bullet)$, and recognizing that $\bar{\mathbf{\Phi}} \equiv \delta\,\mathbf{\Phi} \equiv \bar{\mathbf{R}}\ \mathbf{R}^{\mathbf{T}} \equiv \delta\mathbf{R}\ \mathbf{R}^{\mathbf{T}}$, we have from equation (11.364):

$$\boxed{\begin{aligned} {}^t\bar{\mathbf{\Phi}} &= \frac{2}{\Sigma}\left\{ \underbrace{(\mathbf{S}\,\bar{\mathbf{S}} - \bar{\mathbf{S}}\,\mathbf{S})}_{\text{Commutator or Lie bracket}} \quad + \bar{\mathbf{S}} \right\} \\ &= \frac{2}{\Sigma}\begin{bmatrix} 0 & (-s_1\bar{s}_2 + \bar{s}_1 s_2 - \bar{s}_3) & (-s_1\bar{s}_3 + \bar{s}_1 s_3 + \bar{s}_2) \\ (s_1\bar{s}_2 - \bar{s}_1 s_2 + \bar{s}_3) & 0 & (-s_2\bar{s}_3 + \bar{s}_2 s_3 - \bar{s}_1) \\ (s_1\bar{s}_3 - \bar{s}_1 s_3 - \bar{s}_2) & (s_2\bar{s}_3 - \bar{s}_2 s_3 + \bar{s}_1) & 0 \end{bmatrix} \end{aligned}} \tag{11.367}$$

where a superscripted bar over the variable signifies variation of the variable by the definition: the variation of a along $\bar{a}$ is given by: $\bar{a} = \lim\limits_{\varepsilon \to 0} \dfrac{d}{d\varepsilon}(a + \varepsilon\bar{a})$.

11.12.4.1 Rotational Variational Vector, ${}^t\bar{\varphi} \equiv \delta^t\varphi$

Now, the expressions for the axial rotational variational vector, $\bar{\varphi}$, corresponding to the variation ${}^t\bar{\mathbf{\Phi}} \equiv \delta\,{}^t\mathbf{\Phi}$, follow from equation (11.367), as:

$$\boxed{{}^t\bar{\varphi} = \mathbf{W}_1(\mathbf{s})\ \bar{\mathbf{s}}} \tag{11.368}$$

where we have $\mathbf{W}_1(\mathbf{s})$ as defined before in equation (11.356).

11.12.4.2 $\bar{\varphi}$ in Terms of the Rotation Vector Variation, $\bar{\theta}$

By substituting $\bar{\varphi}$ for $\mathbf{k}_\alpha$, $\alpha = 1, 2$, and $\bar{\theta}$ for $\theta,_\alpha$ in equations (11.361) and (11.265), we get:

$$\boxed{{}^t\bar{\varphi} = \mathbf{W}_1(\mathbf{s})\,\bar{\mathbf{s}} = \mathbf{W}_1\ \mathbf{W}_2\ \bar{\theta} \equiv \mathbf{W}(\theta)\,{}^t\bar{\theta}} \tag{11.369}$$

where $\mathbf{W}(\theta)$, $c_1(\theta)$, $c_2(\theta)$ have been defined earlier in equation (11.265).

11.12.5 Derivatives: $\mathbf{W},_\alpha(\theta)$, $\alpha = 1, 2$ and $\dot{\mathbf{W}}(\theta)$, $\ddot{\mathbf{W}}(\theta)$ and Variation: $\bar{\mathbf{W}}(\theta) \equiv \delta\mathbf{W}(\theta)$ of $\mathbf{W}(\theta)$

For future use, we will need the derivatives: $\mathbf{W},_\alpha(\theta)$, $\alpha = 1, 2$, the rates: $\dot{\mathbf{W}}(\theta)$, $\ddot{\mathbf{W}}(\theta)$, and the variation: $\bar{\mathbf{W}}(\theta) \equiv \delta\mathbf{W}(\theta)$ of the matrix $\mathbf{W}(\theta)$.

Upon differentiating by S^α, $\alpha = 1, 2$ the expression for $\mathbf{W}(\theta)$ as given by equation (11.265), we have:

$$\boxed{\begin{aligned}
\mathbf{W},_\alpha &= c_1\boldsymbol{\Theta},_\alpha + c_2(\boldsymbol{\Theta},_\alpha\boldsymbol{\Theta} + \boldsymbol{\Theta}\boldsymbol{\Theta},_\alpha) + c_1,_\alpha\boldsymbol{\Theta} + c_2,_\alpha\boldsymbol{\Theta}^2\\
&= (c_2 - c_1)(\theta \bullet \theta,_\alpha)\mathbf{I} + c_1\boldsymbol{\Theta},_\alpha + c_2\{(\theta \otimes \theta,_\alpha) + (\theta,_\alpha \otimes \theta)\} + c_1,_\alpha\boldsymbol{\Theta} + c_2,_\alpha(\theta \otimes \theta)\\
c_1,_\alpha &\equiv \frac{1}{\theta^2}(1 - 2c_1 - \theta^2 c_2)(\theta \bullet \theta,_\alpha) = c_3(\theta \bullet \theta,_\alpha),\\
c_2,_\alpha &\equiv \frac{1}{\theta^2}(c_1 - 3c_2)(\theta \bullet \theta,_\alpha) = c_4(\theta \bullet \theta,_\alpha)
\end{aligned}} \tag{11.370}$$

where we have introduced the following definitions.

Definitions

$$\boxed{\begin{aligned}
c_3 &\equiv \frac{1}{\theta^2}(1 - 2c_1 - \theta^2 c_2),\\
c_4 &\equiv \frac{1}{\theta^2}(c_1 - 3c_2)(\theta \bullet \theta,_\alpha)
\end{aligned}} \tag{11.371}$$

and applied the following identities:

$$\boxed{\begin{aligned}
\boldsymbol{\Theta}^2 &= \{(\theta \otimes \theta) - \theta^2\mathbf{I}\}\\
(\boldsymbol{\Theta},_\alpha\ \boldsymbol{\Theta} + \boldsymbol{\Theta}\ \boldsymbol{\Theta},_\alpha) &= \{(\theta \otimes \theta,_\alpha) + (\theta,_\alpha \otimes\ \theta) - 2(\theta \bullet\ \theta,_\alpha)\mathbf{I}\}\\
\theta,_\alpha \equiv \|\theta\|,_\alpha &= \frac{1}{\theta}(\theta \bullet\ \theta,_\alpha)
\end{aligned}} \tag{11.372}$$

Exercise: Prove the identities (11.372).

Solution $\forall \mathbf{h} \in \mathbb{R}^3$,

$$\boldsymbol{\Theta}^2\ \mathbf{h} = \theta \times \theta \times \mathbf{h} = \{(\theta \otimes \theta) - \theta^2\mathbf{I}\}\mathbf{h}$$

and

$$
\begin{aligned}
(\boldsymbol{\Theta},_\alpha \boldsymbol{\Theta} + \boldsymbol{\Theta}\,\boldsymbol{\Theta},_\alpha)\mathbf{h} &= \theta,_\alpha \times \theta \times \mathbf{h} + \theta \times \theta,_\alpha \times \mathbf{h} \\
&= (\theta,_\alpha \bullet \mathbf{h})\theta - (\theta,_\alpha \bullet \theta)\mathbf{h} + (\theta \bullet \mathbf{h})\theta,_\alpha - (\theta \bullet \theta,_\alpha)\mathbf{h} \\
&= \left\{(\theta \otimes \theta,_\alpha) + (\theta,_\alpha \otimes \theta) - 2(\theta \bullet \theta,_\alpha)\mathbf{I}\right\}\mathbf{h}
\end{aligned}
$$

complete the proof.

Exercise: Derive relations given in equation (11.370).

Solution: By taking the derivative with respect to $S^\alpha,\ \alpha = 1,2$ of $\mathbf{H}(\theta)$ in equation (11.265) and using relations in equation (11.372):

$$
\begin{aligned}
\mathbf{W},_\alpha &= c_1\boldsymbol{\Theta},_\alpha + c_2(\boldsymbol{\Theta},_\alpha\boldsymbol{\Theta} + \boldsymbol{\Theta}\boldsymbol{\Theta},_\alpha) + c_1,_\alpha\boldsymbol{\Theta} + c_2,_\alpha\boldsymbol{\Theta}^2 \\
&= c_1\boldsymbol{\Theta},_\alpha + c_2\{(\theta \otimes \theta,_\alpha) + (\theta,_\alpha \otimes \theta) - 2(\theta \bullet \theta,_\alpha)\mathbf{I}\} + c_1,_\alpha\boldsymbol{\Theta} + c_2,_\alpha\{(\theta \otimes \theta) - \theta^2\mathbf{I}\} \\
&= c_1\boldsymbol{\Theta},_\alpha + c_2\{(\theta \otimes \theta,_\alpha) + (\theta,_\alpha \otimes \theta)\} + c_1,_\alpha\boldsymbol{\Theta} + c_2,_\alpha(\theta \otimes \theta) + \{-c_1 + 3c_2 - 2c_2\}(\theta \bullet \theta,_\alpha)\mathbf{I} \\
&= (c_2 - c_1)(\theta \bullet \theta,_\alpha)\mathbf{I} + c_1\boldsymbol{\Theta},_\alpha + c_2\{(\theta \otimes \theta,_\alpha) + (\theta,_\alpha \otimes \theta)\} + c_1,_\alpha\boldsymbol{\Theta} + c_2,_\alpha(\theta \otimes \theta)
\end{aligned}
$$

completes the derivation. Now, replacing $\theta,_\alpha$ by $\dot{\theta}$ and $\boldsymbol{\Theta},_\alpha$ by $\dot{\boldsymbol{\Theta}}$ everywhere in equations (11.370)–(11.372), we have the expression for $\dot{\mathbf{W}}$, the time derivative of $\mathbf{W}$ matrix:

$$
\boxed{
\begin{aligned}
\dot{\mathbf{W}} &= c_1\dot{\boldsymbol{\Theta}} + c_2(\dot{\boldsymbol{\Theta}}\boldsymbol{\Theta} + \boldsymbol{\Theta}\dot{\boldsymbol{\Theta}}) + \dot{c}_1\boldsymbol{\Theta} + \dot{c}_2\boldsymbol{\Theta}^2 \\
&= (c_2 - c_1)(\theta \bullet \dot{\theta})\mathbf{I} + c_1\dot{\boldsymbol{\Theta}} + c_2\{(\theta \otimes \dot{\theta}) + (\dot{\theta} \otimes \theta)\} + \dot{c}_1\boldsymbol{\Theta} + \dot{c}_2(\theta \otimes \theta) \\
\dot{c}_1 &\equiv \frac{1}{\theta^2}(1 - 2c_1 - \theta^2 c_2)(\theta \bullet \dot{\theta}) = c_3(\theta \bullet \dot{\theta}),\ \dot{c}_2 \equiv \frac{1}{\theta^2}(c_1 - 3c_2)(\theta \bullet \dot{\theta}) = c_4(\theta \bullet \dot{\theta})
\end{aligned}
}
\tag{11.373}
$$

Taking the time derivative of $\dot{\mathbf{W}}$ in equation (11.373), we have the expression for $\ddot{\mathbf{W}}$, the second time derivative of $\mathbf{W}$ matrix:

$$
\boxed{
\begin{aligned}
&\ddot{\mathbf{W}} = c_1\ddot{\boldsymbol{\Theta}} + 2\left\{\dot{c}_1\dot{\boldsymbol{\Theta}} + \dot{c}_2(\dot{\boldsymbol{\Theta}}\boldsymbol{\Theta} + \boldsymbol{\Theta}\dot{\boldsymbol{\Theta}})\right\} + c_2(\ddot{\boldsymbol{\Theta}}\boldsymbol{\Theta} + 2\dot{\boldsymbol{\Theta}}^2 + \boldsymbol{\Theta}\ddot{\boldsymbol{\Theta}}) + \ddot{c}_1\boldsymbol{\Theta} + \ddot{c}_2\boldsymbol{\Theta}^2 \\
&\ddot{c}_1 \equiv c_5(\theta \bullet \dot{\theta})^2 + c_3\left\{\theta^2 + (\theta \bullet \dot{\theta})\right\}, \qquad \ddot{c}_2 \equiv c_6(\theta \bullet \dot{\theta})^2 + c_4\left\{\theta^2 + (\theta \bullet \dot{\theta})\right\} \\
&c_5 \equiv -\frac{1}{\theta^2}(2c_2 + 4c_3 + \theta^2 c_4), \qquad c_6 \equiv -\frac{1}{\theta^2}(c_3 - 5c_4),
\end{aligned}
}
\tag{11.374}
$$

where we have used the properties of the constants as given by equations (11.265) and (11.371).

Now, replacing $\theta,_\alpha$ by $\bar{\theta}$ and $\boldsymbol{\Theta},_\alpha$ by $\bar{\boldsymbol{\Theta}}$ everywhere in equations (11.370)–(11.372), we have the expression for $\bar{\mathbf{W}}$, the variation of $\mathbf{W}$ matrix:

$$\begin{aligned}
\bar{\mathbf{W}} &= c_1\bar{\boldsymbol{\Theta}} + c_2(\bar{\boldsymbol{\Theta}}\,\boldsymbol{\Theta} + \boldsymbol{\Theta}\,\bar{\boldsymbol{\Theta}}) + \bar{c}_1\,\boldsymbol{\Theta} + \bar{c}_2\,\boldsymbol{\Theta}^2 \\
&= (c_2 - c_1)\left(\theta \bullet \bar{\theta}\right)\mathbf{I} + c_1\bar{\boldsymbol{\Theta}} + c_2\left\{\left(\theta \otimes \bar{\theta}\right) + \left(\bar{\theta} \otimes \theta\right)\right\} + \bar{c}_1\,\boldsymbol{\Theta} + \bar{c}_2\,(\theta \otimes \theta) \\
\bar{c}_1 &\equiv \frac{1}{\theta^2}(1 - 2c_1 - \theta^2 c_2)(\theta \bullet \bar{\theta}) = c_3(\theta \bullet \bar{\theta}), \bar{c}_2 \equiv \frac{1}{\theta^2}(c_1 - 3c_2)(\theta \bullet \bar{\theta}) = c_4(\theta \bullet \bar{\theta})
\end{aligned} \tag{11.375}$$

11.12.6 Angular Acceleration Vector, $^t\dot{\omega}$, in Terms of the Rotation Vector, $^t\theta$

Taking the time derivative of the angular acceleration given in equation (11.366), we have:

$$^t\dot{\omega} = \dot{\mathbf{W}}(\theta)\,^t\dot{\theta} + \mathbf{W}(\theta)\,^t\ddot{\theta} \tag{11.376}$$

where $\mathbf{W}(\theta)$ is given by equation (11.265), and $\dot{\mathbf{W}}(\theta)$ by equation (11.373).

11.12.6.1 An Important Lemma

We will also need the following important result, relating the derivatives of the variational vector, $^t\bar{\varphi},_\alpha$, and the curvature vectors, $^t\mathbf{k}_\alpha$:

Lemma LD1: Defining $^t\mathbf{k}_\alpha$, $\alpha = 1, 2$ and $^t\bar{\varphi}$ as the axial vectors of $^t\mathbf{K}_\alpha$, $\alpha = 1, 2$ and $^t\bar{\boldsymbol{\Phi}}$, respectively, we have:

$$\begin{aligned}
^t\bar{\boldsymbol{\Phi}},_\alpha &= {^t\bar{\mathbf{K}}_\alpha} + [{^t\mathbf{K}_\alpha}\,{^t\bar{\boldsymbol{\Phi}}} - {^t\bar{\boldsymbol{\Phi}}}\,{^t\mathbf{K}_\alpha}], \quad \alpha = 1, 2 \\
\underbrace{^t\bar{\varphi},_\alpha}_{\text{Effective material curvature vector variation}} &= \underbrace{^t\bar{\mathbf{k}}_\alpha}_{\text{Material curvature vector variation}} + \underbrace{^t\mathbf{k}_\alpha}_{\text{Material curvature vector}} \times \underbrace{^t\bar{\varphi}}_{\text{Material rotation vector variation}}
\end{aligned} \tag{11.377}$$

Proof LD1: Using relations (11.353) and the definition of $^t\bar{\boldsymbol{\Phi}} \equiv \delta\,^t\boldsymbol{\Phi} \equiv \bar{\mathbf{R}}\,\mathbf{R}^\mathbf{T} \equiv \delta\mathbf{R}\,\mathbf{R}^\mathbf{T}$, we have:

$$\bar{\boldsymbol{\Phi}}^t,_\alpha = \bar{\mathbf{R}},_\alpha\,\mathbf{R}^T + ({^t\bar{\boldsymbol{\Phi}}}\,\mathbf{R})\,\mathbf{R}^T,_\alpha = {^t\bar{\mathbf{K}}_\alpha}\,\mathbf{R}\,\mathbf{R}^T + {^t\mathbf{K}_\alpha}\,{^t\bar{\boldsymbol{\Phi}}}\,\mathbf{R}\,\mathbf{R}^T + {^t\bar{\boldsymbol{\Phi}}}\,\mathbf{R}\,\mathbf{R}^T\,({^t\mathbf{K}})^T$$

Now, noting the orthogonality property of a rotation tensor: $\mathbf{R}\,\mathbf{R}^T = \mathbf{I}$, and skew-symmetry property: $({^t\mathbf{K}})^T = -{^t\mathbf{K}}$ completes the proof of the first part. Recognition of the Lie bracket property: for $\mathbf{A}, \mathbf{B}$ skew, and $\mathbf{a}, \mathbf{b}$, the corresponding axial vectors, $[\mathbf{AB} - \mathbf{BA}] = [\mathbf{a} \times \mathbf{b}]$, for the second part, completes the proof.

Exercise: Detail the steps in the LD1 proof.

Solution

$$
{}^t\bar{\boldsymbol{\Phi}},_{\alpha} = {}^t\bar{\mathbf{K}}_{\alpha}\,\mathbf{R}\,\mathbf{R}^T + {}^t\mathbf{K}_{\alpha}\,{}^t\bar{\boldsymbol{\Phi}}\,\mathbf{R}\,\mathbf{R}^T + {}^t\bar{\boldsymbol{\Phi}}\,\mathbf{R}\,\mathbf{R}^T\,({}^t\mathbf{K}_{\alpha})^T = {}^t\bar{\mathbf{K}}_{\alpha} + \underbrace{{}^t\mathbf{K}^t_{\alpha}\bar{\boldsymbol{\Phi}} - {}^t\bar{\boldsymbol{\Phi}}\,{}^t\mathbf{K}_{\alpha}}_{\text{Lie bracket}}
$$

$$
\Rightarrow {}^t\bar{\boldsymbol{\varphi}},_{\alpha} = {}^t\bar{\mathbf{k}}_{\alpha} - {}^t\bar{\boldsymbol{\varphi}} \times {}^t\mathbf{k}_{\alpha}
$$

11.12.6.2 The Central Lemma

We will need the following most important lemma, relating the derivatives, $\boldsymbol{\theta},_{\alpha}$, and the variation, $\bar{\boldsymbol{\theta}}$, or the increment, $\Delta\boldsymbol{\theta}$ (appears in the linearized virtual functional later in the treatment of a Newton-type iterative algorithm), of the rotation vector, $\boldsymbol{\theta}$, for deriving the geometric stiffness operator and proving its symmetry under the conservative system of loading, presented later.

Lemma LD2: Given $\mathbf{W}$, c_1 and c_2 as in equation (11.265), we have:

$$
\boxed{\begin{aligned} \bar{\mathbf{W}}\,\boldsymbol{\theta},_{\alpha} &= \mathbf{X}_{\alpha}\,\bar{\boldsymbol{\theta}}, \quad \alpha = 1,2 \\ \Delta\mathbf{W}\,\boldsymbol{\theta},_{\alpha} &= \mathbf{X}_{\alpha}\,\Delta\boldsymbol{\theta} \end{aligned}} \tag{11.378}
$$

where

$$
\boxed{\begin{aligned} &\mathbf{X}_{\alpha} \equiv -c_1\boldsymbol{\Theta},_{\alpha} + c_2\{[\boldsymbol{\theta}\otimes\boldsymbol{\theta},_{\alpha}] - [\boldsymbol{\theta},_{\alpha}\otimes\boldsymbol{\theta}] + (\boldsymbol{\theta}\bullet\boldsymbol{\theta},_{\alpha})\mathbf{I}\} + \mathbf{V}\,[\boldsymbol{\theta},_{\alpha}\otimes\boldsymbol{\theta}] \\ &\mathbf{V} \equiv -c_2\mathbf{I} + c_3\boldsymbol{\Theta} + c_4\boldsymbol{\Theta}^2, \quad c_3 \equiv \frac{1}{\theta^2}\{(1-\theta^2 c_2) - 2c_1, \quad c_4 \equiv \frac{1}{\theta^2}(c_1 - 3c_2) \end{aligned}} \tag{11.379}
$$

where $\bar{\boldsymbol{\theta}}$ and $\Delta\boldsymbol{\theta}$ are virtual (variational) and incremental rotation vectors, respectively.

Proof LD3: First noting that $\bar{\theta} = \dfrac{1}{\theta}(\boldsymbol{\theta}\bullet\bar{\boldsymbol{\theta}})$ and taking one parameter variation of c_1 and c_2, we have, from equations (11.370), (11.373) and (11.379): $c_1,_{\alpha} = c_3(\boldsymbol{\theta}\bullet\boldsymbol{\theta},_{\alpha})$, $\bar{c}_1 = c_3(\boldsymbol{\theta}\bullet\bar{\boldsymbol{\theta}})$ and $c_2,_{\alpha} = c_4(\boldsymbol{\theta}\bullet\boldsymbol{\theta},_{\alpha})$, $\bar{c}_2 = c_4(\boldsymbol{\theta}\bar{\boldsymbol{\theta}})$. Now, rewriting the variation of $\mathbf{W}$ as defined in the first equality of equation (11.373), and multiplying both sides by $\boldsymbol{\theta},_{\alpha}$, we get:

$$
\bar{\mathbf{W}}\,\boldsymbol{\theta},_{\alpha} = -c_1\bar{\boldsymbol{\Theta}}\,\boldsymbol{\theta},_{\alpha} + c_2(\bar{\boldsymbol{\Theta}}\boldsymbol{\Theta} + \boldsymbol{\Theta}\bar{\boldsymbol{\Theta}})\,\boldsymbol{\theta},_{\alpha} + (c_3\boldsymbol{\Theta} + c_4\boldsymbol{\Theta}^2)(\boldsymbol{\theta}\bullet\bar{\boldsymbol{\theta}})\,\boldsymbol{\theta},_{\alpha} \tag{11.380}
$$

Noting the identities: $\bar{\boldsymbol{\Theta}}\boldsymbol{\Theta} + \boldsymbol{\Theta}\bar{\boldsymbol{\Theta}} = [\boldsymbol{\theta}\otimes\bar{\boldsymbol{\theta}}] + [\bar{\boldsymbol{\theta}}\otimes\boldsymbol{\theta}] - 2(\boldsymbol{\theta}\bullet\bar{\boldsymbol{\theta}})\mathbf{I}$ and $[\boldsymbol{\theta}\otimes\bar{\boldsymbol{\theta}}]\boldsymbol{\theta},_s = [\boldsymbol{\theta}\otimes\boldsymbol{\theta},_s]\bar{\boldsymbol{\theta}}$, we have:

$$
\begin{aligned} \bar{\boldsymbol{\Theta}}\,\boldsymbol{\theta},_{\alpha} &= \bar{\boldsymbol{\theta}}\times\boldsymbol{\theta},_{\alpha} = -\boldsymbol{\theta},_{\alpha}\times\bar{\boldsymbol{\theta}} = -\boldsymbol{\Theta},_{\alpha}\bar{\boldsymbol{\theta}}; \\ (\bar{\boldsymbol{\Theta}}\boldsymbol{\Theta} + \boldsymbol{\Theta}\bar{\boldsymbol{\Theta}})\,\boldsymbol{\theta},_{\alpha} &= \bar{\boldsymbol{\theta}}\times\boldsymbol{\theta}\times\boldsymbol{\theta},_{\alpha} + \boldsymbol{\theta}\times\bar{\boldsymbol{\theta}}\times\boldsymbol{\theta},_{\alpha} \\ &= (\bar{\boldsymbol{\theta}}\bullet\boldsymbol{\theta},_{\alpha})\boldsymbol{\theta} - (\bar{\boldsymbol{\theta}}\bullet\boldsymbol{\theta})\boldsymbol{\theta},_{\alpha} + (\boldsymbol{\theta}\bullet\boldsymbol{\theta},_{\alpha})\bar{\boldsymbol{\theta}} - (\boldsymbol{\theta}\bullet\bar{\boldsymbol{\theta}})\boldsymbol{\theta},_{\alpha} \\ &= \left\{(\boldsymbol{\theta}\otimes\boldsymbol{\theta},_{\alpha}) - (\boldsymbol{\theta},_{\alpha}\otimes\boldsymbol{\theta}) + (\boldsymbol{\theta}\bullet\boldsymbol{\theta},_{\alpha})\mathbf{I}\right\}\bar{\boldsymbol{\theta}} - (\boldsymbol{\theta},_{\alpha}\otimes\boldsymbol{\theta})\bar{\boldsymbol{\theta}}; \end{aligned} \tag{11.381}
$$

Now, inserting relations (11.381) in equation (11.380), we get:

$$
\bar{\mathbf{W}}\,\boldsymbol{\theta},_{\alpha} = \left\{ \begin{aligned} &-c_1\boldsymbol{\Theta},_{\alpha} + c_2\{(\boldsymbol{\theta}\otimes\boldsymbol{\theta},_{\alpha}) - (\boldsymbol{\theta},_{\alpha}\otimes\boldsymbol{\theta}) + (\boldsymbol{\theta}\bullet\boldsymbol{\theta},_{\alpha})\mathbf{I}\} \\ &+(-c_2\mathbf{I} + c_3\boldsymbol{\Theta} + c_4\boldsymbol{\Theta}^2)(\boldsymbol{\theta},_{\alpha}\otimes\boldsymbol{\theta}) \end{aligned} \right\}\bar{\boldsymbol{\theta}} \tag{11.382}
$$

Next, insertion of the definitions for $\mathbf{X}_\alpha$, $\alpha = 1,2$ and $\mathbf{V}$ of equation (11.379) in equation (11.382) completes the proof for the first part of equation (11.378), and replacement of $\overline{(\bullet)}$ terms by the $\Delta(\bullet)$, completes the proof for the increment.

11.12.7 Where We Would Like to Go

We have now introduced various parameters and derived all the necessary relationships anticipating the next step of developing the virtual work equation for the shell reference surface. Clearly, then, we should move on to:

- the computational virtual work principle.

11.13 Computational Virtual Work Equations

11.13.1 What We Need to Recall

- The real configuration space of a shell is characterized by $({}^t\mathbf{d}(S^1,S^2,t), {}^t\mathbf{R}(S^1,S^2,t))$.
- The virtual space for a shell may be described by the virtual displacement vector and the virtual rotational tensor fields, $(\bar{\mathbf{d}}(S^1,S^2) = \mathbf{G}\,{}^t\bar{\mathbf{d}}, \quad \bar{\mathbf{R}}(S^1,S^2) = \mathbf{G}\,{}^t\bar{\mathbf{R}}\,\mathbf{G}^T)$, with the component virtual entities defined as $({}^t\bar{\mathbf{d}} \equiv \delta\,{}^t\mathbf{d},\ {}^t\bar{\mathbf{\Phi}} \equiv \delta\,{}^t\mathbf{\Phi})$ where ${}^t\bar{\mathbf{\Phi}}(S^1,S^2)$, the instantaneous (variational) rotational skew-symmetric matrix describing the one parameter variation ${}^t\bar{\mathbf{R}}(S^1,S^2) \equiv \delta\,{}^t\mathbf{R}(S^1,S^2)$, of the rotation matrix ${}^t\mathbf{R}$ defined by: ${}^t\bar{\mathbf{\Phi}} \equiv {}^t\bar{\mathbf{R}}\,{}^t\mathbf{R}^T$ with ${}^t\bar{\boldsymbol{\varphi}}$ as its axial vector, that is, ${}^t\bar{\mathbf{\Phi}}h = {}^t\bar{\boldsymbol{\varphi}} \times \mathbf{h}, \forall \mathbf{h} \in \mathbb{R}^3$, and ${}^t\bar{\mathbf{\Phi}}\,{}^t\bar{\boldsymbol{\varphi}} = \mathbf{0}$.
- The computational dynamic balance equations in component vector form are:

$$\boxed{\begin{aligned} &\mathbf{G}({}^t\mathbf{F}^\alpha|_\alpha + {}^t\hat{\mathbf{F}}^\alpha - {}^t\dot{\mathbf{L}}) = 0 \\ &\mathbf{G}({}^t\mathbf{M}^\alpha|_\alpha + \mathbf{a}_\alpha \times {}^t\mathbf{F}^\alpha + {}^t\hat{\mathbf{M}}^\alpha - {}^t\dot{\mathbf{A}}) = 0 \end{aligned}} \tag{11.383}$$

or

$$\boxed{\begin{aligned} &\mathbf{G}({}^t\mathbf{F}^\alpha{}_{,\alpha} + \boldsymbol{\kappa}^0_\alpha \times {}^t\mathbf{F}^\alpha + {}^t\hat{\mathbf{F}}^\alpha - {}^t\dot{\mathbf{L}}) = \mathbf{0} \\ &\mathbf{G}({}^t\mathbf{M}^\alpha{}_{,\alpha} + \boldsymbol{\kappa}^0_\alpha \times {}^t\mathbf{M}^\alpha + \{\boldsymbol{\kappa}^0_\alpha \times \mathbf{d} + \mathbf{e}_\alpha\} \times {}^t\mathbf{F}^\alpha + {}^t\hat{\mathbf{M}}^\alpha - {}^t\dot{\mathbf{A}}) = \mathbf{0} \end{aligned}} \tag{11.384}$$

where $\mathbf{G} = [\,\mathbf{G}_1 \quad \mathbf{G}_2 \quad \mathbf{N}\,]$ is the orthonormal matrix with the base vectors as the columns of the matrix such that $\mathbf{G}^T\,\mathbf{G} = \mathbf{I}$.
- The force and moment component vectors in undeformed and current deformed coordinate system are related by the force and moment rotation transformation:

$$\boxed{\begin{aligned} {}^t\mathbf{F}^\alpha &= {}^t\mathbf{R}\,{}^n\mathbf{F}^\alpha, \quad \alpha = 1,2 \\ {}^t\mathbf{M}^\alpha &= {}^t\mathbf{R}\,{}^n\mathbf{M}^\alpha \end{aligned}} \tag{11.385}$$

where the superscript t over an object that reminds us that the object is expressed in the undeformed $\mathbf{G}_1, \mathbf{G}_2, \mathbf{G}_3 \equiv \mathbf{N}$ orthonormal frame, may be dropped occasionally for clarity; a superscript n identifies variables in the rotated frame, $\mathbf{g}_1 = {}^t\mathbf{R}\,\mathbf{G}_1$, $\mathbf{g}_2 = {}^t\mathbf{R}\,\mathbf{G}_2$, $\mathbf{g}_3 = {}^t\mathbf{R}\,\mathbf{G}_3$ where ${}^t\mathbf{R}$ is the current rotation tensor measured in the undeformed coordinate system; the context should help us recall the appropriate frame of reference.

- The covariant derivatives of the displacement and rotation component vectors:

$$\boxed{\begin{aligned} {}^t\mathbf{d}|_\alpha &= {}^t\mathbf{d},_\alpha + \mathbf{k}^0_\alpha \times {}^t\mathbf{d}, \quad \alpha = 1,2 \\ {}^t\boldsymbol{\theta}|_\alpha &= {}^t\boldsymbol{\theta},_\alpha + \mathbf{k}^0_\alpha \times {}^t\boldsymbol{\theta} \end{aligned}} \tag{11.386}$$

- Likewise, we have the time derivatives of the momentum, $\mathbf{L}(S^1, S^2, t)$, and the angular momentum, $\mathbf{A}(S^1, S^2, t)$, in terms of the components measured in the undeformed coordinate system:

$$\boxed{\begin{aligned} \mathbf{L}(S^1,S^2,t) &= \mathbf{G}(S^1,S^2)\,{}^t\mathbf{L}(S^1,S^2,t), \\ {}^t\dot{\mathbf{L}} &= M_0\,{}^t\ddot{\mathbf{d}} \approx (\rho_0 H)\,{}^t\ddot{\mathbf{d}} \\ \mathbf{A}(S^1,S^2,t) &= \mathbf{G}(S^1,S^2)\,{}^t\mathbf{A}(S^1,S^2,t), \\ {}^t\dot{\mathbf{A}} &= I_0\,{}^t\dot{\boldsymbol{\omega}} \approx \left(\rho_0 \frac{H^3}{12}\right)\,{}^t\dot{\boldsymbol{\omega}} \end{aligned}} \tag{11.387}$$

- The vector triple product property: $\mathbf{a} \times \mathbf{b} \bullet \mathbf{c} = \mathbf{b} \times \mathbf{c} \bullet \mathbf{a} = \mathbf{c} \times \mathbf{a} \bullet \mathbf{b}$, $\forall \mathbf{a}, \mathbf{b}, \mathbf{c}$ vectors.
- In what follows notationally, all virtual entities are identified with a superscripted bar over them.

11.13.2 Boundary Conditions

Just as in the reduction of the 3D balance laws of a shell-like body by integration through the thickness direction produced the exact counterparts for a 2D shell surface that are valid for its interior, M, we need to determine the exact statically equivalent boundary conditions for the shell edge from the prescribed boundary conditions on the side surface, M^S, of a 3D shell-like body.

The boundary conditions on a 3D shell-like body can basically be of the following types:

- ***Traction boundary condition***: Tractions are specified on all or part of the enclosing edge (the side-surface) of a shell-like body. If the specification involves the entire edge, some sort of constraint must be imposed against the rigid body movement. For tractions specified partially on some patches of the side surfaces, we collect all the areas and designate the aggregate area as the set $M_f^S \subset M^S$. Let the specified 3D traction, $\hat{\mathbf{t}}(S,\eta), \quad (S,\eta) \in M_f^S \subset M^S$, be given which of course satisfies the relationship: $\hat{\mathbf{t}}(S,\eta) = \mathbf{P}(S,\eta)\ \nu(S,\eta)$ where $\nu(S,\eta)$ is the outward normal at any point (S,η) on the side surface with S as the arc length parameter along the boundary curve defining the side surface with shell reference normal on it. In this case, we may obtain the 2D counterparts by the through-the-thickness integration as:

$$\boxed{\begin{aligned} \mathbf{F} \bullet \nu_0 &= \mathbf{F}^\alpha \otimes \mathbf{G}_\alpha \bullet \nu_{0\alpha} \otimes \mathbf{G}^\alpha = \mathbf{F}^\alpha\ \nu_{0\alpha} = \mathbf{F}_\nu = \hat{\mathbf{F}}_\nu \\ &\equiv \int_{\eta^b}^{\eta^t} \hat{\mathbf{t}}(S,\eta)\ {}^\eta\mu^S\ d\eta, \quad S \in \partial M_f \\ \mathbf{M} \bullet \nu_0 &= \mathbf{M}^\alpha \otimes \mathbf{G}_\alpha \bullet \nu_{0\alpha} \otimes \mathbf{G}^\alpha = \mathbf{M}^\alpha\ \nu_{0\alpha} = \mathbf{M}_\nu = \hat{\mathbf{M}}_\nu \\ &\equiv \int_{\eta^b}^{\eta^t} \mathbf{h}(S,\eta) \times \hat{\mathbf{t}}(S,\eta)\ {}^\eta\mu^S\ d\eta \end{aligned}} \tag{11.388}$$

where $\partial M_f = M \cap M_f^S$ is that part of the shell edge where the shell statically equivalent tractions will be considered; for a thin shell with $\mathbf{h}(S,\eta) = \eta\ \mathbf{N}(S)$, from equation (11.388), we will have:

$$\begin{aligned}
\mathbf{F} \bullet \nu_0 &= \mathbf{F}^\alpha \otimes \mathbf{G}_\alpha \bullet \nu_{0\alpha} \otimes \mathbf{G}^\alpha = \mathbf{F}^\alpha\ \nu_{0\alpha} = \mathbf{F}_\nu = \hat{\mathbf{F}}_\nu \\
&\equiv \int_{\eta^b}^{\eta^t} \hat{\mathbf{t}}(S,\eta)\ {}^\eta\mu^S\ d\eta, \quad S \in \partial M_f \\
\mathbf{M} \bullet \nu_0 &= \mathbf{M}^\alpha \otimes \mathbf{G}_\alpha \bullet \nu_{0\alpha} \otimes \mathbf{G}^\alpha = \mathbf{M}^\alpha\ \nu_{0\alpha} = \mathbf{M}_\nu = \hat{\mathbf{M}}_\nu \\
&\equiv \mathbf{N}(S) \times \int_{\eta^b}^{\eta^t} \eta\ \hat{\mathbf{t}}(S,\eta)\ {}^\eta\mu^S\ d\eta
\end{aligned} \tag{11.389}$$

- ***Kinematic boundary condition***: Displacements are specified on all or part of the side surface. For displacements specified partially on some patches of the side surfaces, we collect all the areas and designate the aggregate area as the set $M_d^S \subset M^S$. We will admit only those displacements that have finite strain energy. Now, if the displacements are homogeneous, we simply set: ${}^t\mathbf{d} = \mathbf{0}$, ${}^t\mathbf{R} = \mathbf{0}$, $({}^t\mathbf{d}, {}^t\mathbf{R}) \in \partial M_d$ where $\partial M_d \subset \partial M$ is that portion of the boundary curve, ∂M, that is also common with M_d^S. For inhomogeneous displacement specifications, we will set: ${}^t\mathbf{d} = \hat{\mathbf{d}}$, ${}^t\mathbf{R} = \hat{\mathbf{R}}$, $({}^t\mathbf{d}, {}^t\mathbf{R}) \in \partial M_d$; the hat superscript means the quantities are prescribed. In this case, the kinematically admissible virtual displacements, $({}^t\bar{\mathbf{d}} \equiv \delta\,{}^t\mathbf{d},\ \bar{\mathbf{R}} \equiv \delta\,{}^t\mathbf{R})$ will be identically zero on ∂M_d. Thus, in the subsequent virtual work equation, we must have the boundary integral on ∂M_d identically vanished. The actual finite element solution for a shell with an inhomogeneous displacement condition will be given as an additive sum of two displacement functions with one part as the solution of the relevant virtual work equation and the other part as an arbitrary fixed admissible displacement function.
- ***Mixed traction and displacement boundary conditions***: In the event that the side surface of the 3D shell-like body is in contact with other elastic bodies such as, say, elements modeled as distributed springs, the boundary conditions will be expressed as a linear combinations of both the traction and displacement vectors with the spring stiffness as given. This condition is essentially a combination of the previous two boundary conditions.

11.13.3 Initial Conditions

For a shell semi-discrete finite element formulation (which we have alluded to in our motivation), the virtual work equation eventually reduces to a matrix second-order of ordinary differential equations. Clearly, the typical initial conditions will be the prescribed initial displacement and velocity at the finite element nodes for calculating the time-evolution of the shell deformations and internal reactive stresses. Now, the initial conditions may be written as:

$$\begin{aligned}
\mathbf{d}(0) &= \mathbf{d}_0. \\
\boldsymbol{\theta}(0) &= \boldsymbol{\theta}_0. \\
\mathbf{v}(0) &\equiv \dot{\mathbf{d}}(0) = \mathbf{v}_0, \\
\boldsymbol{\omega}(0) &\equiv \dot{\boldsymbol{\theta}}(0) = \boldsymbol{\omega}_0
\end{aligned} \tag{11.390}$$

11.13.4 Virtual Work from the Vector Balance Equations

Let us consider a shell reference surface of area M bounded by a smooth closed curve, ∂M. Let $(\bar{\mathbf{d}}, \bar{\boldsymbol{\varphi}})$ be a virtual vector set belonging to the admissible test or virtual space. If $({}^t\bar{\mathbf{d}}, {}^t\bar{\boldsymbol{\varphi}})$ is the corresponding component vectors, then we have $\bar{\mathbf{d}} = \mathbf{G}\,{}^t\bar{\mathbf{d}}$ and $\bar{\boldsymbol{\varphi}} = \mathbf{G}\,{}^t\bar{\boldsymbol{\varphi}}$; note that $({}^t\bar{\mathbf{d}}, {}^t\bar{\boldsymbol{\varphi}})$ is time-independent. Now, with the above background in mind, we multiply equation (11.383) or equation (11.384) by the admissible virtual state vectors $\bar{\mathbf{d}}$ and $\bar{\boldsymbol{\varphi}}$, and integrate over an arbitrary shell reference surface area, Ω, to get the shell virtual work functional, $G(\mathbf{d}, \boldsymbol{\varphi}; \bar{\mathbf{d}}, \bar{\boldsymbol{\varphi}})$, given as:

$$G(\mathbf{d}, \boldsymbol{\varphi}; \bar{\mathbf{d}}, \bar{\boldsymbol{\varphi}}) \equiv -\int_M \left\{ \begin{array}{l} ({}^t\mathbf{F}^\alpha|_\alpha + {}^t\hat{\mathbf{F}}^\alpha - {}^t\dot{\mathbf{L}}) \bullet {}^t\bar{\mathbf{d}} + \\ ({}^t\mathbf{M}^\alpha|_\alpha + (\hat{\mathbf{1}}_\alpha + {}^t\mathbf{d}|_\alpha) \times {}^t\mathbf{F}^\alpha + {}^t\hat{\mathbf{M}}^\alpha - {}^t\dot{\mathbf{A}}) \bullet {}^t\bar{\boldsymbol{\varphi}} \end{array} \right\} dA \qquad (11.391)$$
$$= \mathbf{0}, \quad \forall \text{ admissible } {}^t\bar{\mathbf{d}}, {}^t\bar{\boldsymbol{\varphi}}$$

where we have applied the fact that $\mathbf{G}^T\,\mathbf{G} = \mathbf{I}$. Now, by an application of the Green–Gauss (or divergence) theorem for a surface: $\int_M {}^t\mathbf{F}^\alpha|_\alpha\, dA = \int_M DIV\,{}^f\mathbf{S}\, dA = \oint_{\partial M} {}^t\mathbf{F}^\alpha \otimes \mathbf{G}_\alpha \bullet \nu_{0\beta}\, \mathbf{G}^\beta\, dS = \oint_{\partial M} {}^t\mathbf{F}^\alpha\, \nu_{0\alpha}\, dS \equiv \oint_{\partial M} {}^t\mathbf{F}_\nu\, dS$ where $\boldsymbol{\nu}_0 = \nu_{0\alpha}\,\mathbf{G}^\alpha$ is the outward unit normal on the tangent plane on the boundary curve, ∂M, to the shell reference surface area, M; we have also introduced the following definition.

Definition: ${}^t\mathbf{F}_\nu$ as:

$$\boxed{{}^t\mathbf{F}_\nu \equiv {}^f\mathbf{S} \bullet \boldsymbol{\nu}_0 \Leftarrow {}^t\mathbf{F}^\alpha \otimes \mathbf{G}_\alpha \bullet \nu_{0\beta}\, \mathbf{G}^\beta \Leftarrow {}^t\mathbf{F}^\alpha\, \nu_{0\alpha}} \qquad (11.392)$$

We can have a similar expression for moments as: $\int_M {}^t\mathbf{M}^\alpha|_\alpha\, dA = \oint_{\partial M} {}^t\mathbf{M}^\alpha\, \nu_{0\alpha}\, dS \equiv \oint_{\partial M} {}^t\mathbf{M}_\nu\, dS$ with the:

Definition: ${}^t\mathbf{M}_\nu$ as:

$$\boxed{{}^t\mathbf{M}_\nu \equiv {}^m\mathbf{S} \bullet \boldsymbol{\nu}_0 \Leftarrow {}^t\mathbf{M}^\alpha \otimes \mathbf{G}_\alpha \bullet \nu_{0\beta}\, \mathbf{G}^\beta \Leftarrow {}^t\mathbf{M}^\alpha\, \nu_{0\alpha}} \qquad (11.393)$$

Then, we have: $\int_M ({}^t\mathbf{F}^\alpha \bullet {}^t\bar{\mathbf{d}})|_\alpha\, dA = \oint_{\partial M} ({}^t\mathbf{F}^\alpha \bullet {}^t\bar{\mathbf{d}})\, \nu_{0\alpha} dS = \oint_{\partial M} ({}^t\mathbf{F}^\alpha\, \nu_{0\alpha} \bullet {}^t\bar{\mathbf{d}}) dS = \oint_{\partial M} ({}^t\mathbf{F}_\nu \bullet {}^t\bar{\mathbf{d}}) dS$. Now, by applying the chain rule of differentiation and rearranging terms, we get:

$$-\int_M ({}^t\mathbf{F}^\alpha|_\alpha \bullet {}^t\bar{\mathbf{d}})\, dA = -\int_M ({}^t\mathbf{F}^\alpha \bullet {}^t\bar{\mathbf{d}}|_\alpha)\, dA - \oint_{\partial M} ({}^t\mathbf{F}_\nu \bullet {}^t\bar{\mathbf{d}})\, dS \qquad (11.394)$$

Similarly, for the moments, we have:

$$-\int_M ({}^t\mathbf{M}^\alpha|_\alpha \bullet {}^t\bar{\boldsymbol{\varphi}})\, dA = -\int_M ({}^t\mathbf{M}^\alpha \bullet {}^t\bar{\boldsymbol{\varphi}}|_\alpha)\, dA - \oint_{\partial M} ({}^t\mathbf{M}_\nu \bullet {}^t\bar{\boldsymbol{\varphi}})\, dS \qquad (11.395)$$

Next, recalling the commutation property of vector triple product, and applying equations (11.191), (11.192) and (11.387) in equation (11.190) with boundary conditions given by equation

(11.388), we get:

$$
\begin{aligned}
G_{Dyn}({}^t\mathbf{d}, {}^t\mathbf{R}; {}^t\bar{\mathbf{d}}, {}^t\bar{\varphi}) &\equiv G_{Inertial} + G_{Internal} - G_{External} \\
&= \overbrace{\int_M \left\{ (M_0\, {}^t\ddot{\mathbf{d}} \bullet {}^t\bar{\mathbf{d}} + I_0\, {}^t\dot{\omega} \bullet {}^t\bar{\varphi} \right\} dA}^{\text{Inertial}} \\
&\quad + \overbrace{\int_M \{ ({}^t\mathbf{F}^\alpha \bullet [{}^t\bar{\mathbf{d}}|_\alpha - {}^t\bar{\varphi} \times (\hat{\mathbf{1}}_\alpha + {}^t\mathbf{d}|_\alpha)] + {}^t\mathbf{M}^\alpha \bullet {}^t\bar{\varphi}|_\alpha \} dA}^{\text{Internal}} \\
&\quad - \overbrace{\left\{ \oint_{\partial M_f} ({}^t\hat{\mathbf{F}}_\nu \bullet {}^t\bar{\mathbf{d}} + {}^t\hat{\mathbf{M}}_\nu \bullet {}^t\bar{\varphi})\, dS + \int_M ({}^t\hat{\mathbf{F}} \bullet {}^t\bar{\mathbf{d}} + {}^t\hat{\mathbf{M}} \bullet {}^t\bar{\varphi})\, dA \right\}}^{\text{External}} \\
&= \mathbf{0}, \quad \forall \text{ admissible } {}^t\bar{\mathbf{d}}, {}^t\bar{\varphi}
\end{aligned}
\tag{11.396}
$$

where we have the angular velocity component vector, ${}^t\omega$, related to the Rodrigues' rotation component vector, ${}^t\theta$, and identified the inertial virtual work, the internal virtual work and the external virtual work of the virtual work principle as follows:

Definition: Inertial virtual work, $G_{Inertial}$, as:

$$
G_{Inertial} \equiv \int_M \left\{ (M_0\, {}^t\ddot{\mathbf{d}} \bullet {}^t\bar{\mathbf{d}} + I_0\, {}^t\dot{\omega} \bullet {}^t\bar{\varphi} \right\} dA \tag{11.397}
$$

Internal virtual work, $G_{Internal}$, as:

$$
G_{Internal} \equiv \int_M \{ {}^t\mathbf{F}^\alpha \bullet ({}^t\bar{\mathbf{d}}|_\alpha - {}^t\bar{\varphi} \times \bullet (\hat{\mathbf{1}}_\alpha + {}^t\mathbf{d}|_\alpha)) + {}^t\mathbf{M}^\alpha \bullet {}^t\bar{\varphi}|_\alpha \} dA \tag{11.398}
$$

External virtual work, $G_{External}$, as:

$$
G_{External} \equiv \oint_{\partial M_f} \{ {}^t\hat{\mathbf{F}}_\nu \bullet {}^t\bar{\mathbf{d}} + {}^t\hat{\mathbf{M}}_\nu \bullet {}^t\bar{\varphi} \} dS + \int_M \{ {}^t\hat{\mathbf{F}} \bullet {}^t\bar{\mathbf{d}} + {}^t\hat{\mathbf{M}} \bullet {}^t\bar{\varphi} \} dA \tag{11.399}
$$

Remark

- Although from an apparent energy conjugation in the internal virtual work equation (11.398), the expressions, ${}^t\bar{\mathbf{d}}|_\alpha - {}^t\bar{\varphi} \times (\hat{\mathbf{1}}_\alpha + {}^t\mathbf{d}|_\alpha)$ and ${}^t\bar{\varphi}|_\alpha$ present themselves as the force and moment virtual strains, it turns out that there exist no functions whose variations can possibly be identified with these functions; in other words, these are not truly the virtual strains derivable from some real strains. Fortunately, however, if we rotate the force and moment component vectors by pre-multiplying with ${}^t\mathbf{R}^T$ to obtain forces and moments as the components in the rotating deformed frame, we achieve our goal of establishing the true virtual nature of the newly generated and energy conjugated strain functions; this is precisely what we intend to do next. However, it is important to note that all relevant quantities are measured in the Lagrangian or material coordinate system.

11.13.5 Virtual Translational and Rotational Strain Vectors

Now, we introduce the rotation matrix, ${}^t\mathbf{R}$, to express the internal virtual work in the rotated frame, and by applying equation (11.385) in equation (11.194) and noting that ${}^t\mathbf{R}$ is orthonormal, that is, ${}^t\mathbf{R}^T\,{}^t\mathbf{R} = \mathbf{I}$, we get:

$$
\boxed{
\begin{aligned}
G_{Internal} &= \int_M \left\{ {}^t\mathbf{R}^T\,{}^t\mathbf{F}^\alpha \bullet {}^t\mathbf{R}^T \left({}^t\bar{\mathbf{d}}|_\alpha - {}^t\bar{\boldsymbol{\varphi}} \times (\hat{\mathbf{1}}_\alpha + {}^t\mathbf{d}|_\alpha)\right) + {}^t\mathbf{R}^T\,{}^t\mathbf{M}^\alpha \bullet {}^t\mathbf{R}\,{}^t\bar{\boldsymbol{\varphi}}|_\alpha \right\} dA \\
&= \int_M \left\{ \begin{array}{l} \underbrace{{}^n\mathbf{F}^\alpha}_{\text{Real component force vector along rotated frame}} \bullet \underbrace{\left\{{}^t\mathbf{R}^T\,{}^t\bar{\mathbf{d}}|_\alpha - ({}^t\mathbf{R}^T\,{}^t\bar{\boldsymbol{\Phi}})\,(\hat{\mathbf{1}}_\alpha + {}^t\mathbf{d}|_\alpha)\right\}}_{\text{Virtual translational strain vector}} + \\ \underbrace{{}^n\mathbf{M}^\alpha}_{\text{Real component moment vector along rotated frame}} \bullet \underbrace{{}^t\mathbf{R}^T\,{}^t\bar{\boldsymbol{\varphi}}|_\alpha}_{\text{Virtual rotational strain vector}} \end{array} \right\} dA
\end{aligned}
}
\tag{11.400}
$$

Next, applying the definition of the covariant derivatives as given by equation (11.386) and interpreting ${}^t\bar{\Phi}$ (being skew-symmetric: ${}^t\bar{\Phi}^T = -{}^t\bar{\Phi}$) as the variation of the rotation matrix, ${}^t\mathbf{R}$, related by: ${}^t\bar{\mathbf{\Phi}} \equiv {}^t\bar{\mathbf{R}}\,{}^t\mathbf{R}^T \Rightarrow ({}^t\bar{\mathbf{R}})^T = ({}^t\bar{\mathbf{\Phi}}\,{}^t\mathbf{R})^T = {}^t\mathbf{R}^T\,{}^t\bar{\mathbf{\Phi}}^T = -{}^t\mathbf{R}^T\,{}^t\bar{\mathbf{\Phi}}$, we can rewrite and define the virtual strain vectors as follows.

Definition

- The virtual translational (axial & shear) component strain vector, ${}^n\bar{\boldsymbol{\beta}}_\alpha,\ \alpha = 1, 2$:

$$
\boxed{
\begin{aligned}
{}^n\bar{\boldsymbol{\beta}}_\alpha &\equiv {}^t\mathbf{R}^T\,{}^t\bar{\mathbf{d}}|_\alpha - ({}^t\mathbf{R}^T\,{}^t\bar{\boldsymbol{\Phi}})\,(\hat{\mathbf{1}}_\alpha + {}^t\mathbf{d}|_\alpha) \\
&= {}^t\mathbf{R}^T\,\overline{(\hat{\mathbf{1}}_\alpha + {}^t\mathbf{d}|_\alpha)} + \overline{{}^t\mathbf{R}^T}\,(\hat{\mathbf{1}}_\alpha + {}^t\mathbf{d}|_\alpha) = \overline{{}^t\mathbf{R}^T(\hat{\mathbf{1}}_\alpha + {}^t\mathbf{d}|_\alpha)}
\end{aligned}
}
\tag{11.401}
$$

and

- The virtual rotational (bending) component strain vector, ${}^n\bar{\chi}_\alpha,\ \alpha = 1, 2$:

$$
\boxed{
\begin{aligned}
{}^n\bar{\chi}_\alpha &\equiv {}^t\mathbf{R}^T\,{}^t\bar{\boldsymbol{\varphi}}|_\alpha = {}^t\mathbf{R}^T\{\,{}^t\bar{\boldsymbol{\varphi}}_{,\alpha} + \mathbf{k}^0_\alpha \times {}^t\bar{\boldsymbol{\varphi}}\} \\
&= {}^t\mathbf{R}^T\{\,\bar{\mathbf{k}}_\alpha - {}^t\bar{\boldsymbol{\Phi}}\,(\mathbf{k}^0_\alpha + \mathbf{k}_\alpha)\} \\
&= {}^t\mathbf{R}^T\,\bar{\mathbf{k}}_\alpha + (-{}^t\mathbf{R}^T\,{}^t\bar{\boldsymbol{\Phi}})\,(\mathbf{k}^0_\alpha + \mathbf{k}_\alpha) \\
&= {}^t\mathbf{R}^T\,\overline{(\mathbf{k}^0_\alpha + \mathbf{k}_\alpha)} + \overline{{}^t\mathbf{R}^T}\,(\mathbf{k}^0_\alpha + \mathbf{k}_\alpha) = \overline{{}^t\mathbf{R}^T\,(\mathbf{k}^0_\alpha + \mathbf{k}_\alpha)}
\end{aligned}
}
\tag{11.402}
$$

In equation (11.402), we have used the virtual derivative relation: ${}^t\bar{\boldsymbol{\varphi}}_{,\alpha} = \bar{\mathbf{k}}_\alpha - {}^t\bar{\boldsymbol{\Phi}}\,\mathbf{k}_\alpha$ with $\mathbf{k}_\alpha^0$ as the initial curvature vector, and $\mathbf{k}_\alpha$ as the current curvature vector along the S^α, $\alpha = 1, 2$, direction, respectively. Also, recall that all entities with barred superscript are virtual in nature. Finally, introducing equations (11.397) through (11.402) in equation (11.396), we get the desired shell virtual functional as:

$$
\begin{aligned}
&G_{Dyn}({}^t\mathbf{d}, {}^t\mathbf{R}; {}^t\bar{\mathbf{d}}, {}^t\bar{\boldsymbol{\varphi}}) \equiv G_{Inertial} + G_{Internal} - G_{External} \\
&= \overbrace{\int_M \left\{ (M_0\, {}^t\ddot{\mathbf{d}} \cdot {}^t\bar{\mathbf{d}} + I_0\, {}^t\dot{\boldsymbol{\omega}} \cdot {}^t\bar{\boldsymbol{\varphi}} \right\} dA}^{\text{Inertial}} + \overbrace{\int_M \{ ({}^n\mathbf{F}^\alpha \cdot {}^n\bar{\boldsymbol{\beta}}_\alpha + {}^n\mathbf{M}^\alpha \cdot {}^n\bar{\boldsymbol{\chi}}_\alpha \} \, dA}^{\text{Internal}} \\
&\quad - \overbrace{\left\{ \oint_{\partial M_f} ({}^t\hat{\mathbf{F}}_\nu \cdot {}^t\bar{\mathbf{d}} + {}^t\hat{\mathbf{M}}_\nu \cdot {}^t\bar{\boldsymbol{\varphi}}) \, dS + \int_M ({}^t\hat{\mathbf{F}} \cdot {}^t\bar{\mathbf{d}} + {}^t\hat{\mathbf{M}} \cdot {}^t\bar{\boldsymbol{\varphi}}) \, dA \right\}}^{\text{External}} \\
&= \mathbf{0}, \quad \forall \text{ admissible } {}^t\bar{\mathbf{d}}, {}^t\bar{\boldsymbol{\varphi}}
\end{aligned}
\tag{11.403}
$$

Note that in the internal virtual work, the forces and the moments, and the corresponding conjugate translational and bending virtual strain vectors, although measured in undeformed coordinates, are all expressed as the components in the current deformed (rotated) local coordinate system, that is, the co-spinning coordinate system. One of the main consequences is that the constitutive relationship must also be considered in this deformed coordinate system which is the co-spinning coordinate system: ${}^n\mathbf{G}_1$, ${}^n\mathbf{G}_2$, $\mathbf{N}$, in the undeformed state that becomes $\mathbf{g}_1, \mathbf{g}_2, \mathbf{n}$ in the current deformed configuration.

11.13.6 Static Virtual Work Functional and Virtual Strain Tensors

Just as the virtual strain vectors in the co-rotated coordinates have been represented by the undeformed coordinates, their tensor counterparts, as we shall see, are also most suitably represented in the mixed bases.

Remark

- For this, we will consider only the static part of the invariant form of the balance equations.

Thus, let the static balance equation be:

$$
\begin{aligned}
DIV\, {}^f\mathbf{P} + \hat{\mathbf{F}} &= \mathbf{0} \\
DIV\, {}^m\mathbf{P} + \mathbf{c}_{0,\alpha} \times ({}^f\mathbf{P} \cdot \mathbf{G}^\alpha) + \hat{\mathbf{M}} &= \mathbf{0}
\end{aligned}
\tag{11.404}
$$

As before, we multiply equation (11.404) by any admissible virtual state vectors $\bar{\mathbf{d}}$ and $\bar{\boldsymbol{\varphi}}$, and integrate over an arbitrary shell reference surface area, M, and focus on the shell static virtual work functional, $G_{Static}(\mathbf{d}, \mathbf{R}; \bar{\mathbf{d}}, \bar{\boldsymbol{\varphi}})$, given as:

$$G_{Static}(\mathbf{d}, \mathbf{R}; \bar{\mathbf{d}}, \bar{\boldsymbol{\varphi}}) \equiv -\int_M \left\{\bar{\mathbf{d}} \bullet (DIV^f\mathbf{P} + \hat{\mathbf{F}}) + (DIV^m\mathbf{P} + \bar{\boldsymbol{\varphi}} \bullet \mathbf{c}_{0,\alpha} \times ({}^f\mathbf{P} \bullet \mathbf{G}^\alpha) + \hat{\mathbf{M}})\right\} dA$$
$$= \mathbf{0}, \quad \forall \text{admissible } \bar{\mathbf{d}},\ \bar{\boldsymbol{\varphi}} \tag{11.405}$$

Let us recall the product rule of divergence of a vector times a tensor:

$$\begin{aligned} -\bar{\mathbf{d}} \bullet DIV^f\mathbf{P} &= -DIV(\bar{\mathbf{d}} \bullet {}^f\mathbf{P}) + {}^f\mathbf{P}: GRAD\ \bar{\mathbf{d}} \\ -\bar{\boldsymbol{\varphi}} \bullet DIV^m\mathbf{P} &= -DIV(\bar{\boldsymbol{\varphi}} \bullet {}^m\mathbf{P}) + {}^m\mathbf{P}: GRAD\ \bar{\boldsymbol{\varphi}} \end{aligned} \tag{11.406}$$

where we have, by definition: $GRAD\ \bar{\mathbf{d}} \equiv \bar{\mathbf{d}},_\alpha \otimes \mathbf{G}^\alpha$, and $GRAD\ \bar{\boldsymbol{\varphi}} \equiv \bar{\boldsymbol{\varphi}},_\alpha \otimes \mathbf{G}^\alpha$ with $GRAD$ as the material surface gradient, and $:$ is the usual double contraction for tensors. Let us also recall the surface Green–Gauss or divergence theorem: $\int_M DIV\,(\bullet)\, dA = \oint_{\partial M} (\bullet) \bullet \boldsymbol{\nu}\, dS$ where $\boldsymbol{\nu}$ is the outward unit normal on the shell tangent plane at an enclosing closed boundary, ∂M, of the shell reference surface area, M. Now applying definitions given by equations (11.392) and (11.393), and relationships given by equations (11.406), we have from equation (11.405) with the boundary conditions (11.388):

$$\begin{aligned} &G_{Static}(\mathbf{d}, \mathbf{R}; \bar{\mathbf{d}}, \bar{\boldsymbol{\varphi}}) \\ &= \int_M \left\{{}^f\mathbf{P} : GRAD\ \bar{\mathbf{d}} + {}^m\mathbf{P} : GRAD\ \bar{\boldsymbol{\varphi}} + \bar{\boldsymbol{\varphi}} \bullet \mathbf{c}_{0,\alpha} \times ({}^f\mathbf{P} \bullet \mathbf{G}^\alpha)\right\} dA \\ &\quad - \oint_{\partial M} \{\bar{\mathbf{d}} \bullet ({}^f\mathbf{P} \bullet \boldsymbol{\nu}) + \bar{\boldsymbol{\varphi}} \bullet ({}^m\mathbf{P} \bullet \boldsymbol{\nu})\}\, dS - \int_\Omega \left\{\bar{\mathbf{d}} \bullet \hat{\mathbf{F}} + \bar{\boldsymbol{\varphi}} \bullet \hat{\mathbf{M}}\right\} dA \\ &= \int_M \left\{{}^f\mathbf{P} : GRAD\ \bar{\mathbf{d}} + {}^m\mathbf{P} : GRAD\ \bar{\boldsymbol{\varphi}} + \hat{\mathbf{F}} - (DIV^m\mathbf{P} + \bar{\boldsymbol{\varphi}} \bullet \mathbf{c}_{0,\alpha} \times ({}^f\mathbf{P} \bullet \mathbf{G}^\alpha)\right\} dA \\ &\quad - \oint_{\partial M_f} \{\bar{\mathbf{d}} \bullet \hat{\mathbf{F}}_\nu + \bar{\boldsymbol{\varphi}} \bullet \hat{\mathbf{M}}_\nu\}\, dS - \int_\Omega \left\{\bar{\mathbf{d}} \bullet \hat{\mathbf{F}} + \bar{\boldsymbol{\varphi}} \bullet \hat{\mathbf{M}}\right\} dA = \mathbf{0}, \quad \forall \text{admissible } \bar{\mathbf{d}},\ \bar{\boldsymbol{\varphi}} \end{aligned} \tag{11.407}$$

Now, noting that $\mathbf{c}_{0,\alpha} \times \bar{\boldsymbol{\varphi}} \otimes \mathbf{G}^\alpha = -\bar{\boldsymbol{\varphi}} \times \mathbf{c}_{0,\alpha} \otimes \mathbf{G}^\alpha = -\bar{\boldsymbol{\varphi}} \times GRAD\mathbf{c}_0 = GRAD\mathbf{c}_0 \times \bar{\boldsymbol{\varphi}}$, we have:

$$\begin{aligned} &\bar{\boldsymbol{\varphi}} \bullet \mathbf{c}_{0,\alpha} \times ({}^f\mathbf{P} \bullet \mathbf{G}^\alpha) \\ &= ({}^f\mathbf{P} \bullet \mathbf{G}^\alpha) \bullet (\bar{\boldsymbol{\varphi}} \times \mathbf{c}_{0,\alpha}) = ({}^f\mathbf{P} \bullet \mathbf{G}^\alpha) \otimes \mathbf{G}_\alpha : (\bar{\boldsymbol{\varphi}} \times \mathbf{c}_{0,\alpha}) \otimes \mathbf{G}^\alpha \\ &= (\mathbf{G}_\alpha \otimes \mathbf{G}^\alpha) \otimes {}^f\mathbf{P} : \bar{\boldsymbol{\varphi}} \times GRAD\ \mathbf{c}_0 = -{}^f\mathbf{P} : GRAD\ \mathbf{c}_0 \times \bar{\boldsymbol{\varphi}} \end{aligned} \tag{11.408}$$

Next, using equation (11.408) in equation (11.407), we get:

$$\boxed{\begin{aligned} G_{Static}(\mathbf{d},\mathbf{R};\bar{\mathbf{d}},\bar{\boldsymbol{\varphi}}) &= \underbrace{\int_M \{{}^f\mathbf{P} : (GRAD\ \bar{\mathbf{d}} + GRAD\ \mathbf{c}_0 \times \bar{\boldsymbol{\varphi}}) + {}^m\mathbf{P} : GRAD\ \bar{\boldsymbol{\varphi}}\}\ dA}_{\text{Internal virtual work}} \\ &\quad \underbrace{- \oint_{\partial M_f} \{\bar{\mathbf{d}} \bullet \hat{\mathbf{F}}_\nu + \bar{\boldsymbol{\varphi}} \bullet \hat{\mathbf{M}}_\nu\}\ dS + \int_M \{\bar{\mathbf{d}} \bullet \hat{\mathbf{F}} + \bar{\boldsymbol{\varphi}} \bullet \hat{\mathbf{M}}\}\ dA}_{\text{External virtual work}} \\ &= G_{Internal} - G_{External} = \mathbf{0}, \quad \forall \text{admissible } \bar{\mathbf{d}},\ \bar{\boldsymbol{\varphi}} \end{aligned}} \tag{11.409}$$

Let us note that recalling relation ${}^f\mathbf{P} \equiv \mathbf{F}^\alpha \otimes \mathbf{G}_\alpha$, as before in the translational internal work part in equation (11.403), we have:

$$\begin{aligned} &\int_M \{{}^f\mathbf{P} : (GRAD\ \bar{\mathbf{d}} + GRAD\ \mathbf{c}_0 \times \bar{\boldsymbol{\varphi}})\}\ dA \\ &= \int_M \{\mathbf{F}^\alpha \otimes \mathbf{G}_\alpha : (\bar{\mathbf{d}},_\beta + \mathbf{c}_0,_\beta \times \bar{\boldsymbol{\varphi}}) \otimes \mathbf{G}^\beta\}\ dA \\ &= \int_M \{\mathbf{F}^\alpha \bullet (\bar{\mathbf{d}},_\alpha + \mathbf{c}_0,_\alpha \times \bar{\boldsymbol{\varphi}})\}\ dA = \int_M \{\mathbf{F}^\alpha \bullet (\bar{\mathbf{d}},_\alpha + (\mathbf{G}_\alpha + \bar{\mathbf{d}},_\alpha) \times \bar{\boldsymbol{\varphi}})\}\ dA \\ &= \int_M \{{}^t\mathbf{F}^\alpha \bullet ({}^t\bar{\mathbf{d}}|_\alpha + (\hat{\mathbf{1}}_\alpha + {}^t\bar{\mathbf{d}}|_\alpha) \times {}^t\bar{\boldsymbol{\varphi}})\}\ dA \\ &= \int_M \{\mathbf{R}^T\,{}^t\mathbf{F}^\alpha \bullet \{\mathbf{R}^T\,{}^t\bar{\mathbf{d}}|_\alpha + (-\mathbf{R}^T\bar{\boldsymbol{\Phi}})(\hat{\mathbf{1}}_\alpha + {}^t\bar{\mathbf{d}}|_\alpha)\}\}\ dA \\ &= \int_M \{{}^n\mathbf{F}^\alpha \bullet \overline{\mathbf{R}^T(\hat{\mathbf{1}}_\alpha + {}^t\mathbf{d}|_\alpha)}\}\ dA = \int_M \{{}^n\mathbf{F}^\alpha \bullet {}^n\bar{\boldsymbol{\beta}}_\alpha\}\ dA \end{aligned} \tag{11.410}$$

with all terms and relations defined earlier in the investigation.

Similarly, we can recover the expression for the rotational internal work part of equation (11.403) as:

$$\begin{aligned} \int_M \{{}^m\mathbf{P} : (GRAD\ \bar{\boldsymbol{\varphi}})\}\ dA &= \int_M \{\mathbf{M}^\alpha \otimes \mathbf{G}_\alpha : \bar{\boldsymbol{\varphi}},_\beta \otimes \mathbf{G}^\beta\}\ dA \\ &= \int_M \{\mathbf{M}^\alpha\ \ \bar{\boldsymbol{\varphi}},_\alpha\}\ dA = \int_M \{{}^t\mathbf{M}^\alpha \bullet {}^t\bar{\boldsymbol{\varphi}}|_\alpha\}\ dA \\ &= \int_M \{\mathbf{R}^T\,{}^t\mathbf{M}^\alpha\ \ \mathbf{R}^T\,{}^t\bar{\boldsymbol{\varphi}}|_\alpha\}\ dA = \int_M \{{}^n\mathbf{M}^\alpha \bullet \overline{\mathbf{R}^T(\mathbf{k}_\alpha^0 + \mathbf{k}_\alpha)}\}\ dA \\ &= \int_M \{{}^n\mathbf{M}^\alpha\ \ {}^n\bar{\chi}_\alpha\}\ dA \end{aligned} \tag{11.411}$$

11.13.6.1 Virtual Translational (Axial and Shear) and Rotational (Bending) Strain Tensors

Now we are in a position to define the virtual strain tensors in material bases as follows:

Define: Virtual translational (axial & shear) strain tensor, ${}^{n}\bar{\mathbf{E}}$:

$$\boxed{{}^{n}\bar{\mathbf{E}} \equiv {}^{n}\bar{\boldsymbol{\beta}}_{\alpha} \otimes \mathbf{G}^{\alpha}} \tag{11.412}$$

Virtual rotational (bending) strain tensor, ${}^{n}\bar{\mathrm{K}}$:

$$\boxed{{}^{n}\bar{\mathrm{K}} \equiv {}^{n}\bar{\chi}_{\alpha} \otimes \mathbf{G}^{\alpha}} \tag{11.413}$$

Finally, with the definitions given by equations (11.412) and (11.413), we have from equation (11.409) after applying equations (11.410) and (11.411):

$$\boxed{\begin{aligned} G_{Static}(\mathbf{d}, \mathbf{R}; \bar{\mathbf{d}}, \bar{\varphi}) &= \int_{M} \{\mathbf{R}^{T\,f}\mathbf{P} : {}^{n}\bar{\mathbf{E}} \;+\; \mathbf{R}^{T\,m}\mathbf{P} : {}^{n}\bar{\mathrm{K}}\ \}\, dA \\ &\quad - \left(\oint_{\partial M_f} \{\bar{\mathbf{d}} \bullet \hat{\mathbf{F}}_{\nu} + \bar{\varphi} \bullet \hat{\mathbf{M}}_{\nu}\}\, dS + \int_{M} \{\, \bar{\mathbf{d}} \bullet \hat{\mathbf{F}} + \bar{\varphi} \bullet \hat{\mathbf{M}}\}\, dA \right) \\ &= \mathbf{0}, \quad \forall \,\text{admissible}\ \bar{\mathbf{d}},\ \bar{\varphi} \end{aligned}} \tag{11.414}$$

Remarks

- We can easily see the equivalence of equation (11.414) described by the virtual strain tensors with equation (11.403) modified for the static situation involving the virtual strain vectors.
- The virtual work functional presented above is true and exact (in the sense of our definitions) irrespective of the shell material properties (i.e. the constitutive law) and the loading system.

11.13.7 Where We Would Like to Go

However, let us note that the virtual functional of equations (11.403) is expressed in terms of rotational variation vector, ${}^{t}\bar{\varphi}$, the variational axial vector corresponding to the instantaneous variational matrix, ${}^{t}\bar{\boldsymbol{\Phi}}$, of the rotation matrix, ${}^{t}\mathbf{R}$. Since we parameterize the rotation matrix by the Rodrigues rotation vector, ${}^{t}\theta$, we must express for computational purposes the virtual strains, angular accelerations and the virtual functional as functions of this rotation vector and its virtual counterpart, ${}^{t}\bar{\theta}$. Thus, we must check out:

- computational virtual work: revisited

11.14 Computational Virtual Work Equations and Virtual Strains: Revisited

11.14.1 What We Need to Recall

- The virtual work equation is given as:

$$
\boxed{
\begin{aligned}
&G({}^t\mathbf{d},\,{}^t\mathbf{R};\,{}^t\bar{\mathbf{d}},\,{}^t\bar{\boldsymbol{\varphi}}) \equiv G_{Inertial} + G_{Internal} - G_{External}\\
&= \overbrace{\int_M \left\{ (M_0\, {}^t\ddot{\mathbf{d}} \bullet {}^t\bar{\mathbf{d}} + I_0\, {}^t\dot{\boldsymbol{\omega}} \bullet {}^t\bar{\boldsymbol{\varphi}} \right\} dA}^{\text{Inertial}} + \overbrace{\int_M \{({}^n\mathbf{F}^\alpha \bullet {}^n\bar{\boldsymbol{\beta}}_\alpha + {}^n\mathbf{M}^\alpha \bullet {}^n\bar{\boldsymbol{\chi}}_\alpha\} \, dA}^{\text{Internal}}\\
&\quad \overbrace{-\left\{ \int_{\partial M_f} \left({}^t\hat{\mathbf{F}}_\nu \bullet {}^t\bar{\mathbf{d}} + {}^t\hat{\mathbf{M}}_\nu \bullet {}^t\bar{\boldsymbol{\varphi}} \right) dS + \int_M \left({}^t\hat{\mathbf{F}} \bullet {}^t\bar{\mathbf{d}} + {}^t\hat{\mathbf{M}}\ {}^t\bar{\boldsymbol{\varphi}} \right) dA \right\}}^{\text{External}}\\
&= \mathbf{0}, \quad \forall \text{ admissible } {}^t\bar{\mathbf{d}},\, {}^t\bar{\boldsymbol{\varphi}}
\end{aligned}
}
\tag{11.415}
$$

Definitions

- The translational virtual strain fields, ${}^n\bar{\boldsymbol{\beta}}_\alpha, \quad \alpha = 1, 2$, are given as:

$$
\boxed{
\begin{aligned}
{}^n\bar{\boldsymbol{\beta}}_\alpha &= {}^t\mathbf{R}^T \left({}^t\bar{\mathbf{d}}_{,\alpha} + {}^t\mathbf{k}^0_\alpha \times {}^t\bar{\mathbf{d}} \right) - {}^t\mathbf{R}^T\, {}^t\bar{\boldsymbol{\Phi}} \left({}^t\mathbf{e}_\alpha + {}^t\mathbf{k}^0_\alpha \times {}^t\mathbf{d} \right)\\
&= \overline{{}^t\mathbf{R}^T(\hat{1}_\alpha + {}^t\mathbf{d}|_\alpha)}
\end{aligned}
}
\tag{11.416}
$$

- The bending virtual strain fields, ${}^n\bar{\boldsymbol{\chi}}_\alpha, \quad \alpha = 1, 2$, are given as:

$$
\boxed{ {}^n\bar{\boldsymbol{\chi}}_\alpha \equiv {}^t\mathbf{R}^T\, {}^t\bar{\boldsymbol{\varphi}}|_\alpha = {}^t\mathbf{R}^T\{\, {}^t\bar{\boldsymbol{\varphi}}_{,\alpha} + \mathbf{k}^0_\alpha \times {}^t\bar{\boldsymbol{\varphi}}\} \equiv \overline{{}^t\mathbf{R}^T\,(\mathbf{k}^0_\alpha + \mathbf{k}_\alpha)} }
\tag{11.417}
$$

- The definition of a rotation tensor, its matrix representation, and algebra, geometry and topology of rotation.
- A rotation matrix, ${}^t\mathbf{R}$, corresponding to the 3D special orthogonal matrix group, $SO(3)$, may be parameterized by the Rodrigues vector, $\mathbf{s}$, the axis of the instantaneous rotation.
- The rotation matrix, ${}^t\mathbf{R}(S^1, S^2, t)$, is related, through Cayley's theorem, to the Rodrigues vector, $\mathbf{s}(S^1, S^2, t) = \{\, s_1(S^1, S^2, t) \quad s_2(S^1, S^2, t) \quad s_3(S^1, S^2, t)\,\}^T$, with the components in the

$\mathbf{G}_1, \mathbf{G}_2, \mathbf{N}$ frame at any point, (S^1, S^2), and at a time, $t \in \mathbb{R}_+$, parameterized by the arc lengths along the lines of curvature on a shell reference surface by:

$$\mathbf{R} = \frac{1}{\Sigma}\begin{bmatrix} 1 + s_1^2 - s_2^2 - s_3^2 & 2(s_1 s_2 - s_3) & 2(s_1 s_3 + s_2) \\ 2(s_1 s_2 + s_3) & 1 - s_1^2 + s_2^2 - s_3^2 & 2(s_2 s_3 - s_1) \\ 2(s_1 s_3 - s_2) & 2(s_2 s_3 + s_1) & 1 - s_1^2 - s_2^2 + s_3^2 \end{bmatrix} \quad (11.418)$$

where $\sum \equiv 1 + \| \mathbf{s} \|^2$

- The Rodrigues vector, $\mathbf{s}$, is orientated along the instantaneous axis of rotation at a point of the shell reference surface. The skew-symmetric matrix, $\mathbf{S}$, corresponding to the Rodrigues vector, $\mathbf{s}$, is given by:

$$\mathbf{S} = \begin{bmatrix} 0 & -s_3 & s_2 \\ s_3 & 0 & -s_1 \\ -s_2 & s_1 & 0 \end{bmatrix} \quad (11.419)$$

- Finally, the Rodrigues rotation vector, $\boldsymbol{\theta}$, is the scaled Rodrigues vector related by:

$$\boxed{\mathbf{s} = \frac{1}{\theta}\tan\left(\frac{\theta}{2}\right)\boldsymbol{\theta}} \quad (11.420)$$

where $\theta \equiv \|\boldsymbol{\theta}\| = (\boldsymbol{\theta} \bullet \boldsymbol{\theta})^{\frac{1}{2}}$ is the length measure of rotation vector. Alternatively, defining the unit vector, $\mathbf{n} \equiv \frac{\boldsymbol{\theta}}{\theta}$, and the corresponding skew-symmetric matrix, $\mathbf{N}$, such that $\mathbf{N}\,\mathbf{h} = \mathbf{n} \times \mathbf{h},\ \forall \mathbf{h} \in \mathbb{R}_3$, we have:

$$\boxed{\begin{aligned} &\mathbf{R} = (\mathbf{I} - \mathbf{S})^{-1}\,(\mathbf{I} + \mathbf{S}) \\ &\mathbf{s} = \tan\left(\frac{\theta}{2}\right)\mathbf{n}, \\ &\mathbf{S} = \tan\left(\frac{\theta}{2}\right)\mathbf{N} \end{aligned}} \quad (11.421)$$

- The Rodrigues quaternion and Euler–Rodrigues parameters.
- The finite rotation, $\mathbf{R}$, belongs to the spherical curved space, $SO(3)$, and the instantaneous and infinitesimal virtual rotation, $\bar{\boldsymbol{\Phi}}$, represented by a skew-symmetric matrix, belongs to the tangent space at identity, Lie algebra, and they are related by an exponential map but, computationally, it is more convenient to apply Cayley's formula:

$$\boxed{\mathbf{R} = (\mathbf{I} - \bar{\boldsymbol{\Phi}})^{-1}\,(\mathbf{I} + \bar{\boldsymbol{\Phi}})} \quad (11.422)$$

- For notational uniformity, we denote all variations of entities by a bar superscripted on it. Furthermore all entities measured in $\mathbf{G_1}, \mathbf{G_2}, \mathbf{N}$ frame is superscripted as ${}^t(\bullet)$, and those in the rotating $\mathbf{g_1}, \mathbf{g_2}, \mathbf{g_3}$ coordinate system as ${}^n(\bullet)$. However, for the sake of uncluttered readability, sometimes the superscripts may be dropped, but the context should help determine the appropriate underlying coordinate system.

- The material "curvature" tensors, ${}^t\mathbf{K}_\alpha,\ \alpha = 1,2$, and "curvature" vector, ${}^t\mathbf{k}_\alpha,\ \alpha = 1,2$, as:

$$\boxed{{}^t\mathbf{K}_\alpha \equiv \mathbf{R},_\alpha \mathbf{R}^{\mathbf{T}}, \quad \alpha = 1,2} \tag{11.423}$$

$$\boxed{{}^t\mathbf{k}_\alpha = \mathbf{W}(\boldsymbol{\theta})\ {}^t\boldsymbol{\theta},_\alpha, \quad \alpha = 1,2} \tag{11.424}$$

where we have:

- Definitions:

$$\boxed{\begin{aligned} &\mathbf{W}({}^t\boldsymbol{\theta}) \equiv \mathbf{I} + c_1\ {}^t\boldsymbol{\Theta} + c_2\ {}^t\boldsymbol{\Theta}^2,\\ &c_1({}^t\boldsymbol{\theta}) = \frac{1-\cos\theta}{\theta^2},\\ &c_2({}^t\boldsymbol{\theta}) = \frac{\theta - \sin\theta}{\theta^3} \end{aligned}} \tag{11.425}$$

with ${}^t\boldsymbol{\Theta}$ is the skew-symmetric matrix corresponding to the axial vector, ${}^t\boldsymbol{\theta}$, such that ${}^t\boldsymbol{\Theta}\,{}^t\boldsymbol{\theta} = \mathbf{0}$.

- The material first variation tensor, ${}^t\bar{\boldsymbol{\Phi}} \equiv \delta\,{}^t\boldsymbol{\Phi}$, and first variation vector, ${}^t\bar{\boldsymbol{\varphi}} \equiv \delta\,{}^t\boldsymbol{\varphi}$, as:

$$\boxed{{}^t\bar{\boldsymbol{\Phi}} \equiv \delta\,{}^t\boldsymbol{\Phi} \equiv \bar{\mathbf{R}}\ \mathbf{R}^{\mathbf{T}} \equiv \delta\mathbf{R}\ \mathbf{R}^{\mathbf{T}}} \tag{11.426}$$

$$\boxed{{}^t\bar{\boldsymbol{\varphi}} = \mathbf{W}(\boldsymbol{\theta})\ {}^t\bar{\boldsymbol{\theta}}} \tag{11.427}$$

- The angular velocity component vector, ${}^t\boldsymbol{\omega}$, is related to the Rodrigues rotation component vector, ${}^t\boldsymbol{\theta}$, by:

$$\boxed{{}^t\boldsymbol{\omega} = \mathbf{W}(\boldsymbol{\theta})\,{}^t\dot{\boldsymbol{\theta}}} \tag{11.428}$$

- The angular acceleration component vector, ${}^t\dot{\boldsymbol{\omega}}$, is related to the Rodrigues rotation component vector, ${}^t\boldsymbol{\theta}$, by:

$$\boxed{{}^t\dot{\boldsymbol{\omega}} = \dot{\mathbf{W}}(\boldsymbol{\theta})\,{}^t\dot{\boldsymbol{\theta}} + \mathbf{W}(\boldsymbol{\theta})\,{}^t\ddot{\boldsymbol{\theta}}} \tag{11.429}$$

- Lemma LD1: Defining ${}^t\mathbf{k}_\alpha,\ \alpha = 1,2$ and ${}^t\bar{\boldsymbol{\varphi}}$ as the axial vectors of ${}^t\mathbf{K}_\alpha,\ \alpha = 1,2$ and ${}^t\bar{\boldsymbol{\Phi}}$, respectively, we have:

$$\boxed{\begin{aligned} {}^t\bar{\boldsymbol{\Phi}},_\alpha &= {}^t\bar{\mathbf{K}}_\alpha + [{}^t\mathbf{K}_\alpha\ {}^t\bar{\boldsymbol{\Phi}} - {}^t\bar{\boldsymbol{\Phi}}\ {}^t\mathbf{K}_\alpha], \quad \alpha = 1,2\\ \underbrace{{}^t\bar{\boldsymbol{\varphi}},_\alpha}_{\substack{\text{Effective material}\\ \text{curvature vector}\\ \text{variation}}} &= \underbrace{{}^t\bar{\mathbf{k}}_\alpha}_{\substack{\text{Material}\\ \text{curvature vector}\\ \text{variation}}} + \underbrace{{}^t\mathbf{k}_\alpha}_{\substack{\text{Material}\\ \text{curvature vector}}} \times \underbrace{{}^t\bar{\boldsymbol{\varphi}}}_{\substack{\text{Material}\\ \text{rotation vector}\\ \text{variation}}} \end{aligned}} \tag{11.430}$$

- The central lemma
  We will need the following most important lemma, relating the derivatives, $\boldsymbol{\theta},_{\alpha}$, and the variation, $\bar{\boldsymbol{\theta}}$, or the increment, $\Delta\boldsymbol{\theta}$, (appears in the linearized virtual functional later in the treatment of a Newton-type iterative algorithm), of the rotation vector, $\boldsymbol{\theta}$, for deriving the geometric stiffness operator and proving its symmetry under the conservative system of loading, presented later

Lemma LD2: Given $\mathbf{W}$, c_1 and c_2 as in equation (11.265), we have:

$$\boxed{\begin{aligned} \bar{\mathbf{W}}\,\boldsymbol{\theta},_{\alpha} &= \mathbf{X}_{\alpha}\,\bar{\boldsymbol{\theta}}, \quad \alpha = 1,2 \\ \Delta\mathbf{W}\,\boldsymbol{\theta},_{\alpha} &= \mathbf{X}_{\alpha}\,\Delta\boldsymbol{\theta} \end{aligned}} \tag{11.431}$$

where

$$\boxed{\begin{aligned} \mathbf{X}_{\alpha} &\equiv -c_1\boldsymbol{\Theta},_{\alpha} + c_2\{[\boldsymbol{\theta}\otimes\boldsymbol{\theta},_{\alpha}] - [\boldsymbol{\theta},_{\alpha}\otimes\boldsymbol{\theta}] + (\boldsymbol{\theta}\bullet\boldsymbol{\theta},_{\alpha})\mathbf{I}\} + \mathbf{V}\,[\boldsymbol{\theta},_{\alpha}\otimes\boldsymbol{\theta}] \\ \mathbf{V} &\equiv -c_2\mathbf{I} + c_3\boldsymbol{\Theta} + c_4\boldsymbol{\Theta}^2 \\ c_3 &\equiv \frac{1}{\theta^2}(1-\theta^2 c_2) - 2c_1 \\ c_4 &\equiv \frac{1}{\theta^2}(c_1 - 3c_2) \end{aligned}} \tag{11.432}$$

where $\bar{\boldsymbol{\theta}}$ and $\Delta\boldsymbol{\theta}$ are virtual (variational) and incremental rotation vectors, respectively.

- Notationally, the bracket [•] denotes the skew-symmetric matrix corresponding to the argument, that is, $[\mathbf{a}] \equiv \mathbf{a}\times$ for an axial vector, $\mathbf{a}$, such that $[\mathbf{a}]\,\mathbf{a} = \mathbf{0}$.

11.14.2 Virtual Functional and Virtual Strains Revisited

Here we intend to eliminate ${}^{t}\bar{\boldsymbol{\varphi}}$ from all the virtual strains and the virtual functional, and replace them with ${}^{t}\bar{\boldsymbol{\theta}}$, the Rodrigues rotation vector.

11.14.3 Inertial Virtual Functional Revisited

Using the relationships in equation (11.427) and equation (11.429), we get:

$$\boxed{G_{Inertial} \equiv \int_M \left\{(M_0\,{}^{t}\ddot{\mathbf{d}} \bullet {}^{t}\bar{\mathbf{d}} + I_0\mathbf{W}^T\,(\dot{\mathbf{W}}\,{}^{t}\dot{\boldsymbol{\theta}} + \mathbf{W}\,{}^{t}\ddot{\boldsymbol{\theta}}) \bullet {}^{t}\bar{\boldsymbol{\theta}}\right\} dA} \tag{11.433}$$

11.14.4 Internal Virtual Functional and Virtual Strains Revisited

We would also like to express the translational and the bending virtual strains in computationally suitable matrix form.

11.14.4.1 Computational Translational Virtual Strain in Matrix Form

Using the relationship, ${}^t\bar{\varphi} = \mathbf{W}\,{}^t\bar{\theta}$ from equation (11.427), and using definitions $\mathbf{a}_\alpha \equiv \mathbf{e}_\alpha + \mathbf{k}_\alpha^0 \times \mathbf{d}$ where $\mathbf{e}_\alpha \equiv \mathbf{d},_\alpha + \hat{\mathbf{1}}_\alpha, \quad \alpha = 1,2$ with $\hat{\mathbf{1}}_1 \equiv \{1 \quad 0 \quad 0\}^T$, $\hat{\mathbf{1}}_2 \equiv \{0 \quad 1 \quad 0\}^T$, we have: $-\mathbf{R}^T\,{}^t\bar{\mathbf{\Phi}}({}^t\mathbf{e}_\alpha + {}^t\mathbf{k}_\alpha^0 \times {}^t\mathbf{d}) = \mathbf{R}^T[\mathbf{a}_\alpha]\,{}^t\bar{\varphi} = \mathbf{R}^T[\mathbf{a}_\alpha]\mathbf{W}\,{}^t\bar{\theta}$; then, we can rewrite the expression for the translational virtual strain field, ${}^n\bar{\beta}_\alpha$, $\alpha = 1,2$, of equation (11.416) as:

$$\boxed{\begin{Bmatrix} {}^n\bar{\beta}_1 \\ {}^n\bar{\beta}_2 \end{Bmatrix} = \begin{bmatrix} {}^t\mathbf{R}^T & \mathbf{0} \\ \mathbf{0} & {}^t\mathbf{R}^T \end{bmatrix} \begin{bmatrix} [\mathbf{k}_1^0] & [\mathbf{a}_1]\mathbf{W} & \mathbf{I} & \mathbf{0} & \mathbf{0} & \mathbf{0} \\ [\mathbf{k}_2^0] & [\mathbf{a}_2]\mathbf{W} & \mathbf{0} & \mathbf{0} & \mathbf{I} & \mathbf{0} \end{bmatrix} \begin{Bmatrix} {}^t\bar{\mathbf{d}}_1 \\ {}^t\bar{\theta}_1 \\ {}^t\bar{\mathbf{d}},_1 \\ {}^t\bar{\theta},_1 \\ {}^t\bar{\mathbf{d}},_2 \\ {}^t\bar{\theta},_2 \end{Bmatrix}} \tag{11.434}$$

11.14.4.2 Computational Rotational Virtual Strain in Matrix Form

Using the relationship, ${}^t\bar{\varphi},_\alpha = {}^t\bar{\mathbf{k}}_\alpha + {}^t\mathbf{k}_\alpha \times {}^t\bar{\varphi}$ from the second of equation (11.430), ${}^t\mathbf{k}_\alpha = \mathbf{W}\ \theta,_\alpha$, and the central lemma: $\bar{\mathbf{W}}\,\theta,_\alpha = \mathbf{X}_\alpha\,\bar{\theta}, \quad \alpha = 1,2$ of equation (11.378), we have:

$$\begin{aligned}
{}^n\bar{\chi}_\alpha &= {}^t\mathbf{R}^T\,{}^t\bar{\varphi},_\alpha - {}^t\mathbf{R}^T\,{}^t\bar{\mathbf{\Phi}}\,{}^t\mathbf{k}_\alpha^0 = {}^t\mathbf{R}^T\left\{{}^t\bar{\mathbf{k}}_\alpha + ({}^t\mathbf{k}_\alpha + {}^t\mathbf{k}_\alpha^0) \times {}^t\bar{\varphi}\right\} \\
&= {}^t\mathbf{R}^T\left\{\bar{\mathbf{W}}\,\theta,_\alpha + \mathbf{W}\,\bar{\theta},_\alpha + [{}^t\mathbf{k}_\alpha + {}^t\mathbf{k}_\alpha^0]\mathbf{W}\,\bar{\theta}\right\} \\
&= {}^t\mathbf{R}^T\left\{\mathbf{X}_\alpha\,\bar{\theta} + \mathbf{W}\,\bar{\theta},_\alpha + [{}^t\mathbf{k}_\alpha + {}^t\mathbf{k}_\alpha^0]\mathbf{W}\,\bar{\theta}\right\} \\
&= {}^t\mathbf{R}^T\left\{(\mathbf{X}_\alpha + \ [{}^t\mathbf{k}_\alpha + {}^t\mathbf{k}_\alpha^0]\mathbf{W})\bar{\theta} \ + \mathbf{W}\,\bar{\theta},_\alpha\right\}
\end{aligned} \tag{11.435}$$

Now, introducing the following definition.

Definition

- The total current "curvature" vectors, $\mathbf{k}_\alpha^c, \quad \alpha = 1,2,$

$$\boxed{{}^t\mathbf{k}_\alpha^c \equiv {}^t\mathbf{k}_\alpha + {}^t\mathbf{k}_\alpha^0} \tag{11.436}$$

we have in matrix form:

$$\boxed{\begin{Bmatrix} {}^n\bar{\chi}_1 \\ {}^n\bar{\chi}_2 \end{Bmatrix} = \begin{bmatrix} {}^t\mathbf{R}^T & \mathbf{0} \\ \mathbf{0} & {}^t\mathbf{R}^T \end{bmatrix} \begin{bmatrix} \mathbf{0} & \mathbf{X}_1 + [{}^t\mathbf{k}_1^c]\,\mathbf{W} & 0 & \mathbf{W} & \mathbf{0} & \mathbf{0} \\ \mathbf{0} & \mathbf{X}_2 + [{}^t\mathbf{k}_2^c]\,\mathbf{W} & \mathbf{0} & \mathbf{0} & 0 & \mathbf{W} \end{bmatrix} \begin{Bmatrix} {}^t\bar{\mathbf{d}}_1 \\ {}^t\bar{\theta}_1 \\ {}^t\bar{\mathbf{d}},_1 \\ {}^t\bar{\theta},_1 \\ {}^t\bar{\mathbf{d}},_2 \\ {}^t\bar{\theta},_2 \end{Bmatrix}} \tag{11.437}$$

11.14.4.3 Virtual Strain–Displacement

By introducing the following definitions,

Definitions

- the (18×1) virtual generalized deformation function $\mathbf{H}(\hat{\bar{\mathbf{d}}}) \equiv \{\bar{\mathbf{d}} \quad \bar{\theta} \quad \bar{\mathbf{d}},_1 \quad \bar{\theta},_1 \quad \bar{\mathbf{d}},_2 \quad \bar{\theta},_2\}^T$,
- the total current "curvature" skew-symmetric matrices $[\mathbf{k}_\alpha^c] \equiv [\mathbf{k}_\alpha^0] + [{}^t\mathbf{k}_\alpha], \quad \alpha = 1, 2,$
- the virtual generalized strain vector, $\breve{\bar{\varepsilon}} \equiv \{ {}^n\bar{\beta}_1 \quad {}^n\bar{\chi}_1 \quad {}^n\bar{\beta}_2 \quad {}^n\bar{\chi}_2 \}^T$, and we may express the virtual strain–displacement relations in compact matrix form as:

$$\boxed{\breve{\bar{\varepsilon}}(\mathbf{d}, \theta; \bar{\mathbf{d}}, \bar{\theta}) = \mathbf{E}(\mathbf{d}, \theta)\, \mathbf{H}(\bar{\mathbf{d}}, \bar{\theta})} \tag{11.438}$$

where, we have introduced the following definition,

Definition: The virtual strain–displacement matrix, $\mathbf{E}(\mathbf{d}, \theta)$:

$$\boxed{\mathbf{E}(\mathbf{d}, \theta) = \begin{bmatrix} \mathbf{R}^\mathrm{T} & \mathbf{0} & \mathbf{0} & \mathbf{0} \\ \mathbf{0} & \mathbf{R}^\mathrm{T} & \mathbf{0} & \mathbf{0} \\ \mathbf{0} & \mathbf{0} & \mathbf{R}^\mathrm{T} & \mathbf{0} \\ \mathbf{0} & \mathbf{0} & \mathbf{0} & \mathbf{R}^\mathrm{T} \end{bmatrix} \begin{bmatrix} [\mathbf{k}_1^0] & [\mathbf{a}_1]\mathbf{W} & \mathbf{I} & \mathbf{0} & \mathbf{0} & \mathbf{0} \\ \mathbf{0} & \mathbf{X}_1 + [\mathbf{k}_1^c]\mathbf{W} & \mathbf{0} & \mathbf{W} & \mathbf{0} & \mathbf{0} \\ [\mathbf{k}_2^0] & [\mathbf{a}_2]\mathbf{W} & \mathbf{0} & \mathbf{0} & \mathbf{I} & \mathbf{0} \\ \mathbf{0} & \mathbf{X}_2 + [\mathbf{k}_2^c]\mathbf{W} & \mathbf{0} & \mathbf{0} & \mathbf{0} & \mathbf{W} \end{bmatrix}} \tag{11.439}$$

where, $\mathbf{R}(\theta)$, $\mathbf{W}(\theta)$, $\mathbf{X}_\alpha(\theta)$, $\alpha = 1, 2$ matrices are as defined in equations (11.348) with (11.350), (11.265), and (11.379), respectively, and $\mathbf{I}$, is the 3×3 identity matrix.

11.14.5 External Virtual Functional Revisited

A definitive form of the external virtual work depends on the basic nature of the externally applied loading system. In what follows, we will consider two types of loading system.

11.14.5.1 Quasi-Static Proportional Loading

This type of loading system essentially results in the static analysis of a shell structure in the nonlinear regime; in this case, the linear and the angular momentum of a shell are assumed identically zero, that is, the inertial virtual work vanishes identically: $G_{Inertia} \equiv 0$. Thus, In order that we can produce a concrete solution strategy for the quasi-static analysis of a shell, we need to characterize the external virtual work more precisely.

As indicated earlier, the exact expression for the external virtual work depends on the forces applied to a shell. We may take the distributed forces and the moments along the shell body, and the forces and moments along the edge boundary curve to be purely proportional, with the former being only functions of the shell arc length parameters. In other words, we assume that the prescribed forces and moments are functions of a single real parameter, $\lambda \in \mathbb{R}$. Clearly, within this assumption of one parameter family of forces and moments, various possibilities exist such as the combination of a constant, $\mathbf{b}_1(S^1, S^2, \eta)$, and a proportional distributed load, $\mathbf{b}_2(S^1, S^2, \eta)$, that is, $\mathbf{b}(\lambda, S^1, S^2, \eta) \equiv \mathbf{b}_1(S^1, S^2, \eta) + \mathbf{b}_2(\lambda, S^1, S^2, \eta)$ where λ is the constant of proportionality for the distributed loading. Similarly, displacement state dependent loadings, that is, non-conservative loadings such as the follower forces, and so on, can also be considered through an additional term reflecting the appropriate differentiations. We will only consider conservative loading system for static analysis.

The utility of the choice of a one-parameter family of forces and moments is that by specializing the external virtual work equations for this family, we can nonlinearly trace the configuration path of a shell. With these in mind, we may include the load proportionality parameter, λ, as a formal parameter, and rewrite the external virtual work functional, $G_{External}(\lambda;\, {}^t\bar{\mathbf{d}}, {}^t\bar{\boldsymbol{\theta}})$, as:

$$\boxed{\begin{aligned} &G_{External}(\lambda;\ {}^t\bar{\mathbf{d}},\ {}^t\bar{\boldsymbol{\theta}}) \\ &= \left\{ \iint\limits_{M^f} \left({}^t\bar{\mathbf{d}} \bullet {}^t\hat{\mathbf{F}}(\lambda, S^1, S^2) + \mathbf{W}\,{}^t\bar{\boldsymbol{\theta}} \bullet {}^t\hat{\mathbf{M}}(\lambda, S^1, S^2) \right) dA \right\} \\ &+ \lambda \left\{ \oint\limits_{\partial M^f} \left({}^t\bar{\mathbf{d}} \bullet {}^t\hat{\mathbf{F}}_\nu(\lambda, S) + \mathbf{W}\,{}^t\bar{\boldsymbol{\theta}} \bullet {}^t\hat{\mathbf{M}}_\nu(\lambda, S) \right) dS \right\} = \mathbf{0}, \quad \forall \text{admissible } {}^t\bar{\mathbf{d}},\ {}^t\bar{\boldsymbol{\theta}} \end{aligned}}$$

(11.440)

where the superscript t and n indicate that all quantities are expressed in the undeformed frame and the deformed rotated frame, respectively.

11.14.5.2 Time-dependent Dynamic Loading

This type of loading system essentially results in the full dynamic analysis of a shell structure in the linear and nonlinear regimes. For this case also, we will only consider conservative loading system. For these situations and using the relationships in equation (11.427), we rewrite the external virtual work functional as:

$$\boxed{G_{External}(\lambda;\ {}^t\bar{\mathbf{d}},\ {}^t\bar{\boldsymbol{\theta}}) = \int\limits_{\partial M_f} \left({}^t\hat{\mathbf{F}}_\nu \bullet {}^t\bar{\mathbf{d}} + \mathbf{W}^T\, {}^t\hat{\mathbf{M}}_\nu \bullet {}^t\bar{\boldsymbol{\theta}} \right) dS + \int\limits_{M} \left\{ {}^t\hat{\mathbf{F}} \bullet {}^t\bar{\mathbf{d}} + \mathbf{W}^T\, {}^t\hat{\mathbf{M}} \bullet {}^t\bar{\boldsymbol{\theta}} \right\} dA}$$

(11.441)

11.14.6 Dynamic Computational Virtual Work Functional

Now, we can express the virtual functional of equation (11.415) in terms of the real displacement, ${}^t\mathbf{d}$, and rotation, ${}^t\boldsymbol{\theta}$, and their virtual counterparts for dynamic analysis as:

$$
\begin{aligned}
G({}^t\mathbf{d}, {}^t\theta; {}^t\bar{\mathbf{d}}, {}^t\bar{\theta}) &\equiv G_{Inertial} + G_{Internal} - G_{External} \\
&= \overbrace{\int_M \left\{ (M_0\, {}^t\ddot{\mathbf{d}} \bullet {}^t\bar{\mathbf{d}} + I_0\, \mathbf{W}^T (\dot{\mathbf{W}}\, {}^t\dot{\theta} + \mathbf{W}\, {}^t\ddot{\theta}) \bullet {}^t\bar{\theta} \right\} dA}^{\text{Inertial}} \\
&+ \overbrace{\int_M \{ ({}^n\mathbf{F}^\alpha \bullet {}^n\bar{\beta}_\alpha + {}^n\mathbf{M}^\alpha \bullet {}^n\bar{\chi}_\alpha \} \, dA}^{\text{Internal}} \\
&- \overbrace{\left\{ \int_{\partial M_f} \left({}^t\hat{\mathbf{F}}_\nu \bullet {}^t\bar{\mathbf{d}} + \mathbf{W}^T\, {}^t\hat{\mathbf{M}}_\nu \bullet {}^t\bar{\theta} \right) dS + \int_M \left({}^t\hat{\mathbf{F}} \bullet {}^t\bar{\mathbf{d}} + \mathbf{W}^T\, {}^t\hat{\mathbf{M}} \bullet {}^t\bar{\theta} \right) dA \right\}}^{\text{External}} \\
&= \mathbf{0}, \quad \forall \text{admissible}\, {}^t\bar{\mathbf{d}}, {}^t\bar{\theta}
\end{aligned}
\tag{11.442}
$$

with the initial conditions given as:

$$
\begin{aligned}
\mathbf{d}(0) &= \mathbf{d}_0. \\
\theta(0) &= \theta_0. \\
\mathbf{v}(0) &\equiv \dot{\mathbf{d}}(0) = \mathbf{v}_0, \\
\omega(0) &\equiv \dot{\theta}(0) = \omega_0
\end{aligned}
\tag{11.443}
$$

11.14.7 Static and Quasi-Static Computational Virtual Work Functional

Similarly, we can express the virtual functional of equation (11.415) in terms of the real displacement, ${}^t\mathbf{d}$, and rotation, ${}^t\theta$, and their virtual counterparts for static and quasi-static analysis as:

$$
\begin{aligned}
&G(\lambda, {}^t\mathbf{d}, {}^t\theta; {}^t\bar{\mathbf{d}}, {}^t\bar{\theta}) \\
&\equiv G_{Internal} - G_{External} = \overbrace{\int_M \{ ({}^n\mathbf{F}^\alpha \bullet {}^n\bar{\beta}_\alpha + {}^n\mathbf{M}^\alpha \bullet {}^n\bar{\chi}_\alpha \} \, dA}^{\text{Internal}} \\
&- \overbrace{\left\{ \int_{\partial M_f} \left({}^t\hat{\mathbf{F}}_\nu \bullet {}^t\bar{\mathbf{d}} + {}^t\hat{\mathbf{M}}_\nu \bullet \mathbf{W}\, {}^t\bar{\theta} \right) dS + \int_M \left({}^t\hat{\mathbf{F}} \bullet {}^t\bar{\mathbf{d}} + {}^t\hat{\mathbf{M}} \bullet \mathbf{W}\, {}^t\bar{\theta} \right) dA \right\}}^{\text{External}} \\
&= \mathbf{0}, \quad \forall \text{admissible}\, {}^t\bar{\mathbf{d}}, {}^t\bar{\theta}
\end{aligned}
\tag{11.444}
$$

with appropriate prescribed boundary conditions.

11.14.8 Generalized Definitions

For notational compactness and as a follow-up on our discussion of the shell configuration space, we may combine both the real vector displacement and rotation fields, ${}^t\mathbf{d}$ and ${}^t\boldsymbol{\theta}$, and the virtual vector displacement and rotation fields, ${}^t\bar{\mathbf{d}}$ and ${}^t\bar{\boldsymbol{\theta}}$, respectively, for the following.

Definitions

- (6×1) the generalized real state field, ${}^t\breve{\mathbf{d}}$ as: ${}^t\breve{\mathbf{d}} \equiv \{{}^t\mathbf{d} \quad {}^t\boldsymbol{\theta}\}^T$
- (6×1) the generalized virtual state field, ${}^t\breve{\bar{\mathbf{d}}}$ as: ${}^t\breve{\bar{\mathbf{d}}} \equiv \{{}^t\bar{\mathbf{d}} \quad {}^t\bar{\boldsymbol{\theta}}\}^T$
- (6×1) the generalized real velocity field, ${}^t\dot{\breve{\mathbf{d}}}$ as: ${}^t\dot{\breve{\mathbf{d}}} \equiv \{\mathbf{v} \quad \boldsymbol{\omega}\}^T$
- (6×1) the generalized real inertial force field, ${}^t\breve{\mathbf{F}}_{Iner}$ as: ${}^t\breve{\mathbf{F}}_{Iner} \equiv \{M_0\,{}^t\ddot{\mathbf{d}} \quad I_0\,(\mathbf{W}^T(\dot{\mathbf{W}}\,{}^t\dot{\boldsymbol{\theta}} + \mathbf{W}\,{}^t\ddot{\boldsymbol{\theta}}))\}^T$

Similarly, we have the following.

Definitions

- (12×1) the generalized real strain, ${}^n\breve{\boldsymbol{\varepsilon}}$, as ${}^n\breve{\boldsymbol{\varepsilon}} \equiv \left\{{}^n\boldsymbol{\beta}_1 \quad {}^n\boldsymbol{\chi}_1 \quad {}^n\boldsymbol{\beta}_2 \quad {}^n\boldsymbol{\chi}_2\right\}^{\mathbf{T}}$
- (12×1) the generalized virtual strain, ${}^n\breve{\bar{\boldsymbol{\varepsilon}}}$, as: ${}^n\breve{\bar{\boldsymbol{\varepsilon}}}(\mathbf{d}, \boldsymbol{\varphi}; \bar{\mathbf{d}}, \bar{\boldsymbol{\varphi}}) \equiv \left\{{}^n\bar{\boldsymbol{\beta}}_1 \quad {}^n\bar{\boldsymbol{\chi}}_1 \quad {}^n\bar{\boldsymbol{\beta}}_2 \quad {}^n\bar{\boldsymbol{\chi}}_2\right\}^T$
- (12×1) the generalized reactive force field, ${}^n\breve{\mathbf{F}}$, as: $\underbrace{{}^n\breve{\mathbf{F}}}_{12x1} \equiv \left\{{}^n\mathbf{F}^1 \quad {}^n\mathbf{M}^1 \quad {}^n\mathbf{F}^2 \quad {}^n\mathbf{M}^2\right\}^T$
- (12×1) the generalized external force field, ${}^t\breve{\hat{\mathbf{F}}}$, as: $\underbrace{{}^t\breve{\hat{\mathbf{F}}}}_{12x1} \equiv \left\{{}^t\hat{\mathbf{F}}^1 \quad {}^t\hat{\mathbf{M}}^1 \quad {}^t\hat{\mathbf{F}}^2 \quad {}^t\hat{\mathbf{M}}^2\right\}^T$

Similar generalizations apply for other force and moment terms.

11.14.9 Dynamic Virtual Work Functional in Generalized Form

With the above notational definitions and the load proportionality parameter, λ, we may rewrite the virtual work functional, $G({}^t\mathbf{d}, {}^t\boldsymbol{\theta}; {}^t\bar{\mathbf{d}}, {}^t\underline{\bar{\theta}})$, of equation (11.442) in compact generalized notation as:

$$
\boxed{
\begin{aligned}
&G({}^t\breve{\mathbf{d}};\, {}^t\breve{\bar{\mathbf{d}}}) \equiv G_{Internal} + G_{Inertial} - G_{External} = \mathbf{0}, \quad \forall \text{admissible}\, {}^t\breve{\bar{\mathbf{d}}}\\
&G_{Internal} = \int_M {}^n\breve{\bar{\boldsymbol{\varepsilon}}}\; {}^n\breve{\mathbf{F}}\, dA\\
&G_{Inertial} = \int_M {}^t\breve{\bar{\mathbf{d}}}^T\; {}^t\breve{\mathbf{F}}_{Iner}\, dA\\
&G_{External} = \int_M {}^t\breve{\bar{\mathbf{d}}}^T\; {}^t\breve{\hat{\mathbf{F}}}(S^1, S^2)\, dA + \lambda \int_{\partial M^f} {}^t\breve{\bar{\mathbf{d}}}^T\; {}^t\breve{\hat{\mathbf{F}}}(S)\, dS
\end{aligned}
}
\tag{11.445}
$$

with the applicable initial conditions on the dynamic state variables as:

$$\begin{aligned} \breve{\mathbf{d}}(0) &= \breve{\mathbf{d}}_0. \\ {}^t\dot{\breve{\mathbf{d}}}(0) &= \breve{\mathbf{v}}_0, \end{aligned} \tag{11.446}$$

11.14.10 Static Virtual Work Functional in Generalized Form

With the above notational definitions and the load proportionality parameter, λ, we may rewrite the virtual work functional, $G(\lambda,\, {}^t\mathbf{d},\, {}^t\boldsymbol{\theta};\, {}^t\bar{\mathbf{d}},\, {}^t\overleftarrow{\theta})$, of equation (11.444) in compact generalized notation as:

$$\begin{aligned} &G(\lambda,\, {}^t\breve{\mathbf{d}};\, {}^t\breve{\bar{\mathbf{d}}}) \equiv G_{Internal}({}^t\breve{\mathbf{d}};\, {}^t\breve{\bar{\mathbf{d}}}) - W_{external}(\lambda;\, {}^t\breve{\bar{\mathbf{d}}}) = 0, \quad \forall \text{admissible } {}^t\breve{\bar{\mathbf{d}}} \\ &G_{Internal} = \int_M {}^n\breve{\bar{\boldsymbol{\varepsilon}}}\ {}^n\breve{\mathbf{F}}\ dA \\ &G_{External} = \int_M {}^t\breve{\bar{\mathbf{d}}}^T\ {}^t\breve{\hat{\mathbf{F}}}^B(\lambda, S^1, S^2)\, dA + \lambda \int_{\partial M^f} {}^t\breve{\bar{\mathbf{d}}}^T\ {}^t\breve{\hat{\mathbf{F}}}^S(S)\, dS \end{aligned} \tag{11.447}$$

where the superscripts t and n indicate that all quantities are expressed in the undeformed frame and the deformed frame, respectively.

11.14.11 Where We Would Like to Go

In conformity with our principal interest in computational solid mechanics and structural engineering, we need to bring all the shell equations down to a form where we can easily solve shell problems numerically by our finite element scheme: the c-type finite element method. This, in turn, requires the derivation of the compatible real strain fields along the configuration path from our knowledge of the above virtual strain fields; this endeavor validates our assumption that the derived virtual strains are truly variational, that is, these can be shown to exist as variations of some real strain fields. Our expressions for the virtual strain fields given in equations (11.416) and (11.417) already suggest as to what the real strain fields will be. However, we would like to record formally the expressions for real strain fields through an application of what is known as the Vainberg principle. Accordingly, the most logical route will be:

- the compatible real strains in component form.

Moreover, as we can see from equations (11.445) and (11.447), the virtual functional, being generally highly nonlinear in displacements and rotations, may not be solved in a closed form in an analytical way. The solution, that is, the tracing of the shell motion or configuration path, will require some sort of iterative numerical method. Of all the possible schemes, the one that we intend to pursue is a Newton-type method. This, in turn, requires linearization of the equations at every iteration step. Clearly, then, we will have to further modify the virtual work equation for the ensuing computational strategy. Let us recall, however, that while the translational displacement of the shell reference surface travels on a Euclidean (flat) configuration space, the shell rotation tensor belongs to a Riemannian (curved) configuration space. More specifically, as we will see, we will need to recognize this by resorting to covariant derivatives as opposed to standard Gateaux derivatives of the Euclidean space in our linearization of the virtual work functional. Thus, the next logical place to go from here will be:

- covariant linearization of virtual work functionals

11.15 Computational Real Strains

11.15.1 What We Need to Recall

Now, let us recall that we have defined the virtual strains in terms of the displacement and rotation fields in invariant (defined from the weak form of balance equations) and component vector forms (defined from the virtual work equation) by:

11.15.1.1 Internal Virtual Work Functional

$$\boxed{G_{Internal} \equiv \int_M \left\{ {}^n\mathbf{F}^\alpha \bullet {}^n\bar{\boldsymbol{\beta}}_\alpha + {}^n\mathbf{M}^\alpha \bullet {}^n\bar{\boldsymbol{\chi}}_\alpha \right\} dA} \tag{11.448}$$

11.15.1.2 Component Vector Form

- The translational virtual strain fields, ${}^n\bar{\boldsymbol{\beta}}_\alpha, \quad \alpha = 1, 2$, given as:

$$\boxed{{}^n\bar{\boldsymbol{\beta}}_\alpha = {}^t\mathbf{R}^{\mathbf{T}} \left({}^t\bar{\mathbf{d}},_\alpha + \mathbf{k}^0_\alpha \times {}^t\bar{\mathbf{d}} \right) - {}^t\mathbf{R}^{\mathbf{T}}\, {}^t\bar{\boldsymbol{\Phi}} \left(\mathbf{e}_\alpha + \mathbf{k}^0_\alpha \times {}^t\mathbf{d} \right)} \tag{11.449}$$

Component-wise in spatial representation, that is, in rotated bases:

$$\boxed{{}^n\bar{\boldsymbol{\beta}}_\alpha = \left\{ {}^n\bar{\beta}_{\alpha 1} \quad {}^n\bar{\beta}_{\alpha 2} \quad {}^n\bar{\beta}_{\alpha 3} \right\}^T, \quad \alpha = 1, 2} \tag{11.450}$$

For equation (11.449), we have the following definitions:

$$\begin{aligned} \mathbf{e}_\alpha &\equiv \hat{\mathbf{1}}_\alpha + {}^t\mathbf{d},_\alpha, \quad \alpha = 1, 2, \\ {}^t\mathbf{d}|_\alpha &= {}^t\mathbf{d},_\alpha + \mathbf{k}^0_\alpha \times {}^t\mathbf{d}, \quad \alpha = 1, 2, \end{aligned} \tag{11.451}$$

- The bending virtual strain fields, ${}^n\bar{\boldsymbol{\chi}}_\alpha, \quad \alpha = 1, 2$, given as:

$$\boxed{{}^n\bar{\boldsymbol{\chi}}_\alpha = {}^t\mathbf{R}^{\mathbf{T}}\, {}^t\bar{\boldsymbol{\varphi}},_\alpha - {}^t\mathbf{R}^{\mathbf{T}}\, {}^t\bar{\boldsymbol{\Phi}}\, \mathbf{k}^0_\alpha} \tag{11.452}$$

Component-wise in spatial representation, that is, in rotated bases:

$$\boxed{{}^n\bar{\boldsymbol{\chi}}_\alpha = \left\{ {}^n\bar{\chi}_{\alpha 1} \quad {}^n\bar{\chi}_{\alpha 2} \quad {}^n\bar{\chi}_{\alpha 3} \right\}^T, \quad \alpha = 1, 2} \tag{11.453}$$

Furthermore, we need the following results:

$$\boxed{{}^t\mathbf{k}_\alpha = \mathbf{W}(\theta)\ \theta,_\alpha, \quad \alpha = 1,2} \tag{11.454}$$

where:

$$\boxed{\begin{aligned} &\mathbf{W}(\theta) \equiv \mathbf{I} + c_1\ \mathbf{\Theta} + c_2\ \mathbf{\Theta}^2, \\ &c_1(\theta) = \frac{1-\cos\theta}{\theta^2}, \quad c_2(\theta) = \frac{\theta - \sin\theta}{\theta^3} \end{aligned}} \tag{11.455}$$

Lemma LD1: Defining ${}^t\mathbf{k}_\alpha$, $\alpha = 1,2$ and ${}^t\bar{\varphi}$ as the axial vectors of ${}^t\mathbf{K}_\alpha$, $\alpha = 1,2$ and ${}^t\bar{\mathbf{\Phi}}$, respectively, we have:

$$\boxed{\begin{aligned} {}^t\bar{\mathbf{\Phi}},_\alpha &= {}^t\bar{\mathbf{K}}_\alpha + [{}^t\mathbf{K}_\alpha\ {}^t\bar{\mathbf{\Phi}} - {}^t\bar{\mathbf{\Phi}}\ {}^t\mathbf{K}_\alpha], \quad \alpha = 1,2 \\ \underbrace{{}^t\bar{\varphi},_\alpha}_{\substack{\text{Effective material} \\ \text{curvature vector} \\ \text{variation}}} &= \underbrace{{}^t\bar{\mathbf{k}}_\alpha}_{\substack{\text{Material} \\ \text{curvature vector} \\ \text{variation}}} + \underbrace{{}^t\mathbf{k}_\alpha}_{\substack{\text{Material} \\ \text{curvature vector}}} \times \underbrace{{}^t\bar{\varphi}}_{\substack{\text{Material} \\ \text{rotation vector} \\ \text{variation}}} \end{aligned}} \tag{11.456}$$

11.15.2 *Component Vector Form: Real Strains and the Vainberg Principle*

11.15.2.1 Virtual Strain Symmetry

As indicated, in order for us to apply the Vainberg principle of determining the real strains from the virtual ones, we need to guarantee the symmetry of the virtual strains on a linear vector space which is provided by the tangent space at identity; the components of the virtual strains are expressed in the rotated bases. Noting that $\hat{\bar{\varepsilon}} \equiv ({}^n\bar{\beta}_\alpha, {}^n\bar{\chi}_\alpha)$ belongs to a linear vector space (namely, the tangent space at identity), the following symmetry relations hold:

$$\boxed{\hat{\bar{\varepsilon}}({}^t\mathbf{d}, {}^t\theta; {}^t\tilde{\mathbf{d}}, {}^t\tilde{\theta}; {}^t\bar{\mathbf{d}}, {}^t\bar{\theta}) = \hat{\bar{\varepsilon}}({}^t\mathbf{d}, {}^t\theta; {}^t\bar{\mathbf{d}}, {}^t\bar{\theta}; {}^t\tilde{\mathbf{d}}, {}^t\tilde{\theta})} \tag{11.457}$$

where the parameters ${}^t\theta$, ${}^t\tilde{\theta}$, ${}^t\bar{\theta}$ belong to the tangent space at identity with ${}^t\mathbf{d}$, ${}^t\theta$ as the real parameters, ${}^t\tilde{\mathbf{d}}$, ${}^t\tilde{\theta}$; ${}^t\bar{\mathbf{d}}$, ${}^t\bar{\theta}$ are the two sets of the virtual parameters. In other words, the virtual strains are symmetric for the conservative loading, and thus, by the Vainberg principle, the potential functional exists. In other words, there exist functions (we will call them the real strains) the variation of which will yield the virtual strains.

11.15.2.2 Component Form: Real Translational (Extensional and Shear) Strain Field ${}^n\beta_\alpha$, $\alpha = 1,2$

The real translational strain field, ${}^n\beta_\alpha$, can be derived by an application of the integral expression in the definition of the Vainberg principle as follows (mnemonically, multiply the real variables by $\rho \in [0.1]$ such as $\mathbf{d} \to \rho\mathbf{d}$, and so on, and replace the virtual variables by the corresponding real one such as $\bar{\mathbf{d}} \to \mathbf{d}$, and so on, and integrate over $\rho \in [0.1]$):

$$
\begin{aligned}
{}^n\beta_\alpha &= \int_0^1 {}^n\bar{\beta}_\alpha(\rho\,{}^t\mathbf{d},\,\rho\,{}^t\varphi;\,{}^td,\,{}^t\varphi)\,d\rho = \int_0^1 \{{}^t\mathbf{R}_\rho^T\,{}^t\mathbf{d}|_\alpha + (-{}^t\mathbf{R}_\rho^T\,{}^t\bar{\Phi})(\hat{\imath}_\alpha + \rho\,{}^t\mathbf{d}|_\alpha)\}\,d\rho \\
&= \int_0^1 \frac{d}{d\rho}\{{}^t\mathbf{R}_\rho^T(\hat{\imath}_\alpha + \rho\,{}^t\mathbf{d}|_\alpha)\}\,d\rho = \{{}^t\mathbf{R}_\rho^T(\hat{\imath}_\alpha + \rho\,{}^t\mathbf{d}|_\alpha)\}_{\rho=0}^{\rho=1} \\
&= {}^t\mathbf{R}^T(\hat{\imath}_\alpha + {}^t\mathbf{d}|_\alpha) - \hat{\imath}_\alpha, \quad \alpha = 1,2
\end{aligned}
$$

In the above derivation, we have used relations in equation (11.451), and the following definition.

Definition: ${}^t\mathbf{R}_\rho \equiv {}^t\mathbf{R}(\rho\varphi)$

Facts:

- ${}^t\mathbf{R}_0 \equiv {}^t\mathbf{R}(0) = \mathbf{I} = {}^t\mathbf{R}^T(0) = {}^t\mathbf{R}_0^T$
  $\overline{{}^t\mathbf{R}^T} \equiv \delta({}^t\mathbf{R}^T) = -{}^t\mathbf{R}^T\,{}^t\bar{\Phi}$ (see discussions on the derivatives and variations of rotation)

11.15.2.3 Component Form: Real Bending Strain Field: ${}^n\chi_\alpha$, $\alpha = 1, 2$

Likewise, the real bending strains, ${}^n\chi_\alpha$, can be derived by an application of the integral expression in the definition of the Vainberg principle as in the exercise below, ${}^n\chi_\alpha = {}^t\mathbf{R}^{\mathbf{T}}\,\mathbf{k}_\alpha^c - \mathbf{k}_\alpha^0$, $\alpha = 1,2$ where we have used the following definitions.

Definitions

- The current total curvature vector: $\mathbf{k}_\alpha^c \equiv \mathbf{k}_\alpha^0 + \mathbf{k}_\alpha$, $\alpha = 1,2$, with $\mathbf{k}_\alpha \equiv \mathbf{W}\ \theta,_\alpha$ where $\mathbf{W}(\theta) \equiv I + c_1\,\mathbf{\Theta} + c_2\,\mathbf{\Theta}^2$, $c_1 \equiv \frac{1-\cos\theta}{\theta^2}$, $c_2 \equiv \frac{\theta-\sin\theta}{\theta^3}$ with $\mathbf{W}$ defined in equation (11.455) and introduced in the section on the derivatives and variations of rotation tensor, ${}^t\mathbf{R}$, in which $\mathbf{\Theta}$ is the skew-symmetric tensor corresponding to rotation vector, θ, and θ stands for the length of θ; $\mathbf{k}_\alpha^0$, $\alpha = 1,2$ are the initial curvature vectors of the shell reference surface along the lines of curvature.

Exercise: Following the similar procedure used for the real translational strains, derive the real bending strain as: ${}^n\chi_\alpha = \mathbf{R}^{\mathbf{T}}\,\mathbf{k}_\alpha^c - \mathbf{k}_\alpha^0$, $\alpha = 1,2$.

Solution

$$
\begin{aligned}
{}^n\chi_\alpha &= \int_0^1 {}^n\bar{\chi}_\alpha(\rho\,{}^t\mathbf{d},\,\rho\,{}^t\varphi;\,{}^td,\,{}^t\varphi)\,d\rho \\
&= \int_0^1 \frac{d}{d\rho}\{{}^t\mathbf{R}_\rho^T(\mathbf{k}_\alpha^0 + \rho\,\mathbf{k}_\alpha)\}\,d\rho = \{{}^t\mathbf{R}_\rho^T(\mathbf{k}_\alpha^0 + \rho\,\mathbf{k}_\alpha)\}_{\rho=0}^{\rho=1} \\
&= {}^t\mathbf{R}^T(\mathbf{k}_\alpha^0 + \mathbf{k}_\alpha) - \mathbf{k}_\alpha^0 = {}^t\mathbf{R}^T\,\mathbf{k}_\alpha^c - \mathbf{k}_\alpha^0, \quad \alpha = 1,2
\end{aligned}
\tag{11.458}
$$

where we have recalled the definition for the total curvature vector as: $\mathbf{k}_\alpha^c \equiv \mathbf{k}_\alpha^0 + \mathbf{k}_\alpha$, $\alpha = 1,2$.

11.15.3 The Real Translational and Bending Strains

Thus, we have the real translational (axial and shear) and the bending strains given as:

$$\boxed{\begin{aligned} {}^{n}\boldsymbol{\beta}_{\alpha} &= \mathbf{R}^{\mathbf{T}}\left(\mathbf{e}_{\alpha} + \mathbf{k}^{0}_{\alpha} \times \mathbf{d}\right) - \hat{\mathbf{1}}_{\alpha}, \quad \alpha = 1,2 \\ {}^{n}\boldsymbol{\chi}_{\alpha} &= \mathbf{R}^{\mathbf{T}}\mathbf{k}^{c}_{\alpha} - \mathbf{k}^{0}_{\alpha} \end{aligned}} \tag{11.459}$$

11.15.4 Where We Would Like to Go

So far, we have derived the shell real strain fields by applying the Vainberg principle to the virtual strain fields. Now that both the real and the virtual strain fields are known, we must particularize the virtual functional further by determination of the reactive forces and moments through the application of a relevant constitutive law. Thus, we must check out the special material properties and the loading system for which the analysis will be specialized and pursued henceforth for numerical implementation. Let us also recall that the internal virtual work functional of equation (11.448), has been expressed in terms of the real and virtual displacements and rotations, and the reactive forces and moments. For notational compactness and in favor of uncluttered presentation, it is useful to generalize the expressions by lumping various translational entities with their corresponding rotational counterparts – such as the displacements with the rotations (real and virtual), and so on. Thus, a logical place to go from here should be:

- shell material property (constitutive laws)
- covariant linearization of virtual work equations.

11.16 Hyperelastic Material Property

11.16.1 What We Need to Recall

- The invariant form of the real strain vectors in rotated (spatial) frame, and the components in rotating bases, are given as:

$$\boxed{\boldsymbol{\beta}_{\alpha} = \mathbf{c}_{0},_{\alpha} - \mathbf{g}_{\alpha} = {}^{n}\beta_{\alpha\eta}\, \mathbf{g}^{\eta} + {}^{n}\beta_{\alpha 3}(\equiv \beta_{\alpha})\, \mathbf{g}^{3}, \ \alpha = 1,2} \tag{11.460}$$

$$\boxed{\boldsymbol{\chi}_{\alpha} = \mathbf{k}_{\alpha} = \mathbf{k}^{c}_{\alpha} - \mathbf{R}\,\mathbf{k}^{0}_{\alpha} = {}^{n}\chi_{\alpha\eta}\, \mathbf{g}^{\eta} + {}^{n}\chi_{\alpha 3}(\equiv \chi_{\alpha})\, \mathbf{g}^{3}, \ \alpha = 1,2} \tag{11.461}$$

- The invariant form of the internal virtual work functional is given by:

$$\boxed{\hat{G}_{Internal} \equiv \int_{\Omega} \left\{ (\mathbf{F}^{\alpha} \bullet \overset{\Delta}{\overline{\boldsymbol{\beta}_{\alpha}}} + {}^{n}\mathbf{M}^{\alpha} \bullet \overset{\Delta}{\overline{\boldsymbol{\chi}_{\alpha}}} \right\} dA} \tag{11.462}$$

where $\overset{\Delta}{\overline{\boldsymbol{\beta}_{\alpha}}}$ and $\overset{\Delta}{\overline{\boldsymbol{\chi}_{\alpha}}}$ are the co-rotated or objective variations of the force and moment strain fields, respectively;

- The resistive stress resultant, $\mathbf{F}^\alpha$, is defined by:

$$\boxed{\mathbf{F}^\alpha \equiv \int_{\eta^b}^{\eta^t} {}^\eta\mathbf{P}^\alpha \; {}^\eta\mu^S \, d\eta} \tag{11.463}$$

- The resistive stress couple, $\mathbf{M}^\alpha$, is defined by:

$$\boxed{\mathbf{M}^\alpha(S) \equiv \int_{\eta^b}^{\eta^t} (S,\eta) \times \; {}^\eta\mathbf{P}^\alpha \; {}^\eta\mu^S \, d\eta} \tag{11.464}$$

where (S,η) defines the through-the-thickness deformation of the shell-like body. and component-wise, denoted by:

$$\boxed{\begin{aligned} \mathbf{F}^\alpha &= {}^nF^{\alpha\beta}\;\mathbf{g}_\beta + {}^nF^{\alpha 3}\,\mathbf{g}_3, \;\alpha = 1,2 \\ \mathbf{M}^\alpha &= {}^nM^{\alpha\beta}\;\mathbf{g}_\beta + {}^nM^{\alpha 3}\;\mathbf{g}_3 \end{aligned}} \tag{11.465}$$

- The component form of the real strain vectors, and the components in rotating bases, are given as:

$$\boxed{\begin{aligned} {}^n\boldsymbol{\beta}_\alpha &= \mathbf{R}^\mathbf{T}\left(\mathbf{e}_\alpha + \mathbf{k}^0_\alpha \times \mathbf{d}\right) - \hat{\mathbf{1}}_\alpha, \quad \alpha = 1,2 \\ {}^n\boldsymbol{\chi}_\alpha &= \mathbf{R}^\mathbf{T}\mathbf{k}^c_\alpha - \mathbf{k}^0_\alpha \end{aligned}} \tag{11.466}$$

with components denoted by:

$$\boxed{\begin{aligned} {}^n\boldsymbol{\beta}_\alpha &= \left\{\; {}^n\beta_{\alpha 1} \quad {}^n\beta_{\alpha 2} \quad {}^n\beta_{\alpha 3} \;\right\}^T, \quad \alpha = 1,2 \\ {}^n\boldsymbol{\chi}_\alpha &= \left\{\; {}^n\chi_{\alpha 1} \quad {}^n\chi_{\alpha 2} \quad {}^n\chi_{\alpha 3} \;\right\}^T \end{aligned}} \tag{11.467}$$

where a superscript n signifies that the object has a spatial representation in the rotating bases, $\{\mathbf{g}_i\},\; i = 1,2,3$.

- The internal virtual work functional, $G_{Internal}(\mathbf{d},\mathbf{R};\bar{\mathbf{d}},\bar{\boldsymbol{\varphi}})$, is given as:

$$\boxed{G_{Internal}(\mathbf{d},\mathbf{R};\bar{\mathbf{d}},\bar{\boldsymbol{\varphi}}) \equiv \underbrace{\int_\Omega \left\{ ({}^n\mathbf{F}^\alpha \bullet {}^n\bar{\boldsymbol{\beta}}_\alpha + \; {}^n\mathbf{M}^\alpha \bullet {}^n\bar{\boldsymbol{\chi}}_\alpha \right\} dA}_{G_{Internal}}} \tag{11.468}$$

The shell reactive component or operational force vector, ${}^n\mathbf{F}^\alpha$, and the shell reactive component or operational moment vector, ${}^n\mathbf{M}^\alpha$, are component-wise denoted by:

$$\boxed{\begin{aligned} {}^n\mathbf{F}^\alpha &= \{\; {}^nF^{\alpha 1} \quad {}^nF^{\alpha 2} \quad {}^nF^{\alpha 3} \;\}^T = {}^t\mathbf{R}^T\,{}^t\mathbf{F}^\alpha, \quad \alpha = 1,2 \\ {}^n\mathbf{M}^\alpha &= \{\; {}^nM^{\alpha 1} \quad {}^nM^{\alpha 2} \quad {}^nM^{\alpha 3} \;\}^T = {}^t\mathbf{R}^T\,{}^t\mathbf{M}^\alpha \end{aligned}} \tag{11.469}$$

where the superscript t and n indicate that all quantities are measures and expressed in the undeformed $\mathbf{G}_1, \mathbf{G}_2, \mathbf{G}_3 \equiv \mathbf{N}$ frame, and the deformed and rotated $\mathbf{g}_1, \mathbf{g}_2, \mathbf{g}_3$ frame, respectively.

- The reduced internal energy variation is given as:

$$\boxed{\begin{aligned}\bar{W} \equiv \delta W &= \int_M \left(\int_{\eta^b}^{\eta^t} \mathbf{P} : \bar{\mathbf{F}}\ {}^{\eta}\mu^S\, d\eta \right) dA \\ &= \int_M \left\{ \mathbf{F}^{\alpha} \bullet \overline{\boldsymbol{\beta}_{\alpha}} + \mathbf{M}^{\alpha} \bullet \overline{\boldsymbol{\chi}_{\alpha}} \right\} dA = \bar{U} \equiv \delta U\end{aligned}} \tag{11.470}$$

where $\bar{U} \equiv \delta U$ is the strain energy variation of a two-dimensional shell reference surface identified as the reduced 3D strain energy variation, $\bar{W} \equiv \delta W$, of the corresponding shell-like body.

11.16.2 Elastic Shell Material Property

As mentioned earlier, and in our discussion on the constitutive laws, every constitutive theory is approximate by its very task of modeling and fitting experimental data; the shell constitutive properties will not be an exception and, in fact, will induce additional approximations through the reduction of the 3D theory to an equivalent 2D shell material properties. In any case, as usual, the shell constitutive modeling must also satisfy the four general principles of a consistent 3D theory which we may recall are: the principle of stress determinism, the principle of local action, the principle of material frame indifference, and the principle of material symmetry. As a reminder, we must note that the derived real translational and bending strains as given by equation (11.460) and (11.461), respectively, and the virtual work principle given by equation (11.468) is exact and does not depend on the material properties of a shell.

11.16.2.1 Principle of Equivalence

All the same, we need an equivalent 2D constitutive theory from a corresponding 3D one, that, on the one hand, is suitable for a shell, and replicates the response of its corresponding 3D counterparts, that is, a shell-like body, as close as possible, on the other. One way of measuring this equivalency is to require that the internal virtual work of a 3D shell-like body integrated through the thickness (just like the way we have derived the 2D virtual work principle and the boundary conditions) be as close as possible to the 2D internal virtual work expressed in equation (11.468). This has been achieved elsewhere earlier in the text as shown in equation (11.470). Recall that for an elastic material, by definition, the first Piola–Kirchhoff stress tensor, $\mathbf{P} = \mathbf{P}^i \otimes \mathbf{G}_i$, can be expressed as a function of the deformation gradient tensor, $\mathbf{F}$, by:

$$\boxed{\mathbf{P} = \bar{\mathcal{P}}(\mathbf{F},\ \mathbf{X})} \tag{11.471}$$

where $\mathbf{X}$ denotes a particle on the 3D shell-like body in the undeformed reference state; recall that the imposition of the material frame indifference guarantees that equation (11.471) can be expressed in terms of energy conjugate stress and strain variables and that with $\mathbf{C} \equiv \mathbf{F}^T\, \mathbf{F}$ as the right Cauchy–Green stretch tensor, we can redefine equation (11.471) as:

$$\boxed{\mathbf{P} = \mathcal{P}(\mathbf{C},\ \mathbf{X})} \tag{11.472}$$

An application of the equivalence principle will generally result in the reduction of the 3D strain energy variation, $\bar{W} \equiv \delta W$, and as such establish the existence of the equivalent 2D strain energy variation, $\bar{U} \equiv \delta U$, as:

$$\begin{aligned}
\bar{W} \equiv \delta W &= \int_M \left(\int_{\eta^b}^{\eta^t} \mathbf{P} : \bar{\mathbf{F}}\ {}^{\eta}\mu^S \, d\eta \right) dA \\
&= \int_M \left\{ \mathbf{F}^\alpha \bullet \overline{\boldsymbol{\beta}_\alpha} + \mathbf{M}^\alpha \bullet \overline{\boldsymbol{\chi}_\alpha} \right\} dA \ + \text{ possible error} \\
&= \bar{U} \equiv \delta U
\end{aligned} \tag{11.473}$$

Recall also that under the kinematic assumption of 1D inextensible director, we have already established this reduction exactly as shown by equation (11.470). However, if we generate the stress resultants and the stress couple, and the conjugate strains exactly, as we do in our theory, without any kinematic assumptions, the reduction in the form of equation (11.473) will generally entail possible error which conforms to our design of containing and restricting all the approximations in the constitutive theory of shells.

11.16.2.2 Isotropic, Hyperelastic Material

Now, if we restrict, as we do, our ensuing analysis to only isotropic hyperelastic, or Green-elastic shells, we can recall that, by definition, there exists a strain energy functional, $W = \rho\,\Phi$, per unit of undeformed volume, that depends on the deformation gradient or the strain tensor; ρ is the current density and Φ is the specific strain energy (i.e. strain energy per unit mass), also known as the Helmholtz free energy, such that we have:

$$\mathcal{P}(\mathbf{C}, \mathbf{X}) = \frac{\partial W}{\partial \mathbf{C}} = 2\mathbf{F}\frac{\partial W}{\partial \mathbf{F}} \tag{11.474}$$

Now, considering equation (11.473) with the shell real strain measures given by $\boldsymbol{\beta}_\alpha$, the real translational (axial and shear) spatial strain, and $\boldsymbol{\chi}_\alpha$, the real rotational (bending) strain, the shell strain energy, U, for the 2D surface, must be given by:

$$U \equiv U(\boldsymbol{\beta}_\alpha, \boldsymbol{\chi}_\alpha, S^1, S^2, \eta) = \int_{\eta^b}^{\eta^t} W(\mathbf{C}, \mathbf{X})\ {}^{\eta}\mu^S \, d\eta \ + \text{ possible error} \tag{11.475}$$

where we have included the explicit dependence on the position of a material particle by the arc length parameters, S^1, S^2 along the lines of curvature, and η, the shell thickness. Then, by the postulate of hyperelasticity, there exist two vector valued functions, $\mathcal{F}(\boldsymbol{\beta}_\alpha, \boldsymbol{\chi}_\alpha, S^1, S^2, \eta)$ and $\mathcal{M}(\boldsymbol{\beta}_\alpha, \boldsymbol{\chi}_\alpha, S^1, S^2, \eta)$ such that $\mathbf{F}^\alpha$, the spatial stress resultant, and $\mathbf{M}^\alpha$, the spatial stress couple, are given by:

$$\begin{aligned}
\mathbf{F}^\alpha &= \mathcal{F}(\boldsymbol{\beta}_\alpha, \boldsymbol{\chi}_\alpha, S^1, S^2, \eta) \equiv \frac{\partial U}{\partial \boldsymbol{\beta}_\alpha} \\
\mathbf{M}^\alpha &= \mathcal{M}(\boldsymbol{\beta}_\alpha, \boldsymbol{\chi}_\alpha, S^1, S^2, \eta) \equiv \frac{\partial U}{\partial \boldsymbol{\chi}_\alpha}
\end{aligned} \tag{11.476}$$

11.16.2.3 Homogeneous, Isotropic, Hyperelastic Material

If the material properties are independent of the position of a particle, then we can rewrite equation (11.476) as:

$$\begin{aligned}\mathbf{F}^{\alpha} &= \mathcal{F}(\beta_{\alpha}, \chi_{\alpha}) \equiv \frac{\partial U}{\partial \beta_{\alpha}} \\ \mathbf{M}^{\alpha} &= \mathcal{M}(\beta_{\alpha}, \chi_{\alpha}) \equiv \frac{\partial U}{\partial \chi_{\alpha}}\end{aligned} \tag{11.477}$$

11.16.2.4 Linear, Homogeneous, Isotropic, Hyperelastic Material

In what follows, we restrict the exposition only to linear, homogeneous, isotropic, hyperelastic materials. Let us recall that for such material properties, the strain energy is a quadratic function of the strains and thus, the constitutive law for a 3D body can be represented in tensor form or component-wise by:

$$\begin{aligned} W &= \frac{1}{2}\mathbf{E} : \mathbf{C} : \mathbf{E} \\ &= \frac{1}{2}\left\{E_{ij}\, \mathbf{A}^{i} \otimes \mathbf{A}^{j} : \mathbb{C}^{mnpq}\, \mathbf{A}_{m} \otimes \mathbf{A}_{n} \otimes \mathbf{A}_{p} \otimes \mathbf{A}_{q} : E_{kl}\, \mathbf{A}^{k} \otimes \mathbf{A}^{l}\right\} \\ &= \frac{1}{2}\left\{\delta^{i}_{m}\, \delta^{j}_{n}\, \delta^{k}_{p}\, \delta^{l}_{q}\, \mathbb{C}^{mnpq}\, E_{ij}\, E_{kl}\right\} = \frac{1}{2}\mathbb{C}^{ijkl}\, E_{ij}\, E_{kl} \\ \Rightarrow \mathbf{S} &= \frac{\partial W}{\partial \mathbf{E}} = \mathbf{C} : \mathbf{E} = \left\{\mathbb{C}^{ijkl}\, E_{kl}\right\} \mathbf{A}_{i} \otimes \mathbf{A}_{j}, \\ S^{ij} &= \mathbb{C}^{ijkl}\, E_{kl} \end{aligned} \tag{11.478}$$

where $\mathbf{S} = S^{ij}(S^1, S^2, \eta)\, \mathbf{A}_i(S^1, S^2, \eta) \otimes\ \mathbf{A}_j(S^1, S^2, \eta)$ is the second Piola–Kirchhoff stress tensor, and $\mathbf{E} = E_{kl}\, (S^1, S^2, \eta)\, \mathbf{A}^k(S^1, S^2, \eta) \otimes\ \mathbf{A}^l(S^1, S^2, \eta)$ is the Green–Lagrange strain tensor, $\left\{\mathbf{A}_i(S^1, S^2, \eta)\right\}$, $i = 1, 2, 3$ is the base vector set at any particle point (S^1, S^2, η) in the undeformed reference system. Now, recall that for linear isotropic material with the identity tensors: $\overset{4}{\mathbf{I}} \equiv \mathbf{A}_i \otimes \mathbf{A}_j \otimes \mathbf{A}^i \otimes \mathbf{A}^j$ and $\mathbf{I} \equiv \mathbf{A}_i \otimes \mathbf{A}^i$, the constitutive tensor is given by:

$$\mathbf{C} \equiv 2\mu \left(\overset{4}{\mathbf{I}} + \frac{\nu}{1 - 2\nu}\, \mathbf{I} \otimes \mathbf{I} \right) \tag{11.479}$$

resulting in the coefficients, $\mathbb{C}^{ijkl}$, $i, j, k, l = 1, 2, 3$ of the fourth-order constitutive tensor, $\mathbf{C} = \mathbb{C}^{ijkl}\, \mathbf{A}_i \otimes \mathbf{A}_j \otimes \mathbf{A}_k \otimes \mathbf{A}_l$, in the curvilinear coordinates as given by:

$$\mathbb{C}^{ijkl} \equiv \mu \left(A^{ik}A^{jl} + A^{il}A^{jk} + \frac{2\nu}{1 - 2\nu} A^{ij}A^{kl} \right) \tag{11.480}$$

where $\mu \equiv G = \dfrac{E}{2(1 + \nu)}$ is the shear modulus with E as the elasticity modulus and ν as the Poisson's ratio.

11.16.2.5 First Piola–Kirchhoff Stress Tensor in Mixed Bases

The stress–strain relations expressed in equations (11.478)–(11.488) are all in terms of the second Piola–Kirchhoff stress tensor, **S**, and the Green–Lagrange strain tensor, **E**; however, we are interested in relations in terms of the first Piola–Kirchhoff stress tensor, $\mathbf{P} = P^{ij}\, \mathbf{a}_i \otimes \mathbf{A}_j$, and the deformation gradient tensor, $\mathbf{F}$, defined by $\mathbf{F} \equiv \mathbf{a}_i \otimes \mathbf{A}^j$ where $\{\mathbf{a}_i\}$ and $\{\mathbf{A}_i\}$ are the deformed and the undeformed base vector sets, respectively. Recall that $\mathbf{P} = \mathbf{F}\,\mathbf{S}$. The important thing to note here is that we have represented $\mathbf{P}$ in a mixed basis form, that is, the domain basis in the undeformed reference state and the range basis in the deformed state. Under the mixed basis representation, we have:

$$\mathbf{P} = P^{ij}\ \mathbf{c},_i \otimes \mathbf{A}_j = \mathbf{F}\,\mathbf{S} = \underbrace{\mathbf{c},_i \otimes \mathbf{A}^i}_{\mathbf{F},\ \text{deformation gradient tensor}} \bullet \underbrace{S^{ij}\mathbf{A}_i \otimes \mathbf{A}_j}_{\mathbf{S},\ \text{second Piola-Kirchhoff tensor}} \tag{11.481}$$

Thus, the components, P^{ij}, of $\mathbf{P}$ represented in the mixed basis are identical with the corresponding components, S^{ij}, of $\mathbf{S}$, and hence, symmetric. Now, noting, that $\mathbf{P} = \mathbf{F}\,\mathbf{S}$, we can rewrite the constitutive relation of equation (11.478) as:

$$\begin{aligned}
P^{ij}\ \mathbf{a}_i \otimes \mathbf{A}_j &= \mathbf{P} = \mathbf{F}\,\mathbf{S} = \mathbf{F}\,\mathbb{C} : \mathbf{E}\\
&= \mathbf{a}_i \otimes \mathbf{A}^i \bullet \mathbb{C}^{kjmn}\ \mathbf{A}_k \otimes \mathbf{A}_j \otimes \mathbf{A}_m \otimes \mathbf{A}_n : E_{pq}\ \mathbf{A}^p \otimes \mathbf{A}^q\\
&= \mathbb{C}^{ijmn}\ (E_{mn}\ \mathbf{a}_i \otimes \mathbf{A}_j) = (\mathbb{C}^{ijmn}\ E_{mn})\ \mathbf{a}_i \otimes \mathbf{A}_j\\
&\Rightarrow \boxed{P^{ij} = \mathbb{C}^{ijmn}\ E_{mn}}
\end{aligned} \tag{11.482}$$

From equation (11.482), we see that if the first Piola–Kirchhoff stress tensor is represented in the mixed basis, then for the stress–strain relations, we can continue to use the same constitutive relations, as between the second Piola–Kirchhoff and the Green–Lagrange strain tensor, with the effective strain tensor represented in the mixed form with the coefficients of the Green–Lagrange strain tensor.

11.16.3 Thin, Linear, Homogeneous, Isotropic, Hyperelastic Shell

In what follows, we restrict the exposition only to linear, homogeneous, isotropic, hyperelastic and very thin shells to derive the stress–strain relationship that will be used in our subsequent numerical simulations. Also, for the thin shells, we will assume that $\mathbf{h}(S, \eta) = \eta\ \mathbf{g}_3$ in equation (11.464), keeping the normal to the surface inextensible through the deformation history. Now, let us first recall that for our optimal normal shell coordinate system along the lines of curvature with the arc lengths as the coordinate parameters, for any particle with coordinates (S^1, S^2, η) in the undeformed shell like body, we have:

$$\begin{aligned}
\mathbf{C}(S^1, S^2, \eta) &= \mathbf{C}_0(S^1, S^2) + \eta\ \mathbf{N}(S^1, S^2)\\
\Rightarrow \mathbf{A}_\alpha &= \mathbf{G}_\alpha + \eta\ \mathbf{N},_\alpha = \mathbf{G}_\alpha - \eta\ K_\alpha^\beta\ \mathbf{G}_\beta\\
&= \mathbf{G}_\alpha - \eta\ (G^{\sigma\beta} K_{\alpha\sigma})\ \mathbf{G}_\beta = \mathbf{G}_\alpha - \eta\ \left(\frac{K_{\alpha\alpha}}{G_{\alpha\alpha}}\right)\ \mathbf{G}_\alpha\\
&= \left(1 - \frac{\eta}{R_\alpha}\right)\ \mathbf{G}_\alpha
\end{aligned} \tag{11.483}$$

where R_α is the radius of curvature, $G_{\alpha\beta}$ and $K_{\alpha\beta}$ are the covariant components of the metric tensor and the curvature tensor of the undeformed shell surface with off-diagonal terms identically zero for the coordinate system along the lines of curvature. We have also used the relationship between the covariant and the contravariant coefficients of the metric tensor for the same coordinate system.

11.16.3.1 Thin Shell Material Properties

The assumption of a thin shell may be mathematically expressed by the condition: $\frac{\eta}{R_\alpha} \ll 1$ resulting in $\mathbf{A}_\alpha \simeq \mathbf{G}_\alpha$ for all practical purposes. Thus, we may approximate the coefficients of the constitutive tensor from equation (11.480) to:

$$\mathbb{C}^{ijkl} \equiv \mu \left(G^{ik} G^{jl} + G^{il} G^{jk} + \frac{2\nu}{1-2\nu} G^{ij} G^{kl} \right) \tag{11.484}$$

and in the optimal normal shell coordinates, we can identify and particularize the coefficients of equation (11.484) to get:

$$\begin{aligned}
\mathbb{C}^{\alpha\beta\gamma\delta} &= \mu \left(G^{\alpha\gamma} G^{\beta\delta} + G^{\alpha\delta} G^{\beta\gamma} + \frac{2\nu}{1-2\nu} G^{\alpha\beta} G^{\gamma\delta} \right), \\
\mathbb{C}^{\alpha\beta 33} &= \mu \left(\cancel{G^{\alpha 3} G^{\beta 3}} + \cancel{G^{\alpha 3} G^{\beta 3}} + \frac{2\nu}{1-2\nu} G^{\alpha\beta} \underbrace{G^{33}}_{=1} \right) = \frac{E\nu}{(1+\nu)(1-2\nu)} G^{\alpha\beta}, \\
\mathbb{C}^{\alpha 3\beta 3} &= \mu \left(G^{\alpha\beta} \underbrace{G^{33}}_{=1} + \cancel{G^{\alpha 3} G^{\beta 3}} + \frac{2\nu}{1-2\nu} \cancel{G^{\alpha 3} G^{\beta 3}} \right) = \mu\, G = \frac{E}{2(1+\nu)} G^{\alpha\beta}, \\
C^{3333} &= \mu \left(2 + \frac{2\nu}{(1-2\nu)} \right) = \frac{E(1-\nu)}{(1+\nu)(1-2\nu)}
\end{aligned} \tag{11.485}$$

Now, for the shell-like body, if we continue to stay with the optimal normal coordinate system, that is, coordinates along the lines of curvature with arc length parameterization, and then noting that in this coordinate system:

$$\begin{aligned}
G^{\alpha\beta} &= 0, \quad \alpha \neq \beta \\
G^{\alpha 3} &= 0, \\
G^{33} &= 1
\end{aligned} \tag{11.486}$$

we get:

$$\begin{aligned}
\mathbb{C}^{\alpha\beta\gamma 3} &= \mathbb{C}^{\alpha\beta 3\gamma} = \mathbb{C}^{\gamma 3\alpha\beta} = \mathbb{C}^{3\gamma\alpha\beta} = 0, \\
\mathbb{C}^{\alpha 333} &= \mathbb{C}^{3\alpha 33} = \mathbb{C}^{33\alpha 3} = \mathbb{C}^{333\alpha} = 0
\end{aligned} \tag{11.487}$$

Using the relations given in equation (11.487), we can list the non-trivial stress components from equation (11.478) as:

$$\begin{aligned}
S^{\alpha\beta} &= C^{\alpha\beta\gamma\delta}\, E_{\gamma\delta} + C^{\alpha\beta 33}\, E_{33}, \\
S^{\alpha 3} &= 2C^{\alpha 3\gamma 3}\, E_{\gamma 3}, \\
S^{33} &= C^{33\gamma\delta}\, E_{\gamma\delta} + C^{3333}\, E_{33},
\end{aligned} \tag{11.488}$$

Obviously, by solving for E_{33} in the expression for S^{33}, we can eliminate it from the expression for $S^{\alpha\beta}$ in equation (11.488). For the plane stress assumption on the top and bottom faces, we will place the restriction, $S^{33} \equiv 0$, in which case we will have:

$$E_{33} = -\frac{\mathbb{C}^{33\gamma\delta}}{\mathbb{C}^{3333}} E_{\gamma\delta} \tag{11.489}$$

The coefficients, S^{ij}, of the stress tensor can be similarly particularized. However, before we do so, we may impose the plane stress condition: $S^{33} \equiv 0$, on the top and the bottom faces of the shell. Then, as before we get from equations (11.489) and (11.485): $E_{33} = -\frac{C^{33\gamma\delta}}{C^{3333}} E_{\gamma\delta} = -\frac{\nu}{(1-\nu)} G^{\alpha\beta} E_{\alpha\beta}$. Now, applying this relation to the expressions (10.16.29) for the stress components, we finally get:

$$\begin{aligned}
S^{\alpha\beta} &= \left(C^{\alpha\beta\gamma\delta} - \frac{\nu}{1-\nu} \frac{E\nu}{(1+\nu)(1-2\nu)} G^{\alpha\beta} G^{\gamma\delta} \right) E_{\gamma\delta} \\
&= \mu \left\{ G^{\alpha\gamma} G^{\beta\delta} + G^{\alpha\delta} G^{\beta\gamma} + \left(\frac{2\nu}{1-2\nu} - \frac{2\nu^2}{(1-\nu)(1-2\nu)} G^{\alpha\beta} G^{\gamma\delta} \right) \right\} E_{\gamma\delta} = \hat{H}^{\alpha\beta\gamma\delta}\, E_{\gamma\delta}, \\
S^{\alpha 3} &= 2 C^{\alpha 3\gamma 3}\, E_{\gamma 3}
\end{aligned} \tag{11.490}$$

where we have introduced the definition:

$$\hat{H}^{\alpha\beta\gamma\delta} = \frac{E}{2(1+\nu)} \left(G^{\alpha\gamma} G^{\beta\delta} + G^{\alpha\delta} G^{\beta\gamma} + \frac{2\nu}{1-\nu} G^{\alpha\beta} G^{\gamma\delta} \right) \tag{11.491}$$

and $C^{\alpha 3\gamma 3}$ is as given in equation (11.485).

11.16.3.2 Two-dimensional Strain Energy, $U(\boldsymbol{\beta}_\alpha, \boldsymbol{\chi}_\alpha)$

Before we can formally write down the expression for the 2D strain energy functional, $U(\boldsymbol{\beta}_\alpha, \boldsymbol{\chi}_\alpha)$ per unit of the shell surface area, we need to express the coefficients, E_{ij}, of the Green–Lagrange strain tensors in terms of the coefficients as given by equations (11.460) and (11.461) of $\boldsymbol{\beta}_\alpha$ and $\boldsymbol{\chi}_\alpha$, the spatial translational and bending strain vectors of the shell surface; similarly, the components of the stress resultants, $\mathbf{F}^\alpha$, and those of the stress couple, $\mathbf{M}^\alpha$, as given by equations (11.465) must be expressed in terms of the coefficients, S^{ij}, of the second Piola–Kirchhoff stress tensor. Let us recall some preliminaries: that $\mathbf{A} = \mathbf{A}_i \otimes \mathbf{A}^i = \mathbf{A}_\alpha \otimes \mathbf{A}^\alpha + \mathbf{N} \otimes \mathbf{N}$ is the identity tensor in the undeformed optimal normal coordinate system; that for an arbitrary particle, the current particle position vector is given by $\mathbf{c} = \mathbf{c}_0 + \eta\, \mathbf{g}_3$; that, at an arbitrary particle, the deformed base vector set is given as:

$$\begin{aligned}
\mathbf{a}_\alpha &\equiv \mathbf{c}_{,\,\alpha} = \mathbf{c}_{0,\,\alpha} + \eta\, \mathbf{g}_{3,\,\alpha} \\
\mathbf{a}_3 &\equiv \mathbf{c}_{,\,3} = \mathbf{c}_{,\,\eta} = \mathbf{g}_3
\end{aligned} \tag{11.492}$$

with:

$$
\begin{aligned}
\eta\,\mathbf{g}_{3},_{\alpha} &= \eta \mathbf{k}_{\alpha}^{tot} \times \mathbf{g}_3 = \eta\,\mathbf{k}_{\alpha} \times \mathbf{g}_3 + \eta\,\mathbf{R}\,\mathbf{k}_{\alpha}^{0} \times \mathbf{g}_3 = \eta\,\mathbf{k}_{\alpha} \times \mathbf{g}_3 + \eta\,\mathbf{R}\,\mathbf{k}_{\alpha}^{0} \times \mathbf{R}\,\mathbf{N} \\
&= \eta\,\chi_{\alpha} \times \mathbf{g}_3 + \eta\,\mathbf{R}\,(\mathbf{k}_{\alpha}^{0} \times \mathbf{N}) = \eta\,\chi_{\alpha} \times \mathbf{g}_3 + \eta\,\mathbf{R}\,\mathbf{N},_{\alpha} \\
&= \eta\,\chi_{\alpha} \times \mathbf{g}_3 + \underbrace{\mathbf{R}\frac{\eta}{R_{\alpha}}\mathbf{G}_{\alpha}}_{higher\ order} = \eta\,\chi_{\alpha} \times \mathbf{g}_3
\end{aligned}
\tag{11.493}
$$

where we have used the identity: $\mathbf{R}\,\mathbf{k}_{\alpha}^{0} \times \mathbf{R}\,\mathbf{N} = \mathbf{R}\left(\mathbf{k}_{\alpha}^{0} \times \mathbf{N}\right)\ (det\,\mathbf{R})$ with $det\,\mathbf{R} = 1$; that the deformation gradient, $\mathbf{F}$, is given by:

$$
\begin{aligned}
&\mathbf{F} \equiv \mathbf{a},_{i} \otimes \mathbf{A}^{i} = \mathbf{c},_{\alpha} \otimes \mathbf{A}^{\alpha} + \mathbf{c},_{\eta} \otimes \mathbf{A}^{3} = (\mathbf{c}_{0},_{\alpha} + \eta\,\mathbf{g}_{3},_{\alpha}) \otimes \mathbf{A}^{\alpha} + \mathbf{g}_3 \otimes \mathbf{N} \\
\Rightarrow\ &\mathbf{F}^{T} = \mathbf{A}^{\alpha} \otimes (\mathbf{c}_{0},_{\alpha} + \eta\,\mathbf{g}_{3},_{\alpha}) + \mathbf{N} \otimes \mathbf{g}_3
\end{aligned}
\tag{11.494}
$$

where the deformed shell reference surface spin coordinate base vectors are given as: $\mathbf{g}_i = \mathbf{R}\,\mathbf{G}_i$ with $\{\mathbf{G}_i\}$ as the undeformed shell reference surface base vector set; that noting the strain expressions as given by equations (11.460) and (11.461), we have: $\mathbf{c}_{0},_{\alpha} = \boldsymbol{\beta}_{\alpha} + \mathbf{g}_{\alpha}$; now, considering only the linear terms we get:

$$
\begin{aligned}
\mathbf{c}_{0},_{\alpha} \bullet \mathbf{c}_{0},_{\beta} &= (\boldsymbol{\beta}_{\alpha} + \mathbf{g}_{\alpha}) \bullet (\boldsymbol{\beta}_{\beta} + \mathbf{g}_{\beta}) = \underbrace{\boldsymbol{\beta}_{\alpha} \bullet \boldsymbol{\beta}_{\beta}}_{\text{Nonlinear}} + \boldsymbol{\beta}_{\alpha} \bullet \mathbf{g}_{\beta} + \boldsymbol{\beta}_{\beta} \bullet \mathbf{g}_{\alpha} + \mathbf{g}_{\alpha} \bullet \mathbf{g}_{\beta} \\
&= {}^{n}\beta_{\alpha\beta} + {}^{n}\beta_{\beta\alpha} + \mathbf{G}_{\alpha\beta}
\end{aligned}
\tag{11.495}
$$

where we have used: $\mathbf{g}_{\alpha} \bullet \mathbf{g}_{\beta} = \mathbf{R}\,\mathbf{G}_{\alpha} \bullet \mathbf{R}\,\mathbf{G}_{\beta} = \mathbf{G}_{\alpha\beta}$; now, noting that: $\eta\,\mathbf{g}_{\alpha} \bullet \mathbf{RN},_{\beta} = \eta\,\mathbf{RG}_{\alpha} \bullet \mathbf{RN},_{\beta} = -\eta\,\mathbf{R}(\mathbf{G}_{\alpha} \bullet \frac{1}{R_{\beta}}\mathbf{G}_{\beta}) = -\,\mathbf{R}(\frac{\eta}{R_{\beta}}\mathbf{G}_{\alpha\beta}) \ll 1$, and $\sqrt{G} = 1 \Rightarrow \varepsilon_{\gamma\sigma} = e_{\gamma\sigma}$, we get:

$$
\begin{aligned}
\eta(\mathbf{c}_{0},_{\alpha} \bullet \mathbf{g}_{3},_{\beta}) &= \eta\{(\boldsymbol{\beta}_{\alpha} + \mathbf{g}_{\alpha}) \bullet (\chi_{\beta} \times \mathbf{g}_3 + \mathbf{R}\,\mathbf{N},_{\beta})\} \\
&= \eta\,\{(\boldsymbol{\beta}_{\alpha} + \mathbf{g}_{\alpha}) \bullet (\chi_{\beta} \times \mathbf{g}_3) + \mathbf{RK}_{\beta}^{0}\,\mathbf{N}\} \\
&= \left\{
\begin{aligned}
&\underbrace{\eta\ \boldsymbol{\beta}_{\alpha} \bullet (\chi_{\beta} \times \mathbf{g}_3)}_{\text{Nonlinear}} + \underbrace{\eta\ \boldsymbol{\beta}_{\alpha} \bullet \mathbf{RK}_{\beta}^{0}\,\mathbf{N}}_{\ll 1} \\
&+ \mathbf{g}_{\alpha} \bullet (\chi_{\beta} \times \mathbf{g}_3) + \underbrace{\eta\ \mathbf{g}_{\alpha} \bullet \mathbf{RK}_{\beta}^{0}\,\mathbf{N}}_{\ll 1}
\end{aligned}
\right\} \\
&= \eta\ \mathbf{g}_{\alpha} \bullet (\chi_{\beta\gamma}\ \mathbf{g}_{\gamma} \times \mathbf{g}_3) = \eta\ \mathbf{g}_{\alpha} \bullet (\chi_{\beta\gamma}\ \varepsilon_{\gamma\sigma}\mathbf{g}^{\sigma}) = \eta\ \chi_{\beta\gamma}\ e_{\gamma\alpha}
\end{aligned}
\tag{11.496}
$$

where recall that the surface permutation tensor, $\varepsilon_{\gamma\sigma} = \sqrt{G}\,e_{\gamma\sigma}$ with the surface metric determinant: $\sqrt{G} = 1$ in the optimal normal coordinate system and $e_{\gamma\sigma}$ is the coefficient of the surface permutation matrix: $e = \begin{bmatrix} 0 & 1 \\ -1 & 0 \end{bmatrix}$.

11.16.3.3 Explicit Strain Components

Now, noting that $\mathbf{E} \equiv \frac{1}{2}(\mathbf{C} - \mathbf{A}) = \frac{1}{2}\left(\mathbf{F}^T\,\mathbf{F} - \mathbf{A}\right) = E_{ij}\mathbf{A}^i \otimes \mathbf{A}^j$, with $\mathbf{A}_\alpha \simeq \mathbf{G}_\alpha$, we have in the shell optimal normal coordinates:

$$
\begin{aligned}
E_{\alpha\beta} &= \frac{1}{2}(\mathbf{a}_\alpha \bullet \mathbf{a}_\beta - \mathbf{G}_{\alpha\beta}),\\
E_{\alpha 3} &= \frac{1}{2}(\mathbf{a}_\alpha \bullet \mathbf{a}_3 - \mathbf{G}_{\alpha 3}),\\
E_{33} &= \frac{1}{2}(\mathbf{a}_3 \bullet \mathbf{a}_3 - \mathbf{G}_{33})
\end{aligned}
\tag{11.497}
$$

with $\mathbf{G}_{33} = \mathbf{N} \bullet \mathbf{N} = 1$.

Now, then,

$$
\begin{aligned}
E_{\alpha\beta} &= \frac{1}{2}\{(\mathbf{c}_{0},_{\alpha} + \eta\,\mathbf{g}_{3},_{\alpha}) \bullet (\mathbf{c}_{0},_{\beta} + \eta\,\mathbf{g}_{3},_{\beta}) - \mathbf{G}_{\alpha\beta}\}\\
&= \frac{1}{2}\left\{\mathbf{c}_{0},_{\alpha} \bullet\ \mathbf{c}_{0},_{\beta} + \eta\,(\mathbf{c}_{0},_{\alpha} \bullet\ \mathbf{g}_{3},_{\beta} + \mathbf{c}_{0},_{\beta} \bullet\ \mathbf{g}_{3},_{\alpha}) + \underbrace{\cancel{\eta^2\,\mathbf{g}_{3},_{\alpha} \bullet \mathbf{g}_{3},_{\alpha}}}_{\text{Nonlinear}} - \mathbf{G}_{\alpha\beta}\right\}\\
&= \frac{1}{2}\left\{{}^n\beta_{\alpha\beta} + {}^n\beta_{\beta\alpha} + \cancel{\mathbf{G}_{\alpha\beta}} + \eta\left({}^n\chi_{\beta\gamma}e_{\gamma\alpha} + {}^n\chi_{\alpha\gamma}e_{\gamma\beta}\right) - \cancel{\mathbf{G}_{\alpha\beta}}\right\}\\
&= \frac{1}{2}\left\{{}^n\beta_{\alpha\beta} + {}^n\beta_{\beta\alpha} + \eta\left({}^n\chi_{\beta\gamma}e_{\gamma\alpha} + {}^n\chi_{\alpha\gamma}e_{\gamma\beta}\right)\right\}\\
&= \frac{1}{2}\left({}^n\beta_{\alpha\beta} + {}^n\beta_{\beta\alpha}\right) + \frac{1}{2}\eta\left({}^n\chi_{\beta\gamma}e_{\gamma\alpha} + {}^n\chi_{\alpha\gamma}e_{\gamma\beta}\right) \equiv \overset{C}{E}_{\alpha\beta} + \eta\,\overset{L}{E}_{\alpha\beta}
\end{aligned}
\tag{11.498}
$$

where we have used the relations given by equations (11.495) and (11.496), and introduced the constant variation (related to shear and extensional stresses) of the strain component through the thickness as: $\overset{C}{E}_{\alpha\beta} \equiv \frac{1}{2}\left({}^n\beta_{\alpha\beta} + {}^n\beta_{\beta\alpha}\right)$, and the linear variation (related to bending stresses) strain components as: $\overset{L}{E}_{\alpha\beta} \equiv \frac{1}{2}\left({}^n\chi_{\beta\gamma}e_{\gamma\alpha} + {}^n\chi_{\alpha\gamma}e_{\gamma\beta}\right)$. We get:

$$
\begin{aligned}
E_{\alpha 3} &= \frac{1}{2}\{(\mathbf{c}_{0},_{\alpha} + \eta\,\mathbf{g}_{3},_{\alpha})\ \bullet\ \mathbf{g}_3\} = \frac{1}{2}\left\{\mathbf{c}_{0},_{\alpha}\ \bullet\ \mathbf{g}_3 + \eta\,\underbrace{\cancel{\mathbf{g}_{3},_{\alpha} \bullet \mathbf{g}_3}}_{=0}\right\}\\
&= \frac{1}{2}\{(\boldsymbol{\beta}_\alpha + \mathbf{g}_\alpha)\ \bullet\ \mathbf{g}_3\} = \frac{1}{2}\ {}^n\beta_{\alpha 3} \equiv \overset{C}{E}_{\alpha 3}
\end{aligned}
\tag{11.499}
$$

where we have used the fact that: $\mathbf{g}_{3},_{\alpha} = \mathbf{k}_\alpha^{tot} \times\ \mathbf{g}_3 \Rightarrow \mathbf{g}_{3},_{\alpha}\ \bullet\ \mathbf{g}_3 = 0$. Similarly, $E_{3\alpha} = \frac{1}{2}\ {}^n\beta_{3\alpha}$. Finally, for inextensional normal, we have:

$$
E_{33} = 0 \tag{11.500}
$$

For a thin shell, we will take the mid-surface as the reference shell surface with the thickness defined as: $h \equiv \eta_t + \eta_b$ and the shifter: ${}^n\mu^S \equiv 1 - 2H\eta + K\eta^2 \rightarrow 1$ as $\eta H = \dfrac{\eta}{R_1} + \dfrac{\eta}{R_2} \rightarrow 0$ and

$\eta^2 K = \dfrac{\eta}{R_1}\dfrac{\eta}{R_2} \to 0$. Now, then, the 2D strain energy may be expressed as:

$$
\begin{aligned}
U &= \int_{-\frac{h}{2}}^{\frac{h}{2}} W\ {}^{\eta}\mu^S\, d\eta = \int_{-\frac{h}{2}}^{\frac{h}{2}} \left(S^{\alpha\beta}E_{\alpha\beta} + 2S^{\alpha 3}E_{\alpha 3}\right)\, d\eta \\
&= \int_{-\frac{h}{2}}^{\frac{h}{2}} \left(\hat{H}^{\alpha\beta\gamma\delta}E_{\gamma\delta}E_{\alpha\beta} + 4C^{\alpha 3\gamma 3}\, E_{\gamma 3}E_{\alpha 3}\right)\, d\eta \\
&= \int_{-\frac{h}{2}}^{\frac{h}{2}} \hat{H}^{\alpha\beta\gamma\delta}\left\{\overset{C}{E}_{\gamma\delta}\overset{C}{E}_{\alpha\beta} + \eta\left(\overset{C}{E}_{\gamma\delta}\overset{L}{E}_{\alpha\beta} + \overset{L}{E}_{\gamma\delta}\overset{C}{E}_{\alpha\beta}\right) + \eta^2\ \overset{L}{E}_{\gamma\delta}\overset{L}{E}_{\alpha\beta}\right\}\, d\eta \\
&+ \int_{-\frac{h}{2}}^{\frac{h}{2}} \left(4k_S C^{\alpha 3\gamma 3}\ \overset{C}{E}_{\gamma 3}\overset{C}{E}_{\alpha 3}\right)\, d\eta
\end{aligned}
\tag{11.501}
$$

where we have used definitions for $\overset{C}{E}_{\alpha\beta}$, $\overset{L}{E}_{\alpha\beta}$ as in equation (11.498) and for $\overset{C}{E}_{\alpha 3}$ as equation (11.499); finally, noting that: $\int_{-\frac{h}{2}}^{\frac{h}{2}} d\eta = h$, $\int_{-\frac{h}{2}}^{\frac{h}{2}} \eta\, d\eta = 0$, and $\int_{-\frac{h}{2}}^{\frac{h}{2}} \eta^2\, d\eta = \dfrac{h^3}{12}$, and taking into consideration the definition in equation (11.491), we have:

$$
\boxed{U = \frac{1}{2}\left\{\left(D_E\ \overset{C}{E}_{\gamma\delta}\overset{C}{E}_{\alpha\beta} + D_M\ \overset{L}{E}_{\gamma\delta}\overset{L}{E}_{\alpha\beta}\right)H^{\alpha\beta\gamma\delta} + 2D_S\, C^{\alpha\gamma}\ \overset{C}{E}_{\gamma 3}\overset{C}{E}_{\alpha 3}\right\}}
\tag{11.502}
$$

where we have introduced the following definitions:

$$
\boxed{
\begin{aligned}
&D_M \equiv \frac{Eh}{1-\nu^2} && \text{(Membrane stiffness)}\\
&D_B \equiv \frac{Eh^3}{12(1-\nu^2)} && \text{(Bending stiffness)}\\
&D_S \equiv 2Gh = 2\frac{Eh}{2(1+\nu)} = D_E(1-\nu) && \text{(Shear stiffness)}\\
&H^{\alpha\beta\gamma\delta} \equiv \frac{1-\nu}{2}\left(G^{\alpha\gamma}G^{\beta\delta} + G^{\alpha\delta}G^{\beta\gamma} + \frac{2\nu}{1-\nu}G^{\alpha\beta}G^{\gamma\delta}\right)
\end{aligned}}
\tag{11.503}
$$

From equation (11.503), let us note that:

$$
\begin{aligned}
H^{11\gamma\delta} &\equiv \frac{1-\nu}{2}\left(G^{1\gamma}G^{1\delta} + G^{1\delta}G^{1\gamma} + \frac{2\nu}{1-\nu}G^{11}G^{\gamma\delta}\right)\\
&= \begin{cases}
\dfrac{1-\nu}{2}\left(1+1+\dfrac{2\nu}{1-\nu}\right) = 1, & \gamma=\delta=1\\
\dfrac{1-\nu}{2}\left(\dfrac{2\nu}{1-\nu}\right) = \nu, & \gamma=\delta=2\\
= 0, & \gamma\neq\delta
\end{cases}
\end{aligned}
\tag{11.504}
$$

Similar expressions can be obtained for $H^{11\gamma\delta}$; for $H^{12\gamma\delta}$, we have:

$$H^{12\gamma\delta} \equiv \frac{1-\nu}{2}\left(G^{1\gamma}G^{2\delta} + G^{1\delta}G^{2\gamma} + \frac{2\nu}{1-\nu}\,\cancel{G^{12}G^{\gamma\delta}}\right)$$
$$= \begin{cases} 0, & \gamma = \delta = 1 \\ \dfrac{1-\nu}{2}, & \gamma \neq \delta \\ 0, & \gamma = \delta = 2 \end{cases} \tag{11.505}$$

Using the relations given by equations (11.498) and (11.499), the 2D strain energy, U, can be expressed as functions of the components of the spatial strains, $\boldsymbol{\beta}_\alpha$ and $\boldsymbol{\chi}_\alpha$, and the stress resultants and the stress couples are given by:

$$\begin{aligned} {}^nF^{\alpha\beta} &= \frac{\partial U}{\partial\, {}^n\beta^{\alpha\beta}}, \quad {}^nF^{\alpha 3} = \frac{\partial U}{\partial\, {}^n\beta^{\alpha 3}}, \quad {}^nF^{33} = \frac{\partial U}{\partial\, {}^n\beta^{33}}, \\ {}^nM^{\alpha\beta} &= \frac{\partial U}{\partial\, {}^n\chi^{\alpha\beta}}, \quad {}^nM^{\alpha 3} = \frac{\partial U}{\partial\, {}^n\chi^{\alpha 3}}, \quad {}^nM^{33} = \frac{\partial U}{\partial\, {}^n\chi^{33}} \end{aligned} \tag{11.506}$$

Now, performing the derivative operations on equation (11.502), and using equations (11.498) and (11.499), we finally get:

$$\begin{aligned} {}^nF^{11} &= D_M({}^n\beta^{11} + \nu\, {}^n\beta^{22}), \\ {}^nF^{22} &= D_M({}^n\beta^{22} + \nu\, {}^n\beta^{11}), \\ {}^nF^{12} &= D_M\left(\frac{1-\nu}{2}\right){}^n\beta^{12} = D_{\nu M}\, {}^n\beta^{12}, \\ {}^nF^{21} &= D_M\left(\frac{1-\nu}{2}\right){}^n\beta^{21} = D_{\nu M}\, {}^n\beta^{21}, \\ {}^nF^{\alpha 3} &= D_S\, {}^n\beta^{\alpha 3}, \\ {}^nM^{12} &= D_B({}^n\chi^{12} - \nu\, {}^n\chi^{21}), \\ {}^nM^{21} &= D_B({}^n\chi^{21} - \nu\, {}^n\chi^{12}), \\ {}^nM^{11} &= D_B(1-\nu)\, {}^n\chi^{11} = D_{\nu B}\, {}^n\chi^{11}, \\ {}^nM^{22} &= D_B(1-\nu)\, {}^n\chi^{22} = D_{\nu B}\, {}^n\chi^{22}, \\ {}^nM^{\alpha 3} &= 0, \end{aligned} \tag{11.507}$$

where we have introduced the definitions:

$$\boxed{D_{\nu M} \equiv D_M\left(\frac{1-\nu}{2}\right) \quad \text{and} \quad D_{\nu B} \equiv D_B\,(1-\nu),} \tag{11.508}$$

Remark

- It may be noted that because of the inextensible normal to the shell reference surface, ${}^nM^{\alpha 3} = 0$ for $\alpha = 1, 2$; in other words, the constant normal length constraint reduces the rotation degrees of freedom to only two, and the rotation tensor belongs to 2-sphere, or the S^2 manifold instead of $SO3$ manifold. If we want to continue with the original three degrees of freedom, we may approximate the torsional stress–strain relations for ${}^{n}M^{\alpha 3}$ and ${}^{n}\chi^{\alpha 3}$ by including a very small term such as: ${}^{n}M^{\alpha 3} = \varepsilon\,(1-\nu)\,D_B\,{}^{n}\chi^{\alpha 3} \equiv \varepsilon D_{\nu B}\,{}^{n}\chi^{\alpha 3}$ with a small enough torsional parameter (Chroscielewski *et al.* 1992), say, $\varepsilon \leq 0.01$. Thus, finally, we can express the stress–strain matrix, **D**, of a thin, linear, hyperelastic, homogeneous shell by:

$$\begin{Bmatrix} {}^{n}N^{11} \\ {}^{n}N^{12} \\ {}^{n}N^{13} \\ {}^{n}M^{11} \\ {}^{n}M^{12} \\ {}^{n}M^{13} \\ {}^{n}N^{21} \\ {}^{n}N^{22} \\ {}^{n}N^{23} \\ {}^{n}M^{21} \\ {}^{n}M^{22} \\ {}^{n}M^{23} \end{Bmatrix} = \hat{D} \begin{Bmatrix} {}^{n}\beta_{11} \\ {}^{n}\beta_{12} \\ {}^{n}\beta_{13} \\ {}^{n}\chi_{11} \\ {}^{n}\chi_{12} \\ {}^{n}\chi_{13} \\ {}^{n}\beta_{21} \\ {}^{n}\beta_{22} \\ {}^{n}\beta_{23} \\ {}^{n}\chi_{21} \\ {}^{n}\chi_{22} \\ {}^{n}\chi_{23} \end{Bmatrix} \tag{11.509}$$

where the (12×12) constitutive matrix, $\hat{D}$, is defined as:

$$\hat{D} \equiv \begin{bmatrix} D_M & 0 & 0 & 0 & 0 & 0 & 0 & \nu D_M & 0 & 0 & 0 & 0 \\ 0 & 2D_{\nu M} & 0 & 0 & 0 & 0 & 0 & 0 & 0 & 0 & 0 & 0 \\ 0 & 0 & D_S & 0 & 0 & 0 & 0 & 0 & 0 & 0 & 0 & 0 \\ 0 & 0 & 0 & 2D_{\nu B} & 0 & 0 & 0 & 0 & 0 & 0 & 0 & 0 \\ 0 & 0 & 0 & 0 & D_B & 0 & 0 & 0 & 0 & -\nu D_B & 0 & 0 \\ 0 & 0 & 0 & 0 & 0 & \varepsilon D_{\nu B} & 0 & 0 & 0 & 0 & 0 & 0 \\ 0 & 0 & 0 & 0 & 0 & 0 & 2D_{\nu M} & 0 & 0 & 0 & 0 & 0 \\ \nu D_M & 0 & 0 & 0 & 0 & 0 & 0 & D_M & 0 & 0 & 0 & 0 \\ 0 & 0 & 0 & 0 & 0 & 0 & 0 & 0 & D_S & 0 & 0 & 0 \\ 0 & 0 & 0 & 0 & -\nu D_B & 0 & 0 & 0 & 0 & D_B & 0 & 0 \\ & & & & 0 & 0 & 0 & 0 & 0 & 0 & 2D_{\nu B} & 0 \\ & & & & 0 & 0 & 0 & 0 & 0 & 0 & 0 & \varepsilon D_{\nu B} \end{bmatrix}$$

Note that the components of the stress resultants, the stress couples and the strains are all indicated in the rotated spin bases; the corresponding relations for the engineering components for the stress couples are easily obtained with the help of the simple permutation matrix discussed earlier in the text.

11.16.4 Where We Would Like to Go

Now for the next step of the computational mechanics of shells, as we can see from equation (11.468), the virtual functional in the integral or weak form of equation, being generally highly nonlinear in displacements and rotations, may not be solved in a closed form analytical way. The solution that is – the tracing of the shell motion or the configuration path – will require some sort of iterative numerical method. Of all the possible schemes, the one that we intend to pursue is a Newton-type method. This, in turn, requires linearization of the equations at every iteration step. Clearly, then, we will have to further modify the virtual work equation for computational strategy. Thus, a logical place to go from here will be:

- the covariant linearization of virtual functional.

11.17 Covariant Linearization of Virtual Work

11.17.1 What We Need to Recall

- The generalized virtual work equation for quasi-static proportional loading, for all admissible generalized virtual displacement, ${}^t\breve{\bar{\mathbf{d}}} \equiv \{\,{}^t\bar{\mathbf{d}} \quad {}^t\bar{\boldsymbol{\theta}}\,\}^T$, is given as:

$$
\begin{aligned}
&G_{Static}(\lambda,\, {}^t\breve{\mathbf{d}};\, {}^t\breve{\bar{\mathbf{d}}}) \equiv G_{Internal}({}^t\breve{\mathbf{d}};\, {}^t\breve{\bar{\mathbf{d}}}) - G_{external}(\lambda;\, {}^t\breve{\bar{\mathbf{d}}}) = 0, \quad \forall \text{admissible } {}^t\breve{\bar{\mathbf{d}}} \\
&G_{Internal}({}^t\breve{\mathbf{d}};\, {}^t\breve{\bar{\mathbf{d}}}) = \int_M {}^n\breve{\bar{\varepsilon}}({}^t\breve{\mathbf{d}};\, {}^t\breve{\bar{\mathbf{d}}})^T\, {}^n\breve{\mathbf{F}}({}^t\breve{\mathbf{d}})\, dA \\
&G_{External}(\lambda;\, {}^t\breve{\bar{\mathbf{d}}}) = \int_M {}^t\breve{\bar{\mathbf{d}}}^T\, {}^t\breve{\hat{\mathbf{F}}}(\lambda, S^1, S^2)\, dA \, + \lambda \int_{\partial M^f} {}^t\breve{\bar{\mathbf{d}}}^T\, {}^t\breve{\hat{\mathbf{F}}}_\nu(S)\, dS
\end{aligned}
\tag{11.510}
$$

- The generalized virtual work equation for dynamic loading, for all admissible generalized virtual displacement, ${}^t\breve{\bar{\mathbf{d}}} \equiv \{\,{}^t\bar{\mathbf{d}} \quad {}^t\bar{\boldsymbol{\theta}}\,\}^T$, is given as:

$$
\begin{aligned}
&G_{Dyn}({}^t\mathbf{d}, {}^t\boldsymbol{\theta};\, {}^t\bar{\mathbf{d}}, {}^t\bar{\boldsymbol{\theta}}) \equiv G_{Internal} + G_{Inertial} - G_{External} \\
&G_{Internal}({}^t\breve{\mathbf{d}};\, {}^t\breve{\bar{\mathbf{d}}}) = \int_M {}^n\breve{\bar{\varepsilon}}({}^t\breve{\mathbf{d}};\, {}^t\breve{\bar{\mathbf{d}}})^T\, {}^n\breve{\mathbf{F}}({}^t\breve{\mathbf{d}})\, dA \\
&G_{Inertial}({}^t\breve{\mathbf{d}};\, {}^t\breve{\bar{\mathbf{d}}}) = \int_M {}^t\breve{\bar{\mathbf{d}}}^T\, {}^t\breve{\mathbf{F}}_{Iner}({}^t\breve{\mathbf{d}})\, dA \\
&\qquad = \int_M \left\{ {}^t\bar{\mathbf{d}}^T\, M_0\, {}^t\ddot{\mathbf{d}} + {}^t\bar{\boldsymbol{\theta}}^T\, I_0\, (\mathbf{W}^T\, {}^t\dot{\boldsymbol{\omega}}) \right\} dA \\
&G_{External}({}^t\breve{\bar{\mathbf{d}}}) = \int_M {}^t\breve{\bar{\mathbf{d}}}^T\, {}^t\breve{\hat{\mathbf{F}}}(S^1, S^2)\, dA \, + \int_{\partial M^f} {}^t\breve{\bar{\mathbf{d}}}^T\, {}^t\breve{\hat{\mathbf{F}}}_\nu(S)\, dS \\
&\qquad = \mathbf{0}, \quad \forall \text{admissible } {}^t\bar{\mathbf{d}},\, {}^t\bar{\boldsymbol{\theta}}
\end{aligned}
\tag{11.511}
$$

- ${}^{n}\breve{\varepsilon}$, the generalized virtual strain field is given as:

$$
\begin{aligned}
&{}^{n}\breve{\varepsilon} \equiv \left\{ {}^{n}\bar{\beta}_1 \quad {}^{n}\bar{\chi}_1 \quad {}^{n}\bar{\beta}_2 \quad {}^{n}\bar{\chi}_2 \right\}^T,\\
&{}^{n}\bar{\beta}_\alpha = {}^{t}\mathbf{R}^T \left({}^{t}\bar{\mathbf{d}},_\alpha + {}^{t}\mathbf{k}^0_\alpha \times {}^{t}\bar{\mathbf{d}} \right) - {}^{t}\mathbf{R}^T\, {}^{t}\bar{\boldsymbol{\Phi}} \left({}^{t}\mathbf{e}_\alpha + {}^{t}\mathbf{k}^0_\alpha \times {}^{t}\mathbf{d} \right) = {}^{t}\mathbf{R}^T(\overline{\hat{1}_\alpha + {}^{t}\mathbf{d}|_\alpha}),\\
&{}^{n}\bar{\chi}_\alpha \equiv {}^{t}\mathbf{R}^T\, {}^{t}\bar{\phi}|_\alpha = {}^{t}\mathbf{R}^T \left\{ {}^{t}\bar{\phi},_\alpha + \mathbf{k}^0_\alpha \times {}^{t}\bar{\phi} \right\} = {}^{t}\mathbf{R}^T \left(\overline{\mathbf{k}^0_\alpha + \mathbf{k}_\alpha} \right),\\
&{}^{t}\bar{\phi} = \mathbf{W}\, {}^{t}\bar{\theta}
\end{aligned}
\tag{11.512}
$$

where $\mathbf{W}$ is given as:

$$
\begin{aligned}
&\mathbf{W}({}^{t}\theta) \equiv \mathbf{I} + c_1\, {}^{t}\boldsymbol{\Theta} + c_2\, {}^{t}\boldsymbol{\Theta}^2,\\
&c_1({}^{t}\theta) = \frac{1-\cos\theta}{\theta^2}, \quad c_2({}^{t}\theta) = \frac{\theta - \sin\theta}{\theta^3}
\end{aligned}
\tag{11.513}
$$

with ${}^{t}\boldsymbol{\Theta}$ is the skew-symmetric matrix corresponding to the axial vector, ${}^{t}\theta$, such that ${}^{t}\boldsymbol{\Theta}\,{}^{t}\theta = \mathbf{0}$.

- $\dot{\mathbf{W}}$ is given as:

$$
\begin{aligned}
\dot{\mathbf{W}} &= c_1\, \dot{\boldsymbol{\Theta}} + c_2(\dot{\boldsymbol{\Theta}}\, \boldsymbol{\Theta} + \boldsymbol{\Theta}\, \dot{\boldsymbol{\Theta}}) + \dot{c}_1\, \boldsymbol{\Theta} + \dot{c}_2\, \boldsymbol{\Theta}^2\\
&= (c_2 - c_1)(\theta \bullet \dot{\theta})\mathbf{I} + c_1\, \dot{\boldsymbol{\Theta}} + c_2\{(\dot{\theta} \otimes \theta) + (\theta \otimes \dot{\theta})\} + \dot{c}_1\, \boldsymbol{\Theta} + \dot{c}_2\, (\theta \otimes \theta)\\
\dot{c}_1 &\equiv \frac{1}{\theta^2}(1 - 2c_1 - \theta^2 c_2)(\theta \bullet \dot{\theta}) = c_3(\theta \bullet \dot{\theta}),\ \dot{c}_2 \equiv \frac{1}{\theta^2}(c_1 - 3c_2)(\theta \bullet \dot{\theta}) = c_4(\theta \bullet \dot{\theta}),
\end{aligned}
\tag{11.514}
$$

- $\ddot{\mathbf{W}}$ is given as:

$$
\begin{aligned}
&\ddot{\mathbf{W}} = c_1\, \ddot{\boldsymbol{\Theta}} + 2\left\{ \dot{c}_1\, \dot{\boldsymbol{\Theta}} + \dot{c}_2(\dot{\boldsymbol{\Theta}}\, \boldsymbol{\Theta} + \boldsymbol{\Theta}\, \dot{\boldsymbol{\Theta}}) \right\} + c_2(\ddot{\boldsymbol{\Theta}}\, \boldsymbol{\Theta} + 2\,\dot{\boldsymbol{\Theta}}^2 + \boldsymbol{\Theta}\, \ddot{\boldsymbol{\Theta}}) + \ddot{c}_1\, \boldsymbol{\Theta} + \ddot{c}_2\, \boldsymbol{\Theta}^2\\
&\ddot{c}_1 \equiv c_5(\theta \bullet \dot{\theta})^2 + c_3 \left\{ \dot{\theta}^2 + (\theta \bullet \ddot{\theta}) \right\}, \quad \ddot{c}_2 \equiv c_6(\theta \bullet \dot{\theta})^2 + c_4 \left\{ \dot{\theta}^2 + (\theta \bullet \ddot{\theta}) \right\}\\
&c_5 \equiv -\frac{1}{\theta^2}(2c_2 + 4c_3 + \theta^2 c_4), \quad c_6 \equiv -\frac{1}{\theta^2}(c_3 - 5c_4),
\end{aligned}
\tag{11.515}
$$

- ${}^{n}\breve{\mathbf{F}}$, the generalized reactive force field given as:

$$
{}^{n}\breve{\mathbf{F}} \equiv \left\{ {}^{n}\mathbf{F}_1 \quad {}^{n}\mathbf{M}_1 \quad {}^{n}\mathbf{F}_2 \quad {}^{n}\mathbf{M}_2 \right\}^T
\tag{11.516}
$$

- ${}^{t}\breve{\mathbf{F}}_{Iner}$, the generalized inertial force field given as:

$$
\begin{aligned}
&{}^{t}\breve{\mathbf{F}}_{Iner} \equiv \left\{ M_0\, {}^{t}\ddot{\mathbf{d}} \quad I_0\, (\mathbf{W}^T\, {}^{t}\dot{\omega}) \right\}^T,\\
&{}^{t}\omega \equiv \mathbf{W}\, {}^{t}\dot{\theta}, \quad {}^{t}\dot{\omega} = (\dot{\mathbf{W}}\, {}^{t}\dot{\theta} + \mathbf{W}\, {}^{t}\ddot{\theta})
\end{aligned}
\tag{11.517}
$$

where the angular velocity vector, ${}^t\boldsymbol{\omega}$, is the axial vector corresponding to the skew-symmetric angular velocity matrix, $\boldsymbol{\Omega}$, given by:

$$\boxed{\Omega \equiv \dot{\mathbf{R}}\,\mathbf{R}^T} \tag{11.518}$$

- ${}^t\breve{\hat{\mathbf{F}}}$, the generalized external applied force field given as:

$$\boxed{{}^t\breve{\hat{\mathbf{F}}} \equiv \{\, {}^t\hat{\mathbf{F}}_1 \quad {}^t\hat{\mathbf{M}}_1 \quad {}^t\hat{\mathbf{F}}_2 \quad {}^t\hat{\mathbf{M}}_2 \,\}^T} \tag{11.519}$$

- Equation (11.510) and equation (11.511) are generally nonlinear in ${}^t\breve{\mathbf{d}} \equiv \{\, {}^t\mathbf{d} \quad {}^t\boldsymbol{\theta} \,\}^T$, the generalized displacement state of the shell; similarly, all entities in equation (11.512) to equation (11.519) are functions of generalized state variables, ${}^t\breve{\mathbf{d}} \equiv \{\, {}^t\mathbf{d} \quad {}^t\boldsymbol{\theta} \,\}^T$.
- For the quasi-static case, we have restricted our presentation to a weak form involving only proportional loading system with the assumption that the forces are truly external and hence independent of the displacement state, ${}^t\breve{\mathbf{d}} \equiv \{\, {}^t\mathbf{d} \quad {}^t\boldsymbol{\theta} \,\}^T$, and as such can be expressed with a multiplicative variable of proportionality, λ, as we have indicated in the above equation.
- Briefly, the basic traits of a Newton-type method of tracing a configuration path entail the following: starting from a known state or configuration (which is usually the unloaded and undeformed structure at the first step), we find the tangent to the equilibrium or configuration path (usually, as a rule for every load step for a quasi-static system, or for every time step for a time history dynamic analysis) that helps us set up the instantaneous linear equations to be solved for the incremental and independent variables iteratively for each step; the state variables are updated by the incremental state variables and the process is repeated.
- Finally, the superscript n in the above expressions that helps to identify the relevant entities in the current, rotated or deformed configuration will occasionally be dropped for notational clarity although the context will lend support to consistency.

11.17.2 Incremental State Variables

Thus, for numerical evaluation of the motion of a shell on a digital computer by a typical iterative Newton-type method, we will need to derive for every step of the iteration, the instantaneous linearized form of the virtual equations as a function of the incremental generalized displacement state; we denote it hereon as ${}^t\Delta\breve{\mathbf{d}} \equiv \{\, {}^t\Delta\mathbf{d} \quad {}^t\Delta\boldsymbol{\theta} \,\}^T$.

11.17.2.1 Quasi-static Proportional Loading

For quasi-static proportional load cases, in actual numerical evaluation with the incremental load step, denoted by $\Delta\lambda$, as a preassigned entity in a typical load-control incrementation method, a Newton-type method without any special treatment such as arc length constraint, is inherently disastrous at the limit points or at the bifurcation points of a configuration path. Thus, for computational purpose (i.e. to maintain symmetry of the stiffness matrices), we consider a Newton-type method with an arc length constraint algorithm in which $\Delta\lambda$ also is treated as an additional independent variable as the displacement state variable, ${}^t\breve{\mathbf{d}} \equiv \{\, {}^t\mathbf{d} \quad {}^t\boldsymbol{\theta} \,\}^T$. Thus, the virtual work equation (11.510) must be linearized with respect to both the independent variable displacement sets: ${}^t\breve{\mathbf{d}} \equiv \{\, {}^t\mathbf{d} \quad {}^t\boldsymbol{\theta} \,\}^T$, and the coefficient of proportionality, λ.

11.17.2.2 Dynamic Time History Loading

For dynamic time history load cases, the virtual work equation (11.511) must be linearized with respect to only the independent variable displacement sets: ${}^t\breve{\mathbf{d}} \equiv \{\, {}^t\mathbf{d} \quad {}^t\boldsymbol{\theta}\,\}^T$, but we need to additionally linearize the inertial virtual work functional with particular attention to the angular acceleration, ${}^t\dot{\boldsymbol{\omega}}$.

Finally, in attempting to linearize the virtual equations, we must be alert to the fact that the rotation tensor belongs to a curved space; hence, the linearization must be performed by the covariant differentiation in mind; the practical way to accommodate covariant differentiation is simply to include the differentiation of the associated base vectors (for the vector fields), or the base tensors (for the tensor field).

11.17.3 Linearization of the Virtual Work Equation

As indicated earlier, the virtual work – equation (11.510) for the quasi-static case or equation (11.511) for the dynamic case – must be linearized, that is, expressed in incremental form for an incremental quasi-static load step or a dynamic time step, to apply a Newton-type iteration scheme for tracing the motion of a shell. Let the known – but not necessarily in equilibrium – configuration be given by ${}^t\breve{\mathbf{d}}^0$ at a load level λ^0, that is, $G(\lambda^0, {}^t\breve{\mathbf{d}}^0; {}^t\breve{\bar{\mathbf{d}}})$ may not satisfy equation (11.510) for the quasi-static case (similarly, $G({}^t\breve{\mathbf{d}}^0; {}^t\breve{\bar{\mathbf{d}}})$ for the dynamic analysis). We will try to seek equilibrium in the neighborhood of $(\lambda^0, {}^t\breve{\mathbf{d}}^0)$ for the quasi-static case (similarly, in the neighborhood of ${}^t\breve{\mathbf{d}}^0$ for the dynamic analysis). More specifically, at $(\lambda, {}^t\breve{\mathbf{d}})$ for the quasi-static case, that is, we want $G(\lambda, {}^t\breve{\mathbf{d}}; {}^t\breve{\bar{\mathbf{d}}}) = 0, \forall\, {}^t\breve{\bar{\mathbf{d}}}$ admissible (similarly, at ${}^t\breve{\mathbf{d}}$ for the dynamic analysis, that is, we want $G({}^t\breve{\mathbf{d}}; {}^t\breve{\bar{\mathbf{d}}}) = \mathbf{0}, \forall\, {}^t\breve{\bar{\mathbf{d}}}$ admissible). Expanding by Taylor series around $(\lambda^0, \breve{\mathbf{d}}^0)$ for quasi-static case (similarly, around $\breve{\mathbf{d}}^0$ for the dynamic case) and retaining only the first-order (linear) terms, that is, terms involving $\Delta^t\breve{\mathbf{d}}\ (\equiv {}^t\breve{\mathbf{d}} - {}^t\breve{\mathbf{d}}^0)$ and $\Delta\lambda\ (\equiv \lambda - \lambda^0)$, we have the following.

11.17.3.1 For Quasi-static Proportional Loading

$$
\begin{aligned}
&\tilde{G}(\lambda, {}^t\breve{\mathbf{d}}; {}^t\breve{\bar{\mathbf{d}}}; \Delta\,{}^t\breve{\mathbf{d}}, \Delta\lambda) \equiv G(\lambda^0, {}^t\breve{\mathbf{d}}^0; {}^t\breve{\bar{\mathbf{d}}}) + \underset{\Delta\,{}^t\breve{\mathbf{d}},\Delta\lambda}{\mathcal{L}^{Quasi}}[G(\lambda^0, {}^t\breve{\mathbf{d}}^0; {}^t\breve{\bar{\mathbf{d}}})] = \mathbf{0}\\
&\Rightarrow \underset{\Delta\,{}^t\breve{\mathbf{d}},\Delta\lambda}{\mathcal{L}^{Quasi}}[G(\lambda^0, {}^t\breve{\mathbf{d}}^0; {}^t\breve{\bar{\mathbf{d}}})] = -G(\lambda^0, {}^t\breve{\mathbf{d}}^0; {}^t\breve{\bar{\mathbf{d}}}; \Delta\,{}^t\breve{\mathbf{d}})\\
&\underset{\Delta\,{}^t\breve{\mathbf{d}},\Delta\lambda}{\mathcal{L}^{Quasi}}[G(\lambda^0, {}^t\breve{\mathbf{d}}^0; {}^t\breve{\bar{\mathbf{d}}})] \equiv \nabla_{\lambda^0} G(\lambda^0, {}^t\breve{\mathbf{d}}^0; {}^t\breve{\bar{\mathbf{d}}}) \bullet \Delta\lambda + \nabla_{{}^t\breve{\mathbf{d}}^0} G(\lambda^0, {}^t\breve{\mathbf{d}}^0; {}^t\breve{\bar{\mathbf{d}}}) \bullet \Delta^t\breve{\mathbf{d}}
\end{aligned}
\tag{11.520}
$$

where $\nabla_{{}^t\breve{\mathbf{d}}^0} G$ and $\nabla_{\lambda^0} G$ are the covariant derivatives of G with respect to ${}^t\breve{\mathbf{d}}$ and λ evaluated at ${}^t\breve{\mathbf{d}}^0$ and λ^0, respectively, and the linear part – that is, the first-order terms – of the virtual functional is denoted by $\underset{\Delta\,{}^t\breve{\mathbf{d}},\Delta\lambda}{\mathcal{L}^{Quasi}}[G(\lambda^0, {}^t\breve{\mathbf{d}}^0; {}^t\breve{\bar{\mathbf{d}}})]$. In a non-equilibrium state, $G(\lambda^0, {}^t\breve{\mathbf{d}}^0; {}^t\breve{\bar{\mathbf{d}}})$, known as the

total residual force functional, is non-zero. Noting that for conservative loading, the external loading does not depend on the current configuration, we can express the residual as:

$$G_{Static}(\lambda^0,\ {}^t\breve{\mathbf{d}}^0;\ {}^t\breve{\bar{\mathbf{d}}}) \equiv G_{Internal}({}^t\breve{\mathbf{d}}^0;\ {}^t\breve{\bar{\mathbf{d}}}) - G_{External}(\lambda^0;\ {}^t\breve{\bar{\mathbf{d}}}) \tag{11.521}$$

Similarly, because of the coefficient of proportionality, λ, the linear operator, $\mathcal{L}^{Quasi}_{\Delta^t\breve{\mathbf{d}},\Delta\lambda}[G(\lambda^0,\ {}^t\breve{\mathbf{d}}^0;\ {}^t\breve{\bar{\mathbf{d}}})]$, also consists of two parts: the internal and the external linear virtual functional:

$$\begin{aligned}
&\mathcal{L}^{Quasi}_{\Delta^t\breve{\mathbf{d}},\Delta\lambda}[G(\lambda^0,\ {}^t\breve{\mathbf{d}}^0;\ {}^t\breve{\bar{\mathbf{d}}})] \equiv \Delta G_{Internal}({}^t\breve{\mathbf{d}}^0;\ {}^t\breve{\bar{\mathbf{d}}};\Delta\,{}^t\breve{\mathbf{d}}) + \Delta G_{External}(\lambda^0;\ {}^t\breve{\bar{\mathbf{d}}};\Delta\lambda)\\
&\Delta G_{Internal} \equiv \nabla_{{}^t\breve{\mathbf{d}}^0} G_{Internal}({}^t\breve{\mathbf{d}}^0;\ {}^t\breve{\bar{\mathbf{d}}}) \bullet \Delta^t\breve{\mathbf{d}}\\
&\Delta G_{External} \equiv \nabla_{\lambda^0} G_{External}(\lambda^0;\ {}^t\breve{\bar{\mathbf{d}}}) \bullet \Delta\lambda
\end{aligned} \tag{11.522}$$

11.17.3.2 For Dynamic Time History Loading

$$\begin{aligned}
&\tilde{G}({}^t\breve{\mathbf{d}};\ {}^t\breve{\bar{\mathbf{d}}}) \equiv G({}^t\breve{\mathbf{d}}^0;\ {}^t\breve{\bar{\mathbf{d}}}) + \mathcal{L}_{\Delta^t\breve{\mathbf{d}}}^{Dyn}[G({}^t\breve{\mathbf{d}}^0;\ {}^t\breve{\bar{\mathbf{d}}})] = \mathbf{0}\\
&\Rightarrow \mathcal{L}_{\Delta^t\breve{\mathbf{d}}}^{Dyn}[G({}^t\breve{\mathbf{d}}^0;\ {}^t\breve{\bar{\mathbf{d}}})] = -G({}^t\breve{\mathbf{d}}^0;\ {}^t\breve{\bar{\mathbf{d}}})\\
&\mathcal{L}_{\Delta^t\breve{\mathbf{d}}}^{Dyn}[G({}^t\breve{\mathbf{d}}^0;\ {}^t\breve{\bar{\mathbf{d}}})] \equiv \nabla_{{}^t\breve{\mathbf{d}}^0} G({}^t\breve{\mathbf{d}}^0;\ {}^t\breve{\bar{\mathbf{d}}}) \bullet \Delta^t\breve{\mathbf{d}}
\end{aligned} \tag{11.523}$$

where $\nabla_{{}^t\breve{\mathbf{d}}^0} G$ is the covariant derivative of G with respect to ${}^t\breve{\mathbf{d}}$ evaluated at ${}^t\breve{\mathbf{d}}^0$, and the linear part – that is, the first-order terms – of the virtual functional is denoted by $\mathcal{L}^{Dyn}_{\Delta^t\breve{\mathbf{d}}}[G({}^t\breve{\mathbf{d}}^0;\ {}^t\breve{\bar{\mathbf{d}}})]$. In a non-equilibrium state, $G({}^t\breve{\mathbf{d}}^0;\ {}^t\breve{\bar{\mathbf{d}}})$, known as the ***residual force functional***, is non-zero. Noting that for conservative loading, the external loading does not depend on the current configuration, we can express the residual as:

$$G_{Dyn}({}^t\breve{\mathbf{d}}^0;\ {}^t\breve{\bar{\mathbf{d}}}) \equiv G_{Internal}({}^t\breve{\mathbf{d}}^0;\ {}^t\breve{\bar{\mathbf{d}}}) + G_{Inertial}({}^t\breve{\mathbf{d}}^0;\ {}^t\breve{\bar{\mathbf{d}}}) - G_{External}({}^t\breve{\bar{\mathbf{d}}}) \tag{11.524}$$

The linear operator, $\mathcal{L}^{Dyn}_{\Delta^t\breve{\mathbf{d}}}[G(\lambda^0,\ {}^t\breve{\mathbf{d}}^0;\ {}^t\breve{\bar{\mathbf{d}}})]$, consists only of two parts: the internal and the inertial linear virtual functionals:

$$\begin{aligned}
&\mathcal{L}_{\Delta^t\breve{\mathbf{d}}}^{Dyn}[G({}^t\breve{\mathbf{d}}^0;\ {}^t\breve{\bar{\mathbf{d}}})] \equiv \Delta G_{Internal}({}^t\breve{\mathbf{d}}^0;\ {}^t\breve{\bar{\mathbf{d}}};\Delta\,{}^t\breve{\mathbf{d}}) + \Delta G_{Inertial}({}^t\breve{\mathbf{d}}^0;\ {}^t\breve{\bar{\mathbf{d}}};\Delta\,{}^t\breve{\mathbf{d}})\\
&\Delta G_{Internal} \equiv \nabla_{{}^t\breve{\mathbf{d}}^0} G_{Internal}({}^t\breve{\mathbf{d}}^0;\ {}^t\breve{\bar{\mathbf{d}}}) \bullet \Delta^t\breve{\mathbf{d}}\\
&\Delta G_{Inertial} \equiv \nabla_{{}^t\breve{\mathbf{d}}^0} G_{Inertial}({}^t\breve{\mathbf{d}}^0;\ {}^t\breve{\bar{\mathbf{d}}}) \bullet \Delta^t\breve{\mathbf{d}}
\end{aligned} \tag{11.525}$$

In what follows, we provide a complete description of the various parts just mentioned.

11.17.4 Internal Virtual Functional: Covariant Linearization

Referring to equation (11.510) or equation (11.511), the internal virtual work functional is given as:

$$G_{Internal}({}^t\breve{\mathbf{d}};\, {}^t\breve{\bar{\mathbf{d}}}) = \int_M {}^n\breve{\bar{\varepsilon}}({}^t\breve{\mathbf{d}};\, {}^t\breve{\bar{\mathbf{d}}}) \bullet {}^n\breve{\mathbf{F}}({}^t\breve{\mathbf{d}})\, dA \tag{11.526}$$

Taking the covariant derivative of the expression in equation (11.526), we have:

$$\boxed{\begin{aligned} \Delta G({}^t\breve{\mathbf{d}}^0;\, {}^t\breve{\bar{\mathbf{d}}};\, \Delta\, {}^t\breve{\mathbf{d}}) &\equiv \nabla_{{}^t\breve{\mathbf{d}}} G({}^t\breve{\mathbf{d}}^0;\, {}^t\breve{\bar{\mathbf{d}}}) \bullet\ \Delta {}^t\hat{\mathbf{d}} \\ &= \int_M \breve{\bar{\varepsilon}}^T\ \Delta^n\breve{\mathbf{F}}\, dA + \int_M \Delta^n\breve{\bar{\varepsilon}}^T\ {}^n\breve{\mathbf{F}}\, dA \\ \Delta^n\breve{\mathbf{F}} &\equiv \nabla_{{}^t\breve{\mathbf{d}}}\, {}^n\breve{\mathbf{F}}^T\ \Delta\, {}^t\breve{\mathbf{d}}, \\ \Delta^n\breve{\bar{\varepsilon}} &\equiv \nabla_{{}^t\breve{\mathbf{d}}}\, {}^n\breve{\bar{\varepsilon}}^T\ \Delta\, {}^t\breve{\mathbf{d}} \end{aligned}} \tag{11.527}$$

where $\nabla_{{}^t\breve{\mathbf{d}}} G$ is the covariant derivatives of G with respect to ${}^t\breve{\mathbf{d}}$, and we have introduced the following.

Definitions

The incremental generalized force: $\Delta^n\breve{\mathbf{F}} \equiv (\nabla^n_{{}^t\breve{\mathbf{d}}}\breve{\mathbf{F}})^T\ \Delta\, {}^t\breve{\mathbf{d}}$,

and

The incremental generalized virtual strains: $\Delta\breve{\bar{\varepsilon}} \equiv [(\nabla^n_{{}^t\breve{\mathbf{d}}}\breve{\bar{\varepsilon}})^T\ \Delta\, {}^t\breve{\mathbf{d}}]$,

Now, noting that $\Delta^n\breve{\mathbf{F}} = {}^n\breve{\mathbf{D}}\ \Delta^n\breve{\varepsilon}$, the generalized reactive forces in the current configuration of a linear, isotropic, homogeneous and hyperelastic material with ${}^n\breve{\mathbf{D}}$ as the constitutive matrix, we can rewrite the equation as:

$$\boxed{\nabla_{\breve{\mathbf{d}}^0} G(\lambda^0,\, {}^t\breve{\mathbf{d}}^0;\, {}^t\breve{\bar{\mathbf{d}}}) \bullet \Delta {}^t\breve{\mathbf{d}} = \underbrace{\int_M {}^n\breve{\bar{\varepsilon}}^T\, {}^n\breve{\mathbf{D}}\ \Delta^n\breve{\varepsilon}\, dA}_{\text{Material stiffness-specific}} + \underbrace{\int_M \Delta^n\breve{\bar{\varepsilon}}T\ {}^n\breve{\mathbf{F}}\, dA}_{\text{Geometric stiffness-specific}}} \tag{11.528}$$

Remarks

- It may be noted that the internal virtual functional linearization consists of two parts. The first part, $\int_\Omega {}^n\breve{\bar{\varepsilon}}^T\, {}^n\breve{\mathbf{D}}\ \Delta^n\breve{\varepsilon}\, dA$, on the right-hand side of equation (11.528), involves the virtual strains and the incremental real strains derived from the linearization of the generalized reactive forces.

More specifically, it is the general nature of the nonlinear formulation that the incremental and the first variations of the real strains eventually give rise to what is known as the material stiffness functional, $\mathcal{K}_M$:

Definition: ***Material stiffness functional,*** $\mathcal{K}_M$**:**

$$\boxed{\mathcal{K}_M({}^t\breve{\mathbf{d}},\, {}^t\breve{\bar{\mathbf{d}}};\Delta^t\breve{\mathbf{d}}) \equiv \int_M {}^n\breve{\varepsilon}^T\, {}^n\breve{\mathbf{D}}\ \Delta^n\breve{\varepsilon}\ dA} \tag{11.529}$$

- Likewise, the second part, $\int_\Omega \Delta^n\breve{\bar{\varepsilon}}T\, {}^n\breve{\mathbf{F}}\, dA$, as a general trait of a geometrically nonlinear system, involves the second variations of the real strains or the first variation or increment of the virtual strains, derived from the linearization of the generalized strains and resulting in what is known as the geometric stiffness functional, $\mathcal{K}_G$:

Definition: ***Geometric Stiffness functional,*** $\mathcal{K}_G$**:**

$$\boxed{\mathcal{K}_G({}^t\breve{\mathbf{d}},\, {}^t\breve{\bar{\mathbf{d}}};\Delta\, {}^t\breve{\mathbf{d}}) \equiv \int_M \Delta^n\breve{\bar{\varepsilon}}^T({}^t\breve{\mathbf{d}},\, {}^t\breve{\bar{\mathbf{d}}};\Delta\, {}^t\breve{\mathbf{d}})\ {}^n\breve{\mathbf{F}}\ dA} \tag{11.530}$$

- Clearly, then, if the real strain is linear, the second variation (or the Hessian) and thus, the second part vanish identically, that is, $\mathcal{K}_G \equiv \mathbf{0}$. In other words, for linear shell responses, the geometric stiffness part drops out, with the shell tangent stiffness being purely due to the material stiffness contribution.

Definition: ***Tangent Stiffness functional,*** $\mathcal{K}_T$**:**

$$\boxed{\mathcal{K}_T \equiv \mathcal{K}_M + \mathcal{K}_G} \tag{11.531}$$

As indicated before, if the geometric stiffness functional is absent, as in the case of a linear shell system, the tangent stiffness functional is populated solely by the material stiffness functional.

11.17.5 Inertial Virtual Functional: Covariant Linearization

Referring to equation (11.511), the inertial virtual work functional is given as:

$$\boxed{\begin{aligned} G_{Inertial}({}^t\breve{\mathbf{d}};\, {}^t\breve{\bar{\mathbf{d}}}) &= \int_M {}^t\breve{\bar{\mathbf{d}}}^T\, {}^t\breve{\mathbf{F}}_{Iner}({}^t\breve{\mathbf{d}})\ dA \\ &= \int_M \{\, {}^t\bar{\mathbf{d}}^T\, M_0\, {}^t\ddot{\mathbf{d}} + {}^t\bar{\theta}^T\, I_0\, (\mathbf{W}^T\, {}^t\dot{\omega})\, \}dA \end{aligned}} \tag{11.532}$$

11.17.5.1 Linearized Translational Acceleration Vector, $\Delta\, {}^t\ddot{\mathbf{d}}$

The translational acceleration vector, ${}^t\ddot{\mathbf{d}}$, is trivially linearized as $\Delta\, {}^t\ddot{\mathbf{d}}$, since it is a member of a Euclidean vector space. Now, linearizing the expression in equation (11.532), we have:

$$
\begin{aligned}
\Delta G_{Inertial}({}^t\breve{\mathbf{d}};\, {}^t\bar{\breve{\mathbf{d}}}; \Delta\, {}^t\breve{\mathbf{d}}) &\equiv \{\nabla_{{}^t\breve{\mathbf{d}}}\, G_{Inertial}({}^t\breve{\mathbf{d}};\, {}^t\bar{\breve{\mathbf{d}}})\}^T\, \Delta\, {}^t\breve{\mathbf{d}} = \int_M {}^t\bar{\breve{\mathbf{d}}}^T\, \Delta\, {}^t\mathbf{F}_{Iner}({}^t\breve{\mathbf{d}}; \Delta\, {}^t\breve{\mathbf{d}})\, dA \\
&= \int_M \{\, {}^t\bar{\mathbf{d}}^T\, (M_0\, \Delta\, {}^t\ddot{\mathbf{d}}) + {}^t\bar{\boldsymbol{\theta}}^T\, (I_0\, (\Delta\mathbf{W}^T\, {}^t\dot{\boldsymbol{\omega}} + \mathbf{W}^T\, \Delta\, {}^t\dot{\boldsymbol{\omega}}))\, \}\, dA
\end{aligned}
\tag{11.533}
$$

Before we can generate the computationally efficient matrix form for the covariant linearization of the inertial virtual functional from equation (11.533), we need the following results.

11.17.5.2 Linearized Angular Acceleration Vector, $\Delta\, {}^t\dot{\boldsymbol{\omega}}$

However, the skew-symmetric angular velocity matrix, $\boldsymbol{\omega}$, is related to the rotation matrix, $\mathbf{R}$, that belongs to a curved space; hence, as before, linearization involves additional consideration as follows.

Let $\Delta\boldsymbol{\Psi}$ define the incremental skew-symmetric matrix related to the linearized or incremental rotation matrix, $\Delta\mathbf{R}$, such that:

$$
\boxed{\Delta\boldsymbol{\Psi} \equiv \Delta\mathbf{R}\, \mathbf{R}^T} \tag{11.534}
$$

whereby the axial vector, $\Delta\boldsymbol{\psi}$, to the matrix, $\Delta\boldsymbol{\Psi}$, is related to the incremental rotation vector, $\Delta\boldsymbol{\theta}$, by:

$$
\boxed{\Delta\boldsymbol{\psi} = \mathbf{W}({}^t\boldsymbol{\theta})\, \Delta\, {}^t\boldsymbol{\theta}} \tag{11.535}
$$

Now, starting from the relationship given by equation (11.518), and interchanging the order of the linear operations, the incremental or linearized angular velocity matrix, $\Delta\, {}^t\boldsymbol{\Omega}$, is given by:

$$
\begin{aligned}
\Delta\, {}^t\boldsymbol{\Omega} &= \Delta\, \dot{\mathbf{R}}\, \mathbf{R}^T + \dot{\mathbf{R}}\, \Delta\mathbf{R}^T \\
&= \overbrace{\Delta\boldsymbol{\Psi}\, \mathbf{R}}^{\bullet}\, \mathbf{R}^T + {}^t\boldsymbol{\Omega}\, \mathbf{R}\, \mathbf{R}^T\, \Delta\boldsymbol{\Psi}^T = \Delta\dot{\boldsymbol{\Psi}}\, \mathbf{R}\, \mathbf{R}^T + \Delta\boldsymbol{\Psi}\, \dot{\mathbf{R}}\, \mathbf{R}^T - {}^t\boldsymbol{\Omega}\, \Delta\boldsymbol{\Psi} \\
&= \Delta\dot{\boldsymbol{\Psi}} + \Delta\boldsymbol{\Psi}\, {}^t\boldsymbol{\Omega}\, \mathbf{R}\, \mathbf{R}^T - {}^t\boldsymbol{\Omega}\, \Delta\boldsymbol{\Psi} = \Delta\dot{\boldsymbol{\Psi}} + \left(\Delta\boldsymbol{\Psi}\, {}^t\boldsymbol{\Omega} - {}^t\boldsymbol{\Omega}\, \Delta\boldsymbol{\Psi}\right)
\end{aligned}
\tag{11.536}
$$

where we have used the fact: $\mathbf{R}\,\mathbf{R}^T = \mathbf{I}$, with $\mathbf{I}$ as the 3×3 identity matrix. Noting the commutator or Lie-bracket property (i.e. for any two skew-symmetric matrices, $\mathbf{A}$, $\mathbf{B}$ with $\mathbf{a}$, $\mathbf{b}$ as the corresponding axial vectors in three dimensions, we have: $\mathbf{AB} - \mathbf{BA} = [\mathbf{a} \times \mathbf{b}]$, where the notation $[\mathbf{c}] \equiv \mathbf{c}\times$ for any vector $\mathbf{c}$ is implied) in equation (11.536), the incremental or linearized angular velocity vector, $\Delta\, {}^t\boldsymbol{\omega}$, the axial vector of $\Delta\, {}^t\boldsymbol{\Omega}$, is given by:

$$
\boxed{\Delta\, {}^t\boldsymbol{\omega} = \Delta\dot{\boldsymbol{\psi}} - {}^t\boldsymbol{\omega} \times \Delta\boldsymbol{\psi}} \tag{11.537}
$$

Taking the time derivative of the expression in equation (11.537), for the incremental or linearized angular acceleration, $\Delta^t \dot{\omega}$, we get:

$$\boxed{\Delta^t \dot{\omega} = \Delta\ddot{\psi} - {}^t\omega \times \Delta\dot{\psi} - {}^t\dot{\omega} \times \Delta\psi} \tag{11.538}$$

Now, we first note from equation (11.517) and equation (11.535):

$$\omega \times \Delta\dot{\psi} = \mathbf{W}\dot{\theta} \times (\dot{\mathbf{W}}\,\Delta\theta + \mathbf{W}\,\Delta\dot{\theta}) = (\mathbf{W}\dot{\Theta}\,\dot{\mathbf{W}})\,\Delta\theta + (\mathbf{W}\dot{\Theta}\,\mathbf{W})\,\Delta\dot{\theta} \tag{11.539}$$

Similarly, we have:

$$\dot{\omega} \times \Delta\psi = (\dot{\mathbf{W}}\,\dot{\theta} + \mathbf{W}\,\ddot{\theta}) \times \mathbf{W}\,\Delta\theta = (\dot{\mathbf{W}}\,\dot{\Theta}\,\mathbf{W} + \mathbf{W}\,\ddot{\Theta}\,\mathbf{W})\,\Delta\theta \tag{11.540}$$

Finally, noting that: $\boxed{\Delta\dot{\psi} = \dot{\mathbf{W}}\,\Delta^t\theta + \mathbf{W}\,\Delta^t\dot{\theta}}$ and $\boxed{\Delta\ddot{\psi} = \ddot{\mathbf{W}}\,\Delta^t\theta + 2\dot{\mathbf{W}}\,\Delta^t\dot{\theta} + \mathbf{W}\,\Delta^t\ddot{\theta}}$, and using equations (11.539) and (11.540) in equation (11.538), we have the incremental or linearized angular acceleration, $\Delta^t\dot{\omega}$, in terms of the incremental or linearized rotation vector, $\Delta^t\theta$, the incremental or linearized rotational velocity vector, $\Delta^t\dot{\theta}$, and the incremental or linearized rotational acceleration vector, $\Delta^t\ddot{\theta}$, as:

$$\boxed{\begin{aligned}\Delta^t\dot{\omega} &= \Delta\ddot{\psi} - {}^t\omega \times \Delta\dot{\psi} - {}^t\dot{\omega} \times \Delta\psi \\ &= \underbrace{\mathbf{W}}_{\substack{\text{Mass-}\\\text{specific}}}\;\Delta^t\ddot{\theta} + \underbrace{(2\dot{\mathbf{W}} - \mathbf{W}\,{}^t\dot{\Theta}\,\mathbf{W})}_{\substack{\text{Gyroscopic-}\\\text{specific}}}\,\Delta^t\dot{\theta} \\ &\quad + \underbrace{(\ddot{\mathbf{W}} - \{\mathbf{W}\,{}^t\dot{\Theta}\,\dot{\mathbf{W}} + \dot{\mathbf{W}}\,{}^t\dot{\Theta}\,\mathbf{W}\} - \mathbf{W}\,{}^t\ddot{\Theta}\,\mathbf{W})\,\Delta^t\theta}_{\text{Centrifugal-specific}}\end{aligned}} \tag{11.541}$$

Now, from equation (11.533) and (11.541), we have:

$$\begin{aligned}\mathbf{W}^T\,\Delta\dot{\omega} &= (\mathbf{W}^T\mathbf{W})\Delta\ddot{\theta} + \mathbf{W}^T(2\dot{\mathbf{W}} - \mathbf{W}\,{}^t\dot{\Theta}\,\mathbf{W})\,\Delta\dot{\theta} \\ &\quad + \mathbf{W}^T(\ddot{\mathbf{W}} - \{\mathbf{W}\,\dot{\Theta}\,\dot{\mathbf{W}} + \dot{\mathbf{W}}\,\dot{\Theta}\,\mathbf{W}\} - \mathbf{W}\,{}^t\ddot{\Theta}\,\mathbf{W})\,\Delta\theta\end{aligned} \tag{11.542}$$

Again, from equation (11.533), letting angular acceleration, $\mathbf{a} \equiv \dot{\mathbf{W}}\,\dot{\theta} + \mathbf{W}\,\ddot{\theta}$, we have:

$$\begin{aligned}\Delta\mathbf{W}^T\,\mathbf{a} &= \{-c_1\,\Delta\Theta + c_2\,(\Delta\Theta\,\Theta + \Theta\,\Delta\Theta) + \left(c_3\Theta + c_4\Theta^2\right)(\theta \bullet \Delta\theta)\}\,\mathbf{a} \\ &= \{c_1\,[\mathbf{a}] + \left(-c_2\,\mathbf{I} - c_3\,\Theta + c_4\Theta^2\right)(\mathbf{a} \bullet \theta) + c_2\,((\theta \otimes \mathbf{a}) - (\mathbf{a} \otimes \theta) + (\theta \bullet \mathbf{a})\mathbf{I}\}\,\Delta\theta\end{aligned} \tag{11.543}$$

Exercise: Derive expression in equation (11.543).

Finally, using the results in equation (11.542) and equation (11.543) in equation (11.533), we have the full linearization of the inertial virtual function in matrix form as:

$$\Delta G_{Inertial}({}^t\breve{\mathbf{d}};\ {}^t\breve{\bar{\mathbf{d}}};\Delta\,{}^t\breve{\mathbf{d}}) = \underbrace{\Delta G_M({}^t\breve{\mathbf{d}};\ {}^t\breve{\bar{\mathbf{d}}};\Delta\,{}^t\breve{\mathbf{d}})}_{\text{Mass functional}} + \underbrace{\Delta G_G({}^t\breve{\mathbf{d}};\ {}^t\breve{\bar{\mathbf{d}}};\Delta\,{}^t\breve{\mathbf{d}})}_{\text{Gyroscopic functional}} + \underbrace{\Delta G_C({}^t\breve{\mathbf{d}};\ {}^t\breve{\bar{\mathbf{d}}};\Delta\,{}^t\breve{\mathbf{d}})}_{\text{Centrifugal functional}} \tag{11.544}$$

where

$$\underbrace{\Delta G_M({}^t\breve{\mathbf{d}};\ {}^t\breve{\bar{\mathbf{d}}};\Delta\,{}^t\breve{\mathbf{d}})}_{\text{Mass functional}} = \int_M {}^t\breve{\bar{\mathbf{d}}} \bullet \begin{bmatrix} M_0\,\mathbf{I} & \mathbf{0} \\ \mathbf{0} & I_0\,(\mathbf{W}^T\,\mathbf{W}) \end{bmatrix} \Delta\,{}^t\ddot{\breve{\mathbf{d}}}\ dA$$

$$\underbrace{\Delta G_G({}^t\breve{\mathbf{d}};\ {}^t\breve{\bar{\mathbf{d}}};\Delta\,{}^t\breve{\mathbf{d}})}_{\text{Gyroscopic functional}} = \int_M {}^t\breve{\bar{\mathbf{d}}} \bullet \begin{bmatrix} \mathbf{0} & \mathbf{0} \\ \mathbf{0} & \mathbf{W}^T(2\,\dot{\mathbf{W}} - \mathbf{W}\,\dot{\boldsymbol{\Theta}}\,\mathbf{W}) \end{bmatrix} \Delta\,{}^t\dot{\breve{\mathbf{d}}}\ dA$$

$$\underbrace{\Delta G_C({}^t\breve{\mathbf{d}};\ {}^t\breve{\bar{\mathbf{d}}};\Delta\,{}^t\breve{\mathbf{d}})}_{\text{Centrifugal functional}} = \int_M {}^t\breve{\bar{\mathbf{d}}} \bullet \begin{bmatrix} \mathbf{0} & \mathbf{0} \\ \mathbf{0} & \mathbf{U(a)} + \mathbf{W}^T\{\ddot{\mathbf{W}} - (\dot{\mathbf{W}}\,\boldsymbol{\Theta}\,\dot{\mathbf{W}} + \dot{\mathbf{W}}\,\boldsymbol{\Theta}\,\dot{\mathbf{W}}) - \mathbf{W}\,\ddot{\boldsymbol{\Theta}}\,\mathbf{W}\} \end{bmatrix} \Delta\,{}^t\breve{\mathbf{d}}\ dA \tag{11.545}$$

where, from equation (11.543), we have defined operator, **U**, as:

$$\mathbf{U(a)} \equiv \{c_1\,[\mathbf{a}] + (-c_2\,\mathbf{I} - c_3\,\boldsymbol{\Theta} + c_4\boldsymbol{\Theta}^2)(\mathbf{a}\bullet\theta) + c_2((\theta\otimes\mathbf{a}) - (\mathbf{a}\otimes\theta) + (\theta\bullet\mathbf{a})\mathbf{I}\},$$
$$\mathbf{a} \equiv \dot{\mathbf{W}}\,\dot{\theta} + \mathbf{W}\,\ddot{\theta} \tag{11.546}$$

with $\mathbf{W}$, $\dot{\mathbf{W}}$ and $\ddot{\mathbf{W}}$ are given by equations (11.513), (11.514) and (11.515), respectively.

11.17.6 Quasi-Static External Virtual Functional: Covariant Linearization

Referring to equation (11.510), the quasi-static external virtual work functional for proportional loading is given as:

$$G_{External}(\lambda;\ {}^t\breve{\bar{\mathbf{d}}}) = \int_M {}^t\breve{\bar{\mathbf{d}}}^T\ {}^t\hat{\mathbf{F}}(\lambda, S^1, S^2)\ dA\ + \lambda \int_{\partial M^f} {}^t\breve{\bar{\mathbf{d}}}^T\ {}^t\hat{\mathbf{F}}_\nu(S)\ dS \tag{11.547}$$

Then, we have the following symbolic representation of the first-order terms:

$$\Delta G_{External}(\lambda,\, {}^t\breve{\mathbf{d}};\, {}^t\breve{\bar{\mathbf{d}}};\, \Delta\lambda) \equiv \nabla_\lambda G_{External}(\lambda,\, {}^t\breve{\mathbf{d}};\, {}^t\breve{\bar{\mathbf{d}}}) \bullet \Delta\lambda = -\left\{ \int_M {}^t\breve{\bar{\mathbf{d}}}^T\, {}^t\breve{\hat{\mathbf{F}}}_{,\lambda}(\lambda, S^1, S^2)\, dA + \int_{\partial M^f} {}^t\breve{\bar{\mathbf{d}}}^T\, {}^t\breve{\hat{\mathbf{F}}}_\nu(S)\, dS \right\} \bullet \Delta\lambda \tag{11.548}$$

11.17.7 Quasi-static Linearized Virtual Work Equation

Thus, finally, using equations (11.528) and (11.548) in equation (11.520), we can introduce the following.

Definition

The linearized quasi-static virtual work equations:

$$\overbrace{\int_M {}^n\breve{\bar{\boldsymbol{\varepsilon}}}^T\, {}^n\breve{\mathbf{D}}\, \Delta^n\breve{\boldsymbol{\varepsilon}}\, dA}^{\text{Material stiffness functional}} + \overbrace{\int_M \Delta^n\breve{\bar{\boldsymbol{\varepsilon}}}^T\, {}^n\breve{\mathbf{F}}\, dA}^{\text{Geometric stiffness functional}}$$
$$= \underbrace{-\overbrace{G(\lambda^0,\, {}^t\breve{\mathbf{d}}^0;\, {}^t\breve{\bar{\mathbf{d}}})}^{\text{Unbalanced force functional}} - \overbrace{\left\{ \int_M {}^t\breve{\bar{\mathbf{d}}}^T\, {}^t\breve{\hat{\mathbf{F}}}_{,\lambda}(\lambda^0, S^1, S^2)\, dA + \int_{\partial M^f} {}^t\breve{\bar{\mathbf{d}}}^T\, {}^t\breve{\hat{\mathbf{F}}}_\nu(S)\, dS \right\} \bullet \Delta\lambda}^{\text{Applied force functional}}}_{\text{Residual force functional}} \tag{11.549}$$

Or,

$$\overbrace{\overbrace{\mathcal{K}_M}^{\text{Material stiffness functional}} + \overbrace{\mathcal{K}_G}^{\text{Geometric stiffness functional}}}^{\text{Tangent functional}}$$
$$= \underbrace{-\overbrace{G(\lambda^0,\, {}^t\breve{\mathbf{d}}^0;\, {}^t\breve{\bar{\mathbf{d}}})}^{\text{Unbalanced force functional}} - \overbrace{\left\{ \int_M {}^t\breve{\bar{\mathbf{d}}}^T\, {}^t\breve{\hat{\mathbf{F}}}_{,\lambda}(\lambda^0, S^1, S^2)\, dA + \int_{\partial M^f} {}^t\breve{\bar{\mathbf{d}}}^T\, {}^t\breve{\hat{\mathbf{F}}}_\nu(S)\, dS \right\} \bullet \Delta\lambda}^{\text{Applied force functional}}}_{\text{Residual force functional}} \tag{11.550}$$

where, we have introduced and collected, for convenience, the following.

Definitions

- The ***unbalanced force functional***, $G(\lambda^0, {}^t\breve{\mathbf{d}}^0; {}^t\bar{\breve{\mathbf{d}}})$:

$$\boxed{G(\lambda^0, \breve{\mathbf{d}}^0; \bar{\breve{\mathbf{d}}}) \equiv \int_M {}^n\bar{\breve{\varepsilon}}^T\ {}^n\breve{\mathbf{D}}^n\breve{\varepsilon}\ dA - \left\{ \int_M {}^t\bar{\breve{\mathbf{d}}}^T\ {}^t\breve{\hat{\mathbf{F}}}(\lambda^0, S^1, S^2)\ dA\ + \lambda^0 \int_{\partial M^f} {}^t\hat{\bar{\mathbf{d}}}^T\ {}^t\breve{\hat{\mathbf{F}}}_\nu(S)\ dS \right\}} \tag{11.551}$$

- The ***residual force functional***:

$$\boxed{\begin{aligned} \text{Residual force functional} \equiv &- \overbrace{G(\lambda^0, {}^t\breve{\mathbf{d}}^0; {}^t\bar{\breve{\mathbf{d}}})}^{\text{Unbalanced force functional}} \\ &+ \overbrace{\left\{ \int_M {}^t\bar{\breve{\mathbf{d}}}^T\ {}^t\breve{\hat{\mathbf{F}}}_{,\lambda}(\lambda^0, S^1, S^2)\ dA\ + \int_{\partial M^f} {}^t\bar{\breve{\mathbf{d}}}^T\ {}^t\hat{\mathbf{F}}^{\breve{S}}(S)\ dS \right\}}^{\text{Applied force functional}} \Delta\lambda \end{aligned}} \tag{11.552}$$

Material stiffness functional, $\mathcal{K}_M$:

$$\boxed{\mathcal{K}_\mathbf{M}({}^t\breve{\mathbf{d}}, {}^t\bar{\breve{\mathbf{d}}}; \Delta^t\breve{\mathbf{d}}) \equiv \int_M {}^n\bar{\breve{\varepsilon}}^T\ {}^n\breve{\mathbf{D}}\ \Delta^n\breve{\varepsilon}\ dA} \tag{11.553}$$

Geometric stiffness functional, $\mathcal{K}_G$:

$$\boxed{\mathcal{K}_G({}^t\breve{\mathbf{d}}, {}^t\bar{\breve{\mathbf{d}}}; \Delta\,{}^t\breve{\mathbf{d}}) \equiv \int_M \Delta^n\bar{\breve{\varepsilon}}^T({}^t\breve{\mathbf{d}}, {}^t\bar{\breve{\mathbf{d}}}; \Delta\,{}^t\breve{\mathbf{d}})\ {}^n\breve{\mathbf{F}}\ dA} \tag{11.554}$$

11.17.8 Dynamic Linearized Virtual Work Equation

Thus, finally, using equation (11.528) and equations (11.544)–(11.546) in equation (11.523), we can introduce and collect, for convenience, the following.

Definition

The linearized dynamic virtual work equations:

$$\overbrace{\int_M {}^n\breve{\bar{\varepsilon}}^T\, {}^n\breve{\mathbf{D}}\ \Delta^n\breve{\varepsilon}\ dA}^{\text{Material stiffness functional}} + \overbrace{\int_M \Delta^n\breve{\bar{\varepsilon}}^T\ {}^n\breve{\mathbf{F}}\ dA}^{\text{Geometric stiffness functional}}$$

$$\overbrace{+\underbrace{\Delta G_M({}^t\breve{\mathbf{d}};\ {}^t\breve{\bar{\mathbf{d}}};\Delta\, {}^t\breve{\mathbf{d}})}_{\text{Mass-specific}} + \underbrace{\Delta G_G({}^t\breve{\mathbf{d}};\ {}^t\breve{\bar{\mathbf{d}}};\Delta\, {}^t\breve{\mathbf{d}})}_{\text{Gyroscopic}} + \underbrace{\Delta G_C({}^t\breve{\mathbf{d}};\ {}^t\breve{\bar{\mathbf{d}}};\Delta\, {}^t\breve{\mathbf{d}})}_{\text{Centrifugal}}}^{\text{Inertial functional}} = -\underbrace{\overbrace{G({}^t\breve{\mathbf{d}}^0;\ {}^t\breve{\bar{\mathbf{d}}})}^{\text{Unbalanced force functional}}}_{\text{Residual force functional}}$$

(11.555)

Or,

$$\overbrace{\overbrace{\mathcal{K}_M}^{\text{Material stiffness functional}} + \overbrace{\mathcal{K}_G}^{\text{Geometric stiffness functional}}}^{\text{Tangent functional}}$$

$$\overbrace{+\underbrace{\Delta G_M({}^t\breve{\mathbf{d}};\ {}^t\breve{\bar{\mathbf{d}}};\Delta\, {}^t\breve{\mathbf{d}})}_{\text{Mass-specific}} + \underbrace{\Delta G_G({}^t\breve{\mathbf{d}};\ {}^t\breve{\bar{\mathbf{d}}};\Delta\, {}^t\breve{\mathbf{d}})}_{\text{Gyroscopic}} + \underbrace{\Delta G_C({}^t\breve{\mathbf{d}};\ {}^t\breve{\bar{\mathbf{d}}};\Delta\, {}^t\breve{\mathbf{d}})}_{\text{Centrifugal}}}^{\text{Inertial functional}} = -\underbrace{\overbrace{G({}^t\breve{\mathbf{d}};\ {}^t\breve{\bar{\mathbf{d}}})}^{\text{Unbalanced force functional}}}_{\text{Residual force functional}}$$

(11.556)

where, for convenience, we collect the linearized internal and inertial terms for the following.

Definitions

- The residual force functional, $G({}^t\breve{\mathbf{d}};\ {}^t\breve{\bar{\mathbf{d}}})$:

$$G(\breve{\mathbf{d}};\breve{\bar{\mathbf{d}}}) \equiv \int_M {}^n\breve{\bar{\varepsilon}}^T\ {}^n\breve{\mathbf{D}}\ {}^n\breve{\varepsilon}\ dA - \left\{ \begin{array}{l} \displaystyle\int_M \left({}^t\hat{\mathbf{F}} \bullet {}^t\bar{\mathbf{d}} + {}^t\hat{\mathbf{M}} \bullet \mathbf{W}^t\bar{\theta}\right) dA \\ \displaystyle + \int_{\partial M_f} \left({}^t\hat{\mathbf{F}}_\nu \bullet {}^t\bar{\mathbf{d}} + {}^t\hat{\mathbf{M}}_\nu \bullet \mathbf{W}^t\bar{\theta}\right) dS \end{array} \right\} \tag{11.557}$$

- The linearized mass functional, $\Delta G_M({}^t\breve{\mathbf{d}};\ {}^t\breve{\bar{\mathbf{d}}};\Delta\, {}^t\breve{\mathbf{d}})$:

$$\Delta G_M({}^t\breve{\mathbf{d}};\ {}^t\breve{\bar{\mathbf{d}}};\Delta\, {}^t\breve{\mathbf{d}}) \equiv \int_M {}^t\breve{\bar{\mathbf{d}}} \bullet \begin{bmatrix} M_0\,\mathbf{I} & \mathbf{0} \\ \mathbf{0} & I_0\,(\mathbf{W}^T\,\mathbf{W}) \end{bmatrix} \Delta\, {}^t\ddot{\breve{\mathbf{d}}}\ dA \tag{11.558}$$

- The linearized gyroscopic (damping) functional, $\Delta G_G({}^t\breve{\mathbf{d}};\, {}^t\breve{\bar{\mathbf{d}}};\Delta\, {}^t\breve{\mathbf{d}})$:

$$\boxed{\Delta G_G({}^t\breve{\mathbf{d}};\, {}^t\breve{\bar{\mathbf{d}}};\Delta\, {}^t\breve{\mathbf{d}}) = \int_M {}^t\breve{\bar{\mathbf{d}}} \bullet \begin{bmatrix} \mathbf{0} & \mathbf{0} \\ \mathbf{0} & \mathbf{W}^T(2\,\dot{\mathbf{W}} - \mathbf{W}\,\dot{\boldsymbol{\Theta}}\,\mathbf{W}) \end{bmatrix} \Delta\, {}^t\dot{\breve{\mathbf{d}}}\; dA} \tag{11.559}$$

- The linearized centrifugal (stiffness) functional, $\Delta G_C({}^t\breve{\mathbf{d}};\, {}^t\breve{\bar{\mathbf{d}}};\Delta\, {}^t\breve{\mathbf{d}})$:

$$\boxed{\begin{aligned} &\Delta G_C({}^t\breve{\mathbf{d}};\, {}^t\breve{\bar{\mathbf{d}}};\Delta\, {}^t\breve{\mathbf{d}}) = \int_M {}^t\breve{\bar{\mathbf{d}}} \bullet \begin{bmatrix} \mathbf{0} & \mathbf{0} \\ \mathbf{0} & \mathbf{U}({}^t\dot{\omega}) + \mathbf{W}^T\{\ddot{\mathbf{W}} - (\mathbf{W}\,\dot{\boldsymbol{\Theta}}\,\dot{\mathbf{W}} + \dot{\mathbf{W}}\,\dot{\boldsymbol{\Theta}}\,\mathbf{W}) - \mathbf{W}\,\ddot{\boldsymbol{\Theta}}\,\mathbf{W}\} \end{bmatrix} \Delta\, {}^t\breve{\mathbf{d}}\; dA \\ &\mathbf{U}({}^t\dot{\omega}) \equiv \left\{ \begin{aligned} &c_1\,[{}^t\dot{\omega}] + (-c_2\,\mathbf{I} - c_3\,\boldsymbol{\Theta} + c_4\boldsymbol{\Theta}^2)({}^t\dot{\omega} \bullet \theta) \\ &+c_2((\theta \otimes {}^t\dot{\omega}) - ({}^t\dot{\omega} \otimes \theta) + (\theta \bullet {}^t\dot{\omega})\mathbf{I}) \end{aligned} \right\}, \\ &{}^t\dot{\omega} = (\dot{\mathbf{W}}\; {}^t\dot{\theta} + \mathbf{W}\; {}^t\ddot{\theta}) \end{aligned}} \tag{11.560}$$

with $[\mathbf{h}] \equiv \mathbf{h}\times,\ \forall \mathbf{h} \in \mathbb{R}_3$, and

$$\boxed{\begin{aligned} &\mathbf{W}({}^t\theta) \equiv \mathbf{I} + c_1\,{}^t\boldsymbol{\Theta} + c_2\,{}^t\boldsymbol{\Theta}^2, \\ &c_1({}^t\theta) = \frac{1-\cos\theta}{\theta^2}, \\ &c_2({}^t\theta) = \frac{\theta - \sin\theta}{\theta^3} \end{aligned}} \tag{11.561}$$

$$\boxed{\begin{aligned} \dot{\mathbf{W}} &= c_1\,\dot{\boldsymbol{\Theta}} + c_2(\dot{\boldsymbol{\Theta}}\,\boldsymbol{\Theta} + \boldsymbol{\Theta}\,\dot{\boldsymbol{\Theta}}) + \dot{c}_1\,\boldsymbol{\Theta} + \dot{c}_2\,\boldsymbol{\Theta}^2 \\ &= (c_2 - c_1)(\theta \bullet \dot{\theta})\mathbf{I} + c_1\,\dot{\boldsymbol{\Theta}} + c_2\{(\theta \otimes \dot{\theta}) + (\dot{\theta} \otimes \theta)\} + \dot{c}_1\,\boldsymbol{\Theta} + \dot{c}_2(\theta \otimes \theta) \\ \dot{c}_1 &\equiv \frac{1}{\theta^2}(1 - 2c_1 - \theta^2 c_2)(\theta \bullet \dot{\theta}) = c_3(\theta \bullet \dot{\theta}), \\ \dot{c}_2 &\equiv \frac{1}{\theta^2}(c_1 - 3c_2)(\theta \bullet \dot{\theta}) = c_4(\theta \bullet \dot{\theta}), \end{aligned}} \tag{11.562}$$

and

$$\boxed{\begin{aligned} &\ddot{\mathbf{W}} = c_1\,\ddot{\boldsymbol{\Theta}} + 2\{\dot{c}_1\,\dot{\boldsymbol{\Theta}} + \dot{c}_2(\dot{\boldsymbol{\Theta}}\,\boldsymbol{\Theta} + \boldsymbol{\Theta}\,\dot{\boldsymbol{\Theta}})\} + c_2(\ddot{\boldsymbol{\Theta}}\,\boldsymbol{\Theta} + 2\,\dot{\boldsymbol{\Theta}}^2 + \boldsymbol{\Theta}\,\ddot{\boldsymbol{\Theta}}) + \ddot{c}_1\,\boldsymbol{\Theta} + \ddot{c}_2\,\boldsymbol{\Theta}^2 \\ &\ddot{c}_1 \equiv c_5(\theta \bullet \dot{\theta})^2 + c_3\{\theta^2 + (\theta \bullet \dot{\theta})\}, \quad \ddot{c}_2 \equiv c_6(\theta \bullet \dot{\theta})^2 + c_4\{\theta^2 + (\theta \bullet \dot{\theta})\} \\ &c_5 \equiv -\frac{1}{\theta^2}(2c_2 + 4c_3 + \theta^2 c_4), \quad c_6 \equiv -\frac{1}{\theta^2}(c_3 - 5c_4), \end{aligned}} \tag{11.563}$$

11.17.9 *Where We Would Like to Go*

Clearly, the generalized form of the linearized equation (11.549) or equation (11.555) is presented in functional form and thus is still not suitable for computational purposes; that is, the expressions for the strains hide the sought-after incremental variable, $\Delta\, {}^t\mathbf{d}$. We would like to have a form of the virtual linearized equation that is totally explicit in the incremental variables $\Delta\, {}^t\breve{\mathbf{d}}$ and $\Delta\lambda$ for quasi-static cases, and $\Delta\, {}^t\breve{\mathbf{d}}$, $\Delta\, {}^t\dot{\breve{\mathbf{d}}}$ and $\Delta\, {}^t\ddot{\breve{\mathbf{d}}}$ for dynamic analyses so that it can be easily solved digitally by a Newton-type iterative method. In other words, we would like the matrix forms of the equations that are easily solved on digital computers. For this, we need to make a few more logical stops as follows.

For the matrix form of the incremental equation (11.555) for dynamic time history loading we want:

- finite element formulation: dynamic loading
- For the matrix form of the incremental equation (11.549) for quasi-static loading:
- finite element formulation: quasi-static loading

11.18 c-type FE Formulation: Dynamic Loading

11.18.1 *What We Need to Recall*

- The virtual work functional, $G_{Dyn}({}^t\mathbf{d}, {}^t\boldsymbol{\theta}; {}^t\bar{\mathbf{d}}, {}^t\bar{\boldsymbol{\theta}})$, in compact generalized notation:

$$\boxed{\begin{aligned} &G_{Dyn}({}^t\breve{\mathbf{d}}; {}^t\breve{\bar{\mathbf{d}}}) \equiv G_{Internal} + G_{Inertial} - G_{External} = \mathbf{0}, \quad \forall\, \text{admissible}\, {}^t\breve{\bar{\mathbf{d}}} \\ &G_{Internal} = \int_M {}^n\breve{\bar{\boldsymbol{\varepsilon}}}\ {}^n\breve{\mathbf{F}}\ dA \\ &G_{Inertial} = \int_M {}^t\breve{\bar{\mathbf{d}}}^T\ {}^t\breve{\mathbf{F}}_{Iner}\ dA \\ &G_{External} = \int_M {}^t\breve{\bar{\mathbf{d}}}^T\ {}^t\breve{\hat{\mathbf{F}}}(S^1, S^2)\ dA\ + \lambda \int_{\partial M^f} {}^t\breve{\bar{\mathbf{d}}}^T\ {}^t\breve{\hat{\mathbf{F}}}(S)\ dS \end{aligned}} \tag{11.564}$$

with the applicable initial conditions on the dynamic state variables as:

$$\boxed{\begin{aligned} &\breve{\mathbf{d}}(0) = \breve{\mathbf{d}}_0. \\ &{}^t\dot{\breve{\mathbf{d}}}(0) = \breve{v}_0, \end{aligned}} \tag{11.565}$$

The linearized dynamic virtual work equations: at a time $t \in \mathbb{R}_+$

$$\Delta G_{Inertial} + \Delta G_{Internal} = \text{residual force functional} = R({}^t\breve{\mathbf{d}}(t);\ {}^t\breve{\bar{\mathbf{d}}})$$

$$\overbrace{\underbrace{\int_M {}^n\breve{\bar{\varepsilon}}^T\, {}^n\mathbf{D}\ \Delta{}^n\breve{\varepsilon}(t)\, dA}_{\text{Material stiffness functional}} + \underbrace{\int_M \Delta{}^n\breve{\bar{\varepsilon}}^T\ {}^n\breve{\mathbf{F}}(t)\, dA}_{\text{Geometric stiffness functional}}}^{\text{Linearized internal functional, } \Delta G_{Internal}}$$

$$\overbrace{+\underbrace{\Delta G_M({}^t\breve{\mathbf{d}}(t);\ {}^t\breve{\bar{\mathbf{d}}};\Delta\,{}^t\breve{\mathbf{d}}(t))}_{\text{Inertialmass}} + \underbrace{\Delta G_G({}^t\breve{\mathbf{d}}(t);\ {}^t\breve{\bar{\mathbf{d}}};\Delta\,{}^t\breve{\mathbf{d}}(t))}_{\text{Gyroscopic}} + \underbrace{\Delta G_C({}^t\breve{\mathbf{d}};\ {}^t\breve{\bar{\mathbf{d}}};\Delta\,{}^t\breve{\mathbf{d}}(t))}_{\text{Centrifugal}}}^{\text{Inertial functional, } \Delta G_{Inertial}}$$

$$= -\underbrace{\overbrace{G({}^t\breve{\mathbf{d}}(t);\ {}^t\breve{\bar{\mathbf{d}}})}^{\text{Unbalanced force functional}}}_{\text{Residual force functional}} \tag{11.566}$$

or

$$\overbrace{\overbrace{\mathcal{K}_M(t)}^{\text{Material stiffness functional}} + \overbrace{\mathcal{K}_G(t)}^{\text{Geometric stiffness functional}}}^{\text{Tangent functional}}$$

$$\overbrace{+\underbrace{\Delta G_M({}^t\breve{\mathbf{d}}(t);\ {}^t\breve{\bar{\mathbf{d}}};\Delta\,{}^t\breve{\mathbf{d}}(t))}_{\text{Mass-specific}} + \underbrace{\Delta G_G({}^t\breve{\mathbf{d}}(t);\ {}^t\breve{\bar{\mathbf{d}}};\Delta\,{}^t\breve{\mathbf{d}}(t))}_{\text{Gyroscopic}} + \underbrace{\Delta G_C({}^t\breve{\mathbf{d}}(t);\ {}^t\breve{\bar{\mathbf{d}}};\Delta\,{}^t\breve{\mathbf{d}}(t))}_{\text{Centrifugal}}}^{\text{Inertial functional}} \tag{11.567}$$

$$= -\underbrace{\overbrace{G({}^t\breve{\mathbf{d}}(t);\ {}^t\breve{\bar{\mathbf{d}}})}^{\text{Unbalanced force functional}}}_{\text{Residual force functional}}$$

where the initial conditions on the generalized displacement and velocity are specified as:

$$\breve{\mathbf{d}}(0) \equiv \begin{Bmatrix} {}^t\mathbf{d}(0) \\ {}^t\theta(0) \end{Bmatrix} = \hat{\breve{\mathbf{d}}}_0 \equiv \begin{Bmatrix} {}^t\mathbf{d}_0 \\ {}^t\theta_0 \end{Bmatrix} \tag{11.568}$$

where, for convenience, we collect the linearized internal and inertial terms for the following.

Definitions

- The ***linearized inertial functional***, $\Delta G_{Inertial}({}^t\breve{\mathbf{d}};\, {}^t\breve{\bar{\mathbf{d}}}; \Delta\, {}^t\breve{\mathbf{d}})$:

$$\boxed{\Delta G_{Inertial}({}^t\breve{\mathbf{d}};\, {}^t\breve{\bar{\mathbf{d}}}; \Delta\, {}^t\breve{\mathbf{d}}) = \underbrace{\Delta G_M({}^t\breve{\mathbf{d}};\, {}^t\breve{\bar{\mathbf{d}}}; \Delta\, {}^t\breve{\mathbf{d}})}_{\text{Mass functional}} + \underbrace{\Delta G_G({}^t\breve{\mathbf{d}};\, {}^t\breve{\bar{\mathbf{d}}}; \Delta\, {}^t\breve{\mathbf{d}})}_{\text{Gyroscopic functional}} + \underbrace{\Delta G_C({}^t\breve{\mathbf{d}};\, {}^t\breve{\bar{\mathbf{d}}}; \Delta\, {}^t\breve{\mathbf{d}})}_{\text{Centrifugal functional}}}$$

(11.569)

- The ***linearized (symmetric) mass functional***, $\Delta G_M({}^t\breve{\mathbf{d}};\, {}^t\breve{\bar{\mathbf{d}}}; \Delta\, {}^t\breve{\mathbf{d}})$:

$$\boxed{\Delta G_M({}^t\breve{\mathbf{d}};\, {}^t\breve{\bar{\mathbf{d}}}; \Delta\, {}^t\breve{\mathbf{d}}) \equiv \int_M {}^t\breve{\bar{\mathbf{d}}} \bullet \begin{bmatrix} M_0\,\mathbf{I} & \mathbf{0} \\ \mathbf{0} & I_0\,(\mathbf{W}^T\,\mathbf{W}) \end{bmatrix} \Delta\, {}^t\ddot{\breve{\mathbf{d}}}\; dA} \tag{11.570}$$

- The ***linearized gyroscopic (damping) functional***, $\Delta G_G({}^t\breve{\mathbf{d}};\, {}^t\breve{\bar{\mathbf{d}}}; \Delta\, {}^t\breve{\mathbf{d}})$:

$$\boxed{\Delta G_G({}^t\breve{\mathbf{d}};\, {}^t\breve{\bar{\mathbf{d}}}; \Delta\, {}^t\breve{\mathbf{d}}) = \int_M {}^t\breve{\bar{\mathbf{d}}} \bullet \begin{bmatrix} \mathbf{0} & \mathbf{0} \\ \mathbf{0} & I_0\,\{\mathbf{W}^T(2\,\dot{\mathbf{W}} - \mathbf{W}\,\dot{\boldsymbol{\Theta}}\,\mathbf{W})\} \end{bmatrix} \Delta\, {}^t\dot{\breve{\mathbf{d}}}\; dA} \tag{11.571}$$

- The ***linearized centrifugal (stiffness) functional***, $\Delta G_C({}^t\breve{\mathbf{d}};\, {}^t\breve{\bar{\mathbf{d}}}; \Delta\, {}^t\breve{\mathbf{d}})$:

$$\boxed{\begin{aligned}
&\Delta G_C({}^t\breve{\mathbf{d}};\, {}^t\breve{\bar{\mathbf{d}}}; \Delta\, {}^t\breve{\mathbf{d}}) = \int_M {}^t\breve{\bar{\mathbf{d}}} \bullet \begin{bmatrix} \mathbf{0} & \mathbf{0} \\ \mathbf{0} & I_0\{\mathbf{U}({}^t\,\dot{\boldsymbol{\omega}}) + \mathbf{W}^T\{\ddot{\mathbf{W}} - (\mathbf{W}\,\dot{\boldsymbol{\Theta}}\,\dot{\mathbf{W}} + \dot{\mathbf{W}}\,\dot{\boldsymbol{\Theta}}\,\mathbf{W}) - \mathbf{W}\,\ddot{\boldsymbol{\Theta}}\,\mathbf{W}\}\} \end{bmatrix} \Delta\, {}^t\breve{\mathbf{d}}\; dA \\
&\mathbf{U}({}^t\,\dot{\boldsymbol{\omega}}_S) \equiv \left\{ \begin{aligned} & c_1\,[{}^t\,\dot{\boldsymbol{\omega}}_S] + (-c_2\,\mathbf{I} - c_3\,\boldsymbol{\Theta} + c_4\boldsymbol{\Theta}^2)({}^t\,\dot{\boldsymbol{\omega}}_S \bullet \boldsymbol{\theta}) \\ & + c_2((\boldsymbol{\theta} \otimes {}^t\,\dot{\boldsymbol{\omega}}_S) - ({}^t\,\dot{\boldsymbol{\omega}}_S \otimes \boldsymbol{\theta}) + (\boldsymbol{\theta} \bullet {}^t\,\dot{\boldsymbol{\omega}}_S)\mathbf{I}) \end{aligned} \right\}, \\
&{}^t\,\dot{\boldsymbol{\omega}}_S = (\dot{\mathbf{W}}\,{}^t\dot{\boldsymbol{\theta}} + \mathbf{W}\,{}^t\ddot{\boldsymbol{\theta}})
\end{aligned}}$$

(11.572)

with $[\mathbf{h}] \equiv \mathbf{h}\times, \forall \mathbf{h} \in \mathbb{R}_3$,

- The residual force functional, $R({}^t\breve{\mathbf{d}}(t);\, {}^t\breve{\bar{\mathbf{d}}}) = -G({}^t\breve{\mathbf{d}}(t);\, {}^t\breve{\bar{\mathbf{d}}})$ is given as:

$$
\begin{aligned}
R({}^t\breve{\mathbf{d}}(t);\, {}^t\breve{\bar{\mathbf{d}}}) &= -G({}^t\breve{\mathbf{d}}(t);\, {}^t\breve{\bar{\mathbf{d}}}) = -G({}^t\mathbf{d}(t),\, {}^t\boldsymbol{\theta}(t);\, {}^t\bar{\mathbf{d}},\, {}^t\bar{\boldsymbol{\theta}}) \\
&\equiv -\{G_{Internal} + G_{Inertial} - G_{External}\} \\
G_{Internal}({}^t\breve{\mathbf{d}}(t);\, {}^t\breve{\bar{\mathbf{d}}}) &= \int_M {}^n\breve{\bar{\boldsymbol{\varepsilon}}}({}^t\breve{\mathbf{d}}(t);\, {}^t\breve{\bar{\mathbf{d}}})^T\; {}^n\breve{\mathbf{F}}({}^t\breve{\mathbf{d}}(t))\, dA \\
G_{Inertial}({}^t\breve{\mathbf{d}}(t);\, {}^t\breve{\bar{\mathbf{d}}}) &= \int_M {}^t\breve{\bar{\mathbf{d}}}^T\; {}^t\mathbf{F}_{Iner}({}^t\breve{\mathbf{d}}(t))\, dA \\
&= \int_M \{ {}^t\bar{\mathbf{d}}^T\; M_0\, {}^t\ddot{\mathbf{d}}(t) + {}^t\bar{\boldsymbol{\theta}}^T\; I_0\, (\mathbf{W}^T\, {}^t\dot{\boldsymbol{\omega}}_S(t))\, \} dA \\
G_{External}({}^t\breve{\bar{\mathbf{d}}}) &= \int_M {}^t\breve{\bar{\mathbf{d}}}^T\; {}^t\breve{\hat{\mathbf{F}}}(S^1, S^2)\, dA + \int_{\partial M^f} {}^t\breve{\bar{\mathbf{d}}}^T\; {}^t\breve{\hat{\mathbf{F}}}_\nu(S)\, dS \\
&= \mathbf{0}, \quad \forall \text{ admissible } {}^t\bar{\mathbf{d}},\, {}^t\bar{\boldsymbol{\theta}}
\end{aligned}
\tag{11.573}
$$

where $\mathbf{W}$ is given as:

$$
\begin{aligned}
&\mathbf{W}({}^t\boldsymbol{\theta}) \equiv \mathbf{I} + c_1\, {}^t\boldsymbol{\Theta} + c_2\, {}^t\boldsymbol{\Theta}^2, \\
&c_1({}^t\boldsymbol{\theta}) = \frac{1-\cos\theta}{\theta^2}, \quad c_2({}^t\boldsymbol{\theta}) = \frac{\theta - \sin\theta}{\theta^3}
\end{aligned}
\tag{11.574}
$$

with ${}^t\boldsymbol{\Theta}$ as the skew-symmetric matrix corresponding to the axial vector, ${}^t\boldsymbol{\theta}$, such that ${}^t\boldsymbol{\Theta}\, {}^t\boldsymbol{\theta} = \mathbf{0}$.

- $\dot{\mathbf{W}}$ is given as:

$$
\begin{aligned}
\dot{\mathbf{W}} &= c_1\, \dot{\boldsymbol{\Theta}} + c_2(\dot{\boldsymbol{\Theta}}\, \boldsymbol{\Theta} + \boldsymbol{\Theta}\, \dot{\boldsymbol{\Theta}}) + \dot{c}_1\, \boldsymbol{\Theta} + \dot{c}_2\, \boldsymbol{\Theta}^2 \\
&= (c_2 - c_1)(\boldsymbol{\theta} \bullet \dot{\boldsymbol{\theta}})\mathbf{I} + c_1\, \dot{\boldsymbol{\Theta}} + c_2\{(\boldsymbol{\theta} \otimes \dot{\boldsymbol{\theta}}) + (\dot{\boldsymbol{\theta}} \otimes \boldsymbol{\theta})\} + \dot{c}_1\, \boldsymbol{\Theta} + \dot{c}_2(\boldsymbol{\theta} \otimes \boldsymbol{\theta}) \\
\dot{c}_1 &\equiv \frac{1}{\theta^2}(1 - 2c_1 - \theta^2 c_2)(\boldsymbol{\theta} \bullet \dot{\boldsymbol{\theta}}) = c_3(\boldsymbol{\theta} \bullet \dot{\boldsymbol{\theta}}), \quad \dot{c}_2 \equiv \frac{1}{\theta^2}(c_1 - 3c_2)(\boldsymbol{\theta} \bullet \dot{\boldsymbol{\theta}}) = c_4(\boldsymbol{\theta} \bullet \dot{\boldsymbol{\theta}}),
\end{aligned}
\tag{11.575}
$$

- $\ddot{\mathbf{W}}$ is given as:

$$
\begin{aligned}
\ddot{\mathbf{W}} &= c_1\, \ddot{\boldsymbol{\Theta}} + 2\{\dot{c}_1\, \dot{\boldsymbol{\Theta}} + \dot{c}_2(\dot{\boldsymbol{\Theta}}\, \boldsymbol{\Theta} + \boldsymbol{\Theta}\, \dot{\boldsymbol{\Theta}})\} + c_2(\ddot{\boldsymbol{\Theta}}\, \boldsymbol{\Theta} + 2\, \dot{\boldsymbol{\Theta}}^2 + \boldsymbol{\Theta}\, \ddot{\boldsymbol{\Theta}}) + \ddot{c}_1\, \boldsymbol{\Theta} + \ddot{c}_2\, \boldsymbol{\Theta}^2 \\
\ddot{c}_1 &\equiv c_5(\boldsymbol{\theta} \bullet \dot{\boldsymbol{\theta}})^2 + c_3\{\dot{\theta}^2 + (\boldsymbol{\theta} \bullet \ddot{\boldsymbol{\theta}})\}, \quad \ddot{c}_2 \equiv c_6(\boldsymbol{\theta} \bullet \dot{\boldsymbol{\theta}})^2 + c_4\{\dot{\theta}^2 + (\boldsymbol{\theta} \bullet \ddot{\boldsymbol{\theta}})\} \\
c_5 &\equiv -\frac{1}{\theta^2}(2c_2 + 4c_3 + \theta^2 c_4), \quad c_6 \equiv -\frac{1}{\theta^2}(c_3 - 5c_4),
\end{aligned}
\tag{11.576}
$$

- $^{n}\breve{\boldsymbol{\varepsilon}}$, the generalized virtual strain field is given as:

$$\begin{aligned}
&{}^{n}\breve{\bar{\boldsymbol{\varepsilon}}} \equiv \left\{ {}^{n}\bar{\boldsymbol{\beta}}_1 \quad {}^{n}\bar{\boldsymbol{\chi}}_1 \quad {}^{n}\bar{\boldsymbol{\beta}}_2 \quad {}^{n}\bar{\boldsymbol{\chi}}_2 \right\}^T,\\
&{}^{n}\bar{\boldsymbol{\beta}}_\alpha = {}^{t}\mathbf{R}^T \left({}^{t}\bar{\mathbf{d}}_{,\alpha} + {}^{t}\mathbf{k}^0_\alpha \times {}^{t}\bar{\mathbf{d}} \right) - {}^{t}\mathbf{R}^T\, {}^{t}\bar{\boldsymbol{\Phi}} \left({}^{t}\mathbf{e}_\alpha + {}^{t}\mathbf{k}^0_\alpha \times {}^{t}\mathbf{d} \right)\\
&\qquad = \overline{{}^{t}\mathbf{R}^T(\hat{\mathbf{1}}_\alpha + {}^{t}\mathbf{d}|_\alpha)},\\
&{}^{n}\bar{\boldsymbol{\chi}}_\alpha \equiv {}^{t}\mathbf{R}^T\, {}^{t}\bar{\boldsymbol{\phi}}|_\alpha = {}^{t}\mathbf{R}^T \left\{ {}^{t}\bar{\boldsymbol{\phi}}_{,\alpha} + \mathbf{k}^0_\alpha \times {}^{t}\bar{\boldsymbol{\phi}} \right\} = \overline{{}^{t}\mathbf{R}^T \left(\mathbf{k}^0_\alpha + \mathbf{k}_\alpha \right)},\\
&{}^{t}\bar{\boldsymbol{\phi}} = \mathbf{W}\, {}^{t}\bar{\boldsymbol{\theta}}
\end{aligned} \tag{11.577}$$

- ${}^{n}\breve{\mathbf{F}}$, the generalized reactive force field is given as:

$$ {}^{n}\breve{\mathbf{F}} \equiv \left\{ {}^{n}\mathbf{F}_1 \quad {}^{n}\mathbf{M}_1 \quad {}^{n}\mathbf{F}_2 \quad {}^{n}\mathbf{M}_2 \right\}^T \tag{11.578}$$

- ${}^{t}\breve{\mathbf{F}}_{Iner}$, the generalized inertial force field is given as:

$$\begin{aligned}
&{}^{t}\breve{\mathbf{F}}_{Iner} \equiv \{ M_0\, {}^{t}\ddot{\mathbf{d}} \quad I_0\, (\mathbf{W}^T\, {}^{t}\dot{\boldsymbol{\omega}}_S) \}^T,\\
&{}^{t}\boldsymbol{\omega}_S \equiv \mathbf{W}\, {}^{t}\dot{\boldsymbol{\theta}}, \quad {}^{t}\dot{\boldsymbol{\omega}}_S = (\dot{\mathbf{W}}\, {}^{t}\dot{\boldsymbol{\theta}} + \mathbf{W}\, {}^{t}\ddot{\boldsymbol{\theta}})
\end{aligned} \tag{11.579}$$

where the angular velocity vector, ${}^{t}\boldsymbol{\omega}$, is the axial vector corresponding to the skew-symmetric angular velocity matrix, $\boldsymbol{\omega}_S$, given by:

$$ \boldsymbol{\Omega}_S \equiv \dot{\mathbf{R}}\, \mathbf{R}^T \tag{11.580}$$

The linearized internal functional, $\Delta G_{Internal}$:

$$\begin{aligned}
\Delta G_{Internal} &\equiv \overbrace{\underbrace{\int_M {}^{n}\breve{\boldsymbol{\varepsilon}}^T\, {}^{n}\breve{\mathbf{D}}\, \Delta^{n}\breve{\boldsymbol{\varepsilon}}\, dA}_{\text{Material stiffness functional}} + \underbrace{\int_M \Delta^{n}\breve{\boldsymbol{\varepsilon}}^T\, {}^{n}\breve{\mathbf{F}}\, dA}_{\text{Geometric stiffness functional}}}^{\text{Linearized internal functional}}\\
&= \overbrace{\underbrace{\mathcal{K}_M}_{\text{Material stiffness functional}} + \underbrace{\mathcal{K}_G}_{\text{Geometric stiffness functional}}}^{\text{Tangent functional}}
\end{aligned} \tag{11.581}$$

Definition: Material stiffness functional, $\mathcal{K}_M$:

$$\boxed{\mathcal{K}_M({}^t\breve{\mathbf{d}},\, {}^t\bar{\breve{\mathbf{d}}}; \Delta^t\breve{\mathbf{d}}) \equiv \int_M {}^n\breve{\bar{\varepsilon}}^T\, {}^n\breve{\mathbf{D}}\ \Delta^n\breve{\varepsilon}\, dA} \tag{11.582}$$

$$\boxed{\mathcal{K}_G({}^t\breve{\mathbf{d}},\, {}^t\bar{\breve{\mathbf{d}}}; \Delta\, {}^t\breve{\mathbf{d}}) \equiv \int_M \Delta^n\breve{\bar{\varepsilon}}^T({}^t\breve{\mathbf{d}},\, {}^t\bar{\breve{\mathbf{d}}}; \Delta\, {}^t\breve{\mathbf{d}})\ {}^n\breve{\mathbf{F}}\, dA} \tag{11.583}$$

the virtual strain–displacement relations in compact matrix form is:

$$\boxed{\breve{\bar{\varepsilon}}(\breve{\mathbf{d}}; \bar{\breve{\mathbf{d}}}) = \mathbf{E}(\breve{\mathbf{d}})\ \mathbf{H}(\bar{\breve{\mathbf{d}}})} \tag{11.584}$$

where

$$\boxed{\mathbf{E}(\breve{\mathbf{d}}) = \begin{bmatrix} \mathbf{R^T} & \mathbf{0} & \mathbf{0} & \mathbf{0} \\ \mathbf{0} & \mathbf{R^T} & \mathbf{0} & \mathbf{0} \\ \mathbf{0} & \mathbf{0} & \mathbf{R^T} & \mathbf{0} \\ \mathbf{0} & \mathbf{0} & \mathbf{0} & \mathbf{R^T} \end{bmatrix} \begin{bmatrix} [\mathbf{k}_1^0] & [\mathbf{a}_1]\mathbf{W} & \mathbf{I} & \mathbf{0} & \mathbf{0} & \mathbf{0} \\ \mathbf{0} & \mathbf{X}_1 + [\mathbf{k}_1^c]\mathbf{W} & \mathbf{0} & \mathbf{W} & \mathbf{0} & \mathbf{0} \\ [\mathbf{k}_2^0] & [\mathbf{a}_2]\mathbf{W} & \mathbf{0} & \mathbf{0} & \mathbf{I} & \mathbf{0} \\ \mathbf{0} & \mathbf{X}_2 + [\mathbf{k}_2^c]\mathbf{W} & \mathbf{0} & \mathbf{0} & \mathbf{0} & \mathbf{W} \end{bmatrix}} \tag{11.585}$$

and $\mathbf{I}$, is the 3×3 identity matrix, $\mathbf{H}(\bar{\breve{\mathbf{d}}}) \equiv \{\bar{\mathbf{d}}\ \ \bar{\boldsymbol{\theta}}\ \ \bar{\mathbf{d}},_1\ \ \bar{\boldsymbol{\theta}},_1\ \ \bar{\mathbf{d}},_2\ \ \bar{\boldsymbol{\theta}},_2\}^T$, and from our knowledge of the relevant derivatives and variations, we have:

$$\boxed{\begin{aligned} &\mathbf{W} \equiv \mathbf{I} + c_1\boldsymbol{\Theta} + c_2\boldsymbol{\Theta}^2 \\ &\mathbf{X}_\alpha \equiv -c_1\boldsymbol{\Theta},_\alpha + c_2\{[\boldsymbol{\theta} \otimes \boldsymbol{\theta},_\alpha] - [\boldsymbol{\theta},_\alpha \otimes \boldsymbol{\theta}] + (\boldsymbol{\theta}\boldsymbol{\theta},_\alpha)\mathbf{I}\} \\ &\qquad +\mathbf{V}\,[\boldsymbol{\theta},_\alpha \otimes \boldsymbol{\theta}]), \quad \alpha = 1, 2 \\ &\mathbf{V} \equiv -c_2\mathbf{I} + c_3\boldsymbol{\Theta} + c_4\boldsymbol{\Theta}^2 \end{aligned}} \tag{11.586}$$

$$\boxed{\begin{aligned} c_1 &\equiv (1 - \cos\theta)/\theta^2, \\ c_2 &\equiv (\theta - \sin\theta)/\theta^3, \\ \overline{c_1} \equiv \delta c_1 \equiv c_3 &\equiv \{(1 - \theta^2 c_2) - 2c_1\}/\theta^2, \\ \overline{c_2} \equiv \delta c_2 \equiv c_4 &\equiv (c_1 - 3c_2)/\theta^2 \end{aligned}} \tag{11.587}$$

the (6×1) incremental generalized displacement state vector functions, $\Delta\breve{\mathbf{d}} \equiv (\Delta\,\mathbf{d}\ \ \ \Delta\boldsymbol{\theta}\,)^T$, and finally, ${}^n\breve{\mathbf{D}}$ is the symmetric hyperelastic constitutive matrix. In the expanded form, equation

(11.583) may be expressed as:

$$\mathcal{K}_G \equiv \overbrace{\int_{\Omega} {}^{n}\Delta\bar{\beta}_{\alpha} \bullet {}^{n}\mathbf{F}^{\alpha}\, dA}^{\text{Translational part}} + \overbrace{\int_{\Omega} {}^{n}\Delta\bar{\chi}_{\alpha} \bullet {}^{n}\mathbf{M}^{\alpha}\, dA}^{\text{Rotational part}} \tag{11.588}$$

where Δ, the incremental operations, involve the covariant derivatives of the virtual strain functions, $\underbrace{\mathbf{H}({}^{t}\bar{\mathbf{d}})}_{18\times 1} \equiv \{\, {}^{t}\bar{\mathbf{d}} \quad {}^{t}\bar{\theta} \quad {}^{t}\bar{\mathbf{d}},_1 \quad {}^{t}\bar{\theta},_1 \quad {}^{t}\bar{\mathbf{d}},_2 \quad {}^{t}\bar{\theta},_2 \,\}^T$, and $\underbrace{\mathbf{H}(\breve{\Delta d})}_{18\times 1} \equiv$ $\{\Delta\,{}^{t}\mathbf{d} \quad \Delta\,{}^{t}\theta \quad \Delta\,{}^{t}\mathbf{d},_1 \quad \Delta\,{}^{t}\theta,_1 \quad \Delta\,{}^{t}\mathbf{d},_2 \quad \Delta\,{}^{t}\theta,_2\}^T$ with $(\bullet),_{\alpha}$, $\alpha = 1, 2$ are derivatives with respect to the arc length parameters, S^{α}, $\alpha = 1, 2$, along the lines of curvature.

- Superscript t reminds us that the relevant items are measured in the undeformed coordinate system, $\mathbf{G}_1, \mathbf{G}_2, \mathbf{N}$; superscript n reminds us that the relevant items are measured in the rotated coordinate system, $\mathbf{g}_1(= \mathbf{R}\,\mathbf{G}_1), \mathbf{g}_2(= \mathbf{R}\,\mathbf{G}_2), \mathbf{g}_3 \equiv \mathbf{n}(= \mathbf{R}\,\mathbf{N})$.

Remark

- The expressions given in equation (11.583) or equation (11.588) involve the first variation (incremental) of the virtual rotational strains or, equivalently, the second variation (Hessian) of the real rotational strains. It is this second variation that renders the derivation of the geometric stiffness matrix quite involved, and thus tedious. However, the subsequent reward in terms of high computational accuracy, ease and efficiency in situations of repetitive computations of the geometric stiffness, as we have in a nonlinear numerical analysis, makes it extremely worthwhile.

11.18.2 Material Stiffness Functional Revisited: $\mathcal{K}_M$

As in the previous virtual expression, we group together the incremental state variables and their derivatives with respect to the arc length parameters, (S^1, S^2), into an augmented (18×1) incremental vector function (we can replace $\breve{\mathbf{d}}$ by $\breve{\Delta\mathbf{d}}$ in the definition of $\underbrace{\mathbf{H}(\breve{\mathbf{d}})}_{18\times 1} \equiv$ $\{\bar{\mathbf{d}} \quad \bar{\theta} \quad \bar{\mathbf{d}},_1 \quad \bar{\theta},_1 \quad \bar{\mathbf{d}},_2 \quad \bar{\theta},_2\}^T$) to introduce:

Definition

Incremental generalized displacement function: $\underbrace{\mathbf{H}(\breve{\Delta\mathbf{d}})}_{18\times 1} \equiv \{\Delta\mathbf{d} \quad \Delta\theta \quad \Delta\mathbf{d},_1 \quad \Delta\theta,_1$ $\Delta\mathbf{d},_2 \quad \Delta\theta,_2\}^T$. Now, replacing the variational variables in equation (11.584) with incremental variables, with the above definition in mind, we get a similar expression as follows.

Definition

Incremental strain–displacement relation:

$$\boxed{\Delta\breve{\varepsilon}(\breve{\mathbf{d}};\Delta\breve{\mathbf{d}}) = \mathbf{E}(\breve{\mathbf{d}})\ \mathbf{H}(\Delta\breve{\mathbf{d}})} \tag{11.589}$$

Note that the incremental strain, $\Delta\breve{\varepsilon}(\breve{\mathbf{d}};\Delta\breve{\mathbf{d}})$, is clearly linear in the incremental generalized displacement, $\Delta\breve{\mathbf{d}}$, by being related to the strain–displacement matrix that is independent of $\Delta\breve{\mathbf{d}}$. By applying the virtual strain–displacement relation (11.584) and incremental strain–displacement relation of equation (11.589), we are now ready to transform equation (11.582) as:

$$\boxed{\mathcal{K}_M(\breve{\mathbf{d}},\breve{\bar{\mathbf{d}}};\Delta\breve{\mathbf{d}}) \equiv \int_M \mathbf{H}(\breve{\bar{\mathbf{d}}})^T\ (\mathbf{E}(\breve{\mathbf{d}})^T\ {}^n\hat{\mathbf{D}}\ \mathbf{E}(\breve{\mathbf{d}}))\ \mathbf{H}(\Delta\breve{\mathbf{d}})\ dA} \tag{11.590}$$

Based on equation (11.590), we can redefine and present the material stiffness functional as follows.

Definition: Material stiffness functional, $\mathcal{K}_M$, is:

$$\boxed{\mathcal{K}_M(\breve{\mathbf{d}},\breve{\bar{\mathbf{d}}};\Delta\breve{\mathbf{d}}) \equiv \int_M \mathbf{H}(\breve{\bar{\mathbf{d}}})^T\ \mathbf{K}_M\ \ \mathbf{H}(\Delta\breve{\mathbf{d}})\ dA} \tag{11.591}$$

where the matrix $\mathbf{K}_M$ is defined as:

$$\boxed{\mathbf{K}_M \equiv \mathbf{E}(\breve{\mathbf{d}})^T\ {}^n\mathbf{D}\ \ \mathbf{E}(\breve{\mathbf{d}})} \tag{11.592}$$

Remark

- Clearly, from equation (11.592), the matrix, $\mathbf{K}_M$, is symmetric in state or configuration space variables for a conservative system.

11.18.3 Geometric Stiffness Functional Revisited: $\mathcal{K}_G$

In what follows, in order to avoid undue distraction, first we present the main results, and then we refer the interested reader to the sections on detailed derivations.

- Translational virtual work: linearization and symmetry

  We get as the final result:

$$\boxed{\begin{aligned} {}^n\Delta\beta_\alpha \bullet {}^n\mathbf{F}^\alpha = {} & \bar{\mathbf{d}},_\alpha \bullet ([{}^t\mathbf{F}^\alpha]^{\mathbf{T}}\,\mathbf{W})\Delta\theta + \Delta\mathbf{d},_\alpha \bullet ([{}^t\mathbf{F}^\alpha]^{\mathbf{T}}\,\mathbf{W})\bar{\theta} \\ & + \bar{\mathbf{d}} \bullet ([\mathbf{k}_\alpha^0]\,[{}^t\mathbf{F}^\alpha]\,\mathbf{W})\Delta\theta + \Delta\mathbf{d} \bullet ([\mathbf{k}_\alpha^0]\,[{}^t\mathbf{F}^\alpha]\,\mathbf{W})\bar{\theta} \\ & + \bar{\theta} \bullet \mathbf{Z}_\beta^{Sym}\,\Delta\theta \end{aligned}} \tag{11.593}$$

with

$$\boxed{\begin{aligned}\mathbf{Z}_{\bar{\beta}}^{Sym} = \tfrac{1}{2}\{&\mathbf{W}^{\mathbf{T}}([{}^{t}\mathbf{F}^{\alpha}\otimes\mathbf{a}_{\alpha}]+[\mathbf{a}_{\alpha}\otimes{}^{t}\mathbf{F}^{\alpha}])\,\mathbf{W}+([\mathbf{h}^{f}\otimes\theta]+[\theta\otimes\mathbf{h}^{f}])\}\\ &-({}^{t}\mathbf{F}^{\alpha}\bullet\mathbf{a}_{\alpha})\mathbf{W}^{\mathbf{T}}\,\mathbf{W}+c_2(\theta\ \ \mathbf{b}^{f})\mathbf{I}\end{aligned}}$$

- *Rotational Virtual Work: Linearization and Symmetry*
  We get as the final result:

$$\boxed{\begin{aligned}{}^{n}\Delta\bar{\chi}\bullet{}^{n}\mathbf{M}=\Delta\theta\bullet\{&\mathbf{W}^{\mathbf{T}}[{}^{t}\mathbf{M}^{\alpha}]\,\mathbf{W}+\mathbf{Y}({}^{t}\mathbf{M}^{\alpha})\}\bar{\theta},_{\alpha}\\ +\bar{\theta}\bullet\{&\mathbf{W}^{\mathbf{T}}[{}^{t}\mathbf{M}^{\alpha}]\,\mathbf{W}+\mathbf{Y}({}^{t}\mathbf{M}^{\alpha})\}\Delta\theta,_{\alpha}\\ +\Delta\theta\bullet&\mathbf{Z}_{\bar{\chi}}^{Sym}\,\bar{\theta}\end{aligned}} \tag{11.594}$$

with

$$\boxed{\begin{aligned}\mathbf{Z}_{\bar{\chi}}^{Symm}\equiv\tfrac{1}{2}\{&\mathbf{W}^{\mathbf{T}}([{}^{t}\mathbf{M}^{\alpha}\otimes\mathbf{k}_{\alpha}^{c}]+[\mathbf{k}_{\alpha}^{c}\otimes{}^{t}\mathbf{M}^{\alpha}])\,\mathbf{W}+([\mathbf{h}^{m}\otimes\theta]+[\theta\otimes\mathbf{h}^{m}])\}\\ &-({}^{t}\mathbf{M}^{\alpha}\bullet\mathbf{k}_{\alpha}^{c})\mathbf{W}^{\mathbf{T}}\,\mathbf{W}+c_2(\theta\bullet\mathbf{b}^{m})\mathbf{I}+\mathbf{W}^{\mathbf{T}}[{}^{t}\mathbf{M}^{\alpha}]\mathbf{X}_{\alpha}+\mathbf{X}_{\alpha}^{\mathbf{T}}\,[{}^{t}\mathbf{M}^{\alpha}]^{\mathbf{T}}\mathbf{W}\end{aligned}}$$

and the $\mathbf{Y}(\bullet)$ function is defined later.

Based on equation (11.593) and equation (11.594), we redefine and present here the geometric stiffness functional in its most uncluttered and explicit form.

Definition: The geometric stiffness functional, $\mathcal{K}_G$, is:

$$\boxed{\mathcal{K}_G(\breve{\mathbf{d}},\bar{\mathbf{d}};\Delta\breve{\mathbf{d}})\equiv\int_{\Omega}\left\{\underbrace{\mathbf{H}(\breve{\mathbf{d}})^{T}}_{1\times18}\ \underbrace{\mathbf{G}(\breve{\mathbf{d}},{}^{t}\breve{\mathbf{F}})}_{18\times18}\ \underbrace{\mathbf{H}(\Delta\breve{\mathbf{d}})}_{18\times1}\right\}dA} \tag{11.595}$$

where the matrix $\mathbf{G}(\breve{\mathbf{d}},{}^{t}\breve{\mathbf{F}})$, is given as:

$$\boxed{\mathbf{G}(\breve{\mathbf{d}},{}^{t}\breve{\mathbf{F}})\equiv\begin{bmatrix}\mathbf{0} & \breve{\mathbf{G}}_{12}^{1}+\breve{\mathbf{G}}_{12}^{2} & \mathbf{0} & \mathbf{0} & \mathbf{0} & \mathbf{0}\\ (\breve{\mathbf{G}}_{12}^{1}+\breve{\mathbf{G}}_{12}^{2})^{T} & \breve{\mathbf{G}}_{22}^{1}+\breve{\mathbf{G}}_{22}^{2} & \breve{\mathbf{G}}_{23}^{1} & \breve{\mathbf{G}}_{24}^{1} & \breve{\mathbf{G}}_{23}^{2} & \breve{\mathbf{G}}_{24}^{2}\\ \mathbf{0} & (\breve{\mathbf{G}}_{23}^{1})^{T} & \mathbf{0} & \mathbf{0} & \mathbf{0} & \mathbf{0}\\ \mathbf{0} & (\breve{\mathbf{G}}_{24}^{1})^{T} & \mathbf{0} & \mathbf{0} & \mathbf{0} & \mathbf{0}\\ \mathbf{0} & (\breve{\mathbf{G}}_{23}^{2})^{T} & \mathbf{0} & \mathbf{0} & \mathbf{0} & \mathbf{0}\\ \mathbf{0} & (\breve{\mathbf{G}}_{24}^{2})^{T} & \mathbf{0} & \mathbf{0} & \mathbf{0} & \mathbf{0}\end{bmatrix}} \tag{11.596}$$

where

$$\begin{aligned}
\mathbf{G}_{12}^{\alpha} &\equiv [\mathbf{k}_{\alpha}^{0}][{}^{t}\mathbf{F}^{\alpha}]\,\mathbf{W}, \quad \alpha = 1,2 \text{ and no sum}\\
\mathbf{G}_{22}^{\alpha} &\equiv \tfrac{1}{2}\mathbf{W}^{\mathbf{T}}\{[{}^{t}\mathbf{F}^{\alpha}\otimes\mathbf{a}_{\alpha}] + [\mathbf{a}_{\alpha}\otimes{}^{t}\mathbf{F}^{\alpha}] - 2({}^{t}\mathbf{F}^{\alpha}\bullet\,\mathbf{a}_{\alpha})\}\mathbf{W} + c_2(\boldsymbol{\theta}\bullet\,\mathbf{b}_{\alpha}^{f})\mathbf{I}\\
&\quad + \tfrac{1}{2}([\mathbf{h}_{\alpha}^{b}\otimes\boldsymbol{\theta}] + [\boldsymbol{\theta}\otimes\mathbf{h}_{\alpha}^{b}])\\
&\quad + \tfrac{1}{2}\mathbf{W}^{\mathbf{T}}\{[{}^{t}\mathbf{M}^{\alpha}\otimes\mathbf{k}_{\alpha}^{c}] + [\mathbf{k}_{\alpha}^{c}\otimes{}^{t}\mathbf{M}^{\alpha}] - 2({}^{t}\mathbf{M}^{\alpha}\bullet\,\mathbf{k}_{\alpha}^{c})\}\mathbf{W} + c_2(\boldsymbol{\theta}\bullet\,\mathbf{b}_{\alpha}^{m})\mathbf{I}\\
&\quad + \tfrac{1}{2}([\mathbf{h}_{\alpha}^{m}\otimes\boldsymbol{\theta}] + [\boldsymbol{\theta}\otimes\mathbf{h}_{\alpha}^{m}]) + \mathbf{W}^{\mathbf{T}}[{}^{t}\mathbf{M}^{\alpha}]\,\mathbf{X}_{\alpha} + \mathbf{X}_{\alpha}\,[{}^{t}\mathbf{M}^{\alpha}]^{\mathbf{T}}\,\mathbf{W}\\
\mathbf{G}_{23}^{\alpha} &\equiv \mathbf{W}^{\mathbf{T}}\,[{}^{t}\mathbf{F}^{\alpha}]\\
\mathbf{G}_{24}^{\alpha} &\equiv \mathbf{W}^{\mathbf{T}}\,[{}^{t}\mathbf{M}^{\alpha}]\,\mathbf{W} + \mathbf{Y}({}^{t}\mathbf{M}^{\alpha})
\end{aligned} \tag{11.597}$$

with

$$\begin{aligned}
c_1 &\equiv (1-\cos\theta)/\theta^2,\\
c_2 &\equiv (\theta-\sin\theta)/\theta^3,\\
\bar{c}_1 \equiv \delta c_1 \equiv c_3 &\equiv \{(1-\theta^2 c_2) - 2c_1\}/\theta^2,\\
\bar{c}_2 \equiv \delta c_2 \equiv c_4 &\equiv (c_1 - 3c_2)/\theta^2
\end{aligned} \tag{11.598}$$

$$\begin{aligned}
&\mathbf{a} \equiv \mathbf{e} + [\mathbf{k}^0]\mathbf{d} = \{(\hat{\mathbf{1}} + \mathbf{d},_s) + [\mathbf{k}^0]\mathbf{d}\}\\
&\mathbf{k}^t \equiv \mathbf{W}\,\boldsymbol{\theta},_s\\
&\mathbf{k}^c \equiv \mathbf{k}^0 + \mathbf{k}^t\\
&\mathbf{b}^f \equiv \mathbf{F}^t \times \mathbf{a}\\
&\mathbf{b}^m \equiv \mathbf{M}^t \times \mathbf{k}^c
\end{aligned} \tag{11.599}$$

$$\begin{aligned}
\mathbf{W} &\equiv \mathbf{I} + c_1\boldsymbol{\Theta} + c_2\boldsymbol{\Theta}^2\\
\mathbf{X}_{\alpha} &\equiv -c_1\boldsymbol{\Theta},_{\alpha} + c_2\{[\boldsymbol{\theta}\otimes\boldsymbol{\theta},_{\alpha}] - [\boldsymbol{\theta},_{\alpha}\otimes\,\boldsymbol{\theta}] + (\boldsymbol{\theta}\,\bullet\,\boldsymbol{\theta},_{\alpha})\mathbf{I}\}\\
&\qquad + \mathbf{V}\,[\boldsymbol{\theta},_{\alpha}\otimes\,\boldsymbol{\theta}])\\
\mathbf{V} &\equiv -c_2\mathbf{I} + c_3\boldsymbol{\Theta} + c_4\boldsymbol{\Theta}^2\\
\mathbf{Y}(\mathbf{x}) &\equiv -c_1[\mathbf{x}] + c_2([\mathbf{x}\otimes\boldsymbol{\theta}] - [\boldsymbol{\theta}\otimes\mathbf{x}]) + c_2(\boldsymbol{\theta}\mathbf{x})\mathbf{I}\\
&\qquad + [\boldsymbol{\theta}\otimes\mathbf{V}^{\mathbf{T}}\,\mathbf{x}], \quad \forall\mathbf{x}\in R^3\\
\mathbf{h}_{\alpha}^{f} &\equiv \mathbf{V}^{\mathbf{T}}\,\mathbf{b}_{\alpha}^{f}\\
\mathbf{h}_{\alpha}^{m} &\equiv \mathbf{V}^{\mathbf{T}}\,\mathbf{b}_{\alpha}^{m}
\end{aligned} \tag{11.600}$$

Exercise: Drawing upon the analyses referred to earlier for the translational and rotational geometric stiffness functional, derive the expression for **G** above. (Hint: look at Section 10.18 on the symmetry of the beam geometric stiffness matrix.)

Remark

- Clearly, from equation (11.596), the matrix, **G**, is symmetric in state or configuration space variables for a conservative system.

11.18.3.1 Implementational Note

> Note that, as will be explained later, the computational finite element implementation for the internal work of a shell problem can be achieved essentially by using the corresponding beam formulations along each of the lines of curvature at numerical Gauss quadrature points in a typical Newton-type iteration method of solution under a time-marching formulation.

The material and the geometric stiffness functional parts in equations (11.591) and (11.595), respectively, need to be further modified in the subsequent finite element formulation as the direct sums of the element stiffness matrices for computational purposes. In so doing, we will be able to introduce the computational definition for the shell material and geometric stiffness matrices. Thus, for the final and computationally optimum form of the linearized virtual work, we will focus next on the c-type finite element formulation.

11.18.4 c-type Finite Element Formulation

Under the semi-discrete formulation, and following the standard definition of a finite element discretization, a shell initial surface, Ω, is broken into a disjointed finite number of elemental surfaces such that:

$$\Omega = \bigcup_{e=1}^{e=\text{no. of elements}} \text{Finite element}^e = \bigcup_{e=1}^{e=\text{no. of elements}} \Omega^e \tag{11.601}$$

Then, the shell material and geometric stiffness matrices can be expressed as an assembly of the corresponding elemental matrices; a similar idea applies to the formation of the global unbalanced and the applied forces. As will be seen in our c-type finite element implementation phase, we will use what is known as the direct global stiffness and load assembly of the corresponding elemental entities.

Thus, for now, we can express each of the following properties as a sum over all the corresponding entities belonging to the elements:

- Shell symmetric mass functional:

$$\Delta G_M({}^t\breve{\mathbf{d}};\, {}^t\bar{\breve{\mathbf{d}}};\, \Delta\, {}^t\breve{\mathbf{d}}) = \sum_{elements} \int_{\Omega^e} \underbrace{{}^t\bar{\breve{\mathbf{d}}}^T}_{1\times 6} \underbrace{\begin{bmatrix} M_0\,\mathbf{I} & \mathbf{0} \\ \mathbf{0} & I_0\,(\mathbf{W}^T\,\mathbf{W}) \end{bmatrix}}_{6\times 6} \underbrace{\Delta\,{}^t\ddot{\breve{\mathbf{d}}}}_{6\times 1}\, dA \tag{11.602}$$

- Shell gyroscopic (damping) functional:

$$\Delta G_G({}^t\breve{\mathbf{d}};\, {}^t\bar{\breve{\mathbf{d}}};\, \Delta\, {}^t\breve{\mathbf{d}}) = \sum_{elements} \int_{\Omega^e} \underbrace{{}^t\bar{\breve{\mathbf{d}}}^T}_{1\times 6} \underbrace{\begin{bmatrix} \mathbf{0} & \mathbf{0} \\ \mathbf{0} & I_0\,\{\mathbf{W}^T(2\,\dot{\mathbf{W}} - \mathbf{W}\,\dot{\boldsymbol{\Theta}}\,\mathbf{W})\} \end{bmatrix}}_{6\times 6} \underbrace{\Delta\,{}^t\dot{\breve{\mathbf{d}}}}_{6\times 1}\, dA \tag{11.603}$$

- Shell centrifugal (stiffness) functional:

$$\Delta G_C({}^t\breve{\mathbf{d}};\, {}^t\bar{\breve{\mathbf{d}}};\, \Delta\, {}^t\breve{\mathbf{d}}) = \sum_{elements} \int_{\Omega^e} \underbrace{{}^t\bar{\breve{\mathbf{d}}}^T}_{1\times 6} \underbrace{\begin{bmatrix} \mathbf{0} & \mathbf{0} \\ \mathbf{0} & I_0 \left\{ \begin{array}{l} \mathbf{U}({}^t\,\dot{\boldsymbol{\omega}}) \\ +\mathbf{W}^T\{\ddot{\mathbf{W}} - (\mathbf{W}\dot{\boldsymbol{\Theta}}\dot{\mathbf{W}} + \dot{\mathbf{W}}\dot{\boldsymbol{\Theta}}\mathbf{W}) - \mathbf{W}\ddot{\boldsymbol{\Theta}}\mathbf{W}\} \end{array} \right\} \end{bmatrix}}_{6\times 6} \underbrace{\Delta\,{}^t\ddot{\breve{\mathbf{d}}}}_{6\times 1}\, dA$$

$$\mathbf{U}({}^t\,\dot{\boldsymbol{\omega}}_S) \equiv \left\{ \begin{array}{l} c_1\,[{}^t\,\dot{\boldsymbol{\omega}}_S] + (-c_2\,\mathbf{I} - c_3\,\boldsymbol{\Theta} + c_4\boldsymbol{\Theta}^2)({}^t\,\dot{\boldsymbol{\omega}}_S \bullet \boldsymbol{\theta}) \\ +c_2((\boldsymbol{\theta} \otimes {}^t\,\dot{\boldsymbol{\omega}}_S) - ({}^t\,\dot{\boldsymbol{\omega}}_S \otimes \boldsymbol{\theta}) + (\boldsymbol{\theta} \bullet {}^t\,\dot{\boldsymbol{\omega}}_S))\mathbf{I} \end{array} \right\},$$

$${}^t\,\dot{\boldsymbol{\omega}}_S = (\dot{\mathbf{W}}\,{}^t\,\dot{\boldsymbol{\theta}} + \mathbf{W}\,{}^t\,\ddot{\boldsymbol{\theta}}) \tag{11.604}$$

- Shell material stiffness functional:

$$\mathcal{K}_M(\breve{\mathbf{d}}, \bar{\breve{\mathbf{d}}};\, \Delta\breve{\mathbf{d}}) = \sum_{elements} \int_{\Omega^e} \underbrace{\mathbf{H}(\bar{\breve{\mathbf{d}}})^T}_{1\times 18}\ \underbrace{\mathbf{K}_M(\breve{\mathbf{d}})}_{18\times 18}\ \underbrace{\mathbf{H}(\Delta\breve{\mathbf{d}})}_{18\times 1}\, dA \tag{11.605}$$

- Shell geometric stiffness functional:

$$\mathcal{K}_G(\breve{\mathbf{d}}, \bar{\breve{\mathbf{d}}};\, \Delta\breve{\mathbf{d}}) = \sum_{elements} \int_{\Omega^e} \left\{ \underbrace{\mathbf{H}(\bar{\breve{\mathbf{d}}})^T}_{1\times 18}\ \underbrace{\mathbf{G}}_{18\times 18}\ \underbrace{\mathbf{H}(\Delta\breve{\mathbf{d}})}_{18\times 1} \right\} dA \tag{11.606}$$

- Shell unbalanced residual force functional:

$$
\begin{aligned}
&R({}^t\breve{\mathbf{d}}(t);\, {}^t\bar{\breve{\mathbf{d}}}) = -G({}^t\breve{\mathbf{d}}(t);\, {}^t\bar{\breve{\mathbf{d}}})\\
&G({}^t\breve{\mathbf{d}}(t);\, {}^t\bar{\breve{\mathbf{d}}}) = G_{Internal} + G_{Inertial} - G_{External}\\
&= \sum_{elements} \left(\begin{array}{l} \displaystyle\int_{\Omega^e} {}^n\bar{\breve{\boldsymbol{\varepsilon}}}^T\, {}^n\breve{\mathbf{D}}\, {}^n\breve{\boldsymbol{\varepsilon}}\; dA \\[2ex] \displaystyle + \int_{\Omega^e} {}^t\bar{\breve{\mathbf{d}}}^T\, {}^t\breve{\mathbf{F}}_{Iner}\; dA \\[2ex] \displaystyle - \left\{ \int_{\Omega^e} \bar{\breve{\mathbf{d}}}^{\mathbf{T}}\, \breve{\hat{\mathbf{F}}}(S^1, S^2)\; dA + \int_{\Gamma^e} \bar{\breve{\mathbf{d}}}^{\mathbf{T}}\, \breve{\hat{\mathbf{F}}}_\nu(S)\; dS \right\} \end{array} \right)
\end{aligned}
\tag{11.607}
$$

Based on the assumption that the assembly (i.e. the summation) of the finite elements will be performed by the direct global assembly procedure where the properties of each element enter into the summing operation independent of the all other element data, we can now restrict our discussion to a typical finite element for the shell. Finally, we can access the available library of c-type shell elements of both curved quadrilateral and triangular shape. However, in what follows, we will restrict our discussion to only tensor product quadrilateral elements; similar arguments and ideas apply formally to the triangular elements too. The root element corresponding to a typical finite element is defined over the geometric 2D root parameter plane defined by the parameters $(\hat{\xi}^1, \hat{\xi}^2) \in [0,1] \cup [0,1]$. Next, the parameters, (ξ^1, ξ^2), along the lines of curvature is related to the arc length parameter, (S^1, S^2), of a shell by: $S^1(\xi^1, \xi^2) = \int_0^{\xi^1} \|\, {}^e\mathbf{C}_0(\eta, \xi^2)_{,\eta} \| d\eta$ and $S^2(\xi^1, \xi^2) = \int_0^{\xi^2} \|\, {}^e\mathbf{C}_0(\xi^1, \eta)_{,\eta} \| d\eta$, where ${}^e\mathbf{C}_0(\xi^1, \xi^2)$, $\xi^1 \in [0,1]$, $\xi^2 \in [0,1]$ defines the reference surface position vector at any point (ξ^1, ξ^2) of the eth finite element, that is, the Finite element^e. Let us briefly recall the basic steps of a c-type finite element generation for future use. The coordinate geometry of a shell reference surface is discretized in the mesh generation phase whereby the root element is identified and fixed for an element; the position vector, $\mathbf{C}_0(\hat{\xi}^1, \hat{\xi}^2, t)$, of any point, $(\hat{\xi}^1, \hat{\xi}^2)$, on the element at the initial time $t = 0 \in \mathbb{R}_+$ is given by:

$$
\begin{aligned}
&\mathbf{C}_0(\hat{\xi}^1, \hat{\xi}^2, 0) = \left\{ \begin{array}{l} C_0^1 = X(\hat{\xi}^1, \hat{\xi}^2) \\ C_0^2 = Y(\hat{\xi}^1, \hat{\xi}^2) \\ C_0^3 = Z(\hat{\xi}^1, \hat{\xi}^2) \end{array} \right\} = {}^c\mathbf{T}(\hat{\xi}^1, \hat{\xi}^2)\; {}^c\mathbf{q},\\
&C_0^k(\hat{\xi}^1, \hat{\xi}^2, 0) = \sum_{i=0}^{n_c} \sum_{j=0}^{m_c} B_i^{n_c}(\hat{\xi}^1)\, B_j^{m_c}(\hat{\xi}^2)\; {}^c q_{ij}^k, \quad k = 1, \cdots, 3
\end{aligned}
\tag{11.608}
$$

where ${}^c\mathbf{T}(\hat{\xi}^1, \hat{\xi}^2)$ is the root transformation matrix, $B_i^n(\hat{\xi}^1)$, $B_j^m(\hat{\xi}^2)$ are Bernstein–Bezier polynomials of degree n_c and m_c respectively, and ${}^c\mathbf{q}$, ${}^c q_{ij}^k$ are the generalized coordinate control vector and the component control net for the element, respectively. The discretization of the generalized displacement field, $\breve{\mathbf{d}}(t) \equiv (\mathbf{d}(t), \boldsymbol{\theta}(t))$, $\quad t \in \mathbb{R}_+$, for the finite element, is then performed over the geometric root element except that the degree of the displacement distribution polynomials (i.e. the local basis or shape functions) will be at least equal to that of the geometric coordinate distribution polynomials; in other words, the element is either subparametric or isoparametric.

11.18.5 c-type Shell Elements and State Control Vectors

Let us first recognize that shell geometry and the constituting finite elements are naturally discretized in terms of the parameters $(\hat{\xi}^1, \hat{\xi}^2)$ as tensor product objects for quadrilateral elements. However, for analytical convenience, the linearized virtual work expression as given by equation (11.566) has been derived in terms of the arc length parameters, (S^1, S^2), along the corresponding lines of curvature. Thus, for computational purposes, we must express all our equations back in the $(\hat{\xi}^1, \hat{\xi}^2)$ parameters at the Gauss quadrature points of the numerical integration scheme. This is what is done in the ensuing finite element formulation. Let us, then, introduce a typical c-type shell element displacement interpolation for the semi-discrete finite element formulation as:

$$\boxed{\begin{aligned} &\breve{\mathbf{d}}(\hat{\xi}^1, \hat{\xi}^2, t) = {}^d\breve{\mathbf{T}}(\hat{\xi}^1, \hat{\xi}^2)\, {}^d\breve{\mathbf{q}}(t), \\ &\qquad \breve{\mathbf{d}} = \{\, \breve{d}_1 \ \ \breve{d}_2 \ \ \cdots \ \ \breve{d}_6 \,\}^T \\ &\breve{d}_k(\hat{\xi}^1, \hat{\xi}^2, t) = \sum_{i=0}^{n} \sum_{j=0}^{m} B_i^n(\hat{\xi}^1)\, B_j^m(\hat{\xi}^2)\, {}^d\breve{q}_{ij}^{\,k}(t), \quad k = 1, \ldots, 6 \end{aligned}} \tag{11.609}$$

where ${}^d\breve{\mathbf{T}}(\hat{\xi}^1, \hat{\xi}^2)$ is the root transformation matrix, $B_i^n(\hat{\xi}^1)$, $B_j^m(\hat{\xi}^2)$ are Bernstein–Bezier polynomials of degree n and m respectively, and ${}^d\breve{\mathbf{q}}(t)$, ${}^d\breve{q}_{ij}^{\,k}(t)$ are the generalized time-dependent displacement control vector and the component control net for the element, respectively. Similar distributions can be produced for the incremental generalized displacement, $\Delta\breve{d}$; for the variations in generalized displacements, $\bar{\breve{\mathbf{d}}}$, the distribution is time independent and given as:

$$\boxed{\begin{aligned} &\bar{\breve{\mathbf{d}}}(\hat{\xi}^1, \hat{\xi}^2, t) = {}^d\breve{\mathbf{T}}(\hat{\xi}^1, \hat{\xi}^2)\, {}^d\bar{\breve{\mathbf{q}}}, \\ &\bar{\breve{d}}_k(\hat{\xi}^1, \hat{\xi}^2, t) = \sum_{i=0}^{n} \sum_{j=0}^{m} B_i^n(\hat{\xi}^1)\, B_j^m(\hat{\xi}^2)\, {}^d\bar{\breve{q}}_{ij}^{\,k}, \quad k = 1, \ldots, 6 \end{aligned}} \tag{11.610}$$

where we have introduced the following.

Definitions

- ${}^d\bar{\breve{\mathbf{q}}}$ be the $(6N^2 \times 1)$ generalized time independent virtual displacement control vector, otherwise, similar in the ordering of the components to ${}^d\breve{\mathbf{q}}$.

Let us define N, M as the orders of the polynomials, respectively, given by: $N \equiv n + 1$ and $M \equiv m + 1$. As will be seen later, in actual implementation, we will enforce: $n = m$ for each element. We may see that the tensor product of these polynomials serves as the local basis functions for our c-type shell elements. Thus, a typical n-degree c-type shell element is given by:

$$\boxed{\begin{aligned}
&\underbrace{\breve{\mathbf{d}}(\hat{\xi}^1,\hat{\xi}^2,t)}_{6\times 1} = \underbrace{{}^d\breve{\mathbf{T}}(\hat{\xi}^1,\hat{\xi}^2)}_{6\times 6N^2}\ \underbrace{{}^d\breve{\mathbf{q}}(t)}_{6N^2\times 1},\\
&\underbrace{\dot{\breve{\mathbf{d}}}(\hat{\xi}^1,\hat{\xi}^2,t)}_{6\times 1} = \underbrace{{}^d\breve{\mathbf{T}}(\hat{\xi}^1,\hat{\xi}^2)}_{6\times 6N^2}\ \underbrace{{}^d\dot{\breve{\mathbf{q}}}(t)}_{6N^2\times 1},\\
&\underbrace{\ddot{\breve{\mathbf{d}}}(\hat{\xi}^1,\hat{\xi}^2,t)}_{6\times 1} = \underbrace{{}^d\breve{\mathbf{T}}(\hat{\xi}^1,\hat{\xi}^2)}_{6\times 6N^2}\ \underbrace{{}^d\ddot{\breve{\mathbf{q}}}(t)}_{6N^2\times 1},
\end{aligned}} \tag{11.611}$$

where we have introduced the following.

Definitions

$$\underbrace{{}^d\breve{\mathbf{q}}(t)}_{6N^2\times 1} = \Big\{ \underbrace{q^u_{00}\cdots q^u_{n0}\, q^u_{01}\cdots q^u_{n1}\, q^u_{0n}\cdots q^u_{nn}}_{1\times N^2}\ \underbrace{q^v_{00}\cdots q^v_{n0}\, q^v_{01}\cdots q^v_{n1}\, q^v_{0n}\cdots q^v_{nn}}_{1\times NM}$$

- $$\underbrace{q^w_{00}\cdots q^w_{n0}\, q^w_{01}\cdots q^w_{n1}\, q^w_{0n}\cdots q^w_{nn}}_{1\times N^2}\ \underbrace{q^{\theta_1}_{00}\cdots q^{\theta_1}_{n0}\, q^{\theta_1}_{01}\cdots q^{\theta_1}_{n1}\, q^{\theta_1}_{0n}\cdots q^{\theta_1}_{nn}}_{1\times N^2}$$ is the $(6N^2 \times 1)$
$$\underbrace{q^{\theta_2}_{00}\cdots q^{\theta_2}_{n0}\, q^{\theta_2}_{01}\cdots q^{\theta_2}_{n1}\, q^{\theta_2}_{0n}\cdots q^{\theta_2}_{nn}}_{1\times N^2}\ \underbrace{q^{\theta_3}_{00}\cdots q^{\theta_3}_{n0}\, q^{\theta_3}_{01}\cdots q^{\theta_3}_{n1}\, q^{\theta_3}_{0n}\cdots q^{\theta_3}_{nn}}_{1\times N^2}\Big\}^T$$
generalized time-dependent displacement control vector where all the components listed are time-dependent; a similar definition applies for ${}^d\dot{\breve{\mathbf{q}}}(t)$, the $(6N^2 \times 1)$ generalized time-dependent velocity control vector, and for ${}^d\ddot{\breve{\mathbf{q}}}(t)$, the $(6N^2 \times 1)$ generalized time-dependent acceleration control vector with all the coefficients ${}^dq_{ij}$ replaced by ${}^d\dot{q}_{ij}$ and ${}^d\ddot{q}_{ij}$, respectively; of course, other possibilities exist for sequencing the nodal variables and the degrees of freedoms.
- ${}^d\breve{\mathbf{T}}(\hat{\xi}^1,\hat{\xi}^2)$ is the $(6\times 6N^2)$ interpolation transformation matrix containing the Bernstein–Bezier polynomials, that is,

$${}^d\breve{\mathbf{T}}(\hat{\xi}^1,\hat{\xi}^2) =$$

$$\underbrace{\begin{bmatrix}
B^n_0B^n_0 & B^n_1B^n_0\cdots B^n_nB^n_{n-1} & B^n_nB^n_n & 0 & 0 & \cdots & \cdots & \cdots & 0\\
0 & 0 & 0 & B^n_0B^n_0 & B^n_1B^n_0\cdots & B^n_nB^n_n & 0 & \cdots & 0\\
 & & \cdots & \cdots & & & & & \\
 & & & \cdots & \cdots & & & & \\
0 & 0 & \cdots & \cdots & 0 & B^n_0B^n_0 & B^n_1B^n_0\cdots & B^n_nB^n_n &
\end{bmatrix}}_{6x6N^2}$$

(11.612)

In other words, each row of ${}^{d}\breve{\mathbf{T}}(\hat{\xi}^1, \hat{\xi}^2)$ is identical except that the non-zero contents are shifted by $\left(1 \times 6N^2\right)$ zero entries with increasing row numbers. Based on these definitions, we also have:

$$\boxed{\begin{aligned} \Delta\breve{\mathbf{d}}(t) &= {}^{d}\breve{\mathbf{T}}(\hat{\xi}^1, \hat{\xi}^2)\, \Delta\, {}^{d}\breve{\mathbf{q}}(t), \\ \Delta\, \dot{\breve{\mathbf{d}}}(t) &= {}^{d}\breve{\mathbf{T}}(\hat{\xi}^1, \hat{\xi}^2)\, \Delta\, {}^{d}\dot{\breve{\mathbf{q}}}(t). \\ \Delta\, \ddot{\breve{\mathbf{d}}}(t) &= {}^{d}\breve{\mathbf{T}}(\hat{\xi}^1, \hat{\xi}^2)\, \Delta\, {}^{d}\ddot{\breve{\mathbf{q}}}(t), \end{aligned}} \tag{11.613}$$

where we have similarly introduced the following.

Definitions

- $\Delta {}^{d}\breve{\mathbf{q}}(t)$, $\Delta {}^{d}\dot{\breve{\mathbf{q}}}(t)$ and $\Delta {}^{d}\ddot{\breve{\mathbf{q}}}(t)$, are the $\left(6N^2 \times 1\right)$ time-dependent incremental displacement, velocity and acceleration control vectors, with the components ordered in the same sequence as the components of ${}^{d}\breve{\mathbf{q}}(t)$, ${}^{d}\dot{\breve{\mathbf{q}}}(t)$ and ${}^{d}\ddot{\breve{\mathbf{q}}}(t)$, respectively.

Note that equations (11.611) and (11.612) imply that we have chosen to interpolate both the displacement components and the rotation components with the same degree control net.

Now, let us rewrite $\mathbf{H}({}^{t}\mathbf{d}) \equiv \{\, {}^{t}\mathbf{d} \quad {}^{t}\boldsymbol{\theta} \quad {}^{t}\mathbf{d},_1 \quad {}^{t}\boldsymbol{\theta},_1 \quad {}^{t}\mathbf{d},_2 \quad {}^{t}\boldsymbol{\theta},_2 \,\}^T$ in generalized form as ${}^{S}\mathbf{H}({}^{t}\breve{\bar{\mathbf{d}}}) \equiv \{\, {}^{t}\breve{\mathbf{d}} \quad {}^{t}\breve{\mathbf{d}},_1 \quad {}^{t}\breve{\mathbf{d}},_2 \,\}^T$, with the superscript S reminding us that the derivatives in the expression are with respect to the arc length parameters, $S^{\alpha}, \quad \alpha = 1, 2$.

Next, let us further introduce the following.

Definitions

- ${}^{\hat{\xi}}\breve{\mathbf{H}}(\breve{\bar{\mathbf{d}}}) \equiv \{\, \breve{\bar{\mathbf{d}}} \quad \breve{\bar{\mathbf{d}}},_{\hat{\xi}^1} \quad \breve{\bar{\mathbf{d}}},_{\hat{\xi}^2} \,\}^T$, and similarly, ${}^{\hat{\xi}}\breve{\mathbf{H}}(\Delta\breve{\mathbf{d}}) \equiv \{\, \Delta\breve{\mathbf{d}} \quad \Delta\breve{\mathbf{d}},_{\hat{\xi}^1} \quad \Delta\breve{\mathbf{d}},_{\hat{\xi}^2} \,\}^T$
- ${}^{\hat{\xi}}\breve{\mathbf{B}}(\hat{\xi}^1, \hat{\xi}^2)$ is the $\left(18 \times 6N^2\right)$ augmented interpolation transformation matrix as:

$$\underbrace{{}^{\hat{\xi}}\breve{\mathbf{B}}(\hat{\xi}^1, \hat{\xi}^2)}_{18\times 6N^2} \equiv \begin{bmatrix} \underbrace{{}^{d}\breve{\mathbf{T}}}_{6\times 6N^2} \\ \underbrace{{}^{d}\breve{\mathbf{T}},_{\hat{\xi}^1}}_{6\times 6N^2} \\ \underbrace{{}^{d}\breve{\mathbf{T}},_{\hat{\xi}^2}}_{6\times 6N^2} \end{bmatrix} \tag{11.614}$$

such that

$$\boxed{\begin{aligned}\underbrace{{}^{\hat{\xi}}\mathbf{H}(\breve{\bar{\mathbf{d}}})}_{18x1} &= \underbrace{{}^{\hat{\xi}}\breve{\mathbf{B}}(\hat{\xi}^1,\hat{\xi}^2)}_{18x6N^2}\ \underbrace{{}^{d}\breve{\bar{\mathbf{q}}}}_{6N^2\times 1}\\ \underbrace{{}^{\hat{\xi}}\mathbf{H}(\Delta\breve{\bar{\mathbf{d}}})}_{18x1} &= \underbrace{{}^{\hat{\xi}}\breve{\mathbf{B}}(\hat{\xi}^1,\hat{\xi}^2)}_{18x6N^2}\ \underbrace{\Delta^{d}\breve{\bar{\mathbf{q}}}}_{6N^2\times 1}\end{aligned}} \tag{11.615}$$

with superscript $\hat{\xi}$ in each term reminding us that the derivatives are with respect to parameters, $\hat{\xi}^{\alpha}, \quad \alpha = 1, 2$.

Moreover, suppose the lines of curvature make angles of α_1 and α_2 with the $\hat{\xi}^1$ coordinate direction at any point $(\hat{\xi}^1, \hat{\xi}^2)$ on the shell surface. Then, as indicated before, with the arc length parameters, (S^1, S^2), along the corresponding lines of curvature, we can write the following transformation relationship for the virtual generalized displacement from $(\hat{\xi}^1, \hat{\xi}^2)$ parameters to the arc length parameter, (S^1, S^2):

$$\begin{aligned}\underbrace{{}^{S}\mathbf{H}(\breve{\bar{\mathbf{d}}})}_{18\times 1} &= \underbrace{{}^{S\hat{\xi}}\breve{\mathbf{B}}(\hat{\xi}^1,\hat{\xi}^2)}_{18\times 18}\underbrace{{}^{\hat{\xi}}\mathbf{H}(\breve{\bar{\mathbf{d}}})}_{18\times 1}\\ \underbrace{{}^{S}\mathbf{H}(\Delta\breve{\bar{\mathbf{d}}})}_{18\times 1} &= \underbrace{{}^{S\hat{\xi}}\breve{\mathbf{B}}(\hat{\xi}^1,\hat{\xi}^2)}_{18\times 18}\underbrace{{}^{\hat{\xi}}\mathbf{H}(\Delta\breve{\bar{\mathbf{d}}})}_{18\times 1}\end{aligned} \tag{11.616}$$

with ${}^{S\hat{\xi}}\breve{\mathbf{B}}(\hat{\xi}^1, \hat{\xi}^2)$ defined by:

$$\underbrace{{}^{S\hat{\xi}}\breve{\mathbf{B}}(\hat{\xi}^1,\hat{\xi}^2)}_{18\times 18} \equiv \begin{bmatrix} \underbrace{\mathbf{I}}_{3\times 3} & \mathbf{0} & \mathbf{0} \\ \mathbf{0} & \dfrac{1}{\gamma_1}\cos\alpha_1\,\mathbf{I} & \dfrac{1}{\gamma_1}\sin\alpha_1\,\mathbf{I} \\ \mathbf{0} & -\dfrac{1}{\gamma_2}\sin\alpha_1\,\mathbf{I} & \dfrac{1}{\gamma_2}\cos\alpha_1\,\mathbf{I} \end{bmatrix} \tag{11.617}$$

with $\gamma_{\alpha}(\xi^1, \xi^2) \equiv \dfrac{dS^{\alpha}}{d\xi^{\alpha}} = \|\bar{\mathbf{G}}_{\beta}\|, \quad \alpha = 1, 2$ with $\bar{\mathbf{G}}_{\beta}(\xi^1, \xi^2) \equiv \dfrac{\partial \mathbf{C}_0(\xi^1, \xi^2)}{\partial \xi^{\beta}}$ defined by the tangents to the lines of curvature. Now, from equations (11.615) and (11.616), we have the transformation relations as:

$$\boxed{\begin{aligned}\underbrace{{}^{S}\mathbf{H}(\breve{\bar{\mathbf{d}}})}_{18\times 1} &= \underbrace{\breve{\mathbf{B}}(\hat{\xi}^1,\hat{\xi}^2)}_{18\times 6N^2}\ \underbrace{{}^{d}\breve{\bar{\mathbf{q}}}}_{6N^2\times 1}\\ \underbrace{{}^{S}\mathbf{H}(\Delta\breve{\bar{\mathbf{d}}})}_{18\times 1} &= \underbrace{\breve{\mathbf{B}}(\hat{\xi}^1,\hat{\xi}^2)}_{18\times 6N^2}\ \underbrace{{}^{d}\Delta\breve{\bar{\mathbf{q}}}}_{6N^2\times 1}\end{aligned}} \tag{11.618}$$

where we have defined $\breve{\mathbf{B}}(\hat{\xi}^1, \hat{\xi}^2)$ as:

$$\breve{\mathbf{B}}(\hat{\xi}^1, \hat{\xi}^2) \equiv {}^{S\xi}\breve{\mathbf{B}}(\hat{\xi}^1, \hat{\xi}^2)\, {}^{\xi}\breve{\mathbf{B}}(\hat{\xi}^1, \hat{\xi}^2) \tag{11.619}$$

Finally, we have to express the infinitesimal area, $d\hat{A} = \sqrt{\hat{G}}\, d\hat{\xi}^1\, d\hat{\xi}^2$, of the area integrals where $\sqrt{\hat{G}}$ is the positive square root of the determinant of the surface metric expressed in terms of the geometry coordinate parameters, $\hat{\xi}^\alpha, \quad \alpha = 1, 2$.

11.18.6 Computational Element Symmetric Material Stiffness Matrix, ${}^e\mathbf{K}_M$

By substituting equation (11.618) in equation (11.591), we get for each finite element:

$$\begin{aligned}
&{}^e\mathcal{K}_M(\breve{\mathbf{d}}, \bar{\mathbf{d}}; \Delta\breve{\mathbf{d}}) \\
&\equiv \underbrace{{}^d\breve{\mathbf{q}}^T}_{1\times 6N^2} \underbrace{\left[\int_0^1 \int_0^1 \left\{ \underbrace{\breve{\mathbf{B}}(\hat{\xi}^1, \hat{\xi}^2)^T}_{6N^2\times 18} \underbrace{\overbrace{\mathbf{E}(\breve{\mathbf{d}})^T}^{18\times 12} \overbrace{{}^n\breve{\mathbf{D}}}^{12\times 12} \overbrace{\mathbf{E}(\breve{\mathbf{d}})}^{12\times 18}}_{18\times 18} \underbrace{\breve{\mathbf{B}}(\hat{\xi}^1, \hat{\xi}^2)}_{18\times 6N^2} \right\} \sqrt{\hat{G}}\, d\hat{\xi}^1\, d\hat{\xi}^2 \right]}_{6N^2\times 6N^2} \underbrace{{}^d\Delta\breve{\mathbf{q}}(t)}_{6N^2\times 1} \\
&= \underbrace{{}^d\breve{\mathbf{q}}^T}_{1\times 6N^2} \underbrace{{}^e\mathbf{K}_M}_{6N^2\times 6N^2} \underbrace{{}^d\Delta\breve{\mathbf{q}}(t)}_{6N^2\times 1}
\end{aligned} \tag{11.620}$$

where, we have introduced the following.

Definition

- The ***computational element symmetric material stiffness matrix***, ${}^e\mathbf{K}_M$, as:

$$\underbrace{{}^e\mathbf{K}_M}_{6N^2\times 6N^2} \equiv \underbrace{\int_0^1 \int_0^1 \left\{ \underbrace{\breve{\mathbf{B}}(\hat{\xi}^1, \hat{\xi}^2)^T}_{6N^2\times 18} \underbrace{\overbrace{\mathbf{E}(\breve{\mathbf{d}})^T}^{18\times 12} \overbrace{{}^n\breve{\mathbf{D}}}^{12\times 12} \overbrace{\mathbf{E}(\breve{\mathbf{d}})}^{12\times 18}}_{18\times 18} \underbrace{\breve{\mathbf{B}}(\hat{\xi}^1, \hat{\xi}^2)}_{18\times 6N^2} \right\} \sqrt{\hat{G}}\, d\hat{\xi}^1\, d\hat{\xi}^2}_{6N^2\times 6N^2} \tag{11.621}$$

11.18.7 Computational Element Symmetric Geometric Stiffness Matrix, ${}^e\mathbf{K}_G$

By substituting equation (11.618) into equation (11.595), we get for each finite element:

$$\begin{aligned}
&{}^e\mathcal{K}_G(\breve{\mathbf{d}}, \breve{\bar{\mathbf{d}}}; \Delta\breve{\mathbf{d}}) \\
&\equiv \underbrace{{}^d\breve{\bar{\mathbf{q}}}^T}_{1\times 6N^2} \underbrace{\left[\int_0^1\int_0^1 \left\{\underbrace{\breve{\mathbf{B}}(\hat{\xi}^1,\hat{\xi}^2)^T}_{6N^2\times 18}\ \underbrace{\breve{\mathbf{G}}(\breve{\mathbf{d}}, {}^t\breve{\mathbf{F}})}_{18\times 18}\ \underbrace{\breve{\mathbf{B}}(\hat{\xi}^1,\hat{\xi}^2)}_{18\times 6N^2}\right\} \sqrt{\hat{G}}\, d\hat{\xi}^1\, d\hat{\xi}^2\right]}_{6N^2\times 6N^2} \underbrace{{}^d\Delta\breve{\mathbf{q}}(t)}_{6N^2\times 1} \\
&= \underbrace{{}^d\breve{\bar{\mathbf{q}}}^T}_{1\times 6N^2}\ \underbrace{{}^e\mathbf{K}_G}_{6N^2\times 6N^2}\ \underbrace{{}^d\Delta\breve{\mathbf{q}}(t)}_{6N^2\times 1}
\end{aligned} \tag{11.622}$$

where, we have introduced the following.

Definition

- The ***computational element symmetric geometric stiffness matrix***, ${}^e\mathbf{K}_G$, as:

$$\underbrace{{}^e\mathbf{K}_G}_{6N^2\times 6N^2} \equiv \underbrace{\int_0^1\int_0^1 \left\{\underbrace{\breve{\mathbf{B}}(\hat{\xi}^1,\hat{\xi}^2)^T}_{6N^2\times 18}\ \underbrace{\breve{\mathbf{G}}(\breve{\mathbf{d}}, {}^t\breve{\mathbf{F}})}_{18\times 18}\ \underbrace{\breve{\mathbf{B}}(\hat{\xi}^1,\hat{\xi}^2)^T}_{18\times 6N^2}\right\}}_{6N^2\times 6N^2} \sqrt{\hat{G}} d\hat{\xi}^1 d\hat{\xi}^2 \tag{11.623}$$

with the geometric stiffness matrix, $\mathbf{G}$, defined by equation (11.596).

11.18.8 Computational Element Symmetric Mass Matrix, ${}^e\mathbf{M}$

By substituting expression for the generalized incremental acceleration vector, $\Delta\,\ddot{\breve{\mathbf{d}}}(t)$, and the generalized virtual displacement vector, $\breve{\bar{\mathbf{d}}}$, from equation (11.613) into equation (11.570), we get for each finite element:

$$\Delta G_M({}^t\breve{\mathbf{d}}; {}^t\breve{\bar{\mathbf{d}}}; \Delta {}^t\breve{\mathbf{d}}) \equiv {}^t\breve{\bar{\mathbf{q}}}^{T\,e}\mathbf{M}\Delta\, {}^t\ddot{\breve{\mathbf{d}}} \tag{11.624}$$

where we have introduced the following.

Definition

- The ***computational element symmetric mass matrix***, ${}^e\mathbf{M}$, as:

$$\underbrace{{}^e\mathbf{M}}_{6N^2\times 6N^2} \equiv \int_0^1\int_0^1 \underbrace{\left\{ \underbrace{{}^d\breve{\mathbf{T}}(\hat{\xi}^1,\hat{\xi}^2)^T}_{6N^2\times 6} \underbrace{\begin{bmatrix} M_0\,\mathbf{I} & \mathbf{0} \\ \mathbf{0} & I_0\,(\mathbf{W}^T\,\mathbf{W}) \end{bmatrix}}_{6\times 6} \underbrace{{}^d\breve{\mathbf{T}}(\hat{\xi}^1,\hat{\xi}^2)}_{6\times 6N^2} \right\}}_{6N^2\times 6N^2} \sqrt{\hat{G}}\,d\hat{\xi}^1\,d\hat{\xi}^2 \tag{11.625}$$

11.18.9 Computational Element Gyroscopic (Damping) Matrix, ${}^e\mathbf{C}_G$

By substituting the expression for the generalized incremental acceleration vector, $\Delta\,\ddot{\breve{\mathbf{d}}}(t)$, and the generalized virtual displacement vector, $\breve{\bar{\mathbf{d}}}$, from equation (11.613) into equation (11.571), we get for each finite element:

$$\boxed{\Delta G_G({}^t\breve{\mathbf{d}};\,{}^t\breve{\bar{\mathbf{d}}};\,\Delta\,{}^t\breve{\mathbf{d}}) \equiv {}^t\bar{\mathbf{q}}^T\,{}^e\mathbf{C}_G\,\Delta\,{}^t\dot{\mathbf{q}}} \tag{11.626}$$

where we have introduced the following.

Definition

- The ***computational element gyroscopic matrix***, ${}^e\mathbf{C}_G$, as:

$$\underbrace{{}^e\mathbf{C}_G}_{6N^2\times 6N^2} \equiv \int_0^1\int_0^1 \underbrace{\left\{ \underbrace{{}^d\breve{\mathbf{T}}(\hat{\xi}^1,\hat{\xi}^2)^T}_{6N^2\times 6} \underbrace{\begin{bmatrix} \mathbf{0} & \mathbf{0} \\ \mathbf{0} & \mathbf{W}^T(2\,\dot{\mathbf{W}} - \mathbf{W}\,\dot{\boldsymbol{\Theta}}\,\mathbf{W}) \end{bmatrix}}_{6\times 6} \underbrace{{}^d\breve{\mathbf{T}}(\hat{\xi}^1,\hat{\xi}^2)}_{6\times 6N^2} \right\}}_{6N^2\times 6N^2} \sqrt{\hat{G}}\,d\hat{\xi}^1\,d\hat{\xi}^2 \tag{11.627}$$

11.18.10 Computational Element Centrifugal (Stiffness) Matrix, ${}^e\mathbf{K}_C$

By substituting the expression for the generalized incremental acceleration vector, $\Delta\,\ddot{\breve{\hat{\mathbf{d}}}}(t)$, and the generalized virtual displacement vector, $\breve{\hat{\bar{\mathbf{d}}}}$, from equation (11.613) into equation (11.572), we get for each finite element:

$$\boxed{\Delta G_C({}^t\breve{\mathbf{d}};\, {}^t\bar{\breve{\mathbf{d}}};\, \Delta\, {}^t\breve{\mathbf{d}}) \equiv {}^t\bar{\breve{\mathbf{q}}}^T\, {}^e\mathbf{K}_C\, \Delta\, {}^t\breve{\mathbf{q}}} \tag{11.628}$$

where we have introduced the following.

Definition

- The ***computational element centrifugal stiffness matrix***, ${}^e\mathbf{K}_C$, as:

$$\underbrace{{}^e\mathbf{K}_C}_{6N^2\times 6N^2} \equiv \int_0^1\int_0^1 \underbrace{\left\{ \underbrace{{}^d\breve{\mathbf{T}}(\hat{\xi}^1, \hat{\xi}^2)^T}_{6N^2\times 6} \underbrace{\begin{bmatrix} \mathbf{0} & \mathbf{0} \\ \mathbf{0} & \mathbf{U}({}^t\dot{\boldsymbol{\omega}}_S) + \mathbf{W}^T \left\{ \begin{array}{l} \ddot{\mathbf{W}} - (\mathbf{W}\boldsymbol{\Theta}\dot{\mathbf{W}} \\ + \dot{\mathbf{W}}\boldsymbol{\Theta}\mathbf{W}) - \mathbf{W}\boldsymbol{\Theta}\ddot{\mathbf{W}} \end{array} \right\} \end{bmatrix}}_{6\times 6} \underbrace{{}^d\breve{\mathbf{T}}(\hat{\xi}^1, \hat{\xi}^2)}_{6\times 6N^2} \right\}}_{6N^2\times 6N^2} \sqrt{\hat{G}}\, d\hat{\xi}^1\, d\hat{\xi}^2 \tag{11.629}$$

with

$$\boxed{\begin{array}{l} \mathbf{U}({}^t\dot{\boldsymbol{\omega}}_S) \equiv \left\{ \begin{array}{l} c_1\, [{}^t\dot{\boldsymbol{\omega}}_S] + (-c_2\, \mathbf{I} - c_3\, \boldsymbol{\Theta} + c_4 \boldsymbol{\Theta}^2)({}^t\dot{\boldsymbol{\omega}}_S \bullet \boldsymbol{\theta}) \\ + c_2((\boldsymbol{\theta} \otimes {}^t\dot{\boldsymbol{\omega}}_S) - ({}^t\dot{\boldsymbol{\omega}}_S \otimes \boldsymbol{\theta}) + (\boldsymbol{\theta} \bullet {}^t\dot{\boldsymbol{\omega}}_S)\mathbf{I} \end{array} \right\}, \\ {}^t\dot{\boldsymbol{\omega}}_S = (\dot{\mathbf{W}}\, {}^t\dot{\boldsymbol{\theta}} + \mathbf{W}\, {}^t\ddot{\boldsymbol{\theta}}) \end{array}} \tag{11.630}$$

where $[\mathbf{h}] \equiv \mathbf{h}\times, \forall \mathbf{h} \in \mathbb{R}_3$.

11.18.11 Element Unbalanced Residual Force, $\mathbf{R}(\breve{\mathbf{d}})$

By substituting equation (11.618) into equation (11.573), we get for each finite element:

$$\boxed{\begin{aligned} \overbrace{R(\breve{\mathbf{d}};\bar{\breve{\mathbf{d}}})}^{\text{Residual force}} &= -\overbrace{{}^eG(\breve{\mathbf{d}};\bar{\breve{\mathbf{d}}})}^{\text{Unbalanced force}} \\ &= -{}^d\bar{\breve{\mathbf{q}}}T\, {}^e\breve{\mathbf{R}}(\breve{\mathbf{d}}) \end{aligned}} \tag{11.631}$$

where, we have defined the following.

Definition

$$\underbrace{{}^e\breve{\mathbf{R}}(\breve{\mathbf{d}})}_{6N^2\times 1} \equiv \left\{ \begin{array}{l} \displaystyle\int_0^1\int_0^1 \left(\underbrace{\breve{\mathbf{B}}^T}_{6N^2\times 18} \underbrace{\mathbf{E}^T}_{18\times 12} \underbrace{{}^n\breve{\hat{\mathbf{F}}}}_{12\times 1} + \underbrace{\breve{\mathbf{T}}^T}_{6N^2\times 6} \left\{ \underbrace{{}^t\breve{\mathbf{F}}_{Iner}}_{6\times 1} - \underbrace{{}^t\breve{\hat{\mathbf{F}}})}_{6\times 1} \right\} \right) \sqrt{\hat{G}}\, d\hat{\xi}^1\, d\hat{\xi}^2 \\ \displaystyle - \int_\Gamma \underbrace{\breve{\mathbf{T}}^T}_{6N^2\times 6} \underbrace{{}^t\breve{\hat{\mathbf{F}}}_v(S)}_{6\times 1} \, dS \end{array} \right\} \tag{11.632}$$

11.18.12 Geometrically Exact Element Incremental Dynamic Equation

Now substituting expressions given in equations (11.620), (11.622), (11.625), (11.627), (11.629) and (11.631) in equation (11.566), we have:

$$ {}^d\bar{\mathbf{q}}T\{\, {}^e\mathbf{M}\,\Delta\, {}^d\ddot{\breve{\mathbf{q}}} + {}^e\mathbf{C}_G\,\Delta\, {}^d\dot{\breve{\mathbf{q}}} + \left({}^e\mathbf{K}_M + {}^e\mathbf{K}_G + {}^e\mathbf{K}_C\right)\,\Delta^d\breve{\mathbf{q}}\} = -\,{}^d\bar{\mathbf{q}}T\,{}^e\breve{\mathbf{R}}(\breve{\mathbf{d}}) \tag{11.633}$$

Noting that this equation is true for any arbitrary admissible generalized virtual control displacement, ${}^d\bar{\mathbf{q}}$, we can finally write the $6N^2 \times 1$ geometrically exact fully linearized finite element equation for each element in unknown incremental generalized displacement control vector, $\Delta\,{}^d\breve{\hat{\mathbf{q}}}$, generalized velocity control vector, $\Delta\,{}^d\dot{\breve{\hat{\mathbf{q}}}}$, and generalized acceleration control vector, $\Delta\,{}^d\ddot{\breve{\hat{\mathbf{q}}}}$, for any time $t \in \mathbb{R}_+$, as:

$$\underbrace{{}^e\mathbf{M}}_{6N^2\times 6N^2}\ \underbrace{\Delta\,{}^d\ddot{\breve{\mathbf{q}}}}_{6N^2\times 1} + \underbrace{{}^e\mathbf{C}_G}_{6N^2\times 6N^2}\ \underbrace{\Delta\,{}^d\dot{\breve{\mathbf{q}}}}_{6N^2\times 1} + \underbrace{\left({}^e\mathbf{K}_C + {}^e\mathbf{K}_T\right)}_{6N^2\times 6N^2}\ \underbrace{\Delta\,{}^d\breve{\mathbf{q}}}_{6N^2\times 1} = -\underbrace{{}^e\breve{\mathbf{R}}(\breve{\mathbf{d}})}_{6N^2\times 1} \tag{11.634}$$

where, we have introduced the following.

Definition

- The computational tangent stiffness matrix, ${}^e\mathbf{K}^c_T$:

$$ {}^e\mathbf{K}_T \equiv {}^e\mathbf{K}_M + {}^e\mathbf{K}_G \tag{11.635}$$

of a shell, a concept familiar under quasi-static loading.

11.18.13 Direct Global Assembly

Thus far, we have detailed the development of the incremental second-order ordinary differential equation for a shell element at any time $t \in \mathbb{R}_+$ as given by equation (11.634). The corresponding incremental equation for the entire domain of a shell can be obtained by the standard direct global assembly procedure and is given as:

$$\boxed{\mathbf{M}\ \Delta^d \ddot{\breve{\mathbf{q}}}(t) + \mathbf{C}_G\ \Delta^d \dot{\breve{\mathbf{q}}}(t) + \mathbf{K}\ \Delta^d \breve{\mathbf{q}}(t) = -\breve{\mathbf{R}}(\breve{\mathbf{d}}(t))} \tag{11.636}$$

where the global matrices and vectors are obtained as:

$$\boxed{\begin{aligned}
\mathbf{M} &\equiv \sum_{element} \underbrace{{}^e\mathbf{M}}_{6N^2 \times 6N^2}, \\
\mathbf{C}_G &\equiv \sum_{element} \underbrace{{}^e\mathbf{C}_G}_{6N^2 \times 6N^2}, \\
\mathbf{K} &\equiv \sum_{element} \underbrace{\left({}^e\mathbf{K}_M + {}^e\mathbf{K}_G + {}^e\mathbf{K}_C\right)}_{6N^2 \times 6N^2}, \\
\breve{\mathrm{R}} &\equiv \sum_{element} \underbrace{{}^e\breve{\mathbf{R}}(\breve{\mathbf{d}})}_{6N^2 \times 1}
\end{aligned}} \tag{11.637}$$

An actual global assembly scheme, referred to in the finite element jargon as the direct global assembly, will normally be applied, which we will elaborate upon in the numerical implementation phase next.

11.18.14 Time Integration of the Semi-discrete Equation

So far, we have detailed the development of the global incremental matrix second-order ordinary differential equation as given by equation (11.636) under the semi-discrete c-type finite element formulation. The linearization has been introduced and devised with the understanding that a time-integration algorithm for the ensuing matrix second-order ordinary differential equation will be developed in conjunction with a Newton-type iterative scheme which will require linearization of the nonlinear equations at each iterative step. The time integration must realize that the rotation vector marching and updating must be given special attention as it is constrained to travel over a curved space: the 3-sphere.

Remark

- In this book, we have not implemented the dynamic treatment presented above, so for the following sections, we restrict ourselves to implementation of shell analysis to static and quasi-static loading conditions.
- However, for nonlinear structural dynamic problems involving finite rotations, we now suggest that the way to go may be the rotational acceleration with a spherically linear distribution (see Chapter 4 on rotation tensors) for a predictor–corrector type of numerical analysis; the rotational velocity and rotation distribution may then be obtained by corresponding integration of the acceleration.

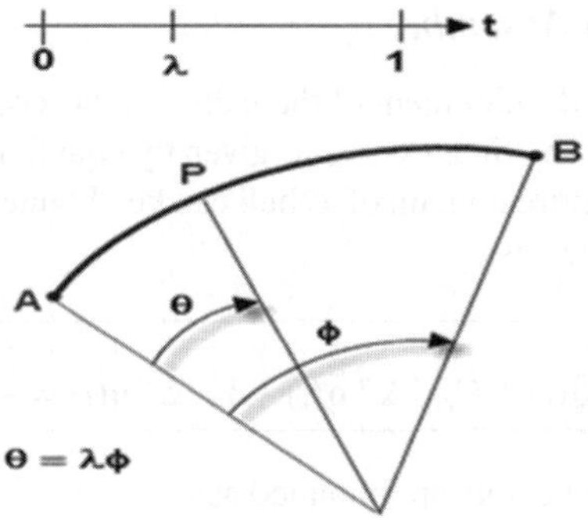

Figure 11.21 Spherical interpolation and approximation.

11.18.14.1 Linear Interpolation or Approximation on 3-Sphere

Let **A**, **B** represent the unit Hamilton–Rodrigues quaternions for rotations at parameters $t = 0$ and $t = 1$, respectively, as shown in Figure 11.21. The rotation from the reference state, **A**, to a final state, **B**, is a taken as a single rotation which is always possible by Euler's theorem. Now, let $\mathbf{P}(t)$, with $\|\mathbf{P}\| = 1$, represent a rotation on this curve at an affine parameter $t = \lambda \in [0, 1]$. We are interested in expressing **P** in terns of **A** and **B** (see Chapter 4 on rotation tensors):

$$\boxed{\mathbf{P} = \frac{\sin(1-t)\,\phi}{\sin\phi}\mathbf{A} + \frac{\sin t\phi}{\sin\phi}\mathbf{B}, \quad t \in [0.1]} \tag{11.638}$$

11.18.15 Where We Would Like to Go

The main goal here was to present the finite element formulation of the dynamic fully linearized variational or virtual work equations. As indicated earlier, shell analysis subjected to dynamic loadings has not been implemented. We will continue with the implementational issues related to static and quasi-static analyses:

- finite element implementation for quasi-static analyses.

11.19 c-type FE Formulation: Quasi-static Loading

11.19.1 What We Need to Recall

The linearized quasi-static virtual work equations:

$$\boxed{\overbrace{\int_M {}^n\breve{\bar{\varepsilon}}^T\, {}^n\breve{\mathbf{D}}\; \Delta^n\breve{\varepsilon}\; dA}^{\text{Material stiffness functional}} + \overbrace{\int_M \Delta^n\breve{\bar{\varepsilon}}^T\; {}^n\breve{\mathbf{F}}\; dA}^{\text{Geometric stiffness functional}} = -\underbrace{\overbrace{G({}^t\breve{\mathbf{d}}^0;\, {}^t\breve{\bar{\mathbf{d}}})}^{\text{Unbalanced force functional}}}_{\text{Residual force functional}}} \tag{11.639}$$

Or,

$$\overbrace{\underbrace{\mathcal{K}_M}_{\text{Material stiffness functional}} + \underbrace{\mathcal{K}_G}_{\text{Geometric stiffness functional}}}^{\text{Tangent functional}} = -\overbrace{\underbrace{G({}^t\breve{\mathbf{d}}^0;\, {}^t\breve{\bar{\mathbf{d}}})}_{\text{Residual force functional}}}^{\text{Unbalanced force functional}} \tag{11.640}$$

The linearized internal functional, $\Delta G_{Internal}$:

$$\begin{aligned}\Delta G_{Internal} &\equiv \overbrace{\underbrace{\int_M {}^n\breve{\bar{\boldsymbol{\varepsilon}}}^T\, {}^n\breve{\mathbf{D}}\; \Delta^n\breve{\boldsymbol{\varepsilon}}\; dA}_{\text{Material stiffness functional}} + \underbrace{\int_M \Delta^n\breve{\bar{\boldsymbol{\varepsilon}}}^T\; {}^n\breve{\mathbf{F}}\; dA}_{\text{Geometric stiffness functional}}}^{\text{Linearized internal functional}} \\ &= \overbrace{\underbrace{\mathcal{K}_M}_{\text{Material stiffness functional}} + \underbrace{\mathcal{K}_G}_{\text{Geometric stiffness functional}}}^{\text{Tangent functional}}\end{aligned} \tag{11.641}$$

Definition: Material stiffness functional, $\mathcal{K}_M$:

$$\boxed{\mathcal{K}_M({}^t\breve{\mathbf{d}},\, {}^t\breve{\bar{\mathbf{d}}};\, \Delta^t\breve{\mathbf{d}}) \equiv \int_M {}^n\breve{\bar{\boldsymbol{\varepsilon}}}^T\, {}^n\breve{\mathbf{D}}\; \Delta^n\breve{\boldsymbol{\varepsilon}}\; dA} \tag{11.642}$$

Definition: Geometric stiffness functional, $\mathcal{K}_G$:

$$\boxed{\mathcal{K}_G({}^t\breve{\mathbf{d}},\, {}^t\breve{\bar{\mathbf{d}}};\, \Delta^t\breve{\mathbf{d}}) \equiv \int_M \Delta^n\breve{\bar{\boldsymbol{\varepsilon}}}^T({}^t\breve{\mathbf{d}},\, {}^t\breve{\bar{\mathbf{d}}};\, \Delta^t\breve{\mathbf{d}})\; {}^n\breve{\mathbf{F}}\; dA} \tag{11.643}$$

the virtual strain–displacement relations in compact matrix form is as follows:

$$\boxed{\breve{\bar{\boldsymbol{\varepsilon}}}(\breve{\mathbf{d}};\, \breve{\bar{\mathbf{d}}}) = \mathbf{E}(\breve{\mathbf{d}})\; \mathbf{H}(\breve{\bar{\mathbf{d}}})} \tag{11.644}$$

where

$$\boxed{\mathbf{E}(\breve{\mathbf{d}}) = \begin{bmatrix} \mathbf{R^T} & \mathbf{0} & \mathbf{0} & \mathbf{0} \\ \mathbf{0} & \mathbf{R^T} & \mathbf{0} & \mathbf{0} \\ \mathbf{0} & \mathbf{0} & \mathbf{R^T} & \mathbf{0} \\ \mathbf{0} & \mathbf{0} & \mathbf{0} & \mathbf{R^T} \end{bmatrix} \begin{bmatrix} [\mathbf{k}_1^0] & [\mathbf{a}_1]\mathbf{W} & \mathbf{I} & \mathbf{0} & \mathbf{0} & \mathbf{0} \\ \mathbf{0} & \mathbf{X}_1 + [\mathbf{k}_1^c]\mathbf{W} & \mathbf{0} & \mathbf{W} & \mathbf{0} & \mathbf{0} \\ [\mathbf{k}_2^0] & [\mathbf{a}_2]\mathbf{W} & \mathbf{0} & \mathbf{0} & \mathbf{I} & \mathbf{0} \\ \mathbf{0} & \mathbf{X}_2 + [\mathbf{k}_2^c]\mathbf{W} & \mathbf{0} & \mathbf{0} & \mathbf{0} & \mathbf{W} \end{bmatrix}} \tag{11.645}$$

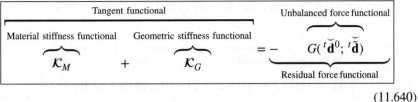

and $\mathbf{I}$, is the 3×3 identity matrix, $\mathbf{H}(\breve{\bar{\mathbf{d}}}) \equiv \{\bar{\mathbf{d}} \quad \bar{\theta} \quad \bar{\mathbf{d}},_1 \quad \bar{\theta},_1 \quad \bar{\mathbf{d}},_2 \quad \bar{\theta},_2\}^T$, and from our knowledge of the relevant derivatives and variations, we have:

$$\boxed{\begin{aligned} &\mathbf{W} \equiv \mathbf{I} + c_1\boldsymbol{\Theta} + c_2\boldsymbol{\Theta}^2 \\ &\mathbf{X}_\alpha \equiv -c_1\boldsymbol{\Theta},_\alpha + c_2\{[\theta \otimes \theta,_\alpha] - [\theta,_\alpha \otimes \theta] + (\theta\theta,_\alpha)\mathbf{I}\} \\ &\qquad +\mathbf{V}[\theta,_\alpha \otimes \theta]), \quad \alpha = 1, 2 \\ &\mathbf{V} \equiv -c_2\mathbf{I} + c_3\boldsymbol{\Theta} + c_4\boldsymbol{\Theta}^2 \end{aligned}} \tag{11.646}$$

$$\boxed{\begin{aligned} c_1 &\equiv (1 - \cos\theta)/\theta^2, \\ c_2 &\equiv (\theta - \sin\theta)/\theta^3, \\ c_1 \equiv \delta c_1 \equiv c_3 &\equiv \{(1 - \theta^2 c_2) - 2c_1\}/\theta^2, \\ c_2 \equiv \delta c_2 \equiv c_4 &\equiv (c_1 - 3c_2)/\theta^2 \end{aligned}} \tag{11.647}$$

the (6×1) incremental generalized displacement state vector functions, $\Delta\breve{\mathbf{d}} \equiv (\Delta\mathbf{d} \quad \Delta\theta)^{\mathbf{T}}$, and finally, ${}^n\breve{\mathbf{D}}$ is the 12×12 symmetric elastic constitutive matrix. In the expanded form, equation (11.643) may be expressed as follows:

$$\boxed{\mathcal{K}_G \equiv \overbrace{\int_M {}^n\Delta\bar{\beta}_\alpha \ {}^n\mathbf{F}^\alpha \, dA}^{\text{Translational part}} + \overbrace{\int_M {}^n\Delta\bar{\chi}_\alpha \ {}^n\mathbf{M}^\alpha \, dA}^{\text{Rotational part}}} \tag{11.648}$$

where Δ, the incremental operations, involve the covariant derivatives of the virtual strain functions.

- The unbalanced force functional, $G(\lambda^0, {}^t\breve{\mathbf{d}}^0; {}^t\breve{\bar{\mathbf{d}}})$:

$$\boxed{G(\lambda^0, \breve{\mathbf{d}}^0; \breve{\bar{\mathbf{d}}}) \equiv \int_M {}^n\breve{\bar{\varepsilon}}^T \ {}^n\breve{\mathbf{D}} \ {}^n\breve{\varepsilon} \, dA - \left\{ \int_M {}^t\breve{\bar{\mathbf{d}}}^T \ {}^t\breve{\hat{\mathbf{F}}}(\lambda^0, S^1, S^2) \, dA + \lambda^0 \int_{\partial M^f} {}^t\breve{\hat{\bar{\mathbf{d}}}}^T \ {}^t\breve{\hat{\mathbf{F}}}_\nu(S) \, dS \right\}}$$

(11.649)

- The residual force functional:

$$\text{Residual force functional} \equiv -\overbrace{G(\lambda^0,\ {}^t\breve{\mathbf{d}}^0;\ {}^t\breve{\bar{\mathbf{d}}})}^{\text{Unbalanced force functional}} + \overbrace{\left\{ \int_M {}^t\breve{\bar{\mathbf{d}}}^T\ {}^t\hat{\mathbf{F}},_\lambda(\lambda^0, S^1, S^2)\, dA + \int_{\partial M^f} {}^t\breve{\bar{\mathbf{d}}}^T\ {}^t\overset{S}{\hat{\mathbf{F}}}\,(S)\, dS \right\} \Delta\lambda}^{\text{Applied force functional}}$$

(11.650)

$\underbrace{\mathbf{H}({}^t\hat{\bar{\mathbf{d}}})}_{18\times 1} \equiv \{\ {}^t\bar{\mathbf{d}}\quad {}^t\bar{\theta}\quad {}^t\bar{\mathbf{d}},_1\quad {}^t\bar{\theta},_1\quad {}^t\bar{\mathbf{d}},_2\quad {}^t\bar{\theta},_2\}^T$, and $\underbrace{\mathbf{H}(\Delta\hat{d})}_{18\times 1} \equiv \{\Delta\,{}^t\mathbf{d}\quad \Delta\,{}^t\theta\quad \Delta\,{}^t\mathbf{d},_1\quad \Delta\,{}^t\theta,_1\quad \Delta\,{}^t\mathbf{d},_2\quad \Delta\,{}^t\theta,_2\}^T$ with $(\bullet),_\alpha$, $\alpha = 1, 2$ are derivatives with respect to the arc length parameters, S^α, $\alpha = 1, 2$, along the lines of curvature.

- Superscript t reminds us that the relevant items are measured in the undeformed coordinate system, $\mathbf{G}_1, \mathbf{G}_2, \mathbf{N}$; superscript n reminds us that the relevant items are measured in the rotated coordinate system, $\mathbf{g}_1(=\mathbf{R}\,\mathbf{G}_1), \mathbf{g}_2(=\mathbf{R}\,\mathbf{G}_2), \mathbf{g}_3 \equiv \mathbf{n}(=\mathbf{R}\,\mathbf{N})$.
- The expressions given in equation (11.643) or equation (11.648) involve the first variation (incremental) of the virtual rotational strains or, equivalently, the second variation (Hessian) of the real rotational strains. It is this second variation that makes the derivation of the geometric stiffness matrix extremely involved, and thus tedious. However, the subsequent reward in terms of high computational accuracy, ease and efficiency in situations of repetitive computations of the geometric stiffness as we have in a nonlinear numerical analysis, makes it well worth while.

11.19.2 Material Stiffness Functional Revisited: $\mathcal{K}_M$

As in the virtual expression above, we group together the incremental state variables and their derivatives with respect to the arc length parameters, (S^1, S^2), into an augmented (18×1) incremental vector function (we can replace $\breve{\mathbf{d}}$ by $\Delta\breve{\mathbf{d}}$ in the definition of $\underbrace{\mathbf{H}(\breve{\mathbf{d}})}_{18\times 1} \equiv \{\breve{\mathbf{d}}\quad \breve{\theta}\quad \breve{\mathbf{d}},_1\quad \breve{\theta},_1\quad \breve{\mathbf{d}},_2\quad \breve{\theta},_2\}^T$) to introduce the following.

Definition

Incremental generalized displacement function:

$$\underbrace{\mathbf{H}(\Delta\breve{\mathbf{d}})}_{18\times 1} \equiv \{\Delta d\quad \Delta\theta\quad \Delta d,_1\quad \Delta\theta,_1\quad \Delta d,_2\quad \Delta\theta,_2\}^T \qquad (11.651)$$

Now, replacing the variational variables in equation (11.644) with incremental variables and keeping the above definition in mind, we get a similar expression as follows:

Definition

Incremental strain–displacement relation:

$$\boxed{\Delta\breve{\varepsilon}(\breve{\mathbf{d}}; \Delta\breve{\mathbf{d}}) = \mathbf{E}(\breve{\mathbf{d}})\ \mathbf{H}(\Delta\breve{\mathbf{d}})} \tag{11.652}$$

Note that the incremental strain, $\Delta\breve{\varepsilon}(\breve{\mathbf{d}}; \Delta\breve{\mathbf{d}})$, is clearly linear in the incremental generalized displacement, $\Delta\breve{\mathbf{d}}$, by being related to the strain–displacement matrix that is independent of $\Delta\breve{\mathbf{d}}$. By applying the virtual strain–displacement relation (11.644) and incremental strain–displacement relation (11.652), we are now ready to transform equation (11.642) as under:

$$\boxed{\mathcal{K}_M(\breve{\mathbf{d}}, \breve{\bar{\mathbf{d}}}; \Delta\breve{\mathbf{d}}) \equiv \int_M \mathbf{H}(\breve{\bar{\mathbf{d}}})^T(\mathbf{E}(\breve{\mathbf{d}})^T\ {}^n\hat{\mathbf{D}}\ \mathbf{E}(\breve{\mathbf{d}}))\mathbf{H}(\Delta\breve{\mathbf{d}})dA} \tag{11.653}$$

11.19.3 Material Stiffness Matrix, $\mathbf{K}_M$

Based on equation (11.653), we can now redefine and present the material stiffness functional as follows:

Definition: Material stiffness functional, $\mathcal{K}_M$ is:

$$\boxed{\mathcal{K}_M(\breve{\mathbf{d}}, \breve{\bar{\mathbf{d}}}; \Delta\breve{\mathbf{d}}) \equiv \int_M \mathbf{H}(\breve{\bar{\mathbf{d}}})^T\ \mathbf{K}_M(\breve{\mathbf{d}})\ \ \mathbf{H}(\Delta\breve{\mathbf{d}})\ dA} \tag{11.654}$$

where, we have introduced:

Definition: Material stiffness matrix, $\mathbf{K}_M$, as:

$$\boxed{\mathbf{K}_M \equiv \mathbf{E}(\breve{\mathbf{d}})^T\ {}^n\mathbf{D}\ \mathbf{E}(\breve{\mathbf{d}})} \tag{11.655}$$

Remark

- Clearly, from equation (11.655), the material stiffness matrix, $\mathbf{K}$, is symmetric in state or configuration space vector, $\breve{\mathbf{d}}$.

11.19.4 Geometric Stiffness Functional Revisited: $\mathcal{K}_G$

Here, in order to avoid undue distraction, first we present the main results, and then refer the interested reader to the sections on detailed derivations.

- Translational virtual work: linearization and symmetry
  We get as the final result:

$$
\begin{aligned}
{}^{n}\Delta\bar{\beta}_{\alpha} \bullet {}^{n}\mathbf{F}^{\alpha} &= \bar{\mathbf{d}},_{\alpha} \bullet([{}^{t}\mathbf{F}^{\alpha}]^{\mathbf{T}})\Delta\theta + \Delta\mathbf{d},_{\alpha} \bullet([{}^{t}\mathbf{F}^{\alpha}]^{\mathbf{T}}\mathbf{W})\bar{\theta} \\
&+ \bar{\mathbf{d}} \bullet ([\mathbf{k}^{0}_{\alpha}][{}^{t}\mathbf{F}^{\alpha}]\mathbf{W})\Delta\theta + \delta\mathbf{d} \bullet ([\mathbf{k}^{0}_{\alpha}][{}^{t}\mathbf{F}^{\alpha}]\mathbf{W})\bar{\theta} \\
&+ \bar{\theta} \bullet \mathbf{Z}^{Sym}_{\beta}\Delta\theta
\end{aligned}
\tag{11.656}
$$

with

$$
\begin{aligned}
\mathbf{Z}^{Sym}_{\bar{\beta}} &= \frac{1}{2}\{\mathbf{W}^{\mathbf{T}}([{}^{t}\mathbf{F}^{\alpha} \otimes \mathbf{a}_{\alpha}] + [\mathbf{a}_{\alpha} \otimes {}^{t}\mathbf{F}^{\alpha}])\,\mathbf{W} + ([\mathbf{h}^{f} \otimes \theta] + [\theta \otimes \mathbf{h}^{f}])\} \\
&- ({}^{t}\mathbf{F}^{\alpha} \bullet \mathbf{a}_{\alpha})\mathbf{W}^{\mathbf{T}}\,\mathbf{W} + c_{2}(\theta \bullet \mathbf{b}^{f})\mathbf{I}
\end{aligned}
\tag{11.657}
$$

- Rotational virtual work: linearization and symmetry
  We get as the final result:

$$
\begin{aligned}
{}^{n}\Delta\bar{\chi} \bullet {}^{n}\mathbf{M} = \Delta\theta \bullet\ & \{\mathbf{W}^{\mathbf{T}}\,[{}^{t}\mathbf{M}^{\alpha}]\,\mathbf{W} + \mathbf{Y}({}^{t}\mathbf{M}^{\alpha})\}\bar{\theta},_{\alpha} \\
&+\bar{\theta} \bullet \{\mathbf{W}^{\mathbf{T}}\,[{}^{t}\mathbf{M}^{\alpha}]\,\mathbf{W} + \mathbf{Y}({}^{t}\mathbf{M}^{\alpha})\}\Delta\theta,_{\alpha} \\
&+\Delta\theta \bullet\ \mathbf{Z}^{Sym}_{\bar{\chi}}\ \bar{\theta}
\end{aligned}
\tag{11.658}
$$

with

$$
\begin{aligned}
\mathbf{Z}^{Symm}_{\bar{\chi}} &\equiv \frac{1}{2}\{\mathbf{W}^{\mathbf{T}}([{}^{t}\mathbf{M}^{\alpha} \otimes \mathbf{k}^{c}_{\alpha}] + [\mathbf{k}^{c}_{\alpha} \otimes {}^{t}\mathbf{M}^{\alpha}])\,\mathbf{W} + ([\mathbf{h}^{m} \otimes \theta] + [\theta \otimes \mathbf{h}^{m}])\} \\
&- ({}^{t}\mathbf{M}^{\alpha} \bullet\ \mathbf{k}^{c}_{\alpha})\mathbf{W}^{\mathbf{T}}\,\mathbf{W} + c_{2}(\theta \bullet\ \mathbf{b}^{m})\mathbf{I} + \mathbf{W}^{\mathbf{T}}[{}^{t}\mathbf{M}^{\alpha}]\mathbf{X}_{\alpha} + \mathbf{X}^{\mathbf{T}}_{\alpha}\,[{}^{t}\mathbf{M}^{\alpha}]^{\mathbf{T}}\mathbf{W}
\end{aligned}
\tag{11.659}
$$

and $\mathbf{Y}(\bullet)$ function is defined later.

11.19.5 *Geometric Stiffness Functional,* $\mathcal{K}_G$ *and Matrix,* $\mathbf{G}(\hat{\mathbf{d}}, \hat{\mathbf{F}})$

Based on equations (11.656) and (11.658), we can now redefine and present below the geometric stiffness functional in its most uncluttered and explicit form.

Definition: Geometric stiffness functional, $\mathcal{K}_G$, is:

$$
\mathcal{K}_G(\breve{\mathbf{d}}, \bar{\breve{\mathbf{d}}}; \Delta\breve{\mathbf{d}}) \equiv \int_M \left\{ \underbrace{\mathbf{H}(\breve{\mathbf{d}})^T}_{1\times 18}\ \underbrace{\mathbf{G}(\breve{\mathbf{d}}, {}^{t}\breve{\mathbf{F}})}_{18\times 18}\ \underbrace{\mathbf{H}(\Delta\breve{\mathbf{d}})}_{18\times 1} \right\} dA
\tag{11.660}
$$

where

Definition: Geometric stiffness matrix, $\mathbf{G}(\breve{\mathbf{d}},\ {}^t\breve{\mathbf{F}})$, is given as:

$$\mathbf{G}(\breve{\mathbf{d}},\ {}^t\breve{\mathbf{F}}) \equiv \begin{bmatrix} \mathbf{0} & \mathbf{G}^1_{12}+\mathbf{G}^2_{12} & \mathbf{0} & \mathbf{0} & \mathbf{0} & \mathbf{0} \\ (\mathbf{G}^1_{12}+\mathbf{G}^2_{12})^T & \mathbf{G}^1_{22}+\mathbf{G}^2_{22} & \mathbf{G}^1_{23} & \mathbf{G}^1_{24} & \mathbf{G}^2_{23} & \mathbf{G}^2_{24} \\ \mathbf{0} & (\mathbf{G}^1_{23})^T & \mathbf{0} & \mathbf{0} & \mathbf{0} & \mathbf{0} \\ \mathbf{0} & (\mathbf{G}^1_{24})^T & \mathbf{0} & \mathbf{0} & \mathbf{0} & \mathbf{0} \\ \mathbf{0} & (\mathbf{G}^2_{23})^T & \mathbf{0} & \mathbf{0} & \mathbf{0} & \mathbf{0} \\ \mathbf{0} & (\mathbf{G}^2_{24})^T & \mathbf{0} & \mathbf{0} & \mathbf{0} & \mathbf{0} \end{bmatrix} \tag{11.661}$$

where

$$\begin{aligned} \mathbf{G}^\alpha_{12} &\equiv \left[\mathbf{k}^0_\alpha\right][{}^t\mathbf{F}^\alpha]\,\mathbf{W}, \quad \alpha = 1,2 \text{ and no sum} \\ \mathbf{G}^\alpha_{22} &\equiv \tfrac{1}{2}\mathbf{W}^\mathbf{T}\{[{}^t\mathbf{F}^\alpha \otimes \mathbf{a}_\alpha] + [\mathbf{a}_\alpha \otimes {}^t\mathbf{F}^\alpha] - 2({}^t\mathbf{F}^\alpha\ \mathbf{a}_\alpha)\}\mathbf{W} + c_2(\boldsymbol{\theta}\ \ \mathbf{b}^f_\alpha)\mathbf{I} \\ &\quad + \tfrac{1}{2}([\mathbf{h}^b_\alpha \otimes \boldsymbol{\theta}] + [\boldsymbol{\theta} \otimes \mathbf{h}^b_\alpha]) \\ &\quad + \tfrac{1}{2}\mathbf{W}^\mathbf{T}\{[{}^t\mathbf{M}^\alpha \otimes \mathbf{k}^c_\alpha] + [\mathbf{k}^c_\alpha \otimes {}^t\mathbf{M}^\alpha] - 2({}^t\mathbf{M}^\alpha\ \mathbf{k}^c_\alpha)\}\mathbf{W} + c_2(\boldsymbol{\theta}\ \ \mathbf{b}^m_\alpha)\mathbf{I} \\ &\quad + \tfrac{1}{2}([\mathbf{h}^m_\alpha \otimes \boldsymbol{\theta}] + [\boldsymbol{\theta} \otimes \mathbf{h}^m_\alpha]) + \mathbf{W}^\mathbf{T}[{}^t\mathbf{M}^\alpha]\,\mathbf{X}_\alpha + \mathbf{X}_\alpha\,[{}^t\mathbf{M}^\alpha]^\mathbf{T}\,\mathbf{W} \\ \mathbf{G}^\alpha_{23} &\equiv \mathbf{W}^\mathbf{T}\,[{}^t\mathbf{F}^\alpha] \\ \mathbf{G}^\alpha_{24} &\equiv \mathbf{W}^\mathbf{T}\,[{}^t\mathbf{M}^\alpha]\,\mathbf{W} + \mathbf{Y}({}^t\mathbf{M}^\alpha) \end{aligned} \tag{11.662}$$

with

$$\begin{aligned} c_1 &\equiv (1-\cos\theta)/\theta^2, \\ c_2 &\equiv (\theta - \sin\theta)/\theta^3, \\ \bar{c}_1 \equiv \delta c_1 \equiv c_3 &\equiv \{(1-\theta^2 c_2) - 2c_1\}/\theta^2, \\ \bar{c}_2 \equiv \delta c_2 \equiv c_4 &\equiv (c_1 - 3c_2)/\theta^2 \end{aligned} \tag{11.663}$$

$$\begin{aligned} \mathbf{a} &\equiv \mathbf{e} + [\mathbf{k}^0]\mathbf{d} = \{(\hat{\mathbf{1}} + \mathbf{d}_{,s}) + [\mathbf{k}^0]\mathbf{d}\} \\ \mathbf{k}^t &\equiv \mathbf{W}\ \boldsymbol{\theta}_{,s} \\ \mathbf{k}^c &\equiv \mathbf{k}^0 + \mathbf{k}^t \\ \mathbf{b}^f &\equiv \mathbf{F}^t \times \mathbf{a} \\ \mathbf{b}^m &\equiv \mathbf{M}^t \times \mathbf{k}^c \end{aligned} \tag{11.664}$$

$$
\begin{aligned}
\mathbf{W} &\equiv \mathbf{I} + c_1\boldsymbol{\Theta} + c_2\boldsymbol{\Theta}^2 \\
\mathbf{X}_\alpha &\equiv -c_1\boldsymbol{\Theta},_\alpha + c_2\{[\boldsymbol{\theta}\otimes\boldsymbol{\theta},_\alpha] - [\boldsymbol{\theta},_\alpha\otimes\ \boldsymbol{\theta}] + (\boldsymbol{\theta}\bullet\boldsymbol{\theta},_\alpha)\mathbf{I}\} \\
&\quad + \mathbf{V}\,[\boldsymbol{\theta},_\alpha\otimes\ \boldsymbol{\theta}]) \\
\mathbf{V} &\equiv -c_2\mathbf{I} + c_3\boldsymbol{\Theta} + c_4\boldsymbol{\Theta}^2 \\
\mathbf{Y}(\mathbf{x}) &\equiv -c_1[\mathbf{x}] + c_2([\mathbf{x}\otimes\boldsymbol{\theta}] - [\boldsymbol{\theta}\otimes\mathbf{x}]) + c_2(\boldsymbol{\theta}\bullet\mathbf{x})\mathbf{I} + [\boldsymbol{\theta}\otimes\mathbf{V}^{\mathrm{T}}\,\mathbf{x}], \quad \forall\mathbf{x}\in R^3 \\
\mathbf{h}^f_\alpha &\equiv \mathbf{V}^{\mathrm{T}}\,\mathbf{b}^f_\alpha \\
\mathbf{h}^m_\alpha &\equiv \mathbf{V}^{\mathrm{T}}\,\mathbf{b}^m_\alpha
\end{aligned}
\tag{11.665}
$$

Exercise: Drawing upon the analyses referred to above for the translational and rotational geometric stiffness functional, derive the expression for **G** above. (Hint: look at Section 10.18 on geometric stiffness and its symmetry property of a beam.)

Remark

- Clearly, from equation (11.661), the geometric stiffness matrix, **G**, is symmetric in state or configuration space variables for a conservative system.
- Because of total similarity in each of the lines of curvature, for derivation of the geometric stiffness and its symmetry property, look up the discussion on the geometric stiffness and its symmetry property of a beam.

11.19.5.1 Implementational Note

Note that, as will be explained later, the computational finite element implementation of a shell problem can be achieved essentially by using the corresponding beam formulations along each of the lines of curvature at numerical Gauss quadrature points in a typical Newton-type iteration method of solution strategy.

The material and the geometric stiffness functional parts in equations (11.653) and (11.660) need to be further modified in the subsequent finite element formulation for computational purposes. In so doing, we will be able to introduce the computational definition for the shell material and the geometric stiffness matrices. Thus, for the final and computationally optimum form of the linearized virtual work, we will focus next on the c-type finite element formulation.

11.19.6 c-type Finite Element Formulation

Following the standard definition of a finite element discretization, the shell surface, Ω, is broken into disjointed finite number of elemental surfaces such that:

$$
\Omega = \bigcup_{e=1}^{e=\text{no. of elements}} \text{Finite element}^e = \bigcup_{e=1}^{e=\text{no. of elements}} \Omega^e \tag{11.666}
$$

Then, the shell material and geometric stiffness matrices can be expressed as an assembly of the corresponding elemental matrices; a similar idea applies to the formation of the global unbalanced and the applied forces. As will be seen in our c-type finite element implementation phase, we will use what is known as the direct stiffness and load global assembly of the corresponding elemental entities.

Thus, for now, we can express each of the following properties as a sum over all the corresponding entities belonging to the elements:

- Shell material stiffness matrix functional:

$$\boxed{\mathcal{K}_M(\breve{\mathbf{d}}, \breve{\bar{\mathbf{d}}}; \Delta\breve{\mathbf{d}}) = \sum_{elements} \int_{\Omega^e} \underbrace{\mathbf{H}(\breve{\bar{\mathbf{d}}})^T}_{1\times18} \; \underbrace{\mathbf{K}_M(\breve{\mathbf{d}})}_{18\times18} \; \underbrace{\mathbf{H}(\Delta\breve{\mathbf{d}})}_{18\times1} \, dA} \tag{11.667}$$

- Shell geometric stiffness matrix functional:

$$\boxed{\mathcal{K}_G(\breve{\mathbf{d}}, \breve{\bar{\mathbf{d}}}; \Delta\breve{\mathbf{d}}) = \sum_{elements} \int_{\Omega^e} \left\{ \underbrace{\mathbf{H}(\breve{\bar{\mathbf{d}}})^T}_{1\times18} \; \underbrace{\mathbf{G}}_{18\times18} \; \underbrace{\mathbf{H}(\Delta\breve{\mathbf{d}})}_{18\times1} \right\} dA} \tag{11.668}$$

- Shell applied force functional:

$$\sum_{elements} \left(\int_{\Omega^e} \breve{\bar{\mathbf{d}}}^{\mathbf{T}} \, \breve{\hat{\mathbf{F}}},_{\lambda}(\lambda, S^1, S^2) \, dA \; + \int_{\Gamma^e} \hat{\bar{\mathbf{d}}}^{\mathbf{T}} \, \breve{\hat{\mathbf{F}}}_{\nu}(S) \, dS \right) \tag{11.669}$$

- Shell unbalanced force functional:

$$G(\lambda^0, \breve{\mathbf{d}}^0; \breve{\bar{\mathbf{d}}}) = \sum_{elements} \left(\int_{\Omega^e} {}^n\breve{\bar{\boldsymbol{\varepsilon}}}^T \; {}^n\breve{\mathbf{D}} \; {}^n\breve{\boldsymbol{\varepsilon}} \; dA - \left\{ \int_{\Omega^e} \breve{\bar{\mathbf{d}}}^T \, \breve{\hat{\mathbf{F}}}(\lambda^0, S^1, S^2) \, dA \; + \lambda^0 \int_{\Gamma^e} \breve{\bar{\mathbf{d}}}^T \, \breve{\hat{\mathbf{F}}}_{\nu}(S) \, dS \right\} \right) \tag{11.670}$$

Based on the assumption that the assembly of the finite elements will be performed by the direct global stiffness assembly procedure where the properties of each element enter into the summing operation independent of the all other element data, we can restrict our discussion here to a typical finite element for the shell. Later, we will discuss the assembly of the global stiffness, and so on in our implementation phase.

Finally, although we can use the available library of c-type shell elements of both curved quadrilateral and triangular shape, in what follows, we will restrict our discussion to only the Cartesian product quadrilateral elements; similar arguments and ideas apply formally to the triangular elements too. The root element corresponding to a typical finite element is defined over the geometric 2D root parameter plane defined by parameters $(\hat{\xi}^1, \hat{\xi}^2) \in [0,1] \cup [0,1]$. Next, the parameters, (ξ^1, ξ^2), along the lines of curvature are related to the arc length parameter, (S^1, S^2), of a shell by: $S^1(\xi^1, \xi^2) = \int_0^{\xi^1} \| {}^e\mathbf{C}_0(\eta, \xi^2),_{\eta} \| d\eta$, and $S^2(\xi^1, \xi^2) = \int_0^{\xi^2} \| {}^e\mathbf{C}_0(\xi^1, \eta),_{\eta} \| d\eta$ where ${}^e\mathbf{C}_0(\xi^1, \xi^2)$, $\xi^1 \in [0,1]$, $\xi^2 \in [0,1]$ defines the reference surface position vector at any point

(ξ^1, ξ^2) of the eth finite element, that is, the Finite element^e. Let us briefly recall the basic steps of the c-type finite element generation for future use. The coordinate geometry of a shell reference surface is discretized in the mesh generation phase whereby the root element is identified and fixed for an element; the position vector, $\mathbf{C}_0(\hat{\xi}^1, \hat{\xi}^2)$, of any point on the element is given by:

$$
\boxed{
\begin{aligned}
&\mathbf{C}_0(\hat{\xi}^1, \hat{\xi}^2) = \begin{Bmatrix} C_0^1 = X(\hat{\xi}^1, \hat{\xi}^2) \\ C_0^2 = Y(\hat{\xi}^1, \hat{\xi}^2) \\ C_0^3 = Z(\hat{\xi}^1, \hat{\xi}^2) \end{Bmatrix} = {}^c\mathbf{T}(\hat{\xi}^1, \hat{\xi}^2)\, {}^c\mathbf{q}, \\
&C_0^k(\hat{\xi}^1, \hat{\xi}^2) = \sum_{i=0}^{n_c} \sum_{j=0}^{m_c} B_i^{n_c}(\hat{\xi}^1)\, B_j^{m_c}(\hat{\xi}^2)\, {}^c q_{ij}^k, \quad k = 1, \ldots, 3
\end{aligned}
}
\tag{11.671}
$$

where ${}^c\mathbf{T}(\hat{\xi}^1, \hat{\xi}^2)$ is the root transformation matrix, $B_i^n(\hat{\xi}^1)$, $B_j^m(\hat{\xi}^2)$ are Bernstein–Bezier polynomials of degree n_c and m_c respectively, and ${}^c\mathbf{q}$, ${}^c q_{ij}^k$ are the generalized coordinate control vector and the component control net for the element, respectively.

The discretization of the generalized displacement field, $\breve{\mathbf{d}}(\lambda) \equiv (\mathbf{d}(\lambda), \Theta(\lambda))$, for the finite element, is then performed over the geometric root element, except that the degree of the displacement distribution polynomials (i.e. the local basis or shape functions) will be at least equal to that of the geometric coordinate distribution polynomials; in other words, the element is either subparametric or isoparametric.

11.19.7 c-type Shell Elements and State Control Vectors

Let us first recognize that the shell geometry and the finite elements are naturally discretized in terms of the parameters $(\hat{\xi}^1, \hat{\xi}^2)$ as tensor product objects for quadrilateral elements. However, for analytical convenience, the linearized virtual work expression as given by equation (11.639) has been derived in terms of the arc length parameters, (S^1, S^2), along the corresponding lines of curvature. Thus, for computational purposes, we must express all our equations back in the $(\hat{\xi}^1, \hat{\xi}^2)$ parameters. This is what is done in the ensuing finite element formulation. Let us, then, introduce a typical c-type shell element displacement interpolation as

$$
\boxed{
\begin{aligned}
&\breve{\mathbf{d}}(\hat{\xi}^1, \hat{\xi}^2, \lambda) = {}^d\breve{\mathbf{T}}(\hat{\xi}^1, \hat{\xi}^2)\, {}^d\breve{\mathbf{q}}(\lambda), \\
&\breve{\mathbf{d}}_k(\hat{\xi}^1, \hat{\xi}^2, \lambda) = \sum_{i=0}^{n} \sum_{j=0}^{m} B_i^n(\hat{\xi}^1)\, B_j^m(\hat{\xi}^2)\, {}^d\breve{q}_{ij}^k(\lambda), \quad k = 1, \cdots, 6
\end{aligned}
}
\tag{11.672}
$$

where ${}^d\breve{\mathbf{T}}(\hat{\xi}^1, \hat{\xi}^2)$ is the root transformation matrix, $B_i^n(\hat{\xi}^1)$, $B_j^m(\hat{\xi}^2)$ are the Bernstein–Bezier polynomials of degree n and m respectively, and ${}^d\breve{\mathbf{q}}$, ${}^d\breve{q}_{ij}^k$ are the generalized displacement control vector and the component control net for the element, respectively. Similar distributions can be produced for the variations in generalized displacements, $\bar{\breve{\mathbf{d}}}$, and the incremental generalized displacement, $\Delta\breve{d}$. Let us define N, M as the orders of the polynomials, respectively, given by: $N \equiv n + 1$ and $M \equiv m + 1$. As will be seen, in actual implementation, we will enforce: $n = m$ for each element. We may see that the Cartesian product of these polynomials serve as the local basis functions for our c-type shell elements. Thus, a n degree typical c-type shell element is given by:

$$\boxed{\underbrace{\breve{\mathbf{d}}(\hat{\xi}^1,\hat{\xi}^2,\lambda)}_{6\times1} = \underbrace{{}^d\breve{\mathbf{T}}(\hat{\xi}^1,\hat{\xi}^2)}_{6\times6N^2}\ \underbrace{{}^d\breve{\mathbf{q}}(\lambda)}_{6N^2\times1}} \tag{11.673}$$

where we have introduced the following.

Definitions

$$\underbrace{{}^d\breve{\mathbf{q}}(\lambda)}_{6N^2\times1} = \Big\{ \underbrace{q^u_{00}\cdots q^u_{n0}\ q^u_{01}\cdots q^u_{n1}\ q^u_{0n}\cdots q^u_{nn}}_{1\times N^2}\ \underbrace{q^v_{00}\cdots q^v_{n0}\ q^v_{01}\cdots q^v_{n1}\ q^v_{0n}\cdots q^v_{nn}}_{1\times NM}$$

$$\underbrace{q^w_{00}\cdots q^w_{n0}\ q^w_{01}\cdots q^w_{n1}\ q^w_{0n}\cdots q^w_{nn}}_{1\times N^2}\ \underbrace{q^{\theta_1}_{00}\cdots q^{\theta_1}_{n0}\ q^{\theta_1}_{01}\cdots q^{\theta_1}_{n1}\ q^{\theta_1}_{0n}\cdots q^{\theta_1}_{nn}}_{1\times N^2}$$

$$\underbrace{q^{\theta_2}_{00}\cdots q^{\theta_2}_{n0}\ q^{\theta_2}_{01}\cdots q^{\theta_2}_{n1}\ q^{\theta_2}_{0n}\cdots q^{\theta_2}_{nn}}_{1\times N^2}\ \underbrace{q^{\theta_3}_{00}\cdots q^{\theta_3}_{n0}\ q^{\theta_3}_{01}\cdots q^{\theta_3}_{n1}\ q^{\theta_3}_{0n}\cdots q^{\theta_3}_{nn}}_{1\times N^2}\Big\}^T$$

- is the $(6N^2 \times 1)$ generalized displacement control vector.
- ${}^d\breve{\mathbf{T}}(\hat{\xi}^1,\hat{\xi}^2)$ is the $(6 \times 6N^2)$ interpolation transformation matrix containing the Bernstein–Bezier polynomials, that is,

$$\boxed{\begin{aligned}&{}^d\breve{\mathbf{T}}(\hat{\xi}^1,\hat{\xi}^2) =\\ &\underbrace{\begin{bmatrix} B^n_0B^n_0 & B^n_1B^n_0\cdots B^n_nB^n_{n-1} & B^n_nB^n_n & 0 & 0 & \cdots & \cdots & \cdots & 0\\ 0 & 0 & 0 & B^n_0B^n_0 & B^n_1B^n_0\cdots & B^n_nB^n_n & 0 & \cdots & 0\\ & & \cdots & \cdots & & & & & \\ & & & \cdots & \cdots & & & & \\ 0 & 0 & \cdots & \cdots & & 0 & B^n_0B^n_0 & B^n_1B^n_0\cdots & B^n_nB^n_n \end{bmatrix}}_{6x6N^2}\end{aligned}} \tag{11.674}$$

In other words, each row of ${}^d\breve{\mathbf{T}}(\hat{\xi}^1, \hat{\xi}^2)$ is identical except shifted by $(1 \times 6N^2)$ zero entries with increasing row numbers. Based on the above definitions, we have:

$$\boxed{\Delta\breve{\mathbf{d}}(\lambda) = {}^d\breve{\mathbf{T}}(\hat{\xi}^1, \hat{\xi}^2)\ \Delta\,{}^d\breve{\mathbf{q}}(\lambda)} \tag{11.675}$$

where we have similarly introduced the following.

Definitions

- $\Delta^d\breve{\mathbf{q}}(\lambda)$ is the $(6N^2 \times 1)$ load-dependent *incremental displacement control vector*, similar in the ordering of components to ${}^d\breve{\mathbf{q}}(\lambda)$.

Note that equations (11.673) and (11.674) imply that we have chosen to interpolate both the displacement components and the rotation components with the same degree control net. Now, let us rewrite $\mathbf{H}({}^t\mathbf{d}) \equiv \{\ {}^t\mathbf{d}\quad {}^t\boldsymbol{\theta}\quad {}^t\mathbf{d},_1\quad {}^t\boldsymbol{\theta},_1\quad {}^t\mathbf{d},_2\quad {}^t\boldsymbol{\theta},_2\ \}^T$ in generalized form as ${}^S\mathbf{H}({}^t\bar{\mathbf{d}}) \equiv \{\ {}^t\breve{\mathbf{d}}\quad {}^t\breve{\mathbf{d}},_1\quad {}^t\breve{\mathbf{d}},_2\ \}^T$, with the superscript S reminding us that the derivatives in the expression are with respect to the arc length parameters, $S^\alpha,\quad \alpha = 1, 2$. Next, let us further introduce the following.

Definitions

- ${}^{\hat{\xi}}\mathbf{H}(\breve{\mathbf{d}}) \equiv \{\ \bar{\mathbf{d}}\quad \breve{\mathbf{d}},_{\hat{\xi}^1}\quad \bar{\mathbf{d}},_{\hat{\xi}^2}\ \}^T$, and similarly, ${}^{\hat{\xi}}\mathbf{H}(\Delta\breve{\mathbf{d}}) \equiv \{\ \Delta\breve{\mathbf{d}}\quad \Delta\breve{\mathbf{d}},_{\hat{\xi}^1}\quad \Delta\breve{\mathbf{d}},_{\hat{\xi}^2}\ \}^T$
- ${}^{\hat{\xi}}\breve{\mathbf{B}}(\hat{\xi}^1, \hat{\xi}^2)$ is the $(18 \times 6N^2)$ augmented interpolation transformation matrix as:

$$\underbrace{{}^{\hat{\xi}}\breve{\mathbf{B}}(\hat{\xi}^1, \hat{\xi}^2)}_{18\times 6N^2} \equiv \begin{bmatrix} \underbrace{{}^d\breve{\mathbf{T}}}_{6\times 6N^2} \\ \underbrace{{}^d\breve{\mathbf{T}},_{\hat{\xi}^1}}_{6\times 6N^2} \\ \underbrace{{}^d\breve{\mathbf{T}},_{\hat{\xi}^2}}_{6\times 6N^2} \end{bmatrix} \tag{11.676}$$

such that

$$\boxed{\begin{aligned} \underbrace{{}^{\hat{\xi}}\mathbf{H}(\breve{\bar{\mathbf{d}}})}_{18\times 1} &= \underbrace{{}^{\hat{\xi}}\breve{\mathbf{B}}(\hat{\xi}^1, \hat{\xi}^2)}_{18\times 6N^2}\ \underbrace{{}^d\breve{\mathbf{q}}}_{6N^2\times 1} \\ \underbrace{{}^{\hat{\xi}}\mathbf{H}(\Delta\breve{\mathbf{d}})}_{18\times 1} &= \underbrace{{}^{\hat{\xi}}\breve{\mathbf{B}}(\hat{\xi}^1, \hat{\xi}^2)}_{18\times 6N^2}\ \underbrace{\Delta^d\breve{\mathbf{q}}}_{6N^2\times 1} \end{aligned}} \tag{11.677}$$

with superscript $\hat{\xi}$ in each term reminding us that the derivatives are with respect to parameters, $\hat{\xi}^\alpha,\quad \alpha = 1, 2$. Moreover, suppose the lines of curvature make an angle of α_1 and α_2 with $\hat{\xi}^1$ coordinate direction at any point $(\hat{\xi}^1, \hat{\xi}^2)$ on the shell surface. Then, as indicated earlier, with the arc length parameters, (S^1, S^2), along the corresponding lines of curvature, we can write the following transformation relationship for the virtual generalized displacement from $(\hat{\xi}^1, \hat{\xi}^2)$ parameters to the arc length parameter, (S^1, S^2):

$$\begin{aligned} \underbrace{{}^{S}\mathbf{H}(\breve{\bar{\mathbf{d}}})}_{18\times1} &= \underbrace{{}^{S\hat{\xi}}\breve{\mathbf{B}}(\hat{\xi}^1, \hat{\xi}^2)}_{18\times18}\,\underbrace{{}^{\hat{\xi}}\mathbf{H}(\breve{\bar{\mathbf{d}}})}_{18\times1} \\ \underbrace{{}^{S}\mathbf{H}(\Delta\breve{\mathbf{d}})}_{18\times1} &= \underbrace{{}^{S\hat{\xi}}\breve{\mathbf{B}}(\hat{\xi}^1, \hat{\xi}^2)}_{18\times18}\,\underbrace{{}^{\hat{\xi}}\mathbf{H}(\Delta\breve{\mathbf{d}})}_{18\times1} \end{aligned} \tag{11.678}$$

with ${}^{S\hat{\xi}}\breve{\mathbf{B}}(\hat{\xi}^1, \hat{\xi}^2)$ defined by:

$$\underbrace{{}^{S\hat{\xi}}\breve{\mathbf{B}}(\hat{\xi}^1, \hat{\xi}^2)}_{18\times18} \equiv \begin{bmatrix} \underbrace{\mathbf{I}}_{3\times3} & \mathbf{0} & \mathbf{0} \\ \mathbf{0} & \dfrac{1}{\gamma_1}\cos\alpha_1\,\mathbf{I} & \dfrac{1}{\gamma_1}\sin\alpha_1\,\mathbf{I} \\ \mathbf{0} & -\dfrac{1}{\gamma_2}\sin\alpha_1\,\mathbf{I} & \dfrac{1}{\gamma_2}\cos\alpha_1\,\mathbf{I} \end{bmatrix} \tag{11.679}$$

with $\gamma_\alpha(\xi^1, \xi^2) \equiv \dfrac{dS^\alpha}{d\xi^\alpha} = \|\bar{\mathbf{G}}_\beta\|$, $\alpha = 1, 2$ with $\bar{\mathbf{G}}_\beta(\xi^1, \xi^2) \equiv \dfrac{\partial \mathbf{C}_0(\xi^1, \xi^2)}{\partial \xi^\beta}$ defined by the tangents to the lines of curvature. Now, from equations (11.677) and (11.678), we have the final transformation relations as:

$$\boxed{\begin{aligned} \underbrace{{}^{S}\mathbf{H}(\breve{\bar{\mathbf{d}}})}_{18\times1} &= \underbrace{\breve{\mathbf{B}}(\hat{\xi}^1, \hat{\xi}^2)}_{18\times6N^2}\,\underbrace{{}^{d}\breve{\bar{\mathbf{q}}}}_{6N^2\times1} \\ \underbrace{{}^{S}\mathbf{H}(\Delta\breve{\mathbf{d}})}_{18\times1} &= \underbrace{\breve{\mathbf{B}}(\hat{\xi}^1, \hat{\xi}^2)}_{18\times6N^2}\,\underbrace{{}^{d}\Delta\breve{\mathbf{q}}z}_{6N^2\times1} \end{aligned}} \tag{11.680}$$

where we have defined $\breve{\mathbf{B}}(\hat{\xi}^1, \hat{\xi}^2)$ as:

$$\breve{\mathbf{B}}(\hat{\xi}^1, \hat{\xi}^2) \equiv {}^{S\hat{\xi}}\breve{\mathbf{B}}(\hat{\xi}^1, \hat{\xi}^2)\,{}^{\hat{\xi}}\breve{\mathbf{B}}(\hat{\xi}^1, \hat{\xi}^2) \tag{11.681}$$

Finally, we must express the infinitesimal area, $d\hat{A} = \sqrt{\hat{G}}\,d\hat{\xi}^1\,d\hat{\xi}^2$, of the area integrals where $\sqrt{\hat{G}}$ is the positive square root of the determinant of the surface metric expressed in terms of the geometry coordinate parameters, $\hat{\xi}^\alpha$, $\alpha = 1, 2$.

11.19.8 Computational Symmetric Element Material Stiffness Matrix, ${}^{e}\mathbf{K}_M^c$

By substituting equation (11.680) in equation (11.667), we get for each finite element:

$$
\begin{aligned}
&{}^{e}\mathcal{K}_M(\breve{\mathbf{d}}, \breve{\bar{\mathbf{d}}}; \Delta\breve{\mathbf{d}}) \\
&\equiv \underbrace{{}^{d}\breve{\bar{\mathbf{q}}}T}_{1\times 6N^2} \underbrace{\left[\int_0^1 \int_0^1 \left\{ \underbrace{\breve{\mathbf{B}}(\hat{\xi}^1, \hat{\xi}^2)^T}_{6N^2\times 18} \underbrace{\overbrace{\mathbf{E}(\breve{\mathbf{d}})^T}^{18\times 12} \overbrace{{}^{n}\breve{\mathbf{D}}}^{12\times 12} \overbrace{\mathbf{E}(\breve{\mathbf{d}})}^{12\times 18}}_{18\times 18} \underbrace{\breve{\mathbf{B}}(\hat{\xi}^1, \hat{\xi}^2)}_{18\times 6N^2} \right\} \sqrt{\hat{G}}\, d\hat{\xi}^1\, d\hat{\xi}^2 \right]}_{6N^2\times 6N^2} \underbrace{{}^{d}\Delta\breve{\mathbf{q}}(\lambda)}_{6N^2\times 1} \\
&= \underbrace{{}^{d}\breve{\bar{\mathbf{q}}}T}_{1\times 6N^2} \underbrace{{}^{e}\mathbf{K}^c_M}_{6N^2\times 6N^2} \underbrace{{}^{d}\Delta\breve{\mathbf{q}}(\lambda)}_{6N^2\times 1}
\end{aligned}
\tag{11.682}
$$

where, we have introduced the following.

Definition

- The computational symmetric element material stiffness matrix, ${}^{e}\mathbf{K}^c_M$, as:

$$
{}^{e}\mathbf{K}^c_M \equiv \underbrace{\int_0^1 \int_0^1 \left\{ \underbrace{\breve{\mathbf{B}}(\hat{\xi}^1, \hat{\xi}^2)^T}_{6N^2\times 18} \underbrace{\overbrace{\mathbf{E}(\breve{\mathbf{d}})^T}^{18\times 12} \overbrace{{}^{n}\breve{\mathbf{D}}}^{12\times 12} \overbrace{\mathbf{E}(\breve{\mathbf{d}})}^{12\times 18}}_{18\times 18} \underbrace{\breve{\mathbf{B}}(\hat{\xi}^1, \hat{\xi}^2)}_{18\times 6N^2} \right\} \sqrt{\hat{G}} d\hat{\xi}^1\, d\hat{\xi}^2}_{6N^2\times 6N^2}
\tag{11.683}
$$

11.19.9 *Computational Symmetric Elem. Geometric Stiffness Matrix,* ${}^{e}\mathbf{K}^c_G$

By substituting equation (11.680) into equation (11.668), we get for each finite element:

$$
\begin{aligned}
&{}^{e}\mathcal{K}_G(\breve{\mathbf{d}}, \breve{\bar{\mathbf{d}}}; \Delta\breve{\mathbf{d}}) \\
&\equiv \underbrace{{}^{d}\breve{\bar{\mathbf{q}}}T}_{1\times 6N^2} \underbrace{\left[\int_0^1 \int_0^1 \left\{ \underbrace{\breve{\mathbf{B}}(\hat{\xi}^1, \hat{\xi}^2)^T}_{6N^2\times 18} \underbrace{\mathbf{G}(\breve{\mathbf{d}}, {}^{t}\mathbf{F})}_{18\times 18} \underbrace{\breve{\mathbf{B}}(\hat{\xi}^1, \hat{\xi}^2)}_{18\times 6N^2} \right\} \sqrt{\hat{G}}\, d\hat{\xi}^1\, d\hat{\xi}^2 \right]}_{6N^2\times 6N^2} \underbrace{{}^{d}\Delta\breve{\mathbf{q}}(t)}_{6N^2\times 1} \\
&= \underbrace{{}^{d}\breve{\bar{\mathbf{q}}}^T}_{1\times 6N^2} \underbrace{{}^{e}\mathbf{K}^c_G}_{6N^2\times 6N^2} \underbrace{{}^{d}\Delta\breve{\mathbf{q}}(t)}_{6N^2\times 1}
\end{aligned}
\tag{11.684}
$$

where, we have introduced the following.

Definition

- The computational symmetric element geometric stiffness matrix, ${}^e\mathbf{G}^c$, as:

$$
{}^e\mathbf{K}_G^c \equiv \int_0^1\int_0^1 \underbrace{\left\{ \underbrace{\breve{\mathbf{B}}(\hat{\xi}^1,\hat{\xi}^2)^T}_{6N^2\times 18}\ \underbrace{\mathbf{G}(\breve{\mathbf{d}},\ {}^t\breve{\mathbf{F}})}_{18\times 18}\ \underbrace{\breve{\mathbf{B}}(\hat{\xi}^1,\hat{\xi}^2)}_{18\times 6N^2} \right\}}_{6N^2\times 6N^2} \sqrt{\hat{G}}\, d\hat{\xi}^1\, d\hat{\xi}^2 \tag{11.685}
$$

with the geometric stiffness matrix, $\mathbf{G}$, defined by equation (11.661).

11.19.10 Element Unbalanced Force, ${}^eG(\lambda^0, \breve{\mathbf{d}}^0; \breve{\bar{\mathbf{d}}})$

By substituting equation (11.680) into equation (11.670), we get for each finite element:

$$
\overbrace{{}^eG(\lambda^0, \breve{\mathbf{d}}^0; \breve{\bar{\mathbf{d}}})}^{\text{Unbalanced force}} = {}^d\breve{\bar{\mathbf{q}}}^T\ {}^e\breve{\mathbf{R}}(\lambda^0, \breve{\mathbf{d}}^0) \tag{11.686}
$$

where, we have defined:

$$
\underbrace{{}^e\breve{\mathbf{R}}(\lambda^0, \breve{\mathbf{d}}^0)}_{6N^2\times 1} \equiv \left\{ \int_0^1\int_0^1 \left(\underbrace{\breve{B}^{\mathbf{T}}}_{6N^2\times 18}\ \underbrace{\mathbf{E}^T}_{18\times 12}\ \underbrace{{}^n\breve{\hat{\mathbf{F}}}}_{12\times 1} + \underbrace{\breve{\mathbf{T}}^T}_{6N^2\times 6}\ \underbrace{{}^t\breve{\hat{\mathbf{F}}}(\lambda^0,\hat{\xi}^1,\hat{\xi}^2)}_{6\times 1} \right) \sqrt{\hat{G}}\, d\hat{\xi}^1\, d\hat{\xi}^2 - \left(\lambda^0 \int_\Gamma \underbrace{\breve{\mathbf{T}}^T}_{6N^2\times 6}\ \underbrace{{}^t\breve{\hat{\mathbf{F}}}_\nu(S)}_{6\times 1}\, dS \right) \right\} \tag{11.687}
$$

11.19.11 Element Incremental Generalized Applied Force, ${}^e\breve{\mathbf{p}}(\lambda)$

By substituting equation (11.673) into equation (11.669), the applied external force for a step can be expressed as:

$$
\overbrace{\left\{ \int_\Omega {}^t\breve{\bar{\mathbf{d}}}^T\ {}^t\breve{\hat{\mathbf{F}}}_{,\lambda}(\lambda^0, S^1, S^2)\, dA + \int_\Gamma {}^t\breve{\bar{\mathbf{d}}}^T\ {}^t\breve{\hat{\mathbf{F}}}_\nu(S)\, dS \right\}}^{\text{Incremental applied force}} \Delta\lambda = {}^d\breve{\bar{\mathbf{q}}}^T\ {}^e\breve{\mathbf{p}}(\lambda)\, \Delta\lambda \tag{11.688}
$$

where, we have defined:

$$\underbrace{{}^e\breve{\mathbf{p}}(\lambda^0)}_{6N^2\times 1} \equiv \int_0^1\int_0^1 [\underbrace{\breve{\mathbf{T}}^T(\hat{\xi}^1,\hat{\xi}^2)}_{6N^2\times 6}\ \underbrace{{}^t\hat{\breve{\mathbf{F}}}_{,\lambda}(\lambda^0,\hat{\xi}^1,\hat{\xi}^2)}_{6\times 1}]\sqrt{\hat{G}}\,d\hat{\xi}^1\,d\hat{\xi}^2 + \int_\Gamma \underbrace{\breve{\mathbf{T}}^T(S)}_{6N^2\times 6}\ \underbrace{{}^t\hat{\breve{\mathbf{F}}}_\nu(S)}_{6\times 1}\,dS \tag{11.689}$$

11.19.12 Geometrically Exact Element Incremental Equation

Now substituting expressions given in equations (11.682), (11.684), (11.686) and (11.688) in equation (11.639), we have:

$$\boxed{{}^d\breve{\mathbf{q}}^T\left\{{}^e\mathbf{K}_M^c + {}^e\mathbf{K}_G^c\right\}\ \Delta^d\hat{\mathbf{q}} = {}^d\breve{\mathbf{q}}^T\{{}^e\breve{\mathbf{p}}(\lambda)\,\Delta\lambda - {}^e\breve{\mathbf{R}}(\lambda_0,\breve{\mathbf{d}}_0)\}} \tag{11.690}$$

Noting that this equation is true for any arbitrary admissible generalized virtual control displacement, ${}^d\breve{\mathbf{q}}$, we can finally write the geometrically exact linearized finite element equation for each element in unknown incremental generalized displacement, $\Delta^d\breve{\mathbf{q}}$, and the incremental load factor, $\Delta\lambda$, as:

$$\begin{aligned} &{}^e\mathbf{K}_T^c\ \Delta^d\breve{\mathbf{q}} - {}^e\breve{\mathbf{p}}(\lambda^0)\,\Delta\lambda = -{}^e\breve{\mathbf{R}}(\lambda^0,\breve{\mathbf{d}}^0) \\ \Rightarrow\quad &{}^e\mathbf{K}_T^c\ \Delta^d\breve{\mathbf{q}} = -{}^e\breve{\mathbf{R}}(\lambda^0,\breve{\mathbf{d}}^0) + {}^e\breve{\mathbf{p}}(\lambda^0)\,\Delta\lambda \end{aligned} \tag{11.691}$$

where, we have introduced the following.

Definition

- The computational tangent control stiffness matrix, ${}^e\mathbf{K}_T^c$:

$$\boxed{{}^e\mathbf{K}_T^c \equiv {}^e\mathbf{K}_M^c + {}^e\mathbf{K}_G^c} \tag{11.692}$$

11.19.13 Where We Would Like to Go

Here we presented the element-level finite element formulation of the linearized variational or virtual work equations. Considering that the shell geometric stiffness can be viewed as the direct sum of those along the direction of the lines of curvature in the beam formulation, we deliberately skipped the lengthy proofs for establishing both the translational and rotational symmetry of the geometric stiffness; for this, we refer to:

- beam geometric stiffness symmetry.

  Also, in our derivation, we relied – without detailing the numerical issues – on various topics such as the determination of the lines of curvature, global formulation of actually gathering the element information, and so on. Thus, the next few logical steps should be the introduction to the numerical transformation of the coordinate frame to the curvilinear frame, a global assembly procedure of the finite element method, the numerical integration of the various stiffnesses and residual force expressions, and an iterative solution including the updating options by a Newton-type algorithm to set out the configuration path for a specific loading condition. Thus, for examination of one of the primary algorithms for a Newton-type solution in the nonlinear regime, we may refresh our memory by reviewing:

- a Newton-type method and the arc-length constraint algorithm.

Finally, for shell element library and practical guidelines on all the relevant shell computational details, and with the details for the general iterative solution methodology behind us, we would like to apply these algorithms to various problems of interest, and to evaluate our shell theoretical and computational formulations; for this, we look into:

- c-type numerical implementation for quasi-static loading and example problems and solutions.

11.20 c-type FE Implementation and Examples: Quasi-static Loading

11.20.1 c-type Shell Element Library

The library consists of the tensor product quadrilateral, and triangular elements; each is determined simply by the degree of the underlying polynomial of interpolation or the so-called shape functions. Within each category, both the non-rational Bernstein–Bezier and the rational Bernstein–Bezier interpolation types may be constructed. For general geometrical shapes, non-rational types are applicable; for quadric surfaces such as spheres and ellipsoids, but not paraboloids, we must use the rational counterparts for accuracy of representation; in fact, for large c-type mega-elements, it is absolutely crucial. For example, it is well known that an acute-angled portion of a spherical surface can be quite "closely" approximated geometrically by a bi-cubic net, but then the subsequent finite element analysis becomes erroneous for such elements of mega size. We will have occasion later to discuss it further.

11.20.1.1 Non-rational Tensor-product Quadrilateral Elements

The root element of a typical n-degree non-rational quadrilateral c-type finite element interpolating the shell geometry is a unit square, and the geometric interpolation of the element itself can be written as:

$$\boxed{\begin{aligned} \mathbf{C}_0(\hat{\xi}^1,\hat{\xi}^2) &= \sum_{i=0}^{L}\sum_{j=0}^{L} \mathrm{B}_i^L(\hat{\xi}^1)\mathrm{B}_j^L(\hat{\xi}^2)\ {}^c q_{ij} \\ {}^t\hat{\mathbf{d}}(\hat{\xi}^1,\hat{\xi}^2) &= \sum_{i=0}^{M}\sum_{j=0}^{M} \mathrm{B}_i^M(\hat{\xi}^1)\mathrm{B}_j^M(\hat{\xi}^2)\ {}^d q_{ij} \end{aligned}} \tag{11.693}$$

where $\boldsymbol{C}(\hat{\xi}^1,\hat{\xi}^2) \equiv \{\, \boldsymbol{X}(\hat{\xi}^1,\hat{\xi}^2) \quad \boldsymbol{Y}(\hat{\xi}^1,\hat{\xi}^2) \quad \boldsymbol{Z}(\hat{\xi}^1,\hat{\xi}^2)\,\}^T$ defines the position vector of a point on the shell element, $\hat{\xi}^1 \in [\boldsymbol{0},\boldsymbol{1}]$, $\hat{\xi}^2 \in [\boldsymbol{0},\boldsymbol{1}]$; $\mathrm{B}_i^L(\hat{\xi}^1)$ and $\mathrm{B}_j^L(\hat{\xi}^2)$, the ith and the jth Bernstein–Bezier interpolation polynomials of degree L, respectively, with the control points ${}^c q_{ij}$ for $i,j \in \{0,1,\ldots,L\}$; the tensor product $\mathrm{B}_i^L(\hat{\xi}^1)\mathrm{B}_j^L(\hat{\xi}^2)$ is the ij^{th} non-rational basis functions, and the control points ${}^c q_{ij}$ for $i,j \in \{0,1,\ldots,L\}$ form the Bernstein–Bezier net of control polygons. For

${}^{t}\hat{\mathbf{d}}(\hat{\xi}^1,\hat{\xi}^2)$, the generalized displacement and rotation interpolations, $\mathrm{B}_i^M(\hat{\xi}^1)$ and $\mathrm{B}_j^M(\hat{\xi}^2)$, the ith and j^{th} Bernstein–Bezier interpolation polynomials of degree M, respectively, with the generalized displacement control points ${}^{d}q_{ij}$ for $i,j \in \{0,1,\ldots,M\}$; the tensor product $\mathrm{B}_i^M(\hat{\xi}^1)\mathrm{B}_j^M(\hat{\xi}^2)$ is the ij^{th} non-rational basis functions, and the control points ${}^{d}q_{ij}$ for $i,j \in \{0,1,\ldots,M\}$ form the Bernstein–Bezier net of generalized displacement control polygons.

Remark

- We must have $L \leq M$ with the equality for isoparametry, and the inequality for subparametry; superparametry, that is, $L > M$ is computationally not recommended.

11.20.1.2 Rational Quadrilateral Elements

For mesh generation, the root element of an n-degree typical rational quadrilateral c-type geometric element interpolating the shell geometry is a unit square, and the geometric interpolation of the element itself can be written as:

$$\boxed{\begin{aligned} \mathbf{C}_0(\hat{\xi}^1,\hat{\xi}^2) &= \sum_{i=0}^{L}\sum_{j=0}^{L} \boldsymbol{F}_{ij}(\hat{\xi}^1,\hat{\xi}^2)\ {}^{c}q_{ij} \\ \boldsymbol{F}_{ij}(\hat{\xi}^1,\hat{\xi}^2) &\equiv \frac{w_{ij}\mathrm{B}_i^L(\hat{\xi}^1)\mathrm{B}_j^L(\hat{\xi}^2)}{\sum_{i=0}^{L}\sum_{j=0}^{L} w_{ij}\mathrm{B}_i^L(\hat{\xi}^1)\mathrm{B}_j^L(\hat{\xi}^2)} \end{aligned}} \tag{11.694}$$

where $\mathbf{C}(\hat{\xi}^1,\hat{\xi}^2) \equiv \{\, X(\hat{\xi}^1,\hat{\xi}^2) \quad Y(\hat{\xi}^1,\hat{\xi}^2) \quad Z(\hat{\xi}^1,\hat{\xi}^2)\,\}^T$, as before, defines the position vector of a point on the shell element, $\hat{\xi}^1 \in [0,1]$, $\hat{\xi}^2 \in [0,1]$; $F_{ij}(\hat{\xi}^1,\hat{\xi}^2)$ for $i,j \in \{0,\ldots,L\}$ is the ij^{th} rational basis functions; w_{ij} for $i,j \in \{0,\ldots,L\}$ are the weights at the control points.

Remarks

- Note that while the rational surface elements are obtained as the projections of the tensor product of root elements, they are themselves not tensor product elements as the basis functions, $\boldsymbol{F}_{ij}(\hat{\xi}^1,\hat{\xi}^2)$, cannot, in general, be expressed as the tensor product of the coordinate basis functions because of the presence of the denominator in the second expression of equation (11.694).
- For generalized displacement interpolation, we still choose the non-rational tensor products as given in the equation (11.693) because, for unknown displacement distribution, we would like to avoid choosing ad hoc estimates for the corresponding required weights.

11.20.1.3 Non-rational Triangular Elements

The root element of a typical n-degree non-rational triangular c-type finite element interpolating the shell geometry can be an equilateral triangle, and the geometric interpolation of the element itself can be written as:

$$\begin{aligned} \mathbf{C}_0(\boldsymbol{r},\boldsymbol{s},\boldsymbol{t}) &= \sum_{i+j+k=M} \mathrm{B}^M_{ijk}(\boldsymbol{r},\boldsymbol{s},\boldsymbol{t})\, {}^c q_{ijk} \\ {}^t\mathbf{d}(\boldsymbol{r},\boldsymbol{s},\boldsymbol{t}) &= \sum_{i+j+k=N} \mathrm{B}^M_{ijk}(\boldsymbol{r},\boldsymbol{s},\boldsymbol{t})\, {}^d q_{ijk} \end{aligned} \tag{11.695}$$

where, $\mathrm{B}^M_{ijk}(\boldsymbol{r},\boldsymbol{s},\boldsymbol{t})$ and $\mathrm{B}^N_{ijk}(\boldsymbol{r},\boldsymbol{s},\boldsymbol{t})$ are the bi-variate Bernstein polynomial of degree M and N, respectively, with $N \geq M$, given by:

$$\begin{gathered} \mathrm{B}^X_{ijk}(\boldsymbol{r},\boldsymbol{s},\boldsymbol{t}) = \begin{pmatrix} \boldsymbol{X} \\ \boldsymbol{ijk} \end{pmatrix} \boldsymbol{r}^i \boldsymbol{s}^j \boldsymbol{t}^k, \quad \boldsymbol{X} = \boldsymbol{M}, \boldsymbol{N} \\ \text{where,}\quad \boldsymbol{i}+\boldsymbol{j}+\boldsymbol{k} = X, \quad \boldsymbol{r}+\boldsymbol{s}+\boldsymbol{t} = 1, \quad \text{and,} \quad \begin{pmatrix} \boldsymbol{X} \\ \boldsymbol{ijk} \end{pmatrix} = \frac{\boldsymbol{X}!}{\boldsymbol{i}\,!\boldsymbol{j}\,!\,\boldsymbol{k}!} \end{gathered} \tag{11.696}$$

Note that because of the constraint: $\boldsymbol{r}+\boldsymbol{s}+\boldsymbol{t} = 1$, the Bernstein–Bezier polynomials effectively depend only on two variables.

Remark

- In what follows, we will restrict our discussion to non-rational tensor product elements, that is, the quadrilateral finite elements; similar considerations for the triangular finite elements can be developed without much difficulty on the basis of the detailed discussion of the triangular geometric elements in the mesh generation phase of patches and surfaces.

11.20.2 Bernstein–Bezier Geometric Element

For a typical non-rational quadrilateral finite element, then, a typical n-degree c-type shell geometry can be written in tensor-product form as:

$$\underbrace{\mathbf{C}_0(\hat{\xi}^1,\hat{\xi}^2)}_{3\times 1} = \underbrace{\sum_{i=0}^{n_c}\sum_{j=0}^{n_c} B_i^{n_c}(\hat{\xi}^1)\, B_j^{n_c}(\hat{\xi}^2)}_{3\times 3N_c^2} \quad \underbrace{{}^c\mathbf{q}_{ij}}_{3N_c^2\times 1}, \tag{11.697}$$

where $\mathbf{C}_0(\hat{\xi}^1,\hat{\xi}^2) \equiv \{\, X(\hat{\xi}^1,\hat{\xi}^2) \quad Y(\hat{\xi}^1,\hat{\xi}^2) \quad Z(\hat{\xi}^1,\hat{\xi}^2)\,\}^T, B_i^{n^c}(\hat{\xi}^1)\, B_j^{n^c}(\hat{\xi}^2)$, ${}^c\mathbf{q}_{ij}$ and n^c are the geometric position vector, the Bernstein–Bezier basis functions, the geometry control vectors and the degree of the geometry element, respectively, at any point $(\hat{\xi}^1,\hat{\xi}^2)$. The equation (11.697) can be rewritten in transformation form as:

$$\underbrace{\mathbf{C}_0(\hat{\xi}^1,\hat{\xi}^2)}_{3\times 1} = \underbrace{{}^c\hat{\mathbf{T}}(\hat{\xi}^1,\hat{\xi}^2)}_{3\times 3N_c^2} \quad \underbrace{{}^c\hat{\mathbf{q}}}_{3N_c^2\times 1}, \tag{11.698}$$

where we have:

$$\underbrace{{}^{c}\hat{\mathbf{q}}}_{3N_c^2\times 1} = \Big\{ \underbrace{q_{00}^{X} \cdots q_{n_c0}^{X}\, q_{01}^{X} \cdots q_{n_c1}^{X}\, q_{0n_c}^{X} \cdots q_{n_cn_c}^{X}}_{1\times N_c^2}\, \underbrace{q_{00}^{Y} \cdots q_{n0}^{Y}\, q_{01}^{Y} \cdots q_{n_c1}^{Y}\, q_{0n_c}^{Y} \cdots q_{n_cn_c}^{Y}}_{1\times N_c^2}$$
$$\underbrace{q_{00}^{Z} \cdots q_{n_c0}^{Z}\, q_{01}^{Z} \cdots q_{n_c1}^{Z}\, q_{0n_c}^{Z} \cdots q_{n_cn_c}^{Z}}_{1\times N_c^2} \Big\}^{T} \tag{11.699}$$

the $(3N_c^2 \times 1)$ geometry control vector, and ${}^{c}\hat{\mathbf{T}}(\hat{\xi}^1, \hat{\xi}^2)$ as the $(3 \times 3N_c^2)$ geometry interpolation transformation matrix containing the Bernstein–Bezier polynomials, that is,

$$ {}^{c}\hat{\mathbf{T}}(\hat{\xi}^1, \hat{\xi}^2) = $$
$$\underbrace{\begin{bmatrix} B_0^{n_c}B_0^{n_c} & B_1^{n_c}B_0^{n_c} & B_{n_c}^{n_c}B_{n_c}^{n_c} & 0\cdots & & \cdots & & & \cdots 0 \\ 0\cdots & & \cdots 0 & B_0^{n_c}B_0^{n_c} & B_1^{n_c}B_0^{n_c} & B_{n_c}^{n_c}B_{n_c}^{n_c} & & & \\ 0\cdots & & & & & \cdots 0 & B_0^{n_c}B_0^{n_c} & B_1^{n_c}B_0^{n_c} & B_{n_c}^{n_c}B_{n_c}^{n_c} \end{bmatrix}}_{3x3N^2} \tag{11.700}$$

Note that $N_c \equiv n_c + 1$ is the order of the geometric element with degree n_c.

Remarks

- It is recommended whenever possible to have the geometry representation at most cubic, that is, $n^c \leq 3$ for ease and efficiency in the computation of tangents, curvatures and twists that are subsequently necessary. So, as an example, for a cubic Bernstein–Bezier geometry net, that is, $n_c = 3$, results in the order of the geometry as: $N = 4$. Thus, we will have 48(i.e. three sets of 4×4 geometry control nets) will be the dimension of the geometry control vector, ${}^{c}\mathbf{q}$, and the interpolation transformation matrix is of dimension 3×16.
- Additionally, whenever conic geometric surfaces (i.e. spheres, ellipsoids, etc.) are involved, the geometric representation must be of rational Bernstein–Bezier form for accuracy.

Example: Let us consider a bilinear element, that is, $n_c = 1$, $N = N_c = 2$. We have:

$$\underbrace{\mathbf{C}_0(\hat{\xi}^1, \hat{\xi}^2)}_{3\times 1} = \underbrace{\sum_{i=0}^{1}\sum_{j=0}^{1} B_i^1(\hat{\xi}^1)\, B_j^1(\hat{\xi}^2)}_{3\times 12}\ \underbrace{{}^{c}\mathbf{q}_{ij}}_{12\times 1}, \tag{11.701}$$

with

$$\begin{aligned} B_0^1(\hat{\xi}^k) &\equiv 1-\hat{\xi}^k, \quad k=1,2 \\ B_1^1(\hat{\xi}^k) &\equiv \hat{\xi}^k \end{aligned} \tag{11.702}$$

implying

$$\boxed{\underbrace{{}^c\hat{\mathbf{q}}}_{12\times1} = \Big\{ \underbrace{q_{00}^X\ q_{10}^X\ q_{01}^X\ q_{11}^X}_{1\times4}\ \underbrace{q_{00}^Y\ q_{10}^Y\ q_{01}^Y\ q_{11}^Y}_{1\times4}\ \underbrace{q_{00}^Z\ q_{10}^Z\ q_{01}^Z\ q_{11}^Z}_{1\times4} \Big\}^T} \tag{11.703}$$

and

$$\begin{aligned} {}^c\hat{\mathbf{T}}(\hat{\xi}^1,\hat{\xi}^2) &= \underbrace{\begin{bmatrix} \underbrace{\mathbf{T}}_{1\times4} & \underbrace{\mathbf{0}}_{1\times4} & \underbrace{\mathbf{0}}_{1\times4} \\ \underbrace{\mathbf{0}}_{1\times4} & \underbrace{\mathbf{T}}_{1\times4} & \underbrace{\mathbf{0}}_{1\times4} \\ \underbrace{\mathbf{0}}_{1\times4} & \underbrace{\mathbf{0}}_{1\times4} & \underbrace{\mathbf{T}}_{1\times4} \end{bmatrix}}_{3x12}, \\ \mathbf{T} &\equiv \Big\{ (1-\hat{\xi}^1)(1-\hat{\xi}^2) \quad \hat{\xi}^1(1-\hat{\xi}^2) \quad (1-\hat{\xi}^1)\hat{\xi}^2 \quad \hat{\xi}^1\hat{\xi}^2 \Big\} \end{aligned} \tag{11.704}$$

11.20.3 Bernstein–Bezier Generalized Displacement Element

For a typical quadrilateral finite element, a typical n-degree c-type shell geometry can be written in tensor-product form as:

$$\boxed{\underbrace{\hat{\mathbf{d}}(\hat{\xi}^1,\hat{\xi}^2)}_{6\times1} = \underbrace{\sum_{i=0}^{n}\sum_{j=0}^{n} B_i^n(\hat{\xi}^1)\, B_j^n(\hat{\xi}^2)}_{6\times\, 6N^2} \quad \underbrace{{}^d\mathbf{q}_{ij}}_{6N^2\times1}} \tag{11.705}$$

where, $\hat{\mathbf{d}}$, $B_i^n(\hat{\xi}^1)\, B_j^n(\hat{\xi}^2)$, ${}^d\mathbf{q}_{ij}$ and n are the generalized (i.e. displacement and rotation) displacement vector, the Bernstein–Bezier basis functions, the generalized displacement control vectors and the degree of the finite element, respectively, at any point $(\hat{\xi}^1,\hat{\xi}^2)$. The equation (11.705) can be rewritten in transformation form as:

$$\boxed{\underbrace{\hat{\mathbf{d}}(\hat{\xi}^1,\hat{\xi}^2)}_{6\times1} = \underbrace{{}^d\hat{\mathbf{T}}(\hat{\xi}^1,\hat{\xi}^2)}_{6\times\, 6N^2} \quad \underbrace{{}^d\hat{\mathbf{q}}}_{6N^2\times1}} \tag{11.706}$$

$$\underbrace{{}^d\hat{\mathbf{q}}}_{6N^2\times 1} = \Big\{ \underbrace{q^u_{00}\cdots q^u_{n0}\; q^u_{01}\cdots q^u_{n1}\; q^u_{0n}\cdots q^u_{nn}}_{1\times N^2}\; \underbrace{q^v_{00}\cdots q^v_{n0}\; q^v_{01}\cdots q^v_{n1}\; q^v_{0n}\cdots q^v_{nn}}_{1\times NM}$$

where we have:

$$\underbrace{q^w_{00}\cdots q^w_{n0}\; q^w_{01}\cdots q^w_{n1}\; q^w_{0n}\cdots q^w_{nn}}_{1\times N^2}\; \underbrace{q^{\theta_1}_{00}\cdots q^{\theta_1}_{n0}\; q^{\theta_1}_{01}\cdots q^{\theta_1}_{n1}\; q^{\theta_1}_{0n}\cdots q^{\theta_1}_{nn}}_{1\times N^2}$$

$$\underbrace{q^{\theta_2}_{00}\cdots q^{\theta_2}_{n0}\; q^{\theta_2}_{01}\cdots q^{\theta_2}_{n1}\; q^{\theta_2}_{0n}\cdots q^{\theta_2}_{nn}}_{1\times N^2}\; \underbrace{q^{\theta_3}_{00}\cdots q^{\theta_3}_{n0}\; q^{\theta_3}_{01}\cdots q^{\theta_3}_{n1}\; q^{\theta_3}_{0n}\cdots q^{\theta_3}_{nn}}_{1\times N^2}\Big\}^T$$

as the $(6N^2 \times 1)$ generalized displacement control vector. Let ${}^d\hat{\bar{\mathbf{q}}}$ be the $(6N^2 \times 1)$ generalized virtual displacement control vector, and ${}^d\Delta\hat{\mathbf{q}}$, be the $(6N^2 \times 1)$ generalized incremental control vector, similar to ${}^d\mathbf{q}$. Then, let ${}^d\hat{\mathbf{T}}(\hat{\xi}^1, \hat{\xi}^2)$ be the $(6 \times 6N^2)$ interpolation transformation matrix containing the Bernstein–Bezier polynomials, that is,

$${}^d\hat{\mathbf{T}}(\hat{\xi}^1, \hat{\xi}^2) =$$

$$\underbrace{\begin{bmatrix} B^n_0B^n_0 & B^n_1B^n_0\cdots B^n_nB^n_{n-1} & B^n_nB^n_n & 0 & 0 & \cdots & \cdots & \cdots & 0 \\ 0 & 0 & 0 & B^n_0B^n_0 & B^n_1B^n_0\cdots & B^n_nB^n_n & 0 & \cdots & 0 \\ & & \cdots & \cdots & & & & & \\ & & \cdots & \cdots & & & & & \\ 0 & 0 & \cdots & \cdots & 0 & B^n_0B^n_0 & B^n_1B^n_0\cdots & & B^n_nB^n_n \end{bmatrix}}_{6x6N^2} \tag{11.707}$$

The derivatives involved in ${}^{\xi}\hat{\mathbf{B}}(\xi^1, \xi^2)$, the $(18 \times 6N^2)$ augmented interpolation transformation matrix:

$$\underbrace{{}^{\hat{\xi}}\hat{\mathbf{B}}(\hat{\xi}^1, \hat{\xi}^2)}_{18\times 6N^2} \equiv \begin{bmatrix} \underbrace{{}^d\mathbf{T}}_{6\times 6N^2} \\ \underbrace{{}^d\mathbf{T},_{\hat{\xi}^1}}_{6\times 6N^2} \\ \underbrace{{}^d\mathbf{T},_{\hat{\xi}^2}}_{6\times 6N^2} \end{bmatrix} \tag{11.708}$$

where ${}^d\mathbf{T},_{\hat{\xi}^1}$ and ${}^d\mathbf{T},_{\hat{\xi}^2}$ are computed by the first derivative recursive formula of the Bernstein basis functions at Gauss points as:

$$\frac{d}{d\hat{\xi}^\alpha}B(\hat{\xi}^\alpha) = n\,(B^{n-1}_{i-1}(\hat{\xi}^\alpha) - B^{n-1}_i(\hat{\xi}^\alpha)), \quad \alpha = 1, 2 \tag{11.709}$$

Finally, we must express the infinitesimal area, $d\hat{A} = \sqrt{\hat{G}}\,d\hat{\xi}^1\,d\hat{\xi}^2$, of the area integrals where $\sqrt{\hat{G}}$ is the positive square root of the determinant of the surface metric expressed in terms of the geometry coordinate parameters, $\hat{\xi}^\alpha$, $\alpha = 1, 2$.

Remarks

- Note that both the ${}^{d}\hat{\mathbf{T}}(\hat{\xi}^1, \hat{\xi}^2)$ and ${}^{\xi}\hat{\mathbf{B}}(\hat{\xi}^1, \hat{\xi}^2)$ matrices for all Gauss points can be computed and saved once and stored away for repeated use in the Newton-type iterative algorithm; of course, in adaptive refinements the locations of the Gauss points change requiring recomputation.
- For convergence, we must have the elements at most isoparametric, that is, $n^c = n$; we must always avoid superparametry, that is, n^c must never allowed to exceed n.
- For computational efficiency such as tangent and curvature computations, a subparametric representation, that is, n^c strictly less than n, with $n^c = 3$, that is, up to cubic spline representation for the non-rational geometry, seems an optimal choice. Similarly, for rational geometry such as spheres, ellipsoids, and so on, the quadratic rational spline is the desired geometric representation.
- It turns out that for shell finite element analysis, the optimum choice for the generalized displacement distribution must be with a quintic Bernstein–Bezier net, that is, $n = 5$, resulting in the order of the element as: $N = 6$. Thus, we will have 216(i.e. six sets of 6×6 control nets) will be the dimension of the generalized displacement control vector, ${}^{d}\mathbf{q}$, and the interpolation transformation matrix is of dimension 6×36.

11.20.4 Gauss Quadrature Points and the Lines of Curvature

Now for a quintic Bernstein–Bezier net, for example, over an element that is, with $n = 5$, we have for the order of the element as: $N = 6$. Thus, for full integration, we will have a set of 6×6 Gauss quadrature points. With the locations of the Gauss quadrature points fixed, our goal is to determine the lines of curvature directions through these points and other desired geometric properties. But, in order to accomplish this, we need to construct a local frame at these points with the triad of base vectors: two tangent vectors along the coordinate iso-curves: $\hat{\mathbf{G}}_\beta \equiv \dfrac{\partial \hat{\mathbf{C}}_0}{\partial \hat{\xi}^\beta}$ $\beta = 1, 2$, and the unit normal to these base vectors, $\mathbf{N} \equiv \hat{\mathbf{N}} \equiv \hat{\mathbf{G}}_3 = \dfrac{\hat{\mathbf{G}}_1 \times \hat{\mathbf{G}}_2}{\|\hat{\mathbf{G}}_1 \times \hat{\mathbf{G}}_2\|}$; these are generally non-orthogonal to each other. So, by taking derivatives of equation (11.697) with respect to the surface parameters, we get:

$$\boxed{\begin{aligned}
\underbrace{\hat{\mathbf{G}}_1(\hat{\xi}^1, \hat{\xi}^2)}_{3\times1} &\equiv \frac{\partial \mathbf{C}_0(\hat{\xi}^1, \hat{\xi}^2)}{\partial \hat{\xi}^1} == n_c \sum_{j=0}^{n_c} \left\{ \sum_{i=0}^{n_c-1} (\Delta^{1,0}\, {}^{c}\mathbf{q}_{ij}) B_i^{n_c-1}(\hat{\xi}^1) \right\} B_j^{n_c}(\hat{\xi}^2). \\
\underbrace{\hat{\mathbf{G}}_2(\hat{\xi}^1, \hat{\xi}^2)}_{3\times1} &\equiv \frac{\partial \mathbf{C}_0(\hat{\xi}^1, \hat{\xi}^2)}{\partial \hat{\xi}^2} == n_c \sum_{i=0}^{n_c} \left\{ \sum_{j=0}^{n_c-1} (\Delta^{0,1}\, {}^{c}\mathbf{q}_{ij}) B_j^{n_c-1}(\hat{\xi}^2) \right\} B_i^{n_c}(\hat{\xi}^1)
\end{aligned}} \tag{11.710}$$

where we have introduced the following first-order forward difference operators, $\Delta^{1.0}$ and $\Delta^{0.1}$ as:

$$\boxed{\begin{aligned}
\Delta^{1,0}\, {}^{c}\mathbf{q}_{ij} &\equiv {}^{c}\mathbf{q}_{(i+1)j} - {}^{c}\mathbf{q}_{ij} \\
\Delta^{0,1}\, {}^{c}\mathbf{q}_{ij} &\equiv {}^{c}\mathbf{q}_{i(j+1)} - {}^{c}\mathbf{q}_{ij}
\end{aligned}} \tag{11.711}$$

Note that the great advantage of using difference operators is that we only need to know and use the differences of original Bezier control points of the geometry. The normal vector is, then, easily obtained as a cross product of the tangent vectors given by equations (11.710) and (11.711); however, note that the normals at four corners are easily obtained as the partials that may be expressed as the difference of boundary points, as for example:

$$\hat{\mathbf{N}}(0,0) = \frac{\Delta^{1,0\,c}\mathbf{q}_{00} \times \Delta^{0,1\,c}\mathbf{q}_{00}}{\|\Delta^{1,0\,c}\mathbf{q}_{00} \times \Delta^{0,1\,c}\mathbf{q}_{00}\|} \tag{11.712}$$

Now, for the determination of the lines of curvature direction, we will additionally need the curvatures at Gauss points, that is, the second derivatives of the position vectors, or equivalently, the first derivatives of the base vectors. This, in turn, will need us to introduce the second-order forward differences. So, by taking second derivatives of equation (11.697) with respect to the surface parameters, we get:

$$\boxed{\begin{aligned}
\mathbf{C}_0(\hat{\xi}^1,\hat{\xi}^2),_{\hat{\xi}^1\hat{\xi}^1} &== \frac{n!}{(n-2)!}\sum_{j=0}^{n_c}\left\{\sum_{i=0}^{n_c-2}(\Delta^{2,0\,c}\mathbf{q}_{ij})B_i^{n_c-2}(\hat{\xi}^1)\right\}B_j^{n_c}(\hat{\xi}^2).\\
\mathbf{C}_0(\hat{\xi}^1,\hat{\xi}^2),_{\hat{\xi}^2\hat{\xi}^2} &== \frac{n!}{(n-2)!}\sum_{i=0}^{n_c}\left\{\sum_{j=0}^{n_c-2}(\Delta^{0,2\,c}\mathbf{q}_{ij})B_j^{n_c-2}(\hat{\xi}^2)\right\}B_i^{n_c}(\hat{\xi}^1)\\
\mathbf{C}_0(\hat{\xi}^1,\hat{\xi}^2),_{\hat{\xi}^1\hat{\xi}^2} &== n^2\sum_{i=0}^{n_c-1}\left\{\sum_{j=0}^{n_c-1}(\Delta^{1,1\,c}\mathbf{q}_{ij})B_j^{n_c-1}(\hat{\xi}^2)\right\}B_i^{n_c-1}(\hat{\xi}^1)
\end{aligned}} \tag{11.713}$$

where we have introduced the second-order forward difference operators as:

$$\boxed{\begin{aligned}
\Delta^{2,0\,c}\mathbf{q}_{ij} &\equiv \Delta^{1,0\,c}\mathbf{q}_{(i+1)j} - \Delta^{1,0\,c}\mathbf{q}_{ij}\\
\Delta^{0,2\,c}\mathbf{q}_{ij} &\equiv \Delta^{0,1\,c}\mathbf{q}_{i(j+1)} - \Delta^{0,1\,c}\mathbf{q}_{ij}\\
\Delta^{1,1\,c}\mathbf{q}_{ij} &\equiv \Delta^{0,1\,c}\mathbf{q}_{(i+1)j} - \Delta^{0,1\,c}\mathbf{q}_{ij}
\end{aligned}} \tag{11.714}$$

Again, note that the great advantage of using difference operators is that we only need to know and use the differences of the original Bezier control points of the geometry. Let us recall that because the normal vector has unit length, and that $\mathbf{N} \bullet \mathbf{C}_0,_{\xi^\alpha} = \mathbf{N} \bullet \hat{\mathbf{G}}_\alpha = 0$ for $\alpha = 1,2$, we have, by differentiation: $\mathbf{C}_0,_{\xi^\alpha} \bullet \mathbf{N},_{\xi^\beta} = -\mathbf{C}_0,_{\xi^\alpha\xi^\beta} \bullet \mathbf{N}$ for $\alpha,\beta = 1,2$. Now, from our discussions on differential geometry of surfaces, with $\lambda \equiv \dfrac{d\hat{\xi}^2}{d\hat{\xi}^1} = \tan\alpha_1$, the curvature is given as a function of λ by:

$$k(\lambda) \equiv \frac{L + 2M\lambda + N\lambda^2}{E + 2F\lambda + G\lambda^2} \tag{11.715}$$

where we have introduced the following definitions:

$$
\boxed{\begin{aligned}
&E \equiv \hat{G}_{11} = \mathbf{C},_{\xi^1} \bullet \mathbf{C},_{\xi^1}\\
&F \equiv \hat{G}_{12} = \mathbf{C},_{\xi^1} \bullet \mathbf{C},_{\xi^2}\\
&G \equiv \hat{G}_{22} = \mathbf{C},_{\xi^2} \bullet \mathbf{C},_{\xi^2}\\
&L \equiv -\mathbf{C},_{\xi^1} \bullet \mathbf{N},_{\xi^1} = \mathbf{C},_{\xi^1\xi^1} \bullet \mathbf{N} \equiv \hat{B}_{11}\\
&M \equiv -\frac{1}{2}(\mathbf{C},_{\xi^1} \bullet \mathbf{N},_{\xi^2} + \mathbf{C},_{\xi^2} \bullet \mathbf{N},_{\xi^1}) = \mathbf{C},_{\xi^1\xi^2} \bullet \mathbf{N} \equiv \hat{B}_{12}\\
&N \equiv -\mathbf{C},_{\xi^2} \bullet \mathbf{N},_{\xi^2} = \mathbf{C},_{\xi^2\xi^2} \bullet \mathbf{N} \equiv \hat{B}_{22}
\end{aligned}} \tag{11.716}
$$

Note that $\hat{B}_{\alpha\beta}$ for $\alpha, \beta = 1, 2$ are the components of the surface curvature tensor. Moreover, also recall that if $\mathbf{C}_0$ is an umbilical point, that is, $L : M : N = E : F : G$, then the curvature is independent of λ; otherwise, we can find the extreme values as the desired curvatures, $\frac{1}{R_1}$ and $\frac{1}{R_2}$ at the roots λ_1, λ_2 (where these roots define the directions of lines of curvature) from vanishing of the following determinant:

$$
\det \begin{bmatrix} \lambda^2 & -\lambda & 1 \\ E & F & G \\ L & M & N \end{bmatrix} = 0 \tag{11.717}
$$

With some manipulation, we can write the expressions of the roots as:

$$
\lambda_{1\,or\,2} = \frac{-(NE - GL) \pm \sqrt{(NE - GL)^2 - 4(FN - GM)(ME - FL)}}{2(FN - GM)} \tag{11.718}
$$

The curvatures, $\frac{1}{R_1}$ and $\frac{1}{R_2}$, themselves are obtained as roots of:

$$
\det \begin{bmatrix} \frac{1}{R}E - L & \frac{1}{R}F - M \\ \frac{1}{R}F - M & \frac{1}{R}G - N \end{bmatrix} = 0 \tag{11.719}
$$

resulting in:

$$
\boxed{\begin{aligned}
&\frac{1}{R_1} \; or \; \frac{1}{R_2} = \frac{-b \pm \sqrt{b^2 - 4ac}}{2a},\\
&a \equiv 2FN - GM\\
&b \equiv 2FM - NE - GL\\
&c \equiv ME - FL
\end{aligned}} \tag{11.720}
$$

Note that if $F \equiv 0$ and $M \equiv 0$ (spherical surfaces, for example), the coordinate iso-curves are already along the lines of curvature, then, $\lambda = 0$, and we get: $\frac{1}{R_1}$ *or* $\frac{1}{R_2} = \frac{N}{G}, \frac{L}{E} = \frac{\hat{B}_{22}}{\hat{G}_{22}}, \frac{\hat{B}_{11}}{\hat{G}_{11}}$. Also, we can compute the determinant of the surface metric tensor necessary for element surface integrals as: $\sqrt{\hat{G}} = \sqrt{EG - F^2}$. Now, the unit base vectors (i.e. with arc length parameterization) along the lines of curvature can be easily related by a rotational matrix to the coordinate base vectors as:

$$\begin{Bmatrix} \mathbf{G}_1 \\ \mathbf{G}_2 \end{Bmatrix} = \begin{bmatrix} \cos\alpha_1 & \sin\alpha_1 \\ -\sin\alpha_1 & \cos\alpha_1 \end{bmatrix} \begin{Bmatrix} \hat{\mathbf{G}}_1 \\ \hat{\mathbf{G}}_2 \end{Bmatrix} \tag{11.721}$$

where α_1 is the angle that the ξ^1- line of curvature makes with the $\hat{\xi}^1$ coordinate curve. Finally, using equation (11.721), we can easily establish the derivative relationship with respect to the arc length parameters and the coordinate parameters as:

$$\begin{Bmatrix} \dfrac{\partial}{\partial S^1} \\ \dfrac{\partial}{\partial S^2} \end{Bmatrix} = \begin{bmatrix} \dfrac{1}{\gamma_1}\cos\alpha_1 & \dfrac{1}{\gamma_1}\sin\alpha_1 \\ -\dfrac{1}{\gamma_2}\sin\alpha_1 & \dfrac{1}{\gamma_2}\cos\alpha_1 \end{bmatrix} \begin{Bmatrix} \dfrac{\partial}{\partial \hat{\xi}^1} \\ \dfrac{\partial}{\partial \hat{\xi}^2} \end{Bmatrix} \tag{11.722}$$

where $\gamma_\alpha \equiv \|\bar{\mathbf{G}}_\alpha(\xi^\alpha)\|, \quad \alpha = 1, 2$ are the lengths of the base vectors along the lines of curvature. Now, note that equation (11.722) can be used to compute $\frac{\partial \gamma_\alpha}{\partial S^\beta}$ for $\alpha, \beta = 1, 2$ with the knowledge that: $\frac{\partial \gamma_\alpha}{\partial \hat{\xi}^\beta} = \frac{\mathbf{G}_\alpha \bullet \mathbf{G}_{\alpha}, _{\hat{\xi}^\beta}}{\gamma_\alpha}$, $\mathbf{G}_1, _{\hat{\xi}^\beta} = \hat{B}_{1\beta}\cos\alpha_1 + \hat{B}_{2\beta}\sin\alpha_1$, and $\mathbf{G}_2, _{\hat{\xi}^\beta} = -\hat{B}_{1\beta}\sin\alpha_1 + \hat{B}_{2\beta}\cos\alpha_1$.

Exercise: Show that $\frac{\partial \gamma_\alpha}{\partial \hat{\xi}^\beta} = \frac{\mathbf{G}_\alpha \bullet \mathbf{G}_{\alpha}, _{\hat{\xi}^\beta}}{\gamma_\alpha}$.

Solution: With $\gamma_\alpha^2 \equiv \|\mathbf{G}_\alpha\|^2 = \mathbf{G}_\alpha \bullet \mathbf{G}_\alpha$, taking derivative: $\not{2}\gamma_\alpha \frac{\partial \gamma_\alpha}{\partial \hat{\xi}^\beta} = \not{2}\mathbf{G}_\alpha \bullet \frac{\partial \mathbf{G}_\alpha}{\partial \hat{\xi}^\beta}$ completes the proof.

With the above background, we can express all the components of both the desired curvature vectors: $\mathbf{k}_1^0 \equiv \begin{Bmatrix} 0 & -\frac{1}{R_1} & -\frac{1}{\gamma_1}\gamma_1, _2 \end{Bmatrix}^T$ and $\mathbf{k}_2^0 \equiv \begin{Bmatrix} \frac{1}{R_2} & 0 & \frac{1}{\gamma_2}\gamma_2, _1 \end{Bmatrix}^T$ in terms of the coordinate parameters, $\hat{\xi}^\alpha, \quad \alpha = 1, 2$. Finally, $\sqrt{\hat{G}} \equiv \sqrt{\hat{G}_{11}\hat{G}_{22} - \hat{G}_{12}^2}$, the positive square root of the determinant of the surface metric is evaluated at all the Gauss quadrature points of integration.

Remark

- Thus, note that the reference surface properties – the initial curvature vectors, $\mathbf{k}_1^0$ and $\mathbf{k}_2^0$ along with the square root of the determinant $\sqrt{\hat{G}}$ of the surface metric – can be computed at all Gauss points and saved once and stored away for repeated use in the Newton-type iterative algorithm; of course, in adaptive refinement (i.e. c_h-, c_p- and c_k- type refinements), the locations of the Gauss points change requiring recomputation.

11.20.5 Algorithmic Code Snippets for Geometric Properties

```
        Subroutine Calc_ReferenceProps(Qc,ri,rj,jCInt)
!       ==================================================
!         n m-1
!       A1hat = m * sum { sum (Dif10*Qij*Bri(ri)) } Brj(rj)
!         j=0 i=0
!       ==================================================
        Implicit Real(kind=8) (a-h,o-z)
!       =================
        include 'SizeVar.h'
        include 'files.h'
        include 'Scratch.h'
        include 'CurParams.h'
!       =================
        Real(kind=8)    Qc
        Dimension              Qc(nDim,nQc1,nQc2)
        Real(kind=8)    X11hat     ,X12hat     ,X22hat
        Dimension              X11hat(nDim),X12hat(nDim),X22hat(nDim)
        Real(kind=8)    A1     ,A2     ,Atem     ,Btem
        Dimension              A1(nDim),A2(nDim),Atem(nDim),Btem(nDim)
        Real(kind=8)    A1c1     ,A1c2     ,A2c1     ,A2c2
        Dimension              A1c1(nDim),A1c2(nDim),A2c1(nDim),A2c2(nDim)
!
        DATA zero/0.D0/,one/1.0D0/,two/2.0D0/,three/3.0D0/
!=========================================================
!       ri = 0.D0
!       rj = 0.D0
!       -----------------------------------------------------------metric Tensor
        nQcm = MAX(nQc1-1,nQc2-1)
!
        call Calc_A1_Hat(Qc,nDim,nQc1,nQc2,nQcm,ri,rj,A1hat)
!       call VectorNorm(A1hat,A,3,Iout)
!       call ScaleVector(A1hat,uA1hat,1.D0/A,3,Iout)
!
        call Calc_A2_Hat(Qc,nDim,nQc1,nQc2,nQcm,ri,rj,A2hat)
!       call VectorNorm(A2hat,A,3,Iout)
!       call ScaleVector(A2hat,uA2hat,1.D0/A,3,Iout)
!       ---------------------------------------------------- unit normal
        call CrossProduct(A1hat,A2hat,xNor,Iout)
        call VectorNorm(xNor,A,3,Iout)
        call ScaleVector(xNor,xNor,1.D0/A,3,Iout)
!       ---------------------------------------------------- RootAhat
        call DotProduct(A1hat,A1hat,E,3,Iout)
        call DotProduct(A1hat,A2hat,F,3,Iout)
        call DotProduct(A2hat,A2hat,G,3,Iout)
!
```

```
        RutAhat = DSQRT(E*G - F*F)
!       -------------------------------------------------- curvature Tensor
        X11hat = 0.D0
        X12hat = 0.D0
        X22hat = 0.D0
!
        if(nQc1.gt.2) then     ! atleast Quadratic
                call Calc_X11_Hat(Qc,nDim,nQc1,nQc2,nQcm,ri,rj,X11hat)
        endif
        if(nQc2.gt.2) then     ! atleast Quadratic
                call Calc_X22_Hat(Qc,nDim,nQc1,nQc2,nQcm,ri,rj,X22hat)
        endif
        if(nQc1.gt.2.AND.nQc2.gt.2) then    ! atleast both Quadratic
                call Calc_X12_Hat(Qc,nDim,nQc1,nQc2,nQcm,ri,rj,X12hat)
        endif
!
        call DotProduct(X11hat,xNor,xL,3,Iout)          ! B11hat
        call DotProduct(X12hat,xNor,xM,3,Iout)          ! B12hat
        call DotProduct(X22hat,xNor,xN,3,Iout)          ! B22hat
!       -------------------------------------------------- Lines of curvatures
        FEG =F/E/G
        if (dabs(FEG).le.1.0D-3.or.xM==0.d0) then    ! already lines of curvatures
               Rho1     = zero                   ! direction
               Rho2     = zero                   ! direction
               Cu1             = xL/E            ! 1/R1
               Cu2             = xN/G   ! 1/R2
!
               ThetaC  = zero
               sT              = zero
               cT              = one
        else
               ENmGL   = E*xN - G*xL
               ENpGL   = E*xN + G*xL
               FNmGM   = F*xN - G*xM
               EMmFL   = E*xM - F*xL
               SQ                 = DSQRT(ENmGL*ENmGL - 4.D0*FNmGM*EMmFL)
               FNGM2   = 2.D0*FNmGM
!
               Rho1               = (-ENmGL + SQ)/FNGM2
               Rho2               = (-ENmGL - SQ)/FNGM2
!
               EGmFsq = E*G - F*F
               TEGmFsq = 2.D0*EGmFsq
               xLNmMsq = xL*xN - xM*xM
               TFMmENpGL= 2.D0*F*xM - ENpGL
               SQ              = DSQRT(TFMmENpGL*TFMmENpGL - 4.D0*EGmFsq*xLNmMsq)
               Cu1             = (-TFMmENpGL + SQ)/TEGmFsq
               Cu2             = (-TFMmENpGL - SQ)/TEGmFsq
!
               ThetaC        = DATAN(Rho1)
               sT              = DSIN(ThetaC)
               cT              = DCOS(ThetaC)
        endif
```

```
!        ------------ Bases: Ao1 & Ao2 & Ao12N Coord Xformation
         call ScaleVector(A1hat,aTem,cT,3,Iout)
         call ScaleVector(A2hat,bTem,sT,3,Iout)
         A1 = aTem + bTem
!
         call ScaleVector(A1hat,aTem,-sT,3,Iout)
         call ScaleVector(A2hat,bTem,cT,3,Iout)
         A2 = aTem + bTem
!        ------------------------------------------- Lamda1 & Lamda2
         call VectorNorm(A1,xLam1,3,Iout)
         call ScaleVector(A1,Ao1,1.D0/xLam1,3,Iout)
         call VectorNorm(A2,xLam2,3,Iout)
         call ScaleVector(A2,Ao2,1.D0/xLam2,3,Iout)
!
         Ao12N(1,1) = Ao1(1)
         Ao12N(2,1) = Ao1(2)
         Ao12N(3,1) = Ao1(3)
!        ------------------------------------------- Xformation:
         Global Cartesian -> Lines of Curvs.
         Ao12N(1,2) = Ao2(1)
         Ao12N(2,2) = Ao2(2)
         Ao12N(3,2) = Ao2(3)
!
         Ao12N(1,2) = Ao2(1)
         Ao12N(2,2) = Ao2(2)
         Ao12N(3,2) = Ao2(3)
!
         Ao12N(1,3) = xNor(1)
         Ao12N(2,3) = xNor(2)
         Ao12N(3,3) = xNor(3)
!        ------------------------------------------- (Lamda_alfa,beta)
         call ScaleVector(X11hat,aTem,cT,3,Iout)
         call ScaleVector(X12hat,bTem,sT,3,Iout)
         A1c1 = aTem + bTem                           ! A1,1
         call DotProduct(A1,A1c1,xLam11hat,3,Iout)    ! Lamda1,1hat
         xLam11hat = xLam11hat/xLam1
!
         call ScaleVector(X12hat,aTem,cT,3,Iout)

         call ScaleVector(X22hat,bTem,sT,3,Iout)
         A1c2 = aTem + bTem                           ! A1,2
         call DotProduct(A1,A1c2,xLam12hat,3,Iout)    ! Lamda1,2hat
         xLam12hat = xLam12hat/xLam1
!
         call ScaleVector(X11hat,aTem,-sT,3,Iout)
         call ScaleVector(X12hat,bTem, cT,3,Iout)
         A2c1 = aTem + bTem                           ! A2,1
         call DotProduct(A2,A2c1,xLam21hat,3,Iout)    ! Lamda2,1hat
         xLam21hat = xLam21hat/xLam2
!
         call ScaleVector(X12hat,aTem,-sT,3,Iout)
         call ScaleVector(X22hat,bTem, cT,3,Iout)
         A2c2 = aTem + bTem                           ! A1,2
         call DotProduct(A2,A2c2,xLam22hat,3,Iout)    ! Lamda2,2hat
```

```
       xLam22hat = xLam22hat/xLam2
!
!      xLam11 = (cT*xLam11hat + sT*xLam12hat)/xLam1        ! lamda1,1
       xLam21 = ( cT*xLam21hat + sT*xLam22hat)/xLam1       ! lamda2,1
       xLam12 = (-sT*xLam11hat + cT*xLam12hat)/xLam2       ! lamda1,2
!      xLam22 = (-sT*xLam12hat + cT*xLam22hat)/xLam2       ! lamda2,2
!      -------------------------------------------Curvatures: ko1 & ko2
       C01(1) = zero
       C01(2) = +Cu1
!      C01(2) = Cu1
       C01(3) = +xLam12/xLam1
!
       C02(1) = -Cu2
       C02(2) = zero
       C02(3) = -xLam21/xLam2
!      ---------------------
!      ---------------------------------- Curvatures: ko1 & ko2 CORRECT ONE
       C01(1) = zero
       C01(2) = -Cu1
!      C01(2) = Cu1
       C01(3) = -xLam12/xLam1
!
       C02(1) = Cu2
       C02(2) = zero
       C02(3) = xLam21/xLam2
!      ============================================= save in tape
       call PutGetRefPropsatGaussPt(iRefPro1,jCInt,1)
!      ============================================= save in tape end
       return
       end

       Subroutine Calc_X11_Hat(Qc,nDim,nQc1,nQc2,nQcm,ri,rj,X11hat)
!      actual degree m = nQc1 -1
!      actual degree n = nQc2 -1
!      Because indices run from 1,...nQc1 instead of i = 0,....m
!      Because indices run from 1,...nQc2 instead of j = 0,....n
!      so, range 0 to m-2 becomes 1 to nQc1-2, etc.
! ==============================================================
!        n m-2
!      X11hat = m*(m-1) * sum { sum (Dif20*Qij*Bri(ri)) } Brj(rj)
!        j=0 i=0
! ==============================================================
       Implicit Real(kind=8) (a-h,o-z)
       Real(kind=8)        X11hat    ,Dif20
       Dimension                     X11hat(nDim),Dif20(nDim)
       Real(kind=8)        Qc        ,Bri   ,Brj
       Dimension                     Qc(nDim,nQc1,nQc2),Bri(nQc1),Brj(nQc2)
!
       nQc1m1  =  nQc1-1 !actual degree      = m
       nQc1m2  =  nQc1-2 !actual degree - 1 = m-1
!      --------------------------------------------Basis Functions
       call Bernstein(Bri,nQc1m2,ri,iOut)
       call Bernstein(Brj,nQc2,rj,iOut)
```

```
!       ------------------------------
        X11hat      = 0.D0
        do 30 j= 1,nQc2
        do 10 i= 1,nQc1m2
!       ===================                                    Dif20 = (Type 2)
        call DifferenceOp_2(Qc,nDim,nQc1,nQc2,i,j, Dif20,2,iOut)
!       ===================
        do 20 k= 1,nDim
        X11hat(k) = X11hat(k) + Dif20(k)*Bri(i)*Brj(j)
   20   continue
   10   continue
   30   continue
!
        X11hat = nQc1m1 * nQc1m2 * X11hat
!       ==========
        return
        end

        Subroutine DifferenceOp_1(Qc,nDim,nQc1,nQc2,i1,j2,Diff,iTyp,iOut)
!       ==================================================
        Implicit Real(kind=8) (a-h,o-z)
!       ===============
        real(kind = 8) Qc
        Dimension               Qc(nDim,nQc1,nQc2)
        real(kind = 8) Diff
        Dimension               Diff(nDim)
        integer                 nQc1,nQc2,iOut
!
        Diff     = 0.D0
        if(iTyp == 1) then
!               --------------------------- Dif10 = Q(i+1,j) - Q(i,j) (Type 1)
                do 30 k= 1,nDim
                Diff(k) = Qc(k,i1+1,j2) - Qc(k,i1,j2)
   30           continue
!               --------------------------- Dif01 = Q(i,j+1) - Q(i,j) (Type 2)
        elseif (iTyp == 2) then
                do 40 k= 1,nDim
                Diff(k) = Qc(k,i1,j2+1) - Qc(k,i1,j2)
   40           continue
        endif
!--------------------------
        return
        end

        Subroutine DifferenceOp_2(Qc,nDim,nQc1,nQc2,i1,j2, Dif2,iTyp,iOut)
!       ==================================================
        Implicit Real(kind=8) (a-h,o-z)
!       ====================
        real(kind = 8)   Qc
        Dimension              Qc(nDim,nQc1,nQc2)
        real(kind = 8)   Diff   ,Dif2
        Dimension              Diff(nDim),Dif2(nDim)
        integer                nQc1,nQc2,iOut
```

```
!-----------------------------------
      Dif2      = 0.D0
!
      if(iTyp == 1) then
!              --------------------- Dif11 = Dif10(i,j+1) - Dif10(i,j) (Type 1)
!              -----------or: Dif11 = Dif01(i+1,j) - Dif01(i,j) (Type 2) SELECTED
              do 20 k= 1,nDim
!     ====================
              call DifferenceOp_1(Qc,nDim,nQc1,nQc2,i1+1,j2, Diff,2,iOut)
              Dif2(k) = Diff(k)
              call DifferenceOp_1(Qc,nDim,nQc1,nQc2,i1,j2, Diff,2,iOut)
              Dif2(k) = Dif2(k) - Diff(k)
   20         continue
      elseif (iTyp == 2) then
!              --------------------- Dif20 = Dif10(i+1,j) - Dif10(i,j) (Type 1)
              do 40 k= 1,nDim
!     ====================
              call DifferenceOp_1(Qc,nDim,nQc1,nQc2,i1+1,j2, Diff,1,iOut)
              Dif2(k) = Diff(k)
              call DifferenceOp_1(Qc,nDim,nQc1,nQc2,i1,j2, Diff,1,iOut)
              Dif2(k) = Dif2(k) - Diff(k)
   40         continue
      elseif (iTyp == 3) then
!              --------------------- Dif02 = Dif01(i,j+1) - Dif01(i,j) (Type 2)
              do 60 k= 1,nDim
              call DifferenceOp_1(Qc,nDim,nQc1,nQc2,i1,j2+1, Diff,2,iOut)
              Dif2(k) = Diff(k)
              call DifferenceOp_1(Qc,nDim,nQc1,nQc2,i1,j2, Diff,2,iOut)
              Dif2(k) = Dif2(k) - Diff(k)
   60         continue
      endif
!--------------------------------------
      return
      end

Subroutine Calc_ReferenceProps_Hem(Qc,tmin,tmax,pmin,pmax, ri,rj,jCInt)
!==================================================================
!HARDCODED for LINES OF CURVATURES ALONG COORD.DIRS. 1 & 2
! For the Hemisphere with Hole Problem in the sequel
!==================================================================
!        n m-1
!     A1hat = m * sum { sum (Dif10*Qij*Bri(ri)) } Brj(rj)
!        j=0 i=0
!     ============================================================
      Implicit Real(kind=8) (a-h,o-z)
!     =====================
      include 'SizeVar.h'
      include 'files.h'
      include 'Scratch.h'
      include 'CurParams.h'
!     =====================
      Real(kind=8)       RutAhatAlt
      Real(kind=8)       Qc
```

```
      Dimension                  Qc(nDim,nQc1,nQc2)
      Real(kind=8)      X11hat        ,X12hat       ,X22hat
      Dimension                  X11hat(nDim),X12hat(nDim),X22hat(nDim)
      Real(kind=8)      A1       ,A2       ,Atem       ,Btem
      Dimension                  A1(nDim),A2(nDim),Atem(nDim),Btem(nDim)
      Real(kind=8)      A1c1      ,A1c2      ,A2c1      ,A2c2
      Dimension                  A1c1(nDim),A1c2(nDim),A2c1(nDim),A2c2(nDim)
!
      DATA zero/0.D0/,one/1.0D0/,two/2.0D0/,three/3.0D0/
!==================================================
      rad = 10.D0                    ! HARDCODED for radius = 10
      the     = tmax - tmin
      phe = pmax - pmin
!     ---------------------------------------Coords along Lines of Curvatures
      cT     = one                   ! needed in A1hat routine etc
      sT     = zero
!     ------------------------------------------ metric Tensor
      tri = tmin + (tmax - tmin)*ri
      prj = pmin + (pmax - pmin)*rj
!
      A1hat(1) = -rad*dcos(prj)*dsin(tri)*the
      A1hat(2) =  rad*dcos(prj)*dcos(tri)*the
      A1hat(3) = zero
!
      A2hat(1) = -rad*dsin(prj)*dcos(tri)*phe
      A2hat(2) = -rad*dsin(prj)*dsin(tri)*phe
      A2hat(3) =  rad*dcos(prj)*phe
!     ------------------------------------------unit normal
      call CrossProduct(A1hat,A2hat,xNor,Iout)
      call VectorNorm(xNor,A,3,Iout)
      call ScaleVector(xNor,xNor,1.D0/A,3,Iout)
!     ------------------------------------------RootAhat
      call DotProduct(A1hat,A1hat,E,3,Iout) ! A11hat
      call DotProduct(A1hat,A2hat,F,3,Iout) ! A12hat
      call DotProduct(A2hat,A2hat,G,3,Iout) ! A22hat
!
      RutAhat = DSQRT(E*G - F*F)
      RutAhatAlt = rad*rad*dcos(prj)*the*phe
!     -------------------------------Bases: Ao1 & Ao2 & Ao12N Coord Xformation
      A1 = A1hat
      A2 = A2hat
!     ------------------------------------------Lamda1 & Lamda2
      call VectorNorm(A1,xLam1,3,Iout)
      call ScaleVector(A1,Ao1,1.D0/xLam1,3,Iout)
      call VectorNorm(A2,xLam2,3,Iout)
      call ScaleVector(A2,Ao2,1.D0/xLam2,3,Iout)
!
      Ao12N(1,1) = Ao1(1)
      Ao12N(2,1) = Ao1(2)
      Ao12N(3,1) = Ao1(3)
!     -------------------------Xformation: Global Cartesian -> Lines of Curvs.
      Ao12N(1,2) = Ao2(1)
      Ao12N(2,2) = Ao2(2)
```

```
      Ao12N(3,2) = Ao2(3)
!
      Ao12N(1,2) = Ao2(1)
      Ao12N(2,2) = Ao2(2)
      Ao12N(3,2) = Ao2(3)
!
      Ao12N(1,3) = xNor(1)
      Ao12N(2,3) = xNor(2)
      Ao12N(3,3) = xNor(3)
!     ----------------------------------------------- (Lamda, alfa, beta)
      X11hat(1) = -rad*dcos(prj)*dcos(tri)*the*the
      X11hat(2) = -rad*dcos(prj)*dsin(tri)*the*the
      X11hat(3) = zero
!
      X22hat(1) = -rad*dcos(prj)*dcos(tri)*phe*phe
      X22hat(2) = -rad*dcos(prj)*dsin(tri)*phe*phe
      X22hat(3) = -rad*dsin(prj)*phe*phe
!
      X12hat(1) =  rad*dsin(prj)*dsin(tri)*the*phe
      X12hat(2) = -rad*dsin(prj)*dcos(tri)*the*phe
      X12hat(3) = zero
!     ============================== Surface Curvatures Tensor
      call DotProduct(xNor,X11hat,B11,3,Iout)       ! B11
      call DotProduct(xNor,X12hat,B12,3,Iout)       ! B12      ! SHOULD BE ZERO
      call DotProduct(xNor,X12hat,B21,3,Iout)       ! B21      ! SHOULD BE ZERO
      call DotProduct(xNor,X22hat,B22,3,Iout)       ! B22
!
      Cu1 = B11/(xLam1*xLam1)           ! B11/A11
      Cu2 = B22/(xLam2*xLam2)           ! B22/A22
!     ================================================ Curvature End
      call DotProduct(A1,X11hat,xLam11hat,3,Iout)      ! Lamda1,1hat w.r.t zi
      xLam11hat = xLam11hat/xLam1                                              !
      call DotProduct(A1,X12hat,xLam12hat,3,Iout)      ! Lamda1,2hat
      xLam12hat = xLam12hat/xLam1                                              !
      call DotProduct(A2,X12hat,xLam21hat,3,Iout)      ! Lamda2,1hat
      xLam21hat = xLam21hat/xLam2
      call DotProduct(A2,X22hat,xLam22hat,3,Iout)      ! Lamda2,2hat
      xLam22hat = xLam22hat/xLam2
!
      xLam21 = xLam21hat/xLam1              ! lamda2,1 w.r.t s1
      xLam12 = xLam12hat/xLam2              ! lamda1,2 w.r.t s2
!     ---------------------------------------------Curvatures: ko1 & ko2
      C01(1) = zero
      C01(2) = Cu1
      C01(3) = xLam12/xLam1
!
      C02(1) = -Cu2
      C02(2) = zero
      C02(3) = -xLam21/xLam2
!     -------------------------------Curvatures: ko1 & ko2 CORRECT ONE
      C01(1) = zero
      C01(2) = -Cu1
      C01(3) = -xLam12/xLam1
```

```
!
        C02(1) = Cu2
        C02(2) = zero
        C02(3) = xLam21/xLam2
!       ========================================================= save in tape
        call PutGetRefPropsatGaussPt(iRefPro1,jCInt,1)
!       ===================================================== save in tape end
        return
        end
```

11.20.6 Stiffness Matrices and Residual Forces

As indicated earlier, the implementation for the shell analysis flows directly along the line of our beam analysis implementation (see Chapter 10 on nonlinear beams). In this sense, the shells can be viewed as beams along the lines of curvature coupled by the constitutive relations and other conditions. Thus, both the instantaneous material stiffness matrix and the geometric stiffness matrix are formed in an additive way with appropriate substitution of various fundamental vectors: $\mathbf{k}_\alpha^0$, $\mathbf{a}_\alpha$, $\mathbf{X}_\alpha$ and so on, for directions given by indices $\alpha = 1, 2$. Moreover, generally for a c-type finite element, the internal Bezier control points can be easily eliminated by static condensation through partial Gauss elimination algorithm.

11.20.6.1 Algorithmic Code Snippets for Material and Geometric Stiffness Matrices

```
         Subroutine Tensor_STIF(RoCOld,CuCOld,RoSav,CKSav,Qd,
        &           Nel,DMat,B,T,Stif,BFors,StifK,nStif,Iout,Trace)
!
!        Notes: Nint should be atleast 2
!
!        Inputs
!               Qd(nStif) = Latest displacement controls: state vector
!               Nel              = Element Number
!               NelType = Element Type:
!               Nint             = Number of Gauss Points
!               E                = Elasticity Modulus
!               PR               = Poisson's ratio
!               Thk              = Thickness
!               nStif            = Row or Col size of Stiffness matrix
!                                           = nDof*nQd1*nQd2
!               Iout             = Error Output file#
!
!        Outputs
!               Stif(nStif,nStif) = Desired Stiffness Matrix
!               bFors(nStif)      = Desired Body force Vector Integrals
!               StifK(nStif)      = Desired kLamda Stiffness last column Vector
Integrals
!               Trace             = trace of the matrix
!
         Implicit Real(kind=8) (a-h,o-z)
!       ==================================
```

```
        include 'Material.h'
        include 'CurParams.h'
        include 'SizeVar.h'
        include 'LoadStepIter.h'
        include 'Scratch.h'
        include 'ShellParams.h'
        include 'LogParams.h'
        include 'GaussIndex.h'
!==================================================
!
!       Input Variables & Arrays
!
        Real(kind=8)    Qd
        Dimension               Qd(nStif)
        Real(kind=8)    RoSav        ,CKSav
        Dimension               RoSav(nCInt),CKSav(nCInt)
        Real(kind=8)    RoCOld       ,CuCOld
        Dimension               RoCOld(nCInt),CuCOld(nCInt)
        Real(kind=8)    DMat
        Dimension               DMat(nCMat,nCMat)
!
!       Working Variables & Arrays
!
!       Real(kind=8)  DD ,RR ,DP1 ,Rp1 ,Dp2 ,Rp2
!       Dimension               DD(3),RR(3),DP1(3),Rp1(3),Dp2(3),Rp2(3)
!
!       Output Variables & Arrays
!
        Real(kind=8)    B          ,T
        Dimension               B(nDof2,nStif),T(nDof,nStif)
        Real(kind=8)    BFors      ,StifK       ,Stif       ,Trace
        Dimension               BFors(nStif),StifK(nStif),Stif(nStif,nStif)
        Real(kind=8)    EDG        ,DB
        Dimension               EDG(nDof2,nDof2),DB(nDof2)
        Real(kind=8)    Add        ,HD          ,Rp
        Dimension               Add(nStif),HD(nDof2),Rp(3)
        Real(kind=8)    BVEC       ,SKLM        ,R0       ,RK
        Dimension               BVEC(nStif),SKLM(nStif),R0(3),RK(3)
        Real(kind=8) AM,BM,CM,TH,Ri,Rj,Si,Wt,sum
        LOGICAL bReset
        DATA zero/0.D0/,one/1.0D0/,onm/-1.0D0/
        DATA NC/3/
!==========================================================
!       Stiffness Matrix (Tangent Operator) Calculation
!
        BFors   = 0.D0  !all
        StifK   = 0.D0  !all
        Stif    = 0.D0  ! all elements = 0.
!==========================================================
!       Rewind CurvatureInfo, B & T Scratch files
        if(Nel == 1) then
                nRec = 0
                rewind iRefPro1
```

```
                rewind iRefPro2
         endif
!        Loop over Gauss Locations
         do 60 iGau = 1,Nint1
         do 60 jGau = 1,Nint2
!        ---------------------------------------- gauss positions
         ri = GaussPtVal(iGau,Nint1,1)
         rj = GaussPtVal(jGau,Nint2,1)
!----------------------------------—- Get Geometric Curvature Vector etc.
         nRec = nRec + 1
         T = zero
         B = zero
         call Get_ReferenceInfoataGaussPt
 &                                          (T,B,nDof,nDof2,nStif,Nel,nRec)
!------------------------------------------------- DisRots & Derivatives
         call Split_Dis_Rot_Deriv(Qd,nStif,B,Iout)
!        ------------------------------------------------------ Update Rotation
!        do 10 i = 1,nC
! 10     RK(i) = R0(i) + RR(i)
         ================================= ED+G(nDof2,nDof2)
         call Form_EtranDEplusGMatrix(DMat,T,B,BVEC,SKLM,EDG,Iout)
!============================================== save
!       if(iNut.LE.1) Then
!              do 10 ii = 1,nC
!              ij = ii + (jGau -1)*nC
!              RoCOld(ij) = RK(ii)
!              CuCold(ij) = CK(ii)
! 10           continue
!              nRec = (Nel-1)*Nint + jGau
!              if(iRotCur == iRotCur1) Then
!                      call PutGetRotCur(iRotCur2,RoCOld,CuCold,nCInt,nRec,1)
!              else
!                      call PutGetRotCur(iRotCur1,RoCOld,CuCold,nCInt,nRec,1)
!              endif
!       endif
!----------------------------------------------------------- Form: Btran_ED+G_B
!       Add to stiffness matrix
        Wt = GaussPtVal(iGau,Nint1,2)*GaussPtVal(jGau,Nint2,2)*RutAhat
        do 50 j = 1,nStif
        do 20 k = 1,nDof2         !
        DB(k) = 0.0
        do 20 l = 1,nDof2         !
        DB(k) = DB(k) + EDG(k,l)*B(l,j)
  20 continue
        do 40 i = j,nStif
        sum = 0.0
        do 30 l = 1,nDof2         !
        sum = sum + B(l,i)*DB(l)
   30   continue
        Stif(i,j) = Stif(i,j) + sum*Wt
   40   continue
   50      continue
!---------------------------------------- Load Vector: Residual & Body force part
```

```
      do 53 i = 1,nStif
      BFors(i) = BFors(i) + Wt*BVec(i)
 53 continue
!------------------------ Form: Last Column of stiffness: KLamda Body Force part
      do 55 i = 1,nStif
      StifK(i) = StifK(i) + Wt*SKLM(i)
 55 continue
 60 continue
!     symmetry: Upper Triangle
      do 70 j = 1,nStif
      do 70 i = j,nStif
      Stif(j,i) = Stif(i,j)
 70 continue
!     Get Trace
      Trace = 0.D0
      do 80 i = 1,nStif
      Trace = Trace + Stif(i,i)
 80 continue
!========================================================
      return
      end

      Subroutine FormEMatrix_1(EMat,nE,nG,nC,iDir,Iout)
!     ---------------------------------------------------
!     Inputs
!             AMat(nC,nC)          = Rotation Matrix
!             nG                   = 18
!             C0i(nC)        = initial Curvature vector = k0i for iCurDir i
!             nC                   = 3 (= nDim)
!             nE                   = 12
!             W(nC,nC)             = W Matrix
!             R1(nC,nC)            = R1 Matrix
!             R2(nC,nC)            = R2 Matrix
!             AAi(nC)        = e + [k0i]d
!             CKi(nC)        = k0i + ki = k0i + W(thetaPrimed)
!
!     Outputs
!             EMat(nE,nG)          = EMat Matrix (12,18)
!     ------------------------------------------
      Implicit Real(kind=8) (a-h,o-z)
!====================================
      include 'Material.h'
      include 'LoadStepIter.h'
      include 'CurParams.h'
      include 'ThetaConstants.h'
      include 'BeamParams.h'
      include 'ShellParams.h'
!=============================================
      Real(kind=8)     EMat
      Dimension              EMat(nE,nG)
      Real(kind=8)     E11          ,E12
      Dimension              E11(nC,nC),E12(nC,nC)
      Real(kind=8)     E13          ,E22          ,E23          ,E24
```

```
      Dimension              E13(nC,nC),E22(nC,nC),E23(nC,nC),E24(nC,nC)
      Real(kind=8)    CoA   ,CoB
      Dimension              CoA(3,3),CoB(3,3)
!
      DATA zero/0.D0/,one/1.0D0/,two/2.0D0/,three/3.0D0/,onem/-1.0D0/
!------------------------------------------------- Get EBlocks for dir 1
!     =========================================
      call FormEBlocks(nC,E11,E12,E13,E22,E24,1,Iout)
!------------------------------------------------------------ Form EMatrix
      EMat = zero
!     --------
      do 10 i = 1,3
      do 10 j = 1,3
      EMat(i  ,j  )    = E11(i,j)         ! [C0]
      EMat(i  ,j+3)    = E12(i,j)         ! [a1]W
      EMat(i  ,j+6)    = E13(i,j)         ! I
      EMat(i+3,j+3)    = E22(i,j)         ! R+[Cn]W
      EMat(i+3,j+9)    = E24(i,j)         ! W
   10 continue
!-----------------------------------------------------------
      return
      End

      Subroutine FormEMatrix_2(EMat,nE,nG,nC,iDir,Iout)
! -----------------------------------------------------------
!     Inputs
!         AMat(nC,nC)              = Rotation Matrix
!         nG                       = 18
!         C0i(nC)            = initial Curvature vector = k0i for iCurDir i
!         nC                       = 3 (= nDim)
!         nE                       = 12
!         W(nC,nC)           = W Matrix
!         R1(nC,nC)                = R1 Matrix
!         R2(nC,nC)                = R2 Matrix
!         AAi(nC)            = e + [k0i]d
!         CKi(nC)            = k0i + ki = k0i + W(thetaPrimed)
!
!     Outputs
!            EMat(nE,nG)          = EMat Matrix (12,18)
! -----------------------------------------------------------
      Implicit Real(kind=8) (a-h,o-z)
!============================================
      include 'Material.h'
      include 'LoadStepIter.h'
      include 'CurParams.h'
      include 'ThetaConstants.h'
      include 'BeamParams.h'
      include 'ShellParams.h'
!=============================================
      Real(kind=8)    EMat
      Dimension              EMat(nE,nG)
      Real(kind=8)    E31   ,E32
      Dimension              E31(nC,nC),E32(nC,nC)
```

```
      Real(kind=8)     E35    ,E42    ,E23    ,E46
      Dimension               E35(nC,nC),E42(nC,nC),E23(nC,nC),E46(nC,nC)
      Real(kind=8)     CoA    ,CoB
      Dimension               CoA(3,3),CoB(3,3)
!
     DATA zero/0.D0/,one/1.0D0/,two/2.0D0/,three/3.0D0/,onem/-1.0D0/
!------------------------------------------ Get EBlocks for dir 2
      E31 = zero
      E32 = zero
      E35 = zero
      E42 = zero
      E46 = zero
!     ==========================================
     call FormEBlocks(nC,E31,E32,E35,E42,E46,2,Iout)
!     ==========================================
!----------------------------------------------------------- Form EMatrix
      do 40 i = 1,3
      do 40 j = 1,3
      EMat(i+6 ,j   )    = E31(i,j)
      EMat(i+6 ,j+3 )    = E32(i,j)
      EMat(i+6 ,j+12)    = E35(i,j)
      EMat(i+9 ,j+3 )    = E42(i,j)
      EMat(i+9 ,j+15)    = E46(i,j)
   40 continue
!-------------------------------------------------
      return
      End

      Subroutine FormEBlocks(nC,E11,E12,E13,E22,E24,iCurDir,Iout)
! ----------------------------------------------------------------------------
!     Inputs
!            C0i(nC)        = initial Curvature vector for dir iCurDir
!            nC                    = 3
!            AMat                  = Rotation Matrix
!            W                     = matrix
!            R                     = matrix
!            AA                    = a Vector for dir iCurDir
!            CK                    = Kc Vector for dir iCurDir
!
!     Outputs
!            E11(nC,nC)            = E11 block
!            E12(nC,nC)            = E12 block
!            E13(nC,nC)            = E13 block
!            E22(nC,nC)            = E22 block
!            E24(nC,nC)            = E24 block
!
!     ------------------------------------------------
      Implicit Real(kind=8) (a-h,o-z)
!
      include 'ThetaConstants.h'
      include 'BeamParams.h'
!     ================================
      Real(kind=8)     E11    ,E12    ,E13    ,E22
      Dimension               E11(nC,nC),E12(nC,nC),E13(nC,nC),E22(nC,nC)
```

```
      Real(kind=8)    TAMat
      Dimension             TAMat(nC,nC)
      Real(kind=8)    W       ,R    ,E24    ,sAA    ,sKc
      Dimension             W(nC,nC),R(nC,nC),E24(nC,nC),sAA(nC,nC),sKc(nC,nC)
!
      DATA zero/0.D0/,one/1.0D0/,two/2.0D0/,three/3.0D0/
!     -----------------------------------------------------------------
      W = WMat
      R = RMat
!-----------------------------------------------------------------AAMatTran
      TAMat = TRANSPOSE(AMat)
!------------------------------------------------------------------------E11Block
      call FormSkew(C0,nC,E11,Iout)                      !skew[k0]
      E11 = MATMUL(TAMat, E11)                !AMatTransposeE11
!------------------------------------------------------------------------E12Block
      call FormSkew(AA,nC,sAA,Iout)                      !skew[a]
      E12 = MATMUL(sAA, W)                               ![a]W
      E12 = MATMUL(TAMat, E12)                !AMatTransposeE12
!------------------------------------------------------------------------E13Block
      E13 = TAMat                                        !AMatTransposeE13
!------------------------------------------------------------------------E22Block
      call FormSkew(CK,nC,sKc,Iout)                      !skew[Kc]
      E22 = R + MATMUL(sKc,W)                 !R + [Kc]W
      E22 = MATMUL(TAMat, E22)                !AMatTransposeE22
!------------------------------------------------------------------------E24Block
      E24 = W
      E24 = MATMUL(TAMat, E24)                !AMatTransposeE24
!-----------------------------------------
      return
      end

      Subroutine FormGMatrix(G,nG,nC,nDof,Iout)
! -----------------------------------------------------------------
!     Inputs
!           HD(nG) = {dis,rot,disPrimed_1,rotPrimed_1,disPrimed_2,rotPrimed_2}
!           nG                          = 18
!           FM(nC)              = {force1,moment1,force2,moment2}
!           nF                          = 12
!           C01(nC)             = initial Curvature vector = k01
!           C02(nC)             = initial Curvature vector = k02
!           nC                          = 3 (= nDim)
!
!     Outputs
!           G22(nC,nC)        = GThetaBarDelThetaBlock
!   -----------------------------------------------------------
      Implicit Real(kind=8) (a-h,o-z)
!==========================================
      include 'LoadStepIter.h'
      include 'CurParams.h'
      include 'ThetaConstants.h'
      include 'BeamParams.h'
      include 'ShellParams.h'
!===========================================
```

```
      Real(kind=8)     G
      Dimension                G(nG,nG)
      Real(kind=8)     G12    ,G22    ,G23    ,G24
      Dimension         G12(3,3),G22(3,3),G23(3,3),G24(3,3)
!     ===================================================
      DATA zero/0.D0/,one/1.0D0/,two/2.0D0/,three/3.0D0/
!    ============
      G = zero
!--------------------------------------------------------------- Form GBLocks_Dir_1
      G12 = zero
      G22 = zero
      G23 = zero
      G24 = zero
!     --------------
      C0            = C01
      Ro            = RR
      Rp            = Rp1
      WMat          = WMat1
      RMat          = RMat1
      CK            = CK1
      AA            = AA1
      FMt           = FMt1
!     -----------------------------------
      call FormGBlocks(nDof,nC,G12,G22,G23,G24,1,Iout)
!----------------------------------------------------- Fill GMatrix for Dir_1
! -------------------------------- Upper Triangle with Diagonals
      do 10 i = 1,3
      do 10 j = 1,3
      G(i,j+3)          = G12(i,j)
      G(i+3,j+3)        = G22(i,j)
      G(i+3,j+6)        = G23(i,j)
      G(i+3,j+9)        = G24(i,j)
   10 continue
!----------------------------------------------------- Form GBLocks_Dir_2
      G12 = zero
      G22 = zero
      G23 = zero
      G24 = zero
!     ------------
      C0             = C02
      Ro             = RR
      Rp             = Rp2
      WMat           = WMat2
      RMat           = RMat2
      CK             = CK2
      AA             = AA2
      FMt            = FMt2
!     -----------------------------------
      call FormGBlocks(nDof,nC,G12,G22,G23,G24,2,Iout)
!---------------------------------------------------------- Fill GMatrix for Dir_2
!     ------------------------ Upper Triangle with Diagonals
      do 40 i = 1,3
      do 40 j = 1,3
```

```
      G(i  ,j+ 3)    = G(i  ,j+ 3)    + G12(i,j)
      G(i+3,j+ 3)    = G(i+3,j+ 3)    + G22(i,j)
      G(i+3,j+12)    = G(i+3,j+12)    + G23(i,j)
      G(i+3,j+15)    = G(i+3,j+15)    + G24(i,j)
  40 continue
!---------------------------------------- Symmetry:Lower Triangle
      do 50 i = 1,17
      do 50 j = i+1,18
      G(j,i) = G(i,j)
  50 continue
!----------------------------------------
      return
      End

      Subroutine FormGBlocks(nF,nC,G12,G22,G23,G24,iCurDir,Iout)
!     ------------------------------------------------
!     Inputs
!           FM(nC)      = {forcei,momenti} for dir iCurDir
!           nF                   = 6
!           C0(nC)      = initial Curvature vector for dir iCurDir
!           nC                   = 3
!
!     Outputs
!           G22(nC,nC)          = GThetaBarDelThetaBlock
!     ------------------------------------------------
          Implicit Real(kind=8) (a-h,o-z)
!
      include 'ThetaConstants.h'
      include 'BeamParams.h'
!     ================================
      Real(kind=8)   G12       ,G22       ,G23      ,G24
      Dimension                G12(nC,nC),G22(nC,nC),G23(nC,nC),G24(nC,nC)
      Real(kind=8)   Fo      ,XM
      Dimension                Fo(nC),XM(nC)
      Real(kind=8)   BB      ,CC
      Dimension                BB(nC),CC(nC)
      Real(kind=8)   W        ,R        ,SM        ,SF        ,SC
      Dimension                W(nC,nC),R(nC,nC),SM(nC,nC),SF(nC,nC),SC(nC,nC)
      Real(kind=8)   FA        ,XMK        ,TA        ,ATA
      Dimension                FA(nC,nC),XMK(nC,nC),TA(nC,nC),ATA(nC,nC)
      Real(kind=8)   WMR        ,TMT        ,SMt
      Dimension                WMR(nC,nC),TMT(nC,nC),SMt(nC,nC)
!
      DATA nV/3/
      DATA zero/0.D0/,one/1.0D0/,two/2.0D0/,three/3.0D0/
! ------------------------------------------------
      W = WMat
      R = RMat
!     ========
! istart = (iCurdir-1)*6
!
!     call VectorExtract_Vector(HD,nG,Ro,nC,4,Iout)         !rotation
!     call VectorExtract_Vector(HD,nG,Rp,nC,istart+10,Iout)!rotationPrimed
```

```
      call VectorExtract_Vector(FMt,nF,Fo,nC,1,Iout)!Force
      call VectorExtract_Vector(FMt,nF,XM,nC,4,Iout)          !Moment
!-----------------------------------------------------------------G22Block
! ---------------------------------------------------- Translational
!     ------------------------------------------ FA = WTran(Fxa + axF -2(F.a)I)W
      call FormF_a_PartMatrix(Fo,AA,FA,3,1,Iout)     !Fxa + axF -2(F.a)I
!
!     call Form_WMatrix(Ro,W,nC,c1,c2,c1b,c2b,Iout)
      FA = MATMUL(FA, W)                               !(Fxa + axF -2(F.a)I)W
      FA = MATMUL(TRANSPOSE(W), FA)       !WTran(Fxa + axF -2(F.a)I)W
!
      call CrossProduct(Fo,AA,BB,Iout)           ! b = F x a
      call DotProduct(Ro,BB,dot,nV,Iout)       ! theta dot b
      C2TdotB = c2*dot                    ! c2*(theta dot b)
! -----------------------------------------.5*FA + c2(theta dot b)I
      FA = 0.5D0*FA
      FA(1,1) = FA(1,1) + C2TdotB
      FA(2,2) = FA(2,2) + C2TdotB
      FA(3,3) = FA(3,3) + C2TdotB
!     ------------------------------------------------------ Rotational
!     ------------------------------------ XMK = WTran(Mxkc + kcxM -2(M.kc)I)W
      call FormF_a_PartMatrix(XM,CK,XMK,3,2,Iout)        !Mxkc + kcxM -2(M.kc)I
!
      XMK = MATMUL(XMK, W)           !(Mxkc + kcxM -2(M.kc)I)W
      !-----------------------------------------WTran(Mxkc + kcxM -2(M.kc)I)W
      XMK = MATMUL(TRANSPOSE(W), XMK)
      call CrossProduct(XM,CK,CC,Iout)           ! c = M x kc
      call DotProduct(Ro,CC,dot,nV,Iout)       ! theta dot c
      C2TdotC = c2*dot                              ! c2*(theta dot c)
!-----------------------------------------------.5*XMK + c2(theta dot c)I
      XMK = 0.5D0*XMK
      XMK(1,1) = XMK(1,1) + C2TdotC
      XMK(2,2) = XMK(2,2) + C2TdotC
      XMK(3,3) = XMK(3,3) + C2TdotC
!      -------------------------------------------------AlfaCxRot + RotxAlfaC
      call FormThetaAlfaTrans_Matrix(Ro,CC,TA,nC,c1,c2,c1b,c2b,Iout)    !RotxAlfa
      ATA = 0.5D0*(TA + TRANSPOSE(TA))
!      -------------------------------------------------AlfaBxRot + RotxAlfaB
      call FormThetaAlfaTrans_Matrix(Ro,BB,TA,nC,c1,c2,c1b,c2b,Iout)  !RotxAlfa
      ATA = ATA + 0.5D0*(TA + TRANSPOSE(TA))
! ------------------------------------------------WTranMR+RTranMTranW
      call FormSkew(XM,nC,SM,Iout)                       !skew[M]
!      call FormR_Matrix(Ro,Rp,R,nC,c1,c2,c1b,c2b,Iout)!R matrix
      WMR = MATMUL(TRANSPOSE(W),SM)      !WTran[M]
      WMR = MATMUL(WMR,R)                                !WTran[M]R
      WMR = WMR + TRANSPOSE(WMR)
!     ---------------------------------------------- G22Block
      G22 = FA + XMK + ATA + WMR
!-------------------------------------------------------G12Block
      call FormSkew(Fo,nC,SF,Iout)            !skew[F]
      call FormSkew(C0,nC,SC,Iout)            !skew[k0]
!
      G12 = MATMUL(SC,SF)                    ![k0][Ft]
```

```
      G12 = MATMUL(G12,W)                  ![k0][Ft]W
!--------------------------------------------------------- G23Block
      G23 = MATMUL(TRANSPOSE(W),SF)       !WTran[Ft]
!---------------------------------------------------------G24Block
      TMT = MATMUL(TRANSPOSE(W),SM)!WTran[Mt]
      TMT = MATMUL(TMT,W)             !WTran[Mt]W
      call FormS_Matrix(Ro,XM,SMt,nC,c1,c2,c1b,c2b,Iout) !S(Mt) Matrix
!
      G24 = TMT + SMt
!---------------------------------------------------------
      return
      end
```

11.20.7 Incremental Equation Solution and Updates

Now, let us recall the geometrically exact linearized finite element equation in unknown incremental generalized displacement, $\Delta\hat{\mathbf{q}}^d$, and the incremental load factor, $\Delta\lambda$, as:

$$\boxed{\mathbf{K}_T^c\ \Delta\hat{\mathbf{q}}^d + \ \hat{\mathbf{p}}(\lambda)\ \Delta\lambda = -\hat{\mathbf{R}}(\lambda_0, \hat{\mathbf{d}}_0)} \tag{11.723}$$

It may be recalled that the linearized equation (11.723) has been formulated to be solved by a Newton-type iterative solution procedure for the configuration path evaluation. The configuration path is sought for a set of discrete load level parameters: $\lambda = \lambda_0, \lambda_1, \lambda_2, \cdots$. Generally, a nonlinear configuration path may consist of both snap-through and snap-back buckling phenomena. A most suitable procedure for capturing these unstable situations is provided by embedding the arc length constraint algorithm in a Newton-type incremental/iterative method. At each step of an iterative solution procedure, both $\mathbf{d}$, the displacement vector, and $\mathbf{R}$, the rotation tensor, need to be updated for use in the next step of iteration, and for the determination of the shell configuration path, $C(\mathbf{d}, \mathbf{R})$, of interest. For the displacement vector, being a member of a vector space, the updating is additively straightforward as given by:

$$\mathbf{d}^{(i+1)} = \mathbf{d}^{(i)} + \Delta\mathbf{d}^{(i)} \tag{11.724}$$

where $\Delta\mathbf{d}^{(i)}$, $\mathbf{d}^{(i)}$, $\mathbf{d}^{(i+1)}$ are the incremental displacement vector obtained as a solution of the linearized incremental equation (11.723), the known starting displacement vector at iteration, i, and the updated displacement vector, respectively.

An additive update for the rotation tensor, being a member of a manifold, that is, a curved space ($SO(3)$, the special orthogonal group), is not allowed. However, we can locally parameterize the rotation tensor by $\boldsymbol{\theta}$, the Rodrigues rotation vector that belongs to the tangent spaces; a tangent space being a vector space, we can have an additive updating scheme for the rotation vector, without leaving the space, similar to that of the displacement update of the reference surface as:

$$\boldsymbol{\theta}^{(i+1)} = \boldsymbol{\theta}^{(i)} + \Delta\boldsymbol{\theta}^{(i)} \tag{11.725}$$

where $\Delta\boldsymbol{\theta}^{(i)}$, $\boldsymbol{\theta}^{(i)}$, $\boldsymbol{\theta}^{(i+1)}$ are the incremental rotation vector obtained as a solution of the linearized incremental equation (11.723), the known starting rotation vector at iteration, i, and the updated rotation vector, respectively. For most structural problems of interest, the additive updating procedure suffices, resulting in extremely accurate approximations close to the exact solution for moderately large rotations.

Remark

- Recalling our section on the rotation tensor, the updating scheme given by equation (11.725) will not work for $\|\theta\| = (2n+1)\pi, \quad n = 0, 1, \cdots$, as the Rodrigues rotation tensor becomes singular at those $\boldsymbol{\theta}$'s.

11.20.8 Singularity-free Update for the Configuration

The update for the translational configuration, that is, the displacement of the shell reference surface, is done as shown earlier by equation (11.724). Here, we are mainly concerned with a singularity-free rotational configuration update. For details on the main ideas about quaternion, and so on, and the numerical algorithm, we refer to Section 10.20 on the numerical implementation for beams.

11.20.8.1 Update for the Rotation Configuration

With the above conversions from the rotation matrix to the quaternion and back, in place, let $\Delta\boldsymbol{\theta}^{(i)}$ be the incremental Rodrigues rotation vector. We will have in storage the current quaternion, $\mathbf{Q}^{(i)}$, for the iteration step i; we can retrieve and convert it to the rotation matrix, $\mathbf{R}^{(i)}$. Then, we can generate the quaternion, $\Delta\mathbf{Q}^{(i)}$, corresponding to $\Delta\boldsymbol{\theta}^{(i)}$ by replacing $\boldsymbol{\theta}$ by $\Delta\boldsymbol{\theta}^{(i)}$, and similarly, the norm; we can construct the incremental rotation matrix, $\Delta\mathbf{R}^{(i)}$ from the quaternion $\Delta\mathbf{Q}^{(i)}$. Now, we get the updated rotation matrix, $\mathbf{R}^{(i+1)}$, as $\mathbf{R}^{(i+1)} = \Delta\mathbf{R}^{(i)}\,\mathbf{R}^{(i)}$. Finally, by application of the algorithm for the conversion of a rotation matrix to the corresponding quaternion, we convert $\mathbf{R}^{(i+1)}$ to $\mathbf{Q}^{(i+1)}$ and store it in the database.

11.20.9 Newton-type Iteration and Arc Length Constraint

Recalling Section 8.5 on Newton-type iteration with arc length constraint, a method can be devised to solve problems and perform stability or buckling analysis with quadratic convergence. For proportional loading, the coefficient of proportionality for the loading, λ, is gradually incremented through a Newton-type iterative method to trace the configuration path of the system. However, a straightforward application of the method fails near, at or beyond the stability limit point. For details of the main ideas about Newton-type method and arc length constraint, and so on, and the numerical algorithm, see Section 10.20 on numerical implementation for beams. Recall from Section 8.5 on Newton-type iteration with arc length constraint that there are generally two fundamentally different scenarios

11.20.10 Numerical Examples

Now that we have some implementation details, we would like to apply these to various problems of interest (that have become industry benchmarks), examine them and interpret the accuracy, robustness and numerical efficiency of the obtained results and compare them with the published solutions by other methods, whenever available. It is important to note that in all of the following

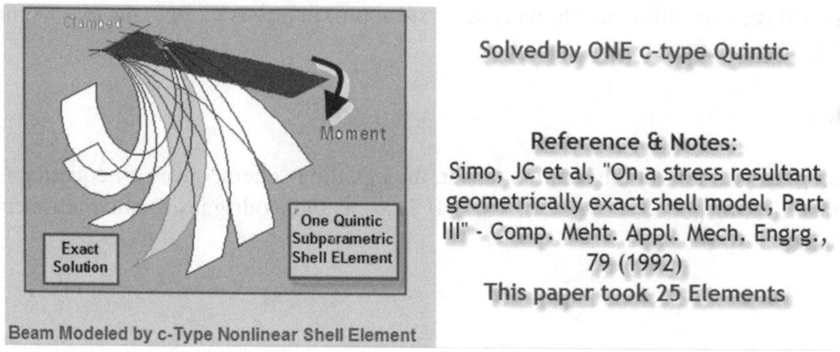

Figure 11.22 Example 1: Flat plate curling problem.

examples full Gauss quadrature points (see Section 10.20 on numerical implementation for beams) were selected with an absolute minimum number of mega c-type shell elements to show that locking is a non-issue and that there is no need for any ad hoc so-called "under integration" and/or "reduced integration"; this was only possible by c-type finite elements that are backed by a balanced exact theoretical formulation and an associated numerical implementation with Bernstein–Bezier polynomials as the basis functions.

Example 1: A 2D straight thin plate coiling to circle

A straight cantilevered plate, reported in the literature of unit cross-sectional depth is subjected to a moment loading at the free end. The plate is modeled as a single degenerate 3D quintic in *X-Y* plane with *Z*-directional constraints. The theoretical solution for a beam results in a complete circular plate of radius, $r = \dfrac{EI}{M}$, where the tip moment, $M = \dfrac{2\pi EI}{L}$, with L, as the length of a beam. The elasticity modulus, E, and the shear modulus, G, are 10^6 and 0.5×10^6, respectively; the length is taken as $L = 10.0$, the depth and width as 1.0 with Poisson's ratio, $\mu = 0$. Finally, the problem is solved subparametrically, with the geometry modeled linearly and the displacement modeled by a c-type quintic shell element. The deformed shapes for various tip moment loading are shown in Figure 11.22. Note that compared to the reported analyses that have used eight elements, a single mega quintic element suffices, and thus points towards the huge reduction in subsequent quality assurance efforts in practical endeavors.

Example 2: Geometrically exact nonlinear curved plate as deep arch

To exemplify unification of geometry modeling and finite element analysis in both formulation and subsequent numerical solution, a geometrically exact nonlinear deep arch problem with shear deformation and extensibility is considered here.

The deformed shapes at various load levels are shown in Figures 11.23(a) and 11.23(b) for two Huddleston compressibility parameters, $c = \dfrac{I}{A(2bd)}$ where, I, is the moment of inertia, A represents cross-sectional area, and for b, d. One model, as in Figure 11.23(a), is analyzed with $c = 2.55\text{x}10^{-6}$, that is, the flexibility is so low that it renders the arch relatively incompressible. The other, as shown in Figure 11.23(b), has $c = .01$, that is, it has very high I, and thus, making the arch substantially compressible. It may be noted that it is solved with only one c-type quintic element to accurately model the problem, while it takes 40 Q-type shell elements of Palazotto and Dennis (1992), to seriously underestimate the limit point and track the wrong equilibrium path.

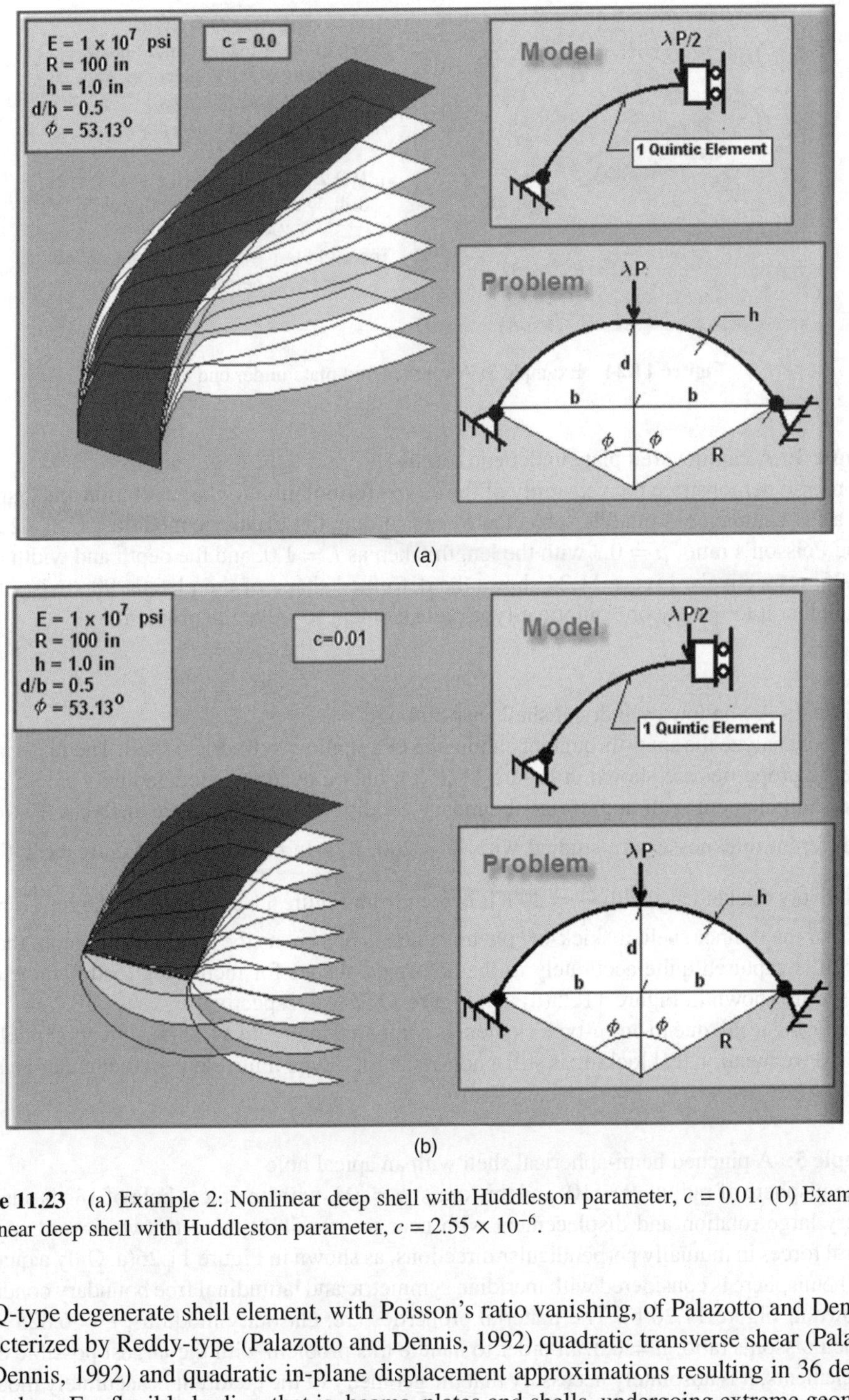

Figure 11.23 (a) Example 2: Nonlinear deep shell with Huddleston parameter, $c = 0.01$. (b) Example 2: Nonlinear deep shell with Huddleston parameter, $c = 2.55 \times 10^{-6}$.

The Q-type degenerate shell element, with Poisson's ratio vanishing, of Palazotto and Dennis is characterized by Reddy-type (Palazotto and Dennis, 1992) quadratic transverse shear (Palazotto and Dennis, 1992) and quadratic in-plane displacement approximations resulting in 36 degrees of freedom. For flexible bodies, that is, beams, plates and shells, undergoing extreme geometric nonlinearity through large rotations and displacement, it seems unwise to compromise theoretical formulation prematurely for numerical manipulations early in the process.

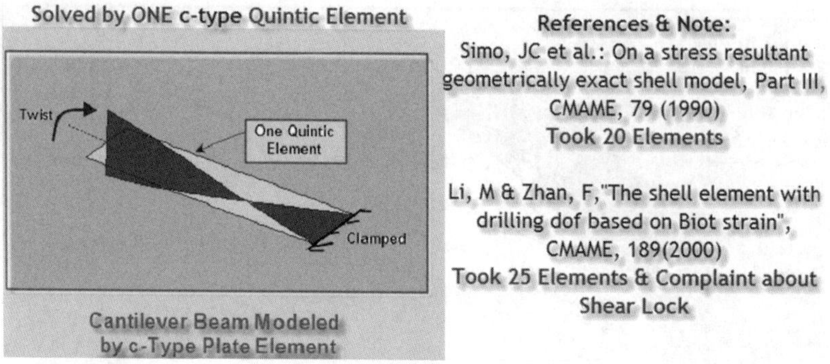

Figure 11.24 Example 3: A cantilevered plate under end torque.

Example 3: A cantilevered plate under end torque

In order to demonstrate the versatility of the c-type formulation for large rotation and displacement, a flat cantilevered plate is subjected to end torque; the elasticity modulus, E, is 12×10^6 and the Poisson's ratio, $\mu = 0.3$ with the length taken as $L = 1.0$, and the depth and width as 0.1 and 0.25, respectively. Figure 11.24 shows the deformed shape of the plate at 90° twist; it may be noted that it took only one quintic c-type plate element to solve the problem.

Example 4: A Shallow cylindrical shell snap-through

Here we analyze the snap-through phenomenon of a shallow cylindrical shell. The material and geometric properties are shown in Figure 11.25(a); only a quadrant modeled by a single c-type quintic shell element with appropriate boundary conditions is used for the analysis. Two cases with different thicknesses are studied with $\frac{R}{h} = 200, 400$. In contrast with Figure 11.25(b) for the relatively thicker case with $\frac{R}{h} = 200$, it is seen from Figure 11.25(c) that for larger $\frac{R}{h} = 400$, that is, for the thinner shell, a "kick-in" phenomenon is present as the sides deform more than the center; it is captured quite accurately as the deformed shapes for increasing load parameter for the cases are shown in Figure 11.25(d) and Figure 11.25(e), respectively.

Finally, the result due to one c-type element is compared in Figure 11.25(f) with other published work and we mention that locking is still a non-issue for c-type finite element method of analysis.

Example 5: A pinched hemi-spherical shell with an apical hole

A hemisphere of radius, $R = 10$, and thickness, $h = .04$, with an apical hole of 18° is analyzed for very large rotation and displacement with near inextensibility due to two inward and two outward forces in mutually perpendicular directions, as shown in Figure 11.26(a. Only a quadrant of the hemisphere is considered with meridian symmetric and latitudinal free boundary conditions as shown in Figure 11.26(b). The material properties are: elasticity modulus, $E = 6.825 \times 10^7$, and the Poisson's ratio, $\mu = 0.3$. In order to handle this problem with the largest possible c-type shell elements, it is absolutely necessary that the geometry of the quadrant is accurately modeled; for example, the cubic non-rational Bernstein–Bezier element are not acceptable. The rational quadratic Bernstein–Bezier elements turned out to be the perfect choice; just a 3×3 set of these elements produced the accurate result; a quadrant with 2×2 set of rational elements is shown in Figure 11.26(c).

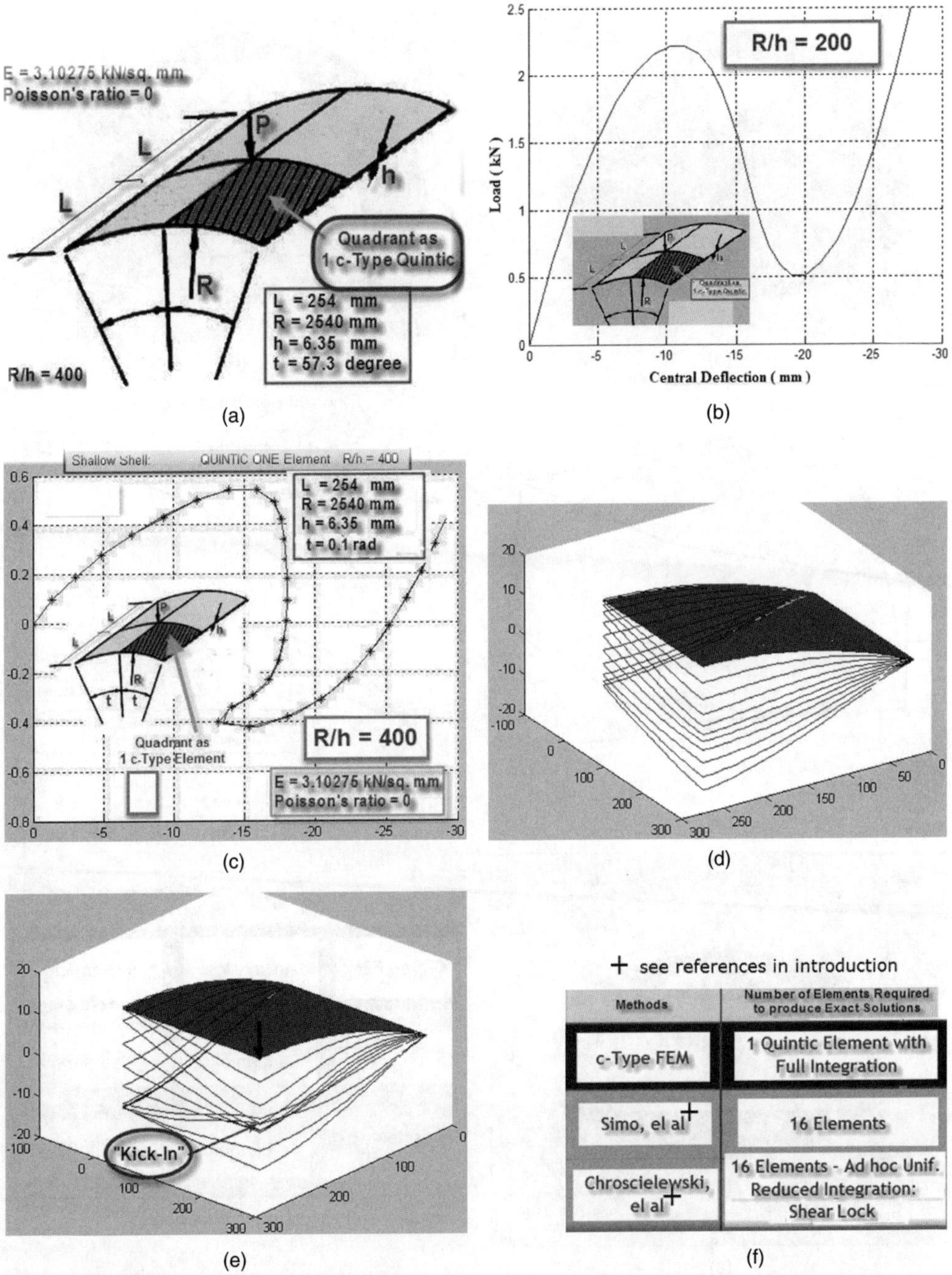

Figure 11.25 (a) Example 4: Shallow shell with one c-type quintic element for a quadrant. (b) Example 4: Shallow shell load deformation at load point for $R/h = 200$. (c) Example 4: Shallow shell load deformation at load point for $R/h = 400$. (d) Example 4: Shallow shell deformation shapes for $R/h = 200$. (e) Example 4: Shallow shell deformation shapes for $R/h = 400$. (f) Example 4: Number of elements comparison.

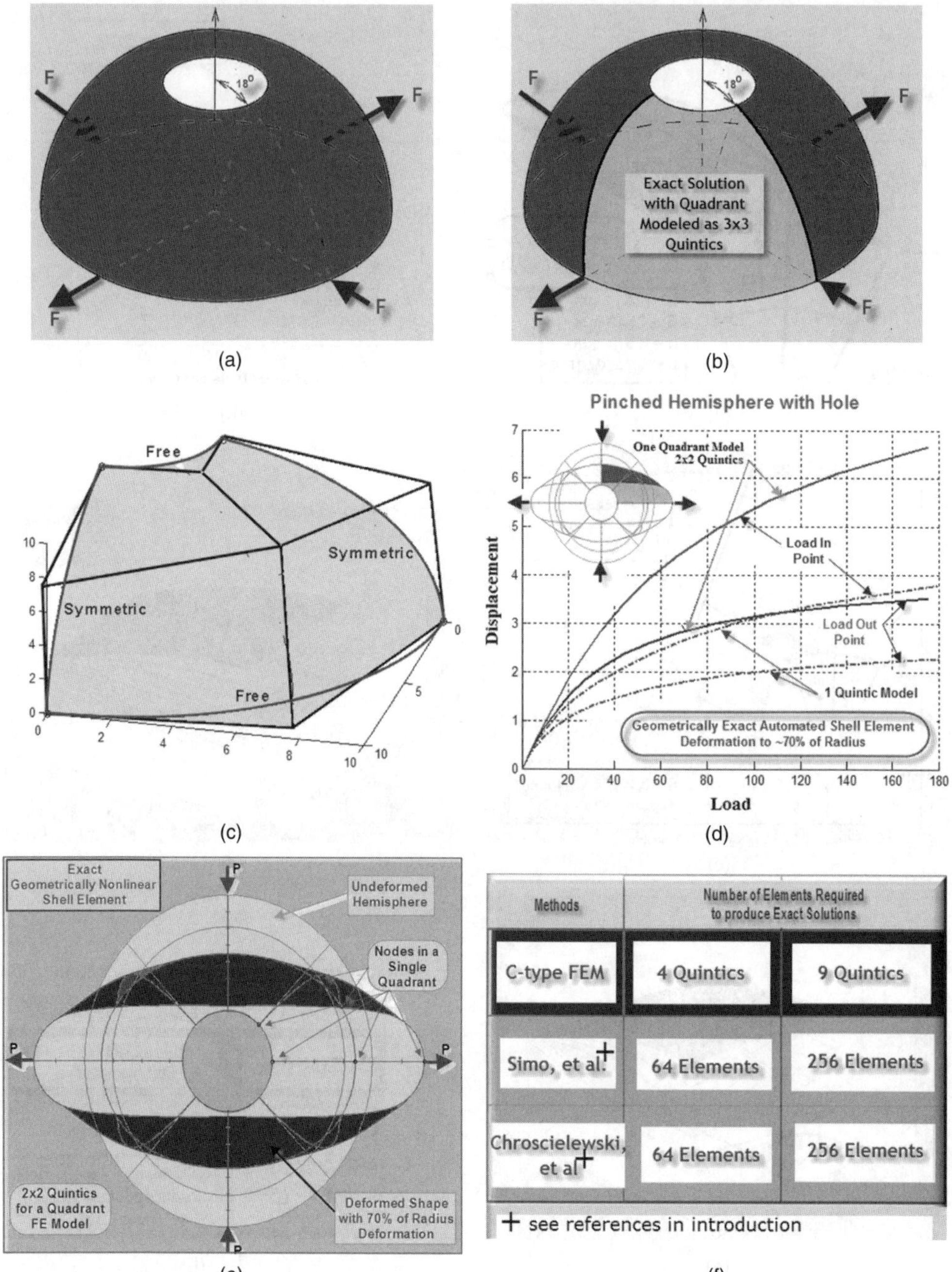

Figure 11.26 **(a)** Example 5: Hemispherical shell with hole geometry. **(b)** Example 5: Hemispherical shell with hole geometry. **(c)** Example 5: Hemispherical shell: rational quadratic Bezier mesh for a quadrant. **(d)** Example 5: Hemispherical shell: load deformation curves under load points for 2 × 2 quintics for a quadrant. **(e)** Example 5: Hemispherical shell: deformed shape. **(f)** Example 5: Hemispherical shell: number of elements comparison.

For the linear range, the displacement inward or outward at the load application point is symmetric and was found to produce the exact value of 0.093. However, in the nonlinear regime, the symmetry disappears with the inward displacements much more than the outward ones for the same load level; load–deformation relationships at load points up to 70 of the radius are shown in Figure 11.26(d); the corresponding deformation shape is shown in Figure 11.26(e). Finally, the comparison of element numbers with published results is shown in Figure 11.26(f).

Index

Computation of Nonlinear Structures: Extremely Large Elements for Frames, Plates and Shells, First Edition. Debabrata Ray.